# Prefixes and Suffixes in Anatomical and Medical Terminology

| Element | Definition and Example |
|---|---|
| mito- | thread: *mitochondrion* |
| mono- | alone, one, single: *monocyte* |
| morph- | form, shape: *morphology* |
| multi- | many, much: *multinuclear* |
| myo- | muscle: *myology* |
| narc- | numbness, stupor: *narcotic* |
| necro- | corpse, dead: *necrosis* |
| neo- | new, young: *neonatal* |
| nephr- | kidney: *nephritis* |
| neuro- | nerve: *neurolemma* |
| noto- | back: *notochord* |
| ob- | against, toward, in front of: *obturator* |
| oc- | against: *occlusion* |
| -oid | resembling, likeness: *sigmoid* |
| oligo- | few, small: *oligodendrocyte* |
| -oma | tumor: *lymphoma* |
| oo- | egg: *oocyte* |
| or- | mouth: *oral* |
| orchi- | testis: *orchiectomy* |
| ortho- | straight, normal: *orthopnea* |
| -ory | pertaining to: *sensory* |
| -ose | full of: *adipose* |
| osteo- | bone: *osteoblast* |
| oto- | ear: *otolith* |
| ovo- | egg: *ovum* |
| par- | give birth to, bear: *parturition* |
| para- | near, beyond, beside: *paranasal* |
| path- | disease, that which undergoes sickness: *pathology* |
| -pathy | abnormality, disease: *neuropathy* |
| ped- | children: *pediatrician* |
| pen- | need, lack: *penicillin* |
| -penia | deficiency: *thrombocytopenia* |
| per- | through: *percutaneous* |
| peri- | near, around: *pericardium* |
| phag- | to eat: *phagocyte* |
| -phil | have an affinity for: *neutrophil* |
| phleb- | vein: *phlebitis* |
| -phobia | abnormal fear, dread: *hydrophobia* |
| -plasty | reconstruction of: *rhinoplasty* |
| platy- | flat, side: *platysma* |
| -plegia | stroke, paralysis: *paraplegia* |
| -pnea | to breathe: *apnea* |
| pneumo(n)- | lung: *pneumonia* |
| pod- | foot: *podiatry* |
| -poiesis | formation of: *hemopoiesis* |
| poly- | many, much: *polyploid* |
| post- | after, behind: *postnatal* |
| pre- | before in time or place: *prenatal* |
| pro- | before in time or place: *prophase* |

| Element | Definition and Example |
|---|---|
| proct- | anus: *proctology* |
| pseudo- | false: *pseudostratified* |
| psycho- | mental: *psychology* |
| pyo- | pus: *pyorrhea* |
| quad- | fourfold: *quadriceps femoris* |
| re- | back, again: *repolarization* |
| rect- | straight: *rectus abdominis* |
| ren- | kidney: *renal* |
| rete- | network: *rete testis* |
| retro- | backward, behind: *retroperitoneal* |
| rhin- | nose: *rhinitis* |
| -rrhagia | excessive flow: *menorrhagia* |
| -rrhea | flow or discharge: *diarrhea* |
| sanguin- | blood: *sanguineous* |
| sarc- | flesh: *sarcoma* |
| -scope | instrument for examining a part: *stethoscope* |
| -sect | cut: *dissect* |
| semi- | half: *semilunar* |
| -sis | process or action: *dialysis* |
| steno- | narrow: *stenosis* |
| -stomy | surgical opening: *tracheostomy* |
| sub- | under, beneath, below: *subcutaneous* |
| super- | above, beyond, upper: *superficial* |
| supra- | above, over: *suprarenal* |
| syn- (sym-) | together, joined, with: *synapse* |
| tachy- | swift, rapid: *tachycardia* |
| tele- | far: *telencephalon* |
| tens- | stretch: *tensor tympani* |
| tetra- | four: *tetrad* |
| therm- | heat: *thermogram* |
| thorac- | chest: *thoracic cavity* |
| thrombo- | lump, clot: *thrombocyte* |
| -tomy | cut: *appendectomy* |
| tox- | poison: *toxemia* |
| tract- | draw, drag: *traction* |
| trans- | across, over: *transfuse* |
| tri- | three: *trigone* |
| trich- | hair: *trichology* |
| -trophy | a state relating to nutrition: *hypertrophy* |
| -tropic | turning toward, changing: *gonadotropic* |
| ultra- | beyond, excess: *ultrasonic* |
| uni- | one: *unicellular* |
| -uria | urine: *polyuria* |
| uro- | urine, urinary organs or tract: *uroscope* |
| vas- | vessel: *vasoconstriction* |
| viscer- | organ: *visceral* |
| vit- | life: *vitamin* |
| zoo- | animal: *zoology* |
| zygo- | union, join: *zygote* |

# HUMAN

## ANATOMY

# &

## physiology

**Book Team**

Editor *Colin H. Wheatley*
Developmental Editor *Kristine Noel*
Production Editor *Jane E. Matthews*
Designer *Jeff Storm*
Art Editor *Mary E. Powers*
Photo Editor *John C. Leland*
Permissions Coordinator *Gail I. Wheatley*

## Wm. C. Brown Publishers
A Division of Wm. C. Brown Communications, Inc.

Vice President and General Manager *Beverly Kolz*
Vice President, Publisher *Kevin Kane*
Vice President, Director of Sales and Marketing *Virginia S. Moffat*
Vice President, Director of Production *Colleen A. Yonda*
National Sales Manager *Douglas J. DiNardo*
Marketing Manager *Craig S. Marty*
Advertising Manager *Janelle Keeffer*
Production Editorial Manager *Renée Menne*
Publishing Services Manager *Karen J. Slaght*
Royalty/Permissions Manager *Connie Allendorf*

## Wm. C. Brown Communications, Inc.

President and Chief Executive Officer *G. Franklin Lewis*
Senior Vice President, Operations *James H. Higby*
Corporate Senior Vice President, President of WCB Manufacturing *Roger Meyer*
Corporate Senior Vice President and Chief Financial Officer *Robert Chesterman*

Copyedited by Ann Mirels

Cover credits:
Hands and ears: © Wm. C. Brown Communications.
Mechanical gears: Photone-Letraset.
Running and jumping series: Courtesy George Eastman House.
Female model: © Butch Martin/The Image Bank.
Illustration of male: Courtesy Francis A. Countway Library of Medicine.
Eye: © Digital Stock.

The credits section for this book begins on page 967 and is considered an extension of the copyright page.

# ANATOMY

# HUMAN

## Concepts of

# & physiology

### fourth edition

**Kent M. Van De Graaff**
*University of Utah*

**Stuart Ira Fox**
*Pierce College*

**Wm. C. Brown Publishers**

Dubuque, IA  Bogota  Boston  Buenos Aires  Caracas  Chicago
Guilford, CT  London  Madrid  Mexico City  Sydney  Toronto

# [ brief contents ]

# [ contents ]

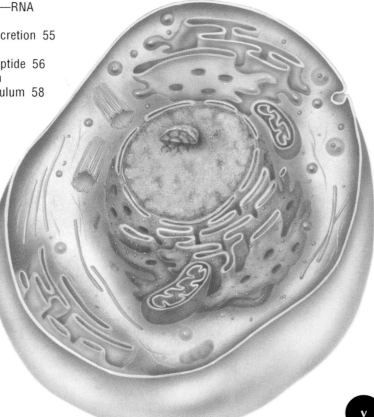

## chapter five

### Membrane Transport and the Membrane Potential 91

## chapter six

### Histology 106

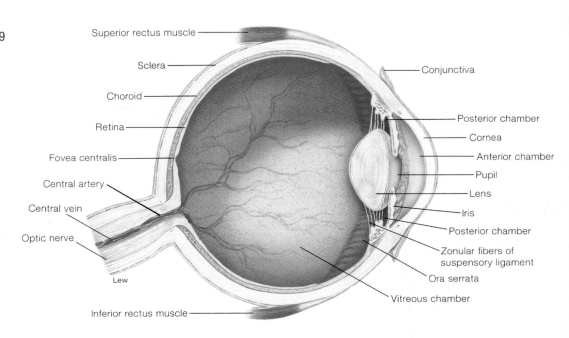

# UNIT 2

### Support and Movement of the Human Body 137

## chapter seven

### Integumentary System 137

## chapter eight

### Skeletal System: Bone Tissue and Development 159

Contents

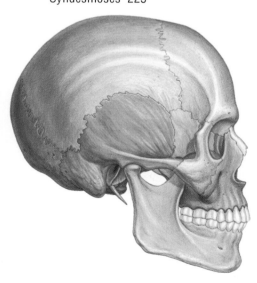

## UNIT 3

## Integration and Control Systems of the Human Body 344

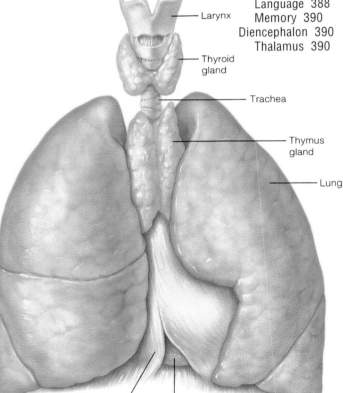

Larynx

Thyroid gland

Trachea

Thymus gland

Lung

Pericardium    Heart

Lew

Contents

## chapter eighteen
### Sensory Organs 464

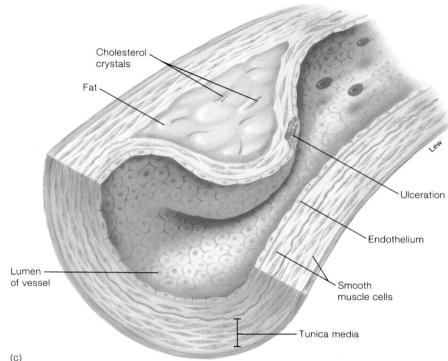

Cholesterol
crystals

Fat

Lew

Ulceration

Endothelium

Lumen
of vessel

Smooth
muscle cells

Tunica media

(c)

## chapter nineteen
### Endocrine System 513

### Regulation and Maintenance of the Human Body 549

## chapter twenty
### Circulatory System: Blood 549

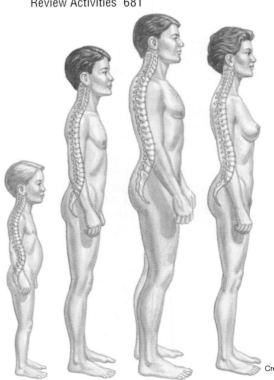

Creek

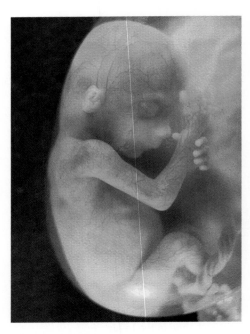

While the fourth edition of *Concepts of Human Anatomy and Physiology* has taken on a fresh, new look and has changed in other significant ways, we have made every effort to retain those features of previous editions that have contributed to the great popularity of this text over the years. Of major importance, the fourth edition is consistent with previous editions in its focus on unifying concepts as a means of integrating factual information. Just as important, a clear and interesting narrative, carefully rendered and attractive illustrations, and numerous pedagogical devices continue to be central in enabling students to assimilate a large body of information and to place what they have learned in a meaningful context.

As in previous editions, the material is organized so that instructors may tailor required text readings to their individual course needs. Because the text is designed for students who do not have extensive science backgrounds but who plan to enter health or other careers that require considerable knowledge of anatomy and physiology, the chapters in the opening unit present basic chemical, cellular, biological, and anatomical concepts. The chapters in the remaining four units then take a detailed approach to the anatomy and physiology of organs and systems. Throughout the text, we continue to promote the view of anatomy and physiology as dynamic sciences that serve as foundations for the health professions.

Having said this, we have no doubt that the fourth edition is the strongest by far. We are confident that it can be of immense value in helping students achieve learning objectives, in fostering in them a love of and respect for the science of human anatomy and physiology, and in persuading them to continue in the field.

# What's New, Revised, or Improved

Followers of previous editions will quickly note the major physical improvements in the fourth edition. These include larger headings and text type, exciting new design elements, and other new features. We have revised the narrative, and in some cases reorganized chapters, in accordance with the helpful suggestions of reviewers and users. We have also added new material in light of recent scientific findings, taking care to connect new developments to basic principles. Listed below are some of the major fourth-edition changes.

1. **New and Revised Illustrations** An already outstanding illustration program has been greatly improved in this edition, which features many new full-color illustrations. In addition, a large number of the illustrations used in previous editions have been substantially revised.

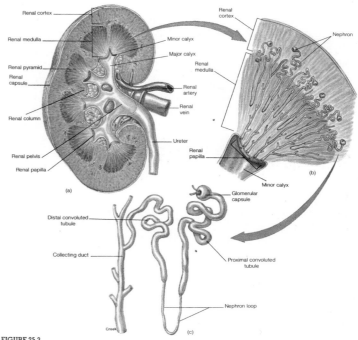

**FIGURE 25.3**
The internal structures of a kidney. (*a*) A coronal section showing the structure of the renal cortex, renal medulla, and renal pelvis. (*b*) A diagrammatic magnification of a renal pyramid and renal cortex to depict the tubules. (*c*) A diagrammatic view of a single nephron and a collecting duct.

In certain places, color coding has been introduced as a technique to aid learning. For example, the bones of the skull in chapter 9 are color coded so that each bone can be readily identified in the many renderings included in this chapter.

2. **Developmental Expositions** These features, entitled "**Under Development**" appear at the end of most systems chapters. Each discussion includes exhibits and explanations of the morphogenic events involved in the development of a body system. Placement at the end of a chapter ensures that the terminology needed to understand the embryonic structures has been introduced.

In a few chapters, an Under Development feature follows the relevant discussion of a specific body part or region; this occurs, for example, in sections on the skull, brain and spinal cord, ear and eye, and pituitary gland.

3. **Interrelationship Charts** New to this edition is a feature entitled **Nexus.** The Nexus page, which appears toward the end in each of the systems chapters, ties the functional aspects of one body system to each of the other systems, underscoring the concept of homeostasis. The gear-mechanism background art is symbolic of the body as a machine, in which all parts are interdependent in maintaining normal functioning.

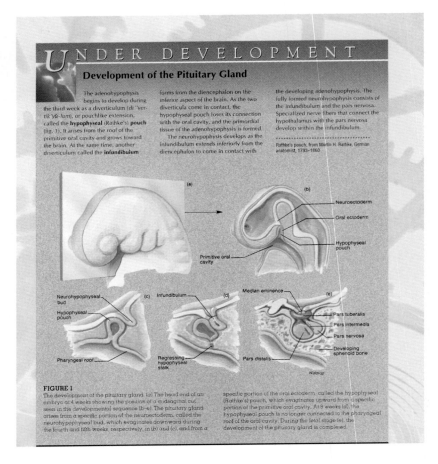

UNDER DEVELOPMENT

**Development of the Pituitary Gland**

The adenohypophysis begins to develop during the third week as a diverticulum (di "vertik'yŭ-lum), or pouchlike extension, called the **hypophyseal** (Rathke's) **pouch** (fig. 1). It arises from the roof of the primitive oral cavity and grows toward the brain. At the same time, another diverticulum called the **infundibulum** forms from the diencephalon on the inferior aspect of the brain. As the two diverticula come in contact, the hypophyseal pouch loses its connection with the oral cavity, and the primordial tissue of the adenohypophysis is formed.

The neurohypophysis develops as the infundibulum extends inferiorly from the diencephalon to come in contact with the developing adenohypophysis. The fully formed neurohypophysis consists of the infundibulum and the pars nervosa. Specialized nerve fibers that connect the hypothalamus with the pars nervosa develop within the infundibulum.

Rathke's pouch, from Martin H. Rathke, German anatomist, 1793–1860

**FIGURE 1**
The development of the pituitary gland. (a) The head end of an embryo at 4 weeks showing the position of a midsagittal cut seen in the developmental sequence (b–e). The pituitary gland arises from a specific portion of the neuroectoderm, called the neurohypophyseal bud, which evaginates downward during the fourth and fifth weeks, respectively, in (b) and (c), and from a specific portion of the oral ectoderm, called the hypophyseal (Rathke's) pouch, which evaginates upward from a specific portion of the primitive oral cavity. At 8 weeks (d), the hypophyseal pouch is no longer connected to the pharyngeal roof of the oral cavity. During the fetal stage (e), the development of the pituitary gland is completed.

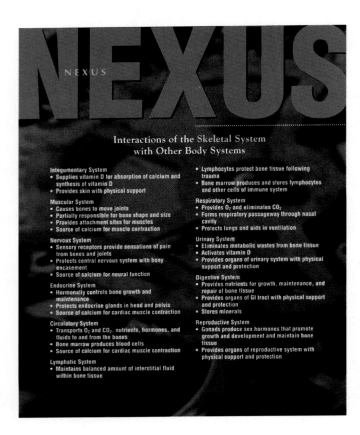

NEXUS

**Interactions of the Skeletal System with Other Body Systems**

**Integumentary System**
- Supplies vitamin D for absorption of calcium and synthesis of vitamin D
- Provides skin with physical support

**Muscular System**
- Causes bones to move joints
- Partially responsible for bone shape and size
- Provides attachment sites for muscles
- Source of calcium for muscle contraction

**Nervous System**
- Sensory receptors provide sensations of pain from bones and joints
- Protects central nervous system with bony encasement
- Source of calcium for neural function

**Endocrine System**
- Hormonally controls bone growth and maintenance
- Protects endocrine glands in head and pelvis
- Source of calcium for cardiac muscle contraction

**Circulatory System**
- Transports $O_2$ and $CO_2$, nutrients, hormones, and fluids to and from the bones
- Bone marrow produces blood cells
- Source of calcium for cardiac muscle contraction

**Lymphatic System**
- Maintains balanced amount of interstitial fluid within bone tissue

**Lymphatic System**
- Lymphocytes protect bone tissue following trauma
- Bone marrow produces and stores lymphocytes and other cells of immune system

**Respiratory System**
- Provides $O_2$ and eliminates $CO_2$
- Forms respiratory passageway through nasal cavity
- Protects lungs and aids in ventilation

**Urinary System**
- Eliminates metabolic wastes from bone tissue
- Activates vitamin D
- Provides organs of urinary system with physical support and protection

**Digestive System**
- Provides nutrients for growth, maintenance, and repair of bone tissue
- Provides organs of GI tract with physical support and protection
- Stores minerals

**Reproductive System**
- Gonads produce sex hormones that promote growth and development and maintain bone tissue
- Provides organs of reproductive system with physical support and protection

4. **Topic Icons** Throughout the text, short commentaries on clinical applications, development, and exercise physiology are set off in a distinct typeface and highlighted by the icons at right.

5. **New Content** A great deal of new information has been incorporated into this edition. The roles of nitric oxide as a regulatory molecule, the significance of NMDA receptors, and the requirements for T cell interactions are just a few examples of the new topics covered.

6. **Improved Flow of Information** Learning Objectives now appear at the beginning of each chapter so as not to interrupt the narrative. For the same reason, we have chosen to delete the Review Activities after each major section and to include the Clinical Investigations in the Instructor's Manual, where they can be easily duplicated and distributed to the class.

7. **Expanded Summaries** All of the end-of-chapter summaries have been reworked to reflect the text changes. These summaries are also more comprehensive, and they now include page references to the major sections.

8. **A More Personal Approach**   It has been our experience that beginning students in anatomy and physiology are often intimidated by a very formal, academic writing style. In this edition, the language has been relaxed to engage the reader and make learning more enjoyable. Simple analogies are frequently used to promote understanding of concepts. The level of difficulty has been carefully controlled, recognizing the wide variation in motivation and background that typifies a broad spectrum of students.

# Learning Aids: A Guide to the Student

The pedagogical devices in this text are designed to help you learn anatomy and physiology. Don't just read this text as you would a novel. Interact with it, using the pedagogical devices as tools. The more you use these tools, the more effective and enjoyable your study will become.

## Chapter Introductions

The opening page of each chapter contains an overview of the contents of the chapter in outline form. Page numbers are indicated to guide you to the major sections. Learning objectives are also included, and should be checked both before and after studying each section of a chapter.

## Concept Statements

One of the unique attributes of this text is the way in which major sections are introduced. Each of these sections is prefaced by a concept statement—a succinct expression of the main idea, or organizing theme, of the information presented in the section. These concept statements will help you gain an overview before encountering the details.

## Terminology Aids

Where each technical term first appears in the narrative, it is set off by boldface or italic type and is often followed by a phonetic pronunciation in parentheses. Pause in your readings to learn the correct pronunciation of a term so that you will be better able to remember it. The phonetic spellings in this edition have been revised, making them easier than ever to interpret.

## Word Derivations

The derivations of some of the terms in the text are provided in footnotes at the bottom of the page on which the term is introduced. Don't skip over these footnotes; they are often interesting in themselves. Furthermore, if you know how a word was derived, it becomes more meaningful and is easier to remember. You can identify the roots of each term by referring to the glossary of prefixes and suffixes on the inside front cover of the text.

## Topical Commentaries

Set off from the text narrative are short paragraphs highlighted by icons that discuss topics of practical value. You will find it enjoyable, as well as instructive, to see how your newly acquired basic science knowledge is applied in real-world situations.

## Illustrations and Tables

This text contains abundant tables and illustrations to support the concepts presented. Carefully studying the tables will enhance your understanding of the text. The summary tables will help you review for examinations. Although many of the figures can be admired for their beauty alone, it should be remembered that they were created for one primary purpose—to illustrate concepts presented in the text. Therefore, refer to the figures and analyze them as you read. Each one has been placed as close as possible to its text reference to spare you from flipping through pages.

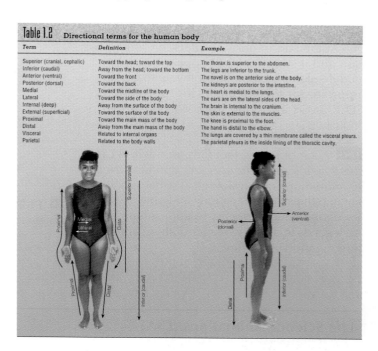

Table 1.2   Directional terms for the human body

| Term | Definition | Example |
|---|---|---|
| Superior (cranial, cephalic) | Toward the head; toward the top | The thorax is superior to the abdomen. |
| Inferior (caudal) | Away from the head; toward the bottom | The legs are inferior to the trunk. |
| Anterior (ventral) | Toward the front | The navel is on the anterior side of the body. |
| Posterior (dorsal) | Toward the back | The kidneys are posterior to the intestine. |
| Medial | Toward the midline of the body | The heart is medial to the lungs. |
| Lateral | Toward the side of the body | The ears are on the lateral sides of the head. |
| Internal (deep) | Away from the surface of the body | The brain is internal to the cranium. |
| External (superficial) | Toward the surface of the body | The skin is external to the muscles. |
| Proximal | Toward the main mass of the body | The knee is proximal to the foot. |
| Distal | Away from the main mass of the body | The hand is distal to the elbow. |
| Visceral | Related to internal organs | The lungs are covered by a thin membrane called the visceral pleura. |
| Parietal | Related to the body walls | The parietal pleura is the inside lining of the thoracic cavity. |

Following chapter 13 is a set of reference plates, including photographs of human cadaver dissections and several full-page newly rendered illustrations of the male and female trunk.

## Chapter Summaries

At the end of each chapter the material is summarized for you in outline form, following the sequence of the text narrative. Review each summary after studying the chapter to be sure that you have not missed any points. In addition, use the chapter summaries in preparing for examinations.

## Review Activities

A series of objective and essay questions follows each chapter summary. Answering these questions will provide you with feedback as to the depth of your learning and understanding. The answers to the objective questions are provided in Appendix A.

## Glossary

The glossary of terms at the end of the text is particularly noteworthy for its comprehensiveness. The definitions for almost all of the terms are accompanied by pronunciation keys, and synonyms are indicated as appropriate. The majority of the terms in the glossary are accompanied by a page number indicating where the term is discussed in the text marrative. (Adjectival terms and general terms are not page referenced.) This useful feature is new to this edition. Look to the glossary as you review to check your understanding of the technical terminology.

## Software Icon

This fourth edition introduces two new and exciting learning tools for the student.

At the end of most chapters, a CD-ROM icon and a statement will appear that lists a module of the *Explorations* software that is appropriate for the chapter. The *Explorations* CD-ROM is available from Wm. C. Brown Publishers.

A set of five videotapes, entitled *WCB Life Science Animations* (*LSA*), contains over 50 animations of physiological processes integral to the study of anatomy and physiology. A videotape icon appears in appropriate figure legends to alert the reader of these animations, as shown in figure 3.1.

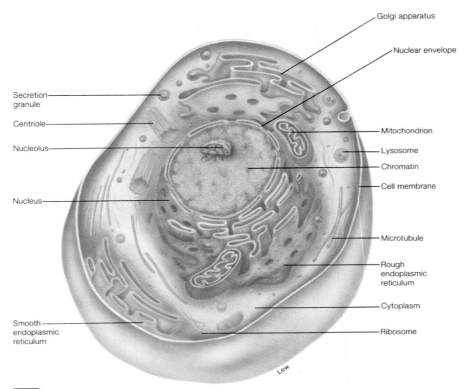

Golgi apparatus
Nuclear envelope
Secretion granule
Centriole
Mitochondrion
Nucleolus
Lysosome
Chromatin
Cell membrane
Nucleus
Microtubule
Rough endoplasmic reticulum
Cytoplasm
Smooth endoplasmic reticulum
Ribosome
Lew

### FIGURE 3.1
A generalized cell and the principal organelles.

## Life Science Animations (LSA)

| | |
|---|---|
| Figure 2.5 LSA 1 | Dissociation of sodium and chlorine |
| Figure 3.1 LSA 2 | Generalized cell |
| Figure 3.4 LSA 3 | Stages of endocytosis |
| Figure 3.19 LSA 16 | RNA synthesis |
| Figure 3.21 LSA 17 | Genetic code |
| Figure 3.23 LSA 17 | Translation of mRNA |
| Figure 3.25 LSA 4 | Golgi apparatus |
| Figure 3.26 LSA 15 | Replication of DNA |
| Figure 3.30 LSA 12 | Stages of mitosis |
| Figure 3.32 LSA 13 | Meiosis |
| Figure 3.33 LSA 14 | Genetic variation |
| Figure 4.13 LSA 11 | Exergonic and endergonic reactions |
| Figure 4.15 LSA 11 | ATP |
| Figure 4.18 LSA 5 | Glycolysis |
| Figure 4.23 LSA 6 | Krebs cycle |
| Figure 4.24 LSA 7 | Electron transport and oxidative phosphorylation |

# Supplementary Materials

The supplementary materials that accompany the text are designed to help students in their learning activities and to guide instructors in planning course work and presentations. Following are brief descriptions of these supplements.

1. A *Laboratory Manual* written by Stuart I. Fox and Kent M. Van De Graaff to accompany the fourth edition of *Concepts of Human Anatomy and Physiology* requires that students read the text for relevant background information before attempting the laboratory exercises. It helps students gain the laboratory experience required to support the lecture portion of the course and includes exercises that have been repeatedly classroom tested.

2. The *Instructor's Manual for the Laboratory Manual* provides the answers to the questions that appear in the laboratory reports in the Laboratory Manual.

3. A *Student Study Guide* written by Kent M. Van De Graaff to accompany the fourth edition of *Concepts of Human Anatomy and Physiology* features the concept statements from the text, focus questions, mastery quizzes, study activities, and answer keys with explanations.

4. An *Instructor's Manual and Test Item File* prepared by Jeffrey and Karianne Prince provides instructional support in the use of the textbook. It also contains a test item file with approximately 70 items for each chapter to aid instructors in constructing examinations.

5. The following slide sets accompany the textbook and are free to adopters:
   a. one hundred histology slides showing basic tissue types representative of all of the human body systems;
   b. seventy clinical application slides, depicting important pathological conditions; and
   c. twenty-five radiographic slides.
   Descriptions of these slides are included in the Instructor's Manual.

6. A set of 200 acetate transparencies is available to instructors who adopt this text. The transparencies were made from text illustrations chosen for their value in reinforcing lecture presentations.

7. All of the illustrations from the transparency set are collected in a *Student Study Art Notebook* available to students. With this notebook at their desk, students no longer have to worry about whether they will be able to see leader lines and labels in a large lecture hall. They can make notes right on the illustrations. Also available is the Pictorial Review Notebook, which contains the same illustrations without the labels.

8. *Knowledge Map of the Human Anatomy Systems* prepared by Craig Gundy is a computer tutorial developed for the Macintosh. The computer disk set (13 diskettes) provides an anatomical review of all of the body systems. A disk icon appears at the end of many chapters that contain material applicable to this software.

9. *Study Cards for Anatomy and Physiology* prepared by Kent M. Van De Graaff, R. Ward Rhees, and Christopher H. Creek is a boxed set of 300 3-by-5-inch cards. It serves as a well-organized and illustrated synopsis of the structure and function of the human body. Clinical information is presented as it applies to specific body organs and systems or physiological processes. The Study Cards offer a quick and effective way for students to review human anatomy and physiology.

10. Microtest—a new computerized testing service for generating examinations and quizzes—is available to instructors who adopt the text.

11. The WCB Anatomy & Physiology Video Series consists of
    a. Human Skeletal Musculature System (90 min., color; ISBN 15210)
    b. Introduction to the Human Cadaver and Prosection (30 min., color; ISBN 11177)
    c. Introduction to Cat Dissection: Musculature (55 min., color; ISBN 11630)
    d. Blood Cell Counting, Identification, & Grouping (30 min., color; ISBN 11629)
    e. Internal Organs and the Circulatory System of the Cat (58 min., color; ISBN 13922)
    f. Review of the Human Skeletal System (51 min., color; ISBN 13243)
    These exceptional videotapes are available free to adopters.

12. *Coloring Review Guide for Anatomy and Physiology* prepared by Robert J. Stone and Judith A. Stone is a plastic comb-bound book emphasizing learning through the process of color association. The Coloring Guide provides a thorough review of anatomical and physiological concepts. (ISBN 17109)

13. *Atlas of the Skeletal Muscles* prepared by Robert J. Stone and Judith A. Stone is a guide to the structure and function of human skeletal muscles. The illustrations help students locate muscles and understand their actions. (ISBN 10618)

14. *Case Histories in Human Physiology* compiled by Donna Van Wynsberge and Gregory M. Colley, M.D., affords students opportunities for integrating their thinking and for problem solving. An answer key is available for instructors. (ISBN 11606)

15. *Laboratory Guide: Human Anatomy and Physiology with Clinical Applications—Cat Version* written by Stuart I. Fox, Kent M. Van De Graaff, and Laurence G. Thouin, Jr., is designed for anatomy and physiology courses that use a body systems' approach to teach structure and function of the human body. (ISBN 05687)

# Acknowledgments

**This book could not have been written without the enduring patience and support of our wives, Karen Van De Graaff and Ellen Fox, to whom this book is gratefully dedicated.**

Many of the improvements in the fourth edition of *Concepts of Human Anatomy and Physiology* came about through comments that we received from the many users of previous editions. Although it would be impossible in this space to acknowledge them individually, we are deeply grateful to each one. As in the past, our colleagues at our respective institutions were very supportive and helpful. In particular, we would like to thank professors Michael J. Shively, Lawrence H. Thouin, R. Ward Rhees, James Rikel, William M. Hess, and Ferron L. Andersen

We also wish to thank physicians who assisted in specific ways. Drs. Eric J. Van De Graaff, J. Phillip Freestone, Douglas W. Hacking, and Charles H. Stewart provided professional advice. Dr. Brent C. Chandler provided many of the radiographs used in the text. Drs. James N. Jones and Paul Urie assisted in updating the clinical information.

Quality illustrations for this text were provided by a number of talented artists. We are especially grateful for their tremendous contributions. Many of the renderings new to this edition were contributed by Christopher H. Creek.

The editorial and production staffs at Wm. C. Brown Publishers inspired, guided, and shaped this enormous project, and they were superb to work with. We owe a large debt of gratitude to Acquisitions Editor Colin Wheatley, Developmental Editor Kristine Noel, Production Editor Jane Matthews, and many other talented professionals at Wm. C. Brown. We are also especially appreciative of Ann Mirels, who laboriously copyedited the manuscript and provided numerous helpful suggestions.

# Reviewers

The forthright criticisms and helpful suggestions of a knowledgeable and hard-working panel of reviewers added immeasurably to the quality of the final draft. The review panel for the fourth edition included

Elizabeth A. Becker
*Elgin Community College*

Idna M. Corbett
*West Chester University*

John G. Dziak
*Community College of Allegheny County, Allegheny Campus*

Kathryn L. Hedges
*Indiana University Northwest*

Stan Hillman
*Portland State University*

Robert L. Knuden
*San Joaquin Delta College*

Florence Levin
*Edison Community College*

Ibrahim Y. Mahmoud
*University of Wisconsin-Oshkosh*

Ben R. Mayne
*Saginaw Valley State University*

Marlene McCall
*Community College of Allegheny County*

John P. Messick
*Missouri Southern State College*

Holly J. Morris
*Lehigh Carbon Community College*

William S. Morrison
*Slippery Rock University of Pennsylvania*

Colleen Nolan
*St. Mary's University*

Thomas E. Oldfield
*Ferris State University*

Justicia Opoku-Edusei
*University of Maryland—College Park*

Charles M. Page
*El Camino College*

Lisa A. Perfetto
*West Chester University*

Pradeep D. Pilakel
*Anson County Hospital*

Marjorie Plummer
*Faulkner University*

David G. Poedel
*Pima Community College*

Mark A. Shoop
*Macon College*

Paul S. Sunga
*Langara College*

P. Stephen Taraskevich
*Nova-Southeastern University*

Charles H. Woodward
*Shepherd College*

It is our sincere hope that students and instructors will enjoy using this text and that it will serve them well. As always, we welcome suggestions and comments for future editions.

[ chapter one ]

# introduction to anatomy and physiology

## objectives

- Discuss the key historical events in the development of anatomy and physiology.

- Explain what is meant by the scientific method.

- Describe the taxonomic classification of humans.

- List the characteristics that identify humans as chordates and mammals.

- Describe the different levels of organization in the human body.

- Identify the planes of reference used to depict structural arrangement.

- Describe the anatomical position and use the descriptive and directional terms that refer to body structures, surfaces, and regions.

- List the regions of the body and the localized areas within each region.

- Identify the body cavities and the organs found within each.

- Define *homeostasis* and describe its significance.

- Describe the components of a negative feedback loop and explain how negative feedback helps to maintain homeostasis.

# The Sciences of Anatomy and Physiology

*Anatomy and physiology are integrated, dynamic sciences with an exciting heritage. These sciences provide the foundation for personal health and clinical applications.*

Human anatomy and physiology are sciences concerned with the structure and function of the body. The term **anatomy** (ă-nat´ŏ-me) is derived from a Greek word meaning "to cut up"; in the past, the word *anatomize* was more commonly used than the word *dissect*. Dissection of human *cadavers* (kă-dav´erz) is the basis for understanding the structure of the human body. **Physiology** (fiz˝e-ol´ŏ-je) is derived from a Greek word meaning the study of nature—the nature of an organism is its function. Physiology attempts to explain how the body functions through physical and chemical processes. Much of the knowledge of physiology is gained through experimentation.

Anatomy and physiology are both subdivisions of the science of *biology,* the study of living organisms. Frequently, anatomy and physiology are studied as separate disciplines, in which case the anatomy of the body is learned before its physiology. However, since anatomical structures are adapted to perform specific functions, a proper understanding of structure and function is best achieved through an integrated study.

## Historical Development

The study of human anatomy and physiology has had a rich, long, and frequently troubled heritage. Its history parallels that of medicine because interest in the structure and function of the body frequently grew out of a desire to understand and treat body dysfunctions.

**Grecian Period**   It was in ancient Greece that anatomy and physiology first found wide acceptance as sciences. *Hippocrates* (460–337 B.C.) was a famous Greek physician who is regarded as the father of medicine (fig. 1.1). Perhaps the greatest contribution of Hippocrates was his attribution of disease to natural causes rather than to the displeasure of the gods. His application of logic and reason to medical study marked the beginning of observational medicine.

The field of medicine at the time of Hippocrates held to the notion that an individual's health depended upon the balance of four body fluids, or **humors.** This so-called

---

anatomy: Gk. *ana,* up; *tome,* a cutting
cadaver: L. *cadere,* to fall
physiology: Gk. *physis,* nature; *logos,* study

**FIGURE 1.1**

A fourteenth-century painting of the famous Greek physician Hippocrates. Hippocrates is referred to as the father of medicine; his creed is immortalized as the Hippocratic oath.

*humoral theory* suggested that, if blood, bile, black bile, and phlegm were balanced, the person would be healthy and have an even disposition. If blood was the predominant humor, one was said to have a *sanguine* personality—courageous and passionate. If there was too much bile, one was *choleric*—angry and mean. A *melancholic* personality—moody and depressed—resulted from an overproduction of black bile. Too much phlegm resulted in a *phlegmatic* personality—sluggish and apathetic. Although medicine has long since abandoned this explanation of health and personality, it is interesting that these terms are still used in our language. The term *sanguine,* however, has evolved to refer simply to the cheerfulness and optimism that accompanied a sanguine personality, and no longer refers directly to the humoral theory.

*Aristotle* (384–332 B.C.) made careful investigations of all kinds of animals, including humans, and pursued a limited type of scientific method in obtaining data (fig. 1.2). He wrote the first known account of embryology, in which he

---

humor: L. *humor,* fluid
sanguine: L. *sanguis,* bloody
choler: Gk. *chole,* bile
melancholic: Gk. *melan,* black; *chole,* bile
phlegm: Gk. *phlegm,* inflammation

described the development of the heart in a chick embryo. His best-known zoological works are *History of Animals, Parts of Animals*, and *Generation of Animals*.

Despite his tremendous accomplishments, Aristotle perpetuated some erroneous theories regarding human anatomy. For example, he disagreed with Plato, who had described the brain as the seat of feeling and thought, and proclaimed the heart to be the seat of intelligence. Aristotle thought that the function of the brain, which was bathed in fluid, was to cool the blood that was pumped from the heart, and thus maintain body temperature.

**FIGURE 1.2**
This Roman copy of a Greek sculpture is believed to be of Aristotle, the famous Greek philosopher.

The Greek scientist *Erasistratus* (about 300 B.C.) was more interested in body functions than structure, and is therefore frequently referred to as the father of physiology. Erasistratus authored a book on the causes of diseases, in which he included observations on the heart, vessels, brain, and cranial nerves. He noted the toxic effects of snake venom on various visceral organs and described changes in the liver resulting from certain metabolic diseases. Although some of the writings of Erasistratus were scientifically accurate, others were based on primitive and mystical concepts. For instance, he thought that the cranial nerves carried animal spirits and that muscles contracted because of distention by spirits.

Both Erasistratus and another Greek philosopher, *Herophilus*, (about 325 B.C.), were greatly criticized later in history for their use of *vivisection* (viv˝ĭ-sek´shun), the dissection of living animals. Herophilus was described as a butcher who had dissected as many as 600 living human beings, some of them in public demonstrations.

**Roman Era**   In many respects, the Roman Empire stifled scientific advancements and set the stage for the Dark Ages. The interest and emphasis of science shifted from the theoretical to the practical under Roman rule. Few dissections of cadavers were performed other than at autopsies in attempts to determine the cause of death in criminal cases. Medicine was not preventive but was limited, almost without exception, to the treatment of soldiers injured in battle.

*Claudius Galen* (A.D. 130–201) was the most famous Roman physician of his time and the most influential writer to date on medical subjects. For nearly 1,500 years, the writings of Galen represented the ultimate authority on anatomy and medical treatment. Galen probably dissected no more than two or three human cadavers during his career, of necessity limiting his anatomical descriptions to animal dissections. He compiled nearly 500 medical papers (of which 83 have been preserved) from earlier works of others, as well as from his personal studies. Galen believed in the humors of the body and perpetuated this concept. He also gave authoritative explanations for nearly all body functions.

Galen's works contain many errors, primarily because of his desire to draw definitive conclusions regarding human body functions on the basis of data obtained largely from nonhuman animals. He did, however, provide some astute and accurate anatomical details that are still regarded as classics. He proved to be an experimentalist, demonstrating that the heart of a pig would continue to beat when the spinal nerve was transected so that nerve impulses could not reach the heart. He showed that the squealing of a pig stopped when the particular nerve that innervated its vocal cords was cut. He also proved that arteries contained blood rather than air.

**Middle Ages**   The Middle Ages (Dark Ages) came with the fall of the Roman Empire in A.D. 476 and lasted nearly 1,000 years. Dissections of cadavers were totally prohibited during this period, and molesting a corpse was a criminal act that was frequently punished by burning at the stake. If mysterious deaths occurred, examinations by inspection and palpation were allowed. During the plague epidemic in the sixth century, however, a few *necropsies* (nek´rop-sēz) and dissections were performed in hopes of determining the cause of this dreaded disease.

**Renaissance**   The period known as the Renaissance was characterized by a rebirth of science. It lasted roughly from the fourteenth through the sixteenth century and was a period of transition between the Middle Ages and the modern age of science. The development of movable type in about

vivisection: L. *vivus*, living; *sectus*, to cut

autopsy: Gk. *autopsia*, seeing with one's own eyes
necropsy: Gk. *nekros*, corpse; *opsy*, view

1450 revolutionized the production of printed books and helped to usher in the Renaissance.

The major advancements in anatomy that occurred during the Renaissance were in large part due to the artistic and scientific ability of *Andreas Vesalius* (1514–64). By the time he was 28 years old, Vesalius had completed the masterpiece of his life, *De Humani Corporis Fabrica*, in which he beautifully illustrated and described the various body systems and individual organs (fig. 1.3). Because of the eventual impact of this book, Vesalius is often called the father of anatomy. His book was especially important because it boldly challenged Galen's erroneous teachings. Vesalius wrote of his surprise upon finding numerous anatomical errors that were being taught as fact, and he refused to accept Galen's explanations on faith. Bitter controversies ensued between Vesalius and the traditional Galenic anatomists, including Vesalius's former teacher, Sylvius. Vesalius became so unnerved by the relentless attacks on him that he destroyed much of his unpublished work and ceased his dissections.

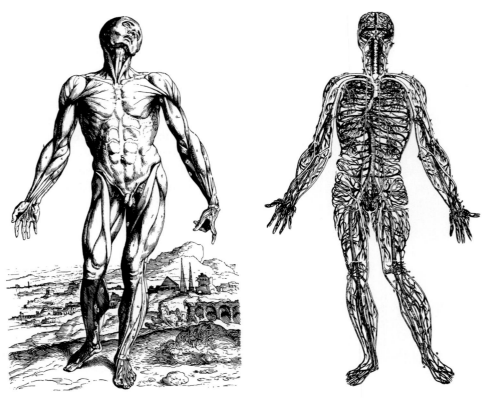

**FIGURE 1.3**
Plates from *De Humani Corporis Fabrica*, which Vesalius completed at the age of 28. This book, published in 1543, revolutionized anatomy and physiology.

**Seventeenth and Eighteenth Centuries**   Two of the most significant contributions to anatomy and physiology of the seventeenth and eighteenth centuries were the explanation of blood flow and the development and use of the microscope.

In 1628, the English physician *William Harvey* (1578–1657) published his pioneering work *On the Movement of the Heart and Blood in Animals*. Not only did this brilliant research establish proof of the continuous circulation of blood within vessels, it also provided a classic example of the scientific method of investigation (fig. 1.4). Harvey is widely regarded as the father of modern physiology, although like Vesalius, he was severely criticized in his time for his departure from Galenic philosophy. The controversy over the circulation of the blood raged for 20 years until other anatomists finally repeated Harvey's experiments and confirmed his observations.

*Antoni van Leeuwenhoek* (1632–1723) was a Dutch lens grinder who so improved the microscope that he achieved a magnification of 270 times. His many contributions included developing techniques for examining tissues and describing blood cells, spermatozoa, and the striated appearance of skeletal muscle.

The development of the microscope added an entirely new dimension to anatomy and physiology and eventually led to explanations of basic body functions. In addition, the

**FIGURE 1.4**
In the early seventeenth century, the English physician William Harvey demonstrated that blood circulates and does not flow back and forth through the same vessels.

improved microscope was invaluable for understanding the etiologies (causes) of diseases, and thus for discovering cures for many of them.

## Nineteenth and Twentieth Centuries

The major contribution in the nineteenth century was the formulation of the biological principle known as the **cell theory** and the implications it had for a clearer understanding of the structure and functioning of the body. According to the cell theory, all living organisms are composed of cells and the products of cells.

*Johannes Müller* (1801–58), a German physiologist and comparative anatomist, is noted for applying the sciences of physics, chemistry, and psychology to the study of the human body. *Claude Bernard* (1813–78), a French physiologist, extended the experimental study of physiology, and is often regarded as the father of experimental medicine. The study of anatomy and physiology during the twentieth century has become highly specialized, and the research more detailed and complex. In response to the increased technology and depths of understanding, new disciplines and specialties have emerged (fig. 1.5). The explosive

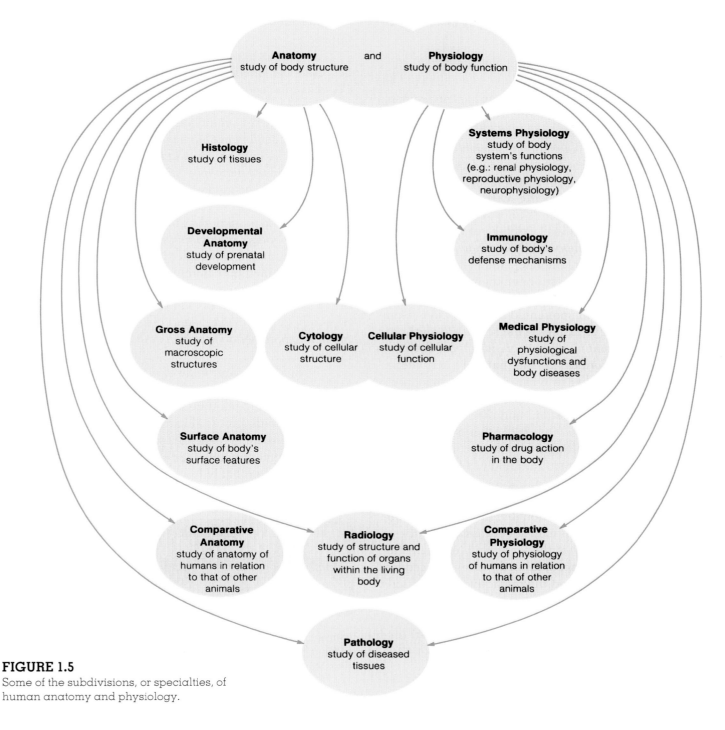

### FIGURE 1.5

Some of the subdivisions, or specialties, of human anatomy and physiology.

growth of knowledge in these specialties in recent times has made it nearly impossible for any one individual to remain expert in them all.

As may be implied from figure 1.5, anatomy and physiology are dynamic, applied sciences that are constantly changing as new discoveries are made. Keeping up with the changes requires a solid and broad understanding of these sciences, hence, an objective of this text is to provide such a foundation for the student. An excellent way to keep up with anatomy and physiology after you complete the formal course is to read scientific magazines like *Science, Scientific American, Discover,* and *Science News.* If you are to be an educated contributor to society, it is essential to become and stay informed.

## Scientific Method

Researchers in anatomy and physiology employ the processes of scientific inquiry in conducting their investigations and making discoveries. Although the boundaries of scientific investigations leave room for considerable creativity, they progress in well-defined, orderly ways. Such a disciplined approach to gaining information (facts) about the world is referred to as the **scientific method.** Simply stated, the scientific method depends on a systematic search for information and a continual checking and rechecking to see whether previous ideas still hold up in the light of new information. Actually, the scientific method may be used in seeking answers to questions encountered in everyday life, as well as in methodical research.

All of the information presented in this text has been gained through the application of the scientific method. Although many different techniques are involved, all share three attributes: (1) confidence that natural phenomena are ultimately explainable in terms we can understand; (2) descriptions and explanations of the natural world that are honestly based on observations and that are subject to modification or refutation as a result of other observations; and (3) humility, or the willingness to accept the fact that we could be wrong. If further study should yield conclusions that refuted all or part of an idea, the idea would have to be modified accordingly. In short, the scientific method is based on a confidence in our rational ability, honesty, and humility. Practicing scientists may not always display these attributes, but the validity of the large body of scientific knowledge that has been accumulated—as evidenced by technological applications and the predictive value of scientific hypotheses—are ample testimony to the fact that the scientific method works.

The scientific method involves specific steps. In the first step, a **hypothesis** is formulated. In order for this hypothesis to be scientific, it must be *testable*; that is, it must be open to possible refutation by experiments or other observations of the natural world. For example, one might hypothesize that people who exercise regularly have a lower resting pulse rate than people who don't. Experiments are conducted, or other observations are made, and the results are analyzed. Conclusions are then drawn as to the validity of the hypothesis. If the hypothesis survives such testing, it might be incorporated into a more general **theory.** Scientific theories are statements about the natural world that incorporate a number of proven hypotheses. They serve as a logical framework by which these hypotheses can be interrelated and provide the basis for predictions that may as yet be untested.

The hypothesis in the preceding example is scientific because it can be tested. The pulse rates of 100 athletes and 100 sedentary people can be measured to see whether there are statistically significant differences. If there are, the statement that athletes, on the average, have lower resting pulse rates than sedentary people is justified *based on this data.* But one must still keep in mind that this conclusion could be wrong. Before the discovery could become generally accepted as fact, other scientists would have to consistently replicate the results. Scientific theories are based on *reproducible data.*

It is quite possible that in attempting to replicate the experiment, other scientists will obtain slightly different results. They may, for example, construct scientific hypotheses that the differences in resting pulse rate also depend on such factors as the nature of the exercise performed, or on other variables. When scientists attempt to test these hypotheses, they will likely encounter new problems, requiring new hypotheses, which then must be tested by additional experiments.

In this way, a large body of highly specialized information is gradually accumulated and a more generalized explanation (a scientific theory) can be formulated. This explanation will almost always be different from preconceived notions. People who follow the scientific method will then appropriately modify their concepts, realizing that their new ideas will probably have to be changed again in the future as additional experiments are performed.

# Classification and Characteristics of Humans

*Humans are biological organisms belonging to the phylum Chordata within the kingdom Animalia and to the family Hominidae within the class Mammalia and the order Primates.*

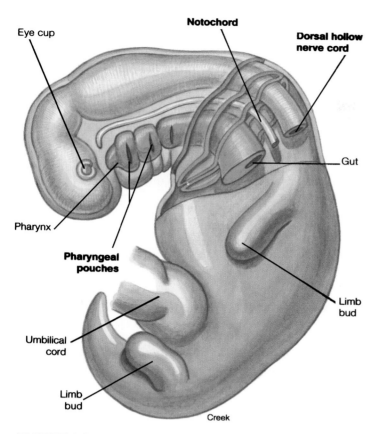

**Eye cup**

**Notochord**

**Dorsal hollow nerve cord**

**Gut**

**Pharynx**

**Pharyngeal pouches**

**Limb bud**

**Umbilical cord**

**Limb bud**

Creek

**FIGURE 1.6**

A schematic diagram of a chordate embryo. The three diagnostic chordate characteristics are indicated in bold type.

Intervertebral discs

Nucleus pulposus

**Intervertebral disc**

Creek

**Vertebral column**

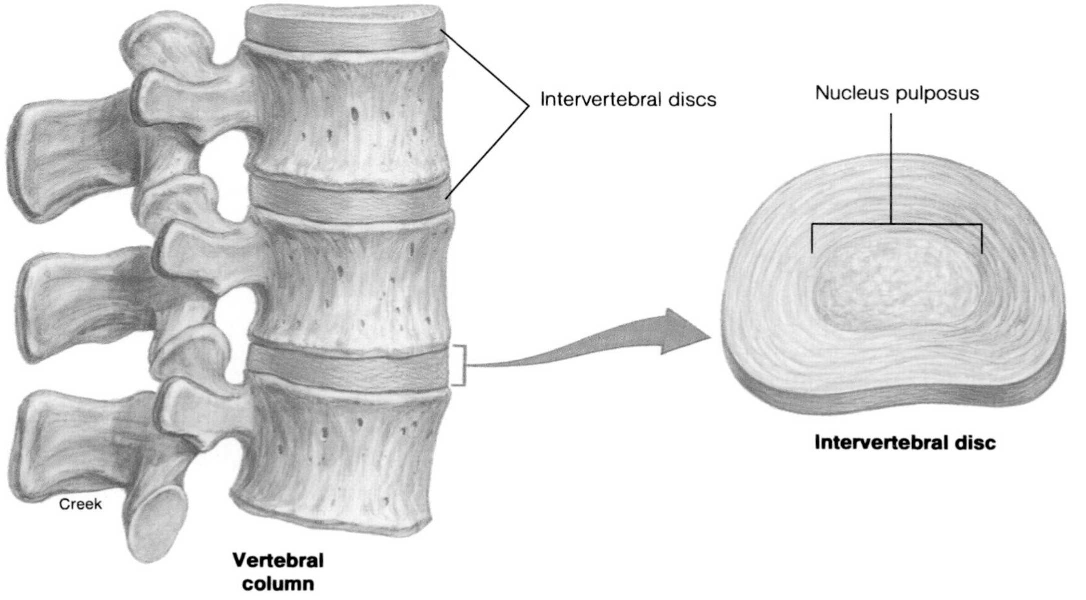

**FIGURE 1.7**

A lateral view of the vertebral column showing the intervertebral discs and, to the right, a superior view of an intervertebral disc showing the nucleus pulposus.

# Taxonomic Scheme

The classification, or taxonomic, scheme has been established by biologists to organize the structural and evolutionary relationships of living organisms. Each category of classification is referred to as a *taxon*. The highest taxon is the kingdom, and the most specific taxon is the species. Humans are species belonging to the **animal kingdom.** *Phylogeny* (*fi-loj´ĕ-ne*) is the evolutionary development and history of species or higher taxa of organisms. A phylogenetic tree of animal taxa can be constructed, much like a family tree, to show the geneology and relationships of different animal groups.

Humans also belong to the **phylum Chordata** (*fi´lum kor-dă-tă*) and **subphylum vertebrata,** along with fish, amphibians, reptiles, birds, and other mammals. All chordates have three structures in common: a **notochord** (*no´tŏ-kord´´*), a **dorsal hollow nerve cord,** and **pharyngeal** (*fă-rin´je-al*) **pouches** (fig. 1.6). These chordate characteristics are well expressed during the embryonic stage of development, and to a certain extent are present in an adult. The notochord is a flexible rod of tissue that extends along the back of an embryo. A portion of the notochord persists in the adult as the *nucleus pulposus*, which is the gelatinous center within each intervertebral disc (fig. 1.7). The dorsal hollow nerve cord is positioned above the notochord and develops into the *brain* and *spinal cord*. Pharyngeal pouches form gill openings in fish and some amphibians. In other chordates, such as humans, embryonic pharyngeal pouches develop, but only one of the pouches persists, becoming the *auditory canal* (or eustachian tube), a connection between the middle-ear cavity and pharynx (*far´ingks*), or throat area.

Humans are in the **class mammalia.** Mammals are vertebrates that possess hair and mammary glands. Hair is a protective covering for most mammals, and mammary glands serve for suckling the young. Other characteristics of mammals include three small auditory ossicles (ear bones), a fleshy outer ear, (auricle), heterodont dentition (teeth, such as incisors and molars, that are shaped differently), a temporomandibular joint (a joint between the lower

## Table 1.1 Classification scheme of human beings

| Taxon | Designated grouping | Characteristics |
|---|---|---|
| Kingdom | Animalia | Eucaryotic cells that lack walls, plastids, and photosynthetic pigments |
| Phylum | Chordata | Dorsal hollow nerve cord; notochord; pharyngeal pouches |
| Subphylum | Vertebrata | Vertebral column |
| Class | Mammalia | Mammary glands; hair |
| Order | Primates | Well-developed brain; prehensile hands |
| Family | Hominidae | Large cerebrum; bipedal locomotion |
| Genus | Homo | Flattened face; prominent chin and nose with inferiorly positioned nostrils |
| Species | sapiens | Largest cerebrum |

jaw and skull), usually seven cervical vertebrae, an attached placenta, well-developed facial muscles, a muscular diaphragm, and a four-chambered heart with a left aortic arch.

Humans are in the **order primates,** along with monkeys and apes. Members of this order have prehensile hands (with the digits modified for grasping) and relatively large, well-developed brains.

Humans are the sole living members of the **family Hominidae.** *Homo sapiens* is included within this family, to which all ethnic groups of humans belong. The taxonomic classification of humans is presented in table 1.1.

### Human Characteristics

Human beings have a few anatomical characteristics that are so specialized that they are diagnostic in separating them from other animals and even from other closely related mammals. Humans also have characteristics that are equally well developed in other animals, but when these function with the human brain, they provide remarkable capabilities. The anatomical characteristics of humans include the following:

**1 A large, well-developed brain.** The average human brain weighs between 1,350 and 1,400 grams (about 3 pounds). This gives humans a large brain-to-body-weight ratio. But more

important is the development of portions of the brain. Certain extremely specialized regions and structures within the brain account for emotion, thought, reasoning, memory, and precise, coordinated movement.

**2 Bipedal locomotion.** Because humans stand and walk on two appendages, their style of locomotion is said to be *bipedal.* Upright posture imposes certain other diagnostic structural features such as the *sigmoid* (S-shaped) *curvature* of the spine, the anatomy of the hip and thighs, and arched feet.

**3 An opposable thumb.** The human thumb joint is structurally adapted for great versatility in grasping objects. Most primates have opposable thumbs.

**4 Well-developed vocal structures.** Humans, like no other animals, have developed articulated speech. The anatomical structure of the vocal organs and the well-developed brain have made this possible.

**5 Stereoscopic vision.** Although this characteristic is well developed in several other animals, it is also keen in humans. Human eyes are directed forward so that when focused upon an object, it is viewed from two angles. Stereoscopic vision gives us depth perception, or a three-dimensional image.

Humans also differ from other animals in the number and arrangement of vertebrae (vertebral formula), the kinds and number of teeth (tooth formula), the degree of development of facial muscles (allowing for a wide range of facial expression), and certain distinguishing features of various body organs.

## Body Organization

*There are different levels of structure and function in the human body, with each level contributing to the total organism.*

### Cellular Level

The **cell** is the basic structural and functional component of life. Humans are multicellular organisms composed of 60 to 100 trillion cells. It is at the cellular level (fig. 1.8) that such vital functions of life as metabolism, growth, irritability (the ability to respond to stimuli), repair, and reproduction are carried on. All cells contain a semifluid substance called **protoplasm** (pro´tŏ-plaz´em). Certain protoplasmic structures and molecules are arranged into small functional units called **organelles** (or˝gă-nelz´). Each organelle carries out a specific function within the cell.

Primates: L. *primas,* first
prehensile: L. *prehensus,* to grasp

bipedal: L. *bi,* two; *pedis,* foot
cell: L. *cella,* small room
protoplasm: Gk. *protos,* first; *plassein,* to mold

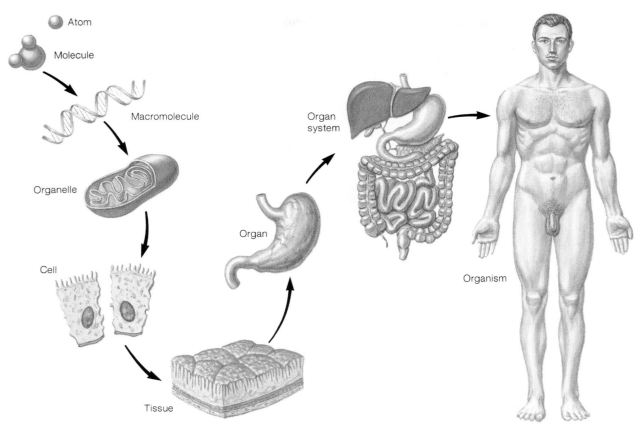

**FIGURE 1.8**
Levels of structural organization and complexity within the human body.

The body contains many distinct kinds of cells, each specialized to perform specific functions. Examples of specialized cells are bone cells, muscle cells, fat cells, blood cells, and nerve cells. Each of these cell types has a unique structure directly related to its function.

## Tissue and Organ Levels

**Tissues** are layers or aggregations of similar cells that perform specific functions. The entire body is composed of only four primary tissue types: epithelial tissues, connective tissues, muscular tissues, and nervous tissues (chapter 6). The outer layer of skin, for example, is a tissue (epithelium) because it is composed of similar cells, bound together, that perform specific functions.

An **organ** is an aggregate of two or more tissues that are integrated to perform a particular function. Organs occur throughout the body and vary greatly in size and function. Examples of organs are the heart, spleen, pancreas, ovary, skin, and even any of the bones within the body. Most organs contain all four primary tissues. In the stomach, for example, the inside epithelial lining performs the functions of secretion and absorption. The wall of the stomach, however, also contains muscle tissue (for contractions of the stomach), nervous tissue (for regulation), and connective tissue, which binds the other tissues together.

## System Level

The **systems** of the body constitute the next level of structural organization. A body system consists of various organs that have similar or related functions. Examples of systems are the circulatory system, nervous system, digestive system, and endocrine system. Certain organs may serve several systems. The pancreas, for example, functions with both the

tissue: Fr. *tissu,* woven; from L. *texo,* to weave
organ: Gk. *organon,* instrument

system: Gk. *systema,* being together

endocrine and digestive systems. All of the systems of the body are interrelated and function together, constituting the **organism.**

A *systemic approach* to studying anatomy and physiology emphasizes the functional relationships of various organs within a system. For example, the functional role of the digestive system can be better understood when all of the organs within that system are studied together. Another approach to anatomy, the *regional approach*, has merit in professional schools because the structural relationships of portions of several systems can be observed simultaneously. This is important for surgeons, who must be familiar with all the systems within a particular region. Dissections of cadavers are usually conducted on a regional basis. By means of radio-graphs (images produced by X rays) (fig. 1.9) and newer medical technologies, including computerized tomography (CT) and magnetic resonance imaging (MRI) scans (fig. 1.10), the organs within different body regions can be safely visualized in a patient.

This text uses a systemic approach to anatomy and physiology. In the chapters that follow, you will become acquainted system by system with the structural and functional aspects of the entire body.

# Planes of Reference and Descriptive Terminology

*All of the anatomical planes of reference and terms of direction are made with respect to the body in anatomical position. Many anatomical terms are derived from Greek and Latin.*

## Planes of Reference

In order to visualize and study the structural arrangements of various organs, the body may be sectioned (cut) and diagrammed according to planes of reference. Three fundamental planes, *midsagittal*

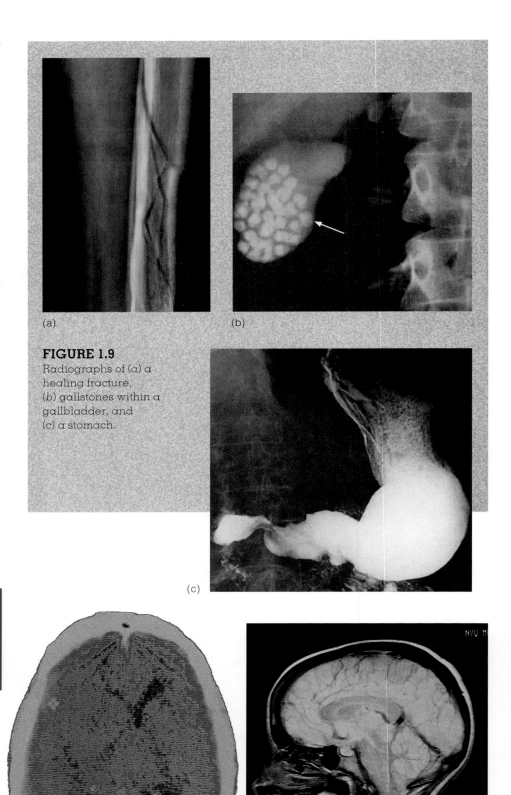

(a)　　　　　　　　(b)

**FIGURE 1.9**
Radiographs of (a) a healing fracture, (b) gallstones within a gallbladder, and (c) a stomach.

(c)

(a)　　　　　　　　(b)

**FIGURE 1.10**
Some newer technologies: (a) a CT scan through the head and (b) an MRI image through the head.

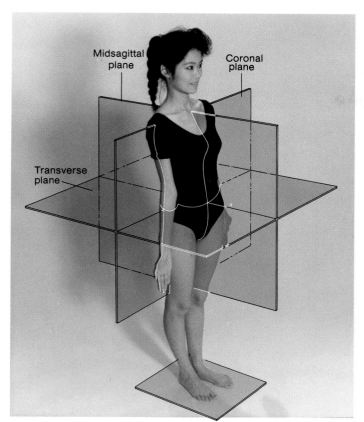

**FIGURE 1.11**
Planes of reference through the body.

(*mid-saj´ ĭ-tal*), *coronal*, and *transverse*, are frequently used to depict structural arrangement (fig. 1.11).

A **midsagittal plane** passes lengthwise through the midplane of the body, dividing it into equal right and left halves. Other *sagittal planes* extend parallel to the midsagittal plane, but off-center, and divide the body into unequal right and left portions. **Coronal,** or **frontal, planes** also pass lengthwise and divide the body into front and back portions. **Transverse planes,** also called **horizontal,** or **cross-sectional, planes,** divide the body into superior (upper) and inferior (lower) portions.

 The value of the computerized tomographic X-ray (CT) scan is that it displays an image along a transverse plane similar to that which could otherwise be obtained only in an actual section through the body. Prior to the development of this X-ray technique, conventional radiographs (X-ray images) were on a vertical plane, and the dimensions of body irregularities were difficult, if not impossible, to ascertain.

## Anatomical Position and Directional Terms

All terms of direction that describe the relationship of one body part to another are made in reference to the **anatomical position.** In the anatomical position, the body is erect, the feet are parallel to one another and flat on the floor, the eyes are directed forward, and the arms are at the sides of the body with the palms of the hands turned forward (fig. 1.12).

Directional terms are used to locate the position of structures, surfaces, and regions of the body. These terms are always relative to the anatomical position. For example, if a person is in the anatomical position, the thumb is always lateral to the little finger and the nose is anterior to the ears. A summary of directional terms is presented in table 1.2.

**FIGURE 1.12**
In the anatomical position, the body is erect, the feet are parallel, the eyes are directed forward, and the arms are to the sides with the palms directed forward and the fingers pointed straight down.

## Word Derivations

Analyzing anatomical and physiological terminology can be a rewarding experience. Not only is an understanding of the roots of words of academic interest, but a familiarity with technical terms reinforces the learning process. The majority of scientific terms are of Greek or Latin derivation, but some are German, French, and Arabic derivatives. Unfortunately, some anatomical and medical terms have been coined in honor of various anatomists or physicians. Such terms have no descriptive basis and must simply be memorized.

Many Greek and Latin terms were coined more than 2,000 years ago. It is exciting to decipher the meanings of these terms and gain a glimpse into our medical heritage. Many terms referred to common plants or animals. Thus, the term *vermis* means worm; *cochlea*, snail shell; *cancer*, crab; and *uvula*, grape. Other terms provide a clue to the warlike environment of the Greek and Latin era. *Thyroid*, for example, means shield; *xiphos* (*zi´fos*), sword; and *thorax*, breastplate. *Sella* means saddle and *stapes* (*sta´pēz*) means stirrup. Various tools or instruments were referred to in early anatomy. The malleus and anvil, for example, resemble miniatures of a blacksmith's implements.

Certain clinical procedures are important in determining body structure and function. *Palpation* is feeling with firm pressure for surface landmarks, lumps, tender spots, or pulsations. *Percussion* is tapping sharply at points on the

## Table 1.2 Directional terms for the human body

| Term | Definition | Example |
|------|-----------|---------|
| Superior (cranial, cephalic) | Toward the head; toward the top | The thorax is superior to the abdomen. |
| Inferior (caudal) | Away from the head; toward the bottom | The legs are inferior to the trunk. |
| Anterior (ventral) | Toward the front | The navel is on the anterior side of the body. |
| Posterior (dorsal) | Toward the back | The kidneys are posterior to the intestine. |
| Medial | Toward the midline of the body | The heart is medial to the lungs. |
| Lateral | Toward the side of the body | The ears are on the lateral sides of the head. |
| Internal (deep) | Away from the surface of the body | The brain is internal to the cranium. |
| External (superficial) | Toward the surface of the body | The skin is external to the muscles. |
| Proximal | Toward the main mass of the body | The knee is proximal to the foot. |
| Distal | Away from the main mass of the body | The hand is distal to the elbow. |
| Visceral | Related to internal organs | The lungs are covered by a thin membrane called the visceral pleura. |
| Parietal | Related to the body walls | The parietal pleura is the inside lining of the thoracic cavity. |

thorax or abdomen to determine fluid concentrations and organ densities. *Auscultation* is listening to the sounds that various organs make as they perform their functions.

You will encounter many new terms throughout your study of anatomy and physiology. It will be easier to learn these terms if you understand the prefixes and suffixes of the new words. Use the glossary of prefixes and suffixes (on the inside front cover) as an aid in learning new terms. A guide to the relationship between the singular and plural forms of words is presented in table 1.3. Pronouncing these terms as you learn them will also help you recall them later.

## Body Regions and Body Cavities

*The human body is divided into regions and specific areas that are identifiable on the surface. The head and trunk contain internal organs housed in distinct body cavities.*

Learning the terminology used in reference to the body regions now will help you learn the names of underlying structures later. The major body regions are the **head, neck, trunk, upper extremity,** and **lower extremity.** The trunk is frequently divided into the thorax and abdomen. Figure 1.13 presents an outline of the major body regions.

# PUMP UP YOUR GRADES!

## TRIM YOUR STUDY TIME

### Student Study Guide for *Concepts of Human Anatomy and Physiology*
*Kent Van De Graaff and Stuart Fox*
ISBN 0-697-16078-5

For each chapter in this text, there is a corresponding study guide chapter—including questions for review, objective questions, and essay questions—that review major chapter concepts.

### Coloring Guide to Anatomy and Physiology
*by Judith Stone and Robert Stone*
ISBN 0-697-17109-4

This helpful manual serves as a framework for learning human anatomy and physiology. By labeling and coloring each drawing, you will easily learn key anatomical and physiological structures and functions.

### Atlas of the Skeletal Muscles
*by Judith Stone and Robert Stone*
ISBN 0-697-10618-7

This inexpensive atlas depicts all of the human skeletal muscles in precise drawings that show their origin, insertion, action, and innervation. Each muscle is presented on a separate page, and the spinal cord level of the nerve fibers is included in parentheses. These features will help you visually understand the action of muscles.

### Study Cards for Human Anatomy and Physiology
*by Kent Van De Graaff et al.*
ISBN 0-697-09731-5

Make studying a breeze with this boxed set of 300, two-sided study cards. Each card provides a complete description of terms, clearly labeled drawings, pronunciation guides, and clinical information on diseases.

 **Wm. C. Brown Publishers**
A Division of Wm. C. Brown Communications, Inc.

### Independent Study Guide to Anatomy and Physiology to Prepare for the ACT/PEP Exam
by Margaret Mueller-Shore
ISBN 0-697-24692-2

Pursuing a career in nursing? Then this study aid is just for you. Every topic currently listed in the ACT/PEP Study Outline *and* some additionally important issues typically addressed in your human anatomy and physiology class are addressed.

### How to Study Science
by Fred Drewes
ISBN 0-697-14474-7

This excellent workbook offers you helpful suggestions for meeting the considerable challenges of a college science course. It offers tips on how to take notes, how to get the most out of labs, as well as how to overcome "science anxiety." The book's unique design helps you develop critical-thinking skills, while facilitating careful note-taking.

### A Life Science Lexicon
by William Marchuk
ISBN 0-697-12133-X

This portable, inexpensive reference will help you quickly master the vocabulary of the life sciences. Not a dictionary, it carefully explains the rules of word construction and derivation and gives complete definitions of all important terms.

### WCB Life Science Animations Videotapes

Each videotape contains colorful, moving visuals that make challenging concepts easier to understand.

#1 *Chemistry, The Cell and Energetics:*
ISBN 0-697-25060-7

#2 *Cell Division, Heredity, Genetics, Reproduction, and Development*
ISBN 0-697-25069-5

#3 *Animal Biology I*
ISBN 0-697-25070-9

#4 *Animal Biology II*
ISBN 0-697-25071-7

# TRIM YOUR STUDY TIME

# PUMP UP YOUR GRADES!

 **Wm. C. Brown Publishers**
A Division of Wm. C. Brown Communications, Inc.

*Concepts of Human Anatomy and Physiology*
by Kent Van De Graaff and Stuart Ira Fox

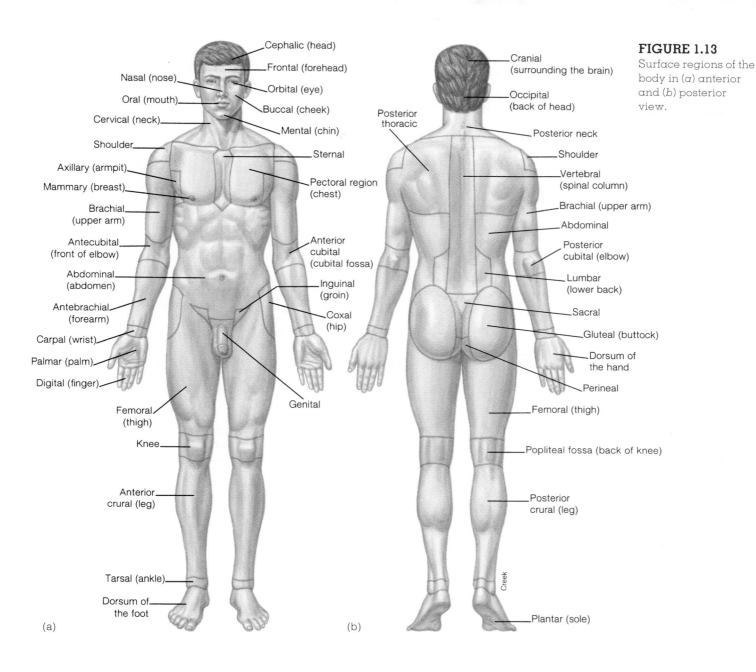

**FIGURE 1.13**

Surface regions of the body in (*a*) anterior and (*b*) posterior view.

**(a)**

- Cephalic (head)
- Frontal (forehead)
- Nasal (nose)
- Orbital (eye)
- Oral (mouth)
- Buccal (cheek)
- Cervical (neck)
- Mental (chin)
- Shoulder
- Sternal
- Axillary (armpit)
- Mammary (breast)
- Pectoral region (chest)
- Brachial (upper arm)
- Antecubital (front of elbow)
- Anterior cubital (cubital fossa)
- Abdominal (abdomen)
- Inguinal (groin)
- Antebrachial (forearm)
- Coxal (hip)
- Carpal (wrist)
- Palmar (palm)
- Digital (finger)
- Femoral (thigh)
- Genital
- Knee
- Anterior crural (leg)
- Tarsal (ankle)
- Dorsum of the foot

**(b)**

- Cranial (surrounding the brain)
- Occipital (back of head)
- Posterior thoracic
- Posterior neck
- Shoulder
- Vertebral (spinal column)
- Brachial (upper arm)
- Abdominal
- Posterior cubital (elbow)
- Lumbar (lower back)
- Sacral
- Gluteal (buttock)
- Dorsum of the hand
- Perineal
- Femoral (thigh)
- Popliteal fossa (back of knee)
- Posterior crural (leg)
- Plantar (sole)
- Creek

## Table 1.3  Examples of singular and plural word endings

| Singular ending | Plural ending | Examples | Singular ending | Plural ending | Examples |
|---|---|---|---|---|---|
| -a | -ae | Axilla, axillae | -on | -a | Mitochondrion, mitochondria |
| -ax | -aces | Thorax, thoraces | -um | -a | Cilium, cilia |
| -en | -ina | Lumen, lumina | -us | -i | Tarsus, tarsi |
| -ex | -ices | Cortex, cortices | -us | -ora | Corpus, corpora |
| -is | -es | Diagnosis, diagnoses | -us | -era | Viscus, viscera |
| -is | -ides | Epididymis, epididymides | -x | -ges | Pharynx, pharynges |
| -ix | -ices | Appendix, appendices | -y | -ies | Ovary, ovaries |
| -ma | -mata | Carcinoma, carcinomata | | | |

Source: Courtesy of Kenneth S. Saladin, Georgia College.

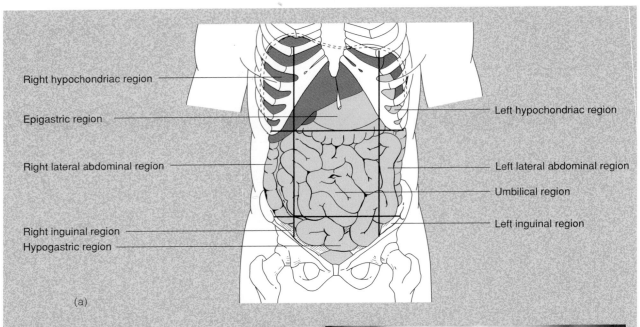

Right hypochondriac region

Epigastric region

Right lateral abdominal region

Right inguinal region
Hypogastric region

Left hypochondriac region

Left lateral abdominal region

Umbilical region

Left inguinal region

(a)

**FIGURE 1.14**

The abdomen is subdivided into nine regions. The vertical planes are positioned just medial to the nipples. The upper horizontal plane is positioned at the level of the rib cage and the lower horizontal plane is even with the upper border of the hipbones.

Midclavicular plane

Subcostal plane

Intertubercular plane

(b)

## *Head and Neck*

The **head** is divided into a **facial region,** which includes the eyes, nose, and mouth, and a **cranial region,** or *cranium* (*kra´ne-um*), which covers and supports the brain. The identifying names for detailed surface regions are based on associated organs—such as the orbital (eye), nasal (nose), oral (mouth), mental (chin), and auricular (ear) regions—or underlying bones—such as the frontal, zygomatic, temporal, parietal, and occipital regions. The neck, referred to as the **cervical region, or** *cervix* (*ser´viks*), supports the head and permits it to move.

## *Thorax*

The **thoracic** (*thŏ-ras´ik*) **region,** or *thorax,* is commonly referred to as the chest. The **mammary region** of the thorax surrounds the nipple and in sexually mature females is enlarged as the breast. Between the mammary regions is the **sternal region.** The armpit is called the **axillary fossa,** or simply **axilla,** and the surrounding area the **axillary region.** Paired **scapular regions** (shoulder blades) can be identified from the back of the thorax. The **vertebral region,** follow-

ing the vertebral column, extends the length of the back. On either lateral side of the thorax are the **pectoral regions.**

 The heart and lungs are contained within the thoracic cavity. Easily identified surface landmarks greatly facilitate physical examination of these organs. A physician must know, for example, where the valves of the heart can best be detected and where to listen for respiratory sounds. The axilla becomes important in examining for infected lymph nodes. When fitting a patient for crutches, a physician will instruct the patient to avoid supporting the weight of the body on the axillary area because of the possibility of damaging the underlying nerves and vessels.

## *Abdomen*

The **abdomen** is located below the thorax. The **navel,** or **umbilicus,** is an obvious landmark on the front and center of the abdomen. The abdomen has been divided into nine regions in order to describe the location of internal organs. Figure 1.14 diagrams the subdivisions of the abdomen.

..........................................

thorax: L. *thorax,* chest

The **pelvic region** forms the inferior portion of the abdomen. Within the pelvic region is the **pubic area,** which is covered with pubic hair in sexually mature individuals. The **perineum** (*per″ĭ-ne-um*) is the region where the external sex organs and the anal opening are located. The center of the back side of the abdomen, commonly called the small of the back, is the **lumbar region.** The **sacral region** is located farther down, at the point where the vertebral column terminates. The large hip muscles form the **buttock,** or **gluteal region.** This region is a common injection site for hypodermic needles.

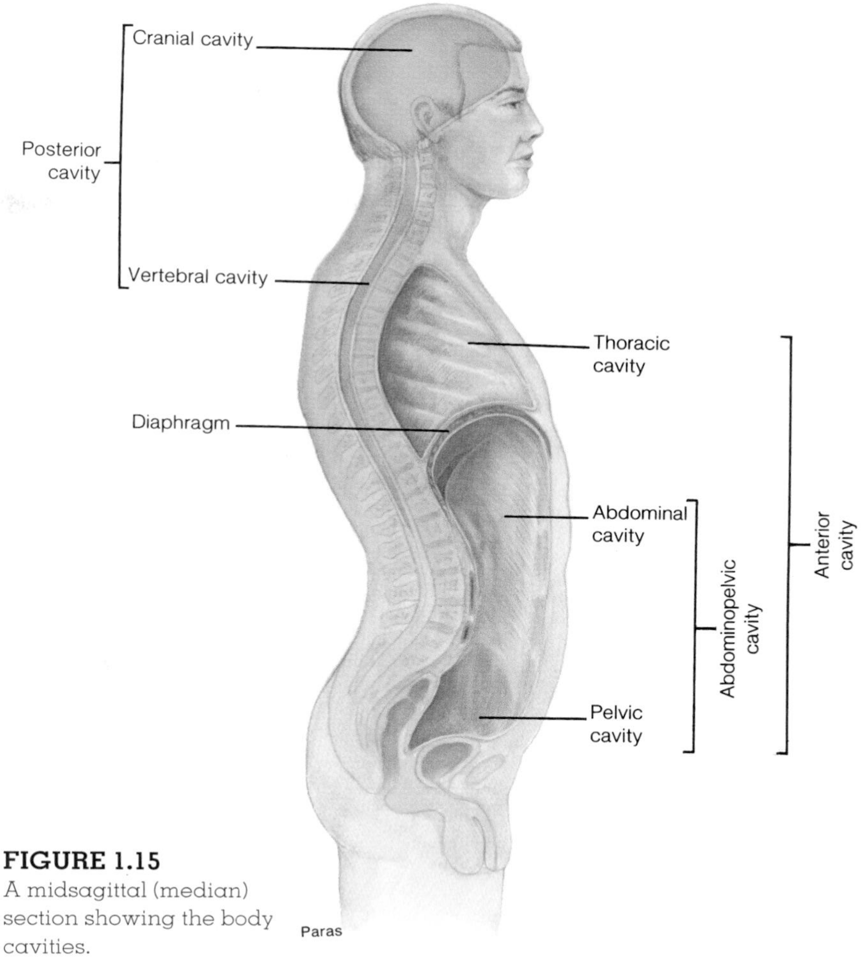

## Upper and Lower Extremities

The upper extremity is anatomically divided into the **shoulder, brachium** (*bra′ke-um*) (upper arm), **antebrachium** (forearm), and **manus** (hand). The shoulder is the region between the pectoral girdle and the brachium in which the shoulder joint is located. The shoulder is referred to as the **omos,** or **deltoid region.** Between the upper arm and forearm is a flexible joint called the *elbow.* The surface area of the elbow is known as the **cubital region.** The front surface of the elbow is also known as the **cubital fossa,** an important site for intravenous injections or the withdrawal of blood. The wrist is the flexible junction between the forearm and the hand. The front of the hand is referred to as the **palm,** or **palmar surface,** and the back of the hand is called the **dorsum of the hand.**

The lower extremity consists of the **thigh, knee, leg,** and **foot.** The thigh is commonly called the upper leg, or **femoral** (*fem′or-al*) **region.** The knee joint has two surfaces: the front surface is the **patellar region,** or kneecap, and the back of the knee is called the **popliteal** (*pop″lĭ-te-al*) **fossa.** The *shin* is a prominent bony ridge extending longitudinally along the **anterior crural region,** and the *calf* is the thickened muscular mass of the **posterior crural region.** The **ankle** is the junction connecting the leg and the foot. The **heel** is the back of the foot, and the **sole** of the foot is referred to as the **plantar surface. The dorsum of the foot** is the top surface.

## Body Cavities

**Body cavities** are confined spaces within the body. They contain organs that are protected, compartmentalized, and supported by associated membranes. There are two principal body cavities: the **posterior body cavity** and the larger

cubital: L. *cubitis,* elbow
coelom: Gk. *koiloma,* cavity

**FIGURE 1.15**
A midsagittal (median) section showing the body cavities.

anterior body cavity. The posterior body cavity contains the brain and the spinal cord.

During development, the anterior cavity forms from a cavity called the **coelom** (*se′lom*). The coelom is a body cavity within the trunk and is lined with a membrane that secretes a lubricating fluid. As development progresses, the coelom is partitioned by the muscular diaphragm into an upper **thoracic cavity,** or chest cavity, and a lower **abdominopelvic cavity** (figs. 1.15 and 1.16). Organs within the coelom are collectively called **viscera,** or **visceral** (*vis′er-al*) **organs** (fig. 1.17). Within the thoracic cavity are two **pleural** (*ploor′al*) **cavities** for the right and left lungs and a **pericardial** (*per″ĭ-kar′-de-al*) **cavity** containing the heart. The area between the two lungs is known as the **mediastinum** (*me″de-ă-sti′num*).

The abdominopelvic cavity consists of an upper **abdominal cavity** and a lower **pelvic cavity.** The abdominal cavity contains the stomach, small intestine, large intestine, liver, gallbladder, pancreas, spleen, and kidneys. The pelvic cavity is occupied by the terminal portion of the large intestine, the urinary bladder, and certain reproductive organs (the uterus, uterine tubes, and ovaries in the female; the seminal vesicles and prostate in the male).

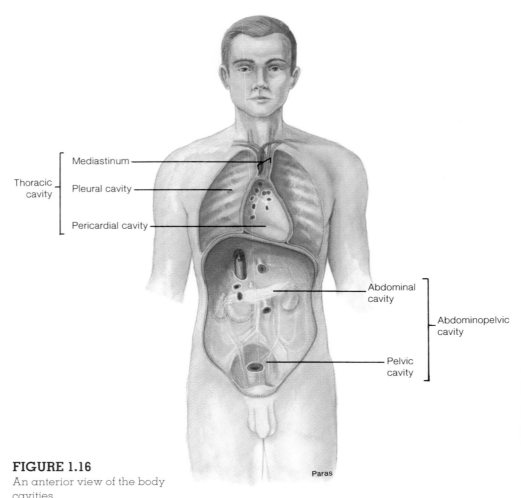

Body cavities serve to confine organs and systems that have related functions. The major portion of the nervous system occupies the posterior cavity; the principal organs of the respiratory and circulatory systems, the thoracic cavity; the primary organs of digestion, the abdominal cavity; and the reproductive organs, the pelvic cavity. Not only do these cavities house and support various body organs, they also effectively compartmentalize them so that infections and diseases cannot spread from one compartment to another. For example, *pleurisy* of one lung membrane does not usually spread to the other, and an injury to the thoracic cavity will usually cause only one lung to collapse rather than both.

**FIGURE 1.16**
An anterior view of the body cavities.

# Homeostasis and Feedback Control

*The regulatory mechanisms of the body can be understood in terms of a single, shared function: that of maintaining a dynamic constancy of the internal environment, or homeostasis. Homeostasis is maintained by effectors, which are regulated by sensory information from the internal environment.*

Over a century ago, the French physiologist Claude Bernard observed that the *milieu interieur* (internal environment) remains remarkably constant despite changing conditions in the external environment. In a book entitled *The Wisdom of the Body*, published in 1932, Walter Cannon coined the term **homeostasis** to describe this internal constancy. He suggested that mechanisms of physiological regulation exist for one purpose—the maintenance of internal constancy.

The concept of homeostasis has been of immense value in the study of anatomy and physiology because it allows diverse regulatory mechanisms to be understood in terms of their "why" as well as their "how." For example, we can study the physiological mechanisms that regulate the blood

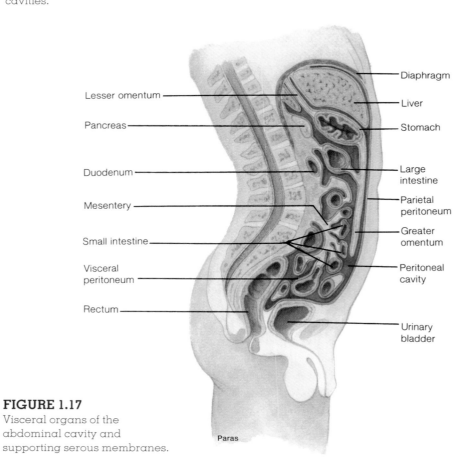

**FIGURE 1.17**
Visceral organs of the abdominal cavity and supporting serous membranes.

| Table 1.4 | Approximate normal ranges for selected blood measurements |
|---|---|
| **Measurement** | **Normal range** |
| Arterial pH | 7.35–7.43 |
| Bicarbonate | 21.3–28.5 mEq/L |
| Sodium | 136–151 mEq/L |
| Calcium | 4.6–5.2 mEq/L |
| Oxygen content | 17.2–22.0 ml/100 ml |
| Urea | 12–35 mg/100 ml |
| Amino acids | 3.3–5.1 mg/100 ml |
| Protein | 6.5–8.0 g/100 ml |
| Total lipids | 350–850 mg/100 ml |
| Glucose | 75–110 mg/100 ml |

glucose concentration and understand that they work that way to maintain homeostasis of blood glucose. The concept of homeostasis also provides a foundation for medical diagnostic procedures. When a particular measurement of the internal environment, such as a blood measurement (table 1.4), deviates significantly from the normal range of values, it can be concluded that homeostasis is not being maintained. A number of such measurements, combined with clinical observations, may allow the particular defective mechanism to be identified.

## Negative Feedback Loops

In order for internal constancy to be maintained, the body must have *sensors* that are able to detect deviations from a *set point*. The set point is analogous to the temperature set on a house thermostat. In a similar manner, there is a set point for body temperature, blood glucose concentration, the tension on a tendon, and so on. When a sensor detects a deviation from a particular set point, it must relay this information to an *integrating center*, which usually receives information from many different sensors. The integrating center is often a particular region of the brain or spinal cord, but in some cases it can also be cells of endocrine glands. The relative strengths of different sensory inputs are weighed in the integrating center, and, in response, the integrating center either increases or decreases the activity of particular *effectors*, which are generally muscles or glands.

In response to sensory information about a deviation from a set point, therefore, effectors act to promote a reverse change in the internal environment. If the body temperature exceeds the set point of 37° C, for example, effectors (such as sweat glands) act to lower the temperature. If, as another example, the blood glucose concentration falls below normal, the effectors (endocrine glands) act to increase the blood glucose. Notice that the change produced by the effectors acts to counter, or negate, the original deviation from

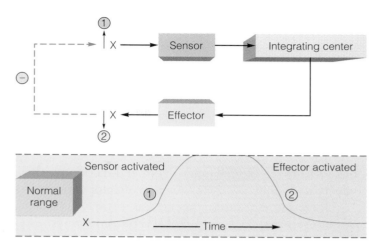

**FIGURE 1.18**

A rise in some factor of the internal environment (↑x) is detected by a sensor. Acting through an integrating center, this caused an effector to produce a change in the opposite direction (↓x). The initial deviation is thus reversed, completing a negative feedback loop (shown by the dashed arrow and negative sign). The numbers indicate the sequence of changes.

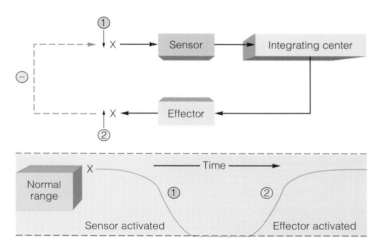

**FIGURE 1.19**

A negative feedback loop that compensates for a fall in some factor of the internal environment (↓x). (Compare this figure with figure 1.18.)

the set point. Since the activity of the effectors is influenced by the effects they produce, and since this regulation is in a negative, or reverse, direction, this type of control system is known as a **negative feedback loop** (fig. 1.18).

It is important to realize that this negative-feedback-loop activity is continuous. Thus, a particular nerve fiber, which is part of an effector mechanism, may always display some activity, and a particular hormone, which is part of another effector mechanism, may always be present in the blood. The nerve activity and hormone concentration may decrease in response to deviations of the internal environment in one direction (fig. 1.18) or they may increase in response to deviations in the opposite direction (fig. 1.19).

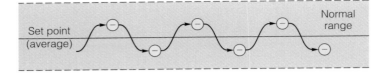

**FIGURE 1.20**

Negative feedback loops (indicated by negative signs) maintain a state of dynamic constancy within the internal environment.

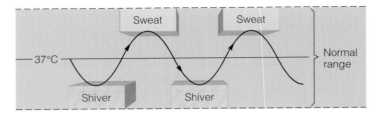

**FIGURE 1.21**

A simplified scheme by which body temperature is maintained within the normal range (with a set point of 37° C) by two antagonistic mechanisms—shivering and sweating. Shivering is induced when the body temperature falls too low and gradually subsides as the temperature rises. Sweating occurs when the body temperature is too high and diminishes as the temperature falls. Most aspects of the internal environment are regulated by the antagonistic actions of different effector mechanisms.

Changes from the normal range in either direction are thus compensated for by reverse changes in effector activity.

Homeostasis is best conceived as a state of *dynamic constancy* rather than as a state of absolute constancy. The values of particular measurements of the internal environment fluctuate above and below the set point, which can be taken as the average value within the normal range of measurements (fig. 1.20). This state of dynamic constancy results from a greater or a lesser degree of activation of effectors in response to sensory feedback and from the competing actions of antagonistic effectors.

**Antagonistic Effectors**  Most factors in the internal environment are controlled by several effectors, which often display antagonistic activity. Control by antagonistic effectors is sometimes described as push-pull, where the increasing activity of one effector is accompanied by decreasing activity of an antagonistic effector. This affords a finer degree of control than could be achieved by simply switching one effector on and off. Normal body temperature, for example, is maintained at a set point of about 37° C by the antagonistic effects of sweating, shivering, and other mechanisms (fig. 1.21).

The blood concentrations of glucose, calcium, and other substances are regulated by negative feedback loops that involve chemical regulators called hormones that produce opposite effects. While insulin, for example, lowers blood glucose, other hormones raise the blood glucose concentration. The heart rate, similarly, is controlled by impulses through nerve fibers that produce opposite effects. Stimulation of one group of nerve fibers increases the heart rate, while stimulation of another group slows the heart rate.

## Positive Feedback

Constancy of the internal environment is maintained by effectors that act to compensate for changes that served as stimuli for their activation; in short, by negative feedback loops. A thermostat, for example, maintains a constant temperature by increasing heat production when it is cold and decreasing heat production when it is warm. The opposite occurs during **positive feedback**—in this case, the action of effectors *amplifies* those changes that stimulated the effectors. A thermostat that worked by positive feedback, for example, would increase heat production in response to a rise in temperature.

It is clear that homeostasis must ultimately be maintained by negative rather than by positive feedback mechanisms. The effectiveness of some negative feedback loops, however, is increased by positive feedback mechanisms that amplify the actions of a negative feedback response. Blood clotting, for example, occurs as a result of a sequential activation of clotting factors. The activation of one clotting factor results in activation of many in a positive feedback, avalanchelike, manner. In this way, a single change is amplified to produce a blood clot. Formation of the clot, however, can prevent further loss of blood, and thus represents the completion of a negative feedback loop.

## Neural and Endocrine Regulation

The effectors of most negative feedback loops include the actions of nerves and hormones. In both neural and endocrine regulation, particular chemical regulators released by nerve fibers or endocrine glands stimulate target cells by interacting with specific receptor proteins in these cells. The mechanisms by which this regulation is achieved will be described in later chapters.

Homeostasis is maintained by two general categories of regulatory mechanisms: (1) those that are *intrinsic*, or "built-in," to the organs that produce them and (2) those that are *extrinsic*, as in regulation of an organ by the nervous and endocrine systems.

The endocrine system functions closely with the nervous system in regulating and integrating body processes and maintaining homeostasis. The nervous system controls the secretion of many endocrine glands, and some hormones in turn affect the function of the nervous system. Together, the nervous and endocrine system regulate the activities of most of the other systems of the body.

Regulation by the endocrine system is achieved by the secretion of chemical regulators called **hormones** into the blood. Since hormones are secreted into the blood, they are carried by the blood to all organs in the body. Only specific organs can respond to a particular hormone, however; these are known as the *target organs* of that hormone.

Nerve fibers are said to *innervate* the organs that they regulate. When stimulated, these fibers produce electrochemical nerve impulses that are conducted from the origin of the fiber to the target organ innervated by that fiber. These target organs can be muscles or glands that may function as effectors in the maintenance of homeostasis.

## Feedback Control of Hormone Secretion

We will discuss the details of the nature of the endocrine glands, the interaction of the nervous and endocrine systems, and the actions of hormones in later chapters. For now it is sufficient to describe the regulation of hormone secretion very broadly, since it so superbly illustrates the principles of homeostasis and negative feedback regulation.

Hormones are secreted in response to specific chemical stimuli. A rise in the plasma glucose concentration, for example, stimulates insulin secretion from the pancreatic islets (islets of Langerhans) in the pancreas. Hormones are also secreted in response to nerve stimulation and to stimulation by other hormones.

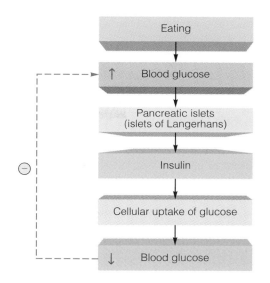

**FIGURE 1.22**
The negative feedback control of insulin secretion and blood glucose concentration.

The secretion of a hormone can be inhibited by its own effects, in a negative feedback manner. Insulin, for example, produces a lowering of blood glucose. Since a rise in blood glucose stimulates insulin secretion, a lowering of blood glucose caused by insulin's action inhibits further insulin secretion. This closed-loop control system is called **negative feedback inhibition** (fig. 1.22).

# Chapter Summary

### The Sciences of Anatomy and Physiology (pp. 2–6)

1. The history of anatomy and physiology parallels that of medicine; Hippocrates, the famous Greek physician, is regarded as the father of medicine.
2. Aristotle established a type of scientific method for obtaining data.
3. Erasistratus is frequently referred to as the father of physiology.
4. The writings of the Roman physician Galen were the ultimate authority on anatomy and medical treatment for nearly 1,500 years.
5. Often regarded as the father of anatomy, Vesalius challenged Galen's teachings and accelerated research in anatomy and physiology during the Renaissance.
6. Two major contributions of the seventeenth and eighteenth centuries were the explanation of blood flow by Harvey and the development and

utilization of the microscope by Leeuwenhoek.
7. The scientific method is a disciplined approach considered necessary for scientific investigation. It includes observation, data collection, and formulation of a hypothesis that is testable.

### Classification and Characteristics of Humans (pp. 6–8)

1. Humans are biological organisms belonging to the phylum Chordata within the kingdom Animalia and to the family Hominidae within the class Mammalia and the order Primates.
2. Humans belong to the phylum Chordata because of the presence of a notochord, a dorsal hollow nerve cord, and pharyngeal pouches during the embryonic period of development.

3. Some of the characteristics of humans include a large, well-developed brain, bipedal locomotion, an opposable thumb, well-developed vocal structures, and stereoscopic vision.

### Body Organization (pp. 8–10)

1. The cell is the structural and functional component of life.
2. Tissues are aggregations of similar cells that perform specific functions.
3. An organ is an aggregate of two or more tissues that performs specific functions.
4. Body systems consist of various organs that have similar or interrelated functions.

### Planes of Reference and Descriptive Terminology (pp. 10–12)

1. In the anatomical position, the subject stands erect and faces forward with arms at the sides and palms turned forward.

2. Directional terms are used to describe the location of one body part with respect to another part.
3. The majority of anatomical and physiological terms are of Greek or Latin derivation.

## Body Regions and Body Cavities (pp. 12–15)

1. The human body is divided into regions, which can be identified on the surface. In each region are internal organs, the

locations of which are anatomically, physiologically, and clinically important.
2. For functional and protective purposes, the viscera are compartmentalized and supported in specific body cavities by connective and epithelial membranes.

## Homeostasis and Feedback Control (pp. 16–19)

1. Homeostasis is the dynamic constancy of the internal environment. This concept is the central theme of anatomy and physiology, and even of medicine.

2. Deviations from a set point are detected by sensors, and changes are instituted by effectors that act to compensate for the deviations in a negative feedback fashion.
3. Neural and endocrine effectors regulate most of the organs of the body. The secretion of hormones is controlled by negative feedback loops.

# Review Activities

## Objective Questions

1. Which of the following men would be most likely to disagree with the concept of body humors?
   a. Galen          b. Hippocrates
   c. Vesalius       d. Aristotle
2. The most important contribution of William Harvey was his research on
   a. the continuous circulation of blood.
   b. the microscopic structure of spermatozoa.
   c. the detailed structure of the kidney.
   d. the striated appearance of skeletal muscle.
3. The taxonomic scheme from specific to general is
   a. species, class, order, phylum.
   b. genus, family, kingdom, phylum.
   c. species, family, class, kingdom.
   d. genus, phylum, class, kingdom.
4. Which of the following is (are) not a principal chordate characteristic?
   a. dorsal hollow nerve cord
   b. distinct head, thorax, and abdomen
   c. notochord
   d. pharyngeal pouches
5. The cubital fossa is located in
   a. the thorax.
   b. the upper extremity.
   c. the abdomen.
   d. the lower extremity.

6. Which of the following is not a fundamental plane?
   a. coronal plane      b. transverse plane
   c. vertical plane     d. midsagittal plane
7. In the anatomical position,
   a. the arms are extended away from the body.
   b. the palms of the hands face posteriorly.
   c. the body is erect and the palms face anteriorly.
   d. the body is in a fetal position.
8. Which of the following statements about homeostasis is true?
   a. The internal environment is maintained absolutely constant.
   b. Negative feedback mechanisms act to correct deviations from a normal range within the internal environment.
   c. Homeostasis is maintained by switching effector actions on and off.
   d. All of the above are true.
9. In a negative feedback loop, the effector organ produces changes that are
   a. similar in direction to that of the initial stimulus.
   b. opposite in direction to that of the initial stimulus.
   c. unrelated to the initial stimulus.
10. A hormone called parathyroid hormone acts to help raise the blood calcium concentration. According to the principles

of negative feedback, an effective stimulus for parathyroid hormone secretion would be
   a. a fall in blood calcium.
   b. a rise in blood calcium.

## Essay Questions

1. What is meant by the humoral theory of body organization? Which great anatomists were influenced by this theory? When was the humoral theory discarded?
2. Discuss the impact of Galen on the advancement of anatomy and physiology and medicine. What ideological circumstances permitted the philosophies of Galen to survive for so long?
3. What role did the development of the microscope play in the advancement of the sciences of anatomy and physiology and medicine? What specialties of anatomical and physiological study have emerged since the introduction of the microscope?
4. Describe the scientific method and comment on its immense importance in anatomical and physiological research.
5. Explain the role of antagonistic negative feedback processes in the maintenance of homeostasis.
6. Explain, using examples, how the secretion of a hormone is controlled by the effects of that hormone's actions.

# [ chapter two ]

# chemical composition of the body

## objectives

- Describe the structure of an atom and the types and relative strengths of different types of chemical bonds.

- Define the terms *acid*, *base*, *pH*, and *ion*.

- Discuss the properties of water and explain why compounds may be either hydrophilic or hydrophobic.

- Describe the structures of some organic molecules and identify different functional groups.

- List the subcategories of carbohydrates and give examples of each.

- Describe dehydration synthesis and hydrolysis reactions and indicate where they occur in the body.

- Identify the subclasses of lipids and explain why they are all classified as lipids.

- Distinguish between saturated and unsaturated fats and describe the chemical reactions in which triglycerides are formed and broken down.

- Describe the structures of phospholipids and prostaglandins and explain their functions in the body.

- Describe the structure of amino acids and explain how one type of amino acid differs from another.

- Describe the primary, secondary, tertiary, and quaternary structure of proteins.

- List some of the functions of different proteins and explain why protein structure is so diverse.

# Atoms, Ions, and Molecules

*In order to understand human anatomy and physiology, a knowledge of basic chemical concepts, including the structure of atoms and molecules, chemical bonding, and the basis of pH measurements, is required.*

The anatomical structures and physiological processes of the body are largely based on the properties and interactions of atoms, ions, and molecules. Water is the major solvent in the body and contributes 65% to 75% of the total weight of an average adult. Of this amount, 30% to 40% is contained within the body cells as the *intracellular compartment;* the remainder is contained outside of the body cells as the *extracellular compartment,* including the blood and tissue fluids. Dissolved in this water are many organic molecules (carbon-containing molecules like carbohydrates, lipids, proteins, and nucleic acids) and inorganic molecules and ions (atoms with a net charge). Before we turn to the structure and function of organic molecules within the body, let's first consider some basic chemical concepts, terminology, and symbols.

## Atoms

**Atoms** are much too small to be seen as individual structures, even with the most powerful microscope. Through the efforts of generations of scientists, however, the structure of atoms is now well understood. At the center of an atom is its *nucleus.* The nucleus contains two types of particles: *protons,* which have a positive charge, and *neutrons,* which are uncharged. The mass of a proton is approximately equal to the mass of a neutron, and the sum of the number of protons and neutrons in an atom is equal to the **atomic mass** of the atom. For example, an atom of carbon, which contains six protons and six neutrons, has an atomic mass of 12 (table 2.1).

The number of protons in an atom is its **atomic number.** Carbon has six protons, and thus has an atomic number of 6. Outside the positively charged nucleus are negatively charged *electrons.* Since the number of electrons in an atom is equal to the number of protons, atoms have a net charge of zero.

Although it is often convenient to think of electrons as orbiting the nucleus like planets orbiting the sun, this view of atomic structure is no longer believed to be correct. A given electron can occupy any position in a certain volume of space surrounding the nucleus. The outer boundary of this volume of space is called the *orbital* of the electron. The orbital is like an energy shell, or barrier, beyond which the electron usually does not pass.

There are potentially several such orbitals around a nucleus, with each successive orbital being farther from the nucleus. The first orbital, closest to the nucleus, can contain only two electrons. If an atom has more than two electrons (as do all atoms except hydrogen and helium), the additional electrons must occupy orbitals that are more distant from the nucleus. The second orbital can contain a maximum of eight electrons; the third can also contain a maximum of eight, and the fourth can contain a maximum of 18. The orbitals are filled from the innermost outward. Carbon, with six electrons, thus has two electrons in its first orbital and four electrons in its second orbital (fig. 2.1). It is always the electrons in the outermost orbital, if this orbital is incomplete, that participate in chemical reactions and form chemical bonds. These outermost electrons are known as the *valence electrons* of the atom.

**Isotopes**   A particular atom with a given number of protons in its nucleus may exist in several forms that differ from each other in their number of neutrons. All of these forms have the same atomic number but a different atomic mass. These different forms of atoms are called **isotopes.** When we refer to a **chemical element,** all of the isotopic

## Table 2.1   Atoms commonly present in organic molecules

| Atom | Symbol | Atomic number | Atomic mass | Orbital 1 | Orbital 2 | Orbital 3 | Number of chemical bonds |
|------|--------|---------------|-------------|-----------|-----------|-----------|--------------------------|
| Hydrogen | H | 1 | 1 | 1 | 0 | 0 | 1 |
| Carbon | C | 6 | 12 | 2 | 4 | 0 | 4 |
| Nitrogen | N | 7 | 14 | 2 | 5 | 0 | 3 |
| Oxygen | O | 8 | 16 | 2 | 6 | 0 | 2 |
| Sulfur | S | 16 | 32 | 2 | 8 | 6 | 2 |

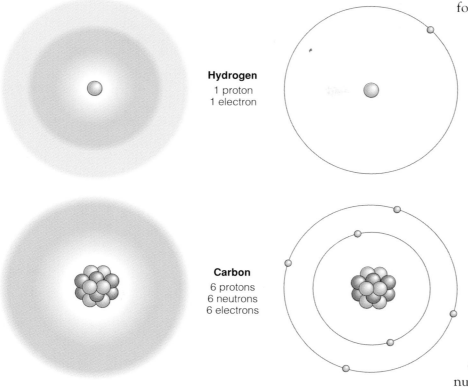

**Hydrogen**
1 proton
1 electron

**Carbon**
6 protons
6 neutrons
6 electrons

Proton ⬤    Neutron ⬤    Electron ◦

**FIGURE 2.1**

Diagrams of the hydrogen and carbon atoms. The electron orbitals on the left are represented by shaded spheres that indicate probable positions of the electrons. The orbitals on the right are represented by concentric circles.

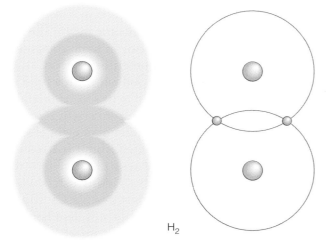

$H_2$

**FIGURE 2.2**

The hydrogen molecule, showing the covalent bonds between hydrogen atoms formed by the equal sharing of electrons.

forms of a given atom are included. The element hydrogen, for example, has three isotopes. The most common of these has a nucleus consisting of only one proton. Another isotope of hydrogen (called *deuterium*) has one proton and one neutron in the nucleus, whereas the third isotope (*tritium*) has one proton and two neutrons. Tritium is a radioactive isotope that is commonly used in physiological research and in many clinical laboratory procedures.

## Chemical Bonds, Molecules, and Ionic Compounds

**Molecules** are formed through interaction of the valence electrons of two or more atoms. These interactions, such as the sharing of electrons, produce *chemical bonds* (fig. 2.2). The number of bonds that each atom can have is determined by the number of electrons needed to complete the outermost orbital. Hydrogen, for example, must obtain only one more electron—and can thus form only one chemical bond—to complete the first orbital of two electrons. Carbon, by contrast, must obtain four more electrons—and can thus form four chemical bonds—to complete the second orbital of eight electrons (fig. 2.3, *left*).

**Covalent Bonds** Covalent bonds result when two or more atoms share electrons. Covalent bonds that are formed between identical atoms, as in oxygen gas ($O_2$) and hydrogen gas ($H_2$), are the strongest because their electrons are equally shared. Since the electrons are equally distributed between the two atoms, these molecules are said to be *nonpolar* and are bonded by nonpolar covalent bonds. When covalent bonds are formed between two different atoms, however, the electrons may be pulled more toward one atom than the other. The side of the molecule toward which the electrons are pulled is electrically negative in comparison to the other side. Such a molecule is said to be *polar* because it has a positive and negative pole. Atoms of oxygen, nitrogen, and phosphorus have a particularly strong tendency to pull electrons toward themselves when they bond with other atoms.

Water is the most abundant molecule in the body and serves as a solvent for body fluids. Water is a good solvent

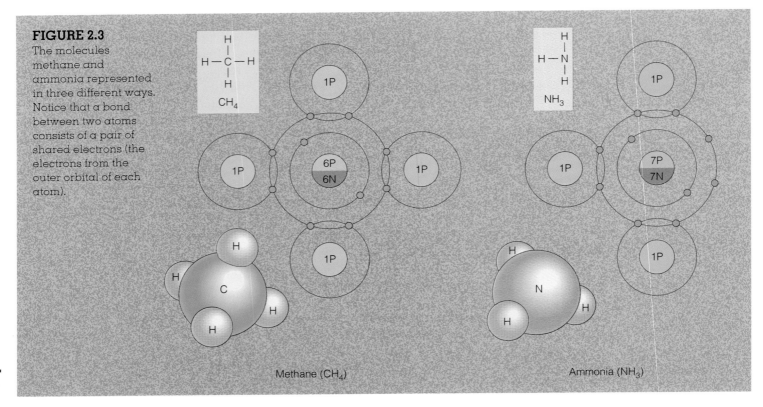

**FIGURE 2.3**

The molecules methane and ammonia represented in three different ways. Notice that a bond between two atoms consists of a pair of shared electrons (the electrons from the outer orbital of each atom).

Methane (CH₄)

Ammonia (NH₃)

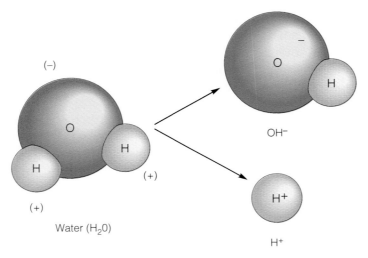

Water (H₂O)

OH⁻

H⁺

H⁺

**FIGURE 2.4**

A model of a water molecule showing its polar nature. Notice that the oxygen side of the molecule is negative, whereas the hydrogen side is positive. Polar covalent bonds are weaker than nonpolar covalent bonds. As a result, some water molecules ionize to form a hydroxyl ion (OH⁻) and a hydrogen ion (H⁺). The H⁺ combines with water molecules to form hydronium (H₃O⁺) ions (not shown).

because it is polar; the oxygen atom pulls its electrons from the two hydrogens toward its side of the water molecule, so that the oxygen side is more negatively charged than the hydrogen side of the molecule (fig. 2.4). In the next section, we will see the significance of the polar nature of water in its function as a solvent.

**Ionic Bonds**  Ionic bonds result when one or more valence electrons from one atom are completely transferred to a second atom. Thus, the electrons are not shared at all. The first atom loses electrons, and with fewer electrons than protons it becomes a positively charged **ion** (*i´on*). Positively charged ions are called *cations* because they move toward the negative pole, or cathode, in an electric field. The second atom now has more electrons than it has protons and becomes a negatively charged ion, or *anion* (so called because it moves toward the positive pole, or a node, in an electric field). The cation and anion attract each other to form an **ionic compound.**

Common table salt, sodium chloride (NaCl), is an example of an ionic compound. Sodium, with a total of 11 electrons, has 2 in its first orbital, 8 in its second orbital, and only 1 in its third orbital. Chlorine, conversely, is one electron short of completing its outer orbital of eight electrons. The lone electron in sodium's outer orbital is attracted to chlorine's outer orbital. This creates a chloride ion (represented as Cl⁻) and a sodium ion (Na⁺). Although table salt is shown as NaCl, it is actually composed of Na⁺Cl⁻ (fig. 2.5).

Ionic bonds are weaker than polar covalent bonds, and therefore ionic compounds easily dissociate when dissolved in water to yield their separate ions. Dissociation of NaCl, for example, yields Na⁺ and Cl⁻. Each of these ions attracts polar water molecules; the negative ends of water molecules are attracted to the Na⁺ and the positive ends are attracted to the Cl⁻ (fig. 2.6). The water molecules that surround these ions in turn attract other molecules of water to form *hydration spheres* around each ion.

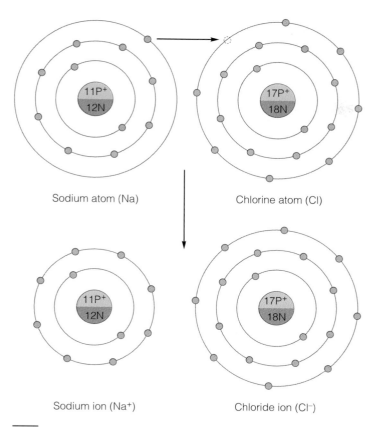

Sodium atom (Na)　　　　Chlorine atom (Cl)

Sodium ion (Na⁺)　　　　Chloride ion (Cl⁻)

**FIGURE 2.5**

The dissociation of sodium and chlorine to produce sodium and chloride ions. The positive sodium ions and the negative chloride ions attract each other to produce the ionic compound sodium chloride (NaCl).

The formation of hydration spheres makes an ion or a molecule soluble in water. Glucose, amino acids, and many other organic molecules are water-soluble because hydration spheres can form around atoms of oxygen, nitrogen, and phosphorous that are joined by polar covalent bonds to other atoms in the molecule. Such molecules are said to be **hydrophilic** (*hi″drŏ-fil′ik*). By contrast, molecules composed primarily of nonpolar covalent bonds, such as the hydrocarbon chains of fat molecules, have few charges, and thus cannot form hydration spheres. They are insoluble in water, and actually avoid it. For this reason, nonpolar molecules are said to be **hydrophobic** (*hi″drŏ-fo′bik*).

**Hydrogen Bonds**　Hydrogen bonds are very weak bonds that help to stabilize the delicate folding and bending of long organic molecules like proteins. When hydrogen forms a polar covalent bond with an atom of oxygen or nitrogen, the hydrogen gains a slight positive charge as the electron is pulled toward the other atom. This other atom is therefore described as being *electronegative*. Since the hydrogen has a slight positive charge, it will have a weak attraction for a second electronegative atom (oxygen or nitrogen) that may be located near it. This weak attraction is called a **hydrogen bond.** Hydrogen bonds are usually represented by

· · · · · · · · · · · · · · · · · · · · · · · · · · · · · · · · · · ·

hydrophilic: Gk. *hydor*, water; *philos*, fond
hydrophobic: Gk. *hydor*, water; *phobos*, fear

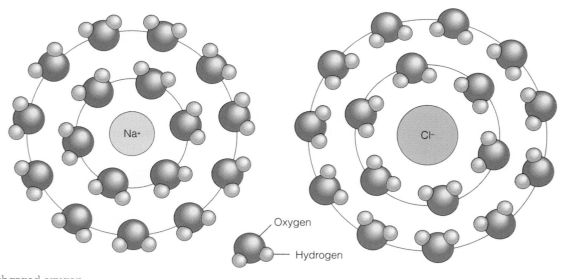

**FIGURE 2.6**

The negatively charged oxygen-ends of water molecules are attracted to the positively charged Na⁺, whereas the positively charged hydrogen-ends of water molecules are attracted to the negatively charged Cl⁻. Other water molecules are attracted to this first concentric layer of water, forming hydration spheres around the sodium and chloride ions.

Oxygen

Hydrogen

**Water molecule**

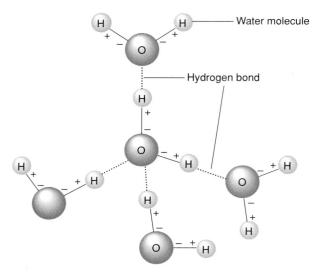

Water molecule

Hydrogen bond

**FIGURE 2.7**
The oxygen atoms of water molecules are weakly joined together by the attraction of the electronegative oxygen for the positively charged hydrogen. These weak bonds are called *hydrogen bonds*.

dotted lines (fig. 2.7) to distinguish them from strong covalent bonds, which are represented by solid lines.

Hydrogen bonds can be formed within the folds of a protein (as we will see in a later section), and between the chains of DNA (chapter 3). They can also be formed between adjacent water molecules, as shown in figure 2.7. The hydrogen bonding between water molecules is responsible for many of the physical properties of water, including its *surface tension* and its ability to be pulled as a column through narrow channels in a process called *capillary action*.

## Acids, Bases, and the pH Scale

The bonds in water molecules joining hydrogen atoms to oxygen atoms are, as previously discussed, polar covalent bonds. Although these bonds are strong, a small proportion of them break as the electron from the hydrogen atom is completely transferred to oxygen. When this occurs, the water molecule ionizes to form a hydroxyl ion ($OH^-$) and a hydrogen ion ($H^+$), which is simply a free proton. A proton released in this way does not remain free for long, because it is attracted to the electrons of oxygen atoms in water molecules. This forms a *hydronium ion,* indicated by the formula $H_3O^+$. For the sake of clarity in the following discussion, however, we will use $H^+$ to indicate the cation resulting from the ionization of water.

Ionization of water molecules produces equal amounts of $OH^-$ and $H^+$. Since only a small proportion of water molecules ionize, the concentration of $H^+$ and $OH^-$ are each equal to only $10^{-7}$ molar. (The term *molar* is a unit of concentration, as described in chapter 5; for hydrogen, 1 molar equals 1 gram per liter.) A solution with this $H^+$ concentration, which

| Table 2.2 | Common acids and bases | | | |
|---|---|---|---|---|
| **Acid** | **Symbol** | **Base** | **Symbol** | |
| Hydrochloric acid | HCl | Sodium hyroxide | NaOH | |
| Phosphoric acid | $H_3PO_4$ | Potassium hydroxide | KOH | |
| Nitric acid | $HNO_3$ | Calcium hydroxide | $Ca(OH)_2$ | |
| Sulfuric acid | $H_2SO_4$ | Ammonium hydroxide | $NH_4OH$ | |
| Carbonic acid | $H_2CO_3$ | | | |

is produced by the ionization of water molecules in which the $H^+$ concentration equals the $OH^-$ concentration, is said to be **neutral.**

A solution that contains a higher $H^+$ concentration than that of water is called an *acidic solution;* a solution with a lower $H^+$ concentration is called a *basic solution*. An **acid** is defined as a molecule that can release protons ($H^+$) to a solution; it is a proton donor. A **base** is a negatively charged ion (anion), or a molecule that ionizes to produce the anion, which can combine with $H^+$ and thus remove the $H^+$ from solution; it is a proton acceptor. Most strong bases release $OH^-$ into a solution, which combines with $H^+$ to form water, and which thus lowers the $H^+$ concentration. Examples of common acids and bases are given in table 2.2.

**pH**  The $H^+$ concentration of a solution is usually indicated in pH units on a scale that runs from 0 to 14. The pH number is equal to the logarithm of 1 over the $H^+$ concentration:

$$pH = \log \frac{1}{[H^+]}$$

where [$H^+$] = molar $H^+$ concentration

Pure water has an $H^+$ concentration of $10^{-7}$ molar at 25° C, and thus it has a pH of 7 (neutral). Because of the logarithmic relationship, a solution with 10 times the hydrogen ion concentration ($10^{-6}$ M) has a pH of 6, whereas a solution with one tenth the $H^+$ concentration ($10^{-8}$ M) has a pH of 8. The pH number is easier to write than the molar $H^+$ concentration, but it is admittedly confusing because it is *inversely related* to the $H^+$ concentration: a solution with a higher $H^+$ concentration has a lower pH number; one with a lower $H^+$ concentration has a higher pH number. A strong acid with a high $H^+$ concentration of $10^{-2}$ molar, for example, has a pH of 2, whereas a solution with only $10^{-10}$ molar $H^+$ has a pH of 10. Acidic solutions, therefore, have a pH of less than 7 (that of pure water), whereas basic solutions have a pH of between 7 and 14 (table 2.3).

**Buffers**  A buffer is a system of molecules and ions that resists changes in $H^+$ concentration, thus serving to stabilize the pH of a solution. In blood plasma, for example, the

## Table 2.3 The pH scale

| | $H^+$ concentration (molar) | pH | $OH^-$ concentration (molar) |
|---|---|---|---|
| Acids | 1.0 | 0 | $10^{-14}$ |
| | 0.1 | 1 | $10^{-13}$ |
| | 0.01 | 2 | $10^{-12}$ |
| | 0.001 | 3 | $10^{-11}$ |
| | 0.0001 | 4 | $10^{-10}$ |
| | $10^{-5}$ | 5 | $10^{-9}$ |
| | $10^{-6}$ | 6 | $10^{-8}$ |
| Neutral | $10^{-7}$ | 7 | $10^{-7}$ |
| Bases | $10^{-8}$ | 8 | $10^{-6}$ |
| | $10^{-9}$ | 9 | $10^{-5}$ |
| | $10^{-10}$ | 10 | 0.0001 |
| | $10^{-11}$ | 11 | 0.001 |
| | $10^{-12}$ | 12 | 0.01 |
| | $10^{-13}$ | 13 | 0.1 |
| | $10^{-14}$ | 14 | 1.0 |

pH is stabilized by the following reversible reaction involving the bicarbonate ion ($HCO_3^-$) and carbonic acid ($H_2CO_3$):

$$HCO_3^- + H^+ \rightleftharpoons H_2CO_3$$

The double arrows indicate that the reaction could go either to the right or to the left; the net direction depends on the concentration of molecules and ions on each side. If an acid (such as lactic acid) should release $H^+$ into the solution, for example, the following reaction would be promoted:

$$HCO_3^- + H^+ \longrightarrow H_2CO_3$$

The preceding reaction serves to decrease the effect of added $H^+$ on the pH of the blood. Bicarbonate is the major buffer of the blood. Under the opposite condition, when the concentration of free $H^+$ in the blood was falling, the reaction previously described would be reversed:

$$H_2CO_3 H^+ \longrightarrow HCO_3^-$$

Here, the dissociation of carbonic acid yields free $H^+$, which helps to prevent an increase in pH. Bicarbonate ions and carbonic acid thus act as a *buffer pair* to prevent either decreases or increases in pH, respectively. This buffering action normally maintains the blood pH at a very stable $7.40 \pm 0.05$.

## Organic Molecules

**Organic molecules** contain the atom *carbon*. Since the carbon atom has four electrons in its outer orbital, it must share four additional electrons by covalent bonding with other atoms to fill its outer orbital with eight electrons. The unique bonding capacity of carbon enables it to join with other carbon atoms to form chains and rings, while still allowing the carbons to bond with hydrogen and other atoms.

Most organic molecules in the body contain hydrocarbon chains and rings, as well as other atoms bonded to carbon. Two adjacent carbon atoms in a chain or ring may share one or two pairs of electrons. If the two carbon atoms share one pair of electrons, they are said to have a *single covalent bond*. This leaves each carbon atom free to bond to as many as three other atoms. If the two carbon atoms share two pairs of electrons, they have a *double covalent bond*, and each carbon atom can only bond to a maximum of two additional atoms (fig. 2.8).

The ends of some hydrocarbons are joined together to form rings. In the shorthand structural formulas of these molecules, the carbon atoms are not shown but are understood to be located at the corners of the ring. Some of these cyclic molecules have a double bond between two adjacent carbon atoms. Benzene and related molecules are shown as a six-sided ring with alternating double bonds. Such compounds are called *aromatic*. Since all of the carbons in an aromatic ring are equivalent, double bonds can be shown between any two adjacent carbons in the ring (fig. 2.9). The hydrocarbon chain or ring of many organic molecules provides a relatively inactive molecular backbone, to which more reactive groups of atoms are attached. These reactive groups are known as *functional groups* of the molecule and they usually contain atoms of oxygen, nitrogen, phosphorous, or sulfur. They are largely responsible for the unique chemical properties of the molecule (fig. 2.10).

Classes of organic molecules can be named according to their functional groups (fig. 2.11). *Ketones* (ke´tōnz) for example, have a carbonyl group within the carbon chain. An organic molecule is an *alcohol* if it has a hydroxyl group at one end of the chain. All *organic acids* (such as acetic acid, citric acids, and others) have a carboxyl (kar-bok´sil) group.

**Stereoisomers** Two molecules may have exactly the same atoms arranged in exactly the same sequence, yet may differ with respect to the spatial orientation of a key functional group. Such molecules are *stereoisomers* of each other. Depending upon the direction in which the key functional group is oriented with respect to the molecule, stereoisomers are called either D-isomers (for *dextro*, or right-handed) or L-isomers (for *levo*, or left-handed). Their relationship is similar to that between a right and left glove—if the palms are both facing forward, the two cannot be superimposed.

These subtle differences in structure are extremely important biologically, since enzymes, which interact with such molecules in a stereo-specific way in chemical reactions, cannot combine with the wrong stereoisomer. The enzymes of all cells (human and others) can only combine

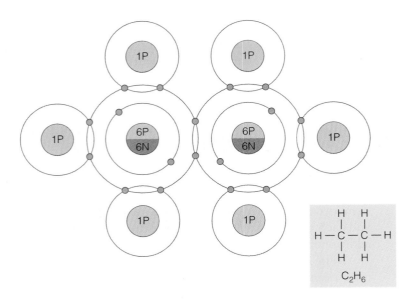

Ethane (C₂H₆)

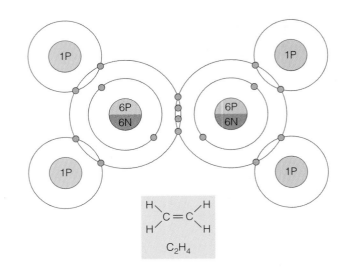

Ethylene (C₂H₄)

**FIGURE 2.8**

Two carbon atoms joined by a single covalent bond (*left*) or by a double covalent bond (*right*). In both cases, each carbon atom shares four pairs of electrons (has four bonds) to complete the eight electrons required to fill its outer orbital.

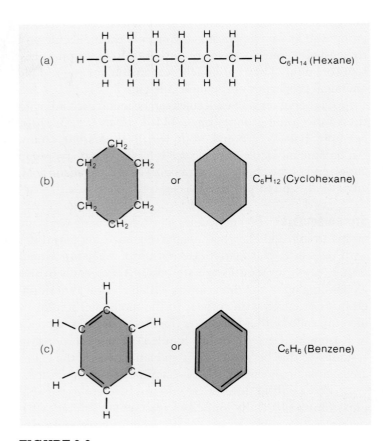

**FIGURE 2.9**

Hydrocarbons that are (*a*) linear, (*b*) cyclic, and (*c*) aromatic rings.

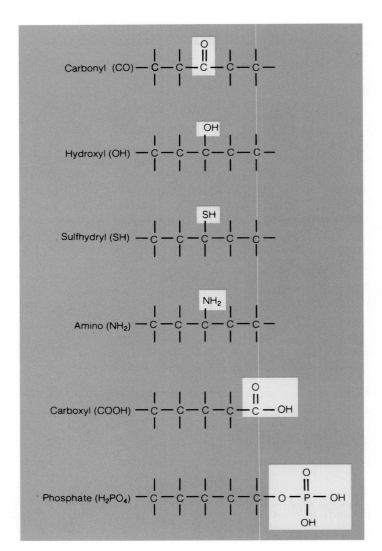

**FIGURE 2.10**

Various functional groups of organic molecules.

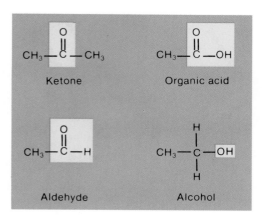

**FIGURE 2.11**
Categories of organic molecules based on functional groups.

with L-amino acids and D-sugars, for example. The opposite stereoisomers (D-amino acids and L-sugars) cannot be used by the body. D-sugars, for example, taste sweet, whereas the L-sugars (produced in a laboratory) are tasteless.

# Carbohydrates and Lipids

*Carbohydrate molecules share a characteristic ratio of carbon, hydrogen, and oxygen atoms, and they are divided into subgroups depending upon the number of simple sugars that each molecule contains. Lipids are a related but distinct group of molecules that share the physical property of being nonpolar, and are thus insoluble in water.*

Carbohydrates and lipids are similar in many ways. Both groups of molecules consist primarily of the atoms carbon, hydrogen, and oxygen, and both serve as major sources of energy in the body (comprising most of the calories consumed in food). Carbohydrates and lipids differ, however, in some important aspects of their chemical structures and physical properties. These differences significantly affect the functions of carbohydrates and lipids in the body.

## Carbohydrates

**Carbohydrates** are organic molecules that contain carbon, hydrogen, and oxygen in the ratio described by their name—*carbo* (carbon, C) and *hydrate* (water, $H_2O$). The general formula of a carbohydrate molecule is thus $CH_2O$; the molecule contains twice the number of hydrogen atoms as it contains carbon or oxygen atoms.

### Monosaccharides, Disaccharides, and Polysaccharides
Carbohydrates include **monosaccharides** (mon´´ŏ-sak´ă-rīdz), or simple sugars, and longer molecules that contain a number of monosaccharides joined together. The suffix *-ose*

..........................................

monosaccharide: Gk. *monos*, single; *sakcharon*, sugar

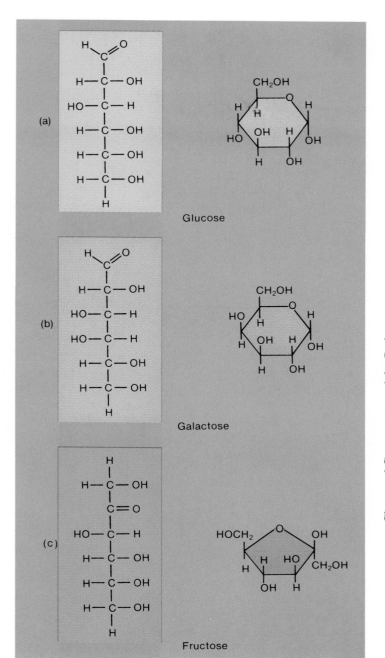

**FIGURE 2.12**
The structural formulas of three hexose sugars: (*a*) glucose, (*b*) galactose, and (*c*) fructose. All three have the same ratio of atoms—$C_6H_{12}O_6$.

denotes a sugar molecule; the term *hexose*, for example, refers to a six-carbon monosaccharide with the formula $C_6H_{12}O_6$. This formula is adequate for some purposes, but it does not distinguish between related hexose sugars, which are *structural isomers* of each other. The structural isomers glucose, galactose, and fructose, for example, are monosaccharides that have the same ratio of atoms arranged in slightly different ways (fig. 2.12).

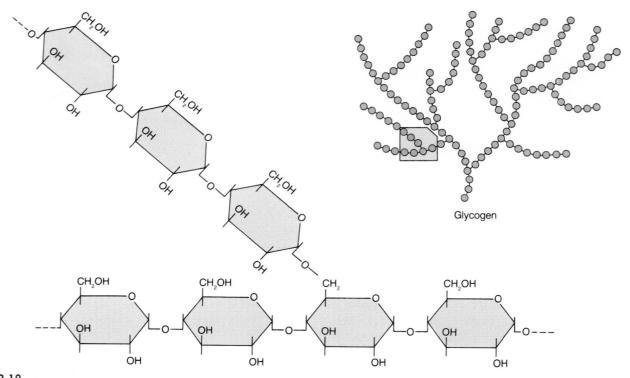

**FIGURE 2.13**

Glycogen is a polysaccharide composed of glucose subunits joined together to form a large, highly branched molecule.

Two monosaccharides can be joined covalently to form a **disaccharide** (*di-sak´ă-rīd*), or double sugar. Common disaccharides include table sugar, or *sucrose* (composed of glucose and fructose); milk sugar, or *lactose* (composed of glucose and galactose); and malt sugar, or *maltose* (composed of two glucose molecules). When many monosaccharides are joined together, the resulting molecule is called a **polysaccharide.** Starch, for example, is a polysaccharide found in many plants and is formed by the bonding together of thousands of glucose subunits. Animal starch, or **glycogen** (*gli´kŏ-jen*), found in the liver and muscles, likewise consists of repeating glucose molecules but differs from plant starch in that glycogen is more highly branched (fig. 2.13).

**Dehydration Synthesis and Hydrolysis**   In the formation of disaccharides and polysaccharides, the separate subunits (monosaccharides) are bonded together covalently by a type of reaction called **dehydration synthesis, or condensation.** In this reaction, which requires the participation of specific enzymes (chapter 4), a hydrogen atom is removed from one monosaccharide and a hydroxyl group (OH) is removed from another. As a covalent bond is formed between the two

monosaccharides, water ($H_2O$) is produced. Dehydration synthesis reactions are illustrated in figure 2.14.

When a person eats disaccharides and polysaccharides, or when the stored glycogen in the liver and muscles is to be used by tissue cells, the covalent bonds that join monosaccharides into disaccharides and polysaccharides must be broken. These *digestion reactions* occur by means of **hydrolysis** (*hi-drol´ĭ-sis*). Hydrolysis is the reverse of dehydration synthesis. A water molecule is split, as implied by the word *hydrolysis*, and the resulting hydrogen atom is added to one of the free glucose molecules as the hydroxyl group is added to the other (fig. 2.15).

When a potato is eaten, the starch within it is hydrolyzed into separate glucose molecules within the small intestine. This glucose is absorbed into the blood and carried to the tissues. Some tissue cells may use this glucose for energy. Liver and muscles, however, can store excess glucose in the form of glycogen by dehydration synthesis reactions in these cells. During fasting or prolonged exercise, the liver can add glucose to the blood through

.........................................................

hydrolysis: Gk. *hydor*, water; *lysis*, break

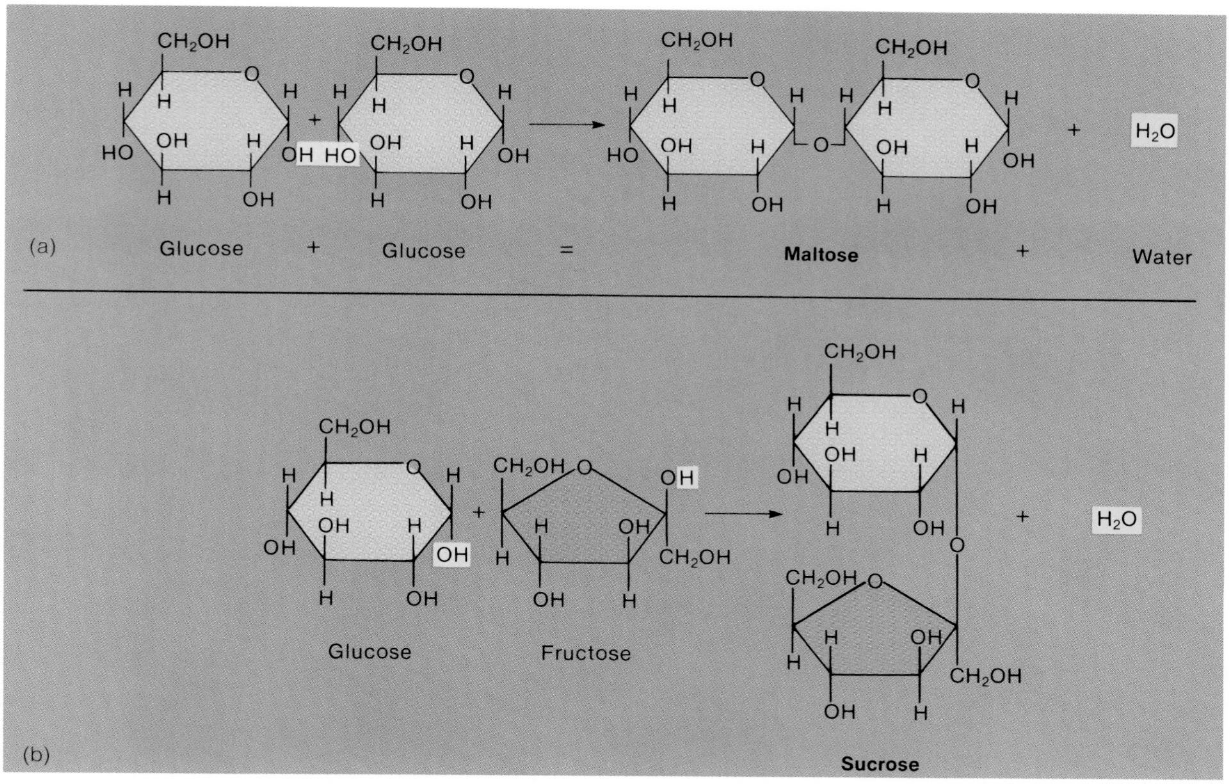

**FIGURE 2.14**
Dehydration synthesis of two disaccharides: (a) maltose and (b) sucrose. Note
that as the disaccharides are formed, a molecule of water is produced.

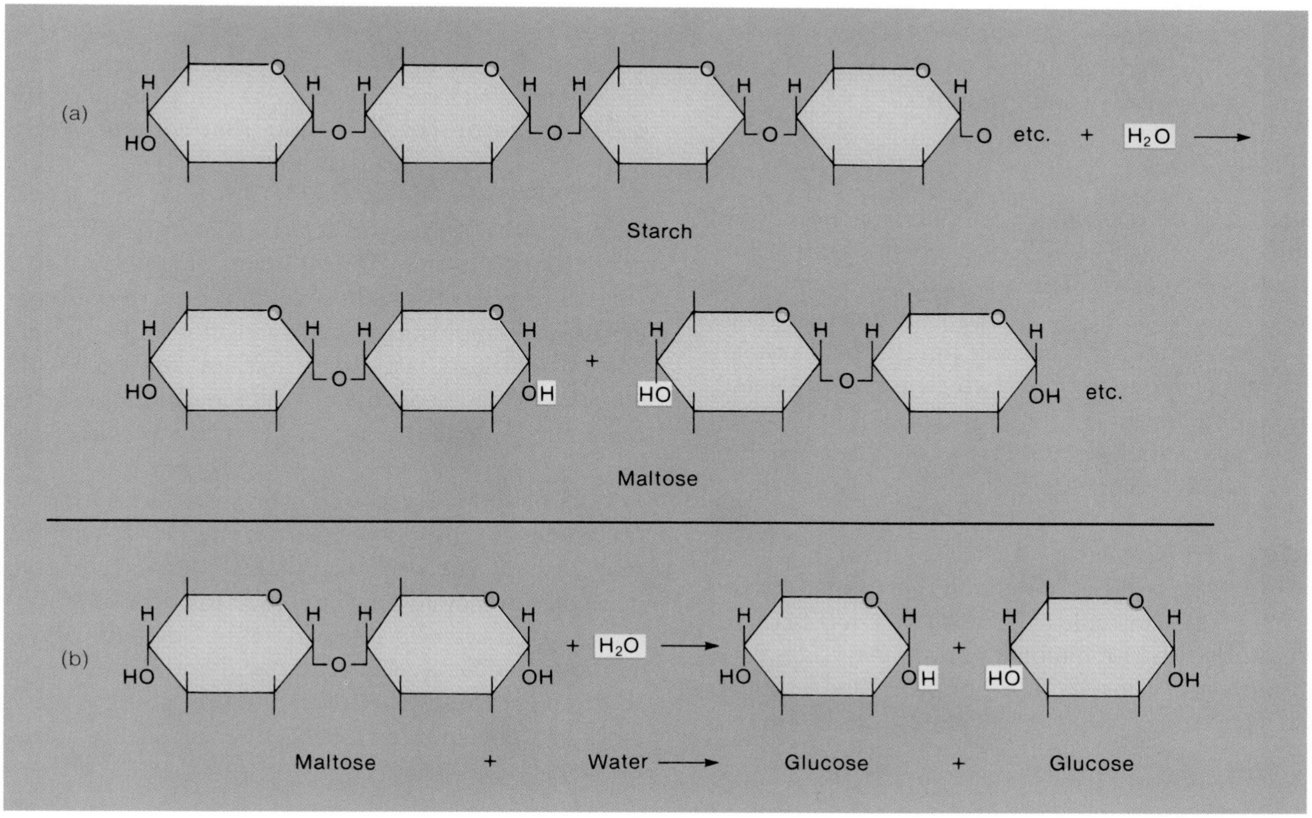

**FIGURE 2.15**
The hydrolysis of starch (a) into disaccharides (maltose) and
(b) into monosaccharides (glucose). Notice that as the covalent
bond between the subunits breaks, a molecule of water is split.

In this way, the hydrogen atom and hydroxyl group from
water are added to the ends of the released subunits.

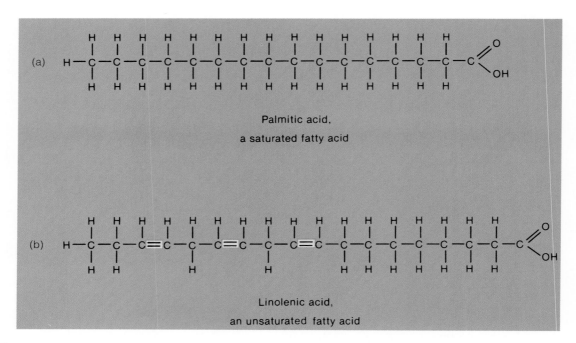

**FIGURE 2.16**
Structural formulas for (a) saturated and (b) unsaturated fatty acids.

hydrolysis of its stored glycogen. Dehydration synthesis and hydrolysis reactions do not occur spontaneously; they require the action of specific enzymes. Similar reactions, in the presence of other enzymes, build and break down lipids, proteins, and nucleic acids. In general, therefore, hydrolysis reactions digest molecules into their subunits, and dehydration synthesis reactions build larger molecules by the bonding together of their subunits.

## Lipids

The category of molecules known as **lipids** includes several types of molecules that vary widely in chemical structure. These diverse molecules are all in the lipid category by virtue of a common physical property—they are all *insoluble in polar solvents* such as water. This is because lipids consist primarily of hydrocarbon chains and rings, which are nonpolar and, thus, hydrophobic. Although lipids are insoluble in water, they can be dissolved in nonpolar solvents, such as ether, benzene, and related compounds.

**Triglycerides** *Triglycerides* are a subcategory of lipids that includes fat and oil. These molecules are formed by the condensation of one molecule of *glycerol* (*glis´ĕ-rol*) (a three-carbon alcohol) with three molecules of *fatty acids*. Each fatty acid molecule consists of a nonpolar hydrocarbon chain with a carboxyl group (abbreviated COOH) on one

end. If the carbon atoms within the hydrocarbon chain are joined by single covalent bonds so that each carbon can also bond to two hydrogen atoms, the fatty acid contains the maximum number of hydrogen atoms and is said to be *saturated*. If there are a number of double covalent bonds within the hydrocarbon chain so that each carbon can bond to only one hydrogen atom, the fatty acid is said to be *unsaturated*. Triglycerides that contain saturated fatty acids are called **saturated fats;** those that contain unsaturated fatty acids are **unsaturated fats** (fig. 2.16).

Within the adipose cells of the body, triglycerides are formed as the carboxylic acid ends of fatty acid molecules condense with the hydroxyl groups of a glycerol molecule (fig. 2.17). Since the hydrogen atoms from the carboxyl ends of fatty acid molecules form water molecules during dehydration synthesis, fatty acids that are combined with glycerol can no longer release $H^+$ and function as acids. For this reason, triglycerides are described as *neutral fats*.

**Ketone Bodies** Hydrolysis of triglycerides within adipose tissue releases *free fatty acids* into the blood. Not only can free fatty acids be used as an immediate source of energy by many organs, they also can be converted by the liver into derivatives called *ketone bodies*. These include four-carbon-long acidic molecules (acetoacetic acid and β-hydroxybutyric acid) and acetone (the solvent in nail polish remover). A rapid breakdown of fat, such as occurs during dieting and

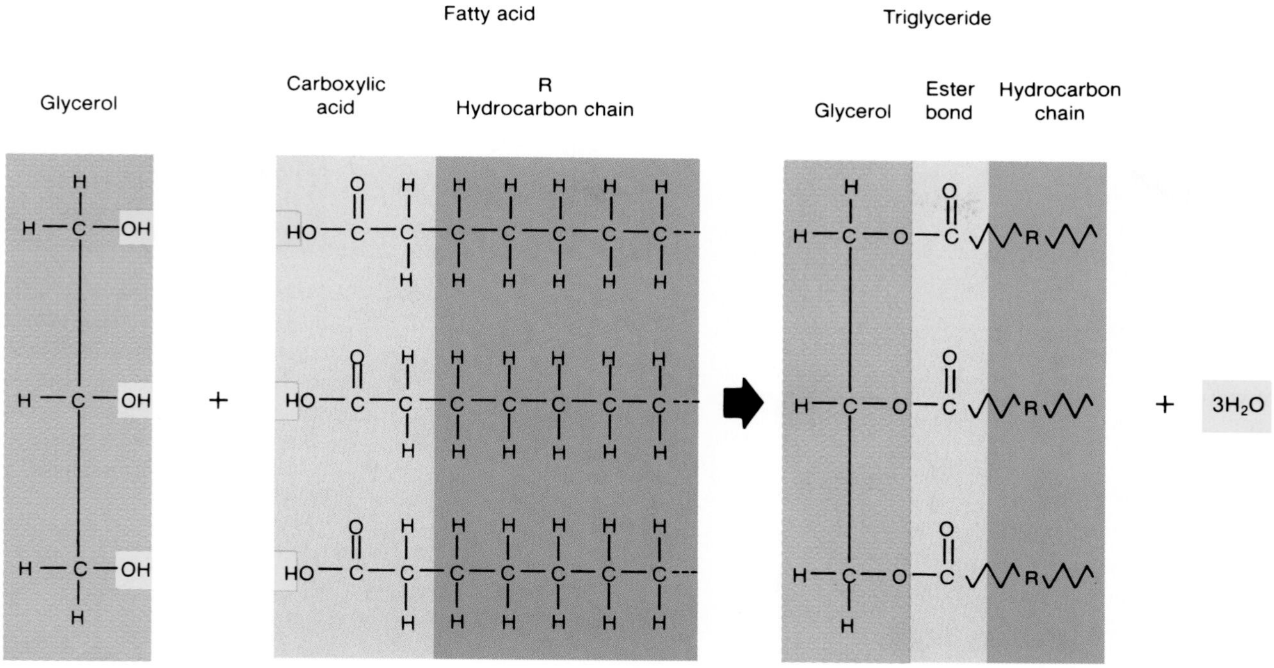

**Fatty acid** — **Glycerol** — **Carboxylic acid** — **R Hydrocarbon chain**

**Triglyceride** — **Glycerol** — **Ester bond** — **Hydrocarbon chain**

$+ \quad 3H_2O$

**FIGURE 2.17**
Dehydration synthesis of a triglyceride molecule from a glycerol and three fatty acids. A molecule of water is produced as an ester bond forms between each fatty acid and the glycerol. Sawtooth lines represent carbon chains, which are symbolized by an R.

in uncontrolled diabetes mellitus, results in elevated levels of ketone bodies in the blood—a condition called **ketosis.** If there are sufficient amounts of ketone bodies in the blood to lower the blood pH, the condition is called **ketoacidosis.** Severe ketoacidosis in uncontrolled diabetes mellitus can lead to coma and death.

**Phospholipids**   The class of lipids known as *phospholipids* includes a number of categories of lipids, all of which contain a phosphate group. In the most common type of phospholipid molecule, the three-carbon glycerol is attached to two fatty acids, the third carbon atom of the glycerol is attached to a phosphate group, and the phosphate group is, in turn, bonded to a nitrogen-containing choline group. The phospholipid molecule thus formed is known as *lecithin* (fig. 2.18). Figure 2.18 shows a simple way of illustrating the structure of a phospholipid. The parts of the molecule capable of ionizing (and thus becoming charged) are shown as a circle, whereas the nonpolar parts of the molecule are represented by lines.

Since the nonpolar ends of phospholipids are hydrophobic, they tend to group together when mixed in water. This allows the hydrophilic parts, which are polar,

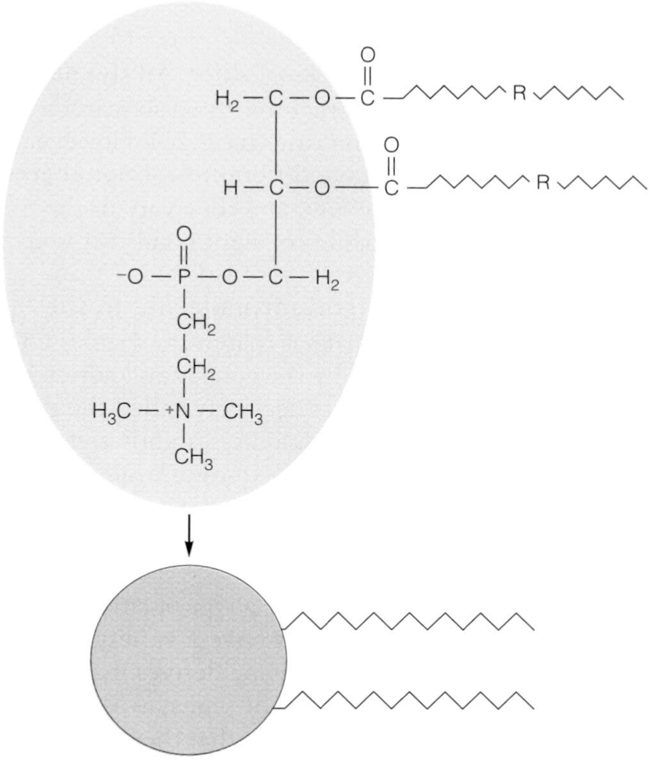

**FIGURE 2.18**
The structure of lecithin, a typical phospholipid (*top*) and its more simplified representation (*bottom*).

to face the surrounding water molecules (fig. 2.19). Such aggregates of molecules are called **micelles** (*mi-selz´*). The dual nature of phospholipid molecules (part polar, part nonpolar) enables them to form the major component of the cell membrane and also to alter the structure of water and decrease its surface tension. This latter function of phospholipids, which makes them *surfactants* (surface-active-agents), prevents collapse of the lungs.

### Steroids

Although the structure of *steroid molecules* is quite different from that of triglycerides or phospholipids, steroids are still included in the lipid category of molecules because they are nonpolar and insoluble in water. All steroid molecules have the same basic structure: three six-carbon rings are joined to one five-carbon ring (fig. 2.20). However, different kinds of steroids have different functional groups attached to this basic structure, and they vary in the number and position of the double covalent bonds between the carbon atoms in the rings.

*Cholesterol* is an important molecule in the body because it serves as the precursor (parent molecule) for the steroid hormones produced by the gonads and adrenal cortex. The testes and ovaries (collectively called the *gonads*) secrete **sex steroids,** which include estradiol and progesterone from the ovaries and testosterone from the testes. The adrenal cortex secretes the **corticosteroids,** including hydrocortisone and aldosterone, among others.

### Prostaglandins

*Prostaglandins* are a type of fatty acid (with a cyclic hydrocarbon group) that have a variety of regulatory functions. Although their name derives from the fact that they were originally noted in the semen as a secretion of the prostate, it has since been shown that they are produced by and active in almost all tissues. Prostaglandins are implicated in regulating the diameter of blood vessels, ovulation, uterine contraction during labor, inflammation reactions,

**FIGURE 2.19**
The formation of a micelle structure by phospholipid molecules.

**FIGURE 2.20**
Cholesterol and some steroid hormones derived from cholesterol.

blood clotting, and many other processes. Some of the different types of prostaglandins are shown in figure 2.21.

## Proteins

*Proteins are large molecules composed of amino acid subunits. Since there are 20 different types of amino acids that can be used in constructing a given protein, the variety of protein structures is immense. This variety allows each type of protein to be constructed so that it can perform very specific functions.*

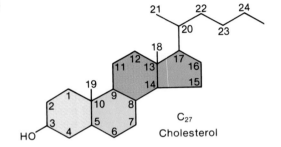

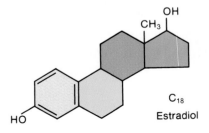

## Structure of Proteins

The enormous diversity of **protein** (pro´tēn´´) structure results from the fact that 20 different building blocks, the **amino acids**, can be used to form a protein. The amino acids are joined together to form a chain that can twist and fold in a specific manner due to chemical interactions between the amino acids. The specific sequence of amino acids in a protein, and thus the specific functional structure of the protein, is determined by genetic information. This genetic information for protein synthesis is contained in another category of organic molecules, the nucleic acids. The structure of nucleic acids, and the mechanisms by which the genetic information they encode directs protein synthesis, is described in chapter 3.

Proteins consist of long chains of subunits called amino acids. As the name implies, each amino acid contains an *amino group* (NH₂) on one end of the molecule and a *carboxylic acid group* (COOH) on another end. As mentioned previously, there are approximately 20 different amino acids with different structures and chemical properties that are used to build proteins. These differences are due to differences in the *functional groups* of these amino acids. The abbreviation for the functional group in the general formula for an amino acid is "R." (fig. 2.22). The R symbol actually stands for the word *residue*, but it can be thought of as indicating the rest of the molecule.

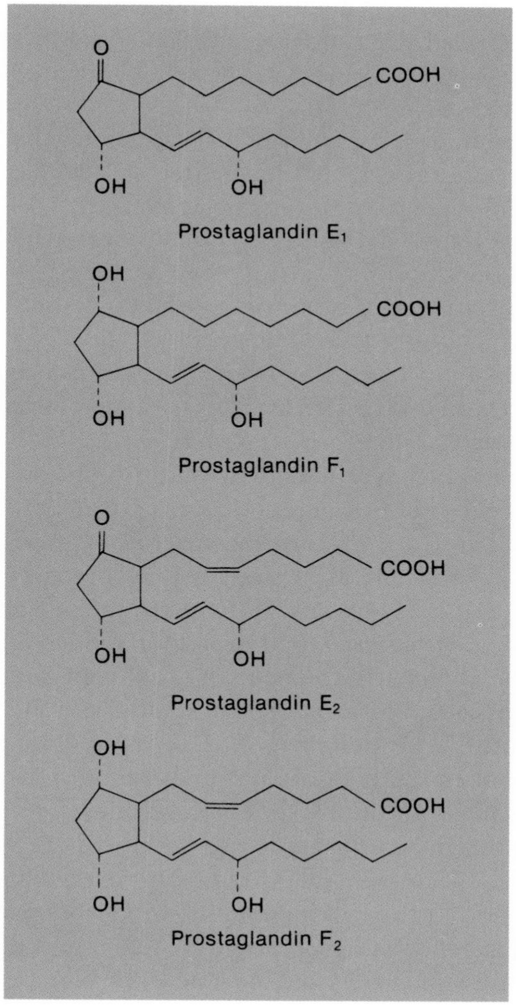

**FIGURE 2.21**
Structural formulas of some prostaglandins.

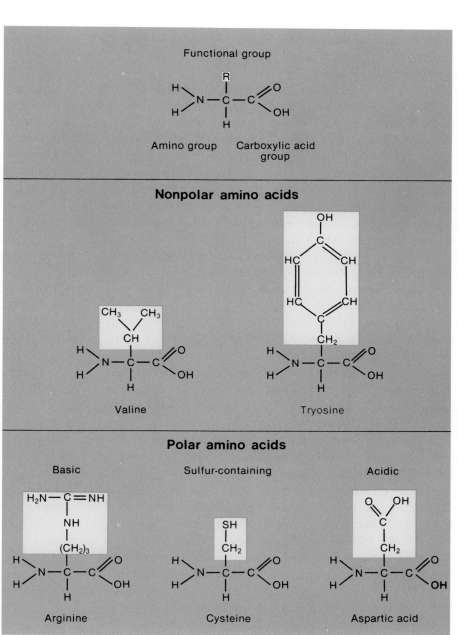

**FIGURE 2.22**
Representative amino acids, showing different types of functional (R) groups.

When amino acids are joined together by dehydration synthesis, the hydrogen from the amino end of one amino acid combines with the hydroxyl group of the carboxylic acid end of another amino acid. As a covalent bond is formed between the two amino acids, water is produced (fig. 2.23). The bond between adjacent amino acids is called a **peptide bond,** and the compound formed is called a *peptide*. When many amino acids are joined in this way, a chain of amino acids, or a **polypeptide,** is produced. The lengths of polypeptide chains vary widely. A hormone called *thyrotropin-releasing hormone*, for example, is only three amino acids long, whereas *myosin* (*mi′ŏ-sin*), a muscle protein, contains about 4500 amino acids. When the length of a polypeptide chain becomes very long (greater than about 100 amino acids), the molecule is called a *protein*.

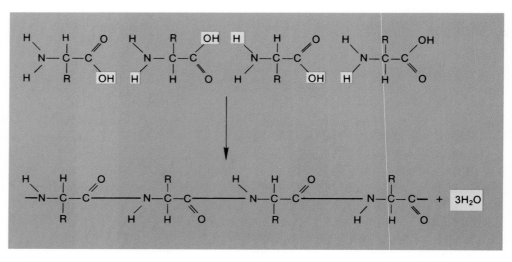

**FIGURE 2.23**
The formation of peptide bonds by dehydration synthesis reactions between amino acids.

The structure of a protein can be described at four different levels. At the first level, the sequence of amino acids in the protein, called the **primary structure** of the protein, is described. Each type of protein has a different primary structure. All of the billions of *copies* of a given type of protein in the body, however, have the same structure, because the structure of a given protein is coded by the genes. The primary structure of a protein is illustrated in figure 2.24*a*.

Weak interactions (such as hydrogen bonds) between amino acids in nearby positions in the polypeptide chain cause this chain to twist into a *helix*. The extent and location of the helical structure is different for each protein because of differences in amino acid composition. A description of the helical structure of a protein is termed its **secondary structure** (fig. 2.24*b*).

Most polypeptide chains bend and fold upon themselves to produce complex three-dimensional shapes, called the **tertiary structure** of the proteins. Each type of protein has its own characteristic tertiary structure. This is because the folding and bending of the polypeptide chain is produced by chemical interactions between particular amino acids that are located in different regions of the chain.

Most of the tertiary structure of proteins is formed and stabilized by weak chemical interactions between the functional (R) groups of widely spaced amino acids. Since most of the tertiary structure is stabilized by weak bonds, this structure can be easily disrupted by high temperatures or by changes in pH. Irreversible changes in the tertiary structure of proteins produced by this means are referred to as **denaturation** (*de-na″chur-a′shun*) of the proteins. The tertiary structure of some proteins is made more stable by strong covalent bonds between sulfur atoms (called *disulfide bonds* and abbreviated S-S) in the functional group of an amino acid known as cysteine (fig. 2.25).

Denatured proteins retain their primary structure (the peptide bonds are not broken) but have altered chemical properties. Cooking a pot roast, for example, alters the texture of the meat proteins—it doesn't result in an amino acid soup. Denaturation is most dramatically demonstrated by frying an egg. Egg albumin proteins are soluble in their native state, in which they form the clear, viscous fluid of a raw egg. When denatured by cooking, these proteins change shape, cross-bond with each other, and by this means form an insoluble white precipitate—the egg white.

Some proteins (such as hemoglobin and insulin) are composed of a number of polypeptide chains covalently bonded together. This is the **quaternary structure** of these proteins. Insulin, for example, is composed of two polypeptide chains—one that is 21 amino acids long and the other, 30 amino acids long. *Hemoglobin* (the protein in red blood cells that carries oxygen) is composed of four separate polypeptide chains, as indicated in table 2.4 along with the compositions of other body proteins.

Many proteins in the body are normally found combined, or *conjugated*, with other types of molecules. **Glycoproteins** are proteins conjugated with carbohydrates. Examples of such molecules include certain hormones and some proteins found in the cell membrane. **Lipoproteins** are proteins conjugated with lipids. These are found in cell membranes and in plasma (the fluid portion of the blood).

## FIGURE 2.24

A polypeptide chain, showing (a) its primary structure and (b) secondary structure.

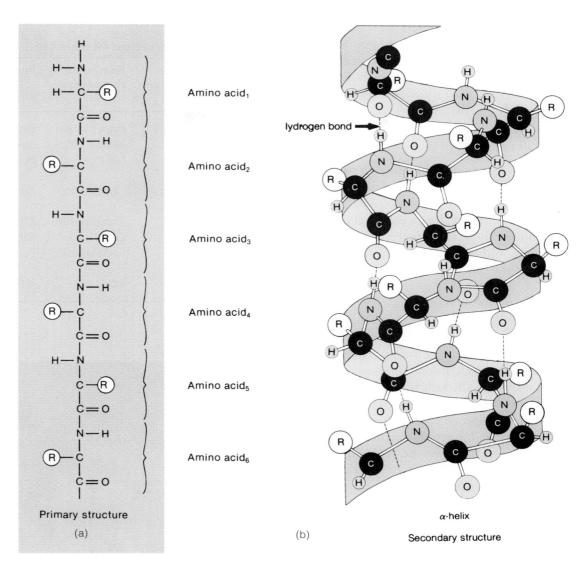

Amino acid₁

Amino acid₂

Amino acid₃

Amino acid₄

Amino acid₅

Amino acid₆

Primary structure

(a)

hydrogen bond →

α-helix

Secondary structure

(b)

## FIGURE 2.25

The tertiary structure of a protein. (a) interactions between functional (R) groups of amino acids result in (b) the formation of complex three-dimensional shapes of proteins.

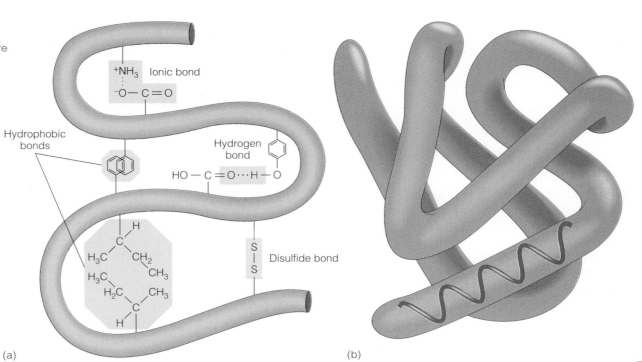

⁺NH₃   Ionic bond

-O—C=O

Hydrophobic bonds

Hydrogen bond

HO—C=O···H—O

S
|
S   Disulfide bond

(a)

(b)

Proteins conjugated with pigment molecules are **chromoproteins.** These include hemoglobin, which transports oxygen in red blood cells, and the cytochromes, which are needed for oxygen utilization and energy production within cells.

## Functions of Proteins

Because of their tremendous structural diversity, proteins can perform a wider variety of functions than any other type of molecule in the body. Many proteins, for example, contribute significantly to the structure of different tissues, and in this capacity play a passive role in the functions of these tissues. Examples of such *structural proteins* are collagen (fig. 2.26) and keratin. Collagen is a fibrous protein that provides tensile strength to connective tissues, such as tendons and ligaments. Keratin is found in the outer layer of dead cells in the epidermis, where it serves to prevent water loss through the skin.

Many proteins play a more active role in the body where specialized structure and function are required. *Enzymes* and *antibodies,* for example, are proteins—no other type of molecule could provide the vast array of different structures needed for their functions. Proteins in cell membranes serve as *receptors* for specific regulator molecules (such as hormones) and as *carriers* that transport specific molecules across the membrane. Because they are so diverse in shape and chemical properties, proteins can provide the specificity these functions require.

..................

chromoprotein: Gk. *chroma,* color; *proteios,* of the first quality

### Table 2.4 Composition of selected body proteins

| Protein | Number of polypeptide chains | Nonprotein component | Function |
|---|---|---|---|
| Hemoglobin | 4 | Heme pigment | Carries oxygen in the blood |
| Myoglobin | 1 | Heme pigment | Stores oxygen in muscle |
| Insulin | 2 | None | Hormone that regulates metabolism |
| Luteinizing hormone | 1 | Carbohydrate | Hormone that stimulates gonads |
| Fibrinogen | 1 | Carbohydrate | Aids in blood clotting |
| Mucin | 1 | Carbohydrate | Forms mucus |
| Blood group proteins | 1 | Carbohydrate | Produces blood types |
| Lipoproteins | 1 | Lipids | Transports lipids in blood |

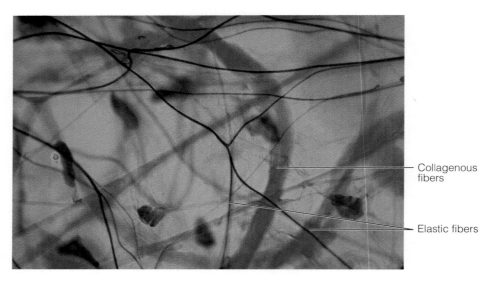

Collagenous fibers

Elastic fibers

**FIGURE 2.26**

A photomicrograph of collagenous fibers within connective tissue.

# Chapter Summary

## Atoms, Ions, and Molecules (pp. 22–29)

1. Covalent bonds are formed by atoms that share electrons. They are the strongest type of chemical bonds.
   a. Electrons are equally shared in nonpolar covalent bonds and unequally shared in polar covalent bonds.
   b. Atoms of oxygen, nitrogen, and phosphorus strongly attract electrons and become electrically negative compared to the other atoms sharing electrons with them.

2. Ionic bonds are formed by atoms that transfer electrons. These weak bonds join atoms together in an ionic compound.
   a. The atom in an ionic compound that takes the electron from another atom gains a net negative charge; the other atom becomes positively charged.

b. Ionic bonds break easily when the ionic compound is dissolved in water. Dissociation of the ionic compound yields charged atoms called ions.

3. When hydrogen is bonded to an electronegative atom, it gains a slight positive charge and is weakly attracted to another electronegative atom. This weak attraction is a hydrogen bond.

4. Acids donate hydrogen ions to a solution, whereas bases lower the hydrogen ion concentration of a solution.
   a. The pH scale is a negative function of the logarithm of the hydrogen ion concentration.
   b. In a neutral solution, the concentration of $H^+$ is equal to the concentration of $OH^-$, and the pH is 7.
   c. Acids raise the $H^+$ concentration, and thus lower the pH below 7. Bases lower the $H^+$ concentration, and thus raise the pH above 7.

5. Organic molecules contain atoms of carbon joined by covalent bonds. Atoms of nitrogen, oxygen, phosphorus, or sulfur may be present as specific functional groups in the organic molecule.

## Carbohydrates and Lipids (pp. 29–34)

1. Carbohydrates contain carbon, hydrogen, and oxygen, usually in a ratio of 1:2:1.
   a. Carbohydrates consist of simple sugars (monosaccharides), disaccharides, and polysaccharides (such as glycogen).
   b. Covalent bonds between monosaccharides are formed by dehydration synthesis, or condensation. Bonds are broken by hydrolysis reactions.

2. Lipids are organic molecules that are insoluble in polar solvents like water.
   a. Triglycerides (fat and oil) consist of three fatty acid molecules joined to a molecule of glycerol.
   b. Ketone bodies are smaller derivatives of fatty acids.
   c. Phospholipids (such as lecithin) are phosphate-containing lipids that have a polar group, which is hydrophilic. The rest of the molecule is hydrophobic.
   d. Steroids (including the hormones of the adrenal cortex and the gonads) are lipids with a characteristic five-ring structure.

e. Prostaglandins are a family of cyclic fatty acids that serve a variety of regulatory functions.

## Proteins (pp. 34–38)

1. Proteins are composed of long chains of amino acids bonded together by covalent peptide bonds.
   a. Each amino acid contains an amino group, a carboxyl group, and a functional group that is different for each of the more than 20 different amino acids.
   b. The polypeptide chain may be twisted into a helix (secondary structure) and bent and folded to form the tertiary structure of the protein.
   c. Proteins that are composed of two or more polypeptide chains are said to have a quaternary structure.
   d. Proteins may be combined with carbohydrates, lipids, or other molecules.
   e. Because they are so diverse structurally, proteins serve a wider variety of specific functions than any other type of molecule.

# Review Activities

## Objective Questions

1. Which of the following statements about atoms is *true?*
   a. They have more protons than electrons.
   b. They have more electrons than protons.
   c. They are electrically neutral.
   d. They have as many neutrons as they have electrons.

2. The bond between oxygen and hydrogen in a water molecule is
   a. a hydrogen bond.
   b. a polar covalent bond.
   c. a nonpolar covalent bond.
   d. an ionic bond.

3. Which of the following is a nonpolar covalent bond?
   a. the bond between two carbons
   b. the bond between sodium and chloride
   c. the bond between two water molecules
   d. the bond between nitrogen and hydrogen

4. Solution A has a pH of 2 and solution B has a pH of 10. Which of the following statements about these solutions is *true?*
   a. Solution A has a higher $H^+$ concentration than solution B.
   b. Solution B is basic.
   c. Solution A is acidic.
   d. All of the above are true.

5. Glucose is
   a. a disaccharide.
   b. a polysaccharide.
   c. a monosaccharide.
   d. a phospholipid.

6. Digestion reactions occur by means of
   a. dehydration synthesis.
   b. hydrolysis.

7. Carbohydrates are stored in the liver and muscles in the form of
   a. glucose.
   b. triglycerides.
   c. glycogen.
   d. cholesterol.

8. Lecithin is
   a. a carbohydrate.
   b. a protein.
   c. a steroid.
   d. a phospholipid.

9. Which of the following lipids have regulatory roles in the body?
   a. steroids
   b. prostaglandins
   c. triglycerides
   d. both a and b
   e. both b and c

10. The tertiary structure of a protein is *directly* determined by
    a. the genes.
    b. the primary structure of the protein.
    c. enzymes that mold the shape of the protein.
    d. the position of peptide bonds.

11. The type of bond formed between two molecules of water is
    a. an ionic bond.
    b. a polar covalent bond.
    c. a nonpolar covalent bond.
    d. a hydrogen bond.

12. The carbon-to-nitrogen bond that joins amino acids together in a protein is called
    a. a glycosidic bond.
    b. a peptide bond.
    c. a hydrogen bond.
    d. a double bond.

**Essay Questions**

1. Compare and contrast nonpolar covalent bonds, polar covalent bonds, and ionic bonds.
2. Define *acid* and *base* and explain how acids and bases influence the pH of a solution.
3. With reference to dehydration synthesis and hydrolysis reactions, describe the relationships between starch in an ingested potato, liver glycogen, and blood glucose.
4. All fats are lipids, but not all lipids are fats. Explain why this is an accurate statement.
5. Describe the relationship between the primary structure of a protein and its secondary and tertiary structures.

# chapter three

# cell structure and genetic regulation

## objectives

- Describe the structure of the cell membrane.

- Discuss the processes of amoeboid motion and phagocytosis.

- Describe the structure and function of cilia and flagella.

- Describe the structure and functions of the cytoskeleton and of lysosomes.

- Describe the structure of mitochondria and the endoplasmic reticulum and discuss their significance.

- Describe the structure of DNA and RNA nucleotides and explain the law of complementary base pairing.

- Discuss the process of RNA synthesis and distinguish between the different types of RNA.

- Explain how RNA directs the synthesis of proteins.

- Describe the functions of mRNA, tRNA, and rRNA.

- Explain how codons and anticodons function in protein synthesis.

- Describe the functions of the rough endoplasmic reticulum and the Golgi apparatus in the packaging and secretion of proteins.

- Explain how DNA replicates itself and state the events that occur during each phase of the cell cycle.

- Describe the different phases of mitosis and the function of mitosis.

- Explain how meiosis differs from mitosis and state the function of meiotic cell division.

# Cell Membrane and Associated Structures

*The cell is the fundamental structural and functional unit in the body. Many of the functions of cells are performed by particular subcellular structures known as organelles. The cell membrane is an extremely important structure that regulates the passage of substances into and out of the cell and participates in cellular movements.*

The cell appears so small and simple when viewed with the ordinary (light) microscope that it is difficult to conceive of each cell as a living entity unto itself. Equally amazing is that the physiology of organs and systems derive from the complex functions of the cells of which they are composed. Complexity of function demands complexity of structure even at the subcellular level.

As the basic functional unit of the body, each cell is a highly organized molecular factory. Cells come in a great variety of shapes and sizes. This great diversity, which is also apparent in the subcellular structures within different cells, reflects the diversity of function of different cells in the body. All cells, however, share certain characteristics; for example, they are all surrounded by a cell membrane and most of them possess the structures listed in table 3.1. Thus, although no single cell can be considered typical, the general structure of cells can be indicated by a single illustration (fig. 3.1).

For descriptive purposes, a cell can be divided into three principal parts:

**1** **The cell (plasma) membrane.** The cell membrane surrounds the cell, gives it form, and separates the cell's internal structures from the extracellular environment. The structure of the cell membrane permits selective communication between the intracellular and extracellular compartments.

**2** **The cytoplasm and organelles** (or˝gă-nelz´). The cytoplasm is the aqueous content of a cell between the nucleus and the cell membrane. Organelles are subcellular structures within the cytoplasm of a cell that perform specific functions.

**3** **The nucleus.** The nucleus is a large, generally spheroid body within a cell. It contains the DNA (genetic material) that directs the cell's activities.

## Structure of the Cell Membrane

Because both the intracellular and extracellular environments (or compartments) are aqueous, a barrier must be present to prevent the loss of such water-soluble cellular molecules as enzymes and nucleotides. Since this barrier cannot itself be composed of water-soluble molecules, it makes sense that the cell membrane should be composed of lipids.

The **cell membrane** (also called the **plasma membrane,** or **plasmalemma**), and indeed all of the membranes surrounding organelles within the cell, are composed primarily of phospholipids and proteins. Phospholipids, as described in chapter 2, are polar on the end that contains the phosphate group and nonpolar (and hydrophobic) throughout

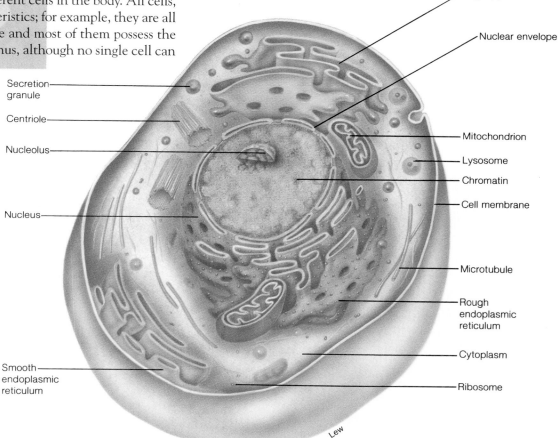

Labels:
Golgi apparatus
Nuclear envelope
Mitochondrion
Lysosome
Chromatin
Cell membrane
Microtubule
Rough endoplasmic reticulum
Cytoplasm
Ribosome
Secretion granule
Centriole
Nucleolus
Nucleus
Smooth endoplasmic reticulum

Lew

■□■ **FIGURE 3.1**
A generalized cell and the principal organelles.

aqueous: L. *aqua*, water

## Table 3.1   Structure and function of cellular components

| Component | Structure | Function |
|---|---|---|
| Cell (plasma) membrane | Membrane composed of phospholipid and protein molecules | Gives form to cell and controls passage of materials in and out of cell |
| Cytoplasm | Fluid, jellylike substance between the cell membrane and the nucleolus in which organelles are suspended | Serves as matrix substance in which chemical reactions occur |
| Endoplasmic reticulum | System of interconnected membrane-forming canals and tubules | Smooth endoplasmic reticulum metabolizes nonpolar compounds and stores $Ca^{++}$ in striated muscle cells; rough endoplasmic reticulum assists in protein synthesis |
| Ribosomes | Granular particles composed of protein and RNA | Synthesize proteins |
| Golgi apparatus | Cluster of flattened, membranous sacs | Synthesizes carbohydrates and packages molecules for secretion; secretes lipids and glycoproteins |
| Mitochondria | Double-walled membranous sacs with folded inner partitions | Release energy from food molecules and transform energy into usable ATP |
| Lysosomes | Single-walled membranous sacs | Digest foreign molecules and worn and damaged cells |
| Peroxisomes | Spherical membranous vesicles | Contain enzymes that produce hydrogen peroxide and use this for various oxidation reactions |
| Centrosome | Nonmembranous mass of two rodlike centrioles | Helps organize spindle fibers and distribute chromosomes during mitosis |
| Vacuoles | Membranous sacs | Store and excrete various substances within the cytoplasm |
| Fibrils and microtubules | Thin, hollow tubes | Support cytoplasm and transport materials within the cytoplasm |
| Cilia and flagella | Minute cytoplasmic extensions from cell | Move particles along surface of cell or move cell |
| Nuclear membrane | Membrane surrounding nucleus composed of protein and lipid molecules | Supports nucleus and controls passage of materials between nucleus and cytoplasm |
| Nucleolus | Dense, nonmembranous mass composed of protein and RNA molecules | Forms ribosomes |
| Chromatin | Fibrous strands composed of protein and DNA molecules | Controls cellular activity for carrying on life processes |

the rest of the molecule. Since there is an aqueous environment on each side of the membrane, the hydrophobic parts of the molecules huddle together in the center of the membrane, leaving the polar ends exposed to water on both surfaces. This results in the formation of a double layer of phospholipids in the cell membrane.

The hydrophobic core of the membrane restricts the passage of water and water-soluble molecules and ions. Certain of these polar compounds, however, do pass through the membrane. The specialized functions and selective-transport properties of the membrane are believed to be due to its protein content. Some proteins are found partially submerged on each side of the membrane; other proteins span the membrane completely from one side to the other. Since the membrane is not solid—phospholipids and proteins are free to move laterally—the proteins within the phospholipid sea are not uniformly distributed, but rather form a constantly changing mosaic pattern. This structure is known as the **fluid-mosaic model** of membrane structure (fig. 3.2).

The proteins found in the cell membrane serve a variety of functions. Some function in structural support, transport of molecules across the membrane, or in the enzymatic control of chemical reactions at the cell surface. Others function as receptors for hormones and other regulatory molecules that arrive at the outer surface of the membrane.

In addition to lipids and proteins, the cell membrane also contains carbohydrates, which are primarily attached to the outer surface of the membrane as glycoproteins and glycolipids. These surface carbohydrates have many negative charges; consequently, they affect the interaction of

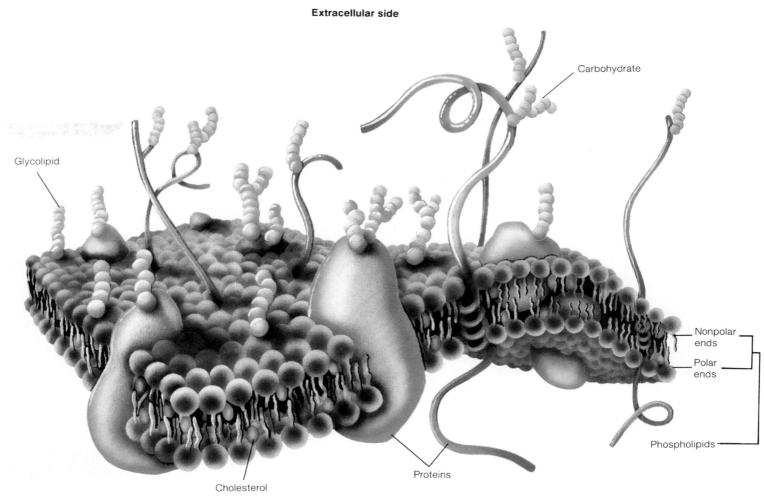

**FIGURE 3.2**
The fluid-mosaic model of the cell membrane. The membrane consists of a double layer of phospholipids, with the phosphates (shown by spheres) oriented outward and the hydrophobic hydrocarbons (wavy lines) oriented toward the center. Proteins may completely or partially span the membrane. Carbohydrates are attached to the outer surface.

regulatory molecules with the membrane. The negative charges at the surface also affect interactions between cells—they help keep red blood cells apart, for example. Stripping the carbohydrates from the outer red blood cell surface results in their more rapid destruction by the liver, spleen, and bone marrow.

## Adaptations of the Cell Membrane

The dynamic nature of the cell membrane and the presence of specialized structures at the cell surface permit the cells to move through their environment and materials to move along the cell surface. Other adaptations of the cell membrane function to improve the transport of materials between the intracellular and extracellular compartments.

**Amoeboid Movement and Phagocytosis**   Some body cells, including certain white blood cells and macrophages in connective tissues, are able to move in the manner of an amoeba (a single-celled organism). This **amoeboid movement** is achieved through the extension of parts of the cytoplasm to form *pseudopods* (soo´dŏ-podz), which attach to a substrate and pull the cell along.

Cells that display amoeboid movement—as well as certain liver cells, which are not mobile—use pseudopods to surround and engulf particles of organic matter (such as bacteria). This process is a type of cellular eating called

pseudopod: Gk. *pseudes*, false; *pod*, foot

 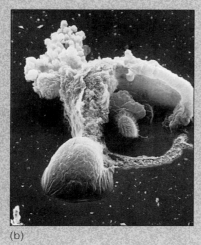

(a)          (b)

**FIGURE 3.3**
Scanning electron micrographs of phagocytosis showing (a) the formation of pseudopods and (b) the entrapment of the prey within a food vacuole.

phagocytosis (*fag″ŏ-si-to´sis*) that serves to protect the body from invading microorganisms and to remove extracellular debris. Phagocytic cells surround their victims with pseudopods, which join together and fuse (fig. 3.3). After the inner membrane of the pseudopods becomes a continuous membrane around the ingested particle, it pinches off from the cell membrane. The ingested particle is now contained in an organelle called a *food vacuole* within the cell. The particle will subsequently be digested by enzymes contained in a different organelle—the lyosome, described in a later section.

Pinocytosis (*pin″ŏ-si-to´sis*) is a related process performed by many cells. Instead of forming pseudopods, the cell membrane invaginates to produce a deep, narrow furrow. The membrane near the surface of this furrow then fuses, and a small vacuole containing the extracellular fluid is pinched off and enters the cell. In this way, a cell can take in large molecules, such as proteins, which may be present in the extracellular fluid.

**Endocytosis**   Phagocytosis and pinocytosis, in which part of the cell membrane invaginates to form first a pouch and then a vacuole, are two means by which extracellular materials can be engulfed by a cell. The general term for this process is **endocytosis.** Another type of endocytosis involves the smallest area of cell membrane, and it occurs only with specific molecules in the extracellular environment. Since the extracellular molecules must bond to very specific *receptor proteins* in the cell membrane, this process is known as *receptor-mediated endocytosis*.

In receptor-mediated endocytosis, the interaction of specific molecules in the extracellular fluid with specific membrane receptor proteins causes the membrane to invaginate, fuse, and pinch off to form a vesicle, which is a small vacuole (fig. 3.4). Vesicles formed in this way contain extracellular fluid and molecules that could not have passed into the cell by other means. Cholesterol attached to specific proteins, for example, is taken up into artery cells by receptor-mediated endocytosis. (This is in part responsible for the vascular disorder, *atherosclerosis,* as described in chapter 21.)

**Exocytosis**   Proteins and other molecules produced within the cell that are destined for export (secretion) are packaged inside of the cell within vesicles by an organelle known as the *Golgi* (*gol´je*) *apparatus.* In the process of *exocytosis,* these secretory vesicles fuse with the cell membrane and release their contents into the extracellular environment (see fig. 3.25). This process adds new membrane material, which replaces what was lost from the cell membrane during endocytosis.

Endocytosis and exocytosis account for only part of the two-way traffic between the intracellular and extracellular compartments. Most of this traffic is due to membrane transport processes—the movement of molecules and ions through the cell membrane (see chapter 5).

**Cilia and Flagella**   Cilia (*sil´e-ă*) are tiny hairlike structures that protrude from the cell and, as in the coordinated action of oarsmen in a boat, stroke in unison. Cilia in the human body are found on the apical surface (the surface facing the lumen, or cavity) of stationary epithelial cells in the respiratory system and female reproductive tract. In the respiratory system, the cilia transport strands of mucus to the throat (pharynx), where the mucus can either be swallowed or expectorated. In the female reproductive tract, ciliary movements in the epithelial lining of the uterine tube draw the egg (ovum) into the tube and move it along toward the uterus.

Spermatozoa are the only cells in the human body that have **flagella** (*flă-jel´ă*). The flagellum is a single, whiplike structure that propels the sperm cell through its environment. Both cilia and flagella are composed of *microtubules* (formed

..........................................
phagocytosis: Gk. *phagein*, to eat; *kytos*, hollow body vacuole: L. *vacuus*, empty
pinocytosis: Gk. *pinein*, to drink; *kytos*, hollow body

..........................................
cilia: L. *cili*, small hair
flagellum: L. *flagrum*, whip

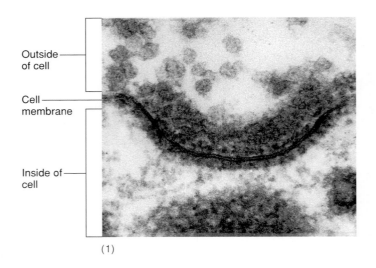

Outside of cell

Cell membrane

Inside of cell

(1)

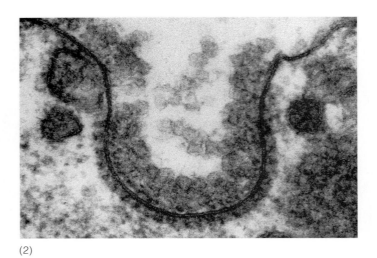

(2)

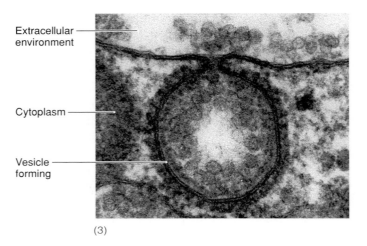

Extracellular environment

Cytoplasm

Vesicle forming

(3)

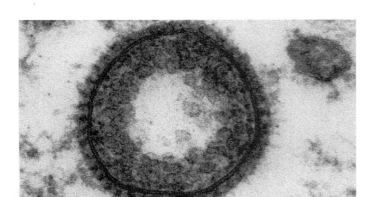

(4)

## FIGURE 3.4

Stages 1–4 of endocytosis, in which specific bonding of extracellular particles to membrane receptor proteins occurs.

from proteins) arranged in a characteristic way. One pair of microtubules in the center of a cilium or flagellum is surrounded by nine other pairs of microtubules to produce what is often described as a "9 + 2" arrangement (fig. 3.5).

**Microvilli** In areas of the body that are specialized for rapid diffusion, the surface area of the cell membranes may be increased by numerous folds. The rapid passage of the products of digestion across the epithelial lining in the small intestine, for example, is aided by this type of structural adaptation. The apical surface of the epithelial cells in the small intestine is increased by many tiny folds that form fingerlike projections called **microvilli** (fig. 3.6). Similar microvilli are found in the epithelium of the kidney tubules, which must reabsorb various molecules that are filtered out of the blood.

microvillus: *mikros,* small; L. *villus,* shaggy hair

## Cytoplasm and Its Organelles

*Many cell functions that are performed in the cytoplasmic compartment result from the activity of specific structures called organelles. Among these are the lysosomes, which contain digestive enzymes, and the mitochondria, where most of the cellular energy is produced. Other organelles participate in the synthesis and secretion of cellular products.*

### Cytoplasm and Cytoskeleton

The jellylike matrix within a cell (exclusive of that within the nucleus) is known as **cytoplasm** (*si´tŏ-plaz´´em*). When viewed in a microscope without special techniques, the cytoplasm appears to be uniform and unstructured. This, however, is not the case. The cytoplasm is actually a

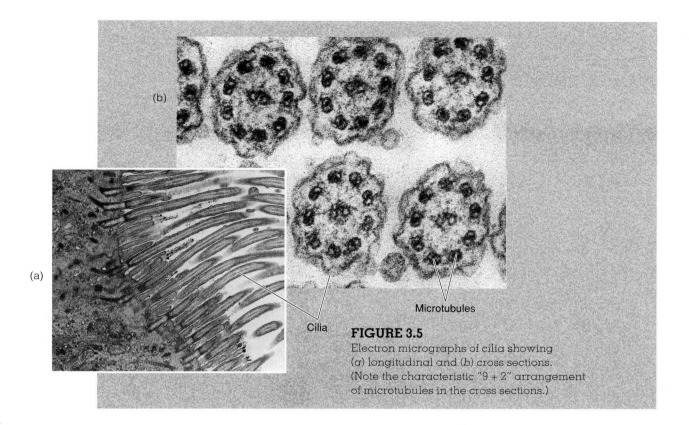

**FIGURE 3.5**
Electron micrographs of cilia showing
(*a*) longitudinal and (*b*) cross sections.
(Note the characteristic "9 + 2" arrangement
of microtubules in the cross sections.)

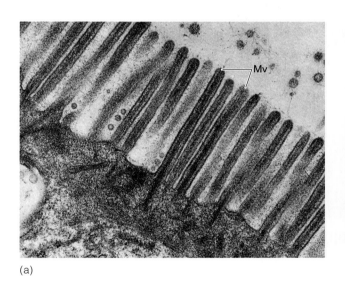

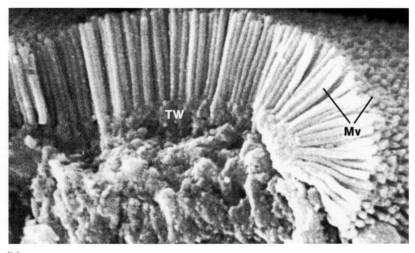

(a)

(b)

**FIGURE 3.6**
Microvilli (labeled Mv) in the small intestine, as seen with the
transmission (*a*) and scanning (*b*) electron microscope. (TW is
the terminal web.)

highly organized structure in which protein fibers, in the
form of *microtubules* and *microfilaments*, are arranged in a
complex lattice. Using fluorescence microscopy, these
structures can be visualized with the aid of antibodies
against their protein components (fig. 3.7). The interconnected microfilaments and microtubules are believed
to provide structural organization for cytoplasmic enzymes
and support for various organelles.

The lattice of microfilaments and microtubules is thus
said to function as a **cytoskeleton** (fig. 3.8). The structure
of this skeleton is not rigid; it has been shown to be capable of quite rapid reorganization. Contractile proteins

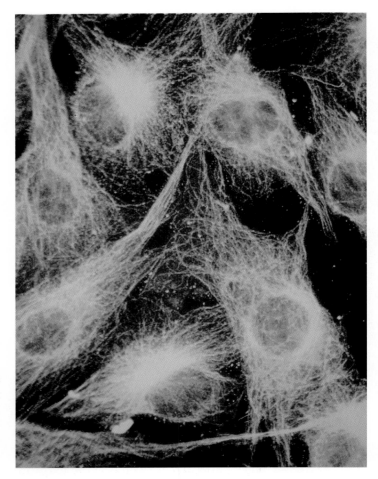

**FIGURE 3.7**

An immunofluorescence photograph of microtubules forming the cytoskeleton of a cell. Microtubules are visualized with the aid of antibodies against tubulin, the major protein component of the microtubules.

(including actin and myosin, which are responsible for muscle contraction) may be able to shorten the length of some microfilaments. The cytoskeleton may thus represent the cellular musculature. Microtubules, for example, form the *spindle apparatus* that pulls chromosomes away from each other in cell division; they also form the central parts of cilia and flagella.

## Lysosomes

After a phagocytic cell has engulfed a particle of food (such as a bacterium), the engulfed particle is still kept isolated from the cytoplasm by the membranes surrounding the food vacuole. The large molecules of proteins, polysaccharides, and lipids within the particle must first be digested into their smaller subunits (amino acids, monosaccharides, and so on) before they can cross the vacuole membrane and enter the cytoplasm.

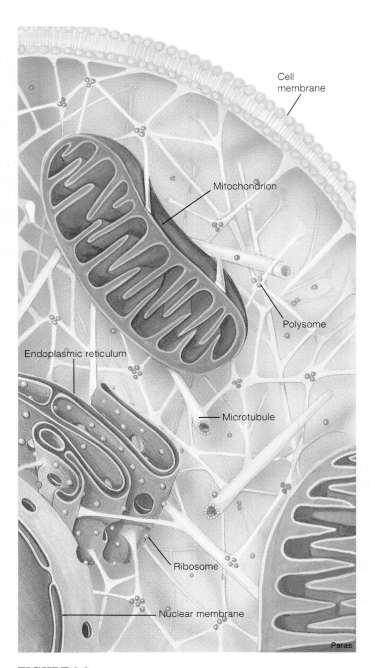

**FIGURE 3.8**

A diagram showing how microtubules form the cytoskeleton.

The digestive enzymes of a cell are isolated from the cytoplasm and concentrated within membrane-bound organelles called **lysosomes** (li´sŏ-somz) (fig. 3.9). A *primary lysosome* contains only digestive enzymes. A primary lysosome may fuse with a food vacuole (or with another cellular organelle) to form a *secondary lysosome*, in which worn-out organelles and the products of phagocytosis can be digested. Thus, a secondary lysosome contains partially digested remnants of other organelles and ingested organic material. A lysosome that contains undigested wastes is

**FIGURE 3.9**

An electron micrograph showing primary and secondary lysosomes. Mitochondria, the Golgi apparatus, and the nuclear envelope are also indicated.

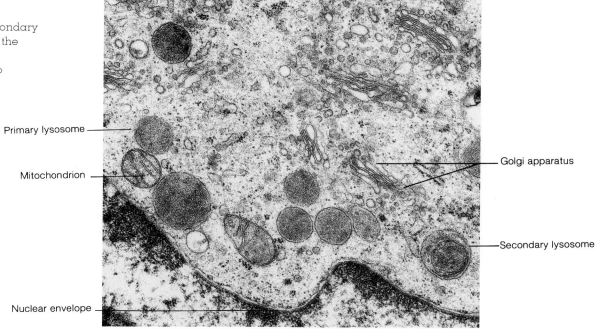

Primary lysosome

Mitochondrion

Golgi apparatus

Secondary lysosome

Nuclear envelope

called a *residual body*. Residual bodies may eliminate their wastes by exocytosis, or the wastes may accumulate within the cell as the cell ages.

The partly digested membranes of various organelles and other cellular debris often observed within secondary lysosomes result from **autophagy,** a process that destroys worn-out organelles so that they can be continuously replaced. Thus, lysosomes are said to constitute the digestive system of the cell.

Lysosomes are also referred to as suicide bags because a rupture in their membranes would release the digestive enzymes, resulting in self-digestion of the cell. This happens normally as part of *programmed cell death*, in which tissues are destroyed in the process of embryological development. It also occurs in white blood cells during an inflammation reaction.

## Mitochondria

All cells in the body, with the exception of mature red blood cells, have from a hundred to a few thousand organelles called **mitochondria** (*mi´´tŏ-kon´dre-ă*). Mitochondria serve as sites for the production of most of

the energy in cells (chapter 4). For this reason, mitochondria are sometimes called the powerhouses of the cell.

Mitochondria vary in size and shape, but all have the same basic structure (fig. 3.10). Each is surrounded by an *outer membrane* that is separated by a narrow space from an *inner membrane*. The inner membrane has many folds, or *cristae*, which extend into the central area, or *matrix*. The cristae and the matrix compartmentalize the space within

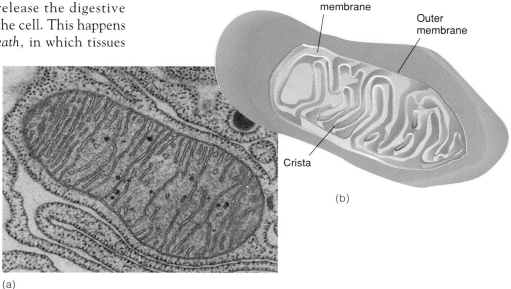

Inner membrane

Outer membrane

Crista

(b)

(a)

**FIGURE 3.10**

(*a*) An electron micrograph of a mitochondrion. The outer membrane and the infoldings of the inner membrane, called the cristae, are clearly seen. The fluid in the center is the matrix. (*b*) A diagrammatic representation of the structure of a mitochondrion.

autophagy: Gk. *autos*, self; *phagein*, to eat
mitochondrion: Gk. *mitos*, a thread; *chondros*, grain

the mitochondrion, and they both have distinct roles in the generation of cellular energy. The detailed structure and functions of mitochondria will be described in the context of cellular metabolism in chapter 4.

Mitochondria are able to migrate through the cytoplasm of a cell, and it is believed that they are able to reproduce themselves. Indeed, mitochondria contain their own DNA! This is a more primitive form of DNA than that found within the cell nucleus. For this and other reasons, many scientists believe that mitochondria evolved from organisms related to bacteria that invaded the ancestors of animal cells and remained in a state of symbiosis.

 An ovum (egg cell) contains mitochondria; the head of a sperm cell contains none. Therefore, all of the mitochondria in a fertilized egg are derived from the mother. The mitochondrial DNA replicates itself and the mitochondria divide, so that all of the mitochondria in the fertilized ovum and the cells derived from it during embryonic and fetal development are genetically identical to those in the original ovum. This represents a unique form of inheritance that is passed only from mother to child. A rare cause of blindness—*Leber's hereditary optic neuropathy*—is believed to be inherited in this manner, as are certain genetically based neuromuscular disorders.

........................................

symbiosis: Gk. *syn*, with; *bios*, life

## Endoplasmic Reticulum

Most cells contain a system of membranes known as the **endoplasmic reticulum** (*en-do-plaz´mik rĕ-tik´yŭ-lum*), or **ER.** The ER may be either of two types: (1) a **rough,** or **granular, endoplasmic reticulum** or (2) a **smooth endoplasmic reticulum** (fig. 3.11). A rough endoplasmic reticulum bears ribosomes on its surface, whereas a smooth endoplasmic reticulum does not. The smooth endoplasmic reticulum serves a variety of purposes in different cells. It provides a site for enzyme reactions in steroid hormone production, for example, and a site for the storage of calcium ions in skeletal muscle fibers. The rough endoplasmic reticulum is found in cells that are active in protein synthesis and secretion, such as those of many exocrine and endocrine glands.

The function of the rough endoplasmic reticulum and that of another organelle called the Golgi apparatus will be described in detail in a later section on protein synthesis. The structure of centrioles and of the spindle apparatus, which are involved in DNA replication and cell division, will also be described in a separate section.

# Cell Nucleus and Nucleic Acids

*The genetic code is based on the structure of DNA. DNA and RNA are composed of subunits called nucleotides, and together these molecules are known as nucleic acids. The sequence of DNA nucleotides constitutes the genetic code and serves to direct the synthesis of RNA molecules. It is through the synthesis of RNA, and eventually of protein, that the genetic code is expressed.*

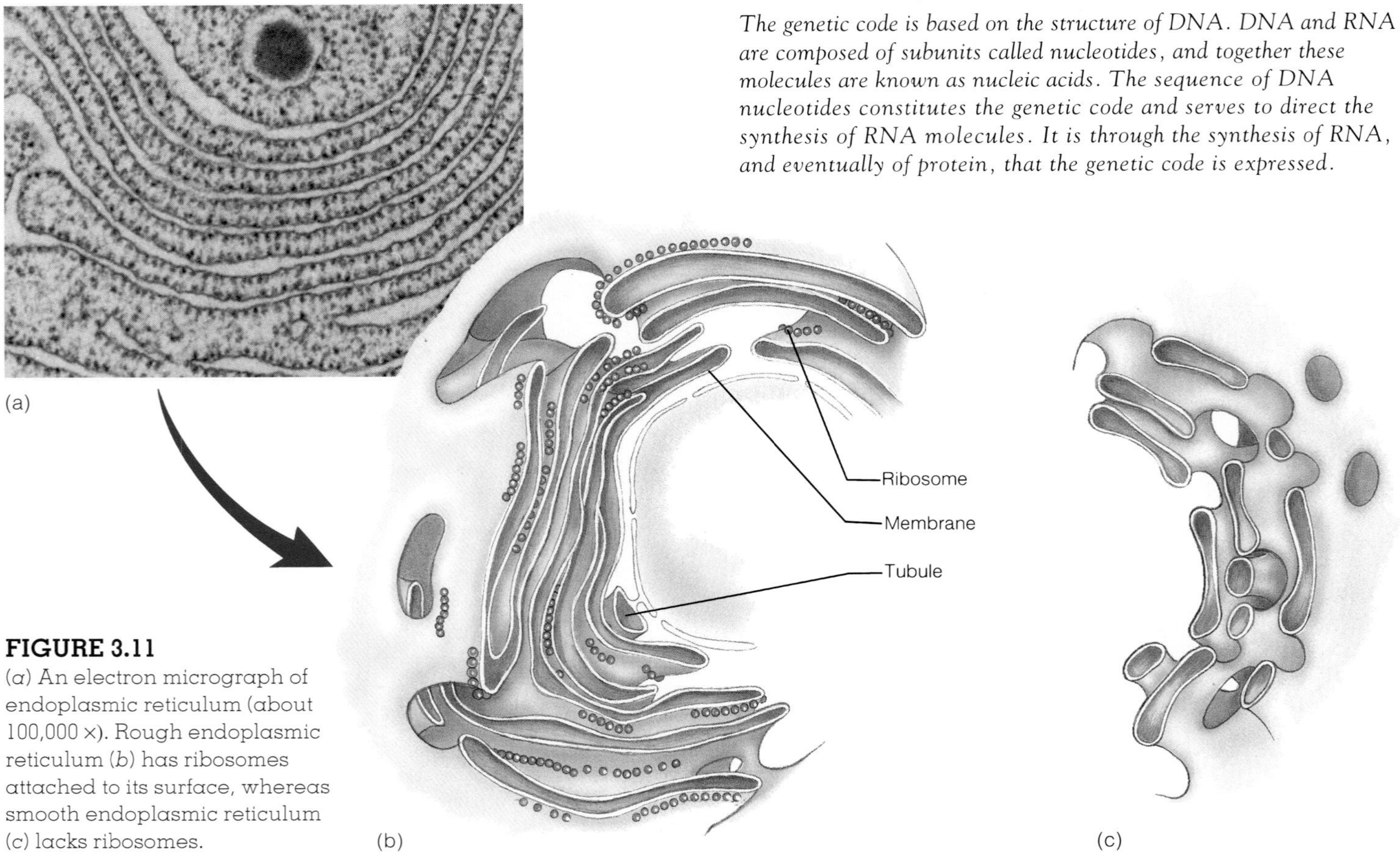

(a)

Ribosome

Membrane

Tubule

**FIGURE 3.11**
(*a*) An electron micrograph of endoplasmic reticulum (about 100,000 ×). Rough endoplasmic reticulum (*b*) has ribosomes attached to its surface, whereas smooth endoplasmic reticulum (*c*) lacks ribosomes.

(b)

(c)

Most cells in the body have a single nucleus, although some, such as skeletal muscle cells, are multinucleate. The nucleus is surrounded by a *nuclear envelope* composed of an inner and an outer membrane. These two membranes fuse together to form thin sacs with openings called *nuclear pores* (figs. 3.12 and 3.13). The nuclear pores are sufficiently large to permit RNA to exit the nucleus and enter the cytoplasm. The larger DNA molecules cannot pass through, however, and remain in the nucleus.

## Nucleic Acids

**Nucleic acids** include the macromolecules of **DNA** and **RNA,** which are critically important in genetic regulation, and the subunits from which these molecules are formed. These subunits are known as **nucleotides.**

Nucleotides function as subunits in the formation of long polynucleotide chains. Each nucleotide, however, is composed of three smaller subunits: a five-carbon sugar, a phosphate group bonded to one end of the sugar, and a *nitrogenous base* bonded to the other end of the sugar (fig. 3.14). The nitrogenous bases are cyclic nitrogen-containing molecules of two kinds: pyrimidines and purines. The *pyrimidines* contain a single ring of carbon and nitrogen, whereas the *purines* have two such rings (fig. 3.15).

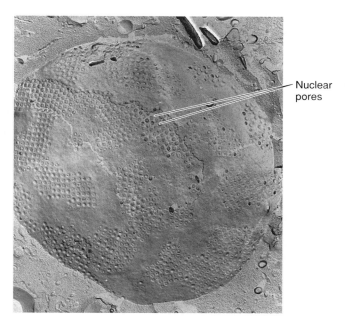

**FIGURE 3.12**
An electron micrograph of a freeze-fractured nuclear envelope, showing the nuclear pores.

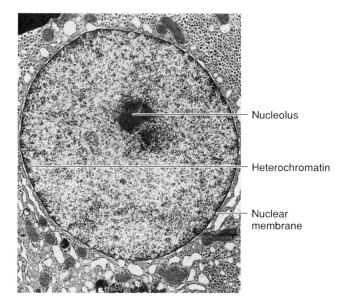

**FIGURE 3.13**
The nucleus of a liver cell, showing the nuclear envelope, heterochromatin, and nucleolus.

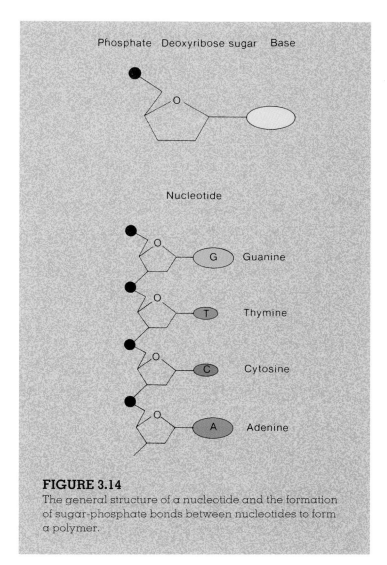

**FIGURE 3.14**
The general structure of a nucleotide and the formation of sugar-phosphate bonds between nucleotides to form a polymer.

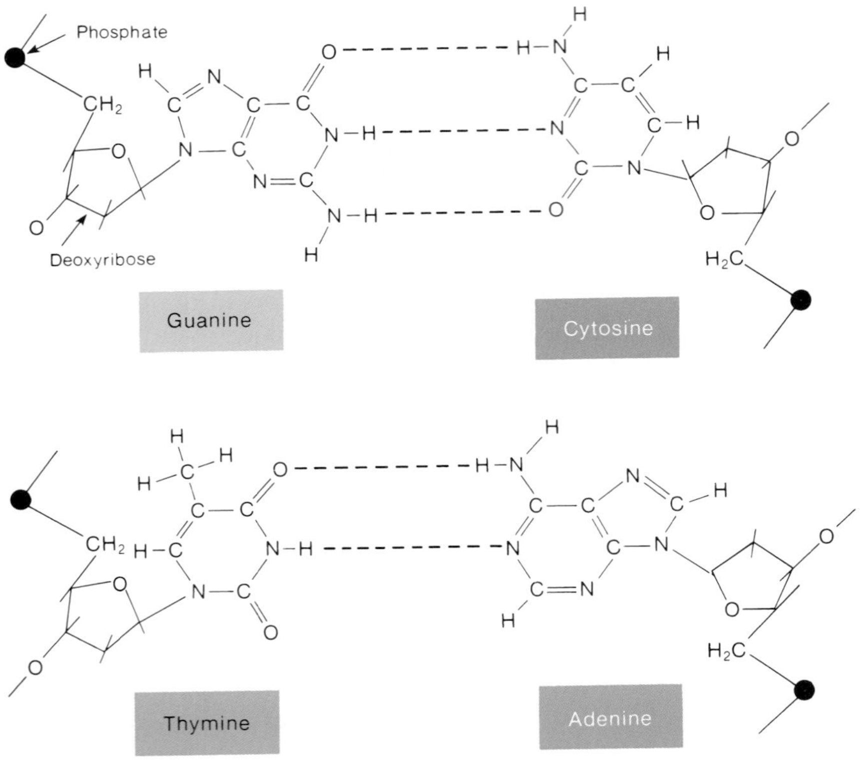

**FIGURE 3.15**

The four nitrogenous bases in deoxyribonucleic acid (DNA). Notice that hydrogen bonds can form between guanine and cytosine and between thymine and adenine.

**Deoxyribonucleic Acid**   The structure of **DNA (deoxyribonucleic acid)** serves as the basis for the genetic code. One might, therefore, expect DNA to have an extremely complex structure. But even though DNA is the largest molecule in the cell, it actually has a simpler structure than that of most proteins. This simplicity of structure deceived some of the early investigators into believing that the protein content of chromosomes, rather than their DNA content, provided the basis for the genetic code.

Sugar molecules in the nucleotides of DNA are a type of pentose (five-carbon) sugar called **deoxyribose** (de-ok″se-ri´bōs). Each deoxyribose can be covalently bonded to one of four possible bases. These bases include the two purines (*adenine* and *guanine*) and the two pyrimidines (*cytosine* and *thymine*). Thus, there are four different types of nucleotides that can be used to produce the long DNA chains.

When nucleotides combine to form a chain, the phosphate group of one nucleotide condenses with the deoxyribose sugar of another. This forms a sugar-phosphate chain as water is removed in dehydration synthesis. Since the nitrogenous bases are attached to the sugar molecules,

the sugar-phosphate chain looks like a backbone from which the bases project. Each of these bases can form hydrogen bonds with other bases, which are part of a different chain of nucleotides. Such hydrogen bonding between bases produces a *double-stranded* DNA molecule; the two strands are like a staircase, with the paired bases as steps.

Actually, the two chains of DNA twist about each other to form a **double helix,** so that the molecule resembles a spiral staircase (fig. 3.16). It has been shown that the number of purine bases in DNA is equal to the number of pyrimidine bases. The reason for this is explained by the **law of complementary base pairing:** adenine can pair only with thymine (through two hydrogen bonds), whereas guanine can pair only with cytosine (through three hydrogen bonds). Applying this rule, we can predict the base sequence of one DNA strand if we know the sequence of bases in the complementary strand.

Although we can be certain of which base is opposite a given base in DNA, we cannot predict which bases will be above or below that particular pair within a single polynucleotide chain. Although there are only four bases, the number of possible base sequences along a stretch of several thousand nucleotides (the length of a gene) is almost infinite. Yet, even with this amazing variety of possible sequences, almost all of the billions of copies of a particular gene in a person are identical. We will see shortly how this is accomplished.

**Ribonucleic Acid**   The genetic information contained in DNA functions to direct the activities of the cell through its production of another type of nucleic acid—**RNA (ribonucleic acid).** Like DNA, RNA consists of long chains of nucleotides joined together by sugar-phosphate bonds. Nucleotides in RNA, however, differ from those in DNA (fig. 3.17) in three ways: (1) a **ribonucleotide** (ri″bo-noo´kle-ŏ-tīd) contains the sugar *ribose* (instead of deoxyribose), (2) the base *uracil* (yoor´ă-sil) is present in place of thymine, and (3) RNA is composed of a single polynucleotide strand (it is not double stranded like DNA).

There are three types of RNA molecules that function in the cytoplasm of cells: *messenger RNA (mRNA), transfer RNA (tRNA),* and *ribosomal RNA (rRNA).* All three types are made within the cell nucleus using information contained in DNA as a guide.

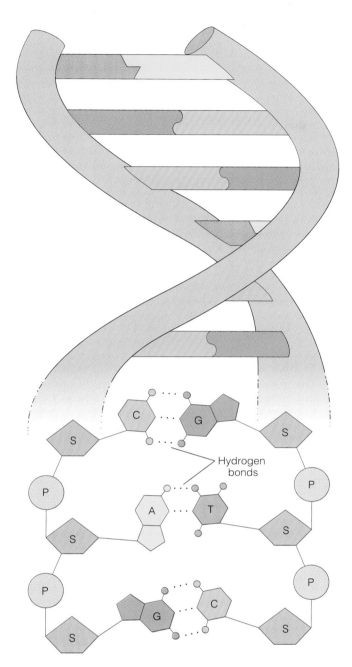

**FIGURE 3.16**
The double helix structure of DNA.

## *Chromatin*

Many granulated threads in the nuclear fluid can be seen with an electron microscope. These threads are called **chromatin** (*kro´mă-tin*) and consist of a combination of DNA

..........................................

chromatin: Gk. *chroma*, color

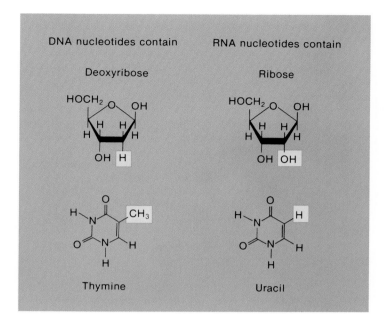

**FIGURE 3.17**
Differences between the nucleotides and sugars in DNA and RNA.

and protein. There are two forms of chromatin. Thin, extended chromatin—or *euchromatin* (*yoo-kro´mă-tin*)—appears to be the active form of DNA in a nondividing cell. Regions of condensed, blotchy-appearing chromatin, known as *heterochromatin,* are believed to contain inactive DNA.

One or more dark areas within each nucleus can be observed. These regions, which are not surrounded by membranes, are called **nucleoli** (*noo-kle´ŏ-li*). The DNA within nucleoli contains genes that code for the production of ribosomal RNA (rRNA), an essential component of ribosomes.

## *Genetic Transcription—RNA Synthesis*

The thin, extended euchromatin is the working form of DNA; the more familiar short, stubby form of chromosomes seen during cell division are inactive packages of DNA. The genes do not become active until the chromosomes unravel. Active DNA directs the metabolism of the cell indirectly through its regulation of RNA and protein synthesis.

*One gene codes for one polypeptide chain.* Each gene is a stretch of DNA that is several thousand nucleotide pairs long. The DNA in a human cell contains 3 to 4 billion base pairs, enough to code for at least 3 million proteins. Since the average human cell contains less than this amount (30,000 to 150,000 different proteins), it follows that only a fraction of the DNA in each cell is used to code

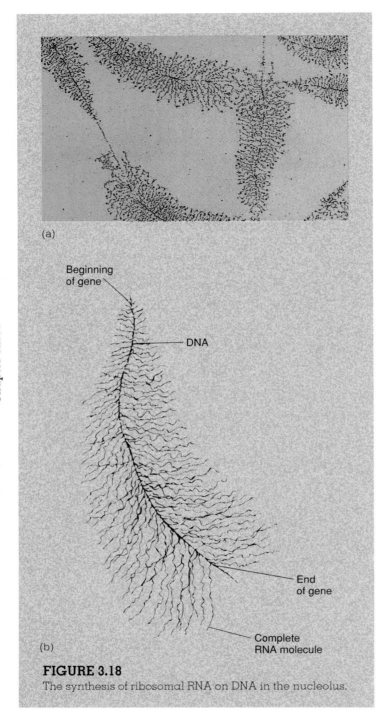

(a)

Beginning
of gene

DNA

End
of gene

Complete
RNA molecule

(b)

**FIGURE 3.18**
The synthesis of ribosomal RNA on DNA in the nucleolus.

does not occur throughout the length of DNA, but only in the regions that are to be transcribed (there are base sequences that code for "start" and "stop"). Double-stranded DNA, therefore, separates in these regions so that the freed bases can pair with the complementary RNA nucleotide bases, which are freely available in the nucleoplasm.

This pairing of bases, like that which occurs in DNA replication (described in a later section), follows the law of complementary base pairing: guanine bonds with cytosine (and vice versa), and adenine bonds with uracil (because uracil in RNA is equivalent to thymine in DNA). Unlike DNA replication, however, only *one* of the two freed strands of DNA serves as a guide for RNA synthesis (fig. 3.19). Once an RNA molecule has been produced, it detaches from the DNA strand on which it was formed. This process can continue indefinitely, producing many thousands of RNA copies of the DNA strand that is being transcribed. When the gene is no longer to be transcribed, the separated DNA strands can then go back together again.

**Types of RNA**    There are four types of RNA produced within the nucleus by genetic transcription: (1) **precursor messenger RNA (pre-mRNA),** which is altered within the nucleus to form mRNA; (2) **messenger RNA (mRNA),** which contains the code for the synthesis of specific proteins; (3) **transfer RNA (tRNA),** which is needed for decoding the genetic message contained in mRNA; and (4) **ribosomal RNA (rRNA),** which forms part of the structure of ribosomes. The DNA that codes for rRNA synthesis is located in the part of the nucleus called the nucleolus. The DNA that codes for pre-mRNA and tRNA synthesis is located elsewhere in the nucleus.

In bacteria, where the molecular biology of the gene is best understood, a gene that codes for one type of protein produces an mRNA molecule that begins to direct protein synthesis as soon as it is formed. This is not the case in higher organisms, including humans. In higher cells, a pre-mRNA is produced that must be modified within the nucleus before it can enter the cytoplasm as mRNA and direct protein synthesis.

Precursor mRNA is much larger than the mRNA that it forms. This large size of pre-mRNA is, surprisingly, not due to excess bases at the ends of the molecule that must be trimmed. Rather, the excess bases are *within* the pre-mRNA. The genetic code for a particular protein, in other words, is split up by stretches of base pairs that do not contribute to the code. These regions of noncoding DNA within a gene are called *introns;* the coding regions are known as *exons.* As a result, pre-mRNA must be cut and spliced to make mRNA. This cutting and splicing can be quite extensive—a single gene may contain up to 50 introns that must be removed from the pre-mRNA in order to convert it to mRNA.

for proteins. The remainder of the DNA may be inactive or redundant, or it may serve to regulate those regions that do code for proteins.

In order for the genetic code to be translated into the synthesis of specific proteins, the DNA code must first be transcribed into an RNA code (fig. 3.18). This is accomplished by DNA-directed RNA synthesis, or **genetic transcription.**

In RNA synthesis, the enzyme *RNA polymerase* breaks the weak hydrogen bonds between paired DNA bases. This

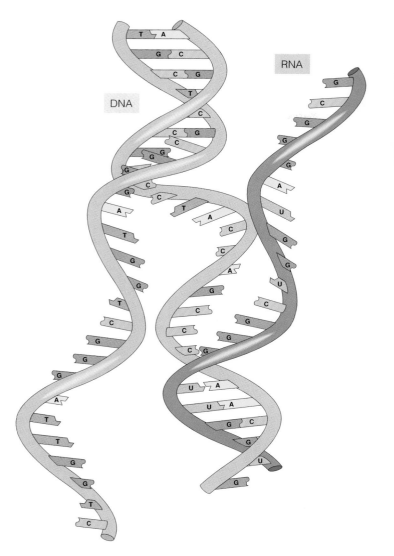

**▣ FIGURE 3.19**

RNA synthesis (genetic transcription). Notice that only one of the two DNA strands is used to form a single-stranded molecule of RNA.

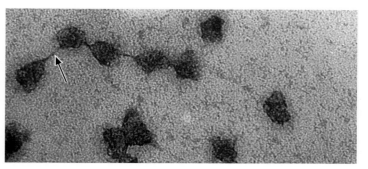

**FIGURE 3.20**

An electron micrograph of polyribosomes. An RNA strand (arrow) joins the ribosomes together.

## Table 3.2 — Selected DNA base triplets and mRNA codons

| DNA triplet | RNA codon | Amino acid |
|---|---|---|
| TAC | AUG | "Start" |
| ATC | UAG | "Stop" |
| AAA | UUU | Phenylalanine |
| AGG | UCC | Serine |
| ACA | UGU | Cysteine |
| GGG | CCC | Proline |
| GAA | CUU | Leucine |
| GCT | CGA | Arginine |
| TTT | AAA | Lysine |
| TGC | ACG | Tyrosine |
| CCG | GGC | Glycine |
| CTC | GAG | Aspartic acid |

# Protein Synthesis and Secretion

*In order for a gene to be expressed, it first must be used as a guide, or template, in the production of a complementary strand of messenger RNA. The mRNA is then used as a guide to produce a particular type of protein whose sequence of amino acids is determined by the sequence of base triplets (codons) in the mRNA.*

When mRNA enters the cytoplasm, it attaches to ribosomes that appear in the electron microscope as numerous small particles. A **ribosome** is composed of three molecules of ribosomal RNA and 52 proteins, arranged to form two subunits of unequal size. The mRNA passes through a number of ribosomes to form a string-of-pearls structure called a *polyribosome* (or *polysome*, for short), as shown in figure 3.20. The association of mRNA with ribosomes is needed for **genetic translation**—the production of specific proteins according to the code contained in the mRNA base sequence.

Each mRNA molecule contains several hundred or more nucleotides, arranged in the sequence determined by complementary base pairing with DNA during genetic transcription (RNA synthesis). Each three-base sequence, or *base triplet*, is a code word—called a **codon**—for a specific amino acid, except for three base triplets (UAA, UAG, and UGA), which serve as "stop signs" in the translation process. Sample codons and their amino acid translations are listed in table 3.2 and illustrated in figure 3.21. As mRNA moves through the ribosome, the sequence of codons is translated into a sequence of specific amino acids within a growing polypeptide chain.

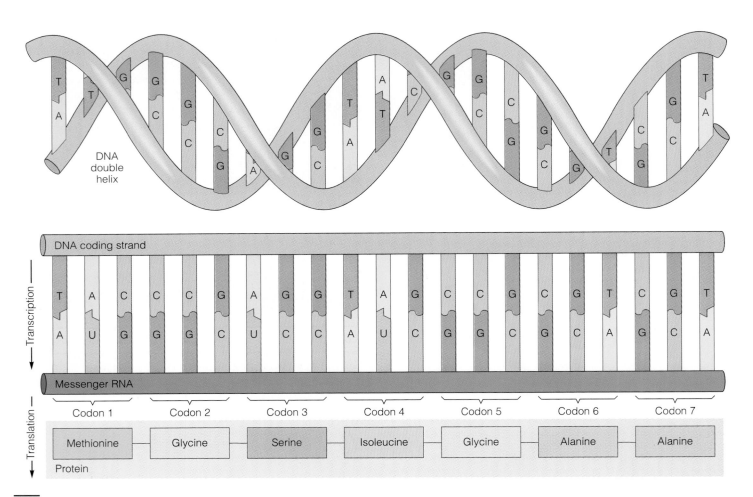

**FIGURE 3.21**

The genetic code is first transcribed into base triplets (codons) in mRNA and then translated into a specific sequence of amino acids in a protein.

## Transfer RNA

Translation of the codons is accomplished by tRNA and particular enzymes. Each tRNA molecule, like mRNA and rRNA, is single stranded. Although tRNA is single stranded, it bends in on itself to form a cloverleaf structure (fig. 3.22a), which is believed to be further twisted into an upside down "L" shape (fig. 3.22b). One end of the "L" contains the **anticodon**—three nucleotides that are complementary to a specific codon in mRNA.

Enzymes in the cell cytoplasm called *aminoacyl-tRNA synthetase enzymes* join specific amino acids to the ends of tRNA, so that a tRNA with a given anticodon is always bonded to one specific amino acid. There are 20 different varieties of synthetase enzymes—one for each type of amino acid. Each synthetase must not only recognize its specific amino acid, it must be able to attach this amino acid to the particular tRNA that has the correct anticodon for that amino acid. The cytoplasm of a cell thus contains tRNA molecules that are each bonded to a specific amino acid and that are each capable of bonding by their anticodon base triplet to a specific codon in mRNA.

## Formation of a Polypeptide

The anticodons of tRNA bond to the codons of mRNA as the mRNA moves through the ribosome. Since each tRNA molecule carries a specific amino acid, the joining together of these amino acids by peptide bonds creates a polypeptide whose amino acid sequence has been determined by the sequence of codons in mRNA.

The first and second tRNA bring the first and second amino acids together, and a peptide bond forms between them. The first amino acid then detaches from its tRNA, so that a dipeptide is linked by the second amino acid to the second tRNA. When the third tRNA bonds to the third codon, the amino acid it brings forms a peptide bond with the second amino acid, which detaches from its tRNA. A tripeptide is attached by the third amino acid to the third tRNA. The polypeptide chain grows as new amino acids are added to its tip (fig. 3.23). This growing polypeptide chain is always attached by means of only one tRNA to the strand of mRNA, and this tRNA molecule is always the one that has added the latest amino acid to the growing polypeptide.

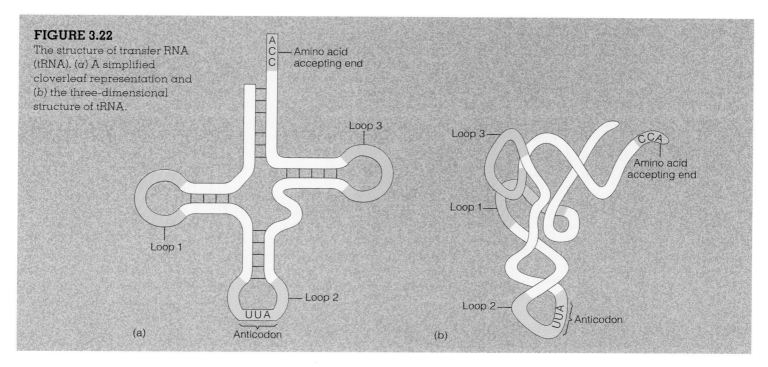

**FIGURE 3.22**

The structure of transfer RNA (tRNA). (a) A simplified cloverleaf representation and (b) the three-dimensional structure of tRNA.

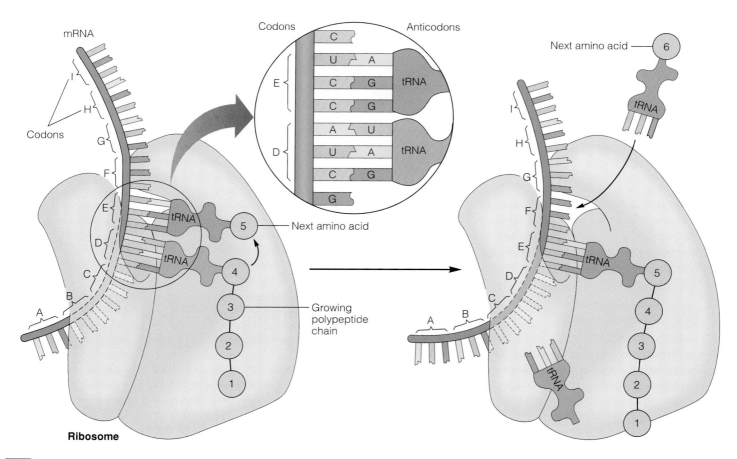

**FIGURE 3.23**

The translation of messenger RNA (mRNA). As the anticodon of each new aminoacyl-tRNA bonds to a codon of the mRNA, new amino acids are joined to the growing tip of the polypeptide chain.

As the polypeptide chain grows in length, interactions between its amino acids cause the chain to twist into a helix (secondary structure) and to fold and bend upon itself (tertiary structure). At the end of this process, the new protein detaches from the tRNA as the last amino acid is added. Many proteins are further modified after they are formed; these modifications occur in the rough endoplasmic reticulum and Golgi apparatus.

## Function of the Rough Endoplasmic Reticulum

Proteins that are to be used within the cell are produced in polyribosomes that are free in the cytoplasm. If the protein is a secretory product of the cell, however, it is produced by mRNA-ribosome complexes located in the rough endoplasmic reticulum. The membranes of this system enclose fluid-filled spaces (cisternae) that may be entered by the newly formed proteins. Once they are within the cisternae, the structure of these proteins is modified in specific ways.

The processing of the hormone insulin can serve as an example of the changes that occur within the endoplasmic reticulum. The original molecule enters the cisterna as a single polypeptide composed of 109 amino acids. This molecule is called preproinsulin. The first 23 amino acids serve as a *leader sequence*; it is attracted to the lipid content of the endoplasmic reticulum membrane and allows the molecule to be injected into the cisterna within the endoplasmic reticulum. The leader sequence is then quickly removed (to produce proinsulin). The remaining chain folds within the cisterna, so that the first and last amino acids in the polypeptide are brought close together. The central region is then enzymatically removed, producing two chains—one of them, 21 amino acids long; the other, 30 amino acids long—which are subsequently joined together by disulfide bonds (fig. 3.24). This is the form of insulin that is normally secreted from the cell.

## Function of the Golgi Apparatus

Secretory proteins do not remain trapped within the rough endoplasmic reticulum; they are transported to another organelle within the cell—the **Golgi apparatus.** In this

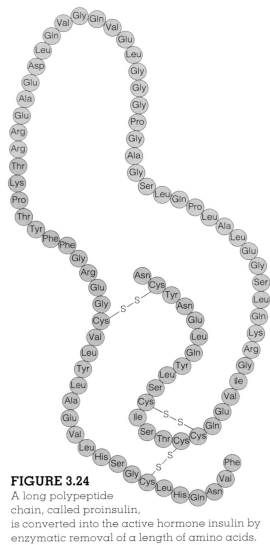

**FIGURE 3.24**

A long polypeptide chain, called proinsulin, is converted into the active hormone insulin by enzymatic removal of a length of amino acids. The insulin molecule produced in this way consists of two polypeptide chains (colored circles) joined by disulfide bonds.

organelle, further modifications of proteins (such as addition of carbohydrates to form *glycoproteins*) occur. Different types of proteins are separated according to their function and destination and the final products are packaged into secretory vesicles. For example, proteins that are to be secreted are separated from those that will be incorporated into the cell membrane and from those that will be introduced into lysosomes and they are packaged into separate membrane-enclosed vesicles.

The Golgi apparatus consists of several flattened sacs. Proteins produced by the rough endoplasmic reticulum are believed to travel in membrane-enclosed vesicles to the sac on one end of the Golgi apparatus. After specialized modifications of the proteins are made within one sac, the modified proteins are passed by means of vesicles to the next sac until the finished products finally leave the Golgi apparatus in vesicles that fuse with the cell membrane or with the membranes of lysosomes (fig. 3.25).

The Golgi apparatus and the rough endoplasmic reticulum contain enzymes that modify the structure of proteins. The action of these organelles, combined with the events that occur during genetic transcription and translation, provide numerous opportunities for the regulation of genetic expression.

# DNA Synthesis and Cell Division

*When a cell is going to divide, each strand of the DNA within its nucleus acts as a template for the formation of a new complementary strand. Organs grow and repair themselves through a type of cell division known as mitosis, in which the two daughter cells produced are genetically identical to one another and to the original parent cell. Gametes contain only half the number of chromosomes as their parent cell and are formed by a type of cell division called meiosis.*

..............................................

Golgi apparatus: from Camilio Golgi, Italian histologist, 1843-1926

Genetic information is required for the life of the cell and for the ability of the cell to perform its functions in the body. Each cell obtains this genetic information from its parent cell through the process of DNA replication and cell division. DNA is the only type of molecule in the body capable of replicating itself, and mechanisms exist within the dividing cell to ensure that the duplicate copies of DNA are properly distributed to the daughter cells.

## DNA Replication

When a cell is going to divide, each DNA molecule replicates itself, and each of the identical DNA copies thus produced is distributed to the two daughter cells. Replication of DNA requires the action of a specific enzyme known as *DNA polymerase* (*pol´ĕ-mĕ-rās*). This enzyme moves along the DNA molecule, breaking the weak hydrogen bonds between complementary bases as it travels. As a result, the bases of each of the two DNA strands become free to bond to new complementary bases, which are part of nucleotides available within the surrounding environment.

According to the rules of complementary base pairing, the bases of each original strand will bond to the appropriate free nucleotides: adenine bases pair with thymine-containing nucleotides, guanine bases pair with cytosine-containing nucleotides, and so on. In this way, two new molecules of DNA, each containing two complementary strands, are formed. The DNA polymerase enzyme links the phosphate groups and deoxyribose sugar groups together to form a second polynucleotide chain in each DNA that is complementary to the first DNA strands. Thus, two new double-helix DNA molecules are produced, each containing the same base sequence as the parent molecule (fig. 3.26).

When DNA replicates, therefore, each copy is composed of one new strand and one strand from the original DNA molecule. Replication is said to be **semiconservative** (half of the original DNA is conserved in each of the new

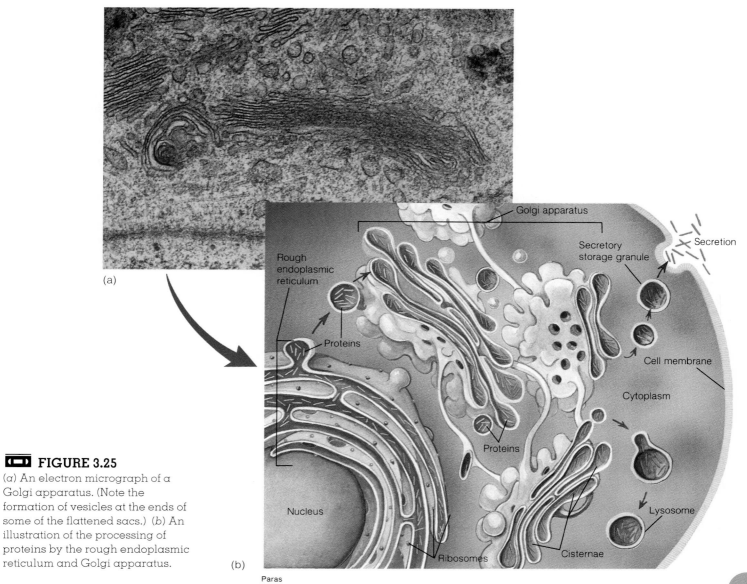

(a)

(b)

Paras

### ▭ FIGURE 3.25

(*a*) An electron micrograph of a Golgi apparatus. (Note the formation of vesicles at the ends of some of the flattened sacs.) (*b*) An illustration of the processing of proteins by the rough endoplasmic reticulum and Golgi apparatus.

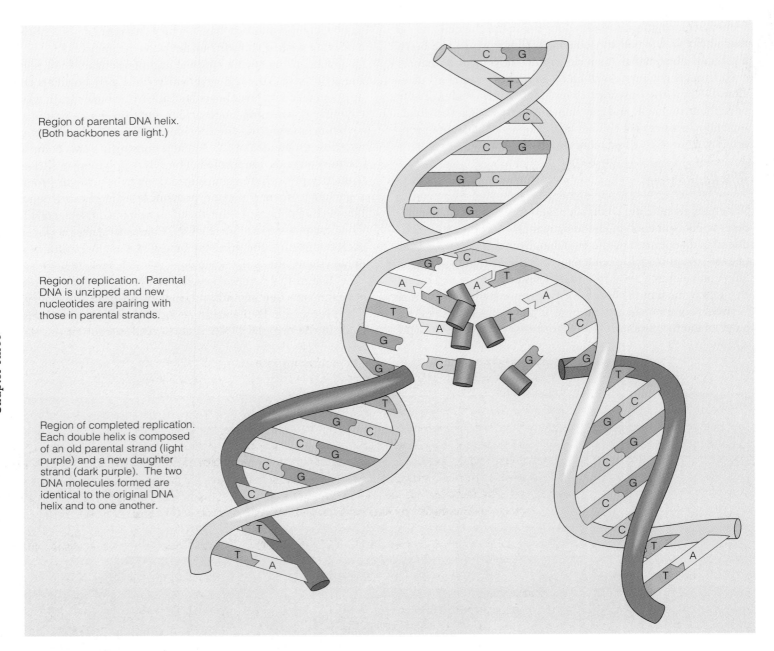

Region of parental DNA helix. (Both backbones are light.)

Region of replication. Parental DNA is unzipped and new nucleotides are pairing with those in parental strands.

Region of completed replication. Each double helix is composed of an old parental strand (light purple) and a new daughter strand (dark purple). The two DNA molecules formed are identical to the original DNA helix and to one another.

**FIGURE 3.26**

The replication of DNA. Each new double helix is composed of one old and one new strand. The base sequences of each of the new molecules is identical to that of the parent DNA because of complementary base pairing.

DNA molecules). Through this mechanism, the sequence of bases in DNA, which is the basis of the genetic code, is preserved from one cell generation to the next.

## Cell Growth and Division

Unlike the life of an organism, which can be viewed as a linear progression from birth to death, the life of a cell follows a cyclical pattern. Each cell is produced as a part of its parent cell; when the daughter cell divides, it in turn becomes two new cells. In a sense, then, each cell is potentially immortal as long as its progeny can continue to divide. Some cells in the body divide frequently; for example, the epidermis of the skin is renewed approximately every 2 weeks and the stomach lining is renewed about every 2 or 3 days. Other cells, however, such as nerve and skeletal muscle cells in the adult, do not divide at all. All cells in the body, of course, live only as long as the person lives. Some cells live longer than others, but eventually all cells die when vital functions cease.

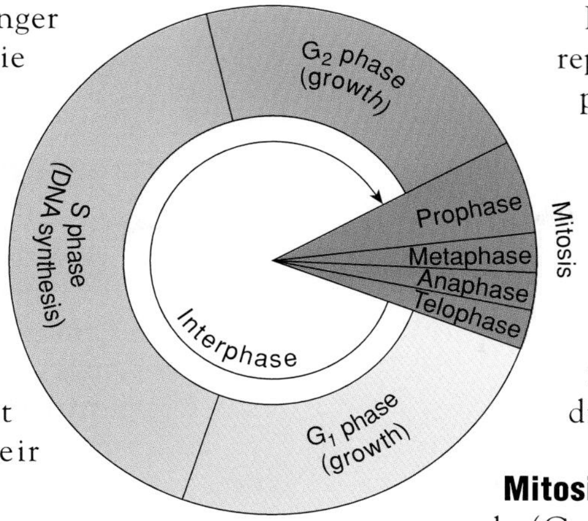

**The Cell Cycle** The nondividing cell is in a part of its life cycle known as **interphase** (fig. 3.27), which is subdivided into $G_1$, S, and $G_2$ phases. The chromosomes are in their extended form (as euchromatin), and their genes actively direct the synthesis of RNA. Through their direction of RNA synthesis, genes control the metabolism of the cell. During this time the cell may be growing, and this part of interphase is known as the $G_1$ *phase*. Although sometimes described as resting, cells in the $G_1$ phase perform the physiological functions characteristic of the tissue in which they are found. The DNA of resting cells in the $G_1$ phase thus produces mRNA and proteins as previously described.

A cell that is going to divide replicates its DNA in a part of interphase known as the *S phase* (S stands for *synthesis*). Once DNA has replicated in the S phase, the chromatin condenses in the $G_2$ *phase* to form short, thick, rodlike structures by the end of $G_2$. This is the more familiar form of chromosomes because they are easily seen in the ordinary (light) microscope. It should be remembered that this form of the chromatin represents a packaged state of DNA—not the extended, threadlike form that is active in directing the metabolism of the cell during the $G_1$ phase.

**Centrosomes** All animal cells capable of cell division have a **centrosome** located on one side of the nucleus. The centrosome is a small, amorphous mass that serves as a factory for the production of microtubules, which are the major component of spindle fibers. At the center of the centrosome are two **centrioles** (*sen´trĭ-ōlz*), positioned at right angles to each other. Each centriole is composed of nine evenly spaced bundles of microtubules, with three microtubules per bundle (fig. 3.28).

**FIGURE 3.27**
The life cycle of a cell.

If a cell is going to divide, the centrosome replicates itself and each centrosome takes a position on the opposite side of the nucleus. The centrosomes then produce **spindle fibers,** which are composed of microtubules. The spindle fibers will eventually pull the duplicated chromosomes to opposite poles of the cell during cell division. Cells that lack centrosomes, such as mature muscle and nerve cells, cannot divide.

**Mitosis** At the end of the $G_2$ phase of the cell cycle ($G_2$ is generally shorter than $G_1$), each chromosome consists of two strands called **chromatids** (*kro´mă-tidz*) that are joined together by a *centromere* (fig 3.29). The two chromatids within a chromosome contain identical DNA base sequences because each is produced by the semiconservative replication of DNA. Each chromatid, therefore,

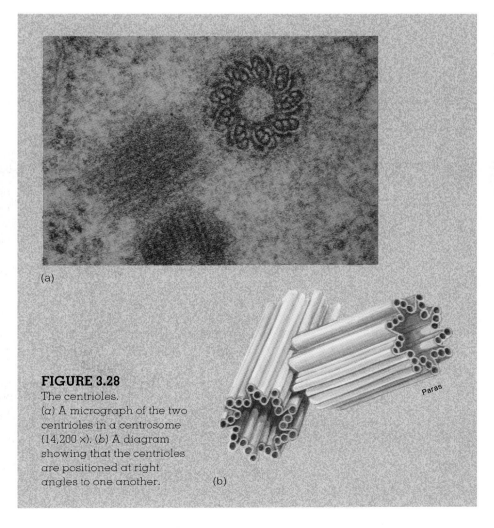

(a)

**FIGURE 3.28**
The centrioles.
(a) A micrograph of the two centrioles in a centrosome (14,200 ×). (b) A diagram showing that the centrioles are positioned at right angles to one another.

(b)

Paras

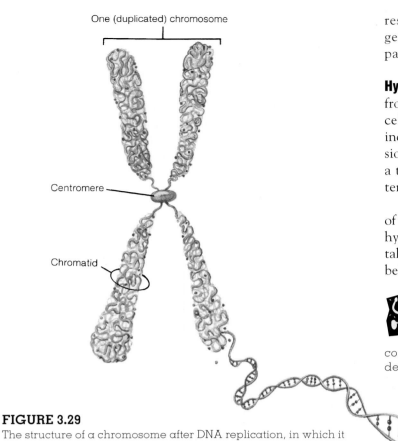

One (duplicated) chromosome

Centromere

Chromatid

**FIGURE 3.29**
The structure of a chromosome after DNA replication, in which it consists of two identical strands, or chromatids.

contains a complete double-helix DNA molecule that is a copy of the single DNA molecule existing prior to replication. Each chromatid will become a separate chromosome once cell division has been completed.

The $G_2$ phase completes interphase. The cell next proceeds through the various stages of cell division, or **mitosis** (the M *phase* of the cell cycle). Mitosis is subdivided into four stages: *prophase, metaphase, anaphase,* and *telophase* (fig. 3.30). In metaphase of mitosis, the chromosomes line up single file along the equator of the cell. This aligning of chromosomes at the equator is believed to result from the action of the spindle fibers that are attached to the centromere of each chromosome (fig. 3.30).

Anaphase begins when the centromeres split apart and the spindle fibers shorten, pulling the two chromatids in each chromosome to opposite poles. Each pole therefore gets one copy of each of the 46 chromosomes. Division of the cytoplasm (*cytokinesis*) during telophase

. . . . . . . . . . . . . . . . . . . . . . . . . . . . . . . . . . . . . .

mitosis: Gk. *mitos,* thread

results in the production of two daughter cells that are genetically identical to each other and to the original parent cell.

**Hyperplasia and Hypertrophy**   The growth of an individual from a fertilized egg into an adult involves an increase in cell number and an increase in cell size. Growth due to an increase in the number of cells results from mitotic cell division and is termed **hyperplasia** (*hi´´per-pla´zhă*). Growth of a tissue or organ due to an increase in the size of cells is termed **hypertrophy** (*hi´´per´trŏ-fe*).

Most growth is due to hyperplasia. A callus on the palm of the hand, for example, involves thickening of the skin by hyperplasia due to frequent abrasion. An increase in skeletal muscle size as a result of exercise, by contrast, is generally believed to be produced by hypertrophy.

Skeletal and cardiac (heart) muscle can grow only by hypertrophy. When this occurs in skeletal muscles in response to an increased workload (during weight training, for example), it is called *compensatory hypertrophy.* The heart muscle may also demonstrate compensatory hypertrophy when its workload increases in response, for example, to hypertension (high blood pressure). The opposite of hypertrophy is *atrophy,* where the cells become smaller than normal. This may result from the disuse of skeletal muscles as occurs in prolonged bed rest, various diseases, and advanced age.

## Meiosis

When a cell is going to divide, either by mitosis or meiosis, the DNA is replicated (forming chromatids) and the chromosomes become shorter and thicker, as previously described. At this point, the cell still has only 46 chromosomes, but each chromosome consists of two duplicate chromatids.

The short, thick chromosomes seen at the end of the $G_2$ phase can be matched into pairs, the members of which appear to be structurally identical. These matched pairs of chromosomes are called **homologous chromosomes** (*hŏmol´ŏ-gus kro´mŏ-sōmz*). One member of each homologous pair is derived from a chromosome inherited from the father, and the other member is a copy of one of the chromosomes inherited from the mother. Homologous chromosomes do not have identical DNA base sequences; for example, one member of the pair may code for blue eyes and the other for brown eyes. There are 22 homologous

. . . . . . . . . . . . . . . . . . . . . . . . . . . . . . . . . . . . . .

meiosis: Gk. *meioun,* lessen

## (a) Interphase

- The chromosomes are in an extended form and seen as chromatin in the electron microscope.
- The nucleus is visible.

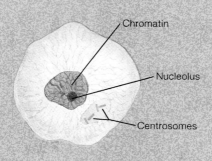
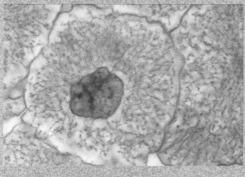

## (b) Prophase

- The chromosomes are seen to consist of two chromatids joined by a centromere.
- The centrioles move apart toward opposite poles of the cell.
- Spindle fibers are produced and extended from each centrosome.
- The nuclear membrane starts to disappear.
- The nucleolus is no longer visible.

Paras

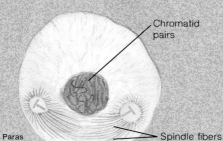

## (c) Metaphase

- The chromosomes are lined up at the equator of the cell.
- The spindle fibers from each centriole are attached to the centromeres of the chromosomes.
- The nuclear membrane has disappeared.

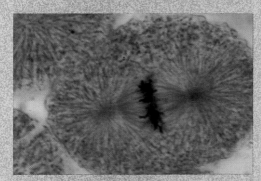

## (d) Anaphase

- The centromeres split, and the sister chromatids separate as each is pulled to an opposite pole.

### (d) Anaphase

The centromeres split, and the sister chromatids separate as each is pulled to an opposite pole.

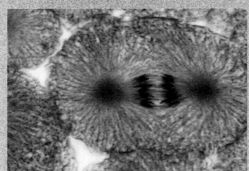

## (e) Telophase

- The chromosomes become longer, thinner, and less distance.
- New nuclear membranes form.
- The nucleolus reappears.
- Cell division is nearly complete.

Paras

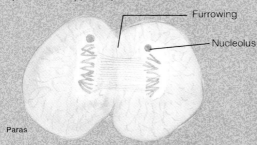
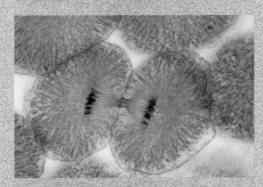

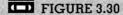 **FIGURE 3.30**

The stages of mitosis.

pairs of *autosomal* (*aw´´to-so´mal*) *chromosomes* and 1 pair of *sex chromosomes*, described as X and Y. Females have two X chromosomes, whereas males have one X and one Y chromosome (fig. 3.31).

**Meiosis** (*mi-o´sis*) is a special type of cell division that occurs only in the gonads (testes and ovaries) and only for the production of *gametes* (sperm cells and ova). In meiosis, the homologous chromosomes line up side by side, rather than single file as in mitosis, along the equator of the cell. The spindle fibers then pull one member of a homologous pair to one pole of the cell and the other member of the pair to the other pole. Each of the two daughter cells thus acquires only one chromosome from each of the 23 homologous pairs contained in the parent. The daughter cells, in other words, contain 23 rather than 46 chromosomes. For this reason, meiosis is also known as **reduction division.**

Meiosis, however, consists of two cell divisions. The need for this is obvious considering that, at the end of the cell division, each daughter cell contains 23 chromosomes, with *each chromosome consisting of two chromatids*. (Since the two chromatids per chromosome are identical, this does not make 46 chromosomes; there are still only 23 *different* chromosomes per cell at this point.) In the second meiotic division, each of the daughter cells from the first cell division then itself divides, with the duplicate chromatids going to each of two new daughter cells. A grand total of four daughter cells can thus be produced from the meiotic cell division of one parent cell. This occurs in the testes, where one parent cell produces four sperm cells, each containing 23 chromosomes. In the ovaries, one parent cell also produces four daughter cells, but three of these die and only one progresses to become a mature egg cell (as described in chapter 29).

The stages of meiosis are subdivided according to whether they occur in the first or the second meiotic cell division. These stages are designated as prophase I, metaphase I, anaphase I, telophase I; and then as prophase II, metaphase II, anaphase II, and telophase II (table 3.3 and fig. 3.32).

The reduction of the chromosome number from 46 to 23 is of obvious necessity for sexual reproduction, where the sex cells join

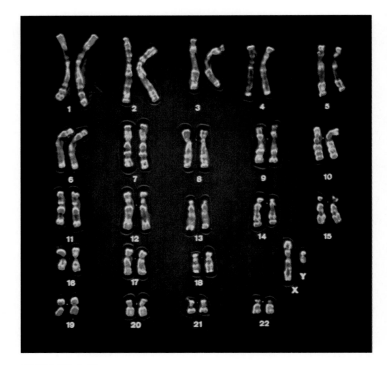

**FIGURE 3.31**
A false-color light micrograph showing the full complement of male chromosomes arranged in numbered homologous pairs.

| Table 3.3 | Stages of meiosis |
|---|---|
| **Stage** | **Events** |
| **First meiotic division** | |
| Prophase I | Chromosomes appear double stranded. Each strand, called a chromatid, contains duplicate DNA joined together by a structure known as a centromere.<br>Homologous chromosomes pair up side by side. |
| Metaphase I | Homologous chromosome pairs line up at the equator.<br>Spindle apparatus is complete. |
| Anaphase I | Homologous chromosomes are separated; the members of a homologous pair move to opposite poles. |
| Telophase I | Cytoplasm divides to produce two haploid cells. |
| **Second meiotic division** | |
| Prophase II | Chromosomes appear, each containing two chromatids. |
| Metaphase II | Chromosomes line up single file along the equator as spindle formation is completed. |
| Anaphase II | Centromeres split and chromatids move to opposite poles. |
| Telophase II | Cytoplasm divides to produce two haploid cells from each of the haploid cells formed at telophase I. |

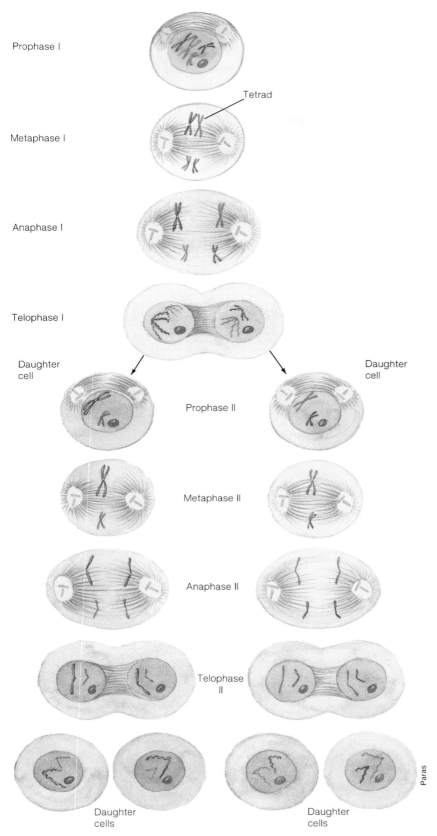

Prophase I

Tetrad

Metaphase I

Anaphase I

Telophase I

Daughter cell

Daughter cell

Prophase II

Metaphase II

Anaphase II

Telophase II

Daughter cells

Daughter cells

Paras

◼▭ **FIGURE 3.32**

Meiosis, or reduction division. In the first meiotic division, the homologous chromosomes of a diploid parent cell are separated into two haploid daughter cells. Each of these chromosomes contain duplicate strands, or chromatids. In the second meiotic division, these chromatids are distributed to two new haploid daughter cells.

(with their content of chromosomes added together) to produce a new individual. The significance of meiosis, however, goes deeper than simply a reduction in chromosome number. At metaphase I, the pairs of homologous chromosomes can line up with either member facing a given pole of the cell. (Recall that each member of a homologous pair came from a different parent.) Maternal and paternal members of homologous pairs are randomly shuffled. When the first meiotic division occurs, each daughter cell will obtain a complement of 23 chromosomes that are randomly derived from the maternal or paternal contribution to the homologous pairs of chromosomes of the parent.

In addition to this shuffling of chromosomes, exchanges of parts of homologous chromosomes can occur at metaphase I. That is, pieces of one chromosome of a homologous pair can be exchanged with the other homologous chromosome in a process called *crossing over* (fig. 3.33). These events together result in **genetic recombination** and ensure that the gametes produced by meiosis are genetically unique. This provides genetic diversity for organisms that reproduce sexually, and genetic diversity has been shown to promote survival of species over evolutionary time.

## Clinical Considerations

## *Functions of Cellular Organelles*

**Lysosomes**   Most, if not all, molecules in the cell have a limited life span. They are continuously destroyed and must be continuously replaced. Glycogen and some complex lipids in the brain, for example, are digested at a particular rate by lysosomes. If a person does not have the proper amount of these lysosomal enzymes because of some genetic defect, the resulting abnormal accumulation of glycogen and lipids could destroy the tissues. Examples of such diseases include *glycogen storage disease*, *Tay–Sach's disease*, and *Gaucher's disease*.

**Endoplasmic Reticulum**   The smooth endoplasmic reticulum in liver cells and other cells contains enzymes used for the inactivation of steroid hormones and many toxic compounds. This inactivation is generally achieved by reactions that convert these compounds to forms that are more

(a) First meiotic prophase

Chromosomes pairing

Chromosomes crossing over

(b) Crossing over

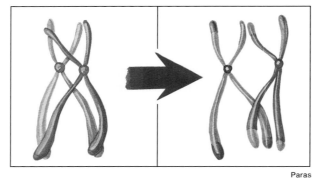

Paras

**FIGURE 3.33**

(a) Genetic variation results from the crossing over of tetrads, which occurs during the first meiotic prophase. (b) A diagram depicting the recombination of chromosomes that occurs as a result of crossing over.

water soluble and less active, and thus more easily excreted by the kidneys. When people take certain drugs (such as alcohol and phenobarbital) for a long period of time, an increasingly large dose is required to produce the effect achieved initially. This phenomenon, called *drug tolerance,* is accompanied by an increase in the smooth endoplasmic reticulum, and thus an increase in the enzymes charged with inactivation of these drugs.

## *Cell Growth and Reproduction*

**Mitosis and Aging**  Certain types of cells can be removed from the body and grown in nutrient solutions (outside the body, or *in vitro*). Under these artificial conditions, the potential longevity of different cell lines can be studied. For unknown reasons, normal connective tissue cells, called fibroblasts, stop dividing in vitro after about 40 to 70 population doublings. Cells that become transformed into cancer, however, apparently do not age and continue dividing indefinitely in culture. It is ironic that these potentially immortal cells may commit suicide by killing their host.

**The Cell Cycle and Cancer**  Mature nerve and muscle cells do not replicate at all; neurons are thus particularly susceptible to damage from alcohol and other drugs. Epithelial cells, by contrast, have very rapid cell cycles that help to replace the continuous loss of cells. *Cancers* have rapid rates of cell division but not necessarily more rapid than normal tissue. The fast growth of some cancers is not due to a rapid rate of cell division but rather to the fact that the rate of cell division far exceeds the rate of cell death. The observation that many normal tissues are at least as rapid in their rates of cell divisions as cancer makes the chemotherapy of cancer much more difficult.

in vitro: L. *in vitro,* in a glass

# Chapter Summary

## Cell Membrane and Associated Structures (pp. 42–46)

1. The structure of the cell, or plasma, membrane is described by a fluid-mosaic model.
   a. The membrane is composed predominately of a double layer of phospholipids.
   b. The membrane also contains proteins, distributed in a mosaic pattern.

2. Some cells move by extending pseudopods. Cilia and flagella protrude from the cell membrane of some specialized cells.
3. Invaginations of the cell membrane, in a process of endocytosis, allow the cells to take up molecules from the external environment.
   a. In phagocytosis, the cell extends pseudopods that eventually fuse together to create a food vacuole;

   pinocytosis involves the formation of a narrow furrow in the membrane that eventually fuses.
   b. Receptor-mediated endocytosis requires the interaction of a specific molecule in the extracellular environment with a specific receptor protein in the cell membrane.
   c. Exocytosis, the reverse of endocytosis, is a process that allows the cell to secrete its products.

## Cytoplasm and Its Organelles
(pp. 46–50)

1. Microfilaments and microtubules produce a cytoskeleton, which aids movements of organelles within a cell.
2. Lysosomes contain digestive enzymes and are responsible for the elimination of structures and molecules within the cell and for digestion of the contents of phagocytic food vacuoles.
3. Mitochondria serve as the major sites for energy production in the cell. They have an outer membrane with a smooth contour and an inner membrane with numerous infoldings called cristae.
4. The endoplasmic reticulum is a system of membranous tubules in the cell.
   a. The rough endoplasmic reticulum is covered with ribosomes and is involved in protein synthesis.
   b. The smooth endoplasmic reticulum provides a site for many enzymatic reactions and, in skeletal muscles, serves to store calcium ions.

## Cell Nucleus and Nucleic Acids
(pp. 50–54)

1. The cell nucleus is surrounded by a nuclear membrane. It contains chromatin, which consists of DNA and protein.
2. Nucleic acids include DNA, RNA, and their nucleotide subunits.
   a. The DNA nucleotides contain the sugar deoxyribose, whereas the RNA nucleotides contain the sugar ribose.
   b. There are four different types of DNA nucleotides that contain one of four possible bases: adenine, guanine, cytosine, and thymine. In RNA, the base uracil substitutes for the base thymine.
   c. DNA consists of two long polynucleotide strands twisted into a double helix. The two strands are held together by hydrogen bonds between specific bases—adenine pairs with thymine, and guanine pairs with cytosine.

   d. RNA is single stranded. Four types are produced within the nucleus: ribosomal RNA, transfer RNA, precursor messenger RNA, and messenger RNA.
3. Active euchromatin directs the synthesis of RNA in a process called genetic transcription.
   a. The enzyme RNA polymerase causes separation of the two strands of DNA along the region of the DNA that constitutes a gene.
   b. One of the two separated strands of DNA serves as a template for the production of RNA. This occurs by complementary base pairing between the DNA bases and ribonucleotide bases.

## Protein Synthesis and Secretion
(pp. 55–58)

1. Messenger RNA leaves the nucleus and attaches to the ribosomes.
2. Each transfer RNA, with a specific base triplet in its anticodon, bonds to a specific amino acid.
   a. As the mRNA moves through the ribosomes, complementary base pairing between tRNA anticodons and mRNA codons occurs.
   b. As each successive tRNA molecule bonds to its complementary codon, the amino acid it carries is added to the end of a growing polypeptide chain.
3. Proteins destined for secretion are produced in ribosomes located in the rough endoplasmic reticulum and enter the cisternae of this organelle.
4. Secretory proteins move from the rough endoplasmic reticulum to the Golgi apparatus, which consists of a stack of membranous sacs.
   a. The Golgi apparatus modifies the proteins it contains, separates different proteins, and packages them in vesicles.

   b. Secretory vesicles from the Golgi apparatus fuse with the cell membrane and release their products by exocytosis.

## DNA Synthesis and Cell Division
(pp. 58–65)

1. Replication of DNA is semiconservative. Each DNA strand serves as a template for the production of a new strand.
   a. The strands of the original DNA molecule gradually separate along their entire length and, through complementary base pairing, form a new complementary strand.
   b. In this way, each DNA molecule consists of one old and one new strand.
2. During the $G_1$ phase of the cell cycle, the DNA directs the synthesis of RNA, and hence that of proteins.
3. During the S phase of the cycle, DNA directs the synthesis of new DNA and replicates itself.
4. After a brief rest ($G_2$), the cell begins mitosis (the M stage of the cycle).
   a. Mitosis consists of the following phases: prophase, metaphase, anaphase, and telophase.
   b. In mitosis, the homologous chromosomes line up single file and are pulled by spindle fibers to opposite poles.
   c. This results in the production of two daughter cells that each contain 46 chromosomes, just like the parent cell.
5. Meiosis is a special type of cell division that results in the production of gametes in the gonads.
   a. The homologous chromosomes line up side by side, so that only one of each pair is pulled to each pole.
   b. This results in the production of two daughter cells, each containing only 23 chromosomes.
   c. The duplicate chromatids in each of the 23 chromosomes go to each of two new daughter cells in the second meiotic cell division.

# Review Activities

## Objective Questions

1. According to the fluid-mosaic model of the cell membrane,
   a. protein and phospholipids form a regular, repeating structure.
   b. the membrane is a rigid structure.
   c. phospholipids form a double layer, with the polar parts facing each other.

   d. proteins are free to move within a double layer of phospholipids.
2. After the DNA molecule has replicated itself, the duplicate strands are called
   a. homologous chromosomes.
   b. chromatids.
   c. centromeres.
   d. spindle fibers.

3. Nerve and skeletal muscle cells in the adult, which do not divide, remain in
   a. the $G_1$ phase.
   b. the S phase.
   c. the $G_2$ phase.
   d. the M phase.

4. The phase of mitosis in which the chromosomes line up at the equator of the cell is called
   a. interphase.
   b. prophase.
   c. metaphase.
   d. anaphase.
   e. telophase.

5. The phase of mitosis in which the chromatids separate is called
   a. interphase.
   b. prophase.
   c. metaphase.
   d. anaphase.
   e. telophase.

6. The RNA nucleotide base that pairs with adenine in DNA is
   a. thymine.
   b. uracil.
   c. guanine.
   d. cytosine.

7. Which of the following statements about RNA is *true?*
   a. It is made in the nucleus.
   b. It is double stranded.
   c. It contains the sugar deoxyribose.
   d. It is a complementary copy of the entire DNA molecule.

8. Which of the following statements about mRNA is *false?*
   a. It is produced as a larger pre-mRNA.
   b. It forms associations with ribosomes.
   c. Its base triplets are called anticodons.
   d. It codes for the synthesis of specific proteins.

9. The organelle that combines proteins with carbohydrates and packages them within vesicles for secretion is
   a. the Golgi apparatus.
   b. the rough endoplasmic reticulum.
   c. the smooth endoplasmic reticulum.
   d. the ribosome.

10. The organelle that contains digestive enzymes is
    a. the mitochondrion.
    b. the lysosome.
    c. the endoplasmic reticulum.
    d. the Golgi apparatus.

11. If four bases in one DNA strand are A (adenine), G (guanine), C (cytosine), and T (thymine), the complementary bases in the RNA strand made from this region are
    a. T,C,G,A.
    b. C,G,A,U.
    c. A,G,C,U.
    d. U,C,G,A.

12. Which of the following statements about tRNA is *true?*
    a. It is made in the nucleus.
    b. It is looped back on itself.
    c. It contains the anticodon.
    d. There are over 20 different types of tRNA.
    e. All of the above are true.

13. The step in protein synthesis during which tRNA, rRNA, and mRNA are all active is known as
    a. transcription.
    b. translation.
    c. replication.
    d. RNA polymerization.

14. The anticodons are located in
    a. tRNA.
    b. rRNA.
    c. mRNA.
    d. ribosomes.
    e. endoplasmic reticulum.

## Essay Questions

1. Give examples of different organelles and describe their functions.

2. Explain how one DNA molecule serves as a template for the formation of another DNA. What do we mean when we say that DNA synthesis is semiconservative?

3. What is the genetic code and how does it affect the structure and function of the body?

4. Why is tRNA thought to be the interpreter of the genetic code?

5. Compare the processing of cellular proteins with that of proteins that are secreted by a cell.

# chapter four

# enzymes, energy, and metabolism

## objectives

- Explain how catalysts function in chemical reactions and how enzymes function as catalysts.

- Describe the effects of pH and temperature on enzyme activity.

- Describe the effects of cofactors and coenzymes on enzyme activity and the effects of substrate and enzyme concentrations.

- Explain how end-product inhibition affects the direction of a branched metabolic pathway.

- Use the first and second laws of thermodynamics to explain why some molecules have more chemical-bond energy than others.

- Describe the coupling of energy-releasing and energy-requiring reactions and discuss the significance of ATP.

- Describe the nature of oxidation-reduction reactions.

- Describe glycolysis in terms of its initial substrate and its products.

- Describe the pathway of anaerobic respiration and discuss the significance of lactic acid formation.

- Define *gluconeogenesis* and discuss its significance.

- Describe the fate of pyruvic acid in aerobic respiration and discuss the nature of the Krebs cycle, naming the products that result from it.

- Explain the function of the electron-transport system and the role of oxygen in aerobic respiration.

- Define *oxidative phosphorylation*, state where it occurs, and discuss its significance.

# Enzymes as Catalysts

*Enzymes are biological catalysts, functioning to increase the rate of chemical reactions. Most enzymes are proteins, and their catalytic action results from their complex structure. The great diversity of protein structure allows enzyme action to be highly specific.*

Although the ability of yeast cells to make alcohol from glucose (a process called *fermentation*) had been recognized since antiquity, by the mid-nineteenth century no chemist had succeeded in duplicating the process in the absence of living yeast. Also, yeast and other living cells could perform a vast array of chemical reactions at body temperature that could not be duplicated in the chemical laboratory without adding a substantial amount of heat energy. These observations led many mid-nineteenth-century scientists to conclude that chemical reactions in living cells were aided by a vital force that operated beyond the laws of the physical world. This vitalist concept was squashed along with the yeast cells when a pioneering biochemist, Eduard Buchner, demonstrated that juice obtained from yeast could ferment glucose to alcohol. The yeast juice was not alive—evidently some chemicals in the cells were responsible for fermentation. Buchner did not know what these chemicals were, so he simply named them **enzymes.**

Chemically, *enzymes are proteins*, although it has recently been learned that RNA can exhibit very specialized enzymatic activity. Biochemists have demonstrated that enzymes act as biological catalysts. A **catalyst** is a chemical that (1) increases the rate of a reaction, (2) is not itself changed at the end of the reaction, and (3) does not change the nature of the reaction or its final result. The same reaction would have occurred to the same degree in the absence of the catalyst, but it would have progressed at a much slower rate.

In order for a given reaction to occur, the reactants must have sufficient energy. The amount of energy required for a reaction to proceed is called the **energy of activation.** By analogy, a match will not burn and release heat energy unless it is first activated by striking the match or by placing it in a flame.

In a large population of reactant molecules, only a small fraction will possess sufficient energy to participate in the reaction. Adding heat will raise the energy level of all the reactant molecules, thus increasing the percentage of the population with the required energy of activation. Heat makes reactions go faster, but it also produces undesirable side effects in cells. Catalysts speed up a reaction at lower temperatures by *lowering the activation energy required*, thus

Buchner, Eduard: German biochemist, 1860–1917
enzyme: Gk. *en*, in; *zyme*, yeast

ensuring that a larger percentage of the population of reactant molecules will have sufficient energy to participate in the reaction (fig. 4.1).

Since a small fraction of the reactants will have the activation energy required for the reaction even in the absence of a catalyst, the reaction could theoretically occur spontaneously at a slow rate. This rate, however, would be much too slow for the needs of a cell. So, from a biological standpoint, the presence or absence of a specific enzyme catalyst acts as a switch—the reaction will occur if the enzyme is present and will not occur if the enzyme is absent.

## Mechanisms of Enzyme Action

The ability of enzymes to lower the activation energy of a reaction is derived from their structure. Enzymes are very large proteins with complex, highly ordered, three-dimensional shapes produced by physical and chemical interactions between their amino acids. Each type of enzyme has a characteristic shape, or *conformation*, with ridges, grooves, and pockets that are lined with specific amino acids. The particular pockets that are active in catalyzing a reaction are called the *active sites* of the enzyme.

The model of how enzymes work is known as the **lock-and-key model** of enzyme activity (fig. 4.2). The reactant molecules, which are the *substrates* of the enzyme, have shapes that allow them to fit into the active sites. The fit may not be perfect at first, but a perfect fit may be induced as the substrate gradually slips into the active site. This induced fit, together with temporary bonds that form between the substrate and the amino acids lining the active sites of the enzyme, weaken the existing bonds within the substrate molecules and allows them to be more easily broken. New bonds are more easily formed as substrates are brought close together in the proper orientation. The *enzyme-substrate complex*, formed temporarily in the course of the reaction, then dissociates to yield *products* and the free, unaltered enzyme.

## Naming of Enzymes

Although an international committee has established a uniform naming system for enzymes, the names that are in common use do not follow a completely consistent pattern. With the exception of the digestive enzymes that were discovered first (including pepsin and trypsin), all enzyme names end with the suffix *-ase.* Classes of enzymes are named according to their job category. *Hydrolases* (hi´drŏ-lās-es), for example, promote hydrolysis reactions. Other enzyme categories include *phosphatases* (fos´fĕ-tās-es), which catalyze the removal of phosphate groups; *synthetases* (sin´thĭ-tās-es), which catalyze dehydration synthesis reactions; and *dehydrogenases* (de´´hi-droj´ĕ-nās-es), which remove hydrogen atoms from their substrates. Enzymes called *isomerases* (i-som´ĕ-rās-es) rearrange atoms within their

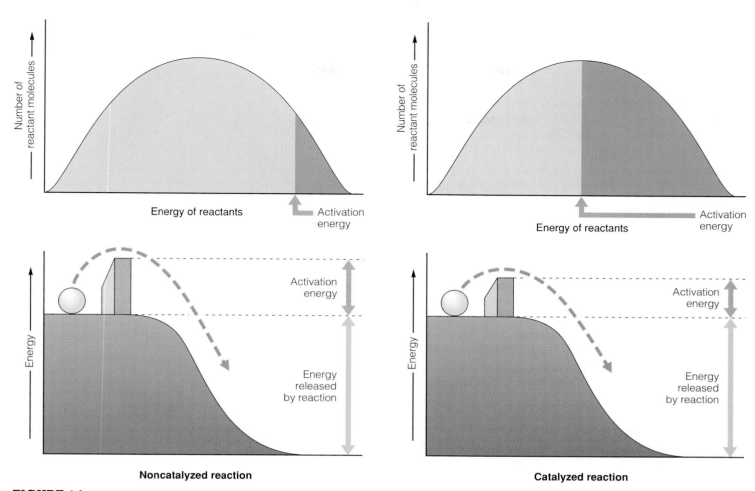

**FIGURE 4.1**

A comparison of a noncatalyzed reaction with a catalyzed reaction. The upper figures compare the proportion of reactant molecules that have sufficient activation energy to participate in the reaction (green). This proportion is increased in the enzyme-catalyzed reaction because enzymes lower the activation energy required for the reaction (shown as a barrier on top of an energy "hill" in the lower figures). Reactants that can overcome this barrier are able to participate in the reaction, as shown by arrows pointing to the bottom of the energy hill.

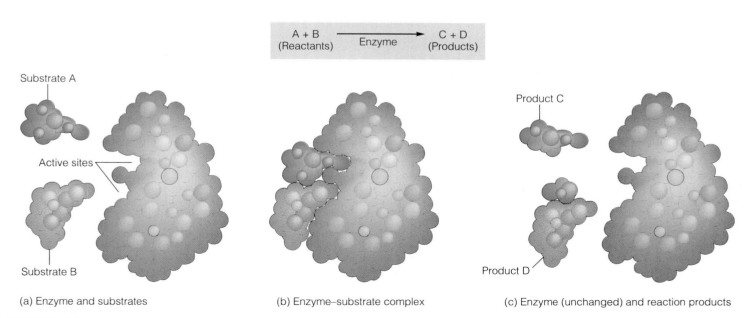

**FIGURE 4.2**

The lock-and-key model of enzyme action. (*a*) Substrate A and B fit into active sites in the enzyme, forming an enzyme-substrate complex (*b*). This then dissociates (*c*), releasing the products of the reaction and the free enzyme.

substrate molecules to form structural isomers. (Examples of such structural isomers include glucose and fructose.)

The names of many enzymes indicate both the substrate of the enzyme and the activity (job category) of the enzyme. Lactic acid dehydrogenase, for example, removes hydrogens from lactic acid. Since enzymes are very specific as to their substrates and activity, the concentration of a specific enzyme in a sample of fluid can be measured relatively easily. This is usually done by measuring the rate of conversion of the enzyme's substrates into products under specified conditions.

Enzymes that do the same job (that catalyze the same reaction) in different organs have the same name, since the name describes the activity of the enzyme. Different organs, however, may make models of the enzyme that vary slightly in one or a few amino acids. These different models of the same enzyme are called **isoenzymes** (*i-so-en´zīmz*). The differences in structure do not affect the active sites of the enzymes (otherwise they would not catalyze the same reaction), but at other locations the structural differences make it possible to separate the different isoenzymatic forms by standard biochemical procedures. These techniques are useful in the diagnosis of diseases, as described at the end of this chapter.

# Control of Enzyme Activity

*The rate of an enzyme-catalyzed reaction depends on numerous factors. Variations in some of these factors control the rate of progress along particular metabolic pathways, and thus help to regulate cellular metabolism.*

The activity of an enzyme, as measured by the rate at which its substrates are converted to products, is influenced by a variety of factors. These include (1) the temperature and pH of the solution; (2) the concentration of cofactors and coenzymes, which are needed by many enzymes to facilitate their catalytic activity; (3) the concentration of enzyme and substrate molecules in the solution; and (4) the stimulatory and inhibitory effects of some products of enzyme action on the activity of the enzymes that helped to form these molecules.

## Effects of Temperature and pH

An increase in temperature will increase the rate of non-enzyme-catalyzed reactions because a larger number of reactant molecules will have the activation energy required. A similar relationship between temperature and reaction rate occurs in enzyme-catalyzed reactions. At a temperature of 0°C, the reaction rate is immeasurably

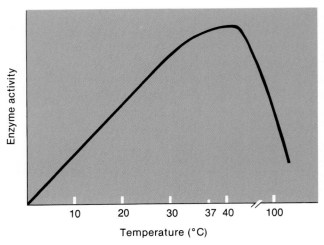

**FIGURE 4.3**

The effect of temperature on enzyme activity, as measured by the rate of the enzyme-catalyzed reaction under standardized conditions.

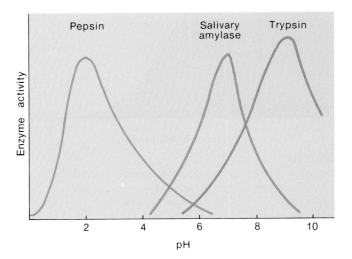

**FIGURE 4.4**

The effect of pH on activity of three digestive enzymes.

slow. As the temperature is raised above 0°C, the reaction rate increases, but only up to a point. At a few degrees above body temperature (which is 37°C), the reaction rate reaches a plateau; further increases in temperature actually *decrease* the rate of the reaction (fig. 4.3). This is because the tertiary structure of enzymes becomes altered at higher temperatures.

A similar relationship is observed when the rate of an enzymatic reaction is measured at different pH values. Each enzyme characteristically has its peak activity in a very narrow pH range, which is the **pH optimum** for the enzyme. If the pH is changed from this optimum, the reaction rate decreases (fig. 4.4). This decreased enzyme activity is due to changes in the conformation of the enzyme and in the charges of the R groups of the amino acids lining the active sites.

The pH optimum of an enzyme usually reflects the pH of the body fluid in which the enzyme is found. The acidic pH optimum of the protein-digesting enzyme *pepsin*, for example, allows it to be active in the strong hydrochloric acid of gastric juice. Similarly, the neutral pH optimum of *salivary amylase* (am´ĭ-lās) and the alkaline pH optimum of *trypsin* (trip´sin) in pancreatic juice allow these enzymes to digest starch and protein, respectively, in other parts of the digestive tract.

Although the pH of other body fluids shows less variation than the fluids of the digestive tract, the pH optima of different enzymes found throughout the body do show significant differences (table 4.1). Some of these differences can be exploited for diagnostic purposes. Disease of the prostate, for example, may cause elevated blood levels of a prostatic phosphatase with an acidic pH optimum (descriptively called acid phosphatase). Bone disease, on the other hand, may produce elevated blood levels of alkaline phosphatase, which has a higher pH optimum than the similar enzyme released from the diseased prostate.

## Cofactors and Coenzymes

Many enzymes are completely inactive when they are isolated in a pure state. Evidently, some of the ions and smaller organic molecules that are removed in the purification procedure play an essential role in enzyme activity. These ions and smaller organic molecules are called **cofactors** and **coenzymes** (ko-en´zīmz).

*Cofactors* are metal ions such as $Ca^{++}$, $Mg^{++}$, $Mn^{++}$, $Cu^{++}$, and $Zn^{++}$. Some enzymes with a cofactor requirement do not have a properly shaped active site in the absence of the cofactor. In these enzymes, the attachment of cofactors causes a conformational change in the protein that allows it to combine with its substrate. The cofactors of other enzymes participate in forming the temporary bonds of the enzyme-substrate complex (fig. 4.5).

*Coenzymes* are organic molecules that are derived from water-soluble vitamins, such as niacin and riboflavin. Coenzymes participate in enzyme-catalyzed reactions by transporting hydrogen atoms and small molecules from one enzyme to another.

Examples of the actions of cofactors and coenzymes in specific reactions will be given in the context of their roles in cellular metabolism later in this chapter.

| Table 4.1 | pH optima of selected enzymes | |
|---|---|---|
| *Enzyme* | *Reaction catalyzed* | *pH optimum* |
| Pepsin (stomach) | Digestion of protein | 2.0 |
| Acid phosphatase (prostate) | Removal of phosphate group | 5.5 |
| Salivary amylase (saliva) | Digestion of starch | 6.8 |
| Lipase (pancreatic juice) | Digestion of fat | 7.0 |
| Alkaline phosphatase (bone) | Removal of phosphate group | 9.0 |
| Trypsin (pancreatic juice) | Digestion of protein | 9.5 |
| Monoamine oxidase (nerve endings) | Removal of amine group from norepinephrine | 9.8 |

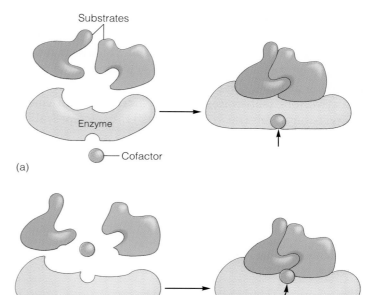

(a)

(b)

**FIGURE 4.5**

The roles of cofactors in enzyme function. In (a) the cofactor changes the conformation of the active site, which allows for a better fit between the enzyme and its substrates. In (b) the cofactor participates in the temporary bonding between the active site and the substrates.

## Substrate Concentration and Reversible Reactions

The rate at which an enzymatic reaction converts substrates into products depends on the enzyme concentration and on the concentration of substrates. When the enzyme concentration is at a given level, the rate of product formation will increase as the substrate concentration increases. Eventually, a point will be reached where additional increases

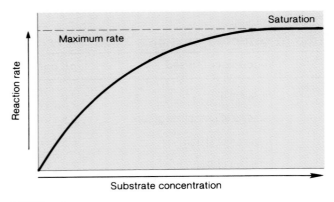

**FIGURE 4.6**
The effect of substrate concentration on the reaction rate of an enzyme-catalyzed reaction. When the reaction rate is maximal, the enzyme is said to be *saturated*.

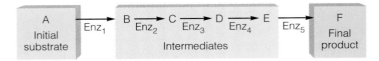

**FIGURE 4.7**
A metabolic pathway, where the product of one enzyme becomes the substrate of the next in a multienzyme system.

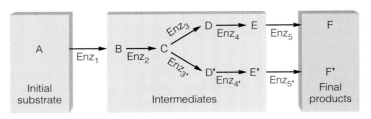

**FIGURE 4.8**
A branched metabolic pathway.

in substrate concentration do not result in comparable increases in the reaction rate. When the relationship between substrate concentration and reaction rate reaches a plateau, the enzyme is said to be *saturated*. If one thinks of enzymes as workers and substrates as jobs, there is 100% employment when the enzyme is saturated; further availability of jobs (substrates) cannot increase employment (conversion of substrate to product). This concept is illustrated in figure 4.6.

Some enzymatic reactions within a cell are reversible, with both the forward and backward reactions catalyzed by the same enzyme. The enzyme *carbonic anhydrase*, for example, is named because it can catalyze the following reaction:

$$H_2CO_3 \longrightarrow H_2O + CO_2$$

The same enzyme, however, can also catalyze the reverse reaction:

$$H_2O + CO_2 \longrightarrow H_2CO_3$$

The two reactions can be more conveniently illustrated by a single equation:

$$H_2O + CO_2 \rightleftharpoons H_2CO_3$$

The direction of a reversible reaction depends, in part, on the relative concentrations of the molecules to the left and right of the arrows. If the concentration of $CO_2$ is very high (as it is in the tissues), the reaction will be driven to the right. If the concentration of $CO_2$ is low (as it is in the lungs), the reaction will be driven to the left. The principle by which reversible reactions are driven from the side of the equation where the concentration is higher to the side where the concentration is lower is known as the **law of mass action.**

Although some enzymatic reactions are not directly reversible, the net effects of the reactions can be reversed by the action of different enzymes. The enzymes that convert glucose to pyruvic acid, for example, are different from those that reverse the pathway and produce glucose from pyruvic acid. Likewise, the formation and breakdown of glycogen (a polymer of glucose) are catalyzed by different enzymes.

## Metabolic Pathways

The many thousands of types of enzymatic reactions within a cell do not occur independently of each other. Rather, they are all linked together in an intricate web of interrelationships, the total pattern of which constitutes cellular metabolism. The part of this web that begins with an *initial substrate*, progresses through a number of *intermediates*, and ends with a *final product* is known as a **metabolic pathway.**

The enzymes in a metabolic pathway cooperate in a manner analogous to workers on an assembly line: each contributes a small part to the final product. In this process, the product of one enzyme in the line becomes the substrate of the next enzyme, and so on (fig. 4.7).

Few metabolic pathways are completely linear, however. Most are branched so that one intermediate at the branch point can serve as a substrate for two different enzymes. Two different products that serve as intermediates of two divergent pathways can thus be formed (fig. 4.8).

**End-Product Inhibition**   The activities of enzymes at the branch points of metabolic pathways are often regulated by a process called **end-product inhibition.** In this process, one of the final products of a divergent pathway inhibits the branch-point enzyme that began the path toward the production of this inhibitor. This inhibition prevents that final product from accumulating excessively and results in a shift toward the final product of the alternate divergent pathway (fig. 4.9).

The mechanism by which a final product inhibits an earlier enzymatic step in its pathway is known as **allosteric** (al″ŏ-ster′ik) **inhibition.** The allosteric inhibitor combines with a part of the enzyme that is distanced from the active site; as a result, the active site changes shape so that it can no longer combine properly with its substrate.

### Inborn Errors of Metabolism

Each enzyme in a metabolic pathway is coded by a different gene. An inherited defect in one of these genes may result in a disease known as an **inborn error of metabolism.** In this type of disease, the quantity of intermediates formed *prior* to the defective enzymatic step *increases* and the quantity of intermediates and final products formed *after* the defective step *decreases*. Diseases may result from a deficiency of the normal end product or from accumulations of toxic levels of intermediates or their alternate derivatives. If the defective enzyme is active at a step that follows a branch point in a pathway, the intermediates and final products of the divergent pathway will increase as a result of the block in the alternate pathway (fig. 4.10). Specific examples of inborn errors of metabolism are discussed at the end of this chapter.

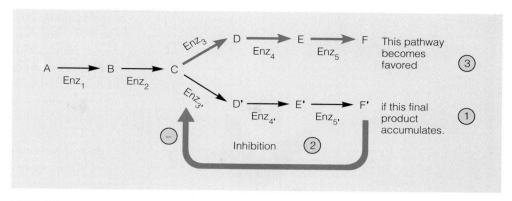

**FIGURE 4.9**

End-product inhibition in a branched metabolic pathway. Inhibition is shown by the arrow in step 2.

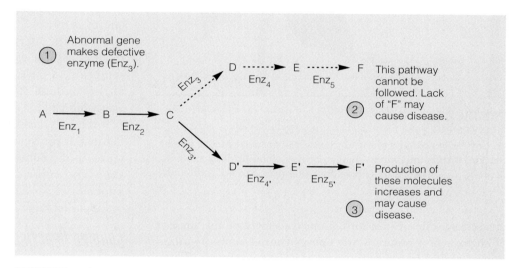

**FIGURE 4.10**

The effects of an inborn error of metabolism on a branched metabolic pathway.

# Bioenergetics

*Living organisms require the constant expenditure of energy to maintain their complex structures and processes. Central to life processes are chemical reactions that are coupled, so that the energy released by one reaction is incorporated into the products of another reaction. The transformation of energy in living systems is largely based on reactions that produce and break down molecules of ATP and on oxidation-reduction reactions.*

*Bioenergetics* refers to the flow of energy in living systems. Organisms maintain their highly ordered structure and life-sustaining activities through the constant expenditure of energy obtained ultimately from the environment. The energy flow in living systems obeys the first and second laws of a branch of physics known as *thermodynamics*.

According to the **first law of thermodynamics,** energy can be transformed (changed from one form to another), but it can neither be created nor destroyed. This is sometimes called the *law of conservation of energy*. As a result of energy transformations, according to the **second law of thermodynamics,** the universe and its parts (including living systems) become increasingly disorganized. The term *entropy* (en′trŏ-pe) is used to describe the degree of disorganization

allosteric: Gk. *allos*, other; *stereos*, position

bioenergetics: Gk. *bios*, life; *energeia*, work
thermodynamics: Gk. *therme*, heat; *dynamis*, force

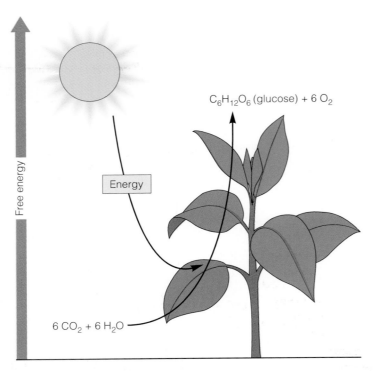

$C_6H_{12}O_6$ (glucose) + 6 $O_2$

Energy

6 $CO_2$ + 6 $H_2O$

Free energy

**FIGURE 4.11**

A simplified diagram of photosynthesis. Some of the sun's radiant energy is captured by plants and used to produce glucose from carbon dioxide and water. As the product of this endergonic reaction, glucose has more free energy than the initial reactants.

of a system. Energy transformations thus increase the amount of entropy of a system. Only energy that is in an organized state—called *free energy*—can be used to do work. Thus, since entropy increases in every energy transformation, the amount of free energy available to do work decreases. As a result of the increased entropy described by the second law, systems tend to go from states of higher free energy to states of lower free energy.

The chemical bonding of atoms into molecules obeys the laws of thermodynamics. Atoms that are organized into complex organic molecules, such as glucose, have more free energy (less entropy) than six separate molecules each of carbon dioxide and water. Therefore, in order to convert carbon dioxide and water to glucose, energy must be added. Plants perform this feat using energy from the sun in the process of *photosynthesis* (fig. 4.11).

## Endergonic and Exergonic Reactions

Chemical reactions that require an input of energy are known as **endergonic** (en´´der-gon´ik) **reactions.** Since energy is added to make these reactions "go," the products

photosynthesis: Gk. *phos,* light; *synthesis,* a putting together
endergonic: Gk. *endon,* within; *ergon,* work

of endergonic reactions must contain more free energy than the reactants. A portion of the energy added, in other words, is contained within the product molecules. This follows from the fact that energy cannot be created or destroyed (first law of thermodynamics) and from the fact that a more-organized state of matter contains more free energy (less entropy) than a less-organized state (as described by the second law).

The fact that glucose contains more free energy than carbon dioxide and water can be proven by the combustion of glucose to $CO_2$ and $H_2O$. This reaction releases energy in the form of heat. Reactions that convert molecules with more free energy to molecules with less—and, therefore, that release energy as they proceed—are called **exergonic reactions.**

As illustrated in figure 4.12, the amount of energy released by an exergonic reaction is the same whether the energy is released in a single combustion reaction or in the many small, enzymatically controlled steps that occur in tissue cells. The energy that the body obtains from the consumption of particular foods, therefore, can be measured as the amount of heat energy released when these foods are combusted.

Heat is measured in units called *calories*. One calorie is defined as the amount of heat required to raise the temperature of one cubic centimeter of water by one degree Celsius. The caloric value of food is usually indicated in kilocalories (1 *kilocalorie* equals 1000 calories), which are often called large calories, designated with a capital C.

## Coupled Reactions: ATP

In order to remain alive, a cell must maintain its highly organized, low-entropy state at the expense of free energy in its environment. Accordingly, the cell contains many enzymes that catalyze exergonic reactions using substrates that come ultimately from the environment. The energy released by these exergonic reactions is used to drive the energy-requiring processes (endergonic reactions) in the cell. Since the cell cannot use heat energy to drive energy-requiring processes, chemical-bond energy that is released in the exergonic reactions must be directly transferred to chemical-bond energy in the products of endergonic reactions. Energy-liberating reactions are thus *coupled* to energy-requiring reactions. Picture, for example, two meshed gears. The turning of one (the energy-releasing, exergonic gear) causes turning of the other (the energy-requiring, endergonic gear). This relationship is illustrated in figure 4.13.

As shown in figure 4.14, the energy released by most exergonic reactions in the cell is used—either directly or indirectly—to drive the formation of **adenosine triphosphate**

exergonic: Gk. *exo,* outside; *ergon,* work
calorie: L. *calor,* heat

# PUMP UP YOUR GRADES!

A+

TRIM .... your study time!

## Computerized Study Guide

for
··· CONCEPTS OF HUMAN ···
ANATOMY AND PHYSIOLOGY, 4/E

by Jeffrey and
Karianne Prince

**Why not call for a copy today and get your grades** PUMPED UP!

### FOCUS YOUR STUDY TIME *ONLY* ON THE AREAS WHERE YOU NEED HELP

*How does it work?*

✔ Take the computerized test for each chapter.
✔ The program presents you with feedback for each answer.
✔ A page-referenced study plan is then created for all incorrect answers.
✔ Look up the answers to the questions you missed.
✔ This easy-to-use **Computerized Study Guide** is **not** a duplicate of the printed study guide that accompanies your text. It's full of all new study aids.

Get your copy of the *Computerized Study Guide for Concepts of Human Anatomy and Physiology.*
Ask your bookstore manager for more information or call toll free **1-800-338-5578**.

The Computerized Study Guide is available on:
✔ IBM PC/PC Compatible 3½" disk (ISBN 28265)
✔ Macintosh 3½" disk (ISBN 28266)

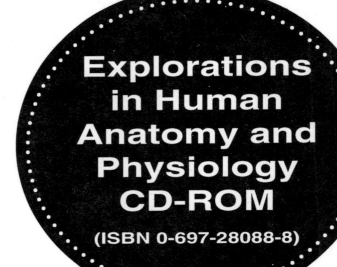

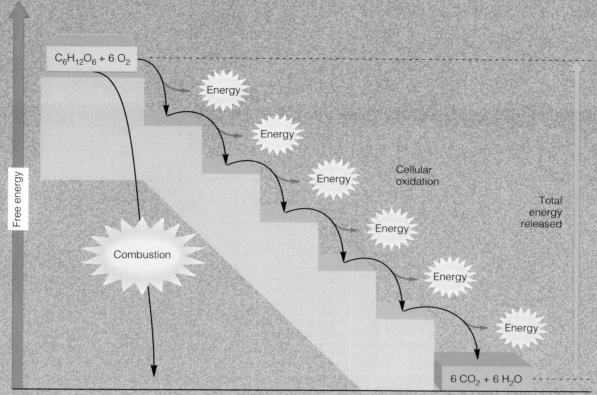

**FIGURE 4.12**
Since glucose contains more free energy than carbon dioxide and water, the combustion of glucose is an exergonic reaction. The same amount of energy is released when glucose is broken down stepwise within the cell.

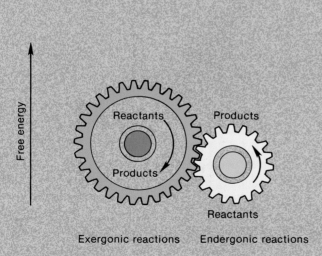

**FIGURE 4.13**
A model of the coupling of exergonic and endergonic reactions. The reactants of the exergonic reaction (represented by the larger gear) have more free energy than the products of the endergonic reaction because the coupling is not 100% efficient—some energy is lost as heat.

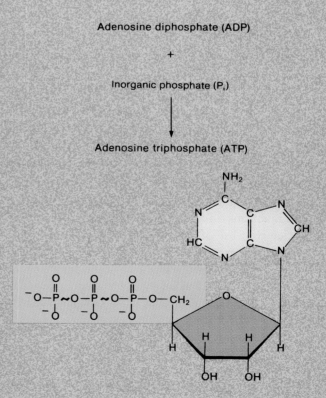

**FIGURE 4.14**
The formation and structure of adenosine triphosphate (ATP).

(ă-den´ŏ-sēn tri-fos´fāt) (**ATP**) from adenosine diphosphate (ADP) and inorganic phosphate ($P_i$). Because the formation of ATP from ADP and $P_i$ requires an input of free energy, it is endergonic.

ATP formation requires the input of a fairly large amount of energy. Since this energy must be conserved (first law of thermodynamics), the bond that is produced by joining $P_i$ to ADP must contain a part of this energy. Thus, when enzymes reverse this reaction and convert ATP to ADP and $P_i$, a large amount of energy is released. Energy released from the breakdown of ATP is used to power the energy-requiring processes in all cells. As the **universal energy carrier,** ATP serves to more efficiently couple the energy released by the breakdown of food molecules to the energy required by the diverse endergonic processes in the cell (fig. 4.15).

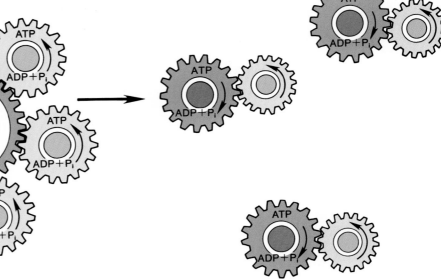

**FIGURE 4.15**

A model of ATP as the universal energy carrier of the cell. Exergonic reactions are shown as gears with arrows going down (reactions produce a decrease in free energy); endergonic reactions are shown as gears with arrows going up (reactions produce an increase in free energy).

## Coupled Reactions: Oxidation-Reduction

When an atom or a molecule gains electrons, it is said to become **reduced;** when it loses electrons, it is said to become **oxidized.** Reduction and oxidation are always coupled reactions: An atom or a molecule cannot become oxidized unless it donates electrons to another, which therefore becomes reduced. The atom or molecule that donates electrons *to* another is a **reducing agent;** the one that accepts electrons *from* another is an **oxidizing agent.** It should be noted that an atom or a molecule may function as an oxidizing agent in one reaction and as a reducing agent in another reaction. It may gain electrons from one atom or molecule and pass them on to another in a series of coupled oxidation-reduction reactions—like a bucket brigade.

Notice that use of the term *oxidation* does not imply that oxygen participates in the reaction. This term is derived from the fact that oxygen has a great tendency to accept electrons; that is, to act as a strong oxidizing agent. This property of oxygen is exploited by cells. Oxygen acts as the final electron acceptor in a chain of oxidation-reduction reactions that provides energy for ATP production.

Oxidation-reduction reactions in cells often involve the transfer of hydrogen atoms rather than free electrons. Since a hydrogen atom contains one electron (and one proton in the nucleus), a molecule that loses hydrogen becomes oxidized, and one that gains hydrogen becomes reduced. In

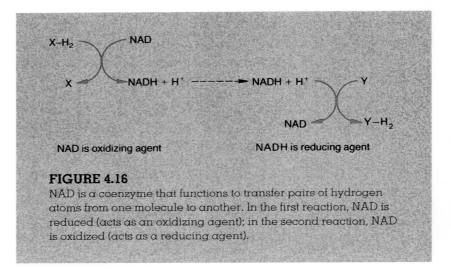

**FIGURE 4.16**

NAD is a coenzyme that functions to transfer pairs of hydrogen atoms from one molecule to another. In the first reaction, NAD is reduced (acts as an oxidizing agent); in the second reaction, NAD is oxidized (acts as a reducing agent).

many oxidation-reduction reactions, pairs of electrons—either as free electrons or as a pair of hydrogen atoms—are transferred from the reducing agent to the oxidizing agent.

Two molecules that serve important roles in the transfer of hydrogens are **nicotinamide** (nik´ŏ-tin´ă-mīd) **adenine dinucleotide (NAD),** derived from niacin (vitamin $B_3$), and **flavin adenine dinucleotide (FAD),** derived from riboflavin (vitamin $B_2$). These molecules are *hydrogen carriers* because they accept hydrogens (becoming reduced) in one enzyme reaction and donate hydrogens (becoming oxidized) in a different reaction (fig. 4.16). The oxidized forms of these molecules may be written as simply NAD and FAD.

Each FAD can accept two electrons and can bind two protons. Therefore, the reduced form of FAD can combine with the equivalent of two hydrogen atoms and may be written as $FADH_2$. Each NAD can also accept two electrons but can bind only one proton. The reduced form of NAD is therefore indicated by $NADH + H^+$ (the $H^+$ represents a free proton). When the reduced forms of these two coenzymes participate in an oxidation-reduction reaction, they transfer two hydrogen atoms to the oxidizing agent (fig. 4.16).

 We need niacin and riboflavin in our diet primarily for the production of the coenzymes NAD and FAD. In the next sections of this chapter, we will describe the role of NAD and FAD in transferring hydrogen atoms in the chemical reactions that provide energy for the body. Niacin and riboflavin do not themselves provide the energy, although this is often claimed in misleading advertisements for health foods. Nor can eating extra amounts of niacin and riboflavin provide extra energy. Once the cells have obtained sufficient NAD and FAD, excess amounts of niacin and riboflavin are simply eliminated in the urine.

# Glycolysis and Anaerobic Respiration

*In cellular respiration, chemical reactions liberate energy, some of which is used to produce ATP. Although the complete combustion of a molecule requires the presence of oxygen, some energy can be obtained in the absence of oxygen by anaerobic respiration. This process produces two ATP molecules per glucose molecule, resulting in the formation of lactic acid.*

All of the reactions in the body that involve energy transformation are collectively termed **metabolism** (*mĕ-tab´ŏ-liz˝em*). Metabolism may be divided into two categories: **anabolism** (*ă-nab´ŏ-liz˝em*) and **catabolism** (*kă-tab´ŏ-liz˝em*). Catabolic reactions release energy, usually by the breakdown of larger organic molecules into smaller molecules. *Anabolic reactions* require the input of energy and include the synthesis of large, energy-storage molecules, such as glycogen, fat, and protein.

The *catabolic reactions* that break down glucose, fatty acids, and amino acids serve as the primary sources of energy for the cellular synthesis of ATP. Collectively, these metabolic pathways refer to the process of *cellular respiration*. When oxygen serves as the final electron acceptor, we use the term **aerobic** (*ă-ro´bik*) **cell respiration.** The final prod-

ucts of aerobic respiration are carbon dioxide, water, and energy (a part of which is trapped in the chemical bonds of ATP). The overall equation for aerobic respiration, therefore, is identical to the equation that describes combustion (fuel + $O_2 \rightarrow CO_2 + H_2O$ + energy).

Notice that the term *respiration* refers to chemical reactions that liberate energy for the production of ATP. The oxygen used in aerobic respiration by tissue cells is obtained from the blood. The blood, in turn, becomes oxygenated in the lungs by the process of breathing. Breathing (also called *ventilation* or *external respiration*) is thus needed for, but is different from, aerobic respiration.

Unlike combustion, the conversion of glucose to carbon dioxide and water within the cells occurs in small, enzymatically catalyzed steps. Oxygen is used only at the last step, as will be described later in this chapter. Since a small amount of the chemical-bond energy of glucose is released at early steps in the metabolic pathway, some cells in the body can obtain energy for ATP production in the temporary absence of oxygen. This process is described next.

## Glycolysis

Both the anaerobic and the aerobic respiration of glucose begin with a metabolic pathway known as **glycolysis** (*gli˝col´ĭ-sis*). Glycolysis is the metabolic pathway by which glucose, a six-carbon (hexose) sugar, is converted into two molecules of *pyruvic (pi-roo´vik) acid*. Even though each pyruvic acid molecule is roughly half the size of a glucose, glycolysis is *not* simply the breaking in half of glucose. Glycolysis is a metabolic pathway involving many enzymatically controlled steps.

Each pyruvic acid molecule contains three carbons, three oxygens, and four hydrogens. The number of carbon and oxygen atoms in one molecule of glucose, $C_6H_{12}O_6$, can be accounted for in the two pyruvic acid molecules. Since the two pyruvic acids together account for only eight hydrogens, however, it is clear that four hydrogen atoms are removed from the intermediates in glycolysis. Each pair of these hydrogen atoms is used to reduce a molecule of NAD. In this process, each pair of hydrogen atoms donates two electrons to NAD, thus reducing it. The reduced NAD binds one proton from the hydrogen atoms, leaving one proton unbound as $H^+$ (as previously described). Starting from one glucose molecule, therefore, glycolysis results in the production of two molecules of NADH and two $H^+$. The $H^+$ will follow the NADH in subsequent reactions, so for simplicity we can refer to reduced NAD simply as NADH.

metabolism: Gk. *metabole,* change
anabolism: Gk. *anabole,* a raising up
catabolism: Gk. *katabole,* a casting down

anaerobic: Gk. *an,* without; *aer,* air; *bios,* life
glycolysis: Gk. *glyco* sugar; *lysis,* breaking

Glycolysis is exergonic, and a portion of the energy that is released is used to drive the endergonic reaction ADP + $P_i$ → ATP. At the end of the glycolytic pathway, there is a net gain of two ATP molecules per glucose molecule, as indicated in the overall equation for glycolysis:

$$\text{Glucose} + 2NAD + 2ADP + 2P_i \rightarrow$$

$$2 \text{ pyruvic acid} + 2NADH + 2ATP$$

Although the overall equation for glycolysis is exergonic, glucose must be activated at the beginning of the pathway before energy can be obtained. This activation requires the addition of two phosphate groups derived from two molecules of ATP. Energy from the reaction ATP → ADP + $P_i$ is therefore consumed at the beginning of glycolysis. This is shown as an "up-staircase" in figure 4.17. In this part of the pathway, the $P_i$ is not shown because the phosphate is not released but instead is added to the intermediate molecules of glycolysis. In later steps, four molecules of ATP are produced (and two molecules of NAD are reduced) as energy is liberated (the "down-staircase" in figure 4.17).

The two molecules of ATP used in the beginning, therefore, represent an energy investment; the net gain of two ATP and two NADH by the end of the pathway represent an energy profit.

The overall equation for glycolysis obscures the fact that this metabolic pathway consists of nine separate steps. The individual steps in this pathway are shown in figure 4.18.

## Anaerobic Respiration

In order for glycolysis to continue, adequate amounts of NAD must be available to accept hydrogen atoms. Therefore, the NADH that is produced in glycolysis must become oxidized by donating its electrons to another molecule. (In aerobic respiration this other molecule is located in the mitochondria and ultimately passes its electrons to oxygen.)

When oxygen is *not* available in sufficient amounts, the NADH (+ $H^+$) produced in glycolysis is oxidized in the cytoplasm by donating its electrons to pyruvic acid.

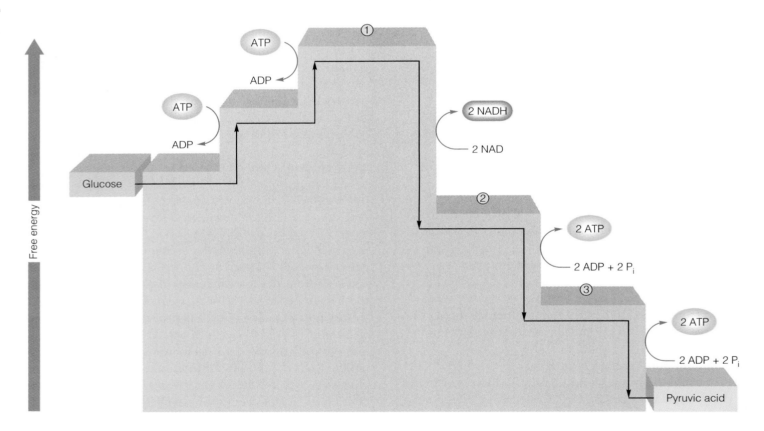

**FIGURE 4.17**

The energy expenditure and gain in glycolysis. Notice that there is a "net profit" of two ATP and two NADH molecules per glucose molecule in glycolysis. Molecules listed by number are (1) fructose 1,6-diphosphate; (2) 1,3-diphosphoglyceric acid; and (3) 3-phosphoglyceric acid (see fig. 4.18).

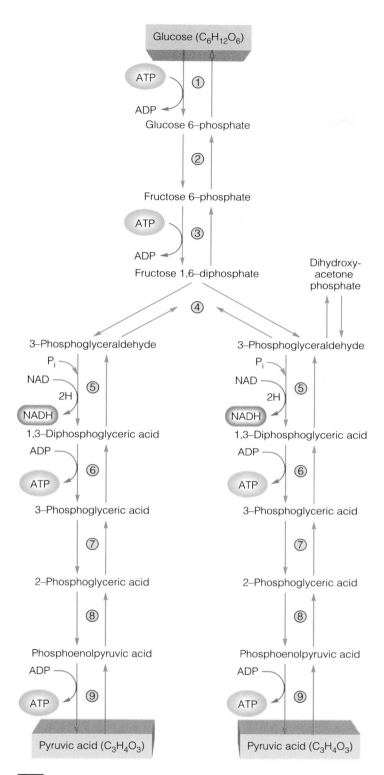

**FIGURE 4.18**

In glycolysis, one glucose molecule is converted into two pyruvic acid molecules in nine separate steps. In addition to two pyruvic acids, these products include two molecules of NADH and four molecules of ATP. Since two ATP molecules were used at the beginning of glycolysis, however, the net gain is two ATP molecules per glucose.

This results in the re-formation of NAD and the addition of two hydrogen atoms to pyruvic acid, which is thus reduced. This addition of two hydrogen atoms to pyruvic acid produces *lactic acid* (fig. 4.19). The metabolic pathway by which glucose is converted to lactic acid may be called **anaerobic respiration.** This term emphasizes the fact that ATP is produced in the absence of oxygen. (The metabolic pathway that produces lactic acid also may be called *lactic acid fermentation*, since it is similar to the pathway by which yeast cells produce alcohol.)

Anaerobic respiration yields a net gain of two ATP molecules (produced by glycolysis) per glucose molecule. A cell can survive anaerobically as long as it can produce sufficient energy for its needs in this way and as long as lactic acid concentrations do not become excessive. Some tissues are better adapted to anaerobic respiration than others; for example, skeletal muscles survive longer than cardiac muscle, which in turn can survive under anaerobic conditions longer than can the brain.

Except for red blood cells, which can respire only anaerobically (thus sparing the oxygen they carry), anaerobic respiration provides only a temporary sustenance for tissues that have energy requirements in excess of their aerobic ability. Anaerobic respiration can occur only for a limited period of time (longest for skeletal muscles, shorter for the heart, and shortest for the brain) when the *ratio of oxygen supply to oxygen need* (related to the concentration of NADH) falls below a critical level. Anaerobic respiration can be viewed, then, as an emergency procedure that provides some ATP until the emergency (oxygen deficiency) has passed. It should be noted, though, that there is no real "emergency" in the case of skeletal muscles, where anaerobic respiration is a normal occurrence that does not harm the tissue or the individual. Anaerobic respiration does not normally occur in the heart, however, and when it does it may signify a potentially dangerous condition.

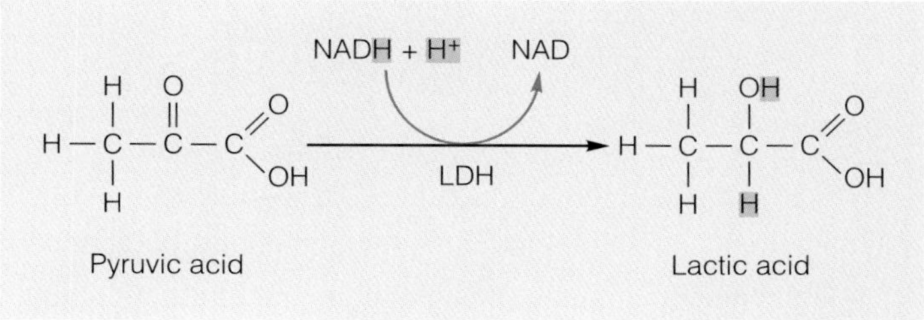

**FIGURE 4.19**

The addition of two hydrogen atoms (colored boxes) from reduced NAD to pyruvic acid produces lactic acid and oxidized NAD. This reaction is catalyzed by lactic acid dehydrogenase (LDH).

*Ischemia* refers to inadequate blood flow to an organ, such that the rate of oxygen delivery is insufficient to maintain aerobic respiration. Inadequate blood flow to the heart, or *myocardial ischemia*, may occur if the coronary blood flow is occluded by atherosclerosis, a blood clot, or by an artery spasm. People with myocardial ischemia often experience *angina pectoris*, severe pain in the chest and left (or sometimes, right) arm area. This pain is associated with increased blood levels of lactic acid, which are produced by anaerobic respiration by the heart muscle. The degree of ischemia and angina can be decreased by vasodilator drugs, such as nitroglycerin and amyl nitrite, which improve blood flow to the heart and also decrease the work of the heart by dilating peripheral blood vessels.

**Gluconeogenesis**   Some of the lactic acid produced by exercising skeletal muscles is delivered by the blood to the liver. The enzyme lactic acid dehydrogenase within liver cells is then able to convert lactic acid to pyruvic acid. In the process, NAD is reduced to NADH + H⁺. Unlike most other organs, the liver contains the enzymes needed to take pyruvic acid molecules and convert them to glucose 6-phosphate in a process that is essentially the reverse of glycolysis. Glucose 6-phosphate in liver cells then can be used as an intermediate for glycogen synthesis, or it can be converted to free glucose that is secreted into the blood.

The conversion of noncarbohydrate molecules (lactic acid, amino acids, and glycerol) through pyruvic acid into glucose is an extremely important process called **gluconeogenesis** (*gloo´´ko-ne´ŏ-jen´i-sis*). In starvation and in prolonged exercise, when glycogen stores are depleted, the formation of new glucose in this way becomes the only means for maintaining constant blood sugar levels. Under these conditions, gluconeogenesis in the liver is the only way that adequate blood glucose levels can be maintained to prevent brain death.

During exercise, some of the lactic acid produced by skeletal muscles may be transformed through gluconeogenesis in the liver to blood glucose. This new glucose can serve as an energy source during exercise, and can be used following exercise in order to replenish the depleted muscle glycogen. This two-way traffic between skeletal muscles and the liver is called the **Cori cycle** (fig. 4.20). Through the Cori cycle, gluconeogenesis in the liver allows depleted skeletal muscle glycogen to be restored within 48 hours.

# Aerobic Respiration

*In the aerobic respiration of glucose, pyruvic acid is formed by glycolysis and then converted into acetyl coenzyme A. This begins the cyclic metabolic pathway of the Krebs cycle. As a result of these pathways, a large number of reduced NAD and FAD molecules are generated. These reduced coenzymes provide electrons for an energy-generating process that drives the formation of ATP.*

Aerobic respiration is equivalent to combustion in terms of its final products ($CO_2$ and $H_2O$) and in terms of the total amount of energy liberated. In aerobic respiration, however, the energy is released in small, enzymatically controlled oxidation reactions, and a portion (38%–40%) of the energy released in this process is captured in the high-energy bonds of ATP.

The aerobic respiration of glucose begins with glycolysis. Glycolysis in both anaerobic and aerobic respiration results in the production of two molecules of pyruvic acid, two molecules of ATP, and two molecules of NADH + H⁺

..........................................

Cori cycle: from Carl F. Cori, American biochemist, 1896–1984
Krebs cycle: from Hans A. Krebs, German biochemist, 1900–1981

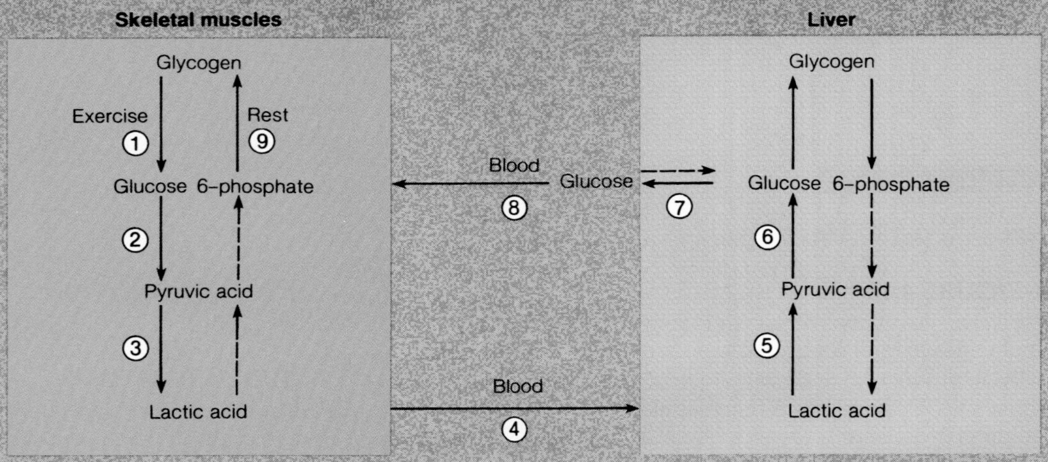

**FIGURE 4.20**
The Cori cycle. The direction of reversible reactions that occur in the Cori cycle is shown by solid arrows; the sequence of steps is indicated by numbers 1 through 9.

Skeletal muscles — Liver

Glycogen
Exercise ① / Rest ⑨
Glucose 6-phosphate ②
Pyruvic acid ③
Lactic acid

Blood — Glucose ⑧ ⑦
Blood ④

Glycogen
Glucose 6-phosphate ⑥
Pyruvic acid ⑤
Lactic acid

Chapter Four

per glucose molecule. In anaerobic respiration, the NADH becomes oxidized by the conversion of pyruvic acid to lactic acid in the cytoplasm. In aerobic respiration, however, the pyruvic acids will move to a different cellular location and undergo a different reaction; the NADH will eventually be oxidized, but that occurs later in the story.

In aerobic respiration, pyruvic acid leaves the cell cytoplasm and enters the interior (the matrix) of mitochondria. Once pyruvic acid is inside a mitochondrion, carbon dioxide is enzymatically removed from each three-carbon-long pyruvic acid to form a two-carbon-long organic acid—acetic acid. The enzyme that catalyzes this reaction combines the acetic acid with a coenzyme (derived from pantothenic acid) called *coenzyme A*. The combination produced is called **acetyl** (*as ˘e-tl*) **coenzyme A,** abbreviated **acetyl CoA** (fig. 4.21).

Glycolysis converts one glucose molecule into two molecules of pyruvic acid. Since each pyruvic acid molecule is converted into one molecule of acetyl CoA and one of $CO_2$, two molecules of acetyl CoA and two molecules of $CO_2$ are derived from each glucose. The acetyl CoA molecules serve as substrates for mitochondrial enzymes in the aerobic pathway; the $CO_2$ is a waste product in this process, which is carried by the blood to the lungs for elimination. (It should be noted that the oxygen in $CO_2$ is derived from pyruvic acid, not from oxygen gas.)

## The Krebs Cycle

Once acetyl CoA is formed, the acetic acid subunit (two carbons long) is combined with oxaloacetic acid (four carbons long) to form a molecule of citric acid (six carbons long). Coenzyme A acts only as a transporter of acetic acid from one enzyme to another (similar to the transport of hydrogen by NAD). The formation of citric acid begins a cyclic metabolic pathway known as the **citric acid cycle,** or **TCA cycle** (for tricarboxylic acid, citric acid having three carboxylic acid groups). Most commonly, however, this cyclic pathway is called the **Krebs cycle,** after its principal discoverer, Sir Hans Krebs. A simplified illustration of this pathway is shown in figure 4.22.

Through a series of reactions involving the elimination of two carbons and four oxygens (as two $CO_2$ molecules) and the removal of hydrogens, citric acid is eventually converted to oxaloacetic acid, which completes the cyclic metabolic pathway. In this process the following events occur: (1) one guanosine triphosphate (GTP) is produced (step 5 of figure 4.23), which donates a phosphate group to ADP to produce one ATP; (2) three molecules of NAD are

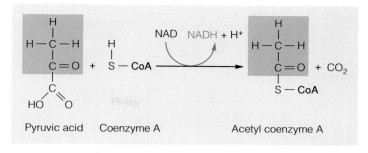

**FIGURE 4.21**

The formation of acetyl coenzyme A in aerobic respiration.

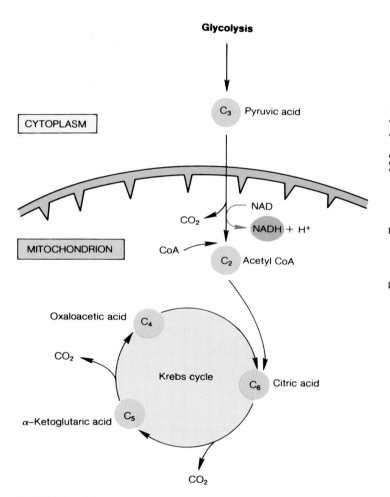

**FIGURE 4.22**

A simplified diagram of the Krebs cycle showing how the original four-carbon-long oxaloacetic acid is regenerated at the end of the cyclic pathway. Only the numbers of carbon atoms in the Krebs cycle intermediates are shown; the numbers of hydrogens and oxygens are not accounted for in this simplified scheme.

Enzymes, Energy, and Metabolism

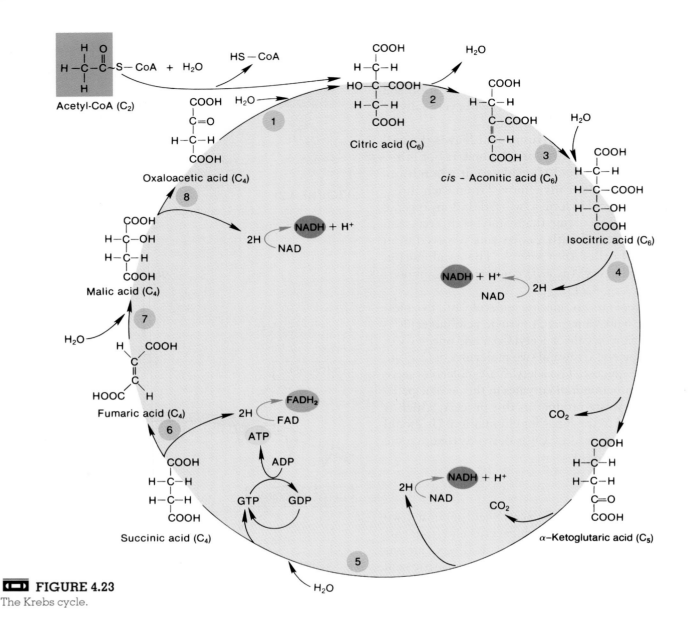

**FIGURE 4.23**
The Krebs cycle.

reduced (steps 4, 5, and 8 of figure 4.23); and (3) one molecule of FAD is reduced (step 6).

The production of reduced NAD and FAD (that is, NADH and $FADH_2$) by each turn of the Krebs cycle is far more significant, in terms of energy production, than the single GTP (converted to ATP) produced directly by the cycle. This is because NADH and $FADH_2$ eventually donate their electrons to an energy-generating process that results in the formation of many molecules of ATP.

## Electron Transport and Oxidative Phosphorylation

Built into the foldings, or cristae, of the inner mitochondrial membrane are a series of molecules that serve in **electron transport** during aerobic respiration. This electron-transport chain of molecules consists of a flavoprotein (derived from riboflavin), coenzyme Q (derived from vitamin E), and a group of iron-containing pigments called *cytochromes*. The last of these cytochromes is cytochrome $a_3$, which donates electrons to oxygen in the final oxidation-reduction reaction (to be described shortly). These molecules of the electron-transport system are fixed in position within the inner mitochondrial membrane in such a way that they can pick up electrons from NADH and $FADH_2$ and transport them in a definite sequence and direction.

In aerobic respiration, NADH and $FADH_2$ become oxidized by transferring their pairs of electrons to the electron-transport system of the cristae. The protons ($H^+$) are not transported together with the electrons; their fate will be described a little later. The oxidized forms of NAD and FAD are thus regenerated and can continue to shuttle electrons from the Krebs cycle to the electron-transport chain. The first molecule of the electron-transport chain becomes reduced when it accepts the electron pair from NADH. When the cytochromes receive a pair of electrons, two ferric ions ($Fe^{+++}$) become reduced to two ferrous ions ($Fe^{++}$). (Notice that the gain of an electron is indicated by the reduction of the number of positive charges.)

The electron-transport chain thus acts as an oxidizing agent for NAD and FAD. Each element in the chain, however, also functions as a reducing agent; one reduced cytochrome transfers its electron pair to the next cytochrome in the chain (fig. 4.24). In this way, the iron ions in each cytochrome alternately become reduced (to ferrous ions) and oxidized (to ferric ions). This is an exergonic process, and the energy derived is used to phosphorylate ADP to ATP. Thus, the production of ATP in this manner is appropriately termed **oxidative phosphorylation** (*ok˝sĭ-da´tiv fos˝for-ĭ-la´shun*).

**Function of Oxygen**   If the last cytochrome remained in a reduced state, it would be unable to accept more electrons. Electron transport would then progress only to the next-to-last cytochrome. This process would continue until all of the elements of the electron-transport chain remained in the reduced state. At this point, the electron-transport system would stop functioning and no ATP could be produced within the mitochondria. With the electron-transport system incapacitated, NADH and $FADH_2$ could not become oxidized by donating their electrons to the cytochrome chain, and through inhibition of Krebs cycle enzymes, no more NADH and $FADH_2$ could be produced in the mitochondria. With cessation of the Krebs cycle, respiration would become anaerobic.

Oxygen, from the air we breathe, allows electron transport to continue by functioning as the **final electron acceptor** of the electron-transport chain. This oxidizes cytochrome $a_3$ so that electron transport and oxidative phosphorylation can continue. At the very last step of aerobic respiration, therefore, oxygen becomes reduced by the two electrons that were passed to the chain from NADH and $FADH_2$. This reduced oxygen binds two protons, and a molecule of water is formed. Since the oxygen atom is part of a molecule of oxygen gas ($O_2$), this last reaction can be shown as follows:

$$O_2 + 4e^- + 4H^+ \rightarrow 2H_2O$$

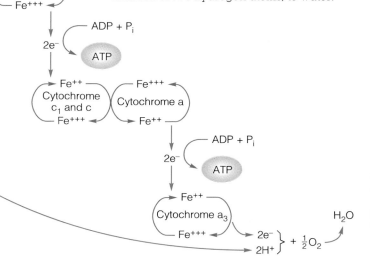

**FIGURE 4.24**
Electron transport and oxidative phosphorylation. Each element in the electron-transport chain alternately becomes reduced and then oxidized as it transports electrons to the next member of the chain. This process provides energy for the formation of ATP. At the end of the electron-transport chain, the electrons are donated to oxygen, which becomes reduced (by the addition of two hydrogen atoms) to water.

*Cyanide* is a fast-acting lethal poison that produces symptoms of rapid heart rate, hypotension, coma, and ultimately death in the absence of quick treatment. The reason cyanide is so deadly is that it has one, very specific action: it blocks the transfer of electrons from cytochrome $a_3$ to oxygen. The effects are thus the same as would occur if oxygen were completely removed—aerobic cell respiration and the production of ATP by oxidative phosphorylation come to a halt.

## ATP Balance Sheet

Each time the Krebs cycle turns, three molecules of NAD are reduced by electrons from three pairs of hydrogens removed from Krebs cycle intermediates. Each NADH donates a pair of electrons to the electron-transport chain. Transport of this pair of electrons to oxygen generates energy for the production of three molecules of ATP through oxidative phosphorylation. Electrons from $FADH_2$ enter the electron-transport chain "down the line" from where the first ATP is produced. Each pair of electrons from $FADH_2$, therefore, produces only two molecules of ATP from oxidative phosphorylation.

Since one NADH provides electrons for the production of three ATP, the three NADH produced per turn of the Krebs cycle results in the production of nine ATP molecules. The single $FADH_2$ per turn of the Krebs cycle results in the production of two ATP. Together with the single ATP made directly by the Krebs cycle, each turn of the Krebs

## Table 4.2  Maximum ATP yield per glucose in aerobic respiration*

| Phase of respiration | ATP produced directly | Reduced coenzymes | ATP from oxidative phosphorylation |
|---|---|---|---|
| Glycolysis (glucose to pyruvic acid) | 2 ATP | 2 NADH + H⁺ | 6 ATP |
| Pyruvic acid to acetyl CoA | — | 1 NADH + H⁺ | 3 ATP |
| Pyruvic acid to acetyl CoA | — | 1 NADH + H⁺ | 3 ATP |
| Krebs cycle | 1 ATP | 3 NADH + H⁺ <br> 1 FADH₂ | 9 ATP <br> 2 ATP |
| Krebs cycle | 1 ATP | 3 NADH + H⁺ <br> 1 FADH₂ | 9 ATP <br> 2ATP |

*One glucose molecule yields two pyruvic acid molecules, and each of these results in two turns of the Krebs cycle.

cycle thus yields a total of twelve ATP molecules. Since one molecule of glucose produces two pyruvic acids, and thus two turns of the Krebs cycle, a total of twenty-four ATP molecules are produced by a single molecule of glucose, taking into account only the ATP made by the Krebs cycle and its NADH and FADH₂ products.

The conversion of pyruvic acid to acetyl CoA, however, also involves the reduction of one NAD (see fig. 4.21). Since two pyruvic acids are produced per glucose, two NADH are formed. And, since each NADH molecule yields three ATP molecules by oxidative phosphorylation, a total of twenty-four plus six, or thirty, ATP molecules are made in the mitochondrion from the steps that occur beyond pyruvic acid.

Now recall that two molecules of NADH are produced in the cytoplasm during glycolysis (conversion of glucose to pyruvic acid). These NADH cannot directly enter the mitochondria; instead, they donate their electrons to other molecules that "shuttle" these electrons into the mitochondria. Depending upon which shuttle is used, either two or three ATP can be produced from each of these cytoplasmic NADH electrons through oxidative phosphorylation. A total of four or six ATP molecules are thus produced. Added to the thirty ATP previously mentioned, this brings the total to thirty-four or thirty-six (depending upon which shuttle is used for the cytoplasmic electrons). Now, when we add the two molecules of ATP produced directly by glycolysis, a grand total of thirty-six to thirty-eight ATP are produced by the aerobic respiration of glucose (table 4.2).

## Energy from Glycogen, Fat, and Protein

Excess ATP is not stored in cells. If more energy (measured in kilocalories) is ingested as food than is needed for the production of ATP, the excess energy is stored primarily as glycogen and fat. Since glycogen is a polysaccharide of glucose, the conversion of glucose to glycogen is relatively straightforward. Glucose can also be changed into fat, using the intermediates of phosphoglyceraldehyde and acetyl CoA (fig. 4.25).

During periods of fasting or exercise, the energy stored as glycogen and fat is available for the production of ATP. The general pathways by which these molecules can fit into the scheme for cell respiration are summarized in figure 4.25. Fatty acids, in particular, are a rich source of energy that can be used to generate many molecules of acetyl CoA, thus initiating many Krebs cycles and producing large amounts of ATP through the electron-transport chain. Proteins can also be metabolized for energy once their amine groups are removed (the amines are incorporated into molecules of urea). Some of the intermediate steps in cell respiration are therefore shared by carbohydrates, lipids, and proteins. These shared intermediates provide a framework by which these different classes of molecules can be interconverted, as shown by the reversible arrows in figure 4.25.

The storage and utilization of carbohydrates, lipids, and proteins for energy are regulated according to the needs of the body by various hormones. A more detailed accounting of the reactions involved, including the hormonal regulation of metabolism, is provided in chapter 27, following a discussion of the digestive system.

## Clinical Considerations

### Clinical Enzyme Measurements

**Assays of Enzymes in Plasma**  When tissues become damaged as a result of diseases, some of the dead cells disintegrate and release their enzymes into the blood. Most of

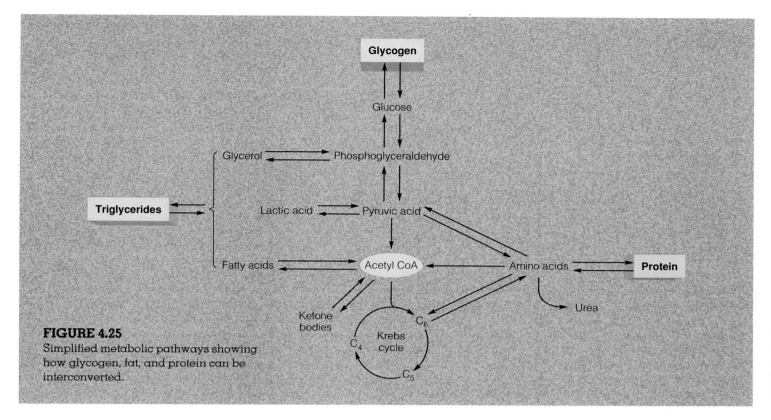

**FIGURE 4.25**
Simplified metabolic pathways showing how glycogen, fat, and protein can be interconverted.

these enzymes are not normally active in the blood because their specific substrates are not available, but their enzymatic activity can be measured in a test tube by the addition of the appropriate substrates to samples of plasma. Such measurements are clinically useful because abnormally high plasma concentrations of particular enzymes are characteristic of certain diseases (table 4.3).

**Identification of Isoenzymes** Different organs, when they are diseased, may liberate different isoenzymatic forms of an enzyme that can be measured in a clinical laboratory. For example, the enzyme **creatine phosphokinase** (abbreviated either **CPK** or **CK**) exists in three isoenzymatic forms. These forms are identified by two letters that indicate two components of this enzyme. One form is identified as MM and is liberated from diseased skeletal muscle; the second is BB, released by a damaged brain; and the third is MB, released from a diseased heart. Newer clinical tests utilize antibodies that can bind to the M and B components to measure only the level of the MB form in the blood when heart disease is suspected. Another enzyme commonly tested in clinical laboratories is **lactate dehydrogenase (LDH),** which is five isoenzymatic forms.

| Table 4.3 | Diseases associated with abnormal plasma concentrations of selected enzymes |
|---|---|
| **Enzyme** | **Associated disease(s)** |
| Alkaline phosphatase | Obstructive jaundice, Paget's disease (osteitis deformans), carcinoma of bone |
| Acid phosphatase | Benign hypertrophy of prostate, cancer of prostate |
| Amylase | Pancreatitis, perforated peptic ulcer |
| Aldolase | Muscular dystrophy |
| Creatine kinase (or creatine phosphokinase-CPK) | Muscular dystrophy, myocardial infarction |
| Lactate dehydrogenase (LDH) | Myocardial infarction, liver disease, renal disease, pernicious anemia |
| Transaminases (GOT and GPT) | Myocardial infarction, hepatitis, muscular dystrophy |

## Metabolic Disturbances

**Phenylketonuria (PKU)** The branched metabolic pathway that begins with phenylalanine as the initial substrate is subject to a number of inborn errors of metabolism (fig. 4.26). When the enzyme that converts this amino acid to tyrosine is defective, the final products of a divergent pathway accumulate and can be detected in the blood and urine. This disease, *phenylketonuria* (fen´´il-kēt´´n-oor´e-ă) (*PKU*), can result in severe mental retardation and a shortened life span. Although inborn errors of metabolism are relatively rare, the incidence of PKU is high enough and the defect is so easy to detect that all newborn babies are routinely tested for it. If this disease is detected early, brain damage can be prevented by placing the child on an artificial diet low in the amino acid phenylalanine.

**Albinism and Other Defects**   One of the conversion products of phenylalanine is a molecule called *DOPA*, which is an acronym for dihydroxyphenylalanine (fig. 4.26). DOPA is a precursor of the pigment molecule melanin, which gives skin, eyes, and hair their normal coloration. An inherited defect in the enzyme that catalyzes the formation of melanin from DOPA results in the lack of normal pigmentation that characterizes an *albino* (al-bi´no). Besides PKU and albinism, there are many other inherited defects of amino acid metabolism, as well as inborn errors in the metabolism of carbohydrates and lipids.

## Endocrine Disorders and Metabolism

Since metabolism is regulated largely by hormones (chemicals secreted by endocrine glands into the blood), endocrine diseases can produce metabolic disorders. *Diabetes mellitus*—characterized by high blood glucose and the presence of

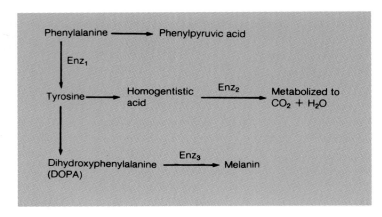

**FIGURE 4.26**

Metabolic pathways for the degradation of the amino acid phenylalanine. Defective enzyme₁ produces phenylketonuria (PKU), defective enzyme₂ produces alcaptonuria (not a clinically significant condition), and defective enzyme₃ produces albinism.

glucose in the urine—for example, results from the inadequate secretion or action of the hormone insulin. This disease may also be associated with the excessive production of ketone bodies, which can alter blood pH and produce *ketoacidosis*. Abnormally low blood glucose, *hypoglycemia*, may be produced by excessive insulin secretion. Other metabolic disorders can result from diseases of the pituitary, thyroid, and adrenal glands, as described in chapter 27.

## Chapter Summary

### Enzymes as Catalysts (pp. 70–72)

1. Enzymes are biological catalysts, acting to increase the rate of chemical reactions.
2. Most enzymes are proteins, and the tertiary structures of proteins grant specificity to the actions of the enzymes.
3. Substrates are the reactant molecules that fit into the active sites of an enzyme. The process by which the enzyme-substrate products are formed leaves the enzyme unaltered, so that it is able to act again.

### Control of Enzyme Activity (pp. 72–75)

1. The activity of an enzyme is affected by a variety of factors, including temperature and pH.
   a. Enzymes function best at one pH, called their pH optimum.
   b. At too high a temperature, the enzyme proteins denature and no longer function effectively.

2. Metabolic pathways involve a number of enzyme-catalyzed reactions, in which enzymes cooperate in a stepwise fashion.
   a. The product of one enzyme becomes the substrate of the next enzyme in the pathway.
   b. If an enzyme is defective, its product is not made; the products formed prior to that step and at branch points in the pathway may accumulate.

### Bioenergetics (pp. 75–79)

1. Reactions that liberate energy may be coupled to those that require energy.
2. ATP is the universal energy carrier of the cells.
   a. Exergonic reactions provide the energy for the formation of ATP, in which some of the liberated energy is trapped in the bond formed between ADP and the last phosphate.

   b. The hydrolysis of ATP provides the energy that powers all of the energy needs of the cells.
3. Oxidation-reduction reactions are coupled and usually involve the transfer of hydrogen atoms.
   a. Atoms or molecules that gain hydrogens (or electrons) are reduced; those that lose hydrogens (or electrons) are oxidized.
   b. In many oxidation-reduction reactions, pairs of electrons are transferred from the reducing agent to the oxidizing agent.

### Glycolysis and Anaerobic Respiration (pp. 79–82)

1. Glycolysis refers to the metabolic pathway that converts glucose to two molecules of pyruvic acid.
   a. Two molecules of ATP are hydrolyzed in the process, but four molecules of ATP are produced, for a net gain of two ATP.

b. Two molecules of the oxidized form of NAD are reduced to NADH + H⁺ during glycolysis.

2. In anaerobic respiration, pyruvic acid is converted into lactic acid.
   a. The two NADH + H⁺ formed during glycolysis are oxidized to NAD when pyruvic acid is reduced to lactic acid.
   b. Anaerobic respiration often occurs during skeletal muscle contraction; the lactic acid thus produced can cause muscle pain and fatigue.

## Aerobic Respiration (pp. 82–86)

1. Pyruvic acid enters into a mitochondrion and, through the loss of carbon dioxide, is converted into a two-carbon molecule that binds to coenzyme A to form acetyl CoA.

2. The Krebs cycle begins when coenzyme A donates acetic acid to an enzyme reaction that combines it with oxaloacetic acid to form citric acid.
   a. As the reactions of the Krebs cycle proceed, carbon dioxide is lost, three NAD are converted to NADH + H⁺, and one FAD is reduced to FADH₂.
   b. Through an intermediate reaction, the Krebs cycle produces one ATP in addition to the NADH and FADH₂.

3. NADH and FADH₂ donate their electrons to an electron-transport chain of molecules, located in the cristae.
   a. Iron ions in the oxidized state ($Fe^{+++}$) gain an electron and are reduced to $Fe^{++}$. They then give up that electron as they pass it to the next molecule in the electron-transport chain.

b. The last cytochrome donates its electron to oxygen, which functions as the final electron acceptor. One atom of oxygen gains two electrons and two protons to form $H_2O$.

4. Electron transport provides energy for the formation of ATP.
   a. The coupling of electron transport to the production of ATP is called oxidative phosphorylation.
   b. The aerobic respiration of 1 glucose molecule can produce 38 ATP molecules. Of these, 2 are produced in the cytoplasm by glycolysis, 2 are produced through two turns of the Krebs cycle, and 34 are produced by oxidative phosphorylation.

# Review Activities

## Objective Questions

1. Which of the following statements about enzymes is *true?*
   a. All proteins are enzymes.
   b. All enzymes are proteins.
   c. Enzymes are changed by the reactions they catalyze.
   d. The active sites of enzymes have little specificity for substrates.

2. Which of the following statements about enzyme-catalyzed reactions is *true?*
   a. The rate of reaction is independent of temperature.
   b. The rate of all enzyme-catalyzed reactions is decreased when the pH is lowered from 7 to 2.
   c. The rate of reaction is independent of substrate concentration.
   d. Under given conditions of substrate concentration, pH, and temperature, the rate of product formation varies directly with enzyme concentration until a point is reached at which the rate cannot be further increased.

3. Which of the following statements about lactate dehydrogenase is *true?*
   a. It is a protein.
   b. It oxidizes lactic acid.
   c. It reduces another molecule (pyruvic acid).
   d. All of the above are true.

4. In an inborn error of metabolism,
   a. a genetic change results in the production of a defective enzyme.
   b. intermediates produced prior to the defective step accumulate.

c. alternate pathways are taken by intermediates at branch points that precede the defective step.
   d. All of the above are true.

5. Which of the following represents an endergonic reaction?
   a. $ADP + P_i \rightarrow ATP$
   b. $ATP \rightarrow ADP + P_i$
   c. glucose + $O_2 \rightarrow CO_2 + H_2O$
   d. $CO_2 + H_2O \rightarrow$ glucose
   e. both a and d
   f. both b and c

6. Which of the following statements about ATP is *true?*
   a. The bond joining ADP and the third phosphate is a high-energy bond.
   b. The formation of ATP is coupled to energy-liberating reactions.
   c. The conversion of ATP to ADP and $P_i$ provides energy for biosynthesis, cell movement, and other cellular processes that require energy.
   d. ATP is the universal energy carrier of cells.
   e. All of the above are true.

7. When oxygen is combined with two hydrogens to make water,
   a. oxygen is reduced.
   b. the molecule that donated the hydrogens becomes oxidized.
   c. oxygen acts as a reducing agent.
   d. both a and b apply.
   e. both a and c apply.

8. The net gain of ATP molecules per glucose molecule in anaerobic respiration is _____ ; in aerobic respiration it is _____ .
   a. 2;4
   b. 2;38
   c. 38;2
   d. 24;30

9. In anaerobic respiration, the oxidizing agent for NADH (that is, the molecule that removes electrons from NADH) is
   a. pyruvic acid.
   b. lactic acid.
   c. citric acid.
   d. oxygen.

10. When organs respire anaerobically, there is an increased blood concentration of
    a. oxygen.
    b. glucose.
    c. lactic acid.
    d. ATP.

11. The conversion of lactic acid to pyruvic acid occurs in
    a. anaerobic respiration.
    b. the heart, where lactic acid is aerobically respired.
    c. the liver, where lactic acid can be converted to glucose.
    d. both a and b.
    e. both b and c.

12. The oxygen in the air we breathe
    a. functions as the final electron acceptor of the electron-transport chain.
    b. combines with hydrogen to form water.
    c. combines with carbon to form $CO_2$.
    d. both a and b apply.
    e. both a and c apply.
13. In terms of the number of ATP molecules directly produced, the major energy-yielding process in the cell is
    a. glycolysis.
    b. the Krebs cycle.
    c. oxidative phosphorylation.
    d. gluconeogenesis.

14. Which of the following organs has an almost absolute requirement for blood glucose as its energy source?
    a. liver
    b. brain
    c. skeletal muscles
    d. heart

## Essay Questions

1. Explain the relationship between the chemical structure and the function of an enzyme and describe how various conditions may alter both the enzyme's structure and function.

2. Discuss the advantages and disadvantages of anaerobic respiration.

3. What purpose is served by the formation of lactic acid during anaerobic respiration? How is this purpose achieved during aerobic respiration?

4. Identify the products of the Krebs cycle and discuss the significance of this cyclic metabolic pathway.

5. Describe the effect of cyanide on oxidative phosphorylation and on the Krebs cycle. Explain why this poison is deadly.

# chapter five

# membrane transport and the membrane potential

## objectives

- Describe how nonpolar molecules and small inorganic ions penetrate the cell membrane and explain how net diffusion occurs.

- Define *osmosis* and describe how the osmotic pressure of solutions affects the direction of osmosis.

- Discuss the significance of osmolality measurements and define the terms *isotonic, hypertonic,* and *hypotonic solutions.*

- Describe the mechanisms that help to maintain a constant plasma osmolality.

- Explain the characteristics of carrier-mediated transport.

- Distinguish between simple diffusion and facilitated diffusion and explain why both are passive transport processes.

- Distinguish between facilitated diffusion and active transport and describe the characteristics of active transport.

- Distinguish between primary and secondary active transport.

- Define the term *membrane potential* and explain how it is determined by the permeability characteristics of the membrane.

- Explain why the true membrane potential is close to, but less than, the theoretical potassium equilibrium potential.

- Describe the role of the $NA^+/K^+$ pumps in the establishment of the membrane potential.

# Diffusion and Osmosis

*Net diffusion of a molecule or ion through a cell membrane always occurs in the direction of its lower concentration. Nonpolar molecules can penetrate the phospholipid barrier and small inorganic ions can pass through channels in the membrane. The net diffusion of water through a membrane is known as osmosis.*

The cell (plasma) membrane separates the intracellular environment from the extracellular environment. Proteins, nucleotides, and other molecules needed for the structure and function of the cell cannot penetrate, or "permeate," the membrane. The cell membrane is, however, **selectively permeable** to certain molecules and many ions. This allows for two-way traffic in nutrients and wastes needed to sustain metabolism and provides electrical currents created by the movements of ions through the membrane.

The mechanisms involved in the transport of molecules and ions through the cell membrane may be divided into two broad categories: (1) transport that requires the action of specific *carrier proteins* in the membrane (*carrier-mediated transport*) and (2) transport through the membrane that is not carrier mediated. Carrier-mediated transport includes *facilitated diffusion* and *active transport*; non-carrier-mediated transport refers to the *simple diffusion* of ions, lipid-soluble molecules, and water through the membrane. The diffusion of water (solvent) through a membrane is called *osmosis (oz-mo´sis)*.

Membrane transport processes may also be categorized on the basis of their energy requirements. Transport in which the net movement is from higher to lower concentration (down a concentration gradient) does not require metabolic energy and is known as **passive transport.** Passive transport includes simple diffusion, osmosis, and facilitated diffusion. Transport that occurs against a concentration gradient (through a membrane to the region of higher concentration) is **active transport.** Active transport requires the expenditure of metabolic energy (ATP) and, as mentioned previously, involves specific carrier proteins.

## Diffusion

Molecules in a gas, as well as molecules and ions dissolved in a solution, are in a constant state of random motion as a result of their thermal (heat) energy. This random motion, called **diffusion,** tends to scatter the molecules evenly, or diffusely, within a given volume of gas or solution. Therefore, whenever a *concentration difference*, or *concentration gradient*, exists between two regions of a solution, random molecular motion tends to eliminate the gradient and to distribute the molecules uniformly (fig. 5.1).

As a result of random molecular motion, molecules in the area of the solution with a higher concentration will enter the area of lower concentration. Molecules will also move in the opposite direction, but not as frequently. As a result, there will be a *net movement* from the region of higher to the region of lower concentration until the concentration difference is eliminated. This net movement is called **net diffusion.** Net diffusion is a physical process that occurs whenever there is a concentration difference. When the concentration difference exists across a membrane, diffusion becomes a type of membrane transport.

## Diffusion through the Cell Membrane

Since the cell membrane consists primarily of a double layer of phospholipids, molecules that are nonpolar (and thus lipid-soluble) can easily pass from one side of the membrane to the other. The cell membrane, in other words, does not present a barrier to the diffusion of nonpolar molecules such as oxygen gas ($O_2$) or steroid hormones. Small organic

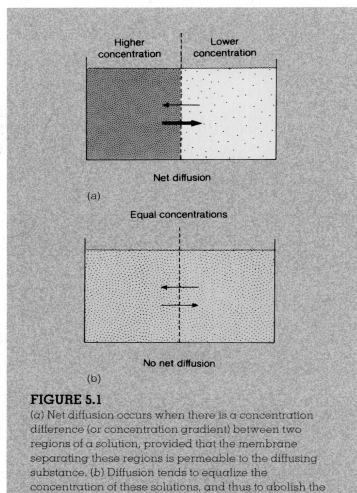

**Higher concentration**     **Lower concentration**

**Net diffusion**

(a)

**Equal concentrations**

**No net diffusion**

(b)

**FIGURE 5.1**

(*a*) Net diffusion occurs when there is a concentration difference (or concentration gradient) between two regions of a solution, provided that the membrane separating these regions is permeable to the diffusing substance. (*b*) Diffusion tends to equalize the concentration of these solutions, and thus to abolish the concentration differences.

molecules that have polar covalent bonds but that are uncharged, such as carbon dioxide ($CO_2$), ethanol, and urea, are also able to penetrate the phospholipid bilayer. Net diffusion of these molecules readily occurs between the intracellular and extracellular compartments when concentration gradients exist.

The oxygen concentration is relatively high, for example, in extracellular fluid because $O_2$ is carried by blood from the lungs to the body tissues. Since $O_2$ is converted to water in aerobic cell respiration, the $O_2$ concentration within the cells is lower than in the extracellular fluid. The concentration gradient for $CO_2$ is in the opposite direction because cells produce $CO_2$. *Gas exchange* occurs by diffusion between the tissue cells and their extracellular environments (fig. 5.2).

Larger polar molecules, such as glucose, cannot pass through the phospholipid bilayer of the membrane, and thus require special *carrier proteins* in the membrane for transport (to be described later). The phospholipid portion of the membrane is similarly impermeable to charged inorganic ions, such as $Na^+$ and $K^+$. Passage of these ions through the cell membrane may take place by means of tiny **ion channels** through the membrane that are too small to be seen even with an electron microscope. These channels are provided by some of the proteins that span the thickness of the membrane (fig. 5.3).

## Rate of Diffusion

The rate of diffusion, measured by the number of diffusing molecules passing through the membrane per unit time, depends on (1) the magnitude of the concentration difference across the membrane (the steepness of the concentration gradient), (2) the degree of permeability of the membrane, and (3) the extent of the surface area of the membrane through which the substances are diffusing.

The magnitude of the concentration difference across the membrane serves as the driving force for diffusion. Regardless of this concentration difference, however, the diffusion of a substance across a membrane will not occur if the membrane is not permeable to that substance. With a given concentration difference, the rate of diffusion through a membrane will vary directly with the degree of permeability. In a resting neuron, for example, the membrane is about 20 times more permeable to $K^+$ than to $Na^+$ and, as a consequence, $K^+$ diffuses much more rapidly than does $Na^+$. Changes in the structure of the membrane channels, how-

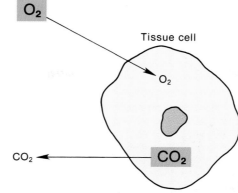

**FIGURE 5.2**

Gas exchange between the intracellular and extracellular compartments occurs by diffusion. The regions of higher concentration are represented by the larger symbols.

ever, can change the permeability of the membrane. This occurs during the production of a nerve impulse (chapter 14), when specific stimulation opens $Na^+$ channels temporarily, resulting in a faster diffusion rate for $Na^+$ than for $K^+$.

In areas of the body that are specialized for rapid diffusion, the surface area of the cell membranes may be increased by numerous folds. The rapid passage of the products of digestion across the epithelial membranes in the small intestine, for example, is aided by such structural adaptations. The surface area of the apical membranes (the part facing the lumen) in the small intestine is increased by many tiny folds that form fingerlike projections called *microvilli* (discussed in chapter 3). Similar microvilli are also found in the epithelium of the kidney tubules, where various molecules that are filtered out of the blood must be reabsorbed.

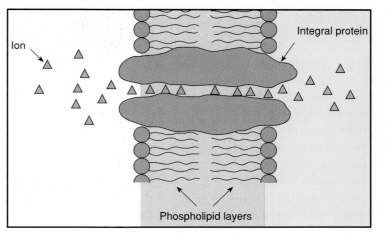

**FIGURE 5.3**

Inorganic ions (such as $Na^+$ and $K^+$) may penetrate the membrane through pores within integral proteins that span the thickness of the double phospholipid layers.

## Osmosis

**Osmosis** is the net diffusion of water (the solvent) across the membrane. In order for osmosis to occur, the membrane must be *semipermeable*; that is, it must be more permeable to water molecules than to solutes. Like the

diffusion: L. *dis-*, apart; *fundere,* to pour

microvilli: Gk. *mikros*, small; L. *villus*, shaggy hair
osmosis: Gk. *osmos*, a thrust

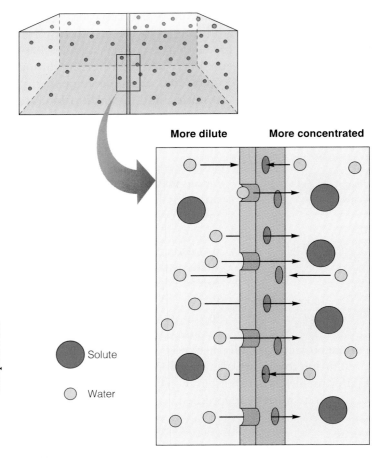

**FIGURE 5.4**

A model of osmosis, or the net movement of water from the solution of lesser solute concentration to the solution of greater solute concentration.

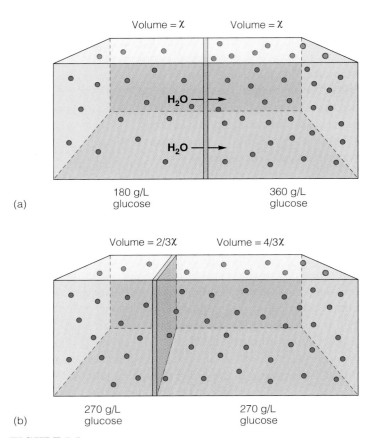

**FIGURE 5.5**

(a) A movable semipermeable membrane (permeable to water but not to glucose) separates two solutions of different glucose concentration. As a result, water moves by osmosis into the solution of greater concentration until (b) the volume changes equalize the concentrations on both sides of the membrane.

diffusion of solute molecules, the diffusion of water occurs when the water is more concentrated on one side of the membrane than on the other side; that is, when one solution is more dilute than the other (fig. 5.4). The more dilute solution has a higher concentration of water molecules and a lower concentration of solute. The principles of osmosis apply to the diffusion of any molecule, but the terminology is backwards because the term *concentration* more frequently is used with reference to the density of solute rather than of solvent molecules.

Imagine a cylinder divided into two equal compartments by a membrane partition that can freely move. One compartment of the cylinder initially contains 180 g/L (grams per liter) of glucose and the other compartment contains 360 g/L of glucose. If the membrane is permeable to glucose, glucose will diffuse from the 360-g/L compartment to the 180-g/L compartment until both compartments contain 270 g/L of glucose. If the membrane is not permeable to glucose but is permeable to water, the same result (270-g/L solutions on both sides of the membrane) is obtained by the diffusion of water. As water diffuses from the 180-g/L compartment to the 360-g/L compartment, the former solution becomes more

concentrated while the latter becomes more dilute. This is accompanied by volume changes, as illustrated in figure 5.5.

 In order for osmosis to occur between two solutions, the two solutions must have different concentrations, and a membrane must be relatively impermeable to the solutes producing the differences in concentration. Those solutes that cannot pass through the membrane are said to be *osmotically active*. Water, for example, returns from tissue fluid to blood capillaries because the protein concentration of blood plasma is higher than the protein concentration of tissue fluid. This is because the plasma proteins, in contrast to other plasma solutes, cannot pass from the capillaries into the tissue fluid. The plasma proteins, in this case, are osmotically active. When clinicians want to expand a patient's blood volume (to raise the blood pressure), they give intravenous infusions of an albumin solution or of plasma, which contains albumin and other proteins. If a person has an abnormally low concentration of plasma proteins, as may occur in liver disease (*cirrhosis*, for example), fluid may accumulate in the tissues and produce edema.

**Osmotic Pressure**  Osmosis and the movement of the membrane partition could be prevented by an opposing force. If one compartment contained 180 g/L of glucose and the other compartment contained pure water, the osmosis of water into the glucose solution could be prevented by pushing against the membrane with a certain force. This is illustrated in figure 5.6.

The force that would have to be exerted to prevent osmosis in the situation just described is the **osmotic pressure** of the solution. This indirect measurement indicates how strongly the solution draws water into it by osmosis. The greater the solute concentration of a solution, the greater its osmotic pressure. Pure water has an osmotic pressure of zero, and a 360-g/L glucose solution has twice the osmotic pressure of a 180-g/L glucose solution.

**Molarity and Molality**   Glucose is a monosaccharide with a molecular weight of 180 (the sum of its atomic weights). Sucrose is a disaccharide of glucose and fructose, which have

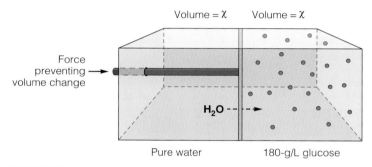

**FIGURE 5.6**

If a semipermeable membrane separates pure water from a 180-g/L glucose solution, water tends to move by osmosis into the glucose solution, thus creating a hydrostatic pressure that pushes the membrane to the left and expands the volume of the glucose solution. The amount of pressure that must be applied to just counteract this volume change is equal to the osmotic pressure of the glucose solution.

molecular weights of 180 each. When glucose and fructose join together by dehydration synthesis to form sucrose (chapter 2), a molecule of water (with a molecular weight of 18) is split off. Therefore, sucrose has a molecular weight of 342 (180 + 180 − 18). Since the molecular weights of sucrose and glucose are in a ratio of 342/180, it follows that 342 grams of sucrose must contain *the same number of molecules* as 180 grams of glucose.

Notice that an amount of any compound equal to its molecular weight in grams must contain the same number of molecules as an amount of any other compound equal to its molecular weight in grams. This unit of weight is called a *mole*, and it always contains $6.02 \times 10^{23}$ molecules (*Avogadro's number*). One mole of solute dissolved in water to make 1 liter of solution is described as a **one-molar solution** (abbreviated 1.0 M). Although this unit of measurement is commonly used in chemistry, it is not altogether desirable in discussions of osmosis because the exact ratio of solute to solvent (water) is not specified. For example, more water is needed to make a 1.0 M NaCl solution (where a mole of NaCl weighs 58.5 grams) than is needed to make a 1.0 M glucose solution because 180 grams of glucose takes up more volume than 58.5 grams of salt.

Since the ratio of solute to water molecules is of critical importance in osmosis, a more useful measurement of concentration is **molality.** In a one-molal solution (abbreviated 1.0 *m*), 1 mole of solute (180 grams of glucose, for example) is dissolved in 1 kilogram of water (equal to one liter at 4° C). A 1.0 *m* NaCl solution and a 1.0 *m* glucose solution both contain a mole of solute dissolved in exactly the same amount of water (fig. 5.7).

Avogadro's number: from Amadeo Avogadro, Italian chemist and physicist, 1766–1856

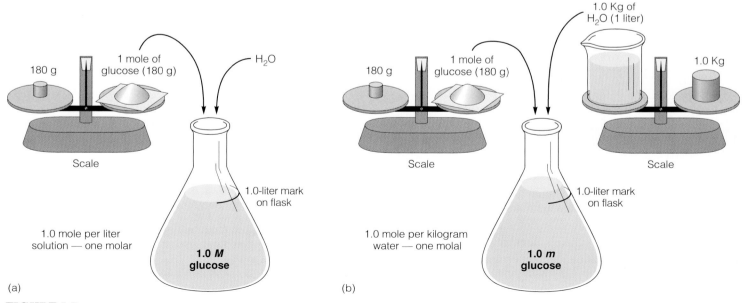

(a)                    (b)

**FIGURE 5.7**

Diagrams illustrating the difference between (*a*) a one-molar (1.0 *M*) and (*b*) a one-molal (1.0 *m*) glucose solution.

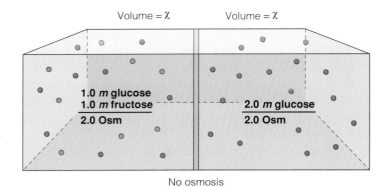

No osmosis

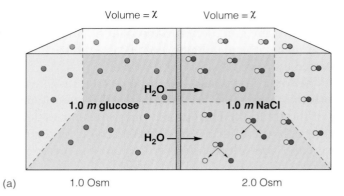

(a)  1.0 Osm                    2.0 Osm

**FIGURE 5.8**

The osmolality (Osm) of a solution is equal to the sum of the molalities of each solute in the solution. If a semipermeable membrane separates two solutions with equal osmolalities, no osmosis will occur.

**Osmolality**   If 180 grams of glucose and 180 grams of fructose were dissolved in the same kilogram of water, the osmotic pressure of the solution would be the same as that of a 360-g/L glucose solution. Osmotic pressure depends on the ratio of solute to solvent, *not* on the chemical nature of the solute molecules. The expression for the total molality of a solution is **osmolality (Osm).** Thus, the solution of 1.0 *m* glucose plus 1.0 *m* fructose has a total molality, or osmolality, of 2.0 osmol/L (abbreviated 2.0 Osm). This is the same as the 360-g/L glucose solution, which is 2.0 *m* and 2.0 Osm (fig. 5.8).

Unlike glucose, fructose, and sucrose, electrolytes such as salt (NaCl) ionize when dissolved in water. One molecule of NaCl dissolved in water yields two ions (Na⁺ and Cl⁻); 1 mole of NaCl ionizes to form 1 mole of Na⁺ and 1 mole of Cl⁻. Thus, a 1.0 *m* NaCl solution has a total concentration of 2.0 Osm. The effect of this on osmosis is illustrated in figure 5.9.

**Tonicity**   A 0.3 *m* glucose solution, which is 0.3 Osm, or 300 milliosmolal (300 mOsm), has the same osmolality and osmotic pressure as blood plasma. The same is true of a 0.15 *m* NaCl solution, which ionizes to produce a total concentration of 300 mOsm. Both of these solutions are used clinically as intravenous infusions, labeled 5% *dextrose* (5 g of glucose per 100 ml, which is 0.3 *m*) and *normal saline* (0.9 g of NaCl per 100 ml, which is 0.15 *m*). Since 5% dextrose and normal saline have the same osmolality as blood plasma, they are said to be **isosmotic** to plasma.

The term *tonicity* is used to describe the effect of a solution on the osmotic movement of water. For example, if an isosmotic glucose or saline solution is separated from blood plasma by a membrane that is permeable to water but

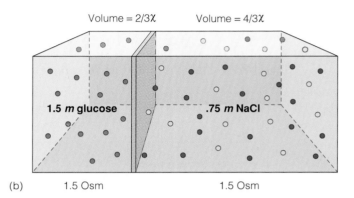

(b)    1.5 Osm                    1.5 Osm

**FIGURE 5.9**

(*a*) If a semipermeable membrane (permeable to water but not to glucose, Na⁺, or Cl⁻) separates a 1.0 *m* glucose solution from a 1.0 *m* NaCl solution, water will move by osmosis into the NaCl solution. This is because NaCl can ionize to yield one-molal Na⁺ plus one-molal Cl⁻. (*b*) After osmosis, the total concentration, or osmolality, of the two solutions is equal.

not to glucose or NaCl, osmosis will not occur. In this case, the solution is said to be **isotonic** to blood plasma. Red blood cells placed in an isotonic solution will neither gain nor lose water.

Solutions that have a lower total concentration of osmotically active solutes and a lower osmotic pressure than plasma are **hypotonic** to plasma. Red blood cells placed in hypotonic solutions gain water and may burst (*hemolysis*). When red blood cells are placed in a **hypertonic** solution (such as seawater), which has a higher osmolality and osmotic pressure than plasma, they shrink due to the osmosis of water out of the cells. In this process, called *crenation*, the cell surface takes on a scalloped appearance (fig. 5.10).

· · · · · · · · · · · · · · · · · · · · · · · · · · · · · · · · · · · · · · ·

isotonic: Gk. *isos*, equal; *tonus*, tension
hypotonic: Gk. *hypo*, under; *tonus*, tension
hypertonic: Gk. *hyper*, over; *tonus*, tension
crenation: L. *crena*, a notch

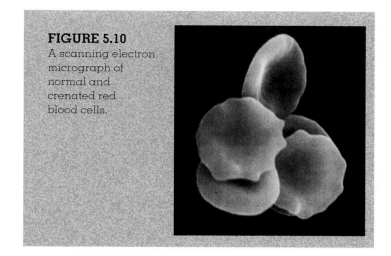

**FIGURE 5.10**
A scanning electron micrograph of normal and crenated red blood cells.

 Fluids delivered intravenously must be isotonic to blood in order to maintain the correct osmotic pressure and prevent cells from either expanding or shrinking due to the gain or loss of water. In addition to normal saline and 5% dextrose, another isotonic solution frequently used in hospitals is *Ringer's lactate*. This solution contains glucose and lactic acid as well as a number of different salts. Isotonic solutions are also used in artificial kidney machines and in heart-lung machines, which take the place of the heart and lungs during open-heart surgery.

## Regulation of Blood Osmolality

The osmolality of the blood plasma is normally maintained within very narrow limits by a variety of regulatory mechanisms. When a person becomes dehydrated, for example, the blood becomes more concentrated as the total blood volume is reduced. The increased blood osmolality and osmotic pressure stimulates *osmoreceptors*, which are neurons located in a part of the brain called the hypothalamus.

As a result of increased osmoreceptor stimulation, the person becomes thirsty and, if water is available, drinks. Along with increased water intake, a person who is dehydrated excretes a lower volume of urine. This occurs as a result of the following sequence of events: (1) increased plasma osmolality stimulates osmoreceptors in the hypothalamus of the brain; (2) the osmoreceptors stimulate the posterior pituitary gland, by means of a tract of nerve fibers, to secrete **antidiuretic** (*an´te-di´´yŭ-ret´ik*) **hormone (ADH);** and (3) ADH acts on the kidneys to promote water retention so that a lower volume of urine is excreted.

A person who is dehydrated, therefore, drinks more and urinates less. This represents a negative feedback loop (fig. 5.11) that acts to maintain homeostasis of the plasma concentration (osmolality) and, in the process, helps to maintain a proper blood volume.

....................................................
Ringer's lactate: from Sidney Ringer, English physiologist, 1835–1910

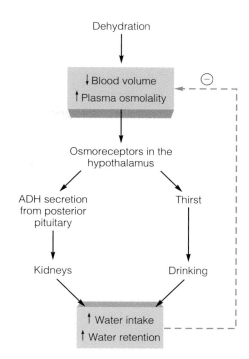

**FIGURE 5.11**
An increase in plasma osmolality (increased concentration and osmotic pressure) due to dehydration stimulates thirst and increased ADH secretion. These effects cause the person to drink more and urinate less. The blood volume, as a result, is increased while the plasma osmolality is decreased. These effects help bring the blood volume back to the normal range and complete the negative feedback loop (indicated by a negative sign).

# Carrier-Mediated Transport

*Molecules such as glucose are transported across the cell membranes by special protein carriers. Carrier-mediated transport in which the net movement is down a concentration gradient, and which is therefore passive, is called facilitated diffusion. Carrier-mediated transport that occurs against a concentration gradient, thus requiring metabolic energy, is called active transport.*

In order to sustain metabolism, cells must be able to take up glucose, amino acids, and other organic molecules from the extracellular environment. Molecules such as these, however, are too large and polar to pass through the lipid barrier of the cell membrane by simple diffusion. The transport of such molecules is mediated by **protein carriers** within the membrane. Although such carriers cannot be directly observed, their presence has been inferred by the observation that this transport has characteristics in common with enzyme activity. These characteristics include (1) *specificity*, (2) *competition*, and (3) *saturation*.

Like enzyme proteins, carrier proteins interact only with specific molecules. Glucose carriers, for example, can interact only with glucose and not with closely related

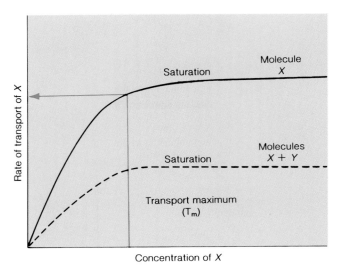

**FIGURE 5.12**
Carrier-mediated transport displays the characteristics of saturation (illustrated by the *transport maximum*) and competition. Molecules *X* and *Y* compete for the same carrier, so that when they are present together the rate of transport of each is lower than when either is present separately.

monosaccharides. As a further example of specificity, particular carriers for amino acids transport some types of amino acids but not others. Two amino acids that are transported by the same carrier compete with each other, so that the rate of transport of each is lower when they are present together than it would be if each were present alone (fig. 5.12).

As the concentration of a transported molecule increases, its rate of transport also increases—but only up to a maximum. Beyond this rate, called the *transport maximum* (or $T_m$), further increases in concentration do not further increase the transport rate. This indicates that the carriers have become saturated (fig. 5.12).

As an example of saturation, imagine a bus stop that is serviced once per hour by a bus that can hold a maximum of 40 people (its transport maximum is thus 40 per hour). If 10 people are waiting at the bus stop, 10 will be transported per hour. If 20 people are waiting at the bus stop, 20 will be transported per hour. This linear relationship will hold, up to a maximum of 40 people; if there are 80 people at the bus stop, the transport rate will still be 40 per hour.

 The kidneys transport a number of molecules from the blood filtrate, which will become urine, back into the blood. Glucose, for example, is normally completely reabsorbed so that urine is normally free of glucose. If the glucose concentration of the blood and filtrate is too high (a condition called *hyperglycemia*), however, the transport maximum will be exceeded. In this case, glucose will be found in the urine (a condition called *glycosuria*). This may result from eating too much sugar or from the inadequate secretion of the hormone *insulin* (in the disease *diabetes mellitus*).

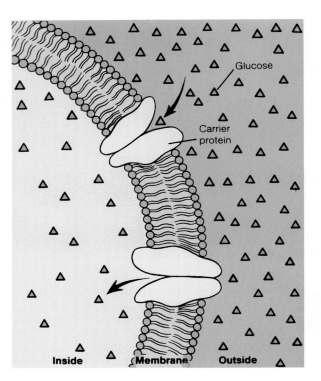

**FIGURE 5.13**
A model of facilitated diffusion in which a molecule is transported across the cell membrane by a carrier protein.

## Facilitated Diffusion

The transport of glucose from the blood across the cell membranes of tissue cells occurs by **facilitated diffusion.** Facilitated diffusion, like simple diffusion, is powered by the thermal energy of the diffusing molecules and involves the net transport of substances through a cell membrane from the side of higher concentration to the side of lower concentration. Active cellular metabolism is not required for either facilitated or simple diffusion.

Unlike simple diffusion of nonpolar molecules, water, and inorganic ions through a membrane, the diffusion of glucose through the cell membrane displays the properties of carrier-mediated transport: specificity, competition, and saturation. The diffusion of glucose through a cell membrane must therefore be mediated by carrier proteins. One conceptual model of the transport carriers is that each may be composed of two protein subunits that interact with glucose in such a way as to create a channel through the membrane (fig. 5.13), thus enabling the movement of the glucose from the higher to the lower concentration.

 The rate of the facilitated diffusion of glucose into tissue cells depends directly on the plasma glucose concentration. When the plasma glucose concentration is abnormally low—a condition called *hypoglycemia*—the rate of transport of glucose into brain cells may be inadequate for the metabolic needs of the brain. Severe hypoglycemia, as may be produced by an overdose of insulin in a person with diabetes, can thus result in loss of consciousness and even death.

## Active Transport

There are some aspects of cell transport that cannot be explained by simple or facilitated diffusion. The epithelial lining of the small intestine and of the kidney tubules, for example, moves glucose from the side of lower concentration to the side of higher concentration (from the lumen to the blood). Similarly, all cells extrude $Ca^{++}$ into the extracellular environment and, by this means, maintain an intracellular $Ca^{++}$ concentration that is 1000 to 10,000 times lower than the extracellular $Ca^{++}$ concentration.

 **Cystic fibrosis** occurs with a frequency of 1:2500 births in the Caucasian population. As a result of a genetic defect, there is abnormal NaCl and water movement across wet epithelial membranes. Where such membranes line the pancreatic ductules and small airways, the secretions of these membranes become more viscous and cannot be properly cleared, leading to pancreatic and pulmonary disorders. The genetic defect involves a particular glycoprotein that forms chloride (Cl⁻) channels in the membrane of the epithelial cells. This protein, known as *CFTR* (for *cystic fibrosis transmembrane conductance regulator*), is formed normally in the endoplasmic reticulum but it doesn't move into the Golgi apparatus for processing. It therefore doesn't get inserted into vesicles that would introduce it into the cell membrane through exocytosis (chapter 3). The gene for CFTR has now been identified and cloned, and research into possible gene therapy of cystic fibrosis is currently ongoing.

The movement of molecules and ions against their concentration gradients, from lower to higher concentrations, requires the expenditure of cellular energy that is obtained from ATP. This type of transport is termed **active transport**. If a cell is poisoned with cyanide (which inhibits oxidative phosphorylation), active transport will be inhibited. On the other hand, passive transport can continue even if metabolic poisons kill the cell by preventing the formation of ATP.

**Primary Active Transport** Primary Active Transport occurs when the hydrolysis of ATP is directly required for the function of the protein carrier. These carriers are proteins that span the thickness of the membrane. The following sequence of events is believed to occur: (1) the molecule or ion to be transported bonds to a specific "recognition site" on one side of the carrier protein; (2) this bonding stimulates the breakdown of ATP, which in turn results in phosphorylation of the carrier protein; (3) as a result of phosphorylation, the carrier protein undergoes a conformational (shape) change; and (4) a hingelike motion of the carrier protein releases the transported molecule or ion on

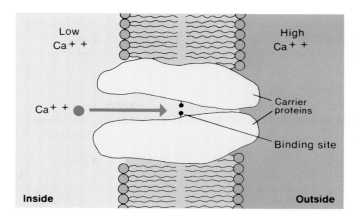

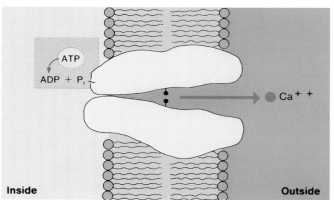

**FIGURE 5.14**
A model of active transport showing the hingelike motion of the integral protein subunits.

the other side of the membrane. This model of active transport is illustrated in figure 5.14.

**The Sodium-Potassium Pump** Primary active transport carriers are often referred to as "pumps." Although some of these carriers transport only one molecule or ion at a time, other carriers exchange one molecule or ion for another. The most important of the latter type of carriers is the **Na⁺/K⁺ pump**. This carrier protein, which is also an ATPase enzyme that converts ATP to ADP and $P_i$, actively extrudes three sodium ions from the cell as it transports two potassium ions into the cell. This transport is energy dependent because the concentration of Na⁺ is greater outside of the cell and that of K⁺ is greater within the cell. Both ions, in other words, are moved against their concentration gradients (fig. 5.15).

All cells have numerous Na⁺/K⁺ pumps that are constantly active. This activity represents an enormous expenditure of energy used to maintain a steep gradient of Na⁺ and K⁺ across the cell membrane. This steep gradient serves three known functions: (1) the steep Na⁺ gradient is used to provide energy for the "cotransport" of other molecules, (2) the activity of the Na⁺/K⁺ pumps can be adjusted (primarily by thyroid hormones) to regulate the resting calorie expenditure and basal metabolic rate of the body, and (3) the $Na^+$ and $K^+$ gradients across the cell membranes of nerve and muscle cells are used to produce electrical impulses.

**Secondary Active Transport (Cotransport)**  In **secondary active transport,** or **cotransport,** the energy needed for the "uphill" movement of a molecule or ion is obtained from

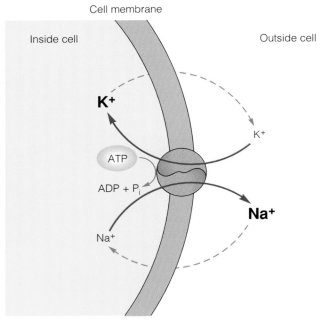

**FIGURE 5.15**
The Na⁺/K⁺ pump actively exchanges intracellular Na⁺ for K⁺. The carrier itself is an ATPase that breaks down ATP for energy. Dashed arrows indicate the direction of passive transport (diffusion); solid arrows indicate the direction of active transport.

the "downhill" transport of Na⁺ into the cell. Hydrolysis of ATP by the action of the Na⁺/K⁺ pumps is required indirectly, in order to maintain low intracellular Na⁺ concentrations. The diffusion of Na⁺ into the cell may power the uphill movement of a different ion or molecule into or out of the cell.

In the epithelial cells of the intestine and renal tubules, for example, glucose is transported against its concentration gradient by a carrier that requires the simultaneous binding of Na⁺ (fig. 5.16). Glucose and Na⁺ move in the same direction (into the cell), as a result of the Na⁺ gradient created by the Na⁺/K⁺ (ATPase) pumps. Because of the distribution of specific carriers in the epithelial cell membrane, the glucose is moved from the lumen of the intestine and renal tubule into the blood (fig. 5.17).

Cotransport can also occur in opposite directions. The uphill extrusion of Ca⁺⁺ from some cells, for example, is coupled to the passive diffusion of Na⁺ into the cell. Cellular energy, obtained from ATP, is not used to move Ca⁺⁺ directly out of the cell in this case, but energy is constantly required to maintain the steep Na⁺ gradient. The net movement of Ca⁺⁺ out of these cells is thus an example of secondary active transport.

# The Membrane Potential

*The selective permeability of the cell membrane, the presence of nondiffusible negatively charged molecules in the cell, and the action of the Na⁺/K⁺ pumps create an unequal distribution of charges across the membrane. As a result, the inside of the cell is negatively charged compared to the outside. This difference in charge, or potential difference, is known as the membrane potential.*

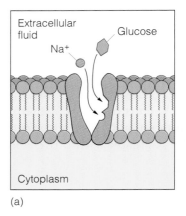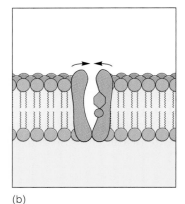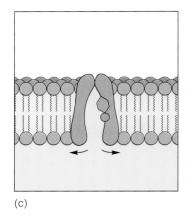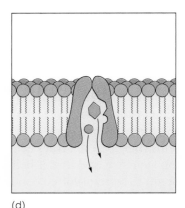

(a)  (b)  (c)  (d)

**FIGURE 5.16**
A model for the cotransport of Na⁺ and glucose into a cell. This is secondary active transport because it is dependent upon the diffusion gradient for Na⁺ created by the Na⁺/K⁺ pumps.

In the preceding section, we considered the action of the $Na^+/K^+$ pumps in the context of active transport, and we noted that these pumps move $Na^+$ and $K^+$ against their concentration gradients. This action alone would create and amplify concentration differences of these ions across the cell membrane. There is, however, another reason for these concentration differences.

Cellular proteins and the phosphate groups of ATP and other organic molecules are negatively charged within the cell cytoplasm. These negative ions (anions) are fixed within the cell because they cannot penetrate the cell membrane. Since these negatively charged organic molecules cannot leave the cell, they attract positively charged inorganic ions (cations) from the extracellular fluid that are small enough to diffuse through the membrane pores. The distribution of small, inorganic cations (mainly $K^+$, $Na^+$, and $Ca^{++}$) between the intracellular and extracellular compartments, in other words, is influenced by the negatively charged fixed ions within the cell.

Since the cell membrane is much more permeable to $K^+$ than to any other cation, $K^+$ builds up within the cell more than the others as a result of its electrical attraction for the fixed anions (fig. 5.18). So, instead of being evenly distributed between the intracellular and extracellular compartments, $K^+$ becomes more highly concentrated within the cell. In the human body, the intracellular $K^+$ concentration is 155 mEq/L compared to an extracellular concentration of 4 mEq/L (mEq = milliequivalents, which is the millimolar concentration multiplied by the valence of the ion—in this case, by one).

As a result of the unequal distribution of charges between the inside and outside of cells, each cell is a tiny battery with the positive pole outside the cell membrane and the negative pole inside. The magnitude of this charge difference is measured in *voltage*. Although the voltage of this battery is very low (less than one-tenth of a volt), it is of critical importance in muscle contraction, the regulation of heartbeat, the generation of nerve impulses, and other physiological events. In order to understand these processes, then, we must first examine the electrical properties of cells.

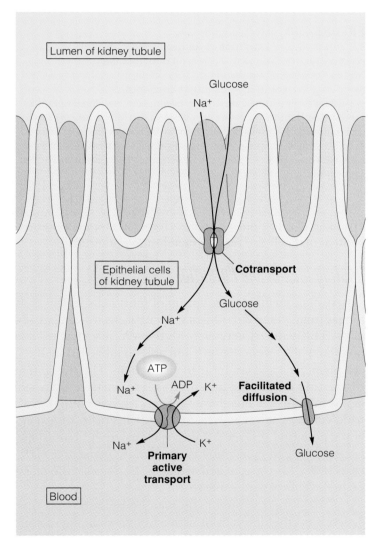

**FIGURE 5.17**
The transport of glucose from the fluid in the kidney tubules through the epithelial cells of the tubule and into the blood. All three types of carrier-mediated transport are used in this process.

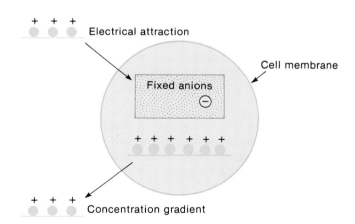

**FIGURE 5.18**
Proteins, organic phosphates, and other organic anions that cannot leave the cell create a fixed negative charge on the inside of the membrane. This attracts positively charged inorganic ions (cations), which therefore accumulate within the cell at a higher concentration than in the extracellular fluid. The amount of cations that accumulate within the cell is limited because a concentration gradient builds up that favors the diffusion of the cations out of the cell.

## Equilibrium Potential

An equilibrium potential is a theoretical voltage that would be produced across a cell membrane if only one ion were able to diffuse through the membrane. Since the membrane is most permeable to $K^+$, we can construct a theoretical approximation to the true situation by considering what would happen if $K^+$ were the *only* ion able to cross the membrane. If this were the case, $K^+$ would diffuse until its concentration inside and outside of the cell became stable, thus establishing an *equilibrium*. In this condition, if a certain amount of $K^+$ were to move inside the cell (by electrical attraction for the fixed anions), an identical amount of $K^+$ would diffuse out of the cell (down its concentration gradient). In other words, at equilibrium the forces of electrical attraction and of the diffusion gradient are equal and opposite.

At this equilibrium, the concentration of $K^+$ would be higher inside the cell than outside. A concentration difference would exist across the cell membrane that was stabilized by the attraction of $K^+$ to the fixed anions. At this point, we could ask, Are the fixed anions neutralized . . . are the charges balanced? The answer depends on how much $K^+$ gets into the cell, which in turn depends on the $K^+$ concentration in the extracellular fluid. At the $K^+$ concentrations found in the body, the answer to our question is no. Not enough $K^+$ is present in the cell to neutralize the fixed anions (fig. 5.19).

At equilibrium, therefore, the inside of the cell membrane has a higher concentration of negative charges than the outside of the membrane. There is a difference in charge, as well as a difference in concentration, across the membrane. The magnitude of the difference in charge, or **potential difference,** on the two sides of the membrane under these conditions is 90 millivolts (mV). This is shown with a negative sign (as –90 mV) to indicate that the inside of the cell is the negative pole. (If we were to write +90 mV, the magnitude of the potential difference would be unchanged, but the inside of the cell would be shown as the positive pole.) The potential difference of –90 mV, which would develop if $K^+$ were the only diffusible ion, is called the **$K^+$ equilibrium potential** (abbreviated $E_K$).

There is another way to look at the equilibrium potential: it is the membrane potential that would *exactly balance* the diffusion gradient and prevent the net movement of a particular ion. A potential difference of –90 mV exactly balances the tendency of $K^+$ to diffuse out of the cell. Similarly, a potential difference of +60 mV (where the inside of the cell is the positive pole) would exactly balance the tendency of $Na^+$ to diffuse into the cell, and is therefore the equilibrium potential for $Na^+$ ($E_{Na}$). Equilibrium potentials are theoretical values because a real cell membrane allows more than one ion to pass through. As a result, no single ion can

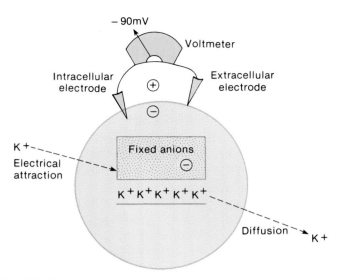

**FIGURE 5.19**

If $K^+$ were the only ion able to diffuse through the cell membrane, it would distribute itself between the intracellular and extracellular compartment until an equilibrium was established. At equilibrium, the $K^+$ concentration within the cell would be higher than outside the cell due to the attraction of $K^+$ for the fixed anions. Not enough $K^+$ would accumulate within the cell to neutralize these anions, however, so the inside of the cell would be 90 millivolts negative compared to the outside of the cell. This membrane voltage is the equilibrium potential ($E_K$) for potassium.

attain an equilibrium. We will call the membrane potential of an actual cell the **resting membrane potential.**

## Resting Membrane Potential

The resting membrane potential of most cells in the body ranges from –65 mV to –85 mV (in neurons it averages –70 mV). This value is very close to the –90 mV that we had predicted if $K^+$ were the only ion able to move through the cell membrane and establish an equilibrium. The reason that the resting membrane potential is less negative than –90 mV is because the membrane is slightly permeable to $Na^+$, which diffuses at a slow rate into the cell. The resting membrane potential, in other words, is closer to $E_K$ than to $E_{Na}$ because the cell membrane is much more permeable to $K^+$ than to $Na^+$. It is not exactly equal to $E_K$ because the membrane is slightly permeable to $Na^+$.

The actual value of the membrane potential thus depends on the differential permeability of the membrane to different ions. When neurons produce nerve impulses, the permeability to $Na^+$ increases dramatically, sending the membrane potential toward the $Na^+$ equilibrium potential (nerve impulses are discussed in chapter 14). This is why the term *resting* is used to describe the membrane potential when it is not producing impulses.

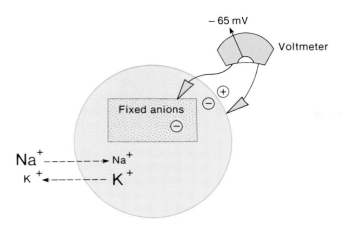

**FIGURE 5.20**
Because some Na⁺ leaks into the cell by diffusion, the actual resting membrane potential is less than the K⁺ equilibrium potential. As a result, some K⁺ diffuses out of the cell, as shown by the dotted lines.

**Role of the Na⁺/K⁺ Pumps**   Since the resting membrane potential is a little less negative than $E_K$ as a result of some Na⁺ entry, some K⁺ leaks out of the cell (fig. 5.20). The cell is *not* at equilibrium with respect to K⁺ and Na⁺ concentrations. Nonetheless, the concentrations of K⁺ and Na⁺ are maintained constant because of the constant expenditure of energy in active transport by the Na⁺/K⁺ pumps. The Na⁺/K⁺ pumps act to counter the leaks and thus maintain the membrane potential.

Actually, the Na⁺/K⁺ pump does more than simply work against the ion leaks. Since it transports *three* sodium ions out of the cell for every *two* potassium ions that it moves in, its action helps to generate a potential difference across the membrane. As a result of all of these activities, a real cell has (1) a relatively constant intracellular concentration of Na⁺ and K⁺ and (2) a constant membrane potential (in the absence of stimulation) in nerves and muscles of –65 mV to –85 mV.

## Clinical Considerations

### Dialysis

Normally functioning kidneys remove waste products from the blood. After the blood is filtered through pores in capillary walls that are large enough to permit the passage of wastes and other molecules, the molecules needed by the body are reabsorbed back into the blood. The wastes generally remain in the filtrate and are excreted in the urine. If the kidneys are not functioning properly, waste molecules can be removed from the blood artificially by a process called

dialysis (*di-al´ĭ-sis*). Dialysis involves the removal of particular molecules from a solution by having them pass, by means of diffusion, through an artificial porous membrane (see fig. 25.30). Since the pores in this dialysis membrane are large enough to permit the passage of some molecules but too small to permit the passage of others (the blood plasma proteins), small waste molecules can be removed from the blood by this technique.

### Inherited Defects in Membrane Carriers

Since membrane transport carriers are proteins that are coded by specific genes, inherited defects in these carriers can result if there is an alteration in the genetic code. Defective protein carriers in the cell membranes of epithelial cells that line the small intestine may produce diseases that result from an inadequate absorption of ingested molecules. *Pernicious anemia*, for example, is due to the inability to absorb vitamin B₁₂. Defects in transport carriers within the epithelial cells of kidney tubules may result in the abnormal excretion of particular molecules in the urine. For example, the inability of kidney tubules to transport glucose is responsible for a rare form of *diabetes mellitus*.

### Hyperkalemia and the Membrane Potential

Although changes in the extracellular concentration of many ions can affect the membrane potential, this potential is particularly sensitive to changes in blood K⁺. Since the maintenance of a particular membrane potential is critical for the generation of electrical events in nerves and muscles (including the heart), the body has a variety of mechanisms that serve to maintain blood K⁺ concentrations within very narrow limits. These mechanisms act primarily through the kidneys, which can excrete K⁺ in the urine or reabsorb it into the blood. The excretion of K⁺ is stimulated by hormones of the adrenal cortex—particularly by aldosterone. Indeed, if the adrenal glands of an experimental animal are removed, the animal may die as a result of an accumulation of K⁺ in the blood. An abnormal increase in the blood concentration of K⁺ is called **hyperkalemia** (*hi´´per-kă-le´me-ă*).

In the event of hyperkalemia, the diffusion gradient that favors the extrusion of K⁺ from the cell is reduced. As a result, more K⁺ can enter the cell and neutralize more of the fixed negative charges. This reduces the membrane potential (brings it closer to zero), thus altering the function of many organs, particularly the heart. For these reasons, the blood electrolyte concentrations are monitored very carefully in patients with heart or kidney disease.

hyperkalemia: Gk. *hyper*, over; L. *kalium*, potash; Gk. *haima*, blood

# Chapter Summary

## Diffusion and Osmosis (pp. 92–97)

1. Diffusion is the net movement of molecules or ions from regions of high concentration to regions of low concentration.
   a. This is a type of passive transport—energy is provided by the thermal energy of the molecules, not by cellular metabolism.
   b. Net diffusion stops when the concentration is equal on both sides of the membrane.
2. The rate of diffusion is dependent on the magnitude of the concentration difference between the two sides of the membrane, the degree of permeability of the cell membrane to the diffusing substance, the extent of the surface area of the membrane, and the temperature of the solution.
3. Simple diffusion is the type of passive transport in which small molecules and inorganic ions, such as $Na^+$ and $K^+$, move through the cell membrane.
   a. Inorganic ions pass through specific channels in the membrane.
   b. Lipids (such as steroid hormones) and dipolar water molecules can pass directly through the phospholipid layers of the membrane by simple diffusion.
4. Osmosis is the simple diffusion of solvent (water) through a membrane that is more permeable to the solvent than it is to the solute.
5. Water moves from the solution that is more dilute to the solution that has a higher solute concentration.
6. Osmosis depends on a difference in total solute concentration, not on the chemical nature of the solute.
   a. The concentration of total solute (in moles) per kilogram (liter) of water is measured in osmolality units.
   b. The solution with the higher osmolality has the higher osmotic pressure.
   c. Water moves by osmosis from the solution of lower osmolality and osmotic pressure to the solution of higher osmolality and osmotic pressure.

7. Solutions that have the same osmotic pressure as plasma (such as 0.9% NaCl and 5% glucose) are said to be isotonic.
   a. Solutions with a lower osmotic pressure are hypotonic; those with a higher osmotic pressure are hypertonic.
   b. Cells in a hypotonic solution gain water and swell; those in a hypertonic solution lose water and shrink (crenate).
8. The osmolality and osmotic pressure of the plasma is detected by osmoreceptors in the hypothalamus of the brain and maintained within a normal range by the action of antidiuretic hormone (ADH) secreted by the posterior pituitary.
   a. Increased osmolality stimulates the osmoreceptors; decreased osmolality inhibits the electrical activity of the osmoreceptors.
   b. Stimulation of the osmoreceptors causes thirst and triggers the secretion of antidiuretic hormone (ADH) from the pituitary.
9. ADH stimulates water retention by the kidneys, which serves to maintain a normal blood volume and osmolality.

## Carrier-Mediated Transport (pp. 97–100)

1. The passage of glucose, amino acids, and other polar molecules through the cell membrane is mediated by carrier proteins in the cell membrane.
   a. Carrier-mediated transport exhibits the properties of specificity, competition, and saturation.
   b. The transport rate of molecules such as glucose reaches a maximum when the carriers are saturated—this maximum rate is called the transport maximum, or $T_m$.
2. The transport of molecules such as glucose from the side of higher to the side of lower concentration by means of membrane carriers is called facilitated diffusion.
   a. Like simple diffusion, this is passive transport—cellular energy is not required.
   b. Unlike simple diffusion, facilitated diffusion displays the properties of specificity, competition, and saturation.

3. The active transport of molecules and ions across a membrane requires the expenditure of cellular energy (ATP).
   a. In active transport, carriers move molecules or ions from the side of lower to the side of higher concentration.
   b. One example of active transport is the action of the $Na^+/K^+$ pump.
4. Sodium is more concentrated on the outside of the cell, whereas potassium is more concentrated on the inside of the cell.
   a. The $Na^+/K^+$ pump helps to maintain these concentration differences by transporting $Na^+$ out of the cell and $K^+$ into the cell.
   b. Three sodium ions are transported out of the cell for every two potassium ions that are transported into the cell.

## The Membrane Potential (pp. 100–103)

1. The cytoplasm of the cell contains negatively charged organic ions (anions) that cannot leave the cell—they are "fixed anions."
   a. These fixed anions attract $K^+$, which is the inorganic ion that can most easily pass through the cell membrane.
   b. As a result of this electrical attraction, the concentration of $K^+$ within the cell is greater than the concentration of $K^+$ in the extracellular fluid.
2. If $K^+$ were the only diffusible ion, the concentrations of $K^+$ on the inside and outside of the cell would reach an equilibrium.
   a. At this point, the rate of $K^+$ entry (due to electrical attraction) would equal the rate of $K^+$ exit (due to diffusion).
   b. At this equilibrium, there would still be a higher concentration of negative charges within the cell (due to the fixed anions) than outside the cell.
   c. At this equilibrium, the inside of the cell would be 90 millivolts negative (−90 mV) compared to the outside of the cell. This potential difference is called the $K^+$ equilibrium potential ($E_K$).

3. The resting membrane potential is less than $E_K$ (usually –65 mV to –85 mV) because some Na+ can also enter the cell.
   a. Na+ is more highly concentrated outside than inside the cell, and the inside of the cell is negative—these forces attract Na+ into the cell.
   b. The rate of Na+ entry is generally slow because the membrane is usually not very permeable to Na+.

4. The slow rate of Na+ entry is accompanied by a slow rate of K+ leakage out of the cell.
   a. The Na+/K+ pump counters this leakage, and thus maintains constant concentrations and a constant resting membrane potential.

   b. There are numerous Na+/K+ pumps in all cells of the body that require a constant expenditure of energy.
   c. The Na+/K+ pump itself contributes to the membrane potential because it pumps more Na+ out than it pumps K+ in (by a ratio of three to two).

# Review Activities

## Objective Questions

1. The movement of water across a cell membrane occurs by
   a. active transport.
   b. facilitated diffusion.
   c. simple diffusion (osmosis).
   d. all of the above.
2. Which of the following statements about the facilitated diffusion of glucose is *true*?
   a. There is a net movement from the region of lower to the region of higher concentration.
   b. Carrier proteins in the cell membrane are required for this transport.
   c. This transport requires energy obtained from ATP.
   d. This is an example of cotransport.
3. If a poison such as cyanide stops the production of ATP, which of the following transport processes would cease?
   a. the movement of Na+ out of a cell
   b. osmosis
   c. the movement of K+ out of a cell
   d. all of the above
4. Red blood cells crenate in
   a. a hypotonic solution.
   b. an isotonic solution.
   c. a hypertonic solution.

5. Plasma has an osmolality of about 300 mOsm. Isotonic saline has an osmolality of
   a. 150 mOsm.
   b. 300 mOsm.
   c. 600 mOsm.
   d. none of the above apply.
6. A 0.5 *m* NaCl solution and a 1.0 *m* glucose solution
   a. have the same osmolality.
   b. have the same osmotic pressure.
   c. are isotonic to each other.
   d. all of the above apply.
7. The diffusible ion that is most important in the establishment of the membrane potential is
   a. K+.
   b. Na+.
   c. Ca++.
   d. Cl−.
8. An increase in blood osmolality
   a. can occur as a result of dehydration.
   b. causes a decrease in blood osmotic pressure.
   c. is accompanied by a decrease in ADH secretion.
   d. all of the above apply.
9. In hyperkalemia, the membrane potential
   a. increases.
   b. decreases.
   c. is not affected.

10. Which of the following statements about the Na+/K+ pump is *true*?
    a. Na+ is actively transported into the cell.
    b. K+ is actively transported out of the cell.
    c. An equal number of sodium and potassium ions are transported with each cycle of the pump.
    d. The pumps are constantly active in all cells.
11. Carrier-mediated facilitated diffusion
    a. uses cellular ATP.
    b. is used for cellular uptake of blood glucose.
    c. is a form of active transport.
    d. none of the above apply.

## Essay Questions

1. Describe the conditions required for osmosis to occur and explain why it occurs under these conditions.
2. Explain how simple diffusion can be distinguished from facilitated diffusion and how active transport can be distinguished from passive transport.
3. Compare the theoretical membrane potential that occurs at K+ equilibrium with the true resting membrane potential. Explain why the resting membrane potential is not exactly equal to $E_K$.
4. Explain how the Na+/K+ pump contributes to the resting membrane potential.

# Explorations ⊛

Two modules of correlating material are available from the Wm. C. Brown CD-ROM *Explorations*. They are #1 Cystic Fibrosis and #2 Active Transport.

Membrane Transport and the Membrane Potential

# histology

## objectives

- Define *tissue* and discuss the importance of histology.

- Describe the functional relationship between cells and tissues.

- Classify the tissues of the body according to their four principal types and list the distinguishing characteristics of each type.

- Compare and contrast the various types of epithelia.

- Define *exocrine gland* and compare and contrast the various types of exocrine glands in the body.

- Describe the general characteristics, locations, and functions of connective tissues.

- List the various ground substances, fiber types, and cells that constitute connective tissue and explain their functions.

- Describe the structure, location, and function of the three types of muscle tissue.

- Describe the basic characteristics and functions of nervous tissue.

- Distinguish between neurons and neuroglia.

# Definition and Classification of Tissues

*Histology is the specialty of anatomy that involves microscopic study of the structure of tissues. Tissues are classified into four basic types on the basis of their cellular composition and histological appearance.*

Although cells are the structural and functional units of the body, the cells of a multicellular organism are so specialized that they do not function independently. *Tissues are aggregations of similar cells and cell products that perform specific functions.* The study of tissues is referred to as **histology** (*his-tol´ŏ-je*). The various types of tissues are established during embryonic development, and as differentiation continues, organs form, each composed of a specific arrangement of tissues. Many adult organs, including the heart and muscles, contain original cells and tissues that were not replaced through mitotic activity during further growth and development. Some functional changes may occur, however, as the tissues of an organ are acted upon by hormones or as their effectiveness diminishes with age.

Although histology is actually microscopic anatomy, it is an essential part of anatomy and physiology because it imparts an understanding of the structure and function of organs at the tissue level. Many diseases profoundly alter the tissues within an affected organ; therefore, by knowing the normal tissue structure, a physician can recognize the abnormal. In medical schools a course in histology is usually followed by a course in *pathology,* which is primarily concerned with identifying diseased tissues.

Although histologists employ a variety of techniques for preparing, staining, and sectioning tissues, basically only two kinds of microscopes are used to view the prepared tissues—light microscopes and electron microscopes. *Light microscopy* (*mi-kros´-kŏ-pe*) is used for the general observation of tissue structure (fig. 6.1) and *electron microscopy* permits observation of the fine details of tissue and cellular structure. Most of the histological photomicrographs in this text are at the light microscopic level. Electron micrographs are used when a student must observe fine structural detail in order to understand a particular function.

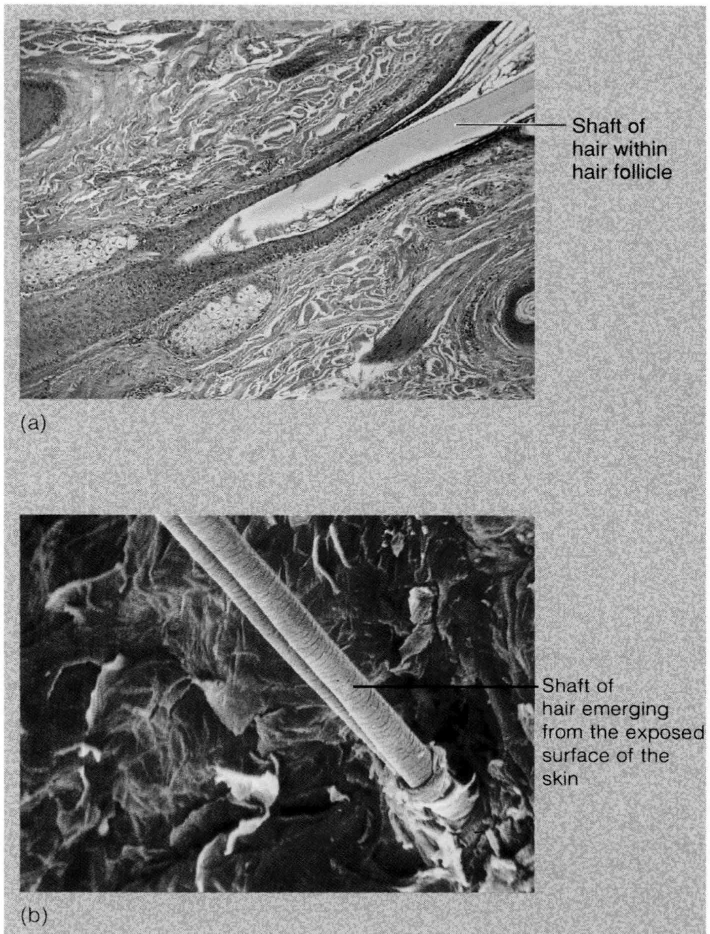

Tissue cells are separated and bound together by a nonliving intercellular *matrix* (*ma´ triks*) that the cells secrete. Matrix varies in composition from one tissue to

(a)

Shaft of hair within hair follicle

(b)

Shaft of hair emerging from the exposed surface of the skin

**FIGURE 6.1**
The appearance of skin (*a*) magnified 25 times, as seen through a compound light microscope, and (*b*) magnified 280 times, as seen through a scanning electron microscope.

another and may take the form of a liquid, semisolid, or solid. Blood, for example, has a liquid matrix, permitting this tissue to flow through vessels. By contrast, bone cells are separated by a solid matrix, permitting this tissue to support the body.

The tissues of the body are classified into four basic kinds on the basis of structure and function: (1) *epithelial* (*ep´´ĭ-the´le-al*) *tissues* cover body and organ surfaces, line body and lumen cavities, and form glands; (2) *connective tissues* bind, support, and protect body parts; (3) *muscle tissues* contract to produce movement; and (4) *nervous tissues* initiate and transmit nerve impulses from one body part to another.

histology: Gk. *histos,* web (tissue); *logos,* study
pathology: Gk. *pathos,* suffering, disease, *logos,* study

matrix: L. *matris,* mother

## Development of Tissues

Human prenatal development is initiated by the fertilization of an ovulated ovum (egg) from a female by a sperm cell from a male. The chromosomes within the nucleus of a **zygote** (*zī′ gōt*) (fertilized egg) contain all the genetic information necessary for the differentiation and development of all body structures.

Within 30 hours after fertilization, the zygote undergoes a mitotic division as it moves through the uterine tube toward the uterus (see fig. 30.6). After several more cellular divisions, the embryonic mass consists of 16 or more cells and is called a **morula** (*mor′yŭ-lă*) (fig. 1). Three or 4 days after conception, the morula enters the uterine cavity where it

remains unattached for about 3 days. During this time, the center of the morula fills with fluid passing in from the uterine cavity. As the fluid-filled space develops inside the morula, two distinct groups of cells form. The single layer of cells forming the outer wall is known as the **trophoblast** and the inner aggregation of cells is known as the **embryoblast.** After

· · · · · · · · · · · · ·

zygote: Gk. *zygotos*, yolked

· · · · · · · · · ·

morula: Gk. *morus*, mulberry

· · · · · · · · · · · · · · ·

trophoblast: Gk. *trophe*, nourishment; *blastos*, germ
embryoblast: Gk. *embryon*, to be full, swell; *blastos*, germ

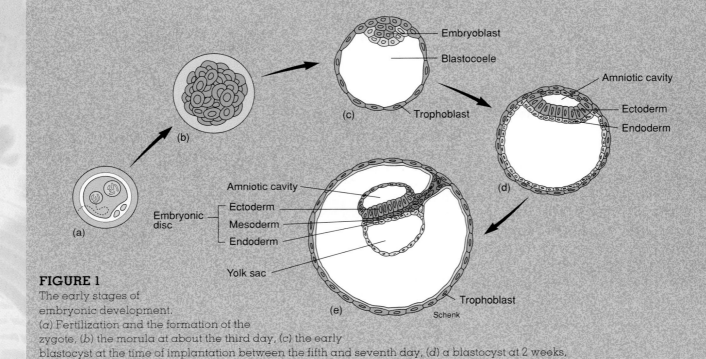

**FIGURE 1**

The early stages of embryonic development.
(*a*) Fertilization and the formation of the zygote, (*b*) the morula at about the third day, (*c*) the early blastocyst at the time of implantation between the fifth and seventh day, (*d*) a blastocyst at 2 weeks, and (*e*) a blastocyst at 3 weeks showing the three primary germ layers that constitute the embryonic disc.

# Epithelial Tissues

There are two major categories of epithelia: membranous and glandular. Membranous epithelia are located throughout the body and form such structures as the outer layer of the skin, the inner lining of body cavities and lumina, and the covering of visceral organs. Glandular epithelia are specialized tissues that form the secretory portion of glands.

· · · · · · · · · · · · · · · · · · · · · · · · · · · · · ·

epithelium: Gk. *epi*, upon; *thelium*, to cover

further development, the trophoblast becomes a portion of the placenta and the embryoblast becomes the embryo. With the establishment of these two groups of cells, the morula becomes known as a **blastocyst** (*blas´tŏ-sist*). Implantation of the blastocyst begins between the fifth and seventh day (see fig. 30.9).

As the blastocyst completes implantation during the second week of development, the embryoblast undergoes marked differentiation. A slitlike space called the **amniotic** (*am´ne-ot-ic*) **cavity** forms within the embryoblast, adjacent to the trophoblast. The embryoblast now consists of two layers: an upper **ectoderm,** which is closer to the amniotic cavity, and a lower **endoderm,** which borders the blastocyst cavity. A short time later, a third layer called the **mesoderm** forms between the endoderm and ectoderm. These three layers constitute the **primary germ layers.**

The primary germ layers are important because all the cells and

..............
ectoderm: Gk. *ecto*, outside; *derm*, skin
endoderm: Gk. *endo*, within; *derm*, skin
mesoderm: Gk. *meso*, middle; *derm*, skin

tissues of the body are derived from them (see fig. 30.10). Ectodermal cells form the nervous system; the outer layer of skin (epidermis), including hair, nails, and skin glands; and portions of the sensory organs. Mesodermal cells form the skeleton, muscles, blood, reproductive organs, dermis of the skin, and connective tissue. Endodermal cells give rise to the lining of the digestive tract, the digestive organs, the respiratory tract and lungs, and the urinary bladder and urethra. In table 6.1, the derivatives of the primary germ layers are listed.

## Table 6.1 Derivatives of the germ layers

| Ectoderm | Mesoderm | Endoderm |
|---|---|---|
| Epidermis of skin and epidermal derivatives: hair, nails, glands of the skin; linings of oral, nasal, anal, and vaginal cavities | Muscle: smooth, cardiac, and skeletal | Epithelium of pharynx, auditory canal, tonsils, thyroid, parathyroid, thymus, larynx, trachea, lungs, gastrointestinal tract, urinary bladder and u rethra, and vagina |
| | Connective tissue: embryonic, connective tissue proper, cartilage, bone, blood | |
| Nervous tissue; sense organs | Dermis of skin; dentin of teeth | Liver and pancreas |
| Lens of eye; enamel of teeth | Epithelium of blood vessels, lymphatic vessels, body cavities, joint cavities | |
| Pituitary gland | Internal reproductive organs | |
| Adrenal medulla | Kidneys and ureters | |
| | Adrenal cortex | |

## *Characteristics of Membranous Epithelia*

**Membranous epithelia** line all body surfaces, cavities, and *lumina* (hollow portions of body tubes) and are specialized for protection and absorption. One side of membranous epithelia is always exposed to a body cavity, lumen, or skin surface. Some membranous epithelia are derived from ectoderm, such as the outer layer of the skin; some from mesoderm, such as the inside lining of blood vessels; and others from endoderm, such as the inside lining of the gastrointestinal, or GI, tract.

Membranous epithelia may be one layer or several layers thick. The upper surface of epithelia may be exposed to gases, as in the case of epithelium in the integumentary and respiratory systems; to liquids, as in the circulatory and urinary systems; or to semisolids, as in the digestive system.

The deep surface of membranous epithelia is bound to underlying supportive tissue by a **basement membrane,** consisting of glycoprotein from the epithelial cells and a meshwork of collagenous and reticular fibers from the underlying connective tissues. Membranous epithelia are avascular (without blood vessels) and must be nourished by diffusion from underlying connective tissues. The cells of membranous epithelia are tightly packed together, and there is little intercellular matrix between them.

Some of the functions of membranous epithelia are quite specific, but certain generalizations apply. Epithelia that cover or line surfaces provide *protection* from pathogens, physical injury, toxins, and desiccation. Epithelia lining the lumen of the GI tract function in *absorption*. The epithelium of the kidneys allows for *filtration,* while the epithelium within the air sacs (alveoli) of the lungs allows for *diffusion*. Highly specialized neuroepithelial cells in the taste buds and in the nasal region respond to chemical molecules and serve as chemoreceptors.

Many membranous epithelia are exposed and, therefore, subject to trauma and destruction. For this reason, epithelial tissues have remarkable regenerative abilities. The mitotic replacement of the outer layer of skin and the lining of the GI tract, for example, is a continuous process.

Membranous epithelia are histologically classified by the number of layers of cells and the shape of the cells along the exposed surface. Epithelial tissues that are composed of a single layer of cells are called *simple;* those that are layered are said to be *stratified*. *Squamous cells* are flattened; *cuboidal cells* are cube-shaped; and *columnar cells* are taller than they are wide.

## Simple Epithelia

**Simple epithelial tissues** are a single layer thick and are located where diffusion, absorption, filtration, and secretion occur. The cells that constitute simple epithelia range from thin, flattened cells to tall, columnar cells, depending on function. These cells may also exhibit surface specializations (for example, cilia and microvilli) to facilitate specific surface functions.

**Simple Squamous Epithelium**    Simple squamous (*skwa´mus*) epithelium is composed of flattened, irregularly shaped cells that are tightly bound together in a mosaiclike pattern (fig. 6.2). Each cell contains an oval centrally located nucleus. Simple squamous epithelium is adapted for diffusion and filtration and occurs in such places as the lining of air sacs (alveoli) within the lungs (where gaseous exchange occurs); a portion of the kidney (where blood is filtered); the inside lining of the walls of blood vessels; the lining of body cavities; and the covering of the viscera (internal

squamos: L. *squamosus,* scaly

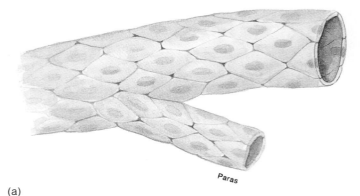
Paras

(a)

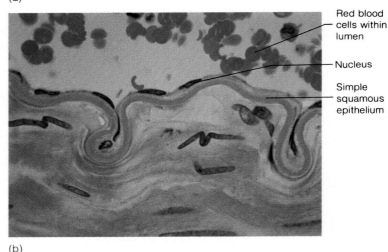

Red blood cells within lumen

Nucleus

Simple squamous epithelium

(b)

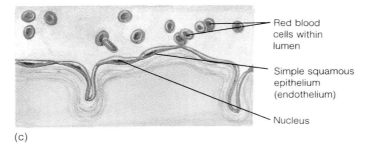

Red blood cells within lumen

Simple squamous epithelium (endothelium)

Nucleus

(c)

**FIGURE 6.2**

(*a*) Simple squamous epithelium lines the lumina of vessels, where it permits diffusion. (*b*) A photomicrograph of this tissue and (*c*) a labeled diagram.

body organs). The simple squamous epithelium lining the lumina of blood and lymphatic vessels is termed **endothelium** (fig. 6.2*c*). That which covers visceral organs and lines body cavities is called **mesothelium.**

**Simple Cuboidal Epithelium**    Simple cuboidal epithelium is composed of a single layer of tightly fitted hexagonal cells (fig. 6.3). This type of epithelium is found lining small ducts and tubules that may have excretory, secretory, or absorptive

endothelium: Gk. *endon,* within; *thelium,* to cover
mesothelium: Gk. *meso,* middle; *thelium,* to cover

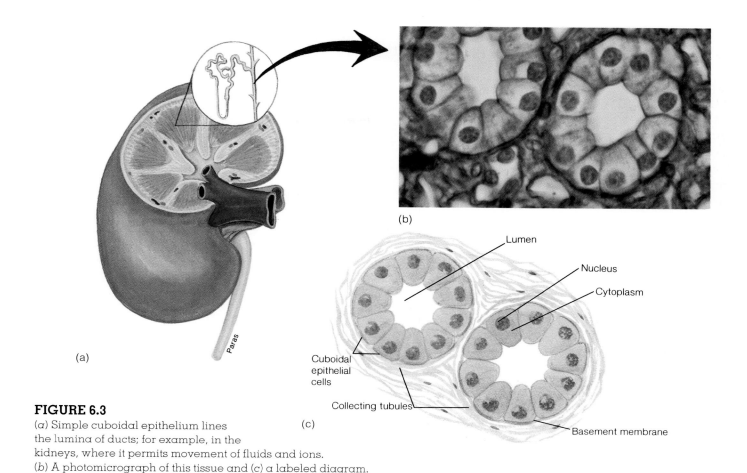

**FIGURE 6.3**

(a) Simple cuboidal epithelium lines
the lumina of ducts; for example, in the
kidneys, where it permits movement of fluids and ions.
(b) A photomicrograph of this tissue and (c) a labeled diagram.

functions. It occurs on the surface of the ovaries, forms a portion of the tubules within the kidney, and lines the ducts of the salivary glands and the pancreas.

**Simple Columnar Epithelium**    Simple columnar epithelium is composed of tall, columnar cells (fig. 6.4). The height of the cells varies, depending on the site and function of the tissue. Each cell contains a single nucleus, which is usually located near the basement membrane. Specialized unicellular glands called **goblet cells** are dispersed throughout this tissue and secrete a lubricative and protective mucus along the free surfaces of the cells. This type of epithelium is found lining the lumen of the stomach and intestine. In the digestive system, simple columnar epithelium forms a highly absorptive surface and also secretes certain digestive substances. Within the stomach, simple columnar epithelium has a tremendous rate of mitotic activity—replacing itself every 2 or 3 days.

**Simple Ciliated Columnar Epithelium**    Simple ciliated columnar epithelium is characterized by the presence of cilia along the free surface (fig. 6.5), whereas the simple columnar type is unciliated. Cilia produce wavelike movements that transport materials through tubes or passageways. As mentioned in chapter 3, this type of epithelium occurs in the uterine tubes of the female, where the currents generated by

the cilia propel the ovum toward the uterus. Furthermore, recent evidence indicates that sperm introduced during sexual intercourse may be moved along the return currents, or eddies, generated by ciliary movement. This greatly enhances the likelihood of fertilization.

**Pseudostratified Ciliated Columnar Epithelium**    As the name implies, this type of epithelium appears stratified. It is actually simple, however, since each cell is in contact with the basement membrane, although not all cells are exposed to the surface (fig. 6.6). The epithelium has a stratified appearance because the nuclei of these cells are located at different levels. Numerous goblet cells and a ciliated, exposed surface are characteristic of this epithelium. The lumina of the trachea and the bronchial tubes are lined with this tissue; hence, it is frequently called *respiratory epithelium*. Its function is to remove foreign dust and bacteria entrapped in mucus from the lower respiratory system.

Coughing, sneezing, or simply "clearing the throat" are protective reflex mechanisms for clearing the respiratory passages of obstructions or inhaled particles that have been trapped in the mucus along the ciliated lining. The coughed-up material consists of the mucus-entrapped particles.

## FIGURE 6.4

(a) Simple columnar epithelium lines the lumen of most of the gastrointestinal tract, where it permits secretion and absorption. (b) A photomicrograph of this tissue and (c) a labeled diagram.

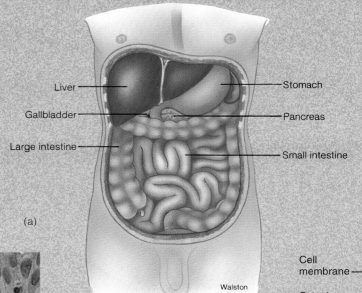

(a)

Liver
Stomach
Gallbladder
Pancreas
Large intestine
Small intestine

Walston

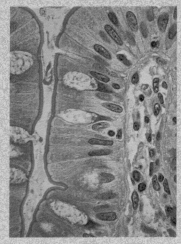

(b)

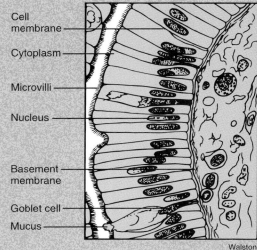

(c)

Cell membrane
Cytoplasm
Microvilli
Nucleus
Basement membrane
Goblet cell
Mucus

Walston

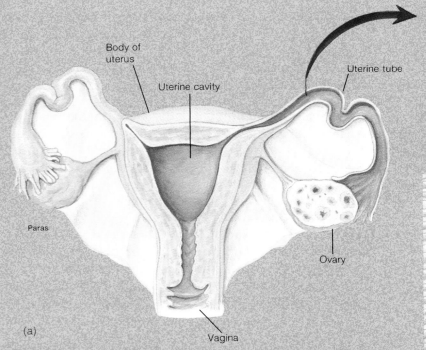

(a)

Body of uterus
Uterine cavity
Uterine tube
Paras
Ovary
Vagina

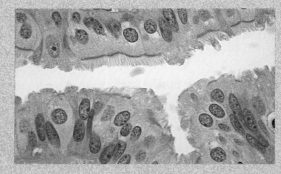

(b)

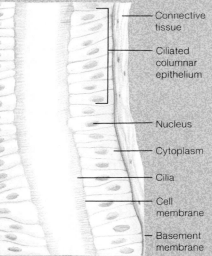

(c)

Connective tissue
Ciliated columnar epithelium
Nucleus
Cytoplasm
Cilia
Cell membrane
Basement membrane

## FIGURE 6.5

(a) Simple ciliated columnar epithelium lines the lumen of the uterine tube, where currents generated by the cilia propel the egg cell toward the uterus. (b) A photomicrograph of this tissue and (c) a labeled diagram.

## FIGURE 6.6

(a) Pseudostratified ciliated columnar epithelium lines the lumen of most of the respiratory tract, where it traps foreign material and moves it away from the alveoli of the lungs. (b) A photomicrograph of this tissue and (c) a labeled diagram.

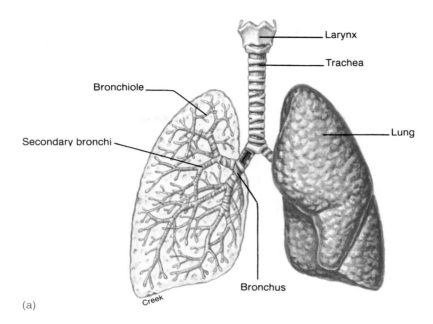

Larynx
Trachea
Bronchiole
Secondary bronchi
Lung
Bronchus
Creek
(a)

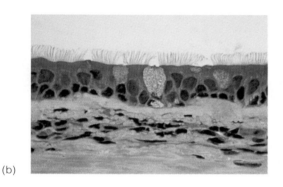

(b)

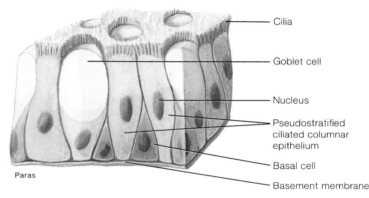

Cilia
Goblet cell
Nucleus
Pseudostratified ciliated columnar epithelium
Basal cell
Basement membrane
Paras
(c)

## Stratified Epithelia

**Stratified epithelia** are tissues consisting of two or more layers of cells. In contrast to simple epithelia, stratified epithelia are poorly suited for absorption and secretion because of their thickness. Stratified epithelia have a primarily protective function that is enhanced by a characteristic rapid mitotic activity. Stratified epithelia are classified according to the shape of the surface layer of cells, since the layer in contact with the basement membrane is cuboidal or columnar in shape.

**Stratified Squamous Epithelium** Stratified squamous epithelium is composed of a variable number of cell layers that tend to flatten near the surface (fig. 6.7). Only at the deepest layer, the **stratum basale,** does mitosis occur. The mitotic rate approximates the rate at which cells are sloughed off at the surface. As the newly produced cells grow, they are pushed toward the surface where they will replace the cells that are sloughed off. Movement of the epithelial cells away from the supportive basement membrane is accompanied by progressive dehydration, flattening, and the production of the protein keratin (ker´ă-tin).

There are two types of stratified squamous epithelia; *keratinized* and *nonkeratinized*. Stratified squamous epithelium that is keratinized forms the outer layer, or *epidermis*, of the skin (see chapter 7). **Keratin** is a protein that strengthens the tissue. This type of epithelium is especially durable and can generally withstand physical abrasion, desiccation, and bacterial invasion. The outer layers of the stratified squamous epithelium of the skin are dead but are kept pliable by local glandular secretions.

Nonkeratinized stratified squamous epithelium lines the oral cavity and pharynx, nasal cavity, vagina, and anal canal. This type of epithelium, called *mucosa*, is well adapted to withstand moderate abrasion but not fluid loss. The cells on the free surface of this tissue remain alive and are always moistened.

. . . . . . . . . . . . . . . . . . . . . . . . . . . . . . . . . . . . . . . . . . . . . .

keratin: Gk. *keras*, horn

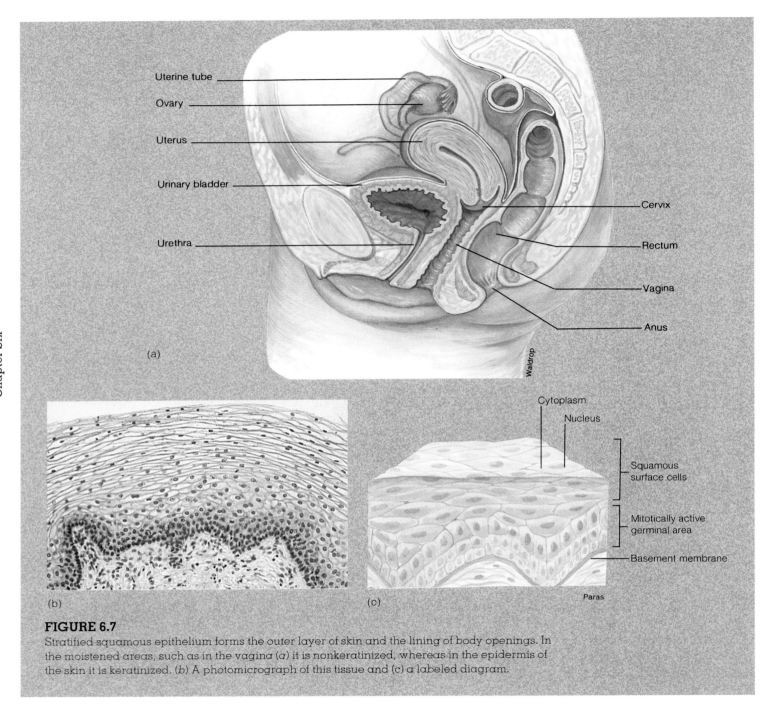

**FIGURE 6.7**
Stratified squamous epithelium forms the outer layer of skin and the lining of body openings. In the moistened areas, such as in the vagina (*a*) it is nonkeratinized, whereas in the epidermis of the skin it is keratinized. (*b*) A photomicrograph of this tissue and (*c*) a labeled diagram.

 Stratified squamous epithelium is the first line of defense against the entry of living organisms into the body. Stratification, along with rapid mitotic activity and keratinization within the epidermis of the skin, are important protective features. An acidic pH along the surfaces of this tissue also helps to prevent disease. The pH of the skin ranges from 4 to 6.8. The pH in the oral cavity ranges from 5.8 to 7.1, which tends to retard the growth of microorganisms. The pH of the anal region is about 6, and the pH along the surface of the vagina is 4 or lower.

**Stratified Cuboidal Epithelium**   Stratified cuboidal epithelium usually consists of only two or three layers of cuboidal cells forming the lining around a lumen (fig. 6.8). This type of epithelium is confined to the linings of the larger ducts of sweat glands, salivary glands, and the pancreas. The stratification of this tissue probably provides a more robust lining than would be afforded by simple epithelium.

**Transitional Epithelium**   Transitional epithelium is similar to nonkeratinized stratified squamous epithelium except that the surface cells of the former are large and round rather than flat, and some may have two nuclei (fig 6.9). Transitional epithelium is located only within the urinary

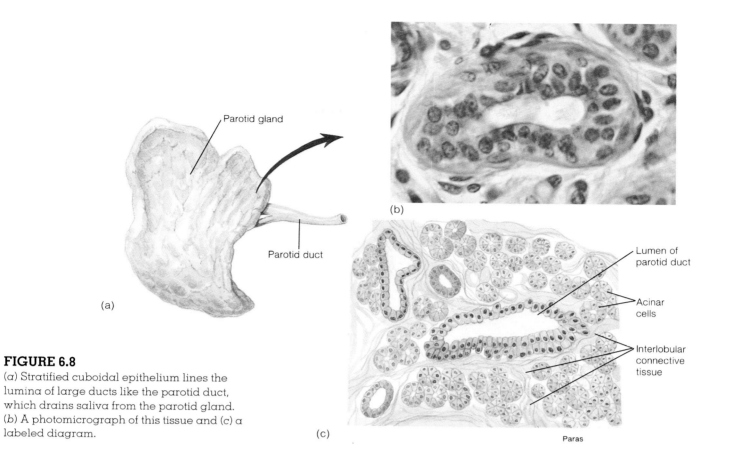

**FIGURE 6.8**

(*a*) Stratified cuboidal epithelium lines the lumina of large ducts like the parotid duct, which drains saliva from the parotid gland. (*b*) A photomicrograph of this tissue and (*c*) a labeled diagram.

Parotid gland

Parotid duct

(a)

(b)

Lumen of parotid duct

Acinar cells

Interlobular connective tissue

(c)

Paras

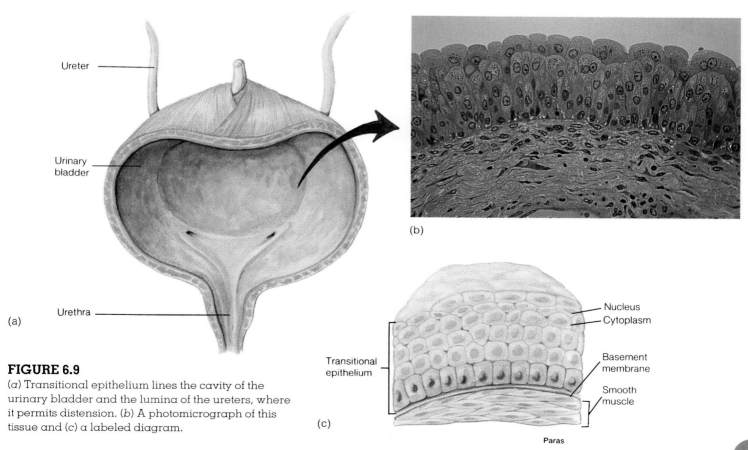

**FIGURE 6.9**

(*a*) Transitional epithelium lines the cavity of the urinary bladder and the lumina of the ureters, where it permits distension. (*b*) A photomicrograph of this tissue and (*c*) a labeled diagram.

Ureter

Urinary bladder

Urethra

(a)

(b)

Nucleus

Cytoplasm

Basement membrane

Smooth muscle

Transitional epithelium

(c)

Paras

## Table 6.2  Summary of membranous epithelial tissues

| Type | Structure and function | Location |
|------|------------------------|----------|
| **Simple epithelia** | Single layer of cells; diffusion and filtration | Covering visceral organs; linings of lumina and body cavities |
| Simple squamous epithelium | Single layer of flattened, tightly bound cells; diffusion and filtration | Capillary walls; alveoli of lungs; covering visceral organs, linings of body cavities |
| Simple cuboidal epithelium | Single layer of cube-shaped cells; excretion, secretion, or absorption | Surface of ovaries; linings of renal tubules, salivary ducts, and pancreatic ducts |
| Simple columnar epithelium | Single layer of nonciliated, tall, columnar-shaped cells; protection, secretion, and absorption | Lining of most of GI tract |
| Simple ciliated columnar epithelium | Single layer of ciliated, columnar-shaped cells; transportive role through ciliary motion | Lining the lumen of the uterine tubes |
| Pseudostratified ciliated columnar epithelium | Single layer of ciliated, irregularly shaped cells, many goblet cells; protection, secretion, ciliary movement | Lining of respiratory passageways |
| **Stratified epithelia** | Two or more layers of cells; protection, strengthening, or distension | Epidermal layer of skin; linings of body openings, ducts, and urinary bladder |
| Stratified squamous epithelium (keratinized) | Numerous layers containing keratin, outer layers flattened and dead; protection | Epidermis of skin |
| Stratified squamous epithelium (nonkeratinized) | Numerous layers lacking keratin, outer layers moistened and alive; protection and pliability | Linings of oral and nasal cavities, vagina, and anal canal |
| Stratified cuboidal epithelium | Usually two layers of cube-shaped cells; strengthening of luminal walls | Larger ducts of sweat glands, salivary glands, and pancreas |
| Transitional epithelium | Numerous layers of rounded, nonkeratinized cells; distension | Luminal walls of ureters and urinary bladder |

system, particularly in the luminal surface of the urinary bladder and the walls of the ureters. This tissue is specialized to permit distension (stretching) of the urinary bladder.

Epithelial tissues are summarized in table 6.2.

## Body Membranes

**Body membranes** are composed of thin layers of epithelial tissue and, in certain locations, epithelial tissue coupled with supporting connective tissue. Body membranes cover, separate, and support visceral organs and line body cavities. There are two basic types of body membranes: **mucous** (*myoo´kus*) **membranes** and **serous** (*se´rus*) **membranes.**

Mucous membranes secrete a thick, viscid substance called *mucus.* Generally, mucus lubricates or protects the associated organs where it is secreted. Mucous membranes line various cavities and tubes that enter or exit from the body, such as the oral and nasal cavities and the tubes of the respiratory, reproductive, urinary, and digestive systems.

Serous membranes line the thoracic and abdominopelvic cavities and cover visceral organs, secreting a watery lubricant called *serous fluid.* **Pleurae** are serous membranes associated with the lungs (see chapter 24). Each pleura (pleura of right lung and pleura of left lung) has two parts. The **visceral pleura** adheres to the outer surface of the lung, while the **parietal** (pă-ri´ĕ-tal) **pleura** lines the thoracic walls and the thoracic surface of the diaphragm. The moistened space between the two pleurae is know as the **pleural cavity.**

**Pericardial membranes** are the serous membranes of the heart (see chapter 21). A thin **visceral pericardium** covers the surface of the heart and a thicker **parietal**

**pericardium** is the durable covering that surrounds the heart. The space between these two membranes is called the **pericardial cavity.**

Serous membranes of the abdominal cavity are called **peritoneal** (*per˝ĭ-tŏ-ne´al*) **membranes** (see chapter 26). The **parietal peritoneum** lines the abdominal wall and the **visceral peritoneum** covers the visceral organs (fig. 6.10). The **peritoneal cavity** is the potential space within the abdominopelvic cavity between the parietal and visceral peritoneal membranes. Certain organs, such as the kidneys, adrenal glands, and a portion of the pancreas, which are within the abdominal cavity, are positioned behind the parietal peritoneum and therefore are said to be **retroperitoneal.** **Mesenteries** (*mes´en-ter˝ēz*) are double folds of peritoneum that connect the parietal to the visceral peritoneum.

## Glandular Epithelia

**Glandular epithelia** are specialized tissues that have secretory functions. During the prenatal development of epithelial tissue, certain epithelial cells invade the underlying connective tissue, forming specialized secretory accumulations called **exocrine** (*ek´šo-krin*) **glands.** Structurally, there are two types of exocrine glands: unicellular and multicellular. *Unicellular exocrine glands* are ductless and their secretion is expelled directly to the surface of a tissue. *Multicellular exocrine glands* retain a connection to the surface in the form of a *duct,* through which secretions flow. Multicellular exocrine glands within the integumentary system include sebaceous (oil) glands, sweat glands, and mammary glands. Multicellular exocrine glands are classified according to the structure of the gland and its means of discharging the secretory product. Exocrine glands should not be confused with endocrine glands (see chapter 19). The latter are ductless and secrete hormones into the blood.

**Unicellular Glands**   Unicellular glands are single-celled exocrine glands interspersed with the various columnar epithelia. A mucus-secreting goblet cell is a good example of a unicellular gland. Goblet cells are found in the epithelial linings of the respiratory, digestive, urinary, and reproductive systems, where the mucus secretion lubricates and protects the surface linings (fig. 6.11).

**Multicellular Glands**   Multicellular glands, as their name implies, are composed of numerous secretory cells in addition to cells that form the walls of the ducts. Multicellular

.........................................................................
exocrine: Gk. *exo*, outside; *krinein*, to separate

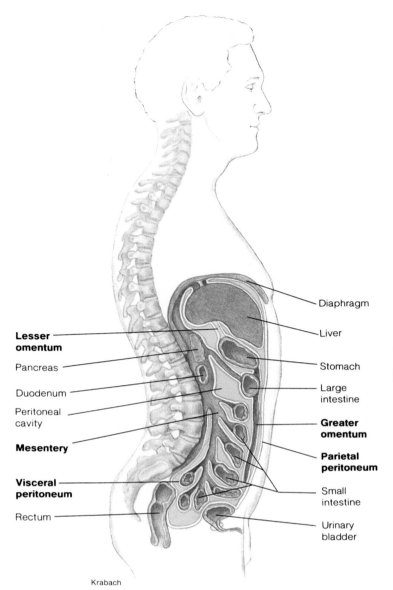

Krabach

**FIGURE 6.10**
Visceral organs of the abdominal cavity and the supporting serous membranes. The serous membranes are labeled in bold face.

exocrine glands are divided into *simple* and *compound glands.* The ducts of the simple glands do not branch or have only a few branches, whereas those of the compound type have multiple branches (fig. 6.12). Multicellular glands are further classified according to the shape of the secretory portion. They are identified as *tubular* if the secretory portion resembles the ductule portion and as *acinar* if the secretory portion is flasklike. Multicellular glands with a secretory portion that resembles both a tube and a flask are termed *tubuloacinar.*

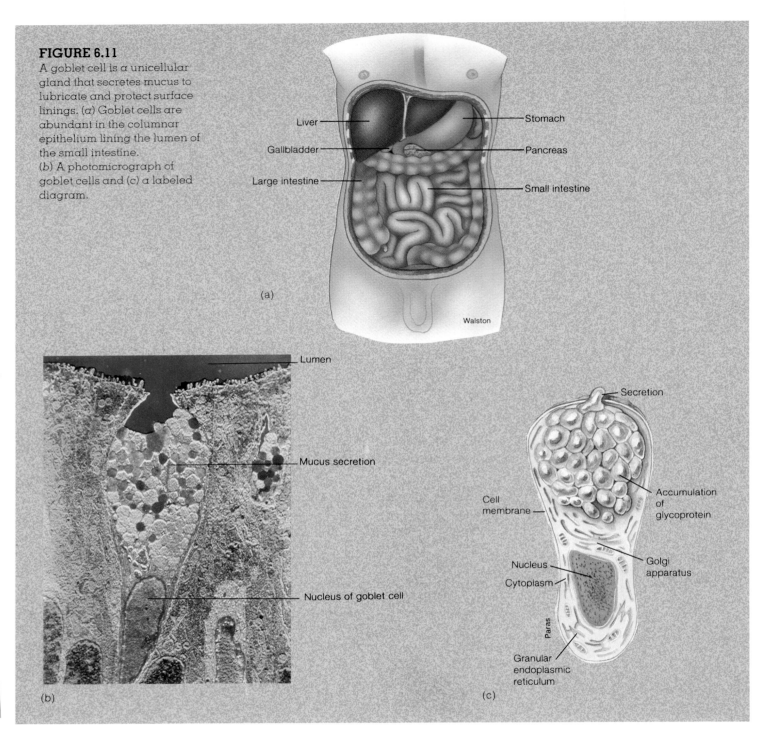

**FIGURE 6.11**
A goblet cell is a unicellular gland that secretes mucus to lubricate and protect surface linings. (*a*) Goblet cells are abundant in the columnar epithelium lining the lumen of the small intestine.
(*b*) A photomicrograph of goblet cells and (*c*) a labeled diagram.

Liver

Gallbladder

Large intestine

Stomach

Pancreas

Small intestine

Walston

(a)

Lumen

Mucus secretion

Nucleus of goblet cell

(b)

Secretion

Cell membrane

Nucleus

Cytoplasm

Paras

Granular endoplasmic reticulum

Accumulation of glycoprotein

Golgi apparatus

(c)

Multicellular exocrine glands are also classified according to the means by which they discharge the secretory product (fig. 6.13). Glands that secrete their products by exocytosis through the cell membrane of the secretory cells are called **merocrine** (*mer´ŏ-krin*) **glands.** Salivary glands, pancreatic glands, and certain sweat glands are of this type. In **apocrine** (*ap´ŏ-krin*) **glands** the secretion accumulates on the surface of the secretory cell, and then a portion of the cell, along with the secretion, is pinched off to be discharged. Mammary glands and certain sweat glands are apocrine glands. In a **holocrine** (*hol´ŏ-krin*) **gland,** the entire secretory cell is discharged

merocrine: Gk. *meros*, part; *krinein*, to separate

apocrine: Gk. *apo*, off; *krinein*, to separate
holocrine: Gk. *holos*, whole; *krinein*, to separate

118

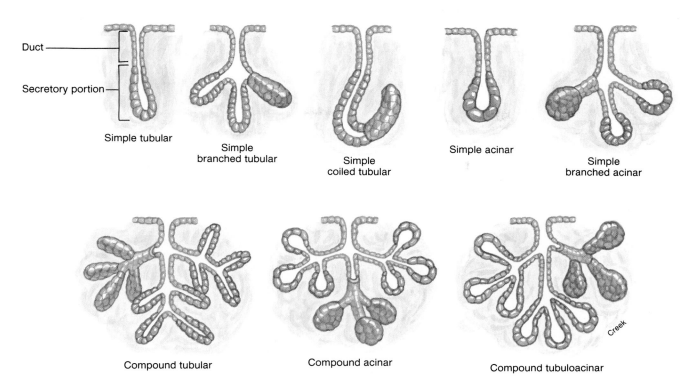

**FIGURE 6.12**

Structural classification of multicellular exocrine glands. The secretory portions of the simple glands either do not branch or have only a few branches, whereas those of the compound type have multiple branches.

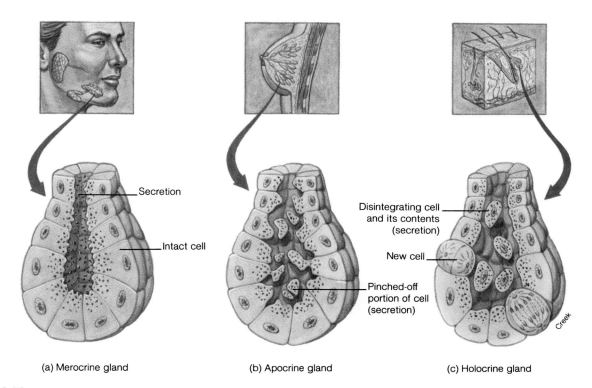

**FIGURE 6.13**

Secretory classification of multicellular exocrine glands: (a) merocrine gland, (b) apocrine gland, and (c) holocrine gland.

## Table 6.3 Summary of glandular epithelia

### Classification of exocrine glands by structure

| Type | Function | Example |
|---|---|---|
| I. Unicellular | Lubricate and protect | Goblet cells of digestive, respiratory, urinary, and reproductive systems |
| II. Multicellular | Protect, cool body, lubricate, aid in digestion, maintain body homeostasis | Sweat glands, digestive glands, mammary glands, sebaceous glands |
|   A. Simple | | |
|     1. Tubular | Aid in digestion | Intestinal glands |
|     2. Branched tubular | Protect, aid in digestion | Uterine glands, gastric glands |
|     3. Coiled tubular | Regulate temperature | Certain sweat glands |
|     4. Acinar | Provide additive for spermatozoa | Seminal vesicle of male reproductive system |
|     5. Branched acinar | Condition skin | Sebaceous skin glands |
|   B. Compound | | |
|     1. Tubular | Lubricate urethra of male, assist body digestion | Bulbourethral gland of male reproductive system, liver |
|     2. Acinar | Provide nourishment for infant, aid in digestion | Mammary gland, salivary gland (sublingual and submandibular) |
|     3. Tubuloacinar | Aid in digestion | Salivary gland (parotid), pancreas |

### Classification of exocrine glands by mode of secretion

| Type | Description of secretion | Example |
|---|---|---|
| Merocrine glands | Watery secretion for regulating temperature or enzymes that promote digestion | Salivary and pancreatic glands, certain sweat glands |
| Apocrine glands | Portion of secretory cell and secretion are discharged; provides nourishment for infant, assists in regulating temperature | Mammary glands, certain sweat glands |
| Holocrine glands | Entire secretory cell with enclosed secretion is discharged; conditions skin | Sebaceous glands of the skin |

along with the secretory product. An example of a holocrine gland is a sebaceous, or oil-secreting, gland of the skin.

Glandular epithelia are summarized in table 6.3.

# Connective Tissues

*Connective tissues are classified according to the characteristics of the matrix that binds the cells. Connective tissues provide structural and metabolic support for other tissues and organs of the body.*

## Characteristics and Classification of Connective Tissues

Connective tissue is found throughout the body and, as the name indicates, supports or binds other tissues and provides for the metabolic needs of all body organs. Certain types of connective tissue store nutritional substances, whereas other types manufacture protective and regulatory materials.

Although connective tissue varies tremendously in structure and function, all connective tissues have similarities. With the exception of mature cartilage, connective

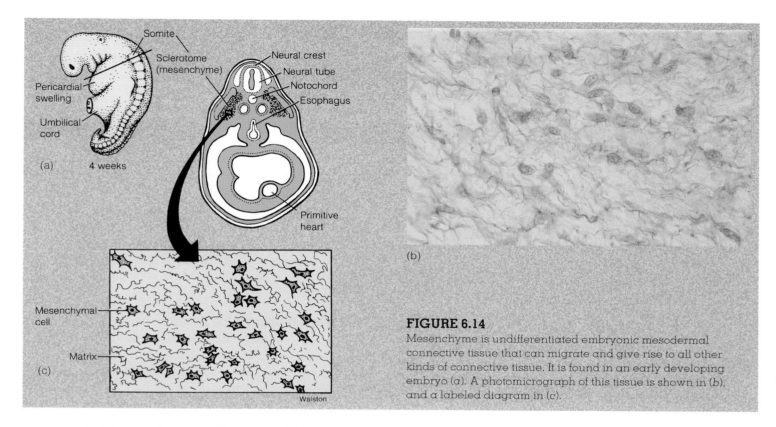

**FIGURE 6.14**

Mesenchyme is undifferentiated embryonic mesodermal connective tissue that can migrate and give rise to all other kinds of connective tissue. It is found in an early developing embryo (*a*). A photomicrograph of this tissue is shown in (*b*), and a labeled diagram in (*c*).

tissues are highly vascular and well nourished. They are able to replicate and, by so doing, are responsible for the repair of body organs. Unlike epithelial tissues, which are composed of tightly fitted cells, connective tissues contain considerably more matrix (intercellular material) than cells. Connective tissues do not occur on free surfaces of body cavities or on the surface of the body, as do epithelial tissues. Furthermore, connective tissues are embryonically derived from mesoderm, whereas epithelial tissues derive from ectoderm, mesoderm, and endoderm.

The classification of connective tissues is not exact, and several schemes have been devised. In general, however, they are named according to the kind and arrangement of the matrix. The following are the basic kinds of connective tissues.

A. Embryonic connective tissue

B. Connective tissue proper
   1. Loose (areolar) connective tissue
   2. Dense regular connective tissue
   3. Dense irregular connective tissue
   4. Elastic connective tissue
   5. Reticular connective tissue
   6. Adipose connective tissue

C. Cartilage
   1. Hyaline cartilage
   2. Fibrocartilage
   3. Elastic cartilage

D. Bone

E. Vascular (blood) tissue

## Embryonic Connective Tissue

The embryonic period of development, which lasts 6 weeks (from the beginning of the third to the end of the eighth week), is characterized by a tremendous amount of tissue differentiation and organ formation. At the beginning of the embryonic period, all connective tissue appears the same and is referred to as **mesenchyme** (*mez´en-kīm*). Mesenchyme is undifferentiated embryonic connective tissue that is derived from mesoderm. It consists of irregularly shaped cells lying in large amounts of a homogeneous, jellylike matrix (fig. 6.14). In certain periods of development, mesenchyme migrates to predisposed sites where it interacts with other tissues to form organs. Before the end of the embryonic period, once mesenchyme has migrated to the appropriate position, it differentiates, and from it all other kinds of connective tissues are formed.

Some mesenchymal-like tissue persists past the embryonic period in certain sites within the body. Good examples are the undifferentiated cells that surround blood vessels and form fibroblasts if the vessels are traumatized. Fibroblasts assist in healing wounds (see chapter 7).

Another kind of prenatal connective tissue exists only in the fetus (the fetal period is from 9 weeks to birth) and is called *mucous connective tissue*, or *Wharton's jelly*. It gives the umbilical cord a turgid consistency.

· · · · · · · · · · · · · · · · · · · · · · · · · · · · · · · · · · · · ·

reticular: L. *rete*, net or netlike
Wharton's jelly: from Thomas Wharton, English anatomist, 1614–73

# Connective Tissue Proper

**Connective tissue proper** has a loose, flexible matrix, frequently called **ground substance.** The most common cell within connective tissue proper is called a **fibroblast** (*fi´bro-blast*). Fibroblasts are large, star-shaped cells that produce collagenous, elastic, and reticular (*rĕ-tik´yŭ-lar*) fibers. **Collagenous** (*kŏ-laj´ĕ-nus*) **fibers** are composed of a protein called *collagen* (*col´ă-jen*); they are flexible, yet they have tremendous strength. **Elastic fibers** are composed of a protein called *elastin*, which provides certain tissues with elasticity. Collagenous and elastic fibers may be either sparse and irregularly arranged, as in loose connective tissue, or tightly packed, as in dense connective tissue. Tissues with loosely arranged fibers generally form packing material that cushions and protects various organs, whereas those that are tightly arranged form the binding and supportive connective tissues of the body.

Resilience in tissues that contain elastic fibers is extremely important for a number of physical body functions. Consider, for example, that elastic fibers are found in the walls of arteries and in the walls of the lower respiratory passageways. As these walls are expanded by blood moving through vessels or by inspired air, the elastic fibers must first stretch and then recoil. This maintains the pressures of the fluid or air moving through the lumina, thus ensuring adequate flow rates and rates of diffusion through capillary and lung surfaces.

*Reticular fibers* reinforce by forming thin, short threads that branch and join to form a delicate lattice or reticulum. Reticular fibers are common in lymphatic glands, where they form a meshlike center called the **stroma.**

Six basic types of connective tissue proper are generally recognized. These tissues are distinguished by the consistency of the ground substance and the type and arrangement of the reinforcement fibers.

### Loose Connective (Areolar) Tissue

Loose connective tissue is distributed throughout the body as a binding and packing material. It binds the skin to the underlying muscles and is highly vascular, providing nutrients to the skin. Loose connective tissue surrounding muscle fibers and muscle groups is known as **fascia** (*fash´e-ă*). It also surrounds blood vessels and nerves, where it provides both protection and nourishment. Specialized cells called **mast cells** are dispersed throughout the loose connective tissue surrounding blood vessels. Mast cells produce *heparin* (*hep´ă-rin*), an anticoagulant that prevents blood from clotting within the vessels. They also produce *histamine*, which is released during inflammation and acts as a powerful vasodilator.

The cells of loose connective tissue are predominantly fibroblasts, with collagenous and elastic fibers dispersed throughout the ground substance (fig. 6.15). The irregular arrangement of this tissue provides flexibility, yet strength, in any direction. It is this tissue layer, for example, that permits the skin to move when a part of the body is rubbed.

Much of the fluid of the body is found within loose connective tissue and is called *tissue fluid.* Sometimes excessive tissue fluid accumulates, causing a swelled condition called edema (*ĕ-de´mă*). Edema is symptomatic of a variety of dysfunctions or disease processes.

### Dense Regular Connective Tissue

Dense regular connective tissue is characterized by large amounts of densely packed collagenous fibers lying parallel to the direction of force placed on this tissue during body movement. Because this tissue is silvery white in appearance, it is sometimes called white fibrous connective tissue.

Dense regular connective tissue occurs where strong, flexible support is necessary (fig. 6.16). **Tendons,** which attach muscles to bones and transfer the forces of muscle contractions, and **ligaments,** which connect bone to bone across articulations, are composed of this type of tissue.

Trauma to ligaments, tendons, and muscles are common sports-related injuries. A *strain* is an excessive stretch of the tissue composing the tendon or muscle, with no serious damage. A *sprain* is a tearing of the tissue of a ligament and may be slight, moderate, or complete. A complete tear of a major ligament is especially painful and disabling. Ligamentous tissue does not heal itself well because it has a poor blood supply. Surgical reconstruction is generally needed for the treatment of a severed ligament.

### Dense Irregular Connective Tissue

Dense irregular connective tissue is characterized by large amounts of densely packed collagenous fibers that are interwoven to provide tensile strength in any direction. This tissue is found in the dermis of the skin, submucosa of the GI tract, and composing the fibrous capsules of organs and joints (fig. 6.17). It also makes up the collagenous matrix of bone, called *osteoid.*

collagen: Gk. *kolla*, glue
elastin: Gk. *elasticus*, to drive
stroma: Gk. *stroma*, a couch or bed
fascia: L. *fascia*, band or girdle

heparin: Gk. *hepatos*, the liver
tendon: L. *tendere*, to stretch
ligament: L. *ligare*, bind

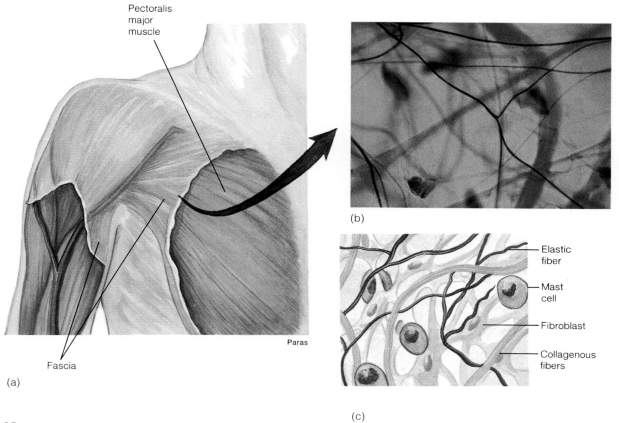

(a)

(b)

(c)

**FIGURE 6.15**

Loose connective tissue is packing and binding tissue that surrounds muscles (*a*), nerves, and vessels and binds the skin to the underlying muscles. (*b*) A photomicrograph of this tissue and (*c*) a labeled diagram.

Labels in (a): Pectoralis major muscle; Fascia; Paras

Labels in (c): Elastic fiber; Mast cell; Fibroblast; Collagenous fibers

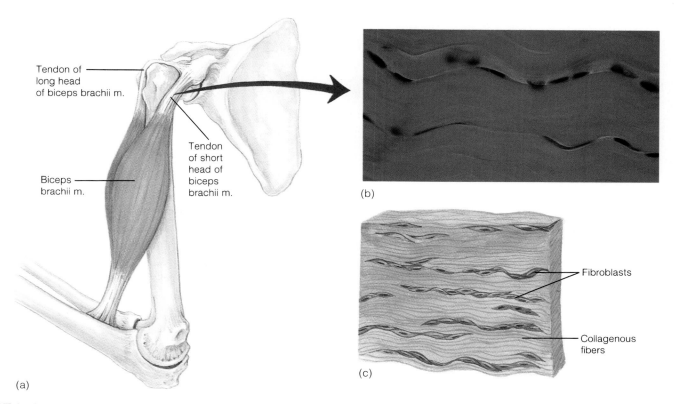

(a)

(b)

(c)

Labels in (a): Tendon of long head of biceps brachii m.; Tendon of short head of biceps brachii m.; Biceps brachii m.

Labels in (c): Fibroblasts; Collagenous fibers

**FIGURE 6.16**

Dense regular connective tissue forms the strong and highly flexible tendons (*a*) and ligaments. (*b*) A photomicrograph of this tissue and (*c*) a labeled diagram.

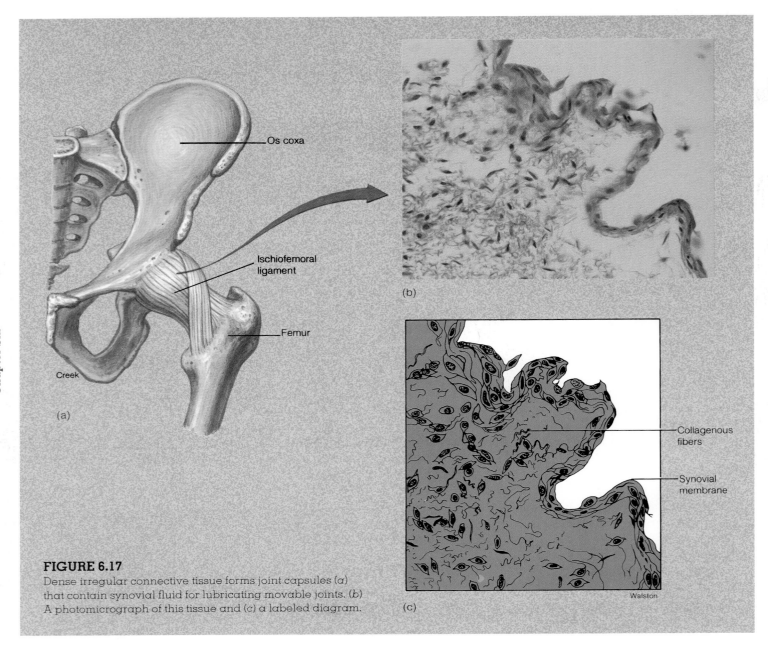

**FIGURE 6.17**
Dense irregular connective tissue forms joint capsules (a) that contain synovial fluid for lubricating movable joints. (b) A photomicrograph of this tissue and (c) a labeled diagram.

**Elastic Connective Tissue**   Elastic connective tissue has a predominance of elastic fibers that are irregularly arranged and yellowish in color (fig. 6.18). They can be stretched to one and a half times their original lengths and will snap back to their former size. Elastic connective tissue is found in the walls of large arteries, in portions of the larynx, and in the trachea and bronchial tubes of the lungs. It is also present between the arches of the vertebrae that make up the vertebral column.

**Reticular Connective Tissue**   Reticular connective tissue is characterized by a network of reticular fibers woven through a jellylike matrix (fig. 6.19). Certain specialized cells within reticular tissue are *phagocytic* (*fag″ŏ-sit´ik*), and therefore ingest foreign materials. The liver, spleen, lymph nodes, and bone marrow contain reticular connective tissue.

**Adipose Connective Tissue**   Adipose tissue is a specialized type of loose fibrous connective tissue that contains large quantities of **adipose cells,** or **adipocytes.** Adipose cells form from mesenchyme and, for the most part, are formed prenatally and during the first year of life. Adipose

. . . . . . . . . . . . . . . . . . . . . . . . . . . . . . . .

adipose: L. *adiposus,* fat

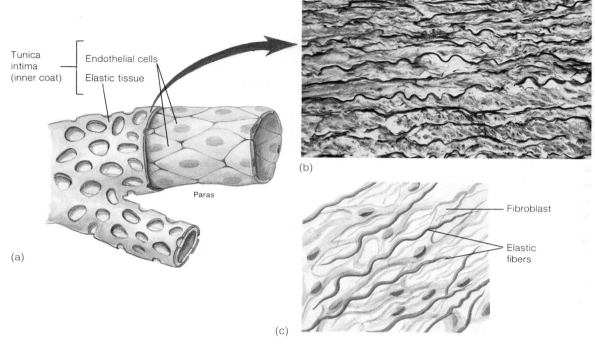

**FIGURE 6.18**

Elastic connective tissue permits stretching of a large artery (*a*) as blood flows through. (*b*) A photomicrograph of this tissue and (*c*) a labeled diagram.

Labels in figure (a): Tunica intima (inner coat), Endothelial cells, Elastic tissue, Paras

Labels in figure (c): Fibroblast, Elastic fibers

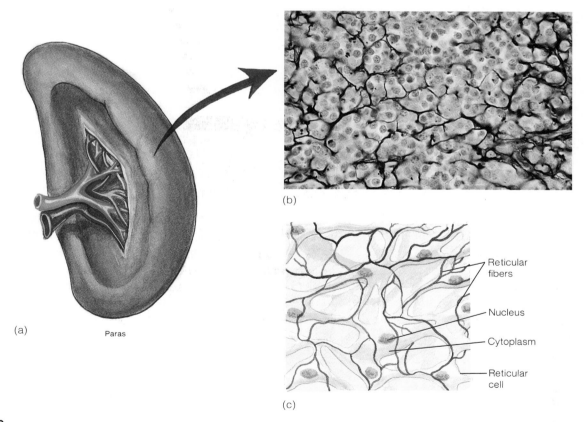

**FIGURE 6.19**

Reticular tissue forms the stroma, or framework, of organs such as the spleen (*a*), liver, thymus, and lymph nodes. (*b*) A photomicrograph of this tissue and (*c*) a labeled diagram.

Labels in figure (a): Paras

Labels in figure (c): Reticular fibers, Nucleus, Cytoplasm, Reticular cell

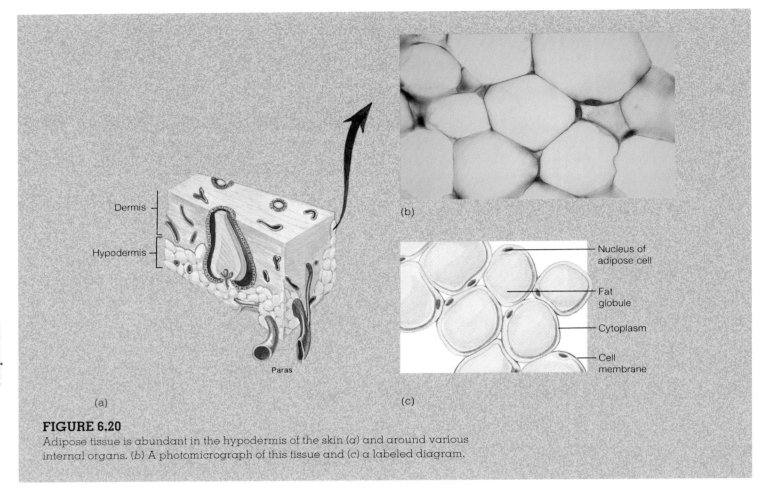

Dermis

Hypodermis

Paras

(a)

(b)

Nucleus of
adipose cell

Fat
globule

Cytoplasm

Cell
membrane

(c)

**FIGURE 6.20**
Adipose tissue is abundant in the hypodermis of the skin (*a*) and around various
internal organs. (*b*) A photomicrograph of this tissue and (*c*) a labeled diagram.

cells store droplets of fat within their cytoplasm, causing
them to swell and forcing their nuclei to one side (fig. 6.20).

Adipose tissue is found throughout the body but is concentrated around the kidney, in the hypodermis of the skin, on
the surface of the heart, surrounding joints, and in the breasts
of sexually mature females. Fat functions not only as a food
reserve but supports and protects various organs. Fat is a good
insulator against cold because it is a poor conductor of heat.

The excessive fat of obesity is a significant risk factor
in cardiovascular disease, in diabetes mellitus, and
in endometrial and breast cancer. For these reasons,
good exercise programs and sensible diets are
extremely important. Adipose tissue can also retain
environmental pollutants that are ingested or absorbed
through the skin. Dieting eliminates the fat stored within the
tissue but not the tissue itself.

The surgical procedure of *suction lipectomy* may be used
to remove small amounts of adipose tissue from certain
localized body areas, such as the breasts, abdomen, buttocks,
or thighs. Suction lipectomy is used for cosmetic purposes
rather than as a treatment for obesity, and the risks for
potentially detrimental side-effects need to be seriously
considered. Potential candidates should be between 30 and
40 years old and only about 15 to 20 pounds overweight. They
should also have good skin elasticity.

The characteristics, functions, and locations of connective tissue proper are summarized in table 6.4.

## Cartilage Tissues

**Cartilage tissue** consists of cartilage cells, or **chondrocytes**
(*kon´dro-sĭtz*), and a semisolid matrix that imparts marked
elastic properties to the tissue. Cartilage is a supportive and
protective connective tissue that is frequently associated with
bone. It forms a precursor of one type of bone and persists at
the articular surfaces on the bones of all movable joints.

The chondrocytes within cartilage may occur singly
but are frequently clustered. Chondrocytes occupy cavities,
called **lacunae** (*lă-kyoo´ne*), within the matrix. Most cartilage tissue is surrounded by a dense regular connective tissue
called **perichondrium** (*per˝ ĭ-kon´dre-um*). Cartilage at the
articular surfaces of bones (articular cartilage) lacks peri-
chondria. Because mature cartilage is avascular, it must
receive nutrients through diffusion from the perichondrium

..............................................................

lacuna: L. *lacuna*, hole or pit

## Table 6.4 Summary of connective tissue proper

| Type | Structure and function | Location |
|------|------------------------|----------|
| Loose connective (areolar) tissue | Predominantly fibroblast cells with lesser amounts of collagen and elastin proteins; binds organs, holds tissue fluids | Surrounding nerves and vessels, between muscles, beneath the skin |
| Dense regular connective tissue | Densely packed collagenous fibers lying parallel to the direction of force; provides strong, flexible support | Tendons, ligaments |
| Dense irregular connective tissue | Densely packed collagenous fibers arranged in a tight interwoven pattern; provides tensile strength in any direction | Dermis of skin, fibrous capsules of organs and joints |
| Elastic connective tissue | Predominantly irregularly arranged elastic fibers; supports, provides framework | Large arteries, lower respiratory tract, between the arches of vertebrae |
| Reticular connective tissue | Reticular fibers forming supportive network; stores, performs phagocytic function | Lymph nodes, liver, spleen, thymus, bone marrow |
| Adipose connective tissue | Adipose cells; protects, stores fat, insulates | Hypodermis of skin, surface of heart, omentum, around kidneys, back of eyeball, surrounding joints |

and the surrounding tissue. For this reason, cartilaginous tissue has a slow rate of mitotic activity and, if damaged, heals with difficulty.

There are three kinds of cartilage: hyaline cartilage, fibrocartilage, and elastic cartilage. Each is distinguished by the type and amount of fibers embedded within the matrix.

**Hyaline Cartilage** Hyaline (hi´ă-lĭn) cartilage has a homogeneous, bluish-white matrix in which the collagenous fibers are so fine that they can be observed only with an electron microscope. When viewed through a microscope, hyaline cartilage has a clear, glassy appearance (fig. 6.21).

Hyaline cartilage is the most abundant cartilage within the body and is commonly called "gristle." It covers the articular surfaces of bones, supports the tubular trachea and bronchi of the respiratory system, reinforces the nose, and forms the flexible bridge, called **costal cartilage,** between the ventral end of each of the first 10 ribs and the sternum. Most of the bones of the body form first as hyaline cartilage and later become bone in a process called *ossification*.

**Fibrocartilage** Fibrocartilage has its matrix reinforced with numerous collagenous fibers (fig. 6.22). It is a durable tissue, adapted to withstand tension and compression. It is

......................................................

hyaline: Gk. *hyalos,* glass

found at the symphysis pubis, where the two pelvic bones articulate, and between the vertebrae as intervertebral discs. It also forms the cartilaginous wedges within the knee joint, called *menisci* (see chapter 11).

 By the end of the day, the intervertebral discs of the vertebral column are somewhat compacted. So a person is actually slightly shorter in the evening than in the morning following a recuperative rest. With aging comes a gradual and irreversible compression of the fibrocartilaginous discs.

**Elastic Cartilage** Elastic cartilage is similar to hyaline except for the presence of abundant elastic fibers that make elastic cartilage very flexible while maintaining its strength (fig. 6.23). The numerous elastic fibers also give it an unstained yellowish appearance. This tissue is found in the outer ear, portions of the larynx, and in the auditory canal.

The structure, function, and location of cartilage is summarized in table 6.5.

## Bone Tissue

**Bone tissue,** or **osseous connective tissue,** is the most rigid of the connective tissues. Unlike cartilage, bone has a rich vascular supply and is the site of considerable metabolic activity. The hardness of bone is largely due to the inorganic calcium phosphate (calcium hydroxyapatite)

## FIGURE 6.21

Hyaline cartilage is the most abundant cartilage within the body. It occurs in such places as the larynx (*a*), trachea, rib cage and embryonic skeleton. (*b*) A photomicrograph of this tissue and (*c*) a labeled diagram.

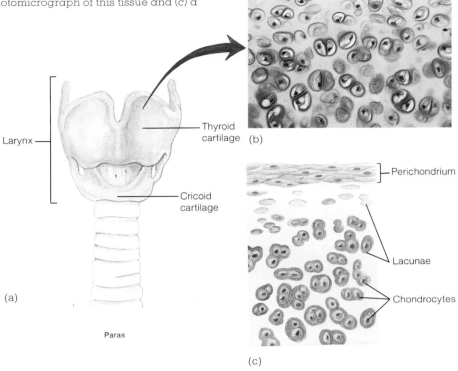

Larynx

Thyroid cartilage (b)

Cricoid cartilage

(a)

Paras

Perichondrium

Lacunae

Chondrocytes

(c)

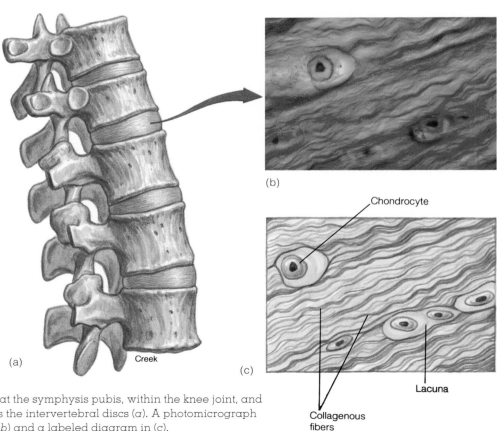

(b)

Chondrocyte

(a)            Creek            (c)

Lacuna

Collagenous fibers

## FIGURE 6.22

Fibrocartilage is located at the symphysis pubis, within the knee joint, and between the vertebrae as the intervertebral discs (*a*). A photomicrograph of this tissue is shown in (*b*) and a labeled diagram in (*c*).

## Table 6.5 Summary of cartilage tissue

| Type | Structure and function | Location |
|------|------------------------|----------|
| Hyaline cartilage | Homogeneous matrix with extremely fine collagenous fibers; provides flexible support, protects, is precursor to bone | Articular surfaces of bones, nose, walls of respiratory passages, fetal skeleton |
| Fibrocartilage | Abundant collagenous fibers within matrix; supports, withstands compression | Symphysis pubis, intervertebral discs, knee joint |
| Elastic cartilage | Abundant elastic fibers within matrix; supports, provides flexibility | Framework of outer ear, auditory canal, portions of larynx |

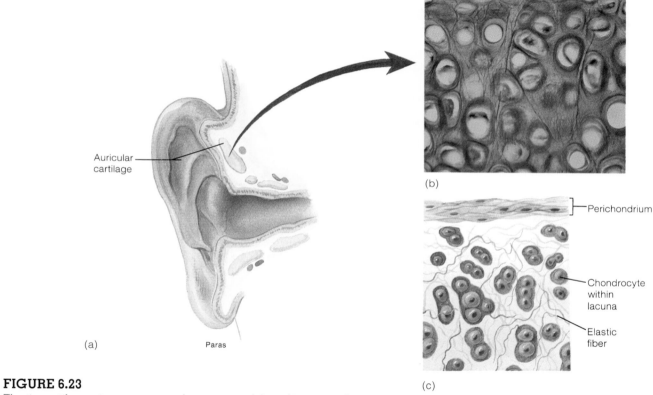

**FIGURE 6.23**

Elastic cartilage gives support to the outer ear (*a*), auditory canal, and parts of the larynx. (*b*) A photomicrograph of this tissue and (*c*) a labeled diagram.

deposited within the intercellular matrix. Numerous collagenous fibers, also embedded within the matrix, give some flexibility to bone.

 When bone is placed in a weak acid, the calcium salts dissolve away and the bone becomes pliable. It retains its basic shape but can be easily bent and twisted. In calcium deficiency diseases, such as *rickets*, the bone tissue becomes pliable and bends under the weight of the body (see fig. 7.8).

The two types of bone tissue are classified according to porosity, and most bones have both types (fig. 6.24). **Compact, or dense, bone tissue** is the hard, outer layer, whereas **spongy,** or **cancellous, bone tissue** is the porous, highly vascular, inner portion. Compact bone tissue is covered by the periosteum. It provides for attachment of muscles and protects and strengthens the bone. Spongy bone tissue makes the bone lighter and provides a space for bone marrow, where blood cells are produced.

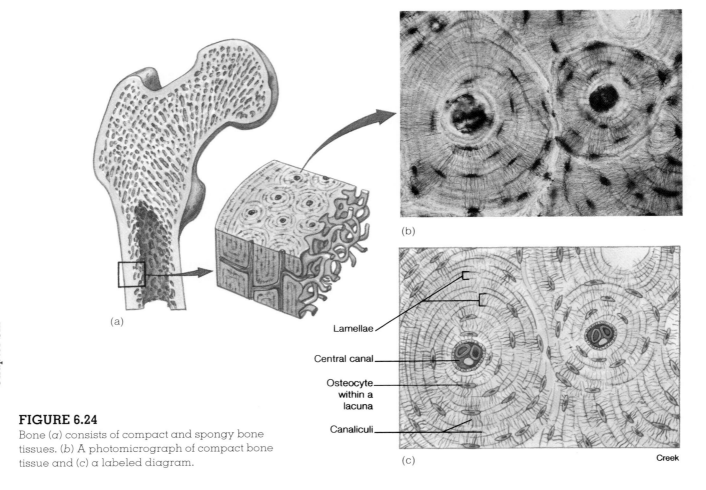

(a)

Lamellae

Central canal

Osteocyte within a lacuna

Canaliculi

(b)

(c)

Creek

**FIGURE 6.24**

Bone (*a*) consists of compact and spongy bone tissues. (*b*) A photomicrograph of compact bone tissue and (c) a labeled diagram.

In compact bone tissue, mature bone cells, called **osteocytes,** are arranged in concentric layers around a **central (haversian) canal,** which contains a vascular and nerve supply. Each osteocyte occupies a cavity called a **lacuna.** Radiating from each lacuna are numerous minute canals, called **canaliculi.** The canaliculi traverse the dense matrix of the bone tissue to adjacent lacunae. Nutrients diffuse through the canaliculi to reach each osteocyte. The inorganic matrix is deposited in concentric layers called **lamellae.** Spongy and compact bone tissues are described further in chapter 8.

## Vascular Connective Tissue

**Blood** is a highly specialized, viscous connective tissue. The cells, or **formed elements,** of vascular connective tissue are suspended in the liquid plasma matrix (fig. 6.25). Blood plays a vital role in maintaining internal body homeostasis. The three types of formed elements found within the blood are **erythrocytes,** (red blood cells), **leukocytes** (white blood

cells), and **thrombocytes** (platelets). Formed elements and the physiology of blood are discussed in detail in chapter 20.

 An injury to a portion of the body may stimulate tissue repair activity, usually involving connective tissue. A minor scrape or cut results in platelet and plasma activity of the exposed blood and the formation of a scab. The epidermis of the skin regenerates beneath the scab. A severe open wound eventuates connective tissue granulation. In this process, collagenous fibers form from surrounding fibroblasts to strengthen the traumatized area. The healed area is known as a *scar.*

# Muscle Tissues

*Muscle tissues are responsible for the movement of materials through the body, the movement of one part of the body with respect to another, and for locomotion. Fibers in the three kinds of muscle tissue are adapted to contract in response to stimuli.*

haversian canal: from Clopton Havers, English anatomist, 1650–1702
erythrocyte: Gk. *erythros*, red; *kytos*, hollow (cell)
leukocyte: Gk. *leukos*, white; *kytos*, hollow (cell)

thrombocyte: Gk. *thrombos*, a clot; *kytos*, hollow (cell)

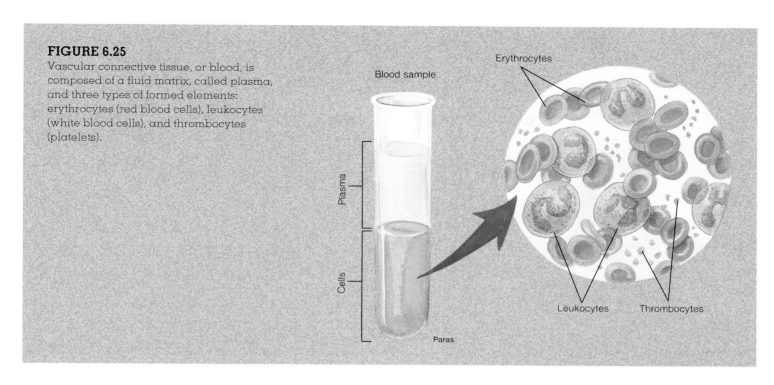

**FIGURE 6.25**
Vascular connective tissue, or blood, is composed of a fluid matrix, called plasma, and three types of formed elements: erythrocytes (red blood cells), leukocytes (white blood cells), and thrombocytes (platelets).

Blood sample

Plasma

Cells

Paras

Erythrocytes

Leukocytes

Thrombocytes

Muscle tissues are unique in possessing the property of *contractility*. The muscle cells, or *fibers*, are elongated in the direction of contraction. Movement is accomplished through the shortening of the fibers in response to a stimulus. Muscle tissue is derived from mesoderm. There are three types of muscle tissue in the body: *smooth, cardiac*, and *skeletal* (fig. 6.26).

**Smooth Muscle**   Smooth muscle tissue is common throughout the body, occurring in many of the systems. For example, in the wall of the GI tract it provides the motive power for the peristaltic movements involved in the mechanical digestion of food. Smooth muscle is also found in the walls of arteries, the walls of respiratory passages, and in the urinary and reproductive ducts. The contraction of smooth muscle is under autonomic (involuntary) nervous control, and is discussed in more detail in chapter 12.

Smooth muscle fibers are long, spindle-shaped cells. Each contains a single nucleus and lacks striations. These cells are usually grouped together in flattened sheets, forming the muscular portion of a wall around a lumen.

**Cardiac Muscle**   Cardiac muscle tissue makes up most of the wall of the heart. This tissue is characterized by bifurcating (branching) fibers, a centrally positioned nucleus, and transversely positioned **intercalated** (*in-ter´kă-lāt-ed*) **discs.** Intercalated discs help to hold adjacent cells together and transmit the force of contraction from cell to cell. Like skeletal muscle, cardiac muscle is striated, but unlike skeletal muscle it experiences rhythmical involuntary contractions. Cardiac muscle is further discussed in chapter 12.

**Skeletal Muscle**   Skeletal muscle tissue attaches to the skeleton and is responsible for voluntary body movements. Each elongated, multinucleated fiber has distinct transverse striations. Fibers of this muscle tissue are grouped into parallel fasciculi (bundles) that can be seen without a microscope in fresh muscle. Both cardiac and skeletal muscle fibers are striated and cannot replicate once tissue formation has been completed shortly after birth. Skeletal muscle tissue is discussed further in chapter 12. The three types of muscle tissue are summarized in table 6.6.

# Nervous Tissues

*Nervous tissue is composed of neurons, which respond to stimuli and conduct impulses to and from all body organs, and neuroglia, which functionally support and physically bind neurons.*

**Neurons**   Although there are several kinds of neurons (*noo´ronz*) in nervous tissue, they all have three principal components: (1) a cell body, or perikaryon; (2) dendrites; and (3) an axon (fig. 6.27*b*). **Dendrites** function to receive stimuli and conduct impulses to the cell body. The **cell body,** or **perikaryon** (*per´´ĭ-kar´e-on*), contains the nucleus and specialized organelles and microtubules. The **axon** is

..........................................................

neuron: Gk. *neuron*, sinew or nerve
perikaryon: Gk. *peri*, around; *karyon*, nut or kernel

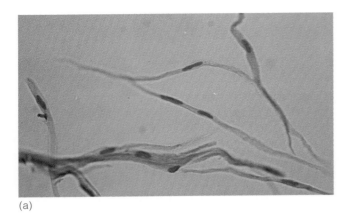
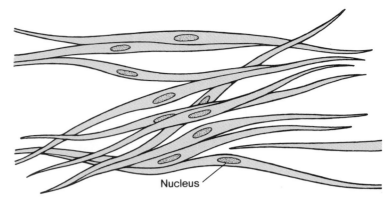

(a)

Nucleus

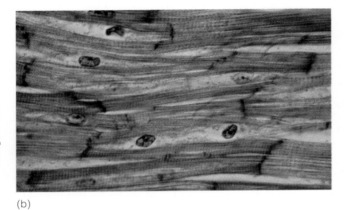
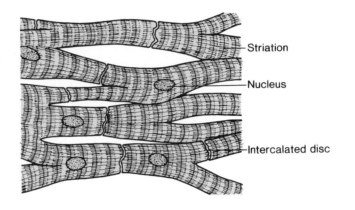

(b)

Striation

Nucleus

Intercalated disc

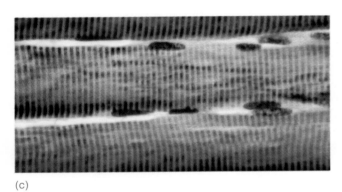
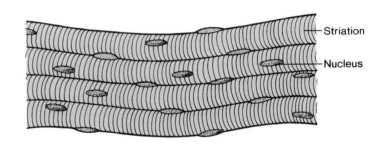

(c)

Striation

Nucleus

## FIGURE 6.26

The three types of muscle tissue: (a) smooth muscle fibers teased apart, (b) cardiac muscle, and (c) skeletal muscle.

## Table 6.6   Summary of muscle tissue

| Type | Structure and function | Location |
|------|------------------------|----------|
| Smooth muscle tissue | Elongated, spindle-shaped fiber with single nucleus; involuntary movements of internal organs | Walls of hollow internal organs |
| Cardiac muscle tissue | Branched, striated fiber with single nucleus and intercalated discs; involuntary rhythmic contraction | Heart muscle |
| Skeletal muscle tissue | Multinucleated, striated, cylindrical fiber that occurs in fasciculi; voluntary movement of skeletal parts | Associated with skeleton; spans joints of skeleton via tendons |

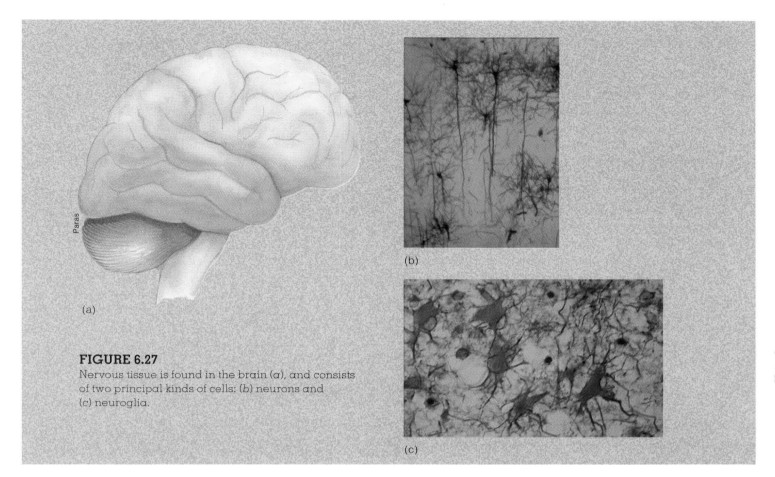

**FIGURE 6.27**
Nervous tissue is found in the brain (a), and consists of two principal kinds of cells: (b) neurons and (c) neuroglia.

a cytoplasmic extension that conducts impulses away from the cell body. The term **nerve fiber** refers to any process extending from the cell body of a neuron.

Neurons derive from ectoderm and are the basic structural and functional units of the nervous system. They are specialized to respond to physical and chemical stimuli, conduct nerve impulses, and perform other functions such as storing memory, thinking, and regulating the activity of other organs or glands. Of all the body's cells, neurons are probably the most specialized. As with muscle cells, the number of neurons is established shortly after birth; thereafter, they lack the ability to undergo mitosis, although under certain circumstances a severed portion can regenerate.

**Neuroglia** In addition to neurons, nervous tissue is composed of neuroglia (*noo-rog´le-ah*)(fig. 6.27c). Neuroglial cells are about five times as abundant as neurons and have limited mitotic abilities. They do not transmit impulses but support and bind neurons together. Certain neuroglial cells are phagocytic; others assist in providing sustenance to the neurons.

Neurons and neuroglia are discussed in detail in chapter 14.

••••••••••••••••••••••••••••••••••

neuroglia: Gk. *neuron*, nerve; *glia*, glue
atrophy: Gk. *a*, without; *trophe*, nourishment

# Clinical Considerations

As discussed at the beginning of this chapter, the study of histology is extremely important in understanding organ and system structure and function. Histology has immense clinical importance as well. Many diseases are diagnosed through microscopic examination of tissue sections. Even in performing an autopsy, an examination of various tissues may be needed to establish the cause of death.

Several sciences are concerned with specific aspects of tissues. *Histopathology* is the study of diseased tissues. *Histochemistry* is concerned with the biochemical physiology of tissues as they function to maintain body homeostasis. *Histotechnology* studies the various ways tissues can be better stained and observed. In all of these disciplines, a thorough understanding of normal, or healthy, tissues is imperative for recognizing altered, or abnormal, tissues.

## Changes in Tissue Composition

Most diseases alter tissue structure *locally* where the disease is prevalent, although some diseases, called *general conditions*, cause changes remote from the locus of the disease. **Atrophy** (wasting of body tissue), for example, may be confined to a particular organ where the disease interferes with the metabolism of that organ; or it may involve an entire limb if nourishment or nerve impulses are decreased or prohibited. *Muscle atrophy,* for example, can be caused by a disease of the nervous system such as polio, or it can be the result of a diminished blood supply to a muscle. *Senescence (sĕ-nes´ens) atrophy,* or simply *senescence,* is the natural aging of tissues and organs within the body. *Disuse atrophy* is a local atrophy that results from the inactivity of a tissue or organ. Muscular dystrophy causes a disuse atrophy that decreases muscle size and strength due to the loss of sarcoplasm within the muscle.

**Necrosis** (nĕ-kro´sis) is cellular or tissue death within the living body. It can be recognized by changes in the dead tissues. Necrosis can be caused by a number of factors, such as severe injury; physical agents (trauma, heat, radiant energy, chemical poisons); or interference with the nutrition of tissues. When examined histologically, the necrotic tissue usually appears opaque, and a whitish or yellow color is assumed. **Gangrene** is a massive necrosis of tissue accompanied by an invasion of microorganisms that live on decaying tissues.

**Somatic death** is the death of the body as a whole. Following somatic death, tissues undergo irreversible changes such as **rigor mortis** (muscular rigidity), clotting of the blood, and cooling of the body. Postmortem changes occur under varying conditions at predictable rates, which is useful in estimating the approximate time of death.

## Tissue Analysis

In diagnosing a disease, it is frequently important to examine tissues from a living person histologically. When this is necessary, a **biopsy** (removal of a section of living tissue) is performed. (The term **biopsy** is also used to refer to the living tissue removed.) There are several techniques for obtaining a biopsy. *Surgical removal* is usually done on large masses or tumors. *Curettage* involves cutting and scraping tissue, as may be done in examining for uterine cancer. In a *percutaneous needle biopsy,* a biopsy needle is inserted through a small skin incision and tissue samples are aspirated. Both normal and diseased tissues are removed for purposes of comparison.

Preparing tissues for examination involves a number of steps. **Fixation** is fundamental for all histological preparation. It is the rapid killing, hardening, and preservation of tissue to maintain its existing structure. **Embedding** the tissue in a supporting medium such as paraffin wax usually

..............................................

necrosis: Gk. *nekros,* corpse
gangrene: Gk. *gangraina,* gnaw or eat

follows fixation. The next step is **sectioning** the tissue into extremely thin slices, followed by **mounting** the specimen on a slide. Some tissues are fixed by rapid freezing and then sectioned while frozen, making embedding unnecessary. Frozen sections enable the pathologist to make a quick diagnosis during a surgical operation.

These are done frequently, for example, in cases of suspected breast cancer. **Staining** is the next step. The hematoxylin and eosin (H & E) stains are routinely used on all tissue specimens. They give a differential blue and red color to the basic and acidic structures within the tissue. Other dyes may be needed to stain for specific structures.

**Examination** is first done with the unaided eye and then with a microscope. Practically all histological conditions can be diagnosed with low magnification (25×). Higher magnification is used to clarify specific details. Further examination may be performed with an electron microscope, which makes visible the cellular structures that are the morphological bases of metabolic processes. Histological observation is the foundation of subsequent diagnosis, prognosis, treatment, and reevaluation.

## Tissue Transplantation

In the last two decades, medical science has made tremendous advancements in tissue transplants. Tissue transplants are necessary for replacing nonfunctional, damaged, or lost body parts. The most successful transplant is one where tissue is taken from one place on a person's body and moved to another place, such as a skin graft from the thigh to replace burned tissue of the hand. This type of transplant is termed an **autotransplant. Isotransplants** are transplants between individuals who are closely related genetically. Identical twins have the best acceptance success in this type of transplant. **Homotransplants** (between individuals of the same species) and **heterotransplants** (between two different species) have low acceptance percentages because of a *tissue-rejection reaction.* When this occurs, the recipient's immune mechanisms are triggered, and the donor's tissue is identified as foreign and is destroyed. The reaction can be minimized by "matching" recipient and donor tissue. Immunosuppressive drugs also may lessen the rejection rate. These drugs act by interfering with the recipient's immune mechanisms. Unfortunately, immunosuppressive drugs may also lower the recipient's resistance to infections. New techniques involving blood transfusions from donor to recipient before transplant are proving successful. In any event, tissue transplants are an important aspect of medical research, and significant breakthroughs are on the horizon.

# Chapter Summary

## Definition and Classification of Tissues (p. 107)

1. Tissues are aggregations of similar cells that perform specific functions. The study of tissues is called histology.
2. Cells are separated and bound together by an intercellular matrix, the composition of which varies from solid to liquid.
3. The four principal types of tissues are epithelial tissues, connective tissues, muscle tissues, and nervous tissues.

## Epithelial Tissues (pp. 108–120)

1. Epithelia derive from all three germ layers and may be one or several layers thick. The lower surface of membranous epithelia is supported by a basement membrane.
   a. Simple epithelia vary in shape and surface characteristics. They are located where diffusion, filtration, and secretion occur.
   b. Stratified epithelia consist of two or more layers of cells and are adapted for protection.
   c. Transitional epithelium lines the urinary bladder and is adapted for distension.
2. The body has two principal types of membranes: mucous membranes, which secrete protective mucus, and serous membranes, which line the thoracic and abdominopelvic cavities and cover visceral organs.
3. Glandular epithelia derive from developing epithelial tissue and function as secretory exocrine glands.

## Connective Tissues (pp. 120–130)

1. Connective tissues derive from mesenchymal mesoderm and, with the exception of cartilage, are highly vascular.
2. Connective tissue proper contains fibroblasts, collagenous fibers, and elastic fibers within a flexible ground substance.
3. Cartilage tissue provides a flexible framework for many organs. It consists of a semisolid matrix of chondrocytes and various fibers.
4. Bone (osseous) tissue consists of osteocytes, collagenous fibers, and a durable matrix of mineral salts.
5. Vascular (blood) tissue consists of formed cellular elements (erythrocytes, leukocytes, and thrombocytes) suspended in a fluid plasma matrix.

## Muscle Tissues (pp. 130–131)

1. Muscle tissues (smooth, cardiac, and skeletal) are responsible for the movement of materials through the body, the movement of one part of the body with respect to another, and for locomotion.
2. Fibers in the three kinds of muscle tissues are adapted to contract in response to stimuli.

## Nervous Tissues (pp. 131–133)

1. Neurons are the functional units of the nervous system. They respond to stimuli and conduct impulses to and from all body organs.
2. Neuroglia support and bind neurons. Some are phagocytic; others provide sustenance to neurons.

# Review Activities

## Objective Questions

1. Which of the following is *not* a principal type of body tissue?
   a. nervous tissue
   b. integumentary tissue
   c. connective tissue
   d. muscular tissue
   e. epithelial tissue
2. Which of the following is a *false* statement regarding tissues?
   a. Tissues are aggregations of similar kinds of cells that perform specific functions.
   b. All tissues are microscopic and are studied within the science of histology.
   c. All tissues are stationary within the body at the location of their developmental origin.
   d. A body organ is composed of two or more kinds of tissues.
3. Connective tissues, muscle tissues, and the dermis of the skin derive from embryonic
   a. mesoderm.
   b. endoderm.
   c. ectoderm.
4. Which statement is *false* regarding epithelia?
   a. They are derived from mesoderm, ectoderm, and endoderm.
   b. They are strengthened by elastic and collagenous fibers.
   c. One side is exposed to the lumen, cavity, or external environment.
   d. They have very few extracellular matrix-binding cells.
5. A gastric ulcer of the stomach would involve
   a. simple cuboidal epithelium.
   b. transitional epithelium.
   c. simple ciliated columnar epithelium.
   d. simple columnar epithelium.
6. Which structural and secretory designation describes mammary glands?
   a. acinar, apocrine
   b. tubular, holocrine
   c. tubular, merocrine
   d. acinar, holocrine
7. Dense regular connective tissue is found in
   a. blood vessels.
   b. the spleen.
   c. tendons.
   d. the wall of the uterus.
8. The phagocytic connective tissue found in the lymph nodes, liver, spleen, and bone marrow is
   a. reticular connective tissue.
   b. areolar connective tissue.
   c. mesenchyme.
   d. elastic connective tissue.

9. Cartilage is slow in healing following an injury because
   a. it is located in body areas that are under constant physical strain.
   b. it is avascular.
   c. its chondrocytes cannot reproduce.
   d. it has a semisolid matrix.
10. Cardiac muscle tissue has
   a. striations.
   b. intercalated discs.
   c. rhythmical involuntary contractions.
   d. all of the above.

## Essay Questions

1. Define *tissue*. What are the differences between cells, tissues, glands, and organs?
2. What physiological functions are epithelial tissues adapted to perform?
3. Identify the epithelial tissue
   a. in the alveoli of the lungs;
   b. lining the lumen of the GI tract;
   c. in the outer layer of skin;
   d. in the urinary bladder;
   e. in the uterine tube; and
   f. lining the lumina of the lower respiratory tract.
   Describe the function of the tissue in each case.
4. Why are both keratinized and nonkeratinized epithelia found within the body?
5. Describe how epithelial glands are classified according to structural complexity and secretory function.
6. Identify the connective tissue
   a. on the surface of the heart and surrounding the kidneys;
   b. within the lumen of the aorta;
   c. forming the symphysis pubis;
   d. supporting the outer ear;
   e. forming the lymph nodes; and
   f. forming the tendo calcaneus.
   Describe the function of the tissue in each case.
7. Compare and contrast the following: reticular fibers, collagenous fibers, elastin, fibroblasts, and mast cells.
8. What is the relationship between adipose cells and fat? Discuss the function of fat and explain the potential danger of excessive fat.
9. Discuss the mitotic abilities of each of the four principal types of tissues.
10. Define the following terms: *atrophy, necrosis, gangrene,* and *somatic death.*

# integumentary system

## objectives

- Explain why the integument is considered an organ and a component of the integumentary system.

- Describe some common clinical conditions of the integument resulting from nutritional deficiencies or body dysfunctions.

- Describe the histological characteristics of each layer of the integument.

- Summarize the transitional events that occur within each of the epidermal layers.

- Discuss the role of the integument in the production of the body from disease and external injury, the regulation of body fluids and temperature, absorption, synthesis, sensory reception, and communication.

- Describe the structure of hair and list the three principal types.

- Describe the structure and function of nails.

- Compare and contrast the structure and function of the three principal kinds of integumentary glands.

# The Integument as an Organ

*The integument (skin) is the largest organ of the body. Together with its epidermal modifications (hair, glands, and nails), it constitutes the integumentary system. It has adaptive modifications in certain body areas that accommodate protective or metabolic functions. The integument is a dynamic interface between the continually changing external environment and the body's internal environment and helps to maintain homeostasis.*

We are more aware of and concerned with our integumentary system than perhaps any other system of our body. One of the first things we do in the morning is examine ourselves in a mirror and assess how to make our appearance presentable. Periodically, we examine our skin for wrinkles and our scalp for gray hairs as signs of aging. We recognize other people to a large extent by features of their skin.

The appearance of our skin frequently determines the initial impression we make on others. Unfortunately, it may also determine whether or not we succeed in gaining social acceptance. For example, social rejection during teenage years, imagined or real, can be directly associated with skin problems such as acne. One's image of oneself and consequent social behavior may be closely associated with physical appearance.

Even clothing styles are somewhat determined by how much skin we, or the designers, want to expose. But our skin is much more than a showpiece; it protects and regulates structures within the body.

The skin, or *integument* (in-teg´yoo-ment), and its associated structures (hair, glands, and nails) constitute the integumentary system. Included in this system are the millions of sensory receptors and a vascular network. The skin is a dynamic interface between the body and the external environment. It protects the body from the environment even as it allows for communication with the environment.

The skin is considered an organ, since it consists of several kinds of tissues that are structurally arranged to function together. It is the largest organ of the body, covering over 7600 sq cm (3000 sq in.) in the average adult, and accounts for approximately 7% of a person's body weight. The skin is of variable thickness, averaging between 1.0 and 2.0 mm. It is thickest on the parts of the body exposed to wear and abrasion, such as the soles of the feet and palms of the hand, where it is about 6 mm. It is thinnest on the eyelids, external genitalia, and tympanum (eardrum), where it is approximately 0.5 mm. Even the texture varies from the rough, callous skin covering the elbows and knuckles to the soft, sensitive areas of the eyelids, nipples, and genitalia.

------------------------------------

integument: L. *integumentum*, a covering

The general appearance and condition of the skin are clinically important because they provide clues to certain body conditions or dysfunctions. Pale skin may indicate shock, whereas red, flushed, overwarm skin may indicate fever and infection. A rash may indicate allergies or local infections. Abnormal textures of the skin may be the result of glandular or nutritional problems (table 7.1). Even chewed fingernails may be a clue to emotional problems.

# Layers of the Integument

*The integument consists of two principal layers. The outer epidermis is stratified into four or five structural and functional layers and the thick and deeper dermis consists of two layers. The hypodermis (subcutaneous tissue) connects the skin to underlying organs.*

## Epidermis

The **epidermis** is the superficial protective layer of the skin and is composed of stratified squamous epithelium that varies in thickness from 0.007 to 0.12 mm. All but the deepest layers of the epidermis are composed of dead cells. The epidermis is composed of either four or five layers, depending on its location within the body (figs. 7.1 and 7.2). The epidermis of the palms and soles has five layers because these areas are exposed to the most friction. The epidermis of all other areas of the body has only four layers. The names and characteristics of the epidermal layers are as follows:

**1 Stratum basale (basal layer).** The stratum basale is composed of a single layer of cells in contact with the dermis. Four types of cells compose the stratum basale: *keratinocytes* (ker˝ă-tin´o-sītz), *melanocytes* (mel´ă-no-sītz), *tactile cells* (Merkel cells), and *nonpigmented granular dendrocytes* (Langerhans cells). With the exception of tactile cells, these cells are constantly dividing mitotically and moving outward to renew the epidermis. It usually takes between 6 and 8 weeks for the cells to move from the stratum basale to the surface of the skin.

**Keratinocytes** are specialized cells that produce **keratin** (ker´ă-tin), which toughens and waterproofs the skin. As keratinocytes are pushed away from the vascular nutrient and oxygen supply of the dermis, their nuclei degenerate, their cellular content is dominated by keratin, and the process of *keratinization* is completed. By the time keratinocytes have reached the surface of the skin, they are scalelike, dead cells filled with keratin enclosed in loose cell membranes. **Melanocytes** are specialized epithelial cells that synthesize the pigment **melanin** (mel´ă-nin), providing a protective barrier to the ultraviolet radiation in sunlight. **Tactile cells** are sparse as compared to keratinocytes and melanocytes.

------------------------------------

epidermis: Gk. *epi*, on; *derma*, skin
stratum: L. *stratum*, something spread out
basale: Gk. *basis*, base
keratinocyte: Gk. *keras*, hornlike; *kytos*, cell
melanocyte: Gk. *melas*, black; *kytos*, cell

## Table 7.1 Conditions of the skin and associated structures indicating nutritional deficiencies or body dysfunctions

| Condition | Deficiency | Comments |
|---|---|---|
| General dermatitis | Zinc | Redness and itching |
| Scrotal or vulval dermatitis | Riboflavin | Inflammation in genital region |
| Hyperpigmentation | Vitamin $B_{12}$, folic acid, or starvation | Dark pigmentation on backs of hands and feet |
| Dry, stiff, brittle hair | Protein, calories, and other nutrients | Usually occurs in young children or infants |
| Follicular hyperkeratosis | Vitamin A, unsaturated fatty acids | Rough skin due to keratotic plugs from hair follicles |
| Pellagrous dermatitis | Niacin and tryptophan | Lesions on areas exposed to sun |
| Thickened skin at pressure points | Niacin | Noted at belt area at the hips |
| Spoon nails | Iron | Thin nails that are concave or spoon-shaped |
| Dry skin | Water or thyroid hormone | Dehydration, hypothyroidism, rough skin |
| Oily skin (acne) | | Hyperactivity of sebaceous glands |

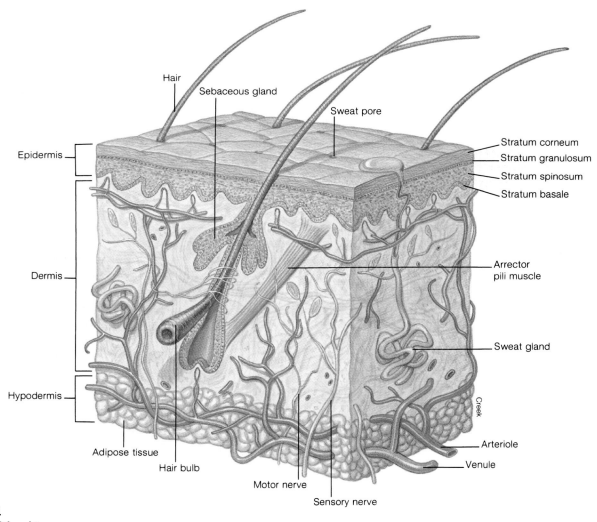

**FIGURE 7.1**

A diagram of the skin.

These cells aid in tactile (touch) reception. **Nonpigmented granular dendrocytes** are scattered throughout the stratum basale. They are protective *macrophagic cells* that ingest bacteria and other foreign debris.

**2** **Stratum spinosum (spiny layer).** The stratum spinosum contains several stratified layers of cells. The spiny appearance of this layer is due to the changed shape of the keratinocytes. Since there is limited mitosis in the stratum spinosum, this layer and the stratum basale are collectively referred to as the **stratum germinativum** (*jer-mĭ″nă-tĭ′vum*).

**3** **Stratum granulosum (granular layer).** This layer consists of only three or four flattened rows of cells. The cells within this layer appear granular due to the process of keratinization.

**4** **Stratum lucidum (clear layer).** The nuclei, organelles, and cell membranes are no longer visible in the cells of the stratum lucidum, and so histologically this layer appears clear. It exists only in the lips and in the thickened skin of the soles and palms.

**5** **Stratum corneum (hornlike layer).** The stratum corneum (*kor′ne-um*) is composed of 25 to 30 layers of flattened, scalelike cells, which are continuously shed as flakelike residues of cells. This surface layer is cornified and is the real protective layer of the skin (fig. 7.3). *Cornification* is brought on by keratinization, and the hardening, flattening process that takes place as the cells die and are pushed to the surface.

Friction at the surface of the skin stimulates additional mitotic activity of the stratum basale, resulting in the formation of a *callus* for additional protection. The specific characteristics of each epidermal layer are described in table 7.2.

Tattooing colors the skin permanently because pigmented dyes are injected below the mitotic basal layer into the dermis. Because of frequent nonsterile conditions, those who administer the tattoo may introduce infections along with the dye.

**Coloration of the Skin**   Normal skin color is caused by the expression of a combination of three pigments; *melanin, carotene,* and *hemoglo-*

•••••••••••••••••••••••••••••••••••
macrophagic: Gk. *makros,* large; *phagein,* to eat
spinosum: L. *spina,* thorn
germinativum: L. *germinare,* sprout or growth
granulosum: L. *granum,* grain
lucidum: L. *lucidus,* light
corneum: L. *corneus,* hornlike
vitiligo: L. *vitiatio,* blemish
carotene: L. *carota,* carrot (referring to orange coloration)

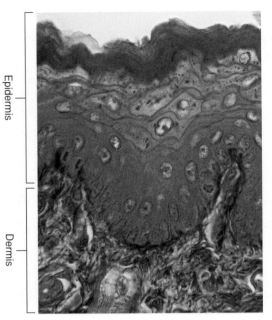

**FIGURE 7.2**
A photomicrograph of the epidermis (250×).

*bin* (*he′mŏ-glo″bin*). **Melanin** is a brown-black pigment produced in the melanocytes of the stratum basale (fig. 7.4). All races have virtually the same number of melanocytes, but the amount of melanin produced and the degree of granular aggregation of the melanin determine whether an individual's skin color is black, brown, red, tan, or white. Melanin is a protective device that guards against the damaging effect of the ultraviolet rays in sunlight. A gradual exposure to the sunlight promotes the increased production of melanin within the melanocytes, and hence tanning of the skin. The skin of a genetically determined albino has the normal complement of melanocytes in the epidermis but lacks the enzyme *tyrosinase,* that converts the amino acid tyrosine to melanin.

Other genetic expressions of melanocytes are more common than albinism. *Freckles,* for example, are caused by aggregated patches of melanin. A lack of melanocytes in localized areas of the skin causes distinct white spots in the condition called *vitiligo* (*vit-ĭ-li′gō*).

**Carotene** is a yellowish pigment found in the epidermal cells and fatty parts of the dermis. Carotene is abundant in the skin of Asians and, together with melanin, accounts for their yellow-tan skin.

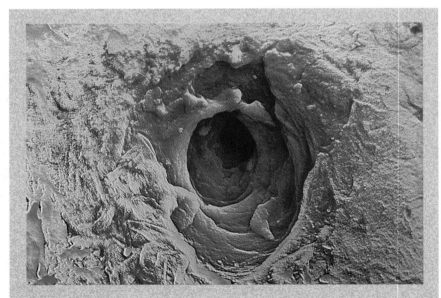

**FIGURE 7.3**
A scanning electron micrograph of the surface of the skin showing the opening of a sweat gland. The fragmented-appearing particles are bacteria, which are present throughout the body on the surface of the skin.

# Table 7.2    Layers of the epidermis

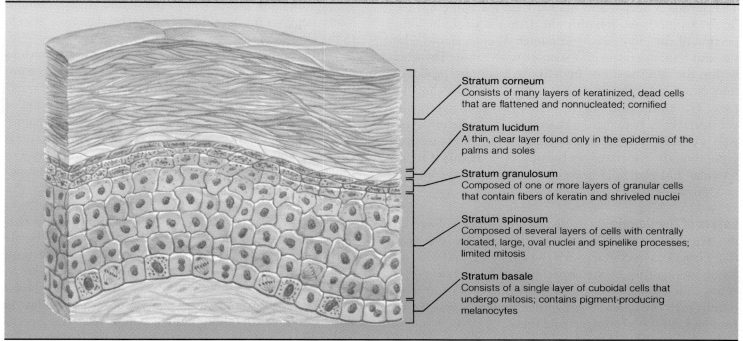

**Stratum corneum**
Consists of many layers of keratinized, dead cells that are flattened and nonnucleated; cornified

**Stratum lucidum**
A thin, clear layer found only in the epidermis of the palms and soles

**Stratum granulosum**
Composed of one or more layers of granular cells that contain fibers of keratin and shriveled nuclei

**Stratum spinosum**
Composed of several layers of cells with centrally located, large, oval nuclei and spinelike processes; limited mitosis

**Stratum basale**
Consists of a single layer of cuboidal cells that undergo mitosis; contains pigment-producing melanocytes

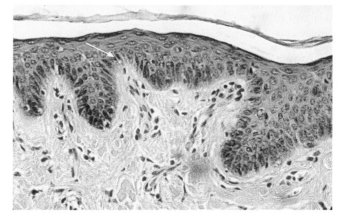

**FIGURE 7.4**
Melanocytes throughout the stratum basale (see arrow) produce melanin.

**Hemoglobin** is not a pigment of the skin; rather, it is the oxygen-binding pigment found in red blood cells. Oxygenated blood flowing through the dermis gives the skin its pinkish tones.

 Certain physical conditions or diseases cause symptomatic discoloration of the skin. *Cyanosis* is a bluish discoloration of the skin that appears in people with certain cardiovascular or respiratory diseases. People also become cyanotic during an interruption of

hemoglobin: Gk. *haima*, blood; *globus*, globe
cyanosis: Gk. *kyanosis, dark-blue color*
*jaundice:* L. galbus, yellow

breathing. In *jaundice*, the skin appears yellowish because of an excess of bile pigment in the bloodstream. Jaundice is usually symptomatic of liver dysfunction and sometimes of liver immaturity, as in a jaundiced newborn. *Erythema* is a redness of the skin generally due to vascular trauma, such as from a sunburn.

**Surface Patterns**   The exposed surface of the skin has recognizable patterns that are either congenital or acquired. Congenital patterns called *fingerprints*, or *friction ridges*, are present on the palms and soles as well as on the finger and toe pads. The designs formed by these lines have basic similarities but are not identical in any two individuals, even identical twins. They are formed by the pull of elastic fibers within the dermis and are well established prenatally. As the name implies, friction ridges function to prevent slippage when grasping objects. Because they are precise and easy to reproduce, fingerprints are customarily used for identifying individuals in the science known as *dermatoglyphics*. All primates have fingerprints, and even dogs have a characteristic "nose print" that is used for identification in the military canine corps.

Acquired lines include the deep *flexion creases* on the palms and the shallow *flexion lines* that can be seen on the knuckles and on the surface of other joints. Furrows on the forehead and face are acquired from continual contraction of facial muscles, such as from smiling or squinting in bright light or against the wind. Facial lines become more strongly delineated as a person ages.

erythema: Gk. *erythros*, red; *haima*, blood

## Dermis

The **dermis** is deeper and thicker than the epidermis (see fig. 7.1). Blood vessels within the dermis nourish the living portion of the epidermis, and numerous collagenous, elastic, and reticular fibers give support to the skin. The fibers within the dermis radiate in definite directions, producing lines of tension on the surface of the skin and providing skin tone. The elastic fibers in the dermis of a young person are considerably more numerous than those in the dermis of one who is elderly. A decreasing amount of elastic fiber is apparently directly associated with aging. The dermis is highly vascular and glandular and contains many nerve endings and hair follicles.

**Layers of the Dermis**   The dermis is composed of two layers. The upper layer, called the **stratum papillarosum** (papillary layer), is in contact with the epidermis and accounts for about one-fifth of the entire dermis. Numerous projections, called *papillae,* extend from the upper portion of the dermis into the epidermis. Papillae form the base for the friction ridges on the fingers and toes.

The deeper and thicker layer of the dermis is called the **stratum reticularosum** (reticular layer). In fact, it is this layer that corresponds to the hide of an animal used to make leather and suede. Fibers within this layer are more dense and regularly arranged to form a tough, flexible meshwork. It is quite distensible, as is evident in pregnant women or obese individuals, but it can be stretched too far, causing "tearing" of the dermis. The repair of a strained dermal area leaves a white streak called a stretch mark, or **linea albicans.** Lineae albicantes frequently develop on the abdomen and breasts of a woman during pregnancy (fig. 7.5).

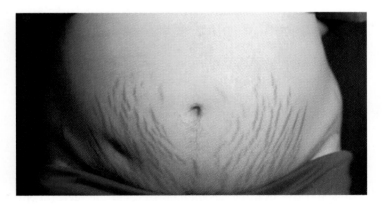

**FIGURE 7.5**
Stretch marks (lineae albicantes) on the abdomen of a pregnant woman. Stretch marks generally fade with time but frequently leave permanent integumentary markings.

papilla: L. *papula,* swelling or pimple

**Innervation of the Skin**   The dermis of the skin has extensive innervation. Specialized integumentary *effectors* consist of muscles or glands within the dermis that respond to motor impulses transmitted from the central nervous system to the skin by autonomic nerve fibers.

Several types of **sensory receptors** respond to various tactile (touch), pressure, temperature, tickle, or pain sensations. Some are exposed nerve endings, some form a network around hair follicles, and some extend into the papillae of the dermis. Certain areas of the body, such as the palms, soles, lips, and external genitalia, have a greater concentration of sensory receptors and are therefore more sensitive to touch. Chapter 18 includes a detailed structural and functional account of the various sensory receptors.

**Vascular Supply of the Skin**   Blood vessels within the dermis supply nutrients to the mitotically active stratum basale of the epidermis and to the cellular structures of the dermis, such as glands and hair follicles. Dermal blood vessels play an important role in regulating body temperature and blood pressure. Autonomic vasoconstriction or vasodilation responses can either shunt the blood away from the superficial dermal arterioles or permit it to flow freely throughout dermal vessels. Fever or shock can be detected by the color and temperature of the skin. Blushing is the result of involuntary vasodilation of dermal blood vessels.

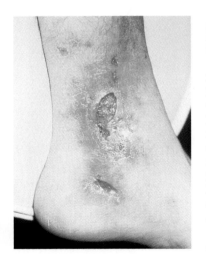

**FIGURE 7.6**
A decubitus ulcer on the medial surface of the ankle. Frequent sites for decubitus ulcers occur most frequently in the skin overlying a bony projection, such as at the hip, ankle, heel, shoulder, or elbow.

 A healthy circulating blood flow in debilitated bedridden patients is important for the prevention of bedsores, or *decubitus ulcers.* When a person lies in one position for an extended period, the dermal blood flow is restricted where the body presses against the bedding. As a consequence, cells die and open wounds develop (fig. 7.6). Changing the position of the patient frequently and periodically massaging the skin to stimulate blood flow help to keep the skin healthy.

decubitus: L. *decumbere,* lie down
ulcer: L. *ulcus,* sore

**Hypodermis** The **hypodermis,** or **subcutaneous tissue,** binds the dermis to underlying organs. The hypodermis is composed primarily of loose fibrous connective tissue and adipose cells interlaced with blood vessels (see fig. 7.1). Collagenous and elastic fibers reinforce the hypodermis— particularly on the palms and soles, where the skin is firmly attached to underlying structures. The amount of adipose in the hypodermis varies with the sex, age, region of the body, and nutritional state of the individual. The hypodermis of females is generally about 8% thicker than that of males. The hypodermis also functions to store lipids, insulate and cushion the body, and regulate temperature.

 The hypodermis is of clinical importance as an injection site for subcutaneous medication. Subcutaneous injections may be administered to patients who are unconscious or uncooperative, and when oral medications are not practical. Slow-release, low-dosage subcutaneous medications are now available. These types of medications may regulate metabolism, inhibit ovulation (for birth control), or mitigate pain. Subcutaneous devices for diabetics, which continuously release small dosages of insulin, are currently in use.

# Physiology of the Integument

*The integument is an extremely dynamic organ. Not only does it protect the body from pathogens and external injury, it also plays a major role in maintaining body homeostasis.*

## Physical Protection

The skin is a physical barrier to most microorganisms, water, and excessive ultraviolet (UV) light. Oily secretions onto the surface of the skin form an acidic (pH 4.0–6.8) protective film that waterproofs the body and retards the growth of most pathogens. The protein keratin in the epidermis also waterproofs the skin. Cornification of the outer layers of the epidermis toughens the mostly dead cells to withstand abrasion and the penetration of microorganisms. Upon exposure to UV light, the melanocytes in the lower epidermal layers are stimulated to synthesize melanin, which in turn absorbs and disperses sunlight. Surface friction causes the epidermis to thicken and form a protective callus.

 Regardless of skin pigmentation, everyone is susceptible to skin cancer if exposure to sunlight is sufficiently intense and continuous. There are an estimated 800,000 new cases of skin cancer yearly in the United States, and approximately 9300 of these are diagnosed as the potentially life-threatening *melanoma* (cancer of melanocytes). Melanomas are usually termed malignant because they may spread rapidly. Sunscreens

• • • • • • • • • • • • • • • • • • • • • • • • • • • • • • • • •
hypodermis: Gk. *hypo,* under; *derma,* skin

are advised for people who must be in direct sunlight for long periods of time.

The skin absorbs two wavelengths of ultraviolet rays from the sun: UV-A and UV-B. The DNA within the basal skin cells may be damaged as the sun's more dangerous UV-B rays penetrate the skin. Although it was once believed that UV-A rays were harmless, recent findings indicate that excessive exposure to UV-A rays may inhibit the DNA repair process that follows exposure to UV-B. Therefore, individuals who are exposed solely to UV-A rays in tanning salons still risk melanomas, since they will later be exposed to UV-B rays of sunlight when they are out-of-doors.

## Hydroregulation

The thickened, keratinized, and cornified epidermis of the skin is adapted for continuous exposure to the air. In addition, the outer layers are dead and scalelike, and a protein-polysaccharide basement membrane adheres the stratum basale to the dermis. Human skin is virtually waterproof, protecting the body from desiccation (dehydration) on dry land and even from water absorption when immersed in water.

## Thermoregulation

The skin plays a crucial role in the regulation of body temperature. Body heat comes from cellular metabolism, particularly in muscle cells as they maintain tone or a degree of tension. A normal body temperature of 37° C is maintained by the antagonistic effects of sweating and shivering, which involve feedback mechanisms. Excess heat is actually lost from the body in three ways, all involving the skin: (1) through radiation from dilated blood vessels, (2) through secretion and the evaporation of perspiration, and (3) through convection and the conduction of heat directly through the skin (fig. 7.7). Sweat secretion increases approximately

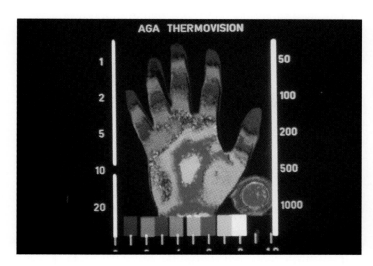

**FIGURE 7.7**
A thermogram of the hand showing differential heat radiation. Hair and body fat are good insulators. Red and yellow indicate the warmest parts of the body. Blue, green, and white indicate the coolest.

100–150 ml/day for each 1-degree elevation in body temperature. Up to 10 L of sweat per day may be secreted to cool the body of a person doing hard physical work out-of-doors in the summertime.

 A serious danger of continued exposure to heat and excessive water and salt loss is *heat exhaustion*, characterized by nausea, weakness, dizziness, headache, and a decreased blood pressure. *Heat stroke* is similar to heat exhaustion, except that sweating is inhibited (for reasons that are not clear) and body temperature rises in heat stroke. Convulsions, brain damage, and death may follow.

Excessive heat loss triggers a shivering response in muscles, which increases cellular metabolism and consequent heat production. Not only do skeletal muscles contract, but tiny smooth muscles called **arrector pili** (ă-rek´tor pi´li), which are attached to hair follicles (see fig. 7.9), contract involuntarily, causing goose bumps.

 When the body's heat-producing mechanisms cannot keep pace with heat loss, *hypothermia* results. A lengthy exposure to temperatures below 20° C and dampness may lead to this condition. This is why it is so important that a hiker, for example, dress appropriately for the weather conditions, especially on cool, rainy spring or fall days. The initial symptoms of hypothermia are numbness, paleness, delirium, and uncontrolled shivering. If the core temperature falls below 32° C (90° F), the heart loses its ability to pump blood and will go into fibrillation (erratic contractions). If the victim is not warmed, extreme drowsiness, coma, and death follow.

## Cutaneous Absorption

Because of the effective protective barriers of the integument already described, cutaneous (through the skin) absorption is limited. Some gases, such as oxygen and carbon dioxide, may pass through the skin and enter the bloodstream. Small amounts of UV light, necessary for synthesis of vitamin D, are absorbed readily. The skin is no barrier to steroid hormones, such as cortisol, and to fat-soluble vitamins (A, D, E, and K). Of clinical consideration is the fact that certain toxins and pesticides enter the body through cutaneous absorption.

. . . . . . . . . . . . . . . . . . . . . . . . . . . . .

pili: L. *pilus*, hair

(a)

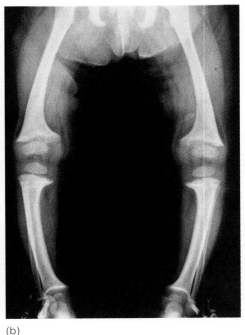

(b)

**FIGURE 7.8**

(*a*) A case of rickets in a child who lives in a village in Nepal, where the people reside in windowless huts. During the 5-to-6-month rainy season, the children are kept indoors. (*b*) A radiograph (X ray) of rickets in a 10-month-old child. Rickets develop from improper diets and also from lack of UV light that is needed to synthesize vitamin D.

## Synthesis

The integumentary system synthesizes melanin and keratin, which remain in the skin, and vitamin D, which is used elsewhere in the body. The integumentary cells contain a compound called *dehydrocholesterol* (de-hi´´dro-kŏ-les´tă-rol), from which they synthesize vitamin D in the presence of UV light. Only small amounts of UV light are necessary for vitamin D synthesis, but these amounts are very important to a growing child (fig. 7.8). Synthesized vitamin D enters the blood and helps regulate the metabolism of calcium and phosphorus, which are important for development of strong and healthy bones. *Rickets* is a disease caused by vitamin D deficiency.

## Sensory Reception

Highly specialized sensory receptors (see chapter 18) that respond to thermal (heat and cold), mechanical (pressure, touch, and vibration), and noxious (pain) stimuli are located throughout the dermis and hypodermis of the integument. These receptors, referred to as **cutaneous receptors,** are abundant in the skin in parts of the face, the palms and fingers of the hands, the soles of the feet, and the genitalia. They are less abundant along the back and on the back of the neck and are sparse in the skin over joints, especially the elbow. Generally speaking, the thinner the skin, the greater the sensitivity.

## Communication

Humans are highly social animals, and the integument plays an important role in communication. Various emotions such as anger or embarrassment may be reflected in changes of skin color. The contraction of specific facial muscles produces facial expressions that convey an array of emotions, including surprise, happiness, sadness, and despair. Secretions from certain integumentary glands frequently elicit subconscious responses from those that detect the odors.

# Epidermal Derivatives

*Hair, nails, and integumentary glands form from the epidermal layer and are therefore of ectodermal derivation. Hair and nails are structural features of the integument and have a limited functional role. By contrast, integumentary glands are extremely important in body defense and the maintenance of homeostasis.*

## Hair

**Hair** is characteristic of all mammals, but its distribution, function, density, and texture varies across mammalian species. Humans are relatively hairless, with only the scalp, face, pubis, and axilla being densely haired. Men and women have about the same amount of hair on their bodies, but it is generally more obvious on men due to male hormones (see chapters 19 and 28). Certain structures and regions of the body are hairless, such as the palms of the hands, soles of the feet, lips, nipples, penis, and labia minora.

 *Hirsutism (her´soo-tiz˝em) is a condition of excessive body and facial hair, especially in women. It may be a genetic expression or occur as the result of a metabolic disorder, usually endocrine in nature. Hirsutism occurs in some women as they experience hormonal changes during menopause (see chapter 29). Various treatments for hirsutism include hormonal injections and electrolysis to permanently destroy selected hair follicles.*

The primary function of hair is protection, even though its effectiveness is limited. Hair on the scalp and eyebrows protects against sunlight. The eyelashes and the hair in the nostrils protect against airborne particles. Hair on the scalp may also protect against mechanical injury. Some secondary functions of hair are to distinguish individuals and to serve as an ornamental sexual attractant.

Each hair consists of a diagonally positioned **shaft, root,** and **bulb** (fig. 7.9). The shaft is the visible but dead portion of the hair projecting above the surface of the skin. The bulb is the enlarged base of the root within the **hair follicle.** Each hair develops from stratum basale cells within

..........................................
hirsutism: L. *hirsutus,* shaggy

the bulb of the hair, where nutrients are received from dermal blood vessels. As the cells divide, they are pushed away from the nutrient supply toward the surface, and cellular death and keratinization occur. In a healthy person, hair grows at the rate of approximately 1 mm every 3 days. As the hair becomes longer, however, it goes through a resting period during which it is anchored in its follicle.

The life span of a hair varies from 3 to 4 months for an eyelash to 3 to 4 years for a scalp hair. Each hair lost is replaced by a new hair that grows from the base of the follicle and pushes the old hair out. Between 10 and 100 hairs are lost each day through replacement. Baldness results when hair is lost and not replaced. This condition may be disease related, but it is generally inherited and most frequently occurs in males as a result of genetic influences combined with the action of the male sex hormone *testosterone* (*testos´tĕ-rōn*). No treatment is effective in reversing genetic baldness; however, flaps or plugs of skin containing healthy follicles from hairy parts of the body can be grafted to hairless regions.

Three layers can be observed in hair that is cut in cross section. An inner **medulla** (*mĕ-dul´ă*) is composed of loosely arranged cells separated by many air cells. The thick median layer, called the **cortex,** consists of hardened, tightly packed cells. A **cuticle** layer covers the cortex and forms the toughened outer portion of the hair. Cells of the cuticle have serrated edges that give a hair a scaly appearance when observed under a dissecting scope.

 People exposed to heavy metals, such as lead, mercury, arsenic, or cadmium, will have concentrations of these metals in their hair that are 10 times as great as those found in their blood or urine. Because of this, hair samples can be extremely important in certain diagnostic tests.

Even evidence of certain metabolic diseases or nutritional deficiencies may be detected in hair samples. For example, the hair of children with cystic fibrosis will be deficient in calcium and display excessive sodium. There is a deficiency of zinc in the hair of malnourished individuals.

Hair color is determined by the type and amount of pigment produced in the stratum basale at the base of the hair follicle. Varying amounts of melanin produce hair ranging in color from blond to brunette to black; the more abundant the melanin, the darker the hair. A pigment with an iron base (trichosiderin) produces red hair. Gray and white hair is the result of a lack of pigment production and air spaces within the layers of the shaft of the hair and generally accompanies aging. The texture of hair is determined by the cross-sectional shape; straight hair is round in cross section, wavy hair is oval, and kinky hair is flat.

..........................................
medulla: L. marrow
cortex: L. bark
cuticle: L. *cuticula,* small skin

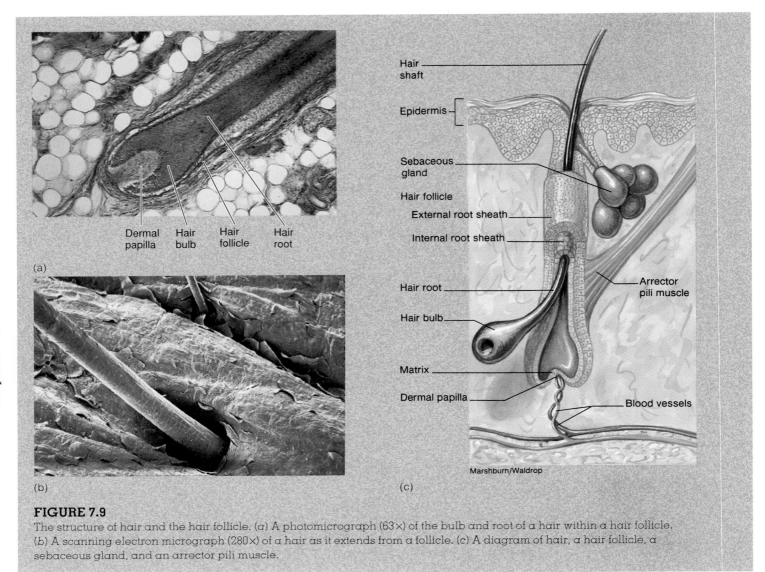

Dermal papilla | Hair bulb | Hair follicle | Hair root

(a)

(b)

Hair shaft

Epidermis

Sebaceous gland

Hair follicle

External root sheath

Internal root sheath

Hair root

Hair bulb

Matrix

Dermal papilla

Arrector pili muscle

Blood vessels

Marshburn/Waldrop

(c)

**FIGURE 7.9**

The structure of hair and the hair follicle. (*a*) A photomicrograph (63×) of the bulb and root of a hair within a hair follicle. (*b*) A scanning electron micrograph (280×) of a hair as it extends from a follicle. (*c*) A diagram of hair, a hair follicle, a sebaceous gland, and an arrector pili muscle.

A sebaceous gland and an arrector pili muscle are attached to the hair follicle (see fig. 7.9c). When the muscle involuntarily contracts due to thermal or psychological stimuli, the hair follicle is pulled into an upright position, causing the hair to "stand on end" and producing goose bumps.

Humans have three distinct kinds of hair.

**1 Lanugo.** Lanugo is a fine, silky fetal hair that appears during the last trimester of development. Lanugo is usually not evident on a baby at birth unless it has been born prematurely.

**2 Angora.** Angora hair grows continuously in length, as on the scalp of males and females and on the face of males.

**3 Definitive.** Definitive hair grows to a certain length and then ceases to grow. It is the most common type of hair. Eyelashes, eyebrows, and pubic and axillary hair are examples.

.........................................

lanugo: L. *lana*, wool

Anthropologists have referred to humans as the "naked apes" because of our relative hairlessness. The clothing that we wear over the exposed surface areas of our bodies functions to insulate and protect us, just as hair or fur does in other mammals. However, the nakedness of our skin does lead to some problems. *Skin cancer* occurs frequently in humans, particularly in regions of the skin exposed to the sun. *Acne*, another problem unique to humans, is partly related to the fact that hair in the affected areas is not sufficiently dense to dissipate the oily secretion from the sebaceous glands.

## Nails

**Nails** are found on the distal dorsum of each of the fingers and toes. Both fingernails and toenails serve to protect the digits, and fingernails also aid in grasping and picking up small objects. Nails form from a hardened, transparent stratum corneum of the epidermis. The hardness of the nail is due to a dense, parallel arrangement of keratin fibrils between the cells.

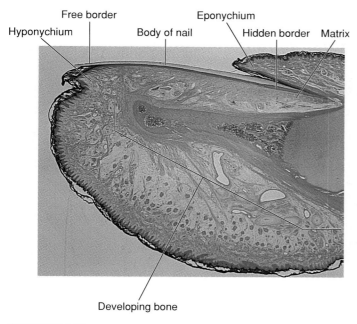

**FIGURE 7.10**

A photomicrograph of a fingertip from a neonatal human showing the nail and associated structures (3.5×).

Each nail consist of a **nail body, free border,** and **hidden border** (fig. 7.10). The platelike body of the nail rests on a **nail bed,** which is actually the stratum spinosum of the epidermis. The nail body and nail bed appear pinkish because of the underlying vascular tissue. The free border is the distal exposed border, which is attached to the undersurface by the **hyponychium** (hi´´pŏ-ni´kē-um). The hidden border of the nail is attached proximally.

An **eponychium** (cuticle) covers the hidden border of the nail. The eponychium frequently splits, causing a hangnail. The growth area of the nail is the **nail matrix.** A small part of the nail matrix, the **lunula,** is seen through the nail body as a white, half-moon-shaped area at the base of the nail. The nail grows by the transformation of the superficial cells of the nail matrix into nail cells. These harder, transparent cells are then pushed forward over the strata basale and spinosum of the nail bed. Fingernails grow at the rate of approximately 1 mm per week. The growth rate of toenails is somewhat slower.

 The condition of nails can be an indication of a person's general health and indicative of his or her personality. Nails should appear pinkish, showing the rich vascular capillaries beneath the translucent nail. A yellowish hue may indicate certain glandular dysfunctions or nutritional deficiencies. Split nails may also be caused by nutritional deficiencies. A prominent bluish tint may indicate improper oxygenation of the blood. Spoon nails (concave body) may be the result of iron-deficiency anemia, and "clubbing" at the base of the nail may be caused by lung cancer. Dirty or ragged nails may indicate poor personal hygiene, and chewed nails may suggest emotional problems.

hyponychium: Gk. *hypo,* under; *onyx,* nail

## Glands

Although they originate in the epidermal layer and are formed of epithelial tissue, all of the glands of the skin are located in the dermis, where they receive physical support and nutritive sustenance. Glands of the skin are referred to as *exocrine* since they excrete substances through ducts. The glands of the skin are of three basic types: *sebaceous, sudoriferous* (soo´dor-if´er-us), and *ceruminous* (sĕ-roo´mĭ-nus).

**Sebaceous**   Sebaceous, or oil, glands are associated with hair follicles, since they develop from the follicular epithelium of the hair. They are simple, branched glands that are connected to hair follicles, where they secrete **sebum** onto the shaft of the hair (fig. 7.11). Sebum, which consists mainly of

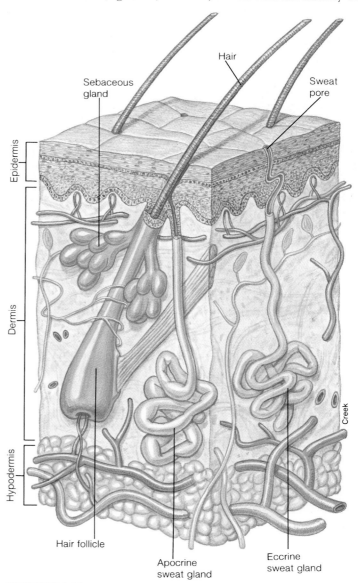

**FIGURE 7.11**

Types of skin glands.

lunula: L. *lunula,* small moon
sebum: L. *sebum,* tallow or grease

# UNDER DEVELOPMENT

## Development of the Integumentary System

Both the ectodermal and mesodermal germ layers (see chapter 6) function in the formation of the structures of the integumentary system. The epidermis and the hair, glands, and nails of the skin develop from the ectodermal germ layer (figs. 1 and 2). The dermis develops from a thickened layer of undifferentiated mesoderm called **mesenchyme** (*mez´en-kīm´*).

By 6 weeks, the ectodermal layer has differentiated into an outer flattened *periderm* and an inner, cuboidal **germinal (basal) layer** in contact with the mesenchyme. The periderm eventually sloughs off, forming the **vernix caseosa** (*ka´se-o´sà*), a cheeselike protective coat that covers the skin of the fetus.

.............................................

mesenchyme: Gk. *mesos*, middle; *enchyma*, infusion
periderm: Gk. *peri*, around; *derm*, skin
vernix caseosa: L. *vernix*, varnish; *caseus*, cheese

By 11 weeks, the mesenchymal cells below the germinal cells have differentiated into the distinct collagenous and elastic connective tissues fibers of the dermis. The tensile properties of these fibers cause a buckling of the epidermis and the formation of dermal papillae. During the early fetal period (about 10 weeks), specialized neural crest cells called **melanoblasts** migrate into the developing dermis and differentiate into **melanocytes.** The melanocytes soon migrate to the germinal layer of the epidermis, where they produce the pigment **melanin,** which colors the epidermis.

Before hair can form, a **hair follicle** must be present. Each hair follicle begins to develop at about 12 weeks (fig. 2), as a mass of germinal cells, called a **hair bud,** proliferates into the underlying mesenchyme. As the hair bud becomes club-shaped, it is referred to as a **hair bulb.** The hair follicle, which physically supports and provides nourishment to the hair, is derived from specialized mesenchyme called the **hair papilla,** which is localized around the hair bulb, and from the epithelial cells of the hair bulb called the **hair matrix.** Continuous mitotic activity in the epithelial cells of the hair bulb results in the growth of the hair.

Sebaceous glands and sweat glands are the two principal types of integumentary glands. Both develop from the germinal layer of the epidermis (fig. 2). Sebaceous glands develop as proliferations from the sides of the developing hair follicle. Sweat glands become coiled as the secretory portion of the developing gland proliferates into the dermal mesenchyme. Mammary glands are modified sweat glands that develop in the skin of the anterior thoracic region.

---

lipids, is dispersed along the shaft of the hair to the surface of the skin, where it lubricates and waterproofs the stratum corneum layer and also prevents the hair from becoming brittle. If the drainage pathway for sebaceous glands becomes blocked for some reason, the glands may become infected, resulting in *acne.* Sex hormones regulate the production and secretion of sebum, and hyperactivity of sebaceous glands can result in serious acne problems, particularly during teenage years.

**Sudoriferous** Sudoriferous, or **sweat glands,** secrete perspiration, or sweat, onto the surface of the skin (fig. 7.11). Sweat glands are most numerous on the palms of the hands

.............................................

sudoriferous: L. *sudorifer*, sweat; *ferre*, to bear

and soles of the feet, in the axillary and pubic regions, and on the forehead. They are coiled and tubular shaped and are of two types.

**1** Eccrine (*ek´rin*) **sweat glands** are widely distributed over the body, especially on the forehead, back, palms, and soles. These glands are formed before birth and function in evaporative cooling in response to thermal or psychological stimuli.

**2** Apocrine (*ap´ŏ-krin*) **sweat glands** are much larger, localized glands found in axillary and pubic regions, where they secrete into hair follicles. Apocrine glands are not functional until puberty, and their odoriferous secretion is thought to act as a sexual attractant.

Perspiration is composed of water, salts, urea, uric acid, and traces of other elements. Perspiration is therefore valuable not only for evaporative cooling but also for the excretion of certain wastes.

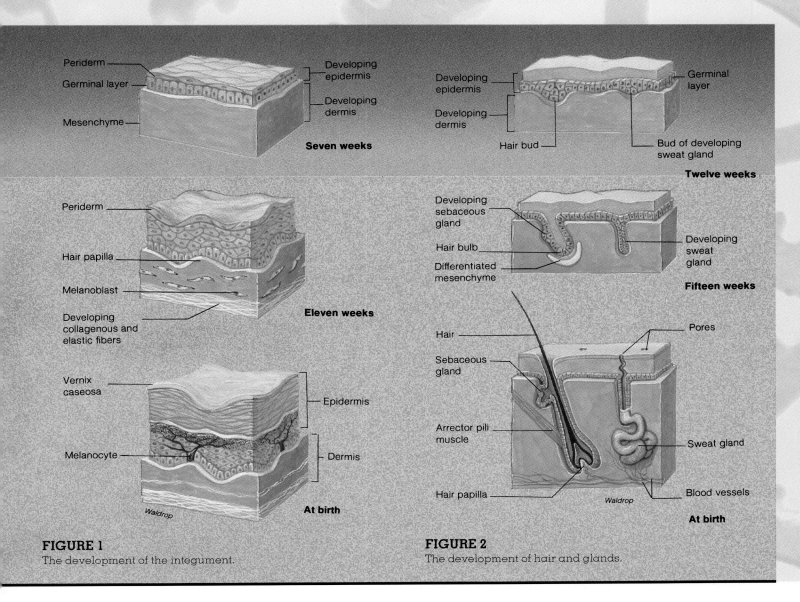

**FIGURE 1**
The development of the integument.

**FIGURE 2**
The development of hair and glands.

**Mammary glands,** found within the breasts, are specialized sudoriferous, or sweat, glands that secrete milk during lactation periods (see chapter 29). The breasts of the human female reach their greatest development during the childbearing years under the stimulus of pituitary and ovarian hormones.

**Ceruminous Glands** These highly specialized glands are found only in the external auditory canal (ear canal). They secrete **cerumen** (sĕ-roo´men), or earwax, which is a water and insect repellent, and which also keeps the tympanum (eardrum) from drying out. Excessive amounts of cerumen may interfere with hearing.

∙∙∙∙∙∙∙∙∙∙∙∙∙∙∙∙∙∙∙∙∙∙∙∙∙∙∙∙∙∙∙∙∙∙∙∙∙∙∙∙∙∙∙∙∙∙∙∙∙∙∙∙∙∙∙∙∙∙

cerumen: L. *cera*, wax

 Good routine hygiene is very important for health and social reasons. Washing away the dried residue of perspiration and sebum eliminates dirt. Excessive bathing, however, can wash off the natural sebum and dry the skin, causing it to crack or itch. The commercial lotions used for dry skin are, for the most part, refined and perfumed lanolin, which is sebum from sheep.

## Clinical Considerations

The skin is a buffer against the external environment and is therefore subject to a variety of disease-causing microorganisms and physical assaults. A few of the many diseases and disorders of the integumentary system are briefly discussed here.

# Inflammatory (Dermatitis) Conditions

Inflammatory skin disorders are caused by immunologic hypersensitivity or infectious agents. Some people are allergic to certain foreign proteins and, because of this inherited predisposition, experience such hypersensitive reactions as asthma, hay fever, hives, drug and food allergies, and eczema. Eczema is characterized by redness, itching, and swollen vascular lesions that become dry, scaly, and crusted. **Lesions,** as applied to inflammatory conditions, are defined as more or less circumscribed pathologic changes in the tissue. Some of the more common inflammatory skin disorders and their usual sites are illustrated in figure 7.12.

There are also a number of *infectious diseases* of the skin, which is not surprising considering that we are highly social and communal animals. Most of these diseases can now be prevented, but too frequently people fail to take appropriate precautionary measures. Infectious diseases involving the skin include childhood viral infections (measles and chicken pox); bacteria, such as staphylococcus (impetigo); sexually transmitted diseases; leprosy; fungi (ringworm, athlete's foot, candida); and mites (scabies).

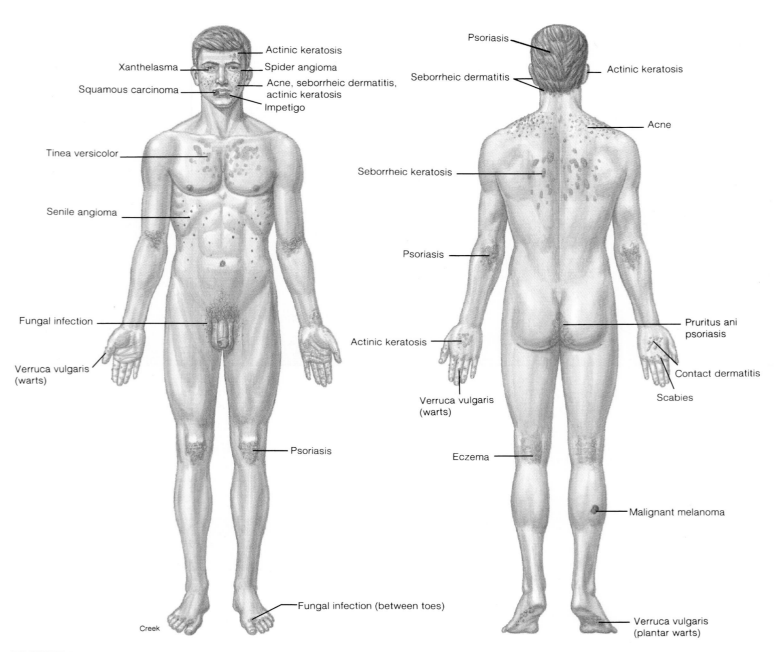

**FIGURE 7.12**

Common inflammatory skin disorders and their usual site of occurrence.

## Neoplasms

Both benign and malignant neoplastic conditions or diseases are common in the skin. Pigmented moles (nevi), for example, are a type of benign neoplastic growth of melanocytes. Dermal cysts and benign viral infections are also common. Warts are virally caused abnormal growths of tissue that occur frequently on the hands and feet. A different type of wart, called a venereal wart, occurs in the anogenital region of sexual partners. Both types of warts are fairly easy to treat by means of excision or various drugs. "Age spots" on elderly people, which appear as pigmented patches on the surface of the skin, are a benign growth of melanin-pigmented basale cells. Usually no treatment is required, unless for cosmetic purposes.

Skin cancer (fig. 7.13) is the most common malignancy in the United States. Except for malignant melanomas, which arise from melanocytes, skin cancer is generally not life threatening. Excessive exposure to ultraviolet light from the sun is a known cause of skin cancer. The preferred treatment for this disease is complete surgical excision of the cancerous portion.

## Burns

A burn is an epithelial injury caused by contact with thermal, radioactive, chemical, or electrical agents. Burns generally occur on the skin, but they can involve the linings of the respiratory and digestive tracts. The extent and location of a burn is frequently less important than the degree to which it disrupts body homeostasis. Burns that have a **local effect** (local tissue destruction) are not as serious as those

........................................

neoplasm: Gk. *neo*, new; *plasma*, something formed
benign: L. *benignus*, good-natured
malignant: L. *malignus*, acting from malice

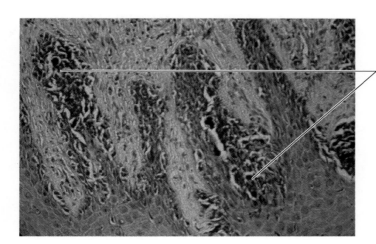

**FIGURE 7.13**
Skin cancer.

that have a **systemic effect.** Systemic effects directly or indirectly involve the entire body and are a threat to life. Possible systemic effects include body dehydration, shock, reduced circulation, and bacterial infections.

Burns are classified as first degree, second degree, and third degree based on their severity (fig. 7.14). In **first-degree burns,** the epidermal layers of the skin are damaged and symptoms are restricted to local effects such as redness, pain, and edema (swelling). A shedding of the surface layers (desquamation) generally follows in a few days. A sunburn is an example. **Second-degree burns** involve both the epidermis and dermis. Blisters appear and recovery is usually complete, although slow. **Third-degree burns** destroy the entire thickness of the skin and frequently some of the underlying connective tissue. The skin appears waxy or charred and is insensitive to touch. As a result, ulcerating wounds develop, and the body attempts to heal itself by forming scar tissue. Skin grafts are frequently used to assist recovery.

As a way of estimating the extent of damaged skin suffered in burned patients, the *rule of nines* (fig. 7.15) is often applied. The surface area of the body is divided into regions, each of which accounts for about 9% (or a multiple of 9%) of the total skin body surface. An estimation of the percentage of surface area damaged is important in treating with intravenous fluid, which replaces the fluids lost from tissue damage.

## Frostbite

Frostbite is a local destruction of the skin resulting from freezing. Like burns, frostbite is classified by its degree of severity: first degree, second degree, and third degree. In **first-degree frostbite** the skin will appear cyanotic (bluish) and swollen. Vesicle formation and hyperemia (engorgement with blood) are symptoms of **second-degree frostbite.** As the effected area is warmed, there will be further swelling, and the skin will redden and blister. In **third-degree frostbite,** there will be severe edema, some bleeding, and numbness, followed by intense throbbing pain and necrosis of the affected tissue. Gangrene will follow untreated third-degree frostbite.

## Skin Grafts

If extensive areas of the stratum basale of the epidermis are destroyed in second-degree or third-degree burns or frostbite, new skin cannot grow back. In order for this type of wound to heal, a skin graft must be performed.

A **skin graft** is a segment of skin that has been excised from a *donor site* and transplanted to the *recipient site*, or *graft bed*. As mentioned

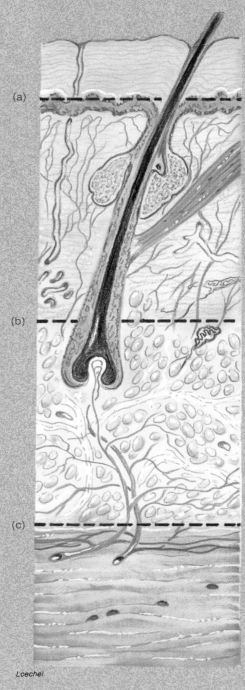

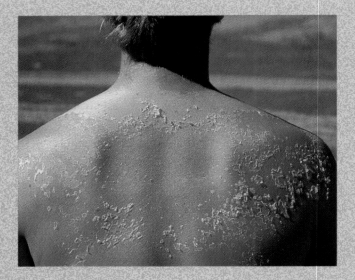

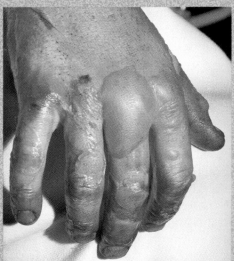

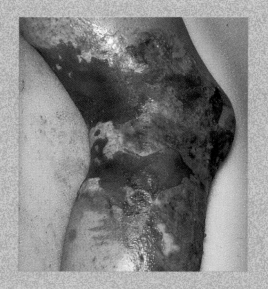

Loechel

## FIGURE 7.14

The classification of burns. (*a*) First-degree burns involve the epidermis and are characterized by redness, pain, and edema—such as with a sunburn; (*b*) second-degree burns involve the epidermis and dermis and are characterized by intense pain, redness, and blistering; (*c*) third-degree burns destroy the entire skin and frequently expose the underlying organs. The skin is charred and numb and does not protect against fluid loss.

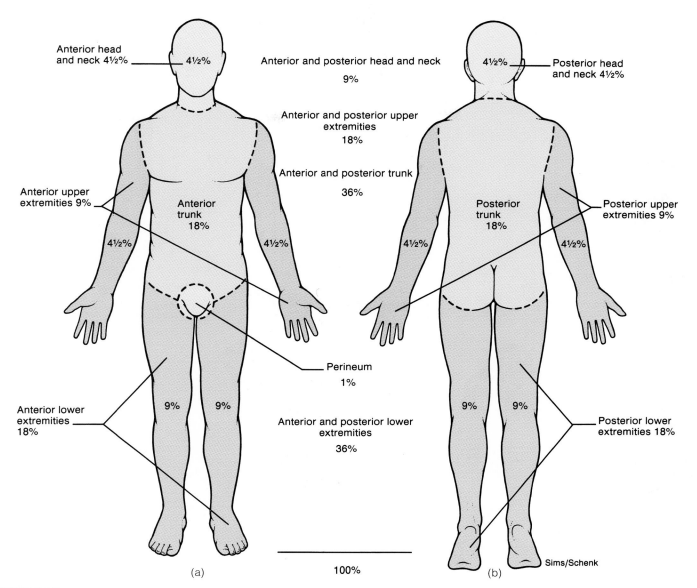

**FIGURE 7.15**
The extent of burns as estimated by the rule of nines. (*a*) Anterior and (*b*) posterior.

in chapter 6, an *autotransplant* is the most successful graft. This type of transplant involves taking a thin sheet of healthy epidermis from a donor site of the burn or frostbite patient and moving it to the recipient site. A *heterotransplant* (between two different species) can serve as a temporary treatment to prevent infection and fluid loss.

Synthetic skin fabricated from animal tissue bonded to a silicone film may be used on a patient who is extensively burned. The process includes seeding the synthetic skin with basal skin cells obtained from healthy areas on the patient's body. This technique and treatment eliminates the problems of skin grafting, such as additional trauma, widespread scarring, and rejection, as in the case of skin obtained from a cadaver.

## Wound Healing

The skin effectively protects against many abrasions, but if a wound does occur, a sequential chain of events promotes rapid healing. The process of wound healing depends on the extent and severity of the injury. Trauma to the epidermal layers stimulates an increased mitotic activity in the stratum basale, whereas injuries that extend to the dermis or subcutaneous tissue elicit activity throughout the body as well as within the wound itself. General body responses include a temporary elevation of temperature and pulse rate.

In an open wound (fig. 7.16), blood vessels are broken and bleeding occurs. Through the action of **blood platelets**

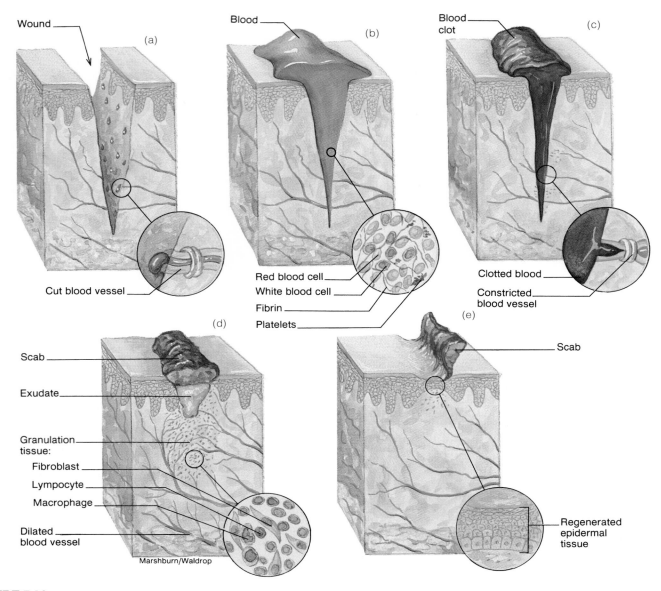

Wound
(a)
Cut blood vessel

Blood
(b)
Red blood cell
White blood cell
Fibrin
Platelets

Blood clot
(c)
Clotted blood
Constricted blood vessel

(d)
Scab
Exudate
Granulation tissue:
Fibroblast
Lympocyte
Macrophage
Dilated blood vessel
Marshburn/Waldrop

(e)
Scab
Regenerated epidermal tissue

**FIGURE 7.16**

The process of wound healing. (*a*) A penetrating wound into the dermis ruptures blood vessels. (*b*) Blood cells, fibrinogen, and fibrin flow out of the wound. (*c*) Vessels constrict and a clot blocks the flow of blood. (*d*) A protective scab is formed from the clot, and granulation tissue forms within the site of the wound. (*e*) The scab sloughs off as the epidermal layers are regenerated.

(*plăt´letz*) and protein molecules, called **fibrinogen** (*fi-brin´ ŏ-jen*), a clot forms and soon blocks the flow of blood and entry of pathogens. A scab forms and covers and protects the damaged area. Mechanisms are activated to destroy bacteria, dispose of dead or injured cells, and isolate the injured area. These responses are collectively referred to as *inflammation* and are characterized by redness, heat, edema, and pain. Inflammation is a response that confines the injury and promotes healing.

The next step in healing is the differentiation of binding **fibroblasts** from connective tissue, forming **fibrin** at the wound margins. Together with new branches from surrounding blood vessels, **granulation tissue** is formed. Phagocytic cells migrate into the wound and ingest dead cells and foreign debris. Eventually the damaged area is repaired and the protective scab is sloughed off.

If the wound is severe enough, the granulation tissue may develop into **scar tissue.** Scar tissue differs from normal skin in that its collagenous fibers are more dense and it has no stratified squamous epidermal layer. Scar tissue has fewer blood vessels and may lack hair, glands, and sensory receptors. The closer together the edges of a wound, the less granulation tissue develops and the less obvious a scar. This is one reason for suturing a large break in the skin.

## Aging

As the skin ages, it becomes thin, dry, and inelastic. Collagenous fibers in the dermis become thicker and less elastic, and the amount of adipose tissue in the hypodermis diminishes, making it thinner. Skinfold measurements indicate that the diminution of the hypodermis begins at about the age of 45. With a loss of elasticity and a reduction in the thickness of the hypodermis, wrinkling, or the permanent infolding of the skin, becomes apparent (fig. 7.17).

During the aging of the skin, the number and activity of hair follicles, sweat glands, and sebaceous glands also diminish. Consequently, there is a marked thinning of scalp hair and hair on the extremities, reduced sweating, and decreased sebum production. Since elderly people cannot perspire as freely, they are more likely to complain of heat and are more subject to heat exhaustion than young people. They also become more sensitive to cold because of the loss of insulating adipose tissue and diminished circulation. A decrease in the production of sebum causes the skin to dry and crack frequently.

The integument of an elderly person is not well protected from the sun because of thinning, and melanocytes that produce melanin gradually atrophy. The loss of melanocytes accounts for graying of the hair and pallor of the skin. After the age of 50, brown, plaquelike growths, called *seborrheic (seb˝ŏ-re´ik) hyperkeratoses,* appear within the skin, particularly on exposed portions. Skin that has been exposed to excessive sunlight tends to develop more cutaneous carcinomas than less exposed skin.

**FIGURE 7.17**
Aging of the skin results in a loss of elasticity and the appearance of wrinkles.

# Important Clinical Terminology

**acne**   An inflammatory condition of sebaceous glands. Acne is affected by gonadal hormones and is therefore most common during puberty and adolescence. Pimples and blackheads on the face, chest, and back are expressions of this condition.

**albinism**   A congenital, genetic deficiency of the pigment of the skin, hair, and eyes due to a metabolic block in the synthesis of melanin (fig. 7.18).

**alopecia**   Loss of hair; baldness. Baldness is usually due to genetic factors and cannot be treated. Baldness may signify anatomical maturity.

**athlete's foot**   (tinea pedis)   A skin-fungus disease of the foot.

**blister**   A collection of fluid between the epidermis and dermis caused by excessive friction or a burn.

**FIGURE 7.18**
The albino individual in this photograph has melanocytes within his skin, but as a result of a mutant gene he lacks the ability to synthesize melanin.

**boil** (furuncle)   A localized bacterial infection originating in a hair follicle or skin gland.

**carbuncle**   A bacterial infection similar to a boil, except that a carbuncle infects the subcutaneous tissues.

**cold sore** (fever blister)   A lesion on the lip or oral mucous membrane caused by type I herpes simplex virus (HSV), transmitted by oral or respiratory exposure.

**comedo**   A plug of sebum and epithelial debris in the hair follicle and excretory duct of the sebaceous gland; also called a blackhead or whitehead.

**corn**   A type of callus that is localized on the foot, usually over toe joints.

**dandruff**   Common dandruff is the continual shedding of epidermal cells of the scalp. Common dandruff can be removed by normal washing and brushing of the hair. Abnormal dandruff may be caused by certain skin diseases, such as seborrhea or psoriasis.

**decubitus ulcer**   A bedsore—an exposed ulcer caused by a continual pressure that restricts dermal blood flow to a localized portion of the skin (see fig. 7.6).

**dermabrasion**   A procedure for removing tattoos or acne scars by high-speed sanding or scrubbing.

**dermatitis**   An inflammation of the skin.

**dermatology**   A specialty of medicine concerned with the study of the skin—its anatomy, physiology, histopathology, and the relationship of cutaneous lesions to systemic disease.

**eczema** (*eg´zĕ-mă*)   A noncontagious inflammatory condition of the skin producing red, itching, vesicular lesions, which may be crusty or scaly.

**erythema** (*er˝ĭ-the´mă*)   Redness of the skin caused by vasodilation from skin injury, infection, or inflammation.

**furuncle**   A boil—a localized abscess resulting from an infected hair follicle.

**gangrene**   Necrosis of tissue due to the obstruction of blood flow. It may be localized or extensive and may secondarily be infected with anaerobic microorganisms.

**impetigo**   A contagious skin infection that results in lesions followed by scaly patches. It generally occurs on the face and is caused by staphylococci or streptococci.

**keratosis**   Any abnormal growth and hardening of the stratum corneum layer of the skin.

**melanoma**   A cancerous tumor originating from proliferating melanocytes within the epidermis of the skin.

**nevus** (*ne´vus*)   A mole or birthmark—a congenital pigmentation of a certain area of the skin.

**papilloma**   A benign epithelial neoplasm, such as a wart or a corn.

**papule**   A small inflamed elevation of the skin, such as a pimple.

**pruritus**   Itching. It may be symptomatic of systemic disorders but is generally due to dry skin.

**psoriasis** (*sŏ-ri˝ă-sis*)   An inherited inflammatory skin disease, usually expressed as circular scaly patches of skin.

**pustule**   A small, localized pus-filled elevation (pimple) of the skin.

**seborrhea** (*seb˝ŏ-re´ă*)   A disease characterized by an excessive activity of the sebaceous glands and accompanied by oily skin and dandruff.

**urticaria** (hives)   A skin eruption of reddish weals, usually with extreme itching. It may be caused by an allergic reaction, stress, or contact with some external or internal precipitating factor.

**wart**   A roughened projection of epidermal cells caused by a virus.

# Chapter Summary

## The Integument as an Organ (p. 138)

1. The skin is considered an organ because it consists of several kinds of tissues that are structurally arranged to function together.
2. The appearance of the skin is clinically important because it provides clues to certain body conditions or dysfunctions.

## Layers of the Integument (pp. 138–143)

1. The stratified squamous epithelium of the epidermis is divisible into five structural and functional layers: the stratum basale, stratum spinosum, stratum granulosum, stratum lucidum, and stratum corneum.
   a. Normal skin color is the result of a combination of melanin and carotene in the epidermis and hemoglobin in the blood of the dermis and hypodermis.
   b. Fingerprints on the surface of the epidermis are individually unique; flexion creases and flexion lines are acquired.
2. The thick dermis of the skin is composed of fibrous connective tissue interlaced with elastic fibers. The two layers of the dermis are the upper papillary layer and the deeper reticular layer.
3. The hypodermis, composed of adipose and fibrous connective tissue, binds the dermis to underlying organs.

## Physiology of the Integument (pp. 143–145)

1. Structural features of the skin protect the body from disease and external injury.
   a. Keratin and an acidic oily secretion on the surface protect the skin from water and microorganisms.
   b. Cornification of the skin protects against abrasion.
   c. Melanin is a barrier to UV light.
2. The skin regulates body fluids and temperatures.
   a. Fluid loss is minimal due to keratinization and cornification.
   b. Temperature regulation is maintained by radiation, convection, and the antagonistic effects of sweating and shivering.
3. The skin permits the absorption of UV light, respiratory gases, steroids, fat-soluble vitamins, and certain toxins and pesticides.
4. The integument synthesizes melanin and keratin, which remain in the skin, and vitamin D, which is used elsewhere in the body.

## Interactions of the Integumentary System with Other Body Systems

**Skeletal System**
- Supports the skin
- Stores minerals needed by the skin
- Covers and protects skeletal system
- Synthesizes vitamin D necessary for calcium absorption and metabolism

**Muscular System**
- Generates body heat to warm the skin
- Covers and protects muscles
- Permits radiant heat loss (sweating) during muscle contractions

**Nervous System**
- Provides autonomic motor impulses to cutaneous vessels and glands
- Houses cutaneous (tactile) receptors that convey sensory sensations to the brain

**Endocrine System**
- Sex hormones cause changes in integumentary features—pubic, axillary, and facial hair and oily skin
- Covers and protects certain endocrine glands

**Circulatory System**
- Transports $O_2$ and $CO_2$, nutrients, and fluids to and from the skin
- Protects circulatory system
- Maintains constant body temperature
- Prevents fluid loss (formation of scabs)

**Lymphatic System**
- Maintains a balanced amount of interstitial fluid within the skin
- Protects against pathogen invasion
- Prevents edema (retention of interstitial fluid)

**Respiratory System**
- Provides $O_2$ and eliminates $CO_2$
- Protects organs of upper respiratory tract

**Urinary System**
- Eliminates metabolic wastes
- Activates vitamin D
- Excretes salts and some nitrogenous wastes

**Digestive System**
- Provides nutrients for growth, maintenance, and repair of skin
- Provides vitamin D necessary for calcium absorption

**Reproductive System**
- Gonads produce sex hormones that promote skin growth, maturation, and maintenance
- Forms scrotum, which covers and protects testes

5. Sensory reception in the skin is provided through cutaneous receptors throughout the dermis and hypodermis. Cutaneous receptors respond to precise sensory stimuli and are more sensitive in thin skin.
6. Certain emotions are reflected in changes in the skin.

## Epidermal Derivatives (pp. 145–149)

1. Hair is characteristic of all mammals, but its distribution, function, density, and texture varies across mammalian species.
   a. Each hair consists of a shaft, root, and bulb. The bulb is the enlarged base of the root within the hair follicle.
   b. The three layers of the hair shaft are the medulla, cortex, and cuticle.
   c. Lanugo, angora, and definitive are the three kinds of human hair.
2. Hardened, keratinized nails are found on the distal dorsum of each digit, where they protect the digits; fingernails aid in grasping and picking up small objects.
   a. Each nail consists of a nail body, free border, and hidden border.
   b. The hyponychium, eponychium, and nail fold support the nail on the nail bed.
3. Integumentary glands are exocrine, since they either secrete or excrete substances through ducts.
   a. Sebaceous glands secrete sebum onto the shaft of the hair.
   b. The two types of sudoriferous (sweat) glands are eccrine and apocrine.
   c. Mammary glands are specialized sudoriferous glands that secrete milk during lactation.
   d. Ceruminous glands secrete cerumen (earwax).

# Review Activities

## Objective Questions

1. Hair, nails, integumentary glands, and the epidermis of the skin are derived from embryonic
   a. ectoderm.          b. mesoderm.
   c. endoderm.          d. mesenchyme.
2. Spoon-shaped nails may result when a person has a dietary deficiency of
   a. zinc.               b. iron.
   c. niacin.             d. vitamin $B_{12}$.
3. The epidermal layer *not* present in the thin skin of the face is the stratum
   a. granulosum.        b. lucidum.
   c. spinosum.          d. corneum.
4. Which of the following does *not* contribute to skin color?
   a. hair papillae      b. melanin
   c. carotene           d. hemoglobin
5. Which of the following is *not* true of the epidermis?
   a. It is composed of stratified squamous epithelium.
   b. As the epidermal cells die, they undergo keratinization and cornification.
   c. Rapid mitotic activity (cell division) within the stratum corneum accounts for the thickness of this epidermal layer.
   d. In most areas of the body, the epidermis lacks blood vessels and nerves.

6. Integumentary glands that empty their secretions into hair follicles are
   a. sebaceous glands.
   b. endocrine glands.
   c. eccrine glands.
   d. ceruminous glands.
7. Fetal hair that is present during the last trimester of development is referred to as
   a. angora.            b. definitive.
   c. lanugo.            d. replacement.
8. Which of these conditions is potentially life threatening?
   a. acne               b. melanoma
   c. eczema             d. seborrhea
9. The skin of a burn victim has been severely damaged through the epidermis and into the dermis. Integumentary regeneration will be slow, with some scarring, but it will be complete. Which kind of burn is this?
   a. first degree
   b. second degree
   c. third degree
10. The technical name for a blackhead or whitehead is
    a. carbuncle.         b. melanoma.
    c. nevus.             d. comedo.

## Essay Questions

1. List the functions of the skin. Which of these functions occurs passively due to the structure of the skin, and which occurs dynamically due to physiological processes?
2. Why is the skin considered an organ? What types of tissues are found in each of the three layers of the skin?
3. Discuss the growth process and regeneration of the epidermis.
4. What are some physical and chemical features of the skin that make it an effective protective organ?
5. Define the following: *lines of tension, friction ridges,* and *flexion lines.* What causes each of these to develop?
6. Distinguish between a hair follicle and a hair. What other accessory structures are associated with hair follicles and hair?
7. Compare and contrast the structure and function of sebaceous sudoriferous, mammary, and ceruminous glands.
8. Discuss the development of the skin and its associated hair and glands. What role do the ectoderm and mesoderm play in integumentary development?
9. Discuss what is meant by an inflammatory lesion. What are some frequent causes of skin lesions?
10. Describe each of the three degrees of burns and discuss the physiological danger of burns.

## Gundy/Weber Software 💾

The tutorial software accompanying Chapter 7 is Volume 1—Introduction, Tissues, Integumentary System.

# skeletal system: bone tissue and bone development

## objectives

- Describe the structural organization of the skeletal system and list the bones of the axial and appendicular portions.

- Discuss the principal functions of the skeletal system and identify the body systems served by these functions.

- Classify bones according to their shapes and give an example of each type.

- Describe the various markings on the surfaces of bones.

- Describe the gross features of a typical long bone and list the functions of each feature.

- Identify the five types of bone cells and list the functions of each.

- Distinguish between spongy and compact bone tissues.

- Describe the process of endochondral ossification as it relates to bone growth.

- Describe the relationships between bone and plasma calcium levels and the importance of maintaining normal blood calcium concentrations.

- Describe how parathyroid hormone and calcitonin secretions are regulated and how these hormones regulate the plasma calcium and phosphate concentrations.

- Explain how vitamin D functions as a prehormone and describe its effects on calcium and phosphate balance.

# Organization of the Skeletal System

*The axial and appendicular components of the skeletal system of an adult human consist of 206 individual bones arranged to form a strong, flexible body framework.*

**Osteology** is the science concerned with the study of bones. Each bone is an organ that plays a part in the total functioning of the skeletal system. The skeletal system of an adult human is composed of approximately 206 bones (fig. 8.1). Actually, the number of bones differs from person to person, depending on age and genetic variations. At birth, the skeleton consists of approximately 270 bones. As further bone development (ossification) occurs during infancy, the number increases. During adolescence, however, the number of bones decreases, as separate bones gradually fuse.

••••••••••••••••••••••••••••••••••••

ossification: Gk. *os*, bone; L. *facio*, to make

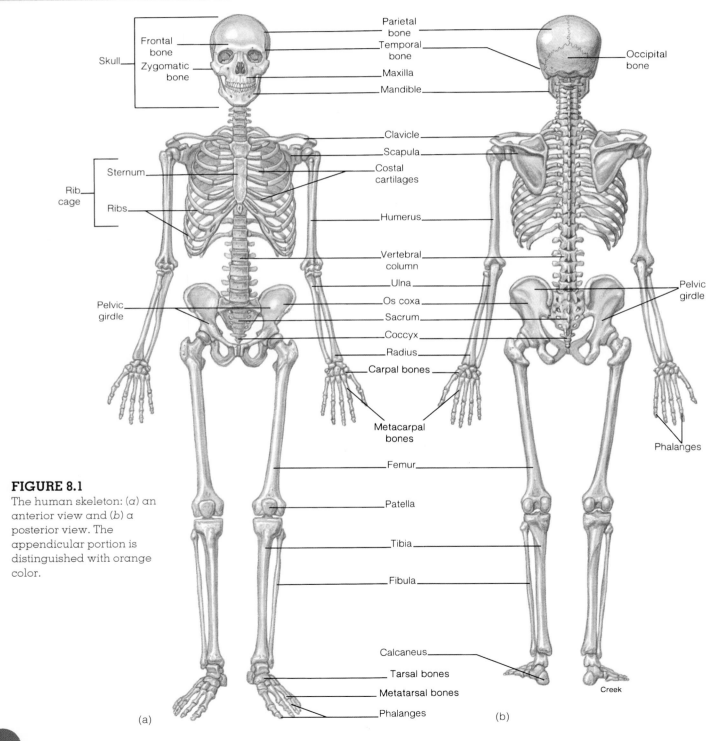

**FIGURE 8.1**

The human skeleton: (*a*) an anterior view and (*b*) a posterior view. The appendicular portion is distinguished with orange color.

Skull
Frontal bone
Zygomatic bone
Parietal bone
Temporal bone
Maxilla
Mandible
Occipital bone

Clavicle
Scapula
Costal cartilages
Sternum
Rib cage
Ribs
Humerus
Vertebral column
Ulna
Os coxa
Sacrum
Coccyx
Radius
Carpal bones
Pelvic girdle
Metacarpal bones
Phalanges
Femur
Patella
Tibia
Fibula
Calcaneus
Tarsal bones
Metatarsal bones
Phalanges

Creek

(a)

(b)

Some adults have extra bones within the joints (sutures) of the skull called **sutural** (*soo´cher-al*) **bones.** Additional bones may develop in tendons in response to stress as the tendons repeatedly move across a joint. Bones formed this way are called **sesamoid** (*ses´ă-moid*) **bones.** Sesamoid bones, like the sutural bones, vary in number. The patellae (kneecaps) are two sesamoid bones all people have.

For the convenience of study, the skeleton is divided into axial and appendicular portions. Anterior and posterior views of the skeleton are shown in figure 8.1. The divisions of the skeleton and the number of bones in each portion are listed in table 8.1.

The **axial skeleton** consists of the bones that form the axis of the body and that support and protect the organs of the head, neck, and trunk.

**1 Skull.** The skull consists of two sets of bones: the *cranial bones* that form the cranium, or braincase, and the *facial bones* that support the eyes, nose, and jaws.

**2 Auditory ossicles.** Three auditory ossicles are present in the middle-ear chamber of each ear and serve to transmit sound impulses.

**3 Hyoid bone.** The hyoid (*hi´oid*) bone is located above the larynx and below the lower jaw. It supports the tongue and assists in swallowing.

**4 Vertebral column.** The vertebral column (backbone) consists of 26 individual bones (*vertebrae*) separated by cartilaginous *intervertebral discs*. In the pelvic region, several vertebrae are fused to form the *sacrum*, which provides attachment for the pelvic girdle. A variable number of terminal vertebrae are fused to form the *coccyx*—the so-called tailbone.

**5 Rib cage.** The rib cage, or thoracic cage, forms the bony and cartilaginous framework of the thorax. The rib cage articulates posteriorly with the thoracic vertebrae and includes the 12 pairs of *ribs*, the flattened *sternum*, and the *costal cartilages* that connect the ribs to the sternum on the anterior side.

........................................

sesamoid: Gk. *sesamon,* like a sesame seed

........................................

ossicle: L. *ossiculum,* little bone

## Table 8.1 Classification of the bones of the adult skeleton

| Axial skeleton | | | | Appendicular skeleton | |
|---|---|---|---|---|---|
| **Skull**—22 bones | | **Auditory ossicles**—6 bones | | **Pectoral girdle**—4 bones | |
| *14 facial bones* | *8 cranial bones* | malleus (2) | | scapula (2) | |
| maxilla (2) | frontal (1) | incus (2) | | clavicle (2) | |
| palatine (2) | parietal (2) | stapes (2) | | | |
| zygomatic (2) | occipital (1) | | | **Upper extremities**—60 bones | |
| lacrimal (2) | temporal (2) | **Hyoid**—1 bone | | humerus (2) | carpal bones (16) |
| nasal (2) | sphenoid (1) | | | radius (2) | metacarpal bones (10) |
| vomer (1) | ethmoid (1) | **Vertebral column**—26 bones | | ulna (2) | phalanges (28) |
| inferior nasal | | cervical vertebra (7) | | | |
| concha (2) | | thoracic vertebra (12) | | **Pelvic girdle**—2 bones | |
| mandible (1) | | lumbar vertebra (5) | | os coxa (2) (each contains 3 fused bones) | |
| | | sacrum (1) (5 fused bones) | | | |
| | | coccyx (1) (3–5 fused bones) | | | |
| | | | | **Lower extremities**—60 bones | |
| | | **Rib cage**—25 bones | | femur (2) | tarsal bones (14) |
| | | rib (24) | | tibia (2) | metatarsal bones (10) |
| | | sternum (1) | | fibula (2) | phalanges (28) |
| | | | | patella (2) | |

The **appendicular skeleton** is composed of the bones of the upper and lower extremities and the bony girdles, which anchor the appendages to the axial skeleton.

**1 Pectoral girdle.** The paired *scapulae* and *clavicles* constitute the pectoral girdle. It is not a complete girdle, having only an anterior attachment to the axial skeleton at the sternum via the clavicles. The primary function of the pectoral girdle is to provide attachment for the muscles that move the brachium and forearm.

**2 Upper extremities.** Each upper extremity contains a proximal *humerus* within the brachium, an *ulna* and *radius* within the forearm, the *carpal bones* of the wrist, and the *metacarpal bones* and *phalanges* of the hand.

**3 Pelvic girdle.** The pelvic girdle is formed by two *ossa coxae* (hipbones) united anteriorly by the *symphysis* (*sim´fĭ-sis*) *pubis* and posteriorly by the sacrum of the vertebral column. The pelvic girdle supports the weight of the body through the vertebral column and protects the lower viscera within the pelvic cavity.

**4 Lower extremities.** Each lower extremity contains a proximal *femur* within the thigh, a *tibia* and *fibula* within the leg, the *tarsal bones* of the ankle, and the *metatarsal bones* and *phalanges* of the foot. In addition, the *patella* (*pă-tel´ă*) is located on the anterior surface of the knee joint between the thigh and leg.

# Functions of the Skeletal System

*The bones of the skeleton perform the mechanical functions of support, protection, and leverage for body movement and the metabolic functions of hemopoiesis and mineral storage.*

The strength of bone comes from its inorganic components, which resist decomposition even after death. Much of what we know of prehistoric animals, including humans, has been determined from preserved skeletal remains. Frequently when we think of bone, we think of a hard, dry structure. In fact, the term *skeleton* comes from a Greek word meaning "dried up." Living bone is not a dry, inert material, however; it is dynamic and adaptable in performing many body functions, including support, protection, leverage for body movement, hemopoiesis, and mineral storage.

**1 Support.** The skeleton forms a rigid framework to which are attached the softer tissues and organs of the body.

**2 Protection.** The skull and vertebral column enclose the central nervous system; the rib cage protects the heart, lungs, great vessels, liver, and spleen; and the pelvic cavity supports and protects the pelvic viscera. Even the site where blood cells are produced is protected within the central portion of certain bones.

**3 Body movement.** Bones serve as anchoring attachments for most skeletal muscles. In this capacity, the bones act as levers, with the joints functioning as pivots, when muscles contract to cause body movement.

**4 Hemopoiesis** (*he´´mŏ-poi-e´sis*). The red bone marrow of an adult produces white blood cells, red blood cells, and platelets. In an infant, the spleen and liver produce red blood cells, but as the bones mature, the bone marrow assumes the performance of this formidable task. It is estimated that an average of 1 million blood cells are produced every second by the bone marrow to replace those that are worn out and destroyed by the liver.

**5 Mineral storage.** The inorganic matrix of bone is composed primarily of the minerals calcium and phosphorus. These minerals give bone its rigidity and account for approximately two-thirds of the weight of bone. About 95% of the calcium and 90% of the phosphorus within the body are deposited in the bones and teeth. Although the concentration of these organic salts within the blood is kept within narrow limits, both of these mineral salts are essential for other body functions. Calcium is necessary for muscle contraction, blood clotting, and the movement of ions and nutrients across cell membranes. Phosphorus is required for the activities of the nucleic acids DNA and RNA, as well as for ATP utilization. If mineral salts are not present in the diet in sufficient amounts, they may be withdrawn from the bones until they are replenished through proper nutrition. In addition to calcium and phosphorus, lesser amounts of magnesium and sodium salts are stored in bone tissue.

In summary, the skeletal system is not an isolated body system. It functions with the muscle system since it stores the calcium needed for muscular contraction and provides an attachment for muscles as they span the movable joints. The skeletal system serves the circulatory system by producing blood cells in protected sites. Also, many of the vessels of the circulatory system are named according to the bones they parallel. The skeletal system supports and protects all of the systems of the body to varying degrees.

# Gross Structure of Bone

*Each bone has a characteristic shape and diagnostic surface features that indicate its functional relationship to other bones, muscles, and to the body structure as a whole.*

The shape and surface features of each bone indicate its functional role in the skeleton (table 8.2). Bones that are long, for example, function as levers during body movement. Bones that support the body are massive and have

## Development of the Skeletal System

Bone formation, or *ossification*, begins about the fourth week of embryonic development, but ossification centers cannot be readily observed until about the tenth week (fig. 1). Bone tissue derives from specialized migratory cells of mesoderm (see chapter 6) known as *mesenchyme*. Some of the embryonic mesenchymal cells will transform into

*chondroblasts* (*kon´dro-blasts*) and develop a cartilage matrix that is later replaced by bone in a process known as **endochondral** (*en˝dŏ-kon´dral*) **ossification.** Most of the skeleton is formed in this fashion—first it goes through a hyaline cartilage stage and then it is ossified as bone.

..........................................

chondroblast: Gk. *chondros*, cartilage; *blastos*, offspring or germ

A smaller number of mesenchymal cells develop directly into bone without first going through a cartilage stage. This type of bone-formation process is referred to as **intramembranous** (*in´tră-mem˝bră-nus*) **ossification.** Facial bones and certain bones of the cranium are formed this way. **Sesamoid bones** are specialized intramembranous bones that develop in tendons. The patella is an example of a sesamoid bone.

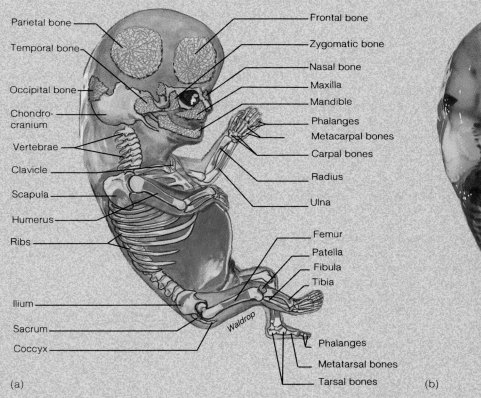

Parietal bone
Temporal bone
Occipital bone
Chondro-cranium
Vertebrae
Clavicle
Scapula
Humerus
Ribs
Ilium
Sacrum
Coccyx

Frontal bone
Zygomatic bone
Nasal bone
Maxilla
Mandible
Phalanges
Metacarpal bones
Carpal bones
Radius
Ulna
Femur
Patella
Fibula
Tibia
Phalanges
Metatarsal bones
Tarsal bones

Waldrop

(a)

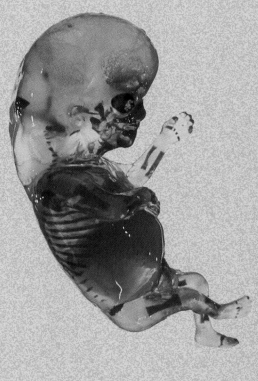

(b)

### FIGURE 1

Ossification centers of the skeleton of a 10-week-old fetus. (a) The diagram depicts endochondrial ossification in red and intramembranous ossification in a stippled pattern. The cartilaginous portions of the skeleton are shown in gray. (b) The photograph shows the ossification centers stained with a red indicator dye.

## Table 8.2 Surface features of bone

| Structure | Description and example |
|---|---|
| **Articulating surfaces** | |
| condyle | A large, rounded, articulating knob (the occipital condyle of the occipital bone) |
| facet | A flattened or shallow articulating surface (the costal facet of a thoracic vertebra) |
| head | A prominent, rounded, articulating end of a bone (the head of the femur) |
| **Depressions and openings** | |
| alveolus | A deep pit or socket (the alveoli for teeth in the maxilla) |
| fissure | A narrow, slitlike opening (the superior orbital fissure of the sphenoid bone) |
| foramen | A rounded opening through a bone (the foramen magnum of the occipital bone) |
| fossa | A flattened or shallow surface (the mandibular fossa of the temporal bone) |
| sinus | A cavity or hollow space in a bone (the frontal sinus of the frontal bone) |
| sulcus | A groove that accommodates a vessel, nerve, or tendon (the intertubercular sulcus of the humerus) |
| **Nonarticulating prominences** | |
| crest | A narrow, ridgelike projection (the iliac crest of the os coxa) |
| epicondyle | A projection above a condyle (the medial epicondyle of the femur) |
| process | Any marked bony prominence (the mastoid process of the temporal bone) |
| spine | A sharp, slender process (the spine of the scapula) |
| trochanter | A massive process found only on the femur (the greater trochanter of the femur) |
| tubercle | A small rounded process (the greater tubercle of the humerus) |
| tuberosity | A large roughened process (the radial tuberosity of the radius) |

facet: Fr. *facette,* little face
trochanter: Gk. *trochanter,* runner
tuberosity: L. *tuberosus,* lump

**FIGURE 8.2**
Shapes of bones.

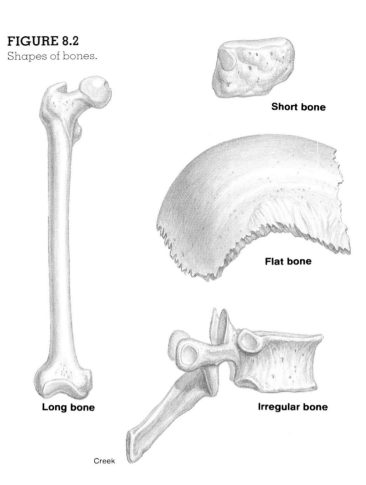

Short bone

Flat bone

Long bone

Irregular bone

Creek

large articular surfaces and processes for muscle attachment. Roughened areas on these bones may serve for the attachment of ligaments, tendons, or muscles. A flattened surface provides a placement site for a large muscle or may provide protection. Grooves around an articular end of a bone are where tendons or nerves pass, and openings through a bone permit the passage of nerves or blood vessels.

## Shapes of Bones

The bones of the skeleton are classified into four principal types on the basis of shape rather than size. The four classes are long bones, short bones, flat bones, and irregular bones (fig. 8.2).

**1 Long bones.** Long bones are longer than they are wide and function as levers. Most of the bones of the upper and lower extremities are of this type (e.g., the humerus, radius, ulna, metacarpal bones, femur, tibia, fibula, metatarsal bones, and phalanges).

**2 Short bones.** Short bones (e.g., the wrist and ankle bones) are somewhat cube-shaped and are found in confined spaces, where they transfer forces.

**3 Flat bones.** Flat bones (e.g., the cranium, ribs, and bones of the shoulder girdle) have a broad, dense surface for muscle attachment or protection of underlying organs.

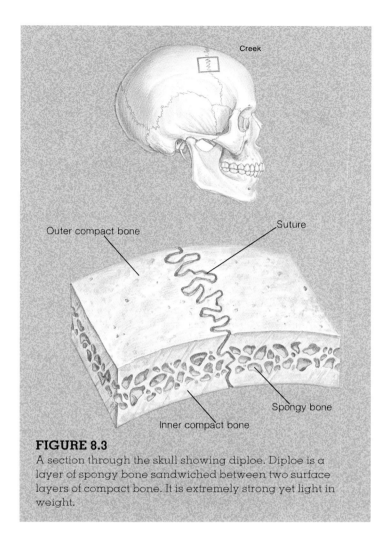

**FIGURE 8.3**
A section through the skull showing diploe. Diploe is a layer of spongy bone sandwiched between two surface layers of compact bone. It is extremely strong yet light in weight.

**4  Irregular bones.** Irregular bones (e.g., the vertebrae and certain bones of the skull) have varied shapes and many surface markings for muscle attachment or articulation.

## Gross Anatomy of a Long Bone

Bone (osseous) tissue is organized as **spongy (cancellous) bone** or **compact (dense) bone,** and most bones have both types. In a flat bone of the skull, for example, the spongy bone is sandwiched between the compact bone and is called a *diploe* (*dip´lo-e*) (fig. 8.3). Because of this protective layering of bone tissue, a blow to the head may fracture the outer compact bone layer without harming the inner compact bone layer and the brain.

In a long bone from an appendage, the bone shaft, or **diaphysis** (*di-af´ĭ-sis*), consists of compact bone forming a cylinder that surrounds a central cavity called the **medullary cavity** (fig. 8.4). The medullary cavity is lined with a thin

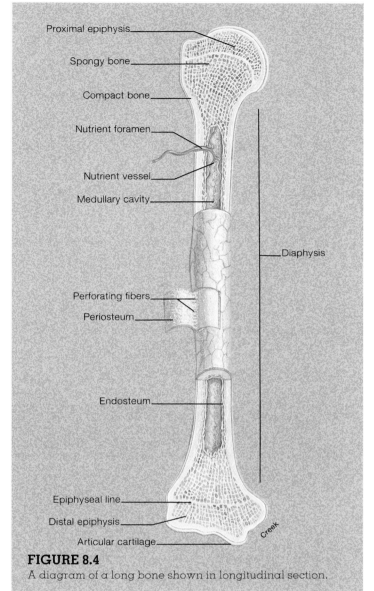

**FIGURE 8.4**
A diagram of a long bone shown in longitudinal section.

layer of connective tissue called the **endosteum** (*en-dos´te-um*) and contains **yellow bone marrow,** so named because of the large amounts of fat it contains. On each end of the diaphysis is an **epiphysis** (*ĕ-pif´ĭ-sis*), consisting of cancellous bone surrounded by a layer of compact bone. **Red bone marrow** is found within the porous chambers of spongy bone. In an adult, erythropoiesis, the production of red blood cells, occurs in the red bone marrow, especially that of the sternum, vertebrae, portions of the ossa coxae, and the proximal epiphyses of the femora and humeri. The red bone marrow is also responsible for the formation of white blood cells and platelets and for the phagocytosis of worn-out red

..........................................
diploe: Gk. *diplous,* double
diaphysis: Gk. *dia,* throughout; *physis,* growth

..........................................
epiphysis: Gk. *epi,* upon; *physis,* growth

blood cells. **Articular cartilage,** which is composed of thin hyaline cartilage, caps each epiphysis and facilitates joint movement. Along the diaphysis are **nutrient foramina**—small openings into the bone that allow for passage of nutrient vessels into the bone for nourishment of the living tissue.

Between the diaphysis and epiphysis is an **epiphyseal** (*ep″ĭ-fiz′e-al*) **plate** of cartilage—a region of mitotic activity that is responsible for *linear bone growth*. As bone growth is completed, an **epiphyseal line** replaces the plate and ossification occurs between the epiphysis and the diaphysis. A **periosteum** (*per″e-os′te-um*) of dense regular connective tissue covers the surface of the bone, except at the articulating surfaces. This highly vascular layer serves as a place for a tendon-muscle attachment and is responsible for *appositional* (width) *bone growth*. The periosteum is secured to the bone by **perforating** (Sharpey's) **fibers** (fig. 8.4) composed of bundles of collagenous fibers.

 Fracture of a long bone in a young person may be especially serious if it results in displacement of an epiphyseal plate. If such an injury is untreated, or treated improperly, linear bone growth may be arrested or retarded, resulting in permanent shortening of the limb.

# Bone Tissue

*Bone tissue is composed of several types of bone cells embedded in a matrix of ground substance, inorganic salts, and collagenous fibers. Bone cells and ground substance give bone flexibility and strength; the inorganic salts give it hardness.*

## Bone Cells

The bone cells of which bone tissue is composed are of five principal types. **Osteogenic** (*os″te-ŏ-jen′ĭk*) **cells** are found in the bone tissues in contact with the endosteum and the periosteum. These cells respond to trauma, such as a fracture, by giving rise to bone-forming cells (*osteoblasts*) and bone-destroying cells (*osteoclasts*). **Osteoblasts** (*os′te-ŏ-blasts*) are bone-forming cells (fig. 8.5) that synthesize and secrete unmineralized ground substance. They are abundant in areas of high metabolism within bone, such as under the periosteum and bordering the medullary cavity. **Osteocytes** (*os′te-ŏ-sĭtz*) are mature bone cells (figs. 8.5 and 8.8) derived from osteoblasts that have secreted bone tissue around themselves. Osteocytes maintain healthy bone tissue by secreting enzymes and influencing bone mineral content. They also regulate the calcium release from bone tissue to blood.

periosteum: Gk. *peri*, around; *osteon*, bone
Sharpey's fibers: from William Sharpey, Scottish physiologist and histologist, 1802–80
osteoblast: Gk. *osteon*, bone; *blastos*, offspring or germ

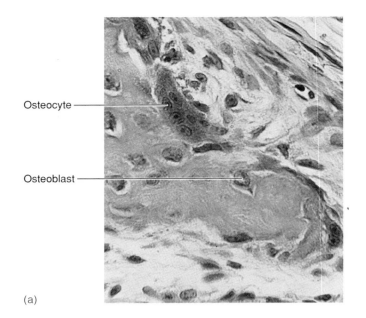

(a)

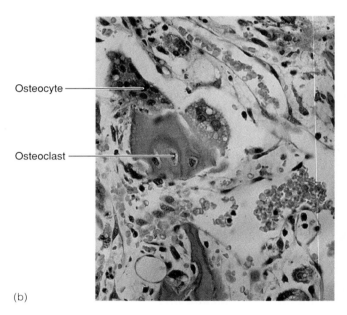

(b)

**FIGURE 8.5**
Types of bone cells. (*a*) Osteoblasts are important in secreting unmineralized ground substance. Osteocytes derive from osteoblasts and play a regulatory role in maintaining bone tissue. (*b*) Osteoclasts are bone-destroying cells that help to maintain the dynamic state of bone tissue.

**Osteoclasts** (*os′te-ŏ-klasts*) are large, multinuclear cells (fig. 8.5) that enzymatically break down bone tissue. These cells are important in bone growth, remodeling, and healing. **Bone-lining cells** are derived from osteoblasts along the surface of most bones in the adult skeleton. These cells are thought to regulate the movement of calcium and phosphate into and out of bone matrix.

osteoclast: Gk. *osteon*, bone; *klastos*, broken

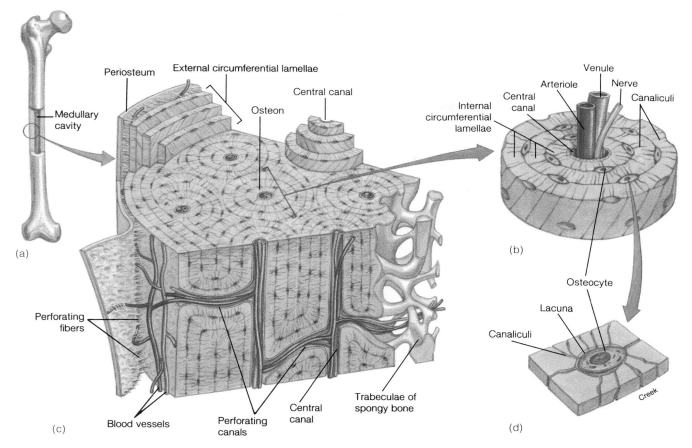

**FIGURE 8.6**

Compact bone tissue. (*a*) A diagram of the femur showing a cut through the compact bone into the medullary cavity. (*b*) The arrangement of the osteons within the diaphysis of the bone.

(*c*) An enlarged view of an osteon showing the osteocytes within lacunae and the concentric lamellae. (*d*) An osteocyte within a lacuna.

## Spongy and Compact Bone Tissues

Both spongy and compact bone tissues are present in most bones (fig. 8.6). **Spongy bone tissue** is deeply located compared to the compact bone tissue, and is quite porous. Minute spikes of bone tissue called **trabeculae** give spongy bone a latticelike appearance. Spongy bone is highly vascular and provides great strength to bone with minimal weight.

**Compact bone tissue** is superficial to the spongy bone tissue, and is very hard and dense. It consists of precise arrangements of microscopic cylindrical structures oriented parallel to the long axis of the bone (fig. 8.6). These column-like structures are the **osteons** (haversian systems) of the bone tissue. The matrix of an osteon is laid down in concentric rings, called **lamellae,** around a **central** (haversian) **canal** (fig. 8.7). The central canal contains minute nutrient vessels and a nerve. Osteocytes within spaces called **lacunae** (fig. 8.8) are regularly arranged between the lamellae. The lacunae are connected by **canaliculi,** through which

nutrients diffuse. Metabolic activity within bone tissue occurs at the osteon level. Between osteons there are incomplete remnants of osteons called **interstitial systems. Perforating** (Volkmann's) **canals** penetrate compact bone connecting osteons with blood vessels and nerves.

## Bone Growth

*The development of bone from embryonic to adult size depends on the orderly processes of mitotic divisions, growth, and the structural remodeling determined by genetics, hormonal secretions, and nutritional supply.*

In most bone development, a cartilaginous model is gradually replaced by bone tissue during **endochondral bone formation**. As the cartilage model grows, the *chondrocytes* (cartilage cells) in the center of the shaft hypertrophy, and minerals are deposited within the matrix in a process called

haversian system: from Clopton Havers, English anatomist, 1650–1702

Volkmann's canal: from Alfred Volkmann, German physiologist, 1800–1877

## FIGURE 8.7

Bone tissue as seen in (a) a scanning electron micrograph, and (b) a photomicrograph. The lacunae (*La*) provide spaces for the osteocytes, which are connected to one another by a canaliculus (*Ca*). Note the divisions between concentric lamellae (arrows). From: *Tissues and Organs: A Text Atlas of Scanning Electron Microscopy* by R. G. Kessel and R. Kardon. © 1979, W. H. Freeman and Company.

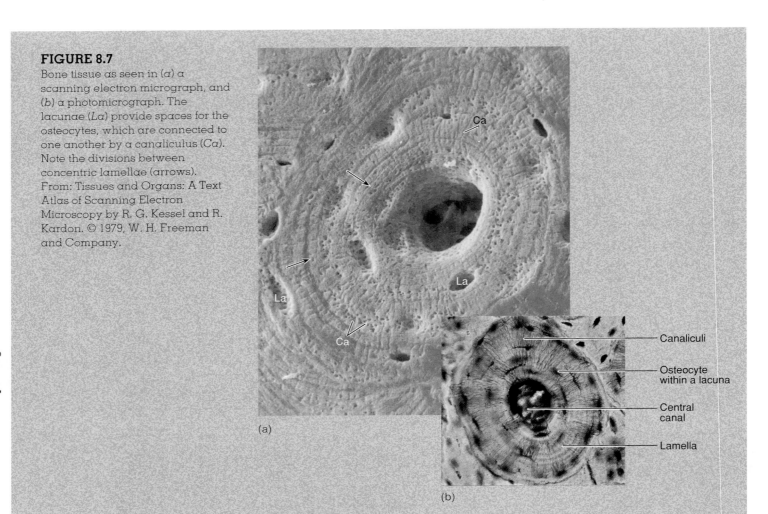

(a)

(b)

Canaliculi

Osteocyte within a lacuna

Central canal

Lamella

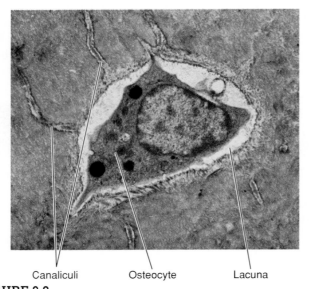

Canaliculi     Osteocyte     Lacuna

## FIGURE 8.8

A photomicrograph of an osteocyte within a lacuna.

*calcification* (fig. 8.9). Calcification restricts the passage of nutrients to the chondrocytes, causing them to die. At the same time, some cells of the perichondrium (dense regular connective tissue surrounding cartilage) differentiate into *osteoblasts* (primordial bone cells) that secrete **osteoid,** the organic component of bone. As the perichondrium calcifies, it gives rise to a thin plate of compact bone called the **periosteal bone collar.** The periosteal bone collar is surrounded by the periosteum.

A **periosteal bud,** consisting of osteoblasts and blood vessels, invades the disintegrating center of the cartilage model from the periosteum. Once in the center, the osteoblasts secrete osteoid, thus establishing a **primary ossification center** from which ossification expands into deteriorating cartilage. This process is repeated in both the proximal and distal epiphyses, forming secondary ossification centers where spongy bone develops.

Once the secondary ossification centers are formed, bone tissue totally replaces cartilage tissue, except at the articular ends of the bone and at the epiphyseal plates. An **epiphyseal plate** contains five *histological zones* (fig. 8.10).

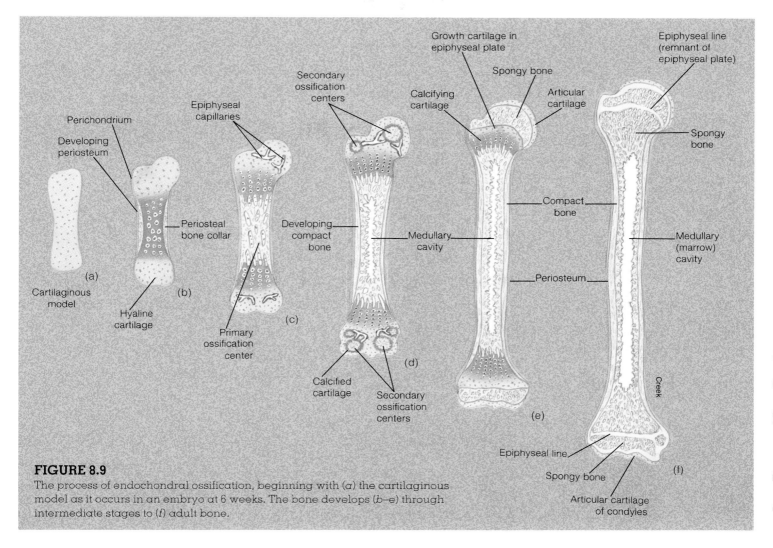

**FIGURE 8.9**

The process of endochondral ossification, beginning with (a) the cartilaginous model as it occurs in an embryo at 6 weeks. The bone develops (b–e) through intermediate stages to (f) adult bone.

The *reserve zone* borders the epiphysis and consists of small chondrocytes that are irregularly dispersed throughout the intercellular matrix. The chondrocytes in this zone anchor the epiphyseal plate to the bony epiphysis. The *proliferating zone* consists of larger, regularly arranged chondrocytes that are constantly dividing. The *hypertrophic zone* consists of very large chondrocytes, arranged in columns. The linear growth of long bones is due to the cellular proliferation at the proliferating zone and the growth and maturation of these new cells within the hypertrophic zone. The *resorption zone* is the area in which mineral content is changing. The *ossification zone* is a region of transformation from cartilage tissue to bone tissue. The chondrocytes within this zone die because the intercellular matrix surrounding them becomes calcified. Osteoclasts then break down the calcified matrix, and the area is invaded by osteoblasts and capillaries from the bone tissue of the diaphysis. As the osteoblasts mature, osteoid is secreted and bone tissue is formed. The result of this process is a gradual increase in the length of bone at the epiphyseal plates.

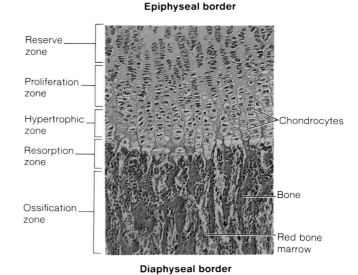

**Epiphyseal border**

**Diaphyseal border**

**FIGURE 8.10**

A photomicrograph from an epiphyseal plate (63×).

| Table 8.3 | Average age of completion of bone ossification | |
|---|---|---|
| **Bone** | | *Chronological age of fusion* |
| Scapula | | 18–20 |
| Clavicle | | 23–31 |
| Bones of upper extremity (brachium, forearm, hand) | | 17–20 |
| Os coxa | | 18–23 |
| Bones of lower extremity (thigh, leg, foot) | | 18–22 |
| Vertebra | | 25 |
| Sacrum | | 23–25 |
| Sternum (body) | | 23 |
| Sternum (manubrium, xiphoid) | | 30+ |

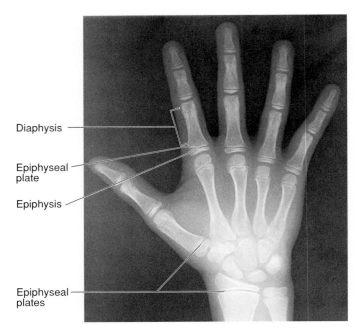

**FIGURE 8.11**
The presence of epiphyseal plates, as seen in a radiograph of a child's hand, indicates that bones are still growing in length.

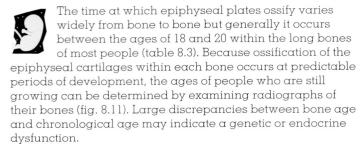 The time at which epiphyseal plates ossify varies widely from bone to bone but generally it occurs between the ages of 18 and 20 within the long bones of most people (table 8.3). Because ossification of the epiphyseal cartilages within each bone occurs at predictable periods of development, the ages of people who are still growing can be determined by examining radiographs of their bones (fig. 8.11). Large discrepancies between bone age and chronological age may indicate a genetic or endocrine dysfunction.

Bone remodeling is a continual process throughout a person's life. The diagnostic processes on the surface of bones develop as stress is applied to the periosteum, resulting in the osteoblastic secretion of osteoid and the formation of new bone tissue. The greater trochanter of the femur, for example, develops in response to forces of stress applied to the periosteum of the bone where the tendons of muscles attach. These processes may continue to change somewhat in people who engage in rigorous physical activity, even though they have stopped growing in height.

As new bone layers are deposited on the outside surface of the bone, osteoclasts dissolve bone tissue adjacent to the medullary cavity. In this way, the size of the cavity keeps pace with the increased growth of the bone.

Bone is highly dynamic and is continually being remodeled in response to mechanical stress or even the absence of stress. The effect of the absence of stress can best be seen in the bones of bedridden or paralyzed individuals. Radiographic examination shows the loss of bone mass, or even osteoporosis. The absence of gravity that accompanies space flight may result in mineral loss from bones if an exercise program is not maintained.

 The movement of teeth in orthodontics involves bone remodeling. The teeth sockets (alveoli) are reshaped through the activity of osteoclast and osteoblast cells as stress is applied through the application of braces. The use of traction in treating certain skeletal disorders has a similar effect.

# Physiology of Bone Tissue

*Bone deposition and bone resorption maintain homeostasis of calcium and phosphate concentrations within the plasma of the blood. These processes are regulated by parathyroid hormone, 1, 25-dihydroxyvitamin D₃, and calcitonin.*

One of the functions of the skeletal system is mineral storage, and bone tissue is highly dynamic in maintaining homeostasis of calcium and phosphate concentrations. Calcium and phosphate are stored in bone as *hydroxyapatite (hidrok´´se-ap´ĭ-tīt) crystals*, which have the formula $Ca_{10}(PO_4)_6(OH)_2$. The calcium phosphate in these hydroxyapatite crystals is derived from the blood by the action of bone-forming cells, or *osteoblasts* (see fig. 8.5). The osteoblasts secrete an organic matrix, composed largely of collagenous fibers, which becomes hardened by deposits of hydroxyapatite. This process is called **bone deposition** or **ossification. Bone resorption** (dissolution), produced by the action of *osteoclasts*, results in the return of bone calcium and phosphate to the blood.

## Table 8.4  Factors affecting bone physiology

| Substance | Effect |
|---|---|
| Growth hormone | Stimulates osteoblast activity and collagen synthesis |
| Thyroid hormones | Stimulate osteoblast activity, collagen synthesis, and formation of ossification centers |
| Parathyroid hormone (PTH) | Stimulates resorption |
| 1,25-Dihydroxyvitamin $D_3$ | Stimulates resorption |
| Calcitonin | Stimulates deposition |
| Sex hormones (especially androgens) | Stimulate osteoblast activity and bone growth |
| Adrenocorticoid hormones | Stimulate osteoclast activity |
| Vitamin A | Promotes chondrocyte function; synthesis of lysosomal enzymes for osteoclast activity |
| Vitamin C | Promotes synthesis of collagen |

The formation and resorption of bone occur constantly at rates determined by a number of physiological factors (table 8.4). Body growth during the first two decades of life occurs because bone formation proceeds at a faster rate than bone resorption. By age 50 or 60, the rate of bone resorption often exceeds the rate of bone deposition. The constant activity of osteoblasts and osteoclasts allows bone to be remodeled throughout life.

Despite the changing rates of bone formation and resorption, the blood plasma concentrations of calcium and phosphate are maintained by hormonal control of the intestinal absorption and urinary excretion of these ions. These hormonal control mechanisms are very effective at maintaining the plasma calcium and phosphate concentrations within narrow limits.

The maintenance of normal blood plasma calcium concentrations is important because of the wide variety of effects that calcium has in the body. In addition to its role in bone formation, calcium is essential for muscle contraction (chapter 12), hormonal action (chapter 19), and maintenance of proper membrane permeability. An abnormally low blood plasma calcium concentration increases the permeability of the cell membranes to $Na^+$ and other ions. Hypocalcemia (low blood plasma calcium concentration) enhances the excitability of nerves and muscles and can result in muscle spasm (tetany).

## Parathyroid Hormone

Whenever the blood plasma concentration of calcium ions begins to fall, the parathyroid glands are stimulated to secrete increased amounts of *parathyroid hormone* (PTH), which acts to raise the blood $Ca^{++}$ back to normal levels. As might be predicted from this action of PTH, people who have their parathyroid glands removed will experience hypocalcemia. This can cause severe muscle tetany (contraction) and serves as a dramatic reminder of the importance of PTH.

Parathyroid hormone helps to raise the blood $Ca^{++}$ concentration primarily by stimulating the activity of osteoclasts to resorb bone. In addition, PTH stimulates the kidneys to retain blood $Ca^{++}$ while promoting $PO_4^{-3}$ (phosphate ion) excretion. This raises blood $Ca^{++}$ levels without promoting the deposition of calcium phosphate crystals in bone. Finally, PTH promotes the formation of 1,25-dihydroxyvitamin $D_3$ (as described in the next section), and so it also helps to raise the blood calcium levels indirectly through the effects of this other hormone.

## 1,25-Dihydroxyvitamin $D_3$

The production of **1,25-dihydroxyvitamin $D_3$** begins in the skin, where vitamin $D_3$ is produced from its precursor molecule (7-dehydrocholesterol) under the influence of sunlight. When the skin does not make sufficient vitamin $D_3$ because of insufficient exposure to sunlight, this compound must be ingested in the diet—that is why it is called a vitamin. Whether this compound is secreted into the blood from the skin or enters the blood after being absorbed from the small intestine, vitamin $D_3$ functions as a *prehormone*; thus, it must be chemically changed in order to be biologically active (see chapter 19).

An enzyme in the liver adds a hydroxyl group (OH) to carbon 25, which converts vitamin $D_3$ into 25-hydroxyvitamin $D_3$. In order to be active, however, another hydroxyl group must be added to the first carbon. Hydroxylation of the first carbon is accomplished by an enzyme in the kidneys, which converts the molecule to 1,25-dihydroxyvitamin $D_3$ (fig. 8.12). The activity of this enzyme in the kidneys is stimulated by parathyroid hormone (fig. 8.13). Increased secretion of PTH, stimulated by low blood $Ca^{++}$, is thus accompanied by the increased production of 1,25-dihydroxyvitamin $D_3$.

The hormone 1,25-dihydroxyvitamin $D_3$ helps to raise the blood plasma concentrations of calcium and phosphate by stimulating (1) the intestinal absorption of calcium and phosphate, (2) the resorption of bones, and (3) the renal reabsorption of calcium and phosphate so that less is excreted in the urine. Notice that 1,25-dihydroxyvitamin $D_3$, but not parathyroid hormone, directly stimulates

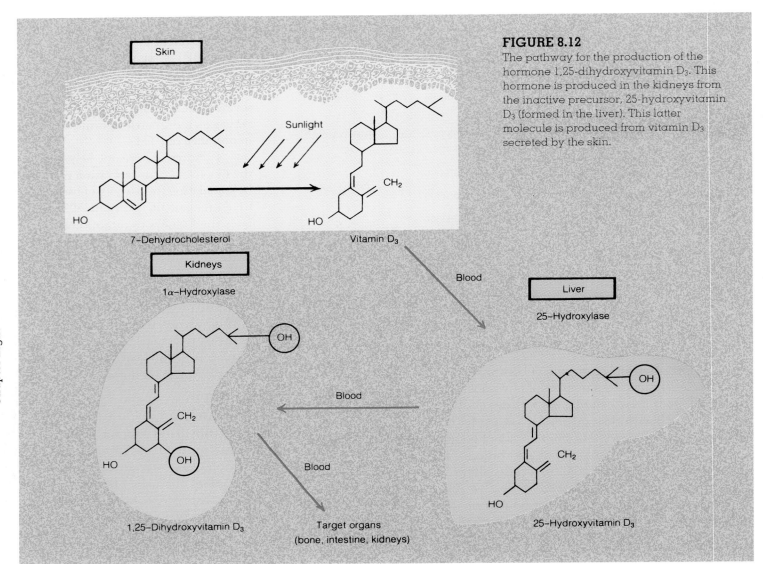

**FIGURE 8.12**

The pathway for the production of the hormone 1,25-dihydroxyvitamin D₃. This hormone is produced in the kidneys from the inactive precursor, 25-hydroxyvitamin D₃ (formed in the liver). This latter molecule is produced from vitamin D₃ secreted by the skin.

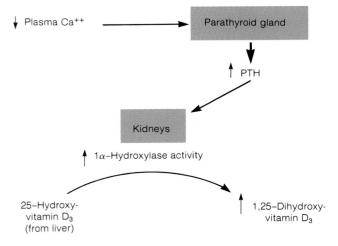

**FIGURE 8.13**

A decrease in plasma Ca⁺⁺ directly stimulates the secretion of parathyroid hormone (PTH). The production of 1,25-dihydroxyvitamin D₃ also rises when Ca⁺⁺ is low because PTH stimulates the final hydroxylation step in the formation of this compound in the kidneys.

intestinal absorption of calcium and phosphate. The effect of simultaneously raising the blood concentrations of $Ca^{++}$ and $PO_4^{-3}$ results in the increased tendency of these two ions to precipitate as hydroxyapatite crystals in bone.

Since 1,25-dihydroxyvitamin D₃ directly stimulates bone resorption, it seems paradoxical that this hormone is needed for proper bone deposition and, in fact, that inadequate amounts of 1,25-dihydroxyvitamin D₃ result in the bone demineralization of osteomalacia and rickets. This apparent paradox may be explained logically by the fact that the primary function of 1,25-dihydroxyvitamin D₃ is stimulation of intestinal $Ca^{++}$ and $PO_4^{-3}$ absorption. When calcium intake is adequate, the major result of 1,25-dihydroxyvitamin D₃ action is the availability of $Ca^{++}$ and $PO_4^{-3}$ in sufficient amounts to promote bone deposition. Only when calcium intake is inadequate does the direct effect of 1,25-dihydroxyvitamin D₃ on bone resorption become significant, acting to ensure proper blood $Ca^{++}$ levels.

## Negative Feedback Control of Calcium and Phosphate Balance

The secretion of parathyroid hormone is controlled by the blood plasma calcium concentrations. Its secretion is stimulated by low-calcium concentrations and inhibited by high-calcium concentrations. Since parathyroid hormone stimulates the final hydroxylation step in the formation of 1,25-dihydroxyvitamin $D_3$, a rise in parathyroid hormone results in an increase in production of 1,25-dihydroxyvitamin $D_3$. Low blood calcium can thus be corrected by the effects of increased parathyroid hormone and 1,25-dihydroxyvitamin $D_3$.

## Calcitonin

Experiments in the 1960s revealed that high blood calcium in dogs may be lowered by a hormone secreted from the thyroid gland. This hormone thus has an effect opposite to that of parathyroid hormone and 1,25-dihydroxyvitamin $D_3$. The calcium-lowering hormone, called **calcitonin,** was found to be a 32-amino-acid polypeptide secreted by the thyroid gland (chapter 19).

The secretion of calcitonin is stimulated by high blood plasma calcium levels and acts to lower calcium levels by (1) inhibiting the activity of osteoclasts, thus reducing bone resorption, and (2) stimulating the urinary excretion of calcium and phosphate by inhibiting their reabsorption in the kidneys.

Although it is attractive to think that calcium balance is regulated by the effects of antagonistic hormones, the significance of calcitonin in human physiology remains unclear. Patients who have had their thyroid gland surgically removed (as for thyroid cancer) are *not* hypercalcemic, as one would expect them to be if calcitonin were needed to lower blood calcium levels. The ability of very large, pharmacological doses of calcitonin to inhibit osteoclast activity and bone resorption, however, is clinically useful in the treatment of *Paget's disease*, in which osteoclast activity causes softening of bone.

## Clinical Considerations

Bone is a dynamic living tissue that is susceptible to hormonal or nutritional deficiency, diseases, and changes brought on by age. Since the development of bone is genetically governed, congenital conditions are possible. The hardness of bone gives it strength, yet it lacks the resiliency to resist fracture if it undergoes excessive trauma. All of these aspects of bone make for some important and interesting clinical considerations.

..............................................
Paget's disease: from Sir James Paget, English surgeon, 1814–99

## Developmental Disorders

Congenital malformations account for several types of skeletal deformities. Certain bones may fail to form during osteogenesis, or they may form abnormally. **Cleft palate** and **cleft lip** are malformations of the palate and face. Cleft palates vary in severity and seem to involve both genetic and environmental factors. **Spina bifida** (*spi´nă bif´ĭ-dă*) is a congenital defect of the vertebral column resulting from a failure of the laminae of the vertebrae to fuse, exposing the spinal cord. The lumbar area is mainly affected, and frequently only a single vertebra is involved.

## Nutritional and Hormonal Disorders

Several bone disorders result from nutritional deficiencies or from excessive or deficient amounts of the hormones that regulate bone development and growth. Vitamin D has a tremendous influence on proper bone structure and function. When there is a deficiency of this vitamin, the body is unable to metabolize calcium and phosphorus. Vitamin D deficiency in children causes **rickets.** The bones of a child with rickets remain soft and are deformed from the weight of the body (see fig. 7.8).

A vitamin D deficiency in the adult causes the bones to demineralize, or to give up stored calcium and phosphorus. This condition is called **osteomalacia** (*os´´te-o-mă-la´shă*). Osteomalacia is prevalent in women who have repeated pregnancies and poor diets and who experience relatively little exposure to sunlight.

The consequences of endocrine disorders are described in chapter 19. Since the impact of hormones on bone development is great, however, a few endocrine disorders will be briefly mentioned here. Hypersecretion of the growth hormone from the pituitary gland leads to **gigantism** in young people, if it starts before ossification of their epiphyseal plates, and to **acromegaly** (*ak´´ro-meg´ă-le*) in adults. Acromegaly is characterized by hypertrophy of the bones of the face, hands, and feet. By contrast, in a child, a hyposecretion of the growth hormone can lead to **dwarfism.**

**Paget's disease** is a disease of disorganized metabolic processes within bone. The activity of osteoblasts and osteoclasts becomes irregular, producing areas with thickened osseous deposits and other areas where too much bone is removed. The etiology of the disease is unknown, but it is a relatively common affliction in people over the age of 50, and it occurs more frequently in males than in females.

## Trauma and Injury

There are a variety of types of trauma to the skeletal system, ranging from injury to the bone itself to damage of the joints in the form of sprains or dislocations. Fractures and the healing of fractures are discussed in chapter 10; joint injuries are discussed in chapter 11.

## Neoplasms of Bone

Malignant bone tumors are three times more common than benign tumors. Pain is the usual symptom of either type of osseous neoplasm, although benign tumors may not have associated pain.

Two types of benign bone tumors are **osteomas,** which are the more frequent and which often involve the skull, and **osteoid osteomas,** which are painful neoplasms of the long bones, usually in children.

**Osteogenic sarcoma** is the most virulent type of bone cancer and frequently metastasizes through the blood to the lungs. This disease usually originates in the long bones and is accompanied by aching and persistent pain.

A **bone scan** (fig. 8.14) is a diagnostic procedure frequently done on a person who has had a malignancy elsewhere in the body that may have metastasized to the bone. The patient receiving a bone scan may be injected with a radioactive substance that accumulates more rapidly in malignant tissue than normal tissue. Entire body radiographs show malignant bone areas as intensely dark dots.

## Aging of the Skeletal System

Senescence affects the skeletal system by decreasing skeletal mass and density and

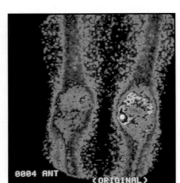

**FIGURE 8.14**
A bone scan of the legs of a patient suffering from arthritis in the left knee joint. In a bone scan, arthritis is depicted as a brighter image than a normal joint.

increasing porosity and erosion (fig. 8.15). Bones become more brittle and susceptible to fracture. Articulating surfaces also deteriorate, contributing to arthritic conditions. Arthritic diseases are second to heart disease as the most common debilitation in the elderly.

**Osteoporosis** (*os´´te-o-pŏ-ro´sis*) is the most prevalent metabolic disorder of bone in elderly people. It is characterized by marked demineralization, which weakens bone. The causes of osteoporosis include aging, prolonged inactivity, malnutrition, and an unbalanced secretion of hormones. It is most common in postmenopausal women. People with osteoporosis are prone to bone fracture, particularly at the pelvic girdle and vertebrae, as the bones become too brittle to support the body. Although there is no known cure for osteoporosis, it can be somewhat prevented in younger adults through proper diet, exercise, and good general health habits. Treatment of the disease in women through dietary calcium, exercise, and estrogens (female sex hormones) has had limited positive results.

Distinct losses in height occur during middle and old age. Between the ages of 50 and 55, there is a decrease of 0.5–2 cm (0.25–0.75 in.) because of compression and shrinkage of the intervertebral discs. Elderly individuals may suffer a further major loss of height because of osteoporosis.

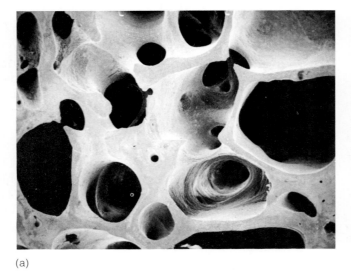

(a)

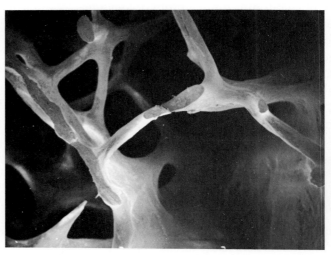

(b)

**FIGURE 8.15**
Scanning electron micrographs of spongy bone biopsy specimens from the iliac crest: (a) is normal and (b) is from a person with osteoporosis.

# Interactions of the Skeletal System with Other Body Systems

**Integumentary System**
- Supplies vitamin D for absorption of calcium and synthesis of vitamin D
- Provides skin with physical support

**Muscular System**
- Causes bones to move joints
- Partially responsible for bone shape and size
- Provides attachment sites for muscles
- Source of calcium for muscle contraction

**Nervous System**
- Sensory receptors provide sensations of pain from bones and joints
- Protects central nervous system with bony encasement
- Source of calcium for neural function

**Endocrine System**
- Hormonally controls bone growth and maintenance
- Protects endocrine glands in head and pelvis
- Source of calcium for cardiac muscle contraction

**Circulatory System**
- Transports $O_2$ and $CO_2$, nutrients, hormones, and fluids to and from the bones
- Bone marrow produces blood cells
- Source of calcium for cardiac muscle contraction

**Lymphatic System**
- Lymphocytes protect bone tissue following trauma
- Bone marrow produces and stores lymphocytes and other cells of immune system

**Respiratory System**
- Provides $O_2$ and eliminates $CO_2$
- Forms respiratory passageway through nasal cavity
- Protects lungs and aids in ventilation

**Urinary System**
- Eliminates metabolic wastes from bone tissue
- Activates vitamin D
- Provides organs of urinary system with physical support and protection

**Digestive System**
- Provides nutrients for growth, maintenance, and repair of bone tissue
- Provides organs of GI tract with physical support and protection
- Stores minerals

**Reproductive System**
- Gonads produce sex hormones that promote growth and development and maintain bone tissue
- Provides organs of reproductive system with physical support and protection

# Chapter Summary

## Organization of the Skeletal System (pp. 160–162)

1. The axial skeleton consists of the skull, auditory ossicles, hyoid bone, vertebral column, and rib cage.
2. The appendicular skeleton consists of the bones within the pectoral girdle, upper extremities, pelvic girdle, and lower extremities

## Functions of the Skeletal System (p. 162)

1. The mechanical functions of bones include the support and protection of softer body tissues and organs; in addition, certain bones function as levers during body movement.
2. The metabolic functions of bones include hemopoiesis and mineral storage.

## Gross Structure of Bone (pp. 162–166)

1. Bone structure includes the shape and surface features of each bone, along with gross internal components.
2. Structurally speaking, bones may be classified as long, short, flat, or irregular.

3. The surface features of bones can be broadly classified into articulating surfaces, nonarticulating prominences, and depressions and openings.
4. A typical long bone has a diaphysis filled with marrow in the medullary cavity, epiphyses, epiphyseal plates for linear bone growth, and a covering of periosteum for appositional bone growth and the attachments of ligaments and tendons.

## Bone Tissue (pp. 166–167)

1. The five types of bone cells are osteogenic cells, in contact with the endosteum and periosteum; osteoblasts (bone-forming cells); osteocytes (mature bone cells); osteoclasts (bone-destroying cells); and bone-lining cells, along the surface of most bones.
2. Compact bone consists of precise arrangements of osteons. The osteons contain osteocytes and lamellae.

## Bone Growth (pp. 167–170)

1. Bone growth from embryonic to adult size is an orderly process determined by genetics, hormonal secretions, and nutritional supply.

2. Most bones develop through endochondral ossification.
3. Bone remodeling is a continual process that involves osteoclasts in bone resorption and osteoblasts in the formation of new bone tissue.

## Physiology of Bone Tissue (pp. 170–173)

1. Bone tissue contains a reserve supply of calcium and phosphate for the blood in the form of hydroxyapatite crystals.
2. Parathyroid hormone stimulates bone resorption and calcium reabsorption in the kidneys, and thus raises the blood calcium concentration.
3. 1,25-dihydroxyvitamin $D_3$ is derived from vitamin D by hydroxylation reactions in the liver and kidneys.
4. A rise in parathyroid hormone, accompanied by the increased production of 1,25-dihydroxyvitamin $D_3$, helps to maintain proper blood levels of calcium and phosphate in response to a fall in calcium levels.
5. Calcitonin is secreted by the thyroid gland and lowers blood calcium by inhibiting bone resorption and stimulating the urinary excretion of calcium and phosphate.

# Review Activities

## Objective Questions

1. A bone is considered to be
   a. a tissue.
   b. a cell.
   c. an organ.
   d. a system.
2. Which of the following statements is *false?*
   a. Bones are important in the synthesis of vitamin D.
   b. Bones and teeth contain about 95% of the body's calcium.
   c. Red bone marrow is the primary site for hemopoiesis.
   d. Most bones develop through endochondral ossification.
3. Bone tissue derives from specialized migratory mesodermal cells called
   a. dermatomes.
   b. mesenchyme.
   c. myotomes.
   d. somites.

4. As a structural feature of certain bones, a fovea is
   a. a rounded opening through a bone.
   b. a small rounded process.
   c. a deep pit or socket.
   d. a small pit or depression.
5. The periosteum is secured to bone by
   a. the epiphyseal plate.
   b. perforating fibers.
   c. interosseous ligaments.
   d. diploe.
6. Columnlike structures within compact bone tissue are called
   a. osteons.
   b. lamellae.
   c. lacunae.
   d. diaphyses.
7. Specialized bone cells that enzymatically reabsorb bone tissue are called
   a. osteoblasts.
   b. osteocytes.
   c. osteons.
   d. osteoclasts.

8. The increased intestinal absorption of calcium is stimulated directly by
   a. parathyroid hormone.
   b. 1,25-dihydroxyvitamin $D_3$.
   c. calcitonin.
   d. all of the above.
9. A rise in blood calcium levels directly stimulates
   a. parathyroid hormone secretion.
   b. calcitonin secretion.
   c. 1,25-dihydroxyvitamin $D_3$ formation.
   d. all of the above.
10. The bone disorder common in elderly people, particularly if they are subject to prolonged inactivity, malnutrition, or an unbalanced secretion of hormones, is
    a. osteitis.
    b. osteonecrosis.
    c. osteoporosis.
    d. osteomalacia.

1. Distinguish between the axial skeleton and the appendicular skeleton. List the bones that compose the pectoral and pelvic girdles.
2. Sketch a typical long bone. Label the diaphysis, epiphyses, articular cartilages, periosteum, and medullary cavity.
3. Define *osteon*. Sketch an osteon and label the osteocytes, lacunae, lamellae, central canal, and canaliculi.
4. Describe how bones grow in length and in circumference. How are these processes similar and how do they differ? Explain how radiographs can be used to determine normal bone growth.
5. Describe the process of endochondral ossification of a long bone. Why is it important that a balance be maintained between osteoblast activity and osteoclast activity?
6. Explain why a proper balance of vitamins, hormones, and minerals is essential in maintaining healthy bone tissue. Give examples of diseases or skeletal conditions that may occur if there is an imbalance of any of these three essential substances.
7. Why is vitamin D considered to be both a vitamin and a prehormone? Explain why people with osteoporosis might be helped by taking controlled amounts of vitamin D.

# Gundy/Weber Software 💾

The tutorial software accompanying Chapter 8 is Volume 3—Skeletal System.

# skeletal system: axial skeleton

## Objectives

- Identify the cranial and facial bones of the skull and describe their structural characteristics.

- Describe the location of each of the bones of the skull and identify the articulations that affix them to each other.

- Identify the bones of the five regions of the vertebral column and describe the characteristic curves of each region.

- Describe the structure of a typical vertebra.

- Identify the parts of the rib cage and compare and contrast the various types of ribs.

# Skull

*The human skull, consisting of 8 cranial and 14 facial bones, contains several cavities that house the brain and sensory organs. Each bone of the skull articulates with the adjacent bones and has diagnostic and functional processes, surface features, and foramina.*

The skull consists of **cranial bones** and **facial bones.** The 8 bones of the cranium join firmly with one another to enclose and protect the brain and associated sense organs. The 14 facial bones form the framework for the facial region and support the teeth. The unique configuration of each human face is largely due to individual differences in the shape and density of the facial bones. With the exception of the bone within the lower jaw, the facial bones are also firmly interlocked with one another and the cranial bones.

..........................................

cranium: Gk, *kranion,* skull

The skull has several cavities. The **cranial cavity** is the largest, with a capacity of about 1300–1350 cc. The **nasal cavity** is formed by both cranial and facial bones and is partitioned into two chambers, or **nasal fossae,** by a **nasal septum** of bone and cartilage. Four sets of **paranasal sinuses** are located within the bones surrounding the nasal area and communicate via ducts into the nasal cavity. **Middle-** and **inner-ear cavities** are positioned inferior to the cranial cavity and house the organs of hearing and balance. The two **orbits** for the eyeballs are formed by facial and cranial bones. The **oral,** or **buccal, cavity** (mouth), which is only partially formed by bone, is completely within the facial region.

The bones of the skull contain numerous foramina to accommodate nerves, vessels, and other structures. A summary of the foramina of the skull is presented in table 9.1. Figures 9.1 through 9.8 show various views of the skull. Radiographs of the skull are shown in figure 9.9.

## Table 9.1    Major foramina of the skull

| Foramen | Location | Structures transmitted |
|---|---|---|
| Carotid canal | Petrous part of temporal bone | Internal carotid artery and sympathetic nerves |
| Greater palatine | Palatine bone of hard palate | Greater palatine nerve and descending palatine vessels |
| Hypoglossal foramen/canal | Anterolateral edge of occipital condyle | Hypoglossal nerve and branch of ascending pharyngeal artery |
| Incisive | Anterior region of hard palate, posterior to incisor teeth | Branches of descending palatine vessels and nasopalatine nerve |
| Inferior orbital fissure | Between maxilla and greater wing of sphenoid bone | Maxillary nerve of trigeminal cranial nerve, zygomatic nerve, and infraorbital vessels |
| Infraorbital | Inferior to orbit in maxilla | Infraorbital nerve and artery |
| Jugular | Between petrous part of temporal and occipital bones, posterior to carotid canal | Internal jugular vein; vagus, glossopharyngeal, and accessory nerves |
| Lacerum | Between petrous part of temporal and sphenoid bones | Branch of ascending pharyngeal artery and internal carotid artery |
| Lesser palatine | Posterior to greater palatine foramen in hard palate | Lesser palatine nerves |
| Magnum | Occipital bone | Union of medulla oblongata and spinal cord, meningeal membranes and accessory nerves; vertebral and spinal arteries |
| Mandibular | Medial surface of ramus of mandible | Inferior alveolar nerve and vessels |
| Mental | Below second premolar on lateral side of mandible | Mental nerve and vessels |
| Nasolacrimal canal | Lacrimal bone | Nasolacrimal (tear) duct |
| Olfactory | Cribriform plate of ethmoid bone | Olfactory nerves |
| Optic | Back of orbit in lesser wing of sphenoid bone | Optic nerve and ophthalmic artery |
| Ovale | Greater wing of sphenoid bone | Mandibular nerve of trigeminal cranial nerve |
| Rotundum | Within body of sphenoid bone | Maxillary nerve of trigeminal cranial nerve |
| Spinosum | Posterior angle of sphenoid bone | Middle meningeal vessels |
| Stylomastoid | Between styloid and mastoid processes of temporal bone | Facial nerve and stylomastoid artery |
| Superior orbital fissure | Between greater and lesser wings of sphenoid bone | Four cranial nerves (oculomotor, trochlear, ophthalmic nerve of trigeminal, and abducens) |
| Supraorbital | Supraorbital ridge of orbit | Supraorbital nerve and artery |
| Zygomaticofacial | Anterolateral surface of zygomatic bone | Zygomaticofacial nerve and vessels |

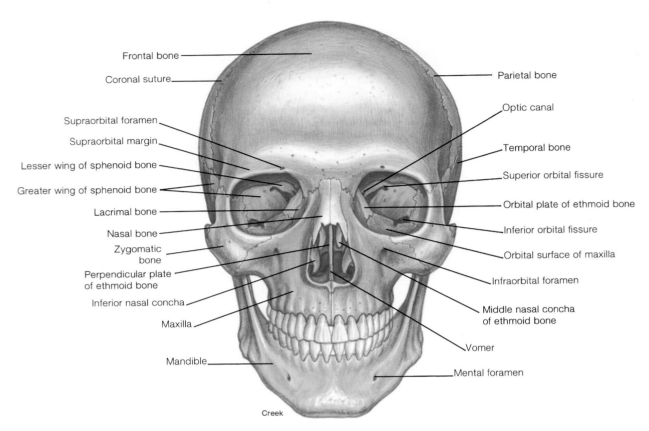

**FIGURE 9.1**
An anterior view of the skull.

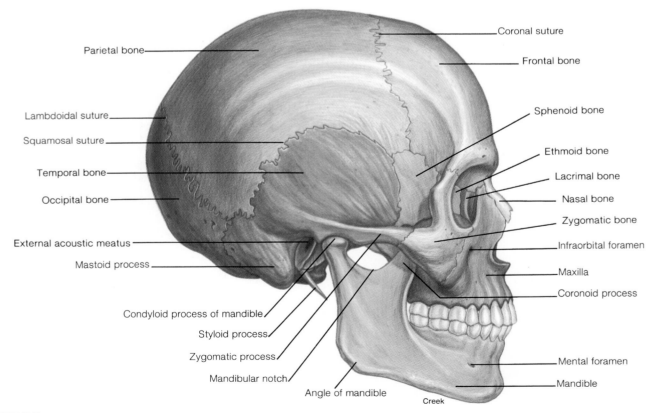

**FIGURE 9.2**
A lateral view of the skull.

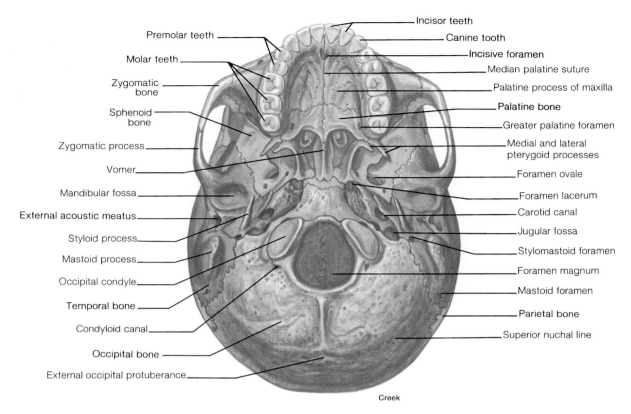

Incisor teeth
Premolar teeth
Canine tooth
Molar teeth
Incisive foramen
Median palatine suture
Zygomatic bone
Palatine process of maxilla
Sphenoid bone
Palatine bone
Zygomatic process
Greater palatine foramen
Vomer
Medial and lateral pterygoid processes
Mandibular fossa
Foramen ovale
External acoustic meatus
Foramen lacerum
Styloid process
Carotid canal
Mastoid process
Jugular fossa
Occipital condyle
Stylomastoid foramen
Temporal bone
Foramen magnum
Condyloid canal
Mastoid foramen
Occipital bone
Parietal bone
External occipital protuberance
Superior nuchal line

Creek

**FIGURE 9.3**

An inferior view of the skull.

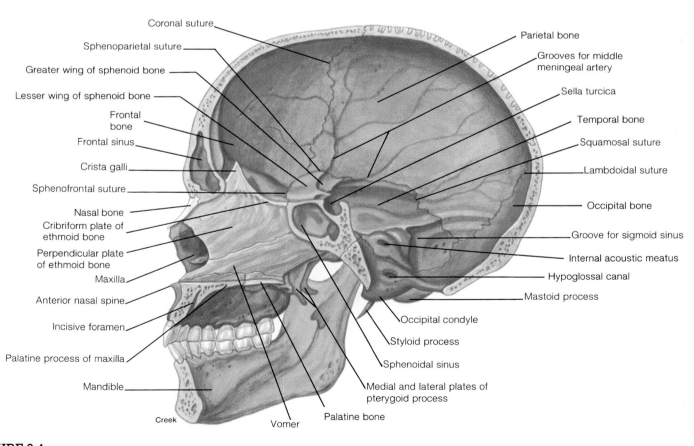

Coronal suture
Parietal bone
Sphenoparietal suture
Grooves for middle meningeal artery
Greater wing of sphenoid bone
Sella turcica
Lesser wing of sphenoid bone
Temporal bone
Frontal bone
Squamosal suture
Frontal sinus
Lambdoidal suture
Crista galli
Sphenofrontal suture
Occipital bone
Nasal bone
Cribriform plate of ethmoid bone
Groove for sigmoid sinus
Perpendicular plate of ethmoid bone
Internal acoustic meatus
Maxilla
Hypoglossal canal
Anterior nasal spine
Mastoid process
Incisive foramen
Occipital condyle
Palatine process of maxilla
Styloid process
Mandible
Sphenoidal sinus
Medial and lateral plates of pterygoid process
Creek
Vomer
Palatine bone

**FIGURE 9.4**

A midsagittal view of the skull.

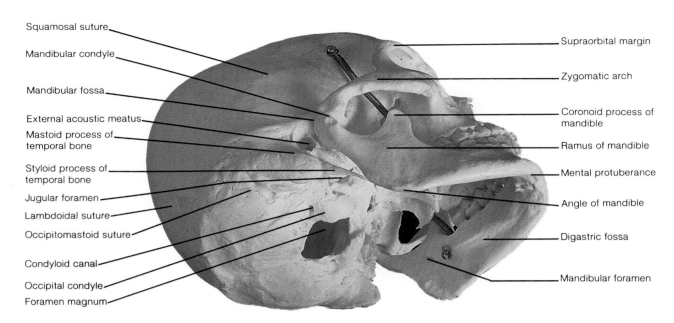

Squamosal suture

Mandibular condyle

Mandibular fossa

External acoustic meatus

Mastoid process of temporal bone

Styloid process of temporal bone

Jugular foramen

Lambdoidal suture

Occipitomastoid suture

Condyloid canal

Occipital condyle

Foramen magnum

Supraorbital margin

Zygomatic arch

Coronoid process of mandible

Ramus of mandible

Mental protuberance

Angle of mandible

Digastric fossa

Mandibular foramen

**FIGURE 9.5**

An inferolateral view of the skull.

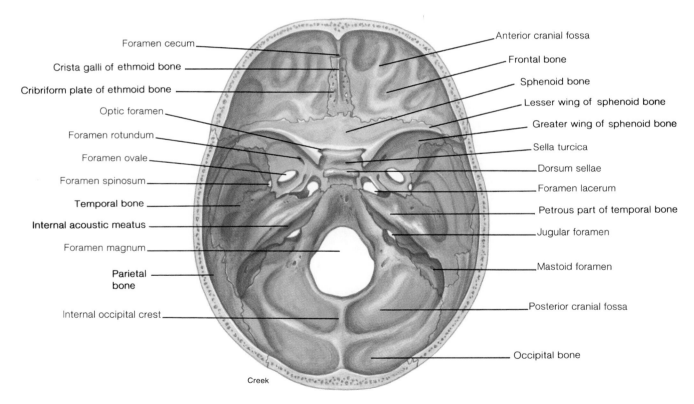

Foramen cecum

Crista galli of ethmoid bone

Cribriform plate of ethmoid bone

Optic foramen

Foramen rotundum

Foramen ovale

Foramen spinosum

Temporal bone

Internal acoustic meatus

Foramen magnum

Parietal bone

Internal occipital crest

Anterior cranial fossa

Frontal bone

Sphenoid bone

Lesser wing of sphenoid bone

Greater wing of sphenoid bone

Sella turcica

Dorsum sellae

Foramen lacerum

Petrous part of temporal bone

Jugular foramen

Mastoid foramen

Posterior cranial fossa

Occipital bone

Creek

**FIGURE 9.6**

The floor of the cranial cavity.

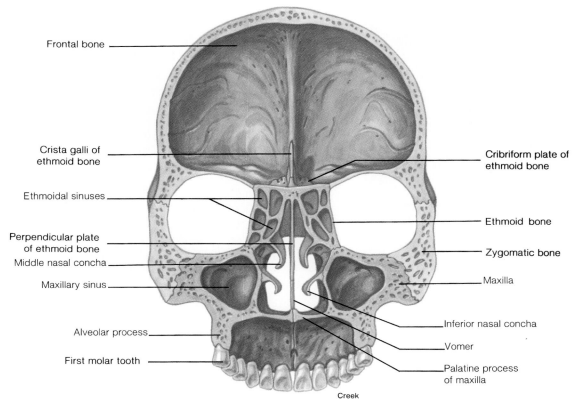

**FIGURE 9.7**

A posterior view of a frontal (coronal) section of the skull.

Frontal bone

Crista galli of
ethmoid bone

Ethmoidal sinuses

Perpendicular plate
of ethmoid bone

Middle nasal concha

Maxillary sinus

Alveolar process

First molar tooth

Cribriform plate of
ethmoid bone

Ethmoid bone

Zygomatic bone

Maxilla

Inferior nasal concha

Vomer

Palatine process
of maxilla

Creek

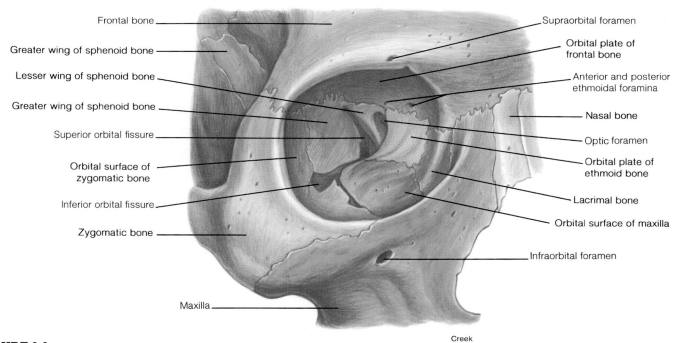

**FIGURE 9.8**

Bones of the orbit of the eye.

Frontal bone

Greater wing of sphenoid bone

Lesser wing of sphenoid bone

Greater wing of sphenoid bone

Superior orbital fissure

Orbital surface of
zygomatic bone

Inferior orbital fissure

Zygomatic bone

Maxilla

Supraorbital foramen

Orbital plate of
frontal bone

Anterior and posterior
ethmoidal foramina

Nasal bone

Optic foramen

Orbital plate of
ethmoid bone

Lacrimal bone

Orbital surface of maxilla

Infraorbital foramen

Creek

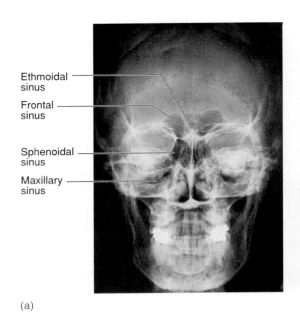

Ethmoidal sinus

Frontal sinus

Sphenoidal sinus

Maxillary sinus

(a)

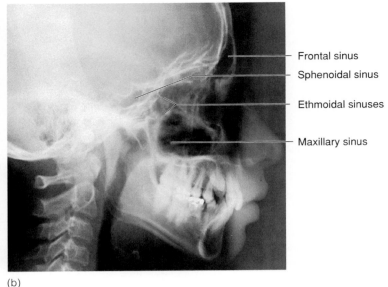

Frontal sinus

Sphenoidal sinus

Ethmoidal sinuses

Maxillary sinus

(b)

**FIGURE 9.9**

Radiographs of the skull showing paranasal sinuses. (*a*) An anteroposterior view, and (*b*) a right lateral view.

Although the hyoid bone and the three paired auditory ossicles are not considered part of the skull, they are located within the axial skeleton and are described immediately following the discussion of the skull.

## Cranial Bones

The cranial bones enclose the brain and consist of one frontal, two parietals, two temporals, one occipital, one sphenoid, and one ethmoid.

**Frontal Bone**   The frontal bone forms the anterior roof of the cranium, the forehead, the roof of the nasal cavity, and the superior arch of the **bony orbits,** which contain the eyeballs. The bones of the orbit are summarized in table 9.2. The frontal bone develops in two halves that grow together. Generally, they are completely fused by age 5 or 6, although a complete suture sometimes persists beyond age 6 and is referred to as a **metopic** (*mĕ-top´ik*) **suture.** The **supraorbital margin** is a prominent bony ridge over the orbit. Openings along this ridge, called *supraorbital foramina,* allow passage of small nerves and vessels.

The frontal bone contains a **frontal sinus,** which is connected to the nasal cavity (fig. 9.9). This sinus, along with the other paranasal sinuses, lessens the weight of the skull and acts as a sound chamber for voice resonance.

........................................

metopic suture: Gk. *metopon,* forehead; L. *sutura,* sew

### Table 9.2   Bones forming the orbit

| Region of the orbit | Contributing bones |
| --- | --- |
| Roof (superior) | Frontal; lesser wing of sphenoid bone |
| Floor (inferior) | Maxilla; zygomatic bone; palatine bone |
| Lateral wall | Zygomatic bone |
| Posterior wall | Greater wing of sphenoid bone |
| Medial wall | Maxilla; lacrimal bone; ethmoid bone |
| Superior rim | Frontal bone |
| Lateral rim | Zygomatic bone |
| Medial rim | Maxilla |

**Parietal Bone**   The **coronal** (*kă-ro´nal*) **suture** separates the frontal bone from the parietals, and the **sagittal** (*saj´ĭ-tal*) **suture** along the superior midline separates the right and left parietals. The parietal bones form the upper sides and roof of the cranium (figs. 9.2 and 9.4). The inner concave surfaces of the parietals and other cranial bones are marked by shallow impressions from convolutions of the brain and vessels serving the brain.

**Temporal Bone**   The two temporal bones form the lower sides of the cranium (figs. 9.2, 9.3, 9.4, and 9.10). Each

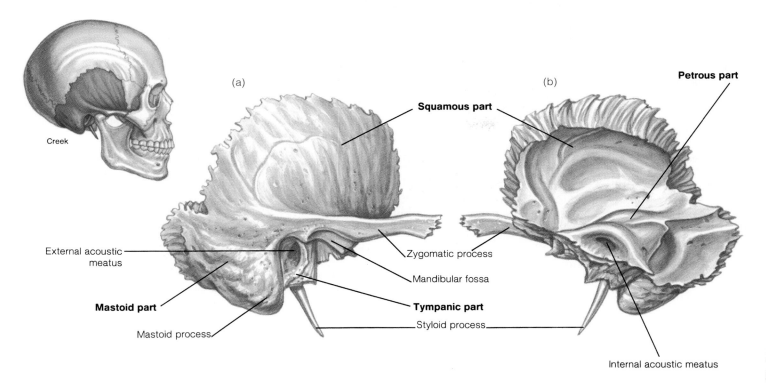

**FIGURE 9.10**
The temporal bone. (*a*) A lateral view, and (*b*) a medial view.

temporal bone is joined to its adjacent parietal bone by the **squamosal** (*skwă-mo´sal*) **suture.** Structurally, each temporal bone has four parts.

**1** **Squamous part.** The squamous part is the flattened plate of bone at the sides of the skull. Projecting forward is a **zygomatic process** that forms the posterior portion of the **zygomatic arch.** On the inferior surface of the squamous part is the **mandibular fossa,** which receives the articular condyle of the mandible. This articulation is referred to as the **temporomandibular** (*tem˝pŏ-ro-man-dib´yŭ-lar*) **joint.**

**2** **Tympanic part.** The tympanic part of the temporal bone contains the **external acoustic meatus,** or ear canal, located immediately posterior to the mandibular fossa. A thin, pointed **styloid process** (figs. 9.3 and 9.4) projects downward from the tympanic portion.

**3** **Mastoid part.** The **mastoid process,** a rounded projection posterior to the external acoustic meatus, accounts for the mass of the mastoid part. The **mastoid foramen** (fig. 9.3) is directly posterior to the mastoid process. The **stylomastoid foramen,** located between the mastoid and styloid processes (fig. 9.3), is the passage for part of the facial nerve.

• • • • • • • • • • • • • • • • • • • • • • • • • • •
zygomatic: Gk. *zygoma*, yolk
styloid: Gk. *stylos*, pillar
mastoid: Gk. *mastos*, breast

**4** **Petrous part.** The petrous (*pet´rus*) part is viewed in figures 9.6 and 9.10*b*. The structures of the middle and inner ear are housed in this dense, bony part of the temporal bone. The **carotid** (*că-rot´id*) **canal** and the **jugular foramen** border on the medial side of the petrous part. The carotid canal allows blood into the brain via the internal carotid artery and the jugular foramen lets blood drain from the brain via the internal jugular vein. Three cranial nerves also pass through the jugular foramen (see table 9.1).

 The mastoid process of the temporal bone can be easily palpated as a bony knob immediately behind the earlobe. This process contains a number of small air-filled spaces called *mastoid cells* that are clinically important because they can become infected in *mastoiditis.* A tubular communication from the mastoid cells to the middle-ear cavity may permit prolonged ear infections to spread to this region.

**Occipital Bone** The occipital bone forms the back and much of the base of the skull. It is fastened to the parietal bones by the **lambdoidal suture** (fig. 9.5). The **foramen magnum** is the large hole in the occipital bone through which the spinal cord attaches to the brain. On each lateral side of the foramen magnum are the **occipital condyles** (fig. 9.3), which

• • • • • • • • • • • • • • • • • • • • • • • • • • •
petrous: Gk. *petra*, rock
magnum: L. *magnum*, great

articulate with the atlas of the vertebral column. At the anterolateral edge of the occipital condyle is the **hypoglossal canal** (fig. 9.4), through which the hypoglossal nerve passes. A **condyloid** (*kon´dĭ-loid*) **canal** lies posterior to the occipital condyle (fig. 9.3). The **external occipital protuberance** is a prominent posterior projection on the occipital bone that can be felt as a definite bump just under the skin. The **superior nuchal** (*noo´kal*) **line** is a ridge of bone extending laterally from the occipital protuberance to the mastoid part of the temporal bone. **Sutural bones** are small clusters of irregularly shaped bones that may be found between the joints of certain cranial bones but that generally occur along the lambdoidal suture.

**Sphenoid Bone**   The sphenoid (*sfe´noid*) bone forms the anterior base of the cranium and can be viewed laterally and inferiorly (figs. 9.2 and 9.3). This bone resembles a butterfly with outstretched wings (fig. 9.11). It consists of a **body** with laterally projecting **greater** and **lesser wings,** which form part of the bony orbit. The body is a wedgelike central portion that contains the **sphenoidal** (*sfe-noi´dal*) **sinuses** and a prominent depression called the **sella turcica** (*sel´ă tur´sĭ-kă*), which supports the pituitary gland. The sella turcica (meaning "Turk's saddle") is seen on the floor of the cranium (figs. 9.4

and 9.6). A pair of **pterygoid** (*ter´ĭ-goid*) **processes (plates)** project inferiorly from the sphenoid bone to help form the lateral walls of the nasal cavity. Several foramina (figs. 9.3, 9.6, and 9.11) are located within the sphenoid bone.

**1** **Optic canal.** A large opening through the lesser wing into the back of the orbit for passage of the optic nerve and the ophthalmic artery.

**2** **Superior orbital fissure.** A triangular opening between the wings of the sphenoid for passage of the ophthalmic nerve off of trigeminal cranial nerve and the oculomotor, trochlear, and abducens cranial nerves.

**3** **Foramen ovale.** An opening at the base of the lateral pterygoid plate, through which the mandibular nerve passes.

**4** **Foramen spinosum.** A small opening at the posterior angle of the sphenoid bone for passage of the middle meningeal vessels.

**5** **Foramen lacerum** (*las´er-um*). An opening between the sphenoid bone and the petrous part of the temporal bone, through which the internal carotid artery and the meningeal branch of the ascending pharyngeal artery pass.

**6** **Foramen rotundum.** An opening located just posterior to the superior orbital fissure at the junction of the anterior and medial portions of the sphenoid bone. The maxillary nerve passes through this foramen.

nuchal: Fr. *nuque,* nape of neck
sphenoid: Gk. *sphenoeides,* wedgelike

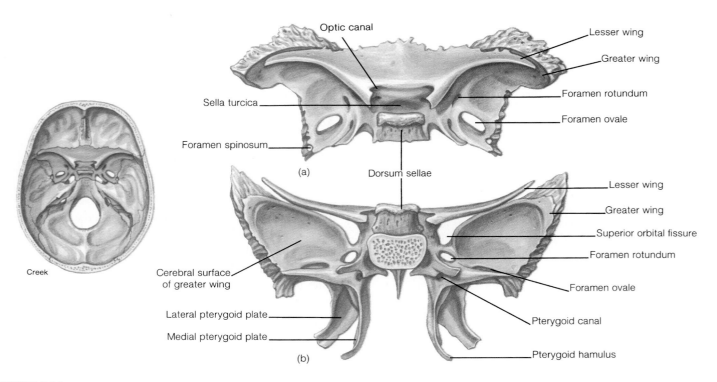

**FIGURE 9.11**
The sphenoid bone. (*a*) A superior view, and (*b*) a posterior view.

Located on the inferior side of the cranium, the sphenoid bone would seemingly be well protected from trauma. Actually just the opposite is true, and in fact the sphenoid is the most frequently fractured bone of the cranium. It has several broad, thin, platelike extensions that are perforated and weakened by numerous foramina. A blow to almost any portion of the skull causes the buoyed, fluid-filled brain to rebound against the vulnerable sphenoid bone, often causing it to fracture.

**Ethmoid Bone**    The ethmoid bone is located in the anterior portion of the floor of the cranium between the orbits, where it forms the roof of the nasal cavity (figs. 9.4, 9.7, and 9.12). An inferior projection of the ethmoid bone, called the **perpendicular plate,** contributes in part to the nasal septum that separates the nasal cavity into two chambers, referred to as **nasal fossae.** A spine of the perpendicular plate, the **crista galli,** projects superiorly into the cranial cavity and serves as an attachment for the meninges covering the brain. On both lateral walls of the nasal cavity are two scroll-shaped plates of the ethmoid bone called the **superior** and **middle nasal conchae** (fig. 9.13). At right angles to the perpendicular plate, within the floor of the cranium, is the **cribriform plate,** which

· · · · · · · · · · · · · · · · · · · · · · · · · · · · · · · ·

ethmoid: Gk. *ethmos*, sieve
crista galli: L. *crista*, crest; *galli*, cock's comb
conchae: L. *conchae*, shells
cribriform: L. *cribrum*, sieve; *forma*, like

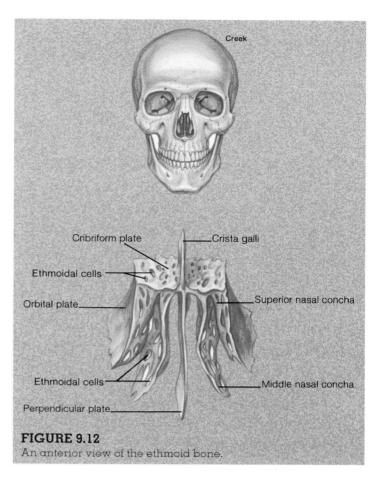

**FIGURE 9.12**
An anterior view of the ethmoid bone.

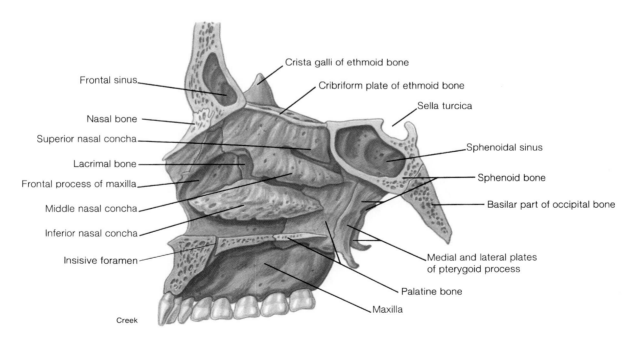

**FIGURE 9.13**
A lateral wall of the nasal cavity.

## Table 9.3    Bones forming the nasal cavity

| Region of nasal cavity | Contributing bones |
|---|---|
| Roof (superior) | Ethmoid bone (cribriform plate); frontal bone |
| Floor (inferior) | Maxilla; palatine bone |
| Lateral wall | Maxilla; palatine bone |
| Nasal septum (medial) | Ethmoid bone (perpendicular plate); vomer; nasal bone |
| Bridge | Nasal bone |
| Conchae | Ethmoid bone (superior and middle conchae); inferior nasal concha |

has numerous perforations for the passage of olfactory nerves from the nasal cavity. The bones of the nasal cavity are summarized in table 9.3.

 The moist, warm vascular lining within the nasal cavity is susceptible to infections, particularly if a person is not in good health. Infections of the nasal cavity can spread to several surrounding areas. The paranasal sinuses connect to the nasal cavity and are especially prone to infection. The eyes may become reddened and swollen during a nasal infection because of the connection of the nasolacrimal duct, through which tears drain from the orbit to the nasal cavity. Organisms may spread via the auditory tube from the nasopharynx to the middle ear. With prolonged nasal infections, organisms may even ascend to the meninges covering the brain, along the sheaths of the olfactory nerves, and through the cribriform plate, resulting in *meningitis*.

## Facial Bones

The 14 bones of the skull not in contact with the brain are called **facial bones.** These bones, together with certain cranial bones (frontal bone and portions of the ethmoid and temporal bones), provide the basic shape of the face. Facial bones also support the teeth and provide attachments for various muscles that move the jaw and act to produce facial expressions. All the facial bones are paired except the vomer and mandible. The articulated facial bones can be seen in figures 9.1 through 9.8.

**Maxilla**    The two maxillae unite at the midline to form the upper jaw, which supports the upper teeth. **Incisors** (*in-si´zorz*), **canines** (cuspids), **premolars,** and **molars** are contained in sockets, or **alveoli,** within the **alveolar** (*al-ve´ŏ-lar*) **process** of the maxilla (fig. 9.14). The palatine process, a horizontal plate of the maxilla, forms the greater portion of the **hard palate** (*pal´it*), or roof of the mouth. The **incisive foramen** (fig. 9.3) is located in the anterior region of the hard palate behind the incisor teeth. An **infraorbital foramen** is located under each orbit and serves as a passageway for the infraorbital nerve and artery to the nose (figs. 9.1, 9.2, 9.8, and 9.14). A final opening within the maxilla is the **inferior orbital fissure.** It is located between the maxilla and the greater wing of the sphenoid bone (fig. 9.1) and is the external opening for the maxillary nerve and infraorbital vessels. The large **maxillary sinus** located within the maxilla is one of the four paranasal sinuses (figs. 9.7, 9.9, and 9.14b).

..........................................

incisor: L. *incidere*, to cut
canine: L. *canis*, dog
molar: L. *mola*, millstone
alveolus: L. *alveus*, little cavity

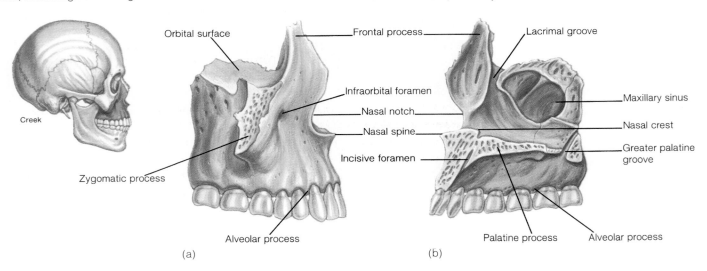

**FIGURE 9.14**
The maxilla. (a) A lateral view, and (b) a medial view.

If the two palatine processes fail to join during early prenatal development (about 12 weeks), a *cleft palate* results. A cleft palate may be accompanied by a *cleft lip* (harelip) lateral to the midline. These conditions can be surgically treated with excellent results. Cleft palate is of immediate concern, however, because a newborn with this condition may have a difficult time swallowing while nursing, being unable to create the necessary suction within the oral cavity.

**Palatine Bone**   The L-shaped palatine bones form the posterior third of the hard palate, a portion of the orbits, and a part of the nasal cavity. The **horizontal plates** of the palatines contribute to the formation of the hard palate (fig. 9.15). On the hard palate of each palatine bone is a large **greater palatine foramen,** which permits the passage of the greater palatine nerve and descending palatine vessels (fig. 9.3). Two or more smaller **lesser palatine foramina** are positioned posterior to the greater palatine foramen. Branches of the lesser palatine nerve pass through these openings.

**Zygomatic Bone**   The two zygomatic bones form the cheekbones of the face. A posteriorly extending **zygomatic process** of this bone unites with that of the temporal bone to form the **zygomatic arch** (figs. 9.3 and 9.5). The zygomatic bone also forms the lateral margin of the orbit. A small **zygomaticofacial** (*zi″gŏ-mat″ĭ-kŏ-fa′shal*) **foramen,** located on the anterolateral surface of this bone, allows passage of the zygomatic nerves and vessels.

**Lacrimal Bone**   The small lacrimals (*lak′rĭ-malz*) are thin bones that form the anterior part of the medial wall of each orbit (fig. 9.8). Each has a **lacrimal sulcus—**a groove that helps to form the **nasolacrimal canal.** This opening permits the tears of the eye to drain into the nasal cavity.

**Nasal Bone**   The small, rectangular nasal bones (fig. 9.1) join in the midline to form the bridge of the nose. The nasal bones support the flexible cartilaginous plates, which are a part of the framework of the nose. Common facial injuries include fractures of the nasal bones or fragmentation of the supporting cartilages.

**Inferior Nasal Concha**   The two inferior nasal conchae are fragile, scroll-like bones that project horizontally and medially from the lateral walls of the nasal cavity (figs. 9.1 and 9.7). They extend into the nasal cavity just below the superior and middle nasal conchae, which are part of the ethmoid bone (fig. 9.12). The inferior nasal conchae are the

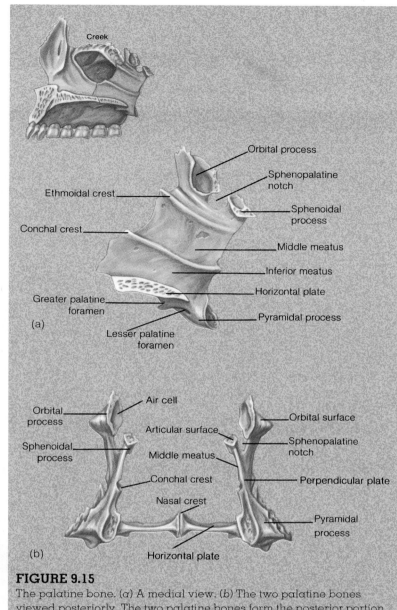

**FIGURE 9.15**

The palatine bone. (*a*) A medial view. (*b*) The two palatine bones viewed posteriorly. The two palatine bones form the posterior portion of the hard palate.

largest of the three conchae, and, like the other two, are covered with mucous membranes to warm, moisten, and cleanse inhaled air.

**Vomer**   The vomer (*vo′mer*) is a thin, elongated bone that forms the lower part of the nasal septum (figs. 9.3, 9.4, and 9.7). The vomer, along with the perpendicular plate of the ethmoid bone, supports the septal cartilage to complete the nasal septum.

..........................................

vomer: L. *vomer*, plowshare

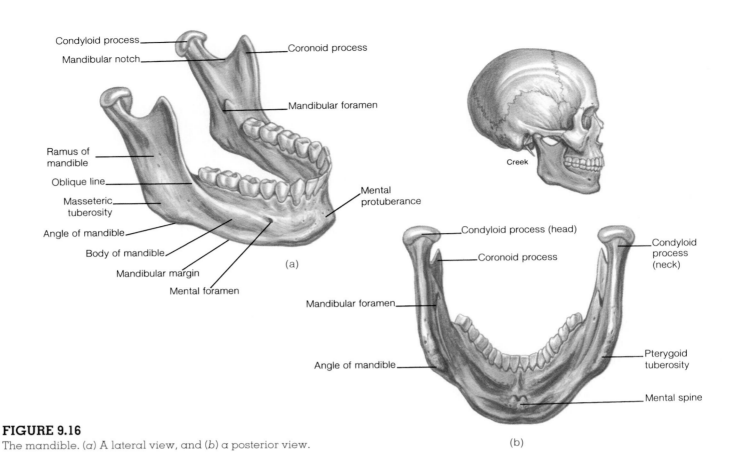

Condyloid process
Mandibular notch
Coronoid process
Mandibular foramen
Ramus of mandible
Oblique line
Masseteric tuberosity
Angle of mandible
Body of mandible
Mandibular margin
Mental foramen
Mental protuberance

(a)

Creek

Condyloid process (head)
Coronoid process
Condyloid process (neck)
Mandibular foramen
Pterygoid tuberosity
Angle of mandible
Mental spine

(b)

**FIGURE 9.16**
The mandible. (a) A lateral view, and (b) a posterior view.

**Mandible** The mandible, or lower jawbone, is attached to the skull by a temporomandibular articulation, and is the only movable bone of the skull. Several muscles that close the jaw extend from the skull to the mandible (see chapter 13). The mandible of an adult supports 16 teeth within alveoli, which occlude with those of the maxilla.

The horseshoe-shaped front and horizontal lateral sides of the mandible are referred to as the **body** (fig. 9.16). Extending vertically from the posterior part of the body are two **rami** (ra´mi; singular, *ramus*). Each ramus has a knoblike **condyloid process,** which articulates with the mandibular fossa of the temporal bone, and a pointed **coronoid process** for the attachment of the temporalis muscle. The depressed area between these two processes is the **mandibular notch.** The angle of the mandible is where the horizontal body and vertical ramus meet at the corner of the jaw.

Two sets of foramina are found on the mandible; the **mental foramen** on the lateral side below the first molar

and the **mandibular foramen** on the medial surface of the ramus. The mental nerve and vessels pass through the mental foramen and the inferior alveolar nerve and vessels are transmitted through the mandibular foramen.

  Dentists use bony landmarks of the facial region to locate the nerves that traverse the foramina so that anesthetics can be injected. For example, the trigeminal cranial nerve is composed of three large nerves, the lower two of which convey sensations from the teeth, gums, and jaws. The mandibular teeth can be desensitized by an injection near the mandibular foramen called a *third-division,* or *lower, nerve block.* An injection near the foramen rotundum of the skull, called a *second-division nerve block,* desensitizes all the upper teeth on one side of the maxilla.

**Hyoid Bone** The U-shaped hyoid bone is located in the neck, just superior to the larynx (voice box). The hyoid is unique in that it does not attach directly to any other bone but is suspended from the styloid processes of the skull by the stylohyoid (sti˝lo-hi´oid) muscles and ligaments. The hyoid has a **body,** two **lesser cornua** extending anteriorly, and two **greater cornua** (fig. 9.17), which project posteriorly to the

mandible: L. *mandere,* to chew
ramus: L. *ramus,* branch
condyloid: L. *condylus,* knucklelike
coronoid: Gk. *korone,* like a crow's beak

cornu: L. *cornu,* horn

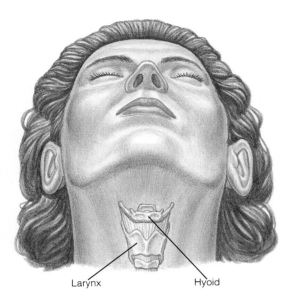

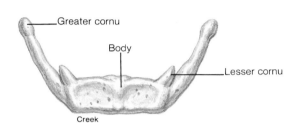

**FIGURE 9.17**
An anterior view of the hyoid bone.

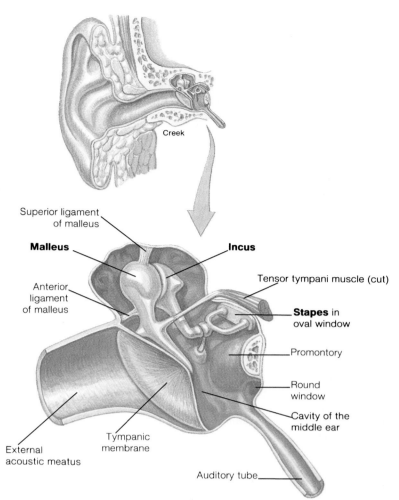

**FIGURE 9.18**
The three auditory ossicles within the cavity of the middle ear.

stylohyoid ligaments. Several neck and tongue muscles attach to the hyoid bone. The hyoid may be palpated by placing a thumb and a finger on either side of the upper neck under the lateral portions of the mandible and firmly squeezing medially.

**Auditory Ossicles**   Three small, paired auditory ossicles—the **malleus, incus,** and **stapes**—are located within the cavities of the middle ears in the petrous part of the temporal bones (fig. 9.18). These bones transfer and amplify sound impulses through the middle ear (see chapter 18).

# Vertebral Column

*The supporting vertebral column consists of vertebrae, separated by fibrocartilaginous intervertebral discs that lend flexibility and absorb the stress of movement. Vertebrae enclose and protect the spinal cord, support the skull and allow for its movement, articulate with the rib cage, and provide for the attachment of trunk muscles.*

· · · · · · · · · · · · · · · · · · · · · · · · · · · · · · · · ·

malleus: L. *malleus,* hammer
incus: L. *incus,* anvil
stapes: L. *stapes,* stirrup

The vertebral column is composed of 33 individual vertebrae. There are 7 **cervical,** 12 **thoracic** (*thŏ-ras´ik*), 5 **lumbar,** 5 fused **sacral,** and 4 or 5 fused **coccygeal** (*kok-sij´e-al*) **vertebrae**; thus, the vertebral column is composed of a total of 26 movable parts. Vertebrae are separated by fibrocartilaginous intervertebral discs and are secured to each other by interlocking processes and binding ligaments. This structural arrangement provides limited movements between vertebrae but extensive movements for the entire vertebral column. Between the vertebrae are openings called **intervertebral** (*in˝ter-ver´tĕ-bral*) **foramina** that permit passage of spinal nerves.

Four curvatures of the vertebral column of an adult can be identified and viewed from the side (fig. 9.19). The **cervical, thoracic,** and **lumbar curves** are identified by the type of vertebrae they include. The **pelvic curve** is formed by the shape of the sacrum and coccyx. The curves of the vertebral column play an important functional role in increasing the

## Development of the Skull

The formation of the skull is a complex process that begins during the fourth week of embryonic development and continues well beyond the birth of the baby. Three factions are involved in the formation of the skull: the chondrocranium, the neurocranium, and the viscerocranium (fig. 1). The **chondrocranium** is the portion of the skull that undergoes endochondral ossification to form the bones supporting the brain. The **neurocranium** is the portion of the skull that develops through membranous ossification to form the bones covering the brain and facial region. The **viscerocranium** (splanchnocranium) is the portion that develops from the embryonic visceral arches and forms the auditory ossicles, the hyoid bone, and specific processes of the skull.

During fetal development and infancy, the bones of the neurocranium covering the brain are separated by fibrous sutures. There are also six large membranous "soft spots" of the skull that provide spaces between the developing bones (fig. 2). Because the baby's pulse can be felt surging in these areas, they are called **fontanels** (*fon´ tă-nelz*), meaning "little fountains." They permit the skull to undergo changes of shape, called *molding,* during parturition (childbirth) and they also allow for rapid growth of the brain during infancy. Ossification of the fontanels is normally

complete by 20 to 24 months of age. A description of the six fontanels follows.

1. **Anterior (frontal) fontanel.** The anterior fontanel is diamond-shaped and is the most prominent of the six. It is located on the anteromedian portion of the skull.

2. **Posterior (occipital) fontanel.** The posterior fontanel is positioned at the back of the skull on the median line.

3. **Anterolateral (sphenoidal) fontanels.** The paired anterolateral fontanels are found on both sides of the skull, lateral to the anterior fontanel.

4. **Posterolateral (mastoid) fontanels.** The paired posterolateral fontanels are located on the posterolateral sides of the skull.

A prominent **sagittal suture** extends the anteroposterior median length of the skull between the anterior and posterior fontanels. A **coronal suture** extends from

chondrocranium: Gk. *chondros*, cartilage; *kranion*, skull
viscerocranium: L. *viscera*, soft parts; Gk. *kranion*, skull
fontanel: Fr. *fontaine*, little fountain

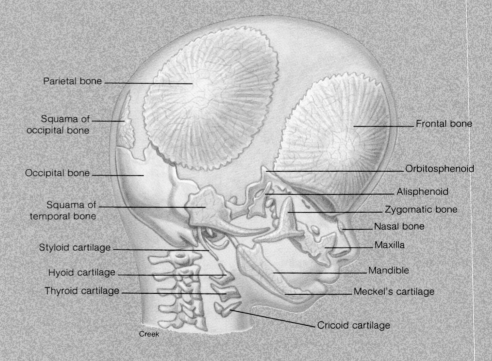

Parietal bone

Squama of occipital bone

Occipital bone

Squama of temporal bone

Styloid cartilage

Hyoid cartilage

Thyroid cartilage

Creek

Frontal bone

Orbitosphenoid

Alisphenoid

Zygomatic bone

Nasal bone

Maxilla

Mandible

Meckel's cartilage

Cricoid cartilage

**FIGURE 1**

The embryonic skull at 12 weeks is composed of bony elements from three developmental sources: the chondrocranium (colored blue-gray), the neurocranium (colored light yellow), and the viscerocranium (colored salmon).

the anterior fontanel to the anterolateral fontanel. A **lambdoidal suture** extends from the posterior fontanel to the posterolateral fontanel. A **squamosal**

......................................

lambdoidal: Gk. *lambda*, letter λ in Greek alphabet

**suture** connects the posterolateral fontanel to the anterolateral fontanel.

During normal parturition, the molding of the fetal skull is such that the occipital bone is usually pressed under the two parietal bones. In addition, one

parietal bone overlaps the other, with the depressed one against the promontory of the mother's sacrum. If a baby is born breech (buttocks first), molding does not occur and delivery is more difficult.

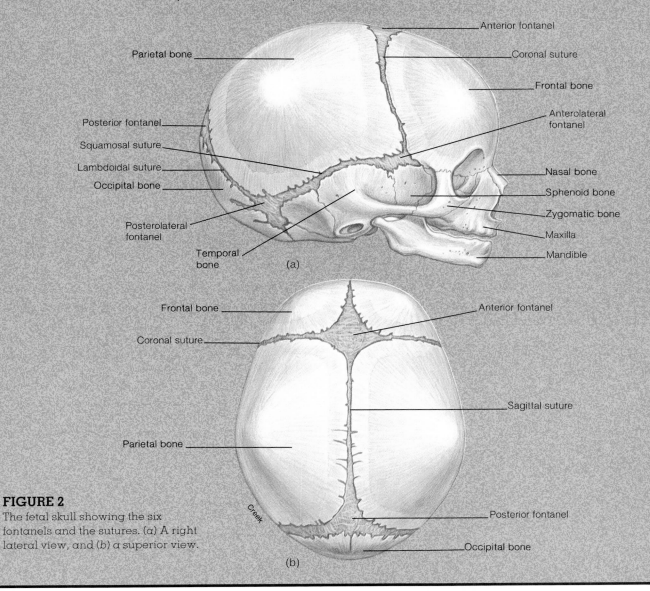

**FIGURE 2**

The fetal skull showing the six fontanels and the sutures. (*a*) A right lateral view, and (*b*) a superior view.

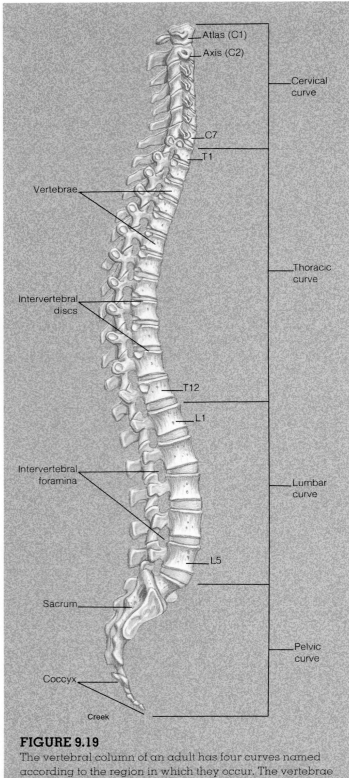

**FIGURE 9.19**

The vertebral column of an adult has four curves named according to the region in which they occur. The vertebrae are separated by intervertebral discs in a structural arrangement that provides flexibility.

strength and maintaining the balance of the upper part of the body; they also make possible a bipedal stance.

The four vertebral curves are not present in a newborn. Instead, the vertebral column is somewhat anteriorly concave, and except for the cervical region, it remains this way even as an infant learns to crawl (fig. 9.20). The cervical curve begins to develop at about 3 months of age, as a baby begins holding up its head, and the curve becomes more pronounced as the baby learns to sit up. The lumbar curve develops as a child begins to walk. The thoracic and pelvic curves are called **primary curves** because they retain the anteriorly concave shape of the fetus. The cervical and lumbar curves are called **secondary curves** because they are modifications of the fetal shape that develop as adaptions to weight bearing as an infant learns to sit up and walk.

The vertebral column is commonly called the "backbone," and together with the spinal cord of the nervous system constitutes the spinal column. The vertebral column has three basic functions:

**1** to support the head and upper extremities while permitting freedom of movement;

**2** to provide attachment for various muscles, ribs, and visceral structures; and

**3** to protect the spinal cord and permit passage of the spinal nerves.

## General Structure of Vertebrae

Vertebrae show similarities in their general structure from one region to another. Figure 9.21 illustrates a typical vertebra. A vertebra is usually composed of an anterior drum-shaped **body,** adapted to withstand compression. The body is in contact with **intervertebral discs** on each end. The **vertebral arch** is affixed to the posterior surface of the body and is composed of two supporting **pedicles** and two arched **laminae.** The hollow space formed by the vertebral arch and body is the **vertebral foramen,** through which the spinal cord passes. Between the pedicles of adjacent vertebrae are the **intervertebral foramina,** through which spinal nerves emerge as they branch off from the spinal cord.

Seven processes arise from the vertebral arch: the **spinous** (*spi′nus*) **process,** two **transverse processes,** two **superior articular processes,** and two **inferior articular**

........................................

pedicle: L. *pediculus,* small foot
lamina: L. *lamina,* thin layer

## FIGURE 9.20

The development of the vertebral curves. An infant is born with the two primary curves but does not develop the secondary curves until it begins sitting upright and walking. (Note the differences in the curves between the sexes.)

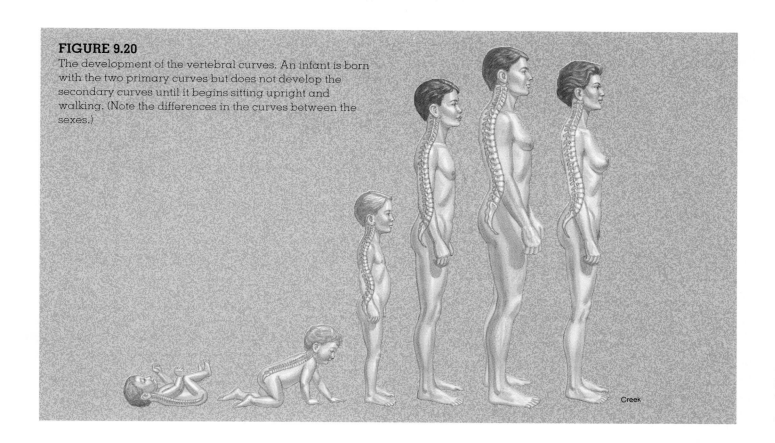

Creek

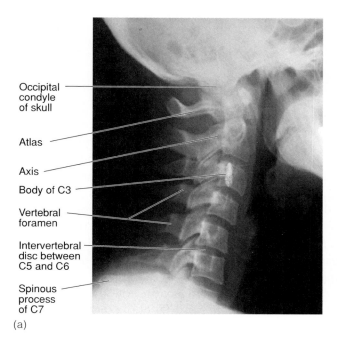

(a)

Occipital
condyle
of skull

Atlas

Axis

Body of C3

Vertebral
foramen

Intervertebral
disc between
C5 and C6

Spinous
process
of C7

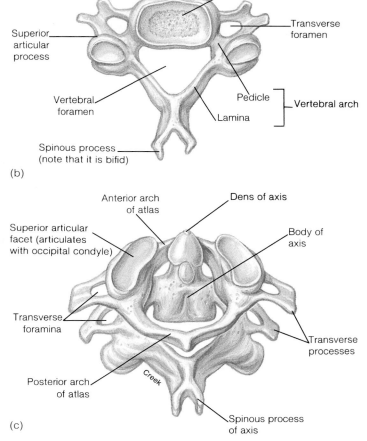

Body

Superior
articular
process

Transverse
foramen

Vertebral
foramen

Pedicle

Lamina

Vertebral arch

Spinous process
(note that it is bifid)

(b)

Anterior arch
of atlas

Dens of axis

Superior articular
facet (articulates
with occipital condyle)

Body of
axis

Transverse
foramina

Transverse
processes

Posterior arch
of atlas

Creek

Spinous process
of axis

(c)

## FIGURE 9.21

Cervical vertebrae. (a) A radiograph of the cervical region, (b) a superior view of a typical cervical vertebra, and (c) the atlas and axis as they articulate.

195

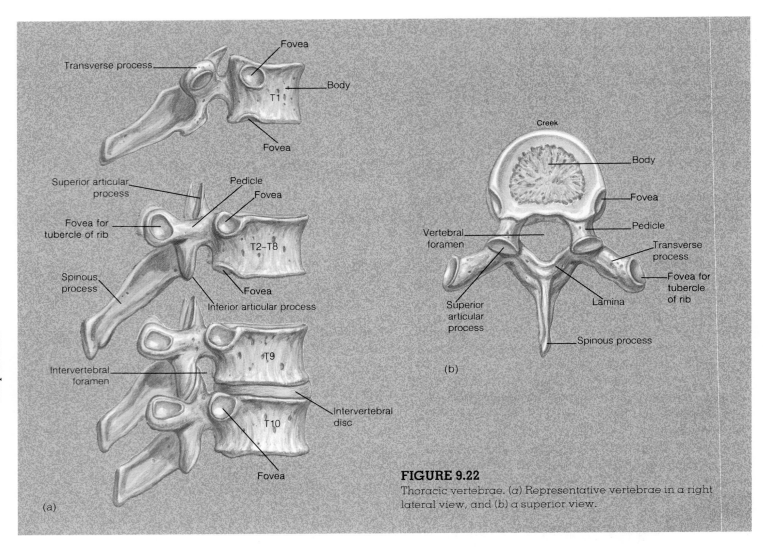

(a)

(b)

**FIGURE 9.22**

Thoracic vertebrae. (a) Representative vertebrae in a right lateral view, and (b) a superior view.

**processes** (fig. 9.22). The first two processes serve for muscle attachment, and the latter two pairs limit twisting of the vertebral column. The spinous process protrudes posteriorly and inferiorly from the vertebral arch. The transverse process extends laterally from each side of a vertebra at the point where the lamina and pedicle join. The superior articular processes of a vertebra have interlocking articulations with the inferior articular processes of the adjacent superior vertebra.

A *laminectomy* is the surgical removal of the spinous processes and their supporting vertebral laminae in a particular region of the vertebral column. A laminectomy may be performed to relieve pressure on the spinal cord or nerve root caused by a blood clot, a tumor, or a herniated disc. It may also be performed on a cadaver to expose the spinal cord and its surrounding meninges.

## Regional Characteristics of Vertebrae

**Cervical Vertebrae**   The seven cervical vertebrae form the flexible framework of the neck region and support the head. The bone tissue of cervical vertebrae is more dense than that found in the other vertebral regions, and, except for those in the coccygeal region, the cervical vertebrae are smallest. Cervical vertebrae are distinguished by the presence of a **transverse foramen** in the transverse process (fig. 9.21). The vertebral arteries pass through this opening as they transfer blood to the brain. The spinous processes of the second through the sixth cervical vertebrae are *bifid* (bi´fid), or notched, for the attachment of the strong **nuchal ligament** that attaches to the back of the skull for added support.

The first cervical vertebra, the **atlas,** is adapted to articulate with the occipital condyles of the skull while supporting the head. The atlas has concave **superior articular surfaces** to articulate with the oval-shaped occipital condyles. This joint permits the nodding of the head in a "yes" movement. The atlas lacks a body. It has a short, rounded spinous process called the **posterior tubercle.**

atlas: from Gk. mythology, Atlas—the Titan who supported the heavens

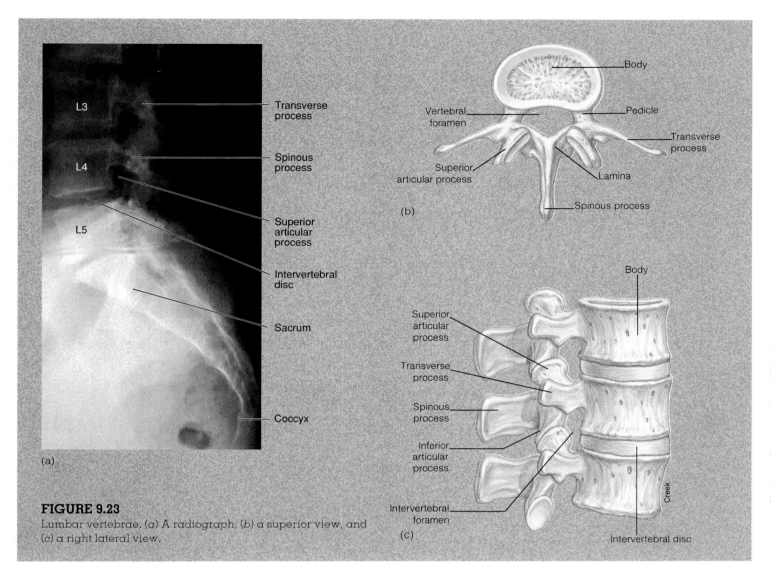

**FIGURE 9.23**

Lumbar vertebrae. (a) A radiograph, (b) a superior view, and (c) a right lateral view.

The second cervical vertebra is called the **axis** and is easily identified by the presence of a peglike projection called the **dens,** or **odontoid process.** This process extends superiorly to provide a pivot for rotation with the atlas. The dens permits rotation, or the turning of the head to the side, as in a "no" movement.

 *"Whiplash"* of the neck is a general term for injury to the cervical region. Muscle, bone, or ligament injury in this portion of the spinal column is relatively common in individuals involved in automobile and sports-related accidents. Joint dislocation without vertebral fracture occurs commonly between the fourth and fifth or fifth and sixth cervical vertebrae, where mobility is greatest. Bilateral dislocations are particularly dangerous because of the probability of spinal cord injury. Compression fractures of the bodies of the first three cervical vertebrae are common and follow abrupt forced flexion of the neck. Fractures of this type may be extremely painful because of pinched spinal nerves.

**Thoracic Vertebrae**   The thoracic vertebrae serve as attachments of the ribs to form the posterior anchor of the rib cage. Thoracic vertebrae are larger than cervical vertebrae and increase in size from superior (T1) to inferior (T12). Each thoracic vertebra has a long spinous process, which slopes obliquely downward, and **fovea** for articulation with the ribs (fig. 9.22).

**Lumbar Vertebrae**   The five lumbar vertebrae are easily identified by their heavy bodies and thick, blunt spinous processes (fig. 9.23) for attachment of powerful back muscles. They are the largest vertebrae of the vertebral column. Their articular processes are also distinctive in that the superior articular processes are directed medially instead of superiorly and the inferior articular processes are directed laterally instead of inferiorly.

axis: L. *axis,* axle
odontoid: Gk. *odontos,* tooth

lumbar: L. *lumbus,* loin

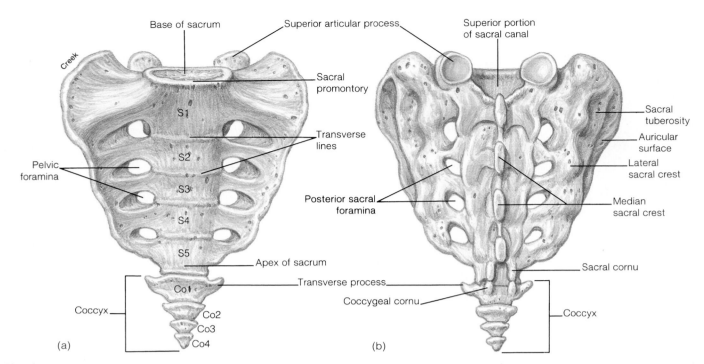

Base of sacrum    Superior articular process    Superior portion of sacral canal

Creek

Sacral promontory

S1

Transverse lines

S2

Pelvic foramina

S3

S4

S5

Apex of sacrum

Co1

Coccyx    Transverse process

Coccygeal cornu

Co2

Co3

Co4

(a)

Sacral tuberosity

Auricular surface

Lateral sacral crest

Posterior sacral foramina

Median sacral crest

Sacral cornu

Coccyx

(b)

**FIGURE 9.24**

The sacrum and coccyx. (*a*) An anterior view, and (*b*) a posterior view.

**Sacrum**    The wedge-shaped sacrum (fig. 9.24) consists of four or five sacral vertebrae, which become fused after age 26. The sacrum is functionally adapted to provide a strong foundation for the pelvic girdle. The sacrum has an extensive **auricular surface** on each side for the formation of a semimovable **sacroiliac** (*sak″ro-il′e-ak*) **joint** with the os coxa. A **median sacral crest** is formed along the posterior surface by the fusion of the spinous processes. **Posterior sacral foramina** on either lateral side of the crest allow the passage of nerves from the spinal cord. The **sacral canal** is the tubular cavity within the sacrum that is continuous with the vertebral canal. Paired **superior articular processes,** which articulate with the fifth lumbar vertebra, arise from the roughened **sacral tuberosity** along the posterior surface.

The smooth anterior surface of the sacrum forms the posterior surface of the pelvic cavity. It has four **transverse lines** denoting the fusion of the vertebral bodies. On both lateral sides of the transverse lines are the paired **pelvic foramina (anterior sacral foramina).** The superior border of the anterior surface of the sacrum, called the **sacral promontory** (*prom′on-tor″e*), is an important obstetric landmark for pelvic measurements.

**Coccyx**    The coccyx (*kok′siks*) is the so-called tailbone. It is composed of four or five fused coccygeal vertebrae, which form a triangular-shaped structure. The first vertebra of the

sacrum: L. *sacris*, sacred
coccyx: Gk. *kokkyx*, like a cuckoo's beak

fused coccyx has two long **coccygeal cornua,** which are attached by ligaments to the sacrum (fig. 9.24). Lateral to the cornua are the transverse processes.

The regions of the vertebral column are summarized in table 9.4.

When a person sits, the coccyx flexes anteriorly somewhat, acting as a shock absorber. An abrupt fall on the coccyx, however, may cause a painful subperiosteal bruising, fracture, or fracture-dislocation of the sacrococcygeal joint. An especially difficult childbirth can even injure the coccyx of the mother. Coccygeal trauma is painful and may require months to heal.

# Rib Cage

*The cone-shaped and flexible rib cage consists of the thoracic vertebrae, 12 paired ribs, costal cartilages, and the sternum. It encloses and protects the thoracic viscera and is directly involved in the mechanics of breathing.*

The **sternum, ribs, costal cartilages,** and the previously described thoracic vertebrae form the **rib cage,** or **thoracic cage,** of the thorax (fig. 9.25). The rib cage is anteroposteriorly compressed and more narrow superiorly than inferiorly. It supports the pectoral girdle and upper extremities, protects and supports the thoracic and upper abdominal viscera, and plays a major role in breathing (see fig. 13.11). Certain

## Table 9.4 Regions of the vertebral column

| Region | Number of bones | Diagnostic features |
|---|---|---|
| Cervical | 7 | Transverse foramina; superior facets of atlas articulate with occipital condyle; dens of axis; spinous processes of third through fifth vertebrae are bifid |
| Thoracic | 12 | Long spinous processes that slope obliquely inferiorly; fovea for articulation with ribs |
| Lumbar | 5 | Large bodies; prominent transverse processes; short, thick spinous processes |
| Sacrum | 4 or 5 fused vertebrae | Extensive auricular surface; median sacral crest; posterior sacral foramina; sacral promontory; sacral canal |
| Coccyx | 4 or 5 fused vertebrae | Small, triangular; coccygeal cornua |

bones of the rib cage contain active sites in the bone marrow for the production of red blood cells.

## Sternum

The **sternum** (breastbone) is an elongated, flattened bony plate consisting of three separate bones; the upper **manubrium,** the central **body,** and the lower **xiphoid** (*zi´foid*)

process. On the lateral sides of the sternum are **costal notches** where the costal cartilages attach. A **jugular notch** is formed at the superior end of the manubrium, and a **clavicular** (*klă-vik´yŭ-lar*) **notch** for articulation with the clavicle is present on both lateral sides of the jugular notch. The manubrium articulates with the costal cartilages of the first and second ribs. The body of the sternum attaches to the costal cartilages of the second through the tenth ribs. The xiphoid process does not attach to ribs but is an attachment for abdominal muscles. The costal cartilages of the eighth, ninth, and tenth ribs fuse to form the *costal margin* of the rib cage. A *costal angle* is formed where the two costal margins come together at the xiphoid process. The **sternal angle** (angle of Louis) may be palpated as an elevation between the manubrium and body of the sternum at the level of the second rib (see fig. 9.25). The costal angle,

• • • • • • • • • • • • • • • • • • • • • • • • • • • • •

sternum: Gk. *sternon*, chest
manubrium: L. *manubrium*, a handle
xiphoid: Gk. *xiphos*, sword
costal: L. *costa*, rib

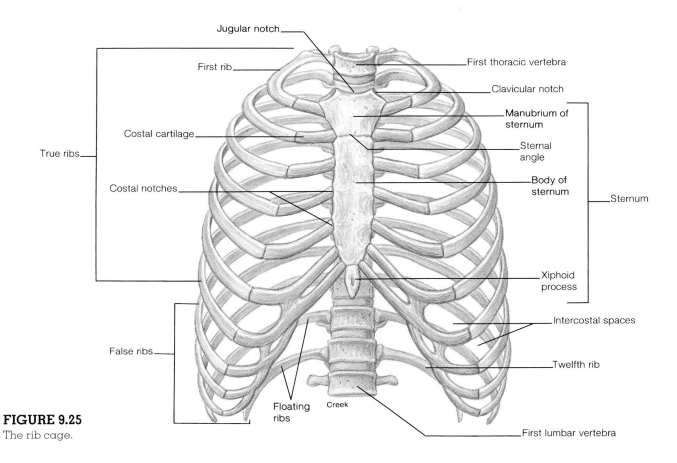

**FIGURE 9.25**
The rib cage.

costal margins, and sternal angle are important surface landmarks of the thorax and abdomen.

## Ribs

There are twelve pairs of **ribs,** each pair being attached posteriorly to a thoracic vertebra. Anteriorly, the first seven pairs are anchored to the sternum by individual **costal cartilages** and are called **true ribs.** The remaining five pairs (ribs 8, 9, 10, 11, and 12) are termed **false ribs.** Because the last two pairs of false ribs do not attach at all to the sternum, they are referred to as **floating ribs.** The four floating ribs are embedded in the muscles of the body wall.

Although the structure of ribs varies, each of the first ten pairs has a **head** and a **tubercle** for articulation with a vertebra. The last two have a head but no tubercle. In addition, each of the twelve pairs has a **neck, angle,** and **body** (fig. 9.26). The head projects posteriorly and articulates with the body of a thoracic

vertebra (fig. 9.27). The tubercle is a knoblike process, just lateral to the head. It articulates with the fovea on the transverse process of a thoracic vertebra. The neck is the constricted area between the head and the tubercle. The body is the main, curved part of the rib. Along the inner surface of the body is a depressed canal called the **costal groove** that protects the costal vessels and nerve. Spaces between the ribs are called **intercostal spaces** and are occupied by the intercostal muscles.

Fractures of the ribs are relatively common injuries and most frequently occur between ribs 3 and 10. The first two pairs of ribs are protected by the clavicles, and the last two pairs move freely and will give with an impact. Little can be done to assist the healing of broken ribs other than binding them tightly to restrict movement.

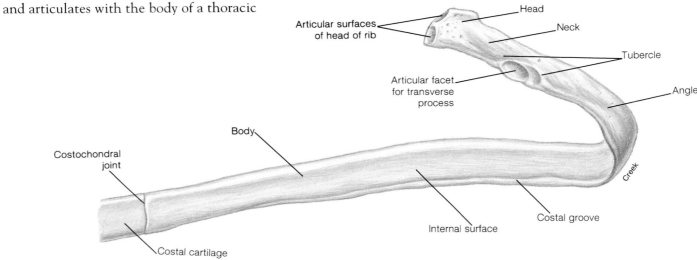

**FIGURE 9.26**
The structure of a rib.

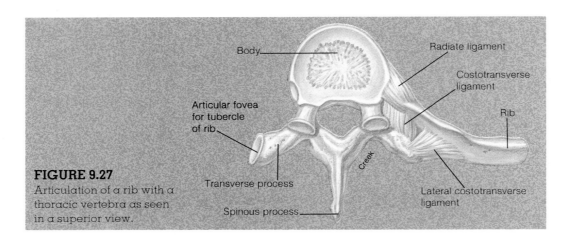

**FIGURE 9.27**
Articulation of a rib with a thoracic vertebra as seen in a superior view.

# Chapter Summary

## Skull (pp. 179–191)

1. The cranium encloses and protects the brain and provides for the attachment of muscles.
   a. Sutures are immovable joints between cranial bones.
   b. The eight cranial bones include the frontal, parietals, temporals, occipital, sphenoid, and ethmoid.
2. Facial bones form the basic shape of the face, support the teeth, and provide for the attachment of the facial muscles.
   a. The 14 facial bones are the nasals, maxillae, zygomatics, mandible, lacrimals, palatines, inferior nasal conchae, and vomer.
   b. The hyoid bone is located in the neck, between the mandible and the larynx.
   c. The three paired auditory ossicles (malleus, incus, and stapes) are located within the middle-ear chambers of the petrous part of the temporal bones.

## Vertebral Column (pp. 191–198)

1. The vertebral column consists of seven cervical, twelve thoracic, five lumbar, four or five fused sacral, and four or five fused coccygeal vertebrae.
2. Cervical vertebrae have transverse foramina; thoracic vertebrae have fovea for articulation with ribs; lumbar vertebrae have heavy bodies; sacral vertebrae are triangularly fused and articulate with the pelvic girdle; and the coccygeal vertebrae form a small triangular bone.

## Rib Cage (pp. 198–200)

1. The sternum consists of a manubrium, body, and xiphoid process.
2. There are seven pairs of true ribs and five pairs of false ribs. The inferior two pairs of false ribs (pairs 11 and 12) are called floating ribs.

# Review Activities

## Objective Questions

Match the following foramina to the correct bone in which it occurs.

1. rotundum
2. mental
3. carotid canal
4. olfactory
5. magnum

a. ethmoid bone
b. occipital bone
c. sphenoid bone
d. mandible
e. temporal bone

6. With respect to the hard palate, which of the following statements is *false?*
   a. The hard palate is composed of two maxillae and two palatine bones.
   b. The hard palate separates the oral cavity (mouth) from the nasal cavity.
   c. The mandible articulates with the posteriolateral angles of the hard palate.
   d. The median palatine suture, incisive fossae, and greater palatine foramina are structural features of the hard palate.
7. The location of the sella turcica is immediately
   a. superior to the sphenoidal sinus.
   b. inferior to the frontal sinus.
   c. medial to the petrous part of the temporal bones.
   d. superior to the perpendicular plate of the ethmoid bone.
8. Which is the most prominent of the six fontanels?
   a. anterior
   b. posterior
   c. anterolateral
   d. posterolateral
9. The parietal bone articulates with the occipital bone at
   a. the coronal suture.
   b. the squamosal suture.
   c. the posterolateral suture.
   d. the lambdoidal suture.
10. Which of the following is *not* a cranial bone?
    a. sphenoid bone
    b. ethmoid bone
    c. vomer
    d. frontal bone
11. Which of the following is *not* one of the four parts of the temporal bone?
    a. squamous part
    b. auricular part
    c. tympanic part
    d. petrous part
    e. mastoid part
12. The mandibular fossa is located in which structural part of the temporal bone?
    a. squamous part
    b. auricular part
    c. tympanic part
    d. petrous part
    e. mastoid part
13. The facial nerve passes through the foramen
    a. stylomastoid.
    b. ovale.
    c. magnum.
    d. spinosum.
14. The crista galli is a structural feature of which bone?
    a. sphenoid bone
    b. ethmoid bone
    c. palatine bone
    d. temporal bone
15. Thoracic vertebrae are distinguished by the presence of
    a. transverse foramina.
    b. bifid spinous processes.
    c. facets.
    d. auricular surfaces.
16. The usual number of false ribs is
    a. two pairs.
    b. three pairs.
    c. five pairs.
    d. seven pairs.

## Essay Questions

1. Describe the development of the skull. What are fontanels, where are they located, and what are their functions?
2. List the bones of the skull that are paired. Which are unpaired? Identify the bones of the skull that can be palpated.
3. Which facial bones contain foramina? What structures traverse these openings?
4. List the bones that form the cranial cavity, the orbit, and the nasal cavity. Describe the location of the paranasal sinuses, the mastoid sinus, and the inner-ear cavity.
5. Describe the curvature of the vertebral column. What is meant by primary curves as compared to secondary curves?
6. List two or more characteristics by which vertebrae from each of the five regions of the vertebral column can be identified.
7. Identify the bones that form the rib cage. What functional role do the bones and the costal cartilages have in respiration?

## Gundy/Weber Software 💾

The tutorial software accompanying Chapter 9 is Volume 3—Skeletal System.

# skeletal system: appendicular skeleton

## objectives

- Describe the bones of the pectoral girdle and the positions of articulations.

- Identify the bones of the upper extremity and list the diagnostic features of each bone.

- Describe the structure of the pelvic girdle and list its functions.

- Describe how the male and female pelves differ structurally.

- Identify the bones of the lower extremity and list the diagnostic features of each bone.

- Describe the structural features and functions of the arches of the foot.

# Pectoral Girdle and Upper Extremity

*The structure of the pectoral girdle and upper extremities is adaptive for freedom of movement and extensive muscle attachment.*

## Pectoral Girdle

The two *scapulae* and two *clavicles* make up the **pectoral girdle** (shoulder girdle). It is not a complete girdle, having only an anterior attachment to the axial skeleton at the sternum. The primary function of the pectoral girdle is to provide attachment for the numerous muscles that move the brachium and antebrachium. The pectoral girdle is not weight bearing and is therefore more delicate in structure than the pelvic girdle.

**Clavicle** The slender S-shaped clavicle (collarbone) binds the shoulder to the axial skeleton and positions the shoulder joint away from the trunk for freedom of movement. The articulation of the medial **sternal extremity** of the clavicle (fig. 10.1) and the manubrium is referred to as the *sternoclavicular joint*. The lateral **acromial** (ă-krō′me-al) **extremity** of the clavicle articulates with the acromion of the scapula. This articulation is referred to as the *acromioclavicular joint*. A **conoid tubercle** is present on the inferior surface of the lateral end, and a **costal tuberosity** is present on the inner surface of the medial end. Both processes serve as points of attachment for ligaments.

The long, delicate clavicle is the most commonly fractured bone in the body. Blows to the shoulder or an attempt to break a fall with an outstretched hand cause the force to be displaced to the clavicle. The most vulnerable area for a fracture of this bone is through its center, immediately proximal to the conoid tubercle. Because the clavicle is subcutaneous and not covered with muscle, a fracture can easily be palpated.

**Scapula** The scapula (shoulder blade) is a large, triangular flat bone positioned on the posterior aspect of the rib cage, overlying ribs 2 to 7. The **spine** of the scapula is a prominent diagonal bony ridge seen on the posterior surface (figs. 10.2 and 10.3). Above the spine is the **supraspinous** (soo″ pră-spi′nus) **fossa,** and below the spine is the **infraspinous fossa.** The spine broadens toward the shoulder as the **acromion** (ă-krō′me-on). The acromion serves for the attachment of several muscles as well as for articulation with the clavicle. Inferior to the acromion is a shallow depres-

clavicle: L. *clavicula,* a small key
conoid tubercle: Gk. *konos,* cone; L. *tuberculum,* a small swelling
scapula: L. *scapula,* shoulder
acromion: Gk. *akros,* peak; *amos,* shoulder

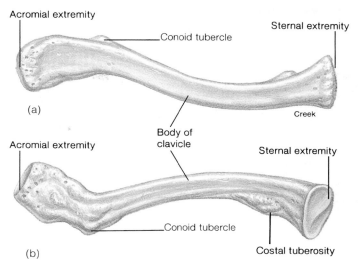

**FIGURE 10.1**
The right clavicle. (*a*) A superior view and (*b*) an inferior view.

sion, the **glenoid cavity,** into which the head of the humerus fits. The **coracoid process** is a thick upward projection that lies superior and anterior to the glenoid cavity. On the anterior surface of the scapula is a slightly concave area known as the **subscapular fossa.**

The scapula has three borders separated by three angles. The superior edge is called the **superior border.** The **medial border** is nearest to the vertebral column, positioned about 5 cm (2 in.) away. The **lateral border** is directed toward the arm. The **superior angle** is located between the superior and medial borders, the **inferior angle** is located between the medial and lateral borders, and the **lateral angle** is located between the superior and medial borders. It is at the lateral angle that the scapula articulates with the head of the humerus. Along the superior border, a distinct depression called the **scapular notch** serves as a passageway for a nerve.

It is important to know the anatomy of the scapula because some 15 muscles attach to its processes and fossae. Clinically, the pectoral girdle is significant because the clavicle and acromion of the scapula are frequently fractured in trying to break a fall. The acromion is palpated when locating the proper site for an intramuscular injection of the arm. This site is chosen because the musculature is quite thick and contains few nerves.

## Brachium (Upper Arm)

The **brachium** extends from the shoulder to the elbow and contains a single bone—the humerus.

**Humerus** The humerus (fig. 10.4) is the longest bone of the upper extremity. It consists of a proximal **head,** which

glenoid: Gk. *glenoeides,* shallow form
coracoid process: Gk. *korakodes,* like a crow's beak

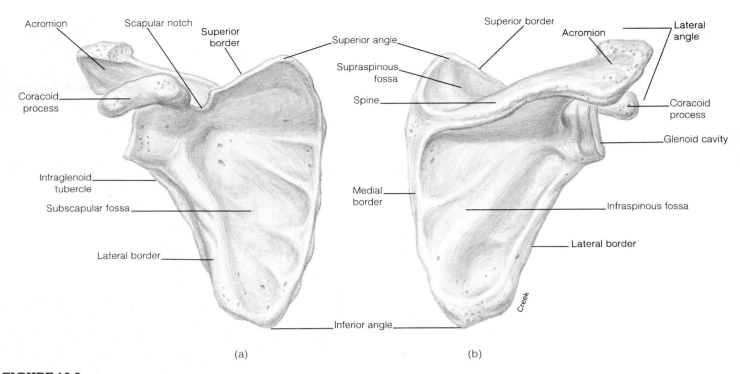

Acromion — Scapular notch — Superior border

Coracoid process

Infraglenoid tubercle

Subscapular fossa

Lateral border

Superior border — Acromion — Lateral angle

Superior angle

Supraspinous fossa

Spine

Coracoid process

Glenoid cavity

Medial border

Infraspinous fossa

Lateral border

Inferior angle

Creek

(a)                                    (b)

**FIGURE 10.2**

The right scapula. (a) An anterior view and (b) a posterior view.

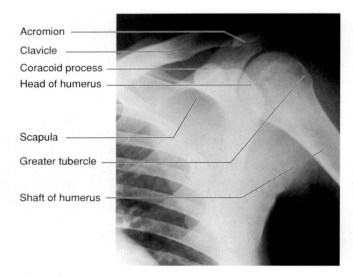

Acromion

Clavicle

Coracoid process

Head of humerus

Scapula

Greater tubercle

Shaft of humerus

**FIGURE 10.3**

A radiograph of the left shoulder shows the articulation of the scapula and humerus forming the shoulder joint.

articulates with the glenoid cavity of the scapula; a **shaft** (body); and a distal end, which is modified to articulate with the two bones of the forearm. Surrounding the margin of the head is a slightly indented groove denoting the **anatomical neck.** The region where the shaft of the humerus begins to taper is referred to as the **surgical neck,** a frequent site of fractures. Lateral to the head is a large eminence, the **greater tubercle.** The **lesser tubercle** is slightly anterior to the greater

and is separated from the greater by an **intertubercular (bicipital) groove,** through which the tendon from the biceps brachii muscle passes.

Along the lateral midregion of the shaft is a roughened area, the **deltoid tuberosity,** for the attachment of the deltoid muscle. Small openings in the bone along the shaft are called **nutrient foramina.**

The distal end of the humerus has two rounded condyloid articular surfaces. The **capitulum** (kă-pich´ŭ-lum) is the lateral rounded condyle that articulates with the radius. The **trochlea** (trok´le-ă) of the humerus is the pulleylike medial surface that articulates with the ulna. On either side above the condyles are the **lateral** and **medial epicondyles.** The large medial epicondyle protects the ulnar nerve that passes posteriorly through a depression on the back of the elbow called the **ulnar sulcus.** The **coronoid fossa** is a depression above the trochlea on the anterior surface. The **olecranon** (o-lek´ră-non) **fossa** is a depression on the distal posterior surface. Both fossae are adapted to receive parts of the ulna during movement of the forearm.

 The medical term for tennis elbow is *lateral epicondylitis,* which means an inflammation of the tissues surrounding the lateral epicondyle of the humerus. Six muscles that control backward

deltoid tuberosity: Gk. *deltoeides,* shaped like the letter Δ
capitulum: L. *caput,* little head
trochlea: Gk. *trochilia,* a pulley
olecranon: Gk. *olene,* ulna; *kranion,* head

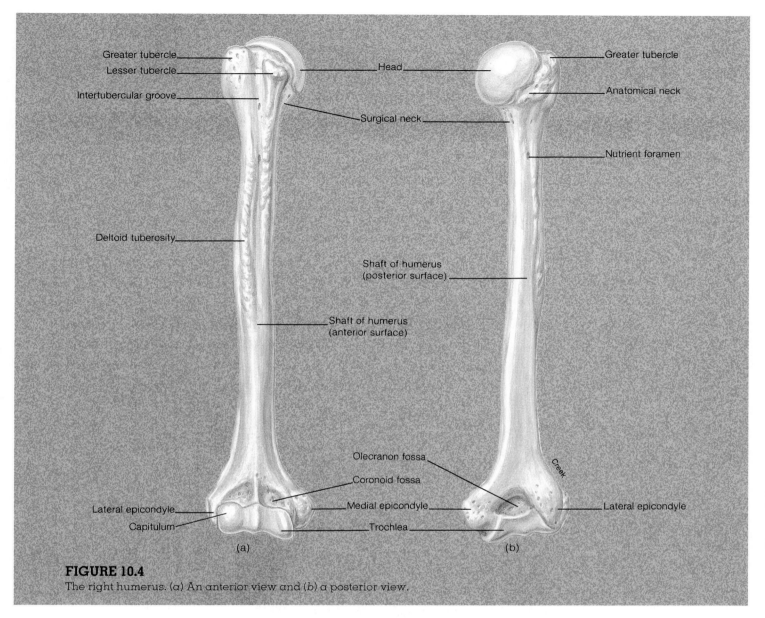

**FIGURE 10.4**
The right humerus. (*a*) An anterior view and (*b*) a posterior view.

(extension) movement of the hand and fingers originate on the lateral epicondyle. Repeated strenuous contractions of these muscles, as in stroking with a tennis racket, may cause a strain on the periosteum and tendinous muscle attachments, resulting in tenderness and pain around the epicondyle. Binding usually eases the pain, but only rest can eliminate the causative factor, and recovery generally follows.

## Antebrachium (Forearm)

The skeletal structures of the antebrachium are the ulna on the medial side and the radius on the lateral (thumb) side (figs. 10.5 and 10.6). The ulna is longer, and it is more firmly connected to the humerus than the radius. The radius, however, contributes more significantly at the wrist joint than does the ulna.

**Ulna** The proximal end of the ulna articulates with the humerus and radius. A distinct depression, the **trochlear notch,** articulates with the trochlea of the humerus. The **coronoid process** forms the anterior lip of the trochlear notch, and the **olecranon** forms the posterior portion, or elbow (fig. 10.5). Lateral and inferior to the coronoid process is the **radial notch,** which accommodates the head of the radius.

The tapered distal end of the ulna has a knobbed part, the **head,** from which the **styloid process** projects posteromedially. The ulna articulates with the radius proximally and distally.

..........................................
styloid process: Gk. *stylos,* pillar; *eidos,* resemblance

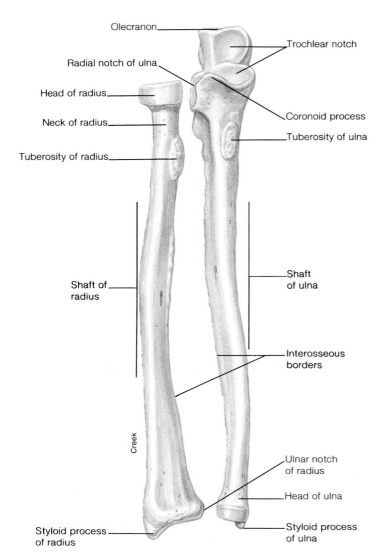

**FIGURE 10.5**
An anterior view of the right radius and ulna.

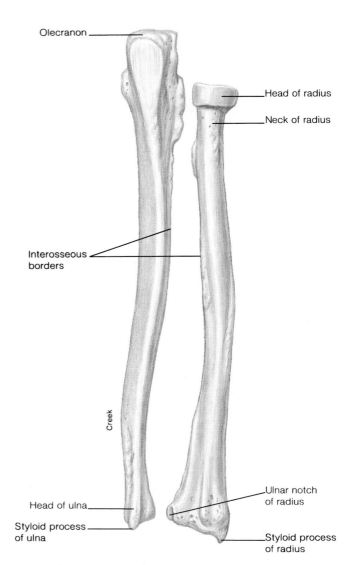

**FIGURE 10.6**
A posterior view of the right radius and ulna.

**Radius**   The radius consists of a **shaft** (body) that has a small proximal end and a large distal end. A proximal disc-shaped **head** articulates with the capitulum of the humerus and the radial notch of the ulna. The prominent **tuberosity of radius** (radial tuberosity) for attachment of the biceps brachii muscle, is on the anteromedial side of the shaft of the radius, just below the head. The distal end of the radius has a double-faceted surface for articulation with the prox-imal carpal bones. The distal end of the radius also has the **styloid process** on the lateral tip and the **ulnar notch** on the medial side that receives the distal end of the ulna. The styloid processes on the ulna and radius provide lateral and medial stability for articulation at the wrist.

When a person falls, the natural tendency is to extend the hand to break the fall. This reflexive movement frequently results in fractured bones. Common fractures of the radius include a fracture of the head as it is driven forcefully against the capitulum, a fracture of

the neck, or a fracture of the distal end (*Colles' fracture*) caused by landing on an outstretched hand.

When falling, it is less traumatic to the body to withdraw the appendages, bend the knees, and let the entire body hit the surface. Athletes learn that this is the safest way to fall.

## Manus (Hand)

The hand contains 27 bones, constituting the carpus, metacarpus, and phalanges (figs. 10.7–10.9).

**Carpus**   The carpus, or wrist, consists of 8 carpal bones arranged in two transverse rows of 4 bones each. The proxi-mal row, naming from the lateral (thumb) to medial side, consists of the **scaphoid (navicular) bone, lunate bone,**

..........................................

carpus: Gk. *karpos*, wrist
scaphoid: Gk. *skaphe*, boat; *eidos*, resemblance
lunate: L. *lunare*, crescent or moon shaped

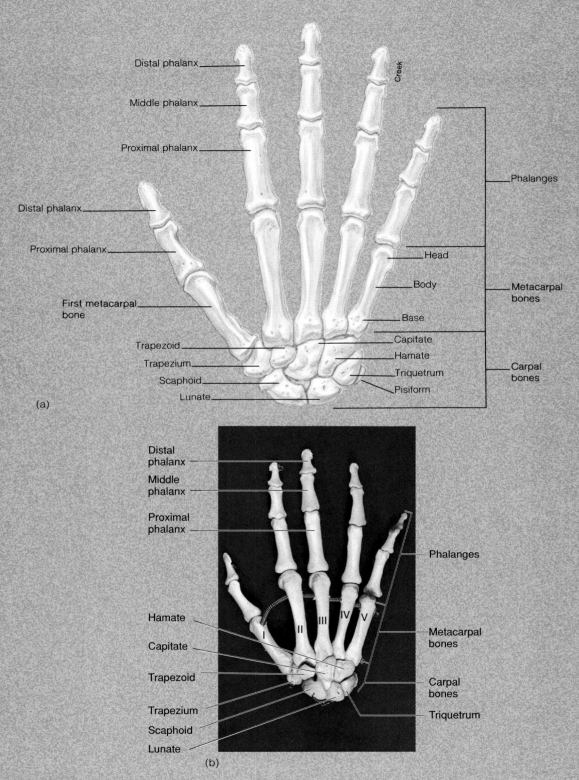

Distal phalanx

Middle phalanx

Proximal phalanx

Distal phalanx

Proximal phalanx

First metacarpal bone

Trapezoid

Trapezium

Scaphoid

Lunate

(a)

Creek

Phalanges

Head

Body

Base

Capitate

Hamate

Triquetrum

Pisiform

Metacarpal bones

Carpal bones

Distal phalanx

Middle phalanx

Proximal phalanx

Hamate

Capitate

Trapezoid

Trapezium

Scaphoid

Lunate

(b)

I

II

III

IV

V

Phalanges

Metacarpal bones

Carpal bones

Triquetrum

**FIGURE 10.7**

A posterior view of the skeleton of the right hand. (a) A drawing and (b) a photograph.

triquetrum (*tri-kwe´trum*) **bone,** and **pisiform** (*pi´si-form*) **bone.** The pisiform bone forms in a tendon as a sesamoid bone. The distal row, from lateral to medial, consists of the **trapezium** (greater multangular) **bone, trapezoid** (lesser multangular) **bone, capitate bone,** and **hamate** (*ham´at*) **bone.** The scaphoid and lunate bones of the proximal row articulate with the distal end of the radius.

**Metacarpus** The metacarpus, or palm of the hand is composed of five metacarpal bones. Each metacarpal bone consists of a proximal **base,** a **shaft** (body), and a distal **head** that is rounded for articulation with the base of each proximal phalanx. The heads of the metacarpal bones are distally located and form the knuckles of a clenched fist. The metacarpal bones are numbered from I to V, the lateral, or thumb, side being I.

**Phalanges** The 14 phalanges are the skeletal elements of the digits. A single finger bone is called a **phalanx** (*fa´langks*). The phalanges of the fingers are arranged in a proximal row, a middle row, and a distal row. The thumb (pollex) has only a proximal and a distal phalanx.

A summary of the bones of the upper extremities is presented in table 10.1.

The hand is a marvel of structural organization that, despite its complexity, can withstand considerable abuse. Other than sprained fingers and dislocation, the most common skeletal injury is a fracture to the scaphoid bone of the wrist (about 70% of carpal fractures occur here). When immobilizing the wrist joint with a plaster cast, the wrist is positioned in the plane of relaxed function. This is the position in which the hand is about to grasp an object between the thumb and index finger.

.............................................
triquetrum: L. *triquetrus*, three cornered
pisiform: Gk. *pisos*, pea
trapezium: Gk. *trapesion*, small table

.............................................
capitate: L. *capitatus*, head
hamate: L. *hamatus*, hook
phalanx: Gk. *phalanx*, finger bone or toe bone

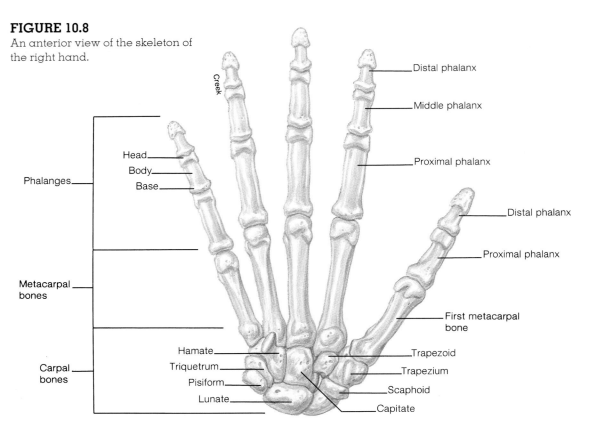

**FIGURE 10.8**

An anterior view of the skeleton of the right hand.

**FIGURE 10.9**

A radiograph of the right wrist and hand shown in an anteroposterior view. (Note the presence of a sesamoid bone at the thumb joint.)

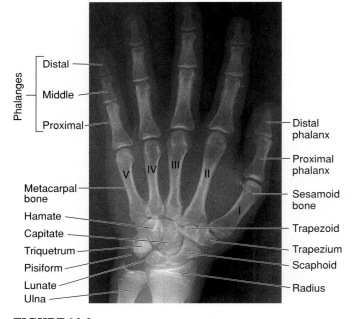

## Table 10.1    Bones of the pectoral girdle and upper extremities

| Name and number | Location | Diagnostic features |
| --- | --- | --- |
| Clavicle (2) | Anterior base of neck, between sternum and scapula | S-shaped; sternal and acromial extremities |
| Scapula (2) | Upper back, forming part of the shoulder | Triangular shaped; spine; acromion and coracoid processes |
| Humerus (2) | Brachium, between scapula and elbow | Longest bone of upper extremity; greater and lesser tubercles; surgical neck; deltoid tuberosity; capitulum; trochlea; coronoid and olecranon fossae |
| Ulna (2) | Medial side of forearm | Trochlear notch; olecranon and styloid processes |
| Radius (2) | Lateral side of forearm | Head; radial tuberosity; styloid process |
| Carpal bone (16) | Wrist | Short bones arranged in two rows of four bones each |
| Metacarpal bone (10) | Palm of hand | Long bones, each aligned with a digit |
| Phalanx (28) | Digits | Three in each digit, except two in thumb |

# Pelvic Girdle and Lower Extremity

*The structure of the pelvic girdle and lower extremities is adaptive for support and locomotion. Extensive processes and surface features on certain bones of the pelvic girdle and lower extremities accommodate massive muscles for posture and locomotion.*

## Pelvic Girdle

The **pelvic girdle,** or **pelvis,** is formed by two **ossa coxae** (hipbones) united anteriorly by the **symphysis pubis** (fig. 10.10). It is attached posteriorly to the sacrum of the vertebral column. The pelvic girdle and its associated ligaments support the weight of the body from the vertebral column. The pelvic girdle also supports and protects the lower viscera, including the urinary bladder, the reproductive organs, and in a pregnant woman, the developing fetus.

Clinically, the basinlike pelvis is frequently divided into a **"greater,"** or **"false,"** **pelvis** and a **"lesser,"** or **"true,"** **pelvis** (see fig. 10.14). These two components are divided by the **pelvic brim,** a curved bony rim passing inferiorly from the sacral promontory to the upper margin of the symphysis pubis. The greater pelvis is the expanded portion of the pelvis superior to the pelvic brim. The pelvic brim not only divides the two portions but surrounds the **pelvic inlet** of the lesser pelvis. The lower circumference of the lesser pelvis

...........................................

coxae: L. *coxae,* hips

bounds the pelvic outlet. During parturition, a child must pass through its mother's lesser pelvis for a natural delivery. *Pelvimetry* measures the dimension of the lesser pelvis to determine whether a cesarean delivery might be necessary. Diameters may be determined by vaginal palpation or by radiographic measurements (fig. 10.11).

Each os coxa actually consists of three separate bones: the **illium,** the **ischium** (*is´ke-um*), and the **pubis** (figs. 10.12 and 10.13). These bones are fused together in the adult. On the lateral surface of the os coxa where the three bones ossify together is a large circular depression, the **acetabulum** (*as´´ĕ-tab´yŭ-lum*), which receives the head of the femur. Although each os coxa is a single bone in the adult, the three components are considered separately for descriptive purposes.

**Ilium**    The ilium is the largest and uppermost of the three pelvic bones. The ilium presents a crest and four angles, or spines, that serve for muscle attachment and as important surface landmarks. The **iliac crest** forms the prominence of the hip. This crest terminates anteriorly as the **anterior superior iliac spine.** Just below this spine is the **anterior inferior iliac spine.** The posterior termination of the iliac crest is the **posterior superior iliac spine,** and just below this is the **posterior inferior iliac spine.**

...........................................

ilium: L. *ilia,* loin
ischium: Gk. *ischion,* hip joint
pubis: L. *pubis,* genital area
acetabulum: L. *acetabulum,* vinegar cup

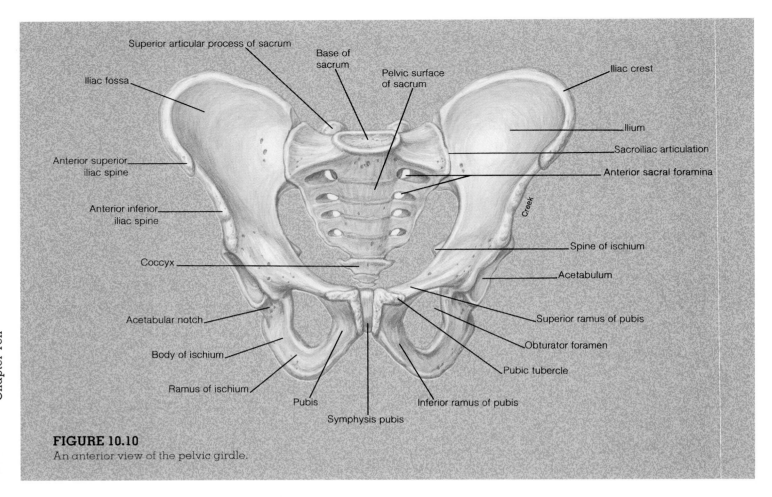

Superior articular process of sacrum

Iliac fossa

Base of sacrum

Pelvic surface of sacrum

Iliac crest

Anterior superior iliac spine

Ilium

Sacroiliac articulation

Anterior sacral foramina

Anterior inferior iliac spine

Spine of ischium

Coccyx

Acetabulum

Acetabular notch

Superior ramus of pubis

Body of ischium

Obturator foramen

Ramus of ischium

Pubic tubercle

Pubis

Inferior ramus of pubis

Symphysis pubis

**FIGURE 10.10**

An anterior view of the pelvic girdle.

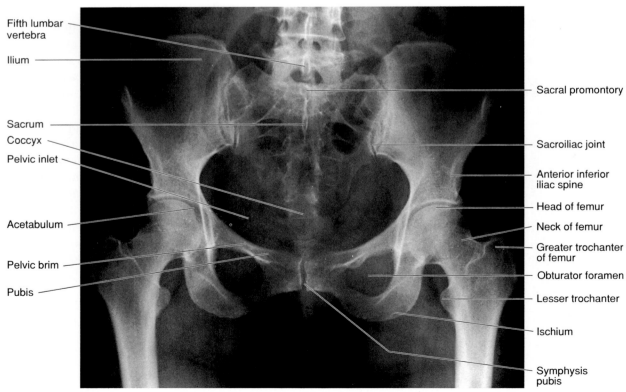

Fifth lumbar vertebra

Ilium

Sacral promontory

Sacrum

Coccyx

Sacroiliac joint

Pelvic inlet

Anterior inferior iliac spine

Acetabulum

Head of femur

Neck of femur

Greater trochanter of femur

Pelvic brim

Obturator foramen

Pubis

Lesser trochanter

Ischium

Symphysis pubis

**FIGURE 10.11**

A radiograph of the pelvic girdle and the articulating femora.

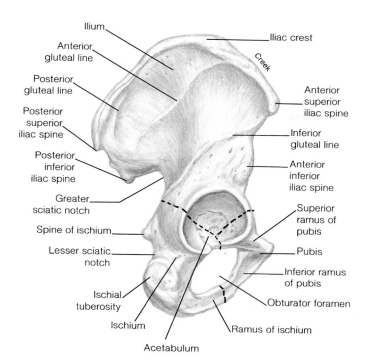

**FIGURE 10.12**
The lateral aspect of the right os coxa.

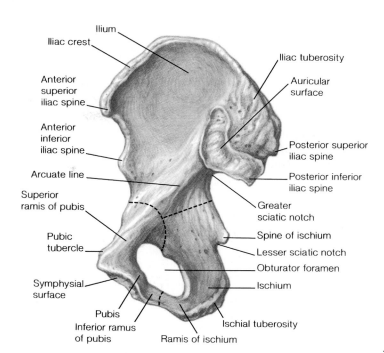

**FIGURE 10.13**
The medial aspect of the right os coxa.

Below the posterior inferior iliac spine is the **greater sciatic** (*si-at´ik*) **notch.** On the medial surface of the ilium is the roughened **auricular surface** that articulates with the sacrum. The **iliac fossa** is the smooth concave surface on the anterior portion of the ilium. The iliacus muscle originates from this fossa. The **iliac tuberosity,** for the attachment of the sacroiliac ligament, is positioned posterior to the iliac fossa. Three roughened ridges are present on the **gluteal surface** of the posterior aspect of the ilium. These ridges serve to attach the gluteal muscles and are the **inferior, anterior,** and **posterior gluteal lines.**

**Ischium**    The ischium is the posteroinferior component of the os coxa. This bone has several significant features. The **spine of the ischium** is the projection immediately posterior and inferior to the greater sciatic notch of the ilium and ischium. Inferior to this spine is the **lesser sciatic notch** of the ischium. The **ischial tuberosity** is the bony projection that supports the weight of the body in the sitting position. A deep **acetabular notch** is present on the inferior portion of the acetabulum. The large **obturator foramen** is formed by the **ramus of the ischium** together with the pubis. The obturator foramen is covered by the obturator membrane, to which several muscles attach.

**Pubis**    The pubis is the anterior component of the os coxa. This bone consists of a **superior ramus** and an **inferior ramus** that supports the **body** of the pubis. The body contributes to the formation of the symphysis pubis—the joint between the two ossa coxae.

**Sex-Related Differences in the Pelvis**    Structural differences between the pelvis of an adult male and that of an adult female (fig. 10.14 and table 10.2) reflect the female's role in pregnancy and parturition.

In addition to the osseous differences listed in table 10.2, the symphysis pubis and sacroiliac joints stretch during pregnancy and parturition.

 The structure of the human pelvis, in its attachment to the vertebral column, permits an upright posture and locomotion on two legs (bipedal) rather than on four legs like other mammals. Although this structural arrangement is well adapted for bipedal locomotion, an upright posture may cause problems. The sacroiliac joint may weaken with age, causing lower back pains. The weight of the viscera may weaken the walls of the lower abdominal area and cause hernias. Some of the problems of childbirth are related to the structure of the mother's pelvis. Finally, the hip joint tends to deteriorate with age, so that many elderly people suffer from fractured hips.

## Thigh

Although the thigh contains only a single bone, the femur, we will also consider the patella, or kneecap, in this section.

**Femur**    The femur (thighbone) is the longest, heaviest, and strongest bone in the body (fig. 10.15). The proximal rounded **head** of the femur articulates with the acetabulum of the

femur: L. *femur,* thigh

os coxa. A shallow pit, called the **fovea capitis femoris** is present in the lower center of the head of the femur. The fovea capitis femoris provides the point of attachment for the ligamentum teres, which helps to support the head of the femur against the acetabulum. The constricted region supporting the head is called the **neck** and is a common site for fractures in the elderly.

The **shaft** (body) of the femur has a slight medial bow so that it converges with the femur of the opposite thigh and brings the knee joints more in line with the body's plane of gravity. The degree of convergence is even greater in the female because of the wide pelvis. The shaft has several important structures for muscle attachment. On the proximolateral side of the shaft of the femur is the **greater trochanter,** and on the medial side is the **lesser trochanter.** Between the trochanters on the anterior side is the **intertrochanteric line.** Between the trochanters on the posterior side is the **intertrochanteric crest.** The **linea aspera** is a vertical ridge on the posterior surface of the shaft.

The distal end of the femur is expanded for articulation with the tibia. The **medial** and **lateral condyles** (*kon'dīlz*) are the articular processes for this joint. The depression between the condyles on the posterior aspect is called the **intercondylar fossa.** The **patellar surface** is located between the condyles on the anterior side. Above the condyles on the lateral and medial sides are the **epicondyles,** which serve for ligament and tendon attachment.

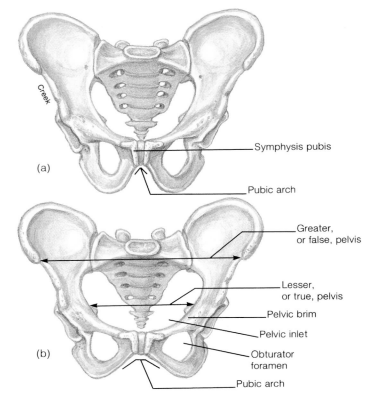

**FIGURE 10.14**

A comparison of (*a*) the male and (*b*) the female pelvic girdle.

**Patella**  The patella (kneecap) is a sesamoid bone positioned on the anterior side of the knee joint (figs. 10.16 and 10.17). It develops in response to strain in the tendon of the quadriceps femoris muscle. The patella is a triangular bone with a broad **base** and an inferiorly pointed **apex. Articular surfaces** on the posterior surface of this bone articulate with the medial and lateral condyles of the femur.

The functions of the patella are to protect the knee joint and to strengthen the tendon of the quadriceps femoris muscle. It also increases the leverage of the quadriceps femoris muscle as it straightens (extends) the knee joint.

 The patella can be fractured by a direct blow. It usually does not fragment, however, because it is confined within the tendon. Dislocations of the patella may result from injury or may be congenital due to underdevelopment of the lateral condyle of the femur.

· · · · · · · · · · · · · · · · · · · · · · · · · · · · ·

linea aspera: L. *linea*, line; *asperare*, rough

| Table 10.2 | Comparison of the male and female pelvic girdle | |
| --- | --- | --- |
| *Characteristics* | *Male pelvis* | *Female pelvis* |
| General structure | More massive; prominent processes | More delicate; processes not so prominent |
| Pelvic inlet | Heart shaped | Round or oval |
| Pelvic outlet | Narrower | Wider |
| Anterior superior iliac spines | Closer together | Farther apart |
| Obturator foramen | Oval | Triangular |
| Acetabulum | Faces laterally | Faces more anteriorly |
| Symphysis pubis | Deeper, longer | Shallower, shorter |
| Pubic arch | Acute (less than 90°) | Obtuse (greater than 90°) |

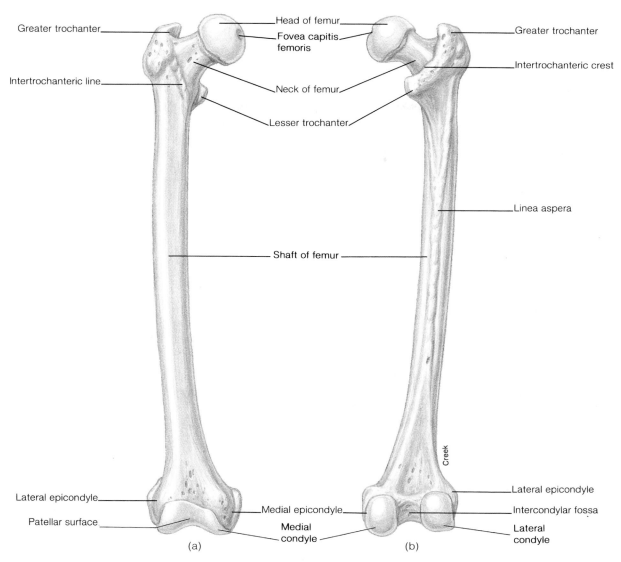

**FIGURE 10.15**

The right femur. (a) An anterior view and (b) a posterior view.

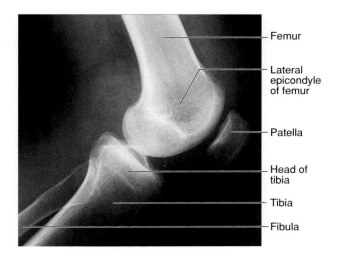

**FIGURE 10.16**

A radiograph of the right knee region.

## *Leg*

The tibia and fibula are the skeletal elements of the leg. The tibia is the larger and more medial of the two bones. The skeletal structure of the leg is illustrated in figure 10.17.

**Tibia** The tibia (shinbone) articulates proximally with the femur at the knee joint to bear the weight of the body. On the distal end, the tibia articulates with the talus of the ankle. Two slightly concave surfaces on the proximal end of the tibia, the **medial** and **lateral condyles,** articulate with the condyles of the femur. Between the condyles, there is a slight upward projection called the **intercondylar eminence.** The **lateral**

..............................................

tibia: L. *tibia,* shinbone, pipe, flute

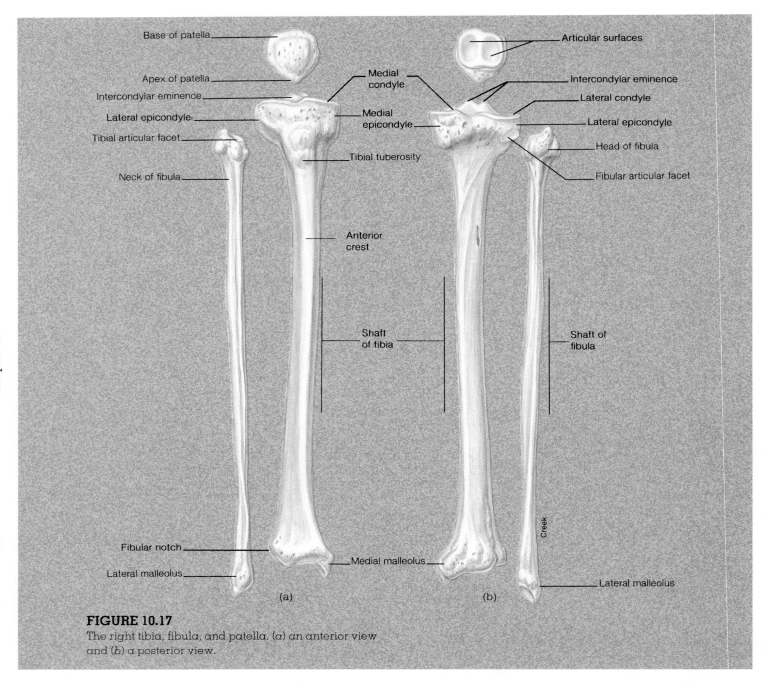

**FIGURE 10.17**
The right tibia, fibula, and patella. (a) an anterior view and (b) a posterior view.

and **medial epicondyles** are located on the proximal end of the tibia on the lateral and medial sides, respectively. The **tibial tuberosity,** for attachment of the patellar ligament, is located on the proximoanterior portion of the shaft. A sharp ridge along the anterior surface of the shaft is called the **anterior crest.**

The **medial malleolus** is a prominent medial knob of bone located on the distal end of the tibia. A **fibular notch,** for articulation with the fibula, is located on the distolateral end.

**Fibula**  The fibula is a long, narrow bone that is more important for muscle attachment than for support. The **head** of the fibula articulates with the proximolateral end of the tibia. The distal end has a prominent knob called the **lateral malleolus.**

The lateral and medial malleoli are positioned on either side of the talus and help stabilize the ankle joint. Both processes can be seen as prominent surface features and are easily palpated. Fractures to either or both malleoli are common in skiers. These fractures, clinically referred to as *Pott's fractures,* result from a shearing force occurring at a vulnerable spot on the leg.

. . . . . . . . . . . . . . . . . . . . . . . . . . . . . . . . . .

malleolus: L. *malleolus,* small hammer

fibula: L. *fibula,* clasp or brooch

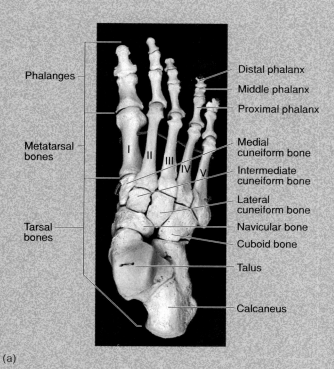

Phalanges

Distal phalanx
Middle phalanx
Proximal phalanx

Metatarsal
bones

Medial
cuneiform bone
Intermediate
cuneiform bone
Lateral
cuneiform bone
Navicular bone
Cuboid bone

Tarsal
bones

Talus

Calcaneus

I  II  III  IV  V

(a)

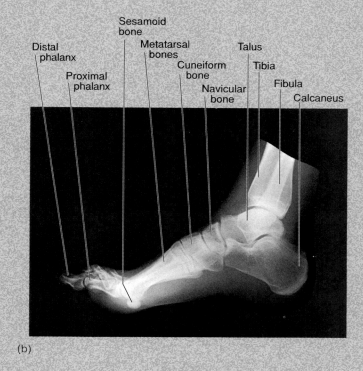

Distal
phalanx
Proximal
phalanx

Sesamoid
bone
Metatarsal
bones
Cuneiform
bone
Navicular
bone

Talus
Tibia
Fibula
Calcaneus

(b)

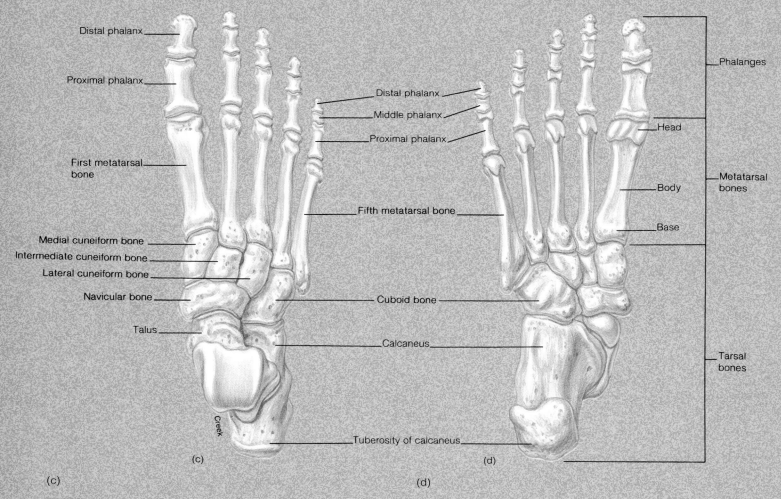

Distal phalanx

Proximal phalanx

First metatarsal
bone

Medial cuneiform bone
Intermediate cuneiform bone
Lateral cuneiform bone
Navicular bone

Talus

Creek

Distal phalanx
Middle phalanx
Proximal phalanx

Fifth metatarsal bone

Cuboid bone

Calcaneus

Tuberosity of calcaneus

(c)

Phalanges

Head

Body

Base

Metatarsal
bones

Tarsal
bones

(d)

(c)                                    (d)

## FIGURE 10.18

(a) A superior view of the right foot. (b) A radiograph of the
right foot. (c) A superior view of the bones of the right foot.

(d) An inferior view of the bones of the right foot. (Note the
presence of a sesamoid bone in (b) at the base of the big toe.)

## Pes (Foot)

The **foot** contains 26 bones constituting the tarsus, metatarsus, and phalanges (figs. 10.18 and 10.19). The bones of the foot are basically like those of the hand. They do, however, have distinct structural differences in order to provide weight support and leverage during walking.

**Tarsus**    There are seven tarsal bones. The **talus** is the tarsal bone that articulates with the tibia and fibula to form the ankle joint. The **calcaneus** (*kal-ka´ne-us*) is the largest of the tarsal bones and provides skeletal support for the heel of the foot. It has a large posterior extension, called the **tuberosity of the calcaneus,** for the attachment of the calf muscles. Anterior to the talus is the block-shaped **navicular** (*nă-vik´yŭ-lar*) **bone.** The remaining four tarsal bones form a distal series that articulate with the metatarsals. They are, from the medial to lateral side, the **medial, intermediate,** and **lateral cuneiform** (*kyoo-ne´ĭ-form*) **bones** and the **cuboid bone.**

**Metatarsus**    Consisting of five bones, the metatarsus forms the skeletal framework for the sole of the foot. The metatarsal bones are numbered from I to V, with the medial, or big toe, side being I. The first metatarsal bone is larger than the others because of its weight-bearing function.

The metatarsal bones each have a **base, shaft** (body), and **head.** The bases of the first, second, and third metatarsal bones articulate proximally with the cuneiform bones. The heads of the metatarsal bones articulate distally with the proximal phalanges. The proximal joints are called *tarsometatarsal joints* and the distal joints are called *metatarsophalangeal joints.* The ball of the foot is formed by the heads of the first two metatarsal bones.

**Phalanges**    The 14 phalanges are the skeletal elements of the toes. As with the fingers of the hand, the phalanges of the toes are arranged in a proximal row, a middle row, and a

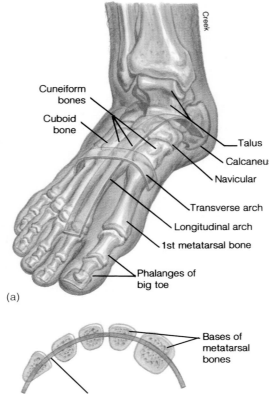

(a)

(b)

Cuneiform bones

Cuboid bone

Talus

Calcaneus

Navicular

Transverse arch

Longitudinal arch

1st metatarsal bone

Phalanges of big toe

Bases of metatarsal bones

Transverse arch

Creek

**FIGURE 10.19**

The arches of the foot. (*a*) A medial view of the right foot showing both arches and (*b*) a transverse view through the bases of the metatarsal bones showing a portion of the transverse arch.

distal row. The great toe (hallux) has only a proximal and a distal phalanx.

**Arches of the Foot**    The foot has two arches that support the weight of the body and provide leverage when walking. These arches are formed by the structure and arrangement of the bones held in place by ligaments and tendons. The arches are not rigid; they yield when weight is placed on the foot and spring back as the weight is lifted (fig. 10.19).

The **longitudinal arch** is divided into medial and lateral parts. The medial, or inner, part is the larger of the two. It is supported by the calcaneus proximally and by the heads of the first three metatarsal bones distally. The wedge, or "keystone," of this part of the longitudinal arch is the talus. The shallower lateral part consists of the calcaneus, cuboid, and fourth and fifth metatarsal bones. The cuboid bone is the "keystone" of this part.

The **transverse arch** extends across the width of the foot and is formed by the distal part of the calcaneus, navicular, and cuboid bones and the proximal portions of all five metatarsal bones.

A weakening of the ligaments and tendons of the foot decreases the height of the longitudinal arch in a condition called *pes planus,* or flatfoot.

A summary of the bones of the lower extremities is presented in table 10.3.

## Clinical Considerations

## Developmental Disorders

Minor defects of the extremities are relatively common malformations. Extra digits, a condition called **polydactyly** (*pol˝e-dak´tĭ-le*) (fig. 10.20), is the most common limb deformity. Usually an extra digit is incompletely formed and does not function. **Syndactyly** (*sin-dak´tĭ-le*), or webbed digits, is

........................................

tarsus: Gk. *tarsos*, flat of the foot
talus: L. *talus*, ankle
calcaneus: L. *calcis*, heel

........................................

polydactyly: Gk. *polys*, many; *daktylos*, finger
syndactyly: Gk. *syn*, together; *daktylos*, finger

likewise a relatively common limb malformation. Polydactyly is inherited as a dominant trait, whereas syndactyly is a recessive trait.

Talipes (*tal´ĭ-pēz*), or clubfoot (fig. 10.21), is a congenital malformation in which the sole of the foot is twisted medially. It is not certain if abnormal positioning or restricted movement in utero causes this condition, but both genetics and environmental factors are involved in most cases.

••••••••••••••••••••••••••••••••••
talipes: L. *talus,* heel; *pes,* foot

## Trauma and Injury

The most common type of bone injury is a **fracture.** A fracture is the cracking or breaking of a bone. Radiographs are often used to diagnose the position and extent of a fracture. Fractures may be classified in several ways, and the type and severity of the fracture varies with the age and the general health of the body. **Spontaneous,** or **pathologic, fractures,** for example, result from diseases that weaken the bones. Most

## Table 10.3  Bones of the pelvic girdle and the lower extremities

| Name and number | Location | Diagnostic features |
|---|---|---|
| Os coxa (2) | Hip, part of the pelvic girdle; composed of three fused bones | Iliac crest, acetabulum, anterior superior iliac spine, ischial tuberosity, obturator foramen |
| Femur (2) | Bone of the thigh, between hip and knee | Head, fovea capitis femoris, neck, greater and lesser trochanters, lateral and medial condyles |
| Patella (2) | Anterior surface of knee | Triangular sesamoid bone |
| Tibia (2) | Medial side of leg, between knee and ankle | Medial and lateral condyles, tibial crest, medial malleolus |
| Fibula (2) | Lateral side of leg, between knee and ankle | Head, lateral malleolus |
| Tarsal bone (14) | Ankle | Large talus and calcaneus to receive weight of leg; five other wedge-shaped bones to help form arches of foot |
| Metatarsal bone (10) | Sole of foot | Long bones, each in line with a digit |
| Phalanx (28) | Digits | Three in each digit; two in big toe |

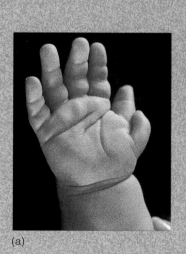
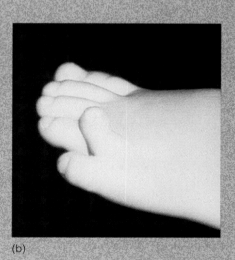
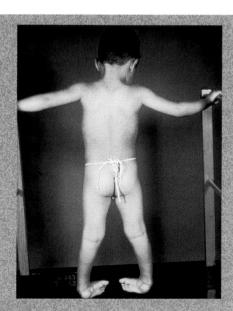

(a)  (b)

### FIGURE 10.20
Polydactyly is the condition in which there are extra digits. It is the most common congenital deformity of the foot, although it also occurs in the hand. Syndactyly is the condition in which two or more digits are webbed together. It is a common congenital deformity of the hand, although it also occurs in the foot. Both conditions can be surgically corrected.

### FIGURE 10.21
Talipes, or clubfoot, is a congenital malformation of a foot or both feet. The condition can be effectively treated surgically if the procedure is done at an early age.

# UNDER DEVELOPMENT

## Development of the Extremities

The development of the upper and lower extremities is initiated toward the end of the fourth week with the appearance of four small elevations called **limb buds** (fig. 1). The anterior pair are the arm buds, which precede the development of the posterior pair of leg buds by a few days. Each limb bud consists of a mass of undifferentiated mesoderm partially covered with a layer of ectoderm. This **apical** (a'pi-kal) **ectodermal ridge** promotes bone and muscle development.

As the limb buds elongate, migrating mesenchymal tissues differentiate into specific cartilaginous bones. Primary ossification centers soon form in each bone, and the hyaline cartilage tissue is gradually replaced by a bony tissue in the process of *endochondral ossification* (see chapter 8).

Initially, the developing limbs are directed caudally, but later there is a lateral rotation in the upper extremity and a medial rotation in the lower extremity. As a result, the elbows are directed backward and the knees directed forward.

**Digital rays** that will form the hands and feet are apparent by the fifth week, and the individual digits separate by the end of the sixth week.

A large number of limb deformities occurred in children born between 1957 and 1962 as a result of mothers ingesting thalidomide during early pregnancy to relieve "morning sickness." It is estimated that 7000 infants were malformed by thalidomide. The malformations ranged from *micromelia* (short limbs) to *amelia* (absence of limbs).

. . . . . . . . . . . .

micromelia: Gk. *mikros*, small; *melos*, limb
amelia: Gk. *a*, without; *melos*, limb

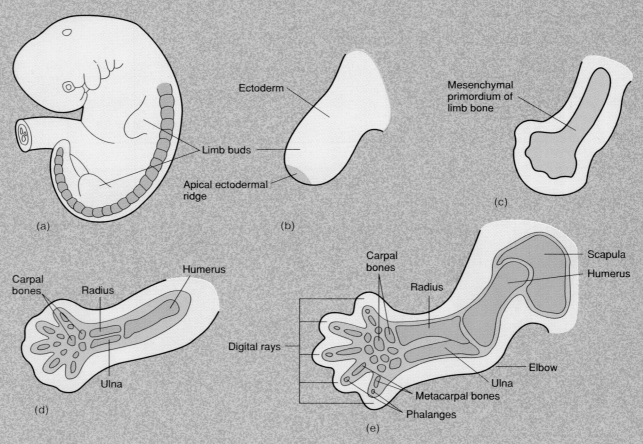

**FIGURE 1**

The development of the appendicular skeleton. (*a*) Limb buds are apparent in an embryo by 28 days and (*b*) an ectodermal ridge is the precursor of the skeletal and muscular structures. (*c*) Mesenchymal primordial cells are present at 33 days. (*d*) Hyaline cartilaginous models of individual bones develop early in the sixth week. (*e*) Later in the sixth week, the cartilaginous skeleton of the upper extremity is well formed.

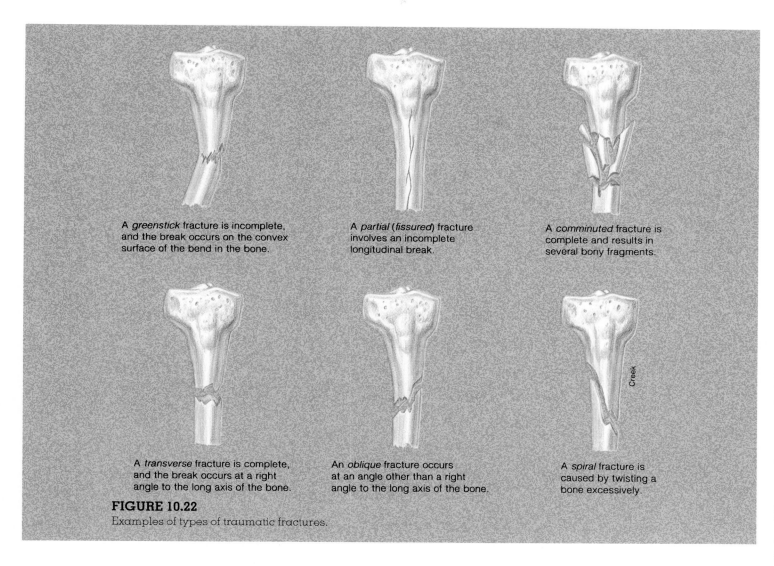

A *greenstick* fracture is incomplete, and the break occurs on the convex surface of the bend in the bone.

A *partial (fissured)* fracture involves an incomplete longitudinal break.

A *comminuted* fracture is complete and results in several bony fragments.

A *transverse* fracture is complete, and the break occurs at a right angle to the long axis of the bone.

An *oblique* fracture occurs at an angle other than a right angle to the long axis of the bone.

A *spiral* fracture is caused by twisting a bone excessively.

**FIGURE 10.22**
Examples of types of traumatic fractures.

fractures, however, are called **traumatic fractures** because they are caused by injuries. The following are descriptions of several kinds of traumatic fractures (fig. 10.22).

**1  Simple,** or **closed.** The fractured bone does not break through the skin.

**2  Compound,** or **open.** The fractured bone is exposed to the outside through an opening in the skin.

**3  Partial (fissured).** The bone is incompletely broken.

**4  Complete.** The fracture has separated the bone into two parts.

**5  Capillary.** A hairlike crack occurs within the bone.

**6  Comminuted** (*kom´ĭ-noot´ed*). The bone is splintered into small fragments.

**7  Spiral.** The fracture line is twisted as it is broken.

**8  Greenstick.** In this incomplete break, one side of the bone is broken and the other side is bowed.

**9  Impacted.** One broken end of a bone is driven into the other.

**10  Transverse.** The fracture occurs across the bone at right angles to the shaft.

**11  Oblique.** The fracture occurs across the bone at an oblique angle to the long axis of the bone.

**12  Colles'.** A fracture of the distal portion of the radius.

**13  Pott's.** A fracture of either or both of the distal ends of the tibia and fibula at the level of the malleoli.

**14  Avulsion.** A portion of a bone is torn off.

**15  Depressed.** The broken portion of the bone is driven inward, as in certain skull fractures.

**16  Displaced.** In this fracture, the bone fragments are not in anatomical alignment.

**17  Nondisplaced.** In this fracture, the bone fragments are in anatomical alignment.

When a bone fractures, medical treatment involves realigning the broken ends and then immobilizing them until new bone tissue is formed and the fracture is healed. The site and severity of the fracture and the age of the patient will determine the type of immobilization. Methods of immobilization include tape, splints, casts, straps, wires, and steel pins. Even with these various methods of treatment, certain fractures heal poorly. Ongoing research into these problems is producing promising results. It has been found, for example, that applying weak electrical currents to fractured bones promotes healing and reduces the time of immobilization by half.

Physicians can realign and immobilize a fracture, but the ultimate repair of the bone occurs naturally within the bone itself. Several steps occur in the repair of a fracture (fig. 10.23).

**1** When a bone is fractured, the surrounding periosteum is usually torn, and blood vessels in both tissues are ruptured. A blood clot called a **fracture hematoma** (*hēm´´ă-to´mă*) soon forms throughout the damaged area. A disrupted blood supply to osteocytes and periosteal cells at the fracture site causes localized cellular death. This is followed by swelling and inflammation.

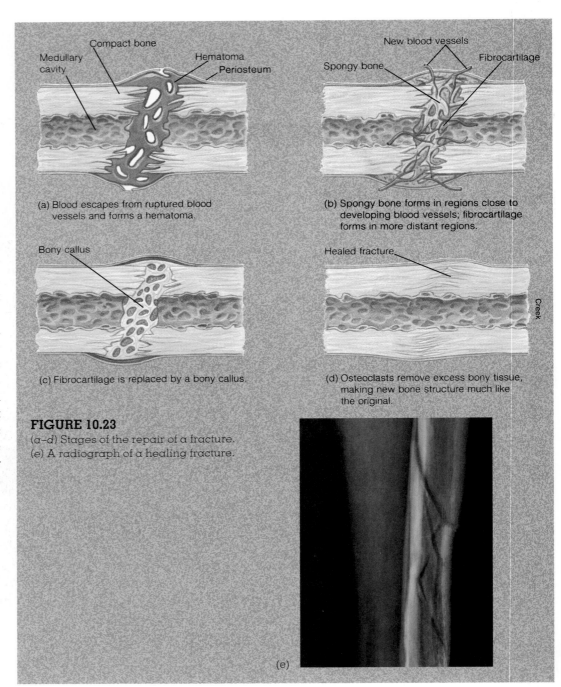

(a) Blood escapes from ruptured blood vessels and forms a hematoma.

Medullary cavity — Compact bone — Hematoma — Periosteum

(b) Spongy bone forms in regions close to developing blood vessels; fibrocartilage forms in more distant regions.

New blood vessels — Spongy bone — Fibrocartilage

(c) Fibrocartilage is replaced by a bony callus.

Bony callus

(d) Osteoclasts remove excess bony tissue, making new bone structure much like the original.

Healed fracture

**FIGURE 10.23**
(a–d) Stages of the repair of a fracture.
(e) A radiograph of a healing fracture.

(e)

Creek

**2** The traumatized area is "cleaned up" by the activity of phagocytic cells within the blood and osteoclasts that resorb bone fragments. As the debris is removed, fibrocartilage fills the gap within the fragmented bone, and a cartilaginous mass, called a **bony callus,** is formed. The bony callus becomes the precursor of bone formation in much the same way that hyaline cartilage is the precursor of developing bone.

**3** The remodeling of the bony callus is the final step in the healing process. The cartilaginous callus is broken down, a new vascular supply is established, and compact bone develops around the periphery of the fracture. A healed fracture line is frequently undetectable in a radiograph, except that the bone in this area is usually slightly thicker.

fracture hematoma: Gk. *hema*, blood; *oma*, tumor
callus: L. *callosus*, hard

# Chapter Summary

## Pectoral Girdle and Upper Extremity (pp. 203–209)

1. The pectoral girdle is composed of two scapulae and two clavicles. The clavicles attach the pectoral girdle to the axial skeleton at the sternum.
   a. Diagnostic features of the clavicle include the conoid tubercle and acromial and sternal extremities.
   b. Diagnostic features of the scapula include the spine, the acromion, and the coracoid process; the supraspinous, infraspinous, and subscapular fossae; the glenoid cavity; superior, medial, and lateral borders; and superior, inferior, and lateral angles.
2. The brachium contains the humerus, which extends from the scapula to the elbow.
   a. Proximally, diagnostic features of the humerus include a rounded head, greater tubercle, anatomical neck, and an intertubercular groove. Distally, they include medial and lateral epicondyles, coronoid and olecranon fossae, a capitulum, and a trochlea.
   b. The head of the humerus articulates proximally with the glenoid cavity of the scapula; distally, the trochlea and capitulum articulate with the ulna and radius, respectively.
3. The antebrachium contains the medial ulna and the lateral radius.
   a. Proximally, diagnostic features of the ulna include the olecranon, coronoid process, and trochlear notch. Distally, they include the styloid process and head of the ulna.
   b. Proximally, diagnostic features of the radius include the head and neck of the radius and the tuberosity of the radius. Distally, they include the styloid process and ulnar notch.
4. The hand contains 27 bones arranged as the carpal bones, metacarpal bones, and phalanges.

## Pelvic Girdle and Lower Extremity (pp. 209–216)

1. The pelvic girdle is formed by two ossa coxae united anteriorly by the symphysis pubis.
2. The pelvis is divided into a greater pelvis, which helps to support the pelvic viscera, and a lesser pelvis, which forms the walls of the birth canal.
3. Each os coxa consists of an ilium, ischium, and pubis. Diagnostic features of the os coxa include an obturator foramen and an acetabulum, the latter of which is the socket for articulation with the head of the femur.
   a. Diagnostic features of the ilium include an iliac crest, iliac fossa, anterior superior iliac spine, and anterior inferior iliac spine.
   b. Diagnostic features of the ischium include the body, ramus, and ischial tuberosity.
   c. Diagnostic features of the pubis include the ramus and pubic tubercle. The two pubic bones articulate at the symphysis pubis.
4. The thigh contains the femur, which extends from the hip to the knee where it articulates with the tibia and the patella.
   a. Proximally, diagnostic features of the femur include the head, neck, and greater and lesser trochanters. Distally, they include the lateral and medial epicondyles, the lateral and medial condyles, and the patellar surface. The linea aspera is a roughened ridge positioned vertically along the posterior aspect of the shaft (body) of the femur.
   b. The head of the femur articulates proximally with the acetabulum of the os coxa and distally with the condyles of the tibia and the articular surfaces of the patella.
5. The leg contains the medial tibia and the lateral fibula.
   a. Diagnostic features of the tibia include the lateral and medial epicondyles, intercondylar eminence, and tibial tuberosity proximally and the medial malleolus distally. The anterior crest is a sharp ridge extending the anterior length of the tibia.
   b. Diagnostic features of the fibula include the head proximally and the lateral malleolus distally.
6. The foot contains 26 bones arranged as the tarsal bones, metatarsal bones, and phalanges.

# Review Activities

## Objective Questions

1. When in anatomical position, the subscapular fossa of the scapula faces
   a. anteriorly.      c. posteriorly.
   b. medially.        d. laterally.

2. The clavicle articulates with
   a. the scapula and humerus.
   b. the humerus and manubrium.
   c. the manubrium and scapula.
   d. the manubrium, scapula, and humerus.

3. Which of the following bones has a conoid tubercle?
   a. scapula          d. clavicle
   b. humerus          e. ulna
   c. radius

4. The "elbow" of the ulna is formed by
   a. the lateral epicondyle.
   b. the olecranon.
   c. the coronoid process.
   d. the styloid process.
   e. the medial epicondyle.

5. Which of the following statements concerning the carpus is *false?*
   a. There are eight carpal bones arranged in two transverse rows of four bones each.
   b. All of the carpal bones are considered sesamoid bones.
   c. The scaphoid and the lunate bones articulate with the radius.
   d. The trapezium, trapezoid, capitate, and hamate bones articulate with the metacarpal bones.

6. In pelvimetry, which of the following is measured?
   a. os coxa
   b. symphysis pubis
   c. pelvic brim
   d. lesser pelvis

7. Which of the following is *not* a structural feature of the os coxa?
   a. obturator foramen
   b. acetabulum
   c. auricular surface
   d. greater sciatic notch
   e. linea aspera

8. A fracture across the intertrochanteric line would involve which bone?
   a. ilium            d. fibula
   b. femur            e. patella
   c. tibia

9. As compared to the male pelvis, the female pelvis
   a. is more massive.
   b. is narrower at the pelvic outlet.
   c. is tilted backward.
   d. has a shallower symphysis pubis.

10. Clubfoot is a congenital malformation that is medically referred to as
    a. talipes.         c. pes planus.
    b. syndactyly.      d. polydactyly.

## Essay Questions

1. Explain the significance of the limb buds, apical ectodermal ridges, and digital rays in limb development. When does limb development begin and when is it completed?

2. Compare the pectoral and pelvic girdles in structure, articulation to the axial skeleton, and function.

3. Explain why the clavicle is more frequently fractured than the scapula.

4. List the processes of the bones of the upper and lower extremities that can be palpated. Why is it important to be able to recognize these bony landmarks?

5. There are basic similarities and specific differences between the bones of the hands and those of the feet. Compare and contrast these appendages, taking into account the functional role of each.

6. Define *bipedal locomotion* and discuss the adaptations of the pelvic girdle and lower extremities that permit this type of movement.

7. What are the structural differences between male and female pelves?

8. What is meant by a congenital skeletal malformation? Give two examples of such abnormalities that occur within the appendicular skeleton.

9. How do spontaneous and traumatic fractures differ? Give some examples of traumatic fractures.

10. How does a fractured bone repair itself? Why is it important that the fracture be immobilized?

## Gundy/Weber Software

The tutorial software accompanying Chapter 10 is Volume 3—Skeletal System.

# articulations

## objectives

- Define *arthrology* and *kinesiology*.

- Compare and contrast the three principal kinds of joints.

- Describe the structure of a suture and indicate where sutures are located.

- Describe the structure of a syndesmosis and indicate where syndesmoses are located.

- Describe the structure and note the location of gomphoses. Also, discuss the importance of these joints to the profession of dentistry.

- Describe the structure of a symphysis and indicate where symphyses occur.

- Describe the structure of a synchondrosis and indicate where synchondroses occur.

- Describe the structure of a synovial joint.

- Discuss the various kinds of synovial joints, noting where they occur and the movements they permit.

- List and discuss the various kinds of movements that are possible at synovial joints.

- Describe the components of a lever and explain the role of synovial joints in lever systems.

- Compare the structure of first-, second-, and third-class levers.

- Describe the structure, function, and possible clinical importance of the following joints: temporomandibular, humeral, elbow, metacarpophalangeal, interphalangeal, coxal, tibiofemoral, and talocrural.

# Classification of Joints

*On the basis of anatomical structure, the articulations between the bones of the skeleton are classified as fibrous joints, cartilaginous joints, or synovial joints. Fibrous joints firmly join skeletal elements with fibrous connective tissue. Cartilaginous joints firmly join skeletal elements with cartilage. Synovial joints are freely movable joints enclosed by joint capsules that contain synovial fluid.*

One of the functions of the skeletal system is to permit body movement. It is not the rigid bones that allow movement but the **articulations,** or **joints,** between the bones. The structure of a joint determines the range of movement it permits. Not all joints are flexible, however, and as one part of the body moves, other joints remain rigid to stabilize the body and maintain balance. The coordinated activity of all of the joints permits the sinuous, elegant movements of a gymnast or ballet dancer, just as it permits all of the commonplace actions associated with walking, eating, writing, and speaking.

**Arthrology** is the science concerned with the study of joints. Generally speaking, an arthrologist is interested in structure, classification, and function of joints, including any dysfunctions that may develop. **Kinesiology** (kĭ-ne´se-ol´ŏ-je) is a more practical and dynamic science concerned with the functional relationship, or biomechanics, of the skeleton, joints, muscles, and innervation as they work together to produce coordinated movement. In a kinetic approach to studying the joints, the various movements permitted at each of the movable joints are demonstrated, with an emphasis on understanding the adaptive advantage as well as the limitations, of each type of movement.

The joints of the body may be classified according to structure or function. Structural classification is based on the presence or absence of a joint cavity and the kind of supportive connective tissue surrounding the joint. In classification by structure, the three types of joints are as follows:

**1** **Fibrous joints.** A fibrous joint lacks a joint cavity, and fibrous connective tissue connects articulating bones.

**2** **Cartilaginous joints.** A cartilaginous joint lacks a joint cavity, and cartilage binds articulating bones.

**3** **Synovial** (sĭ-no´ve-al) **joints.** A synovial joint has a joint cavity, and ligaments help to support the articulating bones.

The functional classification of joints is based on the degree of movement permitted within the joint. Using this type of classification, the three kinds of articulations are as follows:

**1** **Synarthroses** (sin´ar-thro´sēz). Immovable joints.

**2** **Amphiarthroses.** Slightly movable joints.

**3** **Diarthroses** (di˝ar-thro´sēz). Freely movable joints.

In previous editions of this text, the functional classification scheme was used in discussing the various kinds of joints of the body. In accordance with the structural classification of joints presented in the sixth edition of *Nomina Anatomica,* we have revised this chapter using a structural classification of joints.

# Fibrous Joints

*Articulating bones in fibrous joints are tightly bound by fibrous connective tissue. These range from slightly movable joints to joints that are rigid and relatively immovable. Fibrous joints are of three types: sutures, syndesmoses, and gomphoses.*

## Sutures

**Sutures,** one type of fibrous joint, are found only between the flat bones of the skull and are characterized by a thin layer of dense regular connective tissue that binds the articulating bones (fig. 11.1). Sutures form at about 18 months of age and replace the pliable fontanels of an infant's skull (see fig. 2, p. 193).

Sutures can be distinguished on the basis of the appearance of the articulating margin of bone. A **serrate suture** is characterized by interlocking sawlike articulations. This is the most common type of suture, an example being the sagittal suture between the two parietal bones. In a **squamous suture,** the margin of one bone overlaps that of the articulating bone. The squamous suture formed between the temporal and parietal bones is an example (see fig. 9.2). In a **plane suture,** the margins of the articulating bones are fairly smooth. An example is the median palatine suture, where the paired maxillary and palatine bones articulate to form the hard palate (see fig. 9.3).

---

arthrology: Gk. *arthron,* joint; *logos,* study
kinesiology: Gk. *kinesis,* movement; *logos,* study

---

suture: L. *sutura,* sew

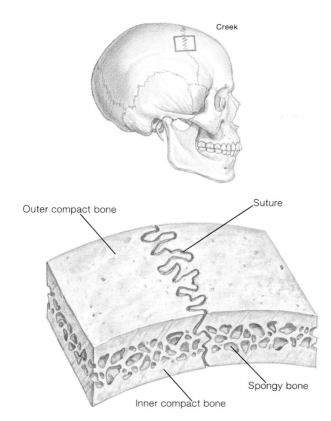

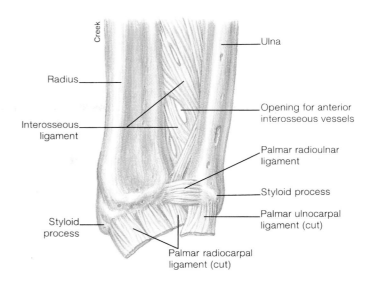

**FIGURE 11.2**
The side-to-side articulation of the ulna and radius forms a syndesmotic joint. An interosseous ligament tightly binds these bones and permits only slight movement between them.

**FIGURE 11.1**
A section across the skull showing a suture.

A **synostosis** (*sin˝os-to´sis*) is a unique sutural joint. It is present during growth of the skull, but in the adult the suture becomes totally ossified. For example, the frontal bone forms as two separate components but the separation becomes obscured in most individuals as the skull completes its growth.

 Fractures of the skull are fairly common in an adult but much less so in a child. The skull of a child is resilient to blows because of the nature of the bone and the layer of fibrous connective tissue within the sutures. The skull of an adult is much like an eggshell in its lack of resilience, and will frequently splinter on impact.

The nomenclature in human anatomy is extensive and precise. There are over 30 named sutures in the skull even though just a few of them are mentioned by name in figures 9.2, 9.3, and 9.4. Review these illustrations and make note of the bones that articulate to form the sutures identified in these figures.

## Syndesmoses

**Syndesmoses** (*sin˝des-mo´sēz*) are fibrous joints found only in the antebrachium (forearm) and leg where adjacent bones are held together by collagenous fibers or *interosseous ligaments*. A syndesmosis is characteristic of the side-to-side joints between the tibia-fibula and the radius-ulna (fig. 11.2). Slight movement is permitted at these joints as the antebrachium or leg is rotated.

## Gomphoses

**Gomphoses** (*gom-fo´sēz*) are fibrous joints that occur between the teeth and the supporting bones of the jaws. More specifically, a gomphosis is located where the root of a tooth is attached to the periodontal ligament of the alveolus (socket) of the bone (see fig. 26.11).

 *Periodontal disease* occurs at gomphoses and is the inflammation and degeneration of the gum, periodontal ligaments, and alveolar bone tissue. With this condition, the teeth become loose and plaque accumulates on the roots. Periodontal disease may be caused by poor oral hygiene, compacted teeth (poor alignment), or local irritants, such as impacted food, chewing tobacco, or cigarette smoke.

• • • • • • • • • • • • • • • • • • • • • • • • • • • • • •

syndesmosis: Gk. *syndesmos*, binding together
gomphosis: Gk. *gompho*, nail or bolt

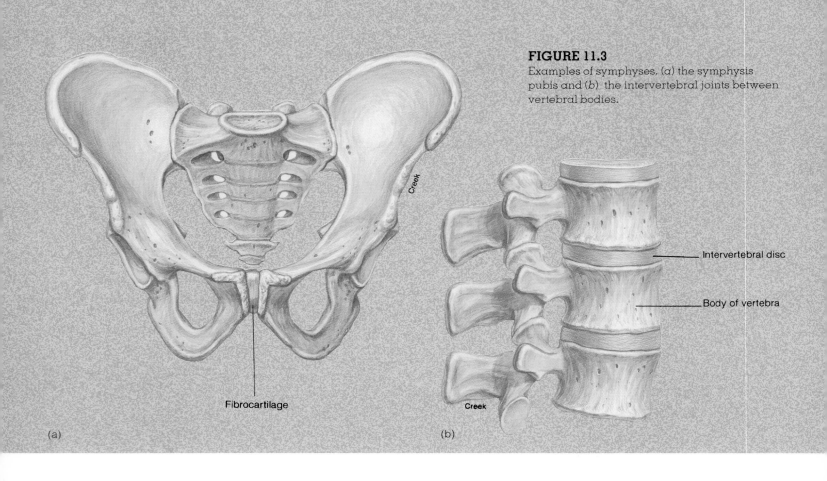

**FIGURE 11.3**

Examples of symphyses. (a) the symphysis pubis and (b) the intervertebral joints between vertebral bodies.

Intervertebral disc

Body of vertebra

Fibrocartilage

Creek

(a)

(b)

# Cartilaginous Joints

*Cartilaginous joints allow limited movement in response to twisting, compression, or stress. The two types of cartilaginous joints are symphyses and synchondroses.*

## Symphyses

The adjoining bones of a **symphysis** (*sim´fĭ-sis*) joint are separated by a pad of fibrocartilage. This pad cushions the joint and allows limited movement. The symphysis pubis and the intervertebral discs (fig. 11.3) are examples of symphyses. Although only limited motion is possible at each intervertebral joint, the combined movement of all the joints of the vertebral column results in extensive movement of the structure as a whole.

## Synchondroses

**Synchondroses** (*sin˝kon-dro´sēz*) are cartilaginous joints that have hyaline cartilage between the bone segments. Some of these joints are temporary, forming the epiphyseal plates (growth lines) between the diaphyses and epiphyses in the long bones of children. When growth is complete,

these synchondrotic joints ossify. A totally ossified synchondrosis may also be referred to as a *synostosis*. Synchondroses can be clearly seen in the radiograph of a long bone of a child in figure 11.4.

 A fracture of a long bone in a child may be extremely serious if it involves the mitotically active epiphyseal plate of a synchondrotic joint. If such an injury is not treated, there likely will be retarded or arrested bone growth, so that the appendage will be shorter than normal.

Synchondroses that do not ossify as a person ages are those between the occipital, sphenoid, temporal, and ethmoid bones of the skull. In addition, the costochondral articulations between the ribs and the sternum are examples of synchondroses. It is interesting that elderly people often exhibit some ossification of costal cartilages of the rib cage. This may restrict movement of the rib cage and obscure an image of the lungs in a thoracic radiograph.

symphysis: Gk. *symphysis*, growing together
synchondrosis: Gk. *syn*, together; *chondros*, cartilage
synostosis: Gk. *syn*, together; *osteon*, bone

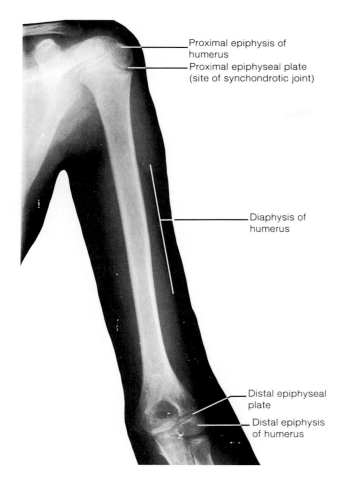

Proximal epiphysis of humerus

Proximal epiphyseal plate (site of synchondrotic joint)

Diaphysis of humerus

Distal epiphyseal plate

Distal epiphysis of humerus

**FIGURE 11.4**
A radiograph of the left humerus of a 10-year-old child showing a synchondrotic joint. In a long bone, this type of joint occurs at both the proximal and distal epiphyseal plates. The mitotic activity at synchondrotic joints is responsible for bone growth in length.

# Synovial Joints

*Synovial joints are freely movable joints enclosed by joint capsules that contain synovial fluid. Based on the shape of the articular surfaces and the kinds of motion they permit, synovial joints are categorized as gliding, hinge, pivot, condyloid, saddle, or ball-and-socket.*

The most obvious type of articulation in the body is the freely movable synovial joint. The function of synovial joints is to provide a wide range of precise, smooth movements, at the same time maintaining stability, strength, and, in certain aspects, rigidity in the body.

Synovial joints are the most complex and varied of the three major types of joints. The range of movement of a synovial joint is limited by three factors: (1) the structure of the bones participating in the articulation (for example, the olecranon of the ulna actually limits the range of motion and "locks" the articulation to prevent overextension of the joint); (2) the strength and tautness of the associated ligaments, tendons, and joint capsule; and (3) the size, arrangement, and action of the muscles that span the joint. Joint motility is characterized by tremendous individual variation, most of which is related to body conditioning. "Double-jointed" is a misnomer because such a joint is not double, although it does permit extreme maneuverability.

 *Arthroplasty* is the surgical repair or replacement of joints. Advancements in arthroplastic procedures continue as devices are sought to provide patients movement capabilities free of pain. A recent advancement in the repair of soft tissues involves the use of *artificial ligaments.* A material consisting of carbon fibers coated with a plastic called polylactic acid is sewn in and around torn ligaments and tendons. This reinforces the traumatized structures while providing a scaffolding on which the body's collagenous fibers can grow. As the healing process continues, the polylactic acid is eventually absorbed and the carbon fibers break down.

## Structure of a Synovial Joint

Synovial joints are enclosed by a fibroelastic **joint capsule** that is filled with lubricating **synovial fluid,** or **synovium** (fig. 11.5). Synovial fluid is secreted by a thin **synovial membrane** that lines the inside of the capsule. Synovial fluid is similar to interstitial fluid (fluid surrounding cells of a tissue) and has a high concentration of hyaluronic acid—a lubricating substance. The bones that articulate in a synovial joint are capped with a smooth **articular cartilage.** The avascular articular cartilage is only about 2 mm thick and depends on the alternating compression and decompression during joint activity for the exchange of nutrients and waste products with the synovial fluid. Ligaments help to bind a synovial joint and may be located within the joint cavity or on the outside of the capsule. Tough, fibrous cartilaginous pads called **menisci**—singular, *meniscus* (mě-nis´kus)—are located within the capsule of certain synovial joints (e.g., the knee joint) and serve to cushion, as well as to guide, the articulating bones.

.........................................

arthroplasty: Gk. *arthron,* joint; *plasso,* to form
meniscus: Gk. *meniskos,* small moon

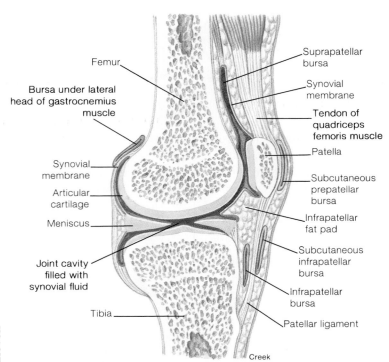

**FIGURE 11.5**

A synovial joint is represented by a lateral view of the knee joint.

 Many people are concerned about the cracking sounds they hear as joints move, or the popping sounds that result from "popping" or "cracking" the knuckles by forcefully pulling on the fingers. These sounds are actually quite normal. When a synovial joint is pulled upon, its volume is suddenly expanded and the pressure of the joint fluid is decreased, causing a partial vacuum within the joint. As the joint fluid is displaced within a vacuum and hits against the articular cartilage, a popping or cracking sound is heard. Similarly, displaced water in a sealed vacuum tube makes the same sound as it hits against the glass wall. Popping one's knuckles does not cause arthritis, but it can lower one's social standing.

Articulating bones of synovial joints do not come in contact with one another. Articular cartilage caps the articular surface of each bone and synovial fluid circulates through the joint during movement. Both of these joint structures serve to minimize friction and cushion the articulating bones. Should trauma or disease render either of these two joint structures nonfunctional, the two articulating bones will come in contact. Bony deposits will then form, and a type of arthritis will develop within the joint.

Located near certain synovial joints are flattened, pouchlike sacs called **bursae**—singular, *bursa (bur´să)*—which are filled with synovial fluid. (fig. 11.6). These closed sacs are commonly located between muscles or within an area where a tendon passes over a bone. The functions of bursae are to cushion certain muscles and to facilitate the movement of tendons or muscles over bony or ligamentous surfaces. A **tendon sheath** is a modified bursa that surrounds and lubricates the tendons of certain muscles, particularly those that cross the wrist and ankle joints.

 Improperly fitted shoes or inappropriate shoes can cause joint-related problems. People who wear high-heeled shoes often have perpetual backaches and leg aches because their posture has to counteract the forward tilt of their bodies when standing or walking. Their knees are excessively bent and the spine is thrust forward at the lumbar curvature in order to maintain balance. Tightly fitted shoes, especially ones with pointed toes, may result in the development of *hallux valgus*. This condition is a deviation of the hallux (big toe) laterally toward the other toes. Hallux valgus is generally accompanied by the formation of a *bunion* at the medial base of the proximal phalanx of the hallux. A bunion is an inflammation and an accompanying callus that develops because of pressure and rubbing of a shoe.

## Kinds of Synovial Joints

Synovial joints are classified into six main types according to the shape of their articular surfaces and the kinds of motion they permit. The six types are gliding, hinge, pivot, condyloid, saddle, and ball-and-socket.

**Gliding** Gliding joints allow only side-to-side and back-and-forth movements with some slight rotation. This is the simplest type of joint movement. The articulating surfaces can be nearly flat, or one may be slightly concave and the other slightly convex (fig. 11.7). The intercarpal and intertarsal joints, the sternoclavicular joint, and the joint between the articular process of adjacent vertebrae are examples.

**Hinge** The structure of a hinge joint permits bending in only one plane, much like the hinge of a door. In this type of articulation, the surface of one bone is always concave and the other, convex (fig. 11.8). Hinge joints are the most common type of synovial joint and include such specific joints as the knee, the humeroulnar articulation within the elbow, and the joints between the phalanges.

..............................................
bursa: Gk. *byrsa*, bag or purse

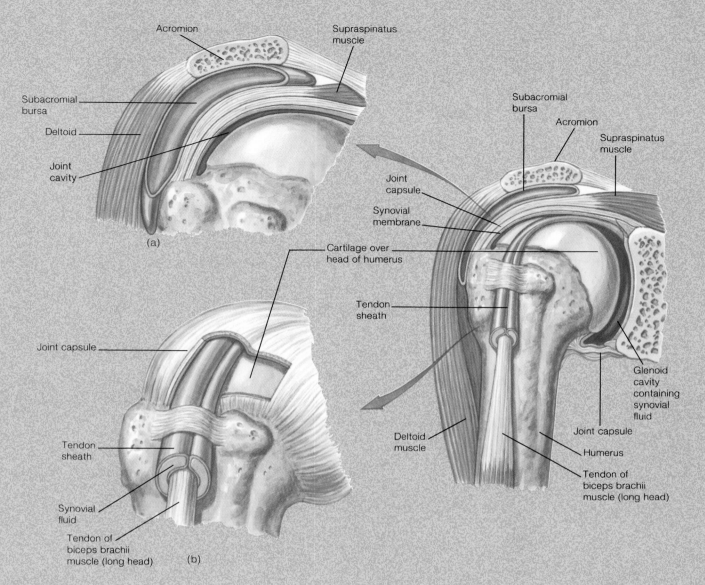

**Acromion**

**Supraspinatus muscle**

Subacromial bursa

Deltoid

Joint cavity

(a)

Joint capsule

Tendon sheath

Synovial fluid

Tendon of biceps brachii muscle (long head)

(b)

Subacromial bursa

Acromion

Supraspinatus muscle

Joint capsule

Synovial membrane

Cartilage over head of humerus

Tendon sheath

Glenoid cavity containing synovial fluid

Deltoid muscle

Joint capsule

Humerus

Tendon of biceps brachii muscle (long head)

**FIGURE 11.6**

Bursae and tendon sheaths are friction-reducing structures found in conjunction with synovial joints. (*a*) A bursa is a closed sac filled with synovial fluid. Bursae are commonly located between muscles or between tendons and articular capsules. (*b*) A tendon sheath encapsulates a tendon through a joint capsule. It is a closed, double-layered sac filled with synovial fluid. One layer of the sac is in contact with the tendon and the other layer is in contact with the joint capsule.

**Pivot**   The movement in a pivot joint is limited to rotation about a central axis. In this type of articulation, the articular surface on one bone is conical or rounded and fits into a depression on another bone (fig. 11.9). Examples are the proximal articulation of the radius and ulna for rotation of the forearm, as in turning a doorknob, and the articulation between the atlas and axis that enables rotational movement of the head.

**Condyloid**   A condyloid articulation is structured so that an oval, convex articular surface of one bone fits into an elliptical, concave depression on another bone (see fig. 11.7). This permits angular movement in two directions (biaxial), as in an up-and-down and side-to-side motion. A condyloid joint does not permit rotational movement. The radiocarpal joint of the wrist is an example.

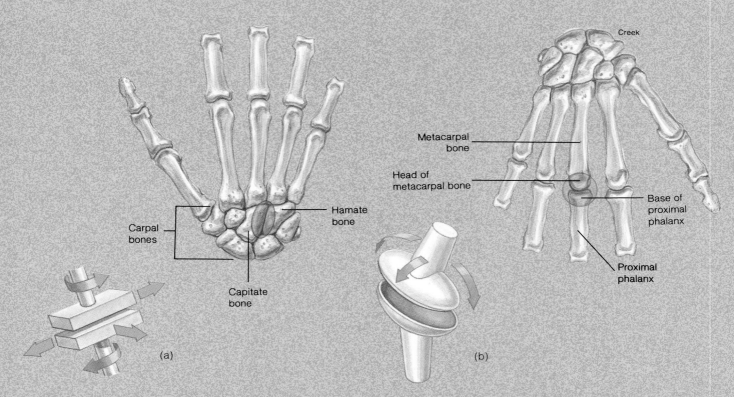

**FIGURE 11.7**

The locations of both gliding and condyloid joints. (a) Gliding joints are located between the proximal and distal carpals and between the individual carpals. (b) Condyloid joints are located between the metacarpal bones and the phalanges. (Note the diagrammatic representation of these joints showing the directions of possible movements.)

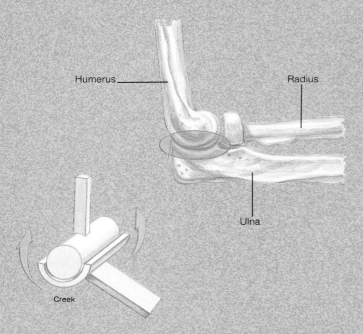

**FIGURE 11.8**

A hinge joint permits only a bending movement (flexion and extension). The hinge joint of the elbow involves the distal end of the humerus articulating with the proximal end of the ulna. (Note the diagrammatic representation of this joint showing the direction of possible movement.)

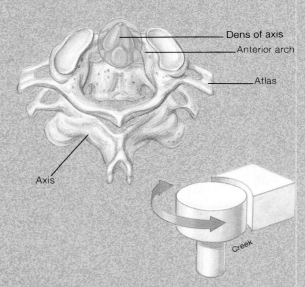

**FIGURE 11.9**

The atlas articulating with the axis forms a pivot joint that permits a rotational movement in one axis. (Note the diagrammatic representation showing the direction of possible movement.) Refer to figure 11.8 and determine which articulating bones of the elbow region form a pivot joint.

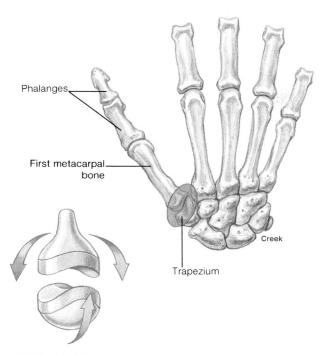

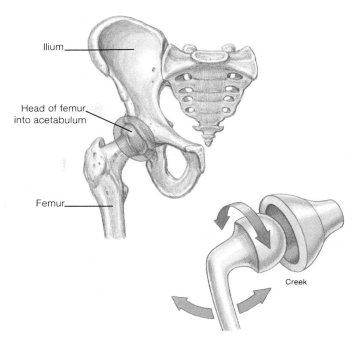

**FIGURE 11.10**

A saddle joint is formed as the trapezium articulates with the base of the first metacarpal bone. (Note the diagrammatic representation of this joint.)

**FIGURE 11.11**

A ball-and-socket articulation illustrated by the hip joint. (Note the diagrammatic representation showing the directions of possible movement.)

**Saddle** Each articular process of a saddle-shaped joint has a concave surface in one direction and a convex surface in another. This unique articulation is a modified condyloid joint that allows a wide range of movement. It is associated only with the thumb and is located at the articulation of the trapezium of the carpus with the first metacarpal bone (fig. 11.10).

**Ball-and-Socket** Ball-and-socket joints are formed by the articulation of a rounded convex surface with a cuplike cavity (fig. 11.11). This type of articulation provides the greatest range of movement of all the synovial joints. Examples are the hip and shoulder joints.

A summary of the various types of joints is presented in table 11.1.

 Synovial fluid within synovial joints lubricates the articular surfaces and provides nourishment to the articular cartilages. Trauma to the joint causes the excessive production of synovial fluid in an attempt to cushion and immobilize the joint. This leads to swelling and discomfort to the joint. The most frequent type of joint injury is a *sprain*, in which the supporting ligaments or the joint capsule are damaged to varying degrees.

# Movements at Synovial Joints

*Movements at synovial joints are produced by the contraction of skeletal muscles that span the joints and attach to or near the bones forming the articulations. In these actions, the bones act as levers, the muscles provide the force, and the joints are the fulcra, or pivots.*

As previously mentioned, the range of movement at a synovial joint is determined by the structure of the individual joint and the arrangement of the associated muscle and bone. The movement at a hinge joint, for example, occurs in only one plane, whereas the structure of a ball-and-socket joint permits movement around many axes. Joint movements are broadly classified as *angular* and *circular*. Within each of these categories are specific types of movements, and certain special movements may involve several of the specific types.

# Table 11.1　Types of articulations

| Type | Structure | Movements | Example |
|---|---|---|---|
| **Fibrous joints** | Skeletal elements joined by fibrous connective tissue | | |
| 1. Suture | Edges of articulating bones frequently serrated; separated by thin layer of fibrous tissue | None | Sutures between bones of the skull |
| 2. Syndesmoses | Articulating bones bound by interosseous ligament | Slightly movable | Joints between tibia-fibula and radius-ulna |
| 3. Gomphoses | Teeth bound into alveoli of bone by periodontal ligament | None | Teeth secured into alveoli (sockets) |
| **Cartilaginous joints** | Skeletal elements joined by fibrocartilage or hyaline cartilage | | |
| 1. Symphyses | Articulating bones separated by pad of fibrocartilage | Slightly movable | Intervertebral joints; symphysis pubis and sacroiliac joint |
| 2. Synchondroses | Mitotically active hyaline cartilage between skeletal elements | None | Epiphyseal plates within long bones |
| **Synovial joints** | Joint capsule containing synovial membrane and synovial fluid | | |
| 1. Gliding | Flattened or slightly curved articulating surfaces | Sliding | Intercarpal and intertarsal joints |
| 2. Hinge | Concave surface of one bone articulates with convex surface of another | Bending motion in one plane | Knee; elbow; joints of phalanges |
| 3. Pivot | Conical surface of one bone articulates with a depression of another | Rotation about a central axis | Atlantoaxial joint; proximal radioulnar joint |
| 4. Condyloid | Oval condyle of one bone articulates with elliptical cavity of another | Biaxial movement | Radiocarpal joint |
| 5. Saddle | Concave and convex surface on each articulating bone | Wide range of movements | Carpometacarpal joint of thumb |
| 6. Ball-and-socket | Rounded convex surface of one bone articulates with cuplike socket of another | Movement in all planes and rotation | Shoulder and hip joints |

## Angular

**Angular movements** increase or decrease the joint angle produced by the articulating bones. The four types of angular movements are flexion, extension, abduction, and adduction.

**Flexion**　Flexion (*flek´shun*) is a movement that decreases the joint angle on an anterior-posterior plane (fig. 11.12). Examples of flexion are the bending of the elbow or knee. Flexion of the elbow joint is a forward movement, whereas flexion of the knee is a backward movement. Flexion in most joints is simple to understand, such as flexion

of the head as it bends forward or the flexing of a digit, but flexion of the ankle and shoulder joints needs further explanation. In the ankle joint, flexion occurs as the dorsum of the foot is elevated. This movement is frequently call **dorsiflexion** (fig. 11.13). Pressing the foot downward (as in rising on the toes) is **plantar flexion.** The shoulder joint is flexed when the arm is brought forward, thus decreasing the joint angle.

**Extension**　In extension (*ik-sten´shun*), which is the reverse of flexion, the joint angle is increased (fig. 11.12). Extension returns a body part to the anatomical position.

flexion: L. *flectere,* to bend

extension: L. *ex,* out, away from; *tendere,* stretch

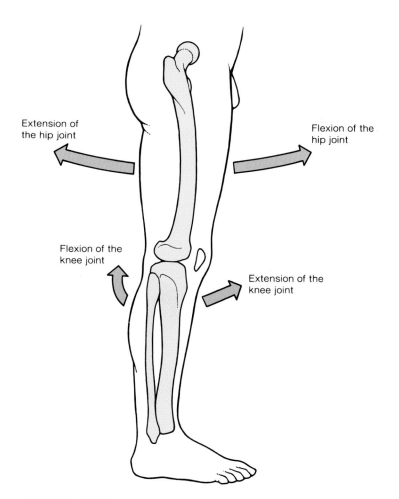

**FIGURE 11.12**
Examples of flexion and extension. Contraction of the posterior thigh muscles with the femur held rigid results in flexion at the knee joint as the leg is moved posteriorly. Contraction of the anterior hip and thigh muscles with the thigh and leg held rigid results in flexion at the hip joint as the lower extremity is moved anteriorly. Contraction of the posterior hip and thigh muscles with the thigh and leg held rigid results in extension at the hip joint as the lower extremity is moved posteriorly.

In an extended joint, the angle between the articulating bones is 180°. The exception to this is the ankle joint, in which there is a 90° angle between the foot and the lower leg in the anatomical position. Examples of extension are the straightening of the elbow or knee joints from flexion positions. **Hyperextension** occurs when a part of the body is extended beyond the anatomical position so that the joint angle is greater than 180°. An example of hyperextension is bending the head backward.

 A common injury in runners is *patellofemoral stress syndrome*, commonly called "runner's knee." This condition is characterized by tenderness and aching pain around or under the patella. During normal knee movement, the patella glides up and down the patellar groove between the femoral condyles. In patellofemoral stress

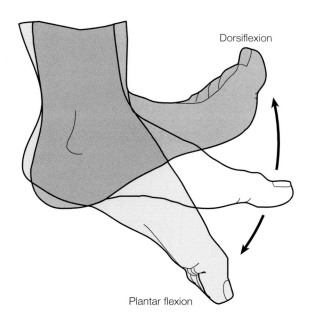

Dorsiflexion

Plantar flexion

**FIGURE 11.13**
Dorsiflexion and plantar flexion at the ankle joint.

syndrome, the patella rubs laterally, causing irritation to the membranes and articular cartilage within the knee joint. Joggers frequently experience this condition from prolonged running on the slope of a road near the curb rather than on a flat surface.

**Abduction**    Abduction is the movement of a body part away from the main axis of the body, or away from the midsagittal plane, in a lateral direction (fig. 11.14). This term usually applies to the arm or leg but can also apply to the fingers or toes, in which case the line of reference is the longitudinal axis of the limb. Examples of abduction are moving the arms sideward and away from the body or spreading the fingers apart.

**Adduction**    Adduction, the opposite of abduction, is the movement of a body part toward the main axis of the body (fig. 11.14). In the anatomical position, the arms and legs have been adducted toward the midplane of the body.

## *Circular*

Joints that permit **circular movement** are composed of a bone with a rounded or oval surface that articulates with a corresponding cup or depression on another bone. The two basic types of circular movements are rotation and circumduction.

......................................

abduction: L. *abducere*, lead away
adduction: L. *adductus*, bring to

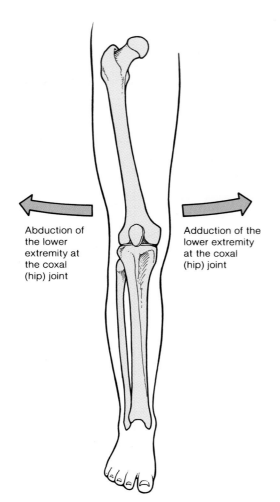

**FIGURE 11.14**
Abduction and adduction of the lower extremity at the coxal (hip) joint reference to the main axis of the body.

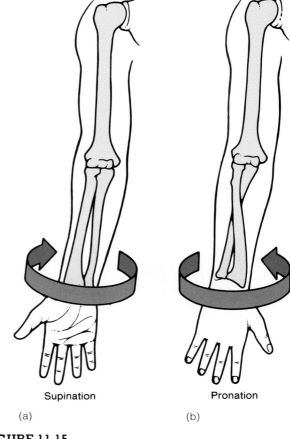

**FIGURE 11.15**
Supination (*a*) and pronation (*b*) of the hand. (Note the relative position of the ulna and radius in both positions.) Pronation requires medial rotation of the forearm relative to the anatomical position.

**Rotation** Rotation is the movement of a body part around its own axis (see fig. 11.9). There is no lateral displacement during this movement. Examples are turning the head from side to side in a "no" motion and moving the forearm from a palm-up position to a palm-down position.

    **Supination** is a specialized rotation of the forearm that results in the palm of the hand being turned forward (anteriorly). In the supine position, the ulna and radius of the forearm are parallel and in the anatomical posi-

rotation: L. *rotare,* a wheel

**FIGURE 11.16**
Circumduction of the shoulder joint.

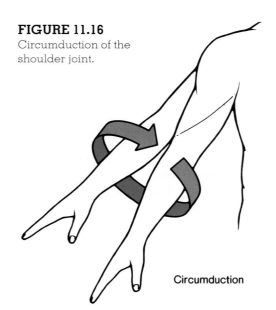

Circumduction

tion. **Pronation** is the opposite of supination (fig. 11.15). It is a rotational movement of the forearm that results in the palm of the hand being directed backward (posteriorly).

**Circumduction** Circumduction is the circular, conelike movement of a body part. The distal extremity forms the circular movement and the proximal attachment forms the pivot (fig. 11.16). This type of motion is possible at the shoulder, wrist, trunk, hip, and ankle joints.

## Special Movements

Because the terms used to describe generalized movements around axes do not apply to the structure of certain joints, other terms must be used to describe the motion of such joints.

**Inversion**   Inversion is the movement of the sole of the foot inward or medially (fig. 11.17). The pivot axes are at the ankle and intertarsal joints.

**Eversion**   Eversion is the opposite of inversion and is the movement of the sole of the foot outward or laterally (fig. 11.17). Both inversion and eversion are clinical terms usually used to describe developmental abnormalities.

**Protraction**   Protraction is the movement of part of the body forward on a plane parallel to the ground. Examples are thrusting out the lower jaw (fig. 11.18) or movement of the shoulder and upper extremity forward.

**Retraction**   Retraction is the pulling back of a protracted part of the body on a plane parallel to the ground (fig. 11.18). Retraction of the mandible brings the lower jaw back in alignment with the upper jaw so that the teeth occlude.

**Elevation**   Elevation is a movement that results in a part of the body being lifted upward. Examples of elevation include elevating the mandible to close the mouth (fig. 11.18) or lifting the scapula to shrug the shoulder.

**Depression**   Depression is the opposite of elevation. Both the mandible (fig. 11.18) and shoulders are depressed when moved downward.

A visual summary of many of the movements permitted at synovial joints is presented in figures 11.19–11.21.

## Biomechanics of Body Movement

A lever is any rigid structure that turns about a fulcrum when force is applied. Because bones are rigid structures that can be moved at synovial joints in response to applied forces, they fit the criteria of levers. There are four basic components to a lever: (1) a rigid bar or other such structure; (2) a pivot or fulcrum; (3) an object or resistance that is moved; and (4) a force that is applied to one portion of the rigid structure.

Levers are generally associated with machines but also apply to other mechanical structures, such as the human body. Synovial joints are usually the fulcra ($F$), the muscles provide the force, or effort ($E$), and the bones are the rigid lever arms that move the resisting object ($R$).

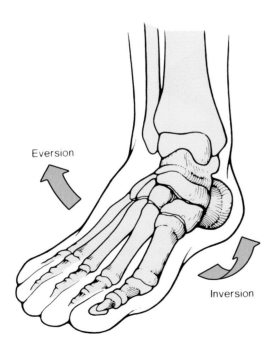

**FIGURE 11.17**
Inversion and eversion of the foot on the intertarsal joints.

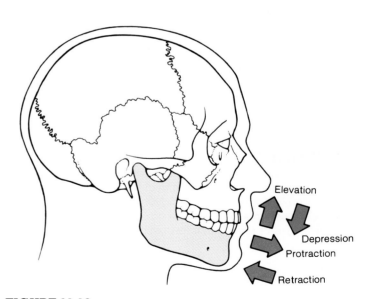

**FIGURE 11.18**
Protraction, retraction, elevation, and depression of the lower jaw.

**FIGURE 11.19**

A photographic summary of joint movements. (a) Adduction of shoulder, hip, and carpophalangeal joints; (b) abduction of shoulder, hip, and carpophalangeal joints; (c) rotation of vertebral column; (d) lateral flexion of vertebral column; (e) flexion of vertebral column; (f) hyperextension of vertebral column; (g) flexion of shoulder, hip, and knee joints of right side of body; extension of elbow and wrist joints; dorsiflexion of right ankle joint; (h) hyperextension of shoulder and hip joints on right side of body; plantar flexion of right ankle joint.

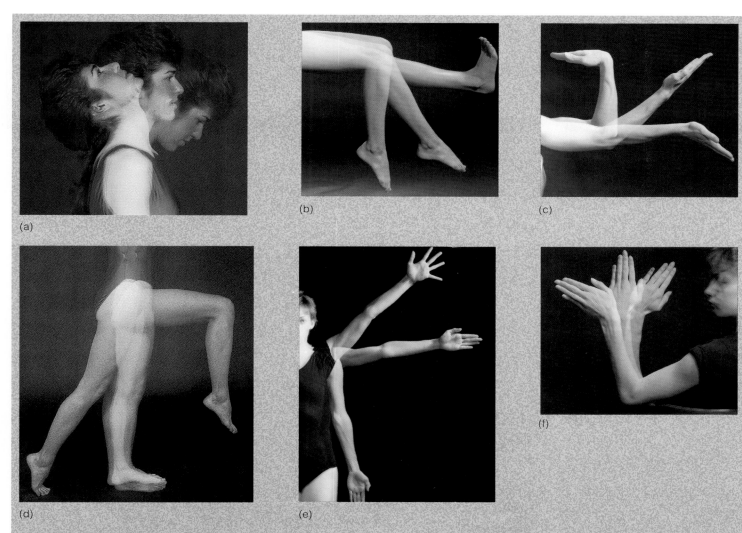

(a)

(b)

(c)

(d)

(e)

(f)

**FIGURE 11.20**

A visual summary of some angular movements at synovial joints. (a) Flexion, extension, and hyperextension in the cervical region; (b) flexion and extension at the knee joint and dorsiflexion and plantar flexion at the ankle joint; (c) flexion, extension, and hyperextension at the wrist joint; (d) flexion, extension, and hyperextension of the hip joint and flexion and extension of the knee joint; (e) adduction and abduction of the arm and fingers; (f) posterior view of abduction and adduction of the hand at the wrist joint. (Note that the range of abduction at the wrist joint is less extensive than the range of adduction as a result of the length of the styloid process of the radius.)

**FIGURE 11.21**

A visual summary of some angular movements at synovial joints. (a) Rotation of the head at the cervical vertebrae—especially at the atlantoaxial joint; (b) rotation of the antebrachium (forearm) at the proximal radioulnar joint.

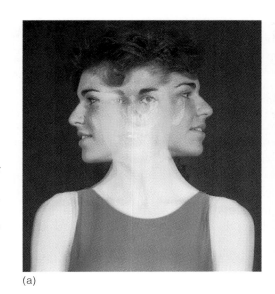

(a)

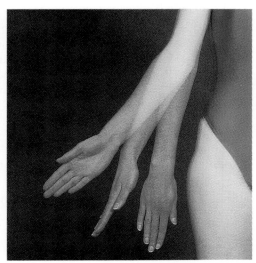

(b)

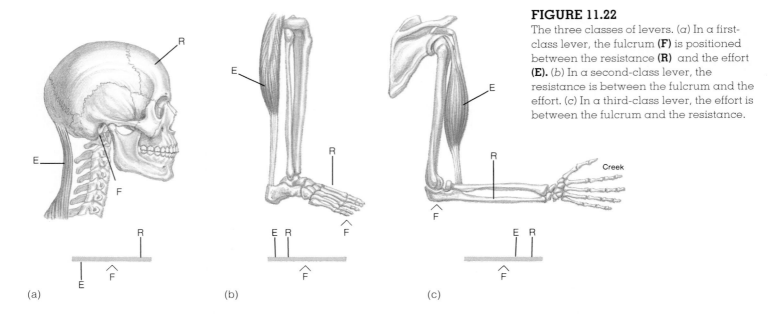

**FIGURE 11.22**

The three classes of levers. (*a*) In a first-class lever, the fulcrum (**F**) is positioned between the resistance (**R**) and the effort (**E**). (*b*) In a second-class lever, the resistance is between the fulcrum and the effort. (*c*) In a third-class lever, the effort is between the fulcrum and the resistance.

(a)　　　　　　　　　　(b)　　　　　　　　　　(c)

There are three kinds of levers, determined by the arrangement of their parts (fig. 11.22).

**1** In a **first-class lever,** the fulcrum is positioned between the effort and the resistance. The sequence of parts in a first-class lever is much like that of a seesaw—a sequence of resistance-pivot-effort. Scissors and hemostats are mechanical examples of first-class levers. In the body, the head at the atlantooccipital (*at-lan´´to-ok-sip´ĭ-tal*) joint is a first-class lever. The weight of the skull and facial portion of the head is the resistance, and the posterior neck muscles that contract to maintain the balance of the head on the joint are the effort.

**2** In a **second-class lever,** the resistance is positioned between the fulcrum and the effort. The sequence of arrangement is pivot-resistance-effort, as in a wheelbarrow or the action of a crowbar when one end is placed under a rock and the other end lifted. Contraction of the calf muscles (*E*) to elevate the body (*R*) on the toes, with the ball of the foot acting as the fulcrum, is another example.

**3** In a **third-class lever,** the effort lies between the fulcrum and the resistance. The sequence of the parts is pivot-effort-resistance, as in the action of a pair of forceps in grasping an object. The third-class lever is the most common type within the body. The flexion at the elbow is an example. The effort occurs as the biceps brachii muscle is contracted to move the resistance of the forearm with the elbow joint forming the fulcrum.

Each bone-muscular interaction at a synovial joint forms some kind of lever system. The specific kind of lever is not always easy to identify. Certain joints are adapted for power at the expense of speed, whereas most are clearly adapted for speed. The specific attachment of muscles that span a joint plays an extremely important role in deter-mining the mechanical advantage. The position of the insertion of a muscle relative to the joint is an important factor in the biomechanics of the contraction. An insertion close to the joint (fulcrum), for example, will produce a faster movement and greater range of movement than an insertion that is farther away from the joint. An attachment far from the joint takes advantage of the lever arm of the bone and increases power at the sacrifice of speed and range of movement.

# Specific Joints of the Body

*Of the numerous joints in the body, some have special structural features that enable them to perform particular functions. Furthermore, these joints are somewhat vulnerable to trauma and are therefore clinically important.*

## Temporomandibular Joint

The **temporomandibular joint** is the only synovial joint in the skull and represents a unique combination of a hinge joint and a gliding joint (fig. 11.23). It is formed by the mandibular condyle of the mandible articulating with the mandibular fossa and the articular tubercle of the temporal bone. An **articular disc** separates the joint cavity into supe-rior and inferior compartments.

Three ligaments support and reinforce the temporo-mandibular joint. The **lateral ligament of the temporo-mandibular joint** is positioned on the lateral side of the joint capsule and is covered by the parotid gland. This

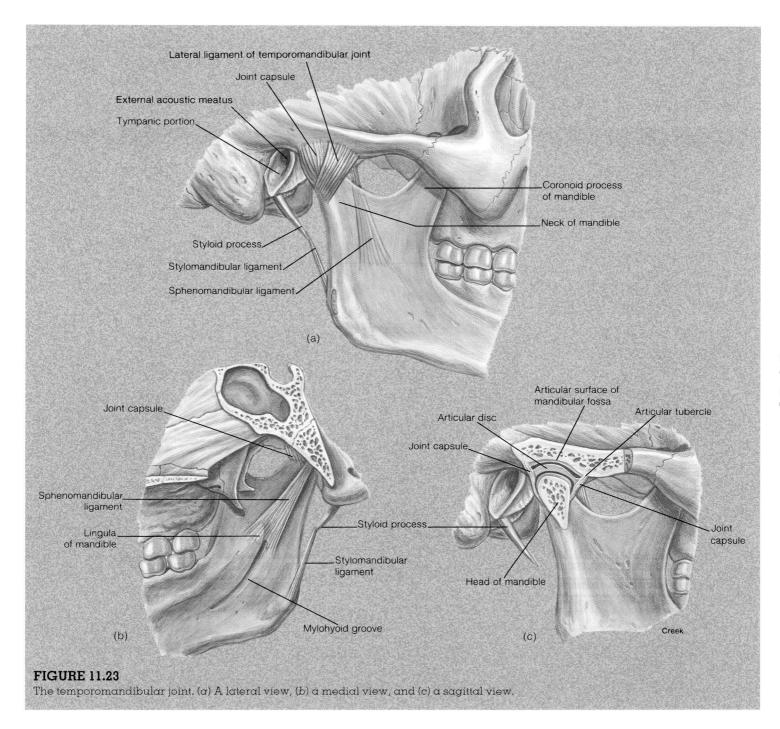

**FIGURE 11.23**

The temporomandibular joint. (*a*) A lateral view, (*b*) a medial view, and (*c*) a sagittal view.

ligament prevents the head of the mandible from being displaced posteriorly and fracturing the tympanic plate when the chin suffers a severe blow. The **stylomandibular ligament** is not directly associated with the joint but extends inferiorly and anteriorly from the styloid process to the posterior border of the ramus of the mandible. A **sphenomandibular** (*sfe´´no-man-dib´yŭ-lar*) **ligament** crosses on the medial side of the joint from the spine of the sphenoid bone to the ramus of the mandible.

The movements of the temporomandibular joint include depression and elevation of the mouth as a hinge joint, protraction and retraction of the jaw as a gliding joint, and lateral rotatory movements. The lateral motion is made possible by the articular disc.

The temporomandibular joint can be easily palpated by applying firm pressure to the area in front of your auricle and opening and closing your mouth. This joint is most vulnerable to dislocation when the mandible is completely depressed, as in yawning. Relocating the jaw is usually a simple task, however, and is accomplished by pressing downward on the molar teeth while pushing the jaw backward.

*Temporomandibular joint (TMJ) syndrome* is a recently recognized ailment that may afflict an estimated 75 million Americans. The apparent cause of TMJ syndrome is a malalignment of one or both temporomandibular joints. The symptoms of the condition vary from moderate and intermittent pain to intense and continuous pain in the head, neck, shoulders, or back. Some vertigo (disorientation of coordination) and tinnitus (ringing in ear) may be experienced.

## Humeral (Shoulder) Joint

The **shoulder joint** is formed by the articulation of the head of the humerus with the glenoid cavity of the scapula (fig. 11.24). It is a ball-and-socket joint and the most freely movable joint in the body. A circular band of fibrocartilage called the **glenoid labrum** passes around the rim of the shoulder joint to deepen the glenoid cavity (fig. 11.24). The shoulder joint is protected from above by an arch formed by the coracoid process and acromion of the scapula and by the clavicle.

......................................

labrum: L. *labrum*, lip

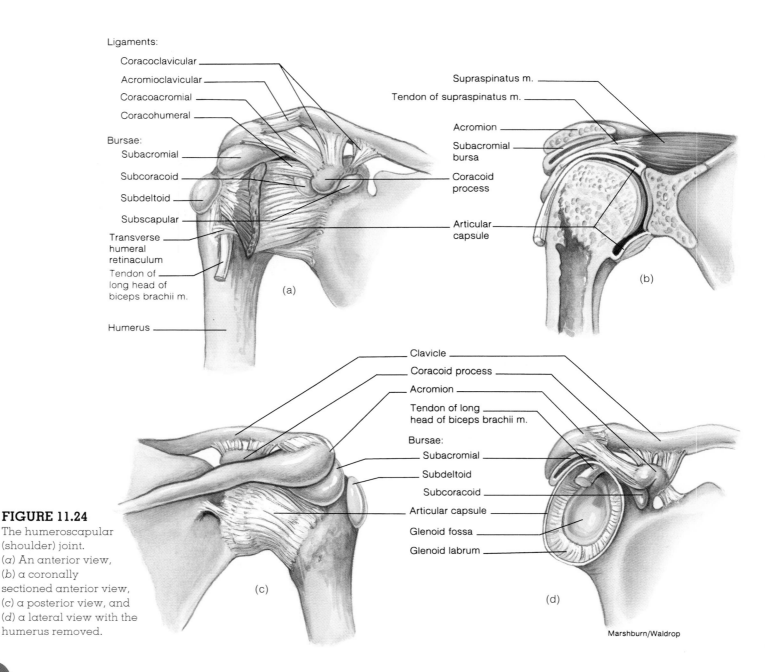

**FIGURE 11.24**

The humeroscapular (shoulder) joint.
(*a*) An anterior view,
(*b*) a coronally sectioned anterior view,
(*c*) a posterior view, and
(*d*) a lateral view with the humerus removed.

Marshburn/Waldrop

Although two ligaments and one retinaculum surround and support the shoulder joint, most of the stability of this joint depends on the powerful muscles and tendons that cross over it. Thus, it is an extremely mobile joint in which stability has been sacrificed for mobility. The **coracohumeral** (*kor″ă-ko-hyoo′mer-al*) **ligament** extends from the coracoid process of the scapula to the greater tubercle of the humerus. The joint capsule is reinforced with three ligamentous bands called the **glenohumeral ligaments.** The final support of the shoulder joint is the **transverse humeral retinaculum,** a thin band that extends from the greater tubercle to the lesser tubercle of the humerus.

 The stability of the shoulder joint is provided only by the tendons of the subscapularis, supraspinatus, infraspinatus, and teres minor muscles, which together form the *musculotendinous (rotator) cuff.* The cuff is fused to the underlying capsule, except in its inferior aspect. Because of the lack of inferior stability, most dislocations (subluxations) occur in this direction. The shoulder is most vulnerable to trauma when the joint is fully abducted and a sudden force from the superior direction is applied to the appendage. Degenerative changes in the musculotendinous cuff produce an inflamed, painful condition known as *pericapsulitis.*

Two major and two minor bursae are associated with the shoulder joint. The larger bursae are the **subdeltoid bursa,** located between the deltoid muscle and the joint capsule, and the **subacromial bursa,** located between the acromion and joint capsule. The **subcoracoid bursa** lies between the coracoid process and the joint capsule and is frequently considered an extension of the subacromial bursa. A small **subscapular bursa** is located between the tendon of the subscapularis muscle and the joint capsule.

 The shoulder joint is vulnerable to dislocations from sudden jerks of the arm, especially in children before strong shoulder muscles have developed. Because of the weakness of this joint in children, parents should be careful not to force a child to follow by yanking on the arm. Dislocation of the shoulder is extremely painful and may cause permanent damage or perhaps muscle atrophy due to disuse.

## Elbow Joint

There are three specific articulations in the elbow region, two of which constitute the **elbow joint** (fig. 11.25). The elbow joint is a hinge joint, formed by the trochlea of the humerus articulating with the trochlear notch of the ulna (humeroulnar joint) and the capitulum of the humerus articulating with the head of the radius (humeroradial joint). Although there are two sets of articulations at the elbow joint, there is only one joint capsule and a large **olecranon bursa** to lubricate this area. A **radial (lateral) collateral ligament** reinforces the elbow joint on the lateral side and an **ulnar (medial) collateral ligament** strengthens the medial side.

 Because so many muscles originate or insert near the elbow, it is a common site of localized tenderness, inflammation, and pain. *Tennis elbow* is a general term for musculotendinous soreness in this area. The structures most generally strained are the tendons attached to the lateral epicondyle of the humerus. The strain is caused by repeated extension of the wrist against some force, as occurs during the backhand stroke in tennis.

## Metacarpophalangeal Joints and Interphalangeal Joints

The **metacarpophalangeal joints** are condyloid joints and the **interphalangeal joints** are hinge joints. These joints are formed as the heads of the metacarpal bones articulate with the proximal phalanges and as the phalanges articulate with one another (fig. 11.26). Each joint in both joint types has three ligaments. A **palmar ligament** spans each joint on the palmar, or anterior, side of the joint capsule. Each joint also has two **collateral ligaments,** one on the lateral side and one on the medial side, to further reinforce the joint capsule. There are no supporting ligaments on the posterior side.

 Athletes frequently jam a finger. It occurs when a ball forcefully strikes a distal phalanx as the fingers are extended, causing a sharp flexion at the joint between the middle and distal phalanges. No ligaments support the joint on the posterior side, but there is a tendon from the digital extensor muscles of the forearm. It is this tendon that is damaged when the finger is jammed. Treatment involves splinting the finger for a period of time. If splinting is not effective, however, surgery will be necessary to avoid a permanent crook in the finger.

## Coxal (Hip) Joint

The ball-and-socket **hip joint** is formed by the head of the femur articulating with the acetabulum of the os coxa (fig. 11.27). It bears the weight of the body and is therefore much stronger and more stable than the shoulder joint. The hip joint is secured by a strong fibrous joint capsule, several ligaments, and a number of powerful muscles.

The primary ligaments of the hip joint are the anterior **iliofemoral** (*il″e-o-fem′or-al*) and **pubofemoral ligaments** and the posterior **ischiofemoral** (*is″ke-o-fem′or-al*) **ligament.**

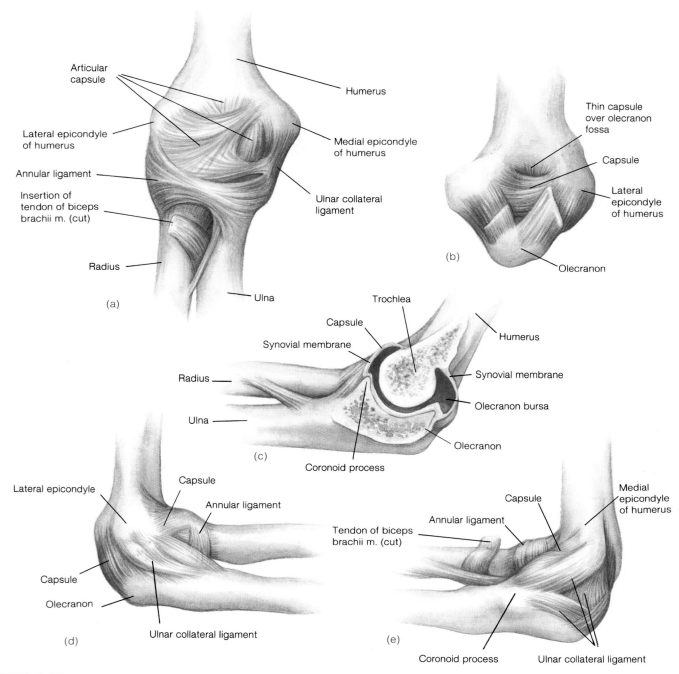

**FIGURE 11.25**

The right elbow region. (*a*) An anterior view, (*b*) a posterior view,
(*c*) a sagittal section, (*d*) a lateral view, and (*e*) a medial view.

The **ligamentum capitis femoris** is located within the articular capsule and attaches the head of the femur to the acetabulum. The **transverse acetabular** (*as˝ĕ-tab´yŭ-lar*) **ligament** crosses the acetabular notch and connects to the joint capsule and the ligamentum capitis femoris. The **acetabular labrum,** a fibrocartilaginous rim that rings the head of the femur as it articulates with the acetabulum, is attached to the margin of the acetabulum.

 *Osteoarthritis* is a degenerative disease of the articular cartilage of synovial joints accompanied by the formation of bony spurs in the joint cavities. Immobility results if the hip joint is severely afflicted with this disease. Fortunately, the entire hip joint can be replaced in a procedure called *hip arthroplasty.* During this surgery, the acetabulum is replaced by a low-friction polyethylene socket, which is fit into the os coxa using bone cement. The head of the femur is replaced by a stainless steel, ball-shaped prosthesis (see fig. 11.31).

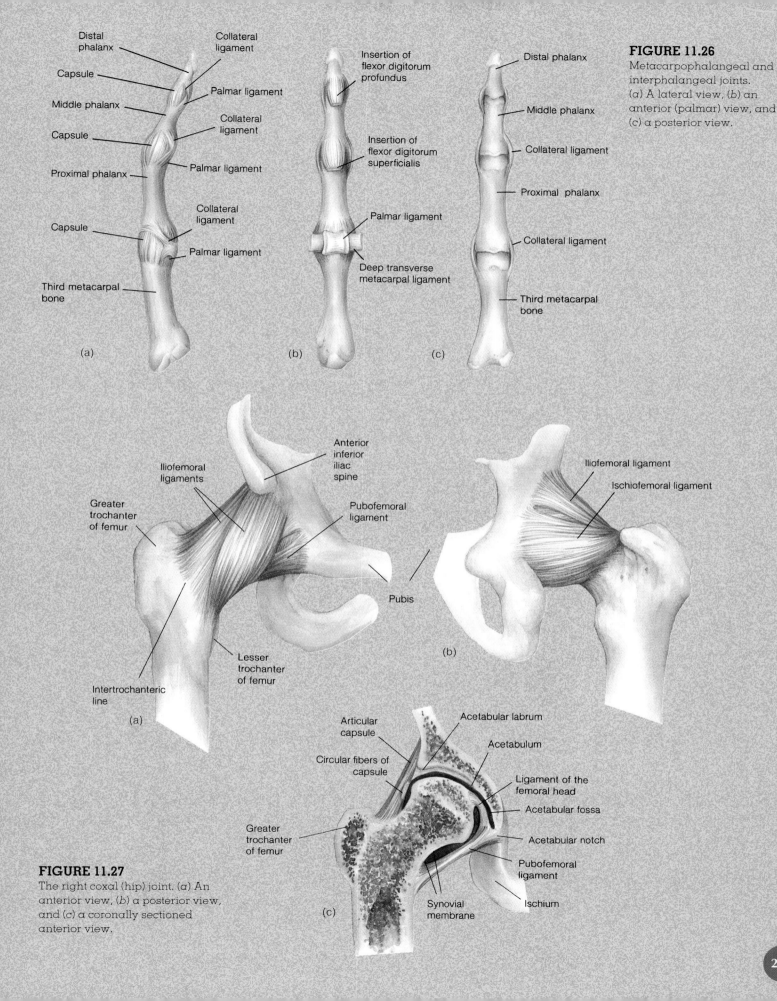

**FIGURE 11.26**
Metacarpophalangeal and interphalangeal joints. (*a*) A lateral view, (*b*) an anterior (palmar) view, and (*c*) a posterior view.

(a)

Distal phalanx
Collateral ligament
Capsule
Palmar ligament
Middle phalanx
Collateral ligament
Capsule
Palmar ligament
Proximal phalanx
Collateral ligament
Capsule
Palmar ligament
Third metacarpal bone

(b)

Insertion of flexor digitorum profundus
Insertion of flexor digitorum superficialis
Palmar ligament
Deep transverse metacarpal ligament

(c)

Distal phalanx
Middle phalanx
Collateral ligament
Proximal phalanx
Collateral ligament
Third metacarpal bone

**FIGURE 11.27**
The right coxal (hip) joint. (*a*) An anterior view, (*b*) a posterior view, and (*c*) a coronally sectioned anterior view.

(a)

Iliofemoral ligaments
Anterior inferior iliac spine
Greater trochanter of femur
Pubofemoral ligament
Pubis
Lesser trochanter of femur
Intertrochanteric line

(b)

Iliofemoral ligament
Ischiofemoral ligament

(c)

Articular capsule
Acetabular labrum
Circular fibers of capsule
Acetabulum
Ligament of the femoral head
Acetabular fossa
Greater trochanter of femur
Acetabular notch
Pubofemoral ligament
Synovial membrane
Ischium

243

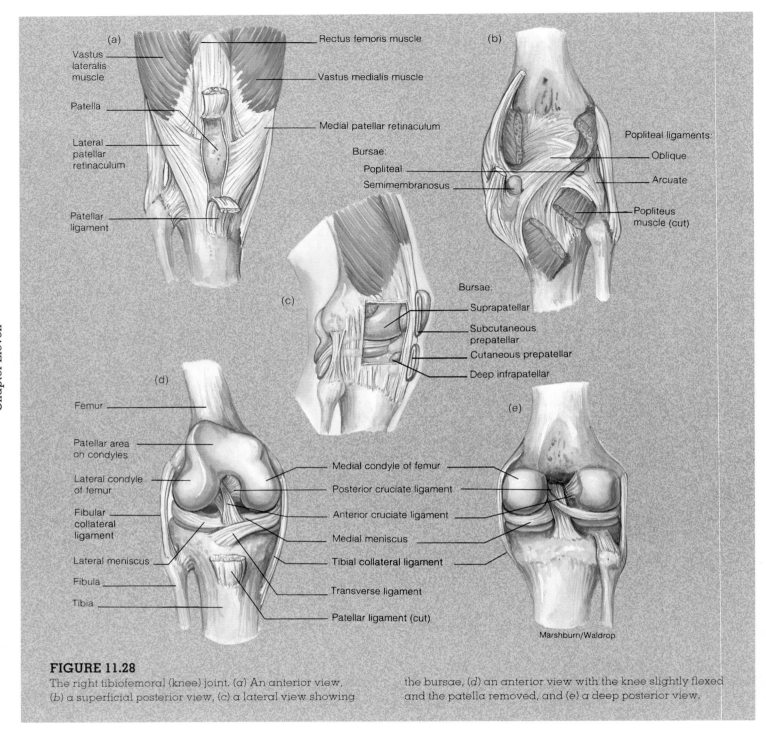

(a)

Vastus lateralis muscle

Patella

Lateral patellar retinaculum

Patellar ligament

Rectus femoris muscle

Vastus medialis muscle

Medial patellar retinaculum

(b)

Popliteal ligaments:

Oblique

Arcuate

Popliteus muscle (cut)

Bursae:

Popliteal

Semimembranosus

(c)

Bursae:

Suprapatellar

Subcutaneous prepatellar

Cutaneous prepatellar

Deep infrapatellar

(d)

Femur

Patellar area on condyles

Lateral condyle of femur

Fibular collateral ligament

Lateral meniscus

Fibula

Tibia

Medial condyle of femur

Posterior cruciate ligament

Anterior cruciate ligament

Medial meniscus

Tibial collateral ligament

Transverse ligament

Patellar ligament (cut)

(e)

Marshburn/Waldrop

**FIGURE 11.28**
The right tibiofemoral (knee) joint. (a) An anterior view,
(b) a superficial posterior view, (c) a lateral view showing
the bursae, (d) an anterior view with the knee slightly flexed
and the patella removed, and (e) a deep posterior view.

# Tibiofemoral (Knee) Joint

The **knee joint** is the largest, most complex, and probably the most vulnerable joint in the body. The knee joint is formed as the femur and the tibia articulate. It is a complex hinge joint that permits limited rolling and gliding movements, in addition to flexion and extension. On the anterior side, the knee joint is stabilized and protected by the patella and the **patellar ligament,** forming a gliding **patellofemoral joint.**

Because of the complexity of the knee joint, only the relative positions of the ligaments, menisci, and bursae will be covered here. Although the detailed attachments and functions will not be discussed, the locations of these structures can be seen in figure 11.28.

In addition to the patella and the patellar ligament on the anterior surface, the tendinous insertion of the quadriceps femoris muscle forms two supportive bands called the **lateral** and **medial patellar retinacula** (*ret″ĭ-nak′yŭ-lă*). Four bursae are associated with the anterior aspect of the knee: the **subcutaneous prepatellar bursa,** the **suprapatellar bursa,** the **cutaneous prepatellar bursa,** and the **deep infra-patellar bursa.**

The posterior aspect of the knee is referred to as the **popliteal fossa.** The broad **oblique popliteal ligament** and the **arcuate** (*ar´kyoo-āt*) **popliteal ligament** are superficial in position, whereas the **anterior** and **posterior cruciate** (*kroo´she-āt*) **ligaments** are deep within the joint. The **popliteal bursa** and the **semimembranosus bursa** are the two bursae associated with the back of the knee.

Strong **collateral ligaments** support both the medial and lateral sides of the knee joint. Two fibrocartilaginous discs, the **lateral** and **medial menisci,** are located within the knee joint interposed between the distal femoral and proximal tibial condyles. The two menisci are connected by a **transverse ligament.** Several other bursae are associated with the knee joint. In addition to the 4 on the anterior side and the 2 on the posterior side, there are 7 bursae on the lateral and medial sides, for a total of 13.

 During normal walking, running, and supporting of the body, the knee joint functions superbly. It can tolerate even moderate stress without tissue damage. However, the knee lacks bony support to withstand sudden forceful stresses, such as commonly occur among professional athletes. Knee injuries frequently require surgery and heal with difficulty due to the avascularity of the cartilaginous tissue. Because this joint is potentially so vulnerable, it is important to appreciate its limitations by understanding its anatomy.

## Talocrural (Ankle) Joint

There are actually two principal articulations within the **ankle joint,** both of which are hinge joints. One articulation is formed as the distal end of the tibia and its medial malleolus articulate with the talus; the other is formed as the lateral malleolus of the fibula articulates with the talus (fig. 11.29).

One joint capsule surrounds the articulations of the three bones, and four ligaments support the ankle joint on the outside of the capsule. The strong **deltoid ligament** is associated with the tibia, whereas the **lateral collateral ligaments, anterior talofibular** (*ta´lo-fib´yŭ-lar*) **ligament, posterior talofibular ligament,** and **calcaneofibular** (*kal-ka´ne-o-fib´yŭ-lar*) **ligament** are associated with the fibula.

The malleoli form a cap over the upper surface of the talus that prohibits side-to-side movement at the ankle joint. Unlike the condyloid joint at the wrist, the movements of the ankle are limited to flexion and extension. Dorsiflexion of the ankle is checked primarily by the tendo calcaneus, whereas plantar flexion, or ankle extension, is checked by the tension of the extensor tendons on the front of the joint and the anterior portion of the joint capsule.

 *Ankle sprains* are a common type of locomotor injury. They vary widely in seriousness but tend to occur in certain locations. The most common cause of ankle sprain is excessive inversion of the foot, resulting in partial tearing of the anterior talofibular ligament and the calcaneofibular ligament. Less commonly, the deltoid ligament is injured by excessive eversion of the foot. Torn ligaments are extremely painful and are accompanied by immediate local swelling. Reducing the swelling and immobilizing the joint are about the only treatments for moderate sprains. Extreme sprains may require surgery and casting of the joint to facilitate healing.

A summary of the principal joints of the body and their movement is presented in table 11.2.

## Clinical Considerations

A synovial joint is a remarkable biologic system that acts as a self-lubricating weight-bearing surface, able to move with almost frictionless precision under tremendous loads and impacts. Under normal circumstances and in most people, the many joints of the body perform without problems throughout life. Joints are not indestructible, however, and are subject to various forms of trauma and disease. Although not all of the diseases of joints are fully understood, medical science has made remarkable progress in the treatment of arthrological problems.

### Trauma to Joints

Joints are well adapted to withstand compression and tension forces. Torsion or sudden impact to the side of a joint, however, can be devastating. These types of injuries frequently occur in athletes.

In a **strained joint,** unusual or excessive exertion stretches the tendons spanning the joint or surrounding muscles, but it causes no serious damage. Strains frequently result from not "warming up," or activating joints and muscles, prior to strenuous activity. A **sprain** is a tearing of the ligaments or tendons that surround a joint. There are various grades of sprains, and the severity will determine the treatment. Severe sprains of the knee joint are frequently accompanied by damage to articular cartilages and menisci that generally requires surgery. Sprains are usually accompanied by **synovitis** (*sin´ŏ-vi´tis*), an inflammation of the joint capsule.

**Luxation,** or **joint dislocation,** is derangement of the articulating bones that compose the joint. Joint dislocation is more serious than a sprain and is usually accompanied by sprains. The shoulder and knee joints are the most vulnerable to dislocation. Self-healing of a dislocated joint may be

Articulations

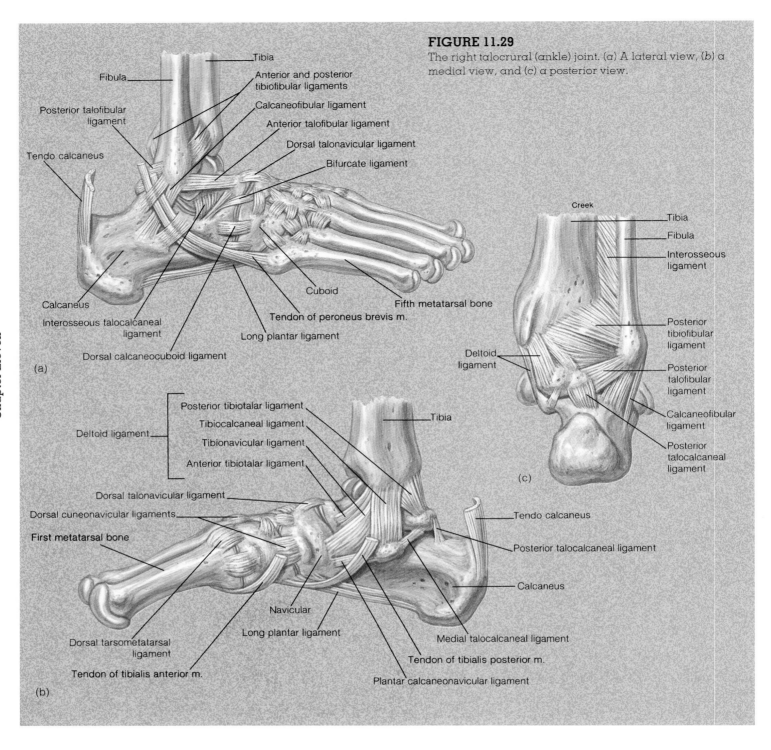

**FIGURE 11.29**

The right talocrural (ankle) joint. (a) A lateral view, (b) a medial view, and (c) a posterior view.

incomplete; for example, leaving the person with a "trick knee" that may unexpectedly give way.

**Subluxation** is partial dislocation of a joint. Subluxation of the hip joint is a common type of birth defect that can be treated by bracing or casting the hip joints to promote suitable bone development.

**Bursitis** (*bur-si´tis*) is an inflammation of the bursa associated with a joint. Because of its close proximity to the joint, bursitis may affect the joint capsule as well. Bursitis may be caused by excessive stress on the bursa from overexertion, or it may be a local or systemic inflammatory process. As the bursa swells, the surrounding muscles become sore and stiff. **Tendonitis** involves the tendon, in or out of a tendon sheath, and usually comes about in the same way as bursitis.

The flexible vertebral column is a marvel of mechanical engineering. Not only do the individual vertebrae articulate one with another, but together they form the portion of the axial skeleton with which the head, ribs, and pelvic girdle

Table 11.2   **Principal articulations**

| Joint | Type | Movement |
|---|---|---|
| Most skull joints | Fibrous (suture) | Immovable |
| Temporomandibular | Synovial (hinge; gliding) | Elevation, depression; protraction, retraction |
| Atlantooccipital | Synovial (condyloid) | Flexion, extension, circumduction |
| Atlantoaxial | Synovial (pivot) | Rotation |
| Intervertebral | | |
|   bodies of vertebrae | Cartilaginous (symphysis) | Slight movement |
|   articular processes | Synovial (gliding) | Flexion, extension, slight rotation |
| Sacroiliac | Cartilaginous (symphysis) | Slight gliding movement; may fuse in adults |
| Costovertebral | Synovial (gliding) | Slight movement during breathing |
| Sternocostal | Synovial (gliding) | Slight movement during breathing |
| Sternoclavicular | Synovial (gliding) | Slight movement when shrugging shoulders |
| Sternal | Cartilaginous (symphysis) | Slight movement during breathing |
| Acromioclavicular | Synovial (gliding) | Protraction, retraction; elevation, depression |
| Humeral (shoulder) | Synovial (ball-and-socket) | Flexion, extension; adduction, abduction; rotation, circumduction |
| Elbow | Synovial (hinge) | Flexion, extension |
| Proximal radioulnar | Synovial (pivot) | Rotation |
| Distal radioulnar (side-to-side) | Fibrous (syndesmosis) | Slight movement |
| Radiocarpal (wrist) | Synovial (condyloid) | Flexion; extension; adduction, abduction; circumduction |
| Intercarpal | Synovial (gliding) | Slight movement |
| Carpometacarpal | | |
|   fingers | Synovial (condyloid) | Flexion, extension; adduction, abduction |
|   thumb | Synovial (saddle) | Flexion, extension; adduction, abduction |
| Metacarpophalangeal | Synovial (condyloid) | Flexion, extension; adduction, abduction |
| Interphalangeal | Synovial (hinge) | Flexion, extension |
| Symphysis pubis | Fibrous (symphysis) | Slight movement |
| Coxal (hip) | Synovial (ball-and-socket) | Flexion, extension; adduction, abduction; rotation; circumduction |
| Tibiofemoral (knee) | Synovial (hinge) | Flexion, extension; slight rotation when flexed |
| Proximal tibiofibular | Synovial (gliding) | Slight movement |
| Distal tibiofibular | Fibrous (syndesmosis) | Slight movement |
| Talocrural (ankle) | Synovial (hinge) | Dorsiflexion, plantar flexion; slight circumduction; inversion, eversion |
| Intertarsal | Synovial (gliding) | Inversion, eversion |
| Tarsometatarsal | Synovial (gliding) | Flexion, extension; adduction, abduction |

Articulations

## Development of Synovial Joints

The sites of developing synovial joints (freely movable joints) are discernible at 6 weeks as mesenchyme becomes concentrated in the areas where precartilage cells differentiate (fig. 1). The future joints at this stage appear as intervals of less concentrated mesenchymal cells. As cartilage cells develop within a forming bone, a thin flattened sheet of cells forms around the cartilaginous model to become the **perichondrium.** These same cells are continuous across the gap between the adjacent developing bone. Surrounding the gap, the flattened mesenchymal cells differentiate to become the **joint capsule.**

During the early part of the third month of development, the mesenchymal cells still remaining within the joint capsule begin migrating toward the epiphyses of the adjacent developing bones. The cleft eventually enlarges to become the **joint cavity.** Thin pads of hyaline cartilage develop on the surfaces of the epiphyses in contact with the joint cavity. These pads become the **articular cartilages** of the functional joint. As the joint continues to develop, a highly vascular **synovial membrane** forms on the inside of the joint capsule and begins secreting a watery *synovial fluid* into the joint cavity.

In certain developing synovial joints, the mesenchymal cells do not migrate away from the center of the joint cavity. Rather, they give rise to cartilaginous wedges called **menisci,** as in the knee joint, or to complete cartilaginous pads, called **articular discs,** as in the sternoclavicular joint.

Most synovial joints have formed completely by the end of the third month. Shortly thereafter, fetal muscle contractions, known as *quickening,* cause movement at these joints. Joint movement enhances the nutrition of the articular cartilage and prevents the fusion of connective tissues within the joint.

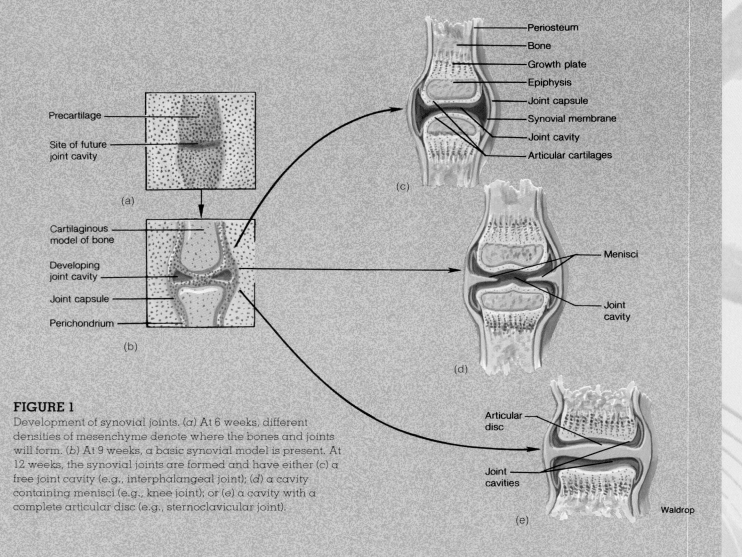

### FIGURE 1

Development of synovial joints. (*a*) At 6 weeks, different densities of mesenchyme denote where the bones and joints will form. (*b*) At 9 weeks, a basic synovial model is present. At 12 weeks, the synovial joints are formed and have either (*c*) a free joint cavity (e.g., interphalangeal joint); (*d*) a cavity containing menisci (e.g., knee joint); or (*e*) a cavity with a complete articular disc (e.g., sternoclavicular joint).

Waldrop

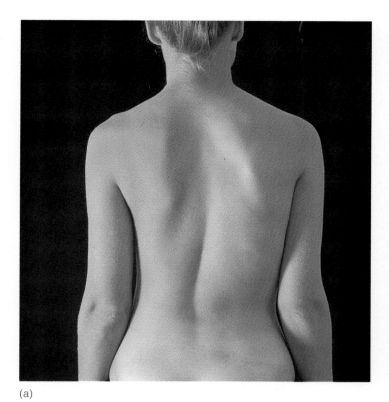

(a)

**FIGURE 11.30**
Scoliosis is a lateral curvature of the spine, usually in the thoracic region. It may be congenital, acquired, or disease related. (a) A posterior view of a 19-year-old woman and (b) a radiograph.

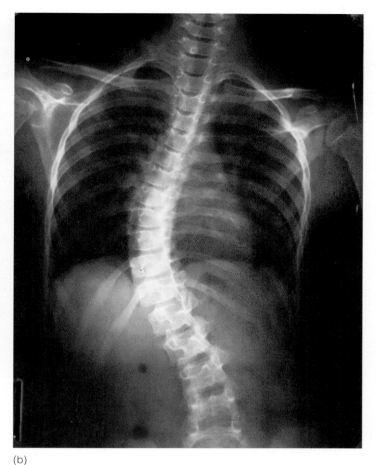

(b)

articulate. The vertebral column also encloses the spinal cord and provides exits for 31 pairs of spinal nerves. Considering all the articulations in the vertebral column and the physical abuse it takes, it is no wonder that back ailments are second only to headaches as our most common physical complaint. Our way of life causes many of the problems associated with the vertebral column. Improper shoes, athletic exertion, sudden stops in vehicles, or improper lifting can all cause the back to go awry. Body weight, age, and general body condition influence a person's susceptibility to back problems.

The most common cause of back pain is *strained muscles*, generally the result of overexertion. The second most frequent back ailment is a *herniated disc*. The dislodged nucleus pulposus of a disc may push against a spinal nerve and cause excruciating pain. The third most frequent back problem is a *dislocated articular facet* between two vertebrae, caused by sudden torsion of the vertebral column. The treatment of back ailments varies from bed rest to spinal manipulation to extensive surgery.

Curvature disorders are another problem of the vertebral column. **Kyphosis** (ki-fo´sis) (hunchback) is an exaggeration of the thoracic curve. **Lordosis** (swayback) is an abnormal anterior convexity of the lumbar curve. **Scoliosis** (crookedness) is an abnormal lateral curvature of the ver-

tebral column (fig. 11.30), which may be caused by one leg being longer than the other or uneven muscular development on the two sides of the vertebral column.

## Diseases of Joints

**Arthritis** is a generalized term for over 50 different joint diseases, all of which have the symptoms of edema, inflammation, and pain. The causes of most kinds of arthritis are unknown, but certain types follow joint trauma or bacterial infections. There is evidence that some types of arthritis are the result of hormonal or metabolic disorders. The most common forms are rheumatoid (roo´mă-toid) arthritis, osteoarthritis, and gouty arthritis.

In **rheumatoid arthritis,** the synovial membrane thickens and becomes tender and synovial fluid accumulates. This change is usually followed by an invasion of fibrous tissue and deterioration of the articular cartilage. When the cartilage is destroyed, the exposed bone tissue is joined by the fibrous tissue and instigates ossification of the joint. It is

..........................................................

kyphosis: Gk. *kyphos,* hunched
lordosis: Gk. *lordos,* curving forward
scoliosis: Gk. *skoliosis,* crookedness
rheumatoid: Gk. *rheuma,* a flowing

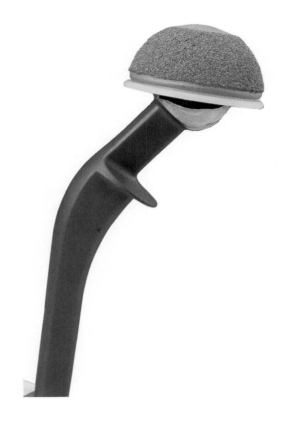

(a)

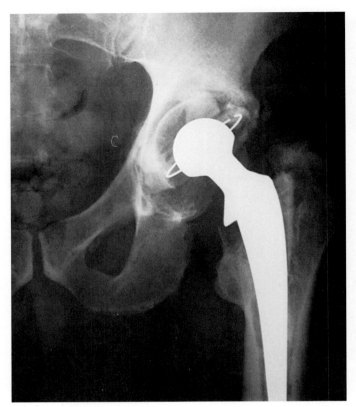

(b)

**FIGURE 11.31**
Two examples of joint prostheses. (*a*, *b*) The coxal (hip) joint and
(*c*, *d*) the tibiofemoral (knee) joint.

the joint ossification that causes the crippling effect of rheumatoid arthritis. This disease affects females slightly more than males and occurs most commonly between the ages of 30 and 50.

**Osteoarthritis** is a degenerative joint disease that results from aging and irritation of the joints. Although osteoarthritis is far more common than rheumatoid arthritis, it is usually less damaging. Osteoarthritis is a slow, progressive disease in which the articular cartilages gradually soften and disintegrate. The affected joints seldom swell, and the synovial membrane is rarely damaged. As the articular cartilage deteriorates, ossified spurs are deposited on the exposed bone, causing pain and restricting the movement of articulating bones. Osteoarthritis most frequently affects the knee, hip, and intervertebral joints.

**Gouty arthritis** results from a metabolic disorder in which an abnormal amount of uric acid is retained in the blood and sodium urate crystals are deposited in the joints. The salt crystals irritate the articular cartilage and synovial membrane, causing swelling, tissue deterioration, and pain. If gout is not treated, the affected joint fuses. Males have a greater incidence of gout than females, and apparently the disease is genetically determined. About 85% of gout cases affect the joints of the

foot and leg. The most common joint affected is the metatarsophalangeal joint of the hallux (great toe).

## Treatment of Joint Disorders

**Arthroscopy** (*ar-thros´kŏ-pe*) is widely used in diagnosing and, to a limited extent, treating joint disorders. Arthroscopic inspection involves making a small incision through the skin and into the joint capsule through which the tubelike arthroscopic instrument is threaded. In arthroscopy of the knee, the articular cartilage, synovial membrane, menisci, and cruciate ligaments can be observed. Samples can be extracted, and pictures taken for further evaluation.

Remarkable advancements have been made in the last 15 years in **joint prostheses** (*pros-the´sēz*) (fig. 11.31). These artificial articulations do not take the place of normal, healthy joints, but they are a valuable option for chronically disabled arthritis patients.

...........................................................

gout: L. *gutta*, a drop (thought to be caused by "drops of viscous humors")
prosthesis: Gk. *pros*, in addition to; *thesis*, a setting down

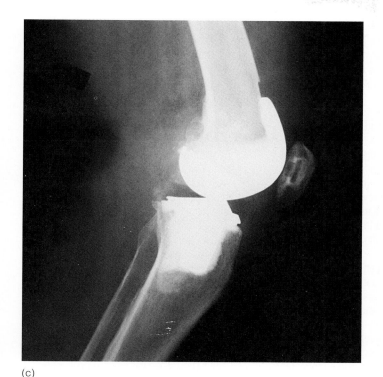

(c)

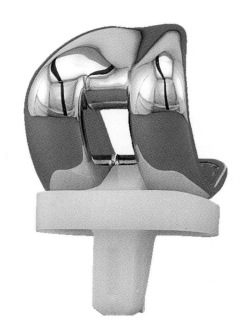

(d)

**FIGURE 11.31** *Continued*

# Important Clinical Terminology

**ankylosis**   Stiffening of a joint resulting in severe or complete loss of movement.

**arthralgia**   (also arthrodynia) Severe pain within a joint.

**arthrolith**   A gouty deposit in a joint.

**arthrometry**   The measurement of the range of movement in a joint.

**arthroncus**   Swelling of a joint due to trauma or disease.

**arthropathy**   Any disease affecting a joint.

**arthroplasty**   The surgical repair of a joint.

**arthrosis**   A joint or an articulation; also, a degenerative condition of a joint.

**arthrosteitis**   An inflammation of the bony structure of a joint.

**chondritis**   An inflammation of the articular cartilage of a joint.

**coxarthrosis**   A degenerative condition of the hip joint.

**hemarthrosis**   An accumulation of blood in a joint cavity.

**rheumatology**   The medical speciality concerned with the diagnosis and treatment of arthritis.

**spondylitis**   An inflammation of one or several vertebrae.

**synovitis**   The inflammation of the synovial membrane lining the inside of a joint capsule.

# Chapter Summary

### Classification of Joints (p. 224)

1. Joints are formed as adjacent bones articulate. Arthrology is the science concerned with the study of joints and kinesiology is the study of movements involving certain joints.
2. Joints can be classified according to structure or function.
   a. Structurally, joints are classified as fibrous, cartilaginous, and synovial types.
   b. Functionally, joints are classified as synarthroses (immovable joints), amphiarthroses (slightly movable joints), and diarthroses (freely movable joints).

### Fibrous Joints (pp. 224–226)

1. Articulating bones in fibrous joints are tightly bound by fibrous connective tissue. They are of three kinds: sutures, syndesmoses, and gomphoses.

2. Sutures are found only in the skull and are classified as serrate, lap, or plane.
3. Syndesmoses are found only in the antebrachium and leg where adjacent bones are held together by interosseous ligaments. Slight movement is permitted at syndesmoses.
4. Gomphoses are found only in the skull where the teeth are bound into their sockets by the periodontal ligament.

### Cartilaginous Joints (pp. 226–227)

1. Cartilaginous joints have fibrocartilage or hyaline cartilage between the adjacent bones. They are of two kinds: symphyses and synchondroses.
2. The symphysis pubis and the intervertebral disc joints are examples of symphyses.
3. Some synchondroses are temporary joints formed in the growth lines (epiphyseal plates) between the diaphyses and epiphyses in the long bones of children. Other synchondroses are permanent, such as the costal cartilages of the rib cage.

### Synovial Joints (pp. 227–231)

1. Synovial joints are freely movable articulations between bones.
2. Each synovial joint contains a joint capsule, articular cartilages, and synovial membranes that produce lubricating synovial fluid. In addition, some synovial joints contain a meniscus or menisci to assist joint movement.
3. The range of movement of a synovial joint is determined by the structure of the articulating bones, the ligaments and tendons, and the muscles that act on the joint.

### Movements at Synovial Joints (pp. 231–238)

1. Movements at synovial joints are produced by the contraction of the skeletal muscles spanning the joints and attaching to or near the bones forming the articulations. In these actions, the bones act as levers, the muscles provide the force, and the joints are the fulcra, or pivots.
2. Angular movements increase or decrease the joint angle produced by the articulating bones.
   a. Flexion decreases the joint angle on an anterior-posterior plane; extension increases the same joint angle.
   b. Abduction is the movement of a body part away from the main axis of the body; adduction is the movement of a body part toward the main axis of the body.
3. Circular movements can occur only in joints that are composed of a bone with a rounded surface articulating with a corresponding depression on another bone.
   a. Rotation is the movement of a bone around its own axis.
   b. Circumduction is a conelike movement of a body part.
4. Special joint movements include inversion and eversion, protraction and retraction, and elevation and depression.
5. Synovial joints can be classified as first-, second-, or third-class levers.
   a. In a first-class lever, the fulcrum is positioned between the effort and the resistance.
   b. In a second-class lever, the resistance is positioned between the fulcrum and the effort.
   c. In a third-class lever, the effort lies between the fulcrum and the resistance.

### Specific Joints of the Body (pp. 238–245)

1. The temporomandibular joint, a combined hinge and gliding joint, is of clinical importance because of temporomandibular joint (TMJ) syndrome.
2. The humeral (shoulder) joint, a ball-and-socket joint, is vulnerable to dislocations from sudden jerks of the arm, especially in children, whose shoulder muscles are still developing.
3. There are two sets of articulations at the elbow as the distal end of the humerus articulates with the proximal ends of the ulna and radius. Strain on the elbow joint is common in certain sports.
4. The ball-and-socket coxal (hip) joint is especially prone to osteoarthritis in elderly people.
5. The hinged tibiofemoral (knee) joint is the largest, most vulnerable joint in the body.
6. There are two hinged articulations within the talocrural (ankle) joint. Ankle sprains are common injuries of this joint.

# Review Activities

**Objective Questions**

1. Which statement regarding joints is *false*?
   a. Joints are the locations where two or more bones articulate.
   b. The structural classification of joints includes fibrous, membranous, and cartilaginous types.
   c. Arthrology is the study of joints; kinesiology is the study of the biomechanics of joint movement.
2. Synchondroses are a type of
   a. fibrous joint.
   b. synovial joint.
   c. cartilaginous joint.
3. An interosseous ligament is characteristic of
   a. a suture.          c. a symphysis.
   b. a synchondrosis.  d. a syndesmosis.
4. Which of the following joint type–function word pairs is *incorrect*?
   a. synchondrosis/growth at the epiphyseal plate
   b. symphysis/movement at the intervertebral joint
   c. suture/strength and stability in the skull
   d. syndesmosis/movement of the jaw
5. Which of the following is a *false* statement?
   a. Synchondroses occur only in children and young adults.
   b. Sutures occur only in the skull.
   c. Saddle joints occur in the thumb and in the neck, where rotational movement is possible.
   d. Syndesmoses occur only in the antebrachium and leg.
6. Which of the following is *not* characteristic of all synovial joints?
   a. articular cartilage
   b. synovial fluid
   c. a joint capsule
   d. a meniscus
7. The atlantoaxial and the proximal radioulnar synovial joints are specifically classified as
   a. hinge.          c. pivotal.
   b. gliding.        d. condyloid.
8. Which of the following joints can be readily and comfortably hyperextended?
   a. interphalangeal joint
   b. coxal joint
   c. tibiofemoral joint
   d. sternocostal joint
9. Which of the following is most vulnerable to luxation?
   a. elbow joint      c. coxal joint
   b. humeral joint   d. tibiofemoral joint
10. A thickening and tenderness of the synovial membrane and the accumulation of synovial fluid are signs of the development of
   a. arthroscopitis.
   b. gouty arthritis.
   c. osteoarthritis.
   d. rheumatoid arthritis.

## Essay Questions

1. What is meant by a structural classification of joints, as compared to a functional classification?
2. Why is the anatomical position so important in explaining the movements that are possible at joints?
3. What are the structural components of a synovial joint that determine the range of movement at that joint?
4. What are the advantages of a hinge joint over a ball-and-socket type? If ball-and-socket joints allow a greater range of movement, why are not all the synovial joints of this type?
5. What is synovial fluid? Where is it produced, and what are its functions?
6. Describe a bursa and discuss its function. What is bursitis?
7. Identify four types of synovial joints found in the wrist and hand region and state the types of movement permitted by each.
8. Discuss the articulations of the pectoral and pelvic regions to the axial skeleton with regard to range of movement, ligamentous attachments, and potential clinical problems.
9. What is meant by a sprained ankle? How does a sprain differ from a strain or a luxation?
10. What occurs within the joint capsule in rheumatoid arthritis? How does rheumatoid arthritis differ from osteoarthritis?

## Activity Question

Refer to figure 11.32 and identify the joints being flexed.

**FIGURE 11.32**

Identify the joints of the body that are being flexed as a person assumes a "fetal position."

## Gundy/Weber Software 💾

The tutorial software accompanying Chapter 11 is Volume 4—Muscle System.

# chapter twelve

# muscle tissue and muscle physiology

## objectives

- Describe the arrangement of muscle fibers within a muscle and the banding pattern of a skeletal muscle fiber.

- Describe the nature of a muscle twitch and explain how summation and tetanus are produced.

- Distinguish between isometric and isotonic contractions and discuss the significance of the series-elastic component of muscles.

- Discuss the relationship between somatic motor neurons and skeletal muscles, noting the significance of motor units.

- Describe the sliding filament mechanism of contraction, noting how the bands in a muscle fiber change during contraction.

- Describe the cross-bridge cycle and the role of ATP in muscle contraction and muscle relaxation.

- Describe the function of actin, myosin, troponin, and tropomyosin in muscle contraction.

- Discuss the role of $Ca^{++}$ in muscle contraction and explain how electrical stimulation influences the availability of $Ca^{++}$.

- Define *maximal oxygen uptake* and *oxygen debt*.

- Describe the role of phosphocreatine in muscle contraction.

- Distinguish between fast-twitch and slow-twitch fibers and explain how muscles adapt to exercise training.

- Compare and contrast smooth muscle and skeletal muscle with respect to structure and contractile mechanisms.

- Explain how the mechanism of contraction is regulated in smooth muscle.

# Structure and Actions of Skeletal Muscles

*Skeletal muscles are composed of individual muscle fibers that contract when stimulated by nerve impulses through motor neurons. Each motor neuron branches to innervate a number of muscle fibers. Activation of varying numbers of motor neurons results in gradations in the strength of contraction of the whole muscle.*

As described in chapter 6 and summarized in table 12.1, there are three types of muscle tissue: skeletal, cardiac, and smooth. Skeletal and cardiac muscle cells are striated, in contrast to the nonstriated cells of smooth muscle. Our focus in this chapter is on muscle structure and function; a detailed description of skeletal muscle anatomy is presented in chapter 13.

## Attachment of Muscles

Skeletal muscles are usually attached to bone on each end by tough connective tissue *tendons.* When a muscle contracts, it shortens, and this places tension on its tendons and attached bones. The muscle tension causes movement of the bones at a joint, where one of the articulating bones generally moves more than the other. The more movable bony attachment of the muscle, known as its *insertion,* is pulled toward its *origin,* which is its less movable attachment.

Specialized tendons are identified by specific names. Flattened, sheetlike tendons, for example, are called **aponeuroses** (*ap˝ŏ-noo-ro´sēz*). An example is the galea aponeurotica over the skull (see fig. 13.5). In certain places, tendons are enclosed by protective tendon sheaths that lubricate the tendons with synovial fluid (see fig. 11.6). In the ankle (see fig. 13.34) and in the wrist, the entire group of tendons is contained in place by a thin but strong band of connective tissue called a **retinaculum** (*ret˝ĭ-nak´yoo-lum*).

## Associated Connective Tissue

Contracting muscle fibers would be ineffective if they worked as isolated units. Each fiber is bound to adjacent fibers to form bundles, and the muscle bundles in turn are bound to other muscle bundles. In this arrangement, the contraction of muscle fibers in one area of a muscle works in conjunction with contracting fibers elsewhere in the muscle. The binding structures within muscles are the *associated connective tissues.*

A fibrous connective tissue, called **fascia** (*fash´e-ă*), is found under the skin and binds adjacent muscles together. Fascia may be categorized as superficial or deep. *Superficial fascia* is the tissue that secures the hypodermis of the skin to the underlying muscles, and it varies in thickness throughout the body. For example, superficial fascia over the buttock and anterior abdominal wall is thick and laced with adipose tissue. By contrast, the superficial fascia under the skin of the dorsum of the hand and facial region is thin. *Deep fascia* is an inward extension of the superficial fascia. It occurs between individual muscles and also surrounds adjacent muscles to bind them into functional groups. Deep fascia generally lacks adipose tissue.

Surrounding each muscle is a connective tissue sheath known as the **epimysium** (*ep˝ĭ-mis´e-um*) (fig. 12.1). The fibers of this sheath are continuous with those of the tendons. Additionally, the connective tissue fibers from the epimysium extend into the body of the muscle, subdividing it into bundles. These subdivisions within the muscle are known as **fasciculi** (*fă-sik´yu-li*), and are the "strings" in stringy meat. Each fasciculus is surrounded by its own connective tissue sheath, known as the **perimysium.**

Dissection of a muscle fasciculus under a microscope reveals that it, in turn, is composed of many **muscle fibers** (or *myofibers*) surrounded by wisps of connective tissue called **endomysium** (fig. 12.1). Since the connective tissue of the tendons, epimysium, perimysium, and endomysium is continuous, muscle fibers do not normally pull out of the tendons when they contract.

## Skeletal Muscle Fibers

The muscle fibers are actually the cells of the muscle. Despite their unusual fibrous shape, muscle fibers have the same organelles that are present in other cells: mitochondria, intracellular membranes, glycogen granules, and others. The most distinctive feature of skeletal muscle fibers, however, is their *striated appearance* when viewed microscopically (fig. 12.2). The striations (stripes) are produced by alternating dark and light bands that appear to cross the width of the fiber.

The dark bands of skeletal muscle fibers are called **A bands** and the light bands are called **I bands.** At high magnification in an electron microscope, thin, dark lines can be seen in the middle of the I bands. These are called **Z lines.** The labels A, I, and Z are useful in describing the functional architecture of muscle fibers and were derived in the course of muscle research. The letters A and I stand for *anisotropic* and *isotropic,* respectively, which indicate the behavior of polarized light as it passes through these regions. The letter Z comes from the German word *Zwischenscheibe,* which translates to "between disc." These derivations are of historical interest only.

........................................

aponeurosis: Gk. *aponeurosis*, change into a tendon
retinaculum: L. *retinere*, to hold back (retain)

fascia: L. *fascia*, a band or girdle
epimysium: Gk. *epi*, upon; *myos*, muscle
fasciculus: L. *fascis*, bundle
perimysium: Gk. *peri*, around; *myos*, muscle

## Table 12.1    Summary of muscle tissue

| Type | Structure and function | Location | |
|------|------------------------|----------|---|
| Smooth | Elongated, spindle-shaped fiber with single nucleus; involuntary movements of internal organs | Walls of hollow internal organs | |
| Cardiac | Branched, striated fiber with single nucleus and intercalated discs; involuntary rhythmic contraction | Heart muscle | |
| Skeletal | Multinucleated, striated, cylindrical fiber; voluntary movement of skeletal parts | Spanning joints and attached to bones of the skeleton | |

## FIGURE 12.1

The relationship between muscle tissue and connective tissue. (*a*) The fascia and tendon attaches a muscle to the periosteum of a bone. (*b*) The epimysium surrounds the entire muscle, and the perimysium separates and binds the fasciculi (muscle bundles). (*c*) The endomysium surrounds and binds individual muscle fibers. (*d*) An individual muscle fiber is composed of myofibrils, which consist of myofilaments of actin and myosin.

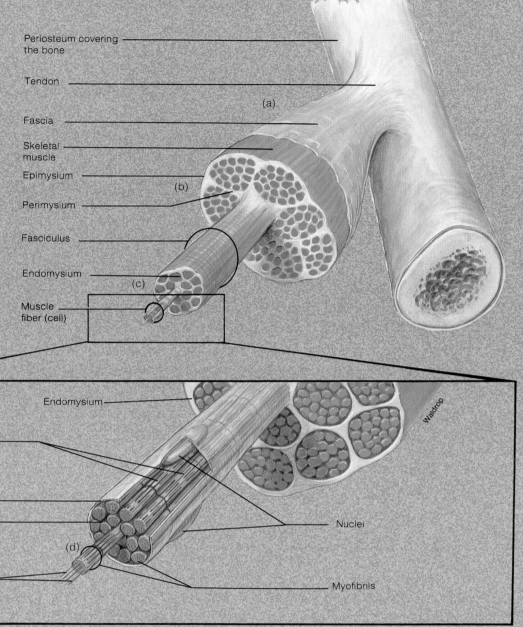

Periosteum covering the bone

Tendon

(a)

Fascia

Skeletal muscle

Epimysium

(b)

Perimysium

Fasciculus

Endomysium

(c)

Muscle fiber (cell)

Endomysium

Striations

Sarcolemma

Sarcoplasm

Nuclei

(d)

Myofilaments

Myofibrils

Waldrop

*Duschenne muscular dystrophy* is the most severe of the muscular dystrophies, afflicting one out of 3,500 boys each year. This disease, inherited as an X-linked recessive trait, involves progressive muscular wasting and usually results in death by the age of 20. The product of the defective gene is a protein named *dystrophin* that is associated with the sarcolemma. Using this information, scientists have recently developed laboratory tests that can detect this disease in fetal cells obtained by amniocentesis. In the future, genetic therapy may be possible. This research is aided by the development of a strain of mice who have an equivalent form of this disease. When the "good genes" for dystrophin are inserted into mouse embryos of this strain, the mice do not develop the disease. Insertion of the gene into large numbers of mature muscle cells, however, is more difficult and has currently met with only limited success.

## Types of Muscle Contractions

The contractile behavior of skeletal muscles is more easily studied in vitro (outside the body) than in vivo (within the body). When a muscle—for example, the gastrocnemius (calf muscle) of a frog—is studied in vitro, it is usually mounted so that one end is fixed and the other is movable. The mechanical force of the muscle contraction is transduced into an electric current, which can be amplified and displayed as pen deflections in a multichannel recorder (fig. 12.3). In this way, the contractile behavior of the whole muscle in response to experimentally administered electric shocks can be studied.

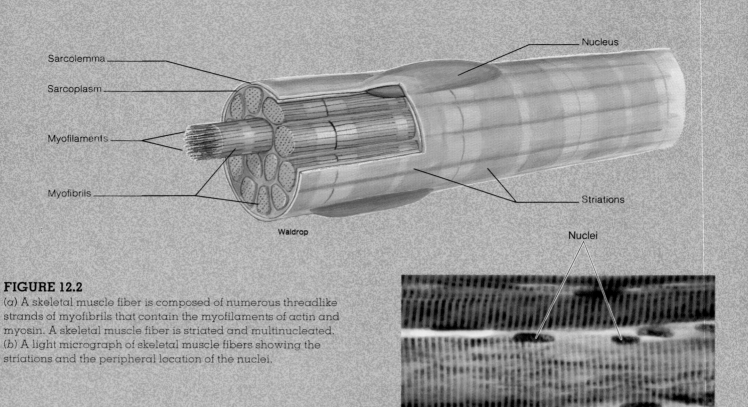

Sarcolemma

Sarcoplasm

Myofilaments

Myofibrils

Nucleus

Striations

Waldrop

Nuclei

**FIGURE 12.2**

(a) A skeletal muscle fiber is composed of numerous threadlike strands of myofibrils that contain the myofilaments of actin and myosin. A skeletal muscle fiber is striated and multinucleated. (b) A light micrograph of skeletal muscle fibers showing the striations and the peripheral location of the nuclei.

(b)

**FIGURE 12.3**

(a) A physiograph recorder. (b) Photograph and (c) illustration of the behavior of an isolated frog gastrocnemius muscle in response to electrical shocks.

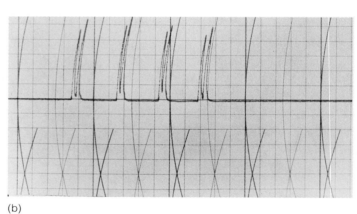

(b)

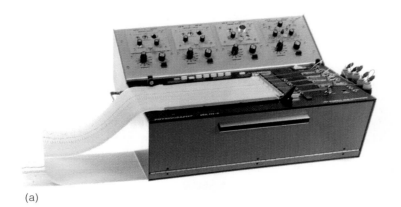

(a)

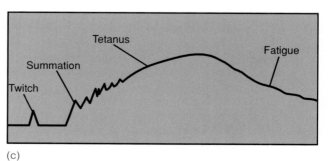

Twitch

Summation

Tetanus

Fatigue

(c)

**Twitch, Summation, and Tetanus** When the muscle is stimulated with a single electric shock of a sufficient voltage, it quickly contracts and relaxes. This response is called a **twitch.** Increasing the stimulus voltage increases the strength of the twitch, up to a maximum. The strength of a muscle contraction can thus be *graded*, or varied. This is an obvious requirement for the proper control of skeletal movements. If a second electric shock is delivered immediately after the first, it will produce a second twitch that may partially ride piggyback on the first. This response is called **summation.**

If the stimulator is set to deliver an increasing frequency of electric shocks automatically, the relaxation time between successive twitches will become shorter and shorter as the strength of contraction increases in amplitude. This effect is known as **incomplete tetanus.** Finally, at a particular fusion frequency of stimulation, there is no visible relaxation between successive twitches (fig. 12.3). Contraction is smooth and sustained, as it is during normal muscle contraction in vivo. This smooth, sustained contraction is called **complete tetanus.** The term *tetanus* should not be confused with the disease of the same name, which is accompanied by a painful state of muscle contracture, or *tetany.*

Stimulation of fibers within a muscle in vitro with an electric stimulator, or in vivo by nerve impulses through motor neurons, usually results in maximal, *all-or-none contractions* of the individual fibers. Stronger muscle contractions are produced by the stimulation of greater numbers of muscle fibers. Skeletal muscles can thus produce **graded contractions,** in which the strength can be varied by variations in the number of muscle fibers stimulated to contract.

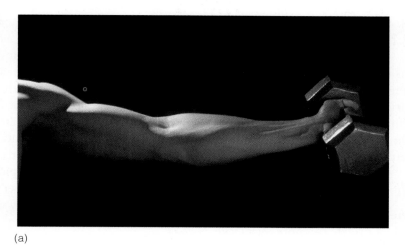

(a)

(b)

**FIGURE 12.4**

(a) Isometric and (b) isotonic contraction.

**Treppe** If the voltage of the electrical shocks delivered to an isolated muscle in vitro is gradually increased from zero, the strength of the muscle twitches will increase accordingly, up to a maximal value at which all of the muscle fibers are stimulated. This demonstrates the graded nature of the muscle contraction. If a series of electrical shocks at this maximal voltage is given to a fresh muscle, each of the twitches evoked will be successively stronger, up to a somewhat higher maximum. This demonstrates **treppe,** or the *staircase effect.* Treppe may represent a warm-up effect, and is believed to be due to an increase in the intracellular Ca++ that is needed for muscle contraction (as discussed in a later section).

**Isotonic and Isometric Contractions** In order for muscle fibers to shorten when they contract, they must generate a force that is greater than the opposing forces that act to prevent movement of the muscle's insertion. Flexion of the elbow joint, for example, occurs against the force of gravity and the weight of the objects being lifted (fig. 12.4). The tension produced by the contraction of each muscle fiber separately is insufficient to overcome these opposing forces, but the combined contractions of many muscle fibers may be sufficient to overcome the opposing forces and flex the elbow. In this case, the muscle and its stimulated muscle fibers shorten in length.

Contraction that results in muscle shortening is called **isotonic contraction,** so-called because the force of contraction remains relatively constant throughout the shortening process. If, however, the opposing forces are

..........................................
isotonic: Gk. *isos,* equal; *tonos,* tension

too great, or if the muscle fibers stimulated are too few in number to shorten the muscle, an **isometric contraction** is produced.

Isometric contractions can be voluntarily produced; for example, by lifting a weight and maintaining the forearm in a partially flexed position. One can then increase the amount of muscle tension (force) until the tension exceeds the load and the muscle begins to shorten; at this point, isometric contraction is converted to isotonic contraction.

## Series-Elastic Component

In order for a muscle to shorten when it contracts, and cause its insertion to move toward its origin, the noncontractile parts of the muscle and the connective tissue of its tendons must first be pulled taut. These structures, particularly the tendons, have elasticity—they resist distension—and when the distending force is released, they tend to spring back to their resting lengths. Since the tendons are in series with the force of muscle contraction, they provide what is called the **series-elastic component** of muscle contraction. The series-elastic component absorbs some of the tension as a muscle contracts, and must be pulled tight before muscle contraction can result in muscle shortening.

When the gastrocnemius muscle was stimulated with a single electric shock as described earlier, the amplitude of the twitch was reduced because some of the force of contraction was used to stretch the series-elastic component. Delivery of a second shock quickly after the first thus produced a greater degree of muscle shortening than the first shock, culminating at the fusion frequency of stimulation with complete tetanus, in which the strength of contraction was much greater than that of individual twitches.

Some of the energy used to stretch the series-elastic component during muscle contraction is released by elastic recoil when the muscle relaxes. This elastic recoil helps the muscles to return to their resting length and is of particular importance for the muscles involved in breathing. As we will see in chapter 24, inspiration is produced by muscle contraction and expiration is produced by the elastic recoil of the thoracic structures that were stretched during inspiration.

## Motor Units

Each muscle fiber receives a single axon terminal from a somatic motor neuron (fig. 12.5). The cell body of a somatic motor neuron, located in the spinal cord, gives rise to a single axon (chapter 6). Each axon, however, can produce a number of collateral branches, with each branch innervating a different muscle fiber. A somatic motor neuron and all of

the muscle fibers that it innervates are referred to collectively as a **motor unit** (fig. 12.6).

Whenever a somatic motor neuron is activated, all of the muscle fibers that it innervates are stimulated to contract with all-or-none twitches. In the body, graded contractions of whole muscles are produced by variations in the number of motor units that are activated. In order to make these graded contractions smooth and sustained, as in complete tetanus, different motor units must be activated at slightly different times in rapid succession.

Fine neural control over the strength of muscle contraction is optimal when many small motor units are involved. In the ocular muscles (see fig. 13.7) that position the eyes, for example, the *innervation ratio* (motor neuron:muscle fibers) of an average motor unit is 1 neuron per 23 muscle fibers. This affords a fine degree of control. By contrast, the innervation ratio of the gastrocnemius muscle (see fig. 13.36) averages 1 neuron per 1000 muscle fibers. Stimulation of these motor units results in more powerful contractions at the expense of finer gradations in contraction strength.

All of the motor units controlling the gastrocnemius are not the same size, however. Innervation ratios vary from 1:100 to 1:2000. A neuron that innervates fewer muscle fibers has a smaller cell body and is stimulated by lower levels of excitatory input than a larger neuron that innervates a greater number of muscle fibers. As a result, the smaller motor units are the ones that are used most often. When contractions of greater strength are required, larger and larger motor units are activated in what is known as **recruitment** of motor units.

# Mechanisms of Contraction

*The bands within each muscle fiber are composed of two kinds of protein filaments, or myofilaments. The A bands are composed of thick myofilaments and the I bands contain thin myofilaments. Movement of cross bridges that extend from the thick to the thin myofilaments causes sliding of the myofilaments, and thus muscle tension and shortening. The activity of the cross bridges is regulated by the availability of Ca++, which is increased by electrical stimulation of the muscle fiber. Electrical stimulation produces contractions of the muscle through the binding of Ca++ to regulatory proteins within the thin myofilaments.*

When muscle fibers are viewed in the electron microscope, each fiber is seen to be composed of many subunits known as **myofibrils** (*mi″ŏ-fi′brilz*) (fig. 12.7). Myofibrils are approximately 1 micrometer (1 μm) in diameter and extend in parallel rows from one end of the muscle fiber to the other. The myofibrils are so densely packed that other organelles, such as mitochondria and intracellular membranes, are restricted to the narrow cytoplasmic spaces that remain between adjacent myofibrils.

isometric: Gk. *isos*, equal; *metron*, measure

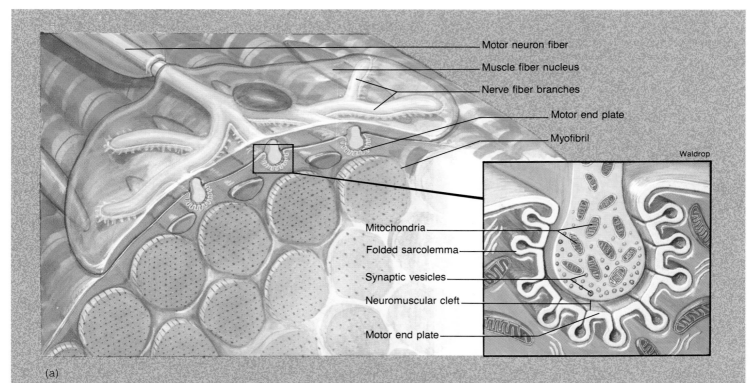

Motor neuron fiber

Muscle fiber nucleus

Nerve fiber branches

Motor end plate

Myofibril

Waldrop

Mitochondria

Folded sarcolemma

Synaptic vesicles

Neuromuscular cleft

Motor end plate

(a)

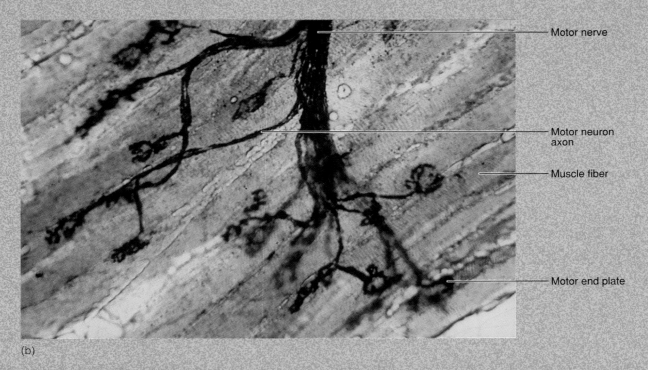

Motor nerve

Motor neuron axon

Muscle fiber

Motor end plate

(b)

## FIGURE 12.5

A motor end plate at the neuromuscular junction. (a) A neuromuscular junction is the site where the nerve fiber and muscle fiber meet. The motor end plate is the specialized portion of the sarcolemma of a muscle fiber surrounding the terminal end of the axon. (Note the slight gap between the membrane of the axon and that of the muscle fiber.) (b) A photomicrograph of muscle fibers and motor end plates. A motor neuron and the muscle fibers it innervates constitute a motor unit.

## FIGURE 12.6

A diagram illustrating the innervation of muscle fibers by different motor units. (Actually, many more muscle fibers would be included in a single motor unit than is shown in this drawing.)

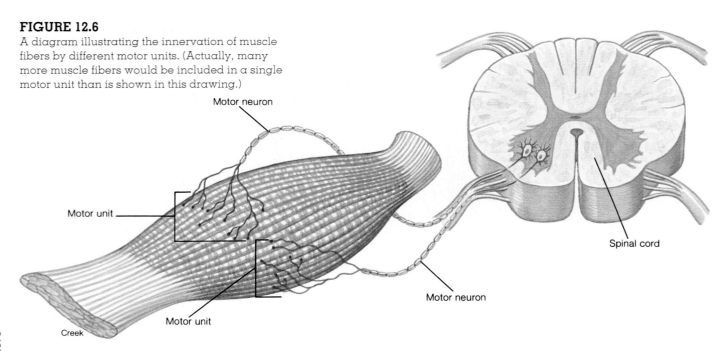

## FIGURE 12.7

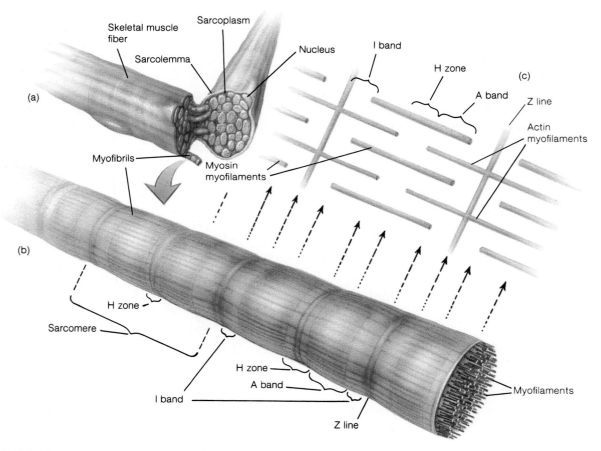

The structure of a myofibril. (a) Each of the many myofibrils of a skeletal muscle fiber is arranged into compartments (b) called sarcomeres. (c) The characteristic striations of a sarcomere due to the arrangement of thin and thick myofilaments, composed of actin and myosin, respectively.

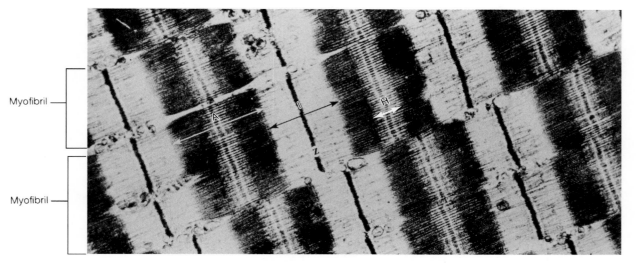

Myofibril

Myofibril

**FIGURE 12.8**
An electron micrograph of a longitudinal section of myofibrils, showing
A, H, and I bands. (Note how the dark and light bands of each myofibril
are stacked in register.)

With the electron microscope, it can be seen that the muscle fiber does not have striations that extend from one side of the fiber to the other. It is the myofibrils that are striated with dark (A) and light (I) bands (fig. 12.8). The striated appearance of the entire muscle fiber when seen with a light microscope is an illusion created by the alignment of the dark and light bands of the myofibrils from one side of the fiber to the other. Since the separate myofibrils are not clearly seen at low magnification, the dark and light bands appear to be continuous across the width of the fiber.

When a myofibril is observed at high magnification in longitudinal section, we see that the banding pattern is produced by an orderly arrangement of myofilaments. The A bands are seen to contain **thick myofilaments** (about 110 angstroms [110 Å] thick; 1 Å = $10^{-10}$m) that are stacked in register. It is these thick myofilaments that give the A band its dark appearance. The lighter I band, by contrast, contains **thin myofilaments** (about 50–60 Å thick). The thick myofilaments are composed of the protein **myosin** (*mi'ŏ-sin*), and the thin filaments are composed primarily of the protein **actin** (*ak'tin*).

The I bands within a myofibril are the lighter areas that extend from the edge of one stack of thick myofilaments to the edge of the next stack of thick myofilaments. They are light in appearance because they contain only thin myofilaments. The thin myofilaments, however, do not end at the edges of the I bands. Instead, each thin myofilament continues partway into the A bands on each side. Since thick and thin myofilaments overlap at the edges of each A band, the edges of the A band are darker in appearance than the central region. The central lighter regions of the A bands are called *H bands* (for *helle*, a German word meaning "bright"). The central H bands contain only thick myofilaments that are not overlapped with thin myofilaments.

In the center of each I band is a thin, dark Z line. The arrangement of thick and thin myofilaments between a pair of Z lines forms a repeating pattern that serves as the basic subunit of striated muscle contraction. These subunits, from Z to Z, are known as **sarcomeres** (*sar'kŏ-mērz*). A longitudinal section of a myofibril presents a side view of successive sarcomeres.

This side view is, in a sense, misleading. There are numerous sarcomeres within each myofibril that are out of the plane of the section (and out of the picture). A better appreciation of the three-dimensional structure of a myofibril can be obtained by viewing the myofibril in transverse section. In this view, it can be seen that the Z lines are actually disc shaped, and that the thin myofilaments that penetrate these Z discs surround the thick myofilaments in a hexagonal arrangement (fig. 12.9c). If one concentrates on a single row of dark thick myofilaments in this transverse section, the alternating pattern of thick and thin myofilaments seen in longitudinal section becomes apparent.

## Sliding Filament Theory of Contraction

When a muscle contracts isotonically, it decreases in length as a result of the shortening of its individual fibers. Shortening of the muscle fibers, in turn, is produced by shortening

myosin: L. *myosin*, within muscle
actin: L. *actus*, motion, doing

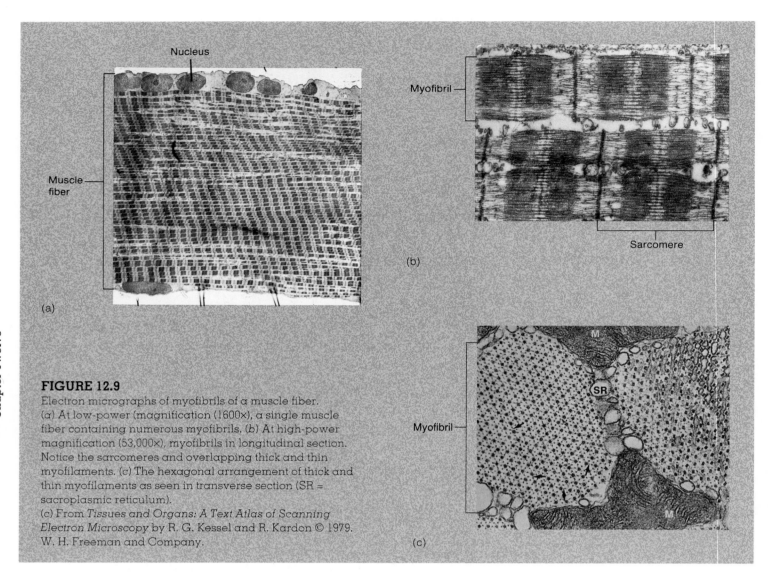

**FIGURE 12.9**
Electron micrographs of myofibrils of a muscle fiber.
(a) At low-power (magnification (1600×), a single muscle
fiber containing numerous myofibrils. (b) At high-power
magnification (53,000×), myofibrils in longitudinal section.
Notice the sarcomeres and overlapping thick and thin
myofilaments. (c) The hexagonal arrangement of thick and
thin myofilaments as seen in transverse section (SR =
sacroplasmic reticulum).
(c) From *Tissues and Organs: A Text Atlas of Scanning
Electron Microscopy* by R. G. Kessel and R. Kardon © 1979.
W. H. Freeman and Company.

of their myofibrils, which occurs as a result of the shortening
of the distance from Z line to Z line. As the sarcomeres
shorten in length, however, the A bands do *not* shorten but
instead move closer together. The I bands, which represent
the distance between A bands of successive sarcomeres,
decrease in length (table 12.2).

The thin actin myofilaments composing the I bands,
however, do not shorten. Close examination reveals that
the thick and thin filaments remain the same length during muscle contraction. Shortening of the sarcomeres is
produced not by shortening of the filaments, but rather
by the *sliding* of thin myofilaments over and between the
thick myofilaments. In the process of contraction, the
thin myofilaments on either side of each A band slide
deeper and deeper toward the center, producing increasing amounts of overlap with the thick myofilaments. The
I bands (containing only thin myofilaments) and H bands
(containing only thick myofilaments) thus get shorter
during contraction (fig. 12.10).

**Cross Bridges**    Sliding of the myofilaments is produced by
the action of numerous **cross bridges** that extend out from
the myosin toward the actin. These cross bridges are part of
the myosin proteins that extend from the axis of the thick
myofilaments to form "arms" that terminate in globular
"heads" (fig. 12.11). The orientation of the cross bridges on
one side of a sarcomere is opposite to that of the cross bridges
on the other side, so that when the myosin cross bridges
attach to actin on each side of the sarcomere they can pull
the actin from each side toward the center.

Isolated muscles in vitro are easily stretched (although
this is opposed in vivo by the stretch reflex, described in
chapter 18), demonstrating that the myosin cross bridges
are not attached to actin when the muscle is at rest. Each
globular head of a cross bridge contains an ATP-binding site
closely associated with an actin-binding site (fig. 12.12).
The globular heads function as **myosin ATPase** ($\bar{a}''te$-$pe'\bar{a}s$)
enzymes, splitting ATP into ADP and $P_i$. This reaction
occurs before the cross bridges combine with actin, and

## Table 12.2 The sliding filament theory of contraction

1. A myofiber, together with all its myofibrils, shortens by movement of the insertion toward the origin of the muscle.

2. Shortening of the myofibrils is caused by shortening of the sarcomeres—the distance between Z lines (or discs) is reduced.

3. Shortening of the sarcomeres is accomplished by sliding of the myofilaments—each myofilament remains the same length during contraction.

4. Sliding of the myofilaments is produced by asynchronous power strokes of myosin cross bridges, which pull the thin myofilaments (actin) over the thick myofilaments (myosin).

5. The A bands remain the same length during contraction, but are pulled toward the origin of the muscle.

6. Adjacent A bands are pulled closer together as the I bands between them shorten.

7. The H bands shorten during contraction as the thin myofilaments from each end of the sarcomeres are pulled toward the middle.

**FIGURE 12.10**

The sliding filament model of contraction. As the filaments slide, the Z lines are brought closer together. (1) Relaxed muscle, (2) partially contracted muscle, and (3) fully contracted muscle.

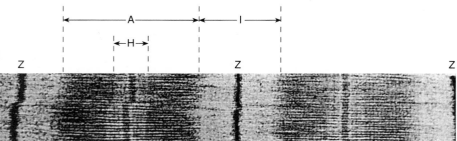
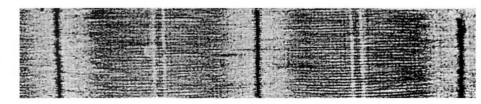

(a)

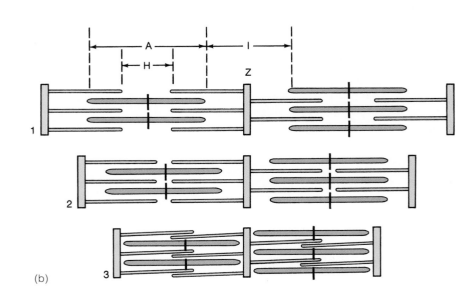

(b)

indeed is required for activating the cross bridges so that they can attach to actin. The ADP and P$_i$ remain bonded to the myosin heads until the cross bridges attach to the actin.

The myosin heads are able to bond to specific attachment sites in the actin subunits. When the cross bridges bond to actin, they undergo a conformation change. This has two effects: (1) ADP and P$_i$ are released and (2) the cross bridges change their orientation, resulting in a *power stroke* that pulls the thin myofilaments toward the center of the A bands. At the end of the power stroke, each cross bridge bonds to a fresh ATP molecule. This bonding of the cross bridge to a new ATP causes the cross bridge to break its bond with actin and resume its resting orientation. The myosin ATPase will then split ATP and become activated as in the previous cycle. Note that the splitting of ATP is required *before* a cross bridge can attach to actin and undergo a power stroke and that the attachment of a *new ATP* is needed for the cross bridge to release from actin at the end of a power stroke.

Because the cross bridges are quite short, a single contraction cycle and power stroke of all the cross bridges in a muscle would shorten the muscle by only about 1% of its resting length. Since muscles can shorten to up to 60% of their resting lengths, it is obvious that the contraction

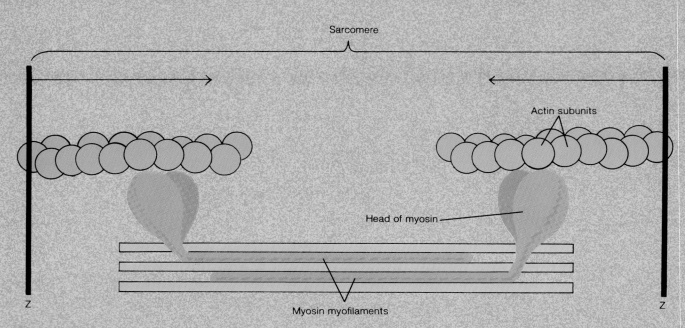

**FIGURE 12.11**

Myosin cross bridges are oriented in opposite directions on both sides of a sarcomere.

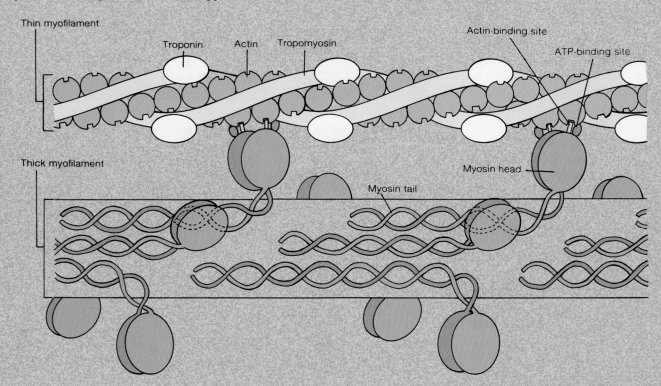

**FIGURE 12.12**

The structure of myosin showing its binding sites for ATP and for actin.

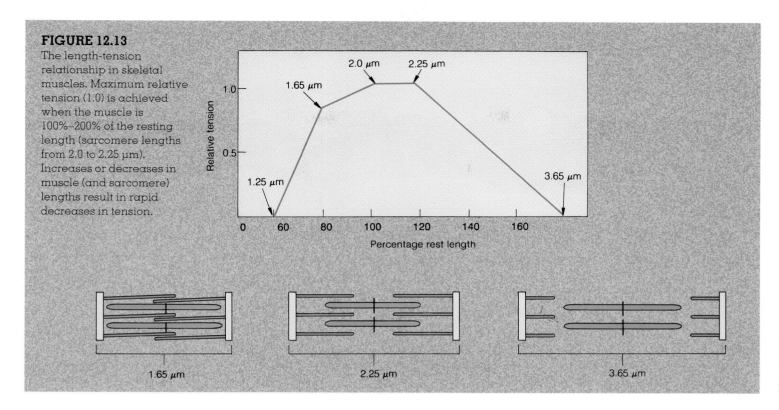

**FIGURE 12.13**
The length-tension relationship in skeletal muscles. Maximum relative tension (1.0) is achieved when the muscle is 100%–200% of the resting length (sarcomere lengths from 2.0 to 2.25 μm). Increases or decreases in muscle (and sarcomere) lengths result in rapid decreases in tension.

cycles must be repeated many times. In order for this to occur, the cross bridges must detach from the actin at the end of a power stroke, reassume their resting orientation, and then reattach to the actin and repeat the cycle.

During normal contraction, however, only about 50% of the cross bridges are attached at any given time. Thus, the power strokes are not in synchrony, as the strokes of a competitive rowing team would be. Rather, they resemble the actions of a team engaged in a tug-of-war, where the pulling action of the members is asynchronous. Some cross bridges are engaged in power strokes at all times during the contraction.

 The detachment of a cross bridge from actin at the end of a power stroke requires that a new ATP molecule bind to the myosin ATPase. The importance of this process is illustrated by the muscular contracture called *rigor mortis* that occurs due to lack of ATP when a muscle dies. This results in the formation of rigor complexes between myosin and actin that cannot detach. In rigor mortis, all of the cross bridges are attached to actin at the same time.

**Length–Tension Relationship**    The strength of a muscle's contraction depends on a number of factors. These include the number of muscle fibers within the muscle that are stimulated to contract, the thickness of each muscle fiber (thicker fibers have more myofibrils and thus can exert more power), and the initial length of the muscle fibers when they are at rest.

There is an ideal resting length for muscle fibers. This is the length at which they can generate maximum force. When the resting length exceeds this ideal, the overlap between actin and myosin is so small that few cross bridges can attach. When the muscle is stretched to the point where there is no overlap of actin with myosin, no cross bridges can attach to the thin myofilaments and the muscle cannot contract. When the muscle is shortened to about 60% of its resting length, the Z lines abut the thick myofilaments so that further contraction cannot occur.

The strength of a muscle's contraction can be measured by the force required to prevent it from shortening. Under these isometric conditions, the strength of contraction, or *tension*, can be measured when the muscle length at rest is varied. Maximum tension is produced when the muscle is at its normal resting length in vivo (fig. 12.13). In other words, if the muscle were any shorter or longer than its normal length, its strength of contraction would be reduced. This resting length is maintained by reflex contraction in response to passive stretching, as described in chapter 18.

## Regulation of Contraction

When the cross bridges attach to actin, they undergo power strokes and cause muscle contraction. In order for a muscle to relax, therefore, the attachment of myosin cross bridges to actin must be prevented. The regulation of cross-bridge attachment to actin is a function of two proteins that are associated with actin in the thin myofilaments.

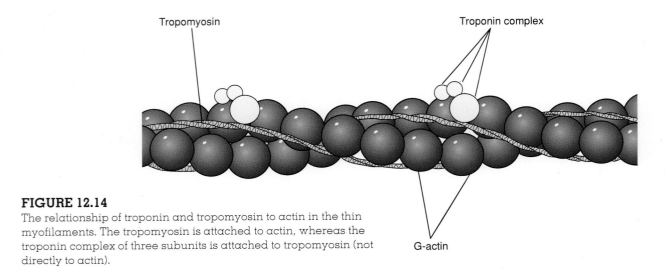

**FIGURE 12.14**
The relationship of troponin and tropomyosin to actin in the thin myofilaments. The tropomyosin is attached to actin, whereas the troponin complex of three subunits is attached to tropomyosin (not directly to actin).

Each actin myofilament is composed of two strands of fibrous actin (*F-actin*)—300 to 400 globular subunits (*G-actin*) arranged in a double row and twisted to form a helix (fig. 12.14). A different type of protein, known as **tropomyosin** (*tro″pŏ-mi″ŏ-sin*), lies within the groove between the F-actin strands. There are 40 to 60 tropomyosin molecules per thin myofilament, with each tropomyosin spanning a distance of approximately seven actin subunits.

Attached to the tropomyosin, rather than directly to the actin, is a third type of protein, called **troponin,** within the thin filaments. Troponin and tropomyosin work together to regulate the attachment of cross bridges to actin, and thus serve as a switch for muscle contraction and relaxation. In a relaxed muscle, the position of the tropomyosin in the thin filaments is such that it physically blocks the cross bridges from bonding to specific attachment sites in the actin. Thus, in order for the myosin cross bridges to attach to actin, the tropomyosin must be moved. This requires the interaction of troponin with $Ca^{++}$, as described in the next section.

**Role of $Ca^{++}$ in Muscle Contraction**   In a relaxed muscle, when tropomyosin blocks the attachment of cross bridges to actin, the concentration of $Ca^{++}$ in the **sarcoplasm** (cytoplasm of muscle cells) is very low. When the muscle fiber is stimulated to contract, certain mechanisms (to be discussed shortly) cause the concentration of $Ca^{++}$ in the sarcoplasm to quickly rise. Some of this $Ca^{++}$ attaches to a subunit of troponin, causing a conformational change that moves the troponin *and* its attached tropomyosin out of the way so that the cross bridges can attach to actin (fig. 12.15). Once the attachment sites on the actin are exposed, the cross bridges can bind to actin, undergo power strokes, and produce muscle contraction.

The position of the troponin-tropomyosin complexes in the thin myofilaments is thus adjustable. When $Ca^{++}$ is not attached to troponin, the tropomyosin is in a position that inhibits attachment of cross bridges to actin, preventing muscle contraction. When $Ca^{++}$ attaches to troponin, the troponin-tropomyosin complexes shift position. The cross bridges can then attach to actin, produce a power stroke, and detach from actin. Moreover, these contraction cycles can continue as long as $Ca^{++}$ is attached to troponin.

**Excitation-Contraction Coupling**   Muscle contraction is turned on when sufficient amounts of $Ca^{++}$ bind to troponin. This occurs when the $Ca^{++}$ concentration of the sarcoplasm rises above $10^{-6}$ molar. In order for muscle relaxation to occur, therefore, the $Ca^{++}$ concentration of the sarcoplasm must be lowered to below this level. Muscle relaxation is produced by the active transport of $Ca^{++}$ out of the sarcoplasm into the **sarcoplasmic reticulum** (fig. 12.16). The sarcoplasmic reticulum is a modified endoplasmic reticulum, consisting of interconnected sacs and tubes that surround each myofibril within the muscle fiber.

Most of the $Ca^{++}$ in a relaxed muscle fiber is stored within expanded portions of the sarcoplasmic reticulum known as *terminal cisternae*. When a muscle fiber is stimulated to contract by either a motor neuron in vivo or by electric shocks in vitro, the stored $Ca^{++}$ is released from the sarcoplasmic reticulum so that it can attach to troponin. When a muscle fiber is no longer stimulated, the $Ca^{++}$ from the sarcoplasm is actively transported back into the sarcoplasmic reticulum. Now, in order to understand how the release and uptake of $Ca^{++}$ is regulated, one more organelle within the muscle fiber must be described.

The terminal cisternae of the sarcoplasmic reticulum are separated by only a very narrow gap from **transverse**

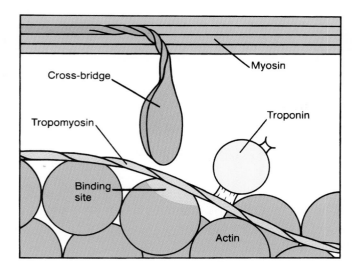 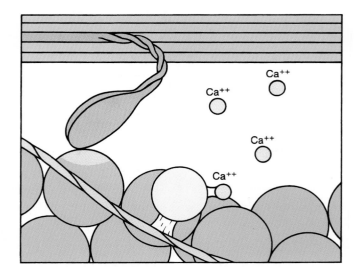

**FIGURE 12.15**

The attachment of Ca⁺⁺ to troponin causes movement of the troponin-tropomyosin complex, which exposes binding sites on the actin. The myosin cross bridges can then attach to actin and undergo a power stroke.

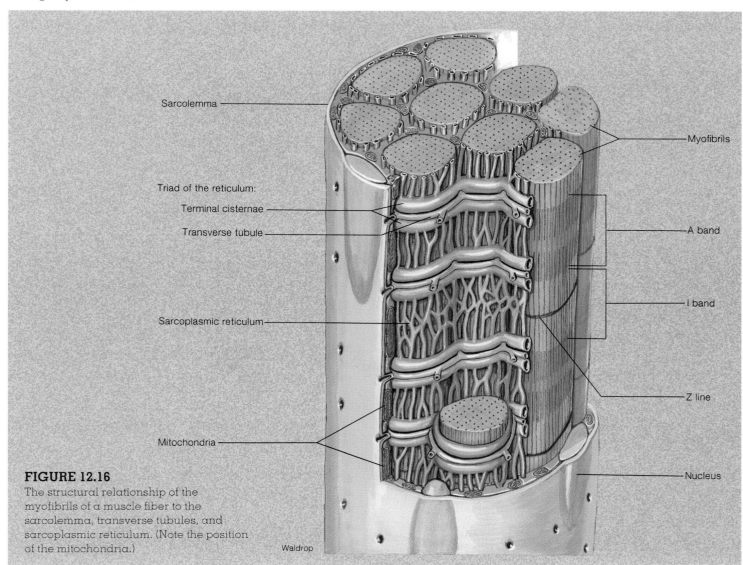

**FIGURE 12.16**

The structural relationship of the myofibrils of a muscle fiber to the sarcolemma, transverse tubules, and sarcoplasmic reticulum. (Note the position of the mitochondria.)

Waldrop

## Table 12.3    Summary of events in excitation-contraction coupling

1. Impulses in a somatic motor nerve cause the release of acetylcholine neurotransmitter at the myoneural junction (one myoneural junction per myofiber).

2. Acetylcholine, through its interaction with receptors in the muscle cell membrane (sarcolemma), produces impulses that are regenerated across the sarcolemma.

3. The membranes of the transverse tubules (T tubules) are continuous with the sarcolemma and conduct impulses deep into the muscle fiber.

4. Impulses in the T tubules, by a mechanism that is poorly understood, stimulate the release of $Ca^{++}$ from the terminal cisternae of the sarcoplasmic reticulum.

5. $Ca^{++}$ released into the sarcoplasm attaches to troponin, causing a change in its structure.

6. The shape change in troponin causes its attached tropomyosin to shift position in the actin myofilament, thus exposing bonding sites for the myosin cross bridges.

7. Myosin cross bridges, previously activated by the hydrolysis of ATP, attach to actin.

8. Once the previously activated cross bridges attach to actin, they undergo a power stroke and pull the thin myofilaments over the thick myofilaments.

9. Attachment of fresh ATP allows the cross bridges to detach from actin and repeat the contraction cycle as long as $Ca^{++}$ remains attached to troponin.

10. When impulses stop being produced, the sarcoplasmic reticulum actively accumulates $Ca^{++}$ and tropomyosin moves again to its inhibitory position.

tubules (or **T tubules**), which are narrow membranous tunnels formed from and continuous with the **sarcolemma** (cell membrane). The transverse tubules open to the extracellular environment through pores in the cell surface and are capable of conducting electrical impulses from the cell membrane. The stage is now set to explain how a motor neuron stimulates a muscle fiber to contract.

Nerve impulses arrive at the axon terminal of a motor neuron and evoke the production of new electrical impulses in the skeletal muscle fibers. These events are discussed in detail in chapter 14, but they can be briefly summarized here as follows: (1) impulses at the axon terminal stimulate the release of a chemical transmitter known as acetylcholine (ACh); (2) ACh diffuses to the muscle fiber and alters the permeability of the cell membrane to specific ions; and (3) the flow of ions (first $Na^+$, then $K^+$) across the membrane produces the electrical impulses in the muscle fiber.

These impulses are conducted along the sarcolemma and into the interior of the fiber by the transverse tubules. Impulses conducted by the transverse tubules then *cause the release of $Ca^{++}$ from the sarcoplasmic reticulum.* The released $Ca^{++}$ binds to troponin, causing the displacement of tropomyosin and allowing the actin to bind to the myosin cross bridges. Muscle contraction is thus stimulated.

As long as electrical impulses continue to be produced, which is as long as the neural stimulation of the muscle is maintained, $Ca^{++}$ will remain attached to troponin and cross bridges will be able to undergo contraction cycles. When neural activity and the production of impulses in the mus-

cle fiber cease, the sarcoplasmic reticulum actively accumulates $Ca^{++}$ and muscle relaxation occurs. Note that the return of $Ca^{++}$ to the sarcoplasmic reticulum involves active transport, and thus requires the hydrolysis of ATP. ATP is therefore needed for muscle relaxation as well as for muscle contraction. Neural regulation of skeletal muscle contraction is mediated by calcium ions. The mechanisms of electrical excitation and the mechanisms of muscle contraction (sliding of the filaments) are coupled through adjustments of the sarcoplasmic $Ca^{++}$ concentration. This is known as **excitation-contraction coupling** and is summarized in table 12.3.

# Energy Requirements of Skeletal Muscles

*Skeletal muscles generate ATP through aerobic and anaerobic respiration and through the use of phosphate groups donated by creatine phosphate. The aerobic and anaerobic abilities of skeletal muscle fibers differ according to muscle fiber type. Slow-twitch (type I) fibers are adapted for aerobic respiration; fast-twitch (type II) fibers are adapted for anaerobic respiration.*

Skeletal muscles at rest obtain most of their energy from the aerobic respiration of fatty acids. During exercise, muscle glycogen and blood glucose are also used as energy sources. Energy obtained by cell respiration is used to make ATP, which serves as the immediate source of energy for

(1) the movement of the cross bridges for muscle contraction and (2) the pumping of $Ca^{++}$ into the sarcoplasmic reticulum for muscle relaxation.

## Metabolism of Skeletal Muscles

Skeletal muscles respire anaerobically for the first 45 to 90 seconds of moderate-to-heavy exercise because the cardiopulmonary system requires this amount of time to sufficiently increase the oxygen supply to the exercising muscles. If exercise is moderate and the person is in good physical condition, aerobic respiration contributes the major portion of the skeletal muscle energy requirements following the first 2 minutes of exercise (fig. 12.17).

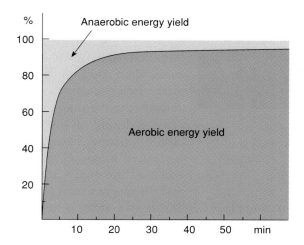

**FIGURE 12.17**
The relative contributions of anaerobic and aerobic respiration to the total energy in a well-trained person performing at maximal effort.

**Maximal Oxygen Uptake and Oxygen Debt** The maximum rate of oxygen consumption (by aerobic respiration) in the body is called the **maximal oxygen uptake,** and it is determined primarily by the person's age, size, and sex. It is about 15%–20% higher for males than for females and is highest at age 20 for both sexes. Some world-class athletes have maximal oxygen uptakes that are twice the average for their age and sex. This appears to be due largely to genetic factors, but training can increase the maximum oxygen uptake by about 20%.

When a person stops exercising, the rate of oxygen uptake does not immediately go back to pre-exercise levels; it returns slowly (the person continues to breathe heavily for some time afterward). This extra oxygen is used to repay the **oxygen debt** incurred during exercise. The oxygen debt includes oxygen that was withdrawn from savings deposits—hemoglobin in blood and myoglobin in muscle (see chapter 24); the extra oxygen required for metabolism by tissues warmed during exercise; and the oxygen needed for the metabolism of the lactic acid produced during anaerobic respiration.

**Phosphocreatine** During sustained muscle activity, ATP may be utilized faster than the rate of ATP production through cell respiration. At these times, the rapid renewal of ATP is extremely important. This is accomplished by combining ADP with phosphate derived from another high-energy phosphate compound called **phosphocreatine** (*fos″fo-kre′ă-tēn*), or **creatine phosphate.**

The phosphocreatine concentration within muscle fibers is more than three times the concentration of ATP and represents a ready reserve of high-energy phosphate that can be donated directly to ADP (fig. 12.18). During times of rest, the depleted reserve of phosphocreatine can be restored by the reverse reaction—phosphorylation of creatine with phosphate derived from ATP.

 The enzyme that transfers phosphate between creatine and ATP is called *creatine kinase*, or *creatine phosphokinase*. Skeletal muscle and heart muscle each have a different form of this enzyme (they have different isoenzymes, as described in chapter 4). The skeletal muscle isoenzyme is elevated in the blood of people with *muscular dystrophy* (a degenerative disease of skeletal muscles). The plasma concentration of the isoenzyme characteristic of heart muscle is elevated in the condition of *myocardial infarction* (damage to heart muscle), and measurements of this enzyme are thus used as a means of diagnosing this condition.

## Slow- and Fast-Twitch Fibers

Skeletal muscle fibers can be divided on the basis of their contraction speed (time required to reach maximum tension) into **slow-twitch,** or **type I, fibers,** and **fast-twitch,** or **type II, fibers.** These differences are associated with different myosin ATPase isoenzymes, designated as "slow" and "fast," by which the two fiber types can be distinguished when they are appropriately stained (fig. 12.19). For example, the ocular muscles that position the eyes (see table 13.3) have a predominance of fast-twitch fibers and reach maximum tension in about 7.3 msec (milliseconds - thousandths of a second); the soleus muscle in the leg, by contrast, has a

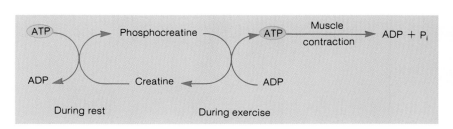

**FIGURE 12.18**
The production and utilization of phosphocreatine in muscles.

predominance of slow-twitch fibers and requires about 100 msec to reach maximum tension (fig. 12.20).

Muscles like the soleus (see fig. 13.34) are *postural muscles* that must be able to sustain a contraction for a prolonged period of time without becoming fatigued. The resistance to fatigue demonstrated by these muscles is aided by several features of slow-twitch fibers that endow them with a high capacity for aerobic respiration. Slow-twitch fibers have a rich capillary supply, numerous mitochondria and aerobic respiratory enzymes, and a high concentration of *myoglobin* pigment. Myoglobin is a red pigment, similar to the hemoglobin in red blood cells, that improves the delivery of oxygen to the slow-twitch fibers. Because of their high myoglobin content, slow-twitch fibers are also called *red fibers*.

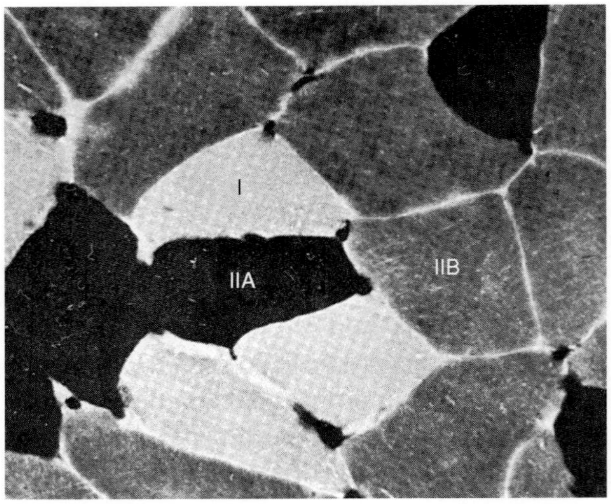

The thicker fast-twitch fibers have fewer capillaries and mitochondria than slow-twitch fibers and not as much myoglobin; hence, these fibers are also called *white fibers*. Fast-twitch fibers are adapted to respire anaerobically by a large store of glycogen and a high concentration of glycolytic enzymes. In addition to the type I (slow-twitch) and type II (fast-twitch) fibers, human muscles may also have an intermediate fiber form. These intermediate fibers are fast-twitch but also have a high aerobic capability. They are sometimes called type IIA, to distinguish them from the anaerobically adapted fast-twitch fibers, which are then labeled IIB. The three fiber types are compared in table 12.4.

Most muscles contain a mixture of fiber types. For example, the gastrocnemius muscle contains both fast- and slow-twitch fibers, although fast-twitch fibers predominate. A given somatic motor axon, however, only innervates muscle fibers of one type. The size of these motor units differ; the motor units composed of slow-twitch fibers tend to be smaller (have fewer fibers) than the motor units of fast-twitch fibers. Since motor units are recruited from smaller to larger when increasing effort is required, the smaller motor units with slow-twitch fibers are used most often in routine activities. Larger motor units with fast-twitch fibers, which can exert a great deal of force but which respire anaerobically and thus fatigue quickly, are used relatively infrequently and for only short periods of time.

## Muscle Fatigue

Muscle fatigue is the inability to maintain a particular muscle tension when the contraction is sustained or to reproduce a particular tension during rhythmic contraction over time. Fatigue during a sustained maximal contraction, when all the motor units are used and the rate of neural firing is maximal (for example, when lifting an extremely heavy weight) appears to be due to an accumulation of extracellular $K^+$. ($K^+$ leaves axons and muscle fibers during the production of electrical impulses.) This reduces the membrane potential of muscle fibers and interferes with their ability to produce impulses. Fatigue under these circumstances is of short duration; maximal tension can again be produced after less than a minute's rest.

Fatigue during moderate exercise occurs as the slow-twitch fibers deplete their reserve glycogen and fast-twitch fibers are increasingly recruited. Fast-twitch fibers obtain their energy through anaerobic respiration, converting glucose to lactic acid with a consequent rise in intracellular $H^+$ and a

**FIGURE 12.19**
Skeletal muscle (of a cat) stained to indicate the activity of myosin ATPase. Type II fibers contain higher ATPase activity than type I fibers.

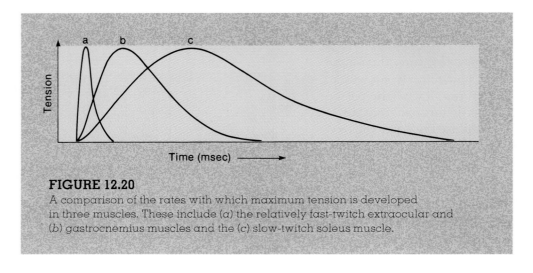

**FIGURE 12.20**
A comparison of the rates with which maximum tension is developed in three muscles. These include (*a*) the relatively fast-twitch extraocular and (*b*) gastrocnemius muscles and the (*c*) slow-twitch soleus muscle.

## Table 12.4 Characteristics of red, intermediate, and white muscle fibers

|  | Red (type I) | Intermediate (type IIA) | White (type IIB) |
|---|---|---|---|
| Diameter | Small | Intermediate | Large |
| Z-line thickness | Wide | Intermediate | Narrow |
| Glycogen content | Low | Intermediate | High |
| Resistance to fatigue | High | Intermediate | Low |
| Capillaries | Many | Many | Few |
| Myoglobin content | High | High | Low |
| Respiration type | Aerobic | Aerobic | Anaerobic |
| Twitch rate | Slow | Fast | Fast |
| Myosin ATPase content | Low | High | High |

## Table 12.5 The effects of endurance training (long-distance running, swimming, bicycling, etc.) on skeletal muscles

1. Improved ability to obtain ATP from oxidative phosphorylation
2. Increased size and number of mitochondria
3. Less lactic acid produced per given amount of exercise
4. Increased myoglobin content
5. Increased intramuscular triglyceride content
6. Increased lipoprotein lipase (enzyme needed to utilize lipids from blood)
7. Increased proportion of energy derived from fat; less from carbohydrates
8. Lower rate of glycogen depletion during exercise
9. Improved efficiency in extracting oxygen from blood

fall in pH. The decrease in muscle pH, in turn, inhibits the activity of key glycolytic enzymes, so that the rate of ATP production is reduced. This may interfere with the ability of the sarcoplasmic reticulum to accumulate $Ca^{++}$ by active transport, and thus interfere with the ability of the muscle to contract in response to electrical stimulation. The decreased cellular pH, in other words, produces fatigue by interfering with excitation-contraction coupling.

## Adaptations to Exercise

Maximal oxygen uptake, obtained during very strenuous exercise, averages 50 ml of $O_2$ per minute per kilogram body weight in males between 20 and 25 years of age (females average 25% lower). For trained endurance athletes (swimmers, long-distance runners), maximal oxygen uptakes can be as high as 86 ml of $O_2$ per minute per kilogram. When exercise is performed at low levels of effort, such that the oxygen consumption rate is below 50% of its maximum, the energy for muscle contraction is obtained almost entirely from aerobic cell respiration. Anaerobic cell respiration, with its consequent production of lactic acid, contributes to the energy requirements as the exercise level rises and more than 60% of the maximal oxygen uptake is required. Highly trained endurance athletes, however, can continue to respire aerobically, with little lactic acid production, at up to 80% of their maximal oxygen uptake. These athletes thus produce less lactic acid at a given level of exercise than most of us; therefore, they are less subject to fatigue than the average person.

Since muscle fiber types are determined by their innervations, endurance training cannot change fast-twitch (type II) fibers to slow-twitch (type I) fibers. All muscle fiber types, however, adapt to endurance training by an increase in myoglobin and aerobic respiratory enzymes. In fact, maximal

oxygen uptake can be increased by as much as 20% through endurance training. In addition to changes in aerobic capacity, muscle fibers show an increase in their content of triglycerides, which serve as an alternate energy source and help to spare their stores of glycogen. A summary of the changes that occur as a result of endurance training is presented in table 12.5.

Although endurance training results in more efficient muscle metabolism, it does not make muscles larger. Muscle enlargement is produced only by frequent periods of high-intensity exercise in which muscles work against a high resistance, as in weight lifting. As a result of resistance training, type II muscle fibers become thicker, and the muscle therefore grows by hypertrophy (an increase in cell size rather than number of cells). The myofibrils within the muscle fiber thicken as a result of the addition of new sarcomeres. After a myofibril attains a certain thickness it may split into two myofibrils, each of which may then become thicker due to the addition of sarcomeres. Muscle hypertrophy, in summary, is associated with an increase in the number and size of myofibrils within the muscle fibers.

# Cardiac and Smooth Muscle

*Cardiac muscle, like skeletal muscle, is striated and contains sarcomeres that shorten by sliding of thin and thick myofilaments. Unlike skeletal muscle, however, cardiac muscle can produce impulses and contract spontaneously. Smooth muscle lacks sarcomeres; hence, it does not have a striated appearance. It does contain actin and myosin that produce contractions in response to a unique regulatory mechanism.*

Unlike skeletal muscles, which are voluntary effectors regulated by somatic motor neurons, cardiac and smooth muscles are involuntary effectors and are regulated by autonomic motor neurons. Although skeletal muscle differs from cardiac and smooth muscle in important ways, all types of muscle have certain features in common. First, all muscle types are believed to contract by means of sliding of thin actin myofilaments over thick myosin myofilaments. Second, the sliding of the myofilaments is produced by the action of myosin cross bridges in all muscle types. Finally, excitation-contraction coupling in all types of muscle involves Ca$^{++}$.

## Cardiac Muscle

Like skeletal muscle fibers, cardiac muscle fibers, or **myocardial cells,** are striated; they contain actin and myosin filaments arranged in the form of sarcomeres and they contract by means of the sliding filament mechanism. The long, fibrous skeletal muscle fibers, however, are structurally and functionally separated from each other, whereas the myocardial cells are short, branched, and interconnected. Adjacent myocardial cells are joined by electrical synapses, or **gap junctions** (described in chapter 14). Gap junctions in cardiac muscle have an affinity for stain that makes them appear as dark lines between adjacent cells when viewed in the light microscope. These dark-staining lines are known as **intercalated discs** (see table 12.1).

Electrical impulses that originate at any point in a mass of myocardial cells, called a *myocardium*, can spread to all cells in the mass that are joined by gap junctions. Because all cells in a myocardium are electrically joined, a myocardium behaves as a single functional unit. Thus, unlike skeletal muscles, which can produce graded contractions in accordance with the number of fibers stimulated, a myocardium contracts with an *all-or-none contraction*. The heart contains two distinct myocardia (within the walls of the atria and the walls of the ventricles), as will be described in chapter 21.

Unlike skeletal muscles, which require external stimulation by somatic motor nerves before they can produce electrical impulses and contract, cardiac muscle is able to produce impulses automatically. Cardiac impulses normally originate in a specialized group of cells called the *pacemaker*. However, the rate of these spontaneously produced impulses, and thus the rate of the heartbeat, is regulated by autonomic innervation.

## Smooth Muscle

Smooth (visceral) muscles are arranged in circular layers around the walls of blood vessels and bronchioles (small air passages in the lungs). Both circular and longitudinal smooth muscle layers occur in the tubular gastrointestinal tract, the ureters (which transport urine), the ductus deferentia (which transport sperm), and the uterine tubes (which transport ova). The alternate contraction of these circular and longitudinal smooth muscle layers produces **peristaltic waves** that propel the contents of the tubes in one direction.

Although smooth muscle cells do not contain sarcomeres, they do contain a great deal of actin and some myosin, which produces a ratio of thin-to-thick myofilaments of about 16:1 (in striated muscles the ratio is 2:1). Unlike striated muscles, in which the thick myofilaments are short and stacked between Z lines in sarcomeres, the myosin myofilaments in smooth muscle cells are quite long.

The long length of myosin myofilaments in smooth muscle and the fact that they are not organized into sarcomeres may be advantageous for the function of smooth muscles. Smooth muscles must be able to contract even when greatly stretched, as in the urinary bladder, for example, where smooth muscle cells may be stretched up to two and a half times their resting length. The smooth muscle cells of the uterus may be stretched up to eight times their original length by the end of a pregnancy. Striated muscles, because of their structure, lose their ability to contract when the sarcomeres are stretched to the point where actin and myosin no longer overlap.

As in striated muscles, the contraction of smooth muscles is triggered by a sharp rise in the Ca$^{++}$ concentration within the cytoplasm of the muscle cells. However, the sarcoplasmic reticulum of smooth muscles is less developed than that of skeletal muscles, and Ca$^{++}$ released from this organelle may account for only the initial phase of smooth muscle contraction. Extracellular Ca$^{++}$ diffusing into the smooth muscle cell through its cell membrane is responsible for sustained contractions. This Ca$^{++}$ enters primarily through membrane channels that open in response to electrical stimulation. The greater the stimulation, the greater the number of calcium ions that can diffuse into the muscle cell and the stronger the contraction.

Drugs such as *verapamil* and related compounds are *calcium channel blockers*. These drugs inhibit the opening of Ca$^{++}$ channels in the smooth muscle membrane, causing the smooth muscle cells of blood vessels to relax. This, in turn, produces vasodilation, which is useful in the treatment of such conditions as *angina pectoris* (pain resulting from heart disease) and *hypertension* (high blood pressure).

The events that follow the entry of Ca$^{++}$ into the cytoplasm are somewhat different in smooth muscles than in striated muscles. Ca$^{++}$ combines with troponin in striated muscles; however, there is no troponin in smooth muscle fibers. In smooth muscles, Ca$^{++}$ combines with a protein called **calmodulin.** Calmodulin is structurally similar to troponin and is also involved in the action of some hormones, as described in chapter 19. The calmodulin-Ca$^{++}$ complex

## Table 12.6 Comparison of skeletal, cardiac, and smooth muscle

| Skeletal muscle | Cardiac muscle | Smooth muscle |
|---|---|---|
| Striated; actin and myosin arranged in sarcomeres | Striated; actin and myosin arranged in sarcomeres | Not striated; more actin than myosin; actin inserts into dense bodies and cell membrane |
| Well-developed sarcoplasmic reticulum and transverse tubules | Moderately developed sarcoplasmic reticulum and transverse tubules | Poorly developed sarcoplasmic reticulum; no transverse tubules |
| Contains troponin in the thin myofilaments | Contains troponin in the thin myofilaments | Contains a $Ca^{++}$ binding protein; may be located in thick myofilaments |
| $Ca^{++}$ released into cytoplasm from sarcoplasmic reticulum | $Ca^{++}$ enters cytoplasm from sarcoplasmic reticulum and extracellular fluid | $Ca^{++}$ enters cytoplasm from extracellular fluid, sarcoplasmic reticulum, and perhaps mitochondria |
| Cannot contract without nerve stimulation; denervation results in muscle atrophy | Can contract without nerve stimulation; action potentials originate in pacemaker cells of heart | Maintains tone in absence of nerve stimulation; visceral smooth muscle produces pacemaker potentials; denervation results in hypersensitivity to stimulation |
| Muscle fibers stimulated independently; no gap junctions | Gap junctions present as intercalated discs | Gap junctions in most smooth muscles |

thus formed combines with and activates an enzyme called *myosin light chain kinase*. This enzyme catalyzes the phosphorylation of (addition of phosphate groups to) the myosin cross bridges. In smooth muscle cells, the cross bridges must be phosphorylated before they can bond to actin, which is not the case in striated muscles.

In the preceding sequence of events, the concentration of $Ca^{++}$ in the smooth muscle cell cytoplasm determines how many cross bridges will combine with actin, and thus determines the strength of contraction. The concentration of $Ca^{++}$ is in turn regulated by the degree of electrical stimulation. Thus, greater degrees of neural stimulation result in greater concentrations of $Ca^{++}$ in the cytoplasm, greater phosphorylation of the cross bridges, and a stronger contraction of the smooth muscle.

In addition to being graded, the contractions of smooth muscle cells are slow and sustained. The slowness of contraction is related to the fact that myosin ATPase in smooth muscle is slower in its action (splitting ATP for the cross-bridge cycle) than it is in striated muscle. The sustained nature of smooth muscle contraction is explained by the theory that cross bridges in smooth muscles can enter a *latch state*.

The latch state allows smooth muscle to maintain its contraction in a very energy-efficient manner, hydrolyzing less ATP than would otherwise be required. This ability is obviously important for smooth muscles, given that they encircle the walls of hollow organs and must sustain contractions for very long time periods. The mechanisms by which the latch state is produced, however, are complex and poorly understood.

Despite their differences, it is currently believed that both smooth muscles and striated muscles contract by means of a sliding filament mechanism, as described earlier in this chapter. The three muscle types—skeletal, cardiac, and smooth—are compared in table 12.6.

**Single-Unit and Multiunit Smooth Muscles** Smooth muscles are often grouped into two functional categories: **single-unit** and **multiunit.** Single-unit smooth muscles have numerous gap junctions (electrical synapses) between adjacent fibers that weld them together electrically; thus, they behave as a single unit, much like cardiac muscle. Most smooth muscles, including those in the gastrointestinal tract and uterus, are single-unit.

Single-unit smooth muscles display *pacemaker* activity, similar to that of cardiac muscle, in which certain fibers stimulate others in the mass. Single-unit smooth muscles also display intrinsic, or *myogenic*, electrical activity and contraction in response to stretch. For example, the stretch induced by an increase in the volume of a small artery or a section of the gastrointestinal tract can stimulate myogenic contraction. Such contraction does not require stimulation by autonomic nerves.

Contraction of multiunit smooth muscles, by contrast, requires nerve stimulation. Multiunit smooth muscles have few, if any, gap junctions and therefore the individual cells must be stimulated separately by nerve endings. Examples of multiunit smooth muscles are the arrector pili muscles in the skin and the ciliary muscles attached to the lens of the eye.

## Development of Skeletal Muscles

The formation of skeletal muscle tissue begins during the fourth week of embryonic development as specialized mesodermal cells, called **myoblasts,** begin rapid mitotic division (fig. 1). The proliferation of new cells continues while the myoblast cells migrate and fuse together into **syncytial** (*sin-sish´al*) **myotubes.** A syncytium is a multinucleated protoplasmic mass formed by the union of originally separate cells. At 9 weeks, primitive myofilaments course through the myotubes and the nuclei of the contributing myoblasts are centrally located. Growth in length continues through the addition of myoblasts.

It is not certain when skeletal muscle is sufficiently developed to sustain contractions, but by week 17 the fetal

movements known as *quickening* are strong enough to be recognized by the mother. The individual muscle fibers have now thickened, the nuclei have moved peripherally, and the myofilaments can be recognized as

............................................

myoblast: Gk. *myos,* muscle; *blastos,* germ

alternating dark and light bands. Growth in length still continues through the addition of myoblasts. Shortly before a baby is born, the formation of myoblast cells ceases. At this time, all of the muscle cells have been determined.

............................................

syncytial: Gk. *syn,* with; *cyto,* cell

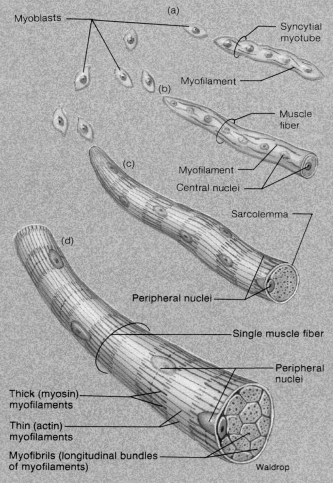

**FIGURE 1**

The development of skeletal muscle fibers. (*a*) At 5 weeks, the myotube is formed as individual cell membranes are broken down. Myotubes grow in length by incorporating additional myoblasts; each adds an additional nucleus. (*b*) Muscle fibers are distinct at 9 weeks, but the nuclei are still centrally located and growth in length continues through the addition of myoblasts. (*c*) At 5 months, thin (actin) and thick (myosin) myofilaments are present and moderate growth in length is still occurring. (*d*) By birth, the striated myofilaments have aggregated into bundles, the fiber has thickened, and the nuclei have shifted to the periphery. Myoblast activity ceases, and all the muscle fibers a person will have are formed.

# Interactions of the Muscular System with Other Body Systems

## Integumentary System
- Covers and protects the body musculature
- Radiates body heat
- Synthesizes vitamin D
- Facial musculature attached to skin produces facial expression when contracted

## Skeletal System
- Stores calcium and phosphate
- Provides attachment sites for muscles
- Joints of skeleton provide levers for movement
- Enables body movement and stabilizes joints
- Muscle contractions maintain the health and strength of bone

## Nervous System
- Coordinates muscle contraction
- Increases cardiac output and respiratory rates during periods of muscle activity
- Sensory receptors monitor body position via autonomic nervous system
- Muscles permit eye movement and speech

## Endocrine System
- Sex hormones promote muscle development and maintenance
- Specific hormones regulate calcium and phosphate concentrations
- Provides protection to certain endocrine glands
- Epinephrine and norepinephrine influence contractions of cardiac and smooth muscles

## Circulatory System
- Transports $O_2$ and $CO_2$, nutrients, and fluids to and from muscles; removes lactic acid and heat
- Tonus and voluntary muscle contractions assist blood movement, particularly within veins
- Cardiac and smooth muscles contribute to function of heart and blood vessels

## Lymphatic System
- Maintains balanced amount of interstitial fluid within muscle tissue
- Lymphocytes provide defense against infection
- Supports and protects superficial lymph nodes

## Respiratory System
- Provides $O_2$ for muscle metabolism and eliminates $CO_2$
- Respiratory muscles are responsible for ventilation of lungs; sound production

## Urinary System
- Eliminates metabolic wastes from muscles
- Assists regulation of calcium and phosphate concentrations
- Muscles of urinary tract surround urinary bladder and form urethral sphincter

## Digestive System
- Provides nutrients for growth, maintenance, and repair of muscles
- Regulates blood glucose
- Supports and protects organs of GI tract
- Muscular sphincters located at openings of GI tract

## Reproductive System
- Testicular androgen promotes growth of skeletal muscle
- Contributes to orgasm in both sexes
- Uterine muscle contractions aid in delivery of fetus

# Chapter Summary

## Structure and Actions of Skeletal Muscles (pp. 255–260)

1. Skeletal and cardiac muscle fibers are striated, whereas smooth muscle fibers are not.
2. Skeletal muscle fibers originate from a number of myoblasts that eventually join together to form the multinucleated skeletal muscle fiber.
3. Skeletal muscles are attached to bones by means of tendons.
   a. Muscles are separated by fascia and each muscle is covered by an epimysium.
   b. Muscle fibers are grouped into fasciculi that are surrounded by a connective tissue perimysium. Each fasiculus is composed of muscle fibers that are surrounded by an endomysium.
4. Muscles in vitro can exhibit twitch, summation, tetanus, and tonus.
   a. The rapid contraction and relaxation of muscle fibers is called a twitch.
   b. The stronger the electric shock, the stronger the muscle twitch. Whole muscles are capable of graded contractions because the number of fibers participating in the contraction varies.
   c. The summation of fiber twitches can occur so rapidly that the muscle produces a smooth, sustained contraction known as tetanus.
   d. When a muscle exerts tension without shortening, the contraction is termed isometric; when shortening does occur, the contraction is isotonic.
5. The contraction of muscle fibers in vivo is stimulated by somatic motor neurons.
   a. Each somatic motor axon branches to innervate a number of muscle fibers.
   b. A motor unit consists of a single motor neuron and the muscle fibers it innervates.

## Mechanisms of Contraction (pp. 260–270)

1. Skeletal muscle fibers are striated because of the arrangement of thick and thin myofilaments.
   a. The A bands contain thick myofilaments composed of the protein myosin; the edges of each A band also contain thin myofilaments overlapped with the thick myofilaments.
   b. The central regions of the A bands contain only thick myofilaments; these regions are called the H bands.
   c. The I bands contain only thin myofilaments, which are composed primarily of the protein actin.
   d. The filaments slide, not shorten, during muscle contraction.
   e. The lengths of the H and I bands decrease during contraction, whereas the A bands stay the same length.
2. Myosin cross bridges extend from the thick myofilaments to the thin myofilaments.
3. The activity of the cross bridges causes the thin myofilaments to slide toward the centers of the sarcomeres.
   a. The cross-bridge heads function as myosin ATPase enzymes.
   b. ATP is split into ADP and $P_i$, activating the cross bridge, prior to attachment of the cross bridge to actin.
4. When the activated cross bridges attach to actin, they undergo a power stroke.
5. At the end of a power stroke, the cross bridge bonds to a new ATP; this allows the cross bridge to detach from actin and repeat the cycle.
6. A protein known as tropomyosin is also located at intervals within the thin myofilaments; another protein, troponin, is attached to the tropomyosin.
7. When a muscle is at rest, the $Ca^{++}$ concentration of the sarcoplasm is very low; cross bridges are prevented from attaching to actin by the position of tropomyosin in the thin myofilaments.
   a. $Ca^{++}$ is actively transported into the sarcoplasmic reticulum when a muscle is at rest.
   b. Electrical impulses, conducted by transverse tubules into the muscle fiber, stimulate the release of $Ca^{++}$ from the sarcoplasmic reticulum.
   c. $Ca^{++}$ binds to troponin, and this causes tropomyosin to shift position so that the myosin cross bridges can bind to actin.
8. When electrical impulses cease, $Ca^{++}$ is removed from the sarcoplasm by active transport into the sarcoplasmic reticulum, and the tropomyosin again inhibits muscle contraction.

## Energy Requirements of Skeletal Muscles (pp. 270–273)

1. Aerobic cell respiration is ultimately required for the production of ATP needed for cross-bridge activity.
2. New ATP can be quickly produced, however, from the combination of ADP with phosphate derived from phosphocreatine.
3. Muscle fibers are of three types.
   a. Slow-twitch red fibers are adapted for aerobic respiration and are resistant to fatigue.
   b. Fast-twitch white fibers are adapted for anaerobic respiration.
   c. Intermediate fibers are fast-twitch but adapted for aerobic respiration.
4. Endurance training increases the aerobic capacity of all muscle fiber types, thus decreasing their reliance on anaerobic respiration and their susceptibility to fatigue.

## Cardiac and Smooth Muscle (pp. 273–275)

1. Cardiac muscle is striated and contains sarcomeres; smooth muscle lacks striations but does contain actin and myosin.
   a. Electrical impulses in cardiac muscle originate in myocardial fibers. These impulses can cross from one myocardial fiber to another through gap junctions.
   b. Electrical impulses are produced spontaneously in the heart.
2. Smooth muscle fibers contain myosin and actin, but not arranged in sarcomeres.
   a. Myosin myofilaments are very long; consequently, smooth muscle fibers can contract even when they are greatly stretched.
   b. When electrically stimulated, $Ca^{++}$ enters smooth muscle fibers and combines with calmodulin. The calmodulin-$Ca^{++}$ complex activates an enzyme that phosphorylates myosin cross bridges.

# Review Activities

## Objective Questions

1. A graded whole muscle contraction is produced in vivo primarily by variations in
   a. the strength of the fiber's contraction.
   b. the number of fibers that are contracting.
   c. both a and b.
   d. neither a nor b.

2. The series-elastic component of muscle contraction is responsible for
   a. increased muscle shortening to successive twitches.
   b. a time delay between contraction and shortening.
   c. the lengthening of muscle after contraction has ceased.
   d. all of the above.

3. Which of the following muscles have motor units with the highest innervation ratio?
   a. leg muscles
   b. arm muscles
   c. muscles that move the fingers
   d. muscles of the trunk

4. When a skeletal muscle shortens during contraction, which of the following events does *not* occur?
   a. The A bands shorten.
   b. The H bands shorten.
   c. The I bands shorten.
   d. The sarcomeres shorten.

5. Electrical excitation of a muscle fiber *most directly* causes
   a. movement of tropomyosin.
   b. attachment of the cross bridges to actin.
   c. release of $Ca^{++}$ from the sarcoplasmic reticulum.
   d. splitting of ATP.

6. The energy for muscle contraction is *most directly* obtained from
   a. phosphocreatine.
   b. ATP.
   c. anaerobic respiration.
   d. aerobic respiration.

7. Which of the following statements about cross bridges is *false?*
   a. They are composed of myosin.
   b. They bond to ATP after they detach from actin.
   c. They contain an ATPase.
   d. They split ATP before they attach to actin.

8. When a muscle is stimulated to contract, $Ca^{++}$ bonds to
   a. myosin.
   b. tropomyosin.
   c. actin.
   d. troponin.

9. Which of the following statements about muscle fatigue is *false?*
   a. It may result when ATP is no longer available for the cross-bridge cycle.
   b. It may be caused by a loss of muscle cell $Ca^{++}$.
   c. It may be caused by the accumulation of extracellular $K^+$.
   d. It may be a result of lactic acid production.

10. Which of the following types of muscle fibers is *not* capable of spontaneous depolarization?
    a. single-unit smooth muscle
    b. multiunit smooth muscle
    c. cardiac muscle
    d. skeletal muscle
    e. both b and d
    f. both a and c

11. Which of the following types of muscle is striated and contains gap junctions?
    a. single-unit smooth muscle
    b. multiunit smooth muscle
    c. cardiac muscle
    d. skeletal muscle

12. In an isotonic muscle contraction,
    a. the length of the muscle remains constant.
    b. the muscle tension remains constant.
    c. both muscle length and tension are changed.
    d. movement of bones does not occur.

## Essay Questions

1. Using the concept of motor units, explain how skeletal muscles in vivo produce graded and sustained contractions.

2. Using the concepts of motor unit recruitment and the series-elastic component of muscles, describe how an isometric contraction can be converted into an isotonic contraction.

3. Why don't the cross bridges attach to the thin myofilaments when a muscle is relaxed? Trace the sequence of events that allows the cross bridges to attach to the thin myofilaments when a muscle is stimulated by a nerve impulse.

4. Using the sliding filament theory of contraction, explain why the contraction strength of a muscle is maximal at a particular muscle length.

5. Explain the role of ATP in muscle contraction and in muscle relaxation.

6. Explain how endurance training raises the level of exercise that can be performed before muscle fatigue occurs.

7. Compare striated muscle and smooth muscle with respect to their mechanism of excitation-contraction coupling.

## Gundy/Weber Software 💾

The tutorial software accompanying Chapter 12 is Volume 4—Muscle System.

## Explorations ✪

A module of correlating material is available from the Wm. C. Brown CD-ROM: *Explorations*. It is #4 Muscle Contraction.

# muscular system

## objectives

- Explain how muscles are described according to their location and cooperative function.

- Explain what is meant by synergistic and antagonistic muscle groups.

- Describe the various ways in which muscle fibers are arranged and discuss the advantage of each of these arrangements.

- Use examples to describe the various ways in which muscles are named.

- Locate the major muscles of the axial skeleton. Describe the action of synergistic and antagonistic muscles, using specific muscles as examples.

- Locate the major muscles of the appendicular skeleton. Identify synergistic and antagonistic muscles and describe their action.

# Organization of the Muscular System

*Skeletal muscles are arranged in functional groups that are adaptive in causing particular movements. Within each muscle, the fibers are arranged in a specific pattern that provides specific functional capabilities.*

Skeletal muscle constitutes its own body system and accounts for approximately 40% of body weight. Over 600 individual muscles make up the skeletal muscular system. The principal superficial muscles are shown in figure 13.1. In describing the various muscles, they are usually grouped according to anatomical location and cooperative function. The *muscles of the axial skeleton* have their attachments to the bones of the axial skeleton and include facial muscles, neck muscles, and the anterior and posterior trunk muscles. The *muscles of the appendicular skeleton* include those that act on the pectoral and pelvic girdles and those that cause movement at the joints of the upper and lower extremities.

## Muscle Groups

Just as individual muscle fibers seldom contract independently, muscles generally do not contract separately but work as functional groups. Muscles that contract together and are coordinated in accomplishing a particular movement are said to be *synergistic muscles* (fig. 13.2). *Antagonistic muscles* perform opposite functions and are generally located on the opposite sides of a limb. The two heads of the biceps brachii muscle together with the brachialis muscle, for example, contract together to *flex* the elbow joint. The triceps brachii muscle, the antagonist to the biceps brachii and brachialis muscles, *extends* the elbow as it is contracted.

Seldom does the action of a single muscle cause a movement at a joint. Utilization of several synergistic muscles rather than one massive muscle allows for a division of labor. One muscle may be an important postural muscle, for example, whereas another may be adapted for rapid, powerful contraction. In terms of total output, the contraction of several small muscles produces more force than the contraction of one large one.

---

synergistic: Gk. *synergein*, cooperate
antagonistic: Gk. *antagonistes*, struggle against

## Muscle Architecture

Skeletal muscles may be classified on the basis of fiber arrangement (fig. 13.3), each type having certain inherent advantages. The major types of fiber arrangements are as follows:

**1** **Parallel.** Parallel muscles have relatively long excursions (contract over a great distance) and good endurance, but they are not especially strong. They are long, straplike muscles, such as the sartorius and rectus abdominis muscles.

**2** **Convergent.** Convergent-fibered muscles are so named because the fibers converge at the insertion point to maximize contraction. They are fan-shaped muscles, such as the deltoid and pectoralis major muscles.

**3** **Pennate.** Pennate-fibered muscles provide dexterity. They have many fibers per unit area, and hence are strong muscles. They have short excursions but generally tire quickly. The three kinds of pennate-fibered muscles are *unipennate, bipennate,* and *multipennate*. Examples of each of the pennate types can be found in the forearm.

**4** **Sphincteral.** Sphincter muscles surround a body opening, or **orifice,** constricting it when contracted. Examples are the orbicularis oculi around the eye and orbicularis oris surrounding the mouth.

Muscle fiber architecture can be observed on a cadaver or other dissection specimen. If you have the opportunity to learn the muscles of the body from a cadaver, observe the fiber architecture of specific muscles and try to determine the advantages afforded to each muscle by its location and action.

# Naming of Muscles

*Skeletal muscles are named on the basis of shape, location, attachment, orientation of fibers, relative position, or function.*

One of the tasks of a myologist is to learn the names of the more than 600 skeletal muscles within the body. This task is not quite as formidable as one might imagine because most of the muscles are paired; that is, the right side is the mirror image of the left. Further simplifying the task are the descriptive names of muscles. In the remaining sections of this chapter, we will describe only the principal muscles.

---

pennate: L. *pennatus*, feather
orifice: L. *orificium*, mouth; *facere*, to make

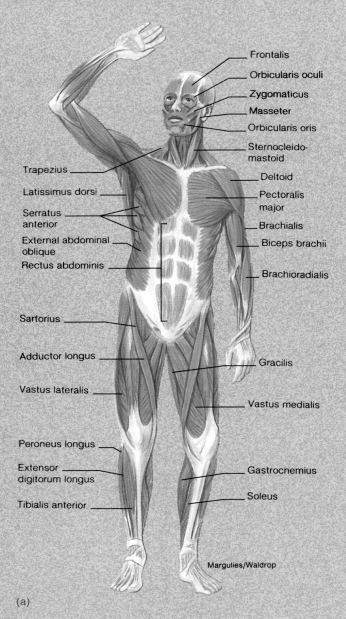

(a)

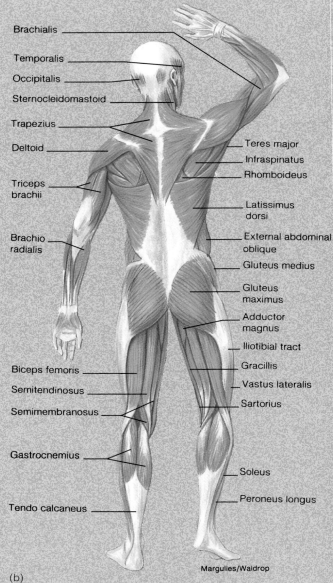

(b)

Margulies/Waldrop

## FIGURE 13.1

An anterior view (*a*) and a posterior view (*b*) of some of the superficial skeletal muscles.

## FIGURE 13.2

Examples of synergistic and antagonistic muscles. The two heads of the biceps brachii muscle and the brachialis muscle are synergistic to each other, as are the three heads of the triceps brachii muscle. The biceps brachii and the brachialis are antagonistic to the triceps brachii, and the triceps brachii is antagonistic to the biceps brachii and the brachialis muscles. When one antagonistic group contracts, the other one must relax; otherwise, movement does not occur.

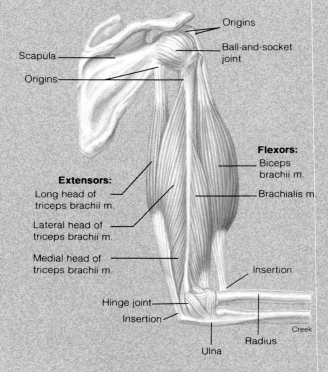

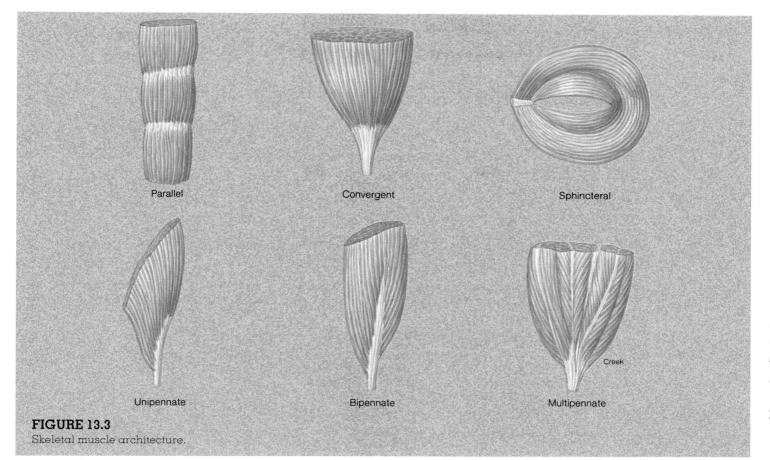

**FIGURE 13.3**
Skeletal muscle architecture.

As you learn the muscles of the body, attend to the derivations of the names and identify the muscles on yourself. Learn them as functional groups so that you will better understand their actions. If you can identify a muscle on yourself, you will be able to contract the muscle and describe its action. Learning anatomy in this way is easiest and most meaningful; moreover, retention is improved.

The following are some examples of how the names of muscles have been logically derived:

**1** **Shape:** rhomboideus (like a rhomboid); trapezius (like a trapezoid); or denoting the number of heads of origin: triceps (three heads), biceps (two heads)

**2** **Location:** pectoralis (in the chest, or pectus); intercostal (between ribs); brachii (upper arm)

**3** **Attachment:** many facial muscles (zygomaticus, temporalis, nasalis); sternocleidomastoid (sternum, clavicle, and mastoid process of the skull)

**4** **Size:** maximus, (larger, largest); minimus (smaller, smallest); longus (long); brevis (short)

**5** **Orientation of fibers:** rectus (straight); transverse (across); oblique (in an oblique direction)

**6** **Relative position:** lateral, medial, internal, and external

**7** **Function:** adductor, flexor, extensor, pronator, and levator (lifter)

## Muscles of the Axial Skeleton

*Muscles of the axial skeleton include those responsible for facial expression, mastication, eye movement, tongue movement, neck movement, and respiration. It also includes those of the abdominal wall, the pelvic outlet, and the vertebral column.*

### Muscles of Facial Expression

Humans have a well-developed facial musculature (figs. 13.4 and 13.5) that provides complex facial expressions as a means of social communication. Indeed, messages are often conveyed by a person without a word spoken.

The muscles of facial expression are in a superficial position located on the scalp, face, and neck. Although highly variable in size and strength, these muscles all originate from the bones of the skull or flat sheetlike tendons and insert onto the hypodermis of the skin (table 13.1). They are all innervated by one of the two facial cranial nerves. The locations and points of attachments of most of the facial muscles are such that, when contracted, they cause movements around the eyes, nostrils, or mouth.

283

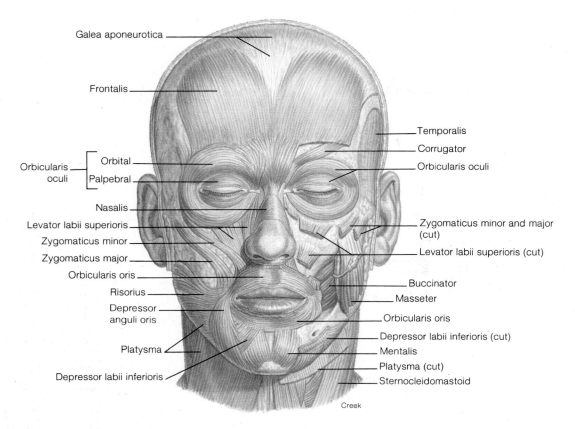

Galea aponeurotica

Frontalis

Orbicularis oculi — Orbital
Palpebral

Nasalis

Levator labii superioris

Zygomaticus minor

Zygomaticus major

Orbicularis oris

Risorius

Depressor anguli oris

Platysma

Depressor labii inferioris

Temporalis

Corrugator

Orbicularis oculi

Zygomaticus minor and major (cut)

Levator labii superioris (cut)

Buccinator

Masseter

Orbicularis oris

Depressor labii inferioris (cut)

Mentalis

Platysma (cut)

Sternocleidomastoid

Creek

**FIGURE 13.4**

An anterior view of the superficial facial muscles involved in facial expression.

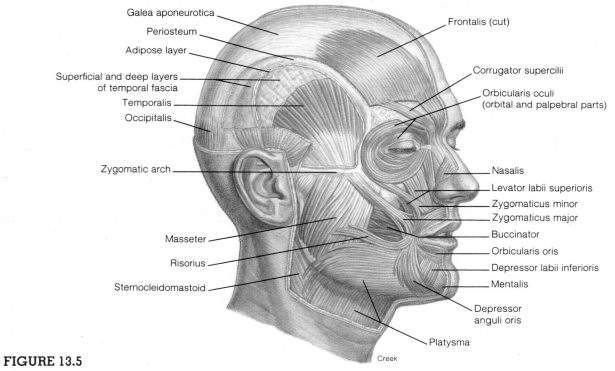

Galea aponeurotica

Periosteum

Adipose layer

Superficial and deep layers of temporal fascia

Temporalis

Occipitalis

Zygomatic arch

Masseter

Risorius

Sternocleidomastoid

Frontalis (cut)

Corrugator supercilii

Orbicularis oculi (orbital and palpebral parts)

Nasalis

Levator labii superioris

Zygomaticus minor

Zygomaticus major

Buccinator

Orbicularis oris

Depressor labii inferioris

Mentalis

Depressor anguli oris

Platysma

Creek

**FIGURE 13.5**

A lateral view of the superficial facial muscles involved in facial expression.

## Table 13.1 Muscles of facial expression

| Muscle | Origin | Insertion | Action |
|--------|--------|-----------|--------|
| Epicranius | Galea aponeurotica and occipital bone | Skin of eyebrow and galea aponeurotica | Wrinkles forehead and moves scalp |
| Frontalis | Galea aponeurotica | Skin of eyebrow | Wrinkles forehead and elevates eyebrow |
| Occipitalis | Occipital bone and mastoid process | Galea aponeurotica | Moves scalp backward |
| Corrugator | Fascia above eyebrow | Root of nose | Draws eyebrows toward midline |
| Orbicularis oculi | Bones of medial orbit | Tissue of eyelid | Closes eyes |
| Nasalis | Maxilla and nasal cartilage | Aponeurosis of nose | Compresses nostrils |
| Orbicularis oris | Fascia surrounding lips | Mucosa of lips | Closes and purses lips |
| Levator labii superioris | Upper maxilla and zygomatic bone | Orbicularis oris and skin above lips | Elevates upper lip |
| Zygomaticus | Zygomatic bone | Superior corner of orbicularis oris | Elevates corner of mouth |
| Risorius | Fascia of cheek | Orbicularis oris at corner of mouth | Draws angle of mouth laterally |
| Depressor anguli oris | Mandible | Inferior corner of orbicularis oris | Depresses corner of mouth |
| Depressor labii inferioris | Mandible | Orbicularis oris and skin of lower lip | Depresses lower lip |
| Mentalis | Mandible (chin) | Orbicularis oris | Elevates and protrudes lower lip |
| Platysma | Fascia of neck and chest | Inferior border of mandible | Depresses lower lip |
| Buccinator | Maxilla and mandible | Orbicularis oris | Compresses cheek |

corrugator: L. *corrugo*, a wrinkle
risorius: L. *risor*, a laughter
mentalis: L. *mentum*, chin

platysma: Gk. *platys*, broad
buccinator: L. *bucca*, cheek

 The muscles of facial expression are of clinical concern for several reasons, each of which involves the facial nerve. Located right under the skin, the many branches of the facial cranial nerve are vulnerable to trauma. Facial lacerations and fractures of the skull frequently damage branches of this nerve. The extensive pattern of motor innervation (see fig. 16.8) becomes apparent in stroke victims and people suffering from *Bell's palsy*. The facial muscles on one side of the face are affected in these people, and that portion of the face appears to sag.

## Muscles of Mastication

The large **temporalis** and **masseter** (*mă-se´ter*) **muscles** (fig. 13.6) are powerful elevators of the mandible in synergy with the **medial pterygoid** (*ter´ĭ-goyd*) **muscle.** The primary function of the medial and lateral pterygoid muscles is to provide grinding movements of the teeth. The **lateral pterygoid muscle** also protracts the mandible (table 13.2). Each of the muscles of mastication is innervated by the mandibular nerve, a branch of the trigeminal cranial nerve.

## Ocular Muscles

The movements of the eyeball are controlled by six extrinsic ocular (eye) muscles (fig. 13.7). Five of these muscles arise from the margin of the optic canal at the back of the orbital cavity and insert on the outer layer (sclera) of the eyeball. Four **rectus muscles** maneuver the eyeball in the direction indicated by their names (**superior, inferior, lateral,** and **medial**), and two **oblique muscles (superior** and **inferior**) rotate the eyeball on its axis. The medial rectus on one side contracts with the medial rectus of the opposite eye when focusing on close objects. When looking to the side, the lateral rectus of one eyeball works with the medial rectus of the opposite eyeball to keep both eyes focused together. The superior oblique muscle passes through a pulleylike cartilaginous loop, the *trochlea*, before attaching to the eyeball. The ocular muscles are innervated by three cranial nerves (table 13.3).

Another muscle, the **levator palpebrae** (*le-ʋa´tor pal´pēbre*) **superioris muscle** (fig. 13.7), is located in the ocular

**FIGURE 13.6**

Muscles of mastication. (*a*) A superficial view, (*b*) a deep view, and (*c*) the deepest view, showing the pterygoid muscles. (The muscles of mastication are labeled in boldface.)

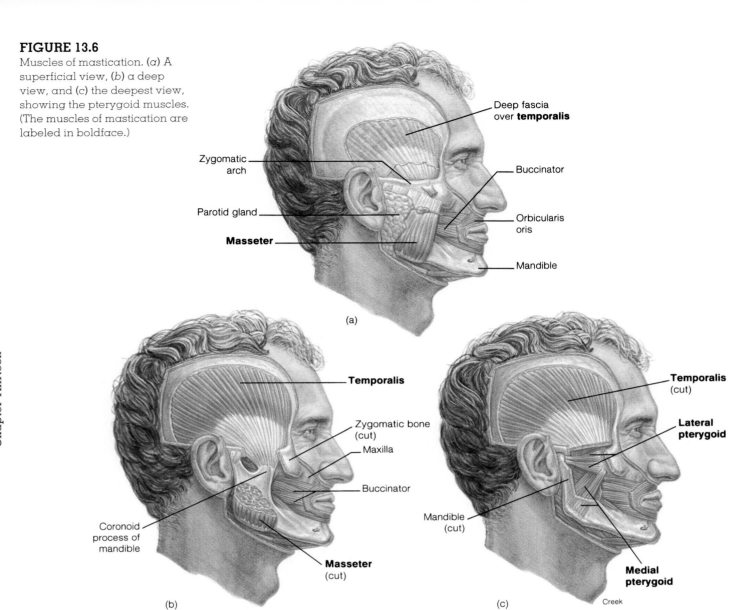

(a)

Deep fascia over **temporalis**

Zygomatic arch

Buccinator

Parotid gland

Orbicularis oris

**Masseter**

Mandible

(b)

**Temporalis**

Zygomatic bone (cut)

Maxilla

Buccinator

Coronoid process of mandible

**Masseter** (cut)

(c)

**Temporalis** (cut)

**Lateral pterygoid**

Mandible (cut)

**Medial pterygoid**

Creek

## Table 13.2    Muscles of mastication

| Muscle | Origin | Insertion | Action |
|---|---|---|---|
| Temporalis | Temporal fossa | Coronoid process of mandible | Elevates mandible |
| Masseter | Zygomatic arch | Lateral part of ramus of mandible | Elevates mandible |
| Medial pterygoid | Sphenoid bond | Medial aspect of mandible | Elevates mandible and moves mandible laterally |
| Lateral pterygoid | Sphenoid bone | Anterior side of mandibular condyle | Protracts mandible |

masseter: Gk. *maseter*, chew
pterygoid: Gk. *pteron*, wing

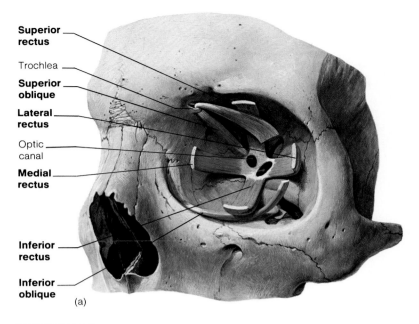

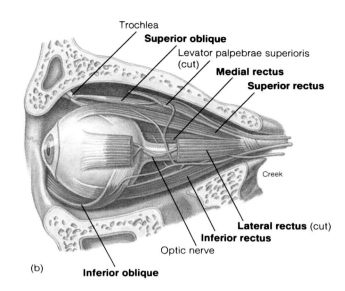

**FIGURE 13.7**

Extrinsic ocular muscles of the left eyeball. (*a*) An anterior view and (*b*) a lateral view. (The extrinsic occular muscles are labeled in boldface.)

| Table 13.3 | Ocular muscles | |
|---|---|---|
| *Muscle* | *Cranial nerve innervation* | *Movement of eyeball* |
| Lateral rectus | Abducens | Lateral |
| Medial rectus | Oculomotor | Medial |
| Superior rectus | Oculomotor | Superior and medial |
| Inferior rectus | Oculomotor | Inferior and medial |
| Inferior oblique | Oculomotor | Superior and lateral |
| Superior oblique | Trochlear | Inferior and lateral |

region but is not attached to the eyeball. It extends into the upper eyelid and raises the eyelid when contracted.

## Muscles That Move the Tongue

The tongue is a highly specialized muscular organ that functions in speaking, manipulating food during mastication, cleansing the teeth, and swallowing. Two groups of skeletal muscles are responsible for tongue movement: intrinsic and extrinsic. The *intrinsic muscles* are confined within the tongue and are responsible for its mobility and changes of shape. *Extrinsic tongue muscles* are those that originate on structures away from the tongue and insert onto it to cause gross tongue movement (see fig. 13.8 and table 13.4). The three paired extrinsic muscles are the **genioglossus, styloglossus,** and **hyoglossus muscles.** Each of the extrinsic tongue muscles is innervated by the hypoglossal cranial

nerve. When the anterior portion of the genioglossus is contracted, the tongue is depressed and thrust forward. If both genioglossus muscles are contracted together along their entire lengths, the surface of the tongue becomes transversely concave. This muscle is extremely important in infants for sucking; the tongue is positioned around the nipple with a concave groove channeled toward the pharynx.

## Muscles of the Neck

Muscles of the neck either support and move the head or are attached to structures within the neck region, such as the hyoid bone and larynx. Only the more obvious neck muscles will be considered in this chapter. These muscles are illustrated in figures 13.9 and 13.10 and are summarized in table 13.5.

**Posterior Muscles** The posterior muscles include the sternocleidomastoid (which originates anteriorly), trapezius, splenius capitis, semispinalis capitis, and longissimus capitis muscles.

As the name implies, the **sternocleidomastoid** (*ster″no-kli″do-mas′toid*) **muscle** originates on the sternum and clavicle and inserts on the mastoid process of the skull (fig. 13.10). When contracted on one side, it turns the head sideways in a direction opposite the side on which the muscle is located. If both sternocleidomastoid muscles are contracted, the head is pulled forward and down. The sternocleidomastoid is covered by the platysma, (see fig. 13.5 and table 13.5).

Although a portion of the **trapezius muscle** extends over the posterior neck region, it is primarily a superficial muscle of the back and will be described later.

**FIGURE 13.8**

Extrinsic muscles of the tongue and deep structures of the neck. (The extrinsic muscles of the tongue are labeled in boldface.)

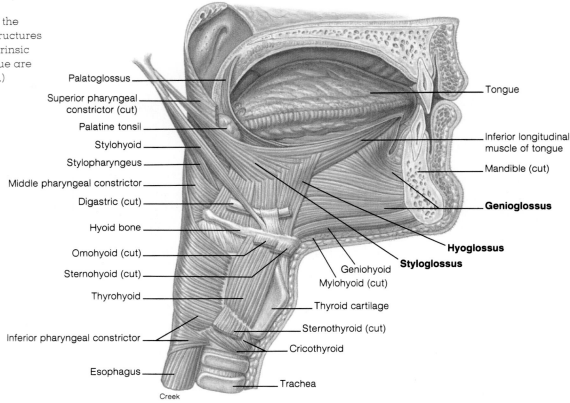

Palatoglossus

Superior pharyngeal constrictor (cut)

Palatine tonsil

Stylohyoid

Stylopharyngeus

Middle pharyngeal constrictor

Digastric (cut)

Hyoid bone

Omohyoid (cut)

Sternohyoid (cut)

Thyrohyoid

Inferior pharyngeal constrictor

Esophagus

Creek

Tongue

Inferior longitudinal muscle of tongue

Mandible (cut)

**Genioglossus**

**Hyoglossus**

**Styloglossus**

Geniohyoid

Mylohyoid (cut)

Thyroid cartilage

Sternothyroid (cut)

Cricothyroid

Trachea

## Table 13.4  Extrinsic tongue muscles

| Muscle | Origin | Insertion | Action |
|--------|--------|-----------|--------|
| Genioglossus | Mental spine of mandible | Undersurface of tongue | Depresses and protracts tongue |
| Styloglossus | Styloid process of temporal bone | Lateral side and undersurface of tongue | Elevates and retracts tongue |
| Hyoglossus | Body of hyoid bone | Side of tongue | Depresses sides of tongue |

genioglossus: L. *geneion*, chin; *glossus*, tongue

The **splenius capitis muscle** is a broad muscle positioned deep to the trapezius (fig. 13.9). It originates on the ligamentum nuchae and the spinous processes of the seventh cervical and first three thoracic vertebrae. It inserts on the back of the skull below the superior nuchal line and on the mastoid process of the temporal bone. When the splenius capitis contracts on one side, the head rotates and extends to one side. Contracted together, these muscles extend the head at the neck. Further contraction causes hyperextension of the neck and head.

The broad, sheetlike **semispinalis capitis muscle** extends upward from the seventh cervical and first six thoracic vertebrae to insert on the occipital bone (fig. 13.9). When the two semispinalis capitis muscles contract together, they extend the head at the neck, along with the splenius capitis muscle. If one of the muscles acts alone, the head is rotated to the side.

The narrow, straplike **longissimus capitis muscle** ascends from processes of the lower four cervical and upper five thoracic vertebrae and inserts on the mastoid process of the temporal bone (fig. 13.9). This muscle extends the head at the neck, bends it to the one side, or rotates it slightly.

**Suprahyoid Muscles**  The group of suprahyoid muscles located above the hyoid bone includes the digastric, mylohyoid, geniohyoid, and stylohyoid muscles (fig. 13.10).

The **digastric muscle** is a two-bellied muscle of double origin that inserts on the hyoid bone. The anterior origin is on the mandible at the point of the chin, and the posterior origin is near the mastoid process of the skull. The digastric can open the mouth or elevate the hyoid.

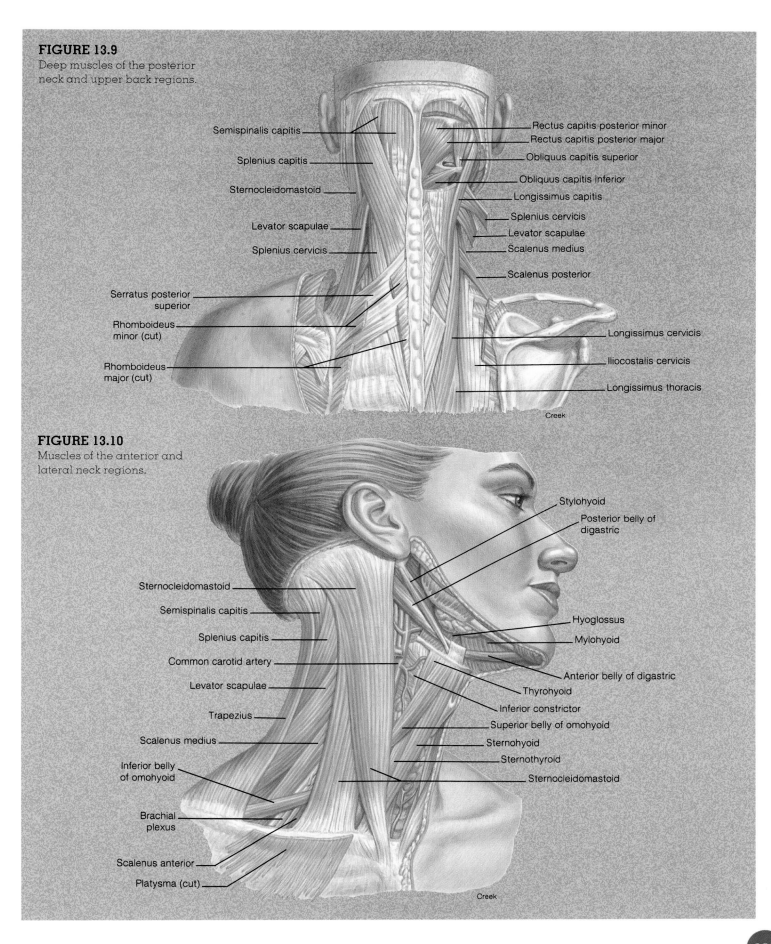

**FIGURE 13.9**
Deep muscles of the posterior neck and upper back regions.

Semispinalis capitis

Splenius capitis

Sternocleidomastoid

Levator scapulae

Splenius cervicis

Serratus posterior superior

Rhomboideus minor (cut)

Rhomboideus major (cut)

Rectus capitis posterior minor

Rectus capitis posterior major

Obliquus capitis superior

Obliquus capitis inferior

Longissimus capitis

Splenius cervicis

Levator scapulae

Scalenus medius

Scalenus posterior

Longissimus cervicis

Iliocostalis cervicis

Longissimus thoracis

Creek

**FIGURE 13.10**
Muscles of the anterior and lateral neck regions.

Sternocleidomastoid

Semispinalis capitis

Splenius capitis

Common carotid artery

Levator scapulae

Trapezius

Scalenus medius

Inferior belly of omohyoid

Brachial plexus

Scalenus anterior

Platysma (cut)

Stylohyoid

Posterior belly of digastric

Hyoglossus

Mylohyoid

Anterior belly of digastric

Thyrohyoid

Inferior constrictor

Superior belly of omohyoid

Sternohyoid

Sternothyroid

Sternocleidomastoid

Creek

# Table 13.5 Muscles of the neck

| Muscle | Origin | Insertion | Action | Innervation |
|---|---|---|---|---|
| Sternocleidomastoid | Sternum; clavicle | Mastoid process of temporal bone | Turns head to side; flexes neck | Accessory n. |
| Digastric | Inferior border of mandible; mastoid groove | Hyoid bone | Opens mouth; elevates hyoid bone | Trigeminal n. (ant. belly); facial n. (post. belly) |
| Mylohyoid | Inferior border of mandible | Body of hyoid bone and median raphe | Elevates hyoid bone and floor of mouth | Trigeminal n. |
| Geniohyoid | Medial surface of mandible at chin | Body of hyoid bone | Elevates hyoid bone | Hypoglossal n. |
| Stylohyoid | Styloid process of temporal bone | Body of hyoid bone | Elevates and retracts tongue | Facial n. |
| Sternohyoid | Manubrium | Body of hyoid bone | Depresses hyoid bone | Hypoglossal n. |
| Sternothyroid | Manubrium | Thyroid cartilage | Depresses thyroid cartilage | Hypoglossal n. |
| Thyrohyoid | Thyroid cartilage | Great cornu of hyoid bone | Depresses hyoid bone; elevates thyroid | Hypoglossal n. |
| Omohyoid | Superior border of scapula | Clavicle; body of hyoid bone | Depresses hyoid bone | Hypoglossal n. |

digastric: L. *di*, two; Gk. *gaster*, belly
mylohyoid: Gk. *mylos*, akin to; *hyoeides*, pertaining to hyoid bone
omohyoid: Gk. *omos*, shoulder

The **mylohyoid muscle** forms the floor of the mouth. It originates on the inferior border of the mandible and inserts on the median raphe and hyoid bone. As this muscle contracts, the floor of the mouth is elevated. It aids swallowing by forcing the food toward the back of the mouth.

The short, straplike **geniohyoid muscle** (see fig. 13.8) is deep to the mylohyoid muscle. The geniohyoid muscle extends from the medial surface of the mandible at the chin to the hyoid bone, which it elevates as it contracts.

The slender **stylohyoid muscle** extends from the styloid process of the skull to the hyoid bone, which it elevates as it contracts. The secondary effect of this muscle on tongue movement has already been described.

**Infrahyoid Muscles** Infrahyoid muscles are paired, thin, straplike muscles located below the hyoid bone. They are individually named on the basis of their origin and insertion and include the sternohyoid, sternothyroid, thyrohyoid, and omohyoid muscles (fig. 13.10).

The **sternohyoid muscle** originates on the manubrium of the sternum and inserts on the hyoid. It depresses the hyoid bone as it contracts.

The **sternothyroid muscle** also originates on the manubrium but inserts on the thyroid cartilage of the larynx. When this muscle contracts, the larynx is pulled downward.

The short **thyrohyoid muscle** extends from the thyroid cartilage to the hyoid bone. It elevates the larynx and lowers the hyoid bone.

The long, thin **omohyoid muscle** originates on the superior border of the scapula and inserts on the clavicle bone and on the hyoid bone. It acts to depress the hyoid bone.

## Muscles of Respiration

The muscles of respiration are skeletal muscles that continually and rhythmically contract, usually involuntarily. Breathing, or *pulmonary ventilation*, is divided into two phases: *inspiration (inhalation)* and *expiration (exhalation)*.

During normal, relaxed inspiration, the important muscles are the **diaphragm** (*di´ă-fram*), the **external intercostal muscles,** and the interchondral portion of the **internal intercostal muscles** (fig. 13.11). A downward contraction of the dome-shaped diaphragm causes a vertical increase in thoracic dimension. A simultaneous contraction of the external intercostal muscles and the interchondral portion of the internal intercostal muscles produces an increase in the lateral dimension of the thorax. In addition, the **sternocleidomastoid muscle** and **scalenes muscle** may assist in inspiration through elevation of the first and second ribs, respectively. The intercostal muscles

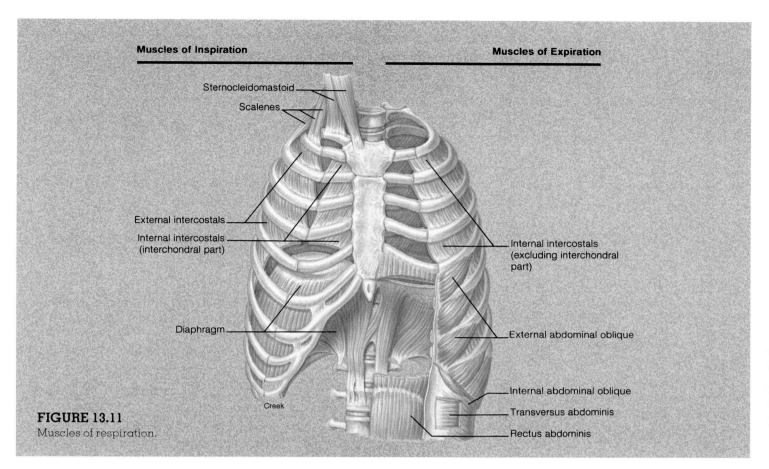

**Muscles of Inspiration**

**Muscles of Expiration**

Sternocleidomastoid

Scalenes

External intercostals

Internal intercostals
(interchondral part)

Internal intercostals
(excluding interchondral
part)

Diaphragm

External abdominal oblique

Creek

Internal abdominal oblique

Transversus abdominis

Rectus abdominis

**FIGURE 13.11**
Muscles of respiration.

are innervated by the intercostal nerves, and the diaphragm receives its stimuli through the phrenic nerves.

Expiration is primarily a passive process, occurring as the muscles of inspiration are relaxed and the rib cage recoils to its original position. During forced expiration, the interosseous portion of the internal intercostal muscles contracts, causing the rib cage to be depressed. This portion of the internal intercostal muscles lies under the external intercostal muscles, and its fibers are directed downward and backward. The **abdominal muscles** may also contract during forced expiration, which increases pressure within the abdominal cavity and forces the diaphragm superiorly, squeezing additional air out of the lungs.

## Muscles of the Abdominal Wall

The anterolateral abdominal wall is composed of four pairs of flat, sheetlike muscles: the external abdominal oblique, internal abdominal oblique, transversus abdominis, and rectus abdominis muscles (fig. 13.12). Each of these muscles is innervated by intercostal nerves. The abdominal muscles support and protect the organs of the abdominal cavity and

aid in breathing. When they contract, the pressure in the abdominal cavity increases, which can aid in defecation and in stabilizing the spine during heavy lifting.

The **external abdominal oblique muscle** is the strongest and most superficial of the three layered muscles of the lateral abdominal wall. Its fibers are directed inferiorly and medially. The **internal abdominal oblique muscle** lies deep to the external abdominal oblique, and its fibers are directed at right angles to the external abdominal oblique. The **transversus abdominis muscle** is the deepest of the abdominal muscles, and its fibers pass directly medially around the abdominal wall. The long, straplike **rectus abdominis muscle** is entirely enclosed in a fibrous sheath formed from the aponeuroses of the other three abdominal muscles. The *linea alba* is a band of connective tissue on the midline of the abdomen that separates the two rectus abdominis muscles. *Tendinous inscriptions* transect the rectus abdominis muscles at several points, causing the surface abdominal anatomy of a well-muscled person to appear segmented.

Refer to table 13.6 for a summary of the muscles of the abdominal wall.

**FIGURE 13.12**
Muscles of the anterolateral neck, shoulder, and trunk regions. The mammary gland is an integumentary structure positioned over the pectoralis major muscle.

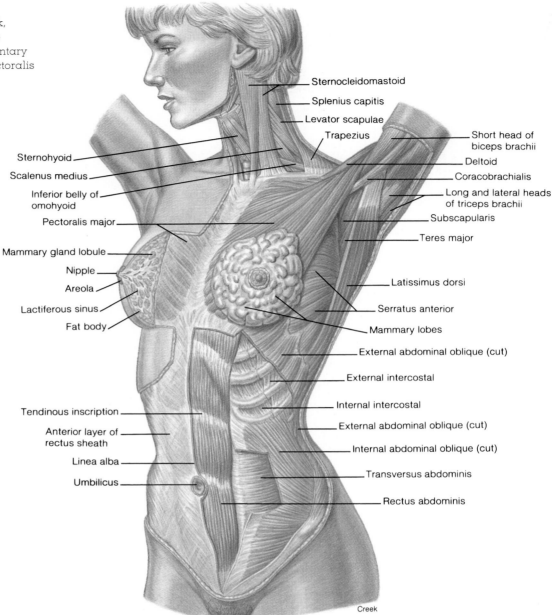

Sternocleidomastoid
Splenius capitis
Levator scapulae
Trapezius
Short head of biceps brachii
Sternohyoid
Scalenus medius
Deltoid
Coracobrachialis
Inferior belly of omohyoid
Long and lateral heads of triceps brachii
Subscapularis
Pectoralis major
Teres major
Mammary gland lobule
Nipple
Areola
Latissimus dorsi
Serratus anterior
Lactiferous sinus
Fat body
Mammary lobes
External abdominal oblique (cut)
External intercostal
Internal intercostal
Tendinous inscription
External abdominal oblique (cut)
Anterior layer of rectus sheath
Internal abdominal oblique (cut)
Linea alba
Transversus abdominis
Umbilicus
Rectus abdominis

Creek

## Muscles of the Pelvic Outlet

The **pelvic outlet,** or **floor of the pelvis,** is the inferiorly positioned muscular wall that supports the pelvic viscera. It consists of the levator ani and coccygeus muscles, collectively called the *pelvic diaphragm.* The openings for the urethra and the rectum, as well as the vaginal opening in the female, pass through the pelvic diaphragm.

The **levator ani muscle** forms a thin sheet of muscle that helps to support the pelvic viscera and constrict the lower part of the rectum, pulling it forward and aiding defecation. The fan-shaped **coccygeus muscle** aids the levator ani in its functions. Either or both of these muscles are occasionally stretched and torn during parturition.

Inferior to the pelvic diaphragm are the *perineal muscles.* These muscles are arranged in a superficial layer, consisting of the bulbocavernosus, ischiocavernosus, and the transversus perinei superficialis muscles, and a deep layer, consisting of the deep transversus perinei muscle and the external anal sphincter (fig. 13.13). Muscles of the deep layer, with a fascia, constitute the *urogenital diaphragm.* The **external anal sphincter muscle** of the deep layer is a funnel-shaped constrictor muscle that surrounds the anal canal. The muscles of the pelvic diaphragm and the urogenital

## Table 13.6  Muscles of the abdominal wall

| Muscle | Origin | Insertion | Action |
|---|---|---|---|
| External abdominal oblique | Lower eight ribs | Illiac crest, linea alba | Compresses abdomen; lateral rotation; draws thorax downward |
| Internal abdominal oblique | Illiac crest, inguinal ligament, and lumbodorsal fascia | Linea alba, costal cartilage of last three or four ribs | Compresses abdomen; lateral rotation; draws thorax downward |
| Transversus abdominis | Illiac crest, inguinal ligament, lumbar fascia, and costal cartilage of last six ribs | Xiphoid process, linea alba, and pubis | Compresses abdomen |
| Rectus abdominis | Pubic crest and symphysis pubis | Costal cartilage of fifth to seventh ribs and xiphoid process of sternum | Flexes vertebral column |

rectus abdominis: L. *rectus,* straplike; abdomino, belly

diaphragm are similar in the male and female, but the perineal muscles of each sex markedly differ.

In males, the **bulbospongiosus muscle** of one side unites with that of the opposite side to form a muscular constriction surrounding the base of the penis. When contracted, the two muscles constrict the urethral canal and assist in emptying the urethra. In females, these muscles are separated by the vaginal orifice, which they constrict as they contract. The **ischiocavernosus muscle** inserts onto the pubic arch and crus of the penis in the male and the pubic arch and crus of the clitoris of the female. This muscle aids the erection of the penis in the male and that of the clitoris in the female.

The levator ani muscle and the transversus perinei muscle are innervated by the pudendal plexus. The coccygeus, bulbospongiosus, and ischiocavernosus muscles are innervated by the perineal branch of the pudendal nerve.

The muscles of the pelvic outlet are illustrated in figure 13.13 and summarized in table 13.7.

## Muscles of the Vertebral Column

The muscles that move the vertebral column are strong and complex because they have to provide support and movement in resistance to the effect of gravity.

The vertebral column can be flexed, extended, abducted, adducted, and rotated. The muscle that flexes the vertebral column, the rectus abdominis, has already been described as a paired, straplike muscle of the anterior abdominal wall. The extensor muscles, located on the posterior side of the vertebral column, have to be stronger than the

flexors because extension (such as lifting an object) is in opposition to gravity. The extensor muscles consist of a superficial group and a deep group. Only some of the muscles of the vertebral column will be described here.

The **erector spinae muscles** constitute a massive superficial muscle group that extends from the sacrum to the skull. It actually consists of three groups of muscles: the **iliocostalis, longissimus,** and **spinalis muscles** (fig. 13.14). Each of these groups, in turn, consists of overlapping slips of muscle. The iliocostalis is the most lateral group, the longissimus is intermediate in position, and the spinalis, in medial position, is positioned in contact with the spinous processes of the vertebrae.

 The erector spinae muscles are frequently strained through improper lifting of objects. A heavy object should not be lifted with the vertebral column flexed; instead, the thighs and knees should be flexed so that the pelvic and leg muscles can aid in the task.

The erector spinae muscles are also frequently strained during pregnancy. Pregnant women will try to counterbalance the effect of the protruding abdomen by hyperextending the vertebral column. This causes an exaggerated lumbar curvature, strained muscles, and a peculiar gait.

The deep **quadratus lumborum muscle** originates on the iliac crest and the lower three lumbar vertebrae. It inserts on the transverse processes of the first four lumbar vertebrae and the inferior margin of the twelfth rib. When the quadratus lumborum muscle contracts on both sides (acts jointly), the vertebral column in the lumbar region extends. Separate contraction causes lateral flexion of the spine.

A summary of the major muscles of the back is presented in table 13.8.

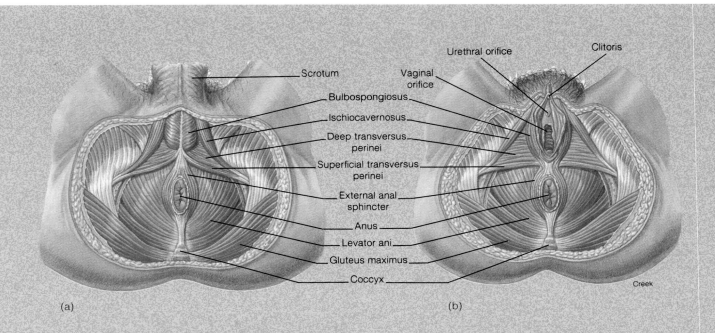

Scrotum

Urethral orifice

Clitoris

Vaginal orifice

Bulbospongiosus

Ischiocavernosus

Deep transversus perinei

Superficial transversus perinei

External anal sphincter

Anus

Levator ani

Gluteus maximus

Coccyx

Creek

(a)

(b)

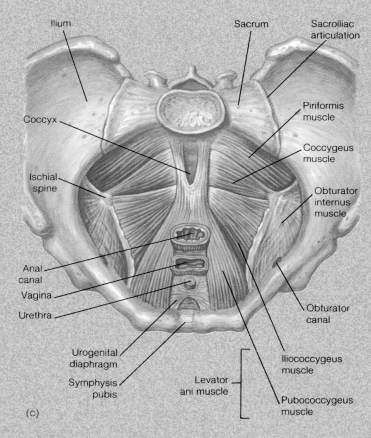

Ilium

Sacrum

Sacroiliac articulation

Coccyx

Piriformis muscle

Coccygeus muscle

Ischial spine

Obturator internus muscle

Anal canal

Vagina

Urethra

Obturator canal

Urogenital diaphragm

Iliococcygeus muscle

Symphysis pubis

Levator ani muscle

Pubococcygeus muscle

(c)

**FIGURE 13.13**

Muscles of the pelvic outlet: (a) male and (b) female.
(c) A superior view of the internal muscles of the female
pelvic outlet.

## Table 13.7 — Muscles of the pelvic outlet

| Muscle | Origin | Insertion | Action |
|---|---|---|---|
| Levator ani | Spine of ischium and pubic bone | Coccyx | Supports pelvic viscera; aids in defecation |
| Coccygeus | Ischial spine | Sacrum and coccyx | Supports pelvic viscera; aids in defecation |
| Transversus perinei | Ischial tuberosity | Central tendon | Supports pelvic viscera |
| Bulbospongiosus | Central tendon | Males: base of penis; females: root of clitoris | Constricts urethral canal; constricts vagina |
| Ischiocavernosus | Ischial tuberosity | Males: pubic arch and crus of the penis; females: pubic arch and crus of the clitoris | Aids erection of penis or clitoris |

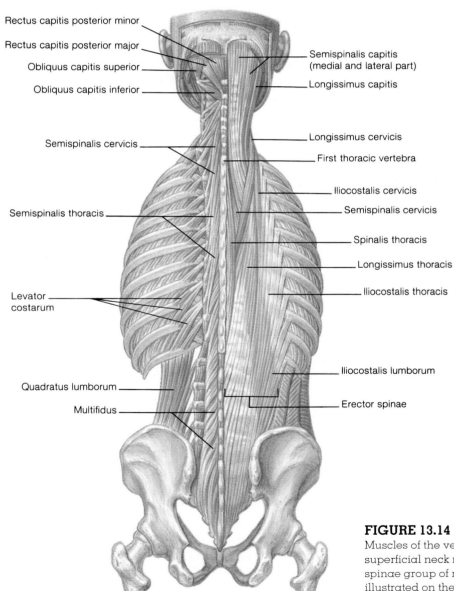

Rectus capitis posterior minor

Rectus capitis posterior major

Obliquus capitis superior

Obliquus capitis inferior

Semispinalis cervicis

Semispinalis thoracis

Levator costarum

Quadratus lumborum

Multifidus

Semispinalis capitis (medial and lateral part)

Longissimus capitis

Longissimus cervicis

First thoracic vertebra

Iliocostalis cervicis

Semispinalis cervicis

Spinalis thoracis

Longissimus thoracis

Iliocostalis thoracis

Iliocostalis lumborum

Erector spinae

Iliocostalis lumborum

Creek

### FIGURE 13.14

Muscles of the vertebral column. The superficial neck muscles and erector spinae group of muscles are illustrated on the right, and the deep neck and back muscles are illustrated on the left.

## Table 13.8    Muscles of the vertebral column

| Muscle | Origin | Insertion | Action | Innervation |
|---|---|---|---|---|
| Quadratus lumborum | Illiac crest and lower three lumbar vertebrae | Twelfth rib and upper four lumbar vertebrae | Extends lumbar region; lateral flexion of vertebral column | Intercostal nerve T12 and lumbar nerves L2–L4 |
| Erector spinae | Consists of three groups of muscles: iliocostalis, longissimus, and spinalis. The iliocostalis and longissimus are further subdivided into three groups on the basis of location along the vertebral column. | | | |
| Iliocostalis lumborum | Crest of illium | Lower six ribs | Extends lumbar region | Posterior rami of lumbar nerves |
| Iliocostalis thoracis | Lower six ribs | Upper six ribs | Extends thoracic region | Posterior rami of thoracic nerves |
| Iliocostalis cervicis | Angles of third to sixth rib | Transverse processes of fourth to sixth cervical vertebrae | Extends cervical region | Posterior rami of cervical nerves |
| Longissimus thoracis | Transverse processes of lumbar vertebrae | Transverse processes of all the thoracic vertebrae and lower nine ribs | Extends thoracic region | Posterior rami of spinal nerves |
| Longissimus cervicis | Transverse processes of upper four or five thoracic vertebrae | Transverse processes of second to sixth cervical vertebrae | Extends cervical region and lateral flexion | Posterior rami of spinal nerves |
| Longissimus capitis | Transverse processes of upper five thoracic vertebrae and articular processes of lower three cervical vertebrae | Posterior margin of cranium and mastoid processes | Extends joint of cervical vertebrae | Posterior rami of middle and lower cervical nerves |
| Spinalis thoracis | Spinous processes of upper lumbar and lower thoracic vertebrae | Spinous processes of upper thoracic vertebrae | Extends vertebral column | Posterior rami of spinal nerves |

# Muscles of the Appendicular Skeleton

*The muscles of the appendicular skeleton include those of the pectoral girdle, arm, forearm, wrist, hand, and fingers and those of the pelvic girdle, thigh, leg, ankle, foot, and toes.*

## Muscles That Act on the Pectoral Girdle

The shoulder is attached to the axial skeleton only at the sternoclavicular joint; therefore, strong, straplike muscles are necessary. Furthermore, muscles that move the brachium originate on the scapula, and during brachial movement the scapula has to be held fixed. The muscles that act on the

pectoral girdle originate on the axial skeleton and can be divided into anterior and posterior groups.

The anterior group of muscles that act on the pectoral girdle includes the **serratus anterior, pectoralis minor,** and **subclavius muscles** (fig. 13.15). The posterior group includes the **trapezius, levator scapulae,** and **rhomboideus muscles** (fig. 13.16). The position of these muscles is such that one muscle does not cause an action on its own, but rather several muscles contract synergistically to result in any movement of the girdle.

 Treatment of advanced stages of *breast cancer* requires the surgical removal of both pectoralis major and pectoralis minor muscles in a procedure called a *radical mastectomy.* Postoperative physical therapy is primarily geared toward strengthening the synergistic muscles of this area. As the muscles that act on the brachium are learned, determine which are synergists with the pectoralis major muscle.

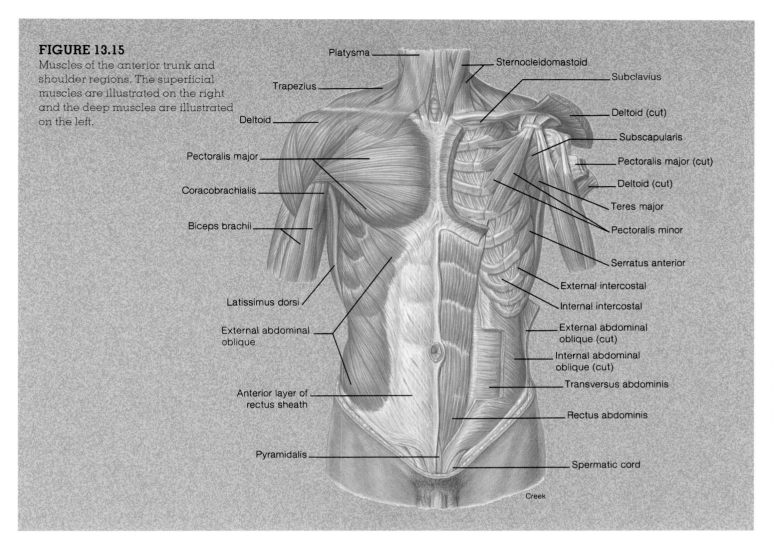

**FIGURE 13.15**
Muscles of the anterior trunk and shoulder regions. The superficial muscles are illustrated on the right and the deep muscles are illustrated on the left.

Platysma
Sternocleidomastoid
Trapezius
Subclavius
Deltoid
Deltoid (cut)
Pectoralis major
Subscapularis
Coracobrachialis
Pectoralis major (cut)
Biceps brachii
Deltoid (cut)
Teres major
Pectoralis minor
Serratus anterior
Latissimus dorsi
External intercostal
Internal intercostal
External abdominal oblique
External abdominal oblique (cut)
Internal abdominal oblique (cut)
Transversus abdominis
Anterior layer of rectus sheath
Rectus abdominis
Pyramidalis
Spermatic cord
Creek

The muscles that act on the pectoral girdle are summarized in table 13.9.

## Muscles That Move the Humerus

Of the nine muscles that span the shoulder joint to insert on the humerus, only two of them, the pectoralis major and latissimus dorsi, do not originate on the scapula. These two are designated as axial muscles, whereas the remaining seven are scapular muscles. The muscles of this region are shown in figures 13.15 and 13.16, and the attachments of all the muscles that either originate or insert on the scapula are shown in figure 13.17.

 In terms of their development, the pectoralis major and the latissimus dorsi muscles are not axial muscles at all. They develop in the forelimb and extend to the trunk secondarily. They are considered axial muscles only because their origins are on the axial skeleton.

**Axial Muscles**   The axial muscles include the pectoralis major and latissimus dorsi muscles.

The **pectoralis major muscle** is a large chest muscle (fig. 13.15) that binds the humerus to the scapula. It is the principal flexor muscle of the arm. The large, flat, triangular **latissimus dorsi muscle** covers the lower half of the thoracic region of the back (see fig. 13.16) and is the antagonist to the pectoralis major muscle. The latissimus dorsi is frequently called the "swimmer's muscle" because it powerfully extends the shoulder joint, drawing the arm downward and backward while it rotates medially. Extension of the arm is in reference to anatomical position and is therefore a backward, retracting (increasing the shoulder joint angle) movement of the arm.

**Scapular Muscles**   The nonaxial scapular muscles include the deltoid, supraspinatus, infraspinatus, teres major, teres minor, subscapularis, and coracobrachialis muscles.

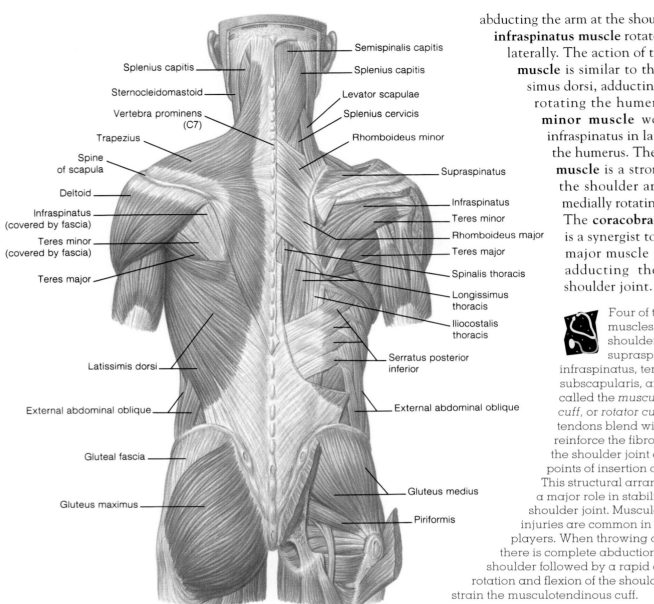

Splenius capitis

Sternocleidomastoid

Vertebra prominens (C7)

Trapezius

Spine of scapula

Deltoid

Infraspinatus (covered by fascia)

Teres minor (covered by fascia)

Teres major

Latissimis dorsi

External abdominal oblique

Gluteal fascia

Gluteus maximus

Semispinalis capitis

Splenius capitis

Levator scapulae

Splenius cervicis

Rhomboideus minor

Supraspinatus

Infraspinatus

Teres minor

Rhomboideus major

Teres major

Spinalis thoracis

Longissimus thoracis

Iliocostalis thoracis

Serratus posterior inferior

External abdominal oblique

Gluteus medius

Piriformis

Creek

**FIGURE 13.16**

Muscles of the posterior neck, shoulder, trunk, and gluteal regions. The superficial muscles are illustrated on the left and the deep muscles are illustrated on the right.

The thick, powerful **deltoid muscle** caps the shoulder joint (figs. 13.18 and 13.19). Although it has several functions (see table 13.10), the principal action of the deltoid muscle is abduction of the arm at the shoulder joint. Functioning together, both the pectoralis major and the latissimus dorsi muscles are antagonists to the deltoid muscle in that they cause adduction of the arm at the shoulder joint. The deltoid muscle is a common site for intramuscular injections.

The remaining six scapular muscles also help stabilize the shoulder and have specific actions at the shoulder joint (see table 13.10). The **supraspinatus muscle** laterally rotates the humerus and is synergistic with the deltoid in

abducting the arm at the shoulder joint. The **infraspinatus muscle** rotates the humerus laterally. The action of the **teres major muscle** is similar to that of the latissimus dorsi, adducting and medially rotating the humerus. The **teres minor muscle** works with the infraspinatus in laterally rotating the humerus. The **subscapularis muscle** is a strong stabilizer of the shoulder and also aids in medially rotating the humerus. The **coracobrachialis muscle** is a synergist to the pectoralis major muscle in flexing and adducting the arm at the shoulder joint.

Four of the nine muscles that cross the shoulder joint, the supraspinatus, infraspinatus, teres minor, and subscapularis, are commonly called the *musculotendinous cuff*, or *rotator cuff*. Their distal tendons blend with and reinforce the fibrous capsule of the shoulder joint en route to their points of insertion on the humerus. This structural arrangement plays a major role in stabilizing the shoulder joint. Musculotendinous cuff injuries are common in baseball players. When throwing a baseball, there is complete abduction of the shoulder followed by a rapid and forceful rotation and flexion of the shoulder, which may strain the musculotendinous cuff.

## Muscles That Act on the Forearm

The powerful muscles of the brachium are responsible for flexion and extension of the elbow joint. These muscles are the biceps brachii, brachialis, brachioradialis, and triceps brachii (figs. 13.18 and 13.19). In addition, a short triangular muscle, the anconeus, is positioned over the distal end of the triceps brachii muscle, near the elbow.

The powerful **biceps brachii muscle,** positioned on the anterior surface of the humerus, is the most familiar muscle of the arm, yet it has no attachments on the humerus. The biceps brachii muscle has a dual origin: a medial tendinous head, the **short head,** arises from the coracoid process of the scapula, and the **long head** originates on the superior tuberosity of the glenoid cavity, passes through the shoulder joint, and descends in the intertubercular groove on the

# Table 13.9  Muscles that act on the pectoral girdle

| Muscle | Origin | Insertion | Action | Innervation |
|---|---|---|---|---|
| Serratus anterior | Upper eight or nine ribs | Medial border of scapula | Pulls scapula forward and downward | Long thoracic n. |
| Pectoralis minor | Sternal ends of third, fourth, and fifth ribs; costal cartilage of first rib | Coracoid process of scapula | Pulls scapula forward and downward | Medial pectoral n. |
| Subclavius | | Inferior surface of acromion of scapula | Draws clavicle downward | Spinal nerves C5, C6 |
| Trapezius | Occipital bone and spines of seventh cervical and all thoracic vertebrae | Clavicle, spine, and acromion of scapula | Elevates scapula, draws head back, adducts scapula, braces shoulder | Accessory nerve and spinal nerves C3, C4 |
| Levator scapulae | First to fourth cervical vertebrae | Superior angle of scapula | Elevates scapula | Dorsal scapular n. |
| Rhomboideus major | Spines of second to fifth thoracic vertebrae | Medial border of scapula | Elevates and adducts scapula | Dorsal scapular n. |
| Rhomboideus minor | Seventh cervical and first thoracic vertebrae | Medial border of scapula | Elevates and adducts scapula | Dorsal scapular n. |

trapezius: Gk. *trapezoeides*, trapezoid shaped
rhomboideus: Gk. *rhomboides*, rhomboid shaped

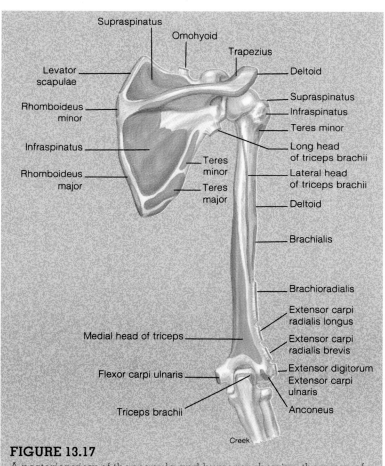

**FIGURE 13.17**

A posterior view of the scapula and humerus showing the areas of attachment of the associated muscles. (Points of origin are color-coded red, and points of insertion are color-coded blue.)

humerus. Both heads of the biceps brachii muscle insert on the radial tuberosity. The **brachialis muscle** is located on the distal anterior half of the humerus, deep to the biceps brachii muscle. It is synergistic to the biceps brachii muscle in flexing the forearm at the elbow joint.

The prominent **brachioradialis muscle** is positioned along the lateral (radial) surface of the forearm. It too flexes the forearm at the elbow joint.

The **triceps brachii muscle,** located on the posterior surface of the brachium, is the muscle antagonistic to the biceps brachii. It has three heads, or origins. Two of the three, the **lateral head** and **medial head,** arise from the humerus, whereas the **long head** arises from the infraglenoid tuberosity of the scapula. A common tendinous insertion attaches the triceps brachii muscle to the olecranon of the ulna. The triceps brachii and the small **anconeus muscles** extend the forearm at the elbow joint.

Refer to table 13.11 for a summary of the muscles that act on the forearm.

## Muscles of the Forearm That Move the Wrist, Hand, and Fingers

The muscles that cause wrist, hand, and most finger movements are positioned along the forearm (figs. 13.20 and 13.21). Several of these muscles act on two joints—the elbow and wrist. Others act on the joints of the wrist, hand, and digits. Still others produce rotational movement

**FIGURE** 13.18
Muscles of the right shoulder and brachium.

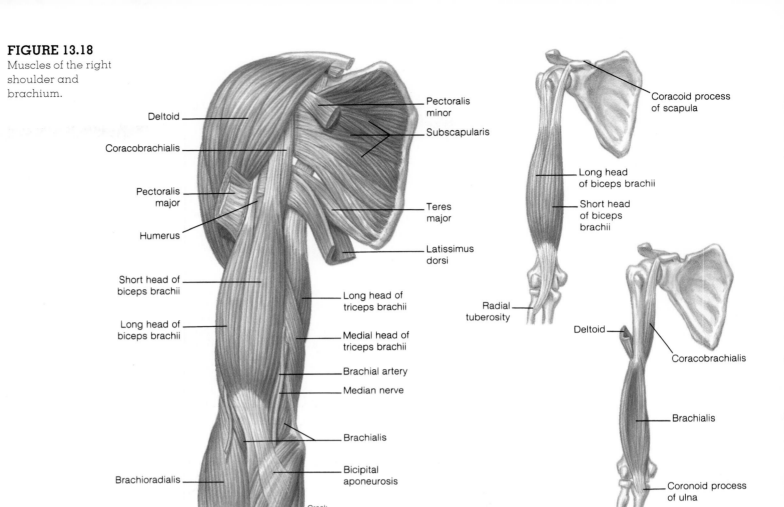

Deltoid

Coracobrachialis

Pectoralis major

Humerus

Short head of biceps brachii

Long head of biceps brachii

Brachioradialis

Pectoralis minor

Subscapularis

Teres major

Latissimus dorsi

Long head of triceps brachii

Medial head of triceps brachii

Brachial artery

Median nerve

Brachialis

Bicipital aponeurosis

Creek

Coracoid process of scapula

Long head of biceps brachii

Short head of biceps brachii

Radial tuberosity

Deltoid

Coracobrachialis

Brachialis

Coronoid process of ulna

## Table 13.10   Muscles that move the humerus

| Muscle | Origin | Insertion | Action | Innervation |
|---|---|---|---|---|
| Pectoralis major | Clavicle, sternum, costal cartilages of second to sixth rib; rectus sheath | Crest of greater tubercle of humerus | Flexes, adducts, and rotates arm medially at shoulder joint | Medial and lateral pectoral nn. |
| Latissimus dorsi | Spines of sacral, lumbar, and lower thoracic vertebrae; iliac crest and lower four ribs | Intertubercular groove of humerus | Extends, adducts, and rotates humerus medially at shoulder joint | Thoracodorsal n. |
| Deltoid | Clavicle, acromion process; spine of scapula | Deltoid tuberosity of humerus | Abducts arm and extends or flexes humerus at shoulder joint | Axillary n. |
| Supraspinatus | Supraspinous fossa of scapula | Greater tubercle of humerus | Abducts and laterally rotates humerus at shoulder joint | Suprascapular n. |

pectoralis major: L. *pectus,* chest
latissimus dorsi: L. *latissimus,* widest
deltoid: Gk. *delta,* triangular

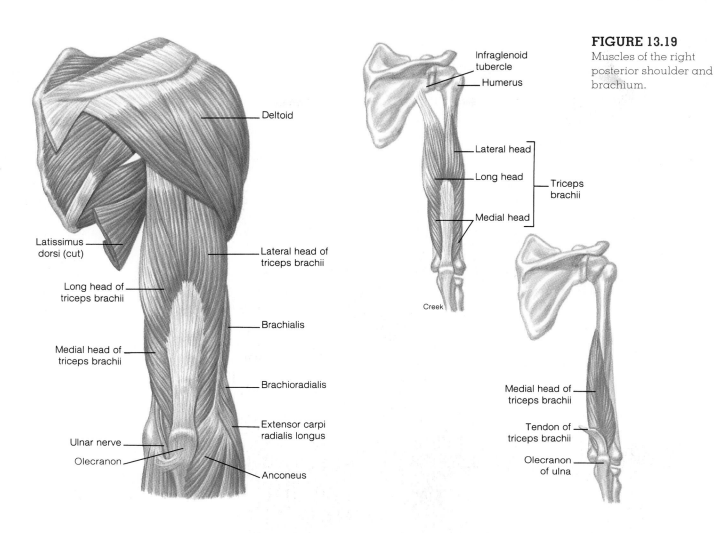

FIGURE 13.19

Muscles of the right posterior shoulder and brachium.

Deltoid

Latissimus dorsi (cut)

Long head of triceps brachii

Medial head of triceps brachii

Ulnar nerve

Olecranon

Lateral head of triceps brachii

Brachialis

Brachioradialis

Extensor carpi radialis longus

Anconeus

Infraglenoid tubercle

Humerus

Lateral head

Long head

Medial head

Triceps brachii

Creek

Medial head of triceps brachii

Tendon of triceps brachii

Olecranon of ulna

Muscular System

| Muscle | Origin | Insertion | Action | Innveration |
|--------|--------|-----------|--------|-------------|
| Infraspinatus | Infraspinous fossa of scapula | Greater tubercle of humerus | Rotates arm laterally at shoulder joint | Suprascapular n. |
| Teres major | Inferior angle and lateral border of scapula | Crest of lesser tubercle of humerus | Extends humerus, or adducts and rotates arm medially at shoulder joint | Lower subscapular n. |
| Teres minor | Lateral border of scapula | Greater tubercle of humerus | Rotates arm laterally at shoulder joint | Axillary n. |
| Subscapularis | Subscapular fossa | Lesser tubercle of humerus | Rotates arm medially at shoulder joint | Subscapular n. |
| Coracobrachialis | Coracoid process of scapula | Body of humerus | Flexes and adducts arm at shoulder joint | Musculocutaneous n. |

teres major: L. *teres*, rounded

# Table 13.11 Muscles that act on the forearm

| Muscle | Origin | Insertion | Action | Innervation |
|---|---|---|---|---|
| Biceps brachii | Coracoid process; tuberosity above glenoid cavity of scapula | Radial tuberosity | Flexes elbow joint; supinates forearm and hand at elbow joint | Musculocutaneous n. |
| Brachialis | Anterior shaft of humerus | Coronoid process of ulna | Flexes elbow joint | Musculocutaneous, median, and radial nn. |
| Brachioradialis | Lateral supracondylar ridge of humerus | Proximal to styloid process of radius | Flexes elbow joint | Radial n. |
| Triceps brachii | Tuberosity below glenoid cavity; lateral and medial surfaces of humerus | Olecranon of ulna | Extends elbow joint | Radial n. |
| Anconeus | Lateral epicondyle of humerus | Olecranon of ulna | Extends elbow joint | Radial n. |

biceps brachii: L. *biceps,* two heads
triceps brachii: L. *triceps,* three heads
anconeus: Gk. *ancon,* elbow

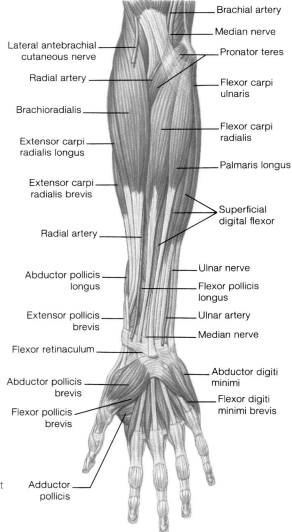

**FIGURE 13.20**
Superficial muscles of the right forearm. (*a*) An anterior view and (*b*) a posterior view.

Labels for part (a):
Brachial artery
Median nerve
Lateral antebrachial cutaneous nerve
Pronator teres
Radial artery
Flexor carpi ulnaris
Brachioradialis
Flexor carpi radialis
Extensor carpi radialis longus
Palmaris longus
Extensor carpi radialis brevis
Superficial digital flexor
Radial artery
Abductor pollicis longus
Ulnar nerve
Flexor pollicis longus
Extensor pollicis brevis
Ulnar artery
Median nerve
Flexor retinaculum
Abductor digiti minimi
Abductor pollicis brevis
Flexor digiti minimi brevis
Flexor pollicis brevis
Adductor pollicis

Labels for part (b):
Brachioradialis
Ulnar nerve
Extensor carpi radialis longus
Anconeus
Extensor carpi ulnaris
Flexor carpi ulnaris
Extensor carpi radialis brevis
Extensor digitorum commun
Extensor digiti minimi
Abductor pollicis longus
Extensor indicis
Extensor pollicis longus
Extensor pollicis brevis
Dorsal branch of ulnar nerve
Superficial branch of radial nerve
Extensor retinaculum
Radial artery
Dorsal interosseous
Creek

(a)     (b)

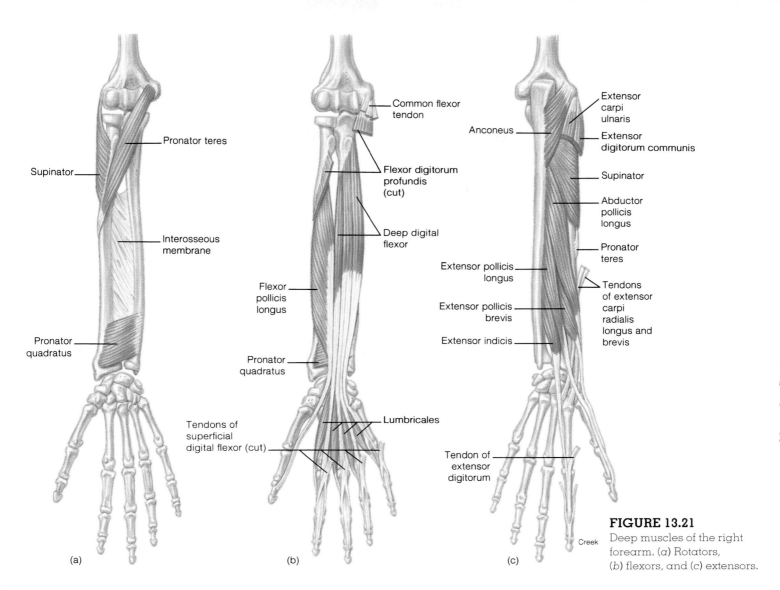

**FIGURE 13.21**
Deep muscles of the right forearm. (*a*) Rotators, (*b*) flexors, and (*c*) extensors.

Creek

(a)  (b)  (c)

at the radioulnar joint. The precise actions of these muscles are complex, so only the basic movements will be described here. Most of these muscles perform four primary actions on the hand and digits: supination, pronation, flexion, and extension. Other actions of the hand include adduction and abduction.

**Supination and Pronation of the Hand**  The **supinator muscle** is positioned around the upper posterior portion of the radius (fig. 13.21), where it works synergistically with the biceps brachii muscle to supinate the hand. Two muscles are responsible for pronating the hand—the pronator teres and pronator quadratus. The **pronator teres muscle** is located on the upper medial side of the forearm, whereas the deep, anteriorly positioned **pronator quadratus muscle** extends between the ulna and radius on the distal fourth of the forearm. These two muscles work synergistically to rotate the palm of the hand posteriorly and position the thumb medially.

**Flexion of the Wrist, Hand, and Fingers**  Six of the muscles that flex the wrist, hand, and fingers will be described from lateral to medial and from superficial to deep (see figs. 13.20 and 13.21). Although four of the six arise from the medial epicondyle of the humerus (table 13.12), their actions on the elbow joint are minimal. The brachioradialis, already described, is an obvious reference muscle for locating the muscles of the forearm that flex the joints of the hand.

The **flexor carpi radialis muscle** extends diagonally across the anterior surface of the forearm, and its distal cord-like tendon crosses the wrist under the *flexor retinaculum*.

The narrow **palmaris longus muscle,** superficial in position on the anterior surface of the forearm, has a long, slender tendon that attaches to the *palmar aponeurosis,* where it assists in flexing the wrist. It is the most variable muscle in the body and is totally absent in approximately 8% of all people. In 4%, it is absent in one or the other forearm. Furthermore, it is absent more often in females than males, and

## Table 13.12 — Muscles of the forearm that act on the wrist, hand, and digits

| Muscle | Origin | Insertion | Action | Innervation |
|---|---|---|---|---|
| Supinator | Lateral epicondyle of humerus and crest of ulna | Lateral surface of radius | Supinates hand | Radial n. |
| Pronator teres | Medial epicondyle of humerus | Lateral surface of radius | Pronates hand | Median n. |
| Pronator quadratus | Distal fourth of ulna | Distal fourth of radius | Pronates hand | Median n. |
| Flexor carpi radialis | Medial epicondyle of humerus | Base of second and third metacarpal bones | Flexes and abducts hand at wrist | Median n. |
| Palmaris longus | Medial epicondyle of humerus | Palmar aponeurosis | Flexes wrist | Median n. |
| Flexor carpi ulnaris | Medial epicondyle and olecranon | Carpal and metacarpal bones | Flexes and adducts wrist | Ulnar n. |
| Superficial digital flexor | Medial epicondyle, coronoid process, and anterior border of radius | Middle phalanges of digits | Flexes wrist and digits at carpophalangeal and interphalangeal joints | Median n. |
| Flexor digitorum profundus | Proximal two-thirds of ulna and interosseous ligament | Distal phalanges | Flexes wrist and digits at carpophalangeal and interphalangeal joints | Median and ulnar nn. |
| Flexor pollicis longus | Shaft of radius, interosseous membrane, and coronoid process of ulna | Distal phalanx of thumb | Flexes joints of thumb | Median n. |
| Extensor carpi radialis longus | Lateral supracondylar ridge of humerus | Second metacarpal bone | Extends and abducts wrist | Radial n. |
| Extensor carpi radialis brevis | Lateral epicondyle of humerus | Third metacarpal bone | Extends and abducts wrist | Radial n. |
| Extensor digitorum communis | Lateral epicondyle of humerus | Posterior surfaces of digits II–V | Extends wrist and phalanges at joints of carpophalangeal and interphalangeal joints | Radial n. |
| Extensor digiti minimi | Lateral epicondyle of humerus | Extensor aponeurosis of fifth digit | Extends joints of fifth digit and wrist | Radial n. |
| Extensor carpi ulnaris | Lateral epicondyle of humerus and olecranon | Base of fifth metacarpal bone | Extends and adducts wrist | Radial n. |
| Extensor pollicis longus | Middle shaft of ulna, lateral side | Base of distal phalanx of thumb | Extends joints of thumb; abducts joints of hand | Radial n. |
| Extensor pollicis brevis | Distal shaft of radius and interosseous ligament | Base of first phalanx of thumb | Extends joints of thumb; abducts joints of hand | Radial n. |
| Abductor pollicis longus | Distal radius and ulna and interosseous ligament | Base of first metacarpal bone | Abducts joints of thumb and joints of hand | Radial n. |

supinator: L. *supin*, bend back
pronator *teres:* L. *pron*, bend forward
palmaris longus: L. *palma*, flat of hand

flexor digitorum profundus: L. *profundus*, deep
flexor pollicis longus: L. *pollex*, thumb

on the left side in both sexes. Because of the superficial position of the palmaris longus muscle, you can readily determine whether or not it is present in your own forearm by tightly clenching your fist and examining for its tendon just proximal to your wrist.

The **flexor carpi ulnaris muscle** is positioned on the medial anterior side of the forearm, where it assists in flexing the wrist and adducting the hand.

The broad **superficial digital flexor muscle** lies directly beneath the three flexors just described (see figs. 13.20 and

13.21). It has an extensive origin, involving all three of the long bones of the brachium and antebrachium (table 13.12). The tendon at the distal end of this muscle is united across the wrist but then splits to attach to the middle phalanx of digits II through V.

The **deep digital flexor muscle** lies deep to the superficial digital flexor muscle. These two muscles flex the wrist, hand, and the second, third, fourth, and fifth digits.

The **flexor pollicis longus muscle** is a deep, lateral muscle of the forearm. It flexes the thumb, assisting the grasping mechanism of the hand.

The tendons of the muscles that flex the hand can be seen on the wrist as a fist is made. These tendons are securely positioned by the *flexor retinaculum* (fig. 13.20a), which crosses the wrist area transversely.

**Extension of the Hand**    The muscles that extend the hand are located on the posterior side of the forearm. Most of the primary extensors of the hand can be seen superficially in figure 13.20b and will be discussed from lateral to medial.

The long, tapered **extensor carpi radialis longus muscle** is located medial to the brachioradialis, where it extends the carpal joints and abducts the hand at the wrist. The **extensor carpi radialis brevis muscle** is positioned immediately medial to the extensor carpi radialis longus muscle and performs approximately the same functions. Its origin and insertion, however, are different (see table 13.12).

The bipennate-fibered **extensor digitorum communis muscle** is positioned in the center of the forearm, along the posterior surface. It originates on the lateral epicondyle of the humerus. Its tendon of insertion divides at the wrist, beneath the extensor retinaculum, into four tendons that attach to the distal tip of the medial phalanges of digits II through V.

The long, narrow **extensor digiti minimi muscle** is located on the ulnar side of the extensor digitorum communis muscle. Its tendinous insertion fuses with the tendon of the extensor digitorum communis muscle going to the fifth digit.

The **extensor carpi ulnaris muscle** is the most medial muscle on the posterior surface of the forearm. It inserts on the base of the fifth metacarpal bone, where it functions to extend and adduct the hand.

The **extensor pollicis longus muscle** extends from the mid-ulnar region, across the lower two-thirds of the forearm, and inserts onto the base of the distal phalanx of the thumb (see fig. 13.21), where it extends the thumb and abducts the hand.

The **extensor pollicis brevis muscle** arises from the lower midportion of the radius and inserts on the base of the proximal phalanx of the thumb (fig. 13.21). The action of this muscle is similar to that of the extensor pollicis longus.

As its name implies, the **abductor pollicis longus muscle** abducts the thumb and hand. It originates on the interosseous ligament, between the ulna and radius, and inserts on the base of the first metacarpal bone.

The muscles that act on the wrist, hand, and digits are summarized in table 13.12.

 Notice that the joints of your hand are partially flexed even when the hand is relaxed. The muscles that extend the hand are not as strong as the muscles that flex it. This is why people who receive strong electrical shocks through the arms will tightly flex their hands and hold on. All the muscles of the arm are stimulated to contract, but the flexors, being stronger, cause the hands to close tightly.

## Muscles of the Hand

The hand is a marvelously complex structure, adapted to permit an array of intricate movements. Flexion and extension movements of the hand and phalanges are accomplished by the muscles of the forearm just described. Precise finger movements that require coordinating abduction and adduction with flexion and extension are the function of the small intrinsic muscles of the hand. These muscles and associated structures of the hand are depicted in figure 13.22. The position and actions of the muscles of the hand are listed in table 13.13.

The muscles of the hand are divided into *thenar* (the´nar), *hypothenar* (hi-poth´ē-nar), and *intermediate groups*. The **thenar eminence** is the fleshy base of the thumb and is formed by three muscles: the **abductor pollicis brevis muscle,** the **flexor pollicis brevis muscle,** and the **opponens pollicis muscle.** The most important of the thenar muscles is the opponens pollicis, which opposes the thumb to the palm of the hand.

The **hypothenar eminence** is the elongated, fleshy bulge at the base of the little finger. It also is formed by three muscles: the **abductor digiti minimi muscle,** the **flexor digiti minimi muscle,** and the **opponens digiti minimi muscle.**

Muscles of the **intermediate group** are positioned between the metacarpal bones in the region of the palm. This group includes the **adductor pollicis muscle,** the **lumbricales** (lum´brī-kalz) **muscles,** and the **palmar** and **dorsal interossei muscles.**

## Muscles That Move the Thigh

The muscles that move the thigh originate from the pelvic girdle and insert on various places on the femur. These muscles stabilize a highly movable hip joint and provide support for the body during bipedal stance and locomotion. Found in this region are the most massive muscles of the body as well as some extremely small muscles. The muscles that move the thigh are divided into anterior, posterior, and medial groups.

**FIGURE 13.22**

Muscles of the hand. (a) An anterior view and (b) a lateral view of the second digit (index finger).

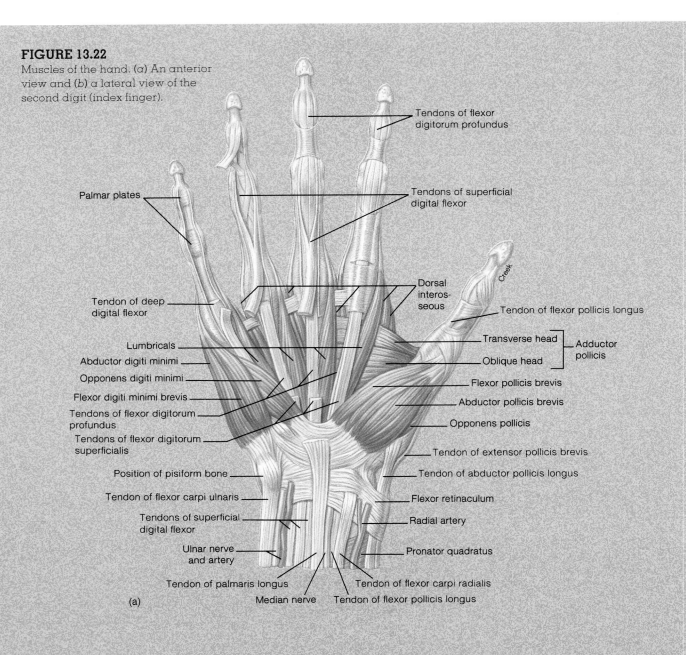

Tendons of flexor digitorum profundus

Palmar plates

Tendons of superficial digital flexor

Dorsal interosseous

Tendon of deep digital flexor

Tendon of flexor pollicis longus

Lumbricals

Transverse head · Adductor pollicis

Abductor digiti minimi

Oblique head

Opponens digiti minimi

Flexor pollicis brevis

Flexor digiti minimi brevis

Abductor pollicis brevis

Tendons of flexor digitorum profundus

Opponens pollicis

Tendons of flexor digitorum superficialis

Tendon of extensor pollicis brevis

Position of pisiform bone

Tendon of abductor pollicis longus

Tendon of flexor carpi ulnaris

Flexor retinaculum

Tendons of superficial digital flexor

Radial artery

Ulnar nerve and artery

Pronator quadratus

Tendon of palmaris longus

Tendon of flexor carpi radialis

Median nerve

Tendon of flexor pollicis longus

(a)

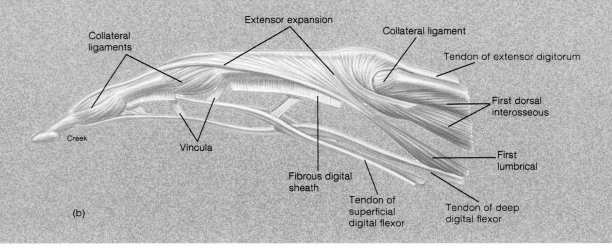

Extensor expansion

Collateral ligaments

Collateral ligament

Tendon of extensor digitorum

First dorsal interosseous

Vincula

First lumbrical

Fibrous digital sheath

Tendon of superficial digital flexor

Tendon of deep digital flexor

(b)

# Table 13.13 Intrinsic muscles of the hand

| Muscle | Origin | Insertion | Action | Innervation |
|---|---|---|---|---|
| **Thenar muscles** | | | | |
| Abductor pollicis brevis | Flexor retinaculum, scaphoid, and trapezium | Proximal phalanx of thumb | Abducts joints of thumb | Median n. |
| Flexor pollicis brevis | Flexor retinaculum and trapezium | Proximal phalanx of thumb | Flexes joints of thumb | Median n. |
| Opponens pollicis | Trapezium and flexor retinaculum | First metacarpal bone | Opposes joints of thumb | Median n. |
| **Intermediate muscles** | | | | |
| Adductor pollicis (oblique and transverse heads) | Oblique head, capitate; transverse head, second and third metacarpal bones | Proximal phalanx of thumb | Adducts joints of thumb | Ulnar n. |
| Lumbricales (4) | Tendons of flexor digitorum profundus | Extensor expansions of digits II–V | Flexes digits at metacarpophalangeal joints; extends digits at interphalangeal joints | Median and ulnar nn. |
| Palmar interossei (3) | Medial side of second metacarpal bone; lateral sides of fourth and fifth metacarpal bones | Proximal phalanges of index, ring, and little fingers and extensor digitorum communis | Adducts fingers toward middle finger at metacarpophalangeal joints | Ulnar n. |
| Dorsal interossei (4) | Adjacent sides of metacarpal bones | Proximal phalanges of index and middle fingers (lateral sides) plus proximal phalanges of middle and ring fingers (medial sides) and extensor digitorum communis | Abducts fingers away from middle finger at metacarpophalangeal joints | Ulnar n. |
| **Hypothenar muscles** | | | | |
| Abductor digiti minimi | Pisiform and tendon of flexor carpi ulnaris | Proximal phalanx of digit V | Abducts joints of digit V | Ulnar n. |
| Flexor digiti minimi | Flexor retinaculum and hook of hamate | Proximal phalanx of digit V | Flexes joints of digit V | Ulnar n. |
| Opponens digiti minimi | Flexor retinaculum and hook of hamate | Fifth metacarpal bone | Opposes joints of digit V | Ulnar n. |

opponens pollicus: L. *opponens,* against

**Anterior Muscles** The anterior muscles that move the thigh are the iliacus and psoas major (figs. 13.23 and 13.24).

The triangular **iliacus muscle** arises from the iliac fossa and inserts on the lesser trochanter of the femur.

The long, thick **psoas major muscle** originates on the bodies and transverse processes of the lumbar vertebrae and inserts, along with the iliacus, on the lesser trochanter (fig. 13.24). The psoas major muscle and the iliacus work synergistically in flexing and rotating the thigh and flexing the vertebral column. They are frequently referred to as a single muscle, the **iliopsoas** (*il´´e-ŏ-so´as*) **muscle.**

**Posterior and Lateral (Buttock) Muscles** The posterior muscles that move the thigh include the gluteus maximus, gluteus medius, gluteus minimus, and tensor fasciae latae muscles.

The large **gluteus maximus muscle** forms much of the prominence of the buttock (fig. 13.25). It is a powerful

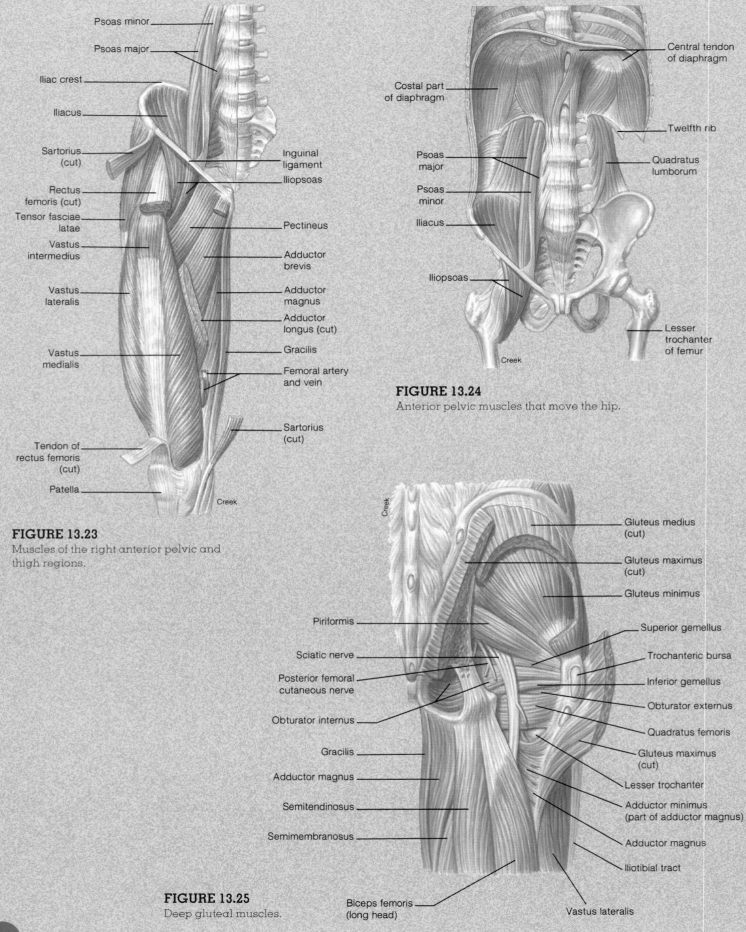

Psoas minor

Psoas major

Iliac crest

Iliacus

Sartorius
(cut)

Rectus
femoris (cut)

Tensor fasciae
latae

Vastus
intermedius

Vastus
lateralis

Vastus
medialis

Tendon of
rectus femoris
(cut)

Patella

Inguinal
ligament

Iliopsoas

Pectineus

Adductor
brevis

Adductor
magnus

Adductor
longus (cut)

Gracilis

Femoral artery
and vein

Sartorius
(cut)

Creek

**FIGURE 13.23**
Muscles of the right anterior pelvic and thigh regions.

Central tendon
of diaphragm

Costal part
of diaphragm

Twelfth rib

Psoas
major

Quadratus
lumborum

Psoas
minor

Iliacus

Iliopsoas

Lesser
trochanter
of femur

Creek

**FIGURE 13.24**
Anterior pelvic muscles that move the hip.

Gluteus medius
(cut)

Gluteus maximus
(cut)

Gluteus minimus

Piriformis

Sciatic nerve

Posterior femoral
cutaneous nerve

Obturator internus

Gracilis

Adductor magnus

Semitendinosus

Semimembranosus

Superior gemellus

Trochanteric bursa

Inferior gemellus

Obturator externus

Quadratus femoris

Gluteus maximus
(cut)

Lesser trochanter

Adductor minimus
(part of adductor magnus)

Adductor magnus

Iliotibial tract

Biceps femoris
(long head)

Vastus lateralis

Creek

**FIGURE 13.25**
Deep gluteal muscles.

# Table 13.14   Anterior and posterior muscles that move the thigh

| Muscle | Origin | Insertion | Action | Innervation |
|---|---|---|---|---|
| Iliacus | Iliac fossa | Lesser trochanter of femur | Flexes and rotates thigh laterally at the hip joint; flexes joints of vertebral column | Femoral n. |
| Psoas major | Transverse processes of all lumbar vertebrae | Lesser trochanter with iliacus | Flexes and rotates thigh laterally at the hip joint; flexes joints of vertebral column | Spinal nerves L2, L3 |
| Gluteus maximus | Iliac crest, sacrum, coccyx, and aponeurosis of the lumbar region | Gluteal tuberosity and iliotibial tract | Extends and rotates thigh laterally at the hip joint | Inferior gluteal n. |
| Gluteus medius | Lateral surface of ilium | Greater trochanter | Abducts and rotates thigh medially at the hip joint | Superior gluteal n. |
| Gluteus minimus | Lateral surface of lower half of ilium | Greater trochanter | Abducts and rotates thigh medially at the hip joint | Superior gluteal n. |
| Tensor fasciae latae | Anterior border of ilium and iliac crest | Iliotibial tract | Abducts thigh at the hip joint | Superior gluteal n. |

psoas major: Gk. *psoa,* loin
gluteus maximus: Gk. *gloutos,* rump

extensor muscle of the thigh and is very important for bipedal stance and locomotion. The gluteus maximus originates on the ilium, sacrum, coccyx, and aponeurosis of the lumbar region. It inserts on the gluteal tuberosity of the femur and the *iliotibial tract,* a thickened tendinous region of the fascia lata extending down the thigh (see fig. 13.27).

The **gluteus medius muscle** is located immediately deep to the gluteus maximus (fig. 13.25). It originates on the lateral surface of the ilium and inserts on the greater trochanter of the femur. The gluteus medius abducts and medially rotates the thigh. The mass of this muscle is of clinical significance as a site for intramuscular injections.

The **gluteus minimus** is the smallest and deepest of the gluteal muscles (fig. 13.25). It also arises from the lateral surface of the ilium and inserts on the lateral surface of the greater trochanter, where it acts synergistically with the gluteus medius to abduct and medially rotate the thigh.

The quadrangular **tensor fasciae latae muscle** is positioned superficially on the lateral surface of the hip (see fig. 13.27). It originates on the iliac crest and inserts on a broad lateral fascia of the thigh called the *fascia lata.* The fascia lata is continuous with the iliotibial tract. The tensor fasciae latae muscle is the principal abductor of the thigh.

A deep group of six lateral rotators of the thigh is positioned directly over the posterior aspect of the hip joint. These muscles are not discussed here but are identified in figure 13.25 from superior to inferior as the **piriformis, superior gemellus, obturator internus, inferior gemellus, obturator externus,** and **quadratus femoris muscles.**

The anterior and posterior group of muscles that move the thigh at the hip joint are summarized in table 13.14.

**Medial, or Adductor, Muscles**   The medial muscles that move the thigh include the gracilis, pectineus, adductor longus, adductor brevis, and adductor magnus muscles (figs. 13.26, 13.27, 13.28, and 13.29).

The long, thin **gracilis** (*gras´il-is*) **muscle** is the most superficial of the medial thigh muscles. It is a two-joint muscle and can adduct the thigh or flex the leg.

The **pectineus muscle** is the uppermost of the medial muscles that move the thigh. It is a flat, quadrangular muscle that flexes and adducts the thigh.

The **adductor longus muscle** is immediately lateral to the gracilis on the upper third of the thigh, and is the most anterior of the adductor muscles. The triangular **adductor brevis muscle** is located behind and largely concealed by the adductor longus and pectineus muscles. The **adductor magnus muscle** is a large, thick muscle, somewhat triangular in shape. It is located deep to the other two adductor muscles. The adductor longus, adductor brevis, and the adductor magnus are synergistic in adducting, flexing, and laterally rotating the thigh.

309

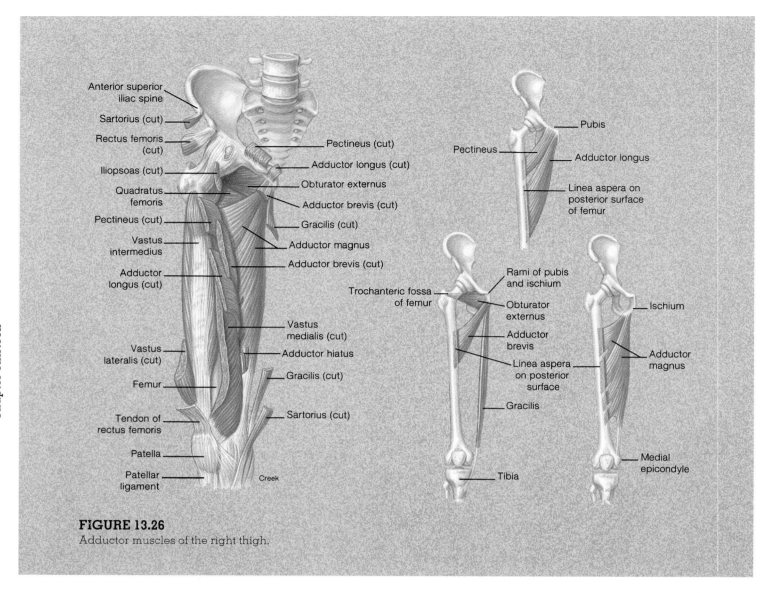

**FIGURE 13.26**
Adductor muscles of the right thigh.

The muscles that adduct the thigh at the hip joint are summarized in table 13.15.

## Muscles of the Thigh That Move the Leg

The muscles that move the leg originate on the pelvic girdle or thigh and are surrounded and compartmentalized by tough fascial sheets that are a continuation of the fascia lata and iliotibial tract. They are divided according to function and position into two groups: anterior extensors and posterior flexors.

**Anterior, or Extensor, Muscles** The anterior muscles that move the leg are the sartorius and quadriceps femoris muscles (figs. 13.26 and 13.27). The femoral nerve innervates these muscles.

The long, straplike **sartorius muscle** obliquely crosses the anterior aspect of the thigh. It can act on both the hip and knee joints to flex and rotate the thigh laterally, and also to assist in flexing the leg and rotating it medially. The sartorius muscle is the longest muscle of the body. It is frequently called the tailor's muscle because it enables one to assume a cross-legged sitting position.

The **quadriceps femoris muscle** is actually a composite of four distinct muscles that have separate origins but a common insertion on the patella via the tendon of the rectus femoris. The tendon of the rectus femoris is continuous over the patella and becomes the *patellar ligament* as it attaches to the head of the tibial tuberosity (fig. 13.27). These muscles function synergistically to extend the leg, as in kicking a

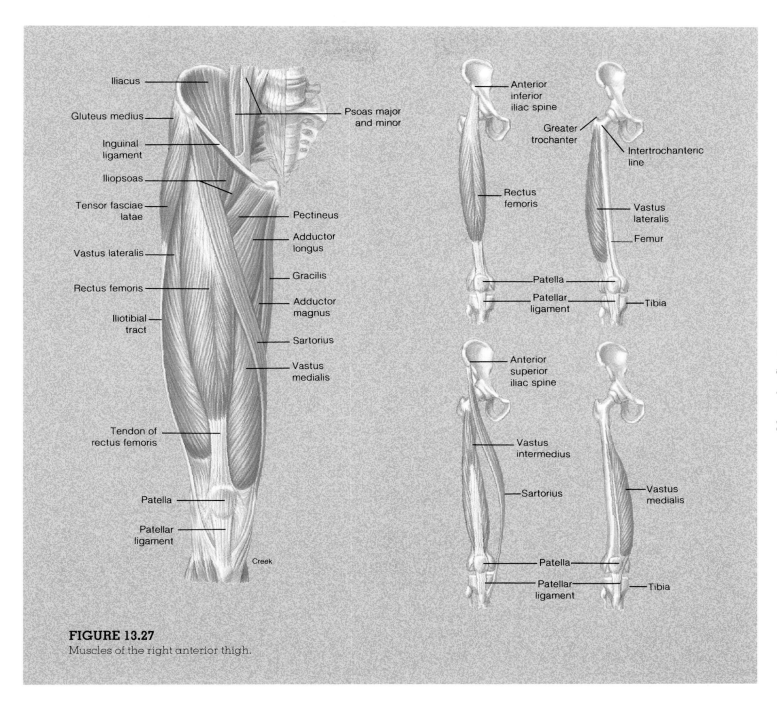

**FIGURE 13.27**
Muscles of the right anterior thigh.

football. The four muscles of the quadriceps femoris muscle are the rectus femoris, vastus lateralis, vastus medialis, and vastus intermedius.

The **rectus femoris muscle** occupies a superficial position and is the only one of the four quadriceps that functions in both the hip and knee joints. The laterally positioned **vastus lateralis muscle** is the largest muscle of the quadriceps femoris. It is a common intramuscular injection site in infants who have small, underdeveloped buttock and shoulder muscles. The **vastus medialis muscle** occupies a medial position along the thigh. The **vastus intermedius muscle** lies deep to the rectus femoris.

The anterior thigh muscles that move the leg at the hip joint are summarized in table 13.16.

**Posterior, or Flexor, Muscles**  There are three posterior thigh muscles antagonistic to the quadriceps femoris muscles in flexing the knee. These muscles, known as the **hamstrings** (fig. 13.28), are innervated by the tibial nerve. Interestingly, this name derives from the old butchers' practice of using the tendons of these muscles to hang the hams of pork for smoke curing.

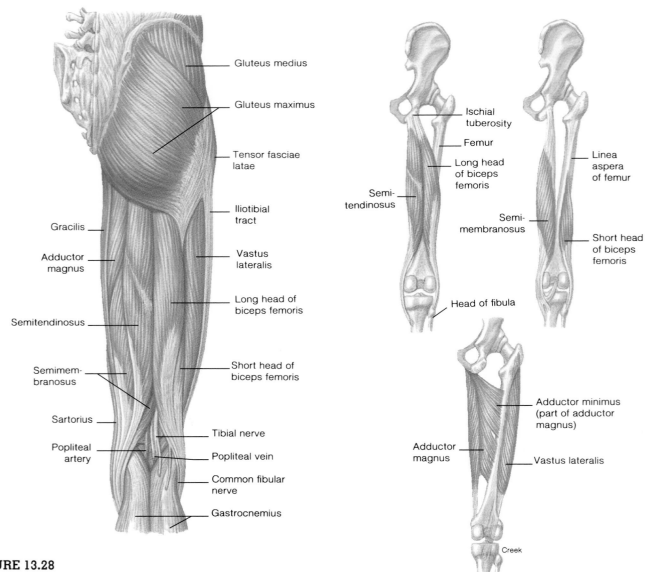

**FIGURE 13.28**
Muscles of the right posterior thigh.

| | | | | |
|---|---|---|---|---|
| **Table 13.15** | **Medial muscles that move the thigh** | | | |
| **Muscle** | **Origin** | **Insertion** | **Action** | **Innervation** |
| Gracilis | Inferior edge of symphysis pubis | Proximal medial surface of tibia | Adducts thigh at hip joint; flexes and rotates leg at knee joint | Obturator n. |
| Pectineus | Pectineal line of pubis | Distal to lesser trochanter of femur | Adducts and flexes thigh at hip joint | Femoral n. |
| Adductor longus | Pubis—below pubic crest | Linea aspera of femur | Adducts, flexes, and laterally rotates thigh at hip joint | Obturator n. |
| Adductor brevis | Inferior ramus of pubis | Linea aspera of femur | Adducts, flexes, and laterally rotates thigh at hip joint | Obturator n. |
| Adductor magnus | Inferior ramus of ischium and pubis | Linea aspera and medial epicondyle of femur | Adducts, flexes, and laterally rotates thigh at hip joint | Obturator n. |

gracilis: Gk. *gracilis,* slender

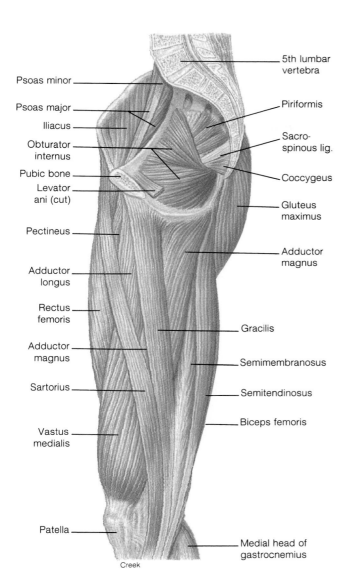

Figure labels:
- 5th lumbar vertebra
- Psoas minor
- Psoas major
- Iliacus
- Obturator internus
- Pubic bone
- Levator ani (cut)
- Pectineus
- Adductor longus
- Rectus femoris
- Adductor magnus
- Sartorius
- Vastus medialis
- Patella
- Piriformis
- Sacro-spinous lig.
- Coccygeus
- Gluteus maximus
- Adductor magnus
- Gracilis
- Semimembranosus
- Semitendinosus
- Biceps femoris
- Medial head of gastrocnemius
- Creek

**FIGURE 13.29**
Muscles of the right medial thigh.

The **biceps femoris muscle** occupies the posterior lateral aspect of the thigh. It has a superficial long head and a deep short head and causes movement at both the hip and knee joints. The superficial **semitendinosus muscle** is fusiform and is located on the posterior medial aspect of the thigh. It also works over two joints. The flat **semimembranosus muscle** lies deep to the semitendinosus muscle on the posterior medial aspect of the thigh.

The posterior thigh muscles that move the leg at the hip joint are summarized in table 13.17. The relationship of the muscles of the thigh is illustrated in figure 13.30.

Hamstring injuries are a common occurrence in some sports. The injury usually occurs when sudden lateral or medial stress to the knee joint tears the muscles or tendons. Because of its structure and the stress applied to it in competition, the knee joint is highly susceptible to injury. To reduce the incidence of knee injury, the rules in contact sports should be altered, or at least additional support and protection provided for this vulnerable joint.

## Muscles of the Leg That Move the Ankle, Foot, and Toes

The muscles of the leg, the **crural muscles,** are responsible for the movements of the foot. There are three groups of crural muscles: anterior, lateral, and posterior. The anteromedial aspect of the leg along the shaft of the tibia lacks muscle attachment.

## Table 13.16  Anterior thigh muscles that move the leg

| Muscle | Origin | Insertion | Action |
|---|---|---|---|
| Sartorius | Anterior superior iliac spine | Medial surface of tibia | Flexes leg and thigh, abducts thigh, rotates thigh laterally, and rotates leg medially at hip joint |
| Quadriceps femoris | | Patella by common tendon, which continues as patellar ligament to tibial tuberosity | Extends leg at knee joint |
| Rectus femoris | Anterior superior iliac spine and lip of acetabulum | | |
| Vastus lateralis | Greater trochanter and linea aspera of femur | | |
| Vastus medialis | Medial surface and linea aspera of femur | | |
| Vastus intermedius | Anterior and lateral surfaces of femur | | |

sartorius: L. *sartor*, a tailor (muscle used to cross legs in a tailor's position)

## Table 13.17  Posterior thigh muscles that move the leg

| Muscle | Origin | Insertion | Action |
|---|---|---|---|
| Biceps femoris | Long head—ischial tuberosity; short head—linea aspera of femur | Head of fibula and lateral epicondyle of tibia | Flexes leg at knee joint; extends and laterally rotates thigh at hip joint |
| Semitendinosus | Ischial tuberosity | Proximal portion of medial surface of shaft of tibia | Flexes leg at knee joint; extends and medially rotates thigh at hip joint |
| Semimembranosus | Ischial tuberosity | Medial epicondyle of tibia | Flexes leg at knee joint; extends and medially rotates thigh at hip joint |

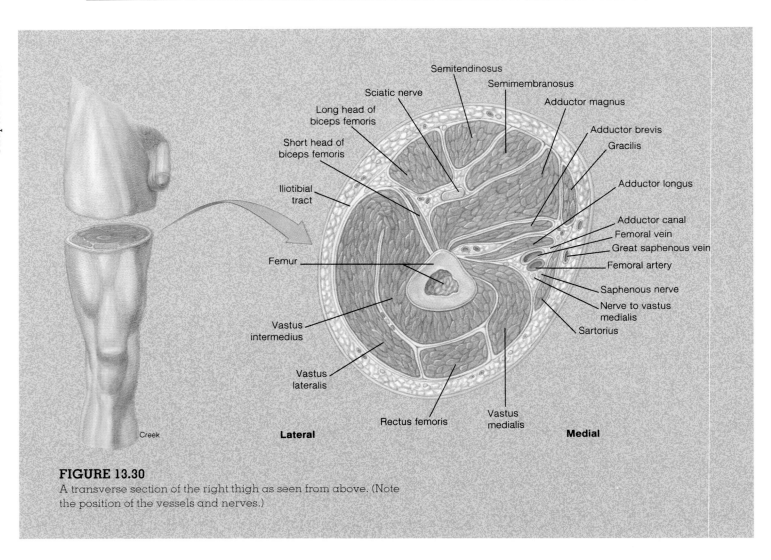

**FIGURE 13.30**

A transverse section of the right thigh as seen from above. (Note the position of the vessels and nerves.)

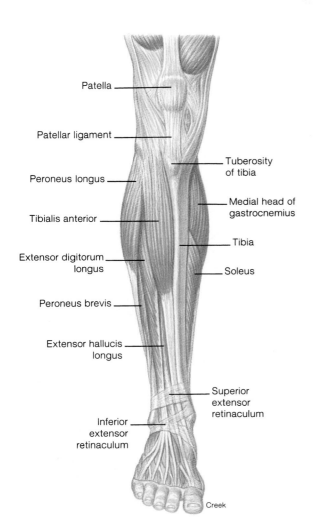

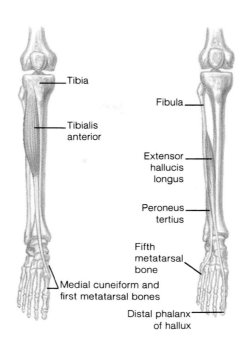

**FIGURE 13.31**
Anterior crural muscles.

**Anterior Crural Muscles**   The anterior crural muscles include the tibialis anterior, extensor digitorum longus, extensor hallucis longus, and peroneus tertius muscles (figs. 13.31, 13.32, and 13.33).

The large, superficial **tibialis anterior muscle** can be easily palpated on the anterior lateral portion of the tibia (fig. 13.31). It parallels the prominent anterior crest of the tibia. The **extensor digitorum longus muscle** lies lateral to the tibialis anterior on the anterolateral surface of the leg. The **extensor hallucis longus muscle** is positioned deep between the tibialis anterior and the extensor digitorum longus. The small **peroneus tertius muscle** is continuous with the distal portion of the extensor digitorum longus muscle.

**Lateral Crural Muscles**   The lateral crural muscles are the peroneus longus and peroneus brevis (figs. 13.31 and 13.32). The long, flat **peroneus longus muscle** is a superficial lateral muscle that overlies the fibula. The **peroneus brevis muscle** lies deep to the peroneus longus muscle and is positioned closer to the foot. These two muscles are synergistic in flexion and eversion of the foot (see table 13.18).

**Posterior Crural Muscles**   The seven posterior crural muscles can be grouped into a superficial and a deep group. The superficial group is composed of the gastrocnemius, soleus, and plantaris muscles (fig. 13.34). The four deep

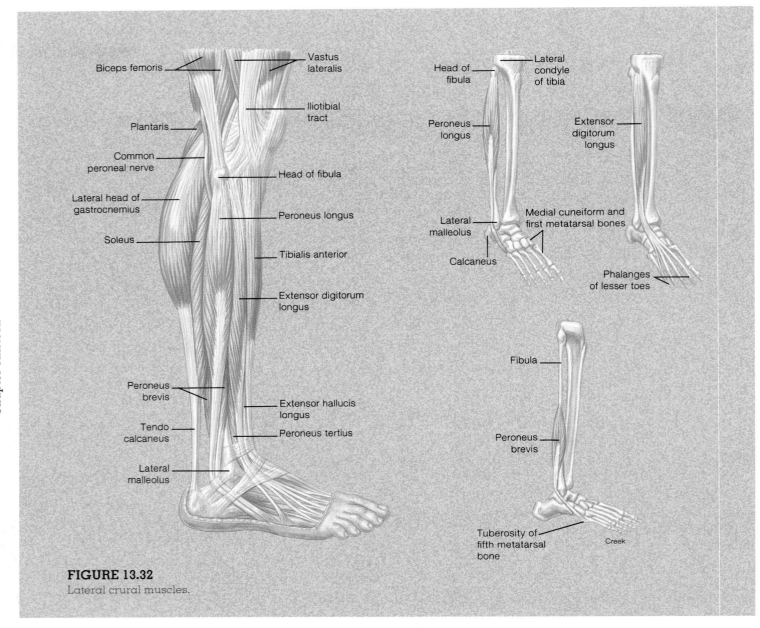

Biceps femoris

Vastus lateralis

Iliotibial tract

Plantaris

Common peroneal nerve

Head of fibula

Lateral head of gastrocnemius

Peroneus longus

Soleus

Tibialis anterior

Extensor digitorum longus

Peroneus brevis

Extensor hallucis longus

Tendo calcaneus

Peroneus tertius

Lateral malleolus

Head of fibula

Lateral condyle of tibia

Peroneus longus

Extensor digitorum longus

Lateral malleolus

Medial cuneiform and first metatarsal bones

Calcaneus

Phalanges of lesser toes

Fibula

Peroneus brevis

Tuberosity of fifth metatarsal bone

Creek

**FIGURE 13.32**
Lateral crural muscles.

posterior crural muscles are the popliteus, flexor hallucis longus, flexor digitorum longus, and tibialis posterior muscles (fig. 13.35).

The large, superficial **gastrocnemius** (*gas″trok-ne′me-us*) **muscle** forms the major portion of the calf of the leg. It consists of two distinct heads, which arise from the posterior surfaces of the medial and lateral condyles of the femur. This muscle and the deeper soleus muscle insert onto the calcaneus bone via the common *tendo calcaneus* (*tendon of* 

*Achilles*). The gastrocnemius acts over two joints to cause flexion of the knee and plantar flexion of the foot.

The **soleus muscle** lies deep to the gastrocnemius muscle. These two muscles are frequently referred to as a single muscle, the **triceps surae muscle.** The soleus muscle has a common insertion with the gastrocnemius muscle but acts on only one joint, the ankle, in plantar flexing the foot.

The small **plantaris muscle** arises just above the origin of the lateral head of the gastrocnemius muscle on the

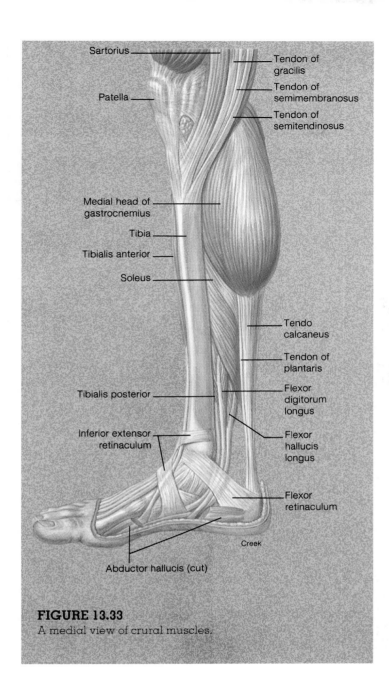

**FIGURE 13.33**
A medial view of crural muscles.

*Labels on figure:*
Sartorius
Tendon of gracilis
Patella
Tendon of semimembranosus
Tendon of semitendinosus
Medial head of gastrocnemius
Tibia
Tibialis anterior
Soleus
Tendo calcaneus
Tendon of plantaris
Tibialis posterior
Flexor digitorum longus
Inferior extensor retinaculum
Flexor hallucis longus
Flexor retinaculum
Creek
Abductor hallucis (cut)

deep to the soleus muscle on the posterolateral side of the leg. It flexes the big toe (hallux) and assists in plantar flexing and inverting the foot.

The **flexor digitorum longus muscle** also lies deep to the soleus muscle, and it parallels the flexor hallucis longus muscle on the medial side of the leg. Its distal tendon passes behind the medial malleolus and continues along the plantar surface of the foot, where it branches into four tendinous slips that attach to the bases of the terminal phalanges of the second, third, fourth, and fifth toes (fig. 13.37). The flexor digitorum longus works over several joints, flexing four of the digits and assisting in plantar flexing and inverting the foot.

The **tibialis posterior muscle** is located deep to the soleus, directly on the interosseous ligament on the posterior surface of the leg. The distal tendon of the tibialis posterior passes behind the medial malleolus and inserts on the plantar surfaces of the navicular, cuneiform and cuboid bones and the second, third, and fourth metatarsal bones (fig. 13.37). The tibialis posterior plantar flexes and inverts the foot and gives support to the arches of the foot.

The crural muscles are summarized in table 13.18.

## Muscles of the Foot

With the exception of one additional intrinsic muscle, the **extensor digitorum brevis muscle,** the muscles of the foot are similar in number and nomenclature to those of the hand. The functions of the muscles of the foot are different, however, because the foot is adapted to provide support while bearing body weight rather than to grasp objects.

The muscles of the foot are topographically arranged into four layers (fig. 13.37) that are difficult to dissociate, even in dissection. The muscles function either to move the toes or to support the arches of the foot through their contraction. Because of their complexity, the muscles of the foot will be presented only in illustrations (see figs. 13.37 and 13.38).

## Clinical Considerations

Compared to the other systems of the body, the muscular system is extremely durable. If properly conditioned, the muscles of the body can adequately serve a person for a lifetime. Muscles are capable of doing incredible amounts of work; through exercise, they can become even stronger.

Clinical considerations include the diagnosis of muscle conditions, functional conditions in muscles, and diseases of muscles.

lateral condyle of the femur. It has a very long, slender tendon of insertion onto the calcaneus bone. The tendon of this muscle is frequently mistaken for a nerve by those dissecting it for the first time. The plantaris is a weak muscle, with limited ability to flex the leg and plantar flex the foot.

The thin, triangular **popliteus muscle** is situated deep to the heads of the gastrocnemius, where it forms part of the floor of the popliteal fossa. The *popliteal fossa* is the depression on the back side of the knee joint (fig. 13.36). The popliteus muscle is a medial rotator of the tibia on the femur. The bipennate **flexor hallucis longus muscle** lies

# Table 13.18  Muscles of the lower leg that move the ankle, foot, and toes

| Muscle | Origin | Insertion | Action | Innervation |
|---|---|---|---|---|
| Tibialis anterior | Lateral epicondyle and body of tibia | First metatarsal bone and first cuneiform | Dorsiflexes ankle and inverts foot at ankle | Deep fibular n. |
| Extensor digitorum longus | Lateral epicondyle of tibia and anterior surface of fibula | Extensor expansions of digits II–V | Extends digits II–V and dorsiflexes foot at ankle | Deep fibular n. |
| Extensor hallucis longus | Anterior surface of fibula and interosseous ligament | Distal phalanx of digit I | Extends joints of big toe and assists dorsiflexion of foot at ankle | Deep fibular n. |
| Peroneus tertius | Anterior surface of fibula and interosseous ligament | Dorsal surface of fifth metatarsal bone | Dorsiflexes and everts foot at ankle | Deep fibular n. |
| Peroneus longus | Lateral epicondyle of tibia and head and shaft of fibula | First cuneiform and metatarsal bone I | Plantar flexes and everts foot at ankle | Superficial fibular n. |
| Peroneus brevis | Lower aspect of fibula | Metatarsal bone V | Plantar flexes and everts foot at ankle | Superficial fibular n. |
| Gastrocnemius | Lateral and medial epicondyle of femur | Posterior surface of calcaneus | Plantar flexes foot at ankle; flexes knee joint | Tibial n. |
| Soleus | Posterior aspect of fibula and tibia | Calcaneus | Plantar flexes foot at ankle | Tibial n. |
| Plantaris | Lateral supracondylar ridge of femur | Calcaneus | Plantar flexes foot at ankle | Tibial n. |
| Popliteus | Lateral condyle of femur | Upper posterior aspect of tibia | Flexes and medially rotates leg at knee joint | Tibial n. |
| Flexor hallucis longus | Posterior aspect of fibula | Distal phalanx of big toe | Flexes joint of distal phalanx of big toe | Tibial n. |
| Flexor joints of digitorum longus | Posterior surface of tibia | Distal phalanges of digits II–V | Flexes joints of distal phalanges of digits II–V | Tibial n. |
| Tibialis posterior | Tibia and fibula and interosseous ligament | Navicular, cuneiforms, cuboid, and metatarsal bones II–IV | Plantar flexes and inverts foot at ankle; supports arches | Tibial n. |

..........................................

extensor hallucis longus: L. *hallus*, great toe
peroneus tertius: Gk. *perone*, fibula, *tertius*, third
gastroenemius: Gk. *gaster*, belly; *kneme*, leg
soleus: L. *soleus*, sole of foot
popliteus: L. *poples*, ham of the knee

## Diagnosis of Muscle Condition

The clinical symptoms of muscle diseases are usually weakness, loss of muscle mass (atrophy), and pain. The most obvious diagnostic procedure is a clinical examination of the patient. Following this, it may be necessary to test muscle function using **electromyography (EMG)** to measure conduction rates and motor unit activity within a muscle. Laboratory tests may include serum enzyme assays or muscle biopsies. Muscle biopsy is perhaps the definitive source of diagnostic information. Progressive atrophy, polymyositis, and metabolic diseases of muscles can be determined through a biopsy.

## Functional Conditions in Muscles

Muscles depend on systematic, periodic contraction to maintain optimal health. Obviously, overuse or disease will cause a change in muscle tissue. The immediate effect of overexertion of muscle tissue is an accumulation of lactic

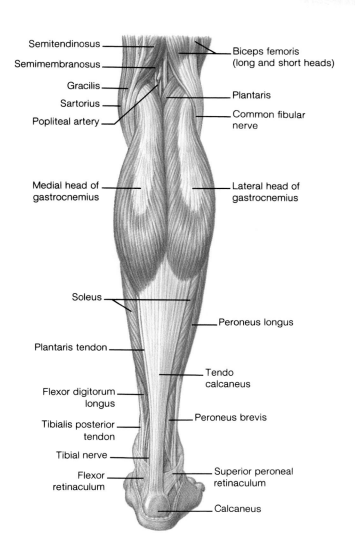

Semitendinosus

Semimembranosus

Gracilis

Sartorius

Popliteal artery

Biceps femoris
(long and short heads)

Plantaris

Common fibular
nerve

Medial head of
gastrocnemius

Lateral head of
gastrocnemius

Soleus

Peroneus longus

Plantaris tendon

Tendo
calcaneus

Flexor digitorum
longus

Tibialis posterior
tendon

Peroneus brevis

Tibial nerve

Flexor
retinaculum

Superior peroneal
retinaculum

Calcaneus

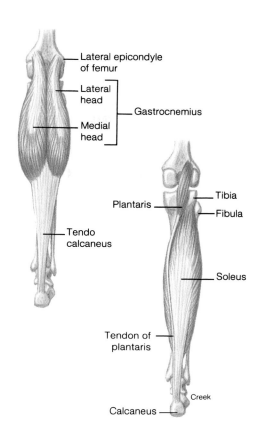

Lateral epicondyle
of femur

Lateral
head

Gastrocnemius

Medial
head

Tendo
calcaneus

Plantaris

Tibia

Fibula

Soleus

Tendon of
plantaris

Creek

Calcaneus

**FIGURE 13.34**

Posterior crural muscles and popliteal region.

acid, resulting in fatigue and soreness. Excessive contraction of a muscle can also damage the fibers or associated connective tissue, resulting in a **strained muscle.**

When skeletal muscles are not contracted, either because the motor nerve supply is blocked or because the limb is immobilized (as when a broken bone is in a cast), the muscle fibers **atrophy,** or diminish in size. Atrophy is reversible if exercise is resumed, as after a healed fracture, but tissue death is inevitable if the nerves cannot be stimulated.

As discussed in chapter 12, the fibers in healthy muscle tissue increase in size, or **hypertrophy,** when a muscle is systematically exercised. This increase in muscle size and strength is due not to an increase in the number of muscle cells but rather to the increased production of myofibrils, accompanied by a strengthening of the associated connective tissue.

A **cramp** within a muscle is an involuntary, painful, and prolonged contraction. Cramps can occur while muscles are in use or at rest. The precise cause of cramps is unknown, but evidence indicates that they may be related to conditions within the muscle (e.g., calcium or oxygen deficiencies) or to stimulation of the motor neurons.

**Rigor mortis** (rigidity of death) affects skeletal muscle tissue several hours after death, as depletion of ATP within the fibers causes a state of muscle contracture and stiffness of the joint. After a few days, however, as the muscle proteins decompose, the rigidity of the corpse disappears.

## Diseases of Muscles

**Fibromyositis** is an inflammation of both skeletal muscular tissue and the associated connective tissue. Its causes are not fully understood. Fibromyositis frequently occurs in the extensor muscles of the lumbar region of the vertebral column, where there are extensive aponeuroses. Fibromyositis of this region is called **lumbago,** or **rheumatism.**

**Muscular dystrophy** is a genetic disease characterized by a gradual atrophy and weakening of muscle tissue. There are several kinds of muscular dystrophy, none of whose etiology is completely understood. The most frequent type affects children and is sex-linked to the male child. As muscular dystrophy progresses, the muscle fibers atrophy and are replaced by adipose tissue.

. . . . . . . . . . . . . . . . . . . . . . . . . . . . . . . . . . .

lumbago: L. *lumbus,* loin

**FIGURE 13.35**
Deep posterior crural
muscles.

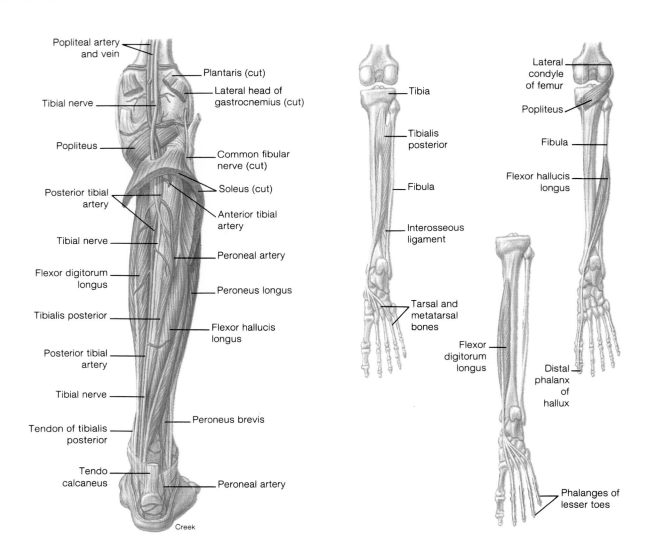

Popliteal artery
and vein

Plantaris (cut)

Tibial nerve

Lateral head of
gastrocnemius (cut)

Popliteus

Common fibular
nerve (cut)

Posterior tibial
artery

Soleus (cut)

Anterior tibial
artery

Tibial nerve

Peroneal artery

Flexor digitorum
longus

Peroneus longus

Tibialis posterior

Flexor hallucis
longus

Posterior tibial
artery

Tibial nerve

Tendon of tibialis
posterior

Peroneus brevis

Tendo
calcaneus

Peroneal artery

Creek

Tibia

Tibialis
posterior

Fibula

Interosseous
ligament

Tarsal and
metatarsal
bones

Flexor
digitorum
longus

Phalanges of
lesser toes

Lateral
condyle
of femur

Popliteus

Fibula

Flexor hallucis
longus

Distal
phalanx
of
hallux

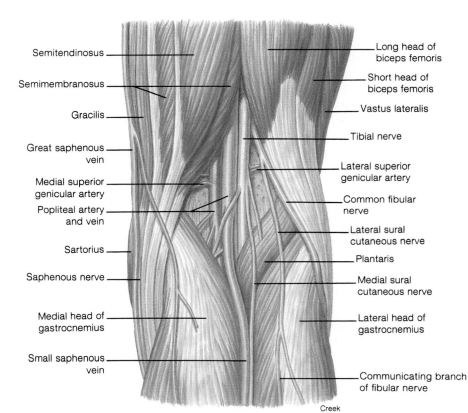

Semitendinosus

Long head of
biceps femoris

Semimembranosus

Short head of
biceps femoris

Gracilis

Vastus lateralis

Great saphenous
vein

Tibial nerve

Medial superior
genicular artery

Lateral superior
genicular artery

Popliteal artery
and vein

Common fibular
nerve

Sartorius

Lateral sural
cutaneous nerve

Saphenous nerve

Plantaris

Medial sural
cutaneous nerve

Medial head of
gastrocnemius

Lateral head of
gastrocnemius

Small saphenous
vein

Communicating branch
of fibular nerve

Creek

**FIGURE 13.36**
Muscles that surround the
popliteal fossa.

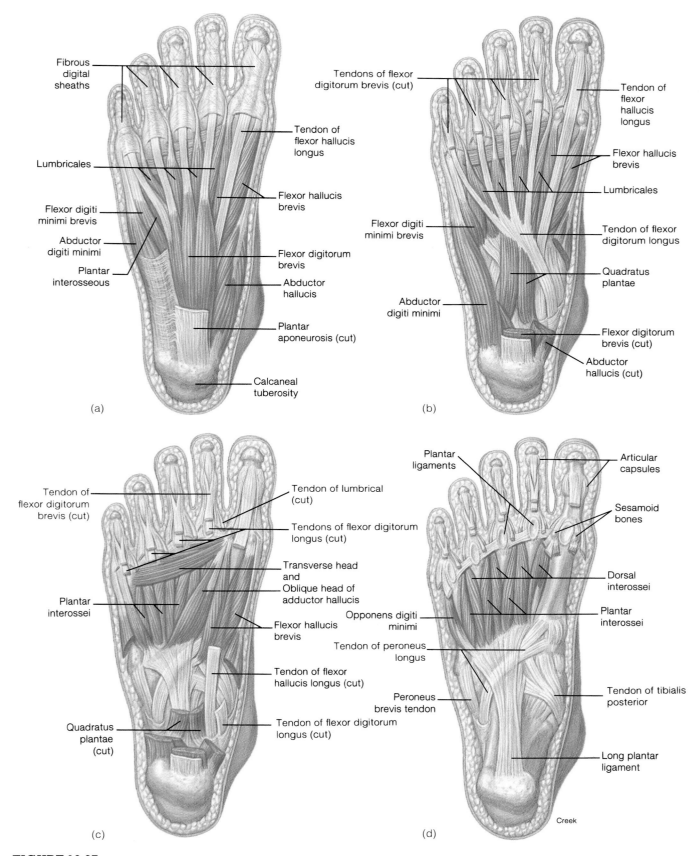

**FIGURE 13.37**

The four musculotendinous layers of the plantar aspect of the foot. (*a*) Superficial layer, (*b*) second layer, (*c*) third layer, and (*d*) deep layer.

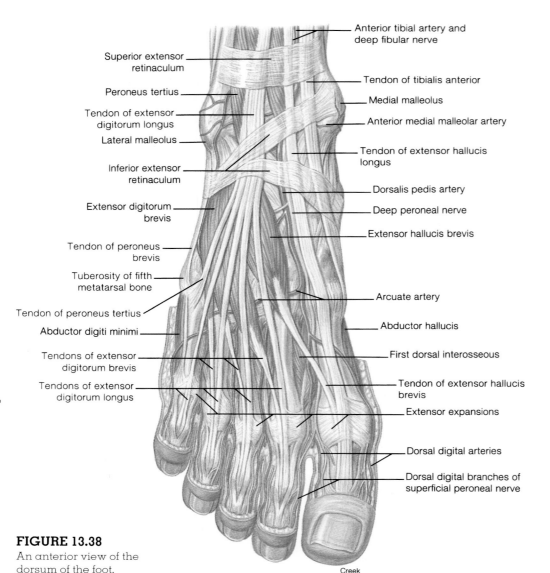

Superior extensor retinaculum

Peroneus tertius

Tendon of extensor digitorum longus

Lateral malleolus

Inferior extensor retinaculum

Extensor digitorum brevis

Tendon of peroneus brevis

Tuberosity of fifth metatarsal bone

Tendon of peroneus tertius

Abductor digiti minimi

Tendons of extensor digitorum brevis

Tendons of extensor digitorum longus

Anterior tibial artery and deep fibular nerve

Tendon of tibialis anterior

Medial malleolus

Anterior medial malleolar artery

Tendon of extensor hallucis longus

Dorsalis pedis artery

Deep peroneal nerve

Extensor hallucis brevis

Arcuate artery

Abductor hallucis

First dorsal interosseous

Tendon of extensor hallucis brevis

Extensor expansions

Dorsal digital arteries

Dorsal digital branches of superficial peroneal nerve

Creek

**FIGURE 13.38**

An anterior view of the dorsum of the foot.

The disease **myasthenia gravis** is characterized by extreme muscle weakness and low endurance. Transmission of impulses at the neuromuscular junction is defective. Myasthenia gravis is believed to be an autoimmune disease, and it typically affects women between the ages of 20 and 40.

**Poliomyelitis** (polio) is actually a viral disease of the nervous system that causes a paralysis of muscles. The viruses are usually localized in the anterior (ventral) horn of the spinal cord, where they affect the motor nerve impulses to skeletal muscles.

**Neoplasms of muscle** are rare, but when they do occur they are usually malignant. **Rhabdomyosarcoma** (*rab″do-mi″ŏ-sar-ko′mă*) is a malignant tumor of the skeletal muscle. It can arise in any skeletal muscle, and most often afflicts young children and elderly people.

myasthenia: Gk. *myos*, muscle; *astheneia*, weakness
poliomyelitis: Gk. *polios*, gray; *myolos*, marrow
rhabdomyosarcoma: Gk. *rhabdos*, rod; *myos*, muscle; *oma*, a growth

# Important Clinical Terminology

**convulsion**   An involuntary, spasmodic contraction of skeletal muscle.

**fibrillation** (*fib-rĭ-la′shun*)   A series of rapid, uncoordinated, and spontaneous contractions of individual motor units within a muscle.

**hernia**   The rupture or protrusion of a portion of the underlying viscera through muscle tissue. The most common hernias occur in the normally weak places of the abdominal wall. Abdominal hernias are of four basic types:

1. **femoral**—viscera passing through the femoral ring

2. **hiatal**—the superior portion of the stomach protruding through the esophageal opening of the diaphragm

3. **inguinal**—viscera protruding through the inguinal canal

4. **umbilical**—a hernia occurring at the navel

**intramuscular injection**   A hypodermic injection at certain heavily muscled areas to avoid damaging nerves. The most common site is the buttock.

**myalgia**   Pain within a muscle resulting from any muscular disorder or disease.

**myokymia**   Continual quivering of a muscle.

**myoma**   A tumor of muscle tissue.

**myopathy**   Any muscular disease.

**myotomy** (*mi-ot′ŏ-me*)   Surgical cutting of muscle tissue.

**myotonia**   A prolonged muscular spasm.

**paralysis**   The loss of nervous control of a muscle.

**shin splints**   Tenderness and pain on the anterior surface of the leg caused by straining the tibialis anterior or extensor digitorium longus muscles.

**torticollis (wryneck)**   A persistent contraction of a sternocleidomastoid muscle, drawing the head to one side and distorting the face. Torticollis may be acquired or congenital.

# Chapter Summary

### Organization of the Muscular System (p. 281)

1. Synergistic muscles contract together. Antagonistic muscles perform in opposition to a particular group of muscles.
2. Muscles may be classified according to fiber arrangement as parallel, convergent, pennate, or sphincteral.
3. Axial muscles include facial muscles, neck muscles, and trunk muscles; appendicular muscles include those that act on the girdles and those that move the segments of the appendages.

### Naming of Muscles (pp. 281–283)

1. Skeletal muscles are named on the basis of shape, location, attachment, size, orientation of fibers, relative position, and function.
2. Most of the muscles are paired; that is, one side of the body is an image of the other.

### Muscles of the Axial Skeleton (pp. 283–295)

The muscles of the axial skeleton include those responsible for facial expression, mastication, eye movement, tongue movement, neck movement, and respiration and those of the abdominal wall, the pelvic outlet, and the vertebral column.

### Muscles of the Appendicular Skeleton (pp. 296–317)

The muscles of the appendicular skeleton include those of the pectoral girdle, humerus, forearm, wrist, hand, and fingers and those of the pelvic girdle, thigh, leg, ankle, foot, and toes.

# Review Activities

### Objective Questions

1. Which of the following muscles is a synergist to the biceps brachii muscle in flexing the elbow joint?
   a. deltoid muscle
   b. triceps brachii muscle
   c. brachialis muscle
   d. anconeus muscle
2. Muscles that are strong, fatigue quickly, and have short excursions are classified as
   a. parallel.
   b. convergent.
   c. pennate.
   d. sphincteral.
3. Which of the following is *not* used as a means of naming muscles?
   a. location
   b. action
   c. shape
   d. attachment
   e. strength of contraction
4. Which of the following muscles originates on the zygomatic arch, inserts on the lateral portion of the mandible, and, when contracted, elevates the mandible?
   a. masseter muscle
   b. lateral pterygoid muscle
   c. temporalis muscle
   d. zygomaticus muscle
5. Which of the following muscles is a flexor of the shoulder joint?
   a. pectoralis major muscle
   b. supraspinatus muscle
   c. teres major muscle
   d. trapezium muscle
   e. latissimus dorsi muscle
6. Which of the following muscles is an antagonist to the longissimus thoracis muscle?
   a. spinalis thoracis muscle
   b. quadratus lumborum muscle
   c. longissimus cervicis muscle
   d. rectus abdominis muscle
7. Which of the following muscles does *not* have either an origin or insertion upon the humerus?
   a. teres minor muscle
   b. biceps brachii muscle
   c. supraspinatus muscle
   d. brachialis muscle
   e. pectoralis major muscle
8. Which of the following muscles in a female has the action of constricting the urethral canal and the vagina?
   a. transversus perinei muscle
   b. bulbocavernosus muscle
   c. coccygeus muscle
   d. ischiocavernosus muscle
   e. levator ani muscle
9. Which of the following muscles does *not* contract over both the hip and knee joints (is *not* a two-joint muscle)?
   a. gracilis muscle
   b. sartorius muscle
   c. rectus femoris muscle
   d. semitendinosus muscle
   e. vastus medialis muscle
10. The genetic disease of muscle that is most common in male children and is characterized by gradual atrophy and weakening of muscle tissue is
    a. poliomyelitis.
    b. myasthenia gravis.
    c. muscular dystrophy.
    d. rhabdomyosarcoma.

### Essay Questions

1. What are the advantages and disadvantages of pennate-fibered muscles?
2. List the muscles of the neck that either originate or insert on the hyoid bone.
3. Diagram a posterior view of the scapula and indicate the locations of the muscles that originate and insert upon it.
4. Discuss the position of flexor and extensor muscles relative to the shoulder, elbow, and wrist joint.
5. Based on function, describe exercises that would strengthen the following muscles: (a) the pectoralis major muscle; (b) the deltoid muscle; (c) the triceps brachii muscle; (d) the pronator teres muscle; (e) the rhomboideus major muscle; (f) the trapezius muscle; (g) the serratus anterior muscle; and (h) the latissimus dorsi muscle.
6. Give three examples of synergistic muscle groups within the lower extremity and identify the antagonistic muscle group of each.
7. Describe the actions of the muscles of inspiration. Which muscles participate in forced expiration?
8. Which muscles of the pelvic outlet support the floor of the pelvic cavity and which are associated with the genitalia?

---

### Gundy/Weber Software 💾

The tutorial software accompanying Chapter 13 is Volume 4—Muscle System.

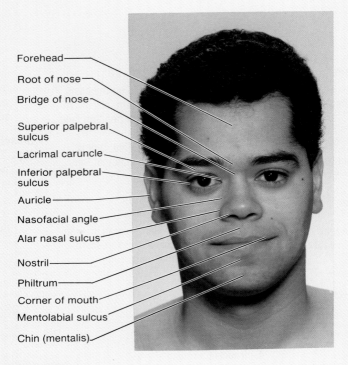

Forehead
Root of nose
Bridge of nose
Superior palpebral sulcus
Lacrimal caruncle
Inferior palpebral sulcus
Auricle
Nasofacial angle
Alar nasal sulcus
Nostril
Philtrum
Corner of mouth
Mentolabial sulcus
Chin (mentalis)

**FIGURE 1**

The surface anatomy of the facial region. (Also refer to figure 13.4 for the underlying muscles.)

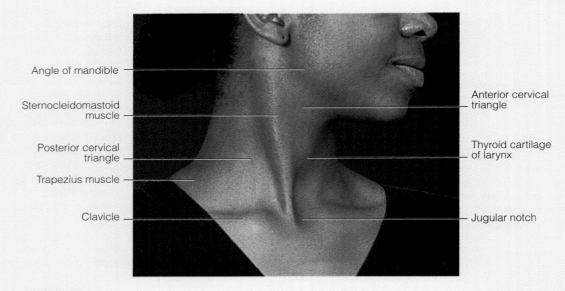

Angle of mandible

Sternocleidomastoid muscle

Posterior cervical triangle

Trapezius muscle

Clavicle

Anterior cervical triangle

Thyroid cartilage of larynx

Jugular notch

**FIGURE 2**

An anterolateral view of the neck. (Also refer to figure 13.10 for the underlying muscles.)

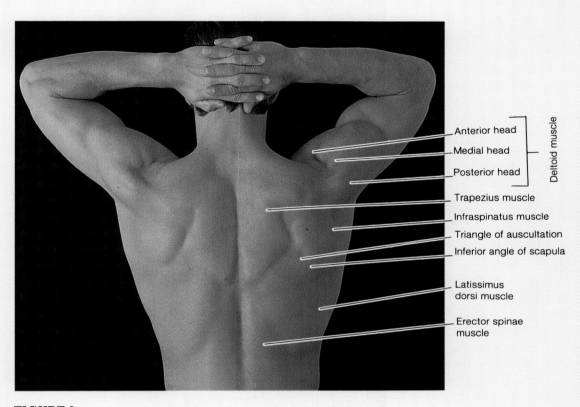

Anterior head

Medial head

Posterior head

Deltoid muscle

Trapezius muscle

Infraspinatus muscle

Triangle of auscultation

Inferior angle of scapula

Latissimus dorsi muscle

Erector spinae muscle

**FIGURE 3**

The surface anatomy of the back. (Also refer to figure 13.17 for the underlying muscles.)

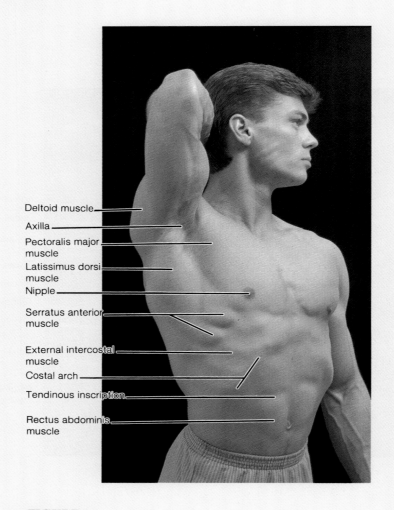

Deltoid muscle

Axilla

Pectoralis major muscle

Latissimus dorsi muscle

Nipple

Serratus anterior muscle

External intercostal muscle

Costal arch

Tendinous inscription

Rectus abdominis muscle

**FIGURE 4**

An anterolateral view of the trunk and axilla. (Also refer to figures 13.12, 13.18, and 13.19 for the underlying muscles.)

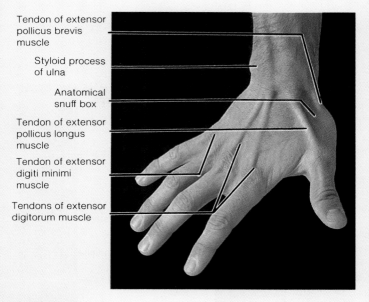

Tendon of extensor
pollicus brevis
muscle

Styloid process
of ulna

Anatomical
snuff box

Tendon of extensor
pollicus longus
muscle

Tendon of extensor
digiti minimi
muscle

Tendons of extensor
digitorum muscle

**FIGURE 5**

The anatomical snuffbox. (Also refer to figure 13.20*b* for the
underlying muscles and tendons.)

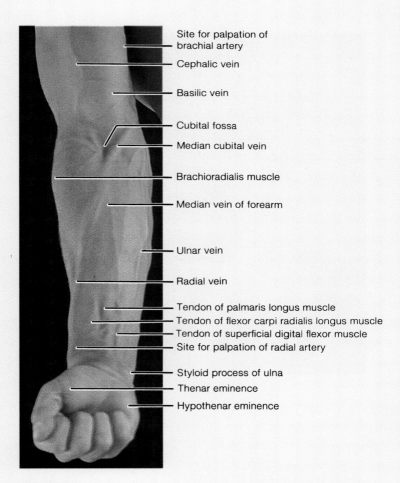

Site for palpation of
brachial artery

Cephalic vein

Basilic vein

Cubital fossa

Median cubital vein

Brachioradialis muscle

Median vein of forearm

Ulnar vein

Radial vein

Tendon of palmaris longus muscle

Tendon of flexor carpi radialis longus muscle

Tendon of superficial digital flexor muscle

Site for palpation of radial artery

Styloid process of ulna

Thenar eminence

Hypothenar eminence

**FIGURE 6**

An anterior view of the forearm and hand. (Also refer to figure 13.20*a*
for the underlying muscles and tendons.)

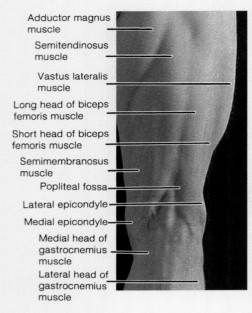

Adductor magnus
muscle

Semitendinosus
muscle

Vastus lateralis
muscle

Long head of biceps
femoris muscle

Short head of biceps
femoris muscle

Semimembranosus
muscle

Popliteal fossa

Lateral epicondyle

Medial epicondyle

Medial head of
gastrocnemius
muscle

Lateral head of
gastrocnemius
muscle

(a)

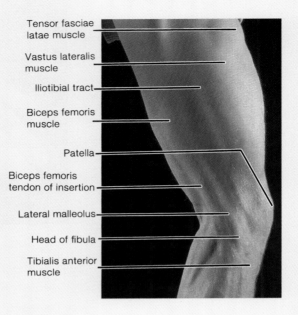

Tensor fasciae
latae muscle

Vastus lateralis
muscle

Iliotibial tract

Biceps femoris
muscle

Patella

Biceps femoris
tendon of insertion

Lateral malleolus

Head of fibula

Tibialis anterior
muscle

(b)

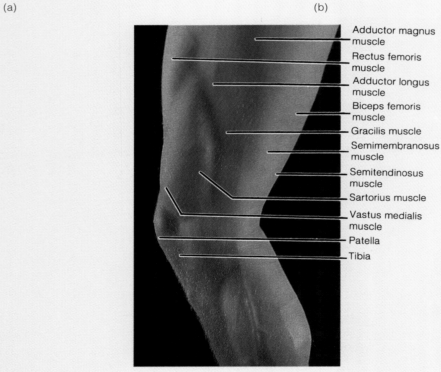

Adductor magnus
muscle

Rectus femoris
muscle

Adductor longus
muscle

Biceps femoris
muscle

Gracilis muscle

Semimembranosus
muscle

Semitendinosus
muscle

Sartorius muscle

Vastus medialis
muscle

Patella

Tibia

(c)

## FIGURE 7

The (a) lateral, (b) posterior, and (c) medial surfaces of the leg. (Also refer to
figures 13.32, 13.33, and 13.34 for the underlying muscles.)

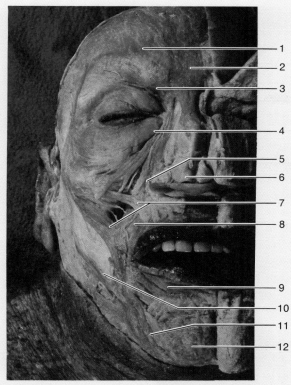

1 Frontalis m.
2 Supratrochlear a.
3 Corrugator m.
4 Orbicularis oculi m.
5 Levator labii superioris m.
6 Alar cartilage
7 Zygomaticus mm.
8 Facial a.
9 Orbicularis oris m.
10 Risorius m.
11 Depressor angularis oris m.
12 Mentalis m.

**FIGURE 8**

An anterior view of the muscles of the head. (Also refer to figures 13.4 and 13.5.)

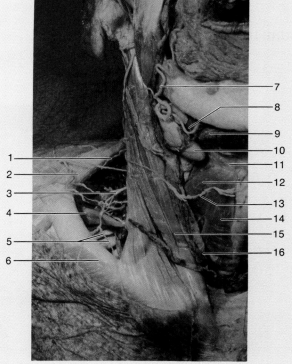

1 Accessory nerve
2 Trapezius m.
3 Supraclavicular nerve
4 Omohyoid m.
5 Brachial plexus
6 Clavicle
7 Facial artery
8 Mylohyoid m.
9 Digastric m.
10 Submandibular gland
11 Hyoid bone
12 Omohyoid m.
13 Transverse cervical nerve
14 Sternohyoid m.
15 Sternocleidomastoid m.
16 External jugular vein

**FIGURE 9**

An anterior view of the right cervical region. (Also refer to figure 13.10.)

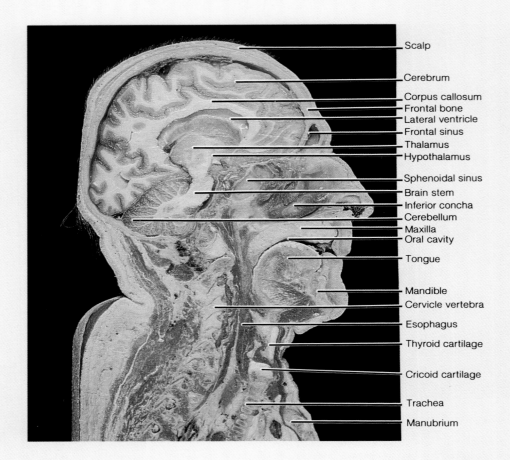

Scalp
Cerebrum
Corpus callosum
Frontal bone
Lateral ventricle
Frontal sinus
Thalamus
Hypothalamus
Sphenoidal sinus
Brain stem
Inferior concha
Cerebellum
Maxilla
Oral cavity
Tongue
Mandible
Cervicle vertebra
Esophagus
Thyroid cartilage
Cricoid cartilage
Trachea
Manubrium

**FIGURE 10**

A sagittal section of the head and neck.

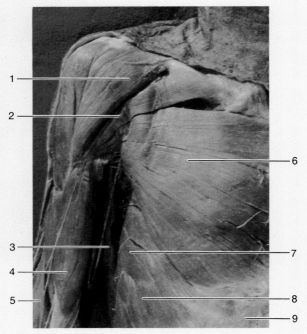

1 Deltoid m.
2 Cephalic vein
3 Latissimus dorsi m.
4 Biceps brachii m.
5 Brachioradialis m.

6 Pectoralis major m.
7 Serratus anterior m.
8 External abdominal oblique m.
9 Rectus sheath

**FIGURE 11**

An anterior view of the right thorax, shoulder, and brachium. (Also refer to figures 13.15 and 13.18.)

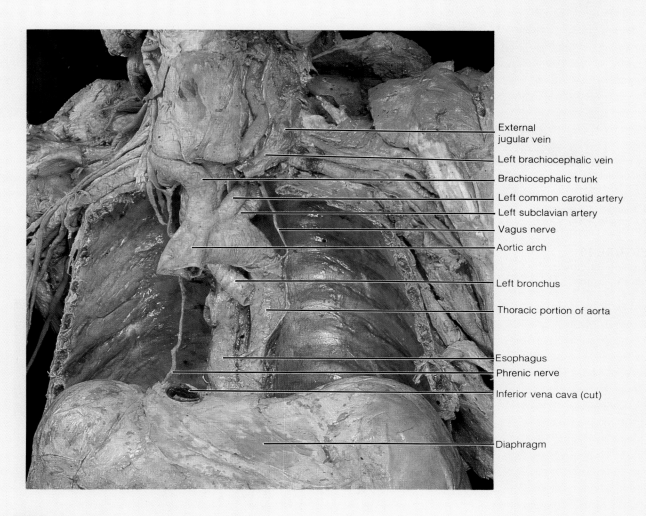

External
jugular vein

Left brachiocephalic vein

Brachiocephalic trunk

Left common carotid artery

Left subclavian artery

Vagus nerve

Aortic arch

Left bronchus

Thoracic portion of aorta

Esophagus

Phrenic nerve

Inferior vena cava (cut)

Diaphragm

**FIGURE 12**

The thoracic cavity with the heart and lungs removed.

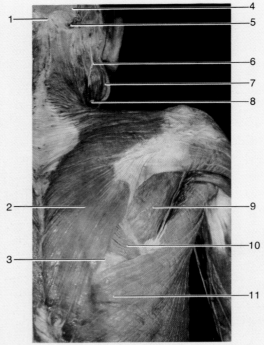

1 External occipital
  protuberance
2 Trapezius m.
3 Triangle of auscultation
4 Occipital artery
5 Greater occipital n.

6 Lesser occipital nerve
7 Sternocleidomastoid m.
8 Great auricular nerve
9 Infraspinatus m.
10 Rhomboideus major m.
11 Latissimus dorsi m.

**FIGURE 13**

A posterior view of the right thorax and neck. (Also refer to figure 13.16.)

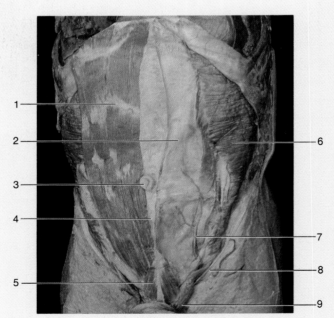

1 Rectus abdominis m.
2 Rectus sheath
3 Umbilicus
4 Linea alba
5 Pyramidalis m.

6 Transverse abdominis m.
7 Inferior epigastric artery
8 Inguinal ligament
9 Spermatic cord

**FIGURE 14**

An anterior view of the muscles of the abdominal wall. (Also refer to figures 13.12 and 13.15.)

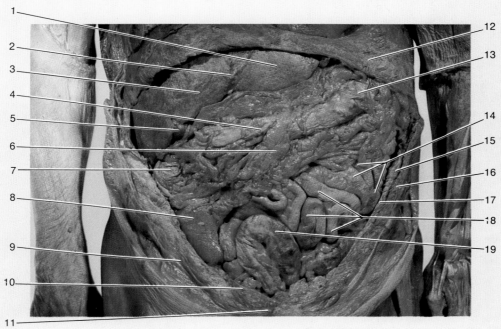

1 Left lobe of liver
2 Falciform ligament
3 Right lobe of liver
4 Transverse colon
5 Gallbladder
6 Greater omentum
7 Hepatic (right colic) flexure
8 Fat deposit within greater omentum
9 Aponeurosis of internal abdominal oblique m.
10 Rectus abdominis m. (cut)
11 Rectus sheath (cut)
12 Diaphragm
13 Splenic (left colic) flexure
14 Jejunum
15 Transversus abdominis m. (cut)
16 Internal and external abdominal oblique mm. (cut)
17 Parietal peritoneum (cut)
18 Ileum
19 Sigmoid colon

**FIGURE 15**

Viscera of the abdomen.

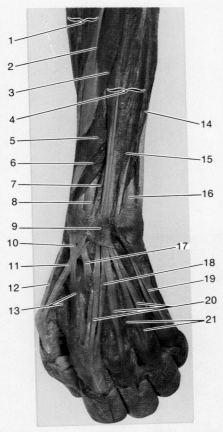

1 Brachioradialis m.
2 Extensor carpi radialis longus tendon
3 Extensor carpi radialis brevis m.
4 Extensor digitorum communis m.
5 Abductor pollicis longus m.
6 Extensor pollicis brevis m.
7 Extensor pollicis longus m.
8 Radius
9 Extensor retinaculum
10 Tendon of extensor carpi radialis longus m.
11 Tendon of extensor pollicis longus m.
12 Tendon of extensor pollicis brevis m.
13 First dorsal interosseous m.
14 Extensor carpi ulnaris m.
15 Extensor digiti minimi m.
16 Ulna
17 Tendon of extensor carpi radialis brevis m.
18 Tendon of extensor indicis m.
19 Tendon of extensor digiti minimi m.
20 Tendons of extensor digitorum m.
21 Intertendinous connections

**FIGURE 16**

A posterior view of the left forearm and hand. (Also refer to figures 13.20 and 13.21 for views of the right forearm.)

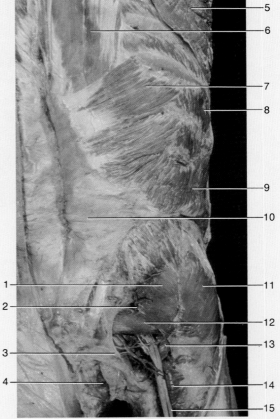

1 Superior gluteal vessels
2 Inferior gluteal vessels
3 Sacrotuberous ligament
4 Levator ani m.
5 Serratus anterior m.
6 Erector spinae m.
7 Serratus posterior m.
8 External intercostal m.
9 Internal abdominal oblique m.
10 Lumbar aponeurosis
11 Gluteus medius m.
12 Piriformis m.
13 Obturator internus m.
14 Quadratus femoris m.
15 Sciatic nerve

**FIGURE 17**

A posterior view of the deep muscles of the right abdominal and gluteal regions. (Also refer to figures 13.14, 13.15, and 13.16.)

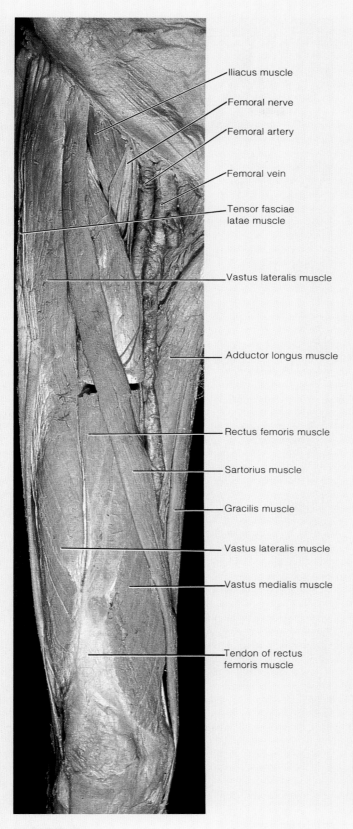

Iliacus muscle

Femoral nerve

Femoral artery

Femoral vein

Tensor fasciae latae muscle

Vastus lateralis muscle

Adductor longus muscle

Rectus femoris muscle

Sartorius muscle

Gracilis muscle

Vastus lateralis muscle

Vastus medialis muscle

Tendon of rectus femoris muscle

**FIGURE 18**
An anterior view of the right thigh. (Also refer to figures 13.23 and 13.27.)

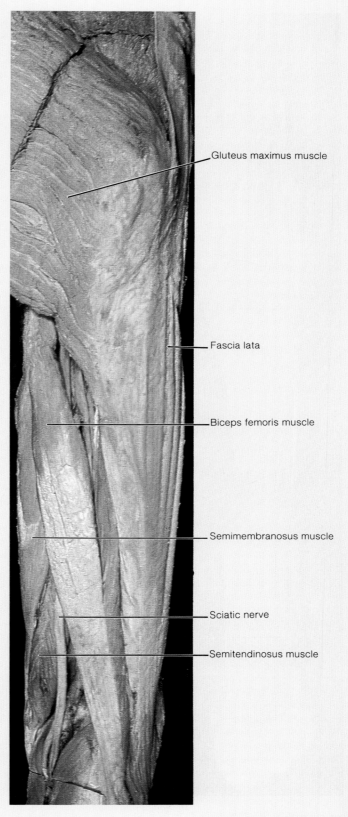

Gluteus maximus muscle

Fascia lata

Biceps femoris muscle

Semimembranosus muscle

Sciatic nerve

Semitendinosus muscle

**FIGURE 19**
A posterior view of the right hip and thigh. (Also refer to figures 13.25 and 13.28.)

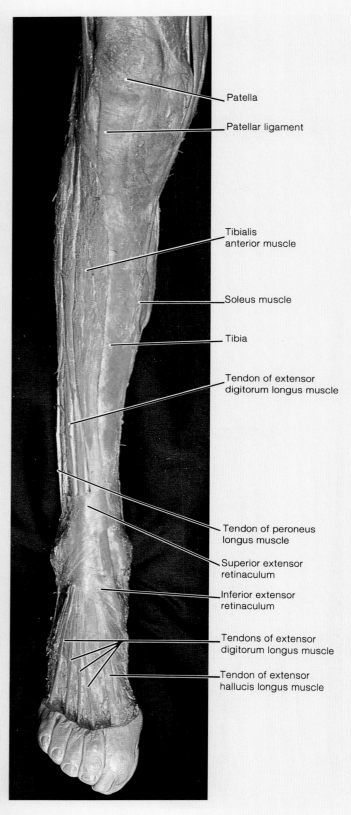

Patella

Patellar ligament

Tibialis
anterior muscle

Soleus muscle

Tibia

Tendon of extensor
digitorum longus muscle

Tendon of peroneus
longus muscle

Superior extensor
retinaculum

Inferior extensor
retinaculum

Tendons of extensor
digitorum longus muscle

Tendon of extensor
hallucis longus muscle

**FIGURE 20**
An anterior view of right leg. (Also refer to figure 13.31.)

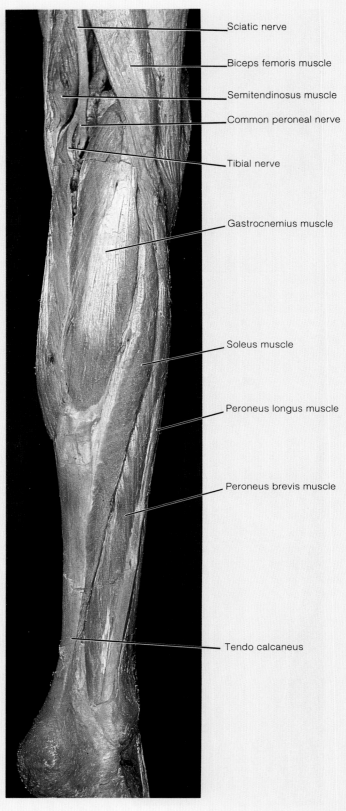

Sciatic nerve

Biceps femoris muscle

Semitendinosus muscle

Common peroneal nerve

Tibial nerve

Gastrocnemius muscle

Soleus muscle

Peroneus longus muscle

Peroneus brevis muscle

Tendo calcaneus

**FIGURE 21**
A posterior view of the right leg. (Also refer to figure 13.34.)

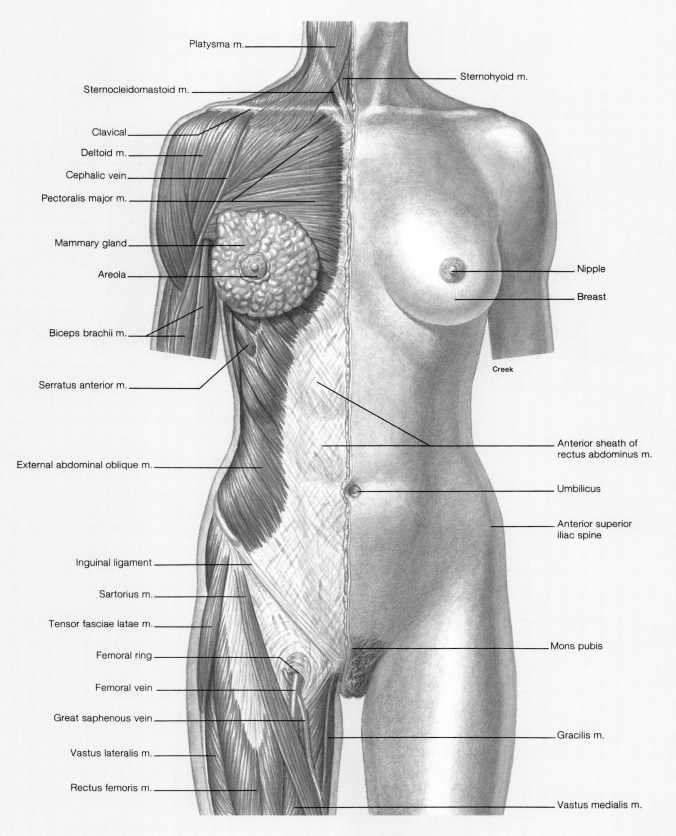

Platysma m.

Sternocleidomastoid m.

Clavical

Deltoid m.

Cephalic vein

Pectoralis major m.

Mammary gland

Areola

Biceps brachii m.

Serratus anterior m.

External abdominal oblique m.

Inguinal ligament

Sartorius m.

Tensor fasciae latae m.

Femoral ring

Femoral vein

Great saphenous vein

Vastus lateralis m.

Rectus femoris m.

Sternohyoid m.

Nipple

Breast

Creek

Anterior sheath of
rectus abdominus m.

Umbilicus

Anterior superior
iliac spine

Mons pubis

Gracilis m.

Vastus medialis m.

**FIGURE 22**

An anterior view of the female trunk with the superficial
muscles exposed on the left side. (*m.* stands for *muscle;*
*v.* stands for *vein.*)

337

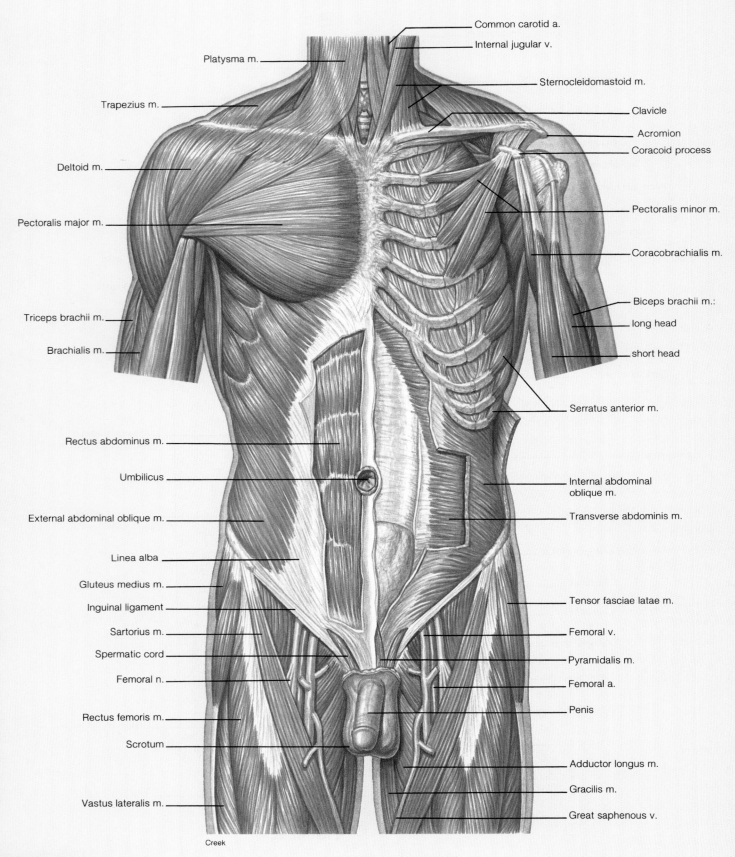

Common carotid a.
Internal jugular v.
Platysma m.
Sternocleidomastoid m.
Trapezius m.
Clavicle
Acromion
Coracoid process
Deltoid m.
Pectoralis minor m.
Pectoralis major m.
Coracobrachialis m.
Biceps brachii m.:
Triceps brachii m.
long head
Brachialis m.
short head
Serratus anterior m.
Rectus abdominus m.
Umbilicus
Internal abdominal oblique m.
External abdominal oblique m.
Transverse abdominis m.
Linea alba
Gluteus medius m.
Inguinal ligament
Tensor fasciae latae m.
Sartorius m.
Femoral v.
Spermatic cord
Pyramidalis m.
Femoral n.
Femoral a.
Rectus femoris m.
Penis
Scrotum
Adductor longus m.
Gracilis m.
Vastus lateralis m.
Great saphenous v.

Creek

**FIGURE 23**
An anterior view of the male trunk with the deeper muscle
layers exposed. (n. stands for *nerve*; a. stands for *artery*.)

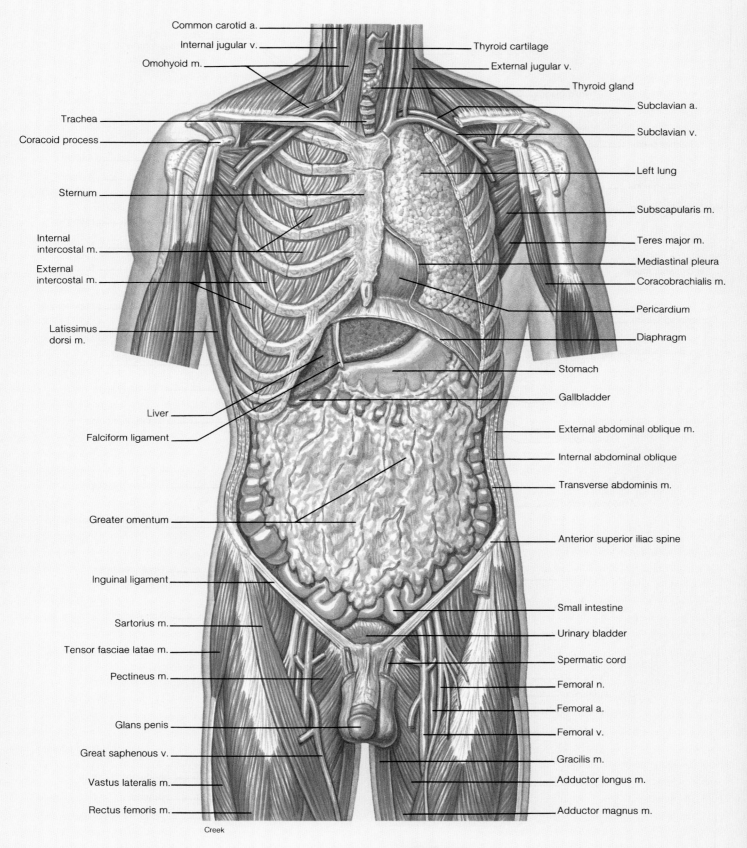

Common carotid a.

Internal jugular v.

Omohyoid m.

Trachea

Coracoid process

Sternum

Internal
intercostal m.

External
intercostal m.

Latissimus
dorsi m.

Liver

Falciform ligament

Greater omentum

Inguinal ligament

Sartorius m.

Tensor fasciae latae m.

Pectineus m.

Glans penis

Great saphenous v.

Vastus lateralis m.

Rectus femoris m.

Creek

Thyroid cartilage

External jugular v.

Thyroid gland

Subclavian a.

Subclavian v.

Left lung

Subscapularis m.

Teres major m.

Mediastinal pleura

Coracobrachialis m.

Pericardium

Diaphragm

Stomach

Gallbladder

External abdominal oblique m.

Internal abdominal oblique

Transverse abdominis m.

Anterior superior iliac spine

Small intestine

Urinary bladder

Spermatic cord

Femoral n.

Femoral a.

Femoral v.

Gracilis m.

Adductor longus m.

Adductor magnus m.

**FIGURE 24**

An anterior view of the male trunk with the deep muscles
removed and the abdominal viscera exposed. (a. stands
for *artery*.)

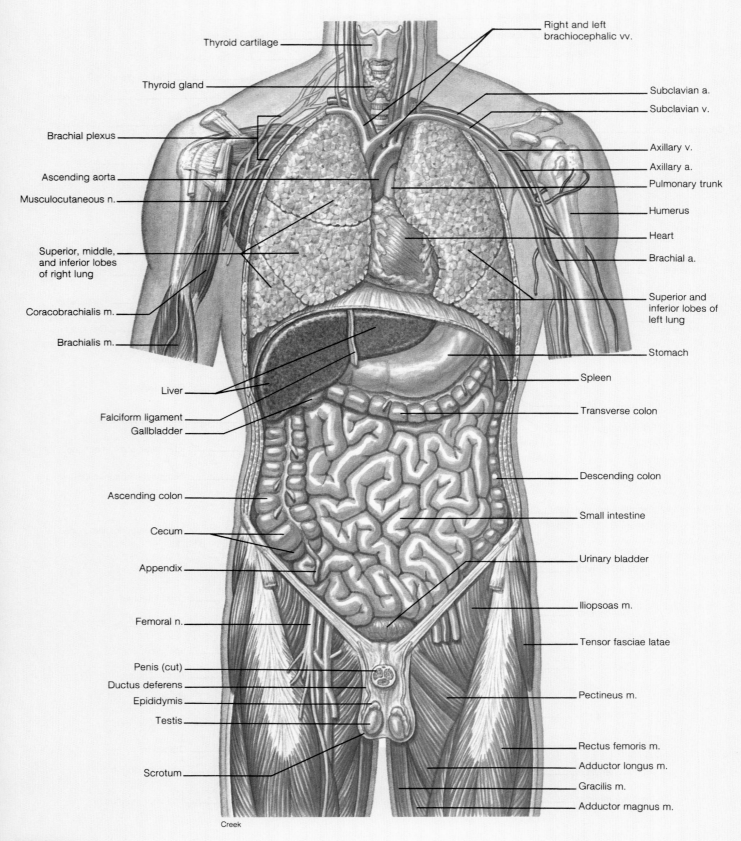

Thyroid cartilage

Thyroid gland

Brachial plexus

Ascending aorta

Musculocutaneous n.

Superior, middle, and inferior lobes of right lung

Coracobrachialis m.

Brachialis m.

Liver

Falciform ligament

Gallbladder

Ascending colon

Cecum

Appendix

Femoral n.

Penis (cut)

Ductus deferens

Epididymis

Testis

Scrotum

Right and left brachiocephalic vv.

Subclavian a.

Subclavian v.

Axillary v.

Axillary a.

Pulmonary trunk

Humerus

Heart

Brachial a.

Superior and inferior lobes of left lung

Stomach

Spleen

Transverse colon

Descending colon

Small intestine

Urinary bladder

Iliopsoas m.

Tensor fasciae latae

Pectineus m.

Rectus femoris m.

Adductor longus m.

Gracilis m.

Adductor magnus m.

Creek

**FIGURE 25**

An anterior view of the male trunk with the thoracic and abdominal viscera exposed.

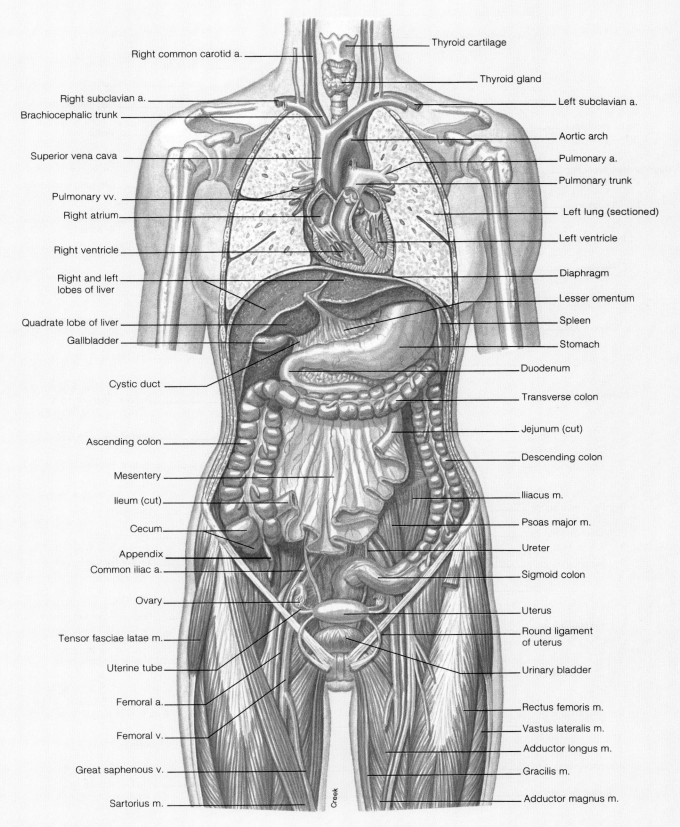

Right common carotid a.
Thyroid cartilage
Thyroid gland
Right subclavian a.
Brachiocephalic trunk
Left subclavian a.
Superior vena cava
Aortic arch
Pulmonary a.
Pulmonary trunk
Pulmonary vv.
Right atrium
Left lung (sectioned)
Right ventricle
Left ventricle
Right and left lobes of liver
Diaphragm
Lesser omentum
Quadrate lobe of liver
Spleen
Gallbladder
Stomach
Duodenum
Cystic duct
Transverse colon
Jejunum (cut)
Ascending colon
Descending colon
Mesentery
Iliacus m.
Ileum (cut)
Psoas major m.
Cecum
Ureter
Appendix
Common iliac a.
Sigmoid colon
Ovary
Uterus
Tensor fasciae latae m.
Round ligament of uterus
Uterine tube
Urinary bladder
Femoral a.
Rectus femoris m.
Vastus lateralis m.
Femoral v.
Adductor longus m.
Great saphenous v.
Gracilis m.
Sartorius m.
Adductor magnus m.

Creek

**FIGURE 26**

An anterior view of the female trunk with the lungs, heart, and small intestine sectioned.

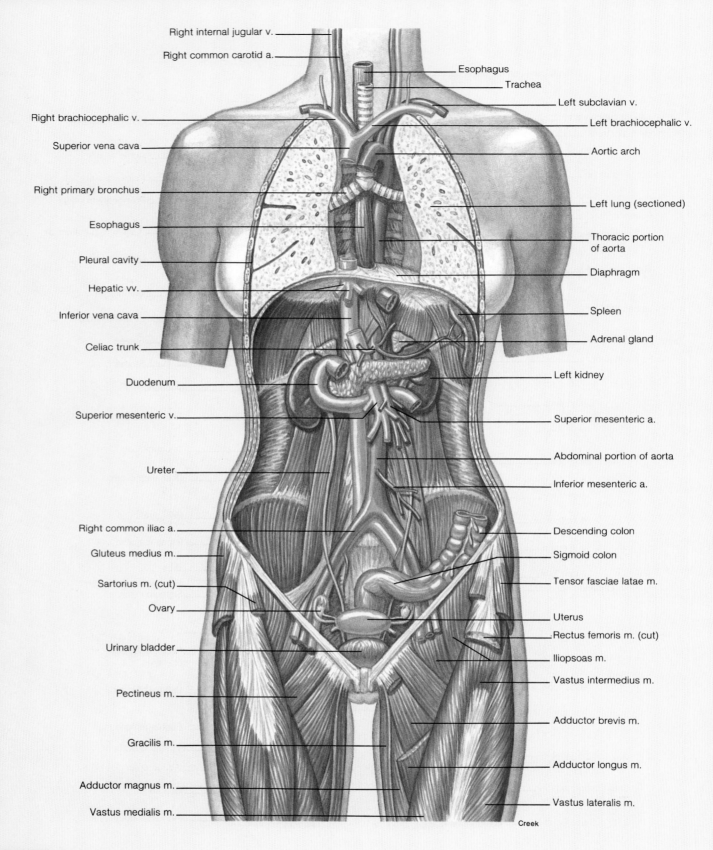

Right internal jugular v.

Right common carotid a.

Esophagus

Trachea

Left subclavian v.

Right brachiocephalic v.

Left brachiocephalic v.

Superior vena cava

Aortic arch

Right primary bronchus

Left lung (sectioned)

Esophagus

Thoracic portion
of aorta

Pleural cavity

Diaphragm

Hepatic vv.

Spleen

Inferior vena cava

Adrenal gland

Celiac trunk

Duodenum

Left kidney

Superior mesenteric v.

Superior mesenteric a.

Abdominal portion of aorta

Ureter

Inferior mesenteric a.

Right common iliac a.

Descending colon

Gluteus medius m.

Sigmoid colon

Sartorius m. (cut)

Tensor fasciae latae m.

Ovary

Uterus

Urinary bladder

Rectus femoris m. (cut)

Iliopsoas m.

Vastus intermedius m.

Pectineus m.

Adductor brevis m.

Gracilis m.

Adductor longus m.

Adductor magnus m.

Vastus lateralis m.

Vastus medialis m.

Creek

## FIGURE 27

An anterior view of the female trunk.

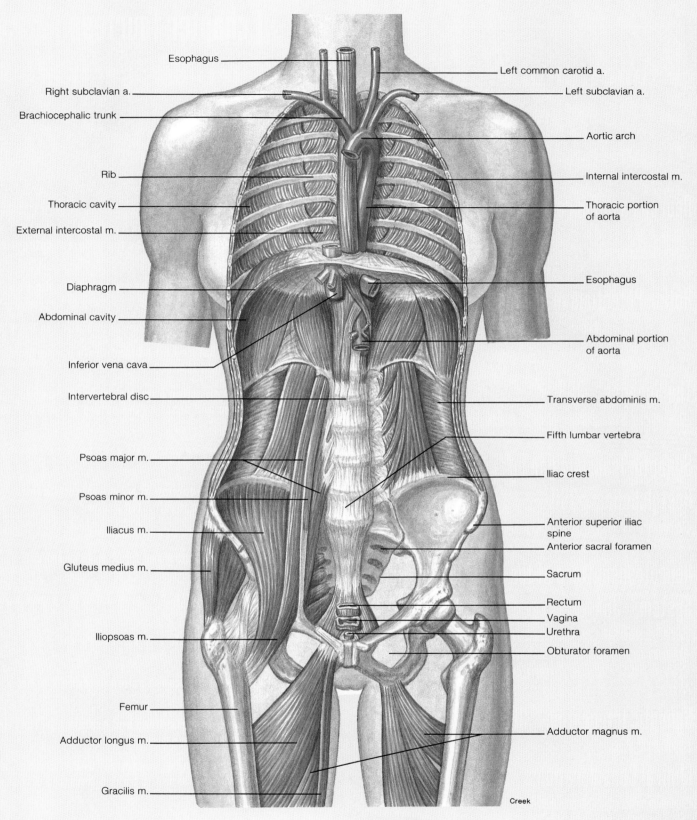

Esophagus

Right subclavian a.

Brachiocephalic trunk

Rib

Thoracic cavity

External intercostal m.

Diaphragm

Abdominal cavity

Inferior vena cava

Intervertebral disc

Psoas major m.

Psoas minor m.

Iliacus m.

Gluteus medius m.

Iliopsoas m.

Femur

Adductor longus m.

Gracilis m.

Left common carotid a.

Left subclavian a.

Aortic arch

Internal intercostal m.

Thoracic portion of aorta

Esophagus

Abdominal portion of aorta

Transverse abdominis m.

Fifth lumbar vertebra

Iliac crest

Anterior superior iliac spine

Anterior sacral foramen

Sacrum

Rectum

Vagina

Urethra

Obturator foramen

Adductor magnus m.

Creek

**FIGURE 28**

An anterior view of the female trunk with the thoracic, abdominal, and pelvic visceral organs removed.

## objectives

- Define the anatomical terms used to describe the nervous system and describe the different categories of neurons.

- Explain how myelin sheaths are formed in the PNS and CNS.

- Describe the process of axon regeneration in the PNS.

- Describe the nature of the blood-brain barrier and discuss its significance.

- Explain why an action potential is an all-or-none phenomenon and describe the nature of the refractory period.

- Explain how action potentials code for the strength of a stimulus.

- Compare the conduction of action potentials in unmyelinated and myelinated axons.

- Describe the nature of gap junctions.

- Describe the structure and function of chemical synapses.

- Explain how acetylcholine stimulates the production of EPSPs.

- Compare the characteristics of EPSPs and action potentials.

- Explain how EPSPs cause action potentials to be produced.

- Identify the catecholamines and explain how they are inactivated at the synapse.

- Describe the actions of dopamine, GABA, glycine, and the endorphins as synaptic transmitters and discuss their significance.

- Describe spatial and temporal summation and explain how EPSPs and IPSPs can interact in the process of postsynaptic inhibition.

- Describe the process of presynaptic inhibition.

# Neurons and Neuroglia

*The nervous system is composed of neurons, which produce and conduct electrochemical impulses, and neuroglia, which support the functions of neurons. Neurons may be classified according to their structure or function; the more abundant neuroglia perform specialized functions.*

The nervous system is divided into the **central nervous system (CNS),** which includes the *brain* and *spinal cord,* and the **peripheral nervous system (PNS),** which includes the *cranial nerves* that arise from the brain and the *spinal nerves* that arise from the spinal cord (table 14.1).

The nervous system is composed of only two principal types of cells—neurons and neuroglia. **Neurons** are the basic structural and functional units of the nervous system. They are specialized to respond to physical and chemical stimuli, conduct electrochemical impulses, and release specific chemical regulators. Through these activities, neurons perform such functions as storing memory, thinking, and controlling muscles and glands.

**Neuroglia,** or **glial** (*glē'al*) **cells,** are supportive cells in the nervous system that aid the function of neurons. Neuroglia are about five times more abundant than neurons and have limited mitotic abilities (brain tumors that occur in adults are usually composed of neuroglia rather than neurons).

## Neurons

Although neurons vary considerably in size and shape, they all have three principal regions: (1) a cell body, (2) dendrites, and (3) an axon (figs. 14.1 and 14.2). Dendrites and axons can be referred to generically as *processes,* or extensions from the cell body.

The **cell body,** or **perikaryon** (*per˝ĭ-kar´e-on*), is the enlarged portion of the neuron that contains the nucleus. It serves as the nutritional center of the neuron, where macromolecules are produced. The cell body also contains **chromatophilic substances** (Nissl bodies) that are not found in the dendrites or axon. These are composed of a granular (rough) endoplasmic reticulum—an organelle in-

| Table 14.1 | Anatomical terms used in describing the nervous system |
|---|---|
| **Term** | **Definition** |
| Central nervous system (CNS) | Brain and spinal cord |
| Peripheral nervous system (PNS) | Nerves, ganglia, and plexuses |
| Association neuron (interneuron) | Multipolar neuron located entirely within the CNS |
| Sensory neuron (afferent neuron) | Neuron that transmits impulses from a sensory receptor into the CNS |
| Motor neuron (efferent neuron) | Neuron that transmits impulses from the CNS to an effector organ; for example, muscle |
| Nerve | Cablelike collection of nerve fibers; may be "mixed" (contain both sensory and motor fibers) |
| Somatic motor nerve | Nerve that stimulates contraction of skeletal muscles |
| Autonomic motor nerve | Nerve that stimulates contraction (or inhibits contraction) of smooth muscle and cardiac muscle and that stimulates glandular secretion |
| Ganglion | Grouping of neuron cell bodies located outside the CNS |
| Nucleus | Grouping of neuron cell bodies within the CNS |
| Tract | Grouping of nerve fibers that interconnect regions of the CNS |

volved in protein synthesis. The cell bodies within the CNS are frequently clustered into groups called **nuclei** (not to be confused with the nucleus of a cell). Cell bodies in the PNS usually occur in clusters called **ganglia** (table 14.1).

**Dendrites** are thin, branched processes that extend from the cytoplasm of the cell body. Dendrites serve as a receptive area for stimuli and transmit electrical impulses to the cell body. The **axon** is a longer process that conducts

neuroglia: Gk. *neuron,* nerve; *glia,* glue
perikaryon: Gk. *peri,* around; *karyon,* nucleus
chromatophilic: Gk. *khroma,* color; *philus,* loving
Nissl body: from Franz Nissl, German neuroanatomist, 1860–1919

ganglion: Gk. *ganglion,* swelling
dendrite: Gk. *dendron,* tree branch
axon: Gk. *axon,* axis

Functional Organization of the Nervous System

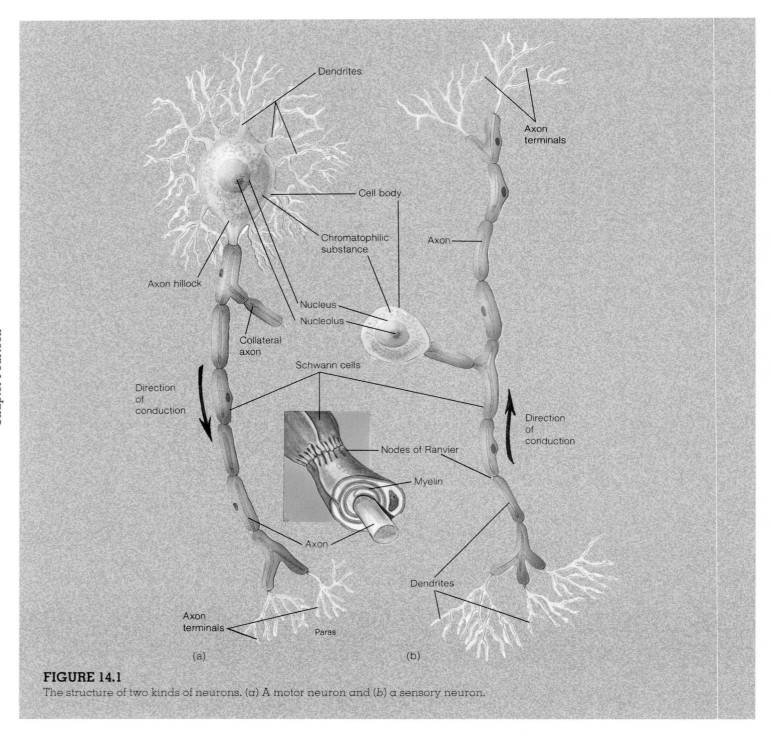

**FIGURE 14.1**
The structure of two kinds of neurons. (*a*) A motor neuron and (*b*) a sensory neuron.

impulses away from the cell body. Axons vary in length from only a millimeter long to up to a meter or more (for those that extend from the CNS to the foot). The origin of the axon near the cell body is called the **axon hillock,** and side branches that may extend from the axon are called **collateral axons,** or *collateral branches*. **Axon terminals** are the bulbular endings of an axon. Axons are frequently referred to as **nerve fibers.**

## Classification of Neurons and Nerves

Neurons may be classified according to their structure or function. The functional classification is based on the direction in which they conduct impulses. **Sensory,** or **afferent, neurons** conduct impulses from sensory receptors into the CNS. **Motor,** or **efferent, neurons** (fig. 14.3) conduct impulses out of the CNS to effector organs (muscles and

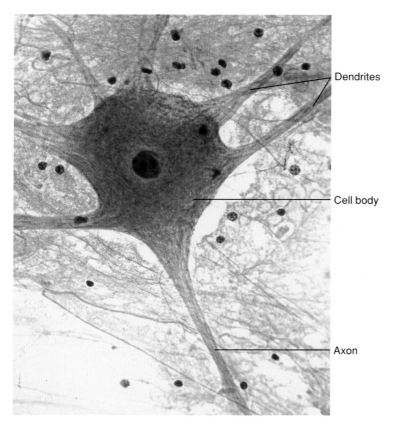

**FIGURE 14.2**
The neuron as seen in a photomicrograph of nerve tissue.

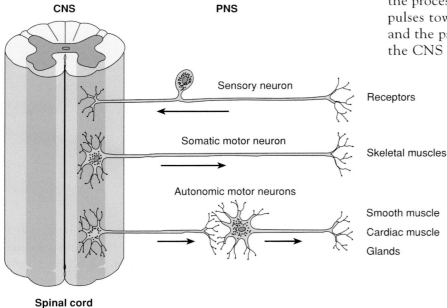

**FIGURE 14.3**
The relationship between sensory and motor fibers of the peripheral nervous system (PNS) and the central nervous system (CNS).

glands). **Association neurons,** or **interneurons,** are located entirely within the CNS and serve the associative, or integrative, functions of the nervous system. The term *innervation* refers to the nerve supply of a structure, which can be either sensory or motor.

There are two types of motor neurons—somatic and autonomic. **Somatic motor neurons** are responsible for both reflex and voluntary control of skeletal muscles. **Autonomic motor neurons** innervate the involuntary effectors—smooth muscle, cardiac muscle, and glands. The cell bodies of the autonomic neurons that innervate these organs are located outside the CNS in autonomic ganglia. There are two subdivisions of autonomic neurons—*sympathetic* and *parasympathetic*. Autonomic motor neurons, together with their central control centers, constitute the *autonomic nervous system* (to be discussed in chapter 17).

The structural classification of neurons is based on the number of processes that extend from the cell body of the neuron (fig. 14.4). **Bipolar neurons** have two processes, one at both ends; this type occurs in the retina of the eye. **Multipolar neurons** are the most common type and are characterized by several dendrites and one axon extending from the cell body. Motor neurons are good examples of this type. A **pseudounipolar neuron** has a single short process that divides like a **T** to form a longer process. Sensory neurons are pseudounipolar—one end of the longer process receives sensory stimuli and produces nerve impulses; the other end delivers these impulses to synapses within the brain or spinal cord. Anatomically, the part of the process that receives sensory stimuli and conducts impulses toward the cell body can be considered a dendrite and the part that conducts impulses from the cell body to the CNS can be considered an axon. Functionally, however, the two constitute a single process (joined at the short stalk of the **T** made near the cell body) that behaves as a long axon. This axon conducts impulses continuously from the receptive "dendritic" branches at its origin to the neurons of the CNS. The cell bodies of the sensory neurons are located outside the CNS in the posterior root ganglia of spinal and cranial nerves.

A **nerve** is a bundle of nerve fibers located outside the CNS. Most nerves are composed of both motor and sensory fibers, and thus are called *mixed nerves.* Some of the cranial nerves, however, contain sensory fibers only. These are the nerves that serve the special senses of smell, sight, hearing, and equilibrium.

## FIGURE 14.4
Three different types of neurons.

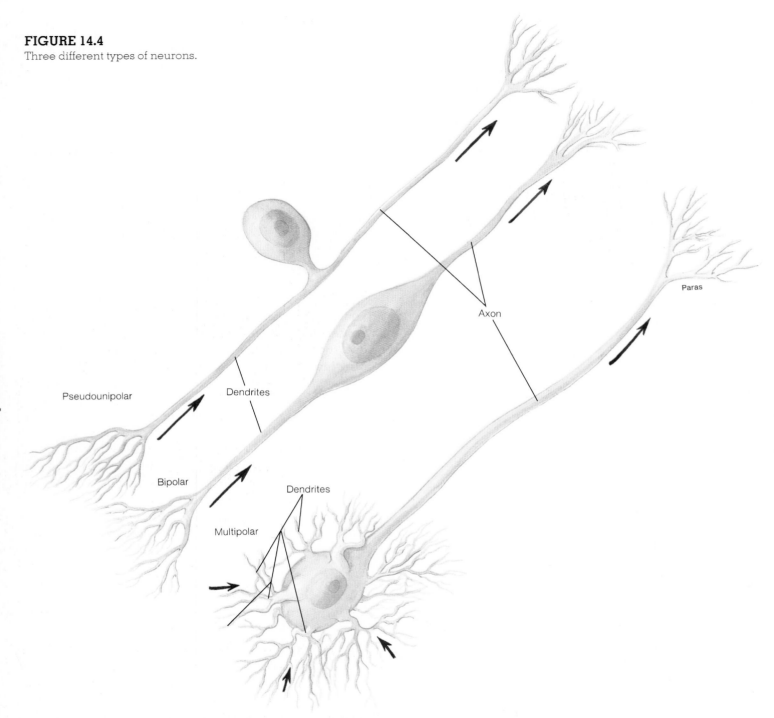

Pseudounipolar

Dendrites

Axon

Paras

Bipolar

Dendrites

Multipolar

# Neuroglia

Unlike other organs that are packaged in connective tissue derived from mesoderm (the middle layer of embryonic tissue), in the nervous system the supporting neuroglial cells are derived from the same embryonic tissue layer (ectoderm) that produces neurons. There are six categories of neuroglial cells: (1) **Schwann cells,** or **neurolemmocytes,** which form

myelin sheaths around peripheral axons; (2) **oligodendrocytes** (*ol″ĭ-go-den′drŏ-sītz*), which form myelin sheaths around axons of the CNS; (3) **microglia,** which are phagocytic cells that migrate through the CNS and remove foreign and degenerated material; (4) **astrocytes,** which help to regulate the passage of molecules from the blood to the brain; (5) **ependymal** (*ĕ-pen′dĭ-mal*) **cells,** which line the

· · · · · · · · · · · · · · · · · · · · · · · · · · · ·

Schwann cell: from Theodor Schwann, German histologist, 1810–1882

· · · · · · · · · · · · · · · · · · · · · · · · · · · ·

oligodendrocyte: Gk. *oligos,* few; L. *dens,* tooth; Gk. *kytos,* hollow (cell)
microglia: Gk. *mikros,* small; *glia,* glue
astrocyte: Gk. *aster,* star; *kytos,* hollow (cell)

## Table 14.2 — Neuroglial cells and their functions

| Neuroglia | Functions |
|---|---|
| Schwann cells | Surround axons of all peripheral nerve fibers, forming a neurolemmal sheath, or sheath of Schwann; wrap around many peripheral fibers to form myelin sheaths; also called **neurolemmocytes** |
| Oligodendrocytes | Form myelin sheaths around axons, producing white matter of the CNS |
| Astrocytes | Vascular processes cover capillaries within the brain and contribute to the blood-brain barrier |
| Microglia | Phagocytize pathogens and cellular debris within the CNS |
| Ependymal cells | Form the epithelial lining of brain cavities (ventricles) and the central canal of the spinal cord; cover tufts of capillaries to form choroid plexuses—structures that produce cerebrospinal fluid |
| Satellite cells | Support ganglia within the PNS; also called *ganglionic gliocytes* |

microglia: Gk. *mikros,* small; *glia,* glue

ventricles of the brain and the central canal of the spinal cord; and (6) **satellite cells,** which support neuron cell bodies within the ganglia of the PNS (table 14.2).

**Sheath of Schwann and Myelin Sheath**    Some axons in the CNS and PNS are surrounded by a myelin sheath and are known as *myelinated axons.* Other axons do not have a myelin sheath and are *unmyelinated axons.*

All axons in the PNS (but not in the CNS) are surrounded by a living sheath of Schwann cells, the **sheath of Schwann,** also known as the *neurolemmal sheath.* The outer surface of this layer of Schwann cells is encased in a glycoprotein *basement membrane,* analogous to the basement membrane that underlies epithelial membranes. The basement membrane and the cell membrane of the Schwann cells is often referred to as the **neurilemma** (*noor˝ĭ-lem´ă*). The axons of the CNS, by contrast, lack a sheath of Schwann (because Schwann cells are only found in the PNS) and also lack a continuous basement membrane. This characteristic is significant in terms of nerve regeneration, as we will describe in a later section.

Axons that are smaller than 2 μm in diameter are usually unmyelinated. Larger axons are generally surrounded by a **myelin sheath,** which is composed of successive wrappings of the cell membrane of Schwann cells (fig. 14.5) or oligodendrocytes. Schwann cells form myelin in the PNS, whereas oligodendrocytes form myelin in the CNS.

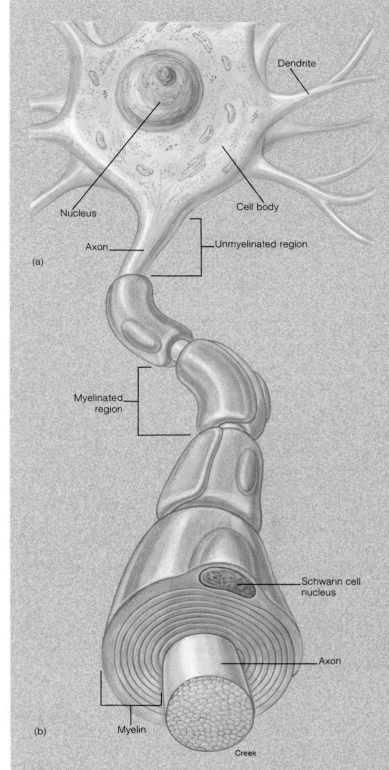

(a)

(b)

**FIGURE 14.5**
A myelinated neuron. A myelin sheath is formed by Schwann cells around the axons of many peripheral neurons. The myelin sheath is composed of wrappings of Schwann cell membrane. The Schwann cell nucleus and most of the cytoplasm, along with an outermost basement membrane, is located to the outside of the myelin sheath.

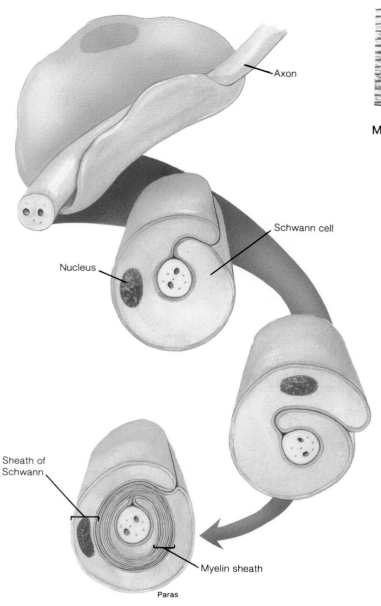

Axon

Schwann cell

Nucleus

Sheath of Schwann

Myelin sheath

Paras

**FIGURE 14.6**
The formation of a myelin sheath in a peripheral axon. The myelin sheath is formed by successive wrappings of the Schwann cell membranes, leaving most of the Schwann cell cytoplasm outside the myelin. The sheath of Schwann is thus located outside the myelin sheath.

In the process of myelin formation in the PNS, Schwann cells roll around the axon, much like electrician's tape is wrapped around a wire. But unlike electrician's tape the wrappings are made in the same spot so that each wrapping overlaps the others. The cytoplasm, meanwhile, becomes squeezed to the outer region of the Schwann cell, much as toothpaste is squeezed to the top of the tube as the bottom is rolled up (fig. 14.6). Each Schwann cell wraps only about 1 mm of axon, leaving gaps of exposed axon between the adjacent Schwann cells. These gaps in

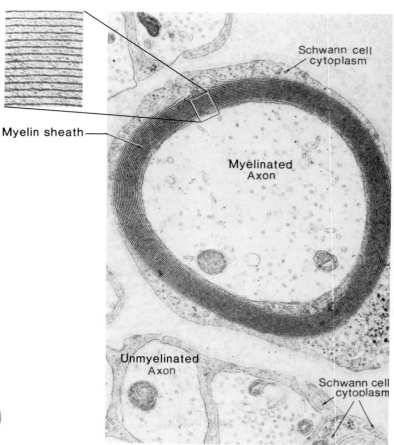

Schwann cell cytoplasm

Myelin sheath

Myelinated Axon

Unmyelinated Axon

Schwann cell cytoplasm

**FIGURE 14.7**
An electron micrograph of unmyelinated and myelinated axons.

the myelin sheath are known as the **nodes of Ranvier** (*ran´ve-a*), also called *neurofibril nodes*. The successive wrappings of Schwann cell membrane serve to insulate the axon, leaving only the nodes of Ranvier exposed to produce nerve impulses.

The Schwann cells remain alive as their cytoplasm is squeezed to the outside of the myelin sheath. As a result, myelinated axons of the PNS, like their unmyelinated counterparts, are surrounded by a living sheath of Schwann, to the outside of which is a continuous basement membrane (fig. 14.7).

The myelin sheaths of the CNS are formed by oligodendrocytes. Unlike a Schwann cell, which forms a myelin sheath around only one axon, each oligodendrocyte has extensions, like the tentacles of an octopus, that form myelin sheaths around several axons (fig. 14.8). Myelinated axons of the CNS, as a result, are not surrounded by a continuous basement membrane. The myelin sheaths around axons of the CNS give this tissue a white color; areas of the CNS

..........................................

nodes of Ranvier: from Louis A. Ranvier, French pathologist, 1835–1922

that contain a high concentration of axons thus form the **white matter.** The **gray matter** of the CNS is composed of high concentrations of cell bodies and dendrites, which lack myelin sheaths. (The only dendrites that have myelin sheaths are those of the sensory neurons of the PNS.)

**Regeneration of a Cut Axon** When an axon in a peripheral nerve is cut, the distal portion of the axon that was severed from the cell body degenerates. The Schwann cells, surrounded by the basement membrane, then form a *regeneration tube*, as the part of the axon that is connected to the cell body begins to grow and exhibit amoeboid movement. The Schwann cells of the regeneration tube are believed to secrete chemicals that attract the growing axon tip, and the regeneration tube helps to guide the regenerating axon to its proper destination. Even a severed major nerve may be surgically reconnected and the function of the nerve largely reestablished if the surgery is performed before tissue death.

Injury in the CNS stimulates growth of axon collaterals, but CNS axons are much more limited in their ability to regenerate than peripheral axons. This may be due in part to the absence of a continuous basement membrane, which precludes the formation of a regeneration tube.

Experiments in vitro suggest that central axons can regenerate if they are provided with the appropriate environment. In a developing fetal brain, chemicals that include *nerve growth factor* promote axon growth, and such chemicals have been shown in experimental animals to promote neuron regeneration in adult brains. Grafts of fetal brain tissue into adult brains have similar effects.

**Astrocytes and the Blood-Brain Barrier** Astrocytes (fig. 14.9) are large, star-shaped cells with numerous cytoplasmic processes that radiate outward. They are the most abundant of the neuroglia in the CNS, constituting up to 90% of the nervous tissue in some areas of the brain.

Astrocytes are known to interact with neurons in two different ways. First, they have been shown to take up potassium ions from the extracellular fluid. Since K$^+$ is released from active neurons during the production of nerve impulses (discussed in a later section), this action of astrocytes may be very important in maintaining a proper ionic environment for the neurons. Second, astrocytes have been shown to take up specific neurotransmitter chemicals that are released from the axon endings, as described in a later section. These neurotransmitters, known as glutamic acid and gamma-aminobutyric acid (GABA), are broken down

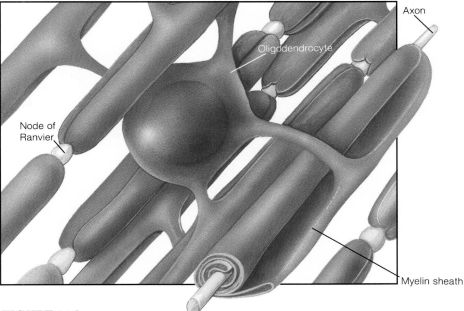

**FIGURE 14.8**
Formation of myelin sheaths in the central nervous system by an oligodendrocyte. One oligodendrocyte forms myelin sheaths around several axons.

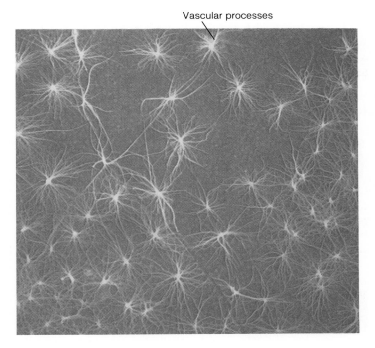

**FIGURE 14.9**
A photomicrograph showing the vascular processes of astrocytes. Astrocyte processes have bulbous ends, or feet, that cover most of the surface area of brain capillaries.

within the astrocytes. The molecule produced from this breakdown—glutamine—is released from the astrocytes and made available to the neurons in order for them to resynthesize these particular neurotransmitters.

Astrocytes have also been shown to interact with blood capillaries within the brain. Indeed, the brain capillaries are almost entirely surrounded by extensions of the astrocytes known as *vascular processes.* This association between astrocytes and brain capillaries has important physiological consequences.

Capillaries in the brain, unlike those of most other organs, do not have pores between adjacent endothelial cells (the cells that compose the walls of capillaries). Instead, the endothelial cells of brain capillaries are joined together by tight junctions. Unlike other organs, therefore, the brain cannot obtain molecules from the blood plasma by a nonspecific filtering process. Instead, molecules within brain capillaries must be moved through the endothelial cells by diffusion, active transport, endocytosis, and exocytosis. This feature of brain capillaries imposes a very selective **blood-brain barrier.** There is evidence to suggest that the development of tight junctions between adjacent endothelial cells in brain capillaries, and thus the development of the blood-brain barrier, results from the effects of astrocytes on the brain capillaries.

The blood-brain barrier presents difficulties in the chemotherapy of brain diseases because drugs that can enter other organs may not be able to enter the brain. In the treatment of *Parkinson's disease,* for example, patients who need a chemical called dopamine in the brain must be given a precursor molecule called levodopa (L-dopa). Dopamine cannot cross the blood-brain barrier, but L-dopa can enter the neurons and be changed to dopamine in the brain.

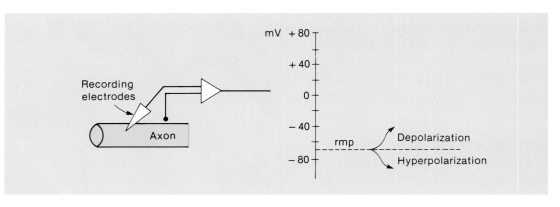

**FIGURE 14.10**

The difference in potential (in millivolts) between an intracellular and extracellular recording electrode is displayed on an oscilloscope screen. The resting membrane potential (rmp) of the axon may be reduced (depolarization) or increased (hyperpolarization).

# Action Potentials in Neurons

*The permeability of the axon membrane to $Na^+$ and $K^+$ is regulated by gates that open in response to stimulation. Net diffusion of these ions occurs in two stages; first $Na^+$ moves into the axon, then $K^+$ moves out. This flow of ions, and the changes in the membrane potential that result, constitute an event known as an action potential.*

All cells in the body maintain a potential difference (voltage) across the membrane, or *resting membrane potential,* in which the inside of the cell is negatively charged in comparison to the outside of the cell (for example, –65 mV). As explained in chapter 5, this potential difference is largely the result of the permeability properties of the cell membrane. The membrane traps large, negatively charged organic molecules within the cell and permits only limited diffusion of positively charged inorganic ions. This differential permeability results in an unequal distribution of sodium and potassium ions across the membrane. The action of the $Na^+/K^+$ pumps also helps to maintain a potential difference because they pump out three sodium ions for every two potassium ions that they transport into the cell. $Na^+$ is thus more highly concentrated in the extracellular fluid than in the cell, whereas $K^+$ is more highly concentrated within the cell.

An increase in membrane permeability to a specific ion results in the diffusion of that ion down its concentration gradient, either into or out of the cell. These *ion currents* occur only across limited patches of membrane (located fractions of a millimeter apart), where specific ion channels are located. Changes in the potential difference across the membrane at these points can be measured by the voltage developed between two electrodes—one placed inside the cell, the other placed outside the cell membrane at the region being recorded. The voltage between these two *recording electrodes* can be visualized by connecting them to an oscilloscope (fig. 14.10).

In an oscilloscope, electrons from a cathode-ray gun sweep across a fluorescent screen, producing a line of light. Changes in the potential difference between the two recording electrodes cause this line to deflect. The oscilloscope can be calibrated so that an upward deflection of the line indicates that the inside of the membrane has become less negative (or more positive) compared to the outside of the membrane. A downward deflection of the line, conversely, indicates that the inside of the cell has become more negative. The oscilloscope can thus function as a fast-responding voltmeter that can display voltage changes as a function of time.

If both recording electrodes are placed outside of the cell, the potential difference between the two will be zero (because there is no charge separation). When one of the two electrodes penetrates the cell membrane, the oscilloscope will indicate that the intracellular electrode is electrically negative with respect to the extracellular electrode; a membrane potential is recorded. If appropriate stimulation causes positive charges to flow into the cell, the line will deflect upward. This change is called **depolarization** because the potential difference between the two recording electrodes is reduced. If, on the other hand, the inside of the membrane becomes more negative as a result of stimulation, the line on the oscilloscope will deflect downward. This change is called **hyperpolarization.**

## Ion Gating in Axons

The permeability of the membrane to $Na^+$, $K^+$, and other ions is regulated by parts of the ion channels through the membrane called **gates.** Gates are believed to be composed of polypeptide chains that can open or close a membrane channel according to specific conditions. When a gated channel for a specific ion is closed, the membrane is not very permeable to that ion; when the gated channel is opened, the permeability to that ion can be greatly increased.

It is believed that there are two types of channels for $K^+$; one type lacks gates and is always open, whereas the other type has gates that are closed in the resting cell. Channels for $Na^+$, by contrast, always have gates, and these gates are closed in the resting cell. The resting cell is thus more permeable to $K^+$ than to $Na^+$. (As described in chapter 5, there is some leakage of $Na^+$ into the cell, which may occur in a nonspecific manner through open $K^+$ channels.) Consequently, the resting membrane potential is close to, but slightly less than, the equilibrium potential for $K^+$ (described in chapter 5).

Whether the gates for the $Na^+$ and $K^+$ channels are open or closed depends on the membrane potential. The gated channels are closed at the resting membrane potential of $-65$ mV, but they open when the membrane is depolarized to a certain threshold level. Since the opening and closing of these gates is regulated by the membrane voltage, the gates are said to be **voltage regulated.**

Depolarization of a small region of an axon can be experimentally induced by a pair of stimulating electrodes that act as if they were injecting positive charges into the axon.

If a pair of recording electrodes are placed in the same region, an upward deflection of the oscilloscope line will be observed as a result of this depolarization. If a certain level of depolarization is achieved (from $-65$ mV to $-55$ mV, for example) by this artificial stimulation, a sudden and very rapid change in the membrane potential will be observed. This is because *depolarization to a threshold level causes the $Na^+$ gates to open.* Now the permeability properties of the membrane are changed, and $Na^+$ diffuses down its concentration gradient into the cell.

A fraction of a second after the $Na^+$ gates open, they close again. At this time, *the depolarization stimulus causes the $K^+$ gates to open.* This makes the membrane more permeable to $K^+$ than it is at rest, and $K^+$ diffuses down its concentration gradient out of the cell. The $K^+$ gates will then close and the permeability properties of the membrane will return to what they were at rest.

## Action Potentials

In this section we will consider the events that occur at one point in an axon when a small region of axon membrane is stimulated artificially and responds with changes in ion permeabilities. The resulting changes in membrane potential at this point are detected by recording electrodes placed in this region of the axon. The nature of the stimulus in vivo (in the body) and the manner by which electrical events are conducted to different points along the axon will be described in later sections.

When the axon membrane has been depolarized to a threshold level, the $Na^+$ gates open and the membrane becomes permeable to $Na^+$. This permits $Na^+$ to enter the axon by diffusion, which further depolarizes the membrane (makes the inside less negative, or more positive). Since the $Na^+$ gates of the axon are voltage regulated, this further depolarization makes the membrane even more permeable to $Na^+$, so that even more $Na^+$ can enter the cell and open even more voltage-regulated $Na^+$ gates. A *positive feedback loop* (fig. 14.11) is thus created, which causes the rate of $Na^+$ entry and depolarization to accelerate in an explosive fashion.

After a slight time delay, depolarization of the axon membrane also causes the opening of voltage-regulated $K^+$ gates and the diffusion of $K^+$ out of the cell. Since $K^+$ is positively charged, the diffusion of $K^+$ out of the cell makes the inside of the cell less positive, or more negative, and acts to restore the original resting membrane potential. This process is called **repolarization** and represents the completion of a *negative feedback loop* (fig. 14.11).

Figure 14.12b illustrates the movement of $Na^+$ and $K^+$ through the axon membrane in response to a depolarization stimulus. Notice that the explosive increase in $Na^+$

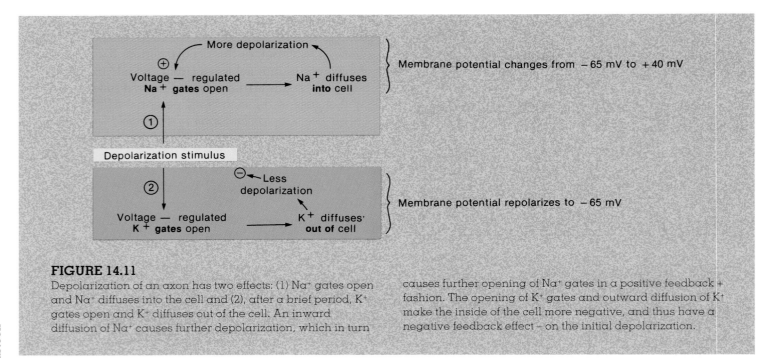

## FIGURE 14.11
Depolarization of an axon has two effects: (1) Na⁺ gates open and Na⁺ diffuses into the cell and (2), after a brief period, K⁺ gates open and K⁺ diffuses out of the cell. An inward diffusion of Na⁺ causes further depolarization, which in turn causes further opening of Na⁺ gates in a positive feedback + fashion. The opening of K⁺ gates and outward diffusion of K⁺ make the inside of the cell more negative, and thus have a negative feedback effect – on the initial depolarization.

diffusion causes rapid depolarization to 0 mV and then *over-shoot* of the membrane potential so that the inside of the membrane actually becomes positively charged (almost +40 mV) compared to the outside (fig. 14.12a). The Na⁺ permeability then rapidly decreases as the diffusion of K⁺ increases, resulting in repolarization to the resting membrane potential. These changes in Na⁺ and K⁺ diffusion and the resulting changes in the membrane potential that they produce constitute an event called the **action potential,** or **nerve impulse.**

Once an action potential has been completed, the Na⁺/K⁺ pumps will extrude the extra Na⁺ that has entered the axon and recover the K⁺ that has diffused out of the axon. This active transport of ions occurs very quickly because the events described occur across only a very small area of membrane, and so only relatively small amounts of Na⁺ and K⁺ actually diffuse through the membrane during the production of an action potential. The total concentrations of Na⁺ and K⁺ in the axon and in the extracellular fluid are not significantly changed during an action potential. Even during the overshoot phase, for

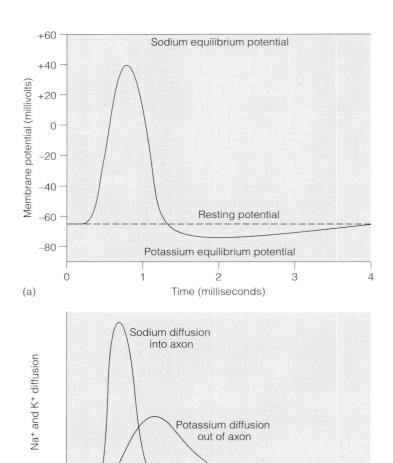

## FIGURE 14.12
An action potential (a) is produced by an increase in sodium diffusion that is followed, with a short time delay, by an increase in potassium diffusion (b). This drives the membrane potential first toward the sodium equilibrium potential and then toward the potassium equilibrium potential.

example, the concentration of Na⁺ remains higher outside the axon; repolarization thus requires the outward diffusion of K⁺, the concentration gradient of which is in the opposite direction to that of the Na⁺.

Notice that active transport processes are not directly involved in the production of an action potential; both depolarization and repolarization are produced by the *diffusion* of ions down their opposing concentration gradients. A neuron poisoned with cyanide, so that it cannot produce ATP, can still produce action potentials for a period of time. After awhile, however, the lack of ATP for active transport by the $Na^+/K^+$ pumps will result in a decline in the ability of the axon to produce action potentials. This demonstrates that the $Na^+/K^+$ pumps are not directly involved, but rather are required to maintain the concentration gradients needed for the diffusion of Na⁺ and K⁺ during action potentials.

**All-or-None Law**   Once a region of axon membrane has been depolarized to a threshold value, the positive feedback effect of depolarization on Na⁺ permeability and of Na⁺ permeability on depolarization causes the membrane potential to shoot toward about +40 mV. It does not normally become more positive because the Na⁺ gates quickly close and the K⁺ gates open. The length of time that the Na⁺ and K⁺ gates stay open is independent of the strength of the depolarization stimulus.

The amplitude of action potentials is therefore **all or none.** When depolarization is below a threshold value, the voltage-regulated gates are closed; when depolarization reaches threshold, a maximum potential change (the action potential) is produced. Since the change from –65 mV to +40 mV and back to –65 mV lasts only about 3 msec, the image of an action potential on an oscilloscope screen looks like a spike. Action potentials are therefore sometimes called *spike potentials*.

Since the gates are open for a fixed period of time, the duration of each action potential is about the same. Likewise, since the concentration gradient for Na⁺ is relatively constant, the amplitude of each action potential is about the same in all axons at all times (from –65 mV to +40 mV, or about 100 mV in total amplitude).

**Coding for Stimulus Intensity**   If one depolarization stimulus is stronger than another, the stronger stimulus is not coded in the nervous system by a greater amplitude of action potentials (because action potentials are all-or-none events). When a stronger stimulus is applied to a neuron, identical action potentials are produced more frequently (more are produced per minute). Therefore, the code for stimulus strength in the nervous system is frequency modulated (FM). This concept is illustrated in figure 14.13.

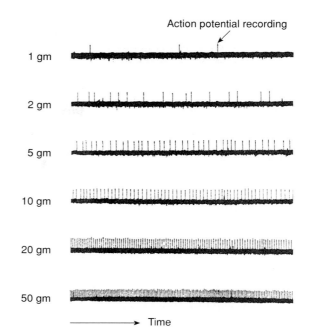

**FIGURE 14.13**

Recordings from a single sensory fiber of the sciatic nerve of a frog stimulated by varying degrees of stretch of the gastrocnemius muscle. Note that increasing amounts of stretch (indicated by increasing weights attached to the muscle) result in an increased frequency of action potentials.

When an entire collection of axons (in a nerve) is stimulated, different axons will be stimulated at different stimulus intensities. A low-intensity stimulus will only activate those few axons with low thresholds, whereas high-intensity stimuli can activate axons with higher thresholds. As the intensity of stimulation increases, more and more axons will become activated. This process, called **recruitment,** represents another mechanism by which the nervous system can code for stimulus strength.

**Refractory Periods**   If a stimulus of a given intensity is maintained at one point along an axon and depolarizes it to threshold, action potentials will be produced at that point at a given frequency (number per minute). As the stimulus strength is increased, the frequency of action potentials produced at that point will increase accordingly. As action potentials are produced with increasing frequency, the delay between successive action potentials will decrease. However this time interval will never be decreased to the point where a new action potential is produced before the preceding one has finished.

During the time that a patch of axon membrane is producing an action potential, it is incapable of responding to further stimulation, and is said to be *refractory*. If a second stimulus is applied, for example, while the Na⁺ gates are open in response to a first stimulus, the second stimulus cannot have any effect (the gates are already open). During the time that the Na⁺ gates are open, therefore, the membrane is in an **absolute refractory period** and cannot respond to any subsequent stimulus. If a second stimulus is applied while the K⁺ gates are open (and the membrane is in the process of repolarizing), the membrane is in a **relative refractory period.** During this time, only a very strong stimulus can depolarize the membrane and produce a second action potential (fig. 14.14).

Because the cell membrane is refractory during the time it is producing an action potential, each action potential remains a separate, all-or-none event. In this way, as a continuously applied stimulus increases in intensity, its strength can be coded strictly by the frequency of the action potentials it produces at each point of the axon membrane.

One might suppose that as an axon produced a large number of action potentials, the relative concentrations of Na⁺ and K⁺ would be changed in the extracellular and intracellular compartments. This is not the case. In a typical mammalian axon that is 1 μm in diameter, for example, only 1 intracellular K⁺ ion in 3000 would be exchanged for a Na⁺. Since a typical neuron has about 1 million Na⁺/K⁺ pumps, which can transport nearly 200 million ions per second, these small changes can be quickly corrected.

**Cable Properties of Neurons**  If a pair of stimulating electrodes produces a depolarization that is too weak to reach threshold (about –55 mV), the change in membrane potential will be *localized* to within 1 to 2 mm of the point of stimulation. For example, if the stimulus causes depolarization from –65 mV to –60 mV at one point, and the recording electrodes are placed only 3 mm away from the stimulus, the membrane potential recorded will remain at –65 mV (the resting potential). The axon is thus a very poor conductor compared to metal wires.

The term *cable properties* is used to describe the ability of a neuron to transmit charges through its cytoplasm. These cable properties are poor because there is a high internal resistance to the spread of charges and because many charges leak out of the axon through its membrane. If an axon had to conduct only through its cable properties, therefore, no axon could be more than a millimeter in length. Some axons are a meter or more in length however, which suggests that the conduction of nerve impulses does not rely on the cable properties of the axon.

## Conduction of Nerve Impulses

When stimulation of electrodes artificially depolarizes one point of an axon membrane to a threshold level, voltage-regulated gates open and an action potential is produced at

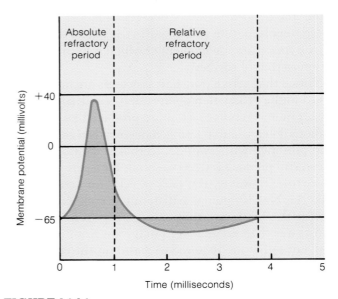

**FIGURE 14.14**
The absolute and relative refractory periods. While a segment of axon is producing an action potential, the membrane is absolutely or relatively resistant (refractory) to further stimulation.

that small region of axon membrane. For about the first millisecond of the action potential, when the membrane voltage changes from –65 mV to +40 mV, a current of Na⁺ enters the cell by diffusion as a result of the opening of the Na⁺ gates. Each action potential thus injects positive charges (sodium ions) into the axon.

These positively charged sodium ions are conducted, by the cable properties of the axon, to an adjacent region that still has a membrane potential of –65 mV. Within the limits of the cable properties of the axon (1 to 2 mm), this conduction helps to depolarize the adjacent region of axon membrane. When this adjacent region of membrane reaches a threshold level of depolarization, it too produces an action potential as its voltage-regulated gates open.

Each action potential thus acts as a stimulus for the production of another action potential at the next region of membrane. In the description of action potentials earlier in this chapter, the stimulus for their production was artificial depolarization, produced by a pair of stimulating electrodes. Now, it can be seen that an action potential at a given point along an axon results from depolarization produced by a preceding action potential. This explains how all action potentials along an axon are produced after the first action potential is generated.

**Conduction in an Unmyelinated Axon**  In an unmyelinated axon, every patch of membrane that contains Na⁺ and K⁺ gates can produce an action potential. Action potentials are thus produced at locations only a fraction of a micrometer apart all along the length of the axon.

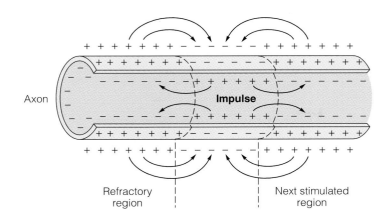

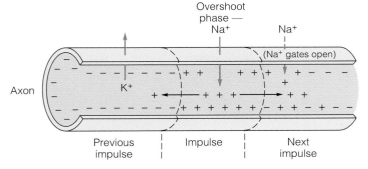

## FIGURE 14.15

The conduction of a nerve impulse (action potential) in an unmyelinated nerve fiber (axon). Each action potential "injects" positive charges that spread to adjacent regions. The region that has previously produced an action potential is refractory. The previously unstimulated region is partially depolarized. As a result, its voltage-regulated Na⁺ gates open, and the process is repeated.

The cablelike spread of depolarization induced by Na⁺ influx during one action potential helps to depolarize the adjacent regions of membrane. This process is also aided by movements of ions on the outer surface of the axon membrane (fig. 14.15). Action potentials are passed *in one direction only* along the length of the axon because the area in the opposite direction has just produced an action potential and cannot produce a new one, being still in its refractory period.

It should be noted that action potentials are not really conducted, although it is convenient to use that word. Each action potential is a separate, complete event that is repeated, or *regenerated*, along the

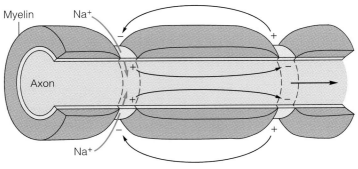

## FIGURE 14.16

The conduction of the nerve impulse in a myelinated nerve fiber. Since the myelin sheath prevents inward Na⁺ current, action potentials can be produced only at the gaps in the myelin sheath called the nodes of Ranvier. This "leaping" of the action potential from node to node is known as saltatory conduction.

axon's length. The action potential produced at the end of the axon is a completely new event that occurs in response to depolarization from the previous action potential. This last action potential has the same amplitude as the first; thus, action potentials are said to be **conducted without decrement** (without decreasing in amplitude).

The spread of depolarization by the cable properties of an axon is fast compared to the time it takes to produce an action potential. Since action potentials are produced at every fraction of a micrometer in an unmyelinated axon, the conduction rate is relatively slow. It is somewhat faster if the unmyelinated axon is thicker. This is because the ability of fibers to conduct charges by cable properties improves with increasing diameter. The conduction rate is substantially faster if the axon is myelinated.

**Conduction in a Myelinated Axon** The myelin sheath provides insulation for the axon, preventing movements of Na⁺ and K⁺ through the membrane. If the myelin sheath were continuous, therefore, action potentials could not be produced. Fortunately, there are interruptions in the myelin—the nodes of Ranvier.

Because the cable properties of axons can only conduct depolarizations over a very short distance (1–2 mm), the nodes of Ranvier must be very close together (they are generally about 1 mm apart). Studies have shown that Na⁺ channels are highly concentrated at the nodes (estimated at 10,000 per square micrometer) and almost absent in the regions of axon membrane between the nodes. Action potentials, therefore, occur only at the nodes of Ranvier (fig. 14.16) and seem to leap from node to node in a type of conduction called **saltatory conduction.**

Since the cablelike spread of depolarization between the nodes is very fast and fewer action potentials need to be produced per given length of axon, saltatory conduction permits a *faster rate of conduction* than is possible in an unmyelinated fiber. Conduction rates in the human nervous system vary from about 1.0 m/sec—in thin, unmyelinated axons that mediate slow, visceral responses—to faster than 100 m/sec (225 mph)—in thick, myelinated axons involved in quick stretch reflexes in skeletal muscles. These differences in conduction rates explain, for example, why the eyeblink reflex is much faster than the reflex that constricts the pupils in response to light.

........................

saltatory: L. *saltatio*, leap

# The Synapse

*Axons end close to, or in some cases at contact with, another cell. Once action potentials reach the end of an axon, they directly or indirectly stimulate the other cell. In specialized cases, action potentials can directly pass from one cell to another. In most cases, however, the action potentials stop at the axon terminal, where they stimulate the release of a chemical neurotransmitter that stimulates the next cell.*

A **synapse** is the functional connection between a neuron and a second cell. In the CNS, this other cell is also a neuron. In the PNS, the other cell may be either a neuron or an *effector cell* within a muscle or a gland. Although the physiology of neuron–neuron synapses and neuron–muscle synapses is similar, the latter synapses are often called **myoneural** (*mi˝ŏ-noor´al*) or **neuromuscular, junctions.**

Neuron–neuron synapses usually involve a connection between the axon terminal of one neuron and the dendrites, cell body, or axon of a second neuron. These are called, respectively, *axodendritic, axosomatic,* and *axoaxonic synapses* (fig. 14.17). In almost all synapses, transmission is in one direction only—from the axon of the first (or presynaptic) cell to the second (or postsynaptic) cell.

In the early part of the twentieth century, most physiologists believed that synaptic transmission was *electrical*—that is, that action potentials were conducted directly from one cell to the next. This was a logical assumption given that axon terminals appeared to touch the postsynaptic cells and that the delay in synaptic conduction was extremely short (about 0.5 msec). Improved histological techniques, however, revealed tiny gaps in the synapses, and experiments demonstrated that the actions of autonomic nerves could be duplicated by certain chemicals. This led to the hypothesis that synaptic transmission might be *chemical*—that the presynaptic axon terminals might release chemical **neurotransmitters** that stimulated action potentials in the postsynaptic cells.

In 1921, a physiologist named Otto Loewi published the results of an experiment suggesting that synaptic transmission was indeed chemical, at least at the junction between a branch of the vagus nerve (chapter 17) and the heart. He had isolated the heart of a frog and, while stimulating the cardiac branch of the vagus that innervates the heart, perfused the heart with an isotonic salt solution. Stimulation of this nerve slowed the heart rate, as expected. More importantly, application of this salt solution to the heart of a second frog caused the second heart to slow its rate of beat.

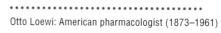

Otto Loewi: American pharmacologist (1873–1961)

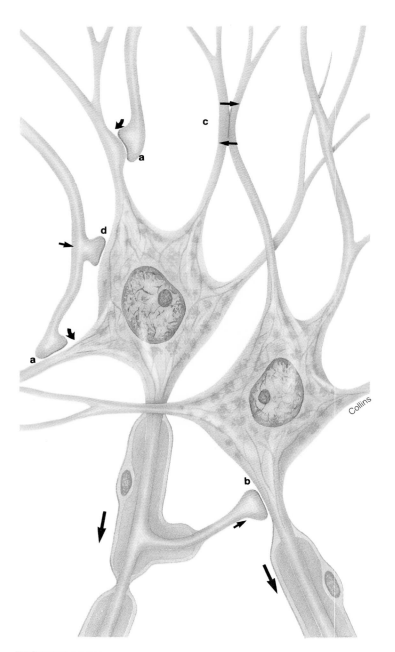

**FIGURE 14.17**

Different types of synapses: (*a*) axodendritic, (*b*) axoaxonic, (*c*) dendrodendritic, and (*d*) axosomatic.

Loewi concluded that the axon terminals of the vagus nerve must have released a chemical, which he called *vagusstoff,* that inhibited the heart rate. This chemical was subsequently identified as **acetylcholine** (*ă-set˝l-ko´lēn*) or **ACh.** In the decades following Loewi's discovery, many other examples of chemical synapses were discovered, and the theory of electrical synaptic transmission fell into disrepute. More recent evidence, ironically, has shown that electrical synapses do exist in the nervous system (although they are the exception), within smooth muscles, and between cardiac cells in the heart.

## Electrical Synapses: Gap Junctions

In order for two cells to be electrically coupled, they must be approximately equal in size and must be joined in areas of low electrical resistance. In this way, impulses can be regenerated from one cell to the next without interruption.

Adjacent cells that are electrically coupled are joined together by **gap junctions.** In gap junctions, the membranes of the two cells are separated by only 2 nanometers (1 nanometer = $10^{-9}$ meter). A surface view of gap junctions in the electron microscope reveals hexagonal arrays of particles that are believed to be channels through which ions and molecules may pass from one cell to the next (fig. 14.18).

Gap junctions are present in smooth and cardiac muscle, where they allow excitation and rhythmic contraction of large masses of muscle cells. Gap junctions have also been observed in various regions of the brain. Although their functional significance in the brain is unknown, it has been speculated that they may allow a two-way transmission of impulses (in contrast to chemical synapses, which are always one-way). Gap junctions have also been observed between neuroglia, which do not produce electrical impulses; these may act as channels for the passage of informational molecules between cells. It is interesting in this regard that many embryonic tissues have gap junctions and that these gap junctions disappear as the tissue becomes more specialized.

## Chemical Synapses

Transmission across the majority of synapses in the nervous system is one-way and occurs through the release of chemical neurotransmitters from presynaptic axon terminals. These presynaptic endings, which are called **axon terminals,** or **terminal boutons** (*boo-tonz´*) because of their swollen appearance, are separated from the postsynaptic cell by a **synaptic cleft** so narrow that it can only be seen clearly with an electron microscope (fig. 14.19).

Neurotransmitter molecules within the axon terminals are contained within many small, membrane-enclosed **synaptic vesicles.** In order for the neurotransmitter within these vesicles to be released into the synaptic cleft, the vesicle membrane must fuse with the axon membrane and release its contents. This process is known as *exocytosis.* The neurotransmitter is released *in quanta*—that is, in multiples of the amount contained in one vesicle. The number of vesicles that undergo exocytosis is directly related to the frequency of action potentials produced at the axon terminals.

The release of neurotransmitters following electrical excitation of the axon terminals accounts for most of the 0.5-msec time delay in synaptic transmission. During this interval there is a sudden, transient inflow of $Ca^{++}$ into the axon terminals. This inflow of $Ca^{++}$ is apparently due to the

......................................

bouton: Fr. *bouton*, button

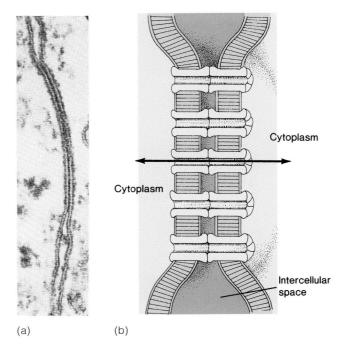

(a)          (b)

**FIGURE 14.18**

Gap junctions. (*a*) An electron micrograph showing that cell membranes of two cells are fused together in the gap junction. (*b*) A graphic representation of gap junctions, with the arrow indicating a channel through which ions and molecules may pass.

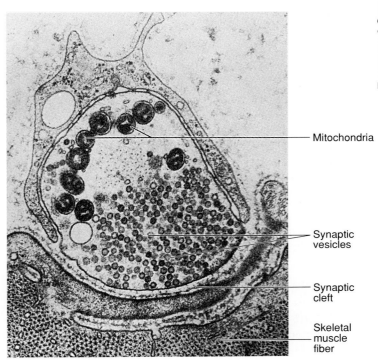

Mitochondria

Synaptic vesicles

Synaptic cleft

Skeletal muscle fiber

**FIGURE 14.19**

An electron micrograph of a chemical synapse showing synaptic vesicles at the end of an axon.

opening of Ca$^{++}$ gates in response to electrical excitation and is required for the release of neurotransmitters. The Ca$^{++}$ activates a regulatory protein within the cytoplasm called **calmodulin,** which in turn activates an enzyme called **protein kinase.** This enzyme phosphorylates (adds a phosphate group to) specific proteins known as **synapsins** in the membrane of the synaptic vesicle. This permits the vesicle to undergo exocytosis and release its content of neurotransmitter molecules. Since regulation by Ca$^{++}$, calmodulin, and protein kinase is also involved in the action of some hormones, it is discussed in more detail in chapter 19.

# Synaptic Transmission by Acetylcholine

*The postsynaptic membrane contains gated channels for Na$^+$ and K$^+$ that open when a neurotransmitter, such as acetylcholine, binds to its receptor in the membrane. Opening of such chemically regulated gates can produce a depolarization. This depolarization, known as an excitatory postsynaptic potential, serves as the stimulus for the production of action potentials in the postsynaptic cell.*

Acetylcholine (ACh) is used as a neurotransmitter by some neurons in the CNS, by somatic motor neurons at the neuromuscular junction, and by certain autonomic nerve endings. The effects of this chemical are excitatory in the first two synapses and either excitatory or inhibitory in the third. In this section, we will consider only the excitatory effects of ACh.

## Chemically Regulated Gates

Axon terminals have synaptic vesicles that contain about 10,000 molecules of ACh each. Once these molecules are released, they quickly diffuse across the narrow synaptic cleft to the membrane of the postsynaptic cell. Here, they chemically bond to **receptor proteins** that are built into the postsynaptic membrane. These receptor proteins combine with ACh in a specific manner, analogous to the specific interaction between transport proteins and their substrates.

Acetylcholine is not transported into the postsynaptic cell after diffusing across the synaptic cleft. Instead, the bonding of ACh to its receptor protein causes changes in the membrane structure that result in the opening of ion channel gates for Na$^+$ and K$^+$. These gates, located only in the postsynaptic membrane, are called **chemically regulated gates** because they open in response to bonding by a chemical (ACh). Unlike the voltage-regulated gates previously described in the axon, the chemically regulated gates do *not* respond to changes in membrane potential.

Also in contrast to voltage-regulated gates, where the outward flow of K$^+$ occurs after the inward flow of Na$^+$, chemically regulated gates permit the simultaneous diffusion of Na$^+$ and K$^+$ (fig. 14.20). The depolarizing effect of Na$^+$ diffusion predominates because the electrochemical gradient for Na$^+$ is greater than that for K$^+$. The outflow of K$^+$, however, does prevent the overshoot characteristic of action potentials—the membrane potential can reach 0 mV but cannot reverse polarity (as occurs in action potentials). Because of the characteristics of chemically regulated gates, neurotransmitters do not directly produce action potentials.

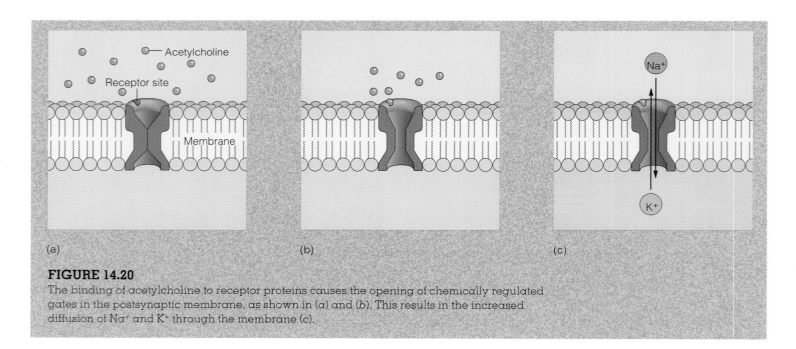

**FIGURE 14.20**
The binding of acetylcholine to receptor proteins causes the opening of chemically regulated gates in the postsynaptic membrane, as shown in (*a*) and (*b*). This results in the increased diffusion of Na$^+$ and K$^+$ through the membrane (*c*).

They can only produce depolarization, which may stimulate the opening of voltage-regulated gates and production of action potentials a short distance away from the site of the synapse.

## Excitatory Postsynaptic Potential (EPSP)

The interaction of ACh with its receptors in the postsynaptic membrane opens chemically regulated gates and depolarizes that region of membrane. Depolarizations produced in response to ACh are known as **excitatory postsynaptic potentials (EPSPs)** when they occur in postsynaptic neurons, or as *end plate potentials* when they occur in skeletal muscle cells. These depolarizations do not overshoot 0 mV for reasons previously discussed; moreover, they differ from action potentials in a number of other respects.

Unlike action potentials, EPSPs have *no threshold*. A single quantum of ACh (released from a single synaptic vesicle) produces a tiny depolarization of the postsynaptic membrane. When more quanta of ACh are released, the depolarization is correspondingly greater. EPSPs are therefore *graded* in magnitude, unlike all-or-none action potentials. Since EPSPs can be graded, they are capable of *summation;* that is, the depolarizations of several different EPSPs can be added together. This is quite different from action potentials, which are prevented from summating by their all-or-none nature and by the occurrence of refractory periods. A comparison of the characteristics of EPSPs and action potentials is provided in table 14.3.

**Acetylcholinesterase**  The bond between ACh and its receptor protein exists for only a brief instant. The ACh-receptor complex quickly dissociates but can be quickly reformed as long as free ACh is in the vicinity. In order for activity in the postsynaptic cell to be controlled, free ACh must be inactivated very soon after it is released. The inactivation of ACh is achieved by means of an enzyme called **acetylcholinesterase** (*ă-set″l-ko″lĭ-nes′tĕ-rās*), or **AChE,** which is present on the postsynaptic membrane or immediately outside the membrane, with its active site facing the synaptic cleft (fig. 14.21).

| Table 14.3 | Comparison of action potentials with excitatory postsynaptic potentials (EPSPs) | |
|---|---|---|
| **Characteristic** | **Action potential** | **Excitatory postsynaptic potential** |
| Stimulus for opening of ionic gates | Depolarization | Acetylcholine (ACh) |
| Initial effect of stimulus | Na⁺ gates open | Na⁺ and K⁺ gates open |
| Production of repolarization | Opening of K⁺ gates | Loss of intracellular positive charges with time and distance |
| Conduction distance | Not conducted—regenerated over length of axon | 1–2 mm; a localized potential |
| Positive feedback between depolarization and opening of Na⁺ gates | Yes | No |
| Maximum depolarization | +40 mV | Close to zero |
| Summation | No summation—all-or-none phenomenon | Summation of EPSPs, producing graded depolarizations |
| Refractory period | Present | Absent |
| Effect of drugs | Inhibited by tetrodotoxin, not by curare | Inhibited by curare, not by tetrodotoxin |

 Nerve gas exerts its odious effects by inhibiting AChE in skeletal muscles. Since ACh is not degraded, it can continue to combine with receptor proteins and continue to stimulate the postsynaptic cell, leading to spastic paralysis. Clinically, cholinesterase inhibitors (such as *neostigmine*) are used to enhance the effects of ACh on muscle contraction when neuromuscular transmission is weak, as in the disease *myasthenia gravis.*

**Stimulation of Skeletal Muscles**  The synapse of a motor axon with a skeletal muscle cell is known as a *motor end plate*. Release of ACh from the motor axon produces an *end plate potential*, which is identical in nature to an EPSP in a postsynaptic neuron. This depolarization opens voltage-regulated gates, and thus serves as the stimulus for the production of action potentials in the muscle fiber. This is significant because electrical excitation of a muscle fiber, through mechanisms discussed in chapter 12, stimulates muscle contraction.

..........................................

myasthenia: Gk. *myos*, muscle; *asthenia*, weakness

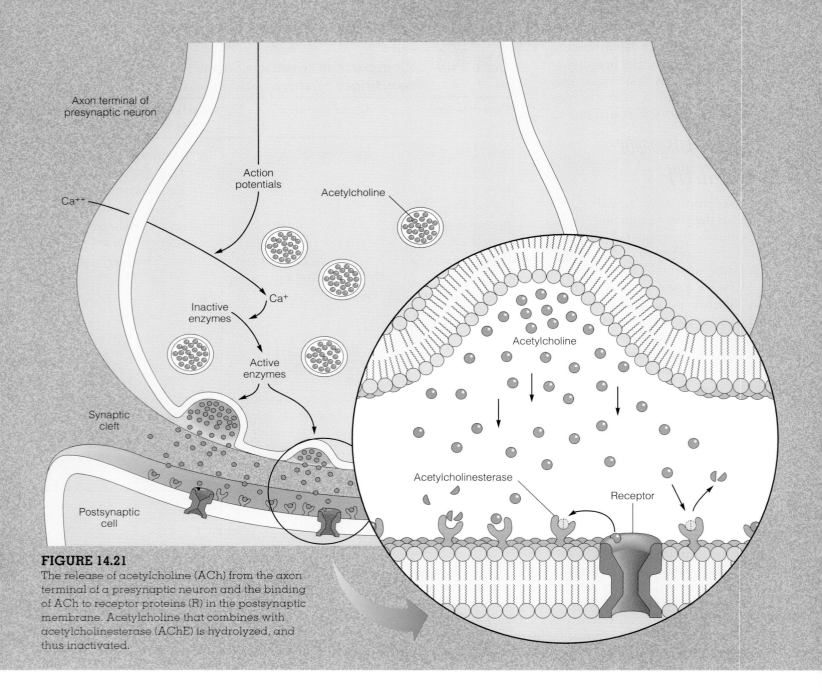

**FIGURE 14.21**
The release of acetylcholine (ACh) from the axon terminal of a presynaptic neuron and the binding of ACh to receptor proteins (R) in the postsynaptic membrane. Acetylcholine that combines with acetylcholinesterase (AChE) is hydrolyzed, and thus inactivated.

Labels within figure: Axon terminal of presynaptic neuron; Action potentials; Acetylcholine; Ca⁺⁺; Ca⁺; Inactive enzymes; Active enzymes; Synaptic cleft; Postsynaptic cell; Acetylcholine; Acetylcholinesterase; Receptor

If any stage in the process of neuromuscular transmission is blocked, muscle weakness—sometimes leading to paralysis and death—may result. The drug *curare* (*koo-rǎ´re*), for example, competes with ACh for attachment to the receptor proteins and reduces the size of the end plate potentials (table 14.4). This drug was first used on poison darts by South American Indians because it produced flaccid paralysis in their victims. Clinically, curare is used as a muscle relaxant during anesthesia and to prevent muscle damage during electroconvulsive shock therapy.

**Stimulation of Neurons**   Within the central nervous system, the axon terminals of one neuron typically synapse with the dendrites or cell body of another. The dendrites and cell body thus serve as the receptive area of the neuron, and it is in these regions that receptor proteins for neurotransmitters and chemically regulated gates are located. The first voltage-regulated gates are located at the beginning of the axon, at the axon hillock. It is here that action potentials are first produced (fig. 14.22).

## Table 14.4  Drugs that affect the neural control of skeletal muscles

| Drug | Derivation | Effect |
| --- | --- | --- |
| Botulinus toxin | Produced by *Clostridium botulinum* (bacteria) | Inhibits release of acetylcholine (ACh) |
| Curare | Resin from a South American tree | Prevents interaction of ACh with the postsynaptic receptor protein |
| α-Bungarotoxin | Venom of *Bungarus* snakes | Binds to ACh-receptor proteins |
| Saxitoxin | Red tide (*Gonyaulax*) protozoa | Blocks voltage-regulated Na$^+$ channels |
| Tetrodotoxin | Pufferfish | Blocks voltage-regulated Na$^+$ channels |
| Nerve gas | Artificial | Inhibits acetylcholinesterase in postsynaptic cell |
| Prostigmine | Nigerian bean | Inhibits acetylcholinesterase in postsynaptic cell |
| Strychnine | Seeds of an Asian tree | Prevents IPSPs in spinal cord that inhibit contraction of antagonistic muscles |

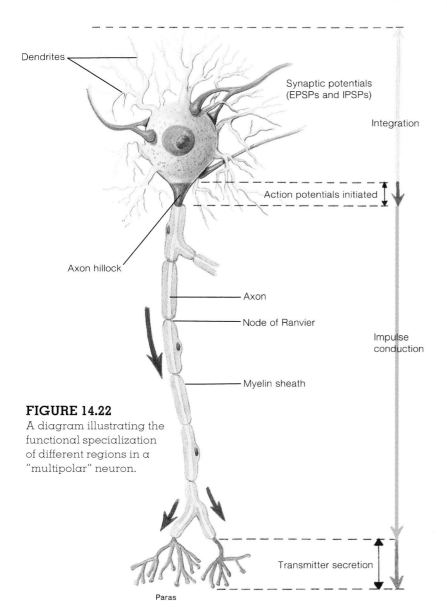

**FIGURE 14.22**

A diagram illustrating the functional specialization of different regions in a "multipolar" neuron.

The cell body and dendrites of multipolar neurons lack voltage-regulated gates. The *initial segment* of the axon, which is the unmyelinated region around the axon hillock, on the other hand, has a high concentration of voltage-regulated gates. Depolarizations in the dendrites and cell body must therefore spread by cable properties to the initial segment of the axon in order to stimulate action potentials.

If the depolarization is at or above threshold by the time it reaches the initial segment of the axon, the EPSP will stimulate the production of action potentials, which can then regenerate themselves along the axon. If, however, the EPSP is below threshold at the initial segment, no action potentials will be produced in the postsynaptic cell (fig. 14.23). Gradations in the strength of the EPSP above threshold determine the frequency with which action potentials will be produced at the axon hillock and at each point in the axon where the impulse is conducted.

Earlier in this chapter, we introduced the concept of action potentials by describing the events that occurred when a depolarization stimulus was artificially produced by stimulating electrodes. Now you know that EPSPs, conducted from the dendrites and cell body, serve as the normal stimuli for the production of action potentials in the axon hillock, and that the action potentials at this point serve as the depolarization stimuli for the next region, and so on. This chain of events ends at the axon terminal, where neurotransmitter is released.

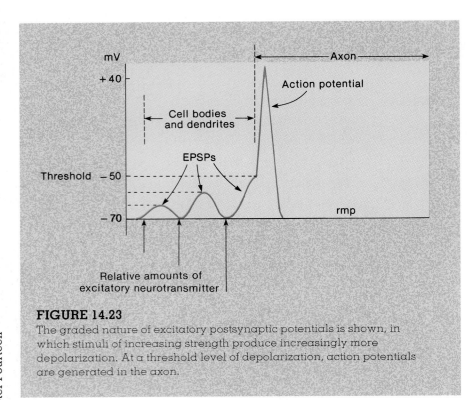

**FIGURE 14.23**
The graded nature of excitatory postsynaptic potentials is shown, in which stimuli of increasing strength produce increasingly more depolarization. At a threshold level of depolarization, action potentials are generated in the axon.

# Neurotransmitters of the Central Nervous System

*A variety of chemicals in the CNS function as neurotransmitters. Among these are ACh, dopamine, norepinephrine, and certain amino acids and polypeptides. A number of brain disorders are associated with particular neurotransmitters.*

Acetylcholine is not only the neurotransmitter of somatic motor neurons and some autonomic neurons of the PNS, it is also an important neurotransmitter within the CNS. In addition to acetylcholine, a wide variety of other molecules serve as neurotransmitters in the CNS.

## Catecholamine Neurotransmitters

**Catecholamines** (kat″ĕ-kol′ă-mēnz) constitute a group of regulatory molecules derived from the amino acid tyrosine. The catecholamines include *dopamine, norepinephrine,* and *epinephrine.* Dopamine and norepinephrine function as neurotransmitters; epinephrine and norepinephrine additionally serve as hormones secreted by the adrenal medulla. Together with a related molecule called *serotonin,* the catecholamines are included in a larger category of molecules called **monoamines.**

Like ACh, catecholamine neurotransmitters are released by exocytosis from presynaptic vesicles and diffuse across the synaptic cleft to interact with specific receptor proteins in the membrane of the postsynaptic cell. The stimulatory effects of these catecholamines, like those of ACh, are quickly inhibited. The inhibition of catecholamine action (fig. 14.24) is due to (1) reuptake of catecholamines into the axon terminals of presynaptic neurons, (2) enzymatic degradation of catecholamines in the axon terminals of presynaptic neuron endings by *monoamine oxidase* (MAO), and (3) their enzymatic degradation in the postsynaptic neuron by *catecholamine O-methyltransferase* (COMT). Drugs that inhibit MAO and COMT thus promote the effects of catecholamine action.

 *Monoamine oxidase* (MAO) is an enzyme in the axon terminals of presynaptic neurons that breaks down catecholamines and serotonin after they have been taken up from the synaptic cleft. Drugs that act as *MAO inhibitors* thus increase transmission at these synapses and have been found to aid people suffering from clinical depression. This suggests that a deficiency in monoamine neural pathways may contribute to severe depression. An MAO inhibitor (*Deprenyl*) has also been used to treat Parkinson's disease by promoting the activity of dopamine as a neurotransmitter.

Drugs that inhibit MAO promote the activity of all of the monoamines, and thus can produce undesired side effects. A newer drug, fluoxetine hydrochloride (*Prozac*), specifically blocks the reuptake of serotonin into axon terminals of presynaptic neurons; thus, it specifically promotes serotonin action and has been found to be effective in the treatment of depression.

**Dopamine and Norepinephrine** Neurons that use **dopamine** as a neurotransmitter and postsynaptic neurons with dopamine receptor proteins in their membranes can be identified in postmortem brain tissue. Through the technique of *positron emission tomography* (PET), the location of dopamine receptor proteins has also been observed in the living brain. These investigations have been spurred by the great clinical interest in the effects of **dopaminergic neurons** (those that use dopamine as a neurotransmitter).

 *Cocaine*—a stimulant related to the amphetamines in its action—is currently widely abused in the United States. Although this drug produces feelings of euphoria and social adroitness at first, continued use leads to social withdrawal, depression, and dependence upon ever-higher dosages to achieve the initial effects. Continued use also leads to serious organic disease that often results in death. The many effects of cocaine on the central nervous system appear to be

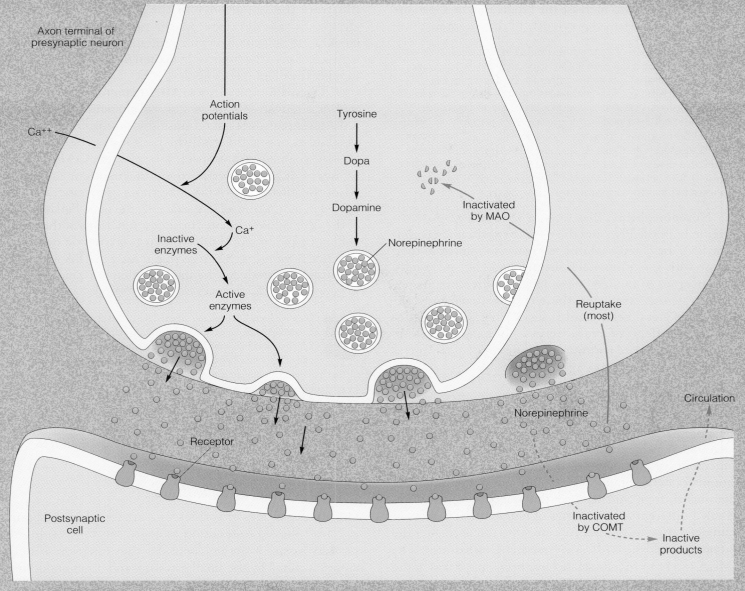

**FIGURE 14.24**

A diagram showing the production, release, and reuptake of catecholamine neurotransmitters from the axon terminal of a presynaptic neuron. The transmitters combine with receptor proteins (R) in the postsynaptic membrane. (COMT = catecholamine O-methyltransferase; MAO = monoamine oxidase.)

mediated by one primary mechanism: cocaine blocks the reuptake of the monoamines into the presynaptic axon endings. This results in overstimulation of those neural pathways that use dopamine and other monoamines as neurotransmitters.

Dopaminergic neurons are highly concentrated in the *substantia nigra* (literally, the "dark substance," so-called because it contains melanin pigment) of the brain. Many neurons in the substantia nigra send fibers to the *basal nuclei*. These large masses of cell bodies deep in the cerebrum are involved in the coordination of skeletal

movements. Medical research findings strongly indicate that **Parkinson's disease** is caused by degeneration of the dopaminergic neurons in the substantia nigra. Parkinson's disease is a major cause of neurological disability in people over the age of 60 and is associated with such symptoms as muscle tremors and rigidity, difficulty in initiating movement and speech, and other severe problems. Patients are

..............................................
Parkinson's disease: from James Parkinson, English physician, 1755–1824

treated with L-dopa to increase the production of dopamine in the brain, as described previously in this chapter.

 A side effect of L-dopa treatment in some patients with Parkinson's disease is the appearance of symptoms characteristic of *schizophrenia*. This effect is not surprising in view of the fact that the drugs used to treat schizophrenic patients (chlorpromazine and related compounds) act as specific antagonists of dopamine receptors. As might be predicted from these observations, schizophrenic patients treated with these drugs often develop symptoms of Parkinson's disease. It seems reasonable to suppose, from this evidence, that schizophrenia may be caused, at least in part, by overactivity of the dopaminergic pathways.

**Norepinephrine,** like ACh, is used as a neurotransmitter in both the PNS and the CNS. Sympathetic neurons of the PNS use norepinephrine as a neurotransmitter at their synapse with smooth muscles, cardiac muscle, and glands. Some neurons in the CNS also appear to use norepinephrine as a neurotransmitter; these neurons seem to be involved in general behavioral arousal. This would help to explain the effects of *amphetamines*, for example, which specifically stimulate pathways that use norepinephrine as a neurotransmitter.

## Amino Acid Neurotransmitters

The amino acids **glutamic acid** and **aspartic acid** function as excitatory neurotransmitters in the CNS. Indeed, glutamic acid (or *glutamate*) is the major excitatory neurotransmitter in the brain. Experiments using chemicals that differ slightly from glutamate but that mimic its actions have revealed different subcategories of glutamate receptors in brain neurons. One of these subcategories is named after the glutamate analogue *NMDA* (*N*-methyl-D-aspartate). The NMDA receptors for glutamate have been implicated in the physiology of memory and in certain types of seizures.

The amino acid **glycine** is inhibitory; instead of depolarizing the postsynaptic membrane and producing an EPSP, it hyperpolarizes the postsynaptic membrane. This hyperpolarization makes the membrane potential even more negative than it is at rest (changing the membrane potential from –65 mV to –85 mV, for example). Such hyperpolarization is known as an **inhibitory postsynaptic potential (IPSP),** and will be discussed in more detail in the next section.

The inhibitory effects of glycine are very important in the spinal cord, where they help in the control of skeletal movements. Flexion of the elbow joint, for example, involves stimulation of the biceps brachii muscle. The motor neurons that innervate this flexor muscle are stimulated in the spinal cord; the motor neurons that innervate the antagonistic extensor muscle (the triceps brachii muscle) are inhibited by IPSPs produced by glycine released from other neurons. The deadly effects of the poison *strychnine*

illustrate the importance of the inhibitory actions of glycine. Strychnine causes spastic paralysis by specifically blocking the glycine receptor proteins. Animals poisoned with strychnine die from asphyxiation due to their inability to relax the diaphragm muscle.

The neurotransmitter **GABA (gamma-aminobutyric acid)** is a derivative of the amino acid glutamic acid. GABA is the most prevalent neurotransmitter in the brain; in fact, as many as one-third of all the neurons in the brain use GABA as a neurotransmitter. Like glycine, GABA is inhibitory—it hyperpolarizes the postsynaptic membrane. Also, the effects of GABA, like those of glycine, are involved in motor control. For example, the large *Purkinje cells* of the cerebellum mediate the motor functions of the cerebellum by producing IPSPs in their postsynaptic neurons. A deficiency in those neurons that release GABA as a neurotransmitter produces the uncontrolled movements seen in people with *Huntington's chorea*.

 *Benzodiazepines* are drugs that act to increase the effectiveness of GABA in the brain and spinal cord. Since GABA inhibits the activity of spinal motor neurons, the intravenous infusion of benzodiazepines acts to inhibit the muscular spasms in *status epilepticus* and seizures resulting from drug overdose and poisons. Probably as a result of its general inhibitory effects on the brain, GABA also appears to function as a neurotransmitter involved in mood and emotion. Benzodiazepines such as *Valium* are therefore used as tranquilizers.

## Polypeptide Neurotransmitters

Many polypeptides of various sizes are found in the brain and are believed to function as neurotransmitters. Interestingly, some of the polypeptides that function as hormones secreted by the small intestine and other endocrine glands are also produced in the brain and may function there as neurotransmitters (table 14.5). For example, *cholecystokinen* (*CCK*) is secreted as a hormone from the small intestine and is released from neurons and used as a neurotransmitter in the brain. Recent evidence suggests that CCK, acting as a neurotransmitter, may promote feelings of satiety in the brain following meals. Another polypeptide found in many organs, *substance P*, functions as a neurotransmitter of pathways in the brain that mediate sensation of pain.

**Synaptic Plasticity** Although some of the polypeptides released from neurons may function as neurotransmitters in the traditional sense—by stimulating the opening of ionic gates and causing changes in the membrane potential—others may have more subtle effects that are poorly understood. **Neuromodulators** has been proposed as a name for compounds with such alternative effects. An exciting recent discovery is that some neurons in both the PNS and

## Table 14.5 Examples of chemicals that are either proven or supposed neurotransmitters

| Category | Chemicals |
|----------|-----------|
| Amines | Acetylcholine |
| | Histamine |
| | Serotonin |
| Catecholamines | Dopamine |
| | Epinephrine |
| | Norepinephrine |
| Amino acids | Aspartic acid |
| | GABA (gamma-aminobutyric acid) |
| | Glutamic acid |
| | Glycine |
| Polypeptides | Glucagon |
| | Insulin |
| | Somatostatin |
| | Substance P |
| | ACTH (adrenocorticotrophic hormone) |
| | Angiotensin II |
| | Endorphins |
| | LHRH (luteinizing hormone-releasing hormone) |
| | TRH (thyrotrophin-releasing hormone) |
| | Vasopressin (antidiuretic hormone) |

these drugs might be due to the stimulation of specific neuron pathways. This implied that opioids (LSD, mescaline, and other mind altering drugs) might resemble neurotransmitters produced by the brain.

The analgesic effects of morphine are blocked in a specific manner by a drug called *naloxone*. In the same year that opioid receptor proteins were discovered, it was found that naloxone also blocked the analgesic effect of electrical brain stimulation. Subsequently, evidence suggested that the analgesic effects of hypnosis and acupuncture could also be blocked by naloxone. These experiments indicated that the brain might be producing its own morphinelike analgesic compounds. These compounds have been identified as a family of chemicals called **endorphins** (for "endogenously produced morphinelike compounds") produced by the brain and pituitary gland. The endorphins include a group of 5-amino acid peptides called *enkephalins* (en-kef´ă-linz), which may function as neurotransmitters, and a 31-amino acid polypeptide called β-*endorphin*, produced by the pituitary gland.

Endorphins have been shown to block the transmission of pain. Current evidence for this effect includes findings from neurophysiological studies—in which endorphins blocked the release of substance P (the chemical transmitter believed to be released by neurons that mediate painful sensations)—and from behavioral studies. The pain threshold of pregnant rats, for example, was found to decrease when they were treated with naloxone.

Endorphins may also provide pleasant sensations and thus mediate reward or positive reinforcement. Overeating in genetically obese mice, for example, appears to be blocked by naloxone. It has also been found that blood levels of β-endorphin increase during exercise. Some people have suggested that the "jogger's high" may thus be due to endorphins. (Naloxone, however, does not appear to block the jogger's high.) Although evidence for this particular effect is poor, endorphins might promote some type of psychic reward system as well as analgesia.

## Nitric Oxide as a Neurotransmitter

**Nitric oxide (NO)** is the first gas known to act as a regulator molecule in the body. Produced by *nitric oxide synthetase* in the cells of many organs from the amino acid *L-arginine*, nitric oxide's actions are very different from those of the more familiar nitrous oxide ($N_2O$), or laughing gas, sometimes used by dentists.

CNS produce both a classical neurotransmitter (ACh or a catecholamine) and a polypeptide neurotransmitter. These are contained in different synaptic vesicles that can be distinguished using the electron microscope. The neuron can thus release either the classical neurotransmitter or the polypeptide neurotransmitter under different conditions.

Discoveries such as the one just described indicate that synapses have a greater capacity for alteration at the molecular level than was previously believed. This attribute has been termed **synaptic plasticity.** Synapses are also more plastic at the cellular level. There is evidence that sprouting of new axon branches can occur over short distances to produce a turnover of synapses, even in the mature CNS. This breakdown and reforming of synapses may occur within a time span of only a few hours. The physiological significance of these interesting discoveries is not yet fully understood.

**Endorphins** The ability of opium and its analogues—that is, the *opioids*—to relieve pain (promote analgesia) has been known for centuries. Morphine, for example, has long been used for this purpose. The discovery in 1973 of opioid receptor proteins in the brain suggested that the effects of

Nitric oxide has different roles in the body. Within blood vessels, it acts as a local tissue regulator that causes the smooth muscles of those vessels to relax, so that the blood vessels dilate. Within macrophages and other cells, nitric oxide helps to kill bacteria. In addition, nitric oxide acts as a neurotransmitter of certain neurons in both the PNS and CNS. It diffuses out of the presynaptic axon and into neighboring cells by simply passing through the lipid portion of the cell membranes.

In the PNS, nitric oxide is released by some neurons that innervate the gastrointestinal tract, penis, respiratory passages, and cerebral blood vessels. These are autonomic neurons that cause smooth muscle relaxation in their target organs. This can produce, for example, the engorgement of the spongy tissue of the penis with blood. In fact, scientists now believe that erection of the penis results from the action of nitric oxide released by specific parasympathetic nerves. Nitric oxide is also released as a neurotransmitter in the brain and has been implicated in the processes of learning and memory. This will be discussed in more detail in a later section.

# Synaptic Integration

*The summation of a number of EPSPs may be needed to produce a depolarization of sufficient magnitude to stimulate the postsynaptic cell. The net effect of EPSPs on the postsynaptic neuron is reduced by hyperpolarization (IPSPs), which is produced by inhibitory neurotransmitters. The activity of neurons within the central nervous system is thus the net result of both excitatory and inhibitory effects.*

Since voltage-regulated Na⁺ and K⁺ gates are absent in dendrites and cell bodies, changes in membrane potential induced by neurotransmitters in these areas do not have the all-or-none characteristics of action potentials. Synaptic potentials are graded and can add together, or summate. **Spatial summation** occurs because many presynaptic axons converge on a single postsynaptic neuron (fig. 14.25). In spatial summation, synaptic depolarizations (EPSPs) produced at different synapses may summate in the postsynaptic dendrites and cell body. In **temporal summation,** the successive activity of a given presynaptic axon, causing successive waves of transmitter release, results in the summation of EPSPs in the postsynaptic neuron. The summation of EPSPs helps to ensure that the depolarization that reaches the axon hillock will be sufficient to generate new action potentials in the postsynaptic neuron (fig. 14.26).

When a presynaptic neuron is experimentally stimulated at a high frequency, for even as short a time as a few seconds, the excitability of the synapse is enhanced—or potentiated—when this neuron pathway is subsequently stimulated. The improved efficacy of synaptic transmission may

last for hours or even weeks and is called **long-term potentiation (LTP).** Long-term potentiation may favor transmission along frequently used neural pathways, and thus may represent a mechanism of neural "learning." It is interesting in this regard that LTP has been observed in the hippocampus of the brain, which is an area implicated in memory storage (chapter 15).

Most of the neural pathways in the hippocampus area of the brain use glutamate as their neurotransmitter. There are two subclasses of glutamate receptors: NMDA (as previously mentioned)

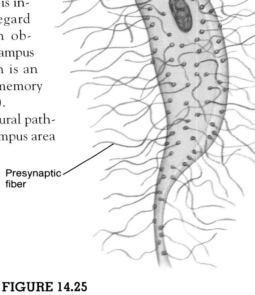

Presynaptic fiber

**FIGURE 14.25**
A diagram illustrating the convergence of large numbers of presynaptic fibers on the cell body of a spinal motor neuron.

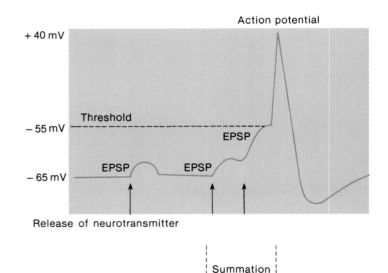

**FIGURE 14.26**
Excitatory postsynaptic potentials (EPSPs) can summate over distance (spatial summation) and time (temporal summation). When summation results in a threshold level of depolarization at the axon hillock, voltage-regulated Na⁺ gates are opened and an action potential is produced.

and non-NMDA. Binding of glutamate to NMDA receptors is not required for normal synaptic transmission. It is the binding of glutamate to non-NMDA receptors that produces the depolarization of the postsynaptic neuron required for the production of action potentials in this neuron. However, when this depolarization occurs at the same time that other glutamate molecules bind to the NMDA receptors, channels for $Ca^{++}$ open in the postsynaptic membrane. The diffusion of $Ca^{++}$ into the postsynaptic neurons may induce LTP by making this neuron more sensitive to subsequent stimulation.

The inward diffusion of $Ca^{++}$ also activates nitric oxide synthetase, causing the postsynaptic neuron to produce and release nitric oxide. The nitric oxide released from the postsynaptic neuron may then act as a "retrograde transmitter" and diffuse to the presynaptic neuron, where it could stimulate the release of more glutamate transmitter. In this way, synaptic transmission during LTP could be strengthened first postsynaptically and then presynaptically. It should be noted that, while much evidence supports the existence of a retrograde transmitter in LTP, the identity of nitric oxide as that transmitter is currently controversial.

## Synaptic Inhibition

Although most neurotransmitters depolarize the postsynaptic membrane (produce EPSPs), some have the opposite effect. These inhibitory neurotransmitters cause *hyperpolarization* of the postsynaptic membrane; they make the inside of the membrane more negative than it is at rest. Since hyperpolarization (from –65 mV to –85 mV, for example) drives the membrane potential farther from the threshold depolarization required to stimulate action potentials, such hyperpolarization inhibits the activity of the postsynaptic neuron. Hyperpolarizations produced by neurotransmitters are therefore called *inhibitory postsynaptic potentials* (IPSPs). The inhibition produced in this way is called **postsynaptic inhibition.**

Excitatory and inhibitory inputs (EPSPs and IPSPs) to a postsynaptic neuron can summate in an algebraic fashion (fig. 14.27). The effects of IPSPs in this way reduce, or may even eliminate, the ability of EPSPs to generate action potentials in the postsynaptic cell. Considering that the nervous system contains approximately $10^{12}$ neurons and that a given neuron may receive as many as 1000 presynaptic inputs, it is apparent that there are tremendous possibilities for synaptic integration.

The neurotransmitters GABA and glycine are exclusively inhibitory. They hyperpolarize central neurons by opening chloride gates. As a result, chloride ions diffuse into the neuron and make the membrane potential more negative on the inside. The physiological significance of these inhibitory effects has already been discussed in the section on amino acid neurotransmitters.

Although acetylcholine produces depolarization at the neuromuscular junction and stimulates skeletal muscle contraction, ACh has the opposite effect—hyperpolarization—in the heart. This is because the combination of ACh with its receptor protein in the myocardial cells causes opening of only $K^+$ gates. The outward diffusion of $K^+$, in turn, produces hyperpolarization. The action of ACh in this way causes a slowing of the heart rate (discussed in chapter 21).

In **presynaptic inhibition** (fig. 14.28), the amount of excitatory neurotransmitter released at an axon terminal is decreased by a second neuron, whose axon synapses with the axon of the first neuron (an axoaxonic synapse). The neurotransmitter released at this axoaxonic synapse partially depolarizes the axon of the first neuron, bringing it closer to threshold and making it easier to "fire" action potentials. The amplitude of these action potentials (subtracting the new lower potential from the +40-mV "top" of the action potential), however, is lower than normal. The smaller action potentials that result cause the release of lesser amounts of excitatory neurotransmitter by the first neuron, so that its effect on its postsynaptic cell is inhibited.

# Clinical Considerations

The clinical aspects of the nervous system are extensive and usually complex. Numerous diseases and developmental problems directly involve the nervous system, and it is indirectly involved with most diseases because of the perception of pain. In this section, we will consider only a few of the diseases involving developmental problems of the nervous system and clinical aspects of nerve conduction and synaptic transmission. Other clinical aspects of neurology will be discussed in later chapters.

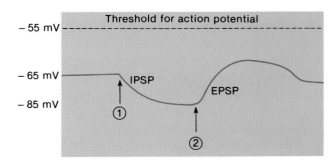

**FIGURE 14.27**
An inhibitory postsynaptic potential (IPSP) makes the inside of the postsynaptic membrane more negative than the resting potential—it hyperpolarizes the membrane. Subsequent or simultaneous excitatory postsynaptic potentials (EPSPs), which are depolarizations, must thus be stronger to reach the threshold required to generate action potentials at the axon hillock.

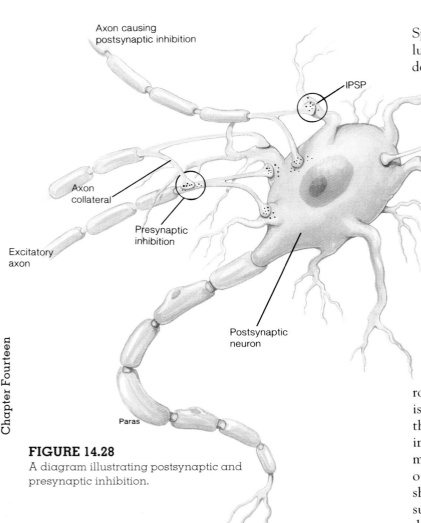

Axon causing
postsynaptic inhibition

IPSP

Axon
collateral

Presynaptic
inhibition

Excitatory
axon

Postsynaptic
neuron

Paras

**FIGURE 14.28**
A diagram illustrating postsynaptic and
presynaptic inhibition.

## Developmental Problems

Congenital malformations of the CNS are common and
frequently involve overlying bone, muscle, and connective
tissue. The more severe abnormalities make life impossible; those that are less severe frequently result in functional disability.

Most congenital malformations of the nervous system
occur during the sensitive embryonic period. Neurological
malformations are generally caused by genetic abnormalities but also may result from environmental factors, such as
anoxia, infectious agents, drugs, and ionizing radiation.

**Spina bifida** is a defective fusion of the vertebral elements and may or may not involve the spinal cord. *Spina bifida occulta* is the most common and least serious type of
spina bifida. This defect usually involves few vertebrae, is
not externally apparent except for perhaps a pigmented spot
with a tuft of hair, and usually does not cause neurological
disturbances. *Spina bifida cystica*, a severe type of spina bifida, is a saclike protrusion of skin and underlying meninges
that may contain portions of the spinal cord and nerve roots.

Spina bifida cystica is most common in the lower thoracic,
lumbar, and sacral regions. The position and extent of the
defect determines the degree of neurological impairment.

**Anencephaly** (*an''en-sef'ă-le*) is a
markedly defective development of the brain
and the surrounding cranial bones. Anencephaly occurs in 1 per 1000 births and
makes sustained extrauterine life impossible. This congenital defect apparently results from the failure of
the neural folds at the cranial portion of the neural plate to fuse and
form the prosencephalon. **Micro-
cephaly** is an uncommon condition
in which brain development is not
completed. If enough neurological tissue is present, the infant will survive
but will be severely mentally retarded.

## Diseases of the Myelin Sheath

**Multiple sclerosis (MS)** is a relatively common neurological disease in people between the ages of 20 and 40. It
is a chronic, degenerating, remitting, and relapsing disease
that progressively destroys the myelin sheaths of neurons
in multiple areas of the CNS. Initially, lesions form on the
myelin sheaths and soon develop into hardened *scleroses,*
or scars (hence the name). Destruction of the myelin
sheaths prohibits the normal conduction of impulses, resulting in a progressive loss of functions. Because myelin
degeneration is widely distributed, MS has a wider variety of
symptoms than any other neurological disease. This characteristic, coupled with remissions, frequently causes mis-
diagnosis of this disease.

During the early stages of MS, many patients are considered neurotic because the symptoms they report vary so
widely and come and go quickly. As the disease progresses,
the symptoms may include double vision (diplopia), spots
in the visual field, blindness, tremor, numbness of the appendages, and locomotor difficulty. Eventually the patient
becomes bedridden, and death may occur anytime from 7 to
30 years after the initial onset of symptoms.

In **Tay–Sachs disease,** the myelin sheaths are destroyed
by the excessive accumulation of one of the lipid components of the myelin. This results from an enzyme defect
caused by the inheritance of genes carried by the parents in
a recessive state. Tay–Sachs disease, which is inherited primarily by individuals of Eastern European Jewish descent,
appears before the infant is a year old. It causes blindness,
loss of mental and motor ability, and ultimately death by

...........................................

multiple sclerosis: L. *multiplus,* many parts; Gk. *skleros,* hardened
Tay–Sachs disease: from Warren Tay, English physician, 1843–1927, and Bernard
Sachs, American neurologist, 1858–1944

the age of 3. Potential parents can tell if they are carriers for this condition by having a special blood test for the defective enzyme.

## Problems of Neuromuscular Transmission

The muscle weakness associated with the disease **myasthenia gravis** results because ACh receptors are blocked by antibodies secreted by the immune system. Paralysis in people who eat shellfish poisoned with *saxitoxin*, produced by the unicellular organisms that cause the red tides, results from inhibition of the chemically regulated $Na^+$ gates at the neuromuscular junction. A similar inhibition and paralysis is produced by *tetrodoxin*, a poison found in the pufferfish.

**Alzheimer's Disease**    *Alzheimer's disease* is the most common cause of senile dementia, which often begins in middle age and produces progressive mental deterioration. The cause of Alzheimer's disease is not known, but there is evidence that it is associated with a loss of neurons that use acetylcholine as a neurotransmitter. These axons terminate in the hippocampus and cerebral cortex of the brain, which are areas concerned with memory storage.

Alzheimer's disease: from Alois Alzheimer, German neurologist, 1864–1915

Alzheimer's is associated with a deficiency in the enzyme responsible for producing acetylcholine from acetyl coenzyme A and choline. Treatment with different drugs that inhibit the activity of acetylcholinesterase has been reported to be fairly effective. By inhibiting this enzyme, the breakdown of ACh in synapses is reduced, so that the action of ACh is improved.

Autopsies of people who have died of Alzheimer's disease reveal "neuritic plaques" that are composed of degenerating axons and deposits of amyloid protein. Similar plaques are seen in the brains of people with Down syndrome, a genetic disease caused by an extra chromosome number 21. Recently, scientists have discovered that the gene that codes for the amyloid protein in people with Alzheimer's disease is located in chromosome number 21, suggesting that Alzheimer's may be caused by genetic defects located on this chromosome.

## Blood-Brain Barrier

Substances that are more lipid-soluble can leave the blood and enter the brain much more readily than can more polar, water-soluble molecules. Injection of a hypertonic solution (of glucose, for example) has been shown to reduce the blood-brain barrier and thus increase the ease with which molecules can enter the brain. Then, chemotherapeutic nerve drugs (for example, drugs that treat disorders such as brain tumors) can more effectively enter the brain.

## Chapter Summary

### Neurons and Neuroglia (pp. 345–352)

1. Every neuron contains a cell body, dendrites, and an axon.
2. On the basis of the number of processes extending from the cell body, neurons can be classified as pseudounipolar, bipolar, or multipolar.
3. Neurons in the PNS that conduct impulses into the CNS are sensory; those that conduct impulses out of the CNS are motor.
4. Motor neurons that innervate skeletal muscles are somatic; those that innervate the heart, smooth muscles, and glands are autonomic.
5. Neuroglia include Schwann cells (neurolemmocytes) and satellite cells in the PNS; in the CNS they include oligodendrocytes, microglia, astrocytes, and ependymal cells.

   a. In the PNS, Schwann cells surround axons to form a sheath of Schwann (neurolemmal sheath) that provides a continuous basement membrane around the axon.
   b. Many axons of the PNS have a myelin sheath, formed by successive wrappings of the Schwann cell membrane.
   c. In the CNS, myelin sheaths are formed by oligodendrocytes.
   d. The sheath of Schwann allows damaged peripheral axons to regenerate; the absence of a sheath of Schwann in the CNS hinders regeneration of central axons.
   e. Astrocytes extend vascular processes that surround capillaries in the brain and help to develop the blood-brain barrier.

### Action Potentials in Neurons (pp. 352–357)

1. The permeability of the axon membrane to $Na^+$ and $K^+$ is regulated by gates at the openings of ion channels.
   a. When the membrane is depolarized to a threshold level, the $Na^+$ gates open first, allowing $Na^+$ to flow into the axon by diffusion.
   b. This is followed quickly by the opening of $K^+$ gates, allowing $K^+$ to diffuse out of the axon.
2. The opening of voltage-regulated gates and the resulting flow of ions produces an action potential.
   a. The inward diffusion of $Na^+$ causes a reversal of the membrane potential from −65 mV to +40 mV.
   b. The opening of $K^+$ gates and outward diffusion of $K^+$ causes the reestablishment of the resting

# Interactions of the Nervous System with Other Body Systems

## Integumentary System
- Supports and protects peripheral receptors
- Influences secretions from integumentary glands and contractions of arrector pili muscles

## Skeletal System
- Supports and protects the brain and spinal cord
- Stores calcium needed for neural function
- Innervates bones and monitors movements within joints

## Muscular System
- Generates body heat to maintain constant temperature for neural function
- Proprioceptors transmit impulses from muscles to brain
- Innervates muscles for autonomic and voluntary muscle contractions
- Monitors muscle activities

## Endocrine System
- Hormones augment and sustain autonomic stimuli to body organs
- Innervates endocrine glands causing rapid, autonomic secretion of hormones

## Circulatory System
- Transports $O_2$ and $CO_2$, nutrients, and fluids to and from the brain and spinal cord
- Innervates the heart and blood vessels to modify heart rate and blood pressure

## Lymphatic System
- Protects against infections within the brain and spinal cord
- Innervates lymphoid organs

## Respiratory System
- Provides $O_2$ and eliminates $CO_2$
- Respiratory centers within brain stem control respiratory rates and depth of respiration

## Urinary System
- Eliminates metabolic wastes
- Regulates pH, body fluids, and electrolyte concentrations
- Innervates organs of urinary system to control urination
- Modifies renal blood pressure

## Digestive System
- Provides nutrients for growth, maintenance, and repair of nervous system
- Innervates digestive organs and autonomically regulates GI tract movements and secretions

## Reproductive System
- Gonads produce sex hormones that influence brain development and sexual behavior
- Innervates reproductive organs to control sexual function

membrane potential in a process called repolarization.

3. Action potentials are all-or-none events.
   a. A depolarization stimulus that is lower than threshold has no effect.
   b. Any depolarization stimulus higher than a threshold level will cause an action potential of maximal amplitude.
   c. Stronger stimuli thus do not produce stronger action potentials; rather, they cause action potentials to be produced at a greater frequency (more per second).
4. The axon membrane is in a refractory period while producing an action potential.
   a. While in the refractory period, the membrane cannot be stimulated (absolute refractory period) or can only be stimulated by a very large stimulus (relative refractory period).
   b. This prevents action potentials from being able to summate or interfere with each other.
5. One action potential serves as the depolarization stimulus for production of the next action potential in the axon.
   a. In unmyelinated axons, action potentials are produced fractions of a micrometer apart.
   b. In myelinated axons, action potentials are produced only at the nodes of Ranvier; this saltatory conduction is faster than conduction in unmyelinated axons.

## The Synapse (pp. 358–360)

1. Gap junctions are electrical synapses and are found in cardiac muscle, smooth muscle, and in some synapses in the CNS.

2. In chemical synapses, neurotransmitters are packaged in synaptic vesicles and released into the synaptic cleft.

### Synaptic Transmission by Acetylcholine (pp. 360–364)

1. The combination of ACh with its receptor protein in the postsynaptic membrane causes chemically regulated gates to open, which produces depolarizations known as excitatory postsynaptic potentials (EPSPs).
   a. Chemically regulated gates allow $Na^+$ and $K^+$ to diffuse simultaneously, resulting in a graded depolarization with a maximum amplitude of 0 volts.
   b. The extent of the depolarization depends on the amount of transmitter released; EPSPs are therefore graded.
   c. EPSPs have no refractory period; they are therefore capable of summation.
   d. Unlike action potentials, EPSPs cannot be self-regenerated; they therefore decrease in amplitude with distance as they are conducted.
2. ACh is inactivated by acetylcholinesterase; this acts to stop the stimulation.
3. EPSPs produced at synapses in the dendrites or cell body travel to the axon hillock.
   a. Here, they stimulate the opening of voltage-regulated gates and generate action potentials in the axon.
   b. Summation of EPSPs, through both spatial and temporal summation, occurs at the axon hillock and influences the frequency of action potentials produced at the axon of the postsynaptic neuron.

### Neurotransmitters of the Central Nervous System (pp. 364–368)

1. Catecholamine neurotransmitters include dopamine and norepinephrine.
   a. These neurotransmitters act via second messengers, such as cyclic AMP.
   b. The catecholamine neurotransmitters are inactivated by the reuptake and degradation of the neurotransmitters by the action of monoamine oxidase.
2. In addition to the classical neurotransmitters, neurons produce many other chemicals that are believed to have a neurotransmitter function.
   a. Proven neurotransmitters include glycine and GABA, which have inhibitory effects via the production of hyperpolarizations, or inhibitory postsynaptic potentials (IPSPs).
   b. Substances believed to act as neurotransmitters include the endorphins and many other polypeptides.

### Synaptic Integration (pp. 368–369)

1. Spatial and temporal summation of EPSPs allows a sufficient depolarization to be produced to cause the stimulation of action potentials in the postsynaptic neuron.
2. Neurotransmitters that cause hyperpolarization of the postsynaptic membrane produce inhibitory postsynaptic potentials (IPSPs).
   a. The production of IPSPs is called postsynaptic inhibition.
   b. IPSPs and EPSPs from different synaptic inputs can summate.
   c. Presynaptic inhibition occurs in an axoaxonic synapse and reduces the amount of neurotransmitter released by the inhibited neuron.

# Review Activities

## Objective Questions

1. The neuroglial cells that form myelin sheaths in the peripheral nervous system are
   a. oligodendrocytes.
   b. satellite cells.
   c. Schwann cells.
   d. astrocytes.
   e. microglia.
2. A collection of neuron cell bodies located outside the CNS is called
   a. a tract.
   b. a nerve.
   c. a nucleus.
   d. a ganglion.

3. Which of the following neurons are pseudounipolar?
   a. sensory neurons
   b. somatic motor neurons
   c. neurons in the retina
   d. autonomic motor neurons
4. Depolarization of an axon is produced by
   a. inward diffusion of $Na^+$.
   b. active extrusion of $K^+$.
   c. outward diffusion of $K^+$.
   d. inward active transport of $Na^+$.
5. Repolarization of an axon during an action potential is produced by
   a. inward diffusion of $Na^+$.
   b. active extrusion of $K^+$.
   c. outward diffusion of $K^+$.
   d. inward active transport of $Na^+$.

6. As the strength of a depolarizing stimulus to an axon is increased,
   a. the amplitude of action potentials increases.
   b. the duration of action potentials increases.
   c. the speed with which action potentials are conducted increases.
   d. the frequency with which action potentials are produced increases.
7. The conduction of action potentials in a myelinated axon is
   a. saltatory.
   b. without decrement.
   c. faster than in an unmyelinated fiber.
   d. all of the above.

8. Which of the following is *not* a characteristic of synaptic potentials?
   a. They are all-or-none in amplitude.
   b. They decrease in amplitude with distance.
   c. They are produced in dendrites and cell bodies.
   d. They are graded in amplitude.
   e. They are produced by chemically regulated gates.

9. Which of the following is *not* a characteristic of action potentials?
   a. They are produced by voltage-regulated gates.
   b. They are conducted without decrement.
   c. Na+ and K+ gates open at the same time.
   d. The membrane potential reverses polarity during depolarization.

10. A drug that inactivates acetylcholinesterase
    a. inhibits the release of ACh from presynaptic endings.
    b. inhibits the attachment of ACh to its receptor protein.
    c. increases the ability of ACh to stimulate muscle contraction.
    d. all of the above apply.

11. Postsynaptic inhibition is produced by
    a. depolarization of the postsynaptic membrane.
    b. hyperpolarization of the postsynaptic membrane.
    c. axoaxonic synapses.
    d. post-tetanic potentiation.

12. Hyperpolarization of the postsynaptic membrane in response to glycine or GABA is produced by the opening of
    a. Na+ gates.
    b. K+ gates.
    c. Ca++ gates.
    d. Cl− gates.

13. The absolute refractory period of a neuron
    a. is due to the high negative polarity of the inside of the neuron.
    b. occurs only during the repolarization phase.
    c. occurs only during the depolarization phase.
    d. occurs during depolarization and the first part of the repolarization phase.

14. The summation of EPSPs from many presynaptic axons converging onto one postsynaptic neuron is called
    a. spatial summation.
    b. long-term potentiation.
    c. temporal summation.
    d. synaptic plasticity.

### Essay Questions

1. Compare the characteristics of action potentials with those of synaptic potentials.
2. Explain how voltage-regulated gates produce an all-or-none action potential.
3. Explain how action potentials are regenerated along an axon.
4. Explain why conduction in a myelinated axon is faster than in an unmyelinated axon.
5. Trace the course of events between the production of an EPSP and the generation of action potentials at the axon hillock. Describe the effect of spatial and temporal summation on this process.
6. Explain how an IPSP is produced and how IPSPs can inhibit activity of the postsynaptic neuron.

## Gundy/Weber Software ⌷

The tutorial software accompanying Chapter 14 is Volume 5—Nervous System.

## Explorations ✪

Two modules of correlating material are available from the Wm. C. Brown CD–ROM: *Explorations*. They are #8 Nerve Condition and #9 Synaptic Transmission.

# chapter fifteen

# central nervous system

## objectives

- Describe the general characteristics of the brain and spinal cord.
- Discuss the basic metabolic demands of the brain.
- Describe the structure of the cerebrum and list the functions of the cerebral lobes.
- Define the term *electroencephalogram* and discuss its clinical importance.
- Describe the locations of the language centers and compare short-term and long-term memory.
- List the autonomic functions of the thalamus and the hypothalamus.
- Describe the location and structure of the pituitary gland.
- List the structures of the mesencephalon and describe their function.

- Describe the location and structure of the pons and cerebellum and list their functions.
- Describe the location and structure of the medulla oblongata and list its functions.
- Define the term *reticular formation* and explain its function.
- Describe the position of the meninges as they protect the CNS.
- Describe the locations of the ventricles of the brain.
- Discuss the formation, function, and flow of cerebrospinal fluid.
- Describe the structure of the spinal cord.
- Describe the arrangement of ascending and descending tracts.

# Characteristics of the Central Nervous System

*The central nervous system, composed of gray and white matter, is covered with meninges and is bathed in cerebrospinal fluid. The tremendous metabolic rate of the brain requires a continuous flow of blood amounting to approximately 20% of the total resting cardiac output.*

The central nervous system (CNS) consists of the brain and spinal cord. The entire delicate CNS is protected by a bony encasement—the cranium surrounding the brain (fig. 15.1) and the vertebral column surrounding the spinal cord. The **meninges** (*mĕ-nin ´jēz*) form a protective membrane between the bone and the soft tissue of the CNS. The CNS is bathed in **cerebrospinal** (*ser ´´ĕ-bro-spi ´nal*) **fluid** that circulates within the hollow **ventricles** of the brain, the **central canal** of the spinal cord, and the **subarachnoid** (*sub ´´ă-rak ´noid*) **space** surrounding the entire CNS.

The CNS is composed of gray and white matter. **Gray matter** consists of either nerve cell bodies and dendrites or bundles of unmyelinated axons and neuroglia. The gray matter of the brain exists as the outer convoluted **cortex layer** of the cerebrum and cerebellum. In addition, specialized gray matter clusters of nerve cells called **nuclei** are found deep within the white matter. **White matter** forms the tracts within the CNS and consists of aggregations of myelinated axons and associated neuroglia.

The brain of an adult weighs nearly 1.5 kg (3–3.5 lb) and is composed of an estimated 100 billion neurons. Neurons communicate with one another by means of innumerable synapses between the axons and dendrites within the brain. Neurotransmission within the brain is regulated by many different neurotransmitter chemicals (chapter 14) that are found in specific brain regions and tracts.

The brain has a tremendous metabolic rate and needs a continuous supply of oxygen and nutrients. The brain accounts for only 2% of a person's body weight, and yet it receives approximately 20% of the total resting cardiac output. This amounts to a flow of about 750 ml of blood per minute. The volume remains relatively constant even during physical or mental activity. This continuous flow is so crucial that a failure of cerebral circulation for as short an interval as 10 seconds causes unconsciousness. The brain is composed of perhaps the most sensitive tissue of the body. Due to its high metabolic rate, not only does the brain require continuous oxygen, but it also requires a continuous nutrient supply and rapid removal of wastes. The brain is also very sensitive to toxins and drugs. The cerebrospinal fluid aids the metabolic needs of the brain through the distribution of nutrients and the removal of wastes. Cerebrospinal fluid also maintains a

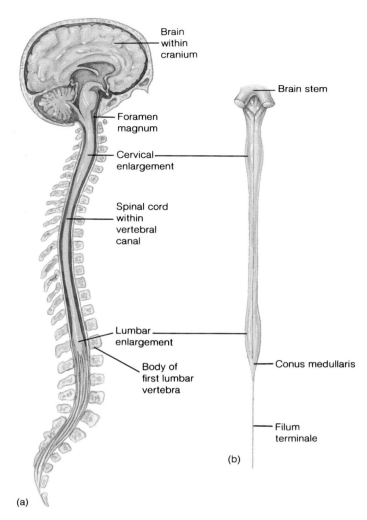

(a)

(b)

**FIGURE 15.1**

The central nervous system consists of the brain and the spinal cord, both of which are covered with meninges and bathed in cerebrospinal fluid. (*a*) A sagittal section showing the brain within the cranium of the skull and the spinal cord within the vertebral canal. (*b*) The spinal cord, shown in a posterior view, extends from the level of the foramen magnum to the first lumbar vertebra (L1).

protective homeostatic environment within the brain. The blood-brain barrier (chapter 14) and the secretory activities of neural tissue also help to maintain homeostasis. The brain has an extensive vascular supply through the paired internal carotid and vertebral arteries that unite at the cerebral arterial circle (circle of Willis) (see chapter 21 and fig. 21.21).

 The brain of a newborn is especially sensitive to oxygen deprivation or to excessive oxygen. If complications arise during childbirth and the oxygen supply from the mother's blood to the baby is interrupted while it is still in the birth canal, the infant may be stillborn or suffer brain damage that can result in cerebral palsy, epilepsy, paralysis, or mental retardation. Excessive oxygen administered to a newborn may cause blindness.

# Table 15.1 Derivation and functions of the major brain structures

| | Region | Structure | Function |
|---|---|---|---|
| Prosencephalon (forebrain) | Telencephalon | Cerebrum | Control of most sensory and motor activities; reasoning, memory, intelligence, etc.; instinctual and limbic functions |
| | Diencephalon | Thalamus | Relay center: all impulses (except olfactory) going into cerebrum synapse here; some sensory interpretation; initial autonomic response to pain |
| | | Hypothalamus | Regulation of renal water flow, body temperature, hunger, heartbeat, etc.; control of secretory activity in anterior pituitary gland; instinctual and limbic functions |
| | | Pituitary gland | Regulation of other endocrine glands |
| Mesencephalon (midbrain) | Mesencephalon | Superior colliculi | Visual reflexes (hand-eye coordination) |
| | | Inferior colliculi | Auditory reflexes |
| | | Cerebral peduncles | Reflex coordination; contain many motor fibers |
| Rhombencephalon (hindbrain) | Metencephalon | Cerebellum | Balance and motor coordination |
| | | Pons | Relay center; contains nuclei (pontine nuclei) |
| | Myelencephalon | Medulla oblongata | Relay center; contains many nuclei; visceral autonomic center (e.g., respiration, heart rate, vasoconstriction) |

Measurable increases in regional blood flow within the brain and in glucose and oxygen metabolism accompany mental functions, including perception and emotion. These metabolic changes can be assessed through the use of *positron emission tomography* (*PET*). The technique of a PET scan (fig. 15.2) is based on injecting radioactive tracer molecules labeled with carbon-11, fluorine-18, and oxygen-15 into the bloodstream and photographing the gamma rays that are emitted from the subject's brain through the skull. PET scans are of value in studying neurotransmitters and neuroreceptors, as well as the substrate metabolism of the brain.

The development of the five basic regions of the brain—telencephalon, diencephalon, mesencephalon, metencephalon, and myelencephalon—is discussed in the boxed developmental material on pages 378 and 379. The distinct functional structures that are formed from these regions (table 15.1) will be considered in greater detail in the following sections.

 Mitotic activity within nervous tissue is completed during prenatal development. Thus, a person is born with all the neurons he or she is capable of producing. However, nervous tissue continues to grow and to specialize after a person is born, particularly in the first several years of postnatal life. The extent to which the nervous tissue is altered during the aging process is not known. It has been estimated that as many as 100,000 neurons die each day

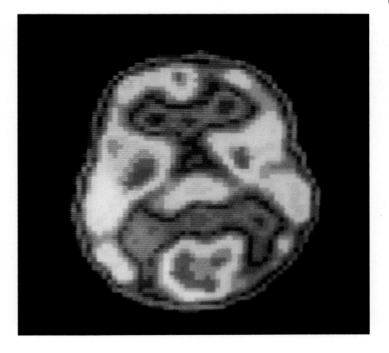

**FIGURE 15.2**

Positron emission tomographic (PET) brain scan (in transverse section) of an unmedicated schizophrenic patient. Red areas indicate high glucose use (uptake of 18–F-deoxyglucose). The scan shows highest glucose uptake in the posterior region, where the brain's visual center is located.

## Development of the Brain

The first indication of nervous tissue development occurs about 17 days following conception, when a thickening appears along the entire dorsal length of the embryo. This thickening, called the **neural plate** (fig. 1), differentiates and eventually gives rise to all of the **neurons** and to most of **neuroglial cells,** which are the supporting cells of the nervous system. As development progresses, the midline of the neural plate invaginates to become the **neural groove.** At the same time, there is a proliferation of cells along the lateral margins of the neural plate, which become the thickened **neural folds.** The neural groove continues to deepen as the neural folds elevate. By day 20, the neural folds have met and fused at the midline and the neural groove has become a **neural tube.**

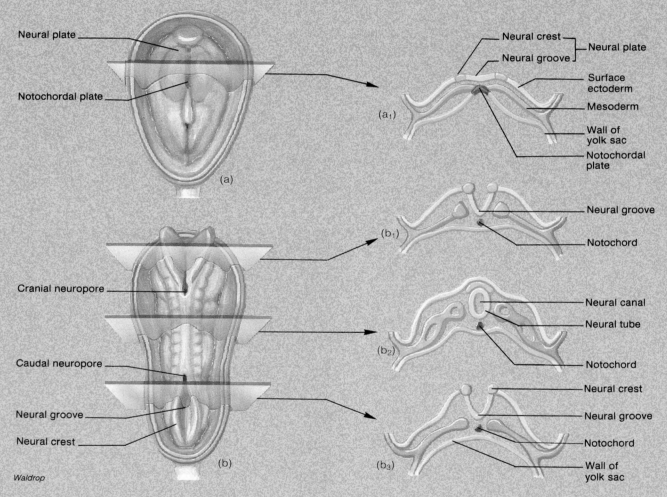

Waldrop

## FIGURE 1

The early development of the nervous system from embryonic ectoderm. (a) A dorsal view of an 18-day-old embryo showing the formation of the neural plate and the position of a transverse cut indicated in (a₁). (b) A dorsal view of a 22-day-old embryo showing cranial and caudal neuropores and the positions of three transverse cuts indicated in (b₁b₃). (Note the amount of fusion of the neural tube at the various levels of the 22-day-old embryo. Note also the relationship of the notochord to the neural tube.)

For a short time, the neural tube is open both cranially and caudally. These openings, called **neuropores,** close during the fourth week. Once formed, the neural tube separates from the surface ectoderm and eventually develops into the central nervous system (brain and spinal cord). The **neural crest** forms from the neural folds as they fuse longitudinally along the dorsal midline.

Most of the peripheral nervous system (cranial and spinal nerves) forms from the neural crest. Some neural crest cells break away from the main tissue mass

and migrate to other locations, where they differentiate into motor nerve cells of the sympathetic nervous system or into Schwann cells (neurolemmocytes), which are a type of neuroglial cell important in the peripheral nervous system.

The brain begins its embryonic development as the cephalic end of the neural tube starts to grow rapidly and to differentiate (fig. 2). By the middle of the fourth week, three distinct swellings are evident: the **prosencephalon** (*pros´en-sef´ă-lon*) (forebrain), the

**mesencephalon** (midbrain), and the **rhombencephalon** (hindbrain). Further development during the fifth week results in the formation of five specific regions: The **telencephalon** and the **diencephalon** (*di´en-sef´ă-lon*) derive from the forebrain, the mesencephalon remains unchanged, and the **metencephalon** and **myelencephalon** form from the hindbrain. The caudal portion of the myelencephalon is continuous with and resembles the spinal cord.

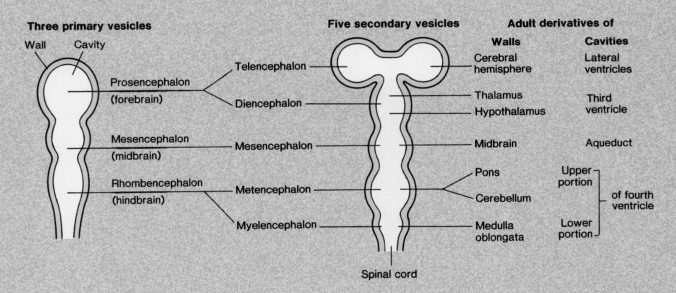

## FIGURE 2

The developmental sequence of the brain. During the fourth week, the three principal regions of the brain are formed. During the fifth week, a five-regioned brain develops and specific structures begin to form.

of our adult life. Recent studies, however, show that such estimates are unfounded. These studies indicate that relatively few neural cells are lost during the normal aging process. Neurons are, however, extremely sensitive and susceptible to various drugs or interruptions of vascular supply, such as those caused by strokes or other cardiovascular diseases.

There is evidence that aging alters neurotransmitters. Age-related conditions such as depression or specific diseases such as *Alzheimer's disease* may be caused by an imbalance of neurotransmitter chemicals. Changes in sleeping patterns in elderly people also probably result from neurotransmitter problems.

# Cerebrum

*The cerebrum, consisting of five paired lobes within two convoluted hemispheres, is concerned with higher brain functions, such as the perception of sensory impulses, the instigation of voluntary movement, the storage of memory, thought processes, and reasoning ability. The cerebrum is also concerned with instinctual and limbic (emotional) functions.*

## Structure of the Cerebrum

The **cerebrum** (*ser´ĕ-brum*) located in the region of the telencephalon, is the largest and most obvious portion of the brain (fig. 15.3). It accounts for about 80% of the mass of the brain and is responsible for the higher mental functions, including memory and reason. The cerebrum consists of the **right** and **left hemispheres,** which are incompletely separated by a **longitudinal cerebral fissure** (fig. 15.4). Portions of the two hemispheres are connected internally by the **corpus callosum** (*că-lo´sum*), a large tract of white matter (see fig. 15.3c). A portion of the meninges, called the **falx** (*falks*) **cerebri** extends into the longitudinal fissure. Each cerebral hemisphere contains a central cavity called the **lateral ventricle** (fig. 15.5), which is lined with ependymal cells and filled with cerebrospinal fluid.

 The two cerebral hemispheres carry out different functions. In most people, the left hemisphere controls analytical and verbal skills, such as reading, writing, and mathematics. The right hemisphere is the source of spatial and artistic kinds of intelligence. The corpus callosum integrates attention and awareness between the two hemispheres and permits a sharing of learning and memory. Severing the corpus callosum is a radical treatment to control severe epileptic seizures. Although this surgery has proven successful, it results in the cerebral hemispheres functioning as separate structures, each with its own memories and thoughts, competing for control. A more recent and effective technique of controlling epileptic seizures is a precise laser treatment of the corpus callosum.

........................................

cerebrum: L. *cerebrum*, brain

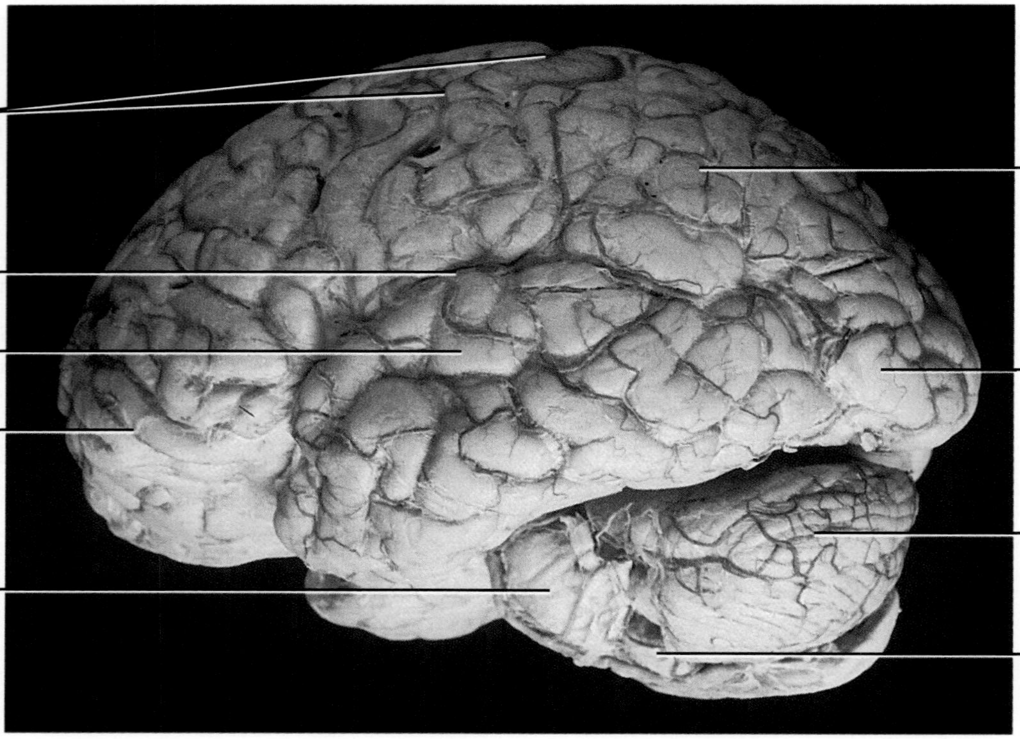

Central sulcus

Lateral sulcus

Temporal lobe

Frontal lobe

Pons

Parietal lobe

Occipital lobe

Cerebellum

Medulla oblongata

(a)

**FIGURE 15.3**
The brain. (*a*) A lateral view, (*b*) an inferior view, (*c*) a sagittal view, and (*d*) a sagittal view of a magnetic resonance image (MRI) of the skull, brain, and cervical portion of the spinal cord.

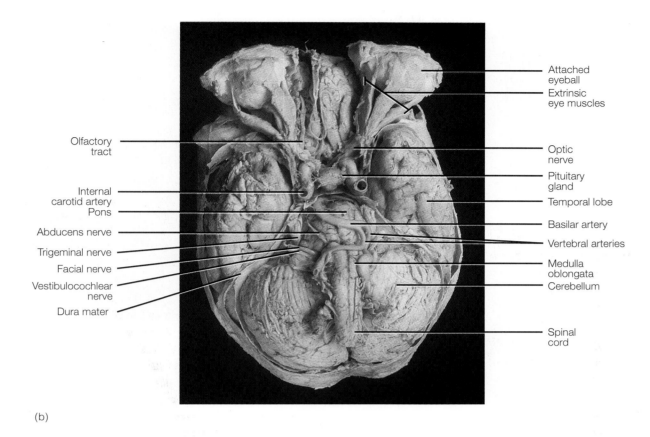

Attached eyeball

Extrinsic eye muscles

Olfactory tract

Optic nerve

Pituitary gland

Internal carotid artery

Temporal lobe

Pons

Basilar artery

Abducens nerve

Vertebral arteries

Trigeminal nerve

Facial nerve

Medulla oblongata

Vestibulocochlear nerve

Cerebellum

Dura mater

Spinal cord

(b)

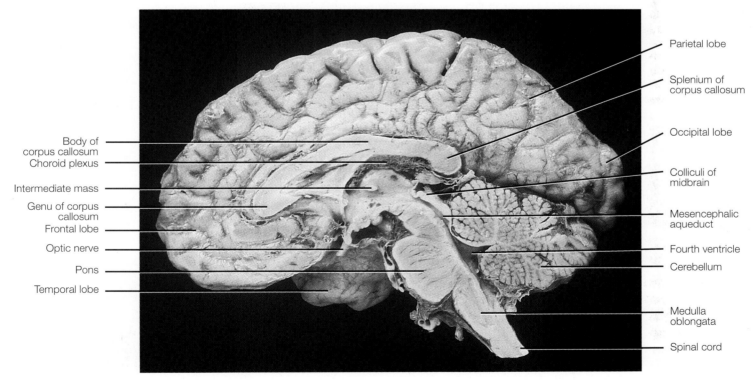

Parietal lobe

Splenium of corpus callosum

Occipital lobe

Body of corpus callosum

Choroid plexus

Colliculi of midbrain

Intermediate mass

Genu of corpus callosum

Mesencephalic aqueduct

Frontal lobe

Optic nerve

Fourth ventricle

Pons

Cerebellum

Temporal lobe

Medulla oblongata

Spinal cord

(c)

**FIGURE 15.3**

*Continued.*

**FIGURE 15.3**
*Continued.*

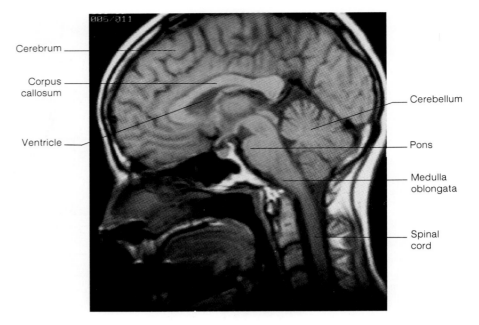

(d)

**FIGURE 15.4**
The cerebrum. (*a*) A lateral view and (*b*) a superior view.

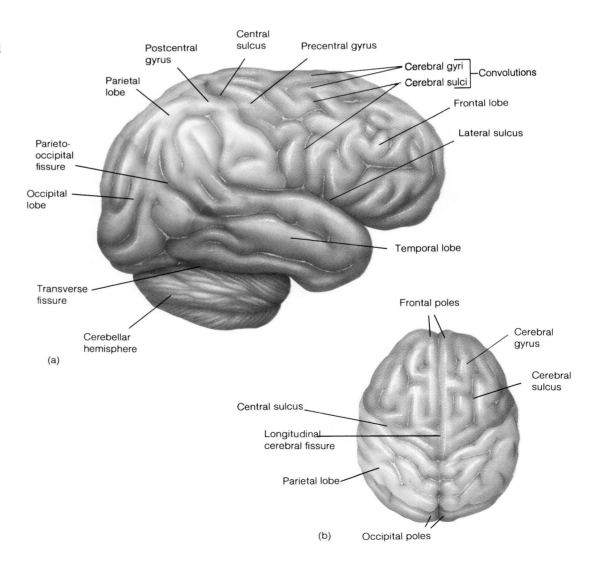

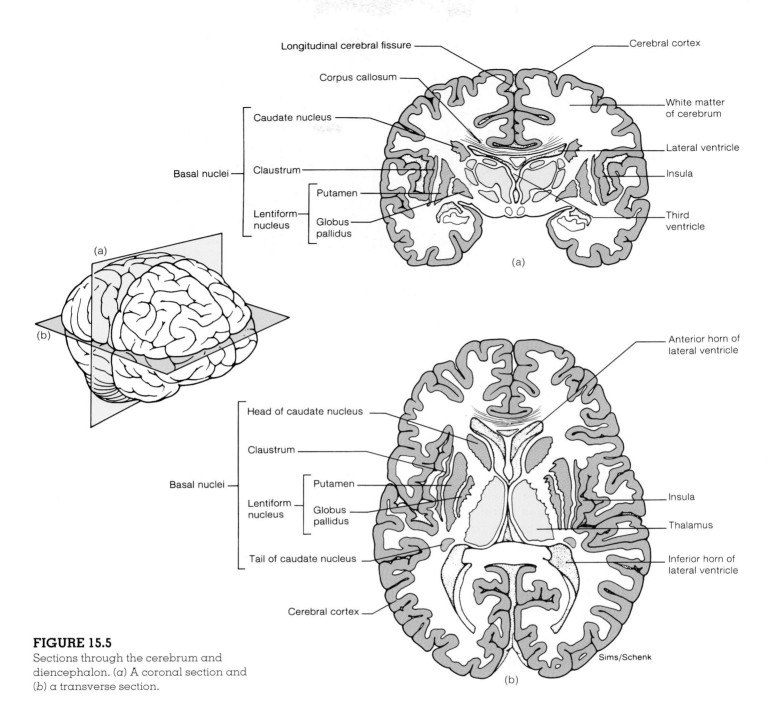

**FIGURE 15.5**
Sections through the cerebrum and diencephalon. (*a*) A coronal section and (*b*) a transverse section.

The cerebrum consists of two layers. The surface layer, referred to as the **cerebral cortex,** is composed of gray matter that is 2–4 mm (0.08–0.16 in.) thick (fig. 15.5). Beneath the cerebral cortex is the thick **white matter** of the cerebrum, which constitutes the second layer. The cerebral cortex is characterized by numerous folds and grooves called **convolutions.** Convolutions form during early fetal development, when brain size increases rapidly and the cortex enlarges out of proportion to the underlying white matter.

The elevated folds of the convolutions are the **cerebral gyri** (singular, *gyrus*), and the depressed grooves are the **cerebral sulci** (*sul´si*) (singular, *sulcus*). The convolutions greatly increase the area of the gray matter, which is composed of neuron cell bodies.

Recent studies indicate that with increased learning, there is an increase in the number of synapses between neurons within the cerebrum. Although the number of neurons is established during prenatal development, the number of synapses is variable depending upon the learning process. The number of cytoplasmic extensions from the cell body of a neuron determines the extent of nerve impulse conduction and the associations that can be made to cerebral areas already containing stored information.

gyrus: Gk. *gyros*, circle
sulcus: L. *sulcus*, a furrow or ditch

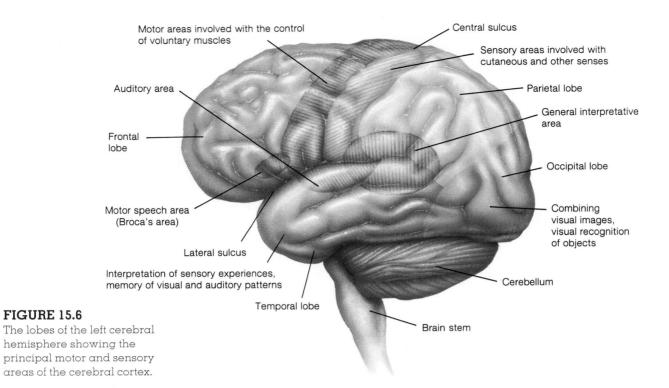

**FIGURE 15.6**
The lobes of the left cerebral hemisphere showing the principal motor and sensory areas of the cerebral cortex.

## *Lobes of the Cerebrum*

Each cerebral hemisphere is subdivided into five lobes by especially deep sulci. Four of these lobes appear on the surface of the cerebrum and are named according to the overlying cranial bones (fig. 15.6). The reasons for the separate cerebral lobes, as well as two cerebral hemispheres, have to do with specificity of function (table 15.2).

**Frontal Lobe** The frontal lobe forms the anterior portion of each cerebral hemisphere (fig. 15.6). A prominent, deep furrow called the **central sulcus** (fissure of Rolando) separates the frontal lobe from the parietal lobe. The central sulcus extends at right angles from the longitudinal fissure to the lateral sulcus. The **lateral sulcus** (fissure of Sylvius) extends laterally from the inferior surface of the cerebrum to separate the frontal and temporal lobes. The **precentral gyrus** (see figs. 15.4 and 15.6), an important motor area, is positioned immediately in front of the central sulcus. Frontal lobe functions include initiating voluntary motor impulses for the movement of skeletal muscles, analyzing sensory experiences, and providing responses relating to personality. The frontal lobes also mediate responses related to memory, emotions, reasoning, judgment, planning, and verbal communication.

.......................................
fissure of Rolando: from Luigi Rolando, Italian anatomist, 1773–1831
fissure of Sylvius: from Franciscus Sylvius de la Boe, Dutch anatomist, 1614–72

## Table 15.2 Functions of the cerebral lobes

| Lobe | Functions |
|------|-----------|
| Frontal | Voluntary motor control of skeletal muscles; personality; higher intellectual processes (e.g., concentration, planning, and decision making); verbal communication |
| Parietal | Somatesthetic interpretation (e.g., cutaneous and muscular sensations); understanding speech and formulating words to express thoughts and emotions; interpretation of textures and shapes |
| Temporal | Interpretation of auditory sensations; storage (memory) of auditory and visual experiences |
| Occipital | Integration of movements in focusing the eye; correlation of visual images with previous visual experiences and other sensory stimuli; conscious perception of vision |
| Insula | Memory; integration of other cerebral activities |

**Parietal Lobe** The parietal lobe is posterior to the central sulcus of the frontal lobe. An important sensory area called the **postcentral gyrus** (see fig. 15.4) is positioned immediately behind the central sulcus. The postcentral gyrus is designated as a *somatesthetic area* because it responds to stimuli from cutaneous and muscular receptors throughout the body.

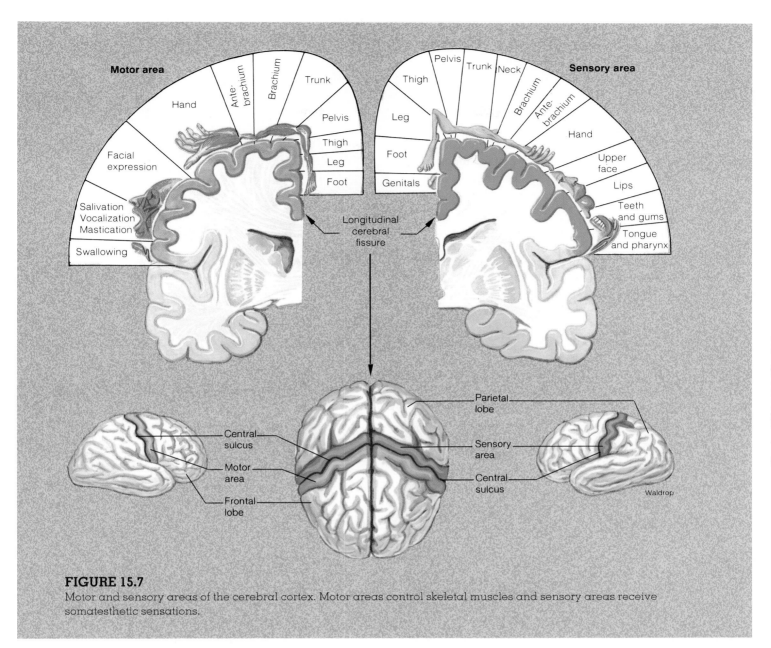

**FIGURE 15.7**
Motor and sensory areas of the cerebral cortex. Motor areas control skeletal muscles and sensory areas receive somatesthetic sensations.

The size of the portions of the precentral gyrus responsible for motor movement and the size of the portions of the postcentral gyrus that respond to sensory stimuli do not correspond to the size of the part of the body being served but rather to the number of motor units activated or the density of receptors (fig. 15.7). For example, because the hand has many motor units and sensory receptors, larger portions of the precentral and postcentral gyri serve it than serve the thorax, even though the thorax is much larger.

In addition to responding to somatesthetic stimuli, the parietal lobe functions in speech comprehension and in verbal articulation of thoughts and emotions. The parietal lobe also interprets the textures and shapes of objects as they are handled.

**Temporal Lobe**  The temporal lobe is located below the parietal lobe and the posterior portion of the frontal lobe. It is separated from both by the lateral sulcus (see fig. 15.6). The temporal lobe contains auditory centers that receive sensory neurons from the cochlea of the ear. This lobe also interprets some sensory experiences and stores memories of both auditory and visual experiences.

**Occipital Lobe**  The occipital lobe forms the posterior portion of the cerebrum and is not distinctly separated from the temporal and parietal lobes (see fig. 15.6). The occipital lobe is superior to the cerebellum and is separated from it by an infolding of the meningeal layer called the **tentorium cerebelli.** The principal functions of the occipital lobe

385

concern vision. The occipital lobe integrates eye movements by directing and focusing the eye. It is also responsible for visual association, correlating visual images with previous visual experiences and other sensory stimuli.

**Insula**   The insula is a deep lobe of the cerebrum that cannot be viewed on the surface (see fig. 15.5). It is deep to the lateral sulcus and is covered by portions of the frontal, parietal, and temporal lobes. Little is known of the function of the insula, though it primarily integrates other cerebral activities and may have some function in memory.

 Because of its size and position, portions of the cerebrum frequently suffer brain trauma. A concussion to the brain may cause a temporary or permanent impairment of cerebral functions; a stroke usually affects cerebral function. Much of what is known about cerebral function comes from observing body dysfunctions when specific regions of the cerebrum are traumatized.

**Brain Waves**   Neurons within the cerebral cortex continuously generate electrical activity, which can be recorded by electrodes attached to precise locations on the scalp, producing an **electroencephalogram (EEG).** An EEG pattern, commonly called *brain waves,* is the collective expression of millions of action potentials from neurons.

Brain waves are first emitted from a developing brain during early fetal development and continue throughout a person's life. The cessation of brain-wave patterns may be a decisive factor in the legal determination of death.

Certain distinct EEG patterns signify healthy mental functions. Deviations from these patterns are of clinical significance in diagnosing trauma, mental depression, hematomas, and various diseases, such as tumors, infections, and epilepsy. Normally, there are four kinds of EEG patterns (fig. 15.8).

**1**  **Alpha waves** are best recorded from the parietal and occipital regions while a person is awake and relaxed but with the eyes closed. These waves are rhythmic oscillations of about 10–12 cycles/second. The alpha rhythm of a child under the age of 8 occurs at a slightly lower frequency of 4–7 cycles/second.

**2**  **Beta waves** are strongest from the frontal lobes, especially the area near the precentral gyrus. These waves are sensory evoked and respond to visual and mental activity. Because they respond to stimuli from receptors and are superimposed on the continuous activity patterns of the alpha waves, they constitute *evoked activity.* The frequency of beta waves is 13–25 cycles/second.

insula: L. *insula,* island

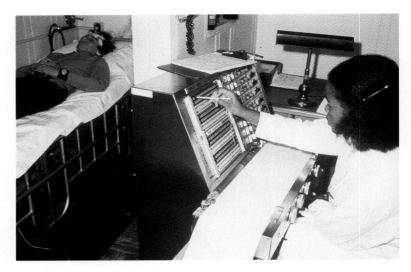

(a)

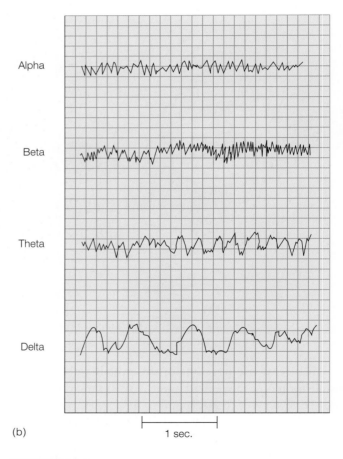

(b)                    1 sec.

**FIGURE 15.8**

Brain waves. (*a*) A technician using an electroencephalograph to take the EEG of a patient. (*b*) Types of EEG waves.

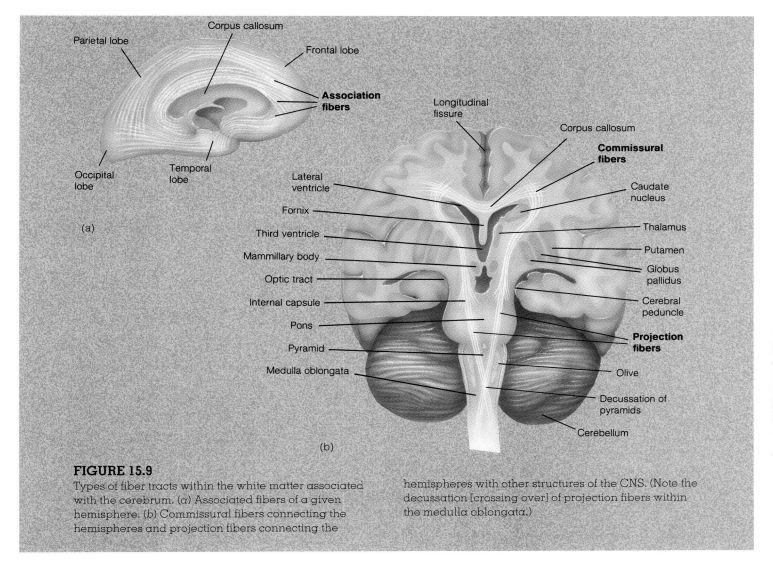

**FIGURE 15.9**
Types of fiber tracts within the white matter associated with the cerebrum. (a) Associated fibers of a given hemisphere. (b) Commissural fibers connecting the hemispheres and projection fibers connecting the hemispheres with other structures of the CNS. (Note the decussation [crossing over] of projection fibers within the medulla oblongata.)

**3** **Theta waves** are emitted from the temporal and occipital lobes. They have a frequency of 5–8 cycles/second and are common in newborn infants. The recording of theta waves in adults generally indicates severe emotional stress and can be a forewarning of a nervous breakdown.

**4** **Delta waves** seem to be emitted in a general pattern from the cerebral cortex. These waves have a frequency of 1–5 cycles/second and are common during sleep and in an awake infant. The presence of delta waves in an awake adult indicates brain damage.

## *White Matter of the Cerebrum*

The thick **white matter** of the cerebrum is deep to the cerebral cortex (see fig. 15.5) and consists of dendrites, myelinated axons, and associated neuroglia. These fibers form the billions of connections within the brain by which information in the form of electrical impulses is transmitted to the appropriate places. The three types of fiber tracts within the white matter are named according to location and the direction in which they conduct impulses (fig. 15.9).

**1** **Association fibers** are confined to a given cerebral hemisphere and conduct impulses between neurons within that hemisphere.

**2** **Commissural** (*kă-mĭ-shur´al*) **fibers** connect the neurons and gyri of one hemisphere with those of the other. The **corpus callosum** and **anterior commissure** (fig. 15.10) are composed of commissural fibers.

**3** **Projection fibers** form the ascending and descending tracts that transmit impulses from the cerebrum to other parts of the brain and spinal cord and from the spinal cord and other parts of the brain to the cerebrum.

387

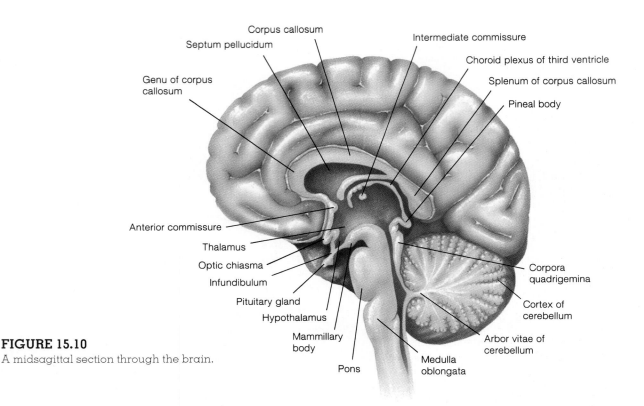

Corpus callosum
Septum pellucidum
Genu of corpus callosum
Intermediate commissure
Choroid plexus of third ventricle
Splenum of corpus callosum
Pineal body
Anterior commissure
Thalamus
Optic chiasma
Infundibulum
Pituitary gland
Hypothalamus
Mammillary body
Pons
Medulla oblongata
Corpora quadrigemina
Cortex of cerebellum
Arbor vitae of cerebellum

**FIGURE 15.10**
A midsagittal section through the brain.

## Basal Nuclei

The **basal nuclei** are specialized paired masses of gray matter located deep within the white matter of the cerebrum (fig. 15.11). The most prominent of the basal nuclei is the **corpus striatum,** so named because of its striped appearance. The corpus striatum is composed of several masses of nuclei. The **caudate nucleus** is the upper mass. A thick band of white matter lies between the caudate nucleus and the next lower two masses, collectively called the **lentiform nucleus.** The lentiform nucleus consists of a lateral portion, called the **putamen** (*pyoo-ta´men*), and a medial portion, called the **globus pallidus**. The claustrum is another portion of the basal nuclei. It is a thin layer of gray matter just deep to the cerebral cortex of the insula.

The basal nuclei are associated with other structures of the brain, particularly within the mesencephalon. The caudate nucleus and the putamen of the lentiform nucleus control unconscious contractions of certain skeletal muscles, such as those of the upper extremities involved in involuntary arm movements during walking. The globus pallidus regulates the muscle tone necessary for specific, intentional body movements. Neural diseases or physical trauma to the basal nuclei generally cause a variety of motor movement dysfunctions, including rigidity, tremor, and rapid and aimless movements.

## Language

Knowledge of the brain regions involved in language has been gained primarily by the study of *aphasias*—speech and language disorders caused by damage to specific language areas of the brain. These areas (fig. 15.12) are generally located in the cerebral cortex of the left hemisphere in both right-handed and left-handed people.

The **motor speech area** (Broca's area) is located in the left inferior gyrus of the frontal lobe. Neural activity in the motor speech area causes selective stimulation of motor impulses in motor centers elsewhere in the frontal lobe, which in turn causes coordinated skeletal muscle movement in the pharynx and larynx. At the same time, motor impulses are sent to the respiratory muscles to regulate air movement across the vocal cords. The combined muscular stimulation translates thought patterns into speech.

**Wernicke's area** is located in the superior gyrus of the temporal lobe and is directly connected to the motor speech area by a fiber tract called the **arcuate fasciculus.** People with *Wernicke's aphasia* produce speech that has been

---

lentiform: L. *lentis,* elongated
putamen: L. *putare,* to cut, prune
globus pallidus: L. *globus,* sphere; *pallidus,* pale

---

aphasia: L. *a,* without; Gk. *phasis,* speech
Broca's area: from Pierre P. Broca, French neurologist, 1824–80
Wernicke's area: from Karl Wernicke, German neurologist, 1848–1905

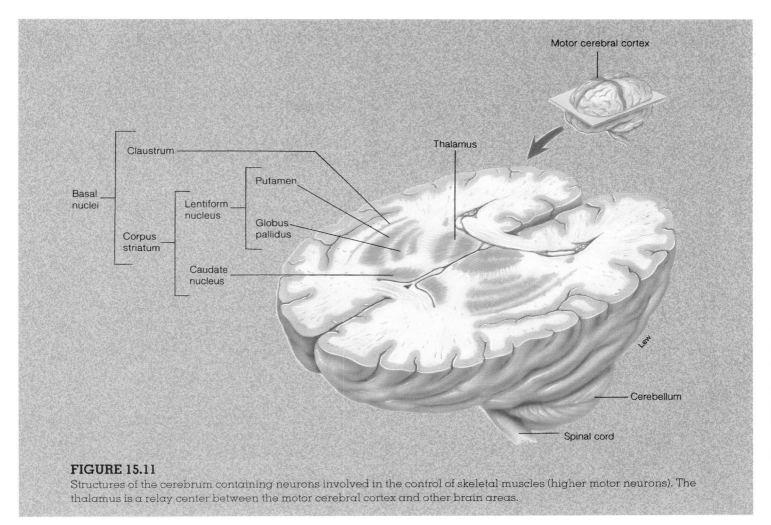

**FIGURE 15.11**

Structures of the cerebrum containing neurons involved in the control of skeletal muscles (higher motor neurons). The thalamus is a relay center between the motor cerebral cortex and other brain areas.

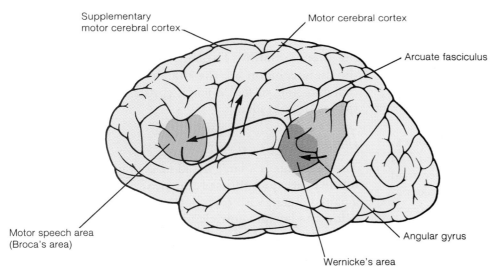

**FIGURE 15.12**

Brain areas involved in the control of speech. Arrows indicate the direction of communication between these areas.

described as a "word salad." The words used may be real words that are chaotically mixed together, or they may be made-up words. Language comprehension is destroyed in people with Wernickes' aphasia; they cannot understand either spoken or written language.

It appears that the concept of words to be spoken originates in Wernicke's area and is then communicated to the motor speech area through the arcuate fasciculus. Damage to the arcuate fasciculus produces *conduction aphasia*, which is fluent but nonsensical speech as in Wernicke's aphasia, even though both the motor speech area and Wernicke's area are intact.

The **angular gyrus,** located at the junction of the parietal, temporal, and occipital lobes, is believed to be a center for the integration of auditory, visual, and somatesthetic information. Damage to the angular gyrus produces aphasias, which suggests that this area projects to Wernicke's area. Some patients with damage to the left angular gyrus can speak and understand spoken language but cannot read or write. Other patients can write a sentence but cannot read it, presumably due to damage to the projections from the occipital lobe (involved in vision) to the angular gyrus.

 Recovery of language ability, by transfer to the right hemisphere after damage to the left hemisphere, is very good in children but decreases after adolescence. Recovery is reported to be faster in left-handed people, possibly because language ability is more evenly divided between the two hemispheres in left-handed people. Some recovery usually occurs after damage to the motor speech area, but damage to Wernicke's area produces more severe and permanent aphasias.

## Memory

Clinical studies of *amnesia* suggest that several different brain regions are involved in memory storage and retrieval. Amnesia has been found to result from damage to the temporal lobe of the cerebral cortex, hippocampus, caudate nucleus (in Huntington's disease), or the dorsomedial thalamus (in alcoholics suffering from Korsakoff's syndrome with thiamine deficiency). Clinical studies also suggest that there are two major categories of memory: **short-term memory** and **long-term memory.** People with head trauma, for example, and patients with suicidal depression who are treated by *electroconvulsive shock* (ECS) *therapy* may lose their memory of recent events but retain their older memories.

The **hippocampus** (see fig. 17.10) appears to be required for short-term memory and for the consolidation of that memory into a long-term form. The surgical removal of the left hippocampus due to the presence of a tumorous growth impairs the consolidation of short-term verbal memories, and removal of the right hippocampus impairs the consolidation of nonverbal memories. The surgical removal of both the left and the right hippocampi leave a patient totally without short-term memory.

Factual information is stored in the cerebral cortex, with verbal memories lateralized to the left hemisphere and visuospatial information to the right hemisphere. Electrical stimulation of various regions in the cerebrum of awake patients often evokes visual or auditory memories that are extremely vivid. Electrical stimulation of specific points in the temporal lobe evokes specific memories in such detail that patients feel as if the events were currently being experienced. Surgical removal of these regions does not, however, eradicate the memory. The amount of memory destroyed by ablation of brain tissue appears to depend more on the amount of brain tissue removed than on the location of the surgery. On the basis of these observations, it appears that memory may be diffusely located in the cerebrum, and that stimulation of the correct location of the cerebral cortex then retrieves the memory.

Since long-term memory is not eradicated by electroconvulsive shock, it seems reasonable to conclude that the consolidation of memory depends on relatively permanent changes in the chemical structure of neurons and their synapses. Experiments suggest that protein synthesis is required for the consolidation of the "memory trace." According to one theory, these proteins may be secreted into the extracellular environment, where they influence synaptic connections. Another theory holds that new receptor proteins in the membrane of the postsynaptic neuron are made available as a result of high-frequency stimulation of the presynaptic neuron. This would help to account for the increased sensitivity of postsynaptic neurons to neurotransmitters, as seen in long-term potentiation (discussed in chapter 14). Much more research is obviously needed in this exciting area of physiology before memory can be fully explained at a cellular and molecular level.

## Diencephalon

*The diencephalon is a major autonomic region of the brain that consists of such vital structures as the thalamus, hypothalamus, epithalamus, and pituitary gland.*

The **diencephalon** is the second subdivision of the forebrain and is almost completely surrounded by the cerebral hemispheres of the telencephalon. The third ventricle (see fig. 15.21) forms a cavity on the median plane within the diencephalon. The most important structures of the diencephalon are the thalamus (*thal´ă-mus*), hypothalamus (*hi´´po-thal´ă-mus*), epithalamus, and pituitary (*pĭ-too´ĭ-ter-e*) gland.

## Thalamus

The **thalamus** is a large ovoid mass of gray matter, constituting nearly four-fifths of the diencephalon. It is actually a paired organ, with each portion positioned immediately below the lateral ventricle of its respective cerebral hemisphere (see figs. 15.3 and 15.5). The principal function of

thalamus: L. *thalamus*, inner room

the thalamus is to act as a relay center for all sensory impulses, except smell, to the cerebral cortex. Specialized masses of nuclei relay the incoming impulses to precise locations within the cerebral lobes for interpretation.

The thalamus also performs some sensory interpretation. The cerebral cortex discriminates pain and other tactile stimuli, but the thalamus responds to general sensory stimuli and provides crude awareness. The thalamus probably plays a role in the initial autonomic response of the body to intense pain and is, therefore, partially responsible for the physiological shock that frequently follows serious trauma.

## Hypothalamus

The **hypothalamus** is a small portion of the diencephalon located below the thalamus, where it forms the floor and part of the lateral walls of the third ventricle (fig. 15.10). The hypothalamus consists of several masses of nuclei interconnected with other parts of the nervous system. Despite its small size, the hypothalamus performs numerous vital functions, most of which relate directly or indirectly to the regulation of visceral activities. It also has emotional and instinctual functions (see also fig. 17.10).

The hypothalamus is an autonomic nervous center because of its role in accelerating or decreasing certain body functions. It secretes eight hormones, including two released from the posterior pituitary gland. These hormones and their functions are discussed in chapter 19. The principal autonomic and limbic (emotional) functions of the hypothalamus are as follows:

**1 Cardiovascular regulation.** Although the heartbeat is automatic, impulses from the hypothalamus cause autonomic acceleration or deceleration of the heart. Impulses from the posterior hypothalamus produce a rise in arterial blood pressure and an increase of the heart rate. Impulses from the anterior portion have the opposite effect. The impulses from these regions do not travel directly to the heart but pass first to the cardiovascular centers of the medulla oblongata.

**2 Body-temperature regulation.** Specialized nuclei within the anterior portion of the hypothalamus are sensitive to changes in body temperature. If the arterial blood flowing through this portion of the hypothalamus is above normal temperature, the hypothalamus initiates impulses that cause heat loss through sweating and vasodilation of cutaneous vessels of the skin. A below-normal blood temperature causes the hypothalamus to relay impulses that result in heat production and retention through the initiation of shivering, the contraction of cutaneous blood vessels, and the cessation of sweating.

**3 Regulation of water and electrolyte balance.** Specialized *osmoreceptors* in the hypothalamus continuously monitor the concentration of the blood. An increased concentration due to lack of water causes antidiuretic hormone (ADH) to be produced by the hypothalamus and released from the posterior pituitary gland. At the same time, a *thirst center* within the hypothalamus produces feelings of thirst.

**4 Regulation of hunger and control of gastrointestinal activity.** The *feeding center* is a specialized portion of the lateral hypothalamus that monitors the blood glucose, fatty acid, and amino acid levels. Low levels of these substances in the blood are partially responsible for a sensation of hunger elicited from the hypothalamus. When sufficient amounts of food have been ingested, the *satiety* (să-ti´ĭ-te) *center* in the midportion of the hypothalamus inhibits the feeding center. The hypothalamus also receives sensory impulses from the abdominal viscera and regulates glandular secretions and the peristaltic movements of the digestive tract.

**5 Regulation of sleeping and wakefulness.** The hypothalamus has both a *sleep center* and a *wakefulness center* that function with other parts of the brain to determine the level of conscious alertness.

**6 Sexual response.** Specialized *sexual center* nuclei within the posterior portion of the hypothalamus respond to sexual stimulation of the tactile receptors within the genital organs. The experience of orgasm involves neural activity with the sexual center of the hypothalamus.

**7 Emotions.** A number of nuclei within the hypothalamus are associated with specific emotional responses, such as anger, fear, pain, and pleasure.

**8 Control of endocrine functions.** The hypothalamus produces neurosecretory chemicals that stimulate the anterior pituitary to release various hormones, which in turn regulate other endocrine glands (see chapter 19). The hypothalamus also produces the two hormones secreted by the posterior pituitary gland.

## Epithalamus

The **epithalamus** is the posterior portion of the diencephalon that includes a thin roof over the third ventricle (see fig. 15.21). The inside lining of the roof consists of a vascular **choroid plexus** where cerebrospinal fluid is produced (see fig. 15.10). A small cone-shaped mass called the **pineal gland** (epiphysis), which has a neuroendocrine function, extends outward from the posterior end of the epithalamus (see fig. 15.10). The posterior commissure, located ventral to the pineal gland, is a tract of commissural fibers that connects the superior colliculi (see fig. 15.16).

........................................

pineal: L. *pinea*, pine cone

**FIGURE 15.13**

The pituitary gland is positioned within the sella turcica of the sphenoid bone and is attached to the brain by the infundibulum.

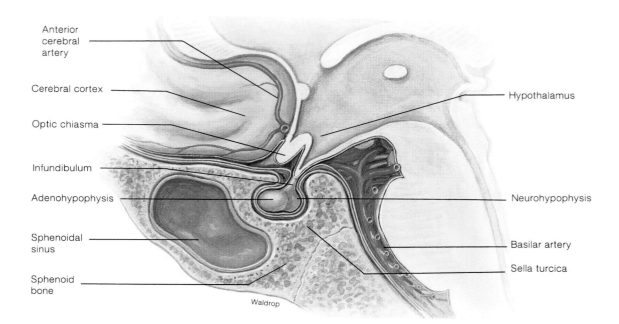

Anterior cerebral artery

Cerebral cortex

Optic chiasma

Infundibulum

Adenohypophysis

Sphenoidal sinus

Sphenoid bone

Hypothalamus

Neurohypophysis

Basilar artery

Sella turcica

Waldrop

## Pituitary Gland

The **pituitary gland,** or **hypophysis** (*hi-pof´ĭ-sis*), is positioned on the inferior aspect of the diencephalon and is attached to the hypothalamus by a stalklike structure called the **infundibulum** (*in˝fun-dib´yŭ-lum*) (see fig. 15.10). The pituitary is a rounded, pea-shaped gland about 1.3 cm (0.5 in.) in diameter. It is covered by the dura mater and is supported by sella turcica of the sphenoid bone (fig. 15.13). The cerebral arterial circle (circle of Willis) (see fig. 21.22*b*) surrounds the highly vascular pituitary gland, providing it with a rich blood exchange. The pituitary, which has an endocrine function, is structurally and functionally divided into an anterior portion, called the **adenohypophysis** (*ad˝ĕ-no-hi-pof´ĭ-sis*), and a posterior portion, called the **neurohypophysis** (see chapter 19).

# Mesencephalon

*The mesencephalon contains the corpora quadrigemina, concerned with visual and auditory reflexes, and the cerebral peduncles, composed of fiber tracts. It also contains specialized nuclei that help to control posture and movement.*

The *brain stem* contains nuclei for autonomic functions of the body and their connecting tracts. It is that portion of the brain that attaches to the spinal cord and includes the midbrain, pons, and medulla oblongata. The **mesencephalon** (*mes˝en-sef´ă-lon*), or **midbrain,** is a short section of the brain stem between the diencephalon and the pons (see fig. 15.16). Within the midbrain is the **mesencephalic aqueduct** (aqueduct of Sylvius) (see fig. 15.21), which connects the third and fourth ventricle. The midbrain also contains the corpora quadrigemina (see fig. 15.10), cerebral peduncles (see fig. 15.9), red nucleus, and substantia nigra.

The **corpora quadrigemina** (*kwad˝rĭ-jem´ĭ-nă*) are the four rounded elevations on the posterior portion of the midbrain (see fig. 15.16). The two upper eminences, the **superior colliculi** are concerned with visual reflexes. The two posterior eminences, the **inferior colliculi,** are responsible for auditory reflexes. The **cerebral peduncles** (*pĕ-dung´k'lz*) are a pair of cylindrical structures composed of ascending and descending projection fiber tracts that support and connect the cerebrum to the other regions of the brain.

The **red nucleus** is deep within the midbrain between the cerebral peduncle and the cerebral aqueduct. The red nucleus is gray matter that connects the cerebral hemispheres and the cerebellum and functions in reflexes concerned with motor coordination and maintenance of posture. Another nucleus called the **substantia nigra** is inferior to the red nucleus. The substantia nigra is thought to inhibit involuntary movements.

. . . . . . . . . . . . . . . . . . . . . . . . . . . . . .

pituitary: L. *pituita*, phlegm (this gland was originally thought to secrete mucus into the nasal cavity)
infundibulum: L. *infundibulum*, funnel

. . . . . . . . . . . . . . . . . . . . . . . . . . . . . .

aqueduct of Sylvius: Sylvius, French anatomist, 1478–1555
corpora quadrigemina: L. *corpus*, body; *quadri*, four; *geminus*, twin
colliculus: L. *colliculus*, small mound

fibers are part of the motor and sensory tracts that connect the medulla oblongata with the tracts of the midbrain.

Scattered throughout the pons are several nuclei associated with specific cranial nerves. The cranial nerves that have nuclei within the pons include the trigeminal (V), which transmits impulses for chewing and sensory sensations from the head; the abducens (VI), which controls certain movements of the eyeball; the facial (VII), which transmits impulses for facial movements and sensory sensations from the taste buds; and the vestibular branch of the vestibulocochlear (VIII), which maintains equilibrium.

Other nuclei of the pons function with nuclei of the medulla oblongata to regulate the rate and depth of breathing. The two respiratory centers of the pons are called the **apneustic** and the **pneumotaxic areas** (fig. 15.14).

## Cerebellum

The **cerebellum** (*ser″ĕ-bel′um*) is the second largest structure of the brain. It is located in the metencephalon and occupies the inferior and posterior aspect of the cranial cavity. The cerebellum is separated from the overlying cerebrum by a **transverse fissure** (see fig. 15.4). A portion of the meninges called the **tentorium cerebelli** extends into the transverse fissure. The cerebellum consists of two **hemispheres** and a central constricted area called the **vermis** (fig. 15.15). The **falx cerebelli** is the portion of the meninges that partially extends between the hemispheres (see table 15.3).

Like the cerebrum, the cerebellum has a thin, outer layer of gray matter, the **cerebellar cortex,** and a thick, deeper layer of white matter. The cerebellum is convoluted into a series of slender, parallel **gyri.** The tracts of white matter within the cerebellum have a distinctive branching pattern called the **arbor vitae** that can be seen in the sagittal view (see fig. 15.10).

Three paired bundles of nerve fibers called **cerebellar peduncles** support the cerebellum and provide it with tracts for communicating with the rest of the brain (fig. 15.16). Following is a description of the cerebellar peduncles.

**1** **Superior cerebellar peduncles** connect the cerebellum with the midbrain. The fibers within these peduncles originate primarily from specialized **dentate nuclei** within the cerebellum and pass through the red nucleus to the thalamus and then to the motor areas of the cerebral cortex. Impulses through the fibers of these peduncles provide feedback to the cerebrum.

**2** **Middle cerebellar peduncles** convey impulses of voluntary movement from the cerebrum through the pons and to the cerebellum.

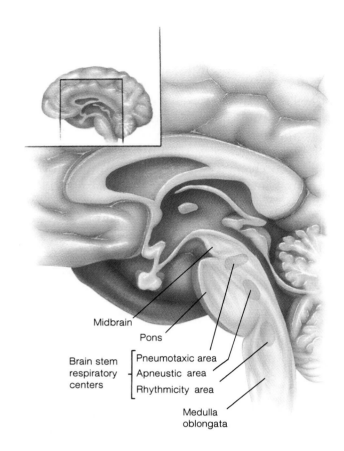

**FIGURE 15.14**
Nuclei within the pons and medulla oblongata that constitute the respiratory center.

# Metencephalon

*The metencephalon contains the pons, which relays impulses, and the cerebellum, which coordinates skeletal muscle contractions.*

The **metencephalon** (*med″en-sef′ă-lon*) is the most superior portion of the hindbrain. Two vital structures of the metencephalon are the pons and cerebellum. The mesencephalic aqueduct of the mesencephalon enlarges to become the **fourth ventricle** (see fig. 15.21) within the metencephalon and myelencephalon.

## Pons

The **pons** can be observed as a rounded bulge on the underside of the brain, between the midbrain and the medulla oblongata (fig. 15.14). The pons consists of white fiber tracts that course in two principal directions. The surface fibers extend transversely to connect with the cerebellum through the middle cerebellar peduncles. The deeper longitudinal

pons: L. *pons*, bridge

cerebellum: L. *cerebellum*, diminutive of *cerebrum*, brain
vermis: L. *vermis*, worm
arbor vitae: L. *arbor*, tree; *vitae*, life
peduncle: L. *peduncle*, diminutive of *pes*, foot

**3** **Inferior cerebellar peduncles** connect the cerebellum with the medulla oblongata and the spinal cord. They contain both incoming vestibular and proprioceptive fibers and outgoing motor fibers.

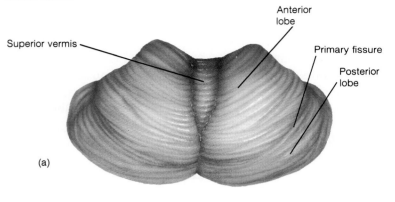

(a)

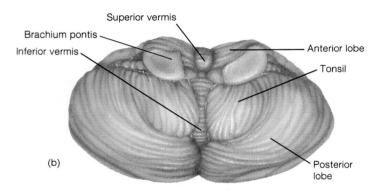

(b)

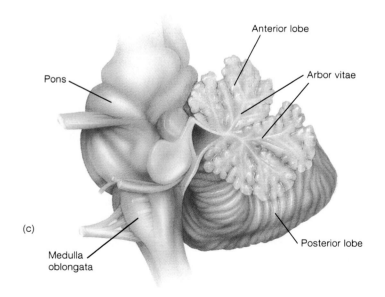

(c)

**FIGURE 15.15**

The structure of the cerebellum: (*a*) a superior view, (*b*) an inferior view, and (*c*) a sagittal view.

The principal function of the cerebellum is coordinating skeletal muscle contractions by recruiting precise motor units within the muscles. Impulses for voluntary muscular movement originate in the cerebral cortex and are coordinated by the cerebellum. The cerebellum constantly initiates impulses to selective motor units for maintaining posture and muscle tone. The cerebellum also adjusts to incoming impulses from **proprioceptors** (*pro″pre-o-sep′torz*) within muscles, tendons, joints, and special sense organs. A proprioceptor is a sensory nerve ending that is sensitive to changes in the tension of a muscle or tendon.

 Trauma or diseases of the cerebellum, such as *cerebral palsy* or a *stroke*, frequently cause an impairment of skeletal muscle function. Movements become jerky and uncoordinated in a condition known as *ataxia*. There is also a loss of equilibrium, resulting in a disturbance of gait. *Alcohol intoxication* causes similar uncoordinated body movements.

# Myelencephalon

*The medulla oblongata, contained within the myelencephalon, connects to the spinal cord and contains nuclei for the cranial nerves and vital autonomic functions.*

## Medulla Oblongata

The **medulla oblongata** is a bulbous structure about 3 cm (1 in.) long that is continuous with the pons anteriorly and the spinal cord posteriorly at the level of the foramen magnum (see figs. 15.9 and 15.10). Externally, the medulla oblongata resembles the spinal cord, except for the two triangular, elevated structures called **pyramids** on the ventral side and an oval enlargement called the **olive** on each lateral surface. The **fourth ventricle,** the space within the medulla oblongata, is continuous posteriorly with the central canal of the spinal cord and anteriorly with the cerebral aqueduct (see fig. 15.21).

The medulla oblongata is composed of vital nuclei and white matter that form all of the descending and ascending tracts communicating between the

medulla: L. *medulla*, marrow

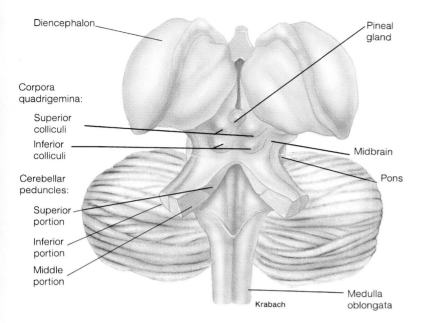

**FIGURE 15.16**

The cerebellar peduncles can be seen when the cerebellum has been removed from its attachment to the brain stem.

Labels for image 1:
Diencephalon
Corpora quadrigemina:
Superior colliculi
Inferior colliculi
Cerebellar peduncles:
Superior portion
Inferior portion
Middle portion
Pineal gland
Midbrain
Pons
Medulla oblongata
Krabach

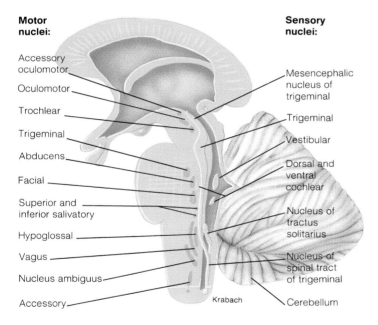

**FIGURE 15.17**

A sagittal section of the medulla oblongata and pons showing the cranial nerve nuclei of gray matter.

Labels for image 2:
Motor nuclei:
Accessory oculomotor
Oculomotor
Trochlear
Trigeminal
Abducens
Facial
Superior and inferior salivatory
Hypoglossal
Vagus
Nucleus ambiguus
Accessory
Sensory nuclei:
Mesencephalic nucleus of trigeminal
Trigeminal
Vestibular
Dorsal and ventral cochlear
Nucleus of tractus solitarius
Nucleus of spinal tract of trigeminal
Cerebellum
Krabach

spinal cord and various parts of the brain. Most of the fibers within these tracts cross over to the opposite side through the pyramidal region of the medulla oblongata, permitting one side of the brain to receive information from and send information to the opposite side of the body (see fig. 15.9).

The gray matter of the medulla oblongata consists of several important nuclei for the cranial nerves, sensory relay, and

for autonomic functions (fig. 15.17). The **nucleus ambiguus** and the **hypoglossal nucleus** are the centers from which arise the vestibulocochlear (VIII), glossopharyngeal (IX), accessory (XI), and hypoglossal (XII) nerves. The vagus nerves (X) arise from **vagus nuclei,** one on each lateral side of the medulla oblongata, adjacent to the fourth ventricle. The **nucleus gracilis** and the **nucleus cuneatus** relay sensory information to the thalamus, and then the impulses are relayed to the cerebral cortex via the thalamic nuclei (not illustrated). The **inferior olivary nuclei** and the **accessory olivary nuclei** of the olive mediate impulses passing from the forebrain and midbrain through the inferior cerebellar peduncles to the cerebellum.

Three other nuclei within the medulla oblongata function as autonomic centers for controlling vital visceral functions.

**1** **Cardiac center.** Both *inhibitory* and *accelerator fibers* arise from nuclei of the cardiac center. Inhibitory impulses constantly travel through the vagus nerves to slow the heartbeat. Accelerator impulses travel through the spinal cord and eventually innervate the heart through fibers within spinal nerves T1–T5.

**2** **Vasomotor center.** Nuclei of the vasomotor center send impulses via the spinal cord and spinal nerves to the smooth muscles of arteriole walls, causing them to constrict or dilate, thus regulating blood pressure and blood flow.

**3** **Respiratory center.** The respiratory center of the medulla oblongata controls the rate and depth of breathing and functions in conjunction with the respiratory nuclei of the pons (see fig. 15.14) to produce rhythmic breathing.

Other nuclei of the medulla oblongata function as centers for nonvital respiratory movements, such as sneezing, coughing, swallowing, and vomiting. Some of these activities (swallowing, for example) may be initiated voluntarily, but once they progress to a certain point they become involuntary and cannot be interrupted.

## Reticular Formation

The **reticular formation** is a complex network of nuclei and nerve fibers within the brain stem that function as the *reticular activating system (RAS)* in arousing the cerebrum. Portions of the reticular formation are located in the spinal cord, pons, midbrain, and parts of the hypothalamus and thalamus (fig. 15.18). The reticular formation contains ascending and descending fibers from most of the structures within the brain.

Nuclei within the reticular formation generate a continuous flow of impulses unless they are inhibited by other parts of the brain. The principal functions of the RAS are to keep the cerebrum in a state of alert consciousness and to selectively monitor the afferent impulses perceived by the cerebrum. The RAS also helps the cerebellum activate

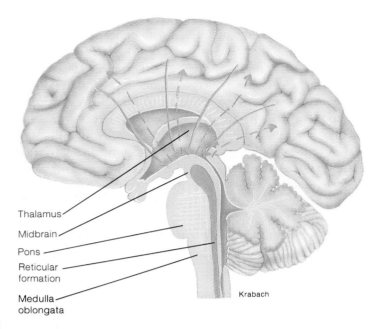

Thalamus

Midbrain

Pons

Reticular formation

Medulla oblongata

Krabach

## FIGURE 15.18

The reticular activating system. The arrows indicate the direction of impulses along nerve pathways that connect with the RAS.

selected motor units to maintain muscle tone and produce smooth, coordinated contractions of skeletal muscles.

 The RAS is sensitive to changes in and trauma to the brain. The sleep response is thought to occur because of a decrease in activity within the RAS, perhaps due to the secretion of specific neurotransmitters. A blow to the head or certain drugs and diseases may damage the RAS, causing unconsciousness. A *coma* is a state of unconsciousness and inactivity of the RAS that even the most powerful external stimuli cannot disturb.

# Meninges of the Central Nervous System

*The CNS is covered by protective meninges; namely a dura mater, an arachnoid mater, and a pia mater.*

The central nervous system is protected by three connective tissue membranous coverings called the **meninges** (figs. 15.19 and 15.20). Individually, from the outside in, they are known as the dura mater, the arachnoid mater, and the pia mater.

## Dura Mater

The **dura mater** is attached to the skull and is composed primarily of tough, fibrous connective tissue. The **cranial dura mater** is a double-layered structure. The thicker, outer **periosteal** (*per″e-os′te-al*) **layer** adheres lightly to the inner surface of the cranium, where it constitutes the periosteum (fig. 15.19). The thinner, inner **meningeal layer** follows the general contour of the brain. The **spinal dura mater** is not double layered but is similar to the meningeal layer of the cranial dura mater.

The two layers of the cranial dura mater are fused and cover most of the brain. In certain regions, however, the layers are separated, enclosing **dural sinuses** (see fig. 15.19) that collect venous blood and drain it to the internal jugular veins of the neck.

In four locations, the meningeal layer of the cranial dura mater forms distinct septa to partition major structures on the surface of the brain and anchor the brain to the inside of the cranial case. These septa are reviewed in table 15.3.

The spinal dura mater forms a tough, tubular **dural sheath** that continues into the vertebral canal and surrounds the spinal cord. There is no connection between the dural sheath and the vertebrae forming the vertebral canal, but instead there is a potential cavity called the **epidural space** (see fig. 15.19). The epidural space is highly vascular and contains loose and adipose connective tissues that form a protective pad around the spinal cord.

## Arachnoid Mater

The **arachnoid mater** is the middle of the three meninges. This delicate, netlike membrane spreads over the CNS but generally does not extend into the sulci or fissures of the brain. The **subarachnoid space,** located between the arachnoid mater and the deepest meninx, the pia mater, contains cerebrospinal fluid. The subarachnoid space is maintained by delicate, weblike strands that connect the arachnoid mater and pia mater (see fig. 15.19).

## Pia Mater

The thin **pia mater** is attached to the surfaces of the CNS and follows the irregular contours of the brain and spinal cord. The pia mater is composed of modified loose connective tissue. It is highly vascular and functions to support the vessels that nourish the underlying cells of the brain and spinal cord. The pia mater is specialized over the roofs of

• • • • • • • • • • • • • • • • • • • • • • • • • • • • •

meninges: L. plural form of *meninx*, membrane

dura mater: L. *dura*, hard; *mater*, mother
arachnoid: L. *arachnoides*, like a cobweb
pia mater: L. *pia*, soft or tender; *mater*, mother

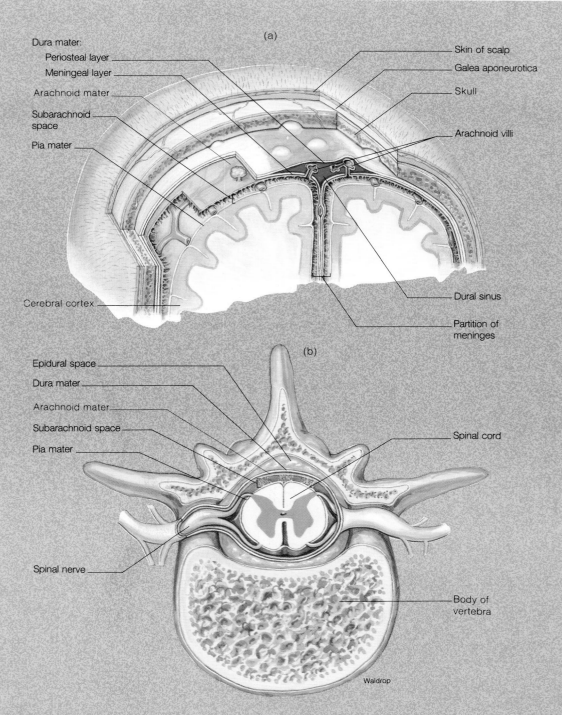

Dura mater:
Periosteal layer
Meningeal layer
Arachnoid mater
Subarachnoid space
Pia mater

Skin of scalp
Galea aponeurotica
Skull
Arachnoid villi

Cerebral cortex

Dural sinus
Partition of meninges

(a)

(b)

Epidural space
Dura mater
Arachnoid mater
Subarachnoid space
Pia mater

Spinal cord

Spinal nerve

Body of vertebra

Waldrop

**FIGURE 15.19**

Meninges and associated structures (a) surrounding the brain and (b) surrounding the spinal cord. The epidural space in the lower lumbar region is of clinical importance as a site for an epidural block that may be administered to facilitate parturition (childbirth).

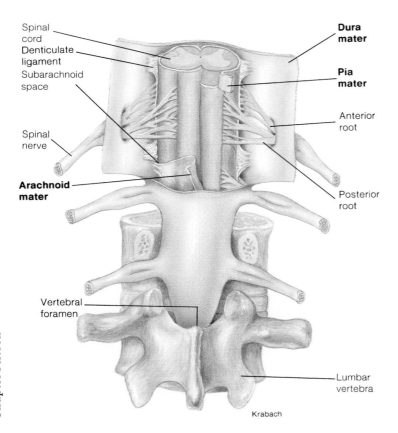

Spinal cord
Denticulate ligament
Subarachnoid space
Spinal nerve
**Arachnoid mater**
Vertebral foramen
Lumbar vertebra
Krabach

**Dura mater**
**Pia mater**
Anterior root
Posterior root

**FIGURE 15.20**

The spinal cord and the meninges. (The meninges are labeled in boldface.)

| Table 15.3 | Septa of the cranial dura mater |
|---|---|
| *Septa* | *Location* |
| Falx cerebri | Extends downward into the longitudinal fissure to partition the right and left cerebral hemispheres; anchored anteriorly to the crista galli of the ethmoid bone and posteriorly to the tentorium |
| Tentorium cerebelli | Separates the occipital and temporal lobes of the cerebrum from the cerebellum; anchored to the tentorium, petrous bones, and occipital bone |
| Falx cerebelli | Partitions the right and left cerebellar hemispheres; anchored to the occipital crest |
| Diaphragma sellae | Forms the roof of the sella turcica |

the ventricles where it contributes to the formation of the choroid plexuses along with the arachnoid mater. Lateral extensions of the pia mater along the spinal cord form the **ligamentum denticulatum,** which attaches the cord to the dura mater.

 *Meningitis* is an inflammation of the meninges, usually caused by bacteria or viruses. The arachnoid mater and the pia mater are the two meninges most frequently affected. Meningitis is accompanied by high fever and severe headache. Complications may cause sensory impairment, paralysis, or mental retardation. Untreated meningitis generally results in coma and death.

# Ventricles and Cerebrospinal Fluid

*The ventricles, central canal, and subarachnoid space contain cerebrospinal fluid, formed by the active transport of substances from blood plasma in the choroid plexuses.*

**Cerebrospinal fluid (CSF)** is a clear, lymphlike fluid that forms a protective cushion around and within the CNS. The fluid also buoys the brain. CSF circulates through the various **ventricles** of the brain, the **central canal** of the spinal cord, and the **subarachnoid space** around the entire CNS. The CSF returns to the circulatory system by draining through the walls of the **arachnoid villi,** which are venous capillaries.

## Ventricles of the Brain

The ventricles of the brain are connected to one another and to the central canal of the spinal cord (fig. 15.21). Each of the two **lateral ventricles** (first and second ventricles) is located in one of the hemispheres of the cerebrum, inferior to the corpus callosum. The **third ventricle** is located in the diencephalon, between the thalami. Each lateral ventricle is connected to the third ventricle by a narrow, oval opening called the **interventricular foramen** (foramen of Monro). The **fourth ventricle** is located in the brain stem, within the pons, cerebellum, and medulla oblongata. The **mesencephalic (cerebral) aqueduct** passes through the midbrain to link the third and fourth ventricles. The fourth

foramen of Monro: from Alexander Monro Jr., Scottish anatomist, 1733–1817

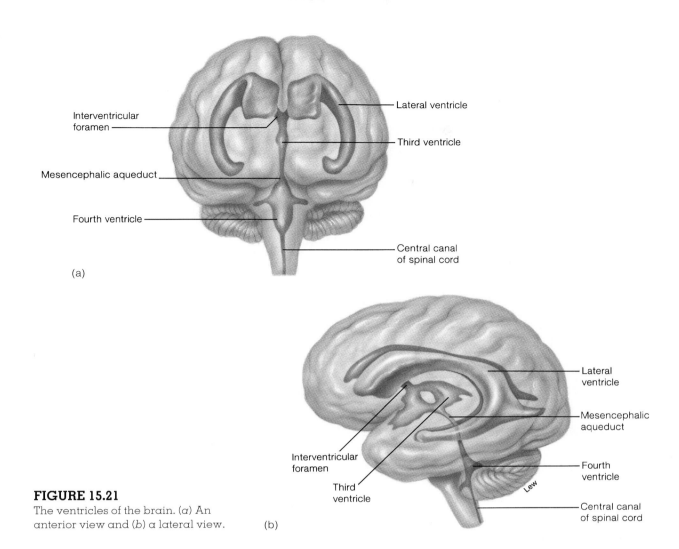

**FIGURE 15.21**
The ventricles of the brain. (*a*) An anterior view and (*b*) a lateral view.

(b)

ventricle also communicates posteriorly with the central canal. CSF exits from the fourth ventricle into the sub-arachnoid space (fig. 15.22) through three foramina: the **median aperture** (foramen of Magendie), a medial open-ing, and two **lateral apertures** (foramina of Luschka) (not il-lustrated). CSF returns to the venous blood through the arachnoid villi.

 *Internal hydrocephalus* is a condition in which CSF builds up within the ventricles of the brain. It is more common in infants whose cranial sutures have not yet strengthened or ossified. If the pressure is excessive, the condition may have to be treated surgically.

*External hydrocephalus*, an accumulation of fluid within the subarachnoid space, usually results from an obstruction of drainage at the arachnoid villi.

foramen of Magendie: from François Magendie, French physiologist, 1783–1855
foramen of Luschka: from Hubert Luschka, German anatomist, 1820–75

## *Cerebrospinal Fluid*

CSF buoys the CNS and protects it from mechanical injury. CSF has a specific gravity of 1.007, which is a density close to that of brain tissue. The brain weight is about 1500 grams, but suspended in CSF its buoyed weight is about 50 grams. This means that the brain has a near neutral buoyancy and can therefore function effectively as a relatively heavy organ. At a true neutral buoyancy, an object does not float or sink but is suspended in its fluid environment.

In addition to buoying the CNS, CSF reduces the dam-aging effect of an impact to the head by spreading the force over a larger area. It also helps to remove metabolic wastes from nervous tissue. Since the CNS lacks lymphatic circu-lation, the CSF moves cellular wastes into the venous re-turn at its places of drainage.

The clear, watery CSF is continuously produced from materials within the blood by masses of specialized capillar-ies called **choroid plexuses** and, to a lesser extent, by secre-tions of the ependymal cells. The ciliated ependymal cells

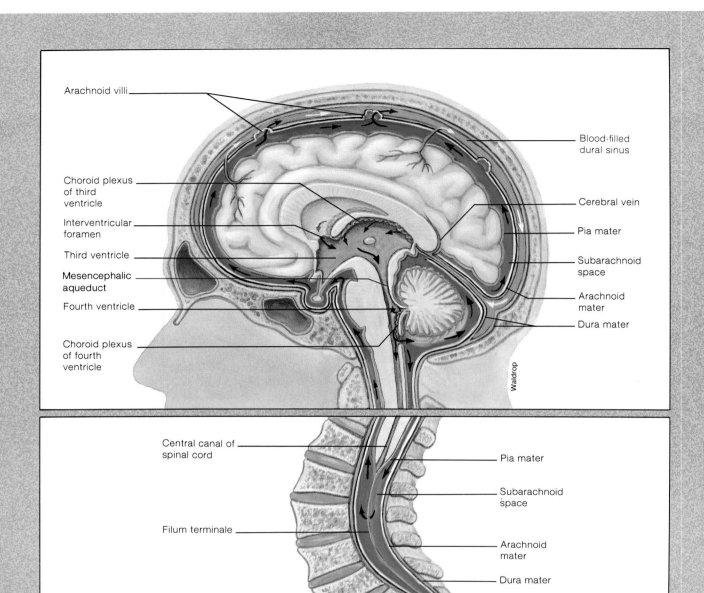

Arachnoid villi

Choroid plexus
of third
ventricle

Interventricular
foramen

Third ventricle

Mesencephalic
aqueduct

Fourth ventricle

Choroid plexus
of fourth
ventricle

Blood-filled
dural sinus

Cerebral vein

Pia mater

Subarachnoid
space

Arachnoid
mater

Dura mater

Waldrop

Central canal of
spinal cord

Filum terminale

Pia mater

Subarachnoid
space

Arachnoid
mater

Dura mater

**FIGURE 15.22**

The flow of cerebrospinal fluid. Cerebrospinal fluid is secreted by choroid plexuses in the ventricular walls. The fluid circulates through the ventricles and central canal, enters the subarachnoid space, and is reabsorbed into the blood of the dural sinuses through the arachnoid villi.

cover the choroid plexuses, as well as line the central canal, and presumably aid the movement of the CSF.

CSF is formed mainly by the active transport and ultrafiltration of substances within the blood plasma. CSF has more sodium, chloride, magnesium, and hydrogen ions than blood plasma and less calcium, potassium, and glucose. In addition, CSF contains some proteins, urea, and white blood cells.

Up to 800 ml of CSF are produced each day, although only 140–200 ml are bathing the CNS at any given moment. A person lying in a horizontal position has a slow but continuous circulation of CSF, with a fluid pressure of about 10 mmHg.

 The homeostatic consistency of the CSF composition is critical, and a chemical imbalance may have marked effects on CNS functions. An increase in glycine (an amino acid) concentration, for example, produces hypothermia and hypotension as temperature and blood pressure regulatory mechanisms are disrupted. A slight change in pH may affect the respiratory rate and depth.

# Spinal Cord

*The spinal cord consists of centrally located gray matter involved in reflexes, and peripherally located ascending and descending tracts of white matter, which conduct impulses to and from the brain.*

The **spinal cord** is the portion of the CNS that extends through the vertebral canal of the vertebral column (fig. 15.23). It is continuous with the brain through the foramen magnum of the skull. The spinal cord has two principal functions.

**1  Impulse conduction.** It provides a means of neural communication to and from the brain through tracts of white matter. **Ascending tracts** conduct impulses from the peripheral sensory receptors of the body to the brain. **Descending tracts** conduct motor impulses from the brain to the muscles and glands.

**2  Reflex integration.** It serves as a center for spinal reflexes. Specific nerve pathways enable some movements to be reflexive rather than initiated voluntarily by the brain. Movements of this type are not confined to skeletal muscles; reflexive movements of cardiac and smooth muscles control heart rate, breathing rate, blood pressure, and digestive activities. Spinal nerve pathways are also involved in swallowing, coughing, sneezing, and vomiting.

## *Structure of the Spinal Cord*

The spinal cord extends inferiorly from the position of the foramen magnum of the occipital bone to the level of the first lumbar vertebra (L1). The spinal cord is somewhat flattened anteroposteriorly, making it oval in cross section. Two prominent enlargements can be seen in a posterior view (see fig. 15.23). The **cervical enlargement** is located between the third

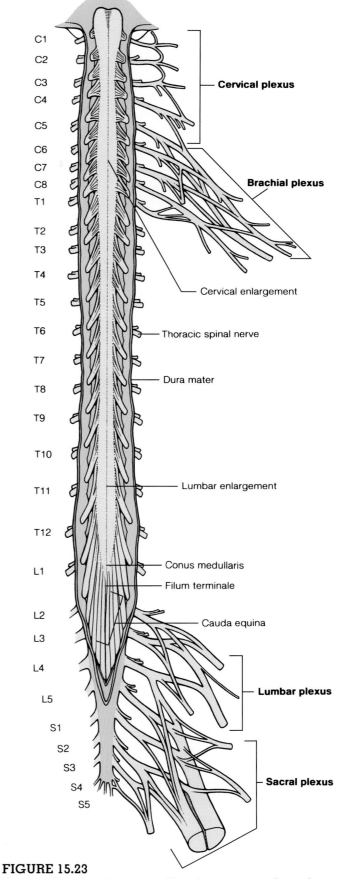

**FIGURE 15.23**

The spinal cord and plexuses. (The plexuses are indicated in boldface.)

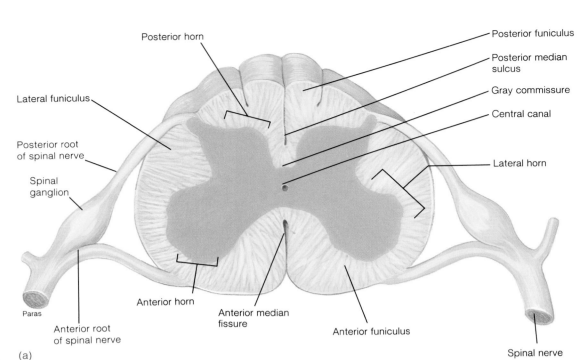

**Posterior horn**

**Lateral funiculus**

**Posterior root of spinal nerve**

**Spinal ganglion**

**Paras**

**Anterior root of spinal nerve**

**Anterior horn**

**Anterior median fissure**

**Anterior funiculus**

**Spinal nerve**

**Posterior funiculus**

**Posterior median sulcus**

**Gray commissure**

**Central canal**

**Lateral horn**

(a)

**FIGURE 15.24**

The spinal cord in cross section. (a) A diagram and (b) a photomicrograph.

cervical (C3) and the second thoracic vertebrae (T2). Nerves emerging from this region serve the upper extremities. The **lumbar enlargement** lies between the ninth and twelfth thoracic vertebrae. Nerves from the lumbar enlargement supply the lower extremities.

The embryonic spinal cord develops more slowly than the associated vertebral column; thus, in the adult, the cord does not extend beyond L1. The tapering, terminal portion of the spinal cord is called the **conus medullaris.** The **filum terminale,** a fibrous strand composed mostly of pia mater, extends interiorly from the conus medullaris at the level of L1 to the coccyx (see fig. 15.22). Nerve roots also radiate inferiorly from the conus medullaris through the vertebral canal. These nerves are collectively referred to as the **cauda equina** because they resemble a horse's tail.

The spinal cord develops as 31 segments, each of which gives rise to a pair of **spinal nerves** that emerge from the cord through the intervertebral foramina. Two grooves, an **anterior median fissure** and a **posterior median sulcus,** extend the length of the spinal cord and partially divide the cord into right and left portions. The spinal cord, like the brain, is protected by three distinct meninges and is

..........................................

filum terminalis: L. *filum,* filament; *terminus,* end
cauda equina: L. *cauda,* tail; *equus,* horse

**Posterior horn**

**Central canal**

**Anterior horn**

**Spinal ganglion**

**Spinal nerve**

(b)

cushioned by cerebrospinal fluid. The pia mater contains an extensive vascular network.

The **gray matter** of the spinal cord is centrally located and surrounded by white matter. It is composed of nerve cell bodies, neuroglia, and unmyelinated association neurons (interneurons). The **white matter** consists of bundles, or tracts, of myelinated fibers of sensory and motor neurons.

The relative size and shape of the gray and white matter varies throughout the spinal cord. The amount of white matter increases toward the brain as the nerve tracts become thicker. More gray matter is found in the cervical and lumbar enlargements where, respectively, innervations from the upper and lower extremities make connections.

The core of gray matter roughly resembles the letter H (fig. 15.24). Projections of the gray matter within the spinal cord are called horns and are named according to the direction in which they project. The paired **posterior horns** extend posteriorly and the paired **anterior horns** project anteriorly. A pair of short **lateral horns** extend to the sides and are located between the other two pairs. Lateral horns are prominent only in the thoracic and upper lumbar regions. The transverse bar of gray matter that connects the paired horns across the center of the spinal cord is called the **gray commissure.** Within the gray commissure is the **central canal.** It is continuous with the ventricles of the brain and is filled with cerebrospinal fluid.

## Spinal Cord Tracts

Impulses are conducted through the ascending and descending tracts of the spinal cord within the columns of white matter. The spinal cord has six columns of white matter called **funiculi** (*fyoo-nik-yŭ-li*), which are named according to their relative position within the cord. The two **anterior funiculi** are located between the two anterior horns of gray matter to either side of the anterior median fissure (fig. 15.24). The two **posterior funiculi** are located between the two posterior horns of gray matter to either side of the posterior median sulcus. Two **lateral funiculi** are located between the anterior and posterior horns of gray matter.

Each funiculus consists of both ascending and descending tracts. The nerve fibers within the tracts are generally myelinated and have specific sites of origin and termination. In fact, the names of the various tracts reflect their origin and termination. The fibers of the tracts either remain on the same side of the brain and spinal cord or cross over within the medulla oblongata or the spinal cord. The crossing over of nerve tracts is referred to as *decussation* (*de˝kŭ-sa´shun*). Figure 15.25 illustrates a descending tract that decussates within the medulla oblongata and figure 15.26 illustrates an ascending tract that decussates within the medulla oblongata.

The principal ascending and descending tracts within the funiculi are presented with their functions in table 15.4 and are illustrated in figure 15.27.

Descending tracts are grouped according to place of origin as either corticospinal or extrapyramidal. **Corticospinal** (pyramidal) **tracts** descend directly, without synap-

tic interruption, from the cerebral cortex to the lower motor neurons. The cell bodies of the neurons that contribute fibers to these tracts are located primarily in the precentral gyrus of the frontal lobe. Most (about 85%) of the corticospinal fibers decussate in the pyramids of the medulla oblongata (see fig. 15.9). The remaining 15% do not cross from one side to the other. The fibers that cross compose the **lateral corticospinal tracts,** and the remaining uncrossed fibers compose the **anterior corticospinal tracts.** Because of the crossing of fibers from higher motor neurons in the pyramids, the right hemisphere primarily controls the musculature on the left side of the body, whereas the left hemisphere controls the right musculature.

The corticospinal tracts are particularly important in voluntary movements that require correlation between the motor cortex and sensory input. Speech, for example, is impaired when the corticospinal tracts are damaged in the thoracic region of the spinal cord, whereas involuntary breathing continues. Damage to the pyramidal motor system can be detected clinically by a positive *Babinski reflex*, in which stimulation of the sole of the foot causes extension (upward movement) of the toes.

The remaining descending tracts are **extrapyramidal tracts** that originate in the brain stem region. Electrical stimulation of the cerebral cortex, the cerebellum, and the basal nuclei indirectly evokes movements because of synaptic connections within extrapyramidal tracts.

The **reticulospinal** (*rĕ-tik˝yŭ-lo-spi´nal*) **tracts** are the major descending pathways of the extrapyramidal system. These tracts originate in the reticular formation of the brain stem. Neurostimulation of the reticular formation by the cerebrum either facilitates or inhibits the activity of lower motor neurons depending on the area stimulated (fig. 15.28).

The basal nuclei, acting through synapses in the reticular formation particularly, appear normally to exert an inhibitory influence on the activity of lower motor neurons. Damage to the basal nuclei thus results in decreased muscle tone. People with such damage display *akinesia* (*a˝kĭ-ne´zha*) (lack of desire to use the affected limb) and *chorea* (sudden and uncontrolled random movements).

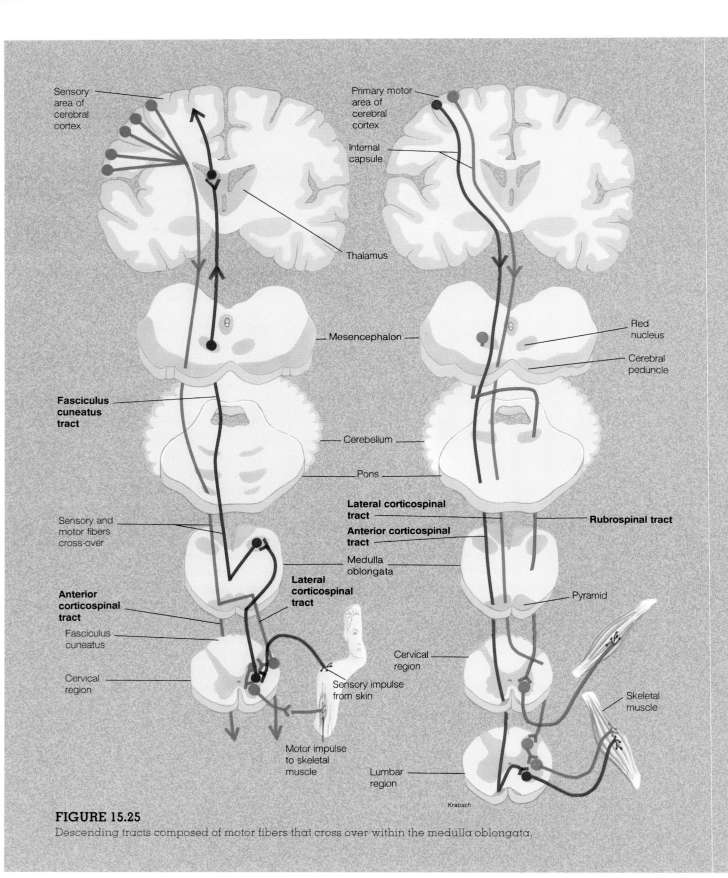

Sensory area of cerebral cortex

Primary motor area of cerebral cortex

Internal capsule

Thalamus

Mesencephalon

Red nucleus

Cerebral peduncle

**Fasciculus cuneatus tract**

Cerebellum

Pons

**Lateral corticospinal tract**

**Anterior corticospinal tract**

**Rubrospinal tract**

Sensory and motor fibers cross-over

Medulla oblongata

**Anterior corticospinal tract**

**Lateral corticospinal tract**

Pyramid

Fasciculus cuneatus

Cervical region

Cervical region

Sensory impulse from skin

Skeletal muscle

Motor impulse to skeletal muscle

Lumbar region

Krabach

## FIGURE 15.25

Descending tracts composed of motor fibers that cross over within the medulla oblongata.

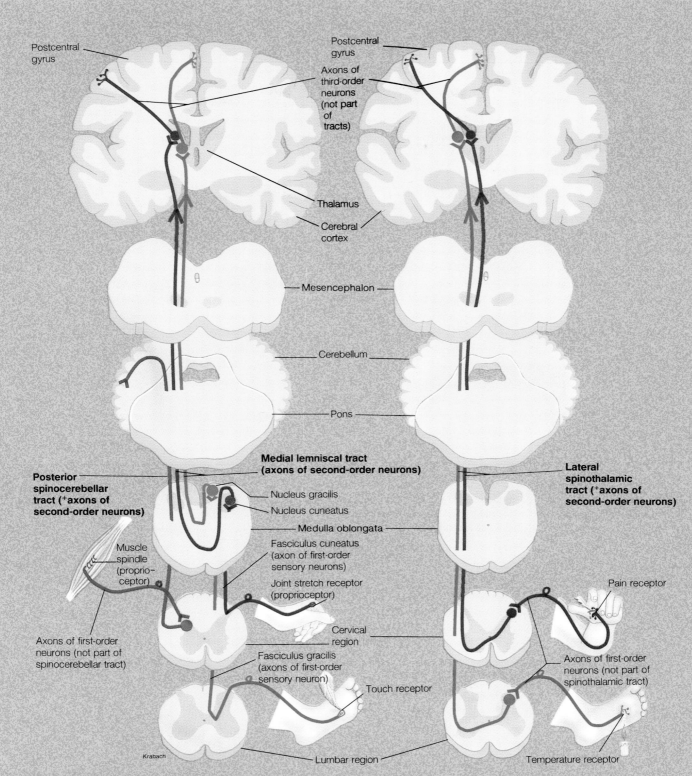

**FIGURE 15.26**

Ascending tracts composed of sensory fibers that cross over within the medulla oblongata.

# Table 15.4 — Principal ascending and descending tracts of the spinal cord

| Tract | Funiculus | Origin | Termination | Function |
|---|---|---|---|---|
| **Ascending tracts** | | | | |
| Anterior spinothalamic | Anterior | Posterior horn on one side of cord but crosses to opposite side | Thalamus, then cerebral cortex | Conducts sensory impulses for crude touch and pressure |
| Lateral spinothalamic | Lateral | Posterior horn on one side of cord but crosses to opposite side | Thalamus, then cerebral cortex | Conducts pain and temperature impulses that are interpreted within cerebral cortex |
| Fasciculus gracilis and fasciculus cuneatus | Posterior | Peripheral sensory neurons; does not cross over | Nucleus gracilis and nucleus cuneatus of medulla oblongata; crosses to the opposite side; eventually thalamus, then cerebral cortex | Conducts sensory impulses from skin, muscles, tendons, and joints, which are interpreted as sensations of fine touch, precise pressures, and body movements |
| Posterior spinocerebellar | Lateral | Posterior horn; does not cross over | Cerebellum | Conducts sensory impulses from one side of body to same side of cerebellum for subconscious proprioception necessary for coordinated muscular contractions |
| Anterior spinocerebellar | Lateral | Posterior horn; some fibers cross, others do not | Cerebellum | Conducts sensory impulses from both sides of body to cerebellum for subconscious proprioception necessary for coordinated muscular contractions |
| **Descending tracts** | | | | |
| Anterior corticospinal | Anterior | Cerebral cortex on one side of brain; crosses to opposite side of cord | Anterior horn | Conducts motor impulses from cerebrum to spinal nerves and outward to cells of anterior horns for coordinated, precise voluntary movements of skeletal muscle |

**FIGURE 15.27**

A transverse section showing the principal ascending and descending tracts within the spinal cord.

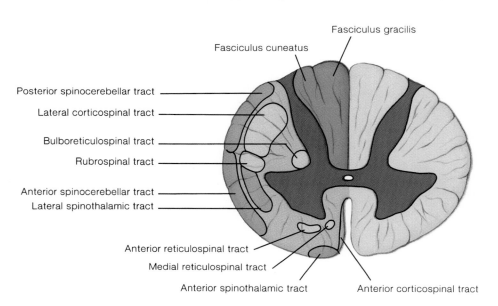

Fasciculus gracilis
Fasciculus cuneatus
Posterior spinocerebellar tract
Lateral corticospinal tract
Bulboreticulospinal tract
Rubrospinal tract
Anterior spinocerebellar tract
Lateral spinothalamic tract
Anterior reticulospinal tract
Medial reticulospinal tract
Anterior spinothalamic tract
Anterior corticospinal tract

## Table 15.4 Continued

| Tract | Funiculus | Origin | Termination | Function |
|---|---|---|---|---|
| **Descending tracts** | | | | |
| Lateral corticospinal | Lateral | Cerebral cortex on one side of brain; crosses in base of medulla oblongata to opposite side of cord | Anterior horn | Conducts motor impulses from cerebrum to spinal nerves and outward to cells of anterior horns for coordinated, precise voluntary movements |
| Tectospinal | Anterior | Mesencephalon; crosses to opposite side of cord | Anterior horn | Conducts motor impulses to cells of anterior horns and eventually to muscles that move head in response to visual, auditory, or cutaneous stimuli |
| Rubrospinal | Lateral | Mesencephalon (red nucleus); crosses to opposite side of cord | Anterior horn | Conducts motor impulses concerned with muscle tone and posture |
| Vestibulospinal | Anterior | Medulla oblongata; does not cross over | Anterior horn | Conducts motor impulses that regulate body tone and posture (equilibrium) in response to movements of head |
| Anterior and medial reticulospinal | Anterior | Reticular formation of brain stem does not cross over | Anterior horn | Conducts motor impulses that control muscle tone and sweat-gland activity |
| Bulboreticulospinal | Lateral | Reticular formation of brain stem; does not cross over | Anterior horn | Conducts motor impulses that control muscle tone and sweat-gland activity* |

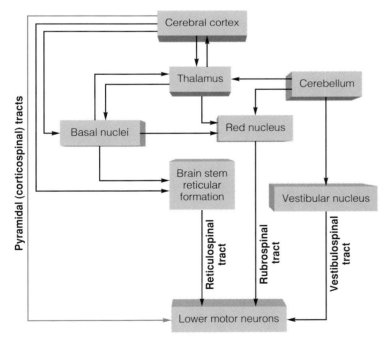

**FIGURE 15.28**

Pathways involved in the higher motor neuron control of skeletal muscles. (The pyramidal [corticospinal] tracts are indicated by the red arrow and the extrapyramidal tracts are indicated by black arrows.)

There are no descending tracts from the cerebellum. The cerebellum can influence motor activity only indirectly, through the vestibular nuclei, red nucleus, and basal nuclei. These structures, in turn, affect lower motor neurons via the **vestibulospinal tracts, rubrospinal tracts,** and reticulospinal tracts. Damage to the cerebellum interferes with the coordination of movements with spatial judgment. Underreaching or overreaching for an object may occur, followed by *intention tremor*, in which the limb moves back and forth in a pendulum-like motion.

# Clinical Considerations

## Neurological Assessment and Drugs

Neurological assessment has become exceedingly sophisticated and accurate in the past few years. In most physical examinations, only the basic aspects such as reflexes and sensory functions are assessed. But if the physician

## Development of the Spinal Cord

The spinal cord, like the brain, develops as the neural tube undergoes differentiation and specialization. Throughout the developmental process, the hollow central canal persists while the specialized white and gray matter forms (fig. 1). Changes in the neural tube become apparent during the sixth week as the lateral walls thicken to form a groove, called the **sulcus limitans** along each lateral wall of the central canal. A pair of **alar plates** forms dorsal to the sulcus limitans and a pair of **basal plates** forms ventrally. By the ninth week, the alar plates have specialized to become the **posterior horns,** containing fibers of the sensory cell bodies, and the basal plates have specialized to form the **anterior** and **lateral horns,** containing motor cell bodies. Sensory neurons of spinal nerves conduct impulses toward the spinal cord, whereas motor neurons conduct impulses away from the spinal cord.

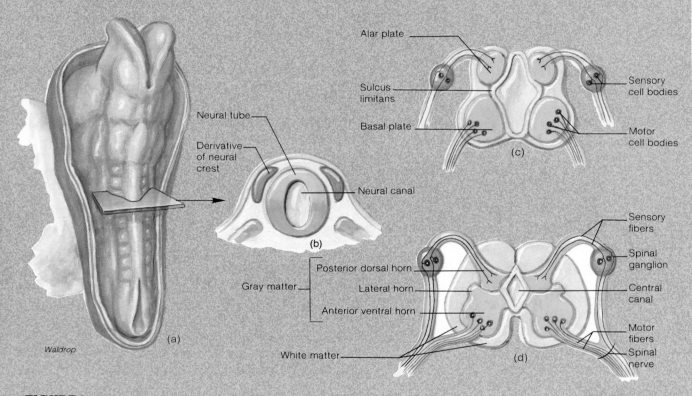

Waldrop

## FIGURE 1

The development of the spinal cord. (*a*) A dorsal view of an embryo at 23 days with the position of a transverse cut indicated in (*b*). (*c*) The formation of the alar and basal plates is evident in a cross section through the spinal cord at 6 weeks. (*d*) The central canal has reduced in size, and functional posterior and anterior horns have formed at 9 weeks.

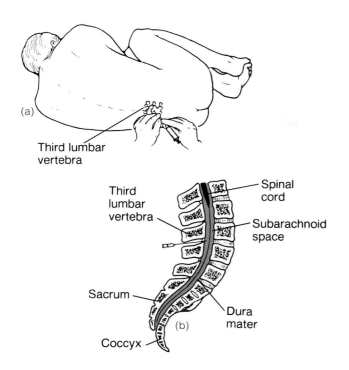

Third lumbar
vertebra

(a)

Third
lumbar
vertebra

Spinal
cord

Subarachnoid
space

Sacrum

Dura
mater

(b)

Coccyx

**FIGURE 15.29**

(a) A lumbar puncture is performed by inserting a needle between the third and fourth lumbar vertebrae (L3–L4) and
(b) withdrawing cerebrospinal fluid from the subarachnoid space.

suspects abnormalities involving the nervous system, further neurological tests are done, employing the following techniques.

A **lumbar puncture** is performed by inserting a fine needle between the third and fourth lumbar vertebrae and withdrawing a sample of CSF from the subarachnoid space (fig. 15.29). A **cisternal puncture** is similar to a lumbar puncture, except that the CSF is withdrawn from the cisterna magna, near the foramen magnum of the skull. The pressure of the CSF, which is normally about 10 mm of mercury, is measured with a *manometer*. Samples of CSF may also be examined for abnormal constituents. In addition, excessive fluids, accumulated as a result of disease or trauma, may be drained.

The condition of the arteries of the brain can be determined through a **cerebral angiogram.** In this technique, a radiopaque substance is injected into the common carotid arteries and allowed to disperse through the cerebral vessels. Aneurysms and vascular constrictions or displacements by tumors may then be revealed on radiographs.

The development of the **CT scanner,** or **computerized axial tomographic scanner,** revolutionized the diagnosis of brain disorders. More recently, the use of **MRI,** or **magnetic resonance imaging,** has immensely improved neurological diagnostic procedures. In MRI, a sharply detailed image of a patient's brain or spinal cord is projected onto a television screen, enabling quick and accurate diagnoses of tumors, aneurysms, blood clots, and hemorrhage. MRI may also be used to detect certain types of birth defects, brain damage, scar tissue, and old or recent strokes.

Another machine that also has greater potential than the CT scanner is the **DSR,** or **dynamic spatial reconstructor.** Like the CT scanner, the DSR is computerized to transform radiographs into composite video images. But the DSR is faster and presents a three-dimensional view. DSR can produce 75,000 cross-sectional images in 5 seconds, whereas CT can produce only one. At that speed, body functions as well as structures may be studied. For example, blood flow through blood vessels of the brain can be observed. This type of data is important in detecting early symptoms leading to a stroke or other disorders.

Certain disorders of the brain may be diagnosed more simply by examining brain-wave patterns using an **electroencephalogram.** Sensitive electrodes placed on the scalp record particular EEG patterns being emitted from evoked cerebral activity. EEG recordings are used to monitor epileptic patients to predict seizures and determine proper drug therapy and also to monitor comatose patients.

The extreme sensitivity of the nervous system to various drugs is fortunate, but is can also be potentially disastrous to a person. *Drug abuse* is a major clinical concern because of the addictive and devastating effect that certain drugs have on the nervous system. Much has been written on drug abuse, and it is beyond the scope of this text to elaborate on the effects of the many drugs that are commonly abused. A positive aspect of drugs is their administration in medicine to temporarily interrupt the passage or perception of sensory impulses. Injecting an anesthetic drug near a nerve, as in dentistry, desensitizes a specific area and causes a *nerve block.* Nerve blocks of a limited extent occur if an appendage is cooled or if a nerve is compressed for a period of time. Before the discovery of pharmacological drugs, physicians would frequently cool an affected appendage with ice or snow before performing surgery. **General anesthetics** affect the brain and render a person unconscious. A **local anesthetic** causes a nerve block by desensitizing a specific area.

## Injuries

Although the brain and spinal cord seem to be well protected within a bony encasement, they are sensitive organs, highly susceptible to injury.

Certain symptomatic terms are used when determining possible trauma within the CNS. **Headaches** are the most common ailment of the CNS. Most headaches are due to dilated blood vessels with the meninges of the brain. Headaches are generally asymptomatic of brain disorders, associated rather with physiological stress, eyestrain, or fatigue. Persistent and intense headaches may indicate a more serious problem such as a brain tumor. A **migraine** is a specific type of headache commonly preceded or accompanied by visual impairments and gastrointestinal unrest. It is not known why only 5%–10% of the population periodically suffer from migraines or why they are more common in women. Fatigue, allergy, or emotional stress tends to trigger migraines.

**Fainting** is a brief loss of consciousness that may result from a rapid pooling of blood in the lower extremities. It may occur when a person rapidly arises from a reclined position, receives a blow to the head, or experiences an intense psychologic stimulus, such as viewing a cadaver for the first time. Fainting is of more concern when it is symptomatic of a particular disease.

A **concussion** is a sudden movement of the brain caused by a violent blow to the head, which may or may not fracture bones of the skull. A concussion usually results in a brief period of unconsciousness followed by mild **delirium,** in which the patient is in a state of confusion. **Amnesia** is a more intense disorientation in which the patient suffers various degrees of memory loss.

A person who survives a severe head injury may be **comatose** for a short or an extended period of time. A coma is a state of unconsciousness from which the patient cannot be aroused, even by the most intense external stimuli. The area of the brain most likely to cause a coma from trauma is the reticular activating system. Although a head injury is the most common cause of coma, chemical imbalances associated with certain diseases (e.g., diabetes) or the ingestion of drugs or poisons may also be responsible.

The flexibility of the vertebral column is essential for body movements, but because of this flexibility the spinal cord and spinal nerves are somewhat vulnerable to trauma. Falls or severe blows to the back are a common cause of injury. A skeletal injury, such as a fracture, dislocation, or compression of the vertebrae, usually traumatizes nervous tissue as well. Other frequent causes of trauma to the spinal cord include gunshot wounds, stabbings, herniated discs, and birth injuries. The consequences of the trauma depend on the location and severity of the injury and the medical treatment the patient receives. If nerve fibers of the spinal cord are severed, motor or sensory functions will be permanently lost.

**Paralysis** is a permanent loss of motor control, usually resulting from disease or a lesion of the spinal cord or specific nerves. Paralysis of both lower extremities is called **paraplegia.** Paralysis of both the upper and lower extremity on the same side is called **hemiplegia,** and paralysis of all four extremities is **quadriplegia.** Paralysis may be flaccid or spastic. **Flaccid** (*flak´sid*) **paralysis** generally results from a lesion of the anterior horn cells and is characterized by noncontractible muscles that atrophy. **Spastic paralysis** results from lesions of the corticospinal tracts of the spinal cord and is characterized by hypertonicity of the skeletal muscles.

**Whiplash** is a sudden hyperextension and flexion of the cervical vertebrae such as may occur during a rear-end automobile collision. Recovery from a minor whiplash (muscle and ligament strains) is generally complete but slow. Severe whiplash (spinal cord compression) may cause permanent paralysis to the structures below the level of injury.

## Disorders of the Nervous System

**Mental Illness**   Mental illness is a major clinical consideration of the nervous system and is perhaps the least understood. The two principal categories of mental disorders are neurosis and psychosis. In *neurosis*, a maladjustment to certain aspects of life interferes with normal functioning, but contact with reality is maintained. An irrational fear (*phobia*) is an example of neurosis. Neurosis frequently causes intense anxiety or abnormal distress that brings about increased sympathetic stimulation. *Psychosis*, a more serious mental condition, is typified by a withdrawal from reality and is usually socially unacceptable. The more common forms of psychosis include *schizophrenia*, in which a person withdraws into a world of fantasy; *paranoia*, in which a person has systematized delusions often of a persecutory nature; and *manic-depressive psychosis*, in which a person's moods swing widely from intense elation to deepest despair.

**Epilepsy**   Epilepsy is a relatively common brain disorder with a strong hereditary basis, but it also can be caused by head injuries, tumors, and childhood infectious diseases. It is sometimes idiopathic (without demonstrable cause). A person with epilepsy may periodically suffer from an *epileptic seizure*, which has various symptoms depending on the type of epilepsy.

---

amnesia: L. *amnesia*, forgetfulness
comatose: Gk. *koma*, deep sleep

paralysis: Gk. *paralysis*, loosening
paraplegia: Gk. *para*, beside; *plessein*, to strike

The most common kinds of epilepsy are petit mal, psychomotor epilepsy, and grand mal. **Petit mal** (*pet´e-mal´*) occurs almost exclusively in children between the ages of 3 and 12. A child experiencing a petit mal seizure loses contact with reality for 5–30 seconds but does not lose consciousness or have convulsions. There may, however, be slight uncontrollable facial gestures or eye movements, and the child will stare, as if in a daydream. During a petit mal seizure, the thalamus and hypothalamus produce an extremely slow EEG pattern of 3 waves per second. Children with petit mal usually outgrow the condition by age 9 or 10 and generally require no medication.

**Psychomotor epilepsy** is often confused with mental illness because of the symptoms characteristic of the seizure. During such a seizure, EEG activity accelerates in the temporal lobes, causing a person to become disoriented and lose contact with reality. Occasionally during a seizure, specific cerebral motor areas will cause involuntary lip smacking or hand clapping. If motor areas in the brain are not stimulated, a person having a psychomotor epileptic seizure may wander aimlessly until the seizure subsides.

**Grand mal** is a more serious form of epilepsy characterized by periodic convulsive seizures that generally render a person unconscious. Grand mal epileptic seizures are accompanied by rapid EEG patterns of 25–30 waves per second. This sudden increase of EEG patterns from the normal of about 10 waves per second may cause an extensive stimulation of motor units and, therefore, uncontrollable skeletal muscle activity. During a grand mal seizure, a person loses consciousness, convulses, and may lose urinary bladder and bowel control. The unconsciousness and convulsions usually last a few minutes, after which the muscles relax and the person awakes but remains disoriented for a short time.

Epilepsy almost never affects intelligence and can be effectively treated with drugs in about 85% of the patients.

**Cerebral Palsy** Cerebral palsy is a condition of motor disorders characterized by paresis (partial paralysis) and lack of muscular contraction. It is caused by damage to the motor areas of the brain during prenatal development, birth, or infancy. During neural development within an embryo, radiation or bacterial toxins (such as from German measles), transferred through the placenta of a pregnant mother, may cause cerebral palsy. Oxygen deprivation due to complications at birth and hydrocephalus in a newborn may also cause cerebral palsy. The three areas of the brain most severely affected by this disease are the cerebral cortex, the basal nuclei, and the cerebellum. The type of cerebral palsy is determined by the particular region of the brain that is affected.

Some degree of mental retardation occurs in 60%–70% of cerebral palsy victims. Partial blindness, deafness, and speech problems frequently accompany this disease. Cerebral palsy is nonprogressive (i.e., these impairments do not worsen as a person ages), but neither are there organic improvements.

**Neoplasms of the CNS** Neoplasms of the CNS are either **intracranial tumors,** which affect cells within or associated with the brain, or they are **intravertebral (intraspinal) tumors,** which affect cells within or near the spinal cord. **Primary neoplasms** develop within the CNS. Approximately half of these are benign, but they may become lethal because of the pressure they exert upon vital centers as they grow. Patients with **secondary,** or **metastatic, neoplasms** within the brain have a poor prognosis because the cancer has already established itself in another body organ, frequently the liver, lung, or breast, and has only secondarily spread to the brain. The symptoms of a brain tumor include headache, convulsions, pain, paralysis, or a change in behavior.

Neoplasms of the CNS are classified according to the tissues in which the cancer occurs. Tumors arising in neuroglial cells are called **gliomas** (*gle-o´maz*) and account for about one-half of all primary neoplasms within the brain. Gliomas are frequently spread throughout cerebral tissue, develop rapidly, and usually cause death within a year after diagnosis. **Astrocytoma** (*as´´tro-si-to´mă*), **oligodendroglioma** (*ol´´ĭ-go-den´´drog-li-o´mă*), *and* **ependymoma** (*ĕ-pen´´dĭ-mo´mă*) are common types of gliomas.

**Meningiomas** arise from meningeal coverings of the brain and account for about 15% of primary intracranial tumors. Meningiomas are usually benign if they can be treated readily.

Intravertebral tumors are classified as **extramedullary** when they develop on the outside of the spinal cord and as **intramedullary** when they develop within the substance of the spinal cord. Extramedullary neoplasms may cause pain and numbness in body structures distant from the tumor as the growing tumor compresses the spinal cord. An intramedullary neoplasm causes a gradual loss of sensory perception and motor function below the spinal-segmental level of the affliction.

Methods of detecting and treating cancers within the CNS have greatly improved in the last few years. Early detection and competent treatment have lessened the likelihood of death from this disease and reduced the probability of physical impairment.

petit mal: L. *pitinnus*, small child; *malus*, bad
epilepsy: Gk. *epi*, upon; *lepsis*, seize

Central Nervous System

**Dyslexia** Dyslexia is a defect in the language center within the brain. In dyslexia, people reverse the order of letters in syllables, of syllables in words, and of words in sentences. The sentence: "The man saw a red dog," for example might be read by the dyslexic as "A red god was the man." Dyslexia is believed to result from the failure of one cerebral hemisphere to respond to written language, perhaps due to structural defects. Dyslexia can usually be overcome by intense remedial instruction in reading and writing.

**Meningitis** The nervous system is vulnerable to a variety of organisms and viruses that may cause abscesses or infections. Meningitis is an infection of the meninges. It may be confined to the spinal cord, in which case it is referred to as **spinal meningitis,** or it may involve the brain and associated meninges, in which case it is known as **encephalitis** or **encephalomyelitis** (*en-sef″ă-lo-mi″ĕ-li-′tis*), respectively. The microorganisms that most commonly cause meningitis are meningococci, streptococci, pneumococci, and tubercle bacilli. Viral meningitis is more serious than bacterial meningitis. Nearly 20% of viral encephalitides are fatal. The organisms that cause meningitis probably enter the body through respiratory passageways.

**Poliomyelitis** Poliomyelitis, or infantile paralysis, is primarily a childhood disease caused by a virus that destroys nerve cell bodies within the anterior horn of the spinal cord, especially those within the cervical and lumbar enlargements. This degenerative disease is characterized by fever, severe headache, stiffness and pain, and the loss of certain somatic reflexes. Muscle paralysis follows within several weeks, and eventually the muscles atrophy. Death results if the virus invades the vasomotor and respiratory nuclei within the medulla oblongata or anterior horn cells controlling respiratory muscles. Poliomyelitis has been effectively controlled with immunization.

**Syphilis** Syphilis is a sexually transmitted disease that, if untreated, progressively destroys body organs. When syphilis causes organ degeneration, it is said to be in the *tertiary stage*

...........................................

dyslexia: Gk,. *dys,* bad; *lexis,* speech

(10 to 20 years after the primary infection). The organs of the nervous system are frequently infected, causing a condition called **neurosyphilis.** Neurosyphilis is classified according to the tissue involved, and the symptoms vary correspondingly. If the meninges are infected, the condition is termed **chronic meningitis. Tabes dorsalis** is a form of neurosyphilis in which there is a progressive degeneration of the posterior funiculi of the spinal cord and posterior roots of spinal nerves. Motor control is gradually lost, and patients eventually become bedridden, unable even to feed themselves.

## Degenerative Diseases of the Nervous System

Degenerative diseases of the CNS are characterized by a progressive, symmetrical deterioration of vital structures of the brain or spinal cord. The etiologies of these diseases are poorly understood, but it is thought that most of them are genetic.

**Cerebrovascular Accident (CVA)** Cerebrovascular accident is the most common disease of the nervous system. It is the third highest cause of death in the United States and perhaps the major cause of disability. The term **stroke** is frequently used synonymously with CVA, but actually a stroke refers to the sudden and dramatic appearance of a neurological defect. Cerebral thrombosis, in which a thrombus, or clot, forms in an artery of the brain, is the most common cause of CVA. Other causes of CVA include intracerebral hemorrhages, aneurysms, atherosclerosis, and arteriosclerosis of the cerebral arteries.

Patients who recover from CVA frequently suffer partial paralysis and mental disorders, such as loss of language skills. The dysfunction depends upon the severity of the CVA and the regions of the brain that were injured. Patients surviving a CVA can often be rehabilitated, but approximately two-thirds die within 3 years of the initial damage.

**Syringomyelia** Syringomyelia (*sĭ-ring″go-mi-e′le-ă*) is a relatively uncommon condition characterized by the appearance of cystlike cavities, called *syringes,* within the gray matter of the spinal cord. These syringes progressively destroy the cord from the inside out. Syringomyelia is a chronic, slow-progressing disease of unknown cause. As the spinal cord deteriorates, the patient experiences muscular weakness and atrophy and sensory loss, particularly of the senses of pain and temperature.

# Chapter Summary

## Characteristics of the Central Nervous System (pp. 376–380)

1. The central nervous system (CNS) consists of the brain and spinal cord and contains gray and white matter. It is covered with meninges and is bathed in cerebrospinal fluid.
2. The tremendous metabolic rate of the 1.5-kg brain requires a continuous flow of blood, amounting to approximately 20% of the total cardiac output.

## Cerebrum (pp. 380–390)

1. The cerebrum, consisting of two convoluted hemispheres, is concerned with higher brain functions, such as the perception of sensory impulses, the instigation of voluntary movement, the storage of memory, thought processes, and reasoning ability.
2. The cerebral cortex of the cerebral hemispheres is convoluted with gyri and sulci.
3. Each cerebral hemisphere contains frontal, parietal, temporal, occipital, and insula lobes.
4. Brain waves generated by the cerebral cortex are recorded as an electroencephalogram and may provide valuable diagnostic information.
5. The white matter of the cerebrum consists of association, commissural, and projection fibers.
6. Basal nuclei are specialized masses of gray matter located within the white matter of the cerebrum.
7. The motor speech area, Wernicke's area, the arcuate fasciculus, and the angular gyrus are the language areas of the brain and are generally located in the cerebral cortex of the left hemisphere.
8. The consolidation of memory requires protein synthesis and probably involves changes in the chemical structure and function of synapses.

## Diencephalon (pp. 390–392)

1. The diencephalon is a major autonomic region of the brain.
2. The thalamus is an ovoid mass of gray matter that functions as a relay center for sensory impulses and responds to pain.
3. The hypothalamus is an aggregation of specialized nuclei that regulate many visceral activities. It also performs emotional and instinctual functions.
4. The epithalamus contains the pineal gland and the vascular choroid plexus over the roof of the third ventricle.

## Mesencephalon (p. 392)

1. The mesencephalon contains the corpora quadrigemina, the cerebral peduncles, and specialized nuclei that help to control posture and movement.
2. The superior colliculi of the corpora quadrigemina are concerned with visual reflexes; the inferior colliculi are concerned with auditory reflexes.
3. The red nucleus and the substantia nigra are concerned with motor activities.

## Metencephalon (pp. 393–394)

1. The pons consists of fiber tracts connecting the cerebellum and medulla oblongata to other structures of the brain. The pons also contains nuclei for certain cranial nerves and the regulation of respiration.
2. The cerebellum consists of two hemispheres connected by the vermis and supported by three paired cerebellar peduncles.
   a. The cerebellum is composed of a white matter tract called the arbor vitae, surrounded by a thin convoluted cortex of gray matter.
   b. The cerebellum is concerned with coordinated contractions of skeletal muscle.

## Myelencephalon (pp. 394–396)

1. The medulla oblongata is composed of the ascending and descending tracts of the spinal cord and contains nuclei for several autonomic functions.
2. The reticular formation functions as the reticular activating system in arousing the cerebrum.

## Meninges of the Central Nervous System (pp. 396–398)

1. The cranial dura mater consists of an outer periosteal layer and an inner meningeal layer. The spinal dura mater is a single layer surrounded by the vascular epidural space.
2. The arachnoid mater is a netlike meninx surrounding the subarachnoid space. The subarachnoid space contains cerebrospinal fluid.
3. The thin pia mater adheres to the contour of the CNS.

## Ventricles and Cerebrospinal Fluid (pp. 398–401)

1. The lateral (first and second), third, and fourth ventricles are interconnected chambers within the brain that are continuous with the central canal of the spinal cord.
2. These chambers are filled with cerebrospinal fluid, which also flows throughout the subarachnoid space.
3. Cerebrospinal fluid is continuously secreted by the choroid plexuses and is absorbed into the blood at the arachnoid villi.

## Spinal Cord (pp. 401–407)

1. The spinal cord is composed of 31 segments, each of which gives rise to a pair of spinal nerves.
   a. It is characterized by a cervical enlargement, a lumbar enlargement, and two longitudinal grooves that partially divide it into right and left halves.
   b. The conus medullaris is the terminal portion of the spinal cord and the cauda equina are nerve roots that radiate inferiorly from that point.
2. Ascending and descending spinal cord tracts are referred to as funiculi.
   a. Descending tracts are grouped as either corticospinal (pyramidal) or extrapyramidal.
   b. Many of the fibers in the funiculi decussate (cross over) in the spinal cord or in the medulla oblongata.

# Review Activities

## Objective Questions

1. Which of the following is *not* a lobe of the cerebrum?
   - a. parietal
   - d. insula
   - b. sphenoid
   - e. occipital
   - c. temporal
2. The principal connection between the cerebral hemispheres is
   - a. the corpus callosum.
   - b. the pons.
   - c. the intermediate mass.
   - d. the vermis.
   - e. the precentral gyrus.
3. Which of the following structures of the brain is most directly involved in the autonomic response to pain?
   - a. pons
   - b. hypothalamus
   - c. medulla oblongata
   - d. thalamus
4. Which statement is *false* concerning the basal nuclei?
   - a. They are located within the cerebrum.
   - b. They regulate the basal metabolic rate.
   - c. They consist of the caudate nucleus, lentiform nucleus, putamen, and globus pallidus.
   - d. They indirectly exert an inhibitory influence on lower motor neurons.
5. In which region of the brain are the corpora quadrigemina, red nucleus, and substantia nigra located?
   - a. diencephalon
   - b. metencephalon
   - c. mesencephalon
   - d. myelencephalon
6. The fourth ventricle is contained within
   - a. the cerebrum.
   - b. the cerebellum.
   - c. the midbrain.
   - d. the metencephalon.
7. The right cerebral cortex controls voluntary movements on the left side of the body because
   - a. most people are right-handed.
   - b. the right hemisphere dominates.
   - c. many of the fibers in the funiculi decussate in the medulla oblongata.
   - d. there is distinct cerebral specialization of hemispheres.
8. A patient experiencing a fluctuating body temperature, lack of hunger and thirst, and psychosomatic disorders may have a malfunctioning
   - a. hypothalamus.
   - b. midbrain.
   - c. cerebellum.
   - d. medulla oblongata.
9. Spinal cord tracts that descend from the cerebral cortex to the lower motor neurons without synaptic interruption are called
   - a. reticulospinal tracts.
   - b. corticospinal tracts.
   - c. rubrospinal tracts.
   - d. vestibulospinal tracts.
10. The disease characterized by the destruction of the myelin sheaths of neurons in the CNS and the formation of plaques is
    - a. syringomyelia.
    - b. neurosyphilis.
    - c. poliomyelitis.
    - d. multiple sclerosis.

## Essay Questions

1. List the types of brain waves recorded on an electroencephalogram, and explain the diagnostic value of each.
2. List the functions of the hypothalamus. Why is the hypothalamus considered a major part of the autonomic nervous system?
3. What structures are found within the midbrain? List the nuclei located in the midbrain, and give the function of each.
4. Describe the location and structure of the medulla oblongata. List the nuclei found within this structure. What are the functions of the medulla oblongata?
5. What is cerebrospinal fluid? Where is it produced and what is its pathway of circulation?
6. What do the following abbreviations stand for? EEG, ANS, CSF, PNS, RAS, CT scan, MS, DSR, and CVA.
7. Describe the various techniques available for conducting a neurological assessment.
8. Define the various psychological terms used to describe mental illness.
9. What is epilepsy? What causes it and how is it controlled?
10. What do meningitis, poliomyelitis, and neurosyphilis have in common? How do these conditions differ?

## Gundy/Weber Software 💾

The tutorial software accompanying Chapter 15 is Volume 5—Nervous System.

# chapter sixteen

# peripheral nervous system

## objectives

- Define *peripheral nervous system* and distinguish between sensory, motor, and mixed nerves.

- List the 12 pairs of cranial nerves and describe the location and function of each.

- Describe the clinical methods for determining cranial nerve dysfunction.

- Explain how the spinal nerves are grouped.

- Describe the general distribution of a spinal nerve.

- List the spinal nerve composition of each of the plexuses arising from the spinal cord.

- List the principal nerves that emerge from the plexuses and describe their general innervation.

- Define *reflex arc* and list its five components.

- Distinguish between the various kinds of reflexes.

# Introduction to the Peripheral Nervous System

*The peripheral nervous system consists of receptors that respond to stimuli and of nerves that convey impulses to and from the central nervous system.*

The **peripheral nervous system (PNS)** is that portion of the nervous system outside the central nervous system. The PNS functions to convey impulses to and from the brain or spinal cord. Sensory receptors within the sensory organs, neurons, nerves, ganglia, and plexuses are all part of the PNS, which serves virtually every part of the body (fig. 16.1). The sensory receptors are discussed in chapter 18.

The nerves of the PNS are classified as cranial nerves or spinal nerves, depending on whether they arise from the brain or the spinal cord. The term *sensory nerve*, *motor nerve*, and *mixed nerve* relate to the direction in which the nerve impulses are being conducted. **Sensory nerves** consist of sensory (afferent) neurons that convey impulses toward the CNS. **Motor nerves** consist of motor (efferent) neurons that

**FIGURE 16.1**

The peripheral nervous system consists of cranial nerves, spinal nerves, plexuses, ganglia (not shown), and peripheral nerves.

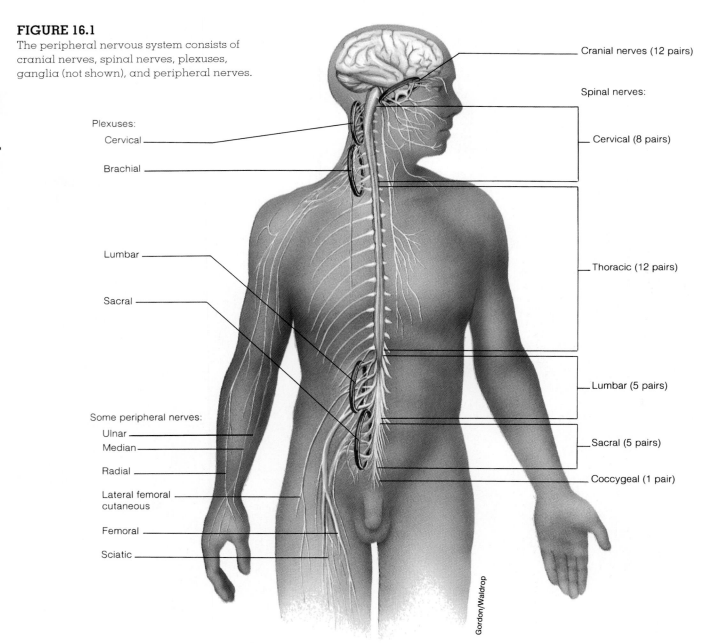

Cranial nerves (12 pairs)

Spinal nerves:

Cervical (8 pairs)

Thoracic (12 pairs)

Lumbar (5 pairs)

Sacral (5 pairs)

Coccygeal (1 pair)

Plexuses:
Cervical
Brachial
Lumbar
Sacral

Some peripheral nerves:
Ulnar
Median
Radial
Lateral femoral cutaneous
Femoral
Sciatic

Gordon/Waldrop

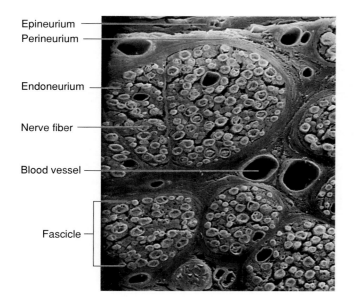

Epineurium
Perineurium

Endoneurium

Nerve fiber

Blood vessel

Fascicle

**FIGURE 16.2**
A scanning electron micrograph of a spinal nerve seen in cross section (about 1000×).
From *Tissues and Organs: A Text-Atlas of Scanning Electron Microscopy*, by R. G. Kessel and R. Kardon. © 1979 W. H. Freeman and Company.

convey impulses away from the CNS. **Mixed nerves** are composed of both sensory and motor neurons and therefore convey impulses in both directions. The reflexes considered in this chapter involve sensory and motor nerves of the PNS and specific portions of the CNS. A cross section of a spinal nerve is depicted in figure 16.2.

# Cranial Nerves

*Twelve pairs of cranial nerves emerge from the inferior surface of the brain and pass through the foramina of the skull to innervate structures in the head, neck, and visceral organs of the trunk.*

## Structure and Function of the Cranial Nerves

Of the 12 pairs of cranial nerves, 2 pairs arise from the forebrain and 10 pairs arise from the midbrain and brain stem (fig. 16.3). The cranial nerves are designated by Roman numerals and names. The Roman numerals refer to the order in which the nerves are positioned from the front of the brain to the back. The names indicate the structures innervated or the principal functions of the nerves. A summary of the cranial nerves is presented in table 16.1, and the locations of the nuclei from which they arise are illustrated in figure 15.17.

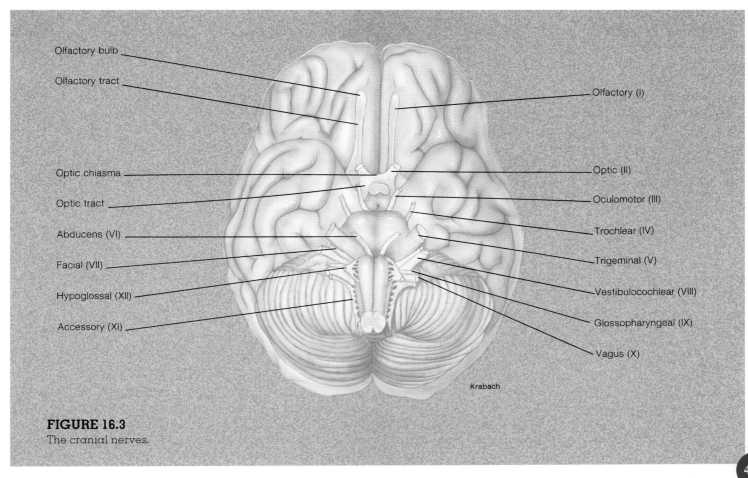

Olfactory bulb
Olfactory tract
Olfactory (I)

Optic chiasma
Optic (II)

Optic tract
Oculomotor (III)

Abducens (VI)
Trochlear (IV)

Facial (VII)
Trigeminal (V)

Hypoglossal (XII)
Vestibulocochlear (VIII)

Accessory (XI)
Glossopharyngeal (IX)

Vagus (X)

Krabach

**FIGURE 16.3**
The cranial nerves.

# Table 16.1    Summary of the cranial nerves

| Number and name | Foramen transmitting | Composition | Location of cell bodies | Function |
|---|---|---|---|---|
| I Olfactory | Foramina in cribriform plate of ethmoid bone | Sensory | Bipolar cells in nasal mucosa | Olfaction |
| II Optic | Optic canal | Sensory | Ganglion cells of retina | Vision |
| III Oculomotor | Superior orbital fissure | Motor | Oculomotor nucleus | Motor impulses to levator palpebrae superioris and extrinsic eye muscles except superior oblique and lateral rectus |
| | | Motor: parasympathetic | | Innervation to muscles that regulate amount of light entering eye and that focus the lens |
| | | Sensory: proprioception | | Proprioception from muscles innervated with motor fibers |
| IV Trochlear | Superior orbital fissure | Motor | Trochlear nucleus | Motor impulses to superior oblique muscle of eyeball |
| | | Sensory: proprioception | | Proprioception from superior oblique muscle of eyeball |
| V Trigeminal Ophthalmic nerve | Superior orbital fissure | Sensory | Trigeminal ganglion | Sensory impulses from cornea, skin of nose, forehead, and scalp |
| Maxillary nerve | Foramen rotundum | Sensory | Trigeminal ganglion | Sensory impulses from nasal mucosa, upper teeth and gums, palate, upper lip, and skin of cheek |
| Mandibular nerve | Foramen ovale | Sensory | Trigeminal ganglion | Sensory impulses from temporal region, tongue, lower teeth and gum, and skin of chin and lower jaw |
| | | Sensory: proprioception | | Proprioception from muscles of mastication |
| | | Motor | Motor trigeminal nucleus | Motor impulses to muscles of mastication and muscle that tenses tympanum |
| VI Abducens | Superior orbital fissure | Motor | Abducens nucleus | Motor impulses to lateral rectus muscle of eyeball |
| | | Sensory: proprioception | | Proprioception from lateral rectus muscle of eyeball |
| VII Facial | Stylomastoid foramen | Motor | Motor facial nucleus | Motor impulses to muscles of facial expression and muscle that tenses the stapes |
| | | Motor: parasympathetic | Superior salivatory nucleus | Secretion of tears from lacrimal gland and salivation from sublingual and submandibular glands |

Although most cranial nerves are mixed, some are associated with special senses and consist of sensory neurons only. The cell bodies of sensory neurons are located in ganglia outside the brain.

Generations of anatomy students have used a mnemonic device to help them remember the order in which the cranial nerves emerge from the brain: "On old Olympus's towering top, a Finn and German viewed a hop."

| Number and name | Foramen transmitting | Composition | Location of cell bodies | Function |
|---|---|---|---|---|
| VII Facial (*continued*) | | Sensory | Geniculate ganglion | Senory impulses from taste buds on anterior two-thirds of tongue; nasal and palatal sensation |
| | | Sensory: proprioception | | Proprioception from muscles of facial expression |
| VIII Vestibulocochlear | Internal acoustic meatus | Sensory | Vestibular ganglion | Sensory impulses associated with equilibrium |
| | | | Spiral ganglion | Sensory impulses associated with hearing |
| IX Glossopharyngeal | Jugular foramen | Motor | Nucleus ambiguus | Motor impulses to muscles of pharynx used in swallowing |
| | | Sensory: proprioception | Petrosal ganglion | Proprioception from muscles of pharynx |
| | | Sensory | Petrosal ganglion | Sensory impulses from taste buds on posterior one-third of tongue, pharynx, middle-ear cavity, and carotid sinus |
| | | Parasympathetic | Inferior salivatory nucleus | Salivation from parotid gland |
| X Vagus | Jugular foramen | Motor | Nucleus ambiguus | Contraction of muscles of pharynx (swallowing) and larynx (phonation) |
| | | Sensory: proprioception | | Proprioception from visceral muscles |
| | | Sensory | Nodose ganglion | Sensory impulses from taste buds on rear of tongue; sensations from auricle of ear; general visceral sensations |
| | | Motor: parasympathetic | Dorsal motor nucleus | Regulate visceral motility |
| XI Accessory | Jugular foramen | Motor | Nucleus ambiguus | Laryngeal movement; soft palate |
| | | | Accessory nucleus | Motor impulses to trapezius and sternocleidomastoid muscles for movement of head, neck, and shoulders |
| | | Sensory: proprioception | | Proprioception from muscles that move head, neck, and shoulders |
| XII Hypoglossal | Hypoglossal canal | Motor | Hypoglossal nucleus | Motor impulses to intrinsic and extrinsic muscles of tongue and infrahyoid muscles |
| | | Sensory: proprioception | | Proprioception from muscles of tongue |

The initial letter of each word in this jingle corresponds to the initial letter of each pair of cranial nerves. A problem with this classic verse is that the eighth cranial nerve represented by *and* in the jingle, which used to be referred to as auditory, is currently recognized as the vestibulocochlear cranial nerve. Hence, the following topical mnemonic: "On old Olympus's towering top, a fat vicious goat vandalized a hat."

**I Olfactory Nerve** Actually, numerous olfactory nerves relay sensory impulses of smell from the mucous membranes of the nasal cavity (fig. 16.4). Olfactory nerves are composed of bipolar neurons that function as *chemoreceptors*, responding to volatile chemical particles breathed into the nasal cavity. The dendrites and cell bodies of olfactory neurons are positioned within the mucosa, primarily that which covers the superior nasal conchae and adjacent nasal septum. The axons of these neurons pass through the cribriform plate of the ethmoid bone to the **olfactory bulb** where synapses are made, and the sensory impulses are passed through the **olfactory tract** to the primary olfactory area in the cerebral cortex.

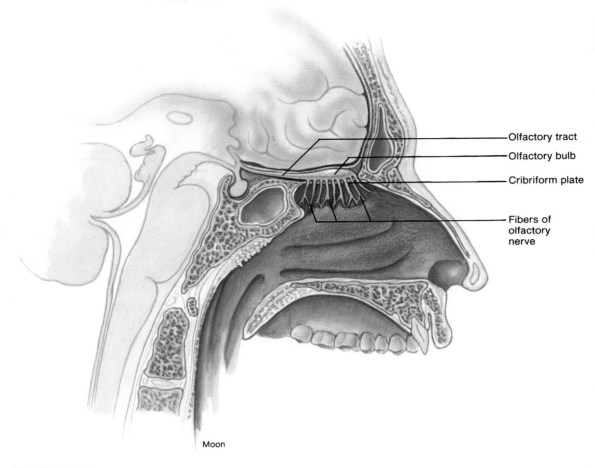

Moon

**FIGURE 16.4**
The olfactory nerve.

**II Optic Nerve** The optic nerve, another sensory nerve, conducts impulses from the *photoreceptors* (rods and cones) in the retina of the eye. Each optic nerve is composed of an estimated 1.25 million nerve fibers that converge at the back of the eyeball and enter the cranial cavity through the optic canal. The two optic nerves unite on the floor of the diencephalon to form the **optic chiasma** (*ki-as′mă*) (fig. 16.5). Nerve fibers that arise from the medial half of each retina cross at the chiasma to the opposite side of the brain, whereas fibers arising from the lateral half remain on the same side of the brain. The optic nerve fibers pass posteriorly from the optic chiasma in the **optic tracts.** The optic tracts lead to the thalamus, where a majority of the fibers terminate within certain thalamic nuclei. A few of the ganglion-cell axons that reach the thalamic nuclei have collaterals that convey impulses to the superior colliculi. Synapses within the thalamic nuclei, however, permit impulses to pass through neurons to the **visual cortex** within the occipital

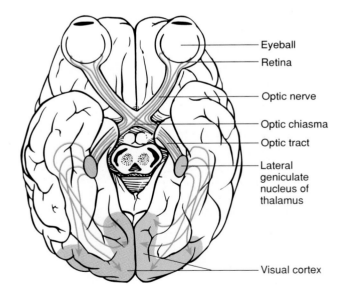

**FIGURE 16.5**
The optic nerve.

olfactory: L. *olfacere*, smell out
optic: L. *optica*, see
chiasma: Gk. *chiasma*, an X-shaped arrangement

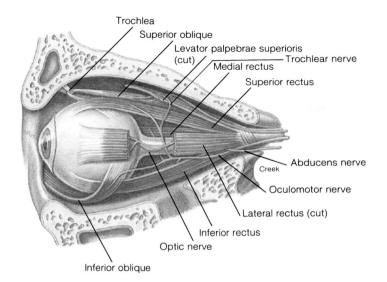

Trochlea
Superior oblique
Levator palpebrae superioris
(cut)
Trochlear nerve
Medial rectus
Superior rectus
Abducens nerve
Creek
Oculomotor nerve
Lateral rectus (cut)
Inferior rectus
Optic nerve
Inferior oblique

**FIGURE 16.6**
The oculomotor, trochlear, and abducens cranial nerves.

lobes. Other synapses permit impulses to reach the nuclei for the oculomotor, trochlear, and abducens nerves, which regulate intrinsic (internal) and extrinsic (from orbit to eyeball) eye muscles. The visual pathway into the eyeball functions reflexively to produce motor responses to light stimuli. If an optic nerve is damaged, the eyeball served by that nerve is blinded.

**III Oculomotor**   Nerve impulses through the oculomotor nerve produce certain extrinsic and intrinsic movements of the eyeball. The oculomotor is primarily a motor nerve that arises from nuclei within the midbrain. It divides into superior and inferior branches as it passes through the superior orbital fissure in the orbit (fig. 16.6). The superior branch innervates the **superior rectus eye muscle,** which moves the eyeball superiorly, and the **levator palpebrae** (*le-va´tor pal´pĕ-bre*) **superioris muscle,** which raises the upper eyelid. The inferior branch innervates the **medial rectus, inferior rectus,** and **inferior oblique eye muscles** for medial, inferior, and superior and lateral movement of the eyeball, respectively. In addition, fibers from the inferior branch of the oculomotor nerve enter the eyeball to supply autonomic motor innervation to the intrinsic smooth muscles of the iris for pupil constriction and to the muscles within the ciliary body for lens accommodation.

A few sensory fibers of the oculomotor nerve originate from proprioceptors within the intrinsic muscles of the eyeball. These fibers convey impulses that affect the position and activity of the muscles they serve. A person whose oculomotor nerve is damaged may have a drooping upper eyelid or a dilated pupil or be unable to move the eyeball in the directions permitted by the four extrinsic muscles innervated by this nerve.

**IV Trochlear**   The trochlear nerve is a very small mixed nerve that emerges from a nucleus within the midbrain and passes from the cranium through the superior orbital fissure of the orbit. The trochlear nerve innervates the **superior oblique muscle** of the eyeball with both motor and sensory fibers (fig. 16.6). Motor impulses to the superior oblique muscle cause the eyeball to rotate downward and away from the midline. Sensory impulses originate in proprioceptors of the superior oblique muscle and provide information about its position and activity. Damage to the trochlear nerve impairs movement in the direction permitted by the superior oblique eye muscle.

**V Trigeminal**   The large trigeminal nerve is a mixed nerve with motor functions originating from the nuclei within the pons and sensory functions terminating in nuclei within the midbrain, pons, and medulla oblongata. Two roots of the trigeminal nerve are apparent as they emerge from the ventrolateral side of the pons (see fig. 16.3). The larger **sensory root** immediately enlarges into a swelling called the **trigeminal** (gasserian) **ganglion,** located in a bony depression on the inner surface of the petrous part of the temporal bone. Three large nerves arise from the trigeminal ganglion (fig. 16.7): the **ophthalmic nerve** enters the orbit through the superior orbital fissure; the **maxillary nerve** extends through the foramen rotundum; and the **mandibular nerve** passes through the foramen ovale. The smaller **motor root** consists of motor fibers of the trigeminal nerve that accompany the mandibular nerve through the foramen ovale and innervate the muscles of mastication and certain muscles in the floor of the mouth. Impulses through the motor portion of the mandibular nerve of the trigeminal ganglion stimulate contraction of the muscles involved in chewing, including the medial and lateral pterygoid, masseter, temporalis, and mylohyoid muscles and the anterior belly of the digastric muscle.

Although the trigeminal is a mixed nerve, its sensory functions are much more extensive than its motor functions. The three sensory nerves of the trigeminal nerve respond to touch, temperature, and pain sensations from the face. More specifically, the ophthalmic nerve consists of sensory fibers from the anterior half of the scalp, skin of the forehead, upper eyelid, surface of the eyeball, lacrimal (tear) gland, side of the nose, and upper mucosa of the nasal cavity. The maxillary nerve is composed of sensory fibers from the lower eyelid, lateral and inferior mucosa of the nasal cavity, palate and portions of the pharynx, teeth and gums of the upper

......................................................

trochlear: Gk. *trochos,* a wheel
trigeminal: L. *trigeminus,* three born together
gasserian ganglion: from Johann L. Gasser, Viennese anatomist, eighteenth century
ophthalmic: L. *ophthalmia,* region of the eye

**FIGURE 16.7**
The trigeminal nerve and its branches.

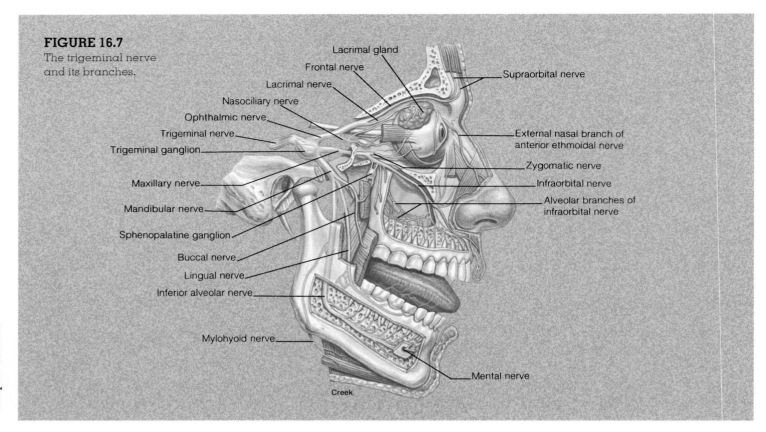

Lacrimal gland
Frontal nerve
Lacrimal nerve
Nasociliary nerve
Ophthalmic nerve
Trigeminal nerve
Trigeminal ganglion
Maxillary nerve
Mandibular nerve
Sphenopalatine ganglion
Buccal nerve
Lingual nerve
Inferior alveolar nerve
Mylohyoid nerve
Supraorbital nerve
External nasal branch of anterior ethmoidal nerve
Zygomatic nerve
Infraorbital nerve
Alveolar branches of infraorbital nerve
Mental nerve
Creek

jaw, upper lip, and skin of the cheek. Sensory fibers of the mandibular nerve transmit impulses from the teeth and gums of the lower jaw, anterior two-thirds of the tongue (but do not serve the sense of taste), mucosa of the mouth, auricle of the ear, and lower part of the face. Trauma to the trigeminal nerve results in a lack of sensation from specific facial structures. Damage to the mandibular nerve impairs chewing.

 The trigeminal nerve is the principal nerve relating to the practice of dentistry. Before teeth are filled or extracted, anesthetic is injected into the appropriate nerve to block sensation. A *maxillary*, or *second-division, nerve block*, performed by injecting near the sphenopalatine ganglion (see fig. 16.7), desensitizes the teeth in the upper jaw. A *mandibular*, or *third-division, nerve block* desensitizes the lower teeth. This is performed by injecting anesthetic into the inferior alveolar branch of the mandibular nerve where it enters the mandible through the mandibular foramen.

**VI Abducens** The small abducens nerve originates from a nucleus within the pons and emerges from the lower portion of the pons and the anterior border of the medulla oblongata. It is a mixed nerve that traverses the superior orbital fissure of the orbit to innervate the **lateral rectus eye muscle** (see fig. 16.6). Impulses through the motor fibers of the abducens nerve cause the lateral rectus eye muscle to contract and the eyeball to move laterally away from the midline. Sensory impulses through the abducens nerve originate in proprioceptors

in the lateral rectus eye muscle and are conveyed to the pons, where muscle contraction is mediated. If the abducens nerve is damaged, not only will the patient be unable to turn the eyeball laterally, but because of the lack of muscle tonus to the lateral rectus, the eyeball will be pulled medially.

**VII Facial** The facial nerve arises from nuclei within the lower portion of the pons, traverses the petrous part of the temporal bone (see fig. 16.9), and emerges on the side of the face near the parotid gland. The facial nerve is mixed. Impulses through the motor fibers cause contraction of the posterior belly of the digastric muscle and the muscles of facial expression, including the scalp and platysma muscles (fig. 16.8). The submandibular and sublingual glands also receive some autonomic motor innervation from the facial nerve, as does the lacrimal gland.

Sensory fibers of the facial nerve arise from taste buds on the anterior two-thirds of the tongue. Taste buds function as *chemoreceptors* because they respond to specific chemical stimuli. The **geniculate ganglion** is the enlargement of the facial nerve just before the sensory portion enters into the pons. Sensory sensations of taste are conveyed into nuclei within the medulla oblongata through the thalamus, and finally to the gustatory (taste) area of the cerebral cortex of the insula.

Trauma to the facial nerve results in inability to contract facial muscles on the affected side of the face and distorts taste perception, particularly of sweets. The affected

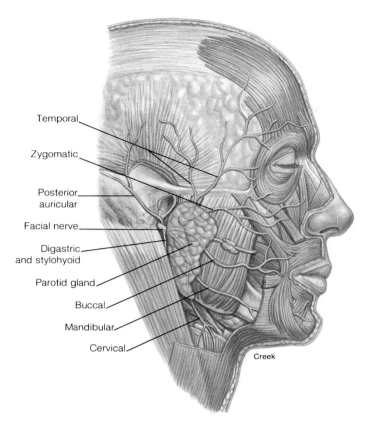

Temporal

Zygomatic

Posterior
auricular

Facial nerve

Digastric
and stylohyoid

Parotid gland

Buccal

Mandibular

Cervical

Creek

**FIGURE 16.8**

The facial nerve and its superficial branches.

side of the face tends to sag because muscle tonus is lost. *Bell's palsy* is a functional disorder (probably of viral origin) of the facial nerve.

**VIII Vestibulocochlear**   The vestibulocochlear nerve, also referred to as the **auditory nerve** or the **acoustic nerve,** serves structures contained within the skull. It is the only cranial nerve that does not exit the cranium through a foramen. A purely sensory nerve, the vestibulocochlear is composed of two nerves that arise within the inner ear (fig. 16.9). The **vestibular nerve** arises from the **vestibular organs** associated with equilibrium and balance. Bipolar neurons from the vestibular organs (saccule, utricle, and semicircular ducts) extend to the **vestibular ganglion,** where cell bodies are contained. From there, fibers convey impulses to the **vestibular nuclei** within the pons and medulla oblongata. Fibers from there extend to the thalamus and the cerebellum.

The **cochlear nerve** arises from the **spiral organ** (organ of Corti) within the cochlea and is associated with hearing. The cochlear nerve is composed of bipolar neurons that convey impulses through the **spiral ganglion** to the **cochlear nuclei** within the medulla oblongata. From there, fibers extend to the thalamus and synapse with neurons that convey the impulses to the auditory areas of the cerebral cortex.

Bell's palsy: from Sir Charles Bell, Scottish physician, 1774–1842
vestibulocochlear: L. *vestibulum,* chamber; *cochlea,* snail shell
organ of Corti: from Alfonso Corti, Italian anatomist, 1822–88

Peripheral Nervous System

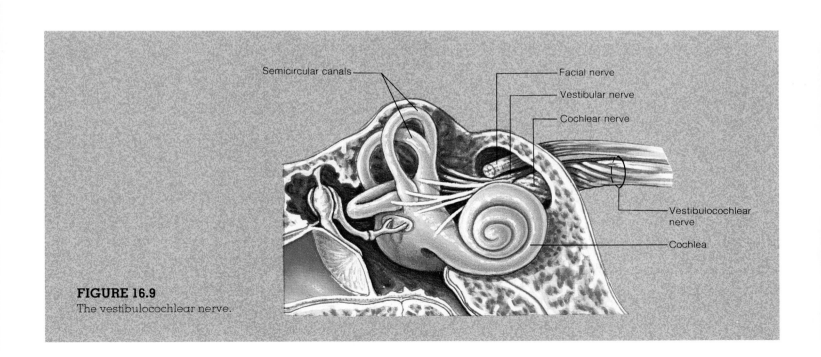

Semicircular canals

Facial nerve

Vestibular nerve

Cochlear nerve

Vestibulocochlear nerve

Cochlea

**FIGURE 16.9**

The vestibulocochlear nerve.

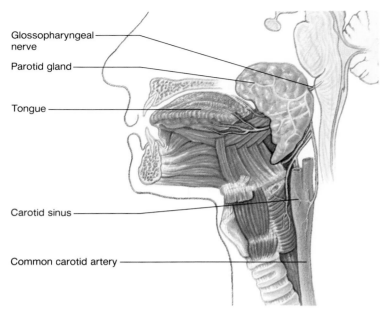

Glossopharyngeal nerve

Parotid gland

Tongue

Carotid sinus

Common carotid artery

**FIGURE 16.10**
The glossopharyngeal nerve.

Injury to the cochlear nerve results in perception deafness, whereas damage to the vestibular nerve causes dizziness and an inability to maintain balance.

**IX Glossopharyngeal**   The glossopharyngeal nerve is a mixed nerve that innervates part of the tongue and pharynx (fig. 16.10). The motor fibers of this nerve originate in a nucleus within the medulla oblongata and pass through the jugular foramen. The motor fibers innervate the muscles of the pharynx and the parotid gland to stimulate the swallowing reflex and the secretion of saliva.

The sensory fibers of the glossopharyngeal nerve arise from the pharyngeal region, the parotid gland, the middle-ear cavity, and the taste buds on the posterior one-third of the tongue. These taste buds, like those innervated by the facial nerve, are *chemoreceptors.* Some sensory fibers also arise from sensory receptors within the carotid sinus of the neck and help regulate blood pressure. Impulses from the glossopharyngeal nerve travel through the medulla oblongata and into the thalamus, where they synapse with fibers that convey the impulses to the gustatory area of the cerebral cortex.

Damage to the glossopharyngeal nerve results in the loss of perception of bitter and sour taste from taste buds on the posterior portion of the tongue. If the motor portion of this nerve is damaged, swallowing becomes difficult.

**X Vagus**   The vagus nerve has motor and sensory fibers that innervate visceral organs of the thoracic and abdominal cavities (fig. 16.11). The motor portion arises from the **nucleus**

**ambiguus** and **dorsal motor nucleus** of the vagus within the medulla oblongata and passes through the jugular foramen. The vagus is the longest of the cranial nerves, and through various branches it innervates the muscles of the pharynx, larynx, respiratory tract, lungs, heart, esophagus, and abdominal viscera, with the exception of the lower portion of the large intestine. One motor branch of the vagus nerve, the **recurrent laryngeal nerve,** innervates the larynx, enabling speech.

Sensory fibers of the vagus convey impulses from essentially the same organs served by motor fibers. Impulses through the sensory fibers relay specific sensations such as hunger pangs, distension, intestinal discomfort, or laryngeal movement. Sensory fibers also arise from proprioceptors in the muscles innervated by the motor fibers of this nerve.

If both vagus nerves are seriously damaged, death ensues rapidly because vital autonomic functions stop. The injury of one nerve causes vocal impairment, difficulty in swallowing, or other visceral disturbances.

**XI Accessory**   The accessory nerve is principally a motor nerve, but it does contain some sensory fibers from proprioceptors within the muscles it innervates. The accessory nerve is unique in that it arises from both the brain and the spinal cord (fig. 16.12). The **cranial root** arises from nuclei within the medulla oblongata (ambiguus and accessory), passes through the jugular foramen with the vagus nerve, and innervates the skeletal muscles of the soft palate, pharynx, and larynx, which contract reflexively during swallowing. The **spinal root** arises from the first five segments of the cervical portion of the spinal cord, passes cranially through the foramen magnum to join with the cranial root, and passes through the jugular foramen. The spinal root of the accessory nerve innervates the sternocleidomastoid and the trapezius muscles that move the head, neck, and shoulders. Damage to an accessory nerve makes it difficult to move the head or shrug the shoulders.

**XII Hypoglossal**   The hypoglossal nerve is a mixed nerve. The motor fibers arise from the hypoglossal nucleus within the medulla oblongata and pass through the hypoglossal canal of the skull to innervate both the extrinsic and intrinsic muscles of the tongue (fig. 16.12). Motor impulses along these fibers account for the coordinated contraction of the tongue muscles necessary for such activities as food manipulation, swallowing, and speech.

The sensory portion of the hypoglossal nerve arises from proprioceptors within the same tongue muscles and conveys impulses to the medulla oblongata regarding the position and function of the muscles.

glossopharyngeal: L. *glossa*, tongue; Gk. *pharynx*, throat

vagus: L. *vagus*, wandering
hypoglossal: Gk. *hypo*, under; L. *glossa*, tongue

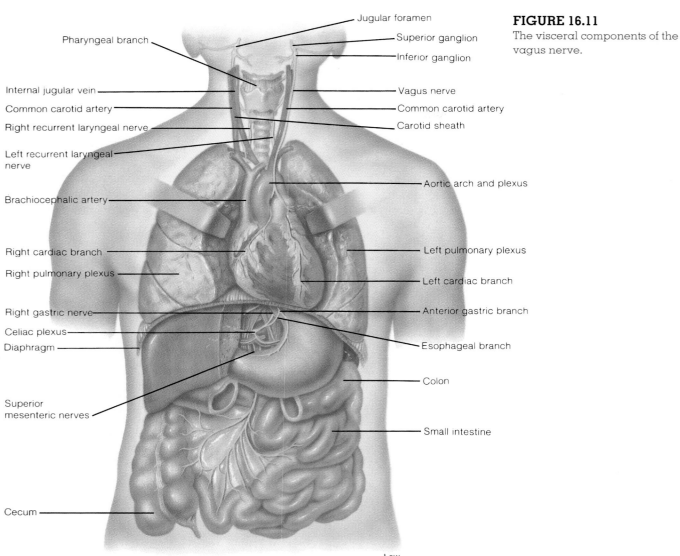

**FIGURE 16.11**

The visceral components of the vagus nerve.

Jugular foramen

Pharyngeal branch

Superior ganglion

Inferior ganglion

Internal jugular vein

Vagus nerve

Common carotid artery

Common carotid artery

Right recurrent laryngeal nerve

Carotid sheath

Left recurrent laryngeal nerve

Aortic arch and plexus

Brachiocephalic artery

Right cardiac branch

Left pulmonary plexus

Right pulmonary plexus

Left cardiac branch

Right gastric nerve

Anterior gastric branch

Celiac plexus

Diaphragm

Esophageal branch

Colon

Superior mesenteric nerves

Small intestine

Cecum

Lew

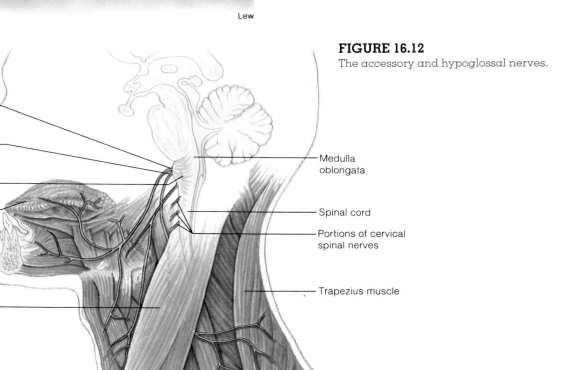

**FIGURE 16.12**

The accessory and hypoglossal nerves.

Hypoglossal nerve

Cranial root

Accessory nerve

Medulla oblongata

Spinal root

Tongue

Spinal cord

Portions of cervical spinal nerves

Trapezius muscle

Sternocleidomastoid muscle

# Table 16.2 Methods of determining cranial nerve dysfunction

| Nerve | Techniques of examination | Comments |
|---|---|---|
| Olfactory | Patient asked to differentiate odors (tobacco, coffee, soap, etc.) with eyes closed. | Nasal passages must be patent and tested separately by occluding the opposite side. |
| Optic | Retina examined with ophthalmoscope; visual acuity tested with eye charts. | Visual acuity must be determined with lenses on, if patient wears them. |
| Oculomotor | Patient follows examiner's finger movement with eyes—especially cross-eyed movement; pupillary change observed by shining light into each eye separately. | Examiner should note rate of pupillary change and coordinated constriction of pupils. Light in one eye should similar pupillary change in other eye, but to a lesser degree. |
| Trochlear | Patient follows examiner's finger movement with eyes—especially lateral and downward movement. | |
| Trigeminal | Motor portion: Temporalis and masseter muscles palpated as patient clenches teeth; patient asked to open mouth against resistance applied by examiner. | Muscles of both sides of the jaw should show equal contractile strength. |
| | Sensory portion: Tactile and pain receptors tested by lightly touching patient's entire face with cotton and then with pin stimulus. | Patient's eyes should be closed and innervation areas for all three nerves branching from the trigeminal nerve should be tested. |
| Abducens | Patient follows examiner's finger movement—especially lateral movement. | Motor functioning of cranial nerves III, IV, and VI may be tested simultaneously through selective movements of eyeball. |
| Facial | Motor portion: Patient asked to raise eyebrows, frown, tightly constrict eyelids, smile, puff out cheeks, and whistle. | Examiner should note lack of tonus expressed by sagging regions of face. |
| | Sensory portion: Sugar placed on each side of tip of patient's tongue. | Not reliable test for specific facial-nerve dysfunction because of tendency to stimulate taste buds on both sides of tip of tongue. |
| Vestibulocochlear | Vestibular portion: Patient asked to walk a straight line. | Not usually tested unless patient complains of dizziness or balance problems. |
| | Cochlear portion: Tested with tuning fork. | Examiner should note ability to discriminate sounds. |
| Glossopharyngeal and vagus | Motor: Examiner notes disturbances in swallowing, talking, and movement of soft palate; gag reflex tested. | Visceral innervation of vagus cannot be examined except for innervation to larynx, which is also served by glossopharyngeal. |
| Accessory | Patient asked to shrug shoulders against resistance of examiner's hand and to rotate head against resistance. | Sides should show uniformity of strength. |
| Hypoglossal | Patient requested to protrude tongue; tongue thrust may be resisted with tongue blade. | Tongue should protrude straight; deviation to side indicates ipsilateral-nerve dysfunction; asymmetry, atrophy, or lack of strength should be noted. |

If a hypoglossal nerve is damaged, a person will have difficulty in speaking, swallowing, and protruding the tongue.

## Neurological Assessment of the Cranial Nerves

Head injuries and brain concussions are common occurrences in automobile accidents. The cranial nerves would seem to be well protected on the inferior side of the brain. But the brain, immersed in and filled with cerebrospinal fluid, is like a water-sodden log; a blow to the top of the head can cause a serious rebound of the brain from the floor of the cranium. Routine neurological examinations involve testing for cranial nerve dysfunction.

Commonly used clinical methods for determining cranial nerve dysfunction are presented in table 16.2.

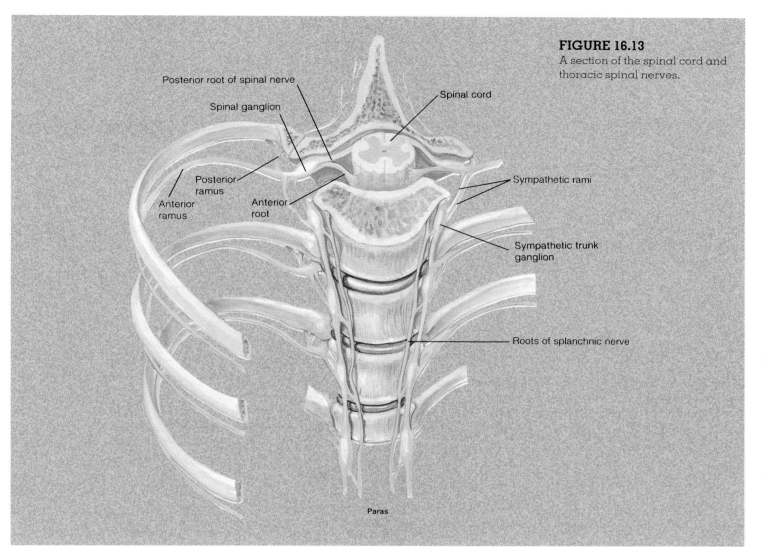

**FIGURE 16.13**
A section of the spinal cord and thoracic spinal nerves.

Posterior root of spinal nerve

Spinal ganglion

Spinal cord

Posterior ramus

Anterior root

Anterior ramus

Sympathetic rami

Sympathetic trunk ganglion

Roots of splanchnic nerve

Paras

# Spinal Nerves

*Each of the 31 pairs of spinal nerves is formed by the union of a posterior and an anterior spinal root that emerges from the spinal cord through an intervertebral foramen to innervate a body dermatome.*

The 31 pairs of **spinal nerves** (see fig. 16.1) are grouped as follows: 8 cervical, 12 thoracic, 5 lumbar, 5 sacral, and 1 coccygeal. With the exception of the first cervical nerve, the spinal nerves leave the spinal cord and vertebral canal through intervertebral foramina. The first pair of cervical nerves emerges between the occipital bone of the skull and the atlas vertebra. The second through the seventh pairs of cervical nerves emerge above the vertebrae for which they are named, whereas the eighth pair of cervical nerves passes between the seventh cervical and first thoracic vertebrae. The remaining pairs of spinal nerves emerge below the vertebrae for which they are named.

A spinal nerve is a mixed nerve attached to the spinal cord by a **posterior (dorsal) root** composed of sensory fibers and an **anterior (ventral) root** composed of motor fibers (fig. 16.13). The posterior root contains an enlargement called the **spinal (sensory) ganglion,** where the cell bodies of sensory neurons are located. The axons of sensory neurons convey sensory impulses through the posterior root and into the spinal cord, where synapses occur with dendrites of other neurons. The anterior root consists of axons of motor neurons that convey motor impulses away from the CNS. A spinal nerve is formed as the fibers from the posterior and anterior roots converge and emerge through an intervertebral foramen.

The disease *herpes zoster*, also known as *shingles*, is a viral infection of the spinal ganglia. Herpes zoster causes painful, often unilateral, clusters of fluid-filled vesicles in the skin along the paths of the affected peripheral sensory neurons. This disease requires no special treatment, as the lesions gradually heal, and is usually not serious except in elderly debilitated patients, who may die from exhaustion.

A spinal nerve divides into several branches immediately after it emerges through the intervertebral foramen. The small **meningeal branch** reenters the vertebral canal to innervate the meninges, vertebrae, and vertebral ligaments. A larger branch, called the **posterior ramus,** innervates the muscles, joints, and skin of the back along the vertebral column (fig. 16.13). An **anterior ramus** of a spinal nerve innervates the muscles and skin on the lateral and anterior side of the trunk. Combinations of anterior rami innervate the limbs.

The **rami communicantes** are two branches from each spinal nerve that connect to a **sympathetic trunk ganglion,** which is part of the autonomic nervous system. The rami communicantes are composed of a **gray ramus,** containing unmyelinated fibers, and a **white ramus,** containing myelinated fibers. This arrangement is described in more detail in chapter 17.

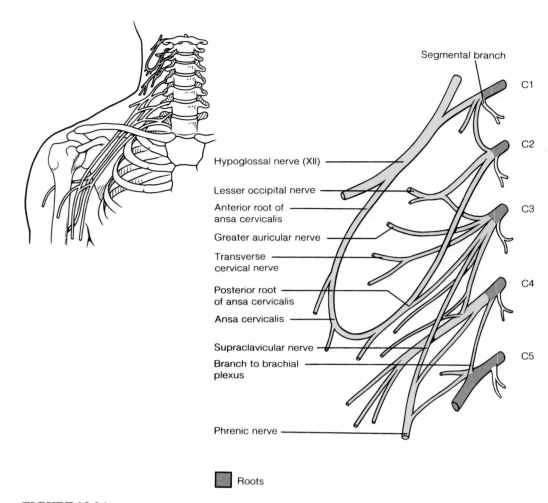

**FIGURE 16.14**
The cervical plexus.

# Nerve Plexuses

*Except in the thoracic nerves T2–T12, the ventral rami of the spinal nerves combine and then split again as networks of nerves referred to as plexuses. There are four plexuses of spinal nerves: the cervical, the brachial, the lumbar, and the sacral. Nerves emerge from the plexuses and are named according to the structures they innervate or the general course they take.*

**Cervical Plexus**   The cervical plexus (*plek´sus*) is positioned deep on the side of the neck, lateral to the first four cervical vertebrae (fig. 16.14). It is formed by the anterior rami of the first four cervical nerves (C1–C4) and a portion

of C5. Branches of the cervical plexus innervate the skin and muscles of the neck and portions of the head and shoulders. Branches of the cervical plexus also combine with the accessory and hypoglossal cranial nerves to supply dual innervation to some specific neck and pharyngeal muscles. Fibers from the third, fourth, and fifth cervical nerves unite to become the **phrenic** (*fren´ik*) **nerve,** which innervates the diaphragm. Motor impulses through the paired phrenic nerves cause the diaphragm to contract, inspiring air into the lungs. The nerves of the cervical plexus are summarized in table 16.3.

**Brachial Plexus**   The brachial plexus is positioned to the side of the last four cervical and first thoracic vertebrae. It is formed by the anterior rami of C5 through T1 with occasional contributions of fibers from C4 and T2. From its emergence, the brachial plexus extends downward and laterally, passes over the first rib behind the clavicle, and enters the axilla. Each brachial plexus innervates the entire upper extremity of one side, as well as a number of shoulder and neck muscles.

## Table 16.3 Branches of the cervical plexus

| Nerve | Spinal component | Innervation |
| --- | --- | --- |
| **Superficial cutaneous branches** | | |
| Lesser occipital | C2,C3 | Skin of scalp above and behind ear |
| Greater auricular | C2,C3 | Skin in front of, above, and below ear |
| Transverse cervical | C2,C3 | Skin of anterior aspect of neck |
| Supraclavicular | C3,C4 | Skin of upper portion of chest and shoulder |
| **Deep motor branches** | | |
| Ansa cervicalis | | |
|   Anterior root | C1,C2 | Geniohyoid, thyrohyoid, and infrahyoid muscles of neck |
|   Posterior root | C3,C4 | Omohyoid, sternohyoid, and sternothyroid muscles of neck |
| Phrenic | C3,C4,C5 | Diaphragm |
| Segmental branches | C1–C5 | Deep muscles of neck (levator scapulae ventralis, trapezius, scalenus, and sternocleidomastoid) |

Structurally, the brachial plexus is divided into *roots, trunks, divisions,* and *cords* (fig. 16.15). The roots of the brachial plexus are simply continuations of the anterior rami of the cervical nerves. The anterior rami of C5 and C6 converge to become the **superior trunk,** the C7 ramus becomes the **middle trunk,** and the ventral rami of C8 and T1 converge to become the **inferior trunk.** Each of the three trunks immediately divides into an **anterior division** and a **posterior division.** The divisions then converge to form three cords. The **posterior cord** is formed by the convergence of the posterior divisions of the upper, middle, and lower trunks, and hence contains fibers from C5 through C8. The **medial cord** is a continuation of the anterior division of the lower trunk and primarily contains fibers from C8 and T1. The **lateral cord** is formed by the convergence of the anterior division of the upper and middle trunk and consists of fibers from C5 through C7.

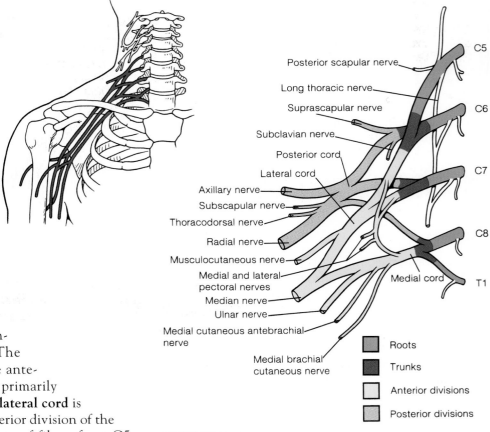

Posterior scapular nerve
Long thoracic nerve
Suprascapular nerve
Subclavian nerve
Posterior cord
Lateral cord
Axillary nerve
Subscapular nerve
Thoracodorsal nerve
Radial nerve
Musculocutaneous nerve
Medial and lateral pectoral nerves
Median nerve
Ulnar nerve
Medial cutaneous antebrachial nerve
Medial brachial cutaneous nerve
Medial cord

C5
C6
C7
C8
T1

Roots
Trunks
Anterior divisions
Posterior divisions

**FIGURE 16.15**
The brachial plexus.

429

## Table 16.4 Selected branches of the brachial plexus

| Nerve | Cord and spinal components | Innervation |
|---|---|---|
| Axillary | Posterior cord (C5,C6) | Skin of shoulder; shoulder joint, deltoid and teres minor muscles |
| Radial | Posterior cord (C5–C8,T1) | Skin of posterior lateral surface of arm, forearm, and hand; posterior muscles of brachium and antebrachium (triceps brachii, supinator, anconeus, brachioradialis, extensor carpi radialis brevis, extensor carpi radialis longus, extensor carpi ulnaris) |
| Musculocutaneous | Lateral cord (C5,C6,C7) | Skin of lateral surface of forearm; anterior muscles of brachium (coracobrachialis, biceps brachii, brachialis) |
| Ulnar | Medial cord (C8,T1) | Skin of medial third of hand; flexor muscles of anterior forearm (flexor carpi ulnaris, flexor digitorum), medial palm, and intrinsic flexor muscles of hand (profundus, third and fourth lumbricales) |
| Median | Medial cord (C6,C7,C8,T1) | Skin of lateral two-thirds of hand; flexor muscles of anterior forearm, lateral palm, and first and second lumbricales |

In summary, the brachial plexus is composed of nerve fibers from spinal nerves C5 through T1 and a few fibers from C4 and T2. Roots are continuations of the anterior rami. The roots converge to form trunks, and the trunks branch into divisions. The divisions in turn form cords, and the nerves of the upper extremity arise from the cords.

Trauma to the brachial plexus sometimes occurs, especially if the clavicle, upper ribs, or lower cervical vertebrae are seriously fractured. Occasionally, the brachial plexus of a newborn is severely strained during a difficult delivery when the baby is pulled through the birth canal. In such cases, the arm of the injured side is paralyzed and eventually withers as the muscles atrophy in relation to the extent of the injury.

The entire upper extremity can be anesthetized in a procedure called a *brachial block* or *brachial anesthesia*. The site for injection of the anesthetic is midway between the base of the neck and the shoulder, posterior to the clavicle. At this point, the anesthetic can be injected in close proximity to the brachial plexus.

Five major nerves, the axillary, radial, musculocutaneous, ulnar, and median, arise from the three cords of the brachial plexus to supply cutaneous and muscular innervation to the upper extremity (table 16.4). The **axillary nerve** arises from the posterior cord and provides sensory innervation to the skin of the shoulder and to the shoulder joint and motor innervation to the deltoid and teres minor muscles (fig. 16.16). The **radial nerve** arises from the posterior cord and extends along the posterior aspect of the brachial region to the radial side of the forearm. The radial nerve provides sensory innervation to the skin of the posterior lateral surface of the upper extremity, including the posterior surface of the hand (fig. 16.17), and motor innervation to all of the extensor muscles of the upper extremity, the supinator muscle, and two muscles that flex the elbow joint.

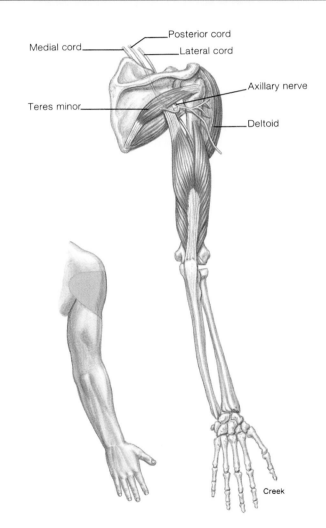

**FIGURE 16.16**
Muscular and cutaneous distribution of the axillary nerve.

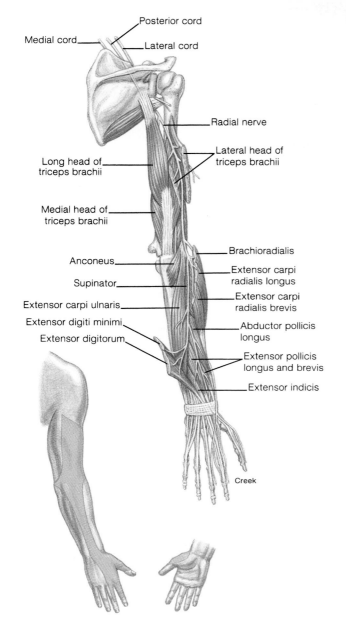

Posterior cord
Medial cord
Lateral cord
Radial nerve
Lateral head of triceps brachii
Long head of triceps brachii
Medial head of triceps brachii
Anconeus
Supinator
Extensor carpi ulnaris
Extensor digiti minimi
Extensor digitorum
Brachioradialis
Extensor carpi radialis longus
Extensor carpi radialis brevis
Abductor pollicis longus
Extensor pollicis longus and brevis
Extensor indicis
Creek

**FIGURE 16.17**

Muscular and cutaneous distribution of the radial nerve.

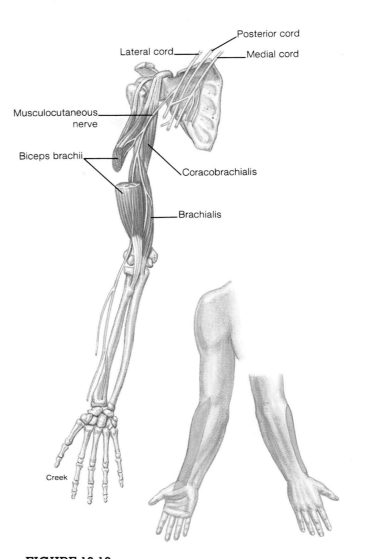

Posterior cord
Lateral cord
Medial cord
Musculocutaneous nerve
Biceps brachii
Coracobrachialis
Brachialis
Creek

**FIGURE 16.18**

Muscular and cutaneous distribution of the musculocutaneous nerve.

 The radial nerve is vulnerable to several types of trauma. *Crutch paralysis* may result when a person improperly supports the weight of the body for an extended period of time with a crutch pushed tightly into the axilla. Compression of the radial nerve between the top of the crutch and the humerus may result in radial nerve damage. Likewise, dislocation of the shoulder frequently traumatizes the radial nerve. Children are particularly at risk as adults yank on their arms. A fracture to the shaft of the humerus may damage the radial nerve that parallels the bone at this point. The principal symptom of radial nerve damage is *wristdrop*, in which the extensor muscles of the fingers and wrist fail to function, and as a result the joints of the fingers, wrist, and elbow are in a constant state of flexion.

The **musculocutaneous nerve** arises from the lateral cord and provides sensory innervation to the skin of the posterior lateral surface of the arm and motor innervation to the anterior muscles of the brachium (fig. 16.18). The **ulnar nerve** arises from the medial cord and provides sensory innervation to the skin on the medial (ulnar side) third of the hand (fig. 16.19). The motor innervation of the ulnar nerve is to two muscles of the forearm and the intrinsic muscles of the hand (except some that serve the thumb).

 The ulnar nerve can be palpated in the groove behind the medial epicondyle. This area is commonly known as the "funny bone" or "crazy bone." Damage to the ulnar nerve may occur as the medial side of the elbow is banged against a hard object. The immediate perception of this trauma is a painful tingling that extends down the ulnar side of the forearm and into the hand and medial two digits. Although common, ulnar nerve damage is generally not serious.

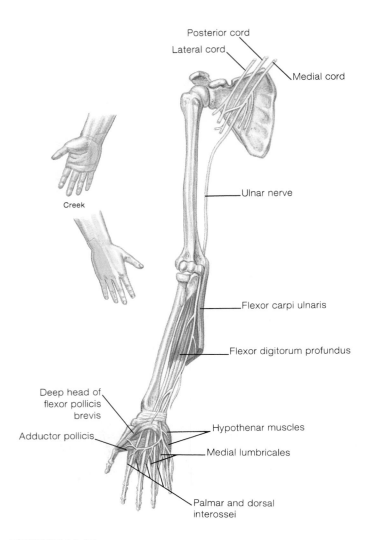

**FIGURE 16.19**
Muscular and cutaneous distribution of the ulnar nerve.

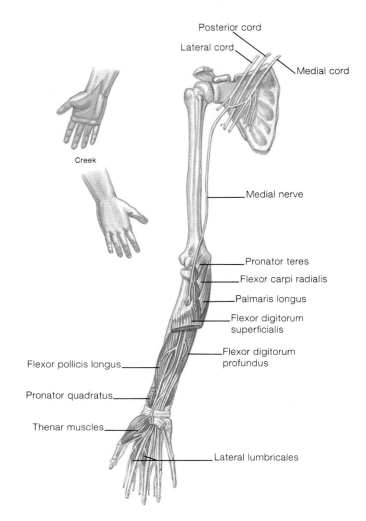

**FIGURE 16.20**
Muscular and cutaneous distribution of the median nerve.

The **median nerve** arises from the medial cord and provides sensory innervation to the skin on the radial portion of the palm of the hand (fig. 16.20) and motor innervation to all but one of the flexor muscles of the forearm and most of the hand muscles of the thumb (thenar muscles).

 The median nerve, which serves the thumb, is the nerve in the forearm most commonly injured by stab wounds or the penetration of glass. If this nerve is severed, the muscles of the thumb are paralyzed and waste away, resulting in an inability to oppose the thumb in grasping.

**Lumbar Plexus** The lumbar plexus is positioned to the side of the first four lumbar vertebrae. It is formed by the anterior rami of spinal nerves L1–L4 and some fibers from T12 (fig. 16.21). The nerves that arise from the lumbar plexus innervate structures of the lower abdomen and anterior and medial portions of the lower extremity. The lumbar plexus is

not as complex as the brachial plexus, having only roots and divisions rather than roots, trunks, divisions, and cords.

Structurally, the **posterior division** of the lumbar plexus passes obliquely outward, deep to the psoas major muscle, whereas the **anterior division** is superficial to the quadratus lumborum muscle. The nerves that arise from the lumbar plexus are summarized in table 16.5. Because of their extensive innervation, the femoral nerve and the obturator nerve are illustrated in figures 16.22 and 16.23, respectively.

The **femoral nerve** arises from the posterior division of the lumbar plexus and provides cutaneous innervation to the anterior and lateral thigh and the medial leg and foot (fig. 16.22). The motor innervation of the femoral nerve is to the anterior muscles of the thigh, including the iliopsoas and sartorius muscles and the quadriceps femoris group.

The **obturator nerve** arises from the anterior division of the lumbar plexus and provides cutaneous innervation to the medial thigh and motor innervation to the adductor muscles of the thigh (fig. 16.23).

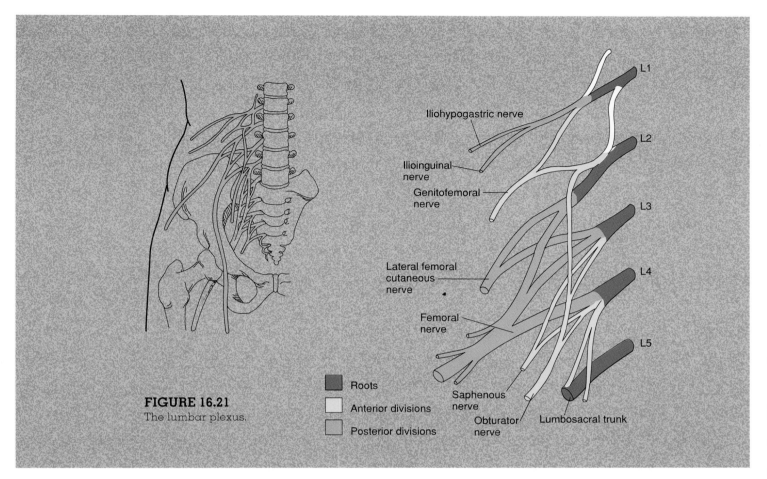

**FIGURE 16.21**
The lumbar plexus.

Roots
Anterior divisions
Posterior divisions

## Table 16.5   Branches of the lumbar plexus

| Nerve | Spinal components | Innervation |
|---|---|---|
| Iliohypogastric | T12–L1 | Skin of lower abdomen and buttock; muscles of anterolateral abdominal wall (external abdominal oblique, internal abdominal oblique, transversus abdominis) |
| Ilioinguinal | L1 | Skin of upper median thigh, scrotum and root of penis in male and labia majora in female; muscles of anterolateral abdominal wall with iliohypogastric nerve |
| Genitofemoral | L1,L2 | Skin of middle anterior surface of thigh, scrotum in male and labia majora in female; cremaster muscle in male |
| Lateral cutaneous femoral | L2,L3 | Skin of anterior, lateral, and posterior aspects of thigh |
| Femoral | L2–L4 | Skin of anterior and medial aspect of thigh and medial aspect of lower extremity and foot; anterior muscles of thigh (iliacus, psoas major, pectineus, rectus femoris, sartorius) and extensor muscles of leg (rectus femoris, vastus lateralis, vastus medialis, vastus intermedius) |
| Obturator | L2–L4 | Skin of medial aspect of thigh; adductor muscles of lower extremity (external obturator, pectineus, adductor longus, adductor brevis, adductor magnus, gracilis) |
| Saphenous | L2–L4 | Skin of medial aspect of lower extremity |

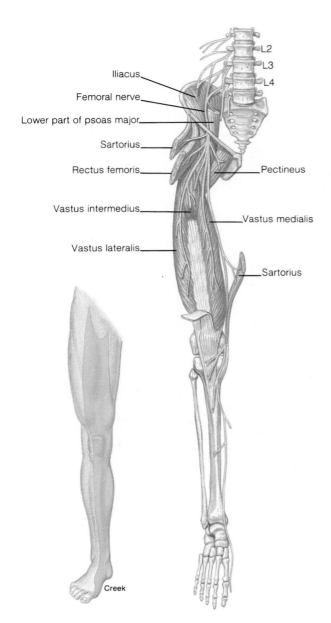

**FIGURE 16.22**
Muscular and cutaneous distribution of the femoral nerve.

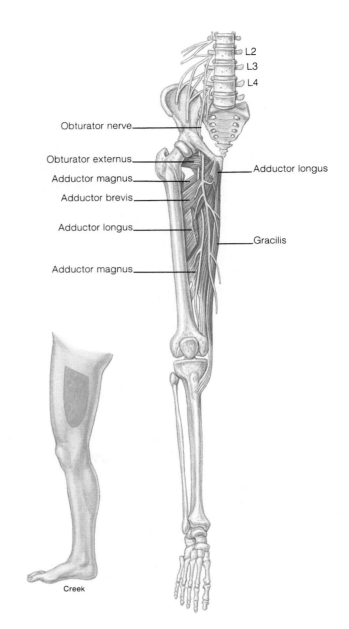

**FIGURE 16.23**
Muscular and cutaneous distribution of the obturator nerve.

**Sacral Plexus**   The sacral plexus is positioned immediately caudal to the lumbar plexus. It is formed by the anterior rami of spinal nerves L4, L5, and S1–S4 (fig. 16.24). The nerves arising from the sacral plexus innervate the lower back, pelvis, perineum, posterior surface of the thigh, anterior and posterior leg, and the dorsum and plantar surface of the foot (table 16.6). Like the lumbar plexus, the sacral plexus consists of *roots* and *anterior* and *posterior divisions* from which nerves arise. Because some of the nerves of the sacral plexus also contain fibers from the nerves of the lumbar plexus through the **lumbosacral trunk,** these two plexuses are frequently described collectively as the **lumbosacral plexus.**

The **sciatic** (*si-at´ik*) **nerve** is the largest nerve arising from the sacral plexus and is the largest nerve in the body. The sciatic nerve passes from the pelvis through the greater sciatic notch of the os coxa and extends down the posterior aspect of the thigh. The sciatic nerve is actually composed of two nerves—the tibial and common fibular nerves, wrapped in a connective tissue sheath.

........................................

sacral: L. *sacris,* sacred
sciatic: L. *sciaticus,* hip joint

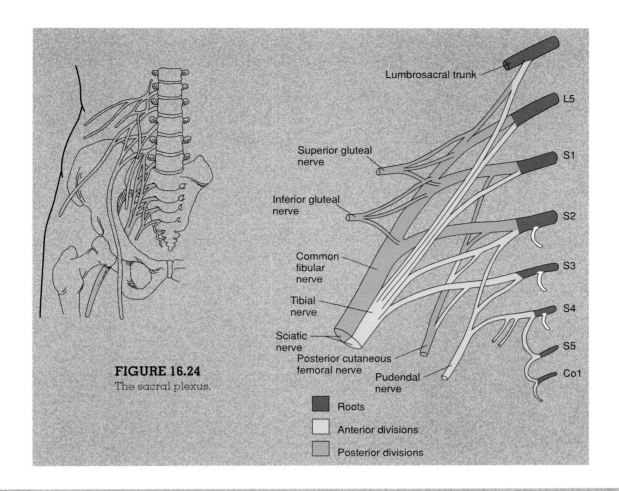

**FIGURE 16.24**
The sacral plexus.

Lumbrosacral trunk
L5
S1
S2
S3
S4
S5
Co1

Superior gluteal nerve

Inferior gluteal nerve

Common fibular nerve

Tibial nerve

Sciatic nerve

Posterior cutaneous femoral nerve

Pudendal nerve

Roots

Anterior divisions

Posterior divisions

## Table 16.6 Branches of the sacral plexus

| Nerve | Spinal components | Innervation |
|---|---|---|
| Superior gluteal | L4,L5,S1 | Abductor muscles of thigh (gluteus minimus, gluteus medius, tensor fasciae latae) |
| Inferior gluteal | L5–S2 | Extensor muscle of hip joint (gluteus maximus) |
| Nerve to piriformis | S1,S2 | Abductor and rotator of thigh (piriformis) |
| Nerve to quadratus femoris | L4,L5,S1 | Rotators of thigh (gemellus inferior, quadratus femoris) |
| Nerve to internal obturator | L5–S2 | Rotators of thigh (gemellus superior, internal obturator) |
| Perforating cutaneous | S2,S3 | Skin over lower medial surface of buttock |
| Posterior cutaneous femoral | S1–S3 | Skin over lower lateral surface of buttock, anal region, upper posterior surface of thigh, upper aspect of calf, scrotum in male and labia majora in female |
| Sciatic | L4–S3 | Composed of two nerves (tibial and common fibular); splits into two portions at popliteal fossa; branches from sciatic in thigh region to "hamstring muscles" (biceps femoris, semitendinosus, semimembranosus) and adductor magnus muscle |
| Tibial (sural, medial, and lateral plantar) | L4–S3 | Skin of posterior surface of leg and sole of foot; muscle innervation includes gastrocnemius, soleus, flexor digitorum longus, flexor hallucis longus, tibialis posterior, popliteus, and intrinsic muscles of the foot |
| Common fibular (superficial and deep fibular) | L4–S2 | Skin of anterior surface of the leg and dorsum of foot; muscle innervation includes peroneus tertius, peroneus brevis, peroneus longus, tibialis anterior, extensor hallucis longus, extensor digitorum longus, extensor digitorum brevis |
| Pudendal | S2–S4 | Skin of penis and scrotum in male and skin of clitoris, labia majora, labia minora, and lower vagina in female; muscles of perineum |

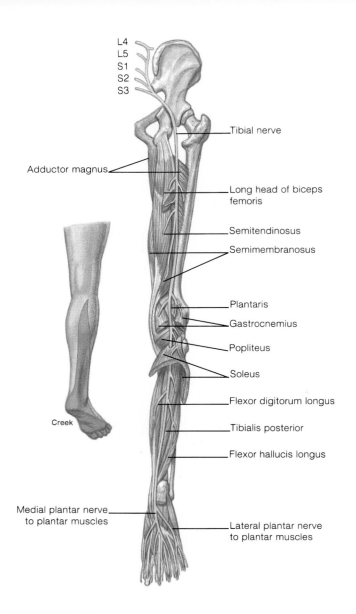

**FIGURE 16.25**

Muscular and cutaneous distribution of the tibial nerve.

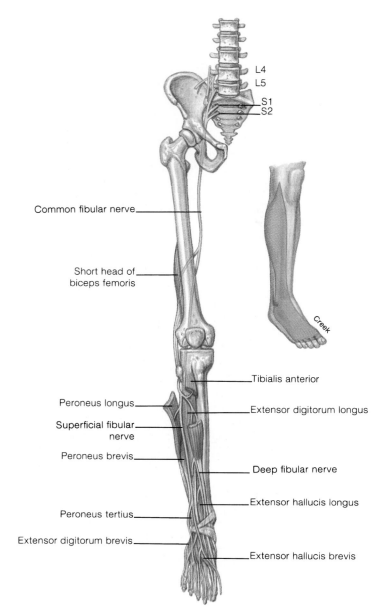

**FIGURE 16.26**

Muscular and cutaneous distribution of the common fibular nerve.

The **tibial nerve** arises from the anterior division of the sacral plexus, extends through the posterior regions of the thigh and leg, and branches in the foot to form the **medial** and **lateral plantar nerves** (fig. 16.25). The cutaneous innervation of the tibial nerve is to the calf of the leg and the plantar surface of the foot. The motor innervation of the tibial nerve is to most of the posterior thigh and leg muscles and to many of the intrinsic muscles of the foot.

The **common fibular nerve** arises from the posterior division of the sacral plexus, extends through the posterior region of the thigh, and branches in the upper portion of the leg into the **deep** and **superficial fibular nerves** (fig. 16.26). The cutaneous innervation of the common fibular nerve and its branches is to the anterior and lateral leg and to the dorsum of the foot. The motor innervation is to the anterior and lateral muscles of the leg and foot.

 The sciatic nerve in the buttock lies deep to the gluteus maximus muscle, midway between the greater trochanter and the ischial tuberosity. Because of its position, the sciatic nerve is of tremendous clinical importance. A posterior dislocation of the hip joint will generally injure the sciatic nerve. A *herniated disc* (fig. 16.27) or pressure from the uterus during pregnancy may damage the roots of the nerves, resulting in a condition called *sciatica* (si-at′ĭ-kă). Sciatica is characterized by a sharp pain in the gluteal region and by pains that extend down the posterior side of the thigh. An improperly administered injection into the buttock may injure the sciatic nerve itself. Even a temporary compression of the sciatic nerve as a person sits on a hard surface for a period of time may result in the perception of tingling throughout the limb as the person stands up. The limb is said to have "gone to sleep."

**FIGURE 16.27**
An MR scan of a herniated disc (arrow) in the lumbar region.

*The conduction pathway of a reflex arc consists of a receptor, a sensory neuron, a motor neuron and its innervation in the PNS, and an association in the CNS. The reflex arc provides the mechanism for a rapid, automatic response to a potentially threatening stimulus.*

Specific **nerve pathways** provide routes by which impulses travel through the nervous system. Frequently, a nerve pathway begins with impulses being conducted to the CNS through sensory receptors and sensory neurons of the PNS. Once within the CNS, impulses may immediately travel back through motor portions of the PNS to activate specific skeletal muscles, glands, or smooth muscles. They also may be sent simultaneously to other parts of the CNS through ascending tracts within the spinal cord.

## Components of the Reflex Arc

The simplest type of nerve pathway is a **reflex arc** (fig. 16.28). A reflex arc implies an automatic, unconscious, protective response to a situation in an attempt to maintain body homeostasis. A reflex arc leads by a short route from sensory to motor neurons and includes only two or three neurons. The five components of a reflex arc are the receptor, sensory neuron, center, motor neuron, and effector. The **receptor** includes the dendrite of a sensory neuron and the place where the electrical impulse is initiated. The **sensory neuron** relays the impulse through the posterior root to the CNS. The **center** is within the CNS and usually involves an association neuron (interneuron). It is here that the arc is made and other impulses are sent through synapses to other parts of the body.

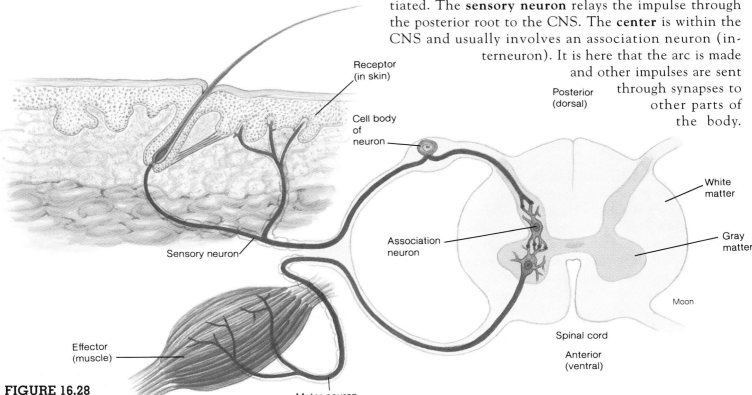

**FIGURE 16.28**
The reflex arc.

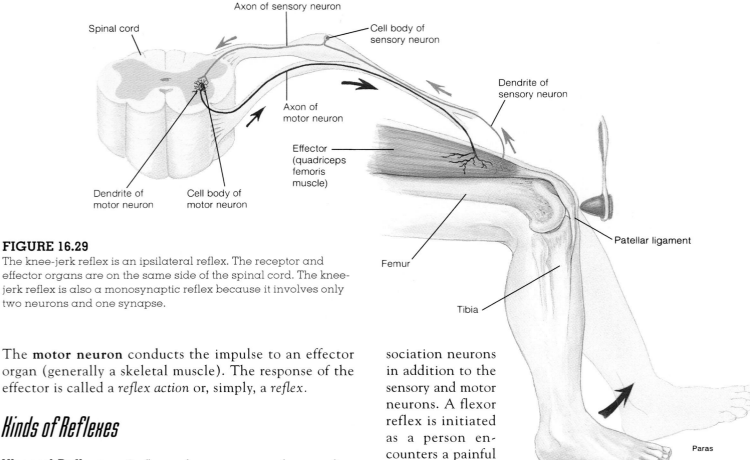

**FIGURE 16.29**

The knee-jerk reflex is an ipsilateral reflex. The receptor and effector organs are on the same side of the spinal cord. The knee-jerk reflex is also a monosynaptic reflex because it involves only two neurons and one synapse.

The **motor neuron** conducts the impulse to an effector organ (generally a skeletal muscle). The response of the effector is called a *reflex action* or, simply, a *reflex*.

## Kinds of Reflexes

**Visceral Reflexes**   Reflexes that cause smooth or cardiac muscle to contract or glands to secrete are **visceral (autonomic) reflexes.** Visceral reflexes help control the body's many involuntary processes, such as heart rate, respiratory rate, blood flow, and digestion. Swallowing, sneezing, coughing, and vomiting may also be reflexive, although they involve the autonomic action of skeletal muscles.

**Somatic Reflexes**   Somatic reflexes are those that result in the contraction of skeletal muscles. The three principal kinds of somatic reflexes are named according to the response they produce.

The **stretch reflex** involves only two neurons and one synapse in the pathway and is therefore called a *monosynaptic reflex arc*. Slight stretching of neuromuscular spindle receptors (described in chapter 18) within a muscle initiates impulses that are conducted along a sensory neuron to the spinal cord. A synapse with a motor neuron occurs in the anterior gray column and a motor unit is activated, causing specific muscle fibers to contract. Since the receptor and effector organs of the stretch reflex involve structures on the same side of the spinal cord, the reflex arc is an *ipsilateral reflex arc*. The knee-jerk reflex is an ipsilateral reflex (fig. 16.29), as are all monosynaptic reflex arcs.

A **flexor reflex,** or **withdrawal reflex,** consists of a *polysynaptic reflex arc* (fig. 16.30). Flexor reflexes involve association neurons in addition to the sensory and motor neurons. A flexor reflex is initiated as a person encounters a painful stimulus, such as a hot or sharp object. As a receptor organ is stimulated, sensory neurons transmit the impulse to the spinal cord where association neurons are activated. There, the impulses are directed through motor neurons to flexor muscles, which contract in response. Simultaneously, antagonistic muscles are inhibited (relaxed) so that the traumatized extremity can be quickly withdrawn from the harmful source of stimulation.

Several additional reflexes may be activated while a flexor reflex is in progress. In an *intersegmental reflex arc*, motor units from several segments of the spinal cord are activated by impulses coming in from the receptor organ. An intersegmental reflex arc produces stimulation of more than one effector organ. Frequently, sensory impulses from a receptor organ cross over through the spinal cord to activate effector organs within the opposite (*contralateral*) limb. This type of reflex is called a **crossed extensor reflex** (fig. 16.31) and is important for maintaining body balance while a flexor reflex is in progress. The reflexive inhibition of certain muscles to contract, called *reciprocal inhibition*, also helps maintain balance while either flexor or crossed extensor reflexes are in progress.

Certain reflexes are important for physiological functions, while others are important for avoiding injury. Some common reflexes are described in table 16.7 and illustrated in figure 16.32.

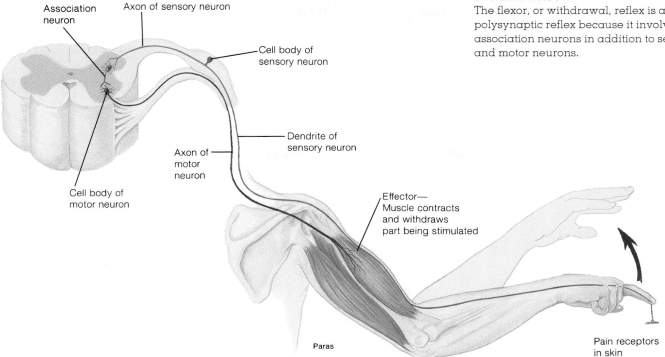

## FIGURE 16.30

The flexor, or withdrawal, reflex is a polysynaptic reflex because it involves association neurons in addition to sensory and motor neurons.

## FIGURE 16.31

A crossed extensor reflex causes a reciprocal inhibition of muscles of the opposite appendage. This type of reflex inhibition is important in maintaining balance.

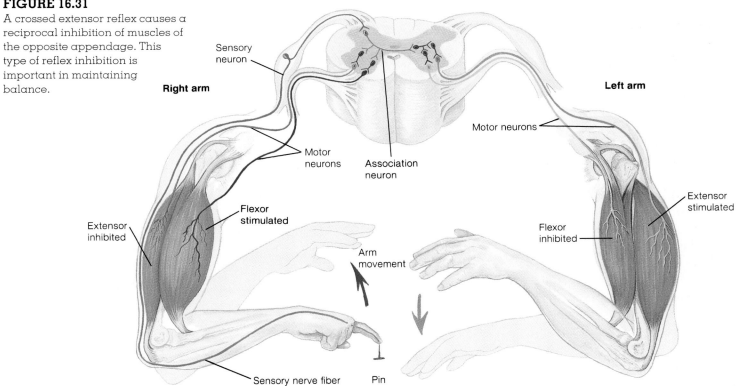

Part of a routine physical examination involves testing a person's reflexes. The condition of the nervous system, particularly the functioning of the synapses, may be determined by examining reflexes. In case of injury to some portion of the nervous system, testing certain reflexes may indicate the location and extent of the injury. Also, an anesthesiologist may try to initiate a reflex to ascertain the effect of an anesthetic.

anesthesia: Gk. *an*, without; *aisthesis*, sensation

# Table 16.7 Selected reflexes of clinical importance

| Reflex | Spinal segment | Site of receptor stimulation | Effector action |
|---|---|---|---|
| Biceps reflex | C5,C6 | Tendon of biceps brachii muscle, near attachment on radial tuberosity | Contracts biceps brachii muscle to flex elbow |
| Triceps reflex | C7,C8 | Tendon of triceps brachii muscle, near attachment on olecranon process | Contracts triceps brachii muscle to extend elbow |
| Supinator or brachioradialis reflex | C5,C6 | Radial attachment of supinator and brachioradialis muscles | Supinates forearm and hand |
| Knee reflex | L2, L3, L4 | Patellar ligament, just below patella | Contracts quadriceps muscle to extend the knee |
| Ankle reflex | S1, S2 | Tendo calcaneus, near attachment on calcaneus | Plantar flexes ankle |
| Plantar reflex | L4, L5, S1, S2 | Lateral aspect of sole, from heel to ball of foot | Plantar flexes foot and flexes toes |
| Babinski reflex* | L4, L5, S1, S2 | Lateral aspect of sole, from heel to ball of foot | Dorsiflexes great toe and fans other toes |
| Abdominal reflexes | T8, T9, T10 above umbilicus and T10, T11, T12 below umbilicus | Sides of abdomen, above and below level of umbilicus | Contracts abdominal muscles and deviates umbilicus toward stimulus |
| Cremasteric reflex | L1, L2 | Upper inside of thigh in males | Contracts cremasteric muscle and elevates testis on same side of stimulation |

*If the Babinski reflex rather than the plantar reflex occurs as the sole of the foot is stimulated, it may indicate damage to the corticospinal tract within the spinal cord. However, the Babinski reflex is present in infants up to 12 months of age because of the immaturity of their corticospinal tracts.

Babinski reflex: from Joseph F. Babinski, French neurologist, 1857–1932

**FIGURE 16.32**
Some reflexes of clinical
importance. (a) Glabellar reflex.
(b) Biceps brachii reflex.
(c) Triceps brachii reflex
(d) Supinator (brachioradialis
reflex).

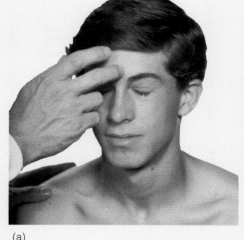

(a)

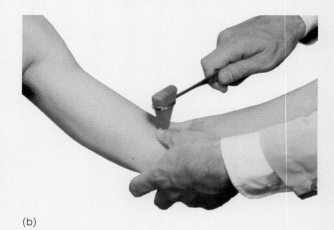

(b)

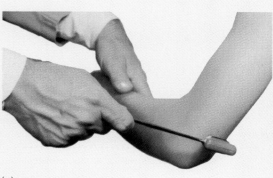

(c)

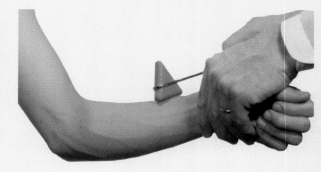

(d)

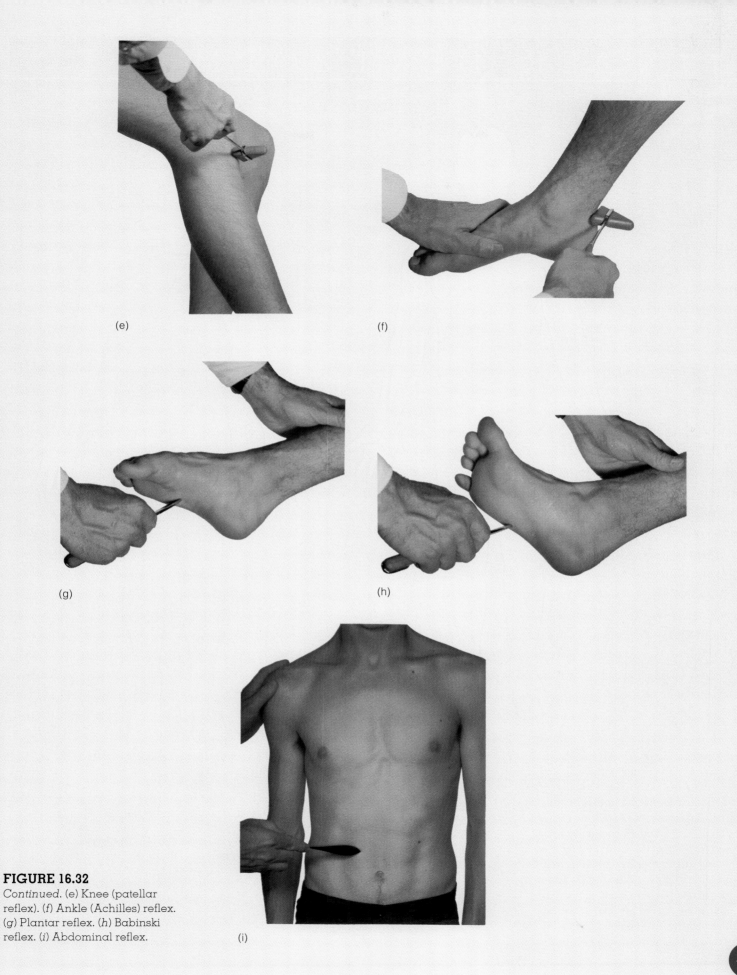

(e)

(f)

(g)

(h)

**FIGURE 16.32**
*Continued.* (e) Knee (patellar reflex). (f) Ankle (Achilles) reflex. (g) Plantar reflex. (h) Babinski reflex. (i) Abdominal reflex.

(i)

## Development of the Peripheral Nervous System

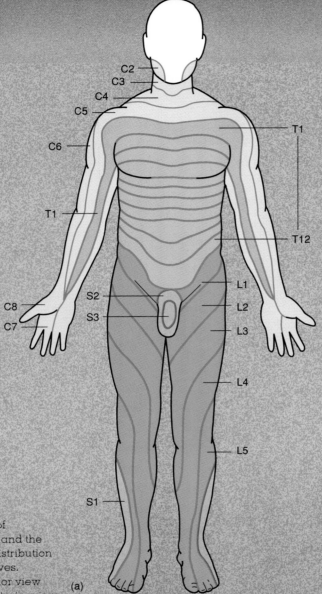

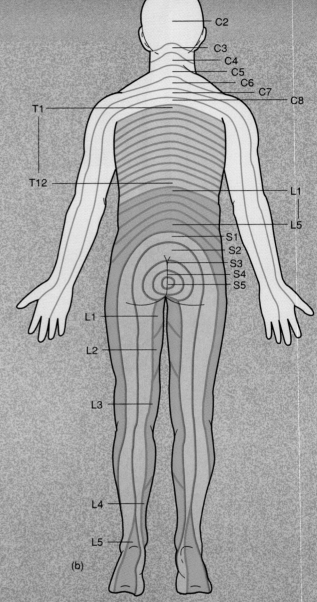

**FIGURE 1**

The pattern of dermatomes and the peripheral distribution of spinal nerves. (a) An anterior view and (b) a posterior view.

Development of the peripheral nervous system produces the pattern of dermatomes within the body (fig. 1). A **dermatome** (*der´mă-tōm*) is an area of the skin innervated by all the cutaneous neurons of a certain spinal or cranial nerve. Most of the scalp and face is innervated by sensory neurons from the trigeminal nerve. With the exception of the first cervical nerve (C1), all of the spinal nerves are associated with specific dermatomes. Dermatomes are consecutive in the neck and trunk regions. In the appendages, however, adjacent dermatome innervations overlap. The apparently uneven dermatome arrangement in the appendages is due to the uneven rate of nerve growth into the limb buds. Actually the limbs are segmented, and dermatomes overlap only slightly.

The pattern of dermatome innervation is of clinical importance when a physician wants to anesthetize a particular portion of the body. Because adjacent dermatomes overlap in the appendages, at least three spinal nerves must be blocked to produce complete anesthesia in these regions. Abnormally functioning dermatomes provide clues about injury to the spinal cord or specific spinal nerves. If a dermatome is stimulated but no sensation is perceived, the physician can infer that the injury involves the innervation to that dermatome.

dermatome: Gk. *derma*, skin; *tomia*, a cutting

# Chapter Summary

## Introduction to the Peripheral Nervous System (pp. 416–417)

1. The peripheral nervous system consists of nerves that convey impulses to and from the central nervous system.
2. The cranial nerves arise from the brain and the spinal nerves arise from the spinal cord.
3. Sensory (afferent) nerves convey impulses toward the CNS, whereas motor (efferent) nerves convey impulses away from the CNS. Mixed nerves are composed of both sensory and motor fibers.

## Cranial Nerves (pp. 417–426)

1. Twelve pairs of cranial nerves emerge from the inferior surface of the brain and, with the exception of the vestibulocochlear nerve, pass through foramina of the skull to innervate structures in the head, neck, and visceral organs of the trunk.
2. The names of the cranial nerves indicate their primary function or the general distribution of their fibers.
3. The olfactory, optic, and vestibulocochlear cranial nerves are sensory only; the trigeminal, glossopharyngeal, and vagus are mixed; and the others are primarily motor, with only the sensory fibers of these being proprioceptive.
4. Some of the cranial nerve fibers are somatic; others are visceral.
5. A test for cranial nerve dysfunction is clinically important in a neurological examination.

## Spinal Nerves (p. 427–428)

1. Each of the 31 pairs of spinal nerves is formed by the union of an anterior (ventral) and posterior (dorsal) spinal root that emerges from the spinal cord through an intervertebral foramen to innervate a body dermatome.
2. The spinal nerves are grouped according to the levels of the spinal column from which they arise, and they are numbered in sequence.
3. Each spinal nerve is a mixed nerve consisting of a posterior root of sensory fibers and an anterior root of motor fibers.
4. Just beyond its intervertebral foramen, each spinal nerve divides into several branches.

## Nerve Plexuses (pp. 428–436)

1. Except in the thoracic nerves T2–T12, the anterior rami of the spinal nerves combine and then split again as networks of nerves called plexuses.
   a. There are four plexuses of spinal nerves: the cervical, the brachial, the lumbar, and the sacral.
   b. Nerves that emerge from the plexuses are named according to the structures they innervate or the general course they take.
2. The cervical plexus is formed by the anterior rami of C1–C4 and by a portion of C5.
3. The brachial plexus is formed by the anterior rami of C5–T1 and occasionally by some fibers from C4 and T2.
   a. The brachial plexus is divided into roots, trunks, divisions, and cords.
   b. The axillary, radial, musculocutaneous, ulnar, and median are the five largest nerves arising from the brachial plexus.
4. The lumbar plexus is formed by the anterior rami of L1–L4 and by some fibers from T12.
   a. The lumbar plexus is divided into roots and divisions.
   b. The femoral and obturator are two important nerves arising from the lumbar plexus.
5. The sacral plexus is formed by the anterior rami of L4, L5, and S1–S4.
   a. The sacral plexus is divided into roots and divisions.
   b. The sciatic nerve, composed of the common fibular and tibial nerves, arises from the sacral plexus.
   c. The lumbar plexus and the sacral plexus are often referred to collectively as the lumbosacral plexus.

## Reflex Arcs and Reflexes (pp. 437–441)

1. The conduction pathway of a reflex arc consists of a receptor, a sensory neuron, a motor neuron and its innervation in the PNS, and an association neuron in the CNS. The reflex arc enables a rapid, automatic response to a potentially threatening stimulus.
2. A reflex arc is the simplest type of nerve pathway.
3. Visceral reflexes cause smooth or cardiac muscle to contract or glands to secrete.
4. Somatic reflexes cause skeletal muscles to contract.
   a. The stretch reflex is a monosynaptic reflex arc.
   b. The flexor reflex is a polysynaptic reflex arc.

# Review Activities

## Objective Questions

1. Which of the following is a *false* statement concerning the peripheral nervous system?
   a. It consists of cranial and spinal nerves only.
   b. It contains components of the autonomic nervous system.
   c. Sensory receptors, neurons, nerves, ganglia, and plexuses are all part of the PNS.
2. An inability to cross the eyes would most likely indicate a problem with which cranial nerve?
   a. optic
   b. oculomotor
   c. abducens
   d. facial
3. Which cranial nerve innervates the muscle that raises the upper eyelid?
   a. trochlear
   b. oculomotor
   c. abducens
   d. facial
4. The inability to walk a straight line may indicate damage to which cranial nerve?
   a. trigeminal
   b. facial

   c. vestibulocochlear
   d. vagus
5. Which cranial nerve passes through the stylomastoid foramen?
   a. facial
   b. glossopharyngeal
   c. vagus
   d. hypoglossal
6. Which of the following cranial nerves does *not* contain parasympathetic fibers?
   a. oculomotor
   b. accessory
   c. vagus
   d. facial
7. Which of the following is *not* a spinal nerve plexus?
   a. cervical
   b. brachial
   c. thoracic
   d. lumbar
   e. sacral
8. Roots, trunks, divisions, and cords are characteristic of
   a. the sacral plexus.
   b. the thoracic plexus.
   c. the lumbar plexus.
   d. the brachial plexus.
9. Which of the following nerve-plexus associations is *incorrect?*
   a. median/sacral
   b. phrenic/cervical

   c. axillary/brachial
   d. femoral/lumbar
10. Extending the leg when the patellar ligament is tapped is an example of
    a. a visceral reflex.
    b. a flexor reflex.
    c. an ipsilateral reflex.
    d. a crossed extensor reflex.

## Essay Questions

1. Explain the structural and functional relationship between the central nervous system, the autonomic nervous system, and the peripheral nervous system.
2. List the cranial nerves, and describe the major function(s) of each. How is each cranial nerve tested for dysfunction?
3. Describe the structure of a spinal nerve.
4. List the roots of each of the spinal plexuses. Describe where each is located and state the nerves that originate from them.
5. Distinguish between monosynaptic, polysynaptic, ipsilateral, stretch, and flexor reflexes.

## Gundy/Weber Software 💾

The tutorial software accompanying Chapter 16 is Volume 5—Nervous System.

# chapter seventeen

# autonomic nervous system

## objectives

- Describe the preganglionic and postganglionic neurons in the motor autonomic pathway.

- Describe the characteristics of the visceral effector organs that are innervated by the autonomic nervous system.

- Identify the location of preganglionic sympathetic neurons and sympathetic ganglia.

- Explain what is meant by the mass activation of the sympathetic division and describe the relationship between the sympathetic division and the adrenal medulla.

- Identify the location of preganglionic parasympathetic neurons and parasympathetic ganglia.

- Identify the neurotransmitters of each of the autonomic neurons.

- Distinguish between the different types of adrenergic receptors and describe sympathetic nerve effects in the body.

- Distinguish between the different types of cholinergic receptors and describe the actions of the drug atropine.

- Describe the antagonistic, complementary, and cooperative actions of sympathetic and parasympathetic nerves on different body organs.

- Explain how the autonomic nervous system is controlled by the brain.

# Neural Control of Involuntary Effectors

*The autonomic nervous system helps to regulate the activities of smooth muscle, cardiac muscle, and glands. In this regulation, impulses are conducted from the CNS by axons that synapse with a second autonomic neuron. It is the axon of this second neuron in the pathway that innervates the involuntary effectors.*

Autonomic motor nerves innervate organs whose functions are not usually under voluntary control. The effectors that respond to autonomic regulation include **cardiac muscle (the heart)**, **smooth** (visceral) **muscles**, and **glands.** These

...........................................
autonomic: Gk. *auto*, self; *nomos*, law
viscera: L. *viscera*, internal organs

are part of the organs of the *viscera* (organs within the thoracic and abdominopelvic cavities) and of blood vessels. The involuntary effects of autonomic innervation contrast with the voluntary control of skeletal muscles by way of somatic motor neurons.

## Autonomic Neurons

As discussed in chapter 16, neurons of the peripheral nervous system (PNS) that conduct impulses away from the central nervous system (CNS) are known as *motor*, or *efferent*, *neurons*. There are two major categories of motor neurons: somatic and autonomic. Somatic motor neurons have their cell bodies within the CNS and send axons to skeletal muscles, which are usually under voluntary control. This was described in chapter 16, in the section on the reflex arc, and is reviewed in the left half of figure 17.1

Unlike somatic motor neurons, which conduct impulses along a single axon from the spinal cord to the neuromuscular

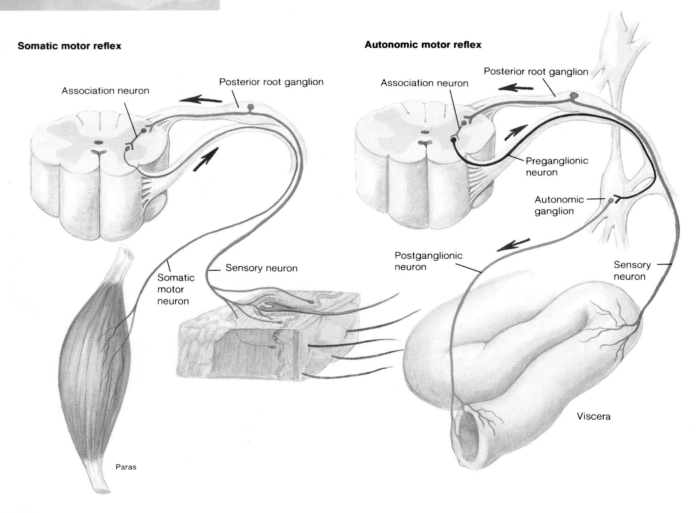

**Somatic motor reflex**

Association neuron

Posterior root ganglion

Somatic motor neuron

Sensory neuron

Paras

**Autonomic motor reflex**

Association neuron

Posterior root ganglion

Preganglionic neuron

Autonomic ganglion

Postganglionic neuron

Sensory neuron

Viscera

**FIGURE 17.1**

A comparison of a somatic motor reflex and an autonomic motor reflex.

## Table 17.1 Comparison of the somatic motor and autonomic motor systems

| Feature | Somatic motor | Autonomic motor |
|---|---|---|
| Effector organs | Skeletal muscles | Cardiac muscle, smooth muscle, and glands |
| Presence of ganglia | No ganglia | Cell bodies of postganglionic autonomic fibers located in paravertebral, prevertebral (collateral), and terminal ganglia |
| Number of neurons from CNS to effector | One | Two |
| Type of neuromuscular junction | Specialized motor end plate | No specialization of postsynaptic membrane; all areas of smooth muscle cells contain receptor proteins for neurotransmitters |
| Effect of nerve impulse on muscle | Excitatory only | Either excitatory or inhibitory |
| Type of nerve fibers | Fast-conducting, thick (9–13 μm), and myelinated | Slow-conducting; preganglionic fibers, lightly myelinated but thin (3 μm); postganglionic fibers, unmyelinated and very thin (about 1.0 μm) |
| Effect of denervation | Flaccid paralysis and atrophy | Muscle tone and function persist; target cells show denervation hypersensitivity |

junction, autonomic motor control involves two neurons in the motor pathway (table 17.1). The first of these neurons has its cell body in the gray matter of the brain or spinal cord. The axon of this neuron does not directly innervate the effector organ but instead synapses with a second neuron within an *autonomic ganglion.* (A ganglion is a collection of cell bodies outside the CNS.) The first neuron is thus called a **preganglionic neuron.** The second neuron in this pathway, called a **postganglionic neuron,** has an axon that extends from the autonomic ganglion and synapses with the cells of an effector organ (fig. 17.1, *right*).

Preganglionic autonomic fibers originate in the midbrain and hindbrain and from the upper thoracic to the fourth sacral levels of the spinal cord. Autonomic ganglia are located in the head, neck, and abdomen; chains of autonomic ganglia also parallel the right and left sides of the spinal cord. The origin of the preganglionic fibers and

.......................................................

ganglion: Gk. *ganglion,* a swelling

the location of the autonomic ganglia help to differentiate the **sympathetic** and **parasympathetic divisions** of the autonomic nervous system, which we will discuss in later sections of this chapter.

## Visceral Effector Organs

Since the autonomic nervous system helps to regulate the activities of glands, smooth muscles, and cardiac muscle, autonomic control is an integral aspect of the physiology of most of the body systems. Autonomic regulation, then, partly explains the functioning of smooth muscle (chapter 12), the endocrine glands (chapter 19), the circulatory system (chapters 21 and 22), and, in fact, of all the remaining systems yet to be discussed. Although we will describe the various functions of the target organs of autonomic innervation in subsequent chapters, at this point we will consider some of the basic features of autonomic regulation.

Unlike skeletal muscles, which enter a state of flaccid paralysis and atrophy when their motor nerves are severed, the involuntary effectors are somewhat independent of their innervation. For example, smooth muscles maintain a resting tone (tension) in the absence of nerve stimulation. In fact, damage to an autonomic nerve makes its target tissue more sensitive than normal to stimulating agents. This phenomenon is called **denervation hypersensitivity.** Such compensatory changes can explain why, for example, the ability of the mucosa of the stomach to secrete acid may be restored after its neural supply from the vagus nerve is severed. (This procedure is called gastric vagotomy, and is sometimes performed as a treatment for ulcers.)

In addition to their intrinsic ("built-in") muscle tone, cardiac muscle and many smooth muscles take their autonomy a step further. These muscles can contract rhythmically, even in the absence of nerve stimulation, in response to electrical waves of depolarization initiated by the muscles themselves. Autonomic innervation simply increases or decreases this intrinsic activity. Autonomic nerves also maintain a resting "tone" in the sense that they maintain a baseline firing rate that can be either increased or decreased. A decrease in the excitatory input to the heart, for example, will slow its rate of beat.

The release of the neurotransmitter (ACh) from somatic motor neurons always stimulates the effector organ (skeletal muscles). By contrast, some autonomic nerve fibers release transmitters that inhibit the activity of their effectors. An increase in the activity of the vagus nerve that supplies inhibitory fibers to the heart, for example, will slow the heart rate, whereas a decrease in this inhibitory input will increase the heart rate.

# Divisions of the Autonomic Nervous System

*Preganglionic neurons of the sympathetic division of the autonomic nervous system originate at the thoracic and lumbar levels of the spinal cord and send axons to sympathetic ganglia, which parallel the spinal cord. Preganglionic neurons of the parasympathetic division, by contrast, originate in the brain and at the sacral level of the spinal cord and send axons to ganglia located in different regions of the body in or near the effector organs.*

The sympathetic and parasympathetic divisions of the autonomic nervous system have some structural features in common: both consist of preganglionic neurons that originate in the CNS and postganglionic neurons that originate outside of the CNS in ganglia. The specific origin of the preganglionic fibers and the location of the ganglia, however, are different in the two divisions of the autonomic nervous system.

## Sympathetic (Thoracolumbar) Division

The sympathetic division is also called the *thoracolumbar division* of the autonomic nervous system because its preganglionic fibers exit the spinal cord from the first thoracic (T1) to the second lumbar (L2) levels. Most sympathetic nerve fibers, however, separate from the somatic motor fibers and synapse with postganglionic neurons within a double row of **sympathetic ganglia,** or **paravertebral ganglia,** one row on each lateral side of the spinal cord (fig. 17.2). Ganglia within each row are interconnected, forming a *sympathetic chain* of ganglia that parallels the spinal cord on either lateral side.

The preganglionic sympathetic fibers are myelinated and thus appear white. These fibers form communicating branches called **white rami communicantes** (singular, *ramus*

**FIGURE 17.2**
The sympathetic chain of paravertebral ganglia showing its relationship to the vertebral column and the spinal cord.

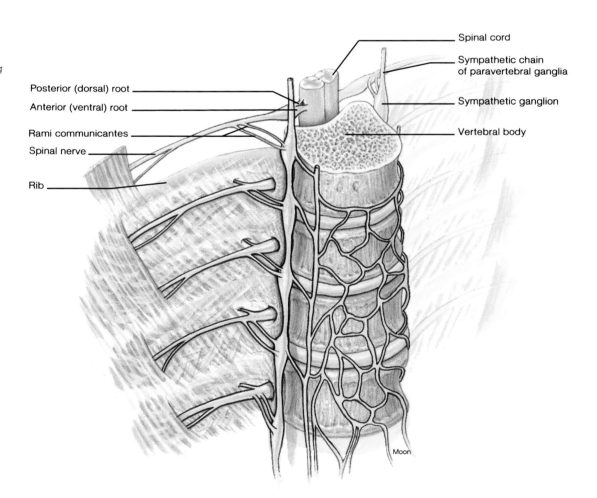

Spinal cord

Sympathetic chain of paravertebral ganglia

Posterior (dorsal) root

Anterior (ventral) root

Sympathetic ganglion

Rami communicantes

Vertebral body

Spinal nerve

Rib

Moon

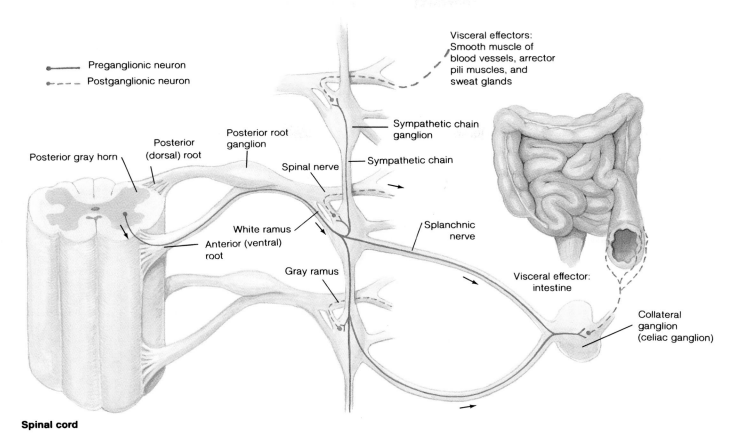

**Preganglionic neuron**
- - - - **Postganglionic neuron**

Posterior gray horn

Posterior (dorsal) root

Posterior root ganglion

Spinal nerve

White ramus

Anterior (ventral) root

Gray ramus

Visceral effectors: Smooth muscle of blood vessels, arrector pili muscles, and sweat glands

Sympathetic chain ganglion

Sympathetic chain

Splanchnic nerve

Visceral effector: intestine

Collateral ganglion (celiac ganglion)

**Spinal cord**

**FIGURE 17.3**

Sympathetic chain ganglia, the sympathetic chain, and rami communicantes of the sympathetic division of the ANS. (Solid lines = preganglionic fibers; dashed lines = postganglionic fibers.)

*communicans*). Some of the preganglionic sympathetic fibers synapse with postganglionic neurons located at their same level in the chain of sympathetic ganglia. Other preganglionic fibers travel up or down within the sympathetic chain before synapsing with postganglionic neurons. Since the fibers of the postganglionic sympathetic neurons are unmyelinated and thus appear gray, they form **gray rami communicantes.** Postganglionic axons in the gray rami extend directly back to the anterior roots of the spinal nerves and travel distally within the spinal nerves to innervate their effector organs (fig. 17.3).

Within the chain of paravertebral ganglia, *divergence* is apparent when a single preganglionic fiber branches to synapse with many postganglionic neurons located at different levels in the chain. *Convergence* is apparent also, when a given postganglionic neuron receives synaptic input from a large number of preganglionic fibers. The divergence of impulses from the spinal cord to the ganglia and the convergence of

impulses within the ganglia usually results in the **mass activation** of almost all of the postganglionic neurons. This explains why the sympathetic division is usually activated as a unit and affects all of its effector organs at the same time.

Many preganglionic fibers that exit the spinal cord in the upper thoracic level travel into the neck, where they synapse in cervical sympathetic ganglia (fig. 17.4). Postganglionic neurons from here innervate the smooth muscles and glands of the head and neck.

**Collateral Ganglia**   Many preganglionic fibers that exit the spinal cord below the level of the diaphragm pass through the sympathetic chain of ganglia without synapsing. Beyond the sympathetic chain, these preganglionic fibers form **splanchnic** (*splangk´nik*) **nerves.** Preganglionic fibers in the splanchnic nerves synapse in **collateral,** or **prevertebral, ganglia.** These include the *celiac, superior mesenteric,* and

ramus: L. *ramus,* a branch

splanchnic: Gk. *splanchono-,* relating to viscera

449

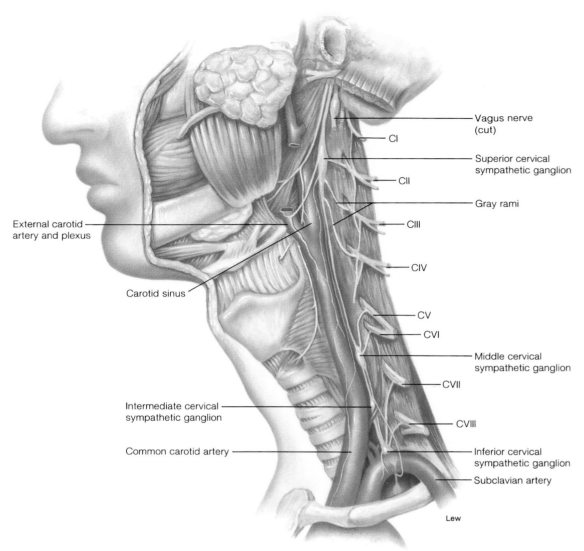

External carotid
artery and plexus

Carotid sinus

Intermediate cervical
sympathetic ganglion

Common carotid artery

Vagus nerve
(cut)

CI

Superior cervical
sympathetic ganglion

CII

Gray rami

CIII

CIV

CV

CVI

Middle cervical
sympathetic ganglion

CVII

CVIII

Inferior cervical
sympathetic ganglion

Subclavian artery

Lew

**FIGURE 17.4**
The cervical sympathetic ganglia.

*inferior mesenteric ganglia* (figs. 17.5 and 17.6). Postganglionic neurons that arise from the collateral ganglia innervate organs of the digestive, urinary, and reproductive systems.

**Adrenal Glands**   The paired adrenal glands are located above each kidney. Each adrenal gland is composed of two parts: an outer **adrenal cortex** and an inner **adrenal medulla.** These two parts are really two functionally different glands with different embryonic origins, different hormones, and different regulatory mechanisms. The adrenal cortex secretes steroid hormones; the adrenal medulla secretes the hormone

· · · · · · · · · · · · · · · · · · · · · · · · · · · · · ·
adrenal: L. *ad*, to; *renes,* kidney
cortex: L. *cortex,* bark
medulla: L. *medulla*, marrow

**epinephrine** (adrenaline) and, to a lesser degree, **norepinephrine,** when it is stimulated by the sympathetic division of the autonomic nervous system.

The adrenal medulla can be likened to a modified sympathetic ganglion; its cells are derived from the same embryonic tissue that forms postganglionic sympathetic neurons. Like a sympathetic ganglion, the cells of the adrenal medulla are innervated by preganglionic sympathetic fibers. The adrenal medulla secretes epinephrine into the blood in response to this neural stimulation. The effects of epinephrine are complementary to those of the neurotransmitter norepinephrine, which is released from postganglionic sympathetic nerve endings. For this reason, and because the adrenal medulla is stimulated as part of the mass activation of the sympathetic system, the two are often grouped together as the **sympathoadrenal system.**

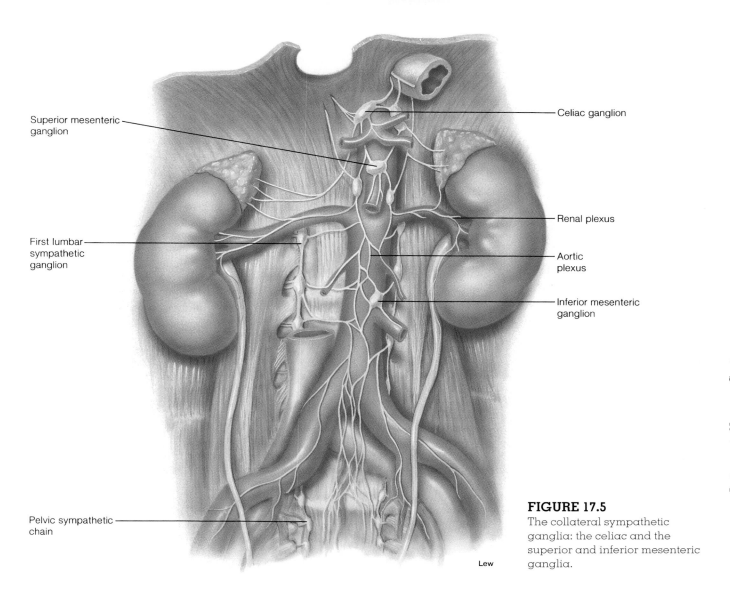

Superior mesenteric ganglion

Celiac ganglion

First lumbar sympathetic ganglion

Renal plexus

Aortic plexus

Inferior mesenteric ganglion

Pelvic sympathetic chain

Lew

**FIGURE 17.5**
The collateral sympathetic ganglia: the celiac and the superior and inferior mesenteric ganglia.

## *Parasympathetic (Craniosacral) Division*

The parasympathetic division is also known as the *craniosacral division* of the autonomic nervous system. This is because its preganglionic neurons originate in the brain (specifically, in the midbrain and the medulla oblongata of the brain stem) and in the second through fourth sacral levels of the spinal cord. These preganglionic parasympathetic fibers synapse in ganglia that are located next to (or actually within) the organs innervated. These parasympathetic ganglia, which are called **terminal ganglia,** supply the postganglionic neurons that synapse with the effector cells.

The comparative structures of the sympathetic and parasympathetic divisions are summarized in tables 17.2 and 17.3. It should be noted that, unlike sympathetic fibers, most parasympathetic fibers do not travel within spinal nerves. As

a result, cutaneous effectors (blood vessels, sweat glands, and arrector pili muscles) and blood vessels in skeletal muscles receive sympathetic but not parasympathetic innervation.

Four of the 12 pairs of cranial nerves (described in chapter 16) contain preganglionic parasympathetic fibers. These are the oculomotor (third cranial), facial (seventh cranial), glossopharyngeal (ninth cranial), and vagus (tenth cranial) nerves. Parasympathetic fibers within the first three of these cranial nerves synapse in ganglia located in the head; fibers in the vagus nerve synapse in terminal ganglia located in many regions of the body.

The oculomotor nerve contains somatic motor and parasympathetic fibers that originate in the oculomotor nuclei of the midbrain. These parasympathetic fibers synapse in the *ciliary ganglion*, whose postganglionic fibers innervate the ciliary muscle and constrictor fibers in the iris of the

## Table 17.2    The sympathetic (thoracolumbar) division

| Parts of body innervated | Spinal origin of preganglionic fibers | Origin of postganglionic fibers |
|---|---|---|
| Eye | C8, T1 | Cervical ganglia |
| Head and neck | T1–T4 | Cervical ganglia |
| Heart and lungs | T1–T5 | Upper thoracic (paravertebral) ganglia |
| Upper extremities | T2–T9 | Lower cervical and upper thoracic (paravertebral) ganglia |
| Upper abdominal viscera | T4–T9 | Celiac and superior mesenteric (collateral) ganglia |
| Adrenal glands | T10, T11 | — |
| Urinary and reproductive systems | T12–L2 | Celiac and inferior mesenteric (collateral) ganglia |
| Lower extremities | T9–L2 | Lumbar and upper sacral (paravertebral) ganglia |

## Table 17.3    The parasympathetic (craniosacral) division

| Effector organs | Origin of preganglionic fibers | Nerve | Location of terminal ganglia |
|---|---|---|---|
| Eye (smooth muscle in iris and ciliary body) | Midbrain (cranial) | Oculomotor (third cranial) nerve | Ciliary ganglion |
| Lacrimal, mucous, and salivary glands | Medulla oblongata (cranial) | Facial (seventh cranial) nerve | Pterygopalatine and submandibular ganglia |
| Parotid gland | Medulla oblongata (cranial) | Glossopharyngeal (ninth cranial) nerve | Otic ganglion |
| Heart, lungs, gastrointestinal tract, liver, pancreas | Medulla oblongata (cranial) | Vagus (tenth cranial) nerve | Terminal ganglia in or near organ |
| Lower half of large intestine, rectum, urinary bladder, and reproductive organs | S2–S4 (sacral) | Pelvic spinal nerves | Terminal ganglia near organs |

eye. Preganglionic fibers that originate in the pons travel in the facial nerve to the *pterygopalatine* (ter˝ ĭ-go-pal´ă-tēn) *ganglion*, which sends postganglionic fibers to the nasal mucosa, pharynx, palate, and lacrimal glands. Another group of fibers in the facial nerve terminate in the *submandibular ganglion*, which sends postganglionic fibers to the submandibular and sublingual glands. Preganglionic fibers of the glossopharyngeal nerve synapse in the *otic ganglion*, which sends postganglionic fibers to the parotid gland.

Other nuclei in the medulla oblongata contribute preganglionic fibers to the very long *vagus (tenth cranial) nerves.* These preganglionic fibers travel through the neck to the thoracic cavity, and through the esophageal hiatus (opening) in the diaphragm to the abdominal cavity (fig. 17.6). In each region, some of these preganglionic fibers branch from the main trunks of the vagus nerves and synapse with postganglionic neurons that are located *within* the innervated organs. The preganglionic vagus fibers are thus quite long; they provide parasympathetic innervation to the heart, lungs, esophagus, stomach, pancreas, liver, small intestine, and the upper half of the large intestine. Postganglionic parasympathetic fibers arise from terminal ganglia within these organs and synapse with effector cells (smooth muscles and glands).

Preganglionic fibers from the sacral levels of the spinal cord provide parasympathetic innervation to the lower half of the large intestine, to the rectum, and to the urinary and reproductive systems. These fibers, like those of the vagus nerves, synapse with terminal ganglia located within the effector organs.

Parasympathetic nerves to the visceral organs thus consist of preganglionic fibers, whereas sympathetic nerves to these organs contain postganglionic fibers. A composite view of the sympathetic and parasympathetic systems is provided in figure 17.7, and the comparisons are summarized in table 17.4.

....................................................

vagus: L. *vagus,* wandering

**FIGURE 17.6**

The vagus nerve and its branches provide parasympathetic innervation to most organs of the internal environment.

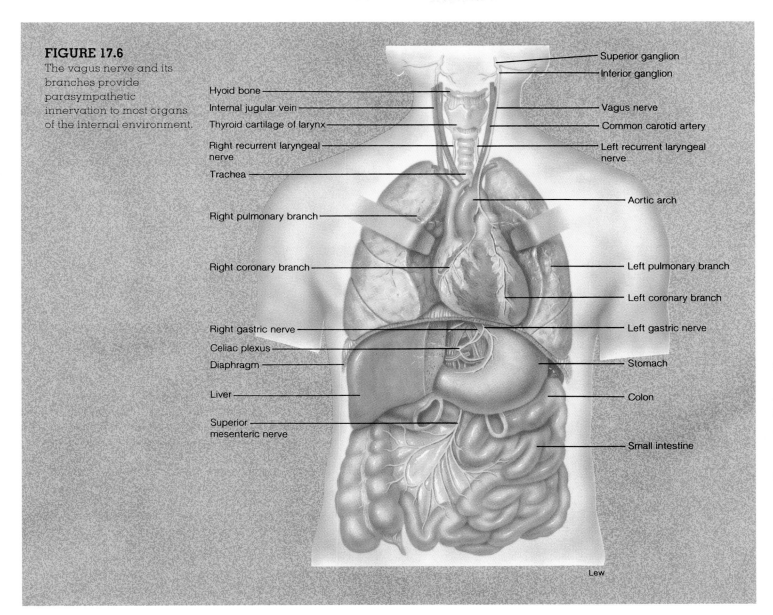

Hyoid bone

Internal jugular vein

Thyroid cartilage of larynx

Right recurrent laryngeal nerve

Trachea

Right pulmonary branch

Right coronary branch

Right gastric nerve

Celiac plexus

Diaphragm

Liver

Superior mesenteric nerve

Superior ganglion

Inferior ganglion

Vagus nerve

Common carotid artery

Left recurrent laryngeal nerve

Aortic arch

Left pulmonary branch

Left coronary branch

Left gastric nerve

Stomach

Colon

Small intestine

Lew

# Functions of the Autonomic Nervous System

*The sympathetic division of the autonomic nervous system activates the body to "fight or flight," largely through the release of norepinephrine from postganglionic fibers and the secretion of epinephrine from the adrenal medulla. The parasympathetic division often produces antagonistic effects through the release of acetylcholine from its postganglionic fibers. The actions of both divisions of the autonomic nervous system must be balanced in order to maintain homeostasis.*

The sympathetic and parasympathetic divisions of the autonomic nervous system affect the visceral organs in different ways. Mass activation of the sympathetic division prepares the body for intense physical activity in emergencies; the heart rate increases, blood glucose rises, and blood is diverted to the skeletal muscles (away from the viscera and skin). These and other effects are listed in table 17.5. The theme of the sympathetic system is aptly summarized in a phrase—**fight or flight.**

The effects of parasympathetic nerve stimulation are often opposite to the effects of sympathetic stimulation. The parasympathetic division, however, is not normally activated as a whole. Stimulation of separate parasympathetic

## Table 17.4  Comparison of the sympathetic and parasympathetic divisions of the ANS

| Feature | Sympathetic | Parasympathetic |
| --- | --- | --- |
| Origin of preganglionic outflow | Thoracolumbar portion of spinal cord | Midbrain, hindbrain, and sacral portion of spinal cord |
| Location of ganglia | Chain of paravertebral ganglia and prevertebral (collateral) ganglia | Terminal ganglia in or near effector organs |
| Distribution of postganglionic fibers | Throughout the body | Mainly limited to the head and viscera |
| Divergence of impulses from pre- to postganglionic fibers | Great divergence (1 preganglionic may activate 20 postganglionic fibers) | Little divergence (1 preganglionic only activates a few postganglionic fibers) |
| Mass discharge of system as a whole | Usually | Not normally |

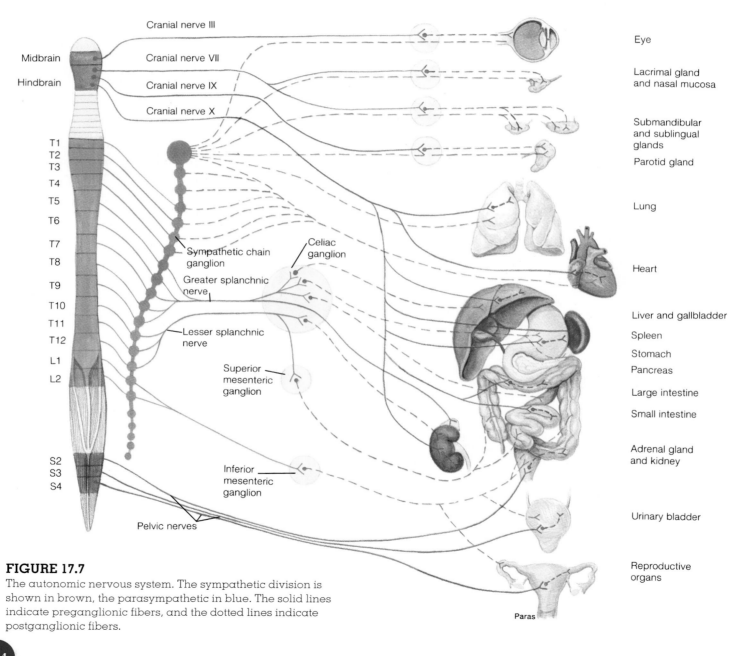

**FIGURE 17.7**

The autonomic nervous system. The sympathetic division is shown in brown, the parasympathetic in blue. The solid lines indicate preganglionic fibers, and the dotted lines indicate postganglionic fibers.

## Table 17.5    Effects of autonomic nerve stimulation on various visceral effector organs

| Effector organ | Sympathetic effect | Parasympathetic effect |
| --- | --- | --- |
| **Eye** | | |
| Iris (pupillary dilator muscle) | Dilation of pupil | — |
| Iris (pupillary sphincter muscle) | — | Constriction of pupil |
| Ciliary muscle | Relaxation (for far vision) | Contraction (for near vision) |
| **Glands** | | |
| Lacrimal (tear) | — | Stimulation of secretion |
| Sweat | Stimulation of secretion | — |
| Salivary | Decreased secretion; saliva becomes thick | Increased secretion; saliva becomes thin |
| Stomach | — | Stimulation of secretion |
| Intestine | — | Stimulation of secretion |
| Adrenal medulla | Stimulation of hormone secretion | — |
| **Heart** | | |
| Rate | Increased | Decreased |
| Conduction | Increased rate | Decreased rate |
| Strength | Increased | — |
| **Blood vessels** | Mostly constriction; affects all organs | Dilation in a few organs (e.g., penis) |
| **Lungs** | | |
| Bronchioles (tubes) | Dilation | Constriction |
| Mucous glands | Inhibition of secretion | Stimulation of secretion |
| **Gastrointestinal tract** | | |
| Motility | Inhibition of movement | Stimulation of movement |
| Sphincters | Closing stimulated | Closing inhibited |
| **Liver** | Stimulation of glycogen hydrolysis | — |
| **Adipocytes (fat cells)** | Stimulation of fat hydrolysis | — |
| **Pancreas** | Inhibition of exocrine secretions | Stimulation of exocrine secretions |
| **Spleen** | Stimulation of contraction | — |
| **Urinary bladder** | Muscle tone aided | Stimulation of contraction |
| **Arrector pili muscles** | Stimulation of hair erection, causing goosebumps | — |
| **Uterus** | If pregnant, contraction; if not pregnant, relaxation | — |
| **Penis** | Ejaculation | Erection (due to vasodilation) |

nerves can result in slowing of the heart, dilation of visceral blood vessels, and increased activity of the gastrointestinal (GI) tract (table 17.5). Visceral organs respond differently to sympathetic and parasympathetic nerve activity because the postganglionic fibers of these two divisions release different neurotransmitters.

## Neurotransmitters of the Autonomic Nervous System

**Acetylcholine (ACh)** is the neurotransmitter of all preganglionic fibers (both sympathetic and parasympathetic). Acetylcholine is also the transmitter released by parasympathetic postganglionic fibers at their synapses with effector cells (fig. 17.8). Transmission at these synapses is thus said to be **cholinergic** (ko″lĭ-ner′jik).

The neurotransmitter released by most postganglionic sympathetic nerve fibers is **norepinephrine** (noradrenalin). Transmission at these synapses is thus said to be **adrenergic** (ad″rĕ-ner′jik). There are a few exceptions, however. Some sympathetic fibers that innervate blood vessels in skeletal muscles and the external genitalia, as well as sympathetic fibers to sweat glands, release ACh (are cholinergic).

In view of the fact that the cells of the adrenal medulla are embryologically related to postganglionic sympathetic neurons, it is not surprising that the hormones released by the adrenal medulla should consist of epinephrine (about 85%) and norepinephrine (about 15%). Epinephrine differs from norepinephrine only in that the former has an additional methyl ($CH_3$) group, as shown in figure 17.9. Epinephrine, norepinephrine, and dopamine (a transmitter within the CNS) are all derived from the amino acid tyrosine and are collectively termed **catecholamines.**

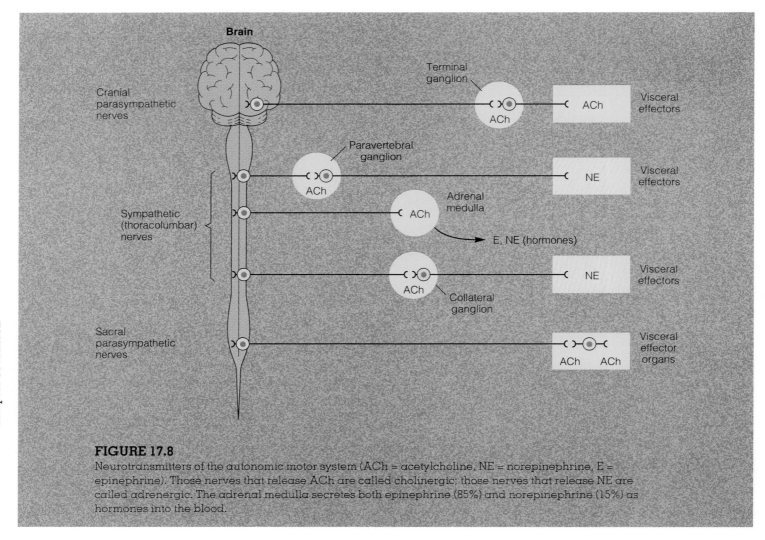

**FIGURE 17.8**

Neurotransmitters of the autonomic motor system (ACh = acetylcholine, NE = norepinephrine, E = epinephrine). Those nerves that release ACh are called cholinergic; those nerves that release NE are called adrenergic. The adrenal medulla secretes both epinephrine (85%) and norepinephrine (15%) as hormones into the blood.

**FIGURE 17.9**

The structure of the catecholamines norepinephrine and epinephrine.

Norepinephrine

Epinephrine

## Responses to Adrenergic Stimulation

Adrenergic stimulation—by epinephrine in the blood and by norepinephrine released from sympathetic nerve endings—has both excitatory and inhibitory effects. The heart, dilatory muscles of the iris, and the smooth muscles of many blood vessels are stimulated to contract. The smooth muscles of the bronchioles and of some blood vessels, however, are inhibited from contracting; adrenergic chemicals, therefore, cause these structures to dilate.

Since excitatory and inhibitory effects can be produced in different tissues by the same chemical, the responses clearly depend on the biochemistry of the tissue cells rather than on the intrinsic properties of the chemical. Included in the biochemical differences among the target tissues for catecholamines are differences in the *membrane receptor proteins* for these chemical agents. The two major classes of these receptor proteins are designated **alpha- ($\alpha$)** and **beta- ($\beta$) adrenergic receptors.** (Recall from chapter 14 the description of the interaction of neurotransmitters and receptor proteins.)

Experiments have revealed that there are two subtypes of each category of adrenergic receptor. These are designated by subscripts: $\alpha_1$ and $\alpha_2$; $\beta_1$ and $\beta_2$. Scientists have developed compounds that selectively bind to one or the other type of adrenergic receptor and, by this means, either promote or inhibit the normal action produced when epinephrine or norepinephrine binds to the receptor. As a result of its binding to an adrenergic receptor, a drug may either promote or inhibit the adrenergic effect. Also, by using these selective compounds, it has been possible to determine which subtype of adrenergic receptor predominates in each organ (table 17.6).

A review of table 17.6 reveals certain generalities about the actions of adrenergic receptors. The stimulation of alpha-adrenergic receptors consistently causes contraction of smooth muscles. One can thus state that the vasoconstrictor effect of sympathetic nerves always results from the activation of alpha-adrenergic receptors. The effects of beta-adrenergic activation are more complex; these receptors stimulate the relaxation of smooth muscles (in the gastrointestinal tract, bronchioles, and uterus, for example), but stimulate contraction of cardiac muscle and promote an increase in cardiac rate.

 The use of drugs that selectively stimulate or block (act as *agonists* or *antagonists)* adrenergic receptors are of great clinical benefit. Many people with hypertension, for example, have received a beta-blocking drug known as *propranolol.* This drug blocks $\beta_1$ receptors, which are located in the heart, and thereby has the desired effect of lowering the cardiac rate and blood pressure. Propranolol, however, also blocks $\beta_2$ receptors, which are located in the bronchioles of the lungs. This reduces the bronchodilation effect of epinephrine, producing bronchoconstriction and asthma in susceptible people. A more specific $\beta_1$ antagonist, *atenolol,* is now used instead to slow the cardiac rate and lower blood pressure. At one time, asthmatics inhaled an epinephrine spray, which stimulates $\beta_1$ receptors in the heart as well as $\beta_2$ receptors in the airways. Now, drugs such as *terbutaline* that selectively function as $\beta_2$ agonists are more commonly used. Drugs that function as $\alpha_1$ agonists, such as *phenylephrine,* are often part of nasal sprays because they promote vasoconstriction in the nasal mucosa.

The diverse effects of epinephrine and norepinephrine can be understood in terms of the fight-or-flight theme. Adrenergic stimulation wrought by activation of the sympathetic division produces an increase in cardiac pumping (a $\beta_1$ effect), vasoconstriction and thus reduced blood flow to the visceral organs (an $\alpha$-adrenergic effect), dilation of pulmonary bronchioles (a $\beta_2$ effect), and so on.

## Responses to Cholinergic Stimulation

Somatic motor neurons, preganglionic autonomic neurons, and postganglionic parasympathetic neurons are cholinergic —they release acetylcholine as a neurotransmitter. The cholinergic effects of somatic motor neurons and preganglionic autonomic neurons are always excitatory. The cholinergic effects of postganglionic parasympathetic fibers are usually excitatory, but there are notable exceptions. The parasympathetic fibers innervating the heart, for example, cause slowing of the heart rate. It is useful to remember that the effects of parasympathetic stimulation are, in general, opposite to the effects of sympathetic stimulation.

There are two known subtypes of cholinergic receptors. The drug *muscarine* (mus´kă-rēn), derived from poisonous mushrooms, stimulates the cholinergic receptor proteins in the target organs of postganglionic parasympathetic nerve fibers (such as in the heart, eye, and digestive system). Muscarine, however, does not stimulate ACh receptor proteins in autonomic ganglia or at the neuromuscular junction of skeletal muscle fibers. The ACh receptors stimulated by muscarine are called **muscarinic receptors,** and the effects produced by parasympathetic nerves in their target organs are called *muscarinic effects* (indicated in table 17.6).

The drug *nicotine*, derived from the tobacco plant, specifically stimulates cholinergic transmission of preganglionic fibers at the autonomic ganglia, as well as activation of the neuromuscular junction of skeletal muscles. These ACh receptors are called **nicotinic receptors** to distinguish them from the muscarinic receptors. The drug *curare*, used clinically to relax skeletal muscles, specifically blocks nicotinic receptors but has little effect on muscarinic receptors.

The muscarinic effects of ACh are specifically inhibited by the drug **atropine,** derived from the deadly nightshade plant (*Atropa belladonna*). Indeed, extracts of this plant were used by women during the Middle Ages to dilate their pupils (atropine inhibits parasympathetic stimulation of the iris), which was thought to enhance their beauty (belladonna—beautiful lady). Atropine is used clinically today to dilate pupils during eye examinations, to dry mucous membranes of the respiratory tract prior to general anesthesia, and to inhibit spasmodic contractions of the lower gastrointestinal tract.

## Other Autonomic Neurotransmitters

Certain postganglionic autonomic axons produce their effects through mechanisms that do not involve either norephinephrine or acetylcholine. This can be demonstrated experimentally by the inability of drugs that block adrenergic and cholinergic effects from inhibiting the actions of those autonomic axons. These axons, consequently, have been termed *nonadrenergic noncholinergic fibers*. Candidate molecules for the role of neurotransmitter by these axons include the gas *nitric oxide* (NO).

The nonadrenergic noncholinergic parasympathetic axons that innervate the blood vessels of the penis cause the smooth muscles of these vessels to relax, hence producing vasodilation and a consequent erection of the penis. These parasympathetic axons have been shown to use nitric oxide (chapter 14) as their neurotransmitter. In a similar manner, nitric oxide appears to function as the autonomic neurotransmitter causing vasodilation of cerebral arteries. In these

# Table 17.6  Adrenergic and cholinergic effects of sympathetic and parasympathetic nerves

| | Effect of | | | |
| | Sympathetic | | Parasympathetic | |
| Organ | Action | Receptor* | Action | Receptor* |
|---|---|---|---|---|
| **Eye** | | | | |
| Iris | | | | |
|   Pupillary dilator muscle | Contracts | α | . . . | . . . |
|   Pupillary sphincter muscle | . . . | . . . | Contracts | M |
| **Heart** | | | | |
| Sinoatrial node | Accelerates | $\beta_1$ | Decelerates | M |
| Contractility | Increases | $\beta_1$ | Decreases (atria) | M |
| **Vascular smooth muscle** | | | | |
| Skin, splanchnic vessels | Contracts | α | . . . | . . . |
| Skeletal muscle vessels | Relaxes | $\beta_2$ | . . . | . . . |
| | Relaxes | M** | . . . | . . . |
| **Bronchiolar smooth muscle** | Relaxes | $\beta_2$ | Contracts | M |
| **Gastrointestinal tract** | | | | |
| Smooth Muscle | | | | |
|   Walls | Relaxes | $\beta_2$ | Contracts | M |
|   Sphincters | Constricts | α | Relaxes | M |
| Secretion | Decreases | α | Increases | M |
| Myenteric plexus | Inhibits | α | . . . | . . . |
| **Genitourinary smooth muscle** | | | | |
| Wall of urinary bladder | Relaxes | $\beta_2$ | Contracts | M |
| Urethral sphincter | Constricts | α | Relaxes | M |
| Uterus, pregnant | Relaxes | $\beta_2$ | . . . | . . . |
| | Contracts | α | . . . | . . . |
| Penis, seminal vesicles | Ejaculation | α | Erection | M |
| **Skin** | | | | |
| Arrector pili muscle | Contracts | α | . . . | . . . |
| Sweat gland activity | | | | |
|   Eccrine | Increases | M | . . . | . . . |
|   Apocrine (stress) | Increases | α | . . . | . . . |

*Adrenergic receptors are indicated as alpha (α) or beta (β); cholinergic receptors are indicated as muscarinic (M).
**Vascular smooth muscle in skeletal muscle has sympathetic cholinergic dilator fibers.

cases, nitric oxide diffuses across the synaptic cleft and promotes relaxation of the postsynaptic smooth muscle cells.

Nitric oxide can produce relaxation of smooth muscles in many organs, including the stomach, small intestine, large intestine, and urinary bladder. There is some controversy, however, about whether the nitric oxide functions in each case as a neurotransmitter, or whether it is produced in the organ itself in response to autonomic stimulation. The latter is a real possibility, because different tissues, such as the endothelium of blood vessels, can produce nitric oxide. Indeed, nitric oxide is a member of a class of local tissue regulatory molecules called *paracrine regulators* (chapter 19). Regulation can therefore be a complex process involving the interacting effects of different neurotransmitters, hormones, and paracrine regulators.

## Organs with Dual Innervation

Most visceral organs receive dual innervation—they are innervated by both sympathetic and parasympathetic fibers. In this condition, the effects of the two divisions may be antagonistic, complementary, or cooperative.

**Antagonistic Effects**    The effects of sympathetic and parasympathetic innervation of the pacemaker region of the heart is the best example of the antagonism of these two systems. In this case, sympathetic and parasympathetic fibers innervate the same cells. Adrenergic stimulation from sympathetic fibers increases the heart rate and cholinergic stimulation from parasympathetic fibers inhibits the pacemaker cells and, thus, decreases the heart rate. A reverse of this antagonism is seen in the gastrointestinal tract, where sympathetic nerves inhibit intestinal movements and secretions and parasympathetic nerves stimulate these movements and secretions.

The effects of sympathetic and parasympathetic stimulation on the diameter of the pupil of the eye are analogous to the reciprocal innervation of flexor and extensor skeletal muscles by somatic motor neurons (chapter 12). This is because the iris contains antagonistic muscle layers. Contraction of the pupillary dilator muscle, which is stimulated by impulses through sympathetic nerve endings, causes dilation; contraction of the pupillary sphincter muscle, which is innervated by parasympathetic nerve endings, causes constriction of the pupil (refer to chapter 18, fig. 18.29).

**Complementary Effects**    The effects of sympathetic and parasympathetic stimulation on salivary gland secretion are complementary. The secretion of watery saliva is stimulated through parasympathetic nerves, which also stimulate the secretion of other exocrine glands in the gastrointestinal tract. Impulses through sympathetic nerves stimulate the constriction of blood vessels throughout the gastrointestinal tract. The resultant decrease in blood flow to the salivary glands causes the production of a thicker, more viscous saliva.

**Cooperative Effects**    The effects of sympathetic and parasympathetic stimulation on the urinary and reproductive systems are cooperative. Erection of the penis, for example, is due to vasodilation resulting from parasympathetic nerve stimulation; ejaculation is due to stimulation through sympathetic nerves. Although the contraction of the urinary bladder is myogenic (independent of nerve stimulation), it is promoted in part by the action of parasympathetic nerves. This *micturition* (*mik''tŭ-rish'un*), or urination, urge and reflex is also enhanced by sympathetic nerve activity, which increases the tone of the urinary bladder muscles. Emotional states that are accompanied by high sympathetic nerve activity (such as extreme fear) may thus result in reflex urination at urinary bladder volumes that are normally too low to trigger this reflex.

## Organs without Dual Innervation

Although most organs are innervated by both sympathetic and parasympathetic nerves, some—including the adrenal medulla, arrector pili muscles, sweat glands, and most blood vessels—receive only sympathetic innervation. In these cases, regulation is achieved by increases or decreases in the tone (firing rate) of the sympathetic fibers. Constriction of blood vessels, for example, is produced by increased sympathetic activity, which stimulates alpha-adrenergic receptors, and vasodilation results from decreased sympathetic nerve stimulation.

The sympathoadrenal system is required for *nonshivering thermogenesis:* animals deprived of their sympathetic division and adrenal glands cannot tolerate cold stress. The sympathetic division is also required for proper thermoregulatory responses to heat. In a hot room, for example, decreased sympathetic stimulation produces dilation of the blood vessels in the surface of the skin, which increases cutaneous blood flow and provides better heat radiation. During exercise, on the other hand, there is increased sympathetic activity, which causes constriction of the blood vessels in the skin of the limbs. This by itself would cause heat retention, but sympathetic fibers also stimulate sweat glands in the trunk. Evaporation of this dilute sweat, secreted in response to cholinergic sympathetic activity, helps to cool the body.

......................................

micturition: L. *micturire,* to urinate

## Table 17.7 — Effects stimulated by sensory input from afferent fibers in the vagus nerves, which transmit this input to centers in the medulla oblongata

| Organs | Type of receptors | Reflex effects |
|---|---|---|
| Lungs | Stretch receptors | Further inhalation inhibited; increase in cardiac rate and vasodilation stimulated |
| | Type J receptors | Stimulated by pulmonary congestion—produces feelings of breathlessness and causes a reflex fall in cardiac rate and blood pressure |
| Aortic arch | Chemoreceptors | Stimulated by rise in $CO_2$ and fall in $O_2$—produces increased rate of breathing, fall in heart rate, and vasoconstriction |
| | Baroreceptors | Stimulated by increased blood pressure—produces a reflex decrease in heart rate |
| Heart | Atrial stretch receptors | Antidiuretic hormone secretion inhibited, thus increasing the volume of urine excreted |
| | Stretch receptors in ventricles | Produces a reflex decrease in heart rate and vasodilation |
| Gastrointestinal tract | Stretch receptors | Feelings of satiety, discomfort, and pain |

## Control of the Autonomic Nervous System by Higher Brain Centers

Visceral functions are largely regulated by autonomic reflexes. In most autonomic reflexes, sensory input is transmitted to brain centers that integrate this information and respond appropriately by modifying the activity of motor preganglionic autonomic neurons. The neural centers that directly control the activity of autonomic nerves are influenced by higher brain areas, as well as by sensory input.

The **medulla oblongata** (chapter 15) of the brain stem is the area that most directly controls the activity of the autonomic nervous system. Almost all autonomic responses can be elicited by experimental stimulation of the medulla oblongata, which contains centers for the control of the circulatory, respiratory, urinary, reproductive, and digestive systems. Much of the sensory input to these centers travels through the sensory fibers of the vagus nerve—a mixed nerve containing both sensory and motor fibers. These reflexes are listed in table 17.7.

Although the medulla oblongata directly regulates the activity of autonomic motor fibers, the medulla oblongata is itself responsive to regulation by higher brain areas. The **hypothalamus** (chapter 15), for example, is a major center for the control of body temperature, hunger, and thirst; regulation of the pituitary gland; and—together with the limbic system and cerebral cortex—various emotional states.

The **limbic system** is a group of fiber tracts and nuclei that form a ring (limbus) around the brain stem. It includes the cingulate gyrus of the cerebral cortex, the hypothalamus, the fornix (a fiber tract), the hippocampus, and the amygdaloid nucleus (fig. 17.10). The limbic system is involved in basic emotional drives, such as anger, fear, sex, and hunger. The involvement of the limbic system with the control of autonomic function is responsible for the visceral responses characteristic of these emotional states. Blushing, pallor, fainting, breaking out in a cold sweat, heart palpitations, and "butterflies in the stomach" are only some of the many visceral reactions that, as a result of autonomic activation, accompany emotions.

The autonomic correlates of motion sickness—nausea, sweating, and cardiovascular changes—are eliminated by cutting the motor tracts of the cerebellum. This demonstrates that impulses from the cerebellum to the medulla oblongata influence activity of the autonomic nervous system. Experimental and clinical observations have also demonstrated that the frontal and temporal lobes of the cerebral cortex influence lower brain areas as part of their involvement in emotion and personality.

One of the most dramatic examples of the role of higher brain areas in personality and emotion is the famous crowbar accident of 1848. A 25-year-old railroad foreman, Phineas P. Gage, was tamping gunpowder into a hole in a rock when it exploded. The rod—three feet, seven inches long and one and one-fourth inches thick—was driven through his left eye, passed through his brain, and emerged from the back of his skull.

After a few minutes of convulsions, Gage got up, rode a horse three-quarters of a mile into town, and walked up a long flight of stairs to see a doctor. He recovered well, with no noticeable sensory or motor deficits. His associates, however, noted striking personality changes. Before the accident, Gage was a responsible, capable, financially prudent man. Afterward, he was much less inhibited socially,

limbic: L. *limbus*, edge or border

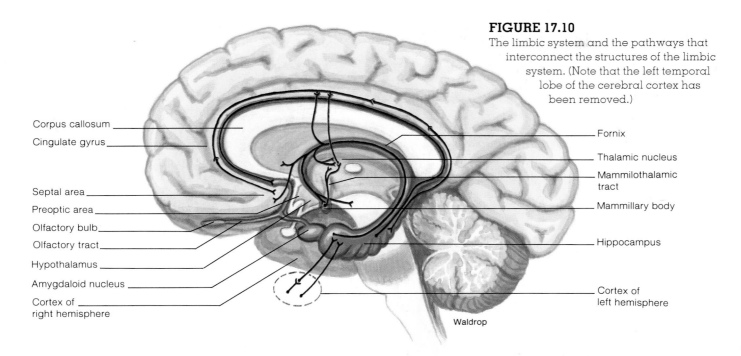

Corpus callosum

Cingulate gyrus

Septal area

Preoptic area

Olfactory bulb

Olfactory tract

Hypothalamus

Amygdaloid nucleus

Cortex of
right hemisphere

Fornix

Thalamic nucleus

Mammilothalamic tract

Mammillary body

Hippocampus

Cortex of
left hemisphere

Waldrop

engaging, for example, in gross profanity (which he had never done previously). He also seemed to be tossed about by chance whims. Eventually Gage was fired from his job, and his old friends remarked that he was "no longer Gage."

# Clinical Considerations

## Autonomic Dysreflexia

Autonomic dysreflexia, a serious condition producing rapid elevations in blood pressure that can lead to stroke (cerebrovascular accident), occurs in 85% of people with quadriplegia and others with spinal cord lesions above the sixth thoracic level. Lesions to the spinal cord first produce the symptoms of spinal shock, characterized by the loss of both skeletal muscle and autonomic reflexes. After a period of time, both types of reflexes return in an exaggerated state. The skeletal muscles may become spastic due to absence of higher inhibitory influences, and the visceral organs experience denervation hypersensitivity. Patients in this state have difficulty emptying their urinary bladders and often must be catheterized.

Noxious stimuli, such as overdistension of the urinary bladder, can result in reflex activation of the sympathetic nerves below the spinal cord lesion. This produces goose bumps, cold skin, and vasoconstriction in the regions served by the spinal cord below the level of the lesion. The rise in blood pressure resulting from this vasoconstriction activates pressure receptors that transmit impulses along sensory nerve

fibers to the medulla oblongata. In response to this sensory input, the medulla oblongata directs a reflex slowing of the heart and vasodilation. Since descending impulses are blocked by the spinal lesion, however, the skin is warm and moist (due to vasodilation and sweat gland secretion) above the lesion but cold below the level of spinal cord damage.

## Pharmacology of the Autonomic Nervous System

Epinephrine, norepinephrine, acetylcholine, and some chemicals that are not normally found in the body can either enhance or block the physiological effects of the autonomic nervous system. Those drugs that promote sympathetic nerve effects are called **sympathomimetic** (sim´´pă-tho-mĭ-met´ik) **drugs,** and those that block sympathetic effects are **sympatholytic** (sim´´pă-tho-lit´ik) **drugs.** In general, drugs that enhance a particular effect are known as agonists; those that block an effect are antagonists.

**Adrenergic Drugs**    Earlier in this chapter, we discussed the use of β-blocking drugs, such as propranolol, to slow the heart and the use of $\alpha_1$ agonist drugs to cause vasoconstriction in mucous membranes. Use of these drugs is easily understood, since stimulation of beta-adrenergic receptors in the heart increases its rate of beat and stimulation of $\alpha_1$ receptors in blood vessels causes vasoconstriction.

Although it may at first seem paradoxical, some people with hypertension are treated with an $\alpha_2$ agonist drug—clonidine. It is currently believed that presynaptic axon terminals in the brain contain $\alpha_2$ receptors, and that stimulation of these receptors inhibits the release of neurotransmitter from the axon terminals. This may represent a negative feedback control of the amount of neurotransmitter released. It is known that clonidine, by stimulating $\alpha_2$ receptors in the CNS, inhibits the central activation of the sympathetic division of the autonomic nervous system, thus serving to decrease the heart rate and lower the blood pressure of hypertensive patients.

**Cholinergic Drugs**   Acetylcholine is used as a neurotransmitter by somatic motor neurons, preganglionic autonomic fibers, all postganglionic parasympathetic nerve fibers, and some sympathetic nerve fibers. Drugs that are similar in structure and action to acetylcholine (methacholine and bethane-

choline) can promote the effects of these neurons. Drugs that inhibit the action of acetylcholinesterase (an enzyme that degrades ACh); for example, physostigmine and neostigmine, enhance the action of ACh. Because of the effects of these drugs on the neuromuscular junction, they are used in the treatment of myasthenia gravis and other muscular disorders.

Some of these drugs are also used as **parasympathomimetics**—drugs that duplicate the action of parasympathetic neurons—in the treatment of glaucoma. Atropine blocks the action of ACh released by postganglionic parasympathetic neurons; it is a **parasympatholytic drug.** Atropine is used clinically to block the actions of the parasympathetic division, as previously described.

## Chapter Summary

### Neural Control of Involuntary Effectors (pp. 446–448)

1. There are two basic categories of neurons in the motor autonomic pathway: preganglionic and postganglionic.
   a. Preganglionic autonomic neurons originate from the brain or spinal cord.
   b. Postganglionic neurons originate from ganglia located outside the CNS.
2. Smooth muscle, cardiac muscle, and glands receive autonomic innervation.

### Divisions of the Autonomic Nervous System (pp. 448–452)

1. Preganglionic neurons of the sympathetic division originate in the spinal cord, between the thoracic and lumbar levels.
   a. Many of these fibers synapse with postganglionic neurons, whose cell bodies are located in a double chain of sympathetic (paravertebral) ganglia outside the spinal cord.
   b. Some preganglionic fibers synapse in collateral or prevertebral, ganglia; these are the celiac, superior mesenteric, and inferior mesenteric ganglia.
   c. Some preganglionic fibers innervate the adrenal medulla, which secretes epinephrine (and some norepinephrine) into the blood in response to stimulation.

2. Preganglionic parasympathetic fibers originate in the brain and in the spinal cord at the sacral levels.
   a. Preganglionic parasympathetic fibers contribute to oculomotor, facial, glossopharyngeal, and vagus cranial nerves.
   b. Preganglionic fibers of the vagus nerve are long and synapse in terminal ganglia located next to or within the innervated organ; short postganglionic fibers then innervate the effector organs.

### Functions of the Autonomic Nervous System (pp. 453–461)

1. The sympathetic division of the autonomic nervous system activates the body to "fight or flight" through adrenergic effects; the parasympathetic division often exerts antagonistic actions through cholinergic effects.
   a. All preganglionic autonomic nerve fibers are cholinergic (use ACh as a neurotransmitter).
   b. All postganglionic parasympathetic fibers are cholinergic.
   c. Most postganglionic sympathetic fibers are adrenergic (use norepinephrine as a neurotransmitter).
2. Adrenergic effects include stimulation of the heart, vasoconstriction in the viscera

and skin, bronchodilation, and glycogenolysis in the liver.
   a. Alpha and beta are the two main groups of adrenergic receptor proteins.
   b. There are two subtypes of alpha receptors ($\alpha_1$ and $\alpha_2$) and two subtypes of beta receptors ($\beta_1$ and $\beta_2$); these subtypes can be selectively stimulated or blocked by therapeutic drugs.
3. In organs with dual innervation, the actions of the sympathetic and parasympathetic divisions can be antagonistic, complementary, or cooperative.
4. The medulla oblongata of the brain stem is the structure that most directly controls the activity of the autonomic nervous system.
   a. The medulla oblongata is in turn influenced by sensory input and by input from the hypothalamus.
   b. The hypothalamus is influenced by input from the limbic system, cerebellum, and cerebrum; these interconnections provide an autonomic component to some of the visceral responses that accompany emotions.

# Review Activities

## Objective Questions

1. Which of the following statements about the superior mesenteric ganglion is *true?*
   a. It is a parasympathetic ganglion.
   b. It is a paravertebral sympathetic ganglion.
   c. It is located in the head.
   d. It contains postganglionic sympathetic neurons.

2. The pterygopalatine, ciliary, submandibular, and otic ganglia are
   a. collateral sympathetic ganglia.
   b. cervical sympathetic ganglia.
   c. parasympathetic ganglia that receive fibers from the vagus nerves.
   d. parasympathetic ganglia that receive fibers from the occulomotor, facial, and glossopharyngeal cranial nerves.

3. When a visceral organ is denervated,
   a. it ceases to function.
   b. it becomes less sensitive to subsequent stimulation by neurotransmitters.
   c. it becomes hypersensitive to subsequent stimulation.

4. Parasympathetic ganglia are located
   a. in a chain parallel to the spinal cord.
   b. in the posterior roots of spinal nerves.
   c. next to or within the organs innervated.
   d. in the brain.

5. The neurotransmitter of preganglionic sympathetic fibers is
   a. norepinephrine.
   b. epinephrine.
   c. acetylcholine.
   d. dopamine.

6. Which of the following results from stimulation of alpha-adrenergic receptors?
   a. constriction of blood vessels
   b. dilation of bronchioles
   c. increased heart rate
   d. sweat gland secretion

7. Which of the following fibers release norepinephrine?
   a. preganglionic parasympathetic fibers
   b. postganglionic parasympathetic fibers
   c. postganglionic sympathetic fibers in the heart
   d. postganglionic sympathetic fibers in sweat glands
   e. all of the above

8. The actions of sympathetic and parasympathetic fibers are cooperative in
   a. the heart.
   b. the reproductive system.
   c. the digestive system.
   d. the eyes.

9. Propranolol is a "beta-blocker." It would therefore be used to
   a. expand the blood vessels.
   b. slow the heart rate.
   c. increase blood pressure.
   d. increase salivary gland secretion.

10. Atropine blocks parasympathetic nerve effects. It would therefore be used to
    a. dilate the pupils.
    b. decrease mucus secretion.
    c. decrease movements of the gastrointestinal tract.
    d. increase the heart rate.
    e. all of the above apply.

11. The area of the brain that is most directly involved in the reflex control of the autonomic nervous system is
    a. the hypothalamus.
    b. the cerebral cortex.
    c. the medulla oblongata.
    d. the cerebellum.

12. The two subtypes of cholinergic receptors are
    a. adrenergic and nicotinic.
    b. dopaminergic and muscarinic.
    c. nicotinic and muscarinic.
    d. nicotinic and dopaminergic.

## Essay Questions

1. Compare the sympathetic and parasympathetic divisions in terms of ganglia location and nerve distribution.

2. Describe the anatomical and physiological relationship between the sympathetic division and the adrenal glands.

3. Compare the effects of adrenergic and cholinergic stimulation on the cardiovascular and digestive systems.

4. Explain how effectors that receive only sympathetic innervation are regulated by the autonomic nervous system.

5. Describe the various types of adrenergic receptors and explain how their differences are clinically exploited.

## Gundy/Weber Software

The tutorial software accompanying Chapter 17 is Volume 5—Nervous System.

# sensory organs

## objectives

- Discuss the different categories of sensory receptors and the selectivity of receptors for specific stimuli.
- Explain the law of specific nerve energies and distinguish between tonic and phasic receptors.
- Describe the structure, function, and location of the tactile receptors and the neural pathway for somatic sensation.
- Describe the receptors and neural pathways for pain and explain what is meant by referred pain and phantom pain.
- Discuss the receptors and neural pathways that mediate proprioception.
- Describe the location and structure of taste buds and the distribution of the different kinds of taste buds.
- Describe the olfactory receptors and the neural pathways for olfaction.
- Distinguish between the membranous and bony labyrinth and describe the structure of the vestibular apparatus.

- Explain how mechanical movements are transduced into nerve impulses in the semicircular canals and in the otolith organs.
- Describe the neural pathways for the sense of equilibrium and explain how the vestibular apparatus can influence eye movements.
- Describe the structure of the outer and middle ear, and explain how they function in hearing.
- Describe the structure of the cochlea and explain how it functions.
- Explain how different pitches of sounds affect the cochlea and how different pitches are coded in the neural pathways of hearing.
- Describe the structures of the eyeball, trace the path of light through the eye, and explain how the focus of the eye is adjusted for viewing at different distances.
- Describe the structure of the retina and compare the structure and function of rods and cones.
- Describe the neural pathways for visual perception.

# Characteristics of Sensory Receptors

*Each type of sensory receptor is most sensitive to a particular modality of environmental stimulus to which it responds by causing the production of action potentials in a sensory neuron. These impulses are conducted to parts of the brain that provide the proper interpretation of sensory perception when that particular neural pathway is activated.*

The sense organs are actually extensions of the nervous system that allow us to perceive our internal and external environments. These sense organs have been described as "windows for the brain" because it is through them that we achieve awareness of the environment. A stimulus must first be received before the sensation can be interpreted and the necessary body adjustments dictated by the central nervous system can be made. Not only do we depend on our sense organs to experience pleasure, they also ensure our very survival. For example, they enable us to hear warning sounds, see dangers, avoid ingesting toxic substances through taste, and perceive sensations of pain, hunger, and thirst.

A *sensation* is the conduction of sensory impulses to the brain. The interpretation of a sensation is referred to as *perception*. In other words, we feel, see, hear, taste, and smell with our brain. In order to perceive a sensation, four conditions are necessary.

**1** A *stimulus* sufficient to initiate a response in the nervous system must be present.

**2** A *receptor* must convert the stimulus to a nerve impulse. A receptor is a specialized, peripheral dendritic ending of a sensory neuron or the specialized receptor cell associated with it.

**3** The *conduction of the nerve impulse* (sensation) must occur from the receptor to the brain along a nervous pathway.

**4** The *interpretation of the sensation* in the form of a perception must occur within a specific portion of the brain.

Only impulses reaching the cerebral cortex of the brain are consciously interpreted. If impulses terminate in the spinal cord or brain stem, they may initiate a reflexive motor activity response but not a conscious awareness. Impulses reaching the higher brain centers travel through nerve fibers composing *sensory,* or *ascending, tracts.* Clusters of neuron cell bodies, called **nuclei,** are synaptic sites along sensory tracts within the CNS. The nuclei that sensory impulses pass through before reaching the cerebral cortex are located in the spinal cord, medulla oblongata, pons, and thalamus.

Through the use of scientific instruments, we have learned that the senses act as energy filters that allow perception of only a narrow range of energy. Vision, for example, is limited to light waves in the visible spectrum. Other types of waves of the same type of energy as visible light, such as X rays, radio waves, and ultraviolet and infrared light, normally cannot excite the sensory receptors in the eyes. Although filtered and distorted by the limitations of sensory functions, perceptions allow us to interact effectively with our environment and are of obvious survival value.

## Categories of Sensory Receptors

Sensory receptors can be categorized on the basis of structure or function. Structurally, sensory receptors may be the dendritic endings of sensory neurons, which are either free (such as those in the skin, which mediate pain and temperature) or are encapsulated within nonneural structures. The photoreceptors in the retina of the eyes (rods and cones) are highly specialized neurons that synapse with other neurons in the retina. In the case of taste buds on the tongue and of hair cells in the inner ears, modified epithelial cells respond to environmental stimuli and activate sensory neurons.

**Functional Categories** Sensory receptors can be grouped according to the type of stimulus energy they transduce. These categories include (1) **chemoreceptors,** such as the taste buds, olfactory epithelium, and the aortic and carotid bodies, which sense chemical stimuli in the environment or the blood; (2) **photoreceptors** (*fo˝to-re-sep´torz*)—the rods and cones in the retina of the eye, which respond to light; (3) **thermoreceptors,** which respond to changes in temperature; and (4) **mechanoreceptors** (*mek˝ă-no-re-sep´torz*), such as the touch and pressure receptors in the skin and the hair cells within the inner ear, which respond to mechanical deformation of the receptor cell membrane. **Nocireceptors,** (*no˝sĭ-re-sep´torz*), or pain receptors, are stimulated by chemicals released from damaged tissue cells and thus are a type of chemoreceptor.

Receptors can also be grouped according to the type of sensory information they deliver to the brain. **Proprioceptors** (*pro˝ pre-o-sep´torz*) include the muscle spindles, neurotendinous receptors (Golgi tendon organs), and joint receptors. These relay information about body position and permit fine control of skeletal movements (as discussed in chapter 12). **Cutaneous receptors** include (1) touch and pressure receptors; (2) hot and cold receptors; and (3) pain receptors. The receptors that mediate sight, hearing, and equilibrium are grouped together as the *special senses.*

Receptors may be classified as *exteroceptors* or *enteroceptors* (*visceroceptors*). **Exteroceptors** (*ek˝stĕ-ro-sep´torz*) are located near the surface of the body and respond to stimuli from the external environment. **Visceroceptors** (*vis˝er-o-sep´torz*)

---

nocireceptor: L. *nocco*, to injure; *ceptus*, taken
proprioceptor: L. *proprius*, one's own; *ceptus*, taken
visceral: L. *viscera*, body organs

## Table 18.1 Classification of receptors based on their normal (or "adequate") stimulus

| Receptor | Normal stimulus | Mechanisms | Examples |
|---|---|---|---|
| Mechanoreceptors | Mechanical force | Deforms cell membrane of sensory dendrites or deforms hair cells that activate sensory nerve endings | Cutaneous touch and pressure receptors; vestibular apparatus and spiral organ in cochlea of ear |
| Pain receptors | Tissue damage | Damaged tissues release chemicals that excite sensory endings | Cutaneous pain receptors |
| Chemoreceptors | Dissolved chemicals | Chemical interaction affects ionic permeability of sensory cells | Smell and taste (exteroreceptors); osmoreceptors and carotid body chemoreceptors (interoreceptors) |
| Photoreceptors | Light | Photochemical reaction affects ionic permeability of receptor cell | Rods and cones in retina of eye |

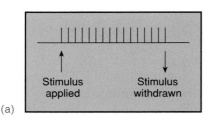

(a)

**Tonic receptor —** slow-adapting

Stimulus applied · Stimulus withdrawn

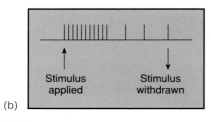

(b)

**Phasic receptor —** fast-adapting

Stimulus applied · Stimulus withdrawn

### FIGURE 18.1

Tonic receptors (a) continue to fire at a relatively constant rate as long as the stimulus is maintained. These produce slow-adapting sensations. Phasic receptors (b) respond with a burst of action potentials when the stimulus is first applied but quickly reduce their rate of firing when the stimulus is maintained. This produces fast-adapting sensations.

produce sensations arising from the viscera, such as internal pain, thirst, hunger, and so on. Specialized visceroceptors within the circulatory system known as *baroreceptors* are sensitive to changes in blood pressure.

**Tonic and Phasic Receptors: Sensory Adaptation** Some receptors respond with a burst of activity when a stimulus is first applied, but then quickly decrease their firing rate—adapt to the stimulus—when the stimulus is maintained. Receptors with this response pattern are called **phasic receptors.** Receptors that produce a relatively constant rate of firing as long as the stimulus is maintained are known as **tonic receptors** (fig. 18.1).

Phasic receptors alert us to changes in sensory stimuli and are partially responsible for the fact that we can cease paying attention to constant stimuli. This ability is called **sensory adaptation.** Odor, touch, and temperature, for example, adapt rapidly; bathwater feels hotter when we first enter it. Sensations of pain, by contrast, adapt little if at all.

## Law of Specific Nerve Energies

Stimulation of a sensory nerve fiber produces only one sensation—touch, cold, pain, and so on. According to the **law of specific nerve energies,** the sensation characteristic of each sensory neuron is that produced by its *normal,* or *adequate, stimulus* (table 18.1). The adequate stimulus for the photoreceptors of the eye, for example, is light. If these receptors are stimulated by some other means—such as by pressure produced by a punch to the eye—a flash of light (the adequate stimulus) may be perceived.

An effect known as *paradoxical cold* provides another example of the law of specific nerve energies. When the tip of a cold metal rod is touched to the skin, the perception of cold gradually disappears as the rod warms to body temperature. Then, when the tip of a rod heated to 45° C is applied to the same spot, the sensation of cold is perceived once again. This paradoxical cold is produced because the heat slightly damages receptor endings, and by this means produces an "injury current" that stimulates the receptor.

Regardless of how a sensory neuron is stimulated, then, only one sensory modality will be perceived. This specificity is due to the synaptic pathways within the brain that are activated by the sensory neuron. The ability of receptors to function as sensory filters and be stimulated by only one type of stimulus (the adequate stimulus) allows the brain to perceive the stimulus accurately under normal conditions.

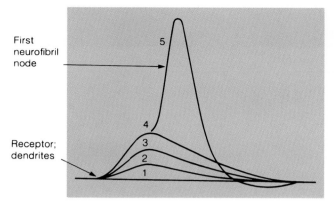

**FIGURE 18.2**
Sensory stimuli result in the production of local, graded potential changes known as receptor, or generator potentials (numbers 1–4). If the receptor potential reaches a threshold value of depolarization, it generates action potentials (number 5) in the sensory neuron.

## Generator (Receptor) Potential

The electrical behavior of sensory nerve endings is similar to that of the dendrites of other neurons. In response to an environmental stimulus, the sensory endings produce local, graded changes in the membrane potential. In most cases these potential changes are depolarizations, analogous to excitatory postsynaptic potentials (EPSPs, as described in chapter 14). In the sensory endings, however, these potential changes in response to environmental stimulation are called **receptor,** or **generator, potentials** because they serve to generate action potentials in response to the sensory stimulation. Since sensory neurons are pseudounipolar (chapter 14), the action potentials produced in response to the generator potential are conducted continuously from the periphery into the CNS.

The *lamellated,* or *pacinian, corpuscle,* a cutaneous receptor for pressure (see table 18.2) can serve as an example of sensory transduction. When a light touch is applied to the receptor, a small depolarization (the generator potential) is produced. Increasing the pressure on the lamellated corpuscle increases the magnitude of the generator potential until it reaches the threshold required to produce an action potential (fig. 18.2). The lamellated corpuscle, however, is a phasic receptor; if the pressure is maintained, the size of the generator potential produced quickly diminishes. It is interesting to note that this phasic response is a result of the onionlike covering around the dendritic nerve ending; if the layers are peeled off and the nerve ending is stimulated directly, it will respond in a tonic fashion.

When a tonic receptor is stimulated, the generator potential it produces is proportional to the intensity of the

stimulus. After a threshold depolarization is produced, increases in the amplitude of the generator potential result in increases in the *frequency* with which action potentials are produced. In this way, the frequency of action potentials that are conducted into the central nervous system serves to code for the strength of the stimulus. As described in chapter 14, this frequency code is needed because the amplitude of action potentials is constant (they are all-or-none events). Acting through changes in action potential frequency, tonic receptors thus provide information about the relative intensity of a stimulus.

# Somatic Senses

*Somatic senses include cutaneous receptors and proprioceptors. There are several types of sensory receptors in the skin, each of which is specialized to be maximally sensitive to one modality of sensation. Proprioceptors provide sensory information about muscles, joints, and tendons.*

## Cutaneous Receptors

**Corpuscles of Touch**    A corpuscle of touch (*Meissner's corpuscle*) is an oval receptor composed of a mass of dendritic endings from two or three nerve fibers enclosed by connective tissue sheaths. These receptors are numerous in the hairless portions of the body, such as the eyelids, lips, tip of the tongue, fingertips, palms of the hands, soles of the feet, nipples, and external genitalia.

Corpuscles of touch lie within the papillary layer of the dermis, where they are especially sensitive to the motion of objects that barely contact the skin (chapter 7). Sensations of fine or light touch are perceived as these receptors are stimulated. They also function when a person touches an object to determine its texture.

**Free Nerve Endings**    Free nerve endings are the least modified and the most superficial of the tactile receptors. These receptors extend into the lower layers of the epidermis, where they end as knobs between the epithelial cells. Free nerve endings are most important as pain receptors, although they also respond to objects that are in continuous contact with the skin, such as clothing.

**Root Hair Plexuses**    Root hair plexuses are a specialized type of free nerve ending. They are coiled around hair follicles, where they respond to movement of the hair.

corpuscle: L. *corpusculum,* diminutive of corpus, body

somatic: Gk. *somatikos,* body
Meissner's corpuscle: from George Meissner, German histologist, 1829–1905

# Table 18.2   Cutaneous receptors

| Type | Location | Function | Sensation |
|------|----------|----------|-----------|
| Corpuscles of touch (Meissner's corpuscles) (mechanoreceptors) | Papillae of dermis; numerous in hairless portions of body (eyelids, fingertips, lips, nipples, external genitalia) | Detect light motion against surface of skin | Fine touch; texture |
| Free nerve endings (pressure receptors; pain receptors) | Lower layers of epidermis | Detect changes in pressure; detect tissue damage | Touch, pressure; pain |
| Root hair plexuses (tactile receptors) | Around hair follicles | Detect movement of hair | Touch |
| Lamellated (pacinian) corpuscles (mechanoreceptors) | Hypodermis; synovial membranes; perimysium, certain visceral organs | Detect changes in pressure | Deep pressure; vibrations |
| Organs of Ruffini (thermoreceptors) | Lower layers of dermis | Detect changes in temperature | Heat |
| Bulbs of Krause (thermoreceptors) | Dermis | Detect changes in temperature | Cold |

Bulb of Krause

Root hair plexus

Lamellated (pacinian) corpuscle

Corpuscle of touch (Meissner's corpuscle)

Free nerve ending

Organ of Ruffini

Creek

## Lamellated Corpuscles

Lamellated (pacinian) corpuscles are large, onion-shaped receptors composed of the dendritic endings of several sensory nerve fibers enclosed by connective tissue layers. They are commonly found within the synovial membranes and connective tissue surrounding joints, in the perimysium of skeletal muscle, and in certain visceral organs. Lamellated corpuscles are also abundant in the skin of the palms and fingers of the hand, soles of the feet, external genitalia, and breasts.

Lamellated corpuscles respond to heavy pressures, generally those that are constantly applied. They can also detect vibrations in tissues and organs.

## Thermoreceptors

**Thermoreceptors** are widely distributed throughout the dermis of the skin but are especially abundant in the lips and the mucous membranes of the mouth and anal regions. There

..........................................
pacinian corpuscle: from Filippo Pacini, Italian anatomist, 1812–1983

are two kinds of thermoreceptors—one that responds to heat and the other to cold.

**Organs of Ruffini**   The organs of Ruffini are heat receptors located deep within the dermis. Heat receptors are elongated, oval structures that are most sensitive to temperatures above 25° C (77° F). Temperatures above 45° C (113° F) elicit impulses through the organs of Ruffini that are perceived as painful, burning sensations.

**Bulbs of Krause**   The bulbs of Krause are receptors for the sensation of cold. They are more abundant than heat receptors and are closer to the surface of the skin. The bulbs of Krause are most sensitive to temperatures below 20° C (68° F). Temperatures below 10° C (50° F) elicit responses through the bulbs of Krause that are perceived as painful, freezing sensations.

The cutaneous receptors are summarized in table 18.2.

## Pain Receptors

The principal receptors for pain are the **free nerve endings.** Several million free nerve endings are distributed throughout the skin and internal tissues. Pain receptors are sparse in certain visceral organs and are absent within the nervous tissue of the brain. Although the free nerve endings are specialized to respond to tissue damage, all of the cutaneous receptors will relay impulses that are interpreted as pain if stimulated excessively.

The protective value of pain receptors is obvious. Unlike other cutaneous receptors, free nerve endings exhibit little accommodation, so impulses are relayed continuously to the CNS as long as the irritating stimulus is present. Although pain receptors can be activated by all types of stimuli, they are particularly sensitive to chemical stimulation. Muscle spasms, muscle fatigue, or an inadequate supply of blood to an organ may also cause pain.

Impulses for pain are conducted to the spinal cord through sensory nerve fibers. The impulses are then conducted to the thalamus along the *lateral spinothalamic tract* of the spinal cord and from there to the cerebral cortex. Although an awareness of pain occurs in the thalamus, the type and intensity of pain is interpreted in the specialized areas of the cerebral cortex.

The sensation of pain can be clinically classified as **somatic pain** or **visceral pain.** Stimulation of the cutaneous pain receptors results in the perception of superficial somatic pain. Deep somatic pain comes from stimulation of receptors in skeletal muscles, joints, and tendons.

Stimulation of the receptors within the viscera causes the perception of visceral pain. Through precise neural pathways, the brain is able to perceive the area of stimulation

organs of Ruffini: from Angelo Ruffini, Italian anatomist, 1864–1929
bulbs of Krause: from Wilhelm J. F. Krause, German anatomist, 1833–1910

and project the pain sensation back to that area. The sensation of pain from certain visceral organs, however, may not be perceived as arising from those organs but from other somatic locations. This phenomenon is known as **referred pain** (fig. 18.3). The sensation of referred pain is consistent from one person to another and is clinically important in diagnosing organ dysfunctions. The pain of a heart attack, for example, may be perceived subcutaneously over the heart and down the medial side of the left arm. Ulcers of the stomach may cause pain that is perceived as coming from the upper central (epigastric) region of the trunk. Pain from problems of the liver or gallbladder may be perceived as localized visceral pain or as referred pain arising from the right neck and shoulder regions.

 The perception of pain has survival value because it alerts the body to an injury, disease, or organ dysfunction. *Acute pain* is sudden, usually short-term, and can generally be endured and attributed to a known cause. *Chronic pain*, however, is long-term and tends to weaken a person as it interferes with the ability to function effectively. Certain diseases, such as arthritis, are characterized by chronic pain. In these patients, relief of pain is of paramount concern. Treatment of chronic pain often requires the use of moderate pain-reducing drugs (analgesics) or intense narcotic drugs. Treatment in severely tormented chronic pain patients may include severing sensory nerves or implanting stimulating electrodes in appropriate nerve tracts.

The phenomenon of the **phantom limb** was first reported by a neurologist during the Civil War, who described a case where a veteran with amputated legs asked for someone to massage his cramped leg muscle. It is now known that such phantom limbs are common in amputees, who may experience complete sensations from the missing limbs. The phantom appears very real to the amputees, especially when their eyes are closed, and seems to move in accordance with the way the limb would naturally move if it were real. This could be useful, as when an amputee is fitted with a prosthetic that the phantom may be perceived to enter. Pain in the phantom is experienced by seventy percent of amputees, however, and the pain can be severe and persistent.

One explanation for phantom limbs is that the nerves remaining in the stump can grow into nodules called *neuromas*, which generate nerve impulses that are transmitted to the brain and interpreted as arising from the missing limb. However, phantom limbs may occur in cases where the limb is not amputated, but the nerves that normally enter from the limb are severed in an accident. Or it may occur in someone with a spinal cord injury above the level of the limb, so that the sensations from the limb do not enter the brain. In these cases, the phantom limb phenomenon requires a different explantion. Current theories propose that the source of the phantom may arise in several brain regions whose activity is somehow changed by the absence of the sensations that would normally arise from the missing limb.

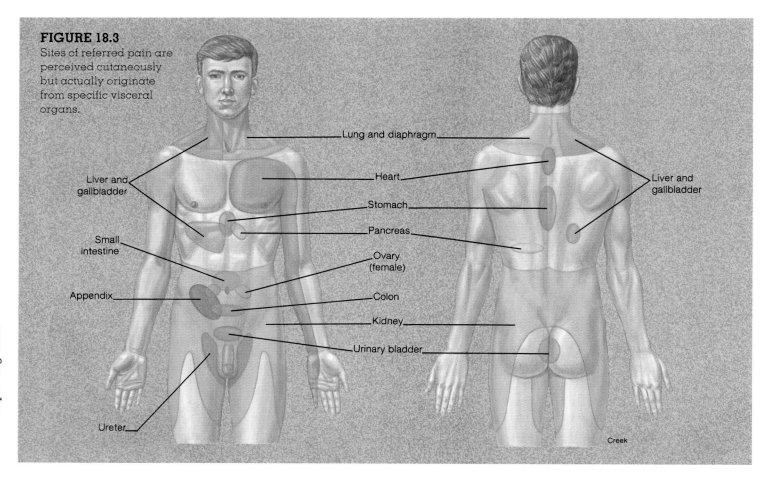

**FIGURE 18.3**
Sites of referred pain are perceived cutaneously but actually originate from specific visceral organs.

Lung and diaphragm

Heart

Liver and gallbladder

Stomach

Pancreas

Small intestine

Ovary (female)

Appendix

Colon

Kidney

Urinary bladder

Ureter

Liver and gallbladder

Creek

## *Receptive Fields and Sensory Acuity*

The **receptive field** of a neuron serving cutaneous sensation is the area of skin whose stimulation results in changes in the firing rate of the neuron. Changes in the firing rate of primary sensory neurons affect the firing of second- and third-order neurons, which in turn affects the firing of those neurons in the postcentral gyrus that receive input from the third-order neurons. Indirectly, therefore, neurons in the postcentral gyrus can be said to have receptive fields in the skin.

The area of each receptive field in the skin varies inversely with the density of receptors in the region. In the back and legs, where a large area of skin is served by relatively few sensory endings, the receptive field of each neuron is correspondingly large. In the fingertips—where a large number of cutaneous receptors serve a small area of skin—the receptive field of each sensory neuron is correspondingly small.

**Two-Point Touch Threshold** The approximate size of the receptive fields serving light touch can be measured by the *two-point touch threshold test*. In this procedure, two points of a pair of calipers are lightly touched to the skin at the same time. If the distance between the points is sufficiently great, each point will stimulate a different receptive field

and a different sensory neuron—two separate points of touch will thus be felt. If the distance is sufficiently small, both points will touch the receptive field of only one sensory neuron, and only one point of touch will be felt (fig. 18.4).

The two-point touch threshold, which is the minimum distance at which two points of touch can be felt, is a measure of the distance between receptive fields. If the distance between the two points of the calipers is less than this minimum distance, only one "blurred" point of touch can be felt. The two-point touch threshold, which varies for different regions of the body, is thus an indication of tactile *acuity*, or the sharpness of touch perception. This threshold ranges from about 2 mm on the highly sensitive tip of the tongue to more than 50 mm on some areas of the back.

The high tactile acuity of the fingertips is exploited in the reading of *Braille*. Braille symbols are formed by raised dots on the page that are separated from each other by 2.5 mm, which is slightly greater than the two-point touch threshold in the fingertips. Experienced Braille readers can scan words at about the same speed that a sighted person can read aloud—a rate of about 100 words per minute.

••••••••••••••••••••••••••••••••••••••••••

acuity: L. *acuo,* sharpen
Braille: from Louis Braille, French teacher of the blind, 1809–52

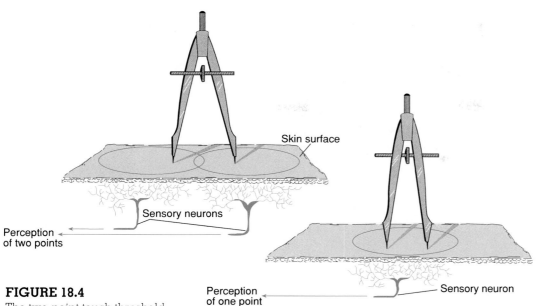

Skin surface

Sensory neurons

Perception of two points

Perception of one point

Sensory neuron

**FIGURE 18.4**
The two-point touch threshold
test. If each point touches the
receptive fields of different sensory neurons, two separate points
of touch will be felt. If both caliper points touch the receptive field
of one sensory neuron, only one point of touch will be felt.

**Lateral Inhibition**   When a blunt object touches the skin,
a number of receptive fields may be stimulated. Those receptive fields in the center areas where the touch is
strongest will be stimulated more than those in neighboring
fields where the touch is lighter. We do not usually feel a
"halo" of light touch surrounding a center of stronger touch,
however. Instead, only a single touch is felt, which is somewhat more defined than the actual shape of the blunt object.
This sharpening of sensation is due to a process called **lateral inhibition.**

Lateral inhibition and the resultant sharpening of sensation occur within the central nervous system. Those sensory
neurons whose receptive fields are stimulated most strongly
inhibit—via association neurons that pass "laterally" within
the CNS—sensory neurons that serve neighboring receptive
fields. Lateral inhibition similarly plays a prominent role in
the ability of the ears and brain to discriminate sounds of different pitch, to be described in a later section.

## Proprioceptors

Proprioceptors are located within skeletal muscle tissue, tendons, and the synovial membranes and connective tissue surrounding joints. Some of the sensory impulses from
proprioceptors reach the level of consciousness as the **kinesthetic sense,** by which the position of the body parts is perceived. Other proprioceptor information is not consciously
interpreted and is used to adjust the intensity and timing of
muscle contractions to provide coordinated movements.
With the kinesthetic sense, the position and movement of

the limbs can be determined
without visual sensations, such
as when dressing or walking in
the dark.

High-speed transmission is
a vital characteristic of the
kinesthetic sense, since rapid
feedback to various body parts
is essential for quick, smooth,
coordinated body movements.
There are three types of proprioceptors: joint kinesthetic receptors, neuromuscular spindles,
and neurotendinous receptors.

### Joint Kinesthetic Receptors
Joint kinesthetic receptors are
located in the connective tissue capsule in synovial joints,
where they are stimulated by changes in position caused by
movement at the joints.

**Neuromuscular Spindles**   The neuromuscular spindles are
located in skeletal muscle, particularly in the muscles of the
extremities. Each neuromuscular spindle (spindle apparatus) contains several thin muscle cells, called **intrafusal
fibers,** packaged within a connective tissue sheath. Like the
stronger and more numerous "ordinary" muscle fibers outside the spindles—the **extrafusal fibers**—the spindles insert into tendons on each end of the muscle. Spindles are
therefore said to be in parallel with the extrafusal fibers.

The extrafusal fibers contain myofibrils along their entire length, but the intrafusal fibers have no contractile apparatus in their central regions. The central, noncontracting
part of an intrafusal fiber contains nuclei. There are two
types of intrafusal fibers. One type, the **nuclear bag fibers,**
have their nuclei arranged in a loose aggregate in the central
regions of the fibers. The other type of intrafusal fibers have
their nuclei arranged in rows and are called **nuclear chain
fibers.** Two types of sensory neurons serve these intrafusal
fibers. **Primary,** or **annulospiral, sensory endings** wrap
around the central regions of the nuclear bag and chain fibers
(fig. 18.5) and **secondary,** or **flower-spray, endings** are located over the contracting poles of the nuclear chain fibers.

Since the spindles are arranged in parallel with the extrafusal muscle fibers, stretching a muscle causes its spindles
to stretch. This stimulates both the primary and secondary
sensory endings. The spindle thus serves as a length detector
because the frequency of impulses produced in the primary
and secondary endings is proportional to the length of the
muscle. The primary endings, however, are most stimulated
at the onset of stretch, whereas the secondary endings respond in a more tonic (sustained) fashion as the stretch is

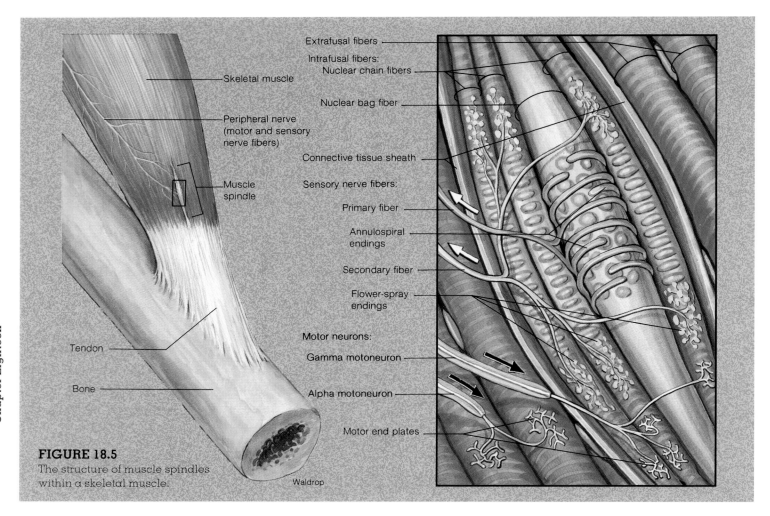

Extrafusal fibers

Intrafusal fibers:
Nuclear chain fibers

Nuclear bag fiber

Skeletal muscle

Peripheral nerve
(motor and sensory
nerve fibers)

Connective tissue sheath

Muscle
spindle

Sensory nerve fibers:

Primary fiber

Annulospiral
endings

Secondary fiber

Flower-spray
endings

Tendon

Motor neurons:

Gamma motoneuron

Bone

Alpha motoneuron

Motor end plates

**FIGURE 18.5**
The structure of muscle spindles
within a skeletal muscle.

Waldrop

maintained. Sudden, rapid stretching of a muscle activates both types of sensory endings, and is thus a more powerful stimulus for the spindles than a slower, more gradual stretching that has less of an effect on the primary sensory endings. Since activation of the sensory endings in neuromuscular spindles produces a reflex contraction, the force of this reflex contraction is greater in response to rapid stretch than to gradual stretch.

In the spinal cord, two types of lower motor neurons innervate skeletal muscles. The motor neurons that innervate the extrafusal muscle fibers are called **alpha motoneurons;** those that innervate the intrafusal fibers are called **gamma motoneurons** (fig. 18.5). The alpha motoneurons are larger and faster conducting (60–90 meters per second) than the thinner, slower conducting (10–40 meters per second) gamma motoneurons. Since only the extrafusal muscle fibers are sufficiently strong and numerous to cause a muscle to shorten, only stimulation by the alpha motoneurons can cause muscle contraction that results in skeletal movements. These are the motor nerve fibers involved in the knee-jerk reflex and other stretch reflexes (fig. 18.6).

The intrafusal fibers of the muscle spindle are stimulated to contract by gamma motoneurons, which represent one-third of all motor fibers in spinal nerves. The intrafusal fibers are too few in number and their contraction is too weak, however, to cause a muscle to shorten. Stimulation by gamma motoneurons thus results in only isometric contraction of the spindles. Since myofibrils are present in the poles but absent in the central regions of intrafusal fibers, the more distensible central region of the intrafusal fiber is pulled toward the ends in response to stimulation by gamma motoneurons. As a result, the spindle is tightened. This effect of gamma motoneurons, which is sometimes termed *active stretch* of the spindles, functions to increase the sensitivity of the spindles when the entire muscle is passively stretched by external forces. The activation of gamma motoneurons thus enhances the stretch reflex and is an important feature in the voluntary control of skeletal movements.

Under normal conditions, the activity of gamma motoneurons is maintained at the level needed to keep the spindles under proper tension while the muscles are relaxed. Undue relaxation of the muscles is prevented by stretch and activation of the spindles, which in turn elicits

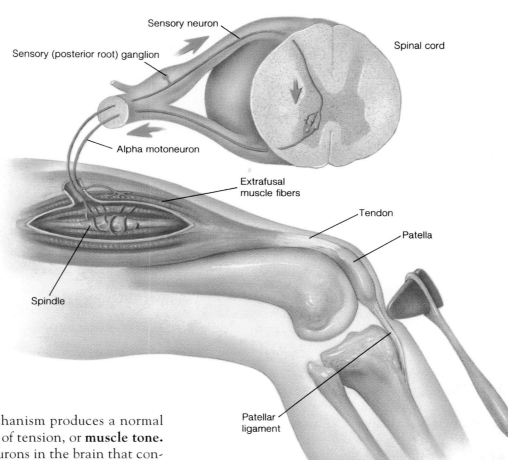

**FIGURE 18.6**
The knee-jerk reflex—an example of a monosynaptic stretch reflex.

a reflex contraction. This mechanism produces a normal resting muscle length and state of tension, or **muscle tone.**

*Higher motor neurons*—neurons in the brain that contribute fibers to descending motor tracts—usually stimulate both alpha and gamma motoneurons simultaneously. Such stimulation is known as **coactivation.** Stimulation of alpha motoneurons results in muscle contraction and shortening; stimulation of gamma motoneurons stimulates contraction of the intrafusal fibers and thus "takes out the slack" that would otherwise be present in the spindles as the muscles shorten. In this way, the spindles remain under tension and provide information about the length of the muscle even while the muscle is shortening.

**Neurotendinous Receptors**   The neurotendinous receptors (Golgi tendon organs) are located where muscles attach to tendons. They continuously monitor tension in the tendons, produced by muscle contraction or passive stretching of a muscle. Sensory neurons from these receptors synapse with association neurons in the spinal cord; these association neurons, in turn, have *inhibitory synapses* (via IPSPs and postsynaptic inhibition—chapter 14) with motor neurons that innervate the muscle (fig. 18.7). This helps to prevent excessive muscle contractions or excessive passive muscle stretching. Indeed, if a muscle is stretched extensively, it will actually relax as a result of the inhibitory effects of the neurotendinous receptors.

 Rapid stretching of skeletal muscles produces very forceful muscle contractions as a result of the activation of primary and secondary endings in the muscle spindles and the monosynaptic stretch reflex. This can result in painful muscle spasms, as may occur, for example, when muscles are forcefully pulled in the process of setting broken bones. Painful muscle spasms may be avoided in physical exercise by stretching slowly and thereby stimulating mainly the secondary endings in the muscle spindles. A slower rate of stretch also allows time for the inhibitory neurotendinous receptor reflex to occur and promote muscle relaxation.

## *Neural Pathways for Somatic Sensations*

The conduction pathways for the somatic senses are shown in figure 18.8. Sensations involving proprioception and pressure are carried by large, myelinated nerve fibers that ascend in the *posterior columns* of the spinal cord on the same (ipsilateral) side. These fibers do not synapse until they reach the medulla oblongata of the brain stem; hence, fibers that carry these sensations from the feet are incredibly long. After synapsing in the medulla oblongata with other, second-order sensory neurons, information in the latter neurons crosses over to

Sensory Organs

473

the contralateral side as it ascends to the thalamus via a fiber tract, called the *medial lemniscus*. Third-order sensory neurons in the thalamus that receive this input in turn project to the postcentral gyrus (the sensory cortex, as described in chapter 15).

Sensations of hot, cold, and pain are carried by thin, unmyelinated sensory neurons into the spinal cord. These synapse with second-order association neurons within the spinal cord, which cross over to the contralateral side and ascend to the brain in the *lateral spinothalamic tract*. Fibers that mediate touch and pressure ascend in the *anterior spinothalamic tract*. Fibers of both spinothalamic tracts synapse with third-order neurons in the thalamus, which in turn project to the postcentral gyrus. Note that, in all cases, somatic information is carried to the postcentral gyrus in third-order neurons. Also, because of crossing over, somatic information from each side of the body is projected to the postcentral gyrus of the contralateral cerebral hemisphere.

All somatic information from the same area of the body projects to the same area of the postcentral gyrus, so that a "map" of the body can be drawn on the postcentral gyrus to represent sensory projection points (see chapter 15, fig. 15.7). This map is very distorted, however, because it shows larger areas of the cerebral cortex devoted to sensation in the face and hands than in other areas of the body. This disproportionately larger area of the cerebral cortex devoted to the face and hands reflects the fact that there is a higher density of sensory receptors in these regions.

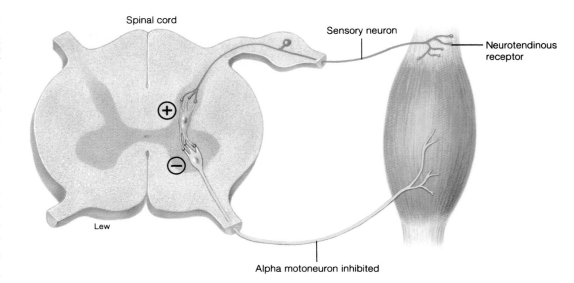

**FIGURE 18.7**

An increase in muscle tension stimulates the activity of sensory nerve endings in the neurotendinous receptor. This sensory input stimulates ⊕ an association neuron, which in turn inhibits ⊖ the activity of a motor neuron innervating that muscle. This is therefore a disynaptic reflex.

chemical changes in the external environment as exteroceptors. Included in the latter category are *taste receptors*, which respond to chemicals dissolved in food or drink, and *olfactory receptors*, which respond to gaseous molecules in the air. This distinction is somewhat arbitrary, however, because odorant molecules in air must first dissolve in fluid within the olfactory mucosa before the sense of smell can be stimulated. Also, the sense of olfaction strongly influences the sense of taste, as can easily be verified by eating an onion with the nostrils pinched together.

## Taste

Taste receptors are specialized epithelial cells that are grouped together into barrel-shaped arrangements called **taste buds.** Taste buds are most numerous on the surface of the tongue but are also present on the soft palate and on the walls of the oropharynx. The cylindrical taste bud is composed of many sensory **gustatory cells** that are encapsulated by supporting cells (fig. 18.9). Projecting from the tip of each gustatory cell is a long microvillus called a **gustatory hair** that passes to the surface through an opening in the taste bud called the **taste pore.** The gustatory hairs compose the sensitive portion of the receptor cells. Saliva provides a moistened environment necessary for a chemical stimulus to activate the gustatory cells.

Taste buds are elevated by surrounding connective tissue and epithelium to form **papillae** (*pă-pil´e*) (fig. 18.9). Three types of papillae can be identified: *vallate, fungiform,* and *filiform* (see chapter 26). Taste buds are found only in the vallate and fungiform papillae. Filiform papillae contain gustatory cells, which are not clustered into taste buds.

## Taste and Olfaction

*The receptors for taste and olfaction respond to molecules that are dissolved in fluid; hence, they are classified as chemoreceptors. Although there are only four basic modalities of taste, they combine in various ways and are influenced by sensations of olfaction, thus allowing for a wide variety of sensory experiences.*

Some chemoreceptors respond to chemical changes in the internal environment as interoceptors; others respond to

····················

gustatory: L. *gustare*, to taste

····················

papilla: L. *papilla*, nipple

## FIGURE 18.8

Pathways that lead from the cutaneous receptors and proprioreceptors into the postcentral gyrus in the cerebral cortex.

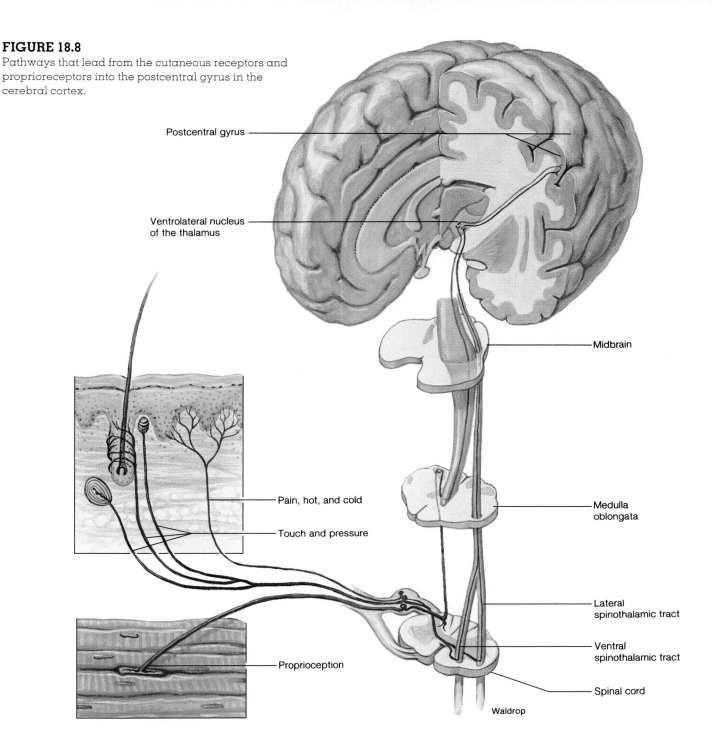

Postcentral gyrus

Ventrolateral nucleus of the thalamus

Pain, hot, and cold

Touch and pressure

Proprioception

Midbrain

Medulla oblongata

Lateral spinothalamic tract

Ventral spinothalamic tract

Spinal cord

Waldrop

Molecules dissolved in saliva at the surface of the tongue interact with receptor molecules in the microvilli of the taste buds. This interaction stimulates the release of a neurotransmitter from the receptor cells, which in turn stimulates sensory nerve endings that innervate the taste buds. Taste buds in the posterior third of the tongue are innervated by the *glossopharyngeal* (*ninth cranial*) *nerve;* those in the anterior two-thirds of the tongue are innervated by the chorda tympani branch of the *facial* (*seventh cranial*) *nerve.*

The four basic modalities of taste are sensed most acutely in particular regions of the tongue. These are *sweet* (tip of the tongue), *sour* (sides of the tongue), *bitter* (back of the tongue), and *salty* (over most of the tongue). This distribution is illustrated in figure 18.10.

The salty taste of food is caused by the presence of $Na^+$ ions. These pass into the sensitive receptor cells through channels in the apical membranes. This depolarizes the cells, causing them to release their transmitter. The anion associated

## FIGURE 18.9

Papillae of the tongue and associated taste buds. (*a*) Numerous taste buds are positioned within each papilla. (*b*) Each gustatory cell and its associated gustatory hair is encapsuled by supporting cells. (*c*) A photomicrograph of some taste buds.

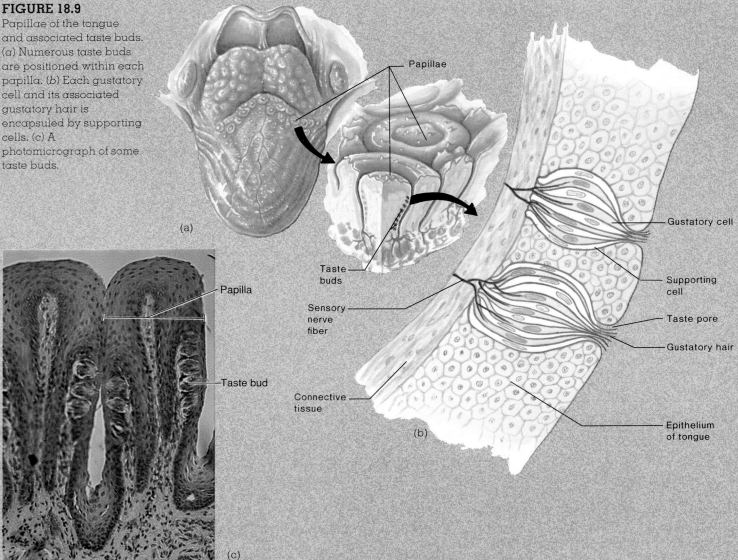

Papillae

Taste buds

Sensory nerve fiber

Connective tissue

(a)

(b)

Gustatory cell

Supporting cell

Taste pore

Gustatory hair

Epithelium of tongue

Papilla

Taste bud

(c)

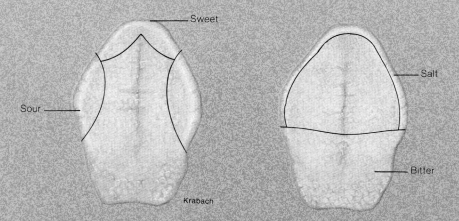

Sweet

Sour

Salt

Bitter

Krabach

## FIGURE 18.10

Patterns of taste receptor distribution on the surface of the tongue.

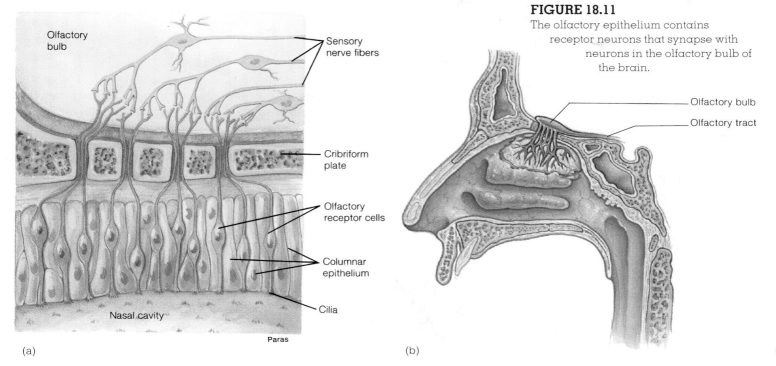

Olfactory bulb

Sensory nerve fibers

Cribriform plate

Olfactory receptor cells

Columnar epithelium

Cilia

Nasal cavity

Paras

(a)

**FIGURE 18.11**

The olfactory epithelium contains receptor neurons that synapse with neurons in the olfactory bulb of the brain.

Olfactory bulb

Olfactory tract

(b)

with the $Na^+$, however, modifies the perceived saltiness to a surprising degree: NaCl tastes much more salty than most other sodium salts (such as sodium acetate) that have been tested. Recent evidence suggests that the anions can pass through the tight junctions between the receptor cells and that the $Cl^-$ anion passes through this barrier more readily than the other anions. This is presumably related to the ability of $Cl^-$ to impart a saltier taste to the $Na^+$ than do the other anions.

Sour taste, like salty taste, is caused by the movement of ions through membrane channels. Sour taste, however, is caused by the presence of hydrogen ions ($H^+$); all acids therefore taste sour. In contrast to the production of salty and sour tastes, the production of sweet and bitter tastes involves interaction of taste molecules with specific membrane receptor proteins. Most organic molecules, particularly sugars, taste sweet to varying degrees. Bitter taste is evoked by quinine and seemingly unrelated molecules. In each case, the stimulated receptor cell activates an associated sensory neuron that transmits impulses to the brain, where they are interpreted as the corresponding taste perception.

## Olfaction

The olfactory receptors are the dendritic endings of the *olfactory (first cranial) nerve*, in association with epithelial supporting cells in the nasal epithelium within the roof of the nasal cavity (fig. 18.11). Unlike other sensory modalities, which are relayed to the cerebrum from the thalamus, the sense of olfaction is transmitted directly to the olfactory bulb and then to the cerebral cortex. This area of the brain is part of the limbic system, which was described in chapter 15 as having an important role in emotion and memory. Perhaps this explains why the smell of a particular odor, more powerfully than other sensations, can evoke emotionally charged memories.

 Certain chemicals activate the trigeminal (fifth) as well as the olfactory (first) cranial nerves and cause particular reactions. Pepper, for example, may cause sneezing; onions cause the eyes to water; and smelling salts (ammonium salts) initiate respiratory reflexes and are used to revive people who are unconscious.

In contrast to the four modalities of taste, many thousands of distinct odors can be distinguished by people who are trained in this capacity (as in the perfume and wine industries). The molecular basis of olfaction is not understood, although various theories have been proposed in an attempt to group families of odors on the basis of similarities in molecular shape and/or charges. Such attempts have been only partially successful. The extreme sensitivity of olfaction is possibly as amazing as its diversity—at maximum sensitivity, only one odorant molecule is needed to excite an olfactory receptor.

A family of genes that codes for the olfactory recepter proteins has been discovered. This is a large family of genes, perhaps as many as a thousand. The large number may reflect the importance of the sense of smell to mammals in general. Even a thousand different genes coding for a thousand differnt receptor proteins, however, cannot account for the fact that humans can distinguish up to 10,000 different odors. Clearly, the brain must integrate the signals from several different receptors and then interpret the pattern as a characteristic "fingerprint" for a particular odor.

# Equilibrium

*The sense of equilibrium is provided by structures in the inner ear, collectively known as the vestibular apparatus. Movements of the head cause fluid within these structures to bend extensions of sensory hair cells, resulting in the production of action potentials.*

The sense of equilibrium, which provides orientation with respect to gravity, is largely due to the function of an organ called the **vestibular apparatus.** The vestibular apparatus and a snail-like structure called the *cochlea*, which is involved in hearing, form the *inner ear* within the temporal bone of the skull. The vestibular apparatus consists of two parts: (1) the *otolith organs*, which include the *utricle* and *saccule*, and (2) the *semicircular canals* (fig. 18.12).

The sensory structures of the vestibular apparatus and cochlea are located within a tubular structure called the **membranous labyrinth** (*lab´ĭ-rinth*), which is filled with a fluid that is similar in composition to intracellular fluid. This fluid is called *endolymph*. The bone surrounding the membranous labyrinth forms a **bony labyrinth** in the inner ear and contains a fluid called *perilymph*, which is similar in composition to cerebrospinal fluid. The membranous labyrinth (fig. 18.13), in other words, is filled with endolymph and is surrounded by perilymph and bone.

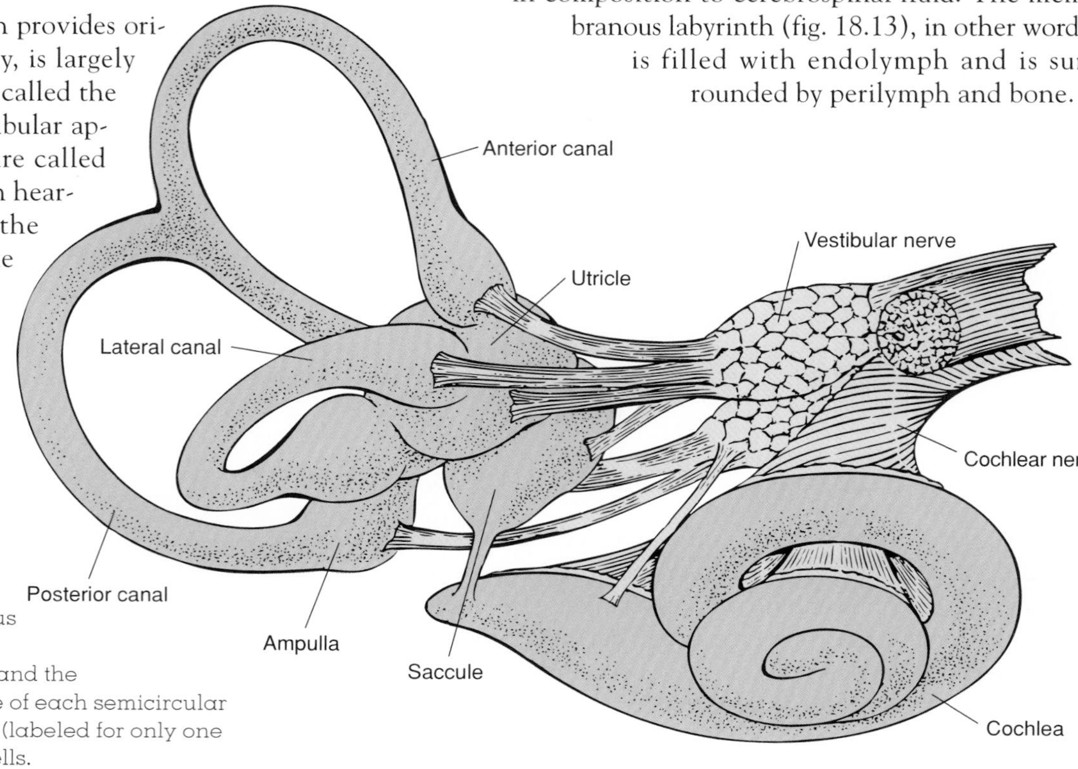

**FIGURE 18.12**
Structures within the inner ear include the cochlea and vestibular apparatus. The vestibular apparatus consists of the utricle and saccule (together called the otolith organs) and the three semicircular canals. The base of each semicircular canal is expanded into an ampulla (labeled for only one canal) that contains sensory hair cells.

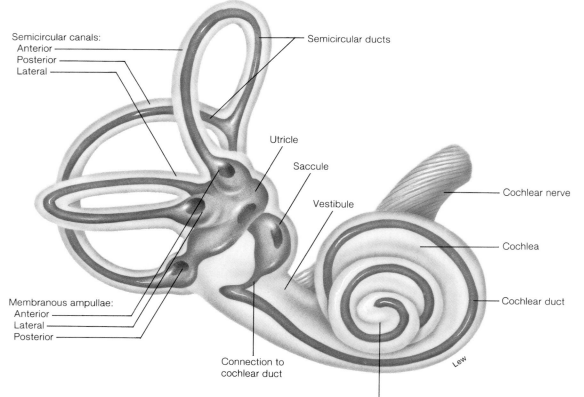

**FIGURE 18.13**
The labyrinths of the inner ear. The membranous labyrinth (darker color) is contained within the bony labyrinth.

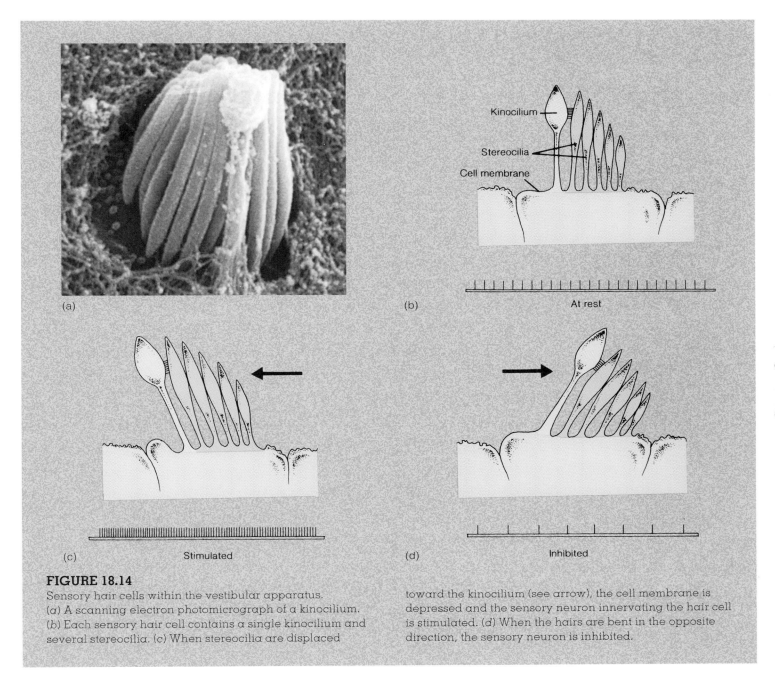

**FIGURE 18.14**
Sensory hair cells within the vestibular apparatus.
(*a*) A scanning electron photomicrograph of a kinocilium.
(*b*) Each sensory hair cell contains a single kinocilium and
several stereocilia. (*c*) When stereocilia are displaced
toward the kinocilium (see arrow), the cell membrane is
depressed and the sensory neuron innervating the hair cell
is stimulated. (*d*) When the hairs are bent in the opposite
direction, the sensory neuron is inhibited.

## *Sensory Hair Cells of the Vestibular Apparatus*

The **utricle** (*yoo´trĭ-k´l*) and **saccule** (*sak´yool*) provide information about *linear acceleration*—changes in velocity
when traveling horizontally or vertically. We therefore have
a sense of acceleration and deceleration when riding in a
car or when skipping rope. A sense of *rotational*, or *angular*,
*acceleration* is provided by the **semicircular canals,** which
are oriented in three planes like the faces of a cube. This
sense helps us maintain balance when turning the head,
spinning, or tumbling.

The receptors for equilibrium are modified epithelial
cells that contain 20 to 50 hairlike extensions. All but one

of these extensions contain filaments of protein and are
known as **stereocilia.** There is one larger extension with the
structure of a true cilium (chapter 3) that is known as a
**kinocilium** (fig. 18.14). When the stereocilia are bent in
the direction of the kinocilium, the hair cell membrane is
depressed and becomes depolarized. This causes the hair cell
to release a synaptic transmitter chemical which stimulates
the dendrites of sensory neurons that are part of the vestibulocochlear (eighth cranial) nerve. When the stereocilia are
bent in the opposite direction, the membrane of the hair cell
becomes hyperpolarized (fig. 18.14) and, as a result, releases
less synaptic transmitter chemical. In this way, the frequency
of action potentials in the sensory neurons that innervate

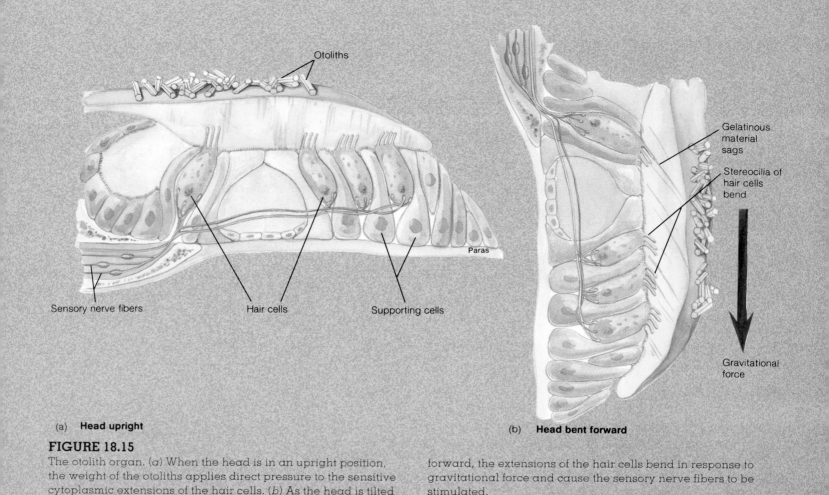

Otoliths

Gelatinous material sags

Stereocilia of hair cells bend

Sensory nerve fibers

Hair cells

Supporting cells

Paras

Gravitational force

(a) **Head upright**

(b) **Head bent forward**

**FIGURE 18.15**

The otolith organ. (a) When the head is in an upright position, the weight of the otoliths applies direct pressure to the sensitive cytoplasmic extensions of the hair cells. (b) As the head is tilted forward, the extensions of the hair cells bend in response to gravitational force and cause the sensory nerve fibers to be stimulated.

the hair cells carries information about movements that cause the hair cell processes to bend.

## Utricle and Saccule

The hair cells of the utricle and saccule protrude into the endolymph-filled membranous labyrinth, with their hairs embedded within a gelatinous **otolith membrane.** The otolith membrane contains microscopic crystals of calcium carbonate called **otoliths** that increase the mass of the membrane and result in a higher *inertia* (resistance to change in movement).

Because of the orientation of their hair cell processes into the otolith membrane, the utricle is more sensitive to horizontal acceleration and the saccule is more sensitive to vertical acceleration. During forward acceleration, the otolith membrane lags behind the hair cells, so the stereocilia of the utricle are pushed backward (fig. 18.15). This is similar to the backward thrust of the body when a car quickly accelerates forward. The inertia of the otolith membrane similarly causes the stereocilia of the saccule to be pushed upward when a person descends rapidly in an elevator. These effects, and the opposite ones that occur when a person accelerates backward

or upward, produce a changed pattern of action potentials in sensory nerve fibers that allows us to maintain our equilibrium with respect to gravity during linear acceleration.

## Semicircular Canals

The three semicircular canals are oriented at right angles to each other. At the base of each canal is a bulge called the **ampulla.** The sensory hair cells are located in an elevated area of the ampulla called the **crista ampullaris.** The processes of these sensory cells are embedded within a gelatinous membrane, the **cupula** (fig. 18.16), which projects into the endolymph of the membranous canals. Like a sail in the wind, the cupula can be pushed in one direction or the other by movements of the endolymph.

The endolymph of the semicircular canals serves a function analogous to that of the otolith membrane—it provides inertia so that the sensory processes will be bent in a direction opposite to that of the angular acceleration. As the head rotates to the right, for example, the endolymph causes the cupula to be bent toward the left, thereby stimulating the hair cells.

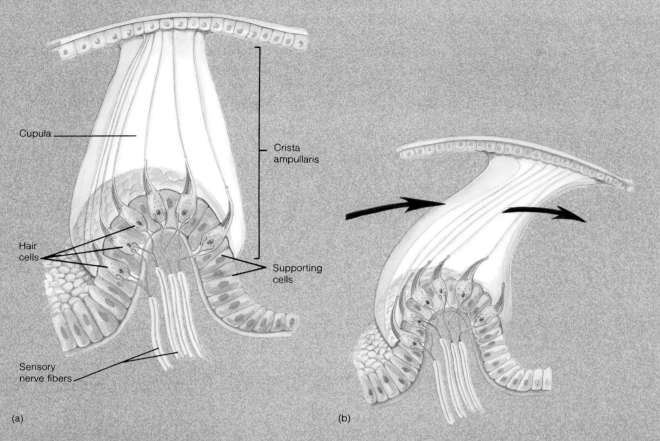

Cupula

Crista
ampullaris

Hair
cells

Supporting
cells

Sensory
nerve fibers

(a)

(b)

**FIGURE 18.16**

(a) The cupula and hair cells within the semicircular canals. (b) Movement of the endolymph during rotation causes the cupula to displace, thus stimulating the hair cells.

**Neural Pathways**    Stimulation of hair cells in the vestibular apparatus activates sensory neurons of the *vestibulocochlear* (*eighth cranial*) *nerve*. These fibers transmit impulses to the cerebellum and to the vestibular nuclei of the medulla oblongata. The vestibular nuclei, in turn, send fibers to the oculomotor center of the brain stem and to the spinal cord (fig. 18.17). Neurons in the oculomotor center control eye movements, and neurons in the spinal cord stimulate movements of the head, neck, and limbs. Movements of the eyes and body produced by these pathways serve to maintain balance and "track" the visual field during rotation.

**Nystagmus and Vertigo**    When a person first begins to spin, the inertia of endolymph within the semicircular canals causes the cupula to bend in the opposite direction. As the spin continues, however, the inertia of the endolymph is overcome and the cupula straightens. At this time, the endolymph and the cupula are moving in the same direction and at the same speed. If movement is suddenly stopped, the greater inertia of the endolymph causes it to continue moving in the previous direction of spin and to bend the cupula in that direction.

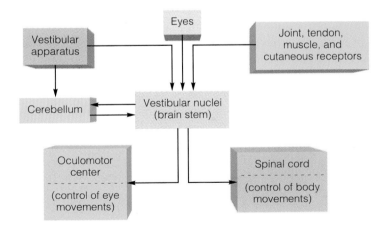

**FIGURE 18.17**

Neural processing involved in the maintenance of equilibrium and balance.

Bending of the cupula after movement has stopped affects muscular control of the eyes and body through the neural pathways previously discussed. The eyes slowly drift in the direction of the previous spin and then are rapidly jerked back to the midline position, producing involuntary oscillations. These movements are called *vestibular nystag-mus,* (*nis-tag´mus*), and people experiencing this effect may feel that they, or the room, are spinning. The loss of equilibrium that results is called *vertigo.* If the vertigo is sufficiently severe, or the person particularly susceptible, the autonomic system may become involved. This can produce dizziness, pallor, sweating, and nausea.

 Vestibular nystagmus is one of the symptoms of an inner-ear disease called *Ménière's disease.* The early symptom of this disease is often "ringing in the ears," or *tinnitus.* Since the endolymph of the cochlea and the endolymph of the vestibular apparatus are continuous through a tiny canal (see fig. 18.12), vestibular symptoms of vertigo and nystagmus often accompany hearing problems in this disease.

## Hearing

*Sound causes vibrations of the tympanic membrane. These vibrations, in turn, produce movements of the auditory ossicles, which press against the oval window. Movements of the oval window produce pressure waves within the fluid of the cochlea, which in turn cause movements of the basilar membrane. Sensory hair cells are located on the basilar membrane, and the movements of this membrane in response to sound result in the bending of the hair cell processes. This stimulates action potentials in sensory fibers that are transmitted to the brain and interpreted as sound.*

Sound waves travel in all directions from their source, like ripples in a pond into which a stone has been dropped. These waves are characterized by their frequency and their intensity. The **frequency,** or distances between crests of the sound waves, is measured in *hertz (Hz),* which is the modern designation for *cycles per second (cps).* The *pitch* of a sound is directly related to its frequency—the greater the frequency of a sound, the higher its pitch.

The **intensity,** or loudness of a sound, is directly related to the amplitude of the sound waves and is measured in units known as *decibels (dB).* A sound that is barely audible—at the threshold of hearing—has an intensity of zero decibels. Every 10 decibels indicates a tenfold increase in sound intensity; a sound is 10 times louder than threshold at 10 dB, 100 times louder at 20 dB, a million times louder at 60 dB, and 10 billion times louder at 100 dB.

The ear of a trained, young individual can hear sound over a frequency range of 20,000–30,000 Hz, yet can distinguish between two pitches that have only a 0.3% difference in frequency. The human ear can detect differences in sound intensities of only 0.1 to 0.5 dB, while the range of

audible intensities covers 12 orders of magnitude ($10^{12}$), from the barely audible to the limits of painful loudness.

### Outer Ear

Sound waves are funneled by the **auricle** (*or´ĭ-kul*), or *pinna* (fig. 18.18), into the external auditory canal, which is the fleshy tube within the bony external acoustic meatus (fig. 18.19). The auricle and external auditory canal constitute the *outer ear.* The external auditory canal channels the sound waves (while increasing their intensity) to the **tympanic** (*tim-pan´ik*) **membrane,** or eardrum. Sound waves in the external auditory canal produce extremely small vibrations of the tympanic membrane; movements of the tympanic membrane during speech (with an average sound intensity of 60 dB) are estimated to be about equal to the diameter of a molecule of hydrogen!

The **external auditory canal** is a slightly S-shaped canal about 2.5 cm (1 in.) in length, extending slightly upward from the auricle to the tympanic membrane (fig. 18.18). The skin that lines the canal contains fine hairs and sebaceous glands near the entrance. Specialized wax-secreting glands, called **ceruminous** (*sĕ-roo´mĭ-nus*) **glands,** are located in the skin, deep within the canal. *Cerumen* (earwax) secreted from ceruminous glands keeps the tympanic membrane soft and waterproof. Cerumen and the hairs also help to prevent small foreign objects from reaching the tympanic membrane. The bitter cerumen is probably an insect repellent as well.

The **tympanic membrane** is a thin partition between the external auditory canal and the middle ear. It is approximately 1 cm in diameter and is composed of an outer concave layer of stratified squamous epithelium and an inner convex layer of low columnar epithelium. The tympanic membrane is extremely sensitive to pain and is innervated by the auriculotemporal nerve (a branch of the mandibular

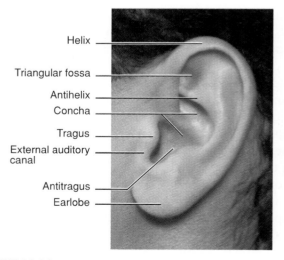

Helix

Triangular fossa

Antihelix

Concha

Tragus

External auditory canal

Antitragus

Earlobe

**FIGURE 18.18**
The surface anatomy of the auricle of the ear.

vertigo: L. *vertigo,* dizziness
Ménière's disease: from Prosper Ménière, French physician, 1799–1862
tinnitus: L. *tinnitus,* ring or tingle

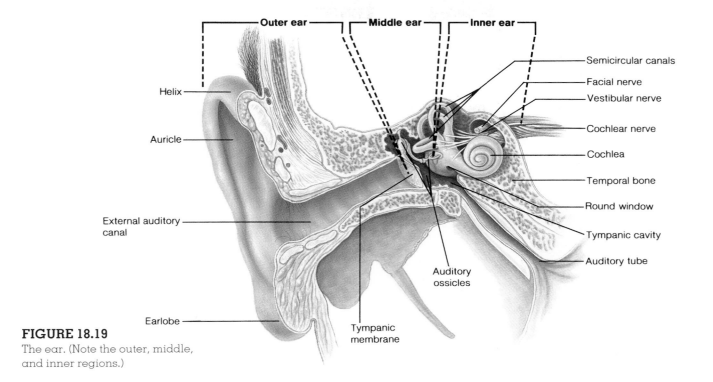

**FIGURE 18.19**
The ear. (Note the outer, middle, and inner regions.)

Outer ear    Middle ear    Inner ear

Helix

Auricle

External auditory canal

Earlobe

Tympanic membrane

Auditory ossicles

Semicircular canals
Facial nerve
Vestibular nerve
Cochlear nerve
Cochlea
Temporal bone
Round window
Tympanic cavity
Auditory tube

nerve of the trigeminal nerve) and the auricular nerve (a branch of the vagus nerve).

Inspecting the tympanic membrane with an otoscope during a physical examination yields significant information about the condition of the middle ear. The color, curvature, presence of lesions, and position of the malleus of the middle ear are features of particular importance. If ruptured, the tympanic membrane can generally regenerate and readily heal itself.

## Middle Ear

The middle ear is the cavity between the tympanic membrane on the outer side and the cochlea on the inner side (fig. 18.20). It is located within the petrous part of the temporal bone and contains three *auditory ossicles*—the **malleus** (hammer), **incus** (anvil), and **stapes** (stirrup). The malleus is attached to the tympanic membrane, so that vibrations of this membrane are transmitted, via the malleus and incus, to the stapes. The stapes, in turn, is attached to a membrane in the cochlea called the **oval (vestibular) window,** which thus vibrates in response to vibrations of the tympanic membrane. When the stapes presses the oval window into the cochlea, another flexible membrane—called the **round (cochlear) window** (fig. 18.20)—bulges outward to relieve the pressure.

The *auditory (eustachian) tube* is a passageway leading from the middle ear to the nasopharynx (a cavity located posterior to the palate of the oral cavity). The auditory tube is usually collapsed, so that debris and infectious agents are prevented from traveling from the oral cavity to the middle ear. In order to open the auditory tube, the *tensor tympani muscle,* attaching to the auditory tube and

the malleus (fig. 18.20), must contract. This occurs during swallowing, yawning, and sneezing. People sense a "popping" sensation in their ears as they swallow when driving up a mountain because the opening of the auditory canal permits air to move from the region of higher pressure in the middle ear to the lower pressure in the nasopharynx.

The fact that vibrations of the tympanic membrane are transferred through three bones instead of just one affords protection. If a sound is too intense, the auditory ossicles may buckle. This protection is enhanced by the action of the *stapedius (stă-pe´de-us) muscle,* which attaches to the neck of the stapes (fig. 18.20). When sound becomes too loud, the stapedius muscle contracts and dampens the movements of the stapes against the oval window. This action helps to prevent nerve damage within the cochlea. If sounds reach high amplitudes extremely rapidly, however—as in gunshots—the stapedius muscle may not respond fast enough to prevent nerve damage.

## Cochlea

Vibrations of the stapes and oval window displace endolymph within a part of the membranous labyrinth of the inner ear known as the **cochlear duct,** or **scala media.** The latter designation indicates that the cochlear duct is the middle part of the snail-shaped cochlea of the inner ear. Like the cochlea as a whole, the cochlear duct coils to form three levels (fig. 18.21), similar to the basal, middle, and apical portions of a

..........................................................
otoscope: Gk. *otikos,* ear; *skopein,* to examine
cochlea: L. *cochlea,* snail shell
scala: Gk. *scala,* staircase

483

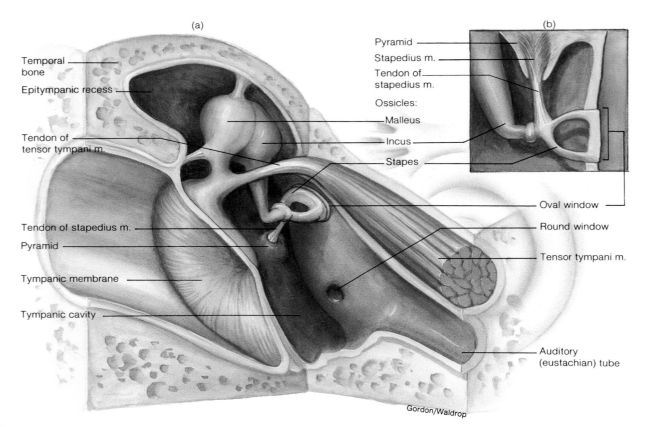

**(a)**

Temporal bone

Epitympanic recess

Tendon of tensor tympani m.

Tendon of stapedius m.

Pyramid

Tympanic membrane

Tympanic cavity

**(b)**

Pyramid

Stapedius m.

Tendon of stapedius m.

Ossicles:

Malleus

Incus

Stapes

Oval window

Round window

Tensor tympani m.

Auditory (eustachian) tube

Gordon/Waldrop

**FIGURE 18.20**

(a) The auditory ossicles and associated structures within the tympanic cavity. (b) The stapedius muscle arises from a bony protrusion called the pyramid.

snail shell. The part of the cochlea above the cochlear duct is called the **scala vestibuli,** and the part below is called the **scala tympani** (fig. 18.21). Unlike the central cochlear duct, which contains endolymph, the scala vestibuli and scala tympani are filled with perilymph.

The perilymph of the scala vestibuli and scala tympani is continuous at the apex of the cochlea because the cochlear duct ends blindly, leaving a small space called the *helicotrema* (hel˝ĭ-kŏ-tre´mă) between the end of the cochlear duct and the wall of the cochlea. Vibrations of the oval window produced by movements of the stapes cause pressure waves within the scala vestibuli, which pass to the scala tympani. Movements of perilymph within the scala tympani, in turn,

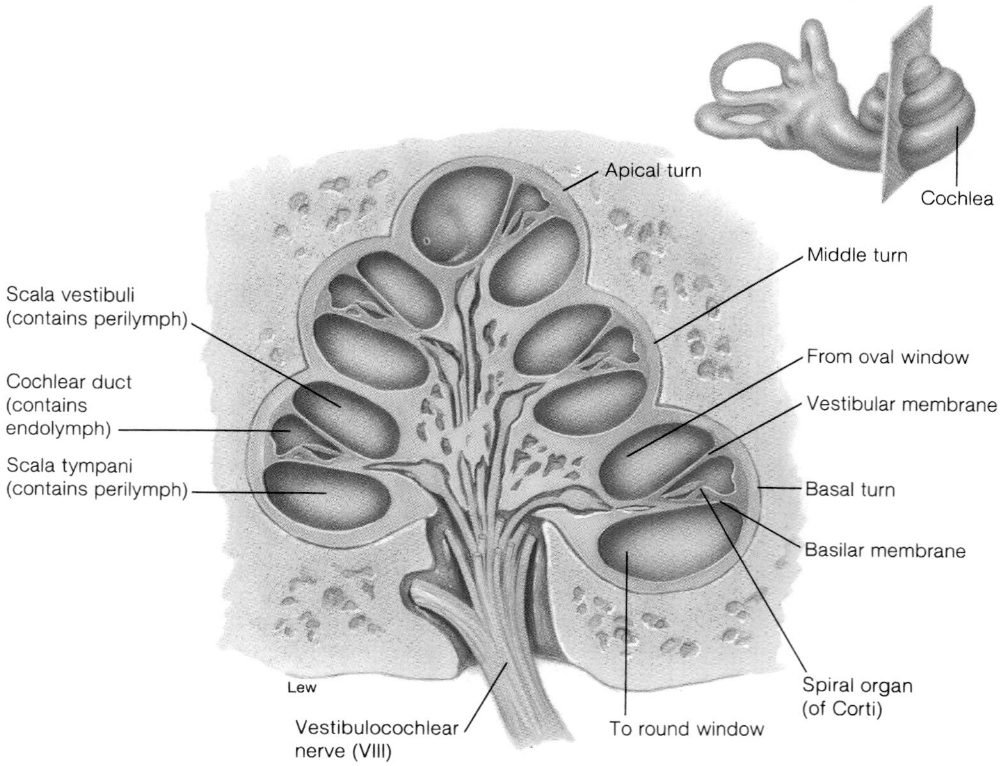

helicotrema: Gk. *helix*, a spiral; *trema*, a hole

Apical turn

Cochlea

Middle turn

From oval window

Vestibular membrane

Scala vestibuli (contains perilymph)

Cochlear duct (contains endolymph)

Scala tympani (contains perilymph)

Basal turn

Basilar membrane

Spiral organ (of Corti)

Lew

Vestibulocochlear nerve (VIII)

To round window

**FIGURE 18.21**

A cross section of the cochlea showing its three turns and its three compartments—the scala vestibuli, cochlear duct (scala media), and scala tympani.

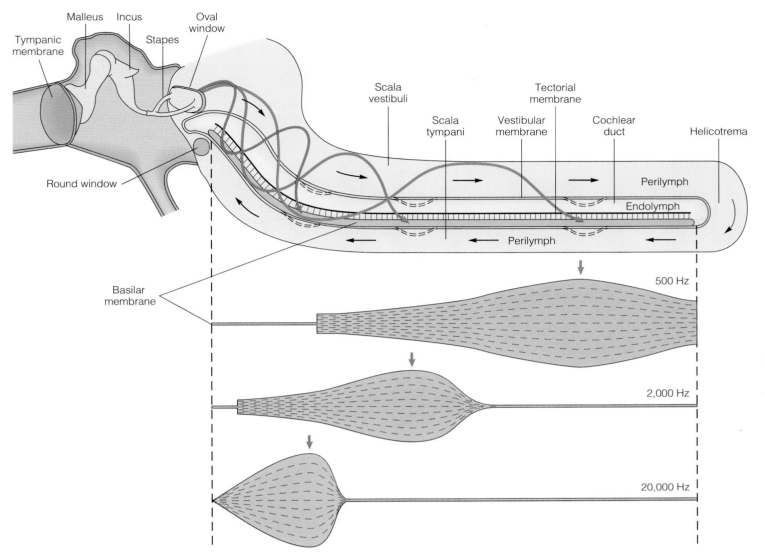

**Malleus**  **Incus**  **Oval window**

**Tympanic membrane**  **Stapes**

**Round window**

**Scala vestibuli**  **Scala tympani**  **Vestibular membrane**  **Tectorial membrane**  **Cochlear duct**  **Helicotrema**

**Perilymph**

**Endolymph**

**Perilymph**

**Basilar membrane**

500 Hz

2,000 Hz

20,000 Hz

*Sensory Organs*

## FIGURE 18.22

Sounds of low frequency cause pressure waves of perilymph to pass through the helicotrema. Sounds of higher frequency cause pressure waves to "shortcut" through the cochlear duct. This causes displacement of the basilar membrane, which is central to the transduction of sound waves into nerve impulses. (The frequency of sound waves is measured in cycles per second, or hertz [Hz].)

travel to the base of the cochlea where they cause displacement of the round window into the middle-ear cavity (see fig. 18.20). This occurs because fluid, such as perilymph, cannot be compressed; an inward movement of the oval window is thus offset by an outward movement of the round window.

When the sound frequency (pitch) is sufficiently low, there is adequate time for the pressure waves of perilymph within the upper scala vestibuli to travel through the helicotrema to the scala tympani. As the sound frequency increases, however, pressure waves of perilymph within the scala vestibuli do not have time to travel all the way to the apex of the cochlea. Instead, they are transmitted through the **vestibular membrane,** which separates the scala vestibuli from the cochlear duct, and through the **basilar membrane,** which separates the cochlear duct from

the scala tympani, to the perilymph of the scala tympani (fig. 18.21). The distance that these pressure waves travel, therefore, decreases as the sound frequency increases.

## Spiral Organ (Organ of Corti)

Movements of perilymph from the scala vestibuli to the scala tympani thus produce displacement of the vestibular membrane and the basilar membrane. Although the movement of the vestibular membrane does not directly contribute to hearing, displacement of the basilar membrane is central to pitch discrimination. The basilar membrane is fixed on the inner side of the cochlear wall to a bony ridge and is supported at its free end by a ligament.

..........................................

organ of Corti: from Alfonso Corti, Italian anatomist, 1822–88

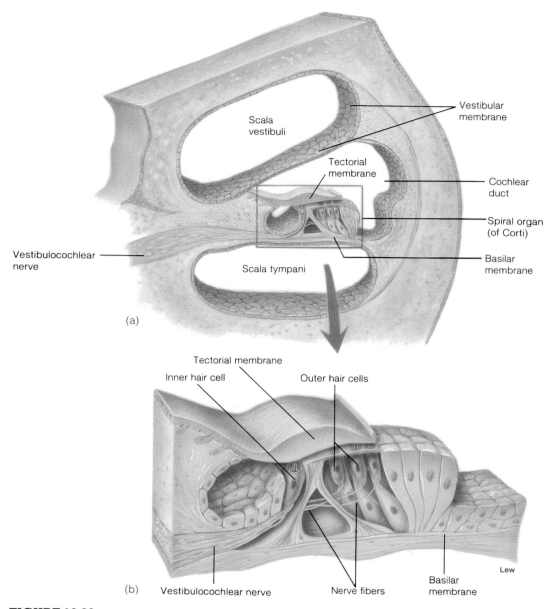

(a)

(b)

**FIGURE 18.23**

The spiral organ (a) within the cochlear duct and (b) shown in greater detail.

ded within a gelatinous **tectorial membrane** that overhangs the hair cells within the cochlear duct (fig. 18.23). The association of the basilar membrane hair cells and their sensory fibers with the tectorial membrane forms a functional unit called the **spiral organ,** or **organ of Corti** (fig. 18.23). When the cochlear duct is displaced by pressure waves of perilymph, a shearing force is created between the basilar membrane and the tectorial membrane. This causes the stereocilia to bend, which in turn produces a generator potential in the sensory nerve endings that synapse with the hair cells.

The greater the displacement of the basilar membrane and the bending of the stereocilia, the greater the generator potential produced, and the greater the frequency of action potentials in the fibers of the cochlear portion of the eighth cranial nerve that synapse with the hair cells. Experiments suggest that the stereocilia need only bend 0.3 nanometers to be detected at the threshold of hearing! A greater bending will result in a higher frequency of action potentials, which will be perceived as a louder sound.

Traveling waves in the basilar membrane reach a peak in different regions, depending on the pitch of the sound. High-pitch sounds produce a peak displacement closer to the base, while those of lower pitch cause peak displacement further toward the apex (see fig. 18.22). Those neurons that originate in hair cells located where the displacement is greatest will be stimulated more than neurons that originate in other regions. This mechanism provides a neural code for **pitch discrimination.**

**Neural Pathways for Hearing** Sensory neurons in the cochlear nerve synapse with neurons in the medulla oblongata (fig. 18.24) that project to the inferior colliculus of the midbrain. Neurons in this area in turn project to the thalamus, which sends axons to the auditory cerebral cortex of the temporal lobe. By means of this pathway, neurons in

Vibrations of the stapes against the oval window set up moving waves of perilymph in the scala vestibuli, which cause displacement of the basilar membrane into the scala tympani. This produces vibrations of the basilar membrane, each region of which vibrates with maximum amplitude to a different sound frequency. Sounds of higher frequency (pitch) cause maximum vibrations of the basilar membrane closer to the stapes, as illustrated in figure 18.22.

The sensory **hair cells** are located on the basilar membrane, with their "hairs" (actually stereocilia) projecting into the endolymph of the cochlear duct. These hair cells are arranged in a single row of inner cells, which extends the length of the basilar membrane, and multiple rows of outer hair cells. The stereocilia of the outer hair cells are embed-

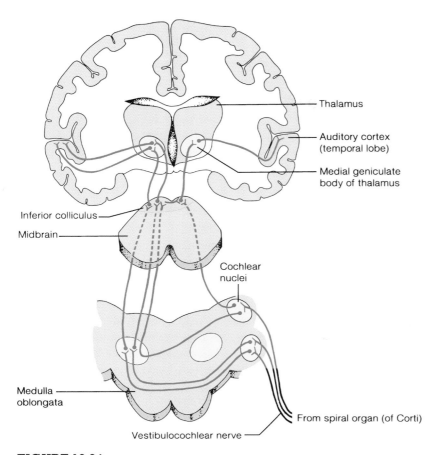

**FIGURE 18.24**
Neural pathways from the spiral ganglia of the cochlea to the auditory cerebral cortex.

different regions of the basilar membrane stimulate neurons in corresponding areas of the auditory cerebral cortex. Each area of the auditory cerebral cortex represents a different part of the basilar membrane and a different pitch.

# Vision

*Light from an observed object is focused by the cornea and lens onto the photoreceptive retina at the back of the eye. The focus is maintained on the retina at different distances between the eyes and the object by variations in the thickness and degree of curvature of the lens.*

The eyes transduce energy in the electromagnetic spectrum (fig. 18.25) into nerve impulses. Only a limited part of this spectrum can excite the photoreceptors—electromagnetic energy with wavelengths between 400 and 700 nanometers (1 nm = $10^9$ m, or one-billionth of a meter) constitute *visible light*. Light of longer wavelengths, which are in the infrared regions of the spectrum, do not have sufficient energy to excite the receptors but are felt as heat. Ultraviolet light, which has shorter wavelengths and more energy than visible light, is filtered out by the yellow color of the eye's lens. Honeybees—and people who have had their lenses removed—can see light in the ultraviolet range.

Sensory Organs

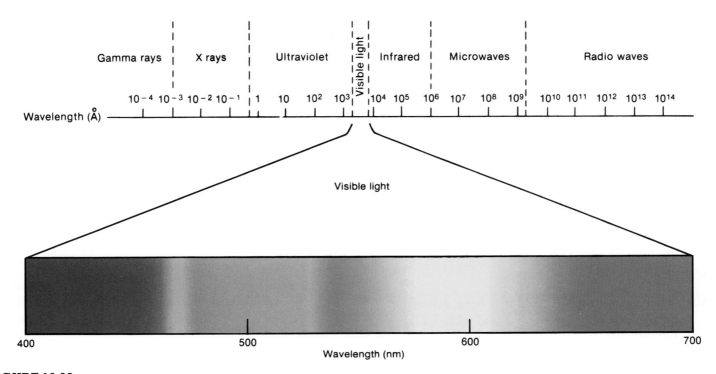

**FIGURE 18.25**
The electromagnetic spectrum (*top*) is shown in Angstrom units (1Å = $10^{-10}$ meter). The visible spectrum (*bottom*) comprises only a small range of this spectrum and is shown in nanometer units (1nm = $10^{-9}$ meter).

## Development of the Ear

The ear begins to develop at the same time as the eye, early during the fourth week. All three embryonic germ layers—ectoderm, mesoderm, and endoderm—are involved in the formation of the ear. Both types of ectoderm (neuroectoderm and surface ectoderm) play a role.

The ear of an adult is structurally and functionally divided into an **outer ear,** a **middle ear,** and an **inner ear.** Although each of these regions has a separate embryonic origin, by the end of the eighth week each of the ear's component parts is in place and formation of the ear is complete. The first indication of ear formation is the appearance of a plate of surface ectoderm called the **otic** (*o'tik*) **placode** lateral to the developing embryonic hindbrain. The otic placode soon invaginates and forms an **otic fovea.** Toward the end of the fourth week, the outer edges of the invaginated otic fovea come together and fuse to form an **otocyst** (fig. 1). The otocyst soon pinches off and separates from the

otic: Gk. *otikos,* ear

surface ectoderm. The otocyst further differentiates to form a posterior **utricular portion** and an anterior **saccular portion.** Structures of the inner ear form from these two portions of the otocyst. Three separate diverticula extend outward from the utricular portion and develop into the **semicircular canals** (not illustrated), which later function in balance and equilibrium. A tubular diverticulum, called the **cochlear duct** (not illustrated), extends in a coiled fashion from the saccular portion and forms the membranous portion of the **cochlea** (*kok'le-ă*), the organ of hearing. The sensory nerves that innervate the inner ear derive from neuroectoderm of the developing brain and grow toward the developing structures of the inner ear so that the nerve tracts will be in place when the rest of the ear has completed its development.

The **auditory ossicles** (bones) have an interesting developmental origin from the first and second pharyngeal arches. Specialized cells known as **mesenchymal condensations** migrate from their sites of origin to a location

just below the developing otocyst. Going first through a cartilaginous stage, they soon ossify to bone and are positioned and structured to amplify sound waves that will pass through the middle ear. The middle-ear chamber is referred to as the **tympanic cavity** and derives from the first pharyngeal pouch. As the tympanic cavity enlarges, it surrounds and encloses the developing auditory ossicles. The connection of the tympanic cavity to the pharynx gradually elongates to develop into the **auditory** (eustachian) **tube,** which remains patent throughout life and is important in maintaining an equilibrium of air pressure between the pharyngeal and tympanic cavities. The **external auditory canal** forms from the surface ectoderm that covers the posterior end of the first branchial groove. This canal permits sound waves to pass from the outer ear to contact the **tympanic membrane.** The tympanic membrane actually derives from the tissues that contributed to the formation of the tympanic cavity and from tissues that contributed to the formation of the external auditory canal.

## Structures Associated with the Eye

Accessory structures of the eye either protect the eyeball or provide eye movement. Protective structures include the bony orbit, eyebrows, facial muscles, eyelids, eyelashes, conjunctiva, and the lacrimal apparatus that produces tears. Eyeball movements are enabled by the actions of the extrinsic eye muscles that arise from the orbit and insert on the outer layer of the eyeball.

**Orbit** Each eyeball is positioned in a bony depression in the skull called the orbit (see fig. 9.1 and table 9.2). Seven bones of the skull (frontal, lacrimal, ethmoid, zygomatic, maxilla, sphenoid, and palatine) form the walls of the orbit that support and protect the eye.

**Eyebrows** Eyebrows consist of short, thick hair positioned transversely above both eyes along the superior orbital ridges of the skull. In this position, they effectively shade the eyes from the sun and prevent perspiration or falling particles from getting into the eyes.

**Eyelids and Eyelashes** Eyelids (palpebrae) develop as reinforced folds of skin with attached skeletal muscle so that they are movable. In addition to the orbicularis oculi muscle attached to the skin that surrounds the front of the eye, the *levator palpebrae* (*lĕ-va'tor pal'pĕ-bre*) *superioris muscle* attached along the upper eyelid, making this eyelid

palpebra: L. *palpebra,* eyelid (related to *palpare,* to pat gently)

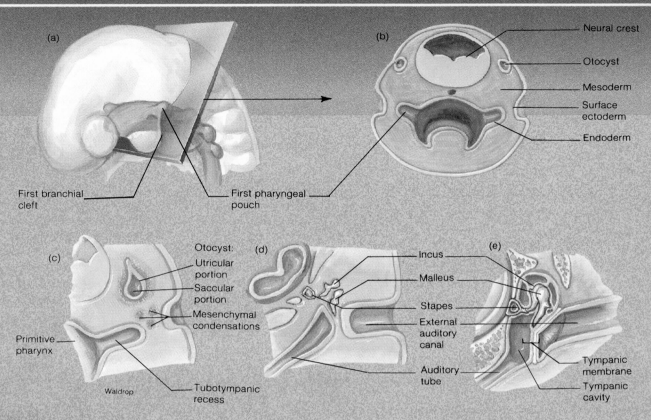

## FIGURE 1

Development of the outer- and middle-ear regions and the auditory ossicles. (a) A lateral view of a 4-week-old embryo showing the position of the cut depicted in the sequential development (b-e). (b) The embryo at 4 weeks illustrating the invagination of the surface ectoderm and the evagination of the endoderm at the level of the first pharyngeal pouch. (c) During the fifth week, mesenchymal condensations are apparent, which will derive the auditory ossicles. (d) Further invagination and evagination at 6 weeks correctly positions the structures of the outer- and middle-ear regions. (e) By the end of the eighth week, the auditory ossicles, tympanic membrane, auditory tube, and external auditory canal have formed.

much more mobile than the lower eyelid. Contraction of the *orbicularis oculi muscle* closes the eyelids over the eye, and contraction of the levator palpebrae superioris muscle elevates the upper eyelid to expose the eye. The eyelids protect the eyeball from desiccation by reflexively blinking about every 7 seconds and moving fluid across the anterior surface of the eyeball. To avoid a blurred image, the eyelid will normally blink when the eyeball moves to a new position of fixation.

The **palpebral fissure** (fig. 18.26) is the interval between the upper and lower eyelids. The **commissures (canthi)** of the eye are the medial and lateral angles where the eyelids come together. The **medial commissure** is broader than the **lateral commissure** and is characterized by a small, reddish, fleshy elevation called the **lacrimal caruncle** (kar´ung-kul) (fig. 18.27), which contains sebaceous and sudoriferous glands. The shape of the palpebral fissure is elliptical when the eyes are open.

Each eyelid supports a row of numerous **eyelashes,** which protect the eye from airborne particles. The shaft of each eyelash is surrounded by a root hair plexus that provides the hair with the sensitivity necessary to elicit a

commissure: L. *commissura*, a joining

caruncle: L. *caruncula*, diminutive of *caro*, flesh

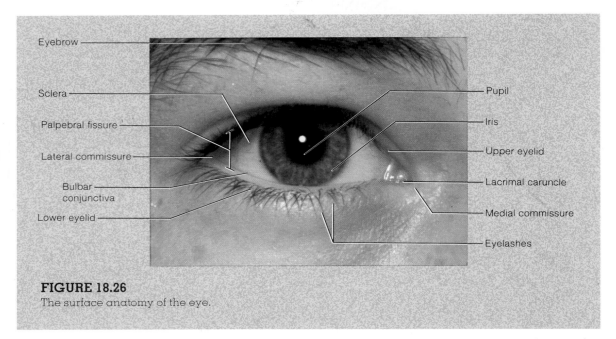

Eyebrow

Sclera

Palpebral fissure

Lateral commissure

Bulbar conjunctiva

Lower eyelid

Pupil

Iris

Upper eyelid

Lacrimal caruncle

Medial commissure

Eyelashes

**FIGURE 18.26**

The surface anatomy of the eye.

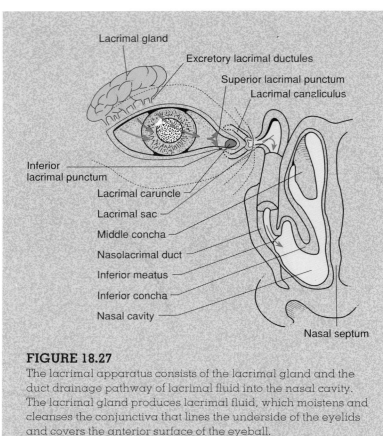

Lacrimal gland

Excretory lacrimal ductules

Superior lacrimal punctum

Lacrimal canaliculus

Inferior lacrimal punctum

Lacrimal caruncle

Lacrimal sac

Middle concha

Nasolacrimal duct

Inferior meatus

Inferior concha

Nasal cavity

Nasal septum

**FIGURE 18.27**

The lacrimal apparatus consists of the lacrimal gland and the duct drainage pathway of lacrimal fluid into the nasal cavity. The lacrimal gland produces lacrimal fluid, which moistens and cleanses the conjunctiva that lines the underside of the eyelids and covers the anterior surface of the eyeball.

reflexive closure of the lids. Eyelashes of the upper lid are long and turn upward, whereas those of the lower lid are short and turn downward.

In addition to the layers of the skin and the underlying connective tissue and orbicularis oculi muscle fibers, each eye-

lid contains a tarsal plate, tarsal glands, and conjunctiva. The **tarsal plates,** composed of dense regular connective tissue, are important in maintaining the shape of the eyelids (fig. 18.28). Specialized sebaceous glands called **tarsal (meibomian) glands** are embedded within the tarsal plates along the exposed inner surfaces of the eyelids. The ducts of the tarsal glands open onto the edges of the eyelids, and their oily secretions help keep the eyelids from adhering to each other. A *chalazion* (kă-la´ze-on) is a tumor or cyst on the eyelid that results from an infection of the tarsal glands. Modified sweat glands called **ciliary glands** are also located within the eyelids, along with additional sebaceous glands at the bases of the hair follicles of the eyelashes. An infection of these sebaceous glands is referred to as a *sty*.

**Conjunctiva** The conjunctiva (con˝jungk-ti´vă) is a delicate mucus-secreting epithelial membrane that lines the interior surface of each eyelid and exposed anterior surface of the eyeball (fig. 18.28). It consists of stratified squamous epithelium that varies in thickness in different regions. The **palpebral conjunctiva** is thick and adheres to the tarsal plates of the eyelids. As the conjunctiva reflects onto the anterior surface of the eyeball, it is known as the **bulbar conjunctiva.** This portion is transparent and especially thin where it covers the cornea. Because the conjunctiva is continuous and reflects from the eyelids to the anterior surface of the eyeball, a space called the **conjunctival sac** exists when the eyelids are closed. The conjunctival sac protects the eyeball by preventing objects (including a contact lens) from passing beyond the confines of the sac. The conjunctiva can repair itself rapidly if it is scratched.

**Lacrimal Apparatus** The lacrimal apparatus consists of the lacrimal gland, which secretes the *lacrimal fluid* (tears), and a series of ducts that drain the secretion into the nasal cavity (see fig. 18.27). The **lacrimal**

tarsal: Gk. *tarsos*, flat basket
meibomian glands: from Heinrich Meibom, German anatomist, 1638–1700
chalazion: Gk. *chalazion*, hail; a small tubercle

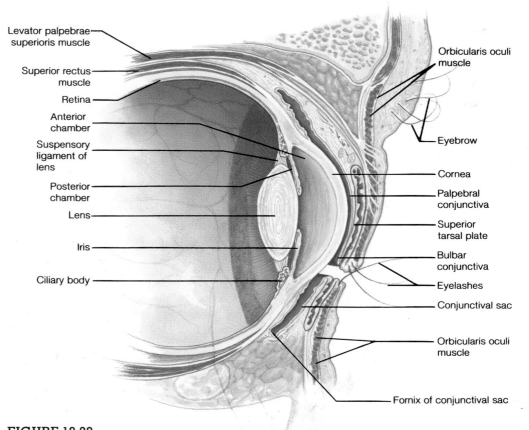

Levator palpebrae superioris muscle
Superior rectus muscle
Retina
Anterior chamber
Suspensory ligament of lens
Posterior chamber
Lens
Iris
Ciliary body

Orbicularis oculi muscle
Eyebrow
Cornea
Palpebral conjunctiva
Superior tarsal plate
Bulbar conjunctiva
Eyelashes
Conjunctival sac
Orbicularis oculi muscle
Fornix of conjunctival sac

**FIGURE 18.28**
Accessory structures of the eyeball as seen in a sagittal section of the eyelids and an anterior portion of the eyeball within the orbit.

Sensory Organs

**gland,** which is about the size and shape of an almond, is located in the superolateral portion of the orbit. It is a compound tubuloacinar gland that secretes lacrimal fluid through several excretory lacrimal ductules into the conjunctival sac of the upper eyelid. With each blink of the eyelids, lacrimal fluid passes medially and downward and drains into two small openings, called **lacrimal puncta** on both sides of the lacrimal caruncle. From here, the lacrimal fluid drains through the lacrimal canaliculus into the lacrimal sac and continues through the nasolacrimal duct to the inferior meatus of the nasal cavity.

Lacrimal fluid is an aqueous, mucus secretion that contains a bactericidal enzyme called *lysozyme*. Lacrimal fluid not only moistens and lubricates the conjunctival sac but also reduces the likelihood of eye infections. Normally about 1 ml of lacrimal fluid is produced each day by the lacrimal gland of each eye. If irritating substances, such as particles of sand or chemicals from onions, come in contact with the conjunctiva, the lacrimal glands are stimulated to oversecrete. The extra lacrimal fluid protects the eye by diluting and washing away the irritating substance.

**Extrinsic Eye Muscles**   The movements of the eyeball are controlled by six extrinsic eye muscles called the **extrinsic ocular muscles** (see fig. 13.7 and table 13.3). Four **recti mus-**

**cles** maneuver the eyeball in the direction indicated by their names (**superior, inferior, lateral,** and **medial**), and two **oblique muscles (superior** and **inferior)** rotate the eyeball on its axis. One of the extrinsic ocular muscles, the superior oblique, passes through a pulleylike cartilaginous loop, the *trochlea* (*trok´le-ă*), before attaching to the eyeball. Although stimulation of each muscle causes a precise movement of the eyeball, most of the movements involve the combined contraction of usually two muscles.

 A physical examination may include an eye movement test. As the patient's eyes follow the movement of a physician's finger, the physician can assess weaknesses in specific muscles or dysfunctions of specific cranial nerves. The patient experiencing *double vision* (*diplopia*) when moving the eyes may be suffering from muscle weakness. Looking laterally tests the abducens cranial nerve, looking inferiorly and laterally tests the trochlear cranial nerve, and crossing the eyes tests the oculomotor and trochlear nerves of both eyes.

*Amblyopia exanopsia,* commonly called "lazy eye," is a condition of ocular muscle weakness causing a deviation of one eye. Because of this, two images are received by the optic cerebral cortex and one is suppressed to avoid diplopia. A person who has amblyopia will experience dimness of vision and partial loss of sight. Young children are tested for amblyopia because little can be done to strengthen the afflicted muscle if it has not been treated before age 6.

## Structure of the Eyeball

The wall of the eyeball contains three layers, or tunics. The outer layer—the **fibrous tunic**—is divided into two regions: the posterior five-sixths is the opaque *sclera* and the anterior one-sixth is the transparent *cornea*. The middle layer is the **vascular tunic,** or **uvea** (*yoo´ve-ă*), and consists of the *choroid, ciliary body,* and the *iris*. The innermost layer of the eyeball is the **internal tunic,** which contains the **retina.** The structures of the eyeball are summarized in table 18.3.

trochlea: Gk. *trochos,* a wheel
amblyopia: Gk. *amblys,* dull; *ops,* vision
uvea: L. *uva,* grape
choroid: Gk. *chorion,* membrane

## Table 18.3  Location and functions of structures of the eyeball

| Tunic and structure | Location | Composition | Function |
|---|---|---|---|
| Fibrous tunic | Outer layer of eyeball | Avascular connective tissue | Gives shape to eyeball |
| Sclera | Posterior, outer layer; white of the eye | Tightly bound elastic and collagen fibers | Supports and protects eyeball |
| Cornea | Anterior surface of eyeball | Tightly packed dense connective tissue—transparent and convex | Transmits and refracts light |
| Vascular tunic (uvea) | Middle layer of eyeball | Highly vascular pigmented tissue | Supplies blood; prevents reflection |
| Choroid | Middle layer in posterior portion of eyeball | Vascular layer | Supplies blood to eyeball |
| Ciliary body | Anterior portion of vascular tunic | Smooth muscle fibers and glandular epithelium | Supports the lens through suspensory ligament and determines its thickness; secretes aqueous humor |
| Iris | Anterior portion of vascular tunic continuous with ciliary body | Pigment cells and smooth muscle fibers | Regulates the diameter of the pupil, and hence the amount of light entering the vitreous chamber |
| Internal tunic | Inner layer of eyeball | Tightly packed photoreceptors, neurons, blood vessels, and connective tissue | Provides location and support for rods and cones |
| Retina | Principal portion of internal tunic | Photoreceptor neurons (rods and cones), bipolar neurons, and ganglion neurons | Photoreception; transmits impulses |
| Lens (not part of any tunic) | Between posterior and vitreous chambers; supported by suspensory ligament of ciliary body | Tightly arranged protein fibers; transparent | Refracts light and focuses onto fovea centralis |

The fibrous tunic of the eye consists of a tough coat of connective tissue called the **sclera,** which can be seen externally as the white of the eyes. The tissue of the sclera is continuous with the transparent **cornea.** Light passes through the cornea to enter the **anterior chamber** of the eye. Light then passes through an opening called the **pupil,** surrounded by a pigmented (colored) smooth muscle known as the **iris.** After passing through the pupil, light enters the **posterior chamber** and then passes through the **lens** as it enters the **vitreous chamber** (fig. 18.29).

The iris is like the diaphragm of a camera; it can increase or decrease the diameter of its aperture (the pupil) to admit more or less light. Constriction of the pupils is produced by contraction of the circularly arranged fibers within the iris; dilation is produced by contraction of the radially arranged fibers. Constriction of the pupils results from parasympathetic stimulation, whereas dilation results from sympathetic stimulation (fig. 18.30). Variations in the diameter of the pupil are similar in effect to variations in the f-stop of a camera.

The posterior part of the iris contains a pigmented epithelium that gives the eye its color. The color of the eye is determined by the amount of pigment—blue eyes have the least pigment, brown eyes have more, and black eyes have the greatest amount of pigment. In the condition of *albinism*—a congenital absence of normal pigmentation due to a defect in the ability to produce melanin pigment—the eyes appear pink because the absence of pigment allows blood vessels to be seen.

Enclosed within a *lens capsule*, the lens is suspended from a muscular process called the **ciliary body,** which is connected to the sclera. Numerous extensions of the ciliary body attach to *zonular fibers*, which in turn attach to the

. . . . . . . . . . . . . . . . . . . . . . . . . . . . . . . .

sclera: Gk. *skleros,* hard
cornea: L. *cornu,* horn
iris: Gk. *irid,* rainbow

. . . . . . . . . . . . . . . . . . . . . . . . . . . . . . . .

zonular fibers: L. *zona,* a girdle

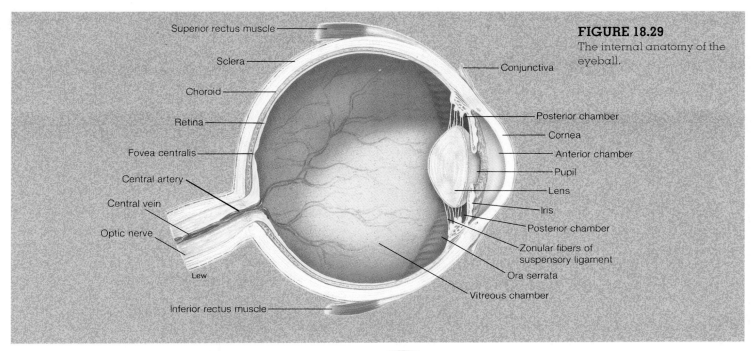

**FIGURE 18.29**
The internal anatomy of the eyeball.

Superior rectus muscle

Sclera

Choroid

Retina

Fovea centralis

Central artery

Central vein

Optic nerve

Lew

Inferior rectus muscle

Conjunctiva

Posterior chamber

Cornea

Anterior chamber

Pupil

Lens

Iris

Posterior chamber

Zonular fibers of suspensory ligament

Ora serrata

Vitreous chamber

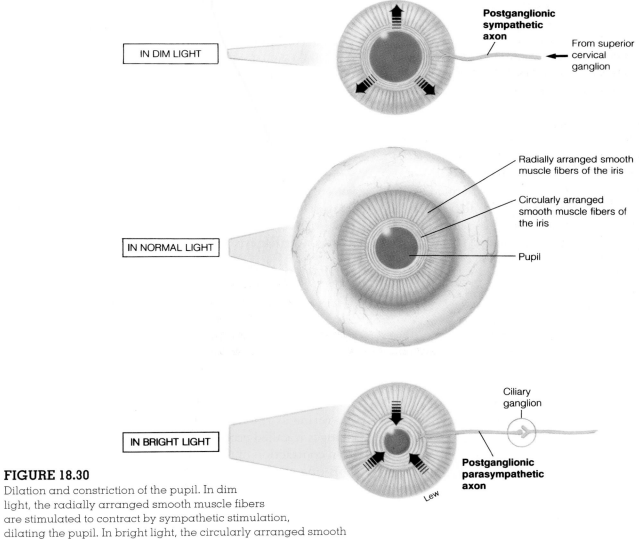

IN DIM LIGHT

**Postganglionic sympathetic axon**

From superior cervical ganglion

IN NORMAL LIGHT

Radially arranged smooth muscle fibers of the iris

Circularly arranged smooth muscle fibers of the iris

Pupil

IN BRIGHT LIGHT

Ciliary ganglion

**Postganglionic parasympathetic axon**

Lew

**FIGURE 18.30**
Dilation and constriction of the pupil. In dim
light, the radially arranged smooth muscle fibers
are stimulated to contract by sympathetic stimulation,
dilating the pupil. In bright light, the circularly arranged smooth
muscle fibers are stimulated to contract by parasympathetic
stimulation, constricting the pupil.

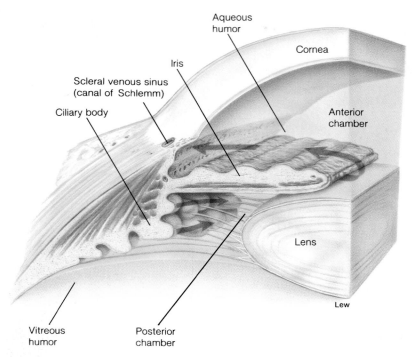

Aqueous humor

Cornea

Iris

Scleral venous sinus (canal of Schlemm)

Ciliary body

Anterior chamber

Lens

Lew

Vitreous humor

Posterior chamber

**FIGURE 18.31**

Aqueous humor maintains the intraocular pressure within the anterior and posterior chambers. It is secreted into the posterior chamber, flows through the pupil into the anterior chamber, and drains from the eyeball through the scleral venous sinus (canal of Schlemm).

lens capsule. Collectively, the zonular fibers constitute the **suspensory ligament.** The space between the cornea and iris is the *anterior chamber,* and the space between the iris and the ciliary body and lens is the *posterior chamber* (fig. 18.31).

The anterior and posterior chambers are filled with a watery fluid called the **aqueous humor.** This fluid is secreted by the ciliary body into the posterior chamber and passes through the pupil into the anterior chamber. From here, it drains into the **scleral venous sinus** (canal of Schlemm), which returns this fluid to the venous blood (fig. 18.31). Inadequate drainage of aqueous humor can lead to excessive accumulation of fluid, which in turn results in increased intraocular pressure. This condition, called *glaucoma,* may seriously damage the retina and cause loss of vision.

The portion of the eye located behind the lens is filled with a clear gel known as the **vitreous humor.** Light passing through the lens and vitreous humor enters the retina. While passing through the retina, some of this light stimulates photoreceptors, which in turn activate other neurons. Neurons in

..........................................................
canal of Schlemm: from Friedrich S. Schlemm, German anatomist, 1795–1858
vitreous: L. *vitreus,* glassy

the retina contribute fibers that are gathered together at a region called the **optic disc** (fig. 18.32) to exit the retina as the optic nerve. This region lacks photoreceptors, and is thus a blind spot. The optic disc is also the site of entry and exit of blood vessels.

## Refraction

Light is *refracted,* or bent, as it passes from a medium of one density into a medium of a different density. The degree of refraction depends on the comparative densities of the two media, as indicated by their *refractive index.* The refractive index of air is set at 1.00; the refractive index of the cornea, by comparison, is 1.38, and the refractive index of the lens is 1.45. Since the greatest difference in refractive index occurs at the air-cornea interface, light is refracted most at the cornea.

The degree of refraction also depends on the curvature of the interface between two media. The curvature of the cornea is constant, however, while the curvature of the lens can be varied. The refractive properties of the lens can thus provide fine control for focusing light on the retina. As a result of light refraction, the image formed on the retina is upside down and right to left (fig. 18.33).

The *visual field*—which is the part of the external world projected onto the retina—is thus reversed in each eye. The cornea and lens focus the right part of the visual field on the left half of the retina of each eye, while the left half of the visual field is focused on the right half of each retina (fig. 18.34). The medial (or nasal) half-retina of the left eye therefore receives the same image as the lateral (or temporal) half-retina of the right eye. The nasal half-retina of the right eye receives the same image as the temporal half-retina of the left eye.

## Accommodation

When a normal eye views an object, parallel rays of light are refracted to a point, or *focus,* on the retina. If the degree of refraction remained constant, movement of the object closer to or farther from the eye would cause corresponding movement of the focal point, so that the focus would either be behind or in front of the retina.

The ability of the eyes to keep an image focused on the retina as the distance between the eyes and object changes is called *accommodation.* Accommodation results from contraction of the ciliary muscles of the ciliary body. This ring of smooth muscle fibers is like a sphincter muscle that can vary its aperture (fig. 18.35). When the ciliary

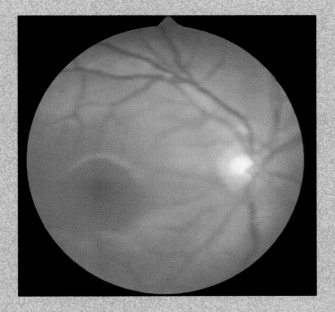

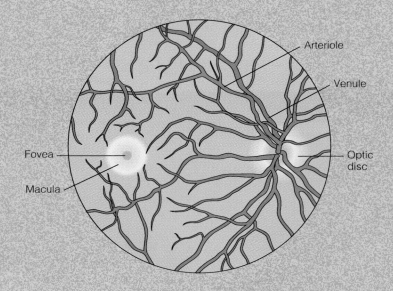

Arteriole

Venule

Optic disc

Fovea

Macula

**FIGURE 18.32**

A view of the retina as seen with an ophthalmoscope. Optic nerve fibers leave the eyeball at the optic disc to form the optic nerve. (Note the blood vessels that can be seen entering the eyeball at the optic disc.)

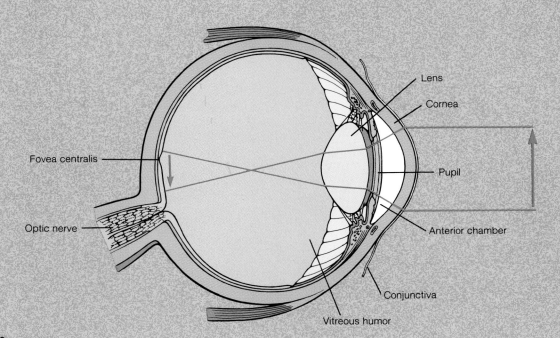

Lens

Cornea

Fovea centralis

Pupil

Optic nerve

Anterior chamber

Conjunctiva

Vitreous humor

**FIGURE 18.33**

The refraction of light waves within the eyeball causes the image of an object to be inverted on the retina.

muscles are relaxed, the aperture is wide. Relaxation of the ciliary muscles thus places tension on the zonular fibers and pulls the lens taut. These are the conditions that prevail when viewing an object that is 20 feet or more from a normal eye; the image is focused on the retina and the lens is in

its most flat, least convex form. As the object moves closer to the eyes, the muscles of the ciliary body contract. This muscular contraction narrows the aperture of the ciliary body and reduces the tension on the zonular fibers that suspend the lens. When the tension is reduced, the lens becomes more rounded and convex as a result of its inherent elasticity (fig. 18.36).

 The ability of a person's eyes to accommodate can be measured by the *near-point-of-vision test*, which is the minimum distance from the eyes at which an object can be maintained in focus. This distance increases with age, and indeed accommodation in almost everyone over the age of 45 is significantly impaired. Loss of accommodating ability with age is known as *presbyopia*. This loss appears to have a number of causes, including thickening of the lens and a forward movement of the attachments of the zonular fibers to the lens. As a result of these changes, the zonular fibers and lens are pulled taut even when the ciliary muscles contract. The lens is thus not able to thicken and increase its refraction when, for example, a printed page is brought close to the eyes.

## Retina

The **retina** is the principal structure of the internal tunic. The retina consists of a thin outer **pigmented layer** (in contact with the choroid of the vascular tunic) and a thick inner **nervous layer.** The retina is principally in the posterior part of

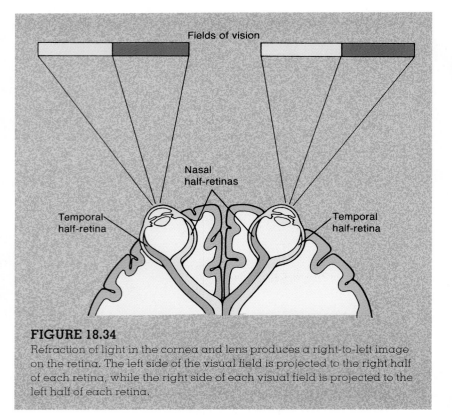

**FIGURE 18.34**
Refraction of light in the cornea and lens produces a right-to-left image on the retina. The left side of the visual field is projected to the right half of each retina, while the right side of each visual field is projected to the left half of each retina.

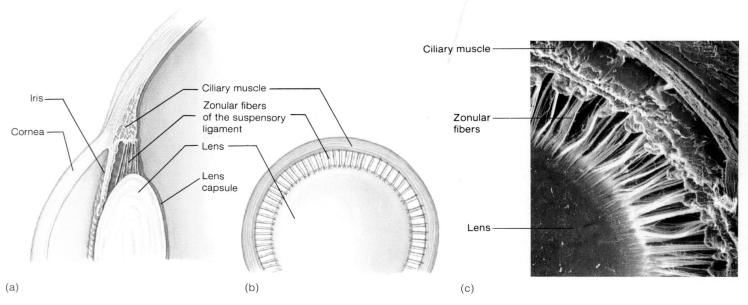

(a)             (b)             (c)

**FIGURE 18.35**
Structure of the anterior portion of the eyeball. (*a*) A sagittal view, (*b*) an anterior view of the lens and supporting structures, and (*c*) a scanning electron micrograph in anterior view showing the relationship between the lens, zonular fibers, and ciliary muscles of the eye.

(*a–b*) From "How the Eye Focuses," by James F. Koretz and George Handleman. Copyright © 1988 by Scientific American, Inc. All rights reserved.

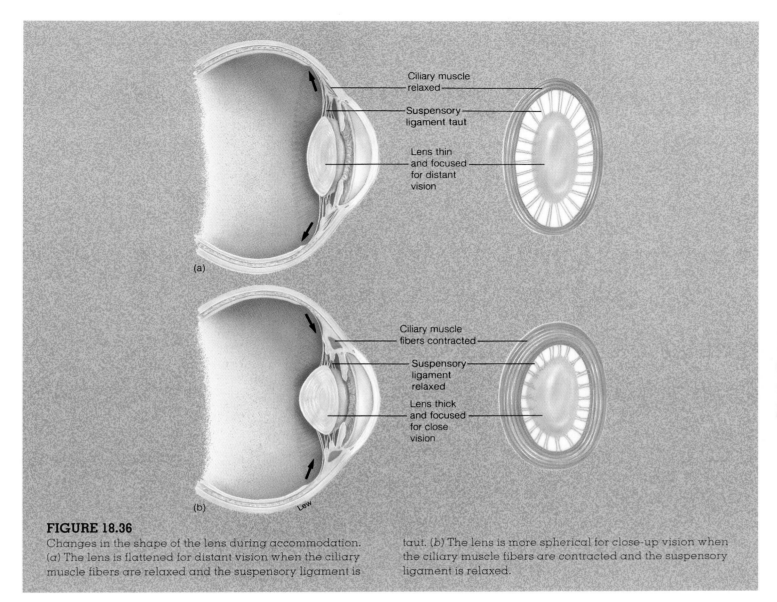

**FIGURE 18.36**

Changes in the shape of the lens during accommodation. (a) The lens is flattened for distant vision when the ciliary muscle fibers are relaxed and the suspensory ligament is taut. (b) The lens is more spherical for close-up vision when the ciliary muscle fibers are contracted and the suspensory ligament is relaxed.

the eyeball (see fig. 18.29). The visual layer of the retina terminates in a jagged margin near the ciliary body called the **ora serrata.** The pigmented layer extends anteriorly over the back of the ciliary body and iris.

 The pigmented and nervous layers of the retina are not attached to each other, except surrounding the optic nerve and at the ora serrata. Because of this loose connection, the two layers may become separated as a *detached retina.* Such a separation can be corrected by fusing the layers with a laser.

The nervous layer of the retina is composed of three principal layers of neurons (fig. 18.37). The neurons located most superficially are called **ganglion cells,**

..........................................

ora serrata: L. *ora*, margin; *serra*, saw

**FIGURE 18.37**

A photomicrograph showing the layers of the retina. The retina is inverted, so that light must pass through various layers of nerve cells before reaching the photoreceptors (rods and cones).

**497**

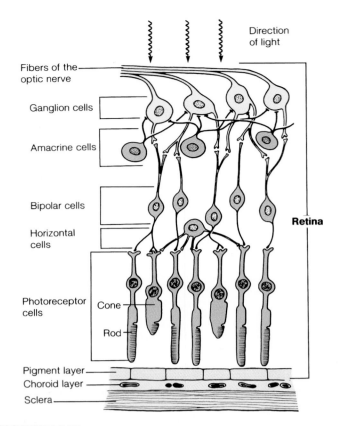

**FIGURE 18.38**

A diagram showing the layers of the retina and the direction of light that travels through it. Nerve impulses are transmitted in the opposite direction.

and their axons form the optic nerve. These neurons receive synaptic input from **bipolar cells** underneath, which in turn receive input from the photoreceptors—the **rods** and **cones.** In addition to the flow of information from photoreceptors to bipolar cells to ganglion cells, neurons called **horizontal cells** synapse with several photoreceptors (and possibly also with bipolar cells) and neurons called **amacrine cells** synapse with several ganglion cells.

The architecture of the retina is the reverse of what might be expected. Light traveling through the vitreous body strikes the ganglion cell layer of the retina first and then must pass through the bipolar cell layer before reaching the photoreceptors (fig. 18.38). Neural information then passes in the reverse direction.

**Effect of Light on the Rods**   The photoreceptor rods and cones, shown in figure 18.39, are activated when light produces a chemical change in molecules of pigment contained within the membranous lamellae of the outer segments of these cells. Rods contain a purple pigment known as

rhodopsin: Gk. *rhodon,* rose; *ops,* eye

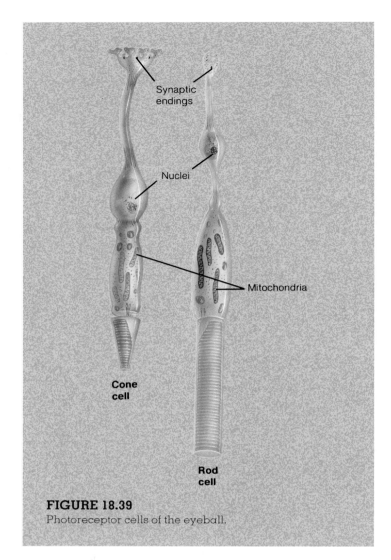

**FIGURE 18.39**

Photoreceptor cells of the eyeball.

**rhodopsin.** The pigment appears purple (a combination of red and blue) because it transmits light in the red and blue regions of the spectrum while absorbing light energy in the green region. The wavelength of light that is absorbed best—the *absorption maximum*—is about 500 nm (a green-colored light).

Green cars (and other objects that are green) are easier to see at night (when rods are used for vision) than those that are red. This is because red light is not well absorbed by rhodopsin, and only absorbed light can produce the photochemical reaction that results in vision. In response to absorbed light, rhodopsin dissociates into its two components: a pigment called **retinene** (or retinaldehyde), derived from vitamin A, and a protein called **opsin.** This reaction is known as the *bleaching reaction.*

Retinene can assume two configurations (shapes): the all-*trans* form or the 11-*cis* form (fig. 18.39). The all-*trans* form

opsin: Gk. *ops,* eye

is more stable, but only the 11-*cis* form is found attached to opsin. In response to absorbed light energy, the 11-*cis* retinene is converted to the all-*trans* form, causing it to dissociate from the opsin. This dissociation reaction in response to light initiates changes in the ionic permeability of the rod cell membrane and ultimately results in the production of nerve impulses in the ganglion cells. As a result of these effects, rods provide black-and-white vision under conditions of low light intensity (as described in a later section).

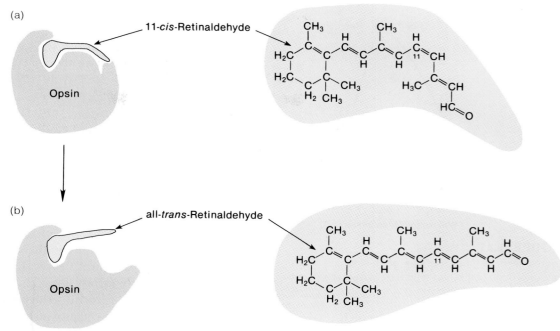

Scientists have recently discovered the genetic basis for blindness in the disease *dominant retinitis pigmentosa*. People with this disease inherit a gene for the opsin protein in which a single base change in the gene (substitution of adenosine for cytosine) causes the amino acid histidine to be substituted for proline at a specific point in the polypeptide chain. This abnormal opsin leads to degeneration of the photoreceptors.

**FIGURE 18.40**

The photopigment rhodopsin consists of the protein opsin combined with 11-*cis* retinaldehyde (*a*). In response to light, the retinaldehyde is converted to a different form, called all-trans, and dissociates from the opsin (*b*). This photochemical reaction induces changes in ionic permeability, ultimately resulting in stimulation of ganglion cells in the retina.

The bleaching reaction that occurs in response to absorbed light results in a lowered amount of rhodopsin in the rods and lowered amounts of visual pigments in the cones. Therefore, when a light-adapted person first enters a darkened room, sensitivity to light is low and vision is poor. A gradual increase in photoreceptor sensitivity, known as **dark adaptation,** then occurs, reaching maximal sensitivity in about 20 minutes. The increased sensitivity to low light intensity is due partly to increased amounts of visual pigments produced in the dark. Increased pigments in the cones produce a slight dark adaptation in the first 5 minutes. Increased rhodopsin in the rods produces a much greater increase in sensitivity to low light levels and is partly responsible for the adaptation that occurs after about 5 minutes in the dark. In addition to the increased concentration of rhodopsin, other more subtle (and less well-understood) changes occur in the rods that ultimately result in a 100,000-fold increase in light sensitivity in dark-adapted as compared to light-adapted eyes.

**Electrical Activity of Retinal Cells**   The only neurons in the retina that produce all-or-none action potentials are ganglion cells and amacrine cells. The photoreceptors, bipolar cells, and horizontal cells instead produce only graded depolarizations or hyperpolarizations, analogous to EPSPs and IPSPs.

In the dark, the photoreceptors have a resting membrane potential that is less negative (closer to zero) than that of most other neurons. This property results from a constant current of $Na^+$ into the cell, called a *dark current*, through special $Na^+$ channels. Light causes a blockage of these $Na^+$ channels; consequently, the photoreceptors become less depolarized than they are in the dark. Put another way, light causes the photoreceptors to become *hyperpolarized* in comparison to their membrane potential in the dark. Since light ultimately must have a stimulatory effect on the optic nerve, this hyperpolarization (which is associated with inhibition—chapter 14) is certainly surprising.

The translation of the effect of light on photoreceptors to the production of nerve impulses may be explained in the following manner. Photoreceptors in the dark release a neurotransmitter chemical at a constant rate at their synapses with bipolar cells. Hyperpolarization of the photoreceptors in response to light decreases their release of a neurotransmitter. In some cases, this neurotransmitter is inhibitory to the bipolar cell. If the neurotransmitter is inhibitory, the secretion of lower amounts of this chemical will stimulate the bipolar cells. These neurons in turn activate ganglion cells, and action potentials will thus be produced on fibers of the optic nerve in response to light.

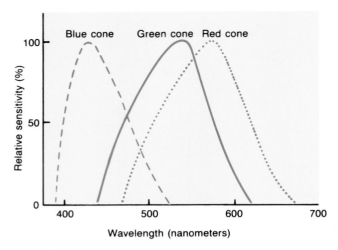

**FIGURE 18.41**
There are three types of cones, each of which contains retinaldehyde combined with a different type of protein.

### Cones and Color Vision

Cones are less sensitive than rods to light, but cones provide color vision and greater visual acuity, as described in the next section. During the day, therefore, the high light intensity bleaches out the rods, and color vision with high acuity is provided by the cones. According to the **trichromatic theory of color vision,** our perception of a multitude of colors is due to stimulation of only three types of cones. Each type of cone contains retinene, as in rhodopsin, but the retinene in the cones is associated with proteins other than opsin. The protein is different for each of the three cone pigments; thus, each of the pigments absorbs light of a given wavelength (color) to a different degree. The three types of cones are designated blue, green, and red, according to the region of the visible spectrum in which each cone pigment absorbs light maximally (fig. 18.41). Our perception of any given color depends on the relative degree to which each cone is stimulated by any given wavelength of visible light.

Suppose a person has become dark adapted in a photographic darkroom over a period of 20 minutes or longer but needs more light to examine some prints. Since rods do not absorb red light but red cones do, a red light in a photographic darkroom allows vision (because of the red cones), but does not cause bleaching of the rods. When the light is turned off, therefore, the rods will still be dark adapted and the person will still be able to see.

 *Color blindness* is due to a congenital lack of one or more types of cones. People with normal color vision are *trichromats;* people with only two types of cones are *dichromats.* Dichromats may be missing red cones (have *protanopia*), or green cones (have *deuteranopia*), or blue cones (have *tritanopia*). They may have difficulty, for example, distinguishing red from green. People who are *monochromats* have only one cone system and can only see black, white, and shades of gray. Color blindness is a trait carried on the X chromosome; since men have only one X chromosome per cell, whereas women have two X chromosomes (see chapter 30), men are far more likely to be color blind than women (who can carry this trait in a recessive state).

### Visual Acuity and Sensitivity

While reading or similarly focusing visual attention on objects in daylight, each eye is oriented so that the image falls within a tiny area of the retina called the **fovea centralis.** The fovea centralis is a pinhead-sized pit within a yellow area of the retina called the *macula lutea.* The depression within the fovea centralis is formed as a result of the displacement of neural layers around the periphery; therefore, light falls directly on photoreceptors in the center (fig. 18.42). Light falling on other areas, by contrast, must pass through several layers of neurons, before reaching the photoreceptors.

There are approximately 120 million rods and 6 million cones in each retina, but only about 1.2 million nerve fibers enter the optic nerve of each eye. This gives an overall convergence ratio of photoreceptors on ganglion cells of about 105:1. This ratio is misleading, however, because the degree of convergence is much lower for cones than for rods, and the ratio is 1:1 in the fovea centralis.

The photoreceptors are distributed in such a way that the fovea centralis contains only cones, whereas more peripheral regions of the retina contain a mixture of rods and cones. Approximately 4000 cones in the fovea centralis provide input to approximately 4000 ganglion cells; each ganglion cell in this region, therefore, has a private line to the visual field. Each ganglion cell thus receives input from an area of retina corresponding to the diameter of one cone (about 2 μm). Peripheral to the fovea centralis, however, many rods synapse with a single bipolar cell, and many bipolar cells synapse with a single ganglion cell. A single ganglion cell outside the fovea centralis thus may receive input from large numbers of rods, corresponding to an area of about 1 square millimeter on the retina (fig. 18.43).

Since each cone in the fovea centralis has a private line to a ganglion cell, and since each ganglion cell receives input from only a tiny region of the retina, visual acuity is greatest and sensitivity to low light is poorest when light falls on the fovea centralis. In dim light only the rods are activated, and vision is best out of the corners of the eye when the image falls away from the fovea centralis. Under these conditions, the convergence of many rods on

fovea: L. *fovea,* small pit
macula lutea: L. *macula,* spot; *luteus,* yellow

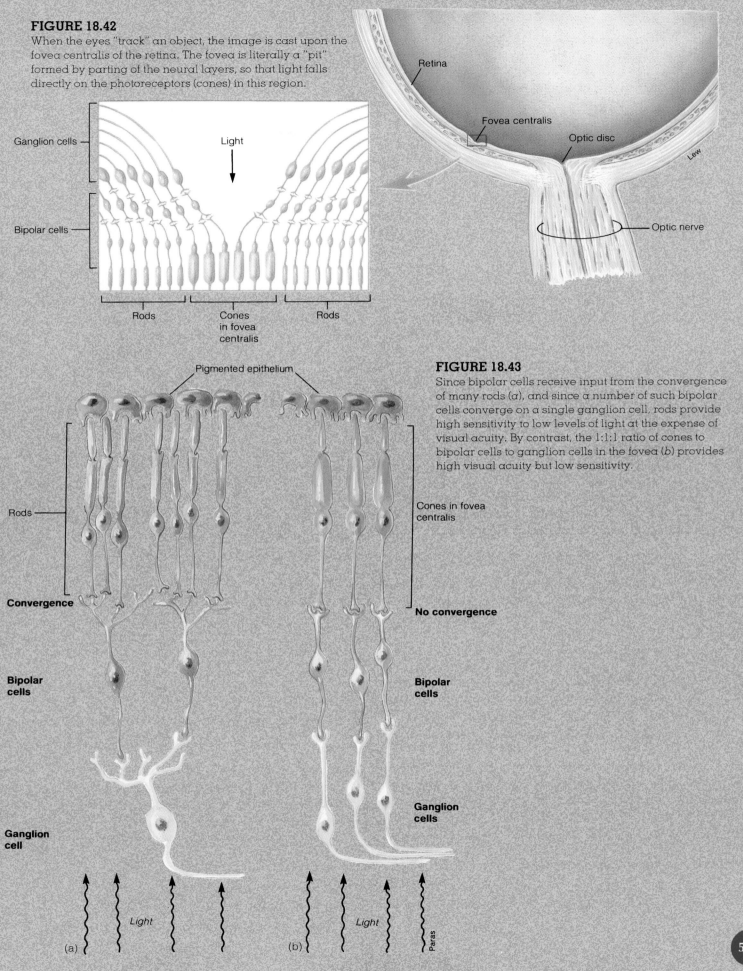

**FIGURE 18.42**

When the eyes "track" an object, the image is cast upon the fovea centralis of the retina. The fovea is literally a "pit" formed by parting of the neural layers, so that light falls directly on the photoreceptors (cones) in this region.

Retina

Fovea centralis

Optic disc

Lew

Optic nerve

Ganglion cells

Light

Bipolar cells

Rods

Cones in fovea centralis

Rods

Pigmented epithelium

**FIGURE 18.43**

Since bipolar cells receive input from the convergence of many rods (*a*), and since a number of such bipolar cells converge on a single ganglion cell, rods provide high sensitivity to low levels of light at the expense of visual acuity. By contrast, the 1:1:1 ratio of cones to bipolar cells to ganglion cells in the fovea (*b*) provides high visual acuity but low sensitivity.

Rods

Cones in fovea centralis

**Convergence**

**No convergence**

**Bipolar cells**

**Bipolar cells**

**Ganglion cells**

**Ganglion cell**

*Light*

*Light*

(a)

(b)

Paras

a single bipolar cell and the convergence of many bipolar cells on a single ganglion cell increase sensitivity to dim light at the expense of visual acuity. Night vision is therefore less distinct than day vision.

## Neural Pathways from the Retina

As a result of light refraction by the cornea and lens, the right half of the visual field is projected to the left half of the retina of both eyes (the temporal half of the left retina and the nasal half of the right retina). The left half of the visual field is projected to the right half of the retina of both eyes. The temporal half of the left retina and the nasal half of the right retina therefore see the same image. Axons from ganglion cells in the left (temporal) half of the left retina pass to the left **lateral geniculate body** of the thalamus. Axons from ganglion cells in the nasal half of the right retina cross (decussate) in the **optic chiasma** to synapse also in the left lateral geniculate body. The left lateral geniculate, therefore, receives input from both eyes that relates to the right half of the visual field (fig. 18.44).

The right lateral geniculate body, similarly, receives input from both eyes relating to the left half of the visual field. Neurons in both lateral geniculate bodies of the thalamus in turn project to the **striate cortex** of the occipital lobe in the cerebral cortex (fig. 18.44).

Approximately 70% to 80% of the axons from the retina pass to the lateral geniculate bodies and to the striate cortex. This **geniculostriate system** is involved in perception of the visual field. Put another way, the geniculostriate system is needed to answer the question, What is it? Approximately 20% to 30% of the fibers from the retina, however, follow a different path to the **superior colliculus** of the midbrain (also called the **optic tectum**). Axons from the superior colliculus activate motor pathways leading to eye and body movements. The **tectal system,** in other words, is needed to answer the question, Where is it?

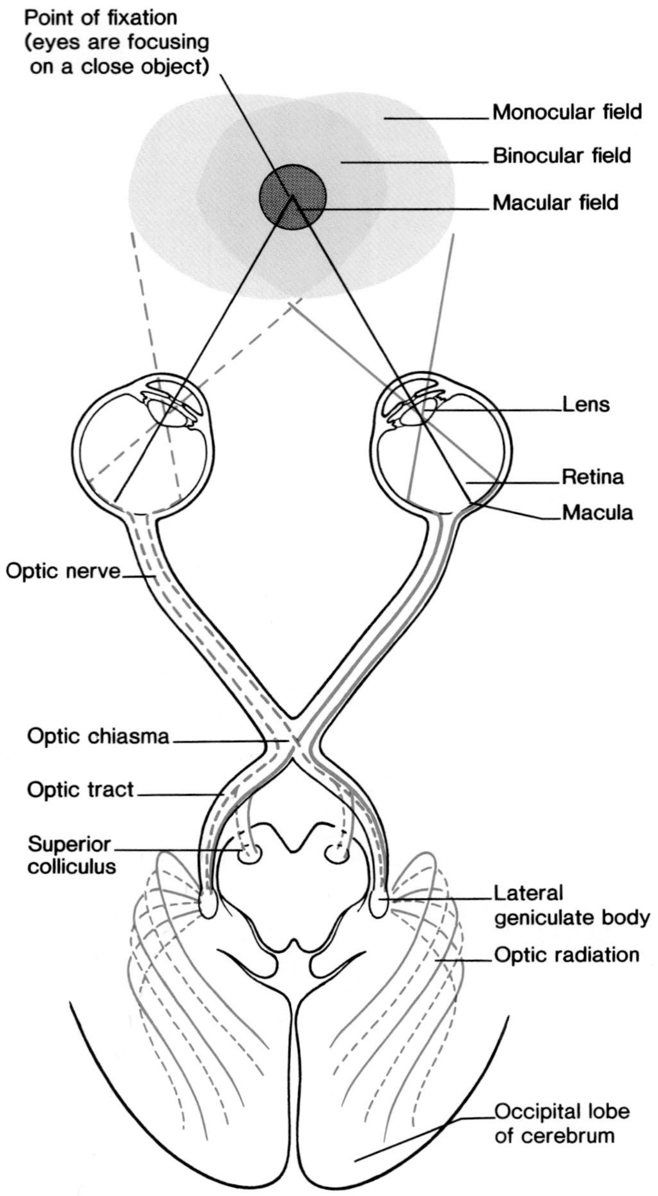

**Superior Colliculus and Eye Movements**   Neural pathways from the superior colliculus to motor neurons in the spinal cord help mediate the startle response to the sight of an unexpected intruder. Other nerve fibers from the superior colliculus stimulate the extrinsic eye muscles, which, it may be recalled, are the striated muscles that move the eyes.

Two types of eye movements are coordinated by the superior colliculus. **Smooth pursuit movements** track moving objects and keep the image focused on the fovea centralis. **Saccadic eye movements** are quick (lasting 20 to 50 msec), jerky movements that occur while the eyes appear to be still. Saccadic movements continuously move the image to different photoreceptors; if they were

····································
saccadic: Fr. *saccade*, a sudden check or jerk

**FIGURE 18.44**
Visual fields of the eyes and neural pathways for vision. An overlapping of the visual field of each eye provides binocular vision, which is the ability to perceive depth.

to stop, the image would disappear as the photoreceptors became bleached.

The tectal system is also involved in the control of the intrinsic eye muscles—the iris and the muscles of the ciliary body. Shining a light into one eye stimulates the **pupillary reflex** in which both pupils constrict. This is caused by activation of parasympathetic neurons by fibers from the superior colliculus. Postganglionic neurons in the ciliary ganglia behind the eyes, in turn, stimulate constrictor fibers in the iris. Contraction of the ciliary body during accommodation also involves stimulation of the superior colliculus.

**Processing Visual Information**   Light cast on the retina directly affects the activity of photoreceptors and indirectly affects the neural activity in bipolar and ganglion cells. The part of the visual field that affects the activity of a particular ganglion cell can be considered its **receptive field.**

Studies of the electrical activity of ganglion cells have yielded some surprising results. In the dark, each ganglion cell discharges spontaneously at a slow rate. When the room lights are turned on, the firing rate of many (but not all) ganglion cells increases slightly. A small spot of light directed at the center of the receptive fields of some ganglion cells, however, stimulates a large increase in firing rate. A small spot of light can thus be a more effective stimulus than wider areas of light.

When the spot of light is moved only a short distance away from the center of the receptive field, the ganglion cell responds in the opposite manner. The ganglion cell that was stimulated with light at the center of its receptive field is inhibited by light in the periphery of its field. The responses produced by light in the center and by light in the "surround" of the visual field are *antagonistic*. Those ganglion cells that are stimulated by light at the center of their visual fields are said to have **on-center fields;** those that are inhibited by light in the center and stimulated by light in the surround have **off-center fields.**

The reason wide illumination of the retina has less effect than pinpoint illumination is now clear; diffuse illumination gives the ganglion cell conflicting orders—on and off. Because of the antagonism between the center and surround of ganglion cell receptive fields, the activity of each ganglion is determined by the *difference in light intensity* between the center and surround of its visual field. This is a form of **lateral inhibition** that helps to accentuate the contours of images and improve visual acuity.

Impulses produced by ganglion cells are transmitted to neurons in the lateral geniculate nuclei. Projections of nerve fibers from the lateral geniculate to the occipital lobe form the **optic radiation** (see fig. 18.44). Because these fiber projections give this area a striped or striated appearance, it is also known as the striate cortex. The cortical neurons that receive input from neurons in the lateral geniculate, in turn, project to other cortical neurons, which in turn send impulses to still others.

Experiments have shown that the dotlike information passed from the retina to the lateral geniculate nuclei to the cerebral cortex is transformed within the cortex into information about edges—their position, length, orientation, and movement. Although this information is highly abstract, the visual association areas of the occipital lobe probably play an early role in the integration of visual information.

Other areas of the brain receive input from the visual association areas and give meaning to visual perception.

# Clinical Considerations

Numerous disorders and diseases afflict the sensory organs. Some of these occur during the sensitive period of prenatal development and are the result of hereditary influences. Other sensory impairments, some of which are avoidable, can be acquired at any time of life. Still others result from changes associated with the natural aging process. The loss of a sense is traumatic and frequently involves a long adjustment period. Fortunately, however, when a sensory function is impaired or lost, the other senses seem to become keener to lessen the extent of the handicap. A blind person, for example, compensates somewhat for the loss of sight by developing a remarkable hearing ability.

Entire specialties within medicine are devoted to treating the disorders of specific sensory organs. It is beyond the scope of this text to attempt a comprehensive discussion of the numerous diseases and dysfunctions of these organs. We will comment generally, however, on the diagnosis of sensory disorders and on developmental problems that can affect the ears and eyes. In addition, we will discuss the more common diseases and dysfunctions of these organs.

## *Diagnosis of Sensory Organs*

**Ear**   Otorhinolaryngology (o˝to-ri˝no-lar˝ing-gol´ŏ-je) is the specialty of medicine dealing with the diagnosis and treatment of diseases or conditions of the ear, nose, and throat. *Audiology* is the study of hearing, particularly assessment of the ear and its functioning.

There are three common instruments or techniques used to examine the ears to determine auditory function: (1) an *otoscope* is an instrument used to examine the tympanic membrane of the ear; abnormalities of this membrane are informative when diagnosing specific auditory problems, including middle-ear infections; (2) *tuning fork tests* are useful in determining hearing acuity and especially for discriminating the various kinds of hearing loss; and (3) *audiometry* is a functional examination for hearing sensitivity and speech discrimination.

**Eye**   There are two distinct professional specialties concerned with the structure and function of the eye. *Optometry* is the paramedical profession concerned with

## Development of the Eye

The development of the eye is a complex, rapid process starting early in the fourth week and involving the precise interaction of neuroectoderm, surface ectoderm, and mesoderm. The initial differentiation involves neuroectoderm forming a lateral diverticulum on each side of the prosencephalon (forebrain). As the diverticulum increases in size, the distal portion dilates to become the **optic vesicle,** and the proximal portion constricts to become the **optic stalk** (fig. 1). Once the optic vesicle has formed, the overlying surface ectoderm thickens and invaginates. The thickened portion is the **lens placode** and the invagination is the **lens fovea.**

During the fifth week, the lens placode is depressed and eventually cut off from the surface ectoderm, causing the formation of the **lens vesicle.** Simultaneously, the optic vesicle invaginates and differentiates into the two-layered **optic cup.** A groove called the **optic fissure** appears along the inferior surface of the optic cup in

which the **hyaloid artery** and **hyaloid vein** traverse to serve the developing eyeball. The walls of the optic fissure eventually close so that the hyaloid vessels are within the tissue of the optic stalk. They become the **central vessels of the retina** of the mature eye. The optic stalk eventually becomes the **optic nerve,** composed of sensory axons from the retina.

By the early part of the seventh week, the optic cup has differentiated into two sheets of epithelial tissue that become the sensory and pigmented layers of the **retina.** Both of these layers also line the entire vascular coat, including the **ciliary body, iris,** and the **choroid.** A proliferation of cells in the lens vesicle leads to the formation of the lens. The **lens capsule** forms from the mesoderm surrounding the lens, as does the **vitreous humor.** Mesoderm surrounding the optic cup differentiates into two distinct layers of the developing eyeball. The inner layer of mesoderm becomes the vascular **choroid;** the outer layer becomes the

toughened **sclera** posteriorly and the transparent **cornea** anteriorly. Once the cornea has formed, additional surface ectoderm gives rise to the thin **conjunctiva** covering the anterior surface of the eyeball. Epithelium of the **eyelids** and the **lacrimal glands** and **duct** develop from surface ectoderm, whereas the **extrinsic eye muscles** and all connective tissues associated with the eye develop from mesoderm. These accessory structures of the eye gradually develop during the embryonic period and into the fetal period as late as the fifth month.

### FIGURE 1

The development of the eye. (a) An anterior view of the developing head of a 22-day-old embryo and the formation of the optic vesicle from the neuroectoderm of the prosencephalon (forebrain). (b) The development of the optic cup. The lens vesicle is formed (c) as the ectodermal lens placode invaginates during the fourth week. The hyaloid vessels become enclosed within the optic nerve ($c_1$ and $e_1$) as there is fusion of the optic fissure. (d) The basic shape of the eyeball and the position of its internal structures are established during the fifth week. The successive development of the eye is shown at 6 weeks (e) and at 20 weeks (f), respectively. (g) The eye of the newborn.

---

optic: L. *optica*, see

hyaloid: Gk. *hyalos*, glass; *eiodos*, form

---

assessing vision and treating visual problems. An *optometrist* prescribes corrective lenses or visual training but is not a medical doctor and does not treat eye diseases. *Ophthalmology* (of˝thal-mol´ŏ-je) is the speciality of medicine concerned with diagnosing and treating eye diseases.

Although the eyeball is an extremely complex organ, it is quite accessible to examination. Various techniques and instruments are used during an eye examination, but the following are used most frequently: (1) a *cycloplegic drug*, which

is instilled into the eyes to dilate the pupils and temporarily inactivate the ciliary muscles; (2) a *Snellen's chart*, which is used to determine the visual acuity of a person standing 20 feet from the chart (a reading of 20/20 is considered normal for the test); (3) an *ophthalmoscope*, which contains a light, mirrors, and lenses to illuminate and magnify the interior of the eyeball so that the structures within may be examined; and (4) a *tonometer*, used to measure ocular tension, which is important in detecting glaucoma.

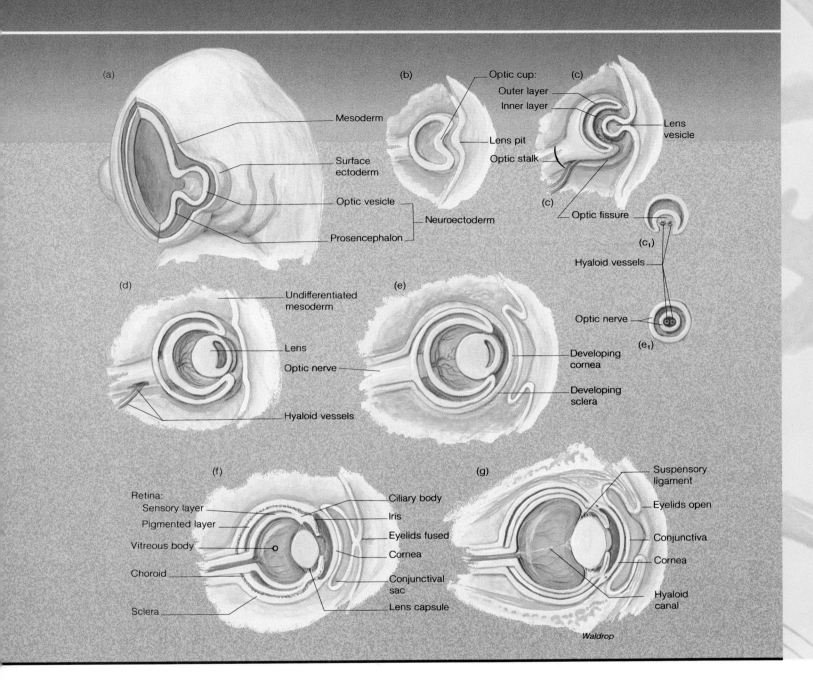

(a)

Mesoderm

Surface ectoderm

Optic vesicle

Neuroectoderm

Prosencephalon

(b)

Optic cup:
Outer layer
Inner layer

Lens pit

Optic stalk

(c)

Lens vesicle

(c)

Optic fissure

(c₁)

Hyaloid vessels

Optic nerve

(e₁)

(d)

Undifferentiated mesoderm

Lens

Optic nerve

Hyaloid vessels

(e)

Developing cornea

Developing sclera

(f)

Retina:
Sensory layer
Pigmented layer

Vitreous body

Choroid

Sclera

Ciliary body

Iris

Eyelids fused

Cornea

Conjunctival sac

Lens capsule

(g)

Suspensory ligament

Eyelids open

Conjunctiva

Cornea

Hyaloid canal

*Waldrop*

## *Developmental Problems of the Ears and Eyes*

Although there are many congenital abnormalities of the eyes and ears, most of them occur rarely. The sensitive period of development for these organs is from about 24 to 45 days after conception. Indeed, 85% of newborns suffer anomalies if infected during this developmental time period. Most congenital disorders of the eyes and ears are caused by genetic factors or intrauterine infections such as *rubella virus*.

If a pregnant woman contracts rubella (German measles), there is a 90% probability that the embryo or fetus will contract it also. An embryo afflicted with rubella is 30% more likely to be aborted, stillborn, or congenitally deformed than one that is not afflicted. Rubella interferes with the mitotic process and thus causes underdeveloped organs. An embryo with rubella may suffer from a number of physical deformities, *cataracts* and *glaucoma* being common deformities of the eye.

505

**Ear** Congenital deafness is generally caused by an auto-somal recessive gene but may also be caused by a maternal rubella infection. The actual functional impairment is generally either a defective set of auditory ossicles or improper development of the neurosensory structures of the inner ear.

Although the shape of the auricle varies widely, **auricular abnormalities** are not uncommon, especially in infants with chromosomal syndromes causing mental deficiencies. In addition, the external auditory canal frequently does not develop in these children, producing a condition called **atresia** (*ă-tre´zhă*) of the external auditory canal.

**Eye** Most **congenital cataracts** are hereditary, but they may also be caused by maternal rubella infection during the critical fourth to sixth week of eye development. In this condition, the lens is opaque and frequently appears grayish white.

**Cyclopia** is a rare condition in which the eyes are partially fused into a median eye enclosed by a single orbit. Other severe malformations, which are incompatible with life, are generally expressed with this condition.

## Infections, Diseases, and Functional Impairments of the Ear

Disorders of the ear are common and may affect both hearing and the vestibular functions. The ear is afflicted by numerous infections and diseases, some of which can be prevented.

**Infections and Diseases** **External otitis** is a general term for infections of the outer ear. The causes of external otitis range from dermatitis to fungal and bacterial infections.

**Acute purulent otitis media** is a middle-ear infection. Pathogens of this disease usually enter through the auditory tube, most often following a cold or tonsillitis. Children frequently have middle-ear infections because of their susceptibility to infections and their short and straight auditory tubes. As a middle-ear infection progresses to the inflammatory stage, the auditory tube closes and drainage is prohibited. An intense earache is a common symptom of a middle-ear infection. The pressure from the inflammation may eventually rupture the tympanic membrane to permit drainage.

Repeated middle-ear infections, particularly in children, usually call for an incision of the tympanic membrane known as a **myringotomy** (*mir˝ing-got´ŏ-me*) and the implantation of a tiny tube within the tympanic membrane (fig. 18.45) to assist the patency of the auditory tube. The tubes, which are eventually sloughed out of the ear, permit the infection to heal and help prohibit further infections by keeping the auditory tube open. **Perforation of the tym-**

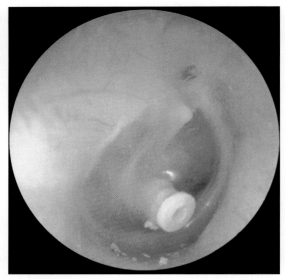

**FIGURE 18.45**
An implanted ventilation tube in the tympanic membrane following a myringotomy.

**panic membrane** may occur as the result of infections or trauma. Sudden, intense noise can rupture the membrane. Spontaneous perforation of the membrane usually heals rapidly, but scar tissue may form and lessen the sensitivity to sound vibrations.

**Otosclerosis** is a progressive deterioration of the normal bone of the bony labyrinth and its replacement with vascular spongy bone. This frequently causes hearing loss as the auditory ossicles are immobilized. Surgical scraping of the bone growth and replacing the stapes with a prosthesis is the most frequent treatment for otosclerosis.

**Ménière's disease** afflicts the inner ear and may cause hearing loss as well as equilibrium disturbance. The causes of Ménière's disease vary and are not completely understood, but they are thought to be related to a dysfunction of the autonomic nervous system that causes a vasoconstriction within the inner ear. The disease is characterized by recurrent periods of **vertigo** (dizziness and a sensation of rotation), **tinnitus** (*tĭ-ni´tus*) (ringing in the ear), and progressive deafness in the affected ear. Ménière's disease is chronic and affects both sexes equally. It is more common in middle-aged and elderly people.

**Auditory Impairment** Loss of hearing results from disease, trauma, or developmental problems involving any portion of the auditory apparatus, cochlear nerve and auditory pathway, or areas of auditory perception within the brain. Hearing impairment varies from slight disablement, which may or may not worsen, to total deafness. Some types of hearing impairment, including deafness, can be mitigated through hearing aids or surgery.

Auditory impairments fall into two broad categories. **Conduction deafness** is caused by an interference with the sound waves through the outer or middle ear. Conduction problems include impacted cerumen (wax), a ruptured tympanic membrane, a severe middle-ear infection, or adhesions of one or more auditory ossicles (*otosclerosis*). Conductive deafness can be successfully treated.

**Perceptive (sensorineural) deafness** results from disorders that affect the inner ear, the cochlear nerve or nerve pathway, or auditory centers within the brain. Perceptive impairment ranges in severity from the inability to hear certain frequencies to deafness. Such deafness may be caused by a number of factors, including diseases, trauma, and genetic or developmental problems. Hearing impairments of this type, which are common in elderly persons, are related to age. The ability to perceive high-frequency sounds is generally affected first. Hearing aids may help patients with perceptive deafness. This type of deafness is permanent, however, because it involves sensory structures that cannot heal themselves through regeneration or replication.

## Functional Impairments of the Eye

Few people have perfect vision. Slight variations in the shape of the eyeball or curvature of the cornea or lens result in an imperfect focal point of light waves onto the retina. Most variations are slight, however, and the error of refraction goes unnoticed. Severe deviations that are not corrected may cause blurred vision, fatigue, chronic headaches, and depression.

The primary clinical considerations associated with defects in the refractory structures or general shape of the eyeball are myopia, hyperopia, presbyopia, and astigmatism. **Myopia** (nearsightedness) is an elongation of the eyeball that causes light waves to focus at a point in the vitreous humor in front of the retina (fig. 18.46). Only light waves from close objects can be focused clearly on the retina; distant objects appear blurred. **Hyperopia** (farsightedness) is a condition in which the eyeball is too short; consequently, light waves have a focal point behind the retina. Although visual accommodation aids a hyperopic person, it generally does not help enough for the person to clearly see objects that are very close or far away. **Presbyopia** is a condition in which the lens tends to lose its elasticity and ability to accommodate. It is relatively common in elderly people. In order to read print on a page, a person with presbyopia must hold the page farther from the eyes. **Astigmatism** is a condition in which an irregular curvature of the cornea or lens of the eye distorts the refraction of light waves. If a person with astigmatism views a circle, the image will not appear clear in all 360 degrees; the parts of the circle that appear blurred can be used to map the astigmatism.

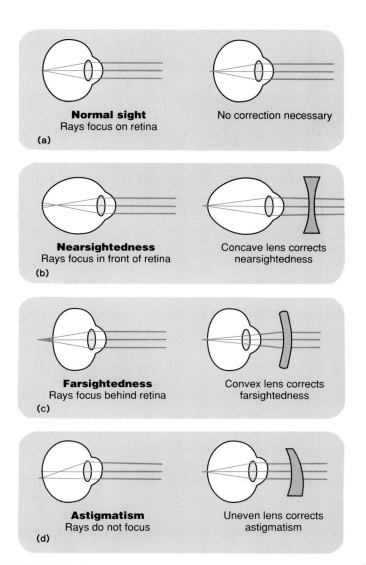

**FIGURE 18.46**

In a normal eye (*a*), parallel rays of light are brought to a focus on the retina by refraction in the cornea and lens. If the eye is too long, as in myopia (*b*), the focus is in front of the retina. This can be corrected by a concave lens. If the eye is too short, as in hyperopia (*c*), the focus is behind the retina. This is corrected by a convex lens. In astigmatism (*d*), light refraction is uneven due to an abnormal shape of the cornea or lens.

Various glass or plastic lenses frequently benefit people with the visual impairments just described. Myopia may be corrected with a biconcave lens; hyperopia with a biconvex lens; and presbyopia with bifocals, or a combination of two lenses adjusted for near and distant vision. Correction for astigmatism requires a careful assessment of the irregularities and a prescription of a specially ground corrective lens.

The condition of myopia may also be treated by a surgical procedure called **radial keratotomy.** In this technique, 8 to 16 microscopic slashes, like the spokes of a wheel, are made in the cornea from the center to the edge. The ocular pressure inside the eyeball bulges the weakened cornea and flattens its center, changing the focal length of the eyeball. Side effects may include vision fluctuations, sensitivity to glare, and incorrect corneal alteration.

**Cataracts**   A cataract is a clouding of the lens that leads to a gradual blurring of vision and the eventual loss of sight. A cataract is not a growth within or upon the eye but a chemical change in the protein of the lens caused by injury, poisons, infections, or age degeneration.

Cataracts are the leading cause of blindness. A cataract can be removed surgically, however, and vision restored by implanting a tiny intraocular lens that either clips to the iris or is secured into the vacant lens capsule. Special contact lenses or thick lenses for glasses are other options.

**Detachment of the Retina**   Retinal detachment is a separation of the nervous or visual layer of the retina from the underlying pigment epithelium. It generally begins as a minute tear in the retina that gradually extends as vitreous humor accumulates between the layers. Retinal detachment may result from hemorrhage, a tumor, degeneration, or trauma from a violent blow to the eye. A detached retina may be repaired by using laser beams, cryoprobes, or intense heat to destroy the tissue beneath the tear and rejoin the layers.

**Strabismus**   Strabismus is a condition in which both eyes do not focus upon the same axis of vision. This prevents stereoscopic vision, and the individuals afflicted will have varied visual impairments. Strabismus is usually caused by a weakened extrinsic eye muscle.

Strabismus is assessed while the patient attempts to look straight ahead. If the afflicted eye is turned toward the nose, the condition is called convergent strabismus (**esotropia**). If the eye is turned outward, it is called divergent strabismus (**exotropia**). Disuse of the afflicted eye causes a visual impairment called **amblyopia.** Visual input from the normal eye and the eye with strabismus results in **diplopia,** or double vision. A normal, healthy person who has excessively consumed alcohol may experience diplopia.

# Infections and Diseases of the Eye

**Infections**   Infections and inflammation can occur in any of the associated structures of the eye and in structures within or on the eyeball itself. The causes of infections are usually microorganisms, mechanical irritation, or sensitivity to particular substances.

**Conjunctivitis** (inflammation of the conjunctiva) may result from sunburn or an infection by organisms such as staphylococci, viruses, or streptococci. Bacterial conjunctivitis is commonly called "pinkeye."

**Keratitis** (inflammation of the cornea) may develop secondarily from conjunctivitis or be caused by diseases such as tuberculosis, syphilis, mumps, or measles. Keratitis is painful and may cause blindness if untreated.

A **chalazion** is a cyst caused by an infection and a subsequent blockage in the ducts of the sebaceous glands along the free edge of the eyelids.

**Styes (hordeola)** are relatively common, but mild, infections of the follicle of an eyelash or the sebaceous gland of the follicle. Styes may spread readily from one eyelash to another if untreated. Poor hygiene and the excessive use of cosmetics may contribute to development of styes.

**Diseases**   Trachoma (*tră-ko´mă*) is a highly contagious bacterial disease of the conjunctiva and cornea. Although rare in the United States, it is estimated that over 500 million people are afflicted by this disease. Trachoma may be treated readily with sulfonamides and some antibiotics, but if untreated it will spread progressively until it covers the cornea. At this stage, vision is lost and the eye undergoes degenerative changes.

**Glaucoma** is the second leading cause of blindness and is particularly common in underdeveloped countries. Although it can afflict individuals of any age, 95% of the cases involve people over the age of 40. Glaucoma is an abnormal increase in the intraocular pressure of the eyeball. Aqueous humor does not drain through the scleral venous sinus as quickly as it is produced. Accumulation of fluid causes compression of the blood vessels in the eyeball and compression of the optic nerve. Retinal cells die and the optic nerve may atrophy, producing blindness.

．．．．．．．．．．．．．．．．．．．．．．．．．．．．

glaucoma: Gk. *glaukos*, gray

# Chapter Summary

## Characteristics of Sensory Receptors (pp. 465–467)

1. Receptors may be classified as chemoreceptors, mechanoreceptors, thermoreceptors, or photoreceptors; they may be further classified as exteroreceptors or enteroreceptors.
   a. Receptors that fire only when a stimulus changes are phasic; those that continue to fire as long as the stimulus is maintained are tonic.
   b. Stimuli produce graded generatory potentials in the sensory organs, analogous to EPSPs.

## Somatic Senses (pp. 467–474)

1. Somatic senses are those in which the receptors are localized within the body wall. They include cutaneous receptors and proprioceptors.
   a. Corpuscles of touch, free nerve endings, and root hair plexuses are tactile receptors responding to light touch.
   b. Lamellated corpuscles are pressure receptors in the dermis.
   c. Thermoreceptors include the organs of Ruffini (heat) and bulbs of Krause (cold).
   d. Free nerve endings are the principal pain receptors, serving to protect the body from injury.
   e. Joint kinesthetic receptors are proprioceptors that are sensitive to changes in stretch and tension.
2. The acuity of sensation depends on the density of the receptors in the stimulated receptive field and on lateral inhibition.
3. Neuromuscular spindles elicit the muscle stretch reflex and provide information about muscle length.
4. Motor neurons that stimulate extrafusal (outside the neuromuscular spindle) fibers to contract are called alpha motoneurons; those that cause the intrafusal muscle fibers of the spindles to contract are called gamma motoneurons.
   a. Only stimulation by the alpha motoneurons can cause muscle contraction that results in skeletal movements.
   b. Activation of gamma motoneurons enhances the stretch reflex.

5. Neurotendinous receptors (Golgi tendon organs) monitor tension in tendons. When activated, they inhibit contracting muscles, causing them to relax.

## Taste and Olfaction (pp. 474–477)

1. Taste buds are located on the papillae of the tongue and respond to either sweet, sour, bitter, or salty tastes, depending upon their location on the tongue.
2. The olfactory receptors are the dendritic endings of the olfactory nerve and transmit their impulses directly through the olfactory bulb to the cerebral cortex.

## Equilibrium (pp. 478–482)

1. The sense of equilibrium is provided by the vestibular apparatus of the inner ear; the vestibular apparatus includes the otolith organs (utricle and saccule) and the three semicircular canals.
   a. The sensory structures for equilibrium and hearing are contained within a membranous labyrinth filled with endolymph.
   b. The membranous labyrinth is contained within a bony cavity called the bony labyrinth, which is filled with perilymph.
2. Sensory hair cells within the fluid-filled membranous labyrinth provide the sense of equilibrium; in the otolith organs, the hair processes protrude into an otolith membrane.
3. In the semicircular canals, the hair processes are embedded in a gelatinous cupula, which is bent by movements of the head along the axis of each canal; bending of the hairs results in the production of nerve impulses.

## Hearing (pp. 482–487)

1. The outer ear consists of the auricle and the external auditory canal, and is bounded by the tympanic membrane.
2. The middle-ear cavity contains the malleus, incus, and stapes; the malleus is attached to the tympanic membrane, and the stapes is attached to the oval (vestibular) window of the cochlea.
3. Vibrations of the oval window in response to sound produces displacements of fluid within the cochlear duct, which is a part of the membranous labyrinth of the inner ear.

4. Displacements of fluid within the cochlea in response to sound cause the lower part of the cochlear duct, known as the basilar membrane, to vibrate.
   a. These vibrations cause hair processes of hair cells on this membrane to bend.
   b. The basilar membrane, hair cells, and the tectorial membrane into which the hair processes are embedded, together compose the spiral organ, or organ of Corti.
   c. Bending of hair in the spiral organ produces impulses that are transmitted to the brain in the eighth cranial nerve and interpreted as sound.
   d. The pitch of the sound is determined by the location of the stimulated hair cells on the basilar membrane.

## Vision (pp. 487–503)

1. Accessory structures of the eyes include the eyebrows, eyelids, eyelashes, conjunctiva, and lacrimal apparatus.
2. The fibrous tunic of the eyeball includes the white sclera and the transparent cornea, which admits light into the eye.
   a. The middle layer, or vascular tunic, consists of the choroid, ciliary body, and iris.
   b. The iris is composed of two layers of muscle that can dilate or constrict the pupil to admit more or less light into the eye.
3. The ciliary body contains muscles that, through attachments of the suspensory ligament to the lens, can alter the degree to which the lens is stretched and thus can vary the refractive power of the lens.
   a. Contraction of the ciliary muscles releases tension on the lens, allowing it to increase its convexity and focusing power as an object is brought closer to the eyes.
   b. Relaxation of the ciliary muscles places tension on the lens and flattens it, allowing an image to remain in focus as an object moves farther away from the eyes.
   c. The ability of the eye to maintain a focus on the retina as the distance to an object is varied is called accommodation.
4. Light passes through the lens and vitreous body to reach the inner layer of the eye—the retina.

## Interactions of the Sensory System with other Body Systems

### Integumentary System
- Protects the body from pathogens and helps to maintain body temperature
- Provides cutaneous sensations of touch, pressure, pain, and heat and cold

### Skeletal System
- Provides protection and support for the eye, ear, and other sensory organs
- Allows body movement
- Provides sensory information about joint movement and the tension of tendons

### Muscular System
- Cardiac and smooth muscles help deliver blood to sensory organs
- Skeletal muscles protect some sense organs
- Provides sensory information about the degree of stretch of muscles and the tension that muscles exert at their tendons

### Nervous System
- Conducts impulses from sensory organs, integrates that information within the central nervous system, and directs responses to the sensory information through activation of effector organs
- Responds to stimuli in the internal or external environment, transduces those stimuli into changes in membrane potential, and activates sensory neurons

### Endocrine System
- Hormones maintain metabolic health of sensory organs
- Sensory stimuli from the internal environment provide stimuli for secretion of appropriate hormones

### Circulatory System
- Delivers oxygen and nutrients to sensory organs and removes metabolic wastes
- Sensory stimuli from the heart and blood vessels provide information for neural regulation of the circulatory system

### Lymphatic System
- Protects against infections of the sensory organs
- Pain sensations may arise from swollen lymph nodes, alerting us to infection

### Respiratory System
- Provides oxygen for transport by the blood and provides for elimination of carbon dioxide from the blood
- Chemoreceptors in the aorta, carotid, and in the medulla oblongata provide sensory information for the regulation of breathing

### Urinary System
- Regulates the volume, pH, and electrolyte balance of the blood and eliminates wastes
- Stretch receptors in the renal tubules and in the atria of the heart provide information for the endocrine regulation of kidney function

### Digestive System
- Provides nutrients for the sensory organs
- Sensory stimuli provide information for neural and endocrine regulation of the digestive system

### Reproductive System
- Gonads produce sex hormones that influence sensations
- Provides sensory information for erection and orgasm, as well as for other aspects of reproductive functioning

5. The retina consists of a pigmented layer and a neural layer; the latter includes photoreceptor rods and cones.
   a. Light passes through a ganglion cell layer, then a bipolar cell layer, before reaching the photoreceptors in the retina.
   b. Impulses are conducted from the photoreceptors to bipolar neurons, then to ganglion neurons that provide axons to the optic nerve, which exits the eye at the optic disc.

6. Cones provide daylight color vision and are responsible for visual acuity.
   a. They are most concentrated in the fovea centralis, where they have 1:1 synapses with bipolar neurons.
   b. There are three types of cones, described by their maximal sensitivity as red, green, and blue.

7. Rods respond to dim light for black-and-white vision.
   a. They are stimulated by the photodissociation of rhodopsin, the retinene part of which is derived from vitamin A.
   b. Rods converge onto bipolar cells, thus providing high visual sensitivity at low levels of illumination.

8. Visual information is transmitted from the eyes to the lateral geniculate bodies and then to the striate cortex of the occipital lobes of the cerebral cortex.

# Review Activities

## Objective Questions

1. Which of the following cutaneous receptors is sensitive to deep pressure?
   a. root hair plexus
   b. lamellated corpuscle
   c. organ of Ruffini
   d. free nerve ending
2. Which of the following is an avascular ocular tissue?
   a. sclera
   b. cornea
   c. ciliary body
   d. iris
3. The middle ear is separated from the inner ear by
   a. the round window.
   b. the tympanic membrane.
   c. the oval window.
   d. both a and c.
   Match the vestibular organ on the left with its correct component on the right.
4. utricle and saccule      a. cupula
5. semicircular canals      b. ciliary body
6. cochlea                  c. basilar membrane
                            d. otolith membrane
7. The dissociation of rhodopsin in the rods in response to light causes
   a. the Na$^+$ channels to become blocked.
   b. the rods to secrete less neurotransmitter.
   c. the bipolar cells to become either stimulated or inhibited.
   d. all of the above.
8. Tonic receptors
   a. are fast-adapting.
   b. do not fire continuously to a sustained stimulus.
   c. produce action potentials at a greater frequency as the generator potential is increased.
   d. all of the above apply.

9. Cutaneous receptive fields are smallest in
   a. the fingertips.
   b. the back.
   c. the thighs.
   d. the arms.
10. The process of lateral inhibition
    a. increases the sensitivity of receptors.
    b. promotes sensory adaptation.
    c. increases sensory acuity.
    d. prevents adjacent receptors from being stimulated.
11. The receptors for taste are
    a. naked sensory nerve endings.
    b. encapsulated sensory nerve endings.
    c. modified epithelial cells.
12. The utricle and saccule
    a. are otolith organs.
    b. are located in the middle ear.
    c. provide a sense of linear acceleration.
    d. both a and c apply.
    e. both b and c apply.
13. Since fibers of the optic nerve that originate in the nasal halves of each retina cross at the optic chiasma, each lateral geniculate receives input from
    a. both the right and left sides of the visual field of both eyes.
    b. the ipsilateral visual field of both eyes.
    c. the contralateral visual field of both eyes.
    d. the ipsilateral field of one eye and the contralateral field of the other eye.
14. When a person with normal vision views an object from a distance of at least 20 feet
    a. the ciliary muscles are relaxed.
    b. the suspensory ligament is taut.
    c. the lens is in its least convex shape.
    d. all of the above apply.

15. Glasses with concave lenses help correct
    a. presbyopia.
    b. myopia.
    c. hyperopia.
    d. astigmatism.
16. Parasympathetic nerves that stimulate constriction of the iris (in the pupillary reflex) are activated by neurons in
    a. the lateral geniculate.
    b. the superior colliculus.
    c. the inferior colliculus.
    d. the striate cortex.
17. The ability of the lens to increase its curvature and maintain a focus at close distances is called
    a. convergence.
    b. accommodation.
    c. astigmatism.
    d. amblyopia.
18. Which of the following sensory modalities is transmitted directly to the cerebral cortex, without being relayed through the thalamus?
    a. taste
    b. sight
    c. smell
    d. hearing
    e. touch

## Essay Questions

1. Define the term *lateral inhibition* and give examples of its effects in three sensory systems.
2. Describe the nature of the generator potential and explain how it relates to stimulus intensity and to frequency of action potential production.

3. Draw a muscle spindle within a muscle and show the sensory and motor innervations. Explain the steps involved in a knee-jerk reflex and then compare the functions of alpha and gamma motoneurons.

4. Diagram the structure of the eyeball and label the sclera, cornea, choroid, macula lutea, ciliary body, suspensory ligament, lens, iris, pupil, retina, optic disc, and fovea centralis.

5. Diagram the ear and label the structures of the outer, middle, and inner ear.

6. Trace a sound wave through the structures of the ear and explain how the mechanism of hearing works.

7. Explain the mechanism by which equilibrium is maintained and the role played by the two kinds of receptor information.

8. Define *accommodation* and explain how it is accomplished. Why is it more of a strain on the eyes to look at a small object that is close to the eyes than at large distant objects?

9. Describe the effects of light on the photoreceptors and explain how these effects influence the bipolar cells.

10. Explain why images that fall on the fovea centralis are seen more clearly than images that fall on the periphery of the retina. Why are the "corners of the eyes" more sensitive to light than the fovea?

## Gundy/Weber Software 💾

The tutorial software accompanying Chapter 18 is Volume 6—Special Senses.

# endocrine system

## objectives

- Describe the nature of endocrine glands and the chemical classification of hormones.
- Explain how hormones may act as neurotransmitters and describe the interactions that can occur between different hormones.
- Describe the structure of the pituitary gland and identify the hormones of the adenohypophysis and neurohypophysis.
- Describe the actions of the anterior pituitary hormones.
- Explain how the secretions of the posterior pituitary and anterior pituitary are regulated by the hypothalamus.
- Explain how the secretions of the hypothalamus and anterior pituitary are regulated by negative feedback inhibition.
- Relate the embryonic origin of the adrenal medulla to the hormones it secretes and the regulation of its secretions.
- State the different categories of corticosteroids and the part of the adrenal cortex that secretes each. Explain how the secretion of these hormones is regulated.
- Explain how the secretions of the adrenal cortex are related to stress.

- Describe the structure of the thyroid, the formation of thyroxine and triiodothyronine, and the regulation of thyroid function.
- Describe the location of the parathyroid glands and discuss their significance.
- Describe the structure of the pancreatic islets and the actions of insulin and glucagon.
- Describe the location and endocrine function of the pineal and thymus glands.
- Identify the hormones of the gonads and placenta.
- Describe the mechanism of action of steroid hormones and thyroxine.
- Describe the mechanism of action of hormones that use cAMP as a second messenger and explain how $Ca^{++}$ can act as a second messenger.
- Discuss the concept of prehormones and the influence of hormone concentration on tissue responsiveness.
- Explain why regulation by prostaglandins is called autocrine rather than endocrine.
- Describe the effects of different prostaglandins and explain how nonsteroidal anti-inflammatory drugs work.

# Endocrine Glands and Hormones

*Hormones are regulatory molecules secreted into the blood by endocrine glands. Categories of hormones include steroids, catecholamines, and polypeptides. Interactions among the various hormones produce effects described as synergistic, permissive, or antagonistic.*

In contrast to exocrine glands whose secretions are transported through ducts to their respective destinations, **endocrine glands** are ductless. The endocrine glands secrete biologically active chemicals called **hormones** directly into the blood. Many endocrine glands are discrete organs (fig. 19.1) whose primary functions are the production and secretion of hormones. The pancreas functions as both an exocrine and an endocrine gland; the endocrine portion of the pancreas is composed of microscopic structures called the pancreatic islets, or islets of Langerhans (fig. 19.1b). The concept of the endocrine system, however, must be extended beyond these organs. In recent years, it has been discovered that many other organs in the body secrete hormones. When these hormones can be demonstrated to have

endocrine: Gk. *endon*, within; *krinein*, to separate
hormone: Gk. *hormon*, to set in motion

significant physiological functions, the organs that produce them may be categorized as endocrine glands, although they serve other functions as well. A partial list of the endocrine glands (table 19.1) should thus include the heart, liver, hypothalamus, and kidneys.

Hormones affect the metabolism of their target organs and, by this means, help regulate (1) total body metabolism, (2) growth, and (3) reproduction. The effects of hormones on body metabolism and growth are discussed in chapter 27; the regulation of reproductive functions by hormones is included in chapters 28, 29, and 30.

## Chemical Classification of Hormones

Hormones secreted by different endocrine glands vary with respect to chemical structure. All hormones, however, can be grouped into three broad chemical categories: (1) **catecholamines** (epinephrine and norepinephrine); (2) **polypeptides** and **glycoproteins,** including shorter chain polypeptides such as antidiuretic hormone and insulin and large glycoproteins such as thyroid-stimulating hormone (table 19.2); and (3) **steroids,** such as cortisol and testosterone.

Steroid hormones, which are derived from cholesterol (fig. 19.2), are lipids and thus are not water-soluble. The gonads—testes and ovaries—secrete *sex steroids*; the adrenal cortex secretes *corticosteroids*, including cortisol and aldosterone among others.

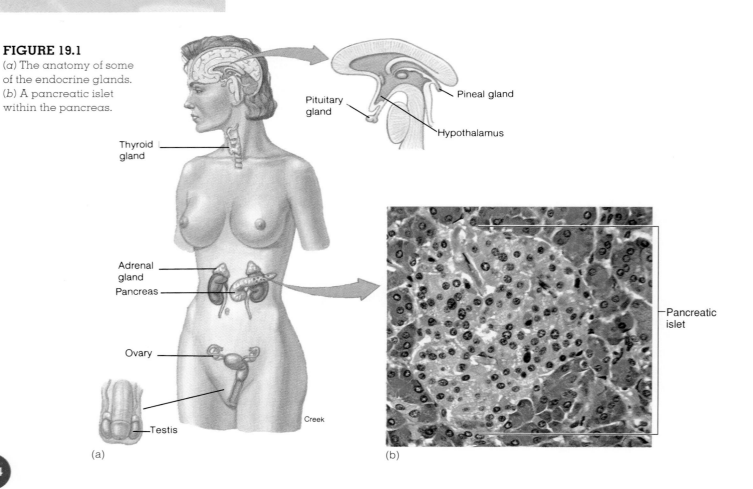

**FIGURE 19.1**
(*a*) The anatomy of some of the endocrine glands. (*b*) A pancreatic islet within the pancreas.

Pituitary gland

Pineal gland

Hypothalamus

Thyroid gland

Adrenal gland

Pancreas

Ovary

Testis

Creek

Pancreatic islet

(a)

(b)

# Table 19.1 A partial list of the endocrine glands

| Endocrine gland | Major hormones | Primary target organs | Primary effects |
|---|---|---|---|
| Adrenal cortex | Glucocorticoids<br>Aldosterone | Liver, muscles<br>Kidneys | Glucocorticoids influence glucose metabolism; aldosterone promotes Na$^+$ retention, K$^+$ excretion |
| Adrenal medulla | Epinephrine | Heart, bronchioles, blood vessels | Causes adrenergic stimulation |
| Heart | Atrial natriuretic hormone | Kidneys | Promotes excretion of Na$^+$ in the urine |
| Hypothalamus | Releasing and inhibiting hormones | Anterior pituitary | Regulates secretion of anterior pituitary hormones |
| Small intestine | Secretin and cholecystokinin | Stomach, liver, and pancreas | Inhibits gastric motility and stimulate bile and pancreatic juice secretion |
| Pancreatic islets | Insulin<br>Glucagon | Many organs<br>Liver and adipose tissue | Insulin promotes cellular uptake of glucose and formation of glycogen and fat; glucagon stimulates hydrolysis of glycogen and fat |
| Kidneys | Erythropoietin | Bone marrow | Stimulates red blood cell production |
| Liver | Somatomedins | Cartilage | Stimulates cell division and growth |
| Ovaries | Estradiol-17β and progesterone | Female reproductive tract and mammary glands | Maintains structure of reproductive tract and promotes secondary sex characteristics |
| Parathyroid glands | Parathyroid hormone | Bone, small intestine, and kidneys | Increases Ca$^{++}$ concentration in blood |
| Pineal gland | Melatonin | Hypothalamus and anterior pituitary | Affects secretion of gonadotrophic hormones |
| Pituitary, anterior | Trophic hormones | Endocrine glands and other organs | Stimulates growth and development of target organs; stimulates secretion of other hormones |
| Pituitary, posterior | Antidiuretic hormone<br>Oxytocin | Kidneys, blood vessels<br>Uterus, mammary glands | Antidiuretic hormone promotes water retention and vasoconstriction; oxytocin stimulates contraction of uterus and mammary secretory units |
| Skin | 1,25-Dihydroxyvitamin D$_3$ | Small intestine | Stimulates absorption of Ca$^{++}$ |
| Stomach | Gastrin | Stomach | Stimulates acid secretion |
| Testes | Testosterone | Prostate, seminal vesicles, other organs | Stimulates secondary sexual development |
| Thymus | Thymosin | Lymph nodes | Stimulates white blood cell production |
| Thyroid gland | Thyroxine (T$_4$) and triiodothyronine (T$_3$) | Most organs | Promotes growth and development and stimulates basal rate of cell respiration (basal metabolic rate or BMR) |

## Table 19.2 Some examples of polypeptide and glycoprotein hormones

| Hormone | Structure | Gland | Primary effects |
| --- | --- | --- | --- |
| Antidiuretic hormone | 8 amino acids | Posterior pituitary | Water retention and vasoconstriction |
| Oxytocin | 8 amino acids | Posterior pituitary | Uterine and mammary contraction |
| Insulin | 21 and 30 amino acids (double chain) | β cells in pancreatic islets | Cellular glucose uptake, lipogenesis, and glycogenesis |
| Glucagon | 29 amino acids | α cells in pancreatic islets | Hydrolysis of stored glycogen and fat |
| ACTH | 39 amino acids | Anterior pituitary | Stimulation of adrenal cortex |
| Parathyroid hormone | 84 amino acids | Parathyroid glands | Increase in blood $Ca^{++}$ concentration |
| FSH, LH, TSH | Glycoproteins | Anterior pituitary | Stimulation of growth, development, and secretion of target glands |

Each of the two major thyroid hormones is composed of two derivatives of the amino acid tyrosine bonded together, and both of these hormones contain iodine, as shown in figure 19.3. When the hormone contains four iodine atoms, it is called *tetraiodothyronine* ($T_4$), or *thyroxine*, when it contains three atoms of iodine, it is called *triiodothyronine* ($T_3$). Although these hormones are not steroids, they are like steroids in that they are relatively small, nonpolar molecules. Steroid and thyroid hormones are active when taken orally (as a pill). Sex steroids are administered as contraceptive pills and thyroid hormone pills are taken by people whose thyroid is deficient (who are hypothyroid). Other types of hormones cannot be taken orally because they would be digested into inactive fragments before being absorbed into the blood.

## Prohormones and Prehormones

Hormone molecules that affect the metabolism of target cells are often derived from less active "parent," or *precursor, molecules*. In the case of polypeptide hormones, the precursor may be a longer chained **prohormone** that is cut and spliced together to make the hormone. Insulin, for example, is produced from *proinsulin* within the endocrine beta cells of the pancreatic islets. Moreover, the prohormone itself may be derived from an even larger precursor molecule; in the case of insulin, this molecule is called *pre-proinsulin*. The term *prehormone* is commonly used to indicate such precursors of prohormones.

In some cases, the molecules secreted by the endocrine gland (and considered to be the hormone of that gland) are actually inactive in the target cells. In order to become active, the target cells must modify the chemical structure of the secreted hormone. Thyroxine ($T_4$), for example, must be changed into $T_3$ within the target cells to affect the metabolism of the target cells. Similarly, testosterone (secreted by the testes) and vitamin $D_3$ (secreted by the skin) are con-

verted into more active molecules within their target cells (table 19.3). In this text, we will use the term **prehormone** to designate those molecules secreted by an endocrine gland that are inactive until they are converted by their target cells into the active forms.

## Common Aspects of Neural and Endocrine Regulation

One might conclude that since endocrine regulation is chemical in nature, it therefore is fundamentally different from neural control systems. This assumption is incorrect. As described in chapter 14, nerve impulses are chemical events produced by the diffusion of ions through the neuron cell membrane. Interestingly, the action of some hormones (such as insulin) is accompanied by ion diffusion and electrical changes in the target cells, so changes in membrane potential are not unique to the nervous system. Also, most neurons stimulate the cells they innervate through the release of a chemical neurotransmitter. Neurotransmitters do not travel in the blood as do hormones; instead, they diffuse only a very short distance across a synapse. In other respects, however, the actions of neurotransmitters and hormones are very similar.

Indeed, many polypeptide hormones, including those secreted by the pituitary gland and by the gastrointestinal tract, have been discovered in the brain. In certain locations in the brain, some of these compounds are produced and secreted as hormones. In other brain locations, some of these compounds apparently serve as neurotransmitters. The discovery of some of these polypeptides in unicellular organisms, which of course lack a nervous and endocrine system, suggests that these regulatory molecules appeared early in evolution and were incorporated into the function of nervous and endocrine tissue as these systems evolved.

Regardless of whether a particular chemical is acting as a neurotransmitter or as a hormone, certain conditions

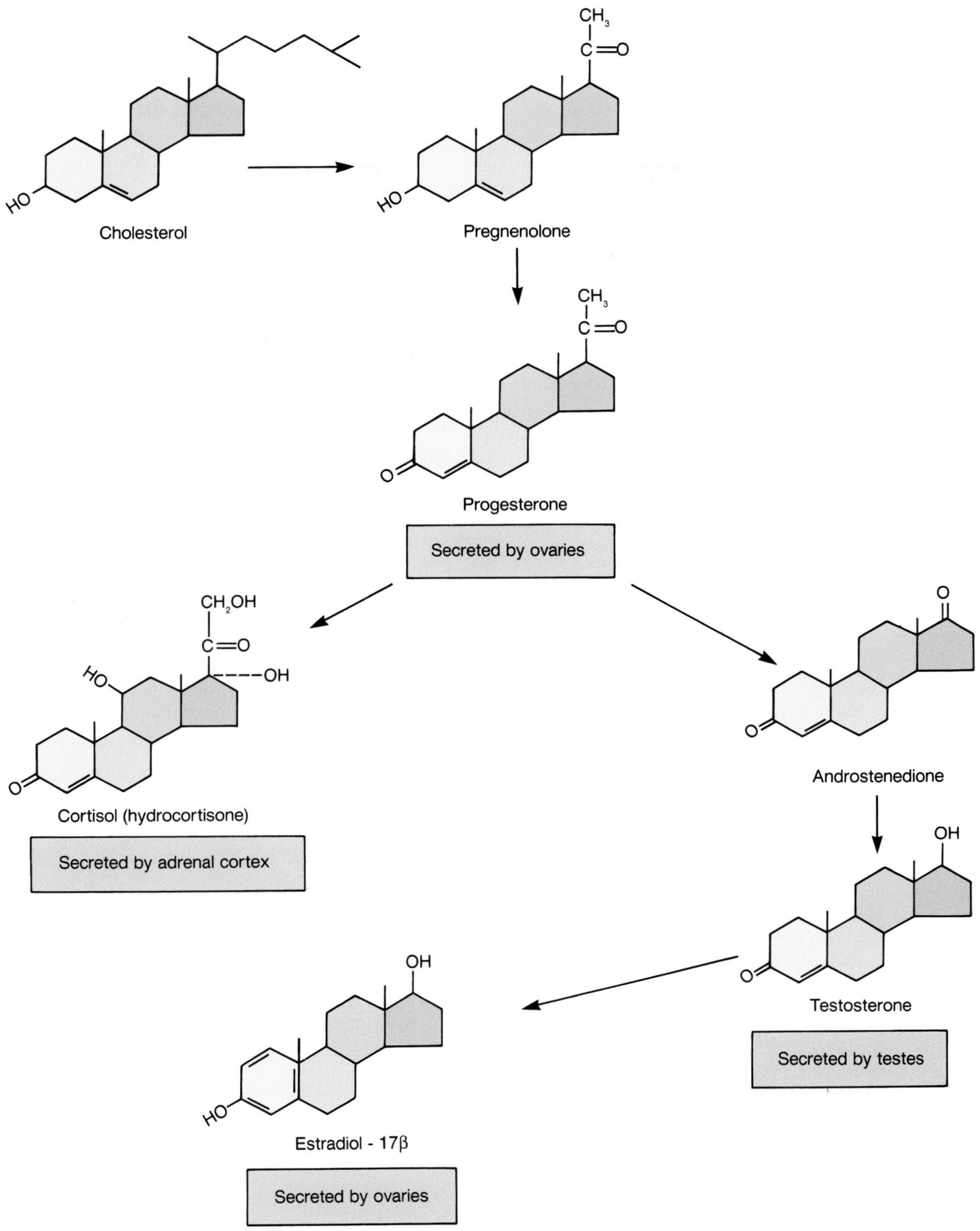

**FIGURE 19.2**

Simplified biosynthetic pathways for steroid hormones. Notice that progesterone (a hormone secreted by the ovaries) is a common precursor in the formation of all other steroid hormones and that testosterone (the major androgen secreted by the testes) is a precursor in the formation of estradiol-17β, the major estrogen secreted by the ovaries.

must apply in order for it to function in physiological regulation: (1) target cells must have specific **receptor proteins** that combine with the chemical; (2) the combination of the regulator molecule with its receptor proteins must cause a specific sequence of changes in the target cells; and (3) there must be a mechanism to quickly turn off the action of the regulator. This mechanism, which involves rapid removal and/or chemical inactivation of the regulator molecules, is essential because without an "off-switch," physiological control would be impossible.

## Hormone Interactions

A given target tissue is usually responsive to a number of different hormones. These hormones may antagonize each other or work together to produce effects that are additive or complementary. The responsiveness of a target tissue to a particular hormone is thus affected not only by the concentration of that hormone, but also by the effects of other hormones on that tissue. Terms used to describe hormone interactions include *synergistic*, *permissive*, and *antagonistic*.

**Synergistic and Permissive Effects**   Synergistic effects occur when two or more hormones work together to produce a particular result. These effects may be *additive* or *complementary*. The action of epinephrine and norepinephrine on the heart is a good example of an additive effect. Each of these hormones, acting alone, produces an increase in cardiac rate; acting together in the same concentrations, they stimulate an even greater increase in cardiac rate. The synergistic action of FSH and testosterone is an example of a complementary effect; each hormone separately stimulates a different stage of spermatogenesis during puberty, so that both hormones together are needed at that time to complete sperm development. Likewise, the ability of mammary glands to produce and secrete milk requires the synergistic action of many hormones—estrogen, cortisol, prolactin, oxytocin, and others.

A hormone is said to have a **permissive effect** on the action of a second hormone when it enhances the responsiveness of a target organ to the second hormone or when it increases the activity of the second hormone. Prior exposure of the uterus to estrogen, for example, induces the formation of receptor proteins for progesterone, which improves the response of the uterus when it is subsequently exposed to progesterone. Estrogen thus has a permissive effect on the responsiveness of the uterus to progesterone. Glucocorticoids (a class of corticosteroids including cortisol) exert permissive effects on the actions of catecholamines (epinephrine and norepinephrine). In the absence of these permissive effects due to abnormally low glucocorticoids, the catecholamines will not be as effective as they are normally. One symptom of this condition may be an abnormally low blood pressure.

Vitamin $D_3$ is a prehormone that must first be converted by enzymes in the kidneys and liver, where two hydroxyl (OH) groups are added to form the active hormone

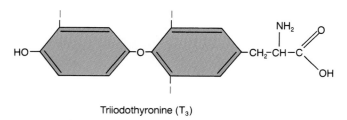

**FIGURE 19.3**

The thyroid hormones thyroxine ($T_4$) and triiodothyronine ($T_3$) are secreted in a ratio of 9:1.

| Table 19.3 | Conversion of prehormones into biologically active derivatives | | |
|---|---|---|---|
| **Endocrine gland** | **Prehormone** | **Active products** | **Comments** |
| Skin | Vitamin $D_3$ | 1,25-Dihydroxyvitamin $D_3$ | Hydroxylation reactions occur in the liver and kidneys. |
| Testes | Testosterone | Dihydrotestosterone (DHT) | DHT and other 5α-reduced androgens are formed in most androgen-dependent tissue. |
| | | Estradiol-17β ($E_2$) | $E_2$ is formed in the brain from testosterone, where it is believed to affect both endocrine function and behavior; small amounts of $E_2$ are also produced in the testes. |
| Thyroid gland | Thyroxine ($T_4$) | Triiodothyronine ($T_3$) | Conversion of $T_4$ to $T_3$ occurs in almost all tissues. |

1,25-dihydroxyvitamin $D_3$. This hormone helps to raise blood calcium levels. Parathyroid hormone (PTH) has a permissive effect on the actions of vitamin $D_3$ because it stimulates the production of the hydroxylating enzymes in the kidneys and liver. By this means, an increased secretion of PTH has a permissive effect on the ability of vitamin $D_3$ to stimulate the intestinal absorption of calcium.

**Antagonistic Effects** In some situations, the actions of one hormone antagonize the effects of another. Lactation during pregnancy, for example, is prevented because the high concentration of estrogen in the blood inhibits the secretion and action of prolactin. Another example of antagonism is the action of insulin and glucagon (two hormones from the pancreatic islets) on adipose tissue. The formation of fat is promoted by insulin, whereas glucagon promotes fat breakdown.

# Pituitary Gland

*The pituitary gland is structurally and functionally divided into the anterior pituitary and the posterior pituitary. The posterior pituitary secretes hormones that are actually produced by the hypothalamus, whereas the anterior pituitary produces and secretes its own hormones. The anterior pituitary is under the control of the hypothalamus, however, by way of releasing hormones secreted by the hypothalamus into a special system of blood vessels. Feedback influences from many glands usually act to inhibit the secretions of the anterior pituitary.*

## Structure of the Pituitary Gland

The **pituitary gland,** or **hypophysis** (*hi-pof´ĭ-sis*), is located on the inferior aspect of the brain in the region of the diencephalon, as described in chapter 15. It is a rounded, pea-shaped gland about 1.3 cm (0.5 in.) in diameter and is attached to the hypothalamus by a stalklike structure called the *infundibulum* (*in´fun-dib´yŭ-lum*) (fig. 19.4). The pituitary gland is structurally and functionally divided into an anterior lobe, or **adenohypophysis** (*ad´n-o-hi-pof´ĭ-sis*), and a posterior lobe called the **neurohypophysis.**

The adenohypophysis consists of three parts: (1) the **pars distalis** is the rounded portion, and is the major endocrine part of the gland; (2) the **pars tuberalis** is the thin extension in contact with the infundibulum; and (3) the **pars intermedia** is located between the anterior and posterior parts of the pituitary.

.........................................

pituitary: L. *pituita*, phlegm (this gland was originally thought to secrete mucus into the nasal cavity)
infundibulum: L. *infundibulum,* a funnel
adenohypophysis: Gk. *adeno,* gland; *hypo,* under; *physis,* a growing

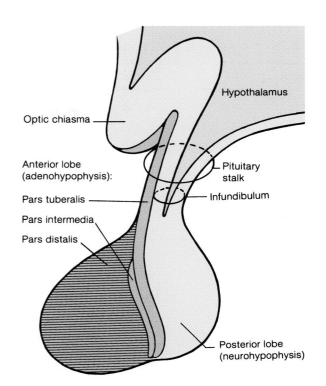

**FIGURE 19.4**
The structure of the pituitary gland as seen in the sagittal view.

The neurohypophysis is the neural part of the pituitary gland. It consists of the **pars nervosa,** which is in contact with the pars intermedia of the adenohypophysis, and the **infundibulum.** Nerve fibers extend through the infundibulum, along with small neuroglia-like cells called **pituicytes** (*pĭ-too´ĭ-sītz*).

## Pituitary Hormones

The hormones secreted by the **anterior pituitary** (the pars distalis of the adenohypophysis) are called **trophic hormones.** The term *trophic* means "food." Although the anterior pituitary hormones are not food for their target organs, this term is used because high levels of the anterior pituitary hormones make their target organs hypertrophy, whereas low amounts cause their target organs to atrophy. In designating the hormones of the anterior pituitary, therefore, "trophic" (conventionally shortened to *tropic,* meaning "attracted to") is incorporated into the names. Also, the shortened forms of the names for the anterior pituitary hormones end in the suffix *-tropin.*

The hormones of the pars distalis (table 19.4) are as follows:

**1 Growth hormone (GH,** or **somatotropin).** This hormone promotes the movement of amino acids into tissue cells and the incorporation of these amino acids into tissue proteins, thus stimulating growth of organs.

# Table 19.4  Anterior pituitary hormones

| Hormone | Target tissue | Stimulated by hormone | Regulation of secretion |
|---|---|---|---|
| ACTH (adrenocorticotropic hormone) | Adrenal cortex | Secretion of glucocorticoids | Stimulated by CRH (corticotropin-releasing hormone); inhibited by glucocorticoids |
| TSH (thyroid-stimulating hormone) | Thyroid gland | Secretion of thyroid hormones | Stimulated by TRH (thyrotropin-releasing hormone); inhibited by thyroid hormones |
| GH (growth hormone) | Most tissue | Protein synthesis and growth; lipolysis and increased blood glucose | Inhibited by somatostatin; stimulated by growth hormone-releasing hormone |
| FSH (follicle-stimulating hormone) and LH (luteinizing hormone) | Gonads | Gamete production and sex steroid hormone secretion | Stimulated by GnRH (gonadotropin-releasing hormone); inhibited by sex steroids |
| Prolactin | Mammary glands and other accessory sex organs | Milk production; controversial actions in other organs | Inhibited by PIH (prolactin-inhibiting hormone) |
| LH (luteinizing hormone) | Gonads | Sex hormone secretion; ovulation and corpus luteum formation | Stimulated by GnRH |

**2** **Thyroid-stimulating hormone** (**TSH,** or **thyrotropin**). This hormone stimulates the thyroid gland to produce and secrete thyroxine (tetraiodothyronine, or T$_4$).

**3** **Adrenocorticotropic hormone** (**ACTH,** or **corticotropin**). This hormone stimulates the adrenal cortex to secrete the glucocorticoids, such as hydrocortisone (cortisol).

**4** **Follicle-stimulating hormone** (**FSH,** or **folliculotropin**). This hormone stimulates the growth and secretion of ovarian follicles in females and the production of sperm in the testes of males.

**5** **Luteinizing hormone** (**LH,** or **luteotropin**). This hormone and FSH are collectively called **gonadotropic** (*go-nad˝ŏ-trop´ik*) **hormones.** In females, LH stimulates ovulation and the conversion of the ovulated ovarian follicle into an endocrine structure called a corpus luteum. In males, LH (which is sometimes also called *interstitial-cell-stimulating hormone,* or *ICSH*) stimulates the secretion of male sex hormones (mainly testosterone) from the interstitial cells in the testes.

**6** **Prolactin.** This hormone is secreted in both males and females. Its best known function is the stimulation of milk production by the mammary glands of women after parturition. Prolactin plays a supporting role in the regulation of the male reproductive system by the gonadotropins (FSH and LH) and acts on the kidneys to help regulate water and electrolyte balance.

The pars intermedia of the adenohypophysis in an adult human is not well defined, and its function, if any, is poorly understood. It produces different forms of **melanocyte-stimulating hormone (MSH),** which in lower vertebrates

(fish, amphibians, and reptiles) cause a darkening of the skin to provide camouflage against a dark background. This reaction involves a cell type that is absent in human skin. Although the exogenous administration of MSH to humans stimulates melanin pigment production in the melanocytes of skin, MSH does not appear to have this effect in normal physiology. It is interesting that ACTH, secreted from the pars distalis, contains the amino acid sequence of MSH as part of its structure, and abnormally high amounts of ACTH secretion (as in Addison's disease) cause a darkening of the skin. The pars intermedia also produces large amounts of β-endorphin (chapter 14), as do other structures, but the physiological significance of this is not understood.

The **posterior pituitary,** or pars nervosa, secretes only two hormones, both of which are produced in the hypothalamus and merely stored in the posterior lobe of the pituitary:

**1** **Antidiuretic** (*an˝te-di˝yŭ-ret´ik*) **hormone** (**ADH,** or **vasopressin**). Antidiuretic hormone stimulates the kidneys to retain water so that less water is excreted in the urine and more water is retained in the blood. This hormone also causes vasoconstriction in experimental animals, although the significance of this effect in humans is controversial.

**2** **Oxytocin.** In females, oxytocin stimulates contractions of the uterus during labor and for this reason is needed for *parturition* (childbirth). Oxytocin also stimulates contractions of the mammary gland alveoli and ducts, which result in the milk-ejection reflex in a lactating woman. In men, a rise in oxytocin at the time of ejaculation has been measured, but the physiological significance of this hormone in males remains to be demonstrated.

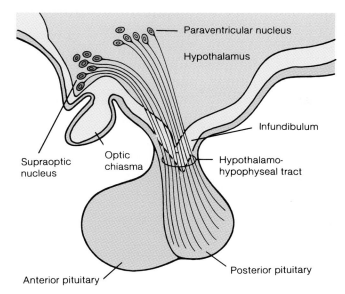

**FIGURE 19.5**

The posterior pituitary, or neurohypophysis, stores and secretes hormones (vasopressin and oxytocin) produced in neuron cell bodies within the supraoptic and paraventricular nuclei of the hypothalamus. These hormones are transported to the posterior pituitary by nerve fibers of the hypothalamo-hypophyseal tract.

 Injections of oxytocin may be given to a woman to induce labor if the pregnancy is overdue or if the fetal membranes have ruptured and there is a danger of infection. Labor may also be induced by injections of oxytocin if there is pregnancy-induced hypertension (*preeclampsia*). Oxytocin is commonly administered after delivery, when, by stimulating contractions of the uterine muscle, it promotes the regression in the size of the uterus and squeezes the blood vessels, thus minimizing the danger of hemorrhage.

## Hypothalamic Control of the Neurohypophysis

Although the posterior pituitary (pars nervosa of the neurohypophysis) secretes two hormones—antidiuretic hormone (ADH) and oxytocin—these hormones are actually produced in neuron cell bodies of the *supraoptic nuclei* and *paraventricular nuclei* of the hypothalamus. These nuclei within the hypothalamus are endocrine glands; the hormones they produce are transported along axons of the **hypothalamo-hypophyseal tract** (fig. 19.5) to the posterior pituitary, which stores and later releases these hormones. The posterior pituitary is thus more a storage organ than a true gland.

The secretion of ADH and oxytocin from the posterior pituitary is controlled by **neuroendocrine reflexes.** In nursing mothers, for example, the stimulus of sucking acts via sensory nerve impulses to the hypothalamus to stimulate the reflex secretion of oxytocin. The secretion of ADH is stimulated by osmoreceptor neurons in the hypothalamus in response to a rise in blood osmotic pressure (chapter 5); its secretion is in-hibited by sensory impulses from stretch receptors in the heart in response to a rise in blood volume. These reflexes are discussed in more detail in later chapters.

## Hypothalamic Control of the Adenohypophysis

At one time, the anterior pituitary was called the "master gland" because it secretes hormones that regulate some other endocrine glands (fig. 19.6 and table 19.4). Adrenocorticotropic hormone (ACTH), thyroid-stimulating hormone (TSH), and the gonadotropic hormones (FSH and LH) stimulate the adrenal cortex, thyroid, and gonads, respectively, to secrete their hormones. The anterior pituitary hormones also have a "trophic" effect on their target glands in that the structure and health of the target glands depend on adequate stimulation by anterior pituitary hormones. The anterior pituitary, however, is not really the master gland, since secretion of its hormones is in turn controlled by hormones secreted by the hypothalamus.

**Releasing and Inhibiting Hormones**   Since axons do not enter the anterior pituitary, hypothalamic control of the anterior pituitary is achieved through hormonal rather than neural regulation. Neurons in the hypothalamus produce releasing and inhibiting hormones that are transported to axon endings in the basal portion of the hypothalamus. This region, known as the **median eminence,** contains blood capillaries that are drained by venules in the stalk of the pituitary.

The venules that drain the median eminence deliver blood to a second capillary bed in the anterior pituitary. Since this second capillary bed receives venous blood from the first (is located "downstream" from the first), there is a vascular link between the median eminence and anterior pituitary. This infrequently occurring arrangement of blood vessels (two capillary beds in series) is described as a *portal system* (see chapter 21). The vascular link between the hypothalamus and the anterior pituitary is thus called the **hypothalamo-hypophyseal portal system.**

Polypeptide hormones are secreted into the hypothalamo-hypophyseal portal system by neurons of the hypothalamus. These hormones regulate the secretions of the anterior pituitary (fig. 19.7 and table 19.5). Thyrotropin-releasing hormone **(TRH)** stimulates the secretion of TSH; corticotropin-releasing hormone **(CRH)** stimulates the secretion of ACTH from the anterior pituitary. A single releasing hormone, gonadotropin-releasing hormone, or **GnRH,** appears to stimulate the secretion of both gonadotropic hormones (FSH and LH) from the anterior pituitary. The secretion of prolactin and of growth hormone from the anterior pituitary is regulated by hypothalamic inhibitory hormones, known as **PIH** (prolactin-inhibiting hormone) and **somatostatin,** respectively.

Recently, a specific **growth hormone–releasing hormone (GHRH)** that stimulates growth hormone secretion

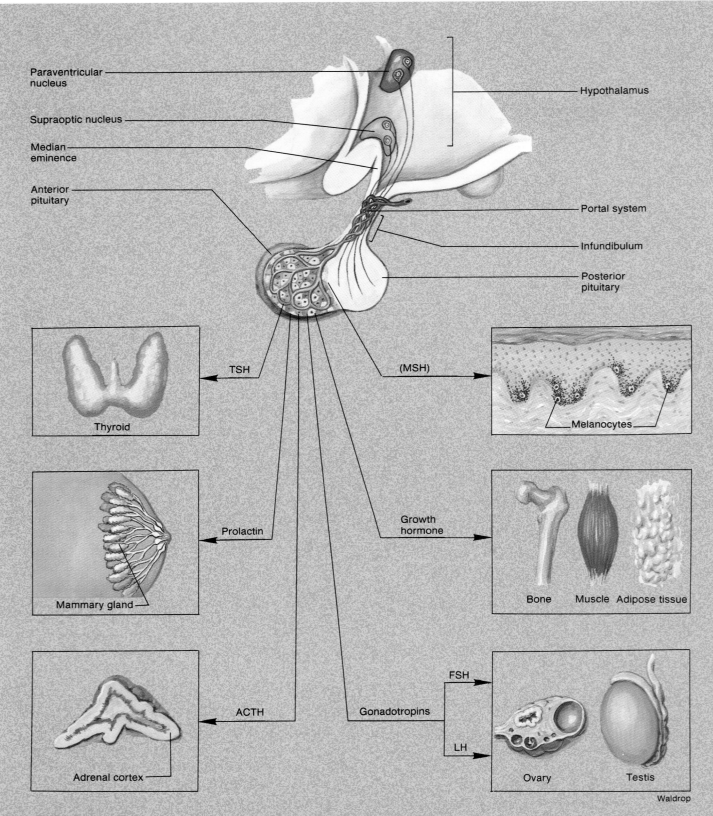

Paraventricular nucleus

Supraoptic nucleus

Median eminence

Anterior pituitary

Hypothalamus

Portal system

Infundibulum

Posterior pituitary

TSH

Thyroid

(MSH)

Melanocytes

Prolactin

Mammary gland

Growth hormone

Bone    Muscle  Adipose tissue

ACTH

Adrenal cortex

Gonadotropins

FSH

LH

Ovary    Testis

Waldrop

## FIGURE 19.6

The hormones secreted by the anterior pituitary and the target organs for these hormones.

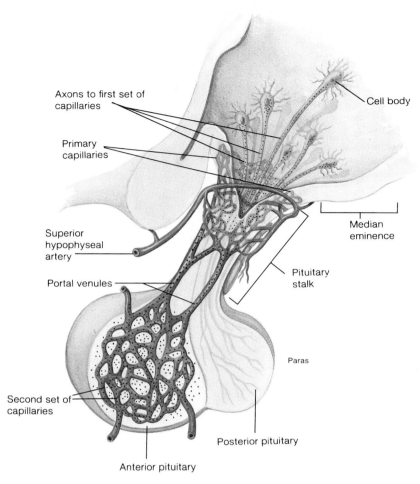

**FIGURE 19.7**

Neurons in the hypothalamus secrete releasing hormones (shown as dots) into the blood vessels of the hypothalamo-hypophyseal portal system. These releasing hormones stimulate the anterior pituitary to secrete its hormones into the general circulation.

*Image labels:* Axons to first set of capillaries; Cell body; Primary capillaries; Median eminence; Superior hypophyseal artery; Pituitary stalk; Portal venules; Paras; Second set of capillaries; Posterior pituitary; Anterior pituitary

has been identified as a polypeptide consisting of 44 amino acids. Experiments suggest that a releasing hormone for prolactin may also exist, but no such specific releasing hormone has yet been discovered.

## Feedback Control of the Adenohypophysis

In view of its secretion of releasing and inhibiting hormones, the hypothalamus might be considered the "master gland." The chain of command, however, is not linear; the hypothalamus and anterior pituitary are controlled by the effects of their own actions. In the endocrine system, by way of analogy, the private takes orders from the general, but the general also takes orders from the private. The hypothalamus and anterior pituitary are not master glands because their secretions are controlled by the target glands they regulate.

Anterior pituitary secretion of ACTH, TSH, and the gonadotropins (FSH and LH) is controlled by **negative feedback inhibition** from the target gland hormones. Secretion of ACTH is inhibited by a rise in corticosteroid secretion, for example, and TSH is inhibited by a rise in the secretion of thyroxine from the thyroid. These negative feedback relationships are easily demonstrated by removal of the target glands. Castration (surgical removal of the gonads), for example, produces a rise in the secretion of FSH and LH. In a similar manner, removal of the adrenals or the thyroid results in an abnormal increase in ACTH or TSH secretion from the anterior pituitary.

## Table 19.5   Hypothalamic hormones involved in the control of the anterior pituitary

| Hypothalamic hormone | Structure | Effect on anterior pituitary | Action of anterior pituitary hormone |
|---|---|---|---|
| Corticotropin-releasing hormone (CRH) | 41 amino acids | Stimulates secretion of adrenocorticotropic hormone (ACTH) | Stimulates secretions of adrenal cortex |
| Gonadotropin-releasing hormone (GnRH) | 10 amino acids | Stimulates secretion of follicle-stimulating hormone (FSH) and luteinizing hormone (LH) | Stimulates gonads to produce gametes (sperm and ova) and secretes sex steroids |
| Prolactin-inhibiting hormone (PIH) | Dopamine | Inhibits prolactin secretion | Stimulates production of milk in mammary glands |
| Somatostatin | 14 amino acids | Inhibits secretion of growth hormone | Stimulates anabolism and growth in many organs |
| Thyrotropin-releasing hormone (TRH) | 3 amino acids | Stimulates secretion of thyroid-stimulating hormone (TSH) | Stimulates secretion of thyroid gland |
| Growth hormone–releasing hormone (GHRH) | 44 amino acids | Stimulates growth hormone secretion | Stimulates anabolism and growth in many organs |

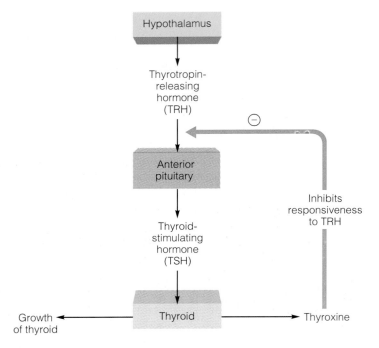

**FIGURE 19.8**

The secretion of thyroxine from the thyroid is stimulated by the thyroid-stimulating hormone (TSH) from the anterior pituitary. The secretion of TSH is stimulated by the thyrotropin-releasing hormone (TRH) secreted from the hypothalamus into the hypothalamo-hypophyseal portal system. This stimulation is balanced by the negative feedback inhibition of thyroxine, which decreases the responsiveness of the anterior pituitary to stimulation by TRH.

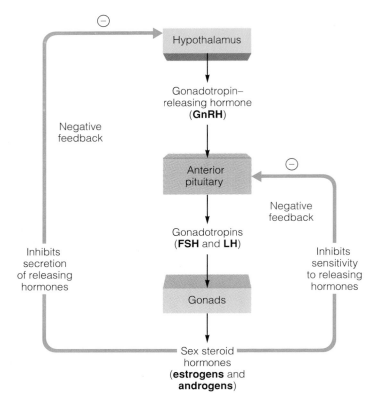

**FIGURE 19.9**

Negative feedback control of gonadotropin secretion.

The effects of removing the target glands demonstrate that, under normal conditions, these glands exert an inhibitory effect on the anterior pituitary. This inhibitory effect can occur at either of two levels: (1) the target gland hormones can act on the hypothalamus and inhibit the secretion of releasing hormones or (2) the target gland hormones can act on the anterior pituitary and inhibit its response to the releasing hormones. Thyroxine, for example, appears to inhibit the response of the anterior pituitary to TRH and thus acts to reduce TSH secretion (fig. 19.8). Sex steroids, by contrast, reduce the secretion of gonadotropins by inhibiting both GnRH secretion and the ability of the anterior pituitary to respond to stimulation by GnRH (fig. 19.9).

In addition to negative feedback control of the anterior pituitary, there is an example of a hormone from a target organ whose action actually stimulates the secretion of an anterior pituitary hormone. Toward the middle of the menstrual cycle, the rising secretion of estradiol from the ovaries stimulates the anterior pituitary to secrete a "surge" of LH, which results in ovulation. This effect is commonly referred to as a *positive feedback* to distinguish it from the more usual negative feedback inhibition of target gland hormones on anterior pituitary secretion. Interestingly, higher levels of estradiol at a later stage of the menstrual cycle exert the opposite effect—negative feedback inhibition—on LH secretion. The control of gonadotropin secretion is discussed in more detail in chapter 29.

## Higher Brain Function and Pituitary Secretion

The relationship between the anterior pituitary and a particular target gland is described as an *axis*. The pituitary-gonad axis, for example, refers to the action of gonadotropic hormones on the testes and ovaries. This axis is stimulated by GnRH from the hypothalamus, as previously described. Since the hypothalamus receives neural input from "higher brain centers" (chapter 15), however, it is not surprising that the pituitary-gonad axis can be affected by emotions. Indeed, alteration of the timing of ovulation or of menstruation as a result of intense emotions is well known. Psychological stress also stimulates the pituitary-adrenal axis, as we will now describe.

The effect of stress on the pituitary-adrenal axis is another prime example of the influence of higher brain centers on pituitary function. Stressors, as described later in this chapter, produce an increase in CRH secretion from the hypothalamus, which in turn results in elevated ACTH and corticosteroid secretion. In addition, the influence of higher brain centers produces *circadian* ("about a day") *rhythms* in

## Development of the Pituitary Gland

The adenohypophysis begins to develop during the third week as a diverticulum (*di´ver-tik´yŭ-lum*), or pouchlike extension, called the **hypophyseal** (Rathke's) **pouch** (fig. 1). It arises from the roof of the primitive oral cavity and grows toward the brain. At the same time, another diverticulum called the **infundibulum** forms from the diencephalon on the inferior aspect of the brain. As the two diverticula come in contact, the hypophyseal pouch loses its connection with the oral cavity, and the primordial tissue of the adenohypophysis is formed.

The neurohypophysis develops as the infundibulum extends inferiorly from the diencephalon to come in contact with the developing adenohypophysis. The fully formed neurohypophysis consists of the infundibulum and the pars nervosa. Specialized nerve fibers that connect the hypothalamus with the pars nervosa develop within the infundibulum.

Rathke's pouch: from Martin H. Rathke, German anatomist, 1793–1860

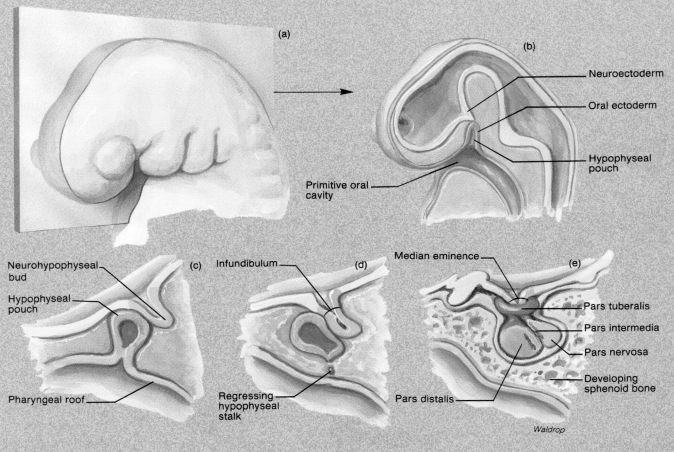

**FIGURE 1**

The development of the pituitary gland. (*a*) The head end of an embryo at 4 weeks showing the position of a midsagittal cut seen in the developmental sequence (*b–e*). The pituitary gland arises from a specific portion of the neuroectoderm, called the neurohypophyseal bud, which evaginates downward during the fourth and fifth weeks, respectively, in (*b*) and (*c*), and from a specific portion of the oral ectoderm, called the hypophyseal (Rathke's) pouch, which evaginates upward from a specific portion of the primitive oral cavity. At 8 weeks (*d*), the hypophyseal pouch is no longer connected to the pharyngeal roof of the oral cavity. During the fetal stage (*e*), the development of the pituitary gland is completed.

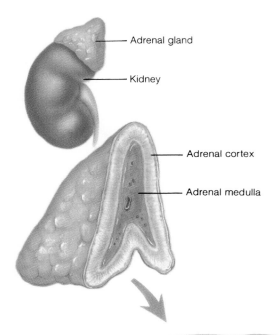

- Adrenal gland
- Kidney
- Adrenal cortex
- Adrenal medulla

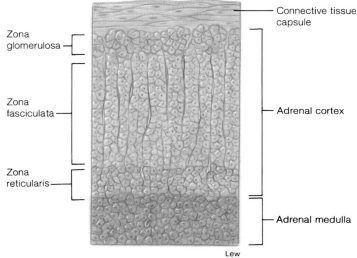

Zona glomerulosa
Zona fasciculata
Zona reticularis

Connective tissue capsule
Adrenal cortex
Adrenal medulla

Lew

**FIGURE 19.10**
The structure of the adrenal gland showing the three zones of the adrenal cortex.

the secretion of many anterior pituitary hormones. The secretion of growth hormone, for example, is highest during sleep and decreases during wakefulness, although its secretion is also stimulated by the absorption of particular amino acids following a meal.

 The influences of higher brain centers on the pituitary-gonad axis help to explain the so-called dormitory effect—that is, a tendency for the menstrual cycles of female roommates to synchronize. This synchronization will not occur in a new roommate if her nasal cavity is plugged with cotton, suggesting that the dormitory effect is due to the action of *pheromones*. Pheromones are chemicals excreted by an individual that act through the olfactory sense and modify the physiology or behavior of another member of the same species. Pheromones are important regulatory molecules in the urine, vaginal fluid,

and other secretions of most mammals, and help to regulate their reproductive cycles and behavior. Although pheromones have a more limited function in human physiology and behavior, their significance is difficult to assess because of our frequent cleansing, use of deodorants, and applications of artificial scents.

# Adrenal Glands

*The adrenal cortex and adrenal medulla are structurally and functionally different. The adrenal medulla secretes catecholamine hormones that complement the sympathetic division of the ANS in the fight-or-flight reaction. The adrenal cortex secretes steroid hormones that participate in the regulation of mineral balance, energy balance, and reproductive function.*

## Structure of the Adrenal Glands

The **adrenal glands** (also called *suprarenal glands*) are paired organs that cap the superior borders of the kidneys (fig. 19.10). Each adrenal gland consists of an outer **adrenal cortex** and inner **adrenal medulla** that function as separate glands. The differences in function of the adrenal cortex and adrenal medulla are related to their different embryonic derivations. The adrenal medulla is derived from embryonic neural crest ectoderm (the same tissue that produces the sympathetic ganglia), whereas the adrenal cortex is derived from embryonic mesoderm.

As a consequence of its embryonic derivation, the adrenal medulla secretes catecholamine hormones (mainly epinephrine, with lesser amounts of norepinephrine) into the blood in response to stimulation by preganglionic sympathetic nerve fibers. The adrenal cortex does not receive neural innervation; therefore, it must be stimulated hormonally (by ACTH secreted from the anterior pituitary).

The adrenal cortex makes up the bulk of the adrenal gland and is histologically subdivided into three zones: an outer **zona glomerulosa** (glo-mer″yoo-lo´să), an intermediate **zona fasciculata** (fă-sik″yoo-lă´tă), and an inner **zona reticularis.** The adrenal medulla is composed of tightly packed clusters of **chromaffin** (kro-maf´in) **cells.**

## Functions of the Adrenal Cortex

The adrenal cortex secretes steroid hormones called **corticosteroids** (kor″tĭ-ko-ster´oidz), or **corticoids,** for short. There are three functional categories of corticosteroids: (1) **mineralocorticoids,** which regulate Na⁺ and K⁺ balance (by acting on the kidneys); (2) **glucocorticoids,** which regulate the metabolism of glucose and other organic molecules; and

..........................................
adrenal: L. *ad*, to; *renes*, kidney

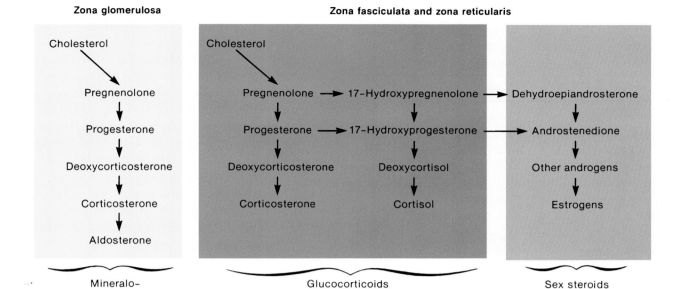

**Zona glomerulosa**

Cholesterol
↓
Pregnenolone
↓
Progesterone
↓
Deoxycorticosterone
↓
Corticosterone
↓
Aldosterone

⎵ Mineralo-
corticoids

**Zona fasciculata and zona reticularis**

Cholesterol
↓
Pregnenolone → 17-Hydroxypregnenolone → Dehydroepiandrosterone
↓                    ↓                              ↓
Progesterone → 17-Hydroxyprogesterone → Androstenedione
↓                    ↓                              ↓
Deoxycorticosterone   Deoxycortisol            Other androgens
↓                    ↓                              ↓
Corticosterone        Cortisol                  Estrogens

⎵ Glucocorticoids          ⎵ Sex steroids

### FIGURE 19.11

Simplified pathways for the synthesis of steroid hormones in the adrenal cortex. The adrenal cortex produces steroids that regulate Na⁺ and K⁺ balance (mineralocorticoids), steroids that regulate glucose balance (glucocorticoids), and small amounts of sex steroid hormones.

(3) **sex steroids**, which are weak androgens (and lesser amounts of estrogens) that supplement the sex steroids secreted by the gonads. These hormones are secreted by the different zones of the adrenal cortex.

**Aldosterone** is the most potent mineralocorticoid. The mineralocorticoids are produced in the zona glomerulosa (fig. 19.11). The predominant glucocorticoid in humans is **cortisol** (*hydrocortisone*), which is secreted by the zona fasciculata and perhaps also by the zona reticularis. The secretion of cortisol by the zona fasciculata is stimulated by ACTH (fig. 19.12). The secretion of aldosterone is controlled by other mechanisms related to blood volume and electrolyte balance (see chapter 22).

### Stress and the Adrenal Gland

In 1936, a Canadian physiologist, Hans Selye, discovered that injections of a cattle ovary extract into rats (1) stimulated growth of the adrenal cortex; (2) caused atrophy of the lymphoid tissue of the spleen, lymph nodes, and thymus; and (3) produced bleeding peptic ulcers. At first Selye thought that the ovarian extracts contained a specific hormone that caused these effects. He later discovered that injections of a variety of substances, including foreign chemicals such as formaldehyde, could produce the same effects. Indeed, the same pattern occurred when he subjected rats to cold environments or when he dropped them into water and made them swim until they were exhausted.

The specific pattern of effects produced by these procedures suggested that these effects were the result of something that the procedures had in common. Selye reasoned that all of the procedures were stressful. *Stress*, according to

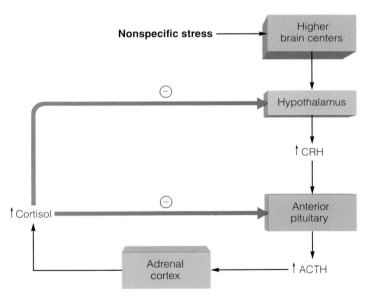

### FIGURE 19.12

The activation of the pituitary-adrenal axis by nonspecific stress.

Selye, is the reaction of an organism to stimuli called *stressors*, which may produce damaging effects. The pattern of changes he observed represented a specific response to any stressful agent. He later discovered that stressors produce these effects because they stimulate the *pituitary-adrenal axis*. Under stressful conditions, there is increased secretion of ACTH from the anterior pituitary and thus there is increased secretion of glucocorticoids from the adrenal cortex.

On this basis, Selye has stated that there is "a non-specific response of the body to readjust itself following any

| Table 19.6 | Comparison of adrenal medullary hormones | |
|---|---|
| **Epinephrine** | **Norepinephrine** |
| Elevates blood pressure because of increased cardiac output and peripheral vasoconstriction | Elevates blood pressure because of generalized vasoconstriction |
| Accelerates respiratory rate and dilates respiratory passageways | Similar effect but to a lesser degree |
| Increases efficiency of muscular contraction | Similar effect but to a lesser degree |
| Increases rate of glycogen breakdown into glucose, so level of blood glucose rises | Similar effect but to a lesser degree |
| Increases rate of fatty acid released from fat, so level of blood fatty acids rises | Similar effect but to a lesser degree |
| Increases release of ACTH and TSH from the adenohypophysis of the pituitary gland | No effect |

demand made upon it." A rise in the plasma glucocorticoid levels results from the demands of the stressors. Selye termed this nonspecific response the **general adaptation syndrome (GAS).** Stress, in other words, produces GAS. There are three stages in the response to stress: (1) the *alarm reaction,* when the adrenal glands are activated; (2) the *stage of resistance,* in which readjustment occurs; and (3) if the readjustment is not complete, the *stage of exhaustion* may follow, leading to sickness and possibly death.

Selye's concept of stress has been refined by subsequent research. These investigations demonstrate that the sympathoadrenal system becomes activated, with increased secretion of epinephrine and norepinephrine, in response to stressors that challenge the organism to respond physically. This is the "fight or flight" reaction described in chapter 18. Different emotions, however, are accompanied by different endocrine responses. The pituitary-adrenal axis, with rising levels of glucocorticoids, becomes more active when the stress is more of a chronic nature and the person feels less in control and is more passive.

## Functions of the Adrenal Medulla

The chromaffin cells of the adrenal medulla produce **epinephrine** and **norepinephrine** in an approximate ratio of 4:1, respectively. These hormones are classified as amines

(more specifically, as catecholamines) and are derived from the amino acid tyrosine.

The effects of these hormones are similar to those produced by stimulation of the sympathetic nervous system, except that the hormonal effects last about 10 times longer. The hormones from the adrenal medulla increase cardiac output and heart rate, dilate coronary blood vessels, enhance mental alertness, increase the respiratory rate, and elevate the metabolic rate. The effects of epinephrine and norepinephrine are compared in table 19.6.

The adrenal medulla is innervated by sympathetic nerve fibers. Many stressors, therefore, activate the adrenal medulla as well as the adrenal cortex. Activation of the adrenal medulla together with the sympathetic nervous system prepares the body for greater physical performance— the *fight-or-flight response.*

 Glucocorticoids, such as hydrocortisone, can inhibit the immune system. Indeed, these steroids are often given medically for this reason to treat various inflammatory diseases and to suppress the immune rejection of a transplanted organ. It seems reasonable, therefore, that the elevated glucocorticoid secretion that can accompany stress may inhibit the ability of the immune system to protect against disease. Indeed, studies show that prolonged stress results in an increased incidence of cancer and other diseases.

# Thyroid and Parathyroid Glands

*The thyroid gland secretes thyroxine ($T_4$) and triiodothyronine ($T_3$). These hormones are essential for proper growth and development and are responsible for determining the basal metabolic rate (BMR). The parathyroid glands secrete parathyroid hormone, which helps to raise the blood calcium concentration.*

## Structure of the Thyroid Gland

The **thyroid gland** is located just inferior to the larynx (fig. 19.13). This gland consists of two lobes positioned on either lateral side of the trachea that are connected anteriorly by a medial tissue mass called the *isthmus.* The thyroid is the largest of the endocrine glands, weighing between 20 and 25 g.

On a microscopic level, the thyroid gland consists of many spherical hollow sacs called **thyroid follicles** (fig. 19.14). These follicles are lined with a simple cuboidal epithelium composed of follicle cells that synthesize the principal thyroid hormones. The interior of the follicles contains **colloid,** a protein-rich fluid. Between the follicles are epithelial cells called **parafollicular cells;** these cells produce a hormone called *calcitonin* (or *thyrocalcitonin*).

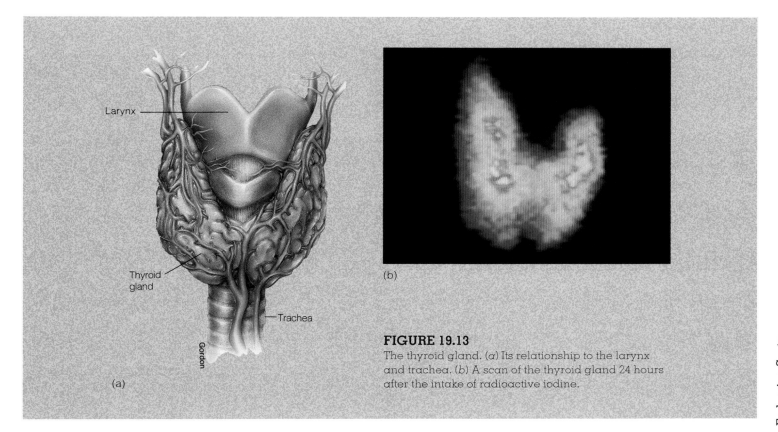

**FIGURE 19.13**

The thyroid gland. (*a*) Its relationship to the larynx and trachea. (*b*) A scan of the thyroid gland 24 hours after the intake of radioactive iodine.

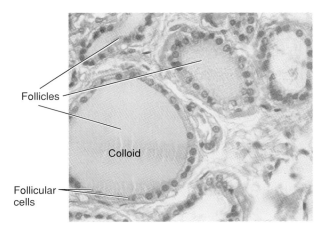

**FIGURE 19.14**

A photomicrograph (250×) of a thyroid gland showing numerous thyroid follicles. Each follicle consists of follicular cells surrounding the fluid known as colloid, which contains thyroglobulin.

## Production and Action of Thyroid Hormones

The thyroid follicles actively accumulate iodide ($I^-$) from the blood and secrete it into the colloid. Once the iodide is in the colloid, it is oxidized to iodine and attached to specific amino acids (tyrosines) within the polypeptide chain of a protein called **thyroglobulin.** The attachment of one iodine to tyrosine produces *monoiodotyrosine* (*MIT*); the attachment of two iodines produces *diiodotyrosine* (*DIT*).

Within the follicle cells, enzymes modify the structure of MIT and DIT and couple them together (fig. 19.15). When two DIT molecules that are appropriately modified are coupled together, a molecule of **tetraiodothyronine** (*tet″ră-i-o″do-thi′rŏ-nēn*)—**T₄,** or **thyroxine**—is produced. The combination of one MIT with one DIT forms **triiodothyronine** (*tri″i-o″do-thi′rŏ-nēn*), or **T₃.** Note that within the colloid, T₄ and T₃ are still attached to thyroglobulin. Upon stimulation by TSH, the cells of the follicle take up a small volume of colloid by pinocytosis, hydrolyze the T₃ and T₄ from the thyroglobulin, and secrete the free hormones into the blood.

Most of the thyroid hormone molecules that enter the blood become attached to plasma carrier proteins. Only the very small percentage of thyroxine that is free in the plasma can enter the target cells, where it is converted to triiodothyronine and attached to nuclear receptor proteins, as described earlier in this chapter. Through the activation of genes, thyroid hormones stimulate protein synthesis, promote maturation of the nervous system, and increase the rate of energy utilization by the body.

The development of the central nervous system is particularly dependent on thyroid hormones, and a deficiency of these hormones during development can cause serious mental retardation. The basal metabolic rate (BMR)—which is the minimum rate of caloric expenditure by the body—is determined largely by the level of thyroid hormones in the blood. The physiological functions of thyroid hormones are described in more detail in chapter 27.

529

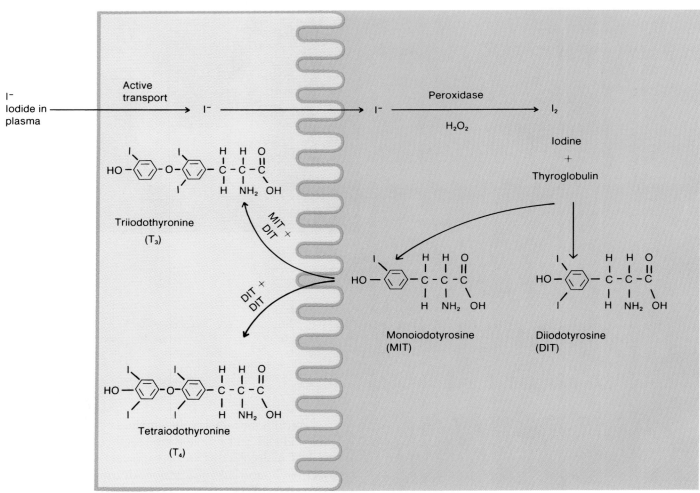

**Follicular cell**          **Colloid**

**FIGURE 19.15**

Stages in the formation and secretion of thyroid hormones. Iodide is actively accumulated by the follicular cells. In the colloid it is converted into iodine and attached to tyrosine amino acids within the thyroglobulin protein. Pinocytosis of iodinated thyroglobulin, coupling of MIT and DIT, and the release of thyroid hormones are stimulated by TSH from the anterior pituitary.

 Thyroid-stimulating hormone (TSH) from the anterior pituitary stimulates the thyroid to secrete thyroxine and exerts a trophic effect on the thyroid gland. This trophic effect shows up dramatically in people who develop an *iodine-deficiency (endemic) goiter* (see fig. 19.26). In the absence of sufficient dietary iodine, the thyroid cannot produce adequate amounts of $T_4$ and $T_3$. The resulting lack of negative feedback inhibition causes abnormally high levels of TSH secretion, which in turn stimulates the abnormal growth of the thyroid (a goiter). These events are summarized in figure 19.16.

## Parathyroid Glands

The small, flattened **parathyroid glands** are embedded in the posterior surfaces of the lateral lobes of the thyroid gland (see fig. 19.17). There are usually four parathyroid glands: a superior and an inferior pair. Each parathyroid gland is a small, yellowish-brown body 3 to 8 mm (0.1 to 0.3 in.) long, 2 to 5 mm (0.07 to 0.2 in.) wide, and about 1.5 mm (0.05 in.) deep.

The parathyroid glands secrete one hormone called **parathyroid hormone (PTH).** This hormone promotes a rise in blood calcium levels by acting on the bones, kidneys, and small intestine (fig. 19.18). Regulation of calcium balance is described in more detail in chapter 26.

# Pancreas and Other Endocrine Glands

*The pancreatic islets secrete two hormones, insulin and glucagon. Insulin promotes the lowering of blood glucose and the storage of energy in the form of glycogen and fat. Glucagon exerts an antagonistic effect by raising the blood glucose concentration. Additionally, many other organs secrete hormones that help to regulate digestion, metabolism, growth, immune function, and reproduction.*

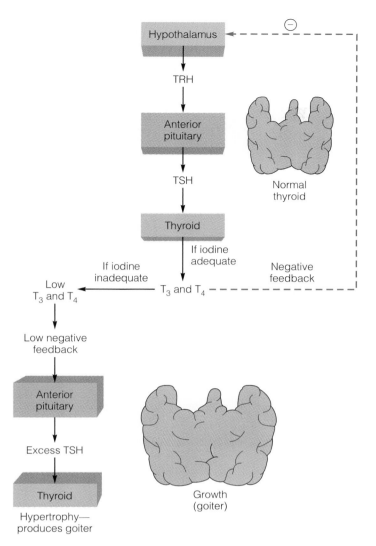

**FIGURE 19.16**
Lack of adequate iodine in the diet interferes with the negative feedback control of TSH secretion, resulting in the formation of an endemic goiter.

The **pancreas** is both an endocrine and an exocrine gland. The gross structure of this gland and its exocrine functions in digestion are described in chapter 17. The endocrine portion of the pancreas consists of scattered clusters of cells called the **pancreatic islets** (islets of Langerhans). These endocrine structures are most common in the body and tail of the pancreas (fig. 19.19).

## *Pancreatic Islets (Islets of Langerhans)*

On a microscopic level, the most conspicuous cells in the islets are the *alpha* and *beta cells.* The alpha cells secrete the hormone **glucagon,** and the beta cells secrete **insulin.**

••••••••••••••••••••••••••••••••••••

islets of Langerhans: from Paul Langerhans, German anatomist, 1847–88
insulin: L. *insula,* island

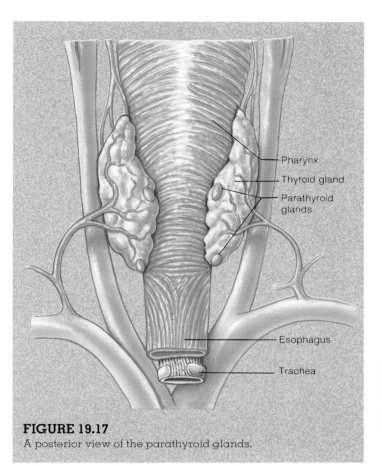

**FIGURE 19.17**
A posterior view of the parathyroid glands.

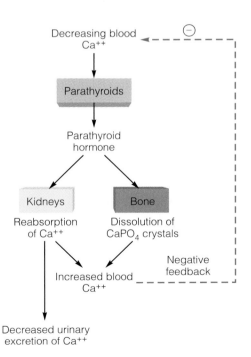

**FIGURE 19.18**
The actions of parathyroid hormone and the negative feedback control of its secretion. An increased level of parathyroid hormone causes the bones to release calcium and the kidneys to conserve calcium that is lost through the urine.

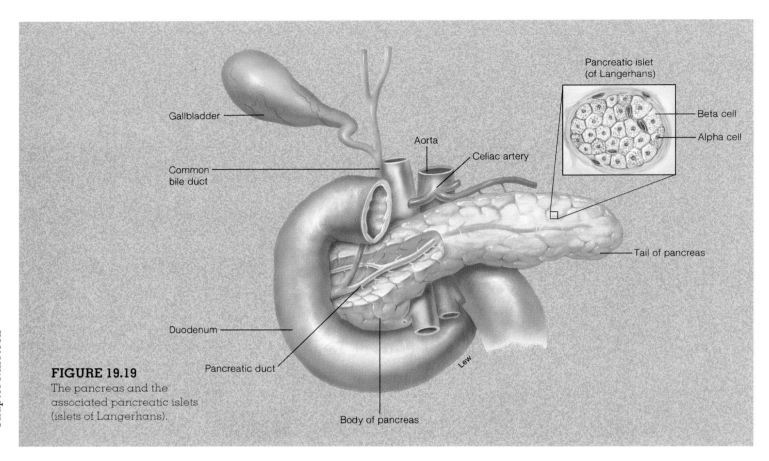

**FIGURE 19.19**
The pancreas and the associated pancreatic islets (islets of Langerhans).

Alpha cells secrete glucagon in response to a fall in the blood glucose concentrations. Glucagon stimulates the liver to hydrolyze glycogen to glucose (*glycogenolysis*), which causes the blood glucose level to rise. This effect represents the completion of a negative feedback loop. Glucagon also stimulates the hydrolysis of stored fat (*lipolysis*) and the consequent release of free fatty acids into the blood. This effect helps to provide energy substrates for the body during fasting, when blood glucose levels fall. Glucagon, together with other hormones, also stimulates the conversion of fatty acids to ketone bodies, which can be secreted by the liver into the blood and serve as an energy source for many organs. Glucagon is thus a hormone that helps to maintain homeostasis during times of fasting, when the body's energy reserves must be utilized.

Beta cells secrete insulin in response to a rise in the blood glucose concentrations. Insulin promotes the entry of glucose into tissue cells, and the conversion of this glucose into energy storage molecules of glycogen and fat. Insulin also aids the entry of amino acids into cells and the production of cellular protein. The actions of insulin and glucagon are thus antagonistic. After a meal, insulin secretion is increased and glucagon secretion is decreased; fasting, by contrast, causes a rise in glucagon and a decline in insulin secretion. The metabolic effects of insulin and glucagon are discussed in more detail in chapter 27.

## Pineal Gland

The small, cone-shaped **pineal gland** is located in the roof of the third ventricle, near the corpora quadrigemina, where it is encapsulated by the meninges covering the brain. The pineal gland of a child weighs about 0.2 g and is 5 to 8 mm (0.2 to 0.3 in.) long and 9 mm wide. The gland begins to regress in size at about age 7, and in the adult appears as a thickened strand of fibrous tissue. Although the pineal gland lacks direct nervous connections to the rest of the brain, it is highly innervated by sympathetic nerve fibers from the superior cervical ganglion.

The principal hormone of the pineal gland is **melatonin.** Production and secretion of this hormone is stimulated by activity of the *suprachiasmatic nucleus (SCN)* in the hypothalamus of the brain via activation of sympathetic neurons to the pineal gland. Activity of the SCN, and thus secretion of melatonin, is highest at night. During the day, neural pathways from the retina of the eyes depress the activity of the SCN, reducing sympathetic stimulation of the pineal gland and thus decreasing melatonin secretion.

..................................................

pineal: L. *pinea,* pine cone

Since cycles of light and dark directly influence the levels of melatonin secretion, other daily (circadian) rhythms of the body may be the result of the ebb and flow of melatonin. Derangements of such cycles (in *jet-lag*, for example) have been tentatively ascribed to the effects of melatonin. Also, phototherapy used in the treatment of *seasonal affective disorder* (SAD), or "winter depression," might be effective because such treatment inhibits melatonin secretion.

It has long been suspected that melatonin inhibits the pituitary-gonad axis. Indeed, a decrease in melatonin secretion in many lower vertebrates is responsible for maturation of the gonads during their reproductive season. Excessive melatonin secretion in humans is associated with a delay in the onset of puberty. Melatonin secretion is highest in children of ages 1–5 and decreases thereafter, reaching its lowest levels at the end of puberty, when concentrations are 75% lower than during early childhood. Therefore, the secretion of melatonin is believed by some researchers to play an important role in the onset of puberty, but this possibility is highly controversial.

## Thymus

The **thymus** is a bilobed organ positioned in the superior mediastinum, in front of the aortic arch and behind the manubrium of the sternum (fig. 19.20). Although the size of the thymus varies considerably from person to person, it is relatively large in newborns and children and sharply regresses in size after puberty. Besides decreasing in size, the thymus of adults becomes infiltrated with strands of fibrous and fatty connective tissue.

The thymus produces **T cells** (thymus-dependent cells), which are the lymphocytes involved in cell-mediated immunity (see chapter 23). In addition to producing T cells, the thymus secretes a number of hormones that are believed to stimulate T cells after they leave the thymus.

## Gastrointestinal Tract

The **stomach** and **small intestine** secrete a number of hormones that act on the gastrointestinal tract (GI tract) itself and on the pancreas and gallbladder, as discussed in chapter 25. The effects of these hormones, in conjunction with regulation by the autonomic nervous system, act to coordinate the activities of different regions of the GI tract and the secretions of pancreatic juice and bile.

## Gonads and Placenta

The **gonads** (**testes** and **ovaries**) secrete sex steroids. These include male sex hormones, or *androgens*, and female sex hormones—*estrogens* and *progestogens*. The principal hormones in each of these categories are *testosterone, estradiol-17β*, and *progesterone*, respectively.

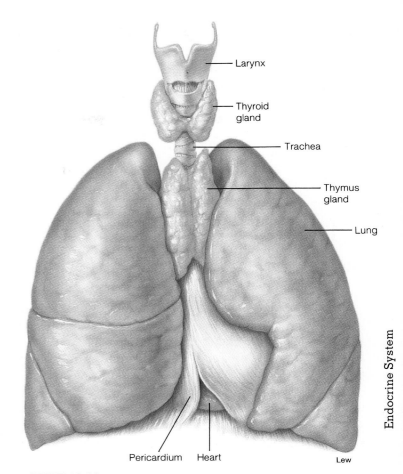

**FIGURE 19.20**

The thymus is a bilobed organ within the mediastinum of the thorax.

The testes contain tightly convoluted *seminiferous tubules*, which produce sperm. Between the convolutions of the tubules are specialized endocrine cells called **interstitial cells (cells of Leydig),** which secrete testosterone. Testosterone is needed for the development and maintenance of the primary and secondary sex organs of the male, as well as for the development of male secondary sex characteristics.

During the first half of the menstrual cycle, estrogen is secreted by many small structures within the ovary called *ovarian follicles*. These follicles contain the egg cell, or *ovum*, and **granulosa cells** that secrete estrogen. By about midcycle, one of these follicles grows very large and, in the process of ovulation, extrudes its ovum from the ovary. The empty follicle, under the influence of luteinizing hormone (LH) from the anterior pituitary, then becomes a new endocrine structure called a **corpus luteum.** The corpus luteum secretes progesterone as well as estradiol-17β.

• • • • • • • • • • • • • • • • • • • • • • • • • • • • • • •

cells of Leydig: from Franz von Leydig, German anatomist, 1821–1908

## Table 19.7 — Functional categories of hormones, based on the location of their receptor proteins and mechanisms of action

| Hormone type | Secreted by | Location of receptors | Effects of hormone–receptor protein interaction |
|---|---|---|---|
| Thyroxine ($T_4$) | Thyroid | Nucleus of target cells | After conversion to triiodothyronine ($T_3$), activates specific genes |
| Steroids | Adrenal cortex, testes, ovaries | Cytoplasm of target cells | Stimulates translocation of hormone-receptor complex to nucleus and activation of specific genes |
| Catecholamines, polypeptides, glycoproteins | All glands except adrenal cortex, gonads, and thyroid | Outer surface of cell membrane | Stimulates production of intracellular "second messenger," which activates previously inactive enzymes |

The **placenta** is the organ responsible for nutrient and waste exchange between the fetus and mother. The placenta is also an endocrine gland; it secretes large amounts of estrogens and progesterone, as well as a number of polypeptide and protein hormones that are similar to some hormones secreted by the anterior pituitary gland. These latter hormones include **human chorionic gonadotropin (hCG),** which is similar to LH, and **somatomammotropin,** which is similar in action to growth hormone and prolactin. The physiology of the placenta and other aspects of reproductive endocrinology are covered in chapter 30.

# Mechanisms of Hormone Action

*Each hormone exerts its characteristic effects on a target organ through its actions on the cells of these organs. The mechanisms of action are similar for hormones that have similar chemical natures. Lipid-soluble hormones pass through the target cell membrane and act directly within the target cell; those that are polar act on the target cell membrane to cause the production of intracellular second-messenger molecules. The effects of a given hormone are influenced by its concentration in the blood and by interactions with other hormones.*

Although each hormone exerts its own characteristic effects on specific target cells, hormones that are in the same chemical category have similar mechanisms of action. These similarities involve the location of cellular receptor proteins and the events that occur in the target cells after the hormone has combined with its receptor protein.

Hormones are delivered by the blood to every cell in the body, but only the **target cells** are able to respond to these hormones. In order to respond to any given hormone, a target cell must have specific receptor proteins for that hormone. Receptor protein–hormone interaction is highly specific, much like the interaction of an enzyme with its substrate. Receptor proteins are not enzymes, however, and

since they cannot be detected by the techniques used to assay enzymes, other methods must be used. These methods are based on the observation that hormones bind to receptors with a *high affinity* (high bond strength) and with a *low capacity.* The latter characteristic refers to the possibility of saturating receptors with hormones because of the limited number of receptors per target cell (usually a few thousand). Notice that the characteristics of hormone–target cell specificity and saturation of receptor proteins are similar to the characteristics of enzyme and carrier proteins discussed in previous chapters.

The location of a hormone's receptor proteins in its target cells depends on the chemical nature of the hormone. Based on the location of the receptor proteins, hormones can be grouped into three categories: (1) receptor proteins within the nucleus of target cells—*thyroid hormones,* (2) receptor proteins within the cytoplasm of target cells—*steroid hormones,* and (3) receptor proteins on the outer surface of the target cell membrane—*catecholamine* and *polypeptide hormones.* This information is summarized in table 19.7.

## Mechanisms of Steroid and Thyroid Hormone Action

Steroid and thyroid hormones are similar in size; moreover, both groups are nonpolar, and thus are not very water-soluble. Unlike other hormones, therefore, steroids and thyroid hormones (primarily thyroxine) do not travel dissolved in the aqueous portion of the plasma but instead are transported to their target cells attached to plasma carrier proteins. These hormones then dissociate from the carrier proteins in the blood and easily pass through the lipid component of the target cell's membrane.

Steroid and thyroid hormones bind to receptor proteins within the target cells. These receptor proteins may be originally located in either the cytoplasm or nucleus, depending on the specific receptor. In response to binding to its hormone, the receptor undergoes a change that allows it

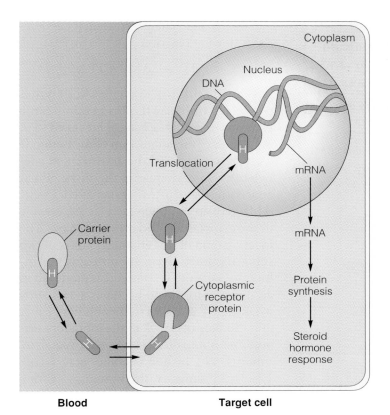

**FIGURE 19.21**
The mechanism of the action of a steroid hormone (H) on the target cells.

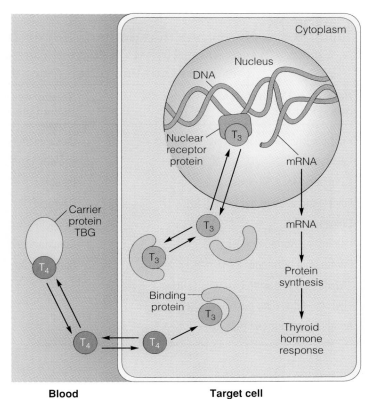

**FIGURE 19.22**
The mechanism of the action of $T_3$ (triiodothyronine) on the target cells.

also to bind to DNA. All of these receptor proteins then promote genetic transcription, or RNA synthesis. The synthesis of specific proteins coded by messenger RNA follows these events and produces the effects of the hormone within the target cell.

**Steroid Hormones**   Once through the cell membrane, most steroid hormones attach to *cytoplasmic receptor proteins* in the target cells. The steroid hormone–receptor protein complex then *translocates* to the nucleus and attaches by means of the receptor proteins to the chromatin. The sites of attachment in the chromatin, termed **acceptor sites,** are specific for the target tissue. This specificity is believed to be determined by acidic (nonhistone) proteins in the chromatin (chapter 3). According to one theory, part of the receptor binds to an acidic protein while a different part of the receptor binds to DNA.

The attachment of the receptor protein–steroid complex to the acceptor site "turns on" genes. Specific genes become activated by this process and produce nuclear RNA, which is then processed into messenger RNA (mRNA). This new mRNA enters ribosomes and codes for the production of new proteins. Since some of these newly synthesized proteins may be enzymes, the metabolism of the target cell is changed in a specific manner (fig. 19.21).

**Thyroxine**   As previously discussed, the major hormone secreted by the thyroid gland is thyroxine, or tetraiodothyronine ($T_4$). Like steroid hormones, thyroxine travels in the blood attached to carrier proteins (primarily attached to *thyroxine-binding globulin,* or *TBG*). The thyroid also secretes a small amount of triiodothyronine ($T_3$). The carrier proteins have a higher affinity for $T_4$ than for $T_3$, however, and as a result, the amount of unbound (or "free") $T_3$ in the plasma is about 10 times greater than the amount of free $T_4$.

Approximately 99.96% of the thyroxine in the blood is attached to carrier proteins in the plasma; the rest is free. Only the thyroxine and $T_3$ that is free can enter target cells; the protein-bound thyroxine serves as a reservoir of this hormone in the blood (this is why it takes a couple of weeks after surgical removal of the thyroid gland for the symptoms of hypothyroidism to develop). Once the free thyroxine passes into the target cell cytoplasm, it is enzymatically converted into $T_3$. As discussed in a later section, it is the $T_3$ rather than $T_4$ that is active within the target cells.

Inactive $T_3$ receptor proteins are already in the nucleus attached to chromatin. They remain inactive until $T_3$ enters the nucleus from the cytoplasm. The attachment of $T_3$ to the chromatin-bound receptor proteins activates genes and results in the production of new mRNA and new proteins. This sequence of events is summarized in figure 19.22.

## Table 19.8 Hormones that activate adenylate cyclase and use cAMP as a second messenger and hormones that use other second messengers

| Hormones that activate adenylate cyclase | Hormones that use other second messengers |
|---|---|
| Adrenocorticotropic hormone (ACTH) | Catecholamines (α-adrenergic) |
| Calcitonin | Growth hormone (GH) |
| Catecholamines (β-adrenergic) | Insulin |
| Follicle-stimulating hormone (FSH) | Oxytocin |
| Glucagon | Prolactin |
| Luteinizing hormone (LH) | Somatomed |
| Parathyroid hormone (PTH) | Somatostatin |
| Thyrotropin-releasing hormone (TRH) | |
| Thyroid-stimulating hormone (TSH) | |
| Antidiuretic hormone (ADH) | |

## Mechanisms of Catecholamine and Polypeptide Hormone Action

Catecholamines (epinephrine and norepinephrine) and polypeptide hormones cannot pass through the lipid barrier of the target cell membrane. Although some of these hormones may enter the cell by pinocytosis, most of their effects are believed to result from their interaction with receptor proteins on the outer surface of the target cell membrane. Since they do not have to enter the target cells to exert their effects, other molecules must mediate the actions of these hormones within the target cells. If you think of hormones as "messengers" from the endocrine glands, the intracellular mediators of the hormone's action can be called **second messengers**.

**Cyclic AMP as a Second Messenger** Cyclic adenosine monophosphate (abbreviated **cAMP**) was the first "second messenger" to be discovered and is the best understood. The effects of epinephrine and norepinephrine induced by binding to beta-adrenergic receptors (chapter 17) are due to cAMP production within the target cells. It also has been demonstrated that the effects of many (but not all) polypeptide hormones are likewise mediated by cAMP (table 19.8).

The binding of these hormones to their membrane receptor proteins activates an enzyme called **adenylate cyclase** (ă-den´l-it si´klās). This enzyme is built into the cell membrane and, when activated, it catalyzes the following reaction:

$$ATP \rightarrow cAMP + PP_i$$

Adenosine triphosphate (ATP) is thus converted into cAMP plus two inorganic phosphates (*pyrophosphate*, abbreviated $PP_i$). As a result of the interaction of the hormone with its receptor and the activation of adenylate cyclase, therefore, the intracellular concentration of cAMP is increased. Cyclic AMP activates a previously inactive enzyme in the cytoplasm called **protein kinase.** The inactive form of this enzyme consists of two subunits: a catalytic subunit and an inhibitory subunit. The enzyme is produced in an inactive form and becomes active only when cAMP attaches to the inhibitory subunit. Binding of cAMP to the inhibitory subunit causes it to dissociate from the catalytic subunit, which then becomes active (fig. 19.23). In summary, the hormone—acting through an increase in cAMP production—causes an increase in protein kinase enzyme activity within its target cells.

Active protein kinase catalyzes the attachment of phosphate groups to different proteins in the target cells. As a result, some enzymes are activated and others are inactivated. Cyclic AMP, acting through protein kinase, thus modulates the activity of enzymes that are already present in the target cell. This alters the metabolism of the target tissue in a manner characteristic of the actions of that specific hormone (table 19.9).

Like all biologically active molecules, cAMP must be rapidly inactivated for it to function effectively as a second messenger in hormone action. This function is served by an enzyme within the target cells called **phosphodiesterase** (fos´´fo-di-es´tĕ-rās), which hydrolyzes cAMP into inactive fragments. Through the action of phosphodiesterase, the stimulatory effect of a hormone that uses cAMP as a second messenger depends on the continuous generation of new cAMP molecules, and thus depends on the level of secretion of the hormone.

 Drugs that inhibit the activity of phosphodiesterase thus prevent the breakdown of cAMP, resulting in increased concentrations of cAMP within the target cells. The drug *theophylline* and its derivatives, for example, are used clinically to raise cAMP levels within bronchiolar smooth muscle. This duplicates and enhances the effect of epinephrine on the bronchioles (producing dilation) in people who suffer from asthma. *Caffeine*, a compound related to theophylline, is also a phosphodiesterase inhibitor, and thus exerts its effects by raising the cAMP concentrations within tissue cells.

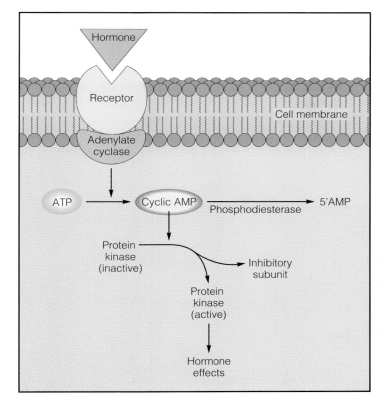

**FIGURE 19.23**
Cyclic AMP (cAMP) as a second messenger in the action of
catecholamine and polypeptide hormones.

**Table 19.9** **Sequence of events involving cyclic AMP as a second messenger**

1. The hormones combine with their receptors on the outer surface of target cell membranes.
2. Hormone-receptor interaction stimulates activation of adenylate cyclase on the cytoplasmic side of the membranes.
3. Activated adenylate cyclase catalyzes the conversion of ATP to cyclic AMP (cAMP) within the cytoplasm.
4. Cyclic AMP activates protein kinase enzymes that were already present in the cytoplasm in an inactive state.
5. Activated cAMP-dependent protein kinase transfers phosphate groups (phosphorylates) to other enzymes in the cytoplasm.
6. The activity of specific enzymes is either increased or inhibited by phosphorylation.
7. Altered enzyme activity mediates the target cell's response to the hormone.

**Ca⁺⁺ as a Second Messenger** The concentration of $Ca^{++}$ in the cytoplasm is kept very low by the action of active transport carriers—calcium pumps—in the cell membrane. Through the action of these pumps, the concentration of calcium in the cytoplasm is 5000 to 10,000 times lower in the cytoplasm than in the extracellular fluid. In addition, the endoplasmic reticulum (chapter 3) of many cells contains calcium pumps that actively transport $Ca^{++}$ from the cytoplasm into the cisternae of the endoplasmic reticulum. The steep concentration gradient for $Ca^{++}$ that results allows various stimuli to evoke a rapid, though brief, diffusion of $Ca^{++}$ into the cell, which can serve as a signal in different control systems.

At the axon terminals, for example, the influx of $Ca^{++}$ serves as a signal for the release of neurotransmitters (chapter 14). Similarly, when skeletal muscles are stimulated to contract, the release of $Ca^{++}$ from the endoplasmic reticulum couples electrical excitation of the muscle cell to the mechanical processes that result in contraction (chapter 13). Additionally, it is now known that $Ca^{++}$ serves as a second messenger in the action of particular hormones.

When epinephrine stimulates its target organs, it must first bind to adrenergic receptor proteins in the membrane of its target cells. As discussed in chapter 17, there are two types of adrenergic receptors—alpha and beta. Stimulation of the beta-adrenergic receptors by epinephrine results in activation of adenylate cyclase and the production of cAMP. By contrast, stimulation of alpha-adrenergic receptors by epinephrine activates the target cell via $Ca^{++}$ as a second messenger (fig. 19.24).

When epinephrine combines with its alpha-adrenergic receptor, an enzyme known as *phospholipase C*, located in the membrane, is activated. The substrate of this enzyme is a type of membrane phospholipid that is split by the active enzyme into **inositol triphosphate (IP₃)** and another product, diacylglycerol. IP₃ leaves the cell membrane and diffuses through the cell cytoplasm to the endoplasmic reticulum. The membrane of the endoplasmic reticulum contains receptor proteins for IP₃, so that the IP₃ is a second messenger in its own right, carrying the hormone's message from the cell membrane to the endoplasmic reticulum. Binding of IP₃ to its receptors causes specific $Ca^{++}$ channels to open, so that $Ca^{++}$ diffuses out of the endoplasmic reticulum and into the cytoplasm.

As a result, there is a rapid and transient rise in the cytoplasmic $Ca^{++}$ concentration. This signal is augmented by the opening of $Ca^{++}$ channels in the cell membrane, presumably due to the action of yet a different (and currently unknown) messenger sent from the endoplasmic reticulum to the cell membrane. $Ca^{++}$ entering the cytoplasm from the endoplasmic reticulum, or from the extracellular fluid, binds to a cytoplasmic protein called **calmodulin.** Once Ca$^{++}$ binds to calmodulin, the now-active calmodulin in turn activates specific protein kinase enzymes (those that add phosphate groups to proteins), which modify the actions of other enzymes in the cell. Activation of specific calmodulin-dependent enzymes is analogous to the activation of enzymes by cAMP-dependent protein kinase. The steps of the Ca$^{++}$ second messenger system are summarized in table 19.10.

Different hormones can act on the same target cell and produce different, and even antagonistic, effects. For example, insulin stimulates the synthesis of fat while the hormone glucagon stimulates hydrolysis of fat in adipose cells. Clearly, these two hormones cannot both use the same second messenger system. The Ca$^{++}$ system, which is stimulated by insulin, promotes lipogenesis, while the cAMP system, which is stimulated by glucagon, promotes lipolysis.

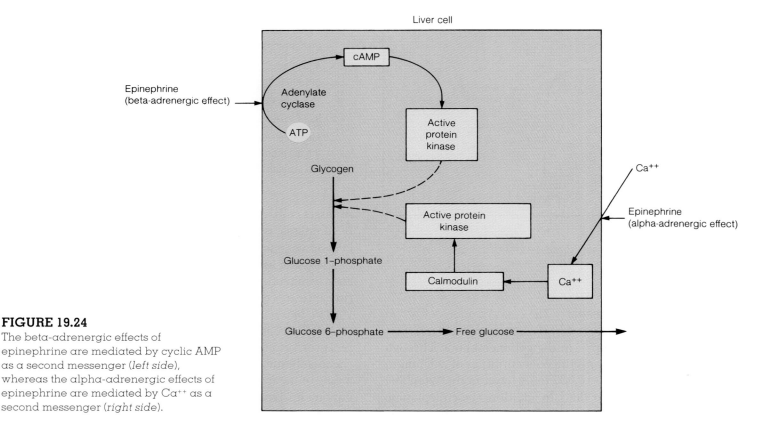

**FIGURE 19.24**

The beta-adrenergic effects of epinephrine are mediated by cyclic AMP as a second messenger (*left side*), whereas the alpha-adrenergic effects of epinephrine are mediated by Ca$^{++}$ as a second messenger (*right side*).

## Table 19.10  The sequence of events by which a hormone stimulates its target cell through the use of the Ca$^{++}$ second messenger system

1. The hormone binds to its receptor on the outer surface of the target cell membrane.

2. Hormone-receptor interaction stimulates the activity of a membrane enzyme, phospholipase C.

3. Phospholipase C catalyzes the conversion of particular phospholipids in the membrane to inositol triphosphate (IP$_3$) and another derivative, diacylglycerol.

4. Inositol triphosphate enters the cytoplasm and diffuses to the endoplasmic reticulum, where it binds to its receptor proteins and causes the opening of Ca$^{++}$ channels.

5. Since the endoplasmic reticulum accumulates Ca$^{++}$ by active transport, there exists a steep Ca$^{++}$ concentration gradient favoring the diffusion of Ca$^{++}$ into the cytoplasm.

6. Ca$^{++}$ that enters the cytoplasm binds to and activates a protein called calmodulin.

7. Activated calmodulin, in turn, activates protein kinase, which phosphorylates other enzyme proteins.

8. Altered enzyme activity mediates the target cell's response to the hormone.

## Effects of Hormone Concentrations on Tissue Response

The concentration of hormones in the blood primarily reflects the rate of secretion by the endocrine glands. Hormones do not generally accumulate in the blood because they are rapidly removed by target organs and by the liver. The **half-life** of a hormone, which is the time required for the plasma concentration of a given amount of a hormone to be reduced to half its reference level, ranges from minutes to hours for most hormones (thyroid hormone, however, has a half-life of several days). Hormones removed from the blood by the liver are converted by enzymatic reactions into less active products. Steroids, for example, are converted to more polar derivatives. These less active, more water-soluble polar derivatives are released into the blood and are excreted in the urine and bile.

The effects of hormones are very concentration dependent. Normal tissue responses are produced only when the hormones are present in the blood within their normal range, or *physiological range*, of concentrations. When some hormones are taken in abnormally high amounts, or *pharmacological concentrations* (as when they are taken as drugs), their effects may be atypical. This may be partly due to the fact that abnormally high concentrations of a hormone sometimes cause the hormone to bind to tissue receptor proteins for different but related hormones. Also, since some steroid hormones can be converted by their target cells into products that have different biological effects (such as the conversion of androgens into estrogens), the administration of large quantities of one steroid can result in the production of a significant quantity of other steroids to which tissues and organs respond differently.

Pharmacological doses of hormones, particularly of steroids, can thus have widespread and often damaging side effects. People with inflammatory diseases who are treated with high doses of cortisone over long periods of time, for example, may develop characteristic changes in bone and soft tissue structure. When oral contraceptives were first introduced, they contained high levels of sex steroids. The potential side effects could not have been predicted at the time "the pill" was first introduced.

*Anabolic steroids* are synthetic androgens (male hormones) that promote protein synthesis in muscles and other organs. Use of these drugs by body builders and other athletes became widespread in the 1960s and, although prohibited by most athletic organizations, the practice is still common today. Although administration of exogenous androgens does promote muscle growth, it can also cause a number of undesirable side effects. Since the liver and adipose cells can change androgens into estrogens, male athletes who take exogenous androgens often develop *gynecomastia*—the development of femalelike mammary tissue. This abnormal tissue must be surgically removed. Damaging effects attributed to anabolic steroids include liver cancer, shrinkage of the testes and temporary sterility, stunted growth in teenage users, masculinization in female users, and antisocial behavior. Anabolic steroids may also raise the blood levels of cholesterol and LDL, thus predisposing users to atherosclerosis of the coronary arteries and heart disease (chapter 21).

**Priming Effects**   Variations in hormone concentration within the normal range can affect the responsiveness of target cells. This is partly due to the effects of polypeptide and glycoprotein hormones on the number of their receptor proteins in target cells. More receptors may be formed in the target cells in response to increasingly higher levels of those hormones to which they respond. Small amounts of gonadotropin-releasing hormone (GnRH), secreted by the hypothalamus, for example, increase the sensitivity of anterior pituitary cells to further GnRH stimulation. This is a priming effect, sometimes called *upregulation*. Subsequent stimulation by GnRH thus causes a more vigorous response from the anterior pituitary.

**Desensitization and Downregulation**   With prolonged exposure to high concentrations of polypeptide hormones, target cells may become desensitized. Subsequent exposure to the same concentration of the same hormone will then produce less of a target tissue response. This desensitization may be partially due to the fact that high concentrations of these hormones cause a decrease in the number of receptor proteins in their target cells—a phenomenon called *downregulation*. Such desensitization and downregulation of receptors has been shown to occur in adipose cells exposed to high concentrations of insulin and in testicular cells exposed to high concentrations of luteinizing hormone (LH).

Desensitization does not occur under normal conditions because polypeptide and glycoprotein hormones are secreted in pulses rather than continuously. The *pulsatile secretion* of GnRH and LH is an important aspect, for example, in the hormonal control of the reproductive system. When these hormones are artificially presented in a continuous fashion, they produce a decrease (rather than the normal increase) in gonadal function. This effect has important clinical implications, as will be described in chapter 28.

# Autocrine and Paracrine Regulation

*Many regulatory molecules are produced throughout the body and act within the organ in which they are produced. These molecules may regulate different cells within one tissue, or they may be produced within one tissue and regulate a different tissue within the same organ.*

In chapters 14 and 19, we considered two types of regulatory molecules—neurotransmitters and hormones. These two classes of regulatory molecules cannot be defined simply by differences in chemical structure, since the same molecule (such as norepinephrine) may be in both categories; rather, they must be defined by function. Neurotransmitters are released by axon terminals, travel across a narrow synaptic cleft, and affect a postsynaptic cell. Hormones are secreted into the blood by an endocrine gland and, through transport in the blood, influence the activities of one or more target organs.

There are yet other classes of regulatory molecules, which are distinguished by the fact that they are produced in many different organs and are active within the organ in which they are produced. Molecules of this type are called **autocrine regulators** if they are produced and act within the same tissue of an organ. They are called **paracrine regulators** if they are produced within one tissue and regulate a dif-

ferent tissue of the same organ (table 19.11). For the sake of simplicity, and because the same chemical can function as an autocrine or a paracrine regulator in different situations, this text will use the term *autocrine* in a generic sense to refer to both types of local regulation.

## Examples of Autocrine Regulation

Many autocrine regulatory molecules are also known as **cytokines,** particularly if they regulate different cells of the immune system, and **growth factors,** if they promote growth and cell division in any organ. This distinction is somewhat blurred, however, because some cytokines may also function as growth factors. Cytokines produced by lymphocytes (the type of white blood cell involved in specific immunity; see chapter 23) are also known as **lymphokines,** and the specific molecules involved are called *interleukins.* The terminology can be confusing because new regulatory molecules, and new functions for previously named regulatory molecules, are being discovered at a rapid pace.

Cytokines secreted by macrophages (a type of phagocytic cell found in connective tissues) and lymphocytes stimulate proliferation of specific cells involved in the immune response. **Neurotrophins,** such as *nerve growth factor,* guide regenerating peripheral neurons that have been injured (chapter 14). Nitric oxide, which can function as a neuro-

## Table 19.11  A partial list of autocrine and paracrine regulators and some examples of their origin and actions

| Autocrine or paracrine regulator | Major sites of production | Major actions |
|---|---|---|
| Insulin-like growth factors (somatomedins) | Many organs, particularly the liver and cartilages | Stimulates growth and cell division |
| Nitric oxide | Endothelium of blood vessels; neurons of CNS; macrophages | Dilation of blood vessels; neural messenger; anti-bacterial |
| Endothelins | Endothelium of blood vessels; other organs | Constriction of blood vessels; other effects |
| Platelet-derived growth factor | Platelets; macrophages; vascular smooth muscle cells | Cell division within blood vessels |
| Epidermal growth factors | Epidermal tissues | Cell division in wound healing |
| Neurotrophins | Neuroglial cells, Schwann cells, neurons | Regeneration of peripheral nerves |
| Bradykinin | Endothelium of blood vessels | Dilation of blood vessels |
| Interleukins | Macrophages; lymphocytes | Regulates immune system |
| Prostaglandins | Many tissues | Many actions (see text) |

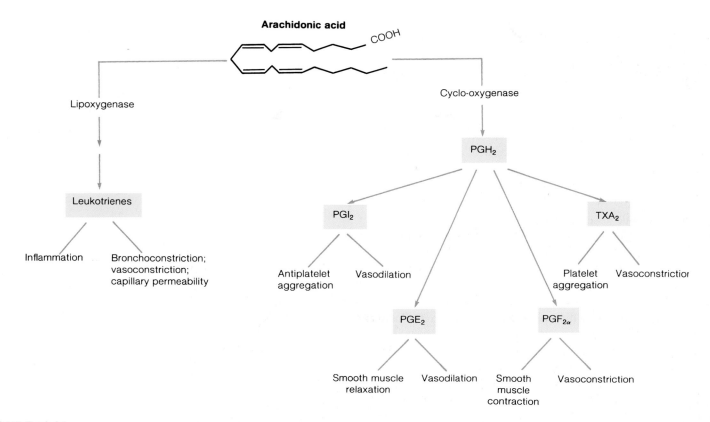

**FIGURE 19.25**
Formation and actions of leukotrienes and prostaglandins
(PG = prostaglandin; TX = thromboxane).

transmitter involved in memory and other functions, is also produced by the endothelium of blood vessels. In this context, it is a paracrine regulator because it diffuses to the smooth muscle layer of the blood vessel and promotes smooth muscle relaxation, leading to dilation of the blood vessel. In this action, nitric oxide functions as the regulator previously known as *endothelium-derived relaxation factor.* Neural and paracrine regulation interact in this case, because autonomic axons that release acetylcholine in blood vessels cause dilation of those vessels by stimulating the synthesis of nitric oxide.

The endothelium of blood vessels also produces other paracrine regulators. These include the *endothelins* (specifically *endothelin-1* in humans), which directly promote vasoconstriction, and *bradykinin,* which promotes vasodilation. These regulatory molecules are thus very important in the control of blood flow and blood pressure (chapter 22). They are also involved in the development of atherosclerosis, the leading cause of heart disease and stroke (chapter 21). In addition, endothelin-1 is produced by the epithelium of the airways and may be important in the embryological development and function of the respiratory system.

All autocrine regulators in some ways control gene expression in their target cells. This is very clearly the case with the various growth factors, such as *platelet-derived growth factor, epidermal growth factor,* and the *insulin-like growth factors* that stimulate cell division and proliferation of their target cells. The latter group of regulators interact with the endocrine system in a number of ways, as described in chapter 27.

## Prostaglandins

The most diverse group of autocrine regulators are the **prostaglandins.** A prostaglandin is a twenty-carbon-long fatty acid that contains a five-membered carbon ring. This molecule is derived from the precursor molecule *arachidonic acid,* which can be released from phospholipids in the cell membrane under hormonal or other stimulation. Arachidonic acid can the enter one of two possible metabolic pathways. In one case, the arachidonic acid may be converted by the enzyme *cyclo-oxygenase* into a prostaglandin, which can then be changed by other enzymes into other prostaglandins. In the other case, arachidonic acid may be converted by the enzyme *lipoxygenase* into **leukotrienes,** which are compounds that are closely related to the prostaglandins (fig. 19.25).

Prostaglandins are produced in almost every organ and have been implicated in a wide variety of regulatory functions. The study of prostaglandin function can be confusing, in part because a given prostaglandin may have opposite effects in different organs. Prostaglandins of the E series (PGE), for example, cause relaxation of smooth muscle in the urinary bladder, bronchioles, GI tract, and uterus, but the same molecules cause contraction of vascular smooth muscle. A different prostaglandin, designated $PGF_{2\alpha}$, has exactly the opposite effects.

The antagonistic effects of prostaglandins on blood clotting make good physiological sense. Blood platelets, which are required for blood clotting, produce *thromboxane* $A_2$. This prostaglandin promotes clotting by stimulating platelet aggregation and vasoconstriction. The endothelial cells of blood vessels, in contrast, produce a different prostaglandin, known as $PGI_2$, or *prostacyclin*, which has the opposite effects—it inhibits platelet aggregation and causes vasodilation. These antagonistic effects help to promote clotting while at the same time they ensure that clots do not normally form on the walls of intact blood vessels.

The following are some of the regulatory functions proposed for prostaglandins in different organs and systems:

**1** **Immune system.** Prostaglandins promote many aspects of the inflammatory process, including the development of pain and fever. Drugs that inhibit prostaglandin synthesis help to alleviate these symptoms.

**2** **Reproductive system.** Prostaglandins may play a role in ovulation and corpus luteum function in the ovaries and in contraction of the uterus. Excessive prostaglandin production may be involved in premature labor, endometriosis, dysmenorrhea (painful menstrual cramps), and other gynecological disorders.

**3** **Digestive system.** The stomach and small intestine produce prostaglandins, which are believed to inhibit gastric secretions and influence intestinal motility and fluid absorption. Since prostaglandins inhibit gastric secretion, drugs that suppress prostaglandin production may predispose a patient to peptic ulcers.

**4** **Respiratory system.** Some prostaglandins cause constriction whereas others cause dilation of blood vessels in the lungs and of bronchiolar smooth muscle. The leukotrienes are potent bronchoconstrictors, and these compounds, together with some prostaglandins, may cause respiratory distress and contribute to bronchoconstriction in asthma.

**5** **Circulatory system.** Some prostaglandins are vasoconstrictors, whereas others are vasodilators. In a fetus, $PGE_2$ is believed to promote dilation of the ductus arteriosus—a short vessel that connects the pulmonary trunk with the aortic arch. After birth, the ductus arteriosus normally closes as a result of a rise in blood oxygen when the baby breathes. If the ductus remains patent (open), however, it can be closed by the administration of drugs that inhibit prostaglandin synthesis.

Prostaglandins also play a role in blood clotting. Thromboxane $A_2$, produced by blood platelets, promotes platelet aggregation and vasoconstriction. Prostacyclin, produced by vascular endothelial cells, inhibits platelet aggregation and promotes vasodilation.

**6** **Urinary system.** Prostaglandins are produced in the renal medulla of the kidneys and cause vasodilation, resulting in increased renal blood flow and increased excretion of water and electrolytes in the urine.

# Clinical Considerations

## *Disorders of the Pituitary Gland*

**Panhypopituitarism** A reduction in the activity of the pituitary gland, called *hypopituitarism* (hi˝po-pĭ-too´ĭ-tă-rizm), can result from intracranial hemorrhage, a blood clot, prolonged steroid treatments, or a tumor. Total pituitary impairment, termed *panhypopituitarism*, brings about a progressive and general loss of hormonal activity. For example, the gonads stop functioning and the person suffers from amenorrhea (lack of menstruation) or aspermia (no sperm production) and loss of pubic and axillary hair. The thyroid and adrenal glands also eventually stop functioning. People with this condition, and those who have had their pituitary gland surgically removed (a procedure called *hypophysectomy*) receive thyroxine, cortisone, growth hormone, and gonadal hormones throughout life to maintain normal body function.

**Abnormal Growth Hormone Secretion** Inadequate growth hormone secretion during childhood causes **pituitary dwarfism.** Hyposecretion of growth hormone in an adult produces a rare condition called **pituitary cachexia** (kă-kek´se-ă) (*Simmonds' disease*). One of the symptoms of this disease is premature aging caused by tissue atrophy. By contrast, oversecretion of growth hormone during childhood causes **gigantism.** Excessive growth hormone secretion in an adult does not cause further growth in height because the epiphyseal plates of the long bones have ossified. Hypersecretion of growth hormone in an adult causes **acromegaly** (ak˝ro-meg´ă-le), in which the person's appearance gradually changes as a result of thickening of bones and growth of soft tissues, particularly in the face, hands, and feet.

**Inadequate ADH Secretion** A dysfunction of the neurohypophysis results in a deficiency in ADH secretion, causing a condition called **diabetes insipidus.** Symptoms of this disease include polyuria (excessive urination), polydipsia (excessive thirst), and severe ionic imbalances. Diabetes insipidus is treated by injections of ADH.

## Disorders of the Adrenal Glands

**Tumors of the Adrenal Medulla** Tumors of the chromaffin cells of the adrenal medulla are referred to as **pheochromocytomas** (*fe-o-kro´´mo-si-to´maz*). These tumors cause hypersecretion of epinephrine and norepinephrine, which produce an effect similar to continuous sympathetic nervous stimulation. The symptoms of this condition are hypertension, elevated metabolism, hyperglycemia and sugar in the urine, nervousness, digestive problems, and sweating. It does not take long for the body to become totally fatigued under these conditions, making the patient susceptible to other diseases.

**Addison's Disease** This disease is caused by inadequate secretion of both glucocorticoids and mineralocorticoids, which results in hypoglycemia, sodium and potassium imbalance, dehydration, hypotension, rapid weight loss, and generalized weakness. A person with this condition who is not treated with corticosteroids will die within a few days because of the severe electrolyte imbalance and dehydration. Another symptom of this disease is darkening of the skin. This is caused by excessive secretion of ACTH and possibly MSH (because MSH is derived from the same parent molecule as ACTH) as a result of lack of negative feedback inhibition of the pituitary by corticosteroids.

**Cushing's Syndrome** Hypersecretion of corticosteroids results in Cushing's syndrome. This is generally caused by a tumor of the adrenal cortex or by oversecretion of ACTH from the adenohypophysis. Cushing's syndrome is characterized by changes in carbohydrate and protein metabolism, hyperglycemia, hypertension, and muscular weakness. Metabolic problems give the body a puffy appearance and can cause structural changes characterized as "buffalo hump" and "moon face." Similar effects are also seen when people with chronic inflammatory diseases receive prolonged treatment with corticosteroids, which are given to reduce inflammation and inhibit the immune response.

**Adrenogenital Syndrome** Usually associated with Cushing's syndrome, this condition is caused by hypersecretion of adrenal sex hormones, particularly the androgens. Adrenogenital syndrome in young children causes premature puberty and enlarged genitals, especially the penis in males and the clitoris in females. An increase in body hair and a deepening of the voice are other characteristics. This condition in a mature woman can cause growth of a beard.

## Disorders of the Thyroid and Parathyroid Glands

**Hypothyroidism** The infantile form of hypothyroidism is known as **cretinism** (*kre´tĭ-nizm*). The clinical symptoms of cretinism are stunted growth, thickened facial features, abnormal bone development, mental retardation, low body temperature, and general lethargy. If cretinism is diagnosed early, it may be successfully treated by administering thyroxine.

**Myxedema** Hypothyroidism in an adult causes **myxedema** (*mik´´sě-de´mă*). This disorder affects body fluids, causing edema and increasing blood volume, hence increasing blood pressure. Symptoms of myxedema include a low metabolic rate, lethargy, and a tendency to gain weight. This condition is treated with thyroxine or with triiodothyronine, which are taken orally (as pills).

**Endemic Goiter** A *goiter* is an abnormal growth of the thyroid gland. When this condition results from inadequate dietary intake of iodine, it is termed *endemic goiter* (fig. 19.26). In this case, growth of the thyroid is due to excessive TSH secretion, which results from low levels of thyroxine secretion. Endemic goiter is thus associated with hypothyroidism.

**Graves' Disease** Graves' disease, also called **toxic goiter,** involves growth of the thyroid associated with hypersecretion of thyroxine. This hyperthyroidism is produced by antibodies that act like TSH and stimulate the thyroid; it is an autoimmune disease. As a consequence of high levels of thyroxine secretion, the metabolic rate and heart rate increase, the person loses weight, and the autonomic nervous system induces excessive sweating. In about half of the cases, **exophthalmos** (*ek´´sof-thal´mos*) (bulging of the eyes) also develops (fig. 19.27) because of edema in the tissues of the eye sockets and swelling of the extrinsic eye muscles.

Endocrine System

......................................

diabetes: Gk. *diabetes,* to pass through a siphon
Addison's disease: from Thomas Addison, English physician, 1793–1860
Cushing's syndrome: from Harvey Cushing, American physician, 1869–1939

......................................

myxedema: Gk. *myxa,* mucus; *oidema,* swelling
Graves' disease: from Robert James Graves, Irish physician, 1796–1853
exophthalmos: Gk. *ex,* out; *opthalmos,* eyeball

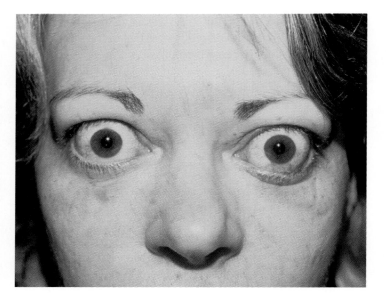

**FIGURE 19.27**
Hyperthyroidism is characterized by an increased metabolic rate, weight loss, muscular weakness, and nervousness. The eyes may also protrude.

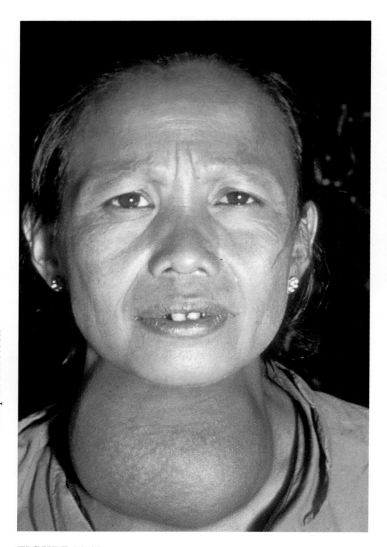

**FIGURE 19.26**
A simple or endemic goiter is caused by insufficient iodine in the diet.

**Disorders of the Parathyroid Glands**    Surgical removal of the parathyroid glands sometimes unintentionally occurs when the thyroid is removed because of a tumor or the presence of Graves' disease. The resulting fall in parathyroid hormone (PTH) causes a decrease in plasma calcium levels, which can lead to severe muscle tetany. Hyperparathyroidism is usually caused by a tumor that secretes excessive amounts of PTH. This stimulates demineralization of bone, which makes the bones soft and raises the blood levels of calcium and phosphate. As a result of these changes, bones are subject to deformity and fracture, and stones (renal calculi) composed of calcium phosphate are likely to develop in the urinary tract.

## Disorders of the Pancreatic Islets

**Diabetes Mellitus**    Diabetes mellitus is characterized by fasting hyperglycemia and the presence of glucose in the urine. There are two forms of this disease. **Type I,** or **insulin-dependent diabetes mellitus,** is caused by destruction of the beta cells and the resulting lack of insulin secretion. **Type II,** or **noninsulin-dependent diabetes mellitus** (which is the more common form) is caused by decreased tissue sensitivity to the effects of insulin, so that increasingly large amounts of insulin are required to produce a normal effect. Both types of diabetes mellitus are also associated with abnormally high levels of glucagon secretion. Diabetes mellitus is discussed in more detail in chapter 27.

**Reactive Hypoglycemia**    People with a genetic predisposition for type II diabetes mellitus often first develop reactive hypoglycemia. In this condition, the rise in blood glucose that follows the ingestion of carbohydrates stimulates excessive secretion of insulin, which in turn causes the blood glucose levels to fall below the normal range. This can result in weakness, changes in personality, and mental disorientation.

## Inhibitors of Prostaglandin Synthesis

*Aspirin* is the most widely used member of a class of drugs known as **nonsteroidal anti-inflammatory drugs.** Other members of this class are *indomethacin* and *ibuprofen*. These drugs produce their effects because they specifically inhibit the cyclo-oxygenase enzyme that is needed for prostaglandin synthesis. Through this action, the drugs inhibit inflammation but produce some unwanted side effects, including gastric bleeding, possible kidney problems, and prolonged clotting time.

It is now known that there are two isoenzyme forms (chapter 4) of cyclo-oxygenase. The *type I* form is produced constitutively (that is, in a constant fashion) by cells of the stomach and kidneys, and by blood platelets, which are cellular structures involved in blood clotting (chapter 20). The *type II* form of the enzyme is induced in a number of cells in response to cytokines involved in inflammation, and the prostaglandins produced by this isoenzyme promote the inflammatory condition.

When aspirin and the other drugs of its class inhibit the type I isoenzyme of cyclo-oxygenase, the synthesis of prostacyclin is inhibited. Since prostacyclin protects the stomach lining, inhibition of its synthesis with aspirin may be responsible for the stomach irritation caused by aspirin and indomethacin, the two most potent inhibitors of the type I isoenzyme. Inhibition of the type I isoenzyme is thus responsible for the negative side effects of these drugs. Inhibition of the type II isoenzyme is responsible for the intended anti-inflammatory benefits of the drugs, and research is currently underway to develop new drugs that more selectively inhibit the type II isoenzyme of cyclo-oxygenase.

There is, however, one important benefit derived from the inhibition of the type I isoenzyme by aspirin. The type I isoenzyme is the form of cyclo-oxygenase present in blood platelets, where it is needed for the production of thromboxane $A_2$. Since this prostaglandin is needed for the ability of platelets to aggregate, inhibition of its synthesis by aspirin reduces the ability of the blood to clot. While this can be a negative effect, low doses of aspirin have been shown to significantly reduce the risk of heart attacks and strokes (chapter 21) by reducing platelet function. It should be noted that this beneficial effect is produced by lower doses of aspirin than are commonly taken to reduce inflammation.

........................................................................................................

# Chapter Summary

### Endocrine Glands and Hormones (pp. 514–519)

1. Hormones are chemicals, including steroids, catecholamines, and polypeptides, that are secreted into the blood by endocrine glands.
2. Precursors of active hormones may be called either prohormones or prehormones.
   a. Prohormones are relatively inactive precursor molecules made in the endocrine cells.
   b. Prehormones are the normal secretions of an endocrine gland that, in order to be active, must be converted to other derivatives by target cells.
3. Hormones can interact in permissive, synergistic, or antagonistic ways.

### Pituitary Gland (pp. 519–526)

1. The pituitary gland secretes 8 hormones.
   a. The anterior pituitary secretes growth hormone, thyroid-stimulating hormone, adrenocorticotropic hormone, follicle-stimulating hormone, luteinizing hormone, and prolactin.
   b. The posterior pituitary secretes antidiuretic hormone (also called vasopressin) and oxytocin.
2. The hormones of the posterior pituitary are produced in the hypothalamus and transported to the posterior pituitary by the hypothalamohypophyseal nerve tract.
3. Secretions of the anterior pituitary are controlled by hypothalamic hormones that stimulate or inhibit secretions of the anterior pituitary.
   a. Hypothalamic hormones include TRH, CRH, GnRH, PIH, somatostatin, and a growth hormone–releasing hormone.
   b. These hormones are carried to the anterior pituitary by the hypothalamohypophyseal portal system.
4. Secretions of the anterior pituitary are also regulated by the feedback (usually negative feedback) of hormones from the target glands.
5. Higher brain centers, acting through the hypothalamus, can influence pituitary secretion.

### Adrenal Glands (pp. 526–528)

1. The adrenal cortex secretes mineralocorticoids (mainly aldosterone), glucocorticoids (mainly cortisol), and sex steroids (primarily weak androgens).
   a. The glucocorticoids help to regulate energy balance; they also can inhibit inflammation and suppress immune function.

# Interactions of the Endocrine System with Other Body Systems

## Integumentary System
- Protects the body from pathogens and helps maintain body temperature
- Sex hormones activate sebaceous glands and promote development of apocrine sweat glands

## Skeletal System
- Supports and protects certain endocrine glands (e.g., the pituitary gland)
- Stores calcium needed for endocrine function
- Anabolic hormones, including growth hormone, stimulate bone development
- Parathyroid hormone and calcitonin regulate calcium deposition and resorption in bones

## Muscular System
- Skeletal muscles provide protection for certain endocrine glands
- Cardiac and smooth muscles help deliver blood to endocrine system
- Anabolic hormones promote muscle growth and development; catabolic hormones promote breakdown of muscle proteins

## Nervous System
- Hypothalamus secretes hormones that control pituitary gland
- Autonomic nerves regulate secretion of adrenal medulla and other endocrine glands
- Sex hormones from the gonads, melatonin from the pineal gland, and many other hormones regulate different aspects of brain function

## Circulatory System
- Transports oxygen and nutrients to endocrine glands and removes wastes
- Transports hormones from endocrine glands to target cells
- Epinephrine and norepinephrine from adrenal medulla stimulate heart
- Thyroxine and other hormones have permissive effects on autonomic regulation of cardiovascular system

## Lymphatic System
- Protects against infections that could damage endocrine system
- Hormones from thymus gland help regulate lymphocyte
- Adrenal corticosteroids have suppressive effects on immune system

## Respiratory System
- Provides oxygen for aerobic respiration of endocrine glands and allows carbon dioxide to be eliminated
- Thyroxine and epinephrine stimulate cell respiration
- Epinephrine promotes bronchodilation

## Urinary System
- Eliminates metabolic wastes
- Releases renin and erythropoietin when local blood pressure declines
- Antidiuretic hormone, aldosterone, and atrial natriuretic hormone regulate functions of kidneys

## Digestive System
- Provides nutrients to endocrine system for metabolism and synthesis of hormones
- Hormones of stomach and small intestine help to coordinate activities of different regions of GI tract

## Reproductive System
- Gonadal hormones influence the function of the pituitary and other endocrine glands
- Several hormones regulate reproductive

b. The pituitary-adrenal axis is stimulated by stress as part of the general adaptation syndrome.
2. The adrenal medulla secretes epinephrine and lesser amounts of norepinephrine; these hormones complement the action of the sympathetic nervous system.

## Thyroid and Parathyroid Glands (pp. 528–530)

1. The thyroid follicles secrete tetraiodothyronine ($T_4$, or thyroxine) and lesser amounts of triiodothyronine ($T_3$).
   a. These hormones are formed within the colloid of the thyroid follicles.
   b. The parafollicular cells of the thyroid secrete the hormone calcitonin, which may act to lower blood calcium levels.
2. The parathyroids are small structures embedded within the thyroid gland; the parathyroids secrete a hormone that promotes a rise in blood calcium levels.

## Pancreas and Other Endocrine Glands (pp. 530–534)

1. Beta cells in the pancreatic islets (islets of Langerhans) secrete insulin; alpha cells secrete glucagon.
   a. Insulin lowers blood glucose and stimulates the production of glycogen, fat, and protein.
   b. Glucagon raises blood glucose by stimulating the breakdown of liver glycogen; it also promotes lipolysis and the formation of ketone bodies.
   c. The secretion of insulin is stimulated by a rise in blood glucose following meals; the secretion of glucagon is stimulated by a fall in blood glucose during periods of fasting.
2. The pineal gland, located on the roof of the third ventricle of the brain, secretes melatonin; this hormone may play a role in regulating reproductive function.
3. The thymus is the site of T cell lymphocyte production and secretes a number of hormones that may help to regulate the immune system.
4. The gastrointestinal tract secretes a number of hormones that help to regulate functions of the digestive system.
5. The gonads secrete sex steroid hormones.
   a. Interstitial (Leydig) cells of the testes secrete testosterone and other androgens.
   b. Granulosa cells of the ovarian follicles secrete estrogen.
   c. The corpus luteum of the ovaries secretes progesterone, as well as estrogen.
6. The placenta secretes estrogen, progesterone, and a variety of polypeptide hormones that have actions similar to some anterior pituitary hormones.

## Mechanisms of Hormone Action (pp. 534–540)

1. Steroid and thyroid hormones enter their target cells, where they bind to receptor proteins.
   a. Thyroid hormones attach to chromatin-bound receptors located in the nucleus.
   b. Steroid hormones bind to cytoplasmic receptor proteins and translocate to the nucleus.
   c. Attachment of the hormone–receptor protein complex to the chromatin activates genes and thereby stimulates RNA and protein synthesis.
2. Amine, polypeptide, and glycoprotein hormones bind to receptor proteins on the outer surface of the target cell membrane.
   a. In many cases, this leads to the intracellular production of cyclic AMP, which serves as a second messenger in the action of these hormones.
   b. In other cases, $Ca^{++}$ and other substances may serve as a second messenger in the action of the hormone.
3. The effects of a hormone in the body depend on its concentration.
   a. Abnormally high levels of a hormone in the blood can result in atypical effects.
   b. Target tissues can become desensitized by high hormone concentrations.

## Autocrine and Paracrine Regulation (pp. 540–542)

1. Autocrine and paracrine regulators act within the same organ in which they are produced.
2. Prostaglandins are special 20-carbon-long fatty acids produced by many different organs. They usually have regulatory functions within the organ in which they are produced.

······································································································

# Review Activities

## Objective Questions

Match the gland to its embryonic origin.

1. Adenohypophysis    a. endoderm of pharynx
2. Neurohypophysis
3. Adrenal medulla    b. diverticulum from brain
   c. endoderm of foregut
   d. neural crest ectoderm
   e. hypophyseal pouch

4. Hypothalamic releasing hormones
   a. are secreted into capillaries in the median eminence.
   b. are transported by portal veins to the anterior pituitary.
   c. stimulate the secretion of specific hormones from the anterior pituitary.
   d. all of the above apply.
5. The hormone primarily responsible for setting the basal metabolic rate and for promoting the maturation of the brain is
   a. cortisol.
   b. ACTH.
   c. TSH.
   d. thyroxine.
6. Which of the following statements about the adrenal cortex is *true?*
   a. It is not innervated by nerve fibers.
   b. It secretes some androgens.
   c. The zona granulosa secretes aldosterone.
   d. The zona fasciculata is stimulated by ACTH.
   e. all of the above are true.
7. The hormone insulin
   a. is secreted by alpha cells in the pancreatic islets.
   b. is secreted in response to a rise in blood glucose.
   c. stimulates the production of glycogen and fat.
   d. both a and b apply.
   e. both b and c apply.

Match the hormone with the primary agent that stimulates its secretion.

8.  Epinephrine
9.  Thyroxine
10. Corticosteroids
11. ACTH

a.  TSH
b.  ACTH
c.  growth hormone
d.  sympathetic nerves
e.  CRH

12. Steroid hormones are secreted by
    a.  the adrenal cortex.
    b.  the gonads.
    c.  the thyroid.
    d.  both a and b.
    e.  both b and c.

13. The secretion of which hormone would be *increased* in a person with endemic goiter?
    a.  TSH
    b.  thyroxine
    c.  triiodothyronine
    d.  all of the above apply.

14. Which hormone uses cAMP as a second messenger?
    a.  testosterone
    b.  cortisol
    c.  insulin
    d.  epinephrine

15. Which of the following terms best describes the interactions of insulin and glucagon?
    a.  synergistic
    b.  permissive
    c.  antagonistic
    d.  cooperative

## Essay Questions

1.  Explain how the regulation of the neurohypophysis and adrenal medulla relates to their embryonic origins.
2.  Compare steroid and polypeptide hormones in terms of their mechanism of action in target organs.
3.  Discuss the significance of the term *trophic* with respect to the actions of anterior pituitary hormones.
4.  Suppose a drug blocks the conversion of $T_4$ to $T_3$. Describe the effects this drug would have (a) on TSH secretion, (b) on thyroxine secretion, and (c) on the size of the thyroid gland.
5.  Explain why the anterior pituitary is sometimes referred to as the "master gland" and why this reference is misleading.
6.  Suppose a person's immune system produced antibodies against insulin receptor proteins. Describe the possible effect of this condition on carbohydrate and fat metabolism.

## Gundy/Weber Software 💾

The tutorial software accompanying Chapter 19 is Volume 7—Endocrine System.

## Explorations ✪

A module of correlating material is available from the Wm. C. Brown CD–ROM: *Explorations*. It is #11 Hormone Action.

# circulatory system: blood

## objectives

- Describe the major components of the circulatory system.

- List the functions of the circulatory system.

- Describe the composition of plasma and the functions of its constituents.

- Discuss the origin, structure, function, and life span of erythrocytes, leukocytes, and platelets.

- Discuss the origin and function of erythropoietin and the lymphokines.

- Discuss the ABO and Rh blood systems, comment on their clinical significance, and explain how the agglutination reaction occurs.

- Describe the platelet release reaction and the formation of a platelet plug.

- Compare the extrinsic and intrinsic clotting pathways.

- Describe the mechanisms of action of different anticoagulants.

- Describe the physiological mechanisms that promote the dissolution of a blood clot.

# Functions and Components of the Circulatory System

*Blood serves numerous functions, including the transport of respiratory gases, nutritive molecules, metabolic wastes, and hormones. Blood circulates through the body in a system of vessels from and back to the heart.*

A unicellular organism can provide for its own maintenance and continuity by performing the wide variety of functions needed for life. By contrast, the complex human body is composed of trillions of specialized cells that demonstrate a division of labor. Cells of a multicellular organism depend on one another for the very basics of their existence. The majority of the cells of the body are firmly implanted in tissues and are incapable of procuring food and oxygen on their own, or even moving away from their own wastes. Therefore, a highly specialized and effective means of transporting materials within the body is needed.

The blood serves this transportation function. An estimated 60,000 miles of vessels throughout the body of an adult ensure that continued sustenance reaches each of the trillions of living cells. However, the blood can also serve to transport disease-causing viruses, bacteria, and their toxins. To guard against this, the circulatory system has protective mechanisms—the white blood cells and lymphatic system. In order to perform its various functions, the circulatory system works together with the respiratory, urinary, digestive, endocrine, and integumentary systems in maintaining homeostasis.

## Functions of the Circulatory System

The many functions of the circulatory system can be grouped into three broad areas: transportation, regulation, and protection.

**1  Transportation.** All of the substances essential for cellular metabolism are transported by the circulatory system. These substances can be categorized as follows:

a  *Respiratory.* Red blood cells, or **erythrocytes** (ĕ-rith′rŏ-sītz), transport oxygen to the tissue cells. In the lungs, oxygen from the inhaled air attaches to hemoglobin molecules within the erythrocytes and is transported to the cells for aerobic respiration. Carbon dioxide produced by cell respiration is carried by the blood to the lungs for elimination in the exhaled air.

b  *Nutritive.* The digestive system is responsible for the breakdown of food so that it can be absorbed through the intestinal wall and into the blood vessels of the circulatory system. The blood then carries these absorbed products of digestion through the liver and to the cells of the body.

c  *Excretory.* Metabolic wastes, excessive water and ions, as well as other molecules in plasma (the fluid portion of blood) are filtered through the capillaries of the kidneys and excreted in urine.

**2  Regulation.** The blood carries hormones and other regulatory molecules from their site of origin to distant target tissues. Diversion of blood from deeper vessels to surface vessels in the skin aids in the regulation of body temperature.

**3  Protection.** The circulatory system protects against injury and foreign microbes or toxins introduced into the body. The clotting mechanism protects against blood loss when vessels are damaged, and white blood cells, or **leukocytes** (loo′kŏ-sītz), render the body immune to many disease-causing agents. Leukocytes may also protect the body through phagocytosis, as described in chapter 3 (fig. 3.3).

## Major Components of the Circulatory System

The circulatory system is frequently divided into the **cardiovascular system,** which consists of the heart and blood vessels, and the **lymphatic system,** which consists of lymph vessels, lymph nodes, and lymphoid organs (spleen, thymus, and tonsils).

The **heart** is a four-chambered double pump. Its pumping action creates the pressure head needed to push blood in the vessels to the lungs and body cells. At rest, the heart of an adult pumps about 5 liters of blood per minute. It takes about 1 minute for blood to be circulated to the most distal extremity and back to the heart.

**Blood vessels** form a tubular network that permits blood to flow from the heart to all the living cells of the body and then back to the heart. *Arteries* carry blood away from the heart, whereas *veins* return blood to the heart. Arteries and veins are continuous with each other through smaller blood vessels.

Arteries branch extensively to form a "tree" of decreasingly smaller vessels. Those that are microscopic in diameter are called *arterioles* (ar-tir′e-olz). Blood passes from the arterial to the venous system in *capillaries* (kap′y-lar-ēz), which are the thinnest and most numerous of the blood vessels. All exchanges of fluid, nutrients, and wastes between the blood and tissues occur across the walls of capillaries. Blood flows through capillaries into microscopic-sized vessels, called *venules* (ven′yoolz), which deliver blood into increasingly larger veins that eventually return the blood to the heart.

As blood plasma passes through capillaries, the hydrostatic pressure of the blood forces some of this fluid out of the capillary walls. Fluid derived from plasma that passes out of capillary walls into the surrounding tissues is called *tissue fluid,* or *interstitial* (in″ter-stish′al) *fluid.* Some of this fluid returns directly to capillaries, and some enters into **lymph vessels** located in the connective tissues around the blood vessels. Fluid in lymph vessels is called *lymph (limf)*.

This fluid is returned to the venous blood at specific sites. **Lymph nodes,** positioned along the way, cleanse the lymph prior to its return to the venous blood. The lymphatic system is discussed as part of the immune system in chapter 23.

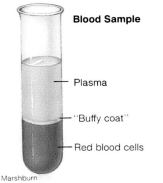

**Blood Sample**

- Plasma
- "Buffy coat"
- Red blood cells

Marshburn

# Composition of the Blood

*Blood consists of formed elements that are suspended and carried in the plasma. These formed elements and their major functions include erythrocytes (oxygen transport), leukocytes (immune defense), and platelets (blood clotting). Plasma contains different types of proteins and many water-soluble molecules.*

The average-sized adult has about 5 liters of blood, constituting about 8% of the total body weight. Blood leaving the heart is referred to as *arterial blood*. Arterial blood, with the exception of that going to the lungs, is bright red in color due to a high concentration of oxyhemoglobin (the combination of oxygen and hemoglobin) in the erythrocytes. *Venous blood* is blood returning to the heart and, except for the venous blood from the lungs, has a darker color (due to hemoglobin that is no longer combined with oxygen).

Blood is composed of a cellular portion, called **formed elements,** and a fluid portion, called **plasma.** When a blood sample is centrifuged, the heavier formed elements are packed into the bottom of the tube, leaving plasma at the top (fig. 20.1). The formed elements constitute approximately 45% of the total blood volume; the plasma accounts for the remaining 55%. The *hematocrit (HCT)* is the percentage of red blood cells per given volume of blood and is an important indicator of the oxygen carrying capacity of blood.

## Plasma

**Plasma** is a straw-colored liquid consisting of water and dissolved solutes. The major solute of the plasma in terms of its concentration is $Na^+$. In addition to $Na^+$, plasma contains many other salts and ions, as well as such organic molecules as metabolites, hormones, enzymes, antibodies, and

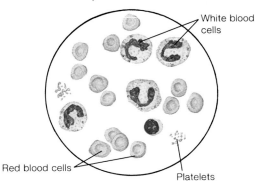

**Peripheral Blood Smear**

White blood cells

Red blood cells

Platelets

**FIGURE 20.1**

Blood cells become packed at the bottom of the test tube.

other proteins. The values of some of these constituents of plasma are indicated in table 20.1.

**Plasma Proteins**  Plasma proteins constitute 7%–9% of the plasma. The three types of proteins are albumins, globulins, and fibrinogen. **Albumins** (*al-byoo´minz*) account for most (about 60%) of the plasma proteins and are the smallest in size. They are produced by the liver and provide the osmotic pressure needed to draw water from the surrounding tissue fluid into the capillaries. This action is needed to maintain blood volume and pressure. **Globulins** (*glob´yoo-linz*) make up about 36% of the plasma proteins and are grouped into three subtypes: **alpha globulins, beta globulins,** and **gamma globulins.** The alpha and beta globulins are produced by the liver and transport lipids and fat-soluble vitamins in the blood. Gamma globulins are antibodies produced by lymphocytes (one of the formed elements found in blood and lymphoid tissues) and function in immunity. **Fibrinogen** (*fi-brin´ŏ-jen*), which accounts for only about 4% of the total plasma proteins, is an important clotting factor produced by the liver. During the process of clot formation (described later in this chapter), fibrinogen is converted into insoluble threads of *fibrin*. The fluid from clotted blood, which is called **serum,** thus does not contain fibrinogen but is otherwise identical to plasma.

**Plasma Volume and pH**  A number of regulatory mechanisms in the body maintain homeostasis of the plasma volume. If the body should lose water, the remaining plasma becomes excessively concentrated (its osmolality increases). This condition is detected by osmoreceptors in the hypothalamus, resulting in a sensation of thirst and an increase in the secretion of antidiuretic hormone (ADH) from the posterior pituitary (chapter 19). This hormone promotes water retention by the kidneys, which—together with increased fluid consumption—helps to compensate for the dehydration and lowered blood volume. This regulatory mechanism is very important in maintaining blood pressure (to be discussed in chapter 22).

The plasma pH is maintained within a very narrow range (7.35–7.45) through numerous mechanisms. Acids in

hematocrit: Gk. *haima*, blood; *krino*, to separate

albumin: L. *albumen*, white
globulin: L. *globulus*, small globe
serum: L. *serum*, liquid

## Table 20.1  Representative normal plasma values

| Measurement | Normal range |
|---|---|
| Blood volume | 80–85 ml/kg body weight |
| Blood osmolality | 280–296 mOsm |
| Blood pH | 7.35–7.45 |
| **Enzymes** | |
| Creatine phosphokinase (CPK) | Female: 10–79 U/L |
| | Male: 17–148 U/L |
| Lactic dehydrogenase (LDH) | 45–90 U/L |
| Phosphatase (acid) | Female: 0.01–0.56 Sigma U/ml |
| | Male: 0.13–0.63 Sigma U/ml |
| **Hematology Values** | |
| Hematocrit | Female: 37%–48% |
| | Male: 45%–52% |
| Hemoglobin | Female: 12–16 g/100 ml |
| | Male: 13–18 g/100 ml |
| Red blood cell count | 4.2–5.9 million/mm$^3$ |
| White blood cell count | 4,300–10,880/mm$^3$ |
| **Hormones** | |
| Testosterone | Male: 300–1,100 ng/100 ml |
| | Female: 25–90 ng/100 ml |
| Adrenocorticotropic hormone (ACTH) | 15–70 pg/ml |
| Growth hormone | Children: over 10 ng/ml |
| | Adult male: below 5 ng/ml |
| Insulin | 6–26 µU/ml (fasting) |
| **Ions** | |
| Bicarbonate | 24–30 mmol/l |
| Calcium | 2.1–2.6 mmol/l |
| Chloride | 100–106 mmol/l |
| Potassium | 3.5–5.0 mmol/l |
| Sodium | 135–145 mmol/l |
| **Organic Molecules (Other)** | |
| Cholesterol | 120–220 mg/100 ml |
| Glucose | 70–110 mg/100 ml (fasting) |
| Lactic acid | 0.6–1.8 mmol/l |
| Protein (total) | 6.0–8.4 g/100 ml |
| Triglyceride | 40–150 mg/100 ml |
| Urea nitrogen | 8–25 mg/100 ml |
| Uric acid | 3–7 mg/100 ml |

Exerpted from material originally appearing in *The New England Journal of Medicine*, "Case Records of the Massachusetts General Hospital," Vol. 302, No. 1, pp. 37–48, January 3, 1980 and "Case Records of the Massachusetts General Hospital," Vol. 314, pp. 39–49. Copyright © 1980, 1986 by *The New England Journal of Medicine*. Reprinted by permission.

the blood are buffered by bicarbonate in the plasma, and blood pH is maintained by the actions of the lungs and kidneys. The lungs function in acid-base balance by eliminating carbon dioxide (see chapter 24), which regulates the amount of carbonic acid in the blood. The kidneys function in acid-base balance by excreting hydrogen ions and retaining plasma bicarbonate (see chapter 25).

## Formed Elements of Blood

The formed elements of blood include two types of blood cells: **erythrocytes** or red blood cells, and **leukocytes,** or white blood cells. Erythrocytes are by far the more numerous of the two: a cubic millimeter of blood contains 5.1 million to 5.8 million erythrocytes in males and 4.3 million to 5.2 million erythrocytes in females. The same volume of blood, by contrast, contains only 5000 to 9000 leukocytes. **Platelets,** a third type of formed element, are cell fragments of extremely large cells found in bone marrow.

**Erythrocytes**   Erythrocytes are flattened, biconcave discs, about 7µm in diameter and 2.2 µm thick (fig. 20.2). Their unique shape relates to their function of transporting oxygen and provides an increased surface area through which gas can diffuse. Erythrocytes lack nuclei and mitochondria (they get energy from anaerobic respiration). Because of these deficiencies, erythrocytes have a circulating life span of only about 120 days before they are destroyed by phagocytic cells in the liver, spleen, and bone marrow.

Each erythrocyte contains approximately 280 million *hemoglobin* molecules, which give blood its red color. Each hemoglobin molecule consists of four polypeptide chains, called *globins*, and four iron-containing, ringlike *heme* groups. Each heme group contains one atom of iron, which is able to combine with oxygen in the lungs and release oxygen in the tissues. The role of hemoglobin in gas transport is described in more detail in chapter 24.

*Anemia* refers to any condition in which there is an abnormally low hemoglobin concentration and/or erythrocyte count. The most common type is *iron-deficiency anemia*, caused by a deficiency in iron, which is an essential component of the hemoglobin molecule. In *pernicious anemia*, there is a deficiency of vitamin B$_{12}$, which is needed for red blood cell production. In most cases, this results from atrophy of the glandular mucosa of the stomach, which normally secretes a substance called *intrinsic factor*. In the absence of that intrinsic factor, vitamin B$_{12}$ cannot be absorbed by intestinal cells. *Aplastic anemia* is due to destruction of the bone marrow, which may be caused by chemicals (including benzene and arsenic) or by X rays.

erythrocytes: Gk. *erythros*, red; *kytos*, hollow (cell)
leukocytes: Gk. *leukos*, white; *kytos*, hollow (cell)

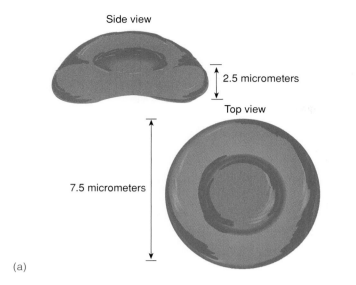

Side view

2.5 micrometers

Top view

7.5 micrometers

(a)

(b)

**FIGURE 20.2**

Erythrocytes. (*a*) A diagram and (*b*) a scanning electron micrograph showing erythrocytes clinging to the tip of a hypodermic needle.

**Leukocytes**  Leukocytes, or white blood cells, differ from erythrocytes in several ways. Leukocytes contain nuclei and mitochondria and can move in an amoeboid fashion (erythrocytes are not able to move independently). Because of their ameboid ability, leukocytes can squeeze through pores in capillary walls and move to a site of infection, whereas erythrocytes usually remain confined within blood vessels. The movement of leukocytes through capillary walls is called *diapedesis*.

· · · · · · · · · · · · · · · · · · · · · · · · · · · · · ·

diapedesis: Gk. *dia*, through; *pedester*, on foot

Leukocytes, which are almost invisible under the microscope unless they are stained, are classified according to their stained appearance. Those leukocytes that have granules in their cytoplasm are called **granular leukocytes;** those that do not are called **agranular** (or nongranular) **leukocytes.**

The stain used to identify leukocytes is usually a mixture of a pink-to-red stain called eosin and a blue-to-purple stain called a "basic stain." Granular leukocytes with pink-staining granules are therefore called **eosinophils** (*e″ŏ-sin′ŏ-filz*), and those with blue-staining granules are called **basophils.** Those with granules that have little affinity for either stain are **neutrophils.** Neutrophils are the most abundant type of leukocyte, accounting for 50% to 70% of the leukocytes in the blood. Because of their oddly shaped nuclei, with lobes and strands, they are sometimes called **polymorphonuclear** (*pol″e-mor″fo-noo′kle-ar*) **neutrophils (PMN).**

There are two types of agranular leukocytes: **lymphocytes** and **monocytes.** Lymphocytes are usually the second most numerous type of leukocyte in the blood. They are small cells with round nuclei and little cytoplasm. Monocytes, by contrast, are the largest of the leukocytes and generally have kidney- or horseshoe-shaped nuclei. In addition to these two cell types, there are smaller numbers of cells, derived from lymphocytes, called **plasma cells.** Plasma cells produce and secrete large amounts of antibodies. The immune functions of the different white blood cells is described in more detail in chapter 23.

**Platelets**  Platelets, or **thrombocytes,** are the smallest of the formed elements. As mentioned previously, they are actually fragments of *megakaryocytes*—large cells found in bone marrow—hence the term *formed elements* rather than *blood cells* as a collective designation for erythrocytes, leukocytes, and platelets. The fragments that enter the circulation as platelets lack nuclei but, like leukocytes, are capable of amoeboid movement. The platelet count per cubic millimeter of blood is 250,000 to 450,000. Platelets survive for about 5 to 9 days before being destroyed by the spleen and liver.

Platelets play an important role in blood clotting. They constitute the major portion of the mass of a clot, and phospholipids in their cell membranes serve to activate the clotting factors in plasma that result in threads of fibrin, which reinforce the platelet plug. Platelets that attach together in a blood clot also release a chemical called *serotonin*, which stimulates constriction of the blood vessels, thus reducing the flow of blood to the injured area.

· · · · · · · · · · · · · · · · · · · · · · · · · · · · · ·

thrombocytes: Gk. *thrombos*, clot; *kytos*, hollow (cell)

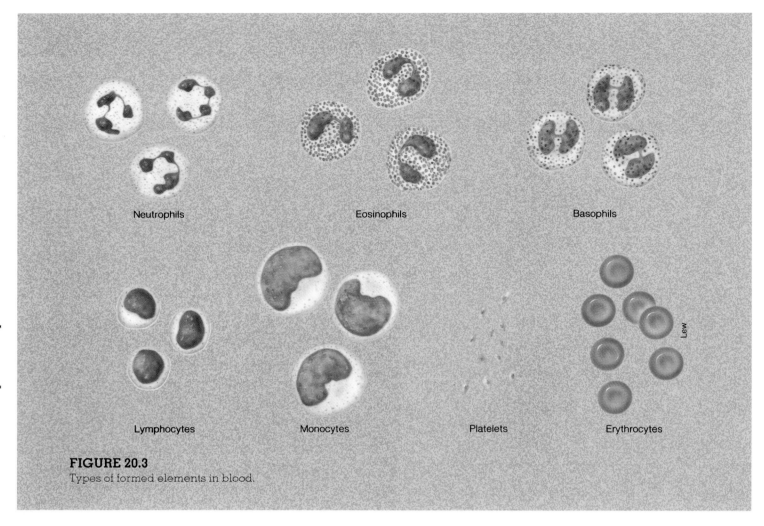

Neutrophils     Eosinophils     Basophils

Lymphocytes     Monocytes     Platelets     Erythrocytes

**FIGURE 20.3**
Types of formed elements in blood.

The appearance of the formed elements of the blood is shown in figure 20.3, and a summary of the characteristics of these formed elements is presented in table 20.2.

 Blood cell counts are an important source of information in assessing a person's health. An abnormal increase in erythrocytes, for example, is termed *polycythemia* (pol″e-si-the′me-ă) and is indicative of several dysfunctions. An abnormally low red blood cell count is termed *anemia*. An elevated leukocyte count, called *leukocytosis*, is often associated with localized infection. A large number of immature leukocytes within a blood sample is diagnostic of the disease *leukemia*.

## Hemopoiesis

Blood cells are constantly being formed through a process called **hemopoiesis** (he″mo-poi-e′sis). The hematopoietic *stem cells*—those that give rise to blood cells—originate in the yolk sac of the human embryo and then migrate to the liver. Hemopoiesis thus occurs in the liver of a fetus. The

hemopoiesis: Gk. *haima*, blood; *poiesis*, production

stem cells then migrate to the bone marrow and, shortly after birth, the liver ceases to be a source of blood cell production.

The term *erythropoiesis* refers to the formation of erythrocytes, and *leukopoiesis* to the formation of leukocytes. These processes occur in two classes of tissues after birth. **Myeloid tissue** is the red bone marrow of the long bones, ribs, sternum, bodies of vertebrae, and portions of the skull. **Lymphoid tissue** includes the lymph nodes, tonsils, spleen, and thymus. The bone marrow produces all of the different types of blood cells, and the lymphocytes are also produced in the lymphoid tissue.

Erythropoiesis is an extremely active process. It is estimated that about 2.5 million erythrocytes are produced every second in order to replace those that are continuously destroyed by the spleen and liver. (Recall that the life span of an erythrocyte is approximately 120 days.) During the destruction of erythrocytes, iron is salvaged and returned to the red bone marrow, where it is used again in the formation of erythrocytes. Agranular leukocytes remain functional for 100 to 300 days under normal conditions. Granular leukocytes, by contrast, have an extremely short life span of 12 hours to 3 days.

# Table 20.2 Formed elements of blood

| Component | Description | Number present | Function |
|---|---|---|---|
| Erythrocyte (red blood cell) | Biconcave disc without nucleus; contains hemoglobin; survives 100–120 days | 4,000,000 to 6,000,000/mm$^3$ | Transports oxygen and carbon dioxide |
| Leukocytes (white blood cells) | | 5000 to 10,000/mm$^3$ | Aid in defense against infections by microorganisms |
| Granular leukocytes | About twice the size of red blood cells; cytoplasmic granules present; survive 12 hours to 3 days | | |
| 1. Neutrophil | Nucleus with 2–5 lobes; cytoplasmic granules stain slightly pink | 54%–62% of white cells present | Phagocytic |
| 2. Eosinophil | Nucleus bilobed; cytoplasmic granules stain red in eosin stain | 1%–3% of white cells present | Helps to detoxify foreign substances; secretes enzymes that break down clots |
| 3. Basophil | Nucleus lobed; cytoplasmic granules stain blue in hematoxylin stain | Less than 1% of white cells present | Releases anticoagulant heparin |
| Agranular leukocytes | Cytoplasmic granules absent; survive 100–300 days | | |
| 1. Monocyte | 2–3 times larger than red blood cell; nuclear shape varies from round to lobed | 3%–9% of white cells present | Phagocytic |
| 2. Lymphocyte | Only slightly larger than red blood cell; nucleus nearly fills cell | 25%–33% of white cells present | Provides specific immune response (including antibodies) |
| Platelet (thrombocyte) | Cytoplasmic fragment; survives 5–9 days | 250,000 to 450,000/mm$^3$ | Aids clotting |

 The major purpose of a *bone marrow transplantation* is to provide competent hematopoietic stem cells to the recipient. If the bone marrow is returned to the same person, the procedure is called an *autotransplantation*. If the donor and recipient are different people, it is an *allogeneic transplantation*.

Hemopoiesis begins the same way in both myeloid and lymphoid tissue (fig. 20.4). Undifferentiated (unspecialized) cells gradually differentiate (specialize) to become **hemocytoblasts.** These are the stem cells that give rise to the blood cells. Hemocytoblasts can duplicate themselves by mitosis, thus ensuring that the parent population never becomes depleted. As the cells become differentiated, they develop membrane receptors for chemical signals that cause further development along particular lines. Hemocytoblasts may develop into *proerythroblasts* (which form erythrocytes), *myeloblasts* (which form granular leukocytes), *lymphoblasts* (which form lymphocytes), *monoblasts* (which form monocytes), or *megakaryoblasts* (which form thrombocytes).

The production of different subtypes of lymphocytes is stimulated by chemicals called **lymphokines,** which are discussed as part of the immune system in chapter 23. The production of erythrocytes is stimulated by a hormone called **erythropoietin,** which is secreted by the kidneys. Erythropoietin, a glycoprotein hormone containing 166 amino acids, stimulates cell division and differentiation of proerythroblasts in bone marrow. The secretion of erythropoietin by the kidneys is stimulated whenever the delivery of oxygen to the kidneys and other organs is lower than normal. Under these conditions—which can occur, for example, when a person lives at high altitude—the increased production of erythrocytes allows the blood to carry a higher concentration of oxygen to the tissues.

Considering that the kidneys produce erythropoietin, and that erythropoietin is needed to stimulate erythrocyte production, it is not surprising that patients with kidney disease may also suffer from anemia. This anemia can now be treated by giving these patients human erythropoietin. This exciting therapeutic breakthrough was made possible by the commercial production of erythropoietin using cultured mammalian cells that had incorporated the human gene for erythropoietin through genetic engineering techniques.

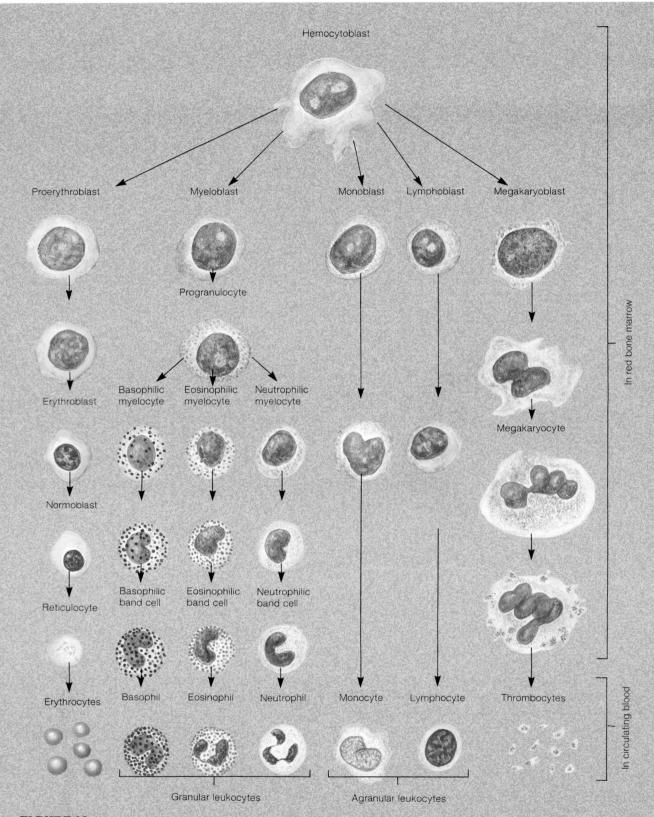

In red bone marrow

In circulating blood

**FIGURE 20.4**

The processes of hemopoiesis. Formed elements begin as hemocytoblasts and differentiate into the various kinds of blood cells in accordance with the needs of the body.

## Erythrocyte Antigens and Blood Typing

All cells in the body contain certain molecules on their surfaces that can be recognized as foreign by the immune system of another individual. These molecules are known as *antigens*. As part of the immune response, particular lymphocytes secrete a class of proteins called **antibodies** that bond in a specific fashion to antigens. The specificity of antibodies for antigens is analogous to the specificity of enzymes for their substrates, and of receptor proteins for neurotransmitters and hormones. A complete description of antibodies and antigens is provided in chapter 23.

**ABO System** The distinguishing antigens on other cells are far more varied than the antigens on red blood cells. Red blood cell antigens, however, are extremely important clinically because their types must be matched between donors and recipients for blood transfusions. Of the several groups of red blood cell antigens, the major group constitutes the **ABO system.** In terms of the antigens present on the red blood cell surface, a person may be *type A* (with only A antigens), *type B* (with only B antigens), *type AB* (with both A and B antigens), or *type O* (with neither A nor B antigens). It should again be noted that the blood type denotes the class of antigens found on the red blood cell surface.

Each person inherits two genes (one from each parent) that control the production of the ABO antigens. The genes for A or B antigens are dominant genes, since O simply means the absence of A and B. A person who is type A, therefore, may have inherited an A gene from each parent (may have the genotype AA), or an A gene from one parent and an O gene from the other (and thus have the genotype AO). Likewise, a person who is type B may have the genotype BB or BO. It follows that a type O person inherited an O gene from each parent (has the genotype OO), whereas a type AB person inherited the A gene from one parent and the B gene from the other (there is no dominant-recessive relationship between A and B).

The immune system is tolerant to its own erythrocyte antigens. A person who is type A, for example, does not produce anti-A antibodies. Surprisingly, however, people with type A blood do make antibodies against the B antigen, and conversely, people with blood type B make antibodies against the A antigen. This is believed to result from the fact that antibodies made in response to some common bacteria cross react with the A or B antigens. These antibodies are absent in the blood until 2 months after birth, lending support to this theory. A person who is type A, therefore, acquires antibodies that can react with B antigens by exposure to these bacteria but does not develop antibodies that can react with A antigens, since this is prevented by tolerance mechanisms.

### Table 20.3 — The ABO system of antigens on erythrocytes

| Antigen on RBCs | Antibody in plasma |
| --- | --- |
| A | Anti-B |
| B | Anti-A |
| O | Anti-A and anti-B |
| AB | Neither anti-A nor anti-B |

People who are type AB develop tolerance to both these antigens and thus do not produce either anti-A or anti-B antibodies. Those who are type O, by contrast, do not develop tolerance to either antigen and, therefore, have both anti-A and anti-B antibodies in their plasma (table 20.3).

**Transfusion Reactions** Before transfusions are performed, a *major crossmatch* is made by mixing serum from the recipient with blood cells from the donor. If the types do not match—if the donor is type A, for example, and the recipient is type B—the recipient's antibodies attach to the donor's erythrocytes and form bridges that cause the cells to clump together, or **agglutinate** (ă-gloot′n-āt) (fig. 20.5). Because of this agglutination reaction, the A and B antigens are sometimes called *agglutinogens,* and the antibodies against them are called *agglutinins*. Transfusion errors that result in such agglutination in the blood can produce a blockage of small blood vessels and cause hemolysis, which may damage the kidneys and other organs.

In emergencies, type O blood has been given to people who are type A, B, AB, or O. Since type O erythrocytes lack A and B antigens, the recipient's antibodies cannot cause agglutination of the donor red blood cells. Type O is, therefore, a *universal donor,* but only as long as the volume of plasma donated is small, because plasma from a type O person would agglutinate type A, type B, and type AB erythrocytes. Likewise, type AB people are *universal recipients* because they lack anti-A and anti-B antibodies and thus cannot agglutinate donor red blood cells. (Donor plasma could agglutinate recipient erythrocytes if the transfusion volume were too large.) Because of the dangers involved, the universal donor and recipient concept in blood transfusions is strongly discouraged.

**Rh Factor** Another group of antigens found in most erythrocytes is the **Rh factor** (Rh stands for rhesus monkey, in which these antigens were first discovered). People who have these antigens are said to be *Rh positive,* whereas those who do not are *Rh negative*. There are fewer Rh negative people because this condition is recessive to Rh positive.

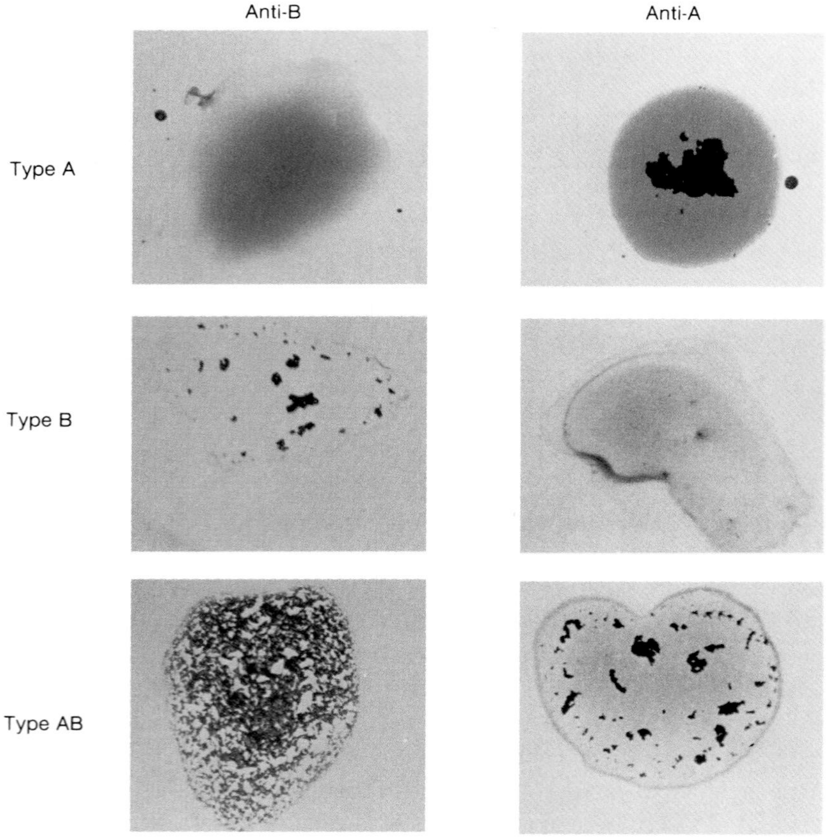

**FIGURE 20.5**

The agglutination (clumping) of red blood cells occurs when cells with A-type antigens are mixed with anti-A antibodies and when cells with B-type antigens are mixed with anti-B antibodies. No agglutination would occur with type O blood (not shown).

The Rh factor is of particular significance when Rh negative mothers give birth to Rh positive babies.

Since the fetal and maternal blood are normally kept separate across the placenta, the Rh negative mother is not usually exposed to the Rh antigen of the fetus during the pregnancy. At the time of birth, however, a variable degree of exposure may occur, and the mother's immune system may become sensitized and produce antibodies against the Rh antigen. This does not always occur, however, because the exposure may be minimal and because Rh negative women vary in their sensitivity to the Rh factor. If the woman does produce antibodies against the Rh factor, these antibodies can cross the placenta in subsequent pregnancies and cause hemolysis of the Rh positive erythrocytes of the fetus. The baby is therefore born anemic, with a condition called *erythroblastosis fetalis,* or *hemolytic disease of the newborn.* If this condition develops, treatment consists of removing the Rh positive blood of the fetus or newborn and infusing Rh negative blood.

Erythroblastosis fetalis can be prevented by injecting the Rh negative mother with an antibody preparation against the Rh factor (a trade name for this preparation is RhoGAM—the GAM is short for gamma globulin, the class of plasma proteins in which antibodies are found) within 72 hours after the birth of each Rh positive baby. This is a type of passive immunization in which the injected antibodies inactivate the Rh antigens and thus prevent the mother from becoming actively immunized to them.

# Blood Clotting

*Trauma to a blood vessel initiates a sequence of events that leads to the formation of a blood clot. These events require the presence of platelets and proteins called clotting factors in the plasma. Ultimately, an active enzyme called thrombin is formed, which catalyzes the conversion of fibrinogen into insoluble threads of fibrin.*

When a blood vessel is injured, a number of physiological mechanisms are activated that promote **hemostasis,** or the cessation of bleeding. Breakage of the endothelium of a vessel exposes collagen proteins from the subendothelial connective tissue to the blood. This initiates three separate, but overlapping, hemostatic mechanisms: (1) vasoconstriction, (2) the formation of a platelet plug, and (3) the production of a web of fibrin proteins around the platelet plug.

## Functions of Platelets

In the absence of vessel damage, platelets are repelled from each other and from the endothelium of vessels. The repulsion of platelets from an intact endothelium is believed to be due to **prostacyclin,** a derivative of prostaglandins, produced within the endothelium. Mechanisms that prevent platelets from sticking to the blood vessels and to each other are obviously needed to prevent inappropriate blood clotting.

Damage to the endothelium of vessels exposes subendothelial tissue to the blood. Platelets are able to stick to exposed collagen proteins that have become coated with a protein (*von Willebrand factor*) secreted by endothelial cells. Platelets contain secretory granules; when platelets stick to collagen, they *degranulate* as the secretory granules release their products. These products include *ADP* (adenosine diphosphate), *serotonin,* and a prostaglandin called *thromboxane A₂.* This event is known as the **platelet release reaction.**

Serotonin and thromboxane $A_2$ stimulate vasoconstriction, which helps to decrease blood flow to the injured vessel. Phospholipids that are exposed on the platelet membrane participate in the activation of clotting factors.

..............................................................

hemostasis: Gk. *haima,* blood; *stasis,* a standing

The release of ADP and thromboxane A$_2$ from platelets that are stuck to exposed collagen makes other platelets in the vicinity "sticky," so that they adhere to those stuck to the collagen. The second layer of platelets, in turn, undergoes a platelet release reaction, and the ADP and thromboxane A$_2$ that are secreted cause additional platelets to aggregate at the site of injury. This produces a **platelet plug** in the damaged vessel, which is strengthened by the activation of plasma-clotting factors.

 In order to undergo a release reaction, the production of prostaglandins by the platelets is required. *Aspirin* inhibits the cyclo-oxygenase enzyme that catalyzes the conversion of arachidonic acid (a cyclic fatty acid) into prostaglandins (see chapter 19), thereby inhibiting the release reaction and consequent formation of a platelet plug. Since platelets are not complete cells, they cannot regenerate new enzymes, and so the inhibited enzymes remain for the life of the platelets. The ingestion of excessive amounts of aspirin can significantly prolong bleeding time for several days, which is why blood donors and women in the last trimester of pregnancy are advised to avoid aspirin. Slight inhibition of platelet aggregation by low doses of aspirin, however, can significantly reduce the risk of atherosclerotic heart disease (described in chapter 21), and is often recommended for patients diagnosed with this condition.

## *Clotting Factors: Formation of Fibrin*

The platelet plug is strengthened by a meshwork of insoluble protein fibers known as **fibrin** (fig. 20.6). Blood clots contain platelets and fibrin, and they usually contain trapped erythrocytes that give the clot a red color (clots formed in arteries, where the blood flow is more rapid, generally lack erythrocytes and thus appear gray). Finally, contraction of the platelets in the process of **clot retraction** forms a more compact and effective plug (fig. 20.7). Fluid squeezed from the clot as it retracts is called **serum,** which is plasma without fibrinogen (the soluble precursor of fibrin).

The conversion of fibrinogen into fibrin may occur via either of two pathways. Blood left in a test tube will clot without the addition of any external chemicals; the pathway that produces this clot is thus called the **intrinsic pathway.** The intrinsic pathway also produces clots in damaged blood vessels when collagen is exposed to plasma. Damaged tissues, however, release a chemical that initiates a "shortcut" to the formation of fibrin. Since this chemical is not part of blood, the shorter pathway is called the **extrinsic pathway.**

The intrinsic pathway is initiated by the exposure of plasma to a negatively charged surface, such as that provided by collagen at the site of a wound or by the glass of a test tube. This activates a plasma protein called factor XII (table 20.4), which is a protein-digesting enzyme (protease). Active factor XII in turn activates another clotting factor, which activates yet another. The plasma-clotting factors are numbered in order of their discovery, which does not reflect the actual sequence of reactions.

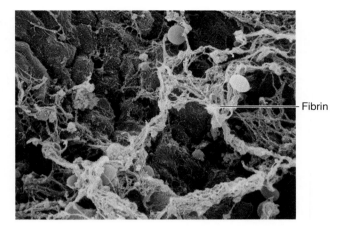

**FIGURE 20.6**
A scanning electron micrograph showing threads of fibrin.

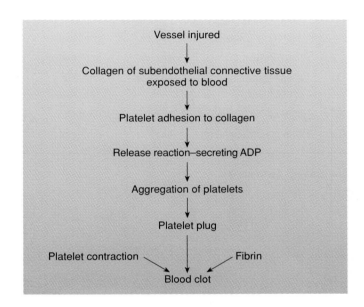

**FIGURE 20.7**
The sequence of events leading to platelet aggregation and the formation of a blood clot.

The next steps in the reaction sequence require the presence of phospholipids, which are provided by platelets, and ionic calcium. These steps result in the conversion of an inactive enzyme, called **prothrombin** into the active enzyme **thrombin.** Thrombin converts the soluble protein **fibrinogen** into **fibrin** monomers. These monomers are joined together to produce the insoluble fibrin polymers that form a supporting meshwork for the platelet plug. The intrinsic clotting sequence is shown on the right side of figure 20.8.

Fibrin can be formed more quickly as a result of the release of **tissue thromboplastin** from damaged tissue cells. This extrinsic clotting pathway is shown on the left side of figure 20.8. Notice that the intrinsic and extrinsic clotting pathways overlap; both result in the activation of factor X and then the formation of thrombin, which converts fibrinogen to fibrin.

## Table 20.4    The plasma-clotting factors

| Factor | Name | Function | Pathway |
|---|---|---|---|
| I | Fibrinogen | converted to fibrin | common |
| II | Prothrombin | enzyme | common |
| III | Tissue thromboplastin | cofactor | extrinsic |
| IV | Calcium ions (Ca++) | cofactor | intrinsic, extrinsic, and common |
| V | Proaccelerin | cofactor | common |
| VII* | Proconvertin | enzyme | extrinsic |
| VIII | Antihemophilic factor | cofactor | intrinsic |
| IX | Plasma thromboplastin component; Christmas factor | enzyme | intrinsic |
| X | Stuart-Prower factor | enzyme | common |
| XI | Plasma thromboplastin antecedent | enzyme | intrinsic |
| XII | Hageman factor | enzyme | intrinsic |
| XIII | Fibrin stabilizing factor | enzyme | common |

\* Note: There is no factor VI.

 People with a genetic deficiency in factor VIII are *hemophilic* and bleed severely when a vessel is cut. But why? After all, the extrinsic pathway is still available to them (see fig. 20.8). One explanation is that the intrinsic pathway is more important. Recent evidence suggests why this may be true. The extrinsic pathway, though initiated first, may become blocked by the formation of an inhibitor. When this occurs, additional activated factor X can be produced only through the intrinsic pathway.

## Dissolution of Clots

As the damaged blood vessel wall is repaired, activated factor XII promotes the conversion of another inactive molecule in plasma into the active form called **kallikrein** (ka´lĭ-kre´in). Kallikrein, in turn, catalyzes the conversion of inactive *plasminogen* into the active molecule called **plasmin.** Plasmin is an enzyme that digests fibrin into "split products," thus promoting dissolution of the clot.

 In addition to kallikrein, a number of other plasminogen activators are used clinically to promote dissolution of clots. An exciting recent development in genetic engineering technology is the commercial availability of an endogenous compound, *tissue plasminogen activator (TPA)*, which is the product of human genes introduced into bacteria. *Streptokinase*, a natural bacterial product, is a potent and more widely used activator of plasminogen. Streptokinase and TPA may be injected into the general circulation or injected specifically into a coronary vessel that has become occluded by a thrombus (blood clot).

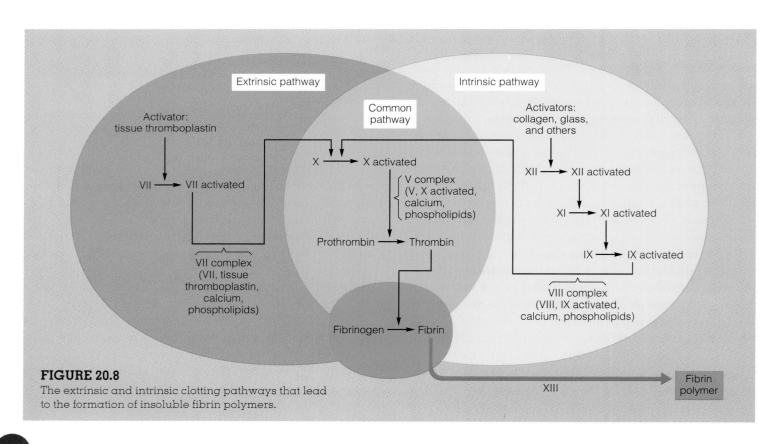

**FIGURE 20.8**
The extrinsic and intrinsic clotting pathways that lead to the formation of insoluble fibrin polymers.

**Anticoagulants** Clotting of the blood in test tubes can be prevented by the addition of *citrate* or *ethylenediaminetetraacetic acid* (*EDTA*), both of which chelate (binds to) calcium. By this means $Ca^{++}$ levels in the blood that can participate in the clotting sequence are lowered, and clotting is inhibited. A mucoprotein called *heparin* can also be added to the tube to prevent clotting. Heparin activates antithrombin III, a plasma protein that combines with and inactivates thrombin. Heparin is given intravenously during certain medical procedures to prevent clotting. Patients may also be given *coumarins* (koo´mă-rinz) as anticoagulants. The coumarins prevent blood clotting by competing with vitamin K.

Vitamin K is needed for the conversion of glutamic acid, an amino acid found within many of the plasma-clotting factors, into a derivative called *gamma-carboxyglutamic acid*. This derivative is more effective than glutamic acid at binding to $Ca^{++}$, and such binding is needed for proper function of clotting factors II, VII, IX, and X. Because of the indirect action of vitamin K on blood clotting, coumarin must be given to a patient for several days to be effective.

 A number of hereditary diseases involve the clotting system. Examples of hereditary clotting disorders include two different genetic defects in factor VIII. A defect in one subunit of factor VIII prevents this factor from participating in the intrinsic clotting pathway. This genetic disease, called *hemophilia A*, is an X-linked recessive trait that is prevalent in the royal families of Europe. A defect in another subunit of factor VIII results in *von Willebrand's disease*. In this disease, rapidly circulating platelets are unable to stick to collagen, and a platelet plug cannot be formed. Some acquired and inherited defects in the clotting system are summarized in table 20.5.

## Arterial Pressure Points and Control of Bleeding

*Because serious bleeding is life threatening, the principal first-aid concern is to stop loss of blood. The following are recommended steps in treating a victim who is hemorrhaging.*

**1** To reduce the chance that the victim will faint, lay the person down and slightly elevate his or her legs. If possible, elevate the site of bleeding above the level of the trunk. To minimize the chance of shock, cover the victim with a blanket.

**2** Without causing further trauma, carefully remove any dirt or debris from the wound. Do not remove any impaling objects. This should be done at the hospital by trained personnel.

**3** Apply direct pressure to the wound with a sterile bandage, clean cloth, or an article of clothing.

**4** Maintain direct pressure until the bleeding stops. Dress the wound with clean bandages or cloth lightly bound in place.

**5** If the bleeding does not stop and continues to seep through the dressing, do not remove the dressing. Rather, place additional absorbent material on top of it and continue to apply direct pressure.

**6** If direct pressure does not stop the bleeding, the pressure point to the wound site may need to be compressed. In the case of a severe wound to the hand, for example, compress the brachial artery against the humerus. This should be done while pressure continues to be applied to the wound itself.

**7** Once the bleeding has stopped, leave the bandage in place and immobilize the injured body part. Get the victim to the hospital or medical treatment center at once.

| Table 20.5 | Acquired and inherited defects in the clotting mechanism and anticoagulant drugs | |
|---|---|---|
| *Category* | *Cause of disorder* | *Comments* |
| Acquired clotting disorders | Vitamin K deficiency | Inadequate formation of prothrombin and other clotting factors in the liver |
| Inherited clotting disorders | Hemophilia A (defective factor VIII$_{AHF}$) | Recessive trait carried on X chromosome; results in delayed formation of fibrin |
| | von Willebrand's disease (defective factor VIII$_{VWF}$) | Dominant trait carried on autosomal chromosome; impaired ability of platelets to adhere to collagen in subendothelial connective tissue |
| | Hemophilia B (defective factor IX), also called Christmas disease | Recessive trait carried on X chromosome; results in delayed formation of fibrin |
| *Anticoagulants* | | |
| Aspirin | Inhibits prostaglandin production, resulting in defective platelet release reaction | |
| Coumarin | Competes with the action of vitamin K | |
| Heparin | Inhibits activity of thrombin | |
| Citrate | Combines with $Ca^{++}$, and thus inhibits activity of many clotting factors | |

von Willebrand's diesease: from E.A. von Willebrand, Finnish physician, 1870–1949

To stop bleeding, apply direct pressure to the wound with a sterile bandage, clean cloth, or an article of clothing.

If direct pressure does not stop the bleeding, apply compression to the arterial pressure point while maintaining direct pressure to the wound.

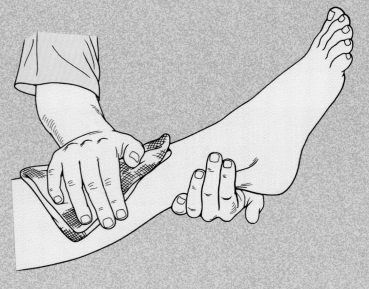

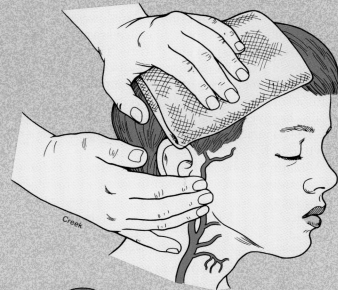

**Arterial pressure points**

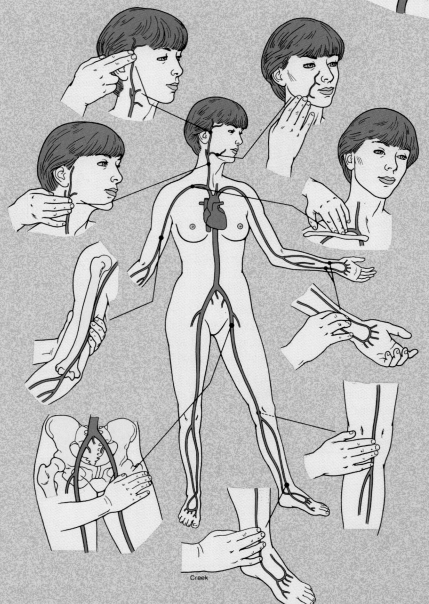

# Recognizing and Treating Victims of Shock

*Shock is defined as the condition of the body as it responds to trauma or severe pain by dilation or relaxation of blood vessels, producing circulatory failure. Symptoms of patients experiencing shock include the following.*

**1** **Skin.** The skin is pale or gray, cool, and clammy.

**2** **Pulse.** The heartbeat is weak and rapid. Blood pressure is reduced, frequently to below measurable values.

**3** **Respiration.** The respiratory rate is hurried, shallow, and irregular.

**4** **Eyes.** The eyes are staring and lusterless. The pupils may be dilated.

**5** **State of Being.** The victim may be conscious or unconscious. If conscious, he or she is likely to feel faint, weak, and confused. Frequently, the victim is anxious and excited.

Most trauma victims will experience some degree of shock, especially if there has been considerable blood loss. Immediate first-aid treatment for shock is essential and includes the following steps.

**1** Get the victim to lie down. Lay the person on his or her back with the feet elevated. This position maintains blood flow to the brain and may relieve faintness and mental confusion. Keep movement to a minimum. If the victim has sustained an injury in which raising the legs causes additional pain, leave the person flat on his or her back.

**2** Keep the victim warm and comfortable. If the weather is cold, place a blanket under and over the person. If the weather is hot, position the person in the shade on top of a blanket. Loosen tight collars, belts, or other restrictive clothing. Do not give the person anything to drink, even if he or she complains of thirst.

**3** Take precautions for internal bleeding or vomiting. If the victim has blood coming from the mouth, or if there is indication that the victim may vomit, position the person on his or her side to prevent choking or inhaling the blood or vomitus.

**4** Treat injuries appropriately. If the victim is bleeding, treat accordingly (see arterial pressure points and control of bleeding). Immobilize fractures and sprains. Always be alert to the possibility of spinal injuries and take the necessary precautions.

**5** See that hospital care is provided as soon as possible.

..................................................

# Chapter Summary

### Functions and Components of the Circulatory System (pp. 550–551)

1. Blood transports oxygen and nutrients to the tissue cells and removes waste products from the tissues; it also serves a regulatory function through its transport of hormones.
   a. Erythrocytes, or red blood cells, transport oxygen.
   b. Leukocytes, or white blood cells, protect the body from disease.
   c. Platelets, or thrombocytes, help to form blood clots.
2. The circulatory system consists of the cardiovascular system (heart and blood vessels) and the lymphatic system.

   a. Veins carry blood to the heart; arteries carry blood away from the heart.
   b. Blood flows from arterioles to capillaries, which transport the blood into venules.

### Composition of the Blood (pp. 551–558)

1. Plasma is the fluid part of the blood, containing dissolved ions and various organic molecules.
   a. Hormones are found in the plasma portion of the blood.
   b. The three types of plasma proteins are albumins; globulins (alpha, beta, and gamma); and fibrinogen.

2. Hemopoiesis is the production of blood cells.
   a. Blood cells form from proerythroblasts, myeloblasts, lymphoblasts, and megakaryoblasts.
   b. Production of erythrocytes is stimulated by the hormone erythropoietin. Development of different kinds of leukocytes is controlled by chemicals called lymphokines.
3. The major blood-typing groups are the ABO system and the Rh system.
   a. Blood type refers to the kind of antigens found on the surface of erythrocytes.

b. When different types of blood are mixed, antibodies against the erythrocytic antigens cause the erythrocytes to agglutinate.
c. A person with type O blood is the universal donor because that person's plasma does not contain antibodies against the A or B antigen. This concept is of limited usefulness, however.
d. Because of potential agglutination reactions, blood types must be matched for blood transfusions.

## Blood Clotting (pp. 558–561)

1. When a blood vessel is damaged, platelets adhere to the exposed subendothelial collagen proteins.
   a. Platelets that stick to collagen undergo a release reaction whereby they secrete ADP, serotonin, and thromboxane $A_2$.
   b. Serotonin and thromboxane $A_2$ cause vasoconstriction; ADP and thromboxane $A_2$ attract other platelets and cause them to adhere to the growing mass of platelets that are stuck to the collagen in the broken vessel.
   c. In the formation of a blood clot, a soluble protein called fibrinogen is converted by the enzyme thrombin into insoluble threads of fibrin.
   d. Thrombin is derived from its inactive precursor prothrombin by either an intrinsic or an extrinsic pathway.
2. Dissolution of the clot eventually occurs by the digestive action of plasmin, which cleaves fibrin into split products.

......................................................................................................................

# Review Activities

## Objective Questions

1. Which of the following is the correct sequence of blood flow?
   a. arteries-capillaries-venules-arterioles-veins
   b. veins-venules-capillaries-arterioles-arteries
   c. capillaries-arteries-arterioles-venules-veins
   d. arteries-arterioles-capillaries-venules-veins
2. Which of the following is the safest to use in a transfusion of a person with type B blood?
   a. type O blood
   b. type A blood
   c. type B blood
   d. type AB blood
3. Which of the following stem cells give rise to granular leukocytes?
   a. myeloblasts
   b. erythroblasts
   c. lymphoblasts
   d. megakaryoblasts
4. Which of the following is an agranular leukocyte?
   a. eosinophil
   b. neutrophil
   c. basophil
   d. monocyte

5. Hemoglobin is normally found in
   a. erythrocytes.
   b. plasma.
   c. granular leukocytes.
   d. agranular leukocytes.
   e. all of the above.
6. Factor X is activated
   a. in the intrinsic pathway only.
   b. in the extrinsic pathway only.
   c. in both the intrinsic and extrinsic pathways.
   d. in neither the intrinsic nor the extrinsic pathway.
7. Platelets
   a. form a plug by sticking to each other.
   b. release chemicals that stimulate vasoconstriction.
   c. provide phospholipids needed for the intrinsic pathway.
   d. all of the above apply.
8. Antibodies against both type-A and type-B antigens are found in the plasma of a person who is
   a. type A.
   b. type B.
   c. type AB.
   d. type O.
   e. all of the above apply.
9. Production of which of the following blood cells is stimulated by a hormone secreted by the kidneys?
   a. lymphocytes
   b. monocytes

   c. erythrocytes
   d. neutrophils
   e. platelets
10. Which of the following statements about plasmin is *true*?
    a. It is involved in the intrinsic clotting system.
    b. It is involved in the extrinsic clotting system.
    c. It functions in fibrinolysis.
    d. It promotes the formation of emboli.

## Essay Questions

1. Describe hemopoiesis in general terms and state where the different types of blood cells are formed.
2. Describe the stages of erythrocyte production and the characteristics of a mature erythrocyte. Explain how erythropoiesis is regulated.
3. Describe the series of events that may culminate in hemolytic disease of the newborn.
4. Explain why agglutination occurs when different blood types are mixed together.
5. Explain how a cut in the skin initiates both the intrinsic and extrinsic clotting pathways. Which pathway finishes first? Why?

## Gundy/Weber Software 💾

The tutorial software accompanying Chapter 20 is Volume 9—Cardiovascular System.

# circulatory system

## objectives

- Describe the location of the heart and its associated membranes. Also, describe the characteristics of the three layers of the heart wall.
- Discuss the nature and significance of the fibrous skeleton of the heart.
- List the chambers, valves, and associated vessels of the heart.
- Trace the flow of blood through the heart and distinguish between the systemic and pulmonary circulations.
- Describe the cycle of contraction and relaxation of the atria and ventricles.
- Explain how pressure changes in the heart affect the closing and opening of its valves and how these events produce the heart sounds.
- Describe the electrical activity of the SA node and explain why it functions as the normal pacemaker.
- Describe the conducting tissue of the heart, noting the pathway of electrical conduction.

- State the events that produce the ECG waves and correlate the ECG waves with the events of the cardiac cycle and the production of the heart sounds.
- Describe the structure of muscular and elastic arteries and the structure of veins. Explain how the structures of arteries and veins are functionally adaptive.
- Describe the structures of different types of capillaries and explain the functions of these capillaries.
- List the arterial branches of the ascending aorta and aortic arch and describe the arterial supply to the brain.
- Describe the arterial pathways that supply blood to the upper extremity, thorax, abdomen, and lower extremity.
- Describe the venous drainage of the head, neck, and upper extremity.
- Describe the venous drainage of the thorax, lower extremity, and abdomen.
- State the vessels involved in the hepatic portal system and explain the significance of this system.

# Structure of the Heart

*The heart contains four chambers: two upper atria, which receive venous blood, and two lower ventricles, which eject blood into arteries. The right ventricle pumps blood to the lungs, where the blood becomes oxygenated, whereas the left ventricle pumps oxygenated blood to the entire body. The proper flow of blood within the heart is aided by two pairs of one-way valves.*

The heart is a hollow, four-chambered muscular organ, roughly the size of a clenched fist. It is located within the thoracic cavity between the lungs, in the region known as the *mediastinum.* About two-thirds of the heart is left of the midline, with its *apex,* or cone-shaped end, pointing downward in contact with the diaphragm. The *base* of the heart is the broad superior end, where the large vessels attach.

The heart is enclosed and protected by a loose-fitting, serous sac of dense regular connective tissue called the **parietal pericardium** (*per˝ĭ-kar´de-um*), or **pericardial sac** (fig. 21.1). The parietal pericardium separates the heart from the other thoracic organs and forms the wall of the

*pericardial cavity,* which contains a watery, lubricating *pericardial fluid.* The parietal pericardium is actually composed of an outer *fibrous layer* and an inner *serous layer.* The serous layer secretes the pericardial fluid.

*Pericarditis* is an inflammation of the parietal pericardium accompanied by an increased secretion of pericardial fluid into the pericardial cavity. Because the tough, fibrous portion of the parietal pericardium is inelastic, an increase in fluid pressure compresses the heart—a condition called *cardiac tamponade*—and impairs the movement of blood into and out of the heart. Some of the pericardial fluid may be withdrawn for analysis by injecting a needle to the left of the xiphoid process to pierce the parietal pericardium.

## Heart Wall

The wall of the heart is composed of three distinct layers (table 21.1). The outer layer is the **epicardium,** also called the *visceral pericardium.* The space between this layer and the parietal pericardium is the pericardial cavity, just described. The thick middle layer of the heart is the **myocardium.** It is composed of cardiac muscle tissue arranged in

**FIGURE 21.1**

The position of the heart and associated serous membranes within the thoracic cavity.

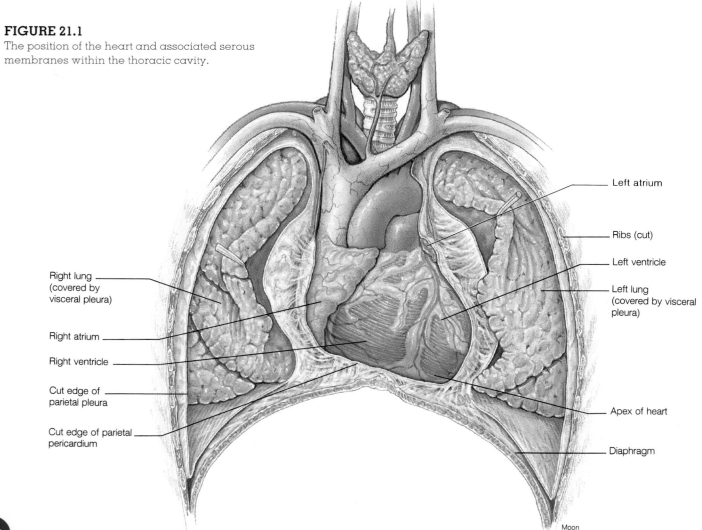

Right lung (covered by visceral pleura)

Right atrium

Right ventricle

Cut edge of parietal pleura

Cut edge of parietal pericardium

Left atrium

Ribs (cut)

Left ventricle

Left lung (covered by visceral pleura)

Apex of heart

Diaphragm

Moon

# Table 21.1    Layers of the heart wall

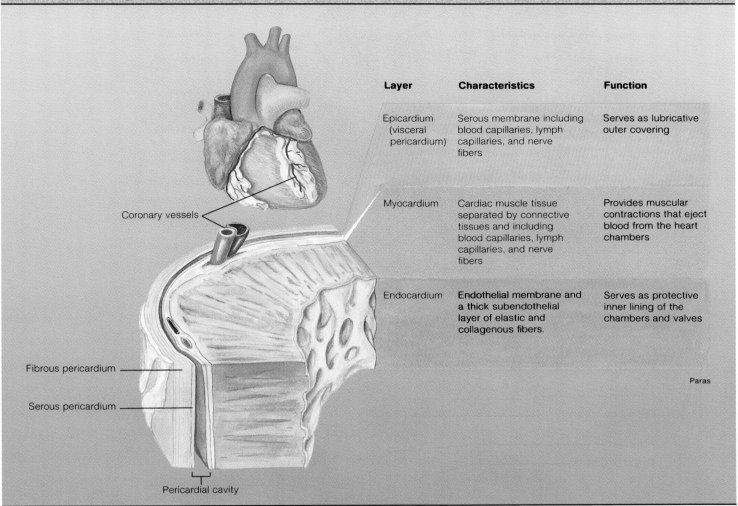

| Layer | Characteristics | Function |
|---|---|---|
| Epicardium (visceral pericardium) | Serous membrane including blood capillaries, lymph capillaries, and nerve fibers | Serves as lubricative outer covering |
| Myocardium | Cardiac muscle tissue separated by connective tissues and including blood capillaries, lymph capillaries, and nerve fibers | Provides muscular contractions that eject blood from the heart chambers |
| Endocardium | Endothelial membrane and a thick subendothelial layer of elastic and collagenous fibers. | Serves as protective inner lining of the chambers and valves |

Coronary vessels

Fibrous pericardium

Serous pericardium

Pericardial cavity

Paras

such a way that the contraction of the muscle bundles results in squeezing or wringing of the heart chambers. The thickness of the myocardium varies in different chambers, depending on the force needed to eject blood. The thickest portion therefore surrounds the left ventricle. The atrial walls are relatively thin. The inner layer of the heart, the **endocardium,** is continuous with the endothelium of blood vessels. The endocardium also forms part of the valves of the heart. Inflammation of the endocardium is called *endocarditis.*

## Chambers and Valves

The interior of the heart is divided into four chambers: upper right and left **atria** (singular, **atrium**) and lower right and left **ventricles.** The atria contract and empty simultaneously into the ventricles (fig. 21.2), which also contract in unison. The atria are separated by the thin muscular **interatrial septum,** while the ventricles are separated by the thick muscular **interventricular septum.** Left and right **atrioventricular valves** are located between the atria and ventricles, and **aortic** and **pulmonary semilunar valves** are present at the bases of the two large vessels leaving the heart. Heart valves maintain the flow of blood in one direction.

Each atrium has a pouchlike extension, or appendage, that is visible when the external surface of the heart is viewed. These appendages of the atria are the right and left **auricles.** Grooved depressions on the surface of the heart indicate the partitions between the chambers and also

⋯⋯⋯⋯⋯⋯⋯⋯⋯⋯⋯⋯⋯⋯⋯

auricle: L. *auricula,* a little ear

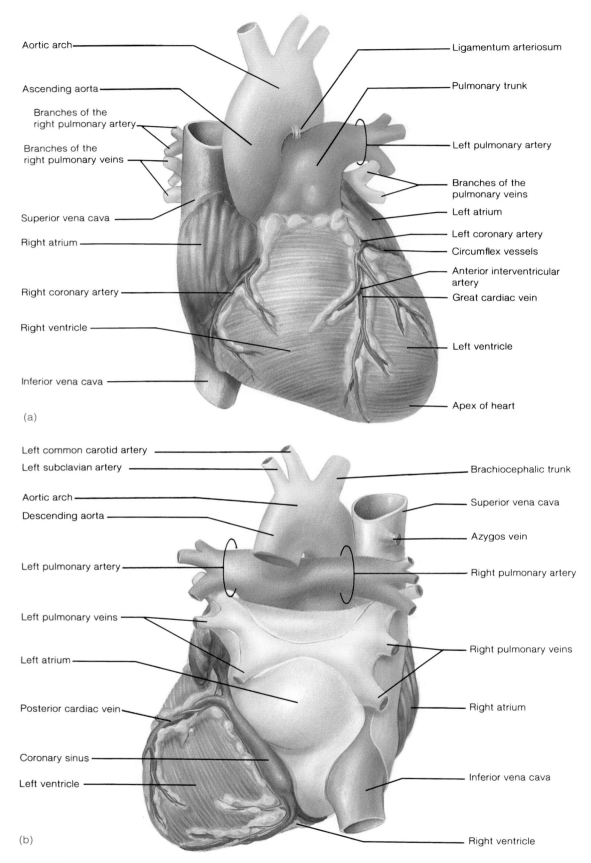

Aortic arch

Ascending aorta

Branches of the right pulmonary artery

Branches of the right pulmonary veins

Superior vena cava

Right atrium

Right coronary artery

Right ventricle

Inferior vena cava

(a)

Ligamentum arteriosum

Pulmonary trunk

Left pulmonary artery

Branches of the pulmonary veins

Left atrium

Left coronary artery

Circumflex vessels

Anterior interventricular artery

Great cardiac vein

Left ventricle

Apex of heart

Left common carotid artery

Left subclavian artery

Aortic arch

Descending aorta

Left pulmonary artery

Left pulmonary veins

Left atrium

Posterior cardiac vein

Coronary sinus

Left ventricle

(b)

Brachiocephalic trunk

Superior vena cava

Azygos vein

Right pulmonary artery

Right pulmonary veins

Right atrium

Inferior vena cava

Right ventricle

**FIGURE 21.2**

The structure of the heart. (*a*) An anterior view, (*b*) a posterior view, and (*c*) an internal view.

contain *cardiac vessels* that supply blood to the wall of the heart. The most prominent groove is the **atrioventricular groove,** or **coronary sulcus,** that encircles the heart and marks the division between the atria and ventricles. The partition between right and left ventricles is denoted by two (anterior and posterior) **interventricular sulci.**

The following discussion follows the sequence in which blood flows through the structures of the atria, ventricles, and valves.

**Right Atrium**    The right atrium receives venous blood from the **superior vena cava,** which drains the upper portion of the body, and from the **inferior vena cava,** which drains the lower portion (fig. 21.2). There is an additional venous opening into the right atrium from the **coronary sinus,** which returns venous blood from the myocardium of the heart itself.

**Right Ventricle**    Blood from the right atrium passes through the **tricuspid valve** to fill the right ventricle. The tricuspid valve is an *atrioventricular* (AV) *valve* that gets its name

.......................................................

chordae tendineae: L. *chorda*, string; *tendere*, to stretch

from its three valve leaflets, or *cusps*. Each cusp is held in position by strong tendinous cords called **chordae tendineae,** which are secured to the ventricular wall by cone-shaped **papillary muscles.** These structures prevent the valves from everting, like an umbrella in a strong wind, when the ventricles contract.

Ventricular contraction causes the tricuspid valve to close and the blood to exit the right ventricle through the **pulmonary trunk** and to enter the lungs through the right and left **pulmonary arteries.** The **pulmonary semilunar valve** lies at the base of the pulmonary trunk, where it prevents the backflow of ejected blood into the right ventricle.

**Left Atrium**    After gas exchange has occurred within the capillaries of the lungs, oxygenated blood is transported to the left atrium through four branches of the **pulmonary veins.**

**Left Ventricle**    The left ventricle receives blood from the left atrium. These two chambers are separated by the **bicuspid,** or **mitral** (*mi´tral*), **valve.** When the left ventricle is relaxed, the valve is open and allows blood to flow from the atrium to the ventricle; when the left ventricle contracts, the valve closes. Closing of the valve during ventricular contraction prevents backflow of blood into the atrium.

The walls of the left ventricle are thicker than those of the right ventricle. The endocardium of both chambers is characterized by distinct ridges called **trabeculae** (*tră-bek´yŭ-le*) **carneae** (see fig. 21.2c) that reinforce the endocardium of the ventricles. Oxygenated blood exits the left ventricle through the **ascending portion of the aorta** (*a-or´tă*). The **aortic**

.......................................................

semilunar: L. *semi*, half; *luna*, moon
mitral: L. *mitra*, like a bishop's mitre
trabeculae carneae: L. *trabecula*, small beams; *carneus*, flesh

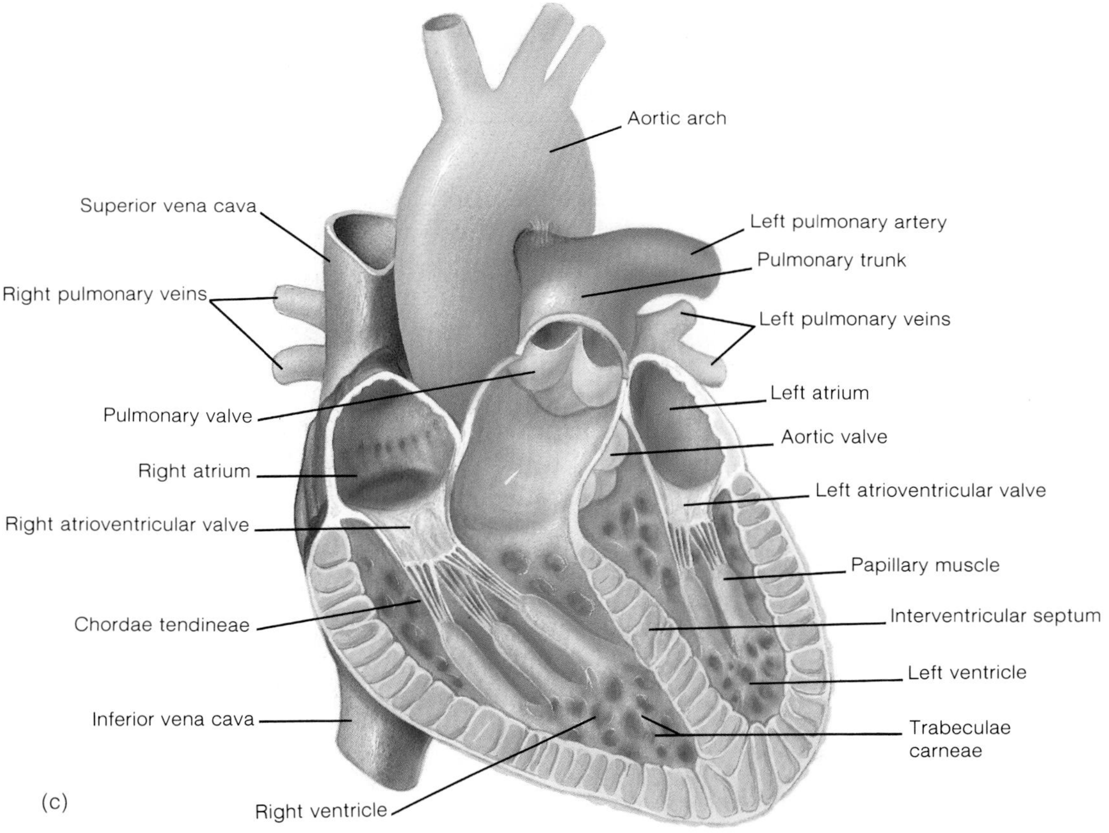

Aortic arch
Superior vena cava
Left pulmonary artery
Right pulmonary veins
Pulmonary trunk
Left pulmonary veins
Pulmonary valve
Left atrium
Aortic valve
Right atrium
Left atrioventricular valve
Right atrioventricular valve
Papillary muscle
Chordae tendineae
Interventricular septum
Left ventricle
Inferior vena cava
Trabeculae carneae
(c)
Right ventricle

**FIGURE 21.2**
*Continued.*

# Table 21.2　Valves of the heart

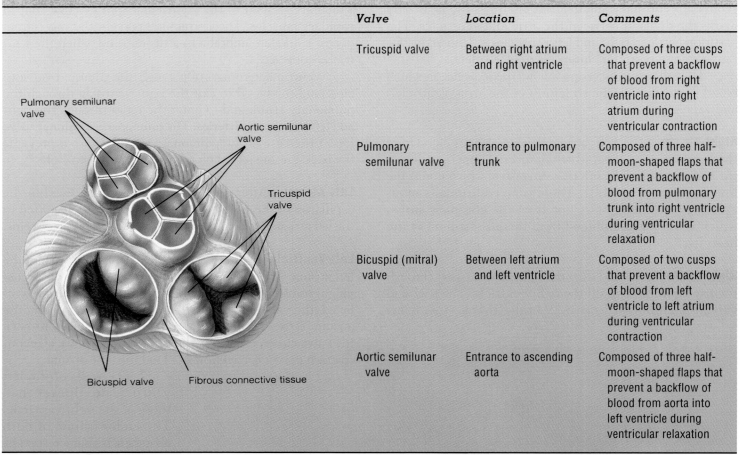

Pulmonary semilunar valve

Aortic semilunar valve

Tricuspid valve

Bicuspid valve

Fibrous connective tissue

| Valve | Location | Comments |
|---|---|---|
| Tricuspid valve | Between right atrium and right ventricle | Composed of three cusps that prevent a backflow of blood from right ventricle into right atrium during ventricular contraction |
| Pulmonary semilunar valve | Entrance to pulmonary trunk | Composed of three half-moon-shaped flaps that prevent a backflow of blood from pulmonary trunk into right ventricle during ventricular relaxation |
| Bicuspid (mitral) valve | Between left atrium and left ventricle | Composed of two cusps that prevent a backflow of blood from left ventricle to left atrium during ventricular contraction |
| Aortic semilunar valve | Entrance to ascending aorta | Composed of three half-moon-shaped flaps that prevent a backflow of blood from aorta into left ventricle during ventricular relaxation |

**semilunar valve,** located at the base of the ascending aorta, closes as a result of the pressure of the blood when the left ventricle relaxes, and thus prevents backflow of blood into the relaxed ventricle.

The valves are illustrated and their actions summarized in table 21.2.

**Fibrous Skeleton**　The layer of dense connective tissue between the atria and ventricles is known as the **fibrous skeleton** of the heart. Bundles of myocardial cells (described in chapter 12) in the atria attach to the upper margin of this fibrous skeleton and form a single functioning unit, or **myocardium.** The myocardial cell bundles of the ventricles attach to the lower margin and form a different myocardium. As a result, the myocardia of the atria and ventricles are structurally and functionally separated from each other so that special conducting tissue is needed to carry action potentials from the atria to the ventricles, as will be described in a later section. The connective tissue of the

fibrous skeleton also forms rings, called **annuli fibrosi,** around the four heart valves, providing a foundation for the support of the valve flaps.

## Circulatory Routes

Figure 21.3 illustrates the circulatory routes of the blood. The principal divisions of the circulatory blood flow are the pulmonary and systemic (*sis-tem´ik*) circulations.

The **pulmonary circulation** consists of blood vessels that transport blood from the right ventricle to the lungs for gas exchange and then to the left atrium of the heart. It includes the pulmonary trunk with its semilunar valve, the pulmonary arteries that transport blood to the lungs, the pulmonary capillaries within each lung, and four pulmonary veins that transport oxygenated blood back to the heart.

The **systemic circulation** is composed of all the remaining vessels of the body that are not part of the pulmonary circulation. It includes the aorta with its semilunar

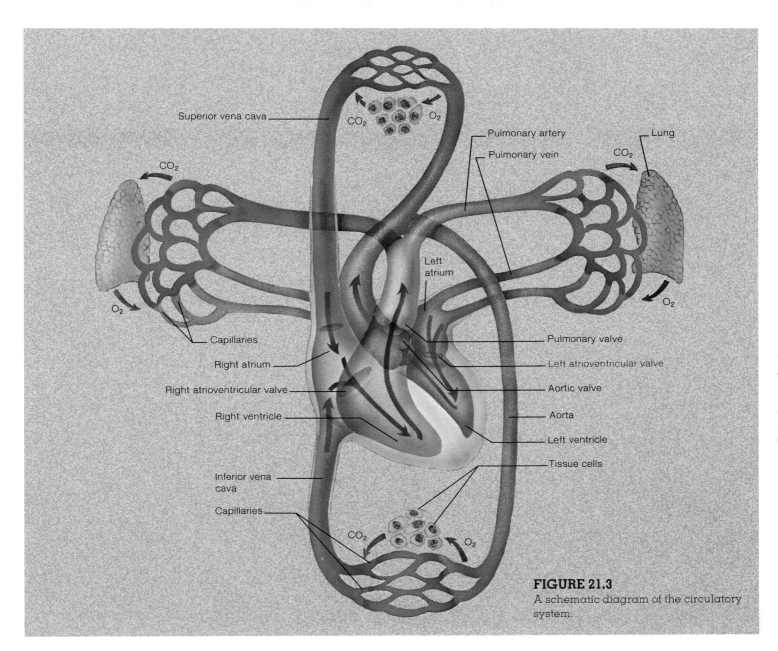

Superior vena cava

$CO_2$   $O_2$

$CO_2$

Pulmonary artery

Pulmonary vein

Lung

$CO_2$

Left atrium

$O_2$

Capillaries

Pulmonary valve

Right atrium

Left atrioventricular valve

Right atrioventricular valve

Aortic valve

Right ventricle

Aorta

Left ventricle

Tissue cells

Inferior vena cava

Capillaries

$CO_2$   $O_2$

**FIGURE 21.3**
A schematic diagram of the circulatory system.

valve, all of the branches of the aorta, all capillaries other than those in the lungs, and all veins other than the pulmonary veins. The right atrium receives all the venous return of oxygen-depleted blood from the systemic veins.

**Coronary Circulation**   The myocardium is supplied with blood by the right and left **coronary arteries** (fig. 21.4a). These two vessels arise from the aorta immediately beyond the aortic semilunar valve and encircle the heart within the atrioventricular sulcus. The left coronary artery gives rise to its major branches, the **anterior interventricular** and **circumflex arteries,** and the right coronary gives rise to the **posterior**

**interventricular** and **marginal arteries** (figs. 21.4a and 21.5). The main trunks of the right and left coronaries anastomose (join together) on the posterior surface of the heart.

From the capillaries in the myocardium, blood enters the **cardiac veins.** The two principal cardiac veins are the **great cardiac vein** and the **middle cardiac vein** (figs. 21.4b and 21.5). These converge into a large venous channel on the posterior surface of the heart called the **coronary sinus** (figs. 21.4b and 21.5). The coronary venous blood then enters the heart through an opening into the right atrium.

⋯⋯⋯⋯⋯⋯⋯⋯⋯⋯⋯⋯⋯⋯⋯⋯⋯⋯⋯

circumflex: L. *circum,* around; *flectere,* to bend

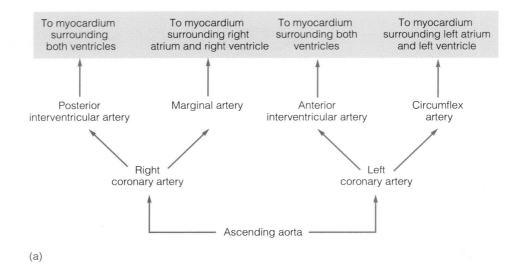

(a)

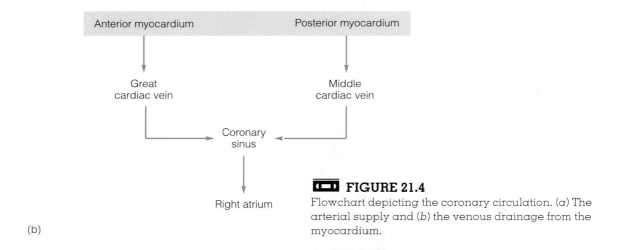

(b)

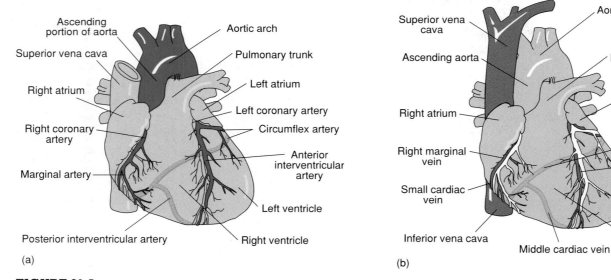

**FIGURE 21.4**

Flowchart depicting the coronary circulation. (a) The arterial supply and (b) the venous drainage from the myocardium.

(a)

(b)

**FIGURE 21.5**

Coronary circulation. (a) An anterior view of the arterial supply to the heart and (b) an anterior view of the venous drainage.

# Cardiac Cycle, Heart Sounds, and the Electrocardiogram

*The two atria fill with blood and then contract simultaneously. This is followed by the simultaneous contraction of both ventricles, which sends blood through the pulmonary and systemic circulations. The AV valves close when the ventricles contract and the semilunar valves close when the ventricles relax, producing the heart sounds. Contraction of the myocardium results from electrical excitation, which can be recorded as an electrocardiogram.*

The *cardiac cycle* refers to the repeating pattern of contraction and relaxation of the heart. The contraction phase is called **systole** and the relaxation phase is called **diastole.** When these terms are used without reference to specific chambers, they refer to contraction and relaxation of the ventricles. It should be noted, however, that the atria also contract and relax. There is an atrial systole and diastole. Atrial contraction occurs toward the end of diastole, when the ventricles are relaxed; when the ventricles contract during systole, the atria are relaxed.

The heart thus has a two-step pumping action. The right and left atria contract almost simultaneously, followed by contraction of the right and left ventricles about 0.1 to 0.2 second later. During the time when both the atria and ventricles are relaxed, the venous return of blood fills the atria. The buildup of pressure that results causes the AV valves to open and blood to flow from atria to ventricles. It has been estimated that the ventricles are about 80% filled with blood even before the atria contract. Contraction of the atria adds the final 20% to the **end-diastolic volume (EDV),** which is the total volume of blood in the ventricles at the end of diastole.

 Interestingly, the blood contributed by contraction of the atria does not appear to be essential for life. Elderly people who have *atrial fibrillation* (a condition in which the atria fail to contract) do not appear to have a higher mortality than those who have normally functioning atria. People with atrial fibrillation, however, become fatigued more easily during exercise because the reduced filling of the ventricles compromises the ability of the heart to sufficiently increase its output and blood flow during exercise. Cardiac output and blood flow during rest and exercise are discussed in chapter 22.

Contraction of the ventricles in systole ejects about two-thirds of the blood that they contain—an amount called the **stroke volume**—leaving one-third of the initial amount

..........................................................

systole: Gk. *systole*, contraction
diastole: Gk. *diastole*, prolonged or expansion

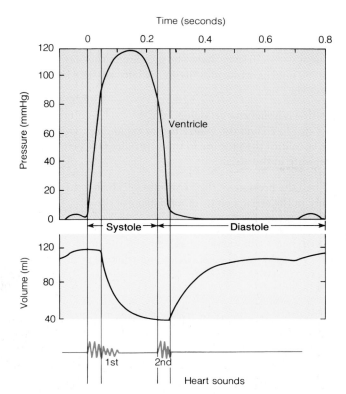

**FIGURE 21.6**

The relationship between the heart sounds and the left intraventricular pressure and volume. Closing of the AV valves occurs during the early part of contraction, when the intraventricular pressure rises prior to ejection of blood. Closing of the semilunar valves occurs at the beginning of ventricular relaxation, just prior to filling. The first and second sounds thus appear during the stages of isovolumetric contraction and isovolumetric relaxation (*iso* = same).

left in the ventricles as the **end-systolic volume.** The ventricles then fill with blood during the next cycle. At an average **cardiac rate** of 75 beats per minute, each cycle lasts 0.8 second; 0.5 second is spent in diastole, and systole takes 0.3 second (fig. 21.6).

## Pressure Changes during the Cardiac Cycle

When the heart is in diastole, pressure in the systemic arteries averages about 80 mmHg (millimeters of mercury). The following events in the cardiac cycle then occur.

**1** As the ventricles begin their contraction the intraventricular pressure rises, causing the AV valves to snap shut. At this time, the ventricles are neither being filled with blood (because the AV valves are closed) nor ejecting blood (because the intraventricular pressure has not risen sufficiently to open the semilunar valves). This is the phase of *isovolumetric contraction.*

**2** When the pressure in the left ventricle becomes greater than the pressure in the aorta, the phase of *ejection* begins as the semilunar valves open. The pressure in the left ventricle and aorta rises to about 120 mmHg (fig. 21.6) as the ventricular volume decreases.

**3** As the pressure in the left ventricle falls below the pressure in the aorta, the back pressure causes the semilunar valves to snap shut. The pressure in the aorta falls to 80 mmHg, while pressure in the left ventricle falls to 0 mmHg.

**4** During *isovolumetric relaxation*, the AV and semilunar valves are closed and the ventricles are expanding. This phase lasts until the pressure in the ventricles falls below the pressure in the atria.

**5** When the pressure in the ventricles falls below the pressure in the atria, a phase of *rapid filling* of the ventricles occurs.

**6** *Atrial contraction* (*atrial systole*) empties the final amount of blood into the ventricles immediately prior to the next phase of isovolumetric contraction of the ventricles.

Similar events occur in the right ventricle and pulmonary circulation, but the pressures are lower. The maximum pressure produced at systole in the right ventricle is 25 mmHg, which falls to a low of 8 mmHg at diastole.

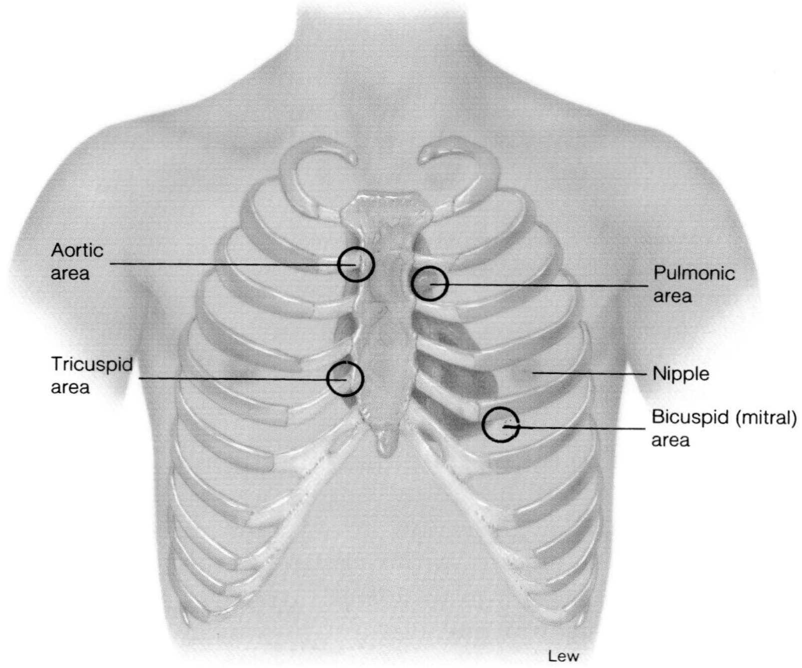

## Heart Sounds

Closing of the AV and semilunar valves produces sounds that can be heard at the surface of the chest with a stethoscope. These sounds are often described as "lub-dub." The "lub," or **first sound,** is produced by closing of the AV valves during isovolumetric contraction of the ventricles. The "dub," or **second sound,** is produced by closing of the semilunar valves when the pressure in the ventricles falls below the pressure in the arteries. The first sound is thus heard when the ventricles contract at systole; the second sound is heard when the ventricles relax at the beginning of diastole.

 Heart sounds are of clinical importance because they provide information about the condition of the heart valves and other heart problems. Abnormal sounds are referred to as *heart murmurs*. Murmurs are caused by valvular leakage or other structural abnormalities that produce turbulence of the blood as it passes through the heart.

The valves of the heart are positioned directly deep to the sternum, which tends to obscure and dissipate valvular sounds. For this reason, a physician will listen with a stethoscope for the heart sounds at locations designated as **valvular auscultatory areas,** which are named according to the valve that can be detected (fig. 21.7). The *aortic area* is at the right second intercostal space near the sternum. The *pulmonic area* is directly across from the aortic area at the left second intercostal space near the sternum. The *tricuspid* and *bicuspid* (mitral) *areas* are both at the fifth intercostal space, with the bicuspid area more laterally placed. Surface landmarks are extremely important in identifying auscultatory areas.

**FIGURE 21.7**
The valvular auscultatory areas are the routine stethoscope positions for listening to the heart sounds.

## Electrical Activity of the Heart

If the heart of a frog is removed from the body and all neural connections are severed, it will still continue to beat as long as the myocardial cells remain alive. The automatic nature of the heartbeat is referred to as *automaticity*. As a result of experiments with isolated myocardial cells and clinical experience with patients who have specific heart disorders, many regions within the heart have been shown to be capable of originating action potentials and functioning as pacemakers.

In a normal heart, however, only one region demonstrates spontaneous electrical activity and by this means functions as a pacemaker. This pacemaker region is called the **sinoatrial node,** or **SA node.** The SA node is located in the right atrium near the opening of the superior vena cava.

A number of abnormal conditions require the surgical implantation of an *artificial pacemaker* under the skin. This battery-powered device, about the size of a locket, stimulates or regulates contractions of the heart. The electrodes from the pacemaker are guided by means of a fluoroscope through a vein to the right atrium, through the tricuspid valve, and into the right ventricle. The electrodes are fixed to the trabeculae carneae and are in contact with the myocardium of the ventricle. When these electrodes deliver shocks, either at a continuous pace or on demand (when the heart's own impulse doesn't arrive on time), both ventricles are depolarized and contract and then repolarize and relax—just as they do in response to endogenous stimulation.

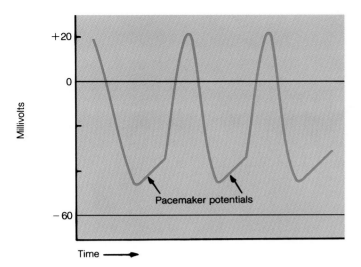

**FIGURE 21.8**
Pacemaker potentials and action potentials in the SA node.

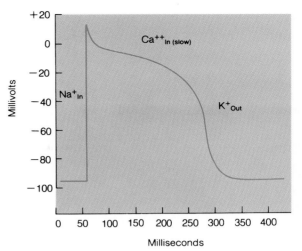

**FIGURE 21.9**
An action potential in a myocardial cell from the ventricles. The plateau phase of the action potential is maintained by a slow inward diffusion of $Ca^{++}$. As a result, the duration of the cardiac action potential is about 100 times longer than that of the "spike potential" of an axon.

The cells of the SA node do not keep a resting membrane potential in the manner of resting neurons or skeletal muscle cells. Instead, during the period of diastole, the SA node exhibits a spontaneous depolarization called the **pacemaker potential.** The membrane potential begins at about –60 mV and gradually depolarizes to –40 mV, which is the threshold for producing an action potential in these cells. This spontaneous depolarization is produced by the diffusion of $Ca^{++}$ through openings in the membrane called *slow calcium channels*. At the threshold level of depolarization, other channels—*fast calcium channels*—open, and $Ca^{++}$ rapidly diffuses into the cells. The opening of voltage-regulated $Na^+$ gates, and the inward diffusion of $Na^+$ that results, may also contribute to the upshoot phase of the action potential in pacemaker cells (fig. 21.8). Repolarization is produced by the opening of $K^+$ gates and outward diffusion of $K^+$. Once repolarization to –60 mV has been achieved, a new pacemaker potential begins, culminating again, at the end of diastole, with a new action potential.

Other regions of the heart, including the area around the SA node and the atrioventricular bundle (see fig. 21.10), can potentially produce pacemaker potentials. The rate of spontaneous depolarization of these regions, however, is slower than that of the SA node. Thus, potential pacemaker cells are stimulated by action potentials from the SA node before they can stimulate themselves through their own pacemaker potentials. If action potentials from the SA node are prevented from reaching these regions (through blockage of conduction), they will generate pacemaker potentials at their own rate and serve as sites for the origin of action potentials—they will function as pacemakers. A pacemaker other than the SA node is called an *ectopic pacemaker,* or an *ectopic focus*. From this discussion, it is clear that the rhythm set by an ectopic pacemaker is usually slower than that normally set by the SA node.

Once another myocardial cell has been stimulated by action potentials originating in the SA node, it produces its own action potential. The majority of myocardial cells have resting membrane potentials of about –90 mV. When stimulated by action potentials from a pacemaker region, these cells become depolarized to threshold, at which point their voltage-regulated $Na^+$ gates open. The upshoot phase of the action potential of nonpacemaker cells is due to the inward diffusion of $Na^+$. Following the rapid reversal of the membrane polarity, the membrane potential quickly declines to about –10 to –20 mV. Unlike the action potential of other cells, however, this level of depolarization is maintained for 200–300 msec before repolarization (fig. 21.9). This *plateau phase* results from a slow inward diffusion of $Ca^{++}$, which balances a slow outward diffusion of cations. Rapid repolarization at the end of the plateau phase is achieved, as in other cells, by the opening of $K^+$ gates and the rapid outward diffusion of $K^+$ that results.

**Conducting Tissue of the Heart** Action potentials that originate in the SA node spread to adjacent myocardial cells of the right and left atria through the gap junctions between these cells. Since the myocardium of the atria are separated from the myocardium of the ventricles by the fibrous skeleton of the heart, however, the impulse cannot be conducted directly from the atria to the ventricles. Specialized conducting tissue, composed of modified myocardial cells, is thus required. These specialized myocardial cells form the AV node, atrioventricular bundle, and conduction myofibers.

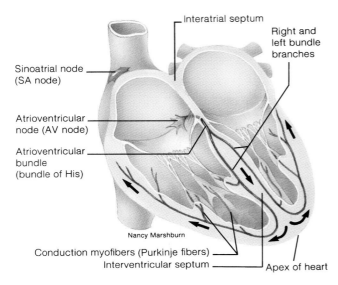

**FIGURE 21.10**
The conduction system of the heart.

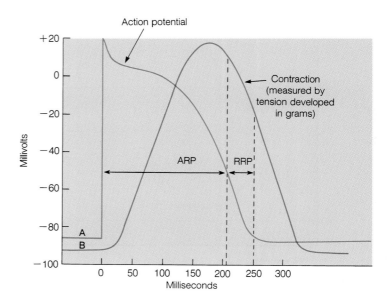

**FIGURE 21.11**
The time course for the myocardial action potential (*A*) is compared with the duration of contraction (*B*). Notice that the long action potential results in a correspondingly long absolute refractory period (ARP) and relative refractory period (RRP). These refractory periods last almost as long as the contraction, so that the myocardial cells cannot be stimulated again until they have completed their contraction from the first stimulus.

Once the impulse spreads through the atria, it passes to the **atrioventricular node (AV node),** which is located on the inferior portion of the interatrial septum (fig. 21.10). The AV node is smaller than the SA node, averaging 5 mm in length, 2 mm in width, and 0.5 mm in depth. From here, the impulse continues through the **atrioventricular bundle** (bundle of His), beginning at the top of the interventricular septum. This conducting tissue pierces the fibrous skeleton of the heart and continues to descend along the interventricular septum. The atrioventricular bundle divides into right and left bundle branches, which are continuous with the **conduction myofibers,** or **Purkinje fibers,** within the ventricular walls. Stimulation of these fibers causes both ventricles to contract simultaneously and eject blood into the pulmonary and systemic circulation.

### Conduction of the Impulse
Action potentials from the SA node spread very quickly—at a rate of 0.8 to 1.0 m/sec—across the myocardial cells of both atria. The conduction rate then slows considerably as the impulse passes into the AV node. Slow conduction of impulses (0.03–0.05 m/sec) through the AV node accounts for over half of the time delay between excitation of the atria and ventricles. After the impulses spread through the AV node, the conduction rate increases greatly in the atrioventricular bundle and reaches very high velocities (5 m/sec) in the conduction myofibers. As a result of this rapid conduction of impulses, ventricular contraction follows the contraction of the atria by only about 0.1 to 0.2 second, as mentioned earlier in the chapter.

bundle of His: from Wilhelm His, Jr., Swiss physician, 1863–1934
Purkinje fibers: from Johannes E. von Purkinje, Bohemian anatomist, 1787–1869

Unlike skeletal muscles, the heart cannot sustain a contraction. This is because the atria and ventricles behave as if each were composed of only one muscle cell. The entire myocardium of each is electrically stimulated as a single unit and contracts as a unit. This contraction, which corresponds in time to the long action potential of myocardial cells and lasts almost 300 msec, is analogous to the twitch produced by a single skeletal muscle fiber (which lasts only 20 to 100 msec in comparison). The heart normally cannot be stimulated again until after it has relaxed from its previous contraction, since myocardial cells have *long refractory periods* (fig. 21.11) that correspond to the long duration of their action potentials. Summation of contractions is thus prevented, and the myocardium must relax after each contraction. The rhythmic pumping action of the heart is thus ensured.

## The Electrocardiogram

A pair of surface electrodes placed directly on the heart will record a repeating pattern of potential changes. As action potentials spread from the atria to the ventricles, the voltage measured between these two electrodes will vary in a way that provides a "picture" of the electrical activity of the heart. By changing the position of the recording electrodes, the observer can gain a more complete picture of the electrical events.

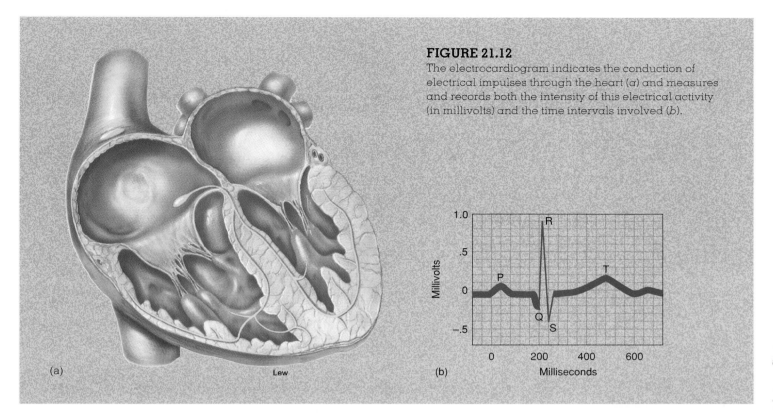

(a)

Lew

(b)

The body is a good conductor of electricity because tissue fluids contain a high concentration of ions that move (creating a current) in response to potential differences. Potential differences generated by the heart are thus conducted to the body surface, where they can be recorded by surface electrodes placed on the skin. The recording is called an **electrocardiogram (ECG or EKG)** (fig. 21.12); the recording device is called an **electrocardiograph.**

Each cardiac cycle produces three distinct ECG waves, designated P, QRS, and T. It should be noted that these waves are not action potentials; they represent changes in potential between two regions on the surface of the heart. These potential differences are produced by the composite effects of action potentials in many myocardial cells. For example, the spread of depolarization through the atria causes a potential difference that is indicated by an upward deflection of the ECG line. When about half the mass of the atria is depolarized, this upward deflection reaches a maximum value because the potential difference between the depolarized and unstimulated portions of the atria is at a maximum. When the entire mass of the atria is depolarized, the ECG returns to baseline because all regions of the atria have the same polarity. The spread of atrial depolarization thus creates the **P wave.**

Conduction of the impulse into the ventricles similarly creates a potential difference that results in a sharp upward deflection of the ECG line, which then returns to the baseline as the entire mass of the ventricles becomes depolarized. The spread of the depolarization into the ventricles is thus represented by the **QRS wave.** The atria repolarize during this period, but the event is hidden by the greater depolarization occurring in the ventricles. Finally, repolarization of the ventricles produces the **T wave** (fig. 21.13).

The time interval from the beginning of the P wave to the beginning of the Q wave, which is the beginning of the QRS complex of waves, is known as the *PR interval.* This is a measure of the time required for the impulse to pass through the AV node. Damage to the AV node, therefore, can cause this interval to be prolonged. A PR interval greater than 0.20 second indicates *first-degree AV node block.* If the PR interval becomes too great, the QRS and T waves may not follow a given P wave, and may only appear with every second or third P wave. This condition is called *second-degree AV node block.* The condition of *third-degree AV node block* occurs when none of the atrial waves enter the ventricles. In this case, the ventricles beat according to a slower rhythm set by an *ectopic pacemaker* in the ventricles.

A depression in the ST segment of the ECG may indicate *myocardial ischemia,* or lack of sufficient blood flow to the heart muscle. This may not be evident at rest but may appear when the metabolism of the heart is increased during exercise. For this reason, an ECG may be performed while a person is walking on a treadmill. Some other abnormal conditions that may be detected by an ECG are described in "Clinical Considerations" at the end of this chapter.

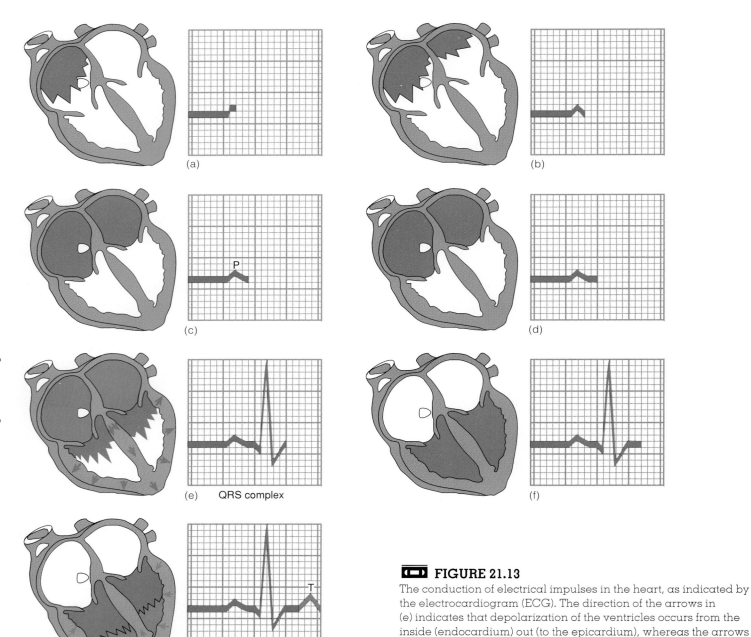

(a)

(b)

(c)

P

(d)

(e)   QRS complex

(f)

(g)

**FIGURE 21.13**

The conduction of electrical impulses in the heart, as indicated by the electrocardiogram (ECG). The direction of the arrows in (e) indicates that depolarization of the ventricles occurs from the inside (endocardium) out (to the epicardium), whereas the arrows in (g) indicate that repolarization of the ventricles occurs in the opposite direction.

**Correlation of the ECG with Heart Sounds**   Depolarization of the ventricles, as indicated by the QRS wave, stimulates contraction by promoting the uptake of $Ca^{++}$ into the regions of the sarcomeres. The QRS wave is thus seen to occur at the beginning systole. The rise in intraventricular pressure that results causes the AV valves to close, so that the first heart sound ($S_1$, or lub) is produced immediately after the QRS wave (fig. 21.14).

Repolarization of the ventricles, as indicated by the T wave, occurs at the same time that the ventricles relax at the beginning of diastole. The resulting fall in intraventricular pressure causes the aortic and pulmonary semilunar valves to close, so that the second heart sound ($S_2$, or dub) is produced shortly after the T wave in an electrocardiogram begins.

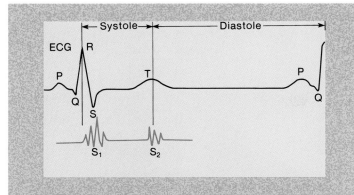

**FIGURE 21.14**
The relationship between the heart sounds and the electrocardiogram during the cardiac cycle. The QRS wave (representing depolarization of the ventricles) occurs at the beginning of systole, while the T wave (representing repolarization of the ventricles) occurs at the beginning of diastole.

# Blood Vessels

*The thick muscle layer of arteries allows them to transmit blood under high pressure ejected from the heart, and the elastic recoil of the large arteries further contributes to blood flow. The thinner muscle layer of veins allows them to distend to accommodate an increasing blood volume, and their one-way valves ensure that blood flows back to the heart. Capillaries are composed of only a single layer of endothelium that allows water and other molecules to move across the capillary walls, thus permitting exchanges between the blood and tissue fluid.*

The walls of arteries and veins are composed of three layers, or tunics. The outermost layer is the **tunica externa,** the middle layer is the **tunica media,** and the inner layer is the **tunica intima.** The tunica externa is composed of loose connective tissue. The tunica media is composed primarily of smooth muscle. The tunica intima consists of three parts: (1) an innermost simple squamous epithelium, the *endothelium (en˝do-the´le-um),* which lines the lumina of all blood vessels; (2) the basement membrane of the endothelium, overlying some connective tissue fibers; and (3) a layer of elastic fibers, or *elastin,* forming an internal elastic lamina.

Although arteries and veins have the same basic structure (fig. 21.15), there are some important differences between the two types of vessels. Arteries have more muscle for their diameters than do comparably sized veins; hence, arteries appear rounder than veins in cross section. Veins are

usually partially collapsed because they are not usually filled to capacity. In addition, many veins have valves, which are absent in arteries.

## Arteries

The aorta and other large arteries contain numerous layers of elastin fibers between smooth muscle cells in the tunica media. These large arteries expand when the pressure of the blood rises as a result of the heart's contraction; they recoil, like a stretched rubber band, when blood pressure falls during relaxation of the heart. This elastic recoil helps to produce a smoother, less pulsatile flow of blood through the smaller arteries and arterioles.

The small arteries and arterioles are less elastic than the larger arteries and have a thicker layer of smooth muscle in proportion to their diameters. Unlike the larger **elastic arteries,** therefore, the smaller **muscular arteries** retain almost the same diameter as the pressure of the blood rises and falls during the heart's pumping activity. Since arterioles and small muscular arteries have narrow lumina, they provide the greatest resistance to blood flow through the arterial system.

Small muscular arteries that are 100 μm or less in diameter branch to form smaller arterioles (20 to 30 μm in diameter). In some tissues, blood from the arterioles can enter the venules directly through **arteriovenous anastomoses** (ă-nas˝tŏ-mo´sēz). In most cases, however, blood from arterioles passes into capillaries (fig. 21.16). Capillaries are the narrowest of blood vessels (7 to 10 μm in diameter), and serve as the "business end" of the circulatory system in which exchanges of gases and nutrients between the blood and the tissues occur.

## Capillaries

The arterial system branches extensively (table 21.3) to deliver blood to billions of capillaries in the body. The extensiveness of these branchings is indicated by the fact that nearly all tissue cells are located within a distance of only 60 to 80 μm of a capillary and by the fact that capillaries provide a total surface area of 1000 square miles for exchanges between blood and tissue fluid.

Despite their large number, capillaries contain only about 250 ml of blood at any time, out of a total blood volume of about 5000 ml (most is contained within veins). The amount of blood flowing through a particular capillary bed is determined in part by the action of the **precapillary sphincter muscles.** These muscles allow only 5% to 10% of

tunic: L. *tunica,* covering or coat
endothelium: Gk. *endo,* within; *thelium,* to cover

anastomosis: Gk. *anastomoo,* to furnish with a mouth (coming together)

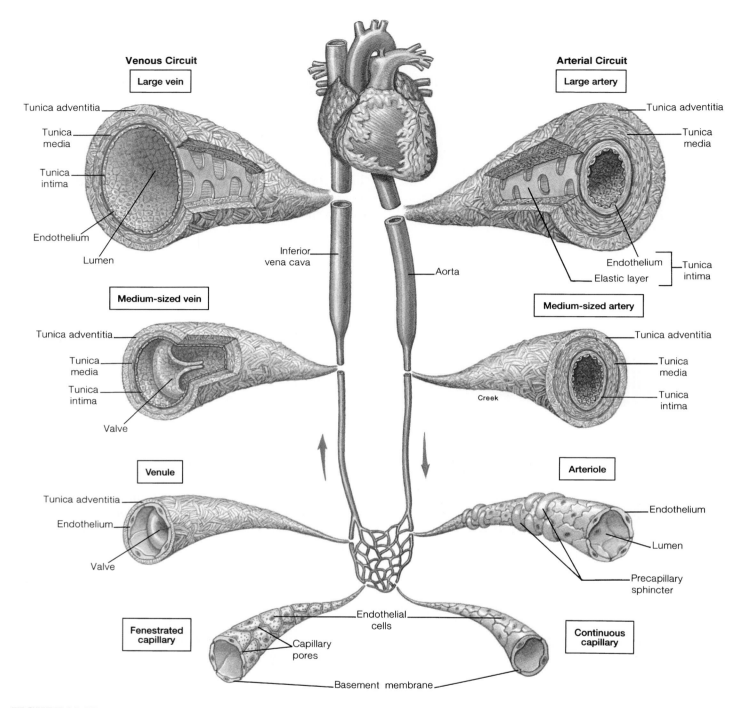

**Venous Circuit**

Large vein

Tunica adventitia

Tunica media

Tunica intima

Endothelium

Lumen

**Arterial Circuit**

Large artery

Tunica adventitia

Tunica media

Endothelium

Elastic layer

Tunica intima

Inferior vena cava

Aorta

**Medium-sized vein**

Tunica adventitia

Tunica media

Tunica intima

Valve

**Medium-sized artery**

Tunica adventitia

Tunica media

Tunica intima

Creek

**Venule**

Tunica adventitia

Endothelium

Valve

**Arteriole**

Endothelium

Lumen

Precapillary sphincter

**Fenestrated capillary**

Endothelial cells

Capillary pores

Basement membrane

**Continuous capillary**

**FIGURE 21.15**

Relative thickness and composition of the tunics in comparable arteries and veins.

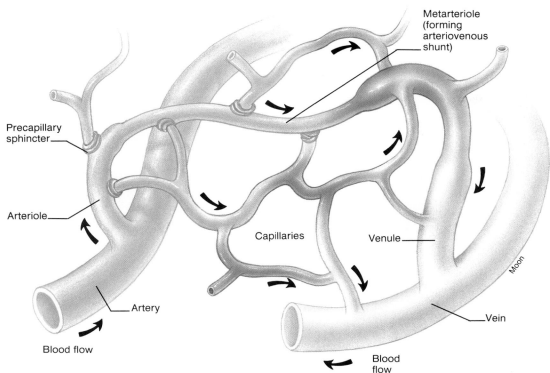

Metarteriole (forming arteriovenous shunt)

Precapillary sphincter

Arteriole

Capillaries

Venule

Artery

Blood flow

Venule

Vein

Blood flow

Moon

**FIGURE 21.16**
The microcirculation. Metarterioles (arteriovenous anastomoses) provide a path of least resistance between arterioles and venules. Precapillary sphincter muscles regulate the flow of blood through the capillaries.

## Table 21.3 Characteristics of the vascular supply to the mesenteries in a dog (pattern is similar in a human)

| Kind of vessel | Diameter (mm) | Number | Total cross-sectional area (cm²) |
|---|---|---|---|
| Aorta | 10 | 1 | 0.8 |
| Large arteries | 3 | 40 | 3.0 |
| Main artery branches | 1 | 600 | 5.0 |
| Terminal branches | 0.6 | 1800 | 5.0 |
| Arterioles | 0.02 | 40,000,000 | 125 |
| Capillaries | 0.008 | 1,200,000,000 | 600 |
| Venules | 0.03 | 80,000,000 | 570 |
| Terminal veins | 1.5 | 1800 | 30 |
| Main venous branches | 2.4 | 600 | 27 |
| Large veins | 6.0 | 40 | 11 |
| Vena cava | 12.5 | 1 | 1.2 |

Source: Data from Malcolm S. Gordon, *Animal Physiology: Principles and Adaptations*, 3d ed., Macmillan Publishing Company. Copyright © 1977 by Malcolm S. Gordon.

the capillary beds in skeletal muscles, for example, to be open at rest. Blood flow to an organ is regulated by the action of these precapillary sphincters and by the degree of resistance to blood flow as determined by the constriction or dilation of the small arteries and arterioles in the organ.

Unlike the vessels of the arterial and venous systems, the walls of capillaries are composed of only an endothelium, which is one cell layer thick (fig. 21.17). The absence of smooth muscle and connective tissue layers permits a more rapid rate of transport of materials between the blood and the tissues.

**Types of Capillaries** Different organs have different types of capillaries, which are distinguished by significant differences in structure. In terms of their endothelial lining, these capillary types include those that are *continuous*, those that are *discontinuous*, and those that are *fenestrated* (fen´ĭ-stra˝tid).

**Continuous capillaries** are those in which adjacent endothelial cells are closely joined together. These are found in muscles, lungs, adipose tissue, and in the central nervous system. The fact that continuous capillaries in the CNS lack intercellular channels contributes to the blood-brain barrier. Continuous capillaries in other organs have narrow intercellular channels (about 40 to 45 Å in width) that permit the passage of molecules other than protein between the capillary blood and tissue fluid.

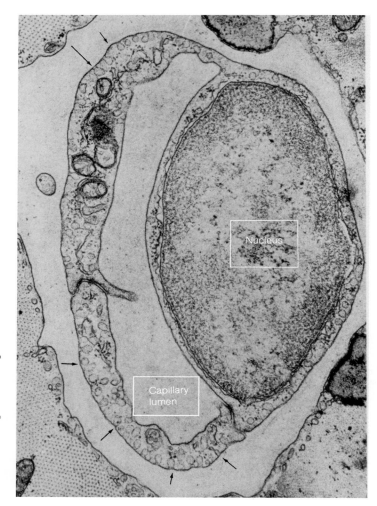

**FIGURE 21.17**

An electron micrograph of a capillary in the heart. Notice the thin intercellular channel (*middle left*) and the capillary wall, composed of only one cell layer. Arrows show some of the many pinocytic vesicles.

Examination of endothelial cells with an electron microscope has revealed the presence of pinocytotic vesicles, which suggests that the intracellular transport of material may occur across the capillary walls. This type of transport appears to be the only mechanism of capillary exchange available within the CNS and may account, in part, for the selective nature of the blood-brain barrier.

The kidneys, endocrine glands, and small intestine have **fenestrated capillaries,** characterized by wide intercellular pores (800 to 1000 Å) that are covered by a layer of mucoprotein, which may serve as a diaphragm. In the bone marrow, liver, and spleen, the distance between endothelial cells is so great that these **discontinuous capillaries** appear as *sinusoids* (little cavities) in the organ.

## *Veins*

Most of the total blood volume is contained in the venous system. Unlike arteries, which provide resistance to the flow

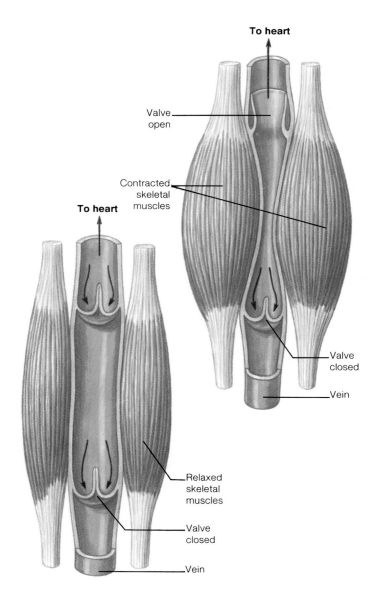

**FIGURE 21.18**

The action of the one-way venous valves. Contraction of skeletal muscles helps to pump blood toward the heart, but is prevented from pushing blood away from the heart by closure of the venous valves.

of blood from the heart, veins are able to expand as they accumulate additional amounts of blood. The average pressure in the veins is only 2 mmHg, compared to a much higher average arterial pressure of about 100 mmHg. These values, expressed in millimeters of mercury, represent the hydrostatic pressure that the blood exerts on the walls of the vessels.

The low venous pressure is insufficient to return blood to the heart, particularly from the lower limbs. Veins, however, pass between skeletal muscle groups that produce a massaging action as they contract (fig. 21.18). As the veins are squeezed by contracting skeletal muscles, a one-way flow of blood to the heart is ensured by the presence of **venous valves.** The ability of these valves to prevent the flow of blood away from the heart was demonstrated in the seventeenth century by William Harvey (see fig. 1.4). After applying a tourniquet to a subject's arm, Harvey found that he could push the blood in a bulging vein toward the heart but not in the reverse direction.

 The effect of the massaging action of skeletal muscles on venous blood flow is often described as the *skeletal muscle pump.* The rate of venous return to the heart is dependent, in large part, on the action of skeletal muscle pumps. When these pumps are less active, as when a person stands still or is bedridden, blood accumulates in the veins and causes them to bulge. When a person is more active, blood returns to the heart at a faster rate and less is left in the venous system.

 The accumulation of blood in the veins of the legs over a long period of time, as may occur in people with occupations that require standing still all day, can cause the veins to stretch to the point where the venous valves are no longer efficient. This can produce *varicose (var'i-kōs) veins.* As a person walks, the movements of the foot activate the soleus muscle pump. This effect can be produced in bedridden people by upward and downward manipulations of the feet.

Action of the skeletal muscle pumps aid the return of venous blood from the lower limbs to the large abdominal veins. Movement of venous blood from abdominal to thoracic veins, however, is aided by an additional mechanism—breathing. When a person inhales, the diaphragm—a dome-shaped muscular sheet separating the thoracic and abdominal cavities—contracts. As it contracts, it moves inferiorly and flattens out, so that it protrudes more into the abdomen. This has the dual effect of increasing the pressure in the abdomen, thus squeezing the abdominal veins, and decreasing the pressure in the thoracic cavity. The pressure difference in the veins created by this inspiratory movement of the diaphragm forces blood into the thoracic veins that return the venous blood to the heart.

## Principal Arteries of the Body

*The aorta ascends from the left ventricle, arches to the left, and descends through the thorax and abdomen. Branches of the aorta carry oxygenated blood to all the cells of the body.*

The principal arteries of the body are shown in figure 21.19. We will describe them by region and identify them in order from largest to smallest, or as the blood flows away from the left ventricle of the heart. The major systemic artery is the **aorta,** from which all the primary systemic arteries arise.

## Aortic Arch

Contraction of the left ventricle forces oxygenated blood into the arteries of the systemic circulation. The systemic vessel that ascends from the left ventricle of the heart is called the **ascending portion of the aorta.** The right and left **coronary arteries** are the only branches that arise from the ascending aorta. The aorta arches to the left and posteriorly over the pulmonary arteries as the **aortic arch** (fig. 21.20). Three vessels arise from the aortic arch: the **brachiocephalic** (*bra´ke-o-sĕ-fal´ik*) **trunk,** the **left common carotid** (*kă-rot´id*) **artery,** and the **left subclavian artery.**

The brachiocephalic trunk, as its name suggests, supplies blood to the arm and head on the right side of the body. It is a short vessel rising superiorly through the mediastinum to a point near the junction of the sternum and the right clavicle. There it bifurcates into the **right common carotid artery,** which extends to the right side of the neck and head, and the **right subclavian artery,** which carries blood to the right upper extremity. Note that the brachiocephalic trunk is unpaired; the **left common carotid artery** and **left subclavian artery,** unlike the right, arise directly from the aortic arch.

## Arteries of the Head and Neck

The common carotid arteries course upward in the neck along either lateral side of the trachea (fig. 21.21). Several small vessels arise from the common carotid artery to supply blood to the larynx, thyroid, anterior neck muscles, and lymph glands of the neck. The common carotid artery bifurcates into the **internal** and **external carotid arteries** slightly below the angle of the mandible. By pressing gently in this area, you can detect your pulse. At the base of the internal carotid artery, near the bifurcation, is a slight dilation called the **carotid sinus.** The carotid sinus contains blood pressure receptors called *baroreceptors.* Surrounding the carotid sinus are *chemoreceptors (ke˝mo-re-sep´torz)* within the **carotid body** that respond to changes in blood gas concentrations.

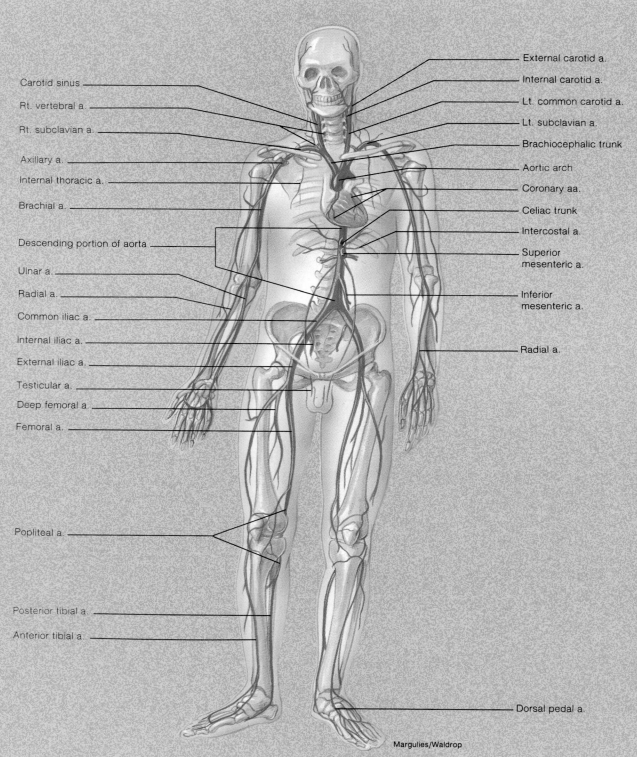

Carotid sinus

Rt. vertebral a.

Rt. subclavian a.

Axillary a.

Internal thoracic a.

Brachial a.

Descending portion of aorta

Ulnar a.

Radial a.

Common iliac a.

Internal iliac a.

External iliac a.

Testicular a.

Deep femoral a.

Femoral a.

Popliteal a.

Posterior tibial a.

Anterior tibial a.

External carotid a.

Internal carotid a.

Lt. common carotid a.

Lt. subclavian a.

Brachiocephalic trunk

Aortic arch

Coronary aa.

Celiac trunk

Intercostal a.

Superior mesenteric a.

Inferior mesenteric a.

Radial a.

Dorsal pedal a.

Margulies/Waldrop

**FIGURE 21.19**

Principal arteries of the body (a. = artery; aa. = arteries).

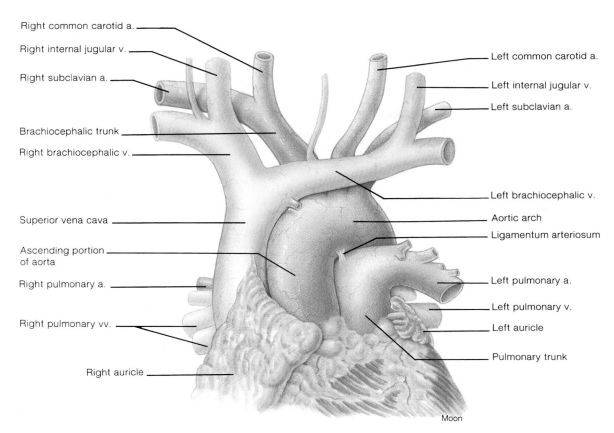

Right common carotid a.

Right internal jugular v.

Right subclavian a.

Brachiocephalic trunk

Right brachiocephalic v.

Superior vena cava

Ascending portion of aorta

Right pulmonary a.

Right pulmonary vv.

Right auricle

Left common carotid a.

Left internal jugular v.

Left subclavian a.

Left brachiocephalic v.

Aortic arch

Ligamentum arteriosum

Left pulmonary a.

Left pulmonary v.

Left auricle

Pulmonary trunk

Moon

**FIGURE 21.20**

The structural relationship between the major arteries and veins to and from the heart (v. = vein; vv. = veins).

**Blood Supply to the Brain**    The brain is supplied with arterial blood that arrives through four vessels that eventually unite on the inferior aspect of the brain surrounding the pituitary gland (fig. 21.22). The four vessels are the paired internal carotid arteries and the paired vertebral arteries. The value of having four separate vessels that anastomose (come together) at one location is that, if one becomes occluded, the three alternate routes may allow for an adequate blood supply to the brain.

The **vertebral arteries** arise from the subclavian arteries at the base of the neck (see fig. 21.21). They pass superiorly through the transverse foramina of the cervical vertebrae and enter the skull through the foramen magnum. Within the braincase, the two vertebral arteries unite to form the **basilar artery** at the level of the pons. The basilar ascends along the inferior surface of the brain stem and terminates by forming two **posterior cerebral arteries,** which supply the posterior portion of the cerebrum. The **posterior communicating arteries** are branches that arise from the posterior cerebral arteries and participate in forming the **cerebral arterial circle** (circle of Willis) around the pituitary gland.

The internal carotid artery branches from the common carotid artery and ascends in the neck until it reaches the base of the skull, where it enters the carotid canal of the temporal bone. Several branches arise from the internal carotid artery once it is on the inferior surface of the brain. Three of the more important ones are the **ophthalmic** (*of-thal´mik*) **artery,** which supplies the eye, and the **anterior** and **middle cerebral arteries,** which provide blood to the cerebrum. The internal carotid arteries are connected to the posterior cerebral arteries at the cerebral arterial circle (fig. 21.23).

**Blood Supply to the Head and Neck**    The external carotid artery gives off several branches as it extends upward along the side of the neck and head (see fig. 21.21). The names of these branches are determined by the areas or structures that they serve. The principal vessels that arise from the external carotid artery are the **superior thyroid, ascending pharyngeal, lingual, facial, occipital,** and **posterior auricular arteries.**

The external carotid artery terminates at a level near the mandibular condyle by dividing into **maxillary** and

Circulatory System

585

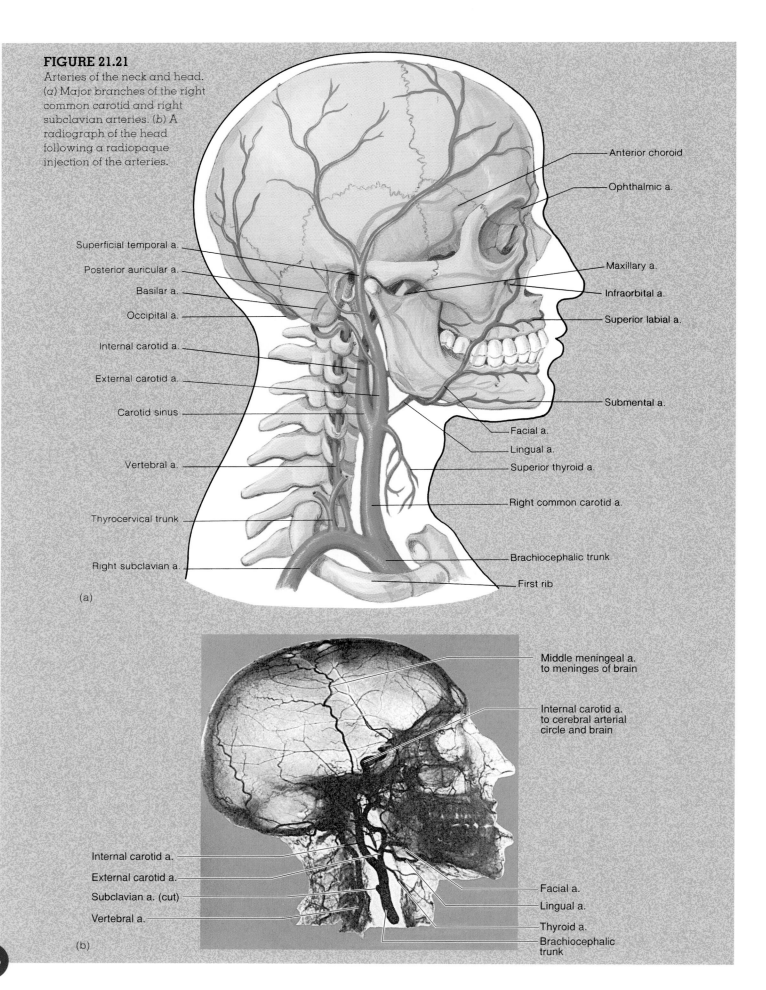

**FIGURE 21.21**

Arteries of the neck and head. (a) Major branches of the right common carotid and right subclavian arteries. (b) A radiograph of the head following a radiopaque injection of the arteries.

Anterior choroid

Ophthalmic a.

Superficial temporal a.

Posterior auricular a.

Basilar a.

Occipital a.

Internal carotid a.

External carotid a.

Carotid sinus

Vertebral a.

Thyrocervical trunk

Right subclavian a.

Maxillary a.

Infraorbital a.

Superior labial a.

Submental a.

Facial a.

Lingual a.

Superior thyroid a.

Right common carotid a.

Brachiocephalic trunk

First rib

(a)

Middle meningeal a. to meninges of brain

Internal carotid a. to cerebral arterial circle and brain

Internal carotid a.

External carotid a.

Subclavian a. (cut)

Vertebral a.

Facial a.

Lingual a.

Thyroid a.

Brachiocephalic trunk

(b)

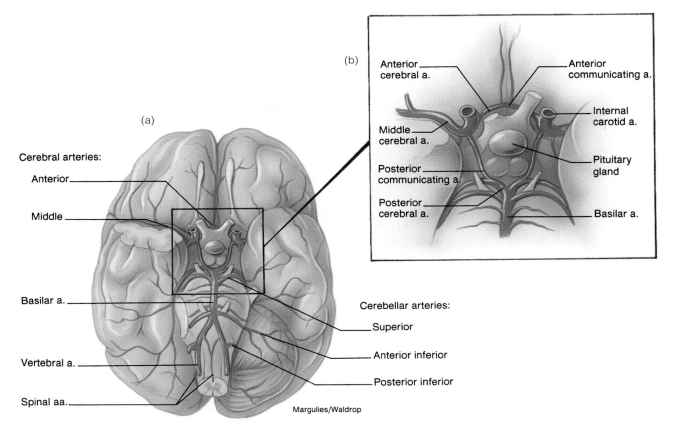

**(a)**

Cerebral arteries:

Anterior

Middle

Basilar a.

Vertebral a.

Spinal aa.

**(b)**

Anterior cerebral a.

Anterior communicating a.

Internal carotid a.

Middle cerebral a.

Pituitary gland

Posterior communicating a.

Posterior cerebral a.

Basilar a.

Cerebellar arteries:

Superior

Anterior inferior

Posterior inferior

Margulies/Waldrop

**FIGURE 21.22**
Arteries that supply blood to the brain. (*a*) An inferior view of the brain and (*b*) a close-up view of the region of the pituitary gland.

**superficial temporal arteries.** Pulsations through the temporal artery can be easily detected by placing your fingertips immediately in front of the ear at the level of the eye. This vessel is frequently used by anesthesiologists to check a patient's pulse rate during surgery.

## Arteries of the Upper Extremity

The **right subclavian artery** branches off the brachiocephalic trunk, and the **left subclavian artery** arises directly from the aortic arch (see fig. 21.19). Each subclavian artery passes laterally deep to the clavicle, carrying blood toward the arm (fig. 21.24). From each subclavian artery arises a **vertebral artery** that carries blood to the brain (already described), as well as a short **thyrocervical trunk** and an **internal thoracic artery.**

The subclavian artery becomes the **axillary** (*ak´sĭ-lar˝e*) **artery** as it passes into the axillary region. Several small branches arise from the axillary artery and supply blood to the tissues of the upper thorax and shoulder region.

The **brachial** (*bra´ke-al*) **artery** is the continuation of the axillary artery through the brachial region. The brachial artery courses on the medial side of the humerus, where it is a major pressure point and the most common site for determining blood pressure. A **deep brachial artery** branches from the brachial artery and curves posteriorly near the radial nerve to supply the triceps brachii muscle. Two additional branches from the brachial, the **anterior** and **posterior humeral circumflex arteries,** form a continuous ring of vessels around the proximal portion of the humerus to supply surrounding muscles.

The brachial artery bifurcates into the **radial** and **ulnar arteries,** which supply blood to the forearm and a portion of the hand and digits. The radial artery is important as a site for recording the pulse near the wrist. Important branches from the radial and ulnar arteries are the **radial recurrent artery** and the **ulnar recurrent artery,** which course superiorly and anastomose with a branch of the brachial artery around the elbow joint.

**587**

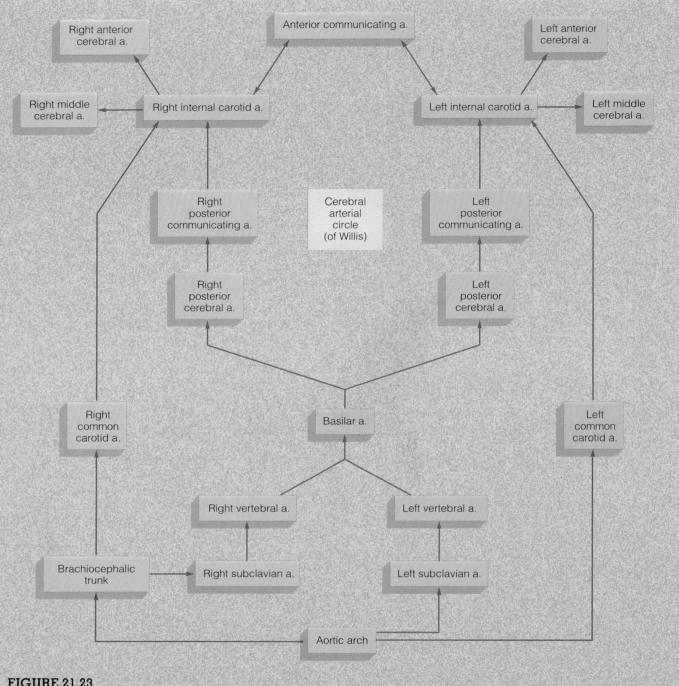

**FIGURE 21.23**
The cerebral arterial circle (of Willis).

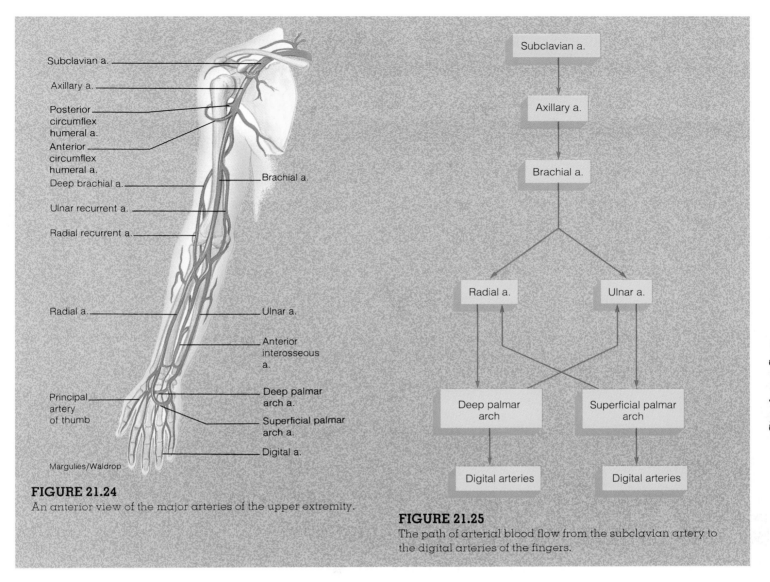

**FIGURE 21.24**
An anterior view of the major arteries of the upper extremity.

Labels in figure 21.24:
Subclavian a.
Axillary a.
Posterior circumflex humeral a.
Anterior circumflex humeral a.
Deep brachial a.
Ulnar recurrent a.
Radial recurrent a.
Radial a.
Principal artery of thumb
Margulies/Waldrop
Brachial a.
Ulnar a.
Anterior interosseous a.
Deep palmar arch a.
Superficial palmar arch a.
Digital a.

**FIGURE 21.25**
The path of arterial blood flow from the subclavian artery to the digital arteries of the fingers.

Labels in figure 21.25:
Subclavian a.
Axillary a.
Brachial a.
Radial a.
Ulnar a.
Deep palmar arch
Superficial palmar arch
Digital arteries
Digital arteries

The radial and ulnar arteries anastomose with each other to form two arches in the hand, the **deep palmar arch** and the **superficial palmar arch.** Branches from the palmar arches form **digital arteries** that extend into the fingers. The path of blood flow from the subclavian artery to the digital arteries is summarized in figure 21.25.

## Branches of the Thoracic and Abdominal Portions of the Aorta

The **thoracic portion of the aorta** is a continuation of the aortic arch as it descends through the thoracic cavity to the diaphragm. This large vessel gives off many branches, including the **pericardial, bronchial, esophageal** (ĕ-sof´´ă-je´al), **posterior intercostal,** and **superior phrenic** (fren´ik) **arteries.** These vessels are summarized according to their location and function in table 21.4.

The **abdominal portion of the aorta** is the segment of the aorta between the diaphragm and the level of the fourth lumbar vertebra. The first branches of the abdominal aorta are the paired **inferior phrenic arteries.** Next, the large **celiac** (se´le-ak) **trunk** arises and divides immediately into three arteries: the **splenic, left gastric,** and the **common hepatic arteries** (fig. 21.26).

Other unpaired arteries are the **superior mesenteric** and the **inferior mesenteric arteries.** Other paired arteries include the **renal, suprarenal,** and **gonadal** (**testicular** or **ovarian**) **arteries.** The organs supplied by these vessels are indicated in table 21.4.

## Arteries of the Pelvis and Lower Extremity

The abdominal aorta terminates in the posterior pelvic area by dividing into the right and left **common iliac** (il´e-ak) **arteries.** These vessels pass downward approximately 5 cm and terminate by dividing into the **internal** and **external iliac arteries.**

## Table 21.4  Segments and branches of the aorta

| Segment of aorta | Arterial branch | General region or organ served |
|---|---|---|
| Ascending portion of aorta | Right and left coronary aa. | Heart |
| Aortic arch | Brachiocephalic trunk Right common carotid a. | Right side of head and neck |
| | Right subclavian a. | Right shoulder and right upper extremity |
| | Left common carotid a. | Left side of head and neck |
| | Left subclavian a. | Left shoulder and left upper extremity |
| Thoracic portion of aorta | Pericardial aa. | Pericardium of heart |
| | Posterior intercostal aa. | Intercostal and thoracic muscles, pleurae |
| | Bronchial aa. | Bronchi of lungs |
| | Superior phrenic aa. | Superior surface of diaphragm |
| | Esophageal aa. | Esophagus |
| Abdominal portion of aorta | Inferior phrenic aa. | Inferior surface of diaphragm |
| | Celiac trunk | |
| | Common hepatic a. | Liver, upper pancreas, duodenum |
| | Left gastric a. | Stomach, esophagus |
| | Splenic a. | Spleen, pancreas, stomach |
| | Superior mesenteric a. | Small intestine, pancreas, cecum, appendix, ascending and transverse colons |
| | Suprarenal aa. | Suprarenal (adrenal) glands |
| | Renal aa. | Kidneys |
| | Gonadal aa. | |
| | Testicular aa. | Testes |
| | Ovarian aa. | Ovaries |
| | Inferior mesenteric a. | Transverse, descending, and sigmoid colons; rectum |
| | Common iliac aa. | |
| | External iliac aa. | Lower extremities |
| | Internal iliac aa. | Genital organs, gluteal muscles |

The internal iliac artery has extensive branches to supply arterial blood to the gluteal muscles and the organs of the pelvic region (fig. 21.27). Branches of the internal iliac artery include the **iliolumbar, lateral sacral, middle rectal, vesicular** (to the urinary bladder), **uterine** and **vaginal, gluteal, obturator,** and **internal pudendal** (to the external genitalia) **arteries.**

The external iliac artery gives off two branches (the **inferior epigastric** and **deep circumflex iliac arteries**) before exiting the pelvic cavity beneath the inguinal ligament (fig. 21.27). Once through the inguinal canal, the external iliac artery becomes the **femoral artery** (fig. 21.28).

The femoral artery passes through an area called the **femoral triangle** on the upper medial portion of the thigh (fig. 21.29). At this point, the femoral artery is close to the surface and is an important arterial pressure point (see page 562). Several vessels arise from the femoral artery, including the **deep femoral** and the **lateral** and **medial femoral circumflex arter-**

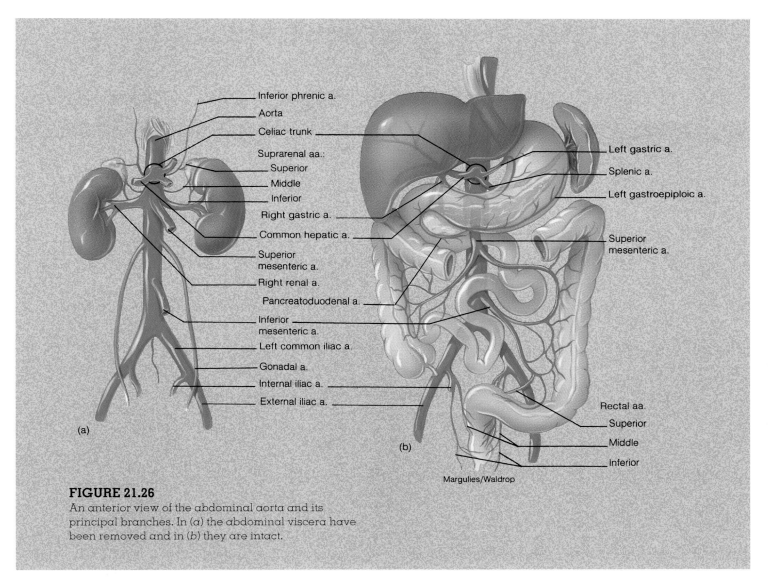

Inferior phrenic a.
Aorta
Celiac trunk
Suprarenal aa.:
  Superior
  Middle
  Inferior
Right gastric a.
Common hepatic a.
Superior
mesenteric a.
Right renal a.
Pancreatoduodenal a.
Inferior
mesenteric a.
Left common iliac a.
Gonadal a.
Internal iliac a.
External iliac a.

Left gastric a.
Splenic a.
Left gastroepiploic a.
Superior
mesenteric a.

Rectal aa.
Superior
Middle
Inferior

(a)

(b)

Margulies/Waldrop

**FIGURE 21.26**
An anterior view of the abdominal aorta and its
principal branches. In (a) the abdominal viscera have
been removed and in (b) they are intact.

**ies.** The femoral artery becomes the **popliteal** (*pop´lĭ-te´al*)
**artery** as it passes across the posterior aspect of the knee.

 *Hemorrhage* can be a serious problem in many
accidents. To prevent a victim from bleeding to death,
it is important to know where to apply pressure to
curtail the flow of blood. The arterial pressure points
for the appendages are the brachial artery on the medial side
of the arm and the femoral artery in the groin (see the section
"Arterial Pressure Points and Control of Bleeding" in chapter
20). Firmly applied pressure to these regions greatly
diminishes blood flow to traumatized areas below. A
tourniquet may have to be applied if bleeding is life
threatening and if other, safer methods have proved
ineffective.

The popliteal artery supplies small branches to the knee
joint and then divides into an **anterior tibial artery** and **posterior tibial artery** (fig. 21.28). At the ankle, the anterior tibial artery becomes the **dorsal pedal artery,** which serves the
foot and then contributes to the formation of the **plantar**

**arch.** The posterior tibial artery gives off the large **peroneal**
(*per´ŏ-ne´al*) **artery** and then, at the ankle, the posterior tibial bifurcates into the **lateral** and **medial plantar arteries.** The
lateral plantar artery anastomoses with the dorsal pedal artery
to form the plantar arch in an arterial arrangement similar to
that of the hand. **Digital arteries** arise from the plantar arch
to supply the toes with blood.

# Principal Veins of the Body

*After systemic blood has passed through the tissues and its oxygen is
depleted, it returns through veins of increasing diameters to the right
atrium of the heart.*

In the venous portion of the systemic circulation, blood
flows from smaller vessels into larger ones, so that a vein receives smaller tributaries instead of giving off branches as

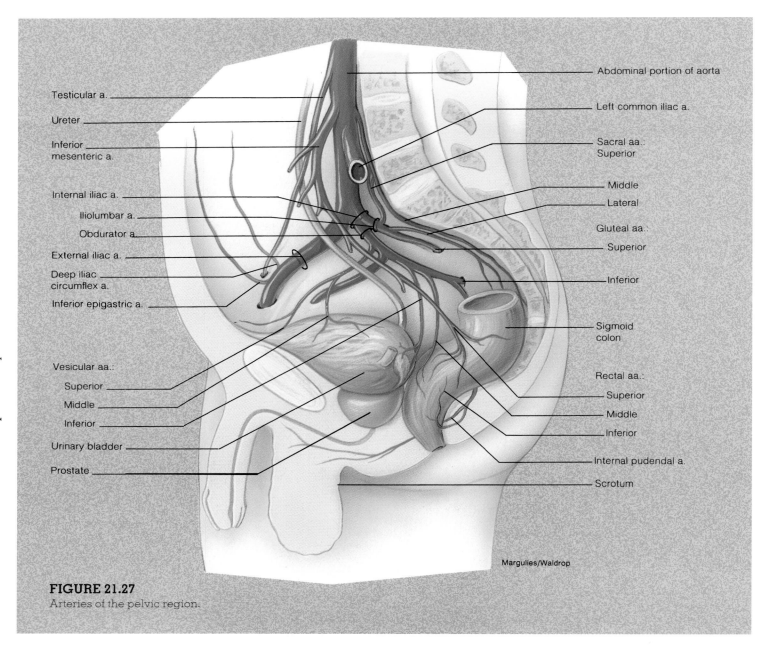

Testicular a.

Ureter

Inferior mesenteric a.

Internal iliac a.

Iliolumbar a.

Obdurator a.

External iliac a.

Deep iliac circumflex a.

Inferior epigastric a.

Vesicular aa.:

Superior

Middle

Inferior

Urinary bladder

Prostate

Abdominal portion of aorta

Left common iliac a.

Sacral aa.:
Superior

Middle

Lateral

Gluteal aa.:

Superior

Inferior

Sigmoid colon

Rectal aa.:

Superior

Middle

Inferior

Internal pudendal a.

Scrotum

Margulies/Waldrop

**FIGURE 21.27**
Arteries of the pelvic region.

an artery does. The veins from all parts of the body converge into two major vessels that empty into the right atrium: the **superior** and **inferior vena cavae.** Veins are more numerous than arteries and are both superficial and deep. Superficial veins generally can be seen just beneath the skin and are clinically important in drawing blood and giving injections. Deep veins are close to the principal arteries and are usually similarly named. As with arteries, veins are named according to the region in which they are found or the organ that they serve (when a vein serves an organ, it drains blood

away from it). The principal systemic veins of the body are illustrated in figure 21.30.

## Veins Draining the Head and Neck

Blood from the scalp, portions of the face, and the superficial neck regions is drained by the **external jugular veins** (fig.

······································

jugular: L. *jugulum,* throat or neck

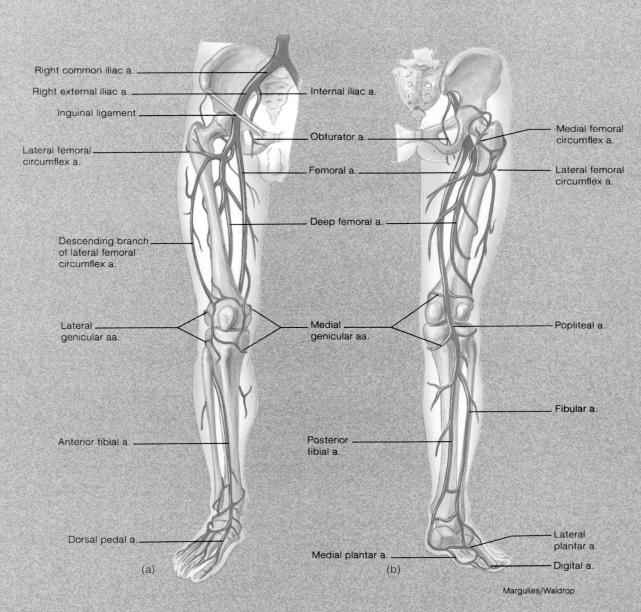

Right common iliac a.

Right external iliac a.

Inguinal ligament

Lateral femoral
circumflex a.

Descending branch
of lateral femoral
circumflex a.

Lateral
genicular aa.

Anterior tibial a.

Dorsal pedal a.

(a)

Internal iliac a.

Obturator a.

Femoral a.

Deep femoral a.

Medial
genicular aa.

Posterior
tibial a.

Medial plantar a.

(b)

Medial femoral
circumflex a.

Lateral femoral
circumflex a.

Popliteal a.

Fibular a.

Lateral
plantar a.

Digital a.

Margulies/Waldrop

**FIGURE 21.28**
Arteries of the right lower extremity. (a) An anterior view and
(b) a posterior view.

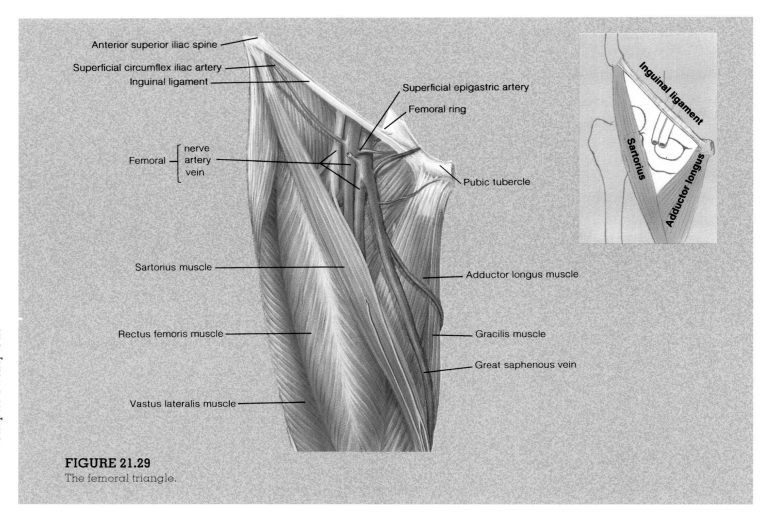

Anterior superior iliac spine

Superficial circumflex iliac artery

Inguinal ligament

Superficial epigastric artery

Femoral ring

Femoral [ nerve
artery
vein ]

Pubic tubercle

Sartorius muscle

Adductor longus muscle

Rectus femoris muscle

Gracilis muscle

Great saphenous vein

Vastus lateralis muscle

Inguinal ligament

Sartorius

Adductor longus

**FIGURE 21.29**
The femoral triangle.

21.31). These vessels descend on either lateral side of the neck and drain into the right and left **subclavian veins,** which are located just behind the clavicles.

The paired **internal jugular veins** drain blood from the brain, meninges, and deep regions of the face and neck. The internal jugular veins are larger and deeper than the external jugular veins. They arise from numerous cranial **venous sinuses,** which constitute a series of both paired and unpaired channels within the dura mater. The venous sinuses, in turn, receive venous blood from the **cerebral,** the **cerebellar,** the **ophthalmic,** and the **meningeal veins.**

The internal jugular vein passes inferiorly down the neck, adjacent to the common carotid artery and the vagus nerve. The internal jugular on each side empties into the subclavian vein, and the union of these two vessels forms the large **brachiocephalic vein** on each side. The two bra-

chiocephalic veins merge to form the **superior vena cava,** which drains into the right atrium of the heart (see fig. 21.30).

## Veins of the Upper Extremity

The upper extremity has both superficial and deep venous drainage (fig. 21.32). The superficial veins are highly variable and form an extensive network just below the skin. The deep veins accompany the arteries of the same region and are given similar names: the **radial, ulnar, brachial,** and **axillary veins** are the major examples.

The main superficial vessels of the upper extremity are the **basilic vein** and the **cephalic vein.** In the cubital fossa of the elbow, the superficial **median cubital vein** connects the cephalic vein on the lateral side with the basilic vein on the medial side. The median cubital vein is commonly punctured to obtain blood samples for clinical tests. Both the

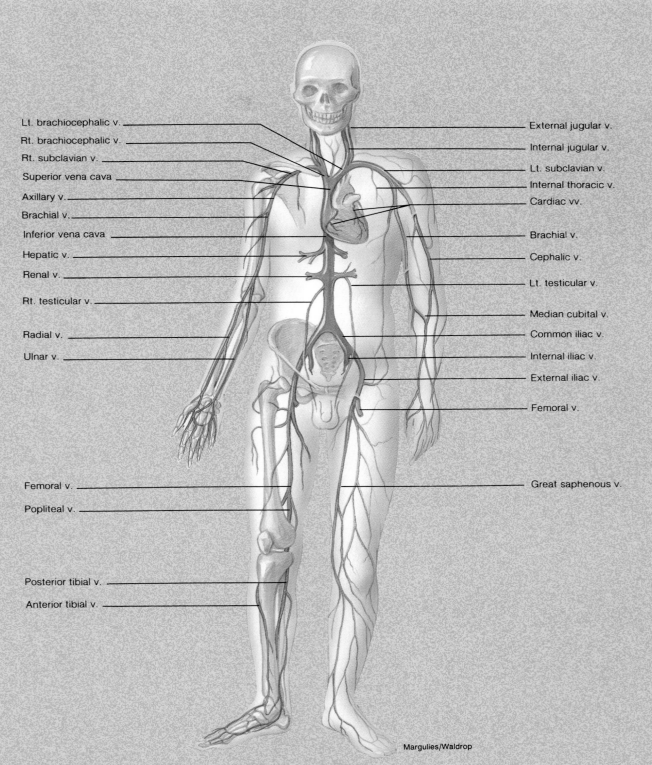

Lt. brachiocephalic v.
Rt. brachiocephalic v.
Rt. subclavian v.
Superior vena cava
Axillary v.
Brachial v.
Inferior vena cava
Hepatic v.
Renal v.
Rt. testicular v.
Radial v.
Ulnar v.
Femoral v.
Popliteal v.
Posterior tibial v.
Anterior tibial v.

External jugular v.
Internal jugular v.
Lt. subclavian v.
Internal thoracic v.
Cardiac vv.
Brachial v.
Cephalic v.
Lt. testicular v.
Median cubital v.
Common iliac v.
Internal iliac v.
External iliac v.
Femoral v.
Great saphenous v.

Margulies/Waldrop

**FIGURE 21.30**
Principal veins of the body. Superficial veins are depicted in the
left extremities and deep veins in the right extremities.

## FIGURE 21.31

Veins that drain the head and neck.

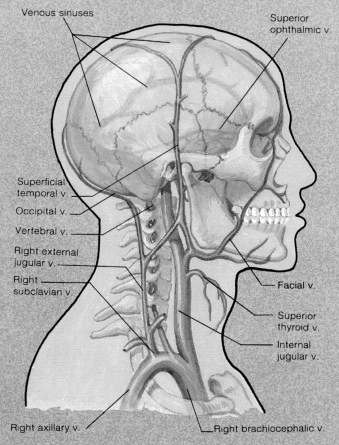

Venous sinuses

Superior
ophthalmic v.

Superficial
temporal v.

Occipital v.

Vertebral v.

Right external
jugular v.

Right
subclavian v.

Facial v.

Superior
thyroid v.

Internal
jugular v.

Right axillary v.

Right brachiocephalic v.

## FIGURE 21.32

An anterior view of the veins that drain
the upper right extremity. (a) Superficial
veins and (b) deep veins.

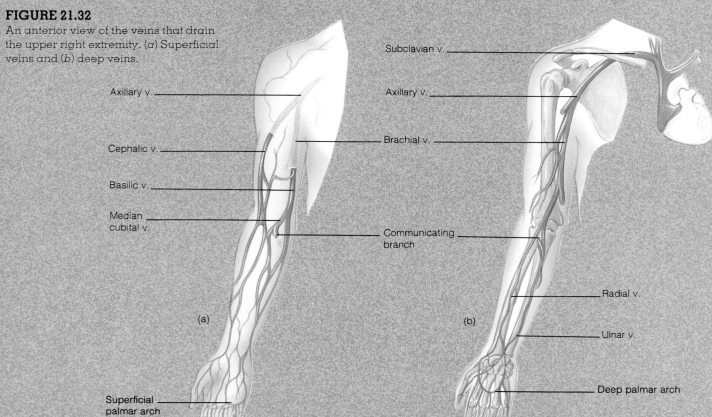

Subclavian v.

Axillary v.

Axillary v.

Cephalic v.

Brachial v.

Basilic v.

Median
cubital v.

Communicating
branch

Radial v.

Ulnar v.

(a)

(b)

Deep palmar arch

Superficial
palmar arch

Margulies/Waldrop

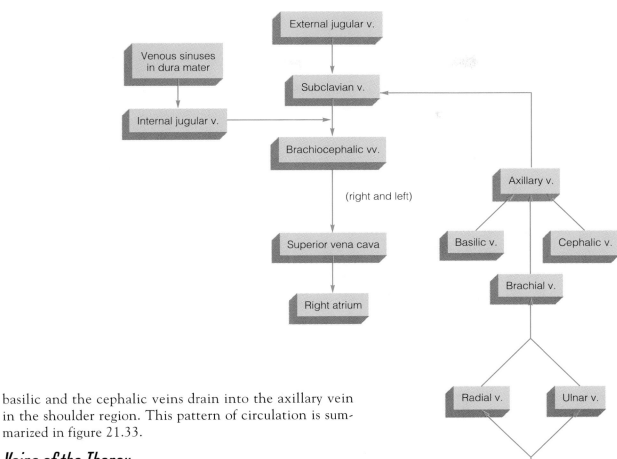

basilic and the cephalic veins drain into the axillary vein in the shoulder region. This pattern of circulation is summarized in figure 21.33.

## Veins of the Thorax

The **superior vena cava,** formed by the union of the two brachiocephalic veins, empties venous blood from the head, neck, and upper extremities directly into the right atrium of the heart. These large vessels lack the valves that are characteristic of most other veins in the body. In addition to receiving blood from the brachiocephalic veins, the superior vena cava collects blood from the *azygos* (*az´ĭ-gos*) *system of veins* arising from the posterior thoracic wall (fig. 21.34). This system includes the **azygos, ascending lumbar, intercostal, accessory hemiazygos,** and **hemiazygos veins.**

## Veins of the Lower Extremity

The lower extremities, like the upper, have both a deep and a superficial group of veins (fig. 21.35). The deep veins accompany corresponding arteries and have more valves than do the superficial veins.

The deep veins include the **posterior** and **anterior tibial veins** that originate in the foot and course upward to the back of the knee, where they merge to form the **popliteal vein.** Just above the knee, this vessel becomes the **femoral**

**FIGURE 21.33**
Venous return of blood from the head and the upper extremity to the heart.

vein. The femoral vein receives blood from the **deep femoral vein** near the groin and then becomes the **external iliac vein** as it passes under the inguinal ligament. The external iliac merges with the **internal iliac vein** at the pelvic and genital regions to form the **common iliac vein.** At the level of the

•••••••••••••••••••••••••••••••••••••••••••

azygos: Gk. *a*, without; *zygon*, yoke

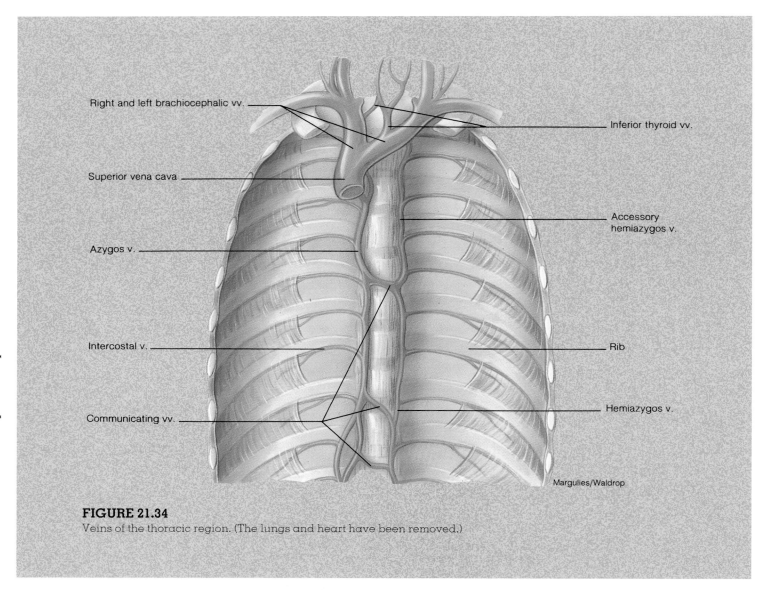

Right and left brachiocephalic vv.

Inferior thyroid vv.

Superior vena cava

Accessory hemiazygos v.

Azygos v.

Intercostal v.

Rib

Communicating vv.

Hemiazygos v.

Margulies/Waldrop

**FIGURE 21.34**
Veins of the thoracic region. (The lungs and heart have been removed.)

fifth lumbar vertebra, the right and left common iliacs unite to form the large **inferior vena cava** (fig. 21.35).

The superficial veins of the lower extremity are the **small** and **great saphenous** (*să-fe´nus*) **veins.** The small saphenous vein arises from the lateral side of the foot and empties into the popliteal vein behind the knee. The great saphenous vein is the longest vessel in the body. It originates at the arch of the foot and ascends superiorly along the medial aspect of the leg and thigh before draining into the femoral vein.

## Veins of the Abdominal Region

The **inferior vena cava** parallels the abdominal aorta on the right side as it ascends through the abdominal cavity to penetrate the diaphragm and enter the right atrium (see fig. 21.30). It is the largest vessel of the body in diameter and is formed by the union of the two common iliac veins drain-

ing the lower extremities. As the inferior vena cava ascends through the abdominal cavity, it receives tributaries from veins that correspond in name and position to arteries previously described.

## Hepatic Portal System

In a *portal system*, one capillary bed is located downstream from another. Venous drainage from the first capillary bed is delivered into the second, and venous drainage from the second is delivered finally into the general circulation. In the hepatic portal system, blood from the intestines, pancreas, spleen, stomach, and gallbladder is drained by veins and delivered to the liver. Blood from the liver, in turn, is

...........................................

saphenous: L. *saphena,* the hidden one

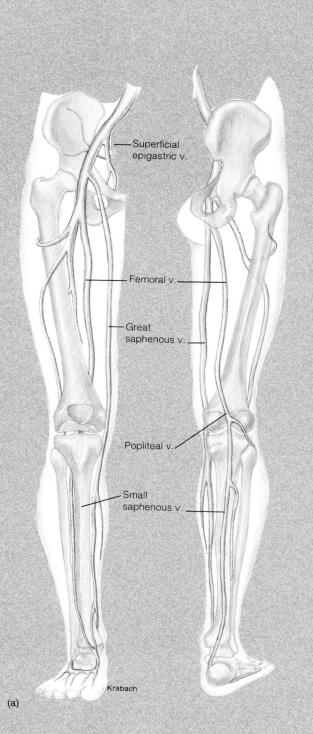

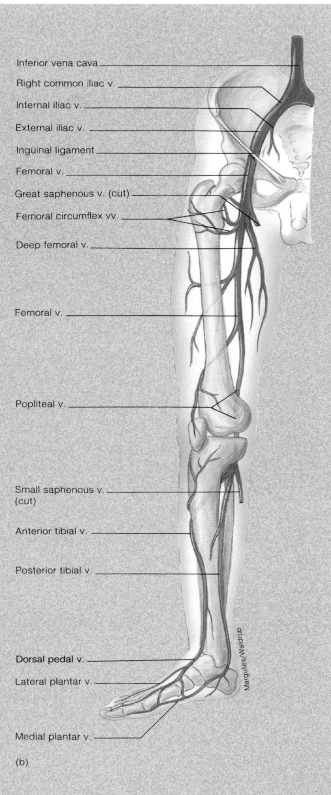

**FIGURE 21.35**

Veins of the lower extremity. (*a*) Superficial veins, medial and posterior aspects and (*b*) deep veins, medial view.

Inferior vena cava

Right common iliac v.

Internal iliac v.

External iliac v.

Inguinal ligament

Femoral v.

Great saphenous v. (cut)

Femoral circumflex vv.

Deep femoral v.

Femoral v.

Popliteal v.

Small saphenous v. (cut)

Anterior tibial v.

Posterior tibial v.

Dorsal pedal v.

Lateral plantar v.

Medial plantar v.

Superficial epigastric v.

Femoral v.

Great saphenous v.

Popliteal v.

Small saphenous v.

Krabach

Margulies/Waldrop

(a)

(b)

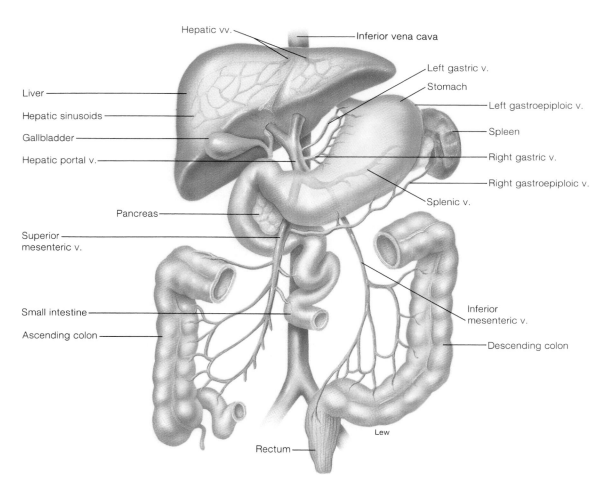

**FIGURE 21.36**
The hepatic portal system.

drained by **hepatic veins** that empty into the inferior vena cava (fig. 21.36). The essential aspect of the hepatic portal system is that the absorbed products of digestion must first pass through the liver to be processed before they can enter the general circulation.

The **hepatic portal vein** is the large vessel that receives blood from the digestive organs. It is formed by a union of the **superior mesenteric vein,** which drains nutrient-rich blood from the small intestine, and the **splenic vein.** The splenic vein drains the spleen but is enlarged because of a convergence of four tributaries: (1) the **inferior mesenteric vein,** (2) the **pancreatic vein,** (3) the **left gastroepiploic vein,** and (4) the **right gastroepiploic vein.** Three additional veins empty directly into the hepatic portal vein. These are the right and left **gastric veins** and the **cystic vein.**

It is important to note that the sinusoids of the liver receive blood from two sources: the hepatic artery and the

hepatic portal vein. The hepatic artery carries oxygen and the hepatic portal vein transports nutrient-rich blood from the small intestine for processing. Liver cells can thus modify the chemical composition of the venous blood that enters the general circulation from the GI tract. The nature of these modifications is discussed in detail in chapter 26.

# Fetal Circulation

*Oxygen and nutrients are supplied to the fetus by the placenta instead of by the fetal lungs and gastrointestinal tract. Fetal circulation is adaptive to these conditions.*

••••••••••••••••••••••••••••••••••••••••

gastroepiploic: Gk. *gastros,* stomach; *epiplein,* to float on (referring to greater omentum)

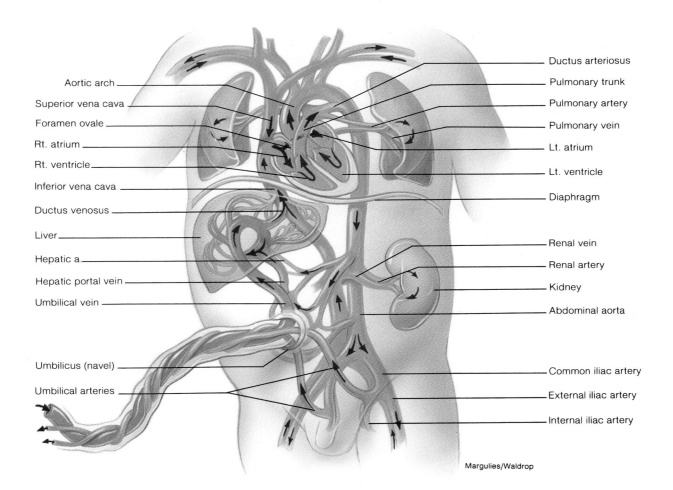

Aortic arch
Superior vena cava
Foramen ovale
Rt. atrium
Rt. ventricle
Inferior vena cava
Ductus venosus
Liver
Hepatic a.
Hepatic portal vein
Umbilical vein

Umbilicus (navel)
Umbilical arteries

Ductus arteriosus
Pulmonary trunk
Pulmonary artery
Pulmonary vein
Lt. atrium
Lt. ventricle
Diaphragm

Renal vein
Renal artery
Kidney
Abdominal aorta

Common iliac artery
External iliac artery
Internal iliac artery

Margulies/Waldrop

**FIGURE 21.37**
Fetal circulation. (Arrows indicate the direction of blood flow.)

The circulation of blood through a fetus is necessarily different from blood circulation in a newborn (fig. 21.37). This is because oxygenation of the blood, the procurement of nutrients, and the elimination of metabolic wastes occur through the placenta instead of through the organs of the fetus.

The **umbilical cord** serves as the connection between the placenta and the fetus. It includes one **umbilical vein** and two **umbilical arteries.** Oxygenated and nutrient-rich blood flows from the placenta to the fetus in the umbilical vein. At the inferior surface of the liver, this vein bifurcates. One branch joins with the hepatic portal vein and the other branch, called the **ductus venosus,** enters the inferior vena cava. Thus, oxygenated blood from the placenta is mixed with venous blood returning from the lower extremities of the fetus before it enters the heart. The umbilical vein, then, is the only vessel of the fetus that carries fully oxygenated blood.

The inferior vena cava empties into the right atrium of the fetal heart. Instead of going into the right ventricle, however, most of the blood passes from the right atrium into the left atrium through the **foramen ovale,** an opening between the two atria. The small amount of blood that does enter the right ventricle and is pumped into the pulmonary trunk is largely diverted by a short **ductus arteriosus** directly into the aortic arch, thereby bypassing the lungs.

The foramen ovale and ductus arteriosus therefore shunt blood away from the pulmonary circulation and into the systemic circulation. This is essential because the function of gas exchange is not performed by the fetal lungs, but rather by the placenta. Blood is delivered to the placenta by the two umbilical arteries, which arise from the internal iliac arteries.

## Development of the Heart

The remarkable development of the heart requires only 6 to 7 days. Heart development is first apparent at day 18 or 19 in the *cardiogenic* (kar´´de-o-jen´ik) *area* of the mesoderm layer. A small paired mass of specialized cells called *heart cords* form here. Shortly after, a hollow center develops in each heart cord, and each structure is then referred to as a *heart tube.* The heart tubes begin to migrate toward each other during day 21 and soon fuse to form a single median *endocardial heart tube.* During this time, the endocardial heart tube undergoes dilations and constrictions so that when fusion is completed during the fourth week, five distinct regions of the heart can be identified. These are the **truncus arteriosus, bulbus cordis, ventricle, atrium,** and **sinus venosus** (fig. 1).

Major changes occur in each of the five primitive dilations of the developing heart during the week-and-a-half embryonic period beginning in the middle of the fourth week. The truncus arteriosus differentiates to form a partition between the aorta and the pulmonary trunk. The bulbus cordis is incorporated in the formation of the walls of the ventricles. The sinus venosus forms the **coronary sinus** and a portion of the wall of the right atrium. The ventricle is divided into the right and left chambers by the growth of the **interventricular septum.** The atrium is partially partitioned into right and left chambers by the **septum secundum.** An opening between the two atria called the **foramen ovale** persists throughout fetal development but normally closes at birth. This opening is covered by a flexible valve, which permits blood to pass from the right to the left side of the heart.

．．．．．．．．．．．．．．．．．．．．．．．

cardiogenic: Gk. *kardia*, heart; *genesis*, be born (origin)
ventricle: L. *ventriculus*, diminutive of *venter*, belly
atrium: L. *atrium*, chamber

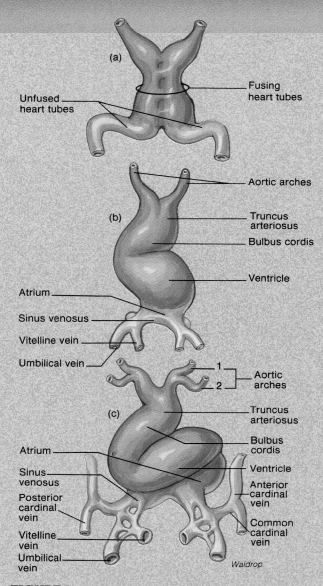

*Waldrop*

## FIGURE 1

Formation of the heart chambers. (*a*) The heart tubes fuse during days 21 and 22. (*b*) The developmental chambers are formed during day 23. (*c*) Differential growth causes folding between the chambers during day 24, and vessels are developed to transport blood to and from the heart. The embryonic heart generally has begun rhythmic contractions and pumping blood by day 25.

## Table 21.5   Cardiovascular structures of the fetus and changes in the neonate

| Structure | Location | Function | Fate in neonate |
|---|---|---|---|
| Umbilical vein | Connects the placenta to the liver; forms a major portion of umbilical cord | Transports nutrient-rich, oxygenated blood from the placenta to the fetus | Forms the round ligament of the liver |
| Ductus venosus | Venous shunt within the liver to connect with the inferior vena cava | Transports oxygenated blood directly into the inferior vena cava | Forms the ligamentum venosum, a fibrous cord in the liver |
| Foramen ovale | Opening between the right and left atria | Acts as a shunt to bypass the pulmonary circulation | Closes at birth and becomes the fossa ovalis, a depression in the interatrial septum |
| Ductus arteriosus | Between pulmonary trunk and aortic arch | Acts as a shunt to bypass the pulmonary circulation | Closes shortly after birth, atrophies, and becomes the ligamentum arteriosum |
| Umbilical arteries | Arise from internal iliac arteries and associated with umbilical cord | Transports blood from the fetus to the placenta | Atrophies to become the lateral umbilical ligaments |

Changes occur in the cardiovascular system at birth. The foramen ovale, ductus arteriosus, ductus venosus, and the umbilical vessels are no longer necessary. The foramen ovale abruptly closes with the first breath of air because the reduced pressure in the right side of the heart causes a flap to cover the opening. The constriction of the ductus arteriosus occurs gradually over a period of about 6 weeks after birth as the vascular muscle cells constrict in response to the higher oxygen concentration in the postnatal blood. The remaining structure of the ductus gradually atrophies and becomes the *ligamentum arteriosum* (see fig. 21.20). The fate of the unique fetal cardiovascular structures is summarized in table 21.5.

## Clinical Considerations

### Electrocardiograph Leads

There are two types of electrocardiograph recording electrodes, or "leads." The **bipolar limb leads** record the voltage between electrodes placed on the wrists and legs. These bipolar leads include lead I (right arm to left arm), lead II (right arm to left leg), and lead III (left arm to left leg). In the **unipolar leads,** voltage is recorded between a single "exploratory electrode" placed on the body and an electrode that is built into the electrocardiograph and maintained at zero potential (ground).

The unipolar limb leads are placed on the right arm, left arm, and left leg, and are abbreviated AVR, AVL, and AVF, respectively. The unipolar chest leads are labeled 1 through 6, starting from the midline position (fig. 21.38).

There are thus a total of 12 standard ECG leads that "view" the changing pattern of the heart's electrical activity from different perspectives (table 21.6). Use of a number of leads is important because certain abnormalities are best seen with some and may not be visible at all with others.

### Arrhythmias Detected by the Electrocardiogram

**Arrhythmias,** or abnormal heart rhythms, can be detected and described by the abnormal ECG patterns they produce. Although the proper clinical interpretation of electrocardiograms requires knowledge of technical information not covered in this chapter, some knowledge of abnormal rhythms is interesting in itself and is useful in understanding normal physiology.

Since a heartbeat occurs whenever a normal QRS complex is seen, and since the ECG chart paper moves at a known speed so that its *x*-axis indicates time, the cardiac rate (beats per minute) can easily be obtained from the ECG recording. A cardiac rate slower than 60 beats per minute indicates **bradycardia** (brad″ĭ-kar′de-ă); a rate faster than 100 beats per minute is described as **tachycardia.**

 Both bradycardia and tachycardia can occur normally. Endurance-trained athletes, for example, commonly have a slower heart rate than that of the general population. This *athlete's bradycardia* occurs as a result of higher levels of parasympathetic inhibition of the SA node and is a beneficial adaptation. Activation of the sympathetic division of the ANS during exercise or emergencies causes a normal tachycardia to occur.

bradycardia: Gk. *bradys*, slow; *kardia*, heart
tachycardia: Gk. *tachys*, rapid; *kardia*, heart

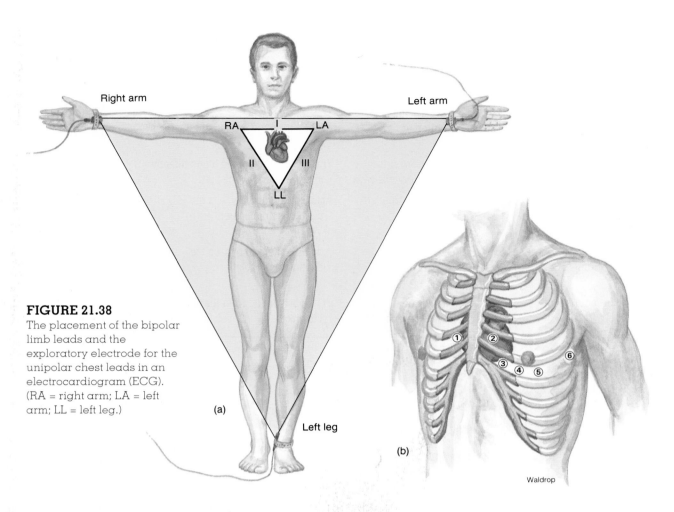

**FIGURE 21.38**
The placement of the bipolar limb leads and the exploratory electrode for the unipolar chest leads in an electrocardiogram (ECG). (RA = right arm; LA = left arm; LL = left leg.)

(a)

Right arm

Left arm

Left leg

(b)

Waldrop

## Table 21.6 Electrocardiograph (ECG) leads

| Name of lead | Placement of electrodes |
| --- | --- |
| Bipolar limb leads | |
| I | Right arm and left arm |
| II | Right arm and left leg |
| III | Left arm and left leg |
| Unipolar limb leads | |
| AVR | Right arm |
| AVL | Left arm |
| AVF | Left leg |
| Unipolar chest leads | |
| $V_1$ | 4th intercostal space right of sternum |
| $V_2$ | 4th intercostal space left of sternum |
| $V_3$ | 5th intercostal space left of sternum |
| $V_4$ | 5th intercostal space in line with the middle of the clavicle |
| $V_5$ | 5th intercostal space to the left of $V_4$ |
| $V_6$ | 5th intercostal space in line with the middle of the axilla |

Abnormal tachycardia occurs when a person is at rest. This may result from abnormally fast pacing by the atria due to drugs, or it may result from the development of abnormally fast *ectopic pacemakers*—cells located outside the SA node that assume a pacemaker function. This abnormal atrial tachycardia thus differs from normal "sinus" (SA node) tachycardia. *Ventricular tachycardia* results when abnormally fast ectopic pacemakers in the ventricles cause them to beat rapidly and independently of the atria (fig. 21.39). This is very dangerous because it can quickly degenerate into a lethal condition known as *ventricular fibrillation*.

**Ventricular Fibrillation** Fibrillation is caused by a continuous recycling of electrical waves through the myocardium. Normally, recycling is prevented because the myocardium enters a refractory period simultaneously at all regions. If some

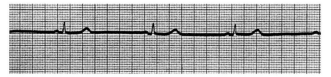

Sinus Bradycardia

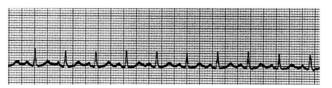

Sinus Tachycardia

(a)

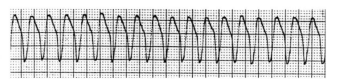

Ventricular tachycardia

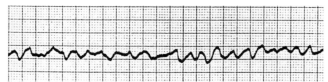

Ventricular fibrillation

(b)

**FIGURE 21.39**

In (a) the heartbeat is paced by the normal pacemaker—the SA node (hence the name sinus rhythm). This can be abnormally slow (bradycardia—46 beats per minute in this example) or fast (tachycardia—136 beats per minute in this example). Compare the pattern of tachycardia in (a) with the tachycardia in (b). Ventricular tachycardia is produced by an ectopic pacemaker in the ventricles. This dangerous condition can quickly lead to ventricular fibrillation, also shown in (b).

cells emerge from their refractory periods before others, however, electrical waves can be continuously regenerated and conducted. The recycling of electrical waves along continuously changing pathways produces uncoordinated contraction and an impotent pumping action. These effects can be produced by damage to the myocardium.

Fibrillation can sometimes be stopped by a strong electric shock delivered to the chest, a procedure called **electrical defibrillation.** The electric shock depolarizes all the myocardial cells at the same time, causing them to enter a refractory state. The conduction of random, recirculating impulses thus stops, and—within a short time—the SA node can begin to stimulate contraction in a normal fashion. Although this does not correct the initial problem that caused the abnormal electrical patterns, it can keep a person alive long enough to take other corrective measures.

## Structural Heart Disorders

Congenital heart problems result from abnormalities in embryonic development and may be attributed to heredity, nutritional problems of the pregnant mother, or viral infections such as rubella. Congenital heart diseases occur in approximately 3 of every 100 births and account for about 50% of early childhood deaths.

**Heart murmurs** are both congenital and acquired. Nearly 10% of all people have heart murmurs, but most are not clinically significant. In general, three basic conditions cause murmurs: (1) *valvular insufficiency,* in which the cusps of the valves do not form a tight seal; (2) *stenosis,* in which the walls surrounding a valve are roughened or constricted; and (3) *turbulence of the blood* moving through the heart during heavy exercise. This last condition produces functional murmurs, which are common in children; they are not considered pathological.

A **septal defect** is the most common type of congenital heart problem. An atrial septal defect, or **patent foramen ovale,** is a failure of the fetal foramen ovale to close after birth. A ventricular septal defect is caused by an abnormal development of the interventricular septum. **Pulmonary stenosis** is a narrowing of the opening into the pulmonary trunk from the right ventricle. It may lead to a pulmonary embolism and is usually recognized by extreme lung congestion. A **patent ductus arteriosus** is a failure of the ductus arteriosus to close after birth, allowing a backflow of blood into the pulmonary circulation from the aortic arch.

The **tetralogy of Fallot** (fig. 21.40) is a combination of four defects within a newborn and immediately causes a cyanotic condition, leading to the newborn being termed a "blue baby." The four characteristics of this anomaly are (1) a ventricular septal defect; (2) an ascending aorta that has shifted in position so that it overrides the interventricular septum and thus receives blood from the right as well as left ventricle; (3) pulmonary stenosis; and (4) right ventricular hypertrophy (fig. 21.40). Pulmonary stenosis obstructs blood flow to the lungs and causes hypertrophy of the right ventricle. Open-heart surgery is required to correct this condition, and the overall mortality rate is about 5%.

Acquired heart diseases include those that result from arterial damage and those that result from bacterial infection. **Bacterial endocarditis** is a disease of the lining of the heart, especially of the cusps of the valves. It is caused by bacteria that enter the bloodstream—most often, the same organisms that cause rheumatic fever.

stenosis: Gk. *stenosis,* a narrowing

tetralogy of Fallot: from Étienne-Louis A. Fallot, French physician, 1850–1922

Circulatory System

## Atherosclerosis

**Atherosclerosis** is the most common form of arteriosclerosis (hardening of the arteries) and, through its contribution to heart disease and stroke, is responsible for about 50% of the deaths in North America, Europe, and Japan. In atherosclerosis, localized plaques, or **atheromas,** protrude into the lumen of the artery and thus reduce blood flow. The atheromas additionally serve as sites for *thrombus* (blood clot) *formation*, which can further occlude the blood supply to an organ (fig. 21.41).

It is currently believed that the process of atherosclerosis begins as a result of damage, or "insult," to the endothelium. Such insults are produced by smoking, hypertension (high blood pressure), high blood cholesterol, and diabetes. The first anatomically recognized change is the appearance of "fatty streaks," which are gray-white areas that protrude into the lumina of arteries, particularly at arterial branch points. These are aggregations of lipid-filled macrophages and lymphocytes within the tunica intima. They are present to a small degree in the aorta and coronary arteries of children aged 10 to 14, but progress to more advanced stages at different rates in different people. In the intermediate stage, the area contains layers of macrophages and smooth muscle cells. The more advanced lesions are called *fibrous plaques* and consist of a cap of connective tissue with smooth muscle cells over accumulated lipid and debris, macrophages that have been derived from monocytes (chapter 23) and lymphocytes.

The process may be instigated by damage to the endothelium, but its progression appears to be a result of a wide variety of cytokines and other autocrine regulators (chapter 19) secreted by the endothelium and by the other participating cells, including platelets, macrophages, and lymphocytes. Some of these regulators attract monocytes and lymphocytes to the damaged endothelium and cause them to penetrate into the tunica intima. The monocytes then become macrophages, engulf lipids, and take on the appearance of "foamy cells." Smooth muscle cells change from a contractile state to a "synthetic" state where they produce and secrete connective tissue matrix proteins. (This is unique; in other tissues, connective tissue matrix is produced by cells

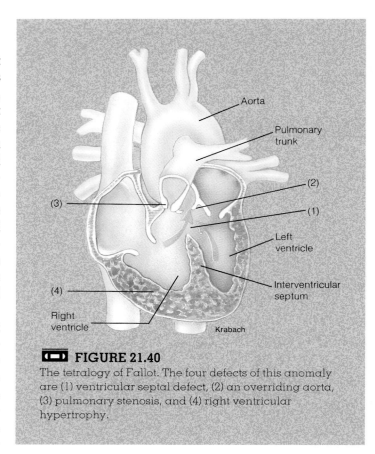

### FIGURE 21.40

The tetralogy of Fallot. The four defects of this anomaly are (1) ventricular septal defect, (2) an overriding aorta, (3) pulmonary stenosis, and (4) right ventricular hypertrophy.

called fibroblasts.) The changed smooth muscle cells respond to chemical attractants and migrate from the tunica media to the tunica intima, where they can proliferate.

Endothelial cells normally prevent this progression by presenting a physical barrier to the penetration of monocytes and lymphocytes, and by secreting autocrine regulators. Hypertension, smoking, and high blood cholesterol, among other risk factors, interfere with this protective function. The role of cholestrol in this process is described in the next section.

## Cholesterol and Plasma Lipoproteins

There is good evidence that high blood cholesterol is associated with an increased risk of atherosclerosis. This high blood cholesterol can be produced by a diet rich in cholesterol and saturated fat, or it may be the result of an inherited condition known as **familial hypercholesteremia.** This condition is inherited as a single dominant gene; individuals who inherit two of these genes have extremely high cholesterol concentrations (regardless of diet) and usually suffer heart attacks during childhood.

Lipids, including cholesterol, are carried in the blood attached to protein carriers, as discussed in detail in chapter 26. Cholesterol is carried to the arteries by plasma proteins called **low-density lipoproteins (LDLs).** These particles, produced by the liver, consist of a core of cholesterol surrounded by a layer of phospholipids (to make the particle water-soluble) and a protein. Cells in various organs contain receptors for the protein in LDL. When LDL attaches to its receptors, the cell engulfs the LDL by receptor-mediated endocytosis (described in chapter 3) and utilizes the cholesterol for different purposes. Most of the LDL in the blood is removed in this way by the liver.

When endothelial cells engulf LDL they oxidize it to a product called oxidized LDL. Recent evidence suggests that oxidized LDL contributes to endothelial cell injury,

lumen: L. *lumen,* opening

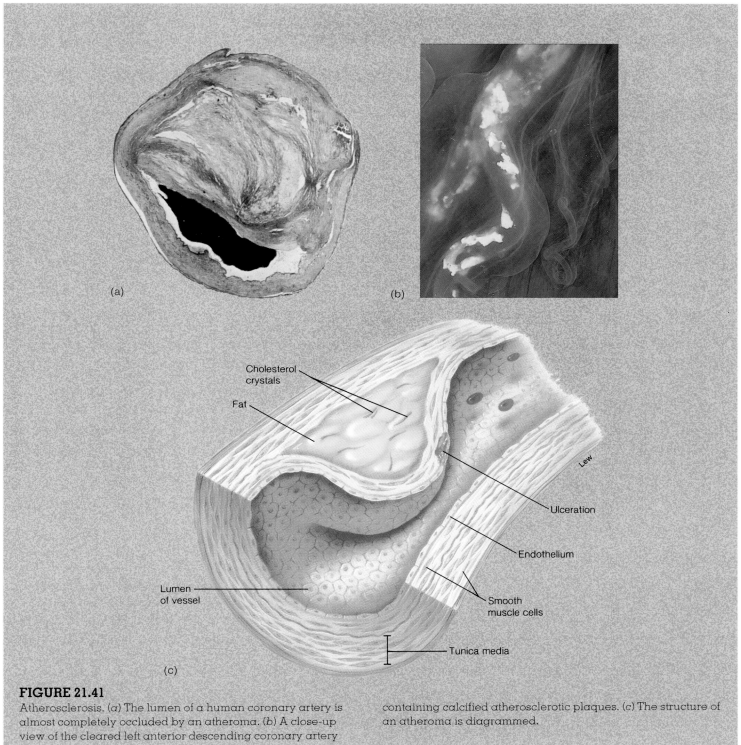

Cholesterol crystals

Fat

Ulceration

Endothelium

Lumen of vessel

Smooth muscle cells

Tunica media

(a)

(b)

(c)

**FIGURE 21.41**

Atherosclerosis. (a) The lumen of a human coronary artery is almost completely occluded by an atheroma. (b) A close-up view of the cleared left anterior descending coronary artery containing calcified atherosclerotic plaques. (c) The structure of an atheroma is diagrammed.

migration of monocytes and lymphocytes into the tunica intima, conversion of monocytes into macrophages, and other events that occur in the progression of atherosclerosis. Since oxidized LDL appears to be so important in the progression of atherosclerosis, it seems logical that anti-oxidant compounds may aid in the prevention or treatment of this condition. The anti-oxidant drug *probucol,* as well as vitamin C, vitamin E, and beta-carotene, which are anti-oxidants (chapter 27), have been shown to be effective in this regard in experimental animals.

People who eat a diet high in cholesterol and saturated fat, and people with familial hypercholesteremia, have a high blood LDL concentration because their livers have a low number of LDL receptors. With fewer LDL receptors, the liver is less able to remove the LDL from the blood and thus more LDL is available to enter the endothelial cells of arteries.

Excessive cholesterol may be released from cells and travel in the blood as **high-density lipoproteins (HDLs),** which are removed by the liver. The cholesterol in HDLs is not taken into the artery wall because these cells lack the membrane receptor required for endocytosis of HDL, and therefore this cholesterol does not contribute to atherosclerosis. Indeed, in contrast to LDL, a high proportion of cholesterol in HDL is beneficial, since it indicates that cholesterol may be traveling away from the blood vessels to the liver. The concentration of HDL-cholesterol appears to be higher and the risk of atherosclerosis lower in people who exercise regularly. The HDL-cholesterol concentration, for example, is higher in marathon runners than in joggers and is higher in joggers than in sedentary individuals. Women in general have higher HDL-cholesterol concentrations and a lower risk of atherosclerosis than men.

Most people can significantly lower their blood cholesterol concentration through a regimen of exercise and diet. Since saturated fat in the diet raises blood cholesterol, such foods as fatty meat, egg yolks, and internal animal organs (liver, brain, etc.) should be eaten only sparingly. The American Heart Association recommends that fat contribute less than 30% to the total calories of a diet, and many experts argue for an even lower percentage. By way of comparison, the typical fast-food meal contains from 40% to 58% of its calories as fat. The single most effective action that smokers can take to lower their risk of atherosclerosis, however, is to stop smoking.

## Ischemic Heart Disease

A tissue is said to be **ischemic** (*ĭ-ske´mik*) when it receives an inadequate supply of oxygen because of an inadequate blood flow. The most common cause of myocardial ischemia is atherosclerosis of the coronary arteries. The adequacy of blood flow is relative—it depends on the metabolic requirements of the tissue for oxygen. An obstruction in a coronary artery, for example, may allow sufficient blood flow at rest but may produce ischemia when the heart is stressed by exercise or emotional conditions.

Myocardial ischemia is associated with increased concentrations of blood lactic acid produced by anaerobic respiration of the ischemic tissue. This condition often causes substernal pain, which may also be referred to the left shoulder and arm, as well as to other areas. This referred pain is called **angina pectoris.** People with angina frequently take nitroglycerin or related drugs that help to relieve the ischemia and pain. These drugs are effective because they stimulate vasodilation, which improves circulation to the heart and decreases the work that the heart must perform to eject blood into the arteries. Myocardial cells are adapted to respire aerobically and cannot respire anaerobically for more than a few minutes. If ischemia and anaerobic respiration continue for more than a few minutes, *necrosis* (cellular death) may occur in the areas most deprived of oxygen. A sudden, irreversible injury of this kind is called a **myocardial infarction,** or **MI.** The lay term "heart attack," though imprecise, usually refers to a myocardial infarction.

Myocardial ischemia may be detected by characteristic changes in the electrocardiogram. The diagnosis of myocardial infarction is aided by measurement of the concentration of enzymes in the blood that are released by the infarcted tissue. Plasma concentrations of *creatine phosphokinase* (*CPK*), for example, increase within 3 to 6 hours after the onset of symptoms and return to normal after 3 days. Plasma levels of *lactate dehydrogenase* (*LDH*) reach a peak within 48 to 72 hours after the onset of symptoms and remain elevated for about 11 days.

## Other Vascular Disorders

An **aneurism** (*an-yŭ-riz-em*) is an expansion or bulging of the heart, aorta, or any other artery. A **coarctation** is a constriction in a segment of a vessel, usually the aorta, and is frequently caused by a remnant of the ductus arteriosus tightening around the vessel. **Varicose veins** are weakened veins that become stretched and swollen. They are most common in the legs because the force of gravity tends to weaken the valves and overload the veins. Varicose veins can also occur in the rectum, in which case they are called **hemorrhoids.** *Vein stripping* is the surgical removal of superficial weakened veins. **Phlebitis** (*flĕ-bi´tis*) is inflammation of a vein. It may develop as a result of trauma or be an aftermath of surgery. Frequently, however, it appears for no apparent reason. Phlebitis interferes with normal venous circulation.

# Chapter Summary

**Structure of the Heart (pp. 566–572)**

1. The right and left sides of the heart pump blood through the pulmonary and systemic circulations.

   a. The right ventricle pumps blood to the lungs; this blood then returns to the left atrium.
   b. The left ventricle pumps blood into the aorta and systemic arteries; this blood then returns to the right atrium.

2. The heart contains two pairs of one-way valves.

   a. The atrioventricular (AV) valves allow blood to go from the atria to the ventricles but not in the reverse direction.

b. The semilunar valves allow blood to exit the ventricles and enter the pulmonary and systemic circulations, but these valves prevent blood from returning from the arteries to the ventricles.

c. Closing of the AV valves produces the "lub" sound at the beginning of systole; closing of the semilunar valves produces the "dub" sound at the beginning of diastole.

3. The pulmonary circulation involves the pulmonary arteries and pulmonary veins; all other arteries and veins in the body are part of the systemic circulation.

a. The pulmonary arteries deliver blood low in oxygen to the lungs; the pulmonary veins deliver oxygen-rich blood from the lungs to the left atrium of the heart.

b. The systemic arteries carry oxygen-rich blood to the body cells for cell respiration.

## Cardiac Cycle, Heart Sounds, and the Electrocardiogram (pp. 573–579)

1. The heart is a two-step pump; first the atria contract, and then the ventricles contract.

a. During diastole, first the atria and then the ventricles fill with blood.

b. The ventricles are about 80% filled before the atria contract and add the final 20% to the end-diastolic volume.

2. When the ventricles contract at systole, the pressure within them first rises sufficiently to close the AV valves and then rises sufficiently to open the semilunar valves.

a. Blood is ejected from the ventricles until the pressure within them falls below the pressure in the arteries; at this point, the semilunar valves close and the ventricles begin relaxation.

b. When the pressure in the ventricles falls below the pressure in the atria, a phase of rapid filling of the ventricles occurs, followed by the final filling caused by contraction of the atria.

3. The electrical impulse begins in the sinoatrial (SA) node and spreads through both atria by electrical conduction from one myocardial cell to another.

a. The impulse then excites the atrioventricular (AV) node, from which it is conducted by the atrioventricular bundle into the ventricles.

b. The conduction myofibers (Purkinje fibers) transmit the impulse into the ventricular muscle and cause it to contract.

4. The regular pattern of conduction in the heart produces a changing pattern of potential differences between two points on the body surface.

a. A recording of the voltage between two points on the surface of the body caused by the electrical activity of the heart is called an electrocardiogram (ECG).

b. The P wave is caused by depolarization of the atria; the QRS wave is caused by depolarization of the ventricles; the T wave is produced by repolarization of the ventricles.

## Blood Vessels (pp. 579–583)

1. Arteries contain three layers, or tunics: the tunica intima, tunica media, and tunica externa.

a. The tunica intima consists of a layer of endothelium that is separated from the tunica media by a band of elastin fibers.

b. The tunica media consists of smooth muscle.

2. Capillaries are the narrowest but the most numerous of the blood vessels; they provide for the exchange of molecules between the blood and the surrounding tissues.

a. Capillaries within most organs of the body have pores between adjacent endothelial cells, so that fluid derived from plasma can be filtered through to produce interstitial, or tissue, fluid.

b. The capillaries in the brain are continuous and lack pores, thus contributing to the blood-brain barrier.

3. Veins have the same three tunics as arteries, but veins generally have a thinner muscular layer than comparably sized arteries.

a. Veins are more distensible than arteries and can expand to hold a larger quantity of blood.

b. Many veins have venous valves that permit a one-way flow of blood to the heart.

c. Contractions of skeletal muscles surrounding veins can squeeze the veins and aid the return of venous blood to the heart. This action is known as the skeletal muscle pump.

## Principal Arteries of the Body (pp. 583–591)

1. Three arteries arise from the aortic arch: the brachiocephalic trunk, the left common carotid artery, and the left subclavian artery. The brachiocephalic trunk divides into the right common carotid artery and the right subclavian artery.

a. The common carotid arteries bifurcate into the internal and external carotid arteries.

b. The subclavian arteries branch to form the vertebral arteries and then continue as the axillary arteries.

2. The head and neck receive an arterial supply from branches of the internal and external carotid arteries and the vertebral arteries.

a. The internal carotid arteries enter the skull through the carotid canal; the vertebral arteries enter through the foramen magnum.

b. Branches of the internal carotid arteries and vertebral arteries form the cerebral arterial circle (circle of Willis), which supplies the brain.

3. The upper extremity is served by the subclavian artery and its derivatives.

4. The abdominal portion of the aorta produces the following branches: the inferior phrenic, celiac trunk, superior mesenteric, renal, suprarenal, testicular (or ovarian), and inferior mesenteric arteries.

5. The abdominal aorta terminates in the posterior pelvic area as it bifurcates into the right and left common iliac arteries. These vessels terminate by dividing into the internal and external iliac arteries, which supply blood to the pelvis and lower extremities.

## Principal Veins of the Body (pp. 591–600)

1. Blood from the head and neck is drained by the external and internal jugular veins; blood from the brain is drained by the internal jugular vein.

2. The upper extremity is drained by superficial and deep veins.

3. In the thorax, the superior vena cava is formed by the union of the two brachiocephalic veins and also collects blood from the azygos vein.

4. The lower extremity is drained by both superficial and deep veins. At the level of the fifth lumbar vertebra, the right and left iliac veins unite to form the inferior vena cava.

5. Blood from capillaries in the GI tract and accessory digestive organs is drained via the hepatic portal vein to the liver.

a. A portal system is one in which there is a second capillary bed downstream from the first; in this case, the liver is downstream from the GI tract.

b. The liver can modify the chemical composition of the blood arriving from the GI tract before this blood goes back to the heart to enter the general circulation.

## Fetal Circulation (pp. 600–603)

1. Fully oxygenated blood is carried only in the umbilical vein, which drains the placenta. This blood is carried via the ductus venosus to the inferior vena cava of the fetus.

2. Partially oxygenated blood is shunted from the right to the left atrium via the foramen ovale and from the pulmonary trunk to the aortic arch via the ductus arteriosus.

a. In this way, blood is diverted away from the lungs, which are not active in oxygenating the blood, to the systemic circulation, where it can be delivered to the placenta via the umbilical arteries.

b. The foramen ovale normally closes immediately following a newborn's first breath; the ductus arteriosus has closed by the sixth week following birth.

# Review Activities

## Objective Questions

1. All arteries in the body contain oxygen-rich blood with the exception of
   a. the aorta.
   b. the pulmonary arteries.
   c. the renal arteries.
   d. the coronary arteries.
2. Most blood from the coronary circulation directly enters
   a. the inferior vena cava.
   b. the superior vena cava.
   c. the right atrium.
   d. the left atrium.
3. The second heart sound immediately follows the occurrence of which event?
   a. P wave
   b. QRS wave
   c. T wave
   d. U wave
4. Which of the following arteries does *not* arise from the aortic arch?
   a. brachiocephalic trunk
   b. coronary artery
   c. left common carotid artery
   d. left subclavian artery
5. Which of the following arteries does *not* supply blood to the brain?
   a. external carotid artery
   b. internal carotid artery
   c. vertebral artery
   d. basilar artery
6. The maxillary and superficial temporal arteries are derived from
   a. the external carotid artery.
   b. the internal carotid artery.
   c. the vertebral artery.
   d. the facial artery.
7. Which of the following statements is *false*?
   a. Most of the total blood volume is contained in veins.
   b. Capillaries have a greater total surface area than any other type of vessel.

c. Exchanges between blood and tissue fluid occur across the walls of venules.
   d. Small arteries and arterioles present great resistance to blood flow.
8. The "lub," or first heart sound, is produced by closing of
   a. the aortic semilunar valve.
   b. the pulmonary semilunar valve.
   c. the tricuspid valve.
   d. the bicuspid valve.
   e. both AV valves.
9. The first heart sound is produced at
   a. the beginning of systole.
   b. the end of systole.
   c. the beginning of diastole.
   d. the end of diastole.
10. Changes in the cardiac rate primarily reflect changes in the duration of
    a. systole.
    b. diastole.
11. The QRS wave of an ECG is produced by
    a. depolarization of the atria.
    b. repolarization of the atria.
    c. depolarization of the ventricles.
    d. repolarization of the ventricles.
12. The cells that normally have the fastest rate of spontaneous diastolic depolarization are located in
    a. the SA node.
    b. the AV node.
    c. the atrioventricular bundle.
    d. the conduction myofibers.
13. Which of the following statements is *true*?
    a. The heart can produce a graded contraction.
    b. The heart can produce a sustained contraction.
    c. All of the myocardial cells in the ventricles are normally in a refractory period at the same time.

14. During the phase of isovolumetric relaxation of the ventricles, the pressure in the ventricles is
    a. rising.
    b. falling.
    c. first rising, then falling.
    d. constant.

## Essay Questions

1. Explain why the beat of the heart is automatic and why the SA node functions as the normal pacemaker.
2. Compare the duration of the heart's contraction with those of the myocardial action potential and refractory period. Explain the significance of these relationships.
3. Describe the pressure changes that occur during the cardiac cycle and relate these changes to the occurrence of the heart sounds.
4. Describe the causes of the P, QRS, and T waves of an ECG and indicate when in the cardiac cycle each of these waves occurs. Explain why the first heart sound occurs immediately after the QRS wave and why the second sound occurs at the time of the T wave.
5. Can a defective valve be detected by an ECG? Can a partially damaged AV node be detected by auscultation with a stethoscope? Explain.
6. Describe the functions of the foramen ovale and ductus arteriosus in a fetus and explain why the newborn is at risk if they remain patent after birth.
7. Trace the flow of blood from the left ventricle to the upper teeth.
8. Trace the flow of blood from the small intestine, to the heart, and back to the small intestine.
9. Define the term *portal system* and explain the functional significance of this system in the liver.

## Gundy/Weber Software ⌷

The tutorial software accompanying Chapter 21 is Volume 9—Cardiovascular System.

## Exploration ⊗

A module of correlating material is avaible from the Wm. C. Brown CD-ROM: *Explorations*. It is #5 Evolution of the Heart.

# [ chapter twenty-two ]

# circulatory system: cardiac output and blood flow

## objectives ...........................................

- Describe how cardiac output is affected by cardiac rate and stroke volume and discuss the effects of autonomic stimulation of the heart.

- Explain the Frank–Starling law of the heart.

- Explain how the venous return of blood to the heart is regulated.

- Describe how tissue fluid is formed and how it returns to blood capillaries.

- Describe the regulation of ADH secretion and the effects of ADH on blood volume.

- Describe the regulation of aldosterone secretion and the effects of aldosterone on blood volume and pressure.

- Explain how blood flow is affected by blood pressure and by vascular resistance.

- Discuss the regulation of vascular resistance by the autonomic nervous system and by intrinsic regulatory mechanisms.

- Describe the relationship between resistance and the radius of a vessel and explain how blood flow can be diverted from one organ to another.

- Explain how resistance and blood flow are regulated by sympathetic and parasympathetic innervation.

- Discuss autoregulation and explain how it is accomplished.

- Describe the mechanisms that control cerebral blood flow.

- Describe the cutaneous circulation and explain how it is regulated.

- Describe the changes in blood pressure as blood passes through the arterial system to capillaries and then to the venous system.

- State the factors that directly influence blood pressure.

- Describe the baroreceptor reflex and comment on its significance.

- Describe the auscultatory method of blood pressure measurements.

# Cardiac Output

*The pumping ability of the heart is a function of the number of beats per minute (cardiac rate) and the volume of blood ejected per beat (stroke volume). The cardiac rate and stroke volume are regulated by autonomic nerves and by mechanisms intrinsic to the cardiovascular system.*

The **cardiac output** is equal to the volume of blood pumped per minute by each ventricle. The average resting **cardiac rate** in an adult is 70 beats per minute; the average **stroke volume** (volume of blood pumped per beat by each ventricle) is 70 to 80 ml per beat. The product of these two variables gives an average cardiac output of about 5.5 liters (L) per minute:

$$\text{Cardiac output} = \text{stroke volume} \times \text{cardiac rate}$$

$$\text{(ml/min)} \qquad \text{(ml/beat)} \qquad \text{(beats/min)}$$

The **total blood volume** is also about 5.5 L. This means that each ventricle pumps the equivalent of the total blood volume each minute under resting conditions. Put another way, it takes about a minute for a drop of blood to complete the pulmonary and systemic circuits. An increase in cardiac output, as occurs during exercise, must thus be accompanied by an increased rate of blood flow through the circulation. This is accomplished by factors that regulate the cardiac rate and stroke volume.

## Regulation of Cardiac Rate

In the complete absence of neural influences, the heart will continue to beat according to the rhythm set by the SA node. This automatic rhythm is produced by the spontaneous decay of the resting membrane potential to a threshold depolarization, at which point voltage-regulated membrane gates are opened and action potentials are produced. As described in chapter 12, Ca$^{++}$ enters the myocardial cytoplasm during the action potential, attaches to troponin, and causes contraction.

Under normal conditions, however, sympathetic innervation to the heart through the cardiac accelerant nerves and parasympathetic innervation through the vagus nerves continuously modify the rate of spontaneous depolarization of the SA node. Norepinephrine, released primarily by sympathetic nerve endings, and epinephrine, secreted by the adrenal medulla, stimulate an increase in the spontaneous firing rate of the SA node. Acetylcholine, released from parasympathetic endings, hyperpolarizes the SA node and thus decreases the rate of its spontaneous firing (fig. 22.1). The actual pace set by the SA node at any time depends on the net effect of these antagonistic influences. Mechanisms

that affect the cardiac rate are said to have a **chronotropic** (*kron´ŏ-trop´ik*) **effect.** Those that increase cardiac rate have a positive chronotropic effect; those that decrease the rate have a negative chronotropic effect.

Autonomic innervation of the SA node represents the major means by which cardiac rate is regulated. However, other autonomic control mechanisms also affect cardiac rate to a lesser degree. Sympathetic stimulation in the musculature of the atria and ventricles increase the strength of contraction and slightly decrease the time spent in systole when the cardiac rate is high (table 22.1).

During exercise, the cardiac rate increases as a result of decreased vagus nerve inhibition of the SA node. Further increases in cardiac rate are achieved by increased sympathetic nerve stimulation. The resting bradycardia (slow heart rate) of endurance-trained athletes is due largely to high

chronotropic: Gk. *chronos*, time; *trope*, turn, change

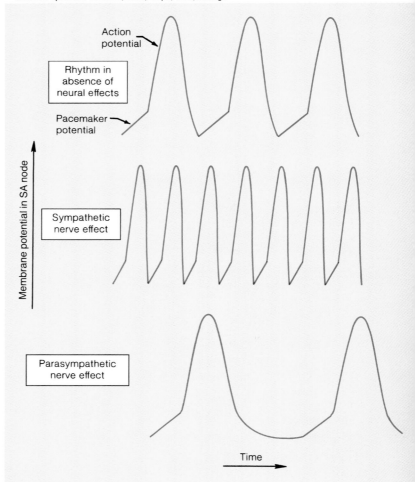

**FIGURE 22.1**

The rhythm set by the pacemaker potentials in the SA node. Sympathetic nerve effects increase the rate of spontaneous depolarization, thus influencing the rate at which action potentials are produced.

| Table 22.1 | Effects of autonomic nerve activity on the heart | |
|---|---|---|
| **Region affected** | **Sympathetic nerve effects** | **Parasympathetic nerve effects** |
| SA node | Increased rate of diastolic depolarization; increased cardiac rate | Decreased rate of diastolic depolarization; decreased cardiac rate |
| AV node | Increased conduction rate | Decreased conduction rate |
| Atrial muscle | Increased strength of contraction | Decreased strength of contraction |
| Ventricular muscle | Increased strength of contraction | No significant effect |

vagus nerve activity. The activity of the autonomic innervation of the heart is coordinated by **cardiac control centers** in the medulla oblongata of the brain stem. The question of whether there are separate cardioaccelerator and cardioinhibitory centers in the medulla oblongata is currently controversial. The cardiac control centers are affected by higher brain areas and by sensory feedback from pressure receptors, or *baroreceptors*, in the aortic arch and the carotid arteries. In this way, a rise in blood pressure can produce a reflex slowing of the heart. We will discuss this *baroreceptor reflex* in more detail later in this chapter, in relation to blood pressure regulation.

## Regulation of Stroke Volume

The **stroke volume** is regulated by three variables: (1) the *end-diastolic volume*, which is the volume of blood in the ventricles at the end of diastole; (2) the *total peripheral resistance*, which is the frictional resistance, or impedance, to blood flow in the arteries; and (3) the *contractility*, or strength, of ventricular contraction.

The **end-diastolic volume (EDV)** is the amount of blood in the ventricles just prior to contraction. Because this workload is imposed on the ventricles before they contract, it is sometimes called a *preload*. The stroke volume is directly proportional to the preload; an increase in end-diastolic volume results in an increase in stroke volume. The stroke volume is also directly proportional to **contractility;** when the ventricles contract more forcefully, they pump more blood.

In order to eject blood, the pressure generated in a ventricle when it contracts must be greater than the pressure in the arteries (since blood flows only from a location of higher to one of lower pressure). The pressure in the arterial system before the ventricle contracts is, in turn, a function of the **total peripheral resistance**—the higher the

peripheral resistance, the higher the pressure. As blood begins to be ejected from the ventricle, the added volume of blood in the arteries causes a rise in pressure against the "bottleneck" presented by the peripheral resistance; ejection of blood stops shortly after the aortic pressure becomes equal to the intraventricular pressure. The total peripheral resistance thus presents an impedance to the ejection of blood from the ventricle, or an *afterload* imposed on the ventricle after contraction has begun.

In summary, the stroke volume is inversely proportional to the total peripheral resistance; that is, the greater the peripheral resistance, the lower the stroke volume. It should be noted that this lowering of stroke volume in response to a raised peripheral resistance occurs only for a few beats. Thereafter, a healthy heart is able to compensate for the increased peripheral resistance by beating more forcefully. This compensation occurs by means of a mechanism called the Frank–Starling law, to be described shortly.

The proportion of the end-diastolic volume that is ejected depends on the strength of ventricular contraction. Normally, contraction strength is sufficient to eject 70 to 80 ml of blood out of a total end-diastolic volume of 110 to 130 ml, making the *ejection fraction* about 60%. More blood is pumped per beat as the end-diastolic volume increases, and thus the ejection fraction remains relatively constant over a range of end-diastolic volumes. In order for this to be true, the strength of ventricular contraction must increase as the end-diastolic volume increases.

**Frank–Starling Law of the Heart**   Two physiologists, Otto Frank and Ernest Starling, demonstrated that the strength of ventricular contraction varies directly with the end-diastolic volume. Even in experiments where the heart is removed from the body (and is thus not subject to neural or hormonal regulation) and where the heart is filled with blood flowing from a reservoir, an increase in end-diastolic volume within the physiological range results in increased contraction strength and, therefore, in an increased stroke volume. This relationship between EDV, contraction strength, and stroke volume is a built-in, or *intrinsic*, property of heart muscle, and is known as the **Frank–Starling law of the heart.**

..........................................................

Frank–Starling law of the heart: From Otto Frank, German physiologist, b. 1865, and Ernest Henry Starling, English physiologist, 1866–1927

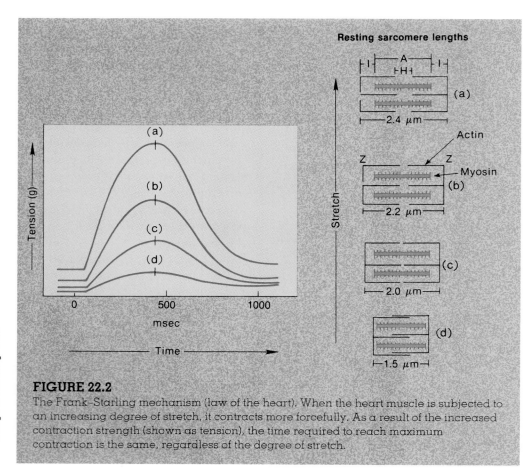

**Resting sarcomere lengths**

**FIGURE 22.2**

The Frank–Starling mechanism (law of the heart). When the heart muscle is subjected to an increasing degree of stretch, it contracts more forcefully. As a result of the increased contraction strength (shown as tension), the time required to reach maximum contraction is the same, regardless of the degree of stretch.

volume), the strength of contraction is intrinsically adjusted by the end-diastolic volume.

The Frank–Starling law explains how the heart can adjust to a rise in total peripheral resistance: (1) a rise in peripheral resistance causes a decrease in the stroke volume of the ventricle so that (2) more blood remains in the ventricle and the end-diastolic volume is greater for the next cycle; as a result, (3) the ventricle is stretched to a greater degree in the next cycle and contracts more strongly to eject more blood. This allows a healthy ventricle to sustain a normal cardiac output. A very important consequence of these events is that the cardiac output (ml/min) of the left ventricle, which pumps blood into the ever-changing resistances of the systemic circulation, can be maintained to match the output of the right ventricle, which pumps blood into the pulmonary circulation. A moment's reflection will reveal that the rate of blood flow through the pulmonary and systemic circulations must be equal in order to prevent fluid accumulation in the lungs and to deliver fully oxygenated blood to the body.

**Intrinsic Control of Contraction Strength**  The intrinsic control of contraction strength and stroke volume is due to variations in the degree to which the myocardium is stretched by the end-diastolic volume. As the end-diastolic volume rises within the physiological range, the myocardium is increasingly stretched and, as a result, it contracts more forcefully.

Stretch can also increase the contraction strength of skeletal muscles. The resting length of skeletal muscles, however, is close to ideal, so that significant stretching decreases contraction strength. This is not true of the heart. Prior to filling with blood during diastole, the sarcomere lengths of myocardial cells are only about 1.5 μm. At this length, the actin filaments from each side overlap in the middle of the sarcomeres, and the cells can only contract weakly (fig. 22.2).

As the ventricles fill with blood the myocardium stretches, so that the actin filaments overlap with myosin only at the edges of the A bands (fig. 22.2). This allows more force to be developed during contraction. Since this more advantageous overlapping of actin and myosin is produced by stretching of the ventricles, and since the degree of stretching is controlled by the degree of filling (the end-diastolic

**Extrinsic Control of Contractility**  The *contractility* is the contraction strength at any given fiber length. At any given degree of stretch, the strength of ventricular contraction depends on the activity of the sympathoadrenal system. Norepinephrine from sympathetic endings and epinephrine from the adrenal medulla produce an increase in contraction strength. This **positive inotropic effect** is believed to result from an increase in the amount of Ca++ available to the sarcomeres.

The activity of the sympathoadrenal system thus affects cardiac output in two ways: (1) through a positive inotropic effect on contractility and (2) through a positive chronotropic effect on cardiac rate (fig. 22.3). Stimulation through the parasympathetic vagus nerves to the heart has a negative chronotropic effect but does not directly affect the contraction strength of the ventricles.

.........................................

inotropic: Gk. *inos*, fiber; *trope*, turn, change

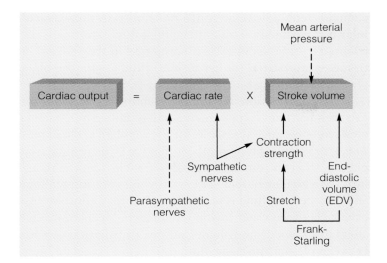

**FIGURE 22.3**
The regulation of cardiac output. Factors that stimulate cardiac output are shown as solid arrows; factors that inhibit cardiac output are shown as dotted arrows.

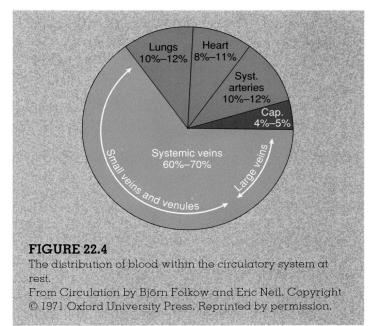

**FIGURE 22.4**
The distribution of blood within the circulatory system at rest.
From Circulation by Björn Folkow and Eric Neil. Copyright © 1971 Oxford University Press. Reprinted by permission.

## Venous Return

The end-diastolic volume—and thus the stroke volume and cardiac output—is controlled by factors that affect the **venous return,** which is the return of blood via veins to the heart. The rate at which the atria and ventricles are filled with venous blood depends on the total blood volume and the venous pressure. The pressure in the veins serves as the driving force for the return of blood to the heart.

Because veins have thinner, less muscular walls than do arteries, they have a higher *compliance*. This means that a given amount of pressure will cause more distension (expansion) in veins than in arteries, so that the veins can hold more blood. Approximately two-thirds of the total blood volume is contained within the veins (fig. 22.4). Veins are therefore called **capacitance vessels,** analogous to the capacitors in electronics that accumulate electrical charges. Muscular arteries and arterioles expand less under pressure (are less compliant), and thus are called **resistance vessels.**

Although veins contain almost 70% of the total blood volume, the mean venous pressure is only 2 mmHg, compared to a mean arterial pressure of 90 to 100 mmHg. The lower venous pressure is due partly to a pressure drop between arteries and capillaries and partly to the high venous compliance.

The venous pressure is highest in the venules (10 mmHg) and lowest at the junction of the venae cavae with the right atrium (0 mmHg). In addition to this pressure difference, the venous return to the heart is aided by (1) sympathetic nerve activity, which stimulates smooth muscle contraction in the venous walls and thus reduces compliance; (2) the skeletal muscle pump, which squeezes veins during muscle contraction; and (3) the pressure difference between the thoracic and abdominal cavities, which promotes the flow of venous blood back to the heart.

Contraction of the skeletal muscles works like a "pump" by virtue of its squeezing action on veins. Contraction of the diaphragm during inhalation also improves venous return, as described in chapter 20. As the diaphragm contracts, it lowers to increase the thoracic volume and decrease the abdominal volume. This creates a partial vacuum in the thoracic cavity and a higher pressure in the abdominal cavity. The pressure difference produced favors blood flow from abdominal to thoracic veins (fig. 22.5).

## Blood Volume

*Fluid in the extracellular environment of the body is distributed between the blood and the tissue fluid compartments by filtration and osmotic forces acting across the walls of capillaries. The function of the kidneys influences blood volume because urine is derived from blood plasma. Through their actions on the kidneys, ADH and aldosterone help to regulate the blood volume.*

Blood volume represents one part, or compartment, of the total body water. Approximately two-thirds of the total body water is contained within cells—in the intracellular

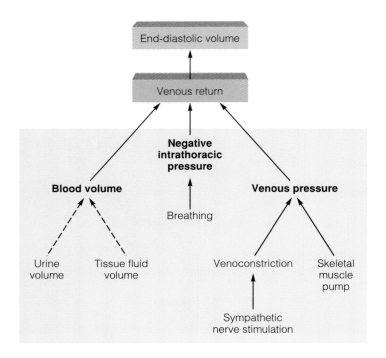

**FIGURE 22.5**

Variables that affect venous return, and thus end diastolic volume. Direct relationships are indicated by solid arrows; inverse relationships are shown with dashed arrows.

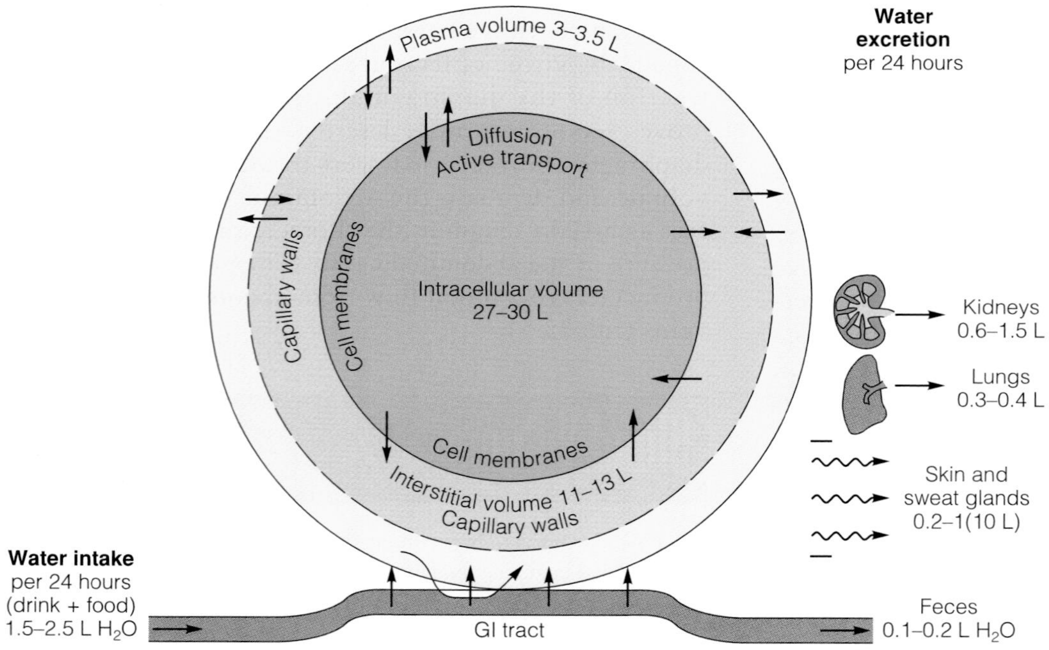

compartment. The remaining one-third is in the **extracellular compartment.** This extracellular fluid is normally distributed so that about 80% is contained in the tissues—as **tissue** or **interstitial** (*in″ter-stish′al*) **fluid**—with the blood plasma accounting for the remaining 20% (fig. 22.6).

The distribution of water between the intracellular and extracellular fluid compartments is determined by a balance between opposing forces acting at the capillaries. Blood pressure, for example, promotes the formation of tissue fluid from plasma, whereas osmotic forces draw water from the tissues into the vascular system. The total volume of intracellular and extracellular fluid is normally maintained constant by a balance between water loss and water gain. Mechanisms that affect drinking, urine volume, and the distribution of water between plasma and tissue fluid thus help to regulate blood volume and, by this means, help to regulate cardiac output and blood flow.

## Exchange of Fluid between Capillaries and Tissues

The distribution of extracellular fluid between the blood plasma and interstitial compartments is in a state of dynamic equilibrium. Tissue fluid is not normally a "stagnant pond" but rather a continuously circulating medium, formed from and returning to the vascular system. The tissue cells receive a continuously fresh supply of glucose and other blood plasma solutes that are filtered through tiny endothelial channels in the capillary walls.

Filtration results from the blood pressure within the capillaries. This hydrostatic pressure, which is exerted against the inner capillary wall, is equal to about 37 mmHg at the arteriolar end of systemic capillaries and drops to about 17 mmHg at the venular end of the capillaries. The **net filtration pressure** is equal to the hydrostatic pressure of the blood in the capillaries minus the hydrostatic pressure of tissue fluid outside the capillaries, which opposes filtration. If, as an extreme example, these two values were

**Water excretion**
per 24 hours

Kidneys
0.6–1.5 L

Lungs
0.3–0.4 L

Skin and
sweat glands
0.2–1(10 L)

Feces
0.1–0.2 L H₂O

**Water intake**
per 24 hours
(drink + food)
1.5–2.5 L H₂O

Plasma volume 3–3.5 L

Diffusion
Active transport

Capillary walls

Cell membranes

Intracellular volume
27–30 L

Cell membranes

Interstitial volume 11–13 L
Capillary walls

GI tract

**FIGURE 22.6**

The distribution of body water between the intracellular and extracellular compartments.

From *Circulation* by Björn Folkow and Eric Neil. Copyright © 1971 Oxford University Press. Reprinted by permission.

equal, there would be no filtration. The magnitude of the tissue hydrostatic pressure varies from organ to organ. With a hydrostatic pressure in the tissue fluid of 1 mmHg, as it is outside the capillaries of skeletal muscles, the net filtration pressure would be 37 − 1 = 36 mmHg at the arteriolar end of the capillary and 17 − 1 = 16 mmHg at the venular end.

Glucose, comparably sized organic molecules, inorganic salts, and ions are filtered along with water through the capillary channels. The concentrations of these substances in tissue fluid are the same as in plasma. The protein concentration of tissue fluid (2 g/100 ml), however, is less than the protein concentration of plasma (6 to 8 g/100 ml). This difference is due to the restricted filtration of proteins by the capillary pores. The osmotic pressure exerted by plasma proteins—called the **colloid osmotic pressure** of the plasma—is therefore much greater than the colloid osmotic pressure of tissue fluid. The difference between these two pressures is called the **oncotic pressure.** Since the colloid osmotic pressure of the tissue fluid is sufficiently low to be neglected, the oncotic pressure is essentially equal to the colloid osmotic pressure of the plasma. This value has been estimated to be 25 mmHg. Since water will move by osmosis from the solution of lower to the solution of higher osmotic pressure (chapter 5), this oncotic pressure favors the movement of water from the tissues into the capillaries.

Whether fluid will move out of or into the capillary depends on the magnitudes of the net filtration pressure, which decreases from the arteriolar to the venular end of the capillary, and on the oncotic pressure, which remains essentially constant. These opposing forces that affect the distribution of fluid across the capillary are known as **Starling forces,** and their effects can be calculated according to the following equation:

Fluid movement is proportional to:

$$(P_c + \pi_i) - (P_i + \pi_p)$$

where:

$P_c$ = hydrostatic pressure in the capillary
$\pi_i$ = colloid osmotic pressure of the interstitial fluid
$P_i$ = hydrostatic pressure of the interstitial fluid
$\pi_p$ = colloid osmotic pressure of the blood plasma

The expression to the left of the minus sign represents the sum of forces acting to move fluid out of the capillary. The expression to the right represents the sum of forces acting to move fluid into the capillary. In the capillaries of skeletal muscles, the values are as follows: at the arteriolar end of the capillary, (37 + 0) − (1 + 25) = 11 mmHg; at the venular end of the capillary, (17 + 0) − (1 + 25) = −9 mmHg. The

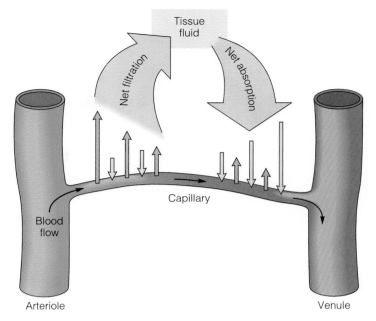

**FIGURE 22.7**
Tissue (interstitial) fluid is formed by filtration (orange arrows) as a result of blood pressures at the arteriolar ends of capillaries and is returned to venular ends of capillaries by the colloid osmotic pressure of plasma proteins (yellow arrows).

positive value at the arteriolar end indicates that the forces favoring the extrusion of fluid from the capillary predominate. The negative value at the venular end indicates that the net Starling forces favor the return of fluid to the capillary. Fluid thus leaves the capillaries at the arteriolar end and returns to the capillaries at the venular end (fig. 22.7).

This "classic" view of capillary dynamics has recently been challenged by some investigators who believe that when capillaries are open, the net filtration force exceeds the force for the osmotic return of water throughout the length of the capillary. They believe that the opposite is true in closed capillaries (capillaries can be opened or closed by the action of precapillary sphincter muscles). The logical conclusion is that net filtration will occur in open capillaries, whereas net absorption of water will occur in closed capillaries.

By either proposed mechanism, blood plasma and interstitial fluid are continuously interchanged. The return of fluid to the vascular system at the venular ends of the capillaries, however, does not exactly equal the amount filtered at the arteriolar ends. According to some estimates, approximately 85% of the capillary filtrate is returned directly to the capillaries; the remaining 15% (amounting to at least 2 L per day) is returned to the vascular system by way of the lymphatic system.

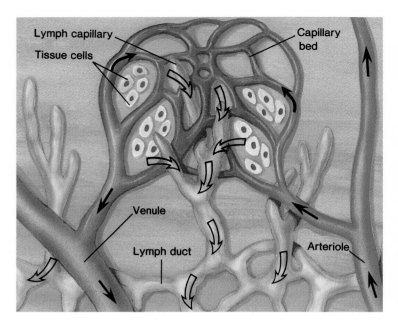

**FIGURE 22.8**
A schematic diagram showing the structural relationship of a capillary bed and a lymph capillary.

**Lymphatic Drainage**   Excessive accumulation of interstitial fluid and filtered proteins is normally prevented by drainage of interstitial fluid into highly permeable, blind-ended **lymphatic capillaries** (fig. 22.8). Interstitial fluid that enters these lymphatic capillaries is known as **lymph.** Lymph is transported by lymphatic ductules into two large lymphatic vessels that drain the lymph into the right and left subclavian veins. In this way, interstitial fluid is ultimately returned to the circulatory system from which it was originally derived. The structure of the lymphatic system is described in detail in chapter 23.

**Causes of Edema**   Excessive accumulation of interstitial fluid is known as **edema** (ĕ-de′mă). This condition is normally prevented by a proper balance between capillary filtration and osmotic uptake of water and by proper lymphatic drainage. Edema may result from (1) *high arterial blood pressure*, which increases capillary pressure and causes excessive filtration; (2) *venous obstruction*—as in phlebitis (where inflammation of a vein causes the formation of a thrombus) or mechanical compression of veins (during pregnancy, for example)—which produces a congestive increase in capillary pressure; (3) *leakage of blood plasma proteins into interstitial fluid*, which causes reduced osmotic flow of water into the capillaries (this occurs during inflammation and allergic reactions as a result of increased capillary permeability); (4) *myxedema*—the excessive production of particular glycoproteins (mucin) in the interstitial spaces

### Table 22.2   Causes and effects of edema

| Cause | Effect |
|---|---|
| Increased blood pressure or venous obstruction | Increases capillary filtration pressure so that more interstitial fluid is formed at the arteriolar ends of capillaries. |
| Increased tissue protein concentration | Decreases osmosis of water into the venular ends of capillaries. Usually a localized tissue edema due to leakage of blood plasma proteins through capillaries during inflammation and allergic reactions. Myxedema due to hypothyroidism is also in this category. |
| Decreased plasma protein concentration | Decreases osmosis of water into the venular ends of capillaries. May be caused by liver disease (which can be associated with insufficient blood plasma protein production), kidney disease (due to leakage of blood plasma protein into urine), or protein malnutrition. |
| Obstruction of lymphatic vessels | Infections by filaria roundworms (nematodes) transmitted by a certain species of mosquito block lymphatic drainage, causing edema and tremendous swelling of the affected areas. |

caused by hypothyroidism; (5) *decreased plasma protein concentration* as a result of liver disease (the liver makes most of the blood plasma proteins) or kidney disease, in which proteins are excreted in the urine; and (6) *obstruction of the lymphatic drainage* (table 22.2).

 In the tropical disease *filariasis* (fil″ă-ri′ă-sis), mosquitoes transmit a parasitic nematode to humans. The larvae of these worms invade lymphatic vessels and block lymphatic drainage. The edema that results can be so severe that the tissues swell to produce an elephantlike appearance, with thickening and cracking of the skin. This condition is thus aptly named *elephantiasis* (fig. 22.9).

**FIGURE 22.9**
Parasitic larvae that block lymphatic drainage cause tissue edema and the tremendous enlargement of the limbs and external genitalia called elephantiasis.

## Regulation of Blood Volume by the Kidneys

The formation of urine begins in the same manner as the formation of interstitial fluid—by filtration of blood plasma through capillary pores. These capillaries are known as *glomeruli*, and the filtrate they produce enters a system of tubules that transports and modifies the filtrate by various mechanisms (to be discussed in chapter 25). The kidneys produce about 180 L per day of blood filtrate, but since there is only 5.5 L of blood in the body, it is clear that most of this filtrate must be returned to the circulatory system and recycled. Only about 1 to 2 L per day of urine is excreted; 98% to 99% of the amount filtered is **reabsorbed** back into the circulatory system.

The volume of urine excreted can be varied by changes in the reabsorption of filtrate. If 99% of the filtrate is reab-

sorbed, for example, 1% must be excreted. Decreasing the reabsorption by only 1%—to 98%—would double the volume of urine excreted (an increase to 2% of the amount filtered). Carrying the logic further, a doubling of urine volume from, for example, 1 to 2 liters would result in the loss of an additional liter of blood volume. The percentage of the glomerular filtrate reabsorbed—and thus the urine volume and blood volume—is adjusted according to the needs of the body by the action of specific hormones on the kidneys. Through their effects on the kidneys, and the resulting changes in blood volume, these hormones serve important functions in the regulation of the cardiovascular system.

## Regulation of Blood Volume by Antidiuretic Hormone (ADH)

One of the major hormones involved in the regulation of blood volume is **antidiuretic** (*an″te-di″yŭ-ret′ik*) **hormone (ADH),** also known as *vasopressin.* As described in chapter 19, this hormone is produced by neurons in the hypothalamus, transported by axons into the neurohypophysis (posterior pituitary), and released from this storage gland in response to hypothalamic stimulation. The secretion of ADH from the posterior pituitary occurs when neurons called **osmoreceptors** in the hypothalamus detect an increase in blood plasma osmolality (osmotic pressure).

An increase in plasma osmolality occurs when the plasma becomes more concentrated (chapter 5). This can occur either through *dehydration* or through *excessive salt intake.* Stimulation of osmoreceptors produces sensations of thirst, leading to increased water intake and increased ADH secretion from the posterior pituitary. Through mechanisms that we will consider in connection with kidney physiology in chapter 25, ADH stimulates water reabsorption from the filtrate. A smaller volume of urine is thus excreted as a result of the action of ADH (fig. 22.10).

A person who is dehydrated or who consumes excessive amounts of salt thus drinks more and urinates less. This raises the blood volume and, in the process, dilutes the plasma to lower its previously elevated osmolality. The rise in blood volume that results from these mechanisms is extremely important in stabilizing the condition of a dehydrated person with low blood volume and pressure.

Drinking excessive amounts of water without excessive amounts of salt does not result in a prolonged increase in blood volume and pressure. The water does enter the blood from the GI tract and momentarily raises the blood volume; at the same time, however, it dilutes the blood. Dilution of the blood decreases the plasma osmolality and thus inhibits ADH secretion. With less ADH, there is less

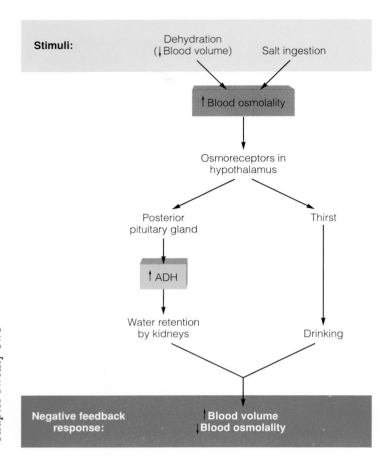

**Stimuli:**

Dehydration (↓Blood volume)    Salt ingestion

↑Blood osmolality

Osmoreceptors in hypothalamus

Posterior pituitary gland    Thirst

↑ADH

Water retention by kidneys    Drinking

**Negative feedback response:**    ↑Blood volume ↓Blood osmolality

**FIGURE 22.10**

The negative feedback control of blood volume and blood osmolality.

reabsorption of filtrate in the kidneys—a larger volume of urine is excreted. Water is therefore a *diuretic* (di˝yŭ-ret´ik)—a substance that promotes urine formation—because it inhibits the secretion of antidiuretic hormone.

In addition to the activity of osmoreceptors, another mechanism operates to inhibit ADH secretion when fluid intake is excessive. An excessively high blood volume stimulates stretch receptors located in the left atrium of the heart. Stimulation of these stretch receptors, in turn, activates a reflex inhibition of ADH secretion, which in turn promotes a lowering of blood volume through increased urine production. This inhibition of ADH secretion is independent of the inhibitory effects that a decrease in plasma osmolality would produce. Experimental stimulation of the stretch receptors by inflation of a balloon in

.........................................

diuretic: Gk. *dia*, through; *ouresis*, urination

the left atrium duplicates the inhibition of ADH secretion produced by high blood volume.

 During prolonged exercise, particularly on a warm day, a substantial amount of water may be lost from the body through sweating (up to 900 ml per hour or more). The lowering of blood volume that results decreases the ability of the body to dissipate heat, and the consequent overheating of the body can cause ill effects and put an end to the exercise. The need for athletes to remain well hydrated is commonly recognized, but drinking pure water may not be the answer. This is because blood sodium is lost in sweat, so that a lesser amount of water is required to dilute the blood osmolality back to normal. When the blood osmolality is normal, the urge to drink is extinguished. For these reasons, endurance athletes should drink solutions containing sodium (as well as carbohydrates for energy), and they should drink at a predetermined rate rather than at a rate determined only by thirst.

## Regulation of Blood Volume by Aldosterone

From the preceding discussion, it is clear that a certain amount of dietary salt is required to maintain blood volume and pressure. Since $Na^+$ and $Cl^-$ are easily filtered in the kidneys, a mechanism must exist to promote the reabsorption and retention of salt when the dietary salt intake is too low. **Aldosterone,** a steroid hormone secreted by the adrenal cortex, stimulates the reabsorption of salt by the kidneys. Aldosterone is a "salt-retaining hormone." Retention of salt indirectly promotes retention of water (in part, by the action of ADH). Aldosterone thus acts to increase the blood volume, but unlike ADH, it does not affect plasma osmolality. This is because aldosterone promotes the reabsorption of salt and water in proportionate amounts, whereas ADH promotes only the reabsorption of water. Thus aldosterone, unlike ADH, does not act to dilute the blood.

The secretion of aldosterone is stimulated during salt deprivation, when blood volume and pressure are reduced. The adrenal cortex, however, is not directly stimulated to secrete aldosterone by these conditions. Instead, a decrease in blood volume and pressure activates an intermediate mechanism that involves the enzyme renin and the plasma protein angiotensin.

Throughout most of human history, salt was in short supply and was therefore highly valued. Moorish merchants in the sixth century traded an ounce of salt for an ounce of gold and salt cakes were used as money in Abyssinia. Part of a Roman soldier's pay was given in salt—it was from this practice that the word salary (*sal* = salt) was derived. The phrase "worth his salt" derives from the fact that Greeks and Romans sometimes purchased slaves with salt.

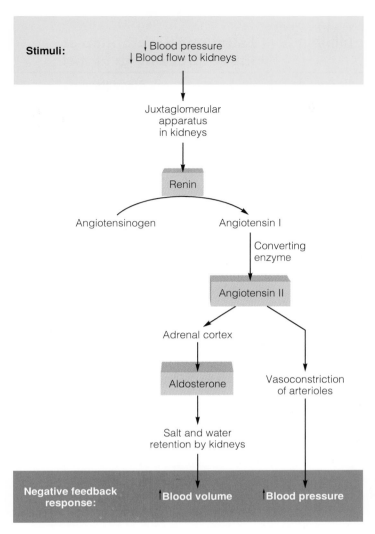

**Stimuli:**

↓ Blood pressure
↓ Blood flow to kidneys

Juxtaglomerular
apparatus
in kidneys

Renin

Angiotensinogen → Angiotensin I

Converting
enzyme

Angiotensin II

Adrenal cortex

Aldosterone

Vasoconstriction
of arterioles

Salt and water
retention by kidneys

**Negative feedback response:**  ↑Blood volume  ↑Blood pressure

**FIGURE 22.11**

The negative feedback control of blood volume and pressure by the renin-angiotensin-aldosterone system.

**Renin-Angiotensin System**   When blood flow and pressure are reduced in the renal artery (as they would be in the low blood volume state of salt deprivation), a group of cells in the kidneys, called the **juxtaglomerular** (*juk˝stă-glo-mer´yŭ-lar*) **apparatus,** secretes the enzyme **renin** into the blood. This enzyme cleaves a 10-amino-acid polypeptide called *angiotensin I* from a plasma protein called *angiotensinogen.* As angiotensin I passes through the capillaries of the lungs and other organs, an *angiotensin-converting enzyme* removes two amino acids. This leaves an 8-amino-acid polypeptide called **angiotensin II** (fig. 22.11). Conditions of salt deprivation, low blood volume, and low blood pressure, in summary, cause increased production of angiotensin II in the blood.

Angiotensin II exerts numerous effects that result in a rise in blood pressure. Vasoconstriction of arterioles and small muscular arteries is produced directly by the effects of angiotensin II on the smooth muscles of these vessels, causing the blood pressure to rise. The increased blood volume is an indirect effect of angiotensin II.

 One of the newer classes of drugs that can be used to treat hypertension (high blood pressure) are the *angiotensin-converting enzyme, or ACE, inhibitors.* These drugs (such as *captopril*) block the formation of angiotensin II and thus promote vasodilation, which decreases the total peripheral resistance. This reduces the afterload of the heart and is thus also beneficial in the treatment of left ventricular hypertrophy and congestive heart failure.

Angiotensin II promotes a rise in blood volume by means of two mechanisms: (1) thirst centers in the hypothalamus are stimulated by angiotensin II, and thus more water is ingested, and (2) secretion of aldosterone from the adrenal cortex is stimulated by angiotensin II, and higher levels of aldosterone cause more salt and water to be retained by the kidneys. The relationship between angiotensin II and aldosterone is sometimes described as the **renin-angiotensin-aldosterone system**.

The renin-angiotensin-aldosterone system can also work in the opposite direction. High-salt intake, leading to high blood volume and pressure, normally inhibits renin secretion. With less angiotensin II formation and less aldosterone secretion, less salt is retained by the kidneys and more is excreted in the urine. Unfortunately, many people with *chronic hypertension* may have normal or even elevated levels of renin secretion. In these cases, the intake of salt must be lowered to match the impaired ability to excrete salt in the urine.

**Atrial Natriuretic Hormone**   As described in the previous section, a fall in blood volume is compensated for by renal retention of fluid through activation of the renin-angiotensin-aldosterone system. An increase in blood volume, conversely, is compensated for by renal excretion of a larger volume of urine. Experiments suggest that the increase in water excretion under conditions of high blood volume is at least partly due to an increase in the excretion of $Na^+$ in the urine, or *natriuresis*.

Increased $Na^+$ excretion may be produced by a decline in aldosterone secretion, but there is evidence for a separate hormone that stimulates natriuresis. This natriuretic (*na˝trĭ-yoo-ret´ik*) hormone would thus be

..........................................

natriuresis: L. *natrium*, sodium; Gk. *ouresis*, urination

antagonistic to aldosterone and would promote Na⁺ and water excretion in the urine in response to a rise in blood volume. The atria of the heart have been shown to produce a hormone with these properties, which is now identified as **atrial natriuretic hormone.** By promoting salt and water excretion in the urine, atrial natriuretic hormone can act to lower the blood volume and pressure. This action is analogous to that of diuretic drugs taken by people with hypertension, as we will describe later in this chapter.

# Vascular Resistance and Blood Flow

*The rate of blood flow to an organ is dependent on the resistance to flow in the small arteries and arterioles that serve the organ. Vasodilation decreases resistance and increases flow, whereas vasoconstriction increases resistance and decreases flow. Vasodilation and vasoconstriction occur in response to intrinsic and extrinsic regulatory mechanisms.*

The amount of blood that the heart pumps per minute is equal to the rate of venous return, and thus is equal to the rate of blood flow through the entire circulatory system. The cardiac output of 5 to 6 L per minute is distributed unequally to the different organs. At rest, blood flow is about 2500 ml per minute through the liver, kidneys, and gastrointestinal tract; 1200 ml per minute through the skeletal muscles; 750 ml per minute through the brain; and 250 ml per minute through the coronary arteries of the heart. The balance of the cardiac output (500 to 1100 ml per minute) is distributed to the other organs (table 22.3).

## *Physical Laws Describing Blood Flow*

The flow of blood through the circulatory system, like the flow of any fluid through a tube, depends on the difference in pressure at the two ends. If the pressure at both ends of a tube is the same, there will be no flow. If the pressure at one end is greater than at the other, blood will flow from the region of higher to the region of lower pressure. The rate of blood flow is proportional to the pressure difference ($P_1 - P_2$) between the two ends of the tube. The term **pressure difference** is abbreviated $\Delta P$, in which the Greek letter $\Delta$ (*delta*) means "change in" (fig. 22.12).

If the systemic circulation is pictured as a single tube leading from and back to the heart (fig. 22.12), blood flow through this system would occur as a result of the pressure difference between the pressure at the beginning of the tube (the aorta) and that at the end of the tube (the junction of

| Organs | Blood flow | |
|---|---|---|
| | **Milliliters per minute** | **Percent total** |
| Gastrointestinal tract and liver | 1400 | 24 |
| Kidneys | 1100 | 19 |
| Brain | 750 | 13 |
| Heart | 250 | 4 |
| Skeletal muscles | 1200 | 21 |
| Skin | 500 | 9 |
| Other organs | 600 | 10 |
| Total organs | 5800 | 100 |

**Table 22.3 Estimated distribution of the cardiac output at rest**

Source: O. L. Wade and J. M. Bishop, *Cardiac Output and Regional Blood Flow.* Copyright © 1962 Blackwell Scientific Publications, Ltd., England.

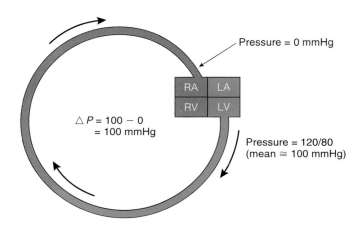

**FIGURE 22.12**

The flow of blood in the systemic circulation is ultimately dependent on the pressure difference ($\Delta P$) between the mean pressure of about 100 mmHg at the origin of the flow in the aorta and the pressure at the end of the circuit—zero mmHg in the vena cava, where it joins the right atrium (RA). (LA = left atrium, RV = right ventricle, LV = left ventricle.)

the venae cavae with the right atrium). The average, or mean, arterial pressure is about 100 mmHg; the pressure at the right atrium is 0 mmHg. The "pressure head," or driving force ($\Delta P$), is therefore about $100 - 0 = 100$ mmHg.

Blood flow is directly proportional to the pressure difference between the two ends of the tube ($\Delta P$) but is *inversely proportional* to the frictional resistance to blood

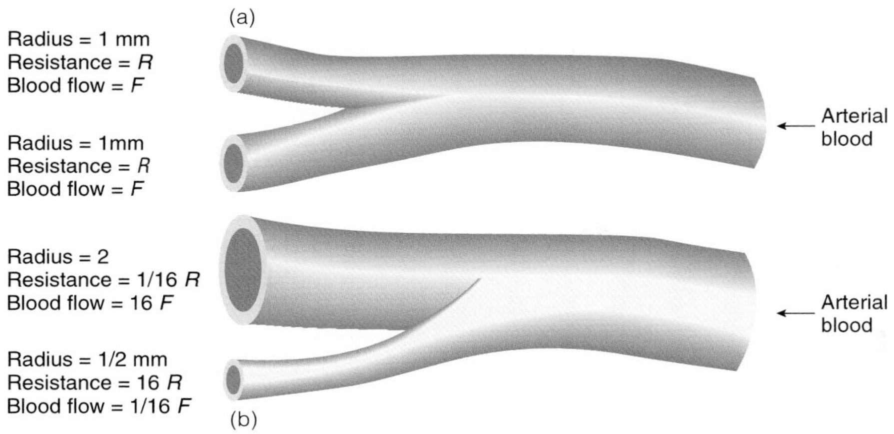

(a)

Radius = 1 mm
Resistance = R
Blood flow = F

Radius = 1mm
Resistance = R
Blood flow = F

Radius = 2
Resistance = 1/16 R
Blood flow = 16 F

Radius = 1/2 mm
Resistance = 16 R
Blood flow = 1/16 F

(b)

← Arterial blood

← Arterial blood

**FIGURE 22.13**

Relationships between blood flow, vessel radius, and resistance.
(a) The resistance and blood flow is equally divided between two branches of a vessel. (b) A doubling of the radius of one branch and halving of the radius of the other produces a sixteenfold increase in blood flow in the former and a sixteenfold decrease of blood flow in the latter.

flow through the vessels. Inverse proportionality is expressed by showing one of the factors in the denominator of a fraction, since a fraction decreases when the denominator increases:

$$\text{Blood flow} \propto \frac{\Delta P}{\text{resistance}}$$

The **resistance** to blood flow through a vessel is directly proportional to the length of the vessel and to the viscosity of the blood (the "thickness," or ability of molecules to "slip over" each other). Vascular resistance is inversely proportional to the fourth power of the radius of the vessel:

$$\text{Resistance} \propto \frac{L\eta}{r^4}$$

where:
L = length of vessel
η = viscosity of blood
r = radius of vessel

For example, if one vessel has half the radius of another and if all other factors are the same, the smaller vessel would have 16 times ($2^4$) the resistance of the larger vessel. Blood flow through the larger vessel, as a result, would be 16 times greater than it would be in the smaller vessel (fig. 22.13).

When physical constants are added to this relationship, the rate of blood flow can be calculated according to **Poiseuille's** (*pwă-zuh´yez*) **law:**

$$\text{Blood flow} = \frac{\Delta P r^4 (\pi)}{\eta L(8)}$$

Vessel length and blood viscosity do not vary significantly in normal physiology, although blood viscosity is increased in severe dehydration and in the polycythemia (high red blood cell count) that occurs as an adaptation to life at high altitudes. The major physiological regulators of blood flow through an organ are the mean arterial pressure (driving the flow) and the vascular resistance to flow. At a given mean arterial pressure, blood can be diverted from one organ to another by variations in the degree of vasoconstriction and vasodilation. Vasoconstriction in one organ and vasodilation in another results in a diversion of blood to the second organ. Since arterioles are the smallest arteries and can become narrower by vasoconstriction, they provide the greatest resistance to blood flow (fig. 22.14). Blood flow to an organ is thus largely determined by the degree of vasoconstriction

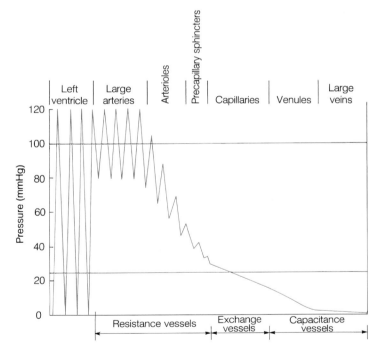

**FIGURE 22.14**

Blood pressure in different vessels of the systemic circulation.

Poiseuille's law: from Jean Poiseuille, French physiologist, 1799–1869

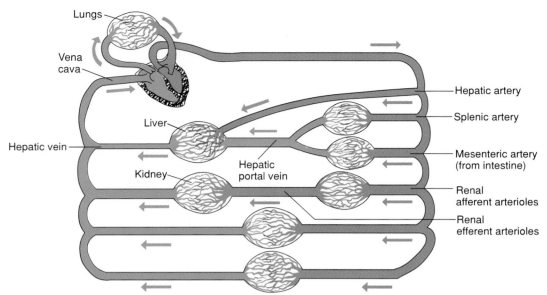

**FIGURE 22.15**

A diagram of the systemic and pulmonary circulations. Notice that with few exceptions (such as the blood flow in the renal circulation) the flow of arterial blood is in parallel rather than in series (arterial blood does not usually flow from one organ to another).

or vasodilation of its arterioles. The rate of blood flow to an organ can be increased by dilation of its arterioles and decreased by constriction of its arterioles.

**Total Peripheral Resistance**   The sum of all the vascular resistances within the systemic circulation is called the **total peripheral resistance.** The arterial supplies to the organs, which contribute to this total, are generally in parallel rather than in series with each other. That is, arterial blood passes through only one set of resistance vessels (arterioles) before returning to the heart (fig. 22.15). Since one organ is not "downstream" from another in terms of its arterial supply, changes in resistance within one organ directly affect blood flow in that organ only.

Vasodilation in a large organ might, however, significantly decrease the total peripheral resistance, and thus decrease the mean arterial pressure. In the absence of compensatory mechanisms, the driving force for blood flow through all organs might be reduced. This situation is normally prevented by an increase in the cardiac output and by vasoconstriction in other areas. During exercise of the large muscles, for example, the arterioles in the exercising muscles

are dilated. This would cause a significant drop in the mean arterial pressure if there were no compensations. But the blood pressure actually rises during exercise because of an increase in the cardiac output and constriction of the arterioles in the viscera and skin.

## Extrinsic Regulation of Blood Flow

The term *extrinsic regulation* refers to control by the autonomic nervous system and endocrine system. **Angiotensin II,** for example, directly stimulates vascular smooth muscle to produce generalized vasoconstriction. Antidiuretic hormone (ADH) also has a vasoconstrictor effect at high concentrations; this is why it is also called **vasopressin.** This vasopressor effect of ADH is not believed to be significant under physiological conditions in humans.

**Regulation by Sympathetic Nerves**   Stimulation of the sympathoadrenal system produces an increase in the cardiac output (as previously discussed) and an increase in total peripheral resistance. The latter effect is due to alpha-adrenergic stimulation (chapter 17) of vascular smooth muscle by norepinephrine and, to a lesser degree, by epinephrine. This produces vasoconstriction of the arterioles in the viscera and skin.

Even when a person is calm, the sympathoadrenal system is somewhat active and helps set the "tone" of vascular smooth muscles. In this case, **adrenergic sympathetic fibers** (those that release norepinephrine) cause a basal level of vasoconstriction throughout the body. During the fight-or-flight reaction, the activity of adrenergic fibers increases so that vasoconstriction is produced in the gastrointestinal tract, kidneys, and skin.

Arterioles in skeletal muscles receive **cholinergic sympathetic fibers,** which release acetylcholine as a neurotransmitter. In the fight-or-flight reaction, the activity of these cholinergic fibers increases, causing vasodilation. Vasodilation in skeletal muscles is also produced by

# Table 22.4   Extrinsic control of vascular resistance and blood flow

| Extrinsic agent | Effect | Comments |
|---|---|---|
| Sympathetic nerves | | |
|    Alpha-adrenergic | Vasoconstriction | The effect occurs throughout the body and is the dominant effect of sympathetic nerve stimulation on the circulatory system. |
|    Beta-adrenergic | Vasodilation | There is some activity in arterioles in skeletal muscles and in coronary vessels, but effects are masked by dominant alpha-receptor-mediated constriction. |
|    Cholinergic | Vasodilation | Effects are localized to arterioles in skeletal muscles and are produced only during defense (fight-or-flight) reaction. |
| Parasympathetic nerves | Vasodilation | Effects are primarily restricted to the gastrointestinal tract, external genitalia, and salivary glands and have little effect on total peripheral resistance. |
| Angiotensin II | Vasoconstriction | A powerful vasoconstrictor produced as a result of secretion of renin from the kidneys, it may function to help maintain adequate filtration pressure in kidneys when systemic blood flow and pressure are reduced. |
| ADH (vasopressin) | Vasoconstriction | Although the effects of this hormone on vascular resistance and blood pressure in anesthetized animals are well documented, the importance of these effects in conscious humans is controversial. |
| Histamine | Vasodilation | It promotes localized vasodilation during inflammation and allergic reactions. |
| Bradykinins | Vasodilation | Bradykinins are polypeptides secreted by sweat glands that promote local vasodilation. |
| Prostaglandins | Vasodilation or vasoconstriction | Prostaglandins are cyclic fatty acids that can be produced by most tissues, including blood vessel walls. Prostaglandin $I_2$ is a vasodilator, whereas thromboxane $A_2$ is a vasoconstrictor. The physiological significance of these effects is presently controversial. |

epinephrine secreted by the adrenal medulla, which stimulates beta-adrenergic receptors. During emergency conditions, therefore, blood flow to the viscera and skin is decreased due to the alpha-adrenergic effects of vasoconstriction in these organs, while blood flow to the skeletal muscles is increased. This diversion of blood flow to the skeletal muscles may give them an "extra edge" in responding to the emergency.

**Parasympathetic Control of Blood Flow**   Parasympathetic endings in arterioles always promote vasodilation. Parasympathetic innervation of blood vessels, however, is limited to the gastrointestinal tract, external genitalia, and salivary glands. Because of this limited distribution, the parasympathetic division of the ANS is less important than the sympathetic division in the control of total peripheral resistance.

The extrinsic control of blood flow is summarized in table 22.4.

## Paracrine Regulation of Blood Flow

Paracrine regulators, as described in chapter 19, are molecules produced by one tissue that help regulate another tissue of the same organ. A blood vessel is an organ that is particularly subject to paracrine regulation. Specifically, the endothelium of the tunica intima produces a number of paracrine regulators that cause the smooth muscle of the tunica media to either relax or contract.

The endothelium produces several molecules that promote smooth muscle relaxation, including *nitric oxide*, *bradykinin*, and *prostacyclin* (chapter 19). Nitric oxide appears to be the *endothelium-derived relaxation factor*, which earlier research had shown to be required for the vasodilation response to nerve stimulation. The ACh released from parasympathetic axons stimulates the opening of $Ca^{++}$ channels in the endothelial cell membrane; the $Ca^{++}$ then binds to and activates calmodulin (chapter 19). Calmodulin, in turn, activates the enzyme nitric oxide synthetase, which converts L-arginine into nitric oxide. Nitric oxide then diffuses into the smooth muscle cells of the vessel to produce the vasodilation response to the nerve stimulation. (Interestingly, the vasodilator drugs often given to treat angina pectoris—nitroglycerin and sodium nitroprusside—promote vasodilation indirectly through their conversion into nitric oxide.) Bradykinin likewise seems to promote vasodilation indirectly by functioning as an autocrine regulator that stimulates nitric oxide synthesis.

The endothelium also produces paracrine regulators that stimulate vasoconstriction. Notable among these is the polypeptide *endothelin-1*. Although the precise physiological role of this regulator is incompletely understood, it is currently believed that it works together with the vasodilator regulators to help maintain normal vessel diameter and blood pressure.

## Intrinsic Regulation of Blood Flow

Extrinsic control mechanisms affect resistance and flow in many regions of the body. In contrast to these more generalized effects, intrinsic (built-in) mechanisms within individual organs provide a more localized regulation of vascular resistance and blood flow. Intrinsic mechanisms are classified as *myogenic* or *metabolic*. Some organs (the brain and kidneys in particular) are capable of utilizing these intrinsic mechanisms to maintain relatively constant flow rates despite wide fluctuations in blood pressure. This ability is termed **autoregulation.**

**Myogenic Control Mechanisms**    If the arterial blood pressure and flow through an organ are inadequate—if the organ is inadequately *perfused* with blood—the metabolism of the organ can only be maintained for a limited period of time. Excessively high blood pressure can also be dangerous, particularly in the brain, because this may cause fine blood vessels to burst (causing cerebrovascular accident—CVA, or stroke).

Changes in systemic arterial pressure are compensated for in the brain and some other organs by the appropriate responses of vascular smooth muscle. A decrease in arterial pressure causes cerebral vessels to dilate, so that adequate rates of blood flow can be maintained despite the decreased pressure. Hypertension, by contrast, causes cerebral vessels to constrict, so that finer vessels downstream are protected

from the elevated pressure. These responses are myogenic; they are direct responses by the vascular smooth muscle to changes in pressure.

**Metabolic Control Mechanisms**    Local vasodilation within an organ can occur as a result of the chemical environment created by the organ's metabolism. The localized chemical conditions that promote vasodilation include (1) *decreased $O_2$ concentrations* that result from an increased metabolic rate; (2) *increased $CO_2$ concentrations*; (3) *decreased tissue pH* (due to $CO_2$, lactic acid, and other metabolic products); and (4) the *release of adenosine or $K^+$* from the tissue cells.

The vasodilation that occurs in response to tissue metabolism can be demonstrated by restricting the blood flow to an area for a short time and then removing the constriction. The constriction allows metabolic products to accumulate by preventing venous drainage of the area. When the constriction is removed and blood flow resumes, the metabolic products that have accumulated cause vasodilation. The tissue thus appears red. This response is called **reactive hyperemia** (*hi″pĕ-re′me-ă*). A similar increase in blood flow occurs in skeletal muscles and other organs as a result of increased metabolism. This is called **active hyperemia.**

Intrinsic control mechanisms are summarized in table 22.5.

# Blood Flow to the Heart and Skeletal Muscles

*Blood flow to the heart and skeletal muscles is regulated by both extrinsic and intrinsic mechanisms. These mechanisms provide increased rates of blood flow when metabolic requirements are increased during exercise.*

Survival requires that the heart and brain receive adequate blood flow at all times. The ability of skeletal muscles to respond quickly in emergencies and to maintain continued high levels of activity may also be critically important for survival. At such times, high rates of blood flow to the skeletal muscles must be maintained without compromising blood flow to the heart and brain. This is accomplished by mechanisms that increase the cardiac output and that direct a higher proportion of the cardiac output to the heart, skeletal muscles, and brain and away from the viscera and skin.

| Table 22.5 | Intrinsic control of vascular resistance and blood flow | | |
|---|---|---|---|
| Category | Agent<br>↑ = increase<br>↓ = decrease | Mechanisms | Comments |
| Myogenic | ↑Blood pressure | Stretching of the arterial wall as the blood pressure rises directly stimulates increased smooth muscle tone pressure (vasoconstriction). | It helps to maintain relatively constant rates of blood flow and pressure within an organ despite changes in systematic arterial pressure (autoregulation). |
| Metabolic | ↓Oxygen<br>↑Carbon dioxide<br>↓pH<br>↑Adenosine<br>↑K+ | Local changes in gas and metabolic concentrations act directly on vascular smooth muscle walls to produce vasodilation in the systemic circulation. The importance of different agents varies in different organs. | It aids in autoregulation of blood flow and also helps to shunt increased amounts of blood to organs with higher metabolic rates (active hyperemia). |

## Aerobic Requirements of the Heart

The coronary arteries supply an enormous number of capillaries that are packed within the myocardium at a density of about 2500 to 4000 per cubic millimeter of tissue. (Fast-twitch skeletal muscles, by contrast, have a capillary density of 300 to 400 per cubic millimeter of tissue.) Each myocardial cell, as a consequence, is within 10 $\mu$m of a capillary (compared to an average distance in other organs of 60 to 80 $\mu$m). The exchange of gases by diffusion between myocardial cells and capillary blood thus occurs very quickly.

Contraction of the myocardium squeezes the coronary arteries. Unlike blood flow in all other organs, flow in the coronary vessels decreases in systole and increases during diastole. The myocardium, however, contains large amounts of *myoglobin*—pigment related to hemoglobin (the molecules in red blood cells that carry oxygen). Myoglobin in the myocardium stores oxygen during diastole and releases its oxygen during systole. In this way, the myocardial cells can receive a continuous supply of oxygen, even though coronary blood flow is temporarily reduced during systole.

In addition to containing large amounts of myoglobin, heart muscle contains numerous mitochondria and aerobic respiratory enzymes. This indicates that—even more than slow-twitch skeletal muscles—the heart is specialized for aerobic respiration. The normal heart always respires aerobically, even during heavy exercise when the metabolic demand for oxygen can rise to five times resting levels. This increased oxygen requirement is met by a corresponding increase in coronary blood flow—from about 80 ml at rest to about 400 ml per minute per 100 g tissue during heavy exercise.

## Regulation of Coronary Blood Flow

Sympathetic nerve fibers, through stimulation of alpha-adrenergic receptors in the coronary arterioles, produce a relatively high vascular resistance in the coronary circulation at rest. Vasodilation of coronary vessels may be produced, in part, by sympathetic nerve activation of beta-adrenergic receptors. Most of the vasodilation that occurs during exercise, however, is due to intrinsic metabolic control mechanisms. As the metabolism of the myocardium increases, local accumulation of carbon dioxide, K+, and adenosine acts directly on the vascular smooth muscle to cause relaxation and vasodilation.

 Under abnormal conditions, the blood flow to the myocardium may be inadequate, resulting in myocardial ischemia (described in chapter 21). The cause may be blockage by atheromas and/or blood clots or muscular spasm of a coronary artery. Occlusion of a coronary artery can be visualized by a technique called *selective coronary arteriography*. In this procedure, a catheter (plastic tube) is inserted into a brachial or femoral artery all the way to the opening of the coronary arteries in the aorta, whereupon radiographic contrast material is injected. The picture obtained is called an *angiogram.*

If an occlusion is sufficiently great, a coronary bypass operation may be performed. In this procedure, a length of blood vessel (usually taken from the great saphenous vein in the leg), is sutured to the ascending aorta and to the coronary artery at a location beyond the site of the occlusion (fig. 22.16).

## Regulation of Blood Flow through Skeletal Muscles

The arterioles in skeletal muscles, like those of the coronary circulation, have a high vascular resistance at rest as a result of alpha-adrenergic sympathetic stimulation. This produces a relatively low rate of blood flow, but because muscles have such a large mass, this still accounts for 20% to 25% of the total blood flow in the body at rest. Also, as in the heart, blood flow in a skeletal muscle decreases when the muscle contracts and squeezes its arterioles, and blood flow stops entirely when the muscle contracts beyond about 70% of its maximum. Pain and fatigue thus occur much more quickly when an isometric contraction is sustained than when rhythmic contractions are performed.

In addition to adrenergic fibers that promote vasoconstriction by stimulation of alpha-adrenergic receptors, there are also sympathetic cholinergic fibers in skeletal muscles. These cholinergic fibers, together with the stimulation

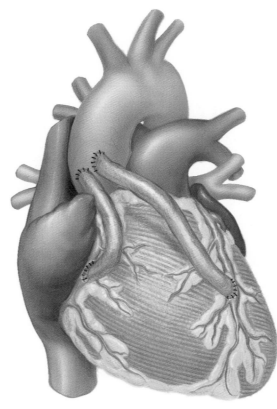

**FIGURE 22.16**
A diagram of coronary artery bypass surgery.

of beta-adrenergic receptors by the hormone epinephrine, stimulate vasodilation as part of the fight-or-flight response to any stressful state, including that existing just prior to exercise (table 22.6). These extrinsic controls have been previously discussed and function to regulate blood flow through muscles at rest and upon anticipation of exercise.

As exercise progresses, the vasodilation and increased skeletal muscle blood flow that occur are almost entirely due to intrinsic metabolic control. The high metabolic rate of skeletal muscles during exercise causes local changes, such as increased carbon dioxide concentrations, decreased pH (due to carbonic acid and lactic acid), decreased oxygen, increased extracellular $K^+$, and the secretion of adenosine. As in the intrinsic control of the coronary circulation, these changes cause vasodilation of arterioles in skeletal muscles, thus decreasing the vascular resistance and increasing the rate of blood flow. This effect is combined with the recruitment of capillaries by the opening of precapillary sphincter muscles (only 5% to 10% of the skeletal muscle capillaries are open at rest). As a result of these changes, skeletal muscles can receive as much as 85% of the total blood flow in the body during maximal exercise.

## Circulatory Changes during Exercise

While the vascular resistance in skeletal muscles decreases during exercise, the resistance to flow through visceral organs

## Table 22.6    Changes in skeletal muscle blood flow under conditions of rest and exercise

| Condition | Blood flow (ml/min) | Mechanism |
|---|---|---|
| Rest | 1000 | High adrenergic sympathetic stimulation of vascular alpha-receptors, causing vasoconstriction |
| Beginning exercise | Increased | Dilation of arterioles in skeletal muscles due to cholinergic sympathetic nerve activity and stimulation of beta-adrenergic receptors by the hormone epinephrine |
| Heavy exercise | 20,000 | Decreased alpha-adrenergic activity<br>Increased sympathetic cholinergic activity<br>Increased metabolic rate of exercising muscles, producing intrinsic vasodilation |

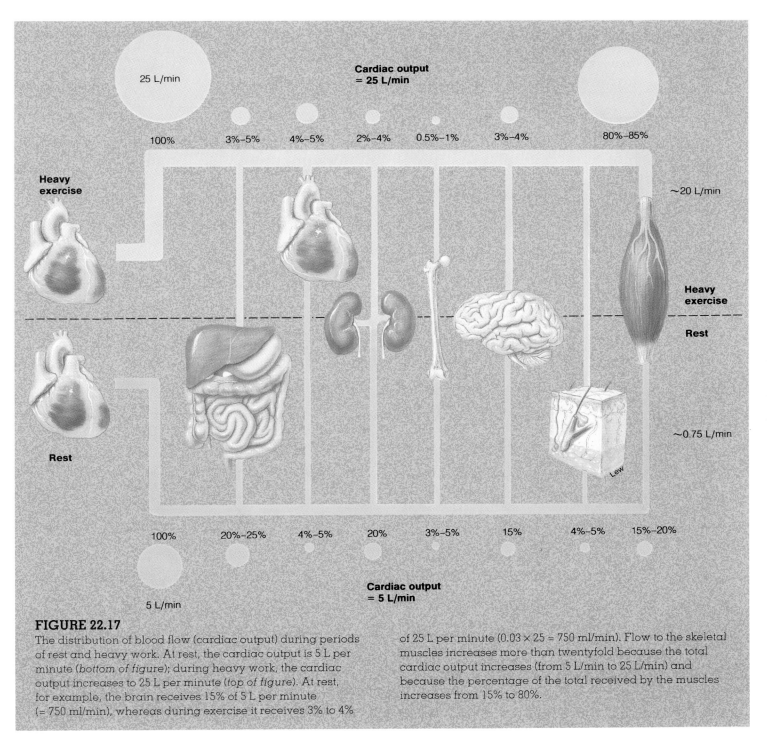

**Cardiac output = 25 L/min**

25 L/min

100%    3%–5%    4%–5%    2%–4%    0.5%–1%    3%–4%    80%–85%

**Heavy exercise**

~20 L/min

**Heavy exercise**

**Rest**

**Rest**

~0.75 L/min

Lew

100%    20%–25%    4%–5%    20%    3%–5%    15%    4%–5%    15%–20%

5 L/min

**Cardiac output = 5 L/min**

**FIGURE 22.17**

The distribution of blood flow (cardiac output) during periods of rest and heavy work. At rest, the cardiac output is 5 L per minute (*bottom of figure*); during heavy work, the cardiac output increases to 25 L per minute (*top of figure*). At rest, for example, the brain receives 15% of 5 L per minute (= 750 ml/minute), whereas during exercise it receives 3% to 4% of 25 L per minute (0.03 × 25 = 750 ml/min). Flow to the skeletal muscles increases more than twentyfold because the total cardiac output increases (from 5 L/min to 25 L/min) and because the percentage of the total received by the muscles increases from 15% to 80%.

and skin increases. This increased resistance occurs because of vasoconstriction stimulated by adrenergic sympathetic fibers and results in decreased rates of blood flow through these organs. During exercise, therefore, the blood flow to skeletal muscles increases because of three simultaneous changes: (1) increased total blood flow (cardiac output), (2) metabolic vasodilation in the exercising muscles, and (3) the diversion of blood away from the viscera and skin.

Blood flow to the heart also increases during exercise, whereas blood flow to the brain does not appear to change significantly (fig. 22.17).

During exercise, the cardiac output can increase five-fold—from about 5 L per minute to about 25 L per minute. This is primarily due to an increase in cardiac rate. The cardiac rate, however, can only increase up to a maximum value (table 22.7), determined mainly by a person's age. In

| Table 22.7 | Relationship between age and average maximum cardiac rate | |
|---|---|
| **Age** | **Maximum cardiac rate** |
| 20–29 | 190 beats/min |
| 30–39 | 160 beats/min |
| 40–49 | 150 beats/min |
| 50–59 | 140 beats/min |
| 60+ | 130 beats/min |

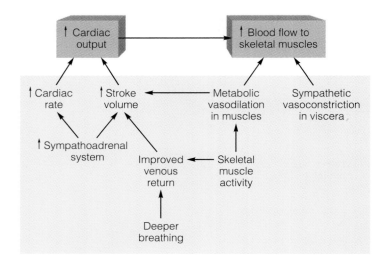

**FIGURE 22.18**

Cardiovascular adaptations to exercise.

athletes who are very well trained, the stroke volume can also increase significantly, allowing these individuals to achieve a cardiac output during strenuous exercise up to six or seven times their resting values.

In most people, the stroke volume can only increase from 10% to 35% during exercise. The fact that the stroke volume can increase at all during exercise may at first be surprising, since the heart has less time to fill with blood between beats when it is pumping faster. Despite the faster beat, however, the end-diastolic volume during exercise is not decreased. This is because the venous return is aided by the improved action of the skeletal muscle pumps and by increased respiratory movements during exercise (fig. 22.18). Since the end-diastolic volume is not significantly changed during exercise, any increase in stroke volume that occurs must be due to an increase in the proportion of blood ejected per stroke.

The proportion of the end-diastolic volume ejected per stroke can increase from 60% at rest to 90% during heavy exercise. This increased ejection fraction is produced by the increased contractility that results from sympathoadrenal stimulation. There also may be a decrease in total peripheral resistance as a result of vasodilation in the exercising skeletal muscles, which decreases the afterload and thus further augments the increase in stroke volume. The cardiovascular changes that occur during exercise are summarized in table 22.8.

Endurance training often results in a lowering of the resting cardiac rate and an increase in the resting stroke volume. The lowering of the resting cardiac rate results from a greater degree of inhibition of the SA node by parasympathetic stimulation through the vagus nerves. The increased resting stroke volume is believed to be due to an increase in blood volume; indeed, studies have shown that the blood volume can increase by about 500 ml after only 8 days of

training. These adaptations enable the trained athlete to produce a larger proportionate increase in cardiac output and achieve a higher absolute cardiac output during exercise. This large cardiac output is the major factor in the improved oxygen delivery to skeletal muscles that occurs as a result of endurance training.

# Blood Flow to the Brain and Skin

*Intrinsic control mechanisms help to maintain a relatively constant blood flow to the brain. Blood flow to the skin, by contrast, can vary tremendously in response to regulation by sympathetic nerve stimulation.*

The examination of cerebral and cutaneous blood flow is a study in contrasts. Cerebral blood flow is regulated primarily by intrinsic mechanisms; cutaneous blood flow is regulated by extrinsic mechanisms. Cerebral blood flow is relatively constant; cutaneous blood flow exhibits more variation than that of any other organ. The brain is the organ that can least tolerate low rates of blood flow; the skin is the organ that can most tolerate low rates.

## Cerebral Circulation

When the brain is deprived of oxygen for only a few seconds, the person loses consciousness; irreversible brain injury may occur after only a few minutes. For these reasons, the

## Table 22.8    Cardiovascular changes during moderate exercise

| Variable | Change | Mechanisms |
|---|---|---|
| Cardiac output | Increased | Cardiac rate and stroke volume increased |
| Cardiac rate | Increased | Increased sympathetic nerve activity; decreased activity of the vagus nerve |
| Stroke volume | Increased | Increased myocardial contractility due to stimulation by sympathoadrenal system; decreased total peripheral resistance |
| Total peripheral resistance | Decreased | Vasodilation of arterioles in skeletal muscles (and in skin when thermoregulatory adjustments are needed) |
| Arterial blood pressure | Increased | Increased systolic and pulse pressure due primarily to increased cardiac output; diastolic pressure rises less due to decreased total peripheral resistance |
| End-diastolic volume (EDV) | Unchanged | Decreased filling time at high cardiac rates is offset by increased venous pressure, increased activity of the skeletal muscle pump, and decreased intrathoracic pressure aiding the venous return |
| Blood flow to heart and muscles | Increased | Increased muscle metabolism produces intrinsic vasodilation; aided by increased cardiac output and increased vascular resistance in visceral organs |
| Blood flow to visceral organs | Decreased | Vasoconstriction in GI tract, liver, and kidneys due to sympathetic nerve stimulation |
| Blood flow to skin | Increased | Metabolic heat produced by exercising muscles produces reflex (involving hypothalamus) that reduces sympathetic constriction of arteriovenous shunts and arterioles |
| Blood flow to brain | Unchanged | Autoregulation of cerebral vessels, which maintains constant cerebral blood flow despite increased arterial blood pressure |

cerebral blood flow is remarkably constant at about 750 ml per minute. This amounts to about 15% of the total cardiac output at rest.

Unlike the coronary and skeletal muscle blood flow, cerebral blood flow is not normally influenced by sympathetic nerve stimulation. Only when the mean arterial pressure rises to about 200 mmHg do sympathetic impulses cause a significant degree of vasoconstriction in the cerebral circulation. This vasoconstriction helps to protect small, thin-walled arterioles from bursting under the pressure, and thus helps to prevent cerebrovascular accident (stroke).

In the normal range of arterial pressures, cerebral blood flow is regulated almost exclusively by intrinsic mechanisms. These mechanisms help to ensure a constant rate of blood flow despite changes in systemic arterial pressure—a process called **autoregulation.** The autoregulation of cerebral blood flow is achieved by both myogenic and metabolic mechanisms.

**Myogenic Regulation**    Myogenic regulation occurs in response to variation in systemic arterial pressure. The cerebral arteries automatically dilate when the blood pressure falls and constrict when the pressure rises. This helps to maintain a constant flow rate during the normal pressure variations that occur during rest, exercise, and emotional states.

The cerebral vessels are also sensitive to the carbon dioxide concentration of arterial blood. When the carbon dioxide concentration rises as a result of inadequate ventilation (hypoventilation), the cerebral arterioles dilate. This is believed to be due to decreases in the pH of cerebrospinal fluid rather than to a direct effect of $CO_2$ on the cerebral vessels. Conversely, when the arterial $CO_2$ falls below normal during hyperventilation, the cerebral vessels constrict. The resulting decrease in cerebral blood flow is responsible for the dizziness experienced during hyperventilation.

**Metabolic Regulation**    The cerebral arterioles are exquisitely sensitive to local changes in metabolic activity, so that those brain regions with the highest metabolic activity get the most blood. Indeed, areas of the brain that control specific processes have been mapped by the changing patterns of blood flow that result when these areas are activated. Visual and auditory stimuli, for example, increase blood flow to the appropriate sensory areas of the cerebral cortex, whereas motor activities, such as movements of the eyes, arms, and organs of speech, result in different patterns of blood flow (fig. 22.19).

The exact mechanisms by which increases in neural activity in a particular area of the brain elicit local vasodilation are not completely understood. There is evidence that local cerebral vasodilation may be caused by $K^+$, which is released from active neurons during repolarization. It has been proposed that astrocytes may take up this extruded $K^+$ near the active neurons and then release the $K^+$ through their vascular processes (chapter 14) surrounding arterioles, thereby causing the arterioles to dilate. Additionally, cerebral vasodilation is produced by nitric oxide from the arterial endothelium, as previously discussed.

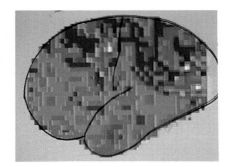
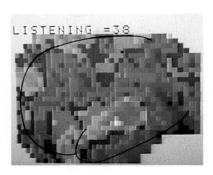
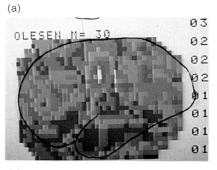
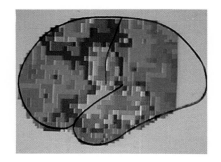

**FIGURE 22.19**
Computerized picture of blood-flow distribution in the brain after injecting the carotid artery with a radioactive isotope. In (a), on the left, the subject followed a moving object with his eyes. High activity is seen over the occipital lobe of the brain. In (a), on the right, the subject listened to spoken words. Notice that the high activity is seen over the temporal lobe (the auditory cortex). In (b), on the left, the subject moved his fingers on the side of the body opposite to the cerebral hemisphere being studied. In (b), on the right, the subject counted to 20. High activity is shown over the mouth area of the motor cortex, the supplementary motor area, and the auditory cortex.

## Cutaneous Blood Flow

The skin is the outer covering of the body and as such serves as the first line of defense against invasion by disease-causing organisms. The skin, as the interface between the internal and external environments, also serves to help maintain a constant deep-body temperature despite changes in the ambient (external) temperature—a process called **thermoregulation.** The thinness and extensiveness of the skin (1.0–1.5 mm thick; 1.7–1.8 square meters in surface area) make it an effective radiator of heat when the body temperature is greater than the ambient temperature. The transfer of heat from the body to the external environment is aided by the flow of warm blood through capillary loops near the surface of the skin.

Blood flow through the skin is adjusted to maintain deep-body temperature at about 37° C (98.6° F). The adjustments are made by variations in the degree of constriction or dilation of ordinary arterioles and of unique **arteriovenous anastomoses** (fig. 22.20). These latter vessels, found predominantly in the fingertips, palms, toes, soles, ears, nose, and lips, shunt blood directly from arterioles to deep venules,

.........................................
anastomosis: Gk. *anastomosis,* opening or outlet

thus bypassing superficial capillary loops. Both the ordinary arterioles and the arteriovenous anastomoses are innervated by sympathetic nerve fibers. When the ambient temperature is low, sympathetic nerves stimulate cutaneous vasoconstriction; cutaneous blood flow is thus decreased, so that less heat will be lost from the body. Since the arteriovenous anastomoses also constrict, the skin may appear rosy because the blood is diverted to the superficial capillary loops. Despite this rosy appearance, the total cutaneous blood flow and rate of heat loss is lower than under usual conditions.

Skin can tolerate an extremely low blood flow in cold weather because its metabolic rate decreases when the ambient temperature decreases. In cold weather, therefore, the skin requires less blood. As a result of exposure to extreme cold, however, blood flow to the skin can be so severely restricted that the tissue dies—a condition known as *frostbite.* Blood flow to the skin can vary from less than 20 ml per minute at maximal vasoconstriction to as much as 3 to 4 L per minute at maximal vasodilation.

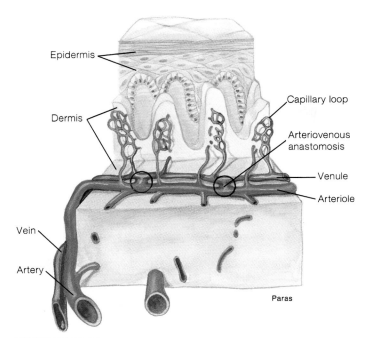

**FIGURE 22.20**
Circulation in the skin showing arteriovenous anastomoses. These vessels function as shunts, allowing blood to be diverted directly from the arteriole to the venule, thus bypassing superficial capillary loops.

As the temperature warms, cutaneous arterioles in the hands and feet dilate as a result of decreased sympathetic nerve stimulation. Continued warming causes dilation of arterioles in other areas of the skin. If the resulting increase in cutaneous blood flow is not sufficient to cool the body, secretion of the sweat glands may be stimulated. Sweat helps to cool the body as it evaporates from the surface of the skin. Also, the sweat glands secrete **bradykinin,** a polypeptide that stimulates vasodilation. This increases blood flow to the skin and to the sweat glands, so that larger volumes of more dilute sweat are produced.

Under the usual conditions of ambient temperature, the cutaneous vascular resistance is high and the blood flow is low when a person is not exercising. In the pre-exercise state of fight or flight, sympathetic nerve stimulation reduces cutaneous blood flow still further. During exercise, however, the need to maintain a deep-body temperature takes precedence over the need to maintain an adequate systemic blood pressure. As the body temperature rises during exercise, vasodilation in cutaneous vessels is accompanied by vasodilation in the exercising muscles. This can produce an even greater lowering of total peripheral resistance. If exercise is performed in hot and humid weather, and if restrictive clothing increases skin temperature and cutaneous vasodilation, a dangerously low blood

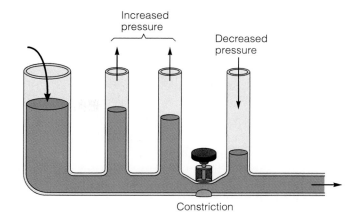

**FIGURE 22.21**
A constriction increases blood pressure upstream (analogous to the arterial pressure) and decreases pressure downstream (analogous to capillary and venous pressure).

pressure may be produced after exercise has ceased and the cardiac output has declined. People have lost consciousness and have even died as a result.

Changes in cutaneous blood flow occur as a result of changes in sympathetic nerve stimulation. Since the autonomic nervous system is controlled by the brain, emotional states, acting through control centers in the medulla oblongata, can affect sympathetic activity and cutaneous blood flow. During fear reactions, for example, vasoconstriction in the skin, along with activation of the sweat glands, can produce a pallor and a "cold sweat." Other emotions may cause vasodilation and blushing.

# Blood Pressure

*The pressure of the arterial blood is regulated by the blood volume, total peripheral resistance, and the cardiac rate. Regulatory mechanisms adjust these factors in a negative feedback manner to compensate for deviations. Arterial pressure rises and falls as the heart goes through systole and diastole.*

Resistance to flow in the arterial system is greatest in the arterioles because these vessels have the smallest diameters. Although the total blood flow through a system of arterioles must be equal to the flow in the larger vessel that gave rise to those arterioles, the narrowness of each arteriole reduces the flow rate in each according to Poiseuille's law. Blood flow rate and pressure are thus reduced in the capillaries, which are located downstream of the high resistance imposed by the arterioles. The blood pressure upstream of the arterioles—in the medium and large arteries—is correspondingly increased (fig. 22.21).

The blood pressure and flow rate within the capillaries are further reduced by the fact that their total cross-sectional area is much greater (due to their large number) than the cross-sectional areas of the arteries and arterioles (fig. 22.22). Thus, although each capillary is much narrower than each arteriole, the capillary beds served by arterioles do not provide as great a resistance to blood flow as do the arterioles.

Variations in the diameter of arterioles due to vasoconstriction and vasodilation thus simultaneously affect blood flow through capillaries and the *arterial blood pressure* "upstream" from the capillaries. An increase in total peripheral resistance due to vasoconstriction of arterioles can raise arterial blood pressure. Blood pressure can also be raised by an increase in the cardiac output. This may be due to elevations in cardiac rate or stroke volume, which in turn are affected by other factors. The three most important variables affecting blood pressure are the **cardiac rate, stroke volume** (determined primarily by the **blood volume**), and **total peripheral resistance.** An increase in any of these variables, if not offset by a decrease in another variable, will result in an increased blood pressure.

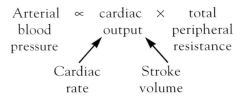

Blood pressure can thus be regulated by the kidneys, which control blood volume, and by the sympathoadrenal system. Increased activity of the sympathoadrenal system can raise blood pressure by stimulating vasoconstriction of arterioles (raising total peripheral resistance) and by promoting an increased cardiac output. Sympathetic stimulation can also affect blood volume indirectly, by stimulating constriction of renal blood vessels and thus reducing urine output.

## Regulation of Blood Pressure

In order to maintain blood pressure within normal limits, specialized receptors for pressure are needed. These **baroreceptors** are stretch receptors located in the *aortic arch* and in the *carotid sinuses.* An increase in pressure causes the walls of these arterial regions to stretch and stimulate the activity of sensory nerve endings. A fall in pressure below

··········································
baroreceptor: Gk. *baros*, pressure; L. *receiver*, to receive

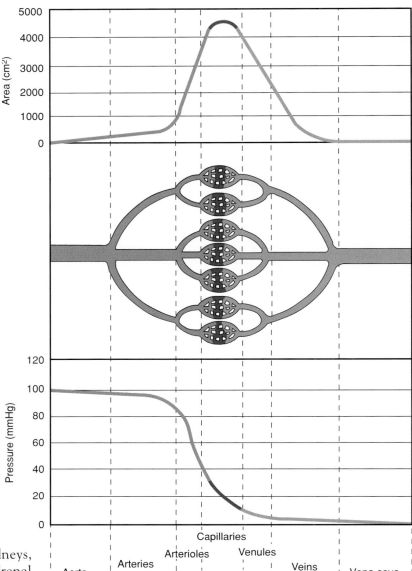

**FIGURE 22.22**

As blood passes from the aorta to the smaller arteries, arterioles, and capillaries, the cross-sectional area increases and the pressure decreases.

the normal range, by contrast, causes a decrease in the frequency of action potentials produced by these sensory nerve fibers.

Sensory nerve activity from the baroreceptors ascends via the vagus and glossopharyngeal nerves to the medulla oblongata, which directs the autonomic nervous system to respond appropriately. **Vasomotor control centers** in the medulla oblongata control vasoconstriction/vasodilation, and hence help regulate total peripheral resistance. **Cardiac control centers** in the medulla oblongata regulate the cardiac rate (fig. 22.23).

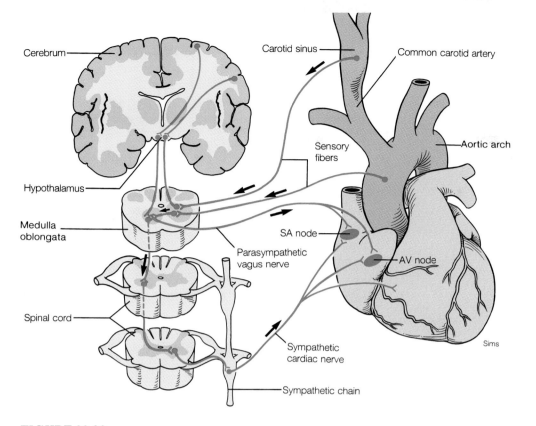

Cerebrum

Carotid sinus

Common carotid artery

Hypothalamus

Sensory fibers

Aortic arch

Medulla oblongata

SA node

Parasympathetic vagus nerve

AV node

Spinal cord

Sims

Sympathetic cardiac nerve

Sympathetic chain

**FIGURE 22.23**

The baroreceptor reflex. Sensory stimuli from baroreceptors in the carotid sinus and the aortic arch, acting via control centers in the medulla oblongata, affect the activity of sympathetic and parasympathetic nerve fibers in the heart.

**Baroreceptor Reflex**   The baroreceptor reflex is activated whenever blood pressure either increases or decreases. The reflex is somewhat more sensitive to decreases in pressure than to increases and is more sensitive to sudden changes in pressure than to more gradual changes. A good example of the importance of the baroreceptor reflex in normal physiology is its activation whenever a person goes from a lying to a standing position.

When a person goes from a lying to a standing position, there is a shift of 500 to 700 ml of blood from the veins of the thoracic cavity to veins in the lower extremities, which expand to contain the extra volume of blood. This pooling of blood reduces the venous return and cardiac output. The resulting fall in blood pressure is almost immediately compensated for by the baroreceptor reflex. A decrease in baroreceptor sensory information, traveling in the glossopharyngeal (ninth cranial) and the vagus (tenth cranial) nerves to the medulla oblongata, inhibits parasympathetic nerve activity and promotes sympathetic

nerve activity, resulting in increased cardiac rate and vasoconstriction. These responses help to maintain an adequate blood pressure upon standing (fig. 22.24).

 Because the baroreceptor reflex may require a few seconds to be fully effective, many people feel dizzy and disoriented if they stand up too rapidly. If baroreceptor sensitivity is abnormally reduced, perhaps by atherosclerosis, an uncompensated fall in pressure may occur upon standing. This condition—called *postural,* or *orthostatic, hypotension* (hypotension = low blood pressure)— can cause a person to feel extremely dizzy or even faint because of inadequate perfusion of the brain.

The baroreceptor reflex can also mediate the opposite response. When the blood pressure rises above an individual's normal range, the baroreceptor reflex causes a slowing of the cardiac rate and vasodilation. Manual massage of the carotid sinus, a procedure sometimes employed by physicians to reduce tachycardia and lower blood pressure, also evokes this reflex. Such carotid massage should be used cautiously, however, because the intense vagus-nerve-induced slowing of the cardiac rate could cause loss of consciousness (as occurs in emotional fainting). Manual massage of both carotid sinuses simultaneously can even cause cardiac arrest in susceptible people.

 *Valsalva's maneuver* is the term used to describe an expiratory effort against a closed glottis (which prevents the air from escaping—see chapter 24). This maneuver, commonly performed during forceful defecation or when lifting heavy weights, increases the intrathoracic pressure. Compression of the thoracic veins causes a fall in venous return and cardiac output, thus lowering the arterial blood pressure. The lowering of arterial pressure stimulates the baroreceptor reflex, resulting in tachycardia and increased total peripheral resistance. When the glottis is finally opened and the air is exhaled, the cardiac output returns to normal but the total peripheral resistance is still elevated, causing a rise in blood pressure. The blood

...........................................................

Valsalva's maneuver: from Antonio Valsalva, Italian anatomist, 1666–1723

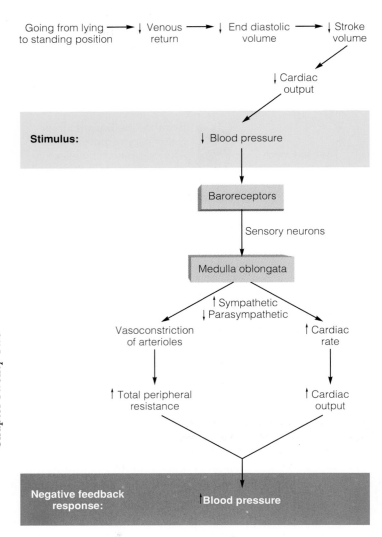

**FIGURE 22.24**
The negative feedback control of blood pressure by the baroreceptor reflex. This reflex helps to maintain an adequate blood pressure upon standing.

pressure is then brought back to normal by the baroreceptor reflex, which causes a slowing of the heart rate. These fluctuations in cardiac output and blood pressure can be dangerous in people with cardiovascular disease. Even healthy people are advised to exhale normally when lifting weights.

**Other Reflexes Controlling Blood Pressure**    The reflex control of ADH secretion by osmoreceptors in the hypothalamus and the control of angiotensin II production and aldosterone secretion by the juxtaglomerular apparatus of the kidneys have been previously discussed. Antidiuretic hormone and aldosterone increase blood volume; angiotensin II stimulates vasoconstriction to cause an increase in blood pressure.

Other reflexes that are important to blood pressure regulation are initiated by **atrial stretch receptors** located in the left atrium of the heart. These receptors are activated by in-

creased venous return to the heart and, in response, stimulate (1) reflex tachycardia, as a result of increased sympathetic nerve activity, (2) inhibition of ADH secretion, resulting in larger volumes of urine excretion and a lowering of blood volume, and (3) increased secretion of atrial natriuretic factor (ANF). The ANF, as previously discussed, lowers blood volume by increasing urinary salt and water excretion.

## Measurement of Blood Pressure

The first documented measurement of blood pressure was accomplished by Stephen Hales, an English physiologist. Hales inserted a cannula into the artery of a horse and measured the height to which blood would rise in the vertical tube. Modern clinical blood pressure measurements are fortunately less direct. The indirect, or **auscultatory** (*aw-skul′tă-tor″e*), **method** of blood pressure measurement is based on the correlation of blood pressure and arterial sounds.

In the auscultatory method, an inflatable rubber bladder within a cloth cuff is wrapped around the upper arm and a stethoscope is applied over the brachial artery (fig. 22.25). The artery is silent before inflation of the cuff because blood normally travels in a smooth laminar flow through the arteries. The term *laminar* means "layered"— blood in the central axial stream moves the fastest and blood flowing closer to the artery wall moves more slowly. There is little transverse movement between these layers that would produce mixing.

The laminar flow that normally occurs in arteries produces little vibration, and is thus silent. When the artery is constricted, however, blood flow becomes turbulent, which causes the artery to vibrate and produce sounds (much like the sounds produced by water through a kink in a garden hose). The tendency of the cuff pressure to constrict the artery is opposed by the blood pressure. In order to constrict the artery, the cuff pressure must be greater than the diastolic blood pressure. If the cuff pressure is also greater than the systolic blood pressure, the artery will be pinched off and silent. Turbulent flow and sounds produced by vibrations of the artery as a result of this flow occur only when the cuff pressure is greater than the diastolic blood pressure but less than the systolic pressure.

Let's say that a person has a systolic pressure of 120 mmHg and a diastolic pressure of 80 mmHg (the average adult normal values). When the cuff pressure is between 80 and 120 mmHg, the artery will be closed during diastole and open during systole. As the artery begins to open with every systole, turbulent flow of blood through

Stephen Hales: English physiologist, 1677–1761
auscultatory: L., *auscultare*, to listen to

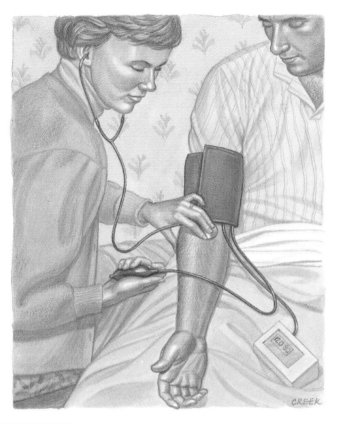

**FIGURE 22.25**
The use of a pressure cuff and a sphygmomanometer to measure blood pressure.

the constriction will create vibrations known as the **sounds of Korotkoff** (kŏ-rot´kof), as shown in figure 22.26. These are usually "tapping" sounds because the artery becomes constricted, blood flow stops, and silence resumes with every diastole. It should be understood that the sounds of Korotkoff are *not* "lub-dub" sounds produced by closing of the heart valves—those sounds can only be heard on the chest, not on the brachial artery.

Initially, the cuff is usually inflated to produce a pressure greater than the systolic pressure so that the artery is pinched off and silent. The pressure in the cuff is read from an attached meter called a *sphygmomanometer* (*sfig´´mo-mă-nom´ĭ-ter*). A valve is then turned to allow the release of air from the cuff, causing a gradual decrease in cuff pressure. When the cuff pressure is equal to the systolic pressure, the *first Korotkoff sound* is heard as blood passes in a turbulent flow through the constricted opening of the artery.

Korotkoff sounds will continue to be heard at every systole as long as the cuff pressure remains greater than the diastolic pressure. When the cuff pressure becomes equal to or drops below the diastolic pressure, the sounds disappear because laminar blood flow is reestablished and the vibrations of

sounds of Korotkoff: from Nicolai S. Korotkoff, Russian physician, 1874–1920
sphygmomanometer: Gk. *sphygmos*, pulse; *manos*, thin; *metro*, measure

No sounds

Cuff pressure = 140

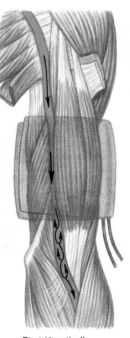

First Korotkoff sounds

Cuff pressure = 120

**Systolic pressure = 120 mmHg**

Sounds at every systole

Cuff pressure = 100

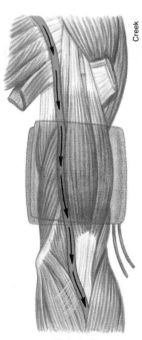

Last Korotkoff sounds

Cuff pressure = 80

**Diastolic pressure = 80 mmHg**

**Blood pressure = 120/80**

**FIGURE 22.26**
Korotkoff sounds are produced by the turbulent flow of blood through the partially constricted brachial artery. This occurs when the cuff pressure is greater than the diastolic pressure but less than the systolic pressure.

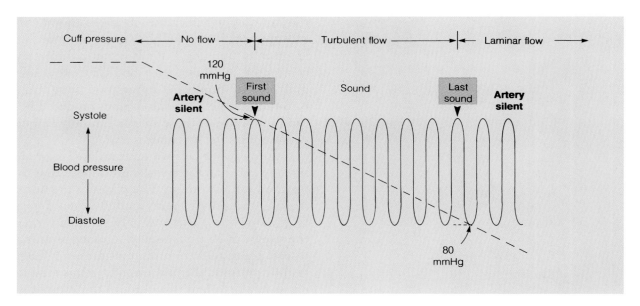

**FIGURE 22.27**
The indirect, or auscultatory, method of blood-pressure measurement. Korotkoff sounds, produced by turbulent blood flow through a constricted artery, occur whenever the cuff pressure is lower than the systolic blood pressure and greater than the diastolic blood pressure. As a result, the first Korotkoff sound is heard when the cuff pressure is equal to the systolic blood pressure, and the last sound is heard when the cuff pressure and diastolic blood pressures are equal.

the artery stop (fig. 22.27). The *last Korotkoff sound* occurs when the cuff pressure is equal to the diastolic pressure.

Different phases in the measurement of blood pressure are identified on the basis of the quality of the Korotkoff sounds (fig. 22.28). In some people, the Korotkoff sounds do not disappear even when the cuff pressure is reduced to zero (zero pressure means that it is equal to atmospheric pressure). In these cases—and often routinely—the onset of muffling of the sounds (phase 4 in fig. 22.28) is used as an indication of diastolic pressure rather than the onset of silence (phase 5). Normal blood pressure values are indicated in table 22.9.

## Pulse Pressure and Mean Arterial Pressure

When someone takes a pulse, he or she palpates an artery (for example, the radial artery) an feels the expansion of the artery occur in response to the beating of

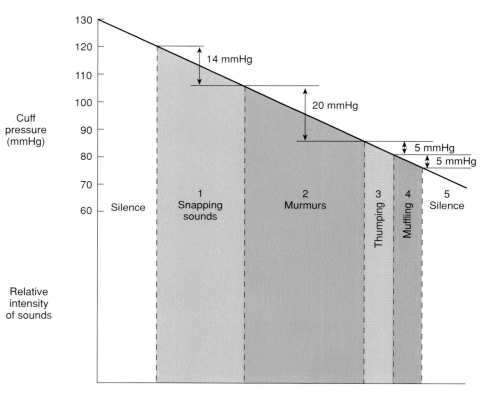

**FIGURE 22.28**
The five phases of blood pressure measurement.

## Table 22.9 — Normal arterial blood pressure at different ages

| Age | Systolic Male | Systolic Female | Diastolic Male | Diastolic Female | Age | Systolic Male | Systolic Female | Dyastolic Male | Dyastolic Female |
|---|---|---|---|---|---|---|---|---|---|
| 1 day | 70 | | | | 16 years | 118 | 116 | 73 | 72 |
| 3 days | 72 | | | | 17 years | 121 | 116 | 74 | 72 |
| 9 days | 73 | | | | 18 years | 120 | 116 | 74 | 72 |
| 3 weeks | 77 | | | | 19 years | 122 | 115 | 75 | 71 |
| 3 months | 86 | | | | 20–24 years | 123 | 116 | 76 | 72 |
| 6–12 months | 89 | 93 | 60 | 62 | 25–29 years | 125 | 117 | 78 | 74 |
| 1 year | 96 | 95 | 66 | 65 | 30–34 years | 126 | 120 | 79 | 75 |
| 2 years | 99 | 92 | 64 | 60 | 35–39 years | 127 | 124 | 80 | 78 |
| 3 years | 100 | 100 | 67 | 64 | 40–44 years | 129 | 127 | 81 | 80 |
| 4 years | 99 | 99 | 65 | 66 | 45–49 years | 130 | 131 | 82 | 82 |
| 5 years | 92 | 92 | 62 | 62 | 50–54 years | 135 | 137 | 83 | 84 |
| 6 years | 94 | 94 | 64 | 64 | 55–59 years | 138 | 139 | 84 | 84 |
| 7 years | 97 | 97 | 65 | 66 | 60–64 years | 142 | 144 | 85 | 85 |
| 8 years | 100 | 100 | 67 | 68 | 65–69 years | 143 | 154 | 83 | 85 |
| 9 years | 101 | 101 | 68 | 69 | 70–74 years | 145 | 159 | 82 | 85 |
| 10 years | 103 | 103 | 69 | 70 | 75–79 years | 146 | 158 | 81 | 84 |
| 11 years | 104 | 104 | 70 | 71 | 80–84 years | 145 | 157 | 82 | 83 |
| 12 years | 106 | 106 | 71 | 72 | 85–89 years | 145 | 154 | 79 | 82 |
| 13 years | 108 | 108 | 72 | 73 | 90–94 years | 145 | 150 | 78 | 79 |
| 14 years | 110 | 110 | 73 | 74 | 95–106 years | 145 | 149 | 78 | 81 |
| 15 years | 112 | 112 | 75 | 76 | | | | | |

Source: *Documenta Geigy Scientific Tables*, edited by K. Diem and C. Lentner, 7th ed. Copyright © 1970 J. R. Geigy S. A., Basle, Switzerland.

the heart; the pulse rate is thus a measure of the cardiac rate. The expansion of the artery with each pulse occurs as a result of the rise in blood pressure within the artery as the artery receives the volume of blood ejected by a stroke of the left ventricle.

The pulse is thus produced by the **pulse pressure,** which is equal to the difference between the systolic and diastolic pressures. If a person has a blood pressure of 120/80 (systolic/diastolic), therefore, the pulse pressure would be 40 mmHg.

Pulse pressure = systolic pressure – diastolic pressure

At diastole in this example the aortic pressure equals 80 mmHg. When the left ventricle contracts, the intraven-

tricular pressure rises above 80 mmHg and ejection begins. As a result, the amount of blood in the aorta increases by the amount ejected from the left ventricle (the stroke volume). Due to the increase in volume, there is an increase in blood pressure. The pressure in the brachial artery, where blood pressure measurements are commonly taken, therefore increases to 120 mmHg in this example. The rise in pressure from diastolic to systolic levels (pulse pressure) is thus a reflection of the stroke volume.

The **mean arterial pressure** represents the average arterial pressure during the cardiac cycle. It is the difference between this pressure and the venous pressure that drives blood through the capillary beds of organs. The mean arterial pressure is not a simple arithmetic average because the period of diastole is longer than the period of systole. Its value can most correctly be approximated by adding one-third of the pulse pressure to the diastolic pressure. If a person has a blood pressure of 120/80, for example, the mean arterial pressure will be 80 + 1/3 (40) = 93 mmHg.

$$\text{Mean arterial pressure} = \text{diastolic pressure} + 1/3 \times \text{pulse pressure}$$

A rise in total peripheral resistance and cardiac rate increases the diastolic pressure more than it increases the systolic pressure. When the baroreceptor reflex is activated by going from a lying to a standing position, for example, the diastolic pressure usually increases by 5 to 10 mmHg, whereas the systolic pressure either remains unchanged or is slightly reduced (as a result of decreased venous return). People with hypertension (high blood pressure) who usually have elevated total peripheral resistance and cardiac rates likewise have a greater increase in diastolic than in systolic pressure. Dehydration or blood loss results in decreased cardiac output and thus also produces a decrease in pulse pressure.

An increase in cardiac output, by contrast, raises the systolic pressure more than it raises the diastolic pressure (although both pressures do rise). This occurs during exercise, for example, when the blood pressure may rise to values as high as 200/100 (yielding a pulse pressure of 100 mmHg).

## Clinical Considerations

### Hypertension

**Hypertension** is blood pressure in excess of the normal range for a person's age and sex. Approximately 20% of all adults in the United States have hypertension. Hypertension that results from (is "secondary to") known disease processes is logically termed **secondary hypertension.** Of the hypertensive population, secondary hypertension accounts for only about 10%. Hypertension that results from complex and poorly understood processes is not so logically termed **primary**, or **essential, hypertension**.

Diseases of the kidneys and arteriosclerosis of the renal arteries can cause secondary hypertension because of high blood volume. More commonly, the reduction of renal blood flow can raise blood pressure by stimulating the secretion of vasoactive chemicals from the kidneys. Experiments in which the renal artery is pinched, for example, produce hypertension that is associated (at least initially) with elevated renin secretion.

**Essential Hypertension** The vast majority of people with hypertension have essential hypertension. An increased total peripheral resistance is a universal characteristic of this condition. Cardiac rate and the cardiac output are elevated in many, but not all, of these cases.

The secretion of renin, which is correlated with angiotensin II production and aldosterone secretion, is likewise variable. In some people with essential hypertension renin secretion is low, but in most cases levels of renin secretion are either normal or elevated. Renin secretion in the normal range is inappropriate for people with hypertension, since high blood pressure should inhibit renin secretion and, through a lowering of aldosterone, result in greater excretion of salt and water. Inappropriately high levels of renin secretion could thus contribute to hypertension by promoting (via stimulation of aldosterone secretion) salt and water retention and high blood volume.

Sustained high stress (acting via sympathetic nerves) and high salt intake appear to act synergistically in the development of hypertension. There is some evidence that $Na^+$ enhances the vascular response to sympathetic stimulation. Further, stimulation through sympathetic nerves can cause constriction of renal blood vessels and thus decrease the excretion of salt and water.

As an adaptive response to prolonged high blood pressure, the arterial wall becomes thickened. This response can lead to arteriosclerosis and result in an even greater increase in total peripheral resistance, thus raising blood pressure still more in a positive feedback fashion.

The interactions between salt intake, sympathetic nerve activity, cardiovascular responses to sympathetic nerve activity, kidney function, and genetics make it difficult to sort out the cause-and-effect sequence that leads to essential hypertension. Many researchers have suggested that there is no single cause-and-effect sequence but rather a web of causes and effects. This view is currently controversial.

**Dangers of Hypertension** If other factors remain constant, blood flow increases as arterial blood pressure increases. The organs of people with hypertension are thus adequately perfused with blood until the hypertension causes vascular damage. Hypertension, as a result, is usually asymptomatic (without symptoms) until a dangerous amount of vascular damage is produced.

Hypertension is dangerous for a number of reasons. First, high arterial pressure increases the afterload, making it more difficult for the ventricles to eject blood. The heart, then, must work harder, which can result in pathological changes in heart structure and function and lead to congestive heart failure. Additionally, high pressure may damage cerebral blood vessels, leading to cerebrovascular accident (stroke). Finally, hypertension contributes to the development of

## Table 22.10  Mechanisms of action of selected antihypertensive drugs

| Category of drugs | Examples | Mechanisms |
|---|---|---|
| Extracellular fluid volume depletors | Thiazide diuretics | Increases volume of urine excreted, and thus lowers blood volume |
| Sympathoadrenal system inhibitors | Clonidine; alpha-methyldopa | Acts on brain to decrease sympathoadrenal stimulation |
| | Guanethidine; reserpine | Depletes norepinephrine from sympathetic nerve endings |
| | Propranolol; Atenolol | Blocks beta-adrenergic receptors, decreasing cardiac output and/or renin secretion |
| | Phentolamine | Blocks alpha-adrenergic receptors, decreasing sympathetic vasoconstriction |
| Direct vasodilators | Hydralazine; sodium nitroprusside | Causes vasodilation by acting directly on vascular smooth muscle |
| Calcium channel blockers | Verapamil | Inhibits diffusion of $Ca^{++}$ into vascular smooth muscle cells, causing vasodilation and reduced peripheral resistance |
| Angiotensin-converting enzyme (ACE) inhibitors | Captopril | Inhibits the conversion of angiotensin I into angiotensin II |

## Table 22.11  Signs of shock

| | Early sign | Late sign |
|---|---|---|
| Blood pressure | Decreased pulse pressure<br>Increased diastolic pressure | Decreased systolic pressure |
| Urine | Decreased $Na^+$ concentration<br>Increased osmolality | Decreased volume |
| Blood pH | Increased pH (alkalosis) due to hyperventilation | Decreased pH (acidosis) due to "metabolic" acids |
| Effects of poor tissue perfusion | Slight restlessness; occasionally warm, dry skin | Cold, clammy skin; "cloudy" senses |

Source: *Principles and Techniques of Critical Care*, edited by R. F. Wilson, Vol. 1. Copyright © 1977 Upjohn Company. Used by permission of F. A. Davis Company, Philadelphia, PA.

atherosclerosis, which can itself lead to heart disease and stroke as previously described.

**Treatment of Hypertension**  The first form of treatment that is usually attempted is modification of lifestyle. This modification includes cessation of smoking, moderation of alcohol intake, and weight reduction, if applicable. It can also include the addition of regular physical exercise and a reduction in sodium intake. People with essential hypertension may have a potassium deficiency, and there is evidence that eating food that is rich in potassium may help to lower blood pressure. There is also evidence that supplementation of the diet with $Ca^{++}$ may be of benefit, but this is more controversial

If lifestyle modifications alone are insufficient, different drugs may be prescribed. Most commonly, these are *diuretics* that increase urine volume, thus decreasing blood volume and pressure. Drugs that block $\beta_1$-adrenergic receptors (such as atenolol) lower blood pressure by decreasing the cardiac rate and are also frequently prescribed. ACE inhibitors, calcium antagonists, and various vasodilators (table 22.10) may also be used in particular situations. Methyldopa, for example, may be given to treat hypertension of a pregnant woman.

## Circulatory Shock

**Circulatory shock** occurs when the tissues receive an inadequate blood supply and/or oxygen utilization by the tissues is adequate. Some of the signs of shock (table 22.11) are a result of inadequate tissue perfusion; other signs of shock are produced by cardiovascular responses that help

## Table 22.12  Cardiovascular reflexes that help to compensate for circulatory shock

| Organ(s) | Compensatory mechanisms |
|---|---|
| Heart | Sympathoadrenal stimulation increases cardiac rate and stroke volume due to "positive inotropic effect" on myocardial contractility |
| Gastrointestinal tract and skin | Decreased blood flow due to vasoconstriction as a result of sympathetic nerve stimulation (alpha-adrenergic effect) |
| Kidneys | Decreased urine production as a result of sympathetic-nerve-induced constriction of renal arterioles; increased salt and water retention due to increased aldosterone and antidiuretic hormone (ADH) secretion |

to compensate for the poor tissue perfusion (table 22.12). When these compensations are effective, they (together with emergency medical care) are able to reestablish adequate tissue perfusion. In some cases, however, and for reasons that are not clearly understood, the shock may progress to an irreversible stage and death may result.

**Hypovolemic Shock**    The term **hypovolemic** (hi″pŏ-vo-le′mik) **shock** refers to circulatory shock due to low blood volume, as might be caused by hemorrhage (bleeding), dehydration, or burns. In response to decreased blood pressure and decreased cardiac output, the sympathoadrenal system is activated by the baroreceptor reflex. As a result, tachycardia is produced and vasoconstriction occurs in the skin, gastrointestinal tract, kidneys, and muscles (table 22.12). Decreased blood flow through the kidneys stimulates renin secretion and activation of the renin-angiotensin-aldosterone system. A person in hypovolemic shock thus has low blood pressure; a rapid, "thready" pulse; cold, clammy skin; and a reduced urine output.

Since the resistance in the coronary and cerebral circulations is not increased, blood is diverted to the heart and brain at the expense of other organs. Interestingly, a similar response has been observed in diving mammals and, to a lesser degree, in Japanese pearl divers during prolonged submersion. These responses help to deliver blood to the two organs that have the highest requirements for aerobic metabolism.

Vasoconstriction in organs other than the brain and heart raises total peripheral resistance, which helps (along with the reflex increase in cardiac rate) to compensate for the drop in blood pressure due to low blood volume. Constriction of arterioles also decreases capillary blood flow and capillary filtration pressure. Less filtrate is formed as a result, while the osmotic return of fluid to the capillaries due

to the plasma colloid osmotic pressure either remains unchanged or is increased (during dehydration). The blood volume is thus raised at the expense of tissue fluid volume. Blood volume is also conserved by the decrease in urine production that results from vasoconstriction in the kidneys and the water-conserving effects of increased amounts of ADH and aldosterone.

**Septic Shock**    Septic shock refers to a dangerously low blood pressure (*hypotension*) that may result from *sepsis*, or infection. This can occur through the action of a bacterial lipopolysaccharide called *endotoxin*. The mortality of septic shock is presently very high, estimated at 50% to 75%. According to recent information, endotoxin activates the enzyme nitric oxide synthetase within macrophages, which are cells that play an important role in the immune response (chapter 23). As previously discussed, nitric oxide synthetase produces nitric oxide, which promotes vasodilation and, as a result, a fall in blood pressure. Septic shock has recently been successfully treated with drugs that inhibit the production of nitric oxide.

**Other Causes of Circulatory Shock**    A rapid fall in blood pressure occurs in **anaphylactic** (ă″nă-fĭ-lak′tik) **shock** as a result of a severe allergic reaction (usually to bee stings or penicillin). This results from the widespread release of histamine, which causes vasodilation and thus decreases total peripheral resistance. The reaction may be fatal if emergency treatment (usually the administration of epinephrine) is not given immediately. A rapid fall in blood pressure also occurs in **neurogenic shock,** in which sympathetic tone is decreased, usually because of upper spinal cord damage or spinal anesthesia. **Cardiogenic shock** results from cardiac failure, as defined by a cardiac output that is inadequate to maintain tissue perfusion.

## Congestive Heart Failure

**Cardiac failure** occurs when the cardiac output is insufficient to maintain the blood flow required by the body. This may be due to heart disease—resulting from myocardial infarction or congenital defects—or to hypertension, which increases the afterload of the heart. The most common causes of left pump failure are myocardial infarction, aortic valve stenosis, and incompetence of the aortic and bicuspid (mitral) valves. Failure of the right pump is usually caused by prior failure of the left pump.

Heart failure can also result from disturbance in the electrolyte concentrations of the blood. Excessive plasma $K^+$ concentration decreases the resting membrane potential of myocardial cells; low blood $Ca^{++}$ reduces excitation-contraction coupling. High blood $K^+$ and low blood $Ca^{++}$ can thus cause the heart to stop in diastole. Conversely, low blood $K^+$ and high blood $Ca^{++}$ can arrest the heart in systole.

The term *congestive* is often used in describing heart failure because of the increased venous volume and pressure that results. Failure of the left pump, for example, raises the left atrial pressure and produces pulmonary congestion and edema. This causes shortness of breath and fatigue; if severe, pulmonary edema can be fatal. Failure of the right pump results in increased right atrial pressure, which produces congestion and edema in the systemic circulation.

People with congestive heart failure are often treated with the drug *digitalis*. Digitalis appears to bind to and inhibit the action of $Na^+/K^+$ pumps in the cell membranes, causing a rise in the intracellular concentrations of $Na^+$. The increased availability of $Na^+$, in turn, stimulates the activity of another membrane transport carrier, which exchanges $Na^+$ for extracellular $Ca^{++}$. As a result, the intracellular concentrations of $Ca^{++}$ are increased, which strengthens the contractions of the heart.

The compensatory responses that occur during congestive heart failure are similar to those that occur during hypovolemic shock. Activation of the sympathoadrenal system stimulates cardiac rate, contractility of the ventricles, and constriction of arterioles. As in hypovolemic shock, renin secretion is increased and urine output is reduced.

As a result of these compensations, chronically low cardiac output is associated with elevated blood volume and dilation and hypertrophy of the ventricles. These changes can themselves be dangerous. Elevated blood volume places a work overload on the heart, and the enlarged ventricles have a higher metabolic requirement for oxygen. These problems are often treated with drugs that increase myocardial contractility (such as digitalis), drugs that are vasodilators (such as nitroglycerin), and diuretic drugs that lower blood volume by increasing the volume of urine excreted.

# Chapter Summary

### Cardiac Output (pp. 612–615)

1. Cardiac rate is increased by sympathoadrenal stimulation and decreased by the effects of parasympathetic fibers that innervate the SA node.
2. Stroke volume is regulated both extrinsically and intrinsically.
   a. The Frank–Starling law of the heart describes the way the end-diastolic volume, through various degrees of myocardial stretching, influences the contraction strength of the myocardium, and thus the stroke volume.
   b. The end-diastolic volume is called the preload. The total peripheral resistance, through its effect on arterial blood pressure, provides an afterload that acts to reduce the stroke volume.
   c. At a given end-diastolic volume, the amount of blood ejected depends on contractility; strength of contraction is increased by sympathoadrenal stimulation.
3. The venous return of blood to the heart is dependent largely on the total blood volume and mechanisms that improve the flow of blood in veins.
   a. The total blood volume is regulated by the kidneys.
   b. The venous flow of blood to the heart is aided by the action of skeletal muscle pumps and the effects of breathing.

### Blood Volume (pp. 615–622)

1. Tissue fluid is formed from and returns to the blood.
   a. The hydrostatic pressure of the blood forces fluid from the arteriolar ends of capillaries into the interstitial spaces of the tissues.
   b. Since the colloid osmotic pressure of plasma is greater than that of tissue fluid, water returns by osmosis to the venular ends of capillaries.
   c. Excess tissue fluid is returned to the venous system by lymphatic vessels.
   d. Edema occurs when there is an excessive accumulation of tissue fluid.
2. The kidneys control the blood volume by regulating the amount of filtered fluid that will be reabsorbed.
   a. Antidiuretic hormone stimulates reabsorption of water from the kidney filtrate, and thus acts to maintain the blood volume.
   b. A decrease in blood flow through the kidneys activates the renin-angiotensin system.

# Interactions of the Circulatory System with Other Body Systems

## Integumentary System
- Protects the body from pathogens and helps to maintain body temperature
- Delivers blood for exchange of gases, nutrients, and wastes
- Provides for clotting and other protective mechanisms if skin is broken

## Skeletal System
- Provides sites (bone marrow) for hemopoiesis
- Protects heart and thoracic vessels
- Delivers blood for exchange of gases, nutrients, and wastes

## Muscular System
- Cardiac and smooth muscles assist blood movement
- Skeletal muscle contractions squeeze veins and thus aid venous blood flow
- Delivers blood for exchange of gases, nutrients, and wastes, including lactic acid

## Nervous System
- Provides autonomic regulation of cardiac output, vascular resistance, blood flow, and blood pressure
- Delivers blood for exchange of gases, nutrients, and wastes
- Cerebral capillaries participate in blood-brain barrier

## Endocrine System
- Epinephrine and norepinephrine from adrenal medulla help to regulate cardiac function and vascular resistance
- Delivers blood for exchange of gases, nutrients, and wastes
- Transports hormones to target organs

## Lymphatic System
- Protects against infections
- Drains tissue fluid and returns it to venous system
- Delivers blood for exchange of gases, nutrients, and wastes
- Provides circulating lymphocytes from bone marrow and lymphoid organs

## Respiratory System
- Provides $O_2$ for transport by blood and provides for elimination of $CO_2$ from blood
- Transports blood to lungs for gas exchange and transports oxygen-rich blood to body tissues
- Transports $CO_2$ from tissues to lungs for elimination

## Urinary System
- Regulates the volume, pH, and electrolyte balance of blood and eliminates wastes
- Delivers blood for exchange of gases, nutrients, and wastes
- Delivers wastes in blood plasma to be excreted in the urine

## Digestive System
- Provides nutrients for blood formation, including iron and B vitamins
- Transports nutrients from GI tract to all tissues in the body

## Reproductive System
- Gonads produce sex hormones for maintenance of blood and blood vessels
- Delivers blood for exchange of gases, nutrients, and wastes
- Erection of penis elicited by vasodilation of blood vessels

c. Angiotensin II stimulates vasoconstriction and the secretion of aldosterone by the adrenal cortex.

d. Aldosterone acts on the kidneys to promote the retention of salt and water.

## Vascular Resistance and Blood Flow (pp. 622–626)

1. According to Poiseuille's law, blood flow is directly related to the pressure difference between the two ends of a vessel and is inversely related to the resistance to blood flow through the vessel.

2. Extrinsic regulation of vascular resistance is provided mainly by the sympathetic nervous system, which stimulates vasoconstriction of arterioles in the viscera and skin.

3. Intrinsic control of vascular resistance allows organs to autoregulate their own blood flow rates.

   a. Myogenic regulation occurs when vessels constrict or dilate in a direct response to a rise or fall in blood pressure.

   b. Metabolic regulation occurs when vessels dilate in response to the local chemical environment within the organ.

## Blood Flow to the Heart and Skeletal Muscles (pp. 626–630)

1. The heart normally respires aerobically because of its extensive capillary supply, myoglobin content, and enzyme content.

2. During exercise, when the heart's metabolism increases, intrinsic metabolic mechanisms stimulate vasodilation of the coronary vessels, and thus increase coronary blood flow.

3. Just prior to exercise and at the start of exercise, blood flow through skeletal muscles increases due to vasodilation caused by stimulation of cholinergic sympathetic nerve fibers. During exercise, intrinsic metabolic vasodilation occurs.

4. Since cardiac output can increase fivefold during exercise, the heart and skeletal muscles receive an increased proportion of a higher total blood flow.

   a. The cardiac rate increases due to decreased activity of the vagus nerve and increased activity of the sympathetic nerves.

   b. The venous return is greater because of greater activity of the skeletal muscle pumps and an increased respiratory movement.

   c. Increased contractility of the heart, combined with a decrease in total peripheral resistance, can result in a higher stroke volume.

## Blood Flow to the Brain and Skin (pp. 630–633)

1. Cerebral blood flow is regulated both myogenically and metabolically.

   a. Cerebral vessels automatically constrict if the systemic blood pressure rises too high.

   b. Metabolic products cause local vessels to dilate and supply more active areas with more blood.

2. The skin has unique arteriovenous anastomoses that can divert blood away from surface capillary loops.

   a. Sympathetic nerve stimulation causes constriction of cutaneous arterioles.

   b. As a thermoregulatory response, there is increased cutaneous blood flow and increased flow through surface capillary loops when body temperature rises.

## Blood Pressure (pp. 633–640)

1. Baroreceptors in the aortic arch and carotid sinuses affect the cardiac rate and the total peripheral resistance via the sympathetic nervous system.

   a. The baroreceptor reflex causes pressure to be maintained when an upright posture is assumed: this reflex can cause a lowered pressure when the carotid sinuses are massaged.

   b. Other mechanisms that affect blood volume help to regulate blood pressure.

2. Blood pressure is commonly measured indirectly by auscultation of the brachial artery when a pressure cuff is inflated and deflated.

   a. The first sound of Korotkoff, caused by turbulent flow of blood through a constriction in the artery, occurs when the cuff pressure equals the systolic pressure.

   b. The last sound of Korotkoff is heard when the cuff pressure equals the diastolic blood pressure.

3. The mean arterial pressure represents the driving force for blood flow through the arterial system.

# Review Activities

## Objective Questions

1. According to the Frank–Starling law, the strength of ventricular contraction is
   a. directly proportional to the end-diastolic volume.
   b. inversely proportional to the end-diastolic volume.
   c. independent of the end-diastolic volume.

2. In the absence of compensations, the stroke volume will decrease when
   a. blood volume increases.
   b. venous return increases.
   c. contractility increases.
   d. arterial blood pressure increases.

3. Which of the following statements about tissue fluid is *false*?
   a. It has the same glucose and salt concentration as blood plasma.
   b. It has a lower protein concentration than blood plasma.
   c. Its colloid osmotic pressure is greater than that of blood plasma.
   d. Its hydrostatic pressure is less than that of blood plasma.

4. Edema may be caused by
   a. high blood pressure.
   b. decreased blood plasma protein concentration.
   c. leakage of blood plasma protein into tissue fluid.
   d. blockage of lymphatic vessels.
   e. all of the above.

5. Both ADH and aldosterone
   a. increase urine volume.
   b. increase blood volume.
   c. increase total peripheral resistance.
   d. all of the above apply.

6. The greatest resistance to blood flow occurs in
   a. large arteries.
   b. medium-sized arteries.
   c. arterioles.
   d. capillaries.

7. If a vessel were to dilate to twice its previous radius and if pressure remained constant, blood flow through this vessel would
   a. increase by a factor of 16.
   b. increase by a factor of 4.
   c. increase by a factor of 2.
   d. decrease by a factor of 2.

8. The sounds of Korotkoff are produced by
   a. closing of the semilunar valves.
   b. closing of the AV valves.
   c. the turbulent flow of blood through an artery.
   d. elastic recoil of the aorta.

9. Vasodilation in the heart and skeletal muscles during exercise is primarily due to the effects of
   a. alpha-adrenergic stimulation.
   b. beta-adrenergic stimulation.
   c. cholinergic stimulation.
   d. products released by the exercising muscle cells.

10. Blood flow in the coronary circulation is
    a. increased during systole.
    b. increased during diastole.
    c. constant throughout the cardiac cycle.

11. Blood flow in the cerebral circulation
    a. varies with systemic arterial pressure.
    b. is regulated primarily by sympathetic stimulation.
    c. is maintained constant within physiological limits.
    d. is increased during exercise.

12. Which of the following organs is able to tolerate the greatest restriction in blood flow?
    a. the brain
    b. the heart
    c. skeletal muscles
    d. the skin

13. Arteriovenous shunts in the skin
    a. divert blood to superficial capillary loops.
    b. are closed when the ambient temperature is very cold.
    c. are closed when the deep-body temperature rises much above 37° C.
    d. all of the above apply.

14. An increase in blood volume will cause
    a. a decrease in ADH secretion.
    b. an increase in $Na^+$ excretion in the urine.
    c. a decrease in renin secretion.
    d. all of the above.

15. The volume of blood pumped by the left ventricle per minute is
    a. greater than the volume pumped by the right ventricle.
    b. less than the volume pumped by the right ventricle.
    c. the same as the volume pumped by the right ventricle.
    d. either less than or greater than the volume pumped by the right ventricle, depending on the strength of contraction.

16. Blood pressure is lowest in
    a. arteries.
    b. arterioles.
    c. capillaries.
    d. venules.
    e. veins.

## Essay Questions

1. Define *contractility, preload,* and *afterload,* and explain how these factors affect the cardiac output.

2. With reference to the Frank–Starling law, explain how the stroke volume is affected (a) by bradycardia and (b) by a "missed beat."

3. Which part of the circulatory system contains the most blood? Which part provides the greatest resistance to blood flow? Which part provides the greatest cross-sectional area? Explain.

4. Explain how the kidneys regulate blood volume.

5. A person who is dehydrated drinks more and urinates less. Explain the mechanisms involved.

6. Using Poiseuille's law, explain how arterial blood flow can be diverted from one organ system to another.

7. Describe the mechanisms that increase the cardiac output during exercise and that increase the rate of blood flow to the heart and skeletal muscles.

8. Explain how an anxious person may develop cold, clammy skin and how the skin becomes hot and flushed on a hot, humid day.

## Gundy/Weber Software 💾

The tutorial software accompanying Chapter 22 is Volume 9—Cardiovascular System.

# [ chapter twenty-three ]

# lymphatic system and immunity

## objectives

- Describe the pattern of lymph flow from the lymphatic capillaries to the venous system.

- Describe the structure of the lymph nodes and lymphoid organs and state where they are located.

- Describe the mechanisms of nonspecific immunity.

- Discuss the nature of antigens and define *hapten* and *antigenic determinant site.*

- Discuss the origin and function of B and T lymphocytes and distinguish between humoral and cell-mediated immunity.

- Describe the structure and origin of antibodies and explain how antibodies promote the destruction of invading pathogens.

- Discuss the complement system and its functions.

- Describe the events that occur in a local inflammation.

- Describe the nature of the primary and secondary immune responses.

- Explain the clonal selection theory and how it relates to the process of active immunization.

- Explain how passive immunizations are performed and discuss the nature of monoclonal antibodies.

- Discuss the role of histocompatibility antigens in the function of T lymphocytes.

- Explain how interaction between T lymphocytes and macrophages leads to stimulation of both cell-mediated and humoral immunity.

- Discuss the nature of immunological tolerance and explain how it might be produced.

- Identify the cells and mechanisms involved in the immunological surveillance against cancer.

- Explain how stress and aging might result in increased susceptibility to cancer.

# Lymphatic System

*The lymphatic system, consisting of lymphatic vessels and lymph nodes, returns interstitial fluid to the bloodstream and helps to protect the body from diseases.*

The lymphatic system has three basic functions: (1) it transports interstitial (tissue) fluid, initially formed as a blood filtrate, back to the bloodstream; (2) it serves as the route by which absorbed fat from the small intestine is transported to the blood; and (3) its cells—called **lymphocytes** (produced in bone marrow, lymphatic organs, and other lymphatic tissues)—help to provide immunological defenses against disease-causing agents.

## Lymph and Lymph Capillaries

The smallest vessels of the lymphatic system are the **lymph capillaries** (see chapter 22, fig. 22.8). Lymph capillaries are microscopic closed-ended tubes that form vast networks in the intercellular spaces within most tissues. Within the villi of the small intestine, for example, lymph capillaries called

lacteal: L. *lacteus*, milk

lacteals (lak´te-alz) transport the products of fat absorption away from the GI tract. Because the walls of lymph capillaries are composed of endothelial cells with porous junctions, interstitial fluid, proteins, microorganisms, and absorbed fat (in the small intestine) can easily enter. Once the interstitial fluid enters the lymphatic capillaries, it is referred to as **lymph.**

## Lymph Ducts

From merging lymph capillaries, the lymph is carried into larger lymphatic vessels, or **lymph ducts.** The walls of lymph ducts are similar to those of veins in that they have the same three layers and contain valves to prevent the backflow of lymph. The pressure that pushes lymph through the lymph ducts comes from the massaging actions produced by skeletal muscle contractions, gravity, intestinal movements, and other body movements. The many valves keep lymph moving in one direction.

Interconnecting lymph ducts eventually empty into one of the two principal vessels: the **thoracic duct** and the **right lymphatic duct** (fig. 23.1). These ultimately drain the lymph

lymph: L. *lympha*, clear water

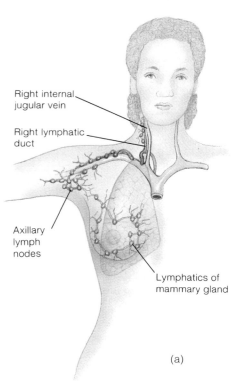

(a)

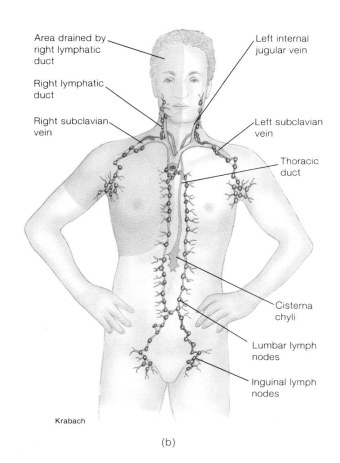

Krabach

(b)

**FIGURE 23.1**
Lymphatic vessels. (*a*) A magnified view of the upper right quadrant showing the lymph drainage of the right breast. (*b*) Major lymph drainage of the body.

into the left and right subclavian veins, respectively, so that it is returned to the circulatory system (fig. 23.2).

The larger thoracic duct drains lymph from the lower extremities, abdomen, left thoracic region, left upper extremity, and left side of the head and neck. The main trunk of this vessel ascends along the spinal column and drains into the left subclavian vein. In the abdominal area is a saclike enlargement of the thoracic duct called the **cisterna chyli** (*sis-ter´nă ki´le*). The shorter right lymphatic duct drains lymph from the right upper extremity, right thoracic region, and right side of the head and neck. The right lymphatic duct empties into the right subclavian vein near the right internal jugular vein.

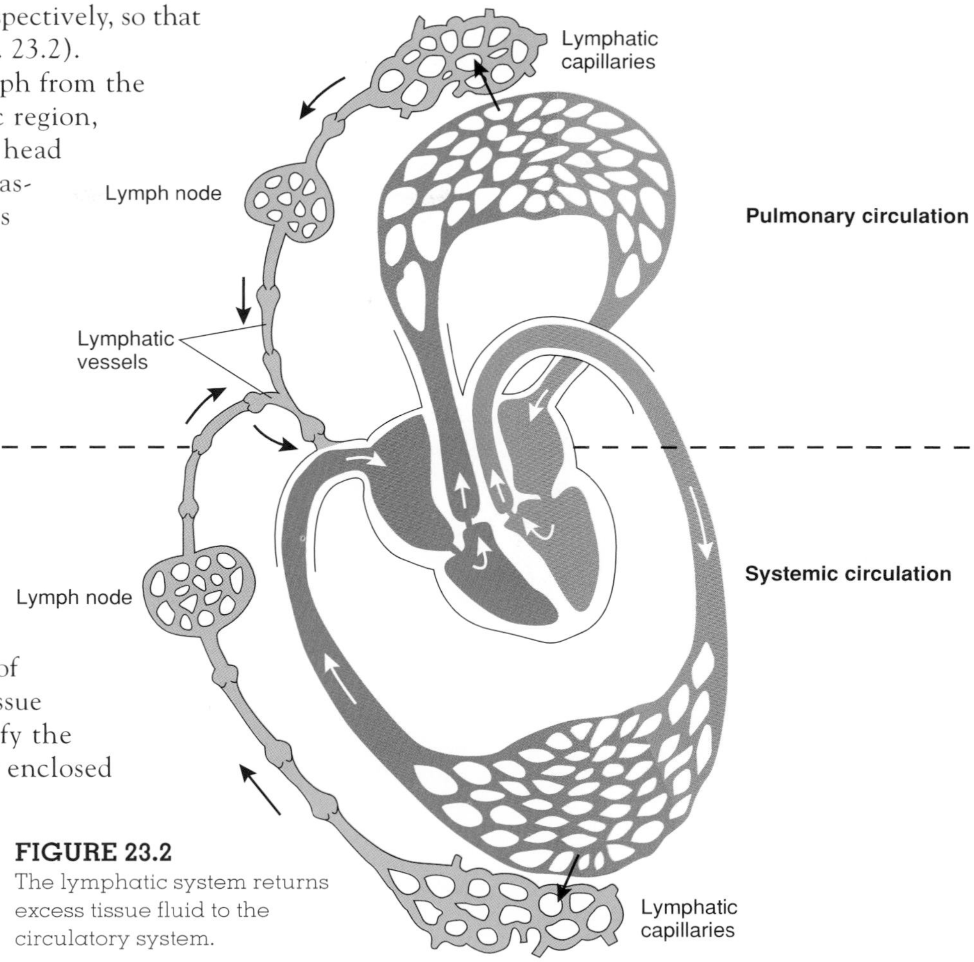

## Lymph Nodes

Lymph filters through the reticular tissue of lymph nodes (fig. 23.3). The reticular tissue contains phagocytic cells that help purify the fluid. A lymph node is a small oval body enclosed within a fibrous connective tissue *capsule*. Specialized connective tissue bands called *trabeculae* divide the node. Afferent

..........................................

cisterna chyli: L. *cisterna*, box; Gk. *chylos*, juice

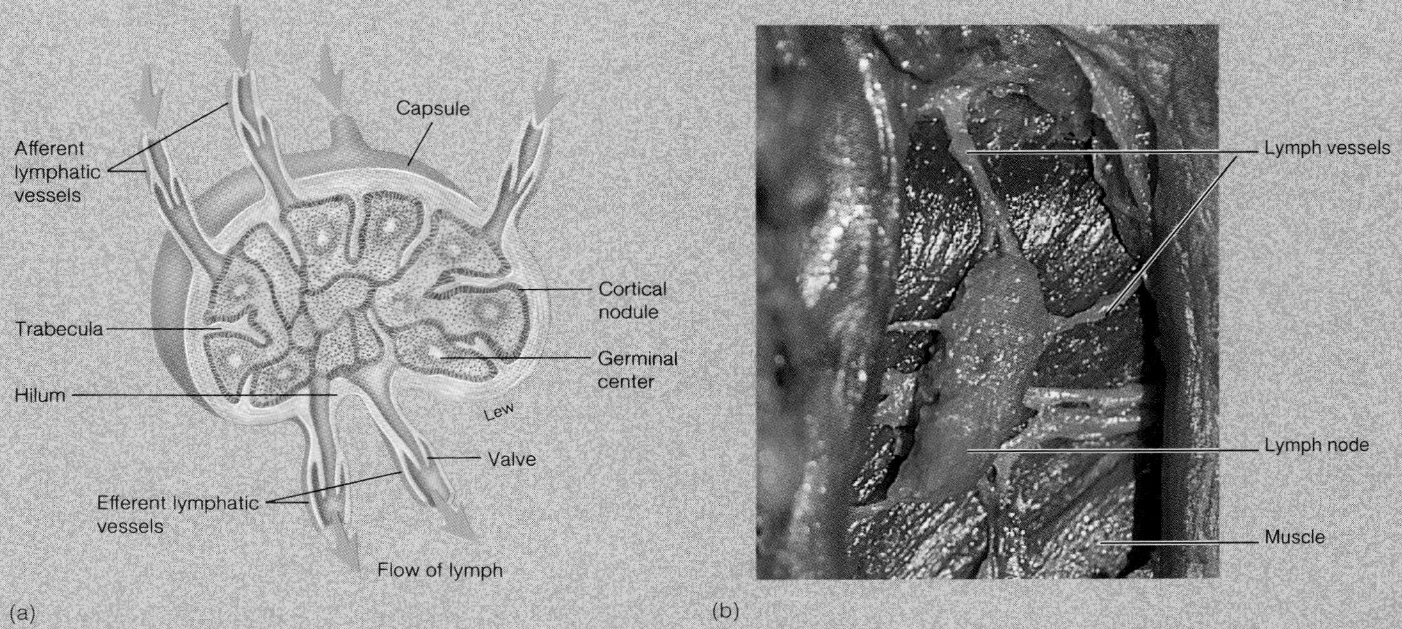

**FIGURE 23.2**
The lymphatic system returns excess tissue fluid to the circulatory system.

**FIGURE 23.3**
The structure of a lymph node. (a) A schematic diagram of a sectioned lymph node and associated vessels and (b) a photograph of a lymph node positioned near a blood vessel.

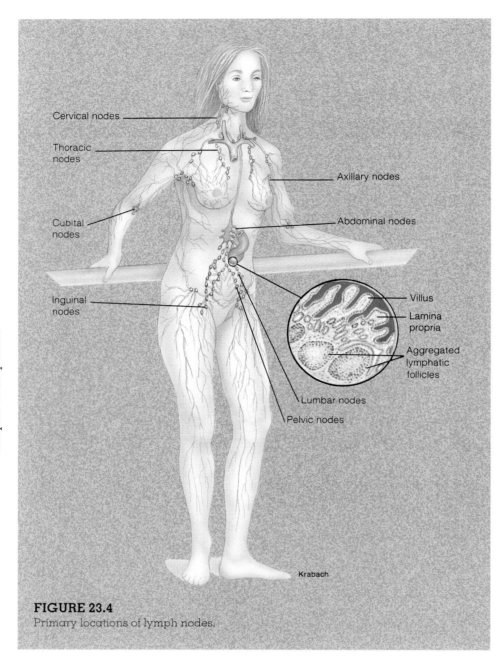

**FIGURE 23.4**
Primary locations of lymph nodes.

Labels in figure:
Cervical nodes
Thoracic nodes
Cubital nodes
Inguinal nodes
Axillary nodes
Abdominal nodes
Villus
Lamina propria
Aggregated lymphatic follicles
Lumbar nodes
Pelvic nodes
Krabach

**thoracic nodes** of the chest, and the **cervical nodes** of the neck. The wall of the small intestine contains numerous scattered lymphocytes and lymphatic nodules and larger aggregations of lymphatic tissue called **aggregated lymphatic follicles,** or **Peyer's patches.**

 Migrating cancer cells (metastases) are especially dangerous if they enter the lymphatic system, which can disperse them widely. On entering the lymph nodes, the cancer cells can multiply and establish secondary tumors in organs far removed from the site of the primary tumor.

## Lymphoid Organs

The *spleen* and the *thymus* are lymphoid organs. The **spleen** is located on the left side of the abdominal cavity, posterior and lateral to the stomach from which it is suspended (fig. 23.5). The spleen is not a vital organ in an adult, but it does assist other body organs in producing lymphocytes, filtering the blood, and destroying old erythrocytes. In an infant, it is an important site for the production of erythrocytes. In an adult, the spleen contains *red pulp*, which serves to destroy old red blood cells, and *white pulp*, which contains germinal centers for the production of lymphocytes.

The **thymus** is located in the anterior thorax, deep to the manubrium of the sternum. Because it regresses in size during puberty, it is much larger in a fetus and child than in an adult. The thymus plays a key role in the immune system, as will be described in a later section.

Table 23.1 summarizes the lymphoid organs of the body.

lymphatic vessels carry lymph into the node, where it is circulated through sinuses in the *cortical tissue*. Lymph leaves the node through the efferent lymphatic vessel, which emerges from the *hilum*—the depression on the concave side. **Germinal centers** within the node are sites of lymphocyte production and are important in the development of an immune response, as we will discuss later in this chapter.

Lymph nodes usually occur in clusters in specific regions of the body (fig. 23.4). Some of the principal groups of lymph nodes are the **popliteal** and **inguinal nodes** of the lower extremity, the **lumbar nodes** of the pelvic region, the **cubital** and **axillary nodes** of the upper extremity, the

 The tonsils, of which there are three pairs, are lymphatic organs of the pharyngeal region. The function of the tonsils is to combat infection of the ear, nose, and throat regions. Because of the persistent infections that some children suffer, however, the tonsils may become so overrun with infections that they

Peyer's patches: from Johann K. Peyer, Swiss anatomist, 1653–1712
spleen: L. *splen*, low spirits (thought to cause melancholy)
thymus: Gk. *thymos*, thyme (compared to the flower of this plant by Galen)

actually become a source of infections that spread to other parts of the body. A *tonsillectomy* may then have to be performed. This operation is not as common as it used to be because of the availability of powerful antibiotics and because the functional value of the tonsils is now appreciated to a greater extent.

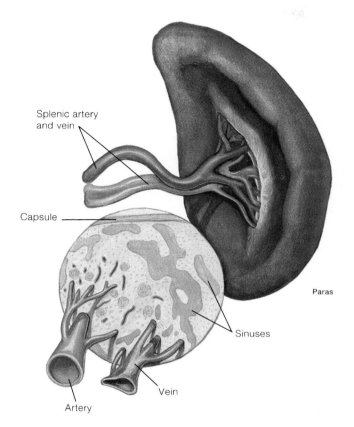

**FIGURE 23.5**
The structure of the spleen.

Labels: Splenic artery and vein; Capsule; Paras; Sinuses; Vein; Artery

## Table 23.1 Lymphoid organs

| Organ | Location | Function |
|---|---|---|
| Lymph nodes | In clusters or chains along the paths of larger lymphatic vessels | Sites of lymphocyte production; house T lymphocytes and B lymphocytes that are responsible for immunity; phagocytes filter foreign particles and cellular debris from lymph |
| Spleen | In upper left portion of abdominal cavity, beneath the diaphragm and suspended from the stomach | Serves as blood reservoir; phagocytes filter foreign particles, cellular debris, and worn erythrocytes from the blood; houses lymphocytes |
| Thymus | Within the mediastinum, behind the manubrium | Important site of immunity in a child; houses lymphocytes; changes undifferentiated lymphocytes into T lymphocytes |

# Defense Mechanisms

*Nonspecific immune protection is provided by such mechanisms as phagocytosis, fever, and the release of interferons. Specific immunity involves the functions of lymphocytes and is directed at specific molecules, or parts of molecules, known as antigens.*

The immune system includes all of the structures and processes that provide a defense against potential pathogens. These defenses can be grouped into *nonspecific* and *specific* mechanisms.

**Nonspecific,** or **innate, defense mechanisms** are inherited as part of the structure of each organism. Epithelial membranes that cover the body surfaces, for example, restrict infection by most pathogens. The strong acidity of gastric juice (pH 1–2) also helps to kill many microorganisms before they can invade the body. These external defenses are backed by internal defenses, such as phagocytosis, which function in both a specific and nonspecific manner (table 23.2).

Each individual can acquire the ability to defend against specific pathogens (disease-causing agents) by a prior exposure to those pathogens. This **specific,** or **acquired, immune response** is a function of lymphocytes. Internal specific and nonspecific defense mechanisms function together to combat infection, with lymphocytes interacting in a coordinated effort with phagocytic cells.

## Nonspecific Immunity

Invading pathogens, such as bacteria, that have crossed epithelial barriers enter connective tissues. These invaders—or chemicals, called *toxins*, released from them—may enter blood or lymphatic capillaries and be carried to other areas of the body. To counter the invasion and spread of infection, nonspecific immunological defenses are first employed. If these defenses are not sufficient to destroy the pathogens, lymphocytes may be recruited, and their specific actions used to reinforce the nonspecific immune defenses.

**Phagocytosis** The three major groups of phagocytic cells are (1) **neutrophils;** (2) the cells of the *mononuclear phagocyte system,* including **monocytes** in the blood and **macrophages** (derived from monocytes) in the connective tissues;

pathogen: Gk. *pathema,* suffering; *gen,* to produce

## Table 23.2 Structures and defense mechanisms of nonspecific immunity

| | Structure | Mechanisms |
|---|---|---|
| External | Skin | Physical barrier to penetration by pathogens; secretions contain lysozyme (enzyme that destroys bacteria) |
| | GI tract | High acidity of stomach; protection by normal bacterial population of colon |
| | Respiratory tract | Secretion of mucus; movement of mucus by cilia; alveolar macrophages |
| | Urinary tract | Acidity of urine |
| | Female reproductive tract | Vaginal lactic acid |
| Internal | Phagocytic cells | Ingest and destroy bacteria, cellular debris, denatured proteins, and toxins |
| | Interferons | Inhibit replication of viruses |
| | Complement proteins | Promote destruction of bacteria and other effects of inflammation |
| | Endogenous pyrogen | Secreted by leukocytes and other cells; produces fever |

## Table 23.3 Phagocytic cells and their locations

| Phagocyte | Location |
|---|---|
| Neutrophils | Blood and all tissues |
| Monocytes | Blood and all tissues |
| Tissue macrophages (histiocytes) | All tissues (including spleen, lymph nodes, bone marrow) |
| Kupffer cells | Liver |
| Alveolar macrophages | Lungs |
| Microglia | Central nervous system |

blood flows through the liver and spleen, foreign chemicals and debris are removed by phagocytosis and chemically inactivated within the phagocytic cells. Invading pathogens are very effectively removed in this manner, so that after a few passes through the liver and spleen, blood is usually sterile. Fixed phagocytes in lymph nodes similarly help to remove foreign particles from the lymph.

Connective tissues contain a resident population of all leukocyte types. Neutrophils and monocytes in particular can be highly mobile within connective tissues as they scavenge for invaders and cellular debris. These leukocytes are recruited to the site of an infection by a process known as **chemotaxis** (ke″mo-tak′sis)—movement toward chemical attractants. Neutrophils are the first to arrive at the site of an infection; monocytes arrive later and can be transformed into macrophages as the battle progresses.

If the infection is sufficiently large, new phagocytic cells from the blood may join those already in the connective tissue. These new neutrophils and monocytes are able to squeeze through the tiny gaps between adjacent endothelial cells in the capillary wall and enter the connective tissues. This process, called **diapedesis** (di″ă-pĕ-de′sis), is illustrated in figure 23.6.

Phagocytic cells engulf particles in a manner similar to the way an amoeba eats. The particle becomes surrounded by cytoplasmic extensions called pseudopods, which ultimately fuse together. The particle thus becomes surrounded by a membrane derived from the cell membrane (fig. 23.7) and is contained within an organelle analogous to a food vacuole in an amoeba. This vacuole then fuses with lysosomes (organelles that contain digestive enzymes), so that the ingested particle and the digestive enzymes remain separated from the cytoplasm by a continuous membrane. Often, however, lysosomal

and (3) **organ-specific phagocytes** in the liver, spleen, lymph nodes, lungs, and brain (table 23.3).

The *Kupffer* (koop′fer) *cells* in the liver, together with phagocytic cells in the spleen and lymph nodes, are **fixed phagocytes.** This term refers to the fact that these cells are immobile ("fixed") in the channels within these organs. As

............................................
Kupffer cell: from Karl Wilhelm von Kupffer, Bavarian anatomist, 1829–1902

............................................
diapedesis: Gk. *dia,* through; *pedesis,* a leaping

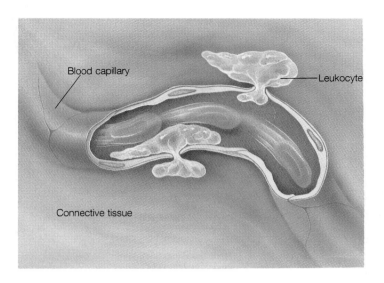

**FIGURE 23.6**
Diapedesis. White blood cells squeeze through openings between capillary endothelial cells to enter underlying connective tissues.

enzymes are released before the food vacuole has completely formed. When this occurs, free lysosomal enzymes may be released into the infected area and contribute to inflammation.

**Fever**    Fever may be a component of the nonspecific defense system. Body temperature is regulated by the hypothalamus, which contains a thermoregulatory control center (a "thermostat") that coordinates skeletal muscle shivering and the activity of the sympathoadrenal system to maintain body temperature at about 37° C. This thermostat is reset upward in response to a chemical called **endogenous pyrogen,** secreted by leukocytes. Endogenous pyrogen secretion is stimulated by a chemical called *endotoxin*, which is released by certain bacteria.

Although high fevers are definitely dangerous, many believe that a mild to moderate fever may be a beneficial response that aids recovery from bacterial infections. There is some evidence to support this view, but the

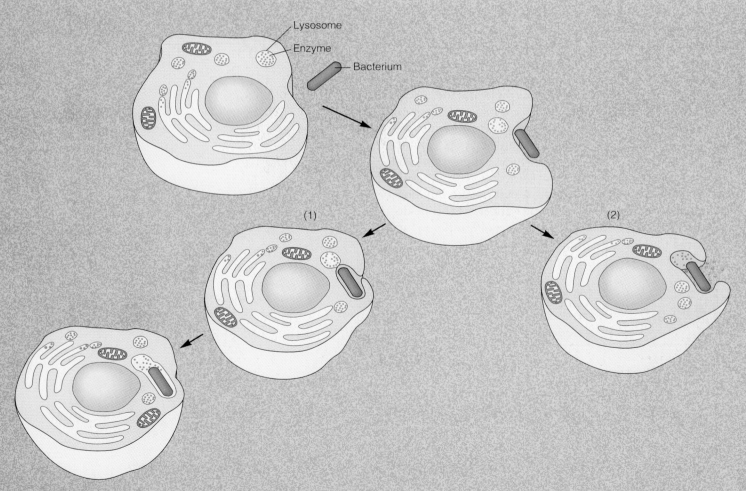

**FIGURE 23.7**
Phagocytosis by a neutrophil or macrophage. A phagocytic cell extends its pseudopods around the object to be engulfed, such as a bacterium. (Blue dots represent lysosomal enzymes.) (*1*) If the pseudopods fuse to form a complete food vacuole, lysosomal enzymes are restricted to the organelle formed by the lysosome and food vacuole. (*2*) If the lysosome fuses with the vacuole before fusion of the pseudopods is complete, lysosomal enzymes are released into the infected area of tissue.

mechanisms involved are not clearly understood. One theory is that elevated body temperature may interfere with the uptake of iron by some bacteria.

**Interferons** It was discovered in 1957 that cells infected with a virus produced polypeptides that interfered with the ability of a second, unrelated strain of virus to infect other cells in the same culture. These **interferons** (in´´ter-fēr´onz), as they were called, thus produced a nonspecific, short-acting resistance to viral infection. This discovery generated a great deal of excitement, but further research in this area was hindered by the fact that human interferons could be obtained only in very small quantities; moreover, animal interferons were shown to have little effect in humans. In 1980, however, a technique called *genetic recombination* made it possible to introduce human interferon genes into bacteria, enabling the bacteria to act as interferon factories.

There are three major categories of interferons: *alpha, beta,* and *gamma interferons.* Almost all cells in the body make alpha and beta interferon. These polypeptides act as messengers that protect other cells in the vicinity from viral infection. The viruses are still able to penetrate these other cells, but the ability of the viruses to replicate and assemble new virus particles is inhibited. Viral infection, replication, and dispersal are shown in figure 23.8, using the virus that causes AIDS (discussed later) as an example. Gamma interferon is only produced by particular lymphocytes and a related type of cell called *natural killer cells.* The secretion of gamma interferon by these cells is part of the immunological defense against infection and cancer, as will be described later. Some of the proposed effects of interferons are summarized in table 23.4.

**FIGURE 23.8**

The life cycle of the human immunodeficiency virus (HIV). This virus, like others of its family, contains RNA instead of DNA. Once inside the host cell, the viral RNA is transcribed by reverse transcriptase into complementary DNA (cDNA). The genes in the cDNA then direct the synthesis of new virus particles.

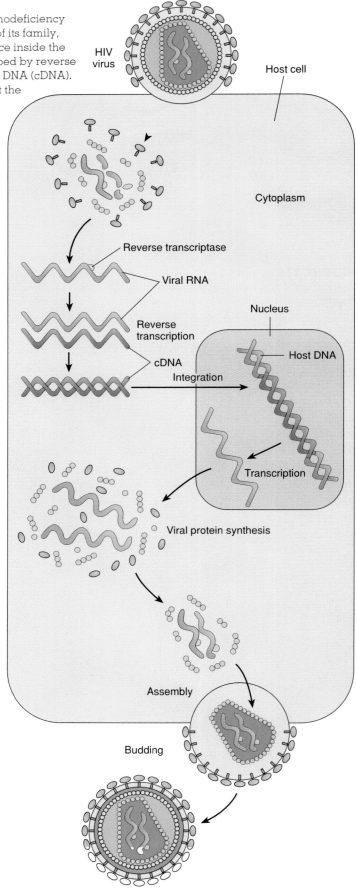

## Table 23.4 Proposed effects of interferons

| Stimulation | Inhibition |
| --- | --- |
| Macrophage phagocytosis | Cell division |
| Activity of cytotoxic (killer) T lymphocytes | Tumor growth |
| Activity of natural killer cells | Maturation of adipose cells |
| Production of antibodies | Maturation of erythrocytes |

The Food and Drug Administration (FDA) has currently approved the use of interferons to treat a number of diseases. Among these are the use of alpha interferon to treat chronic hepatitis A and B, hairy-cell lukemia, virally induced genital warts, and Karposi's sarcoma. The FDA has also approved the use of beta interferon to treat relapsing-remitting multiple sclerosis and the use of gamma interferon to treat chronic granulomatous disease. Interferon treatment of numerous forms of cancer is currently in various stages of clinical trials.

## Specific Immunity

In 1890, a German bacteriologist, Emil Adolf von Behring, demonstrated that a guinea pig that had been previously injected with a sublethal dose of diphtheria toxin could survive subsequent injections of otherwise lethal doses of that toxin. Further, von Behring showed that this immunity could be transferred to a second, nonexposed animal by injections of serum from the immunized guinea pig. He concluded that the immunized animal had chemicals in its serum—which he called **antibodies**—that were responsible for the immunity. He also showed that these antibodies conferred immunity only to subsequent diphtheria infections; the antibodies were *specific* in their actions. It was later learned that antibodies are proteins produced by a particular type of lymphocyte.

**Antigens** Antigens are molecules that stimulate the production of antibodies and combine with these specific antibodies. Antigens are large molecules (such as proteins) with a molecular weight greater than about 10,000, and they are foreign to the blood and other body fluids (although there are exceptions to both descriptions). The ability of a molecule to function as an antigen depends not only on its size but also on the complexity of its structure. Proteins are more antigenic than the simpler structured polysaccharides. Plastics used in artificial implants are composed of large molecules, but they are not very antigenic because of their simple, repeating structures.

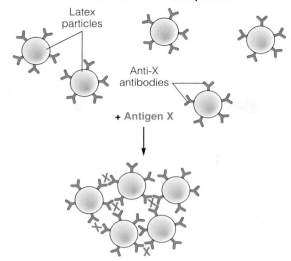

**Antibodies attached to latex particles**

Latex particles

Anti-X antibodies

+ Antigen X

**Agglutination (clumping) of latex particles**

**FIGURE 23.9**

Immunoassay using the agglutination technique. Antibodies against a particular antigen are adsorbed to latex particles. When these are mixed with a solution that contains the appropriate antigen, the formation of the antigen-antibody complexes produces clumping (agglutination) that can be seen with the unaided eye.

A large, complex foreign molecule can have a number of different **antigenic determinant sites,** which are areas of the molecule that stimulate production of and combine with different antibodies. Most naturally occurring antigens have many antigenic determinant sites and stimulate the production of different antibodies with specificities for these sites.

**Haptens** Many small organic molecules are not antigenic by themselves but can become antigens if they bind to proteins (and thus become antigenic determinant sites on the proteins). This discovery was made by Karl Landsteiner, who is also credited with the discovery of the ABO blood groups (chapter 20). By binding these small molecules—which Landsteiner called **haptens**—to proteins in the laboratory, new antigens can be created for research or diagnostic purposes. The binding of foreign haptens to a person's own proteins can also occur in the body; by this means, derivatives of penicillin, for example, that would otherwise be harmless can produce fatal allergic reactions in susceptible people.

**Immunoassays** When the antigen or antibody is attached to the surface of a cell or to particles of latex rubber (in commercial diagnostic tests), the antigen-antibody reaction becomes visible because the particles *agglutinate* (clump) as a result of antigen-antibody bonding (fig. 23.9). These agglutinated particles can be used to assay a variety of antigens, and tests that utilize this procedure are called **immunoassays** (im´´yŭ-no-as´āz). Blood typing (chapter 20) and modern pregnancy tests are examples of such immunoassays. A newly developed latex agglutination test for detecting AIDS using fingertip blood may also soon be available.

## Lymphocytes

Leukocytes, erythrocytes, and blood platelets are all ultimately derived from ("stem from") unspecialized cells in the bone marrow. These *stem cells* produce the specialized blood cells, and they replace themselves by cell division so that the stem cell population is not exhausted.

Emil Adolph von Behring: German bacteriologist, 1854–1917

Karl Landsteiner: Austrian-born American pathologist and immunologist, 1868–1943

## Table 23.5 Comparison of B and T lymphocytes

| Characteristic | B lymphocyte | T lymphocyte |
|---|---|---|
| Site where processed | Bone marrow | Thymus |
| Type of immunity | Humoral (secretes antibodies) | Cell-mediated |
| Subpopulations | Memory cells and plasma cells | Cytotoxic (killer) T lymphocytes, helper cells, suppressor cells |
| Presence of surface antibodies | Yes—IgM or IgD | Not detectable |
| Receptors for antigens | Present—are surface antibodies | Present—are related to immunoglobulins |
| Life span | Short | Long |
| Tissue distribution | High in spleen, low in blood | High in blood and lymph |
| Percent of blood lymphocytes | 10%–15% | 75%–80% |
| Transformed by antigens to | Plasma cells | Small lymphocytes |
| Secretory product | Antibodies | Lymphokines |
| Immunity to viral infections | Enteroviruses, poliomyelitis | Most others |
| Immunity to bacterial infections | *Streptococcus, staphylococcus,* many others | Tuberculosis, leprosy |
| Immunity to fungal infections | None known | Many |
| Immunity to parasitic infections | Trypanosomiasis, maybe to malaria | Most others |

Lymphocytes produced in this manner seed the thymus, spleen, and lymph nodes, producing self-replicating lymphocyte colonies in these organs.

The lymphocytes that become seeded in the thymus become **T lymphocytes.** These cells have surface characteristics and an immunological function that differ from those of other lymphocytes. The thymus, in turn, seeds other organs. About 65% to 85% of the lymphocytes in blood and most of the lymphocytes in lymph nodes are T lymphocytes. These lymphocytes, therefore, came from or had an ancestor that came from the thymus gland.

Most of the lymphocytes that are not T lymphocytes are called **B lymphocytes.** The letter *B* has its origin in immunological research performed in chickens. Chickens have an organ called the *bursa of Fabricius* that processes B lymphocytes. Since mammals do not have a bursa, the *B* is often translated as the "bursa equivalent" for humans and other mammals. It is currently believed that the B lymphocytes in mammals are processed in the bone marrow, which conveniently also begins with the letter *B*.

Both B and T lymphocytes function in specific immunity. The B lymphocytes combat bacterial and some viral infections by secreting antibodies into the blood and lymph. Because blood and lymph are body fluids (humors), the B

.............................................
bursa of Fabricius: from Hieronymus Fabricius, Italian anatomist, 1533–1619

lymphocytes are said to provide **humoral immunity,** although the term **antibody-mediated immunity** is also used. T lymphocytes attack host cells that have become infected with viruses or fungi, transplanted human cells, or cancerous cells. The T lymphocytes do not secrete antibodies; they must come in close proximity or have actual physical contact with the victim cell in order to destroy it. T lymphocytes are therefore said to provide **cell-mediated immunity** (table 23.5).

## Functions of B Lymphocytes

*B lymphocytes secrete antibodies that can bond in a specific fashion with antigens. Binding of these secreted antibodies to antigens stimulates a cascade of reactions whereby a system of proteins in the plasma called complement is activated. Some of these activated complement proteins kill the cells containing the antigen; others promote phagocytosis and other activity, resulting in a more effective defense against pathogens.*

Exposure of a B lymphocyte to the appropriate antigen results in cell growth followed by many cell divisions. Some of the progeny become memory cells, which are indistinguishable from the original cell; others are transformed into **plasma cells** (fig. 23.10). Plasma cells are protein factories that produce about 2000 antibody proteins per second in their brief life span of 5 to 7 days.

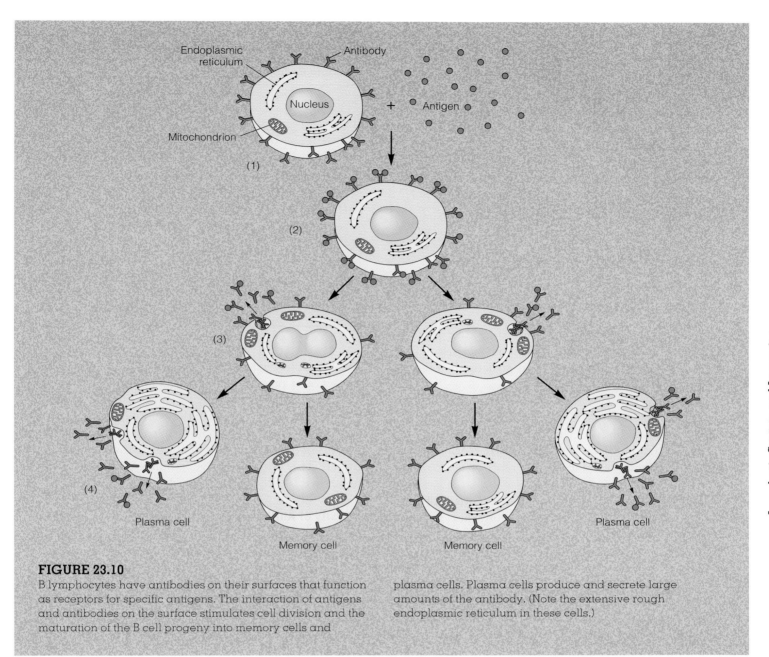

**FIGURE 23.10**

B lymphocytes have antibodies on their surfaces that function as receptors for specific antigens. The interaction of antigens and antibodies on the surface stimulates cell division and the maturation of the B cell progeny into memory cells and plasma cells. Plasma cells produce and secrete large amounts of the antibody. (Note the extensive rough endoplasmic reticulum in these cells.)

The antibodies that are produced by plasma cells when B lymphocytes are exposed to a particular antigen react specifically with that antigen. Such antigens may be isolated molecules, or they may be molecules at the surface of an invading foreign cell. The specific binding of antibodies to antigens serves to identify the enemy and to activate defense mechanisms that lead to the invader's destruction.

## Antibodies

Antibody proteins are also known as **immunoglobulins** (*im″yŭ-no-glob′yŭ-linz*), abbreviated *Ig*. They are mostly found in the gamma globulin class of plasma proteins, as identified by a technique called *electrophoresis* (*ĕ-lek″tro-fŏ-re′sis*). With this technique, classes of plasma proteins are separated by their movement in an electric field (fig. 23.11). The five distinct bands of proteins that appear are albumin, alpha-1 globulin, alpha-2 globulin, beta globulin, and gamma globulin.

The gamma globulin band is wide and diffuse because it represents a heterogeneous class of molecules. Since antibodies are specific in their actions, it follows that different types of antibodies should have different structures. An antibody against smallpox, for example, does not confer

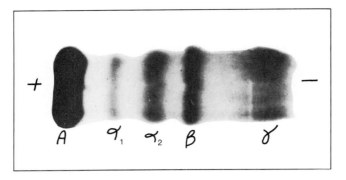

**FIGURE 23.11**
The separation of serum protein by electrophoresis into albumin (A), alpha-1 globulin ($\alpha_1$), alpha-2 globulin ($\alpha_2$), beta globulin ($\beta$), and gamma globulin ($\gamma$).

| Table 23.6 | The immunoglobulins |
|---|---|
| **Immunoglobulin** | **Functions** |
| IgG | Main form of antibodies in circulation; production increased after immunization |
| IgA | Main antibody type in external secretions, such as saliva and mother's milk |
| IgE | Responsible for allergic symptoms in immediate hypersensitivity reactions |
| IgM | Function as antigen receptors on lymphocyte surface prior to immunization; secreted during primary response |
| IgD | Function as antigen receptors on lymphocyte surface prior to immunization; other functions unknown |

immunity to poliomyelitis and, therefore, must have a slightly different structure than an antibody against polio. Despite these differences, antibodies are structurally related and form only a few subclasses.

There are five subclasses of immunoglobulins: **IgG, IgA, IgM, IgD,** and **IgE.** Most of the antibodies in serum are in the IgG subclass, whereas most of the antibodies in external secretions (saliva and milk) are IgA (table 23.6). Antibodies in the IgE subclass are involved in allergic reactions.

**Antibody Structure** All antibody molecules consist of four interconnected polypeptide chains. Two long, *heavy chains* (the *H chains*) are joined to two short, *light chains* (*L chains*). Research has shown that these four chains are arranged in the form of a **Y**. The stalk of the **Y** has been called the "crystallizable fragment" (abbreviated $F_c$), whereas the top of the **Y** is known as the "antigen-binding fragment" ($F_{ab}$). This structure is shown in figure 23.12.

The amino acid sequences of some antibodies have been determined through the analysis of antibodies sampled from people with multiple myelomas. These lymphocyte tumors arise from the division of a single B lymphocyte, which forms a population of genetically identical cells (a clone) that secretes identical antibodies. Clones and the antibodies they secrete differ, however, from one patient to another. Analyses have shown that the $F_c$ regions of different antibodies are the same (are constant), whereas the $F_{ab}$ regions are variable. Variability of the antigen-binding regions is required for the specificity of antibodies for antigens. Thus, it is the $F_{ab}$ region of an antibody that provides a specific site for bonding with a particular antigen (fig. 23.13).

B lymphocytes have antibodies on their cell membrane that serve as **receptors for antigens.** The combination of antigens with these antibody receptors stimulates the B cell to divide and produce more of these antibodies, which are secreted. Exposure to a given antigen thus results in increased amounts of the specific type of antibody that can attack that antigen. This provides active immunity, which we will discuss shortly.

**Diversity of Antibodies** It is estimated that there are about 100 million trillion ($10^{20}$) antibody molecules in each individual, representing a few million different specificities for different antigens. Considering that antibodies to particular antigens can cross-react with closely related antigens to some extent, this tremendous antibody diversity usually ensures that there will be some antibodies that can combine with almost any antigen a person might encounter. These observations evoke a question that has long fascinated scientists: How can a few million different antibodies be produced? A person cannot possibly inherit a correspondingly large number of genes devoted to antibody production.

Two mechanisms have been proposed to explain antibody diversity. First, since different combinations of heavy and light chains can produce different antibody specificities, a person does not have to inherit a million different genes to code for a million different antibodies. If a few hundred genes code for different H chains and a few hundred code for different L chains, various combinations of these polypeptide chains could produce millions of different antibodies. Second, the diversity of antibodies could increase

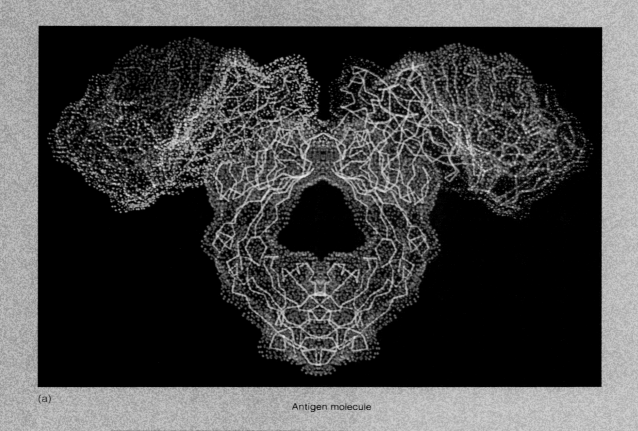

(a)

Antigen molecule

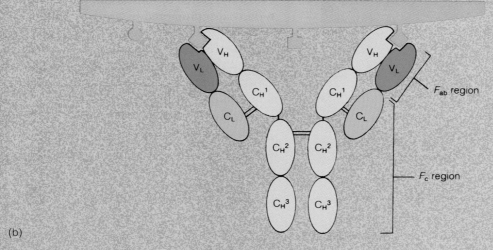

$V_H$

$V_L$

$C_H^1$

$C_L$

$C_H^2$ $C_H^2$

$C_H^3$ $C_H^3$

$V_H$

$V_L$

$C_H^1$

$C_L$

$F_{ab}$ region

$F_c$ region

(b)

### FIGURE 23.12

Antibodies are composed of four polypeptide chains—two are heavy (H) and two are light (L). (a) A computer-generated model of antibody structure. (b) A simplified diagram showing the constant and variable regions. (The variable regions are abbreviated V and the constant regions are abbreviated C.) Antigens combine with the variable regions. Each antibody molecule is divided into an $F_{ab}$ (antigen-binding) fragment and an $F_c$ (crystallizable) fragment.

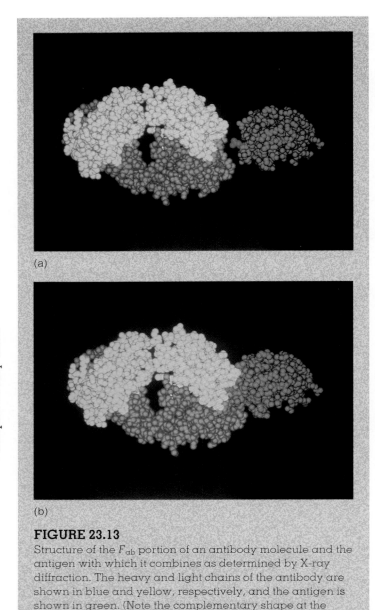

(a)

(b)

**FIGURE 23.13**
Structure of the $F_{ab}$ portion of an antibody molecule and the antigen with which it combines as determined by X-ray diffraction. The heavy and light chains of the antibody are shown in blue and yellow, respectively, and the antigen is shown in green. (Note the complementary shape at the region where the two are joined together in (b).)

during development if, when some lymphocytes divided, the progeny received antibody genes that had been slightly altered by mutations. Such mutations are called *somatic mutations* because they occur in body cells rather than in sperm or ova. Antibody diversity would thus increase as the lymphocyte population increased.

## The Complement System

The combination of antibodies with antigens does not itself cause destruction of the antigens or the pathogenic organisms that contain these antigens. Antibodies, rather, serve to identify the targets for immunological attack and to activate nonspecific immune processes that destroy the

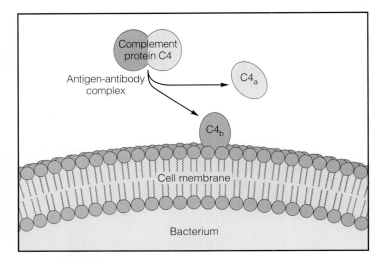

**FIGURE 23.14**
The fixation of complement proteins. The formation of an antibody-antigen complex causes complement protein C4 to be split into two subunits—$C4_a$ and $C4_b$. The $C4_b$ subunit attaches (is fixed) to the membrane of the cell to be destroyed (such as a bacterium). This event triggers the activation of other complement proteins, some of which attach (are fixed) to the $C4_b$ on the membrane surface.

invader. Bacteria that are buttered with antibodies, for example, are better targets for phagocytosis by neutrophils and macrophages. The ability of antibodies to stimulate phagocytosis is termed **opsonization.** Immune destruction of bacteria is also promoted by antibody-induced activation of a system of serum proteins known as *complement*.

In the early part of the twentieth century, it was learned that rabbit antibodies to sheep red blood cell antigens could not lyse (destroy) these cells unless certain protein components of serum were present. These proteins, called **complement,** constitute a nonspecific defense system that is activated by the bonding of antibodies to antigens and by this means is directed against specific invaders that have been identified by antibodies.

The complement proteins are designated C1 (which has three protein components) through C9. These proteins are normally present in plasma and other body fluids in an inactive form and are activated by the attachment of antibodies to antigens. In terms of their functions, the complement proteins can be subdivided into three components: (1) recognition (C1); (2) activation (C4, C2, and C3, in that order); and (3) attack (C5–C9). The attack phase consists of **complement fixation,** in which complement proteins attach to the cell membrane and destroy the victim cell.

Antibodies of the IgG and IgM subclasses attach to antigens on the invading cell's membrane, bind to C1, and by this means activate its enzyme activity. Activated C1 catalyzes the hydrolysis of C4 into two fragments (fig. 23.14), designated $C4_a$ and $C4_b$. The $C4_b$ fragment binds to the cell membrane (is "fixed") and becomes an active enzyme that splits C2 into two fragments, $C2_a$ and $C2_b$. The $C2_a$ becomes attached to $C4_b$ and cleaves C3 into $C3_a$ and $C3_b$. Fragment $C3_b$ becomes attached to the growing complex of complement proteins on the cell membrane. The $C3_b$ converts C5 to $C5_a$ and $C5_b$. The $C5_b$ and, eventually, C6 through C9 become fixed to the cell membrane.

Complement proteins C5 through C9 create large pores in the membrane (fig. 23.15). These pores

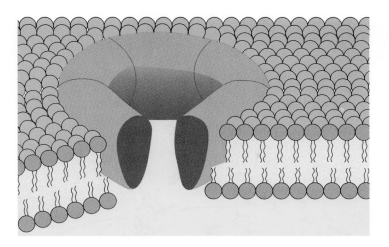

**FIGURE 23.15**

Complement proteins C5 through C9 (illustrated as a doughnut-shaped ring) puncture the membrane of the cell to which they are attached (fixed). This aids destruction of the cell.

permit the osmotic influx of water, so that the victim cell swells and bursts. Note that the complement proteins, not the antibodies directly, kill the cell; antibodies serve only as activators of this process. Other molecules can also activate the complement system in an alternate nonspecific pathway that bypasses the early phases of the specific pathway described here.

$C3_a$ and other complement fragments that are liberated into the surrounding fluid rather than becoming fixed trigger a number of events. These events include (1) *chemotaxis*—the liberated complement fragments attract phagocytic cells to the site of complement activation; (2) *opsonization*—phagocytic cells have receptors for $C3_b$, so that this fragment may form bridges be-

tween the phagocyte and the victim cell, thus facilitating phagocytosis; and (3) *stimulation of the release of histamine* from mast cells (a connective tissue cell type) and basophils by fragments $C3_a$ and $C5_a$. As a result of histamine release, blood flow to the infected area increases due to vasodilation and increased capillary permeability. The latter effect can result in the leakage of plasma proteins into the surrounding tissue fluid, producing local edema.

## Local Inflammation

Aspects of the nonspecific and specific immune responses and their interactions are well illustrated by the events that occur when bacteria enter a break in the skin and produce a local inflammation (table 23.7). The inflammatory reaction is initiated by the nonspecific mechanisms of phagocytosis and complement activation. Activated complement enhances this nonspecific response by attracting new phagocytes to the area and by stimulating their activity.

After some time, B lymphocytes are stimulated to produce antibodies against specific antigens that are part of the invading bacteria. Attachment of these antibodies to antigens in the bacteria greatly amplifies the previously

## Table 23.7  Summary of events in a local inflammation

| Category | Events |
|---|---|
| Nonspecific immunity | Bacteria enter through break in anatomic barrier of skin. |
| | Resident phagocytic cells—neutrophils and macrophages—engulf bacteria. |
| | Nonspecific activation of complement protein occurs. |
| Specific immunity | B lymphocytes are stimulated to produce specific antibodies. |
| | Phagocytosis is enhanced by antibodies attached to bacterial surface antigens. |
| | Specific activation of complement proteins occurs, which stimulates phagocytosis, chemotaxis of new phagocytes to the infected area, and secretion of histamine from tissue mast cells. |
| | Diapedesis allows new phagocytic leukocytes (neutrophils and monocytes) to invade the infected area. |
| | Vasodilation and increased capillary permeability (as a result of histamine secretion) produce redness and edema. |

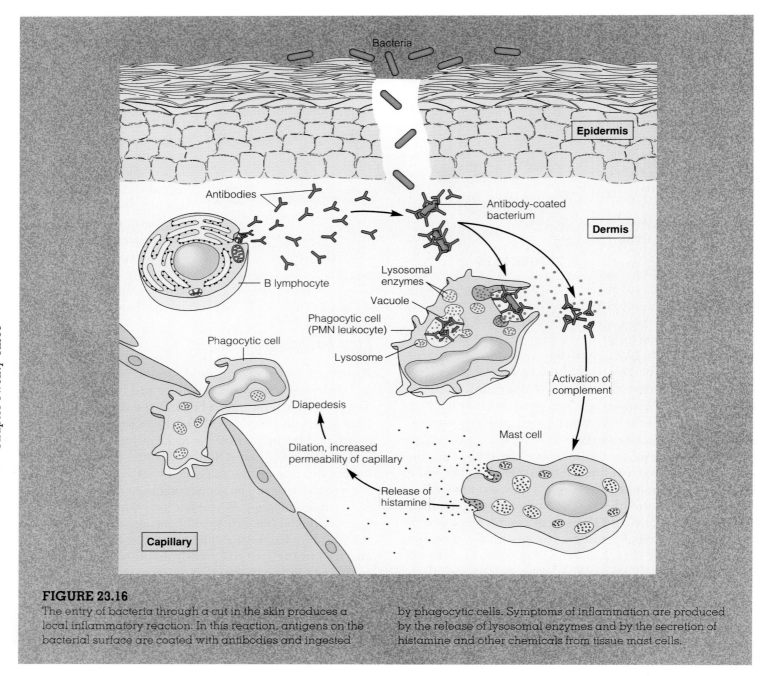

Bacteria

**Epidermis**

Antibodies

**Dermis**

Antibody-coated
bacterium

B lymphocyte

Lysosomal
enzymes

Vacuole

Phagocytic cell
(PMN leukocyte)

Lysosome

Activation of
complement

Phagocytic cell

Mast cell

Diapedesis

Dilation, increased
permeability of capillary

Release of
histamine

**Capillary**

**FIGURE 23.16**
The entry of bacteria through a cut in the skin produces a
local inflammatory reaction. In this reaction, antigens on the
bacterial surface are coated with antibodies and ingested
by phagocytic cells. Symptoms of inflammation are produced
by the release of lysosomal enzymes and by the secretion of
histamine and other chemicals from tissue mast cells.

nonspecific response. This occurs because of greater activation of complement, which directly destroys the bacteria, and which—together with the antibodies themselves—promotes the phagocytic activity of neutrophils, macrophages, and monocytes (fig. 23.16).

As inflammation progresses, the release of lysosomal enzymes from macrophages causes the destruction of leukocytes and other tissue cells. These effects, together with those produced by histamine and other chemicals released from mast cells, produce the characteristic symptoms of a local inflammation: *redness and warmth* (due to vasodilation), *swelling* (edema), and *pus* (the accumulation of dead leukocytes). If

the infection continues, the release of endogenous pyrogen from leukocytes and macrophages may produce a fever.

# Active and Passive Immunity

*When a person is first exposed to a pathogen, the immune response may be insufficient to combat the disease. However, lymphocytes that have specificity for the antigens encountered are stimulated to divide many times and produce a clone. This is active immunity and can protect an individual from getting a disease upon subsequent exposures.*

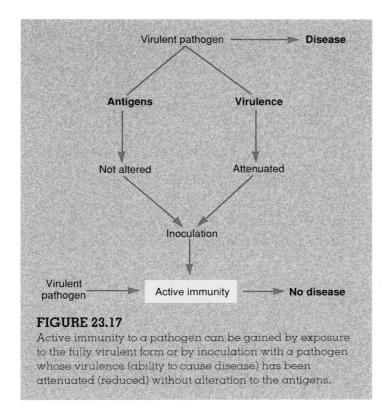

**FIGURE 23.17**
Active immunity to a pathogen can be gained by exposure to the fully virulent form or by inoculation with a pathogen whose virulence (ability to cause disease) has been attenuated (reduced) without alteration to the antigens.

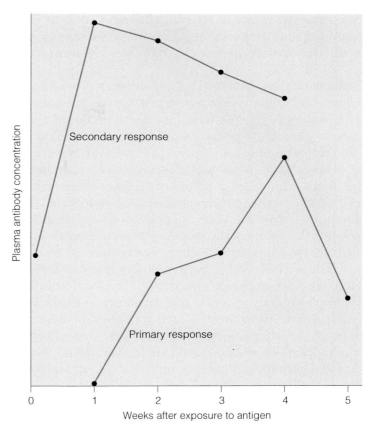

**FIGURE 23.18**
A comparison of antibody production in the primary response (upon first exposure to an antigen) to antibody production in the secondary response (upon subsequent exposure to the antigen). The greater secondary response is believed to be due to the presence of lymphocyte clones, produced during the primary response.

It first became known in Western Europe in the mid-eighteenth century that the fatal effects of smallpox could be prevented by inducing mild cases of the disease. This was accomplished at that time by rubbing needles into the pustules of people who had mild forms of smallpox and injecting those needles into healthy people. Understandably, this method of immunization did not gain wide acceptance.

Acting on the observation that milkmaids who contracted cowpox—a disease similar to smallpox but less *virulent* (pathogenic)—were immune to smallpox, an English physician, Edward Jenner, inoculated a healthy boy with cowpox. When the boy recovered, Jenner inoculated him with what was considered a deadly amount of smallpox, from which he also proved to be immune. (This was fortunate for both the boy—who was an orphan—and Jenner; Jenner's fame spread, and as the boy grew into manhood he proudly gave testimonials on Jenner's behalf.) This experiment, performed in 1796, led to the first widespread immunization program.

A similar, but more sophisticated, demonstration of the effectiveness of immunizations was performed by Louis Pasteur almost a century later. Pasteur isolated the bacteria that cause anthrax and heated them until their ability to cause disease was greatly reduced (their virulence was *attenuated*), although the nature of their antigens was not significantly altered (fig. 23.17). He then injected these

Edward Jenner: English physician, 1749–1823
Louis Pasteur: French chemist and bacteriologist, 1822–95

attenuated bacteria into 25 cows, leaving 25 unimmunized. Several weeks later, before a gathering of scientists, he injected all 50 cows with the completely active anthrax bacteria. All 25 of the unimmunized cows died—all 25 of the immunized animals survived.

## Active Immunity and the Clonal Selection Theory

Upon first exposure to a particular pathogen, there is a latent period of 5 to 10 days before measurable amounts of specific antibodies appear in the person's blood. This sluggish **primary response** may not be sufficient to protect the individual against the disease caused by the pathogen. Antibody concentrations in the blood during this primary response reach a plateau in a few days and decline after a few weeks.

A subsequent exposure of the same individual to the same antigen results in a **secondary response** (fig. 23.18). Compared to the primary response, antibody production during the secondary response is much more rapid. Maximum antibody concentrations in the blood are reached in less than 2 hours and are maintained for a longer time than in the primary response. This rapid rise in antibody production is usually sufficient to prevent the disease.

**Clonal Selection Theory**    The immunization procedures of Jenner and Pasteur were effective because the immune systems of the people who were inoculated produced a secondary rather than a primary response when exposed to the virulent pathogens. The type of protection they were afforded does not depend on accumulations of antibodies in the blood—secondary responses occur even after antibodies produced by the primary response have disappeared. Immunizations, therefore, seem to produce a type of "learning" in which the ability of the immune system to combat a particular pathogen is improved by prior exposure.

The mechanisms by which secondary responses are produced are not completely understood; the **clonal selection theory,** however, appears to account for most of the evidence. According to this theory, B lymphocytes *inherit* the ability to produce particular antibodies (and T lymphocytes inherit the ability to respond to particular antigens). A single B lymphocyte can produce only one type of antibody, with specificity for one antigen. Since this ability is genetically inherited rather than acquired, some lymphocytes can respond to smallpox, for example, and produce antibodies against it even if the person has never been previously exposed to this disease.

The inherited specificity of each lymphocyte is reflected in the antigen receptor proteins on the surface of the lymphocyte's cell membrane. Exposure to smallpox antigens thus stimulates these specific lymphocytes to divide many times until a large population of genetically identical cells—a clone—is produced. Some of these cells become plasma cells that secrete antibodies for the primary response; others become memory cells that can be stimulated to secrete antibodies during the secondary response (fig. 23.19).

Notice that according to the clonal selection theory (table 23.8), antigens do not induce lymphocytes to make the appropriate antibodies. Rather, antigens select lymphocytes (through interaction with surface receptors) that are already able to make antibodies against that antigen. This is analogous to evolution by natural selection. An environmental agent (in this case, antigens) acts on the genetic diversity already present in a population of organisms (lymphocytes) to cause an increase in number of the individuals selected.

**Active Immunity**    The development of a secondary response provides **active immunity** against the specific pathogens. The development of active immunity requires prior exposure to the specific antigens, at which time the

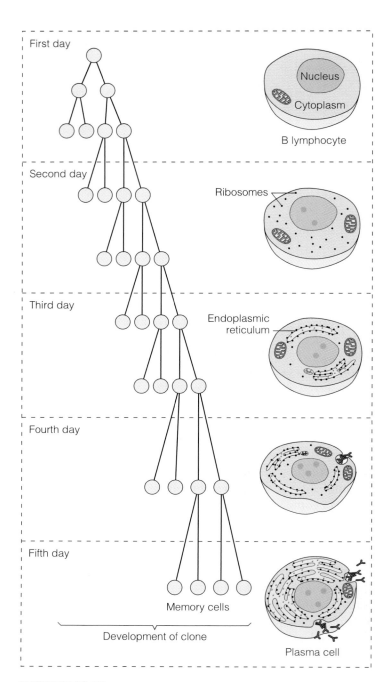

**FIGURE 23.19**

The clonal selection theory as applied to B lymphocytes. Most members of the B lymphocyte clone become memory cells, but some become antibody-secreting plasma cells.

primary response may cause the person to develop symptoms of the disease. Some parents, for example, deliberately expose their children to others who have measles, chicken pox, and mumps so that their children will be immune to these diseases later in life, when they are potentially more serious.

Clinical immunization programs induce primary responses by inoculating people with pathogens whose virulence has been attenuated or destroyed (such as Pasteur's

## Table 23.8    Summary of the clonal selection theory (as applied to B lymphocytes)

| Process | Results |
|---|---|
| Lymphocytes inherit the ability to produce specific antibodies. | Prior to antigen exposure, lymphocytes that can make the appropriate antibodies are already present in the body. |
| Antigens interact with antibody receptors on the lymphocyte surface. | Antigen-antibody interaction stimulates cell division and the development of lymphocyte clones that contain memory cells and plasma cells that secrete antibodies. |
| Subsequent exposure to the specific antigens produces a more efficient response. | Exposure of lymphocyte clones to specific antigens results in greater and more rapid production of specific antibodies. |

heat-inactivated anthrax bacteria) or by using closely re-lated strains of microorganisms that are antigenically similar but less pathogenic (such as Jenner's cowpox inoculations). The name for these procedures—**vaccination** (after the Latin word *vacca*, meaning "cow")—reflects the history of this technique. All of these procedures cause the development of lymphocyte clones that can combat the virulent pathogens by producing secondary responses.

The first successful polio vaccine (the Salk vaccine) was composed of viruses that had been inactivated by treat-ment with formaldehyde. These "killed" viruses were in-jected into the body, in contrast to the oral (Sabin) vaccine in current use. The oral vaccine contains "living" viruses that have attenuated virulence. These viruses invade the epithelial lining of the intestine and multiply but do not invade nerve tissue. The immune system can, therefore, be-come sensitized to polio antigens and produce a secondary response if polio viruses that attack the nervous system are later encountered.

 There is always a danger when vaccines are prepared using attenuated viruses or toxins that some virulence may remain and cause disease in vaccinated people. A case in point is the commonly used vaccine for *pertussis* (whooping cough)—a disease caused by a toxin released from a species of bacteria. Pertussis is responsible for about 1 million infant deaths annually. Although the pertussis vaccine prepared from the bacterial cells or their toxin is relatively effective, it occasionally produces severe side effects. Using genetic engineering techniques, scientists have recently produced a protein subunit of pertussis toxin that appears to confer immunity without the virulent action of the toxin. In the future, production of specific proteins from cloned DNA may provide other vaccines that are safer and more effective than those prepared by traditional methods.

..............................................

Salk vaccine: from Jonas Salk, American immunologist, b. 1914
Sabin vaccine: from Albert B. Sabin, American virologist, b. 1906

## Passive Immunity

The term **passive immunity** refers to the immune protec-tion that can be produced by the transfer of antibodies to a recipient from another person or from an animal. The donor person or animal has been actively immunized, as explained by the clonal selection theory. The recipient of the ready-made antibodies is thus passively immunized to the same antigens. Passive immunity occurs naturally in the transfer of antibodies from mother to fetus during pregnancy. It can also be artificially conferred by injecting antibodies into the individual requiring immunity.

The ability to mount a specific immune response—called **immunological competence**—does not develop until about a month after birth. The fetus, therefore, cannot im-munologically reject its mother. The immune system of the mother is fully competent but does not usually respond to fetal antigens for reasons that are not completely under-stood. Some IgG antibodies from the mother do cross the placenta and enter the fetal circulation, however, and these serve to confer passive immunity to the fetus.

The fetus and the newborn baby are, therefore, im-mune to the same antigens as the mother. However, since the baby did not itself produce the lymphocyte clones needed to form these antibodies, such passive immunity disappears when the infant is about 1 month old. If the baby is breast-fed, it can receive additional antibodies of the IgA subclass in its mother's first milk (the *colostrum*).

Passive immunizations are used clinically to protect people who have been exposed to extremely virulent in-fections or toxins, such as snake venom or tetanus. In these cases, the affected person is injected with *antiserum* (serum containing antibodies), also called *antitoxin*, from an animal that has been previously exposed to the pathogen. The an-imal develops the lymphocyte clones and active immunity and thus has a high concentration of antibodies in its blood. Since the person who is injected with these antibodies does not develop active immunity, he or she is protected only

## Table 23.9  Comparison of active and passive immunity

| Characteristic | Active immunity | Passive immunity |
|---|---|---|
| Injection of person with | Antigens | Antibodies |
| Source of antibodies | The person inoculated | Natural—the mother; artificial—injection with antibodies |
| Method | Injection with killed or attenuated pathogens or their toxins | Natural—transfer of antibodies across the placenta; artificial—injection with antibodies |
| Time to develop resistance | 5 to 14 days | Immediately after injection |
| Duration of resistance | Long (perhaps years) | Short (days to weeks) |
| When used | Before exposure to pathogen | Before or after exposure to pathogen |

for a short time and must again be injected with antitoxin upon subsequent exposures. Active and passive immunity are compared in table 23.9.

## Monoclonal Antibodies

In addition to their use in passive immunity, antibodies are also commercially prepared for use in research and clinical laboratory tests. In the past, antibodies were obtained by chemically purifying a specific antigen and then injecting this antigen into animals. Since an antigen typically has many different antigenic determinant sites, however, the antibodies obtained by in this way were *polyclonal*; they had different specificities. This decreased their sensitivity to a particular antigenic site and resulted in some degree of cross-reaction with closely related antigen molecules.

**Monoclonal antibodies,** by contrast, exhibit specificity for one antigenic determinant only. In the preparation of monoclonal antibodies, an animal (frequently, a mouse) is injected with an antigen and subsequently killed. B lymphocytes are then obtained from the animal's spleen and placed in thousands of different in vitro incubation vessels. These cells soon die, however, unless they are hybridized with cancerous multiple myeloma cells. Cell fusion is promoted by a chemical—polyethylene glycol. The fusion of a B lymphocyte with a cancerous cell produces a hybrid that undergoes cell division and produces a clone called a *hybridoma*. Each hybridoma secretes large amounts of identical *monoclonal antibodies*. From among the thousands of hybridomas produced in this way, the one that produces the desired antibody is cultured for large-scale production (fig. 23.20).

The availability of large quantities of pure monoclonal antibodies has led to the development of much more sensitive clinical laboratory tests (for pregnancy, for example) than had been used previously. These pure antibodies have also been used to pick one molecule (the specific antigen interferon, for example) out of a solution of many molecules and thus isolate and concentrate it.

In the future, monoclonal antibodies against specific tumor antigens may aid the diagnosis of cancer. Even more exciting, antitumor drugs that can kill normal as well as cancerous cells could be aimed directly at a tumor by combining these drugs with monoclonal antibodies against specific tumor antigens. This technique has recently been used in rodent experiments but, as of this writing, has not yet been approved for human tests.

## Functions of T Lymphocytes

*T cells assist all aspects of the immune system, including cell-mediated destruction by killer T cells and supporting roles by helper and suppressor T cells. T cells are activated only by antigens presented to them by macrophages; the activated T cells in turn produce lymphokines, which activate other cells of the immune system.*

The thymus processes lymphocytes in such a way that their functions become quite distinct from those of B cells. Lymphocytes residing in the thymus or originating from the thymus, or those derived from cells that came from the thymus, are all T lymphocytes. These cells can be distinguished from B cells by specialized techniques. Unlike B cells, the T lymphocytes provide specific immune protection without secreting antibodies. This is accomplished in various ways by three subpopulations of T lymphocytes that will be described shortly.

## Thymus

The **thymus** extends from below the thyroid in the neck into the thoracic cavity. This organ grows during childhood but gradually regresses after puberty. Lymphocytes from the

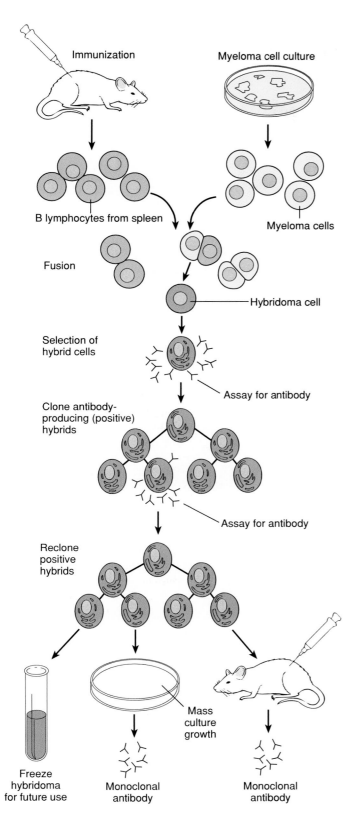

fetal liver and spleen and from the bone marrow postnatally seed the thymus and become transformed into T cells. These lymphocytes then enter the blood and seed lymph nodes and other organs, where they divide to produce new T cells when stimulated by antigens.

Small T lymphocytes that have not yet been stimulated by antigens have very long life spans—months, or perhaps years. Still, new T cells must be continuously produced to provide efficient cell-mediated immunity. Since the thymus atrophies after puberty, this organ may not be able to provide new T cells in later life. Colonies of T cells in the lymph nodes and other organs, however, are apparently able to produce new T cells under the stimulation of various **thymus hormones.**

Two hormones that are believed to be secreted by the thymus—*thymopoietin I* and *thymopoietin II*—may promote the transformation of lymphocytes into T cells. Another thymus hormone, called *thymosin*, may promote the maturation of T lymphocytes.

## Killer, Helper, and Suppressor T Lymphocytes

The **killer,** or **cytotoxic, T lymphocytes** destroy specific victim cells that are identified by specific antigens on their surfaces. In order to effect this *cell-mediated destruction*, the T lymphocytes must be in actual contact with their victim cells (in contrast to B cells, which kill at a distance). Although the mechanisms by which the killer lymphocytes kill their victims are not completely understood, there is evidence that they accomplish this task by secreting certain molecules at the region of contact. Among these molecules, specific polypeptides called *perforins* have been identified that polymerize in the cell membrane of the victim cell and form cylindrical channels through the membrane. These channels are analogous to those formed by complement proteins previously discussed and can result in osmotic destruction of the victim cell.

The killer T lymphocytes defend against viral and fungal infections and are also responsible for transplant rejection reactions and for immunological surveillance against cancer. Although most bacterial infections are fought by B lymphocytes, some are the targets of cell-mediated attack by killer T lymphocytes. This is the case with the *tubercle bacilli* that cause tuberculosis. Injections of some of these bacteria under the skin produce inflammation after a latent period of 48 to 72 hours. This *delayed hypersensitivity reaction* is cell mediated rather than humoral, as shown by the fact that it can be induced in an unexposed guinea pig by an infusion of lymphocytes, but not of serum, from an exposed animal.

**FIGURE 23.20**

The production of monoclonal antibodies that are directed against a specific antigen.

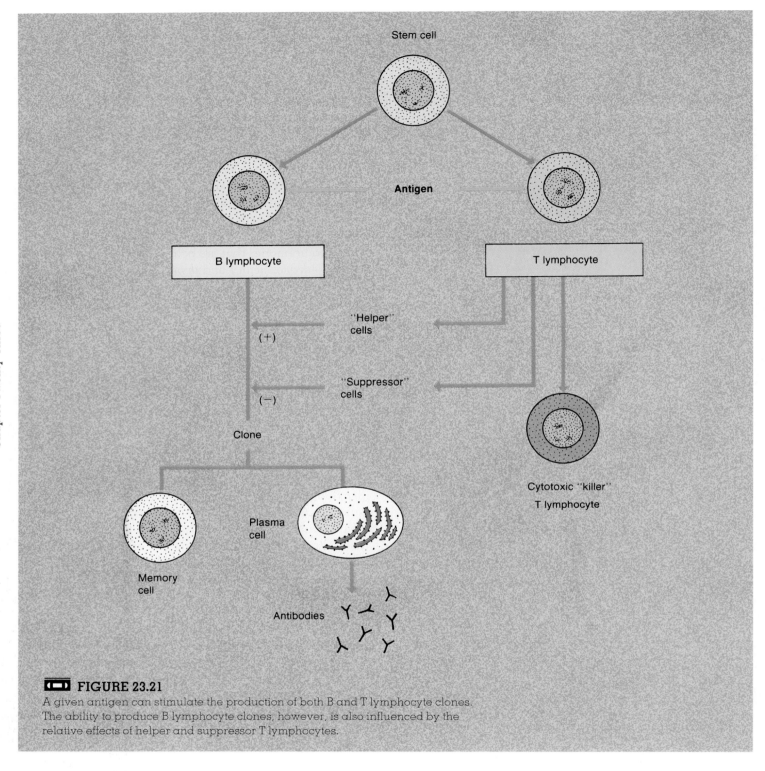

Stem cell

Antigen

B lymphocyte

T lymphocyte

"Helper" cells

(+)

"Suppressor" cells

(−)

Clone

Cytotoxic "killer" T lymphocyte

Memory cell

Plasma cell

Antibodies

▭▭ **FIGURE 23.21**

A given antigen can stimulate the production of both B and T lymphocyte clones. The ability to produce B lymphocyte clones, however, is also influenced by the relative effects of helper and suppressor T lymphocytes.

The **helper T lymphocytes** and **suppressor T lymphocytes** indirectly participate in the specific immune response by regulating the responses of the B cells (fig. 23.21) and the killer T cells. The activity of B cells and killer T cells is increased by helper T lymphocytes and decreased by suppressor T lymphocytes. The amount of antibodies secreted in response to antigens is thus affected by the relative numbers of helper to suppressor T cells that develop in response to a given antigen.

As a result of advances in recombinant DNA technology (genetic engineering) that enable the production of monoclonal antibodies, it is now possible for clinical laboratories to distinguish between the different subcategories of lymphocytes by means of antigen "markers" on

## Table 23.10  The major lymphokines

| Lymphokine | Biological functions | Secreted by |
|---|---|---|
| Interleukin-1 | Activates resting T lymphocytes | Macrophages and others |
| Interleukin-2 | Serves as growth factor for activated T lymphocytes; activates cytotoxic T lymphocytes | Helper T lymphocytes |
| Interleukin-3 | Promotes growth of bone marrow stem cells; serves as growth factor for mast cells | Helper T lymphocytes |
| Interleukin-4  (B lymphocyte-stimulating factor) | Promotes growth of activated B lymphocytes; promotes growth of resting T lymphocytes; enhances activity  of cytotoxic T lymphocytes | Helper T lymphocytes |
| B lymphocyte-differentiating factor | Induces the conversion of activated B lymphocytes into antibody-secreting plasma cells | T lymphocytes and others |
| Colony-stimulating factors | Different colony-stimulating factors stimulate the proliferation of granulocytic leukocytes and macrophages | T lymphocytes and others |
| Interferons | Activate macrophages; augment natural killer cell body activity; exhibit antiviral activity | T lymphocytes and others |
| Tumor necrosis factors | Exert direct cytotoxic effect on some tumor cells; stimulate production of other lymphokines | Macrophages and others |

Source: Adapted from information appearing in C. A. Dinarello and J. W. Mier, *The New England Journal of Medicine*, Vol. 317, p. 940, 1987.

their surfaces. Counting the lymphocytes in each of these subcategories provides far more information about diseases and their causes than was previously available. Tests of this sort have provided valuable information about the effects of the AIDS virus.

 *Acquired immune deficiency syndrome (AIDS)* has caused the deaths of hundreds of thousands of people worldwide. Millions more are infected, and since AIDS has been shown to have a latency period of approximately 8 years, most will display symptoms of the disease in the near future. People at high risk include homosexual and bisexual men (through anal intercourse) and intravenous drug users (through sharing of needles with infected individuals). Intravenous drug users account for one-third of AIDS cases in the United States and Europe; half of the estimated 200,000 intravenous drug users in New York City are believed to be infected. People at lesser risk include those who received blood transfusions prior to 1985 (before blood was tested for AIDS) and the spouses of those at high risk. In Haiti and the countries of central Africa, heterosexual contact is believed to be the primary route of infection.

AIDS is caused by the *human immunodeficiency virus (HIV)* (see fig. 23.8), which specifically destroys the helper T lymphocytes. This results in decreased immunological function and greater susceptibility to opportunistic infections, including *Pneumocystis carinii pneumonia*. Many people with AIDS also develop a previously rare form of cancer known as *Kaposi's sarcoma*.

**Lymphokines**  The T lymphocytes, as well as some other cells such as macrophages, secrete a number of polypeptides that serve in an autocrine fashion (chapter 19) to regulate

many aspects of the immune system. These products are generally called **cytokines;** the term **lymphokines** is often used to refer to the cytokines of lymphocytes. When a cytokine is first discovered, it is named according to its biological activity (e.g., *B cell-stimulating factor*). Since each cytokine has many different actions (table 23.10), however, such names can be confusing. Scientists have thus agreed to use the name *interleukin*, followed by a number, to designate a cytokine once its amino acid sequence has been determined.

*Interleukin-1*, for example, is secreted by macrophages and other cells and can activate the T cell system. B cell-stimulating factor, now called *interleukin-4*, is secreted by T lymphocytes and is required for the proliferation and clone development of B cells. *Interleukin-2* is released by helper T lymphocytes and is required for activation of killer T lymphocytes, among other functions. *Macrophage colony stimulating factor* is secreted by helper T lymphocytes and promotes the activity of macrophages.

Current research has demonstrated that there are two subtypes of helper T lymphocytes, designated $T_H1$ and $T_H2$. Helper T lymphocytes  of the $T_H1$ subtype produce interleukin-2 and gamma interferon. Because they secrete these cytokines, $T_H1$ lymphocytes activate killer T lymphocytes and promote cell-mediated immunity. The $T_H2$ lymphocytes secrete interleukin-4, interleukin-5 and interleukin-10, which stimulates B lymphocytes to promote humoral immunity. Scientists have recently discovered that "uncommitted" helper T lymphocytes are changed into the $T_H1$ subtype in response to a cytokine called interleukin-12, which is secreted by macrophages under appropriate conditions. This process could thus provide a switch from determining how much of the immune response to an antigen will be cell-mediated and how much will be humoral.

**T Cell Receptor Proteins**   Unlike B cells, T cells do not make antibodies; thus, they do not have antibodies on their surfaces to serve as receptors for antigens. The T cells do, however, have specific receptors for antigens on their membrane surfaces, and these T cell receptors have recently been identified as molecules closely related to immunoglobulins. The T cell receptors differ from the antibody receptors on B cells in another, and very important, respect: the T cell receptors *cannot bond to free antigens*. In order for a T lymphocyte to respond to a foreign antigen, the antigen must be presented to the T lymphocyte on the membrane of an **antigen-presenting cell.** The chief antigen-presenting cells are macrophages. Macrophages present the foreign antigen together with other surface antigens, called *histocompatibility antigens*, to the T lymphocytes. Some knowledge of the histocompatibility antigens is thus required before T cell–macrophage interactions and T cell functions can be understood.

**Histocompatibility Antigens**   Tissue that is transplanted from a donor to a recipient contains antigens that are foreign to the recipient. This is because all tissue cells, with the exception of mature red blood cells, are genetically marked with a characteristic combination of **histocompatibility antigens** on the membrane surface. The greater the variance in these antigens between the donor and the recipient in a transplant, the greater will be the chance of transplant rejection. Prior to organ transplantation, therefore, the "tissue type" of the recipient is matched to that of potential donors. Since the person's white blood cells are used for this purpose, histocompatibility antigens in humans are also called **human leukocyte antigens,** abbreviated **HLAs.**

HLAs are proteins that are coded for by a group of genes called the **major histocompatibility complex (MHC),** located on chromosome number 6. These four genes are labeled A, B, C, and D. Each of them can code for only one protein in a given individual, but this protein can be different in different people. Two people, for example, may both have antigen A3, but one might have antigen B17 and the other antigen B21. The closer two people are related, the more similar their histocompatibility antigens will be.

Clinical interest has been generated by the observation that certain diseases are much more common in people who have particular histocompatibility antigens. *Ankylosing spondylitis* (a type of rheumatoid arthritis), for example, is much more common in people who have antigen B27, and *psoriasis* (a skin disorder) is three times more common in people with antigen B17 than in the general population. Other diseases that have a high correlation with particular

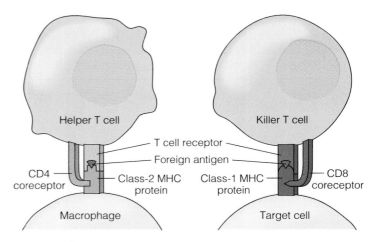

**FIGURE 23.22**

A foreign polypeptide antigen is presented to T lymphocytes in association with MHC proteins. The CD4 and CD8 coreceptors permit the T lymphocyte receptors to interact with only a specific class of MHC protein.

histocompatibility antigens include Hodgkin's disease (a cancer of the lymph nodes), myasthenia gravis, Graves' disease, and type I diabetes mellitus.

## Interactions between Macrophages and T Lymphocytes

The major histocompatibility complex of genes produces two classes of MHC proteins: *class 1* and *class 2*. The class-1 proteins are made by all cells in the body except red blood cells. Class-2 MHC proteins are produced only by macrophages and B lymphocytes and promote the interactions between T cells and these other cells of the immune system.

Killer T lymphocytes can interact only with antigens presented with class-1 MHC proteins, whereas helper T lymphocytes can interact only with antigens presented with class-2 MHC proteins. These MHC restrictions result from the presence of proteins called *coreceptors* that are associated with the T cell receptors. As illustrated in figure 23.22, the coreceptor known as *CD8* is associated with the killer T lymphocyte receptor and interacts with only the class-1 MHC proteins; the coreceptor known as *CD4* is associated with the helper T lymphocyte receptor and interacts with only the class-2 MHC proteins.

When a foreign particle, such as a virus, infects the body, it is taken into macrophages by phagocytosis and partially digested. Within the macrophage, the partially digested virus particles provide foreign antigens that are moved to the surface of the cell membrane. At the membrane, these foreign antigens form a complex with the class-2 MHC proteins. This bonding of the MHC proteins and foreign antigens is required for interaction with the receptors on the surface of helper T cells. The macrophages thus "present" the antigens to the helper T cells, and in this way stimulate

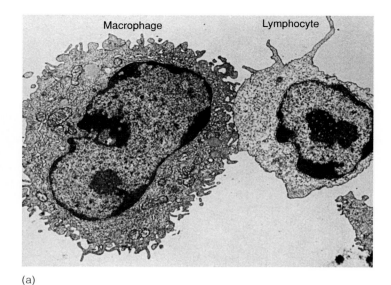

Macrophage   Lymphocyte

(a)

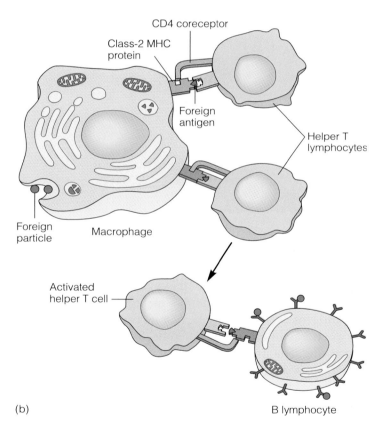

CD4 coreceptor

Class-2 MHC protein

Foreign antigen

Helper T lymphocytes

Foreign particle

Macrophage

Activated helper T cell

(b)

B lymphocyte

**FIGURE 23.23**

(a) An electron micrograph showing contact between a macrophage (*left*) and a lymphocyte (*right*). As illustrated in (b), such contact between a macrophage and a T cell requires that the helper T cell interact with both the foreign antigen and the class-2 MHC protein on the surface of the macrophage.

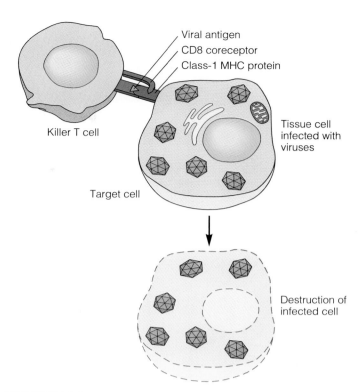

Viral antigen
CD8 coreceptor
Class-1 MHC protein

Killer T cell

Tissue cell infected with viruses

Target cell

Destruction of infected cell

**FIGURE 23.24**

In order for a killer T cell to destroy a tissue cell infected with viruses, the T cell must interact with both the foreign antigen and the class-1 MHC protein on the surface of the infected cell.

activation of the T cells (fig. 23.23). It should be remembered that T cells are "blind" to free antigens; they can respond only to antigens presented to them by macrophages (and some other cells) in combination with class-2 MHC proteins.

The first phase of macrophage–T cell interaction then occurs: the macrophage is stimulated to secrete the lymphokine known as interleukin-1. As previously discussed, interleukin-1 stimulates cell division and proliferation of T lymphocytes. The activated helper T cells, in turn, secrete macrophage colony-stimulating factor and gamma interferon, which promote the activity of macrophages. In addition, interleukin-2 is secreted by the T lymphocytes and stimulates the macrophages to secrete *tumor necrosis factor,* which is particularly effective in killing cancer cells.

Killer T cells can destroy infected cells only if those cells display the foreign antigens together with their class-1 MHC proteins (fig. 23.24). Such interaction of killer T cells with the foreign antigen–MHC class-1 complex also stimulates proliferation of those killer T cells. This proliferation is supported by the interleukin-2 secreted by the helper T cells that were activated by macrophages, as previously described (fig. 23.25).

The network of interaction among different cell types of the immune system now spreads outward. Helper T cells, activated to an antigen by macrophages, can also promote the humoral immune response of B cells. In order to do this, the membrane receptor proteins on the surface of the helper

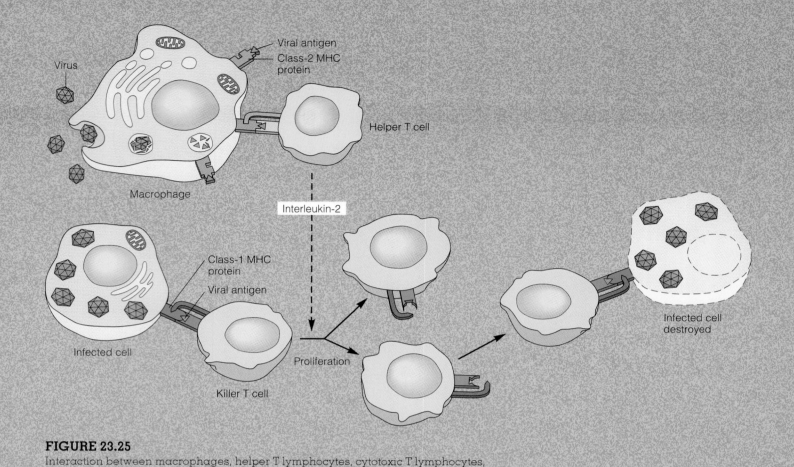

**FIGURE 23.25**
Interaction between macrophages, helper T lymphocytes, cytotoxic T lymphocytes, and infected cells in the immunological defense against viral infections.

T lymphocytes must interact with molecules on the surface of the B cells. This occurs when the foreign antigen attaches to the immunoglobulin receptors on the B cells, so that the B cells can present this antigen together with its MHC class-2 protein to the receptors on the helper T cells (fig. 23.26). This interaction stimulates proliferation of the B cells, their conversion to plasma cells, and their secretion of antibodies against the foreign antigens.

 Glucocorticoids (such as hydrocortisone), secreted by the adrenal cortex, can act to suppress the activity of the immune system and inflammation. This is why *cortisone* and its analogues are used clinically to treat inflammatory disorders and to inhibit the immune rejection of transplanted organs. The immunosuppressive effect of these hormones may result from the fact that they inhibit the secretion of the lymphokines. It is interesting in this regard that interleukin-1 has recently been shown to stimulate ACTH secretion. Rising ACTH, in turn, stimulates glucocorticoid secretion (chapter 19), which inhibits the immune system and suppresses interleukin-1 secretion. Immune control thus involves interaction not only among the cells of the immune system, but also between the immune and endocrine systems.

## Tolerance

The ability to produce antibodies against foreign, **nonself antigens,** while tolerating (not producing antibodies against) **self-antigens** occurs during the first month or so of postnatal life when immunological competence is established. Therefore, if a fetal mouse of one strain receives transplanted antigens from a different strain, it will not recognize tissue transplanted from the other strain as foreign later in life, and, as a result, will not immunologically reject the transplant.

The ability of an individual's immune system to recognize and tolerate self-antigens requires continuous exposure of the immune system to those antigens. If this exposure begins when the immune system is weak—as it is in fetal and early postnatal life—tolerance will be more complete and long lasting than that produced by exposure beginning later in life. Some self-antigens, however, are normally hidden from the blood, such as thyroglobulin within the thyroid gland and lens protein in the eye. An exposure to these self-antigens results in antibody production just as if the proteins were foreign. Antibodies made against self-antigens

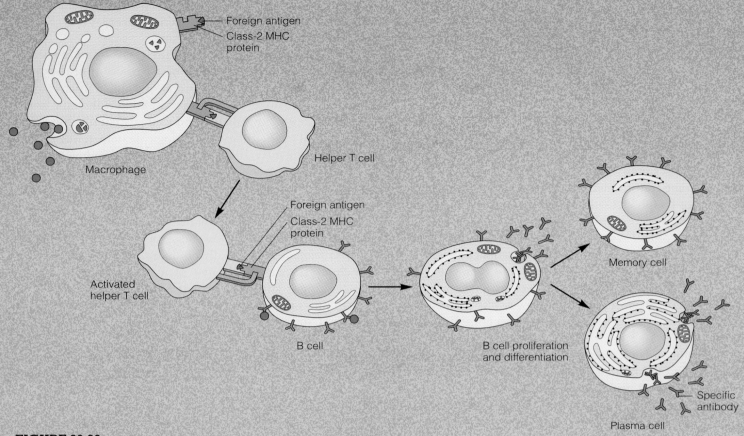

**FIGURE 23.26**
A schematic diagram of the events that are believed to occur in the interactions of macrophages, helper T lymphocytes, and B lymphocytes.

are called **autoantibodies.** Killer T cells that attack self-antigens are called **autoreactive T cells.**

Two general theories have been proposed to account for immunological tolerance: clonal deletion and clonal anergy. According to the **clonal deletion theory,** tolerance to self-antigens is achieved by destruction of the lymphocytes that recognize self-antigens. This occurs primarily during fetal life, when those lymphocytes that have receptors on their surface for self-antigens are recognized and destroyed. There is much evidence for clonal deletion in the thymus, and this mechanism is believed to be largely responsible for T cell tolerance.

**Clonal anergy** occurs when lymphocytes directed against self-antigens are present throughout life but, for complex and poorly understood reasons, do not attack the self-antigens. Clonal anergy may be aided by suppressor T lymphocytes, but evidence suggests that other mechanisms are also involved. Clonal anergy is believed

to be largely responsible for tolerance in B cells, and there is some evidence that it may also contribute to tolerance in T cells.

 A spontaneous mutation in mice leads to the development of a strain that suffers from *severe combined immunodeficiency* (*SCID*). This condition is similar to a rare congenital condition in humans, and is characterized by the absence of both B and T lymphocytes. Grafts in SCID mice are therefore not rejected. This inability to reject transplants has been exploited by reconstituting a human immune system in SCID mice using lymphocytes from peripheral human blood or human fetal liver, thymus, and lymph node grafts. This technique may provide a means for studying the functions of the immune system and diseases of the human immune system. Since the HIV virus, for example, only infects human (and chimpanzee) lymphocytes, the mouse-human chimera may provide an animal model for experimental investigation of AIDS.
........................................
chimera: Gk. *chimaira*, a mythological monster with a lion's head, goat's body, and serpent's tail

# Tumor Immunology

*Tumor cells can reveal antigens that activate an immune reaction that destroys the tumor. When cancers develop, this immunological surveillance system—primarily the function of T lymphocytes and natural killer cells—has failed to prevent the growth and metastasis of the tumor.*

**Oncology** (the study of tumors) has revealed that tumor biology is similar to and interrelated with immune system functioning. Most tumors appear to be clones of single cells that have become transformed, in a process similar to the development of lymphocyte clones in response to specific antigens. Lymphocyte clones, however, are under complex inhibitory control systems—such as those exerted by suppressor T lymphocytes and negative feedback by antibodies. The division of tumor cells, by contrast, is not effectively controlled by normal inhibitory mechanisms. Tumor cells are also relatively unspecialized—they *dedifferentiate*, which means that they become like the less specialized cells of an embryo.

Tumors are described as *benign* when they are relatively slow growing and limited to a specific location (warts, for example). *Malignant tumors* grow more rapidly and **metastasize** (*mĕ-tas´tă-sīz*); that is, cells from the primary growth spread to

..........................................

metastasis: Gk. *metastasis,* a removing

sites elsewhere in the body where they form new tumors called secondary cancer masses. The term **cancer,** as it is generally applied, refers to malignant tumors.

As tumors dedifferentiate, they reveal surface antigens that can stimulate the immune destruction of the tumor cells. Consistent with the concept of dedifferentiation, some of these antigens are proteins produced in embryonic or fetal life that are not normally produced postnatally. Since they are absent at the time immunological competence is established, they are treated as foreign and fit subjects for immunological attack when they are produced by cancerous cells. The release of two such antigens into the blood has provided the basis for laboratory diagnosis of some cancers. *Carcinoembryonic* (*kar´´sĭ-no-em´´bre-on´ik*) *antigen tests* are useful in diagnosing colon cancer, for example, and tests for *alpha-fetoprotein* (normally produced only by the fetal liver) help in the diagnosis of liver cancer.

Tumor antigens activate the immune system, initiating an attack primarily by killer T lymphocytes (fig. 23.27) and natural killer cells. The concept of **immunological surveillance** against cancer was introduced in the early 1970s to describe the proposed role of the immune system in fighting cancer. According to this concept, tumor cells frequently appear in the body but are normally recognized and destroyed by the immune system before they can cause cancer. There is evidence that immunological surveillance does prevent some types of cancer; this explains why, for example, AIDS victims (with a depressed

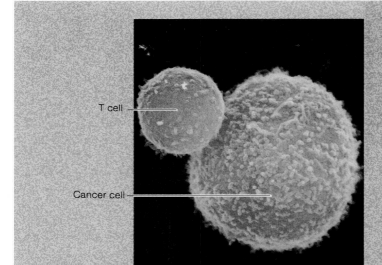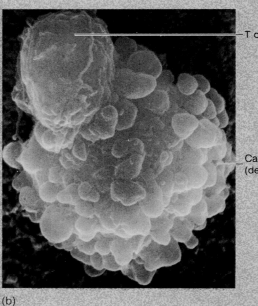

(a)                    (b)

## FIGURE 23.27

A killer T cell (*a*) contacts a cancer cell (the larger cell), in a manner that requires specific interaction with antigens on the cancer cell. The killer T cell releases lymphokines, including toxins that cause the death of the cancer cell (*b*). (Scanning electron micrographs © Andrejs Liepens.)

immune system) have a high incidence of Karposi's sarcoma. It is not clear, however, why all types of cancers do not appear with high frequency in AIDS patients and others whose immune systems are suppressed. For these reasons, the generality of the immunological surveillance system concept is currently controversial.

## Immune Therapy of Cancer

The production of human interferons by genetically engineered bacteria has made large amounts of these substances available for the experimental treatment of cancer. Thus far, interferons have proven to be a useful addition to the treatment of particular forms of cancer, including some types of lymphomas, renal carcinoma, melanoma, Karposi's sarcoma, and breast cancer. They have not, however, proved to be the "magic bullet" against cancer (a termed coined by Paul Ehrlich) as had previously been hoped.

A team of scientists at the National Cancer Institute has pioneered the use of another lymphokine, **interleukin-2 (IL-2)**, also now available through genetic engineering techniques. IL2 activates both killer T lymphocytes and B lymphocytes. These investigators removed some of the blood from cancer patients who could not be successfully treated by conventional means and isolated a population of their lymphocytes. They treated these lymphocytes with IL-2 to produce *lymphokine-activated killer (LAK) cells* and then infused these cells, together with IL-2 and interferons, into the patients. Depending on the combinations and dosages, they obtained remarkable success (but not a complete cure for all cancers) in many of these patients.

The research group next identified a subpopulation of lymphocytes that had invaded solid tumors in mice. These *tumor-infiltrating lymphocyte (TIL) cells* were allowed to replicate in tissue culture, whereupon they were introduced into the mice with excellent results. Recently, these same techniques were used to treat an experimental group of people with metastatic melanoma, a cancer that claims the lives of 6000 Americans annually. First, the patients were given conventional chemotherapy and radiation therapy. Then they were treated with their own TIL cells and IL-2. Some of the preliminary results of this treatment seem very promising. It should be noted, however, that the long-term effects of the treatment are not known at the present time, and more research is needed to clarify the optimum procedures for this therapy.

cancer: L. *cancer*, a crab
Paul Ehrlich: German bacteriologist, 1854–1915

## Natural Killer Cells

In a particular strain of hairless mice, a thymus and T lymphocytes are genetically lacking, yet these mice do not appear to suffer a particularly high incidence of tumor production. This surprising observation led to the discovery of **natural killer (NK) cells,** which are lymphocytes that are related to, but distinct from, T lymphocytes. Unlike killer T cells, NK cells destroy tumors in a nonspecific fashion and do not require prior exposure for sensitization to the tumor antigens. The NK cells thus provide a first line of cell-mediated defense, which is subsequently backed up by a specific response mediated by killer T cells. These two cell types interact, however; the activity of NK cells is stimulated by interferon, released as one of the lymphokines from T lymphocytes.

## Effects of Aging and Stress

Susceptibility to cancer varies widely. The Epstein–Barr virus that causes Burkitt's lymphoma in a few individuals in Africa, for example, can also be found in healthy people throughout the world. Most often the virus is harmless; in some cases, it causes mononucleosis (involving a limited proliferation of white blood cells). Only rarely does this virus cause the uncontrolled proliferation of leukocytes characteristic of Burkitt's lymphoma. The reasons for these different responses to the Epstein–Barr virus and indeed for the variance in degree of susceptibility to other forms of cancer are not well understood.

It is known that cancer risk increases with age. According to one theory, aging lymphocytes gradually accumulate genetic errors that decrease their effectiveness. The age-related atrophy of the thymus is also believed to contribute to an increased susceptibility to disease. The secretion of thymus hormones decreases with age, as does the production of new T lymphocytes. These changes could produce a decline in cell-mediated immune competence, and thus an increased susceptibility to cancer.

Numerous experiments have demonstrated that tumors grow faster in experimental animals subject to stress than in unstressed control animals. Stressed animals, including humans, exhibit increased secretion of corticosteroid hormones that suppress the immune system (which is why cortisone is given to people who receive organ transplants and to people with chronic inflammatory diseases). Some recent experiments, however, suggest that the stress-induced suppression of the immune system may also be due to other factors that do not involve the adrenal cortex. Future advances in cancer therapy may incorporate methods of strengthening the immune system together with methods that directly destroy tumors.

# Clinical Considerations

The ability of the normal immune system to tolerate self-antigens while it identifies and attacks foreign antigens provides a specific defense against invading pathogens. At times, however, this system of defense against invaders is responsible for domestic offenses that may produce diseases ranging in severity from the sniffles to sudden death.

Diseases caused by the immune system can be grouped into three interrelated categories: (1) *autoimmune diseases*, (2) *immune complex diseases*, and (3) *allergy, or hypersensitivity.* It is important to remember that these diseases are not caused by foreign pathogens but by abnormal responses of the immune system.

## Autoimmunity

In an autoimmune disease, the immune system fails to recognize and tolerate self-antigens. As a consequence of this failure, autoreactive T lymphocytes are activated and autoantibodies are produced by B lymphocytes, causing inflammation and organ damage. Over 40 known or suspected autoimmune diseases affect 5% to 7% of the population. These diseases arise through a variety of mechanisms.

**1** **An antigen that does not normally circulate in the blood may become exposed to the immune system.** In *lymphocytic thyroiditis (Hashimoto's disease)*, for example, thyroglobulin protein that is normally trapped within the thyroid follicles stimulates the production of autoantibodies that attack the thyroid (fig. 23.28). Similarly, autoantibodies developed against lens protein in a damaged eye may cause the destruction of a healthy eye, as in *sympathetic ophthalmia.*

**2** **A self-antigen that is otherwise tolerated may be altered by combining with a foreign hapten.** The disease *thrombocytopenia* (low platelet count), for example, can be caused by the autoimmune destruction of thrombocytes (platelets). This occurs when drugs such as aspirin, sulfonamide, antihistamines, digoxin, and others combine with platelet proteins to produce new antigens. The symptoms of this disease usually disappear when the person stops taking these drugs.

**3** **Antibodies may be produced that are directed against other antibodies.** *Rheumatoid arthritis*, for example, is an autoimmune disease associated with the abnormal production of one group of antibodies (of the IgM type) that attack other antibodies (of the IgG type). This contributes to the inflammation reaction of the joints characteristic of the disease.

**4** **Antibodies produced against foreign antigens may cross-react with self-antigens.** Autoimmune diseases of this sort can occur, for example, as a result of *Streptococcus* bacterial infections. Antibodies produced in response to antigens in this bacterium may cross-react with self-antigens in the heart and kidneys. The

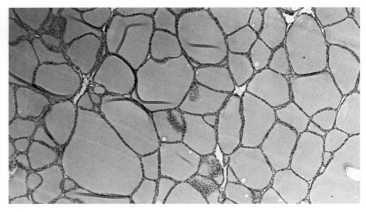

(a)

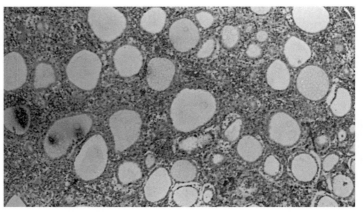

(b)

**FIGURE 23.28**

Autoimmune thyroiditis in a rabbit, induced experimentally by injection with thyroglobulin. Compare the picture of a normal thyroid (*a*) with that of a diseased thyroid (*b*). The grainy appearance of the diseased thyroid is due to the infiltration of large numbers of lymphocytes and macrophages.

inflammation induced by such autoantibodies can produce heart damage (including the valve defects characteristic of *rheumatic fever*) and damage to the glomerular capillaries in the kidneys (*glomerulonephritis*).

**5** **Self-antigens, such as receptor proteins, may be presented to the helper T lymphocytes together with class-2 MHC proteins.** Normally, only macrophages and B lymphocytes produce class-2 MHC proteins. Perhaps as a result of viral infection, however, cells that do not normally produce class-2 HLA antigens may start to do so and, in this way, present a self-antigen to the helper T lymphocytes. In *Graves' disease*, for example, the thyroid cells produce class-2 MHC proteins, and the immune system produces autoantibodies against the TSH receptor proteins in the thyroid cells. These autoantibodies, called *TSAb* for "thyroid-stimulating antibody," interact with the TSH receptors and overstimulate the thyroid gland. Similarly, in type I *diabetes mellitus*, the beta cells of the pancreatic islets abnormally produce class-2 MHC proteins, resulting in autoimmune destruction of the insulin-producing cells.

## Immune Complex Diseases

The term *immune complexes* refers to combinations of antibodies with antigens that are free rather than attached to bacterial or other cells. The formation of such complexes activates complement proteins and promotes inflammation. This inflammation normally is self-limiting because the immune complexes are removed by phagocytic cells. When large numbers of immune complexes are continuously formed, however, they cannot be cleared and the inflammation may be prolonged. Also, the dispersion of immune complexes to other sites can lead to widespread inflammations and organ damage. The damage produced by the inflammatory response to antigens is called **immune complex disease.**

Immune complex diseases can result from infections by bacteria, parasites, or viruses. In hepatitis B, for example, an immune complex that consists of viral antigens and antibodies can cause widespread inflammation of arteries (*periarteritis*). Arterial damage is not caused by the hepatitis virus itself but by the inflammatory process.

Immune complex diseases can also result from the formation of complexes between self-antigens and autoantibodies. This is the case in rheumatoid arthritis, where the inflammation is produced by complexes of altered IgG antibodies (the antigens in this case) and IgM antibodies. Another immune complex disease that has an autoimmune basis is *systemic lupus erythematosus* (*SLE*). People with SLE produce antibodies against their own DNA and nuclear proteins, which can result in the formation of immune complexes throughout the body.

## Allergy

The term *allergy*, often used interchangeably with *hypersensitivity*, refers to particular types of abnormal immune responses to antigens, which in these cases are called *allergens*. There are two major forms of allergy: (1) **immediate hypersensitivity,** which is due to an abnormal B lymphocyte response that produces symptoms within seconds or minutes, and (2) **delayed hypersensitivity,** which is an abnormal T lymphocyte response that produces symptoms within about 48 hours after exposure to an allergen. These two types of hypersensitivity are compared in table 23.11.

**Immediate Hypersensitivity** Immediate hypersensitivity can produce a variety of symptoms, including allergic rhinitis (chronic runny or stuffy nose), conjunctivitis (red eyes), allergic asthma, and atopic dermatitis (urticaria, or hives). These symptoms result from the pro-

**Table 23.11** — **Allergy: Comparison of immediate and delayed hypersensitivity reactions**

| Characteristic | Immediate reaction | Delayed reaction |
|---|---|---|
| Time for onset of symptoms | Within several minutes | Within a period of 1 to 3 days |
| Lymphocytes involved | B lymphocytes | T lymphocytes |
| Immune effector | IgE antibodies | Cell-mediated immunity |
| Allergies most commonly produced | Hay fever, asthma, and most other allergic conditions | Contact dermatitis (such as to poison ivy and poison oak) |
| Therapy | Antihistamines and adrenergic drugs | Corticosteroids (such as cortisone) |

duction of antibodies of the IgE subclass rather than the normal IgG antibodies.

In contrast to IgG antibodies, IgE antibodies do not circulate in the blood. Instead, they attach to tissue mast cells, which have membrane receptors for these antibodies. When the person is again exposed to the same allergen, the allergen bonds to the antibodies attached to the mast cells. This stimulates the mast cells to secrete various chemicals, including **histamine** (fig. 23.29). During this process, leukocytes may also secrete **prostaglandins** and related molecules called **leukotrienes.** These chemicals (table 23.12) produce the symptoms of the allergic reactions. Examination of table 23.12 reveals that histamine stimulates smooth muscle contraction in the respiratory tract but stimulates smooth muscle relaxation in the walls of blood vessels. This seemingly paradoxical effect is because the smooth muscle relaxation in the wall of blood vessels is actually produced by nitric oxide (chapter 22), which is synthesized in response to histamine stimulation.

The symptoms of hay fever (itching, sneezing, watery eyes, runny nose) are produced largely by histamine and can be treated effectively by antihistamine drugs. Food allergies, causing diarrhea and colic, are mediated primarily by prostaglandins and can be treated with aspirin, which inhibits prostaglandin synthesis. (Food allergies are the only allergies that respond positively to aspirin.) Asthma, produced by smooth muscle constriction in the bronchioles in the lungs, is due to the release of leukotrienes. Since there are no antileukotriene drugs presently available, asthma is treated with epinephrine-like compounds (which cause bronchodilation) and corticosteroids.

Immediate hypersensitivity to a particular antigen is commonly tested by injecting various antigens under the skin (fig. 23.30). Within a short time a *flare-and-wheal reaction* is produced if the person is allergic to that antigen. This reaction is due to the release of histamine and other chemical mediators: the flare (spreading flush) is due to vasodilation, and the wheal (elevated area) results from local edema.

Lymphatic System and Immunity

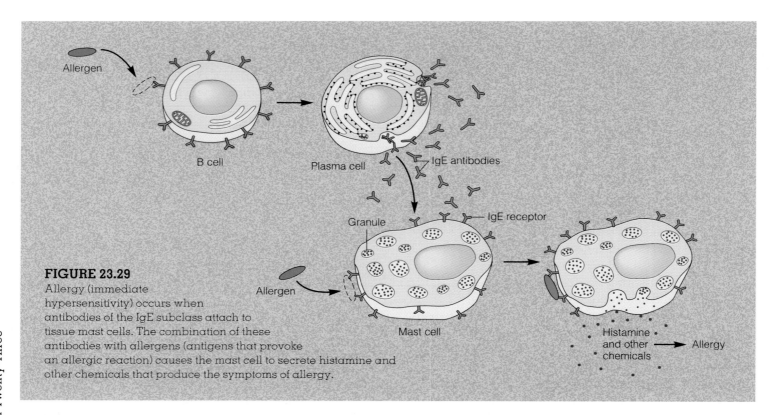

**FIGURE 23.29**
Allergy (immediate hypersensitivity) occurs when antibodies of the IgE subclass attach to tissue mast cells. The combination of these antibodies with allergens (antigens that provoke an allergic reaction) causes the mast cell to secrete histamine and other chemicals that produce the symptoms of allergy.

# Table 23.12  Chemicals that produce allergic reaction symptoms

| Chemical | Derivation | Action |
|---|---|---|
| Histamine | From histidine | Contracts smooth muscles in bronchioles; dilates blood vessels; increases capillary permeability |
| Serotonin | From tryptophane | Contracts smooth muscles |
| Prostaglandins and leukotrienes | Arachidonic acid | Prolonged contraction of smooth muscle; increased capillary permeability |
| Eosinophil chemotactic factor | Polypeptides | Attracts eosinophils |
| Bradykinin and related compounds | Polypeptides | Slow smooth muscle contraction |

Allergens that provoke immediate hypersensitivity include various foods, bee stings, and pollen grains. The most common allergy of this type is seasonal hay fever, which may be provoked by ragweed (*Ambrosia*) pollen grains (fig. 23.31*a*). People with chronic allergic rhinitis and asthma due to an allergy to dust or feathers are usually allergic to a tiny mite (fig. 23.31*b*) that lives in dust and eats the scales of skin that are constantly shed from the body. Actually, most of the antigens from the dust mite are not in its body but rather in its feces, which are tiny particles that can enter the nasal mucosa, much like pollen grains.

**Delayed Hypersensitivity**  In delayed hypersensitivity, as the name implies, symptoms take a longer time (hours to days) to develop than in immediate hypersensitivity. This may be because immediate hypersensitivity is mediated by antibodies, whereas delayed hypersensitivity is a cell-mediated T lymphocyte response. Since the symptoms are caused by the secretion of lymphokines, rather than by the secretion of histamine, treatment with antihistamines provides little benefit. At present, corticosteroids are the only drugs that can effectively treat delayed hypersensitivity.

One of the best-known examples of delayed hypersensitivity is **contact dermatitis,** caused by poison ivy, poison oak, and poison sumac. The skin tests for tuberculosis—the *tine test* and the *Mantoux test*—also rely on delayed hypersensitivity reactions. If a person has been exposed to the tubercle bacillus and consequently has developed T lymphocyte clones, skin reactions appear within a period of a few days after the tubercle antigens have been rubbed into the skin with small needles (tine test) or have been injected under the skin (Mantoux test).

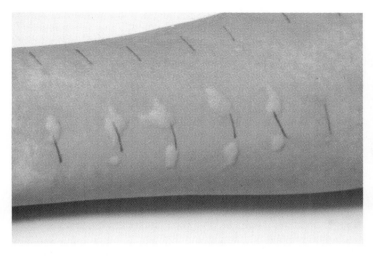
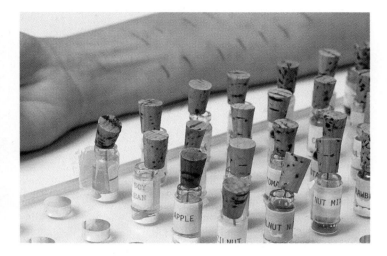

**FIGURE 23.30**

A skin test for allergy. If an allergen is injected into the skin of a sensitive individual, a typical flare-and-wheal response occurs within several minutes.

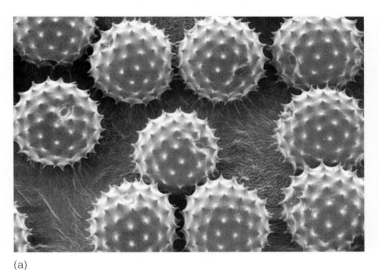

(a)  (b)

**FIGURE 23.31**

(a) A scanning electron micrograph of ragweed (*Ambrosia*), which is responsible for hay fever. (b) A scanning electron micrograph of the house dust mite (*Dermatophagoides farinae*). Dust mites are often responsible for chronic allergic rhinitis and asthma.

(a) From: *Tissues and Organs: A Text Atlas of Scanning Electron Microscopy* by R. G. Kessel and R. Kardon. © 1979 W. H. Freeman and Company.

# Chapter Summary

## Lymphatic System (pp. 648–651)

1. Lymph capillaries drain excess interstitial fluid and are highly permeable to proteins and particles such as microorganisms.
2. Lymph ducts receive lymph from the lymph capillaries and eventually empty into the thoracic duct and right lymphatic duct, which in turn empty into the left and right subclavian veins, respectively.

3. Lymph nodes contain germinal centers for lymphocyte production.
4. The spleen and thymus are lymphoid organs that contain germinal centers.
   a. The spleen also filters the blood and destroys old red blood cells.
   b. The thymus plays a key role in the immune system.

## Defense Mechanisms (pp. 651–656)

1. Nonspecific defense mechanisms include barriers to penetration of the body by pathogens and internal defenses.
2. Specific immune responses are directed against antigens.
   a. Antigens are molecules, or parts of molecules, that are usually large, complex, and foreign.

# Interactions of the Immune and Lymphatic Systems with other Body Systems

**Integumentary System**
- Serves as first line of defense from invasion by pathogens
- Lymphatic vessels drain excess interstitial fluid
- Immune system protects against infection

**Skeletal System**
- Provides sites (bone marrow) for hemopoiesis, including formation of leukocytes involved in immunity
- Lymphatic vessels drain excess interstitial fluid
- Immune system protects against infection

**Muscular System**
- Cardiac and smooth muscles help deliver blood to lymphatic organs
- Lymphatic vessels drain excess interstitial fluid
- Immune system protects against infection

**Nervous System**
- Neural regulation of the pituitary and adrenal glands indirectly influences activity of the immune system
- Nerves regulate blood flow to lymphatic organs
- Lymphatic vessels drain excess interstitial fluid
- Immune system protects against infection

**Endocrine System**
- Pituitary and adrenal glands influence immune function
- Thymus regulates production of T lymphocytes
- Lymphatic vessels drain excess interstitial fluid
- Immune system protects against infection

**Respiratory System**
- Provides oxygen for transport by the blood and provides for elimination of carbon dioxide from the blood
- Lymphatic vessels drain excess interstitial fluid
- Immune system protects against infection

**Urinary System**
- Regulates the volume, pH, and electrolyte balance of the blood and eliminates wastes
- Lymphatic vessels drain excess interstitial fluid
- Immune system protects against infection

**Digestive System**
- Provides nutrients to be transported by the circulatory system to lymphatic organs and tissues
- Lymphatic vessels drain excess interstitial fluid; absorbed fat enters the lacteals
- Immune system protects against infection

**Reproductive System**
- Reproductive hormones may influence immune functioning
- Blood-testes barrier prevents sperm cell antigens from provoking autoimmune response
- Lymphatic vessels drain excess interstitial fluid
- Immune system protects against infection

b. A given molecule can have a number of antigenic determinant sites that stimulate the production of different antibodies.
3. Specific immunity is a function of lymphocytes.
   a. B lymphocytes secrete antibodies and provide humoral, or antibody-mediated, immunity.
   b. T lymphocytes provide cell-mediated immunity.

## Functions of B Lymphocytes (pp. 656–662)

1. There are five subclasses of antibodies, or immunoglobulins.
2. Antigen-antibody complexes activate a system of proteins called the complement system.
3. Specific and nonspecific immune mechanisms cooperate in the development of a local inflammation.

## Active and Passive Immunity (pp. 652–666)

1. A primary response is produced when a person is first exposed to a pathogen; a subsequent exposure results in a secondary response.
2. The secondary response is believed to be due to lymphocyte clones that develop after the first exposure as a result of the antigen-stimulated proliferation of appropriate lymphocytes.

3. Passive immunity is provided by transfer of antibodies from an immune to a nonimmune organism.
   a. Passive immunity occurs naturally in the transfer of antibodies from mother to fetus.
   b. Injections of antiserum provide passive immunity to some pathogenic organisms and toxins.
4. Monoclonal antibodies are made by hybridomas, which are formed artificially by the fusion of B lymphocytes with multiple myeloma cells.

## Functions of T Lymphocytes (pp. 666–673)

1. The thymus processes T lymphocytes and secretes hormones that are believed to be required for the proper function of the immune response of T lymphocytes throughout the body.
2. There are three subcategories of T lymphocytes.
   a. Killer T lymphocytes are responsible for transplant rejection and for the immunological defense against fungal and viral infections, as well as for the defense against some bacterial infections.
   b. Helper and suppressor T lymphocytes stimulate or suppress, respectively, the function of B lymphocytes and killer T lymphocytes.

c. The T lymphocytes secrete a family of compounds called lymphokines, which promote the action of lymphocytes and macrophages.
   d. Receptor proteins on the cell membrane of T lymphocytes must bind to a foreign antigen in combination with a MHC protein in order for the T lymphocyte to become activated.
3. Macrophages partially digest a foreign body, such as a virus, and present the antigens to the lymphocytes on the surface of the macrophage in combination with class-2 MHC proteins.
4. Interleukin-2 stimulates proliferation of killer T lymphocytes that are specific for the foreign antigen.
5. Tolerance to self-antigens may be due to the destruction of lymphocytes that can recognize the self-antigens, or it may be due to suppression of the immune response by the action of specific suppressor T lymphocytes.

## Tumor Immunology (pp. 674–675)

1. Immunological surveillance against cancer is provided mainly by killer T lymphocytes and natural killer cells.
2. Natural killer cells are nonspecific; T lymphocytes are directed against specific antigens on the cancer cell surface.

# Review Activities

### Objective Questions

1. Which of the following offers a nonspecific defense against viral infection?
   a. antibodies
   b. leukotrienes
   c. interferon
   d. histamine

Match the cell type with its secretion.

| | |
|---|---|
| 2. killer T cells | a. antibodies |
| 3. mast cells | b. perforins |
| 4. plasma cells | c. lysosomal enzymes |
| 5. macrophages | d. histamine |

6. Which of the following statements about the $F_{ab}$ portion of antibodies is *true*?
   a. It bonds to antigens.
   b. Its amino acid sequences are variable.
   c. It consists of both H and L chains.
   d. All of the above are true.

7. Which of the following statements about complement proteins $C3_a$ and $C5_a$ is *false*?
   a. They are released during the complement fixation process.
   b. They stimulate chemotaxis of phagocytic cells.
   c. They promote the activity of phagocytic cells.
   d. They produce pores in the victim cell membrane.
8. Mast cell secretion during an immediate hypersensitivity reaction is stimulated when antigens combine with
   a. IgG antibodies.
   b. IgE antibodies.
   c. IgM antibodies.
   d. IgA antibodies.
9. During a secondary immune response,
   a. antibodies are made quickly and in great amounts.
   b. antibody production lasts longer than in a primary response.

c. antibodies of the IgG class are produced.
   d. lymphocyte clones are believed to develop.
   e. all of the above apply.
10. Which of the following cells aids the activation of lymphocytes by antigens?
    a. macrophages
    b. neutrophils
    c. mast cells
    d. natural killer cells
11. Which of the following statements about T lymphocytes is *false*?
    a. Some T lymphocytes promote the activity of B lymphocytes.
    b. Some T lymphocytes suppress the activity of B lymphocytes.
    c. Some T lymphocytes secrete interferon.
    d. Some T lymphocytes produce antibodies.

12. Delayed hypersensitivity is mediated by
    a. T lymphocytes.
    b. B lymphocytes.
    c. plasma cells.
    d. natural killer cells.
13. Active immunity may be produced by
    a. contracting a disease.
    b. receiving a vaccine.
    c. receiving gamma globulin injections.
    d. both a and b.
    e. both b and c

14. Which of the following statements about class-2 MHC proteins is *false*?
    a. They are found on the surface of B lymphocytes.
    b. They are found on the surface of macrophages.
    c. They are required for B lymphocyte activation by a foreign antigen.
    d. They are needed for interaction of helper and killer T lymphocytes.
    e. They are presented together with foreign antigens by macrophages.

**Essay Questions**

1. Explain how antibodies help to destroy invading bacterial cells.
2. Explain how T lymphocytes interact with macrophages and the infected cells in fighting viral infections.
3. Explain how helper and suppressor T lymphocytes may function in (a) defense against infections and (b) tolerance to self-antigens.
4. Describe the clonal selection theory and explain how it accounts for the development of active immunity.
5. Explain the physiological and clinical significance of histocompatibility antigens.

## Explorations ✪

Two modules of correlating material are available from the Wm. C. Brown CD-ROM: *Explorations*. They are #12 Immune Response and #13 AIDS.

# respiratory system

## objectives

- Discuss the functions of the respiratory system and describe and identify the organs associated with respiration.
- State the type of epithelium found in each region of the respiratory tract and describe the location and function of the paranasal sinuses.
- Describe the regions of the pharynx and the structure and functions of the larynx and conducting airways.
- Describe the structure and function of the alveoli.
- Discuss the gross structure of the lungs and the arrangement and significance of thoracic serous membranes.
- Describe the changes in intrapulmonary and intrapleural pressure during breathing and discuss how Boyle's law relates to lung function.
- Define *compliance, elasticity,* and *surface tension* and explain how they affect lung function.
- State the muscles involved in quiet and in forced inspiration and expiration and describe their actions.
- Distinguish between restrictive and obstructive pulmonary disease and give examples of each type.
- Define the terms used to describe lung volumes and capacities and describe how pulmonary function tests help diagnose lung disorders.

- Discuss the calculation of partial pressures of gases in the air and explain how they are measured in the blood.
- Explain the significance of the plasma $P_{O_2}$ measurement.
- Discuss the functions of the pneumotaxic, apneustic, and rhythmicity centers in the brain.
- State the location of the chemoreceptors and explain the chemoreceptor reflex control of breathing.
- Distinguish between the different forms of hemoglobin and describe the loading and unloading reactions.
- Explain how changes in pH, temperature, and 2, 3-DPG influence the function of hemoglobin.
- Describe how carbon dioxide is transported in the blood and discuss the chloride shift.
- Explain how breathing helps to maintain acid-base balance and how changes in ventilation can produce respiratory acidosis and alkalosis.
- Explain the increased ventilation that occurs during exercise and describe the effect of exercise on blood $P_{O_2}$, $P_{CO_2}$, and pH.
- Discuss the effect of endurance training on anaerobic threshold.
- Describe the respiratory system changes associated with acclimatization to a high altitude.

# Functions and Divisions of the Respiratory System

*The respiratory system can be divided anatomically into upper and lower divisions and functionally into a conducting division and a respiratory division.*

The term *respiration* refers to three separate but related functions: (1) **ventilation** (breathing); (2) **gas exchange,** which occurs between the air and blood in the lungs and between the blood and other tissues of the body; and (3) **oxygen utilization** by the tissues in the energy-liberating reactions of cell respiration. Ventilation and the exchange of gases (oxygen and carbon dioxide) between the air and blood are collectively called **external respiration.** Gas exchange between the blood and other tissues and oxygen utilization by the tissues are collectively known as **internal respiration.**

Ventilation is the mechanical process that moves air into and out of the lungs. Since air in the lungs has a higher oxygen concentration than the blood, oxygen diffuses from air to blood. Carbon dioxide, conversely, moves from the blood to the air within the lungs by diffusing down its concentration gradient. As a result of this gas exchange, the inspired air contains more oxygen and less carbon dioxide than the expired air. More importantly, blood leaving the lungs (in the pulmonary veins) has a higher oxygen concentration and a lower carbon dioxide concentration than the blood delivered to the lungs in the pulmonary arteries. This results from the fact that the lungs function to bring the blood into gaseous equilibrium with the air.

The major passages and structures of the respiratory system (fig. 24.1) are the *nasal cavity, pharynx, larynx,* and *trachea,* and the *bronchi, bronchioles,* and *alveoli* within the *lungs.* The respiratory system is frequently divided into the **conducting division** and the **respiratory division.** The conducting division includes all of the cavities and structures that transport gases to the respiratory division. The structures involved in the gas exchange between the air and blood constitute the respiratory division.

## Conducting Division

*Air is conducted through the nasal and oral cavities to the pharynx and then through the larynx to the trachea and bronchial tree. These structures deliver warmed, cleansed, and humidified air to the respiratory division in the lungs.*

........................................

respiration: L. *re,* back; *spirare,* to breathe

## Nose and Pharynx

The **nose** includes an external portion that juts out from the face and an internal **nasal cavity** for the passage of air. The external portion of the nose is supported by paired **nasal bones,** forming the bridge (chapter 9), and pliable cartilage, forming the distal portions. The **septal cartilage** forms the anterior portion of the **nasal septum,** and the paired **lateral cartilages** and **alar cartilages** form the framework around the **nostrils.**

The perpendicular plate of the **ethmoid bone** and the **vomer bone,** together with the septal cartilage, constitute the supporting framework called the **nasal septum,** which divides the nasal cavity into two lateral halves. Each half opens anteriorly through the **nostril** (fig. 24.2) and communicates posteriorly with the **nasopharynx** through the **choana** (ko-a-nă). The roof of the nasal cavity is formed anteriorly by the

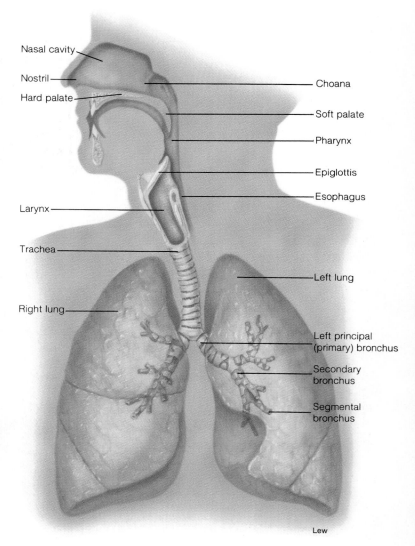

Nasal cavity
Nostril
Hard palate
Choana
Soft palate
Pharynx
Epiglottis
Esophagus
Larynx
Trachea
Left lung
Right lung
Left principal (primary) bronchus
Secondary bronchus
Segmental bronchus

Lew

**FIGURE 24.1**

The basic anatomy of the respiratory system.

frontal bone and paired nasal bones, medially by the cribriform plate of the ethmoid bone, and posteriorly by the sphenoid bone. The palatine and maxillary bones form the floor of the cavity. On the lateral walls are the **superior, middle,** and **inferior conchae** (*kong´ke*). Air passages between the conchae are referred to as **meatuses** (fig. 24.2). The anterior openings of the nasal cavity are lined with stratified squamous epithelium, whereas the conchae are lined with pseudostratified ciliated columnar epithelium (figs. 24.3 and 24.4). Mucus-secreting goblet cells are present in great abundance throughout both regions.

The three functions of the nasal cavity and its contents are as follows:

**1** The nasal epithelium covering the conchae warms, moistens, and cleanses the air. The curved conchae enhance the turbulence of the air and extend the surface area of the nasal epithelium, which is highly vascular. This is important for warming the air but unfortunately also makes humans susceptible to nosebleeds. Nasal hairs called **vibrissae** (*vibris´e*), which often extend from the nostrils, filter macroparticles that might otherwise be inhaled. Fine particles, such as dust, pollen, or smoke, are trapped along the moist mucous membrane lining the nasal cavity.

**2** Olfactory epithelium in the upper medial portion of the nasal cavity responds to inhaled chemicals during olfaction.

**3** The nasal cavity is associated with voice phonetics by functioning as a resonating chamber.

 There are several drainage openings into the nasal cavity (see fig. 24.2). The paranasal ducts drain mucus from the *paranasal sinuses*, located in the frontal, ethmoid, sphenoid, and maxillary bones, and the *nasolacrimal ducts* drain tears from the eyes. An excessive secretion of tears causes the nose to run as the tears drain into the nasal cavity. The auditory tube from the middle ear enters the upper respiratory tract posterior to the nasal cavity in the nasopharynx. With all these accessory connections, it is no

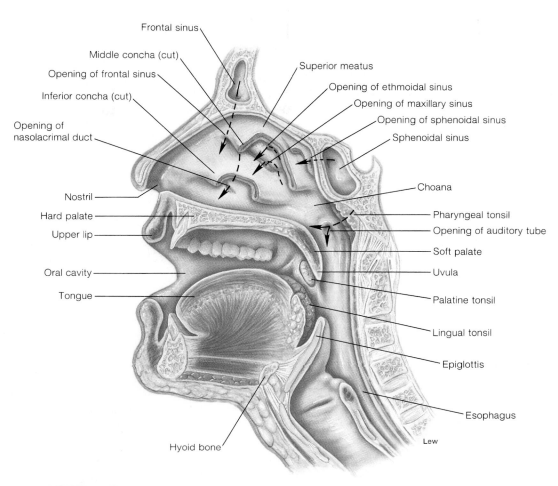

**FIGURE 24.2**
A sagittal section of the head showing the structures of the upper respiratory tract.

wonder that infections can spread so easily from one chamber to another throughout the facial area. To avoid causing damage or spreading infections to other areas, one must be careful not to blow the nose too forcefully.

The **pharynx** (*far´ingks*) is a funnel-shaped passageway, approximately 13 cm (5 in.) long that connects the nasal and oral cavities to the larynx at the base of the skull. The supporting walls of the pharynx are composed of skeletal muscle and the lumen is lined with a mucous membrane. Within the pharynx are several paired lymphoid organs, called **tonsils** (fig. 24.2). The pharynx has both respiratory and digestive functions and is divided on the basis of location and function into three regions: nasal, oral, and laryngeal.

The **nasopharynx** (*na˝zo-far´ingks*) has a respiratory function only. It is the uppermost portion of the pharynx, directly posterior to the nasal cavity and superior to the level of the soft palate. A pendulous **uvula** hangs from the middle lower border of the soft palate. The paired **auditory**

meatus: L. *meatus*, path

pharynx: L. *pharynx*, throat
uvula: L. *uvula*, small grape

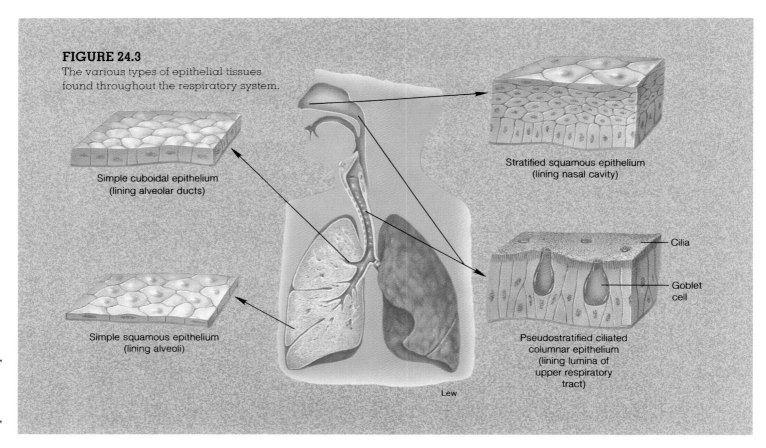

**FIGURE 24.3**
The various types of epithelial tissues found throughout the respiratory system.

Stratified squamous epithelium
(lining nasal cavity)

Simple cuboidal epithelium
(lining alveolar ducts)

Cilia

Goblet cell

Simple squamous epithelium
(lining alveoli)

Pseudostratified ciliated columnar epithelium
(lining lumina of upper respiratory tract)

Lew

**FIGURE 24.4**
A color-enhanced scanning electron micrograph of a bronchial wall showing cilia. In the trachea and bronchi, there are about 300 cilia per cell. The cilia move mucus-dust packages toward the pharynx, where they can either be swallowed or expectorated.

(eustachian) **tubes** connect the nasopharynx with the middle-ear cavities. A collection of lymphoid tissue called the **pharyngeal tonsils,** or **adenoids,** are situated in the posterior wall of this cavity. During the act of swallowing, the soft palate and uvula are elevated to block the nasal cavity and prevent food from entering. Occasionally a person may suddenly exhale air (as with a laugh) while in the process

· · · · · · · · · · · · · · · · · · · · · · · · · · · · · · · ·
adenoid: Gk. *adenoeides,* glandlike

of swallowing fluid. If this occurs before the uvula effectively blocks the nasopharynx, fluid will be discharged through the nasal cavity.

The **oropharynx** (*o″ro-far′ingks*) is the middle portion of the pharynx, between the soft palate and the level of the hyoid bone. The base of the tongue forms the anterior wall of the oropharynx. Paired **palatine tonsils** are located along the posterior lateral wall of the oropharynx, and the **lingual tonsils** are found on the base of the tongue. This portion of the pharynx has both a respiratory and a digestive function.

The **laryngopharynx** (*lă-ring″go-far′ingks*) is the lowermost portion of the pharynx. It extends posteriorly from the level of the hyoid bone to the larynx and opens into the esophagus and larynx. It is at the lower laryngopharynx that the respiratory and digestive systems become distinct. Swallowed food and fluid is directed into the esophagus, whereas inhaled air is moved anteriorly into the larynx.

 During a physical examination, a physician commonly depresses the patient's tongue and examines the condition of the palatine tonsils. Tonsils are pharyngeal lymphoid organs and tend to become swollen and inflamed after persistent infections. As mentioned in chapter 23, tonsils may have to be surgically removed when they become so overrun with pathogens after repeated infections that they themselves become the source of infection. The removal of the palatine tonsils is called a *tonsillectomy,* whereas the removal of the pharyngeal tonsils is called an *adenoidectomy.*

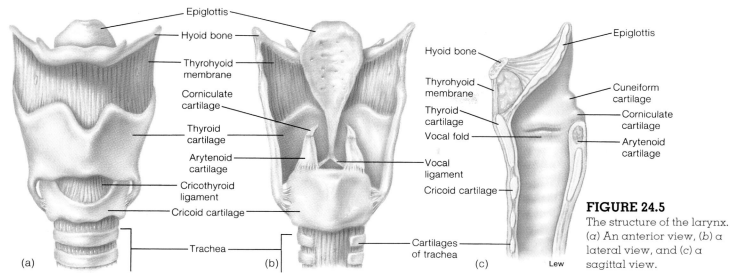

FIGURE 24.5

The structure of the larynx. (a) An anterior view, (b) a lateral view, and (c) a sagittal view.

## Larynx

The **larynx** (*lar´ingks*), or voice box, forms the entrance into the lower respiratory system as it connects the laryngopharynx with the trachea. It is positioned in the anterior midline of the neck at the level of the fourth through sixth cervical vertebrae. The larynx functions to prevent food or fluid from entering the trachea and lungs during swallowing and to permit passage of air while breathing. A secondary function is to produce sounds.

The larynx is shaped like a triangular box (fig. 24.5). It is composed of nine cartilages: three are large, single structures and six are smaller, paired structures. The largest of the unpaired cartilages is the anterior **thyroid cartilage.** The **laryngeal prominence** of the thyroid cartilage, commonly called the "Adam's apple," forms an anterior vertical ridge along the larynx that can be palpated on the midline of the neck. The thyroid cartilage is typically larger and more prominent in males than in females because of the effect of testosterone on the development of the larynx during puberty.

The spoon-shaped **epiglottis** has a cartilaginous framework. It is behind the root of the tongue and aids in closing the **glottis,** or laryngeal opening, during swallowing. The lower end of the larynx is formed by the ring-shaped **cricoid** (*kri´koid*) **cartilage.** This third unpaired cartilage connects the thyroid cartilage above and the trachea below. The paired **arytenoid** (*ar´´ĭ-te´noid*) **cartilages,** located above the cricoid cartilage and behind cartilage the thyroid cartilage, furnish the attachment of the **vocal cords.** The other paired **cuneiform cartilages** and **corniculate** (*kor-nik´yŭ-lāt*) **cartilages** are small accessory cartilages that are closely associated with the arytenoid cartilages (fig. 24.6).

....................................................
thyroid: Gk. *thyreos,* shieldlike
arytenoid: Gk. *arytaina,* ladle or cup-shaped
cuneiform: L. *cuneus,* wedge-shaped
corniculate: L. *corniculum,* diminutive of *cornu,* horn

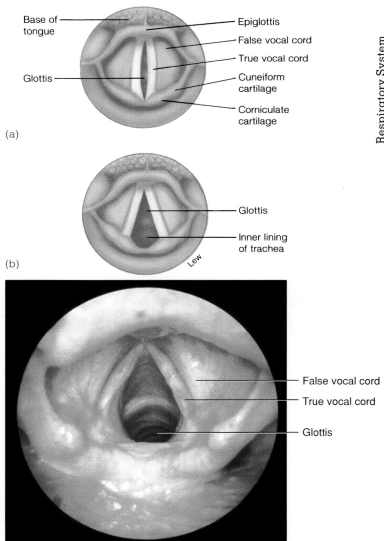

FIGURE 24.6

A superior view of the vocal cords. In (a) the vocal cords are taut; in (b) they are relaxed and the glottis is opened. (c) A photograph through a laryngoscope showing the true and false vocal cords and the glottis.

Two pairs of strong connective tissue bands are stretched across the upper opening of the larynx from the thyroid cartilage anteriorly to the paired arytenoid cartilages posteriorly. These are the **true vocal cords** and the **false vocal cords** (fig. 24.6). The false vocal cords support the true vocal cords and, as their name implies, are not used in sound production. The true vocal cords vibrate in the production of sound.

The **laryngeal muscles** are extremely important in closing the glottis during the act of swallowing and in speech. There are two groups of laryngeal muscles: **extrinsic muscles,** responsible for elevating the larynx during swallowing, and **intrinsic muscles** that, when contracted, change the length, position, and tension of the vocal cords. Various pitches are produced as air passes over the altered vocal cords. If the vocal cords are taut, vibration is more rapid and causes a higher pitch. Less tension on the cords produces lower sounds. Mature males generally have thicker and longer vocal cords than females; therefore, the vocal cords of males vibrate more slowly and produce lower pitches.

During the act of swallowing, the larynx is elevated to close the glottis against the epiglottis. This movement can be noted by cupping the fingers lightly over the larynx and then swallowing. Food may become lodged within the glottis if it is not closed as it should be. In this case, the *abdominal thrust* (Heimlich) *maneuver* can be used to prevent suffocation (see page 721).

## Trachea and Bronchial Tree

The **trachea** (*tra´ke-ă*), commonly called the windpipe, is a rigid tube, approximately 12 cm (4½ in.) long and 2.5 cm (1 in.) in diameter. It is positioned anterior to the esophagus and connects the larynx to the primary bronchi (fig. 24.7).

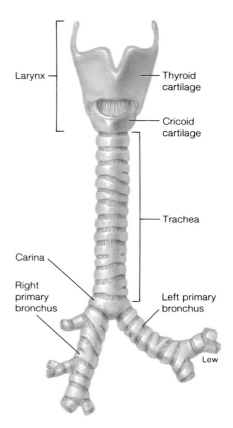

**FIGURE 24.7**
An anterior view of the larynx, trachea, and bronchi.

A series of 16 to 20 **C**-shaped rings of hyaline cartilage form the walls of the trachea (fig. 24.8). The open part of the C is positioned posteriorly and is covered by fibrous connective tissue and smooth muscle. The cartilages provide a rigid but flexible tube that allows the airway to be permanently open.

**FIGURE 24.8**
Histology of the trachea. (a) A photomicrograph showing the relationship of the trachea to the esophagus (3×) and (b) a photomicrograph of tracheal cartilage (63×).

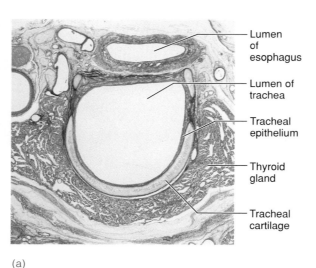

(a)

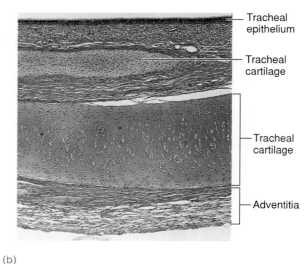

(b)

trachea: L. *trachia*, rough air vessel

The division of the trachea into right and left bronchi is reinforced by the **carina** (kă-ri´nă), a keel-like cartilage plate (see fig. 24.7).

If the trachea becomes occluded, as by aspiration of a foreign object, it may be necessary to create an emergency opening so that ventilation can still occur. A *tracheotomy* is the process of surgically opening the trachea, and a *tracheostomy* is the procedure of inserting a tube into the trachea to permit breathing and to keep the passageway open. A tracheotomy should be performed only by a competent physician because of the great risk of cutting a recurrent laryngeal nerve or the common carotid artery, which is also in this area.

The **bronchial tree** is so named because it is composed of a series of respiratory tubes that branch into progressively narrower tubes as they extend into the lung. The trachea bifurcates into a right and left **primary bronchus** (brong´kus). Each bronchus has hyaline cartilage rings surrounding its lumen to keep it open as it extends into the lung. Because of the more vertical position of the right bronchus, foreign particles are more likely to lodge here than in the left bronchus.

The bronchus divides deeper in the lungs to form **secondary bronchi** and **segmental (tertiary) bronchi.** The

........................................

carina: L. *carina*, keel
bronchus: L. *bronchus*, windpipe

bronchial tree continues to branch into yet smaller tubules called **bronchioles** (brong´ke-olz) (fig. 24.9). There is almost no cartilage in the bronchioles, and the smooth muscle in their walls can constrict or dilate these airways. Bronchioles provide the greatest resistance to air flow in the conducting passages, and thus their function is analogous to that of the arterioles in the vascular system. A simple cuboidal epithelium lines the bronchioles rather than the pseudostratified ciliated columnar epithelium that lines the bronchi (see fig. 24.3). Numerous **terminal bronchioles** mark the end of the air-conducting pathway to the alveoli.

Regardless of the temperature and humidity of the atmosphere, when the inspired air reaches the respiratory zone it is at a temperature of 37° C (body temperature) and it is saturated with water vapor. This ensures that a constant internal body temperature will be maintained and that delicate lung tissue will be protected from desiccation.

Mucus secreted by cells of the conducting zone serves to trap small particles in the inspired air and thereby performs a filtration function. This mucus is moved along at a rate of 1 to 2 centimeters per minute by cilia projecting from the tops of epithelial cells that line the conducting zone (fig. 15.7). About 300 cilia per cell beat in a coordinated fashion to move mucus toward the pharynx, where it can either be swallowed or expectorated.

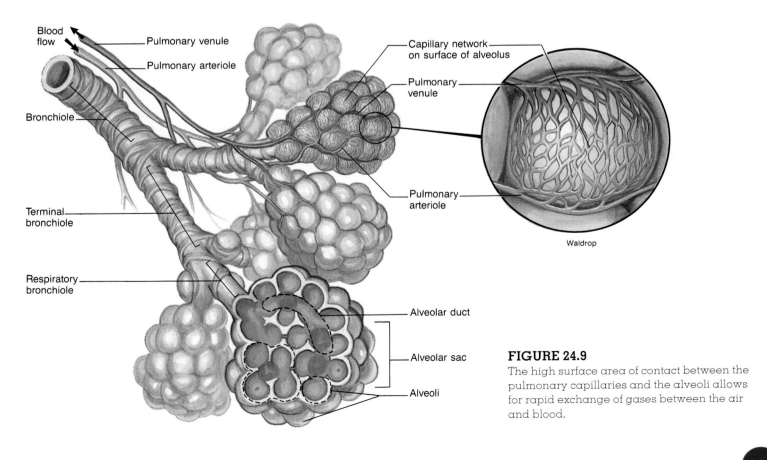

Blood flow
Pulmonary venule
Pulmonary arteriole
Bronchiole
Terminal bronchiole
Respiratory bronchiole
Capillary network on surface of alveolus
Pulmonary venule
Pulmonary arteriole
Waldrop
Alveolar duct
Alveolar sac
Alveoli

**FIGURE 24.9**

The high surface area of contact between the pulmonary capillaries and the alveoli allows for rapid exchange of gases between the air and blood.

As a result of this filtration function, particles larger than about 6 μm do not normally enter the respiratory zone of the lungs. The importance of this function is evidenced by the disease called *black lung*, which occurs in miners who inhale too much carbon dust and therefore develop pulmonary fibrosis (as described in a later section). The alveoli themselves are normally kept clean by the action of macrophages that reside within them. The cleansing action of cilia and macrophages in the lungs has been shown to be diminished by cigarette smoke.

# Alveoli, Lungs, and Pleurae

*Alveoli are the functional units of the lungs, where gas exchange occurs. Right and left lungs are separately contained in pleural membranes.*

## Alveoli

Air from the terminal bronchioles enters the **alveolar ducts.** The alveolar ducts contain individual **alveoli** as outpouchings along their length and open into clusters of alveoli called **alveolar sacs** at their ends (fig. 24.9). The last three structures make up the *respiratory division* of the lungs.

Gas exchange in the lungs occurs across about 300 million tiny (0.25–0.50 mm in diameter) alveoli. The enormous number of these structures provides a large surface area (60–80 square meters, or about 760 square feet) for diffusion of gases. The diffusion rate is further increased by the fact that each alveolus is only one cell layer thick, so that the total "air-blood barrier" is only two cells across (an alveolar cell and a capillary endothelial cell), or about 2 μm. This is an average distance because the type II alveolar cells are thicker than the type I cells (fig. 24.10). Where the basement membranes of capillary endothelial cells fuse with those of type I alveolar cells, the diffusion distance is less than 1 μm.

Alveoli are polyhedral and are usually clustered together, like the units of a honeycomb (fig. 24.11). As mentioned previously, individual alveoli also occur as separate outpouchings along the length of the alveolar ducts. Although the distance between each alveolar duct and its terminal alveoli is only about 0.5 mm, these units together constitute most of the mass of the lungs.

The enormous surface area of alveoli and the short diffusion distance between alveolar air and the capillary blood quickly bring the blood into gaseous equilibrium with the alveolar air. This function is further aided by the fact that each alveolus is surrounded by so many capillaries that they form an almost continuous sheet of blood around each alveolus (see fig. 24.9).

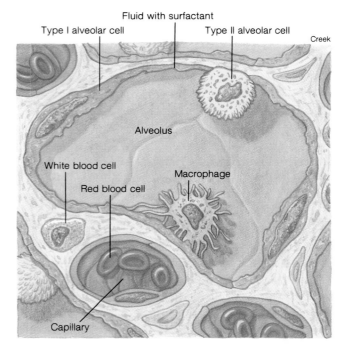

**FIGURE 24.10**
A diagram showing the relationship between lung alveoli and pulmonary capillaries.

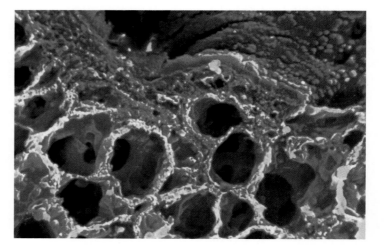

**FIGURE 24.11**
A color-enhanced scanning electron micrograph of lung tissue showing alveolar sacs and a bronchiole.

## Lungs

The **lungs** are large, spongy, paired organs within the thoracic cavity. They lie against the rib cage anteriorly and posteriorly and extend from the diaphragm to a point just above the clavicles. The lungs are separated from one another by the heart and other structures of the **mediastinum** (me˝de-ă-sti´num), as shown in figure 24.12. The mediastinum is the area between

···········································

alveolus: L. diminutive of *alveus*, cavity

···········································

mediastinum: L. *mediastinus*, intermediate

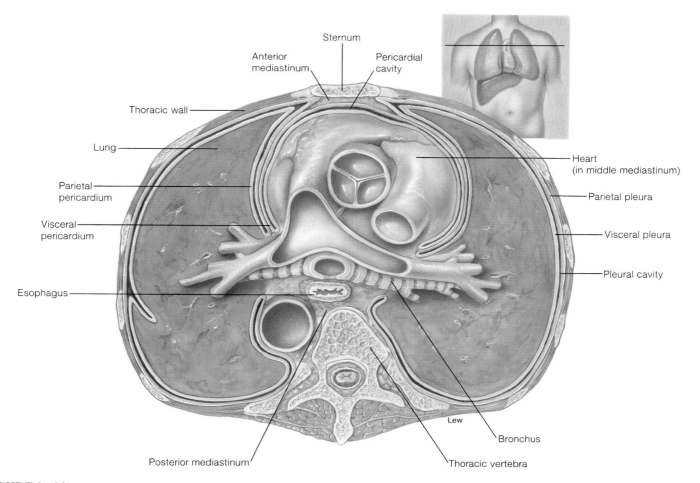

Labels on figure:
Sternum
Anterior mediastinum
Pericardial cavity
Thoracic wall
Lung
Heart (in middle mediastinum)
Parietal pericardium
Parietal pleura
Visceral pericardium
Visceral pleura
Pleural cavity
Esophagus
Lew
Bronchus
Posterior mediastinum
Thoracic vertebra

**FIGURE 24.12**

A cross section of the thoracic cavity showing the mediastinum and pleural membranes.

the lungs. All structures of the respiratory system beyond the primary bronchi, including the bronchial tree and alveoli, are contained in the lungs.

Each lung presents three surfaces that match the contour of the thoracic cavity. The mediastinal (medial) surface of each lung is slightly concave and contains a vertical slit, the **hilum** (*hi´lum*), through which pulmonary vessels, nerves, and bronchi traverse. The inferior border of the lung, called the **base,** is concave as it fits over the convex dome of the diaphragm. Anteriorly, the portion of each lung that extends above the level of the clavicle is called the **apex.** Finally, the broad, rounded surface in contact with the rib cage is called the **costal surface.**

Although the right and left lungs are basically similar, there are some distinct differences. The left lung is somewhat smaller than the right and has a **cardiac notch** on its medial surface to accommodate the heart. The left lung is subdivided into a **superior lobe** and an **inferior lobe** by a

single fissure. The right lung is subdivided by two fissures into **superior, middle,** and **inferior lobes** (fig. 24.13). Each lobe of the lung is divided into many small lobules, which in turn contain the alveoli. Lobular divisions of the lungs compose specific *bronchial segments*. The right lung contains 10 bronchial segments and the left lung contains 8 (fig. 24.13).

## Pleurae

**Pleurae** (*ploor´ē*) are serous membranes surrounding the lungs (see fig. 24.12). The **visceral pleura** adheres to the outer surface of the lung and extends into each of the interlobar fissures. The **parietal pleura** lines the thoracic walls and the thoracic surface of the diaphragm. A continuation of the parietal pleura around the heart and between the lungs forms the boundary of the mediastinum. Between the visceral and parietal pleurae is a moistened space called the **pleural cavity.** An inferiorly extending reflection of the pleural layers

hilum: L. *hilum,* a trifle (little significance)

pleura: Gk. *pleura,* side or rib

691

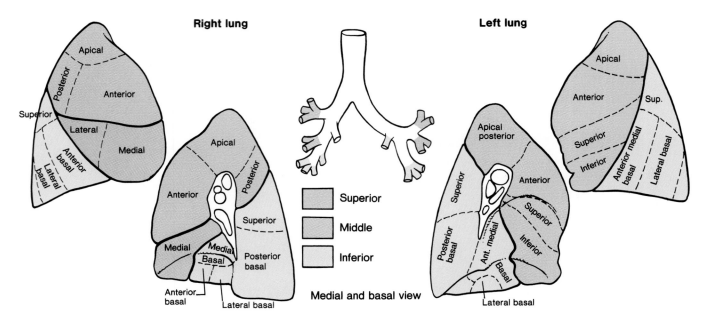

**Right lung**

**Left lung**

Apical

Posterior

Anterior

Superior

Lateral

Medial

Anterior basal

Lateral basal

Apical

Anterior

Posterior

Superior

Medial

Medial

Basal

Posterior basal

Anterior basal

Lateral basal

Apical

Anterior

Sup.

Apical posterior

Superior

Inferior

Superior

Anterior

Inferior

Anterior medial basal

Lateral basal

Posterior basal

Ant. medial

Superior

Inferior

Basal

Lateral basal

Superior

Middle

Inferior

Medial and basal view

**FIGURE 24.13**
Lobes, lobules, and bronchopulmonary segments of the lungs.

around the roots of each lung is called the **pulmonary ligament.** The pulmonary ligaments help to support the lungs. The normal position of the lungs in the thoracic cavity is shown in the radiograph in figure 24.14.

The membranes of the thoracic cavity serve to compartmentalize the different organs. There are four distinct compartments: two pleural cavities (one surrounding each lung); a pericardial cavity in which the heart is situated; and the mediastinum, in which the esophagus, thoracic duct, major vessels, various nerves, and portions of the respiratory tract are located. This *compartmentalization* has protective value because infections are usually confined to one compartment and damage to one organ will not usually involve another. *Pleurisy,* for example, which is an inflamed pleura, is generally confined to one side. A penetrating injury to one side, such as a knife wound, might cause one lung to collapse but not the other.

pulmonary: Gk. *pleumon,* lung

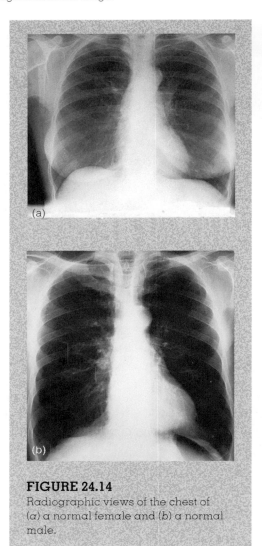

**FIGURE 24.14**
Radiographic views of the chest of (*a*) a normal female and (*b*) a normal male.

# Physical Aspects of Ventilation

*The movement of air into and out of the lungs occurs as a result of pressure differences induced by changes in lung volumes. Ventilation is thus influenced by the physical properties of the lungs, including their compliance, elasticity, and surface tension.*

Movement of air from the conducting division to the terminal bronchioles occurs as a result of the pressure difference between the two ends of the airways. Air flow through bronchioles, like blood flow through blood vessels, is directly proportional to the pressure difference and inversely proportional to the frictional resistance to flow. The pressure differences in the respiratory division are induced by changes in lung volumes. The physical properties of the lungs, including their compliance, elasticity, and surface tension, affect their functioning.

## Intrapulmonary and Intrapleural Pressures

The wet, serous membranes of the visceral and parietal pleurae are normally flush against each other, so that the lungs are stuck to the chest wall in the same manner that two wet pieces of glass stick to each other. The *intrapleural space* between the two wet membranes contains only a thin layer of fluid secreted by the pleural membranes. The pleural cavity in a healthy, living organism is thus potential rather than real; it can become real only in abnormal situations when air enters the intrapleural space. Since the lungs normally remain in contact with the chest wall, they get larger and smaller together with the thoracic cavity during respiratory movements.

Air enters the lungs during inspiration because the atmospheric pressure is greater than the **intrapulmonary,** or **alveolar, pressure.** Since the atmospheric pressure does not usually change, the intrapulmonary pressure must fall below atmospheric pressure to cause inspiration. A pressure below that of the atmosphere is called a *subatmospheric pressure,* or *negative pressure.* During quiet inspiration, for example, the intrapulmonary pressure may decrease to 3 mmHg below the pressure of the atmosphere. This subatmospheric pressure is commonly shown as –3 mmHg. Expiration, conversely, occurs when the intrapulmonary pressure is greater than the atmospheric pressure. During quiet expiration, for example, the intrapulmonary pressure may rise to at least +3 mmHg over the atmospheric pressure.

The lack of air in the intrapleural space produces a subatmospheric **intrapleural pressure** that is lower than the intrapulmonary pressure (table 24.1). Thus, there is a pressure difference across the wall of the lung—the **transpulmonary pressure**—which is the difference between the intrapulmonary pressure and the intrapleural pressure. Since the pressure within the lungs (intrapulmonary pressure) is greater than that outside the lungs (intrapleural pressure), the difference in pressure (transpulmonary pressure) acts to expand the lungs as the thoracic volume expands during inspiration.

**Boyle's Law**   Changes in intrapulmonary pressure occur as a result of changes in lung volume. This follows from **Boyle's law,** which states that the pressure of a given quantity of gas is inversely proportional to its volume. An increase in lung volume during inspiration decreases intrapulmonary pressure to subatmospheric levels; air therefore goes in. A decrease in lung volume raises the intrapulmonary pressure to levels above that of the atmosphere, thus pushing air out. These changes in lung volume occur as a consequence of changes in thoracic volume, as we will describe in a later section on the mechanics of breathing.

...........................................

Boyle's law: from Robert Boyle, Irish-born British physicist, 1627–91

| Table 24.1 | Intrapulmonary and intrapleural pressures in normal, quiet breathing, and the transpulmonary pressure acting to expand the lungs | | |
|---|---|:---:|:---:|
| | | *Inspiration* | *Expiration* |
| Intrapulmonary pressure (mmHg) | | –3 | +3 |
| Intrapleural pressure (mmHg) | | –6 | –3 |
| Transpulmonary pressure (mmHg) | | +3 | +6 |

Note: Pressures indicate mmHg below or above atmospheric pressure. Intrapleural pressure is normally always negative (subatmospheric).

## Physical Properties of the Lungs

In order for inspiration to occur, the lungs must be able to expand when stretched—they must have high *compliance*. In order for expiration to occur, the lungs must get smaller when this stretching force is released—they must have *elasticity*. The tendency to get smaller is also aided by forces produced by *surface tension* within the alveoli.

**Compliance**   The lungs are very distensible, about 100 times more distensible than a toy balloon. Another term for distensibility is **compliance** (*kom-pli´ans*), which is defined as the change in lung volume as a function of change in the transpulmonary pressure. A given transpulmonary pressure, in other words, will cause greater or lesser expansion, depending on the compliance of the lungs.

The compliance of the lungs is reduced by factors that produce a resistance to distension. As an extreme example, if the lungs were filled with concrete, a given transpulmonary pressure would produce no increase in lung volume and no air would enter; the compliance would be zero. The infiltration of lung tissue with connective tissue proteins, a condition called *pulmonary fibrosis*, similarly decreases lung compliance.

**Elasticity**   The term *elasticity* refers to the tendency of a structure to return to its initial size after being distended. A high content of elastin proteins makes the lungs very elastic, so that they resist distension. Since the lungs are normally stuck to the chest wall, they are always in a state of elastic tension. This tension increases during inspiration when the lungs are stretched and is reduced by elastic recoil during expiration. The elasticity of the lungs and of other thoracic structures aids in pushing the air out during expiration.

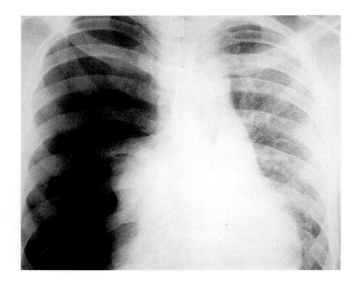

**FIGURE 24.15**

A pneumothorax of the right lung. The right side of the thorax appears uniformly dark because it is filled with air; the spaces between the ribs are also greater than those on the left due to release from the elastic tension of the lungs. The left lung appears denser (less dark) because of shunting of blood from the right to the left lung.

 The elastic nature of lung tissue becomes apparent when air enters the intrapleural space (as a result of an open chest wound, for example). This condition, called a *pneumothorax* (noo″mo-thor′aks), is shown in figure 24.15. As air accumulates in the intrapleural space, the intrapleural pressure rises until it is equal to the atmospheric pressure. When the intrapleural pressure is the same as the intrapulmonary pressure, the lung can no longer expand. Not only does the lung not expand during inspiration, it actually collapses away from the chest wall as a result of elastic recoil. Fortunately, a pneumothorax usually causes only one lung to collapse, since each lung is in a separate pleural compartment.

**Surface Tension**   The forces that act to resist distension include elastic resistance and **surface tension** that is exerted by fluid in the alveoli. Although the alveoli are relatively dry, they do contain a very thin film of fluid, much like soap bubbles. Surface tension is created because water molecules at the surface are attracted more to other water molecules than to air. As a result, the surface water molecules are pulled tightly together by attractive forces from underneath.

The surface tension of an alveolus produces a force that is directed inward and, as a result, creates pressure within the alveolus. As described by the **law of LaPlace,** the pressure created is directly proportional to the surface tension and inversely proportional to the radius of the alveolus (fig. 24.16). According to this law, the pressure in a

..........................................

pneumothorax: Gk. *pneumon*, spirit (breath); L. *thorax*, chest

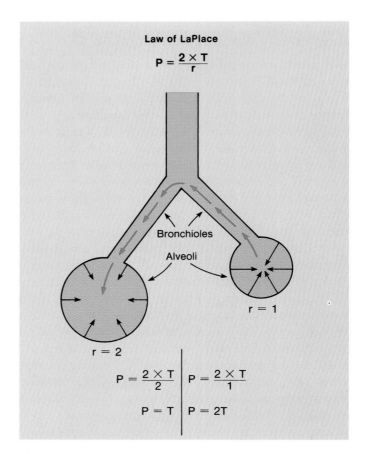

**FIGURE 24.16**

According to the law of LaPlace, the pressure created by surface tension should be greater in the smaller alveolus (*right*) than in the larger alveolus (*left*). This implies that (without surfactant) smaller alveoli would collapse and empty their air into larger alveoli.

smaller alveolus would be greater than that in a larger alveolus if the surface tension were the same in both. The greater pressure of the smaller alveolus would then cause it to empty its air into the larger one (fig. 24.16). This does not normally occur because, as an alveolus decreases in size, its surface tension (the numerator in the equation) is decreased as its radius (the denominator) is reduced. The cause of the reduced surface tension, which prevents the alveoli from collapsing, is described in the next section.

## Surfactant and the Respiratory Distress Syndrome

Alveolar fluid contains a phospholipid known as dipalmitoyl lecithin, probably attached to a protein, which functions to lower surface tension. This compound is called **surfactant** (sur′fak′tant)—a contraction of the term *surface-active agent*. Because of the presence of lung surfactant, the surface tension in the alveoli is lower than would be predicted if surfactant were absent. Further, the ability of lung surfactant to lower surface tension improves as the alveoli get smaller during expiration. This may be because the surfactant

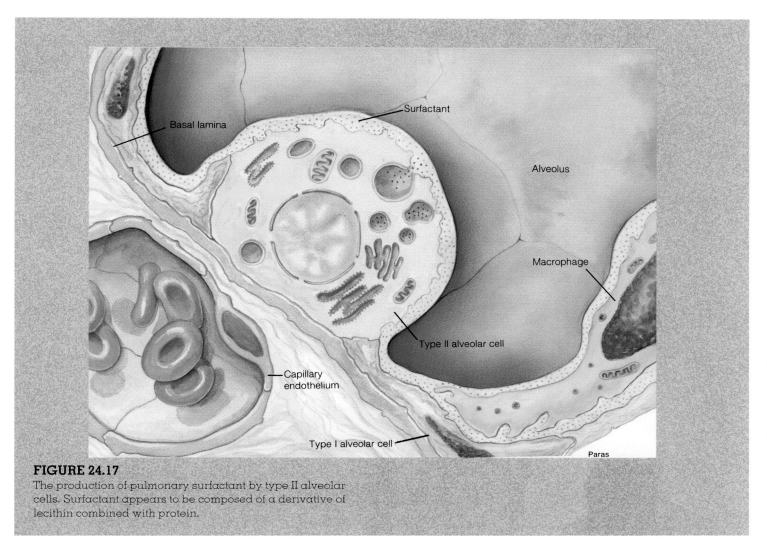

Basal lamina

Surfactant

Alveolus

Macrophage

Type II alveolar cell

Capillary endothelium

Type I alveolar cell

Paras

**FIGURE 24.17**
The production of pulmonary surfactant by type II alveolar cells. Surfactant appears to be composed of a derivative of lecithin combined with protein.

molecules become more concentrated as the alveoli get smaller. Lung surfactant thus prevents the alveoli from collapsing during expiration, as would be predicted from the law of LaPlace. Even after a forceful expiration, the alveoli remain open and a *residual volume* of air remains in the lungs. Since the alveoli do not collapse, less surface tension has to be overcome to inflate them at the next inspiration.

Surfactant is produced by type II alveolar cells (fig. 24.17) in late fetal life. Since surfactant does not start to be produced until about the eighth month, premature babies are sometimes born with lungs that lack sufficient surfactant, and their alveoli are collapsed as a result. This condition is called **respiratory distress syndrome.** It is also called **hyaline membrane disease** because the high surface tension causes plasma fluid to leak into the alveoli, producing a glistening membrane appearance (and pulmonary edema). This condition does not occur in all premature babies; the rate of lung development depends on hormonal conditions (thyroxine and hydrocortisone primarily) and on genetic factors.

 Even under normal conditions, the first breath of life is a difficult one because the newborn must overcome great surface tension forces in order to inflate its partially collapsed alveoli. The transpulmonary pressure required for the first breath is 15 to 20 times that required for subsequent breaths, and an infant with *respiratory distress syndrome* must duplicate this effort with every breath. Fortunately, many babies with this condition can be saved by mechanical ventilators and by exogenous surfactant delivered to the baby's lungs by means of an endotracheal tube. The exogenous surfactant may be a synthetic mixture of phospholipids or it may be surfactant obtained from bovine lungs.

# Mechanics of Breathing

*Normal, quiet inspiration results from muscle contraction, and normal expiration from muscle relaxation and elastic recoil. These actions can be forced by contractions of the accessory respiratory muscles. The amount of air inspired and expired can be measured in a number of ways to test pulmonary function.*

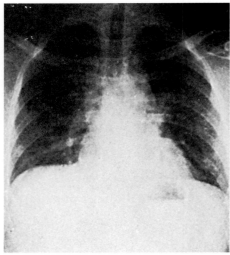

(a)

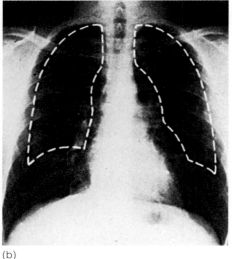

(b)

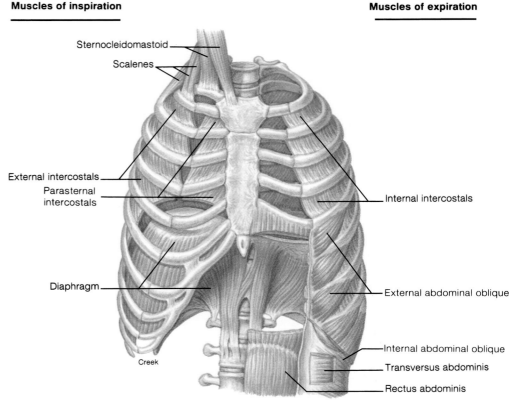

Sternocleidomastoid

Scalenes

External intercostals

Parasternal intercostals

Internal intercostals

Diaphragm

External abdominal oblique

Internal abdominal oblique

Transversus abdominis

Rectus abdominis

Creek

(a)

(b)

**FIGURE 24.19**

The muscles of respiration. (*a*) The principal muscles of inspiration and (*b*) the principal muscles of forced expiration. For the most part, expiration is passive.

**FIGURE 24.18**

A change in lung volume, as shown by radiographs, during expiration (*a*) and inspiration (*b*). The increase in lung volume during full inspiration is shown by comparison with the lung volume in full expiration (*dashed lines*).

The thorax must be sufficiently rigid so that it can protect vital organs and provide attachments for many short, powerful muscles. Breathing, or **pulmonary ventilation** requires that the thorax be flexible to function as a bellows during the ventilation cycle. The rigidity and the surfaces for muscle attachment are provided by the bony composition of the rib cage. The rib cage is pliable, however, because the ribs are separate from one another and because most ribs (the upper 10 of the 12 pairs) are attached to the sternum by resilient costal cartilages. The vertebral attachments likewise provide considerable mobility. The structure of the rib cage and associated cartilages provide continuous elastic tension, so that when stretched by muscle contraction

during inspiration, the rib cage can return passively to its resting dimensions when the muscles relax. This elastic recoil is greatly aided by the elasticity of the lungs.

Pulmonary ventilation consists of two phases, *inspiration* and *expiration*. Inspiration (inhalation) and expiration (exhalation) are accomplished by alternately increasing and decreasing the volumes of the thorax and lungs (fig. 24.18).

## *Inspiration and Expiration*

The thoracic cavity increases in size during inspiration as a result of skeletal muscle contraction. The major expansion of the thoracic volume is produced by the downward contraction of the dome-shaped **diaphragm,** which lowers to a more flattened shape. During *quiet inspiration*, the thoracic volume is also increased by contraction of intercostal muscles, which elevates the ribs and expands the rib cage.

Between the bony portions of the rib cage are two layers of intercostal muscles: the **external intercostal muscles** and the **internal intercostal muscles** (fig. 24.19). Between the costal cartilages, however, there is only one muscle layer and it has fibers oriented in a manner similar to those of

..............................................

diaphragm: Gk. *dia*, across; *phragma*, fence

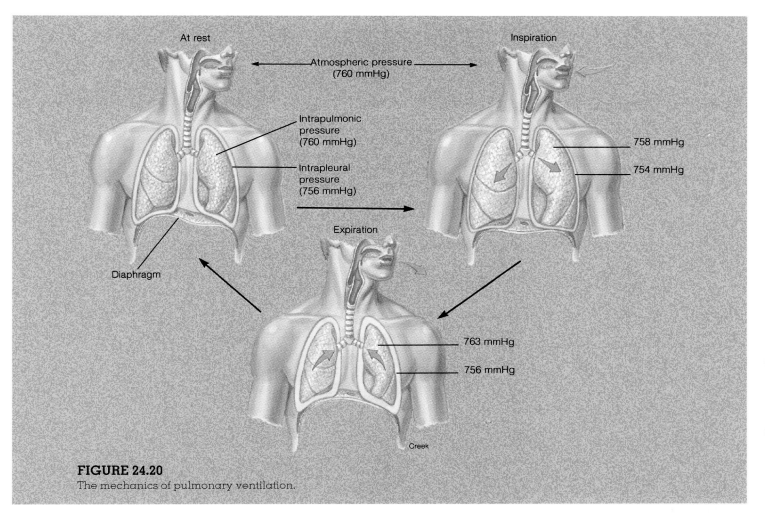

At rest

Atmospheric pressure
(760 mmHg)

Intrapulmonic
pressure
(760 mmHg)

Intrapleural
pressure
(756 mmHg)

Diaphragm

Inspiration

758 mmHg

754 mmHg

Expiration

763 mmHg

756 mmHg

Creek

**FIGURE 24.20**
The mechanics of pulmonary ventilation.

the internal intercostal muscles. These muscles may thus be called the *intercondral part* of the internal intercostal muscles. Another name for them is the **parasternal intercostal muscles.** The parasternal intercostal muscles function along with the external intercostal muscles to expand the rib cage during inspiration.

Other thoracic muscles become involved in *forced* (deep) *inspiration,* including the scalenes and sternocleidomastoid muscles of the neck and the pectoralis minor muscle of the chest. Contraction of these muscles elevates the ribs in an anteroposterior direction; at the same time, the upper rib cage is stabilized so that the intercostal muscles become more effective.

*Quiet expiration* is a passive process. After becoming stretched by contractions of the diaphragm and thoracic muscles, the thorax and lungs recoil as a result of their elastic tension when the respiratory muscles relax. The decrease in lung volume raises the pressure within the alveoli above the atmospheric pressure and pushes the air out.

During *forced expiration,* the internal intercostal muscles (excluding the interchondral part) contract and depress the rib cage. The abdominal muscles may also aid expiration because, when contracted, they force abdominal organs up against the diaphragm, further decreasing the volume of the thorax. By this means, the intrapulmonary pressure within the pleural cavities can rise to 20 or 30 mmHg above atmospheric pressure.

The events that occur during inspiration and expiration are shown in figure 24.20.

## Pulmonary Function Tests

Pulmonary function may be assessed clinically through a technique known as **spirometry.** In this procedure, a subject breathes in a closed system in which air is trapped within a light plastic bell floating in water. The bell moves up when the subject exhales and down when the subject inhales. The movements of the bell cause corresponding movements of a pen, which traces a record of the breathing on a rotating drum recorder (fig. 24.21).

An example of a spirogram is shown in figure 24.22, and the various lung volumes and capacities are defined in table 24.2. A lung capacity is equal to the sum of two or more lung volumes. During quiet breathing, for example, the

spirometry: L. *spiro,* to breathe; Gk. *metron,* measure

**FIGURE 24.21**
A spirometer. With the exception of the residual volume, which is measured using special techniques, this instrument can determine respiratory volumes and capacities.

the tidal volume is 500 ml, the percent of fresh air reaching the alveoli is $350/500 \times 100\% = 70\%$. If the tidal volume is increased to 2000 ml, the percent of fresh air reaching the alveoli is $1850/2000 \times 100\% = 93\%$. An increase in tidal volume can thus be a factor in the respiratory adaptations to exercise and high altitude, as will be described in later sections.

Spirometry is useful in the diagnosis of lung diseases. On the basis of pulmonary function tests, lung disorders can be classified as **restrictive** or **obstructive.** In restrictive disorders, such as *pulmonary fibrosis,* the vital capacity is reduced to below normal. The rate at which the vital capacity can be forcibly exhaled, however, is normal. In disorders that are exclusively obstructive, such as *asthma,* the vital capacity is normal because lung tissue is not damaged. In asthma the bronchioles constrict, and this bronchoconstriction increases the resistance to air flow. Although the vital capacity is normal, the increased airway resistance makes expiration more difficult and take a longer time. Obstructive disorders are thus diagnosed by tests that measure the rate of expiration. One such test is the **forced expiratory volume (FEV),** in which the percentage of the vital capacity that can be exhaled in the first second ($FEV_{1.0}$) is measured (fig. 24.23). An $FEV_{1.0}$ that is significantly less than 75% to 85% suggests the presence of obstructive pulmonary disease.

 *Bronchoconstriction* often occurs in response to the inhalation of noxious agents, such as smoke or smog. The $FEV_{1.0}$ has, therefore, been used by researchers to determine the effects of various components of smog and of passive cigarette smoke inhalation on pulmonary function. These studies have shown that it is unhealthy to exercise on very smoggy days and that inhalation of smoke from other people's cigarettes in a closed environment can measurably affect pulmonary function.

amount of air expired in each breath is the **tidal volume.** The maximum amount of air that can be forcefully exhaled after a maximum inhalation is called the **vital capacity** and is equal to the sum of the **inspiratory reserve volume, tidal volume,** and **expiratory reserve volume** (fig. 24.22). Multiplying the tidal volume at rest by the number of breaths per minute yields a **total minute volume** of about 6 L per minute. During exercise, the tidal volume and the number of breaths per minute increase to produce a total minute volume as high as 100 to 200 L per minute.

It should be noted that not all of the inspired volume reaches the alveoli with each breath. As the fresh air is inhaled, it is mixed with air in the **anatomical dead space** (table 24.3). This dead space comprises the conducting zone of the respiratory system—nose, mouth, larynx, trachea, bronchi, and bronchioles—where no gas exchange occurs. Air within the anatomical dead space has a lower oxygen concentration and a higher carbon dioxide concentration than the external air. Since the air in the dead space enters the alveoli first, the amount of fresh air reaching the alveoli with each breath is less than the tidal volume, But, since the volume of air in the dead space is an anatomical constant, the percentage of fresh air entering the alveoli is increased with increasing tidal volumes. For example, if the anatomical dead space is 150 ml and

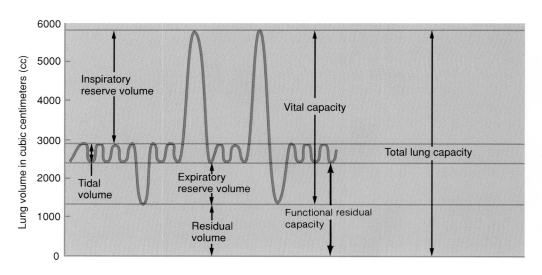

**FIGURE 24.22**
A spirogram showing lung volumes and capacities.

Chapter Twenty-Four

## Table 24.2 — Terminology used to describe lung volumes and capacities

| Term | Definition | Term | Definition |
|---|---|---|---|
| **Lung volumes** | The four nonoverlapping components of the total lung capacity | **Lung capacities** | Measurements that are the sum of two or more lung volumes |
| Tidal volume | The volume of gas inspired or expired in an unforced respiratory cycle | Total lung capacity | The total amount of gas in the lungs at the end of a maximum inspiration |
| Inspiratory reserve volume | The maximum volume of gas that can be inspired during forced breathing in addition to tidal volume | Vital capacity | The maximum amount of gas that can be expired after a maximum inspiration |
| Expiratory reserve volume | The maximum volume of gas that can be expired during forced breathing in addition to tidal volume | Inspiratory capacity | The maximum amount of gas that can be inspired at the end of a tidal expiration |
| Residual volume | The volume of gas remaining in the lungs after a maximum expiration | Functional residual capacity | The amount of gas remaining in the lungs at the end of a tidal expiration |

People with pulmonary disorders frequently complain of *dyspnea* (*disp´-ne-ă*), which is a subjective feeling of "shortness of breath." Dyspnea may occur even when ventilation is normal, however, and may not occur even when total minute volume is very high, as in exercise. Some of the terms used to describe ventilation are defined in table 24.3.

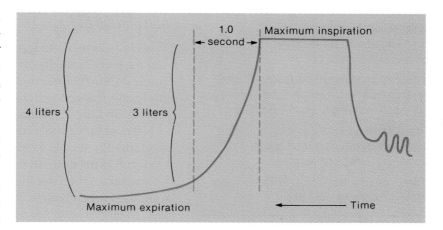

$$FEV_{1.0} = \frac{3L}{4L} \times 100\% = 75\%$$

**FIGURE 24.23**

An illustration of 1-second forced expiratory volume ($FEV_{1.0}$) spirometry test for detecting obstructive pulmonary disorders.

# Gas Exchange in the Lungs

*Gas exchange between the alveolar air and the blood in pulmonary capillaries results in an increased oxygen concentration and a decreased carbon dioxide concentration in the blood leaving the lungs. This blood enters the systemic arteries, where blood gas measurements are taken to assess the effectiveness of lung function.*

The atmosphere is an ocean of gas that exerts pressure on all objects within it. The pressure exerted can be measured using a glass U-tube filled with fluid. One side of the U-tube is open to the atmosphere, while the other side is continuous with a sealed vacuum tube. Since the atmosphere presses on the

open-ended side, but not on the side connected to the vacuum tube, atmospheric pressure pushes fluid in the U-tube up on the vacuum side to a height determined by the atmospheric pressure and the density of the fluid. Water, for example, will be pushed up to a height of 33.9 ft. (10,332 mm) at sea level, whereas mercury (Hg)—which is more dense—will be raised to a height of 760 mm. As a matter of convenience, therefore, devices used to measure atmospheric pressure (barometers) use mercury rather than water. The atmospheric pressure at sea level is thus said to be equal to 760 mmHg (or *760 torr*), which is also described as a pressure of *one atmosphere* (fig. 24.24).

# Table 24.3    Ventilation terminology

| Term | Definition |
|---|---|
| Air spaces | Alveolar ducts, alveolar sacs, and alveoli |
| Airways | Structures that conduct air from the mouth and nose to the respiratory bronchioles |
| Alveolar ventilation | Removal and replacement of gas in alveoli; equal to the tidal volume minus the volume of dead space times the ventilation rate |
| Anatomical dead space | Volume of the conducting airways to the zone gas exchange occurs |
| Apnea | Cessation of breathing |
| Dyspnea | Unpleasant subjective feeling of difficult or labored breathing |
| Eupnea | Normal, comfortable breathing at rest |
| Hyperventilation | Alveolar ventilation that is excessive in relation to metabolic rate; results in abnormally low alveolar $CO_2$ |
| Hypoventilation | An alveolar ventilation that is low in relation to metabolic rate; results in abnormally high alveolar $CO_2$ |
| Physiological dead | Combination of anatomical dead space and space underventilated or underperfused alveoli that do not contribute normally to blood-gas exchange |
| Pneumothorax | Presence of gas in the pleural space (the space between the visceral and parietal pleural membranes) causing lung collapse |
| Torr | Unit of pressure very nearly equal to the millimeter of mercury (760 mmHg = 760 torr) |

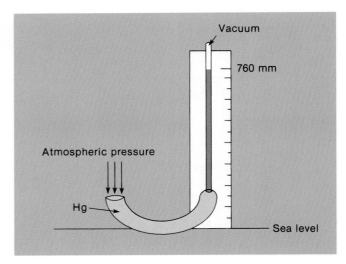

**FIGURE 24.24**

Atmospheric pressure at sea level can push a column of mercury to a height of 760 millimeters. This is also described as 760 torr, or one atmospheric pressure.

## Calculation of $P_{O_2}$

With increasing altitude, the total atmospheric pressure and the partial pressures of the constituent gases decrease (table 24.4). At Denver, for example (5000 feet above sea level), the atmospheric pressure is decreased to 619 mmHg, and the $P_{O_2}$ is therefore reduced to $619 \times 0.21 = 130$ mmHg. At the peak of Mount Everest (at 29,000 feet), the $P_{O_2}$ is only 42 mmHg. As one descends below sea level, as in ocean diving, the total pressure increases by one atmosphere for every 33 feet. At 33 feet therefore, the pressure equals $2 \times 760 = 1520$ mmHg. At 66 feet, the pressure equals three atmospheres.

Inspired air contains variable amounts of moisture. By the time the air has passed into the respiratory zone of the lungs, however, it is normally saturated with water vapor (has a relative humidity of 100%). The capacity of air to contain water vapor depends on its temperature; since the temperature of the respiratory zone is constant at 37° C, its water vapor pressure is also constant (47 mmHg).

Water vapor, like the other constituent gases, contributes a partial pressure to the total atmospheric pressure. Since the total atmospheric pressure is constant (depending only on the height of the air mass), the water vapor "dilutes" the contribution of other gases to the total pressure

$$P_{\text{wet atmosphere}} = P_{N_2} + P_{O_2} + P_{CO_2} + P_{H_2O}$$

When the effect of water vapor pressure is considered, the partial pressure of oxygen in the inspired air is decreased at sea level to

$$P_{O_2} \text{ (sea level)} = 0.21 (760 - 47) = 150 \text{ mmHg}$$

According to **Dalton's law,** the total pressure of a gas mixture (such as air) is equal to the sum of the pressures that each gas in the mixture would exert independently. The pressure that a particular gas in the mixture exerts independently is the **partial pressure** of that gas, which is equal to the product of the total pressure and the fraction of that gas in the mixture. Since oxygen constitutes about 21% of the atmosphere, for example, its partial pressure (abbreviated $P_{O_2}$) is 21% of 760, or about 159 mmHg. Since nitrogen constitutes about 78% of the atmosphere, its partial pressure is equal to $0.78 \times 760 = 593$ mmHg. These two gases thus contribute about 99% of the total pressure of 760 mmHg:

$$P_{\text{dry atmosphere}} = P_{N_2} + P_{O_2} + P_{CO_2} = 760 \text{ mmHg}$$

Dalton's law: from John Dalton, English chemist, 1766–1844

## Table 24.4 — The effect of altitude on $P_{O_2}$

| Altitude (feet above sea level) | Atmospheric pressure (mmHg) | $P_{O_2}$ in air (mmHg) | $P_{O_2}$ in alveoli (mmHg) | $P_{O_2}$ in arterial blood (mmHg) |
|---|---|---|---|---|
| 0 | 760 | 159 | 105 | 100 |
| 2000 | 707 | 148 | 97 | 92 |
| 4000 | 656 | 137 | 90 | 85 |
| 6000 | 609 | 127 | 84 | 79 |
| 8000 | 564 | 118 | 79 | 74 |
| 10,000 | 523 | 109 | 74 | 69 |
| 20,000 | 349 | 73 | 40 | 35 |
| 30,000 | 226 | 47 | 21 | 19 |

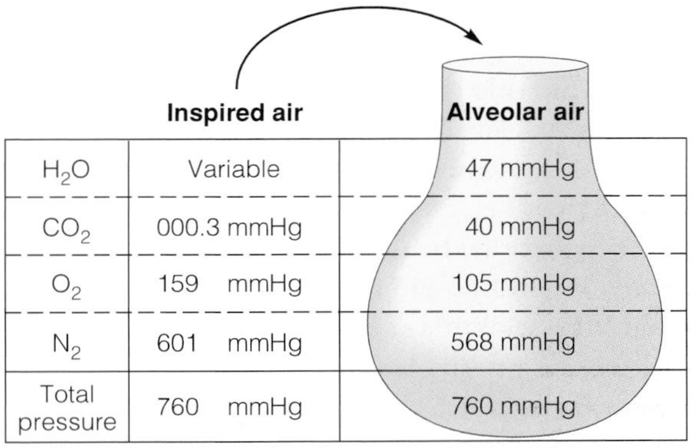

| | Inspired air | Alveolar air |
|---|---|---|
| $H_2O$ | Variable | 47 mmHg |
| $CO_2$ | 000.3 mmHg | 40 mmHg |
| $O_2$ | 159 mmHg | 105 mmHg |
| $N_2$ | 601 mmHg | 568 mmHg |
| Total pressure | 760 mmHg | 760 mmHg |

**FIGURE 24.25**
Partial pressures of gases in the inspired air and the alveolar air.

As a result of gas exchange in the alveoli, the $P_{O_2}$ of alveolar air is further diminished to about 105 mmHg. A comparison of the partial pressures of the inspired air with the partial pressures of alveolar air is shown in figure 24.25.

## Partial Pressures of Gases in Blood

The enormous surface area of alveoli and the short diffusion distance between alveolar air and the capillary blood help to quickly bring the blood into gaseous equilibrium with the alveolar air. When a liquid and a gas—such as blood and alveolar air—are at equilibrium, the amount of gas dissolved in the fluid reaches a maximum value. According to **Henry's law,** this value depends on (1) the solubility of the gas in

..........................................
Henry's law: from William Henry, English chemist, 1775–1837

the fluid, which is a physical constant; (2) the temperature of the fluid—more gas can be dissolved in cold water than warm water; and (3) the partial pressure of the gas. Since the temperature of the blood does not vary significantly, *the concentration of a gas dissolved in a fluid* (such as plasma) *depends directly on its partial pressure in the gas mixture.* When water—or plasma—is brought into equilibrium with air at a $P_{O_2}$ of 100 mmHg, for example, the fluid will contain 0.3 ml $O_2$ per 100 ml of fluid at 37° C. If the $P_{O_2}$ of the gas were reduced by half, the amount of dissolved oxygen would also be reduced by half.

**Blood-Gas Measurements**   Measurement of the oxygen content of blood (in ml of $O_2$ per 100 ml blood) is a laborious procedure. Fortunately, an **oxygen electrode** that produces an electric current in proportion to the amount of *dissolved oxygen* has been developed. If this electrode is placed in a fluid while oxygen is artificially bubbled into it, the current produced by the oxygen electrode will increase up to a maximum value. At this maximum value, the fluid is *saturated* with oxygen—that is, all of the oxygen that can be dissolved at that temperature and $P_{O_2}$ is dissolved. At a constant temperature, the amount dissolved—and thus the electric current—depend only on the $P_{O_2}$ of the gas.

As a matter of convenience, it can now be said that *the fluid has the same $P_{O_2}$ as the gas.* If it is known that the gas has a $P_{O_2}$ of 152 mmHg, for example, the deflection of a needle by the oxygen electrode can be calibrated on a scale at 152 mmHg (fig. 24.26). The actual amount of dissolved oxygen under these circumstances is not particularly important (it can be looked up in solubility tables, if desired); it is simply a linear function of the $P_{O_2}$. A lower $P_{O_2}$ indicates that less oxygen is dissolved; a higher $P_{O_2}$ indicates that more oxygen is dissolved.

If the oxygen electrode is next inserted into an unknown sample of blood, the $P_{O_2}$ of that sample can be read directly from the previously calibrated scale. Let's say, for example, that the blood sample has a $P_{O_2}$ of 100 mmHg as illustrated in figure 24.26. Since alveolar air has a $P_{O_2}$ of about 105 mmHg, this reading indicates that the blood is almost in complete equilibrium with the alveolar air.

The oxygen electrode responds only to oxygen dissolved in water or plasma; it cannot respond to oxygen bound to hemoglobin in red blood cells. Most of the oxygen in blood, however, is located in the red blood cells attached to hemoglobin. The oxygen content of whole blood thus depends on both its $P_{O_2}$ and its red blood cell and hemoglobin content. At a $P_{O_2}$ of about 100 mmHg, whole blood normally contains almost 20 ml $O_2$ per 100 ml blood;

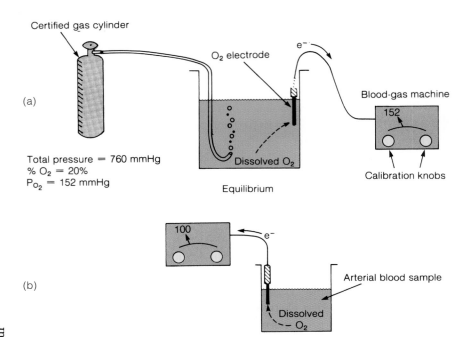

**FIGURE 24.26**
Blood-gas measurements using a $P_{O_2}$ electrode. (a) The electrical current generated by the oxygen electrode is calibrated so that the needle of the blood-gas machine points to the $P_{O_2}$ of the gas with which the fluid is an equilibrium. (b) Once standardized in this way, the electrode can be inserted in a fluid, such as blood, and the $P_{O_2}$ of this solution can be measured.

of this amount, only 0.3 ml $O_2$ is dissolved in the plasma and 19.7 ml $O_2$ is found within the red blood cells. Since only the 0.3 ml $O_2$ per 100 ml blood affects the $P_{O_2}$ measurement, this measurement would be unchanged if the red blood cells were removed from the sample.

## Significance of Blood $P_{O_2}$ and $P_{CO_2}$ Measurements

Since blood $P_{O_2}$ measurements are not directly affected by the oxygen in red blood cells, the $P_{O_2}$ does not provide a measurement of the total oxygen content of whole blood. It does, however, provide a good index of *lung function*. If the inspired air had a normal $P_{O_2}$ but the arterial $P_{O_2}$ was below normal, for example, gas exchange in the lungs would have to be impaired. Measurements of arterial $P_{O_2}$ thus provide valuable information in treating people with pulmonary diseases, in performing surgery (when breathing may be depressed by anesthesia), and in caring for premature babies with respiratory distress syndrome.

When the lungs are functioning properly, the $P_{O_2}$ of systemic arterial blood is only 5 mmHg less than the $P_{O_2}$ of alveolar air. At a normal $P_{O_2}$ of about 100 mmHg, hemoglobin is almost completely loaded with oxygen. An increase in blood $P_{O_2}$—produced, for example, by breathing 100% oxygen from a gas tank—thus cannot significantly increase

the amount of oxygen contained in the red blood cells. It can, however, significantly increase the amount of oxygen dissolved in the plasma because the amount dissolved is directly determined by the $P_{O_2}$. If the $P_{O_2}$ doubles, the amount of oxygen dissolved in the plasma also doubles, but the total oxygen content of whole blood increases only slightly, since so little of the oxygen is in the plasma as compared to the amount in the red blood cells.

Since the oxygen carried by red blood cells must first dissolve in plasma before it can diffuse to the tissue cells, however, a doubling of the blood $P_{O_2}$ means that the *rate of oxygen diffusion* to the tissues would double under these conditions. For this reason, breathing from a tank of 100% oxygen (with a $P_{O_2}$ of 760 mmHg) would significantly increase oxygen delivery to the tissues, although it would have little effect on the total oxygen content of blood.

An electrode that produces a current in response to dissolved carbon dioxide is also used, so that the $P_{CO_2}$ of blood can be measured together with its $P_{O_2}$. Blood in the systemic veins, which is delivered to the lungs by the pulmonary arteries, usually has a $P_{O_2}$ of 40 mmHg and a $P_{CO_2}$ of 46 mmHg. After gas exchange in the alveoli of the lungs, blood in the pulmonary veins and systemic arteries has a $P_{O_2}$ of about 100 mmHg and a $P_{CO_2}$ of 40 mmHg (fig. 24.27). The values in arterial blood are relatively constant and clinically significant because they reflect lung function. Blood-gas measurements of venous blood are not as clinically useful because these values are far more variable. Venous $P_{O_2}$ is much lower and $P_{CO_2}$ much higher after exercise, for example, than at rest, whereas arterial values are not significantly affected by usual changes in physical activity.

# Regulation of Breathing

*The motor neurons that stimulate the respiratory muscles are controlled by two major descending pathways: one from the cerebral cortex, which controls voluntary breathing, and one from the medulla oblongata, which controls involuntary breathing. The unconscious, rhythmic control of breathing is influenced by sensory feedback from receptors sensitive to the $P_{CO_2}$, pH, and $P_{O_2}$ of arterial blood.*

Inspiration and expiration are produced by the contraction and relaxation of skeletal muscles in response to activity in somatic motor neurons in the spinal cord. The activity of these motor neurons, in turn, is controlled by descending tracts from neurons in the respiratory control centers in the medulla oblongata and from neurons in the cerebral cortex.

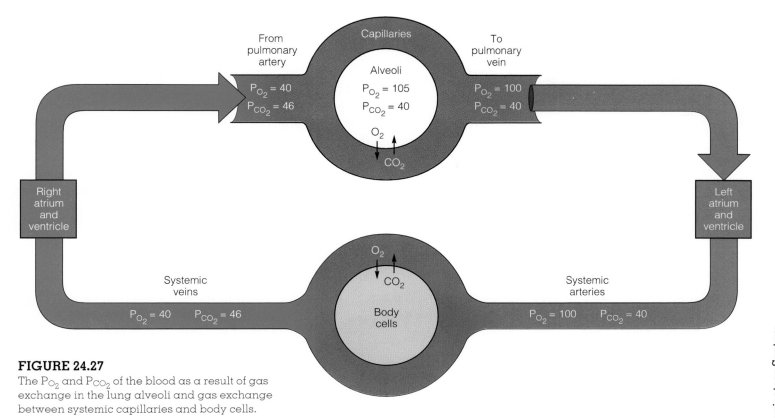

**FIGURE 24.27**
The $P_{O_2}$ and $P_{CO_2}$ of the blood as a result of gas exchange in the lung alveoli and gas exchange between systemic capillaries and body cells.

 The automatic control of breathing is regulated by nerve fibers that descend in the lateral and ventral white matter of the spinal cord from the medulla oblongata. The voluntary control of breathing is a function of the cerebral cortex and involves nerve fibers that descend in the corticospinal tracts (chapter 15). The separation of the voluntary and involuntary pathways that regulate breathing is dramatically illustrated in the condition called *Ondine's curse*. In this condition, neurological damage prevents the automatic but not the voluntary control of breathing. People with Ondine's curse must consciously force themselves to breathe and be put on artificial respirators when they sleep.

## *Brain Stem Respiratory Centers*

A loose aggregation of neurons in the reticular formation of the *medulla oblongata* forms the **rhythmicity** (*rith-mis´ĭ-te*) **center** that controls automatic breathing. The rhythmicity center consists of interacting pools of neurons that respond either during inspiration (*I neurons*) or expiration (*E neurons*). The I neurons project to and stimulate spinal motoneurons that innervate the respiratory muscles. Expiration is a passive process that occurs when the I neurons are inhibited by the activity of the E neurons. The activity of I and E neurons varies in a reciprocal way, so that a rhythmic pattern of breathing is produced. The cycle of inspiration and expiration is thus intrinsic to the neural activity of the medulla oblongata. The rhythmicity center in the medulla oblongata is divided into a posterior group of neurons, which regulates the activity of the phrenic nerves to the diaphragm, and an anterior group, which controls the motor neurons to the intercostal muscles.

The activity of the medullary rhythmicity center is influenced by centers in the *pons*. As a result of research in which the brain stem is destroyed at different levels, two respiratory control centers have been identified in the pons. One area, the **apneustic** (*ap-noo´stik*) **center,** appears to promote inspiration by stimulating the I neurons in the medulla oblongata. The other pons area, called the **pneumotaxic** (*noo˝mŏ-tak´sik*) **center,** seems to antagonize the apneustic center and inhibit inspiration (fig. 24.28). The apneustic center is believed to provide a tonic, or constant, stimulus for inspiration, which is cyclically inhibited by the activity of the pneumotaxic center.

The automatic control of breathing is also influenced by input from receptors sensitive to the chemical composition of the blood. There are two groups of *chemoreceptors* that respond to changes in blood $P_{CO_2}$, pH, and $P_{O_2}$. These are the **central chemoreceptors** in the medulla oblongata and the **peripheral chemoreceptors.** The peripheral chemoreceptors are contained within small nodules associated with the aorta and the carotid arteries and receive blood from these critical arteries via small arterial branches. The peripheral chemoreceptors include the **aortic bodies,** located around the aortic arch, and the **carotid bodies,**

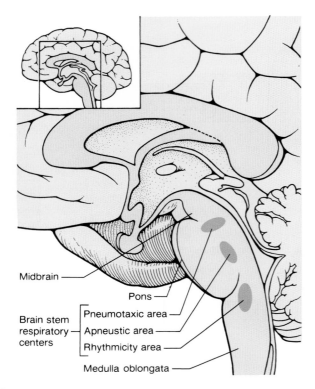

**FIGURE 24.28**

Approximate locations of the brain stem respiratory centers.

Midbrain

Pons

Brain stem respiratory centers
- Pneumotaxic area
- Apneustic area
- Rhythmicity area

Medulla oblongata

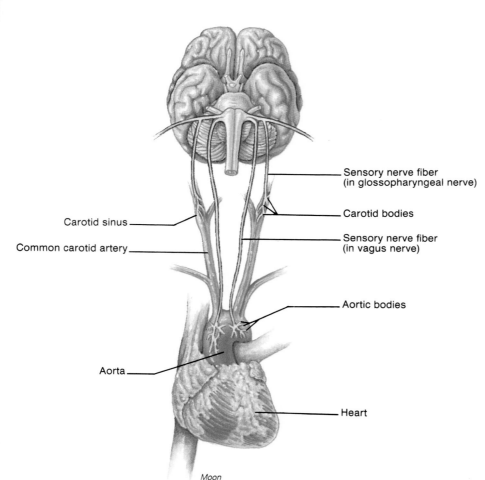

Sensory nerve fiber (in glossopharyngeal nerve)

Carotid bodies

Sensory nerve fiber (in vagus nerve)

Aortic bodies

Heart

Carotid sinus

Common carotid artery

Aorta

*Moon*

located in each common carotid artery just before it branches into the internal and external carotid arteries (fig. 24.29). The aortic and carotid bodies should not be confused with the aortic and carotid sinuses (chapter 22) that are located within these arteries. The aortic and carotid sinuses contain receptors that monitor the blood pressure.

The peripheral chemoreceptors control breathing indirectly via sensory nerve fibers to the medulla oblongata. The aortic bodies send sensory information to the medulla oblongata in the vagus (tenth cranial) nerve; the carotid bodies stimulate sensory fibers in the glossopharyngeal (ninth cranial) nerve.

The neural and sensory control of ventilation is summarized in figure 24.30.

## Effects of Blood $P_{CO_2}$ and pH on Ventilation

Chemoreceptor input to the brain stem modifies the rate and depth of breathing so that, under normal conditions, arterial $P_{CO_2}$, pH, and $P_{O_2}$ remain relatively constant. In the event of hypoventilation (inadequate ventilation), $P_{CO_2}$ quickly rises and pH falls. The pH falls because carbon dioxide can combine with water to form carbonic acid ($H_2CO_3$), which in turn can release $H^+$ to the solution. This is shown in the following equations:

$$CO_2 + H_2O \rightarrow H_2CO_3$$

$$H_2CO_3 \rightarrow HCO_3^- + H^+$$

The oxygen content of the blood decreases much more slowly because there is a large reservoir of oxygen attached to hemoglobin. During hyperventilation, conversely, blood $P_{CO_2}$ quickly falls and pH rises due to the excessive elimination of carbonic acid. The oxygen content of blood, on the other hand, is not significantly increased by hyperventilation (hemoglobin in arterial blood is 97% saturated with oxygen during normal ventilation).

The blood $P_{CO_2}$ and pH are, therefore, more immediately affected by changes in ventilation than is the $O_2$ content. Indeed, changes in $P_{CO_2}$ provide a sensitive index of ventilation, as shown in table 24.5. In view of these facts, it is not surprising that changes in $P_{CO_2}$ provide the most potent stimulus for the reflex control of ventilation. Ventilation, in other words, is adjusted to maintain a constant $P_{CO_2}$; proper oxygenation of the blood occurs naturally as a side product of this reflex control.

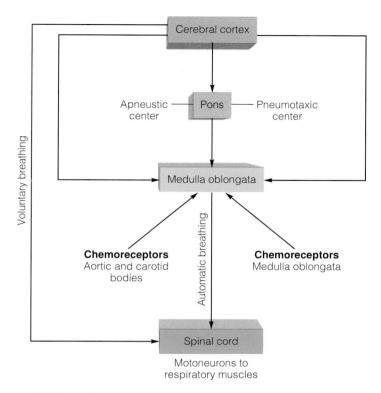

**FIGURE 24.30**
A schematic illustration of the control of ventilation by the central nervous system. The feedback effects of pulmonary stretch receptors and "irritant" receptors are not shown.

**Table 24.5** The effect of ventilation, as measured by total minute volume (breathing rate × tidal volume), on the $P_{CO_2}$ of arterial blood

| Total minute volume | Arterial $P_{CO_2}$ | Type of ventilation |
| --- | --- | --- |
| 2 L/min | 80 mmHg | Hypoventilation |
| 4–5 L/min | 40 mmHg | Normal ventilation |
| 8 L/min | 20–25 mmHg | Hyperventilation |

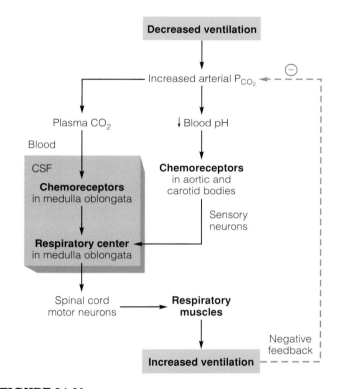

**FIGURE 24.31**
Negative feedback control of ventilation through changes in blood $P_{CO_2}$ and pH. The orange box represents the blood-brain barrier, which allows $CO_2$ to pass into the cerebrospinal fluid while preventing the passage of $H^+$.

Chemoreceptor regulation of breathing in response to changes in $P_{CO_2}$ is illustrated in figure 24.31. Acting through this reflex mechanism, ventilation normally maintains an arterial $P_{CO_2}$ of 40 mmHg. **Hypoventilation,** or inadequate breathing, causes a rise in blood $P_{CO_2}$, which is not being "blown off" as rapidly as it is being produced. Indeed, hypoventilation may be defined as ventilation in which blood carbon dioxide levels are higher than normal—a condition called *hypercapnia* (hi″per-kap′ne-ă). **Hyperventilation,** conversely, is excessive ventilation that results in abnormally low blood carbon dioxide levels, or *hypocapnia.*

**Chemoreceptors in the Medulla Oblongata**  The chemoreceptors most sensitive to changes in the arterial $P_{CO_2}$ are located laterally and anteriorly in the medulla oblongata, near the exit of the glossopharyngeal and vagus (ninth and tenth cranial) nerves. These chemoreceptor neurons are anatomically separate from, but synaptically communicate with, the neurons of the respiratory control center in the medulla oblongata.

An increase in arterial $P_{CO_2}$ causes a rise in the $H^+$ concentration of the blood as a result of increased carbonic acid concentrations. The $H^+$ in the blood, however, cannot cross the blood-brain barrier and, therefore, cannot influence the medullary chemoreceptors. Carbon dioxide in the arterial blood *can* cross the blood-brain barrier and, through the formation of carbonic acid, can lower the pH of cerebrospinal fluid. This fall in cerebrospinal fluid pH directly stimulates the chemoreceptors in the medulla oblongata when there is a rise in arterial $P_{CO_2}$.

The chemoreceptors in the medulla oblongata are ultimately responsible for 70% to 80% of the increased ventilation that occurs in response to a sustained rise in arterial $P_{CO_2}$.

This response, however, takes several minutes. The immediate increase in ventilation that occurs when $P_{CO_2}$ rises is produced by stimulation of the peripheral chemoreceptors.

**Peripheral Chemoreceptors**   The aortic and carotid bodies are not stimulated directly by blood $CO_2$. Instead, they are stimulated by a rise in the $H^+$ concentration (fall in pH) of arterial blood, which occurs when the blood $CO_2$, and thus carbonic acid, is raised. The retention of $CO_2$ during hypoventilation thus stimulates the medullary chemoreceptors through a lowering of cerebrospinal fluid pH and stimulates the peripheral chemoreceptors through a lowering of blood pH.

 People who *hyperventilate* during psychological stress are sometimes told to breathe into a paper bag so that they rebreathe their expired air that is enriched in $CO_2$. This procedure helps to raise their blood $P_{CO_2}$ back up to the normal range, which is necessary because hypocapnia causes cerebral vasoconstriction. In addition to producing dizziness, the cerebral ischemia (inadequate blood flow) that results can lead to acidotic conditions in the brain that, through stimulation of the medullary chemoreceptors, can cause further hyperventilation. Breathing into a paper bag can thus relieve the hypocapnia and stop the hyperventilation.

## Effects of Blood $P_{O_2}$ on Ventilation

Under normal conditions, blood $P_{O_2}$ affects breathing only indirectly by influencing the chemoreceptor sensitivity to changes in $P_{CO_2}$. Chemoreceptor sensitivity to $P_{CO_2}$ is augmented by a low $P_{O_2}$ (so ventilation is increased at a high altitude, for example) and is decreased by a high $P_{O_2}$. If the blood $P_{O_2}$ is raised by breathing 100% $O_2$, therefore, the breath can be held longer because the response to increased $P_{CO_2}$ is blunted.

When the blood $P_{CO_2}$ is held constant by experimental techniques, the $P_{O_2}$ of arterial blood must fall from 100 mmHg to below 50 mmHg before ventilation is significantly stimulated (fig. 24.32). This stimulation is apparently due to a direct effect of $P_{O_2}$ on the carotid bodies. Since this degree of *hypoxemia* (hi″pok-se′me-ă), or low blood oxygen, does not usually occur even in breath holding, $P_{O_2}$ does not usually exert this direct effect on breathing. In emphysema, when there is a chronic retention of carbon dioxide, the chemoreceptors eventually lose their ability to respond to increases in the arterial $P_{CO_2}$. The abnormally high $P_{CO_2}$, however, enhances the sensitivity of the carotid bodies to a fall in $P_{O_2}$. For people with emphysema, breathing may thus be stimulated by a *hypoxic drive* rather than by increases in blood $P_{CO_2}$.

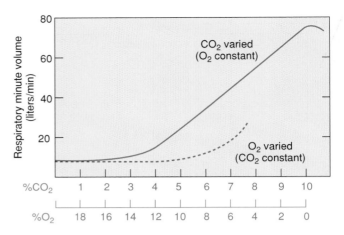

**FIGURE 24.32**

Comparison of the effects of increasing concentrations of $CO_2$ in air with decreasing concentrations of $O_2$ in air on respiration. Notice that respiration increases linearly with increasing $CO_2$ concentration, whereas $O_2$ concentrations must decrease to half the normal value before respiration is stimulated.

The effect of changes in the blood $P_{CO_2}$, pH, and $P_{O_2}$ on chemoreceptors and the regulation of ventilation are summarized in table 24.6.

## Pulmonary Stretch and Irritant Reflexes

The lungs contain various types of receptors that influence the brain stem respiratory control centers via sensory fibers in the vagus nerves. Irritant receptors in the lungs, for example, stimulate reflex constriction of the bronchioles in response to smoke and smog. Similarly, sneezing, sniffing, and coughing may be stimulated by irritant receptors in the nose, larynx, and trachea.

The **Hering–Breuer** (her′ing broy′er) **reflex** is stimulated by pulmonary stretch receptors. The activation of these receptors during inspiration inhibits the respiratory control centers, making further inspiration increasingly difficult. This helps to prevent undue distension of the lungs and may contribute to the smoothness of the ventilation cycles. A similar inhibitory reflex may occur during expiration. The Hering–Breuer reflex appears to be important in the control of normal ventilation in the newborn. Pulmonary stretch receptors in adults, however, are probably not active at normal resting tidal volumes (500 ml per breath) but may contribute to respiratory control at high tidal volumes, as during exercise.

••••••••••••••••••••••••••••••••••••••••
Hering–Breuer reflex: from Ewald Hering, German physiologist, 1834–1918, and Josef Breuer, Austrian physician, 1842–1925

## Table 24.6 — The sensitivity of chemoreceptors to changes in blood gases and pH

| Stimulus | Chemoreceptor | Comments |
|---|---|---|
| $\uparrow P_{CO_2}$ | Medulla oblongata; aortic and carotid bodies | Medullary chemoreceptors are sensitive to the pH of cerebrospinal fluid (CSF). Diffusion of $CO_2$ from the blood into the CSF lowers the pH of CSF by forming carbonic acid. Similarly, the aortic and carotid bodies are stimulated by a fall in blood pH induced by increases in blood $CO_2$. |
| $\downarrow pH$ | Aortic and carotid bodies | Peripheral chemoreceptors are stimulated by decreased blood pH independent of the effect of blood $CO_2$. Chemoreceptors in the medulla oblongata are not affected by changes in blood pH because $H^+$ cannot cross the blood-brain barrier. |
| $\downarrow P_{O_2}$ | Carotid bodies | Low blood $P_{O_2}$ (hypoxemia) augments the chemoreceptor response to blood $P_{CO_2}$ and can stimulate ventilation directly when the $P_{O_2}$ falls below 50 mmHg. |

# Hemoglobin and Oxygen Transport

*Hemoglobin without oxygen, or deoxyhemoglobin, can bond with oxygen to form oxyhemoglobin. This "loading" reaction occurs in the capillaries of the lungs. The dissociation of oxyhemoglobin, or "unloading" reaction, occurs in the tissue capillaries. The bond strength between hemoglobin and oxygen, and thus the extent of the unloading reaction, is adjusted by various factors to ensure an adequate delivery of oxygen to the tissues.*

If the lungs are functioning properly, blood leaving in the pulmonary veins and traveling in the systemic arteries has a $P_{O_2}$ of about 100 mmHg, indicating a plasma oxygen concentration of about 0.3 ml $O_2$ per 100 ml blood. The total oxygen content of the blood, however, cannot be derived from knowing only the $P_{O_2}$ of plasma. The total oxygen content depends not only on the $P_{O_2}$ but also on the hemoglobin concentration. If the $P_{O_2}$ and hemoglobin concentration are normal, arterial blood contains about 20 ml of $O_2$ per 100 ml of blood (fig. 24.33).

## Hemoglobin

Most of the oxygen in the blood is contained within the red blood cells, where it is chemically bonded to **hemoglobin.** Each hemoglobin molecule consists of (1) a protein *globin* part, composed of four polypeptide chains and (2) four nitrogen-containing, disc-shaped organic pigment molecules called *hemes* (fig. 24.34).

The protein part of hemoglobin is composed of two identical *alpha chains*, each 141 amino acids long, and two identical *beta chains*, each 146 amino acids long. Each of the four polypeptide chains is combined with one heme group. In the center of each heme group is one atom of iron, which can combine with one molecule of oxygen ($O_2$). One hemoglobin molecule can thus combine with four molecules of oxygen. Since there are about 280 million hemoglobin molecules per red blood cell, each red blood cell can carry over a billion molecules of oxygen.

Normal heme contains iron in the reduced form ($Fe^{++}$, or ferrous iron). In this form, the iron can share electrons and bond with oxygen to form **oxyhemoglobin.** When oxyhemoglobin dissociates to release oxygen to the tissues, the heme iron is still in the reduced ($Fe^{++}$) form and the hemoglobin is called **deoxyhemoglobin,** or **reduced hemoglobin.** The term *oxyhemoglobin* is thus not equivalent to *oxidized* hemoglobin; hemoglobin does not lose an electron (and become oxidized) when it combines with oxygen. Oxidized hemoglobin, or **methemoglobin,** has iron in the oxidized ($Fe^{+++}$, or ferric) state. Methemoglobin thus lacks the electron it needs to form a bond with oxygen and cannot participate in oxygen transport. Blood normally contains only a small amount of methemoglobin, but certain drugs can increase this amount.

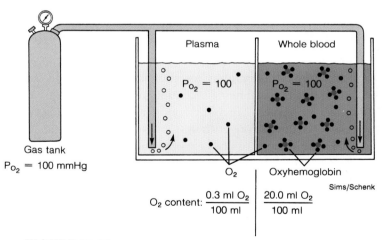

**FIGURE 24.33**
Plasma and whole blood that are brought into equilibrium with the same gas mixture have the same $P_{O_2}$ and thus the same amount of dissolved oxygen molecules (shown as black dots). The oxygen content of whole blood, however, is much higher than that of plasma because of the binding of oxygen to hemoglobin.

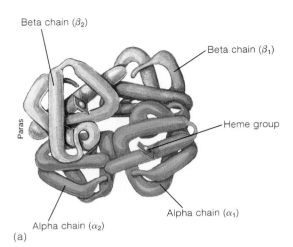

Beta chain ($\beta_2$)

Beta chain ($\beta_1$)

Heme group

Alpha chain ($\alpha_1$)

Alpha chain ($\alpha_2$)

Paras

(a)

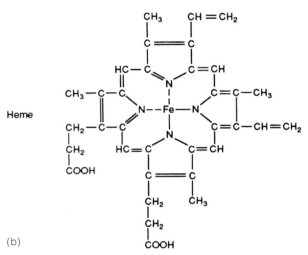

Heme

(b)

**FIGURE 24.34**

An illustration (a) of the three-dimensional structure of hemoglobin in which the two alpha and two beta polypeptide chains are shown. The heme groups are represented as flat structures with spheres in the centers to indicate the iron atoms. (b) The chemical structure of heme.

In **carboxyhemoglobin,** another abnormal form of hemoglobin, the reduced heme is combined with *carbon monoxide* instead of oxygen. Since the bond with carbon monoxide is about 210 times stronger than the bond with oxygen, carbon monoxide tends to displace oxygen in hemoglobin and remains attached to hemoglobin as the blood passes through systemic capillaries. The transport of oxygen to the tissues is thus reduced in carbon monoxide poisoning.

According to federal standards, the percentage of carboxyhemoglobin in the blood of active nonsmokers should not be higher than 1.5%. However, concentrations of 3% in nonsmokers and 10% in smokers have been reported in some cities. Although these high levels may not cause immediate problems in healthy people, long-term adverse effects on health are possible. People with respiratory or cardiovascular diseases would be particularly vulnerable to the negative effects of carboxyhemoglobin on oxygen transport.

**Hemoglobin Concentration**   The *oxygen-carrying capacity* of whole blood is determined by the concentration of normal hemoglobin in the blood. If the hemoglobin concentration is below normal—a condition called **anemia**—the oxygen concentration of the blood is reduced below normal. Conversely, when the hemoglobin concentration is increased to above the normal range—as occurs in **polycythemia** (high red blood cell count)—the oxygen carrying capacity of blood is increased accordingly. This can occur as an adaptation to life at a high altitude.

The production of hemoglobin and red blood cells in bone marrow is controlled by a hormone called **erythropoietin** (ĕ-rith″ro-poi-e′tin), produced primarily by the kidneys (as described in chapter 20). The production of erythropoietin—and the production of red blood cells—is stimulated when the amount of oxygen delivered to the kidneys and other organs is lower than normal. Red blood cell production is also promoted by androgens, which explains why the hemoglobin concentration in men averages 1–2 g per 100 ml higher than in women.

**The Loading and Unloading Reactions**   Deoxyhemoglobin and oxygen combine to form oxyhemoglobin; this is called the **loading reaction.** Oxyhemoglobin, in turn, dissociates to yield deoxyhemoglobin and free oxygen molecules; this is the **unloading reaction.** The loading reaction occurs in the lungs and the unloading reaction occurs in the systemic capillaries.

Loading and unloading can thus be shown as a reversible reaction:

$$\text{Deoxyhemoglobin} + O_2 \underset{\text{(tissues)}}{\overset{\text{(lungs)}}{\rightleftarrows}} \text{Oxyhemoglobin}$$

The extent to which the reaction will go in each direction depends on two factors: (1) the $P_{O_2}$ of the environment and (2) the *affinity*, or bond strength, between hemoglobin and oxygen. High $P_{O_2}$ drives the equation to the right (favors the loading reaction); at the high $P_{O_2}$ of the pulmonary capillaries, almost all of the deoxyhemoglobin molecules combine with oxygen. Low $P_{O_2}$ in the systemic capillaries drives the reaction in the opposite direction to promote unloading. The extent of this unloading depends on how low the $P_{O_2}$ values are.

The affinity (bond strength) between hemoglobin and oxygen also influences the loading and unloading reactions. A very strong bond would favor loading but inhibit unloading; a weak bond would hinder loading but improve unloading. The bond strength between hemoglobin and oxygen is normally strong enough so that 97% of the hemoglobin leav-

| Table 24.7 | The relationship between percent oxyhemoglobin saturation and $P_{O_2}$ (at pH = 7.40 and temperature = 37° C) | | | | | | | | | | |
|---|---|---|---|---|---|---|---|---|---|---|---|
| $P_{O_2}$ (mmHg) | 100 | 80 | 61 | 45 | 40 | 36 | 30 | 26 | 23 | 21 | 19 |
| Percent oxyhemoglobin saturation | 97 | 95 | 90 | 80 | 75 | 70 | 60 | 50 | 40 | 35 | 30 |
| | Arterial blood | | | | Venous blood | | | | | | |

ing the lungs is in the form of oxyhemoglobin, yet the bond is sufficiently weak so that adequate amounts of oxygen are unloaded to sustain aerobic respiration in the tissues.

## Oxyhemoglobin Dissociation Curve

Blood in the systemic arteries, at a $P_{O_2}$ of 100 mmHg, has a *percent oxyhemoglobin saturation* of 97% (which means that 97% of the hemoglobin is in the form of oxyhemoglobin). This blood is delivered to the systemic capillaries, where oxygen diffuses into the tissue cells and is consumed in aerobic respiration. Blood leaving in the systemic veins is thus reduced in oxygen; it has a $P_{O_2}$ of about 40 mmHg and a percent oxyhemoglobin saturation of about 75% (table 24.7). In other words, blood entering the tissues contains 20 ml $O_2$ per 100 ml blood, while blood leaving the tissues contains 15.5 ml $O_2$ per 100 ml blood (fig. 24.35). Thus, 22%, or 4.5 ml of $O_2$ out of 20 ml $O_2$ per 100 ml blood, is unloaded to the tissues.

The **oxyhemoglobin dissociation curve** (fig. 24.35) graphically describes the percent oxyhemoglobin saturation at different values of $P_{O_2}$. The values in this graph are obtained by subjecting samples of blood in vitro (outside the body) to different partial $O_2$ pressures. The percent oxyhemoglobin saturations obtained by this procedure, however, can be used to predict what the unloading percentages would be in vivo (within the body) with given differences in arterial and venous $P_{O_2}$ values.

Figure 24.35 shows the difference between the arterial and venous $P_{O_2}$ and the percent oxyhemoglobin saturation at rest. The relatively large amount of oxyhemoglobin remaining in the venous blood at rest functions as an oxygen reserve. If a person stops breathing, there will be a sufficient reserve of oxygen in the blood to keep the brain and heart alive for approximately 4 to 5 minutes in the absence of cardiopulmonary resuscitation (CPR) techniques. This reserve supply of oxygen can also be tapped when the tissues' requirements for oxygen are raised, as during exercise.

The oxyhemoglobin dissociation curve is S-shaped, or *sigmoidal*. The curve is relatively flat at high $P_{O_2}$ values because changes in $P_{O_2}$ within this range have little effect on

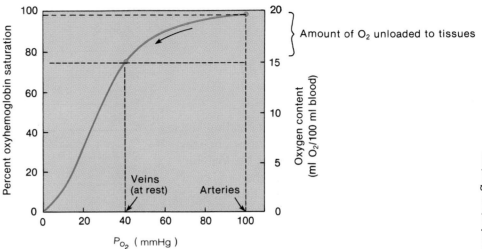

**FIGURE 24.35**

The percentage of oxyhemoglobin saturation and the blood oxygen content are shown at different values of $P_{O_2}$. Notice that there is about a 25% decrease in percent of oxyhemoglobin as the blood passes through the tissue from arteries to veins, resulting in the unloading of approximately 5 ml $O_2$ per 100 ml to the tissues.

the loading reaction. One would have to ascend as high as 10,000 feet, for example, before the oxyhemoglobin saturation of arterial blood would decrease from 97% to 93%. At more common elevations, the percent oxyhemoglobin saturation would not be significantly different from the 97% value at sea level.

At the steep part of the sigmoidal curve, however, small changes in $P_{O_2}$ values produce large differences in percent oxyhemoglobin saturation. A decrease in *venous* $P_{O_2}$ from 40 mmHg to 30 mmHg, as might occur during mild exercise, corresponds to a change in percent saturation from 75% to 58%. Since the *arterial* percent saturation is usually still 97% during exercise, this change in venous percent saturation indicates that more $O_2$ has been unloaded to the tissues. The difference between the arterial and venous percent saturations indicates the unloading percentage. In the preceding example, 97% – 75% = 22% unloading at rest and 97% – 58% = 39% unloading during mild exercise. During heavier exercise, the venous $P_{O_2}$ can drop to 20 mmHg or less, indicating an unloading percentage in excess of 70%.

| Table 24.8 | Effect of pH on hemoglobin affinity for oxygen and unloading of oxygen to the tissues | | | |
|---|---|---|---|---|
| pH | Affinity | Arterial O₂ content per 100 ml | Venous O₂ content per 100 ml | O₂ unloaded to tissues per 100 ml |
| 7.40 | Normal | 19.8 ml $O_2$ | 14.8 ml $O_2$ | 5.0 ml $O_2$ |
| 7.60 | Increased | 20.0 ml $O_2$ | 17.0 ml $O_2$ | 3.0 ml $O_2$ |
| 7.20 | Decreased | 19.2 ml $O_2$ | 12.6 ml $O_2$ | 6.6 ml $O_2$ |

## Effects of pH and Temperature on Oxygen Transport

In addition to changes in $P_{O_2}$, the loading and unloading reactions are influenced by changes in the bond strength, or affinity, of hemoglobin for oxygen. The net effect of such changes is that active skeletal muscles, as a result of their higher metabolism, receive more oxygen from the blood than they do at rest.

The affinity of oxygen for hemoglobin is decreased when the pH is lowered and increased when the pH is raised. This phenomenon is called the **Bohr effect.** When the affinity of hemoglobin for oxygen is reduced, there is slightly less loading of the blood with oxygen in the lungs but greater unloading of oxygen in the tissues. The net effect is that the tissues receive more oxygen when the blood pH is lowered (table 24.8). Since the pH can be decreased by carbon dioxide (through the formation of carbonic acid), the Bohr effect helps to provide more oxygen to the tissues when their carbon dioxide output (and metabolism) is increased.

When the percent oxyhemoglobin saturation at different pH values is graphed as a function of $P_{O_2}$, the dissociation curve is shown to be shifted to the right by a lowering of pH and shifted to the left by a rise in pH (fig. 24.36). If the unloading percentage is calculated by subtracting the percent oxyhemoglobin saturation at given $P_{O_2}$ values for arterial and venous blood, it will be clear that a *shift to the right* of the curve indicates a greater oxygen unloading, whereas a *shift to the left* indicates less unloading but slightly more oxygen loading in the lungs.

When oxyhemoglobin dissociation curves are constructed at constant pH values but at different temperatures, it can be seen that the affinity of hemoglobin for

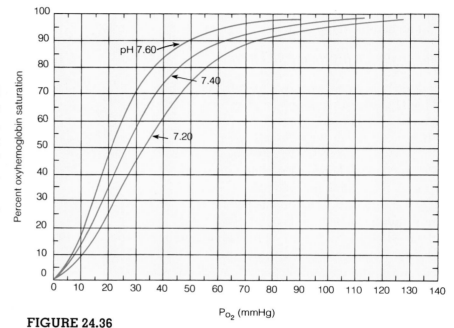

**FIGURE 24.36**

A decrease in blood pH (an increase in H⁺ concentration) decreases the affinity of hemoglobin for oxygen at each $P_{O_2}$ value, resulting in a "shift to the right" of the oxyhemoglobin dissociation curve. A curve that is shifted to the right has a lower percentage of oxyhemoglobin saturation at each $P_{O_2}$, but the effect is more marked at lower $P_{O_2}$ values. This is called the *Bohr effect.*

oxygen is decreased by a rise in temperature. An increase in temperature weakens the bond between hemoglobin and oxygen and thus has the same effect as a fall in pH—the oxyhemoglobin dissociation curve is shifted to the right. At higher temperatures, therefore, more oxygen is unloaded to the tissues than would be the case if the bond strength were constant. This effect can significantly increase the delivery of $O_2$ to muscles that are warmed during exercise.

## Effects of 2,3-DPG on Oxygen Transport

Mature red blood cells lack both nuclei and mitochondria. Without mitochondria they cannot respire aerobically. The very cells that carry oxygen, therefore, are the only cells in the body that cannot use it! Red blood cells must obtain energy through the anaerobic respiration of glucose. At a

..................................................

Bohr effect: From Christian Bohr, Danish physiologist, 1855–1911

certain point in the glycolytic pathway there is a "side reaction" in red blood cells that results in a unique product—**2,3-diphosphoglyceric acid (2,3-DPG).**

The enzyme that produces 2,3-DPG is inhibited by oxyhemoglobin. When the oxyhemoglobin concentration is decreased, therefore, the production of 2,3-DPG is increased. This increase in 2,3-DPG production can occur when the total hemoglobin concentration is low (in anemia) or when the $P_{O_2}$ is low (at a high altitude, for example). The bonding of 2,3-DPG with deoxyhemoglobin, makes the deoxyhemoglobin more stable. At the $P_{O_2}$ values in the tissue capillaries, therefore, a higher proportion of the oxyhemoglobin will be converted to deoxyhemoglobin by the unloading of its oxygen. An increased concentration of 2,3-DPG in red blood cells thus increases oxygen unloading and shifts the oxyhemoglobin dissociation curve to the right.

 The importance of 2,3-DPG within red blood cells is now recognized in *blood banking*. Old, stored red blood cells can lose their ability to produce 2,3-DPG as they lose their ability to metabolize glucose. Modern techniques for blood storage include the addition of energy substrates for respiration and phosphate sources needed for the production of 2,3-DPG.

The effects of 2,3-DPG are also important in the transfer of oxygen from maternal to fetal blood. The mother's hemoglobin molecules are composed of two alpha and two beta chains, as previously described, whereas the fetal hemoglobin contains two alpha and two *gamma chains* in place of beta chains (gamma chains differ from beta chains in 37 of their amino acids). Normal adult hemoglobin in the mother (*hemoglobin A*) is able to bind to 2,3- DPG. Fetal hemoglobin (*hemoglobin F*), by contrast, cannot bond to 2,3-DPG, and thus has a higher affinity for oxygen at a given $P_{O_2}$ than does hemoglobin A. Since hemoglobin F can have a higher percent oxyhemoglobin saturation than hemoglobin A at a given $P_{O_2}$, oxygen is transferred from the maternal to the fetal blood as these two come into close proximity in the placenta.

# Carbon Dioxide Transport and Acid-Base Balance

*Carbon dioxide is transported in the blood primarily in the form of bicarbonate ($HCO_3^-$), which is released when carbonic acid dissociates. Bicarbonate can buffer $H^+$, and thus helps to maintain a normal arterial pH. Hypoventilation raises the carbonic acid concentration of the blood, whereas hyperventilation lowers it.*

Carbon dioxide is carried by the blood in three forms: (1) as *dissolved carbon dioxide*—$CO_2$ is about 21 times more soluble than oxygen in water, and about one-tenth of the total blood $CO_2$ is dissolved in plasma; (2) as *carbaminohemoglobin*—

about one-fifth of the total blood $CO_2$ is carried attached to an amino acid in hemoglobin (carbaminohemoglobin should not be confused with carboxyhemoglobin, the latter being a combination of hemoglobin and carbon monoxide); and (3) as *bicarbonate*, which accounts for most of the $CO_2$ carried by the blood.

Carbon dioxide is able to combine with water to form carbonic acid. This reaction occurs spontaneously in the plasma at a slow rate but occurs much more rapidly within the red blood cells due to the catalytic action of the enzyme **carbonic anhydrase.** Since this enzyme is confined to the red blood cells, most of the carbonic acid is produced there rather than in the plasma. The formation of carbonic acid from $CO_2$ and water is favored by the high $P_{CO_2}$ found in tissue capillaries.

$$CO_2 + H_2O \xrightarrow[\text{high } P_{CO_2}]{\text{carbonic anhydrase}} H_2CO_3$$

## The Chloride Shift

As a result of catalysis by carbonic anhydrase within the red blood cells, large amounts of carbonic acid are produced as blood passes through the systemic capillaries. The buildup of carbonic acid concentrations within the red blood cells favors the dissociation of these molecules into hydrogen ions (protons, which contribute to the acidity of a solution) and $HCO_3^-$ (bicarbonate). The equation describing this reaction has been previously introduced (see chapter 2) and is shown in figure 24.37.

The hydrogen ions released by the dissociation of carbonic acid are largely buffered by their combination with deoxyhemoglobin within the red blood cells. Although the unbuffered hydrogen ions are free to diffuse out of the red blood cells, more bicarbonate diffuses outward into the plasma than does $H^+$. As a result of the trapping of hydrogen ions within the red blood cells by their attachment to hemoglobin and the outward diffusion of bicarbonate, the inside of the red blood cell gains a net positive charge. As a result, negatively charged chloride ions ($Cl^-$) move into the red blood cells as $HCO_3^-$ moves out. This exchange of anions as blood travels through the tissue capillaries is called the **chloride shift** (fig. 24.37).

The ability of the blood to transport carbon dioxide is affected by the transport of oxygen. The unloading of oxygen is increased by the bonding of $H^+$, with oxyhemoglobin. This is the Bohr effect, and results in increased conversion of oxyhemoglobin to deoxyhemoglobin. Now, since deoxyhemoglobin bonds $H^+$ more strongly than does oxyhemoglobin, the act of unloading its oxygen improves the ability of hemoglobin to buffer the $H^+$ released by carbonic acid. Removal of $H^+$ from solution by combining with hemoglobin (through the law of mass action), in turn, favors the continued production of carbonic acid, and thus improves the ability of the blood to transport carbon dioxide.

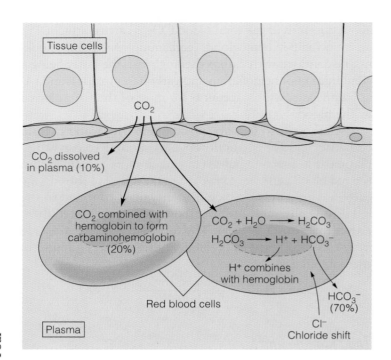

**FIGURE 24.37**

An illustration of carbon dioxide transport by the blood and the "chloride shift." Carbon dioxide is transported in three forms: as dissolved $CO_2$ gas, attached to hemoglobin as carbaminohemoglobin, and as carbonic acid and bicarbonate. Percentages indicate the proportion of $CO_2$ in each of the forms.

When blood reaches the pulmonary capillaries, deoxyhemoglobin is converted to oxyhemoglobin. Since oxyhemoglobin has a lower affinity for $H^+$ than does deoxyhemoglobin, hydrogen ions are released within the red blood cells. This attracts $HCO_3^-$ from the plasma, which combines with $H^+$ to form carbonic acid:

$$H^+ + HCO_3^- \longrightarrow H_2CO_3$$

Under conditions of lower $P_{CO_2}$, as occurs in the pulmonary capillaries, carbonic anhydrase catalyzes the conversion of carbonic acid to $CO_2$ and water:

$$H_2CO_3 \xrightarrow[\text{low } P_{CO_2}]{\text{carbonic anhydrase}} CO_2 + H_2O$$

In summary, the carbon dioxide produced by the tissue cells is converted within the systemic capillaries, mostly through the action of carbonic anhydrase in the red blood cells, to carbonic acid. With the buildup of carbonic acid concentrations in the RBCs, the carbonic acid dissociates into bicarbonate and $H^+$, which results in the chloride shift. A *reverse chloride shift* operates in the pulmonary capillaries to convert carbonic acid to $H_2O$ and $CO_2$ gas, which is eliminated in the expired breath (fig. 24.38). The $P_{CO_2}$, carbonic acid, $H^+$, and bicarbonate concentrations in the systemic arteries are thus maintained relatively constant by normal ventilation.

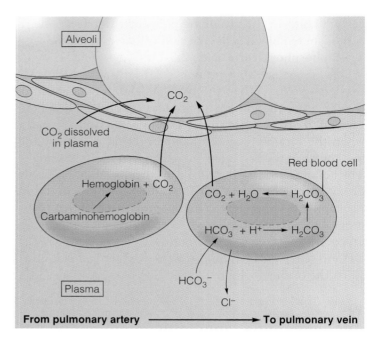

**FIGURE 24.38**

Carbon dioxide is released from the blood as it travels through the pulmonary capillaries. During this time a "reverse chloride shift" occurs, and carbonic acid is transformed into $CO_2$ and $H_2O$.

## Ventilation and Acid-Base Balance

Normal systemic arterial blood has a pH of $7.40 \pm 0.05$. Using the definition of pH described in chapter 2, this means that arterial blood has a $H^+$ concentration of about $10^{-7.4}$ molar. Some of these hydrogen ions are derived from carbonic acid and some are derived from nonvolatile *metabolic acids* (fatty acids, ketone bodies, lactic acid, and others) that cannot be eliminated in the expired breath.

Under normal conditions, the hydrogen ions released by metabolic acids do not affect blood pH because they combine with buffers and are thereby removed from solution. The buffers of the blood include hemoglobin within the red blood cells (as previously described) and proteins and bicarbonate within the plasma. These buffers, however, would eventually become saturated if the body were a closed system that could not eliminate acids. The elimination of acids from the body is accomplished by the lungs, through exhalation of carbon dioxide, and by the kidneys, through excretion of $H^+$ in the urine (see chapter 25).

*Bicarbonate is the major buffer in the plasma* and acts to maintain a blood pH of 7.4 despite the constant production of nonvolatile metabolic acids by the tissues. In this buffering process, some of the $HCO_3^-$ released from the red blood cells during the chloride shift is converted into $H_2CO_3$ in the plasma. Normally, however, there is still a buffer reserve of free bicarbonate that can help protect against unusually large additions of metabolic acids to the blood. These processes are illustrated in figure 24.39.

## Table 24.9    Terminology used to describe acid-base balance

| Term | Definition |
|---|---|
| Acidosis, respiratory | Increased $CO_2$ retention (due to hypoventilation), which can result in the accumulation of carbonic acid and thus a fall in blood pH to below normal |
| Acidosis, metabolic | Increased production of "nonvolatile" acids, such as lactic acid, fatty acids, and ketone bodies, or loss of blood bicarbonate (such as by diarrhea) resulting in a fall in blood pH to below normal |
| Alkalosis, respiratory | A rise in blood pH due to loss of $CO_2$ and carbonic acid (through hyperventilation) |
| Alkalosis, metabolic | A rise in blood pH produced by loss of nonvolatile acids (as in excessive vomiting) or by excessive accumulation of bicarbonate base |
| Compensated acidosis or alkalosis | Metabolic acidosis or alkalosis are partially compensated for by opposite changes in blood carbonic acid levels (through changes in ventilation). Respiratory acidosis or alkalosis are partially compensated for by increased retention or excretion of bicarbonate in the urine. |

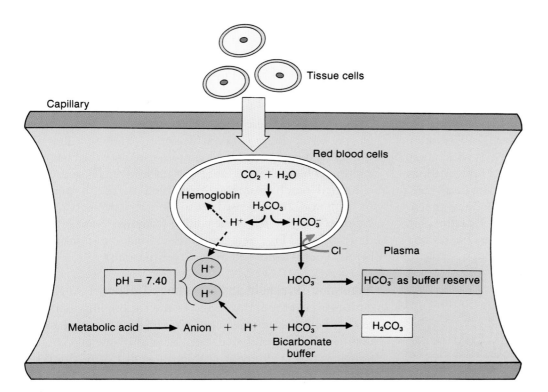

### FIGURE 24.39

Bicarbonate released into the plasma from red blood cells functions to buffer $H^+$ produced by the ionization of metabolic acids (lactic acid, fatty acids, ketone bodies, and others).

Normal plasma, therefore, contains free bicarbonate, carbonic acid, and $H^+$ concentrations indicated by a pH of 7.4. If the $H^+$ concentration of the blood should fall, the carbonic acid produced by the buffering reaction could dissociate and serve as a source of additional $H^+$. If the $H^+$ concentration should rise, bicarbonate could remove this excess $H^+$ from solution. Carbonic acid and bicarbonate are thus said to function as a *buffer pair*.

**Acidosis and Alkalosis**   A fall in blood pH to below 7.35 is called **acidosis** because the pH is to the acid side of normal. Acidosis does not indicate that the blood is acidic (pH of less than 7); a blood pH of 7.2, for example, represents serious acidosis. Similarly, a rise in blood pH to above 7.45 is known as **alkalosis.** There are two components to acid-base balance: *respiratory* and *metabolic*. Respiratory acidosis and alkalosis are due to abnormal concentrations of carbonic acid as a result of abnormal ventilation. Metabolic acidosis and alkalosis result from abnormal amounts of $H^+$ derived from nonvolatile metabolic acids (table 24.9).

Ventilation is normally adjusted to keep pace with the metabolic rate. *Hypoventilation*, as previously described, produces an abnormally high arterial $P_{CO_2}$. This hypercapnia lowers the arterial blood pH and **respiratory acidosis** occurs. In *hyperventilation*, conversely, the arterial $P_{CO_2}$ decreases, causing less formation of carbonic acid than under normal conditions. The depletion of carbonic acid raises the pH, and **respiratory alkalosis** occurs.

**Metabolic acidosis** can occur when the production of nonvolatile acids is abnormally increased. In uncontrolled diabetes mellitus, for example, ketone bodies (derived from fatty acids) may accumulate and produce *ketoacidosis*. In order for metabolic acidosis to occur, however, the buffer reserve of bicarbonate must first be depleted (this is why metabolic acidosis can also be produced by the excessive loss of bicarbonate, as in diarrhea). Until the buffer reserve is depleted, the pH remains normal—ketosis can occur, for example, without ketoacidosis. **Metabolic alkalosis,** a less common condition than metabolic acidosis, can result from loss of acidic gastric juice through vomiting or from excessive intake of bicarbonate (from stomach antacids or from an intravenous solution).

**Compensations for Acidosis and Alkalosis** A change in blood pH, produced by alterations in either the respiratory or metabolic component of acid-base balance, can be partially compensated for by a change in the other component. Metabolic acidosis, for example, stimulates hyperventilation (because the aortic and carotid bodies are sensitive to blood $H^+$) and thus causes a secondary respiratory alkalosis to be produced. The person is still acidotic, but not as much so as would be the case without the compensation. People with partially compensated metabolic acidosis would thus have a low pH, which would be accompanied by a low blood $P_{CO_2}$ as a result of the hyperventilation. Metabolic alkalosis, similarly, is partially compensated for by the retention of carbonic acid due to hypoventilation (table 24.10).

A person with respiratory acidosis would have a low pH and a high blood $P_{CO_2}$ due to hypoventilation. This condition can be partially compensated for by the kidneys, which help to regulate the blood bicarbonate concentration. In short, two organs regulate blood acid-base balance: the lungs (regulating the respiratory component) and the kidneys (regulating the metabolic component). The role of the kidneys in acid-base balance is discussed in more detail in chapter 25.

# Effects of Exercise and High Altitude on Respiratory Function

*Changes in ventilation and oxygen delivery occur during exercise and during acclimatization to high altitudes. These changes help to compensate for the increased metabolic rate during exercise and for the decreased arterial $P_{O_2}$ at high altitudes.*

## Ventilation during Exercise

As soon as a person begins exercise, the rate and depth of breathing increase to produce a total minute volume that is many times the resting value. This increased ventilation, particularly in well-trained athletes, is exquisitely matched to the simultaneous increase in oxygen consumption and carbon dioxide production by the exercising muscles. The arterial blood $P_{O_2}$, $P_{CO_2}$, and pH thus remain remarkably constant as exercise continues (fig. 24.40).

It is tempting to suppose that ventilation increases during exercise as a result of the increased carbon dioxide production by the exercising muscles. Ventilation increases together with increased carbon dioxide production, however, so that blood measurements of $P_{CO_2}$ during exercise are not significantly higher than at rest. The mechanisms responsible for the increased ventilation during exercise must therefore be more complex.

Both *neurogenic mechanisms* and *chemical (humoral) mechanisms* have been proposed to explain the increased ventilation that occurs during exercise. Possible neurogenic mechanisms include the following: (1) sensory nerve activity from the exercising limbs may stimulate the respiratory muscles, either through spinal reflexes or via the brain stem respiratory centers and/or (2) input from the cerebral cortex may stimulate the brain stem centers to modify ventilation. These neurogenic theories help to explain the immediate increase in ventilation that occurs at the beginning of exercise.

Rapid and deep ventilation continues after exercise has stopped, suggesting that chemical factors in the blood may also stimulate ventilation during exercise. Since the $P_{O_2}$, $P_{CO_2}$, and pH of the blood samples from exercising subjects are within the resting range, these chemical theories propose that (1) the $P_{CO_2}$ and pH in the region of the chemoreceptors may be different from these values "downstream" where blood samples are taken and/or (2) cyclic variations in these values that cannot be detected by blood samples stimulate the chemoreceptors. The evidence suggests that both neurogenic and chemical mechanisms are involved in the **hyperpnea** (hi″perp′ne-ă), or increased ventilation, that accompanies exercise. (Note that hyperpnea differs from hyperventilation in that the blood $P_{CO_2}$ remains in the normal range during hyperpnea but is decreased in hyperventilation.)

**Anaerobic Threshold and Endurance Training** At the beginning of exercise, the ability of the cardiopulmonary system to deliver adequate amounts of oxygen to the exercising muscles may be insufficient because of the time lag required to make proper cardiovascular adjustments. During this time, therefore, the muscles respire anaerobically and a "stitch in the side"—probably due to hypoxia of the diaphragm—may develop. After the cardiovascular adjustments have been made, a person may experience a "second wind" when the muscles receive sufficient oxygen for their needs.

Continued heavy exercise can cause a person to reach the **anaerobic threshold,** which is the maximum rate of oxygen consumption that can be attained before blood lactic acid levels rise as a result of anaerobic respiration. This occurs when 50% to 60% of the person's maximal oxygen uptake has been reached. The rise in lactic acid levels is due to the aerobic limitations of the muscles; it is not due to a malfunction of the cardiopulmonary system. Indeed, the arterial oxygen hemoglobin saturation remains at 97% and venous blood draining the muscles contains unused oxygen.

| Table 24.10 | The effect of lung function on blood acid-base balance | | | |
|---|---|---|---|---|
| **Condition** | **pH** | **$P_{CO_2}$** | **Ventilation** | **Cause or compensation** |
| Normal | 7.35–7.45 | 39–41 mmHg | Normal | Not applicable |
| Respiratory acidosis | Low | High | Hypoventilation | Cause of the acidosis |
| Respiratory alkalosis | High | Low | Hyperventilation | Cause of the alkalosis |
| Metabolic acidosis | Low | Low | Hyperventilation | Compensation for acidosis |
| Metabolic alkalosis | High | High | Hypoventilation | Compensation for alkalosis |

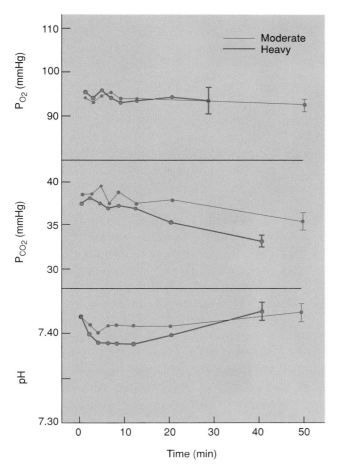

**FIGURE 24.40**

The effect of moderate and heavy exercise on arterial blood gases and pH. Notice that there are no consistent and significant changes in these measurements during the first several minutes of moderate and heavy exercise and that only the $P_{CO_2}$ changes (actually decreases) during more prolonged exercise.

The rise in blood lactic acid that occurs when the anaerobic threshold is exceeded is due to the inability of the exercising muscles to increase their oxygen consumption rate sufficiently to prevent anaerobic respiration. The anaerobic threshold, however, is higher in endurance-trained athletes than it is in other people. Endurance training increases the skeletal muscle content with respect to myoglobin, mitochondria, and Krebs cycle enzymes. These muscles, therefore, are able to utilize more of the oxygen delivered to them by the arterial blood. At a given level of exercise, consequently, the venous blood that drains from muscles in endurance-trained people contains a lower percentage oxyhemoglobin than that in other people. The effects of exercise and endurance training on respiratory function are summarized in table 24.11.

## Acclimatization to High Altitude

When a person who lives near a region near sea level moves to a significantly higher elevation, several adjustments in the respiratory system must be made to compensate for the decreased atmospheric pressure and $P_{O_2}$ at the higher altitude. These adjustments include changes in ventilation, in the hemoglobin affinity for oxygen, and in the total hemoglobin concentration.

Reference to table 24.4 indicates that at an altitude of 8000 feet, for example, the $P_{O_2}$ of arterial blood is 74 mmHg, as compared to 100 mmHg at sea level. Reference to table 24.7 indicates that the percent oxyhemoglobin saturation decreases from 97% at sea level to about 94% at 74 mmHg. The amount of oxygen attached to hemoglobin, and thus the total oxygen content of blood, is decreased. People may, however, experience rapid fatigue even at more moderate elevations (for example 5000 to 6000 feet) at which the oxyhemoglobin saturation is only slightly decreased. Compensations made by the respiratory system gradually reduce the amount of fatigue caused by a given amount of exertion at high altitudes.

**Changes in Ventilation**   Starting at altitudes as low as 1500 m (5000 ft), the decreased arterial $P_{O_2}$ stimulates an increase in ventilation. This is called a *hypoxic ventilatory response*. This response produces hyperventilation, which lowers the arterial $P_{CO_2}$ and thus produces a respiratory alkalosis. The rise in arterial pH then helps to blunt the hyperventilation, which becomes stabilized after a few days at about 2–3 L per minute more than the total minute volume at sea level.

## Table 24.11    Changes in respiratory function during exercise

| Variable | Change | Comments |
|---|---|---|
| Ventilation | Increased | This is not hyperventilation because ventilation is matched to increased metabolic rate. Mechanisms responsible for increased ventilation are not well understood. |
| Blood gases | No change | Blood-gas measurements during light, moderate, and heavy exercise show little change because ventilation is increased to match increased muscle $O_2$ consumption and $CO_2$ production. |
| $O_2$ delivery to muscles | Increased | Although the total $O_2$ content and $P_{O_2}$ do not increase during exercise, there is an increased rate of blood flow to the exercising muscles. |
| $O_2$ extraction by muscles | Increased | Increased $O_2$ consumption lowers the tissue $P_{O_2}$ and lowers the affinity of hemoglobin for $O_2$ (due to the effect of increased temperature). More $O_2$, as a result, is unloaded so that venous blood contains a lower oxyhemoglobin saturation than at rest. This effect is enhanced by endurance training. |

Hyperventilation at high altitude increases tidal volume, thus reducing the proportionate contribution of air from the anatomical dead space and increasing the proportion of fresh air brought to the alveoli. This improves the oxygenation of the blood over what it would be in the absence of the hyperventilation. Hyperventilation, however, cannot increase blood $P_{O_2}$ above that of the inspired air. The $P_{O_2}$ of arterial blood decreases with increasing altitude, regardless of the ventilation. In the Peruvian Andes, for example, the normal arterial $P_{O_2}$ is reduced from 100 mmHg (at sea level) to 45 mmHg. The loading of hemoglobin with oxygen is therefore incomplete, producing an oxyhemoglobin saturation that is decreased from 97% (at sea level) to 81%.

**Hemoglobin Affinity for Oxygen**    Normal arterial blood at sea level only unloads about 22% of its oxygen to the tissues at rest; the percent saturation is reduced from 97% in arterial blood to 75% in venous blood. As a partial compensation for the decrease in oxygen content at high altitude, the affinity of hemoglobin for oxygen is reduced, so that a higher proportion of oxygen is unloaded. This occurs because the low oxyhemoglobin content of red blood cells stimulates the production of 2,3-DPG, which in turn decreases the hemoglobin affinity for oxygen.

At very high altitudes, however, the story becomes more complex. In a 1984 study by J. B. West, the very low arterial $P_{O_2}$ (28 mmHg) of subjects at the summit of Mount Everest stimulated intense hyperventilation, so that the arterial $P_{CO_2}$ was decreased to 7.5 mmHg. The resultant respiratory alkalosis (the arterial pH was greater than 7.7) caused a shift to the left of the oxyhemoglobin dissociation curve (indicating greater affinity of hemoglobin for oxygen) despite the antagonistic effects of increased 2,3-DPG concentrations. It was suggested that the increased affinity of hemoglobin for oxygen caused by the respiratory alkalosis may have been beneficial at such a high altitude, since it increased the loading of hemoglobin with oxygen in the lungs.

**Increased Hemoglobin and Red Blood Cell Production**    In response to tissue hypoxia, the kidneys secrete the hormone erythropoietin. Erythropoietin stimulates the bone marrow to increase its production of hemoglobin and red blood cells. In the Peruvian Andes, for example, people have a total hemoglobin concentration of 19.8 g per 100 ml, as compared to 15 g per 100 ml at sea level. Although the percent oxyhemoglobin saturation is still lower than at sea level, the total oxygen content of the blood is actually greater—22.4 ml $O_2$ per 100 ml compared to a sea level value of about 20 ml $O_2$ per 100 ml. These adjustments of the respiratory system to high altitude are summarized in table 24.12. It should be noted that they are not unalloyed benefits. Polycythemia (high red blood cell count) increases the viscosity of blood, and pulmonary hypertension, which is more common at high altitude, can cause accompanying edema and ventricular hypertrophy, which may lead to heart failure.

 *Acute Mountain Sickness (AMS)* is common in people who arrive at altitudes in excess of 5000 feet. Cardinal symptoms of AMS are headache, malaise, anorexia, nausea, and fragmented sleep. Headache is the most common symptom and may result from changes in blood flow to the brain. Low arterial $P_{O_2}$ stimulates vasodilation of vessels in the pia matter, increasing blood flow and pressure within the skull. The hypocapnia produced by hyperventilation, however, causes cerebral vasoconstriction. The balance between these two antagonistic effects is variable. Pulmonary edema is common at altitudes above 9000 feet and can produce shortness of breath, coughing, and a mild fever. Cerebral edema generally occurs above an altitude of 10,000 feet and can produce mental confusion and even hallucination. Pulmonary and cerebral edema are potentially dangerous and can be alleviated by descent to a lower altitude.

# UNDER DEVELOPMENT

## Embryological Development of the Respiratory System

The formation of the nasal cavity begins at 3½ to 4 weeks of embryonic life. A region of thickened ectoderm called the **olfactory placode** appears on the front and inferior part of the head (fig. 1). The placode invaginates to form the **olfactory pit,** which extends posteriorly to connect with the **foregut.** The foregut, derived of endoderm, later develops into the pharynx.

The mouth, or oral cavity, develops at the same time as the nasal cavity, and for a short time there is a thin **oronasal** (or´o-na´zal) **membrane** separating the two cavities. This membrane ruptures during the seventh week, and a single,

large **oronasal cavity** forms. Shortly thereafter, tissue plates of mesoderm begin to grow horizontally across the cavity. At approximately the same time, a vertical plate develops inferiorly from the roof of the nasal cavity. These plates have completed their formation by 3 months of development. The vertical plate forms the nasal septum, and the horizontal plates form the hard palate. A *cleft palate* forms when the horizontal plates fail to meet in the midline.

The respiratory system begins to form during the fourth week of development as a diverticulum (di´ver-tik´yŭ-lum), or outpouching, called the **laryngotracheal** (lă-ring´go-tra´ke-al) **bud,** from the

ventral surface of endoderm along the lower pharyngeal region. As the bud grows, the proximal portion forms the trachea and the distal portion bifurcates (splits) into a right and left bronchus.

The buds continue to elongate and split until all the tubular network within the lower respiratory tract is formed. As the terminal portion forms air sacs, called **alveoli,** at about 8 weeks of development, the supporting lung tissue begins to form. The complete structure of the lungs, however, is not fully developed until about 26 weeks of fetal development, so premature infants born prior to this time require special artificial respiratory equipment to live.

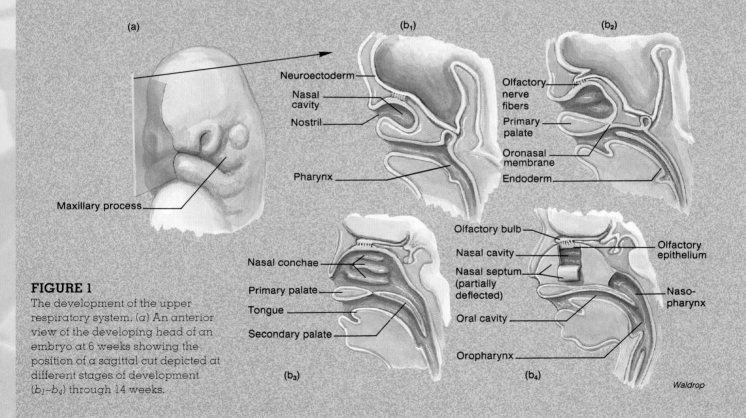

(a)  (b₁)  (b₂)

Neuroectoderm
Nasal cavity
Nostril
Pharynx
Maxillary process

Olfactory nerve fibers
Primary palate
Oronasal membrane
Endoderm

Nasal conchae
Primary palate
Tongue
Secondary palate
(b₃)

Olfactory bulb
Nasal cavity
Nasal septum (partially deflected)
Oral cavity
Oropharynx
(b₄)

Olfactory epithelium
Naso-pharynx

Waldrop

### FIGURE 1

The development of the upper respiratory system. (a) An anterior view of the developing head of an embryo at 6 weeks showing the position of a sagittal cut depicted at different stages of development (b₁–b₄) through 14 weeks.

## Table 24.12  Changes in respiratory function during acclimatization to high altitude

| Variable | Change | Comments |
|---|---|---|
| Partial pressure of $O_2$ | Decreased | Due to decreased total atmospheric pressure |
| Percent oxyhemoglobin saturation | Decreased | Due to lower $P_{O_2}$ in pulmonary capillaries |
| Ventilation | Increased | Due to low arterial $P_{O_2}$; ventilation usually returns to normal after a few days |
| Total hemoglobin | Increased | Due to stimulation by erythropoietin; raises $O_2$ capacity of blood to partially or completely compensate for the reduced partial pressure |
| Oxyhemoglobin affinity | Decreased | Due to increased DPG within the red blood cells; results in a higher unloading percentage of $O_2$ to the tissues, which may partially or completely compensate for the reduced arterial oxyhemoglobin saturation |

# Clinical Considerations

## Developmental Problems

Birth defects, inherited disorders, and premature births commonly cause problems in the respiratory system of infants. A **cleft palate** is a developmental deformity of the hard palate of the mouth in which an opening persists between the oral and nasal cavities, making it difficult, if not impossible, for an infant to nurse. A cleft palate may be hereditary or a complication of some disease (e.g., German measles) contracted by the mother during pregnancy. A **cleft lip** is a genetically based developmental disorder in which the two sides of the upper lip fail to fuse. Cleft palates and cleft lips can be treated very effectively with cosmetic surgery.

**Cystic fibrosis** is the most common fatal inheritable disorder among Caucasians. Approximately 1 out of 20 whites are carriers, and 1 in 2000 children inherit the disease. The defective gene reduces the ability of epithelial cells to actively transport $Cl^-$, causing the production of a very thick, sticky mucus that blocks the airways of the lungs. It also affects the liver and other organs in potentially life-threatening ways.

## Sickle-Cell Anemia and Thalassemia

A number of hemoglobin diseases are produced by inherited (congenital) defects in the protein part of hemoglobin. **Sickle-cell anemia**—a disease carried in a recessive state by 8% to 11% of the black population of the United States—for example, is caused by an abnormal form of hemoglobin called *hemoglobin* S. Hemoglobin S differs from normal hemoglobin A in only one amino acid: valine is substituted for glutamic acid in position 6 on the beta chains. This amino acid substitution is caused by a single base change in the region of DNA that codes for the beta chains.

Under conditions of low blood $P_{O_2}$, hemoglobin S comes out of solution and cross-links to form a "paracrystalline gel" within the red blood cells. This causes the characteristic sickle shape of red blood cells and makes them less flexible and more fragile. The decreased solubility of hemoglobin S in solutions of low $P_{O_2}$ is used to diagnose sickle-cell anemia and sickle-cell trait (the carrier state, in which a person has genes for both hemoglobin A and hemoglobin S).

**Thalassemia** (thal″ă-se′me-ă), is any of a family of hemoglobin diseases found predominantly among people of Mediterranean ancestry. In *alpha thalassemia,* there is decreased synthesis of the alpha chains of hemoglobin, whereas in *beta thalassemia* the synthesis of the beta chains is impaired. One of the compensations for thalassemia is increased synthesis of gamma chains, resulting in the retention of large amounts of hemoglobin F (fetal hemoglobin) into adulthood. Interestingly, a new treatment for sickle-cell disease involves the use of erythropoietin, which stimulates red blood cell production, combined with a drug that stimulates the bone marrow to produce hemoglobin F (as in thalassemia).

Some types of abnormal hemoglobins have been shown to be advantageous in the environments in which they evolved. Carriers for sickle-cell anemia, for example (who therefore have both hemoglobin A and hemoglobin S), are more resistant to malaria than are noncarriers. This is because the parasite that causes malaria cannot live in red blood cells that contain hemoglobin S.

## Trauma or Injury

A collection of air or gas in the pleural cavity that may cause the lung to collapse is referred to as a **pneumothorax.** A pneumothorax can result from an external injury, such as a stabbing, bullet wound, or penetrating fractured rib, or it can occur internally. A severely diseased lung, as in emphy-

........................................

thalassemia: Gk. *thalassa,* sea

sema, can create a pneumothorax as the wall of the lung deteriorates and permits air to enter the pleural cavity.

Choking on a foreign object such as aspirated food is a common serious trauma to the respiratory system. Each day more than eight Americans choke to death on food lodged in their trachea. A simple process termed the **abdominal thrust (Heimlich) maneuver** can save the life of a person who is choking. The technique for performing the abdominal thrust maneuver is presented at the end of this chapter on page 721.

Individuals saved from drowning and shock victims frequently experience apnea (cessation of breathing) and will soon die if not revived by someone performing artificial respiration. The accepted treatment for reviving a person who has stopped breathing is illustrated on page 722.

## Common Respiratory Disorders

A cough is the most common symptom of respiratory disorders. Acute problems may be accompanied by dyspnea or wheezing. Respiratory or circulatory problems may cause **cyanosis,** which is a blue discoloration of the skin caused by blood with a low oxygen content.

Pulmonary disorders are classified as **obstructive** when there is increased resistance to air flow in the bronchioles; they are classified as **restrictive** when alveolar tissue is damaged. Asthma and acute bronchitis are usually just obstructive whereas chronic bronchitis and emphysema are both obstructive and restrictive. Pulmonary fibrosis, by contrast, is a purely restrictive disorder.

**Asthma**    The dyspnea, wheezing, and other symptoms of asthma are produced by the obstruction of air flow through the bronchioles that occurs in episodes or "attacks." This obstruction is caused by inflammation, mucus secretion, and bronchoconstriction. Inflammation of the airways is characteristic of asthma and itself contributes to increased airway responsiveness to agents that promote bronchoconstriction. Bronchoconstriction further increases airway resistance and makes breathing difficult. Constriction of bronchiolar smooth muscles is stimulated by leukotrienes and histamine released by mast cells and leukocytes (chapter 23), which can be provoked by an allergic reaction or by the release of acetylcholine from parasympathetic nerve endings.

Asthma is often treated with glucocorticoid drugs, which inhibit inflammation. Epinephrine and related compounds stimulate beta-adrenergic receptors in the bronchioles and by this means promote bronchodilation. Therefore, epinephrine was used in the past as an inhaled spray to re-

lieve the symptons of an asthma attack. It has since been learned that there are two subtypes of beta receptors for epinephrine and that the subtype in the heart (designated $\beta_1$)is different from the one in the brochioles ($\beta_2$). By utilizing these differences, compounds such as *terbutaline* have been developed that can more selectively stimulate the $\beta_2$-adrenergic receptors and cause bronchodilation, without having as great an effect on the heart as does epinephrine.

**Emphysema**    Alveolar tissue is destroyed in emphysema, resulting in fewer but larger alveoli (fig. 24.41). This reduces the surface area for gas exchange and decreases the ability of the bronchioles to remain open during expiration. Collapse of the bronchioles as a result of the compression of the lungs during expiration produces *air trapping*, which further decreases the efficiency of gas exchange in the alveoli.

Among the different types of emphysema, the most common occurs almost exclusively in people who have smoked cigarettes heavily over a period of years. A component of cigarette smoke apparently stimulates the macrophages and leukocytes to secrete proteolytic (protein-digesting) enzymes that destroy lung tissues. A less common type of emphysema results from a genetic inability to produce a plasma protein called $\alpha_1$-antitrypsin. This protein normally inhibits proteolytic enzymes such as trypsin, and thus normally protects the lungs against the effects of enzymes that are released from alveolar macrophages.

Chronic bronchitis and emphysema, the most common causes of respiratory failure, are together called **chronic obstructive pulmonary disease (COPD).** In addition to the more direct obstructive and restrictive aspects of these conditions, other pathological changes may occur. These include edema, inflammation, hyperplasia (increased cell number), zones of pulmonary fibrosis, pneumonia, pulmonary emboli (traveling blood clots), and heart failure. Patients with severe emphysema may eventually develop *cor pulmonale*—pulmonary hypertension with hypertrophy and the eventual failure of the right ventricle. COPD is the fifth leading cause of death in the United States.

**Pulmonary Fibrosis**    Under certain conditions, for reasons that are poorly understood, lung damage leads to pulmonary fibrosis instead of emphysema. In this condition the normal structure of the lungs is disrupted by the accumulation of fibrous connective tissue proteins. Fibrosis can result, for example, from the inhalation of particles less than 6 μm in size that can accumulate in the respiratory zone of the lungs. This type of fibrosis includes *anthracosis*, or black lung, which is produced by the inhalation of carbon particles from coal dust.

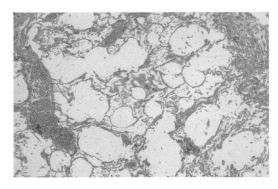

(a)

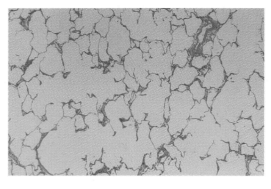

(b)

**FIGURE 24.41**

Photomicrographs of tissue (a) from a normal lung and (b) from the lung of a person with emphysema. In emphysema, lung tissue is destroyed, resulting in the presence of fewer and larger alveoli.

**Other Respiratory Disorders** The **common cold** is the most widespread of all respiratory diseases. Colds occur repeatedly because acquired immunity to one virus does not protect against other viruses that cause colds. Cold viruses cause acute inflammation of the respiratory mucosa, resulting in a flow of mucus, a fever, and often a headache.

Nearly all the structures and regions of the respiratory tract can become infected and inflamed. **Influenza** is a viral disease that causes inflammation of the upper respiratory tract. **Sinusitis** (si-nŭ-si´tis) is an inflammation of the paranasal sinuses. **Tonsillitis** may involve one or all of the tonsils and frequently follows other lingering diseases of the oral or pharyngeal region. **Laryngitis** is inflammation of the larynx, which often produces a hoarse voice and limits the ability to talk. **Tracheobronchitis** and **bronchitis** are infections of the regions for which they are named. Severe inflammation of the bronchioles can cause significant airway resistance and dyspnea.

Diseases of the lungs are common and may be serious. **Pneumonia** is an acute infection and inflammation of lung tissue accompanied by exudation (the accumulation of fluid). It is usually caused by bacteria, most commonly the pneumococcus bacterium. Viral pneumonia is caused by a number of different viruses. **Tuberculosis** is an inflammatory disease of the lungs caused by the presence of tubercle bacilli. Tuberculosis softens and leads to the ulceration of lung tissue. **Pleurisy** is an inflammation of the pleura and is usually secondary to some other respiratory disease. Inspiration may become painful, and fluid may collect within the pleural space.

**Cancer** in the respiratory system is often caused by repeated inhalation of irritating substances, such as cigarette smoke. Cancers of the lips, larynx, and lungs are especially common in smokers over the age of 50.

## Disorders of Respiratory Control

A variety of disease processes can produce cessation of breathing during sleep, or *sleep apnea*. **Sudden infant death syndrome (SIDS)** is an especially tragic form of sleep apnea that claims the lives of about 10,000 babies annually in the United States. Victims of this condition are apparently healthy 2- to-5-month-old babies who die in their sleep without apparent reason—hence, the layperson's term, *crib death*. These deaths seem to be caused by failure of the respiratory control mechanisms in the brain stem and/or by failure of the carotid bodies to be stimulated by reduced arterial oxygen.

Abnormal breathing patterns often appear prior to death from brain damage or heart disease. The most common of these abnormal patterns is **Cheyne–Stokes breathing,** in which the depth of breathing progressively increases and then progressively decreases. These cycles of increasing and decreasing tidal volumes may be followed by periods of apnea of varying durations. Cheyne–Stokes breathing may be caused by neurological damage or by insufficient oxygen delivery to the brain. The latter may result from heart disease or from a brain tumor that diverts a large part of the vascular supply from the respiratory centers.

## Disorders Caused by High Partial Pressures of Gases

The total atmospheric pressure increases by one atmosphere (760 mmHg) for every 10 m (33 ft) below sea level. If a diver descends to 10 m below sea level, therefore, the partial pressures and amounts of dissolved gases in the plasma will be twice those at sea level. At 66 ft they are three times, and at 100 ft they are four times the values at sea level. The increased amounts of nitrogen and oxygen dissolved in the

influenza: L. *influentia*, a flowing in

Cheyne–Stokes breathing: from John Cheyne, Scottish physician, 1777–1836, and William Stokes, Irish physician, 1804–78

blood plasma under these conditions can have serious adverse effects on the body.

**Oxygen Toxicity**    Although breathing 100% oxygen at one or two atmospheres pressure can be safely tolerated for a few hours, higher partial oxygen pressures can be very dangerous. **Oxygen toxicity** develops rapidly when the $P_{O_2}$ rises above about 2.5 atmospheres. This is apparently caused by the oxidation of enzymes and other destructive changes that can damage the nervous system and lead to coma and death. For these reasons, deep-sea divers commonly use gas mixtures in which oxygen is diluted with inert gases such as nitrogen (as in ordinary air) or helium.

Hyperbaric oxygen—oxygen at greater than one atmosphere pressure—is often used to treat conditions such as carbon monoxide poisoning, circulatory shock, and gas gangrene. Before the dangers of oxygen toxicity were realized, these hyperbaric oxygen treatments sometimes resulted in tragedy. Particularly tragic were the cases of **retrolental fibroplasia,** in which damage to the retina and blindness resulted from hyperbaric oxygen treatment of premature babies with hyaline membrane disease.

**Nitrogen Narcosis**    Although nitrogen is physiologically inert at sea level, larger amounts of dissolved nitrogen under hyperbaric conditions have deleterious effects. Since it takes time for the nitrogen to dissolve, these effects usually do not appear until the person has remained submerged for over an hour. **Nitrogen narcosis** resembles alcohol intoxication; depending on the depth of the dive, the diver may experience a euphoria known as "rapture of the deep," or he or she may become so drowsy as to be totally incapacitated.

**Decompression Sickness**    The amount of nitrogen dissolved in the plasma decreases as a diver ascends to sea level as a result of the progressive decrease in the $P_{N_2}$. If the diver surfaces slowly, a large amount of nitrogen can diffuse through the alveoli and be eliminated in the expired breath. If decompression occurs too rapidly, however, bubbles of nitrogen gas ($N_2$) can form in the blood and block small blood channels, producing muscle and joint pain, as well as more serious damage. These effects are referred to as **decompression sickness,** or the bends.

The cabins of airplanes that fly long distances at high altitudes (30,000 to 40,000 ft) are pressurized so that the passengers and crew are not exposed to the very low atmospheric pressures of these altitudes. If a cabin were to become rapidly depressurized at high altitude, much less nitrogen could remain dissolved at the greatly lowered pressure. People in this situation, like divers that ascend too rapidly, would experience decompression sickness.

*The abdominal thrust (Heimlich) maneuver can save the life of a person who is choking. This technique is performed as follows:*

A.    If the victim is standing or sitting:
1. Stand behind the victim or the victim's chair and wrap your arms around his or her waist.
2. Grasp your fist with your other hand and place the fist against the victim's abdomen, slightly above the navel and below the rib cage.
3. Press your fist into the victim's abdomen with a quick upward thrust.
4. Repeat several times if necessary.

B.    If the victim is lying down:
1. Position the victim on his or her back.
2. Face the victim, and kneel on his or her hips.
3. With one of your hands on top of the other, place the heel of your bottom hand on the abdomen, slightly above the navel and below the rib cage.
4. Press into the victim's abdomen with a quick upward thrust.
5. Repeat several times if necessary.

If you are alone and choking, use whatever is available to apply force just below your diaphragm. Press into a table or a sink, or use your own fist.

Respiratory System

721

## Mouth-to-Mouth Method

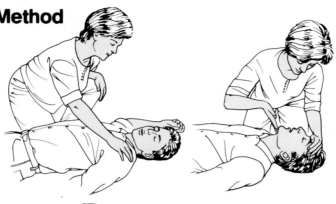

1. **Check for unresponsiveness.**
Gently shake the victim and shout, "Are you okay?" If no response, get the attention of someone who can phone for help. Make sure that the victim is on his or her back.

2. **Open the airway.**
Tilt the victim's head back by pushing on his or her forehead with your hand and lifting the chin with your fingers under his or her jaw. This will open the airway by moving the tongue away from the back of the victim's throat.

3. **Check for breathing.**
Put your ear close to the victim's face to listen and feel for any return of air. At the same time, look to see if there is chest movement. Check for breathing for about 5 seconds.

4. **If no breathing, give two full breaths.**
While maintaining the victim in the head-tilt position, pinch his or her nose to close off the nasal passageway. Take a deep breath, then seal your mouth around the victim's mouth and give two full breaths. (After the first breath, raise your head slightly to inhale quickly and then give the second breath.)

5. **Check for pulse.**
While maintaining head tilt, feel for a carotid pulse for 5 to 10 seconds on the side of the victim's neck.

6. **Continue rescue breathing.**
With the victim in the head-tilt position and his or her nostrils pinched, give one breath every 5 seconds. Observe for signs of breathing between breaths. For an infant, give one gentle puff every 3 seconds.

7. **Recheck for pulse.**
Feel for a carotid pulse at 1-minute intervals. If the victim has a pulse but is not breathing, continue rescue breathing.

## Mouth-to-Nose Method

1. **Open the airway.**
Place the victim in the head-tilt position as described above.

2. **Blow into the victim's nose.**
Using the same sequence described above, blow into the victim's nose while holding his or her mouth closed.

3. **Feel and observe for breathing.**
With the victim's mouth held open, detect for breathing between giving forced breaths.

# Interactions of the Respiratory System with Other Body Systems

## Integumentary System
- Protects respiratory system from pathogens and helps maintain body temperature
- Provides oxygen for cell respiration and eliminates carbon dioxide

## Skeletal System
- Protects lungs within the ribcage
- Produces red blood cells in bone marrow for the transport of oxygen
- Provides oxygen for cell respiration and eliminates carbon dioxide

## Muscular System
- Provides muscular contractions needed for ventilation
- Consumes large amounts of oxygen and produces large amounts of carbon dioxide during exercise
- Provides oxygen for cell respiration and eliminates carbon dioxide

## Nervous System
- Regulates rate and depth of breathing
- Regulates blood flow, and hence delivery of blood to tissues for gas exchange
- Provides oxygen for cell respiration and eliminates carbon dioxide

## Endocrine System
- Epinephrine dilates bronchioles
- Provides oxygen for cell respiration and eliminates carbon dioxide

## Circulatory System
- Delivers oxygen from lungs to body cells and transports carbon dioxide from body tissues to lungs
- Provides oxygen for cell respiration and eliminates carbon dioxide

## Lymphatic System
- Protects against infections that could damage respiratory system
- Provides oxygen for cell respiration and eliminates carbon dioxide
- Tonsils house immune cells

## Urinary System
- Regulates volume and electrolyte balance of the blood
- Participates with lungs in regulation of blood pH
- Provides oxygen for cell respiration and eliminates carbon dioxide

## Digestive System
- Provides nutrients to be used by cells of the lungs and other organs
- Provides oxygen for cell respiration and eliminates carbon dioxide

## Reproductive System
- Promotes changes in respiratory rate and depth during sexual arousal
- Provides oxygen for cell respiration and eliminates carbon dioxide

# Chapter Summary

## Functions and Divisions of the Respiratory System (p. 684)

1. Respiration refers not only to breathing but also to the exchange of gases between the atmosphere, the blood, and individual cells.
2. The respiratory system is divided into a respiratory division, which includes the alveoli of the lungs in which gas exchange occurs, and the conducting division, which includes all the structures that conduct air to the respiratory division.

## Conducting Division (pp. 684–690)

1. The nose is supported by nasal bones and cartilages.
2. The paranasal sinuses are found in the maxillary, frontal, sphenoid, and ethmoid bones.
3. The pharynx is a funnel-shaped passageway that connects the oral and nasal cavities with the larynx.
4. The larynx is composed of a number of cartilages that keep the passageway to the trachea open during breathing and close the respiratory passageway during swallowing.
5. The trachea is a rigid tube, supported by rings of cartilage, that leads from the larynx to the bronchial tree.
6. The bronchial tree includes a right and left primary bronchus, which divides to produce secondary bronchi, tertiary bronchi, and bronchioles; the conducting division ends with the terminal bronchioles, which connect to the alveoli.

## Alveoli, Lungs, and Pleurae (pp. 690–692)

1. Alveoli are the functional units of the lungs where gas exchange occurs; they are numerous, small, thin-walled air sacs.
2. The right and left lungs are separated by the mediastinum. Each lung is divided into lobes and lobules.
3. The lungs are covered by a visceral pleural membrane (visceral pleura), and the thoracic cavity is lined by a parietal pleural membrane (parietal pleura).
   a. A potential space between these two pleural membranes is called the pleural cavity, or intrapleural space.
   b. The pleural membranes package each lung separately and exclude the structures located in the mediastinum.

## Physical Aspects of Ventilation (pp. 692–695)

1. The intrapleural and intrapulmonary pressures vary during ventilation.
   a. The intrapleural pressure is always less than the intrapulmonary pressure.
   b. The intrapulmonary pressure is subatmospheric during inspiration and greater than the atmospheric pressure during expiration.
2. Pressure changes in the lungs are produced by variations in lung volume because, according to Boyle's law, the pressure of a gas is inversely proportional to its volume.
3. The mechanics of ventilation are influenced by the physical properties of the lungs.
   a. The compliance of the lungs refresh the change in lung volume as a function of change in transpulmonary pressure.
   b. The elasticity of the lungs refers to their tendency to recoil after distension.
   c. The surface tension of the fluid in the alveoli exerts a force directed inward, which acts to resist distension.
4. The action of lung surfactant to reduce surface tension in alveoli prevents the alveoli from collapsing during expiration. Babies who lack surfactant suffer from respiratory distress syndrome.

## Mechanics of Breathing (pp. 695–699)

1. Inspiration and expiration are accomplished by contraction and relaxation of skeletal muscles.
   a. During quiet inspiration, the diaphragm and the external intercostal and internal intercostal (interchondral part) muscles contract, increasing the volume of the thorax.
   b. During quiet expiration these muscles relax, and the elastic recoil of the lungs and thorax causes a decrease in thoracic volume.
   c. The lungs are stuck to the wall of the thorax because the intrapulmonary pressure is greater than the intrapleural pressure.
2. Spirometry aids in the diagnosis of a number of pulmonary disorders.
   a. An abnormally low vital capacity indicates restrictive lung disorders, including emphysema and pulmonary fibrosis.
   b. An abnormally low forced expiratory volume indicates obstructive disorders, including asthma and bronchitis.

## Gas Exchange in the Lungs (pp. 699–702)

1. According to Dalton's law, the total pressure of a gas mixture is equal to the sum of the pressures that each gas in the mixture would exert independently.
2. According to Henry's law, the amount of gas that can be dissolved in a fluid is directly proportional to the partial pressure of that gas in contact with the fluid.
3. The $P_{O_2}$ and $P_{CO_2}$ measurements of arterial blood provide information about lung function.
   a. Inadequate ventilation (hypoventilation) causes a rise in arterial $P_{CO_2}$ and a fall in arterial $P_{O_2}$.
   b. Excessive ventilation (hyperventilation) causes a fall in arterial $P_{CO_2}$, but has little effect on arterial $P_{O_2}$.

## Regulation of Breathing (pp. 702–707)

1. The rhythmicity center in the medulla oblongata directly controls the muscles of respiration.
   a. Activity of the inspiratory and expiratory neurons varies in a reciprocal way to produce an automatic breathing cycle.
   b. Activity in the medulla oblongata is influenced by the apneustic and pneumotaxic centers in the pons, as well as by sensory feedback information.
   c. Conscious breathing involves direct control by the cerebral cortex via corticospinal tracts.
2. Breathing is affected by chemoreceptors sensitive to the $P_{O_2}$, pH, and $P_{CO_2}$ of the blood.
   a. The $P_{CO_2}$ of the blood and consequent changes in pH are usually of greater importance than the blood $P_{O_2}$ in the regulation of breathing.
   b. Central chemoreceptors in the medulla oblongata are sensitive to changes in blood $P_{CO_2}$ because of the resultant changes in the pH of cerebrospinal fluid.
   c. The peripheral chemoreceptors in the aortic and carotid bodies are sensitive to changes in blood $P_{CO_2}$ indirectly, because of consequent changes in blood pH.

3. Decreases in blood $P_{O_2}$ directly stimulate breathing only when the blood $P_{O_2}$ is less than 50 mmHg; a drop in $P_{O_2}$ also stimulates breathing indirectly, by making the chemoreceptors more sensitive to changes in $P_{CO_2}$ and pH.

## Hemoglobin and Oxygen Transport (pp. 707–711)

1. Hemoglobin is composed of two alpha and two beta polypeptide chains and four heme groups that contain a central atom of iron.
2. When the iron is in the reduced form and not attached to $O_2$, the hemoglobin is called deoxyhemoglobin; when it is attached to $O_2$, it is called oxyhemoglobin.
3. Deoxyhemoglobin combines with $O_2$ in the lungs (the loading reaction) and breaks its bonds with $O_2$ in the tissue capillaries (the unloading reaction); the extent of each reaction is determined by the $P_{O_2}$ and the affinity of hemoglobin for $O_2$.
4. An oxyhemoglobin dissociation curve is a graph of percent oxyhemoglobin saturation at different values of $P_{O_2}$.

5. The pH and temperature of the blood influence the affinity of hemoglobin for $O_2$ and the extent of loading and unloading.
   a. In the Bohr effect, a fall in pH decreases the affinity of hemoglobin for $O_2$ and a rise in pH increases the affinity of hemoglobin for $O_2$.
   b. A rise in temperature decreases the affinity of hemoglobin for $O_2$.
   c. When the affinity is decreased, the oxyhemoglobin dissociation curve is shifted to the right; this indicates a greater unloading percentage of $O_2$ to the tissues.
6. The affinity of hemoglobin for $O_2$ is also decreased, under conditions of anemia or low blood oxygen, by 2,3-diphosphoglyceric acid (2,3-DPG) within the red blood cells.

## Carbon Dioxide Transport and Acid-Base Balance (pp. 711–714)

1. An enzyme in red blood cells called carbonic anhydrase catalyzes the reversible reaction whereby carbon dioxide and water are used to form carbonic acid.
   a. This reaction is favored by the high $P_{CO_2}$ in the tissue capillaries; as a

result, $CO_2$ produced by the tissues is converted into carbonic acid in the red blood cells.
   b. A reverse reaction occurs in the lungs; in this process, the low $P_{CO_2}$ favors the conversion of carbonic acid to $CO_2$, which can be exhaled.
2. By adjusting the blood concentration of $CO_2$, and thus of carbonic acid, the process of ventilation helps to maintain proper acid-base balance of the blood.
   a. Normal arterial blood pH is 7.40. A pH below 7.35 is termed acidosis; a pH above 7.45 is termed alkalosis.
   b. Hyperventilation causes respiratory alkalosis; hypoventilation causes respiratory acidosis.

## Effects of Exercise and High Altitude on Respiratory Function (pp. 714–716)

1. During exercise there is increased ventilation, or hyperpnea, that can be matched to the increased metabolic rate so that the arterial blood $P_{CO_2}$ remains normal.
2. Acclimatization to a high altitude involves changes that help to deliver $O_2$ more effectively to the tissues, despite reduced arterial $P_{O_2}$.

# Review Activities

## Objective Questions

1. Which is *not* a component of the nasal septum?
   a. the palatine bone
   b. the vomer bone
   c. the ethmoid bone
   d. septal cartilage
2. An adenoidectomy is the removal of
   a. the uvula.
   b. the pharyngeal tonsils.
   c. the palatime tonsils.
   d. the lingual tonsils
3. Which is *not* a paranasal sinus?
   a. the palatime sinus
   b. the ethmiodal sinus
   c. the spheniodal sinus
   d. the frontal sinus
   e. the maxillary sinus
4. Which of the following is *not* characteristic of the left lung?
   a. a cardiac notch
   b. a superior lobe
   c. a single fissure
   d. an inferior lobe
   e. a middle lobe
5. The epithelial lining of the wall of the thorax is called
   a. the parietal pleura.

   b. the pleural peritoneum.
   c. the mediastinal pleura.
   d. the visceral pleura.
   e. the costal pleura.
6. Which of the following statements about intrapulmonary and intrapleural pressure is *true?*
   a. The intrapulmonary pressure is always subatmospheric.
   b. The intrapleural pressure is always greater than the intrapulmonary pressure.
   c. The intrapulmonary pressure is greater than the intrapleural pressure.
   d. The intrapleural pressure equals the atmospheric pressure.
7. If the transpulmonary pressure equals zero,
   a. a pneumothorax has probably occurred.
   b. the lungs cannot inflate.
   c. elastic recoil causes the lungs to collapse.
   d. all of the above apply.
8. Which of the following terms describes the maximum amount of air that can be expired after a maximum inspiration?
   a. tidal volume

   b. forced expiratory volume
   c. vital capacity
   d. maximum expiratory flow rate
9. If the blood lacked red blood cells but the lungs were functioning normally,
   a. the arterial $P_{O_2}$ would be normal.
   b. the $O_2$ content of arterial blood would be normal.
   c. both a and b apply.
   d. neither a nor b applies.
10. If a person were to dive with scuba equipment to a depth of 66 ft, which of the following statements would be *false?*
    a. The arterial $P_{O_2}$ would be three times its normal value.
    b. The $O_2$ content of plasma would be three times its normal value.
    c. The $O_2$ content of whole blood would be three times its normal value.
11. Which of the following would be most affected by a decrease in the affinity of hemoglobin for $O_2$?
    a. arterial $P_{O_2}$
    b. arterial percent oxyhemoglobin saturation
    c. venous oxyhemoglobin saturation
    d. arterial $P_{CO_2}$

12. If a person with normal lung function were to hyperventilate for several seconds, there would be a significant
    a. increase in the arterial $P_{O_2}$.
    b. decrease in the arterial $P_{CO_2}$.
    c. increase in the arterial percent oxyhemoglobin saturation.
    d. decrease in the arterial pH.

13. Erythropoietic factor is produced by
    a. the kidneys.
    b. the liver.
    c. the lungs.
    d. the bone marrow.

14. The affinity of hemoglobin for $O_2$ is decreased under conditions of
    a. acidosis.
    b. fever.
    c. anemia.
    d. acclimatization to a high altitude.
    e. all of the above apply.

15. Most of the $CO_2$ in the blood is carried in the form of
    a. dissolved $CO_2$.
    b. carbaminohemoglobin.
    c. carbonic acid and bicarbonate.
    d. carboxyhemoglobin.

16. The bicarbonate concentration of the blood would be decreased during
    a. metabolic acidosis.
    b. respiratory acidosis.
    c. metabolic alkalosis.
    d. respiratory alkalosis.

17. The chemoreceptors in the medulla oblongata are directly stimulated by
    a. carbon dioxide from the blood.
    b. hydrogen ions from the blood.
    c. hydrogen ions in cerebrospinal fluid, derived from blood $CO_2$.
    d. decreased arterial $P_{O_2}$.

18. The rhythmic control of breathing is produced by the activity of inspiratory and expiratory neurons in
    a. the medulla oblongata.
    b. the apneustic center of the pons.
    c. the pneumotaxic center of the pons.
    d. the cerebral cortex.

19. Which of the following occurs during hypoxemia?
    a. increased ventilation
    b. increased production of 2,3-DPG
    c. increased production of erythropoietin
    d. all of the above occur

20. Which of the following statements pertaining to exercise is *true?*
    a. The arterial percent oxyhemoglobin saturation is decreased.
    b. The venous percent oxyhemoglobin saturation is decreased.
    c. The arterial $P_{CO_2}$ is measurably increased.
    d. The arterial pH is measurably decreased.

### Essay Questions

1. Using a flow diagram to show cause and effect, explain how contraction of the diaphragm produces inspiration.
2. Radiographic (X-ray) pictures show that the rib cage of a person with a pneumothorax is expanded and the ribs are farther apart. Explain why this should be so.
3. Explain, using a flowchart, how a rise in blood $P_{CO_2}$ stimulates breathing. Include both the central and peripheral chemoreceptors in your answer.
4. A person with ketoacidosis may hyperventilate. What is the reason for this hyperventilation and how can it be stopped by an intravenous fluid containing bicarbonate?
5. What blood measurements can be performed to detect (a) anemia, (b) carbon monoxide poisoning, and (c) poor lung function?
6. Explain how measurements of blood $P_{CO_2}$, bicarbonate, and pH are affected by hypoventilation and hyperventilation.
7. Explain how blood pH and bicarbonate concentrations are affected by respiratory and metabolic acidosis.
8. How would an increase in the red blood cell content of 2,3-DPG affect the $P_{O_2}$ of venous blood? Explain your answer.

## Gundy/Weber Software 💾

The tutorial software accompanying Chapter 24 is Volume 10—Respiratory System.

# chapter twenty-five

# urinary system and fluid, electrolyte, and acid-base balance

## objectives

- Describe the position and gross structure of the kidney and the structure of a nephron.
- Explain how the nephrons are positioned in the kidney and trace the path of urine flow from the glomerulus to the renal pelvis.
- Describe the fine structure of the glomerular capillaries and glomerular capsule and discuss the composition of glomerular ultrafiltrate.
- Define the term *glomerular filtration rate* (*GFR*) and explain how the GFR is regulated.
- Describe salt and water reabsorption in the proximal convoluted tubule.
- Describe the transport process in the nephron loop and explain how the countercurrent multiplier effect is produced.
- Discuss the structure and function of the vasa recta.
- Discuss the significance of a hypertonic renal medulla and describe the mechanism of ADH action.

- Define the term *renal plasma clearance* and explain how the processes of reabsorption and secretion affect the clearance of substances.
- Describe how the GFR is measured by the clearance of inulin and explain why total renal blood flow can be measured by the clearance of PAH.
- Describe how the kidneys reabsorb glucose and amino acids and explain how glycosuria is produced.
- Describe the regulation of Na$^+$/K$^+$ balance by aldosterone and explain how aldosterone secretion is regulated.
- Explain the relationship between blood K$^+$ and H$^+$ (pH).
- Distinguish between the respiratory and metabolic components of acid-base balance and discuss the reabsorption of bicarbonate.
- Explain how the kidneys contribute to regulation of acid-base balance and how acid is excreted in the urine.
- Describe the structure of the ureters and of the urinary bladder and explain the process of micturition.
- Compare the male urethra with that of the female.

# Urinary System and Kidney Structure

*Each kidney contains many tiny tubules that empty into a cavity drained by the ureter. Each of the tubules receives a blood filtrate from a capillary bed called the glomerulus. The filtrate is similar to tissue fluid but is modified as it passes through the tubules and is thereby changed into urine. The tubules and associated blood vessels thus form the functioning units of the kidneys, which are known as nephrons.*

The urinary system consists of two kidneys, two ureters, the urinary bladder, and the urethra (fig. 25.1). The primary function of the kidneys is regulation of the extracellular fluid (plasma and interstitial fluid) environment in the body. This function is accomplished through the formation of urine, which is a modified filtrate of plasma. In the process of urine formation, the kidneys regulate (1) the volume of blood plasma (and thus contribute significantly to the regulation of blood pressure); (2) the concentration of waste products in the blood; (3) the concentration of electrolytes ($Na^+$, $K^+$, $HCO_3^-$, and other ions) in the plasma; and (4) the pH of plasma. In order to understand how these functions are performed by the kidneys, a knowledge of kidney structure is required.

## Position and Gross Structure of the Kidney

The reddish-brown **kidneys** lie on each lateral side of the vertebral column, high in the abdominal cavity, between the levels of the twelfth thoracic and the third lumbar vertebrae (fig. 25.2). The right kidney is usually 1.5 to 2.0 cm lower than the left because of the large area occupied by the liver.

The kidneys are *retroperitoneal*, which means that they are positioned behind the parietal peritoneum. Thus, strictly speaking, they are not within the peritoneal cavity. Each adult kidney is a lima-bean-shaped organ about 11.25 cm (4 in.) long, 5.5 to 7.7 cm (2 to 3 in.) wide, and 3.0 cm (1.2 in.) thick. The **hilum** of the kidney is the depression along the medial border through which the **renal artery** and nerves enter and the **renal vein** and **ureter** (*yoo-re´ter*) exit.

Each kidney is embedded in a fatty fibrous pouch consisting of three layers. The **renal capsule** is the innermost layer that forms a strong, transparent fibrous attachment to the kidney. The renal capsule protects the kidney from trauma and the spread of infections. The second layer is formed by a firm protective layer of adipose tissue called the **adipose capsule.** The outermost layer, called the **renal fascia,** is a supportive layer that anchors the kidney to the peritoneum and the abdominal wall.

A coronal section of the kidney shows two distinct regions and a major cavity (fig. 25.3a). The outer **renal cortex,** in contact with the capsule, is reddish brown and granular in

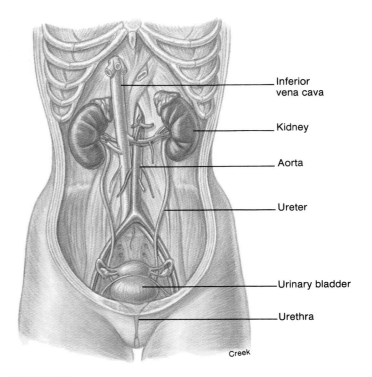

**FIGURE 25.1**

The anatomical locations of the kidneys, ureters, and urinary bladder.

Labels: Inferior vena cava; Kidney; Aorta; Ureter; Urinary bladder; Urethra; Creek

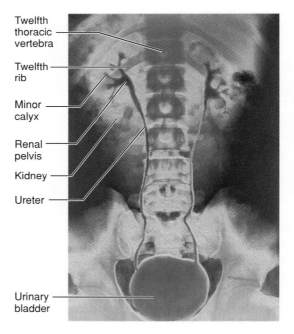

**FIGURE 25.2**

A color-enhanced radiograph of the calyces and renal pelvises of the kidneys, the ureters, and the urinary bladder. (Note the position of the kidneys relative to the vertebral column and ribs.)

Labels: Twelfth thoracic vertebra; Twelfth rib; Minor calyx; Renal pelvis; Kidney; Ureter; Urinary bladder

appearance because of its many capillaries. The deeper region, or **renal medulla,** is darker in color, and the presence of microscopic tubules and blood vessels gives it a striped appearance. The renal medulla is composed of 8 to 15

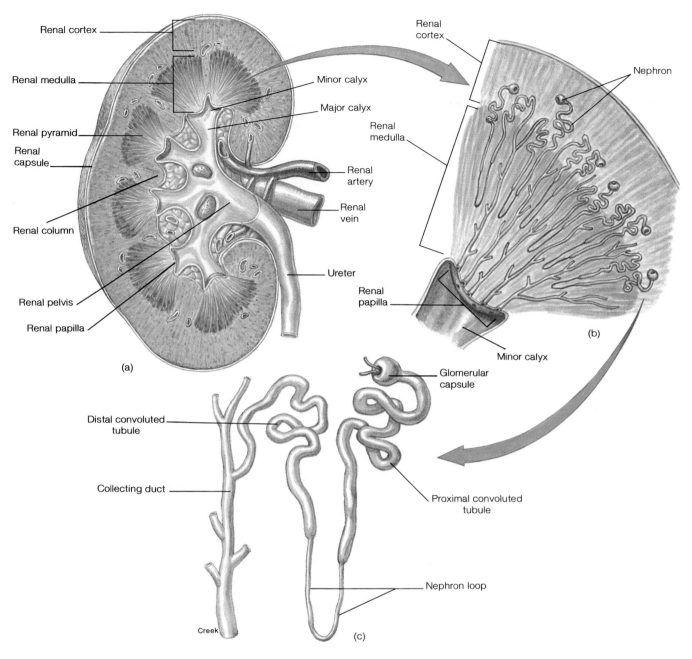

**FIGURE 25.3**

The internal structures of a kidney. (*a*) A coronal section showing the structure of the renal cortex, renal medulla, and renal pelvis. (*b*) A diagrammatic magnification of a renal pyramid and renal

cortex to depict the tubules. (*c*) A diagrammatic view of a single nephron and a collecting duct.

conical **renal pyramids** separated by **renal columns.** The **renal papillae** are the apexes of the renal pyramids. These nipplelike projections are directed toward the large cavity of the kidney called the **renal pelvis**.

The cavity of the kidney collects and transports urine from the kidney to the ureter. It is divided into several

portions. Each papilla of a renal pyramid projects into a small depression called the **minor calyx** (*ka´liks*)—in the plural, *calyces*. Several minor calyces unite to form a **major calyx.** In turn, the major calyces join to form the funnel-shaped renal pelvis. The renal pelvis serves to collect urine from the calyces and transport it to the ureter.

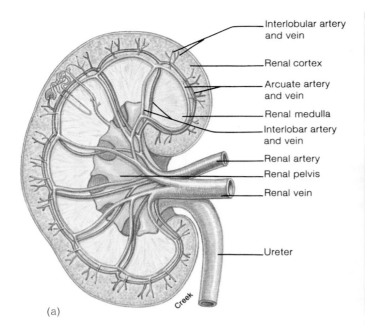

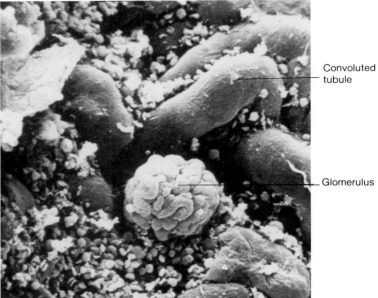

(a)

(b)

**FIGURE 25.4**

The vascular structure of the kidneys. (*a*) An illustration of the major arterial supply and (*b*) a scanning electron micrograph of the glomeruli (300×).

# Microscopic Structure of the Kidney

The **nephron** (*nef´ron*) (fig 25.3c) is the functional unit of the kidney that is responsible for the formation of urine. Each kidney contains more than a million nephrons. A nephron consists of **urinary tubules** and associated small blood vessels. Fluid formed by capillary filtration enters the tubules and is subsequently modified by transport processes. The resulting fluid that leaves the tubules is urine.

**Renal Blood Vessels**    Arterial blood enters the kidney at the hilum through the **renal artery,** which divides into **interlobar** (*in˝ter-lo´bar*) **arteries** (fig. 25.4) that pass between the renal pyramids through the renal columns. **Arcuate** (*ar´kyoo-āt*) **arteries** branch from the interlobar arteries at the boundary of the renal cortex and renal medulla. Many **interlobular arteries** radiate from the arcuate arteries into the renal cortex and subdivide into numerous **afferent arterioles,** which are microscopic in size. The afferent arterioles deliver blood into capillary networks called **glomeruli,** which produce a blood filtrate that enters the urinary tubules. The blood remaining in the glomerulus leaves through an **efferent arteriole,** which delivers the blood into another capillary network, the **peritubular capillaries,** surrounding the tubules (fig. 25.5).

This arrangement of blood vessels, in which a capillary bed (the glomerulus) is drained by an arteriole rather than by a venule and delivered to a second capillary bed located downstream (the peritubular capillaries), is unique. Blood from the peritubular capillaries is drained into veins that parallel the course of the arteries in the kidney. These are the **interlobular veins, arcuate veins,** and **interlobar veins.** The interlobar veins descend between the renal pyramids, converge, and leave the kidney as a single **renal vein** that empties into the inferior vena cava.

 Although the kidneys are generally well protected by being encapsulated retroperitoneally, they may be injured by a hard blow to the lumbar region. The immense vascularity of the kidney makes it highly susceptible to hemorrhage. As a result, such an injury may produce blood in the urine.

**Nephron**    The tubular nephron consists of a glomerular capsule, proximal convoluted tubule, descending limb of the nephron loop (loop of Henle), ascending limb of the nephron loop, and distal convoluted tubule (figs. 25.3c and 25.5).

The **glomerular** (Bowman's) **capsule** surrounds the glomerulus. The glomerular capsule and its associated glomerulus are located in the renal cortex and together constitute the **renal corpuscle.** The glomerular capsule contains an inner visceral layer of epithelium around the glomerular capillaries and an outer parietal layer. The space between these two layers, called the **capsular space,** receives the glomerular filtrate.

....................................................

arcuate: L. *arcuare*, to bend
glomerulus: L. diminutive of *glomus*, ball

....................................................

Bowman's capsule: from Sir William Bowman, English anatomist, 1816–92

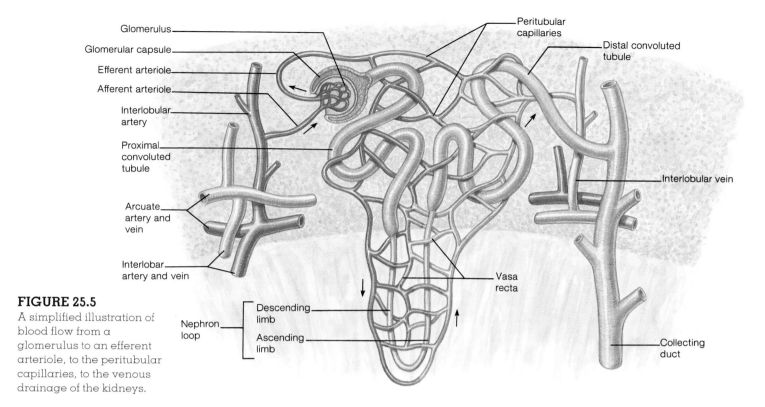

Glomerulus
Glomerular capsule
Efferent arteriole
Afferent arteriole
Interlobular artery
Proximal convoluted tubule
Arcuate artery and vein
Interlobar artery and vein

Peritubular capillaries
Distal convoluted tubule
Interlobular vein
Vasa recta
Collecting duct

Nephron loop
Descending limb
Ascending limb

**FIGURE 25.5**
A simplified illustration of blood flow from a glomerulus to an efferent arteriole, to the peritubular capillaries, to the venous drainage of the kidneys.

Filtrate in the glomerular capsule passes into the lumen of the **proximal convoluted tubule.** The wall of the proximal convoluted tubule consists of a single layer of cuboidal cells containing millions of microvilli; these serve to increase the surface area for reabsorption. In the process of reabsorption, salt, water, and other molecules needed by the body are transported from the lumen, through the tubular cells, and into the surrounding peritubular capillaries.

The glomerulus, glomerular capsule, and proximal convoluted tubule are located in the renal cortex. Fluid passes from the proximal convoluted tubule to the **nephron loop** (loop of Henle). This fluid is carried into the renal medulla in the **descending limb** of the loop and returns to the renal cortex in the **ascending limb** of the loop. Back in the renal cortex, the tubule becomes coiled again and is called the **distal convoluted tubule.** The distal convoluted tubule is shorter than the proximal convoluted tubule and has fewer microvilli. It is the last segment of the nephron and terminates as it empties into a collecting duct.

The two principal types of nephrons are classified according to their position in the kidney and the lengths of their nephron loops. Nephrons that originate in the inner one-third of the renal cortex—called **juxtamedullary nephrons**—have longer loops than the **cortical nephrons** that originate in the outer two-thirds of the renal cortex (fig. 25.6).

The distal convoluted tubules of several nephrons drain into a **collecting duct.** Fluid is then drained by the collecting duct from the renal cortex into the renal medulla as the collecting duct passes through a renal pyramid. This fluid, now called *urine*, passes into a minor calyx. Urine is then funneled through a major calyx to the renal pelvis and out of the kidney in the ureter.

 *Polycystic kidney disease,* a condition inherited as an autosomal dominant trait, affects 1 in 600 to 1000 people. This disease is more common than sickle-cell anemia, cystic fibrosis, or muscular dystrophy, which are also genetic diseases. In 50% of the people who inherit the defective gene, progressive renal failure develops during middle age to the point where dialysis or kidney transplants are required. The cysts that develop are expanded portions of the nephron tubules.

## Glomerular Filtration

*The glomerular capillaries have large pores, and the visceral layer of the glomerular capsule, in contact with the glomerulus, has filtration slits. Water, together with dissolved solutes (but not proteins), can thus pass from the blood plasma to the inside of the capsules and the lumina of the nephron tubules. The volume of this filtrate produced per minute by both kidneys is called the glomerular filtration rate (GFR).*

loop of Henle: from Friedrich G. J. Henle, German anatomist, 1809–85

## FIGURE 25.6

The position of cortical and juxtamedullary nephrons within the kidneys.

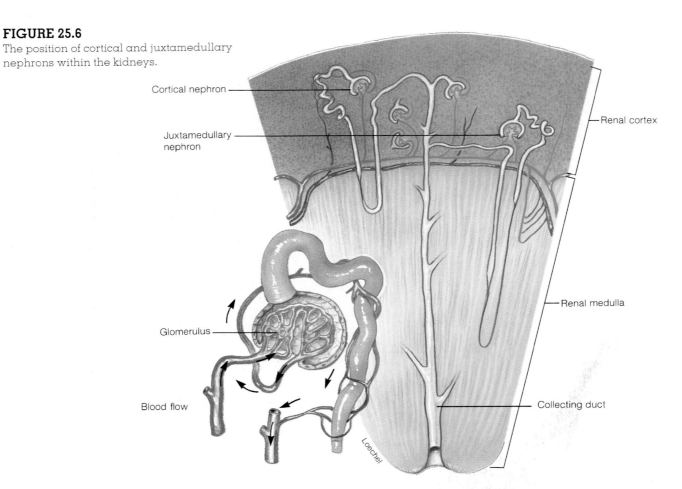

Cortical nephron

Juxtamedullary nephron

Renal cortex

Glomerulus

Renal medulla

Blood flow

Collecting duct

*Loechel*

The glomerular capillaries have extremely large pores (200–500Å in diameter) called **fenestrae,** and are thus said to be *fenestrated.* As a result of these large pores, glomerular capillaries are 100 to 400 times more permeable to plasma, water, and dissolved solutes than are the capillaries of skeletal muscles. Although the pores of glomerular capillaries are large, they are still small enough to prevent the passage of red blood cells, white blood cells, and platelets into the filtrate.

Before the filtrate can enter the interior of the glomerular capsule it must pass through the capillary pores, the basement membrane (a thin layer of glycoproteins immediately outside the endothelial cells), and the inner, visceral layer of the glomerular capsule. The inner layer of the glomerular capsule is composed of unique cells, called **podocytes** (*pod´ŏ-sītz*), with numerous cytoplasmic extensions known as **pedicels** (*ped´ĭ-selz*), or foot processes (fig. 25.7). Pedicels interdigitate, like the fingers of clasped hands, as they wrap around the glomerular capillaries. The narrow slits between adjacent pedicels provide the passageways through which filtered molecules must pass to enter the interior of the glomerular capsule (fig. 25.8).

podocyte: Gk. *pous,* foot; *kytos,* cell
pedicel: L. *peducekkus,* footplate

## FIGURE 25.7

The inner (visceral) layer of the glomerular (Bowman's) capsule is composed of podocytes, as shown in this scanning electron micrograph. Very fine extensions of these podocytes form foot processes, or pedicels, that interdigitate around the glomerular capillaries. Spaces between adjacent pedicels form the filtration slits.

Primary process of podocyte

Pedicel

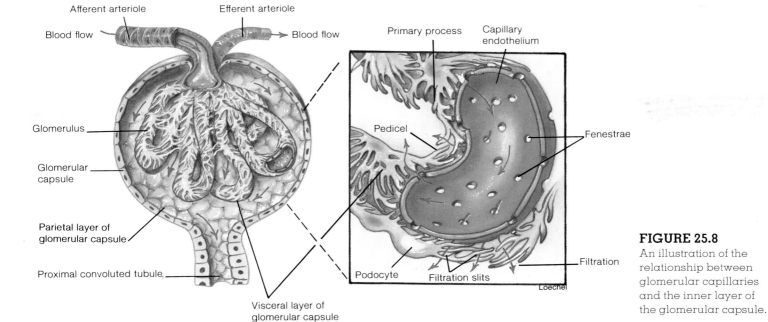

Although the glomerular capillary pores are apparently large enough to permit the passage of proteins, the fluid that enters the capsular space is almost completely free of plasma proteins. This exclusion of plasma proteins from the filtrate is partially a result of their negative charges, which hinder their passage through the negatively charged glycoproteins in the basement membrane of the capillaries (fig. 25.9). The large size and negative charges of plasma proteins may also restrict their movement through the filtration slits between pedicels.

## Glomerular Ultrafiltrate

The fluid that enters the glomerular capsule is called **ultrafiltrate** (fig. 25.10) because it is formed under pressure (the hydrostatic pressure of the blood). This process is similar to the formation of interstitial (tissue) fluid by other capillary beds in the body. The force favoring filtration is opposed by a counterforce developed by the hydrostatic pressure of fluid in the glomerular capsule. Also, since the protein concentration of the tubular fluid is low (less than 2–5 mg per 100 ml) compared to that of plasma (6–8 g per 100 ml), the greater colloid osmotic pressure of plasma promotes the osmotic return of filtered water. When these opposing forces are subtracted from the hydrostatic pressure of the glomerular capillaries, a *net filtration pressure* of approximately 10 mmHg is obtained.

Because glomerular capillaries are extremely permeable and have a large surface area, the modest net filtration pressure produces an extraordinarily large volume of filtrate. The **glomerular filtration rate (GFR)** is the volume of filtrate produced per minute by both kidneys. The GFR aver-

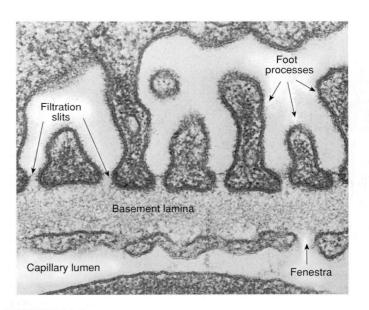

**FIGURE 25.9**
An electron micrograph of the filtration barrier between the lumen of the capillary (*bottom*) and the interior of the glomerular capsule (*top*). Molecules in blood plasma must pass through the capillary.

ages 115 ml per minute in women and 125 ml per minute in men. This is equivalent to 7.5 L per hour or 180 L per day (about 45 gallons)! Since the total blood volume averages about 5.0 L, this means that the total blood volume is filtered into the urinary tubules every 40 minutes. Most of the filtered water must obviously be returned immediately to the circulatory system or a person would literally urinate to death within minutes.

## Regulation of Glomerular Filtration Rate

Vasoconstriction or dilation of afferent arterioles affects the rate of blood flow to the glomerulus and, therefore, the glomerular filtration rate. Changes in the diameter of the afferent arterioles result from both extrinsic (sympathetic innervation) and intrinsic regulatory mechanisms.

**Sympathetic Nerve Effects**   An increase in sympathetic nerve activity, as occurs during the fight-or-flight reaction and exercise, stimulates constriction of afferent arterioles. This is an alpha-adrenergic effect (chapter 17), which helps to preserve blood volume and to divert blood to the muscles and heart. A similar effect occurs during cardiovascular shock, in which sympathetic nerve activity stimulates vasoconstriction. The decreased GFR and the resulting decreased rate of urine formation help to compensate for the rapid blood pressure fall under these circumstances (fig. 25.11).

**Renal Autoregulation**   When the direct effect of sympathetic stimulation is experimentally removed, the effect of systemic blood pressure on the GFR can be observed. Under these conditions, surprisingly, the GFR remains relatively constant despite changes in mean arterial pressure within a range of 70 to 180 mmHg (the normal mean arterial pressure is 100 mmHg). The ability of the kidneys to maintain a relatively constant GFR in the face of fluctuating blood pressures is called **renal autoregulation.**

Renal autoregulation is achieved through the effects of locally produced chemicals on the afferent arterioles (effects on the efferent arterioles are believed to be of secondary importance). When systemic arterial pressure falls toward a mean of 70 mmHg, the afferent arterioles dilate, and when the pressure rises, the afferent arterioles constrict. Blood flow to the glomeruli and the GFR can thus remain relatively constant within the autoregulatory range of blood pressure values. The effects of different regulatory mechanisms on the GFR are summarized in table 25.1.

# Reabsorption of Salt and Water

*Most of the salt and water filtered from the blood into the glomerular capsule is reabsorbed across the wall of the proximal convoluted tubule. The reabsorption of water occurs by osmosis, in which water follows the active extrusion of NaCl from the tubule and into the surrounding peritubular capillaries. Most of the remaining water in the filtrate is reabsorbed across the wall of the collecting duct in the renal medulla. This occurs as a result of the high osmotic pressure of the surrounding tissue fluid, which is produced by transport processes in the nephron loop. Reabsorption of water through the collecting duct is regulated by antidiuretic hormone (ADH).*

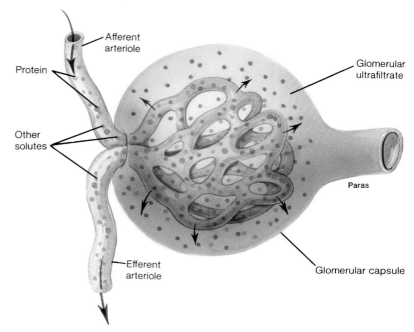

**FIGURE 25.10**

The formation of glomerular ultrafiltrate. Proteins (*green circles*) are not filtered, but smaller plasma solutes (*pink dots*) easily enter the glomerular ultrafiltrate. Arrows indicate the direction of filtration.

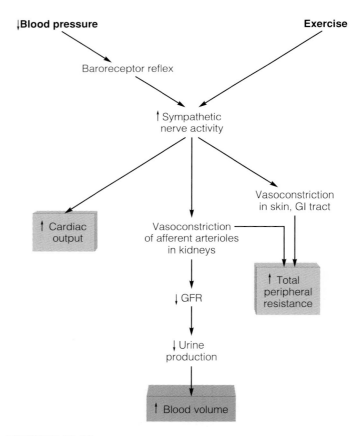

**FIGURE 25.11**

The effect of increased sympathetic nerve activity on the cardiac output, total peripheral resistance, and blood volume.

| Table 25.1 | Regulation of the glomerular filtration rate (GFR) | | |
|---|---|---|---|
| **Regulation** | **Stimulus** | **Afferent arteriole** | **GFR** |
| Sympathetic nerves | Activation by aortic and carotid baroreceptor reflex or by higher brain centers | Constricts | Decreases |
| Autoregulation | Decreased blood pressure | Dilates | No change |
| Autoregulation | Increased blood pressure | Constricts | No change |

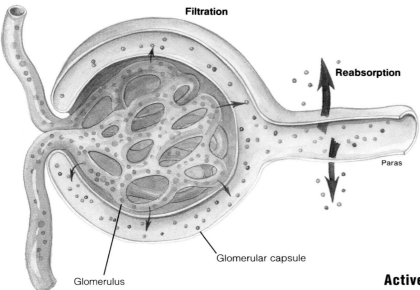

**Filtration**

**Reabsorption**

Paras

Glomerular capsule

Glomerulus

**FIGURE 25.12**

Plasma water and its dissolved solutes (except proteins) enter the glomerular ultrafiltrate, but most of these filtered molecules are reabsorbed. The term *reabsorption* refers to the transport of molecules out of the tubular filtrate back into the blood.

Although about 180 L per day of glomerular ultrafiltrate are produced, the kidneys normally excrete only 1.5 L per day of urine. Approximately 99% of the filtrate must thus be returned to the circulatory system, while 1% is excreted in the urine. The urine volume, however, varies according to the needs of the body. When a well-hydrated person drinks a liter or more of water, urine volume increases to 16 ml per minute (the equivalent of 23 L per day if this were to continue for 24 hours). In severe dehydration, when the body needs to conserve water, only 0.3 ml per minute, or 400 ml per day, of urine are produced. A volume of 400 ml per day of urine is needed to excrete the amount of metabolic wastes produced by the body; this is called the *obligatory water loss*. When water in excess of this amount is excreted, the urine volume is increased and its concentration is decreased.

Regardless of the body's state of hydration, it is clear that most of the filtered water must be returned to the circulatory system to maintain blood volume and pressure. The return of filtered molecules from the tubules to the blood is called **reabsorption** (fig. 25.12). It is important to realize that the transport of water always occurs passively by *osmosis*; there is no such thing as active transport of water. A concentration gradient must thus be created between tubular fluid and blood that favors the osmotic return of water to the circulatory system.

## Reabsorption in the Proximal Convoluted Tubule

Since all plasma solutes, with the exception of proteins, are able to freely enter the glomerular ultrafiltrate, the total solute concentration (osmolality—see chapter 5) of the filtrate is essentially the same as that of plasma. This total solute concentration is equal to 300 milliosmoles per liter (300 mOsm), which is *isosmotic* (chapter 5) to the plasma. Osmosis cannot occur unless the concentration of plasma in the peritubular capillaries and the concentration of filtrate are altered by active transport processes. This is achieved by the active transport of $Na^+$ from the filtrate to the peritubular blood.

**Active and Passive Transport** The epithelial cells that compose the wall of the proximal convoluted tubule are joined together by tight junctions only on their apical sides—that is, the sides of each cell that are closest to the lumen of the tubule (fig. 25.13). Each cell therefore has four exposed surfaces: the apical side, which faces the lumen that contains microvilli; the basal side, which faces the peritubular capillaries; and the lateral sides, which face the narrow clefts between adjacent epithelial cells.

The concentration of $Na^+$ in the glomerular ultrafiltrate—and thus in the fluid entering the proximal convoluted tubule—is the same as that in plasma. The epithelial cells of the tubule, however, have a much lower $Na^+$ concentration. The lower $Na^+$ concentration is due in part to the low permeability of the cell membrane to $Na^+$ and in part to the active transport of $Na^+$ out of the cell by $Na^+/K^+$ pumps, as described in chapter 5. In the cells of the proximal convoluted tubule, the $Na^+/K^+$ pumps are located in the basal and lateral sides of the cell membrane but not in the apical membrane. As a result of the action of these active transport pumps, a concentration gradient is created that favors the diffusion of Na$^+$ from the tubular fluid across the apical cell membranes and into the epithelial cells of the proximal convoluted tubule. The Na$^+$ is then extruded into the surrounding tissue fluid by the Na$^+$/K$^+$ pumps.

The transport of Na$^+$ from the tubular fluid to the interstitial fluid surrounding the epithelial cells of the proximal convoluted tubule creates a potential difference across the wall of the tubule. This electrical gradient favors the passive transport of Cl$^-$ toward the higher Na$^+$ concentration in the tissue fluid. Chloride ions, therefore, passively follow sodium ions out of the filtrate into the interstitial fluid. As a result of the accumulation of NaCl, the osmolality and osmotic pressure values of the interstitial fluid surrounding the epithelial cells increase to above those of the tubular fluid. This is particularly true of the interstitial fluid between the lateral membranes of adjacent epithelial cells, where the narrow spaces permit the accumulated NaCl to achieve a higher concentration.

An osmotic gradient is thus created between the tubular fluid and the interstitial fluid surrounding the proximal convoluted tubule. Since the cells of the proximal convoluted tubule are permeable to water, water moves by osmosis from the tubular fluid into the epithelial cells and then across the basal and lateral sides of the epithelial cells into the interstitial fluid. The salt and water, which were reabsorbed from the tubular fluid, can then move passively into the surrounding peritubular capillaries and in this way be returned to the blood (fig. 25.14).

## Significance of Proximal Convoluted Tubule Reabsorption

Approximately 65% of the salt and water in the original glomerular ultrafiltrate is reabsorbed across the proximal convoluted tubule and returned to the vascular system. The volume of tubular fluid remaining is reduced accordingly, but this fluid is still isosmotic with the blood (has a concentration of 300 mOsm). This is because the cell membranes in the proximal convoluted tubule are freely

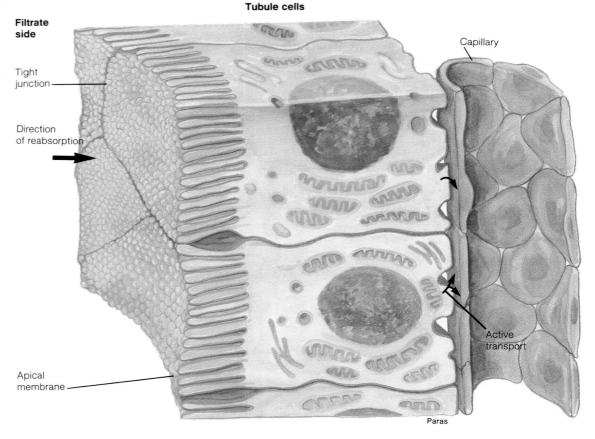

**Tubule cells**

Filtrate side

Tight junction

Direction of reabsorption

Apical membrane

Capillary

Active transport

Paras

### FIGURE 25.13

An illustration of the appearance of tubule cells in the electron microscope. Molecules that are reabsorbed pass through the tubule cells from the apical membrane (facing the filtrate) to the basolateral membrane (facing the blood).

permeable to water, so that water and salt are removed in proportionate amounts.

An additional smaller amount of salt and water is returned to the vascular system by reabsorption in the nephron loop. This reabsorption, like that in the proximal convoluted tubule, occurs constantly, regardless of the person's state of hydration. Unlike reabsorption that occurs in more distal regions of the nephron, it is not subject to hormonal regulation. Approximately 85% of the filtered salt and water is, therefore, reabsorbed in a constant, unregulated fashion in the proximal convoluted tubule and nephron loop. This reabsorption is very costly in terms of energy expenditures, accounting for as much as 6% of the calories consumed by the body at rest.

Since 85% of the original glomerular ultrafiltrate is immediately reabsorbed in the proximal convoluted tubule and nephron loop, only 15% of the initial filtrate remains to enter the distal convoluted tubule and collecting duct. This is still a large volume of fluid—15% × GFR (180 L per day) = 27 L per day—that must be reabsorbed to varying degrees in accordance with the body's state of hydration. In

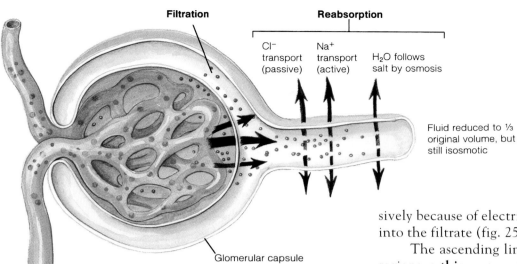

**FIGURE 25.14**
Mechanisms of salt and water reabsorption in the proximal tubule. Sodium is actively transported out of the filtrate and chloride follows passively by electrical attraction. Water follows the salt out of the tubular filtrate into the peritubular capillaries by osmosis.

these regions of the nephron, fine tuning of the percentage of reabsorption and urine volume is accomplished by the action of hormones.

## *The Countercurrent Multiplier System*

Water cannot be actively transported across the tubule wall, and osmosis of water cannot occur if the tubular fluid and surrounding interstitial fluid are isotonic to each other. In order for water to be reabsorbed by osmosis, the surrounding interstitial fluid must be hypertonic. The osmotic pressure of the interstitial fluid in the renal medulla is raised to over four times that of plasma. This results partly because the tubule bends; the configuration of the nephron loop allows for interaction between the descending and ascending limbs. Since the ascending limb is the active partner in this interaction, we will first describe its properties before considering those of the descending limb.

### Ascending Limb of the Nephron Loop

Salt (NaCl) is actively extruded from the ascending limb into the surrounding

interstitial tissue fluid. This is not accomplished, however, by the same process that occurs in the proximal convoluted tubule. Instead, $Na^+$, $K^+$, and $Cl^-$ passively diffuse from the filtrate into the ascending limb cells in a ratio of 1 $Na^+$ to 1 $K^+$ to 2 $Cl^-$. The $Na^+$ is then actively transported across the basolateral membrane to the tissue fluid by the $Na^+/K^+$ pump. $Cl^-$ follows the $Na^+$ passively because of electrical attraction, and $K^+$ diffuses back into the filtrate (fig. 25.15).

The ascending limb is structurally divisible into two regions: a **thin segment,** nearest to the tip of the nephron loop, and a **thick segment** of varying lengths, which carries the filtrate outward into the renal cortex and into the distal convoluted tubule. The thin segment is composed of a simple squamous epithelium, whereas the thick segment is composed of a simple cuboidal epithelium. It is currently believed that only those cells of the thick segments of the ascending limb are capable of actively transporting NaCl from the filtrate into the surrounding tissue fluid.

Although the mechanism of NaCl transport is different in the ascending limb than it is in the proximal convoluted tubule, the net effect is the same: salt (NaCl) is extruded

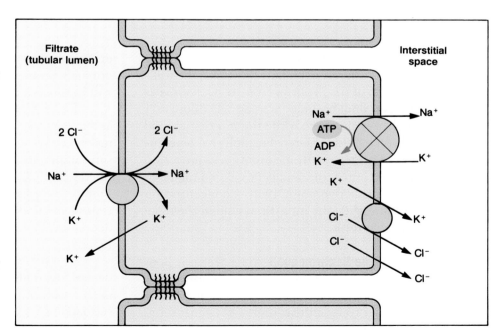

**FIGURE 25.15**
In the thick segment of the ascending limb of the loop, $Na^+$ and $K^+$, together with two $Cl^-$, enter the tubule cells. $Na^+$ is then actively transported out into the interstitial space and $Cl^-$ follows passively. The $K^+$ diffuses back into the filtrate, and some also enters the interstitial space.

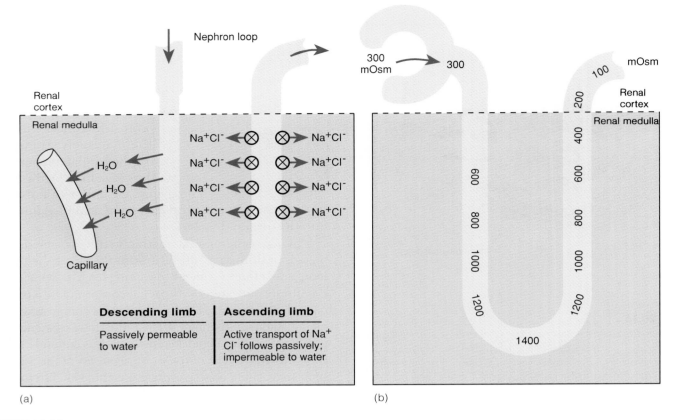

(a)

(b)

**FIGURE 25.16**

The countercurrent multiplier system. The active extrusion of Cl⁻ followed by the Na⁺ from the ascending limb makes the surrounding tissue fluid more concentrated. This concentration is multiplied by the fact that the descending limb is passively permeable so that its fluid increases in concentration as the surrounding tissue fluid becomes more concentrated. The transport properties of the nephron loop and their effect on tubular fluid concentration are shown in (a). The values of these changes in osmolality, together with the effect on surrounding tissue fluid concentration, are shown in (b).

into the surrounding interstitial fluid. Unlike the epithelial walls of the proximal convoluted tubule, however, the walls of the ascending limb of the nephron loop are *not permeable to water*. The tubular fluid becomes increasingly dilute as it ascends toward the renal cortex, while the interstitial fluid around the nephron loops in the renal medulla becomes increasingly more concentrated. By means of these processes, the tubular fluid that enters the distal convoluted tubule in the renal cortex is made hypotonic (with a concentration of about 100 mOsm), while the interstitial fluid in the renal medulla is made hypertonic.

**Descending Limb of the Nephron Loop**  The deeper regions of the renal medulla, around the tips of the loops of juxtamedullary nephrons, reach a concentration of 1200 to 1400 mOsm. In order to reach this high a concentration, the salt pumped out of the ascending limb must accumulate in the interstitial fluid of the renal medulla. This occurs as a result of the properties of the descending limb, to be discussed next, and because blood vessels around the nephron loop do not carry back all of the extruded salt to the gen-

eral circulation. The capillaries in the renal medulla are uniquely arranged to trap NaCl in the interstitial fluid, as will be discussed in a later section.

The descending limb does not actively transport salt, and indeed is believed to be impermeable to the passive diffusion of salt. It is, however, permeable to water. Since the surrounding interstitial fluid is hypertonic to the filtrate in the descending limb, water is drawn out of the descending limb by osmosis and enters blood capillaries. The concentration of tubular fluid is thus increased, and its volume is decreased, as it descends toward the tips of the nephron loops.

As a result of these passive transport processes in the descending limb, the fluid that "rounds the bend" at the tip of the nephron loop has the same osmolality as that of the surrounding tissue fluid (1200–1400 mOsm). There is, therefore, a higher salt concentration arriving in the ascending limb than there would be if the descending limb simply delivered isotonic fluid. Salt transport by the ascending limb is increased accordingly, so that the "saltiness" of the interstitial fluid is multiplied (fig. 25.16).

## Countercurrent Multiplication

Countercurrent flow (flow in opposite directions) in the ascending and descending limbs and the close proximity of the two limbs allow for interaction. Since the concentration of the tubular fluid in the descending limb reflects the concentration of surrounding interstitial fluid, and since the concentration of this interstitial fluid is raised by the active extrusion of salt from the ascending limb, a *positive feedback mechanism* is created. The more salt the ascending limb extrudes, the more concentrated will be the fluid that returns to it from the descending limb. This positive feedback mechanism that multiplies the concentration of interstitial fluid and descending limb fluid is called the **countercurrent multiplier system.**

The countercurrent multiplier system recirculates salt and thus traps some of the salt that enters the nephron loop in the interstitial fluid of the renal medulla. This system results in a gradually increasing concentration of renal tissue fluid from the renal cortex to the inner renal medulla; the osmolality of interstitial fluid increases from 300 mOsm (isotonic) in the renal cortex to 1200 to 1400 mOsm in the deepest part of the renal medulla.

## Vasa Recta

In order for the countercurrent multiplier system to be effective, most of the salt extruded from the ascending limbs must remain in the interstitial fluid of the renal medulla, while most of the water that leaves the descending limbs must be removed by the blood. This is accomplished by thin-walled vessels, called the **vasa recta,** which form long capillary loops that parallel the long nephron loops of the juxtamedullary nephrons (see fig. 25.19).

The vasa recta maintain the hypertonicity of the renal medulla by means of a mechanism known as **countercurrent exchange.** Salt and other solutes (such as urea, described in the next section) that are present at high concentrations in the medullary tissue fluid diffuse into the blood as the blood descends into the capillary loops of the vasa recta, but then they passively diffuse out of the ascending vessels and back into the descending vessels (where the concentration is

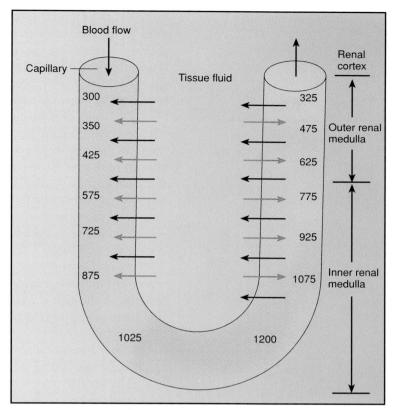

Black arrows = diffusion of NaCl and urea
Blue arrows = movement of water by osmosis

**FIGURE 25.17**

Countercurrent exchange in the vasa recta. The diffusion of salt first into and then out of these blood vessels helps to maintain the "saltiness" (hypertonicity) of the interstitial fluid in the renal medulla (numbers indicate osmolality).

lower). Solutes are thus recirculated and trapped within the renal medulla. Since the walls of the vasa recta are freely permeable to dissolved solutes, the concentrations of these solutes inside the vasa recta and in the surrounding interstitial fluid become the same at each level within the renal medulla. The colloid osmotic pressure within the vasa recta, however, is higher than it is in the interstitial fluid because plasma proteins do not easily pass through the capillary walls. This is similar to the situation in other capillary beds and results in the osmotic movement of water into both the descending and ascending vessels. The vasa recta thus trap salt and urea within the interstitial fluid but transport water out of the renal medulla (fig. 25.17).

## Effects of Urea

Countercurrent multiplication of the NaCl concentration is the mechanism traditionally cited to explain the hypertonicity of the interstitial fluid in the renal medulla. It is currently believed, however, that urea also contributes significantly to the total osmolality of the interstitial fluid.

The role of urea was inferred from experimental evidence showing that active transport of $Na^+$ occurs only in the thick segments of the ascending limbs. The thin segments of the ascending limbs, which are located in the deeper regions of the renal medulla, are not able to extrude salt actively. However, since salt does indeed leave the thin segments, a diffusion gradient for salt must exist, despite the fact that the surrounding interstitial fluid and the tubular fluid have the same osmolality. Investigators have therefore concluded that molecules other than salt—specifically urea—contribute to the hypertonicity of the tissue fluid.

It was later shown that the ascending limb of the nephron loop and the collecting duct are permeable to urea. Urea can thus diffuse out of the collecting duct and into the ascending limb (fig. 25.18). In this way, a certain amount

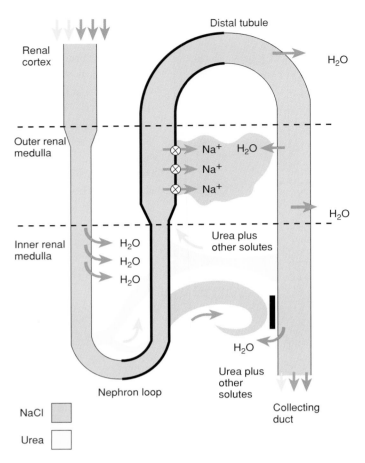

Renal cortex

Distal tubule

$H_2O$

Outer renal medulla

$Na^+$  $H_2O$
$Na^+$
$Na^+$

$H_2O$

Inner renal medulla

$H_2O$
$H_2O$
$H_2O$

Urea plus other solutes

Nephron loop

$H_2O$

Urea plus other solutes

Collecting duct

NaCl

Urea

**FIGURE 25.18**

According to some authorities, urea diffuses out of the collecting duct and contributes significantly to the concentration of the interstitial fluid in the renal medulla. The active transport of $Na^+$ out of the thick segments of the ascending limbs also contributes to the hypertonicity of the renal medulla so that water is reabsorbed by osmosis from the collecting ducts.

transport their fluid through this hypertonic environment in order to empty their contents of urine into the calyces. While the fluid surrounding the collecting ducts in the renal medulla is hypertonic, the fluid that passes into the collecting ducts in the renal cortex is hypotonic as a result of the active extrusion of salt by the ascending limbs of the nephron loops.

The walls of the collecting ducts are *permeable to water but not to salt.* Since the surrounding interstitial fluid in the renal medulla is very hypertonic as a result of the countercurrent multiplier system, water is drawn out of the collecting ducts by osmosis. This water does not dilute the surrounding interstitial fluid because it is transported by capillaries to the general circulation. In this way, most of the water remaining in the filtrate is returned to the circulatory system (fig. 25.19).

Note that the osmotic gradient created by the countercurrent multiplier system provides the force for water reabsorption through the collecting ducts. The rate at which this osmotic movement occurs, however, is determined by the permeability of the collecting duct cell membranes to water. This depends on the number of water channels in the cell membranes of the collecting duct epithelial cells.

The water channels are proteins within the membranes of vesicles that bud from the Golgi apparatus (chapter 3) and enter the cytoplasm. **Antidiuretic hormone (ADH),** acting via cAMP as a second messenger, stimulates the fusion of these vesicles with the cell membrane. This is identical to exocytosis (chapter 3), except that, in this case, no product is secreted. The importance of this process in the collecting duct is that the water channels are incorporated into the cell membrane in response to ADH, so that the collecting duct becomes more permeable to water. When ADH is no longer secreted and no longer binds to its membrane

of urea is recycled through these two segments of the nephron and is trapped in the interstitial fluid.

The transport properties of different tubule segments with respect to the concentrating-diluting mechanisms of the kidney are summarized in table 25.2.

## Collecting Duct: Effect of Antidiuretic Hormone (ADH)

As a result of the recycling of salt between the ascending and descending limbs and of the recycling of urea between the collecting duct and the nephron loop, the medullary interstitial fluid is made very hypertonic. The collecting ducts must

## Table 25.2 Transport properties of different segments of the renal tubules and the collecting ducts

| Nephron segment | Active transport | Passive transport | | |
|---|---|---|---|---|
| | | Salt | Water | Urea |
| Proximal convoluted tubule | $Na^+$ | $Cl^-$ | Yes | Yes |
| Descending limb of nephron loop | None | Maybe | Yes | No |
| Thin segment of ascending limb | None | NaCl | No | Yes |
| Thick segment of ascending limb | $Na^+$ | $Cl^-$ | No | No |
| Distal convoluted tubule | $Na^+$ | No | No | No |
| Collecting duct* | Slight $Na^+$ | No | Yes (ADH) or Slight (no ADH) | Yes |

*The permeability of the collecting duct to water depends on the presence of ADH.

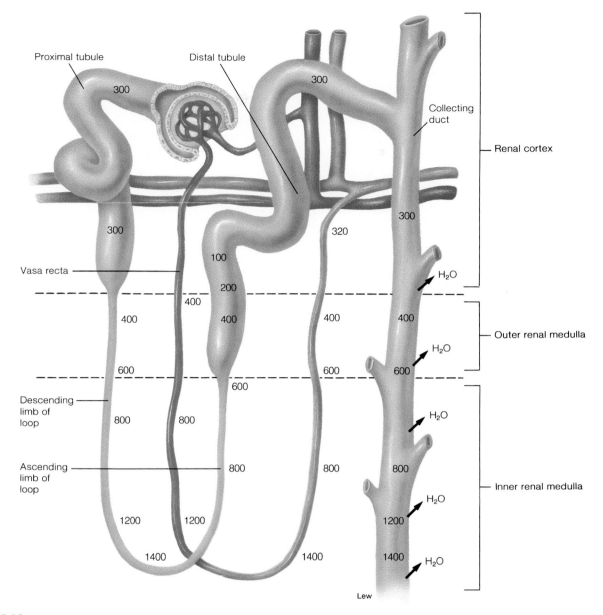

Proximal tubule    Distal tubule

300                300

Collecting duct

Renal cortex

300

300                320

100

H₂O

Vasa recta

200

400

400        400        400        400

Outer renal medulla

600        600        600

H₂O

600

Descending limb of loop

800        800

800

H₂O

Ascending limb of loop

800        800        800

H₂O

1200       1200       1200

H₂O

1400       1400       1400

H₂O

Lew

Inner renal medulla

**FIGURE 25.19**

The countercurrent multiplier system in the loop of the nephron and countercurrent exchange in the vasa recta help create a hypertonic renal medulla. Under the influence of antidiuretic hormone (ADH), the collecting duct is permeable to water so that water is drawn by osmosis out into the hypertonic renal medulla and into the peritubular capillaries.

receptors in the collecing ducts, the water channels are removed from the cell membrane by a process of endocytosis (chapter 3). Endocytosis is the opposite of exocytosis; the cell membrane invaginates to reform vesicles that again contain the water channels. Alternating exocytosis and endocytosis in response to the presence and absence of ADH, respectively, is believed to result in the recycling of water channels within the cell.

When the concentration of ADH is increased, the collecting ducts become more permeable to water, and more

water is reabsorbed. A decrease in ADH, conversely, results in less reabsorption of water and thus in the excretion of a larger volume of more dilute urine. ADH is produced by neurons in the hypothalamus and is screted from the posterior pituitary (chapter 19). The secretion of ADH is stimulated when osmoreceptors in the hypothalamus respond to an increase in blood osmotic pressure. During dehydration, therefore, when the plasma becomes more concentrated, increased secretion of ADH promotes increased permeability of the collecting ducts of water. In severe dehydration, only the

## Table 25.3 — Antidiuretic hormone secretion and action

| Stimulus | Receptors | Secretion of ADH | Effects on | |
|---|---|---|---|---|
| | | | Urine volume | Blood |
| ↑Osmolality (dehydration) | Osmoreceptors in hypothalamus | Increased | Decreased | Increased water retention; decreased blood osmolality |
| ↓Osmolality | Osmoreceptors in hypothalamus | Decreased | Increased | Water loss increases blood osmolality |
| ↑Blood volume | Stretch receptors in left atrium | Decreased | Increased | Decreased blood volume |
| ↓Blood volume | Stretch receptors in left atrium | Increased | Decreased | Increased blood volume |

minimal amount of water needed to eliminate the body's wastes is excreted. This minimum, about 400 ml per day, is limited by the fact that urine cannot become more concentrated than the medullary tissue fluid surrounding the collecting ducts. Under these conditions about 99.8% of the initial glomerular ultrafiltrate is reabsorbed.

A person in a state of normal hydration excretes about 1.5 L per day of urine, indicating that 99.2% of the glomerular ultrafiltrate volume is reabsorbed. Notice that small changes in percent reabsorption translate into large changes in urine volume. Increasing water ingestion—and thus decreasing ADH secretion (table 25.3)—results in the excretion of correspondingly larger volumes of urine.

 *Diabetes insipidus* is a disease associated with the inadequate secretion or action of ADH. The collecting ducts are thus not very permeable to water and, therefore, a large volume (5–10 L per day) of dilute urine is produced. The resulting dehydration causes intense thirst, but a person with this condition has difficulty drinking enough to compensate for the large volumes of water lost in the urine.

# Renal Plasma Clearance

*As blood passes through the kidneys, some of the constituents of the plasma are removed and excreted in the urine. The blood is thus "cleared," to one degree or another, of particular solutes in the process of urine formation. These solutes may be removed from the blood by filtration through the glomerular capillaries or by secretion by the tubular cells into the filtrate. At the same time, certain molecules in the tubular fluid can be reabsorbed back into the blood. The process of reabsorption thus reduces the renal clearance of these substances.*

One of the major functions of the kidneys is the excretion of waste products such as urea, creatinine, and other molecules. These molecules are filtered through the glomerulus into the glomerular capsule along with water, salt, and other plasma solutes. In addition, some waste products can gain access to the urine by a process called **secretion** (fig. 25.20).

Secretion is the opposite of reabsorption. Molecules that are secreted move out of the peritubular capillaries and into the tubular cells, from which they are actively transported into the tubular lumen. In this way, molecules that were not filtered out of the blood in the glomerulus, but instead passed through the efferent arterioles to the peritubular capillaries, can still be excreted in the urine.

Although about 99% of the filtered water is returned to the vascular system by reabsorption, most of the unneeded molecules that are filtered or secreted are eliminated in the urine. The concentration of these substances in the renal vein leaving the kidney is therefore lower than their concentrations in the blood entering the kidney in the renal artery. Some of the blood that passes through the kidneys, in other words, is cleared of these waste products.

## Renal Clearance of Inulin: Measurement of GFR

If a substance is neither reabsorbed nor secreted by the tubules, the amount excreted per minute in the urine will be equal to the amount that is filtered out of the glomeruli. There does not seem to be a single substance produced by the body, however, that is not reabsorbed or secreted to some degree. Plants such as artichokes, dahlias, onions, and garlic, fortunately, do produce such a compound. This compound, a polymer of the monosaccharide fructose, is **inulin.** Once injected into the blood, inulin is filtered by the glomeruli, and the amount of inulin excreted per minute is exactly equal to the amount that was filtered per minute (fig. 25.21).

If the concentration of inulin in urine is measured and the rate of urine formation is determined, the rate of inulin excretion can easily be calculated:

$$\textit{Quantity excreted per minute} = V \times U$$
$$\text{(mg/min)} \quad \left(\frac{\text{ml}}{\text{min}}\right) \left(\frac{\text{mg}}{\text{ml}}\right)$$

where

$V$ = rate of urine formation
$U$ = inulin concentration in urine

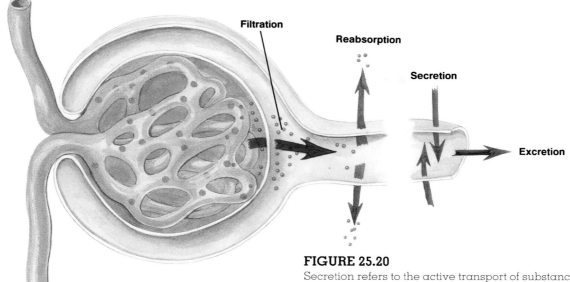

**Filtration**

**Reabsorption**

**Secretion**

**Excretion**

## FIGURE 25.20

Secretion refers to the active transport of substances from the peritubular capillaries into the tubular fluid. This transport is in a direction opposite to that of reabsorption.

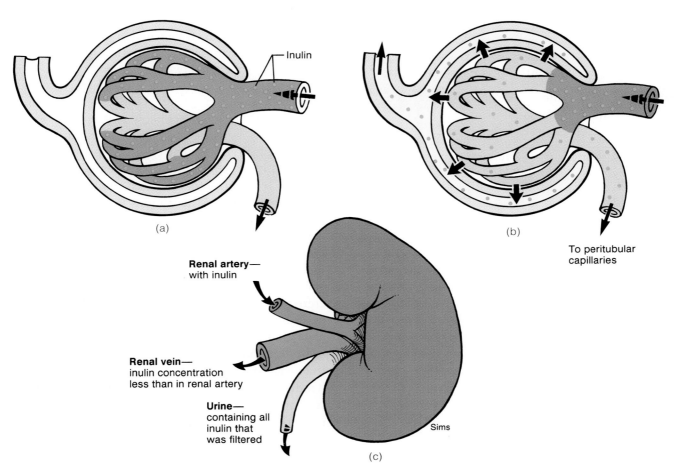

Inulin

(a)

(b)

To peritubular capillaries

**Renal artery—** with inulin

**Renal vein—** inulin concentration less than in renal artery

**Urine—** containing all inulin that was filtered

Sims

(c)

## FIGURE 25.21

The renal clearance of inulin. (*a*) Inulin is present in the blood entering the glomeruli, and (*b*) some of this blood, together with its dissolved inulin, is filtered. All of this filtered inulin enters the urine, whereas most of the filtered water is returned to the vascular system (is reabsorbed). (c) The blood leaving the kidneys via the renal vein thus contains less inulin than the blood that entered the kidneys via the renal artery. Since inulin is filtered but neither reabsorbed nor secreted, the inulin clearance rate equals the glomerular filtration rate (GFR).

## Table 25.4   The effects of filtration, reabsorption, and secretion on renal clearance

| Term | Means | Effect on renal clearance |
|---|---|---|
| Filtered | A substance enters the glomerular ultrafiltrate | Some or all of a filtered substance may enter the urine and be cleared from the blood. |
| Reabsorbed | The transport of a substance from the filtrate, through tubular cells, and into the blood | Reabsorption decreases the rate at which a substance is cleared; clearance rate is less than the glomerular filtration rate (GFR). |
| Secreted | The transport of a substance from peritubular blood through tubular cells and into the filtrate | When a substance is secreted by the nephrons, its clearance rate is greater than the GFR. |

The rate at which a substance is filtered by the glomeruli (in mg per minute) can be calculated by multiplying the ml per minute of plasma that is filtered (the **glomerular filtration rate,** or **GFR**) by the concentration of that substance in the plasma. This is shown in the following equation:

$$\underset{\text{(mg/min)}}{Quantity\ filtered\ per\ minute} = GFR \times P$$
$$\left(\frac{ml}{min}\right)\left(\frac{mg}{ml}\right)$$

where

$$P = \text{inulin concentration in plasma}$$

Since inulin is neither reabsorbed nor secreted, the amount filtered equals the amount excreted:

$$\underset{\text{(amount filtered)}}{GFR \times P} = \underset{\text{(amount excreted)}}{V \times U}$$

If the preceding equation is now solved for the glomerular filtration rate,

$$GFR_{(ml/min)} = \frac{V_{(ml/min)} \times U_{(mg/ml)}}{P_{(mg/ml)}}$$

Let's say, for example, that inulin is infused into a vein and its concentration in the urine and plasma are found to be 30 mg per ml and 0.5 mg per ml, respectively. If the rate of urine formation is 2 ml per minute, the GFR can be calculated as follows:

$$GFR = \frac{(2\ ml/min)(30\ mg/ml)}{0.5\ mg/ml} = 120\ ml/min$$

This equation states that 120 ml of plasma per minute must have been filtered in order to excrete the measured amount of inulin that appeared in the urine. In this example, the glomerular filtration rate is thus 120 ml per minute.

Measurements of the plasma concentration of **creatinine** (kre-at′n-ēn) are often used clinically as an index of kidney function. Creatinine, produced as a waste product of muscle creatine, is secreted in small amounts by the renal tubules so that its excretion rate slightly exceeds that of in-

ulin. Since it is released into the blood at a constant rate and since its excretion is closely matched to the GFR, an abnormal decrease in GFR causes the plasma creatinine concentration to rise. A simple measurement of blood creatinine concentration can thus indicate if the GFR is normal, and thus provide information about the health of the kidneys.

**Clearance Calculations**   The **renal plasma clearance** is the volume of plasma from which a substance is completely removed in one minute by excretion in the urine. Notice that the units for renal plasma clearance are ml/min. In the case of inulin, which is filtered but neither reabsorbed nor secreted, the amount that enters the urine is the same as that contained in the volume of plasma filtered. The clearance of inulin is thus equal to the GFR (120 ml/min in the previous example). This volume of filtered plasma, however, also contains other solutes which may be reabsorbed to varying degrees. If a portion of a filtered solute is reabsorbed, the amount excreted in the urine is less than that which was present in the 120 ml of plasma filtered. Thus, the renal plasma clearance of a substance that is reabsorbed must be less than the GFR (table 25.4).

If a substance is not reabsorbed, all of the filtered amount will be cleared. If this substance is, in addition, secreted by active transport into the renal tubules from the peritubular blood, an additional amount of plasma can be cleared of that substance. The renal plasma clearance of a substance that is filtered and secreted is therefore greater than the GFR (table 25.5). In order to compare the renal handling of various substances in terms of their reabsorption or secretion, the renal plasma clearance is calculated using the same formula used for determining the GFR:

$$Renal\ plasma\ clearance = \frac{V \times U}{P}$$

where

$$V = \text{urine volume per minute}$$
$$U = \text{concentration of substance in urine}$$
$$P = \text{concentration of substance in plasma}$$

## Table 25.5 Renal handling of different plasma molecules

| If substance is | Example | Concentration in renal vein | Renal clearance rate |
|---|---|---|---|
| Not filtered | Proteins | Same as in renal artery | Zero |
| Filtered, not reabsorbed nor secreted | Inulin | Less than in renal artery | Equal to GFR (115–125 ml/min) |
| Filtered, partially reabsorbed | Urea | Less than in renal artery | Less than GFR |
| Filtered, completely reabsorbed | Glucose | Same as in renal artery | Zero |
| Filtered and secreted | PAH | Less than in renal artery; approaches zero | Greater than GFR; up to total plasma flow rate (~625 ml/min) |
| Filtered, reabsorbed, and secreted | $K^+$ | Variable | Variable |

**Clearance of Urea** Urea may be used as an example of how the clearance calculations can reveal the way the kidneys handle a molecule. Urea is a waste product of amino acid metabolism that is secreted by the liver into the blood and filtered into the glomerular capsules. Using the formula for renal clearance previously described and the following sample values, the urea clearance can be obtained:

$$V = 2 \text{ ml/min}$$
$$U = 7.5 \text{ mg/ml of urea}$$
$$P = 0.2 \text{ mg/ml of urea}$$

$$Urea\ clearance = \frac{(2 \text{ ml/min})(7.5 \text{ mg/ml})}{0.2 \text{ mg/ml}} = 75 \text{ ml/min}$$

The clearance of urea in this example is less (75 ml/min) than the clearance of inulin (120 ml/min). Even though 120 ml of plasma filtrate entered the nephrons, only the amount of urea contained in 75 ml of filtrate is excreted per minute. The kidneys must reabsorb some of the urea that is filtered. Although filtered urea is a waste product, a significant portion (between 40% and 60%) is always reabsorbed. This is a passive reabsorption that cannot be avoided because of the high permeability of cell membranes to urea.

## Clearance of PAH: Measurement of Renal Blood Flow

Not all of the blood delivered to the glomeruli is filtered into the glomerular capsules; most of the glomerular blood passes through to the efferent arterioles and peritubular capillaries. The inulin and urea in this unfiltered blood are not excreted but instead return to the general circulation. Blood must make many passes through the kidneys before it can be completely cleared of a given amount of inulin or urea.

In order for compounds in the unfiltered renal blood to be cleared, they must be secreted into the tubules by active transport from the peritubular capillaries. In this way, all of the blood going to the kidneys can potentially be cleared of a secreted compound in a single pass. This is the case for a molecule called **para-aminohippuric acid,** or **PAH,** (fig.

25.22). The clearance (in ml/min) of PAH can be used to measure the *total renal blood flow* (in ml/min). The normal PAH clearance has been found to average 625 ml/min. Since the glomerular filtration rate averages about 120 ml/min, this indicates that only about 120/625, or roughly 20%, of the renal plasma flow is filtered. The remaining 80% passes on to the efferent arterioles.

Since filtration and secretion clear only the molecules dissolved in plasma, the PAH clearance measures the renal plasma flow. In order to convert this measure to the total renal blood flow, the volume of blood occupied by erythrocytes must be taken into account. If the hematocrit (chapter 20) is 45, for example, erythrocytes occupy 45% of the blood volume and plasma accounts for the remaining 55% of the blood volume. The **total renal blood flow** is calculated by dividing the PAH clearance rate by the fractional blood volume occupied by plasma (0.55, in this example). The total renal blood flow in this example is 625 ml/min divided by 0.55, or 1.1 L/min.

 Many antibiotics are secreted by the renal tubules and thus have clearance rates greater than the glomerular filtration rate. Penicillin, for example, is rapidly removed from the blood by renal clearance; hence, large amounts must be administered to be effective. The ability of the kidneys to be visualized in radiographs is improved by the injection of Diodrast. This substance is secreted into the tubules and improves contrast by absorbing X rays. Many drugs and some hormones are inactivated in the liver by chemical transformations and are rapidly cleared from the blood by active secretion in the nephrons.

## Reabsorption of Glucose and Amino Acids

Glucose and amino acids in the blood are easily filtered by the glomeruli into the renal tubules. These molecules, however, are usually not present in the urine. It can be concluded, therefore, that filtered glucose and amino acids are normally completely reabsorbed by the nephrons. This occurs by carrier-mediated active transport processes.

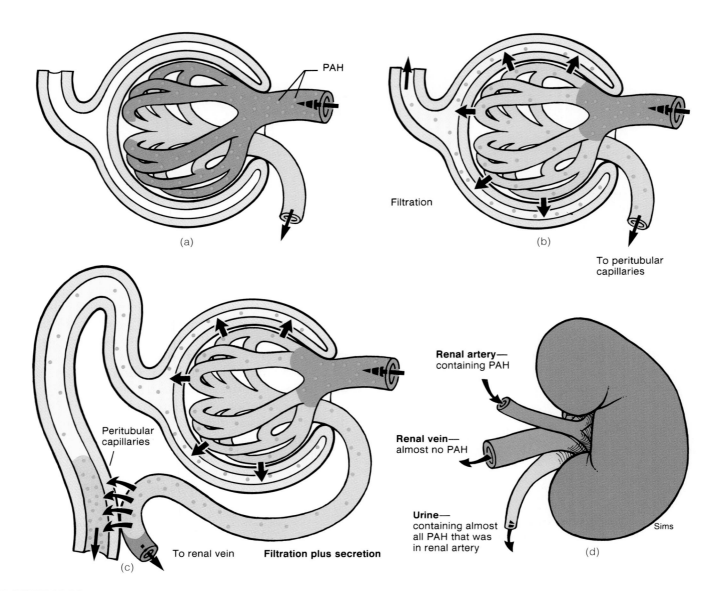

PAH

Filtration

(a)

(b)

To peritubular
capillaries

Peritubular
capillaries

To renal vein    **Filtration plus secretion**

(c)

Renal artery—
containing PAH

Renal vein—
almost no PAH

Urine—
containing almost
all PAH that was
in renal artery

Sims

(d)

**FIGURE 25.22**
Some of the para-aminohippuric acid (PAH) in glomerular blood
(a) is filtered into glomerular capsules (b). The PAH present in the
unfiltered blood is secreted from peritubular capillaries into the
nephron (c), so that all of the blood leaving the kidneys is free of
PAH (d). Therefore, clearance rate of PAH equals the total plasma
flow to the glomeruli.

As described in chapter 5, one of the characteristics
of carrier-mediated transport is *saturation*. This means that
when a transported molecule (such as glucose) is present in
sufficiently high concentrations, all of the carriers are occupied and the transport rate reaches a maximal value. Beyond this rate, called the **transport maximum ($T_m$)**, further
increases in concentration will not increase the transport
rate further.

The carriers for glucose and amino acids in the renal
tubules are not normally saturated and so are able to remove
the filtered molecules completely. The $T_m$ for glucose, for
example, averages 375 mg per minute, which is well above
the rate at which glucose is delivered to the tubules. The
rate of glucose delivery can be calculated by multiplying the
plasma glucose concentration (about 1 mg per ml) by the
GFR (about 125 ml per minute). Approximately 125 mg per
minute are thus delivered to the tubules, whereas a rate of
375 mg per minute is required to reach saturation.

**Glycosuria**    When more glucose passes through the tubules
than can be reabsorbed, glucose appears in the urine in a
condition called *glycosuria (gli″kŏ-soor′e-ă)*. This condition
occurs when the plasma glucose concentration reaches
180–200 mg per 100 ml. Since the rate of glucose delivery is
still below the average $T_m$ for glucose, one must conclude
that some nephrons have considerably lower $T_m$ values than
the average.

The **renal plasma threshold** is the minimum plasma
concentration of a substance that results in the excretion of
that substance in the urine. The renal plasma threshold for

glucose, for example, is 180 to 200 mg per 100 ml. Glucose is normally absent from urine because plasma glucose concentrations remain below this threshold value. Glycosuria occurs only when the plasma glucose concentration is abnormally high (hyperglycemia) and exceeds the renal plasma threshold.

Fasting hyperglycemia is caused by the inadequate secretion or action of insulin. When hyperglycemia results in glycosuria, the disease is called *diabetes mellitus.* A person with uncontrolled diabetes mellitus also excretes a large volume of urine because the excreted glucose carries water with it as a result of the osmotic pressure it generates in the tubules. This condition should not be confused with diabetes insipidus, in which a large volume of dilute urine is excreted as a result of inadequate ADH secretion.

# Renal Control of Electrolyte Balance

*Aldosterone stimulates the reabsorption of $Na^+$ in exchange for $K^+$ in the distal convoluted tubule. Aldosterone thus promotes the renal retention of $Na^+$ and the excretion of $K^+$. Secretion of aldosterone from the adrenal cortex is stimulated directly by a high blood $K^+$ concentration, allowing the completion of a negative feedback loop. A low blood $Na^+$ also stimulates aldosterone secretion, but this is an indirect effect that is mediated by the renin-angiotensin system.*

The kidneys help to regulate the concentrations of plasma electrolytes—sodium, potassium, chloride, bicarbonate, and phosphate—by matching the urinary excretion of these compounds to the amounts ingested. The control of plasma $Na^+$ is important in the regulation of blood volume and pressure; the control of plasma $K^+$ is required to maintain proper function of cardiac and skeletal muscles.

## Role of Aldosterone in $Na^+/K^+$ Balance

Approximately 90% of the filtered $Na^+$ and $K^+$ is reabsorbed by the proximal tubule cells before the filtrate reaches the distal convoluted tubule. This early reabsorption occurs at a constant rate and is not subject to hormonal regulation. The final concentration of $Na^+$ and $K^+$ in the urine is varied according to the needs of the body by processes that occur in the distal convoluted tubule and in the cortical region of the collecting duct (the portion of the collecting duct within the renal medulla does not participate in this regulation). Renal excretion and retention of $Na^+$ and $K^+$ are regulated by **aldosterone,** a steroid hormone secreted by the adrenal cortex.

**Sodium ($Na^+$) Reabsorption** Although 90% of the $Na^+$ present in the filtrate is reabsorbed in the proximal convoluted tubule, the amount left in the filtrate delivered to the distal convoluted tubule is still quite large. In the absence of al-

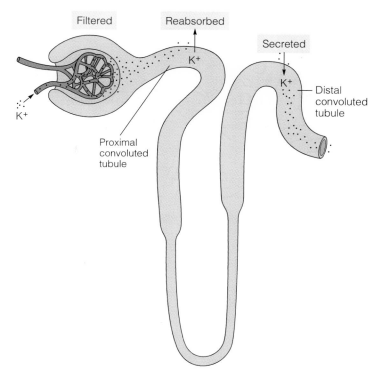

**FIGURE 25.23**
Potassium is almost completely reabsorbed in the proximal tubule, but under aldosterone stimulation, it is secreted into the distal tubule. All of the $K^+$ in urine is derived from secretion rather than from filtration.

dosterone, 80% of this amount is automatically reabsorbed through the wall of the distal convoluted tubule into the peritubular blood; this is 8% of the amount filtered. The amount of $Na^+$ excreted without aldosterone is thus 2% of the amount filtered. Although this percentage seems small, the actual amount of $Na^+$ this represents is an impressive 30 g per day excreted in the urine. By contrast, when aldosterone is secreted in maximal amounts, all of the $Na^+$ delivered to the distal convoluted tubule is reabsorbed. Under these conditions, urine contains no $Na^+$ at all.

**Potassium ($K^+$) Secretion** About 90% of the filtered $K^+$ is reabsorbed in the early regions of the nephron (mainly in the proximal convoluted tubule). When aldosterone is absent, all of the remaining filtered $K^+$ is reabsorbed in the distal convoluted tubule. In the absence of aldosterone, therefore, no $K^+$ is excreted in the urine. The presence of aldosterone stimulates the secretion of $K^+$ from the peritubular blood into the distal convoluted tubule and cortical portion of the collecting duct (fig. 25.23). This aldosterone-induced secretion is the only means by which $K^+$ can be eliminated in the urine. When aldosterone secretion is maximal, as much as 50 times more $K^+$ is excreted in the urine, because of secretion into the distal convoluted tubule, than was originally filtered through the glomeruli.

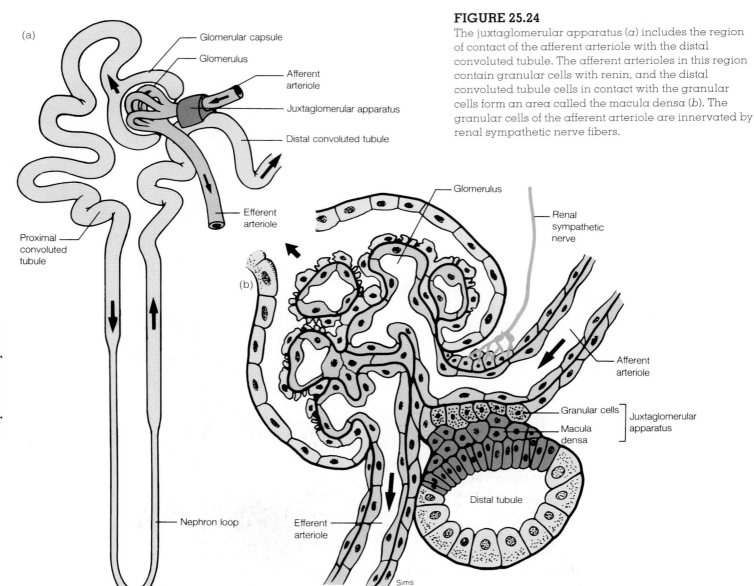

(a)
- Glomerular capsule
- Glomerulus
- Afferent arteriole
- Juxtaglomerular apparatus
- Distal convoluted tubule
- Efferent arteriole
- Proximal convoluted tubule

(b)
- Glomerulus
- Renal sympathetic nerve
- Afferent arteriole
- Granular cells
- Macula densa
- Juxtaglomerular apparatus
- Distal tubule
- Nephron loop
- Efferent arteriole

Sims

**FIGURE 25.24**

The juxtaglomerular apparatus (*a*) includes the region of contact of the afferent arteriole with the distal convoluted tubule. The afferent arterioles in this region contain granular cells with renin, and the distal convoluted tubule cells in contact with the granular cells form an area called the macula densa (*b*). The granular cells of the afferent arteriole are innervated by renal sympathetic nerve fibers.

In summary, aldosterone promotes Na$^+$ retention and K$^+$ loss from the blood by stimulating the reabsorption of Na$^+$ and the secretion of K$^+$ across the wall of the distal convoluted tubules and cortical portions of the collecting ducts. Since aldosterone promotes the retention of Na$^+$, it contributes to an increased blood volume and pressure.

 The body cannot get rid of excess K$^+$ in the absence of aldosterone-stimulated secretion of K$^+$ into the distal convoluted tubules. Indeed, when both adrenal glands are removed from an experimental animal, the *hyperkalemia* (high blood K$^+$) that results can produce fatal cardiac arrhythmias. Abnormally low plasma K$^+$ concentrations, as might result from excessive aldosterone secretion, can also produce arrhythmias, as well as muscle weakness.

## Control of Aldosterone Secretion

Since aldosterone promotes Na$^+$ retention and K$^+$ loss, one might predict (on the basis of negative feedback) that aldosterone secretion will be increased when there is a low Na$^+$ or a high K$^+$ concentration in the blood. This indeed is the case. A rise in blood K$^+$ *directly* stimulates the secretion of aldosterone from the adrenal cortex. Decreases in plasma Na$^+$ concentrations also promote aldosterone secretion, but they do so indirectly.

**Juxtaglomerular Apparatus**   The **juxtaglomerular** (*juk″stă-glo-mer′yŭ-lar*) **apparatus** is the region in each nephron where the afferent arteriole and distal convoluted tubule come into contact (fig. 25.24). The microscopic appearance of the afferent arteriole and distal convoluted tubule in this small region differs from their appearance in other regions. Modified cells in the wall of the afferent arteriole, called **granular cells,** contain the enzyme renin. When the renin is released into the blood, it catalyzes the conversion of *angiotensinogen* (a protein) into *angiotensin I* (a 10-amino-acid polypeptide).

Secretion of renin into the blood results in the formation of angiotensin I, which is converted to **angiotensin II** by *angiotensin-converting enzyme* as blood passes through the lungs and other organs. Angiotensin II, in addition to other effects (described in chapter 22), stimulates the adrenal cortex to secrete aldosterone. Secretion of renin from the granular cells of the juxtaglomerular apparatus is said to initiate the **renin-angiotensin-aldosterone system.** Conditions that result in renin secretion cause increased aldosterone secretion and, by this means, promote the reabsorption of Na$^+$ in the distal convoluted tubules.

**Regulation of Renin Secretion**   A fall in plasma Na$^+$ concentration is always accompanied by a fall in blood volume. This is because ADH secretion is inhibited by the decreased plasma concentration (osmolality). With less ADH, less water is reabsorbed through the collecting ducts and more is excreted in the urine. The fall in blood volume and the fall in renal blood flow that results causes increased renin secretion. Increased renin secretion is believed to be due in part to the direct effect of blood flow on the granular cells, which may function as stretch receptors in the afferent arterioles. Renin secretion is also stimulated by sympathetic nerve activity, which is increased when the blood volume and pressure fall.

An increased secretion of renin, via the increased production of angiotensin II, acts to stimulate aldosterone secretion. Consequently less Na$^+$ is excreted in the urine and more is retained in the blood. This negative feedback system is illustrated in figure 25.25.

**Role of the Macula Densa**   The region of the distal convoluted tubule in contact with the granular cells of the afferent arteriole is called the **macula densa** (see fig. 25.24). There is evidence that this region helps to inhibit renin secretion when the blood Na$^+$ concentration is raised.

According to the proposed mechanism, the cells of the macula densa respond to Na$^+$ within the filtrate delivered to the distal convoluted tubule. When the plasma Na$^+$ concentration is raised, the rate of Na$^+$ delivered to the distal convoluted tubule is also increased. Through an effect on the macula densa, this increase in filtered Na$^+$ may inhibit the granular cells from secreting renin. Aldosterone secretion decreases, and since less Na$^+$ is reabsorbed in the distal

macula densa: L. *macula*, a spot; *densitas*, thick

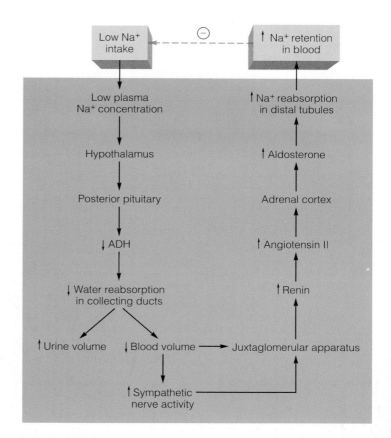

**FIGURE 25.25**

The sequence of events by which a low sodium (salt) intake leads to increased sodium reabsorption by the kidneys. The dotted arrow and negative sign indicate the completion of the negative feedback loop.

convoluted tubule, more Na$^+$ is excreted in the urine. The regulation of renin and aldosterone secretion is summarized in table 25.6.

**Natriuretic Hormone**   Expansion of the blood volume causes increased salt and water excretion in the urine. This is due in part to an inhibition of aldosterone secretion, as previously described. There is much experimental evidence, however, that the increased salt excretion that occurs under these conditions is due not only to the inhibition of aldosterone secretion, but also to the increased secretion of another substance with hormone properties. This other substance is called **natriuretic** (*na˝trĭ-yoo-ret´ik*) **hormone** and is so named because it stimulates salt excretion (in contrast to the inhibitory action of aldosterone). The source and chemical nature of natriuretic hormone remained elusive for many years, but it is now known that the atria of the heart produce a polypeptide that appears to fit the description of the natriuretic hormone proposed by renal physiologists. This polypeptide is called *atrial natriuretic hormone*.

natriuretic: L. *natrium*, sodium; *urina*, urine

## Table 25.6 Regulation of renin and aldosterone secretion

| Stimulus | Effect on renin secretion | Angiotensin II production | Aldosterone secretion | Mechanisms |
|----------|--------------------------|--------------------------|----------------------|------------|
| ↓Na⁺ | Increased | Increased | Increased | Low blood volume stimulates renal stretch receptors; granular cells release renin. |
| ↑Na⁺ | Decreased | Decreased | Decreased | Increased blood volume inhibits stretch receptors; increased Na⁺ in distal convoluted tubule acts via macula densa to inhibit release of renin from granular cells. |
| ↑K⁺ | None | Not changed | Increased | Direct stimulation of adrenal cortex. |
| ↑Sympathetic nerve activity | Increased | Increased | Increased | α-adrenergic effect stimulates constriction of afferent arterioles; β-adrenergic effect stimulates renin secretion directly. |

## Relationship between Na⁺, K⁺, and H⁺

The aldosterone-stimulated reabsorption of $Na^+$ in the distal convoluted tubules creates a large potential difference between the two sides of the tubular wall, with the lumen side being very negative (−50 mV) compared to the basolateral side. The secretion of $K^+$ into the tubular fluid is driven by this electrical gradient. Because of the $Na^+/K^+$ exchange in the distal convoluted tubule, an increase in $Na^+$ reabsorption in the distal convoluted tubule results in an increase in $K^+$ secretion.

Some diuretic drugs inhibit $Na^+$ reabsorption in the nephron loop and, therefore, increase the delivery of $Na^+$ to the distal convoluted tubule. As a result, these diuretics cause an increased reabsorption of $Na^+$ and an increased secretion of $K^+$ in the distal convoluted tubule. People who take diuretics, therefore, tend to have excessive $K^+$ loss in the urine. The actions of different diuretics and their side effects on blood $K^+$ are discussed in the "Clinical Considerations" section of this chapter.

 Complications may arise from the use of diuretics as a result of the $K^+$ loss that occurs. If $K^+$ secretion into the distal convoluted tubules is significantly increased, a condition of *hypokalemia* (low blood $K^+$) may result, which must be compensated for by the increased ingestion of $K^+$. People who take diuretics for the treatment of high blood pressure are usually on a low-sodium diet and often must supplement their meals with potassium chloride (KCl).

The plasma $K^+$ concentration indirectly affects the plasma $H^+$ concentration (pH). Changes in plasma pH likewise affect the $K^+$ concentration of the blood. These effects serve to stabilize the ratio of $K^+$ to $H^+$. When the extracellular $H^+$ concentration increases, for example, some of the $H^+$ moves into the tissue cells and causes cellular $K^+$ to diffuse outward into the extracellular fluid. The plasma concentration of $H^+$ is thus decreased while the $K^+$ increases, helping to reestablish the proper ratio of these ions in the extracellular fluid. A similar effect occurs in the cells of the distal region of the nephron.

In the cells of the late distal convoluted tubule and cortical collecting duct, positively charged ions ($K^+$ and $H^+$) are secreted in response to the negative polarity produced by reabsorption of $Na^+$ (fig. 25.26). When a person has severe acidosis, an increased amount of $H^+$ secretion occurs at the expense of a decrease in the amount of $K^+$ secreted. Acidosis may thus be accompanied by a rise in blood $K^+$. If, on the other hand, hyperkalemia is the primary problem, there is an increased secretion of $K^+$ and a decreased secretion of $H^+$. Hyperkalemia can therefore cause an increase in the blood concentration of $H^+$ and acidosis.

 Aldosterone stimulates the secretion of $H^+$ as well as $K^+$ into the distal convoluted tubules. Therefore, abnormally high aldosterone secretion, as occurs in *primary aldosteronism*, or *Conn's syndrome*, results in both hypokalemia and metabolic alkalosis. Conversely, abnormally low aldosterone secretion, as occurs in *Addison's disease*, can produce hyperkalemia, which is accompanied by metabolic acidosis.

## Renal Control of Acid-Base Balance

*Through their excretion of $H^+$ and their reabsorption of $HCO_3^-$, the kidneys are responsible for the metabolic component of acid-base regulatory systems in the body.*

Normal arterial blood has a pH of 7.35 to 7.45. It should be recalled that the pH number is *inversely* related to the $H^+$

---

hypokalemia: Gk. *hypo*, under; L. *kalium*, potassium

Addison's disease: from Christopher Addison, English anatomist, 1869–1951

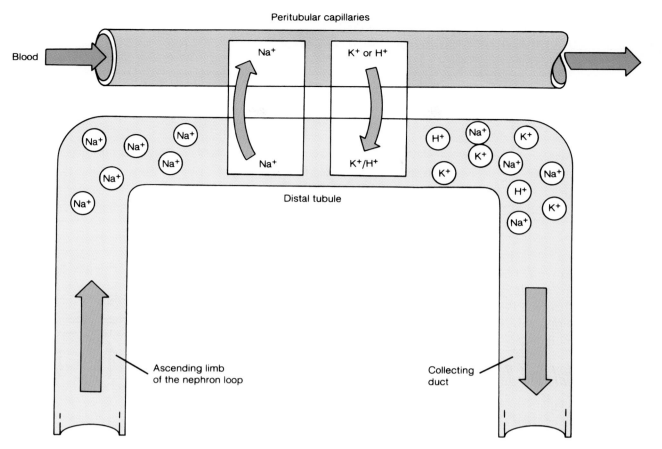

**FIGURE 25.26**
In the distal convoluted tubule, $K^+$ and $H^+$ are secreted in exchange for $Na^+$. High concentrations of $H^+$ may therefore decrease $K^+$ secretion, and vice versa.

concentration. An increase in $H^+$ derived from carbonic acid or the nonvolatile metabolic acids can lower blood pH. With an increase in $H^+$ derived from metabolic acids, however, the blood pH is normally not changed. In the latter case, **bicarbonate buffer** (chapter 24) combines with and thus removes the excess $H^+$ from solution.

## Respiratory and Metabolic Components of the Regulation of Acid-Base Balance

The regulation of acid-base balance has both respiratory and metabolic components. The respiratory component refers to the effect of ventilation on arterial $P_{CO_2}$ and thus on the production of carbonic acid ($H_2CO_3$). The metabolic component refers to the effect of nonvolatile metabolic acids—lactic acid, fatty acids, and ketone bodies—on blood pH. Since these acids are normally buffered by bicarbonate ($HCO_3^-$), the metabolic component can be described in terms of the free $HCO_3^-$ concentration. An increase in metabolic acids uses up free bicarbonate, as $HCO_3^-$ is converted to $H_2CO_3$, and is thus associated with a fall in plasma $HCO_3^-$ concentrations. A decrease in

metabolic acids, conversely, is associated with a rise in free $HCO_3^-$.

Since the respiratory component of acid-base balance is represented by the plasma $CO_2$ concentration and the metabolic component is represented by the free bicarbonate concentration, the study of acid-base balance can be considerably simplified. A normal arterial pH is obtained when both the lungs and kidneys are functioning correctly, and when the ratio of bicarbonate to $CO_2$ is normal. Indeed, given the bicarbonate and $CO_2$ concentrations, the pH can be calculated using the **Henderson–Hasselbalch equation** (the $CO_2$ concentration is obtained indirectly by measuring the plasma $P_{CO_2}$ with an electrode):

$$pH = 6.1 + \log \frac{[HCO_3^-]}{[CO_2]}$$

A normal pH is obtained when the ratio of bicarbonate to $CO_2$ is 20 to 1. A change in this ratio results in an abnormal

∙∙∙∙∙∙∙∙∙∙∙∙∙∙∙∙∙∙∙∙∙∙∙∙∙∙∙∙∙∙∙∙∙∙∙∙∙∙∙∙∙∙∙∙∙∙∙∙∙

Henderson–Hasselbalch equation: from Lawrence Joseph Henderson, American chemist, 1878–1942, and Karl A. Hasselbalch, Dutch scientist, 1874–1962

## Table 25.7 Classification of metabolic and respiratory components of acidosis and alkalosis

| $P_{CO_2}$ | $HCO_3^-$ | Condition | Causes |
|---|---|---|---|
| Normal | Low | Metabolic acidosis | Increased production of nonvolatile acids (lactic acid, ketone bodies, and others) or loss of $HCO_3^-$ in diarrhea |
| Normal | High | Metabolic alkalosis | Vomiting of gastric acid; hypokalemia; excessive steroid administration |
| Low | Low | Respiratory alkalosis | Hyperventilation |
| High | High | Respiratory acidosis | Hypoventilation |

blood pH. Pure respiratory acidosis or alkalosis occurs when the $HCO_3^-$ concentration is normal but the $P_{CO_2}$ and $H_2CO_3$ concentrations are altered. Pure metabolic acidosis or alkalosis occurs when the $P_{CO_2}$ and $H_2CO_3$ are normal but the $HCO_3^-$ concentration is abnormal. This classification is summarized in table 25.7.

## Mechanisms of Renal Acid-Base Regulation

The kidneys help to regulate the blood pH by their excretion of $H^+$ in the urine and by their reabsorption of bicarbonate. These two mechanisms are interdependent—the reabsorption of bicarbonate occurs as a result of the filtration and secretion of $H^+$. The kidneys normally reabsorb all of the filtered bicarbonate and excrete $H^+$. Normal urine, therefore, is free of bicarbonate and is slightly acidic (with a pH range between 5 and 7).

### Reabsorption of Bicarbonate in the Proximal Convoluted Tubule
The apical membranes of the tubule cells (facing the lumen) are impermeable to bicarbonate. The reabsorption of bicarbonate must therefore occur indirectly. When the urine is acidic, $HCO_3^-$ combines with $H^+$ to form carbonic acid. Carbonic acid in the filtrate is then converted to $CO_2$ and $H_2O$ by the action of **carbonic anhydrase.** This enzyme is located in the apical cell membrane of the proximal convoluted tubule in contact with the filtrate. Notice that the reaction that occurs in the filtrate is the same one that occurs within the red blood cells in pulmonary capillaries (discussed in chapter 24).

The tubule cell cytoplasm also contains carbonic anhydrase. Under the conditions of high $CO_2$ that prevail within the cytoplasm, carbonic anhydrase catalyzes the reverse reaction (similar to that which occurs within red blood cells in tissue capillaries). The $CO_2$ that enters the tubule cell is converted to carbonic acid, which in turn dissociates to $HCO_3^-$ and $H^+$ within the tubule cell. The bicarbonate

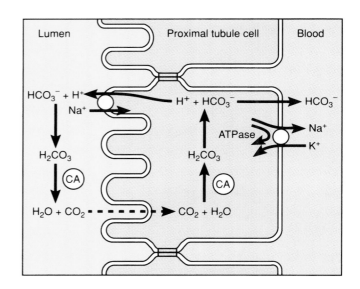

**FIGURE 25.27**
The mechanism of bicarbonate reabsorption. Through this mechanism, the cells of the proximal convoluted tubule can reabsorb bicarbonate while secreting $H^+$. (CA = carbonic anhydrase.)

within the tubule cell can then diffuse through the basolateral membrane and enter the blood (fig. 25.27). Under normal conditions, the same amount of $HCO_3$ passes into the blood as was removed from the filtrate. The $H^+$, which was produced at the same time as $HCO_3^-$ in the cytoplasm of the tubule cell, can either pass back into the filtrate or pass into the blood. Under acidotic conditions, almost all of the $H^+$ goes back into the filtrate and is used to help reabsorb all of the filtered bicarbonate.

During alkalosis, less $H^+$ is secreted into the filtrate. Since the reabsorption of filtered bicarbonate requires the combination of $HCO_3^-$ with $H^+$ to form carbonic acid, less bicarbonate is reabsorbed. This results in urinary excretion of bicarbonate, which helps to partially compensate for the alkalosis.

## Table 25.8  Categories of disturbances in acid-base balance, including those that involve both respiratory and metabolic components

| $P_{CO_2}$ (mmHg) | Bicarbonate (mEq/L) | | |
|---|---|---|---|
| | **Less than 21** | **21–26** | **More than 26** |
| More than 45 | Combined metabolic and respiratory acidosis | Respiratory acidosis | Metabolic alkalosis and respiratory acidosis |
| 35–45 | Metabolic acidosis | Normal | Metabolic alkalosis |
| Less than 35 | Metabolic acidosis and respiratory alkalosis | Respiratory alkalosis | Combined metabolic and respiratory alkalosis |

In this way, disturbances in acid-base balance caused by respiratory problems can be partially compensated for by changes in plasma bicarbonate concentrations. Metabolic acidosis or alkalosis—in which changes in bicarbonate concentrations occur as the primary disturbance—can be similarly compensated for, in part by changes in ventilation. These interactions of the respiratory and metabolic components in the regulation acid-base balance are shown in table 25.8.

When people go to the high elevations of the mountains, they hyperventilate, as discussed in chapter 24. This lowers the arterial $P_{CO_2}$ and produces a respiratory alkalosis. The kidneys participate in this acclimatization by excreting a larger amount of bicarbonate, which helps to partially compensate for the alkalosis and bring the pH back down toward normal. It is interesting in this regard that the drug *acetazolamide*, which inhibits renal carbonic anhydrase, is often used to treat *acute mountain sickness* (*AMS;* see chapter 24). The inhibition of renal carbonic anhydrase causes the loss of bicarbonate and water in the urine, producing a metabolic acidosis and diuresis, which apparently help to alleviate the symptoms of AMS.

**Urinary Buffers**  When a person has a blood pH of less than 7.35 (acidosis), the urine pH almost always falls below 5.5. The nephron, however, cannot produce a urine pH that is significantly less than 4.5. In order for more $H^+$ to be excreted, the acid must be buffered. Actually, even in normal urine most of the $H^+$ excreted is in a buffered form. Bicarbonate cannot serve this function because it is normally completely reabsorbed. Instead, the buffering action of phosphates (mainly $HPO_4^{-2}$) and ammonia ($NH_3$) provide the means for excreting most of the $H^+$ in the urine. Phosphate enters the urine by filtration. Ammonia (whose presence is strongly evident in a diaper pail or kitty litter box) is produced in the tubule cells by deamination of amino acids. These molecules buffer $H^+$ as described in the following equations:

$$NH_3 + H^+ \longrightarrow NH_4^+ \text{ (ammonium ion)}$$

$$HPO_4^{-2} + H^+ \longrightarrow H_2PO_4^-$$

# Ureters, Urinary Bladder, and Urethra

*Urine is channeled from the kidneys to the urinary bladder by the ureters and expelled from the body through the urethra. The mucosa of the urinary bladder permits distension, and the muscles of the urinary bladder and urethra are used in the control of micturition.*

## Ureters

The **ureters,** like the kidneys, are retroperitoneal. Each ureter is a tubular organ about 25 cm (10 in.) long that begins at the renal pelvis and courses inferiorly to enter the urinary bladder at the superior lateral angle of its base. The ureter is thickest—approximately 1.7 cm (0.5 in.) in diameter—near where it enters the urinary bladder.

The wall of the ureter consists of three layers, or tunics. The inner **mucosa** is continuous with the linings of the renal tubules and the urinary bladder. The mucosa consists of transitional epithelium (fig. 25.28). The cells of this layer secrete a mucus that coats the walls of the ureter with a protective film. The middle layer of the ureter is called the **muscularis.** It consists of an inner longitudinal and an outer circular layer of smooth muscle. In addition, the lower third of the ureter contains another longitudinal layer to the outside of the circular layer. Muscular peristaltic waves move the urine through the ureter. The peristaltic waves are initiated by the presence of urine in the renal pelvis, and their frequency is determined by the volume of urine. The waves force urine through the ureter and cause it to spurt into the urinary bladder. The outer layer of the ureter is called the **adventitia.** The adventitia is composed of loose connective tissue that not only covers the ureter but has extensions that anchor it in place.

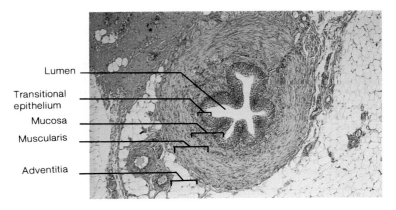

Lumen

Transitional epithelium

Mucosa

Muscularis

Adventitia

**FIGURE 25.28**
A photomicrograph of the ureter in transverse section.

 A *calculus,* or *renal stone,* may obstruct the ureter and greatly increase the frequency of peristaltic waves in an attempt to pass the stone. The pain from a lodged calculus is extreme and extends throughout the pelvic area. A lodged calculus also causes a sympathetic ureterorenal reflex that results in constriction of renal arterioles, thus reducing the production of urine in the kidney on the affected side.

## Urinary Bladder

The **urinary bladder** is a storage sac for urine. It is located posterior to the symphysis pubis and anterior to the rectum. In females, the urinary bladder is in contact with the uterus and vagina. In males, the prostate is positioned below the urinary bladder (fig. 25.29).

The shape of the urinary bladder is determined by the volume of urine it contains. An empty urinary bladder is pyramidal in shape. As the urinary bladder fills, it loses its pyramidal shape and becomes ovoid as the superior surface enlarges and bulges upward into the abdominal cavity. The apex of the urinary bladder is superior to the symphysis pubis and is secured to the **median umbilical ligament** by a fibrous cord called the **urachus.** The base of the urinary bladder receives the ureters along the superolateral angles, and the urethra exits at the neck. The urethra is a tubular continuation of the neck of the urinary bladder.

The wall of the urinary bladder consists of four layers: the mucosa, submucosa, muscularis, and serosa (adventitia). The **mucosa** is composed of transitional epithelium that decreases in thickness as the urinary bladder distends and the cells are stretched. Further distension is permitted by folds of the mucosa, called **rugae,** which can be seen when the urinary bladder is empty. Fleshy flaps of mucosa located where the ureters pierce into the urinary bladder act as valves over the openings

calculus: L. *calculus,* small stone

of the ureters to prevent a reverse flow of urine toward the kidneys as the urinary bladder fills. A triangular area known as the **trigone** (*tri´gōn*) is formed on the mucosa between the two ureter openings and the single urethral opening (fig. 25.29). The internal trigone lacks rugae and is therefore smooth in appearance and remains relatively fixed in position as the urinary bladder changes shape during distension and contraction.

The second layer of the urinary bladder, the **submucosa,** functions to support the mucosa. The **muscularis** consists of three interlaced smooth muscle layers and is referred to as the **detrusor muscle.** At the neck of the urinary bladder, the detrusor muscle is modified to form the upper (the internal) of two muscular sphincters surrounding the urethra. The outer covering of the urinary bladder is the **adventitia.** It appears only on the superior surface of the urinary bladder and is actually a continuation of the peritoneum.

 The urinary bladder becomes infected easily, and because a woman's urethra is so much shorter than a man's, women are particularly susceptible to these infections. A urinary bladder infection, called *cystitis,* may easily ascend from the urinary bladder to the ureters since the mucous linings are continuous. An infection that involves the renal pelvis is called *pyelitis;* if it continues into the nephrons, it is known as *nephritis.*

## Urethra

The tubular **urethra** conveys urine from the urinary bladder to the outside of the body. The urethral wall has an inside lining of mucous membrane surrounded by a relatively thick layer of smooth muscle, the fibers of which are directed longitudinally. Specialized **urethral glands** embedded in the urethral wall secrete mucus into the urethral canal.

Two muscular sphincters surround the urethra (fig. 25.29). The upper, involuntary smooth muscle sphincter is the **internal urethral sphincter,** which is formed from the detrusor muscle of the urinary bladder. The lower sphincter is composed of voluntary, skeletal muscle fibers and is called the **external urethral sphincter.**

The urethra of the female is a simple tube about 4 cm (1.5 in.) long that empties urine through the **urethral orifice** into the vestibule between the labia minora. The urethral orifice is positioned anterior to the vaginal orifice and about 2.5 cm posterior to the clitoris.

The urethra of the male serves both the urinary and reproductive systems. It is about 20 cm (8 in.) long and S-shaped because of the shape of the penis. Three regions can be identified (fig. 25.29). The **prostatic urethra** is the proximal portion, about 2.5 cm long, that passes through the **prostate** located near

trigone: L. *trigonum,* triangle

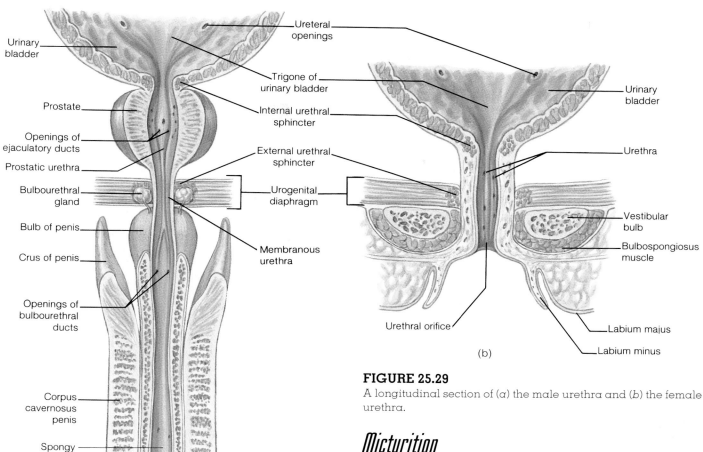

Urinary bladder

Prostate

Openings of ejaculatory ducts

Prostatic urethra

Bulbourethral gland

Bulb of penis

Crus of penis

Openings of bulbourethral ducts

Corpus cavernosus penis

Spongy urethra

Corpus spongiosum penis

Glans penis

Urethral orifice

(a)

Ureteral openings

Trigone of urinary bladder

Internal urethral sphincter

External urethral sphincter

Urogenital diaphragm

Membranous urethra

Urinary bladder

Urethra

Vestibular bulb

Bulbospongiosus muscle

Urethral orifice

Labium majus

Labium minus

(b)

**FIGURE 25.29**

A longitudinal section of (a) the male urethra and (b) the female urethra.

## Micturition

**Micturition** (*mik″tŭ-rish′un*), commonly called urination or voiding, is a reflex action that expels urine from the urinary bladder. It is a complex function that requires a stimulus from the urinary bladder and a combination of involuntary and voluntary nerve impulses to the appropriate muscular structures of the urinary bladder and urethra.

In young children, micturition is a simple reflex action that occurs when the urinary bladder becomes sufficiently distended. Voluntary control of micturition is normally established when a child is 2 or 3 years old. Voluntary control requires the development of inhibitory functioning by the cerebral cortex and a maturing of various portions of the spinal cord. The volume of urine produced by an adult averages about 1200 ml per day, but it can vary from 600 to 2500 ml. The average capacity of the urinary bladder is 700 to 800 ml. A volume of 200 to 300 ml will distend the urinary bladder enough to stimulate stretch receptors and trigger the micturition reflex, creating a desire to urinate.

The micturition reflex center is located in the second, third, and fourth sacral segments of the spinal cord. Following stimulation of this center by impulses arising from stretch receptors in the urinary bladder, parasympathetic nerves that stimulate the detrusor muscle and the internal urethral sphincter are activated. Stimulation of these muscles causes a rhythmic

the neck of the urinary bladder. The **membranous urethra** is the short (0.5 cm) portion of the urethra that passes through the urogenital diaphragm. The **spongy urethra** is the longest portion (15 cm), extending from the outer edge of the urogenital diaphragm to the external urethral orifice on the glans penis. This portion is surrounded by erectile tissue as it passes through the corpus spongiosum of the penis. The paired ducts of the **bulbourethral glands** (Cowper's glands) of the reproductive system attach to the spongy urethra near the urogenital diaphragm.

..........................................

Cowper's glands: from William Cowper, English surgeon, 1666–1709

..........................................

micturition: L. micturire, to urinate

## Development of the Urinary System

The urinary and reproductive systems originate from a specialized elevation of mesodermal tissue called the **urogenital ridge.** The two systems share common structures for part of the developmental period, but by the time of birth two separate systems have formed. The separation in the male is not totally complete, however, since the urethra serves to transport both urine and semen. The development of both systems is initiated during the embryonic stage, but the development of the urinary system starts and ends sooner than that of the reproductive system.

Three successive types of kidneys develop in the human embryo: the *pronephros, mesonephros,* and *metanephros* (fig. 1). The metanephric kidney remains as the permanent kidney.

The **pronephros** (*pro-nef´ros*) develops during the fourth week after conception and persists only through the sixth week. It is the most superior in position on the urogenital ridge of the three kidneys and is connected to the embryonic **cloaca** by the **pronephric duct.** Although the pronephros is nonfunctional and degenerates in humans, most of its duct is used by the mesonephric kidney (fig. 1), and a portion of it is important in the formation of the metanephros.

The **mesonephros** (*mez˝-ŏ-nef´ros*) develops toward the end of the fourth week as the pronephros degenerates. The mesonephros forms from an intermediate portion of the urogenital ridge and functions throughout the embryonic period of development.

Although the **metanephros** (*met˝ă-nef´ros*) begins its formation during the fifth week, it does not become functional until immediately before the start of the fetal stage of development at the end of the eighth week. The paired metanephric kidneys produce urine throughout fetal development. The urine is expelled through the urinary system into the amniotic fluid.

The tubular drainage portion of the kidneys forms as a diverticulum emerges from the wall of the mesonephric duct near the cloaca. This outpouching expands into the metanephrogenic mass to form the drainage pathway for urine (fig. 1*d*). The stalk of the diverticulum develops into the ureter, whereas the expanded terminal portion forms the renal pelvis, calyces, and collecting tubules. Once the metanephric kidneys are formed, they begin to migrate from the pelvis to the upper, posterior portion of the abdomen. The renal blood supply develops as the kidneys become positioned in the posterior body wall.

The urinary bladder develops from the **urogenital sinus,** which is connected to the embryonic umbilical cord by the fetal membrane called the **allantois** (fig. 1*b*). By the twelfth week, the two ureters are emptying into the urinary bladder, the urethra is draining, and the connection of the urinary bladder to the allantois has been reduced to a supporting structure.

• • • • • • • • • • • • • • • • • • • • • • • • • • • • • • • •

pronephros: Gk. *pro,* before; *nephros,* kidney
cloaca: L. *cloaca,* sewer

---

contraction of the urinary bladder wall and a relaxation of the internal urethral sphincter. At this point, a sensation of urgency is perceived in the brain, but there is still voluntary control over the external urethral sphincter. At the appropriate time, the conscious activity of the brain activates the motor nerve fibers (S4) to the external urethral sphincter via the pudendal nerve (S2, S3, and S4), causing the sphincter to relax and urination to occur. The micturition process is summarized in table 25.9.

*Urinary incontinence,* or the inability to void, may occur postoperatively, especially following surgery of the rectum, colon, or internal reproductive organs. The difficulty may be due to nervous tension, the effects of anesthetics, or pain and edema at the site of the operation. If urine is retained beyond 6 to 8 hours, *catheterization* may become necessary. In this procedure, a tube or catheter is passed through the urethra into the urinary bladder so that urine can flow freely.

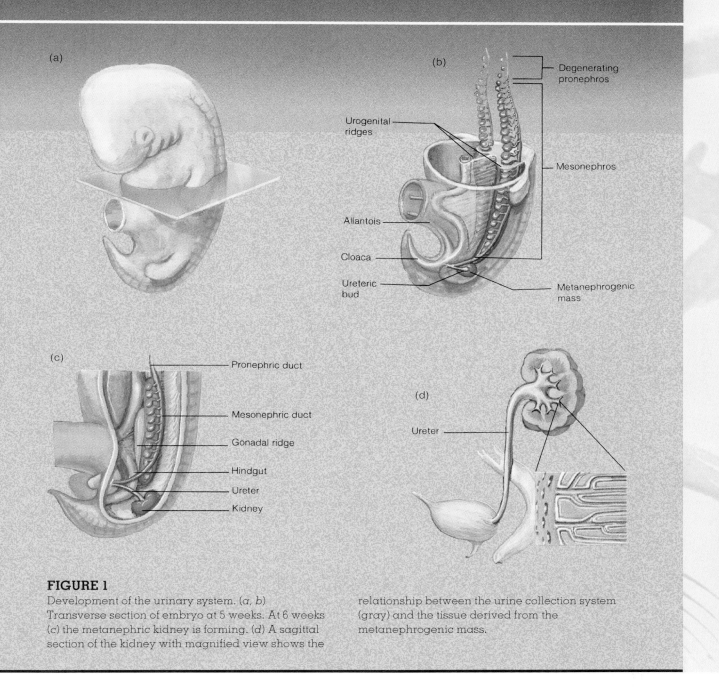

(a)

(b)

Degenerating pronephros

Urogenital ridges

Mesonephros

Allantois

Cloaca

Ureteric bud

Metanephrogenic mass

(c)

Pronephric duct

Mesonephric duct

Gonadal ridge

Hindgut

Ureter

Kidney

(d)

Ureter

**FIGURE 1**

Development of the urinary system. (*a*, *b*) Transverse section of embryo at 5 weeks. At 6 weeks (*c*) the metanephric kidney is forming. (*d*) A sagittal section of the kidney with magnified view shows the relationship between the urine collection system (gray) and the tissue derived from the metanephrogenic mass.

# Clinical Considerations

The importance of kidney function in maintaining homeostasis and the ease with which urine can be collected and used as a mirror of the plasma's chemical composition make the clinical study of renal function and urine composition particularly significant. **Urology** is the medical speciality concerned with dysfunctions of the urinary system. Urinary dysfunctions can be congenital or acquired; they may result from physical trauma or from conditions that secondarily involve the urinary organs.

## Use of Diuretics

People who need to lower their blood volume because of hypertension, congestive heart failure, or edema take medications that increase the volume of urine excreted. Such medications are called **diuretics.** The various diuretic drugs in clinical use act on the nephron in different ways (table 25.10). Based on their chemical structure or aspects of their actions, commonly used diuretics are categorized as carbonic acid inhibitors, loop diuretics, thiazides, osmotic diuretics, or potassium-sparing diuretics.

The most powerful diuretics, inhibiting salt and water reabsorption by as much as 25%, are the drugs that act to inhibit active salt transport out of the ascending limb of the nephron loop. Examples of these nephron loop diuretics include *furosemide* and *ethacrynic acid*. The thiazide diuretics, such as *hydrochlorothiazide*, inhibit salt and water reabsorp-tion by as much as 8% through inhibition of salt transport by the first segment of the distal convoluted tubule. The carbonic anhydrase inhibitors (*acetazolamide*) are much weaker diuretics and act primarily in the proximal convoluted tubule to prevent the water reabsorption that occurs when bicarbonate is reabsorbed.

When extra solutes are present in the filtrate, they increase the osmotic pressure of the filtrate and in this way decrease the osmotic reabsorption of water throughout the nephron. *Mannitol* is sometimes used clinically for this purpose. Osmotic diuresis can occur in diabetes mellitus due to the presence of glucose in the filtrate and urine; this extra solute causes the excretion of excessive amounts of water in the urine and can result in severe dehydration of a person with uncontrolled diabetes.

The previously mentioned diuretics can, as discussed earlier, result in the excessive secretion of $K^+$ into the filtrate and its excessive elimination in the urine. For this reason, potassium-sparing diuretics are sometimes used. *Spironolactones* are aldosterone antagonists that compete with aldosterone for cytoplasmic receptor proteins in the cells of the late distal convoluted tubule. These drugs, therefore, block the aldosterone stimulation of $Na^+$ reabsorption and $K^+$ secretion. *Triamterene* is a different type of potassium-sparing diuretic that appears to act more directly on the $Na^+/K^+$ pumps in the distal convoluted tubule.

## Symptoms and Diagnosis of Urinary Disorders

Normal micturition is painless. **Dysuria** (*dis-yur´e-ă*), or painful urination, is a sign of a urinary tract infection or obstruction of the urethra—as in an enlarged prostate in a male. **Hematuria** means blood in the urine and is usually associated with trauma. **Bacteriuria** means bacteria in the urine, and **pyuria** is the term for pus in the urine, which may result from a prolonged infection. **Oliguria** is an insufficient output of urine, whereas **polyuria** is an excessive output.

### Table 25.9 — Events of micturition

1. The urinary bladder becomes distended as it fills with urine.
2. Stretch receptors in the bladder wall are stimulated, and impulses are sent to the micturition center in the spinal cord.
3. Parasympathetic nerve impulses travel to the detrusor muscle and the internal urethral sphincter.
4. The detrusor muscle contracts rhythmically, and the internal urethral sphincter relaxes.
5. The need to urinate is sensed as urgent.
6. Urination is prevented by voluntary contraction of the external urethral sphincter and by inhibition of the micturition reflex by impulses from the midbrain and cerebral cortex.
7. Following the decision to urinate, the external urethral sphincter is relaxed, and the micturition reflex is facilitated by impulses from the pons and the hypothalamus.
8. The detrusor muscle contracts, and urine is expelled through the urethra.
9. Neurons of the micturition reflex center are inactivated, the detrusor muscle relaxes, and the bladder begins to fill with urine.

### Table 25.10 — Actions of different classes of diuretics

| Category of diuretic | Example | Mechanism of action | Major site of action |
|---|---|---|---|
| Carbonic anhydrase inhibitors | Acetazolamide | Inhibits reabsorption of bicarbonate | Proximal convoluted tubule |
| Loop diuretics | Furosemide | Inhibits $Na^+$ transport | Thick segments of ascending limbs |
| Thiazides | Hydrochlorothiazide | Inhibits $Na^+$ transport | Last part of ascending limb and first part of distal convoluted tubule |
| Potassium-sparing diuretics | Spironolactone | Inhibits action of aldosterone | Last part of distal convoluted tubule and cortical collecting duct |
| | Triamterene | Inhibits $Na^+/K^+$ exchange | Last part of distal convoluted tubule and cortical collecting duct |

Low blood pressure and kidney failure are two causes of oliguria. **Uremia** is a condition in which substances ordinarily excreted in the urine accumulate in the blood. **Enuresis** (*en˝yŭ-re´sis*), or **incontinence,** is the inability to control micturition. It may be caused by psychological factors or by structural impairment.

The palpation and inspection of urinary organs is an important aspect of physical assessment. The right kidney is palpable in the supine position; the left kidney usually is not. The distended urinary bladder is palpable along the superior pelvic rim.

The urinary system may be examined using radiographic techniques. An **intravenous pyelogram (IVP)** permits radiographic examination of the kidneys following the injection of radiopaque dye. In this procedure, the dye that has been injected intravenously is excreted by the kidneys so that the renal pelvises and the outlines of the ureters and urinary bladder can be observed in a radiograph.

**Cystoscopy** (*sĭ-stos´kŏ-pe*) is the inspection of the inside of the urinary bladder by means of an instrument called a cystoscope. With this technique, tissue samples can be obtained, as well as urine samples from each kidney prior to mixing in the urinary bladder. Once the cystoscope is in the urinary bladder, the ureters and pelvis can be viewed through urethral catheterization and inspected for obstructions. A **renal biopsy** is a diagnostic test for evaluating certain types and stages of kidney diseases. The biopsy is performed either through a skin puncture (closed biopsy) or through a surgical incision (open biopsy).

In an **urinalysis,** the voided urine specimen is tested for color, specific gravity, chemical composition, and for the presence of microscopic bacteria, crystals, and *casts*. Casts are accumulations of proteins that leaked through the glomeruli and were pushed through the tubules like toothpaste through a tube. Casts may contain inclusions of red blood cells, white blood cells, bacteria, and other substances. Casts containing red blood cells are diagnostic of glomerulonephritis.

## Infections of Urinary Organs

Urinary tract infections are a significant cause of illness and are also a major factor in the development of chronic renal failure. Females are more predisposed to urinary tract infections than are males, and the incidence of infection increases directly with sexual activity and aging. The higher infection rate in females has been attributed to a shorter urethra, which is in close proximity to the rectum, and to the lack of protection provided by prostatic secretions in males. To reduce the risk of urinary infections, a female should wipe her anal region in a posterior direction, away from the urethral orifice, after a bowel movement.

Infections of the urinary tract are named according to the infected organ. An infection of the urethra is called **ure-thritis** and involvement of the urinary bladder is **cystitis.** Cystitis is frequently a secondary infection from some other part of the urinary tract.

**Nephritis** is inflammation of the kidney tissue. **Glomerulonephritis** (*glo-mer˝yŭ-lo-nĕ-fri´tis*) is inflammation of the glomeruli. Glomerulonephritis frequently occurs following an upper respiratory tract infection because antibodies produced against streptococci bacteria can produce an autoimmune inflammation in the glomeruli. This inflammation may permanently change the glomeruli and figure significantly in the development of chronic renal disease and renal failure.

Any interference with the normal flow of urine, such as from a renal stone or an enlarged prostate in a male, causes stagnation of urine in the renal pelvis and the development of pyelitis. **Pyelitis** is an inflammation of the renal pelvis and its calyces. **Pyelonephritis** is inflammation involving the renal pelvis, the calyces, and the tubules of the nephron within one or both kidneys. Bacterial invasion from the blood or from the lower urinary tract is another cause of both pyelitis and pyelonephritis.

## Trauma to Urinary Organs

A sharp blow to a lumbar region of the back may cause a contusion or rupture of a kidney. Symptoms of kidney trauma include hematuria and pain in the upper abdominal quadrant and flank on the injured side.

Pelvic fractures from accidents may result in perforation of the urinary bladder and urethral tearing. When driving an automobile, it is advisable to stop periodically to urinate because an attached seat belt over the region of a full urinary bladder can cause it to rupture in even a relatively minor accident. Urethral injuries are more common in men than in women because of the position of the urethra in the penis. In a "straddle" injury, for example, a man walking along a raised beam may slip and compress his urethra and penis between the hard surface and his pubic arch, rupturing the urethra.

**Obstruction**   The urinary system can become obstructed anywhere along the tract. Calculi (stones) are the most common cause, but blockage can also come from trauma, strictures, tumors or cysts, spasms or kinks of the ureters, or congenital anomalies. If not corrected, an obstruction causes urine to collect behind the blockage and generate pressure that may cause permanent functional and anatomic damage to one or both kidneys. As a result of pressure buildup in a ureter, a distended ureter, or **hydroureter,** develops. Dilation in the renal pelvis is called **hydronephrosis.**

**Calculi,** or renal stones, are generally the result of infections or metabolic disorders that cause the excretion of

**FIGURE 25.30**
A diagram of a hemodialysis.

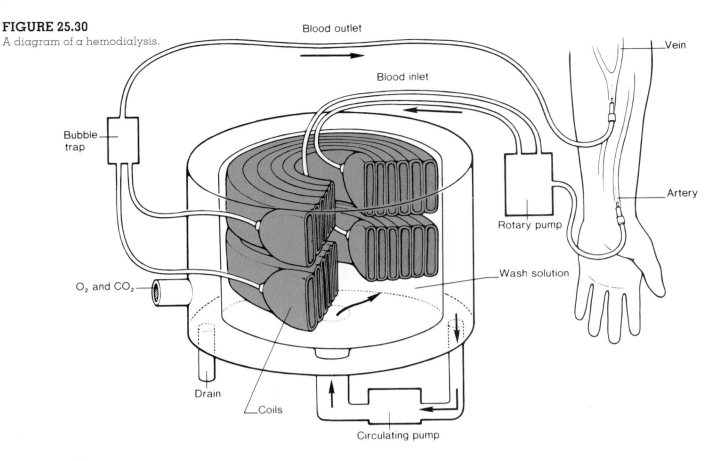

large amounts of organic and inorganic substances. As the urine becomes concentrated, these substances may crystallize and form granules. The granules then serve as cores for further precipitation and development of larger calculi. This becomes dangerous when a calculus grows large enough to cause an obstruction. It also causes intense pain when it passes through the urinary tract.

**Renal Failure**   An output of 50 to 60 cc of urine per hour is considered normal, and an output of less than 30 cc per hour may indicate renal failure. Renal failure is the loss of the kidney's ability to maintain fluid and electrolyte balance and to excrete waste products.

Renal failure can be either acute or chronic. **Acute renal failure** is the sudden loss of kidney function caused by shock and hemorrhage, thrombosis, or other physical trauma to the kidneys. The kidneys may sustain a 90% loss of their nephrons through tissue death and still continue to function without apparent difficulty. If a patient suffering acute renal failure is stabilized, the nephrons have an excellent capacity to regenerate.

A person with **chronic renal failure** cannot sustain life independently. Chronic renal failure is the end result of kidney disease in which the kidney tissue is progressively destroyed. As renal tissue continues to deteriorate, the options for sustaining life are hemodialysis or kidney transplantation.

**Hemodialysis**   Hemodialysis equipment (fig. 25.30) is designed to filter the wastes from the blood of a patient who has chronic renal failure. During hemodialysis, the blood of a patient is pumped through a tube from the radial artery. In the machine, a semipermeable cellophane membrane separates the blood from an isotonic solution containing molecules needed by the body (such as glucose). In a process called **dialysis,** waste products diffuse out of the blood through the membrane while glucose and other molecules needed by the body remain in the blood. After it has been cleansed, the blood is returned to the body through a vein.

More recent hemodialysis techniques include the use of the patient's own peritoneal membranes (which line the abdominal cavity—see chapter 26) for dialysis. Dialysis fluid is introduced into the peritoneal cavity, and then, after a period of time, discarded after wastes have accumulated. This procedure, called *continuous ambulatory peritoneal dialysis* (CAPD), can be performed several times a day by the patients themselves on an outpatient basis.

Patients with kidney failure who are on hemodialysis frequently suffer from anemia. This is due to the lack of the hormone *erythropoietin,* secreted by the normal kidneys, which stimulates red blood cell production in the bone marrow (chapter 20). Such patients are now given recombinant erythropoietin produced by genetic engineering techniques.

# Interactions of the Urinary System with Other Body Systems

## Integumentary System
- Covers and protects the body from excessive fluid loss
- Provides for evaporative water loss
- Maintains blood volume and pressure, blood pH and electrolyte levels, and eliminates metabolic wastes

## Skeletal System
- Supports and protects some organs of urinary system
- Stores calcium and phosphate ions
- Maintains blood volume and pressure, blood pH and electrolyte levels, and eliminates metabolic wastes

## Muscular System
- Supports and protects some organs of urinary system
- Assists storage and voiding of urine
- Maintains blood volume and pressure, blood pH and electrolyte levels, and eliminates metabolic wastes

## Nervous System
- Provides autonomic innervation to urinary system
- Provides motor control of micturition
- Maintains blood volume and pressure, blood pH and electrolyte levels, and eliminates metabolic wastes

## Endocrine System
- Antidiuretic hormone and aldosterone help regulate renal reabsorption of water and electrolytes
- Maintains blood volume and pressure, blood pH and electrolyte levels, and eliminates metabolic wastes
- Kidneys produce the hormone erythropoietin

## Circulatory System
- Transports oxygen and nutrients to urinary system and removes wastes
- Heart secretes atrial natriuretic hormone
- Kidneys filter the blood and remove wastes while regulating blood volume and composition

## Lymphatic System
- Protects urinary tract against infections
- Maintains balance of interstitial fluid
- Kidneys filter the blood and remove wastes while regulating blood volume and composition

## Respiratory System
- Provides oxygen and eliminates carbon dioxide
- Kidneys filter the blood and remove wastes while regulating blood volume and composition

## Digestive System
- Provides nutrients for tissues of urinary system
- Kidneys filter the blood and remove wastes while regulating blood volume and composition

## Reproductive System
- Reproductive organ (penis) provides for passage of urethra in male
- Kidneys filter the blood and remove wastes while regulating blood volume and composition

# Chapter Summary

## Urinary System and Kidney Structure (pp. 728–731)

1. The urinary system consists of two kidneys, two ureters, the urinary bladder, and the urethra.
2. The urinary system maintains the composition and properties of the body fluid, which establishes the internal environment of the body cells. The end product of the urinary system is urine, which is voided from the body through the urethra.
3. The gross structure of the kidney includes the renal pelvis, calyces, renal medulla, and renal cortex.
   a. The renal medulla is composed of the renal pyramids, which are separated by renal columns.
   b. The renal pyramids empty urine into the calyces, which drain into the renal pelvis and out the ureter.
4. Each kidney contains more than a million microscopic functional units called nephrons; nephrons have vascular and tubular components.
   a. A capillary bed, the glomerulus, produces a filtrate that enters the first part of the nephron tubule, known as the glomerular (Bowman's) capsule.
   b. Filtrate from the glomerular capsule enters, in turn, the proximal convoluted tubule, nephron loop (loop of Henle), distal convoluted tubule, and collecting duct.
   c. The glomerulus, proximal convoluted tubule, and distal convoluted tubule are located in the renal cortex; the nephron loop can descend into the renal medulla.
   d. The collecting ducts descend from the renal cortex through the renal medulla to empty their contents of urine into the calyces.

## Glomerular Filtration (pp. 731–734)

1. A filtrate derived from plasma in the glomerulus must pass through the inner layer of the glomerular capsule.
   a. The glomerular ultrafiltrate is formed under the force of blood pressure and has a low protein concentration.
   b. The glomerular filtration rate (GFR) is the volume of filtrate produced per minute by both kidneys; it ranges from 115 to 125 ml per min.
2. The GFR can be regulated by constriction or dilation of the afferent arterioles.

## Reabsorption of Salt and Water (pp. 734–742)

1. Approximately 65% of the filtered salt and water is reabsorbed across the proximal convoluted tubules.
   a. $Na^+$ is actively transported and $Cl^-$ follows, going from the filtrate into the blood in the peritubular capillaries.
   b. Water follows the NaCl by osmosis, so that the volume is reduced but the concentration of the filtrate that remains is unchanged.
2. The reabsorption of most of the remaining water occurs as a result of the action of the countercurrent multiplier system.
   a. Sodium is actively extruded from the ascending limb of the nephron loop followed passively by chloride.
   b. Since the ascending limb is impermeable to water, the remaining filtrate becomes hypotonic.
   c. Because of this salt transport and because of countercurrent exchange in the vasa recta, the tissue fluid of the renal medulla becomes hypertonic.
   d. The hypertonicity of the renal medulla is multiplied by a positive feedback mechanism involving the descending limb, which is passively permeable to water.
3. The collecting duct is permeable to water but not to salt.
   a. As the collecting ducts pass through the hypertonic renal medulla, water leaves by osmosis and is carried away in surrounding capillaries.
   b. The permeability of the collecting ducts to water is stimulated by antidiuretic hormone (ADH).

## Renal Plasma Clearance (pp. 742–747)

1. Inulin is filtered but neither reabsorbed nor secreted; its clearance is thus equal to the glomerular filtration rate.
2. Some of the filtered urea is reabsorbed; its clearance is therefore less than the glomerular filtration rate.
3. The PAH clearance is a measure of the total renal blood flow.
4. Normally, all of the filtered glucose and amino acids are reabsorbed; glycosuria occurs when the transport carriers for glucose become saturated due to hyperglycemia.

## Renal Control of Electrolyte Balance (pp. 747–750)

1. Aldosterone stimulates $Na^+$ reabsorption and $K^+$ secretion in the distal convoluted tubule.
2. Aldosterone secretion is stimulated directly by a rise in blood $K^+$ and indirectly by a fall in blood $Na^+$.
   a. Decreased blood flow through the kidneys stimulates the secretion of the enzyme renin from the juxtaglomerular apparatus.
   b. Renin catalyzes the formation of angiotensin I, which is then converted to angiotensin II.
   c. Angiotensin II stimulates the adrenal cortex to secrete aldosterone.
3. Aldosterone stimulates the secretion of $H^+$ as well as $K^+$ into the filtrate in exchange for $Na^+$.

## Renal Control of Acid-Base Balance (pp. 750–753)

1. The lungs regulate the $P_{CO_2}$ and carbonic acid concentration of the blood, whereas the kidneys regulate the bicarbonate concentration.
2. Filtered bicarbonate combines with $H^+$ to form carbonic acid in the filtrate.
   a. Carbonic anhydrase in the membranes of microvilli in the tubules catalyzes the conversion of carbonic acid to carbon dioxide and water; bicarbonate is thus reabsorbed indirectly.
   b. In addition to reabsorbing bicarbonate, the kidneys excrete $H^+$, which is buffered by ammonium and phosphate buffers.

## Ureters, Urinary Bladder, and Urethra (pp. 753–756)

1. The ureters contain three layers: the mucosa, muscularis, and adventitia.
2. The urinary bladder is lined by a transitional epithelium that is folded into rugae to permit distension.
3. The urethra has an internal urethral sphincter of smooth muscle and an external urethral sphincter of skeletal muscle.
4. Micturition is controlled by reflex centers in the second through fourth segments of the spinal cord.

# Review Activities

## Objective Questions

1. Which of the following statements about metanephric kidneys is *true?*
   a. They become functional at the end of the eighth week.
   b. They are active throughout fetal development.
   c. They are the third pair of kidneys to develop.
   d. All of the above are true.

Match the following descriptions and structures:

2. active transport of sodium; water follows passively
3. active transport of sodium; impermeable to water
4. passively permeable to water
5. passively permeable to water under ADH stimulation

   a. roximal convoluted tubule
   b. descending limb of the nephron loop
   c. ascending limb of the nephron loop
   d. distal convoluted tubule
   e. collecting duct

6. Antidiuretic hormone promotes the retention of water by stimulating
   a. the active transport of water.
   b. the active transport of chloride.
   c. the active transport of sodium.
   d. the permeability of the collecting duct to water.

7. Aldosterone stimulates sodium reabsorption and potassium secretion in
   a. the proximal convoluted tubule.
   b. the descending limb of the nephron loop.
   c. the ascending limb of the nephron loop.
   d. the distal convoluted tubule.
   e. the collecting duct.

8. Substance *X* has a clearance greater than zero but less than that of inulin. What can be concluded about substance *X?*
   a. It is not filtered.
   b. It is filtered, but neither reabsorbed nor secreted.
   c. It is filtered and partially reabsorbed.
   d. It is filtered and secreted.

9. Substance *Y* has a clearance greater than that of inulin. What can be concluded about *Y?*
   a. It is not filtered.
   b. It is filtered but neither reabsorbed nor secreted.
   c. It is filtered and partially reabsorbed.
   d. It is filtered and secreted.

10. About 65% of the glomerular ultrafiltrate is reabsorbed in
    a. the proximal convoluted tubule.
    b. the distal convoluted tubule.
    c. the nephron loop.
    d. the collecting duct.

11. Which of the following statements about the renal pyramids is *false?*
    a. They are located in the renal medulla.
    b. They contain glomeruli.
    c. They contain collecting ducts.
    d. They open by renal papillae into the renal sinus.

12. The detrusor muscle is located in
    a. the kidneys.
    b. the ureters.
    c. the urinary bladder.
    d. the urethra.

13. The internal urethral sphincter is innervated by
    a. sympathetic nerve fibers.
    b. parasympathetic nerve fibers.
    c. somatic motor nerve fibers.
    d. all of the above.

14. Diuretic drugs that act in the nephron loop
    a. inhibit active sodium transport.
    b. result in an increased flow of filtrate to the distal convoluted tubule.
    c. cause an increased secretion of potassium into the tubule.
    d. promote the excretion of salt and water.
    e. all of the above apply.

15. The appearance of glucose in the urine
    a. occurs normally.
    b. indicates the presence of kidney disease.
    c. occurs only when the transport carriers for glucose become saturated.
    d. is a result of hypoglycemia.

16. Reabsorption of water through the tubules occurs by
    a. osmosis.
    b. active transport.
    c. facilitated diffusion.
    d. all of the above.

## Essay Questions

1. Explain how glomerular ultrafiltrate is produced and why it has a low protein concentration.
2. Explain how the countercurrent multiplier system works and discuss its functional significance.
3. Explain how countercurrent exchange occurs in the vasa recta and discuss its functional significance.
4. Explain how the action of diuretic drugs may cause an excessive loss of potassium. Also explain how the potassium-sparing diuretics work.
5. Explain how the structure of the epithelial wall of the proximal convoluted tubule and the distribution of $Na^+/K^+$ pumps in the epithelial cell membranes contribute to the ability of the proximal tubule to reabsorb salt and water.

## Gundy/Weber Software ▪

The tutorial software accompanying Chapter 25 is Volume 12—Urinary System.

# digestive system

## objectives

- Briefly describe the activities of the digestive system and list its structures and regions.
- Locate and describe the serous membranes of the abdominal cavity.
- Describe the four tunics that compose the wall of the GI tract.
- Describe the anatomy of the oral cavity.
- Contrast the deciduous and permanent dentitions and describe the structure of a tooth.
- Describe the histological structure of salivary glands and discuss the functions of saliva.
- Describe the location, gross structure, and functions of the stomach.
- Describe the histological structure of the esophagus and stomach and list the cell types of the gastric mucosa and their secretions.
- Discuss the functions of hydrochloric acid and pepsin and explain how peptic ulcers may be produced.
- State the regions of the small intestine and describe how bile and pancreatic juice are delivered to the small intestine.

- Describe the structure and function of intestinal villi, microvilli, and crypts and explain the nature and significance of the brush border enzymes.
- Describe the different types of intestinal motility.
- State the regions of the large intestine and describe its structure.
- Describe the functions of the large intestine and explain how defecation is accomplished.
- Describe the structure of a liver lobule and trace the flow of blood and bile through a liver lobule.
- Discuss the functions of the liver and describe the enterohepatic circulation.
- Describe the anatomical relationship between the liver and gallbladder.
- Discuss the structure and functions of the pancreas.
- State the regions of the GI tract and list the enzymes involved in the digestion of carbohydrates and proteins.
- Discuss the functions of enzymes and bile in the digestion of lipids and explain how lipids are absorbed.
- Describe the mechanisms that regulate gastric juice secretion.
- Describe the mechanisms that regulate pancreatic juice and bile secretion.

# Introduction to the Digestive System

*Within the lumen of the gastrointestinal (GI) tract, large food molecules are hydrolyzed into their monomers. These monomers pass through the inner layer, or mucosa, of the GI tract to enter the blood or lymph in a process called absorption. Digestion and absorption are aided by specializations of the mucosa and by characteristic movements caused by contractions of the muscle layers of the GI tract.*

Unlike plants, which can form organic molecules using inorganic compounds such as carbon dioxide, water, and ammonia, humans and other animals must obtain their basic organic molecules from food. Some of the ingested food molecules are needed for their energy (caloric) value, which is obtained by the reactions of cell respiration and used in the production of ATP. The balance is used to make additional tissue.

Most of the organic molecules that are ingested are similar to the molecules that form the structure of human tissues. These are generally large molecules (*polymers*), which are composed of subunits (*monomers*). Within the GI tract, **digestion** of these large molecules into their monomers occurs by means of *hydrolysis* reactions (reviewed in fig. 26.1). The monomers formed are transported across the wall of the small intestine into the blood and lymph in a process called **absorption.** Digestion and absorption are the two major functions of the digestive system.

Since the composition of food is similar to the composition of body tissues, enzymes that digest food also are

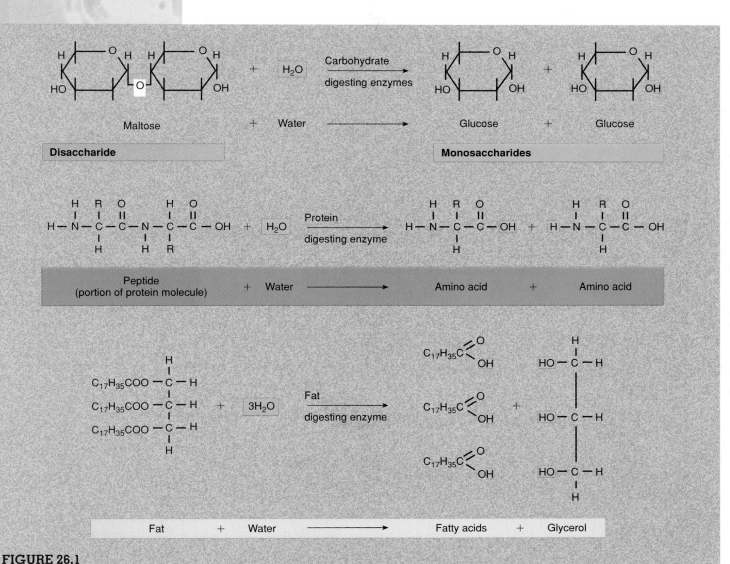

**FIGURE 26.1**
The digestion of food molecules occurs by means of hydrolysis reactions.

capable of digesting a person's own tissues. This does not normally occur, however, because a variety of protective devices inactivate digestive enzymes in the body and separate them from the cytoplasm of tissue cells. The fully active digestive enzymes are normally limited in their location to the lumen (cavity) of the GI tract.

The lumen of the GI tract is continuous with the environment because it is open at both ends (mouth and anus). Indigestible material, such as cellulose from plant walls, passes from one end to the other without crossing the epithelial lining of the GI tract (that is, without being absorbed). In this sense, these indigestible materials never enter the body, and the harsh conditions required for digestion thus occur *outside* the body.

One-way transport in the GI tract is ensured by wavelike muscle contractions called peristalsis (*per˝ĭ-stal´sis*) and by the action of sphincter muscles. This one-way transport permits the specialization of different regions of the GI tract for different functions, as a "dis-assembly line." The principal function of the digestive system is to prepare food for cellular utilization. This involves the following functional activities:

**1** **Motility.** This refers to the movement of food through the digestive tract and includes the following processes.
   a. *Ingestion:* Taking food into the mouth.
   b. *Mastication:* Chewing the food and mixing it with saliva.
   c. *Deglutition:* Swallowing food.
   d. *Peristalsis:* Rhythmic, wavelike contractions that move food through the gastrointestinal tract.

**2** **Secretion.** This includes both exocrine and endocrine secretions.
   a. *Exocrine secretions:* Water, hydrochloric acid, bicarbonate, and many enzymes are secreted into the lumen of the gastrointestinal tract. The stomach alone, for example, secretes 2–3 liters of gastric juice a day.
   b. *Endocrine secretions:* The stomach and small intestine secrete a number of hormones that help regulate the digestive system.

**3** **Digestion.** This refers to the breakdown of food molecules into their smaller subunits, which can be absorbed.

**4** **Absorption.** This refers to the passage of food molecules after their digestion into the blood or lymph.

Anatomically and functionally the digestive system can be divided into a tubular **gastrointestinal (GI) tract,** or *alimentary canal,* and **accessory organs.** The GI tract is

mastication: Gk. *mastichan,* gnash the teeth
deglutition: L. *deglutire,* swallow down
peristalsis: Gk. *peri,* around; *stellain,* compress

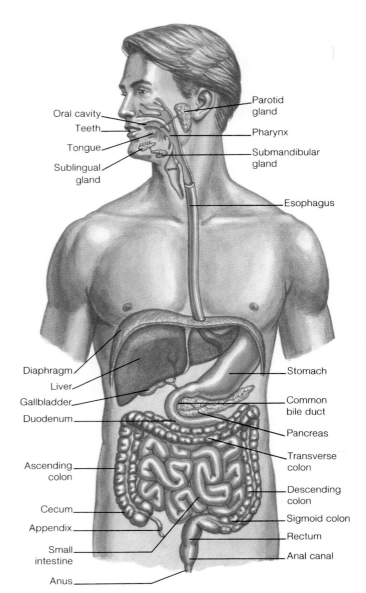

**FIGURE 26.2**
The digestive system consists of the gastrointestinal tract and the accessory digestive organs.

approximately 9 m (30 ft) long and extends from the mouth to the anus. It traverses the thoracic cavity and enters the abdominal cavity at the level of the diaphragm. The anus is located at the inferior portion of the pelvic cavity. The organs of the GI tract include the **oral (buccal) cavity, pharynx, esophagus, stomach, small intestine,** and **large intestine** (fig. 26.2). The accessory digestive organs include the **teeth, tongue, salivary glands, liver, gallbladder,** and **pancreas.** The term *viscera* (*vis´er-ă*) is frequently used to refer to the abdominal organs of digestion, but it also can be used in reference to any of the organs in the thoracic and abdominal cavities. **Gut** is an anatomical term that generally refers to the developing GI tract in the embryo.

## Serous Membranes

Most of the GI tract and abdominal accessory digestive organs are positioned within the **abdominal cavity.** These organs are not firmly embedded in solid tissue but are supported and covered by **serous membranes.** A serous membrane is an epithelial membrane that lines the thoracic and abdominal cavities and covers the organs that lie within these cavities. A serous membrane has a parietal portion lining the body wall and a visceral portion covering the internal organs. The serous membranes associated with the lungs are called pleurae. The serous membranes of the abdominal cavity are called **peritoneal membranes,** or **peritoneum** (*per´´ĭ-tŏ-ne´um*). The peritoneum is the largest serous membrane of the body. It is composed of simple squamous epithelium with portions reinforced by connective tissue.

The **parietal peritoneum** lines the wall of the abdominal cavity (fig. 26.3). Along the posterior, or dorsal, aspect of the abdominal cavity the parietal peritoneum comes together to form a double-layered peritoneal fold called the **mesentery** that supports the GI tract. The mesentery gives the small and large intestines freedom for peristaltic movement and provides a structure through which intestinal nerves and vessels traverse. The **mesocolon** is a specific portion of the mesentery that supports the large intestine (fig. 26.4c,d). The peritoneal covering continues around the intestinal viscera as the **visceral peritoneum.** The **peritoneal cavity** is the space between the parietal and visceral portions of the peritoneum.

Extensions of the parietal peritoneum, located in the peritoneal cavity, serve specific functions (fig. 26.4). The **falciform** (*fal´sĭ-form*) **ligament,** a serous membrane reinforced with connective tissue, attaches the liver to the diaphragm and anterior abdominal wall. The **lesser omentum** (*o-men´tum*) passes from the lesser curvature of the stomach and the upper duodenum to the inferior surface of the liver. The **greater omentum** extends from the greater curvature of the stomach to the transverse colon, forming an apronlike structure over most of the small intestine. Functions of the greater omentum include storing fat, cushioning visceral organs, supporting lymph nodes, and protecting against the spread of infections. In cases of localized inflammation, such as appendicitis, the greater omentum may compartmentalize the inflamed area, sealing it off from the rest of the peritoneal cavity.

Certain of the abdominal organs closely associated with the posterior abdominal wall are not supported by mesentery and are covered only by parietal peritoneum.

mesentery: Gk. *mesos,* middle; *enteron,* intestine
omentum: L. *omentum,* apron

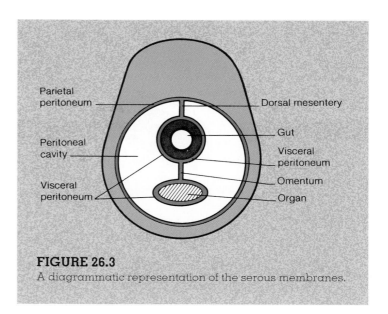

**FIGURE 26.3**
A diagrammatic representation of the serous membranes.

These organs are said to be *retroperitoneal* (behind the peritoneum) and include most of the pancreas, the kidneys, a portion of the duodenum, and the abdominal aorta.

 *Peritonitis* is a bacterial inflammation of the peritoneum. Peritonitis may be caused by trauma, rupture of a visceral organ, or postoperative complications. Peritonitis is usually extremely painful and serious. Treatment involves the injection of massive doses of antibiotics and perhaps peritoneal intubation to permit drainage.

## Layers of the Gastrointestinal Tract

The GI tract from the stomach to the anal canal is composed of four layers, or tunics. Each tunic contains a dominant tissue type that performs specific functions in the digestive process. The four tunics of the GI tract, from the inside out, are the **mucosa, submucosa, muscularis,** and **serosa** (fig. 26.5).

**Mucosa**   The mucosa surrounds the lumen of the GI tract and is the absorptive and major secretory layer. It consists of a simple columnar epithelium, which also contains specialized *goblet cells* that secrete mucus. The epithelium is supported by the **lamina propria,** which is a thin layer of connective tissue. The lamina propria contains numerous lymph nodules, which are important in protecting against disease (fig. 26.5). Deep to the lamina propria is a thin layer of smooth muscle called the **muscularis mucosa.**

**Submucosa**   The relatively thick submucosa is a highly vascular layer of connective tissue serving the mucosa. Absorbed molecules that pass through the columnar epithelial cells of the mucosa enter into blood vessels or lymph ductules of the submucosa. In addition to blood vessels, the

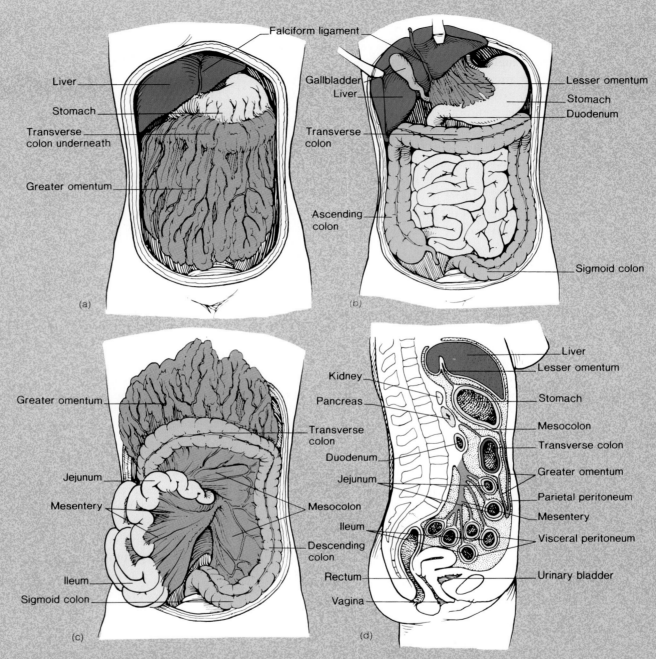

**FIGURE 26.4**

The structural arrangement of the abdominal organs and peritoneal membranes.
(a) The greater omentum, (b) the lesser omentum with the liver lifted, (c) the mesentery
with the greater omentum lifted, and (d) the relationship of the peritoneal membranes
to the visceral organs as shown in a sagittal view.

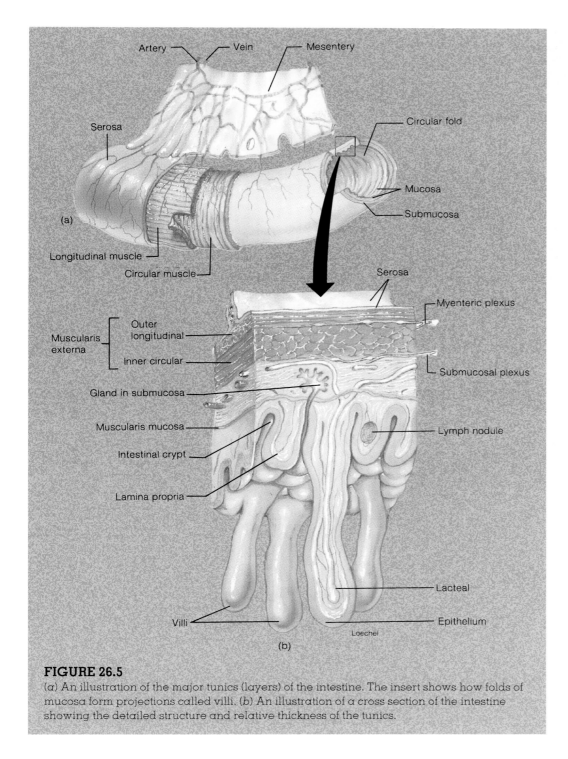

**FIGURE 26.5**

(a) An illustration of the major tunics (layers) of the intestine. The insert shows how folds of mucosa form projections called villi. (b) An illustration of a cross section of the intestine showing the detailed structure and relative thickness of the tunics.

layer of smooth muscle. Contractions of these layers move the food peristaltically through the GI tract and physically pulverize and churn the food with digestive enzymes. The *myenteric* (Auerbach's) *plexus* located between the two muscle layers provides the major nerve supply to the GI tract and includes fibers and ganglia from both the sympathetic and parasympathetic divisions of the autonomic nervous system.

**Serosa**  The outer serosal layer completes the wall of the GI tract. It is a binding and protective layer consisting of loose connective tissue covered with a layer of simple squamous epithelium and subjacent connective tissue. The serosa is actually the visceral peritoneum of the abdominal cavity.

**Innervation**  The GI tract is innervated by the sympathetic and parasympathetic divisions of the autonomic nervous system. The vagus nerves are the source of parasympathetic activity in the esophagus, stomach, pancreas, gallbladder, small intestine, and upper portion of the large intestine. The lower portion of the large intestine receives parasympathetic innervation from spinal nerves in the sacral region. The submucosal plexus and myenteric plexus are the sites where preganglionic fibers synapse with postganglionic fibers that innervate the smooth muscle of the GI tract. Stimulation of the parasympathetic fibers increases peristalsis and the secretions of the GI tract.

Postganglionic sympathetic fibers pass through the submucosal and myenteric plexuses and innervate the GI tract. The effects of sympathetic nerve stimulation reduce peristalsis and secretions and stimulate the contraction of sphincter muscles along the GI tract; therefore, they are antagonistic to the effects of parasympathetic nerve stimulation.

submucosa contains glands and nerve plexuses. The *submucosal* (Meissner's) *plexus* (fig. 26.5), provides autonomic innervation to the muscularis mucosa.

**Muscularis**  The muscularis (also called the muscularis externa) is responsible for segmental contractions and peristaltic movement through the GI tract. The muscularis has an inner circular layer and an outer longitudinal

# Mouth, Pharynx, and Associated Structures

*Ingested food is changed by the mechanical action of teeth and by the chemical activity of saliva into a bolus, which is swallowed in the process of deglutition.*

The **mouth** and associated structures initiate mechanical digestion of food through the process of mastication. The mouth is referred to as the **oral, or buccal** (*buk´al*), **cavity** (fig. 26.6). It is formed by the **cheeks, lips, hard** and **soft palates,** and **tongue.** The **vestibule** of the oral cavity is the depression between the cheeks and lips externally and the gums and teeth internally (fig. 26.7). The opening between the oral cavity and the pharynx is called the **fauces.** The **pharynx** serves as a common passageway for both the respiratory and digestive systems. Both the mouth and pharynx are lined with nonkeratinized stratified squamous epithelium, which is constantly moistened by the secretion of saliva.

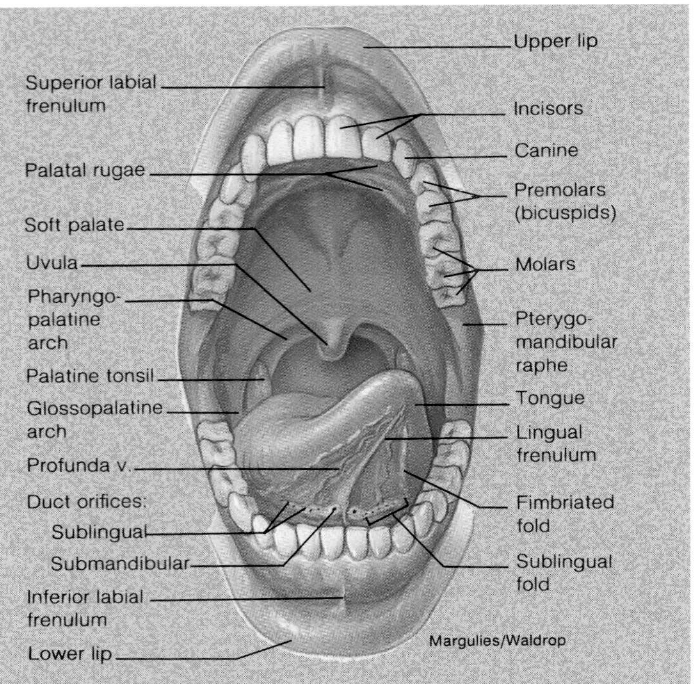

## Cheeks and Lips

The **cheeks** consist of outer layers of skin, subcutaneous fat, facial muscles that assist in manipulating food in the oral cavity, and inner linings of moistened, stratified squamous epithelium. The anterior portion of the cheeks terminates in the superior and inferior lips.

The **lips** are fleshy, highly mobile organs whose principal function in humans is associated with speech. Each lip is attached from its inner surface to the gum by a midline fold of mucous membrane called the **labial frenulum** (*fren´yŭlum*) (fig. 26.6). The lips are formed from the orbicularis oris muscle and associated connective tissue, and they are covered with soft, pliable skin. Between the outer skin

buccal: L. *bucca*, cheek
pharynx: L. *pharynx*, throat

**FIGURE 26.6**
The superficial structures of the oral cavity.

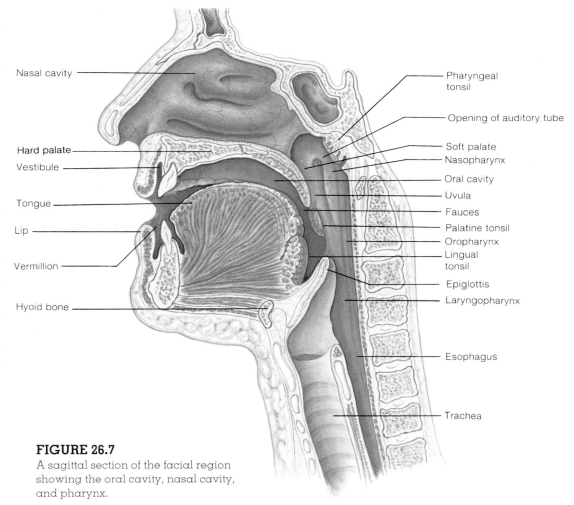

**FIGURE 26.7**
A sagittal section of the facial region showing the oral cavity, nasal cavity, and pharynx.

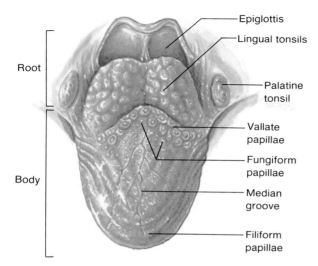

**FIGURE 26.8**
The surface of the tongue.

Labels in figure: Root, Body, Epiglottis, Lingual tonsils, Palatine tonsil, Vallate papillae, Fungiform papillae, Median groove, Filiform papillae

and the mucous membrane of the oral cavity is a transition zone called the **vermilion,** which is reddish-brown in color because of blood vessels close to the surface.

## Tongue

As a digestive organ, the tongue functions to move food around in the mouth during mastication and to assist in swallowing food. It contains **taste buds** (chapter 18) through which various food tastes are sensed, and it is also essential in producing speech. The tongue is a mass of skeletal muscle covered with a mucous membrane. Extrinsic tongue muscles (those that insert upon the tongue) move the tongue from side to side and in and out. Only the anterior two-thirds of the tongue lies in the oral cavity; the remaining one-third lies in the pharynx (fig. 26.7) and is attached to the hyoid bone. Rounded masses of **lingual tonsils** are located on the posterior surface of the tongue (fig. 26.8). The undersurface of the tongue is connected along the midline anteriorly to the floor of the mouth by the vertically positioned **lingual frenulum** (see fig. 26.6).

 When a short lingual frenulum restricts tongue movements, the person is said to be *tongue-tied*. If this developmental problem is severe, an infant may have difficulty suckling. Older children with this problem may have faulty speech. These functional problems can be easily corrected through surgery.

On the surface of the tongue are numerous small elevations called **papillae.** The papillae give the tongue a distinct roughened surface that aids the handling of food. They also contain taste buds that can distinguish sweet, salty, sour, and bitter sensations. The three types of papillae on the

surface of the tongue are **filiform, fungiform,** and **vallate** (fig. 26.8). Filiform papillae, by far the most numerous, have tapered tips and are sensitive to touch. The larger and rounded fungiform papillae are scattered among the filiform type. The few vallate papillae are arranged in a V-shape on the posterior surface of the tongue.

## Palate

The **palate** is the roof of the oral cavity and consists of the bony hard palate anteriorly and the soft palate posteriorly (see figs. 26.6 and 26.7). The hard palate is formed by the palatine processes of the maxillae and the horizontal plates of the palatine bones and is covered with a mucous membrane. Transverse ridges called **palatal rugae** (*roo´je*) are located along the mucous membrane of the hard palate. These structures serve as friction ridges against which the tongue is placed during swallowing. The soft palate is a muscular arch covered with mucous membrane and is continuous anteriorly with the hard palate. Suspended from the middle lower border of the soft palate is a cone-shaped projection called the **uvula.** During swallowing, the soft palate and uvula are drawn upward, closing the nasopharynx and preventing food and fluid from entering the nasal cavity.

Two muscular folds extend downward from both lateral sides of the base of the uvula (see fig. 26.6). The anterior fold is called the **glossopalatine arch** and the posterior fold is the **pharyngopalatine** (*fă-ring´´go-pal´ă-tīn*) **arch.** Between these two arches, toward the posterior lateral portion of the oral cavity, is the **palatine tonsil.**

## Teeth

The **teeth** of humans and other mammals vary in structure and are adapted to handle food in different ways (fig. 26.9). The four pairs (upper and lower jaws) of anteriormost teeth are the **incisors** (*in-si´sorz*). The chisel-shaped incisor teeth are adapted for cutting and shearing food. The two pairs of cone-shaped **canines,** or **cuspids,** are located at the anterior corners of the mouth and are adapted for holding and tearing. Incisor and canine teeth are further characterized by a single root on each tooth. Located behind the canines are the **premolars,** or **bicuspids,** and **molars.** These teeth have two or three roots, and their somewhat rounded, irregular surfaces, or **cusps,** are adapted for crushing and grinding food.

Two sets of teeth develop in a person's lifetime. Twenty **deciduous (milk) teeth** begin to erupt at about 6 months of age (fig. 26.10), beginning with the incisors. All of the

tonsil: L. *toles*, swelling
papilla: L. *papula*, swelling or pimple

fungiform: L. *fungus*, fungus; *forma*, form
incisor: L. *incidere*, to cut
canine: L. *canis*, dog
molar: L. *mola*, millstone
deciduous: L. *deciduus*, to fall away

Digestive System

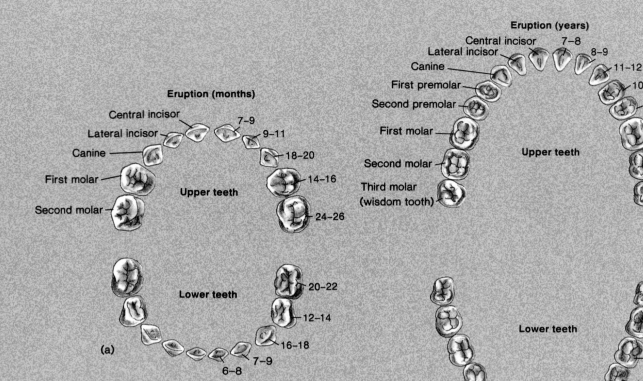

**Eruption (months)**

Central incisor
Lateral incisor
Canine
First molar
Second molar

**Upper teeth**

7–9
9–11
18–20
14–16
24–26

**Lower teeth**

20–22
12–14
16–18
7–9
6–8

(a)

**Eruption (years)**

Central incisor
Lateral incisor
Canine
First premolar
Second premolar
First molar
Second molar
Third molar (wisdom tooth)

**Upper teeth**

7–8
8–9
11–12
10–11
10–12
6–7
12–13
17–25

**Lower teeth**

17–25
11–13
6–7
11–12
10–12
9–10
7–8
6–7

(b)

**FIGURE 26.9**

Dentitions and the sequence of eruptions. (a) Deciduous teeth and (b) permanent teeth.

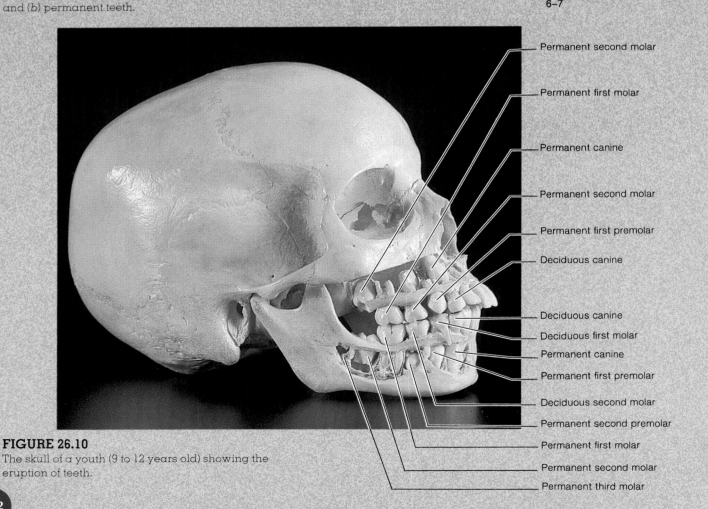

Permanent second molar
Permanent first molar
Permanent canine
Permanent second molar
Permanent first premolar
Deciduous canine
Deciduous canine
Deciduous first molar
Permanent canine
Permanent first premolar
Deciduous second molar
Permanent second premolar
Permanent first molar
Permanent second molar
Permanent third molar

**FIGURE 26.10**

The skull of a youth (9 to 12 years old) showing the eruption of teeth.

deciduous teeth have erupted by the age of 2½. Thirty-two **permanent teeth** replace the deciduous teeth in a predictable sequence. This process begins at about age 6 and continues until about age 17. The **third molars, or wisdom teeth,** are the last to erupt. Eruption of the wisdom teeth is less predictable. If they do erupt, it is between the ages of 17 and 25. Because the jaws are formed by this time and other teeth are in place, the eruption of wisdom teeth may cause serious problems of crowding or impaction.

A **dental formula** is a graphic representation of the types, number, and position of teeth in the oral cavity. Following are the deciduous and permanent dental formulae for humans:

**Formula for deciduous dentition:**
I 2/2, C 1/1, DM 2/2 = 10 × 2 = 20 teeth

**Formula for permanent dentition:**
I 2/2, C 1/1, P 2/2, M 3/3 = 16 × 2 = 32 teeth
(I = incisor; C = canine; P = premolar;
DM = deciduous molar; M = molar)

The cusps of the upper and lower premolar and molar teeth occlude for chewing food (*mastication*), whereas the upper incisors normally form an overbite with the incisors of the lower jaw. An overbite of the upper incisors creates a shearing action as these teeth slide past one another. Masticated food is mixed with saliva, which initiates chemical digestion and facilitates swallowing. The soft mass of chewed food that is swallowed is called a *bolus*.

A tooth consists of an exposed **crown,** supported by a **neck,** anchored firmly into the jaw by one or more **roots** (fig. 26.11). The roots of teeth fit into sockets, called **alveoli,** in the alveolar processes of the mandible and maxillae. Each socket is lined with a connective tissue periosteum, specifically called the **periodontal membrane.** The root of a tooth is covered with a bonelike material called the **cementum;** fibers in the periodontal membrane insert into the cementum and fasten the tooth in its socket. The **gingiva** (*jin ′jĭ-vă*), or **gum,** is the mucous membrane surrounding the alveolar processes in the oral cavity.

The bulk of a tooth consists of **dentin** (*den ′tin*)—a substance similar to bone, but harder. Covering the dentin on the outside and forming the crown is a tough, durable layer of **enamel.** Enamel is composed primarily of calcium phosphate and is the hardest substance in the body. The central region of the tooth contains the **pulp cavity.** The pulp cavity contains the **pulp,** which is composed of connective tissue with blood vessels, lymph vessels, and nerves. A **root canal** is continuous with the pulp cavity and opens to the connective tissue surrounding

..........................................
bolus: Gk. *bolos*, lump
dentin: L. *dens*, tooth

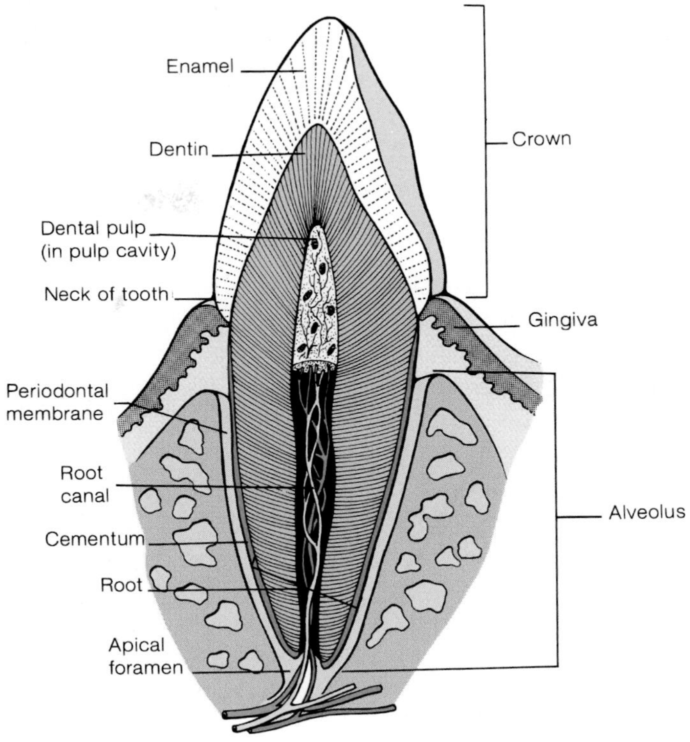

**FIGURE 26.11**
The structure of a tooth shown in a vertical section through a canine tooth.

the root through an **apical foramen** at the tip of the root. The apical foramen permits blood vessels and nerves to enter the pulp.

Although enamel is the hardest substance in the body, it can be weakened due to acidic conditions produced by bacterial activity, resulting in *dental caries (cavities).* These caries must be artificially filled because new enamel is not produced after a tooth erupts. The rate of tooth decay decreases after age 35, but then periodontal diseases may develop. *Periodontal diseases* result from plaque or tartar buildup at the gum line. This buildup wedges the gum away from the teeth, allowing bacterial infections to develop.

## Salivary Glands

**Salivary glands** are accessory digestive glands that produce a fluid secretion called *saliva.* Saliva functions as a solvent in cleansing the teeth and dissolving food molecules so that they can be tasted. Saliva also contains starch-digesting enzymes and mucus that lubricates the pharynx to facilitate swallowing. Numerous minor salivary glands, called **buccal glands,** are located in the mucous membranes of the palatal region of the oral cavity. Most of the saliva, however, is produced by three pairs of salivary glands outside of the oral cavity and is transported to the mouth via salivary ducts.

773

The three major pairs of salivary glands are the parotid, submandibular, and sublingual (fig. 26.12).

The **parotid** (*pă-rot´id*) **gland** is the largest of the salivary glands and is positioned below and in front of the ear, between the skin and the masseter muscle. The **parotid** (Stensen's) **duct** parallels the zygomatic arch across the masseter muscle, pierces the buccinator muscle, and drains into the oral cavity opposite the second upper molar. It is the parotid gland that becomes infected and swollen with the mumps.

The **submandibular gland** is inferior to the body of the mandible about midway along the inner side of the jaw. This gland is covered by the more superficial mylohyoid muscle. The **submandibular** (Wharton's) **duct** empties into the floor of the mouth on either side of the lingual frenulum.

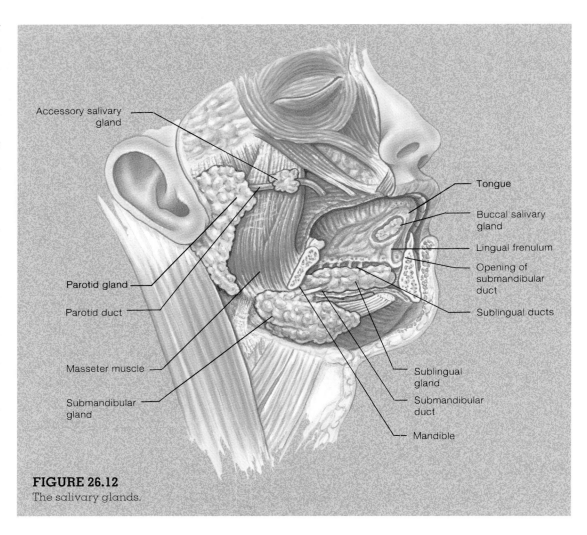

**FIGURE 26.12**
The salivary glands.

The **sublingual gland** lies under the mucosa in the floor of the mouth on the side of the tongue. Associated with each sublingual gland are several small **sublingual** (Rivinus's) **ducts** that empty into the floor of the mouth in an area posterior to the papilla of the submandibular duct.

Two types of secretory cells, *serous* and *mucous cells,* are found in all salivary glands in various proportions. Serous cells produce a watery fluid containing digestive enzymes; mucous cells secrete a thick, stringy mucus. Cuboidal epithelial cells line the lumina of the salivary ducts.

The salivary glands are innervated by both divisions of the autonomic nervous system. Sympathetic impulses stimulate the secretion of small amounts of viscous saliva.

parotid: Gk. *para*, beside; *otos*, ear
Stensen's duct: from Nicholaus Stensen, Danish anatomist, 1638–86
Wharton's duct: from Thomas Wharton, English anatomist, 1614–73
Rivinus's ducts: from Augustus Rivinus, German anatomist, 1652–1723

Parasympathetic stimulation causes the secretion of large volumes of watery saliva. Physiological responses of this type occur whenever a person sees, smells, tastes, or even thinks about desirable food. The amount of saliva secreted daily ranges from 1000 to 1500 ml. Information about the salivary glands is summarized in table 26.1.

## Pharynx

The funnel-shaped **pharynx** (*far´ingks*) is a passageway approximately 13 cm (5 in.) long connecting the oral and nasal cavities to the esophagus and trachea. The external, circular layer of pharyngeal muscles, called *constrictor muscles* (fig. 26.13), serves to compress the lumen of the pharynx involuntarily during swallowing. The **superior constrictor muscle** attaches to bony processes of the skull and mandible and encircles the upper portion of the pharynx. The **middle constrictor muscle** arises from the hyoid bone and stylohyoid ligament and encircles the middle portion of the pharynx. The **inferior constrictor muscle** arises from the cartilages

## Table 26.1  Major salivary glands

| Gland | Location | Entry into oral cavity | Type of secretion |
|---|---|---|---|
| Parotid | Anterior and inferior to auricle; subcutaneous over masseter muscle | Lateral to upper second molar | Watery serous fluid, salts, and enzyme |
| Submandibular | Inferior to the base of the tongue | Papilla lateral to lingual frenulum | Watery serous fluid with some mucus |
| Sublingual | Anterior to submandibular gland; under tongue | Ducts along the base of the tongue | Mostly thick, stringy mucus, salts, and enzyme |

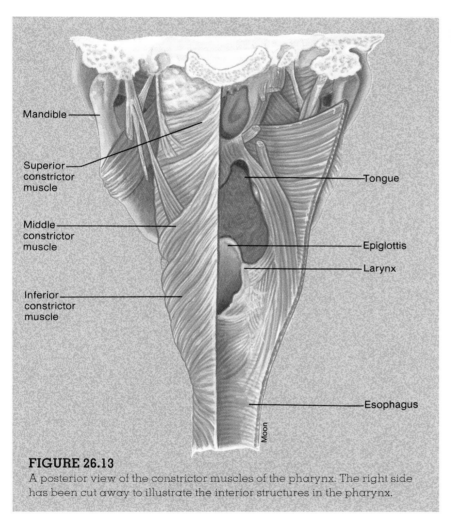

Mandible

Superior constrictor muscle

Middle constrictor muscle

Inferior constrictor muscle

Tongue

Epiglottis

Larynx

Esophagus

Moon

**FIGURE 26.13**

A posterior view of the constrictor muscles of the pharynx. The right side has been cut away to illustrate the interior structures in the pharynx.

# Esophagus and Stomach

*Swallowed food is passed through the esophagus to the stomach by wavelike contractions known as peristalsis. The mucosa of the stomach secretes hydrochloric acid and pepsinogen. Upon entering the lumen of the stomach, pepsinogen is activated to the protein-digesting enzyme known as pepsin. The stomach partially digests proteins and functions to store its contents, called chyme, for later processing by the small intestine.*

## Esophagus

The **esophagus** (ĕ-sof´ă-gus) is that portion of the GI tract that connects the pharynx to the stomach. It is a collapsible muscular tube approximately 25 cm (10 in.) long, located posterior to the trachea within the mediastinum of the thorax. The esophagus passes through the diaphragm by means of an opening called the **esophageal hiatus** (ĕ-sof˝ă-je´al hi-a´tus) before terminating at the stomach. The esophagus is lined with a nonkeratinized stratified squamous epithelium; its walls contain either skeletal or smooth muscle, depending on the location. The upper third of the esophagus contains skeletal muscle; the middle third contains both skeletal and smooth muscle, and the terminal portion contains only smooth muscle.

Swallowing, or **deglutition** (de˝gloo-tish´un), is the complex act of moving food or fluid from the oral cavity through the esophagus to the stomach. Swallowed food is pushed from one end of the esophagus to the other

of the larynx and encircles the lower portion of the pharynx. During breathing, the lower portion of the inferior constrictor muscle is contracted, preventing air from entering the esophagus.

The three regions of the pharynx are described in detail in chapter 23 on the respiratory system.

esophagus: Gk. *oisein*, to carry; *phagema*, food

by a wavelike muscular contraction called **peristalsis** (fig. 26.14). Peristaltic waves are produced by constriction of the lumen as a result of circular muscle contraction, followed by shortening of the tube by longitudinal muscle contraction. These contractions progress from the superior end of the esophagus to the *gastroesophageal junction* at a rate of about 2 to 4 cm per second as they empty the contents of the esophagus into the cardiac region of the stomach.

The lumen of the terminal portion of the esophagus is slightly narrowed because of a thickening of the circular muscle fibers in its wall. This portion is referred to as the **lower esophageal (gastroesophageal) sphincter.** The muscle fibers of this region constrict after food passes into the stomach to help prevent the stomach contents from regurgitating into the esophagus. Regurgitation would occur because the pressure in the abdominal cavity is greater than the pressure in the thoracic cavity as a result of respiratory movements. The lower esophageal sphincter must remain closed, therefore, until food is pushed through it by peristalsis into the stomach.

The lower esophageal sphincter is not a true sphincter muscle that can be identified histologically, and at times it does permit the acidic contents of the stomach to enter the esophagus. This can create a burning sensation commonly called *heartburn*, although the heart is not involved. In infants under a year of age, the lower esophageal sphincter may function erratically, causing them to "spit up" following meals. Some mammals, such as rodents, have a true lower esophageal sphincter and cannot regurgitate, which is why poison grains are effective in killing mice and rats.

## Stomach

The **stomach,** a J-shaped pouch in the left superior portion of the abdomen, is the most distensible part of the GI tract. It is continuous with the esophagus superiorly and empties into the duodenum of the small intestine inferiorly. The functions of the stomach are to store food, to initiate the digestion of proteins, and to move the food into the small intestine as a pasty material called **chyme** (kīm).

The stomach (fig. 26.15) can be divided into four regions: the cardia, fundus, body, and pylorus. The **cardia** is the narrow upper region immediately below the lower

chyme: L. *chymus*, juice
fundus: L. *fundus*, bottom

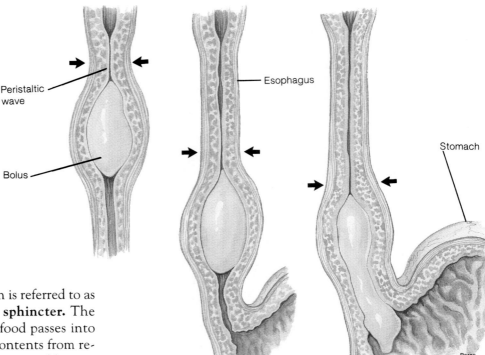

(a)

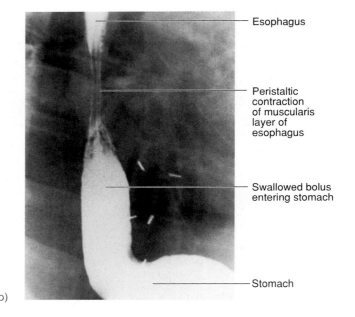

(b)

### FIGURE 26.14

(a) A diagram and (b) a radiograph of peristalsis in the esophagus.

esophageal sphincter. The **fundus** is the dome-shaped portion to the left of the cardia and is in direct contact with the diaphragm. The **body** is the large central portion, and the **pylorus** is the funnel-shaped terminal portion. The pylorus communicates with the duodenum of the small

pylorus: Gk. *pyloros*, gatekeeper

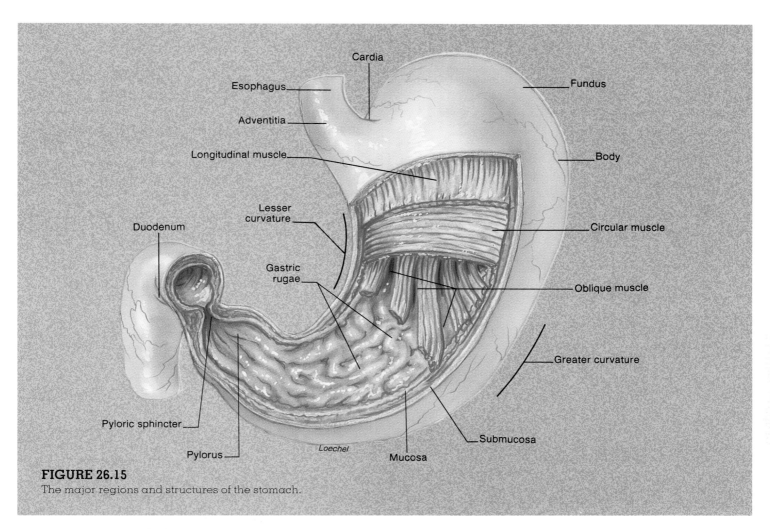

**FIGURE 26.15**
The major regions and structures of the stomach.

intestine through the **pyloric sphincter.** *Pylorus* is a Greek word meaning "gatekeeper," and this junction is just that—regulating the movement of chyme into the small intestine and prohibiting backflow.

The stomach has two surfaces and two borders. The broadly rounded surfaces are referred to as the **anterior** and **posterior surfaces.** The medial concave border is the **lesser curvature** (fig. 26.15), and the lateral convex border is the **greater curvature.** The lesser omentum extends between the lesser curvature and the liver; the greater omentum is attached to the greater curvature.

The wall of the stomach consists of the same four layers found in other regions of the GI tract, with certain modifications. The muscularis is composed of three layers of smooth muscle named according to the direction of fiber arrangement: an outer **longitudinal layer,** a middle **circular layer,** and an inner **oblique layer.** The circular muscle layer is further thickened at the gastroduodenal junction to form the pyloric sphincter.

The inner surface of the stomach is thrown into long folds called **gastric rugae,** which can be seen with the unaided eye. Microscopic examination of the gastric mucosa shows that it is likewise folded. The openings of these folds into the stomach lumen are called **gastric pits.** The cells that line the folds deeper in the mucosa secrete various products into the stomach; these form the exocrine **gastric glands** (figs. 26.16 and 26.17).

Gastric glands have several types of secreting cells: (1) **goblet cells,** which secrete mucus; (2) **parietal cells,** which secrete hydrochloric acid (HCl); (3) **chief (or zymogenic) cells,** which secrete pepsinogen, an inactive form of the protein-digesting enzyme pepsin; (4) **argentaffin** (*ar-jen´tă-fin*) **cells,** which secrete serotonin and histamine as autocrine regulators (chapter 19); and (5) **G cells,**

········································

argentaffin cells: L. *argentum,* silver; *affinis,* attraction (became covered with silver stain)

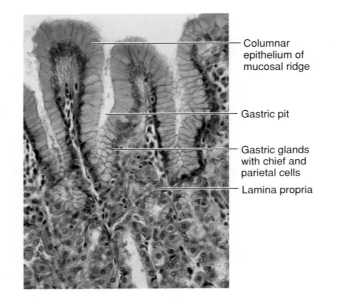

Columnar epithelium of mucosal ridge

Gastric pit

Gastric glands with chief and parietal cells

Lamina propria

**FIGURE 26.16**

Microscopic structures of the mucosa of the stomach.

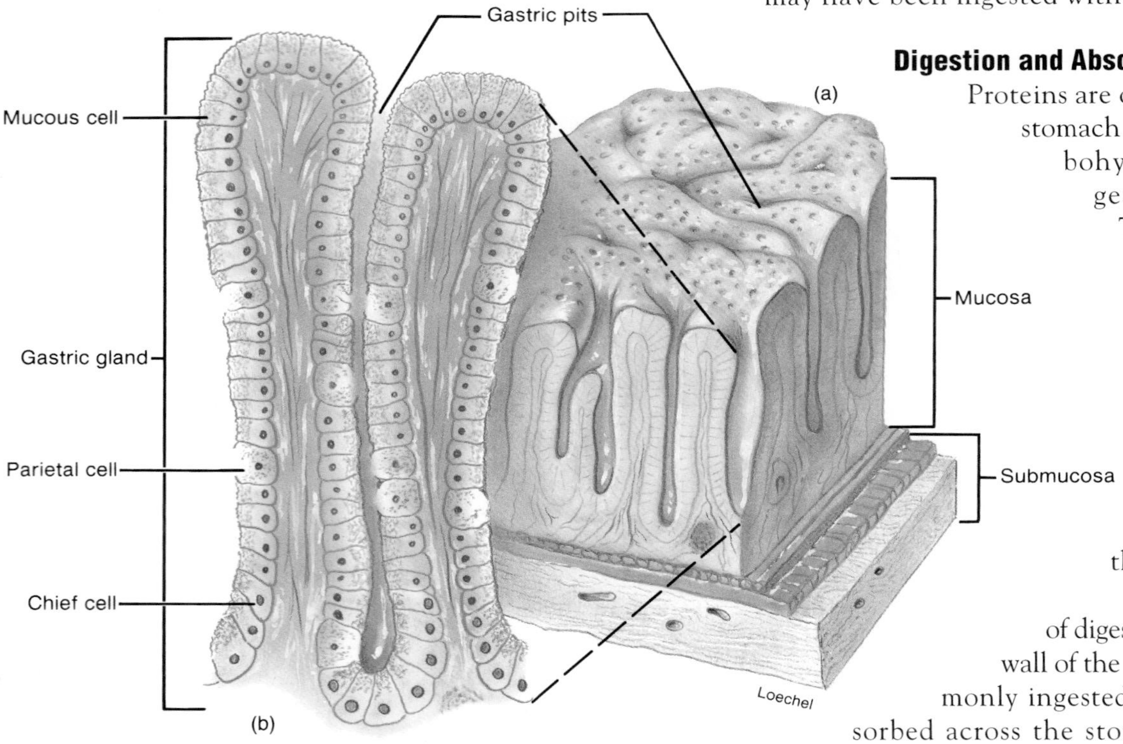

Gastric pits

Mucous cell

Gastric gland

Parietal cell

Chief cell

(b)

(a)

Mucosa

Submucosa

Loechel

**FIGURE 26.17**

Gastric pits and gastric glands of the mucosa. (a) Gastric pits are the openings of the gastric glands. (b) Gastric glands consist of mucous cells, chief cells, and parietal cells, each of which produces a specific secretion.

which secrete the hormone gastrin into the blood. In addition to these products, the gastric mucosa (probably the parietal cells) secretes a polypeptide called *intrinsic factor*, which is required for absorption of vitamin $B_{12}$ in the small intestine.

**Pepsin and Hydrochloric Acid** The secretion of hydrochloric acid by parietal cells makes gastric juice highly acidic, with a pH of less than 2. This acidity serves three digestive functions: (1) ingested proteins are denatured at a low pH—that is, their tertiary structure is altered so that they become more digestible; (2) under acidic conditions, weak pepsinogen enzymes partially digest each other— this frees the active pepsin enzyme as small peptide fragments are removed (fig. 26.18); and (3) pepsin is more active under acidic conditions—it has a pH optimum of about 2.0. The peptide bonds of ingested protein are broken (through hydrolysis reactions) by pepsin under acidic conditions; the HCl itself does not directly digest proteins. The low pH of gastric juice also helps to kill bacteria that may have been ingested with food.

### Digestion and Absorption in the Stomach

Proteins are only partially digested in the stomach by the action of pepsin. Carbohydrates and fats are not digested at all in the stomach. The complete digestion of food molecules occurs later, when chyme enters the small intestine. Patients with partial gastric resections, therefore, and even those with complete gastrectomies (removal of the stomach), can still adequately digest and absorb their food.

Almost all of the products of digestion are absorbed through the wall of the small intestine; the only commonly ingested substances that can be absorbed across the stomach wall are alcohol and aspirin. This occurs as a result of the lipid solubility of these molecules. The passage of aspirin through the gastric mucosa has been shown to cause bleeding, which may be significant if large amounts of aspirin are taken.

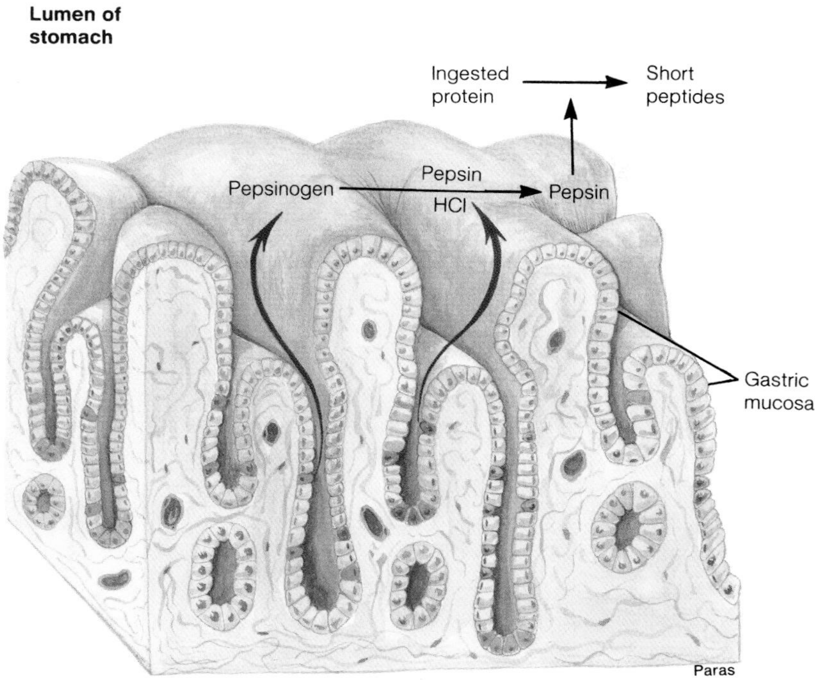

**Lumen of stomach**

Ingested protein → Short peptides

Pepsinogen → [Pepsin / HCl] → Pepsin

Pepsin digests → Short peptides

Gastric mucosa

Paras

**FIGURE 26.18**

The gastric mucosa secretes the inactive enzyme pepsinogen and hydrochloric acid (HCl). In the presence of HCl, the active enzyme pepsin is produced. Pepsin digests proteins into shorter polypeptides.

The only function of the stomach that appears to be essential for life is the secretion of intrinsic factor. This polypeptide is needed for the absorption of vitamin $B_{12}$ in the terminal portion of the ileum in the small intestine, and vitamin $B_{12}$ is required for maturation of red blood cells in the bone marrow. A patient with a gastrectomy has to receive $B_{12}$ orally (together with intrinsic factor) or through injections to prevent the development of *pernicious anemia*.

**Gastritis and Peptic Ulcers**   Peptic ulcers are erosions of the mucous membranes of the stomach or duodenum prodced by the action of HCl. In *Zollinger–Ellison syndrome*, ulcers of the duodenum are produced by excessive gastric acid secretion in response to very high levels of the hormone gastrin, which, in this case, are generally produced by a pancreatic tumor. This is a rare condition, but it does illustrate that excessive gastric acid can cause duodenal ulcers. Ulcers of the stomach, however, are not believed to be due to excessive acid secretion, but rather to mechanisms that reduce the barriers of the gastric mucosa to self-digestion.

Recent experiments demonstrate that the parietal and chief cells of the gastric mucosa are extremely impermeable to the acid in the lumen of the stomach. In fact, they are even impermeable to $CO_2$ and $NH_3$, which are molecules

..........................................

Zollinger–Ellison syndrome: from Robert M. Zollinger, American surgeon, b. 1903, and Edwin H. Ellison, American physician, 1918–70

Brunner's glands: from Johann C. Brunner, Swiss anatomist, 1653–1727

that can freely pass through most other cell membranes. This impermeability of the cell membrane of the gastric epithelium is but one of the mechanisms that work to protect the gastric mucosa. Other protective mechanisms include a layer of alkaline mucus, containing bicarbonate, covering the gastric mucosa; tight junctions between adjacent epithelial cells, preventing acid from leaking into the submucosa; a rapid rate of cell division, allowing damaged cells to be replaced (the entire epithelium is replaced every two to three days); and several protective effects provided by pros-taglandins that are produced by the gastric mucosa. Indeed, a common cause of gastric ulcers is believed to be the use of *non-steroidal anti-inflammatory drugs*. This class of drugs includes aspirin and ibuprofen, which act to inhibit the production of prostaglandins (see chapter 19).

When the gastric barriers to self-digestion are broken down, acid can leak through the mucosa to the submucosa, which causes direct damage and stimulates inflammation. The histamine released from mast cells (chapter 23) during inflammation may stimulate further acid secretion and result in further damage to the mucosa. The inflammation that occurs during these events is called **acute gastritis.**

The duodenum is normally protected from gastric acid by the buffering action of bicarbonate in alkaline pancreatic juice, as well as by secretion of bicarbonate by duodenal (Brunner's) glands in the duodenum. However, people who develop duodenal ulcers produce excessive amounts of gastric acid that are not neutralized by the bicarbonate. People with gastritis and peptic ulcers must avoid substances that stimulate acid secretion, including coffee and wine, and often must take antacids.

It has been known for some time that most poeple who have peptic ulcers are infected with a type of bacterium known as *Helicobacter pylori*, which resides in the GI tract. Also, clinical trials have demonsrated that antibiotics that eliminate this infection appear to help in the treatment of the peptic ulcers. Many people, however, have *H. pylori* infections without having ulcers, so perhaps the infection does not itself cause the ulcer, but only contributes to the weakening of the mucosal barriers to gastric acid damage. Furher research is needed to gain a more complete  understanding of the causes of peptic ulcers.

The secretion of gastric acid is stimulated by acetylcholine (from parasympathetic nerve endings), gastrin (a hormone secreted by the stomach), and histamine (secreted by mast cells in the connective tissue). In response to these stimulants, the parietal cells secrete $H^+$ into the lumen of the stomach by active transport against

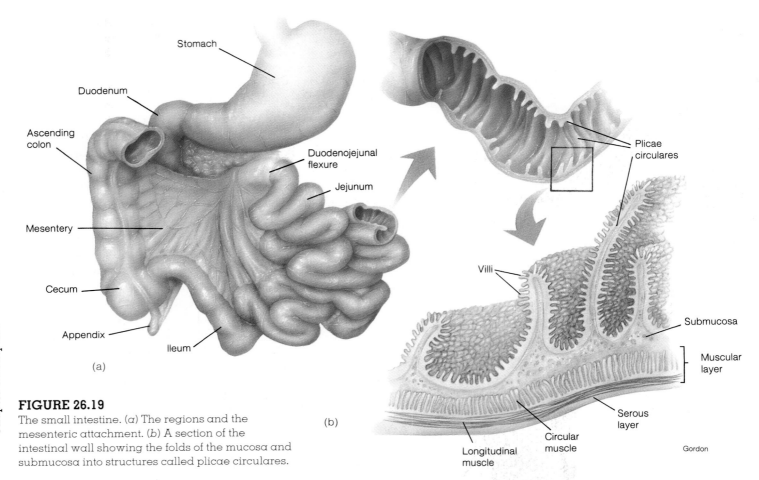

**FIGURE 26.19**

The small intestine. (*a*) The regions and the mesenteric attachment. (*b*) A section of the intestinal wall showing the folds of the mucosa and submucosa into structures called plicae circulares.

---

a million-to-one concentration gradient, using a carrier that functions as a H⁺/K⁺ ATPase pump. People with ulcers, consequently, can be treated with drugs that block ACh, gastrin, or histamine action. Drugs in the last category (for example, Tagamet) are most often used and specifically block the H₂ histamine receptors in the gastric mucosa. This receptor subtype is different from that blocked by antihistamines commonly used to treat cold and allergy symptoms.

# Small Intestine

*The mucosa of the small intestine is folded into villi that project into the lumen. In addition, the cells that line these villi have foldings of their cell membrane called microvilli. This arrangement greatly increases the surface area for absorption. It also improves digestion, since the digestive enzymes of the small intestine are embedded within the cell membrane of the microvilli.*

The **small intestine** is that portion of the GI tract between the pyloric sphincter of the stomach and the ileocecal valve opening into the large intestine. It is positioned in the central and lower portions of the abdominal cavity and is sup-ported, except for the first portion, by mesentery (fig. 26.19). The fan-shaped attachment of mesentery to the small intestine maintains mobility but leaves little chance for the intestine to become twisted or kinked. Enclosed within the mesentery are blood vessels, nerves, and lymphatic vessels that supply the intestinal wall.

The small intestine is approximately 3 m (12 ft) long and 2.5 cm (1 in.) wide in a living person, but it will measure nearly twice this length in a cadaver when the muscle wall is relaxed. It is called the "small" intestine because of its relatively small diameter compared to that of the large intestine. The small intestine serves as the major site of digestion and absorption in the GI tract.

The small intestine is innervated by the **superior mesenteric plexus.** The branches of the plexus contain sensory fibers, postganglionic sympathetic fibers, and preganglionic parasympathetic fibers. The small intestine's arterial blood supply comes through the superior mesenteric artery and small branches from the celiac trunk and the inferior mesenteric artery. Venous drainage is through the superior mesenteric vein, which unites with the splenic vein to form the hepatic portal vein carrying nutrient-rich blood to the liver.

## Regions of the Small Intestine

The small intestine is divided into three regions on the basis of function and histological structure. These three regions are the duodenum, jejunum, and ileum.

The **duodenum** (*doo˝-ŏ-de´num* or *doo-od´ĕ-num*) is a relatively fixed, C-shaped tube measuring approximately 25 cm (10 in.) from the pyloric sphincter of the stomach to the **duodenojejunal** (*doo-od˝-ĕ-no˝jĕ-joo´nal*) **flexure.** The concave surface of the duodenum faces to the left, where it receives bile secretions through the **common bile duct** from the liver and gallbladder and pancreatic secretions through the **pancreatic duct** of the pancreas (fig. 26.20). These two ducts unite to form a common entry into the duodenum called the **hepato-pancreatic ampulla** (*hep˝ă-to-pan˝´kre-at´ik am-pool´ă*) (ampulla of Vater), which pierces the duodenal wall and drains into the duodenum from an elevation called the **duodenal papilla.** It is here that bile and pancreatic juice (described later) enter the small intestine. The papilla can be opened or closed by the action of the **sphincter of papilla** (sphincter of Oddi). The duodenum is retroperitoneal, except for a short portion near the stomach. It differs histologically from the rest of the small intestine by the presence of **duodenal** (Brunner's) **glands** in the submucosa (fig. 26.21). These compound tubuloalveolar glands secrete mucus and are most numerous near the superior end of the duodenum.

The **jejunum** (*jĕ-joo´num*) is approximately 1 m (3 ft) long and extends from the duodenum to the ileum. The lumen of the jejunum is slightly larger than that of the ileum, and the jejunum has more internal folds. The two regions are similar in their histological structure.

The **ileum** (*il´e-um*) makes up the remaining 2 m (6–7 ft) of the small intestine. The terminal portion of the ileum empties into the medial side of the cecum through the **ileocecal** (*il˝´e-ŏ-se´kal*) **valve.** The walls of the ileum have an abundance of lymphatic tissue. Aggregates of lymph nodules called **mesenteric** (Peyer's) **patches** are characteristic of the ileum.

..............................................

duodenum: L. *duodeni*, 12 each (distance of 12 fingers breadth)
ampulla of Vater: from Abraham Vater, German anatomist, 1684–1751
sphincter of Oddi: from Ruggero Oddi, Italian physician, nineteenth century
jejunum: L. *jejunus*, fasting (in dissection it was always found empty)
ilium: Gk. *eileo*, to roll up
Peyer's patches: from Johann K. Peyer, Swiss anatomist, 1653–1712

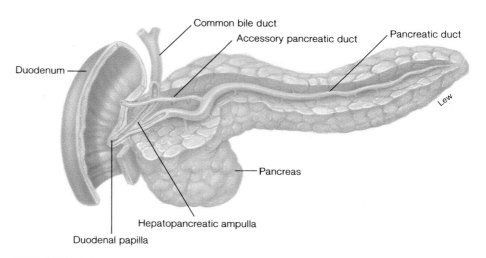

**FIGURE 26.20**
The duodenum and associated structures.

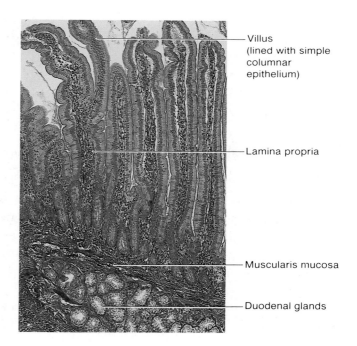

**FIGURE 26.21**
The microscopic structure of the duodenum.

## Structural Modifications for Absorption

The products of digestion are absorbed across the epithelial lining of the intestinal mucosa. Absorption of carbohydrates, lipids, protein, calcium, and iron occurs primarily in the duodenum and jejunum. Bile salts, vitamin B$_{12}$, water, and electrolytes are absorbed primarily in the ileum. The many folds in the small intestine and modifications in the structure of its walls provide a large mucosal surface area; therefore, absorption occurs at a rapid rate. The mucosa and submucosa

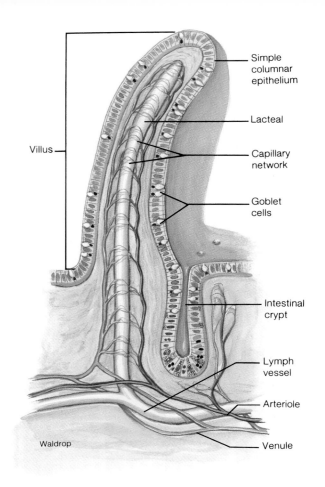

**FIGURE 26.22**
A diagram of the structure of an intestinal villus.

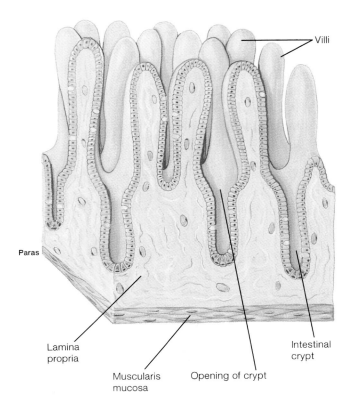

**FIGURE 26.23**
Intestinal villi and intestinal crypts.

form large folds called the **plicae** (*pli´se*) **circulares,** which can be observed with the unaided eye. The surface area is further increased by the microscopic folds of mucosa called **villi** and by the foldings of the apical cell membrane of epithelial cells called **microvilli.**

Each **villus** is a fingerlike fold of mucosa that projects into the lumen of the small intestine (fig. 26.21). The villi are covered with columnar epithelial cells, and interspersed among these are the mucus-secreting goblet cells. The lamina propria forms the connective core of each villus and contains numerous lymphocytes, blood capillaries, and a lymphatic vessel called the **lacteal** (fig. 26.22). Absorbed monosaccharides and amino acids enter the blood capillaries; absorbed fat enters the lacteals.

Epithelial cells at the tips of the villi are continuously exfoliated and replaced by cells that are pushed up from the bases of the villi. The epithelium at the base of each villus invaginates downward at various points to form narrow

pouches that open through pores to the intestinal lumen. These structures are called the **intestinal crypts** (crypts of Lieberkühn) (figs. 26.22 and 26.23). Microvilli, which can be seen clearly only in an electron microscope, are fingerlike projections formed by foldings of the cell membrane. In a light microscope, the microvilli display a somewhat vague **brush border** on the edges of the columnar epithelium. The terms *brush border* and *microvilli* are often used interchangeably in describing the small intestine (fig. 26.24).

## Intestinal Enzymes

In addition to providing a large surface area for absorption, the cell membranes of the microvilli contain digestive enzymes. These enzymes are *not* secreted into the lumen but instead remain attached to the cell membrane with their active sites exposed to the chyme. These **brush border enzymes** hydrolyze disaccharides, polypeptides, and other substrates (table 26.2). One brush border enzyme, *enterokinase* (en´´tĕ-ro-ki´nās), is required for activation of the protein-digesting enzyme *trypsin*, which enters the small intestine in pancreatic juice.

plica: L. *plicatus,* folded
villus: L. *villosus,* shaggy

crypts of Lieberkühn: from Johann N. Lieberkühn, German anatomist, 1711–56

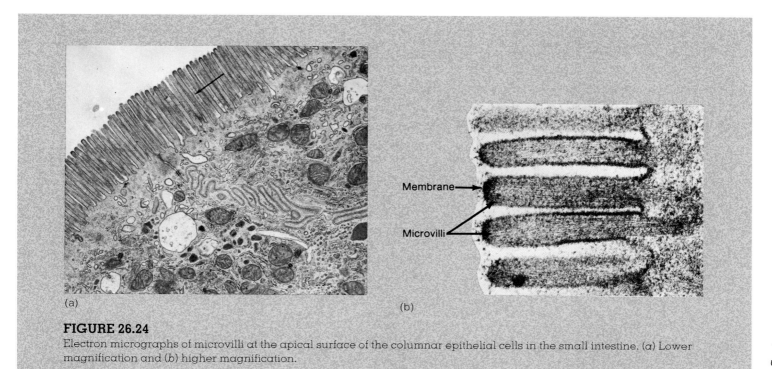

**FIGURE 26.24**
Electron micrographs of microvilli at the apical surface of the columnar epithelial cells in the small intestine. (a) Lower magnification and (b) higher magnification.

| | | | |
|---|---|---|---|
| **Category** | **Enzyme** | **Comments** | |

## Table 26.2 — Brush border enzymes attached to the cell membrane of microvilli in the small intestine

| Category | Enzyme | Comments |
|---|---|---|
| Disaccharidase | Sucrase | Digests sucrose to glucose and fructose; deficiency produces GI disturbances |
| | Maltase | Digests maltose to glucose |
| | Lactase | Digests lactose to glucose and galactose; deficiency produces GI disturbances (lactose intolerance) |
| Peptidase | Aminopeptidase | Produces free amino acids, dipeptides, and tripeptides |
| | Enterokinase | Activates trypsin (and indirectly other pancreatic juice enzymes); deficiency results in protein malnutrition |
| Phosphatase | $Ca^{++}$, $Mg^{++}$—ATPase | Needed for absorption of dietary calcium; enzyme activity regulated by vitamin D |
| | Alkaline phosphatase | Removes phosphate groups from organic molecules; enzyme activity may be regulated by vitamin D |

 The ability to digest milk sugar, or lactose, depends on the presence of a brush border enzyme called *lactase*. This enzyme is present in all children under the age of 4 but becomes inactive to some degree in most adults (with the exception of Scandinavians and some others). This can result in *lactose intolerance*. The presence of large amounts of undigested lactose in the intestine causes diarrhea, gas, cramps, and other unpleasant symptoms. Yogurt is better tolerated than milk because yogurt bacteria produce lactase which, after becoming activated in the duodenum, digests lactose.

## Intestinal Contractions and Motility

The small intestine has two major types of contractions: peristalsis and segmentation. Peristalsis is much weaker in the small intestine than in the esophagus and stomach. Intestinal motility—the movement of chyme through the intestine—is relatively slow and is due primarily to the fact that the pressure at the pyloric end of the small intestine is greater than at the distal end.

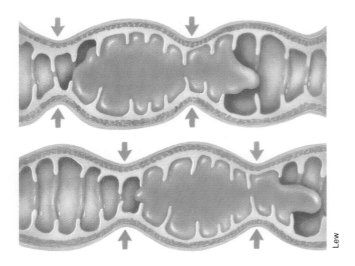

**FIGURE 26.25**

Segmentation of the small intestine. Simultaneous contractions of many segments of the intestine help to mix the chyme with digestive enzymes and mucus.

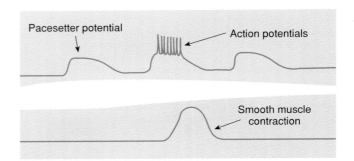

**FIGURE 26.26**

The smooth muscle of the gastrointestinal tract produces and conducts spontaneous pacesetter potentials. As these potential changes reach a threshold level of depolarization, they stimulate the production of action potentials, which in turn stimulate smooth muscle contraction.

The major contractile activity of the small intestine is **segmentation.** This term refers to muscular constrictions of the lumen, which occur simultaneously at different intestinal segments (fig. 26.25). This action serves to mix the chyme more thoroughly.

Like cardiac muscle, intestinal smooth muscle is capable of spontaneous electrical activity and automatic, rhythmic contractions. Spontaneous depolarizations begin in pacemaker smooth muscle cells at the boundary of the circular muscle layer and spread through both the circular and longitudinal muscle layers across *nexuses*. The term *nexus* is used here to indicate an electrical synapse between smooth muscle cells. The spontaneous depolarizations, called **pacesetter potentials,** or *slow waves*, decrease in amplitude as they are conducted from one muscle cell to another, much like excitatory postsynaptic potentials (EPSPs). The pacesetter potentials can stimulate the production of action potentials when they reach a plateau level of depolarization in the smooth muscle cells through which they are conducted (fig. 26.26).

The nexuses conduct the pacesetter potentials, not the action potentials. Action potentials are, therefore, limited to those smooth muscle cells that are depolarized to threshold by the spreading pacesetter potentials. These action potentials therefore stimulate smooth muscle contraction in only limited regions of the small intestine, producing the localized contractions of segementation. Although this

activity is automatic in nature, the excitability of the intestinal smooth muscle cells is increased by acetylcholine released by parasympathetic stimulation and is decreased by impulses through sympathetic nerves. Relaxation of intestinal smooth muscle is promoted by nitric oxide produced within the smooth muscle cells in response to neurotransmitters released by autonomic neurons. The neurotransmitters that cause intestinal relaxation include nitric oxide itself, norepinephrine, and vasoactive intestinal peptide, or VIP.

# Large Intestine

*The large intestine absorbs water and electrolytes from the chyme it receives from the small intestine and, in a process regulated by the action of sphincter muscles, passes undigested waste products out of the GI tract through the rectum and anal canal.*

The **large intestine** is about 1.5 m (5 ft) long and 6.5 cm (2.5 in.) in diameter. It is named the "large" intestine because its diameter is larger than that of the small intestine. The large intestine begins at the terminal end of the ileum in the lower right quadrant of the abdominal cavity. From there, it leads superiorly on the right side to just below the liver, where it crosses to the left and then descends into the pelvis and terminates at the anus. The **mesocolon,** a specialized portion of the mesentery, supports the large intestine along the posterior abdominal wall.

The large intestine has little or no digestive function, but it does absorb water and electrolytes from the remaining chyme. In addition, the large intestine forms, stores, and expels feces from the GI tract.

nexus: L. *nexus*, interconnection

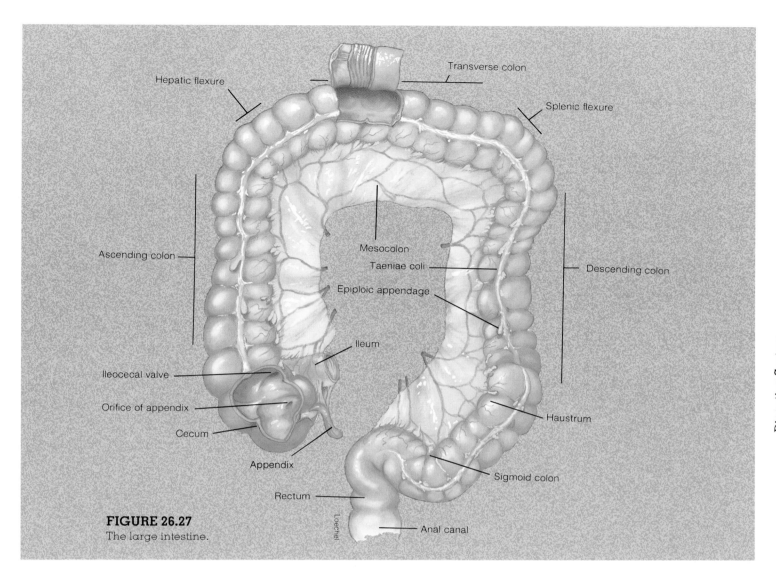

Hepatic flexure

Transverse colon

Splenic flexure

Ascending colon

Mesocolon

Taeniae coli

Descending colon

Epiploic appendage

Ileum

Ileocecal valve

Orifice of appendix

Haustrum

Cecum

Appendix

Sigmoid colon

Rectum

**FIGURE 26.27**
The large intestine.

Loechel

Anal canal

## *Regions and Structures of the Large Intestine*

The large intestine is structurally divided into the cecum, colon, rectum, and anal canal (fig. 26.27).

The **cecum** (*se´kum*) is a dilated pouch that hangs inferiorly, slightly below the ileocecal valve. The ileocecal valve is a fold of mucous membrane at the junction of the small and large intestine that prohibits the backflow of chyme. A fingerlike projection called the **appendix** is attached to the inferior medial margin of the cecum. The 8-cm (3-in.) appendix has an abundance of lymphatic tissue that may serve to resist infection.

The open, superior portion of the cecum is continuous with the colon. The colon consists of ascending, transverse, descending, and sigmoid portions (fig. 26.27). The **ascending**

colon extends superiorly from the cecum along the right abdominal wall to the inferior surface of the liver. Here the colon bends sharply to the left at the **hepatic flexure** and transversely crosses the upper abdominal cavity as the **transverse colon.** At the left abdominal wall, another right-angle bend called the **splenic** (*splen´ik*) **flexure** marks the beginning of the **descending colon.** The descending colon traverses inferiorly along the left abdominal wall to the pelvic region. The colon then angles medially from the brim of the pelvis to form an S-shaped bend known as the **sigmoid colon.**

The terminal 20 cm (7.5 in.) of the GI tract is the **rectum,** and the last 2 to 3 cm of the rectum is referred to as the **anal canal** (fig. 26.28). The **anus** is the external opening of the anal canal. Two sphincter muscles guard the anal opening: the **internal anal sphincter,** composed of smooth muscle fibers, and

cecum: L. *caecum,* blind pouch
appendix: L. *appendix,* attachment

anus: L. *anus,* ring

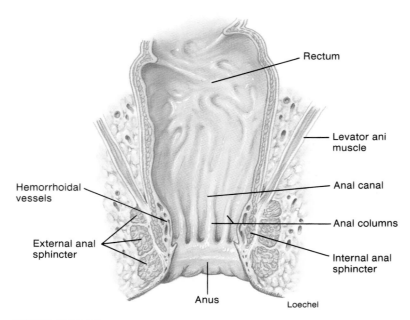

**FIGURE 26.28**
The anal canal.

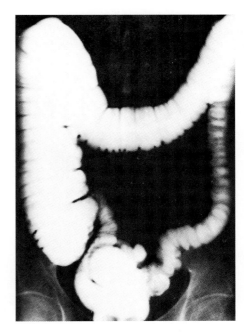

**FIGURE 26.29**
A radiograph after a barium enema showing the haustra of the large intestine.

the **external anal sphincter,** composed of skeletal muscle. The mucous membrane of the anal canal is arranged into highly vascular, longitudinal folds called **anal columns.** Masses of varicose veins in the anal area are known as *hemorrhoids* (*hem´ŏ-roidz*); in reference to the condition in which they occur, they are also called piles.

As in the small intestine, the mucosa of the large intestine contains many scattered lymphocytes and lymphatic nodules and is covered by columnar epithelial cells and mucus-secreting goblet cells. Although this epithelium does form intestinal crypts, there are no villi in the large intestine—the intestinal mucosa therefore appears flat. The outer surface of the colon bulges outwards to form pouches, or **haustra** (*hows´tra*) (figs. 26.27 and 26.29). The longitudinal muscle layer of the muscularis externa forms three distinct muscle bands, called **taeniae coli** (*te´ne-e co´li*) that run the length of the large intestine. Numerous fat-filled pouches called **epiploic appendages** (fig. 26.27) are attached superficially to the taeniae coli in the serous layer.

 A common disorder of the large intestine is inflammation of the appendix, or *appendicitis.* Wastes that accumulate in the appendix cannot be moved easily by peristalsis since the appendix has only one opening. The symptoms of appendicitis include muscular rigidity, localized pain in the lower right quadrant, and vomiting. The chief danger of appendicitis is that the appendix might rupture, resulting in peritonitis.

● ● ● ● ● ● ● ● ● ● ● ● ● ● ● ● ● ● ● ● ● ● ● ● ●
haustrum: L. *haustrum*, bucket or scoop
epiploic: Gk. *epiplein*, to float on

## Intestinal Absorption of Fluid and Electrolytes

Most of the fluid and electrolytes in the lumen of the GI tract are absorbed by the small intestine. Although a person may drink only about 1.5 L/day of water, the small intestine receives 7 to 9 L/day as a result of the fluid secreted into the GI tract by the salivary glands, stomach, pancreas, and liver. The small intestine absorbs most of this fluid and passes about 1.5 to 2.0 L/day of fluid to the large intestine. The large intestine absorbs about 90% of this remaining volume, leaving less than 200 ml/day of fluid to be excreted in the feces.

Absorption of water in the large intestine occurs passively as a result of the osmotic gradient created by the active transport of ions. The epithelial cells of the intestinal mucosa are joined together much like those of the kidney tubules and, like the kidney tubules, contain $Na^+/K^+$ pumps in the basolateral membrane. The analogy with kidney tubules is emphasized by the observation that aldosterone, which stimulates salt and water reabsorption in the renal tubules, also appears to stimulate salt and water absorption in the large intestine.

*Diarrhea* is characterized by excessive fluid excretion in the feces. Three different mechanisms, illustrated by three different diseases, can cause diarrhea. In *cholera,* severe diarrhea results from the action of a chemical called *enterotoxin,* which is released from the infecting bacteria. Enterotoxin stimulates active NaCl transport followed by the osmotic movement of water into the lumen of the large intestine. Diarrhea caused by damage to the large intestinal mucosa occurs in *celiac sprue,* a disease

## Table 26.3  Mechanical activity in the GI tract

| Region | Type of motility | Frequency | Stimulus | Result |
|---|---|---|---|---|
| Oral cavity | Mastication | Variable | Initiated voluntarily, proceeds reflexively | Subdivision, mixing with saliva |
| Oral cavity and pharynx | Deglutition | Maximum of 20 swallows per min | Initiated voluntarily, reflexively controlled by swallowing center | Clears oral cavity of food |
| Esophagus | Peristaltic contractions | Depends on frequency of swallowing | Initiated by swallowing | Movement through the esophagus |
| Stomach | Receptive relaxation | Matches frequency of swallowing | Unknown | Permits filling of stomach |
| | Tonic contractions | 15–20 per min | Autonomic plexuses | Mixing and churning |
| | Peristaltic contractions | 1–2 per min | Autonomic plexuses | Evacuation of stomach |
| | "Hunger contractions" | 3 per min | Low blood sugar level | "Feeding" |
| Small intestine | Peristaltic contractions | 17–18 per min | Autonomic plexuses | Transfer through intestine |
| | Rhythmic segmentation | 12–16 per min | Autonomic plexuses | Mixing |
| | Pendular movements | Variable | Autonomic plexuses | Mixing |
| Large intestine | Peristaltic contractions | 3–12 per min | Autonomic plexuses | Transport |
| | Mass movements | 2–3 per day | Stretch | Fills pelvic colon |
| | Haustral churning | 3–12 per min | Autonomic plexuses | Mixing |
| | Defecation | Variable: 1 per day to 3 per week | Reflex triggered by rectal distension | Defecation |

produced in susceptible people by eating foods that contain gluten (proteins from grains such as wheat). In *lactose intolerance*, diarrhea is produced by the increased osmolarity of the contents of the large intestinal lumen as a result of the presence of undigested lactose.

## Mechanical Activities of the Large Intestine

Chyme enters the large intestine through the ileocecal valve. About 15 ml of pasty material enters the cecum with each rhythmic opening of the valve. The ingestion of food intensifies peristalsis of the ileum and increases the frequency with which the ileocecal valve opens; this is called the **gastroileal reflex.** Material entering the large intestine accumulates in the cecum and ascending colon.

Three types of movements occur throughout the large intestine: peristalsis, haustral churning, and mass movement. **Peristaltic movements** of the colon are similar to those of the small intestine, though they are usually more sluggish. In **haustral churning,** a relaxed haustrum is filled until it reaches a certain point of distension, and then the muscularis layer is stimulated to contract. This contraction not only moves the material to the next haustrum but churns the contents and exposes it to the mucosa, where water and electrolytes are absorbed. **Mass movement** is a very strong peri-

staltic wave, involving the action of the taeniae coli, that moves the colonic contents toward the rectum. Mass movements generally occur only two or three times a day, usually during or shortly after a meal. This response to eating is called the **gastrocolic reflex** and can best be observed in infants who have a bowel movement during or shortly after feeding.

After electrolytes and water have been absorbed, the waste material that is left passes to the rectum, leading to an increase in rectal pressure and the urge to defecate. If the urge to defecate is denied, the feces are prevented from entering the anal canal by the internal anal sphincter. In this case, the feces remain in the rectum and may even back up into the sigmoid colon. The **defecation reflex** normally occurs when the rectal pressure rises to a particular level, determined largely by habit. At this point, the internal anal sphincter relaxes to admit the feces into the anal canal.

During the act of defecation, the longitudinal rectal muscles contract to increase rectal pressure and the internal and external anal sphincter muscles relax. Excretion is aided by contractions of abdominal and pelvic skeletal muscles, which raise the intra-abdominal pressure and help push the feces from the rectum through the anal canal and out the anus.

The mechanical activities of the GI tract are summarized in table 26.3.

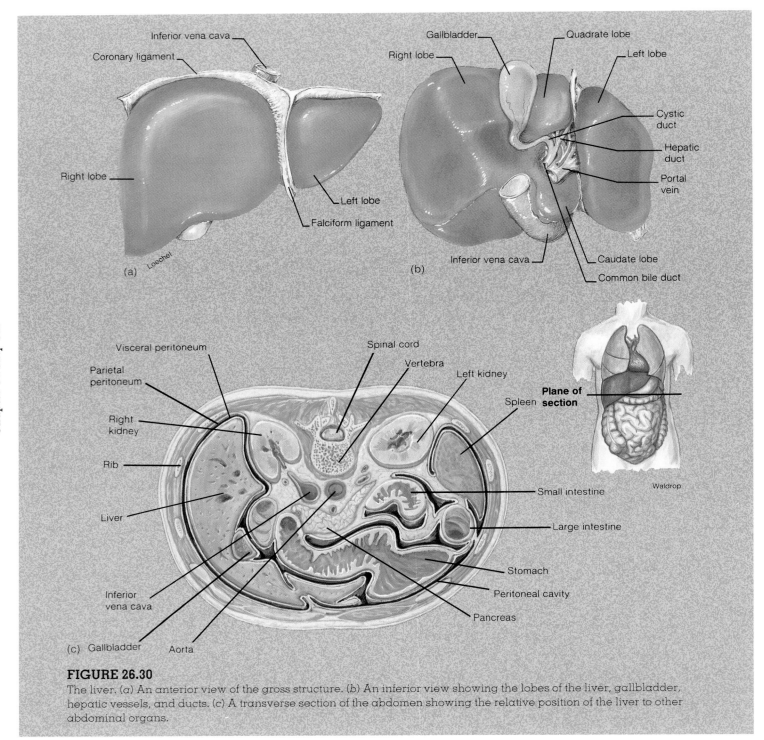

**FIGURE 26.30**

The liver. (*a*) An anterior view of the gross structure. (*b*) An inferior view showing the lobes of the liver, gallbladder, hepatic vessels, and ducts. (*c*) A transverse section of the abdomen showing the relative position of the liver to other abdominal organs.

# Liver, Gallbladder, and Pancreas

*In addition to regulating the chemical composition of the blood in numerous ways, the liver produces and secretes bile, which is stored and concentrated in the gallbladder prior to its discharge into the duodenum. The pancreas produces pancreatic juice, an exocrine secretion containing bicarbonate and important digestive enzymes that is delivered into the duodenum via the pancreatic duct.*

## Structure of the Liver

The **liver** is the largest internal organ of the body, weighing about 1.7 kg (3.5 to 4.0 lbs) in an adult. It is positioned immediately beneath the diaphragm in the right hypochondrium of the abdominal cavity. Its reddish-brown color is due to its great vascularity.

The liver has two major lobes and two minor lobes. Anteriorly, the **right lobe** is separated from the smaller **left lobe** by the **falciform ligament** (fig. 26.30). Inferiorly, the

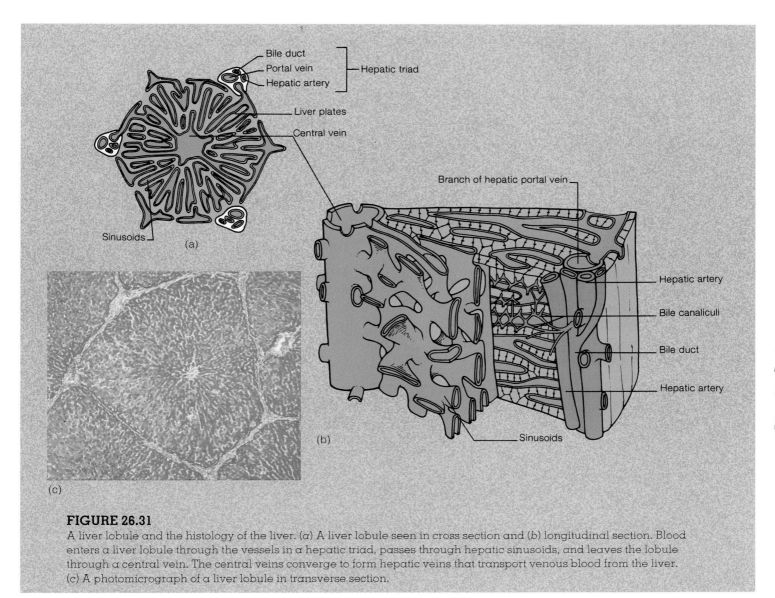

**FIGURE 26.31**
A liver lobule and the histology of the liver. (a) A liver lobule seen in cross section and (b) longitudinal section. Blood enters a liver lobule through the vessels in a hepatic triad, passes through hepatic sinusoids, and leaves the lobule through a central vein. The central veins converge to form hepatic veins that transport venous blood from the liver. (c) A photomicrograph of a liver lobule in transverse section.

caudate lobe is positioned near the inferior vena cava, and the **quadrate lobe** is adjacent to the gallbladder. The falciform ligament attaches the liver to the anterior abdominal wall and the diaphragm. Continuous along the free border of the falciform ligament to the umbilicus is a **ligamentum teres** (round ligament), which is the remnant of the umbilical vein of the fetus. The **porta** of the liver is where the hepatic artery, portal vein, lymphatics, and nerves enter the liver and where the **hepatic ducts** exit.

Although the liver is the largest internal organ, it is, in a sense, only one to two cells thick. This is because the liver cells, or **hepatocytes,** form **hepatic plates** that are one to two cells thick, and the hepatic plates are separated from each other by large capillary spaces called **sinusoids** (figs. 26.31 and 26.32). The sinusoids are lined with phagocytic

**Kupffer** (*koop´fer*) **cells,** but the large intercellular gaps between adjacent Kupffer cells make these sinusoids more permeable than other capillaries. Because of the hepatic plate structure of the liver and the very permeable sinusoids, each hepatocyte is in close contact with the blood.

**Hepatic Portal System** The products of digestion that are absorbed into blood capillaries in the GI tract do not enter the general circulation directly. Instead, this blood is delivered first to the liver. Capillaries in the stomach, small intestine, and large intestine drain venous blood into veins that converge to form the *hepatic portal vein*, which carries blood to capillaries in the liver. It is not until the blood has passed through this second capillary bed that it enters the general circulation through the two *hepatic veins* that drain

hepatic: Gk. *hepatos,* liver

Kupffer cells: from Karl W. Kupffer, Bavarian anatomist, 1829–1902

## FIGURE 26.32

The flow of blood and bile in a liver lobule. Blood flows within sinusoids from a portal vein to the central vein (from the periphery to the center of a lobule). Bile flows within hepatic plates from the center to bile ducts at the periphery of a lobule.

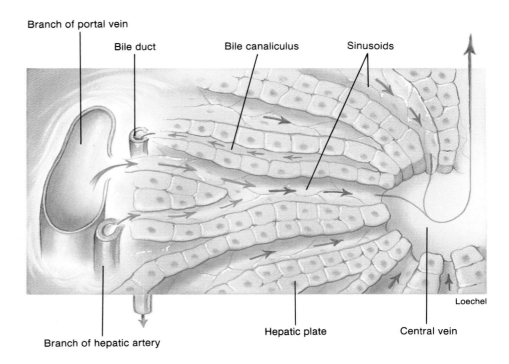

Branch of portal vein

Bile duct

Bile canaliculus

Sinusoids

Loechel

Branch of hepatic artery

Hepatic plate

Central vein

the liver. The term **hepatic portal system** (see fig. 21.33) is used to describe this unique pattern of circulation: capillaries → vein → capillaries → vein. In addition to receiving venous blood from the GI tract, the liver also receives arterial blood from the *hepatic artery.*

**Liver Lobules**   The hepatic plates are arranged into functional units called **liver lobules** (figs. 26.31 and 26.32). In the middle of each lobule is a **central vein,** and at the periphery of each lobule are branches of the hepatic portal vein and of the hepatic artery. These branches open into the spaces *between* hepatic plates. The portal venous blood, containing molecules absorbed in the GI tract, mixes with the arterial blood within the sinusoids as it flows from the periphery of the lobule to the central vein. The central veins of different liver lobules converge to form the two hepatic veins that carry the blood from the liver to the inferior vena cava.

Bile is produced by the hepatocytes and secreted into thin channels called **bile canaliculi** (*kan˝ă-lik´yoo-li*) located *within* each hepatic plate. These bile canaliculi are drained at the periphery of each lobule by **bile ducts,** which in turn drain into **hepatic ducts** that carry bile away from the liver. Since blood travels in the sinusoids and bile travels in the opposite direction within the hepatic plates, blood and bile do not mix in the liver lobules.

 In *cirrhosis,* large numbers of liver lobules are destroyed and replaced with permanent connective tissue and "regenerative nodules" of hepatocytes. These regenerative nodules do not have the platelike structure of normal liver tissue and are therefore less functional. One indication of their depressed function is the entry of ammonia from the hepatic portal blood into the general circulation. Cirrhosis may be caused by chronic alcohol abuse, viral hepatitis, or by other agents that attack liver cells.

### Table 26.4   Compounds excreted by the liver into the bile ducts

|  | Compound | Comments |
|---|---|---|
| Endogenous (naturally occurring) | Bile salts, urobilinogen, cholesterol | High percentage is absorbed and has an enterohepatic circulation[1] |
|  | Lecithin | Small percentage is absorbed and has an enterohepatic circulation |
|  | Bilirubin | No enterohepatic circulation |
| Exogenous (drugs) | Ampicillin, streptomycin, tetracycline | High percentage is absorbed and has an enterohepatic circulation |
|  | Sulfonamides, penicillin | Small percentage is absorbed and has an enterohepatic circulation |

[1]Compounds with an enterohepatic circulation are absorbed to some degree by the intestine and are returned to the liver in the hepatic portal vein.

**Enterohepatic Circulation**   In addition to the normal constituents of bile, a wide variety of exogenous compounds (drugs) are secreted by the liver into the bile ducts (table 26.4). The liver can thus "clear" the blood of particular compounds by removing them from the blood and excreting them into the small intestine with the bile. (The liver can also clear the blood by other mechanisms that will be described in a later section.)

Many compounds that are released with the bile into the small intestine are not excreted with the feces, however.

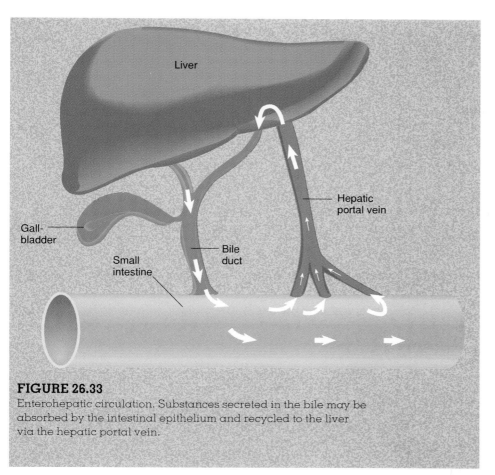

**FIGURE 26.33**
Enterohepatic circulation. Substances secreted in the bile may be absorbed by the intestinal epithelium and recycled to the liver via the hepatic portal vein.

Some of these can be absorbed and enter the hepatic portal blood. These absorbed molecules are carried back to the liver, where they can be again secreted by hepatocytes into the bile ducts. Compounds that recirculate between the liver and small intestine in this way are said to have an enterohepatic circulation (fig. 26.33).

## Functions of the Liver

Because of its very large and diverse enzymatic content, its unique structure, and the fact that it receives food-ladened blood from the small intestine, the liver has a wider variety of functions than any other organ in the body. The major categories of liver function are summarized in table 26.5.

**Bile Production and Secretion** The liver produces and secretes 250 to 1500 ml of bile per day. The major constituents of bile include bile salts, bile pigment (bilirubin), phospholipids (mainly lecithin), cholesterol, and inorganic ions (table 26.6).

## Table 26.5   Major categories of liver function

| Functional category | Actions | Functional category | Actions |
|---|---|---|---|
| Detoxification of blood | Phagocytosis by Kupffer cells | Lipid metabolism | Synthesis of triglyceride and cholesterol |
| | Chemical alteration of biologically active molecules (hormones and drugs) | | Excretion of cholesterol in bile |
| | | | Production of ketone bodies from fatty acids |
| | Production of urea, uric acid, and other molecules that are less toxic than parent compounds | Protein synthesis | Production of albumin |
| | | | Production of plasma transport proteins |
| | Excretion of molecules in bile | | Production of clotting factors (fibrinogen, prothrombin, and others) |
| Carbohydrate metabolism | Conversion of blood glucose to glycogen and fat | Secretion of bile | Synthesis of bile salts |
| | Production of glucose from liver glycogen and from other molecules (amino acids, lactic acid) by gluconeogenesis | | Conjugation and excretion of bile pigment (bilirubin) |
| | Secretion of glucose into the blood | | |

## Table 26.6 — Composition of bile

| Component | Concentration |
|---|---|
| pH | 5.7–8.6 |
| Bile salts | 140–2230 mg/100 ml |
| Lecithin | 140–810 mg/100 ml |
| Cholesterol | 97–320 mg/100 ml |
| Bilirubin | 12–70 mg/100 ml |
| Urobilinogen | 5–45 mg/100 ml |
| Sodium | 145–165 mEq/L |
| Potassium | 2.7–4.9 mEq/L |
| Chloride | 88–115 mEq/L |
| Bicarbonate | 27–55 mEq/L |

Bile pigment, or *bilirubin* (bil″ĭ-roo′bin), is produced in the spleen, liver, and bone marrow from heme groups (minus the iron) derived from hemoglobin. Without the protein part of hemoglobin, the **free bilirubin** is not very water-soluble and thus must be carried in the blood attached to albumin proteins. This protein-bound bilirubin can neither be filtered by the kidneys into the urine nor directly excreted by the liver into the bile.

The liver can take some of the free bilirubin out of the blood and conjugate (combine) it with glucuronic acid. This **conjugated bilirubin** is water-soluble and is secreted into the bile. Once the conjugated bilirubin enters the small intestine, it is converted by bacteria into another pigment, **urobilinogen** (yoo″rŏ-bi-lin′o-jen), which is partially responsible for the color of the feces. About 30% to 50% of the urobilinogen, however, is absorbed by the small intestine and enters the hepatic portal blood. Some is returned to the small intestine in an enterohepatic circulation; the rest enters the general circulation (fig. 26.34). The urobilinogen in plasma, unlike free bilirubin, is not attached to albumin and therefore is easily filtered by the kidneys into the urine, giving urine its characteristic yellow color.

**Bile salts** are derivatives of cholesterol that have two to four polar groups on each molecule. The principal bile salts in humans are *cholic acid* and *chenodeoxycholic* (ke″nŏ-de-ok-sĭ-ko′lik) *acid* (fig. 26.35). In aqueous solutions, these molecules "huddle" together to form aggregates known as **micelles.** The nonpolar parts are located in the central region of the micelle (away from water), whereas the polar groups face water around the periphery of the micelle. Lecithin, cholesterol, and other lipids enter these micelles in a process that aids the digestion and absorption of fats.

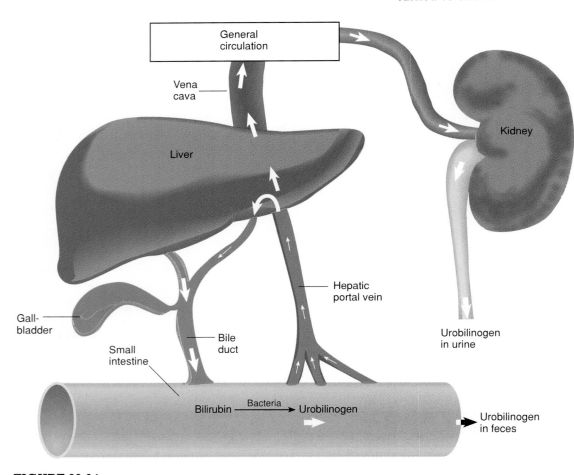

**FIGURE 26.34**

The enterohepatic circulation of urobilinogen. Bacteria in the intestine convert bile pigment (bilirubin) into urobilinogen. Some of this pigment leaves the body in the feces; some is absorbed by the intestine and is recycled through the liver. A portion of the uroglobin that is absorbed enters the general circulation and is filtered by the kidneys into the urine.

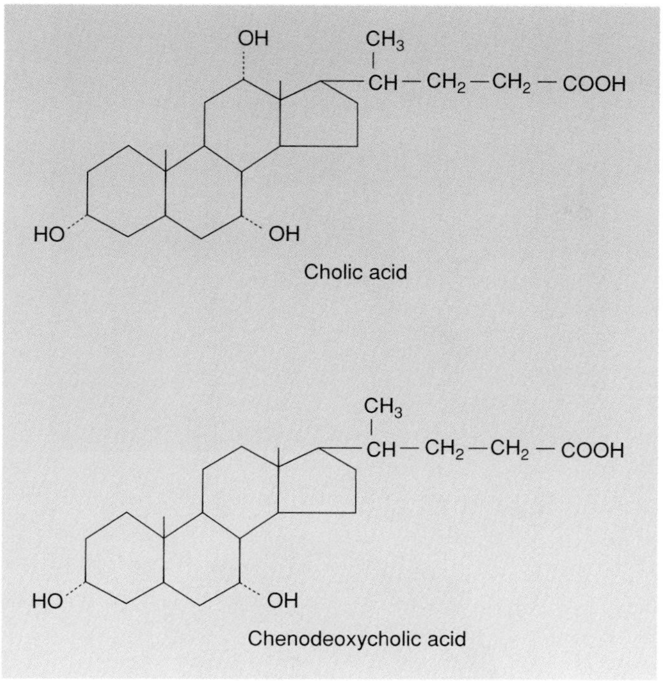

**FIGURE 26.35**
The two major bile acids (which form bile salts) in humans.

**Detoxification of the Blood**   The liver can remove hormones, drugs, and other biologically active molecules from the blood by (1) excretion of these compounds in the bile, (2) phagocytosis by the Kupffer cells that line the sinusoids, and (3) chemical alteration of these molecules within the hepatocytes.

Ammonia, for example, is a very toxic molecule produced by the action of bacteria in the large intestine. Since the ammonia concentration of hepatic portal blood is four to fifty times greater than that of blood in the hepatic vein, it is clear that the ammonia is removed by the liver. The liver has the enzymes needed to convert ammonia into less toxic **urea** molecules. Urea is secreted by the liver into the blood and excreted by the kidneys in the urine. Similarly, the liver converts toxic porphyrins into **bilirubin** and toxic purines into **uric acid.**

Steroid hormones and many drugs are inactivated in their passage through the liver by modifications of their chemical structures. The liver has enzymes that convert these nonpolar molecules into more polar (more water-soluble) forms by *hydroxylation* (the addition of $OH^-$ groups) and by *conjugation* with highly polar groups, such as sulfate and glucuronic acid. Polar derivatives of steroid hormones and drugs are less biologically active and, because of their increased water solubility, are more easily excreted by the kidneys into the urine.

## Secretion of Glucose, Triglycerides, and Ketone Bodies

The liver helps to regulate the blood glucose concentration by either removing glucose from the blood or adding glucose to it, according to the needs of the body. After a carbohydrate-rich meal, the liver can remove some glucose from the hepatic portal blood and convert it into glycogen and triglycerides through the processes of **glycogenesis** (*gli˝kŏ-jen´ĭ-sis*) and **lipogenesis,** respectively. During fasting, the liver secretes glucose into the blood. This glucose can be derived from the breakdown of stored glycogen in a process called **glycogenolysis** (*gli˝kŏ-jĕ-nol´ĭ-sis*), or it can be produced by the conversion of noncarbohydrate molecules (such as amino acids) into glucose in a process called **gluconeogenesis** (*gloo˝ko-ne˝ŏ-jen´ĭ-sis*). The liver also contains the enzymes required to convert free fatty acids into ketone bodies **(ketogenesis),** which are secreted into the blood in large amounts during fasting. These processes are controlled by hormones and are explained in more detail in chapter 27.

**Production of Plasma Proteins**   Plasma albumin and most of the plasma globulins (with the exception of immunoglobulins) are produced by the liver. Albumin constitutes about 70% of the total blood plasma protein and contributes most to the colloid osmotic pressure of the blood. The globulins produced by the liver have a wide variety of functions, including transport of cholesterol and triglycerides, transport of steroid and thyroid hormones, inhibition of trypsin activity, and blood clotting. Clotting factors I (fibrinogen), II (prothrombin), III, V, VII, IX, and XI are all produced by the liver.

## Gallbladder

The **gallbladder** is a saclike organ attached to the inferior surface of the liver. This organ stores and concentrates bile, which drains to it from the liver by way of the bile ducts, hepatic duct, and **cystic duct,** respectively. A sphincter valve at the neck of the gallbladder allows a storage capacity of about 35 to 100 ml. The inner mucosal layer of the gallbladder is arranged in rugae similar to those of the stomach. When the gallbladder fills with bile, it expands to the size and shape of a small pear. Bile is a yellowish-green fluid containing bile salts, bilirubin, cholesterol, and other compounds as previously discussed. Contraction of the muscularis ejects bile from the cystic duct into the **common bile duct,** through which it is conveyed into the duodenum (fig. 26.36).

Bile is continuously produced by the liver and drains through the hepatic and common bile ducts to the duodenum.

..........................................

cystic: Gk. *kystis,* pouch

Digestive System

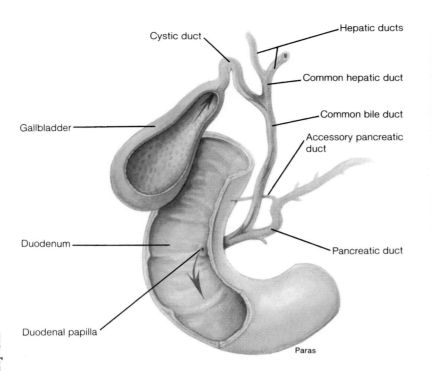

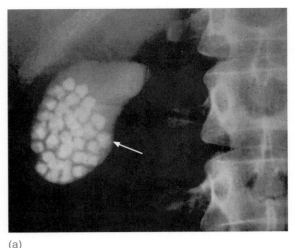

(a)

(b)

**FIGURE 26.36**

The pancreatic duct joins the common bile duct to empty its secretions through the duodenal papilla into the duodenum. The release of bile and pancreatic juice into the duodenum is controlled by the sphincter of ampulla (Oddi).

When the small intestine is empty of food, the **sphincter of ampulla** (sphincter of Oddi) closes, and bile is forced up the cystic duct to the gallbladder for storage.

 Approximately 20 million Americans have *gallstones*—small, hard mineral deposits (calculi) that can produce painful symptoms by obstructing the cystic or common bile ducts. Gallstones usually contain cholesterol as their major component. Cholesterol normally has an extremely low water solubility (20 µg/L), but it can be present in bile at 2 million times its water solubility (40 g/L) because cholesterol molecules cluster together with bile salts and lecithin in the hydrophobic centers of micelles. In order for gallstones to form, the liver must secrete enough cholesterol to create a supersaturated solution, and some substance within the gallbladder must serve as a nucleus for the formation of cholesterol crystals. The gallstone is formed from cholesterol crystals that become hardened by the precipitation of inorganic salts (fig. 26.37). Gallstones may be removed surgically or they may be dissolved by oral ingestion of bile acids. A newer treatment involves fragmentation of the gallstones by high-energy shock waves delivered to a patient immersed in a water bath. This procedure is called *extracorporeal shock-wave lithotripsy.*

## Pancreas

The **pancreas** (see fig. 26.20) is a soft, lobulated, glandular organ that has both exocrine and endocrine functions. The endocrine function is performed by clusters of cells called

**FIGURE 26.37**

(a) A radiograph of a gallbladder that contains gallstones.
(b) A posterior view of a gallbladder that has been removed (cholecystectomy) and cut open to reveal its gallstones (biliary calculi). (A dime is placed in the photo to show relative size.)

the **pancreatic islets** (islets of Langerhans) that secrete the hormones insulin and glucagon into the blood (chapter 19). As an exocrine gland, the pancreas secretes pancreatic juice through the pancreatic duct into the duodenum. The exocrine secretory units of the pancreas are called **acini** (*as´ĭ-ni*). Each acinus (fig. 26.38) consists

..........................................

pancreas: Gk. *pan,* all; *kreas,* flesh
islets of Langerhans: from Paul Langerhans, German anatomist, 1847–88

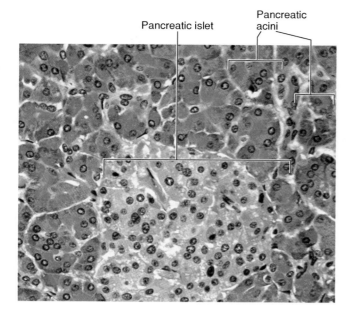

Pancreatic islet

Pancreatic acini

**FIGURE 26.38**
The microscopic structure of the pancreas showing exocrine acini and pancreatic islets.

of a single layer of epithelial cells surrounding a lumen into which the constituents of pancreatic juice are secreted.

The pancreas is positioned horizontally along the posterior abdominal wall, adjacent to the greater curvature of the stomach. It is about 12.5 cm (6 in.) long and 2.5 cm (1 in.) thick and consists of an expanded **head** near the duodenum, a centrally located **body,** and a tapering **tail.** All but a portion of the head is positioned retroperitoneally.

Innervation of the pancreas is provided by branches of the celiac plexus. The glandular portion receives

..........................................

acinus: L. *acinus,* grape

parasympathetic innervation, whereas the pancreatic blood vessels receive sympathetic innervation. The pancreas is supplied with blood by the pancreatic branch of the splenic artery arising from the celiac artery and by the pancreato-duodenal branches arising from the superior mesenteric artery. Venous blood is returned through the splenic and superior mesenteric veins into the hepatic portal vein.

 *Pancreatic cancer* has the worst prognosis of all types of cancer. This is probably because of the spongy, vascular nature of this organ and its vital exocrine and endocrine functions. Pancreatic surgery is a problem because the soft, spongy tissue is difficult to suture.

**Pancreatic Juice** Pancreatic juice contains water, bicarbonate, and a wide variety of digestive enzymes that are secreted into the duodenum. These enzymes include (1) **amylase,** which digests starch; (2) **trypsin,** which digests protein; and (3) **lipase,** which digests triglycerides. Other pancreatic enzymes are indicated in table 26.7. It should be noted that the complete digestion of food molecules in the small intestine requires the action of both brush border enzymes and pancreatic enzymes.

Most pancreatic enzymes are produced as inactive molecules, or **zymogens,** so that the risk of self-digestion within the pancreas is minimized. The inactive form of trypsin, called trypsinogen, is activated within the small intestine by the catalytic action of the brush border enzyme **enterokinase.** Enterokinase converts trypsinogen to active trypsin. Trypsin, in turn, activates the other zymogens of pancreatic juice (fig. 26.39) by cleaving off polypeptide sequences that inhibit the activity of these enzymes.

The activation of trypsin serves as the trigger for the activation of other pancreatic enzymes. Actually, the pancreas does produce small amounts of active trypsin, yet the other

*Digestive System*

## Table 26.7 Enzymes contained in pancreatic juice

| Enzyme | Zymogen | Activator | Action |
|---|---|---|---|
| Trypsin | Trypsinogen | Enterokinase | Cleaves internal peptide bonds |
| Chymotrypsin | Chymotrypsinogen | Trypsin | Cleaves internal peptide bonds |
| Elastase | Proelastase | Trypsin | Cleaves internal peptide bonds |
| Carboxypeptidase | Procarboxypeptidase | Trypsin | Cleaves last amino acid from carboxyl-terminal end of polypeptide |
| Phospholipase | Prophospholipase | Trypsin | Cleaves fatty acids from phospholipids such as lecithin |
| Lipase | None | None | Cleaves fatty acids from glycerol |
| Amylase | None | None | Digests starch to maltose and short chains of glucose molecules |
| Cholesterolesterase | None | None | Releases cholesterol from its bonds with other molecules |
| Ribonuclease | None | None | Cleaves RNA to form short chains |
| Deoxyribonuclease | None | None | Cleaves DNA to form short chains |

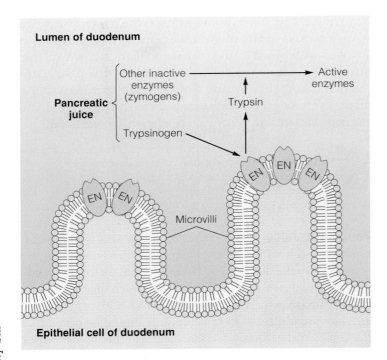

**Lumen of duodenum**

**Pancreatic juice** {
Other inactive enzymes (zymogens) → Active enzymes

Trypsin

Trypsinogen

Microvilli

**Epithelial cell of duodenum**

**FIGURE 26.39**

The pancreatic protein-digesting enzyme trypsin is secreted in an inactive form known as trypsinogen. This inactive enzyme (zymogen) is activated by a brush border enzyme, enterokinase (EN), located in the cell membrane of microvilli. Active trypsin in turn activates other zymogens in pancreatic juice.

enzymes are not activated until the pancreatic juice has entered the duodenum. This is because pancreatic juice also contains a small protein called *pancreatic trypsin inhibitor* that attaches to trypsin and inactivates it in the pancreas.

 Inflammation of the pancreas may result when the various safeguards against self-digestion are insufficient. *Acute pancreatitis* is believed to be caused by the reflux of pancreatic juice and bile from the duodenum into the pancreatic duct. Leakage of trypsin into the blood also occurs, but trypsin is inactive in the blood because of the inhibitory action of two plasma proteins, $\alpha_1$-antitrypsin and $\alpha_2$-macroglobulin. Pancreatic amylase may also leak into the blood, but it is not active because its substrate (glycogen) is not present in blood. Pancreatic amylase activity can be measured in vitro, however, and these measurements are commonly performed to assess the health of the pancreas.

# Digestion and Absorption of Carbohydrates, Lipids, and Proteins

*Polysaccharides and polypeptides are hydrolyzed into their subunits, which enter the epithelial cells of the intestinal villi and are secreted into blood capillaries. Fat is emulsified by the action of bile salts, hydrolyzed*

*into fatty acids and monoglycerides, and absorbed into the intestinal epithelial cells. Once in the cells, triglycerides are resynthesized and combined with proteins to form particles that are secreted into the lymphatic fluid.*

The caloric (energy) value of food is derived mainly from its content of carbohydrates, lipids, and proteins. In the average American diet, carbohydrates account for approximately 50% of the total calories, protein accounts for 11% to 14%, and lipids make up the balance. These food molecules consist primarily of long combinations of subunits (monomers), which must be digested by hydrolysis reactions into free monomers before absorption can occur. The characteristics of the major digestive enzymes are summarized in table 26.8.

## Digestion and Absorption of Carbohydrates

Most of the ingested carbohydrates are in the form of starch, which is a long polysaccharide of glucose in straight chains with occasional branchings. The most commonly ingested sugars are the disaccharides sucrose (table sugar, consisting of glucose and fructose) and lactose (milk sugar, consisting of glucose and galactose). The digestion of starch begins in the mouth with the action of **salivary amylase,** or **ptyalin** (*ti'ă-lin*). Although this enzyme cleaves some of the bonds between adjacent glucose molecules, most people don't chew their food long enough for sufficient digestion to occur in the mouth. The digestive action of salivary amylase stops when the bolus enters the stomach because this enzyme is inactivated at the low pH of gastric juice.

The digestion of starch, therefore, occurs mainly in the duodenum as a result of the action of **pancreatic amylase.** This enzyme cleaves the straight chains of starch to produce the disaccharide *maltose* and the trisaccharide *maltriose*. Pancreatic amylase, however, cannot hydrolyze the bond between glucose molecules at the branch points in the starch. As a result, short, branched chains of glucose molecules, called *oligosaccharides*, are released together with maltose and maltriose by the activity of this enzyme (fig. 26.40).

Maltose, maltriose, and oligosaccharides are hydrolyzed to their monosaccharides by brush border enzymes located on the microvilli of the epithelial cells in the small intestine. The brush border enzymes also hydrolyze the disaccharides sucrose and lactose into their component monosaccharides. These monosaccharides are then moved across the epithelial cell membrane by secondary active transport, in which the glucose shares a common membrane carrier with $Na^+$ (chapter 5). Finally, glucose is secreted from the epithelial cells into blood capillaries within the intestinal villi.

## Digestion and Absorption of Proteins

Protein digestion begins in the stomach with the action of pepsin. Some amino acids are liberated in the stomach, but the major products of pepsin digestion are short-chain

## Table 26.8 Characteristics of the major digestive enzymes

| Enzyme | Site of production | Source | Substrate | Optimum pH | Product(s) |
|---|---|---|---|---|---|
| Salivary amylase | Mouth | Saliva | Starch | 6.7 | Maltose |
| Pepsin | Stomach | Gastric glands | Protein | 1.6–2.4 | Shorter polypeptides |
| Pancreatic amylase | Duodenum | Pancreatic juice | Starch | 6.7–7.0 | Maltose, maltriose, and oligosaccharides |
| Trypsin, chymotrypsin, carboxypeptidase | | | Polypeptides | 8.0 | Amino acids, dipeptides, and tripeptides |
| Pancreatic lipase | | | Triglycerides | 8.0 | Fatty acids and monoglycerides |
| Maltase | | Epithelial membranes | Maltose | 5.0–7.0 | Glucose |
| Sucrase | | | Sucrose | 5.0–7.0 | Glucose + fructose |
| Lactase | | | Lactose | 5.8–6.2 | Glucose + galactose |
| Aminopeptidase | | | Polypeptides | 8.0 | Amino acids, dipeptides, tripeptides |

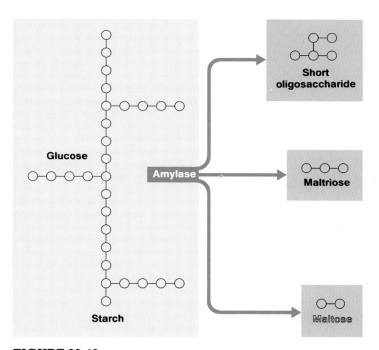

**FIGURE 26.40**

Pancreatic amylase digests starch into maltose, maltriose, and short oligosaccharides containing branch points in the chain of glucose molecules.

polypeptides. Pepsin digestion helps to produce a more homogenous chyme, but it is not essential for the complete digestion of protein that occurs—even in people with total gastrectomies—in the small intestine.

Most protein digestion occurs in the duodenum and jejunum. The pancreatic juice enzymes **trypsin, chymotrypsin** (ki˝mŏ-trip´sin), and **elastase** cleave peptide bonds within the interior of the polypeptide chains. These enzymes are thus classified as *endopeptidases* (en˝do-pep´tĭ-dās˝ēz). Enzymes that remove amino acids from the ends of polypeptide chains, by contrast, are classified as *exopeptidases*. These include the pancreatic juice enzyme **carboxypeptidase,** which removes amino acids from the carboxyl-terminal end of polypeptide chains, and the brush border enzyme **aminopeptidase,** which cleaves amino acids from the amino-terminal end of polypeptide chains.

As a result of the action of these enzymes, polypeptide chains are digested into free amino acids, dipeptides, and tripeptides. The free amino acids are absorbed by cotransport with Na$^+$ into the epithelial cells and secreted into blood capillaries. The dipeptides and tripeptides may enter epithelial cells by a different carrier system, but they are then digested within these cells into amino acids, which are secreted into the blood (fig. 26.41).

Newborns appear to be capable of absorbing a substantial amount of undigested proteins (hence, they can absorb antibodies from their mother's first milk); in adults, however, only the free amino acids enter the hepatic portal vein. Foreign food protein, which would be very antigenic, does not normally enter the blood. An interesting exception is the protein toxin that causes botulism, produced by the bacterium *Clostridium botulinum.* This protein is resistant to digestion and is absorbed into the blood.

## Digestion and Absorption of Lipids

In neonates (newborns), the salivary glands and stomach produce lipases. In adults, however, very little fat digestion occurs until the fat globules in chyme have arrived in the duodenum. Through mechanisms described in the next section, the arrival of fat in the duodenum serves as a stimulus

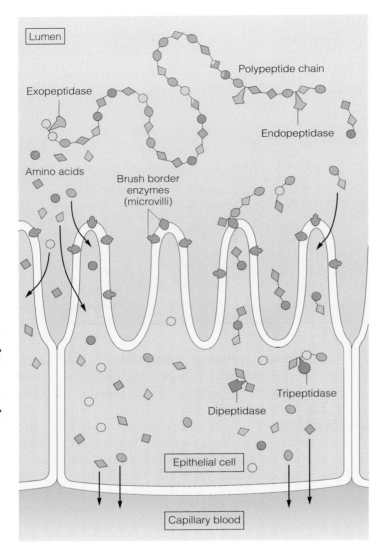

**FIGURE 26.41**

Polypeptide chains are digested into free amino acids, dipeptides, and tripeptides by the action of pancreatic juice enzymes and brush border enzymes. The amino acids, dipeptides, and tripeptides enter duodenal epithelial cells. Dipeptides and tripeptides are hydrolyzed into free amino acids within the epithelial cells, and these products are secreted into capillaries that carry them to the hepatic portal vein.

for the secretion of bile. In a process called **emulsification,** bile salt micelles act to break up the large lipid droplets into tiny *emulsification droplets* of triglycerides. Note that emulsification is not chemical digestion—the bonds joining glycerol and fatty acids are not hydrolyzed by this process.

**Digestion of Lipids** The emulsification of fat aids digestion because the smaller and more numerous emulsification droplets present a greater surface area than the unemulsified fat droplets that originally entered the duodenum. Fat digestion occurs at the surface of the droplets through the

enzymatic action of **pancreatic lipase.** This action is aided by a protein called *colipase*—also secreted by the pancreas—that coats the emulsification droplets and anchors the lipase enzyme to them. Through hydrolysis, lipase removes two of the three fatty acids from each triglyceride molecule and thus liberates *free fatty acids* and *monoglycerides* (fig. 26.42). **Phospholipase A** likewise digests phospholipids, such as lecithin, into fatty acids and lysolecithin (the remainder of the lecithin molecule after two fatty acids have been removed).

Free fatty acids, monoglycerides, and lysolecithin are more polar than the undigested lipids and quickly become associated with micelles of bile salts, lecithin, and cholesterol to form "mixed micelles" (fig. 26.43). These micelles then move to the brush border of the intestinal epithelium, where absorption occurs.

**Absorption of Lipids** Free fatty acids, monoglycerides, and lysolecithin can leave the micelles and pass through the membrane of the microvilli to enter the epithelial cells of the small intestine. There is also some evidence that the micelles may be transported intact into the epithelial cells and that the lipid digestion products may be removed intracellularly from the micelles. In either event, these products are used to *resynthesize* triglycerides and phospholipids within the epithelial cells. This process is different from the absorption of amino acids and monosaccharides, which pass through the epithelial cells without being altered.

Triglycerides, phospholipids, and cholesterol are then combined with protein inside the epithelial cells to form particles called **chylomicrons** (ki´lŏ-mi´kronz). These tiny lipid and protein combinations are secreted into the lymphatic capillaries of the intestinal villi (fig. 26.44). Absorbed lipids pass through the lymphatic system, eventually entering the venous blood by way of the thoracic duct (chapter 23). By contrast, amino acids and monosaccharides enter the hepatic portal vein.

**Transport of Lipids in Blood** Once the chylomicrons are in the blood, their triglyceride content is removed by the enzyme **lipoprotein lipase,** which is attached to the endothelium of blood vessels. This enzyme hydrolyzes triglycerides and thus provides free fatty acids and glycerol for use by the tissue cells. The remaining *remnant particles*, containing cholesterol, are taken up by the liver; this is a process of endocytosis, which requires membrane receptors for the protein part (or *apoprotein*) of the remnant particle.

Cholesterol and triglycerides produced by liver cells are combined with other apoproteins and secreted into the blood as **very-low-density lipoproteins (VLDLs)** that deliver these triglycerides to different organs. Once the triglycerides have been removed, the VLDL is converted to

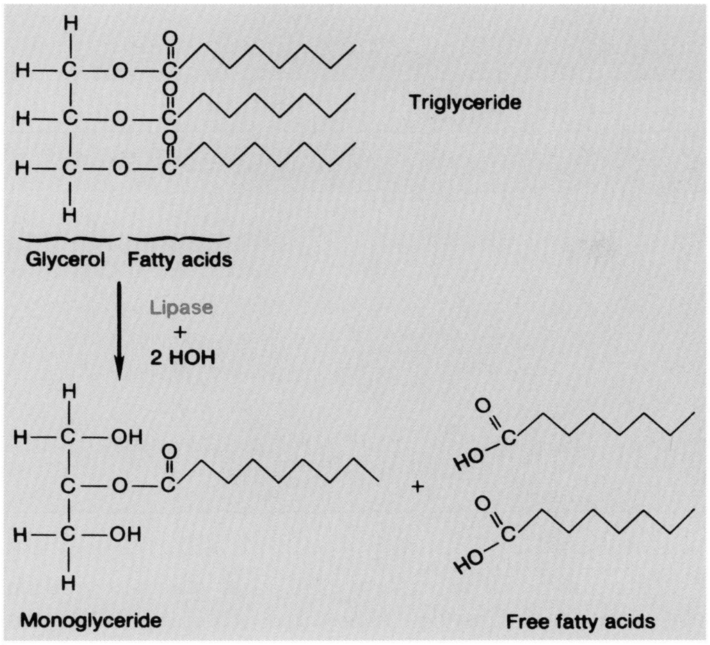

**FIGURE 26.42**

Pancreatic lipase digests fat (triglycerides) by cleaving off the first and third fatty acids. This produces free fatty acids and monoglycerides. Sawtooth structures indicate hydrocarbon chains in the fatty acids.

**FIGURE 26.43**

Steps in the digestion of fat (triglycerides) and the entry of fat digestion products (fatty acids and monoglycerides) into micelles of bile salts secreted by the liver.

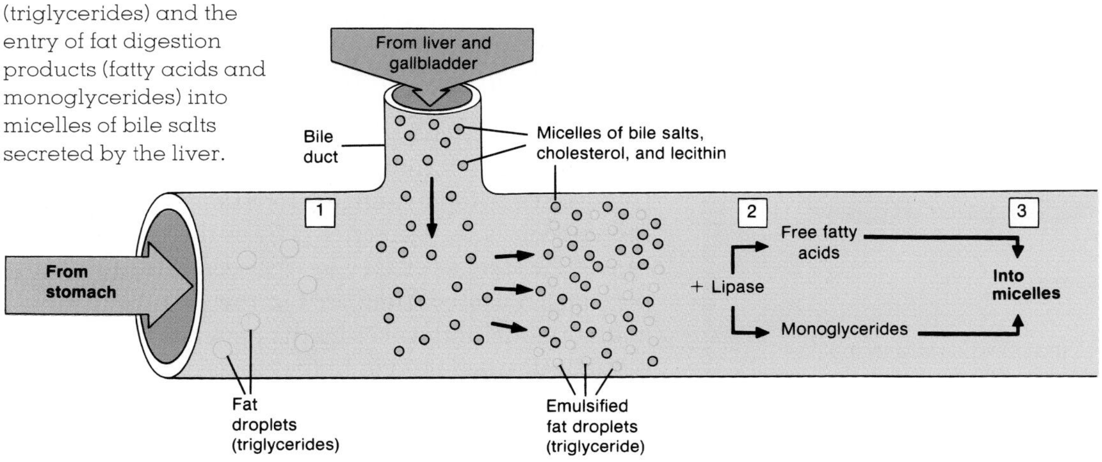

Step 1 Emulsification of fat droplets by bile salts

Step 2 Hydrolysis of triglycerides in emulsified fat droplets into fatty acid and monoglycerides

Step 3 Dissolving of fatty acids and monoglycerides into micelles to produce "mixed micelles"

**low-density lipoproteins (LDLs)** that transport cholesterol to various organs, including blood vessels (chapter 21). Excess cholesterol is returned from these organs to the liver attached to **high-density lipoproteins (HDLs).** A high ratio of HDL-cholesterol to total cholesterol is believed to offer protection against atherosclerosis (chapter 21). The characteristics of these lipoproteins are summarized in table 26.9.

# Neural and Endocrine Regulation of the Digestive System

*The activities of different regions of the GI tract are coordinated by the actions of the vagus nerves and various hormones secreted by the stomach and small intestine. The stomach begins to increase its secretion in anticipation of a meal. In response to the arrival of chyme, it becomes still more active. The entry of chyme into the duodenum stimulates the secretion of hormones that promote contractions of the gallbladder, secretion of pancreatic juice, and inhibition of gastric activity.*

The motility and glandular secretions of the GI tract are, to a large degree, regulated intrinsically. Neural and endocrine control mechanisms, however, can stimulate or inhibit these intrinsic mechanisms to help coordinate the different stages of digestion. The sight, smell, or taste of food, for example, can stimulate salivary and gastric secretions by activation of the vagus nerves, thus helping to "prime" the digestive system in preparation for a meal. Stimulation of the vagus nerves, in this case, originates in the brain and is a conditioned reflex (as Pavlov demonstrated by training dogs to salivate in response to a bell). The vagus nerves are also involved in the reflex control of one part of the digestive system by another—these are "short reflexes," and do not involve the brain.

The GI tract is both an endocrine gland and a target for the action of various hormones. Indeed, the first hormones to be discovered were GI tract hormones. In 1902 two English physiologists, Sir William Bayliss and

Pavlov, Ivan Petrovich: Russian physiologist, 1849–1936
Bayliss, Sir William Maddock: English physiologist, 1860–1924

Digestive System

## Table 26.9 Characteristics of the lipid carrier proteins (lipoproteins) found in plasma

| Lipoprotein class | Origin | Destination | Major lipid(s) | Function |
|---|---|---|---|---|
| Chylomicrons | Intestine | Many organs | Triglycerides, other lipids | Deliver lipids of dietary origin to body cells |
| Very-low-density lipoproteins (VLDLs) | Liver | Many organs | Triglycerides, cholesterol | Deliver endogenously produced triglycerides to body cells |
| Low-density lipoproteins (LDLs) | Intravascular removal of triglycerides from VLDL | Blood vessels, liver | Cholesterol | Deliver endogenously produced cholesterol to various organs |
| High-density lipoproteins (HDLs) | Liver and intestine | Liver and steroid hormone producing glands | Cholesterol | Remove and degrade cholesterol |

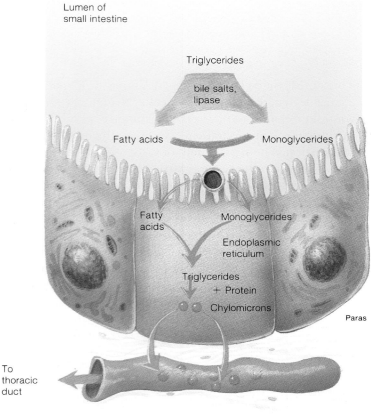

Lumen of small intestine

Triglycerides

bile salts, lipase

Fatty acids — Monoglycerides

Fatty acids — Monoglycerides

Endoplasmic reticulum

Triglycerides + Protein

Chylomicrons

Paras

To thoracic duct

Lacteal

### ▭ FIGURE 26.44

Fatty acids and monoglycerides from the micelles within the small intestine are absorbed by epithelial cells and converted intracellularly into triglycerides. These are then combined with protein to form chylomicrons, which enter the lymphatic vessels (lacteals) of the villi. These lymphatic vessels transport the chylomicrons to the thoracic duct, which empties them into the venous blood.

Ernest Starling, discovered that the duodenum produced a substance that acted as a chemical regulator. They named this substance **secretin** (*sĕ-kreʹtin*) and proposed, in 1905, that it was but one of many yet undiscovered chemical regulators produced by the body. Bayliss and Starling coined the term *hormones* for this new class of regulators. In that same year, other investigators discovered that an extract from the pyloric region of the stomach stimulated gastric secretion. The hormone **gastrin** was thus the second hormone to be discovered.

The chemical structures of gastrin, secretin, and the duodenal hormone **cholecystokinin** (*koʺlĕ-sisʹtŏ-kiʹnin*), or **CCK,** were determined in the 1960s. More recently, a fourth hormone produced by the small intestine, **gastric inhibitory peptide (GIP),** has been added to the list of proven GI tract hormones. The effects of these hormones are summarized in table 26.10.

## Regulation of Gastric Function

Gastric motility and secretion are, to some extent, automatic. Waves of contraction that serve to push chyme through the pyloric sphincter, for example, are initiated spontaneously by pacesetter cells in the greater curvature of the stomach. Likewise, the secretion of HCl and pepsinogen can be stimulated in the absence of neural and hormonal influences by the presence of cooked or partially digested protein in the stomach. The effects of autonomic nerves and hormones are superimposed on this intrinsic activity. The extrinsic control of gastric function is conveniently divided into three phases: the cephalic phase, the gastric phase, and the intestinal phase.

••••••••••••••••••••••••••••••••••••••••••

Starling, Ernest Henry: English physiologist, 1866–1927

## Table 26.10 Physiological effects of the gastrointestinal hormones

| Hormone | Secreted by | Effects |
|---------|-------------|---------|
| Gastrin | Stomach | Stimulates parietal cells to secrete HCl<br>Stimulates chief cells to secrete pepsinogen<br>Maintains structure of gastric mucosa |
| Secretin | Small intestine | Stimulates water and bicarbonate secretion in pancreatic juice<br>Potentiates actions of cholecystokinin on pancreas |
| Cholecystokinin (CCK) | Small intestine | Stimulates contraction of the gallbladder<br>Stimulates secretion of pancreatic juice enzymes<br>Potentiates action of secretin on pancreas<br>Maintains structure of exocrine pancreas (acini) |
| Gastric inhibitory peptide (GIP) | Small intestine | May inhibit gastric emptying<br>May inhibit gastric acid secretion<br>Stimulates secretion of insulin from endocrine pancreas (pancreatic islets) |

**Cephalic Phase** During the **cephalic phase,** the brain regulates gastric gland secretion via the vagus nerves. As previously discussed, various conditioned stimuli can evoke gastric secretion before food reaches the stomach. This conditioning in humans is, of course, more subtle than that exhibited by Pavlov's dogs in response to a bell. In fact, just talking about appetizing food is sometimes a more potent stimulus for gastric acid secretion than the actual sight and smell of food!

Activation of the vagus nerves can stimulate HCl and pepsinogen secretion by two mechanisms: (1) direct vagal stimulation of the gastric parietal and chief cells (the primary mechanism) and (2) vagal stimulation of gastrin secretion by the G cells, which in turn stimulates the parietal and chief cells to secrete HCl and pepsinogen, respectively. This cephalic phase stimulation of gastric secretion continues into the first 30 minutes of a meal but gradually declines in importance as the next phase becomes predominant.

**Gastric Phase** The arrival of chyme into the stomach stimulates the **gastric phase** of regulation. Gastric secretion is stimulated in response to two factors: (1) distension of the stomach, which is determined by the amount of chyme, and (2) the chemical nature of the chyme. While intact proteins have little stimulatory effect, the presence of short polypeptides and amino acids in the stomach stimulates the G cells to secrete gastrin and the parietal and chief cells to secrete HCl and pepsinogen, respectively. Since gastrin also stimulates HCl and pepsinogen secretion, a *positive feedback mechanism* develops: as more HCl and pepsinogen are secreted, more short polypeptides and amino acids are released from the ingested protein, thus stimulating additional secretion of gastrin and, therefore, additional

secretion of HCl and pepsinogen (fig. 26.45). Glucose in the chyme, by contrast, has no effect on gastric secretion, and the presence of fat actually inhibits acid secretion.

Secretion of HCl during the gastric phase is also regulated by a *negative feedback mechanism.* As the pH of gastric juice drops, so does the secretion of gastrin—at a pH of 2.5 gastrin secretion is reduced, and at a pH of 1.0 gastrin secretion shuts off entirely. The secretion of HCl, which is largely under the control of gastrin, declines accordingly. The presence of proteins and polypeptides in the stomach help to buffer the acid and thus to prevent a rapid fall in gastric pH; as a result, more acid can be secreted when proteins are present than when they are absent. The arrival of protein into the stomach thus stimulates acid secretion two ways—by the positive feedback mechanism previously discussed and by inhibition of the negative feedback control of acid secretion. Thus, the amount of acid secreted is closely matched to the amount of protein ingested. As the stomach is emptied, the protein buffers exit, the pH falls, and the secretion of gastrin and HCl is accordingly inhibited.

**Intestinal Phase** The **intestinal phase** of gastric regulation refers to the inhibition of gastric activity when chyme enters the small intestine. Investigators in 1886 demonstrated that the addition of olive oil to a meal inhibits gastric emptying, and in 1929 it was shown that the presence of fat inhibits gastric juice secretion. This inhibitory intestinal phase of gastric regulation is due to both a neural reflex originating from the duodenum and to a chemical hormone secreted by the duodenum.

The arrival of chyme into the duodenum increases its osmolality. This stimulus, together with stretch of the

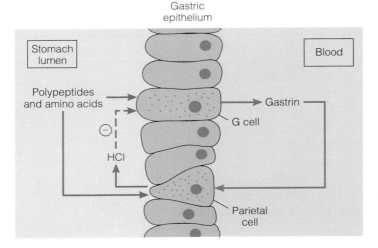

Gastric epithelium

Stomach lumen

Blood

Polypeptides and amino acids

Gastrin

G cell

HCl

Parietal cell

**FIGURE 26.45**

The stimulation of gastric acid (HCl) secretion by the presence of proteins in the stomach lumen and by the hormone gastrin. The secretion of gastrin is inhibited by gastric acidity. This forms a negative feedback loop.

| Table 26.11 | Phases in the regulation of gastric secretion |
|---|---|
| **Phase of regulation** | **Description** |
| **Cephalic phase** | 1. Sight, smell, and taste of food cause stimulation of vagal nuclei in brain<br>2. Vagus nerves stimulate acid secretion<br>  a. Direct stimulation of parietal cells (major effect)<br>  b. Stimulation of gastrin secretion; gastrin stimulates acid secretion (lesser effect) |
| **Gastric phase** | 1. Distension of stomach stimulates vagus nerves; vagus nerves stimulate acid secretion<br>2. Amino acids and peptides in stomach lumen stimulate acid secretion<br>  a. Direct stimulation of parietal cells (lesser effect)<br>  b. Stimulation of gastrin secretion; gastrin stimulates acid secretion (major effect)<br>3. Gastrin secretion inhibited when pH of gastric juice falls below 2.5 |
| **Intestinal phase** | 1. Neural inhibition of gastric emptying and acid secretion<br>  a. Arrival of chyme in duodenum causes distension, increase in osmotic pressure<br>  b. These stimuli activate a neural reflex that inhibits gastric activity.<br>2. In response to fat in chyme, duodenum secretes a hormone that inhibits gastric acid secretion |

duodenum and possibly other stimuli, produces a neural reflex that results in the inhibition of gastric motility and secretion. The presence of fat in the chyme also stimulates the duodenum to secrete a hormone that inhibits gastric function. The general term for such an inhibitory hormone is an *enterogastrone*.

For the past several years, *gastric inhibitory peptide* (*GIP*) was thought to function as an enterogastrone—hence the name for this hormone. Some researchers, however, now believe that a different polypeptide, known as *somatostatin*, may function in this capacity. Somatostatin is produced by the small intestine (as well as by the brain—see chapter 14), where it appears to serve a number of regulatory roles. In addition, the hormone *cholecystokinin* (CCK), which is secreted by the duodenum in response to the presence of chyme, has been found to inhibit gastric emptying. It could be that the only physiological role of GIP is stimulation of insulin secretion from the pancreatic islets in response to the presence of glucose in the intestine. Some scientists therefore propose that the name GIP be retained, but that it serve as an acronym for *glucose-dependent insulinotropic peptide*.

The inhibitory neural and endocrine mechanisms during the intestinal phase of gastric regulation prevent the further passage of chyme from the stomach to the duodenum. This gives the duodenum time to process the load of chyme received previously. Since secretion of the enterogastrone is stimulated by fat in the chyme, a breakfast of bacon and eggs takes longer to pass through the stomach—and makes one feel "fuller" for a longer time—than does a breakfast of pancakes and syrup.

The three phases in the extrinsic control of gastric function are summarized in table 26.11.

## Regulation of Intestinal Function

The submucosal and myenteric plexuses within the wall of the small and large intestine contain preganglionic parasympathetic axons; the ganglion cell bodies of postganglionic parasympathetic neurons; postganglionic sympathetic axons; and sensory neurons. These plexuses also contain association neurons, as does the CNS. Indeed, some scientists refer to the nervous system within the GI tract as an *enteric brain*. Many of the sensory neurons within the intestinal plexuses send impulses all the way to the CNS, but some sensory neurons synapse with the association neurons in the wall of the small and large intestine. This allows for local reflexes that are controlled within the GI tract.

There are several intestinal reflexes controlled both locally and extrinsically. These include (1) the **gastroilial reflex,** in which increased gastric activity causes increased motility of the ileum and increased movement of chyme through the ileocecal sphincter; (2) the **ileogastric reflex,** in which distension of the ileum causes a decrease in gastric motility; and (3) the **intestino-intestinal reflexes,** in which overdistention of one intestinal segment causes relaxation throughout the rest of the intestine.

## Regulation of Pancreatic Juice and Bile Secretion

The secretion of pancreatic juice and bile is stimulated by both neural reflexes initiated in the duodenum and by secretion of the duodenal hormones cholecystokinin (CCK) and secretin.

**Pancreatic Juice**   The secretion of pancreatic juice is stimulated by both secretin and CCK. These two hormones, however, are secreted in response to different stimuli and they have different effects on the composition of pancreatic juice. The secretion of secretin occurs in response to a fall in duodenal pH to below 4.5; this fall in pH occurs for only a short time, however, because the acidic chyme is rapidly neutralized by alkaline pancreatic juice. The secretion of CCK occurs in response to the fat content of chyme in the duodenum.

Secretin stimulates the production of bicarbonate by the pancreas. Since bicarbonate neutralizes the acidic chyme and since secretin is secreted in response to the low pH of chyme, this completes a negative feedback loop in which the effects of secretin inhibit its secretion. Cholecystokinin, by contrast, stimulates the production of pancreatic enzymes, such as trypsin, lipase, and amylase. Secretin and CCK can have different effects on the same cells (the pancreatic acinar cells) because their actions are mediated by different intracellular compounds that act as second messengers. The second messenger of secretin action is cyclic AMP, whereas the second messenger for CCK is $Ca^{++}$.

**Secretion of Bile**   The liver secretes bile continuously, but this secretion is greatly augmented following a meal. The increased bile secretion is due to the release of secretin and CCK from the duodenum. Secretin is the major stimulator of bile secretion by the liver, and CCK enhances this effect. The arrival of chyme in the duodenum also causes the gallbladder to contract and eject bile. Contraction of the gallbladder occurs in response to neural reflexes from the duodenum and in response to stimulation by CCK, but not in response to secretin.

## Pathogens and Poisons

The GI tract presents a suitable environment for an array of microorganisms. Many of these are beneficial, but some bacteria and protozoa can cause diseases. The following discussion includes only a few examples of the pathogenic microorganisms.

**Dysentery** (*dis´en-ter˝e*) is an inflammation of the intestinal mucosa characterized by the discharge of loose stools that contain mucus, pus, and blood. The most common dysentery is **amoebic dysentery,** which is caused by the protozoan *Entamoeba histolytica*. Cysts from this organism are ingested in contaminated food, and after the protective coat has been removed by HCl in the stomach, the vegetative form invades the mucosal walls of the ileum and colon.

**Food poisoning** is a condition that results from ingesting food infected by poisons or by bacteria containing toxins. *Salmonella* is a bacterium that commonly contaminates food. **Botulism** is the most serious type of food poisoning and is caused by ingesting food contaminated with the toxin produced by the bacterium *Clostridium botulinum*. This organism is widely distributed in nature, and the spores it produces are frequently found on food being processed by canning. For this reason food must be heated to 120° C (248° F) before it is canned. It is the toxins produced by the bacteria growing in the food that are pathogenic, rather than the organisms themselves. The poison is a neurotoxin that is readily absorbed into the blood, at which point it can affect the nervous system.

## Disorders of the Liver

The liver is a remarkable organ that has the ability to regenerate itself even if up to 80% has been removed. The most serious diseases of the liver (hepatitis, cirrhosis, and hepatomas) affect the liver throughout, so that it cannot repair itself. **Hepatitis** is inflammation of the liver. Certain chemicals may cause hepatitis, but generally it is caused by infectious viral agents. **Hepatitis A** (infectious hepatitis) is a viral disease transmitted through contaminated foods and liquids. **Hepatitis B** (serum hepatitis) is also caused by a virus and is transmitted in blood plasma during transfusions or by improperly sterilized needles and syringes. Other types of viral hepatitis are designated as hepatitis C, D, and E.

........................................

dysentery: Gk. *dys*, bad; *entera*, intestine

Digestive System

## Development of the Digestive System

The entire digestive system develops from modifications of an elongated tubular structure called the **primitive gut.** These modifications are initiated during the fourth week of embryonic development. The primitive gut is composed solely of endoderm, and for descriptive purposes, can be divided into three regions: the *foregut, midgut,* and *hindgut.*

The **stomodeum** (*sto´mŏ-de´um*), or oral pit, is not part of the foregut but an invagination of ectoderm that breaks through a thin **oral membrane** to become continuous with the foregut and form the oral cavity, or mouth. Structures in the mouth, therefore, are ectodermal in origin. The esophagus, stomach, a portion of the duodenum, the pancreas, liver, and gallbladder are the organs that develop from the foregut (fig. 1). Along the GI tract, only the inside epithelial lining of the lumen is derived from endoderm. The vascular portion and smooth muscle layers are formed from mesoderm that develops from the surrounding mesenchyme. The stomach first appears as an elongated dilation of the foregut. The caudal portion of the foregut and the cranial portion of the midgut form the duodenum. The liver and pancreas arise as small **hepatic** and **pancreatic buds,** respectively, from the wall of the duodenum (fig. 1). The hepatic bud experiences incredible growth to form the gallbladder, associated ducts, and the various lobes of the liver. By the sixth week, the liver is carrying out hemopoiesis (the formation of blood cells), and by the ninth week it has developed to the point where it represents 10% of the total weight of the fetus.

By the fifth week of embryonic development, the midgut has formed a ventral U-shaped loop that projects into the umbilical cord. As development continues, the anterior limb of the midgut loop coils to form most of the small intestine. The posterior limb of the midgut loop expands to form a portion of the small and large intestines.

The hindgut extends from the midgut to the **cloacal membrane.** This membrane is formed in part by the **proctodeum** (*prok´tŏ-de´um*), or **anal pit,** which is a depression in the anal region produced by an invagination of surface ectoderm. A partition develops to divide the cloacal membrane into an anterior **urogenital membrane** and a posterior **anal membrane.** Toward the end of the seventh week, the anal membrane perforates and forms the anal opening, which is lined with ectodermal cells.

....................................

stomodeum: Gk. *stoma*, mouth; *hodaios*, on the way to
cloaca: L. *cloaca*, sewer

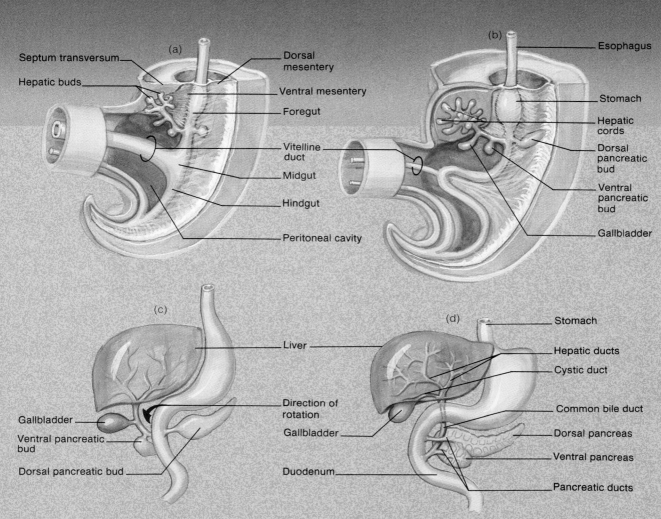

**FIGURE 1**

Progressive stages of development of the foregut to form the stomach, duodenum, liver, gallbladder, and pancreas: (a) 4 weeks, (b) 5 weeks, (c) 6 weeks, and (d) 7 weeks.

Waldrop

**Jaundice** is a yellow staining of the tissue produced by high blood concentrations of either free or conjugated bilirubin. Since free bilirubin is derived from hemoglobin, abnormally high concentrations of heme pigment may result from an unusually high rate of red blood cell destruction. This can occur, for example, as a result of *Rh disease* (*erythroblastosis fetalis*) in an Rh positive baby born to a sensitized Rh negative mother. Jaundice may also occur in otherwise healthy infants because red blood cells are normally destroyed at about the time of birth (hemoglobin concentrations decrease from 19 g per 100 ml to 14 g per 100 ml near the time of birth). This condition is called *physiological jaundice of the newborn* and is not indicative of disease. Premature infants may also develop jaundice because the hepatic enzymes that conjugate bilirubin (a reaction needed for the excretion of bilirubin in the bile) mature late in gestation. Jaundice due to high levels of conjugated bilirubin in the blood is commonly produced in adults when the excretion of bile is blocked by gallstones.

Newborn infants with jaundice are usually treated by *phototherapy*, in which they are placed under blue light in the wavelength range of 400 to 500 nm. This light is absorbed by bilirubin in cutaneous vessels and results in the conversion of bilirubin to a more water-soluble isomer that is soluble in plasma without having to be conjugated with glucuronic acid. The more water-soluble photoisomer of bilirubin can then be excreted in the bile.

## Intestinal Disorders

**Enteritis** is inflammation of the mucosa of the small intestine and is frequently referred to as intestinal flu. Causes of enteritis include bacterial or viral infections, irritating foods or fluids (including alcohol), and emotional stress. The symptoms are abdominal pain, nausea, and diarrhea. **Diarrhea** is the passage of watery, unformed stools. This condition is symptomatic of inflammation, stress, and other body dysfunctions.

A **hernia** is a protrusion of a portion of a visceral organ, usually the small intestine, through a weakened portion of the abdominal wall. Inguinal, femoral, umbilical, and hiatal hernias are the most common. With a **hiatal hernia,** a portion of the stomach pushes superiorly through the esophageal hiatus in the diaphragm and into the thorax. The potential dangers of a hernia are strangulation of the blood supply followed by gangrene, blockage of chyme, or rupture—each of which can threaten life.

**Diverticulosis** is a condition in which the intestinal wall weakens and an outpouching occurs. **Diverticulitis,** or inflammation of a diverticulum, can develop if fecal material becomes impacted in these pockets.

# Other Important Clinical Terminology

**chilitis** Inflammation of the lips.

**colitis** Inflammation of the colon and rectum.

**colostomy** The formation of an abdominal exit from the GI tract by bringing a loop of the colon to the surface of the abdomen. If the rectum is removed because of cancer, the colostomy provides a permanent outlet for the feces.

**cystic fibrosis** An inherited disease of the exocrine glands, particularly the pancreas. Pancreatic secretions that are too thick to drain easily cause inflammation of the ducts and promote connective tissue formation that occludes drainage from the ducts still further.

**gingivitis** Inflammation of the gums. It may result from improper hygiene, poorly fitted dentures, improper diet, or certain infections.

**halitosis** Offensive breath odor. It may result from dental caries, certain diseases, or the ingestion of particular foods.

**heartburn** A burning sensation of the esophagus and stomach. It may result from the regurgitation of gastric juice into the lower portion of the esophagus.

**hemorrhoids** Varicose veins of the rectum and anus.

**jejunoileal bypass** A surgical procedure for creating a bypass of a considerable portion of the small intestine. It reduces the absorptive capacity of the small intestine and is thus used to control extreme obesity.

**nausea** Gastric discomfort and sensations of illness with a tendency to vomit. This feeling is symptomatic of motion sickness and many diseases and may occur during pregnancy.

**pyorrhea** The discharge of pus at the base of the teeth at the gum line.

**regurgitation** The forceful expulsion of gastric contents into the mouth. Nausea and vomiting are common symptoms of almost any dysfunction of the digestive system.

**trench mouth** A contagious bacterial infection that causes inflammation, ulceration, and painful swelling of the floor of the mouth. Generally it is contracted through direct contact by kissing an infected person. Trench mouth can be treated with penicillin and other medications.

**vagotomy** The surgical removal of sections of both vagus nerves where they enter the stomach in order to eliminate nerve impulses that stimulate gastric secretion. This procedure may be used to treat ulcers.

# Chapter Summary

## Introduction to the Digestive System (pp. 765–769)

1. The digestive system functions to break down food into its component monomers (digestion) and to move these monomers into the blood or lymph (absorption).
2. The digestive system consists of a GI tract and accessory digestive organs.
3. Peritoneal membranes line the abdominal wall and cover the visceral organs. The GI tract is supported by a double layer of peritoneum called the mesentery.
4. The layers (tunics) of the abdominal GI tract are, from the inside outward, the mucosa, submucosa, muscularis, and serosa.

## Mouth, Pharynx, and Associated Structures (pp. 770–775)

1. The oral cavity is formed by the cheeks, lips, hard and soft palates, and tongue.
2. The four most anterior pairs of teeth are the incisors and canines, which have one root each; the premolars (bicuspids) and molars have two or three roots.
   a. The tooth is anchored to its bony alveolar socket by a periodontal membrane.
   b. The hard tooth tissues include enamel, dentin, and cementum. The pulp of the tooth receives blood vessels through the apical foramen.
3. The major pairs of salivary glands are the parotid, submandibular, and the sublingual glands.
4. The muscular pharynx is a passageway connecting the oral and nasal cavities to the esophagus and trachea.

## Esophagus and Stomach (pp. 775–780)

1. Peristaltic waves of contraction push food through the lower esophageal sphincter into the stomach.
2. The stomach consists of a cardia, fundus, body, and pylorus; the pylorus terminates at the pyloric sphincter.
   a. The lining of the stomach is thrown into folds, or rugae, and the mucosa is formed into gastric pits and gastric glands.
   b. The parietal cells of the gastric glands secrete HCl; the chief cells secrete pepsinogen.
   c. In the acidic environment of gastric juice, pepsinogen is converted into the active protein-digesting enzyme called pepsin.

## Small Intestine (pp. 780–784)

1. Regions of the small intestine include the duodenum, jejunum, and ileum. The common bile duct and pancreatic duct empty into the duodenum.
2. Fingerlike extensions of mucosa, called villi, project into the intestinal lumen.
   a. Each villus is covered by a columnar epithelium and contains a connective tissue core of lamina propria.
   b. The apical surface (facing the intestinal lumen) of each columnar cell is folded to form microvilli; this brush border of the mucosa increases the absorptive surface area.
3. Digestive enzymes, called brush border enzymes, are located in the membranes of the microvilli.
   a. Brush border enzymes include dipeptidases and disaccharidases, among others.
   b. A deficiency of the disaccharidase known as lactase is responsible for lactose intolerance in many people.

## Large Intestine (pp. 784–787)

1. The large intestine is divided into the cecum, colon, rectum, and anal canal.
   a. The appendix is attached to the inferior medial margin of the cecum.
   b. The colon consists of ascending, transverse, descending, and sigmoid portions.
2. The large intestine absorbs water and electrolytes.

## Liver, Gallbladder, and Pancreas (pp. 788–796)

1. The liver, the largest internal organ, is composed of functional units called lobules.
   a. Liver lobules consist of plates of hepatic cells separated by capillary sinusoids.
   b. Blood flows from the periphery of each lobule, where the hepatic artery and hepatic portal vein empty through the sinusoids and out the central vein.
   c. Bile flows within the hepatocyte plates, in canaliculi, to the bile ducts.
   d. The liver detoxifies the blood and modifies the blood plasma concentration of glucose, triglycerides, ketone bodies, and proteins.
2. The gallbladder stores and concentrates the bile; it releases bile through the cystic duct and common bile duct to the duodenum.
3. The pancreas performs both exocrine and endocrine functions.
   a. The pancreatic islets, which secrete the hormones insulin and glucagon, perform the endocrine function.
   b. The exocrine acini of the pancreas produce pancreatic juice, which contains various digestive enzymes and bicarbonate.

## Digestion and Absorption of Carbohydrates, Lipids, and Proteins (pp. 796–799)

1. The digestion of starch begins in the mouth with the action of salivary amylase.
   a. Pancreatic amylase digests starch into disaccharides and short-chain oligosaccharides.
   b. Complete digestion into monosaccharides is accomplished by brush border enzymes.
2. Protein digestion begins in the stomach with the action of pepsin.
   a. Pancreatic juice contains protein-digesting enzymes, including trypsin, chymotrypsin, and others.
   b. The brush border contains digestive enzymes that help to complete the digestion of proteins into amino acids.
   c. Amino acids, like monosaccharides, are absorbed and secreted into capillary blood entering the hepatic portal vein.

Digestive System

# Interactions of the Digestive System with Other Body Systems

**Integumentary System**
- Synthesizes vitamin D, which indirectly helps to regulate intestinal absorption of $Ca^{++}$
- Stores triglycerides in adipose cells of hypodermis
- Provides nutrients required for health of integumentary system

**Skeletal System**
- Supports and protects certain digestive organs
- Stores calcium phosphate
- Absorbs $Ca^{++}$ and $PO_4^{-3}$ for formation and maintenance of bone tissue

**Muscular System**
- Activity assists chewing and swallowing; also peristalsis and segmention
- Provides nutrients required for function and maintenance of muscles

**Nervous System**
- Autonomic innervation provides neural regulation of digestive system
- Provides nutrients required for function and maintenance of nervous system

**Endocrine System**
- Hormones secreted by stomach and small intestine help to regulate digestive functions
- Provides nutrients required by endocrine system
- Liver removes and metabolizes hormones

**Lymphatic System**
- Protects against infections
- Takes tissue fluid containing absorbed fat from small intestine and conveys it to venous system
- Provides nutrients required for function and maintenance of lymphatic system

**Respiratory System**
- Provides oxygen for metabolism of digestive system and provides for elimination of carbon dioxide
- Provides nutrients required for function and maintenance of respiratory system

**Circulatory System**
- Transports absorbed amino acids, monosaccharides, and other molecules from intestine to liver and then to other organs
- Provides nutrients required for function and maintenance of circulatory system
- Absorbs iron for blood

**Urinary System**
- Eliminates metabolic wastes
- Converts vitamin D to the active form required for calcium absorption
- Provides nutrients required for function and maintenance of urinary system

**Reproductive System**
- Influences metabolic rates through effects of steroids
- Provides nutrients required for function and maintenance of reproductive system
- During pregnancy, provides nutrients for

3. Lipids are digested in the small intestine after being emulsified by bile salts.
   a. After digestion of triglycerides, free fatty acids and monoglycerides are absorbed into the epithelial cells of the small intestine.
   b. Once inside the mucosal epithelial cells, these subunits are used to resynthesize triglycerides.
   c. Triglycerides in the epithelial cells of the small intestine, together with proteins, form chylomicrons; these tiny particles are secreted into the lacteals of the villi.

## Neural and Endocrine Regulation of the Digestive System (pp. 799–803)

1. The regulation of gastric function occurs in three phases.
   a. In the cephalic phase, the activity of higher brain centers, acting via the vagus nerves, stimulates gastric secretion.
   b. In the gastric phase, the secretion of HCl and pepsin is controlled by the gastric contents and by the hormone gastrin, secreted by the gastric mucosa.
   c. In the intestinal phase, the activity of the stomach is inhibited by neural and endocrine reflexes from the duodenum.
2. The secretion of the hormones secretin and cholecystokinin (CCK) regulate pancreatic juice and bile secretion.

....................................................................................................................

# Review Activities

## Objective Questions

1. Which of the following types of teeth are found in the permanent but not in the deciduous dentition?
   a. incisor teeth
   b. canine teeth
   c. premolar teeth
   d. molar teeth
2. The double layer of peritoneum that supports the GI tract is called
   a. the visceral peritoneum.
   b. the dorsal mesentery.
   c. the greater omentum.
   d. the lesser omentum.
3. Which of the following tissue layers in the small intestine contains lacteals?
   a. submucosa
   b. muscularis mucosa
   c. lamina propria
   d. muscularis externa
4. Which of the following statements about gastric secretion of HCl is *false*?
   a. HCl is secreted by parietal cells.
   b. HCl hydrolyzes peptide bonds.
   c. HCl is needed for the conversion of pepsinogen to pepsin.
   d. HCl is needed for maximum activity of pepsin.
5. Intrinsic factor
   a. is secreted by the stomach.
   b. is a polypeptide.
   c. promotes absorption of vitamin $B_{12}$ in the small intestine.
   d. helps prevent pernicious anemia.
   e. all of the above apply.

6. Intestinal enzymes such as lactase are
   a. secreted by the small intestine into the chyme.
   b. produced by the intestinal crypts.
   c. produced by the pancreas.
   d. attached to the cell membrane of microvilli in the epithelial cells of the mucosa.
7. Most digestion occurs in
   a. the mouth.
   b. the stomach.
   c. the small intestine.
   d. the large intestine.
8. Which of the following statements about trypsin is *true?*
   a. Trypsin is derived from trypsinogen by the digestive action of pepsin.
   b. Active trypsin is secreted into the pancreatic acini.
   c. Trypsin is produced in the intestinal crypts.
   d. Trypsinogen is converted to trypsin by the brush border enzyme enterokinase.
9. During the gastric phase, the secretion of HCl and pepsinogen is stimulated by
   a. impulses sent from the brain via the vagus nerves to the stomach.
   b. polypeptides in the gastric lumen and by gastrin secretion.
   c. secretin and cholecystokinin from the duodenum.
   d. all of the above.
10. The secretion of HCl by the stomach mucosa is inhibited by
    a. neural reflexes from the duodenum.
    b. the secretion of gastric inhibitory peptide from the duodenum.
    c. the lowering of gastric pH.
    d. all of the above.

11. Which of the following organs is the first to receive the blood-borne products of digestion?
    a. liver
    b. pancreas
    c. heart
    d. brain
12. Which of the following statements about hepatic portal blood is *true?*
    a. It contains absorbed fat.
    b. It contains ingested proteins.
    c. It is mixed with bile in the liver.
    d. It is mixed with blood from the hepatic artery in the liver.

## Essay Questions

1. Explain how the liver and pancreas are related embryologically and anatomically to the intestine.
2. Explain how the gastric secretion of HCl and pepsin is regulated during the cephalic, gastric, and intestinal phases.
3. Describe how pancreatic enzymes become activated in the lumen of the small intestine and explain the need for these mechanisms.
4. Explain the functions of bicarbonate in pancreatic juice and how peptic ulcers in the duodenum may be produced.
5. Describe the mechanisms that are believed to protect the gastric mucosa from self-digestion, and some proposed reasons for the development of a peptic ulcer in the stomach.
6. Explain why the pancreas is considered to be an exocrine and an endocrine gland. Given this information, predict what effects tying of the pancreatic duct would have on pancreatic structure and function.

7. Explain how jaundice is produced when (a) the person has gallstones, (b) the person has a high rate of red blood cell destruction, and (c) the person has liver disease. In which of these cases would phototherapy for the jaundice be effective? Explain.

8. Describe the steps involved in the digestion and absorption of fat.

9 Distinguish between chylomicrons, very low density lipoproteins, low-density lipoproteins, and high-density lipoproteins.

## Gundy/Weber Software 💾

The tutorial software accompanying Chapter 26 is Volume 11—Digestive System.

# [ chapter twenty-seven ]

## regulation of metabolism

## objectives

- State the factors that influence the metabolic rate and describe how the basal metabolic rate can be measured.

- Describe the nature of the essential amino acids and essential fatty acids and discuss the caloric needs of the body.

- Describe the interrelationship between glucose and glycogen.

- Explain how fat is used as an energy source and how glucose can be converted into fat.

- Explain how protein can be used as an energy source and define the terms *transamination* and *oxidative deamination*.

- Describe the processes involved in the regulation of eating.

- List the hormones that promote anabolism and those that promote catabolism.

- Explain how the secretion of insulin and glucagon is regulated and how plasma concentrations of these hormones change during absorption and fasting.

- Describe the effects of insulin and glucagon on the metabolism of glycogen, fat, and protein and explain the significance of these effects.

- Discuss the metabolic effects of epinephrine and the glucocorticoids.

- Discuss the effects of thyroxine and explain how these effects are produced.

- Describe the metabolic effects of growth hormone and the role of somatomedins in producing these effects.

# Nutritional Requirements

*The body's energy requirements must be met by the caloric value of food to prevent catabolism of the body's own fat, carbohydrates, and protein. Additionally, food molecules—particularly the essential amino acids and fatty acids—are needed to replace molecules in the body that are continuously degraded. Vitamins and elements do not directly provide energy but instead are required for diverse enzymatic reactions.*

Living tissue is maintained by the constant expenditure of energy. This energy is obtained directly from ATP and indirectly from the cell respiration of glucose, fatty acids, ketone bodies, amino acids, and other organic molecules. These molecules are ultimately obtained from food, but they can also be obtained from the glycogen, fat, and protein stored in the body.

The energy value of food is commonly measured in kilocalories (also called "big calories" and spelled with a capital letter [C]). One kilocalorie is equal to 1000 calories; one calorie is defined as the amount of heat required to raise the temperature of 1 gram of water by 1° C (from 14.5° to 15.5° C). As described in chapter 5, the amount of energy released as heat when a quantity of food is combusted in vitro is equal to the amount of energy released within cells through the process of aerobic respiration. This is 4 calories per gram for carbohydrate or protein and 9 calories per gram for fat. When this energy is released by cell respiration, some is transferred to the high-energy bonds of ATP and some is lost as heat.

## Metabolic Rate and Caloric Requirements

The total rate of body metabolism, or the **metabolic rate,** can be measured by either the amount of heat generated by the body or by the amount of oxygen consumed by the body per minute. This rate is influenced by a variety of factors. For example, metabolic rate is increased by physical activity and by eating. The increased rate of metabolism that accompanies the assimilation of food can last more than 6 hours after a meal.

Temperature is also an important factor in determining metabolic rate. The reasons for this are twofold: (1) temperature itself is a determinant of the rate of chemical reactions and (2) the hypothalamus contains *temperature control centers*, as well as temperature-sensitive cells that act as sensors for changes in body temperature. In response to deviations from a set point for body temperature (chapter 1), the control areas of the hypothalamus can direct physiological responses that help to correct the deviations and maintain a constant body temperature. Changes in body temperature are thus accompanied by physiological responses that influence the total metabolic rate.

 During open heart or brain surgery, the core body temperature is often lowered to between 21° and 24° C (a condition of *hypothermia*). Compensatory responses are dampened by the general anesthetic, and the low body temperature drastically reduces the needs of the tissues for oxygen. Under these conditions, the circulation can be stopped and bleeding is significantly reduced.

The metabolic rate of an awake, relaxed person 12 to 14 hours after eating and at a comfortable temperature is known as the **basal metabolic rate (BMR).** The BMR is determined primarily by a person's age, gender, and overall body surface area, but it is also strongly influenced by the level of thyroid secretion. A person with hyperthyroidism has an abnormally high BMR, and a person with hypothyroidism has a low BMR. An interesting finding is that the BMR may be influenced by genetic inheritance, and that at least some families that are prone to obesity may have a genetically determined low BMR.

Differences in energy requirements among most people, however, are due primarily to differences in physical activity. Daily energy expenditures may range from about 1300 to 5000 kilocalories per day. For people not engaged in heavy manual labor but who are active during their leisure time, average values are about 2900 kilocalories per day for men and 2100 kilocalories per day for women. People engaged in office work, the professions, sales, and comparable occupations consume up to 5 kilocalories per minute during work. More physically demanding occupations may require energy expenditures of 7.5 to 10 kilocalories per minute.

When the caloric intake is greater than the energy expenditures, excess calories are stored primarily as fat. This is true regardless of the source of the calories—carbohydrates, proteins, or fats—because these molecules can be converted to fat by various metabolic pathways (to be described). Desirable body weights at ages 25 through 59 are indicated in table 27.1.

Weight is lost when the caloric value of the food ingested is less than the amount required in cell respiration over a period of time. Weight loss, therefore, can be achieved by dieting alone or by dieting combined with an exercise program to raise the metabolic rate. The caloric expenditure associated with different forms of exercise is summarized in table 27.2.

 *Obesity* is a risk factor for cardiovascular diseases, diabetes mellitus, gallbladder disease, and some malignancies (particularly endometrial and breast cancer). The distribution of fat in the body is also important; there is a greater risk of cardiovascular disease when fat produces a high waist-to-hip ratio, or an "apple shape," as compared to a "pear shape." Obesity in childhood is due to an increase in both the size (increased lipid content) and number of adipose cells; weight gain in adulthood is due mainly to an increase in adipose cell size, although the number of these cells may also increase in extreme weight gains. When weight is lost, the size of the adipose cells decreases, but the number of adipose cells does not decrease. It is thus important to prevent further increases in weight in all overweight people but particularly so in children.

## Table 27.1 Desirable body weights

| | Height Feet | Inches | Small frame | Medium frame | Large frame |
|---|---|---|---|---|---|
| **Men** | 5 | 2 | 128–134 | 131–141 | 138–150 |
| | 5 | 3 | 130–136 | 133–143 | 140–153 |
| | 5 | 4 | 132–138 | 135–145 | 142–156 |
| | 5 | 5 | 134–140 | 137–148 | 144–160 |
| | 5 | 6 | 136–142 | 139–151 | 146–164 |
| | 5 | 7 | 138–145 | 142–154 | 149–168 |
| | 5 | 8 | 140–148 | 145–157 | 152–172 |
| | 5 | 9 | 142–151 | 148–160 | 155–176 |
| | 5 | 10 | 144–154 | 151–163 | 158–180 |
| | 5 | 11 | 146–157 | 154–166 | 161–184 |
| | 6 | 0 | 149–160 | 157–170 | 164–188 |
| | 6 | 1 | 152–164 | 160–174 | 168–192 |
| | 6 | 2 | 155–168 | 164–178 | 172–197 |
| | 6 | 3 | 158–172 | 167–182 | 176–202 |
| | 6 | 4 | 162–176 | 171–187 | 181–207 |

Weights at ages 25–59 based on lowest mortality. Weight in pounds according to frame (in indoor clothing weighing 5 lbs., shoes with 1″ heels).

| | Height Feet | Inches | Small frame | Medium frame | Large frame |
|---|---|---|---|---|---|
| **Women** | 4 | 10 | 102–111 | 109–121 | 118–131 |
| | 4 | 11 | 103–113 | 111–123 | 120–134 |
| | 5 | 0 | 104–115 | 113–126 | 122–137 |
| | 5 | 1 | 106–118 | 115–129 | 125–140 |
| | 5 | 2 | 108–121 | 118–132 | 128–143 |
| | 5 | 3 | 111–124 | 121–135 | 131–147 |
| | 5 | 4 | 114–127 | 124–138 | 134–151 |
| | 5 | 5 | 117–130 | 127–141 | 137–155 |
| | 5 | 6 | 120–133 | 130–144 | 140–159 |
| | 5 | 7 | 123–136 | 133–147 | 143–163 |
| | 5 | 8 | 126–139 | 136–150 | 146–167 |
| | 5 | 9 | 129–142 | 139–153 | 149–170 |
| | 5 | 10 | 132–145 | 142–156 | 152–173 |
| | 5 | 11 | 135–148 | 145–159 | 155–176 |
| | 6 | 0 | 138–151 | 148–162 | 158–179 |

Weights at ages 25–59 based on lowest motality. Weight in pounds according to frame (in indoor clothing weighing 3 lbs., shoes with 1″ heels).

Source: Data from *1979 Build Study*, Society of Actuaries and Association of Life Insurance Medical Directors of America, 1980.

## Table 27.2 Energy consumed (in kilocalories per minute) by various activities

| Activity | Weight in pounds | | | |
|---|---|---|---|---|
| | 105–115 | 127–137 | 160–170 | 182–192 |
| **Bicycling** | | | | |
| 10 mph | 5.41 | 6.16 | 7.33 | 7.91 |
| Stationary, 10 mph | 5.50 | 6.25 | 7.41 | 8.16 |
| **Calisthenics** | 3.91 | 4.50 | 7.33 | 7.91 |
| **Dancing** | | | | |
| Aerobic | 5.83 | 6.58 | 7.83 | 8.58 |
| Square | 5.50 | 6.25 | 7.41 | 8.00 |
| **Gardening, weeding, and digging** | 5.08 | 5.75 | 6.83 | 7.50 |
| **Jogging** | | | | |
| 5.5 mph | 8.58 | 9.75 | 11.50 | 12.66 |
| 6.5 mph | 8.90 | 10.20 | 12.00 | 13.20 |
| 8.0 mph | 10.40 | 11.90 | 14.10 | 15.50 |
| 9.0 mph | 12.00 | 13.80 | 16.20 | 17.80 |
| **Rowing, machine** | | | | |
| Easily | 3.91 | 4.50 | 5.25 | 5.83 |
| Vigorously | 8.58 | 9.75 | 11.50 | 12.66 |
| **Skiing** | | | | |
| Downhill | 7.75 | 8.83 | 10.41 | 11.50 |
| Cross-country, 5 mph | 9.16 | 10.41 | 12.25 | 13.33 |
| Cross-country, 9 mph | 13.08 | 14.83 | 17.58 | 19.33 |
| **Swimming, crawl** | | | | |
| 20 yards per minute | 3.91 | 4.50 | 5.25 | 5.83 |
| 40 yards per minute | 7.83 | 8.91 | 10.50 | 11.58 |
| 55 yards per minute | 11.00 | 12.50 | 14.75 | 16.25 |
| **Walking** | | | | |
| 2 mph | 2.40 | 2.80 | 3.30 | 3.60 |
| 3 mph | 3.90 | 4.50 | 5.30 | 5.80 |
| 4 mph | 4.50 | 5.20 | 6.10 | 6.80 |

Regulation of Metabolism

## Table 27.3 Recommended daily allowances

| Category | Age (years) or condition | Weight (kg) | (lb) | Height (cm) | (in) | Protein (g) | Fat-soluble vitamins Vitamin A (μg RE)[1] | Vitamin D (μg)[2] | Vitamin E (mg α-TE)[3] | Vitamin K (μg) |
|---|---|---|---|---|---|---|---|---|---|---|
| Infants | 0.0–0.5 | 6 | 13 | 60 | 24 | 13 | 375 | 7.5 | 3 | 5 |
| | 0.5–1.0 | 9 | 20 | 71 | 28 | 14 | 375 | 10 | 4 | 10 |
| Children | 1–3 | 13 | 29 | 90 | 35 | 16 | 400 | 10 | 6 | 15 |
| | 4–6 | 20 | 44 | 112 | 44 | 24 | 500 | 10 | 7 | 20 |
| | 7–10 | 28 | 62 | 132 | 52 | 28 | 700 | 10 | 7 | 30 |
| Males | 11–14 | 45 | 99 | 157 | 62 | 45 | 1000 | 10 | 10 | 45 |
| | 15–18 | 66 | 145 | 176 | 69 | 59 | 1000 | 10 | 10 | 65 |
| | 19–24 | 72 | 160 | 177 | 70 | 58 | 1000 | 10 | 10 | 70 |
| | 25–50 | 79 | 174 | 176 | 70 | 63 | 1000 | 5 | 10 | 80 |
| | 51+ | 77 | 170 | 173 | 68 | 63 | 1000 | 5 | 10 | 80 |
| Females | 11–14 | 46 | 101 | 157 | 62 | 46 | 800 | 10 | 8 | 45 |
| | 15–18 | 55 | 120 | 163 | 64 | 44 | 800 | 10 | 8 | 55 |
| | 19–24 | 58 | 128 | 164 | 65 | 46 | 800 | 10 | 8 | 60 |
| | 25–50 | 63 | 138 | 163 | 64 | 50 | 800 | 5 | 8 | 65 |
| | 51+ | 65 | 143 | 160 | 63 | 50 | 800 | 5 | 8 | 65 |
| Pregnant | | | | | | 60 | 800 | 10 | 10 | 65 |
| Lactating | 1st 6 months | | | | | 65 | 1300 | 10 | 12 | 65 |
| | 2nd 6 months | | | | | 62 | 1200 | 10 | 11 | 65 |

Reprinted with permission from *Recommended Dietary Allowances*, 10th Edition, © 1989 by the National Academy of Sciences. Published by National Academy Press, Washington, D.C.

[1]Retinol equivalent (1 RE = 1 μg retinol or 6 μg β-carotene)
[2]As cholecalciferol (10 μg cholecalciferol = 400 W of Vitamin D)
[3]α-tocopherol equivalents (1 mg α-tocopherol = 1 α-TE)

## Anabolic Requirements

In addition to providing the body with energy, food also supplies the raw materials for synthesis reactions—collectively termed **anabolism**—that occur constantly within the cells of the body. Anabolic reactions include those that synthesize DNA and RNA, protein, glycogen, triglycerides, and other polymers. These anabolic reactions, in which larger molecules are built from smaller ones, must occur constantly to replace those molecules that are hydrolyzed into their component monomers. The hydrolysis reactions, together with the reactions of cell respiration that ultimately break down the monomers to carbon dioxide and water, are collectively termed **catabolism.**

Acting through changes in hormonal secretion, exercise and fasting increase the catabolism of stored glycogen, fat, and body protein. These molecules are also broken down at a certain rate in a person who is neither exercising nor fasting. Some of the monomers formed (amino acids, glucose, and fatty acids) are used to immediately resynthesize body protein, glycogen, and fat. However, some of the glucose derived from stored glycogen, for example, or fatty acids derived from stored triglycerides, undergo cell respiration for energy. For this reason, new monomers must be obtained from food to prevent a continual decline in the amount of protein, glycogen, and fat in the body.

The *turnover rate* of a particular molecule is the rate at which it is broken down and resynthesized. For example, the average daily turnover for carbohydrates is 250 g/day. Since some of the glucose in the body is reused to form glycogen, the average daily dietary requirement for carbohydrate is less than this amount—about 150 g/day. The average daily turnover for protein is 150 g/day, but since many of the amino acids derived from the catabolism of body proteins can be reused in protein synthesis, a person needs only about 35 g/day of protein in the diet. It should be noted that these are average

| Water-soluble vitamins | | | | | | | Minerals | | | | | | |
|---|---|---|---|---|---|---|---|---|---|---|---|---|---|
| Vitamin C (mg) | Thiamin (mg) | Riboflavin (mg) | Niacin (mg NE)[4] | Vitamin B$_6$ (mg) | Folate (µg) | Vitamin B$_{12}$ (µg) | Calcium (mg) | Phosphorus (mg) | Magnesium (mg) | Iron (mg) | Zinc (mg) | Iodine (µg) | Selenium (µg) |
| 30 | 0.3 | 0.4 | 5 | 0.3 | 25 | 0.3 | 400 | 300 | 40 | 6 | 5 | 40 | 10 |
| 35 | 0.4 | 0.5 | 6 | 0.6 | 35 | 0.5 | 600 | 500 | 60 | 10 | 5 | 50 | 15 |
| 40 | 0.7 | 0.8 | 9 | 1.0 | 50 | 0.7 | 800 | 800 | 80 | 10 | 10 | 70 | 20 |
| 45 | 0.9 | 1.1 | 12 | 1.1 | 75 | 1.0 | 800 | 800 | 120 | 10 | 10 | 90 | 20 |
| 45 | 1.0 | 1.2 | 13 | 1.4 | 100 | 1.4 | 800 | 800 | 170 | 10 | 10 | 120 | 30 |
| 50 | 1.3 | 1.5 | 17 | 1.7 | 150 | 2.0 | 1200 | 1200 | 270 | 12 | 15 | 150 | 40 |
| 60 | 1.5 | 1.8 | 20 | 2.0 | 200 | 2.0 | 1200 | 1200 | 400 | 12 | 15 | 150 | 50 |
| 60 | 1.5 | 1.7 | 19 | 2.0 | 200 | 2.0 | 1200 | 1200 | 350 | 10 | 15 | 150 | 70 |
| 60 | 1.5 | 1.7 | 19 | 2.0 | 200 | 2.0 | 800 | 800 | 350 | 10 | 15 | 150 | 70 |
| 60 | 1.2 | 1.4 | 15 | 2.0 | 200 | 2.0 | 800 | 800 | 350 | 10 | 15 | 150 | 70 |
| 50 | 1.1 | 1.3 | 15 | 1.4 | 150 | 2.0 | 1200 | 1200 | 280 | 15 | 12 | 150 | 45 |
| 60 | 1.1 | 1.3 | 15 | 1.5 | 180 | 2.0 | 1200 | 1200 | 300 | 15 | 12 | 150 | 50 |
| 60 | 1.1 | 1.3 | 15 | 1.6 | 180 | 2.0 | 1200 | 1200 | 280 | 15 | 12 | 150 | 55 |
| 60 | 1.1 | 1.3 | 15 | 1.6 | 180 | 2.0 | 800 | 800 | 280 | 15 | 12 | 150 | 55 |
| 60 | 1.0 | 1.2 | 13 | 1.6 | 180 | 2.0 | 800 | 800 | 280 | 10 | 12 | 150 | 55 |
| 70 | 1.5 | 1.6 | 17 | 2.2 | 400 | 2.2 | 1200 | 1200 | 300 | 30 | 15 | 175 | 65 |
| 95 | 1.6 | 1.8 | 20 | 2.1 | 280 | 2.6 | 1200 | 1200 | 355 | 15 | 19 | 200 | 75 |
| 90 | 1.6 | 1.7 | 20 | 2.1 | 260 | 2.6 | 1200 | 1200 | 340 | 15 | 16 | 200 | 75 |

[4]Niacin equivalent (1 NE = 1 mg of niacin or 60 mg of dietary tryptophan)

figures and will vary in accordance with individual differences in size, gender, age, genetics, and physical activity.

The minimal amounts of dietary protein and fat required to meet the turnover rate are adequate only if they supply sufficient amounts of the essential amino acids and fatty acids. The **essential amino acids** (see table 27.5) are those that must be obtained in the diet because they cannot be made by the body. The **essential fatty acids** are linoleic acid and linolenic acid.

## Vitamins and Minerals

**Vitamins** are small organic molecules that serve as coenzymes in metabolic reactions or that have other, specific functions. They must be obtained in the diet because the body either doesn't produce them, or it produces them in insufficient quantities. (Vitamin D is produced in small quantities by the skin, and the B vitamins and vitamin K are produced by intestinal bacteria.) There are two classes of vitamins: fat-soluble and water-soluble. The *fat-soluble* vitamins include vitamins A, D, E, and K. The *water-soluble* vitamins include thiamine (B$_1$), riboflavin (B$_2$), niacin (B$_3$), pyridoxine (B$_6$), pantothenic acid, biotin, folic acid, vitamin B$_{12}$, and vitamin C (ascorbic acid). Recommended daily allowances for these vitamins are shown in table 27.3.

 β-carotene is a *provitamin*; it is obtained in the diet and converted within the body cells into vitamin A. This conversion occurs when the β-carotene molecules bind free electrons. β-carotene thus serves as an "electron scavenger," or an *antioxidant*. In this role, it may help protect against atherosclerosis and cancer. Other *antioxidants* include the tocopherols (vitamin E) and ascorbic acid (vitamin C).

Derivatives of the water-soluble vitamins serve as coenzymes in the metabolism of carbohydrates, lipids, and proteins. Thiamine, for example, is needed for the activity of the enzyme that converts pyruvic acid to acetyl coenzyme A. Riboflavin and niacin are needed for the production of FAD and NAD, respectively. FAD and NAD serve as coenzymes

## Table 27.4  The Major Vitamins

| Vitamin | Description/comments | Deficiency symptoms | Source |
|---|---|---|---|
| A | Constituent of visual pigment; strengthens epithelial membranes | Night blindness; dry skin | Yellow vegetables and fruit |
| $B_1$ (Thiamine) | Cofactor for enzymes that catalyze decarboxylation | Beriberi; neuritis | Liver, unrefined cereal grains |
| $B_2$ Riboflavin | Part of flavoproteins (such as FAD) | Glossitis; cheilosis | Liver, milk |
| $B_6$ (Pyridoxine) | Coenzyme for decarboxylase and transaminase enzymes | Convulsions | Liver, corn, wheat, and yeast |
| $B_{12}$ (Cyanocobalamin) | Coenzyme for amino acid metabolism; needed for erythropoiesis | Pernicious anemia | Liver, meat, eggs, milk |
| Biotin | Needed for fatty acid synthesis | Dermatitis; enteritis | Egg yolk, liver, tomatoes |
| C | Needed for collagen synthesis in connective tissue | Scurvy | Citrus fruits, green leafy vegetables |
| D | Needed for intestinal absorption of calcium and phosphate | Rickets; osteomalacia | Fish, liver |
| E | Antioxidant | Muscular dystrophy | Milk, eggs, meat, leafy vegetables |
| Folates | Needed for reactions that transfer one carbon | Sprue: anemia | Green leafy vegetables |
| K | Promotes reactions needed for function of clotting factors | Hemorrhage; inability to form clot | Green leafy vegetables |
| Niacin | Part of NAD and NADP | Pellagra | Liver, meat, yeast |
| Pantothenic acid | Part of coenzyme A | Dermatitis; enteritis; adrenal insufficiency | Liver, eggs, yeast |

that transfer hydrogens during cell respiration (chapter 4). Pyridoxine is a cofactor for the enzymes involved in amino acid metabolism. A deficiency of the water-soluble vitamins can thus have widespread effects in the body (table 27.4).

Many fat-soluble vitamins have highly specialized functions. Vitamin K, for example, is required for the production of prothrombin and for clotting factors VII, IX, and X. Vitamin D is converted into a hormone that participates in the regulation of calcium balance (chapter 8). The visual pigments in the rods and cones of the retina are derived from vitamin A. Vitamin A and related compounds, called *retinoids*, also have effects on genetic expression in epithelial cells; these compounds are now used clinically in the treatment of some skin conditions, and researchers are attempting to derive related compounds that may aid the treatment of some cancers.

Elements are needed as cofactors for specific enzymes and for a wide variety of other critical functions. Those that are required daily in relatively large amounts include sodium, potassium, magnesium, calcium, phosphorus, and chlorine (table 27.3). **Trace elements** are required in very small amounts, ranging from 50 μg to 18 mg per day. These include iron, zinc, manganese, fluorine, copper, molybdenum (*mŏ-lib´dĕ-num*), chromium, selenium, and iodine.

# Metabolism of Carbohydrates, Lipids, and Proteins

*Carbohydrates, lipids, and proteins can be used for energy in cell respiration, or they can be interconverted to some degree and stored as energy reserves. The blood concentrations of circulating energy substrates, such as glucose, increase following absorption of a meal.*

## Metabolism of Carbohydrates

The carbohydrates present in food are digested into their component monosaccharides (primarily glucose) and absorbed through the mucosa of the small intestine into the blood. Many organs use this blood glucose as an important energy source. The brain and heart obtain energy (ATP) from glucose via the reactions of aerobic cell respiration. Exercising skeletal muscles can, in addition, obtain energy from glucose via its conversion to lactic acid in the process of anaerobic cell respiration (chapter 4). More glucose may be absorbed following a meal, however, than is immediately needed for energy.

If blood glucose is present in amounts that exceed the body's immediate needs for ATP synthesis, the liver, skeletal muscles, and heart can store carbohydrates in the form of glycogen. The formation of glycogen from glucose is called **glycogenesis.** In this process, glucose is converted to glucose 6-phosphate by utilizing the terminal phosphate group of ATP. Glucose 6-phosphate is then converted into its isomer, glucose 1-phosphate (chapter 4). Finally, the enzyme *glycogen synthetase* removes these phosphate groups as it polymerizes glucose to form glycogen.

The reverse reactions are similar. The enzyme *glycogen phosphorylase* catalyzes the breakdown of glycogen to glucose 1-phosphate. (The phosphates are derived from inorganic phosphate, not from ATP, so glycogen breakdown does not require metabolic energy.) Glucose 1-phosphate is then converted to glucose 6-phosphate. The conversion of glycogen to glucose 6-phosphate is called **glycogenolysis** (gli″kŏ-jĕ-nol´ĭ-sis). In most tissues, glucose 6-phosphate can then be used by the cells for energy (through glycolysis), or for resynthesis of glycogen. Only the liver, however, can convert the glucose 6-phosphate into free glucose for secretion into the blood. This is true for the following reasons.

Organic molecules with phosphate groups cannot cross cell membranes. This has important consequences; the glucose derived from glycogen is in the form of glucose 1-phosphate and then glucose 6-phosphate, so it cannot leak out of the cell. Similarly, glucose that enters the cell from the blood is "trapped" within the cell by conversion to glucose 6-phosphate. Skeletal muscles, which have large amounts of glycogen, can generate glucose 6-phosphate for their own glycolytic needs but cannot secrete glucose into the blood because they lack the ability to remove the phosphate group.

Unlike skeletal muscles, the liver contains an enzyme—known as *glucose 6-phosphatase*—that can remove the terminal phosphate group from the glucose molecule to produce free glucose (fig. 27.1). This free glucose can then be

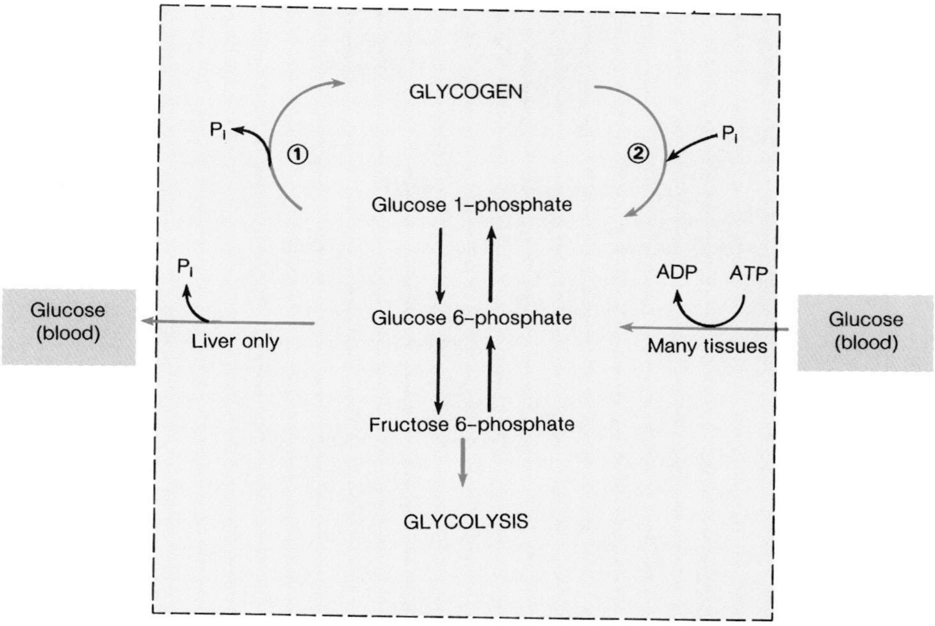

**FIGURE 27.1**

Blood glucose that enters tissue cells is rapidly converted to glucose 6-phosphate. This intermediate can be metabolized for energy in glycolysis or can be converted to glycogen (*1*), a process called *glycogenesis*. Glycogen represents a storage form of carbohydrates, which can be used as a source for new glucose 6-phosphate (*2*), in a process called *glycogenolysis*. The liver contains an enzyme that can remove the phosphate from glucose 6-phosphate; liver glycogen thus serves as a source for new blood glucose.

transported through the cell membrane into the blood. Liver glycogen can thus supply blood glucose for use by other organs, including exercising skeletal muscles that may have depleted much of their own stored glycogen during exercise. Since skeletal muscles lack glucose 6-phosphatase, they cannot secrete glucose into the blood.

## Metabolism of Lipids

It is common experience that eating excessive amounts of carbohydrates increases fat production. It is also well known that foods rich in fat and oil have the highest caloric content. The body obtains 9 kilocalories per gram from fat compared to only 4 kilocalories per gram from either carbohydrates or proteins. Triglycerides (see chapter 2) provide energy primarily through the conversion of fatty acids into acetyl CoA. Acetyl CoA molecules then enter into Krebs cycles and generate ATP through oxidative phosphorylation (fig. 27.2).

**Formation of Fat (Lipogenesis)**  When glucose is to be converted to fat—a process called **lipogenesis** (lip″ŏ-jen´ĕ-sis)—glycolysis occurs and pyruvic acid is converted into acetyl CoA. Instead of entering Krebs cycles, the two-carbon acetic

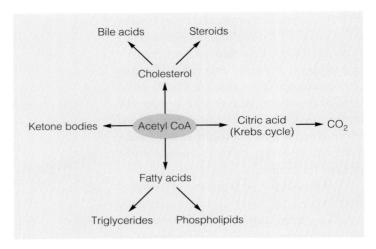

**FIGURE 27.3**

Divergent metabolic pathways for acetyl coenzyme A.

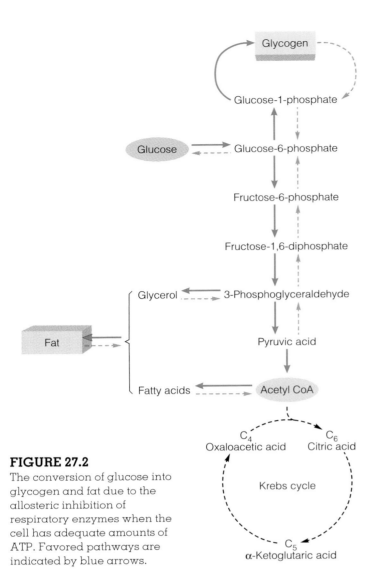

**FIGURE 27.2**

The conversion of glucose into glycogen and fat due to the allosteric inhibition of respiratory enzymes when the cell has adequate amounts of ATP. Favored pathways are indicated by blue arrows.

acid subunits of the acetyl CoA molecules can be used to produce a variety of lipids, including steroids such as cholesterol, and also ketone bodies and fatty acids (fig. 27.3). Acetyl CoA is a branch point from which several metabolic pathways may progress.

In the formation of fatty acids, a number of acetic acid (two-carbon) subunits are joined together to form the fatty acid chain. Six acetyl CoA molecules, for example, will produce a fatty acid that is twelve carbons long. When three of these fatty acids condense with one glycerol (derived from phosphoglyceraldehyde, an intermediate of glycolysis), a triglyceride molecule is formed. Lipogenesis occurs primarily in adipose tissue and in the liver, when the concentration of blood glucose is elevated following a meal.

**Breakdown of Fat (Lipolysis)** When fat stored in adipose tissue is to be used as an energy source, lipase enzymes hy-

drolyze triglycerides into *glycerol* and *free fatty acids* in a process called **lipolysis** (lĭ-pol ́ĭ-sis). These molecules (primarily the free fatty acids) serve as energy sources that can be used by the liver, skeletal muscles, and other organs for aerobic respiration.

A few organs can utilize glycerol for energy by virtue of an enzyme that converts glycerol to phosphoglyceraldehyde. Free fatty acids, however, serve as the major energy source derived from triglycerides. Most fatty acids consist of a long hydrocarbon chain with a carboxylic acid group (COOH) at one end. In a process known as **β-oxidation** (β is the Greek letter *beta*), enzymes remove two-carbon acetic acid molecules from the acid end of a fatty acid. This results in the formation of acetyl CoA, as the third carbon from the end becomes oxidized to produce a new carboxylic acid group. The fatty acid chain is thus decreased in length by two carbons. The process of β-oxidation continues until the entire fatty acid molecule has been converted to acetyl CoA.

A 16-carbon-long fatty acid, for example, yields eight acetyl CoA molecules. Each of these can enter a Krebs cycle and produce twelve ATP per turn of the cycle, so that eight times twelve, or ninety-six, ATP are produced. In addition, each time an acetyl CoA is formed and the end-carbon of the fatty acid chain is oxidized, one NADH and one FADH$_2$ are produced. Oxidative phosphorylation produces three ATP per NADH and two ATP per FADH$_2$. For a 16-carbon-long fatty acid, these five ATP molecules would be formed seven times (producing five times seven, or thirty-five, ATP). Not counting the single ATP used to start β-oxidation, this fatty acid could yield a grand total of 35 + 96, or 131, ATP molecules! Since one triglyceride molecule consists of three fatty acids and one glycerol, the aerobic respiration of one triglyceride molecule yields more than 400 ATP.

**Ketone Bodies**   Even when a person is not losing weight, there is a continuous turnover of triglycerides in adipose tissue. New triglycerides are produced, while others are hydrolyzed into glycerol and fatty acids. This turnover ensures that the blood will normally contain a sufficient level of fatty acids for aerobic respiration by skeletal muscles, the liver, and other organs. When the rate of lipolysis exceeds the rate of fatty acid utilization—as it may do in starvation, dieting, and in diabetes mellitus—the blood concentrations of fatty acids increase.

If the liver cells contain sufficient amounts of ATP so that further production of ATP is not needed, some of the acetyl CoA derived from fatty acids is channeled into an alternate pathway. This pathway involves the conversion of two molecules of acetyl CoA into four-carbon-long acidic derivatives, *acetoacetic acid* and β-*hydroxybutyric acid*. Together with *acetone*, which is a three-carbon-long derivative of acetoacetic acid, these products are known as **ketone bodies.**

 Ketone bodies, which can be used for energy by many organs, are found in the blood under normal conditions. Under conditions of fasting or in the case of *diabetes mellitus*, however, the increased liberation of free fatty acids from adipose tissue results in an elevated production of ketone bodies by the liver. The secretion of abnormally high amounts of ketone bodies into the blood produces *ketosis*, which is one of the signs of fasting or an uncontrolled diabetic state. A person in this condition may have a fruity odor on the breath as acetone vaporizes from the lungs.

## Metabolism of Proteins

Nitrogen is ingested primarily as proteins, enters the body as amino acids, and is excreted mainly as urea in the urine. In childhood, the amount of nitrogen excreted may be less than the amount ingested because amino acids are incorporated into proteins during growth. Growing children are in a state of *positive nitrogen balance*. By contrast, people who are starving or suffering from prolonged wasting diseases are in a state of *negative nitrogen balance*; they excrete more nitrogen than they ingest because they are breaking down their tissue proteins.

Healthy adults maintain a state of nitrogen balance in which the amount of nitrogen excreted is equal to the amount ingested. This does not imply that the amino acids ingested are unnecessary; on the contrary, they are needed to replace the protein that is "turned over" each day. When more amino acids are ingested than are needed to replace proteins, the excess amino acids are not stored as additional protein (one cannot build muscles simply by eating large amounts of protein). Rather, the amine groups can be re-

| Table 27.5 | The essential and nonessential amino acids |
|---|---|
| **Essential amino acids** | **Nonessential amino acids** |
| Lysine | Aspartic acid |
| Tryptophan | Glutamic acid |
| Phenylalanine | Proline |
| Threonine | Glycine |
| Valine | Serine |
| Methionine | Alanine |
| Leucine | Cysteine |
| Isoleucine | Arginine |
| Histidine (children) | Asparagine |
| | Glutamine |
| | Tyrosine |

moved, and the "carbon skeletons" of the organic acids that are left can be used for energy or converted to carbohydrate and fat.

**Transamination**   An adequate amount of all 20 amino acids is required to build proteins for growth and to replace the proteins that are turned over. Fortunately, only 8 (in adults) or 9 (in children) amino acids cannot be produced by the body and so must be obtained in the diet; these are the *essential amino acids* (table 27.5). The remaining amino acids are nonessential only in the sense that the body can produce them if provided with the essential amino acids and with carbohydrates in sufficient amounts.

Pyruvic acid and the Krebs cycle acids are collectively termed *keto acids* because they have a ketone group; these should not be confused with the ketone bodies (derived from acetyl CoA) discussed in the previous section. Keto acids can be converted to amino acids by the addition of an amine ($NH_2$) group. This amine group is usually obtained by cannibalizing another amino acid. A new amino acid is formed as the one that was cannibalized is converted to a new keto acid. This type of reaction, in which the amine group is transferred from one amino acid to form another, is called **transamination** (fig. 27.4).

Each transamination reaction is catalyzed by a specific enzyme (a transaminase) that requires vitamin $B_6$ (pyridoxine) as a coenzyme. The amine group from glutamic acid, for example, may be transferred to either pyruvic acid or oxaloacetic acid. The former reaction is catalyzed by the

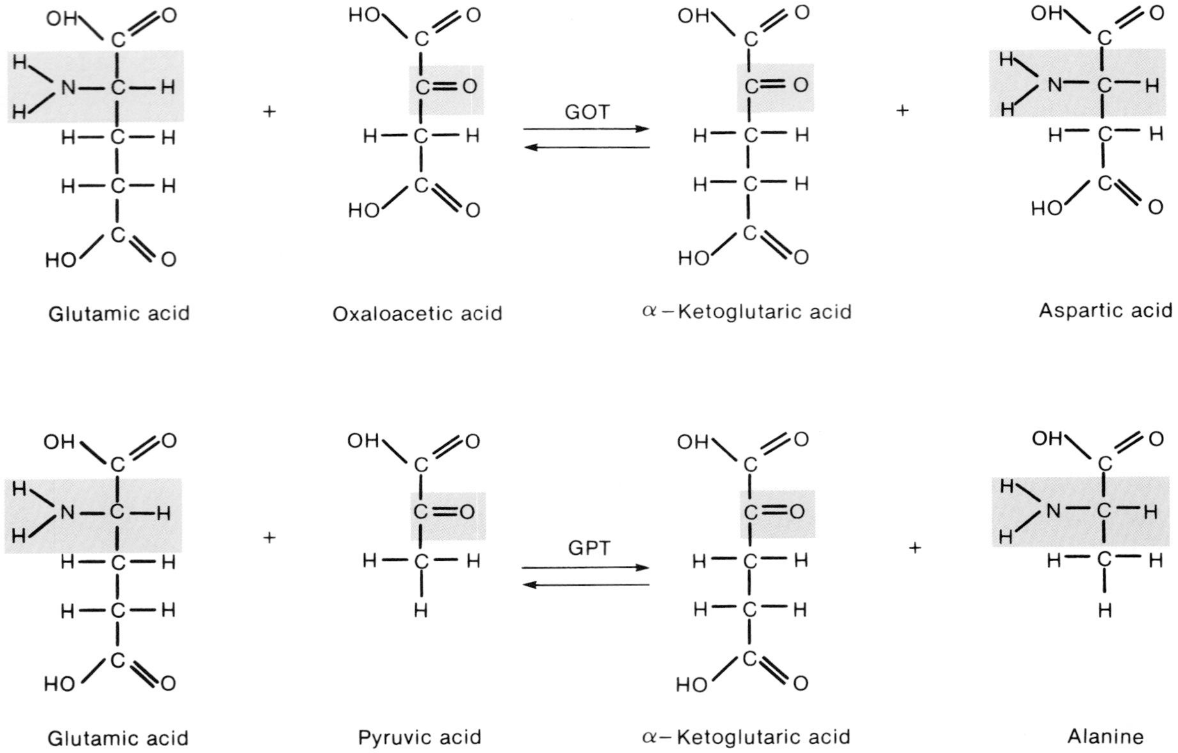

## FIGURE 27.4

The formation of the amino acids aspartic acid and alanine using glutamic acid as the amine donor in transamination (GOT = glutamate oxaloacetate transaminase; GPT = glutamate pyruvate transaminase). The shaded areas show the parts of the molecules that are changed by transamination reactions.

Glutamic acid + Oxaloacetic acid ⇌ (GOT) α–Ketoglutaric acid + Aspartic acid

Glutamic acid + Pyruvic acid ⇌ (GPT) α–Ketoglutaric acid + Alanine

enzyme *glutamate pyruvate transaminase (GPT)*; the latter reaction is catalyzed by *glutamate oxaloacetate transaminase (GOT)*. The addition of an amine group to pyruvic acid produces the amino acid alanine and the addition of an amine group to oxaloacetic acid produces the amino acid known as aspartic acid (fig. 27.4).

**Oxidative Deamination**  As shown in figure 27.5, glutamic acid can be formed through transamination by the combination of an amine group with α-ketoglutaric acid. Glutamic acid is also produced in the liver from the ammonia generated by intestinal bacteria and carried to the liver in the hepatic portal vein. Since free ammonia is very toxic, its removal from the blood and incorporation into glutamic acid is an important function of the healthy liver.

If the body's supply of amino acids exceeds the amount required for protein synthesis, the amine group from glutamic acid may be removed and excreted as *urea* in the urine (fig. 27.5). The metabolic pathway that removes amine groups from amino acids—leaving a keto acid and ammonia (which is converted to urea)—is known as **oxidative deamination** (*ok´´sĭ-da´tiv de-am´´ĭ-na´shun*). A number of amino acids can be converted into glutamic acid by transamination. Since glutamic acid can donate amine groups to urea (through deamination), it serves as a channel through which other amino acids can be used to produce keto acids (pyruvic acid and Krebs cycle acids). These keto acids may then be used in the Krebs cycle as a source of energy (fig. 27.6).

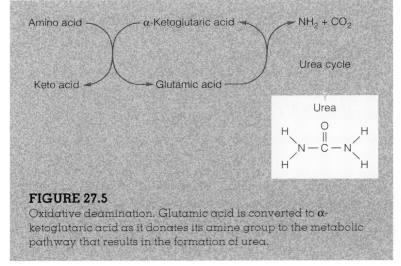

## FIGURE 27.5

Oxidative deamination. Glutamic acid is converted to α-ketoglutaric acid as it donates its amine group to the metabolic pathway that results in the formation of urea.

Depending upon which amino acid is deaminated, the keto acid left over may be either pyruvic acid or one of the Krebs cycle acids. These can be respired for energy, converted to fat, or converted to glucose. In the last case, the amino acids are eventually changed to pyruvic acid, which is used to form glucose. This process, the formation of new glucose from amino acids (or other noncarbohydrate molecules), is called **gluconeogenesis** (*gloo´´ko-ne´´ŏ-jen´ĭ-sis*).

The main substrates for gluconeogenesis are the three-carbon-long molecules of alanine (an amino acid),

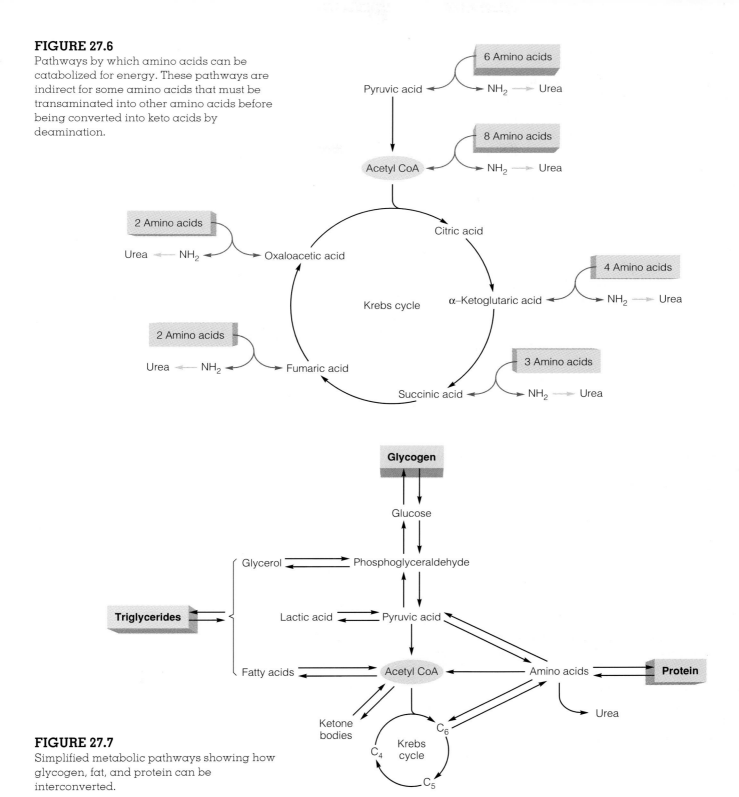

**FIGURE 27.6**

Pathways by which amino acids can be catabolized for energy. These pathways are indirect for some amino acids that must be transaminated into other amino acids before being converted into keto acids by deamination.

**FIGURE 27.7**

Simplified metabolic pathways showing how glycogen, fat, and protein can be interconverted.

lactic acid (derived from glucose during anaerobic respiration), and glycerol (from lipolysis in adipose tissue). The possible interrelationships between amino acids, carbohydrates, and fat are illustrated in figure 27.7. Recent experiments in humans have demonstrated that, even during the early stages of fasting, most of the glucose secreted

from the liver is derived from gluconeogenesis. Findings indicate that hydrolysis of liver glycogen (glycogenolysis) contributes only 36% to the secreted glucose during the early stages of a fast; by 42 hours of fasting, all of the glucose secreted by the liver is being produced by gluconeogenesis.

## Alternative Sources of Energy

The blood serves as a common trough from which all the cells in the body are fed. If all cells used the same energy source, such as glucose, this source would quickly be depleted and cellular starvation would occur. However, normally there are a variety of blood energy sources from which to draw: glucose and ketone bodies from the liver, fatty acids from adipose tissue, and lactic acid and amino acids from muscles. Some organs preferentially use one energy source more than the others, so that each energy source is "spared" for organs with strict energy needs.

The brain uses blood glucose as its major energy source. Under fasting conditions, blood glucose is supplied primarily by the liver through the processes of gluconeogenesis and glycogenolysis. This blood glucose is spared for the brain because skeletal muscles and other organs use fatty acids, ketone bodies, and lactic acid as their major energy sources during fasting. In the case of prolonged fasting, however, the brain too will begin to use ketone bodies as its energy source, in addition to glucose.

The metabolic changes that occur during fasting in a resting person are produced more rapidly in a person who is exercising. Liver (and muscle) glycogen is depleted more rapidly and gluconeogenesis contributes a higher percentage of the hepatic secretion of glucose. Lactic acid produced anaerobically during exercise is also used for energy following exercise. The extra oxygen required to metabolize lactic acid contributes to the *oxygen debt* following exercise (chapter 12).

# Regulation of Energy Metabolism

*Blood plasma contains circulating glucose, fatty acids, amino acids, and other molecules that can be used by body tissues for cell respiration. These circulating molecules may be derived from food or from breakdown of the body's own glycogen, fat, and protein. The building of the body's energy reserves following a meal and the utilization of these reserves between meals, are regulated by the action of hormones that act to promote either anabolism or catabolism.*

The molecules that can be oxidized for energy by the processes of cell respiration may be derived from *energy reserves* of glycogen, fat, or protein. Glycogen and fat serve primarily as energy reserves; for proteins, by contrast, this represents a secondary, emergency function. Although body protein can provide amino acids for energy, it can do so only though the breakdown of proteins needed for muscle contraction, structural strength, enzymatic activity, and other functions. Alternatively, the molecules used for cell

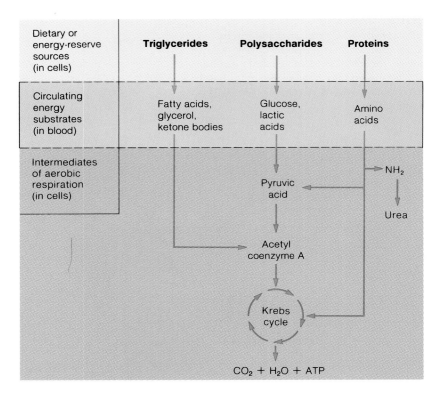

**FIGURE 27.8**
A schematic flowchart of energy pathways in the body.

respiration can be derived from the products of digestion that are absorbed through the small intestine. Since these molecules—glucose, fatty acids, amino acids, and others—are carried by the blood to the tissue cells for use in cell respiration, they can be called **circulating energy substrates** (fig. 27.8).

As discussed in the previous section, different organs have different preferred energy sources. The brain has an almost absolute requirement for blood glucose as its energy source, for example. A fall in the plasma concentration of glucose to below about 50 mg per 100 ml can thus "starve" the brain and have disastrous consequences. Resting skeletal muscles, by contrast, use fatty acids as their preferred energy source. Similarly, ketone bodies (derived from fatty acids), lactic acid, and amino acids can be used to different degrees as energy sources by various organs. The plasma normally contains adequate concentrations of all of these circulating energy substrates to meet the energy needs of the body.

## Eating

Ideally, one should eat the kinds and amounts of foods that provide adequate vitamins, minerals, essential amino acids and fatty acids, and calories. Proper caloric intake maintains energy reserves (primarily fat and glycogen) and results in a body weight within an optimum range for health (see table 27.1).

Body weight tends to be stable despite short-term changes in caloric intake. It has thus been proposed that there may be some regulatory mechanism in the body that is sensitive to the amount of body fat. Although this mechanism is not known, it is clear that there is a relationship between body fat and endocrine function. The secretion of anterior pituitary hormones is affected in a variety of ways. Obese women, for example, may experience menstrual cycle abnormalities and hirsutism (hairiness), whereas women with little body fat—perhaps as the result of extremely strenuous exercise—may experience amenorrhea (cessation of the menstrual cycle). Abnormalities in growth hormone, ACTH, and prolactin secretion have also been observed in obese people.

Eating behavior appears to be at least partially controlled by areas of the hypothalamus. Lesions (destruction) in the ventromedial area of the hypothalamus produce *hyperphagia,* or overeating, and obesity in experimental animals. Lesions of the lateral hypothalamus, by contrast, produce *hypophagia* and weight loss.

Chemical neurotransmitters that may be involved in neural pathways mediating eating behavior are currently being investigated. There is evidence, for example, that endorphins may be involved, since injections of naloxone (a morphine-blocking drug) suppress overeating in rats. There is also evidence that the neurotransmitters norepinephrine and serotonin may be involved; injections of norepinephrine into the brain cause overeating in rats, whereas injections of serotonin have the opposite effect. Interestingly, the intestinal hormone cholecystokinin (CCK) also appears to function as a neurotransmitter in the brain, and it has been shown that injections of CCK cause experimental animals and humans to stop eating. Drugs that block the interaction of CCK with its membrane receptors, conversely, produce overeating in experimental animals.

## Hormonal Regulation of Metabolism

The absorption of energy carriers from the small intestine is not constant; it rises to high levels during a 4-hour period following each meal (the **absorptive state**) and tapers toward zero between meals, after each absorptive state is concluded (the **postabsorptive,** or **fasting, state**). Despite this fluctuation, the plasma concentration of glucose and other energy substrates does not remain high during periods of absorption and does not normally fall below a certain level during periods of fasting. During the absorption of digestion products from the small intestine, energy substrates are removed from the blood and deposited as energy reserves from which withdrawals can be made during times of fasting (fig. 27.9). This ensures an adequate plasma concentration of energy substrates to sustain tissue metabolism at all times.

The rate of deposit and withdrawal of energy substrates into and from the energy reserves and the conversion of one

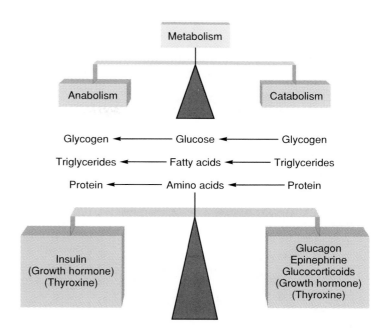

**FIGURE 27.9**

The balance of metabolism can be tilted toward anabolism (synthesis of energy reserves) or catabolism (utilization of energy reserves) by the combined actions of various hormones. Growth hormone and thyroxine have both anabolic and catabolic effects.

type of energy substrate into another are regulated by the actions of hormones. The balance between anabolism and catabolism is determined by the antagonistic effects of insulin, glucagon, growth hormone, thyroxine, and other hormones (fig. 27.9). The specific metabolic effects of these hormones are summarized in table 27.6, and some of their actions are illustrated in figure 27.10.

# Energy Regulation by the Pancreatic Islets

*Insulin secretion is stimulated by a rise in the blood glucose concentration, and insulin promotes the entry of blood glucose into tissue cells. Insulin thus promotes the storage of glycogen and fat while causing the blood glucose concentration to fall. Glucagon secretion is stimulated by a fall in blood glucose, and glucagon acts to raise the blood glucose concentration by promoting glycogenolysis in the liver.*

## Regulation of Insulin and Glucagon Secretion

Insulin and glucagon are hormones secreted by the pancreatic islets (islets of Langerhans), as described in chapter 19. **Insulin** is secreted by the *beta cells* and **glucagon** by the *alpha cells* of the pancreatic islets.

# Table 27.6 Endocrine regulation of metabolism

| Hormone | Blood glucose | Carbohydrate metabolism | Protein metabolism | Lipid metabolism |
|---|---|---|---|---|
| Insulin | Decreased | ↑Glycogen formation<br>↓Glycogenolysis<br>↓Gluconeogenesis | ↑Amino acid transport | ↑Lipogenesis<br>↓Lipolysis<br>↓Ketogenesis |
| Glucagon | Increased | ↓Glycogen formation<br>↑Glycogenolysis<br>↑Gluconeogenesis | No direct effect | ↑Lipolysis<br>↑Ketogenesis |
| Growth hormone | Increased | ↑Glycogenolysis<br>↑Gluconeogenesis<br>↓Glucose utilization | ↑Protein synthesis | ↓Lipogenesis<br>↑Lipolysis<br>↑Ketogenesis |
| Glucocorticoids | Increased | ↑Glycogen formation<br>↑Gluconeogenesis | ↓Protein synthesis | ↓Lipogenesis<br>↑Lipolysis<br>↑Ketogenesis |
| Epinephrine | Increased | ↓Glycogen formation<br>↑Glycogenolysis<br>↑Gluconeogenesis | No direct effect | ↑Lipolysis<br>↑Ketogenesis |
| Thyroxine | No effect | ↑Glucose utilization | ↑Protein synthesis | No direct effect |

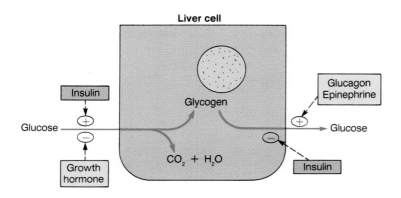

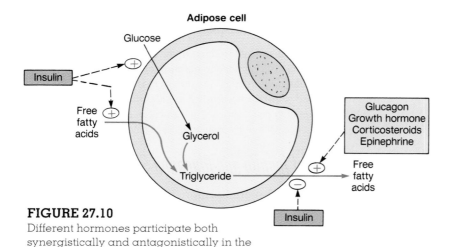

**FIGURE 27.10**

Different hormones participate both synergistically and antagonistically in the regulation of metabolism.

Insulin and glucagon secretion is largely regulated by the plasma concentrations of glucose and, to a lesser degree, of amino acids. The alpha and beta cells act as both the sensors and effectors in this control system. Since the plasma concentration of glucose and amino acids rises during nutrient absorption following a meal and falls during fasting, the secretion of insulin and glucagon likewise fluctuates between the absorptive and postabsorptive states. These changes in insulin and glucagon secretion, in turn, cause changes in blood plasma glucose and amino acid concentrations and thus help to maintain homeostasis via negative feedback loops (fig. 27.11).

**Effects of Glucose** During the absorption of a carbohydrate meal, the blood plasma glucose concentration rises. This rise in plasma glucose (1) stimulates the beta cells to secrete insulin and (2) inhibits the secretion of glucagon from the alpha cells. Insulin acts to stimulate the cellular uptake of plasma glucose. A rise in insulin secretion therefore lowers the blood plasma glucose concentration. Since glucagon has the antagonistic effect of raising the plasma glucose concentration by stimulating glycogenolysis in the liver, the inhibition of glucagon secretion complements the effect of increased insulin during the absorption of a carbohydrate meal. A rise in insulin and a fall in

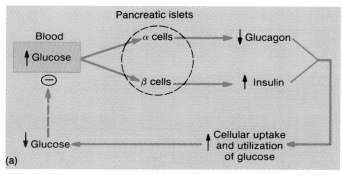

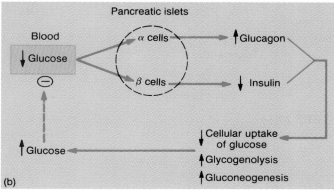

**FIGURE 27.11**

The secretion from the β (beta) cells and α (alpha) cells of the pancreatic islets is regulated to a large degree by the blood glucose concentration. A high blood glucose concentration (a) stimulates insulin and inhibits glucagon secretion. Conversely, a low blood glucose concentration (b) stimulates glucagon and inhibits secretion. The actions of insulin and glucagon provide negative feedback control of the blood glucose concentration.

glucagon secretion thus help to lower the high plasma glucose concentration that occurs during periods of absorption.

During fasting, the plasma glucose concentration falls. At this time, therefore, (1) insulin secretion decreases and (2) glucagon secretion increases. These changes in hormone secretion prevent the cellular uptake of too much blood glucose into the muscles, liver, and adipose tissue and promote the release of glucose from the liver (through the actions of glucagon). A negative feedback loop is therefore completed (fig. 27.11), helping to retard the fall in plasma glucose concentration that occurs during fasting.

The **oral glucose tolerance test** (fig. 27.12) is a measure of the ability of the beta cells to secrete insulin and of the ability of insulin to lower blood glucose. In this procedure, a person drinks a glucose solution and blood samples are taken periodically for plasma glucose measurements. In a normal person, the rise in blood glucose produced by drinking this solution is reversed to normal levels within 2 hours following glucose ingestion.

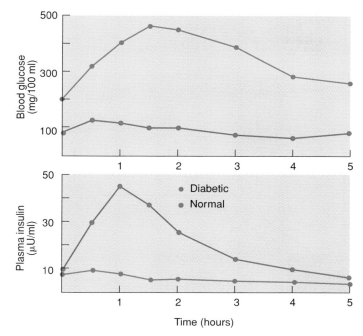

**FIGURE 27.12**

Changes in blood glucose and plasma insulin concentrations after the ingestion of 100 grams of glucose in an oral glucose tolerance test. (Insulin is measured in activity units [U].)

 People with *diabetes mellitus*—due to the inadequate secretion or action of insulin—maintain a state of high plasma glucose concentration (hyperglycemia) during the oral glucose tolerance test (fig. 27.12). People who have *reactive hypoglycemia* (low plasma glucose concentration due to excessive insulin secretion) have lower-than-normal blood glucose concentrations 5 hours following glucose ingestion. These conditions are described in detail under "Clinical Considerations."

**Effects of Autonomic Nerves**  The pancreatic islets receive both parasympathetic and sympathetic innervation. The activation of the parasympathetic system during meals stimulates insulin secretion at the same time that gastrointestinal function is stimulated. The activation of the sympathetic system, by contrast, stimulates glucagon secretion and inhibits insulin secretion. The effects of glucagon, together with those of epinephrine, produce a "stress hyperglycemia" when the sympathoadrenal system is activated.

**Effects of GIP**  Surprisingly, insulin secretion increases more rapidly following glucose ingestion than it does following an intravenous injection of glucose. This is due to the fact that the small intestine, in response to glucose ingestion, secretes a hormone that stimulates insulin secretion before the glucose is absorbed. Insulin secretion thus begins to rise in anticipation of a rise in blood glucose. The intestinal hormone that

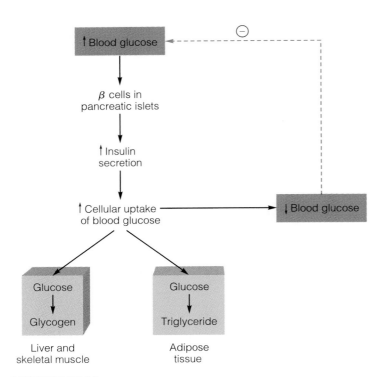

**FIGURE 27.13**

A rise in blood glucose concentration stimulates insulin secretion. Insulin promotes a fall in blood glucose by stimulating the cellular uptake of glucose and the conversion of glucose to glycogen and fat.

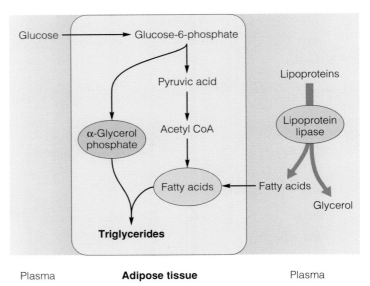

**Adipose tissue**

**FIGURE 27.14**

The synthesis of triglycerides (fat) within adipose cells. Notice that fat can be produced from glucose and from fatty acids released by the hydrolysis of plasma lipoproteins.

mediates this effect is believed to be GIP—gastric inhibitory peptide, or, more appropriately in this context, **glucose-dependant insulinotropic peptide** (chapter 26).

The mechanisms that regulate insulin and glucagon secretion and the actions of these hormones normally prevent the plasma glucose concentration from rising above 170 mg per 100 ml after a meal or from falling below about 50 mg per 100 ml between meals. This regulation is important because abnormally high blood glucose can damage tissue cells (as may occur in diabetes mellitus) and abnormally low blood glucose can damage the brain. The latter effect results from the fact that glucose enters the brain by facilitated diffusion. When the rate of this diffusion is too low due to low plasma glucose concentrations, the supply of metabolic energy for the brain may become inadequate and result in weakness, dizziness, personality changes, and ultimately in coma and death.

## Insulin and Glucagon: Absorptive State

The lowering of blood glucose by insulin is, in a sense, a side effect of the primary action of this hormone. Insulin is the major hormone that promotes anabolism in the body. During absorption of the products of digestion and the subsequent rise in the plasma concentrations of circulating energy substrates, insulin promotes the cellular uptake of glucose and its conversion to glycogen and fat (fig. 27.13). Quanti-

tatively, skeletal muscles are responsible for most of the insulin-stimulated glucose uptake. Insulin also promotes the cellular uptake of amino acids and their conversion to proteins. The stores of large, energy-reserve molecules are thus increased while the plasma concentrations of glucose and amino acids are decreased.

The synthesis of triglycerides (fat) within adipose cells depends upon insulin-stimulated glucose uptake from blood plasma. Once inside the adipose cells, glucose can be converted into α-glycerol phosphate and acetyl CoA. The α-glycerol phosphate is used to form glycerol, and the acetyl CoA is incorporated into fatty acids. The condensation of three fatty acids with glycerol then yields triglycerides (fig. 27.14). Entry of blood glucose into adipose cells—which is directly dependent on insulin—thus determines the rate at which fat is produced.

*Obesity* is a condition of excessive stored fat. A nonobese 70-kg (154-lb) man has approximately 10 kg (about 82,500 kcal) of stored fat. Since 250 g of fat can supply the energy requirements for 1 day, this reserve fuel is sufficient for about 40 days. Glycogen is less efficient as an energy reserve, and less is stored in the body; the glycogen stored in the liver amounts to about 100 g (400 kcal) and, in skeletal muscles, to between 375 and 400 g (1500 kcal). Insulin promotes the cellular uptake of glucose into the liver and muscles and the conversion of glucose into glucose 6-phosphate. In the liver and muscles, this can be changed into glucose 1-phosphate, which serves as the precursor of glycogen. Once the stores of glycogen have been filled, the continued ingestion of excess calories results in the continued production of fat rather than of glycogen.

## Insulin and Glucagon: Postabsorptive State

Glucagon stimulates and insulin suppresses the hydrolysis of liver glycogen, or **glycogenolysis.** Thus, during times of fasting when glucagon secretion is high and insulin secretion is low, liver glycogen is used as a source of additional blood glucose. This process is essentially the reverse of that by which glycogen was formed and results in the liberation of free glucose from glucose 6-phosphate, as previously described.

Since only about 100 g of glycogen are stored in the liver, adequate blood glucose levels could not be maintained for very long during fasting using this source alone. The low levels of insulin secretion during fasting, together with elevated glucagon secretion, however, promote gluconeogenesis. Low insulin allows the release of amino acids from skeletal muscle proteins, while glucagon and cortisol (discussed later) stimulate the production of enzymes in the liver that convert amino acids to pyruvic acid and pyruvic acid to glucose. During prolonged fasting and exercise, gluconeogenesis in the liver using amino acids from muscle proteins may be the only source of blood glucose.

The secretion of glucose from the liver during fasting compensates for the low blood glucose concentrations and helps to provide the brain with the glucose it needs. But, because insulin secretion is low during fasting, skeletal muscles cannot utilize blood glucose as an energy source. Instead, as mentioned previously, the skeletal muscles—as well as the heart, liver, and kidneys—use free fatty acids as their major source of fuel. This helps to spare glucose for the brain.

The free fatty acids are made available by the action of glucagon. In the presence of low insulin levels, glucagon stimulates an enzyme called **hormone-sensitive lipase** in adipose cells. This enzyme catalyzes the hydrolysis of stored triglycerides and the release of free fatty acids and glycerol into the blood. Glucagon also stimulates enzymes in the liver that convert some of these fatty acids into ketone bodies, which

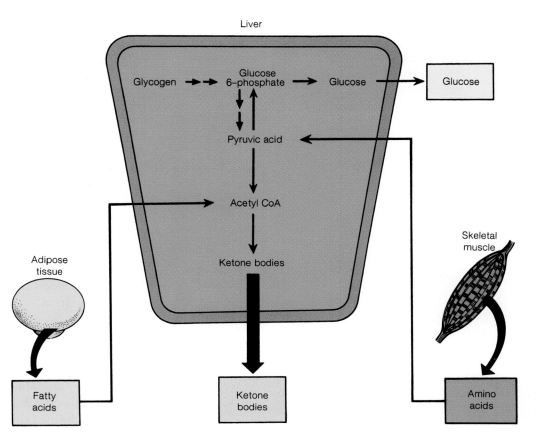

**FIGURE 27.15**

Increased glucagon secretion and decreased insulin secretion during fasting favors catabolism. These hormonal changes result in elevated release of glucose, fatty acids, ketone bodies, and amino acids into the blood. Notice that the liver secretes glucose that is derived both from the breakdown of liver glycogen and from the conversion of amino acids in gluconeogenesis.

are secreted into the blood (fig. 27.15). Several organs in the body can use ketone bodies, as well as fatty acids, as a source of acetyl CoA in aerobic respiration.

Through the stimulation of lipolysis and the formation of ketone bodies, the high glucagon and low insulin levels that occur during fasting provide circulating energy substrates for use by the muscles, liver, and other organs. At the same time, this hormonal state promotes liver glycogenolysis and gluconeogenesis, which provide adequate levels of blood glucose to sustain the metabolism of the brain. The antagonistic action of insulin and glucagon (fig. 27.16) thus promotes appropriate metabolic responses during periods of fasting and periods of absorption.

Regulation of Metabolism

827

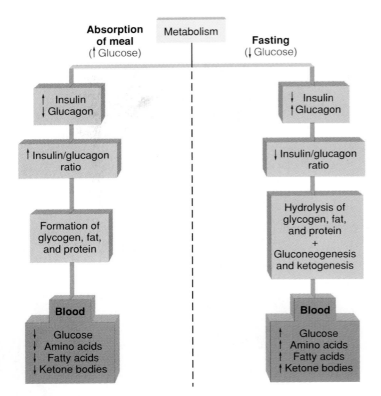

**FIGURE 27.16**
The inverse relationship between insulin and glucagon secretion during the absorption of a meal and during fasting. Changes in the insulin-to-glucagon ratio tilt metabolism toward anabolism during the absorption of food and toward catabolism during fasting.

# Metabolic Regulation by Adrenal Hormones, Thyroxine, and Growth Hormone

*Epinephrine, the glucocorticoids, thyroxine, and growth hormone stimulate the catabolism of carbohydrates and lipids. These hormones are thus antagonistic to insulin in their regulation of carbohydrate and lipid metabolism. Thyroxine and growth hormone, however, stimulate protein synthesis and are needed for body growth and proper development of the central nervous system. These hormones thus have an anabolic effect, which is complementary to that of insulin, on protein synthesis.*

The anabolic effects of insulin are antagonized by glucagon, as previously described, and by the actions of a variety of other hormones. The hormones of the adrenal glands, thyroid, and anterior pituitary (specifically growth hormone) antagonize the action of insulin on carbohydrate and lipid

metabolism. The actions of insulin, thyroxine, and growth hormone, however, can act synergistically in the stimulation of protein synthesis.

## Adrenal Hormones

As described in chapter 19, the adrenal gland consists of two different parts with different embryonic origins that function as separate glands. The two parts secrete different hormones and are regulated by different control systems. The **adrenal medulla** is derived from the neural crest and secretes catecholamine hormones—epinephrine and lesser amounts of norepinephrine—in response to sympathetic nerve stimulation. The **adrenal cortex** is derived from mesoderm and secretes corticosteroid hormones. The corticosteroids are grouped into two functional categories: **mineralocorticoids,** such as aldosterone, which regulate Na$^+$ and K$^+$ balance, and **glucocorticoids** (*gloo˝ko-kor´tĭ-koidz*), such as hydrocortisone (cortisol), which participate in metabolic regulation.

**Metabolic Effects of Epinephrine**   The metabolic effects of epinephrine are similar to those of glucagon. Both stimulate glycogenolysis and the release of glucose from the liver, and both stimulate lipolysis and the release of fatty acids from adipose tissue. These actions occur in response to glucagon during fasting, when low blood glucose stimulates glucagon secretion, and in response to epinephrine during the fight-or-flight reaction to stress. The latter effect provides circulating energy substrates in anticipation of the need for intense physical activity. Glucagon and epinephrine have similar mechanisms of action; the actions of both are mediated by cyclic AMP (fig. 27.17).

**Metabolic Effects of Glucocorticoids**   Hydrocortisone (cortisol) and other glucocorticoids are secreted by the adrenal cortex in response to ACTH stimulation. The secretion of ACTH from the anterior pituitary occurs as part of the general adaptation syndrome in response to stress (chapter 19). Since prolonged fasting or prolonged exercise certainly qualify as stressors, ACTH—and thus glucocorticoid secretion—is stimulated under these conditions. The increased secretion of glucocorticoids during prolonged fasting or exercise supports the effects of increased glucagon and decreased insulin secretion from the pancreatic islets.

Like glucagon, hydrocortisone promotes lipolysis and the formation of ketone bodies. Its major action, however, is the stimulation of *gluconeogenesis* in the liver. Although hydrocortisone stimulates enzyme (protein) synthesis in the liver for gluconeogenesis, it promotes protein breakdown in the muscles. This latter effect increases the blood levels of amino acids and provides the substrates needed by the liver for gluconeogenesis. The release of circulating energy substrates—

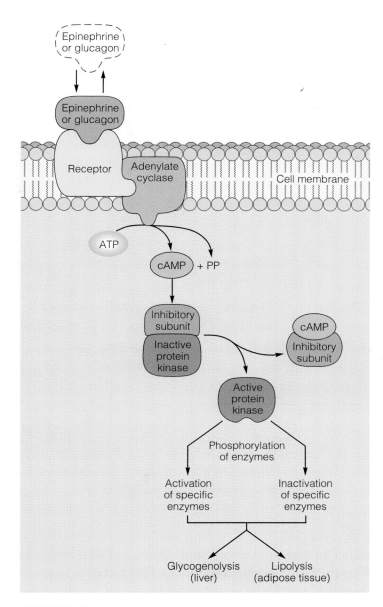

**FIGURE 27.17**
Cyclic AMP (cAMP) serves as a second messenger in the actions of epinephrine and glucagon on liver and adipose tissue metabolism.

amino acids, glucose, fatty acids, and ketone bodies—into the blood in response to hydrocortisone (fig. 27.18) helps to compensate for a state of prolonged fasting or exercise. Whether these metabolic responses are beneficial in other stressful states is open to question.

## *Thyroxine*

The thyroid gland follicles secrete thyroxine, also called tetraiodothyronine ($T_4$), in response to stimulation by thyroid-stimulating hormone (TSH) from the anterior pituitary. Almost all organs in the body are targets of thyroxine action. Thyroxine itself, however, is not the active

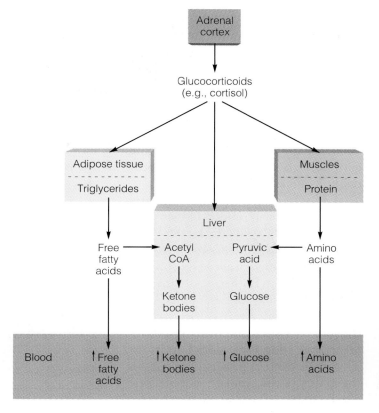

**FIGURE 27.18**
The catabolic actions of glucocorticoids help raise the blood concentration of glucose and other energy-carrier molecules.

form of the hormone within the target cells; thyroxine is a prehormone that must first be converted to triiodothyronine ($T_3$) within the target cells to be active (chapter 19). Acting via its conversion to $T_3$, thyroxine (1) stimulates the rate of cell respiration and (2) contributes to proper growth and development, particularly during early childhood.

**Thyroxine and Cell Respiration**   Thyroxine stimulates the rate of cell respiration in almost all cells in the body. This is believed to be due to thyroxine-induced lowering of cellular ATP concentrations. ATP exerts an end-product inhibition (chapter 4) of cell respiration so that when ATP concentrations increase, the rate of cell respiration decreases. Conversely, a lowering of ATP concentrations, as may occur in response to thyroxine, stimulates cell respiration.

The metabolic rate under carefully controlled resting conditions is known as the basal metabolic rate (BMR), as previously described. Acting through its stimulation of cell respiration, thyroxine acts to "set" the BMR. The BMR can thus be used as an index of thyroid gland function. Indeed, such measurements were used clinically to evaluate thyroid function prior to the development of direct chemical determinations of $T_4$ and $T_3$ in the blood.

The coupling of energy-releasing reactions to energy-requiring reactions is never 100% efficient; a proportion of the energy is always lost as heat. Much of the energy liberated during cell respiration and much of the energy released by the hydrolysis of ATP escapes as heat. Since thyroxine stimulates both ATP consumption and cell respiration, the actions of thyroxine result in the production of metabolic heat.

The heat-producing, or **calorigenic** (*calor* = heat), **effects** of thyroxine are required for cold adaptation. This does not mean that people who are cold-adapted have high levels of thyroxine secretion. Rather, thyroxine levels in the normal range coupled with the increased activity of the sympathoadrenal system and other responses are responsible for cold adaptation. Thyroxine exerts a permissive effect on the ability of the sympathoadrenal system to increase heat production in response to cold stress.

### Thyroxine in Growth and Development

Through its stimulation of cell respiration, thyroxine stimulates the increased consumption of circulating energy substrates, such as glucose, fatty acids, and other molecules. These effects, however, are mediated at least in part by the activation of genes; thyroxine stimulates both RNA and protein synthesis. As a result of its stimulation of protein synthesis throughout the body, thyroxine is considered to be an anabolic hormone like insulin and growth hormone.

Because of its stimulation of protein synthesis, thyroxine is needed for growth of the skeleton and, most importantly, for the proper development of the central nervous system. Recent evidence has demonstrated the presence of receptor proteins for $T_3$ in the neurons and astrocytes of the brain. This need for thyroxine is particularly great when the brain is undergoing its greatest rate of development—from the end of the first trimester of prenatal life to 6 months after birth. Hypothyroidism during this period may result in **cretinism.** Unlike people with dwarfism, who have normal thyroxine secretion but a low secretion of growth hormone, people with cretinism suffer from severe mental retardation. Treatment with thyroxine soon after birth, particularly before 1 month of age, has been found to restore development of intelligence as measured by IQ tests administered 5 years later.

### Hypothyroidism and Hyperthyroidism

As might be predicted from the effects of thyroxine, people who are hypothyroid have an abnormally low BMR and experience weight gain and lethargy. In addition, the ability to adapt to cold stress is diminished in the case of thyroxine deficiency. The most severe form of hypothyroidism is **myxedema** (*mik″sĭ-de′mă*)— a condition characterized by accumulation of mucoproteins in subcutaneous connective tissues. Hypothyroidism has numerous causes, including insufficient thyrotropin-releasing hormone (TRH) secretion from the hypothalamus, insufficient TSH secretion from the pituitary gland, and insufficient iodine in the diet. Hypothyroidism due to lack of iodine is accompanied by excessive TSH secretion, which stimulates abnormal growth of the thyroid gland (goiter). This condition can be reversed by iodine supplements.

A goiter can also be produced by another mechanism. In **Graves' disease,** apparently an autoimmune disease, autoantibodies (chapter 23) exert TSH-like effects on the thyroid. Since the production of these autoantibodies is not controlled by negative feedback, the thyroid gland is stimulated excessively, thus producing the goiter associated with a hyperthyroid state. Hyperthyroidism produces a high BMR accompanied by weight loss, nervousness, irritability, and an intolerance to heat. The symptoms of hypothyroidism and hyperthyroidism are compared in table 27.7.

| Table 27.7 | Comparison of hypothyroidism and hyperthyroidism | |
| --- | --- | --- |
| | *Hypothyroid* | *Hyperthyroid* |
| Growth and development | Impaired growth | Accelerated growth |
| Activity and sleep | Decreased activity; increased sleep | Increased activity; decreased sleep |
| Temperature tolerance | Intolerance to cold | Intolerance to heat |
| Skin characteristics | Coarse, dry skin | Smooth skin |
| Perspiration | Absent | Excessive |
| Pulse | Slow | Rapid |
| Gastrointestinal symptoms | Constipation; decreased appetite; increased weight | Frequent bowel movements; increased appetite; decreased weight |
| Reflexes | Slow | Rapid |
| Psychological aspects | Depression and apathy | Nervousness and emotionality |
| Plasma $T_4$ levels | Decreased | Increased |

Graves' disease: from Robert James Graves, Irish physician, 1796–1853

## Growth Hormone

The anterior pituitary secretes **growth hormone,** also called **somatotropic hormone,** in larger amounts than any other of its hormones. As its name implies, growth hormone stimulates growth in children and adolescents. Growth hormone secretion continues in adults, particularly under the conditions of fasting and other forms of stress. This implies that growth hormone can have important metabolic effects even after the growing years have ended.

**Regulation of Growth Hormone Secretion**   The secretion of growth hormone is inhibited by somatostatin, which is produced by the hypothalamus and secreted into the hypothalamo-hypophyseal portal system (chapter 19). In addition, a hypothalamic-releasing hormone stimulates growth hormone secretion. Growth hormone thus appears to be unique among the anterior pituitary hormones in that its secretion is controlled by both a releasing and an inhibiting hormone from the hypothalamus. The secretion of growth hormone follows a circadian, or 24-hour, rhythm, increasing during sleep and decreasing during periods of wakefulness.

Growth hormone secretion is stimulated by an increase in the plasma concentrations of amino acids and by a decrease in the plasma glucose concentration. These events occur during absorption of a high protein meal, when amino acids are absorbed. The secretion of growth hormone is also increased during prolonged fasting when plasma glucose is low and plasma amino acid concentration is raised by the breakdown of muscle protein.

**Insulin-Like Growth Factors**   **Insulin-like growth factors (IGF)** are polypeptides produced by many tissues that are similar in structure to proinsulin (chapter 2), have insulin-like effects, and also serve as mediators for some of growth hormone's actions. The term **somatomedins** is often used to refer to two of these factors, designated IGF-1 and IGF-2. The liver produces and secretes IGF-1 in response to growth hormone stimulation (fig. 27.19), and this secreted IGF-1 then functions as a hormone in its own right, traveling in the blood to the target tissue. A major target is cartilage, where IGF-1 stimulates cell division and growth. IGF-1 also functions as an autocrine regulator (chapter 19) because the chondrocytes (cartilage cells) themselves produce some IGF-1 in response to growth hormone stimulation. The growth-promoting actions of IGF-1 thus directly mediate the effects of growth hormone on cartilage. These actions are supported by IGF-2, which has more insulin-like actions.

**Effects of Growth Hormone on Metabolism**   The fact that growth hormone secretion is increased during fasting and also during absorption of a protein meal reflects the complex nature of this hormone's action. Growth hormone has both

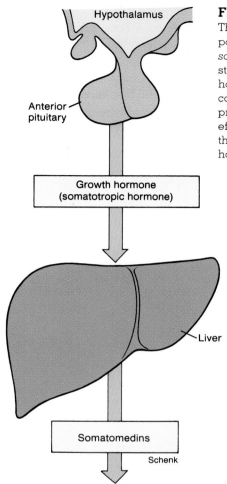

**FIGURE 27.19**
The hepatic production of polypeptides called *somatomedins* is stimulated by growth hormone. These compounds in turn produce the anabolic effects that characterize the actions of growth hormone in the body.

Hypothalamus

Anterior pituitary

Growth hormone (somatotropic hormone)

Liver

Somatomedins

Schenk

Target organs

anabolic and catabolic effects; it promotes protein synthesis (anabolism), and in this respect is similar to insulin. It also stimulates the catabolism of fat and the release of fatty acids from adipose tissue—effects similar to those of glucagon.

In terms of its action on lipid and carbohydrate metabolism, growth hormone is said to have an anti-insulin effect. A rise in the plasma fatty acid concentration induced by growth hormone results in decreased rates of glycolysis in many organs. This inhibition of glycolysis by fatty acids, perhaps along with a more direct action of growth hormone, results in decreased glucose utilization by the tissues. Growth hormone thus acts to raise the blood glucose concentration.

Growth hormone stimulates the cellular uptake of amino acids and protein synthesis in many organs of the body—actions that are useful when eating a protein-rich meal. Amino acids are removed from the blood and used to form proteins, and the blood concentration of glucose and fatty acids is increased to provide alternate energy sources (fig. 27.20). The anabolic effect of growth hormone on protein synthesis is particularly important during the growing years, when it contributes to increases in bone length and in the mass of many soft tissues.

**Effects of Growth Hormone on Body Growth** The stimulatory effects of growth hormone on bone growth results from stimulation of mitosis in the epiphyseal plates of cartilage present in the long bones of growing children and adolescents (see fig. 8.11). This action is mediated by the somatomedins, IGF-1 and IGF-2, which stimulate the chondrocytes to divide and screte more cartilage matrix. Part of this growing cartilage is converted to bone, enabling the

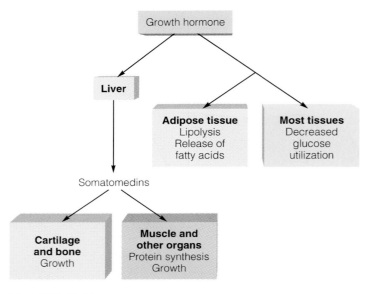

bone to grow in length. Bone growth stops when the epiphyseal plates ossify after the growth spurt during puberty, despite the fact that growth hormone secretion continues throughout adulthood.

An excessive secretion of growth hormone in children can produce **gigantism.** This condition is characterized by an exceptionally rapid growth in height (often to 8 feet tall), at the same time maintaining normal body proportions. Excessive growth hormone secretion that occurs after the epiphyseal plates have sealed, however, cannot produce increases in height. The oversecretion of growth hormone in adults results in an elongation of the jaw and deformities in the bones of the face, hands, and feet. This condition, called **acromegaly,** is accompanied by the growth of soft tissues and coarsening of the skin (fig. 27.21). It is interesting that athletes who (illegally) take growth hormone supplements to increase their muscle mass may also experience body changes similar to those that occur in acromegaly.

An inadequate secretion of growth hormone during the growing years results in **dwarfism.** An interesting variant of this condition is *Laron dwarfism*, in which there is a genetic insensitivity to the effects of growth hormone. This insensitivity is associated with, but may not be caused by, a reduction in the number of growth hormone receptors in the target cells. Genetic engineering has made available recombinant IGF-1, which has recently been approved by the FDA for the medical treatment of Laron dwarfism.

**FIGURE 27.20**

The effects of growth hormone. The growth-promoting, or anabolic, effects of growth hormone are mediated indirectly via stimulation and somatomedin production by the liver.

acromegaly: Gk. *akron*, extremity; *mega*, large
Laron dwarfism: from Zui Laron, Israeli endocrinologist, b. 1927

Age 9

Age 16

Age 33

Age 52

**FIGURE 27.21**

The progression of acromegaly in one individual. The coarsening of features and disfigurement are evident by age 33 and severe at age 52.

An adequate diet, particularly of proteins, is required for the production of IGF-1. This helps to explain why many children are significantly taller than their parents, who may not have had an adequate diet in their youth. Children with protein malnutrition (*kwashiorkor*) have low growth rates and low IGF-1 levels in the blood, even though their growth hormone levels are abnormally elevated. When these children are provided with an adequate diet, IGF-1 levels and growth rates increase.

# Clinical Considerations

Chronic high blood glucose, or *hyperglycemia*, is the hallmark of the disease **diabetes mellitus.** The name of this disease comes from the fact that glucose "spills over" into the urine when the blood glucose concentration is too high (*mellitus* is derived from the Latin word meaning "honeyed" or "sweet"). The hyperglycemia of diabetes mellitus results from either the insufficient secretion of insulin by the beta cells of the pancreatic islets or the inability of secreted insulin to stimulate the cellular uptake of glucose from the blood. Diabetes mellitus, in short, results from the inadequate secretion or action of insulin.

There are two forms of diabetes mellitus. In **insulin-dependent diabetes mellitus (IDDM),** also called *type I diabetes*, the beta cells are progessively destroyed and secrete little or no insulin. This form of the disease accounts for only about 10% of the cases of diabetes in the country. About 90% of the people who have diabetes have **non-insulin-dependent diabetes mellitus (NIDDM),** also called *type II diabetes*. Type I diabetes was once known as *juvenile-onset diabetes*, because this condition is usually diagnosed in

people under the age of twenty. Type II diabetes has also been called *maturity-onset diabetes*, because it is usually diagnosed in people over the age of forty. Some comparisons of these two forms of diabetes mellitus are shown in table 27.8. It should be noted that table 27.8 compares only the early stages of IDDM and NIDDM because, in some people, NIDDM can grade into and later become IDDM.

## Insulin-Dependent Diabetes Mellitus

Insulin-dependent diabetes mellitus results when the beta cells of the pancreatic islets are progressively destroyed by autoimmune attack. Recent evidence in mice suggests that killer T lymphocytes (chapter 23) may target an enzyme known as glutamate decarboxylase in the beta cells. This autoimmune destruction of the beta cells may be provoked by an environmental agent, such as infection by viruses. In other cases, however, the cause is currently unknown. Removal of the insulin-secreting beta cells causes hyperglycemia and the appearance of glucose in the urine. Without insulin, glucose cannot enter the adipose cells; the rate of fat synthesis thus lags behind the rate of fat breakdown and large amounts of free fatty acids are released from the adipose cells.

In a person with uncontrolled IDDM, many of the fatty acids released from adipose cells are converted into ketone bodies in the liver. This may result in an elevated ketone body concentration in the blood (*ketosis*), and if the buffer reserve of bicarbonate is neutralized, it may also result in *ketoacidosis*. During this time, the glucose and excess ketone bodies that are excreted in the urine act as osmotic diuretics and cause the excessive excretion of water in the urine. This can produce severe dehydration, which, together with ketoacidosis and associated disturbances in electrolyte balance, may lead to coma and death (fig. 27.22).

## Table 27.8    Comparison of the two major forms of diabetes mellitus

| Characteristics | Insulin-dependent (Type I) | Non-insulin-dependent (Type II) |
|---|---|---|
| Usual age of onset | Under 20 years | Over 40 years |
| Development of symptoms | Rapid | Slow |
| Percentage of diabetic population | About 10% | About 90% |
| Development of ketoacidosis | Common | Rare |
| Association with obesity | Rare | Common |
| Beta cells of pancreatic islets | Destroyed | Usually not destroyed |
| Insulin secretion | Decreased | Normal or increased |
| Autoantibodies to pancreatic islet cells | Present | Absent |
| Associated with particular HLA antigens | Yes | No |
| Usual treatment | Insulin injections | Diet; oral stimulators of insulin secretion |

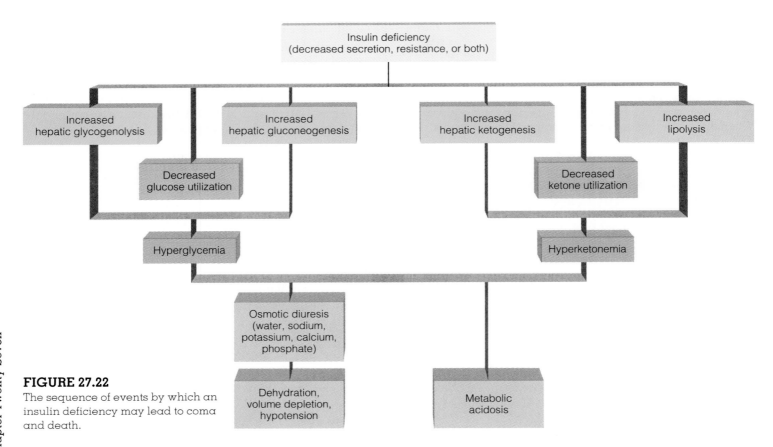

**FIGURE 27.22**

The sequence of events by which an insulin deficiency may lead to coma and death.

In addition to the lack of insulin, people with IDDM have an abnormally high secretion of glucagon from the alpha cells of the pancreatic islets. Glucagon stimulates glycogenolysis in the liver and thus helps to raise the blood glucose concentration. Glucagon also stimulates the production of enzymes in the liver that convert fatty acids into ketone bodies. Some researchers believe that the full symptoms of diabetes result from high glucagon secretion as well as from the absence of insulin. The lack of insulin may be largely responsible for hyperglycemia and for the release of large amounts of fatty acids into the blood. The high glucagon secretion may contribute to the hyperglycemia and be largely responsible for the development of ketoacidosis.

## Non-Insulin-Dependent Diabetes Mellitus

The effects produced by insulin, or any hormone, depend on the concentration of that hormone in the blood and on the sensitivity of the target tissue to given amounts of the hormone. Tissue responsiveness to insulin, for example, varies under normal conditions. For reasons that are not completely understood, exercise increases insulin sensitivity and obesity decreases insulin sensitivity of the target tissues. The pancreatic islets of a nondiabetic obese person, therefore, must secrete high amounts of insulin to maintain the blood glucose concentration in the normal range. Conversely, nondiabetic people who are thin

and exercise regularly require lower insulin secretion to maintain the proper blood glucose concentration.

Non-insulin-dependent diabetes is usually slow to develop, is hereditary, and occurs most often in people who are overweight. Genetic factors are very significant; people at highest risk are those who have both parents with NIDDM and who are members of certain ethnic groups, particularly Mexican-Americans and Pima Indians. Unlike IDDM, people who have NIDDM can have normal or even elevated levels of insulin in the blood. Despite this, people with NIDDM have hyperglycemia, if untreated. This must mean that, even though the insulin levels may be in the normal range, the amount of insulin secreted is inadequate.

Much evidence has been obtained to show that people with NIDDM have an abnormally low tissue sensitivity to insulin, or an *insulin resistance*. This is true even if the person is not obese, but the problem is compounded by the decreased tissue sensitivity that accompanies obesity, particularly of the "apple-shape" variety. There is also evidence that the beta cells are not functioning correctly; whatever amount of insulin they secrete is inadequate to the task. Pre-diabetics often have elevated levels of insulin with hypoglycemia, suggesting insulin resistance. People with established NIDDM have both an insulin resistance and insulin deficiency.

Since obesity decreases insulin sensitivity, people who are genetically predisposed to insulin resistance may develop

# Table 27.9 Comparison of diabetic and hypoglycemic coma

| | Diabetic ketoacidosis | Hypoglycemia |
|---|---|---|
| Onset | Hours to days | Minutes |
| Causes | Insufficient insulin; other diseases | Excess insulin; insufficient food; excessive exercise |
| Symptoms | Excessive urination and thirst; headache, nausea, and vomiting | Hunger, headache, confusion, stupor |
| Physical findings | Deep, labored breathing; breath has acetone odor; blood pressure decreased, pulse weak; skin is dry | Pulse, blood pressure, and respiration are normal; skin is pale and moist |
| Laboratory findings | Urine: glucose present, ketone bodies increased<br>Blood plasma: glucose and ketone bodies increased, bicarbonate decreased | Urine: normal<br>Blood plasma: glucose concentration low, bicarbonate normal |

symptoms of diabetes when they gain weight. Conversely, this type of diabetes mellitus can usually be controlled by increasing tissue sensitivity to insulin through diet and exercise. If this is not sufficient, oral drugs (generically known as the *sulfonylureas*) are available that increase insulin secretion and also stimulate tissue responsiveness to insulin. There is some evidence that overstimulation of the beta cells with sulfonylureas may promote the conversion of NIDDM to IDDM, however, so research into alternative approaches is ongoing.

People with type II diabetes mellitus do not usually develop ketoacidosis. The hyperglycemia itself, however, can be dangerous on a long-term basis. Diabetes is the leading cause of blindness, kidney failure, and amputation of the lower extremities in the United States. People with diabetes frequently have circulatory problems that increase the tendency to develop gangrene and increase the risk of atherosclerosis. The causes of damage to the retina and lens of the eyes and to blood vessels are not well understood. It is believed, however, that these problems may result from a long-term exposure to high blood glucose, which (1) causes water to leave tissue cells by osmosis, and thus produces dehydration of capillary endothelial cells; and (2) results in glycosylation of tissue proteins.

## Hypoglycemia

A person with type I diabetes mellitus depends on insulin injections to prevent hyperglycemia and ketoacidosis. If inadequate insulin is injected, the person may enter a coma as a result of the ketoacidosis, electrolyte imbalance, and dehydration that develop. An overdose of insulin can also produce a coma as a result of the hypoglycemia (abnormally low blood glucose levels) produced. The physical signs and symptoms of diabetic and hypoglycemic coma are sufficiently different (table 27.9) to allow hospital personnel to distinguish between these two types.

Less severe symptoms of hypoglycemia are usually produced by an oversecretion of insulin from the pancreatic

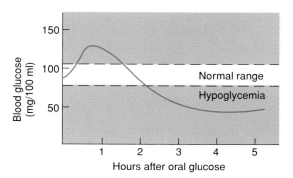

**FIGURE 27.23**

An idealized oral glucose tolerance test in a person with reactive hypoglycemia. The blood glucose concentration falls below the normal range within 5 hours of glucose ingestion as a result of excessive insulin secretion.

islets after a carbohydrate meal. This **reactive hypoglycemia,** caused by an exaggerated response of the beta cells to a rise in blood glucose, is most commonly seen in adults who are genetically predisposed to type II diabetes mellitus. For this reason, people with reactive hypoglycemia must limit their intake of carbohydrates and eat small meals at frequent intervals, rather than two or three meals per day.

The symptoms of reactive hypoglycemia include tremor, hunger, weakness, blurred vision, and impaired mental ability. The appearance of some of these symptoms, however, does not necessarily indicate reactive hypoglycemia; moreover, a given level of blood glucose does not always produce these symptoms. To confirm the diagnosis of reactive hypoglycemia, a number of tests must be performed. In the glucose tolerance test, for example, reactive hypoglycemia is shown when the initial rise in blood glucose produced by the ingestion of a glucose solution triggers excessive insulin secretion, so that the blood glucose levels fall below normal within 5 hours (fig. 27.23).

# Chapter Summary

## Nutritional Requirements (pp. 812–816)

1. Vitamins and minerals serve as cofactors and coenzymes.
   a. Vitamins are classified as fat-soluble (A, D, E, and K) or water-soluble.
   b. Many water-soluble vitamins are needed for the activity of the enzymes involved in cell respiration.
2. Caloric intake must be sufficient to meet the energy expenditures of the body; excessive caloric intake results in obesity.

## Metabolism of Carbohydrates, Lipids, and Proteins (pp. 816–822)

1. When blood glucose enters cells, it combines with a phosphate group and is then used in cell respiration or glycogen synthesis.
2. Triglycerides can be hydrolyzed into free fatty acids and glycerol, which can serve as energy sources.
   a. Fatty acids are converted into molecules of acetyl CoA by the process of β-oxidation; these can then enter Krebs cycles and be used for energy.
   b. Glucose can be converted into fat through the process of lipogenesis; in this process, the intermediates of glycolysis can be used to produce molecules of fatty acid.
3. Amino acids derived from the hydrolysis of proteins can be used for energy.
   a. In the process of oxidative deamination, the amine group of an amino acid is removed and incorporated into urea; the remainder of the amino acid becomes either pyruvic acid or a Krebs cycle intermediate.
   b. In the process of gluconeogenesis, an amino acid may be converted, through pyruvic acid, into glucose.
4. In transamination, a new amino acid is produced by adding the amine group derived from a different amino acid to pyruvic acid or one of the Krebs cycle acids.

## Regulation of Energy Metabolism (pp. 822–823)

1. Circulating energy substrates, which can be used by body cells for cell respiration, may be obtained from either the energy reserves of fat, glycogen, and protein in the body or from the diet.
2. One must eat to obtain an adequate caloric intake that maintains the body's energy reserves and the optimum body weight.
   a. The amount of body fat influences endocrine secretion.
   b. The hypothalamus is involved in the regulation of eating behavior.
3. The metabolism of the body fluctuates as a result of being in either an absorptive state or in a postabsorptive state.
   a. When nutrients are being absorbed from the intestine, anabolism is favored as circulating energy substrates are used to form the energy reserves of triglyceride, glycogen, and protein.
   b. When one is in the postabsorptive, or fasting, state, catabolism is favored as energy reserve molecules are hydrolyzed to yield circulating energy substrates.
   c. Hormones regulate the anabolic and catabolic balance of metabolism.

## Energy Regulation by the Pancreatic Islets (pp. 823–828)

1. A rise in blood glucose concentration, acting on the pancreatic islets, stimulates insulin secretion and inhibits glucagon secretion.
2. During absorption of nutrients following a meal, insulin promotes the uptake of blood glucose into tissue cells and the conversion of that glucose into glycogen and fat.
3. During periods of fasting, insulin secretion decreases and glucagon secretion increases.
   a. Glucagon stimulates glycogenolysis, gluconeogenesis, and ketogenesis, in the liver and lipolysis in adipose cells.
   b. These effects help to maintain adequate levels of blood glucose for the brain and provide alternate energy sources for other organs.

## Metabolic Regulation by Adrenal Hormones, Thyroxine, and Growth Hormone (pp. 828–833)

1. The adrenal hormones involved in energy regulation include epinephrine from the adrenal medulla and glucocorticoids (mainly hydrocortisone) from the adrenal cortex.
   a. The effects of epinephrine are similar to those of glucagon.
   b. Glucocorticoids promote gluconeogenesis by stimulating the breakdown of muscle protein and the conversion of amino acids to glucose in the liver.
2. Thyroxine stimulates the rate of cell respiration in almost all cells in the body.
   a. Thyroxine thus sets the basal metabolic rate (BMR), which is the rate at which energy is consumed by the body under resting conditions.
   b. Thyroxine also promotes protein synthesis which is necessary for proper growth and development, particularly of the central nervous system.
3. The secretion of growth hormone is stimulated by a protein meal and by a fall in plasma glucose concentration during fasting.
   a. Growth hormone stimulates catabolism of lipids and inhibits glucose utilization.
   b. Growth hormone also promotes body growth through protein synthesis.
   c. The anabolic effects of growth hormone, including the stimulation of bone growth in childhood, are believed to be produced indirectly via somatomedins.

# Review Activities

## Objective Questions

Match the following processes with their definitions:

1. lipogenesis
2. lipolysis
3. gluconeogenesis
4. glycogenesis
   a. conversion of glycogen to glucose
   b. hydrolysis of fat
   c. production of glucose from amino acids
   d. conversion of glucose to fat

Match the following states and events:

5. absorption of a carbohydrate meal (absorptive state)
6. fasting (postabsorptive state)
   a. rise in insulin; rise in glucagon
   b. fall in insulin; rise in glucagon
   c. rise in insulin; fall in glucagon
   d. fall in insulin; fall in glucagon

Match the following hormones and effects:

7. growth hormone
8. thyroxine
9. hydrocortisone
   a. increased protein synthesis; increased cell respiration
   b. protein catabolism in muscles; gluconeogenesis in the liver

   c. protein synthesis in muscles; decreased glucose utilization
   d. fall in blood glucose concentrations; increased fat synthesis

10. A lowering of blood glucose concentration
    a. decreases lipogenesis.
    b. increases lipolysis.
    c. increases glycogenolysis.
    d. all of the above apply.

11. Glucose can be secreted into the blood by
    a. the liver.
    b. the muscles.
    c. the liver and muscles.
    d. the liver, muscles, and brain.

12. The basal metabolic rate is determined primarily by
    a. hydrocortisone.
    b. insulin.
    c. growth hormone.
    d. thyroxine.

13. Somatomedins are required for the anabolic effects of
    a. hydrocortisone.
    b. insulin.
    c. growth hormone.
    d. thyroxine.

14. Which of the following hormones stimulates anabolism of proteins and catabolism of fat?
    a. growth hormone
    b. thyroxine

    c. insulin
    d. glucagon
    e. epinephrine

15. Ketoacidosis in untreated diabetes mellitus is due to
    a. excessive fluid loss.
    b. hypoventilation.
    c. excessive eating and obesity.
    d. excessive fat catabolism.

## Essay Questions

1. Trace the metabolic pathways and describe the events by which starch in a chocolate cake can be converted into fat within adipose tissue.
2. Define *glycogenolysis* and *gluconeogenesis* and describe the role of each in maintaining an adequate blood glucose level during prolonged exercise.
3. Compare the metabolic effects of fasting to the state of uncontrolled type I diabetes mellitus and explain their hormonal similarities.
4. Glucocorticoids stimulate the breakdown of protein in muscles but also stimulate the synthesis of protein in the liver. Discuss the significance of these differences.
5. Describe how thyroxine affects cell respiration and explain why a hypothyroid person tends to gain weight and has a reduced tolerance to cold.
6. Compare and contrast the metabolic effects of thyroxine and growth hormone.

Regulation of Metabolism

## Explorations ✪

Two modules of correlating material are available from the Wm. C. Brown CD-ROM: *Explorations*. They are # 7 Diet and Weight Loss and # 11 Hormone Action.

[ chapter twenty-eight ]

# reproduction: development and the male reproductive system

## objectives

- Discuss how genetic sex determines whether testes or ovaries form in the embryo and describe the composition and function of embryonic testes.

- Describe the descent of the testes into the scrotum.

- Describe the hormonal interactions between the hypothalamus, anterior pituitary, and gonads.

- Discuss the mechanisms that regulate the onset of puberty and describe the events that occur during puberty.

- Locate the structures of the male reproductive system and describe the structure and function of the scrotum.

- Describe the structure and function of the two compartments of the testes and discuss their hormonal regulation.

- Discuss the process of spermatogenesis and explain how this process is regulated.

- List the various spermatic ducts and describe the structure of the spermatic cord.

- Describe the location, structure, and function of the seminal vesicles, prostate, and bulbourethral glands.

- Describe the structure and function of the penis.

# Introduction to the Reproductive System

*A gene on the Y chromosome causes the embryonic gonads to differentiate into testes. Females lack a Y chromosome, and the absence of this gene results in the development of ovaries. The embryonic testes secrete testosterone, which triggers the development of male accessory sex organs and external genitalia. The absence of testes (rather than the presence of ovaries) in a female embryo causes the development of female accessory sex organs.*

The incredible complexity of structure and function in living organisms could not be produced in successive generations by chance; mechanisms must exist to transmit the genetic code from one generation to the next. This could be accomplished by either asexual or sexual reproduction. But sexual reproduction, in which genes from two individuals are combined in random and novel ways with each new generation, offers the advantage of introducing great variability into a population. This variability of genetic constitution helps to ensure that some members of a population will survive changes in the environment over evolutionary time.

In sexual reproduction, **germ cells,** or **gametes** (sperm and ova) are formed within the **gonads** (testes and ovaries) by a process of reduction division, or *meiosis* (chapter 3). During this type of cell division, the normal number of chromosomes in most human cells—46—is halved, so that each gamete receives 23 chromosomes. Fusion of a sperm cell (spermatozoon) and an egg cell (ovum) in the act of **fertilization** results in restoration of the original chromosome number of 46 in the fertilized egg (the *zygote*). Growth of the zygote into an adult member of the next generation occurs by means of mitotic cell divisions, as described in chapter 3. When this individual reaches puberty, mature sperm or ova will be formed by meiosis within the gonads so that the life cycle can be continued (fig. 28.1).

## Sex Determination

Each zygote inherits 23 chromosomes from its mother and 23 chromosomes from its father. This does not produce 46 different chromosomes, but rather 23 pairs of **homologous chromosomes.** The members of a homologous pair, with the important exception of the sex chromosomes, appear to be structurally identical and contain similar genes (such as those coding for eye color, height, and so on). Each cell that contains 46 chromosomes (that is, *diploid*) has two number 1 chromosomes, two number 2 chromosomes, and so on

..........................................

gamete: Gk. *gameta*, husband or wife

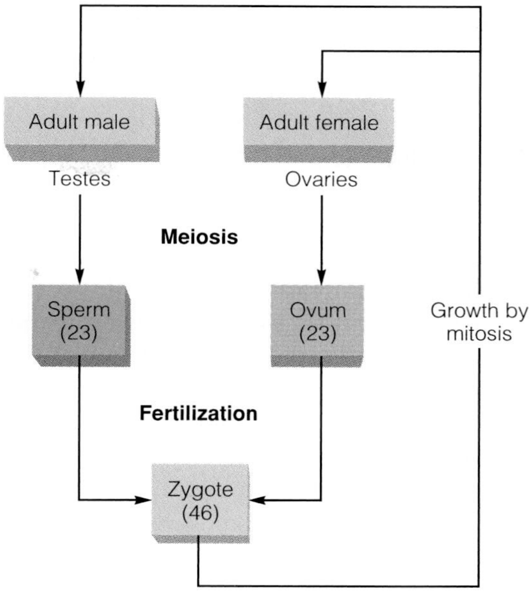

**FIGURE 28.1**
The human life cycle. Numbers in parentheses indicate haploid state (23 chromosomes) and diploid state (46 chromosomes).

through pair number 22. The first 22 pairs of chromosomes are called **autosomal** (*aw″to-so′mal*) **chromosomes.**

The twenty-third pair of chromosomes are the **sex chromosomes.** A normal female has two X chromosomes, whereas a normal male has one X chromosome and one Y chromosome. The X and Y chromosomes look different and contain different genes. This is the exceptional pair of homologous chromosomes mentioned earlier.

When a diploid cell (with 46 chromosomes) undergoes meiotic division, its daughter cells receive only one chromosome from each homologous pair of chromosomes. The gametes are therefore said to be *haploid* (they contain only half the number of chromosomes in the diploid parent cell). Each sperm cell, for example, will receive only one chromosome of homologous pair number 5—either the one originally contributed by the organism's mother, or the one originally contributed by the father. Which of the two chromosomes—maternal or paternal—ends up in a given sperm cell is completely random. This is also true for the sex chromosomes, so that approximately half of the sperm produced will contain an X and approximately half will contain a Y chromosome.

The egg cells (ova) in a woman's ovary will receive a similar random assortment of maternal and paternal chromosomes. Since the body cells of females have two X chromosomes, however, all of the ova resulting from meiosis will normally contain one X chromosome. Because all ova contain one X chromosome, whereas some sperm are X bearing and others are Y bearing, *the chromosomal sex of the zygote is determined by the fertilizing sperm cell.* If a Y-bearing spermatozoon fertilizes the ovum, the zygote will be XY and male; if an X-bearing spermatozoon fertilizes the ovum, the zygote will be XX and female.

### Formation of Testes and Ovaries

The gonads of males and females appear similar for the first 40 or so days of development following conception. During this time, cells that will give rise to sperm (*spermatogonia*) and cells that will give rise to ova (*oogonia*) migrate from the yolk sac to the embryonic gonads. At this stage in development, the gonads have the potential to become either testes or ovaries. The hypothetical substance that promotes their conversion to testes has been called the **testis-determining factor (TDF).**

Though it has long been recognized that male sex is determined by the presence of a Y chromosome and female sex by the absence of the Y chromosome (see fig. 28.2), the genes involved have only recently been located. In rare male babies with XX genotypes, one of the X chromosomes was found to contain a segment of the Y chromosome as a result of an error during the meiotic cell division that formed the sperm. Furthermore, rare female babies with XY genotypes were found to be missing that portion of the Y chromosome erroneously inserted into the X chromosome of XX males.

Through these and other observations, it has been shown that the gene for the testis-determining factor is located on the short arm of the Y chromosome. Evidence suggests that it may be a particular gene known as *SRY* (for "sex-determining region of the Y") which is found in the Y chromosome of all mammals and is highly conserved, meaning that it shows little variation in structure over evolutionary time.

 Notice that it is normally the presence or absence of the Y chromosome that determines whether the embryo will have testes or ovaries. This point is well illustrated by two genetic abnormalities. In *Klinefelter's syndrome,* the affected person has 47 instead of 46 chromosomes because of the presence of an extra X chromosome. This person, with an XXY genotype, develops testes despite the presence of two X chromosomes. Patients with *Turner's syndrome,* who have the genotype XO (and therefore have only 45 chromosomes) have poorly developed ("streak") gonads. Klinefelter's and Turner's syndromes are described more fully under "Clinical Considerations."

The structures that will eventually produce sperm within the testes—the **seminiferous** (*sem″i-nif′er-us*) **tubules**—appear very early in embryonic development—between 43 and 50 days after conception. Although spermatogenesis begins during embryonic life, it is arrested until the onset of puberty. The tubules contain the cells that will eventually produce sperm, generically known as **germinal cells,** and nongerminal cells called **sustentacular cells** (also known as nurse cells, or Sertoli cells). The sustentacular

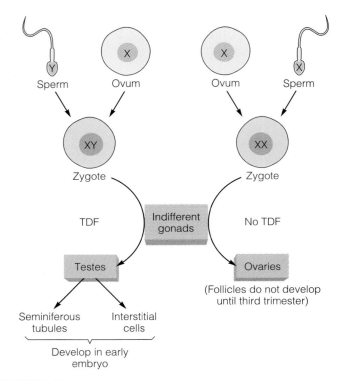

**FIGURE 28.2**

The formation of the chromosomal sex of the embryo and the development of the gonads. The very early embryo has "indifferent gonads," which can develop into either testes or ovaries. The testis-determining factor (TDF) is a gene located on the Y chromosome. In the absence of TDF, ovaries will develop.

cells appear at about day forty-two. At about day sixty-five, the **interstitial cells** (cells of Leydig) appear in the embryonic testes. These cells are located in the *interstitial tissues*, which are outside the seminiferous tubules and between adjacent convolutions of the tubules. The interstitial cells constitute the endocrine tissue of the testes. In contrast to the rapid development of the testes, the functional units of the ovaries—called the **ovarian follicles**—do not appear until the second trimester of pregnancy (at about day 105).

The early appearing interstitial cells in the embryonic testes secrete large amounts of male sex hormones, or **androgens.** The major androgen secreted by these cells is **testosterone** (*tes-tos′t ĕ-rōn*). Testosterone secretion begins as early as 9 to 10 weeks after conception, reaches a peak at 12 to 14 weeks, and thereafter declines so that very low levels are being secreted by the end of the second trimester (at about 21 weeks). During embryonic development, testosterone serves a very important function in males; similarly high levels of testosterone will not appear again in the life of the individual until the time of puberty.

• • • • • • • • • • • • • • • • • • • • • • • • •

spermatogonia: Gk. *sperma*, seed; *gonos*, procreation
oogonia: Gk. *oion*, egg; *gonos*, procreation

• • • • • • • • • • • • • • • • • • • • • • • • •

cells of Leydig: from Franz von Leydig, German anatomist, 1821–1908
androgen: Gk. *andros*, male producing

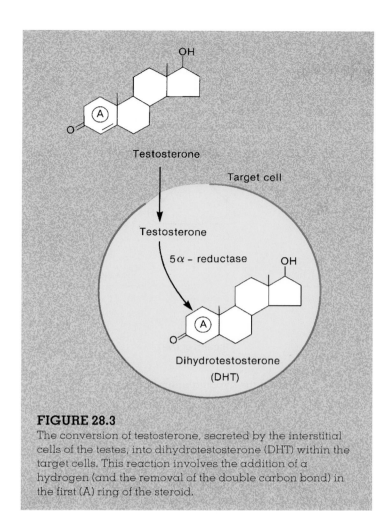

**FIGURE 28.3**
The conversion of testosterone, secreted by the interstitial cells of the testes, into dihydrotestosterone (DHT) within the target cells. This reaction involves the addition of a hydrogen (and the removal of the double carbon bond) in the first (A) ring of the steroid.

Masculinization of the accessory sex organs occurs as a result of testosterone secreted by the embryonic testes. Testosterone itself, however, is not the active agent within these target organs. Once it has entered the target cells, the action of an enzyme called 5α-*reductase* converts the testosterone into the active hormone called **dihydrotestosterone (DHT).** DHT directly mediates the androgenic effect in the accessory sex organs (fig. 28.3).

## Descent of the Testes

Developing on the outside surface of each testis is a fibromuscular cord called the **gubernaculum** (*goo″ber-nak′yŭ-lum*) (fig. 1, p. 860) that attaches to the inferior portion of the testis and extends to the labioscrotal swelling of the same side. At the same time, a portion of the embryonic mesonephric duct (see the discussion of embryonic development later in this chapter) attaches itself to the testis, becomes convoluted, and forms the epididymis. Another portion of the mesonephric duct becomes the ductus deferens.

The external genitalia of a male are completely formed by the end of the twelfth week. The descent of the testes

gubernaculum: L. *gubernaculum,* helm

from the site of development begins between the sixth and tenth week. Descent into the scrotal sac, however, does not occur until about week 28, when paired inguinal canals form in the abdominal wall to provide openings from the pelvic cavity to the scrotal sac. The process by which a testis descends is not well understood, but it seems to be associated with the shortening and differential growth of the gubernaculum, which is attached to the testis and extends through the inguinal canal to the wall of the scrotum. As the testis descends, it passes to the side of the urinary bladder and anterior to the symphysis pubis. It carries with it the ductus deferens, the testicular vessels and nerve, a portion of the internal abdominal oblique muscle, and lymph vessels. All of these structures remain attached to the testis and form what is known as the **spermatic cord.** By the time the testis has taken its position in the scrotal sac, the gubernaculum is no more than a remnant of scarlike tissue.

The temperature of the testes in the scrotum is maintained at about 35° C (95° F), or about 3.6° F below normal body temperature. This lower temperature is needed for proper sperm production and storage, as illustrated by the fact that spermatogenesis does not occur in males with *cryptorchidism*, the condition in which the testes fail to descend.

During the physical examination of a neonatal male child, a physician will palpate the scrotum to determine if the testes are in position. If one or both are not in the scrotal sac, it may be possible to induce descent by administering certain hormones. If this procedure is not effective, surgery is generally performed before the age of 5. Failure to correct the situation may result in sterility and possibly the development of a testicular tumor.

# Endocrine Regulation of Reproduction

*The functions of the testes and ovaries are regulated by gonadotropic hormones secreted by the pituitary gland. The gonadotropic hormones stimulate the gonads to secrete their sex steroid hormones, and these steroid hormones, in turn, have an inhibitory effect on the secretion of the gonadotropic hormones. This interaction between the anterior pituitary and the gonads forms a negative feedback loop.*

During the first trimester of pregnancy the embryonic testes are active endocrine glands, secreting the high amounts of testosterone needed to masculinize the male embryo's external genitalia and accessory sex organs. Ovaries, by contrast, do not mature until the third trimester of pregnancy. During the second trimester of pregnancy, testosterone secretion in the male declines, so that the gonads of both sexes are relatively inactive at the time of birth.

Before puberty, there are equal blood concentrations of *sex steroids*—androgens and estrogens—in both males and females. Apparently, this is not due to deficiencies in the ability of the gonads to produce these hormones, but rather to lack of sufficient stimulation. During puberty, the gonads secrete increased amounts of sex steroid hormones as a result of increased stimulation by **gonadotropic hormones** from the anterior pituitary.

## Interaction between the Hypothalamus, Pituitary Gland, and Gonads

The anterior pituitary (adenohypophysis) produces and secretes two gonadotropic hormones—**FSH (follicle-stimulating hormone)** and **LH (luteinizing hormone).** Although these two hormones are named according to their actions in the female, the same hormones are secreted by the male's pituitary gland. The gonadotropic hormones of both sexes have three primary effects on the gonads: (1) stimulation of spermatogenesis or oogenesis (formation of sperm or ova); (2) stimulation of gonadal hormone secretion; and (3) maintenance of the structure of the gonads (the gonads atrophy if the pituitary gland is removed).

The secretion of both LH and FSH from the anterior pituitary is stimulated by a hormone produced by the hypothalamus and secreted into the hypothalamo-hypophyseal portal vessels (see chapter 19). This releasing hormone is sometimes called *LHRH* (*luteinizing hormone-releasing hormone*). Since attempts to find a separate FSH-releasing hormone have thus far failed, and since LHRH stimulates FSH as well as LH secretion, LHRH is often referred to as **gonadotropin-releasing hormone** (and accordingly abbreviated **GnRH**).

If a male or female animal is castrated (has its gonads surgically removed), the secretion of FSH and LH will increase to much higher levels than those measured in the intact animal. This demonstrates that the gonads secrete products that produce a **negative feedback inhibition** of gonadotropin secretion. This negative feedback is exerted in large part by sex steroids—estrogens and progesterone in the female and testosterone in the male.

The negative feedback effects of steroid hormones are believed to occur by means of two mechanisms: (1) inhibition of GnRH secretion from the hypothalamus, and (2) inhibition of the pituitary gland's response to a given amount of GnRH. In addition to steroid hormones, the testes and ovaries secrete a polypeptide hormone called **inhibin.** Inhibin is secreted by the sustentacular cells of the seminiferous tubules in males and by the granulosa cells of the ovarian follicles in females. This hormone specifically inhibits the anterior pituitary's secretion of FSH, without affecting the secretion of LH.

Figure 28.4 illustrates the process of gonadal regulation. Although hypothalamus-pituitary-gonad interactions are similar in males and females, there are important differences. Secretion of gonadotropins and sex steroids is more or less constant in adult males. Secretion of gonadotropins and sex steroids in adult females, by contrast, is characterized by cyclic variations (during the menstrual cycle).

Studies have shown that secretion of GnRH from the hypothalamus is pulsatile rather than continuous, and thus the secretion of FSH and LH follows this pulsatile pattern. These pulsatile patterns of secretion are required to prevent desensitization and down-regulation of the target glands (discussed in chapter 19). It appears that the frequency of the pulses of secretion, as well as their amplitude (how much hormone is secreted per pulse), affects the target glands' response to the hormone.

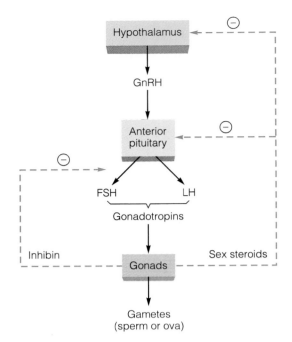

**FIGURE 28.4**

Interactions between the hypothalamus, anterior pituitary, and gonads. Sex steroids secreted by the gonads have a negative feedback effect on the secretion of GnRH (gonadotropin-releasing hormone) and on the secretion of gonadotropins. The gonads may also secrete a polypeptide hormone called *inhibin* that exerts negative feedback control of FSH secretion.

If a powerful synthetic analogue of GnRH (such as *nafarelin*) is administered, the anterior pituitary first increases and then decreases its secretion of FSH and LH. This decrease is contrary to the normal stimulatory action of GnRH, and is due to a desensitization of the anterior pituitary evoked by continuous exposure to GnRH. The decrease in LH causes a decline in testosterone secretion from the testes, or of estradiol secretion from the ovaries. The decreased testosterone secretion is useful in the treatment of men who have *benign prostatic hypertrophy*. In this condition, common in older men, testosterone supports abnormal growth of the prostate. The fall in estradiol secretion in women given synthetic GnRH analogues can be useful in the treatment of *endometriosis*. In this condition, ectopic endometrial tissue from the uterus (dependent on estradiol for growth) is found growing outside the uterus—for example, on the ovaries or on the peritoneum. These treatments, which illustrate why GnRH and the gonadotropins are normally secreted in a pulsatile fashion, are particularly beneficial clinically because their effects are reversible.

## The Onset of Puberty

Secretion of FSH and LH is high in the newborn but falls to very low levels a few weeks after birth. Gonadotropin secretion remains low until the beginning of puberty, which is marked by rising levels of FSH secretion followed by LH secretion. Experimental evidence suggests that this rise in gonadotropin secretion is a result of two processes: (1) maturational changes in the brain that result in increased GnRH secretion by the hypothalamus and (2) decreased sensitivity of gonadotropin secretion to the negative feedback effects of sex steroid hormones.

The maturation of the hypothalamus or other regions of the brain that leads to increased GnRH secretion at the time of puberty appears to be programmed—children without gonads show increased FSH secretion at the normal time. Also during this period of time, a given amount of sex steroids has less of a suppressive effect on gonadotropin secretion than the same dose would have if administered prior to puberty. This suggests that the sensitivity of the hypothalamus and the pituitary gland to negative feedback effects decreases at puberty, which would also help to account for rising gonadotropin secretion at this time.

During late puberty, there is a pulsatile secretion of gonadotropins—FSH and LH secretion increase during periods of sleep and decrease during periods of wakefulness. These pulses of increased gonadotropin secretion during puberty stimulate a rise in sex steroid secretion from the gonads. Increased secretion of testosterone from the testes and of **estradiol–17β** (estradiol is the major *estrogen*, or female sex steroid) from the ovaries during puberty, in turn, produces changes in body appearance characteristic of the two sexes. Such **secondary sex characteristics** (tables 28.1 and 28.2) are the physical manifestations of the hormonal changes occurring during puberty. These changes are accompanied by a growth spurt that begins at an earlier age in females than in males (fig. 28.5).

The age at which puberty begins is related to the amount of body fat and level of physical activity of the individual child. The average age of *menarche*—the first menstrual flow—is later (age 15) in girls who are very active physically than in the general population (age 12.6).

## Table 28.1  Development of secondary sex characteristics and other changes that occur during puberty in girls

| Characteristic | Age of first appearance | Hormonal stimulation |
|---|---|---|
| Appearance of breast bud | 8–13 | Estrogen, progesterone, growth hormone, thyroxine, insulin, cortisol |
| Pubic hair | 8–14 | Adrenal androgens |
| Menarche (first menstrual flow) | 10–16 | Estrogen and progesterone |
| Axillary (underarm) hair | About 2 years after the appearance of pubic hair | Adrenal androgens |
| Eccrine sweat glands and sebaceous glands; acne (from blocked sebaceous glands) | About the same time as axillary hair growth | Adrenal androgens |

# Table 28.2   Development of secondary sex characteristics and other changes that occur during puberty in boys

| Characteristic | Age of first appearance | Hormonal stimulation |
|---|---|---|
| Growth of testes | 10–14 | Testosterone, FSH, growth hormone |
| Pubic hair | 10–15 | Testosterone |
| Body growth | 11–16 | Testosterone, growth hormone |
| Growth of penis | 11–15 | Testosterone |
| Growth of larynx (voice lowers) | Same time as growth of penis | Testosterone |
| Facial and axillary (underarm) hair | About 2 years after the appearance of pubic hair | Testosterone |
| Eccrine sweat glands and sebaceous glands; acne (from blocked sebaceous glands) | About the same time as facial and axillary hair growth | Testosterone |

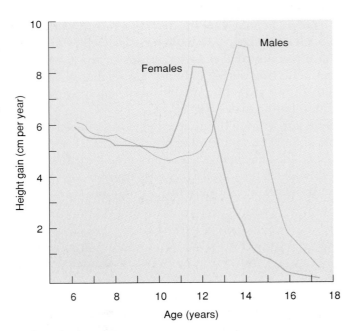

**FIGURE 28.5**

Growth in height of females and males as a function of age. Notice that the growth spurt during puberty occurs at an earlier age in females than in males.

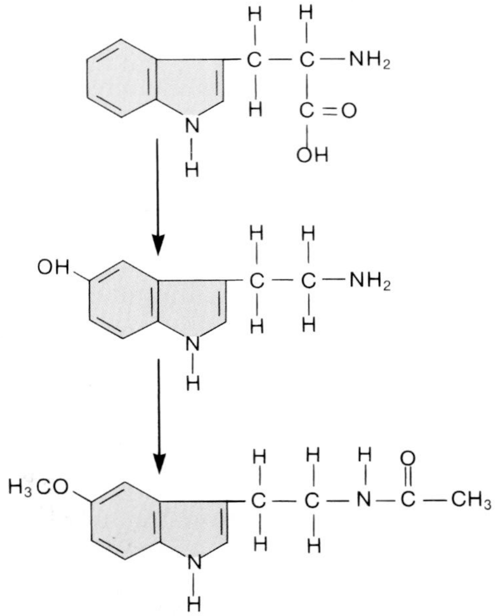

**FIGURE 28.6**

A simplified biosynthetic pathway for the pineal gland hormone melatonin.

Apparently, a minimum percentage of body fat is required for menstruation to begin, which may represent a mechanism favored by natural selection for the ability to successfully complete a pregnancy and nurse the baby. Later in life, women who are very lean and physically active may have irregular cycles and *amenorrhea* (cessation of menstruation). This may also be related to the percentage of body fat. In addition, there is evidence that physical exercise may, through activation of neural pathways involving endorphin neurotransmitters (chapter 14), act to inhibit GnRH and gonadotropin secretion.

## Pineal Gland

The role of the **pineal gland** in human physiology is poorly understood. It is known that this gland, located deep within the brain, secretes the hormone **melatonin** as a derivative of the amino acid tryptophane (fig. 28.6) and that production of this hormone is influenced by light/dark cycles.

The pineal glands of some vertebrates have photoreceptors that are directly sensitive to environmental light. Although no such photoreceptors are present in the pineal glands of mammals, the secretion of melatonin has been shown to increase at night and decrease during daylight. The inhibitory effect of light on melatonin secretion in mammals is indirect. As described in chapter 19, pineal secretion is stimulated by postganglionic sympathetic neurons that originate in the superior cervical ganglion; activity of these neurons, in turn, is inhibited by nerve tracts that are activated by light striking the retina.

There is abundant experimental evidence that melatonin can inhibit gonadotropin secretion in rats, and thus exert an "anti-gonad" effect. In other experimental animals (such as sheep), however, melatonin can stimulate the reproductive system. That melatonin plays a role in the regulation of human reproduction has long been suspected but, because of conflicting and inconclusive data, has not as yet been established.

## Male Reproductive System

*The interstitial cells of the testes are stimulated by LH to secrete testosterone—a potent androgen that acts to maintain the structure and function of the male accessory sex organs and to promote the development of male secondary sex characteristics. The sustentacular (Sertoli) cells in the seminiferous tubules of the testes are stimulated by FSH. This action is needed to initiate the production of sperm and stimulates the sustentacular cells to secrete various products, including inhibin. Spermatogenesis requires the cooperative actions of both FSH and testosterone.*

The structures of the male reproductive system are illustrated in figure 28.7. These include the *primary sex organs* of the male—the testes—and a number of *secondary sex organs*. In this section, we will describe the structure, function, and regulation of the testes. The secondary sex organs are covered in the next section.

### Scrotum

The saclike **scrotum** is suspended immediately behind the base of the penis, anterior to the anal opening, in a region known as the **perineum** (*per˝ĭ-ne´um*). The functions of the scrotum are to support and protect the testes and to regulate their position relative to the pelvic region of the body. The **dartos** (*dar´tos*) **muscle** consists of a layer of smooth muscle fibers in the subcutaneous tissue of the scrotum, and the **cremaster** (*kre-mas´ter*) **muscle** is a thin strand of skeletal muscle associated with the spermatic cord. Both muscles involuntarily contract in response to cold temperatures to move the testes closer to the heat of the body in the pelvic region. The cremaster muscle is a continuation of the internal abdominal oblique muscle and is derived as the testes

descend into the scrotum. Warm temperatures cause the dartos and cremaster muscles to relax and lower the testes away from the body cavity; cold temperatures cause these muscles to contract and elevate the testes. These responses help to maintain the proper temperature required by the testes for spermatogenesis, as previously described.

The scrotum is subdivided into two longitudinal compartments by a fibrous **scrotal septum.** The site of the scrotal septum is apparent on the surface of the scrotum along a median longitudinal ridge called the **scrotal raphe** (*ra´fe*). The purpose of the scrotal septum is to compartmentalize each testis so that trauma or infection in one will not affect the other. Another protective feature is that the left testis is generally suspended lower than the right so that the two are less likely to be compressed forcefully together.

Although uncommon, male infertility may result from an excessively high temperature of the testes over an extended period of time. Tight clothing that keeps the testes close to the body or frequent hot baths or saunas may destroy sperm to the extent that a sperm count of discharged semen will be below that necessary to enable fertilization.

### Structure of the Testes

The **testes** are paired organs about 4 cm (1.5 in.) long and 2.5 cm (1 in.) in diameter. Each testis weighs between 10 and 14 g. Two tissue layers, or tunicas, cover the testis. The outer **tunica vaginalis** is a thin serous sac derived from the peritoneum during the descent of the testis. The **tunica albuginea** (*al˝byoo-jin´e-ă*) is a tough fibrous membrane that directly encapsules the testis (fig. 28.8). Fibrous inward extensions of the tunica albuginea partition the testis into 250 to 300 wedge-shaped **lobules.**

Each lobule of the testis contains one to three tightly convoluted **seminiferous tubules** that may exceed 70 cm (28 in.) in length if uncoiled. It is in the seminiferous tubules that spermatogenesis occurs. Sperm are produced at the rate of thousands per second—more than 100 million per day—throughout the life of a healthy, sexually mature male.

Various stages of spermatogenesis can be observed in a histological section of seminiferous tubules (fig. 28.9). The germinal cells, called **spermatogonia** (*sper-mat˝ĕ-go´ne-ă*), are in contact with the basement membrane. Spermatogonia undergo meiosis to produce, in order of advancing maturity, the primary spermatocytes, secondary spermatocytes, and spermatids (see fig. 28.12). Forming the walls of the seminiferous tubules are **sustentacular** (Sertoli) **cells,** which are sometimes referred to as *nurse cells*. The sustentacular cells

........................................

dartos: Gk. *dartos,* skinned or flayed
cremaster: Gk. *cremaster,* a suspender, to hang

........................................

septum: L. septum, a partition
raphe: Gk. *raphe,* a seam
tunica: L. *tunica,* a coat
vaginalis: L. *vagina,* a sheath
albuginea: L. *albus,* white
Sertoli cells: from Enrico Sertoli, Italian histologist, 1842–1910

(a)

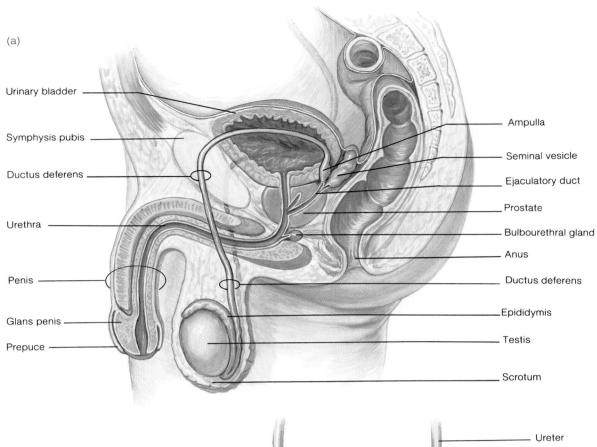

Urinary bladder

Symphysis pubis

Ductus deferens

Urethra

Penis

Glans penis

Prepuce

Ampulla

Seminal vesicle

Ejaculatory duct

Prostate

Bulbourethral gland

Anus

Ductus deferens

Epididymis

Testis

Scrotum

(b)

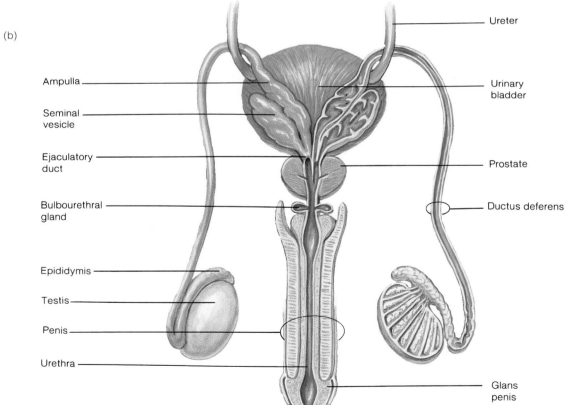

Ampulla

Seminal vesicle

Ejaculatory duct

Bulbourethral gland

Epididymis

Testis

Penis

Urethra

Ureter

Urinary bladder

Prostate

Ductus deferens

Glans penis

**FIGURE 28.7**

Organs of the male reproductive system. (*a*) A sagittal view and (*b*) a posterior view.

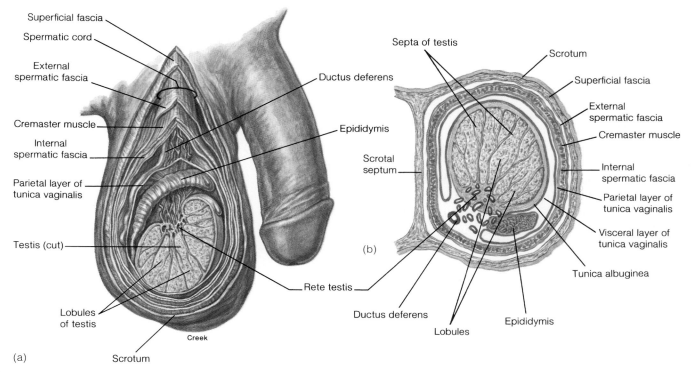

**FIGURE 28.8**

Structural features of a testis and epididymis. (*a*) A longitudinal
view and (*b*) a transverse view.

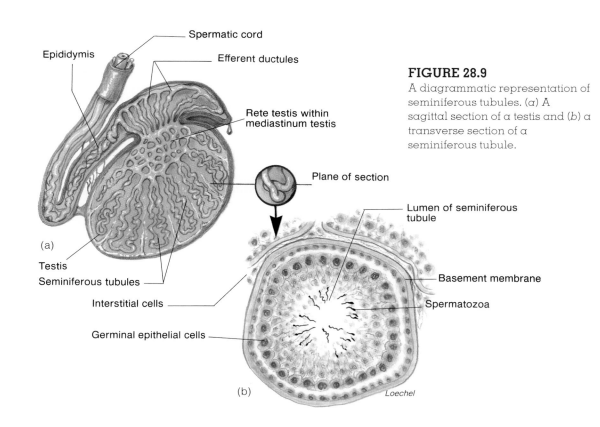

**FIGURE 28.9**

A diagrammatic representation of
seminiferous tubules. (*a*) A
sagittal section of a testis and (*b*) a
transverse section of a
seminiferous tubule.

are critically important to the developing spermatozoa that are embedded between them. The spermatozoa are formed, but not fully matured, by the time they reach the lumina of the seminiferous tubules.

Between the seminiferous tubules are specialized endocrine cells called **interstitial cells** (cells of Leydig). The function of these cells is to produce and secrete the male sex hormones. The testes are considered mixed exocrine and endocrine glands because they produce both sperm and androgens.

Once the sperm have been produced, they move through the seminiferous tubules and enter a tubular network called the **rete** (*re´te*) **testis** for further maturation. Cilia are located on some of the cells of the rete testis, presumably for moving sperm. The sperm are transported out of the testis and into the epididymis through a series of **efferent ductules.**

 The primary cause of male infertility is a condition called *varicocele* (*var´ĭ-ko-sēl*). Varicocele occurs when one or both of the testicular veins draining from the testes becomes swollen, resulting in poor vascular circulation in the testes. A varicocele generally occurs on the left side, because the left spermatic vein drains into the renal vein, where the blood pressure is higher than it is in the inferior vena cava, into which the right testicular vein empties.

## Endocrine Functions of the Testes

With regard to gonadotropin action, the testes are strictly compartmentalized. Cellular receptor proteins for FSH are located exclusively in the seminiferous tubules, where they are confined to the sustentacular cells. LH receptor proteins are located exclusively in the interstitial cells. Secretion of testosterone by the interstitial cells is stimulated by LH but not by FSH. Spermatogenesis in the tubules is stimulated by FSH.

### Control of Gonadotropin Secretion
As mentioned earlier in the chapter, castration of a male animal results in an immediate rise in FSH and LH secretion. This demonstrates that hormones secreted by the testes effect a negative feedback inhibition of gonadotropin secretion. If testosterone is injected into the castrated animal, the secretion of LH can be returned to the previous (precastration) levels. This provides a classical example of negative feedback—LH stimulates testosterone secretion by the interstital cells and testosterone inhibits pituitary secretion of LH (fig. 28.10).

rete: L. rete, a net
efferent ductules: L. *efferre*, to bring out; *ducere*, to lead
varicocele: L. *varico*, a dilated vein; Gk. *kele*, tumor or hernia

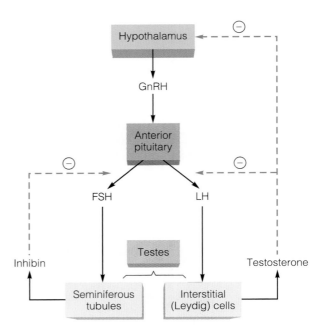

**FIGURE 28.10**
Negative feedback relationships between the anterior pituitary and testes. The seminiferous tubules are the targets of FSH action; the interstitial (Leydig) cells are targets of LH action. Testosterone secreted by the interstitial cells inhibits LH secretion; inhibin secreted by the tubules may inhibit FSH secretion.

The amount of testosterone that will inhibit LH secretion however, is not sufficient to suppress the postcastration rise in FSH secretion that occurs in most experimental animals. In rams and bulls, a water-soluble (and, therefore, a peptide rather than a steroid) product of the seminiferous tubules has been found to specifically suppress FSH secretion. This hormone, produced by the sustentacular cells, is called **inhibin.** There is now good evidence that the seminiferous tubules of the human testes also produce inhibin.

**Secretion and Actions of Testosterone**   Testosterone is by far the most important androgen secreted by the adult testis. This hormone and its derivatives are responsible for initiating and maintaining the body changes associated with puberty in males. Androgens are sometimes called *anabolic steroids* because they stimulate the growth of muscles and other structures (table 28.3). Increased testosterone secretion during puberty is also required for growth of the accessory sex organs—primarily the seminal vesicles and prostate. Removal of androgens by castration results in atrophy of these organs.

Androgens stimulate growth of the larynx (causing a lowering of the voice), hemoglobin synthesis (so that males have higher hemoglobin levels than females), and bone growth. The effect of androgens on bone growth is self-limiting, however, because androgens ultimately cause

| Table 28.3 | Actions of androgens in the male |
|---|---|
| **Category** | **Action** |
| Sex determination | Growth and development of mesenephric ducts into epididymides, ductus deferens, seminal vesicles, and ejaculatory ducts |
| | Development of urogenital sinus and tubercle into prostate |
| | Development of male external genitalia (penis and scrotum) |
| Spermatogenesis | At puberty: completion of meiotic division and early maturation of spermatids |
| | After puberty: maintenance of spermatogenesis |
| Secondary sex characteristics | Growth and maintenance of accessory sex organs |
| | Growth of penis |
| | Growth of facial and axillary hair |
| | Body growth |
| Anabolic effects | Protein synthesis and muscle growth |
| | Growth of bones |
| | Growth of other organs (including larynx) |
| | Erythropoiesis (red blood cell formation) |

conversion of cartilage to bone in the epiphyseal plates, thus "sealing" the plates and preventing further lengthening of the bones (as described in chapter 8).

The negative feedback effects of testosterone and inhibin help to maintain a constant secretion of gonadotropins in males, resulting in relatively constant levels of androgen secretion from the testes. This contrasts with the cyclic secretion of gonadotropins and ovarian steroids in females. Women experience an abrupt cessation in sex steroid secretion during menopause (as described in chapter 29). By contrast, the secretion of androgens declines only gradually and to varying degrees in men over 50 years of age. The causes of this age-related change in testicular function are not currently known.

**Hormonal Interactions within the Testis** Although androgens are by far the major secretory product of the testes, the testes do produce and secrete small amounts of estradiol. There is evidence that both the sustentacular cells of the tubules and the interstitial cells can produce estradiol, although estradiol receptors in the testes appear to be located only in the interstitial cells. Experiments suggest that when LH is present in high amounts and is not secreted in a pulsatile fashion, the desensitization and down-regulation of interstitial cell function that results may be partly mediated by estradiol.

The two compartments of the testes interact with each other in an autocrine fashion (fig. 28.11). Autocrine regulation, as described in chapter 19, refers to chemical regulation

that occurs within an organ. Testosterone from the interstitial cells is metabolized by the tubules into other active androgens and is required for spermatogenesis. The tubules also secrete products that could influence interstitial cell function.

Recent evidence suggests that inhibin secreted by the sustentacular cells in response to FSH can facilitate the interstitial cells' response to LH, as measured by the amount of testosterone secreted. Further, it has recently been shown that the interstitial cells are capable of producing a family of polypeptides previously associated only with the pituitary gland—ACTH, MSH, and β-endorphin. Experiments suggest that ACTH and MSH can stimulate sustentacular cell function, whereas β-endorphin can exert an inhibitory effect. The physiological significance of these fascinating autocrine interactions between the two compartments of the testes remains to be demonstrated.

## Spermatogenesis

The germ cells that migrate from the yolk sac to the testes during early embryonic development become "stem cells," or **spermatogonia** within the outer region of the seminiferous

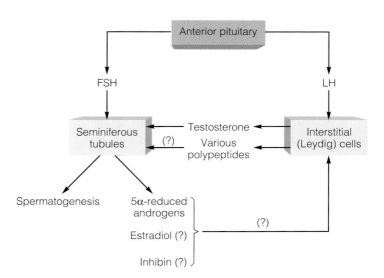

**FIGURE 28.11**
Interactions between the two compartments of the testes. Testosterone secreted by the interstitial cells stimulates spermatogenesis in the tubules. Interstitial cells may also secrete ACTH, MSH, and β-endorphin. Secretion of inhibin by the tubules may affect the sensitivity of the interstitial cells to LH stimulation.

tubules. Spermatogonia are diploid cells (with 46 chromosomes) that ultimately give rise to mature haploid gametes by a process of cell division called meiosis.

As described in chapter 3, *meiosis,* or reduction division, occurs in two parts. In the first part of this process, the DNA duplicates and homologous chromosomes are separated into two daughter cells. Since each daughter cell contains only one of each homologous pair of chromosomes, the cells formed at the end of this first meiotic division contain 23 chromosomes each and are haploid. Each of the 23 chromosomes at this stage, however, consists of two strands (called chromatids) of identical DNA. During the second meiotic division, these duplicate chromatids are separated into daughter cells. Meiosis of one diploid spermatogonia cell therefore produces four haploid cells.

Actually, only about 1000 to 2000 stem cells migrate from the yolk sac into the embryonic testes. In order to produce many millions of sperm throughout adult life, these spermatogonia cells duplicate themselves by mitotic division, and only one of the two cells—now called a **primary spermatocyte**—undergoes meiotic division (fig. 28.12). In this way, spermatogenesis can occur continuously without exhausting the number of spermatogonia.

When a diploid primary spermatocyte completes the first meiotic division (at telophase I), the two haploid cells thus produced are called **secondary spermatocytes.** At the end of the second meiotic division, each of the two secondary spermatocytes produce two haploid **spermatids.** One primary spermatocyte therefore produces four spermatids.

The sequence of events in spermatogenesis is reflected in the cellular arrangement of the wall of the seminiferous tubule. The epithelial wall of the tubule—called the *germinal epithelium*—is indeed composed of germ cells in different stages of spermatogenesis. The spermatogonia and primary spermatocytes are located toward the outer side of the tubule, whereas spermatids and mature spermatozoa are located on the side of the tubule facing the lumen.

At the end of the second meiotic division, the four spermatids produced by meiosis of one primary spermatocyte are interconnected—their cytoplasm does not completely pinch off at the end of each division. Development of these interconnected spermatids into separate, mature spermatozoa (a process called **spermiogenesis**) requires the participation of another type of cell in the tubules, the sustentacular cells (fig. 28.13).

**Sustentacular Cells**   The sustentacular (Sertoli) cells (also called *nurse cells*) are the only nongerminal cell type in the seminiferous tubules. They form a continuous layer, con-

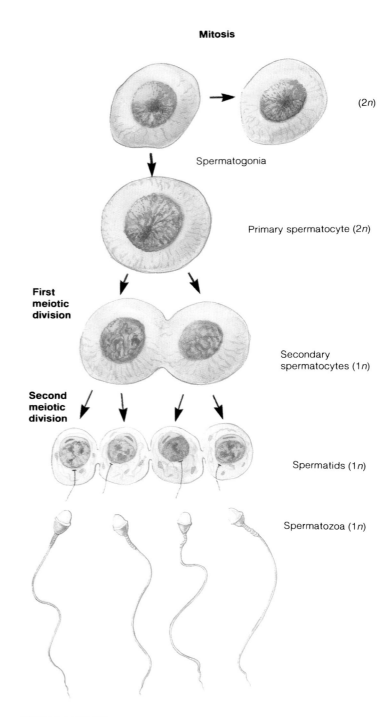

**FIGURE 28.12**

Spermatogonia undergo mitotic division to replace themselves and produce a daughter cell that will undergo meiotic division. This cell is called a primary spermatocyte. Upon completion of the first meiotic division, the daughter cells are called secondary spermatocytes. Each of these completes a second meiotic division to form spermatids. Notice that the four spermatids produced by the meiosis of a primary spermatocyte are interconnected. Each spermatid forms a mature spermatozoan.

## FIGURE 28.13

Seminiferous tubules. (*a*) A transverse section with surrounding interstitial tissue and (*b*) the stages of spermatogenesis within the germinal epithelium of a seminiferous tubule in which the relationship between sustentacular (Sertoli) cells and developing spermatozoa is shown.

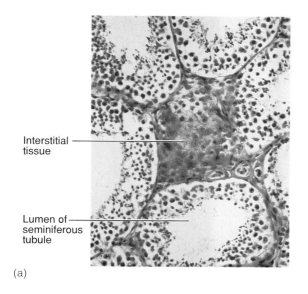

Interstitial tissue

Lumen of seminiferous tubule

(a)

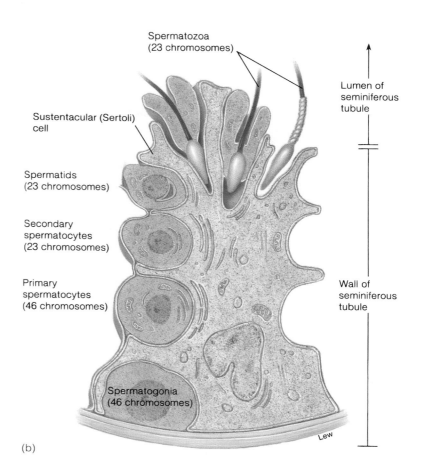

Spermatozoa (23 chromosomes)

Lumen of seminiferous tubule

Sustentacular (Sertoli) cell

Spermatids (23 chromosomes)

Secondary spermatocytes (23 chromosomes)

Primary spermatocytes (46 chromosomes)

Wall of seminiferous tubule

Spermatogonia (46 chromosomes)

Lew

(b)

nected by tight junctions, around the circumference of each tubule. This arrangement of the sustentacular cells creates a *blood-testis barrier* because molecules from the blood must pass through the cytoplasm of these cells before entering germinal cells. Furthermore, this barrier normally prevents the immune system from becoming sensitized to antigens in the developing sperm, and thus prevents autoimmune destruction of the sperm. The cytoplasm of the sustentacular cells extends through the width of the tubule and envelops the developing germ cells, so that it is often difficult to tell where the cytoplasm of the sustentacular cells and that of germ cells is separated.

In the process of *spermiogenesis* (conversion of spermatids to spermatozoa), most of the spermatid cytoplasm is eliminated. This occurs through phagocytosis by sustentacular cells of the "residual bodies" of cytoplasm from the spermatids (fig. 28.14). Many believe that phagocytosis of residual bodies may transmit informational molecules from germ cells to sustentacular cells. The sustentacular cells, in turn, may provide many molecules needed by the germ cells. It is known, for example, that the X chromosome of germ cells is inactive during meiosis. Since this chromosome contains genes

needed to produce many essential molecules, it is believed that these molecules are provided by the sustentacular cells while meiosis is taking place.

Sustentacular cells produce a protein called **androgen binding protein (ABP)** into the lumen of the seminiferous tubules. ABP serves to concentrate testosterone within the tubules. The importance of sustentacular cells in tubular function is further evidenced by the fact that FSH receptors are confined to these cells. The effects of FSH on the tubules, therefore, must be mediated through the action of sustentacular cells. These effects include the FSH-induced stimulation of spermiogenesis and the autocrine interactions between sustentacular cells and interstitial cells previously described.

**Hormonal Control of Spermatogenesis** The very beginning of spermatogenesis—the formation of primary spermatocytes and entry into early prophase I—is apparently somewhat independent of hormonal control and, in fact, is initiated during embryonic development. Spermatogenesis is arrested at this stage, however, until puberty, when

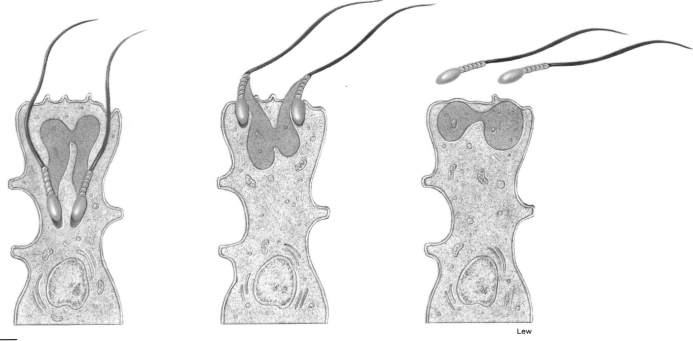

**▣ FIGURE 28.14**

Processing of spermatids into spermatozoa (spermiogenesis). As the spermatids develop into spermatozoa, most of their cytoplasm is pinched off as residual bodies and ingested by the surrounding sustentacular (Sertoli) cell cytoplasm.

Lew

testosterone secretion rises. Testosterone is required for the completion of meiotic division and for the early stages of spermatid maturation. These stages of spermatogenesis probably do not result from the direct action of testosterone, but rather from the action of some of the molecules derived from testosterone in the tubules.

The later stages of spermatid maturation during puberty appear to require stimulation by FSH (fig. 28.15). This FSH effect is mediated by the sustentacular cells, as previously described. During puberty, therefore, both FSH and androgens are needed for the initiation of spermatogenesis. By contrast, within the adult testis it appears that spermatogenesis can be maintained by androgens alone, in the absence of FSH. In other words, FSH apparently is needed to initiate spermatogenesis at puberty, but it may no longer be required once spermatogenesis has begun.

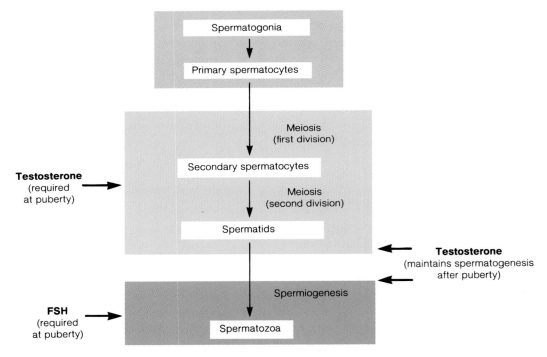

**▣ FIGURE 28.15**

The endocrine control of spermatogenesis. During puberty both testosterone and FSH are required to initiate spermatogenesis. In the adult, however, testosterone alone can maintain spermatogenesis.

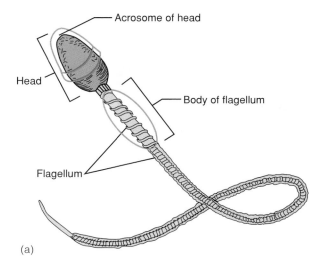

(a)

**FIGURE 28.16**

A human spermatozoon. (*a*) A diagrammatic representation and (*b*) a scanning electron micrograph of a spermatozoan in contact with an egg.

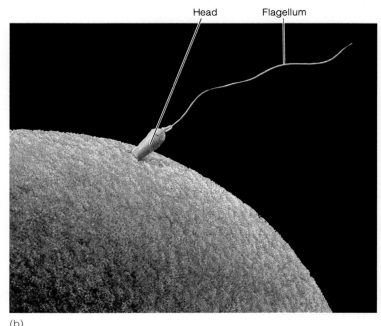

(b)

 Men who have had a *hypophysectomy* (surgical removal of the pituitary gland) experience a cessation of spermatogenesis. Spermatogenesis is restored in these patients by injections of FSH and *hCG* (*human chorionic gonadotropin*, a hormone of the placenta discussed in chapter 30). The hCG has the same biological activity as LH. In this case, it acts like LH to stimulate the interstitial cells to secrete testosterone, which is needed along with FSH to initiate sperm production. After spermatogenesis has been restored, it can be maintained in hypophysectomized patients with hCG injections alone, demonstrating that testosterone can maintain spermatogenesis.

**Structure of Spermatozoa**  At the conclusion of spermiogenesis, spermatozoa are released into the lumina of the seminiferous tubules. A mature sperm cell, or **spermatozoon,** is a microscopic, tadpole-shaped structure about 0.06 mm long (fig. 28.16). It consists of an oval-shaped **head** and an elongated **flagellum.** The head of a sperm cell contains a nucleus with 23 chromosomes. The tip of the head, called the *acrosome*, contains enzymes that help the sperm cell to penetrate the ovum. The flagellum contains numerous mitochondria spiraled around a filamentous core. The mitochondria provide the energy necessary for locomotion. The flagellum propels the sperm with a lashing movement. The maximum unassisted rate of sperm movement is about 3 mm per hour.

. . . . . . . . . . . . . . . . . . . . . . . . . . . . . . . . . . . . . .

spermatozoon: Gk. *sperma*, seed; *zoon*, animal
acrosome: GK. *akros*, extremity; *soma*, body

The life expectancy of ejaculated sperm is between 48 and 72 hours at body temperature. Many of the ejaculated sperm, however, are defective and are of no value. It is not uncommon for sperm to have enlarged heads, dwarfed and misshapen heads, two flagella, or a flagellum that is bent. Sperm such as these are unable to propel themselves adequately.

## Spermatic Ducts, Accessory Glands, and the Penis

*The spermatic ducts store spermatozoa and transport them from the testes to the urethra. The accessory reproductive glands provide additives to the semen, which is discharged from the erect penis during ejaculation.*

### Spermatic Ducts

The male duct system, which stores and transports spermatozoa from the testes to the urethra, includes the epididymides, the ductus deferentia, and the ejaculatory ducts.

**Epididymis**  The **epididymis** (*ep″ĭ-did′ĭ-mis*)—in the plural, *epididymides*—is a long, flattened organ attached to the posterior surface of the testis (see figs. 28.7 and 28.8). The tubular portion of the epididymis is highly coiled and contains millions of sperm in their final stages of maturation (fig. 28.17). It is estimated that if the epididymis were uncoiled,

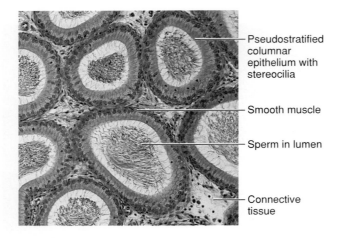

**FIGURE 28.17**
A photomicrograph of the epididymis showing sperm in the lumina (50×).

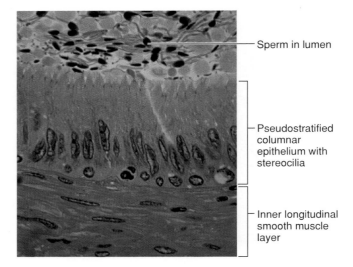

**FIGURE 28.18**
A photomicrograph of the ductus deferens (250×).

it would measure 5 to 6 m (about 17 ft). The upper, expanded portion of the epididymis is the **head,** the tapering middle section is the **body,** and the lower portion is the **tail.** The tail of the epididymis is continuous with the ductus deferens. The time required to produce mature sperm—from meiosis in the seminiferous tubules to storage of motile sperm in the ductus deferens—is approximately 2 months.

**Ductus Deferens**   The **ductus deferens** (*duk´tus def´er-enz*)—in the plural, *ductus deferentia*—is a fibromuscular tube about 45 cm (18 in.) long and 2.5 mm thick (see fig. 28.7) that conveys sperm from the epididymis to the ejaculatory duct. The ductus deferens originates where the tail of the epididymis becomes less convoluted and is no longer attached to the testis. This first portion of the ductus deferens is important for the storage of sperm. The ductus deferens exits the scrotum as it ascends along the posterior border of the testis. From here, it penetrates the inguinal canal, enters the pelvic cavity, and passes to the side of the urinary bladder medial to the ureter. The **ampulla** of the ductus deferens is the terminal portion that joins the ejaculatory duct.

The histological structure of the ductus deferens includes a layer of pseudostratified ciliated columnar epithelium in contact with the tubular lumen and surrounded by three layers of tightly packed smooth muscle (fig. 28.18). Sympathetic nerves from the pelvic plexus serve the ductus deferens. Stimulation through these nerves causes peristaltic contractions of the muscular layer, which forcefully ejects the stored sperm toward the ejaculatory duct.

Much of the ductus deferens is located within a structure known as the **spermatic cord** (see fig. 28.22). The spermatic cord extends from the testis to the inguinal ring and

consists of the ductus deferens, spermatic vessels, nerves, cremaster muscle, lymph vessels, and connective tissue. The portion of the spermatic cord positioned anterior to the pubic bone can be palpated as it is compressed between the skin and the bone.

**Ejaculatory Duct**   The **ejaculatory duct** is 2 cm (0.8 in.) long and is formed by the union of the ampulla of the ductus deferens and the duct of the seminal vesicle. The ejaculatory duct then pierces the capsule of the prostate on its posterior surface and continues through this gland (see fig. 28.7). Both ejaculatory ducts receive secretions from the seminal vesicles and prostate and then eject the sperm with its additives into the prostatic urethra.

## *Accessory Glands*

Accessory reproductive glands include the seminal vesicles, the prostate, and the bulbourethral glands (see fig. 28.7). The contents of the seminal vesicles and the prostate are mixed with the sperm during ejaculation to form *semen* (*seminal fluid*).

**Seminal Vesicles**   The **seminal vesicles,** which are convoluted, club-shaped glands about 5 cm (2 in.) long, are positioned immediately posterior to and at the base of the urinary bladder. They secrete a sticky, slightly alkaline, yellowish substance that serves to enhance sperm movement and longevity. The secretion from the seminal vesicles contains a variety of nutrients, including fructose, a monosaccharide that provides sperm with an energy source. It also

. . . . . . . . . . . . . . . . . . . . . . . . . . . . . . . . . . .

deferens: L. *deferens,* conducting away
ampulla: L. *ampulla,* a two-handled bottle

. . . . . . . . . . . . . . . . . . . . . . . . . . . . . . . . . . .

vesicle: L. *vesicula,* a blister; diminutive of *vesica,* bladder

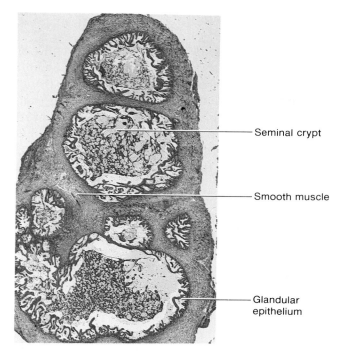

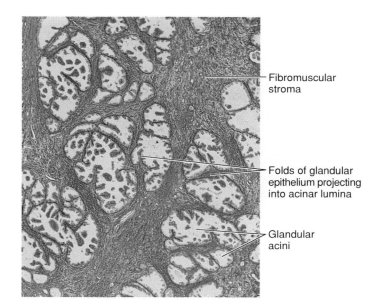

**FIGURE 28.20**
A photomicrograph of the prostate.

Fibromuscular stroma

Folds of glandular epithelium projecting into acinar lumina

Glandular acini

Seminal crypt

Smooth muscle

Glandular epithelium

**FIGURE 28.19**
The histology of the seminal vesicle.

contains citric acid, coagulation proteins, and prostaglandins. The discharge from the seminal vesicles makes up about 60% of the volume of semen.

Histologically, the seminal vesicle looks like a mass of cuts embedded in connective tissue (fig. 28.19). The extensively coiled mucosal layer breaks the lumen into numerous intercommunicating spaces that are lined by pseudostratified columnar and cuboidal secretory epithelia (referred to as glandular epithelium). Sympathetic stimulation causes the contents of the seminal vesicles to empty into the ejaculatory ducts of their respective sides.

**Prostate**  The firm **prostate** (*pros´tāt*) is the size and shape of a horse chestnut. It is about 4 cm (1.6 in.) across and 3 cm (1.2 in.) thick and is positioned immediately below the urinary bladder surrounding the beginning portion of the urethra (see fig. 28.7). It is enclosed by a fibrous capsule and divided into lobules formed by the urethra and the ejaculatory ducts extending through the gland. Twenty to 30 small prostatic ducts from the lobules open into the urethra. Extensive bands of smooth muscular tissue course throughout the prostate to form a meshwork that supports the glandular tissue (fig. 28.20). Contraction of the smooth muscle within the prostate empties the contents from the gland and provides part of the propulsive force needed to ejaculate the semen. The thin, milky-colored prostatic secretion assists sperm motility as a liquefying

agent, and its alkalinity protects sperm in their passage through the acidic environment of the female vagina. The discharge from the prostate makes up about 60% of the volume of semen. (Spermatozoa constitute less than 1% of the volume of semen.) Clotting proteins in the prostatic fluid causes the semen to coagulate after ejaculation, but the hydrolytic action of *fibrinolysin* later causes the coagulated semen to again assume a more liquid form, thereby freeing the sperm. The prostate also secretes the enzyme *acid phosphatase*, which is often measured clinically to assess prostate function.

 A routine physical examination of the male includes rectal palpation of the prostate. Enlargement or overgrowth of the glandular substance of the prostate, a condition called *benign prostatic hypertrophy*, is relatively common in older men. An enlarged prostate may constrict the prostatic urethra and cause difficult micturition. Treatment is usually surgical, and in some cases may be accomplished through the urethral canal. This technique, called a *transurethral prostatic resection* involves cutting and cauterizing the excessive tissue. In another technique, a small balloon is inserted and inflated to keep the prostate from exerting pressure on the prostatic urethra.

**Bulbourethral Glands**  The paired, pea-sized, brownish-colored **bulbourethral** (*bul˝bo-yoo-re´thral*) **glands** (Cowper's glands) are located inferior to the prostate. Each bulbourethral gland is about 1 cm in diameter and drains by a 2.5-cm (1-in.) duct into the urethra (see fig. 28.7). Upon sexual excitement and prior to ejaculation, the bulbourethral glands are stimulated to secrete a mucoid substance. This

---

prostate: Gk. *prostate*, one standing before

---

cauterize: Gk. *kauterion*, a branding iron
Cowper's gland: from William Cowper, English anatomist, 1666–1709

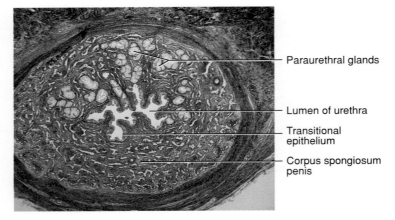

**FIGURE 28.21**
The histology of the urethra (10×).

— Paraurethral glands

— Lumen of urethra

— Transitional epithelium

— Corpus spongiosum penis

secretion coats the lining of the urethra to neutralize the pH of the urine residue and lubricates the tip of the penis in preparation for coitus.

## Urethra

The **urethra** of the male serves as a common tube for both the urinary and reproductive systems. However, urine and semen cannot simultaneously pass through the urethra because the nervous reflex during ejaculation automatically inhibits micturition. The urethra of the male is about 20 cm (8 in.) long and S-shaped due to the shape of the penis. Three regions can be identified—the prostatic urethra, the membranous urethra, and the spongy urethra (see fig. 25.29).

The **prostatic urethra** is the proximal (2.5 cm) portion of the urethra that passes through the prostate. The prostatic urethra receives drainage from the small ducts of the prostate and the two ejaculatory ducts.

The **membranous urethra** is the short (0.5 cm) portion of the urethra that passes through the urogenital diaphragm. The external urethral sphincter muscle is located in this region.

The **spongy urethra** is the longest portion (15 cm), extending from the outer edge of the urogenital diaphragm to the external urethral orifice on the glans penis. This portion is surrounded by erectile tissue as it passes through the corpus spongiosum of the penis. The paired ducts from the bulbourethral glands attach to the spongy urethra near the urogenital diaphragm. The membranous urethra and the spongy urethra are frequently referred to as the *penial urethra.*

The wall of the urethra has an inside lining of mucous membrane, composed of transitional epithelium (fig. 28.21) and surrounded by a relatively thick layer of smooth muscle tissue called *tunica muscularis.* Specialized **urethral glands** are embedded in the urethral wall and function to secrete mucus into the urethral canal.

## Penis

The **penis,** when distended, serves as the copulatory organ of the male reproductive system. It is composed mainly of **erectile tissue**—a spongy network of connective tissue and smooth muscle characterized by vascular spaces that become engorged with blood. The penis is a pendent structure, positioned anterior to the scrotum and attached to the pubic arch. It is divided into a proximally attached root, an elongated tubular body, and a distal glans penis (fig. 28.22).

The **root of the penis** expands posteriorly to form the **bulb of the penis** and the **crus** (*krus*) **of the penis.** The bulb is positioned in the urogenital triangle of the perineum, where it is attached to the undersurface of the perineal membrane and enveloped by the bulbocavernosus muscle (see fig. 13.14). The crus attaches the root of the penis to the pubic arch (ischiopubic ramus) and to the perineal membrane. The crus is positioned superior to the bulb and is enveloped by the ischiocavernosus muscle.

The **body,** or **shaft, of the penis** is composed of three cylindrical columns of erectile tissue that are bound together by fibrous tissue and covered with skin (fig. 28.22). The paired columns that form the dorsum and sides of the penis are named the **corpora cavernosa penis.** The fibrous tissue between the two corpora forms a **median septum.** The **corpus spongiosum penis** lies anterior to the other two and surrounds the spongy urethra. The penis is flaccid and relaxed when the spongelike tissue is not engorged with blood but becomes firm and erect when the spaces are filled.

The **glans penis** is the cone-shaped terminal portion formed from the expanded corpus spongiosum. The opening of the urethra at the tip of the glans penis is called the **urethral orifice** (see fig. 25.29). The **corona glandis** is the prominent posterior ridge of the glans penis. On the undersurface of the glans, a vertical fold of tissue called the **frenulum** (*fren´yŭ-lum*) attaches the skin covering the penis to the glans.

The skin covering the penis is hairless, lacks fat cells, and generally is more darkly pigmented than the other body skin. The skin of the body is loosely attached and is continuous over the glans penis as a retractable sheath called the **prepuce** (*pre´pyoos*), or **foreskin.** The prepuce is commonly removed in male infants by a surgical procedure called *circumcision.*

........................................................

crus: L. *crus*, leg; resembling a leg
cavernous: L. *cavus*, hollow
glans: L. *glans*, acorn
corona: L. *corona*, garland, crown
frenulum: L. diminutive of *frenum*, a bridle
prepuce: L. *prae*, before; *putium*, penis

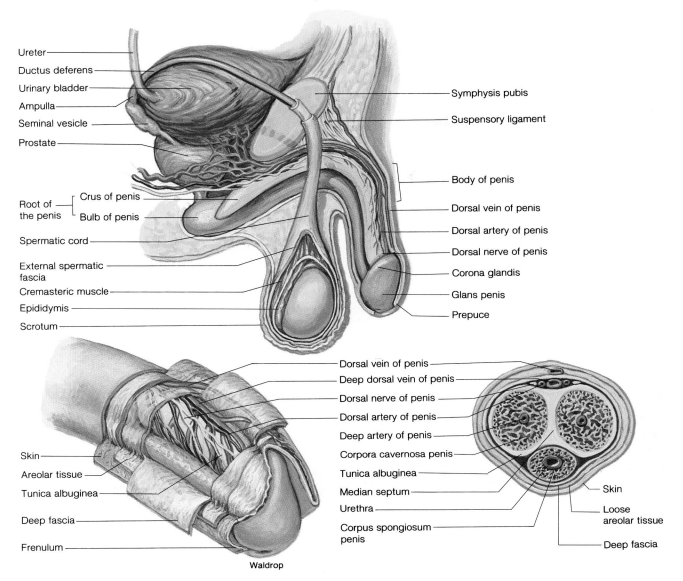

**FIGURE 28.22**

The structure of the penis showing the attachment, blood and nerve supply, and the arrangement of the erectile tissue.

 A *circumcision* is generally performed for hygienic purposes because the glans penis is easier to clean if exposed. A sebaceous secretion from the glans penis, called *smegma*, will accumulate along the border of the corona glandis if good hygiene is not practiced. Smegma can foster bacteria that may cause infections and therefore should be removed through washing. Cleaning the glans penis of an uncircumcised male requires retraction of the prepuce. Occasionally, a child is born with a prepuce that is too tight to permit retraction. This condition is called *phimosis* and necessitates circumcision.

## Erection of the Penis

Erection of the penis depends on the volume of blood that enters the arteries of the penis as compared to the volume that exits through venous drainage. Normally, constant sympathetic stimuli to the arterioles of the penis maintain a partial constriction of smooth muscles within the arteriole walls so that there is an even flow of blood throughout the penis. During sexual excitement, however, parasympathetic impulses cause marked vasodilation within the arterioles of the penis, resulting in more blood entering than venous blood draining. This causes the spongy tissue of the corpora cavernosa and the corpus spongiosum to become distended with blood and the penis to become turgid. Recent evidence has shown that the responses to parasympathetic stimulation that produce erection of the penis are mediated by nitric oxide (NO) as a neurotransmitter.

Erection is controlled by two portions of the central nervous system—the hypothalamus in the brain and the

---

phimosis: Gk. phimosis, a muzzling

turgid: L. turgeo, to swell

**FIGURE 28.23**

The mechanism of emission and ejaculation.

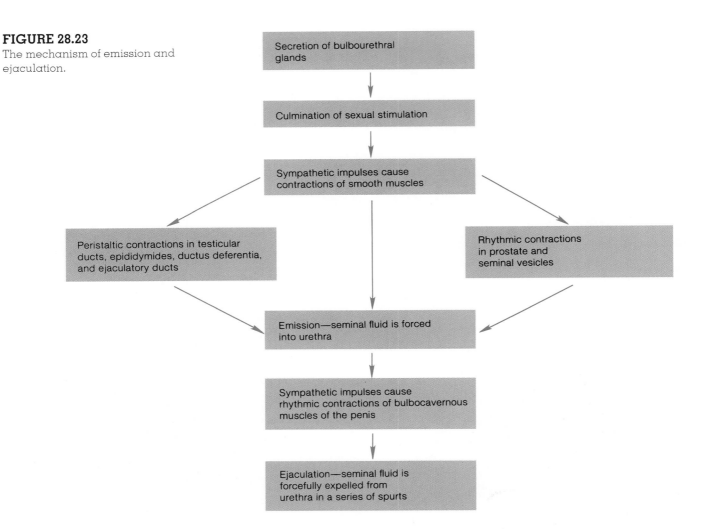

sacral portion of the spinal cord. The hypothalamus controls conscious sexual thoughts that originate in the cerebral cortex. Nerve impulses from the hypothalamus elicit parasympathetic responses from the sacral region that cause vasodilation of the arterioles within the penis. Conscious thought is not required for an erection, however, and stimulation of the penis can cause an erection because of a reflex response in the spinal cord. This reflexive action makes possible an erection in a sleeping male or in an infant—perhaps from the stimulus of a diaper.

## Emission and Ejaculation of Semen

Continued sexual stimulation following erection of the penis causes emission. Emission is the movement of sperm from the epididymides to the ejaculatory ducts and the secretions of the accessory glands into the ejaculatory ducts and urethra in the formation of semen. The first sympathetic response, which occurs prior to ejaculation, is the discharge of fluids from the bulbourethral glands. These fluids are usually discharged before penetration of the penis into the vagina and

..............................................

emission: L. *emittere,* expel or eject

serve to lubricate the urethra and the glans penis. Emission occurs as sympathetic impulses from the pelvic plexus cause a rhythmic contraction of the smooth muscle layers of the testes, epididymides, ductus deferentia, ejaculatory ducts, seminal vesicles, and prostate.

Ejaculation immediately follows emission and is accompanied by *orgasm*, which is considered the climax of the sex act. Ejaculation occurs in a series of spurts of semen from the urethra. This takes place as parasympathetic impulses traveling through the pudendal nerves stimulate the bulbocavernosus muscles at the base of the penis and cause them to contract rhythmically. There is also sympathetic stimulation of the smooth muscles in the urethral wall that peristaltically contract to help eject the semen. Sexual function in the male thus requires the synergistic (rather than antagonistic) action of the parasympathetic and sympathetic nervous systems. The mechanism of emission and ejaculation is summarized in figure 28.23.

Immediately following ejaculation or a cessation of sexual stimuli, sympathetic impulses cause vasoconstriction of the arterioles within the penis, reducing the inflow of blood. At the same time, cardiac output returns to normal, as does venous return of blood from the penis. With the

## Embryonic Development of the Reproductive System

The male and female reproductive systems follow a similar pattern of development, with sexual distinction resulting from the influence of hormones. A significant fact of embryonic development is that the sexual organs for both male and female are derived from the same developmental tissues and are considered *homologous structures.*

The first sign of development of either the male or the female reproductive organs occurs during the fifth week as the medial aspect of each mesonephros (see chapter 25) enlarges to form the **gonadal ridge** (fig. 1). The gonadal ridge continues to grow behind the developing peritoneal membrane. By the sixth week, stringlike masses called **primary sex cords** form within the enlarging gonadal ridge. The primary sex cords in the male will eventually mature to become the seminiferous tubules. In the female, the primary sex cords will contribute to nurturing tissue of developing ova. Each gonad develops near a **mesonephric** (wolffian) **duct** and a **paramesonephric** (müllerian) **duct.**

In the male embryo, each testis connects through a series of tubules to the mesonephric duct. During further development, the connecting tubules become the seminiferous tubules, and the mesonephric duct becomes the *efferent ductules, epididymis, ductus deferens, ejaculatory duct,* and *seminal vesicle.* The paramesonephric duct in the male degenerates without contributing any functional structures to the reproductive system.

In the female embryo, the mesonephric duct degenerates, and the paramesonephric duct contributes in large measure to structures of the female reproductive system. The distal ends of the paired paramesonephric ducts fuse to form the *vagina* and *uterus.* The proximal unfused portions become the *uterine tubes.*

Externally, by the sixth week a swelling called the **genital tubercle** appears anterior to the small embryonic tail (future coccyx). The mesonephric and paramesonephric ducts open to the outside through the genital tubercle. The genital tubercle consists of a *glans*, a *urethral groove*, paired *urethral folds*, and paired *labioscrotal swellings* (fig. 2). As the glans portion of the genital tubercle enlarges, it becomes known as the *phallus*. Early in fetal development (tenth through twelfth week), sexual distinction of the external genitalia becomes apparent. In the male, the phallus enlarges and develops into the *glans* of the penis. The urethral folds fuse around the urethra to form the *body* of the penis. The urethra opens at the end of the glans as the *urethral orifice*. The labioscrotal swellings fuse to form the *scrotum* into which the testes will descend. In the female, the phallus gives rise to the *clitoris*, the urethral folds remain unfused as the *labia minora*, and the labioscrotal swellings become the *labia majora*. The urethral groove is retained as a longitudinal cleft known as the *vestibule*.

### Development of Secondary Sex Organs

In addition to testes and ovaries, which are the **primary sex organs,** or **gonads,** the external genitalia and various internal *accessory sex organs* are needed for reproductive function. These are known as the **secondary sex organs.** Some of the male accessory sex organs are derived from the **mesonephric** (wolffian) **ducts,** and female accessory organs are derived from the **paramesonephric** (müllerian) **ducts** (fig. 2). Interestingly, from about day 25 to about day 50, male and female embryos alike have both duct systems and, therefore, have the potential to form the accessory organs characteristic of either gender.

Experimental removal of the testes (castration) from male embryonic animals results in regression of the mesonephric ducts and development of the paramesonephric ducts into female accessory sex organs: the **uterus** and **uterine** (fallopian) **tubes.** Female accessory sex organs, therefore, develop as a result of the absence of testes (and their secretion of testosterone) rather than the presence of ovaries.

The developing seminiferous tubules within the testes secrete a polypeptide called *müllerian inhibition factor,* that causes regression of the paramesonephric ducts beginning about day 60. The secretion of testosterone by the interstitial cells of the testes subsequently causes growth and development of the mesonephric ducts into male accessory sex organs: the **epididymis, ductus deferens, seminal vesicles,** and **ejaculatory duct.**

Other structures that male and female embryos share in common are the *urogenital sinus, genital tubercle, urethral folds,* and *labiosacral swellings.* The secretions of the testes masculinize these structures to form the **penis, prostate,** and **scrotum.** The genital tubercle that forms the penis in a male will, in the absence of testes, become the **clitoris** in a female. The penis and clitoris are thus said to be *homologous structures.* Similarly, the labiosacral swellings form the scrotum in a male or the **labia majora** in a female; these structures are therefore also homologous (fig. 3).

In summary, the genetic sex is determined by whether a Y-bearing or an X-bearing spermatozoon fertilizes the ovum; the presence or absence of a Y chromosome in turn determines whether the gonads of the embryo will be testes or ovaries; and, finally, the presence or absence of testes determines whether the accessory sex organs and external genitalia will be male or female (table 1).

This regulatory pattern of sex determination makes sense in light of the fact that both male and female embryos develop within an environment high in estrogen, which is secreted by the mother's ovaries and the placenta. If the secretions of the ovaries determined the sex, all embryos would be female.

wolffian ducts: from Kaspar Friedrich Wolff, German embryologist, 1733–94

müllerian ducts: from Johannes Peter Müller, German physiologist, 1801–58

homologous: Gk. *homos*, the same

*Continued*

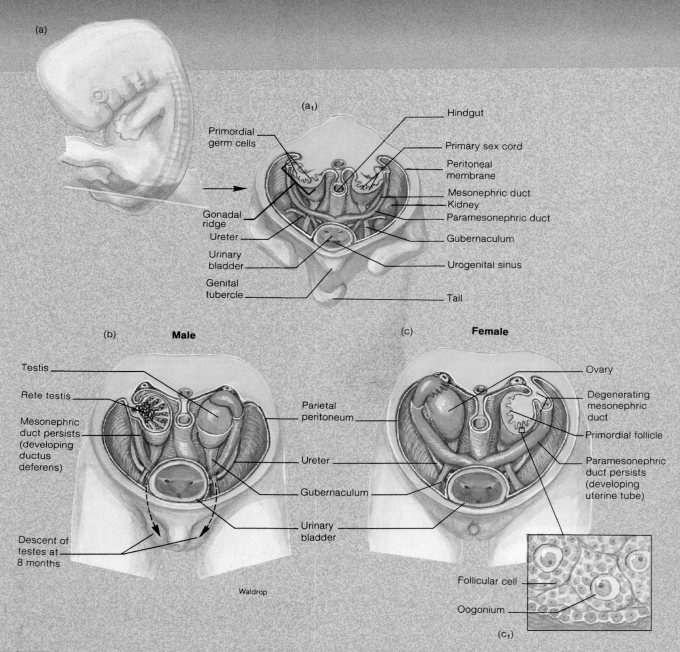

(a)

(a₁)

Primordial germ cells

Gonadal ridge

Ureter

Urinary bladder

Genital tubercle

Hindgut

Primary sex cord

Peritoneal membrane

Mesonephric duct

Kidney

Paramesonephric duct

Gubernaculum

Urogenital sinus

Tail

(b)  **Male**

Testis

Rete testis

Mesonephric duct persists (developing ductus deferens)

Descent of testes at 8 months

Parietal peritoneum

Ureter

Gubernaculum

Urinary bladder

(c)  **Female**

Ovary

Degenerating mesonephric duct

Primordial follicle

Paramesonephric duct persists (developing uterine tube)

Follicular cell

Oogonium

(c₁)

Waldrop

## FIGURE 1

Differentiation of the male and female gonads and genital ducts. (a) An embryo at 6 weeks showing the positions of a transverse cut depicted in (a₁), (b), and (c.) (a₁) At 6 weeks, the developing gonads (primary sex cords) are still indifferent. By 4 months, the gonads have differentiated into male (b) or female (c). The oogonia have formed within the ovaries (c₁) by 6 months.

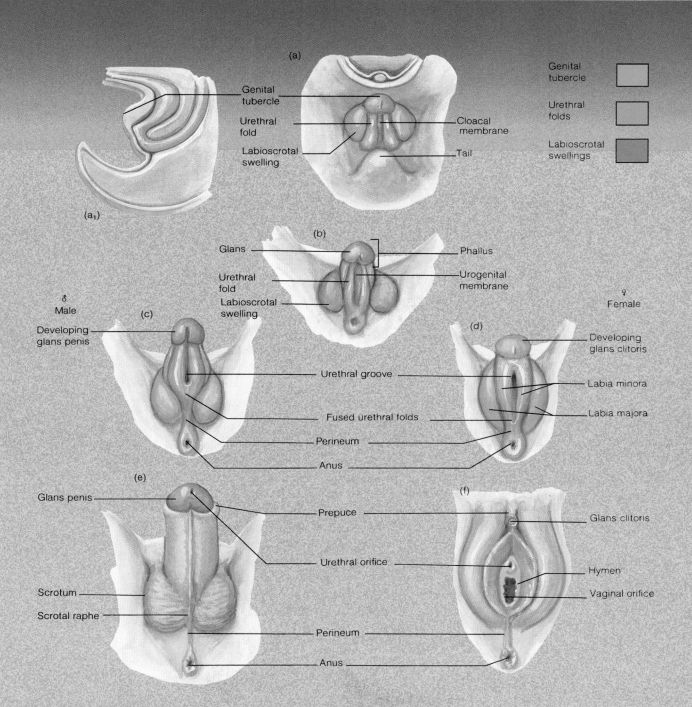

(a)

Genital tubercle

Urethral fold

Labioscrotal swelling

(a₁)

Cloacal membrane

Tail

Genital tubercle

Urethral folds

Labioscrotal swellings

(b)

Glans

Urethral fold

Labioscrotal swelling

Phallus

Urogenital membrane

♂ Male

(c)

Developing glans penis

♀ Female

(d)

Developing glans clitoris

Labia minora

Labia majora

Urethral groove

Fused urethral folds

Perineum

Anus

(e)

Glans penis

Scrotum

Scrotal raphe

Prepuce

Urethral orifice

Perineum

Anus

(f)

Glans clitoris

Hymen

Vaginal orifice

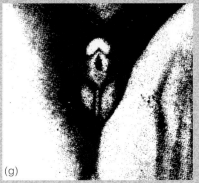

(g)

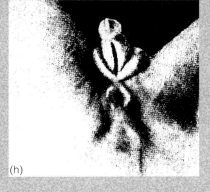

(h)

## FIGURE 2

Differentiation of the external genitalia in the male and female. (a, a₁ [sagittal view]) At 6 weeks, the genital tubercle, urethral fold, and labioscrotal swelling have differentiated from the genital tubercle. (b) At 8 weeks, a distinct phallus is present during the indifferent stage. By week 12, the genitalia have become distinctly male (c) or female (d), being derived from homologous structures. (e, f). At 16 weeks, the genitalia are formed. (g, h) Photographs at week 10 of male and female genitalia, respectively.

*Continued*

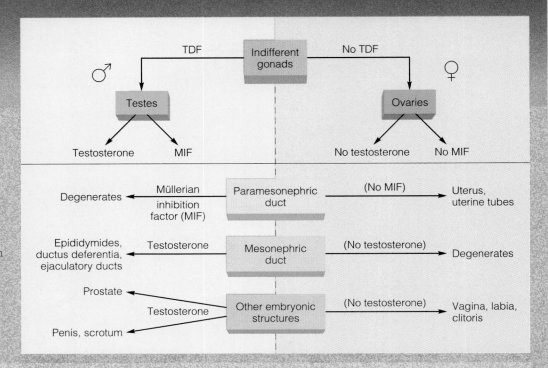

**FIGURE 3**

The embryonic development of male and female accessory sex organs and external genitalia. In the presence of testosterone and müllerian inhibition factor (MIF) secreted by the testes, male structures develop. In the absence of these secretions, female structures develop.

## Table 1  A developmental timetable for the reproductive system

| Approximate time after fertilization | | | Developmental changes | | |
|---|---|---|---|---|---|
| Days | Trimester | Indifferent | Male | Female |
| 19 | First | Germ cells migrate from yolk sac. | | |
| 25–30 | | Mesonephric ducts begin development. | | |
| 44–48 | | Paramesonephric ducts begin development. | | |
| 50–52 | | Urogenital sinus and tubercle develop. | | |
| 43–60 | | | Tubules and sustentacular cells appear. Paramesonephric ducts begin to regress. | |
| 60–75 | | | Interstitial cells appear and begin testosterone production. | Formation of vagina begins. |
| | | | Mesonephric ducts grow. | Regression of mesonephric ducts begins. |
| 105 | Second | | | Development of ovarian follicles begins. |
| 120 | | | | Uterus is formed. |
| 160–260 | Third | | Testes descend into scrotum. Growth of external genitalia occurs. | Formation of vagina complete. |

Source: *Annual Review of Physiology*, vol. 40, p. 279, 1978.

| Table 28.4 | Semen analysis |
|---|---|
| **Characteristic** | **Reference value** |
| Volume of ejaculate | 1.5–5.0 ml |
| Sperm count | 40–250 million/ml |
| Sperm motility | |
| Percentage of motile forms: | |
| 1 hour after ejaculation | 70% or more |
| 3 hours after ejaculation | 60% or more |
| Leukocyte count | 0–2000/ml |
| pH | 7.2–7.8 |
| Fructose concentration | 150–600 mg/100 ml |

Source: July/August 1981 issue of *Diagnostic Medicine*.

return of the normal flow of blood through the penis, it again becomes flaccid. Following an ejaculation of semen from the erect penis, another erection and ejaculation cannot be triggered for a period ranging from 10 minutes to a few hours.

**Semen**    Semen, also called seminal (*sem´ĭ-nal*) fluid, consists of spermatozoa plus the additives from the accessory sex glands, (table 28.4). Generally, the volume of semen ejaculated ranges between 1.5 and 5.0 ml. The bulk of the fluid (about 60%) is produced by the seminal vesicles, and the rest (about 40%) is contributed by the prostate. There are usually between 60 and 150 million sperm per milliliter of ejaculate. In the condition of *oligospermia*, the male ejaculates fewer than 10 million sperm per milliliter and is likely to have fertility problems.

 Human semen can be frozen and stored in sperm banks for future artificial insemination. In this procedure, the semen is diluted with 10% glycerol, monosaccharide, and distilled water buffer and frozen in liquid nitrogen. The freezing process destroys defective and abnormal sperm. For some unknown reason, however, not all human sperm is suitable for freezing.

## Clinical Considerations

Sexual dysfunction is a broad area of medical concern that includes developmental and psychogenic problems as well as conditions resulting from various diseases. Psychogenic problems of the reproductive system are extremely complex, poorly

oligospermia: Gk. *oligos*, few; *sperma*, seed
fertility: L. *fere*, to bear

understood, and beyond the scope of this book. Only a few of the principal developmental conditions, functional disorders, and diseases that affect the physical structure and function of the male reproductive system are included in this section.

### Developmental Abnormalities

The reproductive organs of both sexes develop from similar embryonic tissue that follows a consistent pattern of formation well into the fetal period. Because an embryo has the potential to differentiate into a male or a female, developmental errors can result in various degrees of intermediate sex, or **hermaphroditism** (*her-maf´rŏ-di-tiz˝em*). A person with undifferentiated or ambiguous external genitalia is called a **hermaphrodite.**

True hermaphroditism—in which both male and female gonadal tissues are present, in either the same or opposite gonads—is a rare anomaly. True hermaphrodites usually have a 46, XX chromosome constitution. **Male pseudohermaphroditism** occurs more commonly and generally results from hormonal influences during early fetal development. This condition is caused either by inadequate amounts of androgenic hormones being secreted or by the delayed development of the reproductive organs after the period of tissue sensitivity has passed. These individuals have a 46, XY chromosome constitution and male gonads, but intersexual and variable genitalia. The treatment of hermaphroditism varies, depending on the extent of ambiguity of the reproductive organs. Although people with this condition are sterile, they may engage in normal sexual relations following hormonal therapy and plastic surgery.

Chromosomal anomalies result from the improper separation of the chromosomes during meiosis and are usually expressed in deviations of the reproductive organs. The two most frequent chromosomal anomalies cause Turner's syndrome and Klinefelter's syndrome. **Turner's syndrome** occurs when only one X chromosome is present. About 97% of embryos lacking an X chromosome die; the remaining 3% survive and appear to be females, but their gonads are rudimentary or absent, and they do not mature at puberty. People with **Klinefelter's syndrome** have an XXY chromosome constitution and develop male genitalia, but have underdeveloped seminiferous tubules. These individuals are generally retarded.

A more common developmental problem than genetic abnormalities, and fortunately less serious, is cryptorchidism. **Cryptorchidism** (*krip-tor´kĭ-diz˝em*) means "hidden testis" and is characterized by the failure of one or both testes to descend into the scrotum. A cryptorchid

hermaphrodite: GK. (mythology) Hermaphrodites, son of Hermes (Mercury)
Turner's syndrome: from Henry H. Turner, American endocrinologist, 1892–1970
Klinefelter's syndrome: from Harry F. Klinefelter Jr., American physician, b. 1912
cryptorchidism: Gk. *crypto*, hidden; *orchis*, testis

testis is usually located along the path of descent but can be anywhere in the pelvic cavity (fig. 28.24). It occurs in about 3% of male infants and should be treated before the infant has reached the age of 5 to reduce the likelihood of infertility or other complications.

## Functional Considerations

Functional disorders of the male reproductive system include impotence, infertility, and sterility. **Impotence** (*im´pŏ-tens*) is the inability of a sexually mature male to achieve and maintain penile erection and/or the inability to achieve ejaculation. The causes of impotence may be physical, such as abnormalities of the penis, vascular irregularities, neurological disorders, or certain diseases. Generally, however, the cause of impotence is psychological, and the patient requires skilled counseling by a sex therapist.

**Infertility** is the inability of the sperm to fertilize the ovum. Infertility problems may be due to factors originating in the male or female, or both. The term *impotence* should not be confused with infertility. In males, the most common cause of infertility is the inadequate production of viable sperm. This may result from alcoholism, dietary deficiencies, local injury, varicocele, excessive heat, or exposure to X rays. A hormonal imbalance may also contribute to infertility. Many of the causes of infertility can be treated through proper nutrition, gonadotropic hormone treatment, or microsurgery. If corrective treatment is not possible, however, it may be possible to concentrate the sperm obtained through *masturbation* (in males, self-stimulation to the point of ejaculation) and use this concentrate to artificially inseminate the female.

**Sterility** is similar to infertility, except that sterility is a permanent condition. Sterility may be genetically caused, or it may be the result of degenerative changes in the seminiferous tubules (for example, mumps in a mature male may secondarily infect the testes and cause irreversible tissue damage).

Voluntary sterilization of the male in a procedure called a **vasectomy** is a common technique of birth control and can be performed on an outpatient basis. In this procedure, a small section of each ductus deferens near the epididymis is surgically removed and the cut ends of the ducts are tied (fig. 28.25). A vasectomy interferes with sperm transport but does not directly affect the secretion of androgens from interstitial cells in the interstitial tissue. Since spermatogenesis continues, the sperm cannot be drained from

• • • • • • • • • • • • • • • • • • • • • • • • • • • • •

impotence: L. *im*, not; *potens*, potent
sterility: L. *sterilis*, barren
vasectomy: L. *vas*, vessel; Gk. *ektome*, excision

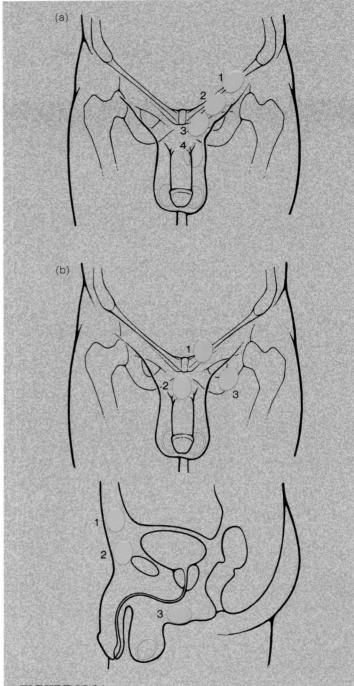

**FIGURE 28.24**
Cryptorchidism. (*a*) Incomplete descent of a testis may involve a region (1) in the pelvic cavity, (2) in the inguinal canal, (3) at the superficial inguinal ring or (4) in the upper scrotum. (*b*) An ectopic testis may be (1) in the superficial fascia of the anterior pelvic wall, (2) at the root of the penis, or (3) in the perineum, in the thigh alongside the femoral vessels.

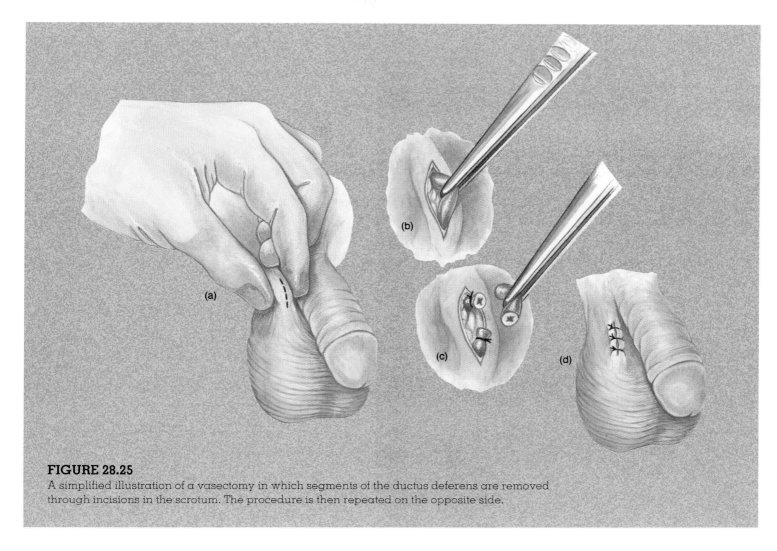

**FIGURE 28.25**
A simplified illustration of a vasectomy in which segments of the ductus deferens are removed through incisions in the scrotum. The procedure is then repeated on the opposite side.

the testes and instead accumulate in the "crypts" that form in the seminiferous tubules and ductus deferens. These crypts present sites of inflammatory reactions in which spermatozoa are phagocytosed and destroyed by the immune system.

## Diseases of the Male Reproductive System

**Sexually Transmitted Diseases**   Sexually transmitted diseases (STDs)—sometimes collectively called "VD" (venereal disease) are contagious diseases that affect the reproductive systems of both the male and the female (table 28.5). STDs are transmitted during sexual activity, and their frequency of occurrence in the United States is regarded by health authorities as epidemic. These diseases have not been eradicated mainly because humans cannot develop immunity to them and because increased sexual promiscuity increases the likelihood of infection and reinfection.

   **Gonorrhea** (*gon´ŏ-re´ă*), commonly called "clap," is caused by the bacterium gonococcus, or *Neisseria gonor-rhoeae*. Males with this disease suffer inflammation of the urethra, accompanied by painful urination and frequently the discharge of pus. In females, the condition is usually asymptomatic, and therefore many women may be unsuspecting carriers of the disease. Advanced stages of gonorrhea in females may infect the uterus and the uterine tubes. A pregnant woman with untreated gonorrhea may transmit the disease to the eyes of her newborn during its passage through the birth canal, possibly causing blindness.

   **Syphilis** (*sif´ĭ-lis*) is caused by the bacterium *Treponema pallidum*. Syphilis is less common than gonorrhea but is the more serious of the two diseases. During the *primary stage* of syphilis, a lesion called a *chancre* develops at the point where contact was made with the infectious syphilitic lesion of a carrier. The chancre—an ulcerated sore that has hard edges—endures for 10 days to 3 months. It is only during the primary stage that syphilis can be spread to another sexual partner. The chancre will heal with time, but if not treated it will be followed by secondary and tertiary stages of

veneral: L (mythology) *Venus*, the goddess of love
gonorrhea: L. *gonos*, seed; *rhoia*, a flow

chancre: Fr. *chancre*, indirectly from L. *cancer*, a crab

## Table 28.5 Sexually transmitted diseases

| Name | Organism | Resulting condition | Treatment |
|---|---|---|---|
| Gonorrhea | *Gonococcus* (bacterium) | Adult: sterility due to scarring of epididymides and tubes; rarely: septicemia; newborn: blindness | Penicillin injections; tetracycline tablets; eyedrops (silver nitrate or penicillin) in newborns as preventative |
| Syphilis | *Treponema pallidum* (bacterium) | Adult: gummas, cardiovascular neurosyphilis; newborn: congenital syphilis (abnormalities, blindness) | Penicillin injections; tetracycline tablets |
| Chancroid (soft chancre) | *Hemophilus ducreyi* (bacterium) | Chancres, buboes | Tetracycline; sulfa drugs |
| Urethritis in men | Various microorganisms | Clear discharge | Tetracycline |
| Vaginitis | *Trichomonas* (protozoan) | Frothy white or yellow discharge | Metronidazole |
| | *Candida albicans* (yeast) | Thick, white, curdy discharge (moniliasis) | Nystatin |
| Acquired immune deficiency syndrome (AIDS) | Human immunodeficiency virus (HIV) | Early symptoms include extreme fatigue, weight loss, fever, diarrhea; increased susceptibility to pneumonia, rare infections, and cancer | Azidothymidine (AZT, or Retrovir); no cure available |
| Chlamydia | *Chlamydia trachomatis* (bacterium) | Whitish discharge from penis or vagina; pain during urination | Tetracycline and sulfonamides |
| Lymphogranuloma venereum (LGV) | Microorganism | Ulcerating buboes; rectal stricture | Tetracycline; sulfa drugs |
| Granuloma venereum (inguinale) | *Donovania granulomatis* | Raw, open, extended sore | Tetracycline |
| Venereal warts | Virus | Warts | Podophyllin; cautery, cryosurgery, or laser treatment |
| Genital herpes | Herpes simplex virus | Sores | Palliative treatment |
| Crabs | Arthropod | Itching | Gamma benzene hexachloride |

syphilis. During the initial contact, the bacteria enter the bloodstream and spread throughout the body. The *secondary stage* of syphilis is expressed by lesions or a rash of the skin and mucous membranes, accompanied by fever. This stage lasts from 2 weeks to 6 months, and the symptoms disappear of their own accord. The *tertiary stage* occurs 10 to 20 years following the primary infection. The circulatory, integumentary, skeletal, and nervous systems are particularly vulnerable to the degenerative changes caused by this disease. The end result of untreated syphilis is blindness, insanity, and eventual death.

**AIDS,** or **acquired immune deficiency syndrome,** is a viral disease that is transmitted primarily through intimate sexual contact and drug abuse (by sharing contaminated syringe needles). Additional information about this fatal disease, for which there is currently no cure, is presented in chapter 23 and table 28.5.

**Disorders of the Prostate** The prostate is subject to several disorders, most of which are common in older men. The four most frequent prostatic problems are acute prostatitis, chronic prostatitis, benign prostatic hypertrophy, and carcinoma of the prostate.

**Acute prostatitis** is common in sexually active young men through infections acquired from a gonococcus bacterium. The symptoms of acute prostatitis are a swollen and tender prostate, painful urination, and in extreme conditions, pus dripping from the penis. It is treated with penicillin, bed rest, and increased fluid intake.

**Chronic prostatitis** is one of the most common afflictions of middle-aged and elderly men. The symptoms of this condition range from irritation and slight difficulty in urinating to extreme pain and urine blockage, which commonly causes secondary renal infections. In this disease, several kinds of infectious microorganisms are believed to be harbored in the prostate and are responsible for inflammations elsewhere in the body, such as in the nerves (neuritis), the joints (arthritis), the muscles (myositis), and the iris (iritis).

**Benign prostatic hypertrophy,** or enlargement of the prostate, occurs in approximately one-third of all males over the age of 60. In this condition, an overgrowth of granular material compresses the prostatic urethra. The cause of prostatic hypertrophy is not known. As the prostate enlarges, urination becomes painful and difficult. If the urinary bladder is not emptied completely, cystitis eventually occurs. People with cystitis may become incontinent and dribble urine continuously. Prostatic hypertrophy is usually treated by the surgical removal of portions of the gland through transurethral curetting (cutting and removal of a small section) or the removal of the entire prostate, called **prostatectomy.**

**Prostatic carcinoma,** or cancer of the prostate, is the second leading cause of death from cancer in males in the United States. When prostatic cancer is confined to the prostate, it is generally small and asymptomatic. But as the cancer grows and invades surrounding nerve plexuses, it becomes extremely painful and is easily detected. The metastases of this cancer to the spinal column and brain are generally what kills the patient.

As prostatic carcinoma develops, it has symptoms nearly identical to those of prostatic hypertrophy—painful urination and cystitis. When examined by rectal palpation with a gloved finger, however, a hard cancerous mass can be detected in contrast to the enlarged, soft, and tender prostate diagnostic of prostatic hypertrophy. Prostatic carcinoma is treated by prostatectomy and frequently by the removal of the testes (called **orchiectomy**) as well. An orchiectomy inhibits metastases by eliminating testosterone secretion.

**Disorders of the Testes and Scrotum**    A **hydrocele** (*hi´dro-sēl*) is a benign fluid mass within the tunica vaginalis that causes swelling of the scrotum. It is a frequent, minor disorder in male infants as well as in adults. The cause is unknown.

An infection in the testes is called **orchitis.** Orchitis may develop from a primary infection from a tubercle bacterium or as a secondary complication of mumps contracted after puberty. If orchitis from mumps involves both testes, it usually causes sterility.

Trauma to the testes and scrotum is common because of their pendent position. The testes are extremely sensitive to pain, and a male responds reflexively to protect the groin area.

# Chapter Summary

## Introduction to the Reproductive System (pp. 839–841)

1. Ova and sperm, collectively called gametes, each contain 23 chromosomes and are haploid.
   a. The gametes are formed by meiosis.
   b. The zygote that is formed as a result of fertilization is diploid, with 46 chromosomes.
   c. The X and Y chromosomes are called the sex chromosomes; a female has the XX and a male has the XY genotype.
2. The indifferent embryonic gonads are converted into testes by the action of a testis-determining factor on the Y chromosome; in the absence of this factor, the gonads become ovaries.
3. Testosterone acts in its target cells through its conversion into derivatives, including dehydrotestosterone (DHT). Testosterone and DHT are responsible for the masculinization of the embryonic tissues to form male accessory sex organs.
4. The testes descend from the body cavity into the scrotum through the action of the gubernaculum.

## Endocrine Regulation of Reproduction (pp. 841–845)

1. The gonads of both sexes are stimulated by FSH and LH, which are secreted by the anterior pituitary in response to stimulation by GnRH from the hypothalamus.
2. At the time of puberty, a rise in the secretion of gonadal steroid hormones causes the development of secondary sex characteristics.
3. The pineal gland, with its secretion of melatonin, is believed by many to play a role in the initiation of puberty.

## Male Reproductive System (pp. 845–853)

1. The testes are contained outside of the body cavity in the scrotum. Through the action of the cremaster and dartos muscles, the scrotum can contract and elevate the testes closer to the body cavity in response to cold temperatures.
2. The testes are partitioned into wedge-shaped lobules; the lobules are composed of seminiferous tubules that produce sperm and of interstitial tissue that produces androgens.

3. LH stimulates the interstitial cells to secrete testosterone and FSH stimulates the tubules to secrete a polypeptide hormone called inhibin; testosterone and inhibin, in turn, exert feedback control of LH and FSH secretion, respectively.
4. Spermatogenesis begins with stem cells called spermatogonia. The diploid spermatogonia undergo meiosis to form haploid secondary spermatocytes at the end of the first division and spermatids at the end of the second division.
   a. The four spermatids formed from the meiotic division of one parent cell are nurtured by the sustentacular cells, which aid the development of spermatids into spermatozoa.
   b. Sustentacular cells are the targets of FSH action, and both FSH and testosterone are required for spermatogenesis at puberty.

## Spermatic Ducts, Accessory Glands, and the Penis (pp. 853–863)

1. Nonmotile sperm pass from the testes to the head of the epididymis, through its

# Interactions of the Reproductive System with Other Body Systems

**Integumentary System**
- Protects the body from pathogens
- Helps to maintain body temperature
- Serves as sexual attractant
- Sex hormones affect distribution of body hair and deposition of subcutaneous fat

**Skeletal System**
- Supports and protects some reproductive organs
- Sex hormones stimulate bone growth and maintenance
- Androgens masculinize skeleten; estrogen feminizes skeleten

**Muscular System**
- Skeletal muscles protect some reproductive organs
- Involuntary action of smooth muscles aids movement of gametes
- Testosterone promotes increase in muscle mass

**Nervous System**
- Provides autonomic regulation of reproductive function
- CNS regulates behavioral aspects of reproduction
- Sex hormones influence brain development and sexual behavior

**Endocrine System**
- Pituitary and gonadal hormones regulate all aspects of reproductive function
- Gonadotropins and GnRH regulate function of gonads

**Circulatory System**
- Transports oxygen and nutrients to reproductive organs and fetus and removes wastes
- Estrogen lowers blood cholesterol levels

**Lymphatic System**
- Protects against infections
- Drains tissue fluid and returns it to venous system

**Respiratory System**
- Provides oxygen for reproductive system and fetus and provides for elimination of carbon dioxide

**Urinary System**
- Regulates the volume, pH, and electrolyte balance of the blood and eliminates wastes
- Male urethra transports semen

**Digestive System**
- Provides nutrients for organ function
- Provides nutrients for embryonic and fetal development

body, and out the tail of the epididymis to enter the ductus deferens as mature sperm.

    a. The ductus deferens exits the scrotum, penetrates through the inguinal canal, and delivers sperm to the ejaculatory duct, which is formed by the union of the ductus deferens and the duct of the seminal vesicle.

    b. The secretions of the seminal vesicles constitute about 60% of the semen; these secretions include fructose, citric acid, and coagulation proteins.

    c. The ejaculatory ducts pass through the prostate, which contributes fluid and chemical agents to the semen. The

secretions of the prostate constitute about 40% of the semen. (Spermatozoa account for less than 1% of the semen content.)

    d. The bulbourethral glands secrete a mucoid substance that coats the urethra and lubricates the tip of the penis.

2. The male urethra, which serves as a common tube for both the urinary and reproductive systems is divided into prostatic, membranous, and spongy portions.

3. The penis contains three long columns of erectile tissue: two dorsal corpora

cavernosa and one ventral corpus spongiosum surrounding the urethra.

    a. Erection is achieved by engorgement of the spongy erectile tissue with blood.

    b. Emission is the movement of sperm into the ejaculatory ducts, together with fluid from the accessory glands.

    c. Ejaculation is the forceful propulsion of semen from the male duct system as a result of muscular contractions of the bulbocavernosus muscles and sympathetic reflexes in the smooth muscles of the reproductive organs.

# Review Activities

## Objective Questions

1. An embryo with the genotype XY develops male accessory sex organs because of
   a. androgens.
   b. estrogens.
   c. the absence of androgens.
   d. the absence of estrogens.
2. Which of the following does *not* arise from the embryonic mesonephric duct?
   a. epididymis.
   b. ductus deferens.
   c. seminal vesicle.
   d. prostate.
3. The external genitalia of a male are completely formed by the end of
   a. the embryonic period.
   b. the ninth week.
   c. the tenth week.
   d. the twelfth week.
4. In the male, FSH
   a. is not secreted by the pituitary.
   b. receptors are located in the interstitial cells.
   c. receptors are located in the spermatogonia.
   d. receptors are located in the sustentacular cells.
5. The secretion of FSH in a male is inhibited by negative feedback effects of
   a. inhibin secreted from the tubules.
   b. inhibin secreted from the interstitial cells.

    c. testosterone secreted from the tubules.
    d. testosterone secreted from the interstitial cells.

6. Which of the following is *not* a spermatic duct?
   a. epididymis
   b. spermatic cord
   c. ejaculatory duct
   d. ductus deferens
7. Spermatozoa are stored prior to emission and ejaculation in
   a. the epididymides.
   b. the seminal vesicles.
   c. the penile urethra.
   d. the prostate.
8. Urethral glands
   a. secrete mucus.
   b. produce nutrients.
   c. secrete hormones.
   d. regulate sperm production.
9. Which statement is *false* regarding erection of the penis?
   a. It is a parasympathetic response.
   b. It may be both a voluntary and an involuntary response.
   c. It has to be followed by emission and ejaculation.
   d. It is controlled by the hypothalamus of the brain and sacral portion of the spinal cord.
10. The condition in which one or both testes fail to descend into the scrotum is
    a. cryptorchidism.
    b. Turner's syndrome.
    c. hermaphroditism.
    d. Klinefelter's syndrome.

## Essay Questions

1. Describe how development of the gonads and of the secondary sex organs is determined by chromosomes and by the secretion of hormones.
2. Explain why a testis is said to be composed of two separate compartments and describe how these compartments may interact.
3. Describe the role of the sustentacular cells in spermatogenesis and explain how spermatogenesis is hormonally controlled.
4. Describe the interactions between the hypothalamus, anterior pituitary, and testes during puberty and discuss the possible role of the pineal gland in puberty.
5. List the structures that constitute the spermatic cord. Where is the inguinal canal? Why are the inguinal canal and inguinal ring clinically important?
6. Compare the seminal vesicles and the prostate in terms of location, structure, and function.
7. Describe the structure of the penis and explain the mechanisms that result in erection, emission, and ejaculation.
8. Distinguish between *impotence, infertility,* and *sterility.*

## Gundy/Weber Software 💾

The tutorial software accompanying Chapter 28 is Volume 13—Reproductive System.

## objectives

- List and briefly describe the primary and secondary
  female sex organs and define *secondary sex
  characteristics.*

- List the functions of the female reproductive system.

- Describe the structures of the uterine tubes, uterus,
  vulva, and vagina.

- Describe the changes that occur in the female
  reproductive structures during sexual excitement and
  coitus.

- Describe the position of the ovaries and of the
  ligaments associated with the ovaries and genital
  ducts.

- Discuss the changes that occur in the ovaries leading
  up to and following ovulation.

- Describe oogenesis and explain why meiosis of one
  primary oocyte results in the formation of only one
  mature ovum.

- Discuss the hormonal secretions of the ovaries during
  an ovarian cycle.

- Describe the hormonal changes that occur during the
  follicular phase and explain the hormonal control of
  ovulation.

- Discuss the formation, function, and fate of the
  corpus luteum.

- Discuss the structural changes that occur in the
  endometrium during the menstrual cycle and explain
  how these changes are hormonally controlled.

- Describe the structure of the breasts and mammary
  glands.

- Discuss the hormonal requirements for mammary
  gland development.

- Describe the action of prolactin and oxytocin on
  lactation and explain how secretion of these
  hormones is regulated.

# Structures and Functions of the Female Reproductive System

*The structures of the female reproductive system include the ovaries; the secondary sex organs (vagina, uterine tubes, uterus, and mammary glands); and the external genitalia. The female reproductive system produces ova, secretes sex hormones, receives sperm from the male, and provides sites for fertilization and development of the embryo and fetus. Parturition follows gestation, and secretions from the mammary glands provide nourishment for the baby.*

The organs of the female and male reproductive systems are considered *homologous* because they develop from the same embryonic structures. The *primary sex organs*, called *gonads*, produce the *gametes*, or *sex cells*. Specifically, the **ovaries** are the gonads in females and the *ova* are the gametes that they produce. The ova of a female are completely formed, but not totally matured, during fetal development of the ovaries. The ova are generally discharged, or *ovulated*, one at a time in a cyclic pattern throughout the reproductive period of the female, which extends from puberty to menopause. *Menstruation* is the discharge of *menses* (blood and solid tissue) from the uterus at the end of each menstrual cycle.

*Menopause* is the termination of ovulation and menstruation. The reproductive period in females generally extends from about age 12 to age 50. The cyclic reproductive pattern of ovulation and the age span during which a woman is fertile are determined by hormonal action.

The functions of the female reproductive system are (1) to produce ova; (2) to secrete sex hormones; (3) to receive the sperm from the male during coitus; (4) to provide sites for fertilization, implantation of the blastocyst (see chapter 30), and embryonic and fetal development; (5) to facilitate *parturition*, or delivery of the baby; and (6) to provide nourishment for the baby through the secretion of milk from the mammary glands in the breasts.

## Secondary Sex Organs and Secondary Sex Characteristics

Secondary sex organs (fig. 29.1) are those structures that are essential for successful fertilization, implantation of the embryo, development of the embryo and fetus, and parturition. The secondary sex organs include the **vagina,** which receives the penis and ejaculated semen during coitus and through

..........................................

vagina: L. *vagina,* sheath or scabbard

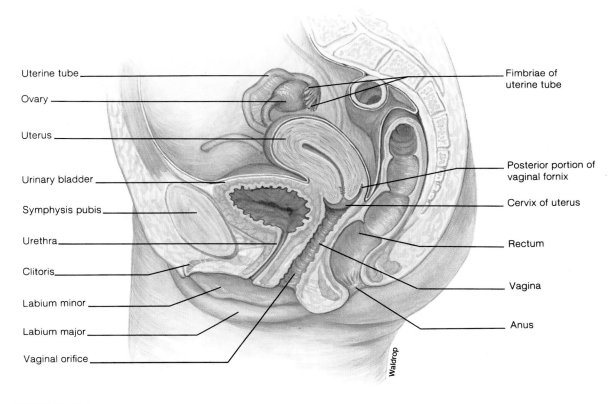

**FIGURE 29.1**

Organs of the female reproductive system seen in sagittal section.

Uterine tube — Fimbriae of uterine tube

Ovary

Uterus

Urinary bladder — Posterior portion of vaginal fornix

Symphysis pubis — Cervix of uterus

Urethra — Rectum

Clitoris

Labium minor — Vagina

Labium major

Vaginal orifice — Anus

Waldrop

which the baby passes during parturition; the **uterine** (fallopian) **tubes,** also called the *oviducts,* through which an ovum is transported toward the uterus after ovulation and in which fertilization normally occurs; and the **uterus** (womb), in which implantation and development occur and whose muscular walls play an active role in parturition. **Mammary glands** are also considered to be secondary sex organs because the milk secreted after parturition provides nourishment for the baby. The structure and function of mammary glands will be discussed in a separate section.

**Secondary sex characteristics** are features that are not essential for the reproductive process but that are considered to be sexual attractants. The female distribution of subcutaneous fat, broad pelvis, body hair pattern, and breast development are examples. Although the breasts contain the mammary glands, large breasts are not essential for nursing the young. Indeed, while all mammals have mammary glands and are capable of nursing, only human females have protruding breasts; the sole purpose of this characteristic, it seems, is to function as a sexual attractant.

 The onset of puberty in females generally occurs between the ages of 12 and 14, varying with the nutritional condition, genetic background, and even sexual exposure of the individual. Generally, girls attain puberty 6 months to 1 year earlier than boys, accompanied by an earlier growth spurt (see chapter 28). Puberty in girls is heralded by the onset of menstruation, or *menarche* (mĕ-nar′ke). Puberty results from the increased secretion of gonadotropins from the anterior pituitary, which stimulates the ovaries to begin their cycles of ova development and sex steroid secretion.

## Uterine Tubes

The paired **uterine** (fallopian) **tubes** transport ova from the ovaries to the uterus. Fertilization normally occurs within the uterine tube. Each uterine tube is approximately 10 cm (4 in.) long and 0.7 cm (0.3 in.) in diameter and is positioned between the folds of the broad ligament of the uterus.

 The term *salpinx* is occasionally used to refer to the uterine tubes. It is a Greek word meaning "trumpet" or "tube" and is the root of such clinical terms as *salpingitis* (sal″pin-ji′tis), or inflammation of the uterine tubes; *salpingography* (radiography of the uterine tubes); and *salpingolysis* (the breaking up of adhesions of the uterine tube to correct female infertility).

The funnel-shaped, open-ended portion of the uterine tube is called the **infundibulum.** Although the infundibulum is close to the ovary, it is not attached. A

fallopian tubes: from Gabriele Fallopius, Italian anatomist, 1523–62
fimbriae: L. *fimbria,* fringe

number of fringed, fingerlike processes, called **fimbriae,** project from the margins of the infundibulum over the lateral surface of the ovary. The fimbriae are covered by ciliated columnar epithelium, which draws the ovum into the lumen of the uterine tube. From the infundibulum, the uterine tube extends medially and inferiorly to open into the superolateral cavity of the uterus at the uterine opening. The **ampulla** (am-pool′ă) is the longest and widest portion of the uterine tube.

The wall of the uterine tube consists of three histological layers (fig. 29.2). The internal **mucosa** lines the lumen and is composed of a ciliated columnar epithelium. The mucosa contains numerous folds that serve to delay the passage of the ovum, thus increasing the chances that fertilization will occur in the upper third of the tube. The **muscularis** is the middle layer, composed of a thick, circular layer of smooth muscle and a thin outer layer of smooth muscle. Peristaltic contractions of the muscularis and ciliary action of the mucosa move the ovum through the lumen of the uterine tube. The outer, lubricative **serous layer** of the uterine tube is part of the visceral peritoneum.

The ovum takes 4 to 5 days to move through the uterine tube. If enough viable sperm are ejaculated into the vagina during coitus, and if there is an oocyte in the uterine tube, fertilization will occur within hours after discharge of the semen. The zygote will move toward the uterus, where implantation occurs. If the developing embryo (called a

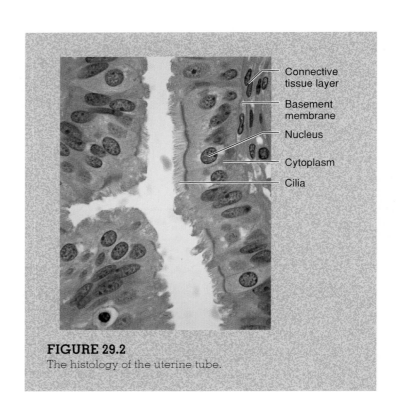

**FIGURE 29.2**
The histology of the uterine tube.

Connective tissue layer
Basement membrane
Nucleus
Cytoplasm
Cilia

*blastocyst*) implants into the uterine tube rather than the uterus, the pregnancy is termed an *ectopic pregnancy*, meaning an implantation of the blastocyst in a site other than the uterus.

Since the infundibulum of the uterine tube is unattached, it provides a potential pathway for pathogens to enter the abdominopelvic cavity. The mucosa of the uterine tube is continuous with that of the uterus and vagina, and it is possible for infectious agents to enter the vagina and cause infections that may ultimately spread to the peritoneal linings, resulting in *pelvic inflammatory disease (PID)*. There is no opening into the abdominopelvic cavity other than through the uterine tubes. The abdominopelvic cavity of a male is totally sealed from external contamination.

## Uterus

The **uterus** is the normal site of implantation for the blastocyst that develops from a fertilized ovum. Prenatal development continues within the uterus until gestation is completed, at which time the uterus plays an active role in the delivery of the baby.

**Structure of the Uterus**   The uterus is a hollow, thick-walled, muscular organ with the shape of an inverted pear. Although the shape and position of the uterus undergo enormous change during pregnancy (fig. 29.3), in its nonpregnant state it is about 7 cm (2.8 in.) long, 5 cm (2 in.) wide (through its broadest region), and 2.5 cm (1 in.) in diameter. The anatomical regions of the uterus include the uppermost dome-shaped portion above the entrance of the uterine tubes, called the **fundus;** the enlarged main portion, called the **body;** and the inferior constricted portion opening into the vagina, called the **cervix** (fig. 29.4). The uterus is located between the urinary bladder anteriorly and the rectum and sigmoid colon posteriorly. The fundus projects anteriorly and slightly superiorly over the urinary bladder. The cervix projects posteriorly and inferiorly, joining the vagina at nearly a right angle (see fig. 29.1).

The **uterine cavity** is the space within the regions of the fundus and body. The lumina of the uterine tubes open into the uterine cavity on the superior-lateral portions. The uterine cavity is continuous inferiorly with the **cervical canal,** which extends through the cervix and opens into the lumen of the vagina. The junction of the uterine cavity with the cervical canal is called the **isthmus of uterus,** whereas the opening of the uterine cavity into the cavity of the vagina is called the **uterine ostium.**

........................................

ectopic: Gk. *ex*, out; *topos*, place
fundus: L. *fundus*, bottom
cervix: L. *cervix*, neck

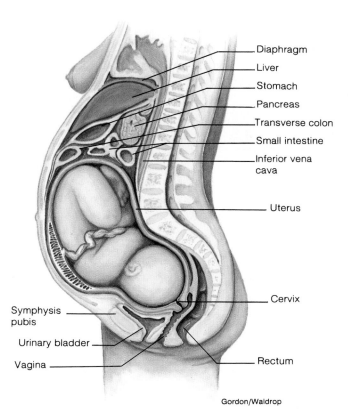

Gordon/Waldrop

**FIGURE 29.3**
The size and position of the uterus in a full-term pregnant woman in sagittal section.

**Support of the Uterus**   The uterus is maintained in position by muscular support and ligaments that extend from the pelvic girdle or body wall to the uterus. Muscles of the perineum, especially the levator ani muscle (see fig. 13.13), provide the principal muscular support. The ligaments that support the uterus undergo marked hypertrophy during pregnancy, regress in size after parturition, and atrophy after menopause.

Four paired ligaments support the uterus in position within the pelvic cavity. The paired **broad ligaments** (fig. 29.4) are folds of the peritoneum that extend from the pelvic walls and floor to the lateral walls of the uterus. The ovaries and uterine tubes are also supported by the broad ligaments. The paired rectouterine folds (not illustrated) are also continuations of peritoneum that curve along the lateral pelvic wall on both sides of the rectum to connect the uterus to the sacrum. The **cardinal (lateral cervical) ligaments** (not illustrated) are fibrous bands within the broad ligament that extend laterally from the cervix and vagina across the pelvic floor, where they attach to the wall of the pelvis. The cardinal ligaments contain some smooth muscle as well as vessels and nerves that supply the cervix and vagina. The fourth paired ligaments are the **round ligaments.** Each round ligament extends from the lateral border of the uterus just below the point where the uterine tube attaches to the lateral

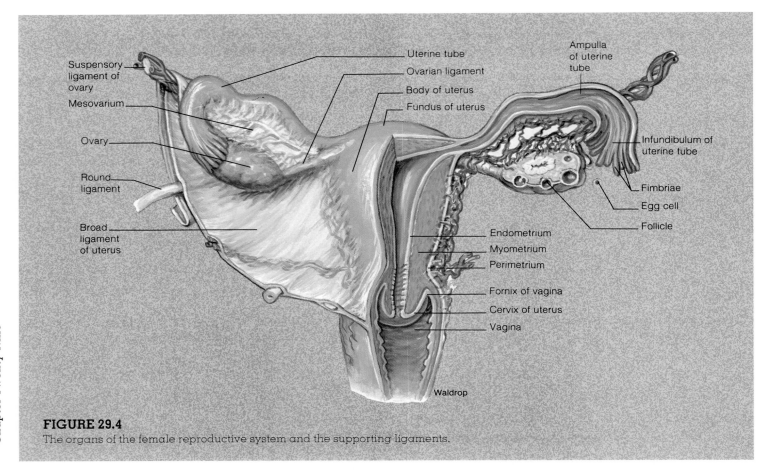

**FIGURE 29.4**

The organs of the female reproductive system and the supporting ligaments.

pelvic wall. Similar to the course taken by the ductus deferentia in the male, the round ligaments continue through the inguinal canals of the abdominal wall, where they attach to the deep tissues of the labia majora.

**Uterine Wall**   The wall of the uterus is composed of three layers: the perimetrium, myometrium, and endometrium (fig. 29.4). The **perimetrium** is the outer, thin serosal covering and a part of the peritoneum. The thick **myometrium** is composed of three poorly defined layers of smooth muscle arranged in longitudinal, circular, and spiral patterns. The myometrium is thickest in the fundus and thinnest in the cervix. During parturition, the muscles of this layer are stimulated to contract forcefully.

The **endometrium** (*en″do-me′tre-um*) is the inner mucosal lining of the uterus. The endometrium has two distinct layers. The superficial **stratum functionale,** composed of columnar epithelium and containing secretory glands, is shed as *menses* during menstruation and built up again under the stimulation of ovarian steroid hormones. The deeper **stratum basale** is highly vascular and serves to regenerate the stratum functionale after each menstruation.

......................................................

menses: L. *menses*, plural of *mensis*, monthly

 The uterus undergoes tremendous change during pregnancy. Its weight increases more than 16 times (from about 60 g to about 1000 g), and its capacity increases from about 2.5 ml to over 5000 ml. The principal change in the myometrium is a marked hypertrophy, or elongation, of the individual muscle cells to as much as 10 times their original length. There is some atrophy of the muscle cells after parturition, but the uterus never returns to its original size.

**Uterine Blood Supply and Innervation**   The uterus is supplied with blood through the **uterine arteries** (fig. 29.5). Each pair of these two vessels anastomose on the upper lateral margin of the uterus. The blood from the uterus returns through uterine veins that parallel the pattern of the arteries.

The uterus receives both sympathetic and parasympathetic innervation from the pelvic and hypogastric plexuses. Both autonomic innervations serve the arteries of the uterus, whereas the smooth muscle of the myometrium receives only sympathetic innervation.

## Vagina

The **vagina** (*vă-ji′nă*) is the organ that receives sperm through the urethra of the erect penis during coitus. It also serves as the birth canal during parturition and provides for

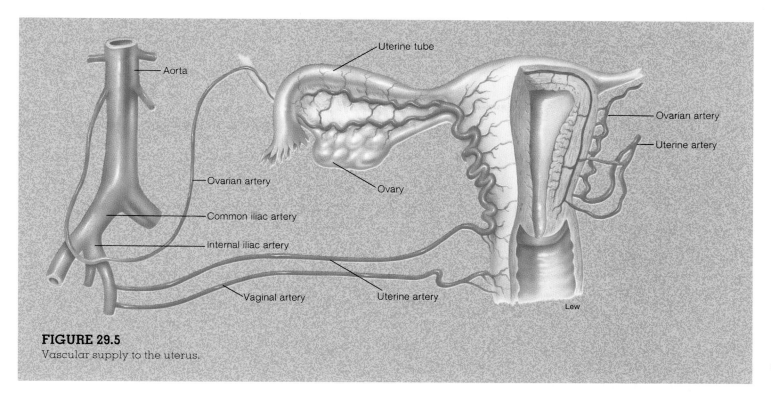

**FIGURE 29.5**
Vascular supply to the uterus.

the passage of menses to the outside. The vagina is a tubular organ about 9 cm (3.5 in.) long, passing from the cervix of the uterus to the vestibule. In its position between the urinary bladder and urethra anteriorly and the rectum posteriorly, it is continuous with the cervical canal of the uterus. The cervix attaches to the vagina at a nearly 90 degree angle. The deep recess surrounding the protrusion of the cervix into the vagina is called the **fornix** (see fig. 29.4). The exterior opening of the vagina, at its lower end, is called the **vaginal orifice.** A thin fold of mucous membrane, called the **hymen,** may partially cover the vaginal orifice.

The vaginal wall is composed of three layers: an inner mucosa, a middle muscularis, and an outer fibrous layer. The **mucosal layer** consists of stratified squamous epithelium, which forms a series of transverse folds called **vaginal rugae** (fig. 29.6). The vaginal rugae provide friction ridges for stimulation of the erect penis during sexual intercourse. They also permit considerable distension of the vagina to facilitate coitus. The mucosal layer contains few glands; the acidic mucus that is present in the vagina comes primarily from glands within the uterus. This acidic environment of the vagina retards microbial growth. The semen, however, temporarily neutralizes the acidity of the vagina to ensure the survival of the ejaculated sperm deposited within the vagina.

The **muscularis layer** consists of longitudinal and circular bands of smooth muscle interlaced with distensible connective tissue. The distension of this layer is especially important during parturition. Skeletal muscle strands near the vaginal orifice, including the levator ani muscle

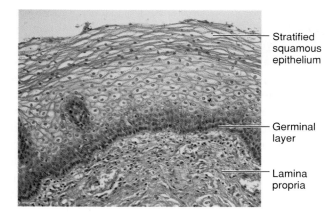

**FIGURE 29.6**
The histology of a vaginal ruga.

(see fig. 13.13), partially constrict this opening. The **fibrous layer** covers the vagina and attaches it to surrounding pelvic organs. This layer consists of dense regular connective tissue interlaced with strands of elastic fibers.

## Vulva

The external genitalia of the female are referred to collectively as the **vulva** (fig. 29.7). The structures of the vulva surround the vaginal orifice and include the mons pubis, labia majora, labia minora, clitoris, vaginal vestibule, vestibular bulbs, and vestibular glands.

.........................................

vulva: L. *volvere,* to roll; wrapper

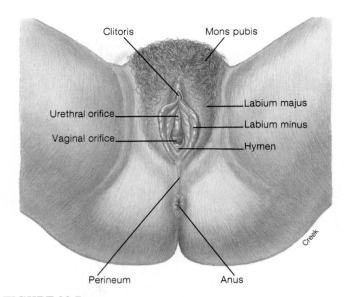

Clitoris  Mons pubis

Urethral orifice

Vaginal orifice

Labium majus

Labium minus

Hymen

Perineum  Anus

Creek

**FIGURE 29.7**

The external female genitalia.

The **mons pubis** is the subcutaneous pad of adipose connective tissue covering the symphysis pubis. At puberty, the mons pubis becomes covered with a pattern of coarse pubic hair that is somewhat triangular, usually with a horizontal upper border. The elevated and padded mons pubis cushions the symphysis pubis and vulva during coitus.

The **labia majora** (*labium majus*, singular) are two thickened longitudinal folds of skin composed of adipose and loose connective tissues, as well as some smooth muscle. The labia majora are continuous anteriorly with the mons pubis, are separated longitudinally by the **pudendal cleft,** and converge again posteriorly on the **perineum** (*per˝ĭ-ne´um*). They are also covered with hair and contain numerous sebaceous and sweat glands. The labia majora are homologous to the scrotum of the male and function to enclose and protect the other organs of the vulva.

Positioned close together between the labia majora are two smaller longitudinal folds called the **labia minora** (*labium minus*, singular). The labia minora are hairless but do contain sebaceous glands. Anteriorly, the labia minora join to form the **prepuce** (*pre´pyoos*), or covering, of the clitoris. These inner folds of skin further protect the vaginal and urethral openings.

The **clitoris** (*klit´or-is*) is a small rounded projection at the upper portion of the pudendal cleft. The clitoris corresponds in structure and origin to the penis in the male; it is, however, smaller and has no urethra. Although most of the clitoris is embedded, it does have an exposed **glans clitoris** of erectile tissue that is richly innervated with sen-

sory endings. The clitoris is about 2 cm (0.8 in.) long and 0.5 cm (0.2 in.) in diameter. The unexposed portion of the clitoris is composed of two columns of erectile tissue called the **corpora cavernosa** that diverge posteriorly to form the **crura** and attach to the sides of the pubic arch.

The **vaginal vestibule** is the longitudinal cleft enclosed by the labia minora. The openings for the urethra and vagina are located in the vaginal vestibule. The external opening of the urethra is about 2.5 cm (1 in.) behind the glans clitoris and immediately in front of the vaginal orifice. The vaginal orifice is lubricated during sexual excitement by secretions from paired major and minor **vestibular** (Bartholin's) **glands** located within the wall of the region immediately inside the vaginal orifice. The ducts from these glands open into the vestibule near the lateral margins of the vaginal orifice. Bodies of vascular erectile tissue, called **vestibular bulbs,** are located immediately below the skin forming the lateral walls of the vestibule. The vestibular bulbs are separated from each other by the vagina and urethra and extend from the level of the vaginal orifice to the clitoris.

The vulva has both sympathetic and parasympathetic innervation, as well as extensive somatic fibers that respond to sensory stimulation. Parasympathetic stimulation causes a response similar to that in the male: dilation of the arterioles of the genital erectile tissue and compression of the venous return.

## *Mechanism of Erection and Orgasm*

The homologous structures of the male and female reproductive systems respond to sexual stimulation in a similar fashion. The erectile tissues of a female, like those of a male, become engorged with blood and swollen during sexual arousal. During sexual excitement, the hypothalamus of the brain sends parasympathetic nerve impulses through the sacral segments of the spinal cord, which cause dilation of arteries serving the clitoris and vestibular bulbs. This increased blood flow causes the erectile tissues to swell. In addition, the erectile tissues in the areola of the breasts become engorged.

Simultaneous with the erection of the clitoris and vestibular bulbs, the vagina expands and elongates to accommodate the erect penis of the male, and parasympathetic impulses cause the vestibular glands to secrete mucus near the vaginal orifice. The vestibular secretion moistens and lubricates the tissues of the vestibule, thus facilitating the penetration of the erect penis into the vagina during coitus. Mucus continues to be secreted during coitus so that the male and female genitalia do not become irritated as they would if the vagina became dry.

mons pubis: L. *mons*, mountain; *pubis*, genital area

vestibule: L. *vestibule*, an entrance, court
Bartholin's glands: from Casper Bartholin Jr., Danish anatomist, 1655–1738

The position of the sensitive clitoris usually allows it to be stimulated during coitus. If stimulation of the clitoris is of sufficient intensity and duration, a woman will experience a culmination of pleasurable psychological and physiological release called *orgasm.*

Associated with orgasm is a rhythmic contraction of the muscles of the perineum and the muscular walls of the uterus and uterine tubes. These reflexive muscular actions are thought to aid the movement of sperm through the female reproductive tract toward the upper end of a uterine tube, where an ovum might be located.

# Ovaries and the Ovarian Cycle

*The ovaries contain a large number of follicles, each of which encloses an ovum. Some of these follicles mature during the ovarian cycle, and the ova they contain progress to the secondary oocyte stage of meiosis. At ovulation, the largest follicle breaks open to extrude a secondary oocyte from the ovary. The empty follicle then becomes a corpus luteum, which ultimately degenerates at the end of a nonfertile cycle.*

The **ovaries** of sexually mature females are solid, ovoid structures about 3.5 cm (1.4 in.) long, 2 cm (0.8 in.) wide, and 1 cm (0.4 in.) thick. On the medial portion of each ovary is a **hilum,** which is the point of entrance for ovarian blood vessels and nerves. The lateral portion of the ovary is positioned near the open ends of the uterine tube (see fig. 29.4).

## Position and Structure of the Ovaries

The ovaries are positioned in the upper pelvic cavity on both lateral sides of the uterus. Each ovary is situated in a shallow depression of the posterior body wall and is secured by several membranous attachments. The principal supporting membrane of the female reproductive tract is the **broad ligament.** The broad ligament is the parietal peritoneum that supports the uterine tubes and uterus. The **mesovarium** (*mes˝ŏ-va´re-um*) is a specialized posterior extension of the broad ligament that attaches to an ovary. Each ovary is additionally supported by an **ovarian ligament** anchored to the uterus and a **suspensory ligament** attached to the pelvic wall (see fig. 29.4).

Each ovary consists of four layers. The **superficial epithelium** (see fig. 29.10) is the thin, outermost layer composed of simple cuboidal epithelium. A collagenous connective tissue layer called the **tunica albuginea** (*al˝byoo-jin´e-ă*) is located immediately below the germinal epithelium. The principal substance of the ovary is divided into an outer **ovarian cortex** and an inner, vascular **ovarian medulla,** although the boundary between these layers is not distinct. The **stroma**—the material of the ovary in which follicles and blood vessels are embedded—lies in both cortical and medullary layers.

Blood is supplied by ovarian arteries that arise from the lateral sides of the abdominal aorta, just below the origin of the renal arteries. An additional supply comes from the ovarian branches of the uterine arteries. Venous return is through the ovarian veins. The right ovarian vein empties into the inferior vena cava, whereas the left ovarian vein drains into the left renal vein.

## Ovarian Cycle

The germ cells that migrate into the ovaries during early embryonic development multiply, so that at about 5 months of gestation (prenatal life) the ovaries contain approximately 6 to 7 million oogonia. The production of new oogonia stops at this point and never resumes again. Toward the end of gestation, the oogonia begin meiosis, at which time they are called **primary oocytes** (*o´ŏ-sītz*) (fig. 29.8). Like spermatogenesis in the prenatal male, oogenesis is arrested at prophase I of the first meiotic division. The primary oocytes are thus still diploid. The number of primary oocytes decreases throughout a woman's reproductive years. The ovaries of a newborn girl contain about 2 million oocytes—all she will ever have. By the time she reaches puberty, this number has been reduced to 300,000 to 400,000. Oogenesis ceases entirely at menopause (the time ovulation and menstruation stop).

Primary oocytes that are not stimulated to complete the first meiotic division are contained within tiny follicles, called **primary follicles.** Immature primary follicles consist of only a single layer of *granulosa cells.* In response to FSH stimulation, some of these oocytes and follicles get larger, and the follicular cells divide to produce numerous layers of granulosa cells that surround the oocyte and fill the follicle. Some primary follicles will be stimulated to grow still bigger and develop a number of fluid-filled cavities, called *vesicles,* at which time they are called **secondary follicles** (fig. 29.8). Continued growth of one of these follicles will be accompanied by the fusion of its vesicles to form a single, fluid-filled cavity called an *antrum.* At this stage, the follicle is known as a **vesicular ovarian,** or **graafian, follicle.**

As the follicle develops, the primary oocyte completes its first meiotic division. This does not form two complete cells, however, because only one cell—the **secondary oocyte**—gets all the cytoplasm. The other cell formed at this time becomes a small *polar body* (fig. 29.9) that eventually fragments and disappears. This unequal division of cytoplasm ensures that the ovum will be large enough to become a viable embryo should fertilization later occur. The

stroma: Gk. *stroma,* a couch or bed
graafian follicle: from Reynier de Graaf, Dutch anatomist and physician

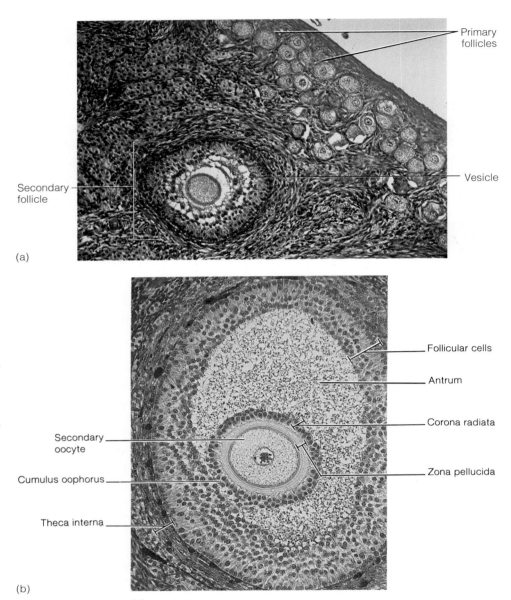

Primary
follicles

Secondary
follicle

Vesicle

(a)

Follicular cells

Antrum

Corona radiata

Secondary
oocyte

Zona pellucida

Cumulus oophorus

Theca interna

(b)

**(b)**

**FIGURE 29.8**

Photomicrographs of (a) primordial and primary follicles and (b) a mature vesicular
ovarian (graafian) follicle.

Under the stimulation of FSH
from the anterior pituitary, the granulosa cells of the ovarian follicles secrete increasing amounts of estrogen
as the follicles grow. Interestingly, the
granulosa cells produce estrogen from
its precursor testosterone, which is
supplied by cells of the *theca interna*
layer, immediately outside the follicle
(see fig. 29.8).

## Ovulation

Usually, by about 10 to 14 days after
the first day of menstruation, only one
follicle has continued its growth to
become a fully mature vesicular ovarian follicle (fig. 29.10); other secondary follicles during that cycle
regress and become *atretic* (ă-tret´ik).
The vesicular ovarian follicle becomes
so large that it forms a bulge on the
surface of the ovary. Under proper
hormonal stimulation, this follicle
will rupture—much like the popping
of a blister—and extrude its secondary
oocyte into the peritoneal cavity near
the opening of the uterine tube in the
process of **ovulation** (fig. 29.11).

The released cell is a secondary
oocyte, surrounded by the zona pellucida and corona radiata. If it is not fertilized, it will degenerate in a couple
of days. If a sperm cell passes through
the corona radiata and zona pellucida
and enters the cytoplasm of the secondary oocyte, the oocyte will then
complete the second meiotic division.
In this process, the cytoplasm is again

secondary oocyte then begins the second meiotic division,
but meiosis is arrested at metaphase II. The second meiotic division is completed only by an ovum that has been fertilized.

The secondary oocyte, arrested at metaphase II, is contained within a vesicular ovarian follicle. The granulosa cells of
the vesicular ovarian follicle form a ring around the circumference of the follicle and a mound that supports the secondary
oocyte. This mound is called the *cumulus oophorus* (o-of´ŏ-rus).
The ring of granulosa cells surrounding the secondary oocyte is
known as the *corona radiata*. Between the oocyte and the corona
radiata is a thin gellike layer of proteins and polysaccharides
called the *zona pellucida* (pě-loo´sĭ-dă) (see fig. 29.8).

not divided equally; most of the cytoplasm remains in the
zygote (fertilized egg), leaving another polar body which, like
the first, degenerates (fig. 29.12).

Changes continue in the ovary following ovulation.
The empty follicle, under the influence of luteinizing hormone from the anterior pituitary, undergoes structural and
biochemical changes to become a **corpus luteum** ("yellow
body"). Unlike the ovarian follicles, which secrete only estrogen, the corpus luteum secretes two sex steroid hormones:
estrogen and progesterone. Toward the end of a nonfertile

cumulus oophorous: L. *cumulus*, a mound; Gk. *oophoros*, egg bearing
zona pellucida: Gk. *zone*, girdle; L. *pellis*, skin; *lucere*, to shine

theca: Gk. *theke*, a box
atretic: Gk. *atretos*, not perforated
corpus luteum: L. *corpus*, body; *luteum*, yellow

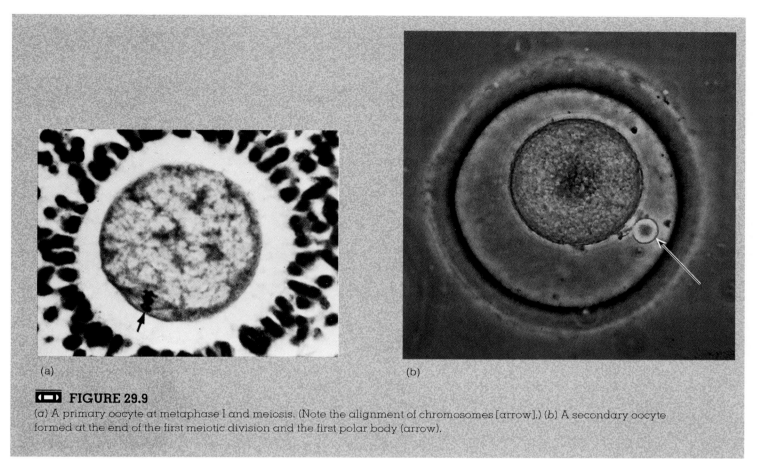

**FIGURE 29.9**

(a) A primary oocyte at metaphase I and meiosis. (Note the alignment of chromosomes [arrow].) (b) A secondary oocyte formed at the end of the first meiotic division and the first polar body (arrow).

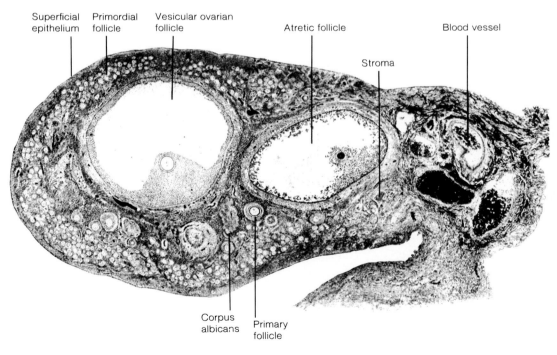

**FIGURE 29.10**

An ovary containing follicles at different stages of development.

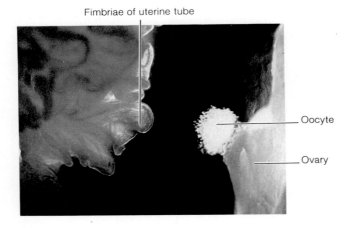

Fimbriae of uterine tube

Oocyte

Ovary

**FIGURE 29.11**
Ovulation from a human ovary. An ovulated oocyte is free in the peritoneal cavity until it enters the lumen of the uterine tube.

cycle, the corpus luteum regresses and is changed into a non-functional **corpus albicans** ("white body"). These cyclic changes in the ovary are summarized in figure 29.13.

## Pituitary-Ovarian Axis

The term *pituitary-ovarian axis* refers to the hormonal interactions between the anterior pituitary and the ovaries. The anterior pituitary secretes two gonadotropic hormones—follicle-stimulating hormone (FSH) and luteinizing hormone (LH)—that promote cyclic changes in the structure and function of the ovaries. The secretion of both gonadotropic hormones, as previously discussed, is controlled by a single releasing hormone from the hypothalamus—called gonadotropin-releasing hormone (GnRH)—and by feedback effects from hormones secreted from the ovaries. The nature of these interactions will be described in detail in the next section.

· · · · · · · · · · · · · · · · · · · · · · · · · · · · · · · · · · · · · · · · · · · ·

albicans: L. *albicare*, to whiten

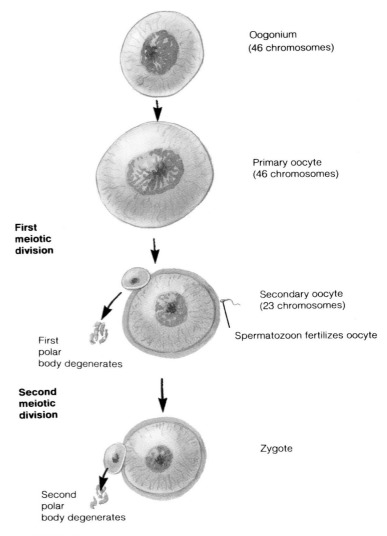

Oogonium
(46 chromosomes)

Primary oocyte
(46 chromosomes)

**First meiotic division**

First polar body degenerates

Secondary oocyte
(23 chromosomes)

Spermatozoon fertilizes oocyte

**Second meiotic division**

Zygote

Second polar body degenerates

**FIGURE 29.12**
A schematic diagram of the process of oogenesis. During meiosis, each primary oocyte produces a single haploid gamete. If the secondary oocyte is fertilized, it forms a secondary polar body and becomes a zygote.

**FIGURE 29.13**
A schematic diagram of an ovary showing the various stages of ovum and follicle development.

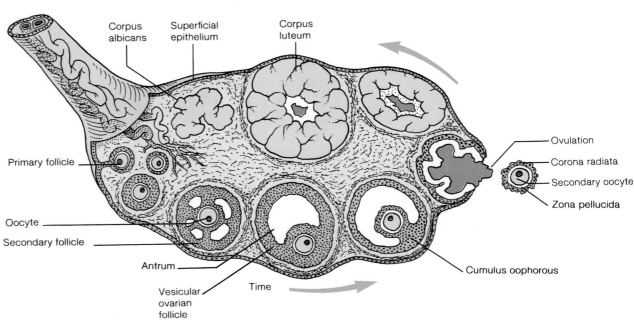

Corpus albicans

Superficial epithelium

Corpus luteum

Ovulation

Corona radiata

Secondary oocyte

Zona pellucida

Primary follicle

Oocyte

Secondary follicle

Antrum

Vesicular ovarian follicle

Time

Cumulus oophorous

Since one releasing hormone can stimulate the secretion of both FSH and LH, one might expect always to see parallel changes in the secretion of these gonadotropins. This, however, is not the case. During an early phase of the menstrual cycle FSH secretion is slightly greater than LH secretion, and just prior to ovulation LH secretion greatly exceeds FSH secretion. These differences are believed to be a result of the feedback effects of ovarian sex steroids, which can change the amount of GnRH secreted, the pulse frequency of GnRH secretion, and the response of the anterior pituitary to GnRH. These complex interactions result in a pattern of hormone secretion that regulates the phases of the menstrual cycle.

# Menstrual Cycle

*Cyclic changes in the secretion of gonadotropic hormones from the anterior pituitary cause the ovarian changes during a monthly cycle. The ovarian cycle is accompanied by cyclic changes in the secretion of sex steroids, which interact with the hypothalamus and pituitary gland to regulate gonadotropin secretion. The cyclic changes in ovarian hormone secretion also cause changes in the endometrium of the uterus during a menstrual cycle.*

Humans, apes, and old-world monkeys have cycles of ovarian activity that repeat at approximately 1-month intervals; hence the name menstrual cycle (*menstru* = monthly). The term *menstruation* is used to indicate the periodic shedding of the stratum functionale of the endometrium, which becomes thickened prior to menstruation under stimulation by ovarian steroid hormones. In primates (other than new-world monkeys), this shedding of the endometrium is accompanied by bleeding. There is no bleeding, by contrast, when other mammals shed the endometrium; their cycles, therefore, are not called menstrual cycles.

In human females and other primates that have menstrual cycles, coitus may be permitted at any time in the cycle. Nonprimate female mammals, by contrast, are sexually receptive only at a particular time in their cycles (shortly before or shortly after ovulation). These animals are therefore said to have *estrous cycles*. Bleeding occurs in some animals (such as dogs and cats) that have estrous cycles shortly before they permit coitus. This bleeding is a result of high estrogen secretion and is not associated with shedding of the endometrium. The bleeding that accompanies menstruation, by contrast, is caused by a fall in estrogen and progesterone secretion.

## Phases of the Menstrual Cycle: Pituitary and Ovarian

The average menstrual cycle has a duration of about 28 days. Since it is a cycle, there is no beginning or end, and the changes that occur are generally gradual. It is convenient, however, to call the first day of menstruation "day 1" of the cycle because menstrual blood flow is the most obvious change to occur. It is also convenient to divide the cycle into phases based on changes that occur in the ovary and in the endometrium. The ovaries are in the **follicular phase** starting on the first day of menstruation and ending on the day of ovulation. After ovulation, the ovaries are in the **luteal phase** until the first day of menstruation. The cyclic changes that occur in the endometrium are called the *menstrual, proliferative,* and *secretory phases* and will be discussed separately.

**Follicular Phase** Menstruation lasts from day 1 to day 4 or 5 of the average cycle. During this time, the secretions of ovarian steroid hormones are at their lowest ebb, and the ovaries contain only primordial and primary follicles. During the *follicular phase* of the ovaries, which lasts from day 1 to about day 13 of the cycle (this is highly variable), some of the primary follicles grow, develop vesicles, and become secondary follicles. Toward the end of the follicular phase, one follicle in one ovary develops a fluid-filled antrum, reaches maturity, and becomes a mature vesicular ovarian follicle. As follicles grow, the granulosa cells secrete an increasing amount of **estradiol** (the principal estrogen), which reaches its highest concentration in the blood at about day 12 of the cycle (2 days before ovulation).

The growth of the follicles and the secretion of estradiol are stimulated by, and dependent upon, FSH secreted from the anterior pituitary. The amount of FSH secreted during the early follicular phase is believed to be slightly greater than the amount secreted in the late follicular phase (fig. 29.14). FSH stimulates the production of FSH receptors in the granulosa cells, so that the follicles become increasingly sensitive to a given amount of FSH. This increased sensitivity is augmented by estradiol, which also stimulates the production of new FSH receptors in the follicles. As a result, the stimulatory effect of FSH on the follicles increases despite the fact that FSH levels in the blood do not increase throughout the follicular phase. Toward the end of the follicular phase, FSH and estradiol also stimulate the production of LH receptors in the vesicular ovarian follicle. This prepares the vesicular ovarian follicle for the next major event in the cycle.

The rapid rise in estradiol secretion from the granulosa cells during the follicular phase acts on the hypothalamus to

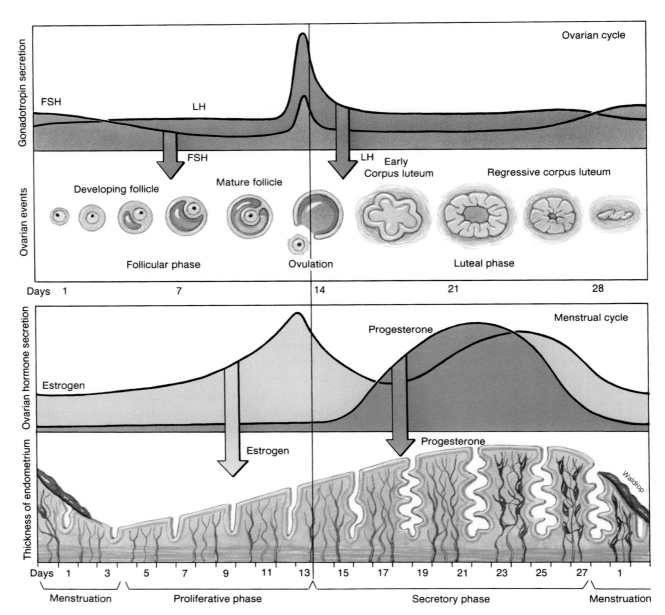

**FIGURE 29.14**

The cycle of ovulation and menstruation.

increase the frequency of GnRH pulses. In addition, estradiol augments the ability of the pituitary gland to respond to GnRH with an increase in LH secretion. As a result of this stimulatory, or **positive feedback,** effect of estradiol on the pituitary gland, there is an increase in LH secretion in the late follicular phase that culminates in an **LH surge** (fig. 29.14).

The LH surge begins about 24 hours before ovulation and reaches its peak about 16 hours before ovulation. It is this surge that acts to trigger ovulation. Since GnRH stimulates the anterior pituitary to secrete both FSH and LH, there is a simultaneous, though smaller, surge in FSH secretion. Some investigators believe that this midcycle peak in FSH acts as a stimulus for the development of new follicles for the next month's cycle.

**Ovulation**   Under the influence of FSH stimulation, the vesicular ovarian follicle grows so large that it becomes a thin-walled "blister" on the surface of the ovary. The growth of the follicle is accompanied by a rapid rate of increase in estradiol secretion. This rapid increase in estradiol, in turn, triggers the LH surge at about day 13. Finally, the surge in LH secretion causes the wall of the mature vesicular ovarian follicle to rupture at about day 14 (fig. 29.14 *top*). In ovulation, a secondary oocyte, arrested at metaphase II of meiosis, is released into the peritoneal cavity. The ovulated oocyte is still surrounded by a zona pellucida and corona radiata as it begins its journey to the uterus.

Ovulation occurs, therefore, as a result of the sequential effects of FSH followed by LH on the ovarian follicles. By means of the positive feedback effect of estradiol on LH

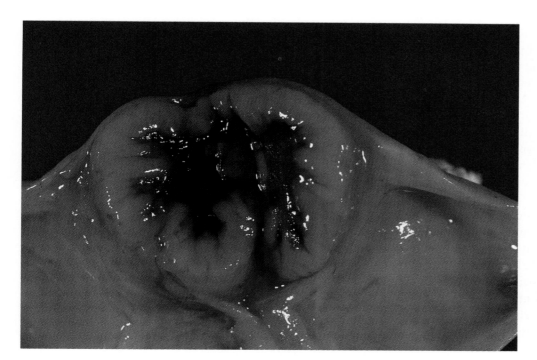

**FIGURE 29.15**
A corpus luteum in a human ovary.

22) because the corpus luteum regresses and stops functioning. In lower mammals, the decline in corpus luteum function is caused by a hormone secreted by the uterus called *luteolysin*. There is evidence to suggest that prostaglandin $F_{2\alpha}$ (see chapter 19) may function as luteolysin in humans, but the exact mechanism of corpus luteum regression in humans is still not well understood. Luteolysis (breakdown of the corpus luteum) can be prevented by high LH secretion, but LH levels remain low during the luteal phase as a result of negative feedback inhibition by ovarian steroids. In a sense, therefore, the corpus luteum causes its own demise.

With the declining function of the corpus luteum, estrogen and progesterone fall to very low levels by day 28 of the cycle. The withdrawal of ovarian steroids causes menstruation and permits a new cycle of ovarian follicle development to progress.

secretion, the follicle in a sense sets the time for its own ovulation. This is because ovulation is triggered by an LH surge, and the LH surge is triggered by increased estradiol secretion that occurs while the follicle grows. In this way, the graafian follicle does not normally ovulate until it has reached the proper size and degree of maturation.

**Luteal Phase**    After ovulation, as mentioned previously, the empty vesicular ovarian follicle is stimulated by LH to become a new structure—the corpus luteum (fig. 29.15). This change in structure is accompanied by a change in function. Whereas the developing follicles secrete only estradiol, the corpus luteum secretes both estradiol and **progesterone.** Progesterone levels in the blood are negligible before ovulation but rise rapidly to reach a peak during the luteal phase, approximately 1 week after ovulation (see fig. 29.14).

The combined high levels of estradiol and progesterone during the luteal phase exert a **negative feedback inhibition** of FSH and LH secretion. This serves to retard development of new follicles, so that further ovulation does not normally occur during that cycle. In this way, multiple ovulations (and possible pregnancies) on succeeding days of the cycle are prevented.

High levels of estrogen and progesterone during the nonfertile cycle do not persist for very long, however, and new follicles do start to develop toward the end of one cycle, in preparation for the next cycle. Estrogen and progesterone levels fall during the late luteal phase (starting about day

## Cyclic Changes in the Endometrium

In addition to a description of the female cycle in terms of the phases of ovarian function, the cycle can also be described in terms of the changes that occur in the endometrium. Three phases can be identified on this basis (fig. 29.14 *bottom*): (1) the proliferative phase, (2) the secretory phase, and (3) the menstrual phase.

The **proliferative phase** of the endometrium occurs while the ovary is in its follicular phase. The increasing amount of estradiol secreted by the developing follicles stimulates growth (proliferation) of the stratum functionale of the endometrium. In humans and other primates, spiral arteries develop in the endometrium during this phase. Estradiol may also stimulate the production of receptor proteins for progesterone at this time, in preparation for the next phase of the cycle.

The **secretory phase** of the endometrium occurs when the ovary is in its luteal phase. In this phase, increased progesterone secretion stimulates the development of mucous glands. As a result of the combined actions of estradiol and progesterone, the endometrium becomes thick, vascular, and spongy in appearance and the uterine glands become engorged with glycogen during the time of the cycle following ovulation. The endometrium is therefore well prepared to accept and nourish an embryo if fertilization should occur.

The **menstrual phase** occurs as a result of the fall in ovarian hormone secretion during the late luteal phase. Necrosis (cellular death) and sloughing of the stratum functionale of the endometrium is produced by constriction of the spiral arteries. This vasoconstriction is believed to be caused by prostaglandin $F_{2\alpha}$, produced in the uterus but prevented from acting by the previously high levels of progesterone. The fall in estrogen and progesterone at the late luteal phase thus sets in motion the events that lead to menstruation. The phases of the menstrual cycle are summarized in figure 29.16 and in table 29.1.

The cyclic changes in ovarian secretion cause other cyclic changes in the female genital ducts. High levels of estradiol secretion, for example, result in cornification of the vaginal epithelium (the upper cells die and become filled with keratin). High levels of estradiol also cause the production of a thin, watery cervical mucus, which can be easily penetrated by spermatozoa. During the luteal phase of the cycle, the high levels of progesterone cause the cervical mucus to become thick and sticky after ovulation has occurred.

 Cyclic changes in ovarian hormone secretion also cause cyclic changes in *basal body temperature*. In the *rhythm method* of birth control, a woman measures her oral basal body temperature upon waking to determine when ovulation has occurred. On the day of the LH peak, when estradiol secretion begins to decline, there is a slight drop in basal body temperature. Starting about 1 day after the LH peak, the basal body temperature sharply rises as a result of progesterone secretion and remains elevated throughout the luteal phase of the cycle (fig. 29.17). The day of ovulation can be accurately determined by this method, making the method useful in increasing fertility if conception is desired. Since the day of the cycle on which ovulation occurs is quite variable in many women, however, the rhythm method is not very reliable for preventing conception by predicting when the next ovulation will occur. The contraceptive pill is a statistically more effective means of birth control.

## Contraceptive Pill

About 10 million women in the United States and 60 million women worldwide are currently using **oral contraceptives.** These contraceptives usually consist of a synthetic estrogen combined with a synthetic progesterone in the form of pills that are taken once each day for 3 weeks after the last day of a menstrual period. This procedure causes an immediate increase in blood levels of ovarian steroids (from the pill), which is maintained for the normal duration of a monthly cycle. As a result of negative feedback inhibition of gonadotropin secretion, *ovulation never occurs*. The entire cycle is like a false luteal phase, with high levels of progesterone and estrogen and low levels of gonadotropins.

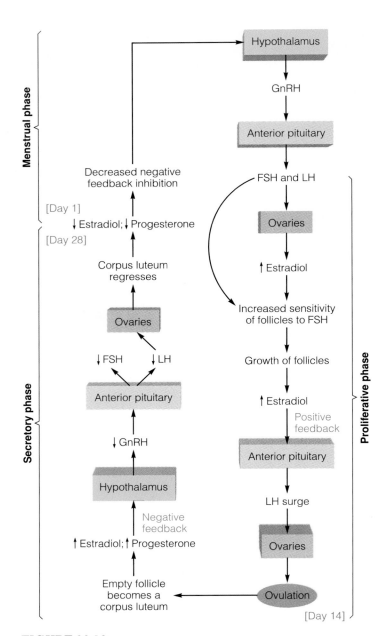

**FIGURE 29.16**

The sequence of events in the endocrine control of the ovarian cycle in context of the phases of the endometrium during the menstrual cycle.

Since the contraceptive pills contain ovarian steroid hormones, the endometrium proliferates and becomes secretory just as it does during a normal cycle. In order to prevent an abnormal growth of the endometrium, women stop taking the steroid pills after 3 weeks (placebo pills are taken during the fourth week). This causes estrogen and progesterone levels to fall, permitting menstruation to occur.

The side effects of earlier versions of the birth control pill have been reduced through a decrease in the content of estrogen and through the use of newer generations of progestogens (analogues of progesterone). The newer contraceptive

# Table 29.1    Phases of the menstrual cycle

| | | Hormonal changes | | Tissue changes | |
|---|---|---|---|---|---|
| *Ovarian* | *Endometrial* | *Pituitary* | *Ovarian* | *Ovarian* | *Endometrial* |
| Follicular (days 1–4) | Menstrual | FSH and LH secretion low | Estradiol and progesterone remain low | Primary follicles grow | Outer two-thirds of endometrium is shed with accompanying bleeding |
| Follicular (days 5–13) | Proliferative | FSH slightly higher than LH secretion in early follicular phase | Estradiol secretion rises (due to FSH stimulation of follicles) | Follicles grow; graafian follicle develops (due to FSH stimulation) | Mitotic division increases thickness of endometrium; spiral arteries develop (due to estradiol stimulation) |
| Ovulatory (day 14) | Proliferative | LH surge (and increased FSH) stimulated by positive feedback from estradiol | Fall in estradiol secretion | Graafian follicle is ruptured and secondary oocyte is extruded into peritoneal cavity | No change |
| Luteal (days 15–28) | Secretory | LH and FSH decrease (due to negative feedback of steroids) | Progesterone and estrogen secretion increase, then fall | Development of corpus luteum (due to LH stimulation); regression of corpus luteum | Glandular development in endometrium (due to progesterone stimulation) |

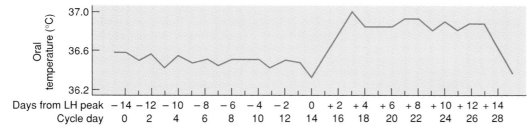

**FIGURE 29.17**

Changes in basal body temperature during the menstrual cycle.

pills are very effective and have a number of beneficial side effects, including a reduced risk endometrial and ovarian cancer, reduced risk of cardiovascular disease, and a reduction in osteoporosis. However, there may be an increased risk of breast cancer, and possibly cervical cancer, with oral contraceptives. The current consensus is that the health benefits of oral contraceptives outweigh the risk.

Newer systems for delivery of contraceptive steroids are designed so that the steroids are not taken orally, and as a result do not have to pass through the liver before entering the general circulation. (All drugs taken orally pass from the hepatic portal vein to the liver before they are delivered to any other organ; see chapter 26.) This permits lower doses of hormones to be effective. Such newer systems include a subdermal implant (see fig. 29.25*h*), which need only be replaced after five years, and vaginal rings, which can be worn for three weeks. The long-term safety of these newer methods is not yet established.

## Menopause

The term *menopause* means literally "pause in the menses" and refers to the cessation of ovarian activity that occurs at about the age of 50. During the postmenopausal years, which account for about a third of a woman's life span, no new ovarian follicles develop and the ovaries stop secreting estradiol. This termination of estradiol secretion is due to changes in the ovaries, not in the pituitary gland; indeed, FSH and LH secretion by the pituitary gland is elevated due to the absence of negative feedback inhibition from estradiol. As in prepubertal boys and girls, the only estrogen found in the blood of postmenopausal women is that formed by conversion of the weak androgen androstenedione, secreted principally by the adrenal cortex, into a weak estrogen called *estrone*.

It is the withdrawal of estradiol secretion from the ovaries that is primarily responsible for the symptoms of menopause. These include vasomotor disturbances and urogential atrophy. Vasomotor disturbances produce the hot flashes of menopause, where a fall in core body temperature is followed by feelings of heat and profuse perspiration.

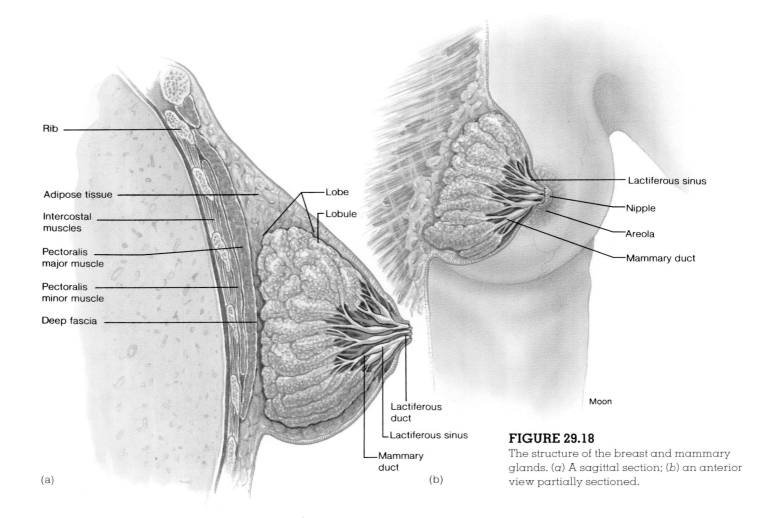

Rib

Adipose tissue

Intercostal muscles

Pectoralis major muscle

Pectoralis minor muscle

Deep fascia

Lobe

Lobule

Lactiferous duct

Lactiferous sinus

Mammary duct

(a)

Lactiferous sinus

Nipple

Areola

Mammary duct

Moon

(b)

**FIGURE 29.18**

The structure of the breast and mammary glands. (*a*) A sagittal section; (*b*) an anterior view partially sectioned.

Atrophy of the urethra, vaginal wall, and vaginal glands occur, with loss of lubrication. There is also increased risk of atherosclerotic cardiovascular disease (see chapter 18) and increased progression of osteoporosis (see chapter 8). These changes can be reversed, to a significant degree, by estrogen treatments.

# Mammary Glands and Lactation

*The structure and function of the mammary glands is dependent on the action of a number of hormones. The secretion of prolactin and oxytocin is directly required for the production and delivery of milk to a suckling infant.*

In structure, the mammary glands, located in the **breasts,** are modified sweat glands and part of the integumentary system (see chapter 7). In function, however, these glands are associated with the reproductive system because they secrete milk for the nourishment of the young. The size and shape of the breasts vary widely from person to person because of differences in genetic makeup, age, and percentage of body fat. At puberty, estrogen from the ovaries

stimulates growth of the mammary glands and the deposition of adipose tissue within the breasts. Mammary glands hypertrophy in pregnant and lactating women and usually atrophy somewhat after menopause.

## Structure of the Breasts and Mammary Glands

Each breast is positioned over ribs 2 through 6 and overlies the pectoralis major muscle and portions of the serratus anterior and external oblique muscles (fig. 29.18). The medial boundary of the breast is over the lateral margin of the sternum, and the lateral margin of the breast is along the anterior border of the axilla. The **axillary process** of the breast extends upward and laterally toward the axilla, where it comes into close relationship with the axillary vessels. This region of the breast is clinically significant because of the high incidence of breast cancer within the lymphatic drainage of the axillary process.

Each mammary gland is composed of 15 to 20 **lobes,** divided by adipose tissue. Each lobe has its own drainage pathway to the outside. The amount of adipose tissue determines the size and shape of the breast but has nothing to do with the ability of a woman to nurse. Each lobe is subdivided into **lobules,** which contain the glandular **alveoli** that secrete the milk of a lactating female. **Suspensory ligaments** (of Cooper) between the lobules extend from the skin to the deep fascia overlying the pectoralis muscle and support the breasts. The clustered alveoli secrete milk into a series of **mammary ducts,** which in turn converge to form **lactiferous** (*lak-tif´er-us*) **ducts.** The lumen of each lactiferous duct expands just deep to the surface of the nipple to form a **lactiferous sinus,** where milk may be stored before it drains at the tip of the nipple.

The **nipple** is a cylindrical projection containing some erectile tissue. A circular pigmented **areola** (*ă-re´ŏ-lă*) surrounds the nipple. The surface of the areola may appear rough because of the presence of sebaceous **areolar glands** close to the surface. The secretions of the areolar glands keep the nipple pliable. The color of the areola and nipple varies according to the complexion of the woman and whether or not she is pregnant. During pregnancy, the areola becomes darker and enlarges somewhat, presumably to become more conspicuous to a nursing infant.

 Lymphatic drainage and the location of lymph nodes within the breast are of considerable clinical importance because of the frequency of *breast cancer* and the high incidence of metastases. About 75% of the lymph drains through the axillary process of the breast into the axillary lymph nodes. Some 20% of the lymph passes toward the sternum to the internal thoracic lymph nodes. The remaining 5% of the lymph is subcutaneous and follows the lymph drainage pathway in the skin toward the back, where it reaches the intercostal nodes near the neck of the ribs.

## Lactation

The changes that occur in the mammary glands during pregnancy and the regulation of lactation provide excellent examples of hormonal interactions and neuroendocrine

.........................................
ligaments of Cooper: from Sir Astley P. Cooper, English anatomist and surgeon, 1768–1841

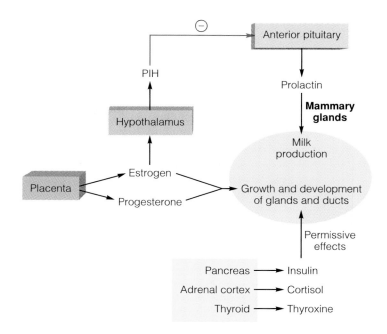

**FIGURE 29.19**

The hormonal control of mammary gland development during pregnancy and lactation. Note that milk production is prevented during pregnancy by estrogen inhibition of prolactin secretion. This inhibition is accomplished by the stimulation of PIH (prolactin-inhibiting hormone) secretion from the hypothalamus.

regulation (table 29.2). Growth and development of the mammary glands during pregnancy requires the permissive actions of insulin, cortisol, and thyroid hormones; in the presence of adequate amounts of these hormones, high levels of progesterone stimulate the development of the mammary alveoli and estrogen stimulates proliferation of the tubules and ducts (fig. 29.19).

The production of milk proteins, including casein and lactalbumin, is stimulated after parturition by **prolactin,** a hormone secreted by the anterior pituitary. The secretion of prolactin is controlled primarily by *prolactin-inhibiting hormone (PIH),* which is believed to be dopamine, produced

## Table 29.2    Hormonal factors affecting lactation

| Hormones | Major source | Effects |
|---|---|---|
| Insulin, cortisol, thyroid hormones | Pancreas, adrenal cortex, and thyroid gland | Permissive effects—adequate amounts of these must be present for other hormones to exert their effects on mammary glands |
| Estrogen and progesterone | Placenta | Growth and development of secretory units (alveoli) and ducts in mammary glands |
| Prolactin | Anterior pituitary | Production of milk proteins, including casein and lactalbumin |
| Oxytocin | Posterior pituitary | Stimulation of milk-ejection reflex |

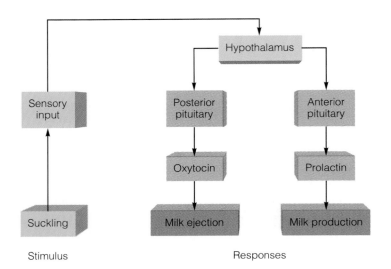

Stimulus           Responses

**FIGURE 29.20**

Lactation occurs in two stages: milk production (stimulated by prolactin) and milk ejection (stimulated by oxytocin). The stimulus of suckling triggers a neuroendocrine reflex that results in increased secretion of oxytocin and prolactin.

by the hypothalamus and secreted into the hypothalamo-hypophyseal portal blood vessels (see fig. 19.7). The secretion of PIH is stimulated by high levels of estrogen. In addition, high levels of estrogen act directly on the mammary glands to block their stimulation by prolactin. During pregnancy, consequently, the high levels of estrogen prepare the breasts for lactation but prevent prolactin secretion and action.

After parturition, when the placenta is eliminated, declining levels of estrogen are accompanied by an increase in the secretion of prolactin. Milk production is therefore stimulated.

If a woman does not wish to breast feed her baby, she may take oral estrogens to inhibit prolactin secretion. A different drug commonly given in these circumstances, and in other conditions in which it is desirable to inhibit prolactin secretion, is *bromocriptine*. This drug binds to dopamine receptors, and thus promotes the action of dopamine. The fact that this action inhibits prolactin secretion offers additional evidence that dopamine may function as the prolactin-inhibiting hormone (PIH).

The act of nursing helps to maintain high levels of prolactin secretion via a *neuroendocrine reflex* (fig. 29.20). Sensory endings in the breast, activated by the stimulus of suckling, relay impulses to the hypothalamus and inhibit the secretion of PIH. There is also indirect evidence that the stimulus of suckling may cause the secretion of a *prolactin-releasing hormone*, but this is controversial. Suckling thus results in the reflex secretion of high levels of prolactin, which promotes the secretion of milk from the alveoli into the ducts. In order for the baby to get the milk, however, the action of another hormone is needed.

The stimulus of suckling also results in the reflex secretion of **oxytocin** from the posterior pituitary. This

hormone is produced in the hypothalamus and stored in the posterior pituitary; its secretion results in the **milk-ejection reflex,** or **milk letdown.** Oxytocin not only stimulates contraction of the lactiferous ducts, but also contraction of the uterus (which explains why women who breast feed regain uterine muscle tone faster than those who do not).

Milk letdown can become a conditioned reflex in response to visual or auditory cues; the crying of a baby can elicit oxytocin secretion and the milk-ejection reflex. On the other hand, this reflex can be suppressed by the adrenergic effects produced in the fight-or-flight reaction. Thus, if a woman becomes nervous and anxious while breast feeding, her milk will be produced but it will not flow (there will be no milk letdown). This can cause increased pressure, intensifying her anxiety and frustration and further inhibiting the milk-ejection reflex. It is therefore important for mothers to nurse their babies in a quiet and calm environment.

Breast feeding, acting through reflex inhibition of GnRH secretion, can inhibit the secretion of gonadotropins from the mother's anterior pituitary and thus inhibit ovulation. Breast feeding is a natural contraceptive mechanism that helps to space births. This mechanism appears to be most effective in women with limited caloric intake who breast feed their babies at frequent intervals throughout the day and night. In the traditional societies of the less industrialized nations, therefore, breast feeding is an effective contraceptive. It has much less of a contraceptive effect in women who are well nourished and who breast feed their babies at more widely spaced intervals.

# Clinical Considerations

Females are more prone to dysfunctions and diseases of the reproductive organs than are males because of cyclic changes in reproductive events, problems associated with pregnancy, and the susceptibility of the female breasts to infections and neoplasms. The termination of reproductive capabilities at menopause can also cause complications due to hormonal alterations. Gynecology is the specialty of medicine concerned with dysfunction and diseases of the female reproductive system, whereas obstetrics is the specialty dealing with pregnancy and childbirth. Frequently a physician will specialize in both obstetrics and gynecology (OBGYN).

A comprehensive discussion of the numerous clinical aspects of the female reproductive system is beyond the scope of this text. Only the most important conditions are discussed in the following sections, along with a description of the more popular methods of birth control.

# UNDER DEVELOPMENT

## Development of the Female Reproductive System

Although the genetic sex is determined at fertilization (XX for females and XY for males), both sexes develop similarly through the indifferent stage of the eighth week. The gonads of both sexes develop from gonadal ridges, and the genital tubercle develops during the sixth week as an external swelling.

The ovaries develop more slowly than do the testes. Ovarian development begins at about the tenth week when **primordial follicles** begin to form within the medulla of the gonads. Each of the primordial follicles consists of an **oogonium** (o˝ŏ-go´ne-um) surrounded by a layer of follicular cells. Mitosis of the oogonia occurs during fetal development, so that thousands of germ cells are formed. Unlike the male reproductive system, in which spermatogonia are formed by mitosis throughout life, all oogonia are formed prenatally and their number continuously decreases after birth.

The uterus and uterine tubes develop from a pair of embryonic tubes called the **paramesonephric** (müllerian) **ducts,** which are so called because they are located to the sides of the mesonephric ducts (which form temporary embryonic kidneys). As the mesonephric kidneys degenerate (chapter 25), the lower portions of the paramesonephric ducts fuse to form the uterus, and the upper portions give rise to the uterine tubes. As

| Table 1 | Homologous reproductive organs and the undifferentiated structures from which they develop | | |
|---|---|---|---|
| **Indifferent stage** | **Male** | | **Female** |
| Gonads | Testes | | Ovaries |
| Urogenital groove | Membranous urethra | | Vestibule |
| Genital tubercle | Glans penis | | Clitoris |
| Urethral folds | Spongy urethra | | Labia minora |
| Labioscrotal swelling | Scrotum | | Labia majora |
| | Bulbourethral glands | | Vestibular glands |

described in chapter 28, if the embryo is male the paramesonephric ducts degenerate due to secretion of *müllerian inhibition factor* from the testes.

A thin membrane called the **hymen** forms to separate the lumen of the vagina from the urethral sinus. The hymen usually is perforated during later fetal development.

The external genitalia of both sexes appear the same during the indifferent stage of the eighth week. A prominent **phallus** (fal´us) forms from the genital tubercle, and a **urethral groove** forms on the ventral side of the phallus. As described in chapter 28, the phallus

becomes the penis in a male and the smaller clitoris in a female. Paired **urethral folds** surround the urethral groove on the lateral sides. In a male, these fuse to form the urethra of the penis; in a female, the urethral folds remain unfused and form the inner labia minora. Similarly, the **labiosacral swellings** in a male fuse to form the scrotum; in a female, these remain unfused and form the prominent labia majora. The male and female structures that share a common embryological origin are said to be homologous structures (table 1).

follicle: L. diminutive of *follis,* bag
oogonium: Gk. *oion,* egg; *gonos,* procreation

hymen: Gk. (mythology) *Hymen,* god of marriage

## Diagnostic Procedures

A gynecological, or pelvic, examination is generally given in a thorough physical examination, especially prior to marriage, during pregnancy, or if problems involving the reproductive organs are suspected. In a gynecological examination, the physician inspects the vulva for irritations, lesions, or abnormal vaginal discharge and palpates the vulva and internal organs. Most of the internal organs can be palpated through the vagina, especially if they are enlarged or tender. Inserting a lubricated *speculum* into the vagina allows visual examination of the cervix and vaginal walls. A speculum is an instrument for opening or distending a body opening to permit visual inspection.

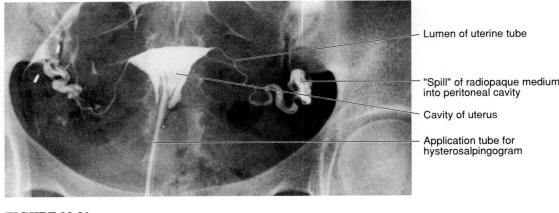

**FIGURE 29.21**
A hysterosalpingogram showing the cavity of the uterus and lumina of the uterine tubes.

Lumen of uterine tube

"Spill" of radiopaque medium into peritoneal cavity

Cavity of uterus

Application tube for hysterosalpingogram

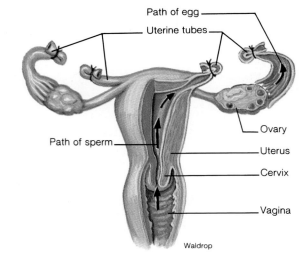

Path of egg

Uterine tubes

Path of sperm

Ovary

Uterus

Cervix

Vagina

Waldrop

**FIGURE 29.22**
Tubal ligation involves removal of a portion of each uterine tube.

In special cases, it may be necessary to examine the cavities of the uterus and uterine tubes by *hysterosalpingography (his˝ter-o-sal˝ping-gog´rǎ-fe)* (fig. 29.21). This technique involves injecting a radiopaque dye into the reproductive tract. The patency of the uterine tubes, irregular pregnancies, and various types of tumors may be detected using this technique. A *laparoscopy (lap˝ǎ-ros´kŏ-pe)* permits in vivo visualization of the internal reproductive organs. The laparoscope may be inserted via the umbilicus, a small incision in the lower abdominal wall, or through the posterior fornix of the vagina into the rectouterine pouch. Although a laparoscope is used primarily in diagnosis, it can be used when performing a *tubal ligation* (fig. 29.22), which is a method of sterilizing a female by tying off the uterine tubes.

hysterosalpingography: Gk. *hystera,* uterus; *salpinx,* trumpet (uterine tube); *graphein,* to record

One diagnostic procedure that should be routinely performed by a woman is a *breast self-examination* (BSE) (see page 895). The importance of a BSE is not to prevent diseases of the breast but to detect any problems before they become serious. A BSE should be performed monthly, 1 week after the cessation of menstruation so that the breast will not be swollen or especially tender.

Another important diagnostic procedure is a *Papanicolaou (Pap) smear.* The Pap smear permits microscopic examination of cells covering the tip of the cervix. Samples of cells are obtained by gently scraping the surface of the cervix with a specially designed wooden spatula. Women should have routine Pap smears for the early detection of cervical cancer.

## Problems Involving the Ovaries and Uterine Tubes

Most **ovarian neoplasms** are nonmalignant ovarian cysts lined by cuboidal epithelium and filled with a serous albuminous fluid. These tumors may frequently be palpated during a gynecological examination and may require surgical removal if they exceed about 4 cm (1.5 in.) in diameter. They are generally removed as a precaution because it is impossible to determine by palpation whether the mass is malignant or benign.

**Ovarian tumors,** which generally occur in women over the age of 60, may reach massive size. Ovarian tumors of 5 kg (14 lbs) are not uncommon, and ovarian tumors of 110 kg (300 lbs) have been reported. Some ovarian tumors produce estrogen and thus cause feminization in elderly women, including the resumption of menstrual periods. The prognosis for women with ovarian tumors varies depending on the type of tumor, whether or not it is malignant, and if it is, the stage of the cancer.

Two frequent problems involving the uterine tubes are salpingitis and ectopic pregnancies. **Salpingitis** is an inflammation of one or both uterine tubes. Infection of the uterine tubes is generally caused by sexually transmitted disease, although secondary bacterial infections from the vagina may also cause salpingitis. Salpingitis may cause sterility if the uterine tubes become occluded.

**Ectopic pregnancy** results from implantation of the blastocyst in a location other than the body or fundus of the

Pap smear: from George N. Papanicolaou, American anatomist and physician, 1883–1962
neoplasm: Gk. neos, new; plasma, something formed

uterus. The most frequent ectopic site is in the uterine tube, where an implanted blastocyst causes what is commonly called a **tubular pregnancy** (see fig. 30.33). One danger of a tubular pregnancy is the enlargement, rupture, and subsequent hemorrhage of the uterine tube where implantation has occurred. A tubular pregnancy is frequently treated by removing the affected tube.

**Infertility,** or the inability to conceive, is a clinical problem that may involve the male or female reproductive system. Based on the number of people who seek help for this problem, it is estimated that 10% to 15% of couples have impaired fertility. Generally, when a male is infertile, it is because of inadequate sperm counts. Female infertility is frequently caused by an obstruction of the uterine tubes or abnormal ovulation.

**Polycystic ovarian syndrome** is characterized by chronic anovulation (lack of ovulation) combined with *hirsutism* (hairiness). The ovulation failure is due to hormone disturbances that originate elsewhere. The hirsutism is produced by abnormally high secretion of androgens from the ovary. Progesterone therapy or birth control pills are generally used for treatment. Fertility drugs, such as Clomid, may also be used to stimulate ovulation in women with polycystic disease who want to become pregnant.

## Problems Involving the Uterus

Abnormal menstruations are among the most common disorders of the female reproductive system. Abnormal menstruations may be directly related to problems of the reproductive organs and pituitary gland or associated with emotional and psychological stress.

**Amenorrhea** (*a-men-ŏ-re´ă*) is the absence of menstruation and can be categorized as normal, primary, or secondary. **Normal amenorrhea** follows menopause, occurs during pregnancy, and in some women may occur during lactation. **Primary amenorrhea** is the failure to have menstruated by the age at which menstruation normally begins. Primary amenorrhea is generally accompanied by failure of the secondary sex characteristics to develop. Endocrine disorders may cause primary amenorrhea and abnormal development of the ovaries or uterus.

**Secondary amenorrhea** is the cessation of menstruation in women who previously have had normal menstrual periods and who are not pregnant and have not gone through menopause. Various endocrine disturbances, as well as psychological factors, may cause secondary amenorrhea. It is not uncommon, for example, for young women who are in the process of making major changes or adjustments in their lives to miss menstrual periods. Secondary amenorrhea is also frequent in female athletes during periods of intense training. A low percentage of body fat may be a contributing factor. Sickness, fatigue, poor nutrition, or emotional stress may also cause secondary amenorrhea.

**Dysmenorrhea** is painful or difficult menstruation that may be accompanied by severe menstrual cramps. The causes of dysmenorrhea are not totally understood but may include endocrine disturbances (inadequate progesterone levels), a faulty position of the uterus, emotional stress, or some type of obstruction that prohibits menstrual discharge.

Abnormal uterine bleeding includes **menorrhagia** (*men˝ŏ-ra´je-ă*), or excessive bleeding during the menstrual period, and **metrorrhagia,** or spotting between menstrual periods. Other types of abnormal uterine bleeding are menstruations of excessive duration, too frequent menstruations, and postmenopausal bleeding. These abnormalities may be caused by hormonal irregularities, emotional factors, or various diseases and physical conditions.

**Uterine neoplasms** are an extremely common problem of the female reproductive tract. Most of the neoplasms are benign and include cysts, polyps, and smooth muscle tumors (leiomyoma). Any of these conditions may provoke irregular menstruations and may cause infertility if the neoplasms are massive.

**Cancer of the uterus** is the most common malignancy of the female reproductive tract. The most common site of uterine cancer is the cervix (fig. 29.23). Cervical cancer is second only to cancer of the breast in frequency of occurrence and is a disease of young women (ages 30 through 50), especially those who have had frequent intercourse with multiple partners during their teens and onward. If detected early through regular Pap smears, the disease can be cured before it metastasizes. The treatment of cervical cancer depends on the stage of the malignancy and the age and health of the woman. In the case of women for whom future fertility is not an issue, a **hysterectomy** (surgical removal of the uterus) is usually performed.

**Endometriosis** is a condition characterized by the presence of endometrial tissues at sites other than the inner lining of the uterus. Frequent sites of ectopic endometrial cells are on the ovaries, outer layer of the uterus, abdominal wall, and urinary bladder. Although it is not certain how endometrial cells become established outside the uterus, it is speculated that some discharged endometrial tissue might be flushed backward from the uterus and through the uterine tubes during menstruation. Women with endometriosis will bleed internally with each menstrual period because the ectopic endometrial cells are stimulated along with the normal endometrium by ovarian hormones. The most common symptoms of endometriosis are extreme dysmenorrhea and a feeling of fullness during each menstrual period. Endometriosis can cause infertility. It is most often treated by suppressing the endometrial tissues with oral contraceptive pills or by surgery. An *oophorectomy*, or removal of the ovaries, may be necessary in extreme cases.

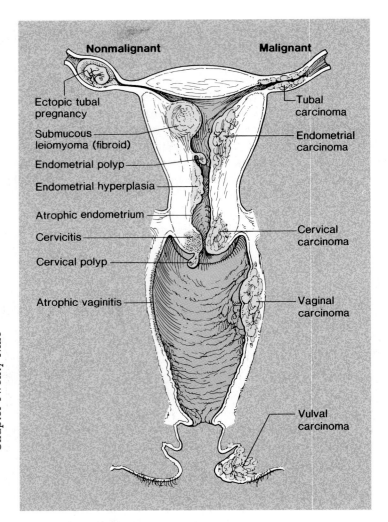

**Nonmalignant**      **Malignant**

Ectopic tubal pregnancy

Submucous leiomyoma (fibroid)

Endometrial polyp

Endometrial hyperplasia

Atrophic endometrium

Cervicitis

Cervical polyp

Atrophic vaginitis

Tubal carcinoma

Endometrial carcinoma

Cervical carcinoma

Vaginal carcinoma

Vulval carcinoma

**FIGURE 29.23**
Sites of various conditions and diseases of the female reproductive tract, each of which could cause an abnormal discharge of blood.

## Diseases of the Vagina and Vulva

**Pelvic inflammatory disease (PID)** is a general term for inflammation of the female reproductive organs within the pelvis. The infection may be confined to a single organ, or it may involve all the internal reproductive organs. The pathogens generally enter through the vagina during coitus, induced abortion, childbirth, or postpartum.

The vagina and vulva are generally resistant to infection because of the acidity of the vaginal secretions. Occasionally, however, localized infections and inflammations do occur; these are termed **vaginitis,** if confined to the vagina, or **vulvovaginitis,** if both the vagina and external genitalia are affected. The symptoms of vaginitis are a discharge of pus (*leukorrhea*) and itching (*pruritus*). The two most common organisms that cause vaginitis are the protozoan *Trichomonas vaginalis* and the fungus *Candida albicans.*

## Diseases of the Breasts and Mammary Glands

The breasts and mammary glands of females are highly susceptible to infections, cysts, and tumors. Infections involving the mammary glands usually follow the development of a dry and cracked nipple during lactation. Bacteria enter the wound and establish an infection within the lobules of the gland. During an infection of the mammary gland, a blocked duct frequently causes a lobe to become engorged with milk. This localized swelling is usually accompanied by redness, pain, and fever. Administering specific antibiotics and applying heat are the usual treatments.

Nonmalignant cysts are the most frequent diseases of the breast. These masses are generally of two types, neither of which is life threatening. **Dysplasia (fibrocystic disease)** is a broad condition involving several nonmalignant diseases of the breast. All dysplasias are benign neoplasms of various sizes that may become painful during or prior to menstruation. Most of the masses are small and remain undetected. Dysplasia affects nearly 50% of women over the age of 30 prior to menopause.

A **fibroadenoma** (*fi˝bro-ad-ĕ-no-m ă*) is a benign tumor of the breast that frequently occurs in women under the age of 35. Fibroadenomas are nontender, rubbery masses that are easily moved about in the mammary tissue. A fibroadenoma can be excised in a physician's office under local anesthetics.

Carcinoma of the breast is the most common malignancy in women. One in nine women will develop breast cancer and one-third of these will die from the disease. Breast cancer is the leading cause of death in women between 40 and 50 years of age. Men are also susceptible to breast cancer, but it is 100 times more frequent in women. Breast cancer in men is usually fatal.

The causes of breast cancer are not known, but women who are most susceptible are those who are over age 35, who have a family history of breast cancer, and who are nulliparous (never having given birth). The early detection of breast cancer is important because the progressed state of the disease will determine the treatment and prognosis.

Confirming suspected breast cancer generally requires *mammography* (fig. 29.24). If the mammogram indicates breast cancer, surgery is performed so that a biopsy can be obtained and the tumor assessed. If the tumor is found to be malignant, surgery is performed, the extent of which depends on the size of the tumor and whether or not metastasis has occurred. The surgical treatment for breast cancer is generally some degree of *mastectomy.* A *simple mastectomy* is removal of the entire breast but not the underlying lymph nodes. A *modified radical mastectomy* is the complete removal of the breast, the lymphatic drainage, and perhaps the pectoralis major muscle. A *radical mastectomy* is similar to a modified except that the pectoralis major muscle is always

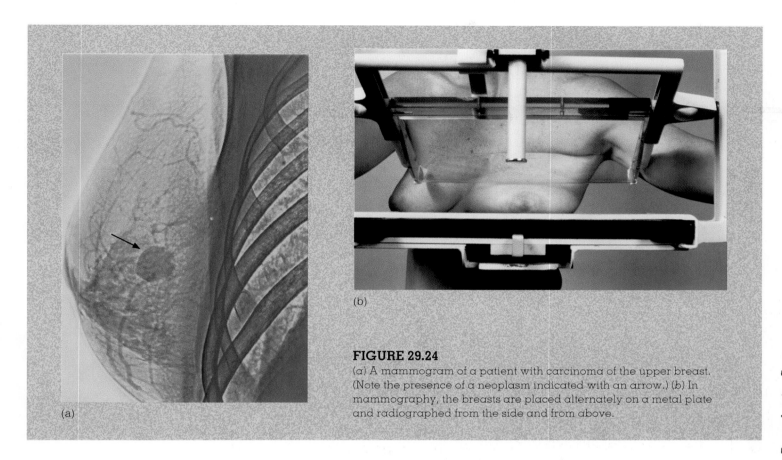

**FIGURE 29.24**

(*a*) A mammogram of a patient with carcinoma of the upper breast. (Note the presence of a neoplasm indicated with an arrow.) (*b*) In mammography, the breasts are placed alternately on a metal plate and radiographed from the side and from above.

removed, as well as the axillary lymph nodes and adjacent connective tissue. A *lumpectomy*—removal of just the lump and a small amount of surrounding breast tissue—is currently used as an option for some small malignancies that are in the beginning stages.

## Methods of Contraception

In addition to the rhythm method and the contraceptive (birth control) pill, which have already been discussed, contraception may be accomplished by sterilization, intrauterine devices (IUDs), and barrier methods—including condoms, diaphragms, sponges, and spermicidal gels and foams (fig. 29.25).

**Sterilization** techniques include *vasectomy* for the male and *tubal ligation* for the female. In the latter technique (which currently accounts for over 60% of sterilization procedures performed in the United States), the uterine tubes are cut and tied. This is analogous to the procedure performed on the ductus deferens in a vasectomy and prevents fertilization of the ovulated ovum. Studies on the long-term effects of these procedures have failed to show deleterious side effects. With current procedures, tubal ligations (as well as vasectomies) should be considered essentially irreversible.

**Intrauterine devices (IUDs)** don't prevent ovulation but instead prevent implantation of the embryo into the uterus should fertilization occur. The mechanisms by which these contraceptive effects are produced are not well understood, but the efficiency of different IUDs appears to be related to their ability to cause inflammatory reactions in the uterus. Uterine perforations are the foremost complication associated with the use of IUDs.

**Barrier methods** of birth control—physical or chemical barriers to keep sperm and ova apart—generally are quite effective for careful users and do not pose serious health risks. The failure rate for condoms—one of the oldest methods of contraception—is 12 to 20 pregnancies per 100 woman years of use, whereas the failure rate for the diaphragm is 12 to 18 pregnancies per 100 woman years. Latex condoms offer an additional benefit: they provide some protection against sexually transmitted diseases, including AIDS.

As an alternative to contraceptive pills, hormonal contraceptives may be delivered to a woman's body by means of **subdermal implants.** Implants are 2-in. rods filled with a synthetic progesterone (progestin) and implanted just under the skin, usually on the upper arm, through a tiny incision. The hormone gradually leaches out through the walls of the rod and enters the bloodstream, preventing pregnancy for at least 5 years.

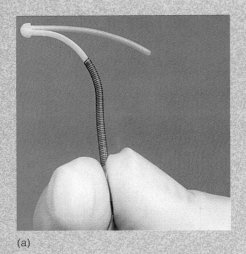

(a)

(b)

(c)

(d)

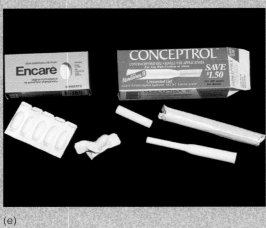

(e)

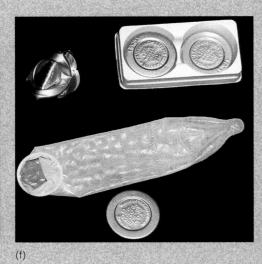

(f)

(g)

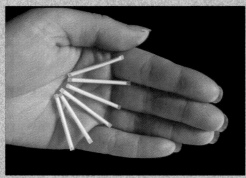

(h)

**FIGURE 29.25**

Various types of birth control devices. (*a*) IUD, (*b*) contraceptive sponge, (*c*) diaphragm, (*d*) birth control pills, (*e*) vaginal spermicide, (*f*) condom, (*g*) female condom, and (*h*) subdermal implants.

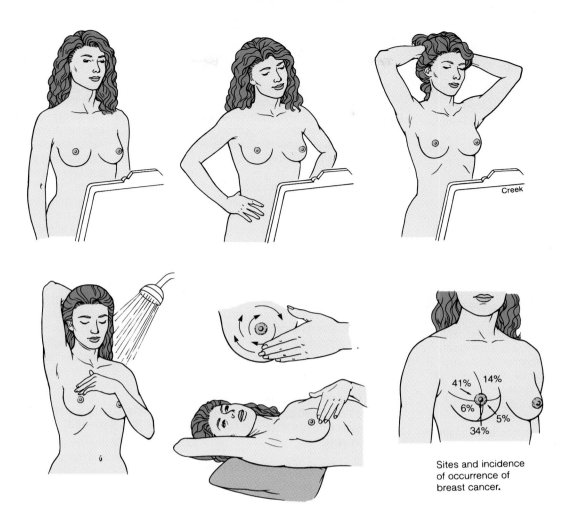

Creek

41%  14%

6%

5%

34%

Sites and incidence
of occurrence of
breast cancer.

# Breast Self-Examination (BSE)

One in nine women will develop breast cancer during her lifetime. Early detection of breast cancer and follow-up medical treatment minimizes the necessary surgical treatment and increases the patient's prognosis. Breast cancer is curable if it is caught early.

A woman should examine her breasts monthly. If she has not yet reached menopause, the ideal time for a BSE is 1 week after her period ends because the breasts are less likely to be swollen and tender at that time. A woman no longer menstruating should just pick a day of the month and do a BSE on that same day on a monthly basis. Visual inspection and palpation are equally important in doing a BSE. The steps involved in this procedure are the following:

**1  Observation before a mirror.** Inspect the breasts with the arms at the sides. Next, raise the arms high overhead. Look for any changes in the contour of each breast—a swelling, dimpling of skin or changes in the nipple. The left and right breast will not exactly match—few women's breasts do. Finally, squeeze the nipple of each breast gently between the thumb and index finger. Any discharge from the nipple should be reported to a physician.

**2  Palpation during bathing.** Examine the breasts during a bath or shower when the hands will glide easily over wet skin. With the fingers flat, move gently over every part of each breast. Use the right hand to examine the left breast, and the left hand for the right breast. Palpate for any lump, hard knot, or thickening. If the breasts are normally fibrous or lumpy (fibrocystic tissue), the locations of these lumps should be noted and checked each month for changes in size and locations.

**3  Palpation while lying down.** To examine the right breast, put a pillow or folded towel under the right shoulder. Place the right hand behind the head—this distributes the breast tissue more evenly on the rib cage. With the fingers of the left hand held flat, press gently in small circular motions around an imaginary clock face. Begin at the outermost top of the right breast for 12 o'clock, then move to 1 o'clock, and so on around the circle back to 12 o'clock. A ridge of firm tissue in the lower curve of each breast is normal. Then move in an inch, toward the nipple, and keep circling to examine every part of the breast, including the nipple. Also examine the armpit carefully for enlarged lymph nodes. Repeat the procedure on the left breast.

If a mammogram indicates breast cancer (see fig. 29.24), surgery is performed so that a biopsy can be obtained and the tumor assessed. The extent of the surgery depends on the size of the tumor and whether or not metastasis has occurred.

# Chapter Summary

## Structures and Functions of the Female Reproductive System (pp. 871–877)

1. The female secondary sex organs—those that are essential for sexual reproduction—include the vagina, uterine tubes, uterus, and mammary glands.
2. The uterine (fallopian) tubes end in fimbriae, which project over the ovary and help to direct a secondary oocyte into the infundibulum of the tube.
3. The uterus consists of a fundus, body, and cervix and is supported by four pairs of ligaments.
   a. The wall of the uterus consists of a perimetrium, a muscular myometrium, and an epithelial lining called the endometrium.
   b. The endometrium is stratified, with the layers divided into a stratum basale and a stratum functionale; the latter is shed during menstruation and rebuilt during the next cycle.
4. The vagina opens to the cervix of the uterus. The structures of the vulva (external genitalia) surround the vaginal orifice and include the mons pubis, labia majora, labia minora, clitoris, vaginal vestibule, vestibular bulbs, and vestibular glands.
5. During sexual excitement, the clitoris and erectile tissue of the areola of the breasts swell with blood. During orgasm, the muscles of the perineum, uterus, and uterine tubes contract rhythmically.

## Ovaries and the Ovarian Cycle (pp. 877–881)

1. The ovaries are supported by the mesovarium, which extends from the broad ligament, and by the ovarian and suspensory ligaments.

2. Primary oocytes, arrested at prophase I of the first meiotic division, are contained within primary follicles in the ovary.
   a. Upon stimulation by gonadotropic hormones, granulosa cells divide and fill the follicle, forming a primary follicle.
   b. Upon further stimulation, a fluid-filled cavity called the antrum begins to form to produce a secondary follicle.
   c. As a secondary follicle develops, the primary oocyte completes its first meiotic division to form a secondary oocyte, arrested at metaphase II, and a polar body.
   d. A single, fully mature follicle called the mature vesicular, or graafian, follicle releases its secondary oocyte at ovulation; the empty follicle then becomes a corpus luteum.
3. The pituitary gland secretes FSH and LH in response to the secretion of GnRH from the hypothalamus; secretion of FSH and LH, as well as GnRH, is modified by feedback from sex steroids secreted by the ovaries.

## Menstrual Cycle (pp. 881–886)

1. The ovarian cycle can be divided into follicular and luteal phases, which are separated by the event of ovulation.
   a. During the follicular phase, FSH stimulates the growth and development of follicles; this is accompanied by increasing secretion of estradiol from the granulosa cells of the follicles.
   b. Estradiol exerts a positive feedback effect on the hypothalamus and pituitary gland, resulting in an LH surge that triggers ovulation.
   c. At ovulation, the mature vesicular ovarian (graafian) follicle ruptures

and its secondary oocyte is released from the ovary.
   d. The empty vesicular ovarian follicle becomes a corpus luteum, which secretes estradiol and progesterone.
   e. Estradiol and progesterone inhibit FSH and LH secretion during the luteal phase.
   f. The corpus luteum regresses at the end of the cycle, and the resulting decline in estradiol and progesterone secretion causes menstruation.
2. In terms of the changes that occur in the endometrium, the cycle can be divided into the proliferative phase (following menstruation), secretory phase (corresponding to the luteal phase of the ovaries), and menstrual phase.
3. The contraceptive pill acts by duplicating the negative feedback inhibition of FSH and LH by estradiol and progesterone that normally occurs during the luteal phase.

## Mammary Glands and Lactation (pp. 886–888)

1. Mammary glands contain secretory alveoli that drain into lactiferous ducts, which in turn open on the surface of the nipple of the breast.
2. Prolactin stimulates the production of milk proteins; oxytocin stimulates contraction of the lactiferous ducts and ejection of milk from the nipple.
   a. Secretion of oxytocin and prolactin occur in response to the stimulus of the baby's suckling, through the activation of a neuroendocrine reflex.
   b. Secretion of prolactin is stimulated by inhibiting the secretion of prolactin-inhibiting hormone from the hypothalamus.

# Review Activities

## Objective Questions

Match the phase/event with its distinguishing feature(s):

1. menstrual phase
2. follicular phase
3. luteal phase
4. ovulation
   a. high estrogen and progesterone; low FSH and LH
   b. low estrogen and progesterone

   c. LH surge
   d. increasing estrogen; low LH and low progesterone
5. The secretory phase of the endometrium corresponds to which of the following ovarian phases?
   a. follicular phase
   b. ovulation
   c. luteal phase
   d. menstrual phase

6. Which of the following statements about oogenesis is *true*?
   a. Oogonia form continuously in postnatal life.
   b. Primary oocytes are haploid.
   c. Meiosis is completed prior to ovulation.
   d. A secondary oocyte is released.

7. The paramesonephric (müllerian) ducts give rise to
   a. the uterine tubes.
   b. the uterus.
   c. the pudendum.
   d. both a and b.
   e. both b and c.

8. In a female, the homologue of the male scrotum is/are
   a. the labia majora.
   b. the labia minora.
   c. the clitoris.
   d. the vestibule.

9. The cervix is a portion of
   a. the vulva.
   b. the vagina.
   c. the uterus.
   d. the uterine tubes.

10. Which of the following is shed as menses?
    a. the perimetrial layer.
    b. the fibrous layer.
    c. the functionalis layer.
    d. the menstrual layer.

11. The transverse folds in the mucosal layer of the vagina are called
    a. perineal folds.
    b. vaginal rugae.
    c. fornices.
    d. labia gyri.

12. Fertilization normally occurs in
    a. the ovary.
    b. the uterine tube.
    c. the uterus.
    d. the vagina.

13. Contractions of the mammary ducts are stimulated by
    a. prolactin.
    b. oxytocin.
    c. estrogen.
    d. progesterone.

14. The suspensory ligaments (of Cooper) support
    a. the ovary.
    b. the uterus.
    c. the uterine tube.
    d. the breast.

15. Uterine contractions are stimulated by
    a. oxytocin.
    b. prostaglandins.
    c. prolactin.
    d. both a and b.
    e. both b and c.

## Essay Questions

1. Explain how the genital ducts develop and why the external genitalia of males and females are considered homologous.

2. Describe the gross and histologic structure of the uterus and explain the significance of the strata functionale and basale of the endometrium.

3. Describe the hormonal interactions that control ovulation and cause it to occur at the proper time.

4. Compare menstrual bleeding and bleeding that occurs during the estrus cycle of a dog in terms of hormonal control mechanisms and the ovarian cycle.

5. "The contraceptive pill tricks the brain into thinking you're pregnant." Interpret this popularized explanation of how birth control pills work in terms of physiological mechanisms.

6. Describe the mechanisms that are thought to trigger lactation in a nursing mother.

## Gundy/Weber Software 💾

The tutorial software accompanying Chapter 29 is Volume 13—Reproductive System.

# chapter thirty

# developmental anatomy and inheritance

## objectives

- Define *morphogenesis*, *capacitation*, and *fertilization*.

- Describe the changes that occur in the spermatozoon and ovum prior to, during, and immediately following fertilization.

- Describe the events of pre-embryonic development that result in the formation of the blastocyst.

- Discuss the role of the trophoblast in the implantation and development of the placenta.

- Explain how the primary germ layers develop and list the structures produced by each layer.

- Define *gestation* and explain how the parturition date is determined.

- Define *embryo* and describe the major events of the embryonic period of development.

- List the embryonic needs that must be met to avoid a spontaneous abortion.

- Describe the structure and function of each of the extraembryonic membranes.

- Describe the development and function of the placenta and umbilical cord.

- Define *fetus* and discuss the major events of the fetal period of development.

- Describe the various techniques available for examining the fetus or monitoring fetal activity.

- Describe the hormonal action that controls labor and parturition.

- Describe the three stages of labor.

- Define *genetics*.

- Discuss the variables that account for a person's phenotype.

- Explain how probability is involved in predicting inheritance and use a Punnett square to illustrate selected probabilities.

# Fertilization

*Upon fertilization of a secondary oocyte by a spermatozoon in the uterine tube, meiotic development is completed and a diploid zygote is formed.*

The structure of the human body forms before birth, or prenatally, through the process of **morphogenesis** (mor˝fo-jen´ĕ-sis). Through morphogenic events, the organs and systems of the body are established in a functional relationship. A comprehensive view of the structure and function of each body system requires an understanding of morphogenesis. Associated with each organ and system are sensitive periods of morphogenesis during which genetic or environmental factors may affect the normal development of a baby. Many clinical problems are congenital in nature and originate during morphogenic development.

The prenatal development of a human is a fascinating and awesome event. It begins with a single fertilized egg and culminates some 38 weeks later with a complex organization composed of billions of cells. Prenatal development can be divided into a *pre-embryonic period*, initiated by the fertilization of a secondary oocyte; an *embryonic period*, during which the body's organ systems are formed; and a *fetal period*, culminating in parturition, or the birth of the baby.

During coitus, a male ejaculates between 100 million and 500 million sperm into the female's vagina. This tremendous number is needed because of the high rate of sperm fatality—only about 100 survive to contact the secondary oocyte in the uterine tube. During passage through the acidic female tract, sperm gain the ability to fertilize a secondary oocyte through a process called **capacitation.** The changes occurring in sperm capacitation are not fully understood. Evidence indicates, however, that when placed in the acidic environment of the female reproductive tract, there is an enzymatic reaction resulting in a molecular alteration in the cell membrane of the sperm.

Experiments confirm that freshly ejaculated sperm are infertile and must be in the female reproductive tract for at least 7 hours before they can fertilize a secondary oocyte. During *in vitro fertilization* (discussed further under "Clinical Considerations"), sperm capacitation is induced artificially by treating the ejaculate with a solution of gamma globulin, free serum, follicular fluid, dextran, serum dialysate, and adrenal gland extract.

A woman usually ovulates one secondary oocyte a month, totaling approximately 400 during her reproductive years. As discussed in chapter 29, each ovulation releases a secondary oocyte arrested at metaphase II. As the secondary oocyte enters the uterine tube, it is surrounded by a thin, transparent layer of protein and polysaccharides, the **zona pellucida,** and a layer of granulosa cells, the **corona radiata** (fig. 30.1).

The head of each spermatozoon is capped by an organelle called an **acrosome** (figs. 30.1 and 30.2). The acrosome contains a trypsin-like protein-digesting enzyme and *hyaluronidase* (hi˝ă-loo-ron´ĭ-dās), which digests hyaluronic acid, an important constituent of connective tissue. When a spermatozoon meets a secondary oocyte in the uterine tube, an *acrosomal reaction* occurs that exposes the acrosome's digestive enzymes and allows the spermatozoon to penetrate through the corona radiata and the zona pellucida.

As a spermatozoon penetrates the zona pellucida, a chemical change in the zona prevents other sperm from entering. Only one spermatozoon, therefore, is allowed to fertilize a secondary oocyte. When the cell membranes of the spermatozoon and the secondary oocyte merge, the secondary oocyte becomes an ovum. As fertilization occurs, the ovum is stimulated to complete its second meiotic division (fig. 30.3). Like the first meiotic division, the second produces one cell, which contains all of the cytoplasm, and one *polar body*. The healthy cell is the mature ovum, and the second polar body, like the first, ultimately fragments and disintegrates.

At fertilization, the entire spermatozoon enters the cytoplasm of the much larger ovum (fig. 30.4). Within 12 hours, the nuclear membrane in the ovum disappears, and the *haploid number* of chromosomes (23) in the ovum is joined by the haploid number of chromosomes from the spermatozoon. A fertilized egg, or zygote (zi´gōt), containing the *diploid number* of chromosomes (46) is thus formed.

A secondary oocyte that is ovulated but not fertilized does not complete its second meiotic division but instead disintegrates 12 to 24 hours after ovulation. Fertilization therefore cannot occur if coitus takes place beyond 1 day following ovulation. Sperm, by contrast, can survive up to 3 days in the female reproductive tract. Fertilization therefore can occur if coitus takes place within 3 days prior to the day of ovulation.

morphogenesis: Gk. *morphe*, form; *genesis*, beginning
capacitation: L. *capacitas*, capable of

zona pellucida: L. *zone*, a girdle; *pellis*, skin
corona radiata: Gk. *korone*, crown; *radiata*, radiate
acrosome: Gk. *akron*, extremity; *soma*, body
haploid: Gk. *haplous*, single; L. *ploideus*, multiple in form
diploid: Gk. *diplous*, double; L. *ploideus*, multiple in form

Developmental Anatomy and Inheritance

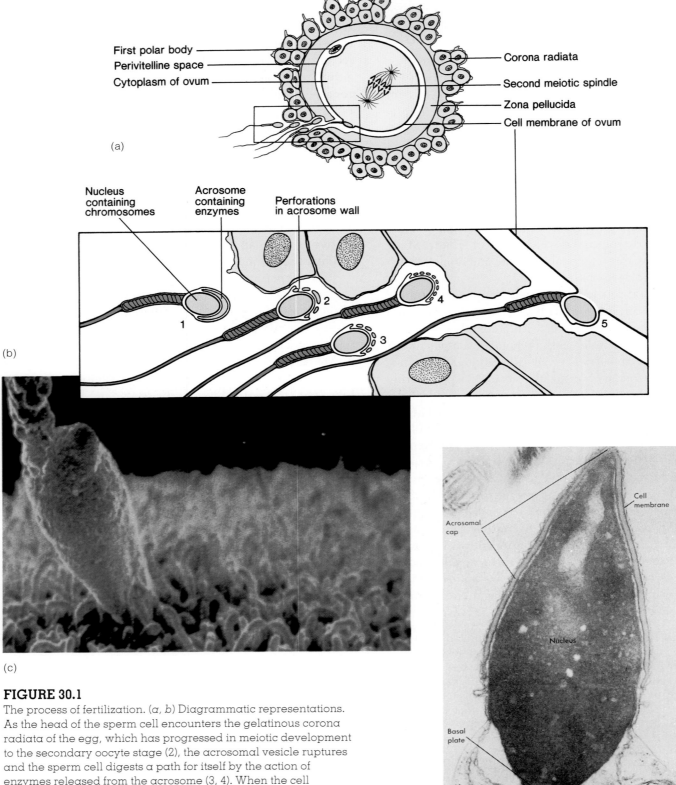

(a)

First polar body
Perivitelline space
Cytoplasm of ovum

Corona radiata
Second meiotic spindle
Zona pellucida
Cell membrane of ovum

Nucleus
containing
chromosomes

Acrosome
containing
enzymes

Perforations
in acrosome wall

Cell
membrane

Acrosomal
cap

Nucleus

Basal
plate

(b)

(c)

**FIGURE 30.1**

The process of fertilization. (*a, b*) Diagrammatic representations. As the head of the sperm cell encounters the gelatinous corona radiata of the egg, which has progressed in meiotic development to the secondary oocyte stage (2), the acrosomal vesicle ruptures and the sperm cell digests a path for itself by the action of enzymes released from the acrosome (3, 4). When the cell membrane of the sperm cell contacts the cell membrane of the egg (5), they become continuous, and the sperm cell nucleus and other contents move into the egg cytoplasm. (c) A scanning electron micrograph of a sperm cell bound to the egg surface.

**FIGURE 30.2**

A scanning electron micrograph showing the head of a human spermatozoon with its nucleus and acrosome.

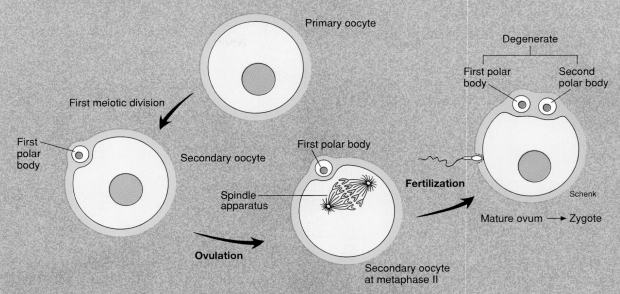

## FIGURE 30.3

A secondary oocyte, arrested at metaphase II of meiosis, is released at ovulation. If this cell is fertilized, it becomes an ovum, completes its second meiotic division, and produces a second polar body.

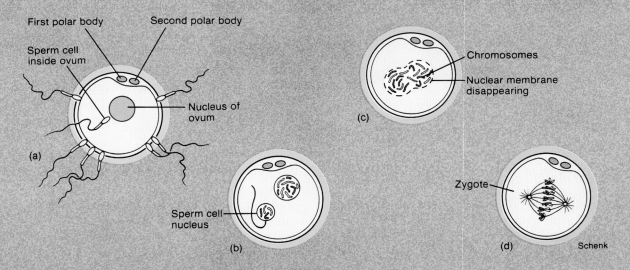

## FIGURE 30.4

Fertilization and the union of chromosomes from the sperm cell and ovum to form the zygote. (a) Sperm cell penetration, (b) the haploid number of chromosomes within the nucleus of each sex cell, (c) degeneration of nuclear membranes, and (d) matching and alignment of chromosomes.

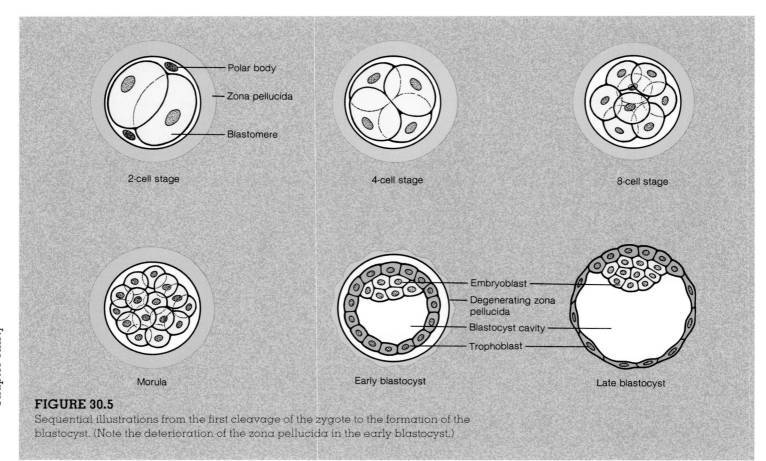

**FIGURE 30.5**

Sequential illustrations from the first cleavage of the zygote to the formation of the blastocyst. (Note the deterioration of the zona pellucida in the early blastocyst.)

# Pre-embryonic Period

*The events of the 2-week pre-embryonic period include fertilization, transportation of the zygote through the uterine tube, mitotic divisions, implantation, and the formation of primordial embryonic tissue.*

## Cleavage and Formation of the Blastocyst

Fertilization occurs within the uterine tube, usually about 12 to 24 hours following ovulation. The fertilized egg (ovum) is referred to as a **zygote.** Within 30 hours, the **cleavage** process begins with a mitotic division that results in the formation of two identical daughter cells called *blastomeres* (fig. 30.5). Several more cleavages occur as the structure passes down the uterine tube and enters the uterus on about the third day. It is now composed of a ball of 16 or more cells called a **morula.** Although the morula has undergone several

mitotic divisions, it is not much larger than the zygote because no additional nutrients necessary for growth have been entering the cells.

The developing structure remains unattached in the uterine cavity for about 3 days. During this time, the center of the morula fills with fluid passing in from the uterine cavity. As the fluid-filled space develops inside the morula, two distinct groups of cells form, and the structure becomes known as a **blastocyst.** The single outer layer forming the wall of the blastocyst is known as the **trophoblast,** whereas the small, inner aggregation of cells is called the **embryoblast,** or *internal cell mass.* With further development, the trophoblast differentiates into a structure called the **chorion,** which will become a portion of the placenta, and the embryoblast will become the embryo. The hollow, fluid-filled center of the blastocyst is called the **blastocyst cavity.** A diagrammatic summary of the ovarian cycle, fertilization, and the morphogenic events of the first week is presented in figure 30.6.

zygote: Gk. *zygotos,* yolked, joined

morula: L. *morus,* mulberry

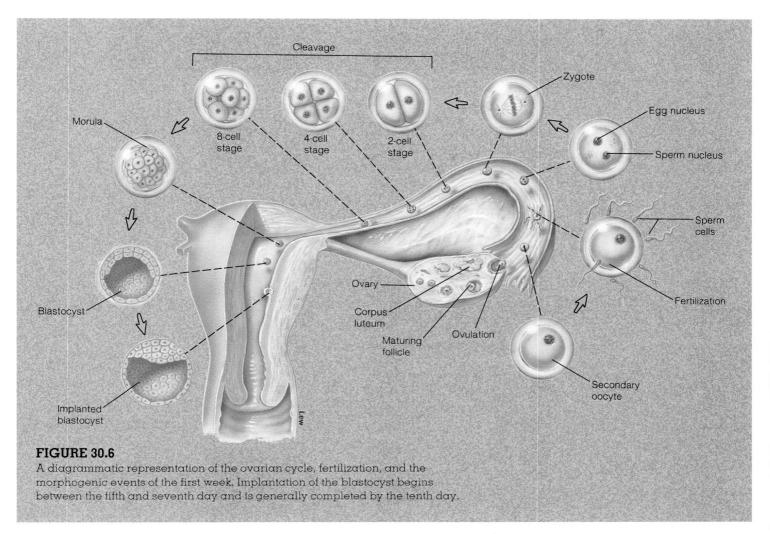

**FIGURE 30.6**
A diagrammatic representation of the ovarian cycle, fertilization, and the morphogenic events of the first week. Implantation of the blastocyst begins between the fifth and seventh day and is generally completed by the tenth day.

## Implantation

The process of **implantation,** or *nidation,* begins between the fifth and seventh day. Attachment is usually upon the posterior wall of the body of the uterus, with the side containing the embryoblast against the endometrium (fig. 30.7). Implantation is made possible by the secretion of *proteolytic enzymes* by the trophoblast, which digest a portion of the endometrium. The blastula sinks into the depression and endometrial cells move back to cover the defect in the wall. At the same time, the part of the uterine wall below the implanting blastocyst thickens, and specialized cells of the trophoblast produce fingerlike projections, called **syncytiotrophoblasts** (*sin-sit″e-ŏ-trof′ŏ-blasts*), into the thickened area. The syncytiotrophoblasts arise from a specific portion of the trophoblast called the **cytotrophoblast,** located at the embryonic pole.

The blastocyst saves itself from being aborted by secreting a hormone that indirectly prevents menstruation. Even before the sixth day when implantation begins, the syncytiotrophoblasts secrete **chorionic gonadotropin** (*kor″e-on′ik go-nad-ŏ-tro′pin*), or **hCG** (the *h* stands for *human*). This hormone is identical to LH in its effects, and therefore is able to maintain the corpus luteum past the time when it would otherwise regress. The secretion of estrogen and progesterone is maintained and menstruation is normally prevented (fig. 30.8).

The secretion of hCG declines by the tenth week of pregnancy. Actually, this hormone is required only for the first 5 to 6 weeks of pregnancy because the placenta itself becomes an active steroid-secreting gland by this time. At the fifth to sixth week, the mother's corpus luteum begins to regress (even in the presence of hCG), but the placenta secretes more than sufficient amounts of steroids to maintain the endometrium and prevent menstruation.

· · · · · · · · · · · · · · · · · · · · · · · · · · · · · · · · ·
implantation: L. *im,* in; *planto,* to plant
nidation: L. *nidus,* nest

All *pregnancy tests* assay for the presence of hCG in the blood or urine because this hormone is secreted by the blastocyst but not by the mother's endocrine glands. Modern pregnancy tests detect the presence of hCG by use of antibodies against hCG or by the use of cellular receptor proteins for hCG.

## Formation of Germ Layers

As the blastocyst completes implantation during the second week of development, the embryoblast undergoes marked differentiation. A slitlike space called the **amniotic** cavity forms between the embryoblast and the invading trophoblast (fig. 30.9). The embryoblast flattens into the **embryonic disc** (see fig. 30.11), which consists of two layers: an upper **ectoderm,** which is closer to the amniotic cavity, and a lower **endoderm,** which borders the blastocyst cavity. A short time later, a third layer called the **mesoderm** forms

........................................................

ectoderm: Gk. *ecto*, outside; *derm*, skin
endoderm: Gk. *endo*, within; *derm*, skin
mesoderm: Gk. *meso*, middle; *derm*, skin

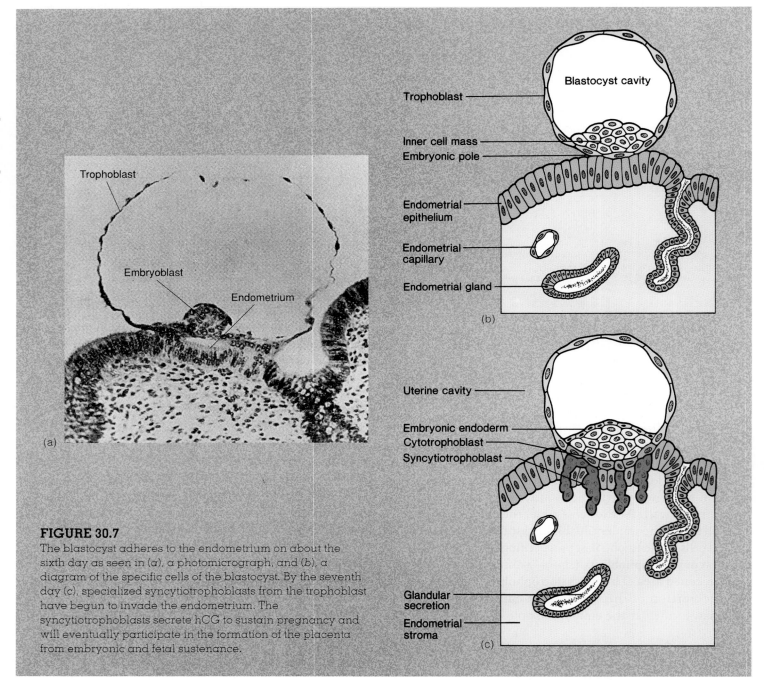

**FIGURE 30.7**
The blastocyst adheres to the endometrium on about the sixth day as seen in (*a*), a photomicrograph, and (*b*), a diagram of the specific cells of the blastocyst. By the seventh day (*c*), specialized syncytiotrophoblasts from the trophoblast have begun to invade the endometrium. The syncytiotrophoblasts secrete hCG to sustain pregnancy and will eventually participate in the formation of the placenta from embryonic and fetal sustenance.

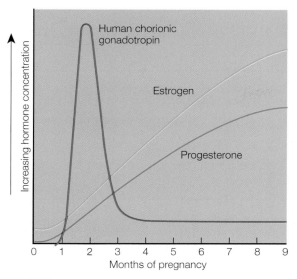

**FIGURE 30.8**

Human chorionic gonadotropin (hCG) is secreted by syncytiotrophoblasts during the first trimester of pregnancy. This hormone maintains the mother's corpus luteum for the first 5½ weeks. After that time the placenta becomes the major sex-hormone-producing gland, secreting increasing amounts of estrogen and progesterone throughout pregnancy.

between the endoderm and ectoderm. These three layers constitute the **primary germ layers.** Once they are formed, at the end of the second week, the pre-embryonic period is complete and the embryonic period begins.

The primary germ layers are important because various cells and tissues of the body are derived from them. Ectodermal cells form the nervous system; the outer layer of skin (epidermis), including hair, nails, and skin glands; and portions of the sensory organs. Mesodermal cells form the skeleton, muscles, blood, reproductive organs, dermis of the skin, and connective tissue. Endodermal cells produce the lining of the GI tract, the digestive organs, the respiratory tract and lungs, and the urinary bladder and urethra.

The events of the pre-embryonic period are summarized in table 30.1 and the derivatives of the three primary germ layers are listed in table 30.2. Refer to figure 30.10 for an illustration of the organs and body systems that derive from each of the primary germ layers.

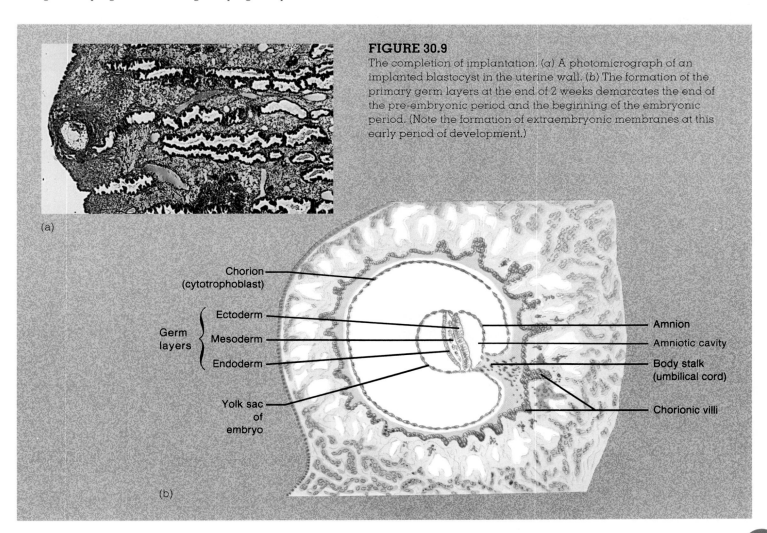

**FIGURE 30.9**

The completion of implantation. (a) A photomicrograph of an implanted blastocyst in the uterine wall. (b) The formation of the primary germ layers at the end of 2 weeks demarcates the end of the pre-embryonic period and the beginning of the embryonic period. (Note the formation of extraembryonic membranes at this early period of development.)

| Table 30.1 | Morphogenic stages and principal events of pre-embryonic development | |
|---|---|---|
| **Stage** | **Time period** | **Principal events** |
| Zygote | 24 to 30 hours following ovulation | Egg is fertilized; zygote has 23 pairs of chromosomes (diploid) from haploid sperm and haploid egg and is genetically unique |
| Cleavage | 30 hours to third day | Mitotic divisions produce increased number of cells |
| Morula | Third to fourth day | Hollow ball-like structure forms, a single layer thick |
| Blastocyst | Fifth day to end of second week | Embryoblast and trophoblast form; implantation occurs; embryonic disc forms, followed by primary germ layers |

| Table 30.2 | Derivatives of germ layers | |
|---|---|---|
| **Ectoderm** | **Mesoderm** | **Endoderm** |
| Epidermis of skin and epidermal derivatives: hair, nails, glands of the skin; linings of oral, nasal, anal, and vaginal cavities | Muscle: smooth, cardiac, and skeletal | Epithelium of pharynx, auditory canal, tonsils, thyroid, parathyroid, thymus, larynx, trachea, lungs, GI tract, urinary bladder and urethra, and vagina |
| Nervous tissue; sense organs | Connective tissue: embryonic, connective tissue proper, cartilage, bone, blood | |
| Lens of eye; enamel of teeth | Dermis of skin; dentin of teeth | Liver and pancreas |
| Pituitary gland | Epithelium of blood vessels, lymphatic vessels, body cavities, joint cavities | |
| Adrenal medulla | Internal reproductive organs | |
| | Kidneys and ureters | |
| | Adrenal cortex | |

The period of prenatal development is referred to as *gestation*. Normal gestation for humans is 9 months. Knowing this and the pattern of menstruation makes it possible to determine the delivery date of a baby. In a typical reproductive cycle, a woman ovulates 14 days prior to the onset of the next menstruation and is fertile for approximately 20 to 24 hours following ovulation. Adding 9 months, or 38 weeks, to the time of ovulation gives one the estimated delivery date, or parturition date.

# Embryonic Period

*The events of the 6-week embryonic period include the differentiation of the germ layers into specific body organs and the formation of the placenta, the umbilical cord, and the extraembryonic membranes. Through these morphogenic events, the needs of the embryo are met.*

The embryonic period lasts from the beginning of the third week to the end of the eighth week. At this period, the developing organism can correctly be called an **embryo.** During the embryonic period, all of the body organs form, as well as

..........................................

gestation: L. *gestatus,* to bear

the placenta, umbilical cord, and extraembryonic membranes. The term **conceptus** refers to the embryo, or fetus, and all of the extraembryonic structures—the products of conception. During parturition, the baby and the remaining portion of the conceptus are expelled from the uterus.

 *Embryology* is the study of the sequential changes in an organism as the various tissues, organs, and systems develop. Chick embryos are frequently studied because of the easy access through the shell and their rapid development. Mice and pig embryos are also extensively studied as mammalian models. Genetic manipulation, induction of drugs, exposure to disease, radioactive tagging or dyeing of developing tissues, and X-ray treatments are some of the commonly conducted experiments that provide information that can be applied to human development and birth defects.

During the pre-embryonic period of cell division and differentiation, the developing structure is self-sustaining. The embryo, however, is not self-supporting and must derive sustenance from the mother. For morphogenesis to continue, certain immediate needs must be met. These needs include: (1) formation of a vascular connection between the uterus of the mother and the embryo so that nutrients and oxygen can be provided and metabolic wastes and carbon

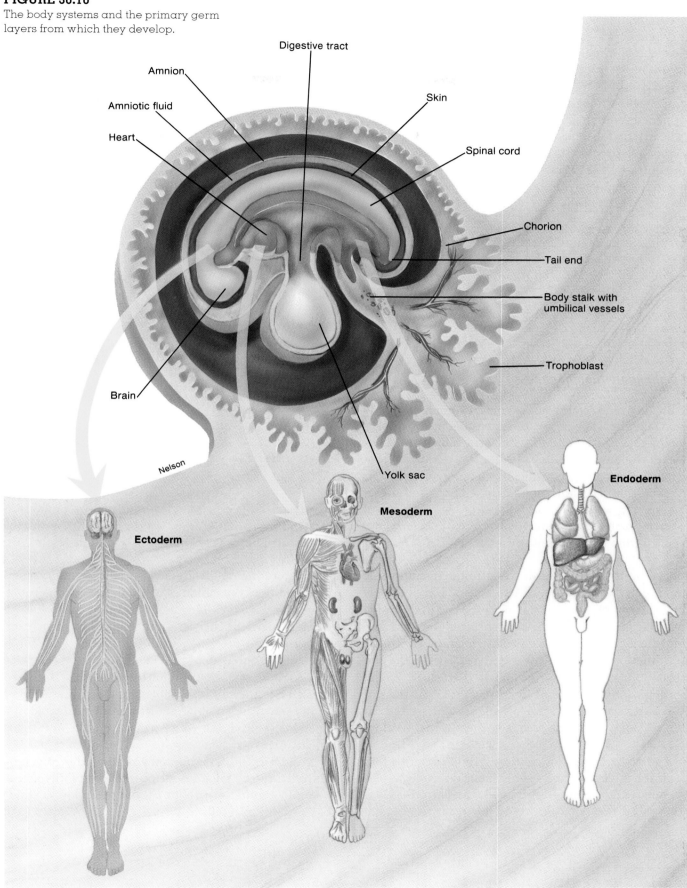

## FIGURE 30.10

The body systems and the primary germ layers from which they develop.

Amnion

Amniotic fluid

Heart

Digestive tract

Skin

Spinal cord

Chorion

Tail end

Body stalk with umbilical vessels

Trophoblast

Brain

Nelson

Yolk sac

Ectoderm

Mesoderm

Endoderm

dioxide can be removed; (2) establishment of a constant, protective environment around the embryo that is conducive to development; (3) establishment of a structural foundation for embryonic morphogenesis along a longitudinal axis; (4) provision for structural support for the embryo both internally and externally; and (5) coordination of the morphogenic events through genetic expression. If these needs are not met, a spontaneous abortion will generally occur.

The first and second of these needs are provided for by extraembryonic structures; the last three are provided for intraembryonically. The extraembryonic membranes, the placenta, and the umbilical cord will be considered separately, prior to a discussion of the development of the embryo.

 Most serious developmental defects cause the embryo to be naturally aborted. About 25% of early aborted embryos have chromosomal abnormalities. Other abortions may be caused by environmental factors, such as infectious agents or teratogenic drugs (drugs that cause birth defects). In addition, an implanting, developing embryo is regarded as foreign tissue by the immune system of the mother, and is rejected and aborted unless maternal immune responses are suppressed.

## Extraembryonic Membranes

While the many intraembryonic events are forming the body organs, a complex system of extraembryonic membranes is developing as well (fig. 30.11). The **extraembryonic membranes** are the amnion, yolk sac, allantois, and chorion. These membranes are responsible for the protection, respiration, excretion, and nutrition of the embryo and subsequent fetus. At parturition, the placenta, umbilical cord, and extraembryonic membranes separate from the fetus and are expelled from the uterus as the *afterbirth*.

**Amnion**   The amnion (*am′ne-on*) is a thin extraembryonic membrane derived from ectoderm and mesoderm. It loosely envelops the embryo, forming an **amniotic sac** that is filled with *amniotic fluid* (fig. 30.12). In later stages of fetal development, the amnion expands to come in contact with the chorion. Amniotic development is initiated early in the embryonic period, at which time its margin is attached around the free edge of the embryonic disc (fig. 30.11). As the amniotic sac enlarges during the late embryonic period (at about 8 weeks), the amnion gradually sheaths the developing umbilical cord with an epithelial covering (fig. 30.13).

As a buoyant medium, amniotic fluid performs four functions for the embryo and subsequent fetus: (1) it permits symmetrical structural development and growth, (2) it cushions and protects by absorbing jolts that the mother

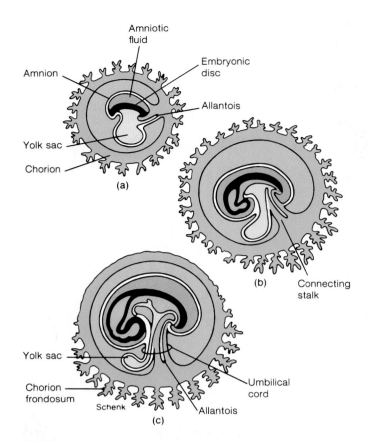

**FIGURE 30.11**

The formation of the extraembryonic membranes during a single week of rapid embryonic development: (*a*) At 3 weeks, (*b*) 3½ weeks, and (*c*) 4 weeks.

may receive, (3) it helps to maintain consistent pressure and temperature, and (4) it allows the fetus to develop freely, which is important for musculoskeletal development and blood flow.

Amniotic fluid is formed initially as an isotonic fluid absorbed from the maternal blood in the endometrium surrounding the developing embryo. Later, the volume is increased and the concentration changed by urine excreted from the fetus into the amniotic sac. Amniotic fluid also contains cells that are sloughed off from the fetus, placenta, and amniotic sac. Since all of these cells are derived from the same fertilized egg, all have the same genetic composition. Many genetic abnormalities can be detected by aspirating this fluid and examining the cells obtained in a procedure called *amniocentesis* (*am′′ne-o-sen-te′sis*).

Amniotic fluid is normally swallowed by the fetus and absorbed in the GI tract. Prior to delivery, the amnion is naturally or surgically ruptured, and the amniotic fluid (bag of waters) is released.

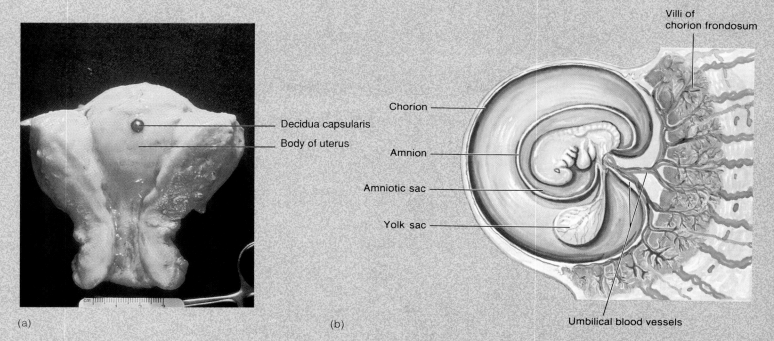

Decidua capsularis

Body of uterus

Villi of chorion frondosum

Chorion

Amnion

Amniotic sac

Yolk sac

Umbilical blood vessels

(a)                                        (b)

**FIGURE 30.12**

An implanted embryo at approximately 4½ weeks. (a) The interior of a uterus showing the implantation site and decidua capsularis elevated by the expanded chorion. (b) The developing embryo, extraembryonic membranes, and the formation of the placenta.

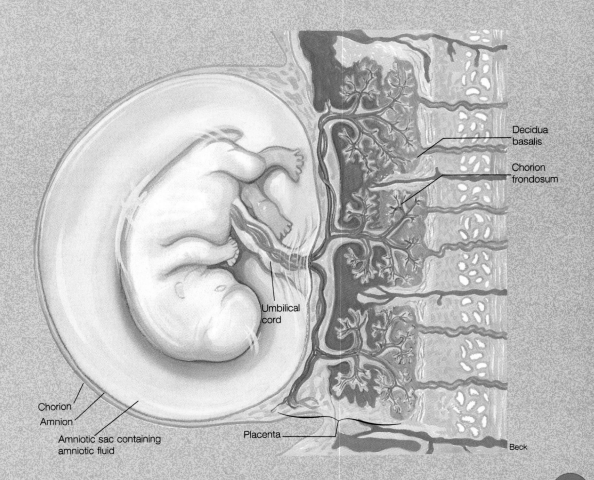

Decidua basalis

Chorion frondosum

Umbilical cord

Chorion

Amnion

Amniotic sac containing amniotic fluid

Placenta

Beck

**FIGURE 30.13**

The embryo, extraembryonic membranes, and placenta at approximately 7 weeks of development. At this time, the amnion and chorion are adherent and are frequently referred to as the amniochorionic membrane. Blood from the embryo is carried to and from the chorion frondosum by the umbilical arteries and vein. The maternal tissue between the chorionic villi is known as the decidua basalis; this tissue, together with the villi, form the functioning placenta.

909

*Amniocentesis* (fig. 30.14) is usually performed at the fourteenth or fifteenth week of pregnancy, when the amniotic sac contains 175–225 ml of fluid. Genetic diseases, such as *Down syndrome* (in which there are three instead of two number 21 chromosomes), can be detected by examining chromosomes. Diseases such as *Tay–Sachs disease*, in which there is a defective enzyme involved in formation of myelin sheaths, can be detected by biochemical techniques.

**Yolk Sac**  The yolk sac is established during the end of the second week as cells from the trophoblast form a thin *exocoelomic (ek″so-sĕ-lo´mik) membrane*. Unlike that of many vertebrates, the human yolk sac contains no nutritive yolk but is an essential structure during early embryonic development. It is attached to the underside of the embryonic disc (figs. 30.11 and 30.12), where it produces blood for the embryo until the liver forms during the sixth week. The dorsal portion of the yolk sac is involved in the formation of the primitive gut. In addition, primordial germ cells form in the wall of the yolk sac and migrate during the fourth week to the developing gonads, where they become the primitive germ cells (spermatogonia or oogonia).

The stalk of the yolk sac usually detaches from the gut by the sixth week. Following this, the yolk sac gradually shrinks as pregnancy advances. Eventually it becomes very small and serves no additional developmental functions.

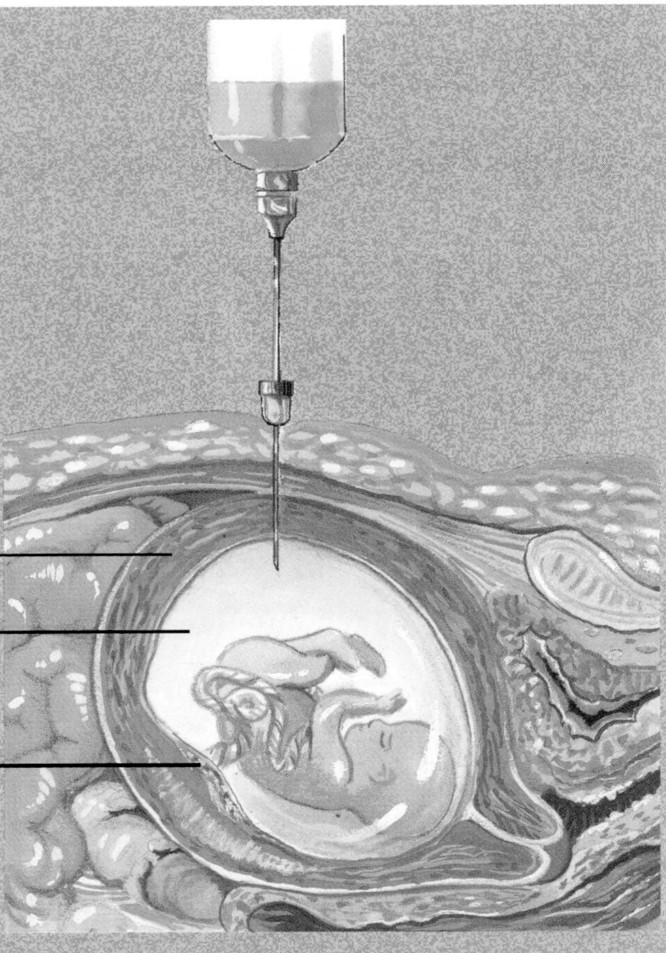

**Allantois**  The allantois forms during the third week as a small outpouching, or diverticulum, from the caudal wall of the yolk sac (see fig. 30.11). The allantois remains small but is involved in the formation of blood cells and gives rise to the fetal umbilical arteries and vein. It also contributes to the development of the urinary bladder.

The extraembryonic portion of the allantois degenerates during the second month. The intraembryonic portion involutes to form a thick urinary tube called the **urachus.** After birth, the urachus becomes a fibrous cord called the median umbilical ligament that attaches to the urinary bladder.

**Chorion**  The chorion is a highly specialized extraembryonic membrane that participates in the formation of the placenta (see fig. 30.11). It is the outermost membrane and originates from the trophoblast of the blastocyst. Numerous small, fingerlike extensions called **villi** (see fig. 30.12) form from the chorion and penetrate deeply into the uterine tis-

Uterine wall

Amniotic fluid

Placenta

**FIGURE 30.14**

Amniocentesis. In this procedure, amniotic fluid containing suspended cells is withdrawn for examination. Various genetic diseases can be detected prenatally by this means.

sue. Initially, the entire surface of the chorion is covered with villi. But those villi on the surface toward the uterine cavity gradually degenerate and produce a smooth, bare area known as the **smooth chorion.** As this occurs, the villi associated with the uterine wall rapidly increase in number and branch out. This portion of the chorion is known as the **villous chorion.** The villous chorion becomes highly vascular, and as the embryonic heart begins to function, blood is pumped in close proximity to the uterine wall.

 *Chorionic villus biopsy* is a technique used to detect genetic disorders much earlier than amniocentesis permits. In chorionic villus biopsy, a catheter is inserted through the cervix to the chorion, and a sample of chorionic villus is obtained by suction or cutting. Genetic tests can be performed directly on the villus sample, since this sample contains much larger numbers of fetal cells than does a sample of amniotic fluid. Chorionic villus biopsy can provide genetic information at 10 to 12 weeks' gestation.

Down syndrome: from John L. H. Down, English physician, 1828–96
allantois: Gk. *allanto*, sausage; *iodos*, resemblance
chorion: Gk. *chorion*, external fetal membrane

villous: L. *villus*, tuft of hair

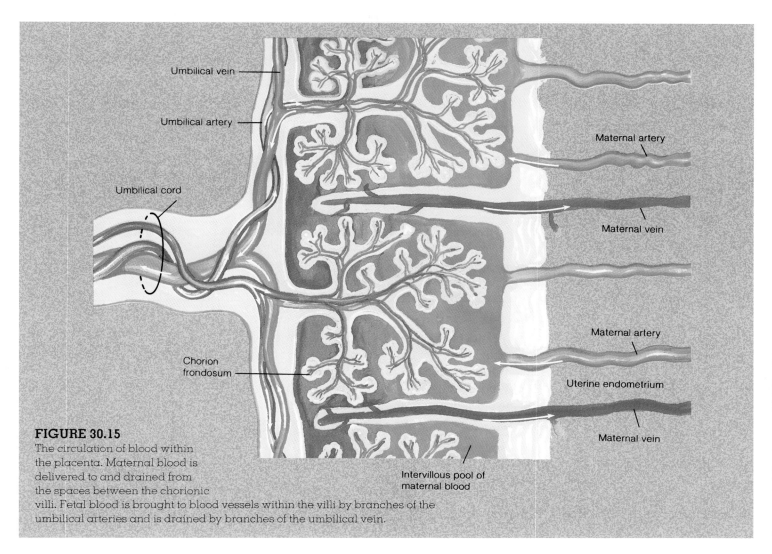

**FIGURE 30.15**

The circulation of blood within the placenta. Maternal blood is delivered to and drained from the spaces between the chorionic villi. Fetal blood is brought to blood vessels within the villi by branches of the umbilical arteries and is drained by branches of the umbilical vein.

Labels in figure: Umbilical vein; Umbilical artery; Umbilical cord; Chorion frondosum; Maternal artery; Maternal vein; Maternal artery; Uterine endometrium; Maternal vein; Intervillous pool of maternal blood

## Placenta

The **placenta** (*plă-cen´tă*) is a vascular structure by which an unborn child is attached to its mother's uterine wall and through which metabolic exchange occurs (fig. 30.15). The placenta is formed in part from maternal tissue and in part from embryonic tissue. The embryonic portion of the placenta consists of the villi of the **chorion frondosum,** whereas the maternal portion is composed of the area of the uterine wall called the **decidua basalis** (see fig. 30.13), into which the villi penetrate. Blood does not flow directly between these two portions, but because of the close membranous proximity, certain substances diffuse readily.

When fully formed, the placenta is a reddish brown oval disc with a diameter of 15 to 20 cm (8 in.) and a thickness of 2.5 cm (1 in.). It weighs 500 to 600 gm, about one-sixth the weight of the fetus.

......................................................

placenta: L. *placenta,* a flat cake

**Exchange of Molecules across the Placenta** The two **umbilical arteries** deliver fetal blood to vessels within the villi of the chorion frondosum of the placenta. This blood circulates within the villi and returns to the fetus via the **umbilical vein** (see fig. 21.37). Maternal blood is delivered to and drained from the cavities within the decidua basalis, which are located between the chorionic villi. In this way, maternal and fetal blood are brought close together but never mix within the placenta.

The placenta serves as a site for the exchange of gases and other molecules between the maternal and fetal blood. Oxygen diffuses from mother to fetus, and carbon dioxide diffuses in the opposite direction. Nutrient molecules and waste products likewise pass between maternal and fetal blood.

The placenta is not merely a passive conduit for exchange between maternal and fetal blood, however. It has a very high metabolic rate, utilizing about one-third of all the oxygen and glucose supplied by the maternal blood. The rate of protein synthesis is, in fact, higher in the placenta

# Table 30.3    Hormones secreted by the placenta

| Hormones | Effects |
| --- | --- |
| **Pituitary-like hormones** | |
| Chorionic gonadotropin (hCG) | Similar to LH; maintains mother's corpus luteum for first 5½ weeks of pregnancy; may be involved in suppressing immunological rejection of embryo; also exhibits TSH-like activity |
| Chorionic somatomammotropin (hCS) | Similar to prolactin and growth hormone; in the mother, hCS acts to promote increased fat breakdown and fatty acid release from adipose tissue and decreased glucose use by maternal tissues (diabetic-like effects) |
| **Sex steroids** | |
| Progesterone | Helps maintain endometrium during pregnancy; helps suppress gonadotropin secretion; promotes uterine sensitivity to oxytocin; helps stimulate mammary gland development |
| Estrogens | Helps maintain endometrium during pregnancy; helps suppress gonadotropin secretion; helps stimulate mammary gland development; inhibits prolactin secretion |

than it is in the liver. Like the liver, the placenta produces a great variety of enzymes capable of converting biologically active molecules (such as hormones and drugs) into less active, more water-soluble forms. In this way, potentially dangerous molecules in the maternal blood are often prevented from harming the fetus.

 Most drugs ingested by a pregnant woman can readily pass through the placenta and may be deleterious to the developing baby. For example, the nicotine taken in by a chain-smoking pregnant woman will stunt the growth of her fetus. Hard drugs, such as heroin, can lead to *fetal drug addiction*. Depressant drugs given to a mother during labor can readily cross the placenta and, if given in high dosages, can cause respiratory depression in the newborn infant.

## Endocrine Functions of the Placenta

The placenta secretes both steroid hormones and protein hormones that have actions similar to those of some anterior pituitary hormones. This latter category of hormones includes **chorionic gonadotropin (hCG)** and **chorionic somatomammotropin (hCS)** (table 30.3). Chorionic gonadotropin has LH-like effects, as previously described; it also has thyroid-stimulating ability, like pituitary TSH. Chorionic somatomammotropin likewise has actions that are similar to two pituitary hormones: growth hormone and prolactin. The placental hormones hCG and hCS thus duplicate the actions of four anterior pituitary hormones.

**Pituitary-Like Hormones from the Placenta**    The importance of chorionic gonadotropin in maintaining the mother's corpus luteum for the first 5½ weeks of pregnancy has been previously discussed. There is also some evidence that hCG may in some way help to prevent immunological rejection of the implanting embryo. Chorionic somatomammotropin synergizes (acts together) with growth hormone from the mother's pituitary to produce a diabetic-like effect in the pregnant woman. The effects of these two hormones include (1) accelerated lipolysis, and therefore increased plasma fatty acid concentrations; (2) decreased maternal utilization of glucose and, therefore increased blood glucose concentrations; and (3) polyuria (excretion of large volumes of urine), thereby producing a degree of dehydration and thirst. This diabetic-like effect in the mother helps to spare glucose for the placenta and fetus that, like the brain, use glucose as their primary energy source.

**Steroid Hormones from the Placenta**    After the first 5½ weeks of pregnancy, when the corpus luteum regresses, the placenta becomes the major sex-steroid-producing gland. The blood concentration of estrogens, as a result of placental secretion, rises to levels more than 100 times greater than those existing at the beginning of pregnancy. The placenta also secretes large amounts of progesterone, changing the estrogen/progesterone ratio in the blood from 100:1 at the beginning of pregnancy to a ratio of close to 1:1 toward full-term.

The placenta, however, is an incomplete endocrine gland because it cannot produce estrogen and progesterone without the aid of precursors supplied to it by both the mother and the fetus. The placenta, for example, cannot produce cholesterol from acetate and so must be supplied with cholesterol from the mother's circulation. Cholesterol, which is a steroid containing 27 carbons, can then be converted by enzymes in the placenta into steroids that contain 21 carbons—such as progesterone. The placenta, however, lacks the enzymes needed to convert progesterone into androgens (which have 19 carbons) and estrogens (which have 18 carbons).

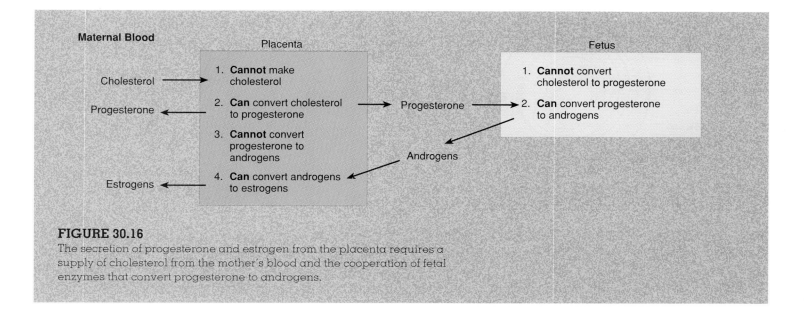

**FIGURE 30.16**
The secretion of progesterone and estrogen from the placenta requires a supply of cholesterol from the mother's blood and the cooperation of fetal enzymes that convert progesterone to androgens.

In order for the placenta to produce estrogens, it needs to cooperate with steroid-producing tissues in the fetus. Fetus and placenta, therefore, form a single functioning system in terms of steroid hormone production. This system has been called the **fetal-placental unit** (fig. 30.16).

The ability of the placenta to convert androgens into estrogen helps to protect the female embryo from becoming masculinized by the androgens secreted from the mother's adrenal glands. In addition to producing estradiol, the placenta secretes large amounts of a weak estrogen called **estriol.** The production of estriol increases tenfold during pregnancy, so that by the third trimester estriol accounts for about 90% of the estrogens excreted in the mother's urine. Since almost all of this estriol comes from the placenta (rather than from maternal tissues), measurements of urinary estriol can be used clinically to assess the health of the placenta.

## Umbilical Cord

The **umbilical cord** forms as the yolk sac shrinks and the amnion expands to envelop the tissues on the underside of the embryo. The formation of the umbilical cord is illustrated in figure 30.17.

The umbilical cord usually attaches near the center of the placenta. When fully formed, it is about 1 to 2 cm (0.5 to 1 in.) in diameter and approximately 55 cm (2 ft) long. The umbilical cord contains two **umbilical arteries,** which carry deoxygenated blood toward the placenta, and one **umbilical vein,** which carries oxygenated blood from the placenta to the embryo (see fig. 21.37). These vessels are surrounded by embryonic connective tissue called **mucoid connective tissue** (Wharton's jelly).

• • • • • • • • • • • • • • • • • • • • • • • • • • • • • • • •
Wharton's jelly: from Thomas Wharton, English anatomist, 1614–73

The umbilical cord has natural twists because the umbilical vein is longer than the arteries. In about one-fifth of all deliveries, the cord is looped once around the baby's neck. If drawn tightly, the cord may cause death or serious perinatal problems.

## Structural Changes of the Embryo by Weeks

**Third Week** Early in the third week a thick linear band called the **primitive line,** appears along the dorsal midline of the embryonic disc (see fig. 30.11). The primitive line is derived from the mesodermal cells, which are specialized from the ectodermal layer of the embryonic mass. The primitive line establishes a structural foundation for embryonic morphogenesis along a longitudinal axis. As the primitive line elongates, a prominent thickening called the **primitive node** appears at its cranial end (fig. 30.18). The primitive node later gives rise to the mesodermal structures of the head and the **notochord.** To give support to the embryo, the notochord forms a midline axis that is the basis of the embryonic skeleton. The primitive line also gives rise to loose embryonic connective tissue called **intraembryonic mesoderm** (mesenchyme), which differentiates into all the various kinds of connective tissue found in the adult. One of the earliest formed organs is the skin, which develops to support and maintain homeostasis within the embryo.

A tremendous amount of change and specialization occurs during the embryonic stage. The factors that cause precise, sequential change from one cell or tissue type to another are not fully understood. It is known, however, that the potential for change is programmed into the genetics of each cell and that under conducive environmental conditions these characteristics become expressed. The process of developmental change is referred

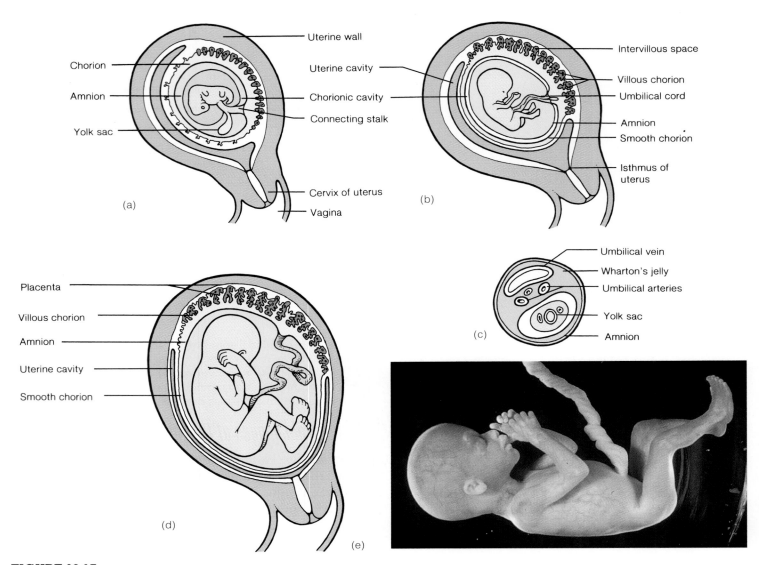

## FIGURE 30.17

The formation of the umbilical cord and other extraembryonic structures as seen in sagittal sections of the gravid uterus from week 4 to week 22. (*a*) A connecting stalk forms as the developing amnion expands around the embryo and finally meets ventrally. (*b*) The umbilical cord begins to take form as the amnion ensheathes the yolk sac. (*c*) A cross section of the umbilical cord showing the embryonic vessels, mucoid connective tissue, and the tubular connection to the yolk sac. (*d*) By week 22, the amnion and chorion have fused and the umbilical cord and placenta have become well-developed structures. (*e*) A 16-week-old fetus.

to as **induction.** Induction occurs when one tissue, called the **inductor tissue,** has a marked effect on an adjacent tissue, causing it to become **induced tissue** and stimulating it to differentiate.

**Fourth Week**  During the fourth week of development, the embryo increases about 4 mm in length. A **connecting stalk,** which is later involved in the formation of the umbilical

cord, is established from the body of the embryo to the developing placenta (fig. 30.19). By this time, the heart is pumping blood to all parts of the embryo. The head and jaws are apparent, and the primordial tissue that will form the eyes, brain, spinal cord, lungs, and digestive organs has developed. The **superior** and **inferior limb buds** are recognizable as small swellings on the lateral body walls.

**Fifth Week**  Embryonic changes during the fifth week are not as extensive as those during the fourth week. The head enlarges, and the developing eyes, ears, and nasal pit are

..........................................

induction: L. *inductus,* to lead in

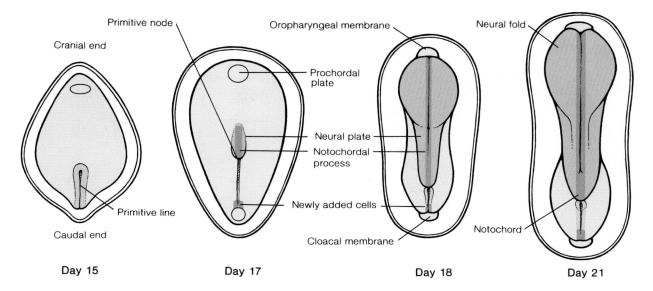

Cranial end

Primitive node

Oropharyngeal membrane

Neural fold

Prochordal plate

Neural plate

Notochordal process

Primitive line

Newly added cells

Caudal end

Cloacal membrane

Notochord

Day 15    Day 17    Day 18    Day 21

**◼ FIGURE 30.18**

The appearance of the primitive line and primitive node along the embryonic disc. These progressive changes occur through the process of induction.

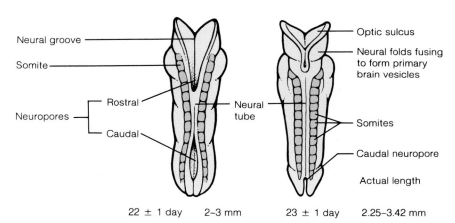

Neural groove

Optic sulcus

Somite

Neural folds fusing to form primary brain vesicles

Neuropores { Rostral / Caudal }

Neural tube

Somites

Caudal neuropore

Actual length

22 ± 1 day    2–3 mm    23 ± 1 day    2.25–3.42 mm

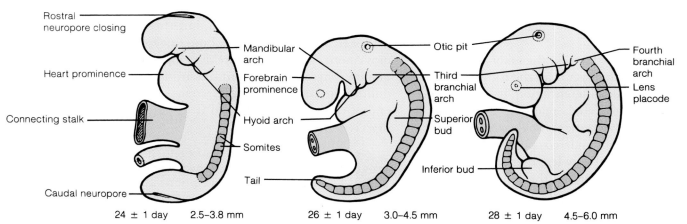

Rostral neuropore closing

Mandibular arch

Otic pit

Fourth branchial arch

Heart prominence

Forebrain prominence

Third branchial arch

Lens placode

Connecting stalk

Hyoid arch

Superior bud

Somites

Caudal neuropore

Tail

Inferior bud

24 ± 1 day    2.5–3.8 mm    26 ± 1 day    3.0–4.5 mm    28 ± 1 day    4.5–6.0 mm

**◼ FIGURE 30.19**

Progressive illustrations of 4-week-old embryos.

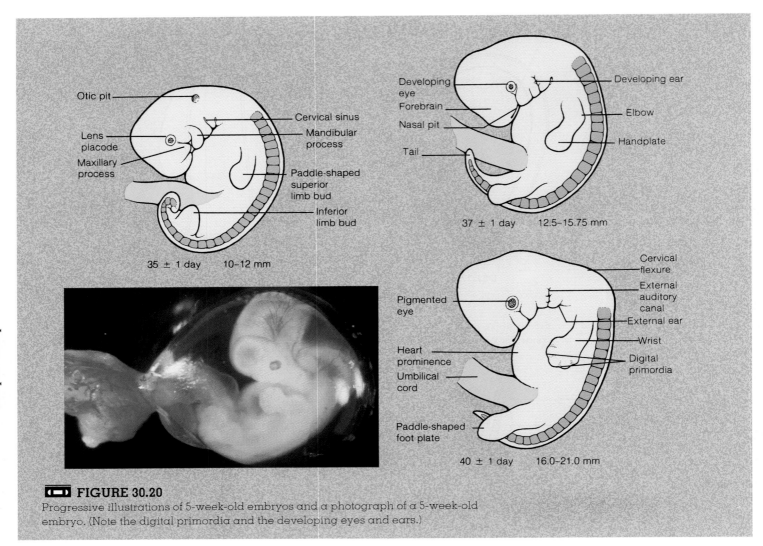

**FIGURE 30.20**

Progressive illustrations of 5-week-old embryos and a photograph of a 5-week-old embryo. (Note the digital primordia and the developing eyes and ears.)

obvious (fig. 30.20). The appendages have formed from the limb buds, and paddle-shaped hand plates develop digital ridges called **finger,** or **digital, primordia.**

**Sixth Week**   During the sixth week, the embryo is 16–24 mm long. The head is much larger than the trunk, and the brain has undergone marked differentiation. It is at this period of development that the vital organs are most vulnerable. An interruption at this critical time can easily cause congenital damage. The limbs undergo considerable change during this week. The forelimbs are lengthened and slightly flexed, and notches appear between the primordia in the hand and foot plates.

**Seventh and Eighth Weeks**   During the last 2 weeks of the embryonic stage, the embryo, which is now 28–40 mm long, has distinct human characteristics (figs. 30.21 and 30.22). The body organs are formed, and the nervous system begins coordinating body activity. The neck region is apparent, and the abdomen is less protuberant. The eyes are well developed, but the lids are stuck together to protect against probing fingers during muscular movement. The nostrils are developed but plugged with mucus. The external genitalia are forming but are still undifferentiated. The body systems are developed by the end of the eighth week, and from this time on the embryo is called a **fetus.**

The most precarious time of prenatal development is during the embryonic period, when there is much tissue differentiation and organ formation. Frequently, however, a woman does not even realize that she is pregnant until she is well into this period of development. For this reason, a woman should consistently take good care of herself and abstain from taking certain drugs (including some antibiotics) if there is even a remote chance that she is pregnant or might become pregnant in the near future.

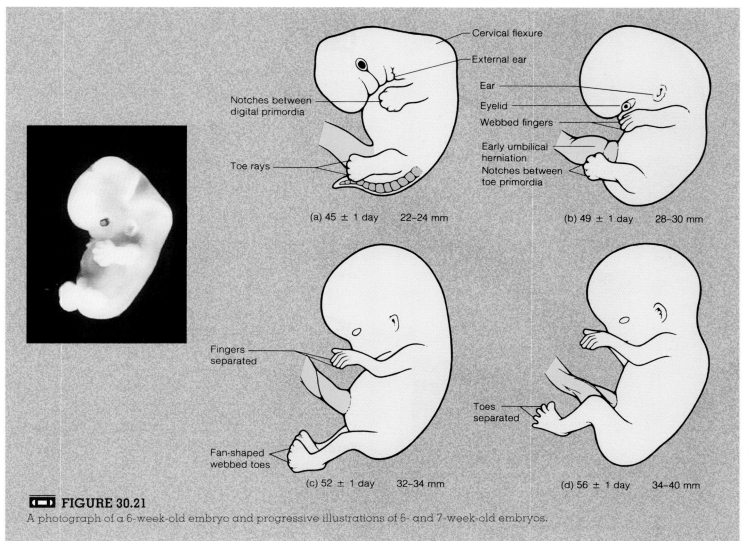

Cervical flexure
External ear
Notches between digital primordia
Toe rays

Ear
Eyelid
Webbed fingers
Early umbilical herniation
Notches between toe primordia

(a) 45 ± 1 day    22–24 mm

(b) 49 ± 1 day    28–30 mm

Fingers separated
Fan-shaped webbed toes

Toes separated

(c) 52 ± 1 day    32–34 mm

(d) 56 ± 1 day    34–40 mm

**FIGURE 30.21**

A photograph of a 6-week-old embryo and progressive illustrations of 6- and 7-week-old embryos.

# Fetal Period

*The fetal period, beginning at week 9 and culminating at birth, is characterized by tremendous growth and the specialization of body structures.*

Since most of the tissues and organs of the body become apparent during the embryonic period, the **fetus** is recognizable as a human being at 9 weeks and is far less vulnerable than the embryo to deformation from viruses, drugs, and radiation. A small amount of tissue differentiation and organ development still occurs during the fetal period but for the most part fetal development is primarily limited to body growth. Changes in external appearance of the fetus from the ninth through the thirty-eighth week are depicted in figure 30.23. The following is a discussion of the weekly structural changes of the fetus as it grows and matures.

**Nine to Twelve Weeks**    At the beginning of the ninth week, the head is as large as the rest of the body. The eyes are widely spaced, and the ears are set low. Head growth slows during the next 3 weeks, whereas lengthening of the body accelerates. Ossification centers appear in most bones during the ninth week. Differentiation of the external genitalia becomes apparent at the end of the ninth week, but the genitalia are not developed to the point of sex determination until the twelfth week. By the end of the twelfth week, the fetus is 87 mm (3.5 in.) long and weighs about 45 g (1.6 oz). It can swallow, digest the fluid that passes through its system, and defecate and urinate into the amniotic fluid. The nervous system and muscle coordination are developed enough so that the fetus will withdraw its leg if tickled. The fetus begins inhaling through its nose but can take in only amniotic fluid. The external appearance of fetuses at 10 and 12 weeks is depicted in figure 30.24.

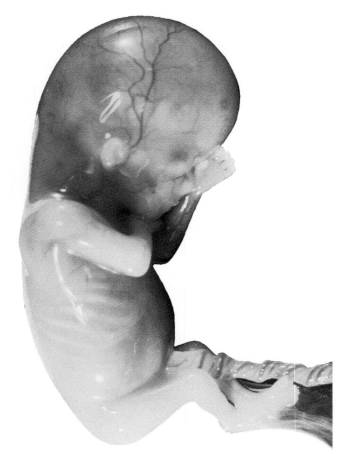

**FIGURE 30.22**
Photograph of an 8-week-old embryo. The body systems have developed by the end of the eighth week, and the embryo is recognizable as human.

 Major structural abnormalities, which may not be predictable from genetic analysis, can often be detected by *ultrasonography* (fig. 30.25). Sound-wave vibrations are reflected from the interface of tissues with different densities—for example, from the interface between the fetus and amniotic fluid—and used to produce an image. This technique is so sensitive that it can be used to detect a fetal heartbeat several weeks before it can be detected by a stethoscope.

**Thirteen to Sixteen Weeks** By the thirteenth week, the facial features are well formed, and epidermal structures such as eyelashes, eyebrows, hair on the head, fingernails, and nipples, begin to develop. The appendages lengthen, and by the sixteenth week the skeleton is sufficiently developed to show up clearly on radiographs. During the sixteenth week, the fetal heartbeat can be heard by applying a stethoscope to the mother's abdomen. By the end of the sixteenth week, the fetus is 140 mm long (5.5 in.) and weighs about 200 g (7 oz).

 After the sixteenth week, fetal length can be determined from radiographs. The reported length of a fetus is generally derived from a straight line measurement from the crown of the head to the developing ischium (crown-rump length). Measurements made on an embryo prior to the fetal stage, however, are not reported as crown-rump measurements but as total length.

**Seventeen to Twenty Weeks** Between the seventeenth and twentieth weeks, the legs achieve their final relative proportions and fetal movements, known as **quickening,** are commonly felt by the mother. The skin is covered with

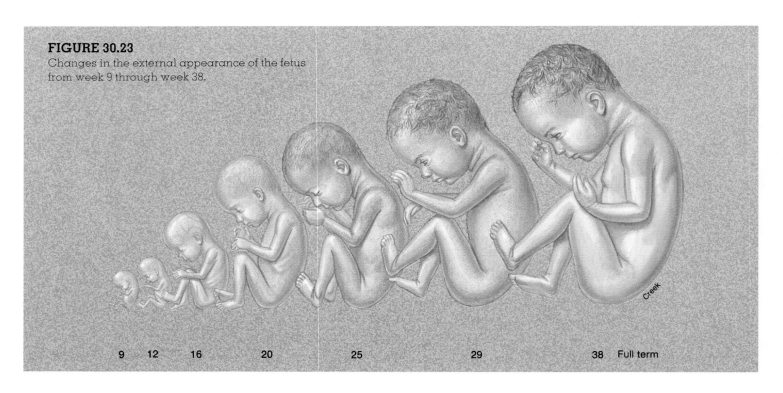

**FIGURE 30.23**
Changes in the external appearance of the fetus from week 9 through week 38.

| 9 | 12 | 16 | 20 | 25 | 29 | 38 Full term |

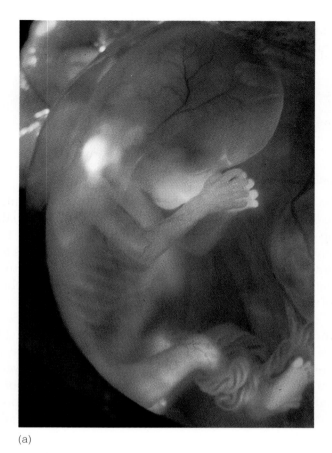

(a)

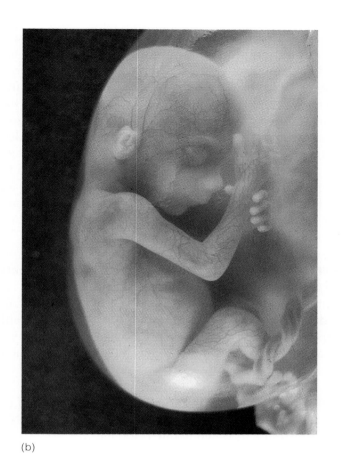

(b)

**FIGURE 30.24**
External appearances of fetuses (*a*) at 10 weeks and (*b*) at 12 weeks.

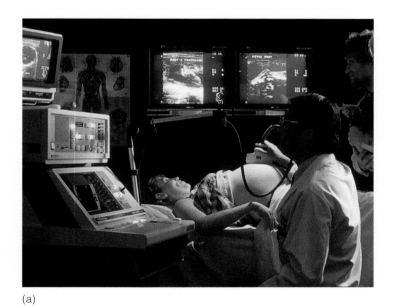

(a)

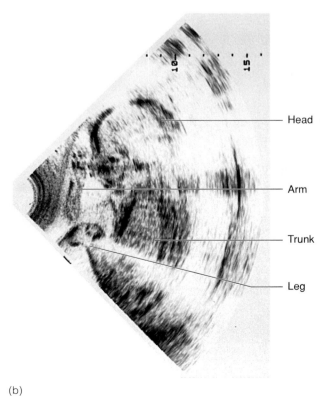

Head

Arm

Trunk

Leg

**FIGURE 30.25**
Ultrasonography. (*a*) Sound-wave vibrations are reflected from the internal tissues of a person's body. (*b*) Structures of the human fetus observed through an ultrasound scan.

(b)

a white, cheeselike material known as **vernix caseosa.** It consists of fatty secretions from the sebaceous glands and dead epidermal cells. The function of vernix caseosa is to protect the fetus while it is bathed in amniotic fluid. Twenty-week-old fetuses usually have fine, silklike fetal hair called **lanugo** (*lă-noo′go*) covering the skin. Lanugo is thought to hold the vernix caseosa on the skin and produce a ciliarylike motion that moves amniotic fluid. The length of a 20-week-old fetus is about 190 mm (7.5 in.) and it weighs about 460 gm (16 oz). Because of cramped space, the fetus develops a marked spinal flexure and is in what is commonly called the fetal position, with the head bent down in contact with the flexed knees.

**Twenty-One to Twenty-Five Weeks**   Between the twenty-first and twenty-fifth weeks, the fetus increases its weight substantially to about 900 gm (32 oz). Body length increases only moderately (to 240 mm), however, so the weight is evenly proportioned. The skin is quite wrinkled and is translucent. Because the blood flowing in the capillaries is now visible, the skin appears pinkish.

**Twenty-Six to Twenty-Nine Weeks**   Toward the end of the 26-to-29-week period, the fetus will be about 275 mm (11 in.) long and will weigh about 1,300 gm (46 oz). A fetus might now survive if born prematurely, but it will need help in adjusting to the outside environment. Its body metabolism cannot yet maintain a constant temperature, and the respiratory muscles have not matured enough to provide a regular respiratory rate. If, however, the premature infant is put in an incubator and a respirator is used to maintain its breathing, it may survive. The eyes open during this period, and the body is well covered with lanugo. If the fetus is a male, the testes should have begun descent into the scrotum. As the time of birth approaches, the fetus rotates to a **vertex position** (fig. 30.26). The head repositions toward the cervix because of the shape of the uterus and because the head is the heaviest part of the body.

**Thirty to Thirty-Eight Weeks**   At the end of 38 weeks, the fetus is considered full-term. It has reached a crown-rump length of 360 mm (14 in.) and weighs about 3,400 gm (7.5 lb). The average total length from crown to heel is 50 cm (20 in.). Most fetuses are plump with smooth skin because of the accumulation of subcutaneous fat. The skin is pinkish-blue, even in fetuses of dark-skinned parents, because melanocytes do not produce melanin until the skin is exposed to sunlight. Lanugo hair is sparse and is generally found on the head and back. The chest is prominent, and

the mammary area protrudes in both sexes. The external genitalia are somewhat swollen.

# Labor and Parturition

*Labor and parturition are the culmination of gestation and require the action of oxytocin, secreted by the posterior pituitary, and prostaglandins, produced in the uterus.*

The time of prenatal development, or the time of pregnancy, is called **gestation.** The human gestational period is usually 266 days or about 280 days from the beginning of the last menstrual period to **parturition,** or birth. Most fetuses are born within 10 to 15 days before or after this time. Parturition is accompanied by a sequence of physiological and physical events called **labor.**

The *onset of labor* is denoted by rhythmic and forceful contractions of the myometrial layer of the uterus (see table 30.4). In *true labor,* the pains from uterine contractions occur at regular intervals and intensify as the interval between contractions shortens. A reliable indication of true labor is dilation of the cervix and a show, or discharge, of blood-containing mucus in the cervical canal and vagina. In *false labor,* abdominal pain is experienced at irregular intervals, and cervical dilation and cervical show are absent.

The uterine contractions of labor are stimulated by two agents: (1) **oxytocin** (*ok″sĭ-to′sin*), a polypeptide hormone

••••••••••••••••••••••••••••••••

gestation: L. *gestatus,* to bear

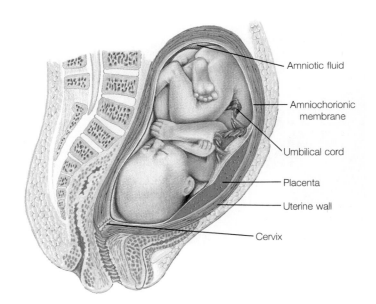

**FIGURE 30.26**

A fetus in vertex position. Toward the end of most pregnancies, the weight of the fetal head causes a rotation of its entire body such that the head is positioned in contact with the cervix of the uterus.

••••••••••••••••••••••••••••••••

vernix caseosa: L. *vernix,* varnish; L. *caseus,* cheese
lanugo: L. *lana,* wool
vertex: L. *vertex,* summit

## Table 30.4    Sequence of events leading to the onset of labor in humans

| Step | Event |
| --- | --- |
| 1 | High estrogen secretion from the placenta stimulates production of oxytocin receptors in the uterus. |
| 2 | Uterine muscle (myometrium) becomes increasingly sensitive to effects of oxytocin during pregnancy. |
| 3 | Oxytocin may stimulate production of prostaglandins in the uterus. |
| 4 | Prostaglandins may stimulate uterine contractions. |
| 5 | Contractions of the uterus stimulate oxytocin secretion from the posterior pituitary. |
| 6 | Increased oxytocin secretion stimulates increased uterine contractions, creating a positive feedback loop and resulting in labor. |

produced in the hypothalamus and released from the posterior pituitary and (2) **prostaglandins** (*pros˝tă-glan´dinz*), a class of fatty acids produced within the uterus itself. Labor can indeed be induced artificially by injections of oxytocin or by the insertion of prostaglandins into the vagina as a suppository.

 The hormone *relaxin*, produced by the corpus luteum, may also be involved in labor and parturition. Relaxin is known to soften the symphysis pubis in preparation for parturition and is thought to also soften the cervix in preparation for dilation. It may be, however, that relaxin does not affect the uterus, but rather that progesterone and estradiol may be responsible for this effect. Further research is necessary to understand the total physiological effect of these hormones.

As illustrated in figure 30.27, labor is divided into three stages:

**1 Dilation stage.** In this period, the cervix dilates to a diameter of approximately 10 cm. Contractions are regular during this stage and the amniotic sac (bag of waters) generally ruptures. If the amniotic sac does not rupture spontaneously, it is done surgically. The dilation stage generally lasts 8 to 24 hours.

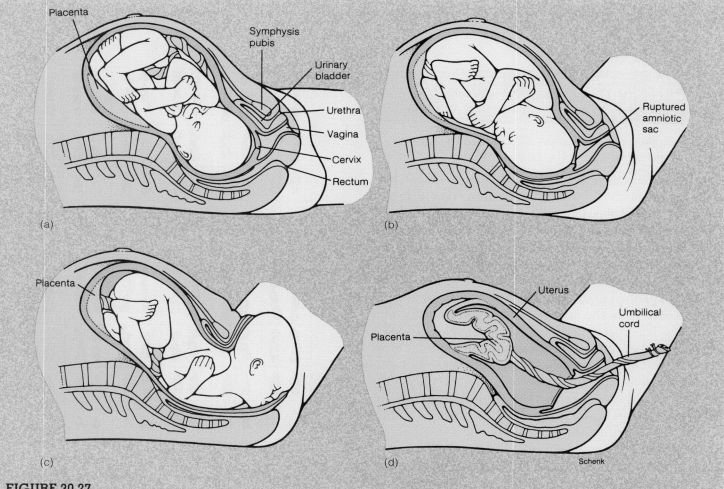

**FIGURE 30.27**
The stages of labor and parturition. (*a*) The position of the fetus prior to labor. (*b*) The ruptured amniotic sac and early dilation of the cervix. (*c*) The expulsion stage, or period of parturition. (*d*) The placental stage.

**2** **Expulsion stage.** This is the period of parturition, or actual childbirth. It consists of forceful uterine contractions and abdominal compressions to expel the fetus from the uterus and through the vagina. This stage may require 30 minutes in a first pregnancy but only a few minutes in subsequent pregnancies.

**3** **Placental stage.** Generally within 10 to 15 minutes after parturition, the placenta is separated from the uterine wall and expelled as the *afterbirth*. Forceful uterine contractions characterize this stage, constricting uterine blood vessels to prevent hemorrhage. In a normal delivery, blood loss does not exceed 350 ml.

A *pudendal nerve block* may be administered during the early part of the expulsion stage to ease the trauma of delivery for the mother and to allow for an episiotomy.

Five percent of newborns are born *breech*. In a breech birth, the fetus has not rotated and the buttocks are the presenting part. The principal concern of a breech birth is the increased time and difficulty of the expulsion stage of parturition. Attempts to rotate the fetus through the use of forceps may injure the infant. If an infant cannot be delivered breech, a *cesarean* (sĭ-zar´e-an) *section* must be performed. A cesarean section is delivery of the fetus through an incision made into the abdominal wall and the uterus.

# Inheritance

*Inheritance is the passage of hereditary traits carried by the genes on chromosomes from one generation to another.*

**Genetics** is the branch of biology that deals with inheritance. Genetics and inheritance are important in anatomy and physiology because of the numerous developmental and functional disorders that have a genetic basis. The knowledge of which disorders and diseases are inherited finds practical application in genetic counseling. The genetic inheritance of an individual begins with conception.

As explained in chapter 28, each zygote inherits 23 chromosomes from the mother and 23 chromosomes from the father. This does not produce 46 different chromosomes; rather, it produces 23 pairs of *homologous chromosomes*. With the important exception of the sex chromosomes, the members of a homologous pair appear to be structurally identical and contain similar genes. These homologous pairs of chromosomes can be **karyotyped** (photographed or illustrated) and identified (as shown in fig. 30.28). Each cell that contains 46 chromosomes (that is *diploid*) has two number 1 chromosomes, two number 2 chromosomes, and so on through chromosomes number 22. The first 22 pairs of chromosomes are called **autosomal chromosomes.** The twenty-third pair are the **sex chromosomes.** In a female these consist of two X chromosomes; in a male there is one X chromosome and one Y chromosome. The X and Y chromosomes look different and contain different genes.

**Genes and Alleles** A **gene** is the portion of the DNA of a chromosome that contains the information needed to synthesize a particular protein molecule. Although each diploid cell has a pair of genes for each characteristic, these genes may be present in a number of alternate forms. Those alternate forms of a gene that affect the same characteristic, but that produce different expressions of that characteristic, are called **alleles** (ă-lēlz). One allele of each pair originates from the female parent and the other from the male. The shape of a person's ears, for example, is determined by the kind of allele received from each parent and how the alleles interact. Alleles are always located on the same spot (called a **locus**) on homologous chromosomes (fig. 30.29).

For any particular pair of alleles in a person, the two alleles are either identical or not identical. If the alleles are identical, the person is said to be **homozygous** (ho´´mo-zi´gus) for that particular characteristic. But if the two alleles are different, the person is **heterozygous** (het´´er-o-zi´gus) for that particular trait.

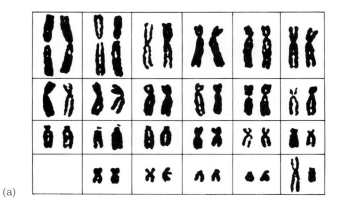

(a)

(b)

**FIGURE 30.28**
A karyotype of homologous pairs of chromosomes obtained from a human diploid cell. The first 22 pairs of chromosomes are called the autosomal chromosomes. The sex chromosomes are (a) XY for a male and (b) XX for a female.

**Genotype and Phenotype**    A person's DNA contains a catalog of genes known as the **genotype** of that person. The expression of those genes results in certain observable characteristics referred to as the **phenotype.**

If the alleles for a particular trait are homozygous, the characteristic expresses itself in a specific manner (two alleles for attached earlobes, for example, results in a person with attached earlobes). If the alleles for a particular trait are heterozygous, however, the allele that expresses itself and the way in which the genes for that trait interact will determine the phenotype. Often one of the alleles expresses itself as the **dominant allele,** while the other does not and is the **recessive allele.** The combinations of dominant and recessive alleles are responsible for a person's hereditary traits (table 30.5).

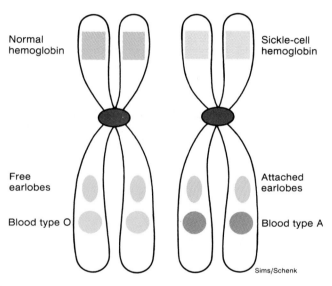

**FIGURE 30.29**

A pair of homologous chromosomes. Homologous chromosomes contain genes for the same characteristic at the same locus.

In describing genotypes, it is traditional to use letter symbols to refer to the alleles of an organism. The dominant alleles are symbolized by uppercase letters and the recessive alleles are symbolized by lowercase. Thus, the *genotype* of a person who is homozygous for free earlobes due to a dominant allele is symbolized *EE*; a heterozygous pair is symbolized *Ee.* In both of these instances, the *phenotypes* of the individuals would be free earlobes due to the presence of a dominant allele in each genotype. A person who inherited two recessive alleles for earlobes would have the genotype *ee* and would have attached earlobes.

Thus, three genotypes are possible when gene pairing involves dominant and recessive alleles. They are *homozygous dominant* (*EE*), *heterozygous* (*Ee*), and *homozygous recessive* (*ee*). Only two phenotypes are possible, however, since the dominant allele is expressed in both the homozygous dominant (*EE*) and the heterozygous (*Ee*) individuals. The recessive allele is expressed only in the homozygous recessive (*ee*) condition. Refer to figure 30.30 for an illustration of how a homozygous recessive trait may be expressed in a child of parents who are heterozygous dominant.

**Probability**    A **Punnett square** is a convenient way to express the probabilities of allele combinations for a particular inheritable trait. In constructing a Punnett square, the male gametes (spermatozoa) carrying a particular trait are placed at the side of the chart, and the female gametes (ova) at the top, as in figure 30.31. The four spaces on the chart represent the possible combinations of male and female gametes that could form zygotes. The probability of an offspring having a particular genotype is 1 in 4 (.25) for homozygous dominant and homozygous recessive and 1 in 2 (.50) for heterozygous dominant.

A genetic study in which a single characteristic (e.g., ear shape) is followed from parents to offspring is referred to as a **monohybrid cross.** A genetic study in which two

| Table 30.5 | Hereditary traits in humans determined by single pairs of dominant and recessive alleles | | |
|---|---|---|---|
| *Dominant* | *Recessive* | *Dominant* | *Recessive* |
| Free earlobes | Attached earlobes | Color vision | Color blindness |
| Dark brown hair | All other colors | Broad lips | Thin lips |
| Curly hair | Straight hair | Ability to roll tongue | Lack of this ability |
| Pattern baldness (♂♂) | Baldness (♀♀) | Arched feet | Flat feet |
| Pigmented skin | Albinism | A or B blood factor | O blood factor |
| Brown eyes | Blue or green eyes | Rh blood factor | No Rh blood factor |

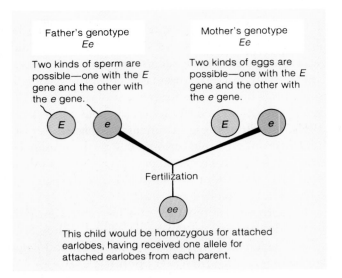

FIGURE 30.30

Inheritance of ear shape. Two parents with free earlobes can have a child with attached earlobes.

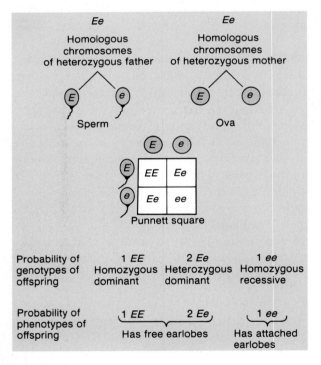

FIGURE 30.31

Inheritance of the shape of earlobes.

characteristics are followed from parents to offspring is referred to as a **dihybrid cross** (fig. 30.32). The term **hybrid** refers to an offspring descended from parents who have different genotypes.

### Sex-Linked Inheritance

Certain inherited traits are located on a sex-determining chromosome and are called **sex-linked characteristics.** The allele for *red-green color blindness,* for

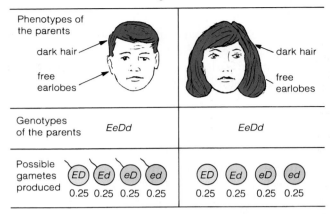

FIGURE 30.32

A dihybrid cross studies the probability of two characteristics at the same time. Any of the combinations of genes that have a *D* and an *E* (nine possibilities) will have free earlobes and dark hair. These are indicated with an asterisk (∗). Three of the possible combinations have two alleles for attached earlobes (*ee*) and at least one allele for dark hair. They are indicated with a dot (•). Three of the combinations have free earlobes and light hair. These are indicated with a square (■). The remaining possibility has the genotype *eedd* for attached earlobes and light hair.

example, is determined by a recessive allele (designated *c*) found on the X chromosome but not on the Y chromosome. Normal color vision (designated C) dominates. The ability to discern red-green colors, therefore, depends entirely on the X chromosomes. The genotype possibilities are

| | |
|---|---|
| $X^CY$ | Normal male |
| $X^cY$ | Color-blind male |
| $X^CX^C$ | Normal female |
| $X^CX^c$ | Normal female carrying the recessive allele |
| $X^cX^c$ | Color-blind female |

In order for a female to be red-green color blind, she must have the recessive allele on both of her X chromosomes. Her father would have to be red-green color blind, and her mother would have to be a carrier for this condition. A male

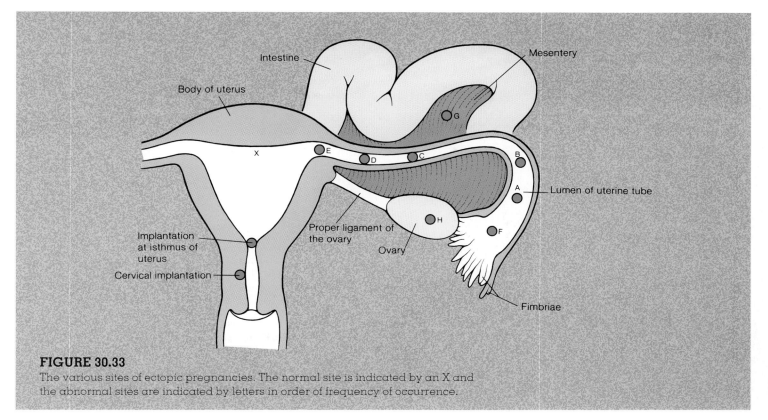

**FIGURE 30.33**
The various sites of ectopic pregnancies. The normal site is indicated by an X and the abnormal sites are indicated by letters in order of frequency of occurrence.

with only one such allele on his X chromosome, however, will show the characteristic. Since a male receives his X chromosome from his mother, the inheritance of sex-linked characteristics usually passes from mother to son.

 *Hemophilia* is a sex-linked condition caused by a recessive allele. The blood in a person with hemophilia fails to clot or clots very slowly after an injury. If *H* represents normal clotting and *h* represents abnormal clotting, then males with $X^HY$ will be normal and males with $X^hY$ will be hemophiliac. Females with $X^hX^h$ will have the disorder.

# Clinical Considerations

Pregnancy and childbirth are natural events in human biology and generally progress smoothly without complications. Prenatal development is amazingly precise, and although traumatic, childbirth for most women in the world takes place without the aid of a physician. Occasionally, however, serious complications arise, and the knowledge of an obstetrician is required. The physician's knowledge of what constitutes normal development and what factors are responsible for congenital malformations ensures the embryo and fetus every possible chance to develop normally. Many of the clinical aspects of prenatal development involve what might be referred to as applied developmental biology.

In clinical terms, gestation is frequently divided into three phases, or **trimesters,** each lasting three calendar months. By the end of the **first trimester** all of the major body systems are formed, the fetal heart can be detected, the external genitalia are developed, and the fetus is about the width of the palm of an adult's hand. During the **second trimester,** fetal quickening can be detected, epidermal features are formed, and the vital body systems are functioning. The fetus, however, would be still unlikely to survive if birth were to occur. At the end of the second trimester, fetal length is about equal to the length of an adult's hand. The fetus experiences a tremendous amount of growth and refinement in system functioning during the **third trimester.** A fetus of this age may survive if born prematurely, and of course, the chances of survival improve as the length of pregnancy approaches the natural delivery date.

Many clinical considerations are associated with prenatal development, some of which relate directly to the female reproductive system. Other developmental problems are genetically related and will be mentioned only briefly. Of clinical concern for developmental anatomy are such topics as ectopic pregnancies, so-called test-tube babies, multiple pregnancy, fetal monitoring, and congenital defects.

## Abnormal Implantation Sites

In an **ectopic pregnancy** the blastocyst implants outside the uterus or in an abnormal site within the uterus (fig. 30.33). About 95% of the time, the ectopic location is within the uterine tube and is referred to as a **tubal pregnancy.** Occasionally, implantation occurs near the cervix, where development of the placenta blocks the cervical opening. This condition, called **placenta previa,** causes serious bleeding. Ectopic pregnancies will not develop normally in unfavorable locations, and the fetus seldom survives beyond the first trimester. Tubal pregnancies are terminated through medical intervention. If a tubal pregnancy is permitted to progress, however, the uterine tube generally ruptures, followed by hemorrhaging. Depending on the location and the stage of development (hence vascularity) of a tubal pregnancy, it may or may not be life-threatening to the woman.

## In Vitro Fertilization and Artificial Implantation

Reproductive biologists have been able to fertilize a human oocyte in vitro (outside the body), culture it to the blastocyst stage, and then perform artificial implantation, leading to a full-term development and delivery. This is the so-called test-tube baby. To obtain the oocyte, a specialized laparoscope (fig. 30.34) is used to aspirate the preovulatory egg from a graafian follicle. The oocyte is then placed in a suitable culture medium, where it is fertilized with sperm. After the zygote forms, the sequential pre-embryonic development continues until the blastocyst stage, at which time implantation is performed. In vitro fertilization with artificial implantation is a means of overcoming infertility problems due to damaged, blocked, or missing uterine tubes in females or low sperm counts in males.

## Multiple Pregnancy

Twins occur about once in 85 pregnancies. They can develop in two ways. **Dizygotic (fraternal) twins** develop from two zygotes resulting from two spermatozoa fertilizing two oocytes in the same ovulatory cycle (fig. 30.35). **Monozygotic (identical) twins** form from a single zygote (fig. 30.36).

previa: L. *previa,* appearing before or in front of

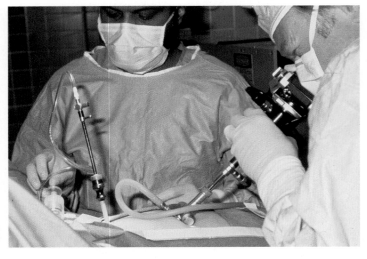

**FIGURE 30.34**

A laparoscope, used for various abdominal operations including the extraction of a preovulatory ovum.

Approximately one-third of twins are monozygotic.

Dizygotic twins may be of the same sex or different sexes and are not any more alike than brothers or sisters born at different times. Dizygotic twins always have two chorions and two amnions, but the chorions and the placentas may be fused.

Monozygotic twins are of the same sex and are genetically identical. Any physical differences in monozygotic twins are caused by environmental factors during morphogenic development (e.g., there might be a differential vascular supply that causes slight differences to be expressed). Monozygotic twinning is usually initiated toward the end of the first week when the embryoblast divides to form two embryonic primordia. Monozygotic twins have two amnions but only one chorion and a common placenta. If the embryoblast fails to completely divide, **conjoined twins** (Siamese twins) may form.

**Triplets** occur about once in 7600 pregnancies and may be (1) all from the same ovum and identical, (2) two identical and the third from another ovum, or (3) three zygotes from three different ova. Similar combinations occur in quadruplets, quintuplets, and so on.

## Fetal Monitoring

Obstetrics has benefited greatly from the advancements made in fetal monitoring in the last two decades. Before these techniques became available, physicians could determine the welfare of the unborn child only by auscultation of the fetal heart and palpation of the fetus. Currently, there are several tests that provide much information about the fetus during any stage of development. Fetal conditions that can now be diagnosed and evaluated include genetic disorders, hypoxia, blood disorders, growth retardation, prematurity, postmaturity, and intrauterine infections. These tests also help determine the advisability of an abortion.

Radiographs of the fetus were once commonly performed but were found harmful and have been replaced by other methods of evaluation that are safer and more informative. **Ultrasonography,** produced by a mechanical vibration of high frequency, produces a safe, high-resolution (sharp) image of fetal structure (fig. 30.37). Ultrasonic imaging is a reliable way to determine pregnancy as early as 6 weeks after ovulation. It can also be used to determine

## FIGURE 30.35

The formation of dizygotic twins. Twins of this type are not identical but fraternal and may have (a) separate or (b) fused placentas. (c) A photograph of fraternal twins at 11 weeks.

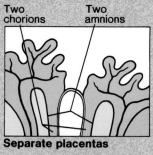

Two chorions   Two amnions

**Separate placentas**

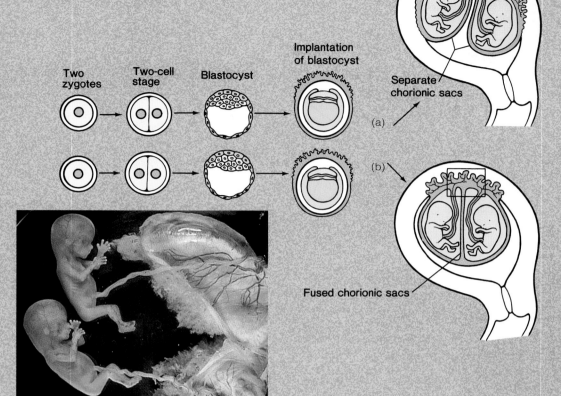

Two zygotes   Two-cell stage   Blastocyst   Implantation of blastocyst

(a)   Separate chorionic sacs

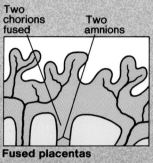

Two chorions fused   Two amnions

**Fused placentas**

(b)   Fused chorionic sacs

(c)

One placenta

One chorionic sac

Two amniotic sacs

Implantation of blastocyst

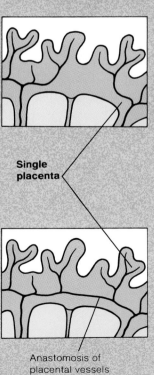

(a)

(b)

Single placenta

Anastomosis of placental vessels

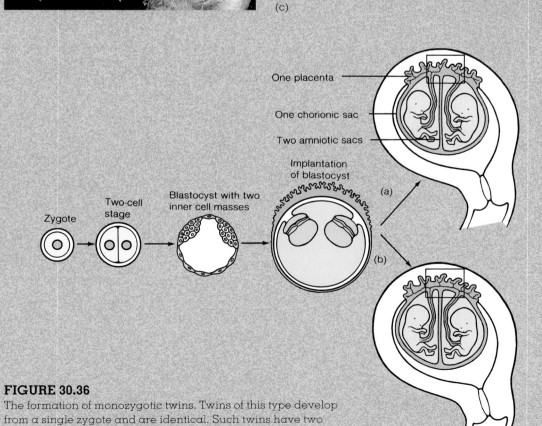

Zygote   Two-cell stage   Blastocyst with two inner cell masses   Implantation of blastocyst

## FIGURE 30.36

The formation of monozygotic twins. Twins of this type develop from a single zygote and are identical. Such twins have two amnions but one chorion and a common placenta.

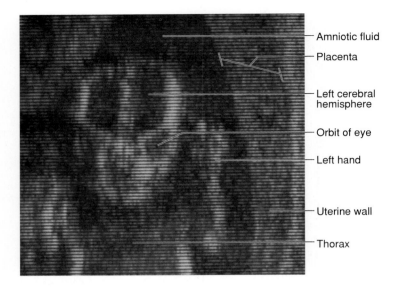

**FIGURE 30.37**
A color-enhanced ultrasonogram of a fetus during the third trimester. The left hand is raised, as if waving to the viewer.

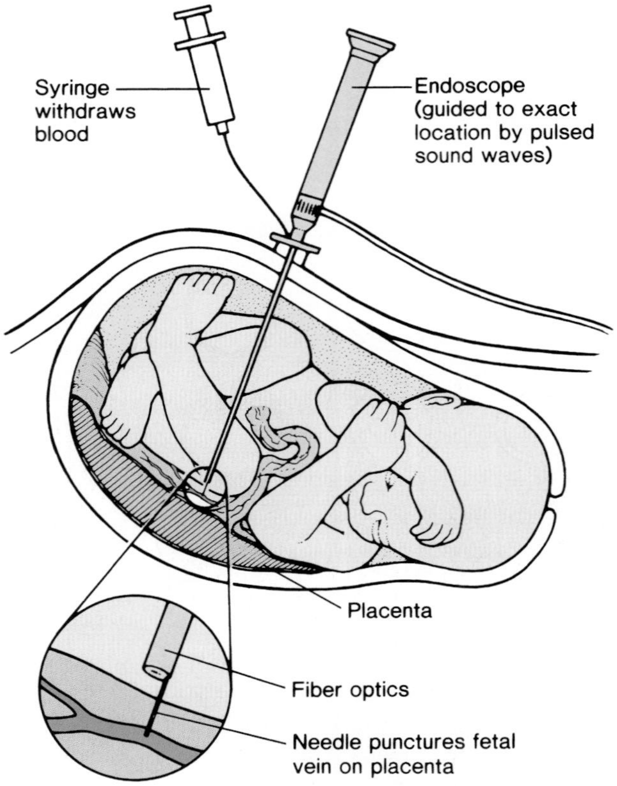

Labels: Amniotic fluid / Placenta / Left cerebral hemisphere / Orbit of eye / Left hand / Uterine wall / Thorax

fetal weight, length, and position, as well as to diagnose multiple fetuses.

**Amniocentesis** is a technique used to obtain a small sample (5–10 ml) of amniotic fluid with a syringe so that the fluid can be assessed (see fig. 30.14). Amniocentesis is most often performed to determine fetal maturity, but it can also help to predict such serious disorders as *Down syndrome* and *Gaucher's disease* (a metabolic disorder).

**Fetoscopy** (fig. 30.38) goes beyond amniocentesis by allowing direct examination of the fetus. Using fetoscopy, physicians scan the uterus with pulsed sound waves to locate fetal structures, the umbilical cord, and the placenta. Skin samples are taken from the head of the fetus and blood samples extracted from the placenta. The principal advantage of fetoscopy is that external features of the fetus (such as fingers, eyes, ears, mouth, and genitals) can be carefully observed. Fetoscopy is also used to determine several diseases, including hemophilia, thalassemia, and sickle-cell anemia cases, 40% of which are missed by amniocentesis.

Most hospitals are now equipped with instruments that monitor fetal heart rate and uterine contractions during labor. This procedure is called Electronic Monitoring of Fetal Heart Rate and Uterine Contractions (FHR-UC Monitoring). The extent of stress to the fetus from uterine

Labels (Figure 30.38): Syringe withdraws blood / Endoscope (guided to exact location by pulsed sound waves) / Placenta / Fiber optics / Needle punctures fetal vein on placenta

**FIGURE 30.38**
Fetoscopy.

contractions can be determined through FHR-UC monitoring (fig. 30.39). Long, arduous deliveries are taxing to both the mother and fetus. If the baby's health and vitality are diagnosed to be in danger because of a difficult delivery, the physician may decide to perform a cesarean section.

## Congenital Defects

Major developmental problems called **congenital malformations** occur in approximately 2% of all newborn infants. The causes of congenital conditions include genetic inheritance, mutation (genetic change), and environmental factors. About 15% of neonatal deaths are attributed to congenital malformations. The branch of developmental biology concerned with abnormal development and congenital malformations is called *teratology*. Many congenital problems have been discussed in previous chapters, in connection with the body system in which they occur.

· · · · · · · · · · · · · · · · · · · · · · · · · · · · · · · · · · · ·

amniocentesis: Gk. *amnion*, lamb (fetal membrane); *kentesis*, puncture
fetoscopy: L. *fetus*, offspring, *skopein*, to view

· · · · · · · · · · · · · · · · · · · · · · · · · · · · · · · · · · · ·

congenital: L. *congenitus*, born with
teratology: Gk. *teras*, monster; *logos*, study of

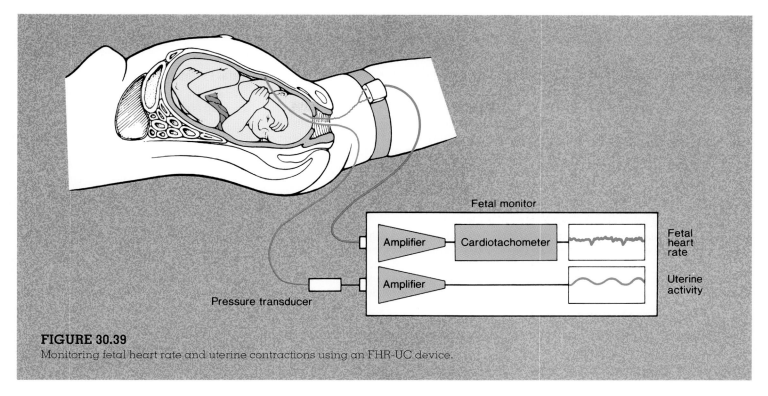

**FIGURE 30.39**
Monitoring fetal heart rate and uterine contractions using an FHR-UC device.

# Genetic Disorders of Clinical Importance

**cystic fibrosis**   An autosomal recessive disorder characterized by the formation of thick mucus in the lungs and pancreas that interferes with normal breathing and digestion.

**familial cretinism**   An autosomal recessive disorder characterized by a lack of thyroid secretion due to a defect in the iodine transport mechanism. Untreated children are dwarfed, sterile, and may be mentally retarded.

**galactosemia**   An autosomal recessive disorder characterized by an inability to metabolize galactose, a component of milk sugar. Patients with this disorder have cataracts, damaged livers, and mental retardation.

**gout**   An autosomal dominant disorder characterized by an accumulation of uric acid in the blood and tissue due to an abnormal metabolism of purines.

**hepatic porphyria**   An autosomal dominant disorder characterized by painful gastrointestinal disorders and neurologic disturbances due to an abnormal metabolism of porphyrins.

**hereditary hemochromatosis**   A sex-influenced, autosomal dominant disorder characterized by an accumulation of iron in the pancreas, liver, and heart, resulting in diabetes, cirrhosis, and heart failure.

**hereditary leukomelanopathy**   An autosomal recessive disorder characterized by decreased pigmentation in the skin, hair, and eyes and abnormal white blood cells. Patients with this condition are generally susceptible to infections and early deaths.

**Huntington's chorea**   An autosomal dominant disorder characterized by uncontrolled twitching of skeletal muscles and the deterioration of mental capacities. A latent expression of this disorder allows the mutant gene to be passed to children before the symptoms develop.

**Marfan's syndrome**   An autosomal dominant disorder characterized by tremendous growth of the extremities, extreme looseness of the joints, dislocation of the lenses, and congenital cardiovascular defects.

**phenylketonuria** (*fen´´il-kēt´´on-oor´e-ă*) **(PKU)**   An autosomal recessive disorder characterized by an inability to metabolize the amino acid phenylalanine. This is accompanied by brain and nerve damage and mental retardation.

**pseudohypertrophic muscular dystrophy**   A sex-linked recessive disorder characterized by progressive muscle atrophy. It usually begins during childhood and causes death in adolescence.

**retinitis pigmentosa**   A sex-linked recessive disorder characterized by progressive atrophy of the retina and eventual blindness.

**Tay–Sachs disease**   An autosomal recessive disorder characterized by a deterioration of physical and mental abilities, early blindness, and early death.

Huntington's chorea: from George Huntington, American physician, 1850–1916

Marfan's syndrome: from Antoine Bernard-Jean Marfan, French physician, 1858–1942
Tay–Sachs disease: from Warren Tay, English physician, 1843–1927, and Bernard Sachs, American neurologist, 1858–1944

# Chapter Summary

## Fertilization (pp. 899–901)

1. Upon fertilization of a secondary oocyte by a spermatozoon in the uterine tube, meiotic development is completed and a diploid zygote is formed.
2. Morphogenesis is the sequential formation of body structures during the prenatal period of human life. The prenatal period lasts 38 weeks and is divided into a pre-embryonic, an embryonic, and a fetal period.
3. A capacitated sperm digests its way through the zona pellucida and corona radiata layers of the secondary oocyte to complete the fertilization process and formation of a zygote.

## Pre-embryonic Period (pp. 902–906)

1. Cleavage of the zygote is initiated within 30 hours and continues until a morula forms; the morula enters the uterine cavity on about the third day.
2. A hollow, fluid-filled space forms within the morula, and it is then called a blastocyst.
3. Implantation begins between the fifth and seventh day and is enabled by the secretion of enzymes that digest a portion of the endometrium.
   a. During implantation, the trophoblast cells secrete human chorionic gonadotrophin (hCG), which prevents the breakdown of the endometrium and menstruation.
   b. The secretion of hCG declines by the tenth week as the developed placenta secretes steroids that maintain the endometrium.
4. The embryoblast of the implanted blastocyst flattens into the embryonic disc, from which the primary germ layers of the embryo develop.
   a. Ectoderm gives rise to the nervous system, the epidermis of the skin and epidermal derivatives, and portions of sensory organs.
   b. Mesoderm gives rise to bones, muscles, blood, reproductive organs, the dermis of the skin, and connective tissue.
   c. Endoderm gives rise to linings of the GI tract, digestive organs, the respiratory tract and lungs, and the urinary bladder and urethra.

## Embryonic Period (pp. 906–917)

1. The events of the 6-week embryonic period include the differentiation of the germ layers into specific body organs and the formation of the placenta, the umbilical cord, and the extraembryonic membranes. These events make it possible for morphogenesis to continue.
2. The extraembryonic membranes include the amnion, yolk sac, allantois, and chorion.
   a. The amnion is a thin membrane surrounding the embryo. It contains amniotic fluid that cushions and protects the embryo.
   b. The yolk sac produces blood for the embryo.
   c. The allantois also produces blood for the embryo and gives rise to the umbilical arteries and vein.
   d. The chorion participates in the formation of the placenta.
3. The placenta, formed from both maternal and embryonic tissue, has a transport role in providing for the metabolic needs of the fetus and in removing its wastes.
   a. The placenta produces steroid and polypeptide hormones.
   b. Nicotine, drugs, alcohol, and viruses can cross the placenta to the fetus.
4. The umbilical cord, containing two umbilical arteries and one umbilical vein, is formed as the amnion envelops the tissues on the underside of the embryo.
5. From the third to the eighth week, the structure of all the body organs, except that of the genitalia, becomes apparent.
   a. During the third week, the primitive node forms from the primitive line, which later gives rise to the notochord and intraembryonic mesoderm.
   b. By the end of the fourth week, the heart is beating; the primordial tissues of the eyes, brain, spinal cord, lungs, and digestive organs are properly positioned; and the superior and inferior limb buds are recognizable.
   c. At the end of the fifth week, the sense organs are formed in the enlarged head and the appendages have developed, with digital primordia evident.
   d. During the seventh and eighth weeks, the body organs, except for the genitalia, are formed and the embryo appears distinctly human.

## Fetal Period (pp. 917–920)

1. A small amount of tissue differentiation and organ development occurs during the fetal period, but for the most part fetal development is primarily limited to body growth.
2. Between weeks 9 and 12, ossification centers appear, the genitalia are formed, and the digestive, urinary, respiratory, and muscle systems show functional activity.
3. Between weeks 13 and 16, facial features are formed and the fetal heartbeat can be detected with a stethoscope.
4. During the 17-to-20-week period, quickening can be felt by the mother, and vernix caseosa and lanugo cover the skin of the fetus.
5. During the 21-to-25-week period, substantial weight gain occurs and the fetal skin becomes wrinkled and pinkish.
6. Toward the end of the 26-to-29-week period, the eyes have opened, the gonads have descended in a male, and the fetus is developed to the extent that it might survive if born prematurely.
7. At 38 weeks, the fetus is full-term; the normal gestation is 266 days.

## Labor and Parturition (pp. 920–922)

1. Labor and parturition are the culmination of gestation and require the action of oxytocin, secreted by the posterior pituitary, and prostaglandins, produced in the uterus.
2. Labor is divided into dilation, expulsion, and placental stages.

## Inheritance (pp. 922–925)

1. Inheritance is the passage of hereditary traits carried on the genes of chromosomes from one generation to another.
2. Each zygote contains 22 pairs of autosomal chromosomes and 1 pair of sex chromosomes—XX in a female and XY in a male.
3. A gene is the portion of a DNA molecule that contains information for the production of one kind of protein molecule. Alleles are different forms of genes that occupy corresponding positions on homologous chromosomes.

4. The combination of genes in an individual's cells constitutes his or her genotype; the appearance of a person is his or her phenotype.
   a. Dominant alleles are symbolized by uppercase letters and recessive alleles are symbolized by lowercase letters.
   b. The three possible genotypes are homozygous dominant, heterozygous, and homozygous recessive.

5. A Punnett square is a convenient means for expressing probability.
   a. The probability of a particular genotype is 1 in 4 (.25) for homozygous dominant and homozygous recessive and 1 in 2 (.50) for heterozygous dominant.

   b. A single trait is studied in a monohybrid cross; two traits are studied in a dihybrid cross.

6. Sex-linked traits like color blindness and hemophilia are carried on the sex-determining chromosome.

# Review Activities

## Objective Questions

1. The pre-embryonic period is completed when
   a. the blastocyst implants.
   b. the placenta forms.
   c. the blastocyst reaches the uterus.
   d. the primary germ layers form.
2. The yolk sac produces blood for the embryo until
   a. the heart is functional.
   b. the kidneys are functional.
   c. the liver is functional.
   d. the baby is delivered.
3. Which of the following is a function of the placenta?
   a. production of steroids and hormones
   b. diffusion of nutrients and oxygen
   c. production of enzymes
   d. all of the above apply.
4. The decidua basalis is
   a. a component of the umbilical cord.
   b. the embryonic portion of the villous chorion.
   c. the maternal portion of the placenta.
   d. a vascular membrane derived from the trophoblast.
5. Which of the following could diffuse across the placenta?
   a. nicotine
   b. alcohol
   c. heroin
   d. all of the above apply.

6. During which week following conception does the embryonic heart begin pumping blood?
   a. fourth week      c. sixth week
   b. fifth week       d. eighth week
7. Twins that develop from two zygotes resulting from the fertilization of two ova by two sperm in the same ovulatory cycle are referred to as
   a. monozygotic twins.
   b. conjoined twins.
   c. dizygotic twins.
   d. identical twins.
8. Match the genotype descriptions in the left-hand column with the correct symbols in the right-hand column.
   homozygous recessive      *Bb*
   heterozygous              *bb*
   homozygous dominant       *BB*
9. An allele that is *not* expressed in a heterozygous genotype is called
   a. recessive.        c. genotypic.
   b. dominant.         d. phenotypic.
10. If the genotypes of both parents are *Aa* and *Aa*, the offspring probably will be
    a. ½ *AA* and ½ *aa*.
    b. all *Aa*.
    c. ¼ *AA*, ½ *Aa*, ¼ *aa*.
    d. ¾ *AA* and ¼ *aa*.

## Essay Questions

1. Describe the implantation of the trophoblast into the uterine wall and its involvement in the formation of the placenta.
2. Explain how the primary germ layers form. What major structures does each germ layer give rise to?
3. Explain why development during the embryonic period is so critical and list the embryonic needs that must be met during the embryonic period for morphogenesis to continue.
4. State the approximate time period (in weeks) for the following occurrences:
   a. appearance of the arm and leg buds.
   b. differentiation of the external genitalia.
   c. perception of quickening by the mother.
   d. functioning of the embryonic heart.
   e. initiation of bone ossification.
   f. appearance of lanugo and vernix caseosa.
   g. survival of the fetus, if born prematurely.
   h. formation of all major body organs completed.
5. State the features of a genetic disorder that would lead one to believe that it was a form of sex-linked inheritance

# Answers to Objective Questions

**Chapter 1**
1. c
2. a
3. c
4. b
5. b
6. c
7. c
8. b
9. b
10. a

**Chapter 2**
1. c
2. b
3. a
4. d
5. c
6. b
7. c
8. d
9. d
10. b
11. d
12. b

**Chapter 3**
1. d
2. b
3. a
4. c
5. d
6. b
7. a
8. c
9. a
10. b
11. d
12. e
13. b
14. a

**Chapter 4**
1. b
2. d
3. d
4. d
5. e

6. e
7. d
8. b
9. a
10. c
11. e
12. d
13. c
14. b

**Chapter 5**
1. c
2. b
3. a
4. c
5. b
6. d
7. a
8. a
9. b
10. d
11. b

**Chapter 6**
1. b
2. c
3. a
4. b
5. d
6. a
7. c
8. a
9. b
10. d

**Chapter 7**
1. a
2. b
3. b
4. a
5. c
6. a
7. c
8. b
9. b
10. d

**Chapter 8**
1. c
2. a

3. b
4. d
5. b
6. a
7. d
8. b
9. b
10. c

**Chapter 9**
1. c
2. d
3. e
4. a
5. b
6. c
7. d
8. a
9. d
10. c
11. b
12. a
13. a
14. b
15. c
16. b

**Chapter 10**
1. a
2. c
3. d
4. b
5. b
6. d
7. e
8. b
9. d
10. a

**Chapter 11**
1. b
2. c
3. d
4. d
5. c
6. d
7. c
8. b
9. b
10. d

**Chapter 12**
1. b
2. d
3. c
4. a
5. c
6. b
7. b
8. d
9. a
10. e
11. c
12. b

**Chapter 13**
1. c
2. c
3. e
4. a
5. a
6. d
7. b
8. b
9. e
10. c

**Chapter 14**
1. c
2. d
3. a
4. a
5. c
6. d
7. d
8. a
9. c
10. c
11. b
12. d
13. d
14. a

**Chapter 15**
1. b
2. a
3. d
4. b
5. c
6. d
7. c

8. a
9. b
10. d

## Chapter 16
1. a
2. b
3. b
4. c
5. a
6. b
7. c
8. d
9. a
10. c

## Chapter 17
1. d
2. d
3. c
4. c
5. c
6. a
7. c
8. b
9. b
10. e
11. c
12. c

## Chapter 18
1. b
2. b
3. d
4. d
5. a
6. c
7. d
8. c
9. a
10. c
11. c
12. d
13. a
14. d
15. b
16. b
17. b
18. c

## Chapter 19
1. e
2. b
3. d
4. d
5. d
6. e
7. e
8. d
9. a
10. b

11. e
12. d
13. a
14. d
15. c

## Chapter 20
1. d
2. c
3. a
4. d
5. a
6. c
7. d
8. d
9. c
10. c

## Chapter 21
1. b
2. c
3. c
4. b
5. a
6. a
7. c
8. e
9. a
10. b
11. c
12. a
13. c
14. b

## Chapter 22
1. a
2. d
3. c
4. e
5. b
6. c
7. a
8. c
9. d
10. b
11. c
12. d
13. b
14. d
15. c
16. e

## Chapter 23
1. c
2. b
3. d
4. a
5. c
6. d
7. d
8. b

9. e
10. a
11. d
12. a
13. d
14. c

## Chapter 24
1. a
2. b
3. a
4. c
5. a
6. c
7. d
8. c
9. a
10. c
11. c
12. b
13. a
14. e
15. c
16. a
17. c
18. a
19. d
20. b

## Chapter 25
1. d
2. a
3. c
4. b
5. e
6. d
7. d
8. c
9. d
10. a
11. b
12. c
13. b
14. e
15. c
16. a

## Chapter 26
1. c
2. b
3. c
4. b
5. e
6. d
7. c
8. d
9. b
10. d
11. a
12. d

## Chapter 27
1. d
2. b
3. c
4. a
5. c
6. b
7. c
8. a
9. b
10. d
11. a
12. d
13. c
14. a
15. d

## Chapter 28
1. a
2. d
3. d
4. d
5. a
6. b
7. a
8. a
9. c
10. a

## Chapter 29
1. b
2. d
3. a
4. c
5. c
6. d
7. d
8. a
9. c
10. c
11. b
12. b
13. b
14. d
15. d

## Chapter 30
1. d
2. c
3. d
4. c
5. d
6. a
7. c
8. *bb*
   *Bb*
   *BB*
9. a
10. c

# Selected Readings

This appendix lists articles and books that may be useful to students who wish to deepen their understanding of particular topics in anatomy and physiology. The sequence of the 10 general content areas into which the references are grouped corresponds to the organization of the text.

## Cell Structure and Function

Afzelius, B. 1986. Disorders of ciliary motility. *Hospital Practice* 21:73.

Allen, R. D. 1987 (February). The microtubule as an intracellular engine. *Scientific American*.

Anderson, W. F. 1992. Human gene therapy. *Science* 256:808.

Bretscher, M. S. 1985 (October). The molecules of the cell membrane. *Scientific American*.

Brown, D. D. 1981. Gene expression in eukaryotes. *Science* 211:667.

Cech, T. R. 1986 (November). RNA as an enzyme. *Scientific American*.

Chambon, P. 1981 (May). Split genes. *Scientific American*.

Collins, F. S. 1993. The molecular biology of cystic fibrosis. *Annual Review of Medicine* 44:133.

Crick, F. 1962 (October). The genetic code. *Scientific American*.

Danielli, J. F. 1973. The bilayer hypothesis of membrane structure. *Hospital Practice* 8:63.

Darnell, J. E., Jr. 1985 (October). RNA. *Scientific American*.

Dautry-Varsal, A., and H. F. Lodish. 1984 (May). How receptors bring proteins and particles into cells. *Scientific American*.

DeDuve, C. 1983 (May). Microbodies in the living cell. *Scientific American*.

Doolittle, R. F. 1985 (October). Proteins. *Scientific American*.

Dustin, P. 1980 (August). Microtubules. *Scientific American*.

Felsenfled, G. 1985 (October). DNA. *Scientific American*.

Fox, C. F. 1972 (February). The structure of cell membranes. *Scientific American*.

Glover, D. M. et al. 1993 (June). The centrosome. *Scientific American*.

Grivell, L. A. 1983 (March). Mitochondrial DNA. *Scientific American*.

Grunstein, M., 1992 (October). Histones as regulators of genes. *Scientific American*.

Hayflick, H. 1980 (January). The cell biology of human aging. *Scientific American*.

Hinkle, P., and R. E. McCarty. 1979 (March). How cells make ATP. *Scientific American*.

Kornfeld, S., and W. S. Sly. 1985. Lysosomal storage defects. *Hospital Practice* 20:71.

Lake, J. A. 1981 (August). The ribosome. *Scientific American*.

Lazarides, E., and J. P. Ravel. 1979 (May). The molecular basis of cell movement. *Scientific American*.

Lienhard, G. E. et al. 1992 (January). How cells absorb glucose. *Scientific American*.

Lodish, H. F., and J. E. Rothman. 1979 (January). The assembly of cell membranes. *Scientific American*.

Mazia, D. 1974 (January). The cell cycle. *Scientific American*.

McKusick, V. A. 1981 The anatomy of the human genome. *Hospital Practice* 16:82.

Miller, A. W., and M. W. Kirschner, 1991 (March). What controls the cell cycle. *Scientific American*.

Miller, O. L., Jr. 1973 (March). The visualization of genes in action. *Scientific American*.

Mulligan, R. C. 1993. The basic science of gene therapy. *Science* 260:926.

Palade, G. 1975. Intracellular aspects of the process of protein synthesis. *Science* 189:347.

Rothman, J. E. 1985 (September). The compartmental organization of the Golgi apparatus. *Scientific American*.

Rothman, J. E., and L. Orci, 1992. Molecular dissection of the secretory pathway. *Nature* 355:409.

Sharon, N., and H. Lis, 1993 (January). Carbohydrates in cell recognition. *Scientific American*.

Singer, S. J. 1973. Biological membranes. *Hospital Practice* 8:81.

Singer, S. J., and G. L. Nicolson. 1972. The fluid mosaic model of the structure of cell membranes. *Science* 175:720.

Sloboda, R. D. 1980. The role of microtubules in cell structure and cell division. *American Scientist* 68:290.

Stein, G., J. S. Stein, and L. J. Kleinsmith. 1975 (February). Chromosomal proteins and gene regulation. *Scientific American*.

Wallace, D. C. 1986. Mitochondrial genes and disease. *Hospital Practice* 21:77.

Weinberg, R. A. 1985 (October). The molecules of life. *Scientific American*.

Wheeler, T. J., and P. C. Hinkle. 1985. The glucose transporter of mammalian cells. *Annual Review of Physiology* 47:503.

White, R., and J. M. Lalouel. 1988 (February). Chromosome mapping with DNA markers. *Scientific American*.

## Integumentary, Skeletal, and Muscular Systems

Astrand, P. O., and K. Rodhal. 1977. *Textbook of Work Physiology: Physiological Basis of Exercise*. New York: McGraw-Hill.

Bluefarb, S. M. 1974. *Dermatology*. Kalamazoo, MI: The Upjohn Company.

Booth, F. W., and B. S. Tseng. 1993. Olympic goal: Molecular and cellular approaches to understanding muscle adaptation. *News in Physiological Sciences* 8:165.

Bourne, G. H., ed. 1973. *The Structure and Function of Muscle*. 2d ed. 4 vols. New York: Academic Press.

Cohen, C. 1975 (November). The protein switch of muscle contraction. *Scientific American*.

Edelson, R. L., and J. M. Fink. 1985 (June). The immunologic function of the skin. *Scientific American*.

Evans, F. G., ed. 1966. *Studies in the Anatomy and Function of Bones and Joints*. New York: Springer-Verlag.

Felig, P., and J. Wahren. 1975. Fuel homeostasis in exercise. *New England Journal of Medicine* 293:1078.

Grinnel, A. D., and M.A.B. Brazier, eds. 1981. *Regulation of Muscle Contraction: Excitation-Contraction Coupling*. New York: Academic Press.

Hall, B. K. 1988 (March–April). The embryonic development of bone. *American Scientist*.

Hoyle, G. 1970 (April). How is muscle turned on and off? *Scientific American*.

Huxley, H. E. 1969. The mechanism of muscle contraction. *Science* 164:1356.

Lamb, G. D., and D. G. Stephenson. 1992. Importance of $Mg^{2+}$ in excitation-contraction coupling in skeletal muscles. *News in Physiological Sciences* 7:270.

Loomis, W. F. 1970 (December). Rickets. *Scientific American*.

Margaria, R. 1972 (March). The sources of muscular energy. *Scientific American*.

Marples, M. J. 1979 (January). Life on the human skin. *Scientific American*.

Merton, P. A. 1972 (May). How we control the contraction of our muscles. *Scientific American.*

Mitlak, B. H., and S. R. Nussbaum. 1993. Diagnosis and treatment of osteoporosis. *Annual Review of Medicine* 44:265.

Moncrief, J. A. 1973. Burns. *New England Journal of Medicine.* 228:444.

Montagna, W. 1969 (June). The skin. *Scientific American.*

Murray, J. H., and A. Weber. 1974 (February). The cooperative action of muscle proteins. *Scientific American.*

Nadel, E. R. 1985. Physiological adaptations to aerobic training. *American Scientist* 73:334.

Pawelek, J. M., and A. M. Korner. 1982 (March–April). The biosynthesis of mammalian melanin. *American Scientist.*

Prosser, C. L. 1992. Smooth muscle: Diversity and rhythmicity. *News in Physiological Sciences* 7:100.

Rasche, P. J., and R. K. Burke. 1978. *Kinesiology and Applied Anatomy: The Science of Human Movement.* 6th ed. Philadelphia: Lea and Febiger.

Ross, R. 1969 (June). Wound healing. *Scientific American.*

Rosse, C., and D. K. Clawson. 1980. *The Musculoskeletal System in Human Health and Disease.* Philadelphia: Harper and Row.

Rushmer, R. L. et al. 1966. The skin. *Science* 154:343.

Sharpe, W. D. 1979. Age changes in human bones: An overview. *Bulletin of the New York Academy of Medicine* 55:757.

Sims, S. M., and L. J. Janssen. 1993. Cholinergic excitation of smooth muscle. *News in Physiological Sciences* 8:207.

Sonstegard, D. A., L. S. Mathews, and H. Kaufer. 1979 (January). The surgical replacement of the human knee joint. *Scientific American.*

Vaughan, J. M. 1981. *The Physiology of Bone.* 3d ed. New York: Oxford University Press.

## Nervous System

Andreasen, N. C. 1988. Brain imaging: Applications in psychiatry. *Science* 239:1381.

Angevine, J. B., Jr., and C. Cottman. 1981. *Principles of Neuroanatomy.* New York: Oxford University Press.

Axelrod, J. 1974 (June). Neurotransmitters. *Scientific American.*

Bailey, C. H., and E. R. Kandel. 1993. Structural changes accompanying memory storage. *Annual Review of Physiology* 55:317.

Barchas, J. D. et al. 1978. Behavioral neurochemistry: Neuroregulators and behavioral states. *Science* 200:964.

Bartus, R. T. et al. 1982. The cholinergic hypothesis of geriatric memory dysfunction. *Science* 217:408.

Benfenati, F., and F. Valtorta. 1993. Synapsins and synaptic transmission. *News in Physiological Science* 8:18.

Blusztain, J. K., and R. J. Wurtman. 1983. Choline and cholinergic neurons. *Science* 221:614.

Brown, A. M. 1992. Ion channels in action potential generation. *Hospital Practice* 27:125.

Catteral, W. A. 1982. The molecular basis of neuronal excitability. *Science* 223:653.

Changeux, J.-P. 1993 (November). Chemical signaling in the brain. *Scientific American.*

Coyle, J. T., D. L. Prince, and M. R. DeLong. 1983. Alzheimer's disease: A disorder of cortical cholinergic innervation. *Science* 219:1184.

Damasio, A. R. 1992. Aphasia. *New England Journal of Medicine* 326:531.

Damasio, A. R., and H. Damasio. 1992 (September). Brain and language. *Scientific American.*

Dunant, Y., and M. Israel. 1985 (April). The release of acetylcholine. *Scientific American.*

Fine, A. 1986 (August). Transplantation in the central nervous system. *Scientific American.*

Frohman, L. A. 1975. Neurotransmitters as regulators of endocrine function. *Hospital Practice* 10:54.

Gershon, E. S., and R. O. Rieder. 1992 (September). Major disorders of mind and brain. *Scientific American.*

Goldman-Rakic, P. S. 1992 (September). Working memory and the mind. *Scientific American.*

Goldstein, G. W., and A. L. Betz. 1986 (September). The blood-brain barrier. *Scientific American.*

Gottlieb, D. I. 1988 (February). GABAergic neurons. *Scientific American.*

Horn, J. P. 1992. The heroic age of neurophysiology. *Hospital Practice* 27:65.

Hubel, D. H. 1979 (September). The brain. *Scientific American.*

Kalin, N. H. 1993 (May). The neurobiology of fear. *Scientific American.*

Kandel, E. R. 1979. Psychotherapy and the single synapse. *New England Journal of Medicine* 301:1028.

Kandel, E. R., and J. H. Schwartz, eds. 1981. *Principles of Neural Science.* New York: Elsevier North Holland.

Keynes, R. D. 1979 (March). Ion channels in the nerve cell membrane. *Scientific American.*

Kimura, D. 1992 (September). Sex differences in the brain. *Scientific American.*

Krieger, D. T. 1983. Brain peptides: What, where, and why? *Science* 222:975.

Kuffler, S. W., and J. G. Nicholls. 1976. *From Neuron to Brain: A Cellular Approach to the Function of the Nervous System.* Sunderland, MA: Sinauer Associates.

Lester, H. A. 1977 (February). The response of acetylcholone. *Scientific American.*

Martin, A. R. 1992. Principles of neuromuscular transmission. *Hospital Practice* 27:147.

Mishkin, M., and T. Appenzeller. 1987 (June). The anatomy of memory. *Scientific American.*

Moncada, S., and A. Higgs. 1993. The l-arginine–nitric oxide pathway. *New England Journal of Medicine* 329:2002.

Morrel, P., and W. Norton. 1980 (May). Myelin. *Scientific American.*

Motulski, J. H., and P. A. Insel. 1982. Adrenergic receptors in man. *New England Journal of Medicine* 307:18.

Nathason, J. A., and P. Greegard. 1977 (August). "Second messengers" in the brain. *Scientific American.*

Neher, E., and B. Sakmann, 1992 (March). The patch clamp technique. *Scientific American.*

Noback, C. E., and R. J. Demerest. 1975. *The Human Nervous System: Basic Principles of Neurobiology.* 2d ed. New York: McGraw-Hill.

Routtenberg, A. 1978 (November). The transport of substances in nerve cells. *Scientific American.*

Selkoe, D. J. 1991 (November). Amyloid protein and Alzheimer's disease. *Scientific American.*

Shashoua, V. E. 1985. The role of extracellular proteins in learning and memory. *American Scientist* 73:364.

Snyder, S. H. 1980. Brain peptides as neurotransmitters. *Science* 209:976.

Snyder, S. H. 1984. Drug and neurotransmitter receptors in the brain. *Science* 224:22.

Springer, S. P., and G. Deutch. 1985. *Left Brain, Right Brain.* Rev. ed. New York: W. H. Freeman.

Squire, L. R. 1986. Mechanisms of memory. *Science* 232:1612.

Squire, L. R., and S. Zola-Morgan. 1991. The medial temporal lobe memory system. *Science* 253:1380.

Stevens, C. F. 1979 (September). The neuron. *Scientific American.*

Sweeney, P. J. 1991. New concepts in Parkinson's disease. *Hospital Practice* 26:84.

Thompson, R. F. 1985. *The Brain.* New York: W. H. Freeman.

Thompson, R. F. 1986. The neurobiology of learning and memory. *Science* 233:941.

Tuomanen, E. 1993 (February). Breaching the blood-brain barrier. *Scientific American.*

Wagner, H. N. 1984. Imaging CNS receptors: The dopaminergic system. *Hospital Practice* 19:187.

Wurtman, R. J. 1982 (April). Nutrients that modify brain function. *Scientific American.*

Wurtman, R. J. 1985 (January). Alzheimer's disease. *Scientific American.*

Young, S. 1993. The body's vital poison: Importance of nitric oxide in biological processes. *New Scientist* 137:36.

## Sensory Organs

Borg, E., and S. A. Counter. 1989 (August). The middle ear muscles. *Scientific American.*

Botstein, D. 1986. The molecular biology of color vision. *Science* 232:142.

Boynton, R. M. 1979. *Human Color Vision.* New York: Holt, Rinehart, and Winston.

Casey, K. L. 1973. Pain: A current view of neural mechanisms. *American Scientist* 61:194.

Cervero, F., and J. M. A. Laird. 1991. One pain or many pains? A new look at pain mechanisms. *News in Physiological Science* 6:268.

Fireman, P. 1987. Newer concepts in otitis media. *Hospital Practice* 22:85.

Freeman, W. J. 1991 (February). The physiology of perception. *Scientific American*.

Freese, A. J. 1977. *The Miracle of Vision*. New York: Harper and Row.

Goldberg, J. M., and C. Fernandez. 1975. Vestibular mechanisms. *Annual Review of Physiology* 37:129.

Green, D. M. 1976. *An Introduction to Hearing*. New York: Lawrence Erlbaum Associates.

Hubel, D. H. 1979. The visual cortex of normal and deprived monkeys. *American Scientist* 67:532.

Hubel, D. H., and T. Wiesel. 1979 (September). Brain mechanisms of vision. *Scientific American*.

Hudspeth, A. J. 1983 (February). The hair cells of the inner ear. *Scientific American*.

Hudspeth, A. J. 1989. How the ear's works work. *Nature* 341:397.

Koretz, J. F., and G. H. Handelman. 1988 (July). How the human eye focuses. *Scientific American*.

Loeb, G. E. 1985 (February). The functional replacement of the ear. *Scientific American*.

Masland, R. H. 1987 (December). The functional architecture of the retina. *Scientific American*.

Melzack, R. 1992 (April). Phantom limbs. *Scientific American*.

Nathans, J. 1989 (February). The genes for color vision. *Scientific American*.

O'Brian, D. F. 1982. The chemistry of vision. *Science* 218:961.

Parker, D. E. 1980 (November). The vestibular apparatus. *Scientific American*.

Pettigrew, J. D. 1972 (August). The neurophysiology of binocular vision. *Scientific American*.

Pfaffmann, C., M. Frank, and R. Norgen. 1979. Neural mechanisms and behavioral aspects of taste. *Annual Review of Physiology* 30:283.

Rhode, W. S. 1984. Cochlear mechanics. *Annual Review of Physiology* 46:231.

Rushton, W. A. H. 1975 (March). Visual pigments and color blindness. *Scientific American*.

Schnapf, J. L., and D. A. Baylor. 1987 (April). How photoreceptor cells respond to light. *Scientific American*.

Stryer, L. 1987 (July). The molecules of visual excitation. *Scientific American*.

Van Essen, D. C. 1979. Visual areas of the mammalian cerebral cortex. *Annual Review of Neurosciences* 2:277.

Van Heyninger, R. 1975 (December). What happens to the human lens in cataract? *Scientific American*.

Von Bekesky, G. 1975 (August). The ear. *Scientific American*.

Zeki, S., 1992 (September). The visual image in mind and brain. *Scientific American*.

## Endocrine System

Axelrod, J., and T. D. Reisine. 1984. Stress hormones: Their interaction and regulation. *Science* 224:452.

Baxter, J. D., and W. J. Funder. 1979. Hormone receptors. *New England Journal of Medicine* 300:117.

Brownstein, M. J. et al. 1980. Synthesis, transport, and release of posterior pituitary hormones. *Science* 207:373.

Carmichael, S. W., and H. Winkler. 1985 (August). The adrenal chromaffin cell. *Scientific American*.

Cohick, W. S., and D. R. Clemmons. 1993. The insulin-like growth factors. *Annual Review of Physiology* 55:131.

Demers, L. M. 1984 (September). The effects of prostaglandins. *Diagnostic Medicine*.

Ebadi, M. et al. (1993). Pineal gland in synchronizing and refining physiological events. *News in Physiological Science* 8:30.

Ganong, W. F., L. C. Alpert, and T. C. Lee. 1974. ACTH and the regulation of adrenocortical secretion. *New England Journal of Medicine* 290:1006.

Gelato, M. C., and G. R. Merriam. 1986. Growth hormone releasing hormone. *Annual Review of Physiology* 48:569.

Gillie, R. B. 1971 (June). Endemic goiter. *Scientific American*.

Henry, J. P. 1993. Biological basis of the stress response. *News in Physiological Science* 8:69.

Katzenellenbogen, B. S. 1980. Dynamics of steroid hormone receptor action. *Annual Review of Physiology* 42:17.

McEwen, B. S. 1976 (July). Interactions between hormones and nerve tissue. *Scientific American*.

O'Malley, B., and W. T. Shrader. 1976 (February). The receptors of steroid hormones. *Scientific American*.

Quinn, S. J., and G. H. Williams. 1988. Regulation of aldosterone secretion. *Annual Review of Physiology* 50:409.

Rasmussen, H. 1986. The calcium messenger system. *New England Journal of Medicine* 314:1094, 1164.

Reisine, T. 1988. Neurohumoral aspects of ACTH release. *Hospital Practice* 23:77.

Reiter, R. J. 1991. Melatonin: That ubiquitously acting pineal hormone. *News in Physiological Science* 6:223.

Reiter, R. J. 1991 (January–February). Pineal gland: Interactions between the photoperiodic environment and the endocrine system. *Trends in Endocrinology and Metabolism*.

Roth, J., and S. I. Taylor. 1982. Receptors for peptide hormones: Alterations in diseases of humans. *Annual Review of Physiology* 44:639.

Schally, A. V. 1978. Aspects of the hypothalamic control of the pituitary gland. *Science* 202:18.

Selye, H. 1973. The evolution of the stress concept. *American Scientist* 61:693.

Thorner, M. O. 1986. Hypothalamic releasing hormones. *Hospital Practice* 21:63.

## Circulatory System

Atlas, S. A. 1986. Atrial natriuretic factor: Renal and systemic effects. *Hospital Practice* 21:67.

Berne, R. M., and M. N. Levy. 1981. *Cardiovascular Physiology*. 4th ed. St. Louis: C. V. Mosby.

Braunwald, E. 1974. Regulation of the circulation. *New England Journal of Medicine* 290:1124, 1420.

Brody, H. J., J. R. Haywood, and K. B. Toun. 1980. Neural mechanisms in hyptertension. *Annual Review of Physiology* 42:441.

Brown, M. S., and J. L. Goldstein. 1984 (November). How LDL receptors influence cholesterol and atherosclerosis. *Scientific American*.

Brown, M. S., and J. L. Goldstein. 1986. A receptor-mediated pathway for cholesterol homeostasis. *Science* 232:34.

Broze, G. J., Jr. 1992. Why do hemophiliacs bleed? *Hospital Practice* 27:71.

Brunner, H. R. 1990. The renin-angiotensin system in hypertension: An update. *Hospital Practice* 25:71.

Cantin, M., and J. Genest. 1986 (February). The heart as an endocrine gland. *Scientific American*.

Conover, M. B. 1980. *Understanding Electrocardiography*. 3d ed. St. Louis: C. V. Mosby.

Del Zoppo, G. J., and L. A. Harker. 1984. Blood/vessel interaction in coronary disease. *Hospital Practice* 19:163.

Donald, D. E., and J. T. Shepard. 1980. Autonomic regulation of the peripheral circulation. *Annual Review of Physiology* 42:429.

Dublin, D. 1981. *Rapid Interpretation of EKG's*. 3d ed. Tampa: Cover Publishing Company.

Eckberg, D. L., and J. M. Fritsch. 1993. How should human baroreflexes be tested? *News in Physiological Sciences* 8:7.

Fulkow, B. 1990. Salt and hypertension. *News in Physiological Sciences* 5:220.

Fulkow, B., and E. Neill. 1971. *Circulation*. London: Oxford University Press.

Fuster, V. et al. 1992. The pathogenesis of coronary artery disease and the acute coronary syndromes. *New England Journal of Medicine* 326:242.

Garcia, R. 1993. Atrial natriuretic factor in experimental and human hypertension. *News in Physiological Science* 8:161.

Gerard, J. M. 1988. Platelet aggregation: Cellular regulation and physiologic role. *Hospital Practice* 23:89.

Gewitz, H. 1991. The coronary circulation: Limitations of current concepts of metabolic control. *News in Physiological Sciences* 6:265.

Glasser, S. P., and R. G. Zobie. 1985. Management of cardiac arrhythmias. *Hospital Practice* 20:127.

Golde, D. W. 1991 (December). The stem cell. *Scientific American*.

Harken, A. H. 1993 (July). Surgical treatment of cardiac arrhythmias. *Scientific American*.

Herd, J. A. 1984. Cardiovascular response to stress in man. *Annual Review of Physiology* 46:177.

Hills, D., and E. Braunwald. 1977. Myocardial ischemia. *New England Journal of Medicine* 296:971, 1033, 1093.

Hilton, S. M., and K. M. Spyer. 1980. Central nervous regulation of vascular resistance. *Annual Review of Physiology* 42:399.

Jagannath, S. et al. 1993. Hematopoietic stem cell transplantation. *Hospital Practice* 28:79.

Katz, A. M. 1987. A physiologic approach to the treatment of heart failure. *Hospital Practice* 22:117.

Kontos, H. A. 1981. Regulation of the cerebral circulation. *Annual Review of Physiology* 43:397.

Laragh, J. H. 1985. Atrial natriuretic hormone, the renin-aldosterone axis, and blood pressure–electrolyte homeostasis. *New England Journal of Medicine* 313:1330.

Lawn, R. M. 1992 (June). Lipoprotein(a) in heart disease. *Scientific American*.

Leon, A. S. 1983. Exercise and coronary heart disease. *Hospital Practice* 18:38.

Little, R. C. 1981. Physiology of the heart and circulation. 2d ed. Chicago: Year Book Medical Publishers.

Nadel, E. R. 1985. Physiological adaptations to aerobic training. *American Scientist* 73:334.

Needleman, P., and J. E. Greenwald. 1986. Atriopeptin: A cardiac hormone intimately involved in fluid, electrolyte, and blood pressure homeostasis. *New England Journal of Medicine* 314:828.

Olsson, R. A. 1981. Local factors regulating cardiac and skeletal muscle blood flow. *Annual Review of Physiology* 43:385.

Oparil, S., and J. M. Wyss. 1993. Atrial natriuretic factor in central cardiovascular control. *News in Physiological Science* 8:223.

Pelleg, A. 1993. Adenosine in the heart: Its emerging roles. *Hospital Practice* 28:71.

Porzig, H. 1991. Signaling mechanisms in erythropoiesis: New insights. *News in Physiological Science* 6:247.

Roberts, H. R., and J. N. Lozier. 1992. New perspectives on the coagulation cascade. *Hospital Practice* 27:97.

Robinson, T. F., S. M. Factor, and E. H. Sonnenblick, 1986 (June). The heart as a suction pump. *Scientific American*.

Ross, R. 1993. The pathogenesis of atherosclerosis: A perspective for the 1990s. *Nature* 362:801.

Said, S. I. 1992. Nitric oxide and vasoactive intestinal peptide: Cotransmitters of smooth muscle relaxation. *News in Physiological Sciences* 7:181.

Smith, J. J., and J. P. Kampine. 1980. *Circulatory Physiology: The Essentials*. Baltimore: Williams and Wilkins Company.

Snyder, S. H., and D. S. Bredt. 1992 (May). Biological roles of nitric oxide. *Scientific American*.

Spear, J. F., and E. N. Moore. 1982. Mechanisms of cardiac arrhythmias. *Annual Review of Physiology* 44:485.

Stephenson, R. B. 1984. Modification of reflex regulation of blood pressure by behavior. *Annual Review of Physiology* 46:133.

Toda, N., and T. Okamura. 1992. Regulation by nitroxidergic nerve of arterial tone. *News in Physiological Science* 7:148.

Vatner, S. F., and E. Braunwald. 1975. Cardiovascular control mechanisms in the conscious state. *New England Journal of Medicine* 293:970.

Weber, K. T., J. S. Janicki, and W. Laskey. 1983. The mechanics of ventricular function. *Hospital Practice* 18:113.

Zellis, R. S., S. F. Flaim, A. J. Liedke, and S. H. Nellis. 1981. Cardiovascular dynamics in the normal and failing heart *Annual Review of Physiology* 43:455.

Zivin, J. A., and D. W. Choi. 1991 (July). Stroke therapy. *Scientific American*.

Zucker, M. B. 1980 (June). The function of blood platelets. *Scientific American*.

## Lymphatic System and Immunity

Acuto, O., and E. Reinhertz. 1985. The human T cell receptor: Structure and function. *New England Journal of Medicine* 312:1100.

Ada, G. L., and G. Nossal. 1987 (August). The clonal selection theory. *Scientific American*.

Alt, F. W., T. K. Blackwell, and G. D. Yancopoulos. 1987. Development of the primary antibody repertoire. *Science* 238:1079.

Baglioni, C., and T. W. Nilsen. 1981. The action of interferon at the molecular level. *American Scientist* 69:392.

Barrett, J. T. 1978. *Textbook of Immunology*. 3d. ed. St. Louis: C. V. Mosby.

Biusseret, P. D. 1982 (August). Allergy. *Scientific American*.

Burnet, F. M. 1976. *Immunology: Readings from Scientific American*. San Francisco: W. H. Freeman.

Burton, D.R. 1992. Human monoclonal antibodies: Achievement and potential. *Hospital Practice* 27:67.

Capra, J. D., and A. B. Edmunson. 1977 (January). The antibody combining site. *Scientific American*.

Cohen, I. R. 1988 (April). The self, the world, and autoimmunity. *Scientific American*.

Cunningham, B. A. 1977 (October). The structure and function of histocompatibility antigens. *Scientific American*.

Dausset, J. 1981. The major histocompatibility complex in man: Past, present, and future concepts. *Science* 213:1469.

DiNome, M. A., and D. E. Young. 1987 (August). The clonal selection theory. *Scientific American*.

Geha, R. S. 1988. Regulation of IgE synthesis in atopic disease. *Hospital Practice* 23:91.

Gleich, G. J. 1988. Current understanding of eosinophil function. *Hospital Practice* 23:137.

Greene, W. C. 1993 (September). AIDS and the immune system. *Scientific American*.

Hamburger, R. N. 1976. Allergy and the immune system. *American Scientist* 64:157.

Herberman, R. B., and J. R. Ortaldo. 1981. Natural killer cells: The role in defense against disease. *Science* 214:24.

Hirsch, M. S., and J. C. Kaplan. 1987 (April). Antiviral therapy. *Scientific American*.

Janeway, C. A., Jr. 1993 (September). How the immune system recognizes invaders. *Scientific American*.

Kapp, J. A., C. W. Pierce, and C. M. Sorensen. 1984. Antigen-specific suppressor T cell factors. *Hospital Practice* 19:85.

Koffler, D. 1980 (July). Systemic lupus erythematosus. *Scientific American*.

Laurence, J. 1985 (December). The immune system in AIDS. *Scientific American*.

Leder, P. 1982 (November). The genetics of antibody diversity. *Scientific American*.

Lichtenstein, L. M. 1993 (September). Allergy and the immune system. *Scientific American*.

Lopate, G., and A. Pestronk. 1993. Autoimmune myasthenia gravis. *Hospital Practice* 28:109.

Marrack, P., and J. Kappler. 1986 (February). The T cell and its receptor. *Scientific American*.

Marrack, P., and J. W. Kappler. 1993 (September). How the immune system recognizes the body. *Scientific American*.

McDevitt, H. O. 1985. The HLA system and its relation to disease. *Hospital Practice* 20:57.

Metcalf, D. 1991. Control of granulocytes and macrophages: Molecular, cellular, and clinical aspects. *Science* 254:529.

Milstein, C. 1980 (October). Monoclonal antibodies. *Scientific American*.

Milstein, C. 1986. From antibody structure to immunological diversification of immune response. *Science* 231:1261.

Nossal, G. J. V. 1987. The basic components of the immune system. *New England Journal of Medicine* 316:1320.

Nossal, G. J. V. 1993 (September). Life, death, and the immune system. *Scientific American*.

Oettgen, H. F. 1981. Immunological aspects of cancer. *Hospital Practice* 16:93.

Old, L. J. 1977 (May). Cancer immunology. *Scientific American*.

Old, L. J. 1988 (May). Tumor necrosis factor. *Scientific American*.

Reichlin S. 1993. Neuroendocrine-immune interactions. *New England Journal of Medicine* 329:1246.

Rennie, J. 1990 (December). The body against itself. *Scientific American*.

Rose, N. R. 1981 (February). Autoimmune diseases. *Scientific American*.

Sachs, L. 1986 (January). Growth, differentiation, and reversal of malignancy. *Scientific American*.

Samuelsson, B. 1983. Leukotrienes: Mediators of immediate hypersensitivity and inflammation. *Science* 220:568.

Schwartz, R. H. 1993 (August). T cell anergy. *Scientific American*.

Steinman, L. 1993 (September). Autoimmune disease. *Scientific American*.

Sutton, B. J., and H. J. Gould. 1993. The human IgE network. *Nature* 366:421.

Tannock, I. F. 1983. Biology of tumor growth. *Hospital Practice* 18:81.

Tonegawa, S. 1985 (October). The molecules of the immune system. *Scientific American*.

Unanue, E. R., and P. M. Allen. 1987. The immunoregulatory role of the macrophage. *Hospital Practice* 22:87.

Vaghan, J. A. 1984. Rheumatoid arthritis: Evidence of a defect in T cell function. *Hospital Practice* 19:101.

Yelton, D. E., and M. D. Scharff. 1980. Monoclonal antibodies. *American Scientist* 63:510.

Young, J. D., and Z. A. Cohen. 1988 (January). How killer cells kill. *Scientific American*.

## Respiratory and Urinary Systems

Alexander, E. 1986. Metabolic acidosis: Recognition and etiologic diagnosis. *Hospital Practice* 21:100E.

Andreson, E. 1977. Regulation of body fluids. *Annual Review of Physiology* 39:185.

Avery, M. E., S. S. Wang, and H. W. Taeusch. 1975 (March). The lung of the newborn infant. *Scientific American*.

Bauman, J. W., and F. P. Chinard. 1975. *Renal Function: Physiological and Medical Aspects*. St. Louis: C. V. Mosby.

Beeuwkes, R., III. 1980. The vascular organization of the kidney. *Annual Review of Physiology* 42:531.

Berger, A. J., R. A. Mitchel, and J. W. Severinghaus. 1977. Regulation of respiration. *New England Journal of Medicine* 297:92, 138, 194.

Bone, R. C. 1993. Bronchial asthma: Diagnosis and treatment issues. *Hospital Practice* 28:45.

Bramble, D. M., and D. R. Carrier. 1983. Running and breathing in mammals. *Science* 219:251.

Brenner, B. M., and R. Beeuwkes, III. 1978. The renal circulation. *Hospital Practice* 13:35.

Brenner, B. M., T. H. Hostetter, and H. D. Humes. 1978. Molecular basis of proteinuria of glomerular origin. *New England Journal of Medicine* 298:826.

Browning, R. J. 1982 (January–February: 39; March–April: 59). Pulmonary disease. Part 1: Back to basics; Part 2: Putting blood gasses to work. *Diagnostic Medicine*.

Buckalew, V. M., Jr., and K. A. Gruber. 1984. Natriuretic hormone. *Annual Review of Physiology* 46:343.

Cherniak, N. S. 1986. Breathing disorders during sleep. *Hospital Practice* 21:81.

Coe, F. L. et al. 1992. The pathogenesis and treatment of kidney stones. *New England Journal of Medicine* 327:1141.

Dantzker, D. R. 1986. Physiology and pathophysiology of pulmonary gas exchange. *Hospital Practice* 121:135.

Decramer, M. 1993. Respiratory muscle interaction. *News in Physiological Science* 8:121.

Dempster, J. A. et al. 1992. The quest for water channels. *News in Physiological Science* 7:172.

Epstein, F. H., and R. S. Brown. 1988. Acute renal failure: A collection of paradoxes. *Hospital Practice* 23:171.

Finch, C. A., and C. Lenfant. 1972. Oxygen transport in man. *New England Journal of Medicine* 286:407.

Flenley, D. C., and P. M. Warren. 1983. Ventilatory response to $O_2$ and $CO_2$ during exercise. *Annual Review of Physiology* 45:415.

Fraser, R. G., and J. A. P. Pare. 1977. *Structure and Function of the Lung*. 2d ed. Philadelphia: W. B. Saunders.

Galla, J. H., and R. G. Luke. 1987. Pathophysiology of metabolic alkalosis. *Hospital Practice* 22:123.

Giebisch, G. H., and B. Stanton. 1979. Potassium transport in the nephron. *Annual Review of Physiology* 41:241.

Glassock, R. J. 1987. Pathophysiology of acute glomerulonephritis. *Hospital Practice* 22:163.

Grantham, J. J. 1992. Polycystic kidney disease: Etiology and pathogenesis. *Hospital Practice* 27:51.

Guz, A. 1975. Regulation of respiration in man. *Annual Review of Physiology* 37:303.

Haddad, G. G., and R. B. Mellins. 1984. Hypoxia and respiratory control in early life. *Annual Review of Physiology* 46:629.

Hays, R. M. 1978. Principles of ion and water transport in the kidneys. *Hospital Practice* 13:79.

Hollenberg, N. K. 1986. The kidney in heart failure. *Hospital Practice* 21:81.

Houston, C. S. 1992 (October). Mountain sickness. *Scientific American*.

Irsigler, G. B., and J. W. Severinghaus. 1980. Clinical problems of ventilatory control. *Annual Review of Medicine* 31:109.

Jacobson, H. R. 1987. Diuretics: Mechanisms of action and uses. *Hospital Practice* 22:129.

Jobe, A. H. 1993. Pulmonary surfactant therapy. *New England Journal of Medicine* 328:861.

Kassirer, J. P., and N. E. Madias. 1980. Respiratory acid-base disorders. *Hospital Practice* 15:57.

Kokko, J. S. 1979. Renal concentrating and diluting mechanisms. *Hospital Practice* 14:110.

Macklem, P. T. 1986. Respiratory muscle dysfunction. *Hospital Practice* 21:83.

McFadden, Jr., E. R., and I. A. Gilbert. 1992. Asthma. *New England Journal of Medicine* 327:1928.

Murray, J. F. 1985. The lungs and heart failure. *Hospital Practice* 20:55.

Naeye, R. L. 1980 (April). Sudden infant death. *Scientific American*.

Peart, W. S. 1975. Renin-angiotensin system. *New England Journal of Medicine* 292:302.

Perutz, M. F. 1978 (December). Hemoglobin structure and respiratory transport. *Scientific American*.

Reid, I. A., B. J. Morris, and W. F. Ganong. 1978. The renin-angiotensin system. *Annual Review of Physiology* 40:377.

Renkin, E. M., and R. R. Robinson. 1974. Glomerular filtration. *New England Journal of Medicine* 290:70.

Rigatto, H. 1984. Control of ventilation in the newborn. *Annual Review of Physiology* 46:661.

Roussos, C., and P. T. Macklem. 1982. The respiratory muscles. *New England Journal of Medicine* 307:786.

Steinmetz, P. R., and B. M. Koeppen. 1984. Cellular mechanisms of diuretic action along the nephron. *Hospital Practice* 19:125.

Their, S. O. 1987. Diuretic mechanisms as a guide to therapy. *Hospital Practice* 22:81.

Tobin, M. J. 1986. Update on strategies in mechanical ventilation. *Hospital Practice*. 21:69.

Vander, A. J. 1980. *Renal Physiology*. 2d ed. New York: McGraw-Hill.

Walker, D. W. 1984. Peripheral and central chemoreceptors in the fetus and newborn. *Annual Review of Physiology* 46:687.

Walker, L. A., and H. Vatlin. 1982. Biological importance of nephron heterogeneity. *Annual Review of Physiology* 44:203.

Warnock, D. G., and F. C. Rector Jr. 1979. Proton secretion by the kidney. *Annual Review of Physiology* 41:197.

Weinberger, S. E. 1993. Recent advances in pulmonary medicine. *New England Journal of Medicine* 328:1389.

West, J. B. 1984. Human physiology at extreme altitudes on Mount Everest. *Science* 223:784.

Whipp, B. J. 1983. Ventilatory control during exercise in humans. *Annual Review of Physiology* 45:393.

## Digestive System and the Regulation of Metabolism

Austin, L. A., and H. Heath III. 1981. Calcitonin: Physiology and pathophysiology. *New England Journal of Medicine* 304:269.

Barret, E. J., and R. A. DeFronzo. 1984. Diabetic ketoacidosis: Diagnosis and treatment. *Hospital Practice* 19:89.

Binder, H. J. 1984. The pathophysiology of diarrhea. *Hospital Practice* 19:107.

Bleich, H. L., and E. S. Boro. 1979. Protein digestion and absorption. *New England Journal of Medicine* 300:659.

Cahill, G. F., and H. O. McDevitt. 1981. Insulin-dependent diabetes mellitus: The initial lesion. *New England Journal of Medicine* 304:454.

Carey, M. C., D. M. Small, and C. M. Bliss. 1983. Lipid digestion and absorption. *Annual Review of Physiology* 45:651.

Cheng, K., and J. Larner. 1985. Intracellular mediators of insulin action. *Annual Review of Physiology* 47:405.

Chou, C. C. 1982. Relationship between intestinal blood flow and motility. *Annual Review of Physiology* 44:29.

Cohen, S. 1983. Neuromuscular disorders of the gastrointestinal tract. *Hospital Practice* 18:121.

Davenport, H. W. 1982. *Physiology of the Digestive Tract*. 5th ed. Chicago: Year Book Medical Publishers.

DeLuca, H. F. 1980. The vitamin D hormonal system: Implications for bone disease. *Hospital Practice* 15:57.

Dockray, G. J. 1979. Comparative biochemistry and physiology of gut hormones. *Annual Review of Physiology* 41:83.

Eisenbarth, G. S. 1986. Type I diabetes mellitus: A chronic autoimmune disease. *New England Journal of Medicine* 314:1360.

Freeman, H. J., and Y. S. Kim. 1978. Digestion and absorption of proteins. *Annual Review of Physiology* 29:99.

Gardner, J. D., and R. T. Jensen. 1986. Receptors and cell activation associated with pancreatic enzyme secretion. *Annual Review of Physiology* 48:103.

Gardner, L. I. 1972 (July). Deprivation dwarfism. *Scientific American.*

Gollan, J. L., and A. B. Knapp. 1985. Bilirubin metabolism and congenital jaundice. *Hospital Practice* 20:83.

Goodman, D. S. 1984. Vitamin A and retinoids in health and disease. *New England Journal of Medicine* 310:1023.

Gray, G. M. 1975. Carbohydrate digestion and absorption: Role of the small intestine. *New England Journal of Medicine* 292:1225.

Grossman, M. I. 1979. Neural and hormonal regulation of gastrointestinal function: An overview. *Annual Review of Physiology* 41:27.

Guengerich, F. P. 1993. Cytochrome P450 enzymes. *American Scientist* 81:440.

Habener, J. F., and J. E. Mahaffey. 1978. Osteomalacia and disorders of vitamin D metabolism. Annual Review of Medicine 29:327.

Hahn, T. J. 1986. Physiology of bone: Mechanisms of osteogenic disorders. *Hospital Practice* 21:73.

Hirsch, J. 1984. Hypothalamic control of appetite. *Hospital Practice* 19:131.

Holt, K. M., and J. I. Isenberg. 1985. Peptic ulcer disease: Physiology and pathophysiology. *Hospital Practice* 20:89.

Isaaksson, O. G., P. S. Eden, and J. O. Jansson. 1985. Mode of action of pituitary growth hormone on target cell. *Annual Review of Physiology* 47:483.

Johnston, D. E., and M. M. Kaplan. 1993. Pathogenesis and treatment of gallstones. *New England Journal of Medicine* 328:412.

Kappas, A., and A. P. Alvarez. 1975 (June). How the liver metabolizes foreign substances. *Scientific American.*

Livingston, E. H., and P. H. Guth. 1992. Peptic ulcer disease. *American Scientist* 80:592.

Lynn, R. B., and L. S. Friedman. 1993. Irritable bowel syndrome. *New England Journal of Medicine* 329:1940.

Marshall, B. J. 1987. Peptic ulcer: An infectious disease? *Hospital Practice* 22:87.

Martin, R. J. et al. 1991. The regulation of body weight. *American Scientist* 79:528.

Masoro, E. J. 1992. A dietary key to uncovering aging process. *News in Physiological Science* 7:157.

McGuigan, J. E. 1978. Gastrointestinal hormones. *Annual Review of Physiology* 29:99.

Mitlak, B. H., and S. R. Nussbaum. 1993. Diagnosis and treatment of osteoporosis. *Annual Review of Medicine* 44:265.

Moller, D. E., and J. S. Flier. 1991. Insulin resistance—mechanisms, syndromes, and implications. *New England Journal of Medicine* 325:938.

Moog, F. 1981 (November). The lining of the small intestine. *Scientific American.*

Notkins, A. L. 1979 (November). The cause of diabetes. *Scientific American.*

Oppenheimer, J. H. 1979. Thyroid hormone action at the cellular level. *Science* 203:971.

Raisz, L. G., and B. E. Kream. 1981. Hormonal control of skeletal growth. *Annual Review of Physiology* 43:225.

Rothman, D. L. et al. 1991. Quantitation of hepatic glycogenolysis and gluconeogenesis in fasting humans with $^{13}$C NMR. *Science* 254:573.

Simon, H. B. 1993. Hyperthermia. *New England Journal of Medicine* 329:483.

Siperstein, M. D. 1985. Type II diabetes: Some problems in diagnosis and treatment. *Hospital Practice* 20:55.

Smith, B. F., and T. Lamont. 1984. The pathogenesis of gallstones. *Hospital Practice* 19:93.

Soll, A., and J. H. Walsh. 1979. Regulation of gastric acid secretion. *Annual Review of Physiology* 41:35.

Tepperman, J. 1980. *Metabolic and Endocrine Physiology.* 4th ed. Chicago: Year Book Medical Publishers.

Unger, R. H., and L. Orci. 1981. Glucagon and the A cell: Physiology and pathophysiology. *New England Journal of Medicine* 304:1518, 1575.

Unger, R. H., and L. Orci. 1981. Insulin, glucagon, and somatostatin secretion in the regulation of metabolism. *Annual Review of Physiology* 40:307.

Van De Graaff, K. M. 1986. Anatomy and physiology of the gastrointestinal tract. *Pediatric Infectious Diseases* 5:S11.

Van Wyk, J., and L. E. Underwood. 1978. Growth hormone, somatomedins, and growth failure. *Hospital Practice* 13:57.

Walsh, J. H., and M. I. Grossman. 1975. Gastrin. *New England Journal of Medicine* 292:1324, 1377.

Weinberg, R. H. 1987. Lipoprotein metabolism: Hormonal regulation. *Hospital Practice* 22:223.

Weisbrodt, N. W. 1981. Patterns of intestinal motility. *Annual Review of Physiology* 43:21.

Williams, J. A. 1984. Regulatory mechanisms in pancreas and salivary acini. *Annual Review of Physiology* 46:361.

Wood, J. D. 1981. Intrinsic neural control of intestinal motility. *Annual Review of Physiology* 43:33.

Wynder, E. L., and D. P. Rose. 1984. Diet and breast cancer. *Hospital Practice* 19:73.

## Reproductive System, Development, and Aging

Baird, D. T., and A. F. Glasier. 1993. Hormonal contraception. *New England Journal of Medicine* 328:1543.

Balinsky, B. I. 1981. *An Introduction to Embryology.* 5th ed. Philadelphia: W. B. Saunders.

Bardin, C. W. 1979. The neuroendocrinology of male reproduction. *Hospital Practice* 14:65.

Bartke, A. A., et al. 1978. Hormonal interaction in the regulation of androgen secretion. *Biology of Reproduction* 18:44.

Beaconsfield, P., G. Birdwood, and R. Beaconsfield. 1980 (July). The placenta. *Scientific American.*

Birnholz, J. C., and E. E. Farrel. 1984. Ultrasound images of human fetal development. *American Scientist* 72:608.

Boyar, R. M. 1978. Control of the onset of puberty. *Annual Review of Medicine* 31:329.

Brann, D. 1993. Progesterone: The forgotten hormone? *Perspectives in Biology and Medicine* 36:642.

Carter, N., ed. 1980. *Development, Growth, and Aging.* London: Croon Helm.

Chervenak, F. A., G. Isaacson, and M. J. Mahoney. 1986. Advances in the diagnosis of fetal defects. *New England Journal of Medicine* 315:305.

Comfort, A. 1979. *The Biology of Senescence.* 3d ed. London: Churchill Livingstone.

D'Alton, M. E., and A. H. DeCherney. 1993. Prenatal diagnosis. *New England Journal of Medicine* 328:114.

Diamond, M. C. 1978 (January–February). The aging brain. *American Scientist.*

Dufau, M. L. 1988. Endocrine regulation and communicating functions of the corpus luteum. *Annual Review of Physiology* 50:483.

England, M. A. 1983. Color atlas of life before birth: Normal fetal development. Chicago: Year Book Medical Publishers.

Epel, D. 1977 (November). The program of fertilization. *Scientific American.*

Fink, G. 1979. Feedback action of target hormones on hypothalamus and pituitary with special reference to gonadal steroids. *Annual Review of Physiology* 41:571.

Frantz, A. G. 1978. Prolactin. *New England Journal of Medicine* 298:112.

Goldzieher, J. W., and A. N. Poindexter. 1987. Medical aspects of contraception. *Hospital Practice* 22:93.

Goldzieher, J. W. 1993. The history of steroidal contraceptive development: The estrogens. *Perspective in Biology and Medicine* 36:363.

Grabowski, C. T. 1983. *Human Reproduction and Development.* Philadelphia: W. B. Saunders.

Grobstein, C. 1979 (March). External human fertilization. *Scientific American.*

Grumbach, M. M. 1979. The neuroendocrinology of puberty. *Hospital Practice* 14:65.

Hatcher, R. A., and A. K. Stewart, 1987. *Contraceptive Technology.* 13th ed. New York: Wiley.

Hayflick L. 1980 (January). The cell biology of human aging. *American Scientist*.

Jackson, L. G. 1985. First trimester diagnosis of fetal genetic disorders. *Hospital Practice* 20:39.

Jones, K. L. et al. 1985. *Dimensions of Human Sexuality*. Dubuque: Wm. C. Brown.

Katchadourian, H. A., and D. T. Lunde. 1985. *Fundamentals of Human Sexuality*. 4th ed. New York: Holt, Rinehart, and Winston.

Keyes, P. L., and M. C. Wiltbank. 1988. Endocrine regulation of the corpus luteum. *Annual Review of Physiology* 50:465.

Lagerkrantz, H., and T. A. Slotkin. 1986 (April). The "stress" of being born. *Scientific American*.

Leong, D. S., L. S. Frawley, and J. D. Neill. 1983. Neuroendocrine control of prolactin secretion. *Annual Review of Physiology* 45:109.

Lipsett, M. B. 1980. Physiology and pathology of the Leydig cell. *New England Journal of Medicine* 303:682.

Marshall, J. C., and R. P. Kelch. 1986. Gonadotropin-releasing hormone: Role of pulsatile secretion in the regulation of reproduction. *New England Journal of Medicine* 315:1459.

Marx, J. L. 1978. The mating game: What happens when sperm meets egg. *Science* 200:1256.

Masters, W. H. 1986. Sex and aging—expectations and reality. *Hospital Practice* 21:175.

Means, A. R. et al. 1980. Regulation of the testis Sertoli cell by follicle stimulating hormone. *Annual Review of Physiology* 42:59.

Naftolin, F. 1981 (March). Understanding the basis of sex differences. *Science* 211:1263.

Nilsson, L., A. Ingelman-Sundbert, and C. Wirsen. 1977. *A Child is Born*. Rev. ed. New York: Dell.

Odell, W. D., and D. L. Moyer. 1971. *Physiology of Reproduction*. St. Louis: C. V. Mosby.

Ojeda, S. R. 1991. The mystery of mammalian puberty: How much more do we know? *Perspectives in Biology and Medicine* 34:365.

Oppenheimer, S. B., and G. Lefevere. 1984. *Introduction to Embryonic Development*. 2d ed. Boston: Allyn and Bacon.

Perone, N. 1993. The history of steroidal contraceptive development: The progestins. *Perspectives in Biology and Medicine* 36:347.

Reiter, E. O., and M. M. Grumbach. 1982. Neuroendocrine control mechanisms and the onset of puberty. *Annual Review of Physiology* 44:595.

Santrock, J. W. 1985. *Adult Development and Aging*. Dubuque: Wm. C. Brown.

Segal, S. J. 1974 (September). The physiology of human reproduction. *Scientific American*.

Short, R. V. 1984 (April). Breast feeding. *Scientific American*.

Simpson, E. R., and P. C. MacDonald. 1981. Endocrine physiology of the placenta. *Annual Review of Physiology* 43:163.

Spitz, I. M., and C. W. Bardin. 1993. Mifepristone (RU 486)—A modulator of progestin and glucocorticoid action. *New England Journal of Medicine* 329:404.

Tamarkin, L. C., C. J. Baird, and O. F. X. Almeida. 1985. Melatonin: A coordinating signal for mammalian reproduction? *Science* 227:714.

Tanner, J. M. 1973 (September). Growing up. *Scientific American*.

Tyson, J. E. 1984 (April). Reproductive endocrinology: New problems call for new solutions. *Diagnostic Medicine*.

Winston, R. M. L., and A. H. Handyside. 1993. New challenges in human in vitro fertilization. *Science* 260:932.

Wilson, J. D. 1978. Sexual differentiation. *Annual Review of Physiology* 40:279.

Wilson, J. D., F. W. George, and J. E. Griffin. 1981. The hormonal control of sexual development. *Science* 211:1278.

Yen, S. S. C. 1979. Neuroendocrine regulation of the menstrual cycle. *Hospital Practice* 14:83.

# Some Laboratory Tests of Clinical Importance

## Common tests performed on blood

| Test | Normal values (adult) | Clinical significance |
|---|---|---|
| Acetone and acetoacetate (serum) | 0.3–2.0 mg/100 ml | Values increase in diabetic acidosis, toxemia of pregnancy, fasting, and high-fat diet. |
| Albumin-globulin ratio or A/G ratio (serum) | 1.5:1–2.5:1 | Ratio of albumin to globulin is lowered in kidney diseases and malnutrition. |
| Albumin (serum) | 3.2–5.5 gm/100 ml | Values increase in multiple myeloma and decrease with proteinuria and as a result of severe burns. |
| Ammonia (plasma) | 50–170 µg/100 ml | Values increase in severe liver disease, pneumonia, shock, and congestive heart failure. |
| Amylase (serum) | 80–160 Somogyi units/100 ml | Values increase in acute pancreatitis, intestinal obstructions, and mumps. They decrease in chronic pancreatitis, cirrhosis of the liver, and toxemia of pregnancy. |
| Bilirubin, total (serum) | 0.3–1.1 mg/100 ml | Values increase in conditions causing red blood cell destruction or biliary obstruction. |
| Blood urea nitrogen or BUN (plasma or serum) | 10–20 mg/100 ml | Values increase in various kidney disorders and decrease in liver failure and during pregnancy. |
| Calcium (serum) | 9.0–11.0 mg/100 ml | Values increase in hyperparathyroidism, hypervitaminosis D, and respiratory conditions that cause a rise in $CO_2$ concentration. They decrease in hypoparathyroidism, malnutrition, and severe diarrhea. |
| Carbon dioxide (serum) | 24–30 mEq/l | Values increase in respiratory diseases, intestinal obstruction, and vomiting. They decrease in acidosis, nephritis, and diarrhea. |
| Chloride (serum) | 96–106 mEq/l | Values increase in nephritis, Cushing's syndrome, and hyperventilation. They decrease in diabetic acidosis, Addison's disease, diarrhea, and following severe burns. |
| Cholesterol, total (serum) | 150–250 mg/100 ml | Values increase in diabetes mellitus and hypothyroidism. They decrease in pernicious anemia, hyperthyroidism, and acute infections. |
| Creatine phosphokinase or CPK (serum) | Men: 0–20 IU/l<br>Women: 0–14 IU/l | Values increase in myocardial infarction and skeletal muscle diseases such as muscular dystrophy. |
| Creatine (serum) | 0.2–0.8 mg/100 ml | Values increase in muscular dystrophy, nephritis, severe damage to muscle tissue, and during pregnancy. |
| Creatinine (serum) | 0.7–1.5 mg/100 ml | Values increase in various kidney diseases. |
| Erythrocyte count or red blood cell count (whole blood) | Men: 4,600,000–6,200,000/cu mm<br>Women: 4,200,000–5,400,000/cu mm<br>Children: 4,500,000–5,100,000/cu mm<br>(varies with age) | Values increase as a result of severe dehydration or diarrhea and decrease in anemia, leukemia, and following severe hemorrhage. |
| Fatty acids, total (serum) | 190–420 mg/100 ml | Values increase in diabetes mellitus, anemia, kidney disease, and hypothyroidism. They decrease in hyperthyroidism. |

# Common tests performed on blood (continued)

| Test | Normal values (adult) | Clinical significance |
| --- | --- | --- |
| Globulin (serum) | 2.5–3.5 gm/100 ml | Values increase as a result of chronic infections. |
| Glucose (plasma) | 70–115 mg/100 ml | Values increase in diabetes mellitus, liver diseases, nephritis, hyperthyroidism, and pregnancy. They decrease in hyperinsulinism, hypothyroidism, and Addison's disease. |
| Hematocrit (whole blood) | Men: 40–54 ml/100 ml<br>Women: 37–47 ml/100 ml<br>Children: 35–49 ml/100 ml<br>(varies with age) | Values increase in polycythemia due to dehydration or shock. They decrease in anemia and following severe hemorrhage. |
| Hemoglobin (whole blood) | Men: 14–18 gm/100 ml<br>Women: 12–16 gm/100 ml<br>Children: 11.2–16.5 gm/100 ml<br>(varies with age) | Values increase in polycythemia, obstructive pulmonary diseases, congestive heart failure, and at high altitudes. They decrease in anemia, pregnancy, and as a result of severe hemorrhage or excessive fluid intake. |
| Iron (serum) | 75–175 $\mu$g/100 ml | Values increase in various anemias and liver disease. They decrease in iron deficiency anemia. |
| Iron-binding capacity (serum) | 250–410 $\mu$g/100 ml | Values increase in iron deficiency anemia and pregnancy. They decrease in pernicious anemia, liver disease, and chronic infections. |
| Lactic acid (whole blood) | 6–16 mg/100 ml | Values increase with muscular activity and in congestive heart failure, severe hemorrhage, and shock. |
| Lactic dehydrogenase or LDH (serum) | 90–200 milliunits/ml | Values increase in pernicious anemia, myocardial infarction, liver diseases, acute leukemia, and widespread carcinoma. |
| Lipids, total (serum) | 450–850 mg/100 ml | Values increase in hypothyroidism, diabetes mellitus, and nephritis. They decrease in hyperthyroidism. |
| Oxygen saturation (whole blood) | Arterial: 94%–100%<br>Venous: 60%–85% | Values increase in polycythemia and decrease in anemia and obstructive pulmonary diseases. |
| ph (whole blood) | 7.35–7.45 | Values increase due to vomiting, Cushing's syndrome, and hyperventilation. They decrease as a result of hypoventilation, severe diarrhea, Addison's disease, and diabetic acidosis. |
| Phosphatase, acid (serum) | 1.0–5.0 King–Armstrong units/ml | Values increase in cancer of the prostate, hyperparathyroidism, certain liver diseases, myocardial infarction, and pulmonary embolism. |
| Phosphatase, alkaline (serum) | 5–13 King–Armstrong units/ml | Values increase in hyperparathyroidism (and in other conditions that promote resorption of bone), liver diseases, and pregnancy. |
| Phospholipids (serum) | 6–12 mg/100 ml as lipid phosphorus | Values increase in diabetes mellitus and nephritis. |
| Phosphorus (serum) | 3.0–4.5 mg/100 ml | Values increase in kidney diseases, hypoparathyroidism, acromegaly, and hypervitaminosis D. They decrease in hyperparathyroidism. |
| Platelet count (whole blood) | 150,000–350,000/cu mm | Values increase in polycythemia and certain anemias. They decrease in acute leukemia and aplastic anemia. |
| Potassium (serum) | 3.5–5.0 mEq/l | Values increase in Addison's disease, hypoventilation, and conditions that cause severe cellular destruction. They decrease in diarrhea, vomiting, diabetic acidosis, and chronic kidney disease. |
| Protein, total (serum) | 6.0–8.0 gm/100 ml | Values increase in severe dehydration and shock. They decrease in severe malnutrition and hemorrhage. |
| Protein-bound iodine or PBI (serum) | 3.5–8.0 $\mu$g/100 ml | Values increase in hyperthyroidism and liver disease. They decrease in hypothyroidism. |

## Common tests performed on blood (continued)

| Test | Normal values (adult) | Clinical significance |
|------|----------------------|----------------------|
| Prothrombin time (serum) | 12–14 sec (one stage) | Values increase in certain hemorrhagic diseases, liver disease, vitamin K deficiency, and following the use of various drugs. |
| Sedimentation rate, Westergren (whole blood) | Men: 0–15 mm/hr<br>Women: 0–20 mm/hr | Values increase in infectious diseases, menstruation, pregnancy, and as a result of severe tissue damage. |
| Sodium (serum) | 136–145 mEq/l | Values increase in nephritis and severe dehydration. They decrease in Addison's disease, myxedema, kidney disease, and diarrhea. |
| Thyroxine or T$_4$ (serum) | 2.9–6.4 $\mu$g/100 ml | Values increase in hyperthyroidism and pregnancy. They decrease in hypothyroidism. |
| Thromboplastin time, partial (plasma) | 35–45 sec | Values increase in deficiencies of blood factors VIII, IX, and X. |
| Transaminases or SGOT (serum) | 5–40 units/ml | Values increase in myocardial infarction, liver disease, and diseases of skeletal muscles. |
| Uric acid (serum) | Men: 2.5–8.0 mg/100 ml<br>Women: 1.5–6.0 mg/100 ml | Values increase in gout, leukemia, pneumonia, toxemia of pregnancy, and as a result of severe tissue damage. |
| While blood cell count, differential (whole blood) | Neutrophils 54%–62%<br>Eosinophils 1%–3%<br>Basophils 0%–1%<br>Lymphocytes 25%–33%<br>Monocytes 3%–7% | Neutrophils increase in bacterial diseases; lymphocytes and monocytes increase in viral diseases; eosinophils increase in collagen diseases, allergies, and in the presence of intestinal parasites. |
| White blood cell count, total (whole blood) | 5000–10,000/cu mm | Values increase in acute infections, acute leukemia, and following menstruation. They decrease in aplastic anemia and as a result of drug toxicity. |

## Common tests performed on urine

| Test | Normal values | Clinical significance |
|------|--------------|----------------------|
| Acetone and acetoacetate | 0 | Values increase in diabetic acidosis. |
| Albumin, qualitative | 0 to trace | Values increase in kidney disease, hypertension, and heart failure. |
| Ammonia | 20–70 mEq/l | Values increase in diabetes mellitus and liver diseases. |
| Bacterial count | Under 10,000/ml | Values increase in urinary tract infection. |
| Bile and bilirubin | 0 | Values increase in melanoma and biliary tract obstruction. |
| Calcium | Under 250 mg/24 hr | Values increase in hyperparathyroidism and decrease in hypoparathyroidism. |
| Creatinine clearance | 100–140 ml/min | Values increase in renal diseases. |
| Creatinine | 1–2 gm/24 hr | Values increase in infections and decrease in muscular atrophy, anemia, leukemia, and kidney diseases. |
| Glucose | 0 | Values increase in diabetes mellitus and various pituitary gland disorders. |
| 17-Hydroxycorticosteroids | 2–10 mg/24 hr | Values increase in Cushing's syndrome and decrease in Addison's disease. |
| Phenylpyruvic acid | 0 | Values increase in phenylketonuria. |
| Urea clearance | Over 40 ml blood cleared of urea/min | Values increase in renal diseases. |
| Urobilinogen | 0–4 mg/24 hr | Values increase in liver diseases and hemolytic anemia. They decrease in complete biliary obstruction and severe diarrhea. |
| Urea | 25–35 gm/24 hr | Values increase as a result of excessive protein breakdown. They decrease as a result of impaired renal function. |
| Uric acid | 0.6–1.0 gm/24 hr. as urate | Values increase in gout and decrease in various kidney diseases. |

# [ glossary ]

Most of the words in this glossary are followed by a phonetic spelling that serves as a guide to pronunciation. The phonetic spellings reflect standard scientific usage and can be easily interpreted following a few basic rules.

Any unmarked vowel that ends a syllable or that stands alone as a syllable has the long sound. Any unmarked vowel that is followed by a consonant has the short sound.

If a long vowel appears in the middle of a syllable (followed by a consonant), it is marked with a macron (¯). Similarly, if a vowel stands alone or ends a syllable but should have a short sound, it is marked with a breve (˘).

Syllables that are emphasized are indicated by stress marks. A single stress mark (´) indicates the primary emphasis; a secondary emphasis is indicated by a double stress mark (˝).

Page references are provided, except in the case of some adjectives and a few general terms.

## A

**abdomen** (ab´dŏ-men, ab-do´men)  The portion of the trunk between the diaphragm and pelvis.  14

**abduction** (ab-duk´shun)  The movement of a body part away from the axis or midline of the body; movement of a digit away from the axis of the limb.  233

**ABO system**  The most common system of classification for red blood cell antigens. On the basis of antigens on the red blood cell surface, individuals can be type A, type B, type AB, or type O.  557

**absorption** (ab-sorp´shun)  The transport of molecules across epithelial membranes into the body fluids.  765

**accessory organs** (ak-ses´ŏ-re)  Organs that assist with the functioning of other organs within a system.  766

**accommodation** (ă-kom˝ŏ-da´shun)  A process whereby the focal length of the eye is changed by automatic adjustment of the curvature of the lens to bring images of objects from various distances into focus on the retina.  494

**acetabulum** (as˝ĕ-tab´yŭ-lum)  A socket in the lateral surface of the hipbone (os coxa) with which the head of the femur articulates.  209

**acetone** (as´ĕ-tōn)  A ketone body produced as a result of the oxidation of fats.  819

**acetyl coenzyme A** (acetyl CoA)  (as´ĕ-tl, ă-set´l)  A coenzyme derivative in the metabolism of glucose and fatty acids that contributes substrates to the Krebs cycle.  83

**acetylcholine** (ACh)  (ă-set´l-ko´lēn)  An acetic acid ester of choline—a substance that functions as a neurotransmitter in somatic motor nerve and parasympathetic nerve fibers.  360

**acetylcholinesterase** (ă-set´l-ko´lĭ-nes´tĕ-rās)  An enzyme in the membrane of postsynaptic cells that catalyzes the conversion of ACh into choline and acetic acid. This enzymatic reaction inactivates the neurotransmitter.  361

**Achilles tendon** (ă-kil´ēz)  *See* tendo calcaneous.  316

**acid** (as´id)  A substance that releases hydrogen ions when ionized in water.  26

**acidosis** (as˝ĭ-do´sis)  An abnormal increase in the $H^+$ concentration of the blood that lowers the arterial pH to below 7.35.  713

**acromegaly** (ak´ro-meg´ă-le)  A condition caused by the hypersecretion of growth hormone from the pituitary gland after maturity and characterized by enlargement of the extremities, such as the nose, jaws, fingers, and toes.  173

**actin** (ak´tin)  A protein in muscle fibers that together with myosin is responsible for contraction.  263

**action potential**  An all-or-none electrical event in an axon or muscle fiber in which the polarity of the membrane potential is rapidly reversed and reestablished.  353

**active immunity** (ĭ-myoo´nĭ-te)  Immunity involving sensitization, in which antibody production is stimulated by prior exposure to an antigen.  662

**active transport**  The movement of molecules or ions across the cell membranes of epithelial cells by membrane carriers. An expenditure of cellular energy (ATP) is required.  92

**adduction** (ad-duk´shun)  The movement of a body part toward the axis or midline of the body; movement of a digit toward the axis of the limb.  223

**adenohypophysis** (ad´n-o-hi-pof´ĭ-sis)  The anterior, glandular lobe of the pituitary gland that secretes FSH (follicle-stimulating hormone), LH (luteinizing hormone), ACTH (adrenocorticotropic hormone), TSH (thyroid-stimulating hormone), GH (growth hormone), and prolactin. Secretions of the adenohypophysis are controlled by hormones produced by the hypothalamus.  519

**adenoids** (ad´ĕ-noidz)  The tonsils located in the nasopharynx; pharyngeal tonsils.  686

**adenylate cyclase** (ă-den´l-it si´klās)  An enzyme found in cell membranes that catalyzes the conversion of ATP to cyclic AMP and pyrophosphate ($PP_1$). This enzyme is activated by an interaction between a specific hormone and its membrane receptor protein.  536

**ADH**  Antidiuretic hormone; a hormone produced by the hypothalamus and released by the posterior pituitary that acts on the kidneys to promote water reabsorption; also known as *vasopressin*.  97

**ADP**  Adenosine diphosphate; a molecule that together with inorganic phosphate is used to make ATP (adenosine triphosphate).  76

**adrenal cortex** (ă-dre´nal kor´teks)  The outer part of the adrenal gland. Derived from embryonic mesoderm, the adrenal cortex secretes corticosteroid hormones (such as aldosterone and hydrocortisone).  526

**adrenal medulla** (mĕ-dul´ă)  The inner part of the adrenal gland. Derived from embryonic postganglionic sympathetic neurons, the adrenal medulla secretes catecholamine hormones—epinephrine and (to a lesser degree) norepinephrine.  526

**adrenergic** (ad´´rĕ-ner´jik)  A term used to describe the actions of epinephrine, norepinephrine, or other molecules with similar activity (as in *adrenergic receptor* and *adrenergic stimulation*).  455

**adventitia** (ad´´ven-tish´ă)  The outermost epithelial layer of a visceral organ; also called *serosa*.  754

**afferent** (af´er-ent)  Conveying or transmitting to.  346

**afferent arteriole** (ar-tir´e-ōl)  A blood vessel within the kidney that supplies blood to the glomerulus.  730

**afferent neuron** (noor´on)  *See* sensory neuron.  346

**agglutinate** (ă-gloot´n-āt)  A clump of cells (usually erythrocytes) formed as a result of specific chemical interaction between surface antigens and antibodies.  557

**agranular leukocytes** (ă-gran´yŭ-lar loo´kŏ-sitz) White blood cells (leukocytes) that do not contain cytoplasmic granules; specifically, lymphocytes and monocytes. 553

**albumin** (al-byoo´min) A water-soluble protein produced in the liver; the major component of the plasma proteins. 551

**aldosterone** (al-dos´ter-ōn) The principal corticosteroid hormone involved in the regulation of electrolyte balance (mineralocorticoid). 527

**alimentary canal** The tubular portion of the digestive tract. *See also* gastrointestinal tract (GI tract). 766

**allantois** (ă-lan´to-is) An extraembryonic membranous sac involved in the formation of blood cells. It gives rise to the fetal umbilical arteries and vein and also contributes to the formation of the urinary bladder. 756

**allergens** (al´er-jenz) Antigens that evoke an allergic response rather than a normal immune response. 677

**allergy** (al´er-je) A state of hypersensitivity caused by exposure to allergens. It results in the liberation of histamine and other molecules with histaminelike effects. 677

**all-or-none principle** The statement of the fact that muscle fibers of a motor unit contract to their maximum extent when exposed to a stimulus of threshold strength. 355

**allosteric** (al´´ŏ-ster´ik) A term used with reference to the alteration of an enzyme's activity as a result of its combination with a regulator molecule. Allosteric inhibition by an end product represents negative feedback control of an enzyme's activity. 75

**alveolar sacs** (al-ve´ŏ-lar) A cluster of alveoli that share a common chamber or central atrium. 690

**alveolus** (al-ve´ŏ-lus) **1.** An individual air capsule within the lung. The alveoli are the basic functional units of respiration. 690 **2.** The socket that secures a tooth (tooth socket). 773

**amniocentesis** (am´´ne-o-sen-te´sis) A procedure in which a sample of amniotic fluid is aspirated to examine suspended cells for various genetic diseases. 910

**amnion** (am´ne-on) A developmental membrane surrounding the fetus that contains amniotic fluid. 908

**amphiarthrosis** (am´´fe-ar-thro´sis) A slightly movable articulation in a functional classification of joints. 224

**amphoteric** (am-fo-ter´ik) Having both acidic and basic characteristics; used to denote a molecule that can be positively or negatively charged, depending on the pH of its environment.

**ampulla** (am-pool´ă) A saclike enlargement of a duct or tube.

**ampulla of Vater** (Fă´ter) *See* hepatopancreatic ampulla. 781

**anabolic steroids** (an´´ă-bol´ik ster´oidz) Steroids with androgenlike stimulatory effects on protein synthesis. 539

**anabolism** (ă-nab´ŏ-liz´´em) A phase of metabolism involving chemical reactions within cells that result in the production of larger molecules from smaller ones; specifically, the synthesis of protein, glycogen, and fat. 814

**anaerobic respiration** (an-ă-ro´bik res´´pi-ra´shun) A form of cell respiration involving the conversion of glucose to lactic acid in which energy is obtained without the use of molecular oxygen. 80

**anal canal** (a´nal) The terminal tubular portion of the large intestine that opens through the anus of the GI tract. 785

**anaphylaxis** (an´´ă-fi-lak´sis) An unusually severe allergic reaction that can result in cardiovascular shock and death. 642

**anastomosis** (ă-nas´tŏ-mo´sis) An interconnecting aggregation of blood vessels or nerves that form a network plexus. 632

**anatomical position** (an´´ă-tom´i-kal) An erect body stance with the eyes directed interior, the arms at the sides, the palms of the hands facing interior, and the fingers pointing straight down. 11

**anatomy** (ă-nat´ŏ-me) The branch of science concerned with the structure of the body and the relationship of its organs. 2

**androgens** (an´drŏ-jenz) Steroids containing 18 carbons that have masculinizing effects; primarily those hormones (such as testosterone) secreted by the testes, although weaker androgens are also secreted by the adrenal cortex. 840

**anemia** (ă-ne´me-ă) An abnormal reduction in the red blood cell count, hemoglobin concentration, or hematocrit, or any combination of these measurements. This condition is associated with a decreased ability of the blood to carry oxygen. 552

**angina pectoris** (an-ji´nă pek´tŏ-ris) A thoracic pain, often referred to the left pectoral and arm area, caused by myocardial ischemia. 608

**angiotensin II** (an´´je-o-ten´sin) An 8-amino-acid polypeptide formed from angiotensin I (a 10-amino-acid precursor), which in turn is formed from cleavage of a protein (angiotensinogen) by the action of renin (an enzyme secreted by the kidneys). Angiotensin II is a powerful vasoconstrictor and a stimulator of aldosterone secretion from the adrenal cortex. 621

**anions** (an´i-onz) Ions that are negatively charged, such as chloride, bicarbonate, and phosphate. 24

**antagonist** (an-tag´ŏ-nist) A muscle that acts in opposition to another muscle. 281

**antebrachium** (an´´te-bra´ke-em) The forearm. 205

**anterior** (ventral) Toward the front; the opposite of *posterior*, or *dorsal*. 12

**anterior pituitary** (pi-too´i-ter-e) *See* adenohypophysis. 519

**anterior root** The anterior projection of the spinal cord, composed of axons of motor neurons. 427

**antibodies** (an´ti-bod´´ēz) Immunoglobin proteins secreted by B lymphocytes that have transformed into plasma cells. Antibodies are responsible for humoral immunity. Their synthesis is induced by specific antigens, and they combine with these specific antigens but not with unrelated antigens. 657

**anticodon** (an´´ti-ko´don) A base triplet provided by three nucleotides within a loop of transfer RNA that is complementary in its base-pairing properties to a triplet (the codon) in mRNA. The matching of codon to anticodon provides the mechanism for translating the genetic code into a specific sequence of amino acids. 56

**antigen** (an´ti-jen) A molecule that can induce the production of antibodies and react in a specific manner with antibodies. 655

**antigenic determinant site** (an-ti-jen´ik) The region of an antigen molecule that specifically reacts with particular antibodies. A large antigen molecule may have a number of such sites. 655

**antiserum** (an´ti-sir´´um) A serum that contains specific antibodies. 665

**anus** (a´nus) The terminal opening of the GI tract. 785

**aorta** (a-or´tă) The major systemic vessel of the arterial system of the body, emerging from the left ventricle. 583

**aortic arch** The superior left bend of the aorta between the ascending and descending portions. 583

**apex** (a´peks) The tip or pointed end of a conical structure. 691

**aphasia** (ă-fa´zhă) Defects in speech, writing, or in the comprehension of spoken or written language caused by brain damage or disease. 388

**apneustic center** (ap-noo´stik) A collection of nuclei (nerve cell bodies) in the brain stem that participates in the rhythmic control of breathing. 703

**apocrine gland** (ap´ŏ-krin) A type of sweat gland that functions in evaporative cooling. It may respond during periods of emotional stress. 148

**aponeurosis** (ap´´ŏ-noo-ro´sis) A fibrous or membranous sheetlike tendon. 255

**appendix** A short pouch that attaches to the cecum. 785

**aqueous humor** (a´kwe-us) The watery fluid that fills the anterior and posterior chambers of the eye. 494

**arachnoid mater** (ă-rak´noid) The weblike middle covering (meninx) of the central nervous system. 396

**arbor vitae** (ar´bor vi´te) The branching arrangement of white matter within the cerebellum. 393

**arm** (brachium) The portion of the upper extremity from the shoulder to the elbow. 203

**arrector pili muscle** (ah-rek´tor pih´le) The smooth muscle attached to a hair follicle that, upon contraction, pulls the hair into a more vertical position, resulting in "goose bumps." 144

**arteriole** (ar-tir´e-ōl) A minute arterial branch. 550

**arteriosclerosis** (ar-tir´´e-o-sklĕ-ro´sis) Any one of a group of diseases characterized by thickening and hardening of the artery wall and in the narrowing of its lumen. 606

**arteriovenous anastomoses** (ar-tir´´e-o-ve´nus ă-nas´´tŏ-mo´sēz) Direct connections between arteries and veins that bypass capillary beds. 632

**artery** (ar´tĕ-re) A blood vessel that carries blood away from the heart. 579

**arthrology** (ar-throl´ŏ-je) The scientific study of the structure and function of joints. 224

**articular cartilage** (ar-tik´yŭ-lar kar´tĭ-lij) A hyaline cartilaginous covering over the articulating surface of the bones of synovial joints. 227

**articulation** (ar-tik´´yŭ-la´shun) A joint. 224

**arytenoid cartilages** (ar´´ĕ-te´noid) A pair of small cartilages located on the superior aspect of the larynx. 687

**ascending colon** (ko´lon) The portion of the large intestine between the cecum and the hepatic flexure. 785

**association neuron** (noor´on) A nerve cell located completely within the central nervous system. It conveys impulses in an arc from sensory to motor neurons; also called *interneuron* or *internuncial neuron*. 347

**astigmatism** (ă-stig´mă-tiz´´em) Unequal curvature of the refractive surfaces of the eye (cornea and/or lens), so that light entering the eye along certain meridians does not focus on the retina. 507

**atherosclerosis** (ath´´ĕ-ro-sklĕ-ro´sis) A common type of arteriosclerosis found in medium and larger arteries in which raised areas within the tunica intima are formed from smooth muscle cells, cholesterol, and other lipids. These plaques occlude arteries and serve as sites for the formation of thrombi. 606

**atomic number** The number of protons in the nucleus of an atom. 22

**atopic dermatitis** (ă-top´ik der´´mă-ti´tis) An allergic skin reaction to agents such as poison ivy and poison oak; a type of delayed hypersensitivity. 687

**ATP** Adenosine triphosphate; the universal energy donor of the cell. 76

**atretic** (ă-tret´ik) Without an opening. Atretic ovarian follicles are those that fail to ovulate. 878

**atrioventricular bundle** (a´´tre-o-ven-trik´yŭ-lar) A group of specialized cardiac fibers that conduct impulses from the atrioventricular node to the ventricular muscles of the heart; also called the *bundle of His* or *AV bundle*. 576

**atrioventricular node** A microscopic aggregation of specialized cardiac fibers located in the interatrial septum of the heart that are a part of the conduction system of the heart; *AV node*. 576

**atrioventricular valve** A cardiac valve located between an atrium and a ventricle of the heart; *AV valve*. 567

**atrium** (a´tre-um) Either of the two superior chambers of the heart that receive venous blood. 567

**atrophy** (at´rŏ-fe) A gradual wasting away or decrease in the size of a tissue or an organ. 133

**atropine** (at´rŏ-pēn) An alkaloid drug obtained from a plant of the species *Belladonna* that acts as an anticholinergic agent. It is used medically to inhibit parasympathetic nerve effects, dilate the pupils of the eye, increase the heart rate, and inhibit intestinal movements. 457

**auditory** (aw´dĭ-tor-e) Pertaining to the structures of the ear associated with hearing. 482

**auditory tube** A narrow canal that connects the middle ear chamber to the pharynx; also called the *eustachian canal*. 483

**auricle** (or´ĭ-kul) **1.** The fleshy pinna of the ear. 482 **2.** An ear-shaped appendage of each atrium of the heart. 567

**autoantibodies** (aw´´to-an´tĭ-bod´´ēz) Antibodies formed in response to, and that react with, molecules that are part of one's own body. 673

**autonomic nervous system** (aw´´tŏ-nom´ik) The sympathetic and parasympathetic portions of the nervous system that function to control the actions of the visceral organs and skin; *ANS*. 446

**autosomal chromosomes** (aw´´to-so´mal kro´mŏ-sōmz) The paired chromosomes; those other than the sex chromosomes. 839

**axilla** (ak-sil´ă) The depressed hollow commonly called the armpit. 14

**axon** (ak´son) The elongated process of a nerve cell that transmits an impulse away from the cell body of a neuron. 345

## B

**ball-and-socket joint** The most freely movable type of synovial joint (e.g., the shoulder or hip joint). 231

**baroreceptor** (bar´´o-re-sep´tor) A cluster of neuroreceptors stimulated by pressure changes. Baroreceptors monitor blood pressure. 634

**basal metabolic rate** (BMR) (ba´sal met´´ă-bol´ik) The rate of metabolism (expressed as oxygen consumption or heat production) under resting or basal conditions (14 to 18 hours after eating). 812

**basal nucleus** (ba´sal noo´kle-us) A mass of nerve cell bodies located deep within a cerebral hemisphere of the brain; also called *basal ganglion*. 388

**base** A chemical substance that ionizes in water to release hydroxyl ions (OH⁻) or other ions that combine with hydrogen ions. 26

**basement membrane** A thin sheet of extracellular substance to which the basal surfaces of membranous epithelial cells are attached; also called the *basal lamina*. 110

**basophil** (ba´sŏ-fil) A granular leukocyte that readily stains with basophilic dye. 553

**B cell lymphocytes** Lymphocytes that can be transformed by antigens into plasma cells that secrete antibodies (and are thus responsible for humoral immunity). The *B* stands for *bursa equivalent*. 656

**belly** The thickest circumference of a skeletal muscle.

**benign** (bĭ-nīn´) Not malignant. 674

**bifurcate** (bi´fur-kāt) Forked; divided into two branches.

**bile** A liver secretion that is stored and concentrated in the gallbladder and released through the common bile duct into the duodenum. It is essential for the absorption of fats. 791

**bilirubin** (bil´´ĭ-roo´bin) Bile pigment derived from the breakdown of the heme portion of hemoglobin. 792

**bipennate** (bi-pen´āt) Denoting muscles that have a fiber architecture coursing obliquely on both sides of a tendon. 281

**blastula** (blas´tyoo-lă) An early stage of prenatal development between the morula and embryonic period. 903

**blood** The fluid connective tissue that circulates through the cardiovascular system to transport substances throughout the body. 551

**blood-brain barrier** A specialized mechanism that inhibits the passage of certain materials from the blood into brain tissue and cerebrospinal fluid. 352

**bolus** (bo´lus) A moistened mass of food that is swallowed from the oral cavity into the pharynx. 773

**bone** A solid, rigid, ossified connective tissue forming an organ of the skeletal system. 127

**bony labyrinth** (lab´ĭ-rinth) A series of chambers within the petrous part of the temporal bone associated with the vestibular organs and the cochlea. The bony labyrinth contains a fluid called perilymph. 478

**Bowman's capsule** (bo´manz kap´sul) *See* glomerular capsule. 730

**brachial plexus** (bra´ke-al plek´sus) A network of nerve fibers that arise from spinal nerves C5–C8 and T1. Nerves arising from the brachial plexuses supply the upper extremities. 428

**bradycardia** (brad´´ĭ-kar´de-ă) A slow cardiac rate; fewer than 60 beats per minute. 603

**bradykinins** (brad´´ĭ-ki´ninz) Short polypeptides that stimulate vasodilation and other cardiovascular changes. 633

**brain** The enlarged superior portion of the central nervous system located in the cranial cavity of the skull. 376

**brain stem** The portion of the brain consisting of the medulla oblongata, pons, and midbrain. 392

**bronchial tree** (brong´ke-al) The bronchi and their branching bronchioles. 689

**bronchiole** (brong´ke-ōl) A small division of a bronchus within the lung. 689

**bronchus** (brong´kus) A branch of the trachea that leads to a lung. 689

**buccal cavity** (buk´al) The mouth, or *oral cavity*. 770

**buffer** A molecule that serves to prevent large changes in pH by either combining with H⁺ or by releasing H⁺ into solution. 26

**bulbourethral glands** (bul´´bo-yoo-re´thral) A pair of glands that secrete a viscous fluid into the male urethra during sexual excitement; also called *Cowper's glands*. 855

**bundle of His** *See* atrioventricular bundle. 576

**bursa** (bur´sa) A saclike structure filled with synovial fluid. Bursae are located at friction points, as around joints, over which tendons can slide without contacting bone. 228

**buttocks** (but´oks) The rump or fleshy masses on the posterior aspect of the lower trunk, formed primarily by the gluteal muscles. 14

## C

**calcitonin** (kal´´sĭ-to´nin) Also called *thyrocalcitonin*. A polypeptide hormone produced by the parafollicular cells of the thyroid and secreted in response to hypercalcemia. It acts to lower blood calcium and phosphate concentrations and may serve as an antagonist of parathyroid hormones. 173

**calmodulin** (kal´´mod-yŭ´lin) A receptor protein for Ca⁺⁺ located within the cytoplasm of target cells. It appears to mediate the effects of this ion on cellular activities. 274

**calorie** (kal´ŏ-re) A unit of heat equal to the amount needed to raise the temperature of one gram of water by 1 C°. 76

**calyx** (ka´liks) A cup-shaped portion of the renal pelvis that encircles a renal papilla. 729

**cAMP** Cyclic adenosine monophosphate; a second messenger in the action of many hormones, including catecholamines, polypeptides, and glycoproteins. It serves to mediate the effects of these hormones on their target cells. 536

**canaliculus** (kan´´ă-lik´yŭ-lus) A microscopic channel in bone tissue that connects lacunae. 130

**canal of Schlemm** (shlem) *See* scleral venous sinus. 494

**cancer** A tumor characterized by abnormally rapid cell division and the loss of specialized tissue characteristics. This term usually refers to malignant tumors. 674

**capacitation** (kă-pas´´ĭ-ta´shun) The process whereby spermatozoa gain the ability to fertilize ova. Sperm that have not have been capacitated in the female reproductive tract cannot fertilize ova. 899

**capillary** (kap´ĭ-lar´´e) A microscopic blood vessel that connects an arteriole and a venule; the functional unit of the circulatory system. 579

**carbonic anhydrase** (kar-bon´ik an-hi´drās) An enzyme that catalyzes the formation or breakdown of carbonic acid. When carbon dioxide concentrations are relatively high, this enzyme catalyzes the formation of carbonic acid from $CO_2$ and $H_2O$. When carbon dioxide concentrations are low, the breakdown of carbonic acid to $CO_2$ and $H_2O$ is catalyzed. These reactions aid the transport of carbon dioxide from tissues to alveolar air. 711

**cardiac muscle** (kar´de-ak) Muscle of the heart, consisting of striated muscle cells. These cells are interconnected into a mass called the myocardium. 273

**cardiac output** The volume of blood pumped per minute by either the right or left ventricle. 612

**cardiogenic shock** (kar´´de-o-jen´ik) Shock that results from low cardiac output in heart disease. 642

**carotid sinus** (kă-rot´id) An expanded portion of the internal carotid artery located immediately above the point of branching from the external carotid artery. The carotid sinus contains baroreceptors that monitor blood pressure. 583

**carpus** (kar´pus) The proximal portion of the hand that contains the eight carpal bones. 206

**carrier-mediated transport** The transport of molecules or ions across a cell membrane by means of specific protein carriers. It includes both facilitated diffusion and active transport. 97

**cartilage** (kar´tĭ-lij) A type of connective tissue with a solid elastic matrix. 126

**cartilaginous joint** (kar´´tĭ-laj´ĭ-nus) A joint that lacks a joint cavity, permitting little movement between the bones held together by cartilage. 226

**cast** An accumulation of proteins molded from the kidney tubules that appears in urine sediment 759.

**catabolism** (kă-tab´o-liz-em) The metabolic breakdown of complex molecules into simpler ones, often resulting in a release of energy. 814

**catecholamines** (kat´´ĕ-kol´ă-mēnz) A group of molecules including epinephrine, norepinephrine, L-dopa, and related molecules with effects similar to those produced by activation of the sympathetic nervous system. 364

**cations** (kat´i-onz) Positively charged ions, such as sodium, potassium, calcium, and magnesium. 24

**cauda equina** (kaw´dă e-kwi´nă) The lower end of the spinal cord where the roots of spinal nerves have a tail-like appearance. 402

**cecum** (se´kum) The pouchlike portion of the large intestine to which the ileum of the small intestine is attached. 785

**cell** The structural and functional unit of an organism; the smallest structure capable of performing all the functions necessary for life. 42

**cell-mediated immunity** (ĭ-myoo´nĭ-te) Immunological defense provided by T cell lymphocytes that come within close proximity of their victim cells (as opposed to humoral immunity provided by the secretion of antibodies by plasma cells). 656

**cellular respiration** (sel´yŭ-lar res´´pĭ-ra´shun) The energy-releasing metabolic pathways in a cell that oxidize organic molecules such as glucose and fatty acids. 79

**cementum** (se-men´tum) Bonelike material that binds the root of a tooth to the periodontal membrane of the bony socket. 773

**central canal** An elongated longitudinal channel in the center of an osteon in bone tissue that contains branches of the nutrient vessels and a nerve; also called a *haversian canal*. 130

**central nervous system** Part of the nervous system consisting of the brain and the spinal cord; *CNS*. 345

**centrioles** (sen´trĭ-olz) Cell organelles that form the spindle apparatus during cell division. 61

**centromere** (sen´trŏ-mēr) The central region of a chromosome to which the chromosomal arms are attached. 61

**centrosome** (sen´trŏ-sōm) A dense body near the nucleus of a cell that contains a pair of centrioles. 61

**cerebellar peduncle** (ser´´ĕ-bel´ar pĕ-dung´k'l) An aggregation of nerve fibers connecting the cerebellum with the brain stem. 393

**cerebellum** (ser´´ĕ-bel´um) The portion of the brain concerned with the coordination of skeletal muscle contraction. Part of the metencephalon, it consists of two hemispheres and a central vermis. 393

**cerebral arterial circle** (ser´ĕ-bral) An arterial vessel that encircles the pituitary gland. It provides alternate routes for blood to reach the brain should a carotid or vertebral artery become occluded; also called the *circle of Willis*. 585

**cerebral peduncles** A paired bundle of nerve fibers along the inferior surface of the midbrain that conduct impulses between the pons and the cerebral hemispheres. 392

**cerebrospinal fluid** (ser´´ĕ-bro-spi´nal) A fluid produced by the choroid plexus of the ventricles of the brain. It fills the ventricles and surrounds the central nervous system in association with the meninges. 376

**cerebrum** (ser´ĕ-brum) The largest portion of the brain, composed of the right and left hemispheres. 380

**ceruminous gland** (sĕ-roo´mĭ-nus) A specialized integumentary gland that secretes cerumen, or earwax, into the external auditory canal. 482

**cervical** (ser´vĭ-kal) Pertaining to the neck or a necklike portion of an organ.

**cervical ganglion** (gang´gle-on) A cluster of postganglionic sympathetic nerve cell bodies located in the neck, near the cervical vertebrae. 452

**cervical plexus** (plek´sus) A network of spinal nerves formed by the anterior branches of the first four cervical nerves. 428

**cervix** (ser´viks) 1. The narrow necklike portion of an organ. 2. The inferior end of the uterus that adjoins the vagina (cervix of the uterus). 873

**chemoreceptor** (ke´mo-re-sep´tor) A neuroreceptor that is stimulated by the presence of chemical molecules. 465

**chemotaxis** (ke˝mo-tak´sis) The movement of an organism or a cell, such as a leukocyte, toward a chemical stimulus. 652

**Cheyne–Stokes respiration** (chān´stōkes´ res˝pĭ-ra´shun) Breathing characterized by rhythmic waxing and waning of the depth of respiration, with regularly occurring periods of apnea (failure to breathe). 720

**chiasma** (ki-as´mă) A crossing of nerve tracts from one side of the CNS to the other; also called a *chiasm*. 420

**choane** (ko-a´ne) The two posterior openings from the nasal cavity into the nasal pharynx; also called the *internal nares*. 684

**cholesterol** (kŏ-les´ter-ol) A 27-carbon steroid that serves as the precursor of steroid hormones. 34

**cholinergic** (ko˝lĭ-ner´jik) Denoting nerve endings that liberate acetylcholine as a neurotransmitter, such as those of the parasympathetic system. 455

**chondrocranium** (kon˝dro-kra´ne-um) The portion of the skull that supports the brain. It is derived from endochondral bone. 192

**chondrocytes** (kon´dro-sītz) Cartilage-forming cells. 126

**chordae tendineae** (kor´de ten-din´e-e) Chordlike tendinous bands that connect papillary muscles to the leaflets of the atrioventricular valves within the ventricles of the heart. 569

**chorea** (kŏ-re-ă) The occurrence of a wide variety of rapid, complex, jerky movements that appear to be well coordinated but that are performed involuntarily. 366

**chorion** An extraembryonic membrane that participates in the formation of the placenta. 910

**choroid** (kor´oid) The vascular, pigmented middle layer of the wall of the eye. 492

**choroid plexus** A mass of vascular capillaries from which cerebrospinal fluid is secreted into the ventricles of the brain. 391

**chromatids** (kro´mă-tidz) Duplicated chromosomes, joined together at the centromere, that separate during cell division. 61

**chromatin** (kro´mă-tin) Threadlike structures in the cell nucleus consisting primarily of DNA and protein. They represent the extended form of chromosomes during interphase. 53

**chromatophilic substances** (kro˝mă-to-fil´ik) Clumps of rough endoplasmic reticulum in the cell bodies of neurons; also called *Nissl bodies*. 345

**chromosomes** (kro´mŏ-sōmz) Structures in the nucleus that contain the genes for genetic expression. 64

**chyme** (kīm) The mass of partially digested food that passes from the pylorus of the stomach into the duodenum of the small intestine. 776

**cilia** (sil´e-ă) Microscopic hairlike processes that move in a wavelike manner on the exposed surfaces of certain epithelial cells. 45

**ciliary body** (sil´e-er˝e) A portion of the choroid layer of the eye that secretes aqueous humor. It contains the ciliary muscle. 492

**circadian rhythms** (ser˝kă-de´an) Physiological changes that repeat at about 24-hour intervals. These are often synchronized with changes in the external environment, such as the day-night cycles. 524

**circle of Willis** See cerebral arterial circle. 585

**circumduction** (ser˝kum-duk´shun) A movement of a body part that outlines a cone, such that the distal end moves in a circle while the proximal portion remains relatively stable. 234

**cirrhosis** (sĭ-ro´sis) Liver disease characterized by loss of normal microscopic structure, which is replaced by fibrosis and nodular regeneration. 790

**clitoris** (klit´or-is, kli´tor-is) A small, erectile structure in the vulva of the female, homologous to the glans penis in the male. 876

**clone** (klōn) 1. A group of cells derived from a single parent cell by mitotic cell division; since reproduction is asexual, the descendants of the parent cell are genetically identical. 62 2. A term used to refer to cells as separate individuals (as in white blood cells) rather than as part of a growing organ. 664

**CNS** See central nervous system. 345

**coccygeal** (kok-sij´e-al) Pertaining to the region of the coccyx; the caudal termination of the vertebral column. 191

**cochlea** (kok´le-ă) The organ of hearing in the inner ear where nerve impulses are generated in response to sound waves. 483

**cochlear window** See round window. 483

**codon** (ko´don) The sequence of three nucleotide bases in mRNA that specifies a given amino acid and determines the position of that amino acid in a polypeptide chain through complementary base pairing with an anticodon in RNA. 55

**coelom** (se´lom) The abdominal cavity. 15

**coenzyme** (ko-en´zīm) An organic molecule, usually derived from a water-soluble vitamin, that combines with and activates specific enzyme proteins. 73

**cofactor** (ko´fak-tor) A substance needed for the catalytic action of an enzyme; generally used in reference to inorganic ions such as $Ca^{++}$ and $Mg^{++}$. 73

**collateral** (kŏ-lat´er-al) A small branch of a blood vessel or nerve fiber.

**colloid osmotic pressure** (kol´oid oz-mot´ik) Osmotic pressure exerted by plasma proteins that are present as a colloidal suspension; also called *oncotic pressure*. 617

**colon** (ko´lon) The first portion of the large intestine. 785

**common bile duct** A tube formed by the union of the hepatic duct and cystic duct that transports bile to the duodenum. 793

**compact bone** Tightly packed bone that is superficial to spongy bone and covered by the periosteum; also called *dense bone*. 165

**compliance** (kom-pli´ans) A measure of the ease with which a structure such as the lung expands under pressure; a measure of the change in volume as a function of pressure changes. 693

**conduction myofibers** Specialized large-diameter cardiac muscle fibers that conduct electrical impulses from the AV bundle into the ventricular walls; also called *Purkinje fibers*. 576

**condyle** (kon´dīl) A rounded process at the end of a long bone that forms an articulation. 164

**cone** A color receptor cell in the retina of the eye. 500

**congenital** (kon-jen´ĭ-tal) Present at the time of birth.

**congestive heart failure** (kon-jes´tiv) The inability of the heart to deliver an adequate blood flow as a result of heart disease or hypertension. This condition is associated with breathlessness, salt and water retention, and edema. 643

**conjunctiva** (kon˝jungk-ti´vă) The thin membrane covering the anterior surface of the eyeball and lining the eyelids. 490

**conjunctivitis** (kon-jungk˝ti´vi-tis) Inflammation of the conjunctiva of the eye, which is sometimes called "pink eye." 508

**connective tissue** One of the four basic tissue types within the body. It is a binding and supportive tissue with abundant matrix. 120

**Conn's syndrome** (konz) Primary hyperaldosteronism; excessive secretion of aldosterone produces electrolyte imbalances. 750

**contralateral** (kon˝tră-lat´er-al) Taking place or originating in a corresponding part on the opposite side of the body. 438

**conus medullaris** (kó nus med˝yŭ-lār´is) The inferior, tapering portion of the spinal cord. 402

**convolution** (kon-vŏ-loo´shun) An elevation on the surface of a structure and an infolding of the tissue upon itself. 383

**cornea** (kor´ne-ă) The transparent, convex, anterior portion of the outer layer of the eyeball. 492

**coronal plane** (kor´ŏ-nal, kŏ-ro´nal) A plane that divides the body into anterior and posterior portions; also called a *frontal plane*. 11

**coronary circulation** (kor´ŏ-nar˝e) The arterial and venous blood circulation to the wall of the heart. 571

**coronary sinus** A large venous channel on the posterior surface of the heart into which the cardiac veins drain. 571

**corpora quadrigemina** (kor´por-ă kwad˝rĭ-jem´ĭ-na) Four superior lobes of the midbrain concerned with visual and auditory functions. 392

**corpus callosum** (kor´pus kă-lo´sum) A large tract of white matter within the brain that connects the right and left cerebral hemispheres. 380

**corpuscle of touch** (kor´pus'l) A touch sensory receptor found in the papillary layer of the dermis of the skin; also called *Meissner's corpuscle*. 467

**cortex** (kor´teks) 1. The outer layer of an internal organ or body structure, as of the kidney or adrenal gland. 526 2. The convoluted layer of gray matter that covers the surface of each cerebral hemisphere. 383

**corticosteroids** (kor´´tǐ-ko-ster´oidz) Steroid hormones of the adrenal cortex, consisting of glucocorticoids (such as hydrocortisone) and mineralocortocoids (such as aldosterone). 34

**costal cartilage** (kos´tal) The cartilage that connects the ribs to the sternum. 198

**cranial** (kra´ne-al) Pertaining to the cranium.

**cranial nerves** One of 12 pairs of nerves that arise from the brain. 417

**cranium** (kra´ne-um) The bones of the skull that enclose or support the brain and the organs of sight, hearing, and balance. 192

**creatine phosphate** (kre´ă-tin fos´fāt) An organic phosphate molecule in muscle cells that serves as a source of high-energy phosphate for the synthesis of ATP; also called *phosphocreatine*. 271

**crenation** (krǐ-na´shun) A notched or scalloped appearance of the red blood cell membrane caused by the osmotic loss of water from these cells. 96

**crest** A thickened ridge of bone for the attachment of muscle.

**cretinism** (krēt´n-iz´´em) A condition caused by insufficient thyroid secretion during prenatal development or the years of early childhood. It results in stunted growth and inadequate mental development. 543

**cricoid cartilage** (kri´koid) A ring-shaped cartilage that forms the inferior portion of the larynx. 687

**crista** (kris´tă) A crest, such as the crista galli that extends superiorly from the cribriform plate of the ethmoid bone. 187

**cryptorchidism** (krip-tor´kǐ-diz´´em) A developmental defect in which one or both testes fail to descend into the scrotum and, instead, remain in the body cavity. 863

**cubital** (kyoo´bǐ-tal) Pertaining to the antebrachium. The cubital fossa is the anterior aspect of the elbow joint. 15

**curare** (koo-ră-re) A chemical derived from plant sources that causes flaccid paralysis by blocking ACh receptor proteins in muscle cell membranes. 362

**Cushing's syndrome** (koosh´ingz) Symptoms caused by the hypersecretion of adrenal steroid hormones as a result of tumors of the adrenal cortex or ACTH-secreting tumors of the anterior pituitary. 543

**cyanosis** (si´ă-no´sis) A bluish discoloration of the skin or mucous membranes due to excessive concentration of deoxyhemoglobin; indicates inadequate oxygen concentration in the blood. 141

**cystic duct** (sis´tik dukt) The tube that transports bile from the gallbladder to the common bile duct. 793

**cytochrome** (si´tŏ-krōm) A pigment in mitochondria that transports electrons in the process of aerobic respiration. 84

**cytokinesis** (si´´to-kǐ-ne´sis) The division of the cytoplasm that occurs in mitosis and meiosis, when a parent cell divides to produce two daughter cells. 62

**cytology** (si-tol´ŏ-je) The science dealing with the study of cells. 5

**cytoplasm** (si´tŏ-plaz´´em) In a cell, the protoplasm located outside of the nucleus. 46

**cytoskeleton** (si´´to-skel´ě-ton) A latticework of structural proteins in the cytoplasm arranged in the form of microfilaments and microtubules. 47

## D

**deciduous** (dǐ-sij´oo-us) Pertaining to something shed or cast off in a particular sequence. Deciduous teeth are shed and replaced by permanent teeth during development. 771

**decussation** (dek´´uh-sa´shun) A crossing of nerve fibers from one side of the CNS to the other. 403

**defecation** (def´ě-ka´shun) The elimination of feces from the rectum through the anal canal and out the anus. 787

**deglutition** (de´´gloo-tish´un) The act of swallowing. 775

**delayed hypersensitivity** An allergic response in which the onset of symptoms may not occur until 2 or 3 days after exposure to an antigen. Produced by T cells, it is a type of cell-mediated immunity. 678

**denaturation** (de-na´´chur-a´shun) Irreversible changes in the tertiary structure of proteins caused by heat or drastic pH changes. 36

**dendrite** (den´drīt) A nerve cell process that transmits impulses toward a neuron cell body. 345

**dentin** (den´tǐn) The main substance of a tooth, covered by enamel over the crown of the tooth and by cementum on the root. 773

**dentition** (den-tish´un) The number, arrangement, and shape of teeth. 773

**depolarization** (de-po´´lar-ǐ-za´shun) The loss of membrane polarity in which the inside of the cell membrane becomes less negative in comparison to the outside of the membrane. The term is also used to indicate the reversal of membrane polarity that occurs during the production of action potentials in nerve and muscle cells. 353

**dermal papilla** (pă-pil´ă) A projection of the dermis into the epidermis. 142

**dermis** (der´mis) The second, or deep, layer of skin beneath the epidermis. 142

**descending colon** The segment of the large intestine that descends on the left side from the level of the spleen to the level of the left iliac crest. 785

**diabetes insipidus** (di´´ă-be´tēz in-sip´ǐ-dus) A condition in which inadequate amounts of antidiuretic hormone (ADH) are secreted by the posterior pituitary. It results in the inadequate reabsorption of water by the kidney tubules and, thus, in the excretion of a large volume of dilute urine. 543

**diabetes mellitus** (mě-li´tus) The appearance of glucose in the urine due to the presence of high plasma glucose concentrations, even in the fasting state. This disease is caused by either lack of

sufficient insulin secretion or inadequate responsiveness of the target tissues to the effects of insulin. 833

**diapedesis** (di´ă-pě-de´sis) The migration of white blood cells through the endothelial walls of blood capillaries into the surrounding connective tissues. 652

**diaphragm** (di´ă-fram) A sheetlike dome of muscle and connective tissue that separates the thoracic and abdominal cavities. 290

**diaphysis** (di-af´ǐ-sis) The shaft of a long bone. 165

**diarrhea** (di´´ă-re´ă) Abnormal frequency of defecation accompanied by abnormal liquidity of the feces. 786

**diarthrosis** (di´´ar-thro´sis) A type of functionally classified joint in which the articulating bones are freely movable; also called a *synovial joint*. 224

**diastole** (di-as´tŏ-le) The sequence of the cardiac cycle during which a heart chamber wall is relaxed. 573

**diencephalon** (di´´en-sef´ă-lon) A major region of the brain that includes the third ventricle, thalamus, hypothalamus, and pituitary gland. 390

**diffusion** (dǐ-fyoo´zhun) The net movement of molecules or ions from regions of higher to regions of lower concentration. 92

**digestion** The process by which larger molecules of food substance are broken down mechanically and chemically into smaller molecules that can be absorbed. 765

**diploe** (dip´lo-e) The spongy layer of bone positioned between the inner and outer layers of compact bone. 165

**diploid** (dip´loid) Denoting cells having two of each chromosome or twice the number of chromosomes that are present in sperm or ova. 839

**disaccharide** (di-sak´ă-rīd) Any of a class of double sugars; carbohydrates that yield two simple sugars, or monosaccharides, upon hydrolysis. 30

**distal** (dis´tal) Away from the midline or origin; the opposite of *proximal*. 12

**diuretic** (di´´yŭ-ret´ik) An agent that promotes the excretion of urine, thereby lowering blood volume and pressure. 620

**DNA** Deoxyribonucleic acid; composed of nucleotide bases and deoxyribose sugar. It is found in all living cells and contains the genetic code. 52

**dopamine** (do´pă-mēn) A type of neurotransmitter in the central nervous system; also is the precursor of norepinephrine, another neurotransmitter molecule. 364

**dorsal** (dor´sal) Pertaining to the back or posterior portion of a body part; the opposite of *ventral*; also called *posterior*. 12

**dorsal root ganglion** See posterior root ganglion. 427

**dorsiflexion** (dor´´sǐ-flek´shun) Movement at the ankle as the dorsum of the foot is elevated. 232

**ductus arteriosus** (duk´tus ar-tir´´e-o´sus) The blood vessel that connects the pulmonary trunk and the aorta in a fetus. 601

**ductus deferens** (def´er-enz) pl. *ductus deferentia* A tube that carries spermatozoa from the epididymis to the ejaculatory duct; also called the *vas deferens* or *seminal duct*. 854

**ductus venosus** (ven-o´sus) A fetal blood vessel that connects the umbilical vein and the inferior vena cava. 601

**duodenum** (doo´ŏ-de´num, doo-od´ĕ-num) The first portion of the small intestine that leads from the pylorus of the stomach to the jejunum. 781

**dura mater** (door´ă ma´ter) The outermost meninx. 396

**dwarfism** A condition in which a person is undersized due to inadequate secretion of growth hormone. 173

**dyspnea** (disp-ne´ă) Subjective difficulty in breathing. 699

## E

**eccrine gland** (ek´rin) A sweat gland that functions in thermoregulation. 148

**ECG** *See* electrocardiogram. 577

**ectoderm** (ek´tŏ-derm) The outermost of the three primary germ layers of an embryo. 904

**ectopic focus** (ek-top´ik) An area of the heart other than the SA node that assumes pacemaker activity. 575

**ectopic pregnancy** Embryonic development that occurs anywhere other than in the uterus (as in the uterine tubes or body cavity). 890

**edema** (ĕ-de´mă) An excessive accumulation of fluid in the body tissues. 618

**EEG** *See* electroencephalogram. 386

**effector** (ĕ-fek´tor) An organ, such as a gland or muscle, that responds to a motor stimulation. 17

**efferent** (ef´er-ent) Conveying away from the center of an organ or structure. 346

**efferent arteriole** (ar-tir´e-ōl) An arteriole of the renal vascular system that conducts blood away from the glomerulus of a nephron. 730

**efferent ductules** (duk´toolz) A series of coiled tubules through which spermatozoa are transported from the rete testis to the epididymis. 848

**efferent neuron** (noor´on) *See* motor neuron. 346

**ejaculation** (ĕ-jak´yŭ-la´shun) The discharge of semen from the male urethra that accompanies orgasm. 858

**ejaculatory duct** (ĕ-jak´yŭ-lă-tor´´-e) A tube that transports spermatozoa from the ductus deferens to the prostatic urethra. 854

**elastic fibers** (ĕ-las´tik) Protein strands that are found in certain connective tissue that have contractile properties. 124

**elbow** The synovial joint between the brachium and the antebrachium. 241

**electrocardiogram** (ĕ-lek´´tro-kar´de-ŏ-gram´´) A recording of the electrical activity that accompanies the cardiac cycle; ECG or EKG. 577

**electroencephalogram** (ĕ-lek´´tro-en-sef´ă-lŏ-gram) A recording of the brain-wave patterns or electrical impulses of the brain from electrodes placed on the scalp; EEG. 386

**electrolytes** (ĕ-lek´tro-lītz) Ions and molecules that are able to ionize and thus carry an electric current. The most common electrolytes in the plasma are Na$^+$, HCO$_3^-$, and K$^+$. 728

**electromyogram** (ĕ-lek´´tro-mi´ŏ-gram) A recording of the electrical impulses or activity of skeletal muscles using surface electrodes; EMG. 318

**electrophoresis** (ĕ-lek´´tro-fŏ-re´sis) A biochemical technique in which different molecules can be separated and identified by their rate of movement in an electric field. 657

**elephantiasis** (el´´ĕ-fan-ti´ă-sis) A disease caused by infection with a nematode worm in which the larvae block lymphatic drainage and produce edema; the lower areas of the body can become enormously swollen as a result. 618

**embryology** (em´´bre-ol´ŏ-je) The study of prenatal development from conception through the eighth week in utero. 906

**EMG** *See* electromyogram. 318

**emphysema** (em´´fĭ-se´mă, em´´fĭ-ze´mă) A lung disease in which the alveoli are destroyed and the remaining alveoli become larger. It results in decreased vital capacity and increased airway resistance. 719

**emulsification** (ĕ-mul´´sĭ-fĭ-ka´shun) The process of producing an emulsion or fine suspension; in the small intestine, fat globules are emulsified by the detergent action of bile. 798

**enamel** (ĕ-nam´el) The outer dense substance covering the crown of a tooth. 773

**endergonic** (en´´der-gon´ik) Denoting a chemical reaction that requires the input of energy from an external source in order to proceed. 76

**endocardium** (en´´do-kar´de-um) The endothelial lining of the heart chambers and valves. 567

**endochondral bone** (en´´dŏ-kon´dral) Denoting bones that develop as hyaline cartilage models first and that are then ossified. 167

**endocrine gland** (en´dŏ-krin) A ductless, hormone-producing gland that is part of the endocrine system. 514

**endocytosis** (en´´do-si-to´sis) A general term for the cellular uptake of particles that are too large to cross the cell membrane. *See also* phagocytosis and pinocytosis. 45

**endoderm** (en´dŏ-derm) The innermost of the three primary germ layers of an embryo. 904

**endogenous** (en-doj´ĕ-nus) Denoting a product or process arising from within the body (as opposed to exogenous products or influences from external sources).

**endolymph** (en´dŏ-limf) A fluid within the membranous labyrinth and cochlear duct of the inner ear that aids in the conduction of vibrations involved in hearing and the maintenance of equilibrium. 478

**endometrium** (en´´do-me´tre-um) The inner lining of the uterus. 874

**endomysium** (en´´do-mis´e-um) The connective tissue sheath that surrounds each skeletal muscle fiber, separating the muscle cells from one another. 255

**endoneurium** (en´´do-nyoo´re-um) The connective tissue sheath that surrounds each nerve fiber, separating the nerve fibers one from another within a nerve. 347

**endoplasmic reticulum** (en-do-plaz´mik rĕ-tik´yŭ-lum) A cytoplasmic organelle composed of a network of canals running through the cytoplasm of a cell. 50

**endorphins** (en-dor´finz) A group of endogenous opiate molecules that may act as a natural analgesic. 367

**endothelium** (en´´do-the´le-um) The layer of epithelial tissue that forms the thin inner lining of blood vessels and heart chambers. 110

**endotoxin** (en´´do-tok´sin) A toxin found within certain types of bacteria that is able to stimulate the release of endogenous pyrogen and produce a fever. 653

**enkephalins** (en-kef´ă-linz) Short polypeptides, containing five amino acids, that have analgesic effects and that may function as neurotransmitters in the brain. The two known enkephalins (which differ in only one amino acid) are endorphins. 367

**enteric** (en-ter´ik) The term referring to the small intestine.

**entropy** (en´trŏ-pe) The energy of a system that is not available to perform work. A measure of the degree of disorder in a system, entropy increases whenever energy is transformed. 75

**enzyme** (en´zīm) A protein catalyst that increases the rate of specific chemical reactions. 70

**eosinophil** (e´´ŏ-sin´ŏ-fil) A type of white blood cell characterized by the presence of cytoplasmic granules that become stained by acidic eosin dye. Eosinophils normally constitute about 2% to 4% of the white blood cells. 553

**epicardium** (ep´´ĭ-kar´de-um) A thin, outer layer of the heart; also called the *visceral pericardium*. 566

**epicondyle** (ep´´ĭ-kon´dīl) A projection of bone above a condyle. 164

**epidermis** (ep´´ĭ-der´mis) The outermost layer of the skin, composed of several stratified squamous epithelial layers. 138

**epididymis** (ep´ĭ-did´ĭ-mis) A highly coiled tube located along the posterior border of the testis. It stores spermatozoa and transports them from the seminiferous tubules of the testis to the ductus deferens. 853

**epidural space** (ep´´ĭ-door´al) A space between the spinal dura mater and the bone of the vertebral canal. 396

**epiglottis** (ep´´ĭ-glot´is) A leaflike structure positioned on top of the larynx. It covers the glottis during swallowing. 687

**epimysium** (ep´´ĭ-mis´e-um) A fibrous outer sheath of connective tissue surrounding a skeletal muscle. 255

**epinephrine** (ep′′ĭ-nef′rin) A hormone secreted from the adrenal medulla resulting in actions similar to those resulting from sympathetic nervous system stimulation; also called *adrenaline*. 528

**epineurium** (ep′′ĭ-nyoo′re-um) A fibrous outer sheath of connective tissue surrounding a nerve. 347

**epiphyseal plate** (ep′′ĭ-fiz′e-al) A hyaline cartilaginous layer located between the epiphysis and diaphysis of a long bone. It functions as a longitudinal growing region. 166

**epiphysis** (ĕ-pif′ĭ-sis) The end segment of a long bone, separated from the diaphysis early in life by an epiphyseal plate but later becoming part of the larger bone. 165

**episiotomy** (ĕ-pe′′ze-ot′ŏ-me) An incision of the perineum at the end of the second stage of labor to facilitate delivery and to avoid tearing the perineum. 922

**epithelial tissue** (ep′′ĭ-the′le-al) One of the four basic tissue types; the type of tissue that covers or lines all exposed body surfaces. 108

**eponychium** (ep′′ŏ-nik′e-um) The thin layer of stratum corneum of the epidermis of the skin that overlaps and protects the lunula of the nail. 147

**EPSP** Excitatory postsynaptic potential; a graded depolarization of a postsynaptic membrane in response to stimulation by a neurotransmitter chemical. EPSPs can be summated but can be transmitted only over short distances. They can stimulate the production of action potentials when a threshold level of depolarization has been attained. 361

**erythroblastosis fetalis** (ĕ-rith′′ro-blas-to′sis fĭ-tal′is) Hemolytic anemia in an Rh positive newborn caused by maternal antibodies against the Rh factor that have crossed the placenta. 558

**erythrocyte** (ĕ-rith′rŏ-sīt) A red blood cell. 552

**esophagus** (ĕ-sof′ă-gus) A tubular portion of the GI tract that leads from the pharynx to the stomach as it passes through the thoracic cavity. 775

**essential amino acids** Those eight amino acids in adults or nine amino acids in children that cannot be made by the human body; therefore, they must be obtained in the diet. 819

**estrogens** (es′tro-jenz) Any of several female sex hormones secreted from the ovarian (graafian) follicle. 533

**estrus cycle** (es′trus) Cyclic changes in the structure and function of the ovaries and female reproductive tract of mammals other than humans, accompanied by periods of "heat" (estrus) or sexual receptivity. Estrus is the equivalent of the human menstrual cycle but differs from the human menstrual cycle in that the endometrium is not shed with accompanying bleeding. 881

**etiology** (e′′te-ol′ŏ-je) The study of cause, especially of disease, including the origin and what pathogens, if any, are involved.

**eustachian canal** (yoo-sta′ke-an) *See* auditory tube. 483

**eversion** (ĕ-ver′zhun) A movement of the foot in which the sole is turned outward. 235

**exergonic** (ek′′ser-gon′ik) Denoting chemical reactions that liberate energy. 76

**exocrine gland** (ek′sŏ-krin) A gland that secretes its product to an epithelial surface, directly or through ducts. 117

**exocytosis** (ek′′so-si-to′sis) The process of cellular secretion in which the secretory products are contained within a membrane-enclosed vesicle. The vesicle fuses with the cell membrane so that the lumen of the vesicle is open to the extracellular environment. 45

**expiration** (ek′′spĭ-ra′shun) The process of expelling air from the lungs through breathing out; also called *exhalation*. 696

**extension** (ek-sten′shun) A movement that increases the angle between parts of a joint. 232

**extensor** A muscle that, upon contraction, increases the angle of a joint.

**external** (superficial) Located on or toward the surface. 12

**external acoustic meatus** (ă-koo′stik me-a′tus) An opening through the temporal bone that connects with the tympanum and the middle-ear chamber and through which sound vibrations pass; also called the *external auditory meatus*. 482

**exteroceptors** (ek′′stĕ-ro-sep′torz) Sensory receptors that are sensitive to changes in the external environment (as opposed to interoceptors). 465

**extraocular muscles** (ek′′stră-ok′yŭ-lar) The muscles that insert into the sclera of the eye and that act to change the position of the eye in its orbit (as opposed to the intraocular muscles, such as those of the iris and ciliary body within the eye). 285

**extrinsic** (eks-trin′sik) Pertaining to an outside or external origin.

**F**

**face** 1. The anterior aspect of the head not supporting or covering the brain. 12 2. The exposed surface of a structure.

**facet** (fas′et) A small, smooth surface of a bone where articulation occurs. 164

**facilitated diffusion** (fă-sil′ĭ-ta′′tid) The carrier-mediated transport of molecules through the cell membrane along the direction of their concentration gradients. It does not require the expenditure of metabolic energy. 98

**FAD** Flavin adenine dinucleotide; a coenzyme derived from riboflavin that participates in electron transport within the mitochondria. 78

**falciform ligament** (fal′sĭ-form lig′ă-ment) An extension of parietal peritoneum that separates the right and left lobes of the liver. 767

**fallopian tube** (fă-lo′pe-an) *See* uterine tube. 872

**false vocal cords** The supporting folds of tissue for the true vocal cords within the larynx. 688

**falx cerebelli** (falks ser′′ĕ-bel′e) A fold of the dura mater anchored to the occipital bone. It projects inward between the cerebellar hemispheres. 398

**falx cerebri** (ser′ĕ-bre) A fold of dura mater anchored to the crista galli of the ethmoid bone. It extends between the right and left cerebral hemispheres. 398

**fascia** (fash′e-ă) A tough sheet of fibrous tissue binding the skin to underlying muscles or supporting and separating muscles. 122

**fasciculus** (fă-sik′yŭ-lus) A small bundle of muscle or nerve fibers. 255

**fauces** (faw′sēz) The passageway between the mouth and the pharynx. 770

**feces** (fe′sēz) Material expelled from the GI tract during defecation, composed of undigested food residue, bacteria, and secretions; also called *stool*. 787

**fertilization** (fer′′tĭ-lĭ-za′shun) The fusion of an ovum and spermatozoon. 899

**fetus** (fe′tus) A prenatal human after 8 weeks of development. 917

**fibrillation** (fib′′rĭ-la′shun) A condition of cardiac muscle characterized electrically by random and continuously changing patterns of electrical activity and resulting in the inability of the myocardium to contract as a unit and pump blood. It can be fatal if it occurs in the ventricles. 604

**fibrin** (fi′brin) The insoluble protein formed from fibrinogen by the enzymatic action of thrombin during the process of blood clot formation. 559

**fibrinogen** (fi-brin′ŏ-jen) A soluble plasma protein that serves as the precursor of fibrin; also called *factor I*. 559

**fibroblast** (fi′bro-blast) An elongated connective tissue cell with cytoplasmic extensions that is capable of forming collagenous fibers or elastic fibers. 122

**fibrous joint** (fi′brus) A type of articulation bound by fibrous connective tissue that allows little or no movement (e.g., a syndesmosis). 225

**filiform papillae** (fil′ĭ-form pă-pil′e) Numerous small projections over the entire surface of the tongue in which taste buds are absent. 771

**filum terminale** (fi′lum ter-mĭ-nal′e) A fibrous, threadlike continuation of the pia mater, extending inferiorly from the terminal end of the spinal cord to the coccyx. 402

**fimbriae** (fim′bre-e) Fringelike extensions from the borders of the open end of the uterine tube. 782

**fissure** (fish′ur) A groove or narrow cleft that separates two parts, such as the cerebral hemispheres of the brain. 380

**flagellum** (flă-jel′um) A whiplike structure that provides motility for sperm. 45

**flare-and-wheal reaction** (hwēl, wēl) A cutaneous reaction to skin injury or the administration of antigens, produced by release of histamine and related molecules and characterized by local edema and a red flare.   679

**flavoprotein**   (fla´´vo-pro´te-in)   A conjugated protein containing a flavin pigment that is involved in electron transport within the mitochondria.   84

**flexion**   (flek´shun)   A movement that decreases the angle between parts of a joint.   232

**flexor**   (flek´sor)   A muscle that decreases the angle of a joint when it contracts.

**fontanel**   (fon´´tă-nel´)   A membranous-covered region on the skull of a fetus or baby where ossification has not yet occurred; commonly called a *soft spot*.   192

**foot**   The terminal portion of the lower extremity, consisting of the tarsal bones, metatarsal bones, and phalanges.   216

**foramen**   (fŏ-ra´men), pl. *foramina*   An opening in an anatomical structure, usually in a bone, for the passage of a blood vessel or a nerve.   164

**foramen ovale**   (o-val´e)   An opening through the interatrial septum of the fetal heart.   601

**forearm**   The portion of the upper extremity between the elbow and the wrist; also called the *antebrachium*.   205

**fornix**   (for´niks)   **1.** A recess around the cervix of the uterus where it protrudes into the vagina.   895   **2.** A tract within the brain connecting the hippocampus with the mammillary bodies.   460

**fossa**   (fos´ă)   A depressed area, usually on a bone.   164

**fourth ventricle**   (ven´trĭ-k'l)   A cavity within the brain, between the cerebellum and the medulla oblongata and the pons, containing cerebrospinal fluid.   398

**fovea centralis**   (fo´ve-ă sen-tra´ lis)   A depression on the macula lutea of the eye, where only cones are located; the area of keenest vision.   500

**frenulum**   (fren´yŭ-lum)   A membranous structure that serves to anchor and limit the movement of a body part.   771

**frontal**   **1.** Pertaining to the region of the forehead.   12   **2.** A plane through the body, dividing the body into anterior and posterior portions; also called the *coronal plane*.   11

**FSH**   Follicle-stimulating hormone; one of the two gonadotropic hormones secreted from the anterior pituitary. In females, FSH stimulates the development of the ovarian follicles; in males, it stimulates the production of sperm in the seminiferous tubules.   520

**fungiform papillae**   (fun´jĭ-form pă-pil´e)   Flattened, mushroom-shaped projections interspersed over the surface of the tongue in which taste buds are present.   771

## G

**GABA**   Gamma-aminobutyric acid; believed to function as an inhibitory neurotransmitter in the central nervous system.   366

**gallbladder**   A pouchlike organ attached to the underside of the liver in which bile secreted by the liver is stored and concentrated.   793

**gamete**   (gam´ēt)   A haploid sex cell; either an egg cell or a sperm cell.   839

**ganglion**   (gang´gle-on)   An aggregation of nerve cell bodies occurring outside the central nervous system.   345

**gastric intrinsic factor**   (gas´trik)   A glycoprotein secreted by the stomach that is needed for the absorption of vitamin $B_{12}$.   779

**gastrin**   (gas´trin)   A hormone secreted by the stomach that stimulates the gastric secretion of hydrochloric acid and pepsin.   778

**gastrointestinal tract**   (GI tract)   (gas´´tro-in-tes´tĭ-nal)   The portion of the digestive tract that includes the stomach and the small and large intestines.   766

**gates**   Structures composed of one or more protein molecules that regulate the passage of ions through channels within the cell membrane. Gates may be chemically regulated (by neurotransmitters) or voltage regulated (in which case they open in response to a threshold level of depolarization).   353

**genetic recombination**   (jĕ-net´ik re´´kom-bĭ-na´shun)   The formation of new combinations of genes, as by crossing-over between homologous chromosomes.   65

**genetic transcription**   (tran-skrip´shun)   The process by which RNA is produced with a sequence of nucleotide bases that is complementary to a region of DNA.   54

**genetic translation**   (trans-la´shun)   The process by which proteins are produced with amino acid sequences specified by the sequence of codons in messenger RNA.   55

**gigantism**   (ji-gan´tiz´´em)   Abnormal body growth as a result of the excessive secretion of growth hormone.   173

**gingiva**   (jin´jĭ-vă)   The fleshy covering over the mandible and maxilla through which the teeth protrude within the mouth; also called the *gum*.   773

**gland**   An organ that produces a specific substance or secretion.

**glans penis**   (glanz pe´nis)   The enlarged, sensitive, distal end of the penis.   856

**gliding joint**   A type of synovial joint in which the articular surfaces are flat, permitting only side-to-side and back-and-forth movements.   228

**glomerular capsule**   (glo-mer´yŭ-lar)   The double-walled proximal portion of a renal tubule that encloses the glomerulus of a nephron; also called *Bowman's capsule*.   730

**glomerular filtration rate**   (GFR)   The volume of filtrate produced per minute by both kidneys.   733

**glomerular ultrafiltrate**   (ul´´tră-fil´trāt)   Fluid filtered through the glomerular capillaries into the glomerular capsule of the kidney tubules.   733

**glomerulonephritis**   (glo-mer´´yŭ-lo-nĕ-fri´tis)   Inflammation of the renal glomeruli,

associated with fluid retention, edema, hypertension, and the appearance of protein in the urine.   759

**glomerulus**   (glo-mer´yŭ-lus)   A coiled tuft of capillaries surrounded by the glomerular capsule that filtrates urine from the blood.   730

**glottis**   (glot´is)   A slitlike opening into the larynx, positioned between the true vocal cords.   687

**glucagon**   (gloo´kă-gon)   A polypeptide hormone secreted by the alpha cells of the pancreatic islets. It acts primarily on the liver to promote glycogenolysis and raise blood glucose levels.   531

**glucocorticoids**   (gloo´´ko-kor´tĭ-koidz)   Steroid hormones secreted by the adrenal cortex (corticosteroids). They affect the metabolism of glucose, protein, and fat and also have anti-inflammatory and immunosuppressive effects. The major glucocorticoid in humans is hydrocortisone (cortisol).   526

**gluconeogenesis**   (gloo´´ko-ne´´ŏ-jen´ĭ-sis)   The formation of glucose from noncarbohydrate molecules, such as amino acids and lactic acid.   82

**glycerol**   (glis´ĕ-rol)   A 3-carbon alcohol that serves as a building block of fats.   32

**glycogen**   (gli´kŏ-jen)   A polysaccharide of glucose—also called *animal starch*—produced primarily in the liver and skeletal muscles. Similar to plant starch in composition, glycogen contains more highly branched chains of glucose subunits than does plant starch.   30

**glycogenesis**   (gli´´kŏ-jen´ĭ-sis)   The formation of glycogen from glucose.   793

**glycogenolysis**   (gli´´kŏ-jĕ-nol´ĭ-sis)   The hydrolysis of glycogen to glucose 1-phosphate, which can be converted to glucose 6-phosphate, which then may be oxidized via glycolysis or (in the liver) converted to free glucose.   793

**glycolysis**   (gli´´kol´ĭ-sis)   The metabolic pathway that converts glucose to pyruvic acid; the final products are two molecules of pyruvic acid and two molecules of reduced NAD, with a net gain of two ATP molecules. In anaerobic respiration, the reduced NAD is oxidized by the conversion of pyruvic acid to lactic acid. In aerobic respiration, pyruvic acid enters the Krebs cycle in mitochondria and reduced NAD is ultimately oxidized to yield water.   79

**glycosuria**   (gli´´kŏ-soor´e-ă)   The excretion of an abnormal amount of glucose in the urine (urine normally only contains trace amounts of glucose).   746

**goblet cell**   A unicellular mucus-secreting gland that is associated with columnar epithelia; also called a *mucous cell*.   111

**Golgi apparatus**   (gol´je)   A network of stacked, flattened membranous sacs within the cytoplasm of cells. Its major function is to concentrate and package proteins for secretion from the cell.   58

**Golgi tendon organ**   A sensory receptor found near the junction of tendons and muscles.   473

**gonad**   (go´nad)   A reproductive organ, testis or ovary, that produces gametes and sex hormones.   839

**gonadotropin hormones**   (go-nad´´ŏ-tro´pin) Hormones of the anterior pituitary that stimulate gonadal function—the formation of gametes and secretion of sex steroids. The two gonadotropins are FSH (follicle-stimulating hormone) and LH (luteinizing hormone), which are essentially the same in males and females.   842

**graafian follicle**   (graf´e-an)   A mature ovarian follicle, containing a single fluid-filled cavity, with the ovum located toward one side of the follicle and perched on top of a hill of granulosa cells.   877

**granular leukocytes**   (loo´kŏ-sītz) Leukocytes with granules in the cytoplasm; on the basis of the staining properties of the granules, these cells are classified as neutrophils, eosinophils, or basophils.   553

**Graves' disease**   A hyperthyroid condition believed to be caused by excessive stimulation of the thyroid gland by autoantibodies; it is associated with exophthalmos (bulging eyes), high pulse rate, high metabolic rate, and other symptoms of hyperthyroidism.   543

**gray matter**   The region of the central nervous system composed of nonmyelinated nerve tissue.   351

**greater omentum**   (o-men´tum)   A double-layered peritoneal membrane that originates on the greater curvature of the stomach. It hangs inferiorly like an apron over the contents of the abdominal cavity.   767

**gross anatomy**   The branch of anatomy concerned with structures of the body that can be studied without a microscope.   5

**growth hormone**   A hormone secreted by the anterior pituitary that stimulates growth of the skeleton and soft tissues during the growing years and that influences the metabolism of protein, carbohydrate, and fat throughout life.   519

**gustatory**   (gus´tă-tor´´e)   Pertaining to the sense of taste.   474

**gut**   The GI tract or a portion thereof; generally used in reference to the embryonic digestive tube, consisting of the foregut, midgut, and hindgut.   766

**gyrus**   (ji´rus)   A convoluted elevation or ridge.   383

## H

**hair**   A threadlike appendage of the epidermis consisting of keratinized dead cells that have been pushed up from a dividing basal layer.   145

**hair cells**   Specialized receptor nerve endings for detecting sensations, such as in the spiral organ (organ of Corti).   479

**hair follicle**   (fol´lĭ-k'l)   A tubular depression in the dermis of the skin in which a hair develops.   145

**hand**   The terminal portion of the upper extremity, containing the carpal bones, metacarpal bones, and phalanges.   206

**haploid**   (hap´loid)   A cell that has one of each chromosome type and therefore half the number of chromosomes present in most other body cells; only the gametes (sperm and ova) are haploid.   839

**haptens**   (hap´tenz)   Small molecules that are not antigenic by themselves, but which—when combined with proteins—become antigenic and thus capable of stimulating the production of specific antibodies.   655

**hard palate**   (pal´it)   The bony partition between the oral and nasal cavities, formed by the maxillae and palatine bones and lined by mucous membrane.   771

**haustra**   (haws´tră)   Sacculations or pouches of the colon.   786

**haversian canal**   (ha-ver´shan)   See central canal.   130

**haversian system**   See osteon.   167

**hay fever**   A seasonal type of allergic rhinitis caused by pollen; it is characterized by itching and tearing of the eyes, swelling of the nasal mucosa, attacks of sneezing, and often by asthma.   677

**head**   The uppermost portion of a human that contains the brain and major sense organs.   12

**heart**   A four-chambered, muscular pumping organ positioned in the thoracic cavity, slightly to the left of midline.   566

**heart murmur**   An auscultatory sound of cardiac or vascular origin, usually caused by an abnormal flow of blood in the heart as a result of structural defects of the valves or septum.   605

**helper T cells**   A subpopulation of T cells (lymphocytes) that helps to stimulate the antibody production of B lymphocytes by antigens.   668

**hematocrit**   (hĭ-mat´ŏ-krit)   The ratio of packed red blood cells to total blood volume in a centrifuged sample of blood, expressed as a percentage.   551

**heme**   (hēm)   The iron-containing red pigment that, together with the protein globin, forms hemoglobin.   552

**hemoglobin**   (he´mŏ-glo´´bin)   The pigment of red blood cells constituting about 33% of the cell volume that transports oxygen and carbon dioxide.   552

**hemopoiesis**   (hem´´ŏ-poi-e´sis)   The production of red blood cells.   554

**heparin**   (hep´ar-in)   A mucopolysaccharide found in many tissues, but most abundantly in the lungs and liver, that is used medically as an anticoagulant.   561

**hepatic duct**   (hĕ-pat´ik)   A duct formed from the fusion of several bile ducts that drain bile from the liver. The hepatic duct merges with the cystic duct from the gallbladder to form the common bile duct.   790

**hepatic portal circulation**   The return of venous blood from the digestive organs and spleen through a capillary network within the liver before draining into the heart.   598

**hepatitis**   (hep´´ă-ti´tis)   Inflammation of the liver.   803

**hepatopancreatic ampulla**   (hep´´ă-to-pan´´kre-at´ik)   A small, elevated area within the duodenum where the combined pancreatic and common bile duct empties; also called the *ampulla of Vater*.   781

**Hering–Breuer reflex**   A reflex in which distension of the lungs stimulates stretch receptors, which in turn act to inhibit further distension of the lungs.   706

**hermaphrodite**   (her-maf´rŏ-dīt)   An organism having both testes and ovaries.   863

**heterochromatin**   (het´´ĕ-ro-kro´mă-tin)   A condensed, inactive form of chromatin.   53

**hiatal hernia**   (hi-a´tal her´ne-ă)   A protrusion of an abdominal structure through the esophageal hiatus of the diaphragm into the thoracic cavity.   806

**hiatus**   An opening or fissure; a foramen.

**high-density lipoproteins** (HDLs)   (lip´´o-pro´te-inz)   Combinations of lipids and proteins that migrate rapidly to the bottom of a test tube during centrifugation. HDLs are carrier proteins for lipids, such as cholesterol, that appear to offer some protection from atherosclerosis.   608

**hilum**   (hi´lum)   A concave or depressed area where vessels or nerves enter or exit an organ; also called *hilus*.   691

**hinge joint**   A type of synovial articulation characterized by a convex surface of one bone fitting into a concave surface of another such that movement is confined to one plane, as in the knee or interphalangeal joint.   228

**histamine**   (his´tă-mēn)   A compound secreted by tissue mast cells and other connective tissue cells that stimulates vasodilation and increases capillary permeability. It is responsible for many of the symptoms of inflammation and allergy.   677

**histology**   (hĭ-stol´ŏ-je)   Microscopic anatomy of the structure and function of tissues.   107

**homeostasis**   (ho´´me-o-sta´sis)   The dynamic constancy of the internal environment, the maintenance of which is the principal function of physiological regulatory mechanisms. The concept of homeostasis provides a framework for understanding most physiological processes.   16

**homologous chromosomes**   (hŏ-mol´ŏ-gus)   The matching pairs of chromosomes in a diploid cell.   62

**horizontal (transverse) plane**   A directional plane that divides the body, organ, or appendage into superior and inferior or proximal and distal portions.   11

**hormone**   (hor´mōn)   A chemical substance produced in an endocrine gland and secreted into the bloodstream to cause an effect in a specific target organ.   514

**humoral immunity**   (hyoo´mor-al ĭ-myoo´nĭ-te)   The form of acquired immunity in which antibody molecules are secreted in response to antigenic stimulation (as opposed to cell mediated immunity); also called *antibody-mediated immunity*.   656

**hyaline cartilage** (hi'ă-lĭn) A cartilage with a homogeneous matrix. It is the most common type, occurring at the articular ends of bones, in the trachea, and within the nose. Most of the bones in the body are formed from hyaline cartilage. 127

**hyaline membrane disease** A disease affecting premature infants who lack pulmonary surfactant, it is characterized by collapse of the alveoli (atelectasis) and pulmonary edema; also called *respiratory distress syndrome*. 695

**hydrocortisone** (hi'drŏ-kor'tĭ-sōn) The principal corticosteroid hormone secreted by the adrenal cortex, with glucocorticoid action; also called *cortisol*. 527

**hydrophilic** (hi''drŏ-fil'ik) Denoting a substance that readily absorbs water; literally, "water loving." 25

**hydrophobic** (hi''drŏ-fo'bik) Denoting a substance that repels, and that is repelled by, water; "water fearing." 25

**hymen** (hi'men) A developmental remnant (vestige) of membranous tissue that partially covers the vaginal opening. 875

**hyperbaric oxygen** (hi''per-bar'ik) Oxygen gas present at greater than atmospheric pressure. 721

**hypercapnia** (hi''per-kap'ne-ă) Excessive concentration of carbon dioxide in the blood. 705

**hyperextension** (hi''per-ek-sten'shun) Extension beyond the normal anatomical position or 180°. 233

**hyperglycemia** (hi''per-gli-se'me-ă) An abnormally increased concentration of glucose in the blood. 98

**hyperkalemia** (hi''per-kă-le'me-ă) An abnormally high concentration of potassium in the blood. 103

**hyperopia** (hi''per-o'pe-ă) A refractive disorder in which rays of light are brought to a focus behind the retina as a result of the eyeball being too short; also called *farsightedness*. 507

**hyperplasia** (hi''per-pla'zha) An increase in organ size due to an increase in cell numbers as a result of mitotic cell division (in contrast to hypertrophy). 64

**hyperpolarization** (hi''per-po''lar-ĭ-za'shun) An increase in the negativity of the inside of a cell membrane with respect to the resting membrane potential. 353

**hypersensitivity** (hi''per-sen''sĭ-tiv'ĭ-te) Another name for *allergy*; abnormal immune response that may be immediate (due to antibodies of the IgE class) or delayed (due to cell-mediated immunity). 677

**hypertension** (hi''per-ten'shun) Elevated or excessive blood pressure. 640

**hypertonic** (hi''per-tion'ik) Denoting a solution with a greater solute concentration and thus a greater osmotic pressure than plasma. 96

**hypertrophy** (hi'per'trŏ-fe) Growth of an organ due to an increase in the size of its cells (in contrast to hyperplasia). 64

**hyperventilation** (hi''per-ven''tĭ-la'shun) A high rate and depth of breathing that results in a decrease in the blood carbon dioxide concentration to below normal. 705

**hypodermis** (hi''pŏ-der'mis) A layer of fat beneath the dermis of the skin. 143

**hyponychium** (hi''pŏ-nik'e-um) A thickened, supportive layer of stratum corneum at the distal end of a digit under the free edge of the nail. 147

**hypothalamic hormones** (hi''po-thal'ă-mik) Hormones produced by the hypothalamus. These include antidiuretic hormone and oxytocin, which are secreted by the posterior pituitary, and both releasing and inhibiting hormones that regulate the secretions of the anterior pituitary. 523

**hypothalamo-hypophyseal portal system** (hi''pŏ-fix'e-al) A vascular system that transports releasing and inhibiting hormones from the hypothalamus to the anterior pituitary. 521

**hypothalamo-hypophyseal tract** The tract of nerve fibers (axons) that transports antidiuretic hormone and oxytocin from the hypothalamus to the posterior pituitary. 521

**hypothalamus** (hi''po-thal'ă-mus) A portion of the forebrain within the diencephalon that lies below the thalamus, where it functions as an autonomic nerve center and regulates the pituitary gland. 391

**hypovolemic shock** (hi''po-vo-le'mik) A rapid fall in blood pressure as a result of diminished blood volume. 642

**hypoxemia** (hi''pok-se'me-ă) A low oxygen concentration of the arterial blood. 706

**I**

**ileocecal valve** (il''e-ŏ-se'kal) A modification of the mucosa at the junction of the small and large intestine that forms a one-way passage and prevents the backflow of food materials. 781

**ileum** (il'e-um) The terminal portion of the small intestine between the jejunum and cecum. 781

**immediate hypersensitivity** (hi''per-sen''sĭ-tiv'ĭ-te) Hypersensitivity (allergy) mediated by antibodies of the IgE class that results in the release of histamine and related compounds from tissue cells. 677

**immunization** (im''yŭ-nĭ-za'shun) The process of increasing one's resistance to pathogens. In active immunity a person is injected with antigens that stimulate the development of clones of specific B or T lymphocytes; in passive immunity a person is injected with antibodies produced by another organism. 663

**immunoassay** (im''yŭ-no-as'a) Any of a number of laboratory or clinical techniques that employ the specific binding between an antigen and its homologous antibody in order to identify and quantify a substance in a sample. 655

**immunoglobulins** (im''yŭ-no-glob'yŭ-linz) Subclasses of the gamma globulin fraction of plasma proteins that have antibody functions, providing humoral immunity. 657

**immunosurveillance** (im''yŭ-no-ser-va'lens) The concept that the immune system recognizes and attacks malignant cells that produce antigens not recognized as "self." This function is believed to be cell mediated rather than humoral. 674

**implantation** (im''plan-ta'shun) The process by which a blastocyst attaches itself to and penetrates into the endometrium of the uterus. 903

**incus** (ing'kus) The middle of three auditory ossicles within the middle-ear chamber; commonly called the *anvil*. 483

**inferior vena cava** (ve'nă ka'vă) A large systemic vein that collects blood from the body regions inferior to the level of the heart and returns it to the right atrium. 592

**infundibulum** (in''fun-dib'yŭ-lum) The stalk that attaches the pituitary gland to the hypothalamus of the brain. 519

**ingestion** (in-jes'chun) The process of taking food or liquid into the body by way of the oral cavity. 766

**inguinal** (ing'gwĭ-nal) Pertaining to the groin region. 12

**inguinal canal** The circular passageway in the abdominal wall through which a testis descends into the scrotum. 841

**inhibin** (in-hib'in) A polypeptide hormone secreted by the testes that is believed to specifically exert negative feedback inhibition of FSH secretion from the anterior pituitary. 842

**inositol** (ĭ-no'sĭ-tol) A sugarlike B-complex vitamin. Inositol triphosphate is believed to act as a second messenger in the action of some hormones. 537

**insertion** The more movable attachment of a muscle, usually more distal. 255

**inspiration** (in''spĭ-ra'shun) The act of breathing air into the alveoli of the lungs; also called *inhalation*. 695

**insula** (in'sŭ-lă) A deep, paired cerebral lobe.

**insulin** (in'sŭ-lin) A polypeptide hormone secreted by the beta cells of the pancreatic islets that promotes the anabolism of carbohydrates, fat, and protein. Insulin acts to promote the cellular uptake of blood glucose and, therefore, to lower the blood glucose concentration; insulin deficiency results in hyperglycemia and diabetes mellitus. 531

**integument** (in-teg'yoo-ment) The skin; the largest organ of the body. 138

**intercalated disc** (in-ter'kă-lāt-ed) A thickened portion of the sarcolemma that extends across a cardiac muscle fiber, indicating the boundary between cells. 131

**intercellular substance** (in''ter-sel'yŭ-lar) The matrix or material between cells that largely determines tissue types. 107

**interferons** (in''ter-fēr'onz) A group of small proteins that inhibit the multiplication of viruses inside host cells and that also have antitumor properties. 654

**internal** (deep) Toward the center, away from the surface of the body. 12

**internal ear** The innermost portion or chamber of the ear, containing the cochlea and the vestibular organs. 483

**interneurons** (in´´ter-noor´onz) Multipolar neurons interposed between sensory (afferent) and motor (efferent) neurons and confined entirely within the central nervous system; also called *association neurons*. 347

**interoceptors** (in´´ter-o-sep´torz) Sensory receptors that respond to changes in the internal environment (as opposed to exteroceptors). 465

**interphase** The interval between successive cell divisions, during which time the chromosomes are in an extended state and are active in directing RNA synthesis. 63

**interstitial cells** (in´´ter-stish´al) Cells located in the interstitial tissue between adjacent convolutions of the seminiferous tubules of the testes; they secrete androgens (mainly testosterone); also called *cells of Leydig*. 848

**intervertebral disc** (in´´ter-ver´tĕ-bral) A pad of fibrocartilage located between the bodies of adjacent vertebrae. 194

**intestinal crypt** A simple tubular digestive gland opening onto the surface of the intestinal mucosa that secretes digestive enzymes; also called the *crypt of Lieberkühn*. 782

**intrafusal fibers** (in´´tră-fyoo´sal) Modified muscle fibers that are encapsulated to form muscle spindle organs, which are muscle stretch receptors. 471

**intramembranous ossification** *See* membranous bone. 163

**intrapleural space** (in´´tră-ploor´al) An actual or potential space between the visceral pleural membrane covering the lungs and the somatic pleural membrane lining the thoracic wall. 693

**intrinsic** (in-trin´zik) Situated within or pertaining to internal origin.

**inulin** (in´yŭ-lin) A polysaccharide of fructose, produced by certain plants, that is filtered by the human kidneys but neither reabsorbed nor secreted. The clearance rate of injected insulin is thus used to measure the glomerular filtration rate. 742

**inversion** (in-ver´zhun) A movement of the foot in which the sole is turned inward. 235

**in vitro** (in ve´tro) Occurring outside the body, in a test tube or other artificial environment. 257

**in vivo** (in ve´vo) Occurring within the body. 257

**ion** (i´on) An atom or group of atoms that has either lost or gained electrons and thus has a net positive or a net negative charge. 24

**ionization** (i-on-ĭ-za´shun) The dissociation of a solute to form ions. 26

**ipsilateral** (ip´´sĭ-lat´er-al) On the same side (as opposed to contralateral). 438

**IPSP** Inhibitory postsynaptic potential; hyperpolarization of the postsynaptic membrane in response to a particular neurotransmitter chemical, which makes it more difficult for the postsynaptic cell to

attain a threshold level of depolarization required to produce action potentials. It is responsible for postsynaptic inhibition. 366

**iris** (i´ris) The pigmented portion of the vascular tunic of the eye that surrounds the pupil and regulates its diameter. 492

**ischemia** (ĭ-ske´me-ă) A rate of blood flow to an organ that is inadequate to supply sufficient oxygen and maintain aerobic respiration in that organ. 82

**islets of Langerhans** (i´letz of lang´er-hanz) *See* pancreatic islets. 794

**isoenzymes** (i´´so-en´zīmz) Enzymes, usually produced by different organs, that catalyze the same reaction but that differ from each other in amino acid composition. 72

**isometric contraction** (i´´sŏ-met´rik) Muscle contraction in which there is no appreciable shortening of the muscle. 260

**isotonic contraction** (i´´sŏ-ton´ik) Muscle contraction in which the muscle shortens in length and maintains approximately the same amount of tension throughout the shortening process. 259

**isotonic solution** A solution having the same total solute concentration, osmolality, and osmotic pressure as the solution with which it is compared; a solution with the same solute concentration and osmotic pressure as plasma. 96

**isthmus** (is´mus) A narrow neck or portion of tissue connecting two structures.

## J

**jaundice** (jawn´dis) A condition characterized by high blood bilirubin levels and staining of the tissues with bilirubin, which imparts a yellow color to the skin and mucous membranes. 806

**jejunum** (jĕ-joo´num) The middle portion of the small intestine, located between the duodenum and the ileum. 781

**joint capsule** The fibrous tissue that encloses the joint cavity of a synovial joint. 227

## K

**keratin** (ker´ă-tin) An insoluble protein present in the epidermis and in epidermal derivatives, such as hair and nails. 113

**ketoacidosis** (ke´´to-ă-sĭ-do´sis) A type of metabolic acidosis resulting from the excessive production of ketone bodies, as in diabetes mellitus. 33

**ketogenesis** (ke´´to-jen´ĭ-sis) The production of ketone bodies. 793

**ketone bodies** (ke´tōn) The substances derived from fatty acids via acetyl coenzyme A in the liver; namely, acetone, acetoacetic acid, and β-hydroxybutyric acid. Ketone bodies are oxidized by skeletal muscles for energy. 32

**ketosis** (ke-to´sis) An abnormal elevation in the blood concentration of ketone bodies that does not necessarily produce acidosis. 33

**kidney** (kid´ne) One of a pair of organs of the urinary system that contains nephrons and that filters wastes from the blood in the formation of urine. 728

**kilocalorie** (kil´ŏ-kal´´ŏ-re) A unit of measurement equal to 1000 calories, which are units of heat (a kilocalorie is the amount of heat required to raise the temperature of 1 kilogram of water by 1 C°). In nutrition, the kilocalorie is called a big calorie (Calorie). 812

**kinesiology** (kĭ-ne´´se-ol´ŏ-je) The study of body movement. 224

**Klinefelter's syndrome** (klīn-fel-terz sin´drōm) An abnormal condition of male sex characteristics due to the presence of an extra X chromosome (genotype XXY). 863

**knee** A region in the lower extremity between the thigh and the leg that contains a synovial hinge joint. 15

**Krebs cycle** (krebz) A cyclic metabolic pathway in the matrix of mitochondria by which the acetic acid part of acetyl CoA is oxidized and substrates provided for reactions that are coupled to the formation of ATP. 83

**Kupffer cells** (koop´fer) Phagocytic cells lining the sinusoids of the liver that are part of the body immunity system. 789

## L

**labial frenulum** (la´be-al fren´yŭ-lum) A longitudinal fold of mucous membrane that attaches the lips to the gum along the midline of both the upper and lower lip. 770

**labia majora** (la´be-ă mă-jor´ă), sing. *labium majus* A portion of the external genitalia of a female consisting of two longitudinal folds of skin extending downward and backward from the mons pubis. 876

**labia minora** (mĭ-nor´ă), sing. *labium minus* Two small folds of skin, devoid of hair and sweat glands, lying between the labia major of the external genitalia of a female. 876

**labyrinth** (lab´ĭ-rinth) An intricate structure consisting of interconnecting passages (e.g., the bony and membranous labyrinths of the inner ear. 478

**lacrimal canaliculus** (lak´rĭ-mal kan´´ă-lik´yŭ-lus) A drainage duct for tears, located at the medial corner of an eyelid. It conveys the tears medially into the nasolacrimal sac. 491

**lacrimal gland** A tear-secreting gland, located on the superior lateral portion of the eyeball underneath the upper eyelid. 490

**lactation** (lak-ta´shun) The production and secretion of milk by the mammary glands. 887

**lacteal** (lak´te-al) A small lymphatic duct associated with a villus of the small intestine. 782

**lactose** (lak´tōs) Milk sugar; a disaccharide of glucose and galactose. 30

**lactose intolerance** A disorder resulting in the inability to digest lactose because of an enzyme, lactase, deficiency. Symptoms include bloating, intestinal gas, nausea, diarrhea, and cramps. 783

**lacuna** (lă-kyoo´nă) A small, hollow chamber that houses an osteocyte in mature bone tissue or a chondrocyte in cartilage tissue. 130

**lambdoidal suture** (lam´doid-al soo´chur) The immovable joint in the skull between the parietal bones and the occipital bone. 185

**lamella** (lă-mel´ă) A concentric ring of matrix surrounding the central canal in an osteon of mature bone tissue. 167

**lamellated corpuscle** (lam´e-la-ted) A sensory receptor for pressure, found in tendons, around joints, and in visceral organs; also called a *pacinian corpuscle.* 467

**lamina** (lam´ĭ-nă) A thin plate of bone that extends superiorly from the body of a vertebra to form either side of the arch of a vertebra. 194

**lanugo** (lă-noo´go) Short, silky fetal hair, which may be present for a short time on a premature infant. 146

**large intestine** The last major portion of the GI tract, consisting of the cecum, colon, rectum, and anal canal. 784

**laryngopharynx** (lă-ring´´go-far´ingks) The inferior or lower portion of the pharynx in contact with the larynx. 686

**larynx** (lar´ingks) The structure located between the pharynx and trachea that houses the vocal cords; commonly called the *voice box.* 687

**lateral** (lat´er-al) Pertaining to the side; farther from the midplane. 12

**lateral ventricle** (ven´trĭ-k'l) A cavity within the cerebral hemisphere of the brain that is filled with cerebrospinal fluid. 398

**L-dopa** Levodopa; a derivative of the amino acid tyrosine. It serves as the precursor for the neurotransmitter molecule dopamine and is given to patients with Parkinson's disease to stimulate dopamine production. 352

**leg** The portion of the lower extremity between the knee and ankle. 213

**lens** (lenz) A transparent refractive organ of the eye positioned posterior to the pupil and iris. 492

**lesion** (le´zhun) A wounded or damaged area. 150

**lesser omentum** (o-men´tum) A peritoneal fold of tissue extending from the lesser curvature of the stomach to the liver. 767

**leukocyte** (loo´kŏ-sīt) A white blood cell; variant spelling, leucocyte. 553

**ligament** (lig´ă-ment) A tough cord or fibrous band of connective tissue that binds bone to bone to strengthen and provide flexibility to a joint. It also may support viscera. 122

**limbic system** (lim´bik) A portion of the brain concerned with emotions and autonomic activity. 460

**linea alba** (lin´e-ă al´bă) A vertical fibrous band extending down the anterior medial portion of the abdominal wall. 291

**lingual frenulum** (ling´gwal fren´yŭ-lum) A longitudinal fold of mucous membrane that attaches the tongue to the floor of the oral cavity. 771

**lipogenesis** (lip´´ŏ-jen´e-sis) The formation of fat or triglycerides. 793

**lipolysis** (lĭ-pol´ĭ-sis) The hydrolysis of triglycerides into free fatty acids and glycerol. 818

**liver** A large visceral organ inferior to the diaphragm in the right hypochondriac region. The liver detoxifies the blood and modifies the blood plasma concentration of glucose, triglycerides, ketone bodies, and proteins. 788

**low-density lipoproteins** (LDLs) (lip´´o-pro´te-inz) Plasma proteins that transport triglycerides and cholesterol. They are believed to contribute to arteriosclerosis. 606

**lower extremity** A lower appendage, including the hip, thigh, knee, leg, and foot.

**lumbar** (lum´bar) Pertaining to the region of the loins. 12

**lumbar plexus** (plek´sus) A network of nerves formed by the anterior branches of spinal nerves L1 through L4. 432

**lumen** (loo´men) The space within a tubular structure through which a substance passes. 109

**lung** One of the two major organs of respiration positioned within the thoracic cavity on either side of the mediastinum. 690

**lung surfactant** (sur-fak´tant) A mixture of lipoproteins (containing phospholipids) secreted by type II alveolar cells into the alveoli of the lungs. It lowers surface tension and prevents collapse of the lungs as occurs in hyaline membrane disease, in which surfactant is absent. 694

**lunula** (loo´nyoo-lă) The half-moon-shaped whitish area at the proximal portion of a nail. 147

**luteinizing hormone** (LH) (loo´te-ĭ-ni´´zing) A hormone secreted by the adenohypophysis (anterior lobe) of the pituitary gland that stimulates ovulation and the secretion of progesterone by the corpus luteum. It also influences mammary gland milk secretion in females and stimulates testosterone secretion by the testes in males. 520

**lymph** (limf) A clear, plasmalike fluid that flows through lymphatic vessels.

**lymphatic system** (lim-fat´ik) The lymphatic vessels and lymph nodes. 618

**lymph node** A small, ovoid mass of reticular tissue located along the course of lymph vessels. 649

**lymphocyte** (lim´fŏ-sīt) A type of white blood cell characterized by agranular cytoplasm. Lymphocytes usually constitute about 20% to 25% of the white blood cell count. 553

**lymphokines** (lim´fŏ-kīns) A group of chemicals released from T cells that contribute to cell-mediated immunity. 669

**lysosomes** (li´sŏ-sōmz) Organelles containing digestive enzymes and responsible for intracellular digestion. 48

## M

**macromolecules** (mak´´ro-mol´ĭ-kyoolz) Large molecules; a term that usually refers to protein, RNA, and DNA.

**macrophage** (mak´rŏ-fāj) A wandering phagocytic cell. 651

**macula lutea** (mak´yŭ-lă loo´te-ă) A yellowish depression in the retina of the eye that contains the fovea centralis, the area of keenest vision. 500

**malignant** Threatening to life; virulent. Of a tumor, cancerous, tending to metastasize. 674

**malleus** (mal´e-us) The first of three auditory ossicles that attaches to the tympanum; commonly called the *hammer.* 483

**mammary gland** (mam´er-e) The gland of the female breast responsible for lactation and nourishment of the young. 886

**marrow** (mar´o) The soft connective tissue found within the inner cavity of certain bones that produces red blood cells. 165

**mast cell** A type of connective tissue cell that produces and secretes histamine and heparin and promotes local inflammation. 122

**mastication** (mas´´tĭ-ka´shun) The chewing of food. 766

**matrix** (ma´triks) The intercellular substance of a tissue. 121

**maximal oxygen uptake** The maximum amount of oxygen that can be consumed by the body per unit time during heavy exercise. 271

**meatus** (me-a´tus) A passageway or opening into a structure. 685

**mechanoreceptor** (mek´´ă-no-re-sep´tor) A sensory receptor that responds to a mechanical stimulus. 465

**medial** (me´de-al) Toward or closer to the midplane of the body. 12

**mediastinum** (me´´de-ă-sti´num) The partition in the center of the thorax between the two pleural cavities. 15

**medulla** (mĕ-dul´ă) The center portion of an organ.

**medulla oblongata** (ob´´long-ga´tă) A portion of the brain stem located between the spinal cord and the pons. 394

**medullary (marrow) cavity** (med´l-er´´e) The hollow core of the diaphysis of a long bone in which marrow is found. 165

**megakaryocyte** (meg´´ă-kar´e-o-sīt) A bone marrow cell that gives rise to blood platelets. 553

**meiosis** (mi-o´sis) A specialized type of cell division by which gametes or haploid sex cells are formed. 62

**Meissner's corpuscle** (mīs´nerz) *See* corpuscle of touch. 467

**melanin** (mel´ă-nin) A dark pigment found within the epidermis or epidermal derivatives of the skin. 140

**melanocyte** (mel´ă-no-sīt) A specialized melanin-producing cell found in the deepest layer of the epidermis. 138

**melanoma** (mel´´ă-no´mă) A dark, malignant tumor of the skin that frequently forms in moles. 143

**melatonin** (mel´´ă-to´nin) A hormone secreted by the pineal gland that produces lightening of the skin in lower vertebrates and that may contribute to the regulation of gonadal function in mammals. Secretion follows a circadian rhythm and peaks at night. 532

**membrane potential** The potential difference or voltage that exists between the inner and outer sides of a cell membrane. It exists in all cells but is capable of being changed by excitable cells (neurons and muscle cells). 102

**membranous bone** (mem´bră-nus) Bone that forms from membranous connective tissue rather than from cartilage. 163

**membranous labyrinth** (lab´ĭ-rinth) A system of communicating sacs and ducts within the bony labyrinth of the inner ear that includes the cochlea and vestibular apparatus. It is filled with endolymph and surrounded by perilymph and bone. 478

**menarche** (mě-nar´ke) The first menstrual discharge. 843

**Ménière's disease** (mān-yarz´) Deafness, tinnitus, and vertigo resulting from a disorder of the labyrinth. 506

**meninges** (mě-nin´jēz), sing. *meninx* A group of three fibrous membranes covering the central nervous system, composed of the dura mater, arachnoid mater, and pia mater. 396

**menisci** (mě-nis´ke) Wedge-shaped fibrocartilages in certain synovial joints. 227

**menopause** (men´ŏ-pawz) The period marked by the cessation of menstrual periods in the human female. 885

**menstrual cycle** (men´stroo-al) The rhythmic female reproductive cycle, characterized by changes in hormone levels and physical changes in the uterine lining. 881

**menstruation** (men´´stroo-a´shun) The discharge of blood and tissue from the uterus at the end of the menstrual cycle. 884

**mesencephalic aqueduct** (mez´´en-sě-fal´ik ak´wě-dukt) The channel that connects the third and fourth ventricles of the brain; also called the *aqueduct of Sylvius*. 398

**mesencephalon** (mes´´en-sef´ă-lon) The midbrain, which contains the corpora quadrigemina and the cerebral peduncles. 392

**mesenchyme** (mez´en-kīm) An embryonic connective tissue that can migrate, and from which all connective tissues arise. 121

**mesenteric patches** (mes´´en-ter´ik) Clusters of lymph nodes on the walls of the small intestine; also called *Peyer's patches*. 650

**mesentery** (mes´en-ter´´e) A fold of peritoneal membrane that attaches an abdominal organ to the abdominal wall. 117

**mesoderm** (mes´ŏ-derm) The middle one of the three primary germ layers. 904

**mesothelium** (mes´´ŏ-the´lium) A simple squamous epithelial tissue that lines body cavities and covers visceral organs; also called *serosa*. 110

**mesovarium** (mes´´ŏ-va´re-um) The peritoneal fold that attaches an ovary to the broad ligament of the uterus. 877

**messenger RNA** (mRNA) A type of RNA that contains a base sequence complementary to a part of the DNA that specifies the synthesis of a particular protein. 54

**metabolism** (mě-tab´ŏ-liz-em) The sum total of the chemical changes that occur within a cell. 79

**metacarpus** (met´´ă-kar´pus) The region of the hand between the wrist and the phalanges, including the five metacarpal bones that support the palm of the hand. 208

**metarteriole** (met´´ar-tir´e-ōl) A small blood vessel that emerges from an arteriole, passes through a capillary network, and empties into a venule. 581

**metastasis** (mě-tas´tă-sis) The spread of a disease from one organ or body part to another. 674

**metatarsus** (met´´ă-tar´sus) The region of the foot between the ankle and the phalanges that includes the five metatarsal bones. 216

**metencephalon** (met´´en-sef´ă-lon) The most superior portion of the hindbrain that contains the cerebellum and the pons. 393

**micelles** (mi-selz´) Colloidal particles formed by the aggregation of many molecules. 34

**microglia** (mi-krog´le-ă) Small phagocytic cells found in the central nervous system. 348

**microvilli** (mi´´kro-vil´i) Microscopic hairlike projections of cell membranes on certain epithelial cells. 46

**micturition** (mik´´tŭ-rish´un) The process of voiding urine; also called *urination*. 755

**midbrain** The portion of the brain between the pons and the forebrain. 392

**middle ear** The middle of the three portions of the ear that contains the three auditory ossicles. 483

**midsagittal plane** (mid-saj´ĭ-tal) A plane that divides the body into equal right and left halves; also called the *median plane* or *midplane*. 11

**mineralocorticoids** (min´´er-al-o-kor´tĭ-koidz) Steroid hormones of the adrenal cortex (corticosteroids) that regulate electrolyte balance. 526

**mitochondria** (mi´´tŏ-kon´dre-ă), sing. *mitochondrion* Cytoplasmic organelles that serve as sites for the production of most of the cellular energy; the so-called powerhouses of the cell. 49

**mitosis** (mi-to´sis) The process of cell division that results in two identical daughter cells, containing the same number of chromosomes. 61

**mitral valve** (mi´tral) The left atrioventricular heart valve; also called the *bicuspid valve*. 569

**mixed nerve** A nerve that contains both motor and sensory nerve fibers. 347

**molal** (mo´lal) Pertaining to the number of moles of solute per kilogram of solvent. 95

**molar** (mo´lar) Pertaining to the number of moles of solute per liter of solution. 95

**mole** (mōl) The number of grams of a chemical that is equal to its formula weight (atomic weight for an element or molecular weight for a compound). 95

**monoclonal antibodies** (mon´´ŏ-klōn´al an´tĭ-bod´´ēz) Identical antibodies derived from a clone of genetically identical plasma cells. 666

**monocyte** (mon´o-sit) A phagocytic type of white blood cell, normally constituting about 3% to 8% of the white blood cell count. 553

**monomer** (mon´ŏ-mer) A single molecular unit of a longer, more complex molecule. Monomers are joined together to form dimers, trimers, and polymers; the hydrolysis of polymers eventually yields separate monomers. 765

**monosaccharide** (mon´´ŏ-sak´ă-rīd) The monomer of the more complex carbohydrates, examples of which include glucose, fructose, and galactose; also called a *simple sugar*. 29

**mons pubis** (monz pyoo´bis) A fatty tissue pad covering the symphysis pubis and covered by pubic hair in the female. 876

**morula** (mor´yu-lă) An early stage of embryonic development characterized by a solid ball of cells. 902

**motile** (mōt´l), mo´tīl) Capable of self-propelled movement.

**motor area** A region of the cerebral cortex from which motor impulses to muscles or glands originate. 388

**motor nerve** A nerve composed of motor nerve fibers. 347

**motor neuron** (noor´on) A nerve cell that conducts action potentials away from the central nervous system and innervates effector organs (muscle and glands). It forms the anterior roots of the spinal nerves; also called an *efferent neuron*. 346

**motor unit** A single motor neuron and the muscle fibers it innervates. 260

**mucosa** (myoo-ko´să) A mucous membrane that lines cavities and tracts opening to the exterior. 113

**mucous cell** (myoo´kus) See goblet cell. 111

**mucous membrane** A thin sheet consisting of layers of visceral organs that include the lining epithelium, submucosal connective tissue, and (in some cases) a thin layer of smooth muscle (the muscularis mucosa). 116

**multipolar neuron** A nerve cell with many processes originating from the cell body. 347

**muscle** (mus´el) A major type of tissue adapted to contract. The three kinds of muscle are cardiac, smooth, and skeletal. 130

**muscle spindles** Sensory organs within skeletal muscles composed of intrafusal fibers. They are sensitive to muscle stretch and provide a length detector within muscles. 471

**muscularis** (mus´´kyŭ-la´ris) A muscular layer or tunic of an organ, composed of smooth muscle tissue. 753

**myelencephalon** (mi´´ĕ-len-sef´ă-lon) The posterior portion of the hindbrain that contains the medulla oblongata. 394

**myelin** (mi´ĕ-lin) A lipoprotein material that forms a sheathlike covering around nerve fibers.

**myelin sheath** A sheath surrounding axons formed by successive wrappings of a neuroglial cell membrane. Myelin sheaths are formed by neurolemmocytes in the peripheral nervous system and by oligodendrocytes within the central nervous system. 349

**myenteric plexus** (mi″en-ter′ik plek′sus) A network of sympathetic and parasympathetic nerve fibers located in the muscularis tunic of the small intestine; also called the *plexus of Auerbach*. 769

**myocardial infarction** (mi″ŏ-kar′de-al in-fark′shun) An area of necrotic tissue in the myocardium that is filled in by scar (connective) tissue. 608

**myocardium** (mi′ŏ-kar′de-um) The cardiac muscle layer of the heart. 566

**myofibril** (mi′ŏ-fi′bril) A bundle of contractile fibers within muscle cells. 260

**myogenic** (mi″ŏ-jen′ik) Originating within muscle cells; used to describe self-excitation by cardiac and smooth muscle cells. 275

**myoglobin** (mi′ŏ-glo′bin) A molecule composed of globin protein and heme pigment. It is related to hemoglobin but contains only one subunit (instead of the four in hemoglobin) and is found in skeletal and cardiac muscle cells where it serves to store oxygen. 272

**myogram** (mi′ŏ-gram) A recording of electrical activity within a muscle. 318

**myology** (mi-ol′ŏ-je) The science or study of muscle structure and function. 281

**myometrium** (mi″o-me′tre-um) The layer or tunic of smooth muscle within the uterine wall. 874

**myoneural junction** (mi′ŏ-noor′al) The site of contact between an axon of a motor neuron and a muscle fiber. 358

**myopia** (mi-o′pe-ă) A visual defect in which objects may be seen distinctly only when very close to the eyes; also called *nearsightedness*. 507

**myosin** (mi′ŏ-sin) A thick myofilament protein that together with actin causes muscle contraction. 263

**myxedema** (mik″sĭ-de′mă) A type of edema associated with hypothyroidism. It is characterized by the accumulation of mucoproteins in tissue fluid. 543

**N**

**NAD** Nicotinamide adenine dinucleotide; a coenzyme derived from niacin that helps to transport electrons from the Krebs cycle to the electron-transport chain within mitochondria. 78

**nail** A hardened, keratinized plate that develops from the epidermis and forms a protective covering on the surface of the distal phalanges of fingers and toes. 146

**naloxone** (nal′ok-sōn, nă-lok′sōn) A drug that antagonizes the effects of morphine and endorphins. 367

**nasal cavity** (na′zal) A mucosa-lined space above the oral cavity, divided by a nasal septum. It is the first chamber of the respiratory system. 684

**nasal concha** (kong′kă) A scroll-like bone extending medially from the lateral wall of the nasal cavity; also called a *turbinate bone*. 189

**nasal septum** (sep′tum) A bony and cartilaginous partition that separates the nasal cavity into two portions. 684

**nasopharynx** (na″zo-far′ingks) The first or uppermost chamber of the pharynx, positioned posterior to the nasal cavity and extending down to the soft palate. 684

**natriuretic** (na″trĭ-yoo-ret′ik) An agent that promotes the excretion of sodium in the urine. Atrial natriuretic hormone has this effect. 749

**neck** 1. Any constricted portion, such as the neck of an organ. 2. The cervical region of the body between the head and thorax.

**necrosis** (nĕ-kro′sis) Cellular death or tissue death due to disease or trauma. 12

**negative feedback** A mechanism in the body for maintaining a state of internal constancy, or homeostasis; effectors are activated by changes in the internal environment, and the actions of the effectors serve to counteract these changes and maintain a state of balance. 17

**neonatal** (ne″o-na′tal) The stage of life from birth to the end of 4 weeks.

**neoplasm** (ne′ŏ-plazm) A new, abnormal growth of tissue, as in a tumor. 151

**nephron** (nef′ron) The functional unit of the kidney, consisting of a glomerulus, convoluted tubules, and a nephron loop. 730

**nerve** A bundle of nerve fibers outside the central nervous system. 347

**neurilemma** (noor″ĭ-lem′ă) A thin, membranous covering surrounding the myelin sheath of a nerve fiber. 349

**neurofibril node** A gap in the myelin sheath of a nerve fiber; also called a *node of Ranvier*. 350

**neuroglia** (noo-rog′le-ă) Specialized supportive cells of the central nervous system. 345

**neurohypophysis** (noor″o-hi-pof′ĭ-sis) The posterior lobe of the pituitary gland derived from the brain. Its major secretions include antidiuretic hormone (ADH), also called vasopressin, and oxytocin, produced in the hypothalamus. 519

**neurolemmocyte** (noor″ŏ-lem′ŏ-sīt) A specialized neuroglia cell that surrounds an axon fiber of a peripheral nerve and forms the neurilemmal sheath; also called a *Schwann cell*. 348

**neuron** (noor′on) The structural and functional unit of the nervous system, composed of a cell body, dendrites, and an axon; also called a *nerve cell*. 345

**neurotransmitter** (noor″o-trans′mit-er) A chemical contained in synaptic vesicles in nerve endings that is released into the synaptic cleft, where it stimulates the production of either excitatory or inhibitory postsynaptic potentials. 358

**neutrons** (noo′tronz) Electrically neutral particles that exist together with positively charged protons in the nucleus of atoms. 22

**neutrophil** (noo′trŏ-fil) A type of phagocytic white blood cell, normally constituting about 60% to 70% of the white blood cell count. 553

**nexus** (nek′sus) A bond between members of a group; the type of bonds present in single-unit smooth muscles. 784

**nidation** (ni-da′shun) Implantation of the blastocyst into the endometrium of the uterus. 903

**nipple** A dark pigmented, rounded projection at the tip of the breast. 887

**Nissl bodies** (nis′l) *See* chromatophilic substances. 345

**node of Ranvier** (ran′ve-a) *See* neurofibril node. 350

**norepinephrine** (nor″ep-ĭ-nef′rin) A catecholamine released as a neurotransmitter from postganglionic sympathetic nerve endings and as a hormone (together with epinephrine) from the adrenal medulla. 528

**notochord** (no′tŏ-kord) A flexible rod of tissue that extends the length of the back of an embryo. 913

**nucleolus** (noo-kle′ŏ-lus) A dark-staining area within a cell nucleus; the site where ribosomal RNA is produced. 53

**nucleoplasm** (noo′kle″o-plaz″em) The protoplasmic contents of the nucleus of a cell. 54

**nucleotide** (noo′kle-ŏ-tīd) The subunit of DNA and RNA macromolecules. Each nucleotide is composed of a nitrogenous base (adenine, guanine, cytosine, and thymine or uracil); a sugar (deoxyribose or ribose); and a phosphate group. 51

**nucleus** (noo′kle-us) A spheroid body within a cell that contains the genetic factors of the cell. 50

**nucleus pulposus** (pul-po′sus) The soft, pulpy core of an intervertebral disc; a remnant of the notochord. 7

**nystagmus** (nĭ-stag′mus) Involuntary oscillary movements of the eye. 481

**O**

**obese** (o-bēs′) Excessively fat. 812

**olfactory** (ol-fak′tŏ-re) Pertaining to the sense of smell. 477

**olfactory bulb** An aggregation of sensory neurons of an olfactory nerve, lying inferior to the frontal lobe of the cerebrum on either lateral side of the crista galli of the ethmoid bone. 420

**olfactory tract** The olfactory sensory tract of axons that conveys impulses from the olfactory bulb to the olfactory portion of the cerebral cortex. 420

**oligodendrocyte** (ol″ĭ-go-den′drŏ-sīt) A type of neuroglial cell concerned with the formation of the myelin of nerve fibers within the central nervous system. 348

**oncology** (on-kol′ŏ-je) The study of tumors. 674

**oncotic pressure** (on-kot′ik) The colloid osmotic pressure of solutions produced by proteins. In plasma, it serves to counterbalance the outward filtration of fluid from capillaries due to hydrostatic pressure. 617

**oocyte** (o´ŏ-sīt) A developing egg cell.

**oogenesis** (o´´ŏ-jen´ĕ-sis) The process of female gamete formation. 877

**opsonization** (op´´sŏ-nĭ-za´shun) The process by which antibodies enhance the ability of phagocytic cells to attack bacteria. 660

**optic** (op´tik) Pertaining to the eye.

**optic chiasma** (ki-az´mă) An X-shaped structure on the inferior aspect of the brain, anterior to the pituitary gland, where there is a partial crossing over of fibers in the optic nerves; also called the *optic chiasm*. 420

**optic disc** A small region of the retina where the fibers of the ganglion neurons exit from the eyeball to form the optic nerve; also called the *blind spot*. 494

**optic tract** A bundle of sensory axons located between the optic chiasma and the thalamus that functions to convey visual impulses from the photoreceptors within the eye. 420

**oral** Pertaining to the mouth.

**ora serrata** The jagged peripheral margin of the retina. 497

**organ** A structure consisting of two or more tissues that performs a specific function. 9

**organelle** (or´´gă-nel´) A minute living structure of a cell with a specific function. 8

**organism** An individual living creature. 10

**organ of Corti** (kor´te) *See* spiral organ. 485

**orifice** (or´ĭ-fis) An opening into a body cavity or tube. 281

**origin** The place of muscle attachment—usually the more stationary point or the proximal bone; opposite the insertion. 255

**oropharynx** (o´´ro-far´ingks) The second portion of the pharynx, located posterior to the oral cavity and extending from the soft palate to the hyoid bone. 686

**osmolality** (oz´´mŏ-lal´ĭ-te) A measure of the total concentration of a solution; the number of moles of solute per kilogram of solvent. 96

**osmoreceptors** (oz´´mŏ-re-cep´torz) Sensory neurons that respond to changes in the osmotic pressure of the surrounding fluid. 97

**osmosis** (oz-mo´sis) The passage of solvent (water) from a more dilute to a more concentrated solution through a membrane that is more permeable to water than to the solute. 93

**osmotic pressure** (oz-mot´ik) A measure of the tendency of a solution to gain water by osmosis when separated by a membrane from pure water. Directly related to the osmolality of the solution, it is the pressure required to just prevent osmosis. 94

**osseous tissue** (os´e-us) Bone tissue. 127

**ossicle** (os´ĭ-kul) One of the three bones of the middle ear; also called the *auditory ossicle*. 483

**ossification** (os´´ĭ-fĭ-ka´shun) The process of bone tissue formation. 168

**osteoblast** (os´te-ŏ-blast) A bone-forming cell. 166

**osteoclast** (os´te-ŏ-klast) A cell that causes erosion and resorption of bone tissue. 166

**osteocyte** (os´te-ŏ-sīt) A mature bone cell. 166

**osteology** (os´´te-ol´ŏ-je) The study of the structure and function of bone and the entire skeleton. 160

**osteomalacia** (os´´te-o-mă-la´shă) Softening of bones due to a deficiency of vitamin D and calcium. 173

**osteon** (os´te-on) A group of osteocytes and concentric lamellae surrounding a central canal, constituting the basic unit of structure in osseous tissue; also called a *haversian system*. 167

**osteoporosis** (os´´te-o-pŏ-ro´sis) Demineralization of bone, seen most commonly in postmenopausal women and patients who are inactive or paralyzed. It may be accompanied by pain, loss of stature, and other deformities and fractures. 174

**otoliths** (o´tŏ-liths) Small, hardened particles of calcium carbonate in the saccule and utricle of the inner ear, associated with the receptors of equilibrium; also called *statoconia*. 480

**outer ear** The outer portion of the ear, consisting of the auricle and the external auditory canal. 482

**oval window** An oval opening in the bony wall between the middle and inner ear, into which the footplate of the stapes fits; also called the *vestibular window*. 483

**ovarian follicle** (o-var´e-an fol´ĭ-kul) A developing ovum and its surrounding epithelial cells. 878

**ovarian ligament** (lig´ă-ment) A cordlike connective tissue that attaches the ovary to the uterus. 877

**ovary** (o´vă-re) The female gonad in which ova and certain sexual hormones are produced. 877

**oviduct** (o´vĭ-dukt) The tube that transports ova from the ovary to the uterus; also called the *uterine tube* or *fallopian tube*. 872

**ovulation** (ov-yu-la´shun) The rupture of an ovarian (graafian) follicle with the release of an ovum. 878

**ovum** (o´vum) A secondary oocyte capable of developing into a new individual when fertilized by a spermatozoon. 839

**oxidative phosphorylation** (ok´´sĭ-da´tiv fos´´for-ĭ-la´shun) The formation of ATP using energy derived from electron transport to oxygen. It occurs in the mitochondria. 85

**oxidizing agent** (ok´sĭ-dīz´ing) An atom that accepts electrons in an oxidation-reduction reaction. 78

**oxyhemoglobin** (ok´´se-he´´mŏ-glo´bin) A compound formed by the bonding of molecular oxygen to hemoglobin. 707

**oxyhemoglobin saturation** The ratio, expressed as a percentage, of the amount of oxyhemoglobin relative to the total amount of hemoglobin in blood. 709

**oxytocin** (ok´´sĭ-to´sin) One of the two hormones produced in the hypothalamus and secreted by the posterior pituitary (the other hormone is vasopressin). Oxytocin stimulates the contraction of uterine smooth muscles and promotes milk ejection in females. 520

**P**

**pacemaker** (pās´ma´´ker) A group of cells that has the fastest spontaneous rate of depolarization and contraction in a mass of electrically coupled cells; in the heart, this is the sinoatrial, or SA, node. 275

**pacinian corpuscle** (pă-sin´e-an) *See* lamellated corpuscle. 467

**PAH** Para-aminohippuric acid; a substance used to measure total renal plasma flow because its clearance rate is equal to the total rate of plasma flow to the kidneys. PAH is filtered and secreted but not reabsorbed by the renal nephrons. 745

**palate** (pal´at) The roof of the oral cavity. 771

**palatine** (pal´ă-tīn) Pertaining to the palate.

**palmar** (pal´mar) Pertaining to the palm of the hand. 15

**palpebra** (pal´pĕ-bră) An eyelid. 488

**pancreas** (pan´kre-as) A mixed organ in the abdominal cavity that secretes pancreatic juices into the GI tract and insulin and glucagon into the blood. 794

**pancreatic duct** (pan´kre-at´ik) A drainage tube that carries pancreatic juice from the pancreas into the duodenum of the hepatopancreatic ampulla. 794

**pancreatic islets** A cluster of cells within the pancreas that forms the endocrine portion and secretes insulin and glucagon; also called *islets of Langerhans*. 794

**papillae** (pă-pil´e) Small, nipplelike projections. 148

**papillary muscle** (pap´ĭ-ler´´e) Muscular projections from the ventricular walls of the heart to which the chordea tendineae are attached. 569

**paranasal sinus** (par´´ă-na´zal si´nus) An air chamber lined with a mucous membrane that communicates with the nasal cavity. 685

**parasympathetic** (par´´ă-sim´´pă-thet´ik) Pertaining to the division of the autonomic nervous system concerned with activities that, in general, inhibit or oppose the physiological effects of the sympathetic nervous system. 447

**parathyroid hormone** (PTH) A polypeptide hormone secreted by the parathyroid glands. PTH acts to raise the blood $Ca^{++}$ levels primarily by stimulating reabsorption of bone. 530

**parathyroids** (par´´ă-thi´roidz) Small endocrine glands embedded on the posterior surface of the thyroid glands that are concerned with calcium metabolism. 530

**parietal** (pă-ri´ĕ-tal) Pertaining to a wall of an organ or cavity. 12

**parietal pleura** (ploor´ă) The thin serous membrane attached to the thoracic walls of the pleural cavity. 116

**Parkinson's disease** (par´kin-sunz) A tremor of the resting muscles and other symptoms caused by inadequate dopamine-producing neurons in the basal nuclei of the cerebrum; also called *paralysis agitans*. 352

**parotid gland** (pă-rot´id) One of the paired salivary glands located on the side of the face over the masseter muscle just anterior to the ear and connected to the oral cavity through a salivary duct. 774

**parturition** (par''tyoo-rish'un) The process of giving birth; childbirth. 920

**passive immunity** (ĭ-myoo'nĭ-te) Specific immunity granted by the administration of antibodies made by another organism. 665

**pathogen** (path'ŏ-jen) Any disease-producing microorganism or substance. 803

**pectoral** (pek'tŏ-ral) Pertaining to the chest region. 14

**pectoral girdle** The portion of the skeleton that supports the upper extremities. 203

**pedicle** (ped'ĭ-k'l) The portion of a vertebra that connects and attaches the lamina to the body. 194

**pelvic** (pel'vik) Pertaining to the pelvis.

**pelvic girdle** The portion of the skeleton to which the lower extremities are attached. 209

**pelvis** (pel'vis) A basinlike bony structure formed by the sacrum and ossa coxae. 209

**penis** (pe'nis) The male organ of copulation, used to introduce sperm into the female vagina and through which urine passes during urination. 856

**pennate** (pen'āt) Pertaining to a skeletal muscle fiber arrangement in which the fibers are attached to tendinous slips in a featherlike pattern. 281

**pepsin** (pep'sin) The protein-digesting enzyme secreted in gastric juice. 778

**peptic ulcer** (pep'tik ul'ser) An injury to the mucosa of the esophagus, stomach, or small intestine due to the action of acidic gastric juice. 779

**perforating canal** A minute duct through compact bone by which blood vessels and nerves penetrate to the central canal of an osteon; also called *Volkmann's canal*. 167

**pericardium** (per''ĭ-kar'de-um) A protective serous membrane that surrounds the heart. 566

**perichondrium** (per''ĭ-kon'dre-um) A toughened connective sheet that covers some kinds of cartilage. 126

**perikaryon** (per''ĭ-kar'e-on) The cell body of a neuron. 345

**perilymph** (per'ĭ-limf) A fluid of the inner ear that provides a liquid-conducting medium for the vibrations involved in hearing and the maintenance of equilibrium. 478

**perimysium** (per''ĭ-mis'e-um) Fascia (connective tissue) surrounding a bundle of muscle fibers. 255

**perineum** (per''ĭ-ne'um) The floor of the pelvis, which is the region between the anus and the symphysis pubis. It is the region that contains the external genitalia. 14

**perineurium** (per''ĭ-noor'e-um) Connective tissue surrounding a bundle of nerve fibers. 347

**periodontal membrane** (per''e-ŏ-don'tal) A fibrous connective tissue lining the dental alveoli. 773

**periosteum** (per''e-os'te-um) A fibrous connective tissue covering the outer surface of bone. 166

**peripheral nervous system** (pĕ-rif'er-al) The nerves and ganglia of the nervous system that lie outside of the brain and spinal cord; *PNS*. 416

**peristalsis** (per''ĭ-stal'sis) Rhythmic contractions of smooth muscle in the walls of various tubular organs by which the contents are forced onward. 766

**peritoneum** (per''ĭ-tŏ-ne'um) The serous membrane that lines the abdominal cavity and covers the abdominal visceral organs. 767

**Peyer's patches** (pi'erz) *See* mesenteric patches. 650

**pH** A measure of the relative acidity or alkalinity of a solution, numerically equal to 7 for neutral solutions. The pH scale in common use ranges from 0 to 14. Solutions with a pH lower than 7 are acidic and those with a higher pH are basic. 26

**phagocytosis** (fag''ŏ-si-to'sis) Cellular eating; the ability of some cells (such as white blood cells) to engulf large particles (such as bacteria) and digest these particles by merging the food vacuole in which they are contained with a lysosome containing digestive enzymes. 45

**phalanx** (fa'langks), pl. *phalanges* A bone of a finger or toe. 208

**pharynx** (far'ingks) The organ of the digestive system and respiratory system located at the back of the oral and nasal cavities that extends to the larynx anteriorly and to the esophagus posteriorly; also called the *throat*. 685

**photoreceptor** (fo''to-re-sep'tor) A sensory nerve ending that responds to the stimulation of light. 465

**physiology** (fiz''e-ol'ŏ-je) The science that deals with the study of body functions. 2

**pia mater** (pi'ă ma'ter) The innermost meninx that is in direct contact with the brain and spinal cord. 396

**pineal gland** (pin'e-al) A small cone-shaped gland located in the roof of the third ventricle. 532

**pinna** (pin'ă) The outer, fleshy portion of the external ear; also called the *auricle*. 482

**pinocytosis** (pin'ŏ-si-to'sis) Cell drinking; invagination of the cell membrane forming narrow channels that pinch off into vacuoles. This allows for cellular intake of extracellular fluid and dissolved molecules. 45

**pituitary gland** (pĭ-too'ĭ-ter-e) A small, pea-shaped endocrine gland situated on the interior surface of the diencephalonic region of the brain, consisting of anterior and posterior lobes; also called the *hypophysis*. 519

**pivot joint** (piv'ut) A synovial joint in which the rounded head of one bone articulates with the depressed cup of another to permit a rotational type of movement. 228

**placenta** (plă-sen'tă) The organ of metabolic exchange between the mother and the fetus. 911

**plantar** (plan'tar) Pertaining to the sole of the foot. 15

**plasma** (plaz'mă) The fluid, extracellular portion of circulating blood. 551

**plasma cells** Cells derived from B lymphocytes that produce and secrete large amounts of antibodies. They are responsible for humoral immunity. 553

**platelets** (plăt-letz) Small fragments of specific bone marrow cells that function in blood coagulation; also called *thrombocytes*. 553

**pleural** (ploor'al) Pertaining to the serous membranes associated with the lungs.

**pleural cavity** The potential space between the visceral pleura and parietal pleura. 691

**pleural membranes** Serous membranes that surround the lungs and provide protection and compartmentalization. 116

**plexus** (plek'sus) A network of interlaced nerves or vessels.

**plexus of Auerbach** (ow'er-bak) *See* myenteric plexus. 769

**plexus of Meissner** (mīs'ner) *See* submucosal plexus. 769

**plicae circulares** (pli'ce sur-kyŭ-lar'ēz) Deep folds within the wall of the small intestine that increase the absorptive surface area. 782

**pneumotaxic area** (noo''mŏ-tak'sik) The region of the respiratory control center located in the pons of the brain. 693

**polar body** A small daughter cell formed by meiosis that degenerates in the process of oocyte production. 877

**polar molecule** A molecule in which the shared electrons are not evenly distributed, so that one side of the molecule is negatively (or positively) charged in comparison with the other side. Polar molecules are soluble in polar solvents, such as water. 23

**polydipsia** (pol''e-dip'se-ă) Excessive thirst.

**polymer** (pol'ĕ-mer) A large molecule formed by the combination of smaller subunits, or monomers. 765

**polymorphonuclear leukocyte** (pol''e-mor''fŏ-noo'kle-ar loo'kŏ-sīt) A granular leukocyte containing a nucleus with a number of lobes connected by thin, cytoplasmic strands. This type includes neutrophils, eosinophils, and basophils. 553

**polypeptide** (pol''e-pep'tīd) A chain of amino acids connected by covalent bonds called peptide bonds. A very large polypeptide is called a protein. 36

**polysaccharide** (pol''e-sak'ă-rīd) A carbohydrate formed by covalent bonding of numerous monosaccharides. Examples include glycogen and starch. 30

**polyuria** (pol''e-yoor'e-ă) Excretion of an excessively large volume of urine in a given period. 758

**pons** (ponz) The portion of the brain stem just above the medulla oblongata and anterior to the cerebellum. 393

**popliteal** (pop''lĭ-te'al, pop-lit'e-al) Pertaining to the concave region on the posterior aspect of the knee. 15

**posterior** (pos-tēr'e-or) Toward the back; also called *dorsal*. 12

**posterior pituitary** (pĭ-too'ĭ-ter-e) *See* neurohypophysis. 519

**posterior root** An aggregation of sensory neuron fibers lying between a spinal nerve and the posterolateral aspect of the spinal cord; also called the *dorsal root* or *sensory root*. 427

**posterior root ganglion** (gang´gle-on) A cluster of cell bodies of sensory neurons located along the posterior root of a spinal nerve. 427

**postganglionic neuron** (pōst´´gang-gle-on´ik) The second neuron in an autonomic motor pathway. Its cell body is outside the central nervous system and it terminates at an effector organ. 447

**postnatal** (pōst-na´tal) After birth.

**postsynaptic inhibition** (pōst´´sĭ-nap´tik) The inhibition of a postsynaptic neuron by axon endings that release a neurotransmitter that induces hyperpolarization (inhibitory postsynaptic potentials). 369

**preganglionic neuron** (pre´´gang-gle-on´ik) The first neuron in an autonomic motor pathway. Its cell body is inside the central nervous system and it terminates on a postganglionic neuron. 447

**pregnancy** A condition in which a female is carrying a developing offspring within the body. 925

**prenatal** (pre-na´tal) Pertaining to the period of offspring development during pregnancy; before birth.

**prepuce** (pre´pyoos) A fold of loose, retractable skin covering the glans of the penis or clitoris; also called the *foreskin*. 856

**presynaptic inhibition** (pre´´sĭ-nap´tik) Neural inhibition in which axoaxonic synapses inhibit the release of neurotransmitter chemicals from the presynaptic axon terminal. 369

**prolactin** (pro-lak´tin) A hormone secreted by the anterior pituitary that, in conjunction with other hormones, stimulates lactation in the postpartum female. It may also participate (along with the gonadotropins) in regulating gonadal function in some mammals. 520

**pronation** (pro-na´shun) A rotational movement of the forearm in which the palm of the hand is turned posteriorly. 234

**proprioceptor** (pro´´pre-o-sep´tor) A sensory nerve ending that responds to changes in tension in a muscle or tendon. 465

**prostaglandin** (pros´´tă-glan´din) Any of a family of fatty acids that have numerous autocrine regulatory functions, including the stimulation of uterine contractions and of gastric acid secretion and the promotion of inflammation. 34

**prostate** (pros´tāt) A walnut-shaped gland surrounding the male urethra just below the urinary bladder that secretes an additive to seminal fluid during ejaculation. 855

**prosthesis** (pros-the´sis) An artificial device to replace a diseased or worn body part. 250

**proton** (pro´ton) A unit of positive charge in the nucleus of atoms. 22

**protoplasm** (pro´tŏ-plaz´´em) A general term for the colloidal complex of protein that constitutes the living material of a cell. It includes cytoplasm and nucleoplasm. 8

**protraction** (pro-trak´shun) The movement of a body part, such as the mandible, forward on a plane parallel with the ground; the opposite of *retraction*. 235

**proximal** (prok´-sĭ-mal) Closer to the midplane of the body or to the origin of an appendage; the opposite of *distal*. 12

**pseudohermaphrodite** (soo´´dŏ-her-maf´rŏ-dīt) An individual with some of the physical characteristics of both sexes, but who lacks functioning gonads of both sexes; a true hermaphrodite has both testes and ovaries. 863

**pseudopods** (soo´dŏ-podz) Footlike extensions of the cytoplasm that enable some cells (with amoeboid motion) to move across a substrate. Pseudopods are also used to surround food particles in the process of phagocytosis. 44

**ptyalin** (ti´ă-lin) An enzyme in saliva that catalyzes the hydrolysis of starch into smaller molecules; also called *salivary amylase*. 796

**puberty** (pyoo´ber-te) The period of development in which the reproductive organs become functional. 843

**pulmonary** (pul´mŏ-ner´´e) Pertaining to the lungs.

**pulmonary circulation** The system of blood vessels from the right ventricle of the heart to the lungs that transports deoxygenated blood and returns oxygenated blood from the lungs to the left atrium of the heart. 570

**pulp cavity** A cavity within the center of a tooth that contains blood vessels, nerves, and lymphatics. 773

**pupil** The opening through the iris that permits light to enter the posterior cavity of the eyeball and be refracted by the lens through the vitreous chamber. 492

**Purkinje fibers** (pur-kin´je) *See* conduction myofibers. 576

**pyloric sphincter** (pi-lor´ik sfingk´ter) A modification of the muscularis tunic between the stomach and the duodenum that functions to regulate the food material leaving the stomach. 777

**pyramid** (pir´ă-mid) Any of several structures that have a pyramidal shape (e.g., the renal pyramids in the kidney and the medullary pyramids on the anterior surface of the brain). 403

**pyrogen** (pi´rŏ-jen) A fever-producing substance. 653

## Q

**QRS complex** The principal deflection of an electrocardiogram that is produced by depolarization of the ventricles. 577

## R

**ramus** (ra´mus) A branch of a bone, artery, or nerve. 211

**raphe** (ra´fe) A ridge or a seamlike structure between two similar parts of a body organ, as in the scrotum. 845

**receptor** (re-sep´tor) A sense organ or a specialized distal end of a sensory neuron that receives stimuli from the environment. 437

**rectum** (rek´tum) The terminal portion of the GI tract, between the sigmoid colon and the anal canal. 785

**red marrow** (mar´o) A tissue that forms blood cells, located in the medullary cavity of certain bones. 165

**red nucleus** (noo´kle-us) An aggregation of gray matter of a reddish color located in the upper portion of the midbrain. It sends fibers to certain brain tracts. 392

**reduced hemoglobin** (he´mŏ-glo´´bin) Hemoglobin with iron in the reduced ferrous state. It is able to bond with oxygen but is not combined with oxygen. Also called *deoxyhemoglobin*. 707

**reducing agent** An electron donor in a coupled oxidation-reduction reaction. 78

**reflex** (re´fleks) A rapid involuntary response to a stimulus. 438

**reflex arc** The basic conduction pathway through the nervous system, consisting of a sensory neuron, an association neuron, and a motor neuron. 437

**regional anatomy** The division of anatomy concerned with structural arrangement in specific areas of the body, such as the head, neck, thorax, or abdomen. 323

**renal** (re´nal) Pertaining to the kidney.

**renal corpuscle** (kor´pus´l) The portion of the nephron consisting of the glomerulus and a glomerular capsule; also called the *malpighian corpuscle*. 730

**renal cortex** The outer portion of the kidney, primarily vascular. 728

**renal medulla** (mě-dul´ă) The inner portion of the kidney, including the renal pyramids and renal columns. 728

**renal pelvis** The inner cavity of the kidney formed by the expanded ureter and into which the calyces open. 729

**renal plasma clearance rate** The milliliters of plasma cleared of a particular solute per minute by the excretion of that solute in the urine. If there is no reabsorption or secretion of that solute by the nephron tubules, the plasma clearance rate is equal to the glomerular filtration rate. 744

**renal pyramid** A triangular structure within the renal medulla composed of nephron loops and the collecting ducts. 729

**repolarization** (re-po´´lar-ĭ-za´shun) The reestablishment of the resting membrane potential after depolarization has occurred. 353

**respiration** (res´´pĭ-ra´shun) The exchange of gases between the external environment and the cells of an organism. 684

**respiratory acidosis** (rĭ-spīr´ă-tor-e as´´ĭ-do´sis) A lowering of the blood pH to below 7.35 due to accumulation of $CO_2$ as a result of hypoventilation. 713

**respiratory alkalosis** (al´´kă-lo´sis) A rise in blood pH to above 7.45 due to excessive elimination of blood $CO_2$ as a result of hyperventilation. 713

**respiratory center** The structure or portion of the brain stem that regulates the depth and rate of breathing. 395

**respiratory distress syndrome** A lung disease of the newborn, most frequently occurring in premature infants, that is caused by abnormally high alveolar surface tension as a result of a deficiency in lung surfactant; also called *hyaline membrane disease*. 695

**respiratory membrane**   A thin, moistened membrane within the lungs, composed of an alveolar portion and a capillary portion, through which gaseous exchange occurs.   690

**rete testis**   (re´te tes´tis)   A network of ducts in the center of the testis associated with the production of spermatozoa.   848

**reticular formation**   (rĕ-tik´yŭ-lar)   A network of nervous tissue fibers in the brain stem that arouses the higher brain centers.   395

**retina**   (ret´ĭ-nă)   The principal portion of the internal tunic of the eyeball that contains the photoreceptors.   496

**retraction**   (re-trak´shun)   The movement of a body part, such as the mandible, backward on a plane parallel with the ground; the opposite of *protraction*.   235

**retroperitoneal**   (ret´´ro-per´´ĭ-tŏ-ne´al)   Positioned behind the parietal peritoneum.   117

**rhodopsin**   (ro-dop´sin)   A pigment in rod cells that undergoes a photochemical dissociation in response to light, and in so doing stimulates electrical activity in the photoreceptors.   498

**rhythmicity area**   (rith-mis´ĭ-te)   A portion of the respiratory control center located in the medulla oblongata that controls inspiratory and expiratory phases.   703

**ribosome**   (ri´bo-sōm)   A cytoplasmic organelle composed of protein and RNA in which protein synthesis occurs.   55

**rickets**   (rik´ets)   A condition caused by a deficiency of vitamin D and associated with an interference of the normal ossification of bone.   173

**right lymphatic duct**   (lim-fat´ik)   A major vessel of the lymphatic system that drains lymph from the upper right portion of the body into the right subclavian vein.   648

**rigor mortis**   (rig´or mor´tis)   The stiffening of a dead body due to the depletion of ATP and the production of rigor complexes between actin and myosin in muscles.   134

**RNA**   Ribonucleic acid; a nucleic acid consisting of the nitrogenous bases adenine, guanine, cytosine, and uracil; the sugar ribose; and phosphate groups. There are three types of RNA found in cytoplasm: messenger RNA (mRNA), transfer RNA (tRNA) and ribosomal RNA (rRNA).   52

**rod**   A photoreceptor in the retina of the eye that is specialized for colorless, dim-light vision.   498

**root canal**   The hollow, tubular extension of the pulp cavity into the root of the tooth that contains vessels and nerves.   773

**rotation**   (ro-ta´shun)   The movement of a bone around its own longitudinal axis.   234

**round window**   A round, membrane-covered opening between the middle and inner ear, directly below the oval window; also called the *cochlear window*.   483

**rugae**   (roo´je)   The folds or ridges of the mucosa of an organ.   754

## S

**saccadic eye movements**   (să-kad´ik)   Very rapid eye movements that occur constantly and that change the focus on the retina from one point to another.   502

**saccule**   (sak´yool)   A saclike cavity in the membranous labyrinth inside the vestibule of the inner ear that contains a vestibular organ for equilibrium.   479

**sacral**   (sa´kral)   Pertaining to the sacrum.

**sacral plexus**   (plek´sus)   A network of nerve fibers that arises from spinal nerves L4 through S3. Nerves arising from the sacral plexus merge with those from the lumbar plexus to form the lumbosacral plexus and supply the lower extremity.   434

**saddle joint**   A synovial joint in which the articular surfaces of both bones are concave in one plane and convex or saddle shaped, in the other plane, such as in the distal carpometacarpal joint of the thumb.   231

**sagittal plane**   (saj´ĭ-tal)   A vertical plane, running parallel to the midsagittal plane, that divides the body into unequal right and left portions.   11

**salivary gland**   (sal´ĭ-ver-e)   An accessory digestive gland that secretes saliva into the oral cavity.   773

**saltatory conduction**   (sal´tă-to´´re)   The rapid passage of action potentials from one node of Ranvier (neurofibril node) to another in myelinated axons.   357

**sarcolemma**   (sar´´kŏ-lem´ă)   The cell membrane of a muscle fiber.   270

**sarcomere**   (sar´´kŏ-mēr)   The portion of a striated muscle fiber between the two adjacent Z lines that is considered the functional unit of a myofibril.   263

**sacroplasm**   (sar´kŏ-plaz´´em)   The cytoplasm within a muscle fiber.   268

**sarcoplasmic reticulum**   (sar´kŏ-plaz´mik rĕ-tik´yŭ-lum)   The smooth or agranular endoplasmic reticulum of skeletal muscle cells. It surrounds each myofibril and stores Ca⁺⁺ when the muscle is at rest.   268

**scala tympani**   (ska´lă tim´pă-ne)   The lower channel of the cochlea that is filled with perilymph.   484

**scala vestibuli**   (vĕ-stib´yŭ-le)   The upper channel of the cochlea that is filled with perilymph.   484

**Schwann cell**   (schwahn)   *See* neurolemmocyte.   348

**sclera**   (skler´ă)   The outer white layer of fibrous connective tissue that forms the protective covering of the eyeball.   492

**scleral venous sinus**   (ve´nus)   A circular venous drainage for the aqueous humor from the anterior chamber; located at the junction of the sclera and the cornea; also called the *canal of Schlemm*.   494

**scrotum**   (skro´tum)   A pouch of skin that contains the testes and their accessory organs.   845

**sebaceous gland**   (sĕ-ba´shus)   An exocrine gland of the skin that secretes sebum.   147

**sebum**   (se´bum)   An oily, waterproofing secretion of the sebaceous glands.   147

**second messenger**   A molecule or ion whose concentration within a target cell is increased by the action of a regulatory compound (e.g., a hormone or neurotransmitter) and which stimulates the metabolism of that target cell in a way that mediates the intracellular effects of that regulatory compound.   536

**secretin**   (sĕ-kre´tin)   A polypeptide hormone secreted by the small intestine in response to acidity of the intestinal lumen. Along with cholecystokinin, secretin stimulates the secretion of pancreatic juice into the small intestine.   800

**semen**   (se´men)   The thick, whitish secretion of the reproductive organs of the male, consisting of spermatozoa and additives from the prostate and seminal vesicles.   863

**semicircular canals**   Tubule channels within the inner ear that contain receptors for equilibrium.   479

**semilunar valve**   (sem´´e-loo´nar)   Crescent- or half-moon-shaped heart valves positioned at the entrances to the aorta and the pulmonary trunk.   567

**seminal vesicles**   (sem´ĭ-nal ves´ĭ-k´lz)   A pair of accessory male reproductive organs lying posterior and inferior to the urinary bladder that secrete additives to spermatozoa into the ejaculatory ducts.   854

**seminiferous tubules**   (sem´ĭ-nif´er-us too´byoolz)   Numerous small ducts in the testes, where spermatozoa are produced.   845

**semipermeable membrane**   (sem´´e-per´me-ă-b´l)   A membrane with pores of a size that permits the passage of solvent and some solute molecules while restricting the passage of other solute molecules.   93

**senescence**   (sĕ-nes´ens)   The process of aging.   134

**sensory area**   A region of the cerebral cortex that receives and interprets sensory nerve impulses.   385

**sensory neuron**   (noor´on)   A nerve cell that conducts an impulse from a receptor organ to the central nervous system; also called an *afferent neuron*.   346

**septum**   (sep´tum)   A membranous or fleshy wall dividing two cavities.

**serous membrane**   (ser´us)   An epithelial and connective tissue membrane that lines body cavities and covers visceral organs within these cavities; also called *serosa*.   116

**Sertoli cells**   (ser-to´le)   *See* sustentacular cells.   845

**serum**   (ser´um)   Blood plasma with the clotting elements removed.   551

**sesamoid bone**   (ses´ă-moid)   A membranous bone formed in a tendon in response to joint stress (e.g., the patella).   161

**sex chromosomes**   The X and Y chromosomes; the unequal pairs of chromosomes involved in sex determination (which is based on the presence or absence of a Y chromosome). Females lack a Y chromosome and normally have the genotype XX; males have a Y chromosome and normally have the genotype XY.   839

**shock**   As it relates to the cardiovascular system, this term refers to a rapid, uncontrolled fall in blood pressure, which in some cases becomes irreversible and leads to death.   641

**shoulder**   The region of the body where the humerus articulates with the scapula.   15

**sickle-cell anemia**   A hereditary, autosomal recessive trait that occurs primarily in people of African ancestry, in which it evolved apparently as a protection  (in the carrier state) against malaria. In the homozygous state, hemoglobin S is made instead of hemoglobin A; this leads to the characteristic sickling of red blood cells, hemolytic anemia, and organ damage.   718

**sigmoid colon**   (sig´moid ko´lon)   The S-shaped portion of the large intestine between the descending colon and the rectum.   785

**sinoatrial node**   (sin´´o-a´tre-al)   A mass of specialized cardiac tissue in the wall of the right atrium that initiates the cardiac cycle; the SA node; also called the *pacemaker*.   574

**sinus**   (si´nus)   A cavity or hollow space within a body organ, such as a bone.

**sinusoid**   (si´nŭ-soid)   A small, blood-filled space in certain organs, such as the spleen or liver.   789

**skeletal muscle**   A specialized type of multinucleated muscle tissue that occurs in bundles, has crossbands of proteins, and contracts in either a voluntary or involuntary fashion.   131

**sleep apnea**   (ap´ne-ă)   A temporary cessation of breathing during sleep, usually lasting for several seconds.   720

**sliding filament theory**   The theory that the thick and thin filaments of a myofibril slide past each other during muscle contraction, while maintaining their initial length.   263

**small intestine**   The portion of the GI tract between the stomach and the cecum whose function is the absorption of food nutrients.   780

**smooth muscle**   A specialized type of nonstriated muscle tissue composed of fusiform, single-nucleated fibers. It contracts in an involuntary, rhythmic fashion within the walls of visceral organs.   131

**sodium/potassium pump**   (so´de-um pŏ-tas´e-um)   An active transport carrier with ATPase enzymatic activity that acts to accumulate K⁺ within cells and extrude Na⁺ from cells, thus maintaining gradients for these ions across the cell membrane.   99

**soft palate**   (pal´at)   The fleshy, posterior portion of the roof of the mouth, from the palatine bones to the uvula.   771

**somatic**   (so-mat´ik)   Pertaining to the nonvisceral parts of the body.

**somatomedins**   (so´´mat´ŏ-med-inz)   A group of small polypeptides believed to be produced in the liver in response to growth hormone stimulation and to mediate the actions of growth hormone on the skeleton and other tissues.   831

**somatostatin**   (so-mat´´ŏ-stat´in)   A polypeptide produced in the hypothalamus that acts to inhibit the secretion of growth hormone from the anterior pituitary.

Somatostatin is also produced in the pancreatic islets, but its function there has not been established.   523

**somatotropic hormone**   (so-mat´´ŏ-trop´ik)   Growth hormone; an anabolic hormone secreted by the anterior pituitary that stimulates skeletal growth and protein synthesis in many organs.   519

**sounds of Korotkoff**   (kŏ-rot´kof)   The sounds heard when pressure measurements are taken. These sounds are produced by the turbulent flow of blood through an artery that has been partially constricted by a pressure cuff.   637

**spermatic cord**   (sper-mat´ik)   The structure of the male reproductive system composed of the ductus deferens, spermatic vessels, nerves, cremaster muscle, and connective tissue. The spermatic cord extends from a testis to the inguinal ring.   841

**spermatogenesis**   (sper-mat´´ŏ-jen´ĭ-sis)   The production of male sex gametes, or spermatozoa.   849

**spermatozoon**   (sper-mat´´ŏ-zo´on), pl. *spermatozoa* or, loosely, *sperm*   A mature male sperm cell, or gamete.   853

**spermiogenesis**   (sper´´me-ŏ-jen´ĕ-sis)   The maturational changes that transform spermatids into spermatozoa.   849

**sphincter**   (sfingk´ter)   A circular muscle that functions to constrict a body opening or the lumen of a tubular structure.   281

**sphincter of ampulla**   The muscular constriction at the opening of the common bile and pancreatic ducts; also called the *sphincter of Oddi*.   794

**sphincter of Oddi**   (o´de)   See sphincter of ampulla.   794

**sphygmomanometer**   (sfig´´mo-mă-nom´ĭ-ter)   A manometer  (pressure transducer) used to measure the blood pressure.   637

**spinal cord**   (spi´nal)   The portion of the central nervous system that extends downward from the brain stem through the vertebral canal.   401

**spinal ganglion**   A cluster of nerve cell bodies on the posterior root of a spinal nerve.   448

**spinal nerve**   One of the 31 pairs of nerves that arise from the spinal cord.   427

**spindle fibers**   (spin´d´l)   Filaments that extend from the poles of a cell to its equator and attach to the chromosomes during the metaphase stage of cell division. Contraction of the spindle fibers pulls the chromosomes to opposite poles of the cell.   62

**spinous process**   (spi´nus)   A sharp projection of bone or a ridge of bone, such as on the scapula.   194

**spiral organ**   The functional unit of hearing, consisting of a basilar membrane supporting receptor hair cells and a tectorial membrane within the endolymph of the cochlear duct; also known as the *organ of Corti*.   485

**spironolactones**   (spi´´rŏ-no-lak´tōnz)   Diuretic drugs that act as an aldosterone antagonist.   758

**spleen**   (splēn)   A large, blood-filled, glandular organ located in the upper left quadrant of the abdomen and attached by mesenteries to the stomach.   650

**spongy bone**   Bone tissue with a latticelike structure; also called *cancellous bone*.   165

**squamous**   (skwa´mus)   Flat or scalelike.   110

**stapes**   (sta´pēz)   The innermost of the auditory ossicles that fits against the oval window of the inner ear; also called the *stirrup*.   483

**steroid**   (ster´oid)   A lipid, derived from cholesterol, that has three 6-sided carbon rings and one 5-sided carbon ring. These form the steroid hormones of the adrenal cortex and gonads.   34

**stomach**   A pouchlike digestive organ located between the esophagus and the duodenum.   776

**stratified**   (strat´ĭ-fīd)   Arranged in layers, or strata.   110

**stratum basale**   (stra´tum bă-să´le)   The deepest epidermal layer, where mitotic activity occurs.   138

**stratum corneum**   (kor´ne-um)   The outer, cornified layer of the epidermis of the skin.   140

**stroke volume**   The amount of blood ejected from each ventricle at each heartbeat.   613

**stroma**   (stro´mă)   A connective tissue framework in an organ, gland, or other tissue.   122

**subarachnoid space**   (sub´´ă-rak´noid)   The space within the meninges between the arachnoid mater and pia mater, where cerebrospinal fluid flows.   396

**sublingual gland**   (sub-ling´gwal)   One of the three pairs of salivary glands. It is located below the tongue and its duct opens to the side of the lingual frenulum.   774

**submandibular gland**   (sub´´man-dib´yŭ-lar)   One of the three pairs of salivary glands. It is located below the mandible and its duct opens to the side of the lingual frenulum.   774

**submucosa**   (sub´´myoo-ko´sa)   A layer of supportive connective tissue that underlies a mucous membrane.   767

**submucosal plexus**   (sub´´myoo-kōs´al plek´sus)   A network of sympathetic and parasympathetic nerve fibers located in the submucosa tunic of the small intestine; also called the *plexus of Meissner*.   769

**substrate**   (sub´strāt)   In enzymatic reactions, the molecules that combine with the amino acids lining the active sites of an enzyme and are converted to products by catalysis of the enzyme.   70

**sulcus**   (sul´kus)   A shallow impression or groove.   164

**superficial**   (soo´´per-fish´al)   Toward or near the surface.   12

**superficial fascia**   (fash´e-ă)   A binding layer of connective tissue between the dermis of the skin and the underlying muscle.   255

**superior**   Toward the upper part of a structure or toward the head; also called *cephalic*.   12

**superior vena cava**   A large systemic vein that collects blood from regions of the body superior to the heart and returns it to the right atrium.   592

**supination**   (soo´´pĭ-na´shun)   Rotation of the arm so that the palm is directed forward or anteriorly; the opposite of *pronation*.   234

**suppressor T cell** A subpopulation of T lymphocytes that acts to inhibit the production of antibodies against specific antigens by B lymphocytes. 668

**surface anatomy** The division of anatomy concerned with the structures that can be identified from the outside of the body. 322

**surfactant** (sur-fak´tant) A substance produced by the lungs that decreases the surface tension within the alveoli. 694

**suspensory ligament** (sŭ-spen´sŏ-re) 1. A portion of the peritoneum that extends laterally from the surface of the ovary to the wall of the pelvic cavity. 877 2. A ligament that supports an organ or body part, such as that supporting the lens of the eye. 492

**sustentacular cells** (sus-ten-tak´yŭ-lar) Specialized cells within the testes that supply nutrients to developing spermatozoa; also called *Sertoli cells* or *nurse cells*. 845

**sutural bone** (soo´chur-al) A small bone positioned within a suture of certain cranial bones; also called a *wormian bone*. 161

**suture** (soo´chur) A type of fibrous joint found between bones of the skull. 224

**sweat gland** A skin gland that secretes a fluid substance for evaporative cooling. 148

**sympathetic** (sim´´pă-thet´ik) Pertaining to the division of the autonomic nervous system concerned with activities that, in general, arouse the body for physical activity; also called the *thoracolumbar division*. 448

**symphysis** (sim´fĭ-sis) A type of cartilaginous joint characterized by a fibrocartilaginous pad between the articulating bones, which provides slight movement. 226

**symphysis pubis** (pyoo´bis) A slightly movable joint located anteriorly between the two pubic bones of the pelvic girdle. 209

**synapse** (sin´aps) A minute space between the axon terminal of a presynaptic neuron and a dendrite of a postsynaptic neuron. 358

**synarthrosis** (sin´´ar-thro´sis) A fibrous joint, such as a syndesmosis or a suture. 224

**synchondrosis** (sin´´kon-dro´sis) A cartilaginous joint in which the articulating bones are separated by hyaline cartilage. 226

**syndesmosis** (sin´´des-mo´sis) A type of fibrous joint in which two bones are united by an interosseous ligament. 225

**synergist** (sin´er-jist) A muscle that assists the action of the prime mover. 281

**synergistic** (sin´´er-jis´tik) Pertaining to regulatory processes or molecules (such as hormones) that have complementary or additive effects. 518

**synovial cavity** (sĭ-no´ve-al) A space between the two bones of a synovial joint, filled with synovial fluid. 227

**synovial joint** A freely movable joint in which there is a synovial cavity between the articulating bones; also called a *diarthrotic joint*. 227

**synovial membrane** The inner membrane of a synovial capsule that secretes synovial fluid into the joint cavity. 227

**system** A group of body organs that function together. 9

**systemic** (sis-tem´ik) Relating to the entire organism rather than to individual parts. 10

**systemic anatomy** The division of anatomy concerned with the structure and function of the various systems. 10

**systemic circulation** The portion of the circulatory system concerned with blood flow from the left ventricle of the heart to the entire body and back to the heart via the right atrium (in contrast to the pulmonary system, which involves the lungs). 570

**systole** (sis´tŏ-le) The muscular contraction of a heart chamber during the cardiac cycle. 573

**systolic pressure** (sis-tol´ik) Arterial blood pressure during the ventricular systolic phase of the cardiac cycle. 636

## T

**tachycardia** (tak´´ĭ-kar´de-ă) An excessively rapid heart rate, usually in excess of 100 beats per minute (in contrast to bradycardia, in which the heart rate is very slow). 603

**tactile** (tak´til) Pertaining to the sense of touch.

**taeniae coli** (te´ne-e ko´li) The three longitudinal bands of muscle in the wall of the large intestine. 786

**target organ** The specific body organ that a particular hormone affects. 514

**tarsal gland** An oil-secreting gland that opens on the exposed edge of each eyelid; also called a *meibomian gland*. 490

**tarsus** (tar´sus) The region of the foot containing the seven tarsal bones. 216

**taste bud** An organ containing the chemoreceptors associated with the sense of taste. 474

**T cell** A type of lymphocyte that provides cell-mediated immunity (in contrast to B lymphocytes, which provide humoral immunity through the secretion of antibodies). There are three subpopulations of T cells: cytotoxic, helper, and suppressor. 666

**tectorial membrane** (tek-to´re-al) A gelatinous membrane positioned over the hair cells of the spiral organ in the cochlea. 486

**telencephalon** (tel´´en-sef´ă-lon) The anterior portion of the forebrain, constituting the cerebral hemispheres and related parts. 379

**tendo calcaneus** (ten´do kal-ka´ne-us) The tendon that attaches the calf muscles to the calcaneous bone; also called the *Achilles tendon*. 316

**tendon** (ten´dun) A band of dense regular connective tissue that attaches muscle to bone. 122

**tendon sheath** A covering of synovial membrane surrounding certain tendons. 228

**tentorium cerebelli** (ten-to´re-um ser´´ĕ-bel´e) An extension of dura mater that forms a partition between the cerebral hemispheres and the cerebellum and covers the cerebellum. 393

**teratogen** (tĕ-rat´ŏ-jen) Any agent or factor that causes a physical defect in a developing embryo or fetus. 928

**testis** (tes´tis) The primary reproductive organ of a male that produces spermatozoa and male sex hormones. 845

**testosterone** (tes-tos´tĕ-rōn) The major androgenic steroid secreted by the interstitial cells of the testes after puberty. 840

**tetanus** (tet´n-us) A smooth contraction of a muscle (as opposed to muscle twitching). 259

**thalamus** (thal´ă-mus) An oval mass of gray matter within the diencephalon that serves as a sensory relay center. 390

**thalassemia** (thal´´ă-se´me-ă) Any of a group of hemolytic anemias caused by the hereditary inability to produce either the alpha or beta chain of hemoglobin. It is found primarily among Mediterranean people. 718

**thigh** The proximal portion of the lower extremity between the hip and the knee in which the femur is located. 211

**third ventricle** (ven´trĭ-k´l) A narrow cavity between the right and left halves of the thalamus and between the lateral ventricles that contains cerebrospinal fluid. 398

**thoracic** (tho-ras´ik) Pertaining to the chest region. 12

**thoracic duct** The major lymphatic vessel of the body that drains lymph from the entire body, except for the upper right quadrant, and returns it to the left subclavian vein. 648

**thorax** (thor´aks) The chest. 12

**threshold stimulus** The weakest stimulus capable of producing an action potential in an excitable cell. 355

**thrombocyte** (throm´bŏ-sīt) A blood platelet formed from a fragmented megakaryocyte. 553

**thrombus** (throm´bus) A blood clot produced by the formation of fibrin threads around a platelet plug. 606

**thymus** (thi´mus) A bilobed lymphoid organ positioned in the upper mediastinum, posterior to the sternum and between the lungs. 650

**thyroid cartilage** (thi´roid kar´tĭ-lij) The largest cartilage in the larynx that supports and protects the vocal cords; commonly called the *Adam's apple*. 687

**thyroxine** (thi-rok´sin) Also called tetraiodothyronine, or T₄. The major hormone secreted by the thyroid gland, which regulates the basal metabolic rate and stimulates protein synthesis in many organs. A deficiency of this hormone in early childhood produces cretinism. 529

**tinnitus** (tĭ-ni´tus) The spontaneous sensation of a ringing sound or other noise without sound stimuli. 506

**tissue** An aggregation of similar cells and their binding intercellular substance, joined to perform a specific function. 9

**tongue** A protrusible muscular organ on the floor of the oral cavity. 771

**tonsil** (ton'sil) A node of lymphoid tissue located in the mucous membrane of the pharynx. 686

**toxin** (tok'sin) A poison. 665

**trabeculae** (tră-bek'yŭ-le) A supporting framework of fibers crossing the substance of a structure, as in the lamellae of spongy bone. 167

**trachea** (tra'ke-ă) The airway leading from the larynx to the bronchi, composed of cartilaginous rings and a ciliated mucosal lining of the lumen; commonly called the *windpipe*. 688

**tract** A bundle of nerve fibers within the central nervous system. 403

**transamination** (trans''am-ĭ-na'shun) The transfer of an amino group from an amino acid to an alpha-keto acid, forming a new keto acid and a new amino acid without the appearance of free ammonia. 819

**transpulmonary pressure** (trans''pul'mŏ-ner''e) The pressure difference across the wall of the lung, equal to the difference between intrapulmonary pressure and intrapleural pressure. 693

**transverse colon** (trans-vers' ko'lon) A portion of the large intestine that extends from right to left across the abdomen between the hepatic and splenic flexures. 785

**transverse fissure** (fish'ur) The prominent cleft that horizontally separates the cerebrum from the cerebellum. 393

**transverse plane** A plane that divides the body into superior and inferior portions; also called a *horizontal*, or *cross-sectional*, *plane*. 11

**tricuspid valve** (tri-kus'pid) The heart valve located between the right atrium and the right ventricle. 569

**trigone** (tri'gōn) A triangular area in the urinary bladder between the openings of the ureters and the urethra. 754

**triiodothyronine** (tri''i-o''do-thi'rŏ-nēn) Abbreviated T₃; a hormone secreted in small amounts by the thyroid; the active hormone in target cells, formed from thyroxine. 529

**trochanter** (tro-kan'ter) A broad, prominent process on the proximolateral portion of the femur. 212

**trochlea** (trok'le-ă) A pulleylike anatomical structure (e.g., the medial surface of the distal end of the humerus that articulates with the ulna). 204

**tropomyosin** (tro''pŏ-mi'ŏ-sin) A filamentous protein that attaches to actin in the thin myofilaments and that acts, together with another protein called troponin, to inhibit and regulate the attachment of myosin cross bridges to actin. 268

**true vocal cords** Folds of the mucous membrane in the larynx that produce sound as they are pulled taut and vibrated. 688

**trunk** The thorax and abdomen together. 12

**trypsin** (trip'sin) A protein-digesting enzyme in pancreatic juice that is released into the small intestine. 795

**tubercle** (too'ber-k'l) A small, elevated process on a bone. 164

**tuberosity** (too'bĭ-ros'ĭ-te) An elevation or protuberance on a bone. 164

**tunica albuginea** (too'nĭ-kă al''byoo-jin'e-ă) A tough, fibrous tissue surrounding the testis. 845

**tympanic membrane** (tim-pan'ik) The membranous eardrum positioned between the external and middle ear. 482

## U

**umbilical cord** (um-bĭ'lĭ-kal) A cordlike structure containing the umbilical arteries and vein and connecting the fetus with the placenta. 601

**umbilicus** (um-bĭ-lĭ-kus) The site where the umbilical cord was attached to the fetus; commonly called the *navel*. 14

**unipolar neuron** (yoo'nĭ-po-lar noor'on) A nerve cell that has a single nerve fiber extending from its cell body. 347

**universal donor** A person with blood type O who is able to donate blood to people with other blood types in emergency blood transfusions. 557

**universal recipient** A person with blood type AB who can receive blood of any type in emergency transfusions. 557

**upper extremity** The appendage attached to the pectoral girdle, consisting of the shoulder, brachium, elbow, antebrachium, and hand. 12

**urea** (yoo-re'ă) The chief nitrogenous waste product of protein catabolism in the urine, formed in the liver from amino acids. 739

**uremia** (yoo-re'me-ă) The retention of urea and other products of protein catabolism as a result of inadequate kidney function. 759

**ureter** (yoo-re'ter) A tube that transports urine from the kidney to the urinary bladder. 753

**urethra** (yoo-re'thră) A tube that transports urine from the urinary bladder to the outside of the body. 754

**urinary bladder** (yoo'rĭ-ner''e) A distensible sac that stores urine, situated in the pelvic cavity posterior to the symphysis pubis. 754

**urobilinogen** (yoo''rŏ-bi-lin'o-jen) A compound formed from bilirubin in the small intestine; some is excreted in the feces, and some is absorbed and enters the enterohepatic circulation, where it may be excreted either in the bile or in the urine. 792

**uterine tube** (yoo'ter-in) The tube through which the ovum is transported to the uterus and the site of fertilization; also called the *oviduct* or *fallopian tube*. 782

**uterus** (yoo'ter-us) A hollow, muscular organ in which a fetus develops. It is located within the female pelvis between the urinary bladder and the rectum; commonly called the *womb*. 873

**utricle** (yoo'trĭ-k'l) An enlarged portion of the membranous labyrinth, located within the vestibule of the inner ear. 479

**uvula** (yoo'vyŭ-lă) A fleshy, pendulous portion of the soft palate that blocks the nasopharynx during swallowing. 771

## V

**vacuole** (vak-yoo'ōl) A small space or cavity within the cytoplasm of a cell. 45

**vagina** (vă-ji'nă) A tubular organ leading from the uterus to the vestibule of the female reproductive tract that receives the male penis during coitus. 874

**vallate papillae** (val'āt pă-pil'e) The largest papillae on the surface of the tongue. They are arranged in an inverted V-shaped pattern at the posterior portion of the tongue. 771

**vasectomy** (vă-sek'tŏ-me, va-zek'tŏ-me) Surgical removal of portions of the ductus deferentia to induce infertility. 864

**vasoconstriction** (va''zo-kon-strik'shun) Narrowing of the lumen of blood vessels due to contraction of the smooth muscles in their walls. 624

**vasodilation** (va''zo-di-la'shun) Widening of the lumen of blood vessels due to relaxation of the smooth muscles in their walls. 624

**vasomotor center** (va''zo-mo'tor) A cluster of nerve cell bodies in the medulla oblongata that controls the diameter of blood vessels. It is therefore important in regulating blood pressure. 395

**vein** (vān) A blood vessel that conveys blood toward the heart. 582

**vena cava** (ve'nă ka'vă) One of two large vessels that return deoxygenated blood to the right atrium of the heart. 592

**ventilation** (ven''tĭ-la'shun) Breathing; the process of moving air into and out of the lungs. 692

**ventral** (ven'tral) Toward the front or facing surface; the opposite of *dorsal*; also called *inferior*. 12

**ventricle** (ven'trĭ-k'l) A cavity within an organ; especially those cavities in the brain that contain cerebrospinal fluid and those in the heart that contain blood to be pumped from the heart. 567

**venule** (ven'yool) A small vessel that carries venous blood from capillaries to a vein. 615

**vermis** (ver'mis) The coiled middle lobular structure that separates the two cerebellar hemispheres. 393

**vertebral canal** (ver'tĕ-bral) The tubelike cavity extending through the vertebral column that contains the spinal cord; also called the *spinal canal*. 401

**vertigo** (ver'tĭ-go) A feeling of movement or loss of equilibrium. 481

**vestibular window** *See* oval window.

**vestibule** (ves'tĭ-byool) A space or cavity at the entrance to a canal, especially that of the nose, inner ear, or vagina.

**villus** (vil'us) A minute projection that extends outward into the lumen from the mucosal layer of the small intestine. 782

**virulent** (vir'yŭ-lent) Pathogenic; able to cause disease.

**viscera** (vis'er-a) The organs within the abdominal or thoracic cavities. 663

**visceral** (vis'er-al) Pertaining to the membranous covering of the viscera. 116

**visceral peritoneum** (per˝ĭ-tŏ-ne´um) A serous membrane that covers the surfaces of abdominal viscera.   117

**visceral pleura** (ploor´ă) A serous membrane that covers the surfaces of the lungs.   116

**visceroceptor** (vis˝er-ŏ-sep´tor) A sensory receptor located within body organs that responds to information concerning the internal environment.   465

**vitreous humor** (vit´re-us hyoo´mer) The transparent gel that occupies the space between the lens and retina of the eyeball.   494

**Volkmann's canal** (fōlk´manz) *See* perforating canal.   167

**vulva** (vul´vă) The external genitalia of the female that surround the opening of the vagina.   875

## W

**white matter** Bundles of myelinated axons located in the central nervous system.   383

**wormian bone** (wer´me-an) *See* sutural bone.   161

## Y

**yellow marrow** (mar´o) Specialized lipid storage tissue within bone cavities.   165

## Z

**zygote** (zi´gōt) A fertilized egg cell formed by the union of a sperm cell and an ovum.   899

**zymogens** (zi´mŏ-jenz) Inactive enzymes that become active when part of their structure is removed by the action of another enzyme or by some other means.   795

## Photographs

### Under Development Box Icon

© Petit Format/Photo Researchers Inc.

### Table of Contents

TOC.4: Kent M. Van De Graaff; TOC.8: © Donald Yeager/Camera M.D. Studios.

### Reference Figures

**Figure 1 page 324:** Kent M. Van De Graaff; **Figure 2 page 324:** Kent M. Van De Graaff; **Figure 3 page 325:** Kent M. Van De Graaff; **Figure 4 page 326:** Kent M. Van De Graaff; **Figure 5 page 327:** Kent M. Van De Graaff; **Figure 6 page 327:** Kent M. Van De Graaff; **Figure 7 A–C page 328:** Kent M. Van De Graaff; **Figure 8 page 329:** From B. Vidic and F. R. Suarez, *Photographic Atlas of the Human Body* (Pl. 27, Pg. 34), 1984, St. Louis, Mosby; **Figure 9 page 329:** From B. Vidic and F. R. Suarez, *Photographic Atlas of the Human Body* (Pl. 34, Pg. 48), 1984, St. Louis, Mosby; **Figure 10 page 330:** © Wm. C. Brown Communications/Karl Rubin, photographer; **Figure 11 page 330:** From B. Vidic and F. R. Suarez, *Photographic Atlas of the Human Body* (Pl. 113, Pg. 182), 1984, St. Louis, Mosby; **Figure 12 page 331:** © Wm. C. Brown Communications/Karl Rubin, photographer; **Figure 13 page 332:** From B. Vidic and F. R. Suarez, *Photographic Atlas of the Human Body* (Pl. 39, Pg. 58), 1984, St. Louis, Mosby; **Figure 14 page 332:** From B. Vidic and F. R. Suarez, *Photographic Atlas of the Human Body* (Pl. 148, Pg. 240), 1984, St. Louis, Mosby; **Figure 15 page 333:** Kent M. Van De Graaff; **Figure 16 page 334:** Kent M. Van De Graaff; **Figure 17 page 334:** From B. Vidic and F. R. Suarez, *Photographic Atlas of the Human Body* (Pl. 206, Pg.

338), 1984, St. Louis, Mosby; **Figures 18 and 19 page 335:** © Wm. C. Brown Communications/Karl Rubin, photographer; **Figures 20 and 21 page 336:** © Wm. C. Brown Communications/Karl Rubin, photographer.

### Chapter 1

**1.1:** Bibliotheque Nationale; **1.2:** Fratelli Alinari; **1.3A, 1.3B:** From the Works of Andrea Vesalius of Brussels by J. Bade, C. M. Saunders and Charley P. O'Malley, Pg. 1096, Dover Publications, Inc.; **1.4:** © Stock Montage; **1.9A:** Courtesy of Kodak; **1.9B:** © Carroll H. Weiss/Camera M.D. Studios; **1.9C:** Courtesy of Utah Valley Regional Medical Center, Dept. of Radiation; **1.10A:** © Lester V. Bergman & Associates, Inc.; **1.10B:** © Hank Morgan/Science Source/Photo Researchers, Inc.; **1.11, 1.12:** © Dr. Sheril D. Burton; **page 12:** Kent M. Van De Graaff; **1.14B:** © Dr. Sheril Burton

### Chapter 2

**2.26:** © Edwin A. Reschke

### Chapter 3

**3.3A, 3.3B:** Kwang W. Jeon; **3.4 (1–4):** M. M. Perry & A. B. Gilbert, *Journal of Cell Science* 39:257–272, 1979; **3.5A:** © Keith R. Porter; **3.5B:** © Richard Chao; **3.6A:** Dr. Carolyn Chambers; **3.6B:** From R. G. Kessel and R. H. Kardon: *Tissues and Organs: A Text-Atlas of Scanning Electron Microscopy*, W. H. Freeman and Company © 1979; **3.7:** © K.G. Murti/Visuals Unlimited; **3.9:** © Richard Chao; **3.10A, 3.11A:** © Keith R. Porter; **3.12:** © E. G. Pollack; **3.13:** © Richard Chao; **3.18A:** © O. L. Miller, B. R. Beatty, D. W. Fawcett/Visuals Unlimited; **3.20:** © Alexander Rich; **3.25A, 3.28A:** © David M. Phillips/Visuals Unlimited;

**3.30A–2, 3.30B–2, 30C–2, 3.30D–2, 3.30E–2:** © Edwin A. Reschke; **3.31:** © CNRI/SPL/Photo Researchers, Inc.

### Chapter 5

**5.10:** © Richard Chao

### Chapter 6

**6.1A:** © Edwin A. Reschke; **6.1B:** © CNRI/SPL/Photo Researchers, Inc.; **6.2B, 6.3B, 6.4B, 6.5B, 6.6B, 6.7B, 6.8B, 6.9B:** © Edwin A. Reschke; **6.11B:** © CNRI/SPL/Photo Researchers, Inc.; **6.14B:** © Biophoto Associates/Science Source/Photo Researchers, Inc.; **6.15B, 6.16B:** © Edwin A. Reschke; **6.17B:** © Bruce Iverson; **6.18B, 6.19B, 6.20B, 6.21B, 6.22B, 6.23B, 6.24B, 6.26A–1, 6.26B–1, 6.26C–1:** © Edwin A. Reschke; **6.27B:** © John D. Cunningham/Visuals Unlimited; **6.27C:** © Martin M. Rotker/Photo Researchers, Inc.

### Chapter 7

**7.2:** © Edwin Reschke/Peter Arnold, Inc.; **7.3:** © J. Burgess/Photo Researchers, Inc.; **7.4:** © Victor B. Eichler, Ph.D.; **7.5:** © Dr. Sheril D. Burton; **7.6:** © James M. Clayton; **7.7:** © Lester V. Bergman & Associates, Inc.; **7.8A:** World Health Organization; **7.8B:** George P. Bogumill, M.D.; **7.9A:** © Michael Abbey/Photo Researchers, Inc.; **7.9B:** Dr. Kerry L. Openshaw; **7.10:** © John D. Cunningham/Visuals Unlimited; **7.13:** © Martin M. Rotker; **7.14A–2:** © Tierbild Okapia/Photo Researchers, Inc.; **7.14B–2:** © J. Stevenson/Photo Researchers, Inc.; **7.14C–2:** © John Radcliff/Photo Researchers, Inc.; **7.17:** (Child): From Science Year, The World Book Science Annual © 1973 Field Enterprise Education Corporation. By

permission of World Book, Inc. (Woman): Black Star Publishing Co.; **7.18:** © Norman Lightfoot/Photo Researchers, Inc.

### Chapter 8

**BOX 8.1B:** Ted Conde; **8.5A, 8.5B:** © Biophoto Associates/Photo Researchers, Inc.; **8.7B:** © Edwin A. Reschke; **8.8:** © Biophoto Associates/Photo Researchers, Inc.; **8.10:** © Edwin A. Reschke; **8.11:** Courtesy of Utah Valley Regional Medical Center, Dept. of Radiology; **8.14:** © CNRI/SPL/Photo Researchers, Inc.; **8.15A, 8.15B:** Raisz, L. G., Dempster, D. W. et al., 1986, J. Bon Miner Res. 1:15–21, Reprinted by permission of the *New England Journal of Medicine*, Vol. 318, No. 13, p. 818.

### Chapter 9

**9.5:** Kent M. Van De Graaff; **9.9A, 9.9B, 9.21A, 9.23A:** Courtesy of Utah Valley Regional Medical Center, Dept. of Radiology.

### Chapter 10

**10.3:** Kent M. Van De Graaff; **10.7B:** © Dr. Sheril D. Burton; **10.9, 10.11, 10.16:** Courtesy of Utah Valley Regional Medical Center, Dept. of Radiology; **10.18A:** © Dr. Sheril D. Burton; **10.18B:** Courtesy of Utah Valley Regional Medical Center, Dept. of Radiology; **10.20A, 10.20B:** © Blayne Hirshche; **10.21:** Kent M. Van De Graaff; **10.23E:** Courtesy of Eastman Kodak Company.

### Chapter 11

**11.4:** Courtesy of Utah Valley Regional Medical Center, Dept. of Radiology; **11.19A, 11.19B, 11.19C, 11.19D, 11.19E, 11.19F, 11.19G, 11.19H:** © Dr. Sheril D. Burton; **11.20A:** Kent

M. Van De Graaff; **11.20B, 11.20C:** © Dr. Sheril D. Burton; **11.20D:** Kent M. Van De Graaff; **11.20E, 11.20F:** © Dr. Sheril D. Burton; **11.21A, 11.21B, 11.30A:** Kent M. Van De Graaff; **11.30B:** © Lester V. Bergman & Associates, Inc.; **11.31A, 11.31B, 11.31C, 11.31D:** SIU, School of Medicine; **11.32:** Kent M. Van De Graaff.

**Chapter 12**

**12.2B:** © Edwin A. Reschke; **12.3A:** International Bio-Medical, Inc.; **12.3B:** Stuart Ira Fox; **12.4A–B:** Kent M. Van De Graaff; **12.5B:** © John D. Cunningham/Visuals Unlimited; **12.8, 12.9A, 12.9B:** © Dr. H. E. Huxley; **12.9C:** From R. G. Kessel and R. H. Kardon: *Tissues and Organs: A Text-Atlas of Scanning Electron Microscopy,* W. H. Freeman and Company © 1979; **12.10A:** © Dr. H. E. Huxley; **12.19:** Hans Hoppler, *Respiratory Physiology* 44:94 (1981).

**Chapter 14**

**14.2:** © Edwin A. Reschke; **14.7:** H. Webster, from John Hubbard, *The Vertebrate Peripheral Nervous System* © 1974, Plenum Press; **14.9:** Andreas Karschin, Heinz Wassle and Jutta Schnitzer, *Scientific American,* Pg. 67, April 1989; **14.13:** Bell et al., *Textbook of Physiology and Biochemistry* , 10th ed. © Churchill Livingstone, Edinburgh; **14.18A:** From Gilula, Ranes and Steinbach, "Metabolic Coupling and Cell Contacts, *Nature* 235:262–265 © McMillan Journals, Ltd.; **14.19:** © John Heuser, Washington University, School of Medicine, St. Louis, MO

**Chapter 15**

**15.1:** Kent M. Van De Graaff; **15.2:** © Monte S. Buchsbaum, M.D.; **15.3A, 15.3B, 15.3C, 15.3D:** Kent M. Van De Graaff; **15.8A:** © Martin M. Rotker; **15.24B:** © Per H. Kjeldsen, University of Michigan, Ann Arbor

**Chapter 16**

**16.2:** From R. G. Kessel and R. H. Kardon: *Tissues and Organs: A Text-Atlas of Scanning Electron Microscopy,* W. H. Freeman and Company © 1979; **16.32A, 16.32B, 16.32C, 16.32D, 16.32E, 16.32F, 16.32G, 16.32H, 16.32I:** All:

Photographed by Dr. Sheril D. Burton assisted by Dr. Douglas W. Hacking

**Chapter 18**

**18.9C:** © Victor B. Eichler, Ph.D.; **18.14A:** © Dean E. Hillman; **18.18:** Kent M. Van De Graaff; **18.26:** © Dr. Sheril Burton; **18.32A:** © Thomas Sims; **18.35B:** P. N. Farnsworth, University of Medicine and Dentistry, New Jersey Medical School; **18.37:** © Per H. Kjeldsen, University of Michigan, Ann Arbor; **18.45:** © Dr. Stephen Clark

**Chapter 19**

**19.1B:** © Edwin A. Reschke; **19.13B:** © SIU/Peter Arnold, Inc.; **19.14:** © Martin M. Rotker; **19.26, 19.27:** © Lester V. Bergman & Associates, Inc.

**Chapter 20**

**20.2B:** Reginald J. Poole/Polaroid International Instant Photomicrography Competition; **20.5:** Stuart Ira Fox (author); **20.6:** © Manfred Kage/Peter Arnold, Inc.

**Chapter 21**

**21.17:** © Don W. Fawcett; **21.21B:** From: Practische Intleedkunde from J. Dankmeyer, H. G. Lambers, and J. M. F. Landsmearr. Bohn, Scheltma and Holkema; **21.39A, 21.39B:** © Richard Menard; **21.41A:** American Heart Association; **21.41B:** © Lewis Lainey

**Chapter 22**

**22.9:** From E. K. Markell and M. Vogue, *Medical Parasitology,* 5th ed, W. B. Saunders; **22.19A, 22.19B:** Courtesy of Niels A. Lassen, Copenhagen, Denmark

**Chapter 23**

**23.12A:** Courtesy Arthur J. Olson, Ph.D., The Scripps Research Institute, Molecular Graphics Laboratory; **23.13A, 23.13B:** From Dr. A. G. Amit, "Three Dimensional Structure of an Antigen Antibody Complex at 2.8 Resolution," *Science* 233:747–753, 15 August 1986, 2 figures © 1986 by AAAS; **23.23A:** From Alan S. Rosenthal, *New England Journal of Medicine* 303:1153, 1980; **23.27A, 23.27B:** © Dr. Andrejs Liepins; **23.28A, 23.28B:** Courtesy of Dr.

Noel Rose; **23.30A–B:** SIU Biomed Comm./Custom Medical Stock; **23.31A:** From R. G. Kessel and C. Y. Shih, *Scanning Electron Microscopy in Biology,* © 1976, Springer-Verlag; **23.31B:** © Catherine Ellis/Photo Researchers, Inc.

**Chapter 24**

**24.4:** © CNRI/SPL/Photo Researchers, Inc.; **24.6C:** © CNRI/Phototake, Inc.; **24.8A:** © John D. Cunningham/Visuals Unlimited; **24.8B:** © Edwin A. Reschke; **24.11:** © CNRI/SPL/Photo Researchers, Inc.; **24.14A, 24.14B, 24.15:** Edward C. Vasquez R. T. C.R.T./Dept. of Radiologic Technology, Los Angeles City College; **24.18A, 24.18B:** From J. H. Comroe, Jr., *Physiology of Respiration,* © 1974, Yearbook Medical Publishers, Inc., Chicago; **24.21:** © SIU/Photo Researchers, Inc.; **24.41A, 24.41B:** © Martin M. Rotker

**Chapter 25**

**25.2:** © SPL/Photo Researchers, Inc.; **25.4B:** © Biophoto Associates/Photo Researchers, Inc.; **25.7:** © F. Spinnelli—D. W. Fawcett/Visuals Unlimited; **25.9:** © Daniel Friend from William Bloom and Don Fawcett, *Textbook of Histology,* 10th, ed, W. B. Saunders, Co.; **25.28:** © Per H. Kjeldsen, University of Michigan, Ann Arbor

**Chapter 26**

**26.10:** Kent M. Van De Graaff; **26.14B:** Courtesy of Utah Valley Regional Medical Center, Dept. of Radiation; **26.16:** © Edwin A. Reschke; **26.21:** © Manfred Kage/Peter Arnold, Inc.; **26.24A, 26.24B:** © Keith R. Porter/Albers, D. H. and Seetharan, D. *New England Journal of Medicine;* **26.29:** From W. A. Sodeman and T. M. Watson, *Pathologic Physiology,* 6th ed., W. B. Saunders Co., 1969; **26.31C:** © Victor B. Eichler, Ph.D.; **26.37A:** © Carroll Weiss/Camera M.D. Studios; **26.37B:** © Dr. Sheril D. Burton; **26.38:** © Edwin A. Reschke

**Chapter 27**

**27.21A, 27.21B, 27.21C, 27.21D:** From "Clinical Pathological Conference Acromegaly, Diabetes,

Hypermetabolism Proteinura and Heart Failure," *American Journal of Medicine* 20:133, 1956

**Chapter 28**

**28.13A:** © Biophoto Associates/Photo Researches, Inc.; **28.16B:** © Francis Leroy, Biocosmos/SPL/Photo Researchers, Inc.; **28.17, 28.18:** © Edwin A. Reschke; **28.19, 28.20:** © Manfred Kage/Peter Arnold, Inc.; **28.21:** © Edwin A. Reschke; **BOX 28.12–G, BOX 28.12–H:** © Dr. Landrum Shettles

**Chapter 29**

**29.2, 29.6, 29.8A:** © Edwin A. Reschke; **29.8B:** © Ed Reschke/Peter Arnold, Inc.; **29.9A:** © Dr. Landrum Shettles; **29.9B:** From R. J. Blandau A *Textbook of Histology,* 10th ed. W. B. Saunders, Co.; **29.10:** From Bloom/Fawcett: *Textbook of Histology,* 10th ed. p. 859, W. B. Saunders; **29.11:** © Dr. Landrum Shettles; **29.15:** © Martin M. Rotker; **29.21:** Courtesy of Utah Valley Regional Medical Center, Dept. of Radiology; **29.24A, 29.24B:** SIU School of Medicine; **29.25A:** © SIU/Visuals Unlimited; **29.25B:** © John Kaprielian/Photo Researchers, Inc.; **29.25C:** © Ray Ellis/Science Source/Photo Researchers, Inc.; **29.25D:** © Science VU-Ortho/Visuals Unlimited; **29.25E, 29.25F:** © M. Long/Visuals Unlimited; **29.25G:** © Scott Camazine/Photo Researchers, Inc.; **29.25H:** © Hank Morgan/Science Source/Photo Researchers, Inc.

**Chapter 30**

**30.1C:** © David Phillips/Visuals Unlimited; **30.2:** Luciano Zabmboni from Greep, Roy and Weiss, Leon, *Histology,* 3rd ed., 1973, McGraw-Hill Book Company; **30.7A, 30.9A:** © Roman O'Rahilly/Carnegie Laboratories of Embryology, University of California; **30.12A:** © Donald Yeager/Camera M.D. Studios; **30.17E:** © Dr. Landrum Shettles; **30.20B:** © Donald Yeager/Camera M.D. Studios; **30.21E:** © Dr. Landrum Shettles; **30.22, 30.24A, 30.24B:** © Donald Yeager/Camera M.D. Studios; **30.25A:** © A. Tsiaras/Photo Researchers, Inc.; **30.25B:** Kent M. Van De Graaff; **30.28:** March of Dimes; **30.34:** © Lester V. Bergman & Associates,

Credits

Credits

## Illustrator Credits

Credits

Index